Proceedings of the Fourteenth Annual Conference of the Cognitive Science Society

July 29 to August 1, 1992
Cognitive Science Program
Indiana University, Bloomington

Psychology Press
Taylor & Francis Group
NEW YORK AND LONDON

First Published by
Lawrence Erlbaum Associates, Inc.
365 Broadway
Hillsdale, New Jersey 07642

Published 2009 by Psychology Press
711 Third Avenue, New York, NY 10017
27 Church Road, Hove, East Sussex BN3 2FA, UK

First issued in paperback 2014

Psychology Press is an imprint of the Taylor and Francis Group, an informa business

ISBN 13: 978-0-8058-1291-6 (hbk)
ISBN 13: 978-0-415-65106-6 (pbk)
ISSN 1047-1316

Publisher's Note
The publisher has gone to great lengths to ensure the quality of this reprint but points out that some imperfections in the original may be apparent.

TABLE OF CONTENTS
The Fourteenth Annual Conference of
the Cognitive Science Society
Indiana University
Bloomington, Indiana 47405
July 29 - August 1, 1992

Connectionist Models of Language I

DAY TWO, SESSION ONE

Vision

Case-Based Reasoning

Similarity and Categories

Cognitive Neuroscience

POSTERS

The Fourteenth Annual Conference of the Cognitive Science Society

This conference is supported in part by funds from the Indiana University College of Arts and Sciences, the Indiana University Research and University Graduate School, and the Indiana University Cognitive Science Program.

Preface

The 1992 Conference of the Cognitive Science Society was held at Indiana University in Bloomington, sponsored in part by the Indiana University Cognitive Science Program, the Indiana University College of Arts and Sciences, and the Indiana University Research and University Graduate School.

Of the 284 papers submitted to this year's Conference, 113 were accepted for oral presentation and 87 for poster presentation. Acceptances were based on reviewers' ratings and on the capacity of the 3-day program. The efforts of the reviewers are gratefully acknowledged; their names appear on the next page.

A paper by Yuri Mashintsev, Moscow State University, entitled "Eliciting implicit theory of personality through processing questionnaires with Boltzmann machine", was accepted for poster presentation, but because of the slow mail between the USA and the former USSR, a camera-ready copy of the paper was not received by press time.

The Conference also featured 7 plenary talks and 9 symposia, the texts of which are not included in these Proceedings. Plenary speakers and talk titles were as follows: Elizabeth Bates (University of California, San Diego), Crosslinguistic studies of language breakdown in aphasia; Daniel Dennett (Tufts University), Problems with some models of consciousness; Martha Farah (Carnegie-Mellon University), Is an object an object? Neuropsychological evidence for domain-specificity in visual object recognition; Douglas Hofstadter (Indiana University), The centrality of analogy-making in human cognition; John Holland (University of Michigan), Must learning precede cognition?; Richard Shiffrin (Indiana University), Memory representation, storage, and retrieval; Michael Turvey (University of Connecticut), Ecological foundations of cognition. Symposium topics and organizers were as follows: Timothy van Gelder (Indiana University) and Beth Preston (University of Georgia), Representation: Who needs it?; Rik Belew (University of California, San Diego), Computational models of evolution as tools for cognitive science; Caroline Palmer (The Ohio State University) and Allen Winold (Indiana University), Dynamic processes in music cognition; Geoffrey Bingham (Indiana University) and Bruce Kay (Brown University), Dynamics in the control and coordination of action; David Leake (Indiana University) and Ashwin Ram (Georgia Institute of Technology), Goal-driven learning; Mary Jo Rattermann (Hampshire College), Similarity and representation in early cognitive development; K. Jon Barwise (Indiana University), Reasoning and visual representations; David Pisoni (Indiana University) and Robert Peterson (Indiana University), Speech perception and spoken language processing; Douglas Hofstadter (Indiana University) and Melanie Mitchell (University of Michigan), Analogy, high-level perception, and categorization. A list of symposium speakers' names was not available at press time. The banquet speaker was Bernhard Flury, Indiana University Department of Mathematics.

REVIEWERS

Robert Allen
Richard Alterman
John Barnden
K. Jon Barwise
William Bechtel
Thomas Bever
Dorrit Billman
Geoffrey Bingham
Daniel Bobrow
Gary Bradshaw
Bruce Britton
Mark Burstein
Richard Catrambone
David Chalmers
Yves Chauvin
Catherine Clement
Axel Cleeremans
Gregg Collins
Cynthia Connine
Paul Cooper
Kenneth Deffenbacher
S. Farnham Diggory
Stephanie Doane
Charles Dolan
Carolyn Drake
Kevin Dunbar
Kurt Eiselt
Martha Evens
Brian Falkenhainer
Martha Farah
Paul Feltovitch
R. James Firby
Ken Forbus
Bob French
Susan Garnsey
Michael Gasser
Dedre Gentner
Morton Gernsbacher
Randy Gobbel
Ashok Goel
Richard Golden
Robert Goldstone
Art Graesser
Jordan Grafman
Richard Granger
Kristian Hammond
Reid Hastie
Geoffrey Hinton
Steven Hirtle
Douglas Hofstadter
Keith Holyoak
William Hoyer
Edwin Hutchins
Susan Kemper
Janet Kolodner
Kenneth Kotovsky
Bruce Krulwich
John Kruschke
John Laird
David Leake
Jill Fain Lehman
Wendy Lehnert
Daniel Levine
Clayton Lewis
Robert Lindsay
Steven Lytinen
Colin MacLeod
Pattie Maes
James Martin
Michael Masson
Richard Mayer
James McClelland
Doug Medin
Melanie Mitchell
Michael Mozer
Paul Munro
Gregory Murphy
Sheldon Nicholl
Donald Norman
Gregg Oden
Stellan Ohlsson
Gary Olson
Judith Orasanu
Alice O'Toole
Christopher Owens
Caroline Palmer
Vimla Patel
Michael Pazzani
Herbert Pick
David Plaut
Kim Plunkett
Robert Port
Bruce Porter
Mitchell Rabinowitz
Ashwin Ram
Michael Ranney
William Rapaport
Michael Redmond
Brian Reiser
Scott Robertson
Paul Rosenbloom
Brian Ross
Roger Schank
Walter Schneider
Alberto Segre
Colleen Seifert
Martin Sereno
Stuart Shapiro
Lokendra Shastri
Richard Shiffrin
Jeff Shrager
Peter Slezak
Steven Small
Linda Smith
Elliot Soloway
Mark St.John
Michael Swain
David Swinney
Katia Sycara
David Touretzky
David Townsend
John Tsotsos
Elise Turner
Roy Turner
Timothy van Gelder
DeLiang Wang
Ed Wisniewski
Dekai Wu
Ivan Yaniv
Ingrid Zukerman

TALKS

Why Intelligent Systems Should Get Depressed Occasionally and Appropriately

Charles Webster
Intelligent Systems Program
University of Pittsburgh
Pittsburgh, PA 15261
cww@pogo.isp.pitt.edu

Abstract

Some researchers suggest that depression may be adaptive. For example, depression may provide an opportunity to assess our capabilities, learn from past failures, trigger personal change, and allocate activity away from futile goals. There are a variety of signature phenomena associated with depression, such as stable, global, and internal styles of failure explanation, a cognitive loop of failure-related rumination, lowered self-esteem and self-efficacy, and increased negative generalization and depressive realism. DEPlanner is presented, a simulated agent that adapts to failure in a simulated environment and exhibits eight targeted signature phenomena of depression.

Introduction

Some types of depression in response to personal failure may be adaptive. Taylor (1989, p. 225) suggests that depression serves as a reality check, an opportunity to take "realistic stock of what one is and where one is going" and make "accurate assessment of his or her capabilities." Williams et al. (1988, p. 183) suggest that depression facilitates coping with long-term problems by "strategic access to previous problem-solving attempts." Flach (1974) and Gut (1989) argue that depression is a normal response to personal failure, and is necessary for personal change. Nesse (1991) hypothesizes that decreased mood reduces activity in situations where effort will not be rewarded, allocating energy away from bad investments. If depression can serve an adaptive function, then perhaps a simulated agent can be designed in such a way as to adapt to failure while exhibiting signature phenomena of depression. DEPlanner, based on ideas discussed in Webster et al. (1988), is such an adaptive agent.

This research was supported by the University of Pittsburgh Alzheimer Disease Research Center.

Overview of DEPlanner Simulation

DEPlanner is a nonlinear planner (Chapman, 1987) that generates sequences of instantiated STRIPS-style operators in response to initial and goal states. Since representing the social environment of a human is so complex, and since something like depression may be generally useful for adapting to some kinds of environmental change, planning and learning occur in a Blocksworld micro-world. DEPlanner's environment is a description of a set of blocks stacked on each other and scattered among locations on a table. Goals are conjunctions of targeted block configurations. A plan is a sequence of block movements. Successes have positive utilities, and failures negative utilities. Interactions between DEPlanner and its environment occur through the use of a problem generator that produces a stream of randomly constructed Blocksworld problems. The problem generator periodically changes internal rules that determine whether one block can be stacked on another block or location. A plan violating such a rule results in failure, and a deduction from DEPlanner's score. DEPlanner's task is to adapt in such a way as to maximize total utility accrued during interactions with the problem generator.

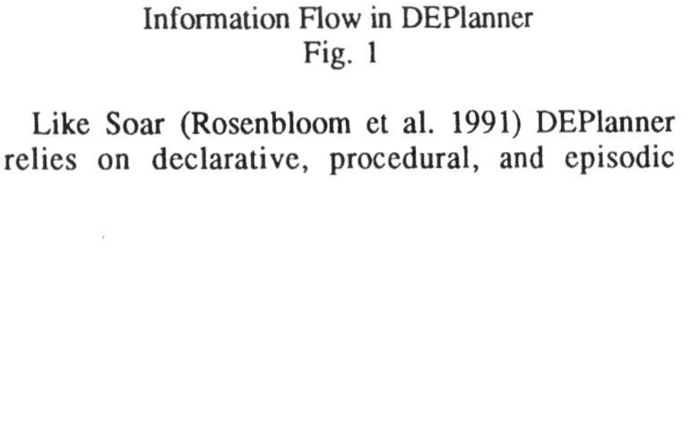

Information Flow in DEPlanner
Fig. 1

Like Soar (Rosenbloom et al. 1991) DEPlanner relies on declarative, procedural, and episodic representations. Fig. 1 depicts the flow of information among them. Declarative knowledge corresponds to representations of which block can be stacked on which block or location. (ASSUME (STACKONABLE B ?X)) means that block A can be stacked anywhere. Fig. 2 illustrates a plan operator. The "SENSE" preconditions match against the problem description provided by the problem generator, while the "ASSUME" precondition matchs against DEPlanner's revisable assumptions in declarative memory. Thus, changing declarative assumptions can result in the production of different plans.

```
#S(OP
  :NAME MOVE        ;move
  :VARS (?X ?Y ?Z)  ;x from y to z
  :PRECONDS
    ((ASSUME (STACKONABLE ?X ?Z))
         ;matches revisable assumption
     (SENSE (CLEAR ?X));matches problem description
     (SENSE (CLEAR ?Z));  "          "         "
     (SENSE (ON ?X ?Y))); "          "         "
  :ADD-LIST ((SENSE (ON ?X ?Z))
             (SENSE (CLEAR ?Y))
  :DEL-LIST ((SENSE (ON ?X ?Y))
             (SENSE (CLEAR ?Z))))
```

DEPlanner Operator
Fig. 2

Procedural knowledge consists of plans like ((MOVE B LOC1) (MOVE A B)) (along with appropriate preconditions and effects). Episodic knowledge is based on a time stamp, problem initial state and goals, the plan used, and outcome (success or failure).

DEPlanner's declarative knowledge changes in response to patterns of failure and success observed in episodic knowledge (A in Fig. 1). Procedural knowledge arises from the application of declarative knowledge (B in Fig. 1), creating a practice effect (DEPlanner becomes faster with experience). In order to change declarative knowledge on the basis of experience, the results of using procedural knowledge are recorded in the form of episodic knowledge (C in Fig. 1). When failures in episodic memory trigger changes in declarative memory, procedural knowledge must also change in order to remain a compiled version of declarative knowledge. DEPlanner's depression is the computationally expensive process of consulting episodic knowledge (retrieving past failures), modifying declarative knowledge (explaining patterns of failure), and compiling new procedural knowledge (creating new plans). Specialized control mechanisms suppress depression unless, and until, future benefits of depression are estimated to outweigh current costs.

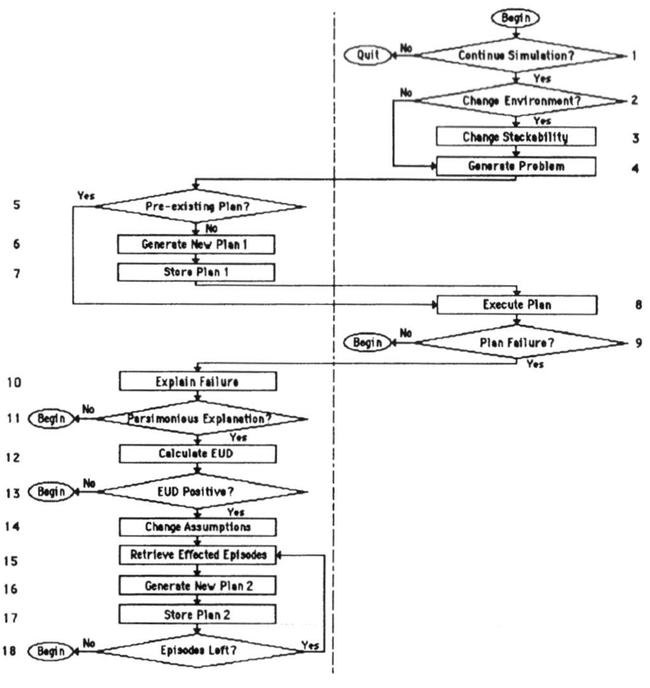

Interaction Between DEPlanner (left) and Problem Generator (right)
Fig. 3

The flowchart in Fig. 3 depicts the flow of information and control between, and within, DEPlanner and its problem generator. The problem generator produces a problem (1-4, Fig. 3). DEPlanner returns a plan (5-7, Fig. 3). The problem generator determines whether the plan is a success (8-9, Fig. 3). And DEPlanner, if certain criteria are met, gets "depressed" (10-18, Fig. 3).

DEPlanner gradually increases its average response speed by reusing plans (5, Fig. 3). Plans can be stored (7, Fig. 3) after generation (6, Fig. 3) in response to problems posed by the problem generator, or they can be regenerated an stored (16 and 17, Fig. 3) under new assumptions in response to memories of old problems (15, Fig. 3). The problem generator periodically changes internal rules about which block can be stacked on which block or location (2 and 3, Fig. 3). If DEPlanner finds a set of assumptions that explains the observed pattern of recent failure (10 and 11, Fig. 3) it calculates a numerical heuristic Expected Utility of Depression (EUD) (12, Fig. 3). If EUD is positive (13, Fig. 3) then DEPlanner changes its assumptions (14, Fig.3) and retrieves previous problem-solving episodes now predicted to result in failure (15, Fig. 3). For each episode (18, Fig. 3) DEPlanner generates a new plan under new assumptions (16, Fig. 3) and stores it (17, Fig. 3).

Signature Phenomena of Depression as Design Constraints on DEPlanner

I used eight signature phenomena, culled from the research literature on depression, as a set of design constraints on the construction of DEPlanner: stable, global, and internal styles of failure explanation (Peterson & Seligman, 1984); a cognitive loop of failure-related rumination (Ingram, 1984); lowered self-esteem (Musson & Alloy, 1988) and self-efficacy (Bandura, 1977; Rehm, 1988); and increased negative generalization (Beck et al. 1979) and depressive realism (Alloy & Abramson, 1988). My task was constructing a simulated agent that adapts to changes in a simulated environment, and exhibits as many signature phenomena of depression as possible.

The first three phenomena (stability, globality, and internality) concern the kinds of events that are most likely to trigger a depression. It is not failure itself, but rather explanation for failure that serves as a trigger. Failure attributions that are stable (continuing to be true in the future), global (affecting many important goals), and internal (having been avoidable) are most likely to precipitate depression (Peterson & Seligman, 1984). Each dimension can be mapped to a different aspect of the formula (Eq. 1) used to trigger or suppress DEPlanner's analog of adaptive depression.

If DEPlanner finds a parsimonious explanation for a recent pattern of failures, in terms of a change in assumptions, DEPlanner uses Eq. 1 to calculate the Expected Utility of Depression (EUD). If EUD is positive for forgoing current opportunities (PGI times PAU), but preparing for future opportunities (PSI times PMU), then DEPlanner gets "depressed." If the EUD is negative then DEPlanner "shrugs off" failure and moves on.

PMU = average Plan Marginal Utility per unit time
(difference in utility accrued per unit time between having and not having a plan)

PSI = average Plan Soundness Interval
(length of time over which a plan will prove useful)

PAU = average Plan Achievable Utility per unit time
(utility accrued per unit time from current repertoire of plans)

PGI = average Plan Generation Interval
(length of time required to generate a plan)

$$EUD = (PMU * PSI) - (PAU * PGI) \qquad \text{Eq. 1}$$

The parameters in Eq. 1 can be associated with the three dimensions of failure attribution in such a way that increasing stability, globality, or internality correspond to increasing EUD and likelihood of depression. The more stable an environment, the longer a period of time over which the benefits of a precomputed plan can be amortized. Therefore stability is associated with the PSI parameter. The more globally important an environmental change, the larger the number of important affected goals. Both PMU and PAU can be associated with globality (PMU positively, and PAU negatively) because they both vary with the total utility affected (in PMU's case) or unaffected (in PAU's case). As for internality, an internal failure attribution is one in which we believe that we could have avoided failure if we had pursued an alternative course of action. In DEPlanner's case an alternative course of action is an alternative plan. The less effort or time required to generate an alternative plan, the more internal should be the attribution. Conversely, impossible plans, which have infinitely long generation times, should result in large PGI's. Therefore internality is mapped to 1.0 / PGI.

When DEPlanner's depression ensues, five more signature phenomena appear: the cognitive loop of failure-related rumination, lowered self-esteem and self-efficacy, and increased negative generalization and depressive realism.

Depressed people experience a cognitive loop of failure-related thoughts that distract them from normally enjoyable activities (Ingram, 1984). During the cognitive loop people think about past failures, construct possible explanations, and consider future implications. When the cognitive loop recedes, people often make important changes in their lives (such as changing their expectations or focusing on different goals). In DEPlanner the cognitive loop consists of retrieving failures from a memory of past problem-solving episodes (accounting for William et al.'s (1988, p. 183) "strategic access to previous problem-solving attempts."), searching for failure explanations (accounting for Taylor's (1989, p. 225) "accurate assessment of his or her capabilities"), and generating new plans under new assumptions. Thus, DEPlanner's analog of the cognitive loop is at the core of DEPlanner's adaptive ability.

William James defined self-esteem to be total successes divided by total pretensions (James, 1890). In DEPlanner's case, self-esteem is the total utility of achievable goals divided by the total utility of all goals in episodic memory. Self-esteem ranges from 0.0 (no goals are possible) to 1.0 (all goals are possible).

Self-efficacy, roughly the subjective probability that a goal can be accomplished, is thought to drop during depression (Rehm, 1988). In DEPlanner self-efficacy is defined in a similar fashion to self-esteem, except that utility is not taken into account: total number of attainable goals divided by total number of goals.

People who are vulnerable to depression are more likely to arrive at negative self-deprecating generalizations in response to small setbacks (Beck et al. 1979). For example a mother whose child complains of a cold breakfast may decide she is a bad mother, rather than she is not a perfect cook. DEPlanner's assumptions, about which block can be stacked on which block or location, are indexed in a hierarchy, with more general assumptions toward the top, and more specific assumptions toward the bottom. (Retracting (ASSUME (STACKONABLE B ?X)) is more general than retracting (ASSUME (STACKONABLE B A)) because retracting (ASSUME (STACKONABLE B ?X)) effectively retracts all assumptions beneath it in the hierarchy.) Given a set of successful and failing past problem-solving attempts, DEPlanner finds positions in the generalization hierarchy that predict the pattern of successes and failures. The more general a failure attribution, the more "ASSUME" preconditions are unsatisfied, and the longer DEPlanner spends precomputing new plans. Thus DEPlanner's behavior is consistent with the correlation between human negative generalization and depression. DEPlanner's measure of negative generalization is the number of retracted assumptions divided by the total number of possible assumptions.

The depressive realism phenomenon is particularly problematic for other theories of depression. In some ways, mildly depressed people appear to be more accurate information processors than nondepressed people (Alloy & Abramson, 1988). For example, nondepressed people seem to over-estimate their chances of success at a variety of tasks, while depressed people are relatively more accurate. A simple measure of DEPlanner's depressive realism is total actual goal utility (according to correct assumptions available in the problem generator) divided by total predicted goal utility (according to DEPlanner's own assumptions and plans). During depression, DEPlanner's measure of depressive realism increases because DEPlanner's assumptions become more accurate.

Comparing DEPlanner's Behavior to Signature Phenomena of Depression

Cognitive modellers aspire to detailed statistical comparisons between the behavior of their simulations and human subjects. However many simulations, especially those that are the first to model a psychological process, do well to qualitatively match a set of signature behaviors obtained from the research literature. DEPlanner is in this latter category. In order to confirm that DEPlanner behaves according to its intended design, an experiment (Fig. 4) was devised.

The value of each number, or change in value, is not important for the purpose of this qualitative assessment. It is the order and direction of change that is relevant. Each of the two decision boxes corresponds to an experimental manipulation. Box 1 corresponds to the presence or absence of a stable, global, and internal environmental change. Box 2 corresponds to the presence or absence of DEPlanner's analog of adaptive depression.

DEPlanner's problem generator has "knobs" that influence the frequency of change (stability), the frequency and total utility of goals affected (globality), and length of time required to construct plans (internality). In order to set these parameters, DEPlanner's assumptions, plans, and past problem-solving attempts are accessed, but not changed. The resulting statistics are used to determine which rules to change in the problem generator, and how often.

Beneath the paths leaving box 1 are tables of numbers representing stability (St), globality (Gl), and internality (In). They are higher on the upper path (see "Compare 1", Fig. 4), consistent with higher stability, globality, and internality. Here the Expected Utility of Depression (EUD) is positive and triggers DEPlanner's adaptive depression (see "Compare 2", Fig. 4). In the low stability, globality, and internality condition EUD is negative (but almost positive) and DEPlanner suppresses its depression.

Box 2 corresponds to enabling (upper path) or disabling (lower path) DEPlanner's adaptive depression. Where adaptive depression is enabled, we see the signature phenomena of decreased self-esteem (Es) and self-efficacy (Ef), and increased negative generalization (Ge) and depressive realism (Re) (see "Compare 3", Fig. 4). Self-esteem and self-efficacy

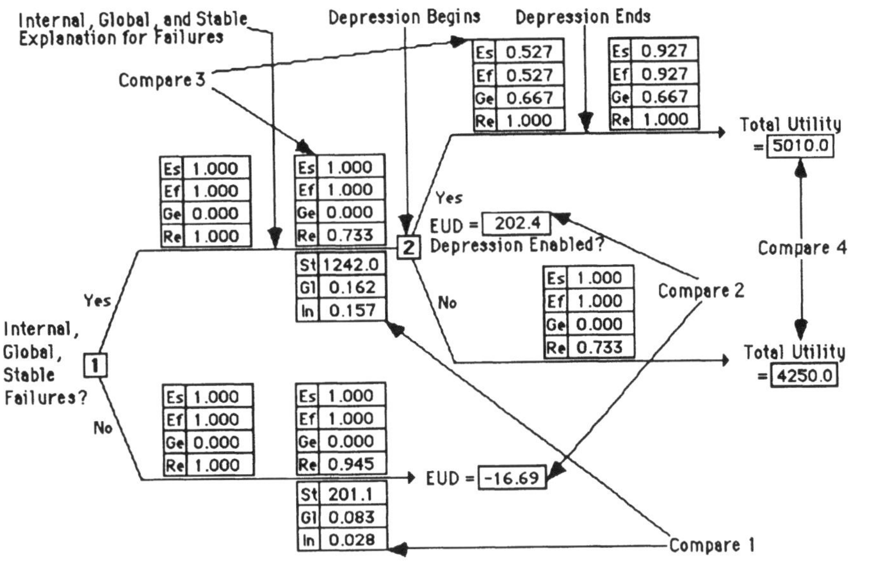

Effects of Environmental and Architectural Change on Signature Phenomena
Fig. 4

have identical values because the current version of DEPlanner's problem generator assigns the same utilities to all problem-solving attempts (5.0 for successes and -5.0 for failures). These numbers would diverge if different problems had different utilities.

Self-esteem and self-efficacy drop because the total utility of unthreatened goals drops. Negative generalization increases in this particular case because DEPlanner retracts (ASSUME (STACKONABLE B ?X)), effectively retracting a large number of assumptions. Correcting DEPlanner's assumptions increases DEPlanner's measure of depression realism.

Eventually all plans that can be precomputed, have been precomputed, and self-esteem and self-efficacy move almost, but not quite, back to their original values. They fail to reach 1.0 because some plans are impossible in light of the change in assumptions. For instance, a goal set that includes (ON B A) is impossible because (ASSUME (STACKONABLE B A) has been retracted.

In either condition, depression-enabled or depression-disabled, the DEPlanner simulation continues until the next change in the environment, whereupon the value of total utility accrued is calculated (see "Compare 4", Fig. 4). When DEPlanner is allowed to get "depressed" the total utility is 5010.0. When DEPlanner's depression is disabled the total utility is 4250.0. Thus DEPlanner achieves a 18 per cent higher total utility in the depression-enabled condition. DEPlanner exhibits analogs for signature phenomena of depression <u>and</u> shows how they may be generated by a functionally useful mechanism.

Conclusion

To become a better model of depression, DEPlanner requires redesigning. Relevant issues include:

1. A more plausible model of goal-driven problem solving and learning should replace the nonlinear planning algorithm. Computer models of knowledge acquisition and compilation in human students will be relevant (e.g., VanLehn et al. 1992).
2. The domain should be changed to reflect concerns of depressed people. Representing changing career and family circumstances will require a more expressive knowledge representation.
3. DEPlanner should be better motivated in terms of concepts like self-esteem and self-schema, as well as general theories of emotional information processing.
4. While DEPlanner is not a model of abnormally triggered, sustained, or intense depression, DEPlanner might be "broken" or "lesioned" in order to provide such an account.
5. The effects of external interventions, analogous to cognitive therapy for depression (Beck et al. 1979), should be modeled with respect to reducing likelihood, intensity, or duration of DEPlanner's depression.
6. Other kinds of cognitive dysfunction, such as Alzheimer Disease, can masquerade as depression, and visa versa (Caine, 81; Merriam et al. 1988). Perhaps DEPlanner can be broken in different ways in order to account for different but related syndromes.

7. DEPlanner uses a heuristic to modulate its cognitive loop. A sophisticated control regime should be based on an explicitly normative metareasoning approach (Horvitz et al. 1991).

8. DEPlanner should be situated with respect to existing systems in the space of cognitive architectures, such as case-based reasoners and explanation-based learners.

9. Extensive sensitivity analysis will be required to understand the complex interactions among DEPlanner's parameters, and between DEPlanner and its environment.

Nevertheless, the current version of DEPlanner is consistent with a relatively large set of signature phenomena associated with depression, and the hypothesis that some forms of depression may be adaptive. Rudimentary computational mechanisms can be assembled into a model of depression that explains a large set of previously unrelated signature phenomena.

Acknowledgements

I thank Kurt VanLehn, Greg Cooper, Gordon Banks, Johanna Moore, Richmond Thomason, Martha Pollack, and two anonymous reviewers, for helpful comments and useful criticisms. Thanks also to Steve Small for access to computing facilities at the Cognitive Modelling Laboratory.

References

Alloy, L. & Abramson, L. (1988). Depressive Realism: Four Theoretical Perspectives. In L. Alloy (Ed.), *Cognitive Processes in Depression*. New York: Guilford Press.

Bandura, A. (1977). Self-Efficacy: Toward a Unifying Theory of Behavioral Change. *Psychological Review*, 84, 191-215.

Beck, A., Rush, A., Shaw, B. & Emery, G. (1979). *Cognitive Therapy for Depression: A Treatment Manual*. New York: Guilford Press.

Caine, E. (1981). Pseudodementia: Current Concepts and Future Directions. *Archives of General Psychiatry*, 38, 1359-1364.

Chapman, D. (1987). Planning for Conjunctive Goals, *Artificial Intelligence*, 32, 333-377.

Flach, F. (1974). *The Secret Strength of Depression*. New York: Lippincott.

Gut, E. (1989). *Productive and Unproductive Depression: Success or Failure of a Vital Process*. New York: Basic Books.

Horvitz. E., Cooper, G. & Heckerman, D. (1989). Reflection and Action Under Scarce Resources: Theoretical Principles and Empirical Study. *11th Int. Journal Conference on Artificial Intelligence*.

Ingram, R. (1984). Toward an Information-Processing Analysis of Depression. *Cognitive Therapy and Research*, 8, 443-478.

James, W. (1890). *The Principles of Psychology, Vol. 1*. Henry Holt & Co.

Merriam, A., Aronson, K., Gaston, P., Wey, S., & Katz, I. (1988). The Psychiatric Symptoms of Alzheimer's Disease. *Journal of the American Geriatric Society*, 36, 7-12.

Musson, R. & Alloy, L. (1988). Depression and Self-Directed Attention. In L. Alloy (Ed.), *Cognitive Processes in Depression*. New York: Guilford Press.

Nesse, R. (1991). What is Mood for? *Psycoloquy*, Vol. 2, Issue 9.2.

Peterson, C. & Seligman, M. (1984). Causal Explanations as a Risk Factor for Depression: Theory and Evidence. *Psychological Review*, 91, 347-374.

Rehm, L. (1988). Self-Management and Cognitive Processes in Depression. In L. Alloy (Ed.), *Cognitive Processes in Depression*. New York: Guilford Press.

Rosenbloom, P., Newell, A. & Laird, J. (1991). Toward the Knowledge Level In Soar: The Role of the Architecture in the Use of Knowledge. In K. VanLehn (Ed.) *Architectures for Intelligence*, Hillsdale, New Jersey: Lawrence Erlbaum.

Taylor, S. (1989). *Positive Illusions: Creative Self-Deception and the Healthy Mind*. Basic Books.

VanLehn, K., Jones, R., & Chi, M. (1992). A Model of the Self-Explanation Effect. In M. Posner (Ed.), *Journal of the Learning Sciences*, 2, 1, 1-59.

Webster, C., Glass, R. & Banks, G. (1988). A Computational Model of Reactive Depression. *Proc. 10th Annual Conf. Cognitive Science Society*, Montreal, Canada: Lawrence Erlbaum.

Williams, J., Watts, F., MacLeod, C. & Mathews, A. (1988). *Cognitive Psychology and Emotional Disorders*. New York: John Wiley & Sons.

Reasoning about Performance Intentions

Michael Freed, Bruce Krulwich, Lawrence Birnbaum, and **Gregg Collins**
Northwestern University, The Institute for the Learning Sciences
1890 Maple Avenue; Evanston, Illinois 60201
Electronic mail: {freed,krulwich,birnbaum,collins}@ils.nwu.edu

Introduction

For an agent to find and repair the faults that underly a planning failure, it must be able to reason about the intended behavior of its planning and decision-making mechanisms. Representations of intended decision-making behaviors, which we refer to as *intentions*, provide a basis for generating testable hypotheses about the source of a failure in the absence of complete information about its cause. Moreover, intentions provide measures by which beneficial modifications to cognitive machinery can be differentiated from harmful or useless ones. This paper presents several examples of these intentions and discusses how they may be used to extend the range of circumstances in which agents can learn.

Since our claim concerns the representations needed to learn from failure, we describe in section 2 a situation in which an agent should be able to learn, and discuss some of the obstacles to doing so. In section 3, we present the idea of *intentions* and show, using the example from the previous section, how they may be used to extend the range of circumstances in which machines can learn. Section 4 presents an implementation of our theory of failure-driven learning and discusses the role of intentions in several stages of the learning process. Finally, in section 5, we relate our claim to the work of other researchers and summarize the argument that the intended behaviors of decision-making mechanisms must be represented explicitly in order to learn from failure.

An every-day example

Consider the situation of a novice driver attempting to traverse a crowded and confusing road feature such as the rotary shown in figure 1. Lacking experience, the novice may end up waiting longer than more experienced drivers before entering the flow of traffic. It is not unreasonable to suppose that the novice notices the undesirably long wait, perhaps with the assistance of impatient drivers waiting behind. The situation implicates some shortcoming in the novice's driving skill, and so warrants an attempt to learn some improvement.

What should be learned?

Drivers can identify *collision-threats* such as road obstacles and moving vehicles, and must take them into account in deciding how and when to enter traffic. For instance, if at some moment traffic is heavy and moving quickly, a driver may notice many collision threats and choose not to enter. Alternately, s/he may employ some plan that avoids or neutralizes all of these threats, and proceed to enter traffic.

For the task of entering the flow of traffic, the category *collision-threat* is useful for constraining the set of plans which may be safely employed. However, the novice in our example may have been employing an overbroad definition of *collision-threat*. As a result, the set of seemingly safe plans will be overconstrained, thereby increasing the average amount of time before some plan seems safe. It is reasonable to blame the category definition for overconstraining the set of plans which can be used to enter traffic safely, thereby preventing timely action. Learning from the failure thus means narrowing the faulty definition of *collision-threat*.

There may of course be many ways in which the definition of collision-threat could be usefully narrowed. For instance, a novice may learn that other drivers tend to stay in their own lanes between exits; thus, the possibility that a vehicle will make a sudden and inexplicable lane change should not be considered a collision-threat.

Tracing the delay to its underlying cause

In our example, the novice's failure to expeditiously enter the flow of traffic stems from a failure to generate a safe plan for entry. This failure stems, in turn,

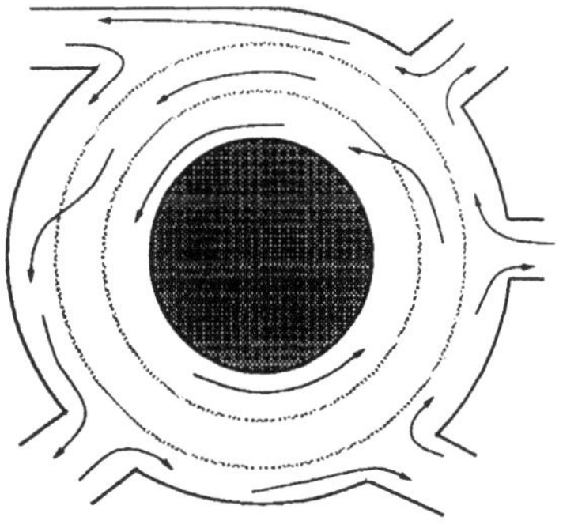

Figure 1: Flow of traffic on a rotary

from having noticed more threats than the novice could plan to counter or avoid. The novice can get better at entering traffic by learning to more accurately differentiate between threats and non-threats.

One way that this diagnosis could proceed is as follows: Suppose the novice receives feedback indicating that a particular perceived threat was not in fact a threat over some time interval. The circumstances during this interval could then be reviewed to show that a safe plan to enter the rotary would have been discovered had the misperceived threat (a "near miss") been ignored. Finally, the misperceived threat could be used to narrow the collision-threat category definition.

The preceding process is unrealistic for several reasons. First, it assumes that the agent can receive specific feedback regarding a *decision* that was made incorrectly, as opposed to an *action* that was incorrectly taken. Moreover, since the novice had no way to know *a priori* which of its decisions should be monitored, it would require that the novice receive continuous feedback on each of its decisions. Finally, it is highly implausible that the novice will be able to receive detailed feedback about particular perceived threats, since novices typically lack the knowledge necessary to evaluate such situations post-hoc [Fitts, 1964; Starkes and Deakin, 1985].

A more realistic diagnostic process begins as the driver of another car honks his horn and thereby leads the novice to question why no safe traffic entry plans have been generated. Hypotheses are developed as to why this might be the case, including for example that either the planning mechanism is inadequate, or its set of collision-threats is faulty. These hypotheses enable the novice to seek specific feedback through experimentation or advice.

Reasoning about agent intentions

As previously argued in [Collins *et al.*, 1991], it is is useful to divide a planner into ***components***, each responsible for a task-independent function such as detecting threats or selecting among competing plans. The effectiveness of a component depends on whether it performs its function quickly, how reliably it attends to relevant input items, whether it avoids pathological interactions with other components, and so on. We call these measures of component effectiveness ***intentions***.

When an agent implemented as a component architecture learns, one or more modifications are made to its constituent components. Component intentions (see figure 2 for examples) are a measure of the value of a modification. For instance, a modification is valuable if it helps avoid pathological interactions between components and harmful to the extent that it aggravates such interactions or introduces new ones. Because the choice of what to learn depends on factors measured by intentions, intentions should be explicitly represented so that learning processes can reason about them. To understand this point, consider again the

Within components	Inter-component
No false positives	Don't flood other components
No false negatives	Don't monopolize resources
Efficient computation	Don't reproduce computation
Output values within acceptable ranges	Don't focus on areas of ultimate irrelevance
	Don't be a bottleneck

Figure 2: Sample planner component intentions

rotary example of section .

Recall that the principal difficulty in speeding traffic-entry performance lay in locating a sample misperceived threat (a near miss) on which to base a refinement of the collision threat classification. The most effective solution was apparently to hypothesize that the category is overbroad and to seek out specific feedback on future threat-detection performance through experimentation or advice. In the absence of an instance of a misclassification, this solution requires some other basis for formulating the fault hypothesis. Intentions provide this alternate basis.

Consider the component intention: ***avoid flooding another component with output***. The negation of this intention represents a situation in which a component is producing too much output; this in turn indicates that some output-definer (category) may be too broad and that fixing the problem requires locating a false positive with which to narrow the category.

A second role of intentions is to evaluate candidate component modification for preventing failure recurrence. In our example, drastically narrowing the collision threat category (so that no collision threats are generated) would solve the problem of flooding the traffic-entry planner with output, but would lead to catastrophically faulty plans. The basis for rejecting this candidate modification is the intention that the threat-detection component's output be free of false-positives.

Implementation and second example

In this section, we describe a system, CASTLE[1], which implements important aspects of our theory. The system, which operates in the domain of chess, detects situations that are contrary to its expectations, and responds to these expectation failures by repairing the faulty planner components which were responsible for the failure. We view this learning as a knowledge-based process, in which the system uses knowledge of its own planning components to learn from events which led to expectation failures. More specifically, the system must reason along different dimensions of intentionality to determine what repairs should be made to its planning

[1] CASTLE stands for *C*oncocting *A*bstract *S*trategies *T*hrough *L*earning from *E*xpectation-failures.

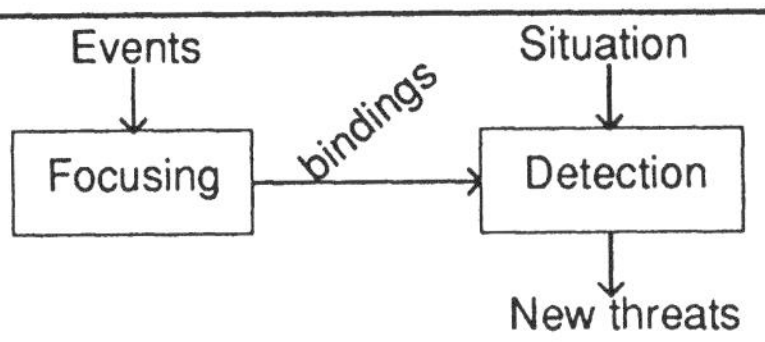

Figure 3: Incremental threat detection

rules.

The knowledge necessary for this repair process is expressed in the form of a planner self-model, which is used to diagnose and repair expectation failures [Davis, 1984; deKleer and Williams, 1987; Simmons, 1988]. More specifically, when the expectation fails, the system first examines an explicit *justification structure* which encodes the the reasoning which let to its belief in the incorrect expectation [deKleer *et al.*, 1977; Doyle, 1979]. This justification is used to isolate the faulty components of its architecture, each of which implements a particular sub-task in the decision-making process [Collins *et al.*, 1991; Krulwich, 1991]. It then uses a *specification* of the faulty components to guide the learning of new rules in response to the failure [Krulwich, 1992]. Each of these information sources must explicitly reference the planner's intentions.

Detection focusing

A central cognitive task in which CASTLE engages is that of noticing threats and opportunities as they become available [Collins *et al.*, 1991]. Rather than recomputing these at each turn, CASTLE maintains a set of active threats and opportunities that is updated over time. To accomplish this incremental threat detection, the system uses a *detection focusing* component, which consists of *focus rules* that specify the areas in which new threats *may have* been enabled. Then, a separate *threat detection* component, consisting of rules for noticing specific types of threats, detects the threats that have in fact been enabled. This relationship between the two components is shown pictorially in figure 3. A sample focus rule is shown in figure 4. This rule embodies the system's knowledge that the most recently moved piece, in its new location, may be a source of new threats. Another focus rule, not shown, specifies that the more recently moved piece can also be a target of newly enabled attacks. Using focus rules such as these, the actual threat detector rules will only be invoked on areas of the board which can potentially contain new threats.

Focusing intentions

What intentions does the system have regarding its detection focusing component? The primary intention that the system has is that there not be any newly enabled threats that are not within the scope of

```
(def-brule focus-new-source
 (focus focus-moved-piece ?player
   (move ?player ?move-type ?piece ?loc1 ?loc2)
   (world-at-time ?time))
<=
 (move-to-make (move ?player ?prev-move-type
                     ?piece ?old-loc ?loc1)
               ?player ?goal (1- ?time)) )
```

Figure 4: Focusing on new moves by a moved piece

the bindings generated by the focusing component. This condition is clearly necessary for the incremental detection scheme to work. A more subtle intention is that the focusing component not generate *too many* bindings in which threats *do not* exist. If this intention is not met, the detection component will be invoked more than is necessary, and in the extreme case the entire point of the detection focusing is lost. Clearly the savings gained by only applying the threat detection rules in constrained ways (and not over the entire board) must be greater than the cost of applying the focusing rules. This will not be the case if the constraints given by the focusing component are too weak. It will also not be the the case if the computational cost of applying the focusing component is too high.

Another planner intention regarding the focusing component is that the division of the tasks shown in figure 3 be enforced. This means that the system should not incorporate information about different types of threats into the focusing rules.

Discovered attacks

To see how CASTLE uses representations of planner intentions in learning, let's first see an example of CASTLE enforcing its simplest intention, that there be no false negatives of its focusing component. Consider, in particular, the example of *discovered attacks* in chess, in which the movement of one piece opens a line of attack for another piece. Novices often fall prey to such attacks, not because they fail to understand the mechanism of the threat (i.e., the way in which the piece can move to

Intention	Application to focusing
No false negatives	Don't let a threat be enabled without detectors being invoked
No false positives	Detectors not over-applied
Efficiency	Incremental scheme shouldn't be less efficient than brute-force
No redundancy	Don't encode information about specific threats in focus rules

Figure 5: Planner intentions in detection focusing

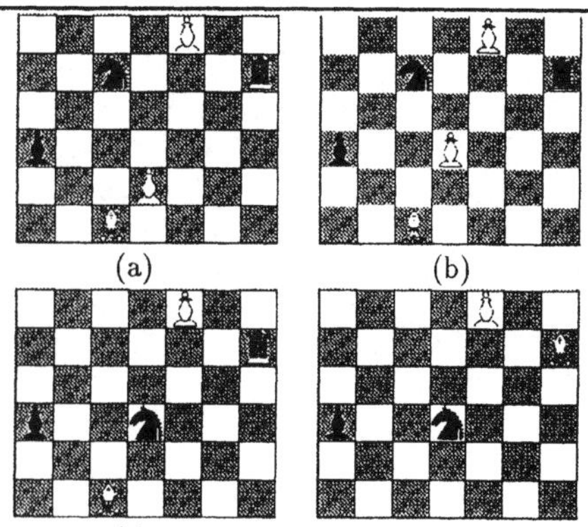

Figure 6: Example: Opponent (white) to move

```
(def-brule learned-focus-method25
  (focus learned-focus-method25 ?player
     (move ?player (capture ?taken-piece)
       ?taking-piece (loc  ?row1 ?col1)
       (loc  ?row2 ?col2)) (world-at-time ?time2))
<=
  (and (move-to-make
          (move ?other-player move ?interm-piece
                (loc ?r-interm ?c-interm)
                (loc ?r-other ?c-other))
          ?player ?goal ?time1)
       (loc-on-line ?r-interm ?c-interm
                    ?row1 ?col1 ?row2 ?col2)
       (at-loc ?player ?taking-piece
               (loc ?row1 ?col1)
               (- gen-time2.24 2)) ))
```

Figure 7: Learned focus rule for *discovered attacks*

make the capture), but rather because they simply fail to consider new threats arising from pieces other than the one just moved. The same is true of CASTLE if it is equipped only with the two focus rules described above.

The example in figure 6 shows the system falling prey to a discovered attack due to its lack of a necessary detection focusing rule. In the situation shown in figure 6(a), the opponent advances its pawn and thereby enables an attack by its bishop on the computer's rook. When the system updates its set of active threats and opportunities, its threat focusing rules will enable it to detect its own ability to attack the opponent's pawn, but it will not detect the threat to its rook. Because of this, when faced with the situation in figure 6(b), the computer will capture the opponent's pawn instead of rescuing its own rook, and it will expect that the opponent's response will be to execute the attack which it believes to the only one available, namely to capture the computer's pawn. Then, in the situation shown in figure 6(c), when the opponent captures the computer's rook, the system has the task of diagnosing and learning from its failure to detect the threat which the opponent executed.

Learning from the failure

To diagnose the failure, CASTLE examines an explicit *justification structure* [deKleer *et al.*, 1977; Doyle, 1979], which record how the planner's expectation was inferred from the rules that constitute its decision-making mechanisms, in conjunction with the policies and underlying assumptions which it has adopted. Diagnosing the failure then involves "backing up" through the justification structure, recursively explaining the failure in terms of faulty rule antecedents [Smith *et al.*, 1985; Simmons, 1988; Birnbaum *et al.*, 1990; Collins *et al.*, 1991]. This diagnosis process will "bottom out" by faulting either an incorrect planner rule or an incorrect assumption that underlies the planning mechanism. In our example, the fault lies in an assumption that the planner could enforce its first intention regarding its focusing component, that it would generate bindings for all enabled threats. CASTLE concludes from this that its set of focusing rules is incomplete and must be augmented.

To construct the new rule, CASTLE retrieves a *component performance specification* for each component. These performance specifications, a form of planner self-knowledge, describe the correct behavior of each component. The specification of the detection focusing component says roughly that *the focusing component will generate bindings that include any capture that is enabled by a given move.* This specification enables CASTLE to focus on the details of the example that are relevant to the component being repaired, by serving as an explanation-based learning target concept [Krulwich, 1991; Krulwich, 1992]. After retrieving the specification, CASTLE invokes its deductive inference engine to construct an explanation of why the possible capture of the rook should have been in the set of constraints generated by the focusing component. This explanation says roughly that the opponent's move should have been generated by the focusing component, *because* the opponent's previous move enabled the attack, *because* it was on a square between the bishop and the rook, *and* there were no other pieces along the line of attack, *and* emptying the line of attack is an enabling condition for the capture to be made. CASTLE then uses explanation-based learning techniques [Mitchell *et al.*, 1986; DeJong and Mooney, 1986] to generalize this explanation and to construct a new detection focusing rule shown in figure 7.

Back to intentionality

In the example of learning *discovered attacks*, the system is able to correct the failure of the detection focusing component to generate bindings that included the new attack. The learning process that we described involves the construction of a new rule to enforce the intention that the focus component not generate any false negatives. Suppose, however, that instead of adding the rule shown in figure 7, the system's learning component added a focusing rule that returned completely unconstrained bindings. This would cause the detection rules to be applied to all board positions in computing the newly enabled threats. This would, of course, enforce the system's intention to have no false negatives, and would also satisfy the component specification, because all threats that could possibly be enabled are by definition within the unconstrained bindings. However, the whole purpose of the incremental threat detection scheme (figure 3) would be undermined, because not only will the system apply the threat detection rules over the entire board, but it will then proceed to integrate the threats that it finds into the set of previously available threats, which is clearly a waste of time. In short, this is a violation of the system's *no-false-positives* intention.

Unfortunately, while the violation of the no-false-negatives intention could be easily noticed by observing an enabled threat that was not in the focus bindings, it is much more difficult to notice the failure of the no-false-positives intention. In our example, after the opponent moves his pawn in figure 6(b), three classes of bindings constraints should be generated: *threats by the moved piece at its new location,* which are generated by the rule in figure 4, *threats against the moved piece at its new location,* and *threats through the square vacated by the moved piece,* generated by the learned discovered attacks rule in figure 7. Two of these in fact reflect new threats that have been enabled, but one of them, threats by the moved piece, in fact do not reflect any new threats.

We can see that this must be the case, because the division of labor between the focusing and detection components requires that no information about the types of threats themselves be present in the focus rules. Since the focusing component is only determining *where* to look for new threats, it is clear that there will be times when a correct place to look for new threats will not in fact contain any.

This complicates the problem of detecting false positives, since there is no direct test to determine whether the focus component was generating false positives. One approach would be for this intention to only come into play when the system is learning new focus rules. The no-false-positives intention could be used to force the learning mechanism to generate the most specific possible rule. Of course, this approach does not allow the system to reason explicitly about this intention.

Another approach would be to have the system generate expectations about the performance of its component that do not relate directly to false positives, but that are good indicators of the system's false positive rate. As we discussed above, the intention not to have any false positives is in service of system efficiency, because the design of the incremental threat detection scheme is based on the focusing component's sufficiently narrowing down the scope of the detection rule application. It follows from this that the system could monitor the computational effort spent on the detection focusing and compare it with the savings in threat detector application. If this tradeoff turned out not to be worthwhile, the system could examine its false positive rate in more detail. This is similar to the example in which the driver was unable to enter the intersection, causing him to examine his detection mechanism for sources of false threats. This requires that the system be able to make utility judgements about different tradeoffs between false positives, false negatives, and efficiency.

A similar learning process could be invoked if the system noticed that too much time was being spent considering pointless opportunities. This could arise if the system found that it was spending too much time considering pawn captures that were always being discarded by the plan selection component. If this were the case, the system could infer that its focusing component should be further constrained not to generate bindings for captures of pawns.

Discussion

We have shown that reasoning about the faults underlying a planning failure requires that an agent explicitly represent performance intentions which describe the desired behavior of its components. When one of its components is faulty, the agent must reason explicitly about its intentions to diagnose the failure and make a repair which is to its overall benefit.

This paper presents several examples of single-component intentions, such as completeness, soundness, and efficiency, as well as intercomponent intentions such as avoiding flooding and competition for global resources. We have discussed aspects of the learning process which require explicit reasoning about these intentions, thereby extending the range of concepts an agent can learn, and allowing it to learn in circumstances in which it could not otherwise learn. This work thus builds on previous research in failure-driven acquisition of new planning knowledge [Hammond, 1989; Birnbaum *et al.*, 1990; Collins *et al.*, 1991].

Previous research has dealt with several of the issues we have discussed. Minton [1988] discussed the need for learned planner rules to be sensitive to the global efficiency of the system. Our work builds on this idea by explicitly modeling and a variety of such intentions. Hunter's system [1989] reasoned about shortcomings in its diagnostic knowledge and explicitly modeled the intentions involved in that task to guide learning. Similarly, Cox and Ram [1991] have modeled several intentions of the case retrieval process for use in the task of understanding. Our research extends these ideas to model intentions for a more general problem solver, as well as modeling the intercomponent intentions. Others have used representations of the system's intentions for planning [Jones, 1991] and understanding [Ram, 1989].

Our previous research has involved extending our model of planning and decision-making to include a variety of tasks and components, all in the domain of competitive games. To date we have developed models of threat detection, counterplanning, schema application, goal regression, lookahead search, and execution scheduling. Future research will elucidate the breadth of planner intentions, and will demonstrate the benefits of explicitly representing them for use in learning.

Acknowledgements: Thanks to Matt Brand, Bill Ferguson, Eric Jones, and Louise Pryor for many discussions on the research presented here. This work was supported in part by the Air Force Office of Scientific Research under grant number AFOSR-91-0341-DEF, and by the Defense Advanced Research Projects Agency, monitored by the Office of Naval Research under contract N-00014-91-J-4092. The Institute for the Learning Sciences was established in 1989 with the support of Andersen Consulting, part of The Arthur Andersen Worldwide Organization. The Institute receives additional support from Ameritech, an Institute Partner, and from IBM.

References

[Birnbaum *et al.*, 1990] L. Birnbaum, G. Collins, M. Freed, and B. Krulwich. Model-based diagnosis of planning failures. In *Proceedings of the Eighth National Conference on Artificial Intelligence*, pages 318–323, Boston, MA, 1990.

[Collins *et al.*, 1991] G. Collins, L. Birnbaum, B. Krulwich, and M. Freed. Plan debugging in an intentional system. In *Proceedings of the Twelfth International Joint Conference on Artificial Intelligence*, pages 353–358, Sydney, Australia, 1991.

[Davis, 1984] R. Davis. Diagnostic reasoning based on structure and function: Paths of interaction and the locality principle. *Artificial Intelligence*, 24(1-3):347–410, 1984.

[DeJong and Mooney, 1986] G. DeJong and R. Mooney. Explanation-based learning: An alternative view. *Machine Learning*, 1:145–176, January 1986.

[deKleer and Williams, 1987] J. deKleer and B.C. Williams. Diagnosing multiple faults. *Artificial Intelligence*, 32(1):97–129, April 1987.

[deKleer *et al.*, 1977] J. deKleer, J. Doyle, G.L. Steele, and G.J. Sussman. Explicit control of reasoning. *SIGPLAN Notices*, 12(8), 1977.

[Doyle, 1979] J. Doyle. A truth maintenance system. *Artificial Intelligence*, 12(3):231–272, 1979.

[Fitts, 1964] P.M. Fitts. Perceptual-motor skill learning. In *Categories of human learning*. Academic Press, New York, 1964.

[Freed, 1991] M. Freed. Learning strategic concepts from experience: A seven-stage process. In *Proceedings of the Thirteenth Annual Conference of The Cognitive Science Society*, pages 132–136, Chicago, IL, 1991.

[Hammond, 1989] K. Hammond. *Case-based planning: Viewing planning as a memory task*. Academic Press, San Diego, CA, 1989.

[Hunter, 1989] L. Hunter. *Knowledge-acquisition planning: Gaining expertise through experience*. PhD thesis, Yale University, 1989.

[Jones, 1991] E. Jones. *The flexible use of abstract knowledge in planning*. PhD thesis, Yale University, 1991.

[Krulwich, 1991] B. Krulwich. Determining what to learn in a multi-component planning system. In *Proceedings of the Thirteenth Annual Conference of The Cognitive Science Society*, pages 102–107, Chicago, IL, 1991.

[Krulwich, 1992] B. Krulwich. *Learning New Methods for Multiple Cognitive Tasks*. PhD thesis, Northwestern University, Institute for the Learning Sciences, 1992. (in preparation).

[Minton, 1988] S. Minton. *Learning Effective Search Control Knowledge: An Explanation-Based Approach*. PhD thesis, Carnegie Mellon University, 1988.

[Mitchell *et al.*, 1986] T.M. Mitchell, R.M. Keller, and S.T. Kedar-Cabelli. Explanation-based generalization: A unifying view. *Machine Learning*, 1(1), January 1986.

[Ram and Cox, 1991] A. Ram and M. Cox. Using introspective reasoning to select learning strategies. In *Proceedings of the First International Workshop on Multistrategy Learning*, 1991.

[Ram, 1989] A. Ram. *Question-driven Understanding: An integrated theory of story understanding, memory, and learning*. PhD thesis, Yale University, 1989.

[Simmons, 1988] R.G. Simmons. A theory of debugging plans and interpretations. In *Proceedings of the Seventh National Conference on Artificial Intelligence*, St. Paul, MN., 1988.

[Smith *et al.*, 1985] Reid G. Smith, H.A. Winston, T. Mitchell, and B.G. Buchanan. Representation and use of explicit justifications for knowledge base refinement. In *Proc. IJCAI-85*, Los Angeles, August 1985.

[Starkes and Deakin, 1985] J.L. Starkes and J.M. Deakin. Perception in sport: A cognitive approach to skilled performance. In *Cognitive Sport Psychology*. Sports Science Associates, Lansing, N.Y., 1985.

Preconditions and appropriateness conditions*

Timothy M. Converse and Kristian J. Hammond
Artificial Intelligence Laboratory
Computer Science Department
University of Chicago
1100 E. 58th St.
Chicago, IL 60637
(312) 702-8584
converse@cs.uchicago.edu

Abstract

Classical plan preconditions implicitly play a dual role, both documenting the facts necessary for a plan to be sound and listing the conditions under which it should be used. As the closed-world assumption is relaxed these two roles begin to diverge, particularly when attempts are made to use plans in situations other than those for which they were originally constructed. Rosenschein and Kaelbling exploit one aspect of the divergence by suggesting that some logical preconditions can be considered in the design phase of building an agent, but "compiled away" so that the agent need not explicitly consider them [Rosenschein and Kaelbling, 1986]. We suggest an alternative view whereby an agent can explicitly reason and learn about which conditions are the best cues for employing standard plans, and discuss the idea in the context of the RUNNER project.

Introduction

Plan execution and the role of preconditions

The closed-world assumption of classical planning allowed the assumption that plans could have a small and explicit set of preconditions which, if true, would ensure that the plan worked. The most straightforward application of this idea to the execution of such plans is that an executor should know or verify the truth of the preconditions before starting the plan. This is particularly unproblematic if, in addition to a closed world assumption, the plan is being generated for exactly the circumstances in which it is to be used; in that case, everything that is known about the state of the world can be taken into account during the synthesis of the plan, and the process of construction can (given the closed-world assumption) itself guarantee the soundness of the plan.

In recent years, greater awareness of the intractability of generative planning [Chapman, 1985], coupled with greater concern about time pressure in activity, has led to attempts to amortize the cost of planning over repeated instances of activity, either by "pre-compiling" action decisions [Rosenschein and Kaelbling, 1986, Drummond, 1989], or by re-using the fruits of previous planning attempts [Hammond, 1989].

At the same time it has been widely acknowledged that the set of logical preconditions for plans in many real-world situations is effectively infinite. This been called the "qualification problem" (defined variously in [McCarthy, 1977, Shoam, 1986, Ginsberg and Smith, 1987]). That is, given any attempt at enumeration of logical statements that need to be true for a given plan to be guaranteed to work, it is usually possible to come up with an additional potential fact that would render the plan unworkable.

Early planning research tried to confront this directly using large numbers of frame axioms. Most generative planning systems since [Fikes and Nilsson, 1971] have used the more optimistic and tractable STRIPS assumption that primitive actions can have associated lists of the facts that are changed by applying them.

Precondition sets for classical plans implicitly play a dual role. They

1. describe the initial conditions under which the plan as described can be expected to be sound

*This work was supported in part by AFOSR grant number AFOSR-91-0112, DARPA contract number F30602-91-C-0028 monitored by Rome Laboratories, DARPA contract number N00014-91-J-4092 monitored by the Office of Naval Research, Office of Naval Research grant number N00014-91-J-1185

(under certain assumptions, and because the facts were used in the plan's construction), and

2. describe the facts that an executor should know to be true before beginning execution of the plan.

Let us keep the existing term of *precondition* for the first sort of fact above, and use the term *appropriateness condition* for the second sort.

Appropriateness conditions

Under assumptions of perfect knowledge and a closed world, there is little divergence between preconditions and appropriateness conditions. When these assumptions are relaxed, however, there are several different ways in which the divergence can become important in plan execution and reuse:

- A precondition can effectively be "always true". This means that the plan may depend upon it for correctness, but an executor will never run into trouble by not worrying about its truth value. This sort of fact should not be an "appropriateness condition", since consideration of it cannot help in the decision whether to use the plan.
- A precondition may be almost always true, and it may be difficult to know or check in advance. If the consequences of an abortive attempt at performing the plan are not too severe, then this sort of fact should not be an appropriateness condition, since the utility of knowing its truth is outweighed by the cost of acquiring the knowledge.
- A precondition may be intermittently true, but may be easily "subgoaled on" in execution, and achieved if false. (This of course depends strongly on representation of plans, and how flexible the execution is.) To the extent this can be handled in "execution", the condition should not be an appropriateness condition, since whether or not the condition holds the plan is likely to succeed.
- A particular condition may not be a precondition *per se*, but may be evidence that the plan will be particularly easy to perform, or will produce results that are preferable to the usual default plan for the relevant goals. This *should* be an appropriateness condition, even though it is not a precondition.

Action nets and appropriateness conditions

Rosenschein and Kaelbling noted the first possibility in the above list, that some preconditions might be "always true", and realized that, while such facts may need to be explicitly considered in the design of an agent for some domain and task, there is no reason why the agent itself need consider them. [Rosenschein and Kaelbling, 1986]. Such facts can essentially be "compiled away" in the design of an agent that will behave appropriately.

To summarize the argument so far:

- The set of facts that an agent should consider before embarking on a given plan is interestingly different from both the (possibly infinite) set of facts that need to be true for the plan to work, and the set of facts explicitly used in the plan's construction. This is true particularly when *de novo* plan construction is impossible or too costly, and plans must be reused.
- One possible approach that recognizes this is to explicitly design an agent so that it only considers the conditions that are actually relevant for action, either by hand-crafting its decision procedure, or by a mixture of hand-crafting and clever compilation of declarative specifications, as in Rosenschein and Kaelbling's work.

And the point we want to make (which will occupy us for the rest of the paper):

There is a large potential middle ground between an approach that requires explicit reasoning about all preconditions on the one hand, and approaches that compile in any needed reasoning of that sort in advance. In particular, even if an agent is assumed to have a largely immutable library of plans and behaviors that will determine its competence, there is still room for learning the particular appropriateness conditions that govern when to invoke particular plans.

An example

To make these distinctions clearer, let's look at a common sense example: the task of making buttered toast, in a well-equipped kitchen, with an electric toaster.

If we start to enumerate the preconditions that we can think of that are associated with this task, the most available ones have to do with the resources we would normally worry about it conjunction with it: possessing bread, possessing butter. Others that come to mind might have to do with available time, instruments (the toaster, a knife), or knowledge about these things (do we know where to find a knife?). As we strain to think of things that are not part of the concerns associated with the plan, we might think of possible "external" problems like an interruption in electric service. Finally, imaginable "preconditions" start to be explicitly counterfactual; what if gravity no longer operated, or heat conduction worked in a different way?

In practice, when deciding whether to make toast, one is probably aware of only the first few considerations, and possibly none are a concern. The technical proposal we would like to make here is that we should trust the introspective availability of these conditions: there is a small set of facts that should be explicitly considered when deciding whether to embark on a given plan. In addition, we argue that these appropriateness conditions should be stored in association with the plan itself, and that learning to refine them is a significant part of the development of expertise in plan use.

The utility of partial plan completion

Agre and Chapman [Agre and Chapman, 1988] have argued against the need by action systems for "plans" in the classical sense, and have pointed out the dangers of confusing the technical and commonsense meanings of the term. The Pengi system [Agre and Chapman, 1987] demonstrated a surprising capacity to exhibit "planful" behavior without explicit representation of plans. What this means is that sequences of actions that were particularly beneficial could be triggered by successive perceptual conditions, without the sequence itself being represented internally in any way.

A good analysis of the sources of domain support for Pengi's behavior can be found in [Chapman, 1990]. An additional one we would like to suggest is that, although the domain rewards successful completion of certain sequences of actions, it does not particularly penalize partial completion. The domain we are investigating in the RUNNER project [Hammond *et al.*, 1990] (performing simple tasks in a simulated kitchen) has the characteristic that many tasks will leave the agent in worse shape if partially completed than if the task had never been started. For example, many tasks require that milk be taken out of the refrigerator, but if that is all that is done, the only result will be sour milk. Note that this contrast is at the level of *domain* analysis, and is orthogonal to the question of how "planful" behavior is generated. We suspect that domains that have this sort of non-additiveness of utility over the course of action sequences may require *some* sort of explicit plan representation from their agents.

So far we have used the term "plan" in a common sense way. What we would like to mean by it is this: the collection of explicitly represented knowledge that is specifically relevant to repeated satisfaction of a given set of goals, and which influences action only when a decision has been made to use the plan. In the RUNNER project this explicitly represented plan serves both as a memory organization point for annotations about the current progress and problems of the use of the plan, and as a hook on which to hang past experiences of its use. Its appropriateness conditions determine whether it should be "active" and hence make suggestions about actions, but it is not the sole determinant of behavior.

A taxonomy of attitudes toward preconditions

Given a plan that requires that certain propositions be true for it to work, there are a limited number of *attitudes* that an agent can take toward those propositions when deciding on whether to employ the plan. Whether explicitly or implicitly, given a certain precondition, the agent can

1. Assume it is always true (because it is always true).

   For example, the assumption of continued gravity in the toast example.

2. Assume it is always true (because the agent *enforces* it).

   For example, the toaster will not work if the power is not on, and the power may not be on if the electric bill is never paid. Nonetheless, most people are probably not aware of thinking of their electric bill when considering making toast. This is because other plans and habits ensure that it is always paid, and therefore electricity does not figure into the appropriateness conditions for making toast. Depending on the extent to which larder-stocking is taken care of by other plans and policies, possession of bread and butter may also be omitted. (For more discussion of the use of enforcement to simplify plan use, see [Hammond and Converse, 1991].)

3. Verify it before plan execution.

   If unbuttered toast is worse than no toast at all, and there is doubt as to the presence of butter in the refrigerator, then a good toast-making plan includes looking in the refrigerator first, and aborting the plan if none is found.

4. Know it to be true.

   "World models" should not be invoked without recognition of the cost of their construction and maintenance, and the cost of memory retrieval of a given fact (even if explicitly represented) may often outweigh the cost of active perceptual verification. Still, in our example, the question of bread and butter could be raised and then answered, for example, by memory of a recent shopping trip.

5. Subgoal on it if necessary.

   If the toaster is found to be unplugged, then it may be easy to plug back in and continue. If this particular state is unpredictable, there still may be no need to include mention of it in the appropriateness conditions for making toast.

We want to include conditions covered by items (3) and (4) above under our rubric of appropriateness conditions; they are the facts that should concern the agent in deciding on the viability of a plan.

Appropriateness Conditions in RUNNER

Our work in the RUNNER project centers around plan use in a commonsense domain. Starting from a case-based planning framework [Hammond, 1989], we believe that an appropriate view of expertise development in many domains is the acquisition and refinement of a library of plans which have been incrementally debugged and optimized for the sets of conjunctive goals that typically recur.

One of the goals of the research is to use plans *flexibly*, where flexibility means that

- Step orderings can be suggested by environmental cues. Where not explicitly constrained, multiple steps can simultaneously suggest actions.
- Particular specializations of plans are environmentally cued.
- Multiple "top-level" plans can be active and each suggest actions, resulting in (unrepresented) interleaving.

The representation structure for RUNNER 's memory, as well as the bulk of the algorithm for marker-passing and activation, is based on Charles Martin's work on the DMAP parser (see [Martin, 1989]). The memory of RUNNER 's agent is encoded in semantic nets representing its plans, goals, and current beliefs. Each node in RUNNER 's plan net has associated with it a (disjunctive) set of *concept sequences*, which are a (conjunctive) listing of states that should be detected before that plan node can be suggested.

Nodes in the plan net become activated in the following ways:

- "Top-level" plans become activated when the goal they subserve is activated, and a concept sequence indicating appropriate conditions is completed.
- Specializations of plans are activated by receiving a *permission marker* from the abstract plan, in addition to the activation of a concept sequence.
- Parts (or steps) of plans are also activated by completion of a concept sequence, and by receiving a permission marker from their parent.

Once activated, many plans have early explicit verification steps which check if other conditions necessary for success are fulfilled, and abort the plan if not.

Passing of permission markers is not recursive, so that the state information indicating an opportunity to perform a sub-plan must be recognized for execution to proceed further. This means that individual subplans must have associated with them concept sequences that indicate opportunities to be performed. (For a fuller explication, see [Hammond *et al.*, 1990].

Appropriateness conditions in RUNNER, then are the union of the concept sequences of plan nodes and the initial verification steps (if any). Since activation of a plan node is sufficient to send activation to its parts and specializations, which can in turn eventually bottom out in primitive actions, it is important that these conditions be well-matched to the conditions under which it is appropriate to invoke the plan.

Implications for Learning

On our view, a large component of expertise in complex domains is the result of the development of a library of conjunctive goal plans, with the simultaneous tuning of the plans, their appropriateness conditions, and the environment itself to maximize the effectiveness of the plans.

Our position is that the state an agent should strive for is one in which

1. its plan library has optimized plans for the different sets of goals that typically recur.
2. each plan (and subplan) has an associated set of appropriateness conditions, which are easily detectable and indicate the conditions under which the plan is appropriate to invoke.
3. when possible, standard preconditions for standard plans are *enforced*, so that they can be assumed true, and do not need to be included as as appropriateness conditions.

Part of the process of refining appropriateness conditions can be taken care of by relatively simple "recategorization" of various conditions in the taxonomy we sketched above, in response to both failure and unexpected success. Here some ways in which this sort of recategorization can be applied.

- Drop appropriateness conditions that turn out to be always true. At its simplest, this is merely a matter of keeping statistics on verification steps at the beginning of plans.
- If a plan fails because some subplan of it fails, and that subplan failed because some appropriateness condition didn't hold, then *promote* that condition to the status of an appropriateness condition for the superordinate plan. That is, make use of the larger plan contingent on finding the condition to be true.

- If a plan is frequently found to have false appropriateness conditions in situations where it is needed, and the conditions are under the agent's control, consider including it in an *enforcement* plan that maintains it, so that it can assumed true by the plan that was failing.

Conclusion

The implicit dual role of classical preconditions should be separated out, into the preconditions that underlie the plan structure and the conditions that signal appropriate use of the plans. This separation opens the door for tuning of the conditions under which an agent will use a given plan. This sort of learning, in conjunction with other methods for improving the match between an environment and a plan library, is a promising method for improving the use and reuse of plans.

Acknowledgements

Daniel Fu gave us helpful comments, as did various anonymous reviewers.

References

[Agre and Chapman, 1987] Phil Agre and David Chapman. Pengi: An implementation of a theory of activity. In *The Proceedings of the Sixth National Conference on Artificial Intelligence*, pages 268–72. AAAI, July 1987.

[Agre and Chapman, 1988] Phil Agre and David Chapman. What are plans for? Memorandum 1050, Massachusetts Institute of Technology Artificial Intelligence Laboratory, 1988.

[Chapman, 1985] David Chapman. Planning for conjunctive goals. Memo AI-802, AI Lab, MIT, 1985.

[Chapman, 1990] David Chapman. On choosing domains for agents. Position paper prepared for the Workshop on Benchmarks and Metrics, NASA Ames, 1990.

[Drummond, 1989] Mark E. Drummond. Situated control rules. In *Proceedings of Conference on Principles of Knowledge Representation*, Toronto, Canada, 1989.

[Fikes and Nilsson, 1971] R. Fikes and N.J. Nilsson. STRIPS: A new approach to the application of theorem proving to problem solving. *Artificial Intelligence*, 2:189–208, 1971.

[Ginsberg and Smith, 1987] Matthew L. Ginsberg and David E. Smith. Possible worlds and the qualification problem. In *Proceedings of AAAI-87*, July 1987.

[Hammond and Converse, 1991] Kristian Hammond and Timothy Converse. Stabilizing environments to facilitate planning and activity: an engineering argument. In *The Proceedings of the 1991 National Conference of Artificial Intelligence*, July 1991.

[Hammond *et al.*, 1990] Kristian Hammond, Timothy Converse, and Charles Martin. Integrating planning and acting in a case-based framework. In *The Proceedings of the 1990 National Conference of Artificial Intelligence*, August 1990.

[Hammond, 1989] Kristian Hammond. *Case-Based Planning: Viewing Planning as a Memory Task*, volume 1 of *Perspectives in Artificial Intelligence.* Academic Press, San Diego, CA, 1989.

[Martin, 1989] Charles E. Martin. *Direct Memory Access Parsing.* PhD thesis, Yale University Department of Computer Science, 1989.

[McCarthy, 1977] John McCarthy. Epistemological problems of artificial intelligence. In *Proceedings of the Fifth International Conference on Artificial Intelligence*, 1977.

[Rosenschein and Kaelbling, 1986] Stanley J. Rosenschein and Leslie Pack Kaelbling. The synthesis of digital machines with provable epistemic properties. In *Proceedings of 1986 Conference on Theoretical Aspects of Reasoning About Knowledge*, March 1986.

[Shoam, 1986] Yoav Shoam. *Reasoning about Change: Time and Causation from the Standpoint of Artificial Intelligence.* PhD thesis, Yale University, 1986. RR #507.

Multi-agent Interactions: A Vocabulary of Engagement*

Patricia Goldweic and Kristian J. Hammond
Artificial Intelligence Laboratory
Computer Science Department
The University of Chicago
Chicago, IL 60637
goldweic@cs.uchicago.edu

Abstract

Our project concerns the definition of a content theory of action appropriate for agents that act in a multi-agent environment and its implementation in a multi-agent system. Such a theory has to explain what agents know and how they use this knowledge; it has to identify what resources are available to the agents when they must decide on an action; it has to allow agents to reason and engage in concrete activity in their domain. More important for our research, and in contrast with numerous works in Distributed Artificial Intelligence, such a vocabulary must provide a basis for agents to decide and learn *when, how* or *with whom* they should cooperate. In this paper we suggest a vocabulary of interactions for intelligent agents. Our vocabulary attempts to do justice to the situated character of action with respect to the disparate but related dimensions of physicality, sociality and experience.

Introduction

Realistic multi-agent environments are characterized by uncertainty, distribution of skills and knowledge, and some degree of unpredictability. Yet many of these environments, like workplaces, can become relatively stable over time, enabling routine patterns of interactions to emerge. Many AI researchers have chosen to develop reactive agent architectures to deal with unpredictability and uncertainty. However, it is striking that almost no multi-agent systems have been built that take advantage of the relative stability of the agents' interactions to learn and improve the agents' behavior over time (Bond and Gasser, 1988; Gasser and Huhns, 1989). Our research attempts to fill this gap.

Our approach is to define a content theory of action appropriate for agents that act in a multi-agent environment and implement it in a multi-agent system. Such a theory has to explain what agents know and how they use this knowledge; it has to identify what resources are available to the agents when they must decide on an action; it has to allow agents to reason and engage in concrete activity in their domain. More important for our research, and in contrast with numerous works in Distributed AI (DAI), such a vocabulary must provide a basis for agents to decide and learn *when, how* or *with whom* they should cooperate.

In this paper we present a vocabulary of interactions for intelligent agents. This vocabulary attempts to do justice to the situated character of action with respect to the disparate but related dimensions of physicality, sociality and experience.

What Supports Action?

To intelligently act in a multi-agent environment, an agent needs to have access to a varied set of resources that serve to influence its actions. We distinguish three main categories of resources.

The first category includes the agent's knowledge about how to act in its domain (e.g. what goals to pursue, how to pursue them, etc.). This knowledge has been frequently called *know-how* or *planning* knowledge.

The second category concerns the social, historical and environmental context of the task at hand.

A third category relates to habitual practices. As opposed to *know-how*, these resources are not idiosyncratic[1]. They include knowledge about standard or recurring interaction patterns (explicit or implicit), social conventions, and dispositions of agents to interact in set ways. Agents who are asocial use resources in the first two categories to decide on action. However, common practice must

*This work was supported in part by AFOSR grant number AFOSR-91-0112, DARPA contract number F30602-91-C-0028 monitored by Rome Laboratories, DARPA contract number N00014-91-J-4092 monitored by the Office of Naval Research, Office of Naval Research grant number N00014-91-J-1185

[1] In fact they represent the *culture* of the agents' community.

orient the agents' behavior if they are to intelligibly act in a community.

But how do agents *use* all these resources? What role do these resources play in the evolution of patterns of interactions? In the rest of this paper we provide example interactions of our project domain that serve to motivate our vocabulary and to suggest answers to these questions. We believe that such a vocabulary, with which agents can reason and engage in concrete situated activity in their domain, is a *must* in any realistic theory of action for intelligent interacting agents.

Interactions and AI

Psychology and MA systems

Recent work in social psychology applied to multi-agent systems (Castelfranchi and Miceli, 1991; Castelfranchi, 1990) has been formally addressing some important points overlooked by the DAI community, mainly the notions of *dependence* and *power* among agents[2]. In fact, the research we describe in this paper is compatible with these authors' notions of autonomy, and with their claim that dependence among agents is an important basis for decision making. We view these concepts as part of the set of resources available to the agents when deciding what to do next.

Interactions in CBR

Recent work in Cased Based Reasoning has been concerned with the indexing of stories that involve multiple agents (Schank and others, 1990). This work has been aimed at defining a universal vocabulary useful for indexing such stories from different perspectives. By covering a large space of possible situations (involving one or more agents), these researchers attempt to contribute to explaining human *reminding* phenomena. Although our vocabularies partly overlap, our main goal is, however, very distinct: instead of developing a *content theory of indexing*, we are trying to develop a *content theory of action* that will support an agent's decision making when actually *participating* in activity. Thus, what we are actually after is a vocabulary that deals *centrally* with issues of *planning* and situated *action*; we expect our indexing vocabulary to derive from it and not vice versa.

Interactions in DAI

A theory of social interactions for artificial agents must be able to account for the moment-to-moment accomplishments of the individual agents. Numerous works in DAI have neglected the development of theories of action and thus concentrated on building centralized systems or systems in which the decisions on what to do next are left to the system's designers (Durfee and Lesser, 1987; Rosenschein, 1988; Georgeff, 1984; Georgeff, 1983). Other work has provided descriptive theories of cooperation which, although insightful, are not readily applicable to agent design (Cohen and Levesque, 1987; Werner, 1989; Lochbaum *et al.*, 1990). For a theory of interactions to be applicable to agent design it *must* support the moment-to-moment decisions and actions of the individual interactants. And it can only do so by explaining how agents can decide upon and engage in meaningful action: what agents know and how they use this knowledge, how the current context determines and shapes action, how standard practice[3] influences the agents' behavior. Without addressing these points, a descriptive theory is only useful to rationalize interactions *a posteriori* of their occurrence, but will never be able to explain the *a priori* process by which agents become socially motivated and act towards common objectives.

Additionally, a theory of interactions, as part of a theory of action, must be able to account for the evolution of multi-agent interactions over time. Most work in AI has completely ignored the fact that interactions do not occur in a vacuum, but that they are historically situated. In our project we attempt to provide a theory that can explain how multi-agent interactions can evolve over time by first acknowledging the fact that agents are situated in the context of their own experiences. Our vocabulary provides a step towards the development of such theory.

The Project

The project's domain is a simulated world in which agents engage in maintenance tasks: they clean floors and windows, move furniture between rooms, and deliver mail within the confines of a unique building. Agents meet when they perform their tasks, either because they happen to be working in the same room or corridors, or because they explicitly decide to interact.

Agents that habitually interact tend to stabilize their relationships over time. Sometimes, the distribution of skills and tasks among the agents, and the dynamics of the activity itself, are such that agents develop cooperative routines to better pursue their goals. For example, two agents that often work in the same rooms cleaning floors and carpets respectively, may soon find out that pushing

[2] It is interesting to note that the example we provide in this paper shows what can occur in an extended interaction when a dependence relationship (the fact that one agent cannot readily push a heavy object by itself) exists.

[3] Although we are concerned with the design of artificial societies (that will nonetheless eventually have to coexist with the human society), we believe there is in fact a need to identify what *practice* in those societies is or can be all about.

heavy furniture is easily done cooperatively, and that a shared wastebasket will be better off in a place where both agents have easy access.

On the other hand, when interactions between agents tend to disrupt individual accomplishments (i.e., agents getting in each other's way), agents are better off if they explicitly coordinate their tasks, or even if they avoid each other. In these cases, they can either individually reorganize their tasks, or explicitly negotiate to avoid the inconveniences.

Perhaps most importantly, agents do not have to continually engage in new interactions from scratch; they can take advantage of their past social experiences to decide on action that is beneficial in the long run. For example, if two agents have skills that complement each other's, we would expect them to mutually cooperate in a way that the agent with the strongest skill helps the weakest party. In these cases, agents can make tacit or explicit deals to more effectively pursue their goals. The rationale is that long term stable relationships are better than beneficial one-time interactions.

An Example: Moving Furniture

In this section we provide three examples of consecutive interactions that motivate our vocabulary. The following are the questions we will be interested in answering:

- What specific resources are **available** to the agents when they intelligently engage in their activities?
- How do those resources interplay to suggest appropriate actions?
- Are there any acceptable or standard behaviors that restrict or help construct the agents' activities?

A first encounter

Consider the following interaction between two parties: Tom and Bob, members of the maintenance team. Assume that both of them have moved heavy objects collaboratively in the past, although not with each other.

> Tom and Bob both want to push a heavy couch. Tom is the first to approach it. He does not see Bob, who just entered the room. Bob realizes that Tom is, with some difficulty, trying to push that same heavy couch and thus offers him help. Tom accepts and they push the couch together to the right place. In doing so, they decide to hold the two opposed ends of the couch and coordinate to push them simultaneously. They communicate briefly in order to coordinate their joint activity.

The agents' resources

What are the resources the agents use to reason and engage in the previous interactions? First, they **know how** to push couches. This knowledge provides them a basis for purposeful action. Second, they know about **common social practices**, in particular about how to cooperate and coordinate to perform concrete activities. This knowledge gives them expectations on what the other party will do or say during the interaction, and restricts their available choices of action. In addition, the agents **perceive** their shared physical environment: they see and hear each other and can see and touch the objects in the room. In this particular interaction, the physical context constitutes a very important part of the common ground of interaction. In fact, the agents' ability to act opportunistically, their appropriate know-how and the fact that they inhabit a shared physical environment, are the basis for successful coordination between the agents themselves and with their physical world.

An agent's perspective on an interaction

The first story exemplifies a common situation in which two agents share a goal: pushing a couch. However simple, little has been done in AI to explain why two parties such as those in the situation above, would realistically come to collaborate in their enterprise. We will attempt to do exactly that.

Agents need to have access to a vocabulary that characterizes different aspects of an interaction. This vocabulary represents part of the common sense knowledge an agent needs in order to engage in meaningful action and in order to learn from it. In what follows, we describe how one participant understands this interaction in terms of such a vocabulary[4].

First, Tom recognizes that Bob is the **initiator** of the interaction. Although apparently trivial, this fact will be important for an *a posteriori* evaluation of the interaction.

Tom did not expect such interaction because he was focusing his **attention** on his activity when Bob interrupted him. Moreover, he recognizes that when Bob initiated the interaction, he had not perceived Bob's presence. This tells Tom that he could not have anticipated that such an interaction was going to take place.

Next, Tom understands that the **rationale** for such an interaction involves:

1. the fact that he cannot readily push the couch by himself, and
2. the fact that Bob has an overlapping goal of moving the same couch, which Bob has communicated to him.

[4] We understand other interpretations are possible. The one we provide seems plausible and does not seem to require complicated reasoning machinery.

Tom also recognizes that if he accepted Bob's offer, he could abandon his current plan for pushing the couch, and more easily achieve his goal with Bob's help. This constitutes Tom's **individual perspective** on the interaction.

From a **social perspective**, Tom recognizes that the interaction would promote a stable relationship between him and Bob. He also recognizes that the social context of his activity would change from a situation of disengagement to one of joint engagement in the couch-pushing task.

After pushing the couch with Bob, Tom is able to evaluate this particular interaction from different perspectives. Because he and Bob were able to successfully move the couch to an agreed upon location, he recognizes that the interaction has done justice to its rationale.

Additionally, Tom will tend to reciprocate Bob's cooperative behavior in the future if the situation provides for doing so. We believe it is important to analyze *how* Tom can reach such decision. Although everybody would agree that reciprocation plays a role in social interactions, it is harder to articulate how an agent can make use of this 'principle' or when to use it. A possible explanation would be that Tom explicitly considers Bob's goals and beliefs and then decides that that is the appropriate thing for him to do. Another explanation, which is the one we prefer, suggests that since the agents are participating in a certain common social situation, they orient to each other in ways that are also part of the common practice. In this particular case, the fact that Tom was taken from a situation in which he was socially disengaged to one in which he is jointly engaged with Bob in an activity he desired, is enough for him to be willing to reciprocate in the future[5].

What can Tom learn from this interaction? He could learn the following:

- Whenever he and Bob have the overlapping goal of pushing a couch, a plan to do it collaboratively benefits both of them and should therefore be suggested if a future situation provides for doing so.
- Whenever Tom is faced with the task of pushing a heavy object, a reasonable plan (or piece of know-how) he could use, involves asking Bob for help.

A second encounter: how experience shapes the activity

Consider the following story:

> Tom and Bob are cleaning a room. Tom is cleaning the floors and Bob the windows. Tom needs to push the heavy couch to clean underneath. He decides to ask Bob for help. Bob accepts and they proceed to move the couch as before.

How does Tom's analysis change with respect to the previous case? Tom anticipates his failure to easily push the couch and he gets reminded of the previous interactive experience. He is now able to use this piece of knowledge as another means to achieve his goals (have the couch be moved by both agents). Since Bob had not shown any interest in moving the couch, he decides to seek Bob's help (thus actively initiating this interaction). He understands that this interaction, building upon a previously established relationship, clearly counts as a favor to him, and thus he will attempt to reciprocate Bob's behavior in the future.

Bob's perspective is richer to analyze. When he remembers the previous interactive experience with Tom, he realizes that he had been able to take advantage of a positive interaction among the two agents (both were pursuing the same goal and had a chance to achieve it jointly). He also reasons that his own goal of moving the couch recurs every time he wants to clean the adjacent window. So in order to take advantage of such an opportunity once again, it would be convenient if his goal of moving the couch were to arise exactly when Tom's does. Clearly, Bob does not have control over Tom's goal generation processes, but he can reasonably behave in one of the following two ways. He could try to make an explicit deal with Tom to fix the schedule of pushing. Alternatively, he could reschedule his own activities so that his couch-pushing goal arises at a convenient time (Hammond and Converse, 1991). Since his desire to push that couch arises as a consequence of his desire to clean the adjacent window, then he could choose to clean that particular window now instead of the one he was currently working on.

Bob chooses the second option because it is socially preferred and because it is not too costly or disruptive for him to switch to working on another window. In general, it is more acceptable that the approached party attempt to cooperate if asked to do so (especially when an ongoing stable relationship among two agents already exists) unless the situation would disrupt his current activities.

A third encounter: developing a more enduring approach

Imagine that the situation we just described repeats itself. How would the interaction between the two agents change? We postulate the following scenario.

Bob is faced with a richer experiential context

[5] This analysis greatly differs from an alternative one based on an explicit consideration of goals and beliefs and expected cost/benefit of a potential interaction. Note that according to such an alternative, Tom might reason that, since the two agents shared the goal of the activity, there wouldn't be anything to reciprocate for.

than ever before. He is now able to recognize other valuable pieces of information. First, he remembers that he rescheduled his activities (previous story) to accommodate for the same sort of situation. This would conceivably make Bob intend to go for a more enduring approach. Moreover, he now recognizes that his desire to push that couch is not only a recurrent goal of his but of Tom's too. These two pieces of knowledge combine to suggest going for an explicit deal this time. Such a deal would include, for example, fixing the schedule for pushing and also for replacing the couch when the activities of the two agents do not depend on the couch's location anymore.

Note that the deal takes long term advantage of a positive inter-agent interaction and reduces the agent's future cognitive requirements (no need to reorganize activities anymore). Moreover, the deal also promotes a relationship by maintaining a cooperative social context while having the agent act within the socially acceptable bounds: because the situation involves a recurrent goal of both agents, it is socially acceptable to try to cut a deal that is convenient to both [6].

The three examples we have described exemplify how a particular multi-agent routine can develop among parties over time that does justice both to individual agency and to a historical process of interactions.

A Closer Look at the Vocabulary

The shared facts

Our vocabulary includes a set of items representing some resources commonly available to the interaction's participants:

- the agents involved
- the interaction's *initiator*
- the physical setting of the task
- the shared past experiences
- knowledge about common social practices

The idiosyncratic resources

Additional vocabulary items represent the following idiosyncratic resources:

- the interaction's *rationale*. An agent may interact with others due to different reasons. Some of these reasons can be traced back to the planning domain's vocabulary for plan failures as applied to any of the participants of the interaction: lack of physical ability, lack of knowledge or skill, or lack of resource. However, other rationales rise out of social considerations. For example, an agent may anticipate inter-agent conflict and seek to avoid it. Interaction can also occur simply because "it's always done that way."
- whether the interaction *satisfied its rationale*.
- whether the interaction was *expected* or not;
- whether the interaction was *desired* or not;
- how the interaction *relates to* the agent's current, suspended or future goals and activities. This vocabulary concerns functional relationships among goals and activities of the same or different agents and extends work in (Wilensky, 1978; Hammond, 1990). Some examples are:
  - The interaction may complete the task in which the agent is engaged, or some part of the task.
  - The interaction may require that the agent suspend work on his task.
  - The interaction may steal a resource needed for the agent's task, or be counter to one of the agent's goals.
- the agent's *perspective on* how the interaction relates to the other agents' activities;

The social perspective

If agents are to decide *when*, *how* or with *whom* they should cooperate, our vocabulary must be able to describe how a particular interaction can affect an inter-agent relationship. This knowledge partly constitutes an agent's social awareness. The fact that agents are socially aware partly explains why they decide to establish long term relationships with others instead of engaging in one-time interactions: relationships may sometimes help agents better (or more cheaply) pursue their goals.

This vocabulary deals with concepts such as whether an interaction *promotes*, *conflicts with* or *stabilizes* an ongoing relationship. Central to this is the notion that an agent holds a certain *attitude* towards the other participants which socially situates himself in the interaction. Thus, our developing vocabulary includes items that represent the following:

- *initial attitude* toward the other participants of the interaction;
- *changes in attitude* towards them as a result of the interaction.

Implementation and Future Work

Our program currently implements the first example we described and some of its variations. The architecture of the agents in our system is based on work on *opportunistic memory* (Hammond, 1989; Hammond *et al.*, 1989) and *agency* (Hammond *et al.*, 1990). We have mostly been concerned with the issues of plan representation (individual or multi-agent) and the *situated use* of those plans. We are currently working towards an implementation of the three stories described in this paper.

[6] Clearly, not every inter-agent relationship will evolve in the same way. The particular evolution (if any) will be based on the particular dynamics of the agents' interaction.

We believe that a theory of interactions is useful only if it can be used to *produce* actual multi-agent behavior. Thus, while we continue improving our vocabulary of engagement we expect to experiment heavily with our multi-agent system and show how evolving patterns of interactions serve to stabilize the agents' interactions over time. We expect our system not only to provide us with feedback useful to constrain and orient the development of our theories, but to help us test their plausibility as well.

Acknowledgements

We would like to thank Mitchell Marks, Paul Crawford, Daniel Fu and an anonymous reviewer for their useful comments on earlier drafts of this paper.

References

P. Agre. *The Dynamic Structure of Everyday Life.* PhD thesis, M.I.T., 1988.

P. Agre. What are plans for? In *New Architectures for Autonomous Agents: Task-level Decomposition and Emergent Functionality.* MIT Press, Cambridge, Massachusetts, 1990.

A. Bond and L. Gasser. An analysis of problems and research in DAI. In A. Bond and L. Gasser, editors, *Distributed Artificial Intelligence*, chapter 1. Morgan Kauffman, 1988.

C. Castelfranchi and M. Miceli. Dependence relations among autonomous agents. In *Proceedings of the Third European Workshop on Modeling Autonomous Agents in a Multi-agent World*, 1991.

C. Castelfranchi. Social power: a point missed in multi-agent, DAI and HCI. In Y. Demazeau and M. J., editors, *Decentralized AI.* Elsevier Science Publishers B.V. (North Holland), 1990.

E. Durfee and V. Lesser. Using partial global plans to coordinate distributed problem solvers. In *Proceedings of the 1987 International Joint Conference on Artificial Intelligence*, 1987.

L. Gasser and M. Huhns. Themes in Distributed Artificial Intelligence research. In L. Gasser and M. Huhns, editors, *Distributed Artificial Intelligence, Volume II.* Morgan Kauffman, 1989.

M. Georgeff. Communication and interaction in multi-agent planning. In *The Proceedings of the Third National Conference on Artificial Intelligence*, 1983.

M. Georgeff. A theory of action for multi-agent planning. In *The Proceedings of the Fourth National Conference on Artificial Intelligence*, 1984.

K. Hammond and T. Converse. Stabilizing environments to facilitate planning and activity: An engineering argument. In *Proceedings of the Ninth National Conference on Artificial Intelligence.* Morgan Kaufmann, July 1991.

K. Hammond, T. Converse, and M. Marks. Learning from opportunity. In *Proceedings of the Sixth International Workshop on Machine Learning*, Ithaca, New York, June 1989. Morgan Kaufmann.

K. Hammond, T. Converse, and C. Martin. Integrating planning and acting in a case-based framework. In *The Proceedings of the 1990 National Conference of Artificial Intelligence*, August 1990.

K. Hammond. *Case-based Planning: An Integrated Theory of Planning, Learning and Memory.* PhD thesis, Yale University, 1986. Technical Report 488.

K. Hammond. Opportunistic memory. In *Proceedings of the Eleventh International Joint Conference on Artificial Intelligence.* IJCAI, 1989.

K. J. Hammond. Explaining and repairing plans that fail. *Artificial Intelligence*, 1990.

W. Hanks. *Referential Practice.* The University of Chicago Press, 1990.

H. Levesque, P. Cohen, and J. H. T. Nunes. On acting together. In *The Proceedings of the Eighth National Conference on Artificial Intelligence*, 1990.

K. E. Lochbaum, B. Grosz, and C. Sidner. Models of plans to support communication: An initial report. In *The Proceedings of the Eighth National Conference on Artificial Intelligence*, 1990.

S. P. Robertson, W. Zachary, and J. B. Black. *Cognition, Computing and Cooperation.* Ablex Publishing Corporation, 1990.

S. Rosenschein. Synchronization of multi-agent plans. In A. Bond and L. Gasser, editors, *Distributed Artificial Intelligence.* Morgan Kauffman, 1988.

R. Schank et al. A content theory of memory indexing. Technical Report 2, Institute for the Learning Sciences, Northwestern University, Evanston, IL, March 1990.

L. Suchman. *Plans and Situated Actions: The Problem of Human-Machine Communication.* Cambridge University Press, 1987.

E. Werner. Cooperating agents: A unified theory of communication and social structure. In L. Gasser and M. Huhns, editors, *Distributed Artificial Intelligence, Volume II.* Morgan Kauffman, 1989.

R. Wilensky. *Understanding Goal-Based Stories.* PhD thesis, Yale University, 1978. Technical Report 140.

A Vocabulary for Indexing Plan Interactions and Repairs

Kristian J. Hammond*
Department of Computer Science
The University of Chicago
1100 East 58th Street
Chicago, IL 60637
hammond@cs.uchicago.edu

Colleen M. Seifert
Department of Psychology
University of Michigan
330 Packard Road
Ann Arbor, MI 48104
seifert@csmil.umich.edu

Abstract

Solving the multiple goals problem has been a major issue in Artificial Intelligence models of planning (Sussman, 1975; Sacerdoti, 1975; Wilensky, 1978; Wilensky, 1980; Wilensky, 1983; Carbonell, 1979); however, most models have assumed that the best plan for a set of goals to be satisfied in conjunction will arise from a simple combination of the best individual plans for each goal. However, human planners seem to possess an ability to look at a set of goals, and characterize them as a whole, instead of as a collection of individual goals (Hayes-Roth and Hayes-Roth, 1979). In this paper, we introduce the notion of indexing complex multiple-goal plans in terms of the *interactions* between the goals that they satisfy. We present the vocabulary requirements for representing the causality behind goal interactions, the general planning strategies used to resolve these interactions, and the specific plans based on these more general resolution strategies that are instantiated in the actual planning problem.

Indexing Plans in Memory

Solving the multiple goals problem has been a major issue in Artificial Intelligence models of planning (Sussman, 1975; Sacerdoti, 1975; Wilensky, 1978; Wilensky, 1980; Wilensky, 1983; Carbonell, 1979); however, most models have assumed that the best plan for a set of goals to be satisfied in conjunction will arise from a simple combination of the best individual plans for each goal. A problem with the "each goal first" planning theories is that they provide no vocabulary capable of characterizing goal and plan interactions in a form that allows access to past cases in memory based on these commonalities. Such goal interactions serve as critical constraints on successful plans, so that taking advantage of these constraints while selecting among and developing plans will not only produce "smarter" plans, but has the advantage of bootstrapping from plans previously developed for similar plan interaction situations.

Three basic requirements of any representational vocabulary used to describe, organize, and index plans are: first, that it characterize abstract patterns of goal interactions that capture relevant similarities between situations; second, that it provide access to general strategies that pertain to resolving the overall goal/plan situation; and finally, that it identifies specific plans that cover the current situation. In the next sections, we present a representational vocabulary that characterizes the causal knowledge behind goals, plans, and their interactions. Human experimental evidence is then presented, along with suggestions about how this proposed paradigm can be extended to encompass a majority of planning situations.

Vocabulary for Goal Interactions

Planners currently use a vocabulary of goals, associated plans, sub- plans, preconditions, and effects (Schank and Abelson, 1977), as well as basic interactions such as *conflict* and *concord* (Wilensky, 1978). The problem of how to describe the similarity between goal situations has been discussed by (Schank, 1982), who introduced abstract memory structures (Thematic Organization Packages or TOPs) to connect episodes in memory on the basis of similarities in the pattern of goals and plans they contain. In planning, such abstract patterns of goal and plan interactions can serve to identify a class of problems where a particular set of resolution strategies are appropriate.

Consider the example (Wilensky, 1978) of "wanting the newspaper from outside on the sidewalk

*This work was supported in part by AFOSR grant number AFOSR-91-0112, DARPA contract number F30602-91-C-0028 monitored by Rome Laboratories, DARPA contract number N00014-91-J-4092 monitored by the Office of Naval Research, Office of Naval Research grant number N00014-91-J-1185

while it is raining," where the planner is trying to achieve a particular goal (getting the newspaper); the chosen plan (carry the paper in) has a particular precondition (be outside); and an existing state (it is raining), in combination with the precondition, results in the violation of an existing preservation goal (stay dry) (Schank and Abelson, 1977). To plan in this situation, the goal conflict must be described in terms of an abstract characterization of the problem that captures the causal chain leading to the violation. This situation can be characterized as *precondition* plus *state* causes *violation* of a preservation goal, or *Plan+State→Violation* (see Figure 1).

```
                                                   GOAL-CONFLICT
PLAN1------planfor------->A-GOAL1                        /
  |                          |                          isa
  |PRECONDITION              |                          /
  |                          +-----interaction---->P+S->V
  V                          |                         \
STATE1-->                    |                          \
        +---result-------->STATE3                         \
STATE2-->                    |                        GENERAL
(Existing STATE)             | VIOLATES              STRATEGIES
                             V
                          P-GOAL1

Goal:  Achievement goal
Plan:  Default plan successful under normal conditions
Precondition:   Precondition for Plan
State:   Existing state of the world, independent of Plan
Violation:    Result of Precondition and State co-occurring,
         causes Preservation Goal failure.
Resolution Strategies:
1. Plan against PRECONDITION
2. Plan against STATE
3. Plan against VIOLATION
```

Figure 1: Representation of the goal interaction situation Plan+State→Violation.

Many other plan/goal conflicts can be characterized in a similar way (see (Hammond *et al.*, 1991; Hammond, 1990)). In this paper, we will concentrate on only one of these, *Plan+State→Violation*, and its associated resolution strategies.

Vocabulary for Resolution Strategies

There are three basic strategies that are designed for use in any *Plan+State→Violation* planning situation. These resolution strategies are plans for altering the causal situation such that the original goal can be achieved while avoiding the preservation goal violation. The three strategies for *Plan+State→Violation* are:

- Plan against precondition: Find a plan for the primary goal that does not require the problematic precondition (find a plan that does not require going outside).
- Plan against state: Alter the existing state such that even when the precondition is met the preservation goal will not be violated (do something to make it stop raining).
- Plan against violation: Add an auxiliary plan that prevents the violation of the preservation goal even in presence of the threatening state and the existing condition (get an umbrella).

One test of these problem resolution strategies is whether or not they apply to other instances of *Plan+State→Violation*. Suppose a planner wants to get a hot pot off of the stove. Like the newspaper and the rain example, the situation fits into *Plan+State→Violation*, and the associated resolution strategies are appropriate here also: the first strategy suggests trying a plan that does not require touching the pot, such as using a stick to push the pot off of the burner; the second suggests waiting for the pot to cool down before moving it; the third suggests that some sort of protection be used, such as a pot holder. While the content of the particular goals is different, the interactions that occur between the goals in both of these situations are similar.

Accessing general strategies through this vocabulary provides the planner with alteration techniques and information as to what parts of the initial causal configuration are appropriate targets of change. In summary, by including a causal analysis of the goal interactions in a situation as part of its representation, it is possible to access in memory the general strategies applicable to the problem, leading to specific plans for the current situation.

Vocabulary for Specific Planning Strategies

Causal relations can be used to organize resolution strategies in terms of the situations for which they are relevant. The resolution strategies indicate *where* a particular causal chain can be effectively altered for a particular planning situation; next we need a more specific vocabulary for characterizing *how* situations can be altered. We will now look at when and how to apply the specific strategies associated with *Plan+State→Violation*.

Specializations of "Plan against precondition." One specialized strategy is *use alternate agent*, appropriate in *Plan+State→Violation* cases such as when a student wants to go into the office and pick up his mail while avoiding his advisor. This strategy is not always appropriate; for example, consider moving the hot pot. To index this strategy so that it will be applied only in appropriate problems, we need a representation of features that identify when this and other particular strategies are relevant. In this case, the distinctive feature lies in the nature of the preservation goal goal being threatened: if it is specific to the planner and not to other agents, then this is a good solution to a *Plan+State→Violation* conflict. Thus, the important features for *use alternate agent* are the commonality of the threatened preservation goal, the nature of the preservation goal, and any special skills or abilities involved in the normal plan to satisfy the initial achievement goal (Schank and Abelson, 1977).

A different specialization of the general strategy to plan against the precondition is to use an alternate plan that does not have the same precondition, or *use alternate achievement goal plan*. For example, to get the hot pot off of the stove, one could pick up the pot without touching it by inserting a stick through its hollow handle. A predictive feature (Johnson and Seifert, 1991) for this strategy is the existence of any alternate plans for the initial goal.

A final specialization of this general strategy of planning against the precondition is to run the initial plan very quickly, the *run fast* strategy. This strategy is effective in those cases where the preservation goal being threatened has degrees of violation linked to a parameter (time or speed) under the agent's control, and when the preservation goal violation is only intermittent (as in the possibility of running into the advisor in the mail room).

Specializations of "Plan against state." As with "plan against precondition," there are three specializations of "plan against state:" *wait out, jump between, and counter plan. Waiting out* the state applied to instances of *Plan+State→Violation* in which the existing state is temporary, such as a hot pot. Use of this strategy depends on the nature of the achievement goal as well as on the nature of the state; if the achievement goal is particularly insistent, then this strategy is inapplicable.

A variation on the *wait it out* strategy is the *run between* strategy, where one runs the initial plan in between fluctuations in the state. This strategy would be indexed under the *Plan+State→Violation* structure, linked to a test concerning the permanence and possible intermittence of the side state being planned against.

Finally, one can *counter plan* against the conflicting state; that is, generate a separate plan to disable the state. For this to work, the planner obviously must have some control over the state. Once again, the indexing vocabulary for all three strategies consists of features related to the practical considerations that the planner has to take into account in order to alter the initial plan: duration of states and existence of specific plans.

Specializations of "Plan against violation." Like the other two general strategies, "plan against violation" has some specialized versions that can be applied in different circumstances. If the preservation goal is relatively minor or short-lived in relation to the achievement goal, then it might make sense to *tolerate* violations of it. To decide if *toleration* is possible, the planner needs to know the relative importance of its goals, and the likely durations of violations.

Like *tolerate*, *run and recover* requires not only that the preservation goal is relatively unimportant, but also that there exists a plan associated with the violation that can be used to recover from it. One can, for example, dry off after fetching the newspaper out of the rain. Like the *tolerate* strategy, this *run and recover* strategy depends on the relative importance of the two goals being planned for, and requires a test for the existence of recovery plans for the violated goal.

A third specialization of "plan against violation" is *counter plan the preservation goal*: generate a concurrent plan in support of the preservation goal, such as using an umbrella. The plan is stored in terms of the causal situation in which it will become relevant, rather than in terms of a specific goal violation.

The specializations of the general strategies apply in some instances of *Plan+State→Violation*, but not others, and therefore must be indexed by their appropriateness conditions. That is, "plan against violation" has three specializations *tolerate* applies when the preservation goal being protected is trivial compared to the achievement goal being satisfied; *run and recover* applies when there is a straightforward recovery plan associated with the violation; and *counter plan the preservation goal* applies when the agent has an existing plan associated with the causal rule leading to the violation, and also when an additional state is required for the violation to occur (such as physical contact).

To summarize this vocabulary, when a situation where a precondition and an existing state cause a preservation goal to be violated, we can respond with one of three possible resolution strategies: plan against precondition, plan against state, and plan against violation. Each of these strategies has several specializations whose appropriateness depends on the pragmatic planning constraints in the situation.

Indices for Retrieving Planning Cases

A plan is proposed, a conflict detected and characterized, a specific strategy chosen. The next step is to search memory for a past instance of that strategy that applies to the current situation; however, we must define the set of indices by which those instances can be recalled at the appropriate time. Every instance of a particular strategy is indexed by the features of that episode that the strategy used or altered to construct the instance, and by implementational causes and effects that are learned through experience. In general, then, the features used to index planning cases are those which have some causal relevance to the way in which that strategy is implemented.

For each of the nine planning strategies proposed above, the following features are likely to lead to useful past instances:

- Indices for ***use alternate agent*** include: The achievement goal, other available agents, the plan itself, the threatened preservation goal, and the state threatening the preservation goal.
- Indices for ***use alternate plan*** include: The achievement goal, the specific plan, possible alternate plans, and the precondition to be avoided.
- Indices for ***run fast*** include: The plan, the state and precondition implicated in the violation of the preservation goal, and the preservation goal being violated.
- Indices for ***wait it out*** include: The achievement goal and the state
- Indices for ***Jump between*** include: The achievement goal, the proposed plan, and the intermittent state.
- Indices for ***counter-plan state*** include: The state, the available plans for that state, and the harmful precondition of the initial plan.
- Indices for ***tolerate the violation*** include: The preservation goal and the violating state.
- Indices for ***run and recover*** include: The violating state and possible plans to recover from the violation.
- Indices for ***counter-plan violation*** include: The rule connecting the precondition and the undesired state, the assumed conditions in that rule, the undesired state, the precondition for the initial plan, and the existing state in the world.

We have presented a set of specific strategies that can be applied in different situations described by *Plan+State→Violation*, each of which has a question or feature that tests for its applicability. The structure *Plan+State→Violation*, then, includes more than just how, in general, to alter the resolution strategies: It also has information about how to ***apply*** specific planning strategies, and in what circumstances the individual strategies are relevant. The features useful for determining the applicability of these strategies are critical to this new vocabulary for describing planning situations.

Evidence for Vocabulary Use

To determine whether humans utilize causal planning factors in selecting among planning strategies, we conducted an experiment in common-sense planning (see (Hammond *et al.*, 1991), for a full description). The planning problems used were six exemplars of the *Plan+State→Violation* structure, all placed in different contexts such as celebrating a sick friend's birthday, jogging after dark, and picking up an exam while avoiding one's professor. Subjects were asked to provide commonsense answers to the planning problems in terms that they would really chose to do in those situations. By examining the types of plans they propose, and how well those plans match the predictions from the vocabulary model, it can be ascertained whether subjects are utilizing these same features in determining plan selection. The responses were then coded using the planning strategies predicted by the vocabulary model. Any responses not fitting one of the categories was coded in a general "other" category. These were:

1. Plan against Precondition: Find a new plan for achieving the primary Goal that does not require the Precondition which threatens the preservation Goal.
   (a) Get alternate agent (if not problematic for other actor)
   (b) Run Plan fast (if limited exposure is acceptable)
   (c) Use alternate plan without Precondition (if available)
2. Plan against Existing State: Alter the Existing State so that even when the Precondition is met, the Preservation Goal will not be violated.
   (a) Waitout Existing-State (if temporary)
   (b) Jump in between phases of State (if intermittent)
   (c) Counterplan against Existing-State (if possible)
3. Plan against Violation (Threatens Preservation-Goal): Add an auxiliary plan that prevents the violation of the preservation Goal even in the presence of Precondition and State.
   (a) Ignore – put up with preservation Goal Violation (if short duration)
   (b) Plan to recover from preservation Goal Violation (if can repair)
   (c) Counterplan Preservation Goal Violation (interrupt connection between Precondition and State)

Overall, the extent to which the responses given fit into the proposed categories support the use of the causal features in commonsense planning, as opposed to other features or plans subjects may potentially generate for the problems.

93% of responses were instances of the planning strategies proposed in the vocabulary theory, while 7% were "other" types of responses. The "other" responses included items such as "see if the dorm has anyone else around and borrow their newspaper" for the newspaper in the rain problem, or "stay home" for the running after dark example. The majority of "other" responses involved abandoning the goal implicated in the goal interaction. This type of response is not predicted by the vocabulary model, which assumes the goal must be satisfied in some way. Overall, it appears the planning strategies for the *Plan+State→Violation* structure were sufficient to account for the plans generated by subjects, with the exception of solutions involving abandoning goal satisfaction.

Among the three resolution strategies, subjects' responses more frequently involved planning against the violation (43%), compared to planning against the precondition (33%) or planning against the existing state (17%). While the model makes no predictions about the use of the three categories beyond which features apply in specific instances, it seems subjects preferred plans that dealt directly with the problematic interaction of precondition and state, rather than attempting to change either separately. In particular, plans to change or work around the existing state of the world were given infrequently compared to other possibilities. This may reflect task demand, in the sense that subjects tried to work within the problem constraints presented, and the states tended to be examples of conditions in the world that are unchangable (such as rain and darkness).

In addition, comparisons of strategies by example indicates high variability in strategy application based on the specific planning constraints in each of the examples. The results show that, while each strategy was used in at least 3% of responses, an uneven pattern of strategy use across examples was evident. Of the specific strategy instantiations, the most frequent was the strategy of "counterplanning against the preservation goal violation," with 27% of responses. Another example of selective use of specific strategy is "selecting an alternate plan without the problematic precondition." This strategy, applied only when such a plan is available, was frequently used in the "driving to Detroit" example, where substituting other means of transportation avoids the faulty brakes in the planner's car. For three of the problems, no responses included substitute plans that avoided the precondition.

In addition, each planning problem showed distinct differences in application of resolution strategies. For example, the most frequently generated plan for the newspaper example was "counterplan against the violation," while for the "driving to Detroit" example, the most frequent plan was "use alternate plan." The reason for these differences rests in the pragmatic features used to determine when a plan is appropriate for application. For example, the use of an alternate plan depends on the existence of such a plan, most obvious in the "driving to Detroit" example where other means of transportation are readily available. Subjects did not perceive many alternate plan for getting the newspaper not involving going outside. Thus, there were many differences in the patterns of plans generated for each example. In general, these patterns fit the predicted categories, such that no plans were generated when the vocabulary tests suggested that the planning strategy was not appropriate. However, there were instances where subjects did not generate plans that could have been expected based on the tests in the planning vocabulary. For example, the strategy of "waiting until the rain lets up" was predicted for "jumping in between" intermittent phases of rain in the newspaper example, but not listed by subjects. However, subjects may have felt information about the state of the rain was lacking, and so avoided using plans based on assumptions about the state not given explicitly in the problem description.

In conclusion, it appears that the proposed planning vocabulary accounts for the set of responses given by subjects to these simple planning problems. Further, there was good evidence that subjects were sensitive to the applicability features associated with each strategy, such that they applied some strategies only in appropriate examples. The vocabulary did not include any prediction of the demonstrated preference for plans against the violation itself, compared to plans against either the precondition or the state separately. The representational scheme also did not account for unsuccessful plan resolution, whereas most of the "other" responses involved subjects' attempts to abandon the goal. Overall, however, the plans subjects generated corresponded extremely well to the causal possibilities laid out in the vocabulary, and few novel intrusions occurred. Further, the application of strategies differentially in the specific problems supports the notion that subjects are sensitive to the features predicting when certain strategies are applicable.

Generality of Vocabulary

The *Plan+State→Violation* vocabulary includes many features that are important for planning in general. It is clearly important for a planner to know the difference between those states that it can plan against and those that it cannot; what the preconditions and effects of its plans are; if it has any other plans for the same goal that has different preconditions and effects; and if the goals it has are only held by it, or also held by others with whom it can share tasks.

The indexing within *Plan+State→Violation* uses the same features a planner needs to detect and monitor in order to plan, and are neither arbitrary nor important only to this structure. Thus, in *Plan+State→Violation*, nine specific strategies are stored under the TOP, indexed by the features of the goals, actions, and states in the structure that determine the applicability of the strategies themselves. The components of structures representing goal interactions should include those of *Plan+State→Violation* (preconditions, existing states, violations) along with many others to describe the prototypical ways in which goals can affect one another (side effect, disable, enable, etc.) Similarly. the set of resolution strategies outline for *Plan+State→Violation* must be extended to capture different modifications to other goal interaction structures. Finally, the specific plan strategies for each situation are greatly affected by the context, and will vary based on the specific features of the planning problems.

With the aid of a complete vocabulary of plan/goal interactions, the planner can, after identifying its situation as an instance of a particular causal structure, apply a few simple tests to select from a set of easily implemented strategies. A refined understanding of the causal pattern underlying the prototypical solutions also allows the generation of alternate solutions when needed; for example, when an umbrella is not available, other materials can serve the same functional purpose.

Conclusion

This paper has presented an outline of a representational scheme for organizing and accessing plans and past episodes relevant to current planning problems. Our argument is that it is the abstract relationship among goals and plans that best constrains what planning choices one might make in a given situation. Therefore, retrieving past plans based on the abstract interaction will assure that the retrieved information will be most useful to the planner.

For any particular planning problem, this representation allows easy access to general resolution strategies and specific past plans related to the goal interaction situation. This allows the planner to search for past plans relevant to its overall situation, rather than building complex plans out of single plans for each of the goals in its current planning situation.

The vocabulary required to support this organization connects three levels of abstraction in planning. In order to identify the particular TOP relevant to any given situation, the planner must be able to characterize its current goal/plan problem in terms of the causal relations between the goals, actions and states included in that episode. This characterization then allows the planner to identify the TOP which packages the general strategies applicable to its current problem. To select among the general strategies, the planner must answer pragmatic questions about its current goal/plan configuration.

Among all possible features in a planning situation, only a limited set of these features – those that are relevant to the way in which the current causal structure can be changed – are used as indices within the TOP.

References

J. Carbonell. *Subjective Understanding: Computer Models of Belief Systems.* PhD thesis, Yale University, 1979.

K. J. Hammond, C. Seifert, and K. Gray. Intelligent encoding of cases in analogical transfer: A hard match is good to find. *The Journal of the Learning Sciences*, January 1991.

K. J. Hammond. Explaining and repairing plans that fail. *Artificial Intelligence Journal*, 1990.

B. Hayes-Roth and F. Hayes-Roth. A cognitive model of planning. *Cognitive Science*, 3(4), 1979.

E. Sacerdoti. A structure for plans and behavior. Technical Report 109, SRI Artificial Intelligence Center, 1975.

R. C. Schank and R. Abelson. *Scripts, Plans, Goals and Understanding.* Lawrence Erlbaum Associates, Hillsdale, New Jersey, 1977.

R. Schank. *Dynamic Memory: A Theory of Reminding and Learning in Computers and People.* Cambridge University Press, 1982.

G. Sussman. *A computer model of skill acquisition*, volume 1 of *Artificial Intelligence Series.* American Elsevier, New York, 1975.

R. Wilensky. *Understanding Goal-Based Stories.* PhD thesis, Yale University, 1978. Technical Report 140.

R. Wilensky. Meta-planning. Technical Report M80 33, UCB College of Engineering, August 1980.

R. Wilensky. *Planning and Understanding: A Computational Approach to Human Reasoning.* Addison-Wesley Publishing Company, Reading, Massachusetts, Reading, Mass, 1983.

Dynamic Construction of Mental Models in Connectionist Networks

Venkat Ajjanagadde*
Wilhelm-Schickard-Institut
Universität Tübingen
Sand 13, W-7400 Tübingen, Germany
email: nnsaj01@mailserv.zdv.uni-tuebingen.de

Abstract

The task of "model construction", which is the one of constructing a detailed representation of a situation based on some clue, forms an important component of a number of cognitive activities. This paper addresses the problem of dynamic model construction from a connectionist perspective. It discusses how to represent models as patterns of activity within a connectionist network, and how dynamic generation of such patterns can be efficiently achieved.

1 Introduction

The subject matter of this paper is what we refer to as the task of "model construction", which forms an important component of a number of cognitive activities. Informally, the task of "model construction" is the one of *constructing a detailed representation of a situation based on some clue*. Consider the problem of language understanding for example. Suppose we are told that "John drove to the supermarket". "Understanding" this sentence involves infering many more things other than what is explicitly stated in the sentence. For example, we would have inferred that "John *went* to the supermarket" using our knowledge that "driving to a place" implies "going to that place". We would have also done "plan recognition" (Charniak & McDermott, 1985), i.e., we would have inferred the most likely reason behind John's supermarket visit (such as "to shop there", "to work there" etc.). We would also have inferred the sequence of actions taking place such as "John reached the supermarket", "He parked his car in the supermarket parking lot", "He got out of the car and grabbed a shopping cart" and so on... In this example, the sentence explicitly provided only a *clue* about the situation; all that it explicitly stated was just *drive-to(john,supermarket1)*. Based on this clue, we inferred many more facts about the situation thereby constructing a detailed representation, or a model, of the situation corresponding to John's supermarket visit.

In the case of other perceptual tasks, the problem is similar. In vision for example, the 2D image on the retina provides the clue about the situation in the world; based on that clue, the problem is that of constructing the representation of object configuration in the 3D world that would give rise to that image.

In addition to perception, "model construction" plays a role in other cognitive tasks as well. For example, Mannes and Kintsch (1991) argue that a number of mundane planning problems are problems of "understanding". But, then, "understanding" in turn is a perceptual task and involves model construction.

It is very natural to expect that other intelligent activities such as solving problems/puzzles, playing games etc. make extensive use of the apparatus that already exist to accomplish perception. Hence, it is not at all surprising that "model construction", which is an important component of perception, has been found to play a major role in tasks such as syllogistic reasoning (Johnson-Laird, 1983) as well.

In this paper, we deal with this all-pervading task of model construction from a connectionist perspective. Specifically, we examine how to represent models in a connectionist network and how efficient, dynamic construction of such models can be achieved. Our treatment is at a general abstract level wherein the details of the individual cognitive activities are suppressed. In their details, there are a number of differences in the model construction process as it takes place in different cognitive activities; thus, for example, low level image processing differs in a number of ways from speech processing. But, when we disregard the details and examine the problem in a rather abstract fashion, there appears a great deal of similarity in the model construction process as it takes place in different cognitive activities such as vision, language processing, problem solving etc. It is

*This work was supported by DFG grant Schr 275/7-1.

at such an abstract level that we treat the problem of model construction in the rest of this paper.

2 Models and their Dynamic Construction

In the previous section, we informally described the task of model construction as the one of constructing a model of a situation based on some clue. In order to proceed, we need to formalize the notions of "model" and "clue".

Our attempt at formalizing the notion of "model" is inspired by the Tarskian semantics of predicate calculus. The idea is to describe the "world" in terms of a set of objects and relations between those objects. Following that scheme, we define a model to be an *explicit* representation of which relations hold between which objects in the "world"[1].

With that definition of "model", the task of model construction can be stated as follows: We are given some of the relation instances that hold in a situation (These constitute the "clue"); the task is to infer all the relation instances that hold in that situation (All those relation instances together constitute the "model").

Thus the task of model construction involves reasoning. What is the nature of that reasoning activity? The answer is that it is an integrated combination of a variety of reasoning that have traditionally been dealt with rather separately within AI. Upon hearing the sentence "John drove to the supermarket", a model constructed in our mind might consist of facts such as "John went to the supermarket", "John used a car", "he would be shopping", etc. Among these, infering that *goto(john,supermarket1)* is an instance of *deductive reasoning* since *drive-to(x,y)* necessarily means that *goto(x,y)* for any x and y. Infering that "John used a car" is an instance of *default reasoning* since with a few exceptions it is usually the case that when one says "x drove to y", the vehicle driven happens to be a car. Infering that the purpose behind John's driving to the supermarket must be one of shopping, is an instance of *abductive reasoning* (Charniak & McDermott, 1985). It involves examining the different possible purposes behind one's driving to a place (such as "to work there", "to shop there", "to meet someone there" etc.) and picking the most likely purpose in the given context. Model construction involves performing all these different kinds of inferences. In order to arrive at a system that dynamically constructs models, it *may not be necessary* that we make these distinctions between different kinds of reasoning. It may be possible to arrive at such a system directly (via learning or via designing) without ever thinking about the differences between the various inferences that constitute the overall model construction process. But, it so happens that the system being presented here was not arrived at directly. It began with a deductive, rule-based, backward reasoning system (Ajjanagadde & Shastri, 1989); later on, a deductive forward reasoning system was developed (Ajjanagadde & Shastri, 1991) and some enhancements in reasoning power were achieved (Shastri & Ajjanagadde, 1990). Then, a system which combines forward and backward chaining was developed. The work was extended to deal with evidential rules and facts, negation, and abductive reasoning in (Ajjanagadde, 1991). Since it is difficult to provide all the relevant details of our system for dynamic model construction here, we will take another approach to describing it. It so happens that our system for dynamic model construction has some important resemblances to Rumelhart et al's (1986) connectionist model of schemata. Since a typical reader can be assumed to be familiar with the work reported in (Rumelhart et al., 1986), it appears that a rough outline of our system can be provided by relating it to the work of Rumelhart et al. We will point out some important similarities and differences between Rumelhart et al.'s schema model and our system. It is hoped that this comparative discussion will provide the reader with a rough understanding of the ideas underlying our system. Details about our system can be found in the publications referred to.

3 Schema Model of Rumelhart et al.

In order to illustrate how schemata can be realized in connectionist networks, Rumelhart et al take as an example the problem of representing knowledge about various kinds of rooms, such as kitchen, bathroom, living room, and bedroom. They select forty microfeatures corresponding to such entities as *sofa, oven, refrigerator, telephone, toilet, television, toaster, bathtub, computer* etc. Corresponding to each microfeature, there exists a node in their network. The nodes corresponding to entities which will be found in the same room have mutual excitatory connections between them; nodes correspond-

[1] To contrast between explicit and implicit representations, let us consider a simple example. Suppose that the objects in the domain are a and b. Now, consider the statement $\forall x \forall y P(x, y)$. This representation *implicitly* represents the relations between the objects in the domain. An equivalent *explicit* representation would consist of the following statements: $P(a,a)$, $P(a,b)$, $P(b,a)$, and $P(b,b)$.

ing to entities which are unlikely to be found in the same room have mutual inhibitory connections between them. Thus, for example, *oven* and *refrigerator* are likely to be found in kitchen. So, there will be mutual excitatory connections between the nodes corresponding to *oven* and *refrigerator*. On the other hand, *bathtub* and *television* are unlikely to be found in the same room; so there exist mutual inhibitory connections between them. Now, the idea is that if we clamp some of the nodes corresponding to items present in a room, the network will settle into a state where the nodes corresponding to the other items in that room will be active and the rest of the nodes will be inactive. Thus, for example, if we clamp the nodes corresponding to *oven* and *refrigerator*, then, in the stable state, the nodes corresponding to other items in the kitchen, such as *toaster* will be active and the nodes corresponding to entities which are unlikely to be in the kitchen, such as *bathtub* will be inactive.

One important similarity between the problem addressed in (Rumelhart et al., 1986) and the problem taken up in this paper must be obvious. In (Rumelhart et al., 1986), the input is a specification of some of the items present in a room. Given that input, the network has to determine what other items are likely to be present in that room. In our case, the input is a specification of some of the relation instances present in the "world"; the network has to determine what other relation instances will be present in that "world".

4 Connectionist Network: Encoding

In (Rumelhart et al., 1986), the building blocks of schemas are microfeatures; a schema is represented by representing which features are present in that schema and which features are absent. We take predicates and objects as building blocks of mental models (Microfeatures can be viewed as special cases corresponding to 0-ary predicates.). A model is represented by representing which relation instances hold between which objects. Particularly, the arguments of these relations can be dynamically bound to objects, to represent the relation instances that hold in a model.

Corresponding to every microfeature, there exists a unique node in the network of Rumelhart et al. (1986). Similarly, corresponding to the different "objects" of interest, there exist unique nodes (referred to as *constant nodes* since they correspond to the "constants" of predicate logic.) in our network. Thus, in the example network of Fig. 2, there are unique nodes (shown as circles) corresponding to the "objects" *mary*, *jack*, *hospiz*, and *super-fries*. Now, consider the other building block of models, i.e., predicates. Corresponding to an *n-ary* predicate, the network has *(n+1)* nodes. Thus, in Fig. 1, corresponding to the tertiary predicate P, there are four nodes. The nodes $a1$, $a2$, and $a3$ correspond to the three arguments of the predicate P. We refer to these nodes as *argument nodes* (shown as diamonds in figures). Also, corresponding to every predicate, there exists a *predicate node* (shown as squares in figures).

In (Rumelhart et al., 1986), the relations between the different microfeatures are indicated by having (excitatory/inhibitory) connections between the corresponding nodes. Similarly, the relationships between the different predicates in our system are represented by connecting the nodes corresponding to different predicates. However, in this case we need to represent the correspondences between the arguments of the predicates as well. Suppose that when $P(x,y,z)$ (for arbitrary x, y, and z) is known to be true, this lends some amount of evidence (say, C) for $Q(y,z,x)$ being true. That is, the knowledge we have here is of the form

$$P(x,y,z) \Rightarrow Q(y,z,x) \text{ (with likelihood } C)$$

As per this rule, the first argument of Q is bound to the same individual that binds the second argument of P. This is denoted by connecting the argument node $a2$ to the argument node $a4$ (Fig. 1). The connection between the other argument nodes are similar. Also, the predicate node of P is connected to the predicate node of Q via a link whose weight is C. The weights of the links connecting the argument nodes is not of significance in our current work; all those links can be assumed to be having the same weight w, where w is some positive constant.

It is typical of practical examples that if knowing proposition A to be true lends positive (negative) evidence to another proposition B, then, knowing B to be true lends positive (negative) evidence to A. Fig. 1 depicts the connections representing that $P(x,y,z)$ lends evidence to $Q(y,z,x)$. To represent that $Q(y,z,x)$ lends evidence to $P(x,y,z)$, we need to also have connections in the direction opposite to that shown in Fig. 1. Thus, similar to the bidirectional connections in (Rumelhart et al., 1986), in our network also, there exist bidirectional connections between the nodes corresponding to different predicates. However, in addition to the difference of representing argument correspondences, there is another difference between the connections in our network and those in (Rumelhart et al., 1986). In the latter, the links running in opposite directions between two nodes have the same weight. In our network, this need not be the case. Consider the relationship between the predicates *have-dinner-at* and *eat-at* in Fig. 2. Knowing that x had dinner at y lends a very high evidence to the proposition that "x ate at y" (In fact, this is a certain implication.). On the other hand, though knowing that "x ate at y" lends a positive evidence to the proposition "x had dinner at y", the magnitude of this evidence is not as high as in the previous case. So, the weight of the link from the predicate node of *eat-at* to the predicate node of *have-dinner-at* is smaller than the weight of the link running in the opposite direction.

A feature of our network for which there is no strict conceptual parallel in (Rumelhart et al., 1986) corresponds to that of *background facts*. Background facts are specific facts present in the agent's memory. Examples of such facts may be "Jack is Mary's brother", "Jim is a computer scientist" etc. Such specific facts already present in the agent's memory significantly influence the model constructed in response to an input. For example, consider the processing of the following two sentences:

> John went to the supermarket.
> Mary went to the supermarket.

These two sentences contain similar information. However, due to the background facts present in the agent's memory, the model constructed in response to the first sentence could be significantly different from the model constructed in response to the second sentence. For example, suppose that the agent knew that "John has run out of groceries" and "Mary is an employee of the supermarket". In that case, the model constructed in response to the first sentence is likely to be the one of John going to the supermarket for shopping there. The model constructed in response to the second sentence is likely to be the one of Mary going to the supermarket to work there.

In the example network of Fig. 2, the encodings of three background facts, namely, *F1: hungry(jack)*, *F2: manager-of(mary,super-fries)* , and *F3: fast-food-shop(super-fries)* are shown (enclosed within hexagonal boxes). Let us skip the details of encoding background facts (Details can be found in (Ajjanagadde, 1991).)[2].

Another set of interconnections in our network for which there are no parallels in (Rumelhart et al., 1986) correspond to the representation of competition between alternative explanatory hypotheses. For example, two of the possible purposes behind one's going to a place are "to work at that place", and "to eat at that place". The competition between these two alternative possibilities is achieved in the network as follows. There is an inhibitory connection from the predicate node of *eat-at* onto the link from the predicate node of *goto* to the predicate node of *work-at* (Fig. 2). Similarly, there is an inhibitory connection from the predicate node of *work-at* onto the link from the predicate node of *goto* to the predicate node of *eat-at*. These inhibitory connections achieve the following winner-take-all kind of effect (Details in (Ajjanagadde, 1991).): If the activity level of the *eat-at* predicate node is higher than that of the *work-at* predicate node, the flow of activity along the link from the *goto* predicate node to the *work-at* predicate node gets cut-off. The reverse would be the situation if the activity level of the *work-at* predicate node is higher than that of the *eat-at* predicate. In effect, this mechanism results in the selection of that hypothesis which acquires maximum evidence.

5 Representation of Mental Models as Patterns of Activity

Previous section discussed the encodings present in our network. In this section, we will discuss how models are represented as patterns of activity in this network.

As mentioned earlier, a model is taken to be an explicit representation of the various relation instances holding in the "world". We will first discuss the pattern of activity representing one relation instance. The overall pattern of activity representing the model is a combination of the individual patterns corresponding to the different relation instances that constitute the model.

Instances of relations are represented in the network by dynamically binding the "objects" to the arguments of relations. The dynamic argument bindings are represented using *phase locked oscillations* (Ajjanagadde & Shastri, 1991). Essentially, the idea is to represent the binding of an object to an argument by the synchrony of activation of the node corresponding to the object and the node corresponding to the argument. Thus, the bindings in the fact *drive-to(jack,super-fries)* will be represented as follows: The argument node $a14$ and the node corresponding to the object *jack* will be active in synchrony. Similarly, the argument node $a15$ and the node corresponding to the object *super-fries* will be active in synchrony. The activity level of the *drive-*

[2] Actually, the network can be extended to encode background facts about classes of individuals instead of just individuals; space limitation precludes the discussion of that aspect here.

to predicate node represents the evidence for the fact *drive-to(jack,super-fries)*.

In order to suggest how models can be represented, the pattern of activity corresponding to the simultaneous representation of three relation instances is shown in Fig. 3.

6 Dynamic Construction of Models

Previous section described how models are represented as patterns of activity in the network. In this section, let us briefly examine how the network constructs a model when the input clue is specified. That is, the process we will be examining is the following: The input proposition(s) (e.g., say, *drive-to(jack,super-fries)*) will be specified to the network by clamping the pattern of activity representing the input proposition(s) onto the network. Now, the network has to generate the patterns of activity corresponding to the other relation instances which hold in the "world". The process is quite similar to the one described in (Rumelhart et al., 1986) wherein some of the nodes in the network are externally clamped and the rest of the nodes in the network settle into appropriate levels of activity. Let us discuss some major differences between the process in (Rumelhart et al., 1986) and in our network.

One main difference is that in our network the additional task of propagating variable bindings has to be done. This aspect has been described in detail in (Ajjanagadde & Shastri, 1991; Ajjanagadde, 1991).

The second main difference between the network of (Rumelhart et al., 1986) and ours is that in addition to the external clamping of input (similar to the external clamping of nodes in (Rumelhart et al., 1986)), there is also what can be viewed as *internal clamping* in our network. This is due to the presence of background facts. In order to clarify this, note that clamping of a node reflects the belief that the proposition represented by that node is true. Thus, in (Rumelhart et al., 1986), clamping of the nodes represents that the items represented by the clamped nodes are *known* to be present in a particular room. The background facts encoded in our connectionist network correspond to the facts the agent already believes to be true. Such background facts should constrain model construction in a fashion similar to the way in which externally clamped nodes do. However, note that not all background facts residing in the agent's memory will be relevant in any given context. Thus, for example, suppose that the natural language sentence being processed is "Mary went to Super-Fries". In this context, a background fact such as *employee-of(mary,super-fries)* will be relevant. That piece of background knowledge makes the agent to conclude that the most likely purpose behind Mary's visit must be the one of working at Super-Fries (rather than, say, eating there). But, when the input proposition is *goto(mary,super-fries)*, a background fact such as *employee-of(lisa,mcdonalds)* will not be relevant. In our network, upon the specification of the input, encodings of those background facts which are relevant in that context automatically get activated. Once activated, they constrain the model construction process in a way similar to externally clamped nodes do. For details of these, please refer to (Ajjanagadde, 1991).

Another difference between the network of Rumelhart et al. (1986) and our system is that the former uses energy minimization technique for reasoning while our network does distributed evidential reasoning by spreading activation. The approach we use is quite similar to that of Pearl (1986). The main reason for our choice is one of reasoning speed. It does not appear that the currently known energy minimization techniques can achieve the kind of reasoning speed we desire. For example, Derthik (1990) reports that with a rather small knowledge base, energy minimization took about 40,000 time steps to settle into the most plausible model. On the other hand, human beings are able to construct and manipulate mental models within fractions of a second. Taking into account the slowness of neurons (Feldman & Ballard, 1982), this means that model construction take place within a few tens to a few hundred time steps. By following the distributed evidential reasoning approach (quite similar to (Pearl, 1986)), it is possible to meet such tight constraints on the number of time steps.

7 Conclusion

The process of dynamic model construction underlies a large number of cognitive tasks. The paper outlined how mental models can be represented as patterns of activity in a massively parallel connectionist network and how can fast, dynamic construction of mental models be efficiently achieved.

Acknowledgments. I would like to thank Seppo Keronen, Vipin Kumar, Uwe Oestermeier and Peter Schroeder-Heister for their comments and suggestions.

References

Ajjanagadde,V. 1991. Abductive reasoning in connectionist networks: Incorporating variables, background knowledge, and structured explananda, Technical Report, WSI 91-7, Wilhelm-Schickard-Institut, Universitaet Tuebingen, Germany.

Ajjanagadde, V., and Shastri, L. 1989. Efficient inference with multi-place predicates and variables in a connectionist system. In Proceedings of the Eleventh Annual Conference of the Cognitive Science Society, 396-403. Hillsdale, NJ: Lawrence Erlbaum.

Ajjanagadde, V., and Shastri, L. 1991. Rules and variables in neural nets. *Neural Computation* 3:121-134.

Charniak, E., and McDermott, D. 1985. *Introduction to Artificial Intelligence.* Reading, MA: Addison Wesley.

Derthik, M. 1990. Mundane reasoning by settling on a plausible model. *Artificial Intelligence* 46:107-157.

Feldman, J.A., and Ballard, D.H. 1982. Connectionist models and their properties. *Cognitive Science* 6:205-254.

Johnson-Laird, P.N. 1983. *Mental Models.* Cambridge, London: Cambridge University Press.

Mannes, S.M., and Kintsch, W. 1991. Routine computing tasks: Planning as understanding. *Cognitive Science* 15:305-342.

Pearl, J. 1986. Fusion, propagation, and structuring in belief networks. *Artificial Intelligence* 29:241-288.

Rumelhart, D., Smolensky, P., McClelland, J., Hinton, G. 1986. Schemata and sequential thought processes in PDP models. In McClelland, J., Rumelhart, D., and the PDP Research Group (eds.), *Parallel Distributed Processing: Psychological and Biological Models.* Cambridge, MA: MIT Press.

Shastri, L., and Ajjanagadde, V. 1990. From simple associations to systematic reasoning: A connectionist representation of rules, variables, and dynamic bindings, Technical Report, MS-CIS-90-05, Department of Computer and Information Science, University of Pennsylvania.

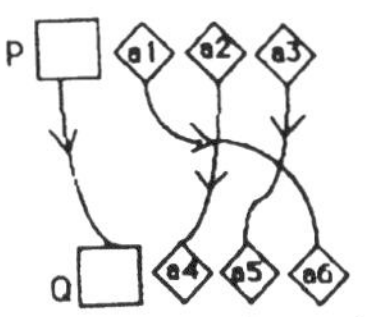

Fig.1 Encoding P(x,y,z) => Q(y,z,x).

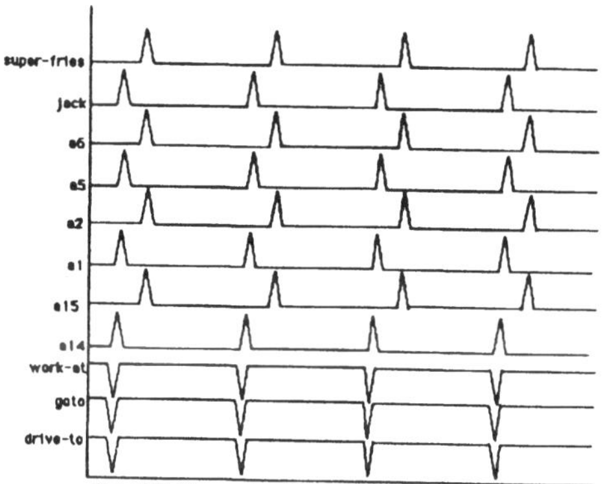

Fig. 3 Pattern of activity representing the facts drive-to(jack,super-fries), goto(jack,super-fries) and work-at(jack,super-fries).

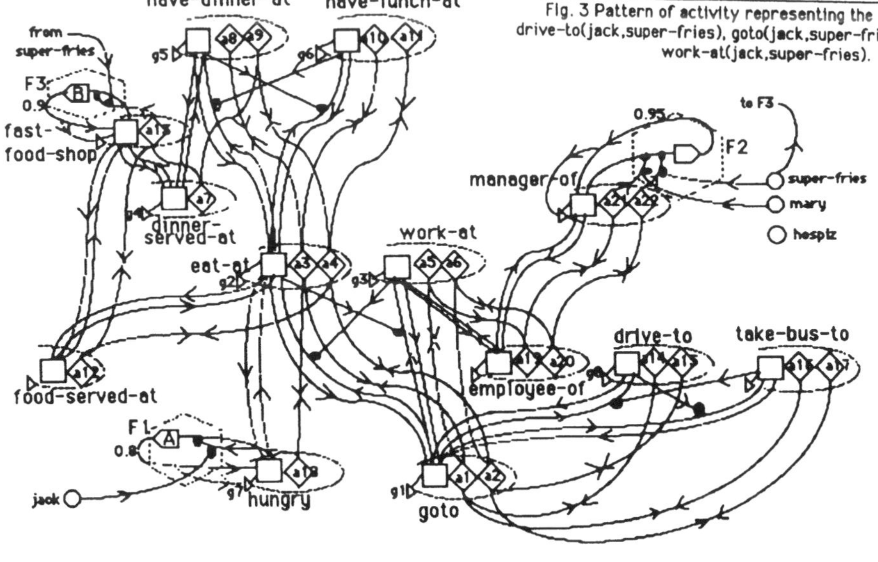

Fig.2 An Example Network.

Learning Relations in an Interactive Architecture

Randall Stark
School of Cognitive and Computing Sciences, University of Sussex
Falmer, Brighton, East Sussex BN1 9QH UK
email: randalls@uk.ac.sussex.cogs

Abstract

This paper presents a connectionist architecture for deriving unknown role fillers in relational expressions. First, a restricted solution to the binding problem is presented which ensures systematicity in principle, and allows for sufficient compositionality so as to enable instantiation of shared variables in conjunctive expressions where the same object may fill a variety of roles in a variety of relations. Next, a more detailed architecture is explicated (an extension of McClelland's 1981 "Interactive Activation Competition" architecture) which allows for systematicity in practice while providing a training procedure for relations. Finally, results of the learning procedure for the Family Tree data set (Hinton, 1990) are used to demonstrate robust generalization in this domain.

1. Introduction

This paper outlines an architecture for connectionist symbol processing. The task driving this architecture is that of **variable instantiation**. This task involves contexts with any number of objects playing any number of roles in any number of relations. Given a subset of the objects, the goal is to simultaneously derive all of the unknown role fillers (Stark, 1992). The focus is on learning relations in a manner compatible with a principled binding strategy (one that allows bindings to be propagated so as to allow role fillers to be derived).

2. An Interactive Binding Strategy

While connectionist networks are good at representing single distinct (or schematic) objects, they do not perform as well when simultaneously representing multiple objects, making it difficult to distinguish which features belong to which objects, or which roles objects are playing in a relation. This is known as the *binding problem* (Hinton, McClelland, & Rumelhart, 1986; Smolensky, 1990).

The variable instantiation task serves to constrain the nature of the binding problem. Rather than being concerned with developing a universal binding scheme to encode arbitrary structures (e.g. Smolensky, 1990; Pollack, 1990), the primary concern here is with providing a connectionist architecture that exhibits *systematicity* (the ability to allow in principle any object to appear in any role of any relation) without having to *a priori* dedicate hardware to allow for all possibilities (cf. Fodor & Pylyshyn, 1988).

Consider an *object-representing network* (Figure 1), viewed as a vector of feature units and a connectivity matrix. Here, the auto-associative network functions as a content-addressable memory that will settle on an appropriate object representation. Such a network can realize a distributed representation of a single object, or a single object schema; the features define a vector space in which points correspond to specific objects or possible schemas.

The problem of simultaneously representing multiple distinct objects can be addressed by utilizing multiple copies of the basic object-representing network (Hinton, McClelland, & Rumelhart, 1986), and arranging them in an interactive architecture. A *context* involving two objects will be computed with a *context network* consisting of two object-representing subnetworks, with additional connections between them governing their interaction. The additional connections are derived from the roles the objects are to play in a given relation in the context network. Functioning as content-addressable memories, the networks can cooperate in simultaneously forming representations of distinct

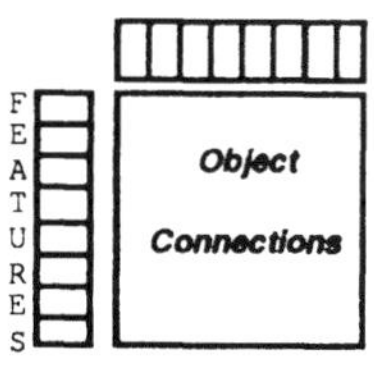

Figure 1: **Object Representing Net**

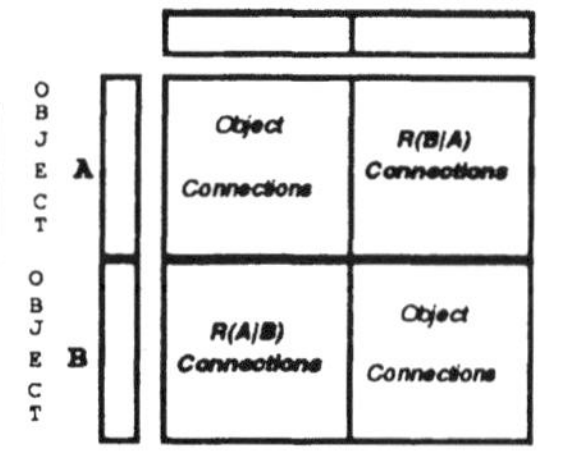

Figure 2: **R(A,B) Context Network**

The work reported in this paper was supported by a grant from the Joint Council Initiative for Cognitive Science (MRC G8920680). The author wishes to thank Dr. Chris Thornton for helpful discussions at various stages of this work.

objects in specific roles, constraining each other's attempts to settle into minima.

A two-object context network operates on a vector made by concatenating two copies of the feature vector associated with an object-representing network, and a connection matrix divided into four equal submatrices. The upper-left and lower-right submatrices (on the main diagonal) are each copies of the object-representing network's connection matrix. The other two submatrices, called *binding matrices*, each contain constraints on one object given the other object.

Figure 2 shows a network for computing a context with two objects (**A** and **B**) that are in role one and role two of the relation **R** respectively. In this network, the "within layer" connections for each object are just exact copies of the basic connection matrix shown in Figure 1. The other two submatrices are the binding matrices. The lower-left binding matrix contains connections representing constraints on object **A**'s feature vector given that it is in the first role of relation **R** with object **B** in the second role (denoted **R(A|B)**), and the upper-right binding matrix contains connections representing constraints on object **B**'s feature vector given that it is in the second role of relation **R** with object **A** in the first role (denoted **R(B|A)**).

Contexts of arbitrary complexity can be created using such binding matrices, for example as in Figure 3, where three distinct objects play roles in two relations, with one object (**B**) playing different roles in different relations. A generative, compositional syntactic description can be used to describe each context (see Figure titles); furthermore, any context describable with this conjunctive, predicate-based syntax has a corresponding network representation.

Systematicity is realized because each object-representing network is in principle capable of representing any object or schema. Since each variable has its own subspace (containing points corresponding to possible bindings), crosstalk problems are brought under control. When crosstalk is desired (as in the case of a variable being bound to a schema), superpositional representation still occurs *within* a variable subspace. Object-representing connection matrices (along the main diagonal of a context) provide mappings *within* these subspaces, while the binding matrices provide mappings *between* variable subspaces.

Most of the work reported in this paper is devoted to describing an advantageous vector representation of objects and explicating procedures for determining the contents of the binding matrices. That such procedures exist can be seen by considering random object vectors and a simple learning procedure (such as that in a Hopfield net). The object-representing network can be trained by applying the Hopfield learning procedure with a training set consisting of all the object vectors. The binding matrices for each relation can be learned by applying the same learning procedure on a network twice the size (as in Figure 2), where the training set for each relation consists of vectors obtained by concatenating the object vectors of each pair of objects observed in the relation. When learning the binding matrices, the connections in the object-representing networks are "frozen", that is, only connections in the upper-right and lower-left quadrants are learned.

Binding is then a *constructive* process in which a context network is generated by creating a unit vector which consists of *n* concatenated copies of the basic object vector, where *n* is the number of objects (or *variables*) in the context. The overall connection matrix can then be constructed dynamically, using the object-representing network's matrix along the main diagonal of the context matrix, and filling in the binding matrices learned for each of the relations, as in Figure 3. Thus contexts involving any configuration of objects in any conjunctive configuration of relations can be modelled. It is this property of the architecture that I refer to as *compositional*.

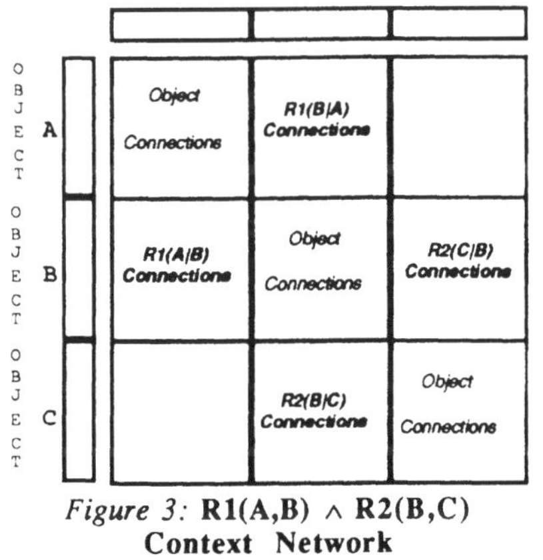

Figure 3: **R1(A,B) $\wedge$ R2(B,C) Context Network**

3. Interactive Representation of Objects

This section looks at one aspect of the more detailed architecture by considering structure *within* the object-representing networks (Figure 1). The architecture used to represent objects is based on McClelland's (1981) "Interactive Activation Competition" (IAC) architecture, best known as the one underlying the "Jets and Sharks" model. The IAC architecture provides both a localist representation with an *instance* subnetwork containing a unit for each object, and a form of distributed representation whereby each object is represented by a pattern of activation over the remaining feature units. These units are further divided into *attribute* subnetworks, each with unique units for each value an attribute may take. Networks using this architecture are able to represent objects and exhibit a number of interesting properties, including the ability to form schemas and function as a content-addressable memory (the network is able to "fill in" an object's attribute values given a subset of them).

When considering the problem of learning relations, however, the *right* features and attributes need be present. The binding scheme described in Section 2 is dependent on features being explicitly represented if they are important in deriving the nature of a relationship. There is nothing inherent in the architecture to guarantee that this condition is met. While not claiming to have found a general solution, it is suggested that certain specific attributes are useful in representing and learning relations.

In particular, attributes associated with the relations themselves can be seen to be of use. If *John loves Mary*, then John has the attribute of loving someone, namely Mary. Likewise, Mary has the attribute of being loved (by John). The IAC architecture offers a simple way of modeling such attributes in the same manner as any other attribute; for a given two-place relation, two additional attribute subnetworks (one for each role of the relation) may be incorporated in the model, each with value units for each object that has been observed in the given role of the given relation. This is a form of *conjunctive encoding* (Hinton, McClelland, & Rumelhart, 1986), since each unit represents the conjunct of a relation, a role, and an object. Thus part of an object's distributed representation will involve features which indicate that object's relationship to other objects in the domain. The remainder of this paper will focus solely on these relational attributes.

Consider an example domain with three individuals (*John, Mary*, and *Sally*), with the five observed facts: `loves(John,Mary)`, `loves(Sally,John)`, `hates(Mary,John)`, `hates(John,Sally)`, `hates(Sally,Mary)`. The corresponding IAC network (in both schematic and vector/matrix form) is shown in Figure 4 (the connection matrix is a detail of the object-representing matrix [Figure 1] duplicated along the main diagonal in Figures 2 and 3).

While this approach allows the realization of simple relational attributes, the limitations of conjunctive coding are well known (Hinton, McClelland, & Rumelhart, 1986; Fodor & Pylyshyn, 1988). As the goal is to be able to compute with contexts of arbitrary complexity, allowing for any configuration of objects in any configuration of relations, a degree of *compositionality* is required that cannot be attained through conjunctive coding alone, if we are to be able to represent not just, e.g., the person that John loves, but also the person who hates the person who John loves, or the mother of the person who hates the person who John loves, etc.

4. Simple Binding

It is to overcome these limitations that the interactive binding strategy was developed. We will first consider a very simple binding procedure (i.e., a method of deriving the connection weights in the binding matrices) to demonstrate the basic principle using the IAC architecture. This procedure does not require any training procedure or learning of weights, as does the full binding procedure presented in the next section.

This simple binding procedure can be seen in terms of Hinton's (1990) discussion of "expanding part/whole hierarchies". An IAC network (the *whole*) consists of a number of subnetworks, some of which represent a specific attribute (a *part*). While the whole is able to represent objects using a distributed representation (which includes all of the attributes), the representation of objects within an attribute network is wholly localist. The effect of the binding procedure is to selectively "expand" some of these "partial" subnetworks into a "whole", enabling a full, distributed representation of its value, which is another unique object (or schema). Thus in a context denoted by *loves(X,Y)*, there will be two copies of the IAC network, one representing *X* and one representing *Y*. The binding procedure will provide a mapping between the *lovee* attribute in the *X* network and the entire *Y* network. This expansion is not strictly hierarchical, as in Hinton's discussion, since there will be an additional mapping between the *lover* subnetwork in the *Y* network and the entire *X* network (i.e., each network is an expansion of a part of the other network).

The first step in determining how to fill in the binding matrices is to map them out in a similar manner to the IAC object-representing network

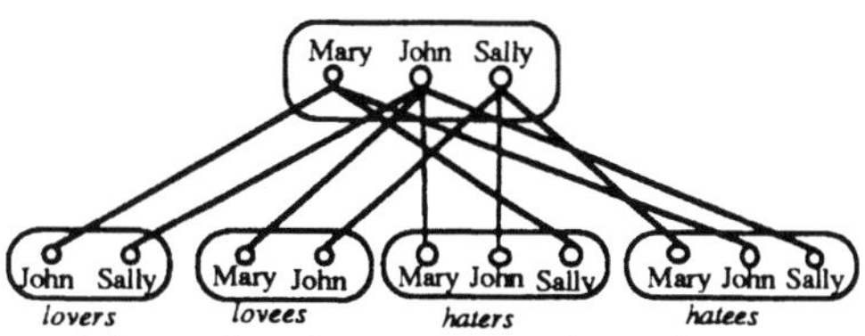

*Figure 4a: **Example Domain IAC Network***

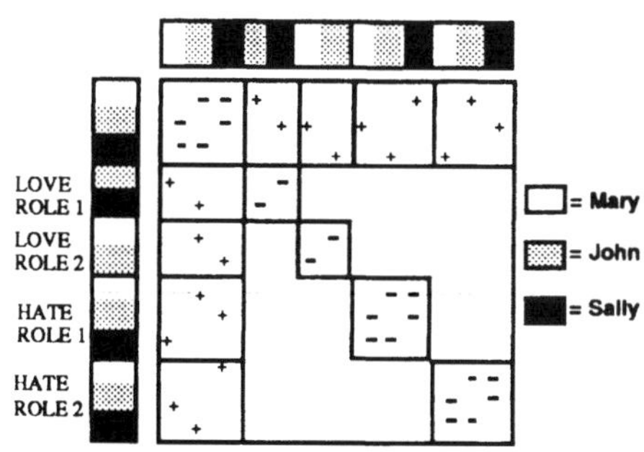

*Figure 4b: **Connection Matrix***

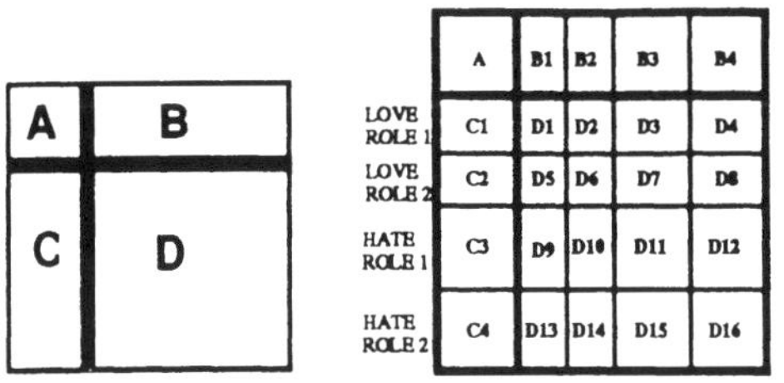

Figure 5 & 6: **Binding Regions & SubRegions**

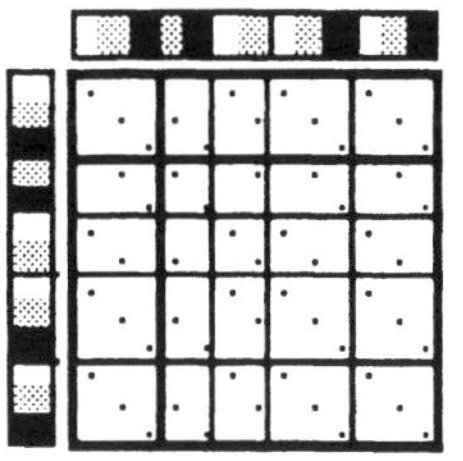

Figure 7: **Binding Connections**

connection matrices (Figure 4). This exposes that the mapping between objects can be seen in terms of mappings between aspects of the representations of objects. Figure 5 shows how each binding matrix can be divided into four "regions": **A** (a mapping between localist representations), **B** and **C** (mappings between localist representations and distributed representations, and *vice versa*), and **D** (a mapping between distributed representations). A further level of structure exposes the individual submappings in each of the regions, involving specific attributes, as shown for the example domain in Figure 6.

Next we consider which of these binding matrix regions and subregions will receive non-zero weights. To accomplish the part/whole expansion of the simple binding procedure, only subregions in the **B** and **C** regions will have non-zero weights. Specifically, two subregions in each binding matrix (one in region **B** and one in region **C**) will be eligible, corresponding to the mappings involving the expanded attribute in these regions. In the example domain (Figure 4), for the context *loves(X,Y)*, subregions **B1** and **C2** will be eligible in the upper-right binding matrix (Figure 2), as will subregions **B2** and **C1** in the lower-left binding matrix.

The issue of which specific connections *within* these subregions will receive non-zero weights is determined by a notion of *co-reference* (motivated in part by Fodor and Pylyshyn's own discussion of "the role of labels in connectionist theories", particularly footnote 12). If we consider two basic sets, one of *units* and one of *labels*, and define a *reference function* that maps from units to labels (such that each unit references a label), then a notion of *co-reference* can be defined as a relation between units which is true iff they reference the same label.

This enables a definition of a rule of co-reference: *Only co-referencing units may be inter-connected at bind time*. Figure 7 shows all of the potential connections in each binding matrix for the example domain. The co-reference rule combined with the subregional breakdown of the binding matrices allows a definition of the simple binding procedure:

Simple Binding Procedure: *To bind two objects in a relation **R**, use two copies of the object-representation network (forming a context network). Positively connect co-referencing units between the instance subnetwork of the first object network and the attribute subnetwork representing the first role of relation **R** in the second object network. Likewise, positively connect co-referencing units between the instance subnetwork of the second object network and the attribute subnetwork representing the second role of relation **R** in the first object network.*

Given a complete description of a domain, contexts of arbitrary complexity (limited only by resources) may be constructed using repeated application of this binding procedure (as in Figure 3). Given information about at least one object in the context, and appropriate network dynamics, the context network will settle in a state whereby all unclamped object-representing subnetworks will represent "solution" objects (or schemas) appropriate for the context. If the set of observed domain facts is complete, *the solutions will be the same as would have been derived by traditional means* (e.g., by Prolog).

5. Training and Generalization

The simple binding procedure, while demonstrating the basic power of the interactive strategy to perform variable instantiation within conjunctively specified contexts, requires complete domain information in order to derive solutions. Although the strategy allows for each object network to represent any object *in principle*, the simple binding procedure will not result in object networks settling on representations of objects that have not been observed in a specified role of a relation. This section outlines a procedure for *learning* binding matrix connection weights and allowing a greater degree of generalization.

The procedure is simple, and follows that outlined near the end of Section 2. The binding matrices are learned using a two-network context (as in Figure 2). A pair of binding matrices are learned for each relation, using as a training set a set of vectors obtained by concatenating the vector representations for each pair of objects observed to be in the relation. Only co-referencing connections are learned (see Figure 7), and the object-representing connections are "frozen" (so as to allow object networks to settle on any object, not just ones observed in the given role of the given relation). Any learning rule may potentially be used.

Generalization is achieved by applying a special "generalization" rule to each resultant matrix. The effect of this rule is to "combine" all of the learned connections in each subregion (Figure 6) into a *single* value, thus forming a *correlation matrix* which represents a generalized version of the binding weights for a relation. Any of a variety of generalization rules may be used, such as taking the mean.

When the binding procedure is invoked, it will supply the weights for each of the potential connections in each of the binding matrices by finding a uniform value for each connection using a normalization formula applied to the values in the appropriate correlation matrix. Thus every eligible connection in a given subregion will have the same weight after binding, even if the learning rule derived different weights for each connection.

Analysis of binding matrices in terms of subregions allows each inter-attribute mapping to be considered separately. As attribute values are mapped at run time, the inherent content-addressable properties of the IAC architecture allow each individual object-representing subnetwork to settle on an object representation that is consistent with the other object representations being derived in the context network.

The co-reference rule provides a means of raising the power of learned mappings by considering not just whether an object in one role is likely to have a specific value for an attribute given that an object in another role has a given attribute value, but whether two objects are likely to have the *same* value for any of their attributes. The correlation matrix subregions can be interpreted in terms of rules governing the objects in the relation. Region **A** contains a single reflexive rule, while regions **B** and **C** encode symmetry rules. Thus in the example domain these regions will encode the rule

`loves(`O_1`,`O_2`) -> hates(`O_2`,`O_1`).`

More complex rules involving third parties are handled in each of the **D** subregions; e.g. subregion **D3** (Figure 6) in the lower-left correlation matrix in the example domain encodes the rule

`loves(X, `O_1`) -> hates(X, `O_2`)`

(which is always true in the example domain when O_1 and O_2 are bound in the relation `loves`, as the rule asserts that if the `lover` [O_1] is loved by someone [X], that person [X] hates the `lovee` [O_2]).

6. Experimental Results

An implementation of this architecture (described in the Appendix) has been used as the basis for an experimental study of various aspects of the architecture, using the "Family Tree" domain of Hinton (1990) (and others, e.g. Quinlan, 1990; Melz & Holyoak, 1991). This domain of kinship relations consists of twenty-four individuals and twelve relations, organized in two isomorphic "family trees". There are a total of 112 "facts" in this domain (when considered as triples). The current architecture is well suited to handle this domain because the kinship relations are all definable in terms of other relations.

Generalization was tested by deriving 400 training sets, such that each set contained between 60% and 100% of the facts in the domain. For each training set, the training procedure was executed, and 172 two-object contexts were constructed, each with one object known (68 contexts in which the object in the first role was known, and 104 contexts in which the second role object was known). After settling, the objects in the missing role were derived by examining the activation of units in the instance subnetwork of the unclamped object network (see the Appendix and Stark, 1992). Perfect performance was indicated by the proper set of 224 objects (112 in each role) being determined by the settled context networks.

Figure 8 show the results of the basic test. The X axis represents the percentage of the domain facts present in each training set, and the Y axis represents the percentage of unknown objects that were derived correctly. The diagonal line indicates expected performance if no generalization took place (i.e., X = Y). In the graph, training sets with the same number of missing facts are grouped together, and their max, min, and mean plotted.

Hinton (1990) reports variable results with 4% of the facts missing, as did Quinlan (1991). The current system performs perfectly on training sets with nearly 20% of the facts missing, and can still retrieve over 90% of missing role fillers in cases where 40% of the facts are missing from the training set. The current experiment, involving 400 different data sets, shows the importance of exactly *which* facts are missing (as can be seen in the variance between the max and min figures for each test set group).

Other experiments show that the effect is quite robust, demonstrating considerable parameter insensitivity and tolerance to "lesions" in the binding matrices. In addition, contexts with up to a dozen objects have been tested and found to perform well, especially when a high percentage of the domain facts are known.

The fact that relational correlations are stored independently of specific objects, coupled with the

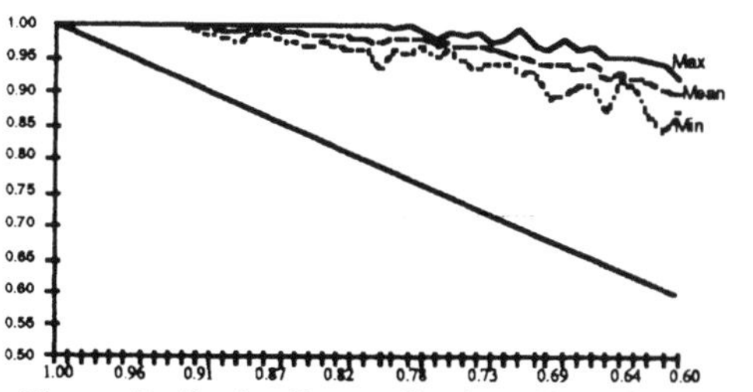

Figure 8: Basic Generalization Test Results

dynamic nature of the binding procedure driven by the co-reference rule, enables a more powerful type of generalization: a set of relations once learned may be applied to a new set of objects *without retraining* provided the regularities governing the relations remain the same, thus achieving effects similar to those reported by Melz & Holyoak (1991).

7. Discussion

This system, and particularly its approach to the binding problem, differs from other similar ones primarily in its goals. Rather than focussing on a universal encoding scheme (such as, e.g. Smolensky, 1990; Pollack, 1990), the emphasis is placed on learning relations and propagating bindings in order to perform variable instantiation in contexts describable by a generative predicate-based syntax.

While Ajjanagadde and Shastri's (1991) temporal binding system does focus on propagation of bindings in a variable instantiation task, and indeed offers provably correct inference in this domain, it does not offer a training procedure, does not address the issue of shared variables (compositionality), and is limited to localist or quasi-localist representation of objects. Hinton, McClelland, and Rumelhart (1986) suggest the possibility of solving the binding problem by making "multiple copies", but express concern about the *implementation* of copies. Temporal binding may indeed provide an implementation mechanism for my scheme, which is dependant on some form of "copies" of a basic network, but the current work focuses not on how copies are made, but rather examines when they are needed and how they should interact.

The current system has a number of important limitations. These include a treatment only of two-place predicates, and the ability to perform only first order bindings; it would be useful to be able to determine what relation two objects are in, given their roles. Perhaps the most important limitation is the lack of hidden units. This lack is partially motivated by an interest in seeing how far one could go in solving problems such as the Family Tree without using hidden units. An extended architecture that exploits hidden units has been developed and will be the subject of future experimentation.

Appendix (Implementation)

Unit activation function:

$$U_j = \max\left(0, \tanh\left(\sum_i U_i W_{ij}\right)\right)$$

Learning Rule:

$$W_{ij} = p(U_j | U_i) = \frac{(U_i \wedge U_j)}{U_i}$$

Generalization Rule:

$C_r = \text{mean}(W_{ij} > 0)$ (for each subregion **r**).

Normalization Rule:

$$W_{ij} = 0.1\, C_r$$

Constant weights in each IAC network were set to **0.1** and **-0.1**. The network was allowed **20** cycles to settle.

The localist competition subnetwork in each object network was assumed to be a "K-winner-take-all" network, where **K** was equal to the number of objects in the solution (this was assumed to be information supplied to the system). The weights for each localist subnetwork were set according to the formula

$$W = \frac{0.3}{K^2}$$

After settling, the **K** units with the highest activations (above zero) were taken as solution. No units with zero activation were considered solutions.

References

Ajjanagadde, V. and Shastri, L. (1991) "Rules and Variables in Neural Nets," *Neural Computation*, 3:121-134.

Fodor, J. A. and Pylyshyn, Z. W. (1988) "Connectionism and Cognitive Architecture: A Critical Analysis," *Cognition*, **28**:3-71.

Hinton, G.E. (1990) "Mapping Part-Whole Hierarchies into Connectionist Networks," *Artificial Intelligence* **46**:47-75

Hinton, G.E. McClelland, J.L., and Rumelhart,D.E. (1986) "Distributed Representations," in Rumelhart, D.E., McClelland, J.L. and the PDP Research Group, eds., *Parallel Distributed Processing*, MIT Press, Cambridge, MA.

McClelland, J.L. (1981): "Retrieving General and Specific Knowledge from Stored Knowledge of Specifics," *Proceedings of the Third Annual Conference of the Cognitive Science Society*, Berkeley, CA.

Melz, E., and Holyoak, K. (1991) "Analogical transfer by constraint satisfaction," *Proceedings of the Thirteenth Annual Conference of the Cognitive Science Society*, Chicago, IL.

Pollack, J.B. (1990) "Recursive Distributed Representations," *Artificial Intelligence* **46**: 77-107.

Smolensky, P. (1990) "Tensor product variable binding and the representation of symbolic structures in connectionist systems," *Artificial Intelligence* **46**:5-46.

Stark, R. (1992) "A Symbolic/Subsymbolic Interface for Variable Instantiation," to appear in *Artificial Neural Networks II: Proceedings of the International Conference on Artificial Neural Networks*, Elsevier.

Quinlan, J.R. (1990) "Learning Logical Definitions from Relations," *Machine Learning* **5**:239-266.

Developing Microfeatures by Analogy

Eric R. Melz
Department of Psychology
University of California, Los Angeles, CA 90024
emelz@cognet.ucla.edu

Abstract*

A technique is described whereby the output of ACME, a localist constraint satisfaction model of analogical mapping (Holyoak & Thagard, 1989) is used to constrain the distributed representations of domain objects developed by Hinton's (1986) multilayer model of propositional learning. In a series of computational experiments, the ability of Hinton's network to transfer knowledge from a source domain to a target domain is systematically examined by first training the model on the full set of propositions representing a source domain together with a subset of propositions representing an isomorphic target domain, and then testing the network on the untrained target propositions. Without additional constraints, basic gradient descent can recover only a negligible proportion of the untrained propositions. Comparison of simulation results using various combinations of the distributed mapping technique and weight decay, indicate that general purpose network optimization techniques may go some ways towards improving the transfer performance of distributed network models. However, performance can be improved substantially more when optimization techniques are combined with the distributed representation mapping technique.

Introduction

Theoretical accounts of analogy posit at least two central stages in the process of analogizing: *mapping*, the establishment of systematic correspondences between objects and relations of a source domain and a target domain, and *transfer*, the importation of knowledge from the source domain into the target domain based on the correspondences established by the mapping phase. Numerous models of analogical transfer have been proposed, each of which explicitly implements the mapping and transfer phases, usually employing traditional techniques of symbolic processing (Hall, 1989; Falkenhainer, Forbus & Gentner, 1989; Holyoak, Novick & Melz, 1992). With the advent of distributed connectionist models, there appears to be some promise for eliminating the cumbersome symbolic machinery of traditional models of analogy. Using a single general purpose learning technique such as backpropagation (Rumelhart, Hinton & Williams, 1986), it may be possible to simply train a model on a set of propositions representing some domain of knowledge, and exploit the generalization capabilities of the network in order to generate useful inferences based on the knowledge the network has obtained.

This paper explores the ability of supervised gradient descent learning to (a) form representations of correspondences between two domains, and (b) use these representations to perform analogical transfer. I first demonstrate that without additional constraints on the learning procedure, gradient descent does not form representations that are optimal for the task of analogical transfer. I then examine two mechanisms which can induce the network to form optimal representations: (1) simple weight decay, and (2) "programming" the network with correspondenceds derived from a connectionist model of analogical mapping. In the next section, I describe the network and domain with which the rest of this paper is conerned.

Hinton's (1986) Family Tree Problem

Hinton (1986) describes a network that learns relationships between people in two isomorphic family trees. The family trees, shown in Fig. 1 can be represented by two sets of 52 propositions of the form *obj1 rel obj2*, where *obj1* and *obj2* are two relatives, and *rel* is the relationship between *obj1* and *obj2* (which may be one of the following: *has_husband, has_wife, has_son, has_daughter, has_mother, has_father, has_brother, has_sister, has_uncle, has_aunt, has_nephew,* and *has_niece*). For example, the fact that Christopher is married to Penelope is represented by the proposition *Christopher has_wife Penelope*. The family tree which contains the English people will henceforth be referred to as the source domain, and the Italian family tree will be referred to as the target domain.

The network which learns the relationships is shown in Fig. 2. The input layer is composed of localist representations of the 24 domain objects (12 English people, and 12 Italians) that can fill the *obj1* slot, and the localist representations of the 12 relationships listed above. The second layer converts the localist representations of the objects and the relations to distributed representations, and the third layer combines these two distributed representations into a distributed representation of the proposition. This layer is transformed into a distributed

* This research was supported by Contract MDA 903-89 K-0179 from the Army Research Institute, and NSF Grant DIR-9024251 to the UCLA Cognitive Science Program.

representation of *obj2* (the penultimate layer), which is finally converted into a localist representation of *obj2* (again consisting of 24 units representing the domain objects). The network is trained by the clamping the appropriate *obj1* and *rel* nodes on the input layer and *obj2* nodes on the output layer for each proposition, and performing backpropagation.

Figure 1: Two isomorphic family trees. The symbol "=" means "married to". (From Hinton, 1986)

To probe the analogical capacity of the network, we train the network on the full set of source propositions and a subset of target propositions. If the network has developed internal representations which allow it to utilize the similarity between the two domains of knowledge, it should be able to activate the correct *obj2* unit on the layer when the *obj1* and *rel* units of an untrained proposition are clamped on the input layer. Of course, success of the network in generating new target propositions does not licence the claim that the network is referring to specific propositions in the source domain. The network may, for example, be using mechanisms more closely related to rule-based inference. For example, if the network has learned that the sister of a person x is a female about the same age as x, it might be able to correctly infer that Sophia is the sister of Alfonso, even without reference to the fact that Charlotte is Colin's sister. However, we can reasonably claim that failure of the network in recreating untrained target propositions indicates that the network is *not* doing analogy. The point is that analogy is a sufficient but not necessary mechanism to make inferences within this problem domain.

The simulations reported throughout paper use Fahlman's (1988) quickprop algorithm, which is essentially a faster version of backpropagation. In standard backpropagation, each weight is adjusted in direct proportion to the magnitude of the partial derivative of the network's cost function with respect to the weight. In contrast, quickprop makes the simplifying assumption that the surface of the cost function is quadratic, and at each epoch, weights are set to the minimum value of the idealized surface by computing the curvature of the cost surface at the current point in weight space. In practical terms, this means that while backprop generally takes small steps in weight space, quickprop is capable of taking large leaps in weight space, which can greatly improve the overall learning times. For example, when the model was trained on all propositions in both domains, 4 runs of backprop using Hinton's (1986) parameters took an average of approximately 15,000 epochs to reduce the cost function to a value below 1.0[1], while 4 runs of quickprop took an average of approximately 500 epochs to meet the same criteria.

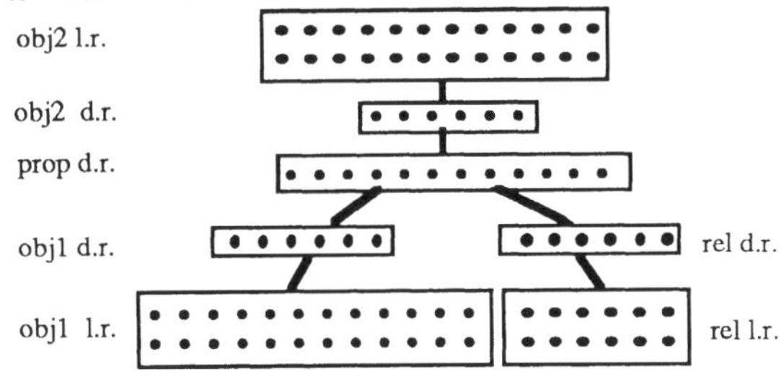

Figure 2: Hinton's (1986) model of propositional learning.

While the specific implementation details of the algorithms are not terribly important with respect to our present concerns, it is important to note that backprop and quickprop are fundamentally similar in that they both implement gradient descent on a cost function. Hence, we can reasonably expect that the internal representations that each algorithm develops will be qualitatively similar, although there may be slight differences as a side effect of the different implementations. The extent to which such differences exist remains an empirical issue, and is not explored further in this paper.

The parameters used for all simulations reported in this paper were $\mu = 1.75$, $\varepsilon = .01$ and *momentum* = 0.9 (see Fahlman 1988 for further explanation of the parameters and the algorithm). Weight decay was set either to 0.0 or 0.0005, depending on the set of simulations. Target values for the output were .8 if the output unit should be on, and .2 if the output should be off. The following cost function was used:

$$C = \Sigma_p \Sigma_o (t_{p,o} - o_{p,o})^2 + {}^1/_2 \lambda \Sigma_i w_i^2,$$

where p is an index over the training patterns, o is an index over the output units, i is an index over the weights, $t_{p,o}$ denotes the target value for unit o on pattern p, $o_{p,o}$ denotes the output value produced by unit o on pattern p, and λ is a weight decay parameter. In addition, one minor modification suggested by Hinton (1986) was implemented: if the network produces

[1] Note that the backprop training times significantly differ from those reported by Hinton. Hinton (personal communication, 1992) confirms that the training times reported in his paper are probably erroneous.

output values more extreme than the target values, the error for that unit is taken to be 0 rather than $(t_{p,o}-o_{p,o})^2$. When testing a training pattern, a liberal criterion is used: a case is considered to have "passed" if all units that should be off have activation values less than .4, and all units that should be on have activations values above .6.

For the first transfer test, the weight decay parameter was set to 0, and the network was trained on all source propositions and all but four randomly chosen target propositions. In the test phase, the network passed on all of the trained propositions, but failed on all of the untrained propositions. A principal components analysis of the distributed representations for *obj1* and *obj2* (i.e., the activations patterns on second and penultimate layers, respectively) yields some insight into the poor performance of the network (Fig. 3). The organization the network develops for the *obj1* representation is fairly sloppy: although the network seems to loosely separate the English people from the Italians, there is no rigid structure apparent within each domain, and the symmetry between the two domains is weak. The network appears to have "memorized" each person without significantly taking into account the relationships people participate in or correspondences between the source and target domains. Likewise, the representations of *obj2* appear to be randomly distributed in activation space. There is little, if any similarity between the organization of the English and the Italian representations. Also, multiple instances of each person seem to be diffusely scattered in the representation space. Ideally, we would like the network to develop a unitary representation for each person, so that the need for a many-to-one mapping between the distributed representation on the penultimate layer and the localist representation on the output layer is obviated. Such many-to-one mappings are bound to degrade the generalization of the network, as we shall see later.

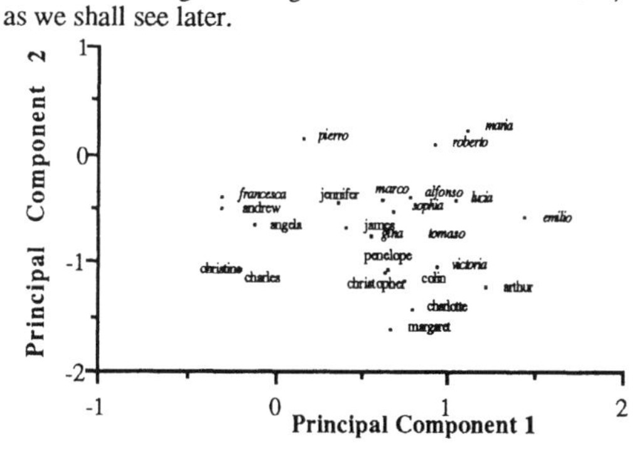

(a)

Clearly, without additional constraints on the development of the network's internal representations, we can expect generalization to be poor. In the next two sections, I examine how the network can be induced to form sensible distributed representations, first by using knowledge of the domain derived from a connectionist model of analogical mapping, and secondly by employing domain- independent network optimization techniques.

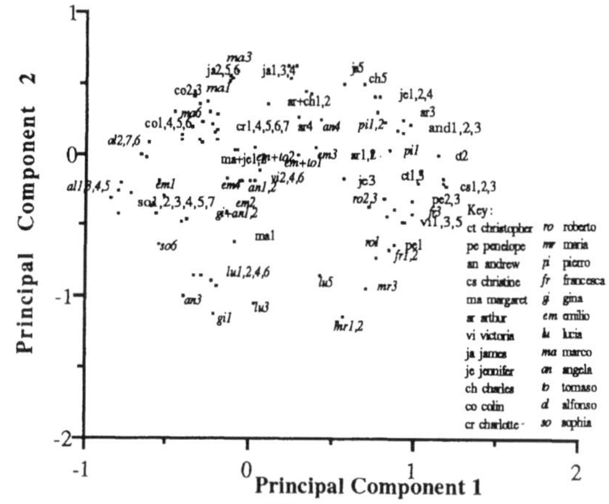

(b)

Figure 3: Principal components analysis of (a) *obj1* and (b) *obj2* for basic gradient descent learning. Each instance of *obj2* is indexed by the order of occurence in the training set. English people are in plain text, and Italians are in italiacs.

Mapping Distributed Representations

Having entertained a couple of probable causes of the network's poor generalization performance, the path to improved generalization seems clear. We would like to be able to "program" the distributed representations to (1) reflect similarity between similar objects and (2) develop singular representations of multiple instances of each object. In this section, I describe how the first requirement can be met by making use of object correspondences produced by a connectionist model of analogical mapping.

In cases where regularities of the training environment are known a priori, hidden units can be "hardwired" to reflect the structure of the domain. For example, Rumelhart et al (1986) describe a network which develops translation-invariant "receptive fields" by constraining the weights of each hidden unit to take equal increments on each epoch. In general, however, it is desirable to do as little handcrafting of the network as is necessary; ideally we would like the network to be capable of flexibly extracting regularities from the environment using general purpose mechanisms. In the domain with which the present paper is concerned with (and in many other domains), analogy is one such mechanism which can assist in defining regularities.

By establishing correspondences between items in the source domain and target domain, we can pressure the network to form similar distributed representations for the mapped objects. However, we can't simply force representations of mapped objects to be identical; some representation of the domain must be present in the representation of each object in order for the network to be able to distinguish objects between domains. Hence, two components of each distributed representation are necessary: a *domain-invariant* component which represents the common aspects of objects between domains, and a *domain-specific* component which encodes the particular domain to which an object belongs. The domain-invariant component of the distributed representation will facilitate transfer between domains, while the domain-specific component will serve as a type of context indicator which will facilitiate specific modes of knowledge processing within each domain.

To create such representations, we first feed the propositional representations of the source and the (possibly incomplete) target domain into ACME (Holyoak & Thagard, 1989), a connectionist model of analogical mapping. ACME sets up a constraint-satisfaction network of mapping hypotheses based on structural, semantic, and pragmatic constraints, and produces a set of object and relation mappings by relaxing the network and returning the hypotheses with the highest activations (see Holyoak & Thagard, 1989, for further details). The object mappings produced by ACME are used to determine corresponding nodes on the input and output layers of Hinton's network (e.g., node 1, the localist representation of Christopher may correspond to node 13, the localist representation of Roberto). Before training the model, we randomize all weights in the network to values between -.3 and .3, with the following restrictions. Weights from corresponding nodes on the input layer projecting to all but one node in the *obj1* distributed representation layer are constrained to be equal. These nodes in the distributed representation layer will represent the domain-invariant component of *obj1*. The remaining second-layer node will encode the domain-specfic component: weights to this node from all source objects are set to +10.0, and weights to this node from all target objects are set to -10.0. Similarly, weights *to* all corresponding nodes on the output layer *from* all but one *obj2* distributed representation layer node are constrained to be equal, and the remaining node has weights of +10.0 to source objects, and weights of -10.0 to target objects. During training, weights that were set to be equal before training are constrained to stay equal by averaging the weight-error derivatives of the corresponding weights before each weight update. Weights that were set to ±10.0 are frozen at these values throughout training.

Note that while this scheme *forces* the network to form distributed representations of *obj1* that are "mapped", the distributed representations of *obj2* are only *pressured* to be mapped, since it is the weights feeding out of, rather than into, the penultimate layer that are constrained. Hence, it might be possible for the network to develop disparate representations of the source and target *obj2*s, although empirically this doesn't seem to occur. Figure 4 shows the distributed representations of *obj1* and *obj2* developed by training the network on the same training set as was used previously (i.e., the entire set of source propositions, and all but 4 target propositions) plotted against the two principal components derived from a principal components analysis. Unsurprisingly, the source and target representations developed for *obj1* are perfectly symmetric. More significantly, the *obj2* representations are also symmetric: the representation of each instance of each English person is adjacent to the corresponding Italian instance, although the instances are still scattered uniformly in representation space. The result of these constraints is a significant improvement in generalization: all four of the untrained target propositions were correctly constructed by the network.

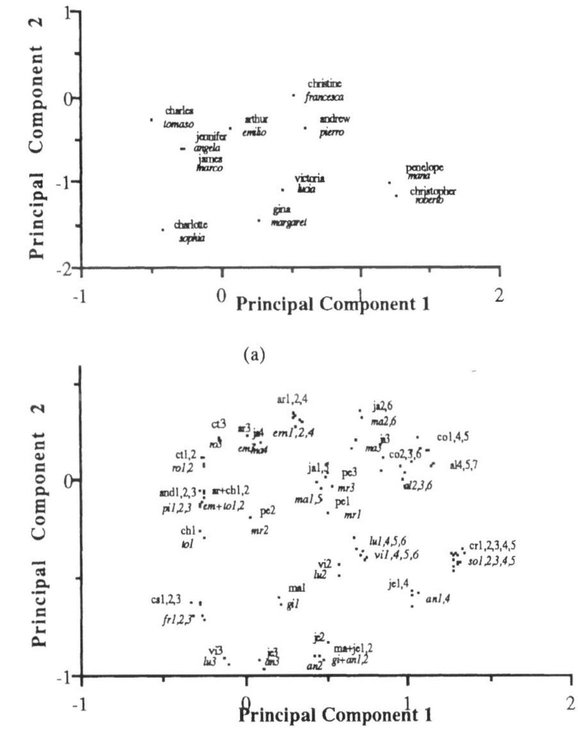

Figure 4: Principal components analysis of (a) *obj1* and (b) *obj2* using the distributed representation mapping technique.

The distributed-representation mapping technique successfuly fulfills our first condition for improved generalization: the network develops representations which reflect similarity between the source and target domains. However, our second condition remains unfulfilled: the network has failed to develop singular representations for multiple instances of objects. To induce the network to develop parsimonious representations, general-purpose network optimization techniques may be successfully employed.

Network Optimization Techniques

Numerous techniques for optimizing the performance of backpropagation have been proposed in the connectionist literature, including weight decay (e.g. Plaut, Nowlan & Hinton 1986; Weigend, Huberman & Rumelhart, 1990; Hanson & Pratt, 1989), eliminating weights (Chauvin, 1989), excising or attenuating hidden units (Kruschke, 1988, Mozer & Smolensky, 1989), limiting the total amount of hidden unit activation (LeCun, 1989), reducing the dimensionality spanned by the hidden weight vectors

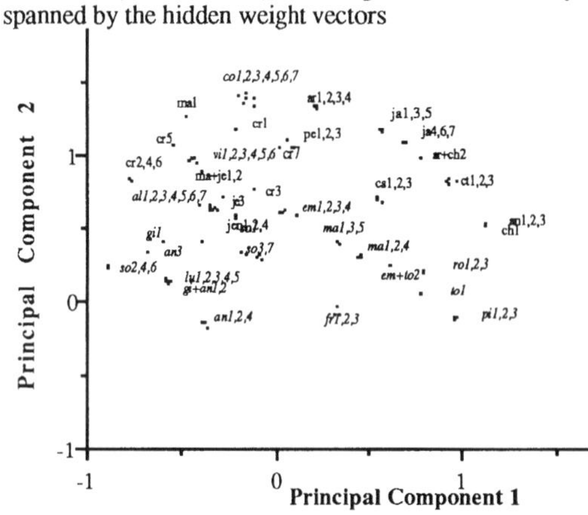

Figure 5: Principal components analysis of *obj2* using weight decay.

(Kruschke, 1988), and forcing hidden layer representations to have uncorrelated node activations (Kruschke & Movellan, 1990). Most of these techniques pressure backprop to make economical use of network resources such as weights, nodes or hidden-unit activation. For example, the second term in eq. 1 (${}^1/_2\lambda\Sigma_i w_i^2$) implements weight decay by pressuring the network to minimize the sum of the squares of all the weights.

Figure 5 shows the results of a principal components analysis of the *obj2* representations when the network was trained on the same set of propositions as in the previous examples, and a weight-decay value of λ = .0005 was used. As expected, multiple instances of each *obj2* are tightly clustered together. The effect of this clustering is a substantial improvement in transfer performance over basic gradient descent: 3 out of the 4 deleted propositions were correctly recovered (compared with recovery levels of 0 for basic gradient descent and 4 for the distributed representation mapping technique). However, the symmetry between the structure of the source and target domains is still imperfect; hence we should not be surprised that knowledge transfer between domains is imperfect. In the next section, I describe a set of computational experiments which systematically test the efficacy of weight decay and the distributed representation mapping technique in improving transfer performance.

Transfer Tests

Four sets of simulations were run, in which the weight decay parameter was set to either 0 or .0005, and the distributed mapping technique was either used or not used. Within each set of simulations, three different amounts of target propositions were omitted from the training set: 4 (8% of the target propositions), 13 (25%), and 26 (50%). Learning was halted when the first term of the cost function dropped below 0.1, or 4000 epochs had elapsed.

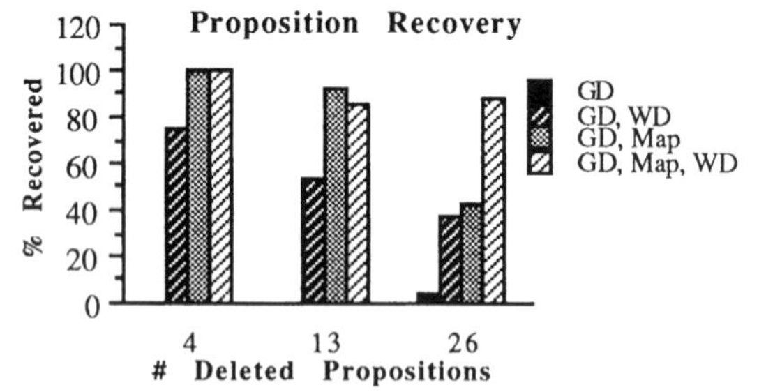

Figure 6: Recovery of Deleted Propositions. "GD" indicates gradient descent, "WD" indicates weight decay, and "Map" indicates the use of the distributed mapping technique.

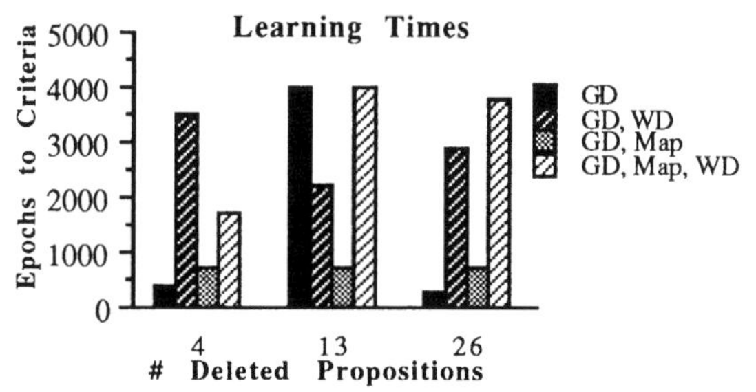

Figure 7: Learning times. Labels are the same as in the previous graph. The second GD run was stuck in a local minimum (2 of the training cases were unlearned at cycle 4000).

The transfer test results (Fig. 6) show that while basic gradient descent is incapable of recovering more than the tiniest fraction of deleted propositions, the distributed representation mapping technique and weight decay both substantially improve transfer, although performance is degraded at higher levels of deletion. The most significant result is that when weight decay is coupled with the distributed representation mapping technique, nearly all of the deleted propositions are recovered, even at the highest level of deletion. This result is obtained because the two conditions of good generalization performance earlier positied are satisified by the two respective techniques: weight decay forms unitary representations of the objects, while the distributed representation mapping technique ensures that similar objects have similar representations.

The order of learning times (Fig. 7) is roughly GD < GD+Map < GD+WD = GD+Map+WD Intuitively, these results are quite sensible: basic gradient descent is the quickest to reach the stopping criteria since it can arive at the closest possible solution without "worrying" about additional solution constraints. Gradient descent with weight decay takes longer to learn than gradient descent with the distributed representation mapping technique because in the case of weight decay, the network must satisfy the conflicting pressures of predicting the training set and driving all weights to 0. In contrast, the distributed representation mapping technique presents two harmonious pressures in the network: the two goals of training set prediction and object representation correspondences can be achieved without conflict.

Additional evidence supports the hypothesis that weight decay and the mapping technique perform complementary optimization functions. For example, computation of the average euclidian distance between each instance (e.g. *emilio1*) in the penultimate layer and the instance's class (e.g. *emilio*) reveals that instances are relatively diffuse when weight decay is not employed. Another important fact is that increasing the weight decay parameter does not substantially improve the generalization performance: as λ is increased, performance on the training set deterioriates as well as performance on the generalization set.

Conclusion

A technique has been described in which high-level domain knowledge produced by a connectionist model of analogical mapping guides the formation of distributed representations of domain objects formed by Hinton's (1986) multilayer model of propositional learning. Simulation results indicate that the use of this technique in isolation can produce substantial improvement in the generalization performance of Hinton's network, and use of this technique in conjunction with weight decay can produce nearly perfect transfer performance. More generally, this paper demonstrates the usefulness, and perhaps even the necessity, of the influence of high level knowledge processing mechanisms on low level subsymbolic learning mechanisms.

Acknowledgements

Keith Holyoak and John Hummel provided helpful comments on an earlier draft.

References

Chauvin, Y. (1989). A back-propagation algorithm with optimal use of hidden units. In D. Touretzky (ed.), *Advances in Neural Information Processing Systems, 1.* San Mateo, CA: Morgan Kaufman.

Fahlman, S. E. (1988). Faster learning variations on back-propagation: an empirical study. In D. Touretzky, G. Hinton, & T. Sejnowski (eds.), *Proceedings of the 1988 Connectionist Models Summer School.* San Mateo, CA: Morgan Kaufmann.

Falkenhainer, B., Forbus, K. D., & Gentner, D. (1989). The structure-mapping engine: Algorithm and examples. *Artificial Intelligence, 41,* 1-63.

Hall, R. P. (1989) Computational approaches to analogical reasoning: a comparative analysis. *Artificial Intelligence, 39,* 39-120.

Hinton, G. E. (1986). Learning distributed representations of concepts. In *Proceedings of the Eighth Annual Conference of the Cognitive Science Society.* Hillsdale, NJ: Erlbaum.

Holyoak, K. J., Novick, L. R., & Melz, E. R. (1992, in press). Component processes in analogical transfer: Mapping, pattern completion, and adaptation. To appear in K. J. Holyoak & J. A. Barnden (eds.), *Advances in Connectionist and Neural Computation Theory, Vol. 2: Analogical Connections.* Norwood, NJ: Ablex.

Holyoak, K. J., & Thagard, P. (1989). Analogical mapping by constraint satisfaction. *Cognitive Science, 13,* 295-355.

Kruschke, J. K. (1989). Creating local and distributed bottlenecks in hidden layers of back propagation networks. In D. Touretzky, G. Hinton, & T. Sejnowski (eds.), *Proceedings of the 1988 Connectionist Models Summer School.* San Mateo, CA: Morgan Kaufmann.

Kruschke, J. K., & Movellan, J. R. (1991). Benefits of gain: Speeded learning and minimal hidden layers in back-propagation networks. *IEEE Transactions on Systems, Man, and Cybernetics, 21 (1).*

LeCun, Y., Denker, J. S., & Solla, J. A. (1989). Optimal Brain Damage. In: D. Touretzky (ed.), *Advances in Neural Information Processing Systems, 1.* San Mateo, CA: Morgan Kaufman.

Mozer, M. C., & Smolensky, P. (1989). Skeletonization: A technique for trimming the fat from a network via relevance assessment. In D. Touretzky (ed.), *Advances in Neural Information Processing Systems, 1.* San Mateo, CA: Morgan Kaufman.

Plaut, D. C., Nowlan, S. J., & Hinton, G. E. (1986). Experiments on learning by back propagation. Carnegie Mellon University Department of Computer Science Technical Report CMU-CS-86-126.

Rumelhart, D. E., Hinton, G. E., & Williams, R. L. (1986). Learning internal representations by error propagation. In D. E. Rumelhart & J. L. McClelland (eds.), *Parallel Distributed Processing: Explorations in the Microstructure of Cognition. Vol. 1.* Cambridge, MA: MIT Press.

Weigend, A. S., Huberman, B. A., & Rumelhart, D. E. (1990). Predicting the future: a connectionist approach. *International Journal of Neural Systems 1 (3),* 193-209.

Direct, Incremental Learning of Fuzzy Propositions

Gregg C. Oden
Departments of Psychology and Computer Science
University of Iowa, Iowa City, IA 52242
oden@cs.uiowa.edu

Abstract

To enable the gradual learning of symbolic representations, a new fuzzy logical operator is developed that supports the expression of negation to degrees. As a result, simple fuzzy propositions become instantiable in a feedforward network having multiplicative nodes and tunable negation links. A backpropagation learning procedure has been straightforwardly developed for such a network and applied to effect the direct, incremental learning of fuzzy propositions in a natural and satisfying manner. Some results of this approach and comparisons to related approaches are discussed as well as directions for further extension.

Introduction

Over the past couple of decades, a wide array of cognitive phenomena have been successfully modeled within a fuzzy propositional theoretical framework[1] (see Massaro, 1987; Oden, 1984, Oden, Rueckl, & Sanocki, 1991 for reviews). The development of this approach to cognitive modeling was motivated by many of the same considerations that underlie current interest in neural information processing systems: in particular, the attainment of robustness and of graceful degradation under duress. In each case, the desired end is achieved largely through reliance on coarse coding, automatic generalization, compensatory information integration, and other consequences of continuous computation. Thus, although the two approaches lie on opposite sides of the symbolic/subsymbolic boundary and might thereby be supposed to be incompatible, they can in fact be seen to be members of the same, more general family of models. From this perspective, it is not surprising that instances of each class of models can be shown to be formally isomorphic under specific common conditions (see Massaro & Cohen, 1987 and Oden, 1988 for two such results). Indeed, facts such as these have been used to support the argument that the two approaches represent separate necessary levels of description of cognitive systems (Oden, 1988; see also Clark, 1989).

The present paper extends this argument by demonstrating how a connectionist learning procedure can be directly and naturally applied within the fuzzy propositional level. This speaks specifically to the common criticism made of symbolic approaches that learning must be an all-or-none process that would require a seemingly magical, external process to wire up new connections.

[1]Hereafter referred to as FuzzyProp for short. The most common FuzzyProp model is the Fuzzy Logical Model of Perception or FLMP (e.g., Massaro & Cohen, 1991; Oden & Massaro, 1978), which is based on the hypothesis of independent evaluation of conjunctive terms.

Learnable Fuzzy Propositions

A fuzzy proposition is a logical expression having component terms that may be more or less true of an object and connectives that are continuous functions of their component terms reflecting the essential logical properties of conjunction, negation, and so on. As applied to the modeling of cognitive processes, fuzzy propositions represent the knowledge that people have about patterns and categories and they provide a basis for evaluating stimuli in a way that fully exploits the information inherent in the systematic continuous variation of stimulus properties. For example, in modeling the identification of handwritten words (Oden & Rueckl, in preparation), the degree to which the initial portion of some stimulus constitutes a lower case letter 'e' involves an evaluation of the proposition that it is a loop that is not too tall:

$$t[loop(x) \wedge \neg tall(x)] = t[loop(x)] \times \{1 - t[tall(x)]\} \qquad (1$$

using (as in all of our work) multiplication to represent fuzzy conjunction[2].

[2]Multiplication is conjunctive in that (a) it yields a value of true (1.0) only if both of its terms are true and a value of false (0.0) if either or both are false, and (b) it has many (arguably the most essential) properties of conjunction such as associativity, commutativity and so on. Importantly, unlike the more common fuzzy conjunction function, t[A ∧ B] = min{t[A], [B]}, multiplication is compensatory, meaning that it allows positive and negative errors to cancel. Much of the robustness of FuzzyProp results from the use of multiplicative conjunction and this is what distinguishes it from most other fuzzy approaches including those that are typically used in constructing fuzzy/neural systems.

It would be advantageous if the knowledge respesented by such propositions could be learned in a gradual or incremental fashion over the course of experience with instances of the concept or relation. To make such propositions incrementally learnable requires some mechanism for gradually converting conjunctions into disjunctions and vice versa. There are many possible ways to do this, but most seem ungainly and ad hoc (e.g., by defining a tunable generic connective as the weighted average of conjunctive and disjunctive expressions). The present approach is to rely on the expressability of disjunctions in conjunctive form through DeMorgan's Law: A ∨ B = ¬(¬A ∧ ¬B). This converts the problem into one of devising a tunable negation operator: a variable connective that allows any term to be negated to some degree. Again, there are many possibilities. For reasons outlined below, the present work makes use of the following rule

$$t[\mathbf{N}_v\, A(x)] = \frac{a^v}{a^v + (1-a)^v} \qquad (2$$

where $a = t[A(x)]$, the degree to which predicate A is true of object x. This operator has a number of attractive properties. The most critical properties, of course, are that of reducing to the identity function for $v = 1$, to standard fuzzy negation for $v = -1$, to a nulling value — a value not dependent on $t[A(x)]$ — for $v = 0$, and to reasonable intermediate functions in between. In addition, as a bonus, the operator yields contrast intensified functions for v values beyond ±1. Thus, the operator can better be thought of as a generalized transfer function that remaps input truth values onto output truth values. Figure 1 plots this function for several representative values of v.

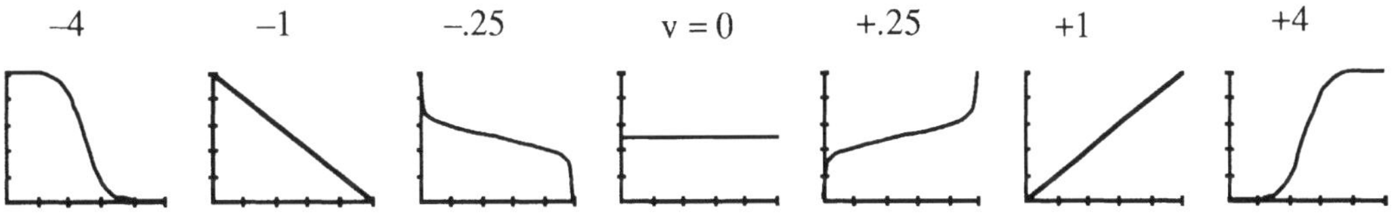

Figure 1. The $\mathbf{N}_v$ operator function for various values of v.

The operator is a natural extension of the version of fuzzy logic employed here (see footnote 2) in the sense that its algebraic structure has a direct natural interpretation in terms of basic fuzzy components (Oden, 1984)[3].

Backprop in FuzzyProp

So far, we have established the instantiability of arbitrary simple fuzzy propositions in a multilayer feedforward network having conjunctive (multiplicative) nodes with tunable generalized (negatable) links connecting nodes of successive layers. Such networks are directly analogous to standard feedforward networks (Rumelhart, Hinton, & Williams, 1986) with the v parameters of the tunable links corresponding to the weights (including those serving as node bias terms) of the standard model. Accordingly, the backward error propagation learning procedure for fuzzy propositions directly follows the form of that laid out by Rumelhart et al: Given a training set of tuples of input and desired output values,

1. evaluate the propositions on the input
2. compute a measure of error between obtained and desired output
3. adjust each v in proportion to the derivative of the error measure with respect to that v, recursively computed.

The calculation of the derivatives is just slightly more complicated in the present case compared to the standard case, in essence because the nonlinearity in the fuzzy propositional system occurs between every pair of nodes from successive layers whereas in standard backprop it occurs just once for each node in the form of the squashing function applied to the output for that node.

[3]As discussed in Oden (1992), it is also algebraically natural in the sense that it is the powering operation for the Abelian (commutative) group defined on [0..1] by the mapping x → x/(x + 1) from the multiplicative group on [0..∞].

Initial tests of this learning procedure demonstrate that it, indeed, performs as it should. For example, when applied to the ever popular test case, XOR, it learns the function forthrightly as indicated by the average learning curve for a representative sample of runs shown in Figure 2.

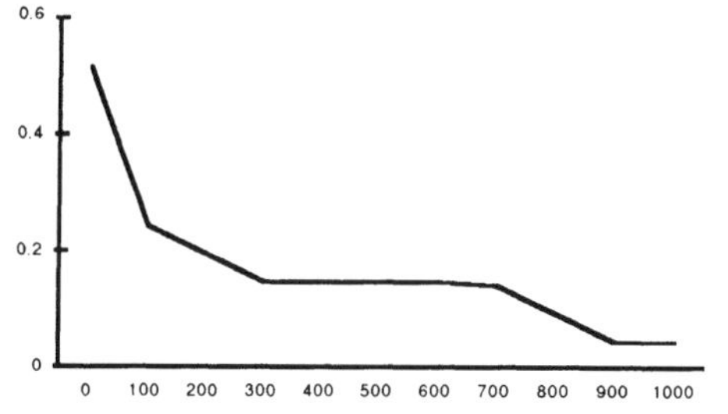

Figure 2. The course of learning XOR for 10 test runs. The root mean squared error is plotted against the number of learning epochs.

The overall form of this curve primarily reflects the fact that most of the test runs in this particular sample resulted in complete learning within a few hundred training epochs or so, a couple required around 800 epochs and one did not complete the learning until 3700 epochs (but then did learn it exactly). The median number of training epochs required to reach a criterion root mean squared error less than .01 was about 250 and the median root mean squared error after 300 epochs was .008. For the sake of comparison, note that these numbers are close to the same number of epochs reported (Rumelhart, Hinton, & Williams, 1986) to be required by standard back propagation to reach a root mean squared error of about .1 The present system needed only about 150 epochs to reach .1 root mean squared deviation.

In these tests, the system is provided with two input nodes, one output node and two intermediate layers having two and one node respectively (2-2-1-1). Thus, altogether the system is allowed seven v values to adjust, precisely the same as the minimum number (and two fewer than the typical number) of free parameters required for this problem in the original Rumelhart et al (1986) paper including both weights and bias terms. This fuzzy propositional network is sufficient for representing XOR as $(A \wedge \neg B) \vee (\neg A \wedge B)$. Sometimes, however, XOR is learnt as $(A \vee B) \wedge \neg(A \wedge B)$ by the system. This is equivalent to the former expression with respect to the input and output values used, which are all (close to) zero and one, but would not be exactly equivalent for intermediate truth values. This form of XOR is actually more compact and only really needs six parameters.

Comparisons and Extensions

The main distinctive feature of the current approach is learned representations that are directly interpretable logical functions of the input variables. In addition, as with other models having multiplicative units such as sigma-pi networks (Rumelhart, Hinton, & McClelland, 1986) or the product unit nets of Durbin and Rumelhart (1989), learning of logical functions may be faster in the present system than with standard backprop. This is due in part to the fact that the v parameters do not have to approach $\pm\infty$ in order to yield outputs close to 0 and 1. Indeed, with respect to feedforward processing in the system, inputs and outputs can take on the values of 0 and 1 exactly. (During the error backpropagation phase, these extreme values must be avoided because the relevant derivatives would be undefined.)

On theoretical grounds, the present approach is interestingly similar to and different from each of the approaches mentioned above in a number of ways. For now, let's just consider the representation of a conjunction of inputs by the standard backpropagation feedforward network in comparison with that of the present system. Figure 3 portrays this relationship in a couple of ways. On the left is a 3D plot and a contour map showing how the standard approach manages to be conjunctive, basically by applying a one dimensional nonlinear cut across the axis corresponding to the sum of the inputs (the diagonal of the 'floor' in the 3D plot). On the right are the corresponding representations for the fuzzy propositional system, which reveal that this approach more directly captures the notion of conjunction as encompassing the vicinity of the (1, 1) corner. This is a direct result of the fact that, as noted above, nonlinearity is more thoroughly ingrained in this system. It is, of course, no accident that the fuzzy propositional system is more naturally conjunctive in this sense, since it is fundamentally logic-based

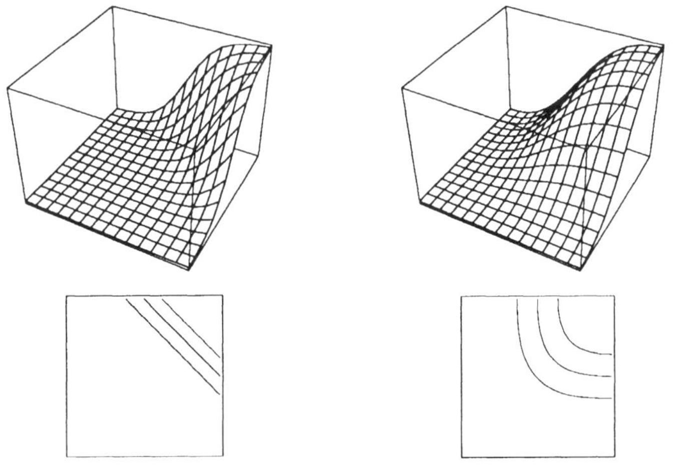

Figure 3. Plots and contour maps of conjunctive functions of two inputs for standard backprop networks (left panels) and the present fuzzy propositional system (right panels).

in structure by design. In contrast, the standard network could be thought of as only emulating the logical functions within a system that does not have an inherently logic-based structure.

The current work is related in quite a different fashion to approaches to the learning of associative relations between fuzzy terms (e.g., Jenison & Oden, 1989; Kosko, 1992). Such techniques are complementary to the one developed here and might serve, for example, to set up the analyses necessary for the evaluation of the primitive terms in the propositions of the present system.

The scheme described here can be variously extended. Inertial terms and other factors analogous to those used with standard backprop can clearly be added and other learning methods from the connectionist literature could be similarly imported into this system. More interestingly, the $\mathbf{N}_v$ operator (Equation 2) can be generalized in at least two reasonable ways: (a) by allowing separate v values for the positive and negative components of the relative ratio expression, and (b) by including an overall exponential weighting factor. Both of these generalizations have natural interpretations in fuzzy terms and yield a significant enrichment of the expressiveness of the resulting fuzzy calculus. Both can be incorporated in the backprop learning scheme with very little complication. Yet another straightforward extension of $\mathbf{N}_v$ (see Oden, 1992) enables it to perform a kind of running average of inputs in order to account for processing dynamics as in Massaro and Cohen (1991).

Conclusion

The overarching moral of this work is that one can have it all. That is, it is not necessary to choose between having symbolic expressions and direct, incremental learning procedures. The two can be combined in a natural and harmonious fashion.

References

Clark, A. 1989. *Microcognition: Philosophy, cognitive science, and parallel distributed processing*. Cambridge, MA: MIT Press.

Durbin, R., and Rumelhart, D. E. 1989. Product units: A computationally powerful and biologically plausible extension to backpropagation networks. *Neural Computation*, 1:133-142.

Jenison, R. L., and Oden, G. C. 1990. Fuzzy implication formation in distributed associative memory. Proceedings of the Twelfth Annual Conference of the Cognitive Science Society, 860–867.

Kosko, B. 1992. *Neural networks and fuzzy systems: A dynamical systems approach to machine intelligence*. Englewood Cliffs, NJ: Prentice-Hall.

Massaro, D. W. 1987. *Speech perception by ear and by eye: A paradigm for psychological inquiry*. Hillsdale, NJ: Erlbaum.

Massaro, D. W., and Cohen, M. M. 1987. Process and connectionist models of pattern recognition. Proceedings of the Ninth Annual Conference of the Cognitive Science Society, Seattle, WA.

Massaro, D. W., and Cohen, M. M. 1991. Integration versus interactive activation: The joint influence of stimulus and context in perception. *Cognitive Psychology*, 23:558-614.

Oden, G. C. 1984. Integration of fuzzy linguistic information in language comprehension. *Fuzzy Sets and Systems*, 14:29-41.

Oden, G. C. 1988. FuzzyProp: A symbolic superstrate for connectionist models. Proceedings of the Second IEEE International Conference on Neural Networks, Vol. 1, 293-300.

Oden, G. C. 1992. An algebra on [0..1], with application to fuzzy values. *Cognotes*, 5.

Oden, G. C., and Massaro, D. W. 1978. Integration of featural information in speech perception. *Psychological Review*, 85:172-191.

Oden, G. C., and Rueckl, J. G. In preparation. Taking language by the hand: Reading handwritten words.

Oden, G. C., Rueckl, J. G., and Sanocki, T. 1991. Making sentences make sense, or words to that effect. In G. B. Simpson (Ed.) *Understanding word and sentence.* Amsterdam: North-Holland.

Rumelhart, D. E., Hinton, G. E., and McClelland, J. L. 1986. A general framework for parallel distributed processing. In D. E. Rumelhart and J. L. McClelland (Eds.), *Parallel distributed processing: Explorations in the microstructure of cognition, Vol. 1.* Cambridge, MA: MIT Press.

Rumelhart, D. E., Hinton, G. E., and Williams, R. J. (1986). Learning internal representations by error propagation. In D. E. Rumelhart and J. L. McClelland (Eds.), *Parallel distributed processing: Explorations in the microstructure of cognition, Vol. 1.* Cambridge, MA: MIT Press.

The Effects of Pattern Presentation on Interference in Backpropagation Networks

Jacob M.J. Murre[1]
Unit of Experimental and Theoretical Psychology
Leiden University

Abstract

This paper reviews six approaches to solving the problem of 'catastrophic sequential interference'. It is concluded that all of these methods function by reducing (or circumventing) hidden-layer overlap. A new method is presented, called 'random rehearsal training', that further explores an approach introduced by Hetherington and Seidenberg (1989). A constant number of patterns, randomly selected from those learned earlier, is rehearsed with every newly learned pattern. This scheme of rehearsing patterns may, perhaps, be compared to the functioning of the 'articulatory loop' (Baddeley, 1986). It is shown that this presentation method may virtually eliminate sequential interference.

Preventing 'Catastrophic Interference'

Both from a psychological and from a practical point of view, standard backpropagation models (Rumelhart, Hinton, and Williams, 1986) suffer from an important weakness: on sequential learning tasks they exhibit strong retroactive interference. Newly learned patterns may erase nearly all existing memories (Grossberg, 1987; McCloskey and Cohen, 1989; Ratcliff, 1990). This behavioral implausibility has become the subject of many studies, usually with reference to the name 'catastrophic interference' coined by McCloskey and Cohen (1989). Several proposals have been made to overcome the strong interference in sequential learning tasks.

A number of studies approaches the issue by enhancing the network architecture or the learning rule. French (1991), for example, uses a method whereby after prolonged learning only *k* nodes in the hidden layer remain active for each pattern. He calls this method '*k*-node sharpening'. Kortge (1990) proposes a new learning rule, called the 'novelty rule'. With this rule, the amount of learning is made dependent on the relative novelty of the input pattern. Sloman and Rumelhart (in press) use a network without hidden units, and with weights that are logically gated by 'episodic units' (i.e., representing the learning context). It seems that networks without hidden units, in general, are less prone to sequential interference (Lewandowsky, 1991; Hetherington, 1990b). Hinton and Plaut (1987) use a network in which the hidden units have either 'fast' or 'slow' weights. Since they focused primarily on the effects of retraining items earlier in the list, this approach does not directly address the problem as posed by either McCloskey and Cohen (1989) or Ratcliff (1990). We might, finally, mention the model by Kruschke (1992) in which the receptive fields of the hidden-layer units are functionally located (restricted) before learning.

Apart from an alteration of the working of the backpropagation algorithm, interference may be reduced by merely changing the representation of the input and output patterns. Some studies have successfully used bipolar pattern features (i.e., values -0.5 and 0.5, or values in this range; Kortge, 1990; Lewandowsky, 1991). Others have argued that normalization of the pattern length may reduce interference (Kruschke, personal communication). It may also be noted that the nature of the patterns used seems to have a strong effect on sequential interference. For example, Brouse and Smolensky (1989) and Hetherington (1990b) have argued that in combinatorial domains (i.e., with a large number of structured patterns, such as words) interference is strongly reduced. Also, Hetherington (1990a) has pointed out that with auto-associative learning one may expect less interference than with hetero-associative learning (i.e., inputs differ from outputs).

As a third general approach we may distinguish between variations in the method of pattern presentation. Hetherington and Seidenberg (1989) trained a network in overlapping blocks (see Table 1, and below), which greatly reduced sequential interference. In this paper, we will focus on another method of presentation, called 'random rehearsal', that is akin to their method. After a brief review of the simulations by McCloskey and Cohen (1989) and Hetherington and Seidenberg (1989), we will describe this new method. In the Discussion, we shall argue that all successful approaches to reducing interference are based on a single underlying factor:

[1] The author is presently employed at the MRC Applied Psychology Unit, 15 Chaucer Road, Cambridge CB2 2EF, UK. E-mail: Jaap.Murre@MRC-APU.CAM.AC.UK.

orthogonalization of hidden-layer representations across subsequent patterns or pattern blocks.

Pattern Presentation and Interference

The study by McCloskey and Cohen (1989) aimed at teaching (by 'rote learning') a model some simple arithmetic: adding, subtracting, dividing, and multiplying numbers in the range zero to nine. During training, two numbers and an arithmetic operator were presented as input patterns, while the correct answer was presented as output. The model consisted of a straightforward, three-layer backpropagation model with fully connected layers. Numbers (single digits) were represented by activating three consecutive nodes in the output or input layer. The number three, for example, was represented as 0 0 0 1 1 1 0 0 0 0 ....., and a zero as 1 1 1 0 0 0 0 ..... The input layer consisted of 28 input nodes (two times twelve nodes for representing the digits, and four additional nodes for the operators), the hidden layer consisted of fifty nodes, and the output layer consisted of 24 nodes (twelve for digits and twelve for tens).

The network could easily be taught all summed digit pairs, as well as all multiplied digit pairs. The pairs were presented for training in blocks with varying random order. When the network was trained on patterns drawn from the entire training set, no problems occurred. But when the network was first taught all additions with one (e.g., [1+1], [2+1], [3+1], ..., and also [1+2], [1+3], [1+4], ...), and only *then* on all additions with two (except [1+2] and [2+1], which had already been learned), the newly learned patterns appeared to have washed out all memory of addition with one. Performance on the ones, dropped from 100% to 57%, after a single run, and to 30%, after two runs on the twos.

The simulations by McCloskey and Cohen (1989) were replicated by Hetherington and Seidenberg (1989) with essentially similar results. A second simulation by these authors, however, indicated that learning the twos does not completely destroy all memory of the ones. It appeared that the ones were relearned faster than a totally novel set of additions (with three). The model thus showed evidence of 'savings' (Ebbinghaus, 1985): the ones were not completely unlearned, which greatly accelerated relearning. Based on these results (also see Hinton and Plaut, 1987), Hetherington and Seidenberg (1989) argue that the catastrophic interference found by McCloskey and Cohen (1989) is primarily dependent on the method of pattern presentation. In particular, they argue that *blocking*
of learning trials (i.e., first a block of ones, then a block of twos) may be an important contributing factor, and

Stage	Pattern sets									
1.	1	1								
2.		1	1	2						
3.			1	2	2	3				
4.			1	2	2	3	3	4		
5.					2	3	3	4	4	5

Table 1. *Training scheme used by Hetherington and Seidenberg (1989). The table shows the sets presented in each stage. A stage lasted for ten epochs. In each epoch, patterns were presented in a different random order.*

that "if, instead of following this strict blocking scheme, there is some minimal retraining on the ones, performance will rapidly improve due to savings." (p.30) Based on this idea they used a training scheme intermediate between both strict blocking and fully concurrent presentation of patterns.

Hetherington and Seidenberg (1989) trained their model in five stages on addition with ones, twos, threes, and fours. For each addition, a set was constructed containing thirteen digit pairs as mentioned above. The sets were constructed, so that they would not overlap (i.e., [1+3] occurred in either the set of ones, or the set of threes, but not in both). The training scheme is shown in Table 1. Presenting two sets of one type corresponds to presenting each element of the set twice, in random order, interleaved with elements from other sets. From the table it becomes clear that the consecutive stages overlap: patterns are trained during a number of consecutive stages. At stage five, the ones are no longer retrained, so that on the basis of the above cited data we might expect considerable interference as a result of training the twos, threes, fours, and fives, while not simultaneously retraining ones. The results, however, indicate that this is not the case. After training on stage 5, the model is still able to correctly reproduce in between twelve and thirteen ones (out of a possible thirteen). The authors further report that after continued training for 35 more epochs following stage four, the mean number of correct responses on the ones was still 91% (11.8 out of 13). Their conclusion, therefore, is that this method of pattern presentation prevents catastrophic interference.

We did a series of simulations to investigate further the effects of pattern presentation schemes on retroactive interference. Our findings indicate that Hetherington and Seidenberg's (1989) results on the detrimental effects of strict blocking do not directly generalize to other models. A method similar to their 'method of overlapping stages', however, appears to work well on

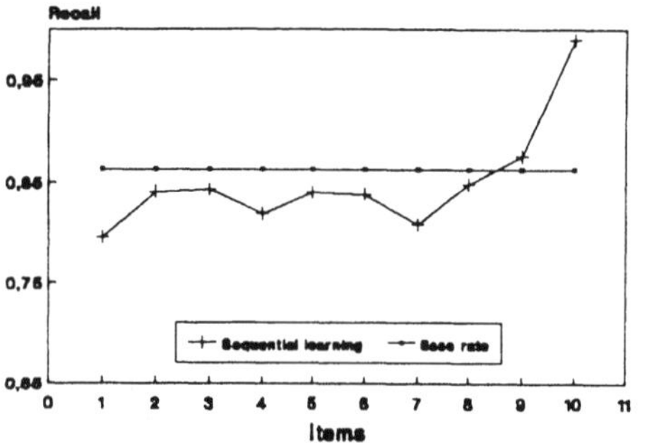

Fig.1. *Interference in backpropagation as a result of strict sequential learning. The results are averaged over 100 replications.*

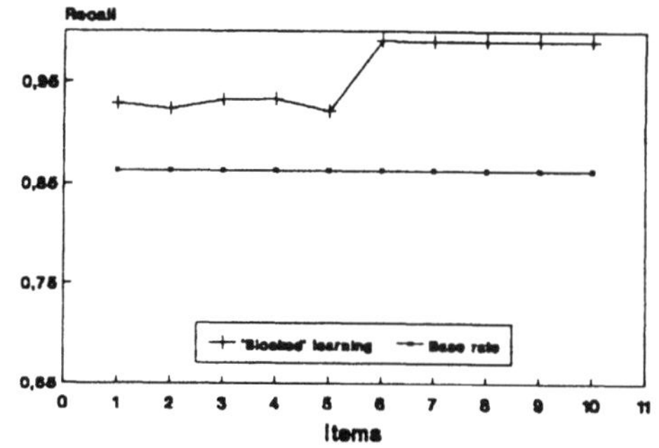

Fig.2. *Interference in backpropagation as a result of a strict blocking scheme. The results are averaged over 100 replications.*

auto-associative learning of random pattern vectors. The patterns used in our simulations consisted of eight elements. Ten new patterns were generated for each simulation (and replication). Each of the pattern elements was assigned a (uniform) random value between zero and one. The length of the vector was normalized to 1.0 (this may have reduced sequential interference, such as between blocks, see discussion below). The model used was a simple three-layer backpropagation network. The size of the input and output layers was eight, the size of the hidden layer was five nodes (simulations indicated that increasing the hidden layer beyond this size did not essentially influence the results, see Figure 6 and the discussion below). Before every simulation, weights were (uniform) randomly initialized in the range [-0.5, 0.5]. The learning rate was 0.5, the momentum parameter was set at 0.9. With these parameters, the networks easily learns ten random patterns to the criterion described below.

Simulation 1. In this simulation, 'strict sequential learning' was used. Each of the patterns was learned until the criterion was reached, and *not repeated thereafter*. The criterion consisted of a correlation coefficient (i.e., cosine of the angle between the two vectors) of more than 0.99 between the (target) pattern and pattern produced at the output layer. The simulation was repeated 100 times. For each replicated simulation both the initial weights and the patterns were generated anew. The averaged results are shown in Figure 1. Recall is represented by the correlations remaining after having learned all patterns. The base rate shown in the figure is the expected correlation of 0.863 between a random pattern and its output before the network has learned anything. It was established by generating 5000 random patterns and averaging the correlations. As can be seen from Figure 1, strict sequential learning causes catastrophic interference to the extent that after learning the network performs actually *worse* than before learning.

Simulation 2. Having established that in this simulation paradigm strict sequential learning gives rise to 'more than catastrophic interference', we trained the network using 'strict blocking' of trials. First, five patterns were simultaneously trained until the criterion was reached (see above), followed by training on patterns six to ten. After these had reached the criterion, the network was tested for recall. The simulation was repeated 100 times. The results are shown in Figure 2. Strict blocking also leads to considerable retroactive interference, although not as bad as strict sequential learning.

Simulation 3. To test whether training in 'overlapping stages', as described by Hetherington and Seidenberg (1989) is a feasible method for reducing interference the following training method was used. A fixed-size 'window' was moved over the ordered pattern set. All patterns in the window were trained to the set criterion. Say, the window can contain three patterns (we will speak of a *depth* of 3). Then, the training stages are as follows. In subsequent stages we train patterns A, B, C, .., as follows: (A), (A,B), (A,B,C); (B,C,D); (C,D,E), etc. Simulations were carried out with windows of depth 1, 2, 3, and 4. Network, patterns, and parameters were as above. For each depth, 100 replications were carried out. The averaged results are shown in Figure 3. A depth of 1 leads to strong retroactive interference, comparable to using zero depth (i.e., strict sequential learning). Depths of 2, 3, and 4, however, lead a to progressively decreasing interference, although even a depth of 4 performs hardly better than strictly blocked learning in this respect.

Simulation 4. According to Hetherington and Seidenberg (1989), the overlapping stages method leads to reduced interference, because old patterns are occasionally retrained. The 'windowing method' of the previous simulations only rehearses the most recent

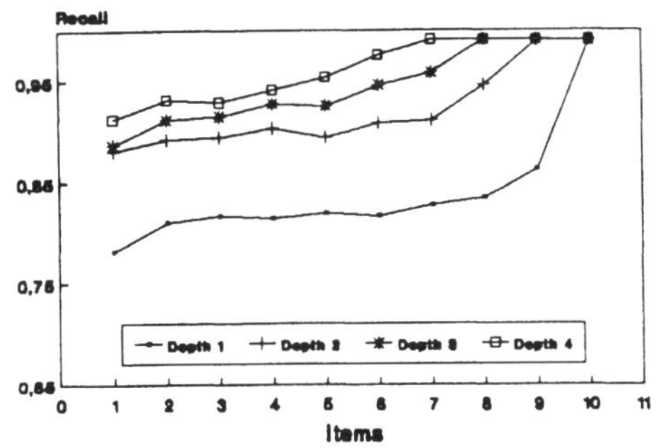

Fig.3. *Interference in backpropagation using a windowed training method. Each point is averaged over 100 replications.*

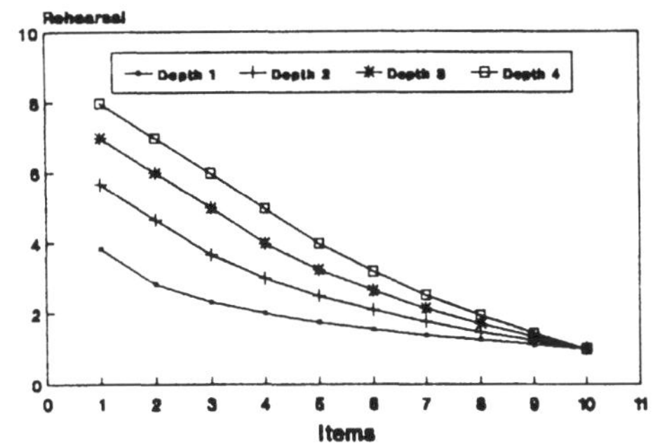

Fig.4. *Expected number of item rehearsals (out of a possible 10) for depths of 1 to 4. See text for an explanation.*

patterns. It would be interesting to see whether an increase in performance (i.e., a reduction in interference) could be achieved by rehearsing a constant number of patterns randomly chosen from the already learned patterns. In the method used, the first patterns have a higher chance of being rehearsed than late patterns in the list. Exact chances of rehearsal with list size 10 are shown in Figure 4. With a depth of two, for example, pattern 3 will on the average be rehearsed about four times (out of a possible ten). The term *depth* is maintained, although here it refers to randomly selected items. We remark, furthermore, that if depth is 3, it implies that the first four items will certainly be rehearsed up until pattern 4.

The results are given in Figure 5. As can be seen, the 'random rehearsal method' works successfully. Especially with depths of 3 and 4 retroactive interference is strongly reduced. Note, that the total number of rehearsals is not greater than that of the windowed training scheme.

Discussion

Simulations 1 to 4 convincingly demonstrate that pattern presentation schemes may considerably influence retroactive interference, from 'more than catastrophic' for strictly sequential learning to 'only slightly' for the random rehearsal method.

As was argued by Hetherington and Seidenberg (1989), their method of presentation may be called plausible from a psychological point of view. Occasional reminders of items are nearly always given during prolonged instruction (e.g., in classroom situations). In our method of 'random rehearsal', reminders are drawn from all items learned, rather that just the most recent ones. This may or may not be a plausible scheme for classroom instruction. We would rather argue, however, that it can be viewed as a partial implementation of the 'articulatory loop' proposed by Baddeley and Hitch (1974; Baddeley, 1986). While learning a list, a fixed number of items is drawn from memory and rehearsed with the new items. As is shown in Figure 4, earlier items have a high chance of being retrained. The articulatory loop may thus be seen as a method whereby occasional 'reminders' are generated for retraining (which occurs quickly due to savings, see above). The proposed method is only a partial implementation, because it presupposes that all items learned are still fully available. In a more complete implementation, rehearsal should be a more self-contained process. Use could, for instance, be made of a recurrent scheme to implement the articulatory loop. Such studies have indeed been carried out recently (Burgess and Hitch, 1991; Nolfi, Parisi, Vallar, and Burani, 1990; Mul, Phaf, and Wolters, in preparation). In these models, early items in the list are recalled *better* than late items. In a recurrent network an 'articulatory loop' may, thus, fully counteract the effects of retrograde interference. More importantly, perhaps, is that this primacy effect is consistent with well-established experimental findings.

One fact that remains surprising in the simulations presented above, is that a seemingly important architectural characteristic such as the size of the hidden layer has a negligible influence on retroactive interference. In Figure 6, for example, a replication of Simulation 1 is shown with hidden layers of 5 and 20 nodes. As can be seen, increasing the size of the hidden layer has only a minor influence on retroactive interference (this was also found by Hetherington and Seidenberg, 1989, see their note 3, p.32). In fact, increasing the hidden layer seems to have a slight *negative* effect on recall.

This effect may be explained with reference to the

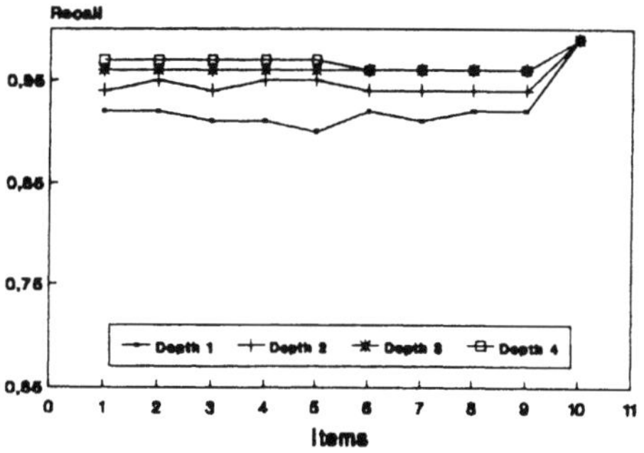

Fig.5. *Interference in backpropagation using a random rehearsal method. Each point is averaged over 100 replications.*

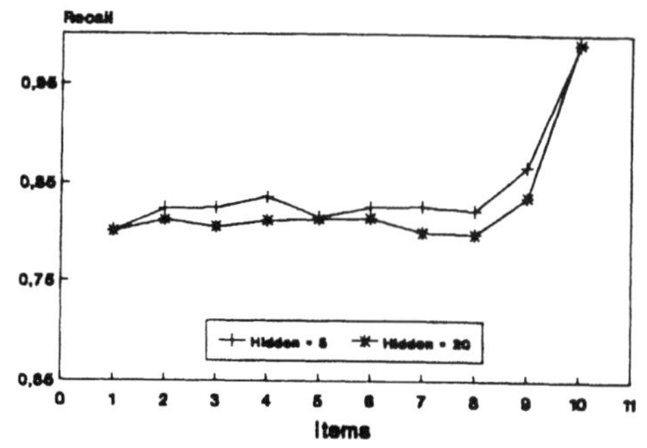

Fig.6. *The negligible effect of increasing the size of the hidden layer on retroactive interference. The results are averaged over 100 replications.*

nature of hidden-layer representations emerging in standard backpropagation networks. A detailed analysis of Ratcliff's (1990) first series of simulations has revealed that nearly all sequential interference can be attributed to overlap of hidden-layer representations (Murre, 1992). Backpropagation is able to develop sufficiently distinct representations for patterns *within* a list (block). The representations ***between*** lists, however, are almost purely random. It can be shown that as few as two overlapping active hidden units (i.e., in two sequentially learned patterns) may cause nearly complete unlearning of the first pattern (Murre, 1992). Normally, roughly half of the hidden-layer will be active for any pattern presented. Chances of an overlap of more than one active unit are, therefore, very high. This is even more so, if the size of the hidden-layer is increased, which may explain the effect of Figure 6.

Reducing hidden-layer overlap will decrease sequential interference (also see French, 1991). In fact, all studies cited above that succeed in reducing interference share this as a common factor:

1. ***Sharpening*** (French, 1990). This method is introduced explicitly to reduce hidden-layer overlap. Fewer active nodes results in a lower chance of overlap. A more detailed analysis of this method, however, shows that in many situations there is a rather high chance that learning will not converge, because within-list representations may overlap (Murre, 1992).
2. ***Novelty rule*** (Kortge, 1990). This method results in emphasizing the between-pattern differences, which gives rise to more distinct hidden-layer representations.
3. ***Restricted receptive fields*** (Kruschke, 1992). Functionally located hidden nodes only respond to a restricted part of the pattern space. This results in a decreased overlap of representations: a unit that responds to a certain pattern, is less likely to respond to another.
4. ***Normalizing patterns*** (Kruschke, personal communication). Normalization results in more restricted receptive fields. This can be most easily seen when considering binary nodes. Suppose, that a certain hidden-unit has weights [1,1] and threshold 1.3. Such a node will be activated by, for example, the pattern (0.8,0.6), but also by all patterns (>1.3,>1.3). Normalization places all patterns on a hypersphere. After normalization, the unit will still be activated by pattern (0.8,0.6), which already has length one. Patterns in the area of (2,0), (0,3), etc., however, will be normalized to (1,0), (0,1), etc., so that the unit is no longer activated by such patterns. Similarly, a linear-sigmoid node can be placed so that it 'carves off' just a small piece of the hypersphere. With normalized patterns such nodes develop more restricted receptive fields, which gives rise to more distinct hidden-layer representations.
5. ***Bipolar pattern features*** (Kortge 1990; Lewandowsky, 1991). With bipolar pattern features, it can easily be shown that for orthogonal patterns, hidden units tend to develop orthogonal representations (also between lists, which is not the case with standard backpropagation, see Ratcliff, 1990).
6. ***Windowed training*** (Hetherington and Seidenberg, 1989). Windowed training is based on providing reminders of recent items. This enables the algorithm to develop representations that are orthogonal over several subsequent lists rather than just within-lists.
7. ***Random rehearsal training*** (Simulation 4, above). This method functions similarly to windowed training. It also provides reminders of items learned earlier. These reminders enable the model to keep representations orthogonal.

We conclude from this brief review that all of these methods derive their succes primarily from making the hidden-layer representations of patterns between-lists more orthogonal. One way of circumventing the problem would be to have a model without hidden layers. This has indeed been found (Hetherington, 1990b, Lewandowsky, 1991, and Sloman and Rumelhart, in press), in particular, if patterns are orthogonal.

To arrive at a more plausible (and perhaps more practical) model of human memory it may be worthwhile to investigate other types of models. Most models based on categorization or other forms of unsupervised learning are able to develop orthogonal representations between lists (e.g., Carpenter and Grossberg, 1987; Grossberg, 1987; Kohonen, 1990; Rumelhart and Zipser, 1985). Elsewhere, we have advocated an approach that combines a modular architecture with intramodular competition. The learning rate in these modules is sensitive to the novelty of the incoming pattern (Murre, Phaf, and Wolters, 1989, 1992; Murre, 1992). This approach combines several partial solutions to the problem of sequential interference outlined above. Once this problem has been fully understood, we may be able to perform detailed simulations of a more challenging phenomenon: the forgetting gradient in retrograde amnesia. Severe disturbance to the brain seems to result in a loss of recently learned patterns, with patterns learned earlier being saved (Squire, 1987). This well-established fact seems to run counter to any neural network model devised thus far.

References

Baddeley, A.D. (1986). *Working memory*. Oxford: Clarendon Press.

Baddeley, A.D., & G.J. Hitch (1974). Working memory. In: G.H. Bower (Ed.) *Recent advances in the psychology of learning and motivation, Vol.VIII*. New York: Academic Press, 47-90.

Brouse, O., & P. Smolensky (1989). Virtual memories and massive generalization in connectionist combinatorial learning. *Proceedings of the Eleventh Annual Conference of the Cognitive Science Society*. Hillsdale, New Jersey: Lawrence Erlbaum Associates, 380-387.

Burgess, N., & G. Hitch (1991). Towards a network model of the articulatory loop. *Journal of Memory and Language*, in press.

Carpenter, G.A., & S. Grossberg (1987). Neural dynamics of category learning and recognition: attention, memory consolidation, and amnesia. In: J. Davis, R. Newburgh & E. Wegman (Eds.) *Brain structure, learning, and memory*. AAAS Symposium Series.

Ebbinghaus, H. (1885). *Über das Gedächtnis*. Leipzig: Dunker.

French, R.M. (1991). Using semi-distributed representations to overcome catastrophic forgetting in connectionist networks. *Proceedings of the Thirteenth Annual Conference of the Cognitive Science Society*. Hillsdale, New Jersey: Lawrence Erlbaum Associates, 173-178.

Grossberg, S. (1987). Competitive learning: from interactive activation to adaptive resonance. *Cognitive Science, 11*, 23-63.

Hetherington, P.A. (1990a). Interference and generalization in connectionist networks: within-domain structure or between-domain correlation? *Neural Network Review*, 4, 27-29.

Hetherington, P.A. (1990b). *The sequential learning problem in connectionist networks*. Unpublished Master's thesis, Department of Psychology, McGill University, Montreal, Canada.

Hetherington, P.A., & M.S. Seidenberg (1989). Is there 'catastrophic interference' in connectionist networks? *Proceedings of the Eleventh Annual Conference of the Cognitive Science Society*. Hillsdale, New Jersey: Lawrence Erlbaum Associates, 26-33.

Hinton, G.E., & D.C. Plaut (1987). Using fast weights to deblur old memories. *Proceedings of the Ninth Annual Conference of the Cognitive Science Society*. Hillsdale, New Jersey: Lawrence Erlbaum Associates, 177-186.

Kohonen, T. (1990). The self-organizing map. *Proceedings of the IEEE*, 78, 1464-1480.

Kortge, C.A. (1990). Episodic memory in connectionist networks. *Proceedings of the Twelfth Annual Conference of the Cognitive Science Society*. Hillsdale, New Jersey: Lawrence Erlbaum Associates, 764-771.

Kruschke, J.K. (1992). ALCOVE: an exemplar-based connectionist model of category learning. *Psychological Review*, 99, 22-44.

Lewandowsky, S. (1991). Gradual unlearning and catastrophic interference: a comparison of distributed architectures. In: W.E. Hockley & S. Lewandowsky (Eds.) *Relating theory and data: essays on human memory in honour of Bennet B. Murdoch*. Hillsdale, New Jersey: Lawrence Erlbaum Associates.

McCloskey, M., & N.J. Cohen (1989). Catastrophic interference in connectionist networks: The sequential learning problem. In: G.H. Bower (Ed.) *The psychology of learning and motivation*. New York: Academic Press.

Mul, N.M., R.H. Phaf, and G. Wolters (in prep.). Sequential recurrent networks in CALM: rehearsal, short-term memory, and ordered recall. Internal Report, Unit of Experimental and Theoretical Psychology, Leiden University.

Murre, J.M.J. (1992). *Categorization and learning in modular neural networks*. Hemel Hempstead: Harvester Wheatsheaf, forthcoming in September 1992.

Murre, J.M.J., R.H. Phaf, & G. Wolters (1989). CALM networks: a modular approach to supervised and unsupervised learning. *Proceedings of the International Joint Conference on Neural Networks Washington DC, 1*, 649-656. (New York: IEEE Press. IEEE Catalog Number 89CH2765-6).

Murre, J.M.J., R.H. Phaf, G. Wolters (1992). CALM: Categorizing And Learning Module. *Neural Networks*, 5, 55-82.

Nolfi, S., D. Parisi, G. Vallar, & C. Burani (1990). Recall of sequences of items by a neural network. In: D.S. Touretzky, J.L. Elman, T.J. Sejnowski, & G.E. Hinton (Eds.) *Proceedings of the 1990 Connectionist Summer School*. San Matteo CA: Morgan Kaufmann.

Ratcliff, R. (1990). Connectionist models of recognition memory: constraints imposed by learning and forgetting functions. *Psychological Review*, 97, 285-308.

Rumelhart, D.E., G.E. Hinton, & R.J. Williams (1986). Learning internal representations by error propagation. In: D.E. Rumelhart & J.L. McClelland (Eds.) *Parallel distributed processing. Volume 1: Foundations*. Cambridge, MA: MIT Press.

Rumelhart, D.E., & D. Zipser (1985). Feature discovery by competitive learning. *Cognitive Science, 9*, 75-112.

Sloman, S.A., & D.E. Rumelhart (in press). Reducing interference in distributed memory through episodic gating. In: A. Healy, S. Kosslyn, & R. Shiffrin (Eds.) *Essays in honor of W.K. Estes*.

Squire, L.R. (1987). *Memory and brain*. Oxford: Oxford University Press.

Developmental Changes in Infants' Perceptual Processing of Biomechanical Motions

Jeannine Pinto
Psychology Department
University of Virginia
Charlottesville, VA 22903
jmp8p@virginia.edu

Jeff Shrager
Xerox
Palo Alto Research Center
Palo Alto, CA 94304
shrager@xerox.com

Bennett I. Bertenthal
Psychology Department
University of Virginia
Charlottesville, VA 22903
bib@virginia.edu

Abstract

In order to process the reduced information in point-light images of human movement, observers rely upon general processing heuristics as well as representations more specific to human gait. This paper explores changes in the perception of structure from motion in young infants. We re-examined data from 17 experiments, involving infants of 3- and 5-months old, to determine which stimulus features of point-light motion infants use to organize percepts, and how perception changes. By combining discrimination and encoding information we provide a picture of developing perceptual processes. Five-month-olds encode the stimuli more quickly than 3-month-olds, while the younger infants discriminate pairs of stimuli more frequently. Infants of both ages use phase information to discriminate displays. Three-month olds discriminate canonical forms from modified forms when the stimuli are organized about a vertical axis, whereas 5-month olds discriminate these forms only when one of the figures take on a human-like configuration. These results support a view in which differential skill in what is encoded characterizes development. Furthermore, this work may help guide the integration of theory-formation models with heuristic and constraint-based models, into a more complete account of perception.

Motion can provide observers with important information about the 3-dimensional world (Gibson, 1950). From motion the visual system can extract information about depth, grouping, and other components of structure. Johansson (1973) introduced moving dot displays of human movement into the exploration of structure from motion. These displays are created by attaching "point lights" to a person's major joints, and filming the person moving in a dimly-lit room. The resulting film provides only the motion information characteristic of human movement; features such as texture, color, and explicit contour have been eliminated. Though a human form is hard to recognize from a single, static frame of a point-light film, Johansson (1973) demonstrated that adult observers can recognize the moving displays as moving persons in as little as 1 second.

Adults almost always identify the point-light displays as depicting human movement, yet the number of possible connections between the 11 point-lights[1] that compose a human form is very large. This suggests that sensory information alone cannot organize a visual array into an object; the visual system organizes the scene with the aid of constraints, or organizing heuristics. Previous research suggests that the visual system exploits the relative motions among point-lights. Human gait cannot be identified in the absolute motion of any individual element. Rather, form emerges as a product of the relative motions of the elements, or motion of the elements in relation to one another. Some relative motions are more salient than others. Wallach & O'Connell (1953), for example, have suggested that, whenever possible, the visual system assumes that relative motions of the elements reflect rigid relations and thus tries to find the simplest interpretation under which the points are rigidly related in 3D. In their account, though non-rigid interpretations are possible, the rigidity assumption privileges one set of interpretations over another.

This work conducted under NICHD grant HD16195 and NICHD Career Development Award HD00678 to the third author.

1. The point-lights appear to be attached to the head, shoulders, elbows, wrists, hips, knees and ankles. In a side view of the walker, one shoulder and hip are always occluded by the rest of the body.

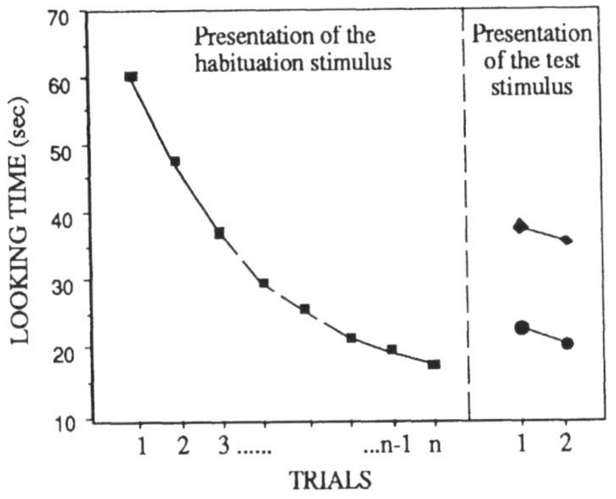

Figure 1. A hypothetical habituation curve. In the infant-control paradigm, the number of trials is variable, determined when the infant reaches approximately half of the initial 3 trials. On the right, two lines illustrate discrimination (diamonds) and lack of discrimination (circles) of the test stimulus.

In addition to general processing heuristics, like the rigidity assumption, evidence from studies of adults suggests that observers also rely upon processes or representations more specific to human gait to help organize point-light walker displays. Adult observers can identify various movements, such as jogging, jumping, or dancing (Johansson, 1975), and the gender of the walker (Cutting, 1978). In so far as this skill is acquired through extensive experience of movement, adults are likely to develop robust strategies and redundant constraints for interpreting human movement. Thus, examinations of the development of the perception of biomechanical motions can provide insights into early perceptual constraints and their interactions with cognitive development.

This paper explores the perception of structure from motion in young infants. Previous work has used measures of discrimination between point-light displays to assess infants' subjective experience. By summarizing a number of discrimination findings, we have developed a picture of the perception of biomechanical motions at 3- and 5- months of age. We have also sought converging evidence for that picture from data about infants' encoding processes as they are presented with point-light displays. We conclude with some general implications of our analyses for computational views of the development of perception.

Tapping Infants' Subjective Experience

The analysis reported here is a re-examination of 17 experiments conducted in the Laboratory for Infant Studies at the University of Virginia, involving 387 infants in two age groups: 3-months old (186 infants) and 5-months old (201 infants)[2]. The goal of these studies was to determine which stimulus features of point-light motion infants use to organize the images, and how this skill develops between 3 and 5 months of age.

Each of the 17 studies used a habituation paradigm to test infants' discrimination of one or more stimulus dimensions. Behavioral habituation is a common tool for studying infant perceptual and cognitive processing. In this paradigm, infants are familiarized, or habituated, to one stimulus and then presented with a second stimulus. The infant's differential response to the two stimuli is used as the measure of pairwise discrimination. The studies reported in this paper were conducted using an infant-control habituation paradigm (Cohen, 1973). In this paradigm, infants are repeatedly presented a visual stimulus (the "habituation" stimulus) and the time they spend looking at this stimulus is measured by trained observers. Gradually, over successive presentations of the same habituation stimulus, infants' looking time per trial declines (Figure 1). When looking time reaches a criterion value, a variant of the habituation stimulus is presented for two trials (the "test" stimulus). The infant's time looking at the test stimulus is then compared to its time looking at the last two presentations of the habituation stimulus. If the infant dishabituates (i.e., looks longer -- with renewed attention) to the test stimulus, the infant is said to discriminate between the two stimuli. Since the stimuli are generally constructed to differ on a single dimension, evidence of discrimination provides evidence of the infant's sensitivity to that dimension, from which we infer that that dimension is involved in processing of the image.

Researchers use pairwise discrimination patterns across many stimulus pairs to piece together theories of what infants organize, perceive, and know, but less attention is generally paid to the habituation data itself. To the extent that performance in the habituation phase of the study is tapping the process of encoding the habituation stimulus, habituation itself can tell us something about perception. In comparisons across stimuli, the habituation trials offer a measure of the relative ease with which stimuli are

2. Each infant was tested only at one age and in one condition.

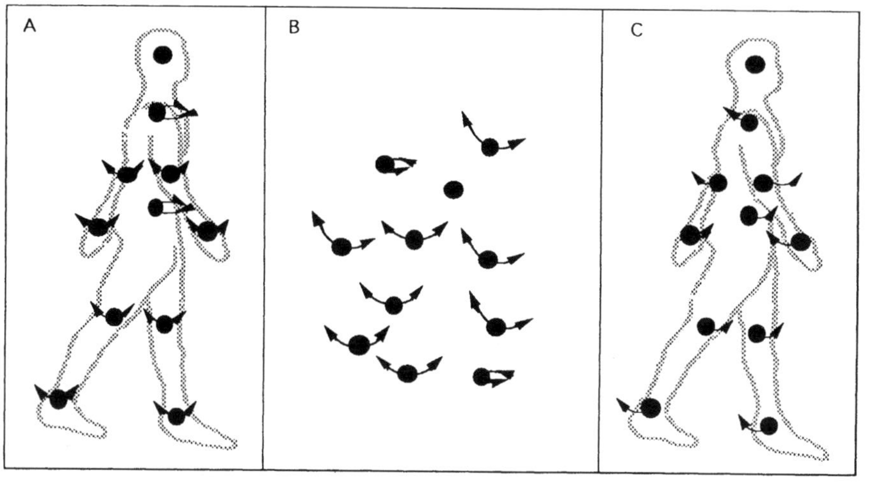

Figure 2. Point-light walker displays. A. Canonical walker. The arrows indicate the pendular motions of each point-light. B. Scrambled walker. The elements preserve only the absolute motions of the canonical walker. C. Phase-shifted walker. While the point-lights on a limb of the canonical walker move together, the point-lights of the phase-shifted walker move asynchronously, sometimes in opposition to one another, as the arrows suggest.

organized during encoding. Ease of encoding, operationalized as the sum of an infant's looking times across habituation trials, is a reflection of the interaction between the infant's organizational processes and the structure of the stimulus. Thus, by the combination of the discrimination information available from the habituation/ dishabituation analyses, and the encoding information available in the habituation trial data, we can develop a fairly detailed picture of the overall processes of perception and its development.

Point-light Displays Of Human Gait

The studies we selected for analysis all involved computer-synthesized point-light walker displays (Cutting, 1978; Bertenthal & Kramer, 1984) and foils in which selected stimulus parameters had been modified. The standard, or canonical, displays (Figure 2a) were comprised of 11 dots that mimic the kinematics of human gait, approximated by the harmonic motions of hierarchically-nested pendula. Adults identify this form as depicting a person walking on a treadmill (Bertenthal & Davis, 1988). Infants show the same pattern of looking during habituation to these synthesized displays as to the displays created using a human model. Foils were constructed varying the relations among the point-light elements, the orientation, and global form.

Relations among the elements: The *scrambled* display (Figure 2b) was comprised of 11 point-lights with individual motion vectors corresponding to those in the canonical display, however, the spatial configuration among elements was random. The *phrase-shifted* display (Figure 2c) maintained the individual motion vectors as well as the hierarchical spatial configuration of the canonical form, but altered the relative motions among the elements. In the canonical displays, point-lights cycled synchronously together or in opposition. In phase-shifted displays, point-lights no longer maintained synchronicity, appearing much like a marionette. Researchers examining motor behavior have suggested that phase relations are an important invariant in human locomotion (Schmidt, 1985). In *non-rigid* displays (not shown), the distance between the point-lights was randomly varied from frame to frame.

Orientation: Displays were oriented upright, inverted, or horizontal in the picture plane.

Global form: In addition to human forms, displays looking much like four-legged "spiders" were created. These forms were comprised of four limbs, identical to human limbs, radiating from a single point.

Each of the 17 studies was designed to test infants' sensitivity to the spatiotemporal properties of point-light walkers across the dimensions just described. Sensitivity was assessed on the basis of habituation/dishabituation patterns obtained from pairwise comparisons made over successive studies. In addition, we obtained each infants' habituation curve, and computed the total looking time (TLT) for each stimulus type. TLT is the sum of the looking times across all habituation trials for a selected stimulus. Alternative measures, such as velocity of decline, peak looking time, or first habituation trial, were correlated with the total looking time but TLT can be more intuitively interpreted. We take it to be a reflection of the ease with which an infant can organize the stimulus during encoding.

Using the TLT measure, we examined the influence of stimulus parameters on encoding times to determine the relative ease of encoding, and by inference, the processing constraints available to the infant. Assessments of discrimination could be made only for stimulus pairs actually tested, but assessments of ease of encoding could be made by way of pseudo-experiments. In these analytic studies, we compiled all habituation curves produced in response to a stimulus exhibiting the property we wished to examine. For example, in order to examine the relative ease with which infants encode canonical versus scrambled forms, we gathered the habituation curves from all infants who had seen either stimulus as the

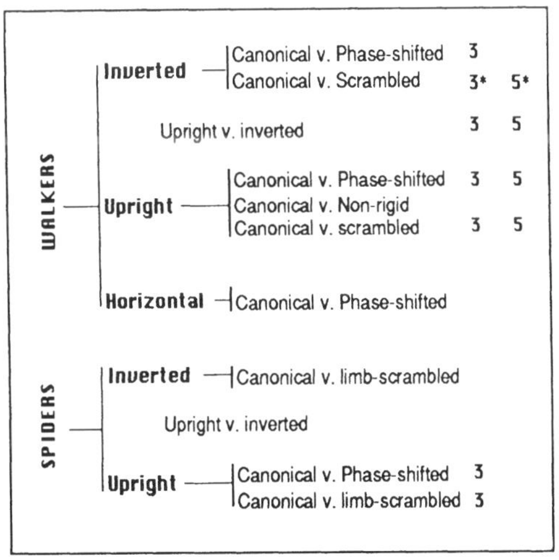

Figure 3. Summary of discrimination analyses. The number to the right of the tree indicates the age group exhibiting discrimination, as indicated when the mean of the test trials is significantly greater than the mean of the last 2 habituation trials. (The * indicates an order effect.)

familiarization display, and compared the total looking times for these data.

Results

Several researchers (e.g., Bronson, 1982; Sokolov, 1960) suggest that infants' visual attention is guided by knowledge and expectancy. If this is the case, 3- and 5- month-olds, who are likely to differ in knowledge, should exhibit different patterns of encoding and discrimination. Moreover, we expected to find a systematic relationship between differential encoding times and discrimination.

Data were organized according to age and stimulus parameters, specifically global form (walker, spider), orientation (upright, inverted, horizontal), and element relations (canonical, scrambled, phase-shifted, nonrigid). The results of a re-analysis of pairwise discrimination by each age group are summarized in Figure 3.

In general, when assessed by either discrimination or encoding behavior, 3- & 5- month-olds respond differently to the point-light displays. As shown in Figure 3, three-month-olds discriminated 7 of the 11 pairs of stimuli, whereas 5-month-olds discriminated only 4. Though differences in encoding times did not mirror discrimination patterns, total looking time also differed systematically by age. Three-month-olds (M=203.1), across stimulus parameters, looked longer during the habituation process than did 5-month-olds (M=120.68), $F(1,385)=59.23$, $p<.01$. This general change in performance may result from maturing neural and motor processes (Johnson, Posner, & Rothbart, 1991), from increased skill or knowledge, or from changes in motivational factors. Though this data cannot directly discern between these accounts, it contributes additional support to positions which posit global change; 5-month-olds habituated faster than did 3-month-olds across all stimuli.

Three-month-olds discriminate forms with differing relations among the elements when those stimuli are organized about a vertical axis. These infants discriminate canonical from phase-shifted forms and canonical from scrambled forms when they are upright or inverted, but not when they are horizontal. Bornstein & Krinsky (1985) report evidence which suggests that 3-month-old infants may show sensitivity to the properties of a form only when it is organized about a vertical axis. Interestingly, at 3 months of age TLT to vertical stimuli (M=208.3) was significantly greater than TLT to horizontal stimuli (M=120.8), $F(1,184)=5.58$, $p<.05$, suggesting a correspondence between higher looking time and subsequent discrimination of stimulus properties.

By 5 months of age, infants are faster to encode stimuli as well as less apt to show discrimination on pairwise comparisons, suggesting that they attend more selectively to information available in the displays. Five-month-old infants showed few differences in encoding times as well. These infants looked significantly longer at out-of-phase walkers (M=153.4) than at in-phase walkers (M=102.5), $F(1,193)=15.6$, $p<.01$. In addition, 5-month-olds looked longer at out-of-phase walkers than at scrambled forms (M=83.83), $F(1,72)=2.99$, $p<.05$. Encoding times to upright canonical and scrambled forms did not differ. Five-month-olds, like 3-month-olds, tend to discriminate stimulus parameters only when they are presented in a particular configural context, or frame. While they encode phase information generally, 5-month olds discriminate canonical and phase-shifted forms only when the figures are presented upright and in the context of a walker-like form (hierarchically organized along a vertical axis of symmetry). Phase-relations in spider-like forms or horizontal orientations are not discriminated. Thus, 5-month-old infants may be responding in part on the basis of some familiarity with the characteristic phase-relations of human gait. Indeed, prior research (Pinto & Davis, 1991; Bertenthal, 1992) suggests that 5-month-olds, but not 3-month-olds, respond on the basis of a category of human movement.

Discussion

We conceive of visual perception as a process whereby information is organized both by general processing heuristics and in accord with knowledge (processes and/or representations) that arises from experience with the domain --in this case, biomechanical motion. We have provided evidence of the development both of general processing heuristics (or the development of their use) and of knowledge that is specific to the perception of biomechanical motion. Both 3- and 5-month old infants use phase information to discriminate point-light displays, but 5-month olds show an encoding difference in phase-shifted vs. canonical displays. Since 3-month-olds discriminate the stimuli along most of the dimensions examined, it is reasonable to interpret their lack of differential encoding times as a reflection of the elevated encoding effort they put into organizing the entire displays. Five-month-olds, on the other hand, may approach the habituation task differently. Rather than encoding every available parameter of the stimulus, 5-month olds may look only long enough to determine whether or not it has been presented previously (Bronson, 1982) in accord with one or a few key features. Thus, though five-month-olds may be sensitive to phase relations in general, perturbations to the characteristic phase relations of human gait may constitute a meaningful difference only in the context of the familiar form of a person.

Differences in what is encoded by children of different ages or experience has been suggested as an explanation for other developmental phenomena, such as skill at making balance-beam discriminations (Siegler, 1989). Our results support a unified view of some forms of perception with these "higher" cognitive phenomena, in which differential skill in what is encoded is a ubiquitous characteristic of human learning and development.

More specific to perception, our results may help to fill out the way in which a computational account of the extraction of structure from biomechanical motion depends upon both general and domain-specific processing heuristics. A number of computational models are able to use general heuristics to organize broad classes of motion-given information by recovering the spatial or geometric properties of the form (Ullman 1983, Webb & Aggarwal, 1982). However, these general algorithms do not yield complete or unique solutions for biomechanical displays nor do they exhibit the robustness of human visual processes (see Proffitt & Bertenthal, 1988, for a discussion of these models). Some integration of a general heuristic account with a theory of schema-driven processing is needed.

Interestingly, one of the early models of habituation was both schema-based, and implied expectation-based differential encoding. Sokolov (1960) proposed a theory of habituation in which the organism builds a model of the stimulus and then tests this model against the stimulus. In Sokolov's view, habituation results from the repression of an orienting response when the perceptual system determines that there is no new information to be gathered from the stimulus. As the organism comes to expect one hypothesis more strongly than another, the value of additional information is reduced significantly, thus resulting in a decline of looking time. Interpreted computationally, Sokolov's theory is very close to computational methods in model-based diagnosis and in theory-formation methods (see the papers in Davis & Hamscher, 1990, and Shrager & Langley, 1991).

Neither Siegler's theory, nor the models of Sokolov, of Ullman, or of Webb & Aggarwal are complete alone; and their integration requires empirical guidance. Some of this guidance may be usefully derived from developmental data such as that provided here. The present work may thus help us sharpen the view of perception as problem-solving (Rock, 1983), by guiding the integration of theory-formation models with general heuristic or constraint-based models, into a more complete account of general perception.

Acknowledgments

We would like to thank Margaret Timmins, Sharon Lambert, and Elizabeth Ong for their assistance with this project. We would also like to thank the reviewers for their helpful comments. Please direct correspondence regarding this paper to the second author at System Sciences Laboratory, Xerox Palo Alto Research Center, 3333 Coyote Hill Road, Palo Alto, CA 94304.

References

Bertenthal, B.I. 1992. Infants' perception of biomechanical motions: Intrinsic image and knowledge-based constraints. In C. Granrud (Ed.), Carnegie *Symposium on cognition: Visual perception and cognition in infancy*. Hillsdale, NJ: Erlbaum. Forthcoming.

Bertenthal, B.I. & Davis, P. 1988. Dynamical pattern analysis predicts recognition and discrimination of biomechanical motions. *Proceedings of the Annual Meeting of the Psychonomic Society*. Chicago, Illinois.

Bertenthal, B.I., & Kramer, S.J. 1984. The TMS 9918a VDP: A new device for generating moving displays on a microcomputer. *Behavior Research Methods, Instruments, & Computers* 16:388-394.

Bornstein, M.H. & Krinsky, S.J. 1985. Perception of symmetry in infancy: The salience of vertical symmetry and the perception of pattern wholes, *Journal of Experimental Child Psychology* 39:1-19.

Bronson, G.W. 1982. *The Scanning Patterns of Human Infants: Implications for visual learning*. New Jersey: Ablex Publishing Co.

Cohen, L.B. 1973. A two process model of infant visual attention. *Merrill-Palmer Quarterly* 19:157-180.

Cutting, J.C. 1978. Generation of synthetic male and female walkers through manipulation of a biomechanical invariant. *Perception* 7:393-405.

Davis, R. & Hamscher, W.C. 1990. Model-based reasoning: Troubleshooting. In P.H. Winston & S.A. Shellard (Eds.), Artificial Intelligence at MIT: Expanding Frontiers. Cambridge, MA: MIT Press, 380-429.

Gibson, J.J. 1950. *Perception of the visual world*. New York: Houghton Mifflin.

Johansson, G. 1973. Visual perception of biological motion and a model for its analysis. *Perception and Psychophysics* 142:201-211.

Johansson, G. 1975. Visual motion perception. *Scientific American* 232(6):76-88.

Johnson, M.H., Posner, M.I., & Rothbart, M.K. 1991. Components of visual orienting in early infancy: Contingency learning, anticipatory looking, and disengaging. *Journal of Cognitive Neuroscience* 3 (4):335-344.

Pinto, J. & Davis, P. 1991. The categorical perception of human gait in 3- & 5- month old infants. *Presented at the biannual meeting of the Society for Research in Child Development*, Seattle, WA.

Proffitt, D.R. & Bertenthal, B.I. 1988. Recovering connectivity from moving point-light displays. In W.N. Martin, & J.K. Aggarwal (Eds.), *Motion Understanding: Robot and Human Vision*. Boston: Kluwer Academic Publishers.

Rock, I. 1983. *The logic of perception*. Cambridge, MA: The MIT Press.

Schmidt, R.A. 1985. The search for invariance in skilled movement behavior. *Research Quarterly for Exercise and Sport* 56(2):188-200.

Shrager, J. & Langley, P. 1991. *Computational Models of Scientific Discovery and Theory Formation*. San Mateo, CA: Morgan Kaufmann.

Siegler, R. S. 1976. Three aspects of cognitive development. *Cognitive Psychology* 8:481-520.

Sokolov, E.N. 1960. Neuronal models and the orienting influence. In M.A.B. Brazier (Ed.), *The central nervous system and behavior: III* New York: Macy Foundation.

Ullman, S. 1983. Recent computational studies in the interpretation of structure from motion. In J.Beck, B. Hope, & A. Rosenfeld (Eds.), *Human and Machine Vision*. New York: Academic Press.

Wallach, H. & O'Connell, D.N. 1953. The kinetic depth effect. *Journal of Experimental Psychology* 45(4):205-217.

Webb, J.A. & Aggarwal, J.K. 1982. Structure from motion of rigid and jointed global forms. *Artificial Intelligence* 19:107-130.

The Role of Measurement in the Construction of Conservation Knowledge

Tony Simon
School of Psychology
Georgia Institute
of Technology
Atlanta, GA 30332-0170
tonys@zunow.gatech.edu

David Klahr
Department of Psychology
Carnegie Mellon University
Pittsburgh, PA 15213
klahr@psy.cmu.edu

Allen Newell
School of Computer Science
Carnegie Mellon University
Pittsburgh, PA 15213
an@centro.soar.cs.cmu.edu

Abstract*

Conservation knowledge and measurement abilities are two central components in quantitative development. Piaget's position is that conservation is a logical pre-requisite of measurement, while Miller's is the reverse. In this paper we illustrate how measurement is employed as an empirical tool in the construction of conservation knowledge. This account predicts the familiar pattern of conservation development from the limits on young children's measurement abilities. We present Q-Soar, a computational model that acquires number conservation knowledge by simulating children's performance in a published conservation training study. This model shows that measurement enables a verification process to be executed which is the basis of conservation learning.

Introduction

Two central conceptual attainments in the development of quantification abilities are conservation knowledge (understanding the behavior of quantities under transformation) and measurement skills (creating quantitative values for bodies of material). Yet, despite the centrality of these two aspects of quantification, relatively little attention has been paid to the developmental roles that they play. Inspection of the literature reveals two incommensurate positions. The view held by Piaget (Piaget, Inhelder and Szeminska, 1960) was that conservation is a logical pre-requisite to the ability to measure. He reasoned that, without an understanding of the essential nature of quantity, measurements in terms of those quantities would mean nothing and would be of no practical use. The opposing view is that measurement is the necessary precursor of conservation (Klahr & Wallace, 1976; Miller, 1984). Measurement is the empirical tool used to gather information about whether or not some dimension of a transformed entity has remained quantitatively invariant. Miller (1984) states that "practical measurement procedures appear not to be late-developing concomitants of a more general understanding of quantity. Instead, the measurement procedures of children embody their most sophisticated understanding of the domain in question. The limitations of these procedures constitute significant limits on children's understanding of quantity ... (p.221)".

Such measurement is not always possible though. The limitations Miller speaks of determine what children can learn about quantity. They are responsible for the pattern in the development of conservation. Number, or discrete quantity, conservation is acquired first. Also, preconservers can reason successfully about transformations of small discrete quantities but not of large ones (Cowan, 1979; Fuson, Secada & Hall 1983; Siegler, 1981). Conservation of continuous quantities such as length, area and volume is acquired a year or two later (Siegler, 1981).

One type of limitation is on processes, i.e. on what kinds of things measurement procedures can be applied to. As Piaget et al (1960) state, "to measure is to take out of a whole, one ... unit, and to transpose this unit on the remainder of the whole". Thus, any material to be measured must afford the measurer some unit which can be used in that process. This characteristic is not present in continuous quantities. Beakers of water or

* This work was supported by contract N00014-86-K-0678 from the Computer Science Division of the Offiice of Naval Research

pieces of string do not exhibit any evident sub-units. Only the employment of special tools such as rulers or measuring cylinders (and the knowledge of how to use them) can create sub-units that can be used for quantification. On the other hand, discrete quantities are defined by collections of individual sub-units of the quantity as a whole. No special tools are needed since quantification abilities are present to some extent in the heads of even the youngest children. Young children appear to be particularly sensitive to the fact that it is at the level of unitary objects, and not subparts of those objects, that quantification of collections should take place (Shipley & Shepperson, 1990). Thus, discrete quantities are clearly much easier to measure.

A second type of limitation is in the abilities of the children who are attempting to use measurement procedures. The children that need to carry out measurements to determine quantitative invariance are those below the age of five. However, their quantification skills are not well developed. They are efficient at subitizing: a fast, accurate perceptual quantification mechanism (Chi & Klahr, 1975; Svenson & Sjoberg, 1983). Subitizing, though, has a limit of about four objects (Atkinson, Campbell & Francis, 1976). Young children's counting is only reliable for collections of about the same size (Fuson, 1988).

The measurement-before-conservation view therefore predicts the learning events that enable the acquisition of quantitative invariance knowledge. It follows that, if measurement is needed to be able to reason about quantity, learning can occur only when the effects of transformations of small collections of objects are evaluated. These quantities will have to be discrete because young children are not capable of creating consistent sub-units from continuous quantities. Gelman (1977) has shown that one-year-olds can reason about some transformations when the number of objects involved is very small. The discrete quantity requirement is supported by Piaget et al's (1960) and Miller's (1984) findings that, given the task of dividing up an object such as a cookie into equal parts, young children created many arbitrarily-sized sub-units. These are unsuitable for quantification because counting them does not produce accurate absolute measures for a single entities or relative measures of multiple entities.

Miller (1989) has further demonstrated the interaction between the use of measurement procedures and the acquisition of quantitative knowledge. Miller tested three- to ten-year-olds on a modified equivalence-conservation task. A variety of transformations were applied to different materials to test number, length and area conservation. He predicted that the effects of transformations would be easy to determine when relevant measurement procedures provided good cues to the actual quantity and vice versa. For example, counting is a good way to determine the resulting quantity of a transformation like spreading objects out. Thus, Miller predicted that the number task would be easier than length or area because it is easiest to measure. But enumeration is a bad method for evaluating the effect of changing the objects' size, since it does not assess their total mass. For this transformation, Miller predicted the number task would produce the worst performance. The results were as predicted, showing that what quantitative knowledge one can learn depends on what measurement procedures one uses.

Our theory (Simon, Newell & Klahr, 1991) follows Klahr (1984) in stating that it is measurement of collections of discrete objects that provides information upon which knowledge about quantitative invariance is built. Conservation knowledge is acquired in situations where invariance can be empirically verified. In other words, learning events occur when the materials allow children to use their measurement capabilities to obtain a numerical measurement for a collection of objects before and after it has been transformed. The two measurements can then be compared and the result attributed to the transformation as its effect.

If the difference is zero, the quantity is unchanged and the transformation is deemed to have a non-quantitative effect for the dimension in question, e.g. it "conserves number". If some difference is detected, the transformation is found to be non-conserving. Such differences can be simply detected by means of discriminations based on subitizing. With sufficient domain knowledge, the direction and magnitude of the change can also be determined. Thus we conclude that the initial learning experiences for invariance knowledge will be based on measurements of small collections of discrete objects within the subitizing range.

Q-Soar

To show that measurements are the stuff of which conservation knowledge is made, we built a computational model, based on the above theory. Q-Soar's foundation is the Soar architecture for intelligent behavior (Laird, Newell & Rosenbloom, 1987) and associated cognitive theory (Newell, 1990) which involves performance organization in terms of problem spaces and goal-oriented, experienced-based learning in terms of chunking. Q-Soar simulated the acquisition of number conservation demonstrated in a training study by Gelman (1982) and thus is the first demonstration that chunking can account for developmental transitions. Gelman's study contained two training conditions of interest; experimental and cardinal-once.

The experimental condition was an equivalence-conservation task where one of two rows of three or four objects was spread out or compressed, leaving the original one untouched. Before and after each transformation, children were required to count each row and state the absolute and relative numerosities of the rows. Gelman designed this condition to help children make use of one-to-one correspondence matching. The cardinal-once condition was an identity conservation condition, so called because the single row involved just one before and after count. The numerical comparison required was between the pre- and post-transformation quantities of the same row. In this condition there was no means of using one-to-one matching and so Gelman predicted that this group would not benefit from training.

Results showed that three- and four-year-olds learned conservation from the experimental condition, since they solved large number tests. The cardinal-once condition produced no learning in three-year-olds but it did benefit the four-year-olds, though they performed less well than their experimental peers. The no-cardinal control group, who saw no transformations, failed the tests.

Q-Soar began with the ability to simulate the pre-training competence of the three- and four-year old children in Gelman's study. The precise details of the model can be found in Simon et al (1991). There were two variants of the model; Q-Soar-3 and Q-Soar-4 (which modelled the three- and four-year-olds respectively). They consisted of a set of problem spaces enabling the execution of the behaviors required in Gelman's conditions. The problem spaces also enabled the model variants to execute a *verification process*. This involved comparing pre- and post-transformation measurements of the numerical aspect of the arrays to determine the effect of the transformation. Thus, after training, the effect of chunking over the verification processing was that Q-Soar now knew the numerical effects of the transformations rather than having to determine them empirically. Untrained versions of each model faced with Gelman's larger number post-tests failed in the characteristic manner. Their quantification abilities were not sufficient to measure the arrays and so they resorted to estimation based on the lengths of the rows. The result was that the wrong answer was always given, just the same as children in Gelman's control group.

The only difference between the two model variants was that Q-Soar-4 always executed the verification process (unless chunks fired to provide the effect of a familiar transformation) whereas Q-Soar-3 did not. The following observations support this. First, Gelman (1977) among others, has shown that young children assume that a set of objects that has undergone no visible transformation will not undergo any alteration in quantity. Second, many experiments (see Donaldson, 1978) have shown that young children assume that the quantitative value of a set of objects will change if it undergoes an obvious visible transformation. These lead to two theoretical assumptions. First, three-year-olds will not attempt to verify the assumed quantitative change resulting from a visible transformation unless presented with conflicting evidence which suggests that the quantity has remained unchanged. Second, four-year-olds will always verify the quantitative effect of a transformation, irrespective of the post-transformation perceptual information.

When the Q-Soar-4 variant undergoes either of the Gelman training conditions it carries out the verification processing which has the effect of allowing it to learn the effects of spreading and compressing transformations. When Q-Soar-3 experiences the cardinal-once training condition, the post-transformation array appears totally consistent with its assumption of quantitative change. There is a single row that is longer or shorter than it was before. There is no reason to check what appears to be an obvious result, i.e. that transforming the row has altered the number of objects in it. Thus Q-Soar-3 in cardinal-once

Gelman, 1982 (% correct)	Experimental	Cardinal-Once	No-Cardinal
3-year-olds	71	9	6
4-year-olds	70	46	15
Q-Soar (response)			
Q-Soar -3	correct	incorrect	incorrect
Q-Soar-4	correct	correct	incorrect

Table 1. Test responses from Gelman's subjects and Q-Soar

does not execute the verification procedure and so does not provide for itself the chance to learn about the conservation effects on number of spreading and compressing transformations.

However, when Q-Soar-3 experiences the experimental condition there are two post-transformation arrays, both within the subitizing range. The transformed row is now assumed to be quantitatively different, but two conflicting types of perceptual input are available. The length of the rows appears to confirm the assumption. However, based on subitizing evidence, the two rows still maintain their original numerical values. This conflict leads Q-Soar-3 to execute the verification procedure and, like Q-Soar-4, learn about the conservation effects of the transformations. This suggests that the conflict is what persuades the three-year-olds to use the verification process when they seek to determine quantitative invariance. Presumably, once stimulated to employ the verification process by such conflict, these three-year-olds will eventually automatically do so, as is the case for four-year-olds.

Results

Q-Soar-3 and -4 underwent the same training and testing as Gelman's subjects. Comparison of test responses is presented in Table 1. Similar performance is evident for Q-Soar and human subjects in the experimental and no-cardinal conditions. The cardinal-once condition produced no learning in Gelman's three-year-olds, as she predicted, but the older children clearly did benefit from the training. The precise reasons for their variable performance are not clear at present. Nevertheless, this counters the prediction that cardinal-once offers no opportunity for conservation learning due to the fact that correspondence matching is not possible. Though Q-Soar-4 learned rather too well from the training, the result suggests that its processes provide the means of learning conservation knowledge.

Conclusions

In this paper we have presented Q-Soar, a computational model, which simulates the acquisition of conservation knowledge in a published training study. Q-Soar implements and extends Klahr's theory that discrete quantities are foundational to conservation learning. It also demonstrates that Soar's chunking mechanism can account for significant developmental acquisitions such as number conservation. Although chunking was originally constructed to model practice effects over many trials (Rosenbloom & Newell, 1986), its application has been extended to a wide range of cognitive tasks (Lewis et al, 1990). This suggests that, as a goal-directed, experience-based learning mechanism within a problem space architecture, chunking may be a sufficient account of human learning.

Q-Soar also predicted learning events that were not consistent with Gelman's theory. Our work shows that conservation knowledge is acquired when young children apply their limited measurement capabilities to empirically verify the quantitative effect of transforming a collection of objects. The result is then bound to the type of transformation observed as its general quantitative effect on the quantity concerned. In other words, measurement enables conservation judgments to be made, while verification enables conservation knowledge to be learned.

This reverses the logical relationship between conservation and measurement in Piaget's formulation, making the empirical process of measurement a prerequisite for necessity judgments about conservation. It also limits the scope of such "logical" conservation judgments from a general principle about all quantities in Piaget's case, to a domain-specific generalization in our case. Number conservation must be learned first, by determining the effects of transformations in terms of number. The transfer to continuous quantities appears to require transfer via representation change, which is presumably why it takes so long after the initial acquisition. We have not yet addressed this issue in detail. Another issue arising from this work is the need to identify the exact processes required to learn conservation knowledge. Q-Soar could complete Gelman's task in a number of ways, not all of which would learn what the children in her experiment did. The data show that the children's learning is not all or none as Q-Soar's is. A parametric analysis of Q-Soar variants is being undertaken. From this we expect to discover the range of necessary knowledge and processes that can be employed to acquire conservation knowledge from this task in the way that human subjects do.

References

Atkinson, J., Campbell, F.W. & Francis, M.R. (1976) The magic number 4 plus or minus 0: A new look at visual numerosity judgments. *Perception* 5, 327-334.

Chi, M.T.H. & Klahr, D. (1975) Span and Rate of Apprehension in Children & Adults. *Journal of Experimental Child Psychology* 19, 434-439.

Cowan, R. (1979) Performance in number conservation tasks as a function of the number of items. *British Journal of Psychology* 73, 77-81.

Fuson, K.C. (1988) *Children's Counting and Concepts of Number*. New York: Springer Verlag.

Fuson, K.C., Secada, W.G. & Hall, J. W. (1983) Matching, counting and conservation of numerical equivalence. *Child Development* 54, 91-97.

Gelman, R. (1977) How young children reason about small numbers. In N.J. Castellan, D.B. Pisoni & G.R. Potts (Eds.) *Cognitive Theory* Vol. 2. Hillsdale, NJ: Erlbaum.

Gelman, R. (1982) Accessing one-to-one correspondence: Still another paper about conservation. *British Journal of Psychology* 73, 209-220.

Klahr, D. (1984) Transition Processes in Quantitative Development. In R.J. Sternberg (Ed.) *Mechanisms of Cognitive Development*. New York, NY: Freeman.

Klahr, D. & Wallace, J.G. (1976) *Cognitive Development: An Information Processing View*. Hillsdale, NJ: Erlbaum.

Laird, J.E., Newell, A. & Rosenbloom, P.S. (1987) *SOAR: An architecture for general intelligence. Artificial Intelligence* 33, 1-64.

Lewis, R.L., Huffman, S.B., John, B.E., Laird, J.E., Lehman, J.F., Newell, A, Rosenbloom, P.S., Simon, T. & Tessler, S.G. (1990) *Soar as a unified theory of cognition: Spring 1990*. Proceedings of the Twelfth Annual Conference of the Cognitive Science Society. Hillsdale, NJ: Erlbaum.

Miller, K.F. (1984) Child as measurer of all things: Measurement procedures and the development of quantitative concepts. In. C. Sophian (Ed.) *The Origin of Cognitive Skills*. Hillsdale, NJ: Erlbaum.

Miller, K.F. (1989) Measurement as a tool for thought: The role of measurement procedures in children's understanding of quantitative invariance. *Developmental Psychology* 25, 589-600.

Newell, A. (1990) *Unified Theories of Cognition*. Cambridge, MA: Harvard University Press.

Piaget, J., Inhelder, B. & Szeminska, A. (1960) *The Child's Conception of Geometry*. London: Routledge & Kegan Paul.

Rosenbloom, P.S. & Newell, A. (1986) The chunking of goal hierarchies: A generalized model of practice. In R.S. Michalski, J.G. Carbonell & T.M. Mitchell *Machine Learning* Vol. 2. Los Altos, CA: Morgan Kaufmann.

Shipley, E.F. & Shepperson, B. (1990) Countable Entities: Developmental Changes. *Cognition* 34, 109-136.

Siegler, R.S. (1981) Developmental sequences within and between concepts. *Monographs of the Society for Research in Child Development.* 46.

Simon, T., Newell, A. & Klahr, D. (1991) A computational account of children's learning about number conservation. In D. Fisher, M. Pazzani & P. Langley (Eds.) *Concept Formation: Knowledge & Experience in Unsupervised Learning.* San Mateo, CA: Morgan Kaufmann.

Svenson, O. & Sjoberg, K. (1983) Speeds of subitizing and counting processes in different age groups. *Journal of Genetic Psychology* 142, 203-211.

An Investigation of Balance Scale Success

William C. Schmidt and Thomas R. Shultz
Department of Psychology
McGill University
1205 Penfield Avenue
Montréal, Québec, Canada H3A 1B1
schmidt@lima.psych.mcgill.ca || shultz@psych.mcgill.ca

Abstract

The success of a connectionist model of cognitive development on the balance scale task is due to manipulations which impede convergence of the back-propagation learning algorithm. The model was trained at different levels of a biased training environment with exposure to a varied number of training instances. The effects of weight updating method and modifying the network topology were also examined. In all cases in which these manipulations caused a decrease in convergence rate, there was an increase in the proportion of psychologically realistic runs. We conclude that incremental connectionist learning is not sufficient for producing psychologically successful connectionist balance scale models, but must be accompanied by a slowing of convergence.

Introduction

Connectionist learning algorithms have successfully acted as transition mechanisms in a number of recent models of cognitive developmental phenomena. McClelland (1988) suggested that gradual, incremental error reduction is a key property of connectionism that is responsible for this success. In the current paper, we focus on McClelland's (1988) connectionist model of cognitive development on the Piagetian balance scale task. We show that variants of the original model perform well psychologically as long as they delay convergence of the back-propagation learning algorithm.

The Balance Scale Task

The balance scale task consists of showing a child a balance scale (Figure 1) supported by blocks so that it stays in the balanced position. A number of weights are placed on one of a number of evenly spaced pegs on each side of the fulcrum, and it becomes the child's task to predict which arm will go down, or whether the scale will balance, once the supporting blocks are removed.

Siegler (1976, 1981) has reported that children's performance on the balance scale progresses through 4 distinct rule based stages: (1) use only weight information to determine if the scale will balance, (2) emphasize weight information but consider distance if weights on either side of the fulcrum are equal, (3) correctly integrate both weight and distance information for simple problems, but respond indecisively when one arm has greater weight and the other greater distance, (4) correctly integrate weight and distance information.

Siegler (1976, 1981) partitioned the set of all possible balance scale problems into 6 distinct problem types. *Balance* problems have equal numbers of weights placed at equal distances from the fulcrum. In *weight* problems, distances on either side of the fulcrum are equal, hence the side with more weights goes down. In *distance* problems, the arm with greater distance goes down since the sides have equal weights. *Conflict* problems have greater weight on one arm and greater distance on the other. The correct response to the problem determines its classification as a *conflict-weight*, *conflict-distance*, or *conflict-balance* problem. Performance in terms of the percentage of correct predictions made on some subset of problems drawn from each of the problem types can be used to classify subjects as conforming to a particular rule. Rules and their predicted performance levels for each of the 6 problem types appear in Figure 1. In order to classify children's performance, Siegler (1981) used 24 testing instances (4 from each of the 6 different problem types). Children scoring 4 or fewer deviations from responses predicted by a given rule were counted as acting in accord with the rule. Additionally, Siegler introduced a number of safeguards to ensure that a child classified at one stage was not actually responding in a manner characterized by another.

McClelland (1988) reported the creation of a connectionist model of the balance scale task with 5 pegs and 5 weights per arm. The network topology, which appears in Figure 2, consisted of two sets of 10 input units, each fully connected to a distinct pair of hidden units. Each pair of hidden units was fully connected to two output units. One set of input units represented, in a localist fashion, the number of weights on each of the balance scale's two arms (1 to 5), while the other represented the distance of the weight from the fulcrum (1 to 5). Activation values of the outputs were interpreted to transform the network's output (2 real numbers between 0.0 and 1.0) into one of three possible prediction responses.

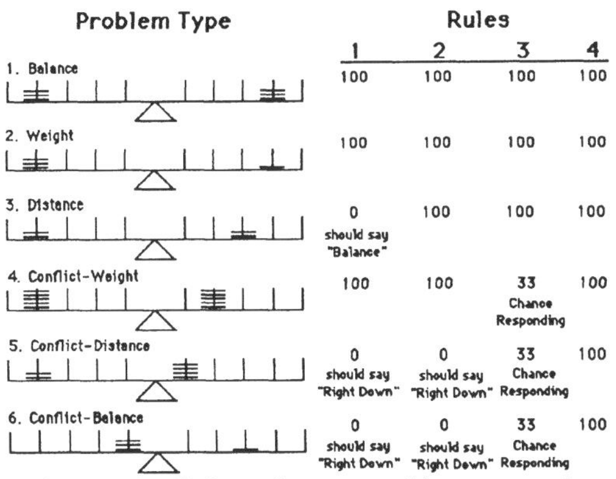

Figure 1. Predictions of percent problems correct for children using different rules.

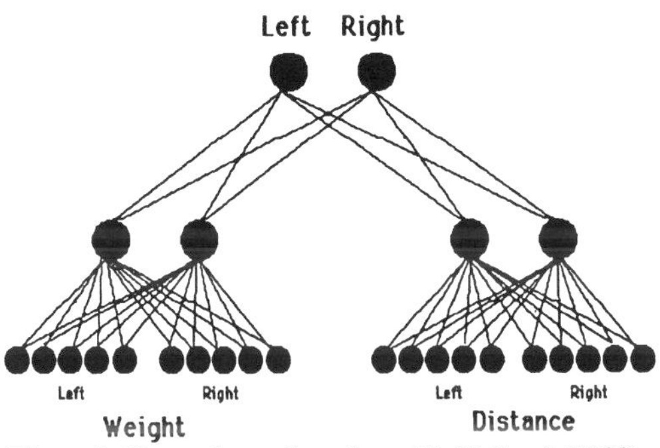

Figure 2. Network topology from McClelland (1988).

The model was trained using the back-propagation learning algorithm.[1] Learning in each epoch was from 100 instances randomly selected from the entire set of 625 possible training problems augmented with a bias for *balance* and *distance* problems (*equal-distance* problems). The bias increased the training set to include 5 times or 10 times the original number of *equal-distance* problems. After each epoch the model's performance was evaluated using Siegler's rule assessment methodology. Other than where noted, all simulations in the current paper assume the network topology, training method, and parameter settings used in McClelland's (1988) original simulations.

Simulation 1: Subset Size and Bias

The first simulation examined the effects of two manipulations. One manipulation varied the size of the bias for *equal-distance* problems. Five different levels of bias were employed. The unadulterated set of 625 possible balance scale problems was augmented with 0, 5, 10, 15 and 20 times the normal number of *equal-distance* problems, resulting in new training corpuses of 625, 1125, 1750, 2375, and 3000 patterns respectively.

The other manipulation varied the size of the subset of training instances randomly selected each epoch from the training corpus. These subsets were selected without replacement, and a permuted batch weight updating method was used as in McClelland's (1988) original model. Since the sizes of training corpuses were unequal across different levels of bias, a percentage of the total number of exemplars belonging to each training corpus was selected. The levels of subset size investigated were 0.25%, 0.5%, 1%, 2%, 3%, 4%, 5%, 15%, 25%, 35%, 45%, 55%, 65%, 75%, 85%, and 95% of the entire training corpus.

Ten runs for each of the 80 groups (5 levels of bias x 16 subset sizes) were carried out. Each run was tested on the entire set of 625 possible training patterns both before training began and after weight updating. The patterns' total sum of squared errors score (TSS) was recorded every epoch, and the network's responses were evaluated for their fit to any of the 4 psychological rules. This longitudinal rule record was assessed to determine whether or not the network passed through all 4 stages. Training continued for 200 epochs each run.

In order to evaluate the style of learning characteristic to each group of runs, a simple linear regression model was fitted to the longitudinal TSS error scores for each run, with epoch predicting error score. This yielded regression equations of the form $error = b_0 + b_1 * log(epoch)$. The log coefficient b_1 assesses the fall off of the learning curve, and hence the rate of convergence. This measure of convergence rate will be more negative for networks which reduce error more slowly. The constant b_0 offsets the learning curve from the abscissa. In the case that there is both a large b_0 term, and a small value of b_1, the network will have failed to converge.

Additionally, for each run, the proportion of error reduction over the 200 training epochs was assessed by dividing the initial TSS error less the average error score for the last 10 epochs of training by the initial level of TSS error. This value can be negative in the event that TSS error increases. Proportion of error reduction was used to assess depth of learning. A network which has a steep convergence rate, but has reduced little error, has failed to solve the problem.

[1] A batch updating method was used in conjunction with permuted presentation of training instances, a learning rate of 0.075, and momentum of 0.9. Weights in the model were initialized randomly in the range of $-0.5 \le w_i \le 0.5$.

A 5 x 16 (bias x subset sizes) ANOVA of the log coefficients was undertaken, revealing main effects for bias ($F(4,720)$ = 148.4, p<.0001), subset sizes ($F(15,720)$ = 237.3, p<.0001), and their interaction ($F(60,720)$ = 16.9, p<.0001). A second ANOVA of learning depth revealed main effects for bias ($F(4,720)$ = 432.5, p<.0001), subset sizes ($F(15,720)$ = 278.5, p<.0001), and their interaction ($F(60,720)$ = 39.6, p<.0001). A third ANOVA of the proportion of networks showing realistic psychological performance showed main effects for bias ($F(4,720)$ = 26.4, p<.0001), subset sizes ($F(15,720)$ = 25.3, p<.0001), and their interaction ($F(60,720)$ = 5.4, p<.0001). In this initial simulation, the average regression captured 57% of available variance (R^2 = .57, sd = .28). After excluding models which failed to learn, this average fit increased to 73% of the variance (R^2 = .73, sd = .16).

A plot of the mean log coefficient as a function of subset size appears in Figure 3, along with the mean proportion of runs demonstrating psychologically realistic stages.[2] The shallowest learning curves occurred for networks trained with subsets of randomly chosen training instances in the range of 1% to 5%. This effect turns around below the 1% level as the learning curves begin to steepen. Could these networks trained on so few instances actually have converged faster? No, since investigation of the amount of error reduced (Figure 4) at these levels indicates that these networks failed to solve the problem at all!

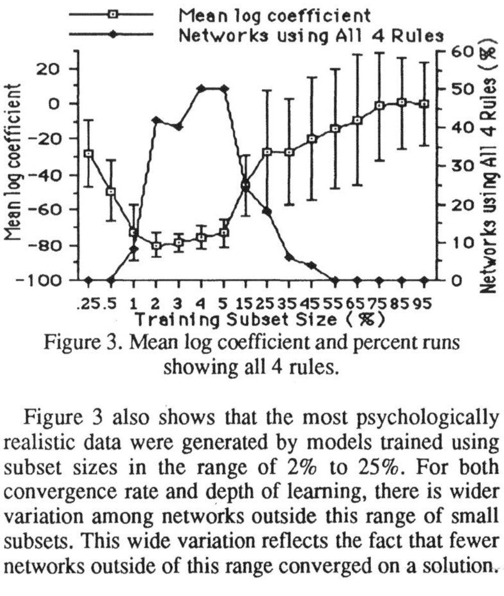

Figure 3. Mean log coefficient and percent runs showing all 4 rules.

Figure 3 also shows that the most psychologically realistic data were generated by models trained using subset sizes in the range of 2% to 25%. For both convergence rate and depth of learning, there is wider variation among networks outside this range of small subsets. This wide variation reflects the fact that fewer networks outside of this range converged on a solution. All networks which failed to converge were trained with subset sizes outside of the range of the range of 0.5% to 15%.

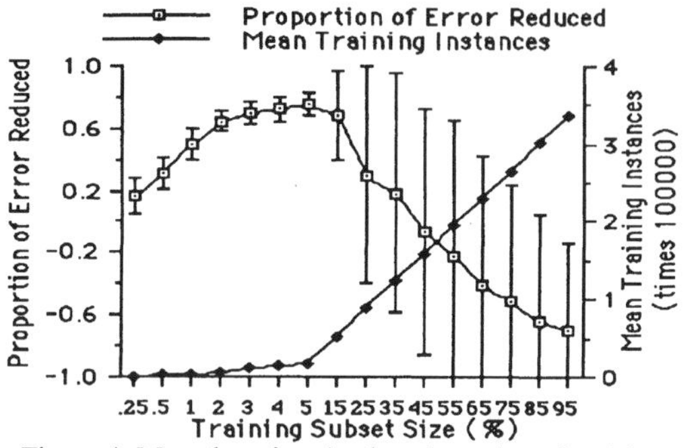

Figure 4. Mean learning depth and number of training presentations.

Investigation of the interaction effects for psychological performance revealed that models trained under levels of 5, 10, 15 and 20 times bias performed realistically while no models run without bias did. This is consistent with McClelland's (1988) two successful models, with subset size 6%, bias 5 times, and subset size 9%, bias 10 times. Here, the ordering of the cell means saw the level of 10 times bias yield the greatest proportion of psychologically realistic subjects, followed by levels of 15, 20, and 5 times bias respectively.

A plot of learning depth appears in Figure 4. Maximum error reduction occurred for networks trained with subsets of instances in the range of 2% to 15%, which overlaps with networks having the slowest convergence rates and the most psychologically realistic performances. In addition, Figure 4 plots the mean number of training instances witnessed in the models' 200 epoch lifetimes.

Comparing across Figures 3 and 4, it is clear that the number of training instances is negatively related to learning depth and positively related to the log coefficient. The negative relation of number of training instances to learning depth is an artifact of the failure of many networks to learn at high levels of subset size. The positive relation of the number of training presentations to log coefficient demonstrates that rate of convergence was generally faster for networks that had more chances to reduce error. Together this reveals that networks seeing a large number of biased training presentations converged quickly on inadequate solutions.

Unfortunately the inequality of the number of training instances across levels of bias prevented us from properly assessing the effect that bias has on convergence rate. A separate experiment controlling for the number of training presentations across all levels of

[2] All error bars in this paper represent 1 standard deviation.

bias revealed that the more bias used in training, the slower the networks were to converge.

In every ANOVA, the interaction effects demonstrated that networks trained with different levels of bias behaved similarly at those psychologically optimal levels of small subset size. For subsets larger than 15%, increasing bias played an increasingly prominent role in preventing convergence. For subsets smaller than 0.5%, there was a gradual drop off in the magnitude of the log coefficient and in the proportion of error reduced. Convergence failed to occur for many runs at the smallest level of subset size.

Thus, the first simulation showed that McClelland's (1988) assumptions of a strongly biased training environment and of a small subset size impeded convergence of the back-propagation learning algorithm. By analyzing a wider range of these variables than were used in his model, we discovered that the most psychologically realistic data were generated by models exhibiting a slow rate of convergence. We also found a failure of back-propagation to learn successfully when trained with a bias and large subset sizes.

Simulation 2: Continuous Weight Updating

The first simulation was surprising in that so many networks failed to converge at all. To determine whether these results might be due to the use of batch weight updating, we repeated the above simulation using a continuous[3] weight updating method. A permuted presentation of training instances was used to prevent any unforeseen side effects due to auto-correlation of the sequence of exemplars.

As before, a 5 x 16 (bias x subset sizes) ANOVA of convergence rate was undertaken, revealing main effects for bias ($F(4,720) = 15.4$, $p<.0001$), subset sizes ($F(15,720) = 1426.4$, $p<.0001$), and their interaction ($F(60,720) = 39.5$, $p<.0001$). The ANOVA of learning depth also revealed main effects for bias ($F(4,720) = 180.4$, $p<.0001$), subset sizes ($F(15,720) = 3891.3$, $p<.0001$), and their interaction ($F(60,720) = 22.1$, $p<.0001$). Finally, the ANOVA predicting psychological performance demonstrated main effects for bias ($F(4,720) = 106.2$, $p<.0001$), subset sizes ($F(15,720) = 33.6$, $p<.0001$), and their interaction ($F(60,720) = 6.9$, $p<.0001$).

The convergence rate and learning depth interaction effects were negligible, and none were of interest. The average regression captured 57% of available variance ($R^2 = .57$, $sd = .24$). After excluding models which failed to converge, this average fit was 58% of the variance ($R^2 = .58$, $sd = .24$). Only 4 of the 800 (.05%) runs failed to converge, and all of these networks were at the lowest level of subset size, having been delivered too few training exemplars.

Figure 5 plots learning depth as a function of subset size for both continuous and batch updating methods. Continuous, but not batch, updating confirmed our intuitions that the amount of error reduced was proportional to the number of observed training instances.

The failure of batch weight updating to learn in a reasonable amount of time may be related to the anecdotal report that highly redundant data sets result in slower convergence on a solution with batch, but not with continuous, weight updating (see connectionists e-mail list exchanges in October 1991).

Psychologically realistic data were generated by continuous runs trained at practically all levels of subset size. However, the interaction effect for both measures of psychological performance showed that the best performance came from models with subset sizes between 1% and 5%. No models trained without bias exhibited realistic performance. A strong linear trend ($F(1,794) = 404.0$, $p<.0001$) in the cell means of bias demonstrated that the larger the bias level used in training, the more likely one was to observe psychologically realistic runs. Additionally, a weaker, but still significant linear trend among the cell means of subset size ($F(1,783) = 32.6$, $p<.0001$) demonstrated that the smaller the subset size, the more psychologically realistic the model.

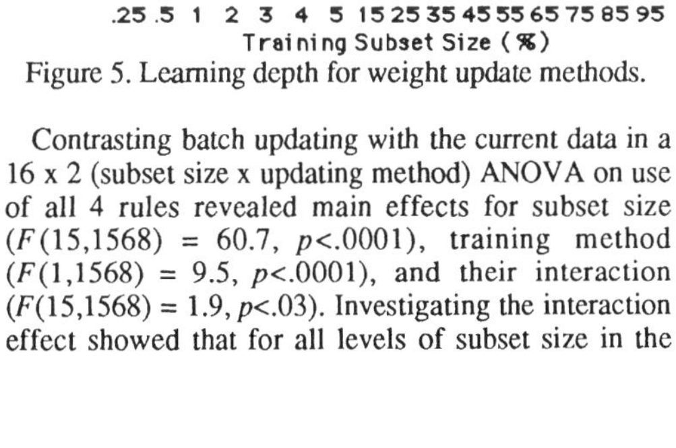

Figure 5. Learning depth for weight update methods.

Contrasting batch updating with the current data in a 16 x 2 (subset size x updating method) ANOVA on use of all 4 rules revealed main effects for subset size ($F(15,1568) = 60.7$, $p<.0001$), training method ($F(1,1568) = 9.5$, $p<.0001$), and their interaction ($F(15,1568) = 1.9$, $p<.03$). Investigating the interaction effect showed that for all levels of subset size in the

[3] Continuous (also known as per-sample, on-line, or pattern) weight updating computes derivatives and weight changes after the presentation of each pattern, as opposed to a batch (also known as per-epoch, or epoch) updating method in which the derivative of the error function summed over all patterns is taken each epoch, before weight changes occur.

segregated network, continuous updating produced more runs which fit the psychological data. Figure 6 plots this interaction. Interestingly, those runs presented with fewer training instances, large biases, and trained with continuous weight updates yielded the greatest proportion of psychologically realistic results, even outperforming McClelland's (1988) original model.

Thus, the second simulation demonstrated that the earlier failures of networks to converge were due to the use of batch weight updating. The first two simulations suggest that a biased training environment and small subset training method slow convergence and enhance psychological realism.

Simulation 3: Fully Connected Nets

The final simulation investigated the effect that segregating the weight and distance dimensions had on producing psychologically realistic performance. This simulation repeated the manipulations of the first two, but without the assumption of segregated hidden units. The network topology had 20 inputs, 4 hidden units, 2 outputs, and was fully connected. Networks were trained under conditions of both continuous and batch weight updating. Since in the previous simulations a lack of training set bias did not result in rule use, this group was dropped. All other details remained unchanged from the earlier simulations.

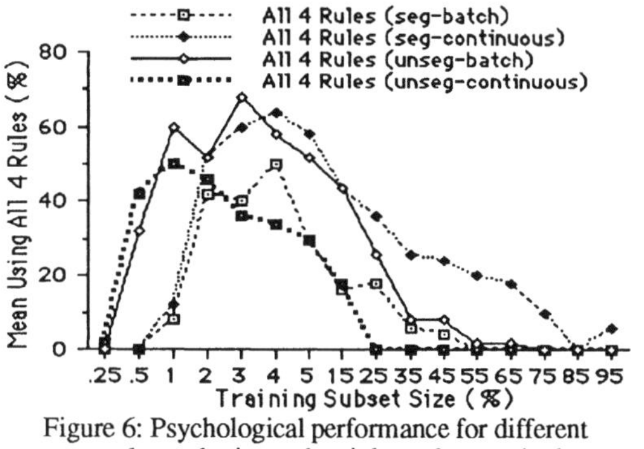

Figure 6: Psychological performance for different network topologies and weight update methods.

For each of the weight update methods (batch and continuous), segregated data generated in earlier simulations were contrasted with the new non-segregated data in 2 x 16 (network topology x subset size) ANOVAs for convergence rate and all 4 psychological rules.

For continuous weight updating, main effects were found for topology ($F(1,1248) = 54.8, p<.0001$), subset size, ($F(15,1248) = 866.8, p<.0001$) and their interaction ($F(15,1248) = 85.3, p<.0001$). Investigation of the interaction revealed that segregated networks converged more slowly than non-segregated for subset levels above 1%. The opposite occurred below 1%. Main effects were observed on all 4 rules for topology ($F(1,1248) = 44.9, p<.0001$), subset size ($F(15,1248) = 31.4, p<.0001$), and their interaction ($F(15,1248) = 12.5, p<.0001$). Investigation of the interaction showed that the segregated networks outperformed the non-segregated networks at subset sizes above 1%. Below 1%, the opposite occurred. Performance corresponded with rate of convergence, in that the interaction effects mirror one another. The group of runs with slowest convergence were also those with highest psychological performance (see Figure 6).

The slower convergence witnessed for the segregated networks seems to be a result of using fewer weights to encode the same amount of information as in the non-segregated networks. More weight changes per epoch in the non-segregated networks speeds convergence.

For batch weight updating, main effects were found for topology ($F(1,1248) = 80.7, p<.0001$), subset size, ($F(15,1248) = 190.7, p<.0001$) and their interaction ($F(15,1248) = 8.1, p<.0001$). Investigation of the interaction showed non-segregated networks converged more slowly than segregated, at all levels of subset size except from 2% to 5%. Main effects were observed on all 4 rules for topology ($F(1,1248) = 70.0, p<.0001$), subset size ($F(15,1248) = 49.4, p<.0001$), and their interaction ($F(15,1248) = 6.8, p<.0001$). Investigation of the interactions showed that the non-segregated batch networks outperformed the segregated batch networks at all levels of subset size (see Figure 6).

Thus, the final simulation showed that, with continuous weight updating, segregated networks converged more slowly than non-segregated networks, and also displayed more realistic psychological performance. With batch weight updating, the opposite effect occurred: non-segregated networks converged more slowly than segregated networks, and also showed more realistic psychological performance. In both cases, whenever network topology impeded convergence of the back-propagation learning algorithm, more realistic psychological performance followed.

The slower convergence for non-segregated batch networks may be due to the failure of so many segregated batch networks to learn. Recall that these nets tended to converge quickly on defective solutions.

Segregated networks do not invariably improve the fit to psychological data. Rather, when segregation slows convergence, as with continuous updating, better psychological performance follows; when segregation speeds convergence, as with batch updating, nets diverge from psychological realism.

Discussion

In these simulations, psychological success of the balance scale models increased as convergence slowed.

Decreasing the number of training presentations in all models caused slower convergence, as did increasing training bias. The precise effects of segregating hidden units depended on the method of weight updating, but the general principle was that psychological realism followed slow convergence.

One ramification of the current findings is that models, like humans, need not have access to all of the information about a problem in order to succeed in finding a solution. Indeed, if models are supplied with complete information, realistic effects do not occur.

Shultz (1991) suggested that stages would emerge whenever network models solve part of the overall problem before solving the range of possible problem types. Among the techniques he listed for encouraging partial solutions (and thus stages) were hidden unit herding, over-generalization, training pattern bias, and hidden unit recruitment. Working on too much of the problem at once may encourage overly rapid convergence on a general solution and thus preclude the appearance of stages. The present findings would suggest that all of these methods slow network convergence. A useful heuristic to apply in the creation of connectionist models of cognitive development may be to consider possible convergence slowing assumptions that bear on the problem domain.

A phenomenon that may be related to the current findings is Elman's (1991) "starting small" effect. Elman reported that recurrent networks had difficulty learning a small grammar unless there was a gradual increase in either the complexity of the training instances or the "working memory capacity" of the network. Here we find analogously that models trained with a reduced number of exemplars perform more realistically, and in the case of batch updating under a heavy bias, often fail to discover a solution unless trained with a small subset of training examples. An important issue for future investigation is whether the staging of the child's environment and her developing cognitive resources work in concert to selectively filter information accessible to learning.

A second result of the current work is that McClelland's (1988) specific set of assumptions is one of several sufficiently capable of producing realistic psychological performance. Although the incremental nature of connectionist learning is crucial for the success of balance scale models (Schmidt, 1991), so is convergence slowing.

It would appear that some assumptions of McClelland's original model are replaceable. The current findings demonstrate balance scale performance without the architectural assumption of a segregated network topology. The bias and subset training assumptions, too, can be replaced by other assumptions which favor one problem dimension over another. Using a generative algorithm, Schmidt (1991) demonstrated that the state of the initial weights can place networks in a position from which they traverse the psychological rules in a realistic fashion. In another simulation, a deliberate patterning of noise added to the training set also achieved the same end.

Another generative connectionist model of balance scale phenomena also demonstrated the disposability of the segregated architectural assumption (Shultz & Schmidt, 1991). In addition, that model showed that a randomly changing environment of training instances could be replaced with a more stable, gradually expanding set of exemplars. An important issue for future research is to examine the plausibility of various sets of balance scale model assumptions and the model's corresponding ability to fit human data.

Acknowledgement

This research was supported by a grant from the Natural Sciences and Engineering Research Council of Canada.

References

Elman, J. L. 1991. Incremental learning, or the importance of starting small. Technical Report 9101, Center for Research in Language, University of California at San Diego.

McClelland, J. L. 1988. Parallel distributed processing: Implications for cognition and development. Technical Report AIP-47, Department of Psychology, Carnegie-Mellon University, Pittsburgh, PA.

Schmidt, W. C. 1991. Connectionist models of balance scale phenomena. Unpublished Honours Thesis, Department of Psychology, McGill University, Montréal, Québec, Canada.

Shultz, T. R. 1991. Simulating stages of human cognitive development with connectionist models. In L. Birnbaum and G. Collins eds., *Machine learning: Proceedings of the Eighth International Workshop*, pp. 105-109. San Mateo, CA: Morgan Kaufman.

Shultz, T. R. and Schmidt, W. C. 1991. A Cascade-Correlation model of balance scale phenomena. In: *Proceedings of the Thirteenth Annual Conference of the Cognitive Science Society, pp. 635-640*. Hillsdale, NJ: Erlbaum.

Siegler, R. S. 1976. Three aspects of cognitive development. *Cognitive Psychology* 8:481-520.

Siegler, R. S. 1981. Developmental sequences between and within concepts. *Monographs of the Society for Research in Child Development* 46:Whole No. 189.

Chunking Processes and Context Effects in Letter Perception

Emile Servan-Schreiber
Department of Psychology, Carnegie Mellon University, Pittsburgh, Pennsylvania 15213
Schreiber@lentil.psy.cmu.edu

Abstract

Chunking is formalized as the dual process of building percepts by recognizing in stimuli chunks stored in memory, and creating new chunks by welding together those found in the percepts. As such it is a very attractive process with which to account for phenomena of perception and learning. Servan-Schreiber and Anderson (1990) demonstrated that chunking is at the root of the "implicit learning" phenomenon, and Servan-Schreiber (1990; 1991) extended that analysis to cover category learning as well. This paper aims to demonstrate the potential of chunking as a theory of perception by presenting a model of context effects in letter perception. Starting from a set of letter segments the model creates from experience chunks that encode partial letters, then letters, then partial words, and finally words. The model's ability to recognize letters alone, or in words, pseudo-words, or strings of unrelated letters is then tested using a backward masking task. The model reproduces the word and pseudoword superiority effects.

To overcome the limited capacity of its short term memory a mind organizes its input into familiar chunks (Miller, 1956). From this fact we can directly derive two more: First, when confronted with a set of input features, a mind will seek to recognize configurations of features, or chunks, that it has stored in its long term memory, and the resulting percept in short term memory will consist of those recognized chunks. Second, additional chunks will be created, and stored in long term memory, by welding together some of the chunks that make up the percept. We have here a general description of an adaptive recognition machine that learns continuously in order to perceive better: A chunking machine.

The facts that minds perceive and learn by chunking have been heavily documented (e.g., Bartram, 1978; Buschke, 1976; Chase & Simon, 1973; Johnson, 1970; Newell & Rosenbloom, 1981), yet most current models of perception, and letter perception in particular, overlook those facts (e.g., McClelland & Rumelhart, 1981; Oden, 1979; Massaro & Sanocki, in press). In this paper I demonstrate that the perceptual advantage of letters in words and pseudowords over letters in unrelated letter strings is a natural characteristic of a chunking machine that has learned, from scratch, to recognize letters and words.

The Chunking Model

Chunks. A chunk is a long term memory hierarchical structure whose constituents are chunks also. There are two kinds of chunks: Elementary chunks are those that the cognitive system never had to create. They are assumed to be the output of an elementary perceptual system. All other chunks are created by welding together lower level chunks. (Any theory of chunking must assume a limit on chunk size, the number of chunks that can be welded together into a new chunk. For simplicity, this theory assumes that it is 2, the lowest number that still enables chunk creation.)

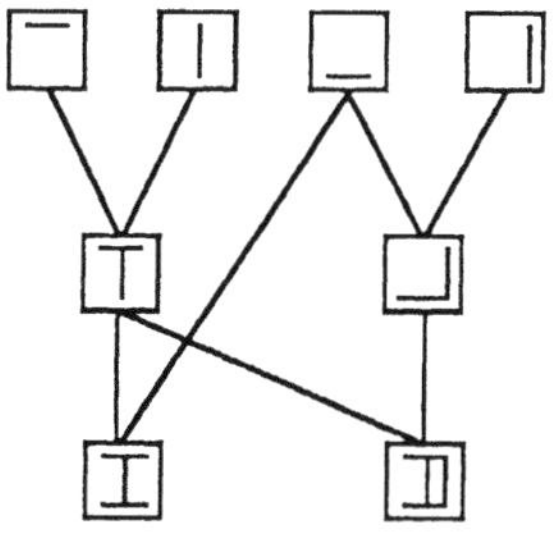

Figure 1. A potential hierarchical network of chunks that encodes the letters T, I, and D in terms of simple line segments. Structurally similar letters may share subchunks, and entire letters may be subchunks of more complex letters.

The letter perception model assumes that simple straight line segments are elementary chunks. As Figure 1 illustrates by showing a potential hierarchical structure of three chunks for the letters T, I, and D, the letters that were used as stimuli in the simulation were made of such simple segments.. Note that, as in this case, structurally similar letters may share subchunks.

Perception. Chunks are used to perceive stimuli, and given the recursive and hierarchical nature of chunks the perception process is necessarily cyclical and bottom-up. Starting with an elementary percept that contains all the elementary chunks present in the stimulus (e.g., letter segments), each cycle of perception seeks to reduce the number of chunks currently in the percept by replacing pairs of chunks in the percept by a chunk that encodes their co-occurrence. This operation is called *encoding*. For example, if the chunks w, x, y, and z are in the percept and there are chunks (w x) and (y z) in memory, then the next cycle of perception puts (w x) and (y z) in the percept in place of their constituent chunks. And if the chunk ((w x) (y z)) is also in memory, then the percept can be encoded further on the following cycle. The process can continue to cycle until the percept cannot be encoded further.

Encoding occurs in parallel on each cycle. As the example above illustrates, two or more chunks can encode the percept simultaneously in one cycle. But there are potential conflict situations: For example, if the percept contains the chunks [w,x,y,z], and there are chunks (w x), (x y), and (y z) in memory, then (w x) and (y z) are compatible encoders while (x y) is incompatible with both of them. To resolve such conflicts, the model first collects the set of all the candidate encoders, then randomly selects a subset of those that are all compatible. Thus the percept could be encoded either as [w,(x y) z] or as [(w x),(y z)].

Note that the choice that is made is not without consequence for the next cycle of perception. If there is a chunk ((w x) (y z)) in memory then it has a chance to encode the percept on the following cycle if it is [(w x),(y z)], but not if it is [w,(x y),z]. As is common with simple hill-climbing procedures, this bottom-up perception process can easily get stuck in a non-optimal encoding of the stimulus.

A simple way to avoid getting stuck in non-optimal encodings is to allow the process to backtrack through a *decoding* operation. To decode a chunk that is in the percept is to remove it and replace it by its subchunks. For example, if the chunk (x y) is decoded in the percept [w,(x y),z] then the resulting percept is [w,x,y,z].

In the model, every perception cycle consists of an encoding stage followed by a decoding stage. Every chunk that is in the percept at the end of an encoding stage has a probability, dp, of being decoded before the onset of the next cycle, and if a chunk is decoded at the end of a cycle, it is forbidden to be an encoder on the immediately following cycle. A chunk's dp is determined throughout the perception process in the following way: Elementary chunks come into the process with an initial probability of being decoded. Then every encoder chunk comes into the percept with a dp that is equal to the average of its subchunks' dps. Finally, and most importantly, whenever an encoding stage has failed to retrieve any encoder the dp of every chunk in the percept is decreased by a small amount (e.g., .01). Thus, the perception process will oscillate between different percepts, but oscillations will become more and more unlikely as the dps of the chunks in the percept decrease. Eventually, the process settles on a stable percept when the dp of each chunk is zero.

This is a straightforward application of simulated annealing to bottom-up encoding. It has the nice property of being likely to settle in one of the more, and often the most, encoded interpretation of the input, as the following example illustrates: Consider Figure 2. It represents the different percepts that the process oscillates between when presented with a D, and given a hypothetical network of chunks. In each network in the figure the chunks that make up the percept are enclosed in bold squares. Thus percept P1 is the elementary percept that consists of the elementary segments of a D, while percept P6 contains only a D chunk. Encoding proceeds from top to bottom on the page, while decoding proceeds from bottom to top following the arrows. There are two possible minima: P5 which represents an "I" and an isolated segment, and P6 which represents a "D". To move from P5 to P6 requires at a minimum 1 decoding followed by 2 encodings, but to move from P6 to P5 requires at a minimum 2 consecutive decodings followed by 1 encoding. Because encoding is guaranteed at every cycle (provided that there exists a pertinent encoder), while decoding is probabilistic, it is easier to encode than to decode. Therefore, given any probability of decoding it is easier to move from P5 to P6 than to move from P6 to P5, and that difference increases as the probability of decoding chunks decreases. So the process will tend to settle on P6 (D) much more often than on P5 (I+ |).

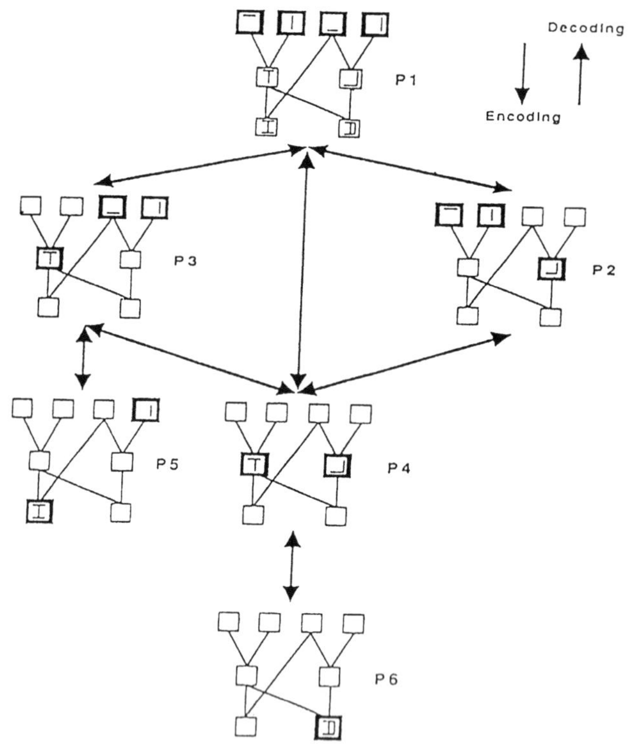

Figure 2. Starting with percept P1 (the segments of the letter "D") the perception process will tend to oscillate between the two possible minima P5 ("I" + "|") and P6 ("D") before settling preferably on P6, the most encoded interpretation of the input. The chunks of each percept are enclosed in bold squares, and arrows indicate how encoding and decoding transform one percept into another.

Learning. Once the perception process has settled on a final percept, a collection of chunks, a new chunk is created by selecting a pair and welding it into a new chunk (if it does not already exist in memory). In cases where the final percept consists of a single chunk, the creation process is not engaged. The selection of the pair of chunks that will be welded into a new chunk is essentially random but may be constrained. For example, when letter segments are welded together into a letter or a partial letter, a constraint may be that the two segments selected must be connected or parallel.

The combination of the perception and chunk creation processes allows the model to continuously grow a network of chunks from its experience with successive stimulus exposures, given only a minimum set of elementary chunks to start with.

Test of the Model

The model was tested with respect to its ability to reproduce several important results in the letter perception literature. They are: (1) The perceptual advantage of letters in words over letters in pronounceable nonwords (also called pseudowords), letters in strings of unrelated letters, and letters presented alone. (2) The perceptual advantage of letters in pseudowords over letters in strings of unrelated letters. (3) The reversal of the advantage of letters in words over letters presented alone at long exposure durations. For a review of those results see McClelland & Rumelhart (1981), or Massaro & Sanocki (in press).

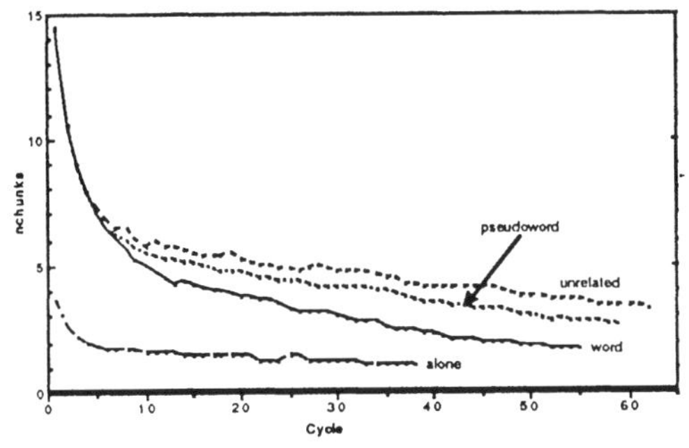

Figure 3. Evolution of the number of chunks in a percept, *nchunks*, with increasing number of perception cycles. There is one curve for each type of stimulus: a single letter (alone), a four letter word (word), a four letter pseudoword (pseudoword), and a string of four unrelated letters (unrelated). Individual curves are plotted up to the cycle where, on average, perception has settled on a stable percept.

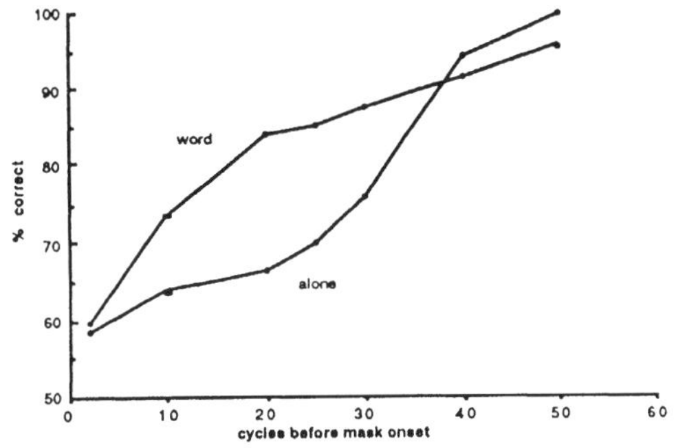

Figure 4. Evolution of the percentage of correct forced-choice recognition of target letters, presented in words or alone, with increasing number of processing cycles before masking.

All these results were obtained in experiments where a high contrast stimulus letter is presented for a short duration in a particular context, followed immediately a high contrast masking stimulus. The subject is then asked to choose among two possible letters which one was in a particular position. For example the subject might see "WORK" quickly followed by a mask "####" and be asked whether "K" or "D" was in the fourth position. As in this example, when the context forms a word with the target, the foil does also, therefore guessing is controlled for.

Before any simulation of those results could be attempted, the model first had to grow a network of chunks to represent letters and words. To start it was given a small set of elementary chunks to represent letter segments of different lengths and orientations. It then learned chunks to recognize the 26 individual letters of the alphabet. Once it could perfectly recognize any letter presented alone, it learned chunks to recognize each word in a sample of 288 four letter words. In the end, from the original 8 elementary segment chunks the model grew a network containing 18 partial-letter chunks, 51 letter chunks, 1665 partial-word chunks, and 1339 word chunks. These numbers indicate a large amount of redundancy in the representation of letters and words.

Time course of encoding. In a first experiment the model was tested for its ability to encode words, pseudowords, single letters, and strings of unrelated letters. Figure 3 plots for 144 stimuli of each type the average number of chunks, *nchunks*, in a percept after a given number of perception cycles. Individual curves are plotted up to the cycle where, on average, perception has settled on a stable percept. For this experiment, and the two that follow, the initial decode-probability was set at .30. Note that the process settles faster when a single letter is presented that when a multi-letter stimulus is presented. For instance, even though the letters in a word are processed in parallel, recognizing a whole word takes longer than recognizing a single letter. Note also that the less related the letters in a multi-letter stimulus are (as evidenced by the final nchunks), the longer it takes to settle on a final percept. These two results can only be attributed to a kind of lateral interference that is an emergent property of the annealing process. To put it simply, the less related the letters are, the more chunks there are in the percept at any cycle, so the more chance of decoding there is, and therefore the longer it takes to encode.

Words vs. single letters. In a second experiment, the recognition of letters in words and letters presented alone was compared given different numbers of cycles before masking. The mask used by the model was the union of the letter segments in "O", "X", and "+", and masking was simulated by adding to a percept those spurious segments of the mask that were not already present in the stimulus.

The model assumed that if perception had settled before the scheduled onset of the mask, then masking could not disrupt perception, that is, responses would be based on the settled percept only. Thus masking could only potentially disrupt performance if it occurred before perception had settled. In those cases, the spurious mask segments were allowed to contaminate the percept for a small number of cycles before a response was made. To choose a response, among two alternatives in a particular position, the model simply checked if either could be found in its percept in that particular position. If yes then it was chosen, else, if neither or both were perceived in that position, then the model chose randomly. (Two or more letters could be recognized in a position because of the spurious segment introduced by the mask.)

Figure 4 plots the results of that experiment involving 144 targets in words and 144 equivalent targets presented alone. This simulation assumed 5 cycles of masking before a response was made. Like the human subjects of Massaro and Klitzke (1979) the model produced an advantage for letters in words that was eventually reversed at late masking onsets.

There are essentially three possible encoding states for the target letter in a percept before masking introduces spurious segments: (1) The target may not be encoded as a single chunk, possibly due to decoding or, more simply, encoding failure. (2) The target may be encoded as a chunk that is not part of a larger chunk, possibly due to the decoding of a larger chunk, or the fact that it was presented alone. (3) The target may be encoded as a chunk that is further encoded in a larger chunk, or a hierarchy of larger chunks. Masking has potentially different effects in any of those three states: (1) If the target is not even encoded as a chunk then the spurious segments of the mask can prevent future encoding of the target by being encoded together with target segments into chunks incompatible with the target's structure. (2) If the target is encoded as a free standing chunk then on each masking cycle there is a probability that this chunk gets decoded with dire consequences as in (1), or that a spurious mask segment gets encoded with the target chunk into a chunk that represents another letter (for instance, the addition of a single letter segment to the chunk for "P" can transform it into an "R"). (3) If the target chunk is well hidden within a further encoded hierarchy, for example in a chunk for a complete word, or part of a word, and these larger chunks resist decoding during the masking cycles, then masking has no harmful effect. But if those larger chunks get decoded, then there may be dire consequences as in (2).

The more cycles of processing there have been before masking, the more unlikely it is that chunks get decoded (the closer to settling perception is). So targets in words, which are likely to be encoded in chunks with other letters, are likely to be quite immune to the effects masking compared to targets encoded in free standing chunks. However, as Figure 3 shows, percepts tend to settle quicker with single letter stimuli than with words. Therefore, because a settled percept is immune to masking, late masking is less likely to disturb the perception of single targets than that of targets presented in words.

Words vs. pseudowords vs. strings of unrelated letters. A final experiment compared the forced-choice recognition of targets in words, pseudowords (e.g., MIPE), and unrelated letter strings (e.g., TCKU). 30 cycles of perception were allowed before 10 cycles of masking. There were 144 stimuli in each condition. The results are in Table 1. Like human subjects the model produced a large advantage for words and a smaller advantage for pseudoword over strings of unrelated letters (e.g., McClelland & Rumelhart, 1981).

The time course of encoding of the different stimulus types plotted in Figure 3 indicated that words are encoded as fewer chunks than pseudowords, themselves encoded as fewer chunks than strings of unrelated letters. This is simply a reflection of the different amounts of relatedness between the letters in the three stimulus types. And as the analysis of the previous experiment demonstrated, more compact encoding directly translates into less adverse effect of masking.

Table 1

Correct forced-choice recognitions of letters in words, pseudowords, and strings of unrelated letters.

Word	Pseudoword	Unrelated
84.4 %	78.6 %	70.3 %

To conclude, briefly, this limited testing of the model demonstrated its potential as a theory of letter perception. Further testing is certainly warranted, but considering that the same chunking analysis was successfully applied elsewhere to "implicit learning" and to category learning (Servan-Schreiber & Anderson, 1990; Servan-Schreiber, 1990; 1991), there is some reason to be confident that chunking processes of the type explored here underly much of human learning and perception. Indeed, one major contribution of this chunking analysis is to show how these apparently unrelated phenomena are in fact deeply related.

References

Bartram, D. J. 1978. Post-iconic visual storage: Chunking in the reproduction of briefly displayed visual patterns. *Cognitive Psychology* 10:324-355.

Buschke, H. 1976. Learning is organized by chunking. *Journal of verbal Learning and Verbal Behavior* 15:313-324.

Chase, W. G., & Simon, H. A. 1973. Perception in chess. *Cognitive Psychology* 4: 55-81.

Johnson, N. F. 1970. The role of chunking and organization in the process of recall. In G. H. Bower (Ed.), *The Psychology of Learning and Motivation* 4:171-247. NY: Academic Press.

Massaro, D. W., & Sanocki, T. (in press). Visual information processing in reading. In D. Willows, R. Kruck, & E. Corcos (Eds.), *Visual processes in reading and reading disabilities.* Hillsdale, NJ: Lawrence Erlbaum.

Miller, G. A. 1956. The magical number seven, plus or minus two: some limits on our capacity for processing information. *Psychological Review* 63:81-97.

Newell, A., & Rosenbloom, P. 1981. Mechanisms of skill acquisition and the law of practice. In J. R. Anderson (Ed.), *Cognitive skills and their acquisition.* Hillsdale, NJ: Lawrence Erlbaum.

Servan-Schreiber, E. 1991. The Competitive Chunking Theory: Models of Perception, Learning, and Memory. Ph.D. diss., Dept. of Psychology, Carnegie-Mellon University.

Servan-Schreiber, E. 1990. Classification of dot-patterns with competitive chunking. In Proceedings of the Twelfth Annual Conference of the Cognitive Science Society, 182-189. Boston, Mass.

Servan-Schreiber, E., & Anderson, J. R. 1990. Learning artificial grammars with competitive chunking. *Journal of Experimental Psychology: Learning, Memory, and Cognition* 16:592-608.

McClelland, J. L., & Rumelhart, D. E. 1981. An interactive activation model of context effects in letter perception, part 1: An account of basic findings. *Psychological Review* 88:375-407.

Oden, G. C. 1979. A fuzzy logical model of letter identification. *Journal of Experimental Psychology: Human Perception and Performance* 5:336-352.

Abstractness and transparency in the mental lexicon

William Marslen-Wilson, Lorraine Komisarjevsky Tyler, Rachelle Waksler, & Lianne Older

Birkbeck College, University of London,
Department of Psychology,
Malet St.,
London WC1E 7HX
Email: ubjta38@cu.bbk.ac.uk

Abstract

This research is concerned with the structure and properties of the mental representations for morphologically complex words in English. In a series of experiments, using a cross-modal priming task, we ask whether the lexical entry for derivationally suffixed and prefixed words is morphologically structured or not, and how this relates to the semantic and the phonological transparency of the relationship between the stem and the affix (*govern* + *ment* is semantically transparent, *depart* + *ment* is not; *happy* + *ness* is phonologically transparent, *vain* + *ity* is not). We find strong evidence for morphological decomposition, at the level of the lexical entry, for semantically transparent prefixed and suffixed forms, independent of the degree of surface transparency in the phonological relationship between the stem and the affix. Semantically opaque forms, in contrast, seem to behave like monomorphemic words. We discuss the implications of this for a theory of lexical representation and the processes of acquisition.

Introduction

To understand language comprehension we need to understand the mental lexicon. This requires us to understand how words are *represented* in the mental lexicon. What are the processing targets of lexical access, and how are representations of lexical form and of lexical content interleaved to permit access, selection, and integration to operate in the rapid and efficient manner documented in earlier research (Marslen-Wilson, 1989; Marslen-Wilson & Tyler, 1987)?

The research we report here is concerned with the properties of the underlying *unit* of lexical representation. Are words represented in the mental lexicon as complete, unanalysed word-forms -- corresponding, more or less, to the words we see on the printed page? Or is the representation broken down into *morphemes* -- traditionally, in linguistic analysis, the smallest meaning-bearing linguistic unit. The word *happy*, for example, is a single morpheme, whereas the words *happiness* and *unhappy* are polymorphemic, being made up, respectively, of the morphemes {happy} + {ness} and {un} + {happy} (where {happy} is the *stem*, and {-ness} and {un-} are morphological *affixes*). In a series of experiments, using the cross-modal repetition priming technique, we explore two basic questions about the representation of morphologically complex words in English: Is their representation phonologically abstract, and what is the role of semantic factors in determining whether or not a word will be represented as morphologically complex?

Preliminaries

In studies of lexical representation, it is crucial to distinguish claims about the *lexical entry* for a given word from claims about its *access representation*. The lexical entry we define as the modality-independent core representation of a word's abstract syntactic, semantic, and phonological properties. The access representation we define as the modality-specific perceptual target for lexical access, constituting the route whereby information in the sensory input is linked to a given lexical entry. Our concern here is with the properties of the lexical entry, and with the role of *semantic* and *phonological transparency* in determining its properties.

These terms refer to two aspects of the surface relationship between stems and affixes in morphological complex words. Semantic transparency refers to whether or not the meaning of a morphologically complex word is synchronically derivable from the meaning of its parts. Words like *happiness* and *unhappy*, for example, are relatively semantically transparent, because their meaning is composable in this way. In contrast, words like *department* or *release* are not semantically transparent -- the meaning of *department* cannot be derived by putting together the meaning of the free stem *depart* with the affix *-ment*. Our research will investigate the role that this factor, of surface semantic interpretability, plays in determining whether or not the lexical entry for a given word-form is morphologically structured.

The second factor, of phonological transparency, refers to the degree to which processes of phonological alternation lead to a change in the phonetic realisation of the stem when it occurs in a morphologically complex word. In English, this applies in particular to the suffixing morphology, as in alternations like *vain/vanity* or *decide/decision*. If pairs like *decide/decision* do share the same stem at the level of the lexical entry -- i.e., the morpheme {decide} -- then this must be represented in a way which abstracts away from the surface phonetic properties of the word-forms in question.

To investigate these issues we will focus on English *derivational* morphology, since this provides the appropriate range of phonological and semantic contrasts. Derivational morphemes in English are both suffixing and prefixing. Suffixing morphemes like *-ness*, *- ment*, *-ence* come after the stem (as in *happiness*, *government*, *defiance*, etc), whereas prefixing morphemes like *re-* , *ex-*, *pre-* (as in *rebuild*, *explain*, *preview*, etc) precede the stem. These are all morphemes which function to change the meaning, and often the grammatical class of the stems to which they are attached.

The experimental task we will use is *cross-modal immediate repetition priming*. This is a task in which the subject hears a spoken prime -- for example, *happiness* -- and immediately at the offset of this word sees a visual probe -- for example, *happy* -- which is related in some way to the prime. The subject makes a lexical decision response to this probe (i.e., judges as quickly as possible whether the string of words presented constitutes a word or not in the language). Response latency relative to a control condition, where listeners respond to the same probe following an unrelated prime, is used to measure any priming effect. Because the task is cross-modal, any priming effects should be attributable to events at the level of the lexical entry, rather than to effects of lower-level overlap at the level of modality-specific access representations.

Abstractness in lexical representation

A theory of lexical representation which claims that morphologically related words share the same stem morpheme in the lexical entry will need to assume that this is a level of representation which abstracts away from surface variation in phonological form. We test this in Experiment 1 by varying the phonological and morphological relationship between the auditory prime and the visual probe.

Note that morphological relatedness is defined here strictly on linguistic and historical grounds. A derived form and a free stem are classified as morphologically related if (a) the derived form has a recognisable affix; (b) when the affix is removed the resulting stem is the same as the paired free stem; (c) the pair of words share the same historical source word (or *etymon*).

In Condition 1 (see Table 1), the auditory prime (always a derived suffixed word) is morphologically related to the visual target (always a free stem) and this relationship is phonologically transparent. Examples of this are pairs like *friendly/friend* or *government/govern*, where the stem is fully contained within the derived word, in a form which is phonologically identical to its realisation as a free form.

If the lexical entries for words like this are morphologically decomposed into stems and affixes, then priming should be obtained in Condition 1 on the basis of shared morphemes in the lexical entry. Hearing *government* should activate the stem morpheme {govern} and the link between this and the suffix {-ment}. If the same stem functions as the lexical entry for the morphologically simple form *govern*, then residual activation of this morpheme after *government* has been heard should facilitate lexical decision responses when *govern* is presented as a visual probe.

Conditions 2 and 3 present the subjects with prime/target pairs which are still morphologically

Table 1: Phonological Transparency and Morphological Relatedness

	Test	*Control*	*Difference*
Condition 1 (friendly/friend)	539	583	-44
Condition 2 (elusive/elude)	563	623	-60
Condition 3 (vanity/vain)	572	608	-36
Condition 4 (termite/term)	647	638	9

related, but where this relationship is no longer phonologically transparent. In Condition 2, we used cases like *tension/tense* or *elusive/elude*, where the phonetic form of the stem is different in isolation from what it is in the derived form. If priming is due to events at the level of the lexical entry, then changes in the surface relationship between forms should not reduce the amount of priming. Condition 3 uses pairs like *vanity/vain* or *gradual/grade*, where not only does the stem have a different phonetic form in isolation, but also the underlying representation of the stem is not identical to its surface form. This has the effect of increasing the abstractness of the relationship between the stem and the phonetic form of the derived word.

The lexical decision responses, given in Table 1, show significant amounts of priming in the three conditions with morphologically related primes and targets. For each of these conditions, responses are significantly faster following the test prime than the control prime, and the size of the facilitation effect does not differ statistically across the three conditions.

There is no sign here that the effectiveness of a prime depends on the surface phonological transparency of the relationship between prime and target. To the contrary, when there is *only* a phonological relationship between prime and target, as in Condition 4, then no priming is obtained. Pairs such as *termite/term* or *planet/plan* are not morphologically related, so that there is no shared morpheme in common. Thus, although the target is transparently contained within the prime, there is no priming at the level of the lexical entry -- which is evidently the level of the system being tapped into by the experimental paradigm.

These results not only support the view that lexical representations are morphologically structured, but also that these representations are abstract. At the level of the lexical entry, representations of lexical form do not simply reflect surface form. If they did, then *decision* would be a much less effective prime of *decide* than *friendly* would be of *friend*, where the surface phonetic overlap is much greater.

Semantic Transparency

What is the role of semantic factors in the priming that we observed in Experiment 1? All of the morphologically related pairs (Conditions 1-3) were semantically transparent, whereas pairs like *planet/plan* (Condition 4) clearly were not. It is possible, therefore, that the lexical relations we are tapping into are semantic in nature and not necessarily morphological at all. The words *government* and *govern*, for example, share many semantic properties, and it may be by virtue of this relationship, rather than any specifically morphological relationship, that priming is obtained. This raises the issue of whether there are grounds for supposing that there is morphological structure in the lexicon independently of semantic structure.

The clearest arguments here are linguistic in nature. Aronoff (1975), for example, argues that morphological relations can be identified which involve morphemes that have no clear semantic interpretation. These are cases like the bound morpheme {-mit}, which only occurs as an element in words like *permit*, *transmit*, and *submit*. Although these words do not share a common meaning, they are linked by a common phonological rule, which generates the forms *permission*, *transmission*, and *submission*, and which is specific to verbs containing the root {-mit}. This suggests, according to Aronoff, that phonetic strings can be identified as morphemes independently of semantic considerations.

Returning to English derivational suffixes, there are plenty of cases where morphological links can be established between pairs of words, but where the relationship is no longer semantically transparent. These are cases like *emergency/ emerge* or *department/depart*, which meet the criteria for morphological relatedness, but where the meaning of the complex form can no longer be derived from the simple composition of the meanings of the stem and the affix. In Experiments 2 and 3 we contrast priming for semantically unrelated but morphologically related

Table 2: Semantic Transparency and Morphological Relatedness

		Test	Control	Difference
Suffixes:	Semantically Related (punishment/punish)	554	595	-41
	Semantically Unrelated (department/depart)	575	574	1
Prefixes:	Semantically Related (insincere/sincere)	503	534	-31
	Semantically Unrelated (restrain/strain)	542	543	-1

pairs with priming for semantically and morphologically related pairs of the type used in Experiment 1, such as *friendly/friend* or *predictable/predict*. Experiment 2 looked at suffixed forms and Experiment 3 at prefixed forms (Table 2).

These contrasts require an assessment of semantic transparency for each test pair. Derived forms are transparent when the meaning of the form is largely determinable from the composition of the stem (or root) with its affix. To determine whether this held synchronically -- that is, for current users of the language -- we used an operational criterion, classifying words as semantically transparent or opaque on the basis of a pre-test, where individuals were asked to judge the relatedness of a derived form and its free stem.

The lexical decision responses (Table 2) show clear effects of semantic relatedness. Although all pairs were morphologically related, according to uniformly applied linguistic and etymological criteria, only those pairs that were also synchronically semantically related showed priming in this task. In other experiments (Marslen-Wilson, Tyler, Waksler, & Older, 1992) we have found the same pattern when the order of prime and target is reversed -- *punish* is a strong prime of *punishment*, but *depart* does not prime *department* (and similarly for prefixed pairs). We also find no evidence of priming for prefixed pairs that share bound stems (of the *include/conclude* type), where there is again morphological but no synchronic semantic relation.

In a final set of experiments (Table 3) we examine the priming effects for suffixed and prefixed pairs where both prime and target are derived forms, sharing the same stem. If priming in this task is due to residual activation of a shared morpheme in the prime and the target, then semantically transparent pairs like *government/governor* should prime each other, for the same reason that *govern* primes *government*, and *vice versa*.

What we find (Table 3) is that although prefixed derived pairs do prime each other, suffixed pairs do not. Semantic relatedness is therefore not enough on its own to ensure priming between morphologically related pairs. We can attribute this absence of priming to inhibitory links between suffixes sharing the same stem. Hearing a transparent suffixed form like *government* will not only activate the stem {govern} but also inhibit other suffixed forms sharing the same stem. This is because forms like *government* and *governor* are mutually exclusive competitors for the same lexical region. The combination of the morpheme {govern} with the affix {-ment} defines a lexical item with a distinct meaning and identity in the language, and this is incompatible with the simultaneous combination of {govern} with a different affix to give a different lexical item.

The fact that prefixed derived pairs (such as *unwind/rewind*) do prime each other (Table 3) is evidence that the lack of priming for suffixed pairs is indeed a competition effect. When the lexical entry is entered *via* a prefix, this does not seem to activate as competitors other prefixed words sharing the same stem. In effect, *rewind* is not in the same cohort as *unwind*, so that these competitors do not need to be inhibited in the same way as suffixed words sharing the same stem.

Table 3: Semantic and Morphological Relatedness: Derived/Derived Pairs.

		Test	Control	Difference
Suffixes	Semantically Related (excitable/excitement)	580	591	-11
	Semantically Unrelated (successful/successor)	611	614	-4
Prefixes	Semantically Related (unfasten/refasten)	576	635	-59
	Semantically Unrelated (express/depress)	576	554	22

Conclusions

This series of experiments allow us to draw three main conclusions:

(i) There is a level of lexical representation which is abstract in nature. Phonetic overlap between primes and targets does not by itself produce priming, and the amount of priming is not affected, for morphologically related forms, by variations in the phonological transparency of the relation between prime and target.

(ii) Semantic relatedness between a prime and a target is a necessary but not sufficient condition for priming to occur. Semantically unrelated pairs, whether morphologically related or not, do not prime reliably.

(iii) The type of morphological relation between a prime and a target, and the stem-affix order within a morphologically complex prime, affect whether or not priming is obtained.

To accomodate these results we need to postulate a model of the mental lexicon which treats separately words like *department* (which are semantically opaque) and words like *punishment* (which are semantically transparent) at the level of the lexical entry. Semantically opaque words will be represented as if they were morphologically simple -- they can enter into combination with other morphemes (as in *interdepartmental*) but they themselves have no internal structure. Synchronically transparent forms, in contrast, will be represented as free stems linked to derivational affixes. Within this system of linked stems and affixes, inhibitory links will need to be set up between suffixes sharing the same stem, but not between prefixes, or, indeed, between prefixes and suffixes.

This proposal has the effect of re-interpreting semantic relatedness in terms of its consequences for the learning process. The structure of the adult lexicon reflects individuals' experience with the language as they learn it. The listener does not mentally represent words as sharing the same stem, and therefore as morphologically related, unless there are semantic grounds for doing so. An item like *department*, although it has a phonetically transparent morphological structure on the surface, will not be analysed during language acquisition into the free stem {depart} plus the affix {-ment} at the level of the lexical entry, since this gives the wrong semantics.

The challenge for learning models of English derivational morphology will be to devise ways of allowing phonological and semantic criteria to interact in the complex but rational ways that this research has begun to uncover.

Acknowledgements

This research was supported by grants to William Marslen-Wilson and Lorraine Komisarjevsky Tyler from the UK SERC, ESRC, and MRC. Rachelle Waksler at now in the Department of Linguistics, San Francisco State University, San Francisco.

References

Aronoff, M. (1976) *Word Formation in Generative Grammar*. Cambridge, Mass.:MIT Press.

Marslen-Wilson, W.D. (1989) Access and integration: Projecting sound onto meaning. In W.D.Marslen-Wilson (Ed.), *Lexical Representation and Process*. Cambridge, Mass.: MIT Press.

Marslen-Wilson, W.D., & Tyler, L.K. (1987) Against modularity. In J.Garfield (Ed.), *Modularity in Knowledge Representation and Natural-Language Understanding*. Cambridge, Mass.:MIT Press

Marslen-Wilson, W.D., Tyler, L.K., Waksler, R., & Older, L. (1992) Morphology and meaning in the English mental lexicon. Manuscript, Birkbeck College.

Polysemy and Lexical Representation: The Case of Three English Prepositions

Sally A. Rice
Dept. of Linguistics
University of Alberta
Edmonton, Alberta T6G 2E7 Canada
userrice@mts.ucs.ualberta.ca

Abstract

This paper is a preliminary analysis from a cognitive linguistics perspective of the meaning of three very high frequency prepositions in English, *at*, *on*, and *in*, which are argued to be inherently polysemous. Although these so-called grammatical morphemes are usually defined in terms of topological relations, the majority of their usages are far too abstract or non-geometric for such spatially-oriented characterizations. Because they seem to sustain a variety of meanings which often overlap, they are exemplary lexical items for testing theories of lexical representation.

Arguments against monosemous accounts center on their inability to formulate schemas which include all appropriate usages while excluding usages of other prepositions. Many of the usages differ only on the basis of variable speaker perspective and construal. A polysemic account is currently being developed and tested experimentally in a series of studies involving how native and non-native speakers of English evaluate and categorize various usages of the different prepositions. Initial results indicate that these naive categorizations reflect a gradient of deviation from a canonical spatial sense. Furthermore, deviant usages tend to form fairly robust clusters consonant with a constrained polysemic analysis.

Monosemy, Polysemy, or Homonymy?

Lexical representations play a central role in theoretical and computational linguistics, as well as in psycholinguistics, in mediating between linguistic and conceptual knowledge (Schreuder & Flores d'Arcais, 1989). The most commonly proposed and thoroughly discussed type of lexical representational system in the cognitive science literature has been the semantic network, a radially or hierarchically structured entity consisting of interconnected nodes representing various facets of meaning from inter-word relations to bundles of semantic features (in the case of lexical decomposition). Many of these network models directly exploit connectionist architecture and are often praised or damned in the same breath as connectionism. Leaving processing issues aside, arguments for and against semantic networks (cf. Johnson-Laird et al., 1984; Evens, 1988) have centered on models built around single lexical items or items selected from a restricted, and often very concrete, semantic field. Rarely have actual lexical entries or semantic networks been worked out in detail, nor have the models been subjected to native speaker validation for what is actually one of the most basic types of human categorization--actual word use. This empirical neglect casts doubt on many theories of lexical meaning, three of which concern us here.

Monosemy is a hypothesis that maintains that the majority of lexical items have a single, highly schematic meaning that extends to all usages of that item. Monosemy is primarily contrasted with two other hypotheses, namely **polysemy** and **homonymy**, each of which allows for the possibility that multiple meanings may be associated with a single lexical form. Polysemy assumes that the multiple meanings constitute a family of related senses and is therefore distinguished from homonymy, in which different meanings are not presumed to have any apparent connection. Given the relative rarity in English of true homonyms, such as *plant* ('vegetative organism' vs. 'factory') or *seal* ('aquatic mammal' vs. 'wax signet'), and our familiarity with multiple but similar dictionary definitions in listings for lexical items, polysemy appears to be the more intuitively plausible alternative to monosemy and will be our focus here.

The monosemy/polysemy debate [cf. Macnamara, 1971; Ruhl, 1989] may be seen as the manifestation at the lexical level of a more generalized issue which subsumes many theoretical controversies in linguistics, that is, whether human language is modular or interactional in nature. This semantic dispute polarizes those who maintain a single meaning for each linguistic form and those that allow for the possibility of multiple but associated meanings. That is, it sets the "meaning minimalists" against the "meaning maximalists," as Ruhl, 1989, and others have called them. Generative approaches, perhaps guided by the exigencies of their computational foundation, have championed the idea that lexical items have stable, if not singular, meanings and syntactic behaviors. In formal terms, this amounts to treating the meanings of lexical items as exclusive disjunctives and defining the items themselves in terms of semantic primitives.

On the other hand, more functionally- or cognitively-oriented frameworks, which place matters of usage before matters of universality and strict parsimony requirements, explicitly recognize the role that convention, pragmatic context, language-specific conceptualization patterns, and speaker construal play in assigning variable meaning and acceptability to particular expressions. As a cognitive linguist, my sympathies tend to fall with the polysemists. As an empirically-minded one, I am committed to justifying each related meaning posited for a single lexical item as well as establishing the nature of the links between them. I contend that, for most lexical items, even for so-called grammatical morphemes like prepositions, polysemy is the norm, but it is also very systematic and more constrained than most monosemists would have us believe.

I argue here that *at*, *on*, and *in*, three very high frequency "contact" prepositions in English, are inherently polysemous. Many researchers in linguistics and AI (cf. Lindkvist, 1978; Hawkins, 1984; Wesche, 1986/87; Herskovits, 1986) have defined these prepositions in terms of highly schematic topological relations (such as the coincidence of a figured entity and a **zero-** or **one-** (in the case of *at*), **two-** (*on*), or **three-** (*in*) **dimensional ground**) or more broadly as predicating relations of **contiguity** (*at*), **support** (*on*), or **containment** (*in*). They have been able to maintain these schemas because they confined the scope of their studies to the prepositions' spatial usages. However, the majority of their usages are far too abstract, non-geometric, or simply non-spatial to sustain such simplistic characterizations. Furthermore, in many cases, the individual meanings of the prepositions overlap, creating a lexical nightmare for anyone trying to represent prepositional meaning on the basis of semantic contrast and a syntactic nightmare for anyone trying to characterize their occurrence on the basis of lexical meaning or grammatical category alone. Given this state of affairs, the English prepositions are exemplary lexical items for testing theories of lexical representation (cf. Lindner, 1981; Brugman, 1981; and, for a similar treatment of the French prepositions, Vandeloise, 1991). As grammatical theories and computational models increasingly rely on lexical knowledge as the impetus for linguistic form and behavior, knowledge about how the lexicon is structured necessarily has profound implications for theories about the mental representation of language.

Why Monosemy Doesn't Work

Basic Usages of the Prepositions

In order to highlight problems facing a monosemic analysis, I present below a small but not unrepresentative sampling of some of the constructions in which these prepositions occur. The analysis is based on an extensive spoken and written database compiled by the author. *In*, *on*, and *at* are three of the earliest and most frequently occurring lexical items in the English language. Some cross-linguistic child language studies suggest that they are among the first five prepositions acquired (notably, Johnston & Slobin, 1979), but Bowerman, 1991 argues that there is no reason to expect all languages to carve up their spatial and conceptual world equivalently or for developmental patterns to be universal. Carroll, Davies, & Richman, 1978 put them among the 20 most frequent morphemes of English.

As prepositions, they are both commonly and technically thought of as having a **spatial or locative function**, serving to place a figured entity or event in relation to one of several variously configured backgrounds such as a 0-D point, a 1-D line, a 2-D surface, or a 3-D container:

(1) *They put the books {**at** the end of the shelf (0 or 1), **on** top of each other (1 or 2), **in** the center of the pile of papers (0 or 2), **in** a row (1), **on** the table (2), **in** the box (3)}.*

Note here that each preposition appears to be compatible with multiply-configured grounds. Or, stated conversely, it appears that zero-dimensionality can be predicated with *at* or *in*; one-dimensionality with *at*, *on*, or *in*; two-dimensionality with *on* or *in*; while three-dimensionality strongly favors *in* alone.

These prepositions also have a **temporal function**, which is not too surprising since time and other abstract domains are regularly conceived of in spatial terms (cf. Lakoff & Johnson, 1980) and time, like space, is routinely segmented into various-sized episodes. Used temporally, *at*, *on*, and *in* serve to situate an event relative to a brief point, short period, or vast expanse of time:

(2) *He died {**at** 6:01, **at** dawn, **at** 6 **on** the dot, **on** Sunday, **on** the 12th, **in** May, **in** summer, **in** 1897, **in** the 19th century}.*

One would be hard-pressed to sustain these geometric notions or even notions of **contiguity**, **support**, or **containment** for the apparent locative usages in (3a) or the event-like usages in (3b):

(3) a. *She met him {**at** UCLA, **on** Guam, **in** Singapore}.*
b. *She met him {**at** the conference, **on** a trip, **in** college}.*

If we maintain, for example, that ***at*** requires a point-like ground, then we need a way of interpreting *UCLA* in (3a) and *the conference* in (3b) in a point-like fashion.

It is easy to concoct data featuring fairly basic usages which further compromise a monosemic analysis. Consider these usages of *on*:

(4) a. *The cat is **on** the mat.*
b. *The handle **on** that mug is chipped.*
c. *What a cruel look **on** his face!*
d. *He turned the light out **on** me.*

What schema or set of abstract features could possibly unite all these senses while excluding the other prepositions? Features that suggest themselves include **contact** and **pressure**, or more abstractly, **support**. But in (4b), a handle is part of a mug, and so not really in contact with it nor does it exert pressure (although the mug does support the handle in a sense). In (4c), a look is an expression that temporarily distorts a face but does not really exert pressure on it (although, the face temporarily supports it). And in (d), the pressure, if there is any, could only be interpreted metaphorically as an act of annoyance. A strictly monosemic analysis is rejected on the grounds that no semantic features are common and exclusive to all usages, even when allowing for context effects. If the meaning assigned to *on* is too schematic, then it will never rule out a usage like **the bottom **on** the jar* or **he stuck his tongue out **on** me*, based on examples like (4b) and (4d). A strictly homonymic analysis is also rejected because native speakers do intuit certain correspondences between these usages. If different usages are treated as instances of homonymy, then how many different lexical *on*'s do English speakers have? Any decision will be arbitrary and will ignore clear commonalities, tenuous though they might first appear to be.

Extended Usages of the Prepositions

By examining a wider assortment of data, we find that these prepositions support a vast array of semantic and syntactic patterns. These extended usages may be far removed from the idealized characterizations given above. In short, the topological relations are, in a sense, too specific to characterize the full range of their linguistic functions adequately. These same relations, if interpreted schematically, would require a serious examination of the real and conceived world knowledge that a speaker brings to bear on the task of determining semantic meaning. No adequate model of the lexicon has the luxury of ignoring the fact that speakers can construe a concept or a lexical item in a variety of ways in order to achieve different semantic effects.

To her credit, Herskovits (1986) tries to provide a formal representation of what speakers know about physical properties of objects in order to model the differential use of *at*, *on*, and *in*, and account for "sense shifts" that allow us to say, for example, both *the water **in** the vase* and *the bird **in** the tree*. She lists criteria such as shape, size, typical physical context and orientation, gravitational properties, conceived geometry and function, characteristic interaction patterns, and most salient subparts. Unfortunately, these attributes are of little help in determining the appropriateness of uses involving subjectively construed or non-spatial relations between abstract entities and events. Even in their purely spatial senses, these prepositions have a **deictic function**, and predicate meanings that are wholly dependent on subjective aspects or expectations of the speaker (e.g., in the proximity or distance of the speaker from the scene) and not on objective properties of the event. In (5), we find contrasts involving minimal pairs that boil down to whether or not the speaker is taking a remote or close-up point of view as in (a) and (b) or a more external or internal point of view as in (c):

(5) a. *He's {**at** the store (distal), **in** the store (proximate)}.*
b. *She's {**at** the beach (distal), **on** the beach (proximate)}.*
c. *I'm still {**at**/**on** Chap. 2 (external) **in** Chap. 2 (internal)} of the book.*

In (6) a similar contrast results from varying degrees of closeness between figure (*tools*) and ground (*hand*):

(6) *When working, it's best to keep all necessary tools {**at** hand (distal), **on** hand, **in** hand (proximate)}.*

Very likely, the proximate/distal contrast is elated to a type of ground canonically but not exclusively associated with each of these prepositions. *At* frequently takes a point-like ground that possibly contrasts with other potential points. *On* often situates a figure with respect to a surface-like ground and *in* to a medium of some sort. In a physical sense, the farther one is from a scene, the more reduced in scope and scale and the more pointlike the scene appears. Conversely, the closer one is, the larger and more enveloping it appears. This experientially-based difference in specificity might motivate usages in which indirect or immediate perspective matters, as in (7):

(7) a. *I was horrified by what happened* ***at*** *Tiananmen Square.*
  b. *Eyewitnesses said tanks ran over people* ***on*** *the square.*
  c. *No one died* ***in*** *Tiananmen Square, reported the government.*

Some contrasts between the prepositions underscore a difference in degree of involvement between figure and ground, which has little to do with spatial coincidence but perhaps a lot to do with cognitive perspective. Take, for example, their **institutional association function**:

(8) *She's {**in** the Physics Department,* ***on*** *the faculty,* ***at*** *MIT}.*

In (8), *in* seems to predicate the most direct, relevant, or local association, while *at* predicates the most generic or global.

Some usages of these prepositions predicate a **cognitive association** between figure and ground. *At* and *on* often appear in two-word verbal expressions where a more superficial **perceptual focus** is involved, as in (9):

(9) a. *They looked* ***at*** *the map.*
  b. *He frowned* ***at*** *her.*
  c. *He focused* ***on*** *the TV.*
  d. *Let's eavesdrop* ***on*** *their conversation.*
  e. *He can speak* ***on*** *any subject.*

while *in* is more likely to be found in similar verbal expressions where a deeper **conceptual focus** is at issue, as in (10):

(10) a. *She's lost* ***in*** *thought.*
  b. *I believe* ***in*** *equal pay for equal work.*
  c. *We take pride* ***in*** *our work.*
  d. *He has tremendous faith* ***in*** *her.*
  e. *He spoke* ***in*** *great detail.*

Of course, these characterizations are not sustainable for all usages. Once the relevant background domain becomes abstract as in the following predications of **cognitive ability**, differences between the prepositions become more a matter of convention and acceptability is ultimately a matter of degree:

(11) a. *He's good {**at** math,* ***in*** *math,* ****on*** *math}.*
  b. *He's having trouble {**in** math,* ***?at*** *math,* ****on*** *math}.*
  c. *He did well {**on** his math test,* ***?at*** *his math test,* ****in*** *his math test}.*

Indeed, if the domain becomes too abstract, as in the following usages with deverbal nominalizations which tend to assume a **pragmatic, summarizing function**, each of the prepositions is acceptable, although subtle semantic differences remain:

(12) a. *{**In** hearing that,* ***On*** *hearing that,* ***At*** *hearing that}, she turned and ran out.*
  b. *She became quite despondent {**in** seeing him again,* ***on*** *seeing him again,* ***at*** *(the idea of) seeing him again}.*

Finally, all three prepositions figure in a host of bare nominal usages that predicate the **state** or **condition** an entity is in or the **manner** in which an activity is carried out. I would maintain that *at, on,* and *in* are still meaningful, though perhaps schematically so, in expressions like those in (13), which might convey abstract notions of **contrast, foundation**, or **medium**, respectively:

(13) a. *The countries are* ***at*** *war.*
  b. *The countries are* ***on*** *a war footing.*
  c. *The countries are* ***in*** *a state of war.*

Although we could link these notions to more spatial notions like **point, surface**, and **container**, the extensions are indirect and non-unique. As the sentences in (14) demonstrate, some usages are more affected by convention than conceptualization, and, consequently, their specific motivation may be more a matter of historical development than contemporary semantics in the minds of the speaker and the linguist:

(14) a. *The man is {**at** risk,* ***at*** *peace,* ***at*** *ease}.*
  b. *The man is {**on** drugs,* ***on*** *good behavior,* ***on*** *duty}.*
  c. *The man is {**in** trouble,* ***in*** *custody,* ***in*** *pain}.*

The discussion so far has downplayed the particular and specialized functions of these prepositions. I have not characterized the precise nature of the necessary semantic extensions so much as argued that these words support a range of meanings that are more or less preserved as they get used in more and more abstract predications. My central claim is that the basic spatial relations that many researchers have posited as their primary meaning do not apply in all circumstances. Highly schematic relations, on the other hand, would also fail to characterize what is unique to each of these prepositions spatially. The examination of these data alone points to the need for both central and peripheral and specific and schematic meanings to be posited for each preposition, while allowing for a certain degree of overlap between usages of different prepositions.

A Constrained Polysemy

It could be argued that, at the very least, these prepositions support **spatial**, **temporal**, and a variety of **abstract** associations between a figured entity or event and a variably-dimensioned ground (the so-called object of the preposition). Most native speakers of English informally queried about the semantic function of these prepositions tend to concur with this rather modest and seemingly obvious claim. However, there are two lexical hypotheses that, in their extreme versions, do not agree with this claim. We might call them the strong monosemy and the strong polysemy views, respectively, although they are effectively equivalent in terms of the degree to which they attribute meaning to the individual prepositions.

A rigidly monosemic analysis assigns a single, indeterminate, and perhaps invariant meaning to each preposition, allowing context to modulate or fill in the remaining information specific to each usage. The preposition is thus like a prism that requires available light in order to transmit semantic color. The infinite polysemic account, on the other hand, like a homonymic account, may attribute a specific meaning to each usage. Such overspecificity and inconstancy essentially strip the lexical item of its peculiar semantic integrity. The preposition is like a chameleon, changing its hue to suit each semantic backdrop. Whether the semantic flexibility is extrinsic or intrinsic is immaterial; the effect is that context supplies or changes the preposition's meaning.

I am proposing, by contrast, a more constrained polysemic account in which each preposition is attributed with a small set of canonical meanings which over time can engender additional meanings that may be either highly schematic or specific in character. Thus, each preposition is represented by a constellation of related senses, some of which are very close and similar while others are rather tenuous and distant. The claim is that these extensions can, for the most part, be motivated, not on the basis of context necessarily, but on the basis of what we know about the plasticity of conceptual perspective and the pervasiveness of metaphor and reasoning by image schemas (cf. Lakoff & Johnson, 1980; Johnson, 1987).

I have only been able to highlight in this limited space the more salient aspects of prepositional meaning not amenable to the purported monosemic characterizations given in the first section. I have also suggested that polysemy is not unbridled and that different lexical items within a semantic field may share the same sorts of extended senses. The conceptual analysis sketched here has been based on certain leading assumptions of cognitive approaches in linguistics (cf. Lakoff, 1987; Langacker, 1987, 1991a and b): i.e., that all lexical items are meaningful in each application; that no single concept necessarily underlies every usage of a lexical item; that there may be schematic concepts that subsume other less central concepts; that the extension from concept to concept is gradual; that there may be multiple motivations for different extended senses; and that the role of convention and speaker construal is very great in assigning meaning to lexical items.

Since the majority of usages of these ubiquitous prepositions are non-spatial, they provide the greatest challenge for a complete semantic analysis. But one needn't sacrifice coherence when one abandons a monosemic solution. A unified account doesn't necessarily depend on finding an overarching schema that sanctions all usages of a polysemous morpheme, but rather on motivating most of the usages as relatively modest extensions from one of several core meanings. I would argue that a constrained polysemic account is thus far the best working hypothesis of lexical meaning for items like the English prepositions. While it is well beyond the scope of this paper to map out all the core and extended senses for even a single preposition, such detailed analysis is being undertaken and will be forthcoming.

Empirical Evidence for Polysemy

The extreme views face other more serious empirical challenges that I hope to address in future work. If a monosemic account is to be sustainable, then we need to ask where the highly abstract schemas come from. How are they built up developmentally from specific instances that seem quite unrelated? I am currently investigating the order of acquisition of the usages of these prepositions by infants and second language learners in order to track lexical extension and schematicization. On the other hand, if an infinite polysemic or homonymic account is to be maintained, then we need to ask why native speakers can group different usages together and even rank order different items within clusters.

A series of experiments is now under way which have been designed to reveal native speaker intuitions about the syntax and semantics of these prepositions. One preliminary study elicited similarity ratings between usages of a single preposition in different sentences. In this study,

60 token sentences containing different usages of a preposition were presented in pairs on a computer screen to 30 native speakers of English who were then asked to rate the similarity between the pairs according to usage. Each rating was indicated by manipulating a cursor via a mouse on an anchored but uncalibrated on-screen scale, whose ends were labelled *completely different* and *absolutely identical*. In each case, subjects rated sentences against the same highly spatial usage. Results show that subjects were able to attend to differences between the sentence pairs such that non-spatial or abstract usages judged a priori to be similar were systematically given similar ratings by subjects. That is, subjects were sensitive to relative deviations from a highly spatial usage and could rate them accordingly. Ratings for like similarities and like deviations tended to form clusters. These clusters were fairly robust within and across subjects and were consonant with a polysemic analysis. The groundwork has now been laid for additional studies using similar methodologies that will focus on fine-tuning the conceptual distance between usages within these small clusters and testing the robustness of the ratings for multiple tokens of a similar usage.

Thus far, I cannot sketch out in greater detail the exact nature of a polysemic account for prepositional meaning that holds across speakers. While I suspect that an integrated network organized around a small number of canonical usages is the best model of prepositional meaning, I do not yet have the independent empirical evidence needed to propose usages which best exemplify the core meanings. Furthermore, I do not reject highly schematic senses out of hand, for I believe that they form part of a well-integrated and mature prepositional network. Although low-level prototypes play the greatest role in motivating productive extension to novel uses, the existence of various higher and intermediate level schemas allows us to recognize graded distinctions between usages. Langacker's (1991b:266-272) approach to lexical networks specifically allows for the coexistence of prototypic and schematic senses. In fact, one might argue that a successful cognitive analysis which is able to demonstrate extensive semantic relatedness between usages would eventually vindicate a monosemic hypothesis of lexical representation, albeit a more relaxed version than is usually conceived, and thus achieve the ultimate in semantic unity.

Acknowledgments

I thank Terry Nearey for help in experimental design and analysis. Mary Hare and John Newman graciously commented on an earlier draft.

References

Bowerman, M. 1991. The origins of children's spatial semantic categories: Cognitive vs. linguistic determinants. Ms.

Brugman, C. 1981. The story of **over**. MA thesis, Dept. of Linguistics, UC Berkeley.

Carroll, J, Davies, P, & Richman, B. 1971. *Word Frequency Book*. Boston: Houghton Mifflin.

Evens, M.W. (ed.). 1988. *Relational Models of the Lexicon: Representing Knowledge in Semantic Networks*. Cambridge: CUP.

Hawkins, B. 1984. The semantics of English spatial prepositions. PhD diss., Dept. of Linguistics, UCSD.

Herskovits, A. 1986. *Language and Spatial Cognition: An Interdisciplinary Study of the Prepositions in English*. Cambridge: CUP.

Johnson, M. 1987. *The Body in the Mind: The Bodily Basis of Meaning, Imagination, and Reason*. Chicago: Univ. of Chicago Press.

Johnson-Laird, P., Hermann, D., and Chaffin, R. 1984. Only connections: A critique of semantic networks. *Psychological Bulletin* 96:292-315.

Johnston, J.R. & Slobin, D.I. 1979. The development of locative expressions in English, Italian, Serbo-Croatian and Turkish. *Journal of Child Language* 6:529-545.

Lakoff, G. 1987. *Women, Fire, & Dangerous Things*. Chicago: Univ. of Chicago Press.

Lakoff, G. & Johnson, M. 1980. *Metaphors We Live By*. Chicago: Univ. of Chicago Press.

Langacker, R. 1987/1991a. *Foundations of Cognitive Grammar*, Vols. I and II. Stanford: Stanford University Press.

-----. 1991b. *Concept, Image, and Symbol: The Cognitive Basis of Grammar*. Berlin: Mouton de Gruyter.

Lindkvist, K.G. 1978a. *At versus On, In, By*. Acta Universitatis Stockholmiensis (Stockholm Studies in English) 49.

Lindner, S. 1981. A lexico-semantic analysis of verb-particle constructions with "up" and "out." PhD diss., Dept. of Linguistics, UCSD.

Macnamara, J. 1971. Parsimony & the lexicon. *Language* 47:359-374.

Ruhl, C. 1989. *On Monosemy: A Study in Linguistic Semantics*. Albany: SUNY Press.

Schreuder, R. & Flores d'Arcais, G. 1989. Psycholinguistic issues in the lexical representation of meaning. In *Lexical Representation and Process*, W. Marslen-Wilson (ed), 409-436. Cambridge, MA: MIT Press.

Vandeloise, C. 1991. *Spatial Prepositions: A Case Study from French*. Chicago: Univ. of Chicago Press.

Wesche, B. 1986/87. At ease with "at". *Journal of Semantics* 5:385-398.

Grammaticality judgment in Chinese-English bilinguals: A gating experiment

Hua Liu **Elizabeth Bates** **Ping Li**
Department of Cognitive Science, University of California, San Diego
La Jolla, CA 92093-0515
liu@cogsci.uscd.edu bates@crl.ucsd.edu liping@crl.ucsd.edu

Abstract

An on-line gating method was used to investigate Chinese-English bilinguals' performance in a grammaticality judgment task. Evidence of different transfer patterns (i.e., *backward* and *forward* transfer in early and late second language acquisition) was found in the data reported here. There were strong and systematic relations between performance on the judgments of grammaticality and a separate sentence interpretation task. However, there is also some evidence that inter-language transfer or interference occurs earlier in acquisition for the judgment task than for the sentence interpretation. Judgments of well-formedness might be one of the first domains to "soften" when one language comes into contact with another. Furthermore, it is possible that Chinese and English are more "inter-penetrable" for both forward and backward transfer between these two languages than has been observed between any two language types to date.

Introduction

Cross-linguistic studies of monolinguals have revealed dramatic differences in the processing strategies that native speakers use to interpret sentences, reflecting differences in the relative information value of lexical and grammatical cues in each language. Drawing on results from studies of sentence processing in a large number of different languages, Bates and MacWhinney (1982, 1989) have constructed a model of sentence processing, known as the Competition Model, to emphasize the extent to which languages can vary in the way that cues compete and converge to determine meaning. A cue, in this context, is a particular piece of information that a speaker or listener can use to determine the relationship between meaning and form. The Competition Model assumes an interactive process in which the mapping between surface forms and underlying meanings is mediated by competitions and collaborations among cues. Cues can be evaluated with respect to their validity, i.e., their information value for the identification of linguistic functions. In any given language, the overall validity of a cue is a joint product of its availability (how often the cue is present when a given interpretation has to be made) and its reliability (when the cue is available, how often it leads to the right answer). Having different cue validities, different cues cooperate and compete in the comprehension process, resulting in different interpretation patterns in different languages.

Most previous studies within the Competition Model have adopted a sentence comprehension task in which native speakers of different languages are presented with simple sentences in their own languages and are asked to identify the agent (actor) of the sentence. In the sentence *The pencil is kissing the elephant*, for example, native English speakers choose the "pencil" as the agent much more often than the "elephant" while native Chinese speakers show the opposite strategy, choosing "elephant" as the actor regardless of word order (Bates & MacWhinney, 1989; Miao, 1981). This finding is compatible with the cue validity principle, because Chinese permits far more word order variation than English, and because the sentence subject (and object) can be omitted in free-standing declarative sentences. Hence word order information is a very strong cue to the agent role in English, but not as strong as animacy in Chinese.

The finding that processing strategies differ

markedly across language types opens up a series of questions concerning the performance of bilingual individuals in each of their languages. Researchers working within the framework of the Competition Model have described four logically possible outcomes that we might expect to find in adult bilinguals: 1) "forward transfer", transfer of the first language strategies in the interpretation of sentences into the second language; 2) "backward transfer", transfer of the second language strategies into the first language; 3) "differentiation", adoption of different strategies for the two languages corresponding to the strategies used by monolingual speakers of each language; 4) the use of a new set of "amalgamated" strategies by bilinguals to both of their languages, which is different from the strategies used by either group of monolingual speakers. All of these patterns have been observed in at least some individuals (Kilborn, 1989; McDonald, 1987). Because they suggest "degrees" of transfer, in more than one direction, they are difficult to explain without invoking the kinds of quantitative, interactive activation principles provided in the Competition Model.

Prior to the study reported here, we conducted a sentence interpretation task to examine the patterns of transfer displayed by Chinese-English and English-Chinese bilinguals (Liu, Bates, & Li, in press). The results showed that novice bilinguals display strong evidence for forward transfer (Chinese-English novices transfer animacy-based strategies to English sentences; English-Chinese novices transfer English-like word order strategies to Chinese). Advanced bilinguals display a variety of transfer patterns, including differentiation (use of animacy strategies in Chinese and word order strategies in English) and backward transfer (i.e., use of L2 processing strategies in L1 – a possible symptom of language loss). These findings were shown to reflect a complex interaction of variables including age of exposure to L2 and patterns of daily language use.

So far, all the bilingual work within the Competition Model has focussed on sentence comprehension, raising concerns for the generalizability of these complex bilingual findings. The present study will examine bilingual processing in a different domain, i.e., grammaticality judgment. By combining results from the judgment and the interpretation tasks, we can learn more about the nature of inter-language transfer, in particular the process by which first language strategies "invade" second language strategies (and vice versa).

Method

In this experiment, a sentence-level "gating" method was used to evaluate judgments of grammaticality for Chinese sentences. This gating task was modeled after the well-known word-level gating paradigm pioneered by Grosjean (1980) and by Marslen-Wilson and Tyler (1980). In the word-level gating task, subjects hear increasingly long fragments of a word, starting with the first fragment (e.g., "S—", followed by a slightly longer fragment (e.g., "ST—"), with progressive expansion on each trial until the subject indicates that he is now sure of the identity of the word (e.g., "STRING"). In our sentence-level adaptation of the gating task, subjects were asked not to guess the identity of the sentence (an impossible task, since the set of sentences in any language is infinite), but to determine whether the sentence is grammatical SO FAR, starting with the first constituent in the sentence, with progressive expansion until the whole sentence has been presented.

Subjects. 33 subjects (19 male and 14 female adults ranging in age from 19 to 44 years) who had participated in sentence comprehension tests in our previous study (see Liu et al., in press) were brought into an on-line Chinese grammaticality judgment task. They were divided into 5 groups: 1) controls – monolingual native speakers of Chinese who had been exposed to English speaking environment for no more than half a year and had received little or no formal training in English when they were in China; 2) native Chinese who were novices in English, with their first exposure to English occurring after 20 years of age; 3) native Chinese who were more advanced in English, with age of first exposure before 20 years of age; 4) native English speakers who were novices in Chinese, and were first exposed to Chinese after 19 years of age; 5) native English speakers who were advanced in Chinese, and were exposed to Chinese from early childhood.

Procedure. The three possible judgments of grammaticality, i.e., "grammatical", "ungrammatical" or "don't know", were represented by three buttons on a button box. Chinese sentences were presented auditorily, ONE PORTION at a time. Each sentence portion was read by a native speaker in a smooth and flat intonation and then digitized into the computer. For example, if the sentence is (the Chinese equivalent of) *The cow kick the horse*, subjects would hear the following in sequence and need to push one of the buttons for their judgments at each point: *The cow*, *The cow kick*, *The cow kick the horse*. All the sentence types were counterbalanced across word order and noun animacy, which yielded the following types of sentences: AVA, AVI, IVA, VAA, VAI, VIA, AAV, AIV, IAV.

Data Analysis. Scoring of the dependent variable was based on the rating of the grammaticality of the phrases. For each phrase, subjects were given a score of 3 for rating the phrase as *grammatical*, a score of 2 for rating the phrase as *unsure*, a score of 1 for rating the phrase as *ungrammatical*. These values were entered as the raw data for subsequent statistic analysis.

Results and Discussion

Although data are available from a broader range of sentence types, we will simplify the present discussion by focusing on the results from sentences that follow canonical word order NVN, under different animacy conditions (Animate-Verb-Inanimate or AVI sentences like *The cow kick the pencil*; Inanimate-Verb-Animate or IVA sentences like *The pencil kick the cow*; semantically neutral Animate-Verb-Animate or AVA sentences like *The cow kick the horse*).

Control data from monolingual Chinese speakers indicate that when word order was consistent with animacy (e.g., AVI items), subjects considered the sentence to be grammatical at all three judgment points: N, NV, and NVN. When there was a competition type between semantics and word order (e.g., IVA items, see Figure 1), subjects would accept the first inanimate noun as grammatical (averaged rating score = 2.83). However, they began to judge the sentence as ungrammatical after IV (averaged rating score = 1.64), and reject it still further at the end (averaged rating score = 1.47). In short, although these monolingual Chinese reliably interpret IVA strings as Object-Verb-Subject (OVS), the judgment task shows that they do not like OVS structures at all, rejecting them as possible structures in their language. ANOVA results show that all the analyses in the control group reached significance, including word order ($F(2,12) = 14.4$, $P < 0.001$), animacy ($F(2,12) = 50.8$, $P < 0.001$), position ($F(2,12) = 31.0$, $P < 0.001$), and two-way as well as higher-level interactions ($P < 0.001$).

The results from our four bilingual groups are also illustrated in Figure 1, where we focus once again on IVA (although a similar story emerges on other sentence types). The most important result was an inverse relationship between sentence interpretation strategies (from animacy-dominance in Chinese to word order dominance in English) and judgments of grammaticality. Ratings of grammaticality for these competition items increase significantly as we move away from our animacy-dominant monolingual Chinese group to our word-order-dominant English-Chinese novices. The difference between the two extreme groups, i.e., the most Chinese group (control group) and the most English group (novice English-Chinese), reached significance ($F(1,12)=252$, $P < 0.001$). To illustrate, consider the strongest competition cell (IVA, as in *The rock hit the cow*). Chinese-dominant bilinguals use semantics to make their interpretation (i.e., *the cow did it*, see Liu et al., in press). This is a semantically plausible reading, but it also means that the sentences has been interpreted as an OVS. OVS is an impossible structure in the Chinese language – a fact that is reflected in left-to-right grammaticality judgments offered by these subjects. English-dominant bilinguals rely on word order to make their interpretations (i.e., *the rock did it*). This is a semantically implausible reading, but it preserves SVO structure. This fact is also reflected in the left-to-right grammaticality judgments made by English-dominant bilinguals, who judge strings like *The rock hit the cow* to be perfectly grammatical.

These findings suggest that forward and backward transfer strategies observed in the sentence interpretation task also have implications for judgments of grammaticality. Particularly strong support for this view comes from those bilingual subjects who stand somewhere "in between" in their use of semantics and word order to interpret sentences. Many of these subjects also display a grammaticality judgment pattern that is "in between", showing a great deal of uncertainty regarding the grammaticality of competition strings. For other Chinese-English

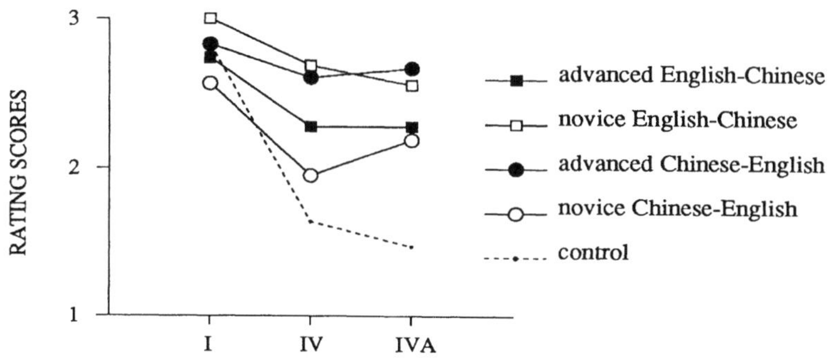

Figure 1: Judgment of grammaticality of IVA sentences

bilinguals, a comparison of the interpretation and judgment tasks suggests that "backward transfer" (i.e., invasion of Language 1 by Language 2) may show up in grammaticality judgments before it appears in sentence interpretation. That is, some of our "in between" subjects (e.g., novice Chinese-English subjects) experience great uncertainty in the judgment task (with judgments persisting in the "don't know" range across the course of the sentence), even though they still display a native-like preference for animacy strategies in the interpretation task.

Summary and Conclusion

One of the main results in the sentence gating task is that novice Chinese-English subjects tend to think that the IVA sentences like *The pencil hit the cow* are relatively LESS grammatical than novice English-Chinese bilinguals. This result maps directly onto results from the comprehension task, where novice Chinese-English subjects show animacy dominance (interpreting IVA strings as semantically plausible but grammatically unacceptable OVS) while novice English-Chinese subjects show word order dominance (interpreting IVA strings as "silly" but grammatically acceptable SVO). In other words, these subjects distinguish very clearly between grammaticality (which is about the well-formedness of sentences) and plausibility (which is about how well the apparent meaning of the sentence fits the real world).

This finding is consistent with Bates et al. (1982)'s interview with English-dominant English-Italian bilinguals, i.e., English subjects interpret a sentence like *the rock kiss the cow* as grammatical but "silly", and they choose the "rock" as the actor. Their Italian-dominant bilinguals report that such sentences are both grammatical and semantically plausible. Chinese is more or less like Italian in the sense that the animacy cue is stronger than word order, but they differ because OVS is a possible (albeit infrequent) structure in Italian while the same order is not possible in Chinese – and yet the Chinese do not "trust" any word order cue enough to override semantics.

A comparison of the interpretation and the judgment tasks also provides interesting information about degrees of "backward" and "forward" transfer. In particular, some of our bilingual subjects display an "in-between" pattern, in both the interpretation and the judgment task. Others show a "softening" of grammaticality judgments while they are still actively using strategies from their first language for sentence interpretation. That is, bilingual speakers may begin to lose sensitivity to the well-formedness of sentences in their first language, even though they are still processing sentences in the normal way.

Finally, the sentence-level gating task provides interesting information about the point at which judgments of grammaticality are made. A comparison of the different bilingual groups in Figure 1 shows that their decisions diverge at the point where a contrast between OVS and SVO becomes clear. However, for those bilingual subjects who are "in between", uncertainty begins at that very same point and persists across the sentence. Evidence from several other sentence types in our data set leads to a similar conclusion. These "in between" patterns of forward and backward transfer are compatible with the interactive activation principles of the Competition Model, and may lead to a deeper understanding of the quantitative and qualitative processes that underlie inter-language transfer.

Reference

Bates, E., & MacWhinney, B. (1982). Functionalist approaches to grammar. In: Wanner, E. & Gleitman, L. (Eds.), *Language acquisition: The state of the art.* New York: Cambridge University Press.

Bates, E., & MacWhinney, B. (1989). Functionalism and the competition model. In: MacWhinney, B. & Bates, E. (Eds.), *The crosslinguistic study of sentence processing.* New York: Cambridge University Press.

Grosjean, F. (1980). Spoken word recognition processes and the gating paradigm. *Perception & Psychophysics*, 28, 267-283.

Liu, H., Bates, E., & Li, P. (In press). Sentence interpretation in bilingual speakers of English and Chinese. *Applied Psycholinguistics.*

Kilborn, K. (1989). Sentence processing in a second language: The timing of transfer. *Language & Speech*, 32, 1-23.

Marslen-Wilsen, W.D., & Tyler, L.K.T. (1980). The temporal structure of spoken language understanding. *Cognition*, 8: 1-71.

McDonald, J. (1987). Sentence interpretation in bilingual speakers of English and Dutch. *Applied Psycholinguistics*, 8: 379-415.

Miao, X. (1981). Word order and semantic strategies in Chinese sentence comprehension. *International Journal of Psycholinguistic Research*, 8: 23-33.

Understanding English Past-Tense Formation: The Shared Meaning Hypothesis

Catherine L. Harris

Department of Psychology
Boston University
Boston, MA 02215
charris@bass.bu-edu

Abstract

It has long been controversial whether language behavior is best described and explained with reference to the constructs of formal linguistic theory, or with reference to information processing concepts and the communicative goals of speakers. Recent work by Kim, Pinker, Prince and Prasada (1991) argues that the vocabulary of formal grammatical theory is essential to psychological explanation. They demonstrate that speakers' evaluation of the well formedness of past-tense forms is sensitive to whether novel verb forms are perceived to be extended from nouns or verbs. I show this pattern of preferences to be a consequence of semantic similarity between the novel sense of the verb and the irregular verb to which it is phonologically related. The data is consistent with the tenets of **functional grammar**: speaker' choice of one linguistic form over another is influenced by perceived communicative gain (Kuno, 1987; Bates & MacWhinney, 1989). The salient task in judging novel verbs phonologically related to irregular verbs is guarding against miscommunication. Dizzy Dean aside, that so few mortals have ever flown out to center field testifies to speakers' success.

Background

Over the last few years, the past-tense has become a virtual crucible for theories of linguistic representation (Rumelhart & McClelland, 1986; Pinker and Prince, 1988; MacWhinney & Leinbach, 1991). Recently, Kim, Pinker, Prince and Presada (1991) used speakers' judgments of the well-formedness of novel past-tense verbs (such as *high-sticked the goalie*) to support the psychological reality of formal grammatical theory. In the current paper, I follow in Kim et al.'s footsteps to probe the processes involved in creating a new lexical entry and extending words into new semantic territory.

If speaking of a person who performed a greater feat than astronaut Sally Ride, we are more likely to exclaim, "*She out-Sally-Rided Sally Ride!*" than "*She out-Sally-Rode Sally Ride!*" (Pinker & Prince, 1988). Similarly, the verb *to grandstand* (impress onlookers) sounds most natural with the past-tense form *grandstanded*, rather than *grandstood*. By contrast, new verbs derived from existing irregular verbs (such as *withstand*) sound more natural if they agree with the past-tense of the verb from which they are derived.

Kim et al.'s stated goal was to show that even so simple a system as English past-tense formation can not be adequately described without reference to the classic descriptive constructs of formal grammatical theory (FGT), such as lexical item, part of speech (noun, verb) and morphological structure. Consider the question of why the base ball sense of *fly* is regular, despite its semantic relation to the more frequent sense of *fly* (to move through the air or before the wind). The irregular marker on the original verb root *fly* is lost when the noun compound *fly ball* is created because only verbs can have irregularity markers (Williams, 1981). Thus, when a new verb is derived from *fly ball*, no irregularity marker is present, and the default past-tense formation rule applies, resulting in *The batter flied out twice to center field.*

To see if these principles stand up in the lab, Kim et al. constructed passages, such as those in (1) and (2), containing either an irregular verb, or a noun polysemous or homophonous to it. Novel semantic extensions of this noun or verb appeared in both the regular and irregular past-tense form, and raters were asked to rate the naturalness of each form on a scale of 1 to 7.

In 18 of 21 passages describing novel nominal and denominal verbs, the predictions of (FGT) were met: subjects preferred the regular past for verbs extended from nouns, and the irregular past for verbs extended from irregular verbs. As an example of the former, raters

From Kim et al. Example of a new verb extended from the noun *shrink:* (Naturalness ratings at right.)

(1) Sam is always acting like a shrink, psychoanalyzing half the people at the table. But last night we had Jonathan over, and he analyzed ALL the people at the table.

He finally out-shrinked Sam. 3.8750 He finally out-shrank Sam. 2.5625

A new verb extended from the verb *shrink*:

(2) My wife Hilda was always washing the clothes at too high a temperature, shrinking them beyond recognition, but we hired a housekeeper who ruined six shirts in one load.

She actually out-shrinked Hilda. 2.5000 She actually out-shrank Hilda. 3.6875

Three exceptions to FGT predictions:

(3) Both boxers managed to land heavy blows... Tyson (*out-blowed* 2.81 / *out-blew* 3.00) his opponent and won ...

(4) Janet was fed up with Sam's recurrent flings with young women... After her fifth willing partner she had actually (*outflinged* 3.31 / *out-flung* 3.62) the guy.

(5) Pitcher Roger allowed the Orioles only three hits ... He (*three-hitted* 3.12 / *three-hit* 4.43) them for the second...

much preferred *He William-Telled the apple* to *He William-Told the apple* when the desired meaning was "He put an apple on his son's head, and tried to pull a William Tell." Subjects slightly preferred *He story-told the children for a solid two hours* to *He story-telled the children,* given the new verb *story-telling.*

Kim et al. point out that FGT can explain not only the 18 cases where the results were in the expected direction, but the three cases that are apparent partial counter-examples. Examples (3)-(6) show that raters still judged the irregular past tense form to be slightly more natural sounding despite the denominals *blows, flings*, and *hits*.

These ratings can be fit to the predictions of FGT if we imagine that raters may have *perceived* that the new usage was derived from the original irregular verb, rather than from the noun. Kim et al. call this addendum **the short circuit theory**, because the normal derivation from a noun is by-passed in favor of derivation from the root verb. Intuitively, raters are most likely to "short-circuit" if they perceive that the new verb retains some semantic similarity to the original. Kim et al. collected ratings of the similarity between each new verb sense and the central sense of the verb root. They found that the semantic similarity ratings for the three exceptions were significantly higher than the mean similarity ratings of the rest of the items, suggesting that this perception of similarity had led previous raters to represent the above three items as deverbal instead of denominal.

If semantic similarity explains raters' preferences in these three cases, one wonders what role it plays in the other cases. In the remainder of this paper, I propose that speakers use shared meaning when judging past-tense well-formedness. I use this account to explain *all* of KIm et al.'s data, embed it in a conception of speakers' processing costs and communicative goals, and generate predictions about how noun/verb category may interact with other aspects of the passage to influence ratings of past-tense well-formedness.

Representation and Function

Of the types of explanations scientists use, Kim et al.'s is "representational": the observed behaviors logically follow from mechanical operations on representational structures. We can call "functional" those theories in which observed behaviors logically follow from plausible theorems about organisms' goals in behaving in one way rather than another. For language, the plausible explanatory parameters are communicative efficacy and effort (Givon, 1979). But because behavior is ultimately causally related to mental structures, there is great interest in developing representational as well as functional explanations.

A frequent shortcoming of representational explanations is that the representational structures are motivated by the data the theorist aims to explain, and there may not be enough data left for an independent test of the explanatory framework. This is not a failing of Kim et al.'s work, as their account of novel past-tense formation appealed to independently motivated linguistic principles. However, because linguistics has historically emphasized representational rather than functional theories (Chomsky, 1957), there are few clues about why the system is set up the way it is. Why is the system set up so that nouns can't inherit and pass on irregularity markers? Why do speakers attend to derivational status when creating a lexical entry for a new verb?

In the next section, I motivate an account of novel past-tense formation that is both communicatively and representationally plausible.

The Shared Meaning Hypothesis

Hopefully the reader agrees that *She out-Sally-Rode Sally Ride*! is not the appropriate way to communicate, "She performed a greater feat than astronaut Sally Ride." The reason for this follows from the basic communicative principle of refraining from knowingly misleading your listener: the new verb "out-Sally-Ride" has no semantic connection with the existing verb *ride*, and to use the past-tense verb *rode* would be incorrectly implying a connection.

I will call the following the **shared meaning hypothesis**:

> Speakers copy the irregular past-tense form of the original verb, rather than use *-ed*, to emphasize shared meaning between the new verb and the original irregular verb. Because *-ed* is the default past-tense and would be used with any completely novel lexical item, use of *-ed* is a strong signal that the meaning of the new word is distinct from that of the original irregular verb.

Because 21 of Kim et al's 29 novel denominal verbs were **homophones** of existing irregular verbs, while *all* of their deverbal verbs were transparently semantically related to existing irregular verbs, this simple version of the shared meaning hypothesis (SMH) accounts for a good part of their data simply by making identical predictions to FGT in these cases. However, the SMH may be able to more closely match the naturalness ratings because both felicity ratings and the degree of shared meaning are continuous, while FGT divides the world of naturalness judgments into two categories.

Below I list some different types of meaning distortions, and describe communicative reasons why speakers will emphasize shared meaning in some cases while disavowing it in others. The SMH can then be further tested by seeing whether past-tense naturalness judgments correlate with these predictions of relative meaning emphasis or disavowal.

The Costs and Benefits of Meaning Extension

If we view speakers' task as one of **minimizing processing costs while maximizing communicative impact** (loosely following Givon, 1989 and others), then we can begin to characterize the costs and benefits of using a new word. Humans, like all animal species, grow accustomed to the commonplace and dishabituate to novelty. The new word has the impact of novelty. But using a novel word has two costs: (a) the encoding cost of inventing a phonological string that the speaker and listener will be able to remember and access later, and (b) the risk of failed communication. (An advantage of polysemy is that it obviates the first problem; Harris, 1992.) In any type of meaning extension, the speaker needs to ensure that the new sense is sufficiently connected to the old sense, but that meaning elements are not incorrectly transferred from the original sense to the new sense. Below I list some types of meaning extensions for which connection of meaning or disavowal might be more or less important.

i. **Metonymy**. New usage picks out one aspect of the meaning conveyed by the original usage. This is typically employed by speakers to reduce processing costs, as the conventional meaning is salient in the discourse or extralinguistic environment, and is thus easier to access than the lexical item that conventionally codes for the intended concept (Deane, 1989). Because metonymy is usually used with contextual support, the risk of comprehension failure may be minimal. **Prediction: emphasize shared meaning.**

ii. **Metaphor**. New usage builds on abstract relations present in the original. The partial mapping of elements is usually thought to be determined by conceptual factors such as highest structural match (Gentner, 1989). This suggests a reason for *not* using a linguistic device to disavow identity between the original meaning and the intended meaning: the intended meaning builds on the original meaning, and may do so in ways that require the original meaning to be available for processing for a significant time period. **Prediction: emphasize shared meaning.**

iii. **Inclusion**. New usage completely contains original usage, but adds to it. **Prediction: emphasize shared meaning.**

iv. **Concatenative compounding**. Like inclusion, but two existing words are joined together, so that the resulting form may be different (as in *oversleep*). Prediction: emphasize shared meaning.

v. **Aspect change**. Central to a verb's meaning is whether it describe an abstract, atemporal relation, or a process. If a process, the verb can refer to a punctate or temporally extended event. Example: *I told the children a story yesterday* (completive aspect). *I story-told the children for two hours* (durative aspect). **Prediction: Mixed.** Aspect is basic to verb meaning, and is part of listeners' automatic inferences about an event described with a certain verb. Therefore, if the new meaning conflicts with the aspect of the original, the speaker might want to disavow a meaning connection. On the other hand, there are verbs in English that have malleable aspect, and many verbs can sound felicitous with a different aspect if the context is right (Langacker, 1987).

vi. **Argument structure change.** Arguments of the original verb are incorporated into the meaning of the new verb or dropped completely. Example: Basic meaning of *fly* assigns to its subject the agent or experiencer of flying (moving through the air). The *base ball* meaning of *fly* incorporates the sense of a ball moving through the air, and assigns to its subject the causative role (initiator of the ball's flight). **Prediction: Mixed, but some meaning disavowal likely.** Maintaining predictable argument structure decreases processing costs: the nouns in a clause can be rapidly assigned the semantic roles defined by the verb, giving the listener has a quick way to find out who is doing what to whom (Bates & MacWhinney, 1987). This rapid (and perhaps automatic) argument assignment is intuitively one good reason to disavow, with whatever devices are at hand, the image of the batter sailing through the sky.

vii. **Denominalization.** One of the strengths of FGT was that is explained the circumlocutive route traveled by an irregular verb, made into a noun, made into a regular verb. It is difficult to make an irregular verb undergo these stages without the eventual meaning differing in aspect and argument structure. **Prediction: disavow shared meaning.** If the meaning *isn't* very different, then the listener may assume the verb is simply the original irregular.

viii. **Homophony.** Although the new meaning may been have extended from the original at one time, there is now very little similarity in meaning left. The two words are likely to be treated as homophones by speakers. **Prediction: Disavow meaning connection.**

Listeners obtain two types of information from the speech signal. **Grammaticized information** is coded by grammatical devices, such as when the duration of an event is signaled by *-ing*, or when the supporting nature of an event is signaled by encoding the event with a subordinate clause. **Inferences** accrue by integrating the linguistic message with non-linguistic material. If we assume that the purpose of grammaticized information is to facilitate processing (Givon, 1989), then meaning distortions involving grammaticized elements (such as aspect and argument structure) are most likely to be disavowed by a linguistic device, such as using the *-ed* past tense.

Analysis of Past Tense Ratings

To test whether the SMH hypothesis can make more fine grained predictions that FGT, Kim et al's 74 passages (37 deverbal, 37 denominals) were categorized according to what meaning relation they had to Webster's entries for the phonologically similar irregular verb.

The categories were defined as follows:

Known/metaphorical. Passage meaning is listed in the on-line version of Webster. All of Kim et al.'s 16 metaphorical deverbals fell into this category. For example, Webster's defines *write off* as *to take off the books: CANCEL.* One of the definitions of *to fly* is *to assail suddenly and violently,* which is similar to Kim et al.'s metaphorical item *he flew off the handle.* Because these are essentially dead metaphors, I am calling them "known."

Concatenative compounds. Passage meaning fully includes a meaning of an irregular verb listed in Webster. Of the 15 passages that fell into this category, all were deverbals.

Argument conflict. Passage meaning is transparently semantically related to an existing irregular, but the semantic role filled by the subject of the new verb is different from that of the existing irregular verb. Of the 14 passages in this category, 6 were deverbal passages, and 7 were denominal. Denominal examples include *I immediately drinked him* (I caused him to be supplied with drinks such that he probably drank a lot) and *Janet out-flinged Sam* (had experiences that led to a state of affairs in which she had more extra-marital affairs than Sam). (It's insightful to see how long a paraphrase one must construct in order to include the original irregular verb -- couldn't achieve this with *fling.*) Deverbal examples include *I'm shaked-out* (I'm experiencing fatigue from the action of shaking) and *He "knew" me once too often* (he has the experience of bumping into me and saying "Don't I know you?").

Homophones. 29 passages were categorized as having no semantic relation to the irregular verb to which they were phonologically similar if they were derived from a proper name (*William-Telled the apple, out-Big-Sleep the Big Sleep*), had spelling differences (*reeded the posts* vs. *read the Captain's mind*), or had very significant and clear meaning differences (*ringed the city*).

The passages were not further divided into a class of aspect change to avoid cross-cutting the other categories. None of the known/metaphorical items involved an aspect change, while many of the novel items did. Aspect change is not a meaningful question for the homophones, since they are already maximally different from their homophonous iregular verb.

Results. Table 1 compares past tense ratings for each of the categories, showing that, as predicted by the SMH, speakers judge irregular past tense forms favorably if the meaning of the new verb incorporates the meaning of an existing irregular, and judge regular past tense forms favorably in the absence of shared meaning, while cases of argument structure conflict yielded intermediate ratings for both deverbals and denominals.

Semantic Similarity Ratings

The row in Table 1 labeled "semantic similarity" contains mean ratings of the degree of similarity holding between Kim et. al's passages and the closest dictionary entry for the relevant irregular verb. 32 Boston University undergraduates compared present-tense-only versions of Kim et al.'s passages to phrases taken from the dictionary. Subjects checked a box marked "unrelated"

Table 1

	Homonyms known	Homonyms novel	Arg Conflict denom	Arg Conflict deverb	Inclusion concat	Inclusion known
Regular	4.41	4.17	3.83	2.86	1.90	1.95
Irreg	1.67	2.23	2.87	4.0	4.45	6.50
Semantic Similarity	5.56	5.14	3.18	2.58	2.12	2.06
Total Items	8	21	8	5	16	16

or placed a hash mark through a scale extending from "closely related" (translated as 1) to "weakly related" (translated as 5). "Unrelated" judgments were coded as a 6. The rank order of the ratings match the predictions made in the previous section.

The correlation between Kim *et al.*'s ratings of past tense well-formedness and the ratings of similarity are summarized in Table 2. The data was fit separately for ratings of regulars and irregulars by multiple regression with four predictors: **derivational status, semantic relatedness,** and two indicator variables: **homonym/polyseme** (argument conflict cases coded as polysemes of an existing irregular), and **known/novel.** The t statistics are the results of partial correlation analysis and show which variables make additional significant contributions after the previous variables have been taken into account.

For the irregular ratings, although derivational status is the predictor most highly correlated with well-formedness ratings, it was not significant when other variables are taken into account. A scatterplot of the iregular ratings (Figure 1) suggests that the variable of known/novel adds with the variable of semantic similarity to render derivational status insignificant. Data points deviating from the linear relationship between well-formedness rating and similarity rating are known deverbals and

Table 2

Predicting Regular Ratings

Variable	r	t(69)	p
derivational status	.78	2.52	0.01
semantic similarity	.71	1.95	0.05
homonym/polyseme	.74	0.64	> .50.
known/novel	.05	1.40	> .15.

Predicting Irregular Ratings

Variable	r	t(69)	p
derivational status	-.76	1.64	0.11
semantic similarity	-.74	2.90	0.005
homonym/polyseme	-.75	1.31	> .15
known/novel	.25	3.40	0.001

novel denominals, with the known items having inflated ratings and the novel items having depressed ratings.

Including in the multiple regression an interaction term for *either* **known X derivational status** or **known X similarity** had the effect of increasing the multiple-R from.83 to .87, increasing the t value of *all* other predictors, including derivational status, $t(68)=2.1$, $p<.05$; and semantic similarity, $t(68)=3.5$, $p<.0001$.

For regression on regular ratings, the same interaction term was significant, $t(68)=2.07$, $p < 0.5$ and its inclusion increased multiple-R from.80 to .82, but known/novel variable alone was *not* significant.

Summary. Contrary to Kim et al's assertions that semantic aspects don't predict well-formedness rating, if known/novel status is taken into account, semantic similarity is a *better* predictor than derivational status (for the irregular ratings).

A Representational Hypothesis

Two ideas will be useful in developing a representational account of what it means to share or disavow a meaning relation. (1) Lexical items do not contain direct pointers to conceptual structure. Words are *conventionalized* form-meaning associations and thus include horizontal co-occurrence statistics, abstractions over which yield argument-structure relations (Langacker, 1987; some ideas about connectionist implementation in Harris, 1991). (2) The "principle of least effort" (Zipf, 1949) applied to meaning extension suggests that new senses "inherit" the meaning associations of the verb from which they are extended, including any association with an irregular past tense. With time, a dead metaphor (such as "I blew him off") may have little meaning relation with the basic sense of blow, but its irregular past has become habitual.

When an irregular verb is extended to become a noun, its association with the irregular form decays due to disuse. If it is *immediately* made into a new verb (as in psychology experiments), we are likely to see the type of indecision displayed in the 8 passages of this type constructed by Kim et al.

Conclusion

Drawing upon the ideas of functionalist grammarians, different types of meaning extension were categorized according to whether the speaker would need to emphasize or disavow shared meaning. Homophony and argument-structure changes were proposed to be the cases most likely to lead to incorrect inferences. Analysis of Kim et al's data showed that *-ed* form was judged more felicitous in these cases than cases of compounding and metaphoric extension. Semantic similarity ratings corroborated and extended these findings. It is concluded that accepting the psychological reality of linguistic constructs doesn't mitigate the importance of functional explanations of these abilities.

References

Bates, E. A. & MacWhinney, B. (1987) Competition, variation, and language learning. In B. MacWhinney (Ed.), *Mechanisms of language acquisition*. Hillsdale, NJ: Lawrence Erlbaum.

Bates, E.A. & MacWhinney, B. (1989) Functionalism and the competition model. In B. MacWhinney & E. A. Bates (Eds.), *The crosslinguistic study of sentence processing*. New York: Cambridge University Press.

Chomsky, N. (1957). *Syntactic structures*. The Hague: Mouton.

Deane, P.D. (1989) Polysemy and cognition. Lingua, 75, 325-361.

Gentner, D. (1989) The mechanisms of analogical learning. In S. Vosniadou & A. Ortony (Eds.) Similarity and analogical reasoning.\R Cambridge: Cambridge University Press.

Givon, T. (1979) On understanding grammar. New York: Academic Press.

Givon, T. (1989) *Mind, code and context: Essays in pragmatics*. Hillsdale, NJ: Erlbaum.

Harris, C.L. (1991) PDP Models and Metaphors for Language and Development. Ph.D. diss, UCSD.

Harris, C.L. (1992) Coarse coding and the lexicon. Paper presented at Universite de Caen, Round Table on 'Continuum in linguistic semantics,' June 22-24, 1992.

Kim, J.J., Pinker, S., Prince, A., & Prasada, S. (1991) Why no mere mortal has ever flown out to center field. *Cognitive Science, 15*, 173-218.

Kuno, S. (1987) *Functional Syntax: Anaphora, Discourse and Empathy*. Chicago: University of Chicago Press.

MacWhinney, B., & Leinbach, J. (1991) Implementations are not conceptualizations: Revising the verb learning model. *Cognition*.

Pinker, S. & Prince, A. (1988) On language and connectionism: Analysis of a parallel distributed processing model of language acquisition. *Cognition, 28*, 73-193.

Rumelhart, D.E., & McClelland, J. L. (1986) Learning the past tense of English verbs. In J. L. McClelland & D.E. Rumelhart (Eds.) *Parallel distributed processing: Vol. 2*. Cambridge, MA: MIT Press.

Williams, E. (1981) On the notions "lexically related" and "head of a word." *Linguistic Inquiry, 12*, 245-274.

Zipf, G.K. (1949) *Human behaviour and the principle of least effort*. Cambridge, MA: Addison-Wesley Press

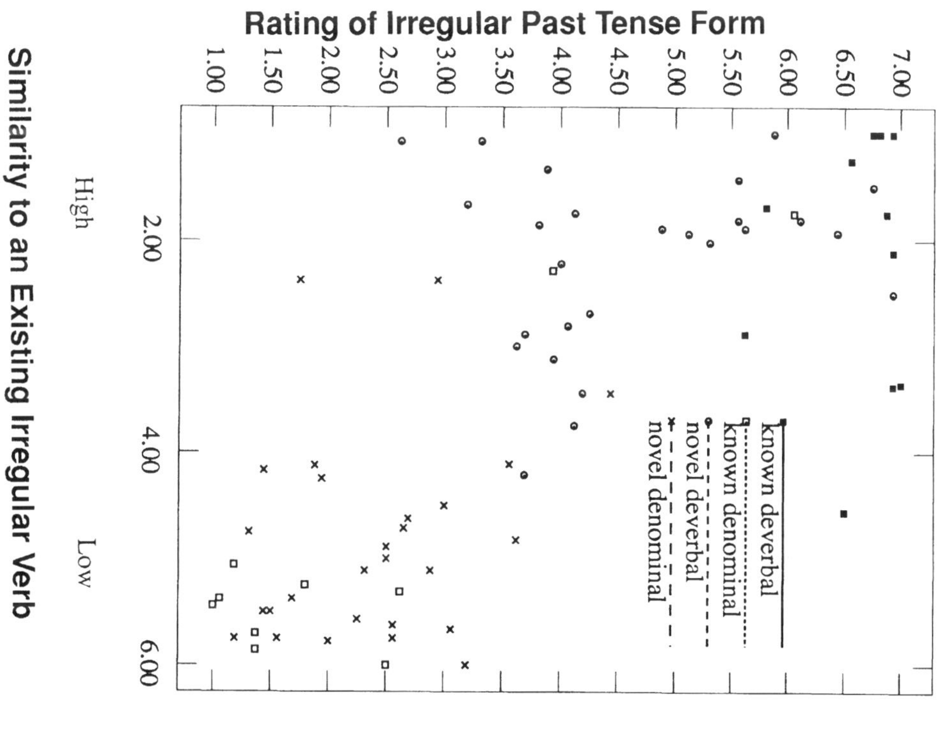

Figure 1

A recognition model of geometry theorem-proving

Tom McDougal
Kristian Hammond
University of Chicago AI Lab
1100 E. 58th Street
Chicago, IL 60637
(312) 702-1571
mcdougal@cs.uchicago.edu
hammond@cs.uchicago.edu

Abstract

This paper describes POLYA, a computer program that writes geometry proofs. POLYA actively collects features from a geometry diagram on the basis of which it recognizes and applies knowledge from known examples. We present a vocabulary of visual targets, results, and actions to support incremental parsing of diagrams. We also show how scripts can be used to organize visual actions into useful sequences. We show how those sequences can be used to parse diagrams and instantiate proofs. Finally, we show how scripts represent the implicit spatial knowledge conveyed by examples.

Introduction

Anyone who has struggled in a math class knows the difference between understanding the solution to a problem and knowing how to come up with that solution. In *How to Solve It* [1957], mathematician and educator George Polya gives advice to the student having trouble deriving solutions. He breaks the problem-solving process into four steps:

This work was supported in part by the University of Chicago Department of Computer Science, AFOSR grant number AFOSR-91-0112, DARPA contract number F30602-91-C-0028, DARPA contract number N00014-91-J-4092 monitored by the Office of Naval Research, Office of Naval Research grant number N00014-91-J-1185, and an internal fellowship from UCSMP.

1. Understand the problem.
2. Devise a plan.
3. Carry out the plan.
4. Look back.

Most of the work occurs in the second step. There Polya recommends drawing on experience: "...It is often appropriate to start the work with the question: *Do you know a related problem?*" [p. 9, italics in original]

How to Solve It is peppered with examples from geometry. However, the history of geometry theorem-proving in AI and Cognitive Science contains little that resembles Polya's four steps (see [Koedinger & Anderson 1990] for a review). Instead, nearly all computer programs which construct geometry theorems do so using forward and backward chaining of if-then rules corresponding to the traditional theorems and axioms of plane geometry. The programs have differed from one another mostly in the heuristics used to make the chaining approach tractable.

This paper describes a computer program, POLYA, which constructs geometry proofs in the way George Polya suggests. It devises a plan by recognizing similarity to known examples; it carries out the plan by mapping significant features of the examples to the new case. POLYA addresses some of the limitations of Koedinger and Anderson's Diagram Configuration model (DC), resulting in a more cognitively plausible model of geometry theorem-proving.

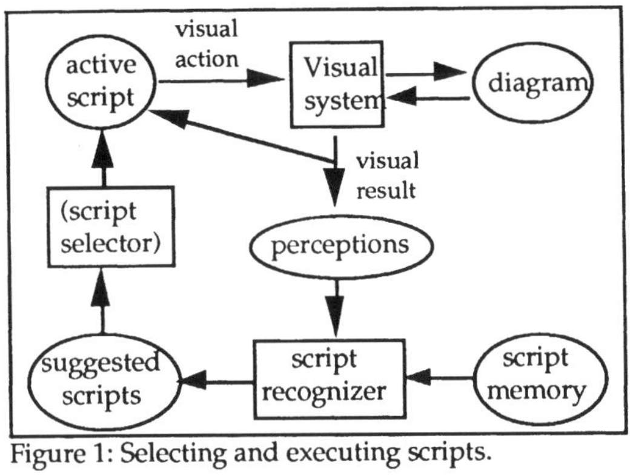

Figure 1: Selecting and executing scripts.

Overview of POLYA

POLYA accepts as input a problem statement and a diagram. The problem statement consists of a short list of "givens" and a goal. The diagram is a bitmap drawing. POLYA marks the diagram to reflect the givens, and then begins to parse the diagram.

POLYA parses the diagram incrementally, using *visual actions* to shift its focus of attention to areas of likely interest in response to what it sees. POLYA's knowledge of how to parse diagrams, as well as its knowledge of how to write proofs, is contained in *scripts*. *Search scripts* are sequences of visual actions for detecting a particular useful pattern or relationship in the diagram. *Proof scripts* contain a sequence of visual actions for verifying the relevance of an example, and another sequence for mapping the example to the current diagram.

POLYA's algorithm for selecting scripts and running them is shown in figure 1. The steps are:

1. Select a script.
2. Perform each action in the script, obtaining visual results. Add each result to a list of perceptions.
3. Compare the visual result to the prediction made by the active script. Also compare the list of perceptions to the triggering sets of inactive scripts.
4. Update the list of suggested scripts, adding ones which have had their triggering set satisfied.
5. Repeat until there are no more scripts or until the proof is complete.

The rest of this paper will describe POLYA's visual system, its visual search knowledge, and its proof-writing knowledge. We will illustrate with a working example how these interact to devise and carry out a plan.

POLYA's visual system

The design of POLYA's visual system is influenced by certain facts about the human visual system. The human visual field has a very small area (about 3 degrees) of high resolution at the fovea with much lower resolution elsewhere. This presents a computational advantage for processing retinal images, but presents challenges for gathering information. Much of the success of human vision is due to our ability to shift the fovea rapidly among areas of likely interest, with additional processing relating those narrow perceptions [Carpenter 1988]. These advantages and challenges have inspired new research in *active vision* [Ballard, Clark, Schwartz, Swain, Tistarelli, etc.].

Focus of attention

Matching configuration schema against a geometry diagram, as DC does in its first problem-solving phase, reduces to NP-complete subgraph isomorphism. The addition of a single irrelevant line segment to a diagram can double the time DC requires to parse it [Koedinger, personal communication]. Koedinger and Anderson [1990] acknowledge in their paper that an accurate model of human problem solving would integrate parsing and schema search.

This is what POLYA does. It parses the diagram opportunistically, using what it sees (*visual results*) to help it decide what to look at next (*visual targets*). Figure 2 contains excerpts from POLYA's vocabulary of visual targets and results. Note that POLYA can focus on pairs of objects, such as two triangles or a point and a segment, as well as on individual objects. In focusing on pairs of objects, POLYA loses specific information about the individual objects while gaining relational information. For example, looking at a pair of triangles yields no information about markings on the individual triangles; for that information POLYA must focus on one triangle at a time.

What POLYA can look at	What POLYA sees	Examples	Visual result
Single point	ray-pattern label no. of angle marks	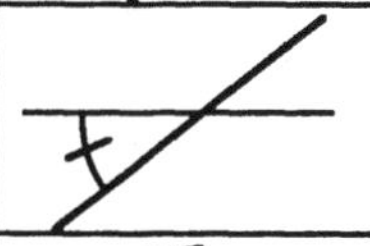	X-HORIZ-UP LABEL-A 1-ANGLE-MARK
pair of segments	rel. lengths rel. extents rel. orientation	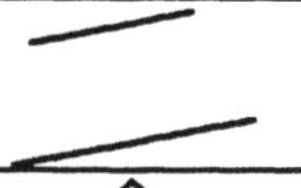	1<2 DISJOINT PARALLEL
triangle	number of sides marked no. of angles marked no. of interior lines basic shape	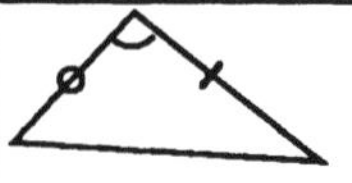	TWO-MARKED-SIDES ONE-MARKED-ANGLE 0-INTERIOR-LINES OTHER-TRIANGLE-SHAPE
triangle pair	symmetry rel. sizes rel. extents	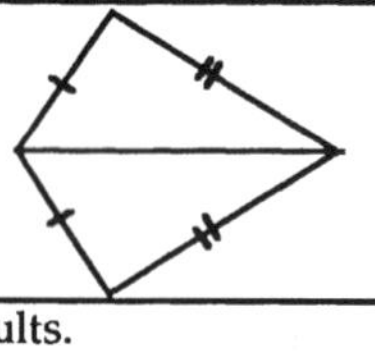	T/B-SYMMETRIC APPROX-SAME-SIZE SHARED-SIDE

Figure 2: Some of POLYA's visual targets and visual results.

Visual actions

Currently, POLYA has three types of actions for shifting its focus from one target to another: FIND, LOOK-AT, and COMPARE. All three shift the focus and return a description (visual result) of the target object(s). FIND shifts the focus to a hypothesized object whose description is (partly) known but whose location is not known. FIND may return nil if no object matching the description exists in the diagram. LOOK-AT shifts the focus to an object whose location is known but whose description is not known. COMPARE shifts the focus to pairs of objects. FIND-MARKED-SEGMENT, LOOK-AT-LOWER-LEFT-VERTEX, and COMPARE-TRIANGLES are instances of the three action types.

As one reasonableness criterion, we intend that POLYA should be able to perform its visual actions directly on a bitmap diagram. Currently, however, POLYA computes its visual results from coordinate listings of points, lines, and segment and angle marks.

Geometric planning knowledge

POLYA's planning knowledge of where to look in a diagram, as well as its knowledge of how to write proofs, is contained in geometry *scripts*. These are modelled after the scripts described in [Schank & Abelson, 1977] and implemented in SAM for understanding newspaper stories [Cullingford 1978]. Scripts store routine actions and sequences of events so that, once a script has been selected, inferencing is tightly controlled.

POLYA has two kinds of scripts: *search scripts* and *proof scripts*. Both kinds of scripts have one or more *triggering sets*, sketchy lists of perceptual features which suggest the relevance of the script. A script becomes *suggested* when the accumulated perceptions contain all elements of the triggering set.

Search scripts

Search scripts direct the focus of attention to potentially salient parts of the diagram. They themselves do nothing with the visual results; their purpose is to gather the features needed to trigger proof scripts. The predictions usually serve only as a check on the success of the visual actions. ISOSC-LEGS search script (figure 3) directs the focus to the legs of a triangle which appears to be isosceles. POLYA has another, similar script which directs the focus to the base angles. The two scripts represent POLYA's knowledge of the important parts of isosceles triangles.

Proof scripts

Proof scripts may correspond to formal geometric facts—axioms, theorems, and properties—or to parts of complete proofs POLYA has seen. They

ISOSC-LEGS search script

Triggering set
DIAGRAM [left-right symmetry]
TRIANGLE [type = ISOSC-UP]

Sequence
1. Action = [look at left side of isosc. triangle]
   Prediction = [generic segment]
2. Action = [look at right side of isosc. triangle]
   Prediction = [generic segment]
3. Action = [compare the two segments]
   Prediction = [segments share common endpoint]

Figure 3: A script to look at isosceles triangles.

are similar to the Diagram Configuration schemas in DC. However, the proof-writing knowledge of POLYA's proof scripts is very specific and uni-directional, whereas DC's schema may package multiple rules for forward or backward inferencing.

Proof scripts have four parts:

1. A triggering set.
2. A *verification sequence.*
3. A *template-filler sequence* for locating objects needed for the proof.
4. A *proof template* .

Once activated by a match against the triggering set, proof scripts operate in three phases. First, the verification sequence checks that the relevant objects in the diagram are in the proper configuration for the proof script to be valid. Second, the template-filler sequence looks again at the diagram to determine variable bindings for the template. Third, the proof script instantiates its template with the variable bindings from the previous step.

The SSS-SHARED-SIDE proof script (figure 4) can prove that two triangles are congruent to each other if two pairs of sides are congruent and if they share a third side (see the triangle-pair example in figure 2). It is triggered on the basis of a triangle visual result in which two of the triangle's sides are marked, and a triangle-pair result in which the two triangles share a side. Because these results may be separated in time, there is no guarantee that they have anything to do with each other, so the first actions of the SSS-SHARED-SIDE script verify that each of the triangles which share a side have two sides marked. Additional actions locate the corresponding pairs of marked sides, and the shared unmarked side.

A complete proof generally requires more than one proof script, and is thus a composite of

sss-shared-side proof script

Triggering set
DIAGRAM [some symmetry]
TRIANGLE [2 sides marked]
TRIANGLE-PAIR [shared-side]

Verification sequence
1. Action = [look at triangle1 of triangle-pair]
   Prediction = [2 sides marked]
2. Action = [look at triangle2 of triangle-pair]
   Prediction = [2 sides marked]
3. Action = [look at shared side of triangle-pair]
   Prediction = [unmarked segment]

Template-filler-sequence
1. Action = [look at triangle1 of triangle-pair]
   Prediction = [2 sides marked]
   Bind-to: ?TRIANGLE1
2. Action = [look at triangle2 of triangle-pair]
   Prediction = [2 sides marked]
   Bind-to: ?TRIANGLE2
3. Action = [find marked-side of ?triangle1]
   Prediction = [marked-segment]
   Bind-to: ?SIDE1A
4. Action = [look at symmetric partner of ?SIDE1A]
   Prediction = [marked-segment]
   Bind-to: ?SIDE2A
5. Action = [compare ?SIDE1A & ?SIDE2A]
   Prediction = [seg-pair similarly marked]
   Bind-to: —

   ....

Proof template
1. Statement = (?SIDE1A = ?SIDE2A)
   Reason = "As marked"
2. Statement = (?SIDE1B = ?SIDE2B)
   Reason = "As marked"
3. Statement = (?SHARED-SIDE = ?SHARED-SIDE)
   Reason = "Reflexive property (shared side)."
4. Statement = (?TRIANGLE1 = ?TRIANGLE2)
5. Reason = "SSS with shared side."

Figure 4: A script for proving two triangles congruent, if they have two sides congruent and share a third side.

several instantiated templates. POLYA currently lacks a mechanism for organizing steps from multiple templates in logical order.

Script selection

When more than one script is suggested, which is frequently the case, POLYA chooses a script essentially at random, except that preference is given to proof scripts. This is adequate for the current model, since the order in which perceptions are gathered does not affect the final solution.

Representing examples with scripts

Scripts capture the visual search and proof-writing knowledge implicit in examples and sample problems. Because this knowledge is rarely, if ever, explicitly taught, we have relied on a careful study of textbook examples to tell us what POLYA's scripts should contain.

When the triangle congruence theorems are introduced in one text [Rhoad et al., 1986], the diagrams in the examples and first several problems emphasize two types of patterns: triangles with two parts marked (sides, angles, or a combination), and triangles with three parts marked. The associated skills are, respectively, to identify what third side or angle would have to be marked for a particular theorem to apply (always two possible theorems), and to identify which one theorem (if any) applies to the given case.

To represent these skills, POLYA has search scripts whose triggering sets consist of triangle-visual-results with two angles marked, or two sides marked, or one side and one angle. They focus attention on the additional side or angle which would be needed for a particular theorem, and compare that object with its symmetric partner in the other triangle. The proof scripts for this section are triggered on the basis of three marks (or two marks and some significant object-pairing). Their verification sequences check that the marked objects are in the proper configuration.

An example

One problem POLYA can recognize is shown in figure 5, a textbook problem from a section which introduces isosceles triangles. To recognize this problem and write its proof requires four search scripts and two proof scripts. The search scripts are a script which looks at the dominant triangle, ISOSC-LEGS and ISOSC-ANGLES, discussed earlier, and CONGRUENT-TRIANGLES. CONGRUENT-TRIANGLES focuses POLYA's attention on the marked triangles in the lower-left and lower-right corners.

The features collected by those four search scripts complete the triggering sets for two proof scripts: SSS-SIMPLE and a proof script specific to this problem (3.6-PROBLEM-3). Of the two proof scripts, POLYA arbitrarily chooses to run 3.6-PROBLEM-3. As its verification sequence, this

Given: $\overline{WS} = \overline{TZ}$
$\overline{WX} = \overline{YZ}$
$\overline{SX} = \overline{TY}$

Prove: $\overline{WR} = \overline{RZ}$

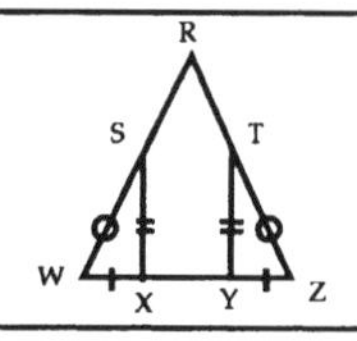

Figure 5: The example.

script COMPAREs the marked triangle with the dominant isosceles triangle, predicting (correctly) that they will share an angle.

The template contains two proof steps related to two geometry rules:

1. Statement = (?BASE-ANGLE1 = ?BASE-ANGLE2)
   Reason = "Corresponding parts of congruent triangles are congruent."
2. Statement = (?SIDE1 = ?SIDE2)
   Reason = "If two angles of a triangle are congruent, then the sides opposite them are congruent."

The sequence of steps for binding the template symbols is:

1. LOOK-AT-BASE-ANGLE1 → ?BASE-ANGLE1
2. (similarly for ?BASE-ANGLE2)
3. LOOK-AT-ISOSC-LEG1 → ?SIDE1
4. (similarly for ?SIDE2)

Step 1 in the template assumes that the smaller triangles are congruent. This has not yet been proven, but soon will be.

Now POLYA runs the SSS-SIMPLE proof script, which represents an iconic example of side-side-side triangle congruence. The template for SSS-SIMPLE looks like this:

1. Statement = (?SIDE1A = ?SIDE2A)
   Reason = "As marked"
2. Statement = (?SIDE1B = ?SIDE2B)
   Reason = "As marked"
3. Statement = (?SIDE1C = ?SIDE2C)
   Reason = "As marked"
4. Statement = (?TRIANGLE1 = ?TRIANGLE2)
   Reason = "SSS"

This template is filled in with the actions LOOK-AT-SIDE1, LOOK-AT-SIDE2, and LOOK-AT-SIDE3 on one of the corner triangles, and using FIND-SYMMETRIC-SEGMENT to locate the corresponding side in the other triangle. This template completes the proof for this problem.

Discussion

POLYA is not really "solving" the problem above, but merely recognizing it as a problem for which it knows the solution. This is the simplest type of reasoning from examples. Simple or not, however, the example shows that POLYA's visual vocabulary is adequate for representing both general patterns and specific solutions. The example also suggests that search scripts can be used for efficient diagram parsing. Finally, the example shows shows how multiple search scripts and proof scripts can interact to recognize and write the proof of a known problem.

Conclusion

This paper has presented a new model of geometry theorem-proving consistent with George Polya's steps of problem-solving. POLYA constructs a proof plan and carries it out through visual search and recognition. To support this behavior we have defined a visual system—a vocabulary of visual targets, results, and actions—for representing and interacting with diagrams. We have shown how search scripts and proof scripts can be used to direct a constrained focus of attention for efficient visual parsing and for writing proofs. We have described an algorithm that allows smooth interaction of search scripts and proof scripts.

POLYA's visual system is still evolving, especially the vocabulary of visual actions. As we add examples we find that new actions are required, or at least new ways of describing them. Generally the new actions stem directly from new geometric concepts. For instance, to handle isosceles triangles requires knowledge of how to focus on the base angles. A full domain knowledge of geometry will comprise a large set of such context-specific visual actions.

We are anxious to test our model on a much larger set of examples with a much larger set of scripts. We expect that additional examples will reveal gaps in our vocabulary of visual actions. We expect the set of visual actions to grow quickly for awhile, then level off to a stable vocabulary of the visual skills required to parse plane geometry diagrams.

Thus far the emphasis of our research has been on reasoning from the diagram; POLYA currently ignores the goal in the problem input. However, we plan to incorporate some goal-directed reasoning in POLYA. Certainly this is something people do, and POLYA will need it to solve more complicated problems.

Finally, we are interested in the expanding POLYA into a case-based system which can learn new cases in response to difficulties experienced during problem-solving.

Acknowledgements

Many thanks to Paul Schiffer for his fine-tooth editing of earlier drafts of this paper.

References

Ballard, D. H. (1991). "Animate vision." *Artificial intelligence*, Vol. 48.

Carpenter, R. (1988) *Movements of the eyes.* Pion.

Clark, J. J. & Ferrier, N. J. (1988). "Modal control of an attentive vision system." *Proceedings,* International Conference on Computer Vision.

Cullingford, R. (1978). *Script application: Computer understanding of newspaper stories.* Ph.D. Dissertation, Research Report #116. Computer Science Dept., Yale University.

Koedinger, K. R. and Anderson, J. R. (1990). "Abstract planning and perceptual chunks: Elements of expertise in geometry." *Cognitive Science*, **14**, 511-550.

McDougal, T. (1988) "A computational model for the structural comparison of secondary plane geometry problems." M.A.T. Thesis, University of Chicago.

Polya, G. (1957). *How to solve it: A new aspect of mathematical method, 2nd Ed.* Princeton University Press.

Rhoad, R., Whipple, R., and Milauskas, G. (1986) *Geometry for enjoyment and challenge.* McDougal, Littell.

Rojer, A.S. and Schwartz, E. L. (1990) "Design considerations for a space-variant visual sensor with complex-logarithmic geometry." *Proceedings*, Int'l Conference on Pattern Recognition.

Schank, R., Abelson, R. (1977). *Scripts, plans, goals, and understanding.* Lawrence Erlbaum.

Swain, M. J. (1991). "Low resolution cues for guiding saccadic eye movements." *SPIE advances in intelligent robot systems.*

Tistarelli, M. & Sandini, G. (1990). "On the estimation of depth from motion using an anthropomorphic visual sensor." *Proceedings*, European Conference on Computer Vision.

Simulating Theories of Mental Imagery

Janice Glasgow and Darrell Conklin
Department of Computing and Information Science
Queen's University, Kingston
Canada, K7L 3N6
Email: {janice,conklin}@qucis.queensu.ca

Abstract

A knowledge representation scheme for computational imagery has previously been proposed. This scheme incorporates three interrelated representations: a long-term memory descriptive representation and two working memory representations, corresponding to the distinct visual and spatial components of mental imagery. It also includes a set of primitive functions for reconstructing and reasoning with image representations. In this paper we suggest that the representation scheme addresses the controversy involved in the imagery debate by providing the computational tools for specifying, implementing and testing alternative theories of mental imagery. This capability is illustrated by considering the representation and processing issues involved in the mental rotation task.

Introduction

A concept of computational imagery has previously been proposed as a reasoning paradigm in artificial intelligence (Papadias & Glasgow, 1991). The knowledge representation scheme developed for computational imagery can also serve as a tool for specifying, implementing and testing cognitive theories of imagery. Unlike Kosslyn's (1980) computational model, which was designed to support a particular theory, the representations and functions of computational imagery can be used to simulate and analyze alternative, and possibly conflicting, theories of mental imagery.

Computational imagery involves techniques for visual and spatial reasoning, where images are generated or recalled from long-term memory and then manipulated, transformed, scanned, associated with similar forms, increased or reduced in size, distorted, etc. In particular, it is concerned with the reconstruction of image representations to facilitate the retrieval of information that was not explicitly stored in long-term memory. The image representations generated to retrieve this information may correspond to real physical scenes or to abstract concepts that are manipulated in ways similar to visual forms.

The knowledge representation scheme for computational imagery separates visual from spatial reasoning and defines independent representations for the two modes (Glasgow & Papadias, 1992). Whereas visual thinking is concerned with *what* an image looks like, spatial reasoning depends more on *where* an object is located relative to other objects in an image. Each of these representations is constructed, as needed, from a descriptive representation stored in long-term memory. Thus, the scheme includes three representations, each appropriate for a different kind of processing. An image is stored in long-term memory as a hierarchically organized, descriptive, *deep representation* that contains all the relevant information about the image. The *spatial representation* of an image denotes the image components using a symbolic array that preserves relevant spatial and topological properties. This array data structure also preserves the hierarchical structure of an image through nested symbolic arrays, which can be used to specify and reason about images at varying levels of the decomposition hierarchy. The *visual representation* depicts the space occupied by an image as an occupancy array. It can be used to retrieve information such as shape, relative distance and relative size. While the deep representation is used as a permanent store for information, the spatial and visual representations act as working (short-term) memory stores for images.

Components of an image may be grouped into features and stored based on their topological relations, such as adjacency or containment, or their spatial relations, such as above, beside, north-of, etc. Because of the relevance of storing and reasoning about such properties of images, the knowledge representation scheme for computational imagery is based on a formal theory of arrays. Similar to set theory, array theory (More, 1979) is concerned with the concepts of nesting and aggregation. It is also concerned with the notion that objects have a location relative to other objects in an array. Several primitive functions, which are used to retrieve, construct and transform representations of images, have been specified in the theory and mapped into the functional programming language Nial (Jenkins, Glasgow & McCrosky, 1986). These functions, along with the primitive functions for computational imagery, provide a general framework for specifying theories of mental imagery.

Computational Imagery

Although no one seems to deny the existence of the phenomenon called imagery, there has been an ongoing debate about the structure and the function of imagery in human cognition. The imagery debate is concerned with whether images are represented as *descriptions* or *depictions*. It has been suggested that descriptive representations contain symbolic, interpreted information, whereas depictive representations contain geometric, uninterpreted information (Finke, Pinker & Farah, 1989). Others debate whether or not images play any causal role in the brain's information processing (Block, 1981).

Pylyshyn (1981), a forceful proponent of the descriptive view, argues that mental imagery simply consists of the use of general thought processes to simulate perceptual events, based on tacit knowledge of how these events happened. He disputes the idea that mental images are stored in a raw uninterpreted form resembling mental photographs and argues for an abstract format of representation called propositional code. Kosslyn's (1980) model of mental imagery is based on a depictive theory which claims that images are quasi-pictorial; that is, they resemble pictures in several ways but lack some of their properties. Hinton disputes the picture metaphor for imagery and claims that images are more like generated constructions (Hinton, 1979). In this approach, as in Marr's (1982) *3D* model, complex images can be represented as a hierarchy of parts.

A primary objective of research in cognitive science is to study and explain how the mind functions. One aspect of work in this area is the theory of computability. If a model is computable, then it is usually comprehensible, complete and available for analysis; implementations of theories can be checked for sufficiency and used to simulate new predictive results. In a discussion of the issues of computability of cognitive theories for imagery, Kosslyn (1980) expresses frustration with existing implementation tools and goes on to state that "The ideal would be a precise, explicit language in which to specify the theory and how it maps into the program." Array theory, combined with the primitive functions and representations for computational imagery, provides such a meta-language. Moreover, it allows us to represent an image either visually or spatially, and provides for the implementation and testing of cognitive theories.

In Kosslyn's computational theory for imagery, images have two components: a surface representation (a quasi-pictorial representation that occurs in a visual buffer), and a deep representation for information stored in long-term memory. Similar to Kosslyn, we consider a separate long-term memory model for imagery, which encodes visual information descriptively. Unlike Kosslyn, we consider the long-term memory to be structured according to the decomposition and conceptual hierarchies of an image domain. Thus we use a semantic network model, implemented using frames, to describe the properties of images. The long-term memory model in Kosslyn's theory is structured as sets of lists of propositions, stored in files.

The surface representation in Kosslyn's theory has been likened to spatial displays on a cathode ray tube screen; an image is displayed by selectively filling in cells of a two-dimensional array. Our scheme for representing images in working memory is richer in three fundamental ways. First, we treat images as inherently three-dimensional, although two-dimensional images can be handled as special cases. As pointed out by Pinker (1988), images must be represented and manipulated as patterns in three-dimensions, which can be accessed using either an object-centered or a world-centered coordinate system. Second, we consider two working memory representations, corresponding to the visual and spatial components of mental imagery. Finally, just as the long-term memory stores images hierarchically, the visual and spatial representations use nested arrays to depict varying levels of resolution or abstraction of an image. While the functionality of many of the primitive operations for computational imagery were initially influenced by the processes defined for Kosslyn's theory, their implementation varies greatly because of the nature of the image representations.

An important distinction between our approach to computational imagery and Kosslyn's computational theory is the underlying motivation behind

a) Visual representation

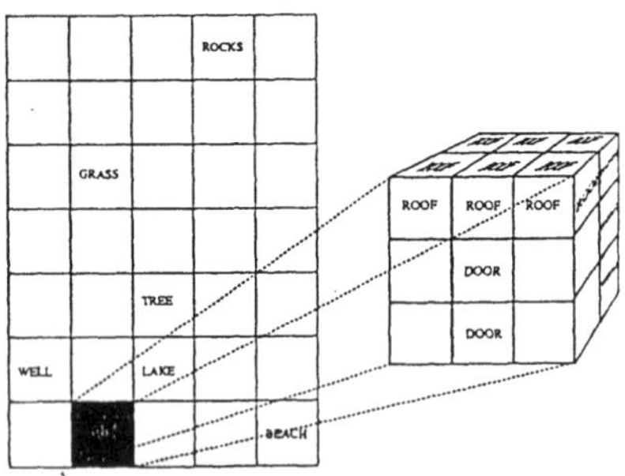

b) Spatial representation

Figure 1: Representations for computational imagery

the two pieces of work. Kosslyn's model was initially developed to simulate and test a particular theory for mental imagery. Although in its initial development the representations and processes of computational imagery were motivated by efficiency and expressiveness concerns, it was also designed to provide the predictive and explanatory power necessary to model alternative theories of cognition.

As an illustration of the visual and spatial representations for computational imagery, consider the island map used by Kosslyn to investigate the processes involved in mental image scanning. Figure 1 presents an occupancy array visual depiction of such a map, as well as a symbolic array spatial representation that preserves the properties of closeness (expressed as adjacency) and relative location of the important features of the island. It can also be used to depict and reason about the hierarchical structure of images (see subimage for hut) using nested data structures. Consider addressing such questions as those involved in mental image scanning and focusing of attention. Although both representations provide access to information about relative location and could be used to simulate scanning processes, they would predict different results in terms of time complexity.

The hypothesis of distinct visual and spatial image components of mental imagery is suggested by results of studies of visually impaired patients (e.g. Kosslyn 1987). It is further supported by the conflicting results of studies dealing with whether size or relative distance is preserved in the representation of a mental image. This controversy could be attributed to the hypothesis of multiple image representations. Thus, as well as testing theories involving individual representations, computational imagery can be used to analyze theories that involve multiple representations.

It is worth noting here that the spatial representation for computational imagery is not just an interpreted low resolution version, or approximation, of the visual representation of an image. The symbolic array may discard, not just approximate, irrelevant visual information. It may also abstract away details by using nested arrays to depict image components at varying levels of the structural hierarchy.

Another approach to visual reasoning was presented by Funt (1980), who represented the state of the world as a diagram, and actions in the world as corresponding actions in the diagram. Similar to Kosslyn, Funt uses two–dimensional arrays to denote his visual images. A more recent model describes how visual information can be represented within the computational framework of discrete symbolic representations in such a way that both mental images and symbolic thought processes can be explained (Chandrasekaran & Narayanan, 1990). While this model allows a hierarchy of descriptions, it is not spatially organized.

The subjective nature of mental imagery has made it a difficult topic to study experimentally; many qualities of an image are not directly observable and may differ from one person to another. In fact, it has been argued by some researchers that it is impossible to resolve the imagery debate experimentally (Anderson, 1978). The difficulties involved in psychological studies emphasize the importance of computer models for mental imagery. While computational imagery does not imply a particular model, it does provide the tools for formally analyzing a variety of theories, and thus may contribute to resolving the imagery debate.

Theories of Rotation

To illustrate the use of computational imagery for studying cognitive theories of imagery, we consider the problem of mental rotation. Although empirical observations suggest that rotation involves an object representation being moved through intermediate orientations (Shepard & Cooper, 1982), an as yet unresolved issue is the actual nature of the representation used. One possibility is a visual depiction of the object which preserves detailed three-dimensional shape information. An alternative approach is one in which the object is represented as vectors corresponding to the major skeletal axes of the object (Just & Carpenter, 1985). This type of representation can be considered as spatial in nature; it preserves connectivity of parts but discards detailed surface information about the image.

A visual representation of an object can be used to test theories of mental rotation where an object is incrementally shifted through some medium. The medium, such as occupancy arrays (Kosslyn, 1980) or concentric rings in a simulated retina structure (Funt, 1983), supports visual inferences and transformations. An object, or a portion of an object, is rotated by shifting its position in the medium. As Funt illustrates, this can be done in parallel using message passing between adjacent cells in a retinal diagram. In Kosslyn's rectangular model, objects will be deformed as they are incrementally rotated, and the granularity of the medium is one of the determining factors of rotation time. This approach can similarly be implemented using a three-dimensional occupancy array in our representation scheme.

The spatial representation of an object preserves important relations among parts, and is pseudo-visual in that it does not necessarily preserve relative distance or metric information. A theory of rotation using symbolic arrays would have an encoding followed by a comparison process. First, an object is rotated into a Cartesian axis-aligned orientation. It is then scanned to extract the skeletal axes, which are labeled and encoded into a symbolic array. Figure 2 shows the partial encoding grammar for simple two-dimensional orthogonal figures. On the left is a visual depiction of such a figure, in the middle its labeled skeletal axes, and on the right the symbolic spatial array. For brevity we do not show the encoding for all four orientations of each stick figure. Encoding visual objects using this grammar

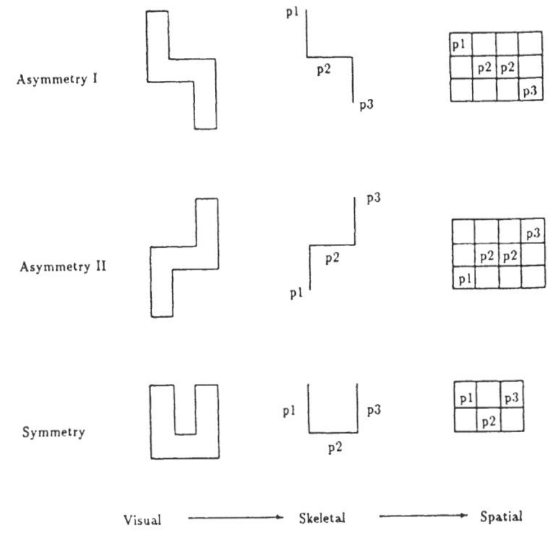

Figure 2: Encoding connected regions of a visual object in a symbolic array.

preserves such important relations as connectivity and ternary symmetry. Figure 3 shows the encodings for more complex visual objects. In this figure, (A) and (B) denote the same image at varying orientations, while (C) is a mirror image of the first two figures.

Two symbolic arrays are considered equivalent under a particular relation if there is a relation homomorphism between them. Informally, this is a function, f, which maps the parts of one symbolic array to the parts of another such that if parts are related in one array, the function maps them to parts which are similarly related. For example, two symbolic arrays $A1$ and $A2$ are equivalent under the relation of connectivity if and only if there exists a one-to-one mapping f between the components of $A1$ and $A2$ such that for all a, b in $A1$, $connected(a, b) \leftrightarrow connected(f(a), f(b))$. Spatial relations are determined simply by computationally inspecting the symbolic array representation. Consider the two symbolic arrays in (A) and (B) of Figure 3. There exists a mapping $f = \{(p1, q5), (p2, q4), (p3, q3), (p4, q2), (p5, q1)\}$ which preserves the relation of connectivity. This mapping also preserves the ternary symmetry relations specified in Figure 2. Although the array in (C) is equivalent under the connectivity relation to the other two, there does not exist a mapping in which the symmetry relations are preserved.

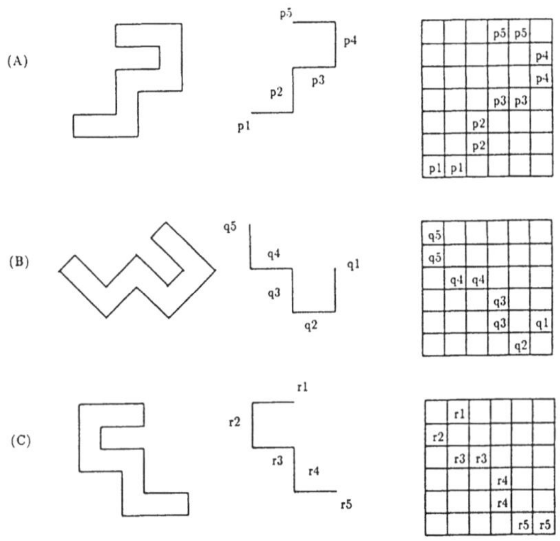

Figure 3: Spatial encodings for visual objects

For this simple class of orthogonal figures, we rephrase the problem of rotation as one of determining whether two images are equivalent under the relations of symmetry and connectivity. The interesting point is that testing two spatial representations for equivalence under these relations involves only identifying a terminal segment for each object, tracing the connectivity of the object in the symbolic array, while checking that symmetry is maintained. It involves no explicit rotation or transformation of the arrays. Note that the symbolic arrays for (A) and (B) in Figure 3 satisfy the conditions for rotation, whereas the array in (C) is not spatially equivalent to the other two arrays.

This spatial rotation method might explain the comparison of *orientation-free* descriptions reported by Just and Carpenter (1985), who hypothesized, based on experimental results, that rotation can sometimes be performed without explicit object manipulation. In their experiments on the comparison of cubes, a subject using an orientation free description strategy exhibited rotation times more or less constant and independent of the amount of rotation involved.

In this section we have presented one example illustrating the use of the spatial representation for computational imagery to explain a mental rotation result. Other issues we could use our knowledge representation scheme to address include whether objects are rotated in whole or in parts, parallel versus sequential process models, axis or task–defined axes for visual objects (Just and Carpenter, 1985), and the use of hierarchical encoding schemes (Anderson, 1983). Similarly, we could examine alternative theories for tasks involving mental scanning, creative imagery and the relationship between imagery and vision.

Discussion

Previous research in imagery has focussed on two issues: the nature of the image representation and the function of imagery in tasks such as rotation, scanning and language processing. The proposed knowledge representation scheme for computational imagery allows us to address both of these interrelated issues; image representations are data structures which can be considered in conjunction with, or distinctly from, the functions that operate on them. The multiple representations of the scheme provides a computational framework for considering images as propositions stored in long–term memory, or as working memory visual or spatial depictions. Using the primitive functions for computational imagery, we can study the relationship between mind and computer by implementing and testing algorithms that simulate theories of mental imagery.

The knowledge representation scheme for computational imagery also provides a basis for implementing programs that involve reconstructing and reasoning with image representations (Glasgow, 1990). In particular, it can be used to develop knowledge–based systems for problems that involve mental imagery when solved by humans. One such system, which is currently under development, is an application to the problem of molecular scene analysis (Glasgow, Fortier & Allen, 1991). This knowledge–based system combines tools form the areas of protein crystallography and molecular–database analysis, through a computational imagery framework. The spatial representation for computational imagery has also been applied to the problem of understanding the use of diagrams and visual analogies in the development of Dalton's atomic theory (Thagard & Hardy, 1992).

Computational imagery is currently being considered in the domain of machine learning (Conklin & Glasgow, 1992). In this research, the spatial representation for imagery is used to develop a theory of subsumption and similarity, which provides a formal model for structure–based classification in the long–term memory model of images.

In conclusion, the underlying mathematics for computational imagery satisfies Kosslyn's (1980) ideal by providing a precise and explicit language for specifying theories of mental imagery. Visual and spatial representations are implemented as arrays and manipulated using the primitive functions of computational imagery, which themselves are expressed as array theory operations. Finally, the primitives of array theory, computational imagery and particular theories of mental imagery can all be directly mapped into programs which run without the "kluges" and "ad hoc manipulations" faced by Kosslyn in the development of his computational theory.

References

Anderson, J.R. 1978. Arguments concerning representations for mental imagery. *Psychological Review,* 85, 249-277.

Block, N., editor 1981. *Imagery.* MIT Press.

Chandrasekaran, B. & Narayanan, N. H. 1990. Integrating imagery and visual representations. *Proceedings of the 12th Annual Conference of the Cognitive Science Society.* Lawrence Erlbaum Associates, 670-677.

Conklin, D. & Glasgow, J.I. 1992. Spatial analogy and subsumption. In Sleeman & Edwards (Eds.), *Machine Learning: Proceedings of the Ninth International Conference (ML92).* Morgan Kaufmann.

Finke, R.A., Pinker, S. & Farah, M.J. 1989. Reinterpreting visual patterns in mental imagery. *Cognitive Science,* 13, 51-78.

Funt, B.V. 1980. Problem-solving with diagrammatic representations. *Artificial Intelligence,* 13, 201-230.

Funt, B.V. 1983. A parallel-process model of mental rotation. *Cognitive Science,* 7, 67-93.

Glasgow, J.I. 1990. Artificial intelligence and imagery. *Proceedings of the Second IEEE Conference on Tools for AI.*

Glasgow, J.I., Fortier S. & Allen F.H. 1991. Crystal and molecular structure determination through imagery, In Hunter (Ed.), *Artificial Intelligence and Molecular Biology,* AAAI Press, Forthcoming.

Glasgow, J.I. & Papadias, D. 1992. Computational Imagery, *Cognitive Science*, Forthcoming.

Hinton, G. 1979. Some demonstrations of the effects of structural descriptions in mental imagery, *Cognitive Science*, 3, 231-250.

Jenkins, M.A., Glasgow, J.I. & McCrosky, C. 1986. Programming styles in Nial, *IEEE Software,* 86, 46-55.

Just, M.A. & Carpenter, P.A. 1985. Cognitive coordinate systems: Accounts of mental rotation and individual differences in spatial ability. *Psychological Review*, 92, 137-172.

Kosslyn, S.M. 1980. *Image and mind.* Harvard University Press.

Kosslyn, S.M. 1987. Seeing and imagining in the cerebral hemispheres: A computational approach. *Psychological Review,* 94, 148-175.

Marr, D. 1982. *Vision: a computational investigation in the human representation of visual information.* San Francisco: Freeman.

More, T. 1979. The nested rectangular array as a model of data. From proc. APL79, *APL Quote Quad,* 9.

Papadias, D. and Glasgow, J.I. 1991. A knowledge representation scheme for computational imagery. *Proceedings of the 13th Annual Conference of the Cognitive Science Society.* Lawrence Erlbaum Associates.

Pinker, S. 1988. A computational theory of the mental imagery medium. In Denis, Engelkamp & Richardson (Eds), *Cognitive and Neuropsychological Approaches to Mental Imagery.* Martinus Nijhorff Publishers, 17-36.

Pylyshyn, Z.W. 1981. The imagery debate: analogue media versus tacit knowledge. *Psychological Review,* 88, 16-45.

Shepard, R.N. & Cooper, L.A. 1982. *Mental images and their transformations.* MIT Press.

Thagard, P. & Hardy, S. 1992. Visual thinking in the development of Dalton's atomic theory, *Proceedings of AI '92.*

Fractal (Reconstructive Analogue) Memory

David J. Stucki and Jordan B. Pollack
Laboratory for Artificial Intelligence Research
Department of Computer and Information Science
The Ohio State University
Columbus, OH 43210
stucki@cis.ohio-state.edu
pollack@cis.ohio-state.edu

Abstract

This paper[1] proposes a new approach to mental imagery that has the potential for resolving an old debate. We show that the methods by which fractals emerge from dynamical systems provide a natural computational framework for the relationship between the "deep" representations of long-term visual memory and the "surface" representations of the visual array, a distinction which was proposed by (Kosslyn, 1980). The concept of an iterated function system (IFS) as a highly compressed representation for a complex topological set of points in a metric space (Barnsley, 1988) is embedded in a connectionist model for mental imagery tasks. Two advantages of this approach over previous models are the capability for topological transformations of the images, and the continuity of the deep representations with respect to the surface representations.

The Imagery Debate

The phenomena of mental imagery is widely disputed among cognitive scientists primarily because it occupies a position in the boundary between perception and cognition. On the one hand, mental images seem to be purely symbolic structural descriptions that are independent of any perceptual mechanisms. In this way images are no different from any other knowledge structures and therefore require no special purpose mechanisms, but can be reasoned about and operated on in the traditional propositional fashion. On the other hand, mental images seem to be represented by a special-purpose cognitive architecture that shares components with the visual perceptual system. Under this approach, additional mechanisms must be proposed for inspecting, transforming, and reasoning about images, providing a means for translating between purely symbolic representations and the "visual buffer".

This latter view that mental imagery is performed in an analogue, "pictorial" medium began to regain acceptance in the early 1970s as a result of empirical studies which indicated both that mental imagery belongs to a different modality than language and that there are cognitive tasks in which mental imagery is brought into play when symbolic reasoning and explicit knowledge is insufficient for solving the problem.

As an example of the former, an experiment designed by Lee Brooks (Brooks, 1968), required subjects to imagine a block letter and report whether successive corners were at the extreme top or bottom of the letter. The experiment showed that visually oriented responses (i.e., pointing to the letters Y or N) took longer than verbal responses (saying 'yes' or 'no') implying that the visual response task was interfering with the imagery task. In a similar experiment (also Brooks, 1968), he asked the subjects to report whether successive words in a sentence were nouns. In this case, verbal responses were slower than visually oriented responses. Brooks' conclusion was that mental imagery is distinct from verbal processes, and shares processing resources with the visual perceptual system.

The most famous experiment illustrating the latter was performed by Roger Shepard and Jacqueline Metzler on the mental rotation (Shepard & Metzler, 1971). When presented with pairs of drawings of three dimensional shapes at differing orientations that were either identical or mirror images, the subjects were to report whether the objects had the same shape, independent of any difference in orientation. They found that the response times varied linearly with the difference of angular rotation of the objects, which implied that the subjects were performing a sort of "mental rotation" in order to solve the problem.

Although extensive contributions have been made by many researchers to the theory of analogue imagery (see Finke, 1989; Chandrasekaran & Narayanan, 1990; Tye, 1991), most of its essential qualities have been incorporated into a single framework described by Stephen Kosslyn (Kosslyn, 1981; Kosslyn, 1980). The primary notion in Kosslyn's theory is that the representations of mental images are quasi-pictorial, or "picture-like". This means that in some way the representation preserves some of the topological or spatial properties of the objects being represented, by embedding these relation-

[1] This research has been partially supported by the Office of Naval Research grant N00014-92-J-1195.

ships in the architectural and functional medium of the representational mechanism. Kosslyn essentially recognizes two structural components and two kinds of processes in any encoding system. The constructs, which he calls the *representation* and the *medium*, are data structures with the medium serving as a host to the representation. For example, when a circular queue is implemented in a computer program as an array, the array is the medium in which the queue is represented. It is also important to note that the word "representation" is being used to denote both the entire encoding system and the data structure for a specific object, the usage being determined from context. The processes which operate within the encoding systems are either for making comparisons between representations, or parts of representations, or for transforming them in various ways (including the generation of new representations in place of old ones).

As in the distinction between sound and meaning in language, there is a distinction between image and meaning in mental imagery. Indeed, Kosslyn borrowed the linguists' terms *surface* and *deep* representations to capture this dichotomy. Surface representations are the analogue, pictorial images which we attribute to the "mind's eye". Deep representations are long term (symbolic) memories that can be used to "display" images on the *visual buffer*, which serves as the viewing screen for the mind's eye. The visual buffer is analogous to the memory a computer uses to store a bitmap for its display monitor. This memory serves as a *functional* coordinate space, since while the mapping from the bits to the pixels must preserve the coordinates—which must also be respected by any processes accessing the structure—there is no constraint that the individual bits be physically contiguous.

Although the nature of the surface representations has been specified in great detail in Kosslyn's model, the details of the deep representations have not been as well developed. The theory suggests two types of deep representation for visual images, called literal and propositional. The literal representations are intended to consist of information about what an object looked like, without any reference to coordinate spaces, but Kosslyn has been unable to formulate an appropriate representation and medium that is relevant to the theory:

> "We have not as yet made any strong claims about the precise format of the underlying literal encodings." (Kosslyn, 1981)

The propositional encodings are simply assertions used to describe the properties and features of an object, which presumably can be manipulated by mechanized logic. These representations are governed by syntax driven rules for interpretation and manipulation that are independent of the semantics of their values. The interpretation of a representation is based on the truth-value assigned to it under these rules.

Zenon Pylyshyn has been the most vocal opponent of the analogue approach to mental imagery (Pylyshyn, 1981; Pylyshyn, 1973). However, rather than suggesting an alternative, new theory, Pylyshyn questions the necessity of abandoning the traditional theory of cognitive processing as a physical symbol system of functional architectures:

> "In my view, however, the central theoretical question in this controversy is whether the explanation of certain imagery phenomena requires that we postulate special types of processes or mechanisms, such as ones commonly referred to by the term *analogue*.... whether certain aspects of cognition, generally (though not exclusively) associated with imagery, ought to be viewed as governed by tacit knowledge...or whether they should be viewed as intrinsic properties of certain representational media or of certain mechanisms that are not alterable in nomologically arbitrary ways by tacit knowledge." (Pylyshyn, 1981)

In order to substantiate his feeling that the answer to the first question is 'no,' Pylyshyn attacks the analogue position on two fronts.

The first attack consists of several specific criticisms of the Kosslyn model. He claims that a theory can not serve as a principled or constrained account of mental imagery if it is not substantive (explanatory), or if it is *ad hoc*, or if it has too many degrees of freedom (free parameters). He asserts that the analogue theory fails on all three counts. He does not claim that the analogue position is wrong in principle, but simply that none of the theories advanced so far have satisfied the conditions necessary for a principled account. The crux of the failure of Kosslyn's model, as he sees it is expressed in the following quote from (Pylyshyn, 1981):

> "Cognitive principles such as those invoked by [Kosslyn] would only be theoretically substantive (i.e., explanatory) if they specified (a) how it was possible to have formal operations that had the desired semantic generalization as their consequence—that is, how one could arrange a formal representation and operations upon it so that small steps in the formal representations corresponded to small steps in the represented domain—and (b) why these particular operations, rather than some other ones that could also accomplish the task, should be used..."

The second front on which Pylyshyn attempts to undermine analogue imagery systems is parsimony. Even the phrasing of the question quoted at the beginning of this section betrays his belief that as long as a propositional attitude can account for all of the empirical evidence on mental imagery, that to advance a theory requiring specialized mechanisms violates Occam's Razor. After all, Pylyshyn might say, if physical symbol systems have succeeded in explaining this much of cognition already, then the simplest thing would be if they could do the whole job.

To provide an example of the error that Pylyshyn perceives in the analogue viewpoint, he briefly discusses the scanning experiments Kosslyn performed with his colleagues in the 1970's (see Kosslyn, 1980). The results of these experiments showed that the further away from the current point of focus in an image a target object was, the longer it took to refocus on the target. Pylyshyn admits that this is clear evidence that inter-object distances are represented in mental images. However, he balks at the conclusion that this implies that the images have spatial extent. He argues here for a distinction between "having" and "representing" dimension or size. Once this distinction is recognized, he insists, the argument of the previous paragraph becomes perfectly natural.

The Dynamical Systems Road to Parsimony

The discovery within the last thirty years that non-linear dynamical systems are capable of exhibiting deterministic but unpredictable behavior and of generating fascinating images has sent a reverberating ripple through the physical sciences that caught the attention of world (Gleick, 1987). As tools have begun to emerge over the last three decades for analyzing, controlling and understanding non-linear systems, the taboo against these non-linearities has diminished, opening the doors for a broader class of the sciences to assimilate dynamical modelling into their theories.

As constituents of the physical universe, it is obvious that brains are subject to physical law and the passage of time, but it was not clear until very recently that there was anything to be gained by viewing cognition, and its various elements, from the physical perspective. For example, the earliest use of dynamical systems as explanatory tools for cognitive functions came from neuroscientific research, such as the work of Walter Freeman on EEGs of the olfactory bulb (Freeman, 1979; Skarda & Freeman, 1987). In the last decade dynamical systems have been applied to coordinated behavior (Kelso & Scholz, 1985; Jordan, 1986), decision processes (Usher & Zakay, 1990), language acquisition (Pollack, 1991; van Gert, 1991) and several other aspects of cognitive and perceptual processing.

The earliest inspirations for our current model can be traced back to a research plan presented in (Pollack, 1989) in which it was proposed that it is within the intersection of AI, Neural Networks, Fractal Geometry and Chaotic Dynamical Systems that various conundrums for cognitive science will be resolved. In this work the relationship between fractals and memories was proposed:

> "Consider something like the Mandelbrot set as the basis for a reconstructive memory. Rather than storing all pictures, one merely has to store the 'pointer' to a picture, and, with the help of a simple function and large computer, the picture can be retrieved...."

A reconstructive memory based upon fractals will require a solution to the "fractal inversion" problem: given a picture within the generative range of some dynamical system, determine the precise parameters that would cause the dynamical system to generate it. Although a very hard problem in general, a mathematician claims to have solved it using the techniques of "Iterated Function Systems" (IFSs) (Barnsley, 1988)[2]. In this approach, the 'pointer' referred to in the above quote would be a single point in a multi-dimensional space of IFS parameters. Although IFSs have primarily received attention for their compact representation of visually complex two dimensional sets (fractals), Barnsley's results can be extended to more classical Euclidean sets, or even to three dimensions. This framework provides a strong mathematical and parsimonious foundation for our contribution to the imagery debate.

Fractal Memory (FRAME)

We have been developing a prototype reconstructive memory system based on fractals, called FRAME. Our encoding system for images is derived from Barnsley's work on IFSs and is commensurate with Kosslyn's dichotomy of deep and surface levels of description. The deep representation of an image is a small set of contractive affine transforms (i.e., linear functions, each of which maps the domain to one of its subsets) over a metric space (e.g. the Euclidean plane). The surface representation is the *attractor* (i.e., fixed point) of the functional union of this set, and can be constructed from the trajectory of a single point in the metric space through random selection and application of the transforms. We have shown that our sequential cascaded networks (SCNs), which are mathematically similar to IFSs, will exhibit state trajectories with complex, fractal properties when randomly stimulated, indicating that a simple neural model can instantiate the mathematical theory of iterated functions (Pollack, 1991; for another approach, see Stark, 1991).

While the images from IFSs are generally thought of as the result of a random infinite sequential process, this iterative process is extremely amenable to massive parallelization, producing rapid visual image reconstruction and even animation from the deep codes. This follows from the fact that the image is the fixed-point attractor of the IFS over the whole space. In other words, since every point in the space follows a trajectory under the IFS that approaches the attractor at an exponential rate, a large number of processors running the same IFS (with different random sequences and initial conditions) in parallel will produce almost instantaneously a surface representation of the image that captures its gross structure. Of course, finer detail will emerge over time.

2. This claim remains unverified, since his solution is being treated as proprietary by his corporation, which is selling digital image data compression systems.

IFS	a	b	c	d	e	f	p
Sierpinski's Gasket	0.5	0.0	0.0	0.5	0.125	0.125	0.333
	0.5	0.0	0.0	0.5	0.25	0.375	0.333
	0.5	0.0	0.0	0.5	0.375	0.125	0.333
A Dragon	0.5	-0.4	0.5	0.5	0.429	0.143	0.5
	0.5	-0.4	0.5	0.5	0.5	-0.214	0.5
Conch Spiral	-0.24	-0.0825	0.125	-0.25	0.428	0.868	0.1
	0.925	-0.225	0.266	0.925	0.15	-0.11	0.9

Figure 1: *Table of coefficients for the affine transformations of three IFSs approximated by FRAME networks*

As mentioned above, one of the most difficult problems is finding the deep representation for a given image. In FRAME, *the fractal inversion problem is refashioned as a neural network learning problem.* While we have not completely solved it, and acknowledge the need for large amounts of domain knowledge, we believe that the emergent complexity of simple nonlinear dynamical systems will provide a computationally feasible solution. The following model is an initial confirmation of this hypothesis.

As a first step towards this goal we used a network with the SCN architecture, which corresponds to the IFS structure. This network models an IFS by tracing the orbit of a point in the unit square under a probabilistically weighted sequence of transformations that are selected by the inputs to the net. The outputs of the network are recurrently connected back to the inputs, to simulate the iterative nature of the IFS. The network is trained in a supervised environment, in which the "teacher" knows the IFS for the image which is being learned. Since the teacher provides target outputs during training, the recurrent connections are only used during performance of the network.

A training set was randomly generated by forming triplets consisting of a point lying on the fractal attractor, an index, and the image of the point under the indexed transformation of the IFS for that attractor. The three sets on which our training sets are based are typical fractal images. These images, while lacking the geometric simplicity of more traditional experimental stimuli, nevertheless mirror the complexity of the stimuli found in nature. The coefficients for the three IFSs we trained on are displayed in the above table, in which each row represents a single affine transformation of the form

$$f(x, y) = (ax + by + e, cx + dy + f) ,$$

and the value of p is the probabilistic weight of the transformation in the reconstruction algorithm. The transformation indices were represented as 1-in-N encodings presented as input to the network, while the output and state of the network represented x-y coordinate values in the unit square. Since the network's task during training is to induce the invariant mathematical relationships in the training set, the training set needs to be large enough to eliminate any bias. On the other hand, a large training set imposes too many simultaneous constraints when using epoch learning. To balance these considerations, the network was only trained on a small subset of the entire training set during each epoch, similar to the independent method of (Cottrell & Tsung, 1991). In order to solve the bias problem, a new subset was randomly chosen at increasing intervals. The network was trained 100 epochs beyond convergence (Pollack, 1991), with the

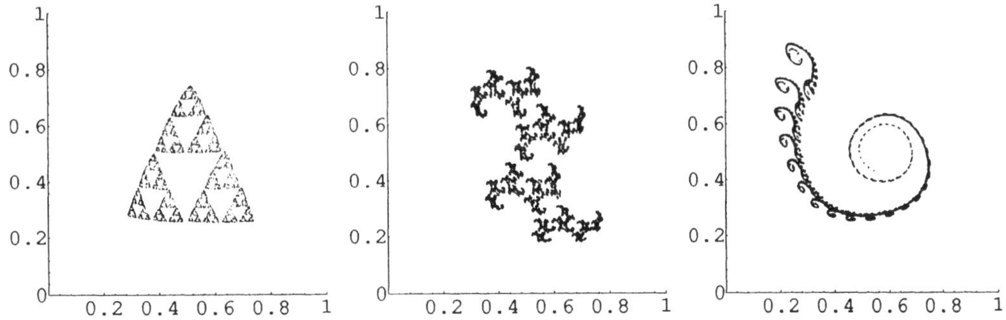

Figure 2: *Reconstructed images generated from FRAME networks that learned the "deep representations" of (a) Sierpinski's Gasket, (b) a Fractal Dragon, and (c) a Conch Spiral.*

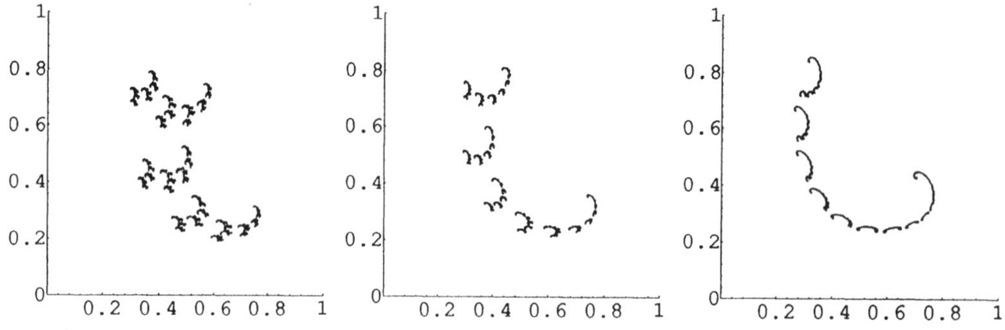

Figure 3: *Reconstructed images from three intermediate stages in the smooth interpolation of the deep network representations for the Fractal Dragon and the Conch Spiral.*

error threshold typically set to a value in the range [0.01, 0.05] and the overtraining usually distributed among several training sets, due to the cycling algorithm. To retrieve the image, the network is started with some random initial point, whose trajectory is plotted for 5000 iterations. The first 50 points in the trajectory (transients) are dropped, allowing it to approach the attractor. The performances of the networks trained on the IFSs from the table are pictured on the previous page. These results show the representational efficacy of this architecture and learnability of the set of transformations of an IFS.

One of the exciting aspects of our fractal representation for images is that *the surface representations vary continuously with the deep representations.* In other words, IFS codes meet Pylyshyn's challenge of small changes in one representation effecting small changes in the other. Furthermore, it varies in a predictable fashion. The affine nature of the transformations of an IFS allow us to decompose a transform into its primitive components: translation, rotation, and scaling. Thus the equation $f(x, y) = (ax + by + e, cx + dy + f)$ becomes

$$f\begin{pmatrix} x \\ y \end{pmatrix} = \begin{pmatrix} (r\cos\theta)\, x + (-s\sin\varphi)\, y + e \\ (r\sin\theta)\, x + (s\cos\varphi)\, y + f \end{pmatrix}$$

where r and s are scaling factors, e and f are translational parameters, and θ and φ are angular rotation. Therefore, any of these transformations of the image can be accomplished by simply manipulating the appropriate parameters in the IFS.

A surprising capability of our model is that *one image can be continuously deformed into another image* merely by linearly interpolating between their deep codes. This technique can be implemented in our cascaded back propagation model by adding an additional cascaded layer to the network that chooses which code to use. The deep representations for the images are loaded into the upper network from long term memory as parallel slices of its weight matrix. The inputs to this network then select a linear combination of these representations as the weights of the middle network, producing an interpolated image (trajectory) on the lower net. Refer to (Pollack, 1987) for architectural details. This is illustrated by the "snapshots" of an interpolation between the spiral and the dragon shown above. By increasing the number of snapshots—by reducing the grainsize of a discrete interpolation—it is possible to produce an animation of the deformation process.

Conclusions

As part of a general theory of reconstructive memory based on fractal inversion, we demonstrate a neural network that is capable of approximating the mathematical theory of iterated function systems. This model addresses the imagery debate in that the weights of the network serve as the deep representation of an image, which is reconstructed using fractal analogue techniques. We believe this substantially addresses the foundational needs of theories of mental imagery, such as Kosslyn's. This model also circumvents Pylyshyn's substantivity criticism of Kosslyn's work. The mathematics in which our system is embedded guarantees that the surface representation of an image will vary continuously with the parameters in the deep representation. In other words, small changes in one will correspond to small changes in the other. This leads to the simple ability to rotate, zoom, and translate mental images by operating on the compact codes. It also leads to a prediction for a cognitive ability to smoothly deform one image to another, which is clearly a task of the "imagination," and which has not been accounted for by any other theories that we are aware of to date.

There are two areas which are incomplete in our fractal memory model. We have begun to solve the problem of inducing deep representations from surface images by constraining the learning task to a supervised environment. However the general problem, which is much more difficult, will be the focus of further research. A second area of further work is in the design of a memory system, containing many deep codes, that serves as a mechanism for the recognition task. If the first problem is solved, recognition of visual images can be done rather quickly using nearest-neighbor techniques.

The complexity attributed to objects in the world is at best dependent on the modelling tools and interpretations that are available. Fractal geometry's first lesson was that the apparent complexity of nature (e.g., the shape of a tree or coastline, the branching structure of rivers and lungs) simply reflected an unsuitable mathematical formalism. Perhaps its second lesson will speak more to the imagination.

Acknowledgments

The authors would like to thank Peter Angeline, John Kolen, and Greg Saunders for criticizing earlier drafts of this paper. Special thanks also goes to John Kolen for many long discussions and suggestions throughout the course of this research.

References

Barnsley, M. (1988). *Fractals Everywhere*. Academic Press.

Brooks, L. (1968). Spatial and verbal components of the act of recall. *Canadian Journal of Psychology*, 22, 349-368.

Chandrasekaran, B. & Narayanan, H. (1990). Towards a theory of commonsense visual reasoning. In K. V. Nori and C. E. Veni Madhavan, (Eds.), *Foundations of Software Technology and Theoretical Computer Science: Proceedings of the Tenth Conference, Lecture Notes in Computer Science, 472*. Springer-Verlag.

Cottrell, G. W. & Tsung, F. S. (1991). Learning simple arithmetic procedures. In John Barnden and Jordan Pollack, (Eds.), *High-level Connectionist Models*, Norwood, N. J.: Ablex.

Finke, R. A. (1989). *Principles of Mental Imagery*. The MIT Press.

Freeman, W. J. (1979). Nonlinear dynamics of paleocortex manifested in the olfactory EEG. *Biological Cybernetics, 35*, 21-34.

Gleick, J. (1987). *Chaos: Making a New Science*. Viking.

Jordan, M. I. (1986). Serial order: A parallel distributed processing approach. ICS Report, La Jolla, CA.

Kelso, J. A. S. & Scholz, J. P. (1985). Cooperative Phenomena in Biological Motion. In Haken, (Ed.), *Complex Systems: Operational Principles in Neurobiology, Physical Systems, and Computers*. Springer-Verlag.

Kosslyn, S. M. (1980). *Image and Mind*. Harvard University Press.

Kosslyn, S. M. (1981). The medium and the message in mental imagery: a theory. *Psychological Review, 88*, 46-66.

Pollack, J. B. (1987). Cascaded back propagation on dynamic connectionist networks. In *Proceedings of the Ninth Conference of the Cognitive Science Society*. Seattle, 391-404.

Pollack, J. B. (1989). Implications of Recursive Distributed Representations. In David S. Touretzky, (Ed.), *Advances in Neural Information Processing Systems*. Los Gatos, California: Morgan Kaufmann.

Pollack, J. B. (1991). The induction of dynamical recognizers. *Machine Learning, 7*, 227-252.

Pylyshyn, Z. (1973). What the mind's eye tells the mind's brain: A critique of mental imagery. *Psychological Bulletin, 80*, 1-24.

Pylyshyn, Z. (1981). The imagery debate: Analogue media versus tacit knowledge. *Psychological Review, 88*, 16-45.

Shepard, R. N. & Metzler, J. (1971). Mental rotation of three-dimensional objects. *Science*, 171, 701-703.

Skarda, C. A. & Freeman, J. (1990). How brains make chaos in order to make sense of the world. In James A Anderson, Andras Pellionisz & Edward Rosenfeld, (Eds.), *Neurocomputing 2*. The MIT Press.

Stark, J. (1991). Iterated function systems as neural networks. *Neural Networks*, 4, 679-690.

Tye, M. (1991). *The Imagery Debate*. The MIT Press.

Usher, M. & Zakay, D. (1990). Multiattribute Decision Processes by Neural Networks. In *Advances in Neural Information Processing Systems 3*. Morgan Kaufmann.

van Gert, P. (1991). A dynamic systems model of cognitive and language growth. *Psychological Review, 98*, 1,3-53.

When Can Visual Images Be Re-Interpreted? Non-Chronometric Tests of Pictorialism.

Peter Slezak
Center for Cognitive Science
University of New South Wales
P.O. Box 1, Kensington 2033 NSW., Australia
peters@spinifex.cs.unsw.oz.au

Abstract

The question of re-interpreting images can be seen as a new focus for the imagery debate since the possibility would appear to be a direct prediction of the pictorial account. Finke, Pinker and Farah (1989) have claimed that their results "refute" the earlier negative evidence of Chambers and Reisberg (1985), while Peterson, Kihlstrom, Rose & Glisky (1992) have used the ambiguous stimuli of Chambers and Reisberg to show that under certain conditions, these images may be re-interpreted after all. By employing newly devised tasks, our own experiments have provided further conflicting evidence concerning the conditions under which images can and cannot be reinterpreted. We consider their bearing on the fundamental 'format' issue which neither Finke et al (1989) nor Peterson et al. (1992) address directly.

New Focus for Imagery Debate

Kosslyn, Finke and other 'pictorialists' take internal representations to be importantly like external ones regarding their 'privileged' spatial properties of depicting and resembling their referents. Thus, Finke (1990) suggests that "perceptual interpretive processes are applied to mental images in much the same way that they are applied to actual physical objects. In this sense, imagined objects can be "interpreted" much like physical objects" (1990, p. 18). Elsewhere he suggests that "The image discoveries which then 'emerge' resemble the way perceptual discoveries can follow the active exploration and manipulation of physical objects" (1990, p. 171).

The impasse in the imagery debate has led some (Anderson 1978) to conclude that the issue between pictorialism and the 'tacit knowledge' alternative is undecidable in principle on the basis of behavioral evidence. On the contrary, however, the experiments reported here show that divergent predictions from the contending theories may be readily formulated and tested with an appropriate experimental paradigm.

At least part of the reason for the persistence of the imagery debate has been the fact that the dispute has centered upon alternative explanations of the *same* body of reaction-time data concerning tasks such as mental rotation and scanning. Since the rival theories make identical predictions for chronometric evidence, adducing new evidence of time-dependent measures, as has repeatedly been done, cannot strengthen the case for a pictorial, spatial medium against the tacit knowledge theory. Thus, experiments are needed on which the contending accounts deliver *different* predictions.

The issue of reinterpreting visual patterns in mental imagery has recently emerged in this way as a new focus for the controversy since the question of whether, and under what conditions, novel information may be discovered from images provides a new means for testing the properties of the conjectured pictorial medium and the claimed parallel between imagery and perception..

Our own evidence concerns perceptual organization tasks which provide unequivocal criteria of the successful rotation, inspection and re-interpretation of images using "recognition processes" and "shape classification" procedures. However, despite the demonstrated ease of our tasks under *perceptual conditions,* naive subjects have generally been unable to succeed in some of the tasks under imagery conditions as would be predicted on the pictorial theory. Just as in the classical "crucial experiment" of Michelson and Morley concerning the speed of light, one might conclude from our null results that the pictorial medium, like the luminiferous ether, does not exist. However, as is familiar from the history of science, the situation is somewhat more complicated, as our own further experiments and those of others have shown.

There can be strictly no such thing as a "crucial experiment", since a falsified prediction can always be blamed on the auxiliary hypotheses on which any theory must depend. Thus, with the shift to non-chronometric investigations of image reinterpretation, the situation in the "imagery debate" has become less a matter of deciding between predictively equivalent theories than a matter of comparing their respective virtues according to the usual criteria of explanatory comprehensiveness, simplicity etc. in relation to the full set of available evidence. Above all, rather than being posed on either side of the debate in a partisan manner as "refuting" the competing theory, the inconsistent data on image reconstrual provide an opportunity for considering the significance of experimental conditions. A step in this direction has been taken by Peterson, Kihlstrom, Rose and Glisky (1992), who have used the ambiguous duck/rabbit figure to show that, under certain conditions, images may be reinterpreted after all, despite the null results of Chambers and Reisberg (1985).

"Equivalence"

Although the recent investigations of image reconstrual have a direct bearing on the deadlocked issue between pictorial and 'tacit knowledge' theories, it is noteworthy that the question is not addressed directly in these studies which focus more narrowly on the issue of reinterpretation itself. However, the possibility of reinterpreting an image follows as a direct implication of the pictorial theory which posits an "equivalence" between imagery and perception, and so the data on image reconstrual has deeper theoretical significance for the vexed question of the nature of mental representations. On the pictorial view, a mental image is conceived to be a "surrogate percept, allowing people to detect some pattern or property in a remembered scene that they did not encode explicitly when they saw the scene initially" (Pinker and Finke 1980, p. 246). Kosslyn (1987, p. 149) explains that one purpose of imagery involves "recognition processes" to discover information which is not stored explicitly in memory and an image "can then be reprocessed as if it were perceptual input (e.g., the shape could be recategorized)" (1987, p.155).

By contrast, the 'tacit knowledge' account would predict that the re-interpretation of images is difficult because it assumes that the mental representations are very abstract output of 'higher' cognitive processes and the encodings of conceptualizations or beliefs and, in this sense, already intrinsically meaningful and not requiring interpretation, - nor susceptible of easy *re*-interpretation (Pylyshyn 1978).

Conflicting Evidence

Consistent with Pylyshyn's explicit predictions, Chambers and Reisberg (1985) found that subjects were uniformly unable to reverse their mental images of the familiar ambiguous figures such as the duck/rabbit and Necker cube. Chambers and Reisberg see their results as supporting the "philosophical" arguments for taking imagery to be conceptual, symbolic representations. However, despite these negative results and earlier skeptical claims, Pinker and Finke (1980) report subjects' ability to "see" novel properties which "can be 'read off' the display" and which should emerge from images after mental rotation. Finke and Slayton (1988) have extended this work, providing further evidence "that people are capable of making unexpected discoveries in imagery" and that novel patterns can "emerge" from within imaged patterns. In the same vein, Finke, Pinker and Farah (1989) have shown that subjects can inspect and reinterpret their images by "applying shape classification procedures to the information in imagery" (1989, p. 51). Most recently, Peterson, Kihlstrom, Rose & Glisky (1992) have shown that, notwithstanding the negative results of Chambers and Reisberg (1985) with ambiguous figures, under suitable conditions subjects can reconstrue these very shapes in imagery.

It is in the light of this clash of experimental results and theoretical claims that our own experiments are to be understood. The findings of Peterson et al. (1992) are restricted to the case of ambiguous shapes, whereas our study seeks to investigate image reconstruals of entirely different kinds - notably, 'mental rotations' among others. Furthermore, our experiments avoid the specific objections by Finke et al. to Chambers and Reisberg.

Peterson et al. (1992) cite certain differences between the experiments of Chambers and Reisberg (1985) and Finke et al. (1989) as relevant to their discrepant outcomes. For example, the role of demonstration figures and hints to subjects might affect the strategies employed and thereby the relative ease of reconstruals. Also cited by Peterson et al. are the differential effects of the fidelity demanded in retention instructions to subjects. Yet another difference is the quality of the stimulus figures regarding their familiarity. However, Peterson et al. do not mention certain crucial methodological differences between the two studies which are likely to be even more important in accounting for the discrepant outcomes. Of particular interest is the fact that in our own experiments we have eliminated the uncertainties which had been introduced as a result of these differences between the basic experimental paradigms of Finke et al. and Chambers and Reisberg. Peterson et al. evidently overlook a crucial difference when they observe that, unlike Chambers & Reisberg, Finke et al. gave their subjects no instructions to remember the picture exactly as it was presented. But this is because, in a significant departure from earlier procedures, Finke et al. did not present subjects with visual stimuli at all. In order to avoid perceptual confounding, Finke et al. resorted to generating images by means of *verbal descriptions* of certain patterns. However, this serves to obscure the precise relevance of their results to the question in dispute. In our own study, by reverting uniformly to visual stimuli and avoiding perceptual confounding in other ways, we are able to make inferences concerning the likely causes of the discrepant results. That is, by keeping experimental paradigm constant, we can reasonably infer that the variability in image reconstrual can be attributed to particular features of the stimuli in question. Of considerable significance is the fact that, among our own stimuli, those which subjects were more readily able to re-interpret have a close resemblance to those of Finke et al. (1989).

Although Peterson et al. (1992) show that reinterpretation of imagined figures is possible, these successful reconstruals do not necessarily count against a 'tacit knowledge' account, since there is no reason to believe that the "reversal strategies" encouraged by the demonstration figures and other experimental conditions are the same as those employed in perception itself. On the contrary, the very need for special conditions which encourage successful strategies suggests precisely the *difference* between imagery and the spontaneous reversals in perception. The issue turns on the precise character of the "reversal strategies" about which presumably nothing is known, though they seem more akin to problem solving heuristics and inferences than perception. In particular, "reference frame" orientation is more likely to be part of the abstract, higher cognitive representations or tacit knowledge of objects and their properties (see Bryant & Tversky, 1992 and Hinton & Parsons, 1981).

Experiment 1a: Mental Rotation.

As reported in detail (Slezak, 1991), when tested on a non-chronometric, perceptual organization task providing unequivocal criteria of successful rotation and recategorization, subjects have been generally unable to perform imagery tasks which have been taken as well established.

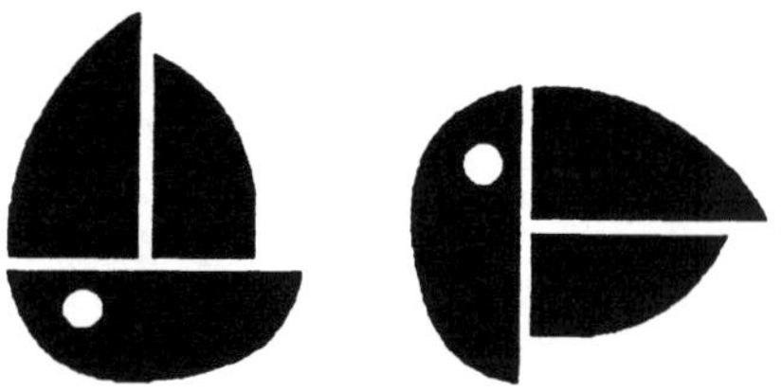

Figure 1.

The stimulus materials have been designed to have two distinct interpretations which are highly orientation specific. Thus, the figures are recognizable as a certain object in one orientation, but are interpretable as an entirely different object when rotated by 90 degrees. These stimuli are variants of the stimuli used by Rock (1973), and are considerably improved in their recognizability. In this respect, the shapes have the important feature that the alternative interpretations are readily obtained by rotation under *perceptual* conditions.

Figure 2

It is important that the task of reinterpretation can be readily accomplished in this way during perception because this makes the conditions for reconstrual under imagery conditions as favorable as possible. Thus, for example, when subjects are shown figure 2 in one orientation, it is immediately recognized as the duckling; then upon rotating the figure by 90 degrees, subjects immediately notice (with frequent expressions of surprise and delight) the alternative interpretation, the rabbit.

The direct expectation of the pictorial medium theory is that the same effect should be obtainable under imagery conditions. That is, if subjects are shown the figures in only one orientation, it would be expected that they could *rotate their image* and *discover* the alternative construal by inspection *from their rotated image*. Of course, the tacit knowledge alternative account takes images to be abstract, intrinsically interpreted conceptualizations and would predict that such reinterpretation would be difficult or impossible for subjects to perform in this way on their rotated images.

At first glance, the experimental results appear somewhat equivocal on the question of reinterpretation since subjects were generally able to reconstrue in imagination about *one third* (35%) of the figures they were presented. Even on these data it is clear that reinterpretation of the rotated image was difficult to perform, even if not always impossible. However, these results across multiple presentations take on a greater significance when the *order of presentation* is taken into account: It is most significant that no subject was able to reconstrue the *first* stimulus presented, which is, of course, the only one for which they did not know in advance that there might be an alternative interpretation. This striking relevance of stimulus order supports our conjecture concerning the effect of loss of naivete regarding the task. Moreover, stimuli were presented in order of decreasing suitability as confirmed by later experiments, and this supports our explanation for the slight improvement in subjects' success rate. In order to clarify this issue, we control for these confounding factors in the follow up experiments reported here, but even without making such allowances the mean success rate overall was still only 35 per cent and surprisingly low given the expectations of the pictorial theory. Typical of the predicted difficulty was the reaction of subjects when pressed to interpret their rotated image of the duckling: just as one would expect on a tacit knowledge account according to which the image is intrinsically bound to its interpretation, many subjects would volunteer that it is a "duckling on its back"!

Expt. 1b. Practice & Perception

The ordering effect in the foregoing data in which subjects showed a slight improvement from their initial failure could be due to practice in the task rather than to perceptual confounding as we had suggested. We controlled for this possibility by giving each subject prior practice with image rotation using Cooper's (1975) random polygon experiment.

At the same time, in order to preserve subjects' naivete on all the stimulus figures, we altered the previous instructions so that the imagery task would not be known until *all* the figures had been viewed and memorized in their initial orientation. Once all stimuli had been memorized in this way subjects were prompted by the brief flash of a figure on the screen, and asked if they were able to recall the shape clearly. Only then were subjects asked to rotate the shape 90 degrees clockwise and asked whether they were able to find an alternative interpretation.

Results: Significantly, prior practice with rotation of images on the Cooper (1975) random polygons had no effect on subjects' performance and this possibility could, therefore, be eliminated as a possible explanation for the ordering effect in the preceding data. Indeed, despite practice in rotating images, there was a dramatic drop in the success rate as a consequence of the new strategy to avoid perceptual confounding. Our data show only 8 successful reconstruals in 100 trials, and these were almost entirely confined to two of the figures whose shapes were said by subjects in debriefing to be a "give-away" due to certain telling clues which led them to speculate about possible alternatives in a way which was clearly not a "perceptual" apprehension, but rather a kind of inferential, searching process.

Experiment 1c. Image quality.

It could be argued that under the new conditions for avoiding perceptual reconstrual, the high failure rate was now due to poor, inaccurate or otherwise degraded images. In order to clarify the possible role of this factor, we altered the conditions in such a way as to maximize the accuracy of encoded shapes in memory.

Subjects were tested on the imagery rotation task now only after being permitted a very long (3 minute) visual presentation of one of the stimuli (following the usual distractors). During this extended viewing time, subjects were encouraged to remember details of the single figure as accurately as possible. When the stimulus was removed, subjects were asked to draw it from memory in order to have some evidence of the accuracy of the image. Subjects were then asked to rotate and reinterpret it in imagination. In addition, D.F. Marks (1973) VVIQ (Vividness of Visual Imagery Questionnaire) was administered for further evidence of image accuracy.

Results and Conclusion.: Despite high scores on the VVIQ averaging 2.5, there were still only 2 successful reconstruals in 23 trials. Above all, the accuracy of the drawings now provided firm grounds for supposing that degraded image quality is not a likely reason for the failure of image reconstrual.

The significance of our negative results derives from the fact that the mental rotation and reinterpretation are not only explicitly predicted by pictorial theorists, but involve precisely the mental transformations which have been classically taken as well established. Of additional importance is the fact that our task is readily performed under perceptual conditions, thereby entitling us to expect it in imagery as well according to the pictorial account. Further favouring reconstrual is the fact that our figures are considerably simpler than the representations of blocks stacked in three dimensions employed by Shepard and Metzler (1971) for which the mental rotation has been claimed, and our own shapes are geometrically no more complex than those of Cooper (1975) for which complexity was specifically found *not* to be a factor in the claimed ease of rotation.

Even if the Finke et al. (1989) study is not problematic in the ways I have suggested (Slezak 1991), their "refutation" is specific to the ambiguous stimuli and, therefore, irrelevant to our entirely different imagery tasks. A pattern of such failures on diverse perceptual phenomena would leave only *ad hoc* ways of avoiding their significance for the pictorial theory of imagery. Therefore, as a follow-up, we have devised additional experiments which attempt to reconstruct yet other perceptual phenomena in imagery.

Expt 2. Figure-Ground Reversal

The shapes illustrated in the left half of figures 3 are such as to encourage perceptual organization into several black objects which may, however, be reversed to become the ground and thereby reveal the letters "EI". Since the reversal in this form is somewhat difficult to achieve in perception itself, the effect can be readily elicited by asking subjects to bring the horizontal lines together to touch the shapes as in the right hand figures, clearly revealing the letters.

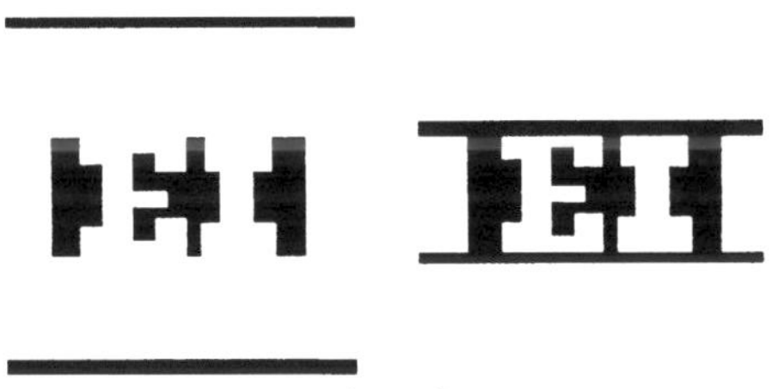

Figure 3

Despite the ease of the imagery task, not a single subject in twenty trials was able to reconstrue their image to reveal the alternative construal as letters. This was despite the fact that subjects' subsequent drawings of their memorized image were highly accurate.

Expt. 3. Kanizsa Illusory Contours

Stimuli of the sort illustrated in figures 4 were designed to produce the familiar illusory contours, but were, of course, not presented to subjects in this form, since the effect would then be created in perception. In order to test the parallel with imagery, circumstances must be contrived which generate the figure only in imagination and, accordingly, the entire figure was not presented at once.

Figure 4

Instead, the black shapes were designed to have good gestalt properties and are such as to discourage any inferences about other shapes of which they might be a part. These shapes were presented one at a time for 30 seconds at their respective positions and then removed from view. Having seen them only one at a time, subjects were instructed to imagine them all together and were then asked whether they were able to detect any other emergent shape, figure or object in their reconstructed image.

Despite research on "creative mental synthesis" which suggests that people can use imagery to mentally assemble the separately presented parts of a pattern (Finke 1990, p.21), only one subject out of thirty trials reported seeing a geometrical shape, correctly identifying the emergent white figure. Again, this overwhelming difficulty with the task was despite the fact that subjects' drawings were highly accurate and they were generally able to notice the emergent shape from their own drawing with frequent expressions of surprise.

Expt. 4. Creative Mental Synthesis

Method: In this experiment again, subjects were tested for their ability to discover something from their images which they had not noticed during perceptual exposure. The shapes illustrated on the *top row* in figure 5 ("M", "heart" and "pot") are such as to be generally unfamiliar in the sense that subjects see them as a whole without noticing that they are composed of a familiar numeral on the right joined to its mirror image on the left. As in each of the foregoing experiments, an essential feature of these stimuli is the fact that the imagery task is one which can be readily accomplished *in perception*. In this case, although people invariably fail to recognize the symmetrical shapes, when partially covered to reveal only the right-hand half, as in the bottom row, this remainder is instantly recognized as the familiar numeral.

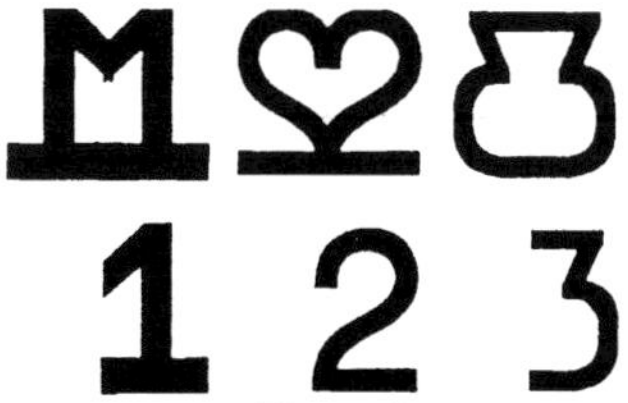

Figure 5

The symmetrical shapes were shown to subjects with the instruction to memorize them as accurately as possible in order to answer certain questions afterwards. To avoid the possibility of perceptual confounding, the task was explained to subjects only after all three shapes had been memorized. This explanation was by means of a demonstration showing a different shape (a square) together with its right half alone. Subjects were then asked to recall each of the stimuli and, in the fashion of the demonstration, to imagine their right halves standing alone. When subjects confirmed that they were visualizing the right half of the figure by itself, they were asked whether it looked like any familiar shape or object.

Results: With these stimuli, the results were strikingly different from those of the previous experiments. In 54 trials, a significant proportion (65 per cent) of subjects were easily able to report the numerals "1", "2" and "3" as a discovery made from their image. In each case they confirmed that they had not noticed the familiar shape when shown the figures and had only realized its meaning from "inspecting" their image.

The ease of reconstrual in these cases is in marked contrast with the earlier experiments, despite the essentially similar logic of the imaginal task. The overall success rate of 65 percent should be qualified by noting an order effect in which there was a consistent improvement from only 50 percent success rate on the first stimulus to 66 percent on the second and 77 percent on the third. From subjects' reports it appeared that if they once discovered a numeral from their image, this provided a strategy which helped to make the subsequent discovery by directing their search in a specific way. Though no less genuine discoveries from imagery, this improvement gives further insight into the scope and limitations of operations with images and supports the insight of Peterson et al (1992) regarding the value of appropriate strategies. The facilitation effect evident here is not a matter of perceptual confounding, but nonetheless suggests that later tests should nevertheless be considered separately from the first naive attempt at discovery in imagery. In the case of this first, naive test, the 50 percent success rate in imagery is still evidence of some significant difficulty of re-interpretation, despite the instantaneous result in perception.

Though qualified in the ways just noted, it remains that our results in this experiment broadly support those reported by Finke et al. (1989). The significant level of image reconstrual needs to be accounted for by the tacit knowledge theory which would predict difficulty in the task. Symmetrically, however, the roughly equal failure rate needs to be explained by the pictorial theory which would predict less difficulty in the imagery task if performed "perceptually".

This situation reinforces the point that there is an acute need for both accounts to develop theoretically motivated explanations for their respective *anomalies* rather than merely adducing supporting evidence. This requirement has evidently not been acknowledged on either side of the controversy between 'tacit knowledge' and pictorial accounts even though, as often in science, the mark of a successful explanatory theory is how well it can accommodate the full range of available empirical evidence.

Our results can be seen as casting further light on the *precise conditions* under which visual imagery transformations are possible and it is perhaps significant that the stimulus figures in this fourth experiment bear a similarity to those of Finke et al. (1989) - namely, line drawings of geometrical shapes which are somehow relevant to their easier reconstrual. On the other hand, our own figures have a more obvious interpretation than those of Finke et al., even if they are highly stylized, and this is not fully consistent with our own supposition that it is the intrinsic meaningfulness of images which makes reconstrual difficult or impossible.

Conclusion.

Notwithstanding the claim by Finke et al. (1989) to have "refuted" Chambers and Reisberg (1985), we have shown that image reconstrual is generally difficult or impossible to perform under certain conditions in which one would have expected it according to the pictorial theory. These negative results, are precisely as one would expect on the tacit knowledge account according to which imagery is highly abstract and cognitive, and does not involve any internal, surrogate, 'objects' to be apprehended by the visual system.

On the other hand, our own positive evidence of image reconstrual, like that of Finke et al. (1989) and Peterson et al. (1992) must be accommodated into our preferred theory, as *prima facie* counter-examples. Peterson et al. showed the importance of appropriate strategies for enhancing the likelihood of image reconstrual and the improved positive results with our own last experiment can be plausibly explained in the same way, as noted.

Finally, it must be acknowledged that, although we have placed emphasis on the semantic interpretability of shapes as a factor in explaining their memory encoding and subsequent difficulty of reconstrual, this kind of factor is not invoked with equal plausibility in the case of all our own results. This factor suggests itself most obviously in the case of our orientation-dependent shapes and even the figure-ground reversals of Experiment 2 can be seen as involving the interpretation of objects. The subjective contours are less obviously cases of this sort, but even here one might suggest that the black shapes presented singly cohere as good gestalt objects which then resist the reinterpretation in imagery to being seen as partially occluded background.

Recent investigations have neglected the the fundamental question of the 'format' of mental representations. Since image reconstrual is clearly not an all-or-none phenomenon, these studies can illuminate this issue by revealing constraints on imagery processes which the contending theories must explain.

References

Anderson, J.R. 1978. Arguments Concerning Representations for Mental Imagery. *Psychological Review* 85:249-277.

Bryant, D.J. and Tversky, B. 1992. Assessing Spatial Frameworks with Object and Direction Probes. *Bulletin of the Psychonomic Society,* 30 (1):29-32.

Chambers, D. and Reisberg, D. 1985. Can Mental Images Be Ambiguous? *Journal of Experimental Psychology: Human Perception and Performance* 11:317-28.

Cooper, L.A. 1975. Mental Rotation of Random Two-Dimensional Shapes. *Cognitive Psychology* 7:20-43.

Finke, R.A. 1990. *Creative Imagery.* NJ: Erlbaum.

Finke, R.A., Pinker, S. and Farah, M.J. 1989. Reinterpreting Visual Patterns in Mental Imagery. *Cognitive Science* 13:51-78.

Finke, R.A., & Slayton, K. 1988. Explorations of Creative Visual Synthesis in Mental Imagery. *Memory and Cognition.* 16:252-257.

Hinton, G.E. and Parsons, L.M. 1981. Frames of Reference and Mental Imagery. In J. Long & A. Baddeley, eds., *Attention and Performance IX.* Hillsdale: Erlbaum.

Kosslyn, S.M. 1987. Seeing and Imagining in the Cerebral Hemispheres. *Psychological Review* 94: 148-175

Peterson, M.A., Kihlstrom, J.F., Rose, P.M., & Glisky, M.L. 1992. Mental Images Can be Ambiguous: Reconstruals and Reference Frame Reversals. *Memory and Cognition,* 20:107-123.

Pinker, S. and Finke, R. 1980. Emergent two-Dimensional Patterns in Images Rotated in Depth, *Journal of Experimental Psychology: Human Perception and Performance* 6(2):244-264.

Pylyshyn, Z. 1978. Imagery and artificial intelligence. In *Minnesota Studies in the Philosophy of Science,* C. Wade Savage ed. Vol. IX. Minnesota: University of Minnesota Press.

Reisberg, D. and Chambers, D. 1991. Neither Pictures Nor Propositions: What Can We Learn from a Mental Image? *Canadian Journal of Psychology,* 45 (3): 336-352.

Rock, I. 1973. *Orientation and Form.* New York: Academic Press.

Shepard, R.N., and Metzler, J. 1971. Mental rotation of three-dimensional objects. *Science* 171:701-703.

Slezak, P. 1991. Can Images Be Rotated and Inspected? *Proceedings of 13th Annual Conference of the Cognitive Science Society,* 55-60. N.J.: Erlbaum.

"Ill-Structured Representations" for Ill-Structured Problems

Vinod Goel
Institute of Cognitive Science
University of California, Berkeley
goel@cogsci.berkeley.edu

Abstract

While the distinction between well-structured and ill-structured problems is widely recognized in cognitive science, it has not generally been noted that there are often significant differences in the external representations which accompany the two problem types. It is here suggested that there is a corresponding distinction to be made between "well-structured" and "ill-structured" representations. Such a distinction is used to further differentiate diagrams into finer-grained types, loosely corresponding to sketches and drafting-type diagrams, and it is argued that ill-structured, open-ended problems, like the preliminary phases of design problem solving, *need* "ill-structured" diagrammatic representations. Data from protocol studies of expert designers are used to support the thesis.

Introduction

Cognitive science has made considerable progress in understanding how certain well-structured problems[1] are solved and the role external representations play in such solutions (Newell & Simon, 1972). A typical example of a well-structured problem is the game of chess. In chess, the start state is specified, as is the goal state and the set of legal transformations (though generating or selecting the "best" transformation at any given point is a non-trivial task). The representation of the task (whether it be in internal or external memory) is also "well structured" in that it is clear what state is being instantiated, by virtue of what it is that state, what states of affairs are being referred to, and what the set of legal transformations from one state to another state are.

While it has been frequently noted that many of the problems that we confront in life are not well structured, it has generally not been appreciated that the representations which often accompany the solutions to such problems are also not "well structured" in the above sense. In fact, the predicates 'ill structured' and 'well structured' have not to my knowledge been applied to representations. The major differentiating criterion for representations has been, and continues to be, the diagrammatic (or pictorial or imagistic) and propositional (or linguistic) dimension (Kosslyn, 1981; Larkin & Simon, 1987; Pylyshyn, 1981; Simon, 1972).

While I do not deny the importance of the diagrammatic and propositional distinction, I will here focus on and argue for an "ill-structured" and "well-structured" distinction. In fact, I will use the "ill-structured" and "well-structured" distinction to further differentiate diagrammatic representations into finer-grained types. Informally, and as a first pass, one might understand the distinction between "ill-structured" and "well-structured" diagrams in terms of the difference between fast freehand sketches and formal, box-like drafting diagrams, where sketches correspond to the former, and drafting diagrams to the latter.[2] A more formal statement follows.

The goal of this paper is to differentiate "well-structured" diagrammatic representations from "ill-structured" diagrammatic representations and to show that some ill-structured problems require "ill-structured" representations to prevent premature crystallization of ideas and to facilitate the generation and exploration of alternate solutions. This is a very brief summary of work reported in full elsewhere (Goel, 1992a; Goel, 1992b).

Differentiating "Well-Structured" & "Ill-Structured" Diagrams

Consider the diagrammatic representation in Figure 1. It depicts two states and a transformation in a game of chess. The representation and the symbol system it belongs to have the following seven properties:

(p1) *Syntactic Disjointness:* Each token belongs to at most one symbol type. Thus for example, no tokens of the type 'rook' belong to the type 'queen'.

(p2) *Syntactic Differentiation:* It is possible to tell which symbol type a token belongs to. So given the types 'queen' and 'rook' and a token of the type 'rook', it is possible to tell which type it does and does not belong to.

[1] A well-structured problem is one in which the information necessary to construct a problem space is specified.

[2] Strictly speaking, this is not true. But it is a useful starting point.

(p3) *Unambiguity:* Every symbol type has the same referent in each and every context in which it appears. Thus no 'bishop' refers to a knight regardless of context.

(p4) *Semantic Disjointness:* The classes of referents are disjoint; i.e., each object referred to belongs to at most one reference-class. So, for example, no pawn belongs to the class of rooks.[3]

(p5) *Semantic Differentiation:* It is possible to tell which class a particular object belongs to. Thus, given a king and two classes of objects, one could determine which class, if any, the king belongs to.

(p6) The rules of transformation of the system are well specified. Thus, for example, there is no question as to what does or does not constitute a legal move for a bishop.

(p7) The legal transformations of the system are such that these properties are preserved at each and every state.

The first five of these properties (p1-p5) are adopted from Goodman (1976). The reader is referred to Goodman (1976), Elgin (1983) and Goel (1992b) for a more complete discussion.

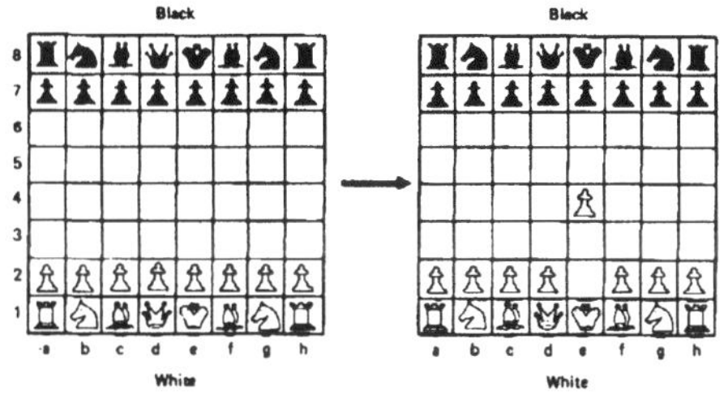

Figure 1: States and transformation from a game of chess (reproduced from Rich (1983, p.65))

Six of these seven properties of symbol systems are actually presupposed by the notion of a computational problem space (Goel, 1991b). The satisfaction of properties p1 and p2 is necessary for there to be a discernable fact of the matter as to what state is being instantiated. Satisfaction of properties p3 and p5 is necessary for there to be a discernible fact of the matter as to what state of affairs is being referred to.[4] Property p6 is necessary to constrain the class of possible transformations, while property p7 is necessary to maintain the above properties during the transformation of one state to the next.

[3]This is, of course, true only in the vocabulary of chess, narrowly defined. In the larger context, a pawn also belongs to the class of chess pieces, the class of material objects, etc. But this is consistent with the point being made here.

[4]Property p4 is necessary to go from the referent, to the referring state. But it is not clear whether this is *necessary* for the notion of a problem space.

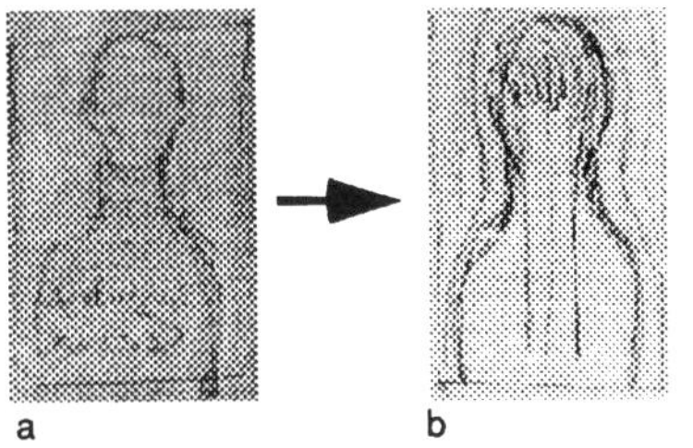

Figure 2: States and transformation from early part of a graphic design session

In contrast, consider the diagrammatic representations in Figure 2 extracted from the early part of a graphic design problem-solving session. They belong to the symbol system of sketching and differ from the representations in Figure 1 with respect to each of the above seven properties. In fact, they fall on the opposite extreme with respect to each of the seven properties (p1-p7):

(p1') *Failure of Syntactic Disjointness:* Each token may belong to many symbol types at the same time. That is, in the absence of any agreement as to the constitutive versus contingent properties of tokens, there may be no fact of the matter as to which equivalence-class they belong to. Thus for example, what equivalence-class does token *a* in Figure 2 belong to? Do tokens *a* and *b* belong to the same equivalence-class? There may be no agreed-upon answers to these questions.

(p2') *Failure of Syntactic Differentiation (through density):* Because the symbol system of sketching allows for a dense ordering of symbol types (i.e., between any two types there is a third), it is not always possible to tell which type a token belongs to. So, for example, even if we agree that the token *a* in Figure 2 belongs to only one equivalence-class, it may not be possible to tell which of several classes it does or does not belong to.

(p3') *Ambiguity:* Symbol types do not have the same referent in each and every context in which they appear. For example, the token *b* in Figure 2 was interpreted as a human head and later reinterpreted as a light bulb.

(p4') *Failure of Semantic Disjointness:* The classes of referents are not disjoint; i.e., each object referred to may belong to many reference-classes. So, for example, the human figure referred to by the symbol-type *a* may belong to the class of humans and the class of students.

(p5') *Failure of Semantic Differentiation:* The system of sketching allows for a dense ordering of reference-classes. When this is the case, it is not possible to tell which class a particular object belongs to. For example, in a perspective drawing of a human figure, every height of the figure would correspond to a different class of heights of humans in the world, and these classes of heights are of course densely ordered. In such a case it would not be possible to tell which class a particular human height belongs to.

(p6') The system of sketching has no well-specified rules for transforming one state into another. There is no transformation of the token *b* which would be "incorrect" or "illegal."

(p7') As the properties p1-p6 are not present to begin with, they are not preserved in the transformation of the system from one state to the next.

Having defined "ill-structured" and "well-structured" representations as such, I will henceforth dispense with the scare quotes.

It makes sense that the representations which underlie well-structured problems (e.g. cryptarithmetic, Moore-Anderson task, Tower of Hanoi, 8-Puzzle problem, checkers, etc.) should be well structured (by virtue of having properties p1-p7). After all, if these properties were absent, then the states, operators, and evaluation functions could not be specified and the problem would not be a well-structured problem.

There is, however, no similar reason for ill-structured problem spaces to be accompanied by representations belonging to well-structured symbol systems. The fact that the problem spaces are ill structured means that states, operators and evaluation functions are not defined thus, there is little need for the information which specifies them to be actually present. In fact, not only is there no compelling reason for representations accompanying ill-structured problems to be well structured, there actually seems to be a case to be made to the effect that they *need* to be ill structured to facilitate certain cognitive processes. This point is argued for in the next sections with some results from design problem solving.

The Role of Ill-Structured Diagrams in Design Problem Solving

Two empirical studies of design problem solving were conducted. The first examined some of the cognitive processes involved in design problem solving while the second focused on the impact on these cognitive processes when the symbol systems the designers were allowed to use were manipulated along the well-structured and ill-structured dimensions.

One result of the first study (Goel, 1991b; Goel & Pirolli, in press) was that the development of design solutions has several distinct phases. Four of these phases are problem structuring, preliminary design, refinement, and detailing.[5] These phases differ with respect to the type of information dealt with, the degree of commitment to generated ideas, the level of detail attended to, and the number and types of transformations engaged in.

What is of interest to us here is the contrast between the preliminary design phase and the refinement and detailing phases. Preliminary design is a classical case of creative, ill-structured problem solving. It is a phase where alternatives are generated and explored. This generation and exploration of alternatives is facilitated by the abstract nature of information being considered, a low degree of commitment to generated ideas, the coarseness of detail, and a large number of lateral transformations. A lateral transformation is one where movement is from one idea to a slightly different idea rather than a more detailed version of the same idea. Lateral transformations are necessary for the widening of the problem space and the exploration and development of kernel ideas (Goel, 1991b).

The refinement and detailing phases are more constrained and structured. They are phases where commitments are made to a particular solution and propagated through the problem space. They are characterized by the concrete nature of information being considered, a high degree of commitment to generated ideas, attention to detail, and a large number of vertical transformations. A vertical transformation is one where movement is from one idea to a more detailed version of the same idea. It results in a deepening of the problem space (Goel, 1991b).

It was also noted that the preliminary design phases were accompanied by ill-structured representations (belonging to the symbol system of sketching), while the refinement and detail phases were accompanied by more well-structured representations, belonging to the system of drafting (Goel, 1991b). A second study was conducted to investigate the role played by ill-structured representations in the preliminary design phase.

In the second protocol study the following four of the seven properties of ill-structured representations were examined and manipulated: (p1') failure of syntactic disjointness, (p2') failure of syntactic differentiation, (p3') ambiguity, and (p5') failure of semantic differentiation. It was predicted that when these properties are absent (i.e., when the symbol system is well structured) the number of lateral transformations is likely to be hampered. The underlying rationale was that properties p1' and p3' facilitate lateral movement by allowing multiple interpretation of symbol-types while properties p2' and p5' facilitate lateral movement by allowing for overlapping (or closely ordered) symbol-

[5] While these categories may seem trivial, they do constitute a significant claim about the design problem space because they are not found in at least some nondesign problem spaces (Goel, 1991b; Goel & Pirolli, in press).

types and ideas (Goel, 1991b; Goel, 1992b). The balance of the paper describes this study.

Subjects & Design: Nine expert designers from the disciplines of industrial design and graphic design were engaged in two (one-hour) problem solving sessions while the symbol systems they were allowed to use were manipulated along the dimensions of ill-structured and well-structured representations. In the one case subjects were allowed to use an ill-structured symbol system with properties p1', p2', p3', and p5'. In the other case they were requested to use a well-structured symbol system with properties p1, p2, p3, and p5.

Manipulation of Symbol Systems: The manipulation of symbol systems was through the manipulation of drawing tools and media. In one session each designer was allowed to use the tools, media, and symbol systems of his/her choice. They invariably chose to use paper and pencils and did a lot of sketching. In the second session they were requested to use a computational interface. Specifically, they were asked to use a subset of the drawing package MacDraw[6] (version 1.9.5; with the freehand tool turned off and the grid turned on) running on a Mac II[7] with a large two-page monitor. MacDraw is not a sketching tool; it is a restrictive subset of a drawing or drafting tool. It only allows one to make precise lines, boxes, and circles. The subjects all used sophisticated computational drawing tools as part of their jobs and so were proficient with MacDraw.

The expectations were that freehand sketching would be used to generate substantially ill-structured representations while the representations generated with MacDraw would be substantially well structured. It was also expected (as noted above) that ill-structured representations would result in more lateral transformations.

Task Descriptions: There were three graphic design tasks and two industrial design tasks. The graphic design tasks were to design (i) a poster for the new cognitive science program at UC-Berkeley, (ii) a poster for the Shakespeare Festival at Stratford-on-Avon, and (iii) a poster promoting the city of San Francisco. The industrial design tasks required the design of (i) a desk time piece to commemorate Earth Day, and (ii) a toy to amuse and educate a 15-month-old toddler.

Informal Overview of Data: Informally, the difference between the two cases seems to be the following: In freehand, when a new idea is generated, a number of variations quickly follow. The variations expand the problem space and are necessary for the reasons noted earlier. One actually gets the sense that the exploration and transformation of ideas is happening on the paper in front of one's eyes as the subject moves from sketch to sketch. Indeed, designers have very strong intuitions to this effect.

When a new idea is generated in MacDraw, its external representation (in MacDraw) serves to fixate and stifle further exploration. Most subsequent effort after the initial generation is devoted to detailing and refining the same idea. One gets the feeling that all the work is being done internally and recorded after the fact, presumably because the external symbol system cannot support such operations.

Hypotheses: It is necessary to measure two things: (1) How are the two symbol systems being used with respect to the ill-structured/well-structured properties? (2) How does this impact the number of lateral transformations and reinterpretations? The hypotheses with respect to (1) are the following:

(H1) Free-hand sketching is syntactically more dense than MacDraw.

(H2) Free-hand sketching is semantically more dense than MacDraw.

(H3) Free-hand sketching is more ambiguous and/or nondisjoint than MacDraw.[8]

The specific hypothesis with respect to (2) is the following:

(H4) Well-structured representations will hamper the exploration and development of alternative solutions (i.e., lateral transformations) and force early crystallization of the design.

Coding Scheme: A coding scheme was developed to measure syntactic and semantic density, ambiguity and/or nondisjointness, and lateral transformations. A few aspects of the scheme are presented here. A full discussion, complete with examples, appears in Goel (1992a).

The protocols were segmented into episodes along the lines of alternative solutions (which correlated with drawings on a one-to-one basis) and analyzed at this level. Syntactic and semantic density were measured in terms of a *variation* relationship between episodes (and the accompanying drawings).

A *variation* rating means that the current drawing is recognizably similar to earlier drawings. At the syntactic level this means that the equivalence-class of marks (i.e., syntactic types) constituting the drawing are closely related to, but distinct from, the equivalence-class of marks constituting one or more previous drawings. A *variation* rating at the semantic level means that the idea or content of the drawing is similar (but not identical) to the ideas or contents of one or more previous drawings.[9]

---

[6]'MacDraw' is a registered trademark of Apple Computer, Inc.

[7]'Mac II' is a registered trademark of Apple Computer, Inc.

[8]Notice here the collapse of the logically distinct notions of unambiguity and disjointness. It was not possible to distinguish between them with the given methodology.

[9] The connection between the variation rating and density can be seen with the aid of the following example. Consider two symbol systems, *SS1* and *SS2*. In *SS1* characters consist of equivalence-classes of line lengths which, when measured in feet, correspond to the integers. So we have lengths of 1', 2', 3', etc. In *SS2* characters

Ambiguity and/or nondisjointness was measured in terms of *reinterpretations* of drawings. Reinterpretations occurred whenever subjects returned to earlier drawings and gave them a different interpretation.

In addition to the relationship between drawings/episodes and the interpretation of drawings, the types of operations which transformed one drawing into another were also coded for. A *lateral transformation* was one which modified a drawing into another related but distinctly different drawing (as opposed to a more detailed version of the same drawing, a totally unrelated drawing, or an identical drawing).

Results: Sequences of episodes which received a *variation* rating were considered to be more densely ordered than those which received some other rating. Measured as such, the ordering of episodes (or alternative solutions) is significantly denser in freehand sketching than in MacDraw. The first row of Table 1 (Syntactic Density) shows the number of densely ordered drawings in freehand sketching versus MacDraw per session. The second row of Table 1 (Semantic Density) indicates that the number of densely ordered ideas per session is also much greater in freehand sketching than in MacDraw.

Table 1: Mean Numbers of Densely Ordered Episodes and Reinterpreted Episodes per Session.

	Free-hand	MacDraw
Syntactic Density	11.2	3.0**
Semantic Density	10.4	4.1**
Reinterpretations	2.4	0.67*

**p<.005, one-tail; *p<.05, one-tail

There was also a significantly greater number of reinterpretations in freehand sketching than in MacDraw (see Table 1, row 3). Thus as predicted, the freehand sketches displayed greater ambiguity and/or lack of syntactic disjointness.

On the basis of these results, and converging verbal evidence, it is concluded that the two symbol systems are indeed being used in the way predicted. That is, the freehand sketches belong to a substantially ill-structured symbol system while MacDraw drawings belong to a substantially well-structured system. Finally, we want to know whether this has the predicted impact on the number and types of transformations.

consist of equivalence-classes of line lengths which, when measured in feet, correspond to the rational numbers. So we have lengths of 1', 2', 3'...; but also lengths of 1.5', 2.5', 3.5'... and 1.25', 2.25', 3.25'... and 1.125', 2.125', 3.125'... and so on. Lines of lengths 1.125' and 1.25' are no more identical than lines of length 1' and 2'; neither of these pairs belongs to the same equivalence-class. However, line lengths of 1.125' and 1.25' are much more "similar" or "closer to each other" -- with respect to length -- than lines of 1' and 2'. Thus the notions of "similarity" or "closeness" seem to be an integral (necessary?) part of density.

Table 2: Mean Numbers of Lateral Transformations per Session

	Free-hand	MacDraw
Syntactic Lateral Transformations	8.9	3.2*
Semantic Lateral Transformations	8.0	3.9*

*p<.05, one-tail

It turns out that there is a statistically significant difference in the number and types of transformations (see Table 2). As predicted we get significantly more lateral transformations, at both the syntactic and semantic levels, with the ill-structured representations (freehand sketching) than with the well-structured representations (MacDraw).

Discussion & Conclusion

Before rejecting the null hypothesis associated with H4, and concluding that well-structured representations hamper the exploration and development of alternative solutions (i.e., lateral transformations) and force early crystallization of the design, it is necessary to examine some alternative interpretations of the data. A rather obvious alternative interpretation is that the behavioral differences have nothing to with the theoretical differences underlying the manipulation, it is simply that freehand sketching is easier to use than MacDraw. One would get a similar hampering of exploration and development if, instead of MacDraw, subjects were forced to draw with a twelve-foot pencil.

Table 3: Mean Numbers of Sessions & Episodes in Minutes, & Mean Number of Episodes & New Solutions per Session.

	Free-hand	MacDraw
Duration of sessions (min)	57.7	53.2
Duration of episodes (min)	2.5	2.8
Number of episodes	16.4	14.4
Number of new solutions (syntactic level)	5.2	4.0
Number of new solutions (semantic level)	5.6	3.9

This interpretation is not, however, supported by the data. The effects of the manipulation are selective, and as predicted. There are no significant differences (F<1) along a number of other important dimensions, including the duration of the sessions, number of episodes per session, the duration of episodes, and the number of new solutions generated per session (Table 3). If the difference was just one of ease vs. difficulty of use, then one would expect significant differences along each of these dimensions.

There is, however, a second alternative interpretation which needs to be taken more seriously. On this account, there are no behavioral differences across the two treatment conditions. What seems like a difference (the hampering of lateral transformations) is just an artifact of the methodology. It is a well-accepted assumption of protocol analysis that a more complete record of internal activity will occur when there is a good match between internal and external symbol systems (Ericsson & Simon, 1984). If, this is the case, and one also assumes that the system of internal representation is ill structured then it follows that the freehand sketching record is more complete than the MacDraw record. So the appearance of behavioral differences is caused by a different degree of completeness in the records.

However, this interpretation -- by assuming that the system of internal representation is ill structured -- violates some very important metatheoretical constraints on the system of internal representation (Goel, 1991a; Goel, 1991b), and leads to a much stronger conclusion. I am postulating that different symbol systems are correlated with different cognitive functions. The alternative interpretation requires one to make an assumption about the structure of internal representations which is very strong, and contrary to much of the literature. It seems more prudent to accept the original interpretation.

On the basis of these results, the failure of alternative interpretations, and the assumption that lateral transformations are desirable,[10] it can be concluded that, at least some ill-structured problems -- like design -- require (or at least benefit from) ill-structured diagrammatic representations during the early, explorative and generative phases of problem solving.

On one reading, this is a rather unremarkable conclusion. Any designer can tell us that sketching is important for preliminary design. Why does this obvious fact need to be established by experiment? What makes the conclusion interesting is that the analysis of symbol systems employed, and the design of the study, allow us to tie the results to certain specific properties of symbol systems, namely density and ambiguity. So the study not only confirms the obvious, it offers an explanation of it in terms of ambiguity and density of symbol systems.

[10]This, I take it, is an unproblematic assumption. It amounts to little more than the claim that better solutions will result if one is allowed to explore the space of solutions and to customize any preexisting solutions to the present context.

Acknowledgements

The author is indebted to Peter Pirolli, Mimi Recker, and Susan Monaghan for helpful discussions and comments. This work has been supported by a Gale Fellowship, a Canada Mortgage and Housing Corporation Fellowship, a research internship at System Sciences Lab at Xerox PARC and CSLI at Stanford University, and an Office of Naval Research Cognitive Science Program grant (# N00014-88-K-0233 to Peter Pirolli).

References

Elgin, C. Z. (1983). *With Reference to Reference.* Indianapolis, IN: Hackett Publishing.

Ericsson, K. A., & Simon, H. A. (1984). *Protocol Analysis: Verbal Reports as Data.* Cambridge, Massachusetts: The MIT Press.

Goel, V. (1991a). Notationality and the Information Processing Mind. *Minds and Machines, 1* (2), 129-165.

Goel, V. (1991b) *Sketches of Thought: A Study of the Role of Sketching in Design Problem Solving and its Implications for the Computational Theory of Mind.* Ph.D. Dissertation, University of California, Berkeley.

Goel, V. (1992a). The Cognitive Role of Sketching in Problem Solving. Manuscript submitted for publication.

Goel, V. (1992b). Specifying and Classifying Representational Systems: A Critique and Proposal for Cognitive Science. In *Proceedings of Conference on Cognition and Representation.* SUNY, Buffalo.

Goel, V., & Pirolli, P. (in press). The Structure of Design Problem Spaces. *Cognitive Science.*

Goodman, N. (1976). *Languages of Art: An Approach to a Theory of Symbols (second edition).* Indianapolis, IN: Hackett Publishing.

Kosslyn, S. M. (1981). The Medium and the Message in Mental Imagery: A Theory. *Psychological Review, 88* (1), 46-66.

Larkin, J. H., & Simon, H. A. (1987). Why a Diagram is (Sometimes) Worth Ten Thousand Words. *Cognitive Science, 11*, 65-99.

Newell, A., & Simon, H. A. (1972). *Human Problem Solving.* Englewood Cliffs, N.J.: Prentice-Hall.

Pylyshyn, Z. W. (1981). The Imagery Debate: Analogue Media Versus Tacit Knowledge. *Psychological Review, 88* (1), 16-45.

Rich, E. (1983). *Artificial Intelligence.* N.Y.: McGraw-Hill Book Company.

Simon, H. A. (1972). What is Visual Imagery?: An Information Processing Interpretation. In L. W. Gregg (Eds.), *Cognition in Learning and Memory.* NY: John Wiley & Sons, Inc.

Sensory Discrimination in a Short-Term Trace Memory

J. Devin McAuley and **Sven E. Anderson** and **Robert F. Port** *
Department of Computer Science
Department of Linguistics
Indiana University
Bloomington, Indiana 47405
mcauley@cs.indiana.edu

Abstract

We propose a fully recurrent neural network to model low-level auditory memory in a task to discriminate intensities of sequentially presented tones across a range of varying interstimulus intervals. In this model, memory represents a *sensory-trace* of the stimulus and takes the form of slow relaxation of a number of units to a globally attractive equilibrium value near zero. The same-different judgment is based on a derivative of the output of the dynamic memory. Gaussian noise added to unit activations was found to improve the resilience of stored information although at the cost of decreased sensitivity. The model exhibits many qualitative properties of human performance on a roving-standard intensity discrimination task.

Introduction: Memory and Comparison

A critical step in the development of a model of auditory processing is the ability to discriminate input stimuli: the capacity to make same-different judgments about simple properties of a pair of stimuli. Since sequential presentation is unavoidable for auditory processing models, some form of memory for the first stimulus is required while the second is presented and a comparison made. If this task is approached from an engineering perspective, the problem is easily solved with a buffer which stores the first item perfectly (for all practical purposes) until the second item is available (Port, 1990). However, as we will show below, there is evidence against storage of the raw stimulus.

The process of serial comparison in relation to the underlying memory mechanism is an important issue. In traditional discrimination procedures, where the intensity level of the standard (or reference) stimulus, I, is held constant across trials and only the ΔI component of the comparison stimulus, $(I + \Delta I)$, changes from trial to trial, the duration of the interstimulus interval is found to have little or no effect. For example, in a single block of trials, a subject might be asked to discriminate between 50 dB and 50 $+\Delta I$ dB. Only a small increase in the Weber ratio is found when the two stimuli are separated by as much as 24 hours (Pollack, 1955). However, if a between-trial roving discrimination task is used, it is found that increasing ISI *does* reduce performance (Berliner & Durlach, 1973) [1].

To explain why roving the level of the standard between trials has such an effect, Durlach and Braida (1969) proposed two modes of memory processing for intensity discrimination: *sensory-trace coding* and *context-coding*. When using *context* mode, subjects are believed to base discrimination on some form of categorical description of the standard stimulus level. Such categorical descriptions are known to be very resistant to changes in ISI. Most studies of same-different comparison in cognitive science are implicitly models of *context-coded*, or categorical, representations—codes that are well-learned and highly resistant to temporal decay (e.g., Liberman et al., 1967; Gasser and Smith, 1991). Since in a roving-level discrimination task, intensity varies randomly, subjects cannot learn to use a categorical representation and must instead rely upon an ephemeral *sensory trace* of the stimulus. Thus, their performance deteriorates with increasing ISI.

The research reported here is an attempt to model the *sensory-trace* processing mode for intensity resolution. We believe that such work is an essential first step in the development of a biologically plausible general model of auditory pattern (category) recognition. Our goal was to construct a system that exhibits the general properties of human performance in roving level discrimination tasks. The evidence from human subjects experiments suggests that this requires performing auditory discrimination without using a special comparison buffer (and the specialized buffer-transfer operations that are implied). Instead, our system measures input signal intensity and then stores that value for the first stimulus while decaying slowly

*This research was supported by ONR grant N00014-91-J1261.

[1] In a roving discrimination task, ΔI is fixed and I varies between trials. For example, the subject might be asked to discriminate between 50 and 52 on one trial and between 40 and 42 dB on the next trial.

toward the equilibrium of the system (located near 0 intensity). When the second stimulus arrives, it re-excites the system to a value corresponding to the intensity of the second stimulus. A differentiation module computes a criterion variable based on information available just after the onset of the second stimulus.

A Model of Memory and Discrimination

The model presented here for intensity discrimination (see Figure 1) consists of two components: (1) a model of auditory short-term memory adapted from a model proposed by (Zipser, 1991) for cortical neurons that incorporates random output fluctuations; and (2) a decision model that generates a criterion variable used to decide when two stimuli differ based on the local change in the output of the memory module.

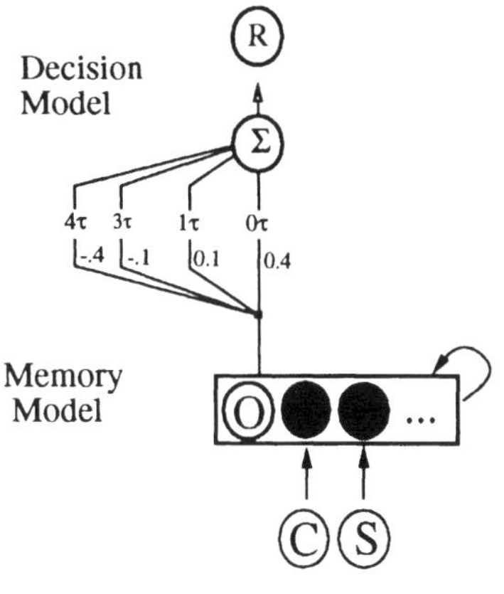

Figure 1: Network architecture. The cue (C) and stimulus (S) input units are shown below. Each unit connects to all units in the recurrent layer. The units of the memory module are fully interconnected and all weights are learned. The response unit R is connected to the output unit O via a set of weighted time-delay connections. The values of the time delays are shown on each connection along with the corresponding weights.

Memory Model. The memory model consists of 2 linear input units and 9 fully connected logistic units representing auditory short-term memory (STM), as shown in Figure 1. The inputs have connections to all STM nodes. The input units are a binary cue input and a real-valued stimulus representing intensity in the range $[0, 1]$. It is known that the rate of neural firing is an important cue for perceived loudness (Moore, 1989) and this corresponds to the activation level of individual units in this model. One of the STM units is the Output unit for the dynamic memory—the only unit trained directly during the learning phase of the simulations. All of the memory units have the activation function

$$y_i(t+1) = \phi(\sum_j w_{ji}y_j(t)+w_{si}x_s+w_{ci}x_c+\theta_i)+X_i(t)$$

where $\phi(x) = (1 + e^{-x})^{-1}$; the cue and stimulus inputs and weights are subscripted with c and s, respectively. X is a random variable drawn, on each time step for each unit, from a Gaussian distribution with mean and standard deviation μ and σ. The random variable X was included during testing trials to simulate random neural excitation of unit activations (Zipser, 1991). It was not included during training. In all of the simulations discussed in this paper, the biases θ_i were fixed at negative values in the range $[-1.0, -2.5]$, as in (Zipser, 1991), to avoid spontaneous unit activity. Also, the dynamics of $\phi(x)$ are only interesting when the biases are negative (McAuley, 1992).

Decision Model. The decision model implements a form of comparison without using buffers. The model consists of a set of several time-delayed connections from the output unit of short-term memory to the response unit. The weights on the connections between the two units are pre-wired and were not adjusted during training. The links effectively implement a low-pass filter that approximates the scaled derivative of the output unit's activation over 5 time steps. We anticipated that if short-term memory is effective in retaining past input values, then a following stimulus will perturb the memory model to the extent it differs from the initial input value. Of course, perfect memory is impeded by the imperfect initial encoding of the sensory stimulus, the internal noise of the trace memory, and the relaxation ("forgetting") of trace memory over time.

Training the Memory Model. The training task, shown in panel A of Figure 2, was to store in short-term memory a cued intensity value for an unspecified duration. During a training sequence, the network was presented with a cue input of 1.0 plus a random stimulus input from the interval $[0, 1]$. The cue may be thought of as representing a signal from some other part of the nervous system indicating that the value of the simultaneous external stimulus should be remembered. The network was trained to autoassociate the current stimulus input for a random number of time steps. For each trial the number of time-steps was drawn from a uniform distribution from 2 to 12. Following the initial stimulus and cue pair, up until the next stimulus and cue pair, the cue unit was set to 0.0 and the stimulus unit varied randomly within the range $[0, 1]$.

The network was trained using the real-time recurrent learning algorithm (Williams & Zipser, 1989) to update weights. All forward weights between the input and recurrent layers and all weights within the recurrent layer were modified during training. Training lasted for 400,000 iterations or

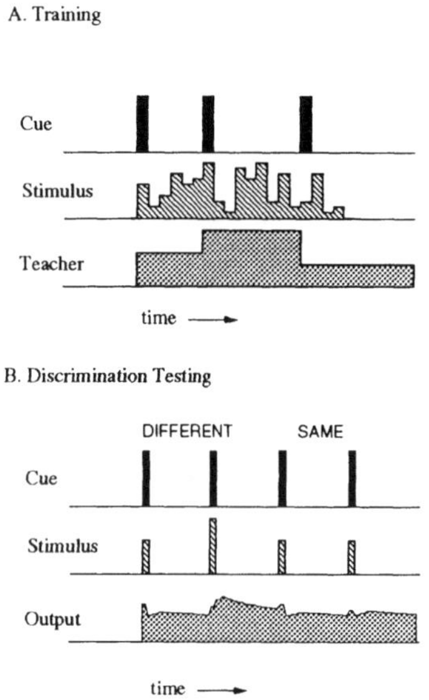

Figure 2: **A.** Cue, stimulus, and teacher values for a hypothetical training sequence. **B.** Actual testing data from a Different followed by a Same trial. The first two stimuli presented are Different (0.6 followed by 1.0). The next two stimuli are the Same (0.6 and 0.6). Note that all input values decay over time and are momentarily affected by introduction of the cue signal.

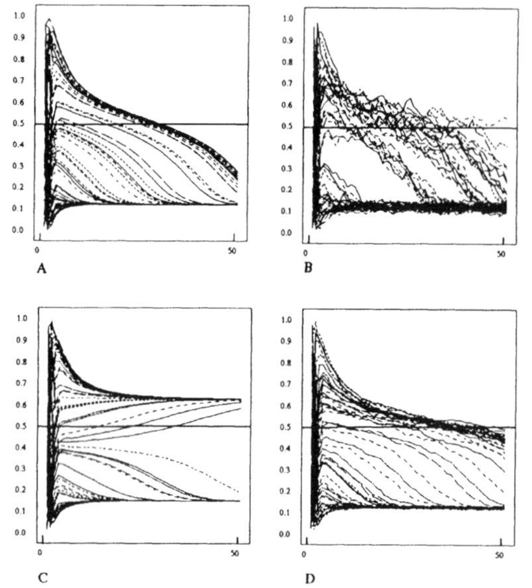

Figure 3: Qualitative dynamics of the model. Each plot shows the activation of the Output node in the range $[0, 1]$ for 50 time steps from 50 random initial conditions. **A.** without noise. **B.** with Gaussian noise, $\mu = 0$, $\sigma = 0.01$, the maximum amount of noise. **C.** with Gaussian noise, $\mu = 0.01$, $\sigma = 0.0$. **D.** with Gaussian noise, $\mu = 0.0025$, $\sigma = 0.005$

approximately 60,000 training trials. Final mean squared error approached 0.01 on the network examined below.

The training task is fairly difficult to learn because the stored input is real-valued and because the interstimulus interval varies between trials. The network has finite capacity, and cannot resolve all the possible real-valued inputs on the unit interval. The randomly varying ISI prevents anticipation of the time of occurrence of the next stimulus.

Simulations

Memory Dynamics. The basis for memory in this model is its very slow relaxation to equilibrium following presentation of the stimulus. The qualitative dynamics of the trained model were explored by setting the cue and stimulus units to 0.0, randomizing the initial activations, and then letting the network run for the 50 randomly chosen initial conditions. Graphs of the results of these tests are shown in Figure 3. All graphs show only one dimension of trace memory, the Output unit (with the range $[0, 1]$) for 50 time steps. Similarities were observed between hidden unit activity and single unit recordings in the auditory cortex of monkeys performing a memory task, as described in (Zipser, 1991). Panel A of Figure 3 shows that memory decay to equilibrium is very slow. For most initial conditions, decay is approximately linear. Even though during training, the network stored stimuli for at most 12 iterations, for some ('high intensity') initial conditions the network is still relatively far away from its asymptote after 50 iterations.

The effect of noise was also explored by adding a Gaussian distributed random variable to the output of each unit on each time step. The variance of this distribution is an important parameter in determining the qualitative dynamics. Panels B and D of Figure 3 represent standard deviations of 0.01 and .005, respectively. Compare the activation levels in these panels at iteration 50 with panel A. Increasing the noise variance slows memory decay, but, as we show below, this results in degraded resolution of the original stimulus intensity. Possibly then, optimal performance is a compromise between sensitivity (improved by less noise) and memory (improved by greater noise variance). Panel D would be a candidate for such a compromise. Panel C shows that the addition of a sufficient amount of 0-variance noise (equivalent to the addition of a sufficiently large constant) creates a second equilibrium point in the system. This suggests that the system achieves a longer memory span (as in Panel B and D) by operating near a bifurcation point.

Because the rate at which the network approaches equilibrium is, at high and low activations, proportional to its overall level of activity, one might suspect that intensity discrimination is better at lower activations, in accordance with Weber's Law ($\Delta I/I = k$). While this is somewhat true for the network, the midrange of the output unit has a rather linear decay, and we found that Weber's law does not hold very well throughout the range of intensities encoded by the memory units of this model.

Memory Span vs Resolution in Intensity Discrimination. Network memory was evaluated using a same-different between-trial roving discrimination task. Panel B of Figure 2 shows two sample trials, one Different and one Same. I varied between 0.1 and 0.9 *between* trials, while ΔI and the interstimulus interval (ISI), measured in discrete time steps, remained fixed. Testing blocks consisted of 1800 trials. Blocks were run for all combinations of ΔI in the set {0.02,0.04,0.06,0.08,0.1} and ISI in the set {3,7,9,11,15,19,29} of time steps. The model's performance on a block of trials was measured by computing hit and false alarm rates for a range of response thresholds applied to the response unit. d' was found to be roughly constant, excluding edge effects, indicating that the response unit approximately obeys the assumptions of signal detection theory (Swets, 1961). The graph in Figure 4 depicts the ΔI required to achieve performance of $d' = 1$ (implying 71% correct with no response bias) as a function of ISI. Four different noise conditions are plotted corresponding to fixed $\mu = 0.0025$ and σ between 0 and 0.01. All four plots show results consistent with human performance on a roving level discrimination task; stimulus sensitivity degrades with increasing ISI. The rate at which sensitivity decreases is inversely related to the amount of noise variance. For large variances (shown with the filled circles in Figure 4), sensitivity is lost for shorter ISIs (where the change in intensity at threshold is .10) yet there is little degradation with increasing ISI. On the other hand, with no noise (shown in the open triangles), an intensity difference of only .04 can be resolved at short ISIs yet performance at longer ISIs is *worse* than the noisier conditions.

In the context of the underlying dynamics shown in Figure 3, these results can be explained. A relatively large internal noise variance can slow memory decay to equilibrium and consequently slow loss of resolution over time. In this model, there is a trade-off between memory and resolution. Memory is improved at the cost of resolving power and improved resolution sacrifices memory performance. Optimal discrimination in a task in which ISI varies within trials might be best achieved by a noise condition which produces a weighted balance between memory span and resolving power.

Discussion

These simulations bear on at least two issues: the nature of human memory for intensity of sensory stimulation, and the role of noise in facilitating the memory function of a dynamic system.

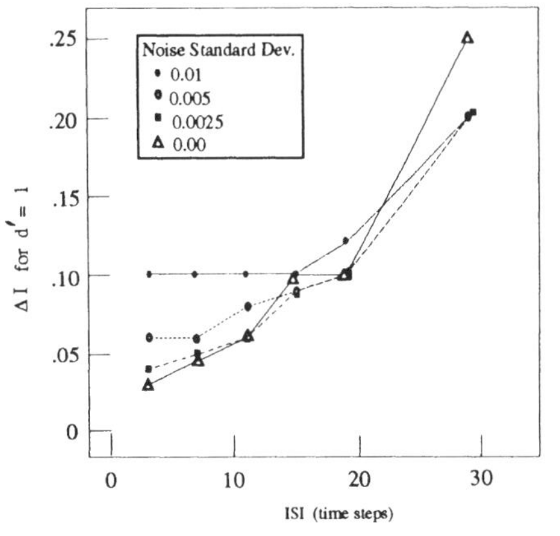

Figure 4: ΔI required for d' of 1 (or maximum percent correct discrimination of 71%) as a function of interstimulus interval. Each line represents a condition with neural noise added, having a mean of ($\mu = 0.0025$) and standard deviation as specified in the legend. The filled circles have the most noise variance and the open triangle has the least variance

Comparison with Human Sensory Memory. We have attempted to model certain properties of human performance on auditory discrimination tasks. Although we only modelled performance for a single tone, our model can easily be extended by creating a bank of identical modules, covering the entire auditory spectrum. The results of these simulations exhibit some critical properties of human performance on analogous roving-standard discrimination tasks. Our model is able to store the intensity of a stimulus for a short while and exhibits decay with accompanying loss of performance. The critical effect of interstimulus interval on our model's performance is analogous to results with human subjects found by Berliner and Durlach (1973).

One aspect of the current model that seems incorrect is that it confounds decay with intensity—inputs stored longer evolve in state-space through representations of less intense inputs, since the fixed point of the system lies near 0 intensity. A more appropriate consequence of decay would seem to be greater uncertainty about intensity—rather than weaker perceived intensity. This may be achievable within our current model by using several trace memory modules with equilibria at different locations in activation space. A stimulus would decay towards lower intensities in some modules, but to higher intensities in others. The mean activity, or population code, of the modules would then represent the stimulus trace, with variance representing the level of stimulus uncertainty.

The Role of Noise in Trace Memory. In our model, a dynamic system has learned to function near a bifurcation point—where a single attractor gives way to a pair of attractors. Adding a constant to unit activations acts as a bifurcation parameter and creates a second attractor. One consequence of this is that the vector field of the system state-space is relatively weak—and the equilibrium near zero is less attractive than it would be farther from the bifurcation point. The effect of adding noise is to 'tease' the memory dynamics by causing vacillation between systems with one and two attractors. Thus relaxation toward equilibrium is slowed and the system behaves as though the vector field were flatter. Of course, the noise also increases uncertainty in the resolution of stimulus intensity.

As suggested for the intensity decay problem, the undesirable effects of noise might be minimized by replication of the memory module, while preserving the benefits (improved performance at long ISI). For example, if several independent modules were used to store the intensity of a single frequency band, their mean would provide a much better estimate of the original intensity than a single module could. It is interesting to hypothesize that the well-documented noisy behavior of real neural systems functions in part to improve memory span in a way similar to that of our model, by slowing relaxation to equilibrium.

Acknowledgments

The authors wish to express their gratitude to Gary R. Kidd, Catherine Rogers, Joseph Stampfli, and Charles S. Watson, for many useful discussions and comments on various drafts of this paper.

References

Berliner, J. E. and Durlach, N. I. 1973. Intensity perception. IV. resolution in roving-level discrimination. *Journal of the Acoustical Society of America*, 53(5):1270–1287.

Berliner, J. E., Durlach, N. I., and Braida, L. D. 1977. Intensity perception. VII. further data on roving-level discrimination and the resolution and bias edge effects. *Journal of the Acoustical Society of America*, 61:1577–1585.

Durlach, N. and Braida, L. 1969. Intensity perception. I. preliminary theory of intensity resolution. *Journal of the Acoustical Society of America*, 46(2):372–383.

Gasser, M. and Smith, L. 1991. The development of the notion of sameness: A connectionist model. In *Proceedings of the Thirteenth Annual Meeting of the Cognitive Science Society*, pages 719–723.

Georgopoulos, A., Schwartz, A., and Kettner, R. 1986. Neuronal population coding of movement direction. *Science*, 233:1416–1419.

Gottlieb, Y., Vaadia, E., and Abeles, M. 1989. Single unit activity in the auditory cortex of a monkey performing a short term memory task. *Experimental Brain Research*, 74:139–148.

Green, D. 1988. *Profile Analysis: Auditory Intensity Discrimination.* Oxford University Press.

Laming, D. 1986. *Sensory Analysis.* London: Academic Press.

Liberman, A., Cooper, F., Shankweiler, D., and Studdert-Kennedy, M. 1967. Perception of the speech code. *Psychological Review*, 74:431–461.

McAuley, J. D. 1992. The dynamics of sensory trace memory. Unpublished manuscript.

Moore, B. C. J. 1989. *An Introduction to Psychology of Hearing.* Harcourt Brace Jovanovich, third edition edition.

Pollack, I. 1955. Long-time differential intensity sensitivity. *Journal of the Acoustical Society of America*, 27:330–381.

Port, R. F. 1990. Representation and recognition of temporal patterns. *Connection Science*, 2:151–176.

Swets, J. A. 1961. Is there a sensory threshold? *Science*, 34:168–177.

Williams, R. and Zipser, D. 1989. A learning algorithm for continually running fully recurrent neural networks. *Neural Computation*, 1(2):270–280.

Zipser, D. 1991. Recurrent network model of the neural mechanism of short-term active memory. *Neural Computation*, 3:179–193.

A Speech Based Connectionist Model of Human Short Term Memory

Dr. Dimitrios Bairaktaris[1] **Prof. Keith Stenning**[2]

Human Communication Research Centre,
University of Edinburgh,
2 Buccleuch Place,
Edinburgh EH8 9LW, Scotland
tel: (+44) 31 6504450, facsimile: (+44) 31 6504587,
dimitris@uk.ac.ed.cogsci, keith@uk.ac.ed.cogsci

Abstract

In recent years connectionist modelling of Short Term Memory (STM) has been a popular subject of research amongst cognitive psychologists. The direct implications in natural language generation and processing, of the speech based phenomena observed in immediate recall STM experiments, make the development of a psychologically plausible STM model very attractive. In this paper we present a connectionist Short Term Store (STS) which is developed using both traditional STM theories of interference and decay trace. The proposed store has all the essential characteristics of human short term memory. It is capable of on-line storage and recall of temporal sequences, it has a limited span, exhibits clear primacy and recency effects, and demonstrates word-length and phonological similarity effects.

Introduction

Short Term Memory (STM) has been a major subject of investigation for cognitive psychologists since the 50's. Initial experiments established that STM has a limited storage capacity, or span (Miller 1956). It was later shown that when span is exceeded, immediate STM recall performance is impaired in a very specific way. Only the first few, primacy, and the last few, recency, memory items can be recalled at some significant level of accuracy (Postman & Philips 1965). Subsequent experiments established that span is affected when memory items are phonologically similar. This feature of STM is known as the Phonological Similarity Effect (PSE) (Baddeley 1966). Later, it was also shown that the time taken to articulate a memory item has a negative effect on span.

[1] Joint Council Initiative in HCI/Cognitive Science Training Research Fellow.

[2] Supported by the Economic and Social Research Council.

Recall performance is reduced when longer words are stored in STM (Baddeley et. al. 1975). This STM feature is known as the Word Length Effect (WLE) and together with PSE suggests that for written verbal material STM access involves some form of speech processing. Further experiments using articulatory suppression, during both STM storage and recall phases (Baddeley 1990), have shown that there exists a route for STM access which involves phonological encoding with the possible use of a lexicon. This immediately relates STM performance to language generation processing tasks and makes the modelling of STM a very attractive task indeed. There are two major STM theories. One claims that span is limited mainly due to interference; more recent memory traces affecting earlier ones. The other claims that forgetting occurs because memory traces decay through time. Interference theory has been supported by a number of mathematical STM models such as (Murdock 1983) (Schweikert 1986) and does conform with psychological data. However, mathematical models fail to provide an account for some of the speech based characteristics of STM recall experiments (Brown & Hulme 1991). Trace decay as applied in the working memory model of STM (Baddeley 1990) provides an account to all speech based aspect of STM recall but fails in at least two ways. It does not provide an explicit computational model of STM which will facilitate testable theoretical predictions, while the articulatory loop rehearsal mechanism is not completely consistent with recent experimental evidence (Baddeley 1986) (Howard & Franklin 1990). Pioneering work on STM connectionist modelling was done by Grossberg (1976) and was further strengthened with the connectionist theory revival of the early 80's resulting in a number of connectionist STM models which conformed, to some degree, with psychological evidence. Some of these models adopted the interference theory (Wang & Arbib 1991) and others the trace decay theory (Brown 1989) (Burgess & Hitch 1992). We discuss these models and some general STM modelling issues next.

Connectionist Models of STM

One particular area in connectionist theory which is closely related to STM modelling is that concerned with the problem of serial order. In its simpler form the problem of serial order manifests itself in list learning and recall, a task very similar to that of immediate recall STM experiments. Is it possible for a connectionist network to learn a sequence of patterns and recall them in their original order? Various solutions have been given to this problem (Jordan 1986) (Elman 1988) (Norris 1990) (Houghton 1989) (Bairaktaris 1992). It is essential for any STM model to perform the serial order task. For example, Brown (1990) uses the solution proposed in Norris (1990) to construct his STM model, Burgess and Hitch use the solution of Houghton (1990) for their sequence storage and generation, while Wang & Arbib (1991) STM model is a improved variation of Houghton's solution.

In general most of the STM models in the literature perform well in the sequence generation task. Performance starts breaking down when they are tested against STM performance criteria such as span, primacy and recency, WLE and PSE. From all the connectionist STM models only Burgess & Hitch's (1992) makes an attempt to address all the above criteria. Their model is based on the trace decay theory of STM but a major part of its dynamical behaviour depends on the existence of random noise rather than decay. In brief, their model does well with span, primacy and WLE, but fails to demonstrate recency, PSE and it is not capable of on-line list learning.

A close study of the Burgess & Hitch model revealed that the majority of the problems with their model were due to the fact that they failed to distinguish between a Short Term Memory and Short Term Store (STS). To identify a separate STS embedded in STM is clearly consistent with the working memory (Baddeley 1986) framework. Such an approach, also favoured by Brown & Hulme (1991), provides the advantage of separating the process of generating the phonological code from the actual storage and retrieval of memory items. This paper describes an STS mechanism which has an accurately defined phonological interface to the main STM mechanism.

The model

We will divide the description of our model in two parts. First we will give an account of its static characteristics and then we will describe its dynamic behaviour as an STS module. The network model of STS is shown in Exhibit 1.

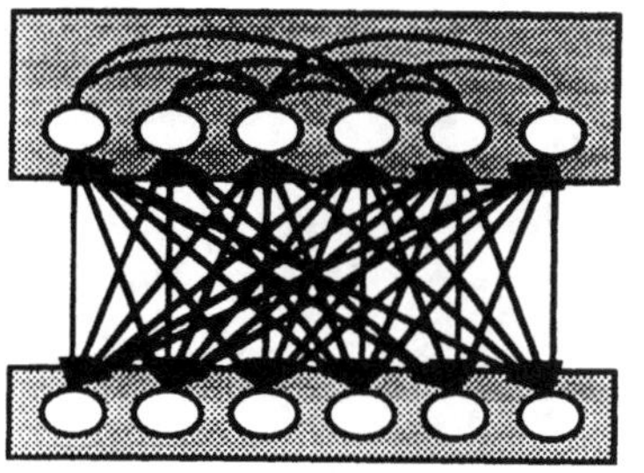

Exhibit 1. Two layer Short Term Store.

It comprises two layers of units, an input layer where the phonological code of the memory items is clamped and the STM layer where the actual memory items are stored. The two layers of units are fully connected with bidirectional connections. The STM layer nodes are also fully interconnected with excitatory connections. Similar STM models, which emphasize different aspects of STM, have been previously described in the literature (Zipser 1991) (Wang et. al 1991). The phonological encoding layer is provided with a sufficient number of nodes in order to store all the significant features of the memory item. It is assumed that a pre-processor is available for the generation of the phonological code. Such a simple pre-processor network is described in Burgess & Hitch (1992). It must be made clear however that the generation of the phonological code is a complex process which possibly requires the generation of intermediate phonetic codes (Besner 1987) as well as the use of a lexicon (Monsell 1987). It should also be mentioned that the quality of the phonological code used by the STM layer is subject to modality effects and various reception conditions such as articulatory suppression (Howard et. al. 1990). We are currently developing a model which deals explicitely with the generation of the phonological code.

Each node in the input layer stores a feature of the item's phonological code in a binary form and in such a way that phonologically similar items have similar codes. The input nodes propagate their values to the STM layer where a different node is allocated to the representation of every memory item. The connections between the input and the STM layer are modified using a hebbian learning rule in order to retain the phonological code of every memory item. It is shown in Bairaktaris (1991, 1992) that using a modifiable threshold technique, a one-to-one correspondence between STM nodes and memory items can be achieved without the use of intra layer inhibitory connections (Grossberg 1976). To avoid limiting the system's capacity artificially, STM nodes are allocated to the memory items dynamically (Bairaktaris 1991). For a network with j input nodes the STM node activation A_i and output O_i are computed as follows:

$$A_i = \Sigma_j W_{ij} P_j$$

$$O_i = A_i \text{ if } A_i > T_i \text{ and } O_i = 0 \text{ if } A_i \leq T_i$$

where **P** is the vector of the activations of the input nodes, **W** is the weight matrix of the connection between the input nodes and STM nodes and **T** is vector of the threshold of the STM nodes. The node allocation, threshold setting and weight modification mechanisms are described in more detail in Bairaktaris (1991, 1992). When an STM node fires its output decays through time as follows:

$$O_i(t+1) = O_i(t) \frac{\delta}{e^{(\lambda - O_i(t))}}, \text{ where } \delta, \lambda \text{ are constants}$$

The effect of the above decay rule is shown in Exhibit 2 for $O_i(0) = 1$, $\delta = \lambda = 0.6$.
The weight Zij on the connection between nodes i and j in the STM layer is modified as follows:

$$Zij(t+1) = Zij(t) + O_i(t)\, O_j(t) \quad (1)$$

During the recall phase, where there no activation propagated from the input layer to the STM layer, STM nodes compute their activation solely on feedback from other STM nodes as follows:

$$A_i = \Sigma_j Z_{ij} O_j$$

It is assumed that every node receives a constant amount of activation from background noise. However not all the nodes fired at the presence of background noise. It is only the node which represents the first memory item, and has the lowest threshold, which will fire thus initiating the recall phase:

$A_1(1) = \kappa$ where κ is constant; typically $\kappa = 0.15$
The training regime between the input layer and the STM layer guarantees that there is a one-to-one correspondence between the memory items and the STM nodes. In recall mode however, more than one STM node can be active at any time. This means that the system cannot decide about the exact recall sequence of the memory items. The relative output of node i (trace decay) against the sum of output of all the nodes (interference) in STM is the probability (Pi) assigned to the hypothesis of the system recalling memory item i at time t:

$$Pi(t) = \frac{Oi(t)}{\Sigma_j Oj(t)}$$

The proposed network architecture is very similar to the model of the articulatory loop described in Baddeley (1986). There are however two major differences between the Baddeley approach and our model.

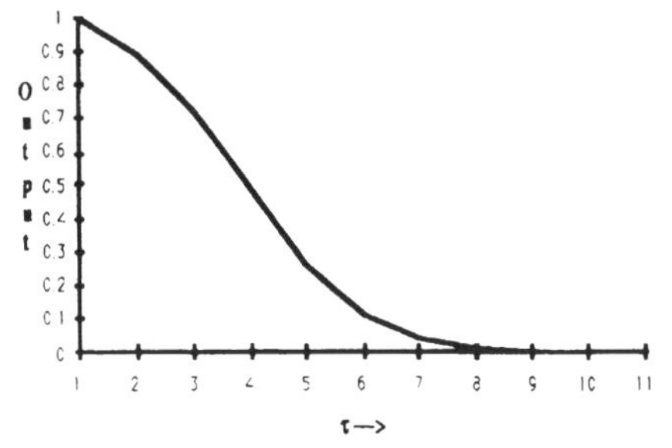

Exhibit 2. Output from STM nodes decays through time.

Baddeley proposes that memory items are rehearsed and dynamically stored in the loop between the phonological encoding and the STM layer and that span is limited purely due to trace decay. Our model stores the memory items in the STM nodes while the item sequence is stored, maintained and reproduced by the recurrent connections in the STM layer. Rehearsal is possible, but not essential (Howard et. al. 1990), via the bidirectional connections between the input layer and STM. Furthermore, recall of the memory items depends on both trace decay of individual memory nodes and interference from other memories.

We will now describe the dynamic behaviour of the proposed model. To simplify the description, we will focus on the dynamic behaviour of the STM layer assuming that the input layer provides all the appropriate phonological memory traces. As is described above when a node in the STM is allocated to the representation of a memory item its output is set to 1 and it decays thereafter. The longer it takes for the second memory item to be registered in the STM layer the weaker the recurrent output signal from the previous memory item becomes. When the second memory item is allocated a node in the STM layer, the connection between the previously active STM node and the current one is modified as shown in (1). Modifications on the STM recurrent connections occur in the same way every time a new memory item it added. A close inspection of Exhibit (3) reveals that by the time the ninth memory item is registered in STM the output of the first item has diminished to zero. This shows that the output decay mechanism applied on the STM nodes, imposes an implicit limit on span which is very close to the empirical 7±2 observation made in Miller (1956).

During recall the node representing the first memory item becomes active due to background noise and initiates the sequence generation process. Immediately after the first node all the other nodes in the layer receive varying degrees of activation depending on their original position in the sequence. Depending on whether their activation exceeds their preset threshold, they activate themselves or not. A wave of nodes firing is spread through the layer.

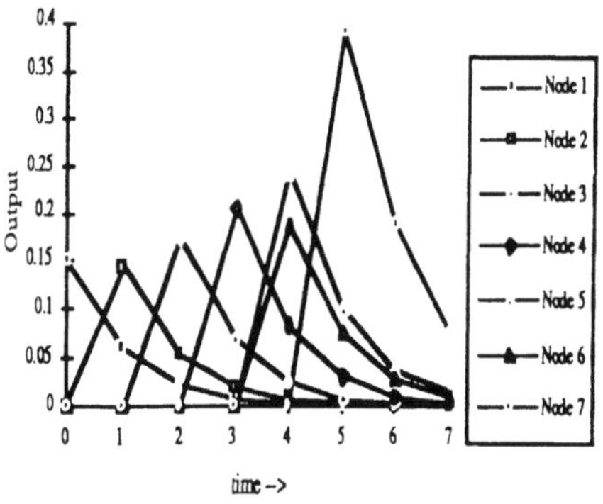

Exhibit 3. Output of STM nodes during recall.

Exhibit 3 shows the output of 7 STM layer nodes through time. The original encoding corresponds to a sequence of 7 memory items presented at equal time intervals. In the example shown $\delta = \lambda = 0.6$ and $\kappa = 0.15$. Exhibit 3 demonstrates a case where more than one node fires at the same time. At time $t = 4$, nodes 5 and 6 fire simultaneously, but node 5 has a relatively stronger output than node 6. Therefore the relative probability of the system recalling memory item 5 at time 4 is higher than the probability of recalling item 6. In general, the probability of recalling a memory item X at a particular time Y, is equivalent to the relative output of the node representing item X at time Y, over the sum of the output from all the nodes in STM at the same time Y. To place this into the context of STM immediate recall experiments, the probability of recalling item 1 at time 0, item 2 at time 1, item 3 at time 2 and so on is equivalent to the probability of correctly recalling all the memory items in their original positions in the sequence. Exhibit 4 shows the recall probabilities for every memory item, for the same sequence of items and same parameter settings of exhibit 3.

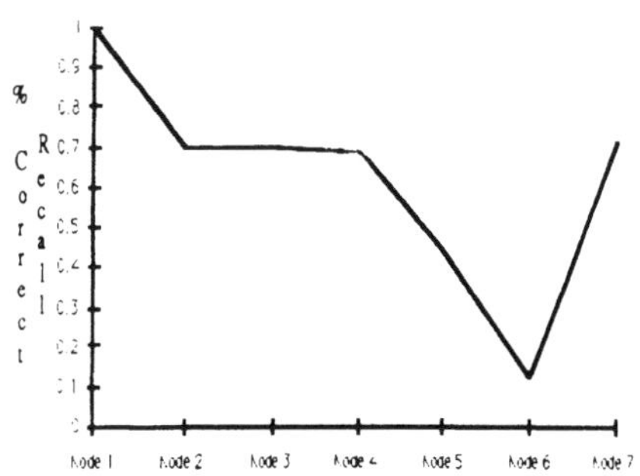

Exhibit 4. The probabilities of correctly recalling 7 memory items in their original ordering.

The above interpretation of the network's dynamics is used throughout the simulation results presented in the following section.

Simulation Results

The proposed STM store was simulated for a variety of different parameter settings before the results reported below were achieved. Setting the κ parameter of the network proved to be a difficult task, but at the same time a number of 'interesting" network behaviours emerged from the simulation process. These behaviours are currently analyzed within the context of neuropsychological, 'patient specific', STM evidence. Here we will only refer to the simulations results that are relevant to our task; to demonstrate that the proposed model conforms to psychological evidence. In all the results presented below, $\kappa = 0.15$ and $\delta = \lambda = 0.6$.

Span-Primacy - Recency
Exhibit 4 shows that the model demonstrates clear primacy and recency effects. For a list of 7 items the network is capable of recalling all the items in their original order, with greater confidence for the first and last list items and smaller confidence for the intermediate items. As is shown in Exhibit 5, the network also demonstrates primacy and recency effects for list lengths of 10 and 20. When span is exceeded, the probability of correctly recalling intermediate list items is effectively 0.

Phonological Similarity
In the introduction of the paper it was mentioned that when the memory items are phonologically similar immediate recall success rates are decreased. Phonologically similar memory items will produced phonologically similar codes at the input layer of our model and because of the hebbian learning algorithm between the input and the STM layer activation will pass not only to the node allocated to the current memory item but also to the nodes which represent phonologically similar items.

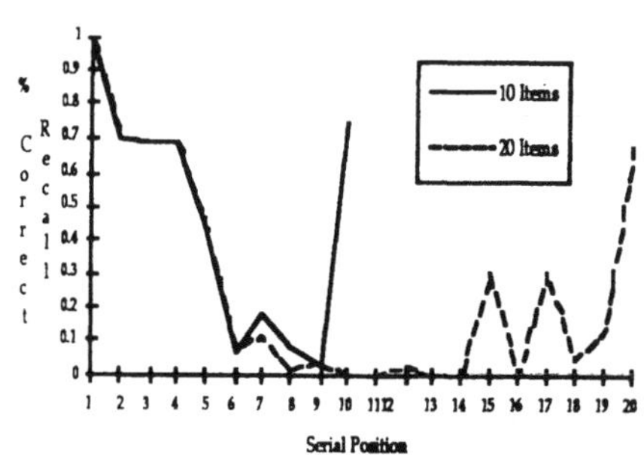

Exhibit 5. Primacy and recency effects for lists of 10 and 20 items

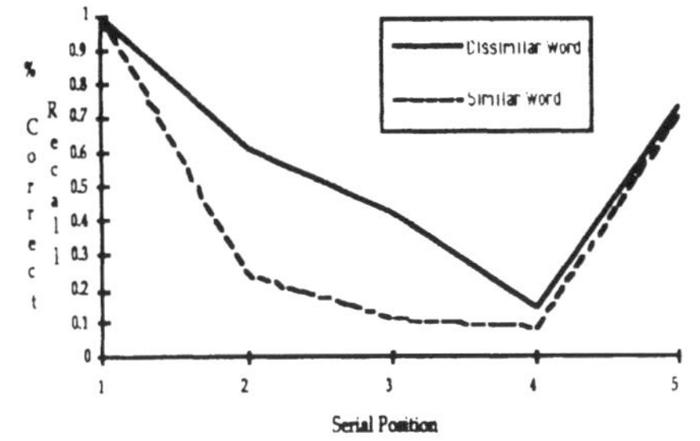

Exhibit 6. Phonological Similarity Effect when recalling a lists of 5 phonologically similar and dissimilar items.

The effect of phonologically similar memory items on the model's recall rates is shown in Exhibit 6 for a list of 5 similar and 5 dissimilar items. The phonologically similar list has lower recall rates that the dissimilar list, still both lists maintain the characteristic primacy and recency effects. An close of node outputs in the STS revealed that items with similar phonological representation are more likely to be recalled in their reverse list order.

Word Length Effect
In the description of our STS model it was mentioned that for the generation of the phonological code pre-processing of the raw data has to be made. It is reasonable to assume that the time taken to articulate a memory item is proportional to the pre-processing time required to generate the phonological code. This means that for longer words it will take longer before our model is provided with its phonological code, and for shorter words the generation of the equivalent phonological code will be shorter. In order to simulate the time taken to articulate a word in our model, we modified the number of time steps taken before two consecutive memory items are clamped at the input units. In all the simulations described above a new memory item was encoded in the STM at every time step.

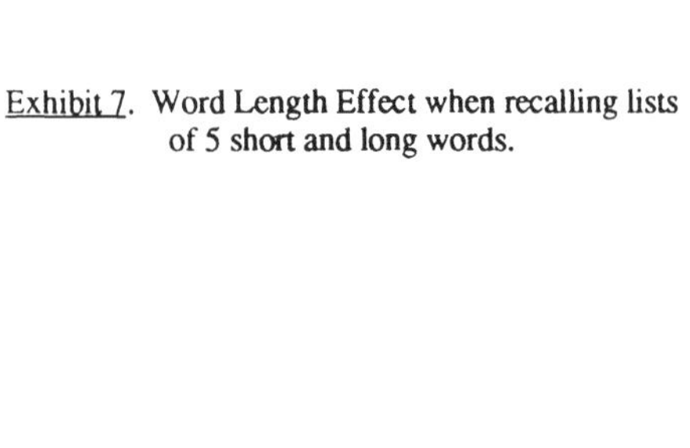

Exhibit 7. Word Length Effect when recalling lists of 5 short and long words.

Exhibit 7 shows recall rates for a list of 5 items, for both the standard case of registering a new memory item at every step (Short Word) and for the case where a new memory item was registered every three time steps instead of one (Long Word). Exhibit 7 shows that recall rates are worse for long words list than the short words list. Primacy effect is present in both cases, but recency is only present for the short word list case. Absence of primacy for the long word list case looks somewhat inconsistent with the psychological evidence. However, looking back at the original word length experiments (Baddeley et. al 1975) the recency effect in their experimental results is not very strong either. In fact the graph of Exhibit 7 is extremely similar to the equivalent word length effect graph in (Baddeley et. al. 1975).

Conclusions

A connectionist network model of a short term memory store was presented. The proposed network architecture comprises a fully interconnected layer of nodes which interacts with the core of the Short Term memory using a layer of input units where the phonological code of the memory item is clamped. The model uses a constructive learning algorithm which combined with a hebbian-type synaptic modification rule allows on-line storage of memory items. The proposed network is different to earlier STM models, in that the interpretation of its dynamics incorporates both the decay trace and the interference STM theories. Simulation results demonstrated that the model conforms to some of the major STM psychological evidence. The basic span, primacy and recency STM effects, are faithfully reproduced by the network model. These are the standard benchmark STM effects that have to be met by all STM models. It is in the interpretation of the main speech based STM effects, such as word length and phonological similarity, that our model makes a significant contribution. It provides an explicit computational account of the above effects by accurately reproducing the psychological data. Furthermore, it can explain some more subtle speech based STM effects, such as phonemic confusion, where non-adjacent phonologically similar list items are transposed during recall.
The proposed model is currently augmented with the development of a network model for the generation of the phonological code. This is intended to provide a computation account for some of the lexical access STM effects such rhyme and pseudo-homophone judgement.

References

Baddeley, A.D., 1966. Short-term memory for word sequences as a function of acoustic, semantic and

formal similarity. Quarterly Journal of Experimental Psychology, 18:302-309.
Baddeley, A.D., 1986. Working Memory. London: Oxford University Press, Oxford Psychology Series, 11:75-107.
Baddeley, A.D., Thomson, N. Buchanan, M., 1975. Word Length and the Structure of Short Term Memory. Journal of Verbal Learning and Verbal Behaviour 14:575-589.
Bairaktaris, D., 1991. Adaptive Pattern Recognition in a Real-World Environment. PhD. Thesis, Dept. of Computational Science, University of St.Andrews.
Bairaktaris, D.,1992. Discovering temporal structure using Hebbian learning. In Beale, R. and Finlay, J. (eds), Neural Networks and Pattern Recognition in Human Computer Interaction. Ellis Howard, U.K.
Besner, D., 1987. Phonology, Lexical Access and Reading and Articulatory Suppresion: A critical review. The Quarterly Journal of Experimental Psychology, 39A:467-478.
Brown, G.D.A, & Hulme, C., 1991. Connectionist Models of Human Short-Term Memory. In Progress in Neural Networks. ed.O. Omidvar, publ. Ablex, Norwood NJ, in press.
Brown, G.D.A, 1990. A Neural Net Model of Human Short-term Memory Development. In Lecture Notes in Computer Science: Proceedings of EURASIP Workshop on Neural Networks. Heidelberg: Springer Verlag.
Brown, G.D.A., 1989, A Connectionist Model of Phonological Short Term Memory, In Proceedings. of the 11th Annual Conference of the Cognitive Science Society, 572-579. Hillsdale, NJ: Lawrence Erlbaum Associates.
Burgess, N., Hitch, G., 1992. Towards a network model of the articulatory loop. Journal of Memory and Language.
Burgess, N., Moore, M.A., Shapiro, J.L.,1990. Human-Like Forgetting in Neural Network Models of Memory, Technical Report, Dept. of Theoretical Physics, University of Manchester.
Elman, J, 1998. *Finding Structure in Time*, CLR Technical Report 8801, Centre for Research in Language, University of California, San Diego.
Grossberg, S., 1976. Adaptive Pattern Classification and Universal Recoding: I Parallel Development and Coding of Neural Feature Detectors. Biological Cybernetics, 23:121-134.
Hinton, G.E., Plaut, D.C., 1987. Using fast weights to deblur old memories. In Proceedings. of the Ninth Annual Conference of the Cognitive Science Society. Hiilsdale, NJ: Lawrence Erlbaum Associates.
Houghton, G., 1989. The problem of serial order: A neural network model of sequence learning and recall. In Proceedings 2nd European Workshop on Language Generation, Edinburgh.
Howard, D., Franklin, S., 1990. Memory Without Rehearsal, In Neuropsychological Impairments of Short Term Memory:287-320. eds. Vallar & Shallice, Cambridge University Press.
Jordan, M.I., 1986. Serial Order: A Parallel Distributed Processing Approach. Technical Report 8604. Institute for Cognitive Science, University of California, San Diego.
Miller, G.A., 1956. The magical Number Seven, Plus or Minus Two: Some limits on Our Capacity for Processing Information. Psychological Review, 63:81-97.
Monsell, S., 1987. On the relation between lexical input and output pathways for speech, Language Perception and Production:273-311. Academic Press Inc.
Murdock, B.B., 1983. A distributed Model for Serial-Order Generation, Psychological Review, 90:316-338.
Norris, D., 1990. Dynamic net model of human speech recognition. In G.T. Altman (Ed.) Cognitive Models of Speech Processing: Psycholinguistic and Computational Perspectives, MIT Press, in press.
Postman, L., & Phillips, L.W., 1965. Short-term temporal changes in free recall. Quarterly Journal of Experimental Psychology, 17:135-.
Schweickert R., & Boruff, B., 1986. Short-Term Memory Capacity: Magic Number or Magic Spell?. Journal of Experimental Psychology, Learning, Memory, and Cognition: 419-425.
Wang, D., Arbib, A.M., 1991, A Neural Model of Temporal Sequence Generation with Interval Maintenance. In Proceedings of the 13th Annual Meeting of Cognitive Science Society, 944-948.
Waugh, N.C., & Norman, D.A., 1965. Primary Memory. Psychological Review:89-104.
Zipser, D., 1991. Recurrent Network Model of the Neural Mechanism of Short-Term Active Memory. Neural Computation, 3:179-193.

Does Memory Activation Grow with List Strength and/or Length?

David E. Huber, Heidi E. Ziemer, and Richard M. Shiffrin
Indiana University
Bloomington, IN 47405
dhuber@ucs.indiana.edu
hziemer@ucs.indiana.edu
shiffrin@ucs.indiana.edu

Kim Marinelli
Northwestern University
Evanston, IL 60208
kim@nugm.psych.nwu.edu

Abstract

Recognition of an item from a list is typically modeled by assuming that the representations of the items are activated in parallel and combined or summed into a single measure (sometimes termed 'familiarity' or 'degree-of-match') on which a recognition decision is based. The present research asks whether extra items (length), or extra repetitions (strength), increase this activation measure. Activation was assessed through examining hits and false alarms as the length or strength of word categories were varied. The use of a categorized list insured that response criteria were not changed across the length and strength manipulations. The results demonstrated that: 1) The activation does not change with an increase in the strength of presented items other than the test item; and 2) The activation is increased by an increase in the number of presented items in a category. The results provide important constraints for models of memory, because most models predict or assume either that activation grows with both length and strength, or grows with neither. In fact, the only extant model that can predict both the length and strength findings is the differentiation version of the SAM model (Shiffrin, Ratcliff, & Clark, 1990).

This research was supported by grants AFOSR90-0215 to the Institute for the Study of Human Capabilities, and by grants NIMH 12717 and AFOSR 870089 to Richard M. Shiffrin.

Introduction

The explosion of interest in memory models associated with the advent of neural net and connectionist frameworks has called into focus certain fundamental assumptions about memory representation and retrieval. Many models assume that the result of probing memory is the activation of all of memory. Especially for recognition tests, it is typically assumed that the recognition decision ('old' or 'new') is based on a single number, representing total activation (or 'familiarity', or 'degree of match'). In the present research we ask whether activation is increased by extra items added to memory (i.e. by list-length) and whether activation is increased by stronger or repeated items added to memory (i.e. by list-strength). The answers provide critical constraints for modelers of memory.

To help in understanding the experiments and the models, we review briefly the way in which performance in recognition tasks is measured. The subject studies a list of items and is then given items to judge as 'old' or 'new'. Responses of 'old' to items from the list (targets) are correct and termed 'hits'; responses of 'old' to new items (distractors) are incorrect and termed 'false alarms'. These data are usually related to theory in the following way: The subject is assumed to base the recognition judgment upon a measure representing 'familiarity', 'summed activation', 'degree of match', or some similar statistic. It is assumed that targets have a distribution of familiarity with a somewhat higher mean than the distribution for distractors, as shown in Figure 1. A given trial results in a sample from either the target or distractor distribution: An old response is given if the sampled value is higher than a subject-chosen criterion. Thus, the hit probability is the area under the target curve above the criterion, and the false alarm probability is the area under the distractor curve above that criterion.

If an experimental manipulation is carried out on a list (such as increasing its length or strength), it is of course possible that the subject will adjust the criterion to a new position, changing the hit rates and false alarm rates. For this reason, in most studies the concern is not with the absolute levels of hits and false alarms but rather a comparison between them that can be used to measure sensitivity of performance. The measure **d'** is theoretically independent of the placement of the criterion and is defined as the difference between the means of the target and distractor distributions, divided by the (common) standard deviation:

$$d' = \frac{\mu_t - \mu_d}{\sigma}$$

This measure is simply calculated from the standard normal (z) transforms of the hit and false alarm rates.

In the present research, however, we are interested in the placements of the distributions and their movements, information not available from **d'**. We therefore developed an experimental procedure in which it is reasonable to expect the criterion to remain fixed over the conditions of interest. Under these circumstances, various models can be discriminated by the values of the hit and false alarm rates.

Experiments

We embedded categories of words in a single long list for study. Words from a given category (but never the prototype) were spaced randomly throughout the list, disguising the category structure. The length and strength of different categories were varied. Following presentation of the study list, items were randomly chosen and tested. The distractors that were tested included non-studied exemplars from each category, non-studied category prototypes and items from none of the categories. A studied item (target) from each category was also tested.

The experiment consisted of eight conditions: 4 length manipulations and 4 strength manipulations. The L-1, L-2, L-6, and L-10 conditions were comprised of 1, 2, 6, or 10 words chosen from a given category. Each of these words was presented once. The 'Pure' strength conditions consisted of 6 words, each of which was presented the same number of times; for different categories the number of repetitions were 1, 2, and 3. Unlike these Pure conditions, the 'Mixed' strength condition consisted of 6 words of which 2 were presented once, 2 were presented twice and 2 were presented three times. The selected words from all 8 conditions were randomly placed into a study list of 335 presentations, and presented for 3 seconds each. The test list consisted of 170 words for which the subject gave 'old' or 'new' judgments.

There were two separate types of word categories used in the experiment: Semantic and orthographic-phonemic. For example, the semantic category of prototype 'butterfly' contained the words 'moth', 'nectar', 'fragile', 'cocoon', 'monarch', 'flutter', 'metamorphosis', 'dragonfly', 'flitting', 'caterpillar', and 'camouflage'. In general, the semantic categories consisted of long words with relatively low natural language frequency (both prototypes and exemplars); the exemplars were all related to the prototype but not so closely that the presentation of an exemplar would likely call to mind the prototype. An example of an orthographic-phonemic category is that formed for the prototype 'sip'; it contained the words 'tip', 'lip', 'hip', 'sir', 'sin', 'sit', 'dip', 'rip', 'six', 'big', and 'fib'. In general, an orthographic-phonemic category contained short words with relatively high natural language frequency; the exemplars all shared vowels with the prototype, and one but not both of the starting and ending consonant clusters.

Predictions

Some patterns predicted for our study by typical models are illustrated in Figure 1. It is important to note that the predictions are somewhat different from those for studies in which variables are manipulated one list at a time. In list studies, manipulations are often predicted to alter the variance of the distributions of activations. However, for essentially all models in which activations of all items participate in the resultant distributions, the mixing of so many categories of different types means that the variances of the distributions are not predicted to differ noticeably for different conditions.

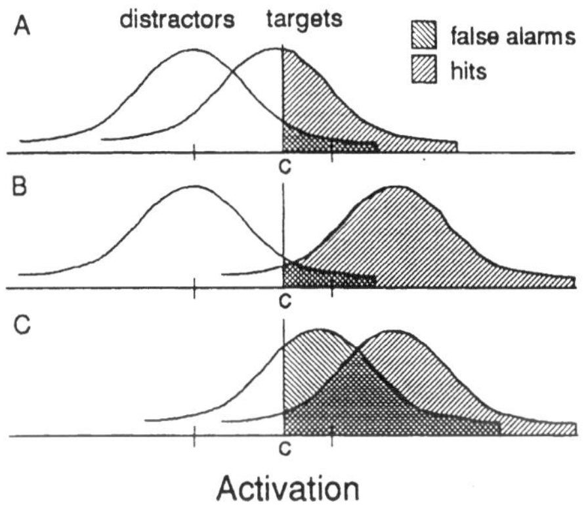

Figure 1. Probability distributions for activation due to targets and distractors. The panels represent predictions for experimental manipulations for various models (see text).

Given this, panel B illustrates the situation when a target is tested of increased strength relative to panel A, but where all items other than the target have unchanged strength-- the target distribution increases; false alarms are unchanged, hits rise, and **d'** rises (virtually all models). The category-length predictions vary with the model. For models in which non-target items contribute activations with greater than zero mean (e.g. the SAM model of Gillund and Shiffrin, 1984; the Matrix model of Pike, 1984), panel C illustrates the situation-- both distributions increase; hits and false alarms increase, and **d'** does not change. However, for models in which non-target items contribute zero mean activations (e.g. the TODAM model of Murdock, 1982; the CHARM model of Metcalfe, 1985; various feed-forward connectionist networks), the distributions do not move. This situation is given by panel A-- hits, false alarms and **d'** do not change.

For virtually all models, the category-strength predictions are at least qualitatively the same as the category-length predictions just discussed. That is, repetitions of an item should affect other items more or less as would an equivalent number of new items. However, Shiffrin, Ratcliff, and Clark (1990) discussed two alternative models. Their <u>differentiation</u> model utilized a tradeoff (discussed below) that causes summed activation to remain constant; this situation is represented by panel A. The category-length predictions are still those illustrated in panel C. The other model discussed by Shiffrin et al (1990) posited that both distributions increase with category strength, so panel C would illustrate these models' predictions for both length and strength. (For a list experiment this last model would predict a variance increase for length but not strength, but in the present category study, the variance differences "wash out").

Finally, we can consider the case in which <u>all</u> list items are increased in strength (in effect confounding the effect of target strength and the effect of strength of other category items). All models predict the same patterns as they do for strength of *other* items in the category (which for most models are also the predictions for category length), with the exception that the target distribution should increase, increasing the hit rates.

Results and Discussion

The results for performance (d') were as follows: Increasing strength led to a sizable increase in performance, increasing length led to a slight decrease in performance, and there was no appreciable difference in performance between the mixed and pure conditions for the 1, 2, and 3, times presented items. The latter result indicates that category strength had little effect, since performance for an item of fixed strength (repetitions) is being compared when the other category members are varied in strength (repetitions). This finding is consistent with many earlier list studies (e.g. Ratcliff,

Clark, & Shiffrin, 1990; Murnane & Shiffrin, 1991). As contrasted with the earlier list studies, however, all these **d'** data are consistent with the predictions of almost all models.

These performance results were expected, but the goal of the present research involved the separate hit and false alarm results. We assume that our procedure led the subjects to use a single criterion, regardless of the category, or category type, being tested. The pattern of results below is certainly consistent with this position.

Figure 2 shows the effect of increasing the strengths of all items in a category (solid lines). As predicted by all models, hit rates for targets rise; $t(888) = 4.871$, $p < .001$. Most importantly, the distractor false alarms clearly do not rise with strength of category, whereas the prototype false alarms show an upward trend that does not reach statistical significance.

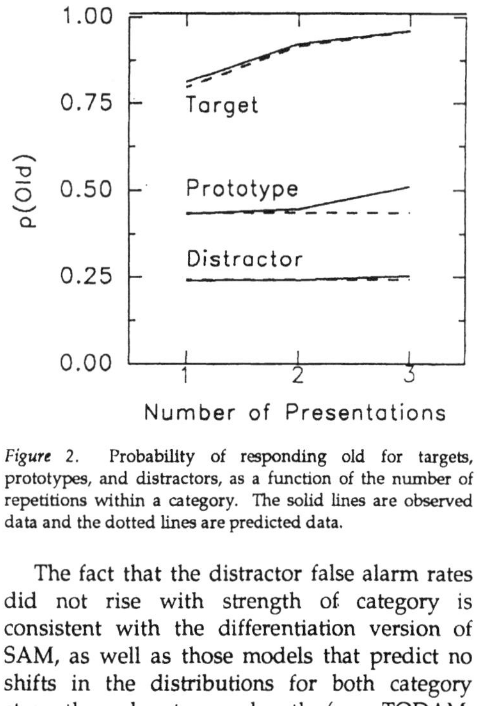

Figure 2. Probability of responding old for targets, prototypes, and distractors, as a function of the number of repetitions within a category. The solid lines are observed data and the dotted lines are predicted data.

The fact that the distractor false alarm rates did not rise with strength of category is consistent with the differentiation version of SAM, as well as those models that predict no shifts in the distributions for both category strength and category length (e.g. TODAM, CHARM, and some feed-forward connectionist nets).

The overall increased false alarm rate for prototypes would be predicted by almost all models on the basis of similarity; the prototypes should, on the average, be more similar to the words presented within the category than is the average distractor from that category. If the slight increase in the prototypes with strength is real, it would require explanation. It may be that subjects occasionally think of the prototype during the study list (especially when the words are repeated many times), and this occasionally leads the prototype to behave as a target at test.

In summary, the key result is the flat distractor function as strength increases.

Figure 3 shows the effect of extra items in the category. The probability of responding 'old' rises with category length for targets, $t(1184) = 2.54$, $p < .05$, for prototypes, $t(1184) = 6.840$, $p < .001$, and for distractors, $t(1184) = 3.357$, $p < .001$. Of the models consistent with the strength results, only the differentiation version of SAM predicts this pattern (since the other models predict similar patterns for length and strength). Thus the critical finding here is the contrast with the strength results. This contrast is reinforced by the category-strength effect that we turn to next.

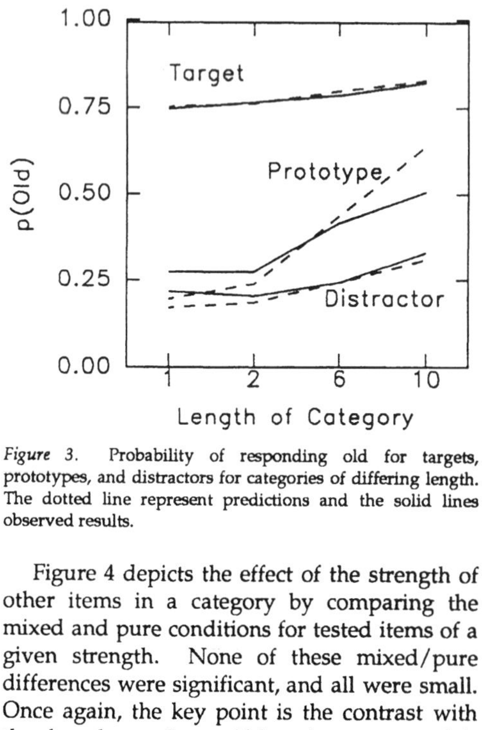

Figure 3. Probability of responding old for targets, prototypes, and distractors for categories of differing length. The dotted line represent predictions and the solid lines observed results.

Figure 4 depicts the effect of the strength of other items in a category by comparing the mixed and pure conditions for tested items of a given strength. None of these mixed/pure differences were significant, and all were small. Once again, the key point is the contrast with the length results. Although some models predict no effect of category strength, they also predict no effect of category length. Of the models we have considered and know of, only the differentiation version of the SAM model predicts this pattern.

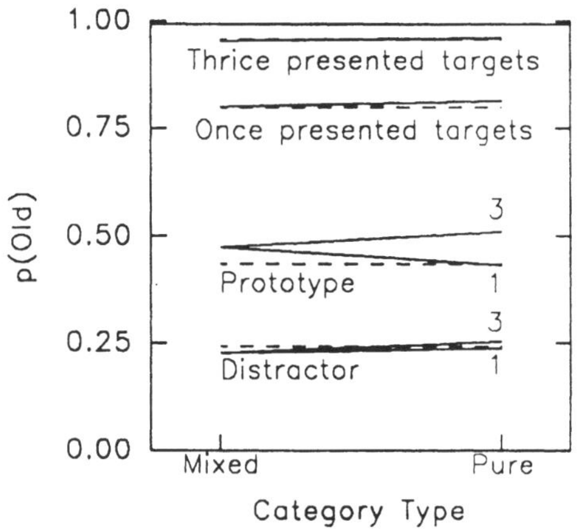

Figure 4. Probability of responding old for three times presented targets, once presented targets, prototypes, and distractors across the Mixed and Pure conditions. For both the prototypes and distractors there are two Pure points, representing tests from categories of differing strengths.

Modeling

To model the data, a simplified version of SAM was employed (see also Murnane and Shiffrin, 1991). In the SAM model introduced by Shiffrin et al (1990), it is assumed that all repetitions of an item are stored in the same trace; whenever an item is repeated, the new trace is appended to the pre-existing trace. At recognition, familiarity is determined by summing activations over all the traces. Activation of a given trace by two cues, a test item and a context cue, is posited to be the product of the separate activation tendencies for these two cues. An increase in repetitions for an item surely increases the activation tendency for the context cue. However, according to the differentiation hypothesis, an increase in the strength of a trace will cause it to mismatch more strongly the (different) test item. Thus, the activation tendency of the item cue will decrease. The model assumes that there is an approximate tradeoff of these two opposing factors. Hence activation of trace *i* will not change with the repetitions of item *i*, when item *j* and the context cue are used as memory probes. Of course, when item *i* is activating trace *i* (tests of a target), then differentiation does not operate, and activation rises with strength. Finally, it is assumed that the standard deviation of activation of an image is proportional to the mean activation. The activations of all images are summed and compared to the criterion in order to make a decision.

The effect of similarity is dealt with in the following manner: Items outside the category of the probe have one level of activation, and items within the category have another. Items within the category when the probe is a prototype have a third level of activation.

For computational ease, parameter estimation was performed through the fitting of predicted z-scores to the z-transforms of the hit and false alarm rates. The following is an example of the predicted formula for a target test from a Pure-2 category:

$$Z = \frac{C_T - [N_o S_o + N_i S_i + S_2]}{\sqrt{\alpha [N_o S_o^2 + N_i S_i^2 + S_2^2]}}$$

where C_T is the criterion, N_o is the number of out-of-category traces, S_o is the out-of-category strength, N_i is the number of different in-category traces, S_i is the in-category strength, S_2 is the strength of a twice repeated item that matches the probe, and α is a proportionality constant. The same type of formula applies to all cases; one must simply use, where needed, the remaining parameters: S_1 (once presented target strength), S_3 (three times presented target strength), and S_p (the in-category prototype strength). It is clear from the figures that the resultant predictions capture the main features of the data.

Conclusion

Under the assumption, partially supported by the data, that the subjects in our study utilize the same recognition criterion for all test items, a number of results concerning activation can be drawn: 1) Increasing the number of presentations of an item causes greater activation when that item is tested; 2) Increasing the number of presentations of some items in a category does not cause more activation when some other item in that category is tested (either target or distractor); and 3) Additional items presented in a category cause more activation when any item in that category is tested (either target or distractor). These results are consistent with just one of the models we know of-- the differentiation version of the SAM model proposed by Shiffrin et al (1990).

It is interesting to note that the differentiation version of SAM was proposed to explain the lack of an effect upon d' of strengthening other items in a list for recognition. This lack of list-strength effect for recognition result was used to argue against models that are sufficiently composite in nature to produce storage interference. The present results further constrain the class of models that might be proposed for recognition memory: When an item is tested, increasing the number of presentations of some other item does not increase activation (i.e. familiarity, degree of match), even though additional items do cause such an increase.

References

Gillund, G., & Shiffrin, R. M. (1984). A retrieval model for both recognition and recall. *Psychological Review, 91,* 1-67.

Metcalfe Eich, J. (1985). Levels of processing, encoding specificity, elaboration, and CHARM. *Psychological Review, 92,* 1-38.

Murdock, B. B., Jr. (1982). A theory for the storage and retrieval of item and associative information. *Psychological Review, 89,* 609-626.

Murnane, K., & Shiffrin, R. M. (1991). Word repetitions in sentence recognition. *Memory & Cognition, 19(2),* 119-130.

Pike, R. (1984). Comparison of convolution and matrix distributed memory systems for associative recall and recognition. *Psychological Review, 91,* 281-294.

Ratcliff, R., Clark, S., & Shiffrin, R. M. (1990). The list-strength effect: I. Data and discussion. *Journal of Experimental Psychology: Learning, Memory, & Cognition, 16,* 163-178.

Shiffrin, R. M., Ratcliff, R., & Clark, S. (1990). The list-strength effect. II. Theoretical mechanisms. *Journal of Experimental Psychology: Learning, Memory, & Cognition, 16,* 179-195.

Misinformed and Biased: Genuine Memory Distortions or Artifactual Phenomena?

Rüdiger F. Pohl

FB I - Psychology, University of Trier, Postfach 3825, D-5500 Trier, Germany

e-mail: pohl@cogpsy.uni-trier.dbp.de

Abstract

In the present study, two cognitive phenomena until now treated apart were compared to each other: *hindsight bias* and *misinformation effect*. Both phenomena result from the same basic retroactive-interference procedure focussing on how memory of originally encoded material is distorted by the encoding of subsequent, conflicting information. The results showed that subjects' recollections of the original information were similarly distorted under both conditions, that is, the amount of hindsight bias was as large as the misinformation effect. More fine-grained analyses, however, revealed important differences. With the additional results of a probability mixture model it was found that only hindsight subjects suffered from memory impairment and that, moreover, their recollections included genuine blends. The misinformation effect, on the other hand, turned out to be an artifact of averaging across two different sets of recollections. These results represent compelling data with respect to the ongoing discussion about the existence of genuine memory blends.

The reported research was partly supported by grant We 498/14 from the Deutsche Forschungsgemeinschaft.

Introduction

Since the early times of pro- and retro-active inhibition it has been shown repeatedly that two pieces of information tied to the same memory node may interfere with one another, thus leading to an impaired memory performance. More recently, two paradigms have studied these interference processes at some length: The *hindsight-bias* paradigm (see, e.g., Christensen-Szalanski & Fobian Willham, 1991; Hawkins & Hastie, 1990) and the *eyewitness-misinformation* paradigm (see, e.g., Loftus, Korf, & Schooler, 1989). In both areas, researchers found that presentation of "new", conflicting information can distort memory for "old", original information.

Hindsight bias. A typical hindsight-bias experiment proceeds in the following way: First, subjects are asked to answer difficult almanac-type questions, for example: "How high is the Statue of Liberty?" Later they receive the solution and finally they are asked to remember their original answers. Typically the recollections lie closer to the solution than the original answers did (see, e.g., Hell et al., 1988; Pohl, 1990). This effect has been referred to as "hindsight bias" or "knew-it-all-along" effect (Fischhoff, 1977; Wood, 1978). The hindsight bias seems to be independent from a variety of experimental manipulations and has, moreover, been observed for rather different materials (see Hawkins & Hastie, 1990, for a recent overview).

Most theoretical explanations favor the final rejudgment process as the point where biasing occurs. According to these, the hindsight bias is a necessary and unavoidable by-product of collecting evidence in the judgment process. The automatic memory distortion comes about because during the rejudgment process subjects are apparently unable to ignore outcome knowledge (cf. Tversky & Kahneman, 1974).

Misinformation effect. In eyewitness-misinformation studies, subjects typically view a sequence of slides depicting some complex event (e.g., a car accident). In a following questionnaire, the experimenter hides information contradictory to some of the details in the slides. For example, a stop sign in the slides is now labeled as a yield sign. Finally, in the test phase, subjects are asked to choose the original information (as seen in the slides) from a pair of two alternatives. In the above example these alternatives would include the stop sign and the yield sign. Subjects who were misled in the questionnaire typically tend to choose the correct alternative less often than subjects who were not misled (see, e.g., Loftus, Miller, & Burns, 1978). This result has been labeled "misinformation effect".

The effect was found over a variety of materials (see Loftus et al., 1989) and has resisted a number of experimental manipulations that were devised to eliminate it (summarized in Loftus, 1979). Accordingly, Loftus strongly argued for automatic updating processes suggesting that misinformation erased the memory representation of original information rendering it irretrievable. This argument is known as the "substitution hypothesis". But later research showed that both original and misinformation may coexist in memory (see, e.g., Bekerian & Bowers, 1983). The misinformation effect then can be understood as being caused by differences in retrievability (Christiaansen & Ochalek, 1983; Morton, Hammersley, & Bekerian, 1985). McCloskey and Zaragoza (1985), though, challenged both views (substitution and coexistence) and criticized the misinformation effect as an artifact of the testing procedure used. In their opinion, the misinformation effect does not result from memory impairment, but arises because different memory and guessing states are unjustly summed across.

The theoretical interpretation of the misinformation effect is still a matter of debate (Chandler, 1991; Metcalfe & Bjork, 1991; Schooler & Tanaka, 1991). Besides the introduction of different new testing procedures, a more promising approach has focussed on the use of material with continuous features (instead of whole objects). Results of these experiments typically showed some form of "memory blends" (Chandler, 1991; Loftus, 1977) paralleling the typical results found in hindsight-bias studies.

Memory blends. Memory distortion is understood (in this paper) as a general term denoting that in some way memory was affected by an experimental condition. Memory blends are recollections that show integration of features from different items. Following Schooler and Tanaka (1991) it appears useful to further distinguish between "representation blends" (that result from encoding or storage processes) and "recollection blends" (that emerge from retrieval processes). While these two types refer to the point where blending occurs, a further distinction denotes the content of blending: A "composite blend" integrates separable objects from different sources, for example, adding a stop sign (from a verbal description) to a traffic scene (from a slide sequence; Loftus, Miller, & Burns, 1978). A "compromise blend", on the other hand, integrates different feature values into one value (by some sort of averaging process), for example, an originally blue car, which was later referred to as being green, is remembered as a bluish green one (Loftus, 1977). Similarly, Metcalfe and Bjork (1991, p. 203) defined a "positive blend" (Metcalfe, 1990) as one where "subjects' recollections of the original event must show a unimodal shift along some dimension in the direction of the misleading event."

But still, the critique of McCloskey and Zaragoza (1985) applies. Metcalfe and Bjork (1991) referred to this situation in their "summed distribution hypothesis": If one type of recollection was centered around the original information, the other one around the misinformation, and if both original and misinformation were not too far apart from each other on their underlying dimension, then the summed recollections may show a unimodal distribution. The mean recollection would then mimick a typical misinformation effect (or hindsight bias) despite the absence of any genuine (memory or recollection) blends.

Summarizing this discussion, it again seems urgent to separate genuine blends from statistical averaging "effects". To this end, a probability mixture model was devised that is described next.

Probability mixture model. (Because of space constraints, only a general outline will be given here; cf. Pohl & Gawlik, 1992.) The model assumes as latent states two knowledge-retrieval cases for control items and five cases for experimental items. Each case is considered to be true for an unknown proportion of all recollections and is associated with different probabilities that a recollection falls within one out of five recollection classes. The latent states reflect which information is available to the subject at the time of retrieval. The recollection classes are defined as follows: (1) recollections shifted away from the solution/misinformation, (2) perfect recollections, (3) recollections shifted towards the solution/misinformation, (4) erroneous recall of the solution/misinformation itself, and (5) recollections shifted beyond the solution/misinformation.

The model was varied according to two major questions. First, the percentage of perfect recollections was either allowed to differ between control and experimental items or not. (For *experimental* items, the solution/misinformation was provided to the subject, while for *control* items, this was not the case.) The McCloskey and Zaragoza (1985) argument stated that—despite the presence of a misinformation effect—the percentages of correctly known (not guessed) recollections could be equal for both types of items. In that case no *memory impairment* occurred. This can be tested by comparing the fits of the corresponding model versions.

Second and more important, the case of *positive blends* was either included in the model or not. Again, the presence or absence of such blends can be concluded from the resulting model fits. It should be noted that both questions (memory impairment and positive blends) can each be accessed independently from one another.

Method

Subjects. 40 students of the University of Trier were randomly assigned to the *hindsight* group or to the *misinformation* group.

Material. A catalogue of 20 objects to-be-sold at a fictitious auction at Sothebys was prepared. Each page of the catalogue contained the picture of one of the objects together with a short description of it. Each object contained one critical (numerical) information. The auction catalogue was prepared in two versions. In the hindsight version, the values of the critical information were replaced by an empty box with a question mark. In the misinformation version, all values were filled in and not marked as critical in any way.

Procedure. Each subject was run individually in two sessions one week apart. The hindsight subjects were run first in order to allow matching of the misinformation subjects. In Session 1, all subjects received the auction catalogue. Each hindsight subject was given the hindsight version and asked to estimate the left-out values. Each misinformation subject was given the misinformation version (with the estimates of a hindsight subject filled in as original information) and instructed to read all descriptions carefully. In Session 2, all subjects first received a feedback list (with solutions/misinformations to some of the items). Then, all subjects again received the auction catalogue, now all in the hindsight version. Subjects of the hindsight group were asked to remember their estimates given in the first session. Misinformation subjects were asked to remember the original information from the first presentation of the catalogue.

Results

The level of significance was set at $\alpha = .05$ for all statistical analyses.

Shift in recollections. In order to allow comparison across numerically dissonant items, all data were z-transformed separately for each item. Then, the absolute z-score distances between original estimate/information and solution/ misinformation (i.e., the *original distance*) and between recollection and solution/misinformation (i.e., the *recollection distance*) were computed. For experimental items, the mean changes from original distances to recollection distances reached -.41 z and -.34 z for hindsight and misinformation subjects, respectively. The corresponding figures for control items were .07 z and .12 z. The analyses of variance revealed one effect: With experimental items, recollection distances were significantly shorter than original distances ($F(1,9) = 14.42$) implying that on an average recollections were closer to the solution/misinformation than the original estimate/information had been. But with control items, the distance did not change remarkably ($F(1,9) = 2.91$, $\alpha > .10$). The group variable (hindsight vs. misinformation) produced neither main nor interaction effects (all $F < 1$).

Perfect recollections. In the hindsight group, the proportion of perfect recollections reached 33% with control items and dropped to 11% with experimental ones ($\chi^2(1) = 11.56$). The misinformation group produced 11% perfect recollections with control items and 8% with experimental ones ($\chi^2(1) < 1$).

Model fitting. The numbers of recollections for each of the five recollection classes were submitted to the model fit. A least-squares iteration procedure was run separately for the hindsight and for the misinformation group with all four versions of the probability mixture model (as described in the Introduction). The results are presented in Table 1. For the hindsight group, the four versions produced model fits of clearly diverging quality, suggesting that both memory impairment and positive blending occurred. For the misinformation group, though, all versions of the model fitted the data equally well, suggesting that neither memory impairment nor positive blending occurred.

Discussion

At first glance, the results seemed to be clear and simple. When looking at the *mean shift of recollections* as compared to original estimate/ information, both hindsight and misinformation group showed the same effect: While there was no memory distortion with control items, both groups revealed an equal amount of memory distortion with experimental items, that is, the hindsight bias was as large as the misinformation effect. The further, more fine-grained analysis, though, proved this conclusion to be premature.

Table 1.
Best Fits (χ^2) of the Four Versions of the Probability Mixture Model for Hindsight and Misinformation Subjects

Model version		*df*	Hindsight	Misinformation
Without memory impairment.				
Without blends	(1)	4	18.39 *	1.36 +
With blends	(2)	3	12.78 *	1.03 +
With memory impairment				
Without blends	(3)	3	4.77	.69 +
With blends	(4)	2	.14 +	.39 +

* $\alpha < .01$, + $\alpha > .70$.

First of all, consider the *percentages of perfect recollections*. The percentages of perfect recollections of experimental items (as compared to control items) was severely diminished in the hindsight group, but not in the misinformation group. Following the argument of McCloskey and Zaragoza (1985), this observation implies that only the memory of hindsight subjects was impaired by the items' solutions, while the memory of misinformation subjects was unaffected by the misinformation.

Fitting the four versions of our *probability mixture model* to the data revealed quite different results for the two groups. In the hindsight group, the model fitted the data perfectly only when two parameters were used to account for memory impairment and positive blends. In the misinformation group, though, all versions produced the same good fit of the model. This, once more, suggests that only hindsight subjects suffered from memory impairment and that they, moreover, produced a significant number of bias recollections (positive blends), while misinformation subjects apparently produced none.

In the light of these observations, the conclusion drawn earlier receives strong support: Only the recollections of hindsight-subjects include cases of genuine positive blends, while the misinformation effect appears as an artifact of statistical averaging. This interpretation corroborates the considerations of McCloskey and Zaragoza (1985) and dismisses their opponents' attempt to rescue the misinformation effect as a memory phenomenon (Loftus, Schooler, & Wagenaar, 1985). The use of experimental material with continuous features (instead of whole objects) and recall as test method (instead of recognition) proved to be a sensible way to detect the suspected artifact.

Now that no cognitive account seems to be necessary for the misinformation effect, what remains to be explained is how the *hindsight bias* arose. If the subject knew roughly both sources of information, that is, she remembered the numerical region of her estimate and the relation between it and the solution, she was hampered by knowing the solution when she tried to reconstruct her estimate. As a consequence, recollections of this sort should fall between estimate and solution reflecting the genuine hindsight-bias case. This situation was described by Tversky and Kahneman (1974) as one of their heuristics named "anchoring and adjustment", where the anchor prevents full adjustment.

With respect to past as well as future research on hindsight bias, misinformation effect, or similar phenomena, recollection data should be analyzed more carefully. The proposed probability mixture model—or generally, any multinomial model (Riefer & Batchelder, 1988)—should be able to detect statistical "effects" of the McCloskey-Zaragoza type, thereby avoiding artifactual explanations.

References

Bekerian, D. A.; and Bowers, J. M., 1983. Eyewitness testimony: Were we misled? *Journal of Experimental Psychology: Learning, Memory, and Cognition* 9(1):139-145.

Chandler, C. C. 1991. How memory for an event is influenced by related events: Interference in modified recognition tests. *Journal of Experimental Psychology: Learning, Memory, and Cognition* 17(1):115-125.

Christensen-Szalanski, J. J. J.; and Fobian Willham, C., 1991. The hindsight bias: A meta-analysis. *Organizational Behavior and Human Decision Processes* 48:147-168.

Christiaansen, R. E.; and Ochalek, K., 1983. Editing misleading information from memory: Evidence for the coexistence of original and postevent information. *Memory and Cognition* 11(5):467-475.

Fischhoff, B. 1977. Perceived informativeness of facts. *Journal of Experimental Psychology: Human Performance and Perception* 3(2):349-358.

Hawkins, S. A.; and Hastie, R., 1990. Hindsight: Biased judgments of past events after the outcomes are known. *Psychological Bulletin* 107(3):311-327.

Hell, W.; Gigerenzer, G.; Gauggel, S.; Mall, M.; and Müller, M., 1988. Hindsight bias: An interaction of automatic and motivational factors? *Memory and Cognition* 16(6):533-538.

Loftus, E. F. 1977. Shifting human color memory. *Memory and Cognition* 5(6):696-699.

Loftus, E. F. 1979. *Eyewitness testimony*. Cambridge, MA: Harvard University Press.

Loftus, E. F.; Korf, N. L.; and Schooler, J. W., 1989. Misguided memories: Sincere distortions of reality. In J. C. Yuille (Ed.), *Credibility assessment* (pp. 155-173). Dordrecht: Kluwer.

Loftus, E. F.; Miller, D. G.; and Burns, H. J., 1978. Semantic integration of verbal information into a visual memory. *Journal of Experimental Psychology: Human Learning and Memory* 4(1):19-31.

Loftus, E. F.; Schooler, J. W.; and Wagenaar, W. A., 1985. The fate of memory: Comment on McCloskey & Zaragoza. *Journal of Experimental Psychology: General* 114(3):375-380.

McCloskey, M.; and Zaragoza, M., 1985. Misleading postevent information and memory for events: Arguments and evidence against memory impairment hypotheses. *Journal of Experimental Psychology: General* 114(1):1-16.

Metcalfe, J. 1990. A composite holographic associative recall model (CHARM) and blended memories in eyewitness testimony. *Journal of Experimental Psychology: General* 119(2):145-160.

Metcalfe, J.; and Bjork, R. A., 1991. Composite models never (well, hardly ever) compromise: Reply to Schooler and Tanaka (1991). *Journal of Experimental Psychology: General* 120(2): 203-210.

Morton, J.; Hammersley, R. H.; and Bekerian, D. A., 1985. Headed records: A model for memory and its failures. *Cognition* 20(1):1-23.

Pohl, R. 1990. *Changing one's memory: Influences of interest and pre-experimental knowledge on the hindsight bias of male and female persons.* Paper presented at the 4th Conference of the European Society for Cognitive Psychology, Como/Italy.

Pohl, R.; and Gawlik, B., 1992. Hindsight bias and misinformation effect: Two cases of memory distortion or artifactual phenomena? (Paper submitted for publication)

Riefer, D. M.; and Batchelder, W. H., 1988. Multinomial modeling and the measurement of cognitive processes. *Psychological Review* 95(3):318-339.

Schooler, J. W.; and Tanaka, J. W., 1991. Composites, compromises, and CHARM: What is the evidence for blend memory representations? *Journal of Experimental Psychology: General* 120(1):96-100.

Tversky, A.; and Kahneman, D., 1974. Judgment under uncertainty: Heuristics and biases. *Science*, 185:1124-1131.

Wood, G. 1978. The "knew-it-all-along" effect. *Journal of Experimental Psychology: Human Perception and Performance* 4(2):345-353.

Neurally motivated constraints on the working memory capacity of a production system for parallel processing: Implications of a connectionist model based on temporal synchrony*

Lokendra Shastri
Department of Computer and Information Science
University of Pennsylvania
Philadelphia, PA 19104, USA
shastri@cis.upenn.edu

Abstract

The production system formulation plays an important role in models of cognition. However, there do not exist neurally plausible realizations of production systems that can support fast and automatic processing of productions involving variables and n-ary relations. In this paper we show that the neurally plausible model for rapid reasoning over facts and rules involving n-ary predicates and variables proposed by Ajjanagadde and Shastri can be interpreted as such a production system. This interpretation is significant because it suggests neurally motivated constraints on the capacity of the working memory of a production system capable of fast parallel processing. It shows that a large number of rules — even those containing variables — may fire in parallel and a large number of facts may reside in the working memory, provided no predicate is instantiated more than a small number of times (≈ 3) and the number of distinct *entities* referenced by the facts in the working memory remains small (≈ 10).

Introduction

Understanding language is a complex task and involves, among other things, recognizing words, accessing lexical items, disambiguating word senses, parsing, and carrying out inferences to establish referential and causal coherence, recognize speaker's plans and make predictions.[1] Nevertheless we can understand written language at the rate of *several hundred words per minute* (Carpenter & Just 1977). In view of the complexity of the language understanding task, the rapid rate at which we can understand language has strong implications and poses a challenge to computational models of cognition. In particular, it suggests that certain kinds of inferences can be drawn within a few hundred milliseconds and significant syntactic processing can occur in a similar time frame. The speed and spontaneity with which we understand language also highlights our ability to perform a class of inferences automatically and without conscious effort — as though they are a *reflex* response of our cognitive apparatus. In view of this we have described such reasoning as *reflexive* (Shastri 1990).[2]

Motivated by a concern for explaining reflexive (rapid) reasoning, Ajjanagadde and Shastri have proposed a connectionist model — let us call it SHRUTI — that can encode a large body of specific *facts*, general *rules* involving *n-ary predicates* and *variables*, as well as *IS-A* relationships between concepts, and perform a range of reasoning with extreme efficiency (Shastri & Ajjanagadde 1990; Ajjanagadde & Shastri 1991; Mani & Shastri 1991). The system performs a class of inferences in time that is independent of the size of the 'knowledge base' and is only proportional to the *length* of the shortest chain of reasoning leading to the conclusion. The reasoning system solves the dynamic (variable) binding problem (Feldman 1982; Malsburg 1986) in a neurally plausible manner: It maintains and propagates variable bindings using temporally synchronous firing of appropriate nodes. This computational model has also been used by Henderson (1991) to design a parser for English. The parser's speed is independent of the size of the lexicon and the grammar, and it offers a natural explanation for certain center embedding phenomena.

In this paper we interpret SHRUTI as a production system and examine the functional properties of the resulting production system. Such an interpretation is motivated by several factors. First, it leads to a production system with novel and interesting working memory characteristics. Second, it points the way to a neurally plausible realization of production systems. Third, it helps relate the working memory *capacity* of such a system and the time taken by each production cycle, to basic biological parameters. The interpretation also helps specify the syntactic properties of productions that can participate in reflexive processing. This aspect, however, is not discussed in this paper. The interested reader may refer to (Shastri & Ajjanagadde 1990; Shastri 1992).

A number of cognitive models are based on the

*This work was supported by NSF grant IRI 88-05465 and ARO grant DAAL 03-89-C-0031.

[1]Empirical data suggests that inferences required to establish referential and causal coherence occur rapidly and automatically during text understanding (see e.g., McKoon & Ratcliff 1980; McKoon & Ratcliff 1981; Keenan, Baillet, and Brown 1984). The evidence for the automatic occurrence of *elaborative* or predictive inferences however, is mixed (see e.g., Kintsch 1988; Potts, Keenan, and Golding 1988).

[2]A formal characterization of reflexive reasoning in terms of time and space complexity is given in (Shastri 1992): Reflexive reasoning occurs in time that is independent of the size of the long-term knowledge base and is proportional only to the length of the shortest chain of inference leading to a conclusion. Also the number of nodes required to encode a long-term knowledge base should be at most *linear* in the size the knowledge base.

production system formalism; two of the most comprehensive being ACT* (Anderson 1983) and SOAR (Newell 1990). Neurally plausible realizations of these models, however, have not been proposed. Although several aspects of ACT* such as its use of levels of activation, weighted links and decay of activation had neural underpinnings, it had not been shown how certain critical aspects of the model could be realized in a neurally plausible manner. For example, ACT* represented productions with variables, but Anderson did not suggest a neurally plausible account of how variable bindings are propagated and matched. In his exposition of SOAR, Newell has used the time course of neural processes to estimate how long various SOAR operations should take, but he has not suggested how a SOAR-like system may be realized in a neurally plausible manner (see p. 440 Newell, 1990). Although a complete mapping of comprehensive systems such as SOAR and ACT* to a neurally plausible architecture still remains an open problem, SHRUTI does provide a a concrete basis for a neurally plausible realization of production systems. Of particular significance are the specific and biologically motivated constraints SHRUTI suggests on the capacity of the working memory of a production system capable of supporting rapid 'knowledge level' parallelism.

Other researchers have proposed connectionist production systems. However, the functional characteristics of SHRUTI when interpreted as a production system are quite distinct from these connectionist models (for a detailed discussion refer to (Shastri & Ajjanagadde 1990)). For example, DCPS the distributed connectionist production system (Touretzky & Hinton 1988) only deals with productions containing a single variable. DCPS is also serial at the knowledge level and it can only apply one rule at a time. Thus in terms of efficiency, DCPS is like a traditional (serial) production system and must deal with the combinatorics of search and the associated problem of backtracking. TPPS a production system based on the tensor product encoding (Dolan & Smolensky 1989), and Conposit a system based on relative position encoding (Barnden & Srinivas 1991), are also serial at the knowledge level. Hence these systems are inappropriate for modeling reflexive processing. A connectionist system that does support knowledge level parallelism is ROBIN (Lange & Dyer 1989). However, the variable binding mechanism incorporated by ROBIN does not lead to the sort of biologically motivated constraints on working memory suggested by SHRUTI.

A Brief Overview of SHRUTI

Refer to the schematic representation of some predicates and individual concepts shown in Fig. 1. Nodes drawn as circles are what we call ρ-**btu** nodes. These nodes have the following idealized behavior: On receiving a periodic spike train, a ρ-btu node produces a periodic spike train that is *in-phase* with the driving input. Thus oscillatory activity in a ρ-btu node can lead to synchronous activity in a ρ-btu node connected to it. We assume that ρ-btu nodes can respond in this manner as long as the period of oscillation, π, lies in the interval $[\pi_{min}, \pi_{max}]$, where π_{min} and π_{max} correspond to the highest and lowest frequencies, respectively, at which ρ-btu nodes can sustain oscillatory activity. [3]

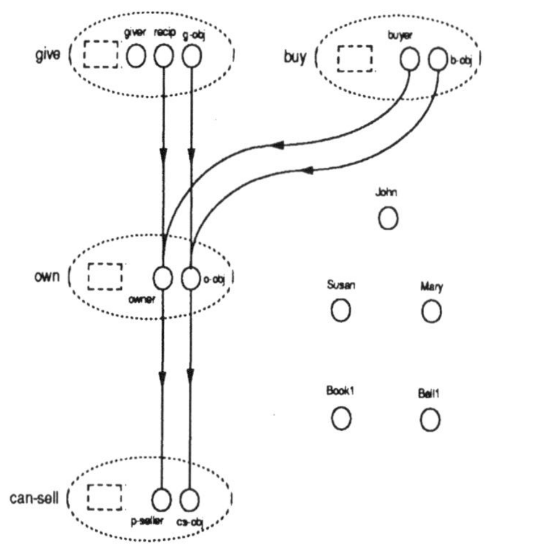

Figure 1: Encoding of predicates, individual concepts, and the rules: $\forall x, y, z\ [give(x, y, z) \Rightarrow own(y, z)]$, $\forall x, y\ [own(x, y) \Rightarrow can\text{-}sell(x, y)]$, and $\forall x, y\ [buy(x, y) \Rightarrow own(x, y)]$.

An n-ary predicate is represented by a cluster of n ρ-btu nodes (the rectangular 'nodes' shown in Fig. 1 are not relevant to our discussion). Nodes such as *John* and *Mary* are also ρ-btu nodes and correspond to *focal* nodes of the complete representations of the individuals 'John' and 'Mary' (Shastri 1988; Feldman 1989). A rule is encoded by linking the arguments of the antecedent and consequent predicates in accordance with the correspondence between arguments specified in the rule. For example, the rule $give(x, y, z) \Rightarrow own(y, z)$ is encoded by connecting the arguments *recip* and *g-obj* of *give* to the arguments *owner* and *o-obj* of *own*, respectively.

SHRUTI represents dynamic bindings using *synchronous* — i.e., *in-phase* — firing of the appropriate argument and concept nodes. With reference to the nodes in Fig. 1, the dynamic representation of the bindings *(giver=John,recip=Mary,g-obj=Book1)* (i.e., the *dynamic* fact *give*(*John*, *Mary*, *Book*1)) will be represented by the *rhythmic* pattern of activity shown in Fig. 2. Observe that while *John*, *Mary* and *Book1* are firing in distinct phases, *giver* is firing in synchrony with *John*, *recip* in synchrony with *Mary*, and *g-obj* in synchrony with *Book1*.

By virtue of the interconnections between argument nodes of the predicates *give*, *own*, and *can-sell*, the state of activation described by the rhythmic pattern shown in Fig. 2 will lead to the rhythmic activation pattern shown in Fig. 3, where the firing pattern of nodes corresponds to the dynamic bindings *(giver=John, recip=Mary, g-*

[3] We can generalize the behavior of a ρ-btu node to account for weighted links by assuming that a node will fire if and only if the weighted sum of synchronous inputs is greater than or equal to n.

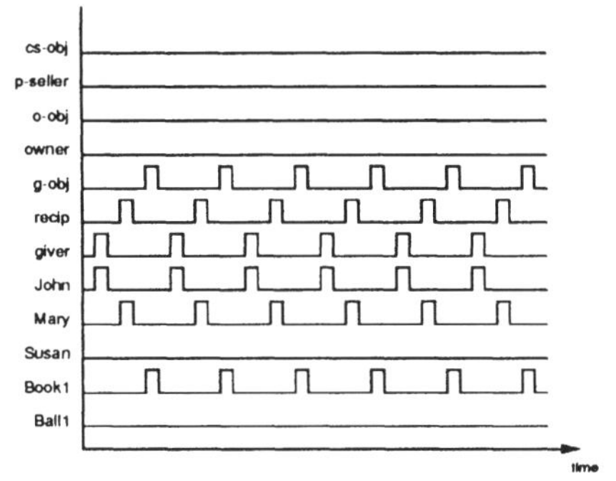

Figure 2: Rhythmic pattern of activation representing the dynamic bindings (*giver* = *John*, *recipient* = *Mary*, *give-object* = *Book*1).

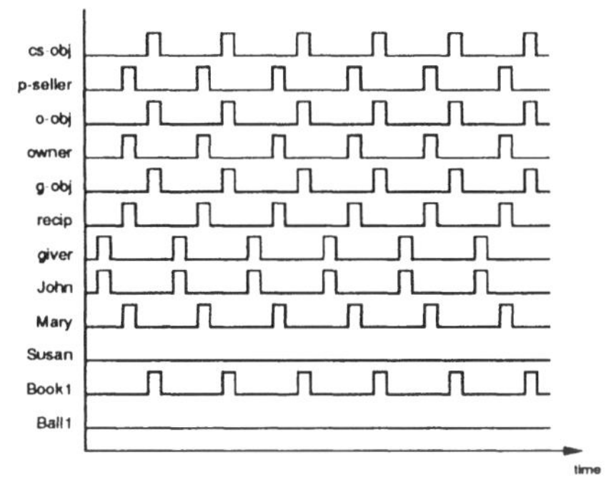

Figure 3: Pattern of activation representing the dynamic bindings (*giver* = *John*, *recipient* = *Mary*, *give-object* = *Book*1, *owner* = *Mary*, *own-object* = *Book*1, *potential-seller* = *Mary*, *can-sell-object* = *Book*1).

obj=Book1, owner=Mary, o-obj=Book1, p-seller= Mary, cs-obj=Book1) which encode the dynamic facts *give(John,Mary,Book1)*, *own(Mary,Book1)*, and *can-sell(Mary,Book1)*. In other words, given 'John gave Mary Book1', the network has inferred 'Mary owns Book1' and 'Mary can sell Book1' by using the 'rules' $give(x,y,z) \Rightarrow own(y,z)$ and $own(u,w) \Rightarrow can\text{-}sell(u,w)$.

Observe that the (multiple) bindings between *Mary* and the arguments *recip*, *owner*, and *p-seller* are represented by these argument nodes firing in-phase with *Mary*.

Conceptually, the rule application process corresponds to a *parallel* breadth-first traversal of a directed *inferential dependency graph* and a large number of rules may fire in parallel. In general, the time taken to generate a chain of inference is independent of the total number of rules and facts and is just equal to $l*\pi*\alpha$ where l equals the *length* of the chain of inference, π equals the period of oscillation of the nodes, and α is the number of cycles required for a ρ-btu node to synchronize with a connected ρ-node.

The system allows a large number of bindings — and hence, dynamic facts — to be represented simultaneously, and it also allows a large number of rules to fire simultaneously. The number of distinct entities involved in simultaneous dynamic bindings, however, is bounded by the ratio π_{max}/ω, where π_{max} is the period corresponding to the *lowest* frequency at which ρ-btu nodes can sustain and propagate oscillations, and ω is the width of the window of synchrony (i.e., two nodes firing with a lead or lag of $\omega/2$ can be considered to be firing in synchrony).

Other Representational Aspects of SHRUTI SHRUTI can also encode *long-term* facts. The encoding of a long-term fact encodes the static bindings pertaining to the fact and rapidly recognizes that the static bindings it encodes, match the dynamic bindings existing in the system's state of activation. Given that SHRUTI represents dynamic bindings as temporal patterns, the encoding of a long-term fact behaves like a *temporal pattern matcher* and is described in (Shastri & Ajjanagadde 1990). With the encoding of long-term facts, SHRUTI can answer queries that follow from the encoded long-term facts and rules.

The network in Fig. 1 can only represent one dynamic instance per predicate and concept. The encoding may be extended to represent a bounded number of instantiations of each predicate and concept (for details see (Mani & Shastri 1992)). This allows SHRUTI to deal with 'bounded recursion'. However, a significant cost has to be paid for encoding multiple instantiations and as discussed below, this has implications on the working memory capacity of the associated production system.

SHRUTI can be combined with the representation of a type (IS-A) hierarchy (Mani & Shastri 1991). Such an integration allows the occurrence of types (categories) as well as instances in rules, facts, and queries. The resulting system can combine rule-based reasoning with type inheritance. For example, the system can infer 'Tweety is scared of Sylvester', based on the generic fact 'Cats prey on birds', the rule 'If x preys on y then y is scared of x' and the *IS-A* relations 'Sylvester is a Cat', 'Tweety is a Canary', and 'Canaries are birds'. The integrated system can also use type information to specify restrictions on the types of argument fillers and encode context sensitive rules such as: $\forall x : animate, y : solid\text{-}obj\ walk\text{-}into(x,y) \Rightarrow hurt(x)$. This rule will fire only if the first and second arguments of 'walk-into' are bound to fillers of the type 'animate' and 'solid-object', respectively.

Finally, SHRUTI can also encode rules involving multiple antecedents, thus it can encode a rule such as $\forall x,y,z \quad love(x,y) \wedge love(y,z) \wedge notequal(x,z) \Rightarrow jealous(x,z)$.

Biological Plausibility of SHRUTI and Neurally Plausible Parameter Values

The potential of synchronous oscillation in neural representation has long been recognized (Hebb 1949; Freeman 1981; Malsburg 1981; Sejnowski 1981; Abeles 1982; Damasio 1989). Compelling evidence of the existence of synchronous activity in the brain comes from recent findings of stimulus-related synchronous oscillations in the cat visual cortex (Eckhorn et al. 1988; Gray et al. 1989). The data suggests that synchronous and rhythmic activity occurs in the brain and the time course of such activity is consistent with the requirements of reflexive reasoning. We summarize some relevant aspects of the data: i) Synchronous oscillations have been observed in the frequency range of 35 – 80 Hz (Eckhorn et al. 1988) and 35 – 65 Hz (Gray et al. 1991). Thus the observed oscillatory activity has periods ranging from about 12 to 28 msecs.; ii) Synchronization of neural activity can occur within a few (sometimes even one) periods of oscillations (Gray et al. 1991); iii) In a large number of cases synchronization occurs with precise phase-locking (zero time lag) and in most cases it occurs with a lag or lead of less than 3 msec. (Gray et al. 1991); and iv) Once achieved, synchrony may last several hundred msecs. (Gray et al. 1991).

The above data provides a basis for making coarse estimates of neurally plausible values of some of SHRUTI's parameters. The data indicates that a plausible estimate of the maximum period of oscillation, π_{max}, may be 28 msecs. and a conservative estimate of ω, the width of the window of synchrony, may be 6 msecs.

SHRUTI as a Production System

As may be evident, there exists a correspondence between SHRUTI and the production system formulation. The correspondence for the declarative memory and the production memory of a production system is straightforward: the declarative memory corresponds to the collection of long-term facts and the production memory corresponds to the collection of rules encoded in SHRUTI (each rule is a production).

Observe that dynamic bindings, and hence, dynamic (active) facts are represented in SHRUTI as a rhythmic pattern of activity over nodes in the network. In functional terms, this transient state of activation temporarily holds information during an episode of reflexive reasoning and corresponds to SHRUTI's *working memory*: a production fires if its antecedents match the contents of the working memory and introduces facts into the working memory. Observe that SHRUTI is a parallel production system that allows a large number of rules — including rules with variables — to fire in parallel as long as the capacity of the working memory is not exceeded (explained below). Furthermore, the time taken to compute a result is independent of the size of the declarative and production memory, and only depends upon the length of the sequence of productions required to produce the result.

Functional Characteristics of the Production System Implied by SHRUTI

Estimates of the working memory capacity of production system models range from very small (about 7 elements) to essentially unconstrained. SHRUTI predicts that the capacity of the *working memory underlying reflexive reasoning* (WMRR) is very large, *but* constrained in critical ways. The number of dynamic facts that can be present in the working memory at any given time is $k_2 p$, where k_2 is a system parameter (see below) and p is the number of predicates represented in the system. Thus the number of dynamic facts that may potentially be present in the working memory is very high. But as discussed below, there exist constraints that limit the number of dynamic facts that may *actually* be present in the working memory at any given time.

Before moving on, let us clarify that the dynamic facts represented in the WMRR during an episode of reflexive reasoning should not be confused with the small number of short-term facts an agent may *overtly* keep track of during *reflective* processing and problem solving. In particular, the WMRR should not be confused with the short-term memory implicated in various memory span tasks (Baddeley 1986).

A Bound on the Number of Distinct Entities Referenced in the Working Memory During an episode of reflexive reasoning, each entity involved in dynamic bindings occupies a distinct phase in the rhythmic pattern of activity. Hence the number of distinct entities[4] that can occur as argument-fillers in the dynamic facts represented in the working memory cannot exceed $\lfloor \pi_{max}/\omega \rfloor$, where π_{max} is the maximum period at which ρ-btu nodes can sustain oscillations and ω equals the width of the window of synchrony.

As pointed out above, a neurally plausible value of π_{max} is about 28 and a conservative estimate of ω is around 6. This suggests that as long as the number of distinct entities referenced by the dynamic facts in the working memory is five or less, there will essentially be no cross-talk among the dynamic facts. If more entities occur as argument-fillers in dynamic facts, the window of synchrony ω would have to shrink appropriately in order to accommodate all the entities. For example, the value of ω would have to shrink to 4 msecs. in order to accommodate 7 entities. As ω shrinks, the possibility of cross-talk between dynamic bindings would increase until eventually, the cross-talk would become excessive and disrupt the system's ability to perform systematic reasoning. The exact bound on the number of distinct entities that may fill arguments in dynamic facts would depend on the smallest feasible value of ω. Given the noise and variation indicated by the data on synchronous activity, it appears unlikely that ω can be less than 3 msecs. Hence we predict that a neurally plausible *upper bound* on the number of distinct entities that can occur in the dynamic facts represented in the working memory is about 10.

It is remarkable that the bound on the number of entities that may be referenced by the dynamic facts in the working memory relates so well to 7 ± 2, the robust measure of short-term memory capacity (Miller 1956). This unexpected coincidence suggests that temporal synchrony may also underlie other short-term and dynamic representations.

In a large system made up of several SHRUTI-like modules, the bounds on the number of distinct entities referenced by the working memory of one mod-

[4] Note that 'Tweety', 'Tweety the Canary', 'Tweety the bird', and 'Tweety the animal' may be active simultaneously and all these count as only *one* entity.

ule is independent of similar bounds on the working memories of other modules. As we discuss in (Shastri & Ajjanagadde 1990), dynamic structures in the working memory of other subsystems may refer to different sets of entities using phase distributions local to those subsystems. Aaronson (1991) has described a connectionist interface that allows two SHRUTI-like modules, each with its own phase structure, to exchange binding information in a consistent and rapid manner.

A Bound on the Multiple Instantiation of Predicates The capacity of the working memory is also limited by the constraint that it may only contain a bounded number of dynamic facts pertaining to each predicate. This constraint follows directly from the limitation that each predicate can only be instantiated a bounded number (k_2) times. The cost of maintaining multiple instantiations of a predicate is significant in terms of space and time. The number of nodes required to represent a predicate and associated long-term facts is proportional to k_2 while the number of nodes required to encode a rule for backward reasoning is proportional to the square of k_2.[5] Thus a system that can represent three dynamic instantiations of each predicate will have anywhere from three to nine times as many nodes as a system that can only represent one instantiation per predicate. Furthermore, the worst case time required for propagating multiple instantiations of a predicate also increases by a factor of k_2 . In view of the additional space and time costs associated with multiple instantiation, and given the necessity of keeping these resources within bounds in the context of reflexive reasoning, we predict that the value of k_2 is quite small, perhaps no more than 3.

Bound on the Number of Rule Firings SHRUTI implies a production system in which any number of rules — even those containing variables — may fire in parallel as long as no relation (predicate) is instantiated more than k_2 times (where k_2 is ≈ 3) and the number of distinct entities referenced by the active facts in the working memory remains less than $\lfloor \pi_{max}/\omega \rfloor$ (≈ 10). This may be compared with Newell's suggestion (1980) that while productions without variables can be executed in parallel, productions with variables may have to be executed in a serial fashion.

Some Typical Retrieval and Inference Timings If the values of appropriate system parameters are set to the neurally plausible values identified in Section 3.1, SHRUTI performs systematic reasoning within a few hundred milliseconds. Note that we are only referring to the time taken by the internal (reflexive) reasoning process, and not considering the time taken by other perceptual, linguistic and motor processes.[6]

We choose π to be 20 msecs., assume that ρ-btu nodes can synchronize within two periods of oscillations (i.e., α equals 2), and pick the bound on the maximum number of instantiations per predicate to be 3 (i.e., k_2 equal to 3). The system takes 320 msecs. to infer 'John is jealous of Tom' after being given the dynamic facts 'John loves Susan' and 'Susan loves Tom' (this involves the production 'if x loves y and y loves z then x is jealous of z). The system takes 260 msecs. to infer 'John is a sibling of Jack' given 'Jack is a sibling of John' (this involves the production 'if x is a sibling of y then y is a sibling of x). Similarly, the system takes 320 msecs. to infer 'Susan *owns* a car' after its internal state is initialized to represent 'Susan *bought* a Rolls-Royce' (using the production 'if x buys y then x owns y' and the *IS-A* relation, 'Rolls-Royce is a car').

If SHRUTI's declarative memory includes 'John bought a Rolls-Royce', SHRUTI will take 140 msecs., 420 msecs., and 740 msecs., respectively, to answer 'yes' to the queries 'Did John buy a Rolls-Royce', 'Does John own a car?' and 'Can John sell a car?' (the last query also makes use of the production 'if x owns y then x can sell y). Note that while the first query amounts to recognizing an existing long-term fact, the second and third queries also involve inferences using other productions and *IS-A* relations in SHRUTI's declarative or production memory.

The above times are independent of the sizes of the declarative or production memories and do not increase when additional productions, facts, and *IS-A* relationships are added. If anything, these times may decrease if a new rule is added as a result of chunking.

Conclusion

We have shown that the neurally plausible model for rapid reasoning over facts and rules involving n-ary predicates and variables proposed by Ajjanagadde and Shastri can be interpreted as a production system. This interpretation leads to neurally motivated constraints on the capacity of the working memory of a production system engaged in fast parallel (reflexive) processing and helps in the estimation of the time it would take to perform such processing.

References

Aaronson, J. (1991) Dynamic Fact Communication Mechanism: A Connectionist Interface. *Proceedings of the Thirteenth Conference of the Cognitive Science Society*. Lawrence Erlbaum.

Abeles, M. (1982) *Local Cortical Circuits: Studies of Brain Function* vol. 6. Springer Verlag.

Ajjanagadde, V. G. & Shastri, L. (1991). Rules and variables in neural nets. *Neural Computation*, **3**:121-134.

Anderson, J. R. (1983) *The Architecture of Cognition*. Harvard University Press.

Baddeley, A. (1986) *Working Memory*. Clarendon Press.

Barnden, J., & Srinivas, K. (1991) Encoding Techniques for Complex Information Structures in Connectionist Systems. *Connection Science*, Vol. 3, No. 3, 269-315.

[5] A detailed discussion of the relation between k_2 and the number of nodes required to encode rules appears in (Mani & Shastri 1992).

[6] The following results were obtained using the simulator for SHRUTI described in (Mani 1992).

Carpenter, P. A. & Just, M. A. (1977) Reading Comprehension as Eyes See It. In: *Cognitive Processes in Comprehension.* ed. M. A. Just & P. A. Carpenter. Lawrence Erlbaum.

Damasio, A. R. (1989) Time-locked multiregional retroactivation: A systems-level proposal for the neural substrates of recall and recognition. *Cognition*, 33, 25-62.

Dolan, C. P. & Smolensky, P. (1989) Tensor product production system: a modular architecture and representation. *Connection Science*, 1, 53–68.

Eckhorn, R., Bauer, R., Jordan, W., Brosch, M., Kruse, W., Munk, M., & Reitboeck, H.J. (1988) Coherent oscillations: A mechanism of feature linking in the visual cortex? Multiple electrode and correlation analysis in the cat. *Biol. Cybernet.* **60** 121-130.

Feldman, J. A. (1989) Neural Representation of Conceptual Knowledge. In *Neural Connections, Mental Computation* ed. L. Nadel, L.A. Cooper, P. Culicover, & R.M. Harnish. MIT Press.

Feldman, J. A. (1982) Dynamic connections in neural networks, *Bio-Cybernetics*, 46:27-39.

Freeman, W.J. (1981) A physiological hypothesis of perception. In *Perspectives in Biology and Medicine*, 24(4), 561–592. Summer 1981.

Gray, C. M., Koenig, P., Engel, A. K., & Singer, W. (1989) Oscillatory responses in cat visual cortex exhibit inter-columnar synchronization which reflects global stimulus properties. *Nature.* Vol. 338, 334-337.

Gray, C. M., Engel, A. K., Koenig, P., & Singer, W. (1991) Properties of Synchronous Oscillatory Neuronal Interactions in Cat Striate Cortex. In *Nonlinear Dynamics and Neural Networks*, ed. H. G. Schuster & W. Singer. Weinheim.

Hebb, D.O. (1949) *The Organization of Behavior.* Wiley.

Henderson, J. (1991) *A connectionist model of real-time syntactic parsing in bounded memory.* Dissertation proposal. Department of Computer and Information Science, University of Pennsylvania.

Keenan, J. M., Baillet, S. D., & Brown, P. (1984) The Effects of Causal Cohesion on Comprehension and Memory. *Journal of Verbal Learning and Verbal Behavior*, 23, 115-126.

Kintsch, W. (1988) The Role of Knowledge Discourse Comprehension: A Construction-Integration Model. *Psychological Review*, Vol. 95, 163-182.

Lange, T. E., & Dyer, M. G. (1989) High-level Inferencing in a Connectionist Network. *Connection Science*, Vol. 1, No. 2, 181-217.

von der Malsburg, C. (1981) The correlation theory of brain function. Internal Report 81-2. Department of Neurobiology, Max-Planck-Institute for Biophysical Chemistry, Gottingen, FRG. 1981.

von der Malsburg, C. (1986) Am I thinking assemblies? In *Brain Theory*, ed. G. Palm & A. Aertsen. Springer-Verlag.

Mani, D. R. (1992) *Using the Connectionist Rule-Based Reasoning System Simulator.* Version 12.

Mani, D. R. & Shastri, L. (1991) Combining a Connectionist Type Hierarchy with a Connectionist Rule-Based Reasoner, *Proceedings of the Thirteenth Conference of the Cognitive Science Society.* Lawrence Erlbaum.

Mani, D. R. & Shastri, L. (1992) Multiple Instantiations of Predicates in a Connectionist Rule-Based Reasoner. Technical report MS-CIS-92-05. Department of Computer and Information Science, University of Pennsylvania.

McKoon, G., & Ratcliff, R. (1980) The Comprehension Processes and Memory Structures Involved in Anaphoric Reference. *Journal of Verbal Learning and Verbal Behavior*, 19, 668-682.

McKoon, G., & Ratcliff, R. (1981) The Comprehension Processes and Memory Structures Involved in Instrumental Inference. *Journal of Verbal Learning and Verbal Behavior*, 20, 671-682.

Miller, G.A. (1956) The magical number seven, plus or minus two: Some limits on our capacity for processing information, *The Psychological Review*, 63(2), pp. 81-97.

Newell, A. (1990) *Unified Theories of Cognition.* Harvard University Press.

Newell A. (1980) Harpy, production systems and human cognition. In *Perception and production of fluent speech*, ed. R. Cole. Lawrence Erlbaum.

Potts, G. R., Keenan, J. M., & Golding, J. M. (1988) Assessing the Occurrence of Elaborative Inferences: Lexical Decision versus Naming. *Journal of Memory and Language*, 27, 399-415.

Sejnowski, T.J. (1981) Skeleton filters in the brain. In *Parallel models of associative memory*, ed. G.E. Hinton & J.A. Anderson. Lawrence Erlbaum.

Shastri, L. (1988) *Semantic networks : An evidential formulation and its connectionist realization*, Pitman London/ Morgan Kaufman Los Altos. 1988.

Shastri, L. (1990) Connectionism and the Computational Effectiveness of Reasoning. *Theoretical Linguistics*, Vol. 16, No. 1, 65–87, 1990.

Shastri, L. & Ajjanagadde, V. G. (1990). A connectionist representation of rules, variables and dynamic bidings. Technical Report MS-CIS-90-05, Department of Computer and Information Science, Univ. of Pennsylvania. (Revised January 1992). To appear in *Behavioral and Brain Sciences.*

Shastri, L. (1992) A computational model of tractable reasoning — taking inspiration from cognition. Proceedings of the Workshop on Tractable Reasoning, AAAI-92, San Jose, CA. To appear.

Touretzky, D. S. & Hinton, G. E. (1988) A Distributed Connectionist Production System. *Cognitive Science*, 12(3), pp. 423-466.

Psychological Responses to Anomalous Data

Clark A. Chinn and William F. Brewer
Center for the Study of Reading
University of Illinois
51 Gerty Drive Champaign, IL 61820
chinn@vmd.cso.uiuc.edu
wbrewer@s.psych.uiuc.edu

Abstract

A crucial aspect of understanding knowledge acquisition and theory change is understanding how people respond to anomalous information. We propose that there are seven fundamental responses that people make to anomalous information. We provide evidence from the history of science and from psychology for each of these responses, and we present the results of a study that explores some of the factors that determine these responses.

Anomalies play a pivotal role in the process of knowledge change. On the one hand, anomalous data can force the learner to realize that a current theory must be changed because it is inconsistent with the real world. On the other hand, people often distort or explain away anomalous data so as to protect their favored theories. Thus, the process of theory change appears to be mediated by the way in which a person evaluates anomalous data. In order to understand the process of theory change, we need a more complete understanding of *how* people respond to anomalous data and *why* they respond as they do.

The disciplines that have been most interested in theory change have not provided detailed accounts of people's response to anomalous data. In artificial intelligence, for example, anomalous data are usually treated as correct and unimpeachable (Tweney, 1990). Most scientific discovery and theory revision systems assume that any empirical data that conflict with the current theory are correct; it is the theory that must be changed (e.g., O'Rorke, Morris, & Schulenburg, 1989; Rajamoney, 1989).

In cognitive and developmental psychology, there is widespread recognition that people often discount anomalous data in some way (e.g., Dunbar, 1989; Kuhn, 1989; Piaget, 1980). However, there has been little work on analyzing the specific ways in which people discount anomalous data. Nor has there been a systematic attempt to delineate the factors that affect the way people respond to anomalous data.

The history and philosophy of science contain many insights relevant to the process of responding to anomalous data (Kuhn, 1962; Lakatos, 1970; Laudan, 1977). However, most of these insights are asides given in the course of analyzing particular cases in the history of science.

We propose that in order to understand the process of responding to anomalous data, one needs answers to two questions:

1. What are the different categories of response a person can make to anomalous data?
2. What are the factors that converge to produce each of the different responses? For example, what factors lead an individual to reject anomalous data in one instance but accept anomalous data in another instance?

In the remainder of this paper, we will address these two questions. In the first section, we present a classification of seven forms of response to anomalous data. We present evidence for our classification from the history of science and from psychology. In the second section, we discuss factors that we hypothesize will influence how people respond to anomalous data, and we present the results of an experiment designed to test our hypotheses.

Responses to Anomalous Data

Suppose that a person holding theory A encounters anomalous data that is inconsistent with theory A. The anomalous data may be accompanied by an alternative theory B, which is a competitor of theory A.

We propose that there are seven ways in which a person can respond to the anomalous data. The person can (1) ignore the data, (2) reject the data, (3) exclude the data from the domain of theory A, (4) hold the data in abeyance and retain theory A, (5) reinterpret the data and retain theory A, (6) reinterpret the data but make peripheral changes to theory A, or (7) accept the data and change theory A, perhaps adopting theory B. We think that this is close to an exhaustive set of the possible responses to anomalous data.

These seven responses vary along three dimensions. The first dimension is whether the individual accepts the anomalous data as valid. The second is whether the individual offers an explanation for why he or she has accepted or not accepted the data. And the third is whether the individual changes his or her theory in any way. As we present each of the seven responses, we will discuss their values for each of the three dimensions. We will also briefly present evidence from the history of science and from empirical studies in psychology for the validity of each form of response.

1. Ignoring

A person who ignores data does not accept the data as valid. No explanation for the data is offered, nor is theory A changed at all. The person gives no indication of having been exposed to the data.

History of Science. According to Osborne (1979), the fact that hot water freezes faster than cold water was known to scientists through the writings of Aristotle, Descartes, and Bacon. But after the development of thermodynamics, this fact vanished from the scientific literature until it was rediscovered by a Tanzanian high school student.

Psychology. The typical psychology experiment is designed so as to make ignoring contradictory data very difficult for the subject. There are a few studies that state that subjects appear to be ignoring anomalous data (e.g., Klahr & Dunbar, 1988), but the experimental situations do not provide enough information for us to be sure these data fit our criteria for ignoring.

2. Rejection

Like a person who ignores data, the person who rejects data does not accept the data as valid. But unlike the person who ignores data, the person who rejects data does generate an explanation for why the data are invalid. This explanation can range from a detailed critique of experimental methodology to a vague claim that something must be wrong with the experiment. With rejection, there is no change at all in theory A.

History of science. In the dispute between Millikan and Ehrenhaft over the nature of charge on the electron, each rejected the other's data on methodological grounds (see Holton, 1978). Ehrenhaft believed that Millikan had illegitimately discarded data in order to support his view that electron charge was unitary. Millikan's rejoinder was to argue that Ehrenhaft was mixing bad data with good data in order to achieve results that appeared to support the case against unitary charge.

Psychology. Champagne, Gunstone, and Klopfer (1985) report a study in which students who believed that heavy objects fall faster than light objects subsequently watched the teacher attempt to refute their beliefs by dropping two blocks of different weights from a common height. Although the blocks appeared to strike the ground simultaneously, two middle school students "reasoned that the blocks had, in fact, fallen at different rates, but that the difference in descent times was too small to be observed over the short distance (approximately one meter) used in the original demonstration" (p. 65). These students rejected the experiment as methodologically too insensitive to detect the predicted effect.

3. Exclusion

Another response to anomalous data is to assert that the data are not relevant to theory A, i.e., that theory A is not intended to account for this data (see Kuhn, 1962; Laudan, 1977). The person who excludes data from the domain of theory A can either accept the anomalous data or remain agnostic about the validity of the data. Like the person who ignores data, the person who excludes data does not offer any account of the data. And once again, there is no change in theory A.

History of science. According to Laudan (1977), most theorists in the nineteenth century excluded Brownian motion from their theories. At various times, Brownian motion was regarded as a biological problem, as a chemical problem, as a problem of electrical conductivity, and as a problem in heat theory. "So long as the problem remained unsolved, any theorist could conveniently choose to ignore it simply by saying that it was not a problem which theories in *his* field had to address" (Laudan, 1977, pp. 19-20, italics in original).

Psychology. Karmiloff-Smith and Inhelder (1975; Karmiloff-Smith, 1988) investigated children attempting to balance blocks on a narrow metal support. Some of the blocks were ordinary blocks that balanced at their geometric centers, but other blocks had a weight hidden at one end so that they did not balance at their geometric center. Some children developed the theory that things balance in the middle, but with this theory they couldn't get the weighted blocks to balance. Instead of changing their theory, these children declared that the uneven blocks were impossible to balance and did not worry about them further, which suggests that the children excluded the data from their theory. Their theory was intended to cover only normal blocks, and they felt no need to develop a theory that encompassed all of the blocks.

4. Abeyance

A common response to anomaly is to hold it in abeyance. In this case, the individual faced with anomalous data cannot explain the data but is confident that it will eventually be given an account within the theoretical current framework (Kuhn, 1962). Abeyance is different from the previous forms of response in that the person accepts the anomaly as valid data that his or her theory should be able to explain. But the person cannot, at the present time, provide an explanation for the data.
History of science. An example of abeyance comes from Ampère's assessment of contrary evidence during the period when he was developing his theory of electrodynamics. Ampère was unable to explain one anomalous experiment that he himself had conducted. This anomaly was held in abeyance for over two years, until he was able to make a modification in his theory that could account for his data (Hofmann, 1988).
Psychology. Brewer and Chinn (1991) had undergraduate subjects read about experiments supporting several principles of quantum mechanics. These principles violated certain deeply-entrenched beliefs held by most of the subjects. In response to the data, one subject held the data in abeyance, confident that physicists would eventually solve the paradoxes of quantum mechanics so that he would not need to give up his commitment to realism. In response to one question, he wrote, "Not sure--I'll tell you in 20 years," indicating his belief that scientists will eventually resolve the anomaly within the realist framework.

5. Reinterpretation

When a person reinterprets anomalous data, he or she accepts the data as valid, at least at some level. The person also offers an explanation to account for the data, but the explanation is such that the person need not change theory A at all. In effect, the person acknowledges the data but claims to be able to explain them without altering theory A at all.
History of science. When Alvarez proposed the meteor impact theory of Cretaceous extinctions, his main evidence was an anomalously high concentration of iridium in the K-T boundary (the clay separating the Cretaceous and Tertiary sediments). Some scientists reinterpreted the iridium anomaly as normal seepage of trace amounts of iridium from layers of limestone above the K-T boundary (Raup, 1986). These scientists did not deny the iridium anomaly, but they reinterpreted it as having a terrestrial source rather than an extraterrestrial source.
Psychology. Piaget (1980, Chapter 6) asked children to predict what would happen when equal weights were put in each pan of a balance scale. Most young children predicted that one pan would go down and the other up, like a seesaw. After watching the experimenter place one weight in each pan and finding that nothing happened, a six year old hesitated and scrutinized the scale closely. Then the child declared that the pans were in the same place because both weights were light. The child did not reject the data; there was no attempt to deny that the pans were level. But the data were reinterpreted to show that it was only because the weights were too light that the seesaw effect did not occur.

6. Peripheral theory change

Lakatos (1970) argued that a theorist can always preserve favorite hypotheses in a theory by changing less central, auxiliary hypotheses. When a person makes a peripheral changes to theory A in response to anomalous data, the individual accepts the data as valid and attempts to explain the data. However, the data can be explained only by modifying one or more hypothesis in theory A.
History of science. Galileo's critics denied that there were mountains on the moon because they believed the moon was a perfect sphere. When one of Galileo's opponents looked through Galileo's telescope, he conceded that he saw mountains, but he declared that the mountains were embedded in a perfectly transparent crystal sphere (Drake, 1980). In this way, he protected his core belief that the moon was a

perfect sphere by adding an additional hypothesis.

Psychology. Vosniadou and Brewer (in press) have found that most young children (ages 4-6 years) have a flat earth theory of the shape of the earth. When young children are told by adults that "the earth is round," they are faced with anomalous data. Some of the children account for this anomalous data by making peripheral changes in their flat earth view. For example, some children interpret the data from the adults to indicate that the earth is a flat disc. This approach accounts for the anomalous data about the earth being round but leaves the basic flat earth belief intact.

7. Theory change

A person may be so convinced by the anomalous data that he or she changes theory A, perhaps adopting theory B instead. In this case, the anomalous data are accepted, and they are explained, but they are explained only by giving up core ideas from theory A.

History of Science. The chemical revolution is a good example of theory change. Driven by more than a decade of active empirical research, much of which was anomalous for the phlogiston theory, almost all chemists abandoned phlogiston theory in favor of Lavoisier's oxygen theory (Musgrave, 1976).

Psychology. Even those psychological experiments that demonstrate that many people distort and discredit data also find that some people do change their theories. For example, in the Karmiloff-Smith and Inhelder (1975) research, older children eventually do use the anomalies to develop an improved theory of balancing.

Factors that Influence How People Respond to Anomalous Data

From the point of view of understanding the process of theory change, the crucial question is why a person chooses one response over another. We have begun a series of studies designed to answer this question, and we will present some preliminary data in the present paper.

In the study reported here, we chose two factors that the literature suggested might be particularly powerful influences on how people respond to anomalous data. The first factor is how entrenched a person's beliefs are. An entrenched belief is a belief that has a great deal of evidentiary support and participates in a broad range of explanations in different sub-domains. The more entrenched a belief is, the more it should resist change, and the more likely it should be that anomalous data will be ignored, rejected, excluded, held in abeyance, or reinterpreted.

The second factor is specific background knowledge related to the anomalous data. We hypothesize that the availability of pertinent background knowledge should strongly influence the likelihood of rejection or reinterpretation. For example, background knowledge or beliefs about proper procedures for conducting experiments should lead a person to accept data that has been gathered according to those procedures but to reject data that has not been gathered according to those procedures. In the present study, we decided to focus on background knowledge that might raise the likelihood of rejecting or reinterpreting data.

Method

Domain. The domain in this study was the mass extinction at the end of the Cretaceous period.

Subjects. The subjects were 54 undergraduates enrolled in an introductory psychology course at the University of Illinois at Urbana-Champaign.

Design. Degree of entrenchment and type of background knowledge were manipulated in a 2 X 3 factorial design. There were two levels of entrenchment: entrenched versus non-entrenched. There were three levels of background knowledge: provision of background knowledge for rejecting anomalous data, provision of background knowledge for reinterpreting the anomalous data, and provision of no background knowledge.

Procedure. Each subject began by reading a version of the meteor impact theory. Subjects in the entrenched condition read a 5-page text containing a broad array of evidence supporting the theory. The evidence included the iridium anomaly in the K-T boundary, the discovery of a crater that is the appropriate size and age, anomalous isotope ratios, and several other pieces of evidence. Subjects in the non-entrenched condition read only about the iridium anomaly.

Embedded in one third of these texts was a piece of background knowledge that could be used to reject anomalous data that would be encountered later (e.g., some texts asserted that small-scale laboratory models of global events are not very reliable). Embedded in another third of the texts was a piece of background knowledge that could be used to reinterpret anomalous data that would be encountered later (e.g., some texts explained different ways in which some species might survive 18 months of darkness). The remaining third of the texts contained no background information that could be used to reject or reinterpret the anomalous data.

After reading these texts, subjects rated their belief in the impact theory on a 0 to 10 Likert scale. Then they were provided with a piece of contradictory evidence (e.g., some students read that based on extrapolations from a small-scale laboratory model, a scientist had concluded that a meteor striking the earth would produce so much dust and debris that all sunlight would be blocked for 18 months, effectively killing all life; the scientist argued that since we know that all life was not exterminated, a meteor could not have struck the earth). Subjects rated their belief in this evidence and wrote an explanation for their rating. They re-rated their belief in the impact theory, explaining any change in their belief.

Then all subjects read a brief description of an alternative theory that explains the iridium anomaly by positing that a prolonged period of intense volcanic activity on earth caused the iridium anomaly as well as the extinctions (iridium is also contained in some volcanic magma). Then subjects rated their belief in the volcano theory and the impact theory, explaining their ratings.

Finally, the subjects were provided with a second piece of evidence that contradicted the impact theory (and supported the volcano theory). Subjects rated their belief in the evidence, in the impact theory, and in the volcano theory, again explaining their beliefs and any changes.

Results and Discussion

Effects of entrenchment. Entrenchment clearly influenced whether subjects changed theories. Subjects in the entrenched condition firmly maintained their belief in the meteor theory. Even after being presented with two pieces of contradictory evidence and a plausible alternative theory, entrenched subjects' mean belief in the meteor impact theory was much stronger than their mean belief in the volcano theory (7.2 versus 4.3). This positive difference was significant [$t(24) = 4.61$, $p < .001$]. By contrast, after being presented with the two pieces of contradictory evidence and the volcano theory, non-entrenched subjects preferred the volcano theory. Mean belief in the meteor impact theory was only 4.5, while mean belief in the volcano theory was 6.1 [$t(21) = -2.56$, $p < .05$].

Thus, our manipulation of entrenchment clearly affected subjects' belief in the alternative theories.

Table 1
Effects of Entrenchment
on Responses to the Evidence:
Frequency and Percentage of Each Response

	Entrenched		Non-Entrenched	
Ignoring	0	(0%)	0	(0%)
Rejection	27	(57%)	25	(54%)
Abeyance	3	(6%)	1	(2%)
Exclusion	0	(0%)	0	(0%)
Reinterpretation	6	(13%)	2	(4%)
Theory change	11	(23%)	18	(39%)

There was also an indication that entrenchment affected the distribution of responses to anomalous data. Table 1 presents the distribution of responses to the two pieces of evidence in the entrenched and non-entrenched conditions. The pattern was not as pronounced as we had expected.

Effects of background knowledge. Our manipulation of background knowledge failed to influence the likelihood that subjects would reject or reinterpret data. Only two subjects whose texts included experimentally-provided information that could be used to reject the anomalous data used that information as grounds for rejecting data. Similarly, only two subjects who read texts that provided information that could be used to reinterpret the anomalous data used that information to reinterpret the data.

The reason for the subjects' failure to use the background information provided in the text appears to be that they instead used background knowledge that they brought with them to the experiment. For example, some subjects used their background knowledge about experimentation and about the physical world to reject the data that declared that a meteor would kill everything because it would block all sunlight for 18 months: "The earth is much too big and different to be correctly represented by some small rock in a laboratory. Things like atmosphere, spin, gravity, etc. play a part as well." Others used their background knowledge not to reject the claim that the earth would be dark for 18 months but to reinterpret it to show that all life would not be killed: "<u>Some</u> things could survive with no light, for example, anaerobic respirators." Others appeared to rely on background intuitions to ratify the study's conclusions: "I think he knows what he's talking about. If a large meteor did hit the earth, a lot of debris would be thrown in the air."

It appears, then, that background knowledge is very important in the response to anomalous data. But it may be difficult for background knowledge supplied by the text to compete with background knowledge already possessed by students. The background knowledge in the text may be less accessible than students' own prior knowledge. (The alert reader will note that this paragraph is an example of reinterpretation on our part!)

Summary and Conclusions

We have argued that understanding how people respond to anomalous data is crucial to understanding the process of knowledge change. We have proposed a taxonomy of seven forms of response to anomalous data, and we believe that this taxonomy can provide a framework for systematic investigation into the factors that influence how people respond to anomalous data. In the study reported here, we found that the entrenchment of beliefs and the availability of background knowledge influenced how undergraduates responded to anomalous data.

References

Brewer, W. F., and Chinn, C. A. 1991. Entrenched beliefs, inconsistent information, and knowledge change. In *Proceedings of the 1991 International Conference of the Learning Sciences*, 67-73. Charlottesville, VA: Association for the Advancement of Computing in Education.

Champagne, A. B.; Gunstone, R. F.; and Klopfer, L. E. 1985. In L. H. T. West and A. L. Pines (eds.), *Cognitive structure and conceptual change* (pp. 61-90). Orlando, FL: Academic Press.

Drake, S. 1980. *Galileo*. New York: Hill and Wang.

Dunbar, K. 1989. Simulating reasoning strategies in a simulated molecular genetics environment. In *Program of the Eleventh Annual Conference of the Cognitive Science Society*, 426-433. Hillsdale, NJ: Lawrence Erlbaum Associates.

Hofmann, J. R. 1988. Ampère's electrodynamics and the acceptability of guiding assumptions. In A. Donovan, L. Laudan, and R. Laudan (eds.), *Scrutinizing science: Empirical studies of scientific change* (pp. 201-217). Dordrecht: Kluwer Academic Publishers.

Holton, G. 1978. Subelectrons, presuppositions, and the Millikan-Ehrenhaft dispute. *Historical Studies in the Physical Sciences* 9:161-224.

Karmiloff-Smith, A. 1988. The child is a theoretician, not an inductivist. *Mind & Language* 3:183-195.

Karmiloff-Smith, A., and Inhelder, B. 1975. "If you want to get ahead, get a theory". *Cognition* 3:195-212.

Klahr, D., and Dunbar, K. 1988. Dual space search during scientific reasoning. *Cognitive Science* 12:1-48.

Kuhn, D. 1989. Children and adults as intuitive scientists. *Psychological Review* 96:674-689.

Kuhn, T. S. 1962. *The structure of scientific revolutions*. Chicago: University of Chicago Press.

Lakatos, I. 1970. Falsification and the methodology of scientific research programmes. In I. Lakatos and A. Musgrave (eds.), *Criticism and the growth of knowledge* (pp. 91-196). London: Cambridge University Press.

Laudan, L. 1977. *Progress and its problems*. Berkeley: University of California Press.

Musgrave, A. 1976. Why did oxygen supplant phlogiston? Research programmes in the chemical revolution. In C. Howson (ed.), *Method and appraisal in the physical sciences: The critical background to modern science, 1800-1905* (pp. 181-209). Cambridge: Cambridge University Press.

O'Rorke, P.; Morris, S.; and Schulenburg, D. 1989. Abduction and world model revision. In *Program of the Eleventh Annual Conference of the Cognitive Science Society*, 789-796. Hillsdale, NJ: Lawrence Erlbaum Associates.

Osborne, D. G. (1979). Mind on ice. *Physics Education* 14:414-417.

Piaget, J. 1980. *Experiments in contradiction* (trans. D. Coltman). Chicago: University of Chicago Press.

Rajamoney, S. A. 1989. Exemplar-based theory rejection: An approach to the experience consistency problem. In *Proceedings of the Sixth International Workshop on Machine Learning*, 284-289. San Mateo, CA: Morgan Kaufmann Publishers.

Raup, D. M. 1986. *The Nemesis Affair: A story of the death of dinosaurs and the ways of science*. New York: W. W. Norton & Company.

Tweney, R. D. 1990. Five questions for computationalists. In J. Shrager and P. Langley (eds.), *Computational models of scientific discovery and theory formation* (pp. 471-484). San Mateo, CA: Morgan Kaufmann Publishers.

Vosniadou, S., and Brewer, W. F. (in press). Mental models of the earth: A study of conceptual change in childhood. *Cognitive Psychology*.

Modelling Inductive And Deductive Discovery Strategies In Galilean Kinematics[1]

Peter C-H. Cheng
Department of Psychology
Carnegie Mellon University
Pittsburgh, PA 15213
pc2n+@andrew.cmu.edu

Abstract

This paper investigates how different strategies affect the success and efficiency of scientific discovery, by examining different approaches in Galilean kinematics. Computational models with biases for inductive or deductive approaches to discovery were constructed to simulate the processes involved in finding coherent and empirically correct sets of laws. The performance of the models shows that the best overall strategy is to begin with an inductive bias and then perform tight cycles of law generation and experimental testing. Comparison of the models with previous findings indicates that the best overall strategy for discovery depends on the relative ease of search in hypothesis and experiment spaces.

1 Introduction

Scientific discovery is an important and growing area of research in the cognitive sciences. A major issue in the area concerns what strategies scientists use to make discoveries and the effectiveness and efficiency of those strategies. Much of the empirical work has focussed on the biases that seem to exist in the process of seeking evidence to assess hypotheses, Gorman (1992). The work of Klahr and Dunbar (1988) and Klahr *et. al* (1990) addresses discovery strategies from a wider context. Their studies employed a simulated discovery environment consisting of a toy robot, BigTrak, controlled using a LOGO-like programming language. The task was to determine how a mystery programming key functioned by writing programs incorporating the key and observing the subsequent behavior of BigTrak. It was found that subjects could be classified as either experimenters or theorists according their preference to search the space of experiments or the space of hypothesis. Theorists were more successful and faster, because they generated new hypotheses using relevant prior knowledge. Experimenters were less effective and efficient, because they laboriously performed experiments and attempted to generalize the results into hypotheses. Klahr and Dunbar characterize scientific discovery as the dual search of the hypothesis and experiment spaces. The best discovery strategy for this task is initially to generate several hypotheses and then test them experimentally.

In contrast to the work of Klahr and colleagues, this paper describes computational work that models different strategies in Galileo's discovery of the laws of free fall. A large number of computational systems already exist that model many aspects of scientific discovery (see Cheng, 1992a, for a review). The conventional computational approach attempts to demonstrate the acceptability of a complete model by simulating one or more episodes of discovery. Two relevant examples are: Kulkarni and Simon's (1988) KEKEDA system, which demonstrates that prompt investigation of surprising experimental outcomes is a good strategy; and, Cheng's (1990, 1991) STERN system that has previously modelled Galilean kinematic discoveries. However, the approach adopted here is different. The modelling will focus on a particular factor that is important in discovery by constructing models that differ with respect to the factor but are otherwise as similar as possible. Thus the difference in the models' performance will directly demonstrate how the factor affects the success and efficiency in a particular discovery task. This approach was previously used to demonstrate the computational benefits of using diagrams in discovery, by comparing models using diagrammatic and conventional mathematical representations (Cheng, 1992b; Cheng & Simon, 1992). Here the models have been given biases for either inductive or deductive approaches to discovery, to examine how the approaches influence the overall success and efficiency of discovery in the Galilean domain.

We will begin by considering Galileo's kinematic discoveries.

[1]Thanks should go to Herbert Simon for the discussions that helped to shape this research. This work was supported by a UK Science and Engineering Research Council postdoctoral fellowship.

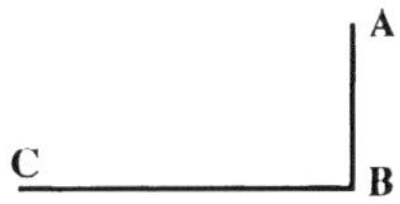

Figure 1

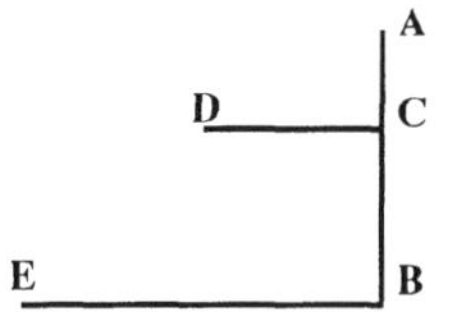

Figure 2

2 Galilean Kinematic Discovery

Although historians of science do not agree upon the precise manner in which Galileo found the laws governing the motion of bodies in free fall, there is reasonable agreement on the main stages of the episode (e.g., Drake, 1973, 1975, 1978; Drake & MacLachlan, 1975; Hill, 1988; Naylor, 1974). We take up the story from the point where Galileo had rejected the Aristotelian views of motion but incorrectly believed the speed of a naturally accelerated body to be in proportion to the distance travelled from rest, which may be expressed as an equation thus;

$$v_{ab}/v_{ac} = d_{ab}/d_{ac}\,, \qquad \ldots (1)$$

where *v* and *d* are speed and distance, respectively, and *ab* and *ac* indicate two different falls from rest. Galileo typically stated quantitative laws as sentences referring to ratios of similar variables, but for ease of comprehension equations will be used here, without affecting the claims being made. Galileo eventually found the correct laws of motion;

$$v_{ab}/v_{ac} = t_{ab}/t_{ac}\,, \qquad \ldots (2)$$

$$d_{ab}/d_{ac} = t_{ab}^{2}/t_{ac}^{2}\,, \qquad \ldots (3)$$

and, $$v_{ab}/v_{ac} = d_{ab}^{1/2}/d_{ac}^{1/2}\,, \qquad \ldots (4)$$

where *t* is time. The precise manner of their discovery is uncertain, the historians have conjectured many different paths, but it is clear that Galileo used a combination of deductive and inductive methods. They are considered in turn.

Galileo (1974) published his kinematic findings in the Third and Fourth Days (sections) of the *Two New Sciences*, TNS hereafter. The Third Day has two subsections. The first concerns constant speed motion, and presents laws relating speed, time and distance when acceleration is absent. For example, the fourth and sixth propositions are, respectively;

$$d_{ef}/d_{gh} = v_{ef}/v_{gh} \cdot t_{ef}/t_{gh}\,, \qquad \ldots (5)$$

and, $$v_{ef}/v_{gh} = d_{ef}/d_{gh} \,/\, t_{ef}/t_{gh}\,, \qquad \ldots (6)$$

where the variables have the same meanings as above, the values of *v* are constants, and the subscripts denote two different bodies or paths. The second subsection of the Third Day concerns the accelerated motion of bodies under many different circumstances. Beginning with a single postulate, a definition, and the laws of constant speed motion, 38 propositions are derived. The definition happens to be Equation 2, and the second proposition (TNS-III-2) and its corollaries are equivalent to Equation 3. TNS-III-2 is derived from TNS-III-1, which states that the mean speed of a body falling from rest is one half its maximum speed. As the speeds of two bodies are in proportion to their times, Equation 2, their mean speeds will thus be in proportion to the times. However, as the mean speeds are both constant by definition, Equation 5 holds, so times can be substituted for speeds to give Equation 3. Another proposition of TNS-III is the *double distance* law, TNS-III-9. Referring to Figure 1, if a body falls from rest through distance *AB* and is turned through a right angle so that it travels horizontally along *BC*, then the distance *BC* will be double that of *AB* when the time for *AB* equals the time for *BC*. This proposition is used in the derivation of Equation 4, considered next.

In TNS-IV-3 Galileo derives the relation for speed and distance, Equation 4. For vertical falls from rest over two distances *AC* and *AB*, Figure 2, the double distance law can be applied twice to distances *CD* and *BE*. As the speeds along *CD* and *BE* are constant their ratio is given by Equation 6,

$$v_{cd}/v_{be} = d_{cd}/d_{bfe} \,/\, t_{cd}/t_{be}. \qquad \ldots (7)$$

However, from the double distance rule we know: (i) the ratio of the times *CD* to *BE* is equal to the ratio of the times *AC* to *AB*; (ii) distance *CD* is double *AC* and distance *BE* double *AB*; and, (iii) the speed at end of the fall *AC* equals the speed along *CD*, and similarly for AB and *BE*. Therefore, the times, distances and speeds of *AC* and *AB* can be substituted into Equation 7;

$$v_{ab}/v_{ac} = d_{ab}/d_{ac} \,/\, t_{ab}/t_{ac}. \qquad \ldots (8)$$

Now, Equation 8 relates distances and times in free fall, so the times can be eliminated using Equation 3 giving Equation 4, so completing the derivation of the set of three laws[2].

The classes of motion experiments that historians are sure Galileo had at his disposal were pendulums, inclined planes or ramps, and projectiles. In the inclined plane experiments Galileo rolled a ball down the plane and measured how the distance along the plane varied with time. To perform experiments on projectiles, Galileo used an inclined plane with a lip at its end to launch the ball horizontally into the air. This combined inclined plane and projectile experiment allowed Galileo to determine speeds using Equation 6, because the horizontal speed of a projectile is constant and the time of fall is also constant when the vertical distance is fixed. Cheng (1991) describes the experiments in more detail. The

[2] Why Galileo did not just substitute for t in Equations 2 and 3, to derive Equation 4, is a mystery.

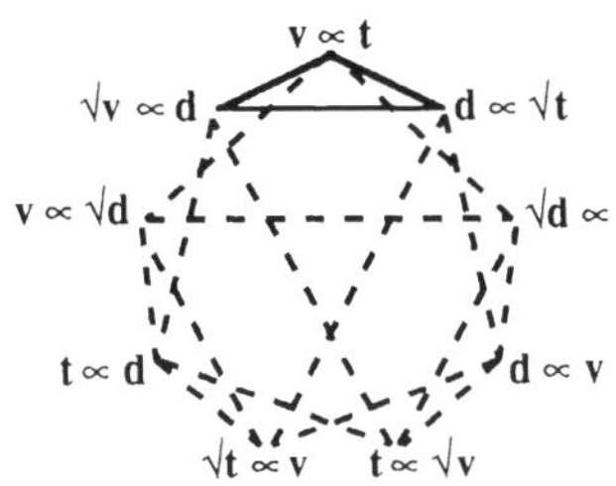

Figure 3

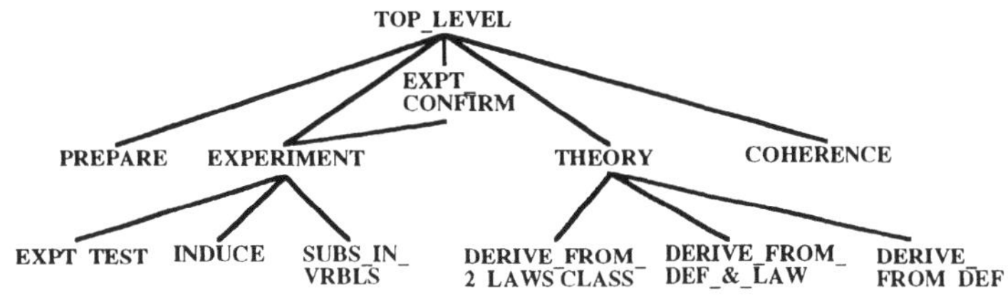

Figure 4

specific purposes behind the experiments are disputed because Galileo (1979) only left terse and cryptic records of them, but is possible that they could have been performed in either an inductive manner or a confirmatory mode. Galileo was a competent mathematician so was able to find expressions describing experimental data or to judge how well numerical predictions and data matched.

We will now define more precisely the discovery task to be modelled.

3 Simplified Discovery Task

A simplified version of the Galilean episode is considered for two reasons. First, it is necessary to reduce the complexity of the problem to make the construction of models and the performance analysis a practical proposition. Second, the simplification will focus on the aspects of the discovery with the greatest historical certainty.

The discovery task is to find an acceptable set of laws relating the speed, distance and time of a body in accelerated motion, given the laws of constant speed motion. The laws to be found will be power function of pairs of variables. A set of three laws is necessary to cover all the combinations of pairs of variables and a set is acceptable if it is coherent and empirically correct. A set is coherent if from any two laws the third can be validly inferred by elimination of a shared variable; which is the case for Equations 2, 3 and 4. Laws are empirically correct when they match experimental data with sufficient accuracy.

To clarify the nature of the discovery task, consider a subset of the space of laws in which the variables are linearly related or one variable is in proportion the square-root of the other. There are nine laws in the subset, so 84 $(=9!/(6!\cdot3!))$ combinations of sets of three laws are possible. Figure 3 shows the nine laws (using proportionality signs rather than ratios of variables). The task is to find an acceptable set from amongst all the (84) possible sets. The triangles indicate the six sets of coherent laws. The uppermost triplet, with the solid triangle, is the empirically correct set. This characterization of the task suggests that enumeration of all the coherent sets may be an effective way to begin tackling the problem, because of the substantial reduction in the search space (from 84 to 6, in the example). However, the models will take a more historically constrained approach, employing methods that Galileo would have had at his disposal.

4 Program and Models

The basis for the models is a single production system that can be given a bias for either an inductive or a deductive approach to discovery. Care has been taken in the specification of the knowledge structures and productions to ensure a close match to the inferences steps seen in the historical material. The declarative knowledge representations will be considered first followed by a description of the various inference processes of the model.

The knowledge representations are simple predicate-argument-like structures, that can be considered at three levels. Information on the highest level relates to the overall task of discovery and includes: a list of the relevant variables (e.g., speed distance time); the permitted indices of the power functions (e.g., 1, 1/2, 2); the type of phenomenon (i.e., free fall accelerated motion); and, the identifiers of sets of laws that have been considered (nil initially). These top level items are given as run-time inputs. The second level of knowledge has information required for inferences involving sets of laws, including: the identifier of the set under active consideration; the pairs of variables for each law; and, a statement about a set's coherence, when tested. The lowest level concerns individual laws and associated information, including: the law's power equation; and records of the status of the deductive and inductive inferences made.

Figure 4 shows the hierarchy of classes of rules used by the program. The TOP_LEVEL rules control the overall operation of program by invoking the one of five classes on the next level. These classes are briefly described before returning to TOP_LEVEL

The PREPARE rules place second and third level knowledge structures into working memory for use by other processes, so it is the first class to be invoked by TOP_LEVEL when a simulation begins. It is also invoked whenever a set of rules is rejected to initialize a new set of structures. PREPARE may generate a new law as a definition based on a pair of preferred variables and an index not previously considered with the variables.

The EXPERIMENT class attempts to find laws by performing experiments. Descriptions of experimental tests, given as inputs to the program, specify which properties can be employed as *manipulated input* and *measured output* variables (Cheng, 1991). For example, time is the manipulated input and distance the measured output in an inclined plane experiment. Given a pair of variables EXPT_TEST is called to find an experiment test with matching input and output variables. When there is no direct match an alternative variable may be used as the input or output, if there is a known relationship between it and the given variable. For example, horizontal distance in the combined inclined plane and projectile experiment can be used to measure speed. When a suitable test is found, a special rule places the experimental results in working memory, as if the experiment had been performed. The INDUCE class is then engaged to seek a simple power law that accounts for the numerical data. Finally, the SUBS_IN_VRBLS class is employed to substitute variables in to the power law when necessary; e.g., speed for distance in the inclined plane and projectile experiment.

The function of EXPT_CONFIRM is to test experimentally a law found deductively. EXPT_CONFIRM simply invokes EXPERIMENT, described above, to induce a law that is compared with the derived law to see if they are the same. Galileo would have made a quantitative prediction from the derived law and compared this with the experimental data, rather than generalizing the data and comparing laws. However, the difference is minor and does not have a significant bearing on the conclusions to be drawn; so for simplicity it is acceptable.

The THEORY class is the deductive module of the program and calls three other classes. DERIVE_FROM_DEF and DERIVE_FROM_DEF_&_LAW attempt to find a law from a definition, or from a definition and an existing law, respectively. The two classes include rules instantiating constant speed laws and generalized versions of the inferences that Galileo employed in the TNS. The rules are general in two ways: (i) they employ variables for predicates and arguments; and, (ii) they consider any power function rather than a particular relation. General rules mean that the correct accelerated motion laws are not implicitly built into the deductive mechanisms of the system. For example, TNS-III-1 states that the mean speed is one half the maximum speed of a uniformly accelerated body, as the speed increases linearly with time. The equivalent rule is (translated into pseudo-English):

If the set of laws G is being considered,
  and the law M is being considered,
  and M is a definitional law of the form
    $SSxy^p/SSxz^p=TTxy^r/TTxz^r$,
  and G's context is free fall motion,
  and SSxy and TTxy are known,
Then SSab is the mean of SSxy,
  and SSab is constant with respect to TTab,
  and TTxy=TTab,
  and SSab=n.SSxy where n=p/(p+r),
  and note SSab & TTab can be determined.

G and M are names; SS and TT stand for predicates; n, p, and r are numbers; and, a, b, x, y and z are points on paths. This rule determines the mean of SS as a function of its maximum, as it increases as some power of TT.

DERIVE_FROM_2_LAWS attempts to find a new law from two given laws by substitution. All three derivation classes invoke the MATHS rules (not shown in Figure 4) to make simple mathematical inferences.

The remaining class, COHERENCE, tests whether a set of three laws is coherent by examining the values of the indices in the equations. Galileo would probably have performed the test by inspection or possibly by substituting values into the equations.

TOP_LEVEL calls various rule classes under different circumstances. PREPARE is invoked whenever a new set of laws is to be considered. COHERENCE is invoked whenever three laws have been found in the active set. EXPT_CONFIRM follows the COHERENCE class when a coherent set of laws has been found. Either EXPERIMENT or THEORY can be invoked when a set of laws is incomplete depending on the relative priority of the two classes set by the user. When EXPERIMENT has the higher priority it is always invoked in preference to the THEORY class, so making the program a model with an inductive bias. A model with a deductive bias is obtained when THEORY has the higher priority. TOP_LEVEL includes rules to evaluate the acceptability of a set of laws. The program halts when an acceptable set is discovered but a new set is sought when an unacceptable set is found.

That completes the overview of the program that can be employed as models with inductive or deductive biases. The following section considers the performance of the two models.

5 Performance of the Models

The two models have been run with different input

conditions. One inductive and two deductive simulations are described. In all three simulations the inclined plane, and the combined inclined plane and projectile experiments, were given as inputs.

The first simulation employed the inductive model and no initial law was given as a definition. Once PREPARE had set up the various knowledge structures, EXPERIMENT was invoked because of the inductive bias. Pairs of variables were considered in turn and suitable experimental tests were found for two of the three pairs. The inclined plane was used to find data relating distance and time; and the combined inclined plane and projectile experiment used to find data relating speed and distance. The data were generalized into two equations, Equations 3 and 4, by INDUCE. No experiments were available to find the relation between speed and time, so THEORY was invoked to find the third law by combining the induced laws, resulting in Equation 2. COHERENCE then determined the laws were coherent, and as two laws matched the experimental data, an acceptable set had been found.

Included amongst the input of the first of the two deductive simulations was one of the correct laws of motion, Equation 2. The system was not told that the law was acceptable and it was used as the basis for the inferences made by THEORY, following the steps in the TNS. First, the relation between distance and time was found using the mean speed relation. Then, the law relating speed and distance was inferred by applying: the definitional law; the law just derived; and, the double distance scenario twice. The set of laws was found to be coherent so EXPT_CONFIRM was called to find whether the laws were compatible with experimental data, which they were. Thus an acceptable set was found. Approximately two thirds more computation, in terms of numbers of productions fired, was required for this simulation than the first, even though one of the correct laws was given as an input.

The second deductive simulation had the same inputs as the first except the given definition was Galileo's earlier incorrect definition, Equation 1. Again two further laws were derived;

$$t_{ab}/t_{ac} = d_{ab}^2/d_{ac}^2 \,, \qquad \ldots (9)$$

and,

$$v_{ab}/v_{ac} = t_{ab}^{1/2}/t_{ac}^{1/2} \,. \qquad \ldots (10)$$

The two laws and the definition form a coherent set, but during experimental testing Equation 9 did not match the data. This demonstrates that the deductive approach can generate a coherent set of laws that are not empirically acceptable. The whole set was rejected and PREPARE invoked to begin the search for a new set of laws. PREPARE generated a new law that happened to be Equation 2, so the simulation then followed the course taken by the first deductive run. The effort to generate the first but incorrect set of laws increased the amount of computation over the previous deductive simulation by approximately 70%.

The implications of the performance of the models will be considered next.

6 Discussion

In the present discovery task both inductive and deductive approaches are necessary for successful discovery, irrespective of whether the model has an inductive or deductive bias. This is consistent with the historical picture. Further, and more interestingly, the performance of the models indicates that the most effective strategy is initially to adopt an inductive or experimental approach that covers as many of the pairs of variables as is possible. When deductive inferences have to be made, it is preferable to employ tight cycles of law generation and experimental testing. Three related aspects of the task explain why this is the best strategy. First, the experiments will have to be performed no matter which approach is taken, because empirical data will be required to determine whether laws in a coherent sets are empirically correct. It is a waste of effort to derive several laws under the deductive approach only then to perform experiments from which the laws could have been more easily inferred, by inductive generalization, in the first place. Second, laws induced from experiments rule out a larger part of the search space of laws than those found by deductive inferences. Two experimental laws will uniquely identify a coherent and empirically correct set, whereas two deductive laws only define a coherent set without any indication of empirical acceptability. Third, the amount of work required to derive each law is substantial, and as the derivation does not guarantee the acceptability of the law, it is best to assess the empirical acceptability of a law as soon as it is found to avoid making further inferences based on an unacceptable law. In summary, the strategy aims to minimize the amount of computation that might be wasted in generating laws that are coherent but not empirically correct.

The overall strategy for the kinematic domain contrasts with Klahr and Dunbar's best strategy for the BigTrak discovery context (see §1). They contend that the best strategy is to maintain a bias for hypothesis space search and initially to think of hypotheses before conducting any experiments. Comparison of the tasks and models reveals the reasons for the differences in the strategies. Many of Klahr and Dunbar's (1988) subjects found it relatively easy to think of novel hypotheses, and the correct hypothesis was often among the small number generated. When the correct hypothesis was not one proposed, the initial search of the hypothesis space made it easier for subjects to induce the correct hypothesis from experimental outcomes. In the Galilean task, however, deductively generating sets of laws is difficult, and existing sets do not help in the generalization of experimental data into laws. This suggests that the experimentally lead strategy could be the most effective in the BigTrak task under certain circumstances; specifically, when the correct hypothesis is not a member of the small set of seemingly reasonable hypotheses, but is highly unexpected[3]. Subjects will not be able to think of the hypothesis initially, and considering seemly reasonable but irrelevant hypotheses will be unlikely to help subjects find the correct hypothesis when the experiments are conducted.

To summarize, it seems best to adopt an inductive experimental approach when the search of the hypothesis space for a particular domain is likely to be relatively difficult and computationally expensive. Alternatively, when the relative amount of effort to perform experiments is likely to be great, it is best initially to generate hypotheses before conducting the experiments.

7 Conclusions

The best strategy for the discovery task considered here is initially to take an inductive approach followed by tight cycles of law generation and experimental testing. The comparison of this finding with Klahr and Dunbar's work indicates that the most effective and efficient strategy will depend on the relative ease with which hypothesis and experiment spaces can be searched. How, or even whether, scientists make such evaluations, to aid the selection of the most appropriate strategy for a particular discovery task, remains as an important issue to be addressed by future research. The kinematic domain will be a good starting point for this work as Galileo's knowledge of Euclidean geometry and his ability as an experimenter could have been the basis for such reasoning.

References

Cheng, P.C-H. 1990. Modelling Scientific Discovery. Technical Report No.65, Human Cognition Research Laboratory, The Open University, Milton Keynes, England.

Cheng, P.C-H. 1991. Modelling experiments in scientific discovery. *Proceedings of the 12th International Joint Conference on Artificial Intelligence.* Mountain View, CA: Morgan Kaufmann.

Cheng, P.C-H. 1992a. Approaches, models and issues in computational scientific discovery. In M.T. Keane & K. Gilhooly. *Advances in the Psychology of Thinking* Vol. 1. Hemel Hempstead, Hertfordshire, UK: Harvester-Wheatsheaf. In press.

Cheng, P.C-H. 1992b. Diagrammatic reasoning in scientific discovery: Modelling Galileo's Kinematic Diagrams. In *Working notes of the AAAI Spring Symposium Series: Reasoning with diagrammatic representations.* Stanford, March 1992.

Cheng, P.C-H. & Simon, H.A. 1992. The right representation for discovery: Finding the conservation of momentum. In D. Sleeman and P. Edwards eds. *Machine Learning: Proceedings of the Ninth International Conference (ML92)*, San Mateo, CA; Morgan Kaufmann.

Drake, S. 1973. Galileo's discovery of the law of free fall. *Scientific American*, **228**:85-92.

Drake, S. 1975. The role of music in Galileo's experiments. *Scientific American*, **232**:98-104.

Drake, S. 1978. *Galileo at Work.* Chicago: University of Chicago Press.

Drake, S. & MacLachlan, J. 1975. Galileo's Discovery of the parabolic trajectory. *Scientific American* **232**(3):102-110.

Galileo Galilei. 1974. *Two new sciences.* S. Drake trans. Madison, Wisconsin: University of Wisconsin Press.

Galileo Galilei. 1979. *Galileo's Notes On Motion.* S. Drake. ed. Annali dell'Instituto e Museo di Storia della Scienza. Florence: Italy.

Gorman, M.E. 1992. Using experiments to determine the heuristic value of falsification. In M.T. Keane & K. Gilhooly. *Advances in the Psychology of Thinking* Vol. 1. Hemel Hempstead, Hertfordshire, UK: Harvester-Wheatsheaf. In press.

Hill, D.K. 1988. Galileo's early experiments on projectile motion and the law of free fall. *ISIS*, **79**,646-68.

Klahr, D. & Dunbar, K. 1988. Dual space search during scientific reasoning. *Cognitive Science*, **12**:1-48.

Klahr, D., Dunbar, K. & Fay, A.L. 1990. Designing good experiments to test bad hypotheses. In J. Shrager & P. Langley eds. *Computational Models of Scientific Discovery and Theory Formation*, 356-402. San Mateo, CA: Morgan Kaufmann.

Kulkarni, D. & Simon, H.A. 1988. The processes of scientific discovery: the strategy of experimentation. *Cognitive Science* **12**:139-75.

Naylor, R.H. 1974. Galileo and the problem of free fall. *British Journal For The History Of Science*, 7(26):105-34.

[3]E.g., the mystery key might function in different ways when its numeric parameter is, or is not, a prime.

Complexity Management in a Discovery Task.[1]

Christian D. Schunn
Department of Psychology
Carnegie Mellon University
Pittsburgh, PA 15213
schunn@cmu.edu

David Klahr
Department of Psychology
Carnegie Mellon University
Pittsburgh, PA 15213
klahr@cmu.edu

Abstract

Previous psychological research about scientific discovery has often focused on subjects' heuristics for discovering simple concepts with one relevant dimension or a few relevant dimensions with simple two-way interactions. This paper presents results from an experiment in which subjects had to discover a concept involving complex three-way interactions on a multi-valued output by running experiments in a computerized microworld. Twenty-two CMU undergraduates attempted the task, of which sixteen succeeded, in an average of 85 minutes. The analyses focus on three strategies used to regulate task complexity. First, subjects preferred depth-first to breadth-first search, with successful subjects regulating the number of features varied from experiment to experiment most effectively. Second, subjects systematically regulated the length of their experiments. Third, a new explicit search heuristic (Put Upon Stack Heuristic) used by successful subjects is described.

One of the most complex cognitive tasks that humans face is scientific discovery. It combines the mystery of creativity with the rigor of experimentation and hypothesis testing. By studying the psychological processes involved in scientific discovery, one can test the scalability of psychological theories developed using simpler tasks, and develop theories of the integration process of numerous subprocesses, which does not occur in simple tasks.

The first psychological investigations of scientific discovery used simple rule discovery tasks (e.g., Bruner, Goodnow, & Austin, 1956; Wason, 1960). Subjects had to discover a rule for classifying instances as either members or nonmembers of a concept which was an arbitrary concatenation of a few simple features. Simon & Lea (1974) proposed that rule discovery could be viewed as search in two problem spaces: the rule space (the set of all rules for classification of instances) and the instance space (the set of all instances to be examined). Rule discovery is comprised of the set of processes, algorithms, and heuristics for searching each of the two spaces, and integrating the search between the spaces.

Klahr and Dunbar (1988) extended the dual search idea by proposing that scientific discovery can be understood in terms of search in two problem spaces: the hypothesis space and the experiment space. A primary difference between scientific discovery and rule discovery is that the relationship between the hypothesis and the experiment is very straight-forward in rule discovery tasks, whereas it is very complex in the real scientific discovery process (Klahr & Dunbar, 1988). To study the psychological processes in situations with this more complex relationship, some researchers have used computerized microworlds (e.g., Mynatt, Doherty, & Tweney,1977; Klahr & Dunbar, 1988; Dunbar, 1989, 1992).

As the complexity of the domain grows, it becomes increasingly necessary to use heuristics to simplify the search in the two spaces. Klahr & Dunbar (1988) describe two heuristics associated with two subgroups of subjects in their experiments: their "experimenters" limited their search mainly to the experiment space and their "theorists" limited their search mainly to the hypothesis space. Klahr, Dunbar, & Fay (1990) further identified several search heuristics used by subjects for searching within each of the two spaces. Example heuristics identified for searching the experiment space included designing experiments which maintain easy observability, and exploiting surprising results.

Although these heuristics were identified in contexts involving much more complexity than the classic rule discovery tasks, there remains a question as to whether the heuristics found will generalize to even more complex situations. That is, as the task becomes more complex, will subjects use other strategies to deal with increased complexity in order

[1] This research was funded by a scholarship from la Formation de Chercheurs et l'Aide à la Recherche to the first author, and by grants from the National Institute of Child Health and Human Development (R01-HD25211) and the A.W. Mellon Foundation to the second author.

to make the task more tractable? This paper reports a study designed to investigate how subjects discover a complex concept. The analysis will focus on the strategies used by subjects to deal with difficulties encountered in discovering complex concepts.

Method

Overview. Subjects were shown the function of all but one command of a device. The subjects designed, conducted and evaluated experiments with the device to discover how the mystery command works. The mystery command was a complex sort operator that took three arguments.

Subjects. Twenty-two Carnegie Mellon University undergraduates took part in the experiment for course credit and eight dollars. Twenty-one of the subjects had taken at least one programming course, and all had used a computer before.

The Computer Interface. Subjects worked in a complex microworld in which a "milk truck" could execute a sequence of actions associated with a dairy delivery route. At any of 6 different locations along its route, it could toot its horn, deliver milk or eggs, or receive money or empties. The route of the milk truck was programmed using the keypad to enter a sequence of action-location parirs (see figure 1). As subjects entered their programs, the steps were displayed on the screen in the program listing. After the route had been entered, subject pressed 'RUN' and the milk truck executed its route on the screen. The milk truck went to each location on the programmed route in the order that it was programmed. The milk truck stopped at the location, and the subjects were shown by way of animated icons what transpired at the location. Also, as the route was being completed, a trace listing displayed in program format what transpired during that run (see figure 1).

When the mystery command, δ (delta), was not used, the trace listing was identical to the program listing. However, the δ command could change the order of delivery, and the resultant trace would then be discrepant from the program listing. Subjects could also look over the program and trace listings of their old programs.

The effect of the δ command was to reorder the execution sequence of part of the program according to the values of its three arguments, a number (1-6), a triangle (white or black), and a Greek letter, (α or β). Table 1 describes the effects of the delta command. The first and second programs in figure 1 show the effects of the δ with white triangle and α, and with black triangle and β.

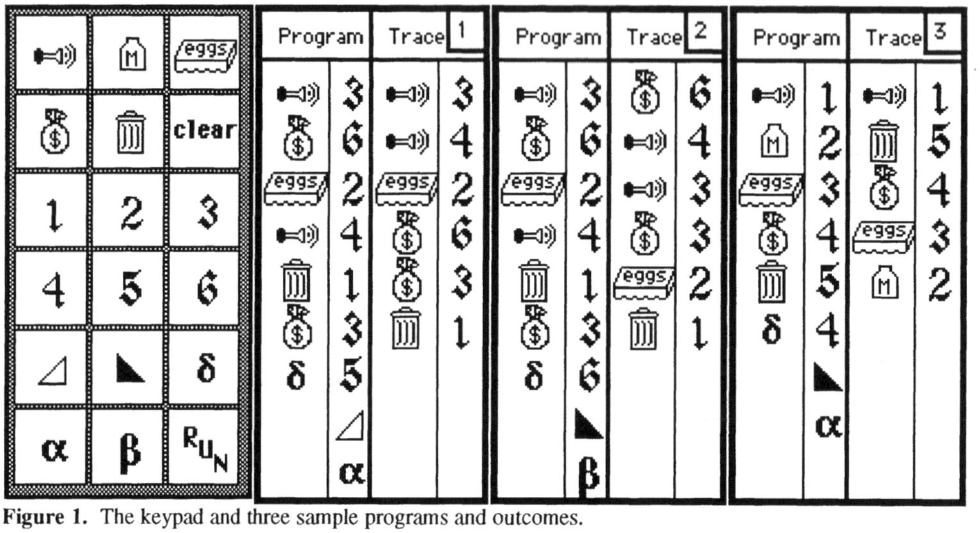

Figure 1. The keypad and three sample programs and outcomes.

For the last N steps in the program, δ reorders the execution sequence of the program by...

	White triangle (increasing)	Black triangle (decreasing)
α (item)	...items in keypad order.	...items in reverse keypad order.
β (number)	...increasing house number order.	...decreasing house number order.

Table 1. The function of the arguments to the δ command.

Materials. The interface was run on a Macintosh IIcx. Subjects were given a pen and scratch paper to take any notes that they wished during the task. A small audio tape-recorder and lapel microphone were used to record the verbal protocols.

Procedure. Subjects worked on the problem in two separate sessions on consecutive days. The first day consisted of an introduction to the task and 30 minutes of problem solving. The introduction to the task presented each aspect of the interface incrementally, and allowed the subjects to try a sample program before being introduced to the δ command. After being introduced to the δ command and its three arguments, subjects began experimentation with the explicit goal of discovering what the δ command and its three arguments did.

For the second session, subjects worked at the task until they had either solved it or had given up. If a subject falsely accepted an incorrect hypothesis, a counter-example was presented (see program 1 in figure 1). The same counter-example was used in all such cases. It was designed so that no subject would be likely to predict the actual outcome on the basis of an incorrect hypothesis.

Results

Overview. The 22 subjects produced 1103 total experiments over 33 hours. Analyses were conducted at three levels of detail: final solutions, computer protocols, and verbal protocols. The first two levels of analysis were carried out for all 22 subjects. Verbal protocols were carried out for only the first five subjects. These five subjects adequately represent the range of subjects: the second fastest solver, the second slowest solver, an average solver, a solver with counter-example, and a non-solver.

Subjects were grouped according to the success of their solutions. Three solution groups will be used for the analyses: subjects who solved without a counter-example (Solve); subjects who solved after receiving the counter-example (Challenge); and subjects who quit without solving (Quit). All subjects who received a counter-example eventually solved the task. The counter-example was usually given in the middle of the second day (mean program number 34.6, s=17.6). Table 2 presents the mean number of experiments and time on task for each group. The task was difficult but solvable: solution rates ranged from 30 to 179 minutes. Note that the Quit subjects did not simply give up prematurely: they ran slightly more experiments ($F(2,19)<1$) over a slightly longer period of time ($F(2,19)=2.6$, $p<.1$) than the other subjects (Quit vs. Solve Bonferonni/Dunn $t(19)=2.94$, $p<.05$).

Group	n	Experiments	Total time
Solve	11	49.5 (18.6)	78.6 (40.9)
Challenge	5	47.8 (17.2)	97.2 (34.2)
Quit	6	54.5 (17.6)	118.8 (17.8)
Total	22	50.4 (17.4)	93.8 (37.4)

Table 2. Mean number of experiments and time on task for each solution group (and standard deviation).

The MilkTruck domain was significantly more difficult for the subjects than the original BigTrak studies reported in Klahr & Dunbar (1988): the MilkTruck subjects ran 6 times as many programs over 2.5 times as much time.

Subjects varied a small number of experiment features. In the experiment space framework, an experiment can be viewed as a vector of features. One way to characterize a subject's search in the experiment space is to measure the number of features that change from one experiment to the next. In choosing how many features to vary, the subject must strike a balance between depth and breadth of search. If a subject varies few features, valid inferences can be made by comparing effects across programs. However, few feature changes between experiments may prevent the subject from reaching certain very informative experiments in finite time. On the other hand, if a subject varies many features, a broader range of experiments are tried at the cost of being able to make valid comparisons across experiments. These two main strategies have been described in concept attainment as "Conservative Focusing" and "Focus Gambling" (Bruner, Goodnow, & Austin, 1956).

For the following analyses four program features were used: the programmed route, and the three delta parameters. Since the subjects had access to more than just the last program, there are several ways of conducting the analyses, of which two have been chosen: comparing each program to the previous program, and comparing each program to the previous seven programs and choosing the lowest feature variation. Seven programs back represents the number of programs on the screen at all times. Comparison with the last program best represents the breadth of search. Comparison with the last seven programs is a measure of the validity of comparisons across experiments. To set a benchmark of comparison, the scores on these two measures by a completely random search in the experiment space were computed using a Monte Carlo simulation[2]. The base rates for the random model were 2.49 features varied for comparing to the last program, and 1.45 features varied for comparing to the last seven programs. Subjects varied a mean of 1.77 features with respect to the last program, and 1.33 features with respect to the last seven programs. Both of these means were significantly lower than the random model base rates ($t(21)=-15.8$, $p<.0001$, and $t(21)=-3.52$, $p<.002$ respectively), suggesting that subjects preferred to do depth-first search. However, both measures were significantly greater than 1 ($t(21)=17.08$, $p<.0001$, and $t(21)=10.01$, $p<.0001$ respectively), indicating that subjects often chose programs which prevented them from being able to make valid comparisons with other programs.

Successful subjects varied features most effectively. To investigate the changes in the number of features varied across time, the two measures of feature changes were taken for each quartile of the total number of programs each subject wrote. An ANOVA was done on the subject means for each quartile for each solution group. Since the analyses of the two measures produced all the same effects, only the analyses of the measure of feature changes with respect to the last program are reported. Overall, the differences between the solution groups were not significant ($F(2,19)<1$). The quartile differences were significant ($F(3,57)=5.99$, $p<.003$). Post-hoc tests revealed that only the decrease from the first to second quartiles was significant (Scheffé $F(1,57)= 11.54$, $p<.001$). The interaction was marginally significant ($F(6,57)=1.86$, $p<.10$), with the Solvers tending to vary the same number of features across the quartiles, and the Challenge and Quit subjects showing a decrease in the number of features varied (See figure 2).

Since these data are correlational, the causal link between feature changes and solution group is ambiguous. However, one plausible account for the differences is that the early changing of many features in the Challenge and Quit subjects was one source of their problems. i.e., changing too many features at once produced ambiguous effects. For the differences in late features changes, the use of fewer feature changes in the Quit subjects was probably a symptom of their difficulties. i.e., they might be stuck investigating one particular situation. One would expect the Challenge subjects not to have this problem, since they all solved the task, and hence should look like the Solve subjects at the end of the task. Using the mean of the last five programs, the Challenge and Quit groups did differ significantly ($t(9)=2.17$, $p<.06$), with means of 1.88 and 1.43 feature changes respectively.

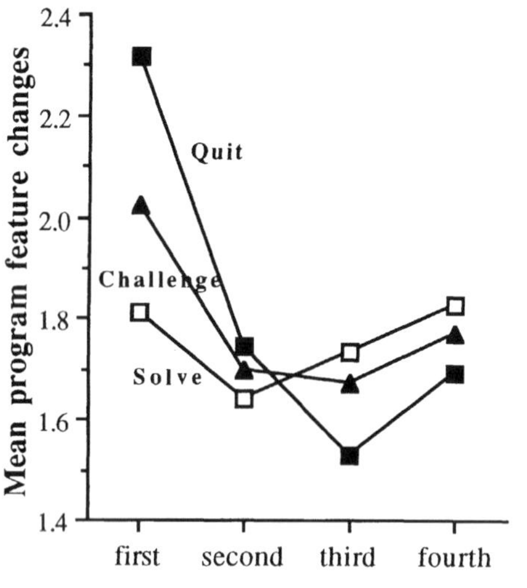

Figure 2. Mean number of program features varied with respect to the last program for each solution group for each quartile of total experiments.

The analyses of the changes in features emphasized the changes in δ parameters at the cost of treating all changes in the base program the same. However, subjects did some systematic variation of the basic program as well, as is shown in the following analysis.

Subjects systematically varied program length. Analyses of the program length revealed that most of the subjects used a strategy of gradually increasing the program length. When the length of each program was regressed against program number, 21 of the 22 subjects showed a positive slope, with a mean r of .56 ($\sigma=.22$). To further investigate the nature of this strategy, each subject's set of programs was divided into four equal parts, and an ANOVA of the means for each quartile for each solution group were calculated. The main effect for Solution group was not significant ($F(2,19)<1$). There were strong effects of quartile ($F(3,57)=9.4$, $p<.0001$). The quartiles increased in program length from the first to the fourth quartile, with significant increases between the first and second quartiles (Scheffé $F(3,57)=10.89$, $p<.05$), and the third and four quartiles (Scheffé $F(3,57)=7.43$, $p<.05$).

[2]A simulation of 100 random subjects was run having 50 programs each of mean length 5. For each program, a length was chosen randomly. For each step in the program, a delivery item and delivery location was chosen randomly. Similarly the delta parameters were chosen randomly. The random choices were all based upon a uniform distribution.

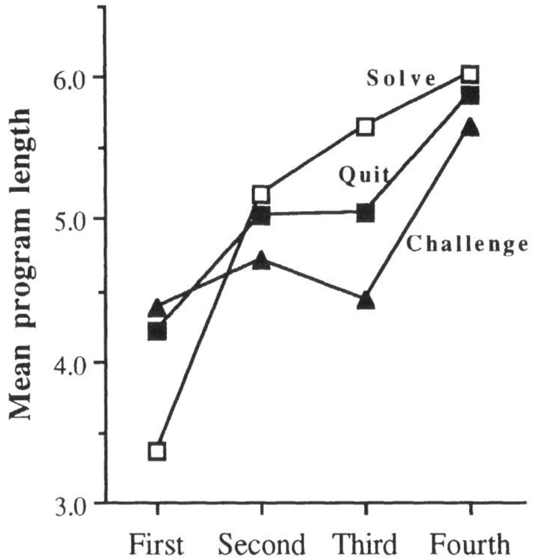

Figure 3. Mean program length for each quartile of the task for each solution group.

The interaction with Solution group was nonsignificant ($F(6,57)=1.41$, $p<.23$). As can be seen in figure 3, it is not simply the case that Solve subject ran longer programs which contained more information—although the differences were not significant, Solve actually started with slightly shorter programs, and were more consistent in their gradual increase of program length.

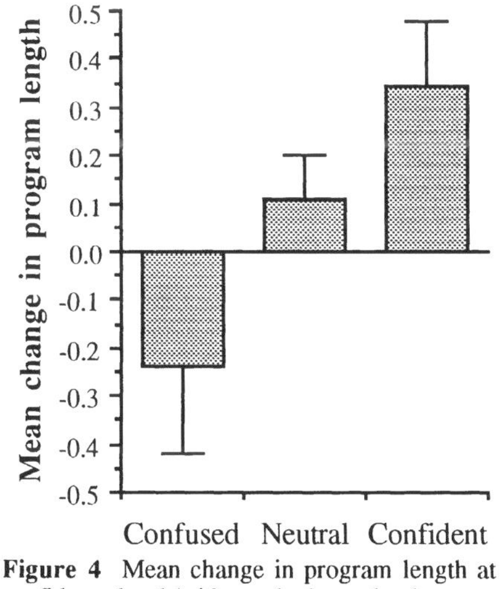

Figure 4 Mean change in program length at each confidence level (with standard error bars).

It is possible that program length could also be related to confidence levels. Since program length increases during the task, and presumably subjects are becoming more confident later in the task, one would expect that subjects were trying longer, more complex programs when they were confident. To test for such a relationship, the post program outcome statements from the full verbal protocols were coded for confidence levels on a 3 point scale: confused, neutral, and confident. The rater was blind to the actual program being run. An ANOVA was done on the change in program length from one program to the next for each confidence level (See figure 4). The effect of confidence level was significant ($F(2,255)=4.05$, $p<.02$), with post-hoc tests revealing a significant difference between the Confused and Confident means ($F(1,255)= 8.06$, $p<.02$).

Subjects used an explicit search strategy. When confronted with great difficulty while investigating a particular situation, subjects explicitly switched to investigating a different situation with the goal of returning to the confusing situation later. Upon completing their investigations of the new situation, subjects returned directly to the old situation, rather than a third, new situation. The time spent away from the original problem proved useful as subjects successfully then solved the old situation as well. This heuristic of putting a problem on hold and returning to it as soon as a different problem is solved was labelled the Put Upon Stack Heuristic (PUSH).

Of the five protocol subjects, only the three Solve subjects displayed evidence of using PUSH. Each of these subjects made explicit comments about being confused by the particular situation that they were currently investigating, and wanting to return to that situation later. And this is, in fact, what each of them did.

PUSH can be useful in three different ways. First, by enabling the subject to work on a different problem, PUSH allows new ideas to become activated and the activation of old ideas to decay, thereby reducing set effects and affecting the hypothesis space search in the old situation.

Second, the investigation of a different problem can suggest new operators which may be applied to the old situation, thereby improving the experiment space search. One subject's protocol provided evidence for this use of PUSH. This subject indicated that he was going to use PUSH: "Yeah. I'm just baffled right now. Ok, let's take a different tack. See if we can figure something out from a different angle. Let's go at it, looking at these other, what this other triangle does." Then, for the first time, he ran two programs which directly contrasted α with β, i.e., varied only α/β between two programs. This new operator proved successful in producing useful information to the subject, and lead the subject to say: "Let's try the same combination with other triangle again. I have a feeling that this might be what I need to be doing all along. The white triangle.

We'll start with α." The subject then went on using this operator, among others, to successfully solve the task.

Third, in inducing a complex concept involving an interaction such as the one used in the current experiment, discoveries about one part of the concept facilitate discoveries about another part of the concept. Thus, as the predictive power of hypotheses improve, the easier it is to resolve the remaining ambiguities. One subject seemed to use PUSH in this way.

Discussion

This experiment has demonstrated three strategies used by subjects in the induction of a complex concept. First, subjects tended to vary one or two features at a time between programs, with those varying the fewest features early more likely to solve. For strictly valid comparisons, scientific methodology states that only one feature may be varied at a time. If one were to take this strict view, then few subjects would be said to have used the strategy since the mean feature change for most subjects was well above one. However, there are two reasons why the number of feature changes in between program comparisons actually used by subjects may be lower than the simple numbers reported here: 1) the structure of the changes within a single program provides a large portion of the information; 2) subjects can design a sequence of programs which do not have to follow from the previous sequence.

Second, subjects start with short programs and gradually increased their length. Longer programs tend to be more informative, since they usually produce more step changes than short programs, and distinguish between more hypotheses. For example, a two step program has only two possible outcomes: the same order, or the reverse order. Thus, one would expect the solvers to have used longer programs. However, the use of long programs involves a high cognitive load in program design, interpretation, and memory. Thus, it is not necessarily better to always use longer programs. Indeed, the fastest solver used programs of a mean length of only 2.6 steps (1.9 standard deviations below the mean), indicating that long programs are not necessary for solution. Therefore, it is reasonable for subjects to start with short programs when they are relatively unfamiliar with the task, and use longer programs later in the task, when they need to test subtle differences between competing hypotheses. In fact, subjects did use shorter programs when they were less confident.

Third, subjects successfully made use of PUSH. Previous research has identified the Investigate Surprising Phenomena (ISP) strategy (Kulkarni & Simon, 1990). On the surface, the two strategies would seem to be incompatible: in the face of difficulty, one strategy (PUSH) advises breadth-first search, whereas the other strategy (ISP) advises depth-first search. However, the two apply to slightly different situations. When the surprising phenomenon has some unique, salient characteristics or features which can be tested in follow-up experiments, then ISP applies. On the other hand, when all the salient possible reasons for the surprising phenomena have been investigated, then PUSH applies. The strategy is to delay investigation until new information has been gathered. From the described mechanisms that may underlie the effectiveness of PUSH, it can be seen that PUSH may be related to incubation phenomena, especially those described in the history of science literature.

In sum, by extending the complexity of the microworld domain, several new heuristics used in scientific discovery have been revealed. Future studies need to address the applicability of these heuristics in other contexts, as well as assess the effects of the increased domain complexity on the use and effectiveness of previously described discovery heuristics.

References

Bruner, J. S., Goodnow, J. J. & Austin, G. A. (1956). *A study of thinking*. New York: Wiley.

Dunbar, K. (1989). Scientific reasoning strategies in a simulated Molecular genetics environment. In *the proceedings of the 11th annual meeting of the Cognitive Science society*, MI Ann Arbor, 426-433.

Dunbar, K. (1992). Concept Discovery in a scientific domain. Manuscript submitted for publication to *Cognitive Science*.

Klahr, D. & Dunbar, K. (1988). Dual space search during scientific reasoning. *Cognitive Science*, **12**, 1-48.

Klahr, D., Dunbar, K., & Fay, A. (1990). Designing Good Experiments to Test "Bad" Hypotheses. In J. Shrager & P. Langley (Eds.), *Computational Models of Discovery and Theory Formation*. Hillsdale, NJ: Lawrence Erlbaum Associates.

Kulkarni, D. & Simon, H.A. (1988). The process of Scientific Discovery: The strategy of Experimentation. *Cognitive Science*, 12, 139-176.

Simon, H. A., & Lea, G. (1974). Problem solving and rule induction: A unified view. In L. W. Gregg (Ed.), *Knowledge and cognition*. Hillsdale, NJ: Lawrence Erlbaum Associates.

Scientific Induction: Individual versus Group Processes and Multiple Hypotheses

Eric G. Freedman
Department of Humanities
Michigan Technological University
Houghton, MI 49931-1295
freedman@mtus5.bitnet

Abstract

It has been suggested that groups can evaluate multiple hypotheses better than individuals. The present study employed Wason's (1960) 2-4-6 task to examine the effects of multiple hypotheses in scientific induction. Subjects worked either individually or in four-member interacting groups. Subjects were also instructed to test either a single or a pair of hypotheses. The results indicate that groups perform significantly better than individuals. When testing multiple hypotheses, groups were more likely to determine the target hypothesis than individuals. Interacting groups generated more positive tests that received negative feedback and received more disconfirmation than individuals. When multiple hypotheses were tested, interacting groups used greater amounts of diagnostic tests than individuals. Interacting groups appear to search their experiment space and evaluate the evidence received better than individuals.

Introduction

In science as well as in everyday induction, people have been shown to rely on a positive-test strategy, that is, they tend to generate tests intended to confirm their hypotheses (see, Klayman & Ha, 1987, for review). Although a negative-test strategy (i.e., disconfirmatory) has been assumed to facilitate induction, subjects are often unable to benefit from this strategy (Freedman, 1991a; Gorman & Gorman, 1984; Tweney et al., 1980). Farris and Revlin (1989) have suggested that subjects may not benefit from a negative-test strategy as a result of an inability to consider alternate hypotheses.

In previous research, subjects have typically worked individually to test a single hypothesis (Gorman & Gorman, 1984; Gorman, Gorman, Latta, & Cunningham, 1984; Hacker, Freedman, Gorman, & Isaacson, 1990; Wason, 1960). Platt (1964) has claimed that scientific induction is facilitated when several hypotheses are tested simultaneously and particular hypotheses are eliminated through experimentation. In fact, Klayman and Ha (1987) have suggested that the evaluation of multiple hypotheses remains an important area for further research. Unfortunately, studies, which encouraged subjects to consider multiple hypotheses, have produced mixed results. Whereas Tweney et al. (1980, Experiment 2) found that encouraging subjects to use multiple hypotheses reduced performance, Klahr and Dunbar (1988, Experiment 2) and Klayman and Ha (1989) found that asking subjects to consider alternative hypotheses improved performance. Yet, McDonald (1990) found that multiple hypotheses were effective only when the target hypothesis was a subset of subjects' initial hypothesis. Freedman (1991b) found that multiple hypotheses improved performance only when used in conjunction with a negative-test strategy. Moreover, Freedman (1991a, 1991b) and Klahr, Dunbar, and Fay (1990) found that subjects employing multiple hypotheses generated significantly fewer experiments than subjects testing a single hypothesis. Thus, multiple hypotheses may reflect a more efficient strategy than single hypotheses.

Whereas individuals have difficulty testing more than one hypothesis at a time, Gorman (1986) has hypothesized that groups "can keep track of several hypotheses at once" (p. 93). Laughlin and Futoran (1985) found that groups performed better than individuals. Gorman et al. (1984, Experiment 1) found that interacting groups determined the target hypothesis as well as the best members of non-interacting groups. Laughlin and Futoran found that groups did not form better hypotheses than individuals; however, groups did evaluate hypotheses better than individuals. Nevertheless, Freedman (1991a) found that even though encouraging individual members of interacting groups to consider alternative hypotheses did not facilitate induction, successful groups who tested multiple hypotheses conducted significantly fewer experiments than successful groups who evaluated a single hypothesis at a time.

In short, whereas several studies seem to favor the testing of multiple hypotheses over single hypotheses, the overall results are inconclusive. It is therefore important to determine more precisely when multiple hypotheses facilitate scientific induction (Klayman & Ha, 1987). Multiple hypotheses are assumed to be effective when they permit a more extensive search of the hypothesis and experiment problem spaces (Klahr, Dunbar, & Fay, 1990). The present experiment used Wason's (1960) 2-4-6 task to investigate two dimensions simultaneously: single versus multiple hypotheses and four-member interacting groups versus individuals working alone. Multiple hypotheses have not been shown to facilitate induction, in part, because multiple-hypotheses studies have typically been run on individual subjects working alone. Therefore, this study examined whether groups can evaluate multiple hypotheses better than individuals. Another reason why multiple hypotheses may not routinely enhance induction is due to the fact that subjects have difficulty mentally representing more than one hypothesis at a time (Freedman, 1991a). In order to make alternate hypotheses more concrete, subjects were required to state a pair of hypotheses on each trial. Consistent with previous group problem-solving research, interacting groups are predicted to perform better than individuals. If interacting groups are better than individuals at evaluating alternative hypotheses, groups should be more likely to determine the target hypothesis when testing multiple hypotheses.

Method

Subjects. One hundred-twenty students enrolled in introductory psychology classes at Michigan Technological University participated in this study. All subjects received course credit for participation.

Procedure. Each group of three to five people was randomly assigned to either the individual-subjects condition or the four-member interacting group condition and to either the single-hypothesis or the multiple-hypotheses conditions. In the single-hypothesis (**SH**) condition, subjects proposed a single hypothesis and a number sequence to test it. In the multiple-hypothesis (**MH**) condition, subjects generated a pair of hypotheses and a number sequence to test them. They were read instructions very similar to those used in previous research (Gorman & Gorman, 1984). Subjects were told that the sequence, 2-4-6, is an instance of a target hypothesis and that they had to determine the target hypothesis by proposing other number sequences. The target hypothesis was—any three different numbers.

For each trial, subjects recorded their hypothesis or hypotheses, number sequence, whether their sequence conformed to their hypothesis and the experimenter's feedback. Subjects were given feedback regarding whether their number sequence was consistent with the target hypothesis but they were not told whether their hypotheses were correct. Prior to the main task, subjects were given a four-trial practice problem. Finally, subjects terminated the experiment when they believed they had determined the target hypothesis. However, a 25-minute time limit was imposed on the main task.

Results

A 2 X 2 (Number of Hypotheses X Number of Subjects) ANOVA was computed on each dependent variable. Differences between subjects who did and did not determine the target hypothesis were also analyzed.

Solutions. The proportion of subjects (based on 12 subjects/cell) who successfully discovered the target hypothesis is presented in Figure 1. Although neither the main effect of Number of Hypotheses or the two-way interaction reached significance, the main effect of the Number of Subjects indicated that interacting groups were more likely to determine the target hypothesis than individuals, $\underline{F}(1,44) = 19.57$, $\underline{p} < .0001$. When multiple hypotheses were evaluated, groups were more likely than individuals to discover the target hypothesis.

Experiments. Because subjects terminated the task when they believed they had discovered the target hypothesis, the total amount of number sequences proposed was measured. SH subjects conducted more experiments than MH subjects ($\underline{M} = 16.13$ vs. 10.49), $\underline{F}(1,44) = 15.37$, $\underline{p} < .0005$. Subjects testing a single hypothesis determined the target hypothesis as often as subjects testing multiple hypotheses because SH subjects gathered more information than MH subjects. In addition, for successful subjects (i.e. subjects who determined the target hypothesis), MH subjects announced the target hypothesis earlier than SH

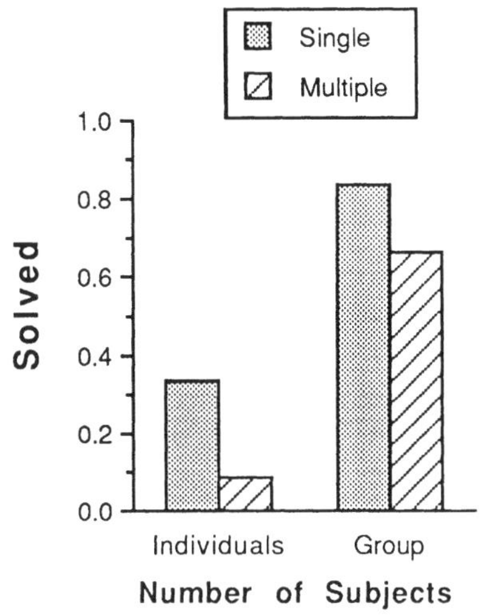

Figure 1
Probability of determining the target hypothesis

subjects ($\underline{M}$ = 9.56 vs. 12.57), $\underline{F}(1,19) = 4.70$, $\underline{p} < .05$. Thus, these results provide further support for the idea that a MH strategy appears to be more efficient than a SH strategy.

Hypotheses. Because subjects explicitly stated a hypothesis or a pair of hypotheses on each trial, the total number of different hypotheses proposed was measured. MH subjects generated significantly more unique hypotheses than SH subjects ($\underline{M}$ = 9.08 vs. 5.38), $\underline{F}(1,44) = 11.03$, $\underline{p} < .002$. This finding provides some preliminary support for the idea that testing multiple hypotheses led to an increased search of the hypothesis space. The average number of tests conducted for each hypothesis was also measured. SH subjects also conducted a greater number of tests per hypothesis than MH subjects ($\underline{M}$ = 3.93 vs. 1.20), $\underline{F}(1,44) = 23.34$, $\underline{p} < .0001$. Thus, subjects testing a single hypothesis maintain their hypotheses longer than those testing multiple hypotheses. It also suggests that subjects testing a single hypothesis rely more on a search of the experiment space than subjects testing multiple hypotheses.

Test Strategy. The number and percentage of sequences that conformed (i.e., positive tests) or did not conform (i.e., negative tests) to subjects' hypotheses was gathered. For the number and the percentage of positive and negative tests, none of the effects reached statistical significance. However, successful subjects proposed a significantly greater amount of negative tests than unsuccessful subjects ($\underline{M}$ = 8.96 vs. 6.28), $\underline{F}(1,44) = 8.90$, $\underline{p} < .005$. Apparently, successful subjects in the present study are able to benefit from the use of a negative-test strategy.

Experimenter's Feedback. Number sequences which were either consistent (i.e., positive feedback) or inconsistent (i.e., negative feedback) with the target hypothesis were recorded. SH subjects received greater amounts of positive feedback than MH subjects ($\underline{M}$ = 13.92 vs. 9.29), $\underline{F}(1,44) = 7.86$, $\underline{p} < .01$. SH subjects also received a greater percentage of positive feedback than MH subjects ($\underline{M}$ = .828 vs. .455), $\underline{F}(1,44) = 152.58$, $\underline{p} < .0001$. Individuals received a greater percentage of positive feedback than interacting groups ($\underline{M}$ = .695 vs. .587), $\underline{F}(1,44) = 12.80$, $\underline{p} < .001$. MH subjects also received significantly less negative feedback than SH subjects ($\underline{M}$ = 1.00 vs. 2.88), $\underline{F}(1,44) = 10.94$, $\underline{p} < .0005$. Interacting groups received more negative feedback than individuals ($\underline{M}$ = 3.00 vs. 0.875), $\underline{F}(1,44) = 14.05$, $\underline{p} < .002$. Furthermore, successful subjects received greater amounts of negative feedback than unsuccessful subjects ($\underline{M}$ = 3.91 vs. 0.12), $\underline{F}(1,44) = 42.17$, $\underline{p} < .0001$. This pattern of results suggests that subjects testing multiple hypotheses and interacting groups conduct a more extensive search of the experiment space.

Confirmation. The amount of confirmation was calculated by combining positive tests that received positive feedback and negative tests that received negative feedback. The amount of disconfirmation was calculated by combining positive tests that received negative feedback and negative tests that received positive feedback. Although no significant differences were observed in the amounts or percentages of confirmation received, a Success X Number of Subjects interaction indicated that successful interacting groups and unsuccessful individuals received a relatively greater percentage of confirmation than the other groups, $\underline{F}(1,44) = 5.59$, $\underline{p} < .05$. A Success X Number of Hypotheses interaction indicated that successful SH subjects and unsuccessful MH subjects received a relatively greater percentage of confirmation than the other groups, $\underline{F}(1,44) = 5.18$, $\underline{p} < .05$. Successful interacting groups may benefit from additional confirmation because of their superior hypothesis evaluation abilities.

A Number of Subjects X Number of Hypotheses interaction indicated that MH interacting groups received greater amounts of disconfirmation than the other conditions (see Figure 2), $F(1,44) = 4.25$, $p < .05$. Subjects in the MH condition received significantly more disconfirmation than subjects in the SH condition ($M = 8.25$ vs. 5.50), $F(1,44) = 11.58$, $p < .002$. A Number of Subjects X Number of Hypotheses interaction indicated that MH interacting groups received a greater percentage of disconfirmation than the other conditions, $F(1,44) = 4.17$, $p < .05$. Thus, one reason why interacting groups may be able to evaluate multiple hypotheses better than individuals is that interacting groups receive more disconfirmation than individuals. Disconfirmation may help interacting groups to eliminate incorrect alternate hypotheses.

Individuals conducted a significantly greater percentage of positive tests resulting in positive feedback than interacting groups ($M = .586$ vs. $.486$), $F(1,44) = 5.29$, $p < .05$. Successful subjects received a smaller percentage of positive feedback to positive tests compared with unsuccessful subjects ($M = .449$ vs. $.616$), $F(1,44) = 11.88$, $p < .002$. A Success X Number of Subjects interaction indicated that unsuccessful individuals received a relatively higher percentage of this type of confirmation than the other subjects, $F(1,44) = 4.28$, $p < .05$. According to Klayman and Ha (1987), this type of test allows subjects to determine the sufficiency of their hypotheses. Thus, individuals may be less successful than interacting groups because individuals focus on tests of the sufficiency rather than the necessity of their hypotheses.

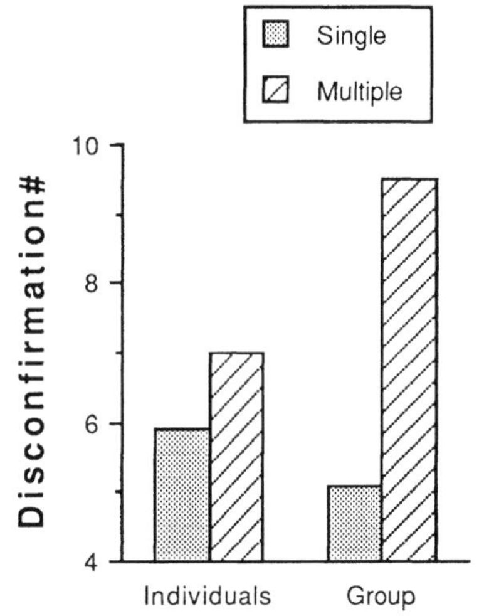

Figure 2
Mean amount of disconfirmation

Interacting groups received greater amounts of negative feedback to negative tests than individuals ($M = 2.50$ vs. 0.75), $F(1,44) = 10.35$, $p < .005$. Interacting groups also received a significantly greater percentage of negative feedback to negative tests than individuals ($M = .134$ vs. $.043$), $F(1,44) = 8.52$, $p < .01$. SH subjects received a greater percentage of negative feedback to negative tests than MH subjects ($M = .128$ vs. $.049$), $F(1,44) = 6.41$, $p < .02$. Successful subjects received greater amounts of negative feedback to negative tests than unsuccessful subjects ($M = 3.39$ vs. 0.00). Thus, not all types of confirmation are detrimental to scientific induction. This type of confirmation may allow subjects to determine the limits of the target hypothesis.

For positive tests that received negative feedback, interacting groups received significantly greater amounts of this type of information than individuals ($M = 1.38$ vs. 0.25), $F(1,44) = 8.92$, $p < .01$. Interacting groups also received a greater percentage of negative feedback to positive tests than individuals ($M = .064$ vs. $.018$), $F(1,44) = 6.88$, $p < .02$. Furthermore, successful subjects received greater amounts of negative feedback to positive tests than unsuccessful subjects ($M = 1.57$ vs. 0.12), $F(1,44) = 8.04$, $p < .01$. Successful subjects also received a significantly greater percentage of negative feedback to positive tests than unsuccessful subjects ($M = .075$ vs. $.010$), $F(1,44) = 7.33$, $p < .01$. As Hoenkamp (1989) has suggested, groups may be able to use positive tests more effectively than individuals because groups generate experiments that help them to decide between their hypotheses and the target hypothesis. For negative tests which received positive feedback, MH subjects generated more of this type of this type of test than SH subjects ($M = 7.25$ vs. 4.71), $F(1,44) = 13.20$, $p < .001$.

Diagnostic Tests. Because subjects can not know beforehand whether their hypotheses will be disconfirmed, the only sure way to disconfirm a hypothesis, when multiple hypotheses are tested, is to conduct a diagnostic test. In the MH condition, diagnostic tests were measured by counting each test that was an instance of one hypothesis and was not an instance of the other one. Clearly, subjects appreciated the importance of diagnostic tests. Diagnostic tests were employed on 62% of the trials. More diagnostic tests were generated by interacting groups than by individuals ($\underline{M}$ = 7.92 vs. 4.50), $\underline{F}(1,22) = 18.81$, $\underline{p} < .01$. Interacting groups also generated a significantly greater percentage of diagnostic tests than individuals ($\underline{M}$ = .754 vs. .481), $\underline{F}(1,22) = 10.47$, $\underline{p} < .005$. Successful subjects produced greater amounts of diagnostic tests than unsuccessful subjects ($\underline{M}$ = 8.33 vs. 4.93), $\underline{F}(1,22) = 5.79$, $\underline{p} < .05$. Thus, another reason why groups may evaluate multiple hypotheses better than individuals is because groups conduct more diagnostic tests.

Discussion

Interacting groups were able to determine the target hypothesis more often than individuals. When subjects employed multiple hypotheses, groups performed better than individuals. The results of the present study provide further support for the view that individuals may have difficulty forming a mental representation of alternate hypotheses and therefore they are not able to benefit from the presence of multiple hypotheses (Freedman, 1991a). Consistent with my previous research (Freedman, 1991a, 1991b), testing multiple hypotheses does not increase the overall likelihood of determining the target hypothesis, but MH subjects generate significantly fewer number sequences than subjects in the SH condition. Additionally, successful MH subjects announce the target hypothesis sooner than successful SH subjects. Thus, testing multiple hypotheses enables discovery of the target hypothesis more efficiently than testing a single hypothesis. As Hoenkamp (1989) has suggested, rather than emphasizing the use of confirmatory or disconfirmatory strategies, the optimal strategy may be one which minimizes the number of experiments conducted.

Evaluating multiple hypotheses may be an efficient strategy during scientific induction because it promotes the elimination of incorrect hypotheses. This conclusion was supported by the finding that MH subjects proposed significantly fewer tests per hypothesis compared with SH subjects. MH subjects also received more disconfirmation and generated more negative tests that received positive feedback than SH subjects. Thus, the presence of alternate hypotheses appears to make the possibility of disconfirmation more salient because it forces subjects to consider the necessity as well as the sufficiency of their hypotheses. When testing a single hypothesis, subjects tend to focus on the sufficiency of their hypotheses as reflected in a relatively greater reliance on positive tests that receive positive feedback. Higher levels of negative feedback in the presence of multiple hypotheses suggests that multiple hypotheses result in a more extensive search of the experiment space. This conclusion is based on the fact that it is harder to obtain negative feedback when a general target hypothesis is employed. Interacting groups may be able to evaluate multiple hypotheses better than individuals because groups receive a greater amount and a higher percentage of disconfirmation. This disconfirmation allows groups testing multiple hypotheses to eliminate incorrect hypotheses. Thus, interacting groups do benefit from receiving disconfirmation. Furthermore, a greater reliance on a diagnostic strategy may facilitate interacting groups' utilization of multiple hypotheses because this type of test provides the best strategy to eliminate incorrect hypotheses.

The results of the present study indicate that interacting groups do not generate greater amounts of number sequences, positive tests, and negative tests. Thus, groups do not receive more information than individuals. Nor, do groups differ in their *intended* strategies (i.e., positive- versus negative-test strategies). In other words, interacting groups do not seek more confirmation or disconfirmation than individuals. Rather, interacting groups are superior in evaluating the information they receive. Nevertheless, it may be the case that groups generate more informative tests (Hoenkamp, 1989) than individuals. Indeed, as Hoenkamp has suggested, a formal analysis of the informativeness of particular tests may yield further insights into the evaluation of multiple hypotheses by individuals and interacting groups. Still, the fact that interacting groups received more negative feedback, less positive feedback, more negative feedback to negative tests, more negative feedback to positive tests, and a smaller percentage of positive feedback to positive tests than individuals indicates that the information interacting groups receive differs from the information individuals receive. Once more, with a broad target hypothesis, negative feedback to various types of tests requires that groups conduct a more extensive search of the experiment space. Interacting groups may conduct a more extensive search of the hypothesis space than individuals because group members typically propose several number sequences from which the most informative sequence is chosen. Often, the most informative number sequence is the one that diverges from previous tests. Furthermore, a greater reliance on tests which lead to disconfirmation may help interacting groups to abandon incorrect alternate hypotheses.

The results of the present study suggest that the way in which subjects search through the experiment and hypothesis space is more important than whether subjects seek confirmation or disconfirmation of their hypotheses. When multiple hypotheses are evaluated, an extensive search of the hypothesis and experiment space may allow these subjects to discover the target hypothesis more expeditiously because they can determine the boundaries of the target hypotheses as well as eliminate incorrect alternative hypotheses. Still, the present study did not attempt to influence the types of hypotheses proposed or the experiments conducted. It is quite possible that encouraging subjects to propose maximally different hypotheses and to conduct diagnostic tests would facilitate scientific induction. While this study alone can not provide a comprehensive explanation of the conditions under which multiple hypotheses facilitate scientific induction, the present study does provide further evidence that when the cost of conducting experiments is high, use of multiple hypotheses may be preferable to the use of a single hypothesis.

References

Farris, H., and Revlin, R. 1989. Sensible reasoning in two tasks: Rule discovery and hypothesis evaluation. *Memory and Cognition* 17:221-232.

Freedman, E. G. 1991a. Role of multiple hypotheses during collective induction. Paper presented at the 61st Annual Meeting of the Eastern Psychological Association New York, NY.

Freedman, E. G. 1991b. Scientific induction: Multiple hypotheses and test strategy. Paper presented at the 32nd Annual Meeting of the Psychonomic Society, San Francisco, CA.

Gorman, M. E. 1986. How the possibility of error affects falsification on a task that models scientific problem solving. *British Journal of Psychology* 77:85-96.

Gorman, M. E., and Gorman, M. E. 1984. A comparison of disconfirmatory, confirmatory and a control strategy on Wason's 2, 4, 6 task. *Quarterly Journal of Experimental Psychology* 36:629-648.

Gorman, M. E., Gorman, M. E., Latta, M., and Cunningham, G. 1984. How disconfirmatory, confirmatory and combined strategies affect group problem solving. *British Journal of Psychology* 75:65-79.

Hacker, K. L., Freedman, E. G., Gorman, M. E., and Isaacson, R. 1990. The emergence of task representations in small-group simulations of scientific reasoning. *Journal of Social Behavior and Personality* 5:175-186.

Hoenkamp, E. 1989. 'Confirmation bias' in rule discovery and the principle of maximum entropy. In Proceedings of the Eleventh Annual Conference of the Cognitive Science Society, 551-558. Hillsdale, N.J.: Lawrence Erlbaum.

Klahr, D., and Dunbar, K. 1988. Dual space search during scientific reasoning. *Cognitive Science* 12:1-48.

Klahr, D., Dunbar, K., and Fay, A. L. 1990. Designing good experiments to test bad hypotheses. In J. Shrager and P. Langley Eds. *Computational models of scientific discovery and theory formation* (pp. 356-402). Palo Alto, CA: Morgan Kaufmann.

Klayman, J., and Ha, Y-W. 1987. Confirmation, disconfirmation, and information in hypothesis testing. *Psychological Review* 94:211-228.

Klayman, J., and Ha, Y-W. 1989. Hypothesis testing in rule discovery: Strategy, structure, and content. *Journal of Experimental Psychology: Learning, Memory, and Cognition* 15:317-330.

Laughlin, P. R., and Futoran, G. C. 1985. Collective induction: Social combination and sequential transition. *Journal of Personality and Social Psychology* 48:608-613.

McDonald, J. 1990. Some situational determinants of hypothesis-testing strategies. *Journal of Experimental Social Psychology* 26:255-274.

Platt, J. R. 1964. Strong inference. *Science* 146:347-353.

Tweney, R. D., Doherty, M. E., Worner, W. J., Pliske, D. B., Mynatt, C. R., Gross, K. A., and Arkkelin, D. L. 1980. Strategies of rule discovery in an inference task. *Quarterly Journal of Experimental Psychology* 32:109-123.

Wason, P. C. 1960. On the failure to eliminate hypotheses in a conceptual task. *Quarterly Journal of Experimental Psychology* 12:129-140.

Team Cognition in the Cockpit: Linguistic control of shared problem solving

Judith Orasanu
NASA-Ames Research Center
M/S 262-4
Moffett Field, CA 94035-1000
jorasanu@eos.arc.nasa.gov

Ute Fischer
NAS/NRC Research Associate
NASA-Ames Research Center
M/S 262-4
Moffett Field, CA 94035-1000
ute@eos.arc.nasa.gov

Abstract

Communication of professional air transport crews (2- and 3-member crews) in simulated inflight emergencies was analyzed in order to determine (1) whether certain communication features distinguish high-performing from low-performing crews, and (2) whether crew size affects communication used for problem solving. Analyses focused on metacognitively explicit talk; i.e., language used to build a shared understanding of the problem, goals, plans and solution strategies. Normalized frequencies of utterances were compared during normal (low workload) and abnormal (high workload) phases of flight. High-performing captains, regardless of crew size, were found to be more metacognitively explicit than low-performing captains, and effective captains in 3-member crews were found to be most explicit. First officers' talk complemented their captains' talk: First officers in low-performing crews tended to be more explicit than first officers in high-performing crews.

A string of recent disasters (e.g., the Vincennes incident in which an Iranian passenger jet was shot down, mistaken for a military plane; the Exxon Valdez oil spill in Alaska; and the Avianca jet crash in New York due to fuel exhaustion) has stimulated public concern for team problem solving and performance. They have also sensitized the research community to how little we know about team problem solving, decision making, and performance. A large literature exists on group problem solving (Steiner, 1972; McGrath, 1984), but much of this work was based on college students performing laboratory tasks in ad hoc groups (see Orasanu & Salas, in press, for a review; Ilgen et al., 1991). The relevance of much of this early group problem solving research to present applied problems has been called into question by recent findings on the role of expertise in <u>individual</u> problem solving performance (Chi, Glaser, & Farr, 1988). Yet we know little about the role of knowledge in expert <u>team</u> problem solving.

Recent efforts in the area of socially-shared cognition have focused on the role of shared knowledge in coordinating actions (see Resnick et al., 1991). For example, Hutchins and Klausen (1991) have analyzed 2 minutes of talk in a simulated air transport cockpit and have shown how shared knowledge gave meaning to ambiguous and cryptic gestures, glances, and utterances. Shared knowledge evidently provided the basis for mutual expectations and interpretations that allowed a 3-member crew to function as a single cognitive unit.

Teams of professionals in any area share considerable background knowledge--for their tasks, systems, artifacts, procedures, and the roles of various team members (Canon-Bowers, Salas, Converse 1990). Thus, when a problem arises, they should be well primed to cope with it. But the fact that tragic accidents occur means that something more is needed besides shared background knowledge. Orasanu (1990) suggested that a team must create a shared model for the current problem situation so that all members have the same understanding of what the problem is, what environmental cues mean, what solutions might be tried, and what is expected of various team members. The elements of the shared model are the same elements that research on metacognition has identified for effective individual problem solvers (e.g., Brown, Armbruster, & Baker, 1986; Flavell, 1981). That is, the shared situation model should include a clear goal, should be based on an accurate interpretation of the situation, should specify appropriate strategies to reach that goal, and should be updated by continuous progress monitoring. Knowledge and problem models held by individual crew members do not contribute to the common team effort unless they are shared. Team members need to communicate to each other how they understand the situation. Through communication they build a common model of the problem situation.

We propose that the degree to which a team establishes a shared mental model for a problem and the degree to which it is made explicit in communication will determine the team's effectiveness in coping with the problem. The fundamental issue addressed in this study was whether differences in how well crews coped with complex problems were associated with systematic differences in language use. Specifically, we sought to determine whether the language of more effective crews was more metacognitively explicit.

Explicitness is a slippery concept, because following Grice (1957), one does not want to say more than is useful. One can be explicit at the lexical level--that is, by using concrete referents rather than implicit ones or various forms of ellipsis. One can assure all participants share common terms of reference (c. f., Garrod & Anderson, 1987). One can also be explicit at a more abstract level, letting others know one's views of a problem and approach to its solution. The second aspect is addressed in this paper. Certain language functions were identified as relevant to metacognitive aspects of problem solving. They included goals, plans, strategies, predictions or warnings, and explanations. These were identified using a speech act theory approach (Austin, 1960; Searle, 1969). Utterances were coded in terms of the most prevalent function they performed within the context of preceding utterances and the given state of the task environment. For example, "Can you get me some ride reports," in the context of turbulence was taken as a comment on the deteriorating state of the weather, not simply as a question or even as a request. Obviously, utterances could be coded at many levels.

The present study examined the language used by teams of experts (cockpit crews) as they solved dynamic problems in a familiar work environment. Crew performance was observed in a high fidelity full-mission simulator. This environment is highly realistic and offers at the same time experimental control. Each crew can be presented with exactly the same scenario, pre-flight information, en route conditions (e.g., turbulence) and system malfunctions. This presents an opportunity to examine how each crew handles realistic emergencies in a safe environment, but one that is familiar to them and calls upon their domain-relevant knowledge.

In order to establish the generality of our findings, two different simulation studies were compared, one involving 2-person crews, the other 3-person crews. This between-study replication enabled us to address the following two hypotheses: (1) High-performing groups, regardless of crew size, show higher levels of explicit metacognitive talk than low-performing crews. (2) Three-member crews use more explicit metacognitive talk than 2-member crews. The second hypothesis is based on the fact that 2-member crews share a visual space and little ambiguity exists as to what is being addressed. The third crew member in a cockpit sits behind the other two and faces a different direction. This difference in visual field increases the need for explicitness.

Method

This study used videotapes from two simulator studies run at the NASA-Ames Man-Vehicle Systems Research Facility (Foushee et al., 1986; Chidester et al., 1990). These studies were selected because both involved the same scenario during one flight leg, with minor differences reflecting the aircraft systems (B-727 and B-737). The scenario involved a 1 to 1 1/2 hour flight characterized by a normal take-off and cruise to the destination. Weather deteriorated at the destination, which required crews to abort the landing. They had to decide whether to try again to land at the original destination, to go to their designated alternate, or to choose another alternate. During climb out, one of the hydraulic systems failed. This initiated a high workload period during which the decision had to be made about the alternate, now complicated by constraints imposed by the system failure. In addition, crews had to lower the gear and flaps manually, a time consuming and low frequency task.

There were 18 2-member crews (captain and first officer), who flew B-737s; 24 3-member crews flew B-727s, an older plane. The third crew member was the second officer, whose main duty is to monitor systems and attend to navigation. Communications of the second officers were not included in this study in order to maintain comparability with the 2-member crews. The captain designates himself or the first officer as pilot-flying, whose duty it is to fly the plane and watch out for traffic. The pilot-not-flying handles radio communication, checklists, and other procedures as needed.

As this was an exploratory study, the five highest performing and 5 lowest performing crews from each simulator study were selected for comparison. Performance was judged on the basis of operational errors, mainly dealing with adherence to standard procedures and aircraft handling (e.g., failures to obtain clearances, to complete checklists, or altitude deviations). Crews were <u>not</u> selected on the basis of their problem solving performance. Performance judgments were made by a check pilot (whose usual job is to evaluate pilots' cockpit performance). This was done on-line during the simulator run, and then two other observers made identical judgments from the videotapes. Videotapes were made of all crews in the simulator. Their talk was transcribed and served as the primary data for the present analyses.

Coding. A coding scheme that characterized the cognitive functions of cockpit discourse was developed (Orasanu, 1990). In brief, that scheme distinguished among three types of communication in the cockpit: Standard Operating Procedure talk (SOP), which is standard formulaic communication that is part of flying the plane; Housekeeping talk, which is sharing non-system information and allocation of tasks; and Metacognitive/Problem Solving talk, which is used to talk about problems. It is not formulaic or prescribed and is most variable across crews.

Because of our interest in the explicitness between 2- and 3-member crews, only certain functions were examined in this study. These include the following types of Metacognitive/ Problem Solving utterances: Goals, Plans/Strategies, Predictions, and Explanations. In addition, two other categories were included that are inherently explicit: Task Allocation and Commands. These served as anchors for the other measures.

Utterances were coded by assistants who were knowledgeable about aircraft systems and procedures, but who were blind to the conditions or performance levels of the crews. A reliability check indicated .86 agreement on scoring, which involved deciding on the unit of analysis, as well as the code.

Design. Separate analyses were run on data for each crew position (captains and first officers) using the same 2 (B) x 2 (B) x 2 (W) design. High and low-performing crews were compared within each crew size (2-member and 3-member). The within-crew variable was phase of flight: Normal vs. Abnormal. The abnormal phase began when the hydraulic system failed, causing the problem solving and coordination demands to escalate. Utterance frequencies for each crew member were normalized as rates per minute by dividing the observed frequencies by the amount of time the crew spent in the normal and abnormal phases, as appropriate. This normalization allowed comparisons across crew sizes and phases. All results reported here are significant at the $p<.05$ level or better, unless indicated otherwise.

Results

1. Do high and low-performing groups differ in the metacognitive explicitness of their talk? First, the total amount of task relevant talk was computed for each group. High- and low-performing captains did not differ in amount of total talk. Similarly, no significant effect of performance on first officers' task relevant talk was observed. This pattern forms the baseline against which to examine differences in types of talk.

Overall, captains of high-performing crews stated more plans and strategies than those in low-performing crews (see Figs. 1 and 2). In contrast, first officers in low-performing crews suggested more plans (M_{low} = .40) than those in high-performing crews (M_{high} = .31), seeming to compensate for the lack of planning by their captains. High-performing captains issued more commands than their low-performing counterparts, but only in 3-member crews. Commands may be considered the ultimate in explicitness. High- and low-performing captains in 2-member crews did not differ in commands. A marginally significant main effect suggests that high-performing captains also are more explicit in allocating tasks. The only other significant difference involving crew performance level was a 3-way interaction with crew size and phase of flight, which will be discussed in the next section.

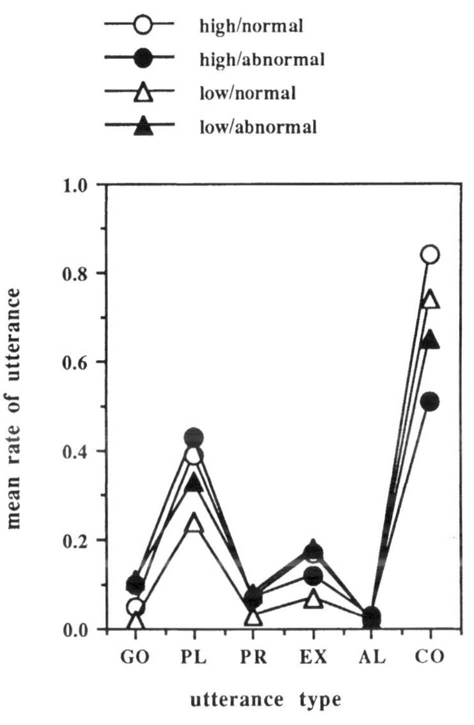

Figure 1: Mean rate of utterances by Captains of 2-member crews

Legend:	
GO	Goal utterances
PL	Planning utterances
PR	Predictions
EX	Explanations
AL	Task allocating utterances
CO	Commands

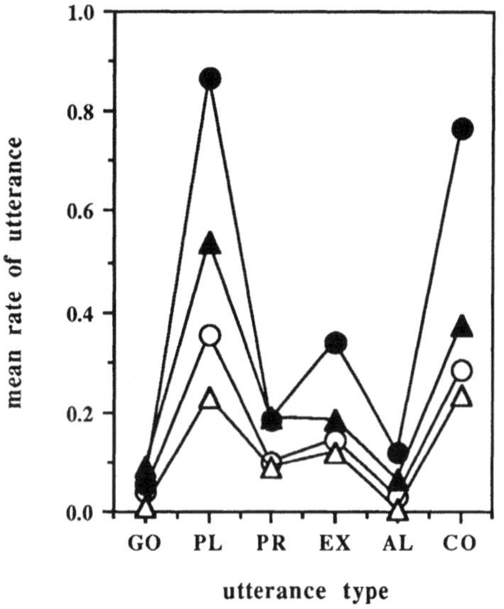

Figure 2: Mean rate of utterances by Captains in 3-member crews

2. Does crew size make a difference in the use of explicitness features in crew problem solving talk? First, it should be noted that the overall rate of talk in the abnormal phase was greater than in the normal phase for both 2- and 3-member captains. However, this effect was exclusively due to high-performing 3-member and low-performing 2-member captains. No significant phase effect on rate of talk was observed for high-performing 2-member and low-performing 3-member captains. In effective 3-member crews, captains' task-relevant talk rate almost doubled in the abnormal phase. For first officers, the reverse pattern was found: high-performing 3-member and low-performing 2-member first officers did not change their rate of talk significantly in the abnormal phase. High-performing 2-member and low-performing 3-member first officers, in contrast, talked more in the abnormal than in the normal phase.

Several main effects and interactions indicate that captains of 3-member crews were more explicit than captains of 2-member crews. In 3-member crews, captains more explicitly assigned tasks, particularly during the abnormal phase when workload was high. The significant phase effect for task assignment was due exclusively to the 3-member crews. They also were more explicit in alerting other crew members about things to watch out for or to expect. Essentially, they were the metacognitive crew monitors. The main effect of crew size was marginally significant for planning, with 3-member captains stating more plans and strategies, especially during the abnormal phase. Regardless of crew size, all captains stated more explicit goals during the abnormal phase than in the normal phase, as might be expected. The picture is somewhat complicated for commands, our bellwether of explicitness, reflecting different patterns associated with crew size. In 2-member crews, high-performing captains gave fewer commands during the abnormal phase than in the normal phase. The reverse was true for low-performing captains: they gave more commands in the abnormal phase. In 3-member crews, high-performing captains issued many more commands during the abnormal phase than during the normal phase. For low-performing crews there was no difference across phases.

First officers' utterances in many ways complemented those of their captains. Like the captains, they made more plans during the abnormal than the normal phase of flight, when coordination demands were high. Overall, first officers in 2-member crews made more plans than first officers in 3-member crews. A marginally significant interaction suggests they do this during the abnormal phase. No other effects were significant for first officers, although several were marginally significant and all of those were in the direction of supporting more explicit contributions by the first officers during the abnormal phase, specifically: stating goals, giving explanations, and allocating tasks.

Discussion

This pattern of findings paints a picture that shows some overriding commonalities in language use among high-performing crews. Two aspects of captains' talk cut across crew size: More effective captains explicitly state their plans and explicitly allocate tasks among crew members. By stating their plans, they let all crew members know what they want to accomplish. This allows other crew members to offer contributions and take actions that are consistent with the captain's intentions.

Other effective behavior patterns reflect the size of the crew. An effective captain communication strategy in a 2-member crew is different from what is effective in a 3-member crew. These differences appear to reflect possible task allocations, mainly with respect to who is flying the plane. In 2-member crews the captain has to make do with fewer cognitive resources. Effective captains exploit low-workload periods to do contingency planning and rehearsal for events that might occur (like the aborted landing). This enables the crew to be prepared when difficulties in fact strike; they are primed and are not just beginning to think about what they might do. Preplanning appears to save cognitive energy. The overall level of high-performing captains' talk drops in the abnormal phase. He is the one flying the plane under difficult circumstances. After he states his goals, plans, and priorities, he turns over the troubleshooting and checklist review to the first officer, allowing him to manage his own workload. He does not micromanage with frequent commands. In contrast, low-performing captains in 2-member crews do little preplanning and find themselves overwhelmed by work during the abnormal phase. They issue many commands and are reactive to problems, but do little overall planning.

In 3-member crews the patterns are quite different, because of the presence of the third person. Effective captains immediately shift flying responsibilities to the first officer, freeing themselves to manage the problems. This results in a high level of talk by effective captains, including many plans, commands, explanations, and explicit allocation of workload. In contrast, captains in low-performing crews fly the plane themselves and assign the first officer to work on trouble shooting and other manual tasks. However, the captain is still in charge of managing the problem and prioritizing tasks. From these captains we see few commands, little planning, and overall a lower level of talk.

First officers appear to be reactive to the style of their captains. If the captain has everything in hand, the first officer just does his job, which in the abnormal phase is to trouble shoot, monitor systems, and to assess progress. If the first officer is flying the plane, that's basically all he does, and talks very little. If the captain has not taken charge of the situation, the first officer is more likely to suggest plans and strategies, as shown by the higher level of plan utterances made by first officer in 2-member low-performing crews.

This pattern of findings suggests that differences in overall crew effectiveness, measured by commission of errors, reflects two things: how well the captain has managed the crew's workload by distributing tasks among himself and his first and second officers, and how explicit he is in letting the crew know what he wants to do. Interestingly, first officers' and captains' levels of explicit metacognitive talk seem to work inversely. If the captain's is high, the first officer's is low and vice versa. It is not the case that some crews appear to use categorically more metacognitive talk. The captain seems to set the lead. The finding that first officers in low-performing crews do not succeed in their efforts to compensate for their captains' lack of metacognitive talk needs further investigation. We suspect that low-performing captains do not sufficiently acknowledge the validity of first officers' suggestions. This interpretation is in line with previous research by Torrance (1954), who found that suggestions and solutions offered by low-status crew members were frequently rejected even when correct. In contrast, high-status members' contributions were accepted even when wrong. Also, Goguen, Linde, & Murphy (1986) showed that first officers' speech is more mitigated and perceived as less forceful than captains'.

Our analysis demonstrates that it is possible to identify classes of utterances that are related to overall quality of crew performance. Crews that use more explicit metacognitive talk when faced with in-flight emergencies perform more effectively than those whose talk is less explicit. Exactly how performance differences are brought about cannot be discerned from this study, which is a post hoc analysis. But the relations suggest directions for future efforts to tease out causal links between crew metacognition, communication and performance effectiveness.

References

Austin, J. L. 1960. *How to do things with words.* Oxford: Clarendon Press.

Brown, A. L., Armbruster, B. B., & Baker, L. 1986. The role of metacognition in reading and studying. In J. Orasanu ed., *Reading Comprehension: From research to practice.* Hillsdale, NJ: Lawrence Erlbaum Associates.

Cannon-Bowers, J. A., Salas, E., & Converse, S. 1990. Cognitive psychology and team training: Training shared mental models of complex systems. *Human Factors Society Bulletin* 33(12):1-4.

Chi, M. T. H., Glaser, R., & Farr, M. J. 1988. *The nature of expertise.* Hillsdale, NJ: Lawrence Erlbaum.

Chidester, T. R., Kanki, B. G., Foushee, H. C., Dickinson, C. L., & Bowles, S. V. 1990. Personality factors in flight operations: Volume I. Leadership characteristics and crew performance in a full-mission air transport simulation. Technical Memo, 259, NASA Ames Research Center.

Flavell, J. H. 1981. Cognitive monitoring. In W. P. Dickson ed., *Children's oral communication skills.* New York, NY: Academic Press.

Foushee, H.C., Lauber, J.K., Baetge, M.M., & Acomb, D.B. 1986. Crew factors in flight operations: III. The operational significance of exposure to short-haul air transport operations. Technical Memo, 88322, NASA-Ames Research Center.

Garrod, S., & Anderson A. 1987. Saying what you mean in dialogue: A study in conceptual and semantic co-ordination. *Cognition*(27):181-218.

Goguen, J., Linde, C., & Murphy, M. 1986. Crew communication as a factor in aviation accidents. Technical Memo, 88254, NASA Ames Research Center.

Grice, H. P. 1957. Meaning. *Philosophical Review*, 67.

Hutchins, E., & Klausen, T. 1991. Distributed cognition in an airline cockpit. Unpublished manuscript. University of California, San Diego, CA.

Ilgen, D. R., Major, D. A., Hollenbeck, J. R., & Sego, D. J. 1991. Decision making in teams: Raising an individual decision making model to the team level. Technical Report, 91-2. Industrial/Organizational Psychology and Organizational Behavior, Michigan State University, East Lansing.

McGrath, J. E. 1984. *Groups: Interaction and performance.* Englewood Cliffs, NJ: Prentice Hall.

Orasanu, J. (1990). Shared mental models and crew performance. Technical Report, 46. Cognitive Science Laboratory, Princeton University.

Orasanu, J., & Salas, E. (in press). Team decision making in complex environments. In G. Klein, J. Orasanu, R. Calderwood, & C. Zsambok eds., *Decision making in action: Models and methods.* Norwood, NJ: Ablex Publishers.

Resnick, J. Levine, & S. Behrend eds. 1991, *Shared Cognition: Thinking as social practice.* Washington, D.C.: American Psychological Association.

Searle, J. 1969 *Speech Acts.* London: Cambridge University Press.

Steiner, I. 1972. *Group process and productivity.* New York: Academic Press.

Torrance, E. P. 1954. Some consequences of power differences on decision making in permanent and temporary 3-man groups. *Research Studies, State College of Washington* 22:130-140.

A Unified Process Model of Syntactic and Semantic Error Recovery in Sentence Understanding

Jennifer K. Holbrook
Department of Psychology
Albion College
Albion, Michigan 49224
jholbrook@albion.bitnet

Kurt P. Eiselt
College of Computing
Georgia Institute of Technology
Atlanta, Georgia 30332-0280
eiselt@cc.gatech.edu

Kavi Mahesh
College of Computing
Georgia Institute of Technology
Atlanta, Georgia 30332-0280
mahesh@cc.gatech.edu

Abstract

The development of models of human sentence processing has traditionally followed one of two paths. Either the model posited a sequence of processing modules, each with its own task-specific knowledge (e.g., syntax and semantics), or it posited a single processor utilizing different types of knowledge inextricably integrated into a monolithic knowledge base. Our previous work in modeling the sentence processor resulted in a model in which different processing modules used separate knowledge sources but operated in parallel to arrive at the interpretation of a sentence. One highlight of this model is that it offered an explanation of how the sentence processor might recover from an error in choosing the meaning of an ambiguous word: the semantic processor briefly pursued the different interpretations associated with the different meanings of the word in question until additional text confirmed one of them, or until processing limitations were exceeded. Errors in syntactic ambiguity resolution were assumed to be handled in some other way by a separate syntactic module.

Recent experimental work by Laurie Stowe strongly suggests that the human sentence processor deals with syntactic error recovery using a mechanism very much like that proposed by our model of semantic error recovery. Another way to interpret Stowe's finding that two significantly different kinds of errors are handled in the same way is this: the human sentence processor consists of a single unified processing module utilizing multiple independent knowledge sources in parallel. A sentence processor built upon this architecture should at times exhibit behavior associated with modular approaches, and at other times act like an integrated system. In this paper we explore some of these ideas via a prototype computational model of sentence processing called COMPERE, and propose a set of psychological experiments for testing our theories.

Overview

Most models of human language processing enforce a separation of language levels either through an assumption of individual modules each devoted to a different level of language or, de facto, by focusing on only one aspect of language processing (e.g., lexical disambiguation, theta-role assignment, or syntactic structure building). In contrast, our ongoing research has focused on finding ways to integrate language processing using as few assumptions of separate processes as possible. However, we have always been cognizant of the fact that theories of modular processes have support in the literature, and we have found it convenient in our own work to focus on lexical and pragmatic disambiguation during sentence processing in a modular fashion.

Our current work represents a meeting of theoretical intent with computational instantiation. In this new model, a unified processor is able to generate multiple inferences and make decisions among these inferences at all levels of language processing. Currently, our model encompasses lexical, syntactic, semantic, and pragmatic processes. The model is also able to make the kinds of inferential errors that people do and to recover from them automatically, as people do. Finally, this model, although a single processor, unites two schools of thought regarding the modularity of language. Our model is able to exhibit seemingly modular processing behavior that matches the results of experiments showing different levels of language processing (e.g., Forster, 1979; Frazier, 1987) but is also able to display seemingly integrated processing behavior that matches the results of experiments showing semantic influences on syntactic structure assignment (e.g., Crain & Steedman, 1985; Tyler & Marslen-Wilson, 1977).

Background

ATLAST (Eiselt, 1989) was a model of unified lexical and pragmatic disambiguation and error recovery. The model included lexical and world knowledge; it also included some amount of syntactic knowledge. The syntactic information was processed separately, using an ATN parser. The model achieved disambiguation using multiple access of meanings for lexical items and pragmatic situations, choosing the meaning that matched previous context, and deactivating but retaining all other meanings. If later context proved the initial disambiguation decision incorrect, the retained meanings could be reactivated without reaccessing the lexicon or world knowledge. ATLAST proved to have great psychological validity for lexical and pragmatic processing—its use of multiple access was well grounded in psychological literature (e.g., Tanenhaus, Leiman, & Seidenberg, 1979), and, more importantly, it made psychological predictions about the retention of unselected meanings that were experimentally validated (Eiselt & Holbrook, 1991; Holbrook, 1989).

ATLAST was not intended to model syntactic disambiguation and error recovery, but we believed that the principles embodied in the model should extend to syntactic knowledge as well (Granger, Eiselt, & Holbrook, 1984): that syntactic disambiguation and error recovery would follow the same pattern of multiple access, selection based on previous context, deactivation and retention of unselected structures, and reactivation of unselected structures should an error be discovered. At last year's meeting of the Cognitive Science Society, Stowe presented the finding that syntactic information and semantic information interact as the knowledge structure is built. Stowe's work (1991; Holmes, Stowe & Cupples, 1989) has lent credence to the prediction that syntactic knowledge is processed just like other language knowledge sources. Particularly relevant to the work presented herein is Stowe's conclusion that in cases of syntactic ambiguity, the sentence processor accesses all possible syntactic structures simultaneously and, if the structure preferred for syntactic reasons conflicts with the structure favored by the current semantic bias, the competing structures are maintained and the decision is delayed. Furthermore, the work suggests an interaction of the various knowledge types, as in some cases semantic information influences structure assignment or triggers reactivation of unselected structures. The new psychological evidence inspired us to extend ATLAST to include syntactic knowledge as an integral part of a unified language processor.

The New Theoretical Model

We propose that the human sentence processor can best be described as a single unified language processor which operates on distinct knowledge sources. These knowledge sources correspond to what are typically labeled *syntax* and *semantics*. While these sources contain different types of knowledge, the same process is used to manipulate and integrate each type of information into a coherent and plausible interpretation. The single processor allows inferences about the interpretation to be generated uniformly, regardless of the type of inference that must be made. Thus, an ambiguous word, an ambiguous parse tree, an ambiguous thematic role assignment, and an ambiguous semantic representation are all disambiguated by the processor in the same way.

This model of sentence processing attempts to explain several different phenomena. For example, lexical/semantic disambiguation and error recovery are accounted for by the approach first postulated in the ATLAST sentence processing model. Since we are using a single processor for all processing in our new model, the approach used by ATLAST is now applied to syntactic disambiguation and error recovery. As a result, we have a plausible process account of Stowe's (1991) findings.

Additionally, because the knowledge sources are modular while the processing is unified, we predict that this new model will sometimes exhibit ambiguity resolution behavior like that expected from a strong modular, autonomous process approach to sentence processing (e.g., Forster, 1979), and at other times it will exhibit behavior more like that expected from a strong interactive approach (e.g., Tyler & Marslen-Wilson, 1977). These differences in behavior will depend on whether the information available from the different knowledge sources is sufficient to resolve the specific ambiguities at hand at any given time. In short, this model should account for the wide range of data accumulated by the opposing camps in this ongoing debate.

Implementation

To explore how well ATLAST's approach to lexical/semantic disambiguation and error recovery would actually work when applied to the resolution of syntactic ambiguity, we constructed a prototype computational model called COMPERE (Cognitive Model of Parsing and Error Recovery) to serve as a testbed. This computational model follows closely the spirit of the theoretical model described above, but diverges slightly in actual implementation. The divergence appears in the processor itself: the theoretical model has a single processor, while the prototype computational model has two nearly-identical processors—one for syntax and one for semantics—which share identical control structures but are duplicated for convenience because each processor must work with information encoded in slightly different formats. Because we intended only to explore syntactic ambiguity and error recovery with this initial computational model, the distinction between two identical processors and one unified processor is unimportant. As we expand the scope of our investigations, however, we will need to unify the two processing components completely.

Both types of knowledge are represented as networks of structures. Syntactic knowledge is represented as a network in which each node holds all the knowledge about a particular syntactic category necessary for parsing a sentence into its surface structure. A node in the network representing semantic knowledge stands for a concept. The structure of the node (i.e., its slots) represents the relationships between the node and other concepts. These relationships can include world knowledge in the form of selectional restrictions.

In addition to syntactic and semantic knowledge, COMPERE has a lexicon which provides

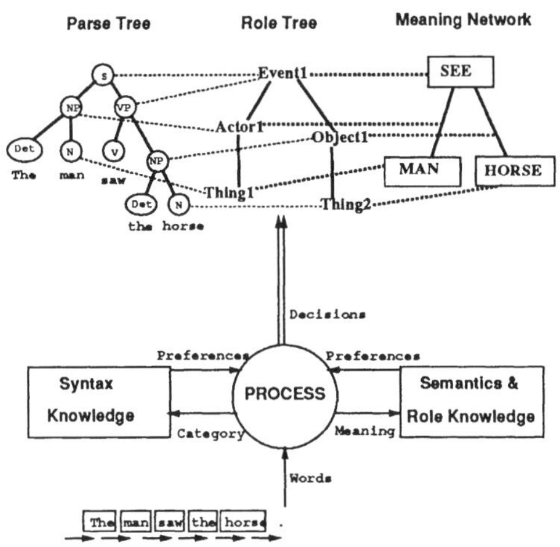

Figure 1: Architecture of COMPERE

the syntactic categories and subcategories of words as well as the meanings of words represented as pointers to nodes in the semantic network. The semantic component also has knowledge of thematic roles which helps bridge syntax and semantics. For instance, it knows that a noun phrase has a primitive role called **THING** which can evolve *in context* to an **ACTOR** or an **OBJECT** role. The representations of the different bodies of knowledge and the flow of information between them is shown in Figure 1.

The Process

Words are read from left to right, and their lexical entries are retrieved. The syntactic categories are passed to the syntax processor; at the same time, the pointers to corresponding meanings are passed to the semantic processor. The semantic processor builds a tree of thematic roles, as well as a network of instances of meaning structures.

As explained above, the control structures of the syntax and semantic processors are identical, though they process different kinds of knowledge. The processors interact many times in processing each word as they build the trees. The syntax processor first builds the basic node for the category of the word which will be a leaf of the parse tree. The semantic processor builds a node for the primitive role the word plays (if any) and also instantiates the meaning structure for the word. For instance, on reading the verb "saw," the syntax processor builds a verb node (V) to be added to the parse tree of the current sentence. The semantic processor builds nodes for an **EVENT** role and an instance of the **SEE** structure. These structures must be connected to other role and meaning structures already built for the sentence. The processors now try to connect the new nodes with the partial trees built earlier. When the syntactic structure of a sentence is successfully parsed, the meaning of the sentence is available as the meaning attached to the root node (S) of the parse tree.

Whenever the syntactic processor connects a node to its parent, it communicates with the semantic processor. The semantic processor tries to find corresponding relationships in the meanings associated with the two nodes by way of connecting their roles in the role tree. Thus the meanings associated with the nodes move up along the syntactic structure. When they meet at a common node, the semantic processor tries to bind them together through their roles. For example, consider the following sentence:

Text 1: The man saw the horse.

The structures that exist after reading "The man saw" are shown in Figure 2.

Figure 2: COMPERE's output for "The man saw."

Now, after reading "the horse," the system creates a noun phrase (NP) node to be connected to the above parse tree, a **THING** role to be connected to the above role tree, and a **HORSE1** structure to be connected to the meaning structures above. Syntactic processing could propose a connection from the new NP to the verb phrase (VP) in the tree, making "the horse" the syntactic object. The semantic processor finds corresponding links between the **HORSE1** node and the **SEE1** node through its **OBJECT** slot. This results from specializing the **THING** role of "the horse" to an **OBJECT** role which can now be connected to the **EVENT1** role. This process can be viewed as the meaning of "horse" propagating up the parse tree to meet the meaning of "see" at the VP node where the corresponding semantic connections are found. The structures built at the end of the sentence are shown in Figure 3.

Though the syntactic and semantic processors interact with each other, they are functionally independent; each can do its job should the other fail. If the syntactic processor fails to build a parse structure for a sentence, the semantic processor connects the primitive role for a word with the role tree (or a set of subtrees) built thus far. The processor can make decisions based on preferences coming only from one source of knowledge (such as syntax or semantics) if other sources fail to provide any preferences. Such a failure of the other sources could be either due to a lack of

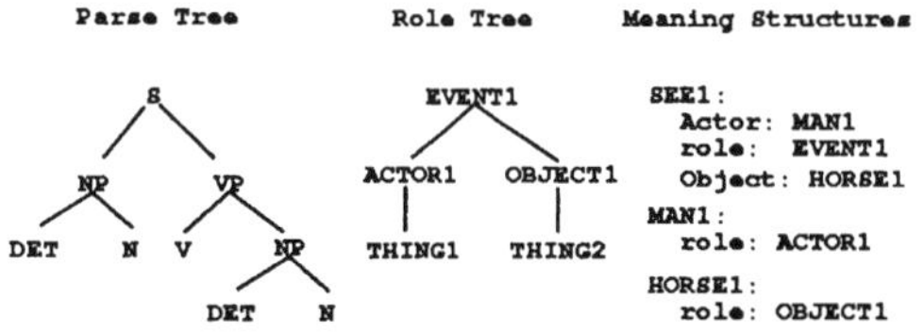

Figure 3: COMPERE's output for Text 1.

knowledge or due to a lesion in the communication pathways. Functionally independent behavior of this kind would not have been possible if the system had a single integrated source of knowledge together with a unified processor as seen in other models (e.g., Jurafsky, 1991; Lebowitz, 1983).

Ambiguity Resolution

Structural ambiguities in a sentence can be resolved through semantic or syntactic processing. For instance, if Text 1 were changed to:

> **Text 2:** The man saw the woman with the horse.

there would be at least two possible interpretations from a syntactic point of view—attaching the prepositional phrase (PP) to the VP or to the object NP—but only one of them is supported by semantics. The NP-attachment interpretation with its "woman together with the horse" meaning is acceptable whereas the VP-attachment interpretation with its "saw using the horse as an instrument" is not acceptable since it violates the constraint that the **INSTRUMENT** slot of the event **SEE** must be filled by an optical instrument.

On the other hand, consider the following sentence:

> **Text 3:** The officers taught at the military academy were very demanding.

The verb "taught" is interpreted as the main verb of the sentence since that would satisfy the expectation of a VP at that point in processing. In other words, we would rather use the verb to begin the VP that is required to complete the sentence structure, instead of treating it as the verb in a reduced relative clause which would have left the expectation of a VP unsatisfied. This behavior is the same as the one explained by the "first analysis" models of Frazier and colleagues (Frazier, 1987) using a minimal-attachment preference.

Error Recovery

When choices are made to resolve structural ambiguities, the alternatives that were not selected are retained for possible recovery from erroneous decisions. When it is not possible to attach a structure to the existing tree(s), the previously retained alternatives are examined to see if choosing another alternative at an earlier point provides a way to attach the current structure. If so, the tree is repaired accordingly to recover from the error. Since the subtree that was originally misplaced is merely attached at a different point, error recovery does not amount to reprocessing the structure of the phrase that corresponds to the subtree.

In Text 3, until seeing the word "were," the verb "taught" is treated as the main verb since it satisfies the expectation of a VP that is required to complete the sentence. However, at this point, the structure is incompatible with the remaining input. The processor now tries the other way of attaching the VP as a reduced relative clause so that there will still be a place for a main verb. In doing so, it did not have to process the PP that was part of the VP for the verb "taught."

In resolving the structural ambiguity in Text 3, semantic preferences did not play a significant role. In other situations, semantic preferences could influence the decisions that the processor makes in resolving syntactic ambiguities. Such behavior would be the same as the ones explained by models which argue for the early effects of semantic and contextual information in syntactic processing (e.g., Crain & Steedman, 1985; Tyler & Marslen-Wilson, 1977). COMPERE is intended to demonstrate that the range of behaviors that these models account for, and the behaviors that the "first analysis" models (e.g., Frazier, 1987) account for, can be explained by a unified model with a single processor operating on multiple independent sources of knowledge.

COMPERE has been implemented on a Symbolics workstation in the Common Lisp language with the Common Lisp Object System. It can process both the syntax and semantics of simple sentences (including all examples used in this paper) and uses semantic information in resolving structural ambiguity. Recovery from errors in resolving structural ambiguity has been implemented in the syntax processor alone; recovery from lexical/semantic errors has not yet been implemented in this model, but it will require very little effort to adapt the mechanism already used successfully by the ATLAST (Eiselt, 1989) system.

Proposed Psychological Studies

To test the validity of our psychological claims, we must answer the following questions: (1) How do we show that there is a single processing architecture which applies to multiple knowledge sources to make language decision, as opposed to multiple, non-identical processors? (2) How can we show error recovery occurring automatically and on-line for lexical, syntactic, semantic and other types of errors?

Answering Question 1

Recent experiments (e.g., Holmes, Stowe, & Cupples, 1989) have focussed on manipulating the information processed, but not the act of processing itself. By varying the type of task assigned to the subject, we can manipulate the processing style that is being executed. We have created materials that make processing more (or less) syntactic or semantic, by giving a task that biases the processor toward any given level. In one experiment, we are using two sets of materials, one semantically weighted and the other syntactically weighted. We have manipulated the level of processing by changing the task that subjects must perform. We are comparing the time it takes for subjects to make word-by-word completion decisions: either a decision on whether a sentence can still be completed grammatically, or whether a sentence can still be completed semantically. We are looking at the kinds of comprehension errors that are made for syntactically versus semantically weighted sentences, as well as at how the reaction time curve changes for the stimuli depending on the level of processing. Thus, in this experiment, we are able to assess the separate effects of the processor and the type of information processed on parsing decisions. Both processing models make empirical predictions. The single-processor model predicts uniform processing errors when we manipulate the processing environment but not the information processed. The multiple processor model predicts that processing errors will be different when we manipulate the processing environment.

A second point of comparison between single and multiple processor models is that the single processor model assumes interaction between lexical information and syntax and semantics, while the multiple processor model assumes that these would be separable. One point at which the information sources may interact is when lexical items are recognized. Some words are syntactically ambiguous, such that more than one part of speech (and probably meaning as well) must be called up.

Seidenberg, Tanenhaus, Leiman, and Bienkowski (1982) looked at ambiguous words that each had a meaning which lexically subcategorized as a noun and a meaning which lexically subcategorized as a verb. Their results showed that even when subcategorization information is available, it does not immediately restrict the processor from viewing all possible meanings of a word any more than other aspects of the word's meaning do. This is evidence that, at the place where meaning and structure are first constructed, the information is extracted in the same manner. We are conducting a similar experiment to that of Seidenberg et al., the main difference being that in our study, the ambiguous word is embedded within the sentence instead of at the end. This is because active suppression of alternate meanings is more likely to occur at the end of materials than within them (Holbrook, 1989). Seidenberg et al.'s results suggest support for the single processor model over the multiple processor model, but only at 0 msec. We are testing to see the time course of disambiguation due to subcategorization information, and the extent to which subcategorization information is relied upon exclusively for disambiguation. If the single processor model is correct, the subcategorization information should be useful but not always deterministic. A multiple processor model would predict that the subcategorization information will be an early and unassailable determiner of meaning choice.

Answering Question 2

Error recovery ought to act differently for a single processor system than for a multiple processor system. A unified process ought to make the task easy, and multiple processes ought to make it hard. The single processor model predicts that error recovery is uniform, no matter at what level of processing the error occurs. The same elements will be brought to bear to fix the error at the lexical, syntactic, and semantic levels. Our previous experiments (e.g., Eiselt & Holbrook, 1991; Holbrook, 1989) have validated the mechanism for lexical ambiguity, but have not validated it for other types of errors. Evidence from similar experiments by Holmes, Stowe and Cupples (1989) showed similar findings for syntactic subcategorization: as in our experiments, one interpretation was chosen and then discarded when later information negated this decision. To tie these two sets of experiments together, we are running the variations on the Holmes et al. experiments described above. To look at error recovery, we will look for priming effects for both meanings of the ambiguous word and for evidence of reinstantiation of a discarded structure.

Conclusion

A model that unifies separate processing mechanisms can only be considered successful if it is able to explain apparently different types of output, such as syntactic and semantic output. In this paper we have developed a model that is able to do so by uniformly processing different types of information. The advantages to this model are that processing errors are usually avoided; many of the processing errors that still occur can be corrected immediately and unconsciously, so that processing can remain automatic and unconscious. The emphasis on different information types allows our model to remain consistent with work that suggests modularity at various levels of processing; the modularity lies in the division of the information types. However, the single processor simplifies the task of building compatible syntactic and semantic structures and allows for their interaction as the meaning of the text is evolved from the separate types of information. Hence, we can explain apparently anomalous psychological findings (e.g, Frazier, 1987; Tyler & Marslen-Wilson, 1977) within a single perspective.

References

Crain, S., and Steedman, M. 1985. On not being led up the garden path: The use of context by the psychological syntax processor. In D.R. Dowty, L. Kartunnen, and A.M. Zwicky (Eds.), *Natural language parsing: Psychological, computational, and theoretical perspectives* (pp. 320-358). Cambridge, England: Cambridge University Press.

Eiselt, K.P. 1989. *Inference processing and error recovery in sentence understanding* (Technical Report 89-24). Doctoral dissertation. Irvine: University of California, Department of Information and Computer Science.

Eiselt, K.P., and Holbrook, J.K. 1991. Toward a unified theory of lexical error recovery. *Proceedings of the Thirteenth Annual Conference of the Cognitive Science Society*, 239-244. Hillsdale, NJ: Lawrence Erlbaum.

Forster, K.I. 1979. Levels of processing and the structure of the language processor. In W.E. Cooper and E.C.T. Walker (Eds.), *Sentence processing: Psycholinguistic studies presented to Merrill Garrett* (pp. 27-85). Hillsdale, NJ: Lawrence Erlbaum.

Frazier, L. 1987. Theories of sentence processing. In J.L. Garfield (Ed.), *Modularity in Knowledge Representation and Natural-Language Understanding* (pp. 291-307). Cambridge, MA: MIT Press.

Granger, R.H., Eiselt, K.P., and Holbrook, J.K. 1984. The parallel organization of lexical, syntactic, and pragmatic inference processes. *Proceedings of the First Annual Workshop on Theoretical Issues in Conceptual Information Processing*, 97-106.

Holbrook, J.K. 1989. *Studies of inference retention in lexical ambiguity resolution.* Unpublished doctoral dissertation. Irvine: University of California, School of Social Sciences.

Holmes, V.M., Stowe, L.A., and Cupples, L. 1989. Lexical expectations in parsing complement-verb sentences. *Journal of Memory and Language, 28*, 668-689.

Jurafsky, D. 1991. An on-line model of human sentence interpretation. *Proceedings of the Thirteenth Annual Conference of the Cognitive Science Society*, 449-454. Hillsdale, NJ: Lawrence Erlbaum.

Lebowitz, M. 1983. Memory-based parsing. *Artificial Intelligence, 21*, 363-404.

Seidenberg, M.S., Tanenhaus, M.K., Leiman, J.M., and Bienkowski, M. 1982. Automatic access of the meanings of ambiguous words in context: Some limitations of knowledge-based processing. *Cognitive Psychology, 14*, 489-537.

Stowe, L.A. 1991. Ambiguity resolution: Behavioral evidence for a delay. *Proceedings of the Thirteenth Annual Conference of the Cognitive Science Society*, 257-262. Hillsdale, NJ: Lawrence Erlbaum.

Tanenhaus, M., Leiman, J., and Seidenberg, M. 1979. Evidence for multiple stages in processing of ambiguous words in syntactic contexts. *Journal of Verbal Learning and Verbal Behavior, 18*, 427-440.

Tyler, L.K., and Marslen-Wilson, W.D. 1977. The on-line effects of semantic context on syntactic processing. *Journal of Verbal Learning and Verbal Behavior, 16*, 683-692.

What You Infer Might Hurt You – A Guiding Principle for a Discourse Planner

Ingrid Zukerman
Department of Computer Science
Monash University
Clayton, VICTORIA 3168, AUSTRALIA
ingrid@bruce.cs.monash.edu.au

Abstract

Most Natural Language Generation systems developed to date assume that a user will learn only what is explicitly stated in the discourse. This assumption leads to the generation of discourse that states explicitly all the information to be conveyed, and that does not address further inferences from the discourse. The content planning mechanism described in this paper addresses these problems by taking into consideration the inferences the user is likely to make from the presented information. These inferences are modeled by means of inference rules, which are applied in a prescriptive manner to generate discourse that conveys the intended information, and in a predictive mode to draw further conclusions from the presented information. In addition, our mechanism minimizes the generated discourse by presenting only information the user does not know or about which s/he has misconceptions. The domain of our implementation is the explanation of concepts in high school algebra.

Introduction

It has been widely accepted that much of what is intentionally conveyed during language use is not explicitly expressed [Grice 1978]. The recognition of this fact has significantly influenced research in Natural Language Understanding, e.g., [Norvig 1989], and in Plan Recognition, e.g., [Allen and Perrault 1980], where extensive inferences are drawn from a piece of discourse. A few Natural Language Generation (NLG) systems have addressed particular types of inferences which can be made from statements issued by a system [Joshi et al. 1984, Reiter 1990, Zukerman 1990, Cawsey 1991]. However, most NLG systems developed to date, e.g., [Appelt 1982, McKeown 1985, Paris 1988, Hovy 1988, Moore and Swartout 1989, Cawsey 1990], assume that a user[1] will learn only what is explicitly stated in the discourse.

[1] The terms hearer/user/addressee are used interchangeably in this paper.

This assumption may result in the generation of discourse that on one hand is *overly explicit*, i.e., includes information that could have been easily inferred by the hearer from the presented information, and on the other hand *fails to address possible indirect inferences drawn by the user*, i.e., is not complemented by information that contradicts possible erroneous inferences. For example, a possible erroneous indirect inference from the statement "wallabies are like kangaroos" is that wallabies are the same size as kangaroos. A system that is sensitive to a user's inferences should complement this statement with the disclaimer "but smaller."

In this paper, we present a content planning mechanism which addresses these problems. Our mechanism generates *Rhetorical Devices (RDs)*, such as Descriptions, Instantiations and Similes. The generation of RDs is performed by applying inference rules which relate RDs to beliefs. These rules are applied in two different ways during the discourse planning process: *forward reasoning* and *backward reasoning*.

Forward reasoning reasons from the RDs to their possible effects. This process accounts for the generation of the disclaimer for the above Simile between wallabies and kangaroos, by first applying a similarity-based inference rule which conjectures that the hearer will transfer what s/he knows about kangaroos to wallabies, and then blocking the transference of features which are not correct with respect to wallabies. This reasoning mechanism was used in [Zukerman 1990] for the generation of Contradictions and Revisions of a user's possible inferences.

Backward reasoning reasons from a communicative goal to the RDs that may be used to achieve it. For instance, the concept of a stack may be conveyed to a student by means of a Definition, an Analogy (say to a stack of plates in a cafeteria), an Example, or a combination of these RDs. This reasoning mechanism has been widely used in NLG systems, e.g., [Appelt 1982, Hovy 1988, Moore and Swartout 1989, Cawsey 1990]. However, these systems do not represent explicitly the inferential process that allows a hearer to deduce a belief from an RD. In our system, this process is modeled by the inference rules.

In the following section, we describe briefly the rules of inference used by our mechanism. In the remainder of this paper, we describe the tasks performed by the content planner.

The Rules of Inference

Our mechanism takes into consideration three types of inferences: (1) direct inferences, (2) indirect inferences, and (3) uniqueness implicatures.

Direct inferences reproduce directly the content of the discourse. The likelihood that a hearer will understand a statement by means of a direct inference is influenced by the complexity and abstractness of the information in the statement and by the addressee's ability to understand abstract explanations, such as stand-alone descriptions or definitions. The abstract-understand inference rule assesses this likelihood.

Indirect inferences produce inferences that are further removed from what was said. These inferences are not always sound. The indirect inference rules considered at present in our model are based on the ones described in [Zukerman 1990], namely: generalization, specialization, similarity and applicability. However, they draw inferences from RDs, rather than from already acquired beliefs. The first three rules reflect student behaviour observed in [Matz 1982]. The generalization rule was also postulated in [Van Lehn 1983, Sleeman 1984]. Applicability is a simple deductive reasoning rule. It states that if we can apply the first set of steps of a procedure to an object, then we can apply the entire procedure to this object. The likelihood of acquiring a message through indirect inferences from an RD depends on the hearer's confidence in the corresponding inference rules and on the strength of the beliefs which participate in the inference process. For instance, in the above wallabies-kangaroos example, the likelihood that the hearer will infer the desired features of wallabies from the Simile "wallabies are like kangaroos" depends on his/her knowledge about kangaroos and on his/her confidence in the similarity inference rule. Finally, in the current system, the application of indirect inference rules emulates a behaviour observed in [Sleeman 1984], whereby good students retain more correct conclusions than incorrect ones, while the opposite happens for mediocre students.

Given a proposition $P(O)$, **uniqueness implicatures** license the inference that P is true *only* with respect to O. For example, upon hearing the statement "Joe has one leg," most people will probably infer that Joe has one leg *only* [Hirschberg 1985][2]. Although the occurrence of uniqueness implicatures is mainly influenced by the wording of the discourse, their impact on a hearer's understanding of the discourse is affected by the hearer's beliefs and by the manner in which the hearer processes the incoming information. That is, people usually expect to add the incoming information to their knowledge pool [Zukerman 1991b]. However, if other information is already in place, a conflict ensues due to the uniqueness implicature resulting from the normal wording of the discourse. For instance, if a speaker says "Bracket Simplification applies to Like Algebraic Expressions," the uniqueness implicature will license the inference that Bracket Simplification applies *only* to Like Algebraic Expressions. Now, if the user believes that Bracket Simplification applies to Numbers, the uniqueness implicature will cause a conflict with this belief.

The domain of our inference rules is RDs, and their range is beliefs. That is, their format is: $Inference(RD) \xrightarrow{pr} Belief$, where pr is the probability that when applied to the RD in the antecedent, the rule will produce the belief in the consequent. For example, if the RD is an Instantiation and the rule is generalization, then pr is the probability that the hearer will generalize the intended belief from the Instantiation. Thus, our rules allow our system to conjecture the effect of an RD on a hearer's beliefs, and act accordingly, i.e., omit information that may be inferred from this RD, and add information that addresses possible incorrect inferences from the RD.

Operation of the Content Planner

Our content planner receives as input a **concept** to be conveyed to the hearer (e.g., Distributive Law), a list of **aspects** that must be conveyed about this concept (e.g., operation and domain), and a **communicative goal** which states the degree to which these aspects must be known (e.g., know well). The output of the content planner is a set of RDs, where each RD is composed of a rhetorical action, such as Assert and Instantiate, applied to a proposition.

In order to convey the intended aspects of a concept, our mechanism first determines the information to be presented, and then proposes RDs to convey this information. However, it is possible that the hearer does not understand the concepts mentioned in a particular RD well enough to understand this RD. Therefore, the generation process is repeated with new communicative goals and aspects with respect to the concepts mentioned in the proposed RDs, in order to add information about these concepts, if necessary. Other discourse planning tasks, such as organizing the generated messages, and selecting a set of RDs among a number of candidate sets of RDs which convey the intended information, are the subject of future research.

Throughout this section, we will use the following sample input to illustrate the operation of the content planner: (Bracket-Simplification, {*domain*,*operation*}, **KNOW**). In this input, the communicative goal is for the hearer to know the domain and operation of the Bracket Simplification procedure.

[2] Uniqueness implicatures differ from the implicatures discussed in [Reiter 1990], since they pertain to propositions, rather than to concepts.

Table 1: Propositions Relevant to the Sample Aspects	
Aspect	Domain Predicate
domain	[Bracket-Simplification apply-to Like-Algebraic-Expressions] [Bracket-Simplification apply-to Numbers]
operation	[Bracket-Simplification use-1 ±] [Bracket-Simplification use-2 ×]

Deciding which Information to Present

In this step, our system produces a list of propositions that must be conveyed in order to satisfy a given communicative goal with respect to specified aspects of a given concept. Our system caters for Grice's Maxim of Quantity [Grice 1975] in that the generated propositions contain only information that the user does not know or about which the user has misconceptions. This feature is particularly useful in situations such as the ones described in [Sleeman 1984], where the user knows most of the steps in a procedure, and needs to be instructed only with respect to a few of them.

Our system first retrieves from a knowledge base the propositions relevant to the given aspects. Next, based on consultation with a model of the hearer's beliefs [Zukerman 1992], the propositions already known by the user are filtered out, and propositions which address misconceptions held by the user are added. Propositions that are weakly believed by the hearer are presented, but they must be prefixed with a Meta Comment which credits the hearer with the belief in question [Zukerman 1991b], e.g., "*As you probably know*, Bracket Simplification applies to Numbers."

For instance, in order to satisfy the aspects in our sample input, the first step determines that the propositions in Table 1 must be known by the hearer[3]. Now, consider a situation where the hearer has the following beliefs with respect to the aspects in question:

[Bracket-Simplification apply-to Algebraic-Expressions]
[Bracket-Simplification apply-to Numbers]
[Bracket-Simplification use-2 ×]

In this case, the propositions [Bracket-Simplification use-2 ×] and [Bracket-Simplification apply-to Numbers] are omitted from the propositions to be conveyed, and the negation of the wrongly believed proposition [Bracket-Simplification apply-to Algebraic-Expressions] is added. This process results in the propositions in Table 2.

Proposing Rhetorical Devices

In this step, the content planner proposes RDs to convey the set of propositions produced in the previous step. To this effect, it takes into consideration inferences a hearer is likely to perform based on these RDs. Our procedure is based on the tenet that while processing a piece of discourse in an interactive setting, a hearer will draw immediate inferences from the discourse, but will perform further reaching inferences after the entire discourse has been processed. Thus, in order to address these immediate inferences, our algorithm draws one round of inferences from a proposed RD. Each inference rule that is applicable to this RD may be instantiated more than once during a round of inferences. This process is carried out by the procedure *Propose-RDs*.

Propose-RDs(propositions-to-be-conveyed)

1. Select a set of propositions that pertain to a particular aspect to be conveyed.
2. Apply inference rules in ***backward reasoning*** mode in order to propose a set of RDs which convey these propositions. Each RD in this set constitutes a different alternative for conveying the propositions in question.
3. For each alternative RD in the set of RDs, apply inference rules in ***forward reasoning*** mode in order to draw the inferences that can de made from this RD.
   (a) Update the list of propositions to be conveyed as follows:
      i. If an inference is correct and it corresponds to one of the propositions to be conveyed, then if the inference is strong enough, the proposition no longer has to be said, and it is deleted from the list of propositions to be conveyed. Otherwise, the inference has had some effect on the proposition to be conveyed, but this effect is not sufficient to determine that the proposition will be believed by the addressee. (A correct inference that does not correspond to one of the propositions to be conveyed has no effect on the discourse[4].)
      ii. If an inference is incorrect, then if it does not correspond to any of the propositions to be conveyed, its negation is added to the list of propositions to be conveyed. (If it corresponds to a

[3] The relationships use-1 and use-2 indicate the temporal ordering of a mathematical operation.

[4] [Zukerman 1990] describes a mechanism which produces discourse that addresses such inferences if they are weak.

Table 2: Propositions to be Conveyed	
Aspect	Domain Predicate
domain	[Bracket-Simplification apply-to Like-Algebraic-Expressions] ¬[Bracket-Simplification (always) apply-to Algebraic-Expressions]
operation	[Bracket-Simplification use-1 ±]

proposition to be conveyed, it is already being addressed.)

(b) Update the model of the hearer with the above inferences.

(c) If the updated list of propositions to be conveyed is not empty, then add the RDs produced by Propose-RDs(updated-propositions-to-be-conveyed) to the RD proposed in this alternative.

To illustrate the workings of this algorithm, let us return to our Bracket Simplification example. For our discussion, we assume that the addressee is able to understand abstract explanations, i.e., the *pr* of the rule *Abstract-understand(Assertion)* $\xrightarrow{pr}$ *Belief* is quite high. Now, the aspects to be conveyed with respect to Bracket Simplification are domain and operation. In the current implementation, we select operation first, since the inferences from the RDs generated to convey this aspect tend to affect other propositions to be conveyed. Next, we apply rules of inference in backward reasoning mode to generate RDs that can convey the proposition [Bracket-Simplification use-1 ±]. This step yields the RDs {Assertion} and {Assertion + Instantiation}, where both RDs have a sufficiently high probability of conveying the intended proposition. In both alternatives, the relationship use-1 in the Assertion is conveyed by a descriptor such as "before multiplying" which identifies the position of the ± operation in the Bracket Simplification procedure.

Let us first consider the alternative initiated by {Assertion}. In this case, the application of the inference rules in forward reasoning mode does not affect any of the other propositions to be conveyed. Hence, we update the model of the hearer to reflect the fact that s/he has been informed of the first step of Bracket Simplification, and re-activate our algorithm with respect to the propositions in the aspect domain.

During the backward reasoning step, our mechanism determines that the proposition [Bracket-Simplification apply-to Like-Algebraic-Expressions] may also be conveyed either by an Assertion or by an Assertion accompanied by an Instantiation. In both cases, during the forward reasoning stage, the following inferences may be drawn from the Assertion: (1) a similarity-based inference based on the hearer's belief that Numbers are similar to Like Algebraic Expressions; (2) a generalization based on the belief that Like Algebraic Expressions are a subset of Algebraic Expressions; and (3) a uniqueness implicature. The similarity-based inference corroborates the hearer's correct belief that Bracket Simplification applies to Numbers; the generalization corroborates his/her incorrect belief in the applicability of Bracket Simplification to Algebraic Expressions; and the uniqueness implicature concludes that Bracket Simplification applies *only* to Like Algebraic Expressions, and hence not to Numbers or to Algebraic Expressions.

The uniqueness implicature, which conflicts with the similarity-based inference and with the user's belief that Bracket Simplification applies to Numbers, may be prevented by prefixing the proposed Assertion with information that corroborates the user's belief, e.g., "*In addition to Numbers*, Bracket Simplification applies to Like Algebraic Expressions." At first glance, it appears that information that was omitted in the filtering process (see preceding section) is now being reinstated. However, the generation of this preamble links the new information to an existing belief held by the hearer, rather than informing the hearer that Bracket Simplification applies to Numbers.

The generalization, which conflicts with the uniqueness implicature and corroborates the hearer's erroneous belief that Bracket Simplification applies to Algebraic Expressions, is already being addressed by the second domain proposition in Table 2. Hence, nothing needs to be added to the list of propositions to be conveyed. However, the fact that the generalization can be inferred from the proposed Assertion supports the generation of an expectation violation Meta Comment [Zukerman 1991b], such as "but" or "however," which links this Assertion with the RD(s) that will be generated to convey the second domain proposition.

The generation of RDs for the second domain proposition in Table 2 is performed similarly. This yields the output in Table 3 for the alternative where an Assertion was generated for the first and third proposition in Table 2, and a Negation for the second proposition. Our current implementation produces the names of the RDs and the propositional representation. The English text has been added for illustrative purposes.

We conclude this discussion by considering briefly the alternative headed by {Assertion + Instantiation} of the proposition [Bracket-Simplification use-1 ±]. This alternative will result in a discourse which is markedly different from the one in Table 3, if the proposition is instantiated with respect to a Like Algebraic Expression, such as $2(2x + 3x)$, and thereafter, in the forward inference step, the generalization inference rule

Table 3: Sample RDs Produced during Content Planning	
Mention	[Bracket-Simplification apply-to Numbers] "In addition to Numbers,
Assert	[Bracket-Simplification apply-to Like-Algebraic-Expressions] Bracket Simplification applies to Like Algebraic Expressions,
Negate	[Bracket-Simplification (always)apply-to Algebraic-Expressions] but it does not always apply to Algebraic Expressions.
Assert	[Bracket-Simplification use-1 ±] Before multiplying, we add or subtract the terms inside the brackets."

produces the inference [Bracket-Simplification apply-to Like-Algebraic-Expressions] from this Instantiation. In this case, this proposition will be deleted from the list of propositions to be conveyed.

Conveying the Concepts in an RD

At this point in the content planning process, we have a number of candidate sets of RDs, where each set conveys the specified aspects of the intended concept. For each of these sets, we now have to ascertain that the hearer understands the concepts mentioned in its RDs well enough to understand these RDs. To this effect, for each of these concepts, the content planner performs the following actions: (1) it determines the aspects of the concept which are relevant to the understanding of the proposition which contains the concept, (2) it determines a communicative goal for these aspects, and (3) it regresses to generate RDs that accomplish this communicative goal with respect to the selected aspects of the concept. This process generalizes the mechanism described in [Zukerman 1991a].

The determination of the aspects the hearer must know about a concept in order to understand a proposition which contains this concept is based on the main predicate of the proposition and on the role of the concept with respect to this predicate. For example, in order to understand the Assertion [Bracket-Simplification apply-to Like-Algebraic-Expressions] proposed above, the hearer must know what Like-Algebraic-Expressions are and what they look like. Hence, the system returns the aspects *membership-class* and *structure*.

The determination of a communicative goal with respect to the selected aspects of a concept is based on the relevance of this concept to the original communicative goal. That is, the more relevant the concept is to this communicative goal, the better it should be known by the addressee, and vice versa. This consideration is implemented by lowering the expertise requirements with respect to a concept as the recursion becomes deeper. In this manner, we preclude the elaboration of concepts which are far removed from the main concept to be conveyed, while at the same time, ensuring a minimal level of competence with respect to these concepts.

Conclusion

The content planning mechanism presented in this paper generates RDs by taking into consideration the inferences a hearer is likely to draw from the presented information. To this effect, our mechanism applies inference rules both in backward and in forward reasoning mode. Although these inference rules are generally applicable, the conditions for the application of the direct and indirect inference rules and for the acquisition of the conclusions they draw vary for different types of users. Uniqueness implicatures, on the other hand, are influenced by expectations which are common to all users, in addition to the wording of the discourse.

Our mechanism minimizes the generated discourse by presenting only information that the user does not know or about which s/he has a misconception, and by omitting information which the hearer is likely to infer from the presented information. The inference mechanism that supports the latter capability also enables our mechanism to address possible incorrect inferences from the discourse. To perform these tasks, our mechanism requires a model of a user's beliefs and skills, and of his/her possible inferences. The former may be acquired with the help of a diagnostic system, such as the ones described in [Sleeman 1982, Burton 1982], and the latter is based on research by [Matz 1982, Van Lehn 1983, Sleeman 1984] about mathematical inferences commonly drawn by students.

A prototype of our content planning mechanism is in advanced stages of implementation. The implementation of the generation of Assertions, Negations and Instantiations in the framework of the algorithm Propose-RDs has been completed. The generation of Analogies and Similes is currently being implemented. Once the system is fully operational, it will be evaluated by presenting the texts generated for different types of students to the corresponding target audiences. The response of the students to these texts will be compared with their response to texts from algebra textbooks and texts produced by the traditional NLG approach.

Finally, an interesting line of investigation for further work consists of activating the system in a reflective mode after a session with the user has been completed. In this mode, the system would draw further reaching inferences from the generated discourse. Typically, these inferences would interact with each other, thereby requiring a processing mechanism that combines the inferences until the beliefs in the user model reach quiescence. The result of this process would then be the starting point of the next interaction with the user.

References

Allen, J.F. and Perrault, C.R. (1980), Analyzing Intention in Utterances. In *Artificial Intelligence* 15, pp. 143-178.

Appelt, D.E. (1982), Planning Natural Language Utterances to Satisfy Multiple Goals. Technical Note 259, SRI International, March 1982.

Burton, R.R. (1982), Diagnosing Bugs in a Simple Procedural Skill. In *Intelligent Tutoring Systems*, Sleeman, D. and Brown, J.S. (Eds.), London: Academic Press, pp. 157-183.

Cawsey, A. (1990), Generating Explanatory Discourse. In *Current Research in Natural Language Generation*, Dale, R., Mellish, C. and Zock, M. (Eds.), Academic Press, pp. 75-102.

Cawsey, A. (1991), Using Plausible Inference Rules in Description Planning. In *Proceedings of the Fifth Conference of the European Chapter of the ACL*.

Grice, H.P. (1975), Logic and Conversation. In *Syntax and Semantics, Volume 3: Speech Acts*, Cole, P.J. and Morgan, J.L. (Eds.), Academic Press, pp. 41-58.

Grice, H.P. (1978), Further Notes on Logic and Conversation. In *Syntax and Semantics, Volume 9: Pragmatics*, Cole, P.J. and Morgan, J.L. (Eds.), Academic Press.

Hirschberg, J.B. (1985), A Theory of Scalar Implicature. Doctoral Dissertation, The Moore School of Electrical Engineering, University of Pennsylvania, Philadelphia, Pennsylvania.

Hovy, E.H. (1988), Planning Coherent Multisentential Text. In *Proceedings of the Twenty-Sixth Annual Meeting of the Association for Computational Linguistics*, Buffalo, New York, pp. 163-169.

Joshi, A., Webber, B.L., and Weischedel, R.M. (1984), Living Up to Expectations: Computing Expert Responses. In *AAAI-84, Proceedings of the National Conference on Artificial Intelligence*, Austin, Texas, pp. 169-175.

Matz, M. (1982), Towards a Process Model for High School Algebra Errors. In *Intelligent Tutoring Systems*, Sleeman, D. and Brown, J.S. (Eds.), London: Academic Press, pp. 25-50.

McKeown K.R. (1985), Discourse Strategies for Generating Natural Language Text. In *Artificial Intelligence* 27, pp. 1-41.

Moore, J.D. and Swartout, W. (1989), A Reactive Approach to explanation. In *Proceedings of the Eleventh International Joint Conference on Artificial Intelligence*, Detroit, Michigan, pp. 1504-1510.

Norvig, P. (1989), Marker Passing as a Weak Method for Text Inferencing. In *Cognitive Science* 13, pp. 569-620.

Paris, C.L. (1988), Tailoring Object Descriptions to a User's Level of Expertise. In *Computational Linguistics* 14(3), pp. 64-78.

Reiter, E. (1990), Generating Descriptions that Exploit a User's Domain Knowledge. In *Current Research in Natural Language Generation*, Dale, R., Mellish, C. and Zock, M. (Eds.), Academic Press, pp. 257-285.

Sleeman, D. (1982), Inferring (mal) rules from pupil's protocols. In *ECAI-82, Proceedings of the European Artificial Intelligence Conference*, pp. 160-164.

Sleeman, D. (1984), Mis-Generalization: An Explanation of Observed Mal-rules. In *Proceedings of the Sixth Annual Conference of the Cognitive Science Society*, Boulder, Colorado, pp. 51-56.

Van Lehn, K. (1983), Human Procedural Skill Acquisition: Theory, Model and Psychological Validation. In *AAAI-83, Proceedings of the National Conference on Artificial Intelligence*, Washington, D.C., pp. 420-423.

Zukerman, I. (1990), A Predictive Approach for the Generation of Rhetorical Devices. In *Computational Intelligence* 6(1), pp. 25-40.

Zukerman, I. (1991a), Avoiding Mis-communication in Concept Explanations. In *Proceedings of the Thirteenth Annual Conference of the Cognitive Science Society*, Chicago, Illinois, pp. 406-411.

Zukerman, I. (1991b), Using Meta Comments to Generate Fluent Text in a Technical Domain. In *Computational Intelligence* 7(4), Special Issue on Natural Language Generation, pp. 276-295.

Zukerman, I. (1992), Content Planning based on a Model of a User's Beliefs and Inferences. In *Proceedings of the Third International Workshop on User Modeling*, Dagstuhl, Germany.

Theme Construction from Belief Conflict and Resolution

John F. Reeves
Artificial Intelligence Laboratory
3531 Boelter Hall, Computer Science Department
University of California, Los Angeles
Los Angeles, California 90024
reeves@cs.ucla.edu

Abstract

Story themes are generalized advice that a story contains, and theme recognition provides a way for a system to show that it has understood the story. THUNDER is a story understanding system that implements a model of theme construction from belief conflicts and resolutions. A belief conflict is conflicting evaluative beliefs regarding a story character's plan. When execution of the plan results in a realized success or failure for the character, a resolution to the conflict is recognized from the additional reasons that the realization provides for the evaluative beliefs in conflict. The theme of the story is generated by reasoning about how the resolution shows the beliefs in conflict to be correct or incorrect, and produces a statement of generalized advice about reasons for evaluation. Two types of advice are generated by THUNDER: (1) reason advice about the reasons for evaluation that the story shows to be correct, and (2) avoidance advice about how failures that occur as the result of erroneous evaluations could be avoided. The algorithms for constructing both type of advice and examples of THUNDER constructing themes are presented.

Introduction

Two general problems for story understanding systems are (1) knowing when a story has been understood, and (2) showing that the story has been understood. One solution is to recognize the 'theme' of the story; a story is understood when the theme is recognized, and the system shows its understanding by answering questions about what was learned from the story. The problems then become how story themes are defined, represented, constructed, and used. Additionally, a theory of theme representation and recognition identifies the components (both processes and knowledge structures) that are used in complex understanding. A computational implementation of the theory provides a independent and testable formulation of the theory and an experimental tool for creating and extending the theory.

THUNDER (THematic UNDerstanding From Ethical Reasoning) (Reeves, 1988, 1991) is a computer program that reads short narratives and answers ethical and thematic questions. For THUNDER, themes are the *generalized advice* that are constructed from conflicts and resolutions in the story. THUNDER recognizes conflicts in evaluative judgments between the reader and story characters. The resolution of a belief conflict is an event in the story that provides additional reasons for one of the beliefs in conflict, and 'shows' the 'correctness' of the belief.

To construct themes from conflicts and resolutions, THUNDER contrasts and generalizes the reasons that led to the conflict to the reasons provided by the resolution. THUNDER constructs two types of themes: (1) *reason advice* about why evaluative beliefs are correct or incorrect, and (2) *avoidance advice* about how planning failures can be avoided.

One of the stories that THUNDER reads is:

Hunting Trip

> Two men on a hunting trip captured a live rabbit. They decided to have some fun by tying a stick of dynamite to the rabbit. They lit the fuse and let it go. The rabbit ran for cover under their truck.

From *Hunting Trip*, THUNDER recognizes the following themes:

THE THEME IS THAT YOU SHOULD NOT PLAY

WITH DYNAMITE BECAUSE YOU WOULD NOT LIKE BAD THINGS TO HAPPEN TO YOU.

THE THEME IS THAT YOU SHOULD NOT EXECUTE PLANS THAT CAUSE BAD THINGS TO HAPPEN TO OTHERS FOR YOUR ENTERTAINMENT BECAUSE YOUR ENTERTAINMENT IS LESS IMPORTANT THAN BAD THINGS HAPPENING TO YOU.

THE THEME IS THAT YOU SHOULD NOT EXECUTE PLANS THAT CAUSE BAD THINGS TO HAPPEN TO OTHERS BECAUSE YOU WOULD NOT LIKE BAD THINGS TO HAPPEN TO YOU.[1]

All three of the themes that are recognized in *Hunting Trip* contain reason advice. The first theme is based on a pragmatic expectation and resolution. One of the reasons that THUNDER believed that the hunters were wrong to blow up the rabbit was because they could get hurt playing with dynamite. When the hunters' truck blows up because they were playing with dynamite, the advice in the theme is generalized from "hurting yourself" and "damaging your possessions" to "bad things happening." The second theme is based on THUNDER's reasoning that the hunters' plan was wrong because they believed that their entertainment was more important than the rabbit's life. The third theme is based on THUNDER's reasoning that the hunters' plan was wrong was because they were going to kill the rabbit. When the hunters suffer at the end of the story, they do not like it just as THUNDER did not like them blowing up the rabbit.

An example of avoidance advice is generated by THUNDER from the following *Twilight Zone* story(Day, 1985):

Four O'Clock

Political fanatic Oliver Crangle is convinced that people who do not agree with his political views are evil. He keeps detailed files on people, makes threatening phone calls, and sends letters discrediting his 'evil' political enemies. One day, he finds a book of black magic and casts a spell to shrink every evil person in the world to a height of two feet tall at exactly four o'clock. But when the time rolls around, it is *he* who becomes two feet tall!

[1] The themes are generated in English by THUNDER when they are constructed. All I/0 is verbatim from the program.

THE THEME IS THAT YOU SHOULD JUDGE YOURSELF BEFORE JUDGING OTHERS BECAUSE YOU WOULD NOT LIKE TO BE PUNISHED.

The theme is generated from the belief conflict over Oliver's plan to punish his political enemies, and the resolution of Oliver's becoming two feet tall. Avoidance advice is constructed by reasoning about how the planner could have avoided the failure. In *Four O'Clock*, Oliver was shrunk because he was guilty of the crime he was punishing others for, and the reason that he was guilty was because he was punishing others unjustly. If Oliver had evaluated his own plan, he would have avoided the failure.

THUNDER Overview

During story understanding, THUNDER's primary task is to create *evaluative beliefs* about story characters' plans. An evaluative belief is a belief that is evaluated in terms of "goodness", in contrast to a factual belief which is evaluated in terms of truth. Positive and negative evaluations of plans (called *obligation beliefs*) correspond to beliefs that the plan should or should not be used, respectively. There are two type of reasons for an evaluative belief: (1) *pragmatic* reasons, reasons about the consequences of the plan for the planner, and (2) *ethical* reasons, reasons about the consequences of the plan for others. These reasons can be broken down into two components: (1) a set of factual beliefs about the plan, and (2) a pragmatic or ethical *judgment warrant* that is used to derive an evaluative belief from factual beliefs. To generate appropriate factual beliefs about a character's plan, THUNDER has to reason about (1) plan availability, or what other options were available to the planner (2) the goal importance of successes and failures caused by the plan, both for the planner and others, and (3) the intention of the planner, and if the planner realizes that he is causing goal failures for himself or others.

THUNDER's evaluative knowledge is organized in an ideology. The representation for ideology has two components: (1) a value system, to represent what is believed to be 'good', and (2) planning strategies, to represent good ways for the goals to be achieved (Carbonell, 1980). The value system represents the goals that THUNDER believes that characters *should* try to achieve, and try not to violate. THUNDER infers the evaluative beliefs of story characters from their actions. For example, in *Hunting Trip* THUNDER infers that the hunters believe that their entertainment is more important than the life of the rabbit. THUNDER's inference rules for evaluative belief are given in (Reeves, 1989).

From THUNDER's inferences about the beliefs of characters, THUNDER recognizes belief conflict patterns (BCPs). There are three types of belief conflicts: (1) *plan execution BCPs*, where the evaluator makes an ethical judgment that a character's plan is wrong, (2) *evaluation BCPs*, where the evaluator makes an ethical judgment that a character's reward or punishment is undeserved, and (3) *expectation BCPs*, where a character violates the evaluator's evaluative expectations. The BCP that is recognized in *Hunting Trip* is the plan execution BCP BCP:Inhumane. BCP:Inhumane represents the situation where an actor's plan is judged to be ethically wrong and (1) the plan causes non-recoverable health goal failures for another, (2) the goal failure is an integral part of the actor's plan, and (3) the goal failure is more important than the actor's goal success. The evaluation BCP BCP:No-Crime is recognized in *Four O'Clock*. BCP:No-crime represents the conflict between THUNDER's and Oliver's evaluation of Oliver's plan: Oliver believes that his plan is ethically right because punishing all evil people will protect society, while THUNDER believes that the plan is ethically wrong because he is punishing people for something that they should not be punished for.

BCPs are recognized when the plan is evaluated. Resolutions to belief conflicts are recognized when the plan is executed, and unexpected consequences (such as the hunters' truck blowing up, or Oliver shrinking) occur. The resolution provides an additional reason for the plan's evaluation. There are two types of resolutions: (1) *positive resolutions*, which are goal failures that provide additional reasons that the evaluator's belief was correct, and (2) *negative resolutions*, which are goal success that show that the planners belief was correct. The hunters' truck blowing up in *Hunting Trip*, and Oliver's shrinkage in *Four O'Clock* are instances of positive resolutions. Stories where the hunters blew up the rabbit, had a good laugh and went home, or Oliver succeeded in shirking his political opponents would have negative resolutions.

For a detailed discussion of THUNDER's natural language parser and generator, knowledge representation, inference rules, judgment warrants, and the complete set of BCPs, see (Reeves, 1991).

Theme Construction

The theme of a story is the controlling idea, central insight, unifying generalization, and purpose of the story (Perrine, 1974). To recognize the theme of the story, the reader has to identify the advice that is contained in the story, or what the story is designed to teach. The advice in the story can be an insight about life, how the world works, how to get along with others, or the reasons for or against certain courses of action.

THUNDER represents themes as (1) an abstract plan failure situation, (2) the reason for the situation's evaluation, and (3) the mistake state that led to the failure. THUNDER's representation of themes is designed to capture the following characteristics: (1) *advice* for the reader about how to plan or reason about plans so that the reader's performance will be improved, (2) *generality* so that the advice can be applied to situations that have an abstract similarity to the situation in the story, and (3) *content* that specifies the situation where the advice is to be applied and the reasons for applying the advice in the situation.

Themes in THUNDER are classified by (1) the two types of advice that the theme provides: reason advice and avoidance advice, and (2) the two types of reasons that are used to construct the theme: pragmatic and ethical. *Pragmatic themes* are advice about how to avoid planning mistakes that result in failures for the planner, and *ethical themes* are advice about why plans should not be executed because of the consequences for others.

Reason Theme Construction

A reason theme is a lesson about why a plan should not be executed. Reason themes are constructed by matching and generalizing two reasons for plan evaluation: (1) the evaluator's (THUNDER's) reason from *before* the plan is executed, and (2) the actor's reason *after* the plan is executed. The BCP recognized in the story contains the evaluator's belief and reason, and the resolution contains the actor's belief and reason.

THUNDER's reason theme construction process can be illustrated by considering what happens when the hunters' truck blows up in *Hunting Trip*. In constructing a theme, THUNDER is trying to answer the questions: (1) what does the hunters' goal failure say about why it is wrong to blow up rabbits for entertainment? and (2) how is the truck blowing up a confirmation of THUNDER's belief that blowing up the rabbit was wrong?

When the truck blows up, the hunters evaluation of their plan changes from positive to negative because the (unexpected) loss of their truck is more important than their entertainment. The hunters' reason structure is similar to THUNDER's belief that the plan was wrong, as shown in figure 1.[2]

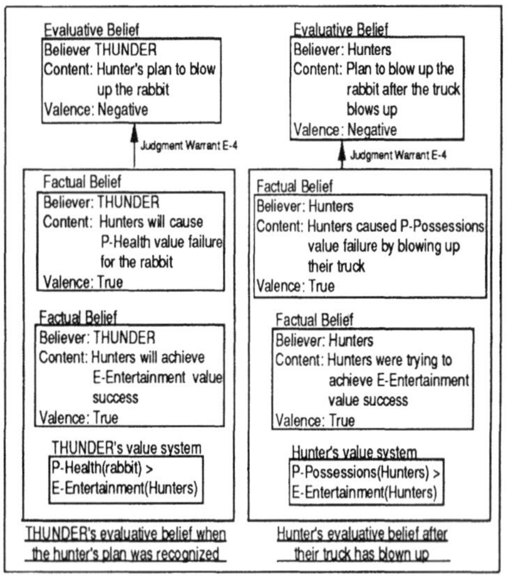

Figure 1: Beliefs used to construct the theme of *Hunting Trip*

The left side of the figure is THUNDER's belief that the hunters' are wrong to blow up the rabbit, supported by the instantiated schema for judgment warrant E-4 (If plan P1 achieves goal G while intentionally causing goal failure GF and the object of G is less important than the object of GF, then P1 is negatively evaluated). THUNDER's belief was generated when the plan was recognized after reading the second sentence of the story, and was used to recognize BCP:Inhumane. The right side of the figure shows the hunters' negative evaluation of their plan after the truck has blown up, which is also supported by judgment warrant E-4.

The matching and generalization process identifies differences in the variable bindings of the two beliefs, and generalizes the differences until the structures unify. Some variables have assigned generalizations (i.e. the believer is generalized to "you" and plan schemas are generalized to "plans"), while other types of generalizations are found by searching the item's is-a hierarchy for a common parent. If the instantiations are equal, then the instantiation is returned. For example, when the two beliefs in figure 1 are matched, the believer slot is generalized from "THUNDER" and "Hunters" to "you", and the value failure type in the top factual belief is generalized from "P-Health" and "P-Possessions" to "bad things."

THUNDER constructs the pragmatic theme from *Hunting Trip* in a similar manner. When the hunters' plan was evaluated, the TAU (Dyer, 1983) TAU:Dangerous-object was recognized that represents the potential planning failure associated with playing with dynamite ("If you play with fire, you're going to get burned"). When the reason that it is wrong to play with dynamite is matched against that hunters' pragmatic reason that the plan was wrong because their truck was destroyed, the following theme is generated:

> THE THEME IS THAT YOU SHOULD NOT PLAY WITH DYNAMITE BECAUSE YOU WOULD NOT LIKE BAD THINGS TO HAPPEN TO YOU.

The content of the belief ("playing with dynamite") is common to both THUNDER's and the hunters' beliefs. The reason in the theme ("bad things happening") is produced by generalizing from the THUNDER's expected health goal failure in TAU:Dangerous-object and the hunters' realized possessions goal failure when the truck was destroyed.

Avoidance Theme Construction

Avoidance themes are constructed by THUNDER from a *planning failure* on the part of the planner, and the planner's evaluative belief that led to the planning failure. A planning failure is a schema that is instantiated from realized goal failures, containing the action that caused the plan to fail, the action's intended effect, and the action's realized effect.

The algorithm for constructing avoidance themes is:

1. Given a BCP, resolution, and a planning failure, identify a *mistaken belief*. The mistaken belief is the belief that was used in their orig-

[2]The notation for goals is based on Schank and Abelson's goal primitives [1977]. In the notation, the goal type is signified by the letter preceding the goal name. Preservation goals (P) are motivations to keep valued acquisitions or social positions, and enjoyment (E) goals are motivation from the happiness that comes from a plan.

inal evaluation of the plan that is shown to be incorrect by the planning failure.

2. From the mistaken belief, find the part of the plan where the mistaken belief should have been recognized, and generate a new plan where the mistaken belief is checked for.

3. From the new plan, identify the failure that would have been avoided.

4. Generalize the new plan by matching the new plan to the plan executed, and generalize the reason for executing the new plan from the failure that occurred and the failure that would have been avoided.

5. Construct the avoidance theme from the generalized plan and the failure that would have been avoided.

The avoidance theme construction process is used in *Four O'Clock*. The input is (1) the BCP BCP:No-crime, (2) the resolution schema GF:Injury representing the health damage from being shrunk, and (3) Oliver's plan failure of casting the spell and shrinking himself. Since BCP:No-crime is an evaluation BCP, and THUNDER believed that Oliver's evaluation was in error, Oliver's belief that his political enemies are evil is identified as the mistaken belief. From Oliver's goal failure in the resolution, THUNDER knows that Oliver's planning error was that he failed to evaluate himself before executing the plan. By putting the step of evaluating the planner before evaluating others and generalizing the resulting structure, THUNDER generates the theme:

> THE THEME IS THAT YOU SHOULD JUDGE YOURSELF BEFORE JUDGING OTHERS BECAUSE YOU WOULD NOT LIKE TO BE PUNISHED.

The avoidance theme is recognized by backtracking to find planning errors based on evaluative beliefs, and specifying the judgments that need to be made to avoid the error.

Related Work

THUNDER's model of story understanding is based on the *explanation-based* story understanding systems BORIS (Dyer, 1983) and PAM (Wilensky, 1983a). BORIS and PAM both implemented thematic knowledge structures (TAUs in BORIS, Story Points (Wilensky, 1982; Wilensky, 1983b) in PAM) to explain the events in a story in terms of what the reader learns from the story. Story points also represented the distinction between problem and solution components in stories, which is generalized in THUNDER to conflict and resolution components. THUNDER extends BORIS and PAM by (1) constructing thematic structures instead of instantiating existing schema, and (2) using evaluative belief and belief relationships as the building blocks of theme, instead of relying exclusively on planning knowledge. The representation for belief and belief relationships for ethical knowledge in THUNDER is based on the types of belief used in the OpEd system (Alvarado, 1990) for beliefs about economic plans.

The most closely related current research to THUNDER is the AQUA program (Ram, 1989). AQUA is a story understanding system that models comprehension as a goal-directed task by generating and answering questions. AQUA uses Schank's (1986) anomaly → explanation-question → explanation-pattern model where problems in understanding motivate the application of general explanations. BCPs are similar to explanation questions in that they index planning advice and potential resolutions, and motivate and direct processing. BCPs are a special class of anomaly involving evaluation that motivates explanation. AQUA takes an open-ended view of story understanding, generating many explanation questions and explanation patterns and uses the explanation patterns to learn from the story. In contrast, THUNDER takes a stratified view of story understanding where ethical judgments are used to control the understanding process. Instead of generating many explanations, THUNDER models the recognition of specific types of story themes, and uses the themes to explain the purpose of story. THUNDER and AQUA model opposing ends of the story understanding spectrum: AQUA models the inquisitive reader that ponders a story for all of its nuances, while THUNDER models the reader that recognizes one type of 'point' of the story.

Conclusions

THUNDER is a story understanding system that constructs themes from stories, based on (1) the difference between ethical and pragmatic reasons for belief conflicts and how resolutions in the story show those reasons to be correct or incorrect, and (2) the different types of advice that can be constructed from the conflict and resolution. Ethical reasons for the belief conflict are used to generate ethical themes about how the resolution shows the plan to be right or wrong because of the consequences for others, while pragmatic reasons are used to construct pragmatic themes about the plan's consequences for the planner. THUNDER implements procedures to construct two types of advice: (1) reason advice about the reasons for evaluation that the story shows to be correct, and (2) avoidance advice about how failures that occur as the result of erroneous evaluations could be avoided.

Recognition of a belief conflict in a story is a judgment by THUNDER that something is wrong in the story. Finding a positive resolution to the belief conflict supports the evaluative belief that led to the original belief conflict. Because the story provides support for the reader's evaluation, the resolution to the belief conflict is thematic.

Recognition of a story theme provides a way for THUNDER (1) to show that it has understood the story, and (2) to know that the story has been completely processed. Identification of a theme is identification of the advice that the story was written to teach. *Thematic learning* is a two step process: (1) the theme is identified, and (2) the theme is incorporated into memory and used to improve future planning and reasoning. THUNDER accomplishes the first task, but not the second. The process of theme incorporation in THUNDER would involve adding stories and themes to episodic memory indexed by the BCP that was used to identify the theme. If the theme provides new advice, the theme can be associated with the BCP for use in future understanding. However, the theme construction algorithms provide the knowledge that will be learned.

References

Alvarado, S. J. 1990. *Understanding Editorial Text: A Computer Model of Argument Comprehension.* Kluwer Academic, Norwell, MA.

Carbonell, Jr., J. G. 1980. Towards a process model of human personality traits. *Artificial Intelligence*, 15:49–74.

Day, P. 1985. Four o'clock. In Greenberg, M. H., Matheson, R., and Waugh, C. G., editors, *The Twilight Zone: The Original Stories.* Avon Books, New York. Originally published in 1958.

Dyer, M. G. 1983. *In-Depth Understanding: A Computer Model of Integrated Processing for Narrative Comprehension.* MIT Press, Cambridge, MA.

Perrine, L. 1974. *Literature: Structure, Sound, and Sense.* Harcourt Brace Jovanovich, New York, second edition.

Ram, A. 1989. *Question-driven Understanding: An Integrated Theory of Story Understanding, Memory, and Learning.* PhD thesis, Department of Computer Science, Yale University, New Haven CT. Technical report YALEU/CSD/RR 710.

Reeves, J. F. 1988. Ethical understanding: Recognizing and using belief conflict in narrative understanding. In *Proceedings of the Sixth National Conference on Artificial Intelligence (AAAI-88)*, St Paul, MN.

Reeves, J. F. 1989. Computing value judgements during story understanding. In *Proceedings of the Eleventh Annual Conference of the Cognitive Science Society (CogSci-89)*, Ann Arbor, MI.

Reeves, J. F. 1991. *Computational Morality: A Process Model of Belief Conflict and Resolution for Story Understanding.* PhD thesis, UCLA Artificial Intelligence Laboratory, University of California, Los Angeles.

Schank, R. C. 1986. *Explanation Patterns.* Lawrence Erlbaum, Hillsdale, NJ.

Schank, R. C. and Abelson, R. P. 1977. *Scripts, Plans, Goals, and Understanding.* Lawrence Erlbaum, Hillsdale, NJ.

Wilensky, R. 1982. Points: A theory of the structure of stories in memory. In Lehnert, W. G. and Ringle, M. H., editors, *Strategies for Natural Language Processing*, pages 345–374. Lawrence Erlbaum, Hillsdale, NJ.

Wilensky, R. 1983a. *Planning and Understanding: A Computational Approach to Human Reasoning.* Addison-Wesley, Reading, MA.

Wilensky, R. 1983b. Story grammars versus story points. *Behavioral and Brain Sciences*, 6:579–623.

Communicating Abstract Advice: the Role of Stories*

Eric K. Jones
Victoria University of Wellington
P.O. Box 600, Wellington, New Zealand
eric.jones@comp.vuw.ac.nz

Abstract

People often give advice by telling stories. Stories both recommend a course of action and exemplify general conditions in which that recommendation is appropriate. A computational model of advice taking using stories must address two related problems: determining the story's recommendations and appropriateness conditions, and showing that these obtain in the new situation. In this paper, we present an efficient solution to the second problem based on caching the results of the first. Our proposal has been implemented in BRAINSTORMER, a planner that takes abstract advice.

Introduction

People often use stories to give advice. Advice in the form of a story frequently seems more convincing than an unadorned list of instructions [Schank, 1991]. Why should this be so? While there are probably a number of factors at work, it is important not to overlook the obvious: stories are often more convincing in large part because they are more informative. A story not only recommends a course of action, it also exemplifies conditions in which this recommendation is appropriate.

Unfortunately for an advice taker, a story's recommendations and appropriateness conditions may not be explicitly mentioned; even if they are, they may not always be identified as such. It follows that stories present a harder problem of inference than other kinds of advice such as lists of instructions, in which these features are explicit. In addition to the usual problems of operationalization [Mostow, 1983], a story-based advice taker must face the problem of working out which features of the story are relevant to the problem at hand.

To obtain a better understanding of this problem, we have studied a particularly simple class of naturally-occurring stories: familiar, advice-giving proverbs. People often give abstract advice using proverbs; many proverbs are worded in terms of a little story or vignette of a domain-specific situation in which the point of the proverb applies. Examples include *make hay while the sun shines*, *a bird in the hand is worth two in the bush*, *a stitch in time saves nine*, and *far from eye, far from heart*.

Proverbs are conventionally associated with an abstract point, which it is the proverb's function to communicate. The abstract point of *make hay while the sun shines*, for example, can be paraphrased as *take advantage of opportunities while they exist.* Knowing a proverb at a minimum requires grasping its abstract point. Proverbs' stories are easier to interpret as advice than than many other kinds of stories, because their abstract points have been computed in advance and stored in memory as part of the representation of the proverb.

Proverbs with stories, however, often contain more information than just their abstract point: their story may also provide a comprehensible exemplar of a specific situation in which the abstract point of the proverb applies. Most people who understand a proverb expressed in terms of a story also understand at least in part how the story illustrates the proverb's abstract point. For example, most people familiar with *make hay while the sun shines* seem to understand that hay making is some kind of goal-directed activity for which sunshine is a precondition, and that the opportunity the proverb refers to is exemplified by sunshine. This understanding is also cached as part of their representation of the proverb.

Even a partial understanding of a proverb's story is usually sufficient to suggest reasons why the abstract point of the proverb might plausibly apply in new situations. Consider, for example, the proverb *far from eye, far from heart*, whose abstract point is that long-distance relationships tend to weaken or dissolve. The proverb's metonymic reference to visual perception (*eye*) suggests the following causal chain in support of its abstract point: if you don't *see* someone regularly, then you aren't reminded of them, so your attachment to them weakens.

In summary, proverbs with stories come with their abstract points precomputed, together with some appropriateness conditions for their recommendations. It follows that the stories of proverbs are somewhat easier to make sense of than other kinds of stories: their interpretation is more tightly con-

*The research described here was conducted at the Institute for the Learning Sciences at Northwestern University, and was supported in part by the Air Force Office of Scientific Research. The Institute for the Learning Sciences was established in 1989 with the support of Andersen Consulting, part of The Arthur Andersen Worldwide Organization.

```
positive-evaluation
   object  scheduling-decision
               object plan =p
               time   time-interval =t
   context (opportunity
                plan   =p
                object =obj
                          time-interval =t)
```

Figure 1: Abstract point of *make hay* ...

```
positive-evaluation
   object  scheduling-decision
               object hay-making-plan =p
               time   time-interval =t
   context (precondition-opportunity
                plan   =p
                object sunshine
                          time-interval  =t)
```

Figure 2: Story of *make hay* ...

strained. An account of how proverbs with stories can be represented and used for advice taking is a first step towards developing a system that can extract advice from a wider range of stories. This paper presents such an account.

Our approach is implemented in BRAINSTORMER, a planner linked to an advice taker [Jones, 1991a; Jones, 1992]. The advice taker elicits advice from a user in the form of a representation of a proverb, and transforms it into an operational planner data structure that help the planner resolve a current problem-solving difficulty. BRAINSTORMER operates in the domain of political and military policy as it relates to terrorism.

Representing Proverbs with Stories

The key problem of representing proverbs with stories is encoding them in ways that highlight their functionally significant features, by which we mean their recommendations and any appropriateness conditions that constrain their application. It is useful to divide these functionally significant features into two categories: essential features and inessential features. Essential features of a proverb's story are aspects of the story that must hold in every situation in which the proverb applies; the union of these features constitutes the proverb's abstract point. Inessential features, when true in a new situation, lend weight to a proverb's recommendation, but need not obtain for the proverb to apply.

We offer a two-part solution to the problem of representing proverbs with stories. First is a notation for highlighting essential features of stories. Second is a way to encode non-standard but functionally significant abstractions of components of stories' representations.

Relating Stories and Abstract Points

Our approach to highlighting essential features can be summarized as follows. We view proverbs as consisting of a story and an abstract point that are structurally related to each other. The essential features of a proverb's story are those shared by its abstract point; therefore, by encoding this structural relationship, we highlight the story's essential features.

Consider, for example, the proverb *make hay while the sun shines*. As shown in figure 1, the proverb's abstract point is the recommendation ***take advantage of opportunities when they exist.***[1] The story of *make hay while the sun shines*, on the other hand, concerns a particular kind of opportunity: an opportunity afforded by the satisfaction of a precondition of a hay-making plan. The representation of this story is shown in figure 2.

While the proverb's abstract point must obtain in a new situation for the proverb to apply, the same is not true of its story. Obviously, the new situation need not be about sunshine; perhaps less obviously, it need not involve fortuitous satisfaction of a precondition of a plan. ***Make hay while the sun shines*** could, for example, be used to advise selling a stock after its price increases; the higher price signifies a good opportunity not because high prices are a precondition for selling, but because selling when prices are high yields a greater profit.

To capture the structural relationship between a proverb's story and its abstract point, we represent the proverb in terms of its story, but with its abstract point superimposed. That is, we begin with representations of its story and its abstract point, and then embed the abstract point in the story. Viewing representations as graph structures [Sowa, 1984], we make the assumption that the abstract point can be represented as a ***subgraph*** of the story, where possibly some of the nodes in the subgraph are more general than the corresponding nodes in the story. Under this assumption, each frame in the abstract point corresponds to a frame in the story.

To build up a representation of the proverb, we start with a representation of its story, and annotate each frame in the representation as follows:

1. If there is no corresponding frame in the abstract point, we postfix the frame with a ⇑. Thus, for example, a `resource` frame would become `resource⇑`.
2. If there is a corresponding frame in the abstract point, but its type label differs from the story (in

[1] Brainstormer uses a frame-based representation system with a slot-filler notation:

```
<frame>
     <slot1> <filler1>
     <slot2> <filler2>
     ...
```

Equality relationships between frames are encoded using the notation `=<symbol>`.

```
positive-evaluation
 object  scheduling-decision
            object  hay-making-plan↑plan =p
            time    time-interval =t
 context (precondition-opportunity↑opportunity
            plan   =p
            object sunshine↑entity
                      time-interval  =t
```

Figure 3: Embedding an abstract point in a story.

which case, it must be more general), then we annotate the frame in the story with this more general type label: **sunshine↑entity**, for example.

3. Otherwise, the frame in the story corresponds to a frame in the abstract point, and the type labels of the two frames are the same. In that case, we do nothing.

Figure 3 illustrates the result of applying this process to the proverb *make hay while the sun shines.*

We represent proverbs in this fashion because we want to glean as much information as possible from their stories, yet be free to generalize away inessential parts that do not hold when the proverb is used in a new situation. The abstract point of a proverb is a representation of its essential core claim that must be true in every situation in which it applies. Our chosen representation has the advantage that it can be easily interpreted as instructions to that effect:

- Those parts of the story that overlap with the abstract point are annotated with "↑"s, which can be thought of as instructions about how far they can be generalized. If there is no annotation at all, the story coincides with the abstract point, so no generalization is permitted. If the annotation is of the form $x \uparrow y$, then story component x corresponds to more general component y in the abstract point, so it can be generalized as far as y but no further.
- Those parts of the story that do not overlap with the abstract point at all (annotated with a ⇑) can be freely generalized or even deleted.

Encoding Non-Standard Abstractions

So far, we have presented a notation for distinguishing essential features of proverb's stories from inessential ones. By itself, however, this notation is not enough. Many proverbs have stories containing features that although inessential to the proverb's application, are nevertheless functionally significant.

The story of *make hay while the sun shines*, for example, involves features of his kind. As we have seen, the abstract point of this proverb can be paraphrased as *take advantage of opportunities when they exist*, while its story talks about hay making and sunshine. There is an important intermediate level of representation, however, which is missed if only the proverb's abstract point and story are represented. In particular, the choice of *sunshine* for the opportunity is significant. Sunshine is a resource that is only intermittently and unpredictably available, but that is essentially free when available; moreover, the availability of sunshine is not under the planner's control, so it cannot be planned for. These features of sunshine are all very relevant for planning: it is particularly important to take immediate advantage of opportunities afforded by the availability of resources that have some or all of these properties.

This description of sunshine as a resource is a generalization of BRAINSTORMER's usual representation for sunshine. Sunshine can be classified as a resource of this kind, but so can any number of other things, including rain and the generosity of kings. Moreover, these features of sunshine do not have to obtain in a new situation for the proverb to apply, although their presence constitutes good evidence that the proverb's recommendation is appropriate.

Encoding this abstraction presents a potential difficulty, however. Concepts in BRAINSTORMER are arranged in an abstraction hierarchy, which supports "property inheritance" of necessary properties of concepts. Unfortunately, it impossible to use the abstraction hierarchy to represent resources as an abstraction of sunshine. Almost any entity can serve as resource in some situations, so if **resource** were to be encoded as an abstraction of **sunshine**, it would also have to be an abstraction of most other concepts in memory. In that case, every represented object would inherit various properties of resources. This would be wasteful, and in many instances, incorrect: an object can be a resource for some purposes and in some situations, but not in others. It follows that **resource** cannot be a generalization of **sunshine** in the abstraction hierarchy.

While it is not *always* useful or correct to think of sunshine as a resource, it is certainly desirable in *some* circumstances. For example, in representing *make hay while the sunshine*, we want to encode that sunshine as an unpredictably available resource whose availability is beyond the planner's control.

Fortunately, there is a way out of this dilemma. BRAINSTORMER is able to reason with multiple descriptions of given objects, using a process called *redescription inference* [Jones, 1991a; Jones, 1991b]. Redescription inference is used to transform instances of one concept into co-referential instances of another. Redescription inference leaves traces in the form of *views* or co-reference links with attached justifications. We use views to explicitly encode important but non-standard abstractions of components of stories that are not represented in the abstraction hierarchy. Thus, we represent the proverb as it would look had **sunshine** been *redescribed* as the appropriate kind of resource. Figure 4 illustrates the resulting representation of the proverb (omitting details of the internal structure of views). Sunshine is described in terms of both the concepts **sunshine** and **resource**. **Resource** frames index information

```
positive-evaluation
 object scheduling-decision
         object hay-making-plan↑plan =p
         time   time-interval =t
 context (precondition-opportunity↑opportunity
          plan =p
          object sunshine↑entity =s
                 time-interval =t)

VIEWS:
 (=s 1. sunshine↑entity
     2. resource⇑
          availability (intermittent⇑
                        unpredictable⇑
                        not-plan-for⇑)
          unit-cost zero⇑)
```

Figure 4: Full representation of *make hay* ...

relevant to planning regarding the availability and cost of objects.

In summary, we have described two notations that together allow us to explicitly represent functionally significant features of proverbs' stories. In the next section, we describe how these representations assist BRAINSTORMER's advice taker.

The Advice-Taking Process

BRAINSTORMER consists of a planner with an associated advice taker. The planner is handed goals having to do with countering terrorism and attempts to come up with plans of action. If the planner gets into trouble, it issues a query for information sufficient to resolve its difficulty. A user then presents advice in the form of a representation of proverb, which the advice taker attempts to transform into an answer to the query. Answers must take the form of operational planner data structures that match (abductively unify) with the query [Charniak, 1988].

Suppose, for example, that the planner is currently considering whether or not to carry out a preemptive raid against a terrorist organization. It therefore issues a query for information that would allow it to make a principled decision. Let us additionally suppose that the planner knows that public opinion is currently running high against this terrorist organization; the planner, however, has not considered the implications of this fact for its decision.

In that case, the proverb *make hay while the sun shines* could be aptly used with the intent of recommending *carry out the raid now, in the window of opportunity afforded by temporarily favorable public opinion.* The task of advice taking is to generate this specific recommendation from the initial, generic representation of the proverb.

The advice-taking process has three phases. The first involves inferring an operational planner data structure from the proverb that can match the planner's query. As we explain in [Jones, 1992], this transformation requires reasoning with an explicit model of the planning process. In this paper, however, we focus our attention on the latter phases of advice taking, because this is where the impact of highlighting functionally significant features of proverbs' stories is most acutely felt.

In the second phase of advice taking, the advice taker generates a common abstraction of the proverb's story and the planner's current problem that matches the planner's query. Here the common abstraction recommends running a plan on the basis of an unspecified opportunity afforded by an unpredictably available resource not under the planner's control. This common abstraction is then matched to the query, yielding a hypothetical answer to the query: *carry out the raid now, because there currently exists an opportunity afforded by a resource like sunshine that is unpredictably available and not under the planner's control.*

The system has now been provided with a recommendation, but thus far it has no reason to believe it particularly plausible, except that it originated with a user who presumably intends to be helpful. In the third and final phase of advice taking,[2] the system attempts to flesh out this provisional answer into a complete and well-justified recommendation. In the current example, this involves positing that the object of the opportunity is in fact the favorable state of public opinion, and then attempting to verify that public opinion can indeed be described as an unpredictably available resource not under the planner's control.

The central task of the two later phases of advice-taking is computing a common abstraction of the proverb's story and the planner's situation that can be fleshed out into a well-justified answer to the planner's query. We want this abstraction to be as useful to the planner as possible, so in fact we desire a *maximally specific* abstraction of the story in terms of features that are *functionally significant* for planning. To reiterate, a feature is functionally significant if it is part of a proverb's recommendation or if it helps specify an appropriateness condition for that recommendation.

BRAINSTORMER computes this abstraction using a knowledge-intensive matching process that compares the proverb's story with the planner's query. Abstractions are computed by incrementally generalizing components of the proverb's story in response to local inconsistencies that the matcher detects when attempting to fit these components to the planner's query. Hay making, for example, cannot be consistently described as a preemptive attack on terrorists, so this aspect of the proverb's story must be generalized. Generalization is only attempted when an inconsistency is detected; it follows that the resulting abstraction will be maximally specific.

[2]In fact, the second and third phases of advice taking are to some extent interleaved.

```
(QUERY-MATCH story query):
 IF the type labels of story and query match
 THEN For each slot of query,
         Recursively QUERY-MATCH corresponding slot
           fillers of story and query
 ELSE IF (REDESCRIBE story and query)
      THEN return success
 ELSE IF the type label of story can be generalized
      THEN LET gen be the generalization of story
                obtained by replacing its type label
                with the next more general type in
                the abstraction hierarchy.
           Generalize or delete subparts of story
           justified only by property inheritance
           from the old type label;
           (QUERY-MATCH gen query)
 ELSE IF the type label of story is entity and
         story can be deleted
      THEN delete it and return success
```

Figure 5: Finding maximally-specific abstractions

Two Problems

The matcher faces two problems in generating this abstraction, both of which are substantially mitigated by having precomputed and cached all of the functionally significant features of the proverb's story. The first problem is to ensure that the common abstractions constructed by the matcher are useful to the planner. There are a great many possible generalizations of the proverb's story, most of which do not focus on functionally significant commonalities. When the matcher faces a local inconsistency, which generalization should it prefer?

Our representation of proverbs' stories is very helpful in this regard. All of the features of the story that are conceivably useful to the planner are either implicit in the abstraction hierarchy, or are precomputed and cached with the proverb. Moreover, those features that can be generalized or deleted without invalidating the proverb's recommendation have been explicitly tagged as such. It follows that generalizing a component of a proverb's story can be accomplished by simply replacing it with an instance of a more general concept in the abstraction hierarchy (or perhaps deleting it altogether); see figure 5. BRAINSTORMER's abstraction hierarchy admits no upward branching, so this process is quite efficient.

A second problem the matcher faces is determining which components of the planner's problematic situation should be matched against candidate answers to a query. Some of these components are mentioned in the query, in which case a correspondence is established automatically by the query-matching process. Further components, however, must be determined during the third phase of advice taking, to justify this candidate answer.

For example, we have seen that a candidate answer to a query can be constructed from *make hay while the sun shines* that mentions sunshine. To justify this answer, the system has to find something in the planner's problematic situation that has something in common with sunshine. If the proverb did not encode functionally significant abstractions of sunshine, it would be necessary to search for an instance of an arbitrary abstraction of sunshine. However, sunshine is tagged as an unpredictably available resource not under the planner's control, so the advice taker can instead engage in a more focused search for a hitherto unnoticed *resource* of the appropriate kind.

Of course, this search may not always succeed in identifying a suitable resource. If the search succeeds, however, it will succeed more quickly than a search for an instance of an arbitrary generalization of sunshine. Moreover, if a resource is found that is in fact unpredictably available or not under the planner's control, then the system can have greater confidence that the proverb's recommendation actually applies.

Discussion and Related Work

Taking advice in the form of a story can be viewed as a process of analogical reasoning, or more specifically, exploiting an "analogical hint" [Greiner, 1988]. Proverbs' stories help both to construct recommendations and to determine aspects of the planner's problematic situation that justify these recommendations. We have seen that both can be accomplished by computing a common abstraction of the story and the current problem in terms of functionally significant features of the proverb's story. In our chosen representation, all such features are cached in advance with the story. These cached features guide generalization towards a suitable abstraction, and simultaneously constrain search for aspects of the planner's problem that justify the proverb's recommendation.

A knowledge-intensive approach to analogical inference is central to BRAINSTORMER's success. The system infers commonalities between a proverb's story and a planning problem on the basis of explicit representations of the story's functionally significant features. Stuart Russell likewise advocates a knowledge-intensive approach [Russell, 1989]. His DBAR system relies on knowledge about functional significance encoded as explicit *determinations*. A determination $P \succ Q$ (pronounced "P determines Q") licenses analogical inference that Q holds for a target, provided that Q also holds for a source that shares properties P with the target. For example, logos on running shoes determine their manufacturer.

Russell notes that determinations guide inference of commonalities between a source and a target. $P \succ Q$ explicitly encodes that P are the features that source and target must share to license analogical inference of Q. Trying to demonstrate that P holds of a source and target is a considerably less open-ended task than an arbitrary search for common features of a source and target. Similarly, BRAINSTORMER explicitly encodes the features of proverbs' stories that if true of a target support the proverb's recommendation, and uses these features to guide inference.

BRAINSTORMER's advice taker relates proverbs to queries using a knowledge-intensive matching process. Knowledge-intensive matching is motivated by the following realization about the nature of analogy: once it is accepted that objects, properties, and relations can be described in different ways using different vocabularies, no purely syntactic criterion for assessing similarity of a source and target could conceivably be adequate. This is not to say, of course, that no syntactic criterion intervenes at *any* stage of analogical reasoning. Rather, it is to emphasize that the major part of a theory of analogical reasoning must consist in specifying (1) which features of the source and target can be meaningfully compared, and (2) the knowledge needed to compute these features efficiently if they are not already explicit. Once this is done, *identity* of functionally relevant features may suffice as a criterion for similarity.

BRAINSTORMER accomplishes (1) by explicit encodings of functionally significant features, and (2) in terms of viewing schemas for redescription inference. As explained in [Jones, 1991a], redescription inference derives ultimately from MERLIN [Moore and Newell, 1973], and is related to the idea of "views" in Jacob's ACE system [Jacobs, 1987].

In contrast, most existing work in analogy ignores one or both of these aspects of analogical reasoning. (Russell's DBAR is a notable exception.) Greiner sidesteps (2) altogether [Greiner, 1988]. Gentner's theory of structure mapping [Falkenhainer *et al.*, 1986] advocates a purely syntactic approach to assessing similarity of source and target, and thus does not address either (1) or (2). However, Falkenhainer's PHINEAS system [Falkenhainer, 1989], while ostensibly an implementation of structure mapping, in fact adopts a more knowledge-intensive approach.

What are the limitations of BRAINSTORMER's approach to advice taking? As currently implemented, there is no way to specify that a feature of a proverb's story is functionally *insignificant*, so currently the system attempts to transfer *all* features of proverbs' stories to the planning problem at hand, even seemingly irrelevant features such as *hay*. While this slows down the inference process a little, it has not proven problematic in practice. (Tagging features of stories as functionally insignificant of course immediately raises the question of why they should be kept around at all. One reason is to allow for the possibility of future learning. A feature of a proverb that currently appears to be irrelevant to its abstract point may later be discovered to be important.)

A more fundamental limitation of the approach is the assumption that all functionally significant features of stories are computed in advance. While this assumption is reasonable for many proverbs, it is less reasonable for arbitrary stories. A story about a given episode can be used to make a variety of very different points on different occasions [Schank, 1991]. It is possible, however, that the following extension of BRAINSTORMER's approach might be adequate. Stories are usually laden with linguistic cues designed to highlight the aspects of the story that the speaker intends the hearer to focus on. Perhaps these cues can be used by an advice taker to help infer candidate abstractions of the story ready for matching to a problem. This is a direction for future research.

References

[Charniak, 1988] Charniak, E. 1988. Motivation analysis, abductive unification, and nonmonotonic equality. *Artificial Intelligence* 34:275–295.

[Falkenhainer *et al.*, 1986] Falkenhainer, B., Forbus, D.K., and Gentner, D. 1986. The structure-mapping engine. In *Proceedings AAAI-86*, Philadelphia, PA. AAAI. 272–277.

[Falkenhainer, 1989] Falkenhainer, B. 1989. *Learning from Physical Analogies: A Study of Analogy and the Explanation Process.* Ph.D. Dissertation, University of Illinois at Urbana-Champaign.

[Greiner, 1988] Greiner, R.D. 1988. Leaning by understanding analogies. *Artificial Intelligence* 35:81–126.

[Jacobs, 1987] Jacobs, P.S. 1987. Knowledge-intensive natural language generation. *Artificial Intelligence* 33:325–378.

[Jones, 1991a] Jones, E.K. 1991a. Adapting abstract knowledge. In *Proceedings of the Thirteenth Annual Conference of the Cognitive Science Society*, Chicago, IL. Lawrence Erlbaum Associates.

[Jones, 1991b] Jones, E.K. 1991b. *The Flexible Use of Abstract Knowledge in Planning.* Ph.D. Dissertation, Yale University.

[Jones, 1992] Jones, E.K. 1992. Model-based case adaptation. In *Proceedings AAAI-92 Tenth National Conference on Artificial Intelligence*, San Jose, CA. Morgan Kaufmann.

[Moore and Newell, 1973] Moore, J. and Newell, A. 1973. How can MERLIN understand? In Gregg, L. W., editor 1973, *Knowledge and Cognition.* Lawrence Erlbaum Associates, New Jersey.

[Mostow, 1983] Mostow, D.J. 1983. Machine transformation of advice into a heuristic search procedure. In Michalski, R.S., Carbonell, J.G., and Mitchell, T.M., editors, *Machine Learning: An Artificial Intelligence Approach.* Tioga Publishing Company, Cambridge, MA. 367–404.

[Russell, 1989] Russell, S.J. 1989. *The Use of Knowledge in Analogy and Induction.* Morgan Kaufmann, San Mateo, CA.

[Schank, 1991] Schank, R.C. 1991. *Tell Me a Story: A New Look at Real and Artificial Intelligence.* Simon and Schuster, New York.

[Sowa, 1984] Sowa, J.F. 1984. *Conceptual Structures: Information Processing in Mind and Machines.* Addison-Wesley, Reading, MA.

Primacy Effects and Selective Attention in Incremental Clustering

David M. Thau
University of Michigan, Ann Arbor
330 Packard Rd.
Ann Arbor, Michigan 48104
Phone: (313) 764-0318
E-mail: thau@engin.umich.edu

Abstract

Incremental clustering is a type of categorization in which learning is unsupervised and changes to category structure occur gradually. While there has been little psychological research on this subject, several computational models for incremental clustering have been constructed. Although these models provide a good fit to data provided by some psychological studies, they overlook the importance of selective attention in incremental clustering. This paper compares the performance of two models, Anderson's (1990) rational model of categorization, and Fisher's (1987) COBWEB, to that of human subjects in a task which stresses the importance of selective attention. In the study, subjects were shown a series of pictorial stimuli in one of two orders. The results showed that subjects focussed their attention on the first extreme feature they saw, and later used this feature to classify ambiguous stimuli. Both models fail to predict human performance. These results indicate the need for a selective attention mechanism in incremental clustering as well as provide one constraint on how such a mechanism might work.

Introduction

Imagine trying to acclimate yourself to a city you have never visited before. As you wander through the streets, you may begin to notice similarities and differences between the styles of some of the houses. Each new house may remind you of a few others, leading you to group them together. Eventually, you may form fairly well defined categories. The process through which these categories are devised is called incremental clustering. Incremental clustering may be characterized by two qualities. First, learning is unsupervised. In the example above, the houses were divided into categories without feedback from a teacher. Second, changes to the category representation are made incrementally Each new exemplar is incorporated into an already existing category structure. This is in contrast to non-incremental categorization, in which the entire category structure is reconsidered whenever a new exemplar is encountered.

Surprisingly, incremental clustering has received little attention from the cognitive psychology community. While both supervised, incremental category learning (e.g. Posner & Keele, 1968, Smith & Medin, 1981), and unsupervised, non-incremental category learning (e.g. Ahn & Medin, in press, Bersted, Brown & Evans, 1969) have been studied in detail, there have been few experiments on unsupervised, incremental category learning (Fried & Holyoak, 1984, Homa & Cultice, 1984).

In the machine learning literature, on the other hand, incremental clustering has received a good deal of attention. The combinatoric explosions that result in computer systems that try to organize categories in a non-incremental fashion have lead machine learning researchers to study incremental learning. This, coupled with the need for systems that learn without constant and consistent feedback, has lead to several models of incremental clustering. I will briefly describe two of the more recent computational models of incremental clustering, Anderson's (1990) rational model of categorization, and Fisher's (1987) COBWEB model. These descriptions will be followed by a study that demonstrates a flaw shared by these models.

The first model I will describe is Anderson's rational model of categorization. Anderson provides a Bayesian analysis of category structure goodness. When presented with a new stimulus, the model calculates the goodness of the whole category scheme for each possible categorization of the new stimulus. For example, if the model has already constructed three categories, it determines the goodness of four different category structures: one structure for when the new item is placed in each of the already existing categories, and one structure for when a new category containing only the new item is created.

Although the actual formula Anderson uses to determine category structure goodness is not important for the purposes of this paper, it is important to note that goodness is determined by feature counts within each category. Information

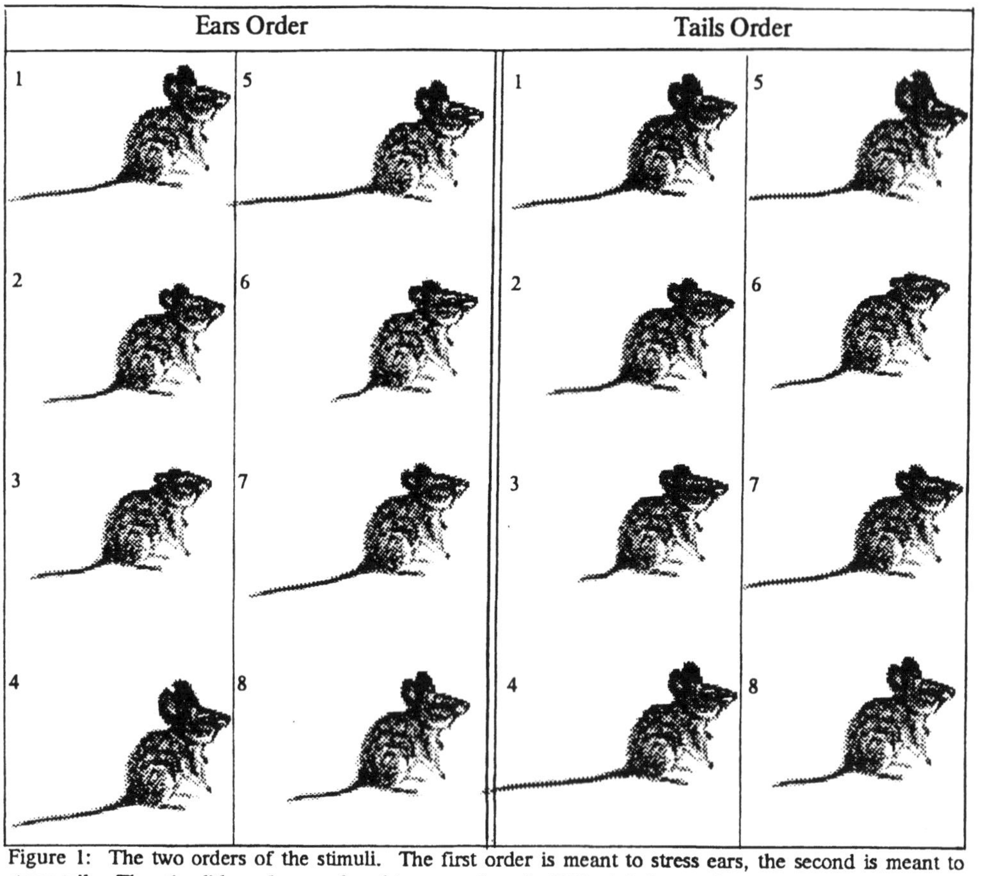

Figure 1: The two orders of the stimuli. The first order is meant to stress ears, the second is meant to stress tails. The stimuli have been reduced to approximately 30% of their actual size.

about the order in which the items were classified is lost. Information about why an item was classified the way it was is also lost. Another interesting limitation of Anderson's model is that it cannot change its partitionings. Once an item is classified, it cannot be reclassified unless it is seen again

While Fisher's COBWEB does allow for reclassification of stimuli, and uses a different category structure goodness function, it is in other ways very similar to Anderson's model. When COBWEB is shown a new stimulus, it, like Anderson's model, considers the goodness of placing the item in each of its categories, or a new category. In addition, COBWEB considers either merging the two best categories, and placing the item in the merged category, or splitting the single best category into two, and placing the item into one of them. Although the model can reclassify items, it shares the information reduction limitations of Anderson's model. All information other than the present membership of each category is lost.

People, however, may represent information beyond category membership. They may, for example, remember why they categorized an exemplar the way they did. Consider two people who have sorted a set of items into the same categories, but have done so for different reasons. Although these different reasons may not have manifested themselves yet, there may be a point in the future where these two people will react differently to a new exemplar. Consider, on the other hand, two runs of COBWEB or Anderson's model. Because these models only represent the current membership of each category, two runs which have formed the same categories will categorize a new exemplar the same way.

The simulations and experiment that follow will demonstrate that this limitation of the models is indeed a problem. Eight pictures of mice with different sized tails and ears were used as stimuli. The mice were actually eight pictures of the same mouse, scanned into a Macintosh computer, and modified using a graphics program. In six out of the eight mice, the ear size and tail size were positively correlated; mice with big ears also had big tails, mice with small ears had small tails. Of these six, four had one relatively extreme feature, either ears that were very big or very small, or a tail that was very big or very small. In the remaining two mice, ear size and tail length were negatively correlated; one mouse had big ears and a small tail, the other had small ears and a big tail. See Figure 1 for a graphical representation of the stimuli.

The mice were put into two orders, each of which is represented in the Figure. The first two mice (mice 1 and 2) were the same in both orders. These were the two mice that maintained a positive correlation between ear size and tail length, but had moderately sized features. The next four mice were either ordered such that the two mice with the extreme ear sizes came next, or the two mice with the extreme tail sizes came next. The next two mice in each order were the other two mice with extreme features. The final mice (mice 7 and 8) were those in which the features violated the positive correlation. One of these mice had a long tail but short ears, the other had a short tail and long ears.

The stimuli were ordered to induce human subjects to pay attention to ears in one condition and tails in the other. The hypothesis was that subjects would attend to the first extreme feature they saw and then focus most of their attention on that feature throughout the sorting. During the sorting of the first six mice, however, this attention weighting has no effect on performance. Because the features of these mice were positively correlated, the sorting will be the same regardless of which feature was more important; mice with big ears and big tails will be sorted into one category, mice with small ears and small tails will be sorted into the other. In terms of the models described above, both orders will produce the same category structure.

The test of the models comes during the sorting of the final two mice. Because the features of these mice violate the positive correlation, the way in which they are sorted provides important information about the subject's sorting strategy. If the subject thinks ear size is more important, he or she will put the mouse with small ears, but a large tail, into the category of mice with small ears and small tails. If tail size is more important, the subject will put the same mouse into the other category. However, the two models can not account for this result. Because the models base their sortings entirely on the current category structure, the order by which that structure was created has no effect. Therefore, because the two orderings produce the same categories, the different ordering of the stimuli should have no effect on the models' sortings.

The preceding observation was supported by simulations using Fisher's and Anderson's models[1]. Both models took two features as input, the size of each mouse's ears (in cm^2) and tail (in cm). The stimuli were presented to the models in the two different orders described above. As can be seen in Table 1, neither Fisher's COBWEB, nor Anderson's rational model showed any effect of order. COBWEB[2] sorted the last two mice according to tail length regardless of the order. Anderson's model[3], which outputs the probability of sorting a stimulus into each possible category, provided the same probabilities regardless of the order.

In summary, neither model was affected by the different orderings of the stimuli. The remainder of the paper compares these results with those of human subjects.

Method

Thirty-five University of Michigan undergraduates participated in the study as part of an introductory psychology course requirement. The subjects, who were tested individually, were told that they would be shown pictures of different mice. They were told that their task was to sort the mice into two different kinds, but that it was up to them to decide how to divide them. They were also told that at any point they could reclassify any of the mice. This reclassification was permitted for two reasons. First, in early stages of categorization, reclassification should be expected (Fried and Holyoak, 1984). Second, the models being tested both take into account the need for reclassification. Fisher's COBWEB model explicitly allows reclassification through its merging and dividing operations. Anderson's rational model requires that the mean and variance of each feature of the stimuli be predetermined, reducing the amount of reclassification necessary.

The experiment proceeded with the experimenter presenting the mice to the subjects one by one. After each mouse was shown, subjects classified it by verbally responding either A or B. The mouse was then placed in front of the subject in a way that allowed subjects to see how each mouse had been classified. Subjects were permitted to see all their

[1]Versions of both models were kindly provided by their authors.

[2]Instead of COBWEB, Fisher's CLASSIT program was used. CLASSIT is a version of COBWEB which allows for features with real values. COBWEB only allows nominal values. Acuity in this simulation was set to 0.5.

[3]The coupling parameter in these runs was set to 0.3.

Table 1
Percentage of Simulated Subjects Who Sorted by Ear Size, Tail Length, or Something Else

Fisher's COBWEB

Order	Sorted by: Ears	Tail	Other
Ears	0	100	0
Tail	0	100	0

Anderson's Rational Model

Order	Sorted by: Ears	Tail	Other
Ears	27	23	50
Tail	27	23	50

classifications in order to diminish reliance on memory. The models of incremental clustering being tested both assume that the system has full memory of the stimuli it has seen. If subjects were not allowed to see the mice they had classified, they would have been working with less information than the models being tested. After all the mice from one order had been presented and classified, the experimenter asked the subject if he or she was satisfied with the sorting, and then asked the subject to describe the categories that were formed.

Results

Table 2 shows the percentage of subjects in each order condition who sorted the mice by ear size, tail length, or something else. As described in the introduction, the stimuli were constructed such that one sorting clearly indicated that the subject was sorting by tail length, and another sorting clearly indicated that the subject was sorting by ear size. Only six of the 35 subjects provided sortings different from the two that were anticipated. These subjects are represented in the 'other' column of Table 2. Most of the subjects who fell into this condition put a mouse that had an extreme feature (such as the mouse with the smallest ears) into one category and the rest of the mice in the other category.

The manipulation clearly had the expected effect, $X^2(2, N=35) = 10.6$, $p < 0.005$. Subjects who were presented with the stimuli that were ordered to emphasize ear size did in fact sort by ear size. Those who were presented with the stimuli that were ordered to emphasize tail length were more likely to sort by tail length.

Discussion and Conclusion

The results summarized above indicate that models of incremental clustering need to take the role of selective attention into account. Neither Fisher's COBWEB, nor Anderson's rational model provide mechanisms by which different features can become

Table 2
Percentage of Subjects in Each Order Who Sorted by Ear Size, Tail Length, or Something Else

Order	Sorted by: Ears	Tail	Other
Ears	72	0	28
Tail	41	53	6

more or less important in the midst of a categorization task apart from the straight accumulation of features. Although both models can preset the salience of the different features, neither can change feature weights on-line.

The results further provide one constraint on how a selective attention mechanism should work. Subjects in this task focussed on the first feature that clearly differentiated between two categories and later used this feature when classifying ambiguous examples.[4]

How this feature weighting occurs in people has yet to be determined. One simple explanation for the results of this study is that once the subjects found a salient feature along which to classify, they ignored all other features. If, for example, ear size seemed like a diagnostic feature, a subject could simply look at the ears of each stimulus and ignore the other features. This mechanism might be implemented as a rule-like system (if big ears, then category A), which would obviate the need for the Anderson and Fisher models once a satisfactory feature was discovered.

Although this approach to selective attention would fit the results of this study, it is unlikely that subjects completely ignore all but the most diagnostic features. If this were true, people could not adapt to changing circumstances. In addition, there is evidence (Medin, Wattenmaker and Michalski, 1987) that people include redundant information when devising classification rules. An alternative approach would involve learning feature weights across dimensions. While there has been a great deal of work on feature weighting models when feedback is immediate, few models (cf. Gennari, 1991, Kohonen, 1982, Rumelhart and Zipser, 1985, Grossberg, 1987) have been developed which apply to unsupervised learning. Future research will involve applying these models to the present task, and extending them or positing new models where necessary.

In conclusion, current models of incremental clustering must be extended to take into account on-line learning of feature weighting. Although the mechanisms involved in this weighting are still uncertain, this study has provided one constraint; salient differences between features are weighted more heavily when they occur in early examples.

---

[4]Recently, Gennari (1991) described CLASSWEB, an extension of COBWEB that includes a selective attention mechanism. CLASSWEB's selective attention mechanism, however, mainly acts to focus attention away from irrelevant features. In the study described here, both features are relevant. Consequently, CLASSWEB does not predict the results provided by human subjects. Instead, in this situation, CLASSWEB behaves exactly like COBWEB.

Acknowledgements

I would like to thank Douglas Medin for his help throughout all phases of this study. I would also like to thank Woo-Kyoung Ahn, Evan Heit and Joushua Rubinstein for reading an earlier draft of this paper.

References

Ahn, W-K., & Medin, D. L. (In Press). A two stage model of free sorting. *Cognitive Science.*

Anderson, J. R. (1990). *The adaptive character of thought.* Hillsdale, NJ: Erlbaum.

Bersted, C. T., Brown, B. R., & Evans, S. H. (1969). Free sorting with stimuli clustered in a multidimensional space. *Perception & Psychophysics, 6,* 409-414.

Fisher, D. H. (1987). Knowledge acquisition via incremental conceptual clustering. *Machine Learning, 2,* 139-172.

Fried, L. S. & Holyoak, K. J. (1984). Induction of category distributions: A framework for classification learning. *Journal of Experimental Psychology: Learning, Memory, and Cognition, 10,* 234-257.

Gennari, J. H. (1991). Concept formation and attention. Proceedings of the Thirteenth Annual Congference of the Cognitive Science Society (pp. 724-728). Chicago, Illinois: Earlbaum.

Grossberg, S. (1987). Competitive learning: from interactive activation to adaptative resonance. *Cognitive Science, 11,* 23-63.

Homa, D., & Cultice, J. (1984). Role of feedback, category size, and stimulus distortion on the acquisition and utilization of ill-defined categories. *Journal of Experimental Psychology: Learning, Memory, and Cognition, 10,* 83-94.

Kohonen, T. (1982). Self-organized formation of topologically correct feature maps. *Biological Cybernetics, 43,* 59-60.

Medin, D. L., Wattenmaker, W. D., & Michalski, R. S. (1987). Constraints and preferences in inductive learning: An experimental study of human and machine performance. *Cognitive Science, 11,* 299-339.

Posner, M. L., & Keele, S. W. (1968). On the genesis of abstract ideas. *Journal of Experimental Psychology, 83,* 304-308.

Rumelhart, D. E. & Zipser, D. (1985). Feature discovery by competitive learning. *Cognitive Science, 9,* 75-112.

Smith, E., & Medin, D. L. (1981). *Categories and concepts.* Cambridge: Harvard University Press.

Some Epistemic Benefits of Action: Tetris, a Case Study

David Kirsh Paul Maglio
dkirsh@ucsd.edu pmaglio@ucsd.edu
Cognitive Science Department
University of California, San Diego
La Jolla, CA 92093-0515

Abstract

We present data and argument to show that in Tetris—a real-time interactive video game—certain **cognitive** and **perceptual** problems are more quickly, easily, and reliably solved by performing actions in the world rather than by performing computational actions in the head alone. We have found that some translations and rotations are best understood as using the world to improve cognition. They are not being used to implement a plan, or to implement a reaction. To substantiate our position we have implemented a computational laboratory that lets us record keystrokes and game situations, as well as allows us to dynamically create situations. Using the data of over 30 subjects playing 6 games, tachistoscopic tests of some of these subjects, and results from our own successful efforts at building expert systems to play Tetris, we show why knowing how to use one's environment to enhance speed and robustness are important components in skilled play.

Introduction

In this paper we present data and argument to show that in Tetris—a real-time interactive video game—certain **cognitive** and **perceptual** problems are more quickly, easily, and reliably solved by performing actions in the world rather than by performing computational actions in the head alone.

In Tetris, there are only four actions a player can take: translate right, translate left, rotate, drop. Tetrazoids—henceforth zoids—enter from the upper boundary of a rectangular playing field at a fixed speed which increases as the game proceeds, leaving the player with less and less time to make the judgements involved in choosing and executing a placement. Because all actions move the current zoid one way or another, every action the player takes has the effect of bringing the current zoid closer to its final position or farther from it. See figure 1.

Owing to the pace of the game it is not surprising that players in the earliest phase make moves before they can know where they wish to place the current piece. We have found that often these moves are best understood as having an *epistemic* function. They are not intended to achieve the *pragmatic* end of bringing a piece closer to its goal position. They are being used to change the world so as to help the agent acquire vital information early on.

Surprisingly, such epistemic functions are not confined to the earliest phases. Some translations and rotations occuring in later phases of decision and execution are also best understood as using the world to improve cognition. Some, for instance, seem designed to help the player *identify* a piece, or to *verify* that a particular action is a good one to take, or to *minimize the mental rotation* necessary to decide on a placement. We see the general function of these actions to be that of improving cognition by:

1. reducing the space complexity of mental computation;
2. reducing the time complexity of mental computation;
3. reducing the unreliability of mental computation.

These are not easy claims to defend in a game as complex as Tetris. We have implemented a computational laboratory that lets us record keystrokes and game situations, as well as allows us to dynamically create situations. Using the data of over 30 subjects playing 6 games, tachistoscopic tests of some of these subjects, and results from our own successful efforts at building expert systems to play Tetris, we will try to defend our conclusion that knowing how to use one's environment to enhance the speed and robustness of mental computation are important components in skilled play.

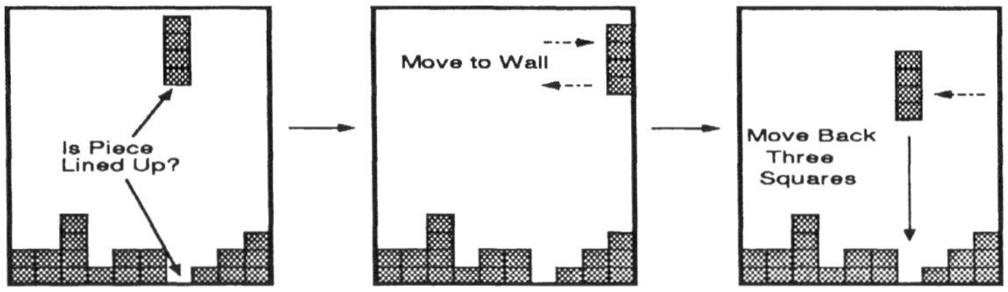

Figure 1: **Translating for verification**

In figure 1 we have an example of the basic game of Tetris. As each piece descends from the ceiling, the player must choose a region in which to place the piece. When a row of squares fills up, it disappears and all the rows above it drop down. As the game speeds up, achieving good placements becomes increasingly difficult.

One technique most players strike on to reduce error is to translate zoids to the wall. In figure 1, we see an instance of how is regularly translated to the outer right wall and back again before it is dropped. The explanation we prefer is that the subject confirms that the column of the zoid is correct, relative to his or her intended placement, by quickly moving the zoid to the wall and then with eyes on the contour tapping out the number of squares to the edge of the intended placement.

The idea that external actions can simplify mental computation is commonplace when symbol manipulation is involved. The activities of adding, accounting, composing, navigating (Hutchins, 1990), etc., are more difficult if agents must rely on their own memory without aid from external supports. Writing reduces the space complexity of the mental computations involved. When symbol manipulation is not involved, however, especially in tasks requiring quick response, it is less widely appreciated that certain non-perceptual actions can simplfy mental computation.

For instance, there is a tacit belief among planning theorists that intelligent behavior is either *reactive* or *planned* (Tate, Hendler & Drummond, 1990). In environments where an agent has time for reflection or forethought, planning can occur, and the agent may benefit from the advantages of previewing possibilities, hence mental backtracking is possible and local minima can be avoided. In rapidly changing environments, where there is not enough time to formulate a planned response—as is typical of arcade video games—the advantage lies with agents who have precompiled plans into reactions (Agre & Chapman, 1987). Where time is scarce, reactive systems, based on reliable statistical models of contingencies plus rapid sensing of environmental conditions, can be expected to score higher than systems which plan, unless, of course, there is enough time between actions to combine elements of both planning and reaction (Georgeff & Lansky, 1987). In each case, though, the assumption is that actions either are perceptual or should, if possible, bring the system physically closer to its goals.

A significant percentage of non-perceptual actions in Tetris actually take the agent physically farther from its ultimate goals. These costs are worth incurring because they are more than made up for by the epistemic or computational benefits they provide. They are rational actions if seen to be directed at transforming the agent's state, rather than the world's.

The idea that real-time systems must act so as to intelligently regulate their intake of environmental information is, at present, a topic of considerable interest (Simmons et al., 1992). But whereas existing inquiries have tended to focus on *control of attention*—the selection of elements within an image for further processing—or *control of gaze*—the orientation and resolution of a sensor—as the means of selecting information, our concern in this paper is with *control of activity*. We wish to know how an agent can use ordinary actions—not sensor actions—to unearth valuable information that is currently unavailable, hard to detect, or hard to compute.

Early and Late Epistemic Actions

Let us call the phase spanning the period when pieces are being identified, and the onset of the phase when plans are implemented, the *identification phase* (see figure 3). The duration of this phase varies with piece, subject, and the speed of the game. For instance, for and , when the game is proceeding at average speed, the identification phase probably begins around 600 ms and ends around 1800 ms, whereas for and the identification phase begins around 800 ms and ends around 2200

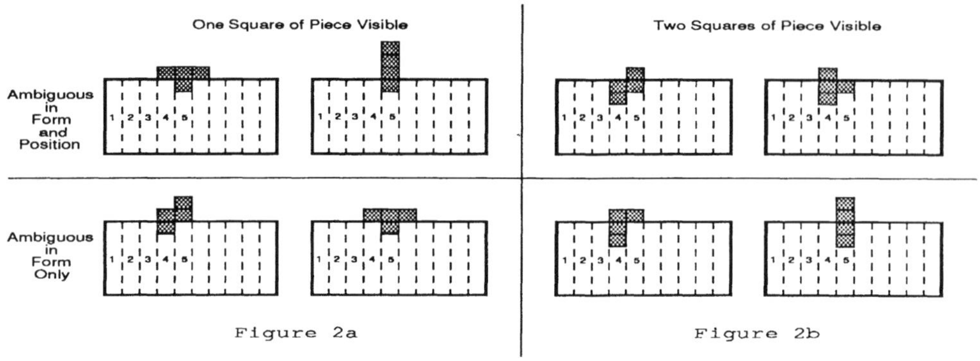

Figure 2: **Rotating for early discovery**

Rotating certain zoids very soon after they appear is a practical method for getting extra information about pieces. Here we see zoids as they first enter the playing field, in 2a they are one square in, in 2b they are two squares in. The upper portion of both 2a and 2b show zoids that look identical at this stage, both in position and in form. The bottom portions show zoids that look identical in form alone. Careful examination reveals that they are in different columns. Players are not explicitly aware of this column difference. The data show that players do not come out rotating, as we originally thought, but rather have a great burst once they are two rows out. At this point they show considerable sensitivity to column difference. Players have a much greater tendency to rotate zoids ambiguous in both form and position (such as those seen in the upper portion of 2b) than they have of rotating zoids that are ambiguous in form alone. By rotating ambiguous zoids early, players are able to make faster identifications, thereby either setting up the conditions for testing candidate placements early or setting early constraints on a candidate generator.

ms. The period before this phase we call the *pre-identification phase*, and the period after it, the *post-identification phase*.

As figures 2 and 1 illustrate, players at the intermediate and expert level regularly perform unambiguously epistemic actions in the pre- and post-identification phases.

1. Very early in the pre-identification phase players often rotate certain zoids before they have competely emerged, as if trying to disambiguate the zoid from all others as soon as possible.
2. In the post-identification phase players often drop certain zoids only after translating them to the nearest outer wall and then back again, as if to verify the column of placement.

The value of these actions is easy to appreciate. The first procedure, **rotate early**, serves to unearth facts otherwise hidden until later. When a zoid first enters the playing field and only a fraction of its total form is visible, the player must rely on subtle clues to disambiguate it. See figure 2. To be sure, players need not follow a strategy that requires them to disambiguate zoids as quickly as possible. But, in fact, we have noted that the more perfectly ambiguous a piece is, the more it is rotated early. The simplest explanation is that early rotation is for fact finding. By rotating a partially hidden piece, a player un-occludes part of it, thereby scaring up new information. The faster this may be done, the sooner an unambiguous image of the piece can be formed. The value of gaining this information early presumably outweighs the cost of possibly rotating a piece beyond its "goal" position.

The second procedure, **translate zoid to an edge and translate back again**, serves to confirm the column which the piece is currently in. After having chosen a spot to place the current piece, and having implemented a plan to direct the piece to that spot, a player may wish to confirm that he or she has succeeded in moving the piece to its intended column. This further phase is most useful when the piece is still high above the active contour and about to be dropped. See figure 1. This action cannot be confused with a pragmatic action, for by definition it requires moving the piece away from the currently intended column. On the occasions when it is performed, the pragmatic cost is more than offset by the benefit of reducing possible error.

Epistemic Actions During the Identification Phase

Actions performed in the identification phase are more difficult to classify, particularly since an ac-

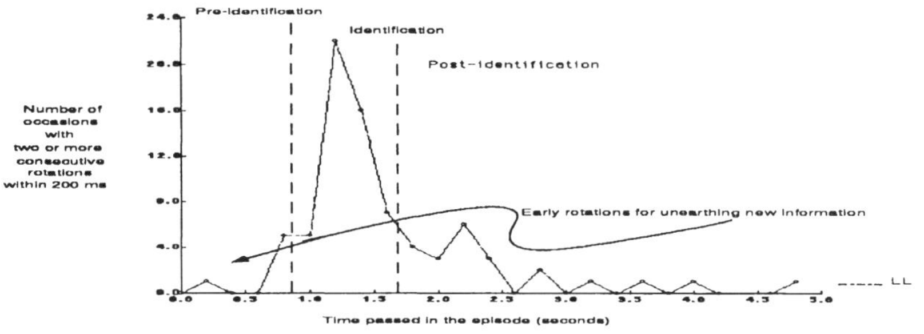

Figure 3: **Rotating to help identification**

When are the fastest rotations performed? Here we see subject PD's time course of double rotations for L's. By a double rotation we mean two rotations in very quick sequence (i.e., 200 ms or less). Two points should be noted. First, the greatest number of double rotations occurs in the region of 800 ms to 1600 ms: a period that corresponds nicely with the period in which pieces are being fully identified, as shown in figure 4. Second, PD had one double rotation at 200 ms, well before he could have identified the piece. PD also had 6 single rotations before 200 ms, a fact we interpret as confirming our conjecture that very early rotations serve to dig up information that otherwise would be hidden for another 400 ms. Similar results hold for all the subjects we have examined so far.

tion may serve both epistemic and pragmatic functions simultaneously. For instance, a zoid rotated in the direction needed for final placement, may, at one and the same time, help the player make an identification, while advancing the cause physically. The two functions—epistemic and pragmatic—are logically distinct, though it is hard to prove which function a given action subserves. Three epistemic functions an action may perform in the identification phase are:

1. help to *identify* a piece's type;
2. help to *verify* the identity of a piece once it is typed, i.e., reduce probability of misidentification;
3. help to *generate* candidate placements.

As can be seen in figure 3, subjects are more prone to have a burst of rapid rotations in the identification phase than at any other time. These actions of rotating pieces in the world take far less time than rotating pieces mentally. A natural conjecture is that they are being used to either facilitate identification or to reduce the probability of misidentification during this crucial period. This becomes more convincing when we consider how players decide where to place pieces.

Although we do not yet know exactly how a player selects where to place a piece, we have good reason to believe that some matching of piece shape to potential placement location must occur. To make this idea more precise, we need to introduce some terminology.

Each piece **type**, except the square, has two or four different orientations, called piece **tokens**. An L type, for instance, has four tokens: . Tokens are structures in the world. The mental image corresponding to a token is an **icon**; and the time required to create an icon is the **iconifying period**. We assume the iconifying period lasts 50-80 ms, the time required to flood V1, primary visual cortex, with a retinal image (Hillyard, 1985). The pieces already sitting on the board have an upper boundary called the current **contour**. The process of comparing an icon to small regions of the current contour we call running an iconic **mask** over the contour and **envisioning** a placement. The measure of how snugly a placement fits into its neighboring pieces is called its **local fitness**. On the basis of experiments with Robotetris, see figure 5, we have discovered that aiming to maximize the local fitness of placements is an important factor in player longevity.[1]

With these terms in mind we state two different methods (with variants) for determining placement, and consider how permitting epistemic actions can reduce their space-time complexity, and probability of error.

[1] In Robotetris the decision concerning where to place a piece is determined by a judicious weighting of such features as "how many holes would this placement create", "how many rows would this placement eat", "how flat is the resulting contour", in addition to "how close to the globally maximum fitness is this placement".

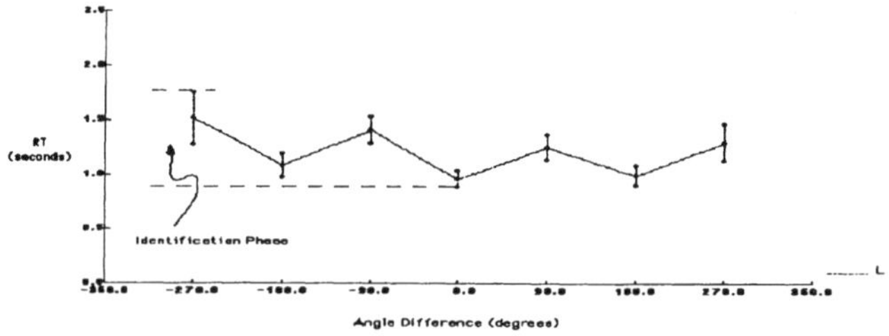

Figure 4: **Mental Rotation Task**
In this figure we display the findings of PD's mental rotation tests for L's. The horizontal line marks the mean times for successful recognitions; the vertical lines mark the 90% confidence interval. The region within this interval we call the identification phase. We found that the identification phase for an L under its various rotations was nearly linear, as suggested by Tarr & Pinker (1989). More precisely, given our data at this stage we can only report that we haven't disconfirmed the constant time hypothesis.

Type-based Method

1. Iconify token;
2. identify piece type, automatically creating a stack of appropriate icons;
3. (a) *computationally intensive version*:
   for each icon in the stack
      i. run its corresponding mask over the current contour,
      ii. envision the result of a placement,
      iii. compute that envisioned placement's local fitness, and
      iv. store the information about (place, score) in a list;

   or

   (b) *memory intensive version*:
   for each icon in the stack
      i. run an orientation-neutral mask over the current contour,
      ii. look-up in an associative list the best orientation for each point on the current contour, and
      iii. store the information about (place, score, orientation) in a list;
4. choose the placement that best maximizes local fitness as well as certain other weighted features.

Token-based Method

1. Iconify token;
2. create icon mask;
3. same as the steps in 3a and 3b above (i.e., without iteration);
4. generate a new icon by
   (a) *physically* rotating the current token (go to 1); or
   (b) *mentally* rotating current icon, (go to 2)
5. choose the placement that best maximizes local fitness as well as certain other weighted features.

If the token-based method resembles the human process of selecting placements, physical rotation is likely to be valued as a means of reducing both the time and effort of mental computation occurring in step 4. Pieces can be physically rotated in less than 100 ms whereas we estimate that mental rotation takes in the neighborhood of 800 to 1200 ms, based on pilot data, such as that displayed in figure 4. This may be misleading if we assume that because of priming effects, second and subsequent rotations are faster than first rotations.

If the type-based method resembles the human process, on the other hand, physical rotation is not especially helpful in enhancing the speed of computing local fitness. We assume that once a piece has been correctly identified, one may have access to its shape under all rotations, since it may be stored in this multiple perspective form. If physical rotation is useful in this type-based method, it will be because it abbreviates the time needed for Step 1: identify type of piece. For example, suppose it takes 1200 ms to identify a piece type from a presentation of a single token, whereas it takes 1000 ms to identify a type if shown one token for 600 ms immediately followed by another token for 400 ms. In such cases, it seems natural to conclude that rapid presentation of multiple perspectives of a piece stimulate retrieval of all perspectives faster than presentation of a single perspective.

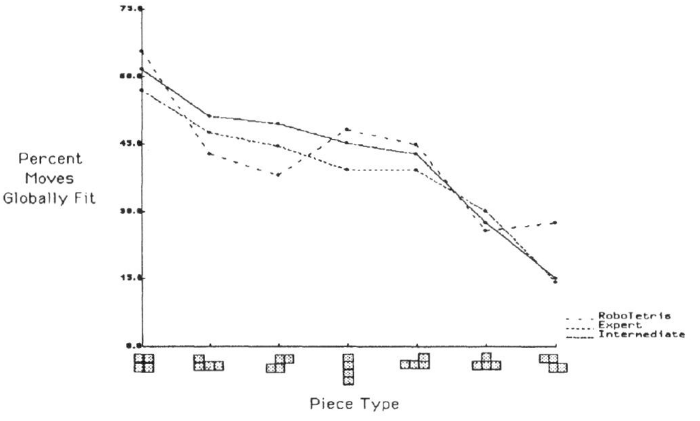

Figure 5: **Global Fitness**

In this figure we compare the tendency of human players and Robotetris to choose locations on the basis of global fitness. As the results show, human subjects vary in how strongly they weight global fitness. Intermediates (mean score 36 rows) place pieces in the globally fit place 41% of the time; experts (mean score 93 rows) 39%. In the version of Robotetris considered here (moderate performance with mean score of 876 rows) 41% of placements are globally fit, lumping Robotetris with intermediates in global fitness, though far above experts in performance.

Conclusion

We have argued that standard state transforming actions are, at times, best understood as serving an epistemic rather than a pragmatic purpose. The point of a particular action may seem to be that of bringing an agent physically closer to its goals, yet upon more careful analysis the real point of that action may be to increase the reliability of a judgement, or to reduce the space-time resources needed to compute it. Most thoughtful theorists of action now agree that a natural part of planning and acting is gathering information. Characteristically, however, this has been interpreted to mean that planners should have an active hand in controlling sensor actions. The thrust of our account of epistemic actions in the game of Tetris is that the scope of epistemic activity is much wider than sensor related activity. Verification and experimentation are the simplest of epistemic functions. There are countless others in every natural form of intelligent activity. It is axiomatic that adaptive creatures would strike on such strategies for augmenting their cognitive abilities.

References

Agre, P. & Chapman, D. (1987). Pengi: An implementation of a theory of activity. In *Proceedings of the Sixth National Conference on Artificial Intelligence*, pages 196–201.

Georgeff, M. & Lansky, A. (1987). Reactive reasoning and planning. In *Proceedings of the Sixth National Conference on Artificial Intelligence*, pages 677–682.

Hillyard, S. (1985). Electrophysiology of human selective attention. *Trends in Neuroscience*, *8*, 400–405.

Hutchins, E. (1990). The technology of team navigation. In J. Galegher, R. Kraut, & C. Egido (Eds.), *Intellectual teamwork: Social and technical bases of collaborative work*. Hillsdale, NJ: Lawrence Erlbaum.

Simmons, R., Ballard, D., Dean, T. & Firby, J. (Eds.). (1992, March). *Control of selective perception*. AAAI Spring Symposium Series. Stanford University.

Tarr, M. & Pinker, S. (1989). Mental rotation and orientation-dependence in shape recognition. *Cognitive Psychology*, *21*, 233–282.

Tate, A., Hendler, J., & Drummond, M. (1990). A review of AI planning techniques. In J. Allen, J. Hendler, & A. Tate (Eds.), *Readings in Planning*, pages 26–49. Morgan Kaufman.

Reference features as guides to reasoning about opportunities

Louise Pryor and Gregg Collins*
Northwestern University
The Institute for the Learning Sciences
1890 Maple Avenue, Evanston IL 60201
email: pryor@ils.nwu.edu, collins@ils.nwu.edu

Abstract

An intelligent agent acting in a complex and unpredictable world must be able to both plan ahead and react quickly to changes in its surroundings. In particular, such an agent must be able to react quickly when faced with unexpected opportunities to fulfill its goals. We consider the issue of how an agent should respond to perceived opportunities, and we describe a method for determining quickly whether it is rational to seize an opportunity or whether a more detailed analysis is required. Our system uses a set of heuristics based on *reference features* to identify situations and objects that characteristically involve problematic patterns of interaction. We discuss the recognition of reference features, and their use in focusing the system's reasoning onto potentially adverse interactions between its ongoing plans and the current opportunity.

1. Introduction

An intelligent agent acting in a complex and unpredictable world must be able to both plan ahead and react quickly to changes in its surroundings. In AI, agent models have generally exhibited one or the other, but not both, of these capabilities. In particular, two opposing schools of thought have arisen: classical planning, in which a sequence of actions that the agent intends to execute is produced ahead of time (*e.g.* Newell & Simon 1963, Fikes & Nilsson 1971, Sacerdoti 1977, Tate 1977, Wilkins 1988), and reactive planning, in which the agent simply responds to its surroundings at any given moment, instead of following an explicit plan[1] (*e.g.* Brooks 1986, Agre & Chapman 1987, Beer *et al* 1990, Kaelbling & Rosenschein 1990). It seems clear that a competent agent model must combine elements of both approaches (*c.f.* Georgeff & Lansky 1987, Firby 1989, Hammond *et al* 1990, Simmons 1990, McDermott 1991).

Classical planning has principally concerned itself with the construction of models that are complete and sound (*e.g.* Chapman 1987, McAllester & Rosenblitt 1991), but proofs of these formal properties depend on unrealistic assumptions. For example, it must be assumed that the agent has full knowledge of the conditions in which the plan will be executed, that all actions have perfectly predictable results, and that no unpredictable changes will occur through causes other than the agent's actions. Such an approach leads naturally to models in which issues of plan execution are ignored, since without unpredictability nothing can happen that has not been foreseen, and plan execution will simply consist of performing the preordained steps. The assumption of perfect foresight severely reduces the practicality of classical planners.

Reactive systems tend to the opposite extreme, performing no lookahead, and concomitantly constructing no plans. A reactive system is instead directed by a set of rules that specify how to react in any given situation, and its competence thus depends entirely upon the extent to which its rules are able to specify the precise action to take in the particular situation in which it finds itself. This approach leads to models in which projection is ignored, as a reactive system is incapable of making use of a predictive model of its world; it does not use projection to determine whether a contemplated action is in fact a good one to take.

A competent agent should fall somewhere between the extremes of classical and reactive planning, making use of projection where possible, yet being able to react with minimal forethought when necessary. In particular, such an agent must be able to react quickly in the face of unexpected opportunities to fulfill its goals[2], even in situations in which it lacks the time or the information necessary to construct a detailed plan before proceeding (*e.g.* Birnbaum 1986, Hammond *et al* 1988, Brand & Birnbaum 1990). The issue of engineering a compromise between classical and reactive planning thus comes down to the problem of responding to opportunities: when an opportunity arises, the

*This work was supported in part by the AFOSR under grant number AFOSR-91-0341-DEF, and by DARPA, monitored by the ONR under contract N00014-91-J-4092. The Institute for the Learning Sciences was established in 1989 with the support of Andersen Consulting, part of The Arthur Andersen Worldwide Organization. The Institute receives additional support from Ameritech, an Institute Partner, and from IBM.

[1]There are obvious similarities to behaviorism (Skinner 1974).

[2]Or to unexpected threats against its goals, but for our purposes these can be regarded as being the same thing.

agent should put only as much effort as is rationally justified into projecting the consequences of pursuing that opportunity. In this paper, we shall concentrate on the issue of how an agent should respond to perceived opportunities, and we shall introduce a method for determining quickly whether it is rational to seize an opportunity without first acquiring more information.

2. Effective independence

We are building a system, PARETO[3], that operates a simulated robot delivery truck in the TRUCKWORLD domain[4] of Firby and Hanks (1987). PARETO can, for example, recognize that a sack of cement mix sitting by the side of the road presents an opportunity to achieve the goal of satisfying a customer who has asked for cement, and can reason about whether this particular opportunity should be pursued. PARETO might choose not to take advantage of an opportunity if doing so would be detrimental to other goals it is currently pursuing. For example, it might not want to pick up a sack of cement that it has come across if it is low on fuel, or if it is late in making another delivery.

The decision to pursue an opportunity depends on an analysis of the costs and benefits of doing so. PARETO must therefore have a way of determining what the costs and benefits are, and how they compare. For opportunities, the benefits can be measured in terms of goal achievement and beneficial side effects on other goals, and costs in terms of forgoing other opportunities and harmful side effects on other goals. A theory of expected utility can be used to compare the results of taking the different courses of action that are available to the agent (Von Neumann & Morgenstern 1944, Feldman & Sproull 1977). There is a well-defined theory of how to arrive at the expected utilities, given certain information; in particular, the agent requires the prior and conditional probabilities from which it can calculate the probabilities of the various outcomes, and the utility values of the outcomes. Unfortunately, precise values are often not available, and indeed in most real-world situations the planner only has access to at most crude approximations of the necessary probabilities and utilities (Haddawy & Hanks 1990).

Even if the necessary information were available, the calculations of the expected utilities for all possible courses of action and all possible outcomes would in general be extremely complex and time-consuming (Hanks 1990). For decisions about whether to take advantage of an opportunity, aspects that might be relevant include anything that might bear on how the pursuit of an opportunity will interact with ongoing plans that are intended to achieve other goals. This covers a great deal of territory. For example, the decision of whether to pick up a sack of cement might depend on whether there is a gas station nearby, in the case where the truck is low on gas; it might depend on whether there is a bridge with a low weight limit on the route that the truck plans to take, if the load is a heavy one; it might depend on whether cement thieves have been reported in the vicinity recently; and so on. If a system considers all the information that could potentially be relevant to a decision, it will be unlikely to complete the reasoning in time for it to be of any use: there are simply too many ways in which plans can interact with each other. The calculation of expected utility is thus not by itself an adequate theory of how an intelligent agent should react to opportunities. Intelligent agents must have a quick and easy way to decide whether the detailed reasoning will be worthwhile.

What is needed is *focus*: in addition to determining whether the pursuit of an opportunity is likely to interact significantly with ongoing plans, the system must identify the areas in which such interactions are likely to occur. These decisions must often be made rapidly if an opportunity is to be seized in a timely fashion. Because of this it is impractical to attempt to make such a determination analytically; instead, the system must reason heuristically. A major simplification that would significantly reduce the complexity of the reasoning required would be to assume that the agent's various goals are *independent*, *i.e.* the pursuit of one goal does not in any way interact with the pursuit of any other goals. Unfortunately, this assumption would deny the possibility of recognizing those circumstances in which the likelihood of adverse interactions should suggest that the opportunity not be pursued.

PARETO therefore uses a weaker version of the independence assumption. It assumes that its various goals are *effectively independent* of each other, *i.e.* that there are no *significant* interactions between them, unless it can infer otherwise (Pryor & Collins 1991). So, in our example, in the absence of evidence to the contrary the system would assume that picking up the cement would have no adverse effects on any other deliveries the truck might be making. If it is valid to assume effective independence of the agent's goals, the decision about whether to pursue an opportunity becomes much simpler, since all insignificant interactions with other goals can simply be ignored. However, if this assumption is to be used we need to be able to recognize potential violations of effective independence quickly and easily. PARETO uses heuristics that indicate potential violations to focus its attention on those aspects of the decision that are likely to repay more detailed analysis.

PARETO pursues plans[5] in order to achieve its delivery goals, and while pursuing them may notice oppor-

[3]Planning and Acting in Realistic Environments by Thinking about Opportunities.

[4]TRUCKWORLD simulates a world in which items can react in a rich variety of ways and can change state with the passing of time. The actions performed by the truck can fail for a variety of reasons, including chance, and other random events can occur.

[5]PARETO is based on Firby's (1990) RAPs system, and thus uses a hierarchy of sketchy plans. At any time one of these plans is active, and others are dormant, awaiting execution.

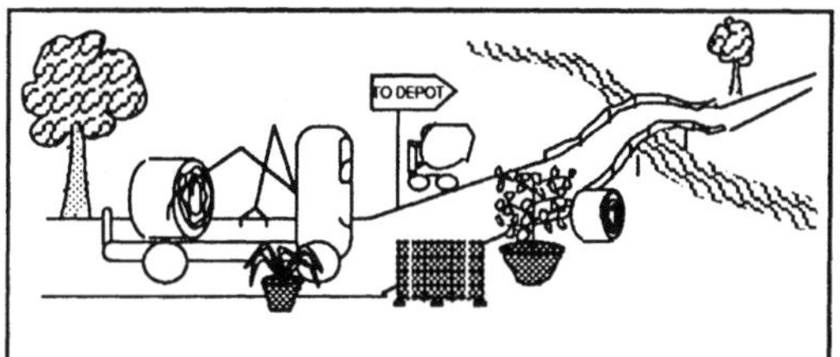

Figure 1: The truck has an opportunity to pick up some cinder blocks

tunities to achieve currently dormant goals. When an opportunity is recognized, a two-stage process is employed to determine whether or not that opportunity should be pursued. The first stage of the process involves the use of heuristics which flag potential violations of effective independence. If there are potential violations, PARETO moves on to the second stage and performs a more detailed analysis of the adverse interactions indicated by the heuristics.

2.1 An example

The types of decisions that PARETO faces when deciding whether to take advantage of an opportunity are illustrated in the following example. Suppose that the robot delivery truck is in the process of delivering some rolls of insulating material, the plan for which requires the truck to cross a bridge. Suppose further that there are some cinder blocks near the truck (figure 1), and that the truck has a currently inactive goal to deliver cinder blocks to a customer. The presence of the cinder blocks thus represents an opportunity to pursue an existing goal, and the agent must determine whether to pursue that opportunity.

In deciding whether or not to pursue the opportunity, the agent might in principle consider any number of possible interactions between its existing plans and the pursuit of the opportunity. The truck would have to stop by the side of the road to pick the cinder blocks up which might involve problems with passing traffic; the other objects by the side of the road might obstruct the truck during the loading process; cinder blocks are heavy, and the load might exceed the truck's weight capacity; they take up space in the truck, which might not be available; they are both hard and abrasive, and might damage other objects in the truck's load; other objects such as acid or iron beams might damage the cinder blocks; if they are not securely fastened to a pallet, they will be difficult to handle; loose cinder blocks might get damaged in transit; the time taken to pick them up might cause other delivery deadlines to be missed; picking them up might use more fuel than the truck has available; their weight might affect the fuel consumption and speed of the truck, and hence the truck's ability to make other deliveries; and so on. The number of possible interactions is enormous.

Unfortunately, any one of these interactions could actually constitute a serious threat. For example, let us suppose that in the situation depicted above the combined weight of the truck and the cinder blocks is greater than the bridge can bear. The problem confronting the agent is how to spot this problematic interaction without considering all the possible interactions in detail.

3. Reference features

An agent operating on the assumption that its various goals are effectively independent must be able to recognize when this assumption is inappropriate. This involves the ability to pick out the few genuinely problematic interactions from the potentially enormous number of harmless interactions in a given situation, and doing so quickly enough that the opportunity is not lost before the computation is complete. As the example above makes clear, this is by no means a trivial task. One approach is to tag elements of situations that are frequently involved in problematic interactions, and then to concentrate resources on detecting interactions involving the tagged elements.

This strategy can be seen as a simple application of common sense. For example, we might expect a nursery school teacher to take note of a child who is often involved in fights and disagreements, and to mark this child as a potential troublemaker to be watched closely in the future. In a similar way, our agent can tag potentially problematic elements in its planning environments, and use these labels to help it spot potential problems. By using different labels to designate different types of potential problems, the agent can in addition use these tags to focus subsequent analysis aimed at determining whether the problem will actually arise in the current situation. For instance, if objects made of a certain substance frequently break when they are involved in impacts with other objects, the agent can take note of that fact and mark such objects as fragile. When handling fragile objects, the agent should recognize that breakages are likely. Similarly, heavy objects often cause supporting structures to collapse, and bulky objects fill large volumes of space.

Objects are not the only elements of situations that can lead to unwanted interactions. For instance, when it is important that a goal be achieved within a short time period, there are often time conflicts with other tasks. By marking such goals as urgent, the agent can use that knowledge to avoid undertaking tasks that will interfere with their timely achievement.

We use the term *reference features* to denote tags such as disruptive, fragile, and urgent that help to direct an agent's attention to interesting functional aspects of the situation. In this paper we are primarily concerned with their use in indicating problematic interactions,

Object	Description	Reference Features
insmat-2	insulating-material	bulky
cblocks-6	cinder-block	heavy rough
bridge-23		rickety
customer-A		impatient
road-57		bumpy

Figure 2: Some reference features

but they also facilitate detecting a specified object, and can be used in planning to achieve goals.

PARETO uses heuristics based on reference features to indicate potentially problematic interactions involved in the pursuit of opportunities. There are several requirements that must be met in order for this strategy to be effective:

- The reference features must be easily recognizable.
- The agent must be able to determine quickly which elements of the situation may have reference features that indicate potentially adverse interactions.
- The agent must be able to use the reference features to indicate the *type* of the potentially problematic interaction.

There are many ways of meeting these requirements. In the current implementation of PARETO we are experimenting with a simple algorithm that looks for reference features indicating similar interaction types.

3.1 Availability

Reference features are useful only insofar as they provide cheap heuristics that indicate the desirability of more detailed reasoning. Reference features must therefore be easily inferable in most situations in which they are applicable, and must be inferable in few of the situations in which they are not. PARETO can link reference features to individual objects (it may know, for example, that a specific bridge is rickety), to descriptions that may apply to objects (the description cinder-block has the reference feature heavy attached to it), to sketchy plans and actions (which may be, for example, lengthy), and to goals (*e.g.* urgent). In TRUCKWORLD it is easy[6] to observe, for example, that an object is a stack of cinder blocks: the fact that the object has the reference feature heavy can then be inferred. Figure 2 shows some of the reference features in our example.

3.2 Situation elements

Since reference features are associated with elements of the situation in which PARETO finds itself, PARETO must be able to determine which elements of the situation are relevant. PARETO uses sketchy plans (Firby 1989) that comprise, among other things, a list of actions to be executed and the goal that the plan serves. Action descriptions consist of an action predicate applied to a set of objects (see figure 3). The set of situational elements associated with a given plan thus consists of all the objects that play a role in any primitive action, the actions themselves, and the goal that the plan serves.

Goal	(deliver ?item ?dest)	(travel-to ?dest)
Plan steps	(travel-to ?item-loc) (load ?item) (travel-to ?dest) (unload ?item)	(traverse ?road1) (traverse ?road2) (traverse ?road3)
Current plan Variables	(?item insmat-2) (?destination cust-A)	(?destination cust-A) (?road1 road-42) (?road2 bridge-23) (?road3 road-31)
Opportunity Variables	(?item cblocks-6) (?destination cust-B)	(?destination cust-B) (?road1 road-45) (?road2 road-57) (?road3 road-76)

Figure 3: Sketchy plans

When PARETO is considering whether to pursue an opportunity, it examines both the plan it is currently executing and the plan that would be used to pursue the opportunity, collecting the relevant situational elements from each. It then checks to see whether any of these elements is associated with a reference feature, and, if so, it flags that element.

In our example, PARETO's current plan is to deliver some insulating material to customer-A. An outline of the sketchy plan for this task is shown in figure 3. The sketchy plan (travel-to cust-A), for example, consists of the three steps: (traverse road-42), (traverse bridge-23) and (traverse road-31). Similarly, the plan for pursuing the opportunity , involving the goal (deliver cblocks-6 cust-B), is a different instantiation of the same sketchy plan. The reference features of bridge-23 (rickety) and cblocks-6 (heavy, rough) are therefore among those that are relevant to the decision of whether to pursue the opportunity to deliver the cinder blocks.

3.3 Focusing reasoning

In addition to flagging potential violations of effective independence, reference features play a role in focusing the agent's reasoning onto the particular aspects of the situation that should be considered in determining whether the violation will actually occur. In the example described above, for instance, there is a potentially problematic interaction involving the rickety bridge and the heavy cinder blocks. PARETO's analysis should concentrate on the question of whether the bridge is likely to collapse under the weight of the blocks.

The knowledge that PARETO needs in order to guide the analysis process is associated with reference fea-

[6]Perception in TRUCKWORLD grounds out at the level of object descriptions, and thus ignores the many important problems of object recognition.

REFERENCE FEATURES OF TASKS	INTERACTION TYPE	ROLE OF TASK
lengthy	time fuel	consumer consumer
urgent	time	requirer

REFERENCE FEATURES OF TERRAINS	INTERACTION TYPE	ROLE OF TERRAIN
bumpy	impact	cause
narrow	volume-capacity	limiter
rickety	load-bearing	base

REFERENCE FEATURES OF OBJECTS	INTERACTION TYPE	ROLE OF OBJECT
bulky	volume-capacity	consumer
explosive	fire	igniter
flammable	fire	burner
fragile	impact time	hit consumer
hard	impact	hitter
heavy	fuel load-bearing	consumer load
rough	surface-damage	abrader

Figure 4: Reference features and interactions

tures themselves. Each reference feature predicts a particular type of problematic interaction (see figure 4), which is represented by an *interaction description* consisting of three items: the configuration that is required in order for the interaction associated with the feature to occur, the potentially problematic outcome of the interaction, and a list of variables designating the elements that play a role in the interaction (see figure 5). When a situational element is flagged with a reference feature, PARETO must determine whether the interaction description associated with that feature applies to that element in the current situation.

In order to determine whether an interaction description applies, PARETO must first determine which other elements of the situation could be involved in such an interaction with the flagged element. For instance, the pile of cinder blocks in the example creates the potential for an interaction in which an object supporting the blocks collapses, but this does not tell us which, if any, specific objects are in danger of collapsing. One approach might be to examine every element of the situation to determine if any are likely to be involved in this interaction; in effect, this would be like asking, for every element, whether it is ever likely to support the cinder blocks, and, if so, whether it can bear their weight. While this is possible, it involves a somewhat unfocused search.

A more focused solution is based on the observation that different reference features may be used to designate objects that play different roles in the same type of interaction. For example, heavy means that an object is likely to make the object that supports it collapse, while rickety means that an object is likely to collapse under an object it is supporting. The features heavy and rickety thus form a natural pair. This knowledge can be used to focus the analysis by considering interactions only when at least one object has been flagged for each role in the interaction. In our example, PARETO considers the possibility that something will collapse only when it recognizes that both a heavy object—the blocks—and a rickety object—the bridge—have been flagged.

Interaction type:	load-bearing	surface-damage
Roles:	?base ?load	?abrader ?hurt
Configuration:	(bears ?base ?load)	(rubs ?abrader ?hurt)
Outcome:	(collapse ?base)	(scratched ?hurt)

Figure 5: Interactions

Once PARETO knows the type of problematic interaction it is looking for, and the elements that are involved in that interaction, it must determine whether the interaction will actually occur. This involves two steps: determining whether the *configuration* described in the interaction description will arise, and determining whether the problematic *outcome* will result if it does. PARETO must therefore be able to perform inference over the causal theory of the planning environment, which must include a theory of action projection in addition to the causal relationships among the objects in the world. We are assuming that this inference is performed by a general purpose query system since we have not yet addressed the problem of a a more specialized system for plan projection.

In our example, then, PARETO should query whether, if the opportunity is pursued, the cinder blocks will at some stage be supported by the bridge. In this case it will discover that this configuration condition will indeed occur. PARETO must therefore decide whether the pursuit of the opportunity would result in the problematic outcome of the interaction, or if that outcome can be avoided.

In order to answer this query, PARETO applies inference rules describing the causality of the domain, for example that the bridge will collapse if the total load on the bridge is greater than its weight limit. In order to use this particular rule, PARETO will need to know the weight limit on the bridge, and the weight of the total load on the bridge at the time when the configuration conditions are met. Further inference rules will enable it to recognize that the total load on the bridge will consist of the truck, its current contents, and the cinder blocks, whose weights must be added together. These individual weights are therefore pieces of information that will be useful to PARETO in making the decision. However, there may be costs involved in acquiring this information (for further discussion of this point see Pryor & Collins 1992). PARETO must therefore consider the value and acquisition costs of each piece of information. For example, if the weight limit on the bridge is very high compared to the weight of any load the truck would carry, information about the weight of the truck's load would not change the decision as to whether to pick up the cinder blocks, and so is not worth acquiring.

4. Discussion

We have described the design of a system, PARETO, that reasons efficiently about whether to pursue an opportunity. The keys to PARETO's approach are, first, that it assumes that its various goals are effectively independent of each other, and, second, that it uses reference features to flag situations in which this assumption of effective independence is likely to be violated. By using reference features to indicate potentially adverse interactions between an opportunity and its current plans, PARETO is able to ignore the many insignificant interactions that may be present.

It is important to note that reference features are not infallible: clearly there may be a problematic interaction that is not indicated by any reference features. As a result, PARETO will occasionally produce incorrect plans. We believe that such an outcome is unavoidable for any planner that is intended to operate in a complex, and unpredictable environment. the upshot of this is that such a system must be able to recover from errors and unforeseen failures, and to learn from its mistakes. One way in which such a system might be expected to learn from mistakes is by positing new reference features indicating problematic interactions that the system has observed. Reference features thus form a natural basis for a theory of learning to plan. We will pursue this issue further in future work.

Acknowledgments: Thanks to Larry Birnbaum, Matt Brand, Will Fitzgerald, Mike Freed, and Bruce Krulwich for many useful discussions.

References

Agre, P. E., & Chapman, D. (1987). Pengi: An implementation of a theory of activity. Proceedings of AAAI 87.

Allen, J., Hendler, J., & Tate, A. (Eds) (1990). Readings in Planning. Morgan Kaufman.

Beer, R. D., Chiel, H. J., & Sterling, L. S. (1990). A Biological Perspective on Autonomous Agent Design. Robotics and Autonomous Systems 6, 169-186.

Birnbaum, L. (1986). Integrated processing in planning and understanding. Technical Report YALEU/CSD/RR #489, Dept. of Computer Science, Yale University, New Haven CT.

Brand, M., & Birnbaum, L. (1990). Noticing Opportunities in a Rich Environment. Proceedings of the Twelfth Annual Conference of the Cognitive Science Society, Cambridge MA.

Brooks, R. (1986). A robust layered control system for a mobile robot. IEEE Journal of Robotics and Automation, vol. 2, no 1

Chapman, D. (1987). Planning for Conjunctive Goals. Artificial Intelligence, 32, 333-337. Also in (Allen *et al* 1990).

Feldman, J. A., & Sproull, R. F. (1977). Decision Theory and Artificial Intelligence II: The Hungry Monkey. Cognitive Science 1, 158-192, and in (Allen *et al.* 1990).

Fikes, R. E., & Nilsson, N. J. (1971). STRIPS: A new approach to the application of theorem proving to problem solving. Artificial Intelligence 2, 189-208, and in (Allen *et al.* 1990).

Firby, R. J. (1989). Adaptive execution in complex dynamic worlds. Technical Report YALEU/CSD/RR #672, Dept. of Computer Science, Yale University, New Haven CT.

Firby, R. J., & Hanks, S. (1987). The simulator manual. Technical Report YALEU/CSD/RR #563, Dept. of Computer Science, Yale University, New Haven CT.

Georgeff, M., & Lansky, A. (1987). Reactive Reasoning and Planning. Proceedings of AAAI 1987, and in (Allen *et al* 1990).

Haddawy, P., & Hanks, S. (1990). Issues in Decision-Theoretic Planning: Symbolic Goals and Numeric Utilities. Innovative Approaches to Planning, Scheduling and Control: Proceedings of a Workshop. DARPA, San Diego.

Hammond, K. J., Converse, T., & Marks, M. (1988). Learning from opportunities: Storing and reusing execution-time optimizations. Proceedings of AAAI 1988.

Hammond, K., Converse, T., & Martin, C. (1990). Integrating Planning and Acting in a Case-Based Framework. Proceedings of the Eighth National Conference on Artificial Intelligence, Boston,

Hanks, S. J. (1990). Projecting plans for uncertain worlds. YALEU/CSD/RR #756, Yale University, New Haven CT.

Kaelbling, L. P., & Rosenschein, S. J. (1990). Action and Planning in Embedded Agents. Robotics and Autonomous Systems 6 35-48.

McAllester, D., & Rosenblitt, D. (1991). Systematic Nonlinear Planning. Proceedings of the Ninth National Conference on Artificial Intelligence, Anaheim, CA. pp. 634-639.

McDermott, D. (1991). Robot planning. Technical Report YALEU/CSD/RR#861, Dept. of Computer Science, Yale University, New Haven CT.

Newell, A., & Simon, H. A. (1963). GPS, a program that simulates human thought. In E.A. Feigenbaum & J. Feldman (Eds), Computers and Thought, R. Oldenbourg K.G., and in (Allen *et al* 1990).

Pryor, L., & Collins, G. (1991). Information gathering as a planning task. In Proceedings of the Thirteenth Annual Conference of the Cognitive Science Society. Chicago, IL.

Pryor, L., & Collins, G. (1992). Planning to perceive: A utilitarian approach. In Working Notes of the AAAI Spring symposium on Selective Perception.

Sacerdoti, E. (1977). A Structure for Plans and Behavior. American Elsevier, New York.

Simmons, R. (1990). An architecture for coordinating planning, sensing, and action. DARPA Planning Workshop, San Diego, CA.

Skinner, B. F. (1974). About Behaviorism. Random House, New York.

Tate, A. (1977). Generating project networks. IJCAI 1977, and in (Allen *et al* 1990).

Von Neumann, J., & Morgenstern, O. (1944). The Theory of Games and Economic Behavior. Princeton University Press.

Wilkins, D. E. (1988). Practical Planning: Extending the Classical AI Planning Paradigm. Morgan Kaufman, CA.

The Evolutionary Induction of Subroutines

Peter J. Angeline and Jordan B. Pollack
Laboratory for Artificial Intelligence Research
Computer and Information Science Department
The Ohio State University
Columbus, Ohio 43210
pja@cis.ohio-state.edu
pollack@cis.ohio-state.edu

Abstract

In this paper[1] we describe a genetic algorithm capable of evolving large programs by exploiting two new genetic operators which construct and deconstruct parameterized subroutines. These subroutines protect useful partial solutions and help to solve the scaling problem for a class of genetic problem solving methods. We demonstrate our algorithm acquires useful subroutines by evolving a modular program from "scratch" to play and win at Tic-Tac-Toe against a flawed "expert". This work also amplifies our previous note (Pollack, 1991) that a phase transition is the principle behind induction in dynamical cognitive models.

Introduction

While complex processes of cognition require some form of modularity, learning this modularity has been problematic. It is ignored by simple learning systems (which cannot learn complex processes) or built into the architectural "bias" of more complex learning systems (begging the origin of such complexity). Thus, the issue of inducing modularity from a complex task in order to perform that task has not been addressed, although a few connectionists are beginning this research (Saunders et al., 1992; Angeline & Pollack, 1991; Jacobs & Jordan, 1991; Jacobs et al., 1991; Nowlan & Hinton, 1991).

In this paper we describe a genetic algorithm capable of evolving large programs by exploiting two new genetic operators which construct and deconstruct parameterized subroutines. These subroutines protect useful partial solutions and help to solve the scaling problem for a class of genetic problem solving methods. After a brief introduction to genetic algorithms, we show that our system is able to learn how to play and win at Tic-Tac-Toe from "scratch" against an imperfect "expert" player. We discuss the formation and tuning of the subroutines and the reasons why their acquisition addresses the scaling problem within this framework. Analysis of the frequency of subroutine calls shows an exponential growth or decay of subroutine usage as they are induced or expelled from the language, leading us to name this phenomenon *evolutionary induction*, an amplification of our earlier principle of "induction by phase transition" (Pollack, 1991).

1. This research is supported by the Office of Naval Research under contract #N00014-89-J1200.

Genetic Algorithms Background

The genetic algorithm (Holland, 1975; Goldberg 1989a) is a form of problem solving search analogous to natural selection, and is a surprisingly adept search method in even very large ill-formed problem spaces. A simple genetic algorithm typically operates by reproducing and altering a population of fixed-length binary strings. A *fitness function* interprets the strings as task solutions and scores their ability to solve the task. Novel strings are added to the population by a process akin to biological reproduction using a collection of genetic *operators*. One such operator, the *crossover* operator, takes two "parent" strings selected for their fitness and returns a "child" string which is a complementary collection of components from both parents. The *point mutation* operator alters the value of a single position of a single parent string to create offspring.

The schema theorem (Holland, 1975), often called the Fundamental Theorem of Genetic Algorithms, illustrates the power behind these search methods. Holland defines a *schema* to be a class of binary strings which share a collection of subsequences. We use "#" to indicate *don't care* positions in the schema, i.e. positions which can be either a 1 or 0. For instance, the string '100101' is a member of the schema '10##01' as is '100001'. The intuition behind schemata is that certain combinations of bits will have a larger contribution to the fitness for a particular string than others. The schema notation allows us to talk about such desirable organizations concisely. A schema's *defining length* is the number of positions at which if we divided the schema into two parts, some of the defined positions (i.e. ones not '#') would be separated. For instance, dividing the schema '#1.0.0##' at any position marked with a '.' will separate some of the defined components giving it a defining length of 2. Similarly, the defining length of '1.0.#.#.0.1' is 5. The schema theorem proves that above average schemata with small defining lengths will be copied an exponential number of times in the generations subsequent to their appearance (Holland, 1975). For a more detailed introduction to genetic algorithms, the schema theorem and its implications see (Goldberg, 1989a).

While simple genetic algorithms can evolve solutions to a wide range of tasks two problems prevent their scaling to more interesting tasks. The first limitation arrises due to the fixed-length nature of the string representation. Because of the closed nature of the representation, the maximum complexity needed to solve the task must be anticipated before the search takes place. The second problem is that each bit of the string representation generally stands for the presence or absence of a specific feature of the interpretation. This positional encoding of the binary representation requires that every possible interpretation also be anticipated prior to the search. This amounts to nothing more than using the string as a pointer to a table of predefined interpretations.

In order to increase the amount of available complexity in simple genetic algorithms, some researchers have devised elaborate interpretation routines (e.g. Belew, McInerney & Schraudolph, 1992 and Dawkins, 1987). Essentially, this approach removes the complexity from the jurisdiction of the representation and places it into the interpretation. Unfortunately, rather than address the representation of complexity problem in simple genetic algorithms this approach merely shifts the problem to a new component. By placing an undue amount of design into the interpretation of the representation these researchers beg the question of evolving complexity since they have provided the complexity a priori.

Recently, Koza has described an exciting advance in genetic algorithms. In his Genetic Programming Paradigm (GPP), Koza uses a hierarchy of primitive functions rather than a fixed-length string to represent potential solutions (Koza, 1992, Koza, 1990). These hierarchies are interpreted as programs written in a language defined by the primitive functions which when executed compute the solution to the task. Koza's genetic operators exchange subtrees of the hierarchies rather than substrings.

Although Koza's dynamic representation alleviates both the fixed-length and positional encoding limitations of simple genetic algorithms, it also suffers from a malady which prevents scaling. Consider that a dynamic representation will eventually grow large enough to encompass the complexity necessary to solve the desired problem. At some point in the learning of a very complex task, the structure will be quite large and the chance of breaking up desirable portions of the program with the crossover operator will overwhelm the chance of improving the program. In other words, as the defining length of a desirable schema increases it becomes more likely that we will consistently break it apart rather than improve upon it. We call this the *defining length problem*.[2] As an empirical indication of this problem, we note that the largest evolved program Koza reports is only 48 nodes.

These scaling difficulties call to mind Simon's parable of the two watchmakers Tempus and Hora (Simon, 1969). In this parable, the two watchmakers build products of similar complexity (1000 parts) using differing design philosophies. Tempus constructs the entire watch directly from the primitive components, much like GPP constructs programs. Consequently, if he is interrupted before completing a watch, say by a customer calling on the phone, the intermediate state is lost and he must rebuild the entire watch from the individual components. Hora's method of construction, on the other hand, uses stable intermediate modules which are individually created, assembled into larger and larger modules and eventually into the completed product. When Hora is interrupted, only the work for the module currently being constructed is lost. The lesson from Simon's parable is clear: in the development of complex systems it is prudent to build incrementally and modularly.

The Genetic Library Builder (GLiB)

The Genetic Library Builder (GLiB) is a genetic algorithm environment based on the ideas forged by Koza in GPP but with provisions for the evolution and evaluation of program subroutines. As in GPP, GLiB uses an expression tree of primitive functions as its representation for potential solution programs. The essential difference between GPP and GLiB is the addition of two new genetic operators. The first operator, called *compression*, creates subroutines from subtrees of individuals in the current population and introduces the subroutines into the "genetic library". This library is simply the collection of subroutines which appear in the programs of population and thus are available for constructing task solutions. Once in the library, the usefulness of a newly constructed subroutine is evaluated by the extent it is used in future generations. The second operator, called *expansion*, replaces compressed subroutines with their original definition. In the following sections we describe these operators and their implementations in detail.

Creation of Subroutines in GLiB

The compression operator in GLiB, the sole method of subroutine definition in the system, works as follows. During the construction of each new generation of programs, the compression operator is applied to a percentage of the population selected by relative fitness. The compression operator is asexual, like the point mutation

2. Because they exploit positional encodings, simple genetic algorithms do not suffer from this problem.

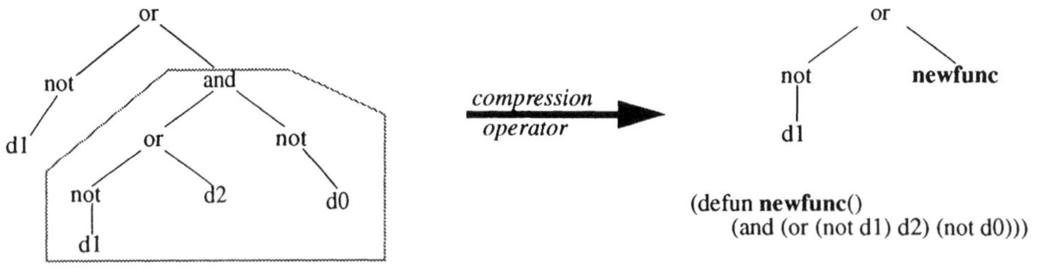

***Figure 1**: Creation of a new subroutine from a randomly selected subtree of an individual in the current population.*

operator described above, so only a single "parent" program is selected and copied. The copy serves as the "child" of this parent in the coming generation. A node in the interior of the child's expression tree is then randomly selected and designated the root of the subtree which will become the newly compressed subroutine. Next, a maximum depth for the subtree is randomly selected from a user defined range. When none of the branches of the subtree exceeds this maximum depth we have the instance of subroutine creation depicted in Figure 1. Here, the entire subtree is removed from the offspring and used as the body of a new LISP function definition with no parameters. Once the new subroutine is defined, the expression tree of the offspring is altered replacing the extracted subtree with the equivalent LISP function call. This compression of the subtree into the name of the equivalent subroutine call introduces the new subroutine into the genetic library.

Occasionally, some branches of the selected subtree will have a depth greater than the allowed maximum depth for the subroutine being created. In this event, we replace each branch of the subtree at the point where it exceeds the maximum depth with a unique variable. When the LISP function is defined, the variables introduced into the subtree are used as parameters to the new subroutines. When we then compress the expression tree of the child, the portions of the subtree which exceeded the maximum depth are not removed but serve as the values for the parameters in the subroutine call. This instance of modularization in GLiB is depicted in Figure. 2. *Note that invariably when a compression takes place the semantics of the program are not altered, only the manner in which the program is expressed.*

Unfortunately, while the compression operator supplies a method to create subroutines from the population during GLiB's genetic search, it also serves to remove unique subtrees from the population, lowering the diversity of the population. For a genetic search to work, there must be sufficient genetic material in the population so that combinations of promising candidates from the current generation can be recombined into novel organizations. By lowering the diversity of the population and consequently the number of novel combinations, we limit the distance from the current state that a genetic search can look.

In order to balance the undesirable effects of the compression operator, we have also added an expansion operator which restores the genetic material from the compressed subtrees. This operator searches the offspring's expression tree for a call to an evolved subroutine. If one is found, it is expanded from its atomic reference back into the full subtree and thus replaces the genetic material previously removed.

The complementary nature of the compression and expansion operators implements a form of iterative refinement. The random selection of a subtree for compression provides no guarantee that the selected subtree will be an above average schema. It is more likely that it will be either a portion of a useful schema or simply of no import at all. By periodically replacing a copy of the

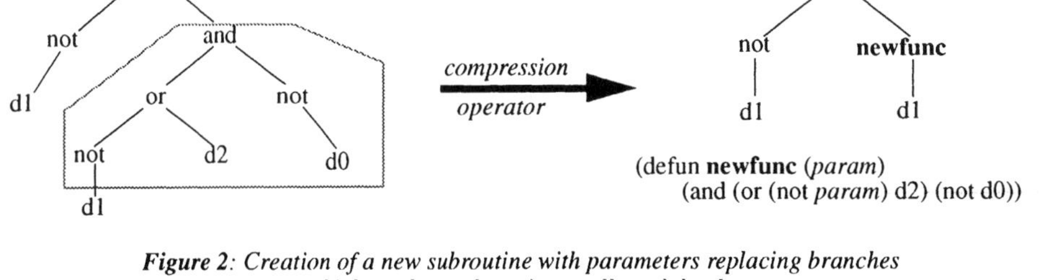

***Figure 2**: Creation of a new subroutine with parameters replacing branches which are beyond maximum allowed depth.*

pos00	*pos01*	*pos02*
pos10	*pos11*	*pos12*
pos20	*pos21*	*pos22*

pos00 .. pos22 - board positions

and - binary LISP "and"

or - binary LISP "or"

if - if <test> then <arg1> else <arg2>

open - returns <arg> if unplayed else NIL

mine - returns <arg> if player's else NIL

yours - returns <arg> if opponent's else NIL

play-at - places player's mark at <arg>

***Figure 3**: Primitives used in evolving modular programs to play Tic Tac Toe.*

compressed subtree back into the population, we provide the chance to capture a better version of the schema at a later time.

Evaluation of Subroutine Performance

Now that we have a method of extracting potentially useful subroutines from the evolved programs, we need a method for evaluating their contribution. We suggest that an appropriate measure of success for a particular evolved subroutine should be the number of times it is put into use by the population in the course of solving the task. If many members of the population are using the subroutine at some point in the genetic search, then it is likely that the subroutine provided some consistent advantage in earlier generations. When this occurs, we say that the subroutine is *evolutionarily viable*.

Our task now is to insure that good subroutines will be copied generously into subsequent generations while inappropriate ones will be suppressed. The "enlightening" guidelines provided for genetic algorithm design in (Goldberg, 1989b) suggest one should never be too clever when dealing with genetic algorithms as a "frontal assault" to the solution of a design problem usually defeats the inherent non-linear interactions. Thus, one should practice prudence when possible.

Appropriately enough, the genetic search which evolves programs to solve the task, automatically evaluates the worth of the subroutines without any additional intervention. The logic of this is straightforward. Initially, when a new subroutine is created there is only one member of the population which has a reference to it. If this program is comparatively fit, then, by the schema theorem, the call to the subroutine will be copied into several offspring in the next generation. If those individuals are also relatively fit then each of them will have multiple offspring which contain the subroutine call as well. Eventually, the subroutine will spread throughout a significant portion of the population. On the other hand, if the program is comparatively unfit, possibly due to one of its subroutines being more of a hinderance than a help, it will have little or no chance to create offspring. This results in a decrease in the number of calls to the subroutine from generation to generation until virtually no member of the population relies upon it. In other words, if a subroutine presents no advantage to the individuals which use it, it will in time go the way of the human appendix. Once the subroutine is no longer used by the population, it is no longer in the genetic library. Thus the genetic search process at the level of the overall task implicitly determines the fitness of evolved subroutines and allows only those that are useful to be propagated.

Learning to Play Tic-Tac-Toe

In order to illustrate our form of subroutine acquisition at work, we used GLiB to evolve programs to play Tic-Tac-Toe (TTT). The primitive language used for this experiment is shown in Figure 3. The first collection of primitives, ***pos00*** to ***pos22***, are the data points to be used in the program which represent the nine positions on the TTT board. This set of data points serves as the leaves of the expression tree. For the remaining primitives, the return value is either one of the positions or NIL, which represents FALSE in LISP, the current language in which GLiB is implemented. For instance, the binary ***and*** operator takes two arguments and when both are non-NIL returns the second. If either argument is NIL then NIL is returned. The ***play-at*** primitive takes a single argument. If the argument is a position and no player has placed a mark there, then the current player's mark is placed at that position and their turn is halted. Otherwise, ***play-at*** returns whatever it is passed. Finally, the operators ***mine***, ***yours*** and ***open*** take a position and return that position when the mark on the playing board in that position fits the test. Otherwise, they return NIL. We have purposely made these functions as general as possible to cover any number of games rather than just TTT. *Note there is no guarantee that a random program in this language will observe the rules of TTT or even place a single mark on a TTT board.* If the program does not make a valid move, then its turn is forfeited. We consider legal moves to be apart of the complexity of the task and consequently should be induced by GLiB.

An "expert" TTT algorithm constructed in LISP served as the opponent for all of the evolved programs. This expert was designed in such a way that it could not lose a game unless the opponent it was playing against had forked it, i.e. created a situation where the addition of a single mark by the opponent resulted in more than one possible winning play on its next turn. In addition, the expert was slightly adaptive such that when no clear best move was available it would select a position known to be frequented by the program it was currently

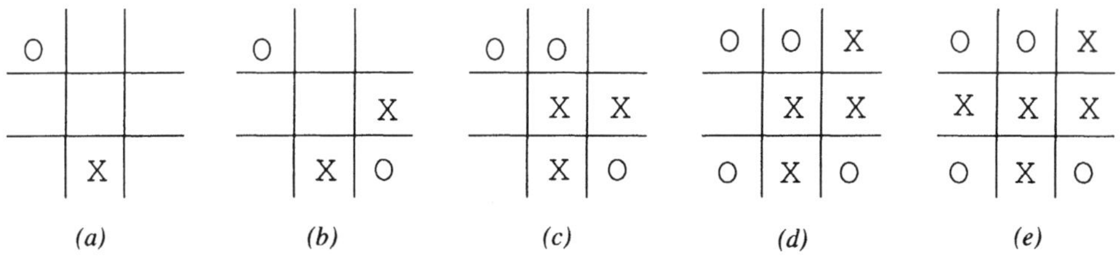

Figure 4: *A sample game from the described run. The evolved program is "X" and is playing first against the expert. Opening moves in (a) lead to first fork setup in (b). Evolved program completes first fork in (c), sets up second fork in (d) and wins in (e). The program received a score of 20 points for this win.*

playing against. The intention of this feature was to increase the apparent complexity of the expert's actions forcing the evolved programs to be more robust. The generality of the primitives combined with the level of play of the expert make this quite a formidable environment for learning TTT.

In order to rate the performance of an evolved program against the expert, a scoring function assigned points for various moves. First, since the semantics of the primitive language do not guarantee a program will actually make a legal play, a point was awarded for every legal move made. An additional point was awarded if a move blocked the expert from winning on its next turn. If the game ended in a draw or a win, the accumulated score of the evolved program was increased by 4 or 12 points respectively. It is important to note that the score for the program was a lump sum and provided no indication of which actions were being rewarded. The same results would be achieved if only the final state of the board were scored rather than the individual moves.

We ran GLiB with a population size of 1000 using the described expert and scoring method as the fitness function. In this run we applied the compression operator to 10 percent of the population each generation. All other parameters were as set in (Koza, 1990) for the "ant" experiment. The best evolved program after 200 generations had an average score of 16.5 points for the 4 games it played against the expert to determine its fitness. This score suggests that while the program was able to beat the flawed expert more than once, the best it could do after the expert had adapted to its playing strategy was to get a draw.

The evolved program had 60 nodes, a maximum depth of 13 nodes, and used 15 evolved subroutines at its top-level. As expected, expanding the definition of these subroutines back into their original subtrees revealed additional subroutine calls in their bodies. In all a total of 43 distinct subroutines were used by this evolved program in 89 subroutine calls making the virtual size of the program 477 nodes with a virtual maximum depth of 39. Two of the subroutines had a total of 9 separate calls each in the fully expanded tree. Note that this evolved program is almost 10 times the size of the largest program reported by Koza.

Figure 4 shows the first game played between the evolved program and the expert. There is an interesting point to be made about the apparent strategy of the evolved program. Notice that it was able to establish a fork by its third move (Figure 4c) but did not win the game until 2 turns later (Figure 4e). While this seems an odd strategy, recall that the evolved program receives points for each move it makes and additional points if it blocks the expert. Its strategy, then, is to maximize its total point score by forking the expert not once but *twice* in the same game! If the program had won the game on its fourth move it would have received 3 less points. By extending the game it actually increases its score without the possibility of losing.

The analysis of the created subroutines is equally interesting. Overall in the run there were 16,852 subroutines created by the compression operator with only 257 in use during the final generation. Figure 5 shows the number of calls per generation for three of the evolved subroutines. Each of these has a distinct period during the run where its number of calls per generation rises extremely quickly. In Figure 5a the sharp increase happens relatively soon after the subroutine is defined, showing it posed an immediate advantage. Figure 5b shows an example of a subroutine which was extremely useful shortly after its creation but whose use fell off dramatically. Finally, Figure 5c shows a subroutine which was present in the population for almost 100 generations before being recognized as being useful.

Discussion

The dramatic shapes of the calls per generation curves for the subroutines shown in Figure 5 are interesting for two reasons. First, it is apparent that we have been able to capture useful schemata in our subroutines by the exponential-like rises in the subroutine call counts. Second this work amplifies our previous note (Pollack, 1991) that a phase transition is the principle behind "induction" in dynamical cognitive models. We call this method of random selection and evolutionary evaluation of subroutines *evolutionary induction*.

But there is more to the story than a simple attachment to Holland's powerful theorem. We also believe

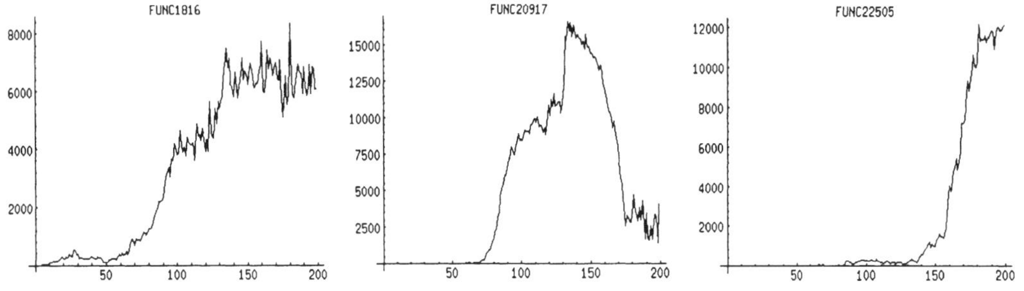

(a) (b) (c)

Figure 5: *Graphs showing number of calls (y-axis) per generation (x-axis) for three evolved subroutines. See text for explanation. Note graphs are not equally scaled.*

that evolutionary induction impacts the defining length problem for dynamic genetic algorithms, although we have as yet no formal verification of this claim. Our reasons are as follows: By compressing random subtrees from the program into representational atoms, we literally reduce the defining length of that subtree to zero. Because we know by the schema theorem that above average schemata will be copied more readily than below average schemata, our chances of compressing a useful schema increases each generation. Once we compress a useful schema and reduce its defining length to zero, it can be used to create more complex structures which are still small enough to be propagated intact to future generations. The end result is a complex modular program with nested subroutine calls and an overall structure similar to Hora's watches.

Acknowledgments

Thanks to Greg Saunders, John Kolen and Dave Stucki for reading and commenting on various drafts of this paper.

References

Angeline, P., and Pollack, J., 1991, "Hierarchical RAAMs: A Uniform Modular Architecture", Technical Report No. 91-PA-HRAAMS, Laboratory for Artificial Intelligence Research, The Ohio State University.

Belew, R., McInerney, J. and Schraudolph, N., 1991, "Evolving Networks: Using the Genetic Algorithm with Connectionist Learning", In *Artificial Life II*, C. Langton, C. Taylor, J. D. Farmer and S. Rasmussen (eds.), Reading, MA: Addison-Wesley Publishing Company, Inc.

Dawkins R., 1987, *The Blind Watchmaker*, New York, W. W. Norton and Co.

Goldberg, D., 1989a, *Genetic Algorithms in Search, Optimization, and Machine Learning*, Reading, MA: Addison-Wesley Publishing Company, Inc.

Goldberg, D., 1989b, "Zen and the Art of Genetic Algorithms", In *Proceedings of the Third International Conference on Genetic Algorithms*, J. Schaffer (ed), Los Altos, CA: Morgan Kaufmann Publishers, Inc.

Holland, J., 1975, *Adaptation in Natural and Artificial Systems*, Ann Arbor, MI: The University of Michigan Press.

Jacobs, R., Jordan, M., Nowlan, S., & Hinton, G., 1991, "Adaptive Mixtures of Local Experts", *Neural Computation*, 3, 79 - 87.

Jacobs, S., & Jordan, M., 1991, "A Competitive Modular Connectionist Architecture", In *Advances in Neural Information Processing 3*, R. Lippmann, J. Moody, D. Touretzky (eds.), San Mateo CA: Morgan Kaufmann Publications, Inc.

Koza, J., 1992, "Genetic Evolution and Co-Evolution of Computer Programs", In *Artificial Life II*, C. Langton, C. Taylor, J. Farmer and S. Rasmussen (eds.), Reading, MA: Addison-Wesley Publishing Company, Inc.

Koza, J., 1990, "Genetic Programming: A Paradigm for Genetically Breeding Populations of Computer Programs to Solve Problems", Technical Report No. STAN-CS-90-1314, Computer Science Department, Stanford University.

Nowlan, S., & Hinton, G., 1991, "Evaluation of Adaptive Mixtures of Competing Experts", In *Advances in Neural Information Processing 3*, R. Lippmann, J. Moody, D. Touretzky (eds.), San Mateo CA: Morgan Kaufmann Publications, Inc.

Pollack, J., 1991, "The Induction of Dynamical Recognizers", *Machine Learning* (7), 227 - 252.

Saunders, G., Kolen, J., Angeline, P. & Pollack, J., 1992, "Additive Modular Learning in Preemptrons", In *Proceedings of the Fourteenth Annual Conference of the Cognitive Science Society*, Bloomington, Indiana, Hillsdale, NJ: Lawrence Erlbaum Associates, Inc.

Simon, H. A., 1969, *The Sciences of the Artificial*, Cambridge, MA: MIT Press.

Learning Several Lessons from One Experience

Bruce Krulwich, Lawrence Birnbaum, and **Gregg Collins**
Northwestern University, The Institute for the Learning Sciences
1890 Maple Avenue, Evanston, IL 60201
Telephone: (708) 491-3500
Electronic mail: `krulwich@ils.nwu.edu`

Abstract

The architecture of an intelligent agent must include components that carry out a wide variety of cognitive tasks, including perception, goal activation, plan generation, plan selection, and execution. In order to make use of opportunities to learn, such a system must be capable of determining which system components should be modified as a result of a new experience, and how lessons that are appropriate for each component's task can be derived from the experience. We describe an approach that uses a self-model as a source of information about each system component. The model is used to determine whether a component should be augmented in response to a new example, and a portion of the model, *component performance specifications,* are used to determine what aspects of an example are relevant to each component and to express the details of the lessons learned in vocabulary that is appropriate to the component. We show how this approach is implemented in the CASTLE system, which learns strategic concepts in the domain of chess.

Cognitive tasks and components

In the course of pursuing its goals, an intelligent agent must notice opportunities, devise plans of action, and select among such plans. In domains that involve interactions with other agents, including such games as chess (in which our system operates), an agent must additionally notice threats posed by the other agents and develop plans to respond to them. It is useful for a variety of reasons to model such an agent as a collection of components, each of which is responsible for one of these planning tasks. In particular, such an approach to modeling the agent is useful in learning [Collins *et al.*, 1991c; Krulwich, 1991]. In this view of a problem-solving agent, learning involves three steps. The first is recognizing situations in which there is a lesson to be learned, such as when the system experiences an expectation failure. The second is determining which component is implicated in the lesson. The third is determining how that component should be modified.

Of course, the tasks in which an agent engages are not completely disjoint: Generating goals requires the agent to reason about its own plans and abilities as well as other agents; plan generation requires the agent to reason about its perceptual and mechanical abilities, as well as reasoning about its future decision-making processes; selecting among plans requires the agent to reason about its ability to execute them. The interrelations between these tasks require that the system's components be correspondingly intertwined. This interconnection of the system's components in turn requires the agent's learning module to be able to reason about interactions between components in formulating new concepts.

Reasoning about such interactions requires that the agent have a degree of *self-knowledge* in order to properly assimilate new knowledge. This paper investigates a *model-based* approach to handling these issues [Collins *et al.*, 1991a]. Our system, named CASTLE,[1] uses an explicit model of its decision-making mechanism to diagnose planning errors [Birnbaum *et al.*, 1990] and to repair its faulty components [Krulwich, 1991; Krulwich, 1992].

An every-day example

Consider the case of a person cooking rice pilaf for the first time. The last step in the directions says to "cover the pot and cook for 25-30 minutes." Suppose the person starts the rice cooking and then goes off to do something else—say, clean up the house. In the interim, the pot boils over. When the person returns to the kitchen a half-hour later, the rice pilaf is ruined.

What should be learned from this sequence of events? This depends on which of the generic decision-making tasks involved in planning the agent chooses to modify. For each task there will be a concept that operationalizes the idea of pots boiling over, in terms that are meaningful in the context of carrying out that task. Figure 1 summarizes the following three ways of operationalizing the problem posed by pots boiling over:

1. Whenever a covered pot containing liquid is on the stove, keep an ear peeled for the sound of the lid bouncing or the sound of the water bubbling.
2. Do not put a covered pot with liquid in it over a high flame, because it will boil over. The flame should be turned down or the pot lid should be left ajar.

[1] CASTLE stands for *C*oncocting *A*bstract *S*trategies *T*hrough *L*earning from *E*xpectation-failures.

System task	What to learn
Perception	Listen for the bubbling sound that warns of the pot boiling over
Planning	Leave the lid ajar or turn down the flame
Plan execution and scheduling	Don't execute any other plans that involve leaving the kitchen

Figure 1: Learned concepts in the *rice pilaf* example

3. When cooking liquid in a covered pot, stay in the kitchen, because it's hard to hear a pot boiling over from the other rooms.

Which of these concepts the agent should learn depends on its perceptual and plan execution abilities, the plans that it typically generates, and the constraints under which it operates. If the agent would in general be able to hear the pot boiling over from the other room, but simply had not attended to the soft sounds that it heard in this instance, then tuning its perceptual attention apparatus when a pot is on the stove is a good way to adapt to the new concept of pots boiling over. This is the first lesson listed above. If the agent would not be able to hear the pot boiling over however hard it listened, another lesson must be learned, either to prevent pots from boiling over by changing the parameters of the cooking process (e.g., by turning down the flame), or to avoid leaving the kitchen when a pot is on the stove. If the recipe will work properly with the flame turned very low, as is the case with rice, or with the pot lid off (which is not usually the case with rice but is with other foods such as spaghetti), then the second lesson in figure 1 will suffice for the agent to plan properly in the future. If the recipe cannot be cooked uncovered or over a low flame, the third lesson is the one that should be learned, that it should not leave the kitchen while a pot is on the stove.

We see, then, that the agent could learn several things in response to the rice pilaf boiling over. The first lesson, that the problem can be averted if the lid can be left ajar or if the flame is lowered, should be learned regardless of the agent's abilities, but should only be applied as appropriate. Which of the other lessons the agent should learn, the idea of staying in the kitchen, or of tuning its perceptual apparatus, depend on the agent's knowledge of its hearing abilities. Other possible lessons, such as the need to compute the ideal height for the flame under the pot, are ruled out due to the inability of the system to perform this computation accurately.

Constraints on concept formulation

Our discussion of learning about pots boiling over while cooking rice pilaf demonstrates several elements of the agent's self-knowledge that affect the formulation of learned concepts:

- *Knowledge of the agent's components:* Different aspects of the example will be relevant to different components in the agent's decision-making architecture (e.g., the agent could learn concepts relating to planning, plan execution/scheduling, and perception).
- *Knowledge of the agent's physical abilities:* Some formulations of the concept will not be effective due to limitations in the agent's physical abilities (e.g., whether to learn to listen harder depends on the physical capability of the agent to hear the bubbling from the other room).
- *Knowledge of the agent's cognitive abilities:* Limitations could also be cognitive (e.g., the agent could attempt to learn to compute the precisely optimal flame height, but limitations in the agent's ability to perform this reasoning make this untenable).
- *Knowledge of typical planning situations:* The possible alternative plans (e.g., lowering the flame and leaving the lid ajar) must be selected based on the situations in which the agent expects to find itself.

Each of these is a type of self-knowledge that an intelligent agent must possess in order to assimilate learned knowledge effectively. This self-knowledge will enable the agent to relate new concepts to relevant components of its decision-making architecture.

The CASTLE system

Our research is an investigation of a *failure-driven* approach to acquiring new planning knowledge [Birnbaum *et al.*, 1990; Collins *et al.*, 1991c]. Our system, CASTLE, detects situations that are contrary to its expectations, and responds to these expectation failures by repairing the faulty planner components which were responsible for the failure. We approach this learning task in a knowledge-intensive fashion, in which the system uses knowledge of its own planning components to assimilate events which led to expectation failures. This knowledge is expressed in the form of a planner self-model, which is used to diagnose and repair expectation failures [Davis, 1984; deKleer and Williams, 1987]. More specifically, the system first examines an explicit *justification structure* that encodes the the reasoning that led to its belief in the incorrect expectation [deKleer *et al.*, 1977; Doyle, 1979]. This justification is used to isolate the components of its architecture that are responsible for the failure [Collins *et al.*, 1991b]. It then uses a *specification* of the faulty components to guide the learning of new rules to embody the concept which must be learned in response to the failure [Krulwich, 1991].

The CASTLE system carries out the tasks we have been discussing in the domain of chess. CASTLE is broken up into a number of components, which reflect a functional decomposition of the decision-making process [Collins *et al.*, 1991]. Each component is dedicated to a particular cognitive task, and is implemented as a set of rules which provide different methods for performing

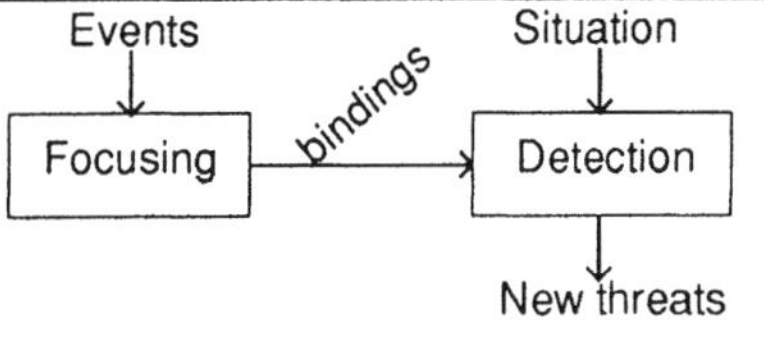

Figure 2: Incremental threat detection

the task. These rules in turn invoke other components as necessary.

We will now examine more specifically some of the knowledge that CASTLE has regarding its components. One cognitive task in which CASTLE engages is that of noticing threats and opportunities as they arise. Rather than recomputing these at each turn, CASTLE maintains a set of active threats and opportunities that is updated over time. To accomplish this incremental threat detection, the system uses a *detection focusing* component, which consists of *focus rules* that specify the areas in which new threats may have been enabled. Then, a separate *threat detection* component, consisting of rules for noticing specific types of threats, detects the threats that have in fact been enabled. The relationship between the two components is shown pictorially in figure 2. A sample focus rule is shown in figure 3, which embodies the system's knowledge that the most recently moved piece, in its new location, may be a source of new threats. Another focus rule, not shown, specifies that the most recently moved piece can also be a target of newly enabled attacks. Using focus rules such as these, the actual threat detector rules will only be invoked on areas of the board that can possibly contain new threats.

Another task in which CASTLE engages is *plan generation*. One of the system's components for doing plan generation is a *schema applier*. This component retrieves schemata, which are generalized sequences of actions, that will achieve a particular goal in the current situation. In CASTLE these schemata are called *offensive strategies,* and the system's offensive strategy component consists of rules that encode the actions in a schema, its conditions of applicability, and the goal which it satisfies. A strategy rule for the classic chess

```
(def-brule focus-new-source
 (focus focus-moved-piece ?player
   (move ?player ?move-type ?piece ?loc1 ?loc2)
   (world-at-time ?time))
<=
 (move-to-make (move ?player ?prev-move-type
                      ?piece ?old-loc ?loc1)
              ?player ?goal (1- ?time)) )
```

Figure 3: Focusing on new moves by a moved piece

```
(def-brule strategy-fork-sample
  (strategy fork ?player (world-at-time ?time)
            (goal (capture ?target2))
    (plan (move ?player non-capture ?piece
                ?loc1 ?loc2 ?time)
      (next (move ?player (capture ?target2)
                  ?piece ?loc2 ?loc4 (1+ ?time)))))
<=
 (and (at-loc ?player ?piece ?loc1 ?time)
      (at-loc ?opponent ?target1 ?loc3 ?time)
      (at-loc ?opponent ?target2 ?loc4 ?time)
      (not (at-loc ?anyone ?any-piece ?loc2 ?time))
      (move-legal ?player ?piece ?loc1 ?loc2)
      (move-legal ?player ?piece ?loc2 ?loc3)
      (move-legal ?player ?piece ?loc2 ?loc4)
      (> (value ?target1) (value ?target2))
    (no (and (counterplan ?cp-meth1 ?opponent
               (goal-capture ?target ?loc3
                 (move ?player (capture ?target1)
                       ?piece ?loc2 ?loc3))
               ?time ?counterplan)
           (counterplan ?cp-meth2 ?opponent
             (goal-capture ?target ?loc4
               (move ?player (capture ?target2)
                     ?piece ?loc2 ?loc4))
             ?time ?counterplan))) ))
```

Figure 4: A strategy rule: The *fork*

strategy the *fork* is shown in figure 4.[2] This rule says roughly that *one way to capture an opponent piece is to find a piece that can move to a location from which it can capture two opponent pieces, if the opponent will have no one counterplan against both the attacks.* Other such strategies are *pin* and the *sactifice.* Issues in acquiring such strategies have been discussed previously [Birnbaum *et al.*, 1990; Freed, 1991].

Learning focusing and a new strategy

Consider the partial chess situations shown in figure 5. Initially CASTLE (playing black) uses its offensive strategy component for plan recognition [Schank and Abelson, 1975; Cullingford, 1978], that is, to see if the opponent can be expected to have any good strategies to apply. Since none of its strategy rules apply from the perspective of the opponent, CASTLE assumes that the opponent will not be able to make a move that will enable a guaranteed capture on the following turn. In other words, the opponent would presumably like to make a situation in which no matter what CASTLE does, the opponent will capture a piece on the next turn. CASTLE believes that since none of its strategy rules apply for the opponent, the opponent will not be able to create such a situation.

[2]In practice this rule is specialized to use geometric reasoning. Alternatively, it could evaluate the results of a projection engine without duplicating its computation.

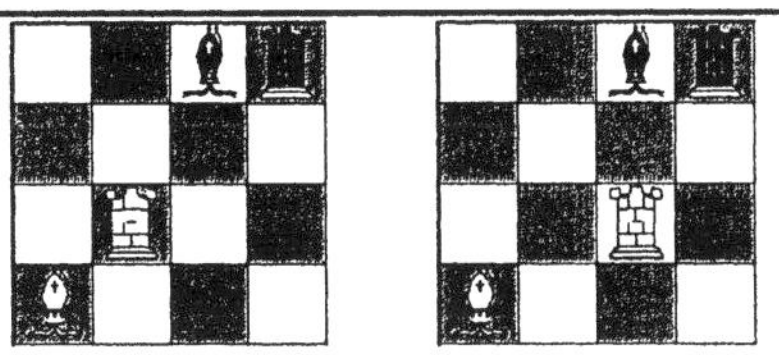

(a) (b)

Figure 5: Example: Opponent (white) to move

The opponent (playing white) then moves its rook, resulting in the situation in figure 5(b). By making this move, the opponent has enabled two attacks simultaneously, the bishop attack on the computer's rook, and the rook attack on the computer's bishop, and one of the attacks is sure to succeed. CASTLE's lack of an offensive strategy rule for this type of *simultaneous attacks* resulted in its being unable to counterplan early enough. CASTLE should learn a new strategy rule as a result of the loss.

The planning failure in our example is also relevant to another of CASTLE's components, namely the detection focusing component. Initially CASTLE is only equipped with the two focus rules discussed earlier, for the new threats by and against the most-recently moved piece, and does not have a rule for focusing on the *discovered attack* that was enabled in our example by moving the rook out of the line of attack between the bishop and the computer's rook [Collins *et al.*, 1991c]. Because of this, CASTLE is at first unable to detect the threat against its rook, and believes that the threat against its bishop is the only threat on the board. Because of this error, in figure 5(b) it moves its bishop, and thinks that all of its pieces are safe as a consequence. When the opponent executes the capture of CASTLE's rook, the computer realizes the extent of its error.

These two concepts, *simultaneous attacks* and *discovered attacks,* should both be learned from the sequence of events that we have seen. Each concept involves a different aspect of the situation, and each must be characterized in a way that can be effectively used by the relevant components of the agent's architecture.

Component	Learned concept
Planning	Make a successful attack by moving a piece off a line of attack to a new location which can also make a second attack
Perception	Threats can be enabled by moving a piece off of a line of attack that is otherwise open

Figure 6: Concepts in the chess example

Developing these characterizations requires the system to use knowledge of the functions and interactions of its components. A characterization of simultaneous attacks that can be used by the offensive strategy component must mention the simultaneous enablement of the two attacks, one through the vacated square and one from the new location of the moved piece, and must encode the fact that the opponent must be unable to react to both attacks in a single move. Additionally, it must be predictive, because it will be invoked before any move has been made, and so it must refer not to moves that have already been made, but rather to moves that can potentially be made. A characterization of discovered attacks that can be used effectively to focus the detection rules must generate a set of constraints describing all the possible moves through a vacated square, without referring to the legal moves themselves which will be checked subsequently by the detection component (see figure 2). This rule must be expressed in terms that can be applied after the enabling move has been made, but before the discovered attack is made.

CASTLE detects the opportunity to learn by observing an expectation failure when its rook is captured by the opponent's bishop. As we have discussed above, CASTLE uses a model-based reasoning approach to diagnosing expectation failures, in which the system diagnoses the failure by examining an explicit justification structure which encodes the basis for its belief in its expectation that its pieces were safe. The system's diagnosis engine traverses this justification, which is shown in figure 7, to find the underlying beliefs of the system that were responsible for the failure. In our example there were two such incorrect assumptions: that *the system's set of threat detection focusing rules is complete,* and *the system's set of strategy schemata is complete.* The task now at hand is for the system to repair its planning

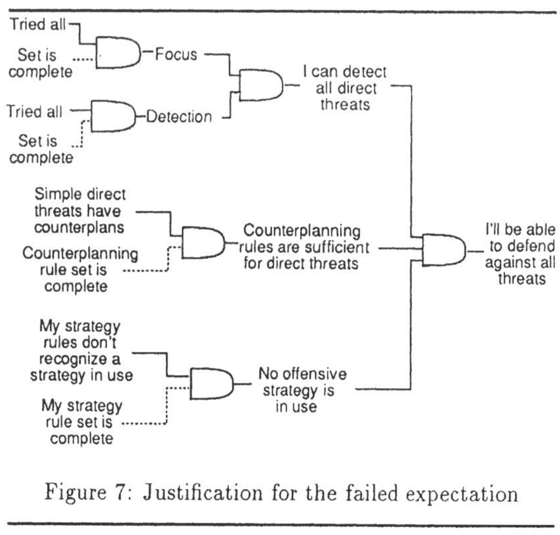

Figure 7: Justification for the failed expectation

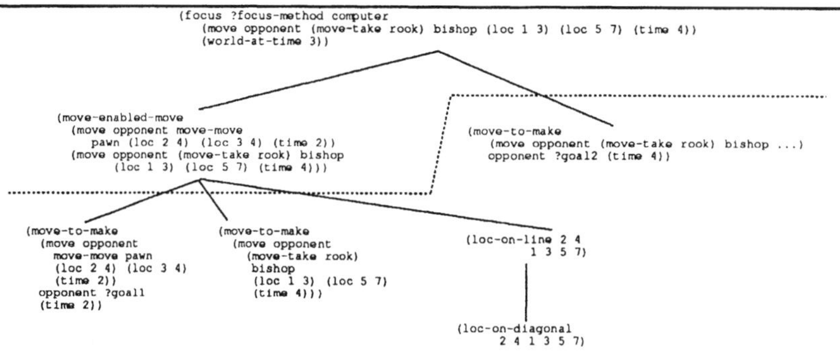

Figure 8: Explanation of desired detection focusing performance

mechanism. Each of the two faults can be repaired by augmenting a corresponding rule set, one for the focusing component and one for the strategy component.

To construct the new rules, CASTLE retrieves a *component performance specification* for each component [Krulwich, 1991]. These performance specifications, a form of planner self-knowledge, describe the correct behavior of each component. These specifications can be used to recognize correct behavior that was not produced by the component's rule sets. The specification of the detection focusing component says roughly that *the focusing component will generate bindings that include any capture that is enabled by a given move.* This specification enables CASTLE to focus on the details of the example that are relevant to the component being repaired, by serving as an explanation-based learning target concept. After retrieving the specification, CASTLE invokes its deductive inference engine to construct an explanation of why the possible capture of the rook should have been in the set of constraints generated by the focusing component. This explanation, shown in figure 8, says roughly that the opponent's move should have been generated by the focusing component, *because* the opponent's previous move enabled the attack, *because* it was on a square between the bishop and the rook, *and* there were no other pieces along the line of attack, *and* emptying the line of attack is an enabling condition for the capture to be made. CASTLE then uses explanation-based learning [Mitchell *et al.*, 1986; DeJong and Mooney, 1986] to generalize this explanation and to construct the new detection focusing rule shown in figure 9.

The same mechanism is used to construct the new offensive strategy rule. A specification of the offensive strategy component is retrieved, which says roughly *the offensive strategy component will generate any plans which are sure to result in a successful attack.* CASTLE then explains why the opponent's move resulted in a certain capture. This explanation says roughly that the opponent's move was a good offensive strategy, *because* it enabled one attack through the vacated square, *and* it enabled a second attack from the new location of the moved piece, *and* there was no counterplan for the opponent that could disable both attacks. The crucial inference in constructing this explanation is that the existence of two attacks, in a situation where they cannot both be counterplanned against simultaneously, means that one of them will necessarily succeed. One approach is for this knowledge to be built in, as has been done implicitly by others. Our approach is for this knowledge to be inferred from more primitive axioms of plan execution [Birnbaum *et al.*, 1990]. After this explanation is constructed it is generalized to form a rule for simultaneous attacks.

Discussion

In section we saw several types of reasoning in which an agent might engage in the course of learning from a sequence of events. In our example in section we saw

```
(def-brule learned-focus-method25
  (focus learned-focus-method25 ?player
    (move ?player (capture ?taken-piece)
      ?taking-piece (loc  ?row1 ?col1)
      (loc  ?row2 ?col2))
    (world-at-time ?time2))
<=
  (and (move-to-make
         (move ?other-player move ?interm-piece
               (loc ?r-interm ?c-interm)
               (loc ?r-other ?c-other))
         ?player ?goal ?time1)
       (loc-on-line ?r-interm ?c-interm
                    ?row1 ?col1 ?row2 ?col2)
       (at-loc ?player ?taking-piece
               (loc ?row1 ?col1)
               (- gen-time2.24 2)) ))
```

Figure 9: Learned focus rule for *discovered attacks*

how CASTLE performs some of these types of reasoning, in particular:

- CASTLE determined which components should be repaired *(e.g., perception, planning)*
- CASTLE formulated a concept for each component being repaired *(e.g., focus rule, strategy schema)*

CASTLE is capable of carrying out several other types of reasoning about the state of its knowledge. Consider, for example, a situation in which the system applied its *simultaneous attacks* strategy against an opponent, and the opponent is able to counterplan against both attacks in a way that CASTLE does not know about. One such counterplanning method might, for example, be to move one of the attacked pieces to a square along the line of attack against the second piece. Since CASTLE does not initially have a rule for counterplanning by interposing pieces, it would think that its simultaneous attack would be successful. When the strategy is seen to be unsuccessful, CASTLE must determine whether the fault lies with the strategy rule or with another component. In this case CASTLE should realize that the strategy is sound, but that it needs to learn a new counterplanning rule for interposition.

This learning process requires that CASTLE have a *self-model* which describes the functions and interactions of its components. This self-model consists of several forms of self-knowledge. *Belief justifications* relate expectations and other beliefs to the reasons that CASTLE believes them to be true. *Implicit assumptions* describe the assumptions that CASTLE is making (such as rule set completeness) that underly the validity of its decision-making mechanisms. *Performance specifications* describe the performance that CASTLE expects from each of its components. Using these forms of self-knowledge, CASTLE can reason about the state of its knowledge and abilities in order to effectively assimilate new knowledge.

Our research has involved extending our model of planning and decision-making to include a variety of tasks and components. To date we have developed models of threat detection, counterplanning, schema application, goal regression, lookahead search, and execution scheduling. Future research will determine the degree to which our theory of model-based knowledge assimilation through diagnosis and learning applies to other decision-making tasks and to planning in other domains.

Acknowledgements: Thanks go to Matt Brand, Michael Freed, Menachem Jona, Eric Jones, and Louise Pryor for discussions on this paper and on the research presented. This work was supported in part by the Air Force Office of Scientific Research under grant number AFOSR-91-0341-DEF, and by the Defense Advanced Research Projects Agency, monitored by the Office of Naval Research under contract N-00014-91-J-4092. The Institute for the Learning Sciences was established in 1989 with the support of Andersen Consulting, part of The Arthur Andersen Worldwide Organization. The Institute receives additional support from Ameritech, an Institute Partner, and from IBM.

References

[Birnbaum *et al.*, 1990] L. Birnbaum, G. Collins, M. Freed, and B. Krulwich. Model-based diagnosis of planning failures. In *Proceedings of the Eighth National Conference on Artificial Intelligence*, pages 318–323, Boston, MA, 1990.

[Collins *et al.*, 1991a] G. Collins, L. Birnbaum, B. Krulwich, and M. Freed. A model-based approach to learning from planning failures. In *Notes of the AAAI Workshop on Model-Based Reasoning*, Anaheim, CA, 1991.

[Collins *et al.*, 1991b] G. Collins, L. Birnbaum, B. Krulwich, and M. Freed. Model-based integration of planning and learning. *SIGART Bulletin*, 2(4):56–60, 1991. Originally in Working Notes of the AAAI Spring Symposium on Integrated Intelligent Architectures.

[Collins *et al.*, 1991c] G. Collins, L. Birnbaum, B. Krulwich, and M. Freed. Plan debugging in an intentional system. In *Proceedings of the Twelfth International Joint Conference on Artificial Intelligence*, pages 353–358, Sydney, Australia, 1991.

[Cullingford, 1978] R. Cullingford. *Script Application: Computer Understanding of Newspaper Stories.* PhD thesis, Yale University, 1978. Technical Report 116.

[Davis, 1984] R. Davis. Diagnostic reasoning based on structure and function: Paths of interaction and the locality principle. *Artificial Intelligence*, 24(1-3):347–410, 1984.

[DeJong and Mooney, 1986] G. DeJong and R. Mooney. Explanation-based learning: An alternative view. *Machine Learning*, 1:145–176, January 1986.

[deKleer and Williams, 1987] J. deKleer and B.C. Williams. Diagnosing multiple faults. *Artificial Intelligence*, 32(1):97–129, April 1987.

[deKleer *et al.*, 1977] J. deKleer, J. Doyle, G.L. Steele, and G.J. Sussman. Explicit control of reasoning. *SIGPLAN Notices*, 12(8), 1977.

[Doyle, 1979] J. Doyle. A truth maintenance system. *Artificial Intelligence*, 12(3):231–272, 1979.

[Freed, 1991] M. Freed. Learning strategic concepts from experience: A seven-stage process. In *Proceedings of the Thirteenth Annual Conference of The Cognitive Science Society*, pages 132–136, Chicago, IL, 1991.

[Krulwich, 1991] B. Krulwich. Determining what to learn in a multi-component planning system. In *Proceedings of the Thirteenth Annual Conference of The Cognitive Science Society*, pages 102–107, Chicago, IL, 1991.

[Krulwich, 1992] B. Krulwich. *Learning New Methods for Multiple Cognitive Tasks.* PhD thesis, Northwestern University, Institute for the Learning Sciences, 1992. (in preparation).

[Mitchell *et al.*, 1986] T.M. Mitchell, R.M. Keller, and S.T. Kedar-Cabelli. Explanation-based generalization: A unifying view. *Machine Learning*, 1(1), January 1986.

[Schank and Abelson, 1975] R. C. Schank and R. P. Abelson. *Scripts, Plans, Goals, and Understanding.* Lawrence Erlbaum Associates, Hillsdale, NJ, 1975.

Energy Minimization and Directionality in Phonological Theories

David S. Touretzky and **Xuemei Wang**
School of Computer Science
Carnegie Mellon University
Pittsburgh, PA 15213

Abstract

Goldsmith (1990, 1991) and Lakoff (in press) have both proposed phonological theories involving parallel constraint satisfaction, and making explicit reference to Smolensky's (1986) harmony theory. We show here that the most straightforward implementation of phonological constraint satisfaction models as spin glasses does not work, due to the need for directionality in constraints. Imposing directionality negates some of the advantages hoped for from such a model. We have developed a neural network that implements a subset of the operations in the Goldsmith and Lakoff phonological theories, but proper behavior requires asymmetric connections and essentially feed-forward processing. After describing the architecture of this network we will move on to the issue of whether spin glass models are really an appropriate metaphor for phonological systems.

Introduction

Goldsmith (1990, 1991) and Lakoff (in press) have both proposed phonological theories involving parallel constraint satisfaction, and making explicit reference to Smolensky's (1986) harmony theory. These proposals have been criticized by Touretzky and Wheeler (Touretzky, 1989; Touretzky & Wheeler, 1990a; Wheeler & Touretzky, in press) as computationally infeasible. They offer an alternative theory using deterministic rules, with no reapplication, that is implementable by purely feed-forward circuitry (Touretzky & Wheeler, 1991). Goldsmith (personal communication) has in turn criticized this theory on the grounds that it may not be powerful enough to handle the complexity of rule interactions found in some human languages. An answer to this challenge awaits further work by Wheeler and Touretzky.

Smolensky's harmony theory is equivalent to the Boltzmann machine model of Hinton & Sejnowski (1986), and thus falls into the category of spin glass energy minimization models (Hopfield, 1982). Here we wish to address the original criticism directed at the constraint satisfaction proposal: that it is not obvious how such models could be implemented in a spin glass architecture, and even if they could, the computation required to find a maximum harmony/minimum energy state would be excessive.

In this paper we show that the most straightforward implementation of phonological constraint satisfaction models as spin glasses does not work, due to the need for directionality in constraints. Imposing directionality negates some of the advantages hoped for from such a model. We have developed a neural network that implements a subset of the operations in the Goldsmith and Lakoff phonological theories, but proper behavior requires asymmetric connections and essentially feed-forward processing. After describing the architecture of this network we will move on to the issue of whether spin glass models are really an appropriate metaphor for phonological systems.

The Goldsmith and Lakoff Models

Goldsmith and Lakoff utilize similar three-level models, where the levels are labeled M (underlying, or morphophonemic), W (word), and P (surface, or phonetic).[1] See Figure 1. The initial underlying form of a word is denoted M_1. Free reapplication of unordered intra-level (M,M) rules produces a new string, M_n. Interaction-free, parallel application of inter-level (M,W) rules then produces the initial W-level representation, W_1. The process continues with free reapplication of unordered (W,W) rules, parallel application of (W,P) rules, and finally, free reapplication of unordered (P,P) rules.

A sample derivation in this formalism, taken from Lakoff (in press), is shown below. The data comes from Yawelmani, an American Indian dialect from California (Kenstowicz & Kisseberth, 1979). Yawelmani has an epenthesis process that inserts an /i/ after the first member of a triconsonantal cluster. It also has a vowel harmony process[2] in which round vowels cause all contiguous

[1] In the original version of his paper Lakoff called these levels M, P, and F, and Touretzky and Wheeler used these labels in their earlier papers. Subsequently, they and Lakoff adopted Goldsmith's notation.

[2] Not to be confused with Smolensky's *harmony theory*, an unrelated use of the word.

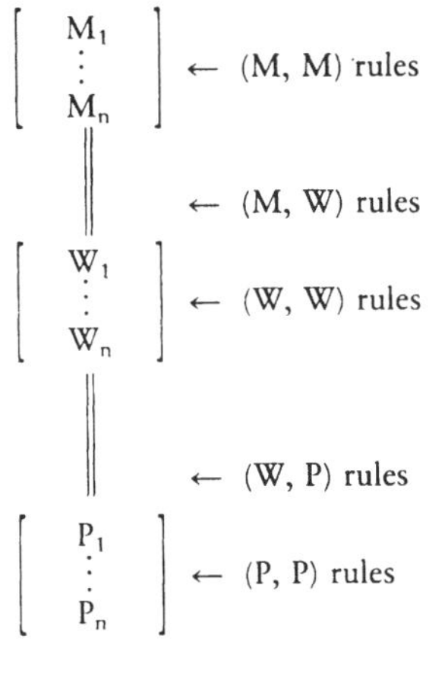

Figure 1: Structure of Goldsmith's model, taken from (Goldsmith, 1990:324).

succeeding vowels of the same height to become round, back, and non-low. The harmony rule can be stated as a (W, W) constraint as follows:

W: If [+syl,+rd,αhigh] C_0 X,
then if X = [+syl,αhigh],
then X = [+rd,+back,−low]

In /du:ll+hin/ "climbs," the (M, W) epenthesis process inserts an /i/ after the first /l/. This epenthetic vowel is not only subject to the (W, W) harmony process, in rounding it becomes the trigger for harmony applying again, to the vowel in /hin/. After an independent lowering process at (W, P), the surface form is [do:lul-hun].

M:	d u: l l - h i n
W:	d u: l u l h u n
P:	d o: l u l h u n

Goldsmith and Lakoff see constraints as maximizing the linguistic "harmony" or well-formedness of a string. Rules act as "repair strategies" triggered by constraint violations (Sommerstein, 1974); they alter the string in language-specific ways so as to best satisfy all applicable constraints. Spin glass models also perform constraint satisfaction, and thus, spin glasses are an appealing metaphor for phonological processing. But this analogy is not without problems.

Criticisms of the Constraint Satisfaction Model

Lakoff, in introducing his theory of "cognitive phonology," argues that classical linguistic derivations are too lengthy to be taken literally as mental theories of phonological processing. However, allowing free reapplication of unordered intra-level rules leads to the same sort of lengthy derivations, at least in Goldsmith's version of the theory, where each representational level goes through a sequence of forms (e.g., $M_1, \ldots, M_n$ in Figure 1) due to serial effects of constraints. Goldsmith makes an analogy to traveling downhill in constraint violation space. (He also suggests in (Goldsmith, 1991) that an actual implementation would use units with graded activations that settle into a minimum energy state in a continuous dynamical system sort of way, which seems at odds with the notion of well-defined symbolic intermediate states.)

Our interpretation of Lakoff is that he takes a nondeterministic approach to derivations, such that there is only one form at each level rather than a sequence of forms. The two requirements of, say, a W-level form are that it must satisfy any applicable constraints at that level, and that its points of difference from M-level must all be "licensed" by some combination of (M, W) rules and (W, W) constraints. For example, rather than positing an intermediate W_1 form /du:lil+hin/ resulting from epenthesis in the Yawelmani derivation of [do:lulhun], Lakoff goes directly to the final W-level form /du:lul+hun/, which is licensed by a combination of (M, W) epenthesis, (W, W) harmony applying to the epenthetic vowel, and (W, W) harmony reapplying to the vowel of the suffix /hin/. This abstract notion of licensing of derived forms avoids adding sequentiality to the theory, but at a cost: deducing the surface form from the underlying form of a string requires either nondeterminism (not available in actual physical computing devices) or search.

Structure of Our Model

A straightforward way to encode constraints in a connectionist net is as propositional implications. Pinkas (1991) showed that any sentence in propositional logic can be encoded directly in a Hopfield net with n-ary (higher order) connections, such that the global energy minima of the network correspond exactly to those truth assignments that satisfy the sentence. For example, the implication **abc** $\rightarrow$ **d** would be encoded as shown in Figure 2 by a third-order connection among **a, b,** and **c** with weight -1, plus a fourth-order connection among **a, b, c,** and **d** with weight $+1$. All units have 0/1 states and zero thresholds. The minimum energy states of this network are those where either at least one of **a, b,** or **c** is in state 0, or all four nodes are in state 1. If **a, b,** and **c** are clamped on, then after settling, **d** will be on. If **a** is clamped on and **d** is clamped off, then after settling, at least one of **b** and **c** will be off.

Figure 3 shows the structure of a portion of our model. A segment (phoneme), depicted as a rectangle in the figure, is represented as a binary feature vector. By default, each M-level segment is mapped to an identical W-level segment; this is accomplished by weak one-to-one connections between corresponding elements. (M, W) rules may alter this mapping in one of three ways: they can alter individual features of a segment, they can cause the segment to be deleted by turning on its deletion bit at W-level, and they can cause a new segment to be inserted to the right of the current segment. The last is achieved

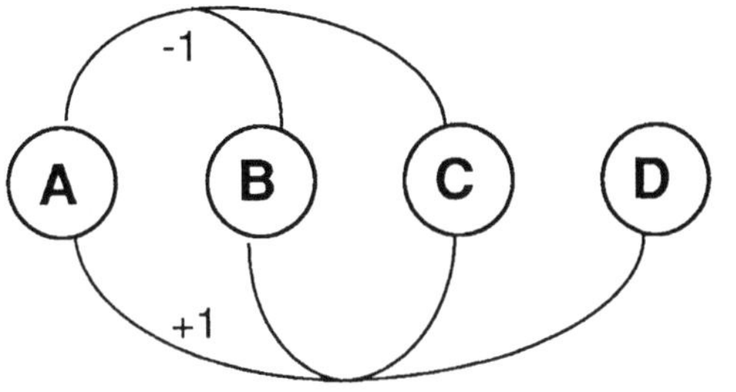

Figure 2: Encoding of **abc** → **d** in a Hopfield net with higher-order connections.

by turning on the segment's insertion bit and writing the binary feature pattern into the associated insertion slot, which appears below and to the right of the segment in the figure.

In order to permit these changes to affect the environment of subsequent rule applications, each segment is augmented with dynamically computed left and right neighbor slots, shown as diamonds in the figure. Let Del_i and Ins_i stand for the deletion and right insert bits associated with segment i. Then the left neighbor of segment i is computed as shown in Figure 4.

A similar computation is used to find the nearest vowels to the left and right of a segment (required for harmony rules, which operate on the "vowel tier".) Insert slots also have left and right neighbor and left and right vowel slots.

This rather ad hoc solution places some limits on multiple insertion and deletion processes. It is not possible to insert more than one segment at W-level between segments that were adjacent at M-level, since there is only a single insertion slot in each position. In order for left and right neighbor computations to work correctly. it is forbidden to delete adjacent segments, though this restriction could be relaxed by increasing the complexity of the neighbor computations. Finally, segments that have been inserted by (M, W) rules can be modified by (W, W) rules, but not deleted, since the insertion slot does not have its own deletion bit.

After the W-level has settled, the string is copied into a fresh buffer, removing any segments marked for deletion and making room for new segments where an insertion bit was turned on. Finally, the mechanism for mapping strings from W to P level is employed; it is identical to the M-to-W mechanism just described.

Rules in the model are replicated at each buffer position, as in (Touretzky, 1989). Intra-level rules can refer to segments, their left and right neighbors, and the nearest vowel to the left and right. Inter-level rules, since they cannot reapply, do not need to consider insertion and deletion bits; they therefore can refer to more distant neighbors with no increase in circuit complexity.

Directionality

Cross-level rules have an inherent directionality. For example, if one clamps the M-level representation and lets the W-level portion of the network settle, the applicable (M, W) rules will exert their influence and modify the W-level representation as specified, just as clamping **a**, **b**, and **c** in Figure 2 will cause **d** to turn on. However, a great deal of the power of the Goldsmith and Lakoff models comes from *intra*-level constraints. The (W, W) harmony constraint is a case in point. With intra-level constraints none of the relevant parts of the representation can be clamped; the entire W-level string is moving downhill in constraint violation space to find a more harmonious state. The problem is that there is usually more than one way to resolve a constraint violation. Constraints stated as implications have an implied directionality of inference, but the straightforward encoding of constraints as symmetric higher order connections does not retain this directionality. Specifically, if nodes **a**, **b**, **c**, of Figure 2 happen to be on while node **d** is off, the network can resolve the constraint violation equally well by turning **d** on or by turning one or more of **a**, **b**, or **c** off.

When we implemented the Yawelmani (W, W) harmony rule as a logical implication in a higher-order Hopfield net version of our model, we found that the lack of directionality could indeed produce this undesirable behavior.

It is of course possible to replace a nondeterministic search for satisfying truth assignments for propositional sentences with the deterministic application of directed inference rules. We did this by switching to a model with asymmetric connections, so that rules were implemented by essentially feed-forward circuitry. The resulting model succesfully performed the Yawelmani derivations in Lakoff's paper, including reapplication of the vowel harmony rule in the example cited above, and with a slight modification, also replicated Lakoff's example of a Mohawk derivation that involves six rule applications.

Reconsidering Constraint Satisfaction

What does the asymmetric model just described really say about constraint satisfaction? It is, after all, merely applying rules in a deterministic, feed-forward fashion, though it employs some clever tricks to make this happen in a connectionist framework. We have several observations to make about this.

First, many phonological processes appear quite compatible with a deterministic model, and do not require true constraint satisfaction. We would not claim that all of phonology fits this mold; Goldsmith and others are emphatic that it does not. But even in areas like syllabification, where one can easily imagine multiple constraints competing to mark a consonant as an onset or a coda, Touretzky & Wheeler (1990b) have shown that there are perfectly adequate parallel but deterministic accounts. So we remain skeptical about the need for constraint satisfaction as a computational mechanism.

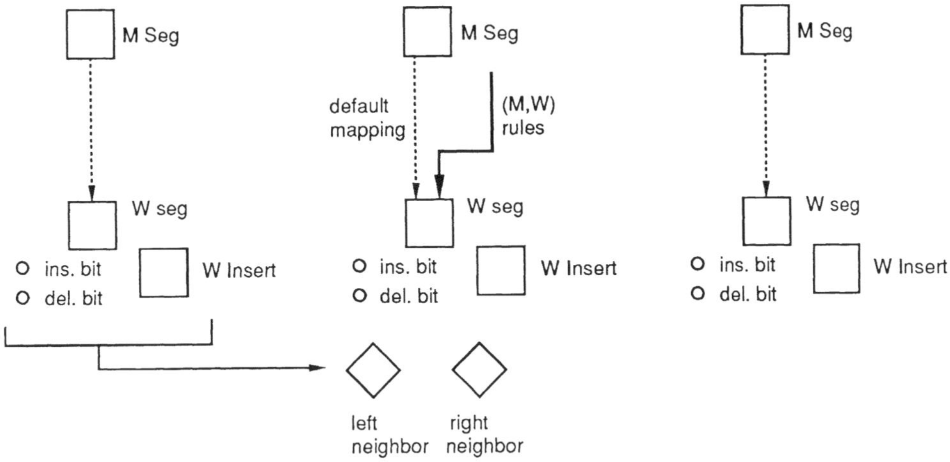

Figure 3: A portion of the circuitry of our phonological model.

$$\text{Left neighbor of } i = \begin{cases} \text{insert slot } i-1 & \text{if Ins}_{i-1} = 1 \\ \text{segment } i-1 & \text{if Ins}_{i-1} = 0 \text{ and Del}_{i-1} = 0 \\ \text{insert slot } i-2 & \text{if Ins}_{i-1} = 0 \text{ and Del}_{i-1} = 1 \text{ and Ins}_{i-2} = 0 \\ \text{segment } i-2 & \text{if Ins}_{i-1} = 0 \text{ and Del}_{i-1} = 1 \text{ and Ins}_{i-2} = 0 \end{cases}$$

Figure 4: Formula for the left neighbor of a segment.

Further investigation is required on this point.

A second point, though, is that there is a tension between the computational flavor of spin glass models (symmetric connections, nondeterministic behavior) and the directed way phonological systems resolve constraint violations. This brings us back to the traditional separation of constraints from repair strategies. The linguistic motivation for this has been parsimony: the same repair strategy may be used to fix several constraint violations, but it should only have to be stated once.[3] The computational motivation for such a separation has received less attention, perhaps because phonological theories are rarely implemented. But it now seems that if all segments are potentially subject to modification (as is the case with intra-level constraints), and rules can freely reapply (so clamping of rule antecedents is not possible), then there can be no straightforward mapping of linguistic contraints into an energy minimization model due to the loss of directionality.

Compare this situation with Touretzky & Hinton's (1988) distributed connectionist production system (DCPS), a Boltzmann machine that used simulated annealing to match production rules against working memory. Rules could reapply in DCPS, but there were two crucial differences from the phonological model. Only one rule could apply at a time, and after each rule application the state of working memory was latched to provide a steady input for the next rule match. One could of course do the same thing in the phonological model, which would bring us back once again to sequential rule application. But it is the potential for simultaneous application and interaction of multiple rules that makes the constraint satisfaction model interesting.

[3] Sommerstein (1974), cited in Goldsmith (1990:321) shows five phonotactic constraints in Latin that share a repair strategy.

Conclusions

We have shown how to implement a limited version of the Goldsmith and Lakoff proposals using connectionist style hardware with asymmetric connections. However, the real contribution of this paper is a negative result, namely, that when you look closely at the details, energy minimization in spin glass models is not as apt a metaphor for linguistic constraint satisfaction as previously thought. In phonology, directionality is important.

This not to say that constraint satisfaction does not exist in phonology, but we must decide which features of the proposal are most important to preserve. If we want constraints to direct the application of repair strategies, then we may have to replace the notion of derivation as massively parallel settling into enery minima denoting well-formed strings with a more sequential form of derivation by succesive repairs. This would appear to undermine Lakoff's (and our) goal of a model fast enough to be cognitively pausible.

Alternatively, we could give up the linguist's goal of

parsimony and abandon the separation of constraints and repair strategies. Constraints would become much more complicated structures that include knowledge of not just individual repair strategies, but also desired rule interactions. In this way we could maintain efficiency of derivations – at the cost of more complex rules.[4]

Can the spin glass approach be salvaged? It is conceivable that a more complex encoding of phonological constraints into a spin glass model could preserve directionality as an additional type of constraint. This would seem to require that points of application of a particular constraint be explicitly represented as part of the network's final state, resulting in a derivation that was annotated with licensing information.[5] We think this idea deserves further study.

Acknowledgements

This work was supported in part by funding from Hughes Aircraft Company. We thank Deirdre Wheeler for valuable discussions.

References

[1] Goldsmith, J. A. (1990) Autosegmental and Metrical Phonology. Oxford, UK: Basil Blackwell.

[2] Goldsmith, J. A. (1991) Phonology as an intelligent system. In D. J. Napoli and J. A. Kegl (eds.), *Bridges between Psychology and Linguistics: A Swarthmore Festschrift for Lila Gleitman*. Hillsdale, NJ: Erlbaum.

[3] Hinton, G. E., and Sejnowski, T. J. (1986) Learning and relearning in Boltzmann machines. In D. E. Rumelhart and J. L. McClelland (eds.), Parallel Distributed Processing: Explorations in the Microstructure of Cognition, vol. 1, pp. 282-317. Cambridge, MA: The MIT Press.

[4] Hopfield, J. J. (1982) Neural networks and physical systems with emergent collective computational abilities. *Proceedings of the National Academy of Sciences, USA*, 81, 6871-6874.

[5] Kenstowicz, M., and Kisseberth, C. (1979) Generative Phonology: Description and Theory. San Diego, CA: Academic Press.

[6] Lakoff, G. (to appear) Cognitive Phonology. In J. Goldsmith (ed.), *The Last Phonological Rule: Reflections on Constraints and Derivations in Phonology*. Chicago: University of Chicago Press.

[7] Pinkas, G. (1991) Symmetric neural networks and propositional logic satisfiability. Neural Computation 3(2):282-291

[8] Smolensky, P. (1986) "Information processing in dynamical systems: foundations of harmony theory." In D. E. Rumelhart and J. L. McClelland (eds.), Parallel Distributed Processing: Explorations in the Microstructure of Cognition, volume 1, pp. 194-281. Cambridge, MA: The MIT Press.

[9] Sommerstein, A. H. (1974) On phonotactically motivated rules. *Journal of Linguistics* 10: 71-94.

[10] Touretzky, D. S. (1989) Towards a connectionist phonology: the "many maps" approach to sequence manipulation. Proceedings of the Eleventh Annual Conference of the Cognitive Science Society, pp. 188-195. Hillsdale, NJ: Erlbaum.

[11] Touretzky, D. S., and Hinton, G. E. (1988) A distributed connectionist production system. *Cognitive Science* 12:(3):423-466.

[12] Touretzky, D. S., and Wheeler, D. W. (1990a) A computational basis for phonology. In D. S. Touretzky (ed.), *Advances in Neural Information Processing Systems* 2, pp. 372-379. San Mateo, California: Morgan Kaufmann.

[13] Touretzky, D. S., and Wheeler, D. W. (1990b) Two derivations suffice: the role of syllabification in cognitive phonology. In C. Tenny (ed.), The MIT Parsing Volume, 1989-1990, pp. 21-35. MIT Center for Cognitive Science, Parsing Project Working Papers 3.

[14] Touretzky, D. S., and Wheeler, D. W. (1991) Sequence manipulation using parallel mapping networks. *Neural Computation* 3, 98-109.

[15] Wheeler, D. W., and Touretzky, D. S. (to appear) A connectionist implementation of cognitive phonology. In J. Goldsmith (ed.), *The Last Phonological Rule: Reflections on Constraints and Derivations in Phonology*. Chicago: University of Chicago Press. An ealier version is available as technical report CMU-CS-89-144, Carnegie Mellon University School of Computer Science, 1989.

[4] Chunking is a well-known technique for incrementally constructing efficient systems with complex rules, by observing the interactions of simple rules.

[5] Such a derivation would resemble a constructive proof.

Integrating Category Acquisition with Inflectional Marking: A Model of the German Nominal System

Prahlad Gupta
Department of Psychology
Carnegie Mellon University
Pittsburgh, PA 15213
prahlad@cs.cmu.edu

Brian MacWhinney
Department of Psychology
Carnegie Mellon University
Pittsburgh, PA 15213
brian@andrew.cmu.edu

Abstract

Linguistic *categories* play a key role in virtually every theory that has a bearing on human language. This paper presents a connectionist model of grammatical category formation and use, within the domain of the German nominal system. The model demonstrates (1) how categorical information can be created through co-occurrence learning; (2) how grammatical categorization and inflectional marking can be integrated in a single system; (3) how the use of co-occurrence information, semantic information and surface feature information can be usefully combined in a learning system; and (4) how a computational model can scale up toward simulating the full range of phenomena involved in an actual system of inflectional morphology. This is, to our knowledge, the first connectionist model to simultaneously address all these issues for a domain of language acquisition.

Introduction

In virtually every model of language processing, the notion of linguistic category plays a key role. For example, syntactic categories such as *noun* and *verb* are the stuff of which sentence processing is thought to be made; grammatical categories such as *gender* and *person* are essential to the co-ordination of conjugational and declensional paradigms in many languages. Linguistic categorization has thus usually been a cornerstone of thinking about language.

This paper presents a connectionist account of how grammatical categories could be formed and usefully incorporated into processing. The phenomenon we model is learning of grammatical gender, within the German nominal system. This domain involves coordination of case, number, and gender information, and for this reason has often been regarded as a challenge to models of language acquisition (Maratsos and Chalkley, 1980; Maratsos, 1982; Pinker, 1984). We therefore chose this domain as an excellent test-bed for proposals about cue-driven learning and categorization.

Aims and Relation to Previous Work

A number of models of linguistic category acquisition have previously been proposed (MacWhinney, 1978; Maratsos and Chalkley, 1980; Pinker, 1984). The similar accounts in (MacWhinney, 1978) and (Pinker, 1984) both involve row- and column-splitting algorithms that operate on a data structure representing the paradigm for the German definite article. However, these matrix-manipulation operations are rather *ad hoc* in nature; problems with these accounts are discussed in more detail in (MacWhinney, 1991). The account in (Maratsos and Chalkley, 1980) and (Maratsos, 1982), while intuitively appealing, has not been specified in computationally precise form.

The aim of the present research was to provide a computational account of the formation of the grammatical category of gender in German, and of how this categorical information could be usefully employed in language processing and acquisition, without reliance on the kinds of *ad hoc* mechanisms specified in the earlier MacWhinney-Pinker account. We aimed, moreover, to make this computational investigation within a connectionist framework.

Previous work by the second author and colleagues has presented a computational model of the acquisition of the German definite article (MacWhinney, Leinbach, Taraban and McDonald, 1989; Taraban, McDonald and MacWhinney, 1989). As will be discussed in more detail in the final section, the present work achieves several significant advances over the earlier model, while also replicating the earlier results.

The German Nominal System

The system of *grammatical gender* in German assigns every noun to one of three gender categories: *masculine, feminine*, or *neuter*. The grammatical gender assigned to a noun will in general have little to do with the sex of its referent. For example, the noun *Fräulein*, meaning "young lady", has neuter gender, while the noun *Polizei*, meaning "police", has feminine gender.

Case	Singular			Plur
	Masc	Fem	Neut	
Nominative	der	die	das	die
Genitive	des	der	des	der
Dative	dem	der	dem	den
Accusative	den	die	das	die

Table 1: Gender, number and case paradigm for the German definite article.

As shown in Table 1, the correct definite article for use with a given noun depends on the gender of the noun, and on the case and number in which the noun is used. Potentially, this leads to 24 cells in the paradigm (4 case possibilities x 3 gender possibilities x 2 number possibilities). However, gender is not relevant in the plural number, and so there are only 16 cells in the paradigm. As there are only six distinct definite articles (*der, die, das, des, dem, den*), a particular article obviously can and does appear in more than one cell.

The stem of a noun undergoes various inflectional modifications according to the case and number context in which it is used, and also depending on its gender. Possible inflectional changes include *umlauting* of a vowel in the stem, and various *suffixation* processes, with *voicing* of a final consonant accompanying certain suffixes. These changes are discussed in more detail in (Mugdan, 1977).

The Model

The architecture of the model is shown in Figure 1. Essentially, this is a connectionist architecture, though with some departures from what is most typical of such models. The overall system consists of three networks, described below: a *categorization network*, an *article-learning network*, and a *stem-modification-learning network*.

Categorization Network

The *categorization network* is shown in the region marked 1 in Figure 1. It constitutes a mechanism that learns to *categorize* articles, based on their co-occurrence with case and number information. This takes the form of a *competitive learning* network (Rumelhart and Zipser, 1986) whose inputs are the representations of case, number, and the article, and whose output response is a pattern over the "Winner-Take-All" layer that identifies that case-number-article combination.

We have assumed that there is a "lexicon", consisting of "lexical representations" of noun stems[1]. For our current purposes, a "lexical entry" comprises information about both the phonology of the noun and the co-occurrence relations in which the noun has participated. In the present case, this latter information is limited to co-occurrences with particular articles. We assume that the categorization responses of the competitive learning network shape the part of the lexical representation of the noun that stores co-occurrence information. Over time, this lexical information comes to be a trace of which articles have occurred with the noun in which case and number. These encodings constitute the noun's *co-occurrence history*.

There are fourteen possible distinct combinations of Case, Number and Article that can occur. These are: Nom-Sing-der (Nominative-Singular-der), Gen-Sing-des, Dat-Sing-dem, Acc-Sing-den, Nom-Sing-die, Gen-Sing-der, Dat-Sing-der, Acc-Sing-die, Nom-Sing-das, Acc-Sing-das, Nom-Plur-die, Gen-Plur-der, Dat-Plur-den, and Acc-Plur-die.

Competitive learning results in single, specific units in the Winner-Take-All layer responding to each possible combination. Note that the Winner-Take-All layer consists, not of exactly fourteen predetermined units, but of an arbitrary number of units (we used 50). Nevertheless, the unsupervised competitive learning algorithm results in there being fourteen units that come to "recognize" the fourteen possible combinations[2].

Only certain combinations of case, number and article will co-occur with a noun of a particular gender. For example, for a Feminine noun such as *Frau*, only the combinations Nom-Sing-die, Gen-Sing-der, Dat-Sing-der, Acc-Sing-die, Nom-Plur-die, Gen-Plur-der, Dat-Plur-den, and Acc-Plur-die will be observed; Feminines will not co-occur with Nom-Sing-der or Gen-Sing-der. Thus, a certain set of combinations of case, number and article will co-occur with Feminine nouns, a different set with Masculine nouns, and a different set for Neuter nouns.

It is important to note that articles are homophonous. For example, *der* is used with both Masculine and Feminine nouns. Occurrence of a particular article with a particular noun therefore does not provide sufficient information to determine the noun's gender (except for the article *das*). The *set* of all articles that can occur with a particular noun does provide sufficient information to encode gender uniquely. So also does the set of all possible combinations of case, number and article. However, if only *part* of the paradigm for a noun has been ob-

[1]Although, for convenience, we have depicted the lexicon as an array-like data structure, we envisage it as a collection of topographically organized maps. We have not attempted to implement this lexical organization; however, work by Mikkulainnen has demonstrated how such a *distributed lexicon* could be formed (Miikkulainen, 1990).

[2]The classification is sometimes into thirteen rather than fourteen categories, with the combinations Nom-Sing-das and Acc-Sing-das being grouped into a single category. However, this does not affect the usefulness of the categorizations to be discussed in the section on "Simulations and Results".

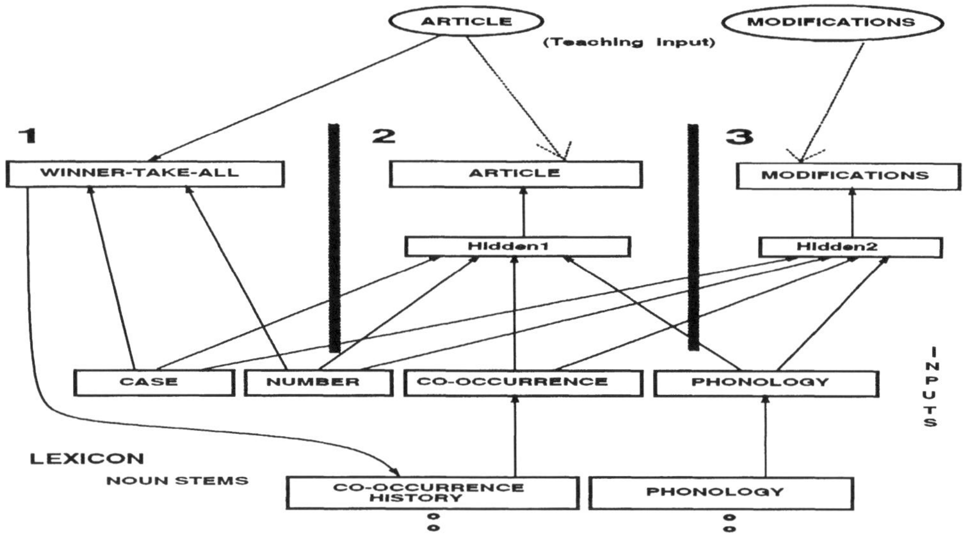

Figure 1: Architecture of the model used in simulations.

served, then a record of the observed case-number-article combinations is a more robust encoding of gender than a record of only the observed articles.

In the model, the co-occurrence history for a particular noun stem is formed in the following way. The categorization responses for each case-number-article combination observed with the stem are additively encoded in the "co-occurrence history". This additive encoding involves the arithmetic addition of the pattern of activation evoked over the Winner-Take-All layer to the co-occurrence history part of the noun stem's lexical representation. As successive categorization responses are added to a particular noun's co-occurrence history, additional units in the co-occurrence history come to be active. Recall that the sets of case-number-article combinations that can co-occur with nouns of different gender are different. Therefore, different sets of units will come to be active in the co-occurrence history of stems of different gender. In other words, the lexical co-occurrence history comes to form a distributed representation of the *grammatical gender* of the stem.

Article-learning and Modification-learning Networks

The *article-learning* and *modification-learning* networks are shown in the regions marked 2 and 3 in Figure 1. These networks together model the process by which the child could learn to use the cues of Case, Number, the phonology of the noun, and its co-occurrence history, to predict the correct article, as well as to produce the corrected inflected form of the noun stem. In what follows, we will sometimes refer to the combination of the article-learning network and the modification-learning network as the *inflectional system*.

Each of these two networks is a typical three-layer connectionist architecture, whose inputs are representations of the noun's case, number, phonology and co-occurrence history. Case is represented by an 8-bit vector in which each of the four case possibilities is coded for by two bits. Number is represented by a 4-bit vector in which each of the two number possibilities is encoded in two bits. The phonological input is a 216-bit vector consisting of phonological distinctive feature representations of each phone in the noun stem; for further details of the phonological representation, the reader is referred to (MacWhinney et al., 1989). The hidden layer of each of these two networks comprises 60 units.

The output of the article-learning network is a representation of the correct article. This representation is a 12-bit vector in which two bits encode each of the six possible articles.

The outputs of the modification-learning network are the appropriate modifications that must be made to a noun stem, for a particular case, number and gender. The nine possible stem modifications are: umlauting of a vowel; addition of one or more of the suffixes *-e*, *-n*, *-s*, *-r*, *-ina*, *-se*, and *-ien*; and voicing of the final consonant in certain cases of suffixation. The output is represented as a 9-bit vector with one bit encoding each of the nine modifications.

Note that more than one of these modifications may be applicable to a particular noun stem in a particular case and number[3]. The primary determinants of the correct set of nominal markings given a particular case and number include (i) gender, (ii) the details of the phonological form of the stem, and (iii) a variety of semantic features which are not included in the present model. A complete linguistic analysis of this system can be found in (Mugdan, 1977).

As an example of training, suppose that the phrase *die Männer*, meaning "the men" (nominative plural), has been "heard". The inputs to both the article-learning network and the modification-learning network are patterns of activation representing Nominative case, Plural number, the phonology of the noun stem *Mann*, and the co-occurrence history of articles with the stem *Mann*. The article-learning network is trained to associate these items of information with the article it has observed (*die*). At the same time, the modification-learning network is trained to associate these same inputs with the inflectional changes that must be made to the stem *Mann*, viz., umlauting of the vowel, and suffixation of *-er*.

Simulations and Results

In the absence of detailed information about the linguistic input available to children learning German, we have based our data sets on a corpus of over 80,000 words from adult German usage (Wangler, 1963). From this corpus, we selected (on the basis of frequency) 2,094 inflected forms of 1,234 noun stems as the training data set, and another 315 inflected forms as a test data set.

Each *trial* involved presentation of input representing one of the 2,094 training patterns to the categorization, article-learning and modification-learning networks[4]. One *epoch* consisted of a trial for each of the 2094 words in the training set.

During training, the article-learning network was trained to produce the article appropriate for the presented word, while the modification-learning network was trained to produce the stem modifications

Epoch	% errors in: NOM	GEN	DAT	ACC
5	1 %	31 %	3 %	10 %
10	0 %	22 %	1 %	1 %
15	0 %	10 %	0 %	0 %
20	0 %	6 %	0 %	0 %

Table 2: Percentage of errors made by the article-learning network in various case contexts over the first 20 epochs of training. NOM=Nominative, GEN=Genitive, DAT=Dative, ACC=Accusative.

appropriate for the presented word. In both cases, training was via the back-propagation learning algorithm (Rumelhart et al., 1986a). Synchronously, on each trial, the categorization network was trained to categorize the co-occurrence of Case, Number and Article, via the competitive learning algorithm (Rumelhart and Zipser, 1986). This categorization response was additively encoded in the lexical representation of the noun stem, as described in the section discussing the categorization network. As a result, on next access of the lexical representation of this stem, the modified co-occurrence history became available.

Simulation 1 was run exactly as described above. The article-learning network learned to produce the correct article for all 2094 patterns in the training set in 66 epochs of training. The modification-learning network learned to produce the correct stem inflections in 68 epochs of training.

The types of errors made by the article-learning network at early stages in learning (over the first 20 epochs) parallel those made by German children learning this paradigm. First, the network learned all nominative forms within 5 epochs of training (see Table 2), which corresponds to childrens' early acquisition of the nominative. Second, the network made errors on an average 17% of genitive forms per epoch over the first 20 epochs, which corresponds to childrens' delayed acquisition of the genitive. Both of these results can be explained in terms of the fact that our training set incorporated approximately the real-world percentages of occurrence of various cases (40% for nominatives, 10% for genitives). Third, the response produced by the network was often below threshold for any of the possible articles, which corresponds to childrens' omission of articles. Fourth, the most common error was production of *der* for *des* for masculine and neuter nouns in the genitive singular, which would have been correct had the noun been of feminine gender (see Table 1). This can be interpreted as paralleling the child's overgeneralizations of a particular gender. These aspects of childrens' errors on the definite article are discussed in (MacWhinney, 1978) and (Mills, 1986).

To test generalization abilities, we examined the

[3]Note also that, although the total number of possible modifications is small, selection of the appropriate *set* of modifications for a given stem in each of the the eight cells of the declension (i.e., in each of the eight possible case-number combinations) involves a complex set of conditions. German has a large number of declensional classes with different assignments across these eight cells, with each class composed of many subgroups, partial regularities, and lists of exceptions.

[4]As noted previously, these inputs were representations of the Case, Number, stem phonology, and stem co-occurrence history. During training, the correct article and stem modifications were also presented, whereas during testing, they were not presented.

responses of the networks to patterns on which they had not been trained. The testing set of 315 forms consisted of 175 forms representing stems the networks had been trained on in other case-number contexts (*familiar-stem tests*), and 140 forms representing stems the network had not been exposed to at all (*novel-stem tests*). Once the article-learning network had learned the training set with 100% accuracy, it produced an incorrect article on only 7 of the 175 familiar-stem test forms (4% error rate), and on only 14 of the 140 novel-stem test forms (10% error rate)[5]. Similarly, once the modification-learning network had learned the training set to criterion, it produced correct modifications on 257 of the 315 generalization test forms (82% correct generalization). Thus, both the article-learning and modification-learning networks exhibited a substantial capacity for both kinds of generalization.

Co-occurrence information was created as described in the section discussing the categorization network; it categorized stems according to gender. To examine the *usefulness* of this information, we ran a simulation (Simulation 2) in which the co-occurrence information was not provided to the article-learning and modification-learning networks. This simulation was in every other respect identical to the one previously described.

In Simulation 2, it took 800 epochs for the article-learning network to learn to produce the correct article for all items in the training set. This is significantly worse performance than that in Simulation 1 (error-free production of the article in 66 epochs of training). Furthermore, the errors made by the article-learning network at early points in training during Simulation 2 were mostly on nominative forms. This is quite unlike the developmental course observed in children, and also unlike Simulation 1. In Simulation 2, it took 117 epochs for the modification-learning network to learn to produce stem modifications correctly for the entire training set. This compares with 68 epochs in Simulation 1.

These comparisons between Simulations 1 and 2 demonstrate that the categorical grammatical gender information that develops is genuinely useful for processing, and highlights the fact that the explicit re-representation of information may be an important technique for models developed within the overall connectionist framework.

Discussion

The use of a separate competitive learning network appears to capture important categorization effects in the process of learning the German article. The question arises, however, of whether such processing has applicability outside the present domain. In this connection, it is interesting to note that the hippocampus has been hypothesized to create orthogonalized episodic encodings (McClelland et al., 1992), which is very similar to the notion of encoding co-occurrences in the present categorization network. The same general categorization mechanisms potentially also provide a basis for the encoding of various regularities and sub-groupings. For example, for the German nominal inflection system, such a mechanism could lead to lexical encodings of the *pluralization paradigm class* of the noun stem. Similarly, for a language such as Hungarian, co-occurrences could lead to encoding of the *vowel harmony class* of the stem. Thus mechanisms very similar to what we propose may, in fact, play an important and quite general role in learning and memory processes.

We have hard-wired the categorization network to receive only exactly the inputs that were expected to be powerfully predictive of gender, namely, Case, Number and Phonology. At present we do not have a satisfactory answer to this criticism, except to note that this criticism is probably partially applicable to almost any model that makes assumptions about input and output information. Further work would be needed to determine the performance of the categorization network under conditions of noisy and extraneous data.

As mentioned at the beginning of this paper, previous work by the second author (MacWhinney et al., 1989) has addressed some of the same issues as the present model. This earlier work presented a computational model that learned the definite article in German without rules, and which matched the developmental data. The present model also uses a cue-driven system to match the developmental sequence of article learning observed in German children.

However, in achieving our aim of modeling the formation and utilization of grammatical gender in German, we feel we have made the following additional, significant, demonstrations, none of which was addressed by the (MacWhinney et al., 1989) model.

First, we have demonstrated how categorical information can be *created* through co-occurrence learning, *made available* in explicit distributed form, and usefully *utilized* by other parts of the processing system. The categorizations created by the competitive learning network in our model in effect construct the paradigm for the German definite article, but without reliance on the problematic row- and column-splitting mechanisms in the MacWhinney-Pinker account (MacWhinney, 1978; Pinker, 1984). We have shown how the encoding of these categorizations in the lexicon can lead to classification of nouns by gender. We have also shown that this

[5]Non-erroneous responses consisted of either production of the correct article (78% and 58% for familiar- and novel-stem generalization, respectively), or omission of the article altogether (18% and 32% for familiar- and novel-stem generalization, respectively).

gender/co-occurrence information is useful in the processes of learning the inflectional system.

Second, we have combined the task of learning the German definite article, examined previously in (MacWhinney et al., 1989), with the task of learning to inflect the noun stem that the article accompanies. This is a significant extension over the earlier model, both in terms of *coverage of linguistic phenomena*, and in terms of *integration of different kinds of processing* (grammatical categorization, and inflectional marking). We are not aware of any previous computational model of the present domain that combines these processes.

Third, we have demonstrated the integration of various *types* of information that have been regarded as important in language learning, viz., co-occurrence information (co-occurrence of Case, Number and Article), semantic information (the semantically based notions of Case and Number), and surface features (phonological information), and we have shown how these types of information can be usefully combined in a learning system. In effect, we have devised a computational implementation of the type of learning proposed in (Maratsos and Chalkley, 1980). To the best of our knowledge, such an implementation has not previously been constructed.

Fourth, we believe that it is vital for cognitive modeling of language to scale up to dealing with realistically sized data sets, because it is only then that linguistic regularities and sub-regularities really emerge. Our simulations used over 1200 noun stems in over 2000 inflected forms. We feel that this steps beyond the realm of a toy-sized model, and thus constitutes the beginnings of an important demonstration of realistic robustness. It also represents a substantial scaling up from the model in (MacWhinney et al., 1989), which used a training corpus consisting of 305 inflected forms of 102 noun stems[6].

In conclusion, the present work offers the first computational account of the synthesis of various kinds of information that have been regarded as important in language leaning. It also suggests how grammatical categories could develop and constitute useful processing information. Finally, this research begins to address questions about the ability of models of language acquisition to scale up to dealing with more realistic data.

[6]We have limited our data set to approximately 2,000 training forms, in order to reduce the time required to run a simulation. (Larger training sets would mean more stimuli per epoch, but would not affect the computational *tractability* of the simulation). However, it is not clear at what training set size all the basic regularities and patterns will be represented in the input set. We therefore consider it important to examine the effect of further increases in training set size.

Acknowledgements

We thank Jay McClelland, Dave Plaut and Dave Touretzky for helpful discussion, and Jay McClelland for use of computing facilities. Jared Leinbach developed formatting programs for network inputs and outputs. Of course, the present authors remain responsible for any errors in this work.

References

MacWhinney, B. (1978). The acquisition of morphophonology. *Monographs of the Society for Research in Child Development*, 43.

MacWhinney, B. (1991). Connectionism as a framework for language acquisition theory. In Miller, J., editor, *Research on Child Language Disorders*. Pro-Ed, Austin, TX.

MacWhinney, B., Leinbach, J., Taraban, R., and McDonald, J. L. (1989). Language learning: Cues or rules? *Journal of Memory and Language*, 28:255–277.

Maratsos, M. (1982). The child's construction of grammatical categories. In Wanner, E. and Gleitman, L., editors, *Language Acquisition: The State of the Art*. Cambridge University Press, New York.

Maratsos, M. and Chalkley, M. (1980). The internal language of children's syntax: The ontogenesis and representation of syntactic categories. In Nelson, K., editor, *Children's Language*, volume 2. Gardner, New York.

McClelland, J., McNaughton, B. L., O'Reilly, R. C., and Nadel, L. (1992). Complementary roles of hippocampus and neocortex in learning and memory. Society for Neuroscience abstracts, submitted.

Miikkulainen, R. (1990). A distributed feature map model of the lexicon. In *Proceedings of the Twelfth Annual Conference of the Cognitive Science Society*. Lawrence Erlbaum, Hillsdale, NJ.

Mills, A. E. (1986). *The Acquisition of Gender*. Springer-Verlag, Berlin.

Mugdan, M. (1977). *Flexionsmorphologie und Psycholinguistik*. Gunter Narr, Tubingen.

Pinker, S. (1984). *Language Learnability and Language Development*. Harvard University Press, Cambridge, MA.

Rumelhart, D., Hinton, G., and Williams, R. (1986a). Learning internal representations by error propagation. In (Rumelhart et al., 1986b).

Rumelhart, D. E., McClelland, J. L., and the PDP Research Group (1986b). *Parallel Distributed Processing*, volume 1: Foundations. MIT Press, Cambridge, MA.

Rumelhart, D. E. and Zipser, D. (1986). Feature discovery by competitive learning. In (Rumelhart et al., 1986b).

Taraban, R., McDonald, J. L., and MacWhinney, B. (1989). Category learning in a connectionist model: Learning to decline the German definite article. In Corrigan, R., Eckman, F., and Noonan, M., editors, *Linguistic Categorization*. Benjamins, New York.

Wangler, H. H. (1963). *Rangworterbuch hochdeutscher Umgangsprache*. Elwert, Marburg, Germany.

Rules or Connections? The Past Tense Revisited

Kim Daugherty
Hughes Aircraft Company
and
University of Southern California
kimd@gizmo.usc.edu

Mark S. Seidenberg
University of Southern California
marks@neuro.usc.edu

Abstract

We describe a connectionist model of the past tense that generates both regular and irregular past tense forms with good generalization. The model also exhibits frequency effects that have been taken as evidence for a past tense rule (Pinker, 1991) and consistency effects that are not predicted by rule-based accounts. Although not a complete account of the past tense, this work suggests that connectionist models may capture generalizations about linguistic phenomena that rule-based accounts miss.

Introduction

A seemingly minor aspect of linguistic knowledge—the past tense of verbs—has generated considerable debate over the role of connectionist models in explaining language. Whereas Rumelhart & McClelland (1986) claimed that their model of past tense learning illustrated a new way of explaining linguistic phenomena, Pinker & Prince's (1988) critique of this work suggested that connectionism had little to add to standard linguistic accounts. Several developments in subsequent years have moved this debate forward. Pinker (1991), for example, now agrees that connectionist networks are needed in order to account for facts about irregular past tenses (e.g., SING-SANG) and generalization. Moreover, the models developed by MacWhinney & Leinbach (1991), Plunkett & Marchman (1989), and Cottrell & Plunkett (1991) show that connectionist networks can exhibit many of the characteristics of the past tense that were lacking in the original Rumelhart & McClelland (1986) work. Thus, there has been considerable movement toward theories of the past tense in which connectionist networks play a central explanatory role. Importantly, however, Pinker (1991) and Kim et al. (1991) retain the idea that a proper account of the past tense will have to include a rule governing regular forms such as LIKE-LIKED. This conclusion is based on a mass of behavioral and other types of evidence thought to implicate this rule. Insofar as connectionist models do not, by definition, incorporate this type of knowledge representation, connectionism cannot provide a complete account of the past tense. Pinker therefore opts for a mixed model employing both a rule and a net.

It would be important to determine whether a connectionist network can explain all of the relevant facts about the past tense or whether, as Pinker suggests, it will have to be supplemented by a rule, since these represent very different claims about linguistic knowledge. Although we cannot present a complete account of the past tense in this paper, we will describe a simple connectionist model that learns the past tense and use it to develop two central points. The first is that the kinds of data that Pinker (1991) sees as evidence for a rule of past tense formation may reflect very simple properties of connectionist nets. As such, these phenomena cannot be taken as uniquely compatible with the rule-based account. This argument is mostly negative: it says that the kinds of evidence that Pinker takes to differentiate between network and rule-based accounts are consistent with both approaches. Our second point, however, is that the network exhibits behaviors that are not predicted by the traditional (Pinker & Prince, 1988) or modified traditional (Pinker, 1991) linguistic approaches. Moreover, these behaviors are also observed in psycholinguistic studies of people (Seidenberg, in press; Seidenberg & Bruck, 1990). Therefore, the network model is not merely an alternative to a rule-based account; rather, it is to be preferred because it captures generalizations that rule-based accounts miss.

Architecture of the Model

The model is a simple feed-forward network. The input layer represents the present tense of a monosyllabic verb; the output layer, its past tense. The phonological representation is similar to one used by MacWhinney et al. (1989). There were 120 phonological units on each layer, representing a CCCVVCCC template for monosyllables in English. This is the maximum structure a syllable can take; the double V is needed to represent dipthongs. Each phonemic segment is represented by 15 binary articulatory features, using a scheme developed by Plunkett and Prince and modified by Mary Hare. The features are: back, tense, low, medium, high, glide, sonorant, fricative, stop, labial, coronal, velar, nasal, sibilant, and voiced. This scheme represents a plausible

```
C C C V V C C C
    b l a   k
      b a   k
```

Figure 1: Phonological Representation of a Syllable

compromise among various proposals within phonetics. If a feature exists for a segment, its value is 1.0; if not, its value is 0.0.

The phonological representation is centered on the nucleus of the syllable as shown in Figure 1. Hence, the rimes of the words **BLACK** and **BACK** receive the same representation. Aligning the representations on the rimes was thought to be desirable because of the perceptual salience of these units in English (e.g., Pinker & Prince, 1988; Seidenberg & McClelland, 1989). For unused segments in a word representation, the units are set to 0.0. There were 200 hidden units. Figure 2 shows the architecture of the entire model.

The phonological form of a present tense stem (e.g., BAKE) is activated on the input units and the model's task is to generate the phonological form of the past tense (e.g., BAKED) on the output units (we use orthography here to represent these phonological codes for typographical convenience). The model was presented with present-past tense pairs during training, with frequency of exposure determined by the logarithm of the verb's Francis & Kucera (1982) frequency. Weight correction was determined by standard back-propagation (Rumelhart, Hinton, & Williams, 1986). In scoring the model's performance, we determined for each phonemic segment whether the best fit to the computed output was provided by the correct target. The output pattern was scored as correct only if the correct targets provided the best fit for all segments in a word. We also calculated the total sum of squared error as a measure of goodness of fit.

Simulation 1: Initial Corpus

Two simulations using the above architecture but different training sets were performed. In the first simulation, the training set consisted of all monosyllabic present-past tense pairs with Francis and Kucera

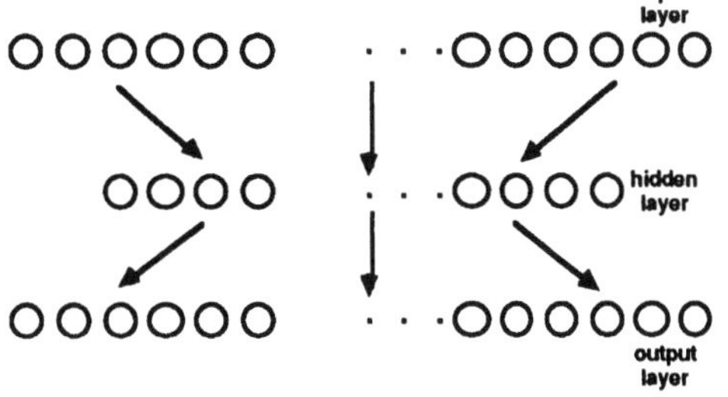

Figure 2: Architecture of the Model

frequencies greater than 1. This included 309 verbs with regular past tenses ("regular verbs") and 104 verbs with irregular past tenses. 112 regular verbs with frequency = 1 were reserved for testing the trained network's capacity to generalize on novel items. The present/past pairs were probabalistically presented during training according to their frequency. The most frequent pairs were presented once per epoch; the least frequent once per 100 epochs. The model was trained on this corpus for 400 epochs, at which point learning approached asymptote. The results below were averaged over three training sessions with random initial weights.

In terms of overall performance, the model learned all (100%) of the regular past tenses and 86/102 (84.3%) of the irregular forms. Errors on the irregulars included regularizations (FALL-FALLED), no change errors (GET-GET, analogous to HIT-HIT), and vowel errors (HIDE-HED).

After training, the model's performance on 112 additional verbs was used to assess its capacity to generalize. The model produced correct output for 84/112 (76.8%) of these items. The two most frequent errors were no change (PEEK-PEEK) and assimilation with phonologically-similar irregular past tenses in the training set (e.g., SEEP-SEPT, which is similar to SWEEP-SWEPT, and WRITHE-WROTHE, which is similar to WRITE-WROTE).

We then examined two theoretically-important phenomena, consistency and frequency effects. Consistency effects have been identified in previous work on spelling-sound correspondences and were simulated in the Seidenberg & McClelland (1989) model of word pronunciation. Briefly, networks trained using learning algorithms such as backpropagation (Rumelhart, Hinton, & Williams, 1986) pick up on the consistency of the mapping between input and output codes. The mapping between the present and past tenses is highly consistent in English because most past tenses obey the regular rule. However, the mapping is not entirely predictable because of irregular cases such as TAKE-TOOK and SIT-SAT. Standard accounts such as Pinker & Prince's (1988) distinguish between rule-governed cases and exceptions. However, connectionist models predict the existence of intermediate cases, so-called "regular but inconsistent" patterns such as BAKE-BAKED and FLIT-FLITTED, which obey the rule but have inconsistent "neighbors" (Seidenberg, in press). Thus, even though BAKE-BAKED is rule-governed, performance may be impaired because the model must also encode MAKE-MADE and TAKE-TOOK. Specifically, it should be worse than on a completely regular pattern such as LIKE-LIKED (all of the -IKE verbs have regular past tenses). Thus, the standard theory predicts that BAKE-BAKED should pattern with LIKE-LIKED, because both are rule governed. However, a backprop net might be expected to perform more poorly on inconsistent

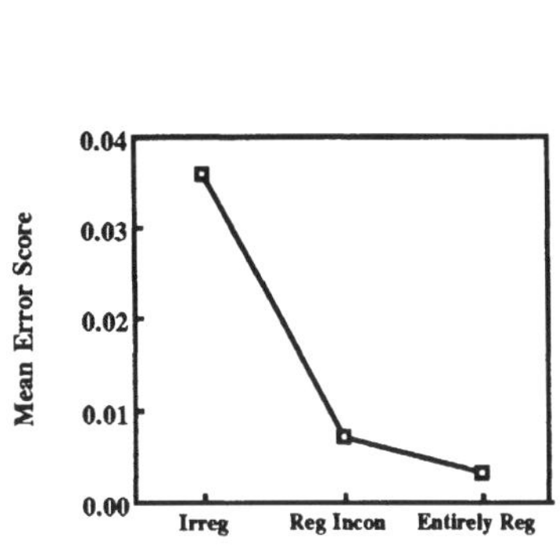

Figure 3: Performance on Matched Subsets of Items

items such as BAKE-BAKED, owing to TAKE-TOOK and MAKE-MADE.

To test this prediction, we identified sets of 60 entirely regular, 60 regular inconsistent, and 60 irregular verbs, equated in terms of frequency Performance on these words is shown in Figure 3. As predicted, the entirely regular verbs yielded better performance than inconsistent verbs, which in turn generated better performance than irregular verbs. Analogous results have been reported in the domain of spelling-sound correspondences (e.g., MUST is regular, HAVE is irregular, GAVE is regular but inconsistent) and simulated by the Seidenberg & McClelland (1989) model. Importantly, Seidenberg & Bruck (1990) and Seidenberg (in press) observed these effects in a study of past tense generation. The subjects in their experiment (college students) were presented with a present tense stem on each trial and had to generate the past tense. Response latencies followed the pattern illustrated in Figure 3: Irregular >> Regular Inconsistent > Entirely Regular. Thus, the model picks up on an important fact about peoples' knowledge of verbs: they encode the degree of consistency in the mapping between present and past tenses. It is important to recognize that these intermediate, regular-but-inconsistent cases are not predicted by either the traditional (Pinker & Prince, 1988) or modified traditional (Pinker, 1991) theories of the past tense. They are, however, a consequence of learning in a multi-layer backprop net. The reason for this is simple. The traditional accounts have the regular and irregular verbs processed by separate mechanisms. Hence there is no basis for predicting that they will interfere with each other. Our network, however, encodes both regular and irregular past tenses in the same set of connection weights. Hence the processing of a "rule-governed" item is affected by whether it has an irregular neighbor or not.

We also examined a second important phenomenon, frequency effects. Pinker (1991) has accumulated several types of evidence thought to support the existence of a past tense rule. The question that arises is whether such phenomena could also be captured by a connectionist net. This issue can be illustrated as follows. Prasada, Pinker & Snyder (1990) observed that the frequency of a past tense form (how often it is used in the language) affects the generation of irregular past tenses, but not regulars. TOOK, for example, is higher in frequency than RANG and takes longer for subjects to generate. However, there is no frequency effect for regular past tenses; BIKED (low frequency) is as easy to generate as LIKED (high frequency). Pinker (1991) interprets this pattern as follows. Regular past tenses are generated by rule; hence they are not affected by frequency. All that matters is how long it takes to recognize the present-tense stem and apply the rule. Irregular pasts are different, however. Either they have to be looked up in a list (traditional theory) or generated by a connectionist net (modified traditional theory). Both processes are thought to be affected by frequency. Thus, the interaction between frequency and regularity of the past tense was thought to implicate two separate mechanisms, a rule and a net.

We thought it likely, however, that our net would also produce this interaction, mainly because we observed the same effect in the Seidenberg & McClelland (1989) model. In that model, frequency has a bigger effect on words with irregular pronunciations (e.g., DEAF, SHOE) than words with regular, rule-governed pronunciations (LIKE, MALE). The explanation for this effect is also simple. Regular, "rule-governed" words contain patterns that occur repeatedly in the corpus. The weights reflect exposure to all these patterns. Mastering a rule-governed instance does not depend very much on its frequency because performance benefits from exposure to neighbors that contain the same pattern. Mastering an irregular instance, however, is highly sensitive to frequency; performance on DEAF (irregular pronunciation) or TAKE-TOOK (irregular past tense) depends on how often the model is exposed to these patterns because the correct output cannot be derived from exposure to neighbors. Thus, we expected that at least one of the behavioral phenomenon that Pinker takes as evidence for a rule would be exhibited by our model.

To test this prediction, we examined the 50 highest frequency regular verbs, the 50 lowest frequency

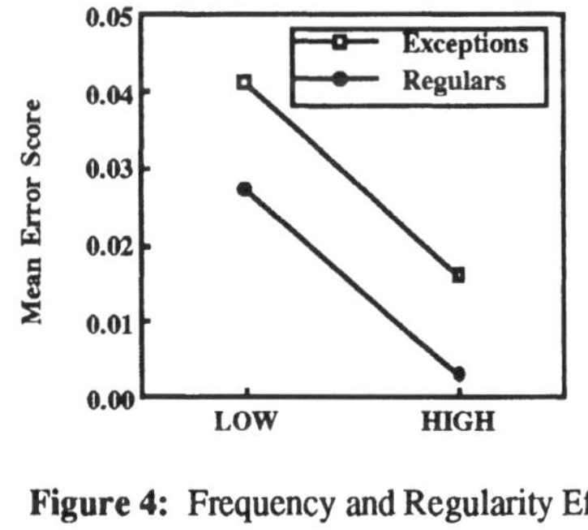

Figure 4: Frequency and Regularity Effects

regular verbs, the 50 highest frequency irregular verbs, and the 50 lowest frequency irregular verbs from the training set. The model's performance on these items is summarized in Figure 4. The results are not as predicted. There are main effects of verb type and frequency but no interaction between the two.

To summarize, the results of the initial simulation were mixed. The model was able to learn all of the regular items in the training set and a high proportion of the irregular items; it produced plausible errors and correct output on most generalization trials. It also showed the consistency effects seen in the behavioral studies, which are not predicted by traditional accounts. However, the model's performance is problematical in two respects. First, it performed more poorly than people on irregular items and generalization trials. Second, it did not exhibit the interaction between frequency and regularity.

At this point, we noticed several hints that the defects in the model's performance were principally due to the large number of irregular items in the training corpus. The model failed to master all of the irregulars in the training corpus, and produced unexpected frequency effects on regular items. Moreover, many of the errors on the generalization trials seemed to occur because the model was affected by phonologically-similar irregular past tenses in the training corpus. In order to assess the effects of the proportion of irregular items on performance, we conducted simulations in which we varied the number of irregular verbs in the training set, while keeping the number of regular verbs constant. The results are shown in Figure 5. As seen in the figure, the number of errors on generalization trials was related to the number of irregular verbs in the training set. The stimuli used in the generalization tests all require the regular past tense; however, some of them are entirely regular with respect to the training corpus whereas some are regular but inconsistent. Figure 6 presents the results for these two types of generalization trials separately. The number of errors on entirely regular novel verbs remained largely invariant as the number of irregular verbs was increased. However, the number of errors on regular inconsistent novel verbs was affected by the number of irregular verbs in the training set. This indicates that irregular verb neighbors create interference for regular inconsistent verbs. Entirely regular verbs are unaffected because they do not have irregular verb neighbors.

Together, these findings suggested that the large number of irregular verbs in the training set was adversely affecting performance.

Simulation 2: A More Realistic Corpus

We then compared the type and token frequencies in our corpus to those in the language at large. An analysis of the Francis & Kucera (1982) sample revealed that irregular verbs comprise 5% of all verb types listed there and 22% of the verb tokens. In the corpus employed in the first simulation, 25% of the verbs were irregular and they accounted for 65% of the tokens presented during training. Thus, irregular items were overrepresented in our training corpus compared to the language as a whole. Other factors also contributed to the overrepresentation of irregulars as well. The model's phonological representation only permitted the use of monosyllabic words, and the proportion of irregular verbs is higher among the monosyllabic words than the multisyllabic words (this is because most irregular verbs are monosyllabic). Finally, regular verbs predominate in the lower frequency range; the training corpus was restricted to items with frequency > 1, meaning that the many regular but very low frequency verbs were excluded. In sum, the large number of irregular items in the training corpus had a negative impact on the model's performance, but these words were overrepresented in the training corpus.

A new training set was then constructed with the goal of maintaining more realistic proportions of regular and irregular verbs. We also attempted to represent the different types of irregular verbs accurately. Pinker & Prince identified 25 classes and sub-classes of irregular verbs which we collapsed into 5 major classes reflecting the most important subtypes. The classes were no

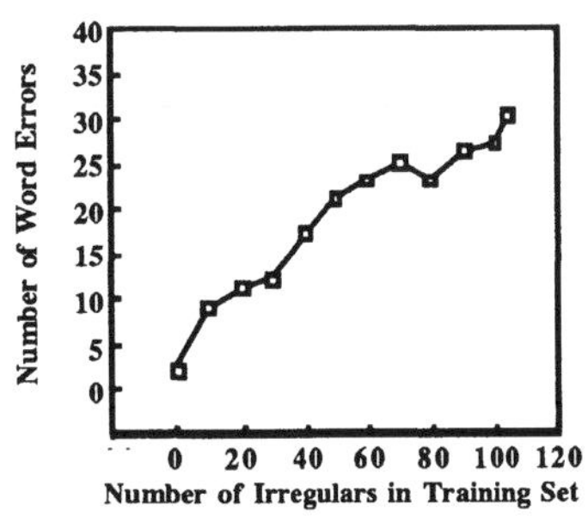

Figure 5: Generalization Errors vs. Number of Irregulars in Training Set

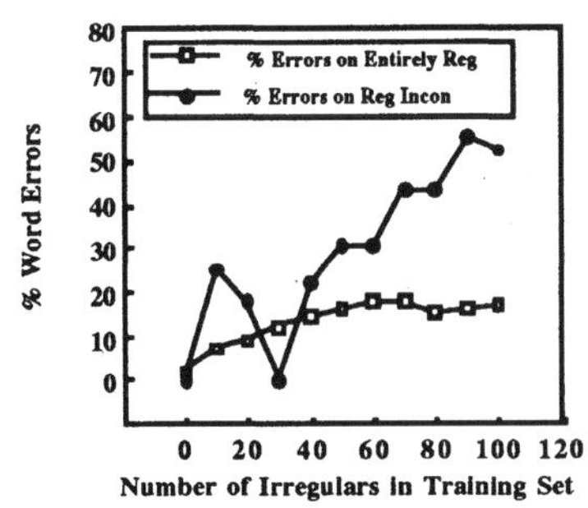

Figure 6: Generalization Errors vs. Number Irregulars in Training Set

change (HIT-HIT), vowel internal change (MEET -MET), vowel internal change plus consonant (LEAVE-LEFT), suppletion (GO-WENT), and consonant change (SEND-SENT). We then devised the training set so that the proportions of these subtypes matched the proportions in the Francis & Kucera corpus. The new training set included 309 regular verbs and 24 irregular verbs. The number of irregular verbs had to be relatively small in order to maintain the correct proportions while keeping the overall size of the training corpus within manageable limits.

The model was trained as in the previous simulation for 500 epochs, at which point performance approached asymptote. The following results reflect averages over three training sessions with random initial weights. All regular verbs in the training set were learned, as before. 22 of the 24 irregular verbs in the training set (92%) were learned, better than in the previous simulation. The 2 errors on irregular verbs were FALL-FELLED an overregulaization error and WIN-WAUN a vowel error.

On the generalization trials, 105 of 112 regular past tenses (94%) were correctly generated, an improvement over the first simulation and a rate that compares well with people. The 7 past tenses that were incorrectly generated were MERGE-MERGT, WAKE-WAKED, WHINE-WOOND, CLINK-CLANGT, WANE-WONE, MEW-VIEW, BROOK-BROOK. The first reflects a substitution of /t/ for /d/ (i.e., incorrect voicing) and the second is an overregularization error. The others are a variety of vowel, consonant and no change errors. As before, some of these can be described as assimilation with irregular verbs in the training set. WHINE-WHOUND is similar to WIND-WOUND, CLINK-CLANGT is similar to CLING-CLANG. Subjects in behavioral experiments produce some of these responses as well (e.g., Bybee & Moder, 1983).

We then re-examined the consistency and frequency effects described earlier. For the consistency effects, we constructed sets of 20 entirely regular, 20 regular inconsistent, and 20 irregular verbs from the training set equated in terms of frequency. The model's performance on the three types is given in Figure 7. As in the previous simulation, the model showed the graded

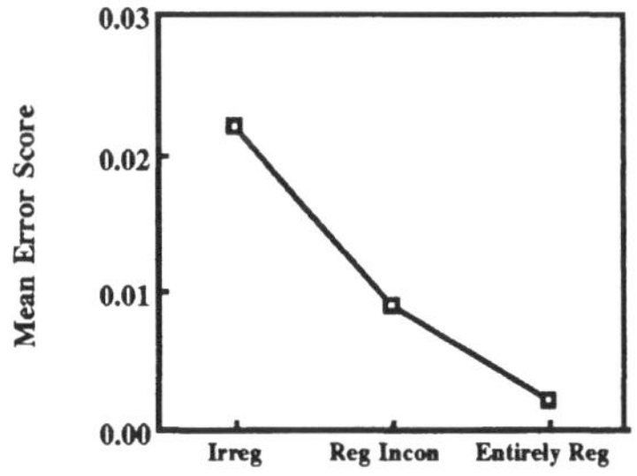

Figure 7: Performance on Matched Subsets of Items

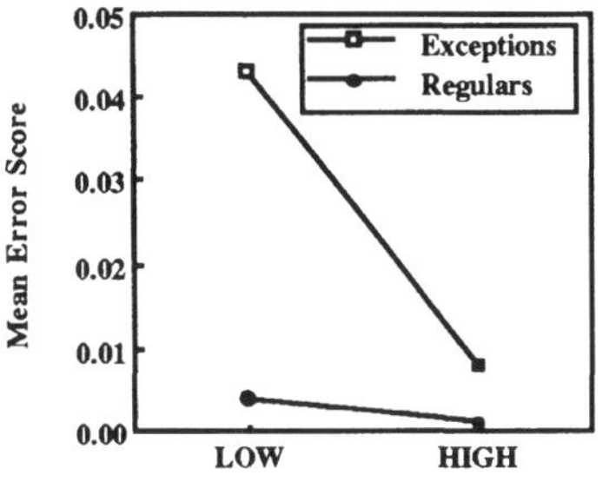

Figure 8: Frequency and Regularity Effects

effects of the consistency of the mapping between present and past tense (the difference between the two types of "regular" verbs) that is not predicted by rule-based accounts. The consistency effect was actually larger than in the previous simulation. In addition, the model now exhibits the predicted interaction of frequency and regularity. We constructed sets of the 10 highest frequency regular verbs, 10 lowest frequency regular verbs, 10 highest frequency irregular verbs, and 10 lowest frequency irregular verbs from the training set. The model's performance on these items is shown in Figure 8. For regular verbs, frequency has little effect on performance. For irregular verbs, performance is better on high frequency items than on low frequency items. This is the pattern that was reported by Pinker (1991) and Prasada et al. (1990) and taken as evidence for a rule-based mechanism.

Thus, changing the corpus so that it better reflected the facts about the distribution of verb types in the language yielded better simulation results. The model continued to master the regular items and over 90% of the irregulars; there was better generalization on novel verbs; and the consistency effect and frequency by regularity interaction seen in behavioral studies were obtained.

Discussion

Our results, like those of MacWhinney & Leinbach (1991), Plunkett & Marchman (1989), and Cottrell & Plunkett (1991), suggest that connectionist models can exhibit the phenomena that Pinker & Prince (1988) see as central to an understanding of the past tense. We are in no way restricted to the mechanisms employed by Rumelhart & McClelland (1986) and are not doomed to the same failures. Two aspects of our models contributed to their relatively better performance. First, the phonological representation that we employed addresses many of the concerns that Pinker & Prince expressed concerning the Wickelphonology that Rumelhart & McClelland had used. Our representation of segments is motivated by articulatory constraints and the slot positions in the representation are motivated by independent evidence concerning the salience of the rime. This representation is by no means complete; however, it utilizes plausible featural, segmental, and syllabic representations, and avoids some of the problems with earlier approaches. Second, the simulations highlight the importance of using a realistic training regime (see also Hetherington & Seidenberg, 1989). Our first simulation clearly overrepresented the number of irregular types and tokens in the training set and its performance was inadequate. Once the training set was modified to be more realistic, performance improved greatly. This result suggests the interesting possibility that for English, the past tense is learnable only if the proportion of irregular pasts is limited.

By understanding the nature of the input representation, the learning algorithm, the phenomena we were trying to capture, and the architecture of the model, we were able to make predictions about the difficulty of learning different types of verbs. In the second simulation, we observed the expected consistency and frequency effects. These effects have also been observed in experiments with human subjects and replicate results that have been obtained in another domain (spelling-sound correspondences; Seidenberg & McClelland, 1989). The frequency effects (i.e. the fact that frequency only affects irregular pasts, not regular pasts) indicate that the kind of phenomena that Pinker (1991) cites as evidence for a rule may be simply captured within connectionist nets. Of course, it remains to be seen whether all of the phenomena he cites can be accommodated in the same way.

The consistency effects are not predicted by the earlier theories and strongly implicate the connectionist alternative. Of course, it might be possible to modify the traditional (Pinker & Prince, 1988) or modified traditional (Pinker, 1991) theories to accommodate these results. Pinker does not present implemented computational models; he describes models and assigns computational properties to them by stipulation. Working at this level of description, it might be possible to formulate a new theory that preserves the idea of a rule but accommodates the new behavioral phenomena. For example, one might introduce ad hoc assumptions about interactions between the rule and network mechanisms. If that were the case, however, the rule-based theory could be said to be useful insofar as it is able to "implement" the connectionist model. Thus, whereas Pinker & Prince (1988) presented a view in which connectionist accounts of language are parasitic on symbolic accounts, we view the present results as suggesting the opposite relation. Of course, it will be necessary to develop our models further, so as to account for the entire range of verb-related phenomena. We take the present results as one step along the route toward an explanatory computational theory.

References

Bybee, J.L. & Moder, C. (1983). Morphological classes as natural categories. *Language, 59*, 251-270.

Cottrell, G. & Plunkett, K. 1991. Using a recurrent net to learn the past tense. In Proceedings of the Cognitive Science Society Conference. Erlbaum.

Francis, W.N. & Kucera, H. 1982. *Frequency Analysis of English Usage*. Houghton-Mifflin.

Hetherington, P., & Seidenberg, M.S. 1989. Is there "catastrophic interference" in connectionist networks? In Proceedings of the Cognitive Science Society Conference. 26-33. Erlbaum.

Kim, J.J., Pinker, S., Prince, A. & Prasada, S. 1991. Why no mere mortal has ever flown out to center field. *Cognitive Science*, 15:73-218.

MacWhinney, B. & Leinbach, J. 1991. Implementations are not conceptualizations: Revising the verb learning model. *Cognition*, 40:121-157.

MacWhinney, B., Leinbach, J., Taraban, R., & McDonald, J. (1989). Language learning: Cues or rules? *Journal of Memory & Language, 28*, 255-277

Pinker, S. 1991. Rules of language. *Science*, 253:530-534.

Pinker, S. & Prince, A. 1988. On language and connectionism. *Cognition*, 28:73-194.

Plunkett, K & Marchman, V. 1991. U-shaped learning and frequency effects in a multi-layered perceptron. *Cognition*, 39:43-102.

Prasada, S., Pinker, S., & Snyder, W. 1990. Some evidence that irregular forms are retrieved from memory but regular forms are rule generated. Psychonomic Society meeting, November.

Rumelhart, D., Hinton, G., & Williams, R.J. 1986. Learning internal representations by error propagation. In Rumelhart, D. & McClelland, J. (Eds.), *Parallel Distributed Processing, vol. 1*. MIT Press.

Rumelhart, D. & McClelland, J.L. 1986. On learning the past tenses of English verbs. In Rumelhart, D.E. & McClelland, J.L. (Eds.), *Parallel Distributed Processing, vol. 2*. MIT Press.

Seidenberg, M.(In press). Connectionism without tears. In S. Davis (Ed.), *Connectionism: Theory and Practice*. Oxford University Press.

Seidenberg, M., & Bruck, M. 1990. Frequency and consistency effects in the past tense. Psychonomic Society meeting, November.

Seidenberg, M.S. & McClelland, J.L. 1989. A distributed, developmental model of word recognition and naming. *Psychological Review*, 96:523-568.

A connectionist account of English inflectional morphology: Evidence from language change

Mary Hare
Department of Psychology, Birkbeck College, University of London
hare@crl.ucsd.edu

Jeffrey L. Elman
Department of Cognitive Science, University of California, San Diego
elman@crl.ucsd.edu

Abstract

One example of linguistic productivity that has been much discussed in the developmental literature is the verb inflection system of English. Opinion is divided on the issue of whether regular/irregular distinctions in surface behavior must be attributed to an underlying distinction in the mechanisms of production. Looking at the course of historical development in English, the current paper evaluates potential shortcomings of two competing approaches. Two sets of simulations are presented. The first argues that a single-mechanism model offers a natural account of historical facts that would be problematic for a dual mechanism approach. The second addresses a potential problem for a single-mechanism account, the question of default behavior, and demonstrates that even in the absence of superior type frequency a network is capable of developing a "default" category. We conclude that the single network account offers a more promising mechanism for explaining English verb inflection.[1]

Introduction

A crucial fact about natural language is its productivity, and any successful model of language must offer an explanation for how this is accomplished. One example that has been much discussed in the developmental literature is the verb inflection system in English. Although the great majority of English verbs form the past tense through the regular and productive process of adding the suffix -d to the verb stem, English includes approximately 160 verbs which mark the past in some other way. Irregular verbal inflection has interesting properties that suggest it may be qualitatively different from the regular system.

There are two current proposals on how to account for these facts. One takes the difference in behavior between the regular and irregular systems to reflect an underlying difference in the mechanisms by which they are produced. On this account the regular verbs are the product of a symbolic system utilizing rules in the traditional sense, while irregularly inflected verbs result from an analogical network that identifies particular verbs as taking irregular past tense inflection, blocking the application of the regular rule (Pinker and Prince 1988). This view contrasts with the position taken by Rumelhart and McClelland (1986) and more recently by Plunkett and Marchman (1990, 1991). According to this work, both regular and irregular inflection result from the same mechanism, with differences in behavior attributed to other factors. The mechanism involved is a single connectionist network operating on largely analogical principles. The goal of the current paper is to consider these proposals in the light of other linguistically relevant data, having to do with historical change.

Issues

Both approaches deal only with the contemporary language facts. However, languages change over time, and it is important to ask how each account would explain historical change. The view of language as carried out by two competing subsystems raises interesting questions about the mechanisms of change. One reasonable assumption, given this approach, is that change involves the loss of a specific "blocking" effect: an exceptional inflection may be lost, allowing the regular rule to apply to a formerly irregular form. Hence irregulars may get regularized, but no mechanism is in place that would account for change in the opposite direction, or for analogical change among the regular verbs themselves. We believe historical change to be more complex than such a simple scenario suggests, and will report on the interim progress of a long-term project in this area. As we hope to demonstrate, a single-process model offers a natural account of facts that might be problematic for a dual mechanism approach and so appears to provide a more promising mechanism for historical change.

We will also address a potential problem for the network account, the need to learn a "default" inflection. It is important not to confound the phenomenon of a default with category size (cf Plunkett & Marchman 1991). Ear-

[1] This work was supported in part by a grant from the ESRC, a grant from the SERC Computational Science Initiative, and by a contract from U.S. Army Avionics (Ft. Monmouth, NJ).

ly work (Rumelhart and McClelland 1986) might be interpreted as suggesting that a network can learn to treat a category as an "elsewhere" case only if it is numerically superior to other categories. This interpretation arises from the belief that a network is a simple analogical system, capable of generalizing only on the basis of frequency and surface similarity. If this were true, a novel form would be treated as a member of the default category only if it resembles another form already learned in that category. Consequently, a true "default" could arise only if the category were well-populated and spanned the entire phonological space. We believe this to be an overly simplistic view of network dynamics, and will demonstrate that default behavior can be learned even in the absence of superior type frequency.

Historical data

Early Old English (ca 870) had approximately three classes of weak verbs, and seven classes of strong verbs. The strong verbs were the descendents of the IndoEuropean vowel change or ablaut series of verbal tense/number inflection. These classes decreased continually in size over the Old and Middle English periods, although remnants of the system appear among the irregular verbs even today (Stark, 1982, Wright 1925, Flom 1930).

The weak verb classes were also in a state of transition during this period. In early Old English, weak Class I, once the most productive weak class, had splintered into three subclasses and various exceptions. Class III was no longer a class at all, but rather four exceptional and very common verbs (live, have, think, and say). Class II, the largest and most productive, was the only class to exhibit a consistent paradigm. Over the OE period Class II attracted new members not only from the strong verb classes, but also from the two smaller subclasses of weak Class I.

One question of interest in the current paper is the motivation for this transfer from one weak class to another. Our claim is that the process of analogical attraction, which is known to affect the strong verbs of English, motivated the movement among the weak verbs as well. If true, the fact that the same process affected both categories of verbs is clear evidence in favor of a single mechanism account. Before continuing, however, we must justify an assumption that is basic to this argument: All classes of OE weak verbs were regular in the sense outlined in the Introduction. Lacking this, one might propose that a dual-mechanism account could easily explain the analogical behavior of the weak verbs by assuming that the verbs of Class I were irregular.

The clearest argument for the regular status of the weak verbs comes from Pinker and Prince's description of the dual-mechanism approach (1988). There the authors explicitly state that denominal verbs cannot be irregular:

"...irregularity is a property of verb roots. Nouns and adjectives by their very nature do not classify as irregular (or regular) with respect to the past tense.... Such verbs [denomincal and de-adjectival] can receive no special treatment and are inflected in accord with the regular system, regardless of any phonetic resemblance to strong roots."

All OE weak verb classes were made up of derived verbs, and in particular the members of both Class I and Class II were predominantly denominal (Flom 1930, Stark 1982). Thus, by established criteria, the dual-mechanism account defines these verbs as necessarily regular.

Ths paper offers a network account of why instability should arise in the first place, and why the resulting change took the direction it did. Since this account relies on the formal (phonological) similarity between members of the various weak verb classes, we will begin with a sketch of the relevant data (data based on Stark 1982).

Change in the Weak verb system

Early Old English

Proto-germanic weak verbs are classified according to the derivative suffix they took between the stem and the preterit suffix. Class I verbs were those which originally took the suffix *-j-*. This segment triggered various phonological changes in the verb stem, eventually resulting in three distinctive subclasses. In the indicative voice, typical EOE West Saxon Class I paradigms are as follows:

	Ia de:man 'judge'	Ib fremman 'do'	Ic nerjan 'save'
present:			
1st sg	*de:me*	*fremme*	*nerje*
2nd sg	*de:mst*	*fremest*	*nerest*
plural	*de:ma?*	*fremmad*	*nerja?*
past:			
1st sg	*de:mde*	*fremede*	*nerede*
plural	*de:mdon*	*fremedon*	*neredon*
pst. part.	*de:med*	*fremed*	*nered*

The subclasses of Class I can be distinguished by the form of their stems. Ia (*de:man*-type verbs) is made up of long-stem verbs, with either a long vowel or a final consonant cluster. Members of Ib (e.g., *fremman*) were originally short-stem verbs whose final consonant geminated under certain circumstances, resulting in a stem alternation between CVC and CVCC. Ic (e.g., *nerjan*) consisted of a small group of short-stem verbs ending in *r*, which did not geminate. Verbs of this third sub-class also had a high front glide (*j*) stem-finally in certain parts of the paradigm.

Class II verbs, unlike those of the Class I subgroups, had no formal criterion of membership. Stems were consistently CVC, with no alternation with CVCC forms, and no requirement that the stem end in a specific consonant or contain a long vowel. A typical Class II paradigm is given below:

	lufian 'love'	
	present:	past:
1st sg	*lufige*	*lufode*
2nd sg	*lufast*	*lufodest*
plural	*lufiad*	*lofodon*
	p. part: *lufod*	

The distinctions of note are these: In Class I, the short-stem verbs (Ib and Ic) take the suffix vowel *e* both in the personal endings of the present singular and as the "medial vowel" between the stem and suffix in the past. The long-stem verbs of Ia take no medial vowel except in the past participle. Verbs of Class II take the suffix vowel *a* in the present, and the medial vowel *-o-*. Furthermore, the high front vowel *i* appears in the Class II paradigm in all the forms where the corresponding glide *j* appears in Ic. II and Ia are large and internally coherent classes of verbs. Although Ib is smaller, it is still a good-sized class. Ic, on the other hand, is quite small.

Developments during Old English

Two phonological changes affecting the language as a whole had interesting consequences for the OE verbal system. First, throughout this period English developed a strong tendency toward glide vocalization. Both the *j* of the Ic (*nerjan*) verbs and the *ig* of the Class II 1sg present went to *i*. As a result, these two groups closely resembled each other, differing only in their medial vowel. At the same time, and arguably as a result of this increased formal similarity, the two classes collapsed into one. Verbs of the small Ic subclass adopted the medial vowel *-o-* of Class II, becoming indistinguishable from members of that class.

It was also during this period that English began to simplify its geminate consonants. Recall that one major distinction of the Ib (*fremman*) subclass was its alternation between geminate and non-geminate stems. This distinction disappeared by late OE. Interestingly, most of the verbs of the *fremman* subclass then adopted the Class II paradigm as well. At the same time a very small number of these verbs drifted into Class Ia.

The long-stem verbs of Class Ia continue unaffected throughout old English. By late OE there remained essentially two weak verb classes, Class II and the long-stemmed (*de:man*-type) subclass Ia. Over the course of Middle English the picture simplified further. Vowel reduction in unstressed syllables (and eventual deletion of the unstressed medial vowel) eliminated most remaining distinctions between conjugation classes.

The result of these changes is the regular past inflection of modern English. Thus from a historical perspective the regular past can be said to result from the operation of an analogical system, although this is obscured when one looks only at the synchronic data. In the next section we will demonstrate that a single network model is capable of accounting for this analogical change.

Network account of weak-verb change

Plunkett and Marchman (1991) analyze the conditions under which competing stem to inflectional mappings are learned in a single network. This work suggests that any mapping with low type and token frequency will be difficult to learn and is likely to be lost. Learning is aided, however, if the network is able to exploit phonological regularities in the relationship between the stem and inflected form. This is consistent with the weak-verb facts. At one stage in the language the smaller weak classes exhibited a certain amount of formal coherence. This made the learning task easier, since it allowed the learner to exploit information about the phonological characteristics of each class. It was as general phonological change eroded these characteristics that membership shifted.

In the simulations that follow we will demonstrate that this combination of factors leads to the correct behavior in the network. Part of the training procedure involves training multiple 'generations' of networks; the targets for each new generation are the outputs (after learning) from the previous generation. Although this training regimen does not claim to exactly mimic the time-course of language change across generations of speakers, it does capture the gradual nature of such change, and the causal role played by inaccurate transmission. In this way we hope to model one of the mechanisms underlying historical change.

Architecture and stimuli

The problem was modeled with a feedforward network implementing the back-propagation learning algorithm. The input bank consisted of 480 units standing for individual verbs, and 6 units standing for individual tense/number inflections. At each training iteration, one "verb" unit and one "inflection" unit were activated simultaneously, and the task of the network was to produce a representation of the inflected verb over the output units. The output was designed to represent the formal features that distinguished the various classes of weak verbs. For each 21-element output there were 12 units dedicated to these features, followed by a 9-element random pattern intended both to mark each verb as unique and to allow the network to treat each set of six inflected forms as individual manifestations of the same verb. The 12 "inflection" units represented the following information: (1) presence and identity of medial (or inflectional) vowel; (2)presence and identity of stem-final high segment; (3) presence or absence of gemination; and (4) presence or absence of a long vowel.

There were 480 "verbs", divided into four subclasses: 32 in class Ic, 64 in class Ib, 128 in class Ia, and 256 in class II. Each verb was learned in six inflected forms: 1st sg, 2nd sg, and plural present, 1st sg and plural preterit, and past participle. These specific forms were chosen because in combination they illustrate the significant distinctions among the four subclasses of verbs.

Training and results

A training regimen was designed to test the hypothesis that low frequency mappings are difficult for the network to learn unless they are formally distinct. The first input-output mapping was based on the canonical OE weak verb system as described in Section III. A network was trained for 30 passes through the set of verbs in each of the six forms. After training the three largest classes (II, Ia, and Ib) were all produced correctly. The small subclass Ic showed the effect of attraction to Class II, producing a vowel rather than a glide for the stem-final high segment. The medial vowel in these verbs, however, is still that of Class I.

At this point a new network was set up and taught to produce the classes as formed after the application of glide vocalization and degemination. The second net was trained for 10 sweeps through the data set. Once more the two large and distinctive classes II and Ia are learned well. Error in class Ib is also low with one verb (out of 64) showing a tendency to adopt the high vowel of class II while a second shows an equally weak tendency to adopt the long vowel of Class Ib. The Ic verbs, as expected, show more interference. Three (out of 32) verbs tend toward Class II vowels. Still, the great majority of verbs of this class remain firmly Class I at this point.

A third network was then built and given as its teacher the output of the previous net. As a result, any errors in learning are propagated on to the next "generation", leading to increasing difficulty in learning those patterns. This training regimen was repeated for 5 subsequent networks, with each daughter network trained for 10 epochs to reproduce the output of its parent net. In each generation error on classes Ib and Ic increased as larger numbers of these verbs failed to learn the correct mapping and were produced on the model of Class II instead. Looking at the output of the sixth generation, we see that the Class Ic verbs are almost identical to those of Class II. No *nerjan*-type verb shows any interference from Class Ia, although at least two from the *fremman* class continue to be pulled very strongly in that direction. The majority of Ib verbs, however, have by this point merged with Class II.

Discussion

Consider the state of the weak verb system in early Old English. Class I had splintered into three groups, each of which had some strong formal criterion of membership. Class Ia, the long stem subclass, had the further advantage of high type frequency. Ic, on the other hand, had extremely low type frequency without any particular token frequency (Wright 1925). If a pattern of this kind is to be learned at all, a network account predicts that it must retain its formal characteristic. As predicted, it is when the distinctive form in the present tense is lost that Ic verbs begin to be modeled on Class II in the past tense as well.

A similar situation holds for the verbs of class Ib. The stem-final gemination was an identifying feature of these verbs, but once degemination had applied this information was no longer available. Class Ib, while not huge, did have a much higher type frequency than did Ic. This was an advantage, for although the loss of the geminate makes the class more difficult to learn, frequency of occurrence partially compensates. The class still eroded gradually, however. The more difficult learning task caused some members of this class to be mis-classified, and as this continued type frequency decreased, leading eventually to a situation not unlike that of Class Ic.

It might be asked, however, why the *fremman*-type verbs assimilated to Class II rather than class Ia. Part of the answer has to do with class size, for Class II was by far the largest class at this point. However, part of the answer also has to do with phonological form: Class II is the least restricted in terms of the phonological structure of its members. A small number of verbs, both in the network and in the real-language data, did in fact assimilate to Class Ia. Those that did so appear to have been drawn by surface similarity, suggesting that only the restricted set of Class Ib verbs sufficiently similar to those of Class Ia were able to merge with that class. No such constraint operated on assimilation to Class II.

While the two short-stem subclasses drifted into Class II, Ia remained unchanged. This is also the outcome observed in the network model. Ia not only had strong type frequency, but was also the one Class I subtype to maintain its formal characteristic. The combination of these two factors results in successful learning of the patterns involved.

The problem of the default

We now turn to a separate question, which is whether the default category can be learned even when it contains few members. Plunkett and Marchman (1991) raise this issue in a discussion of the Arabic plural system, in which the "sound" plural, the default, is of relatively low type and token frequency. They suggest that a crucial fact of this system is that "the numerous exceptions to the default mapping,... tend to be clustered around sets of relatively well-defined features."

The inflectional situation in earlier stages of English was parallel in many respects. As discussed in detail above, the weak preterit was treated as the default inflection even at a point when it enjoyed low type and token frequency. Furthermore, Vowel-change (strong) inflection applied to sets of verbs that exhibited internal phonological coherence. Each strong class had its own vowel series by definition, as well as other formal features by which the classes were distinguished. The weak verb classes as a whole had no such criteria for membership. Although certain weak sub-classes could be formally characterized at different points in their development, this was not consistently true.

The hypothesis is that phonological information will be exploited by a network in following way. In learning to produce the exceptions, the net must learn to respond to the phonological characteristics that are typical of each class. To a network, this absence of phonological basis for default classification can be equally informative, allowing all patterns *not* meeting the criteria for membership in an exceptional class to be classified together, regardless of surface dissimilarity. The goal of the next section is to test this hypothesis and demonstrate that the learning requirement, far from being a stumbling block, leads to an insightful account of default behavior.

Simulation of default data

In the simulation that follows, a connectionist network is trained to perform a classification task based on OE data. The results of this simulation show that even when the type frequency of the default class is artificially constrained, a network is able to generalize appropriately to novel patterns.

Architecture and training

The network used a feed-forward architecture with 50 input, 18 hidden, and 6 output units. The net was trained using the back-propagation of error algorithm. Inputs were 50-element vectors, each representing a word in which a subpart is a particular VC or VCC pattern, defined over distinctive features. There were six output nodes, one for each of six categories. The task of the network was to learn to respond to each input pattern by activating the appropriate category label on the output.

Five of the six classes were made up of patterns based on the "characteristic" by which the OE strong verb classes were distinguished. In the network, these are the following:

1. *i* + any one consonant
2. *e* + one stop or fricative
3. *e* + a consonant cluster
4. *i* + a nasal+stop cluster
5. *a* + one consonant
6. any other VC or VCC

The goal was to show that the network could learn to treat the sixth class as an elsewhere case by reason of its phonological diversity. However, if this diversity allowed the class to sample the entire phonological space of the language, then generalization of new patterns need not result not from the special status of class six, but from similarity to some other member of the class. For this reason, the default class (6) contained a minority of the total forms to be learned. In training, the network was shown 32 randomly generated members of each class. Each was presented once per pass through the data set. The network was trained for 20 such passes, at which point error was extremely low.

Results

Generalization tests were then applied to assess the success of the network. An additional 32 members of each category were chosen from the initial random set, and given to the fully-trained network as input. All novel forms for classes 1-5 were categorized into the appropriate class. Of the 32 novel members of Class 6 (the default class), three were of the form "æ+C" and were most strongly classified as Class 5. A fourth which included the string "irn" was ambiguously categorized between classes 3 and 4. All others novel patterns were placed in Class 6. A second generalization test was then run, in which the network as tested on 63 patterns that did not match any subtype seen during the training phase. The vast majority were treated as members of the sixth (default) class. As in the first test, the net "mis-categorized" certain strings of the form "æ + C" into class 5, and often treated patterns with the vowel *i* followed by a CC cluster as being in either 3 or 4.

Discussion

These results are consistent with an account which claims that the network generalizes membership in classes 1-5 to novel patterns on the basis of surface similarity, but treats class 6 as the elsewhere category. This account explains what might have appeared to be errors in generalization. For example, Class 5 is made up of patterns including the string "a + C" and in the code used as input, *a* and *æ* differ in only one feature. Therefore the "æ + C" patterns are sufficiently similar to those of Class 5 to be attracted to that class. Note also that Class 4 patterns have *i* followed by a NC cluster, while Class 3 patterns have CC clusters after the vowel. Again, the similarity to previously learned patterns leads to predictable overgeneralization.

However, the results as stated do not rule out a second possibility: the network may be generalizing *all* novel patterns on the basis of perceived similarity to known patterns. To argue convincingly that the network developed a true default, we must eliminate this possibility. Two further tests were devised to make certain that membership in Class 6 was not based on similarity to known forms.

Similarity tests

In the first of these tests, we carried out a hierarchical clustering analysis on the input vectors of the training and initial test items. We found that for each of the five well-defined classes, the training data were clustered into the five appropriate groups. Members of class 6 (the default) were scattered randomly through the cluster tree. This is as expected, since there is no similarity basis for class 6 membership.

The initial test items for classes 1-5 were also clustered with the appropriate groups. Many of the class 6 test items also fell near the class 6 training items, since in the first test many of the patterns were indeed similar to the training data. Others, however, were clustered in a major branch by themselves, and dissimilar to any of the other forms (including the previously learned class 6 members). These truly novel forms were also classified by the network as class 6.

Finally, we constructed a third generalization test set which contained patterns that were not only dissimilar from any that had been seen during training, but deviated in various ways from the well-formedness criteria by which the training data were generated. This dissimilarity was apparent in a clustering analysis. When these new items were presented to the network, all were classified as members of class 6, the default.

General discussion

Certain synchronic facts about English are compatible with both the dual- and the single-mechanism approaches to inflectional morphology. First, in the current state of the language there is a single "weak" or *d*-suffixed conjugation. Second, this suffixed past is by far the most common form of English past tense inflection. The weak past tense is the default in that it applies productively as the elsewhere case to any verb not previously learned as irregular: Borrowings from other languages, verbs derived from nouns, adjectives, and other verbs, and neologisms all routinely take the *-d* suffix.

Old English and its immediate predecessors differed from the modern language in two crucial respects. First, there was not a unique *d*-affixed conjugation in early Old English. In proto-Germanic there were four such classes, and while these were distinct from each other, all were productive, applying to the many borrowings and denominal verbs that are characteristic of the language. By early Old English these four classes had been effectively reduced to two, and over the course of the OE period many verbs of the first class began to take on the characteristics of the second. This shift presents a difficulty for an account which disallows phonological information in the application of the regular rule, for as discussed above the migration from one weak class to another was largely governed by phonologically-based analogy.

Second, the weak preterit was an innovation in proto-Germanic. At that stage in the language it had no statistical edge over the more established ablaut series of past tense formation, yet even at an early point it appears to have been treated as the default, used to inflect borrowings and derived verbs. This points to the need to separate out the properties of the default from the effects of frequency and class size.

In this work we have tried to demonstrate the importance of considering historical change when trying to understand the language mechanism which underlies morphological processes. We believe historical facts may usefully constrain hypotheses about the nature of the language mechanism in cases where synchronic facts alone admit multiple hypotheses. The current work, although preliminary, suggests that a single network account of English verbal morphology may have advantages over dual-mechanism accounts. The current work also acknowledges that default categories in language need not require large type frequency, and suggests that this property can be successfully modeled in a connectionist network.

References

Flom, G. 1930. *Introductory Old English Grammar and Reader*. Boston: Heath and Co.

Pinker, S., and Prince, A. 1988. On language and connectionism: Analysis of a Parallel Distributed Processing model of language acquisition. *Cognition* 28:73-193.

Plunkett, K., and Marchman, V. 1991. U-shaped learning and frequency effects in a multi-layered perceptron: Implications for child language acquisition. *Cognition* 38: 3-102.

Rumelhart, D., and McClelland, J. 1986. On Learning the Past Tense of English Verbs. In Rumelhart, D., and McClelland, J., eds. *Parallel Distributed Processing, Vol. II*. Cambridge, MA: MIT Press.

Stark, D. 1982. *Old English Weak Verbs*. Turbingen: Niemeyer.

Wright, J., and Wright, E. 1925. *Old English Grammar*. Oxford: Oxford University Press.

Learning Language in the Service of a Task

Mark F. St. John
Department of Cognitive Science
University of California, San Diego
La Jolla, CA 92093-0515
stjohn@cogsci.ucsd.edu

Abstract

For language comprehension, using an easily specified task instead of a linguistic theoretic structure as the target of training and comprehension ameliorates several problems, and using constraint satisfaction as a processing mechanism ameliorates several more: namely, 1) stipulating an *a priori* linguistic representation as a target is no longer necessary, 2) meaning is grounding in the task, 3) constraints from lexical, syntactic, and task-oriented information is easily learned and combined in terms of constraints, and 4) the dramatically informal, "noisy" grammar of natural speech is easily handled. The task used here is a simple jigsaw puzzle wherein one subject tells another where to place the puzzle blocks. In this paper, only the task of understanding to which block each command refers is considered. Accordingly, the inputs to a recurrent PDP model are the consecutive words of a command presented in turn and the set of blocks yet to be placed on the puzzle. The output is the particular block referred to by the command. In a first simulation, the model is trained on an artificial corpus that captures important characteristics of subjects' language. In a second simulation, the model is trained on the actual language produced by 42 subjects. The model learns the artificial corpus entirely, and the natural corpus fairly well. The benefits of embedding comprehension in a communicative task and the benefits of constraints satisfaction are discussed.

Introduction

Understanding language, and particularly learning to understand language, is a tricky task. A variety of imposing practical and theoretical problems stand in the way. This paper addresses four of these obstacles and shows ways to surmount them and grasp a better understanding of language learning and understanding along the way.

The four obstacles are 1) how to specify a satisfactory representation of the meaning of the language input, 2) how to ground the semantics of concepts used in the communication 3) how to combine lexical, syntactic, and task-oriented information to produce comprehension, and 4) how to handle incomplete and grammatically "noisy" language such as natural spoken language.

The message of this paper is that using constraint satisfaction as a processing mechanism and employing the fact that language is learned and used in the service of performing a task goes a long way toward surmounting these obstacles. First, I'll briefly discuss each obstacle in turn and show how these two ideas can address them. Then I'll present two simulations that show these two ideas at work. This paper concentrates on the first two obstacles and points to preliminary work to address the remainder.

Obstacles

The first obstacle is the need to stipulate a linguistic-theoretic representation as the target or result of comprehension. One well known and useful representation is thematic case roles, such as agent and patient (Fillmore 1968). Unfortunately, case roles are known to either proliferate in number or to become inexact as situations become more diverse and complicated. Specifying the features that characterize concepts is similarly difficult.

Additionally, all but the most trivial language requires something like embedded propositions to specify the relations between concepts. Such symbolic representations impose strong assumptions about the representation of the results of comprehension and place a particularly heavy burden on Parallel Distributed Processing (PDP) models since propositions are virtually impossible to represent in a single vector of units.

Several PDP models have attempted to handle this problem. Miikkulainen and Dyer (1991), St. John (1992), and Touretzky and Hinton (1985) represented multiple propositions concurrently in a hidden layer. Individual propositions could be pulled out one at a time for inspection. Not only is this pull out scheme awkward, but individual propositions still cannot contain embeddings since there is still no way to represent the case where one argument in a proposition is a whole other proposition. Pollack's RAAM model (1988) allowed distributed representations in the output layer, but training still required these representations to be unpacked into their fundamental components.

The solution proposed here is to define the target to be some easily represented task. For example, imagine two people solving a jigsaw puzzle where one person commands the other where to place the blocks. The listener's job is to understand the language input and then move a block. The target can be a simple representation of the updated state of the puzzle after each command. The benefit is that the task is easy to represent, and the complex linguistic structure of the language is hidden inside the listener. If we make the listener a PDP model, the output can be the task, and any linguistic structures required to compute the output reside internally in the hidden layers.

The second obstacle is the need to define the meaning of concepts in the language. Again, the researcher may be required to specify this information, for example, by coding a set of semantic features for each concept. In contrast, Allen (1987), Elman (1990), Miikkulainen and Dyer (1991), and St. John and McClelland (1990) showed that the task itself can be used to specify the semantics. That is, the needed semantics can be learned by a model in the service of solving some task. These models learned which concepts are seen with which other concepts. Semantics, in these models, is defined by the co-occurrence statistics of the concepts in the corpus of training examples.

The approach taken in the puzzle task is to have the model learn the meanings of words like "big" and "blue" and the ramifications of syntactic forms in terms of their ability to help determine to which block the speaker is referring. Thus, words and syntax are learned to be defined in terms of their communicative functions. This idea of language meaning as language use is developed much more fully by Clark (1985). Having the target actually be some performed task, as is the case in this paper, makes this point especially clear.

The third obstacle is the need to combine information from words, syntax, and the task to understand the command specified by the speaker. Speakers often rely on the situation to convey important information. Constraint satisfaction is a powerful mechanism for performing this process. Each piece of information from any source is viewed as a separate constraint. These constraints are combined to compute a single, coherent interpretation (cf. St. John, 1992).

The fourth obstacle is the need to handle the grammatical informality of natural speech. Informal constructions such as repetitions, restarts, and ellipses are so common that they really are the rule and not the exception. As such, they must be treated within the normal course of processing. Constraint satisfaction is well suited to this task. When the language is viewed as a set of constraints, the important factor is that there be sufficient constraints to specify the communication, not that the communication correspond to a specific grammatical form.

Recent models in the literature that handle informal grammatical forms (Lehman, 1990) and ellipsis (Frederking 1988) first assume that input language will be grammatical, and only when the normal processes fail do they perform a time-consuming search for corrections or additions that will produce grammatical and sensible parses. The constraint satisfaction alternative is to use the available constraints to compute an interpretation and only concern itself with the grammar to the extent that it informs that computation.

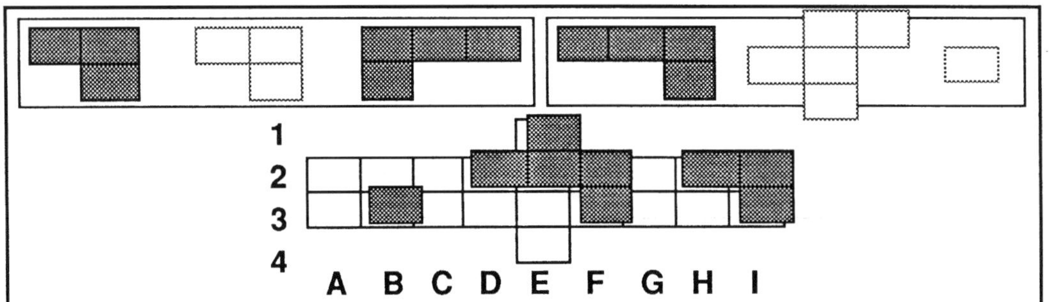

Figure 1. The jigsaw puzzle. Pieces of varying shapes, sizes, and colors are place on an outlined figure.

Task

The task to be use here is the simple puzzle task introduced above. Subjects, UCSD undergraduates, were asked to solve a series of three simple jigsaw puzzles. Each puzzle consisted of six wooden blocks of varying shapes, sizes, and colors that had to be placed on an outlined figure. Unplaced blocks were arranged on two mats above the puzzle figure (see figure 1). The trick was that the subjects were required to sit on their hands and tell the experimenter where to move the blocks. The figure was labeled with Cartesian coordinates to facilitate this task. Subjects' spoken commands were recorded and transcribed.

Since there is considerable linguistic complexity in just referring to which block was to be moved, I will concentrate on just this block reference task in this paper.

Simulation 1 - Artificial Corpus

Rather than proceed directly to the natural corpus of commands produced by subjects, I will begin with a corpus constructed by hand. An artificial corpus is a good place to start the investigation because features of the language are easily controlled. As a further simplification, this corpus contains only first moves in the puzzle task rather than whole series of consecutive moves. This restriction will be removed in the second simulation.

Commands were composed from a small set of adjectives (small, big, square, red, blue, and green), and phrase types, such as simple noun phrases, relative clauses, and prepositional phrases. The commands also made use of "left", "right", and "middle" that require paying attention to word order, for example, "the left block," "the right block that is on the left page," and "the left block on the left page." Finally, blocks could be referenced in terms of other blocks, for example, "the block on the right of the big blue block."

All together, there were 138 commands. Several provisions were made to insure that the corpus was combinatoric in the sense that roughly equal numbers of commands referring to each of the six blocks, and adjectives and relative terms applied to each relevant block in roughly equal numbers. A combinatoric corpus insures that the model will learn the pure meaning of each term without becoming muddied by biases in frequency. Of course the real world may not be so kind, and the second simulation addresses this issue.

Example Commands

the big blue block that is on the right of the left page
the big block that is on the right of the left page
the big red block that is on the left of the right page
the small blue block
the small red block that is on the left of the big blue block

Architecture. The architecture is a simple recurrent network (Elman 1990) wherein the activation in the internal hidden layer is copied back to the input layer on each consecutive step. A recurrent network is useful because it allows each word to be processed sequentially by the same set of weights, yet allows information from previous words to be carried forward. The model cycles through a command one word at a time, with activations from the hidden layer being copied back to the input on each consecutive step (see figure 2). The input is a command, a representation of the blocks yet to be placed on the puzzle, and the activations from the recurrent hidden layer. One unit in the input layer was used to represent each possible word, and as the model worked through a command, the current word was activated in the input layer and the previous word was removed. While the input layer is therefore local in its representation of individual words, the internal hidden layer is free to develop more efficient distributed representations.

The representation of to-be-placed blocks was simply to activate one unit for each remaining block. There was no feature description of any block. Because this simulation deals only with the first command in the puzzle, all six blocks were activated for each command.

The output and target for the task was to activate one unit for the block referred to by the command. The target is specified after each word.

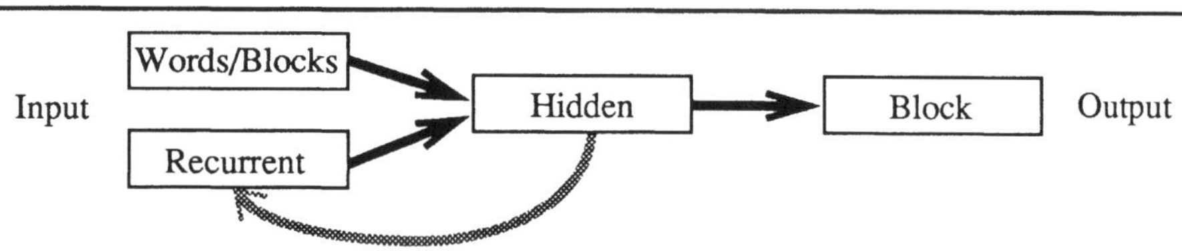

Figure 2. The architecture of the network. See text for details.

That is, the model is asked to produce the correct target even after just the first word of a command is presented. Error between the target and the model's response is used to change the weights of the network via the backpropagation algorithm (Rumelhart, Hinton, and Williams, 1986). This training procedure requires the model to extract the most information from each incoming word.

There were 17 units for words and 6 units for blocks in the input layer, 40 units in both the recurrent and hidden layers, and 6 units for blocks in the output layer.

Results and Discussion. The model was trained for 2000 epochs (trips through the training corpus) with a learning rate of .01. The model mastered the corpus entirely by activating only the correct block for each command in the corpus.

This mastery demonstrates a number of points. First, the model is able to learn and correctly combine a number of partial constraints to activate the correct block. For example, both "big" and "red" refer to two blocks, but together they specify only one block. Each adjective acts as a constraint on the specification of a block. The process of constraint satisfaction embodied in the network works well to combine these constraints.

Second, the model learns to process relative clauses, prepositional phrases, and word order correctly. Commands like "the small block that is on the left of the big blue block" describe two blocks, yet the model picks the correct one as the referent. Commands like "the block that is on the left of the right page" and "the block that is on the right of the left page" refer to different blocks and so require handling the order of left and right correctly.

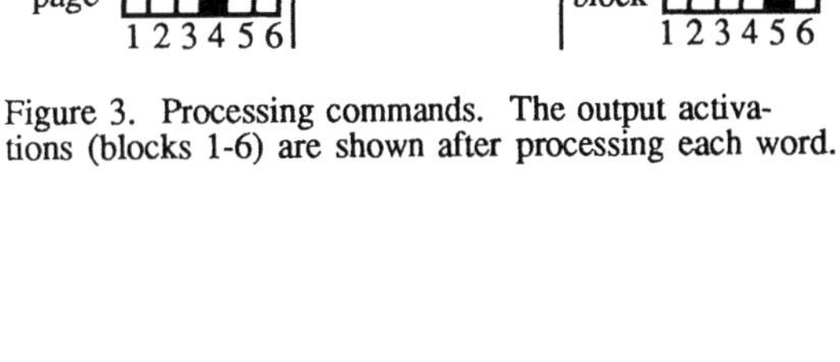

Figure 3. Processing commands. The output activations (blocks 1-6) are shown after processing each word.

A simple way to observe what the network has learned, and to confirm that it has not simply memorized the training set, is to observe the activation of the output units as a command is processed. If the activations throughout the command reflect the constraints on meaning imposed by each new word, we can have greater confidence that the model has learned those constraints. Figure 3 provides three examples. The output activations for the six blocks are shown in black after each word of a command is processed.

Third, the model's understanding of words derives from the task - the words function to differentiate the blocks and determine to which one a command refers. The semantics of the terms, then, is grounded in their functions in the puzzle task.

Fourth, neither the semantics nor any linguistic-theoretic structures had to be specified in the input or target. The input was the sequence of words, and the target was simply the block referenced by the command. To whatever extent case roles, embedded propositions, or tree structures needed to be computed for the task, they were represented and computed internally in the hidden layer of the network.

Simulation 2 - Natural Corpus

A fresh network was trained on a large corpus of commands produced by actual subjects solving the puzzle task. The results from this simulation are preliminary, but they demonstrate important points.

A natural corpus is interesting in a number of ways. Foremost, the grammar used by actual subjects is highly informal and the language is frequently vague since subjects can rely to a large extent on the puzzle situation to convey information. Nearly every command contained repetitions of words or phrases, restarts, or ellipses.

A second interesting aspect of the natural corpus is that subjects solved entire puzzles, and these sequences of block commands demonstrated important task constraints. Most importantly, there was a rough standard order for placing blocks on the puzzles: essentially from the most constraining block to the least constraining block. There was substantial variability in the ordering, but statistically, an ordering is evident. If the comprehension system can utilize these constraints, it can resolve otherwise ambiguous language and it can generally ease its comprehension task when these constraints are redundant with the language of a command. For example, if one of the two red blocks has already been placed, the otherwise ambiguous reference "the red block" becomes clear.

Finally, subjects were allowed to change the positions of blocks and start over. Therefore, consecutive references to the same blocks could occur. Subjects almost invariably used relative terms like "it" or "those blocks." Another relative term is "the block on the left." After the first block on the left is placed, the second block becomes the new block on the left. The model must bind these relative terms to the correct block on each occasion.

Architecture. The architecture was identical to the previous architecture except that many more words were used by subjects, 613, and therefore more words were represented in the input layer. The one difference in training was that the representation of to-be-placed blocks was updated from command to command throughout the course of solving a puzzle. Between commands, the activation of the hidden layer was copied to the input layer, as within commands. The recurrent layer was only reset to zero activations between puzzles. The corpus contained 42 speakers and a total of 1495 commands.

Results and Discussion. The model was trained for 400 epochs with a learning rate that gradually dropped from .005 to .002. The model activated the correct block most strongly on 78% of the commands. The trained model was also tested on a corpus of 3 novel subjects. It performed correctly on 53% of the 134 novel commands. These figures are certainly not great, but an examination of the model's successes and failures is illuminating.

First, the model handles the informality of the grammar very well. Ellipsis of nouns and entire phrases, restarts, and repetitions cause no trouble for the model. For example, commands like "the blue, the big blue, no, yeah, the big" are processed correctly.

The model also handles the relative terms "it" and "left" correctly, and uses the task constraints to understand otherwise ambiguous references like "the red block" discussed above.

The model acquires the rough standard order of block placements readily, and even too well. That is, on many occasions, a subject will violate the standard order. Depending on the violation, the model will either follow the language or follow the standard order. More egregious violations of order, for example choosing the smallest, least constraining block first, are quite rare in the corpus. In these cases, the model will override the language input and activate the standard first block.

To some extent this effect is reasonable, though overly strong. A number of researchers have found that semantic constraints can override the language input. A telling example is to ask subjects "how many animals of each kind did Moses take on the ark?" Most subjects respond "two" without noticing that Moses is not the correct biblical figure (Erickson and Mattson, 1981; and Reder and Cleeremans, 1990). This effect can be viewed in terms of constraint satisfaction as a case where conflicting constraints are present. The task constraints representing the standard order are stronger than the language-based constraints, and the understander goes with the stronger set of constraints.

In fact, it is difficult to really know how often task constraints play an important role in everyday comprehension. It seems reasonable to believe that they actually play a rather large role, and that tell-tale cases of conflict between task constraints and language constraints are just rare. This overriding of the language input, however, is far more frequent in the model than for the experimenter listening to the subjects. For this reason, it seems necessary to reduce the effect of the task constraints.

The question is what changes to the corpus are needed to switch the relative strength of the task and language constraints? One solution is suggested by the artificial corpus. Namely, reduce the strength of the task constraints by making the frequency of different commands similar. Reducing the regularities and increasing the combinatorics of the language in the corpus will force the model to learn strong language constraints. This solution strategy was pursued by St. John (1992) in a simulation study of text comprehension.

Another potential solution is to provide the model with a pre-training task in which there are no regularities aside from the language itself. The model would learn strong language constraints that it could then transfer to the puzzle task.

In general, a more combinatoric corpus with few semantic constraints produces essentially context free language constraints: each word means what it says and little more. Elman (1992) has suggested that the wide range of language contexts provided to children as they learn their native language serves to decorrelate the language from any specific context. This condition produces a virtually combinatoric corpus for children to learn, and underlies their ability to understand unexpected language input in the face of possible task constraints - just like the experimenter in the puzzle task. On the other hand, the relatively weaker task constraints available in any given context can still facilitate comprehension when they cooperate with the language constraints.

Conclusions

The simulation of the natural corpus is preliminary and more work is required to make the model effective. In particular, some method must be found to change the relative strength of the language and task constraints. The strength of the task constraints does, however, demonstrate the powerful ability of the model to acquire and then use task constraints for comprehension.

More generally, constraint satisfaction provides a useful framework for learning and using constraints from different sources, whether lexical, syntactic, or task-oriented.

Constraint satisfaction is also a boon to processing informal language such as natural speech. The model does not attempt to match an input to a known grammatical form. Instead, all that is required of any input is that it contain sufficient constraints to compute the correct message, and those constraints can come either from the language or the task itself.

Using an easily represented task as the target of training provides other important advantages. First, it provides the technical advantage of relieving the experimenter of the burden of creating a linguistic theoretic representation of thematic case roles, propositions, or the like. The experimenter needs only to specify the task, and the model is required to learn whatever representation it needs to perform that task. This idea has the potential to significantly advance the science of PDP models of language comprehension.

A potentially limiting condition is the requirement of finding an adequate task for whatever language is desired to be learned. In this paper the language only pertained to referencing wooden blocks. However, with some creativity, the range of possible tasks and language may expand considerably.

One other advantage of using a task as the target of training is that the meaning of concepts and words do not have to be provided *a priori* in either the input or the target. The model acquires exactly those meanings necessary to perform the task. In this way the semantics of the language are grounded out in the task itself: language meaning as language use.

References

Allen, R. B., (1987). Several studies on natural language and back-propagation. *Proceedings of the International Conference on Neural Networks, Vol. 2*, p. 335-341, June 21-24, 1987, San Diego, CA.

Clark, H. H. (1985). Language use and language users. In G. Lindzey & E. Aronson (Eds.), *The handbook of social psychology, 2, (3rd ed.)*. New York: Harper & Row.

Elman, J. L. (1990). Finding structure in time. *Cognitive Science, 14*, 179-212.

Elman, J. L. (1992). Personal communication.

Erickson, T. D. & Mattson, M. E., (1981). From words to meaning: A semantic illusion, *Journal of Verbal Learning and Verbal Behavior, 20*, 540-551.

Fillmore, C. J. (1968). The case for case. In E. Bach & R. T. Harms (Eds.), *Universals in linguistic theory*. New York: Holt, Rinehart, & Winston.

Frederking, R. E. (1988). *Integrated natural language dialogue: A computational model*. Boston: Kluwer Academic Publishers.

Lehman, J. (1990). Adaptive parsing: Self-extending natural language interfaces. *Proceedings of the 12th Annual Meeting of the Cognitive Science Society*. Hillsdale, NJ: Lawrence Erlbaum.

Miikkulainen, R. & Dyer, M. G. (1991). Natural language processing with modular PDP networks and a distributed lexicon. *Cognitive Science, 15*, 343-400.

Pollack, J. (1988). Recursive auto-associative memory: Devising compositional distributed representations. *Proceedings of the 10th Annual Conference of the Cognitive Science Society*. Hillsdale, NJ: Lawrence Erlbaum.

Reder, L. M., & Cleeremans, A. (1990). The role of partial matches in comprehension: The Moses illusion revisited. In A. C. Graesser and G. H. Bower (Eds.) *Inferences and text comprehension., The psychology of learning and motivation, vol. 25*. San Diego, CA: Academic Press.

Rumelhart, D. E., Hinton, G. E., & Williams, R. J. (1986). Learning internal representations by error propagation. In D. E. Rumelhart, J. L. McClelland, and the PDP Research Group (Eds.), *Parallel distributed processing: Explorations in the microstructure of cognition, Volume 1*. Cambridge, MA: MIT Press.

St. John, M. F. (1992). The story gestalt: A model of knowledge intensive processes in text comprehension. *Cognitive Science, 16*, 271-306.

St. John, M. F. & McClelland, J. L. (1990). Learning and applying contextual constraints in sentence comprehension. *Artificial Intelligence, 46*, 217-257.

Touretzky, D. & Hinton, G. E. (1985). Symbols among the neurons: Details of a connectionist inference architecture. *Proceedings of the 9th International Joint Conference on Artificial Intelligence*.

On the Unitization of Novel, Complex Visual Stimuli

Nancy Lightfoot and Richard M. Shiffrin
Indiana University
Bloomington, IN 47405
lightfoo@ucs.indiana.edu
shiffrin@ucs.indiana.edu

Abstract

We investigated the degree to which novel conjunctions of features come to be represented as perceptual wholes. Subjects were trained in a visual search task using novel, conjunctively-defined stimuli composed of discrete features. The stimulus sets were designed so that successful search required identification of a conjunction of at least two features. With extended training, the slope of the search functions dropped by large amounts. Various transfer tasks were used to rule out the possibility that the organization of sequential search strategies involving simple features could account for this result. The perceptual discriminability or confusibility of the stimuli exerted an important influence on the rate of unitization. The nature of the perceptual unit appears to depend on the subset of features which are diagnostic for carrying out a particular discrimination task. The results provide important constraints for models of visual perception and recognition.

What are the mechanisms by which representations of individual features are bound together and processed as perceptual wholes? One solution to this binding problem involves pre-specifying a representation for all possible conjunctions of features, in addition to the features themselves (Hummel & Biederman, 1990). This approach has been criticized, however, as being relatively inefficient and implausible when applied to higher level conjunctions (Hummel & Biederman, 1990; Hinton, McClelland & Rumelhart, 1986).

We conducted a number of experiments to examine the nature of processing for conjunctively defined stimuli, and the way in which such stimuli might become unitized during training. Our specific interest was in examining the transition from the processing of unfamiliar stimuli at the level of individual features, to the processing of those same stimuli, after familiarization, as conjunctive perceptual wholes. Our results suggest that perceptual unitization occurs after prolonged exposure to particular stimuli. These findings argue against the notion that conjunctive representations are pre-specified. In conditions under which conjunctive representations have developed, however, they may form the basic functional units on which attentional processes operate.

The processing of visual information is often investigated using visual search tasks, in which the subject attempts to locate a target among multiple distractors, typically responding 'target present' or 'target absent'. The slope of the function relating response time to display size provides a convenient way of assessing capacity demands associated with increasing stimulus and display complexity. This slope is often assumed to reflect the time taken for a single comparison between a subject's representation of the target and a stimulus appearing in the display. Slopes on positive (target present) trials are typically half as steep as slopes on negative (target absent) trials. This pattern of results has been interpreted as evidence for a self-terminating search process, since subjects will on average identify the target half-way through their search of the display on positive trials. The increase in reaction time as a function of display size is assumed to reflect the operation of a limited-capacity search mechanism.

Many well-known theories of visual search assume that stimuli are processed at the level of

This research was supported by grants AFOSR90-0215 to the Institute for the Study of Human Capabilities, and by grants NIMH 12717 and AFOSR 870089 to the second author.

primitive visual features such as line orientation, curvature, or color (eg. Triesman & Gelade, 1980; Fisher, 1986). Treisman and her colleagues (eg. Treisman & Gelade, 1980; Treisman & Gormican, 1988), for example, propose that subjects first parse the visual field into individual feature maps, in which the presence of visual features is coded without information as to their location, or as to the objects to which they belong. According to Treisman, it is only through limited-capacity attentional processing that features are conjoined into coherent objects.

Triesman argues that the distinction between pre-attentive processing, in which the simple presence of features is coded, and attentive processing, in which these features are conjoined into coherent wholes, accounts for differences in efficiency in visual search tasks. Targets which are distinguishable from distractors on the basis of a single primitive feature will be detected automatically and without capacity limitations, whereas targets which require identification of a conjunction of features to be uniquely identified will require attentional processing.

An important limitation of Treisman's approach is its difficulty in accounting for the well-documented effects of training in visual search tasks. Two types of training paradigms have been studied extensively in visual search: Consistent mapping (CM) and varied mapping (VM) training (eg. Shiffrin & Schneider, 1977; Schneider & Shiffrin, 1977). In CM training, stimuli are designated either "targets" or "distractors" and never change roles across trials. In VM training, on the other hand, stimuli change roles randomly from trial to trial, appearing as targets on some trials, and distractors on others. In some search tasks, there is a large advantage for CM training, which is well accounted for by automatic attention attraction to targets (Shiffrin & Schneider, 1977; Schneider & Shiffrin, 1977). In other visual search tasks, however, typically where search is more difficult, slopes appear to decrease consistently across days of training in both CM and VM training paradigms, although there is generally at least some degree of CM advantage.

Fisher (1986) has proposed a feature-based model which accounts for a wide pattern of training effects found in visual search (see Shiffrin & Schneider, 1977; Schneider & Shiffrin, 1977 for an alternative explanation). According to Fisher's model, stimuli are decomposed and processed at the level of individual features, as Treisman has suggested. In visual search tasks, subjects sequentially compare the individual features of a target with those of the distractors. Display items not having the first feature are rejected, remaining items are tested on the next feature, and so on, until search terminates.

In Fisher's (1986) model, as in Treisman's model, targets distinguishable on the basis of a primitive feature will be easily identified. In the case of conjunctively defined targets, however, search efficiency depends on the particular order in which the features are compared. With increasing experience in searching for a given target among all of its possible combinations of distractor elements, subjects learn to organize a maximally efficient feature comparison strategy. The development of this search strategy gives rise to the observed effects of training.

There are several critical implications of feature-based theories. If all stimuli are assumed to be processed at the level of individual features, and all stimuli require limited-capacity resources to be conjoined into coherent wholes, then the basic units of information and the nature of processing should be roughly similar for stimuli with high and low levels of familiarity. This same prediction holds in the case where conjunctions of features form the basic unit of processing, but these conjunctive representations are pre-specified within the visual processing system. With regard to training effects, if stimulus sets are designed in which all feature-based search strategies require approximately the same number of comparisons to uniquely identify the target, there should be greatly attenuated effects for training.

Experiment 1

We tested these assumptions by training subjects on sets of novel stimuli in which featural overlap was carefully controlled. Figure 1 shows the two stimulus sets assigned to each subject. For each subject, one set was assigned to CM training and one to VM training. The stimuli were composed of an external frame and three internal line segments. The stimulus sets were designed so that within each set, all feature comparison sequences would lead to approximately comparable performance when averaged over displays composed of different distractor elements.

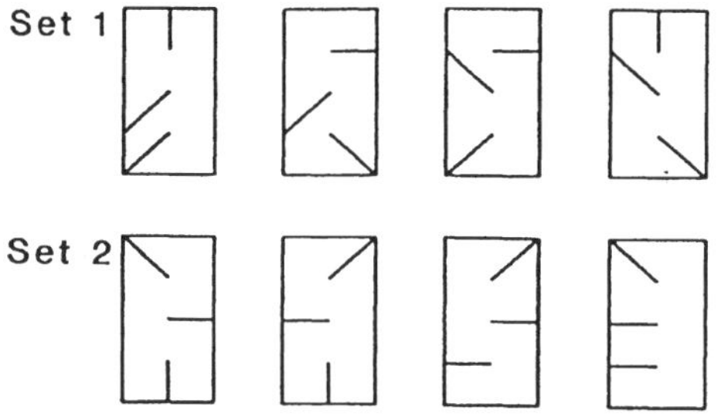

Figure 1. Novel stimulus sets for Experiment 1.

We assumed that the three internal line segments within each character formed the functional features in the search task. Based on this assumption, each target shared a feature with exactly one other stimulus in the set. Each target could be uniquely identified only by examining a conjunction of two features, but any two features would, averaged across displays of different distractor compositions, work equally well in identifying the target. There was no featural overlap between the two stimulus sets.

Under these conditions, training effects would be unlikely to be due to learning optimal feature comparison strategies. Furthermore, to the extent to which all stimuli are decomposed and processed at the level of individual features, performance on these characters should be comparable to that for familiar stimuli. As a result, extended practice using these stimuli should produce greatly attenuated training effects under the assumptions of feature-based models of visual search.

Subjects were run in CM and VM conditions for a period of 50 days. In the CM condition, one stimulus was designated as the target and the other stimuli were designated as distractors over the entire course of training. In the VM conditions, all stimuli within the set appeared equally often as targets, and served as distractor elements on remaining trials. In order to equate training on particular combinations of targets and distractors in the two training conditions, there were four times as many trials in VM as in CM training. Half of the trials in each condition were positive and half were negative. Display sizes varied from one to eight. For display sizes larger than three, displays were filled in a pseudo-random manner, with as few repetitions of each distractor stimulus as possible.

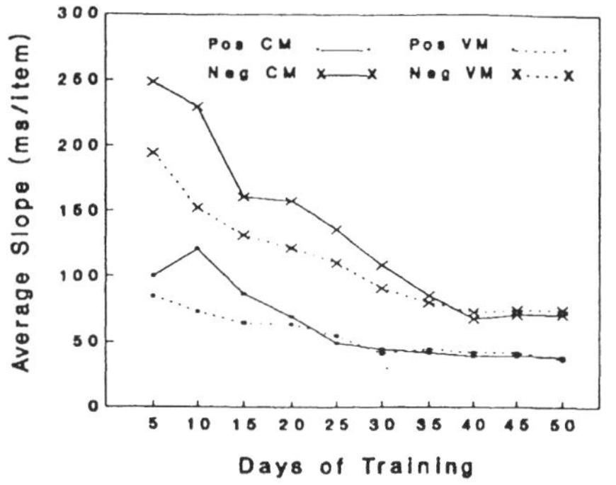

Figure 2. Training effects for Experiment 1.

Figure 2 shows the average slopes across days of training for positive and negative trials, using CM and VM training paradigms. These data are remarkable in several respects. First of all, the training effects are unusually large, with comparison times dropping from a high of 240 to an asymptote of 80 msec per item in the negative CM condition. The effects of training are also unusually protracted. Search performance does not asymptote for 35 to 40 sessions using these novel stimuli. A third unusual aspect of the data is that performance is initially better for VM than for CM training trials. Although CM and VM training can lead to comparable performance in visual search, an advantage for VM training is unprecedented.

The huge effects for training found with these novel stimuli suggest that there is an important role for familiarity in the processing of visual stimuli. This hypothesis is also supported by the fact that performance on CM trials was initially worse than on VM trials. Because there are four times as many trials in VM as in CM, a plausible explanation for this finding is that subjects are simply gaining more familiarity with the VM stimuli early in training. This argument is strengthened by the finding that the training paradigm itself has little influence on asymptotic search performance: reaction time functions at asymptote, graphed in Figure 3, are virtually identical for the two training conditions.

The nature of these familiarity effects is somewhat open to question. One explanation is that through repeated exposure, subjects are learning to unitize these novel stimuli and treat them as perceptual wholes. If subjects learn to perform comparisons at the level of the entire

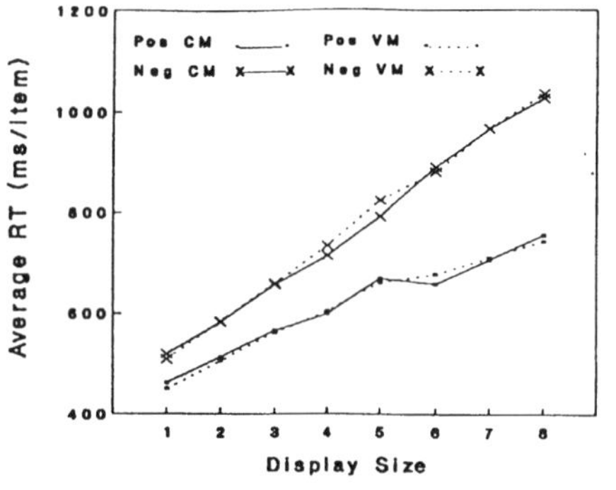

Figure 3. Display size effects at asymptote.

stimulus, instead of at the level of individual features, this greatly reduces the number of comparisons required to identify a target. This reduction in the actual number of comparisons offers a plausible explanation for the large training effects found in Experiment 1.

An alternative possibility relies on the high similarity among the characters within each search set. Both the large degree of featural overlap and the addition of the redundant external frames made the stimuli within a set highly similar, and guaranteed that search would be extremely difficult. It is possible that under these conditions, subjects were initially confused and grossly inefficient in developing appropriate feature search strategies for these displays. Subjects may have initially done a great deal of re-checking, for example, or may have attended to irrelevant or non-diagnostic features, such as the line segments in the external frame, before organizing a more coherent search strategy.

Experiment 2

To further investigate the nature of these familiarity effects, we trained the same subjects who had participated in our first experiment in an additional search task, in which the featural overlap of the targets and distractors was reduced. In particular, we made the items from the former VM set consistent targets, and the distractors from the former CM set consistent distractors. Targets and distractors now shared no features, so a feature based search should always terminate with the first comparison. Thus, if subjects were using a feature based search in Experiment 1, we should see an immediate, marked, and discrete jump in search performance when featural overlap is reduced. If subjects were using whole characters units in their search, however, the shift in stimuli should produce little change until subjects learn to switch to a feature based approach. Figure 4 shows a sample re-pairing for this condition.

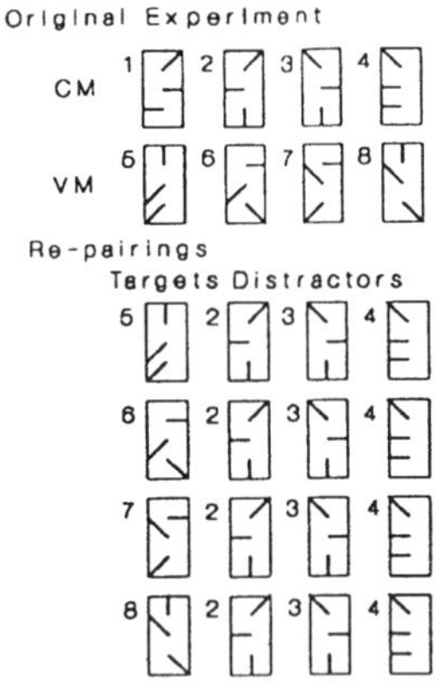

Figure 4. Sample stimulus sets, Experiment 2.

Subjects trained using these new stimulus sets for 15 sessions. Figure 5 shows five days of baseline performance using the original stimulus sets, followed by the data from the stimulus re-pairing. The results are not very compatible with the notion that subjects were employing a feature-based search. There was no sudden jump in performance after re-pairing. Instead slopes showed a moderate and continuous decline over the full course of training in the new condition, possibly reflecting a gradual switch to a more efficient feature-based search.

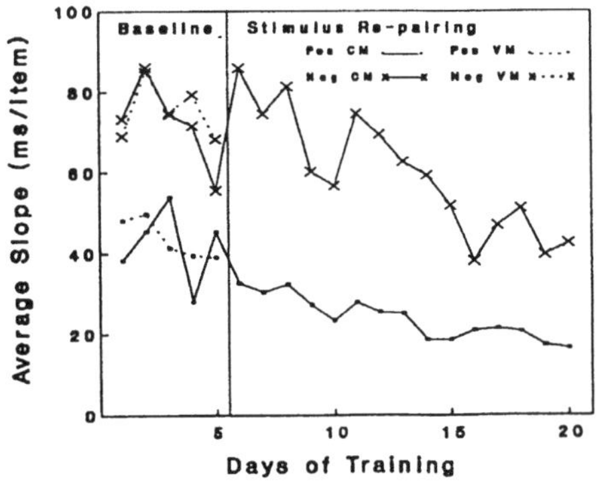

Figure 5. Results of stimulus re-pairing.

Experiment 3

An alternative explanation for these data is that subjects may be using a feature-based search, but using features other than those which we had anticipated. It is possible, for example, that subjects were using emergent features as the basis for their feature comparison process, such as corners or angles formed between the internal line segments and the external stimulus frames.

To test this hypothesis, we decided to continue training on the original search sets with the external stimulus frames removed. Most models of visual search would predict an improvement in search performance after removing the external frames, since search is much more efficient when the featural overlap between targets and distractors is reduced. If, on the other hand, subjects are relying on emergent features as the basis of their search strategy, removing the external stimulus frames should lead to an elimination of the emergent features and a disruption in search performance.

Figure 6 shows the slopes for five days of baseline training followed by the removal of the external stimulus frames. To establish a baseline, the subjects from the first two experiments were re-trained using the original stimulus sets (conjunctively-defined, framed stimuli). After re-training, the external frames were removed, and search with the unframed stimuli began.

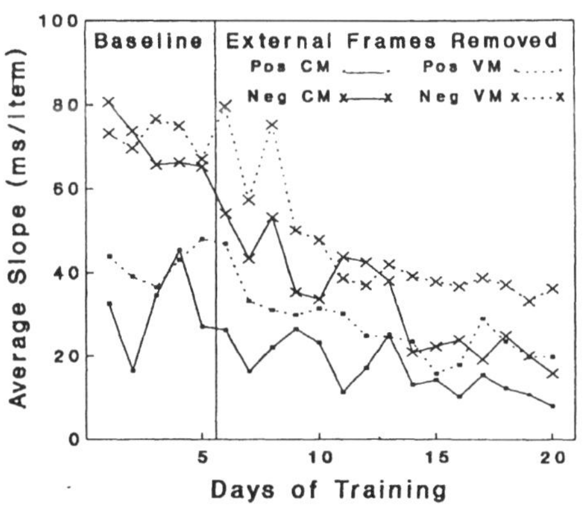

Figure 6. Results of removing external frames.

There was no evidence for disruption in search performance following the removal of the external frames. In fact, search gradually improved, as would be expected based on the decreased similarity between targets and distractors after the removal of the redundant line segments. These data offer no support for the idea that subjects are using emergent features as the basis for a strategic feature search. Given this demonstration that the internal stimulus features appear to form the basis for the search process, the results of Experiment 2 suggest that this process operates at the level of the character or feature conjunction, rather than at the level of individual features.

Experiment 4

This last study raises a puzzle, however. If search is based on unitized character representations, why does removal of a large part of the character not disrupt performance? One possibility is that the functional representation of these characters does not include the boundaries. Perhaps a good deal of the training involves learning to ignore the redundant and confusing external frames. If the learned unit consists of the arrangement of internal line segments, then the large degree of transfer seen in Experiment 3 would be expected. If this reasoning is correct, then initial training on unframed characters should be fast and easy, but subsequent transfer to framed stimuli should be very poor. We therefore trained new subjects in just this way. Experiment 1 was repeated using unframed stimuli. When subjects reached asymptote, the frames were added and training continued. The results are shown in Figure 7.

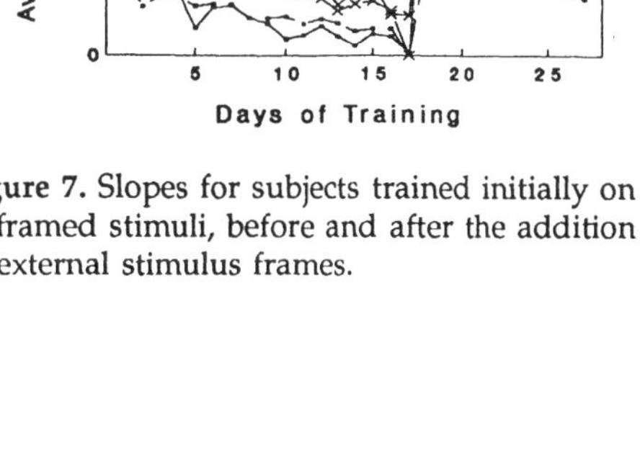

Figure 7. Slopes for subjects trained initially on unframed stimuli, before and after the addition of external stimulus frames.

The results are remarkable: Slopes began at the level indicative of feature search, but dropped quickly to a low asymptotic level. When the frames were added, no transfer was seen -- performance reverted to a level at least as poor as that seen at the start of training in both this experiment and in Experiment 1. This was followed by an additional gradual reduction in slopes, presumably reflecting the subjects' learning to remove the external frames from their perceptual representations.

This complete asymmetry of transfer, depending on the order in which the external frames are added or deleted, further reveals the nature of perceptual unitization. The learned units are apparently not the items themselves, but rather the parts of the items that are useful for making required discriminations.

Taken together, our data provide evidence for a process of perceptual unitization with increasing exposure to novel stimuli. Although it is logically possible to generate a feature-based explanation for the huge training effects seen in Experiment 1, this explanation is incompatible with the results of Experiments 2 and 3. However the exact nature of this perceptual unitization is somewhat ambiguous. Subjects may be learning to unitize the character as a whole (though without the external frames), or they may be learning to unitize only a simple diagnostic conjunction of two features.

It seems clear that these perceptually unitized representations do not act very much like the primitive visual features which Treisman has proposed as the functional units in perceptual processing. Search based on the distinction between basic features such as color or shape appears to be unlimited in capacity, generating flat slopes for both positive and negative trials. In contrast, slopes for our framed novel stimuli remained high over the course of 50 days of training, with the two-to-one slope ratio for negative versus positive trials characteristic of a limited-capacity, self-terminating search mechanism. Thus highly similar, unitized stimuli appear to be dealt with by an attentive, limited-capacity search process. The more discriminable characters without frames, however, showed considerably lower asymptotic slopes. It is possible that sufficiently discriminable conjunctions may come to exhibit characteristics similar to those for simple features.

We are clearly able to encode and recognize novel combinations of features, though with some difficulty, probably due to the necessity of processing such stimuli one feature at a time. With increasing familiarity with conjunctions of features, however, we apparently develop an alternative form of conjunctive or unitized representation of whole perceptual objects. Evidence for this type of perceptual unitization calls into question both the assumption that familiar stimuli are processed simply at the level of basic features, and the assumption that conjunctions of features are somehow pre-specified in the visual system. The fundamental differences in the processing of stimuli based on primitive visual features, novel combinations of features, and familiar visual wholes place strong constraints on models of visual processing and perceptual binding.

References

Fisher, D. L. (1986). Hierarchical models of visual search: Serial and parallel processing. Paper presented at the annual meetings of the Society for Mathematical Psychology, Cambridge, MA.

Hummel, J. E., & Biederman, I. (1990). Dynamic Binding in a Neural Network for Shape Recognition. Ph.D. Diss., Dept. of Psychology, University of Minnesota, Minneapolis, MN.

Hinton, G. E., McClelland, J. L., & Rumelhart, D. E. (1986). Distributed representations. In Rumelhart, D. E., & McClelland, J. L., and the PDP Research Group, *Parallel Distributed Processing: Explorations in the Microstructure of Cognition. Volume 1: Foundations*. Cambridge, MA: MIT Press/Bradford Books.

Schneider, W. & Shiffrin, R.M. (1977). Controlled and automatic human information processing: I. Detection, search, and attention. *Psychological Review, 84*, 1-66.

Shiffrin, R.M., & Schneider, W. (1977). Controlled and automatic human information processing: II. Perceptual learning, automatic attending, and a general theory. *Psychological Review, 84*, 127-190.

Triesman, A., & Gelade, G. (1980). A feature integration theory of attention. *Cognitive Psychology, 12*, 97-136.

Treisman, A., & Gormican, S. (1988). Feature analysis in early vision: Evidence from search asymmetries. *Psychological Review, 95*, 15-48.

Discovering and using perceptual grouping principles in visual information processing

Michael C. Mozer
Department of Computer Science &
Institute of Cognitive Science
University of Colorado
Boulder, CO 80309–0430

Richard S. Zemel
Department of Computer Science
University of Toronto
Toronto, Ontario M5S 1A4

Marlene Behrmann
Department of Psychology & Rotman
Research Institute of Baycrest Centre
University of Toronto
Toronto, Ontario M5S 1A1

Abstract

Despite the fact that complex visual scenes contain multiple, overlapping objects, people perform object recognition with ease and accuracy. Psychological and neuropsychological data argue for a *segmentation* process that assists in object recognition by grouping low-level visual features based on which object they belong to. We review several approaches to segmentation/recognition and argue for a bottom-up segmentation process that is based on feature grouping heuristics. The challenge of this approach is to determine appropriate grouping heuristics. Previously, researchers have hypothesized grouping heuristics and then tested their psychological validity or computational utility. We suggest a basic principle underlying these heuristics: they are a reflection of the structure of the environment. We have therefore taken an adaptive approach to the problem of segmentation in which a system, called MAGIC, *learns* how to group features based on a set of presegmented examples. Whereas traditional grouping principles indicate the conditions under which features should be bound together as part of the same object, the grouping principles learned by MAGIC also indicate when features should be segregated into different objects. We describe psychological studies aimed at determining whether limitations of MAGIC correspond to limitations of human visual information processing.

Recognizing an object in a visual scene involves matching a collection of visual features in the scene that correspond to the object against stored object models. In scenes that contain multiple objects, the matching process alone is insufficient for recognition because it presumes that the features are partitioned by object. Consequently, a complete model of scene recognition requires the ability to *group* features of an object together, or equivalently, to *segment* the scene into regions corresponding to different objects. Psychophysical and neuropsychological evidence suggests that the human visual system possesses such an ability (Duncan, 1984; Farah, 1990; Kahneman & Henik, 1981; Treisman, 1982).

Models of scene recognition can be divided roughly into three classes (Figure 1). *Interactive* models are based on the observation that the scene cannot be properly segmented until object identities are known, yet objects cannot be properly identified until they are segmented. Consequently, segmentation and matching form an iterative cycle in which the matching system can propose refinements of the initial segmentation, which in turn refines the output of the matching system, and so forth (Hinton, 1981; Hinton, Williams, & Revow, 1992; Hanson & Riseman, 1978; Waltz, 1975). The problem with this approach is that it involves a simultaneous search for a good segmentation and a good interpretation of the data. We are skeptical about the computational feasibility of such massive combinatorial searches; they are slow and often lead to local optima in the search space (e.g., Hinton & Lang, 1985).

Bottom-up models are based on the premise that matching processes can be devised that do not require a precise segmentation (e.g., Mozer, 1992). Consequently, segmentation can be viewed as an early heuristic process that depends solely on low-level features. The results of segmentation are fed to the matching system, but the matching system does not directly influence segmentation. Although the heuristics used to group features will not be infallible, the hope is that they will suffice for most recognition tasks (Enns & Rensink, 1992). In cases where recognition fails the first time around, the segmentation can be adjusted and the process restarted. Although this restarting procedure is iterative, iteration is the exception, in contrast to the interactive model which intrinsically relies on an iterative constraint-satisfaction process to perform segmentation and matching jointly. The difficulty with the bottom-up approach is that an adequate set of grouping heuristics is required. We return to this issue later.

*This research was supported by NSF PYI award IRI–9058450, grant 90–21 from the James S. McDonnell Foundation, and DEC external research grant 1250 to MM, and by a National Sciences and Engineering Research Council Postgraduate Scholarship to RZ. Our thanks to Chris Williams, Paul Smolensky, Radford Neal, Geoffrey Hinton, and Jürgen Schmidhuber for helpful comments regarding this work.

Interactive and bottom-up models attempt to achieve object-based segmentation of the scene. That is, features of an object are collected together even if the features are noncontiguous in space and overlap with features of other objects. An alternative approach, *location-based* segmentation, simply determines a coherent region of space that is sufficiently large to be assured of containing all features of a single object, even if the features of other objects are present in that region. The hope is then to devise a matching process that can ignore irrelevant context surrounding the object of interest. It would seem quite difficult to achieve this robust a matching process. Recently, however, Rumelhart (1992; Keeler & Rumelhart, 1992) have proposed such a system using neural net learning techniques. The claim is that learning will find cues reliably indicating the presence of an object regardless of the context in which it is embedded. Even if such cues exist for real-world scenes—and of this we are skeptical—this class of model is inconsistent with the previously mentioned data indicating that people perform object-based grouping of featural information.

Note that the interactive and bottom-up models do not deny the possibility of location-based selection. Indeed, prior to the operation of these models, a spatial focus of attention may well be applied to select the general region of interest. Such a preselection stage would simplify the object-based segmentation task.

Our conviction is that the interactive and location-based models have serious complications both in terms of computational feasibility and psychological validity. We have thus turned to the bottom-up model and attempted to overcome its limitations. The primary concern is whether, based on information from the scene alone, a set of grouping heuristics exist that can determine which features belong together.

Gestalt psychologists have suggested a variety of grouping principles that govern human perception. In exploring how people group elements of a display, evidence has been found for grouping of elements that are close together in space or time, that appear similar, that move together, or that form a closed figure (Rock & Palmer, 1990). There is a long history of attempts by the computer vision community to turn these principles into grouping heuristics, with a fair degree of success (e.g., Lowe & Binford, 1982). The degree of success depends on the ingenuity of the researchers in proposing an adequate set of heuristics.

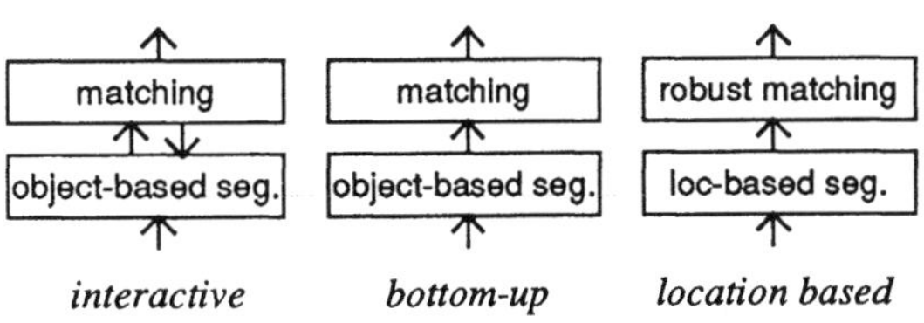

Figure 1: Three classes of object recognition models

We believe there is a more basic principle that underlies these heuristics and that can be used to suggest better heuristics. Namely, these grouping heuristics are a reflection of the structure of the environment. Preliminary evidence from neurobiology suggests that experience can indeed affect the strength of synaptic connections that may play a role in perceptual grouping (Löwel & Singer, 1992).

We have therefore taken an *adaptive* approach to the problem of segmentation in which a system learns how to group features based on a set of examples. We call our system MAGIC, an acronym for multiple-object adaptive grouping of image components. In many cases MAGIC discovers grouping heuristics similar to those proposed in earlier work, but it also has the capability of finding nonintuitive structural regularities in scenes. Hummel and Biederman (1992) have also discussed the possibility of discovering grouping heuristics based on environmental regularities.

MAGIC is trained on a set of presegmented scenes containing multiple objects. By "presegmented", we mean that each feature is labeled as to which object it belongs. MAGIC learns to detect configurations of the scene features that have a consistent labeling in relation to one another across the training examples. Identifying these configurations then allows MAGIC to label features in novel, unsegmented scenes in a manner consistent with the training examples.

This training procedure is a form of supervised learning. Of course, the real world does not directly provides such examples to a learner. However, there is a wealth of information in the environment that can supply the supervision. Perhaps the most important piece of information is motion. A rigid object moving in the plane perpendicular to the line of sight will have the property that all of its features travel across the visual field with the same velocity vector. Thus, by designing the learning system to treat velocity information as a training signal, the system can discover grouping principles that will also apply to stationary objects. Evidence from developmental psychology indeed suggests that the representation of object unity is initially derived from motion (Spelke, 1990).

The Domain

Our initial work has been conducted in the domain of two-dimensional geometric contours. The contours are constructed from four primitive feature types—oriented line segments at 0°, 45°, 90°, and 135°—and are laid out on a 25 × 25 grid. At each location on the grid are units, called *feature units*, that represent each of the four primitive feature types. In our present experiments, scenes contain two contours. We exclude scenes in which the two contours share a common edge. This permits a unique labeling of each feature. Examples of several randomly generated scenes containing rectangles and diamonds are shown in Figure 2. Although the scenes we have tested are composed only of

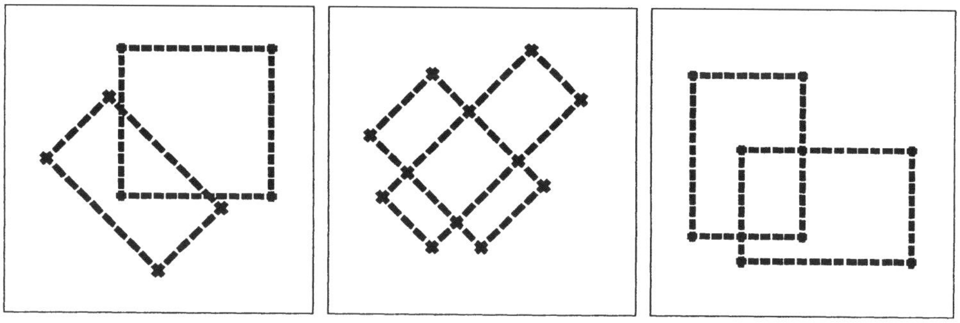

Figure 2: Examples of randomly generated two-dimensional geometric contours

shape features, MAGIC could be directly used to learn grouping principles based on color, texture, etc.

Representing Feature Labelings

Before describing MAGIC, we must first discuss a representation that allows for the labeling of features. Von der Malsburg (1981; von der Malsburg & Schneider, 1986), Gray et al. (1989), Eckhorn et al. (1988), and Strong and Whitehead (1989), among others, have suggested a biologically plausible mechanism of labeling through temporal correlations among neural signals, either the relative timing of neuronal spikes or the synchronization of oscillatory activities in the nervous system. The key idea here is that each processing unit conveys not just an activation value—average firing frequency in neural terms—but also a second, independent value which represents the relative phase of firing. The dynamic grouping or *binding* of a set of features is accomplished by aligning the phases of the features.

In MAGIC, the activity of a feature unit is a complex value with *amplitude* and *phase* components. The phase represents a labeling of the feature, and the amplitude represents the confidence in that labeling. The amplitude ranges from 0 to 1, with 0 indicating a complete lack of confidence and 1 indicating absolute certainty. There is no explicit representation of whether a feature is present or absent in a scene. Rather, absent features are clamped off—their amplitudes are forced to remain at 0—which eliminates their ability to influence other units, as will become clear when the activation dynamics are presented later.

The Architecture

When a scene is presented to MAGIC, units representing features absent in the scene are clamped off and units representing present features are set to a small amplitude and random initial phases. MAGIC's task is to assign appropriate phase values to the units. Thus, the network performs a type of pattern completion.

The network architecture consists of two layers of units, as shown in Figure 3. The lower (input) layer contains the feature units, arranged in spatiotopic arrays with one array per feature type. The upper layer contains hidden units that help to align the phases of the feature units; their response properties are determined by training. There are interlayer connections, but no intralayer connections. Each hidden unit is reciprocally connected to the units in a local spatial region of all feature arrays. We refer to this region as a *patch*; in our current simulations, the patch has dimensions 4 × 4. For each patch there is a corresponding fixed-size *pool* of hidden units. To achieve uniformity of response across the scene, the pools are arranged in a spatiotopic array in which neighboring pools respond to neighboring patches and the patch-to-pool weights are constrained to be the same at all locations in the array.

The feature units activate the hidden units, which in turn feed back to the feature units. Through a relaxation process, the system settles on an assignment of phases to the features. One might consider an alternative architecture in which feature units were directly connected to one another (Hummel & Biederman, 1992). However, this architecture is in principle not as powerful as the one we propose because it does not allow for higher-order contingencies among features.

Once MAGIC reaches equilibrium, grouped features can be passed on to an object matching system (Figure 1, middle panel). Essentially, this involves considering all phases in a particular range as belonging to a single object. A filter situated between the segmentation system and the matching system permits only features having phases in this range to pass through. The determination of how many objects are present and their range of phases can easily be made using Hough transforms (Ballard, 1981).

Network Dynamics

We summarize here the activation dynamics and learning algorithm. Further justification and intuitions underlying each are presented in Mozer, Zemel, Behrmann, & Williams (1992).

The response of each feature unit i, x_i, is a complex

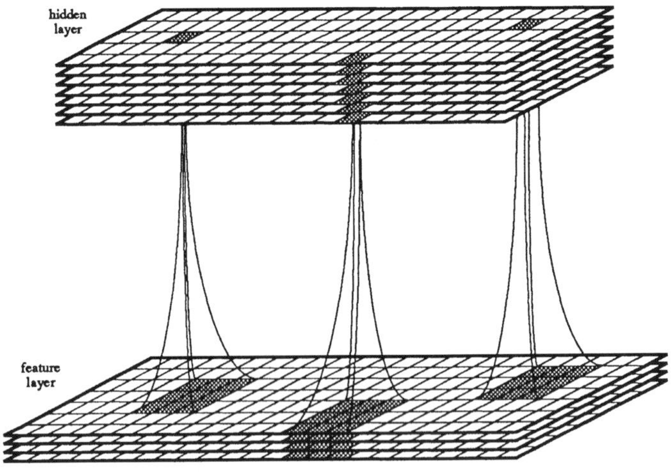

Figure 3: The architecture of MAGIC. The lower (input) layer contains the feature units; the upper layer contains the hidden units. Each layer is arranged in a spatiotopic array with a number of different feature types at each position in the array. Each plane in the feature layer corresponds to a different feature type. The grayed hidden units are reciprocally connected to all features in the corresponding grayed region of the feature layer. The lines between layers represent projections in both directions.

value in polar form, (a_i, p_i), where a_i is the amplitude and p_i is the phase. Similarly, the response of each hidden unit j, y_j, has components (b_j, q_j). The weight connecting unit i to unit j, w_{ji}, is also complex valued, having components (ρ_{ji}, θ_{ji}). The activation rule we propose is a generalization of the dot product to the complex domain. For a particular time step t,

$$net_j(t+1) = \mathbf{x}(t) \cdot \mathbf{w}_j = \sum_i x_i(t) w_{ji}^*$$

where net_j is the net input to hidden unit j and the asterisk denotes the complex conjugate. The net input is passed through a squashing nonlinearity that maps the amplitude of the response from the range $0 \rightarrow \infty$ to $0 \rightarrow 1$ but leaves the phase unaffected. The flow of activation from the hidden layer to the feature layer follows the same dynamics as the flow from the feature layer to the hidden layer. Note that updates are sequential by layer: the feature units activate the hidden units, which then activate the feature units.

In MAGIC, the weight matrix is Hermitian, i.e., $w_{ji} = w_{ij}^*$. This form of weight symmetry ensures that MAGIC will converge to a fixed point (Zemel, Williams, & Mozer, 1992).

Learning Algorithm

During training, we would like the hidden units to learn to detect configurations of features that reliably indicate phase relationships among the features. For instance, if the contours in the scene contain extended horizontal lines, one hidden unit might learn to respond to a collinear arrangement of horizontal segments. Because the unit's response depends on the phase pattern as well as the activity pattern, it will be strongest if the segments all have the same phase value.

The algorithm we have used is a generalization of back propagation. It involves running the network for a fixed number of iterations and, for each iteration, using back propagation to adjust the weights so that the feature phase pattern better matches a target phase pattern. Each training trial proceeds as follows:

1. A training example is generated at random. This involves selecting two contours and instantiating them in a scene. The features of one contour have *target* phase $0°$ and the features of the other contour have target phase $180°$.
2. The training example is presented to MAGIC by setting the initial amplitude of a feature unit to 0.1 if its corresponding scene feature is present, or clamping it at 0.0 otherwise. The phases of the feature units are set to random values in the range $0°$ to $360°$.
3. Activity is allowed to flow from the feature units to the hidden units and back to the feature units.
4. The new phase pattern over the feature units is compared to the target phase pattern (see step 1), and a measure of error is computed. This measure attempts to minimize the difference between the target and actual phases, and to maximize the confidence in the response. The error measure factors out any constant difference between the target and actual phases. See Mozer et al. (1992) for details.
5. Using a generalization of back propagation to complex valued units, error gradients are computed for

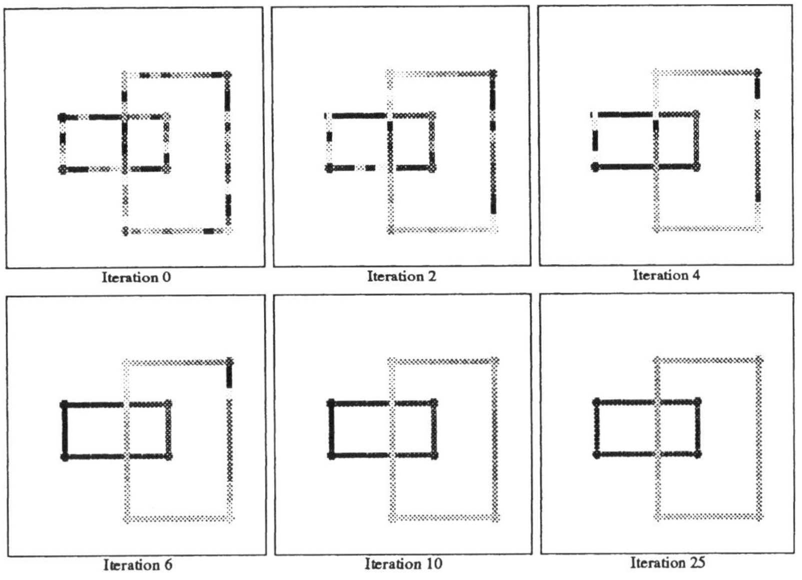

Figure 4: An example of MAGIC segmenting a scene. The "iteration" refers to the number of times activity has flowed from the feature units to the hidden units and back. The phase value of a feature is represented by a gray level. The cyclic phase continuum can only be approximated by a linear gray level continuum, but the basic information is conveyed nonetheless.

the feature-to-hidden and hidden-to-feature weights.

6. Steps 3–5 are repeated for a maximum of 30 iterations. The trial is terminated if the error increases on five consecutive iterations.
7. Weights are updated by an amount proportional to the average error gradient over iterations. Weight constraints are enforced to ensure that $w_{ji} = w_{ij}^*$ and that hidden units of the same "type" responding to different regions of the scene have the same weights.

Simulation Results

We trained a network with 20 hidden units per pool on examples like those shown in Figure 2. Each hidden unit attempts to detect and reinstantiate activity patterns that match its weights. One clear and prevalent pattern in the weights is a collinear arrangement of segments of a given orientation, all having the same phase value. When a hidden unit having weights of this form responds to a patch of the feature array, it tries to align the phases of the patch with the phases of its weight vector. By synchronizing the phases of features, it acts to group the features. Thus, one can interpret the weight vectors as the rules by which features are grouped.

Whereas traditional grouping principles indicate the conditions under which features should be bound together as part of the same object, the grouping principles learned by MAGIC also indicate when features should be segregated into different objects. For example, the weights of the vertical and horizontal segments are generally 180° out of phase with the diagonal segments. This allows MAGIC to segregate the vertical and horizontal features of a rectangle from the diagonal features of a diamond (see Figure 2, left panel). We had anticipated that the weights to each hidden unit would contain two phase values at most because each scene patch contains at most two objects. However, some units make use of three or more phases, suggesting that the hidden unit is performing several distinct functions. As is the usual case with hidden unit weights, these patterns are difficult to interpret.

Figure 4 presents an example of the network segmenting a scene. The scene contains two rectangles. The top left panel shows the features of the rectangles and their initial random phases. The succeeding panels show the network's response during the relaxation process. The lower right panel shows the network response at equilibrium. Features of each object have been assigned a uniform phase, and the two objects are 180° out of phase. The task here may appear simple, but it is quite challenging due to the illusory rectangle generated by the overlapping rectangles.

Empirical Tests of the Model

We are currently conducting psychological experiments to examine the role of feature grouping in human visual processing. Our experiments include the following:

- Previous studies have shown that judgements of two features of a single object (e.g., size, texture) can be made without loss of accuracy or speed whereas a cost is incurred when the features to be judged are drawn from two different objects (Duncan, 1984; Vecera and Farah, 1992). Based on this rationale, we might expect subjects to identify two elements of a single contour (similar to those used with MAGIC) more rapidly and accurately than elements of disparate contours. This paradigm provides a means of determining whether people group in the same way as MAGIC and what the limitations of grouping are. For instance, we are currently conducting experiments to examine whether a contour is processed as a single entity even when its features are spatially distant and it is partially occluded by a second contour.
- The bottom-up and interactive segmentation models presented in Figure 1 make divergent predictions about the recognition process. In the bottom-up model, segmentation is guided by low-level cues and is not influenced by object knowledge per se. Hence, familiarity should not influence segmentation performance. This is a challenge to test empirically because of the difficulty in measuring segmentation performance directly. The paradigm we are considering involves a search for unfamiliar targets embedded in—and difficult to segment from—a background of distractors. The familiarity of the distractors is manipulated. The interactive model suggests that targets should be easier to identify among familiar distractors. The bottom-up model predicts no effect of distractor familiarity.
- If indeed there is a distinct stage of information processing at which segmentation occurs, then it might be possible to find a neurological patient who has an impairment in segmentation. There are now two such reports in the literature in which patients are unable to bind individual features from disparate locations simultaneously (Grailet et al., 1990; Riddoch and Humphreys, 1987). We are currently studying the feature binding abilities of a visually agnosic subject, CK, and believe that he too has an impairment at this stage of processing.

References

Ballard, D. H. (1981). Generalizing the Hough transform to detect arbitrary shapes. *Pattern Recognition, 13(2)*, 111–122.

Duncan, J. (1984). Selective attention and the organization of visual information. *Journal of Experimental Psychology: General, 113*, 501–517.

Eckhorn, R., Bauer, R., Jordan, W., Brosch, M., Kruse, W., Munk, M., & Reitboek, H. J. (1988). Coherent oscillations: A mechanism of feature linking in the visual cortex? *Biological Cybernetics, 60*, 121–130.

Enns, J. T., & Rensink, R. A. (1992). A model for the rapid interpretation of line drawings in early vision. In D. Brogan (Ed.), *Visual search II*. London: Taylor and Francis. In press.

Farah, M. J. (1990). *Visual agnosia*. Cambridge, MA: MIT Press/Bradford Books.

Grailet, J. M., Seron, X., Bruyer, R., Coyette, F., & Frederix, M. (1990). Case report of a visual integrative agnosia. *Cognitive Neuropsychology, 7*, 275–310.

Gray, C. M., Koenig, P., Engel, A. K., & Singer, W. (1989). Oscillatory responses in cat visual cortex exhibit intercolumnar synchronization which reflects global stimulus properties. *Nature (London), 338*, 334–337.

Hanson, A. R., & Riseman, E. M. (1978). *Computer vision systems*. New York: Academic Press.

Hinton, G. E. (1981). A parallel computation that assigns canonical object-based frames of reference. In *Proceedings of the Seventh International Joint Conference on Artificial Intelligence* (pp. 683–685). Los Altos, CA: Morgan Kaufmann.

Hinton, G. E., & Lang, K. (1985). Shape recognition and illusory conjunctions. In *Proceedings of the Ninth International Joint Conference on Artificial Intelligence* (pp. 252–259). Los Angeles, California: (null).

Hinton, G. E., Williams, C. K. I., & Revow, M. D. (1992). Adaptive elastic models for hand-printed character recognition. In J. E. Moody, S. J. Hanson, & R. P. Lippman (Eds.), *Advances in neural information processing systems IV*. San Mateo, CA: Morgan Kaufmann.

Hummel, J. E., & Biederman, I. (1992). Dynamic binding in a neural network for shape recognition. *Psychological Review*. In Press.

Kahneman, D., & Henik, A. (1981). Perceptual organization and attention. In M. Kubovy & J. R. Pomerantz (Eds.), *Perceptual organization* (pp. 181–211). Hillsdale, NJ: Erlbaum.

Keeler, J. D., & Rumelhart, D. E. (1992). Self-organizing segmentation and recognition neural network. In J. E. Moody, S. J. Hanson, & R. P. Lippman (Eds.), *Advances in neural information processing systems IV*. San Mateo, CA: Morgan Kaufmann.

Lowe, D. G., & Binford, T. O. (1982). Segmentation and aggregation: An approach to figure-ground phenomena. *Proceedings of the DARPA IUS Workshop* (pp. 168–178). Palo Alto, CA.

Löwel, S., & Singer, W. (1992). Selection of intrinsic horizontal connections in the visual cortex by correlated neuronal activity. *Nature, 255*, 209–211.

Mozer, M. C. (1991). *The perception of multiple objects: A connectionist approach*. Cambridge, MA: MIT Press/Bradford Books.

Mozer, M. C., Zemel, R. S., Behrmann, M., & Williams, C. K. I. (1992). Learning to segment images using dynamic feature binding. *Neural Computation*. In Press.

Riddoch, M. J., & Humphreys, G. W. (1987). A case of integrative visual agnosia. *Brain, 110*, 1431–1462.

Rock, I., & Palmer, S. E. (1990). The legacy of Gestalt psychology. *Scientific American, 263*, 84–90.

Rumelhart, D. E. (1992). Script handwritten word recognition in a neural network. Colloquium presented at the Institute of Cognitive Science, University of Colorado.

Spelke, E. S. (1990). Principles of object perception. *Cognitive Science, 14*, 29–56.

Strong, G. W., & Whitehead, B. A. (1989). A solution to the tag-assignment problem for neural networks. *Behavioral and Brain Sciences, 12*, 381–433.

Treisman, A. (1982). Perceptual grouping and attention in visual search for features and objects. *Journal of Experimental Psychology: Human Perception and Performance, 8*, 194–214.

Vecera, S., & Farah, M. J. (1992). Visual attention can select from spatially invariant object representations. Submitted for publication.

von der Malsburg, C. (1981). *The correlation theory of brain function* (Internal Report 81-2). Goettingen: Department of Neurobiology, Max Planck Institute for Biophysical Chemistry.

von der Malsburg, C., & Schneider, W. (1986). A neural cocktail-party processor. *Biological Cybernetics, 54*, 29–40.

Waltz, D. A. (1975). Generating semantic descriptions from drawings of scenes with shadows. In P. H. Winston (Ed.), *The psychology of computer vision* (pp. 19–92). New York: McGraw-Hill.

Zemel, R. S., Williams, C. K. I., & Mozer, M. C. (1992). Adaptive networks of directional units. Submitted for publication.

The Role of Genericity in the Perception of Illusory Contours

Marc Albert*
Department of Information and Computer Science
University of California, Irvine
Irvine, CA 92717
email: albert@ics.uci.edu

Abstract

Visual images are ambiguous. Any given image, or collection of images, is consistent with an infinite number of possible states of the external world. Yet, the human visual system seems to have little difficulty in reducing this potential uncertainty to one, or perhaps a few perceptual interpretations. Many vision researchers have investigated what sort of constraints—assumptions about the external world and the images formed of it—that the visual system might be using to arrive at its perceptions. One important class of constraints are those based on *genericity* or *general position*.

We propose a theory of illusory contours in which general position assumptions are used to infer certain necessary conditions for the occurrence of illusory figures that appear to occlude their inducers. Experiments with human subjects are described. The results of these experiments suggest an important role for general position assumptions in understanding the perception of illusory contours. It is also demonstrated that *parallelism* of contours of "blob" type inducers is an important determinant of illusory contour strength.

Introduction

Illusory contours are contours that are perceived in regions of the visual field where there are, in fact, no physical contours, i.e., where there are no sharp gradients in any image property. For example, in Figure 1 most observers perceive a rectangular illusory surface that is brighter than the surrounding white area, and is partially occluding the black elements in the display. The theory of illusory contour perception presented in this paper has its roots in other theories proposed in the literature on human and machine vision. These theories have all used "general position" or "generic view" assumptions to understand some aspect of human perception, or to constrain the design of a computer vision algorithm. Roughly speaking, these assumptions are satisfied when the eye of the observer (or TV camera), and the physically independent objects in a scene are placed "randomly" with respect to each other, so that the image received by the eye is not in any way qualitatively "special" or improbable. These assumptions are closely related to the theory of perceptual preference proposed by Rock (1983).

Previous theories of illusory contours (IC's) fall into three main categories; peripheral, central and Gestaltist. Theorists in the peripheral group (Brigner and Gallagher, 1974; Frisby and Clatworthy, 1975) believe that IC's can be accounted for primarily in terms of peripheral neurobiological processes in the visual system. Theorists of the central group (Gregory, 1972; Rock and Anson, 1979; Coren, 1972) have pointed out that many of the properties of IC's do not fit with purely peripheral explanations. They claim that IC's are created higher up in the visual system, and that a "cognitive" sort of explanation is more appropriate. Our theory perhaps best fits in this category. Finally, the Gestaltists (Kanizsa, 1955, 1974) believe that the phenomena are best understood in terms of the Gestalt laws of perceptual organization.

Transversality

The Transversality Principle is central to the field of differential topology in mathematics (see, for example, Guillemin and Pollack, 1974). For our purposes we can state a special case of that principle as follows: If two differentiable curves in $\Re^2$ (i.e., the plane) are independently and randomly selected, then the probability that the derivatives of those curves will agree at any intersection point of the curves is zero. In other words, *generically* at all points where they inter-

*Partially supported by ONR contract N-00014-88-K-0345.

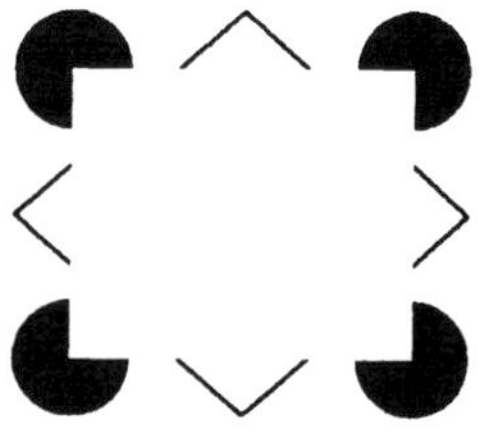

Figure 1

Figure 2

Figure 3

sect they intersect transversely.

The Transversality Principle is used in the work of Hoffman and Richards (1984) on the decomposition of physical objects into parts. For occlusion the Transversality Principle implies that, generically, the tangents to the curves that bound the objects in an image differ at all intersection points of those curves. In displays with "blob" inducers, we can use this observation to derive a necessary condition for the occurrence of a special class of IC's, viz., IC's that appear to partially occlude some or all of their inducers (ICO's).

The standard examples of IC's consist of some black regions on a white background in which an illusory "whiter than white" surface appears to partially occlude the black regions (e.g., Figure 2). By the Transversality Principle we can conclude that if the illusory surface occludes the black regions in a generic way, then each point of intersection of the IC with the contour of a partially occluded black region is a point of transverse intersection. We now make the following

General position assumption for illusory contours: ICO's are generated by the visual system only if the occlusion is generic.

If this assumption is correct, then a necessary condition for the occurrence of an illusory white surface that appears to partially occlude black inducers is the presence of ***convex discontinuities*** in the tangent lines to the contours of the inducers (also see Brady and Grimson, 1981; Kellman and Shipley, 1991).

Using magnitude estimation, 25 naive human subjects rated the IC in Figure 2 to be much stronger than in Figure 3, and they all said that the black elements appeared to be occluded in Figure 2, whereas 22 out of 25 said that they did not in Figure 3. Typically, subjects described the inducers in Figure 3 as pushed up against or crowded around the illusory square.

Line Drawing Interpretation

A number of theories of line drawing interpretation have been proposed by researchers in human and machine vision. Many of these theories have used the generic viewpoint assumption. In particular, Binford (1981) showed that a lot of information about the relative depth of curves in an image can be inferred by applying rules based on this assumption. These rules can explain, for example, why "special" viewpoints on a Necker cube lead to 2-D rather than 3-D interpretations. Two rules that will be useful for us are: 1) If three or more curves intersect at a common point in an image, then their preimages intersect at a common point in space and, 2) If two or more curves terminate at a common point in an image, then their preimages terminate at a common point in space.

Suppose we take a display that generates a strong ICO using line-end inducers, such as Figure 4. For each inducing line ending, we add to that display another line that terminates at a common point with it, as in Figure 5. If the human visual system generated an IC in Figure 5, then, treating the IC on an equal footing with the real lines in the display, and applying rule 1, it would conclude that the intersection point of the IC and the inducers in the image must correspond to an intersection of their preimages in space. So the IC and the lines would appear to be at the same depth where they meet, and the illusory surface would not appear to be occluding the inducers (also see Kennedy, 1978). Our experiments show that human subjects perceive the IC in Figure 5 to be much weaker, or nonexistent compared to that in Figure 4, and the appearance of occlusion is gone. This is true despite the fact that the inducers in Figure 5 are just as "well aligned" as in Figure 4 (see Rock and Anson 1979). In fact, the additional lines might lead one to expect a ***stronger*** IC in Figure 5 than in Figure 4. Thus, genericity seems to be important not only for the interpretation of ordinary line drawings, but also in determin-

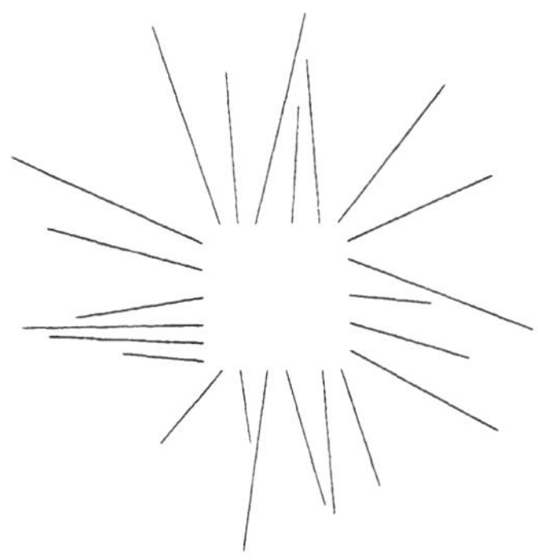

Figure 4

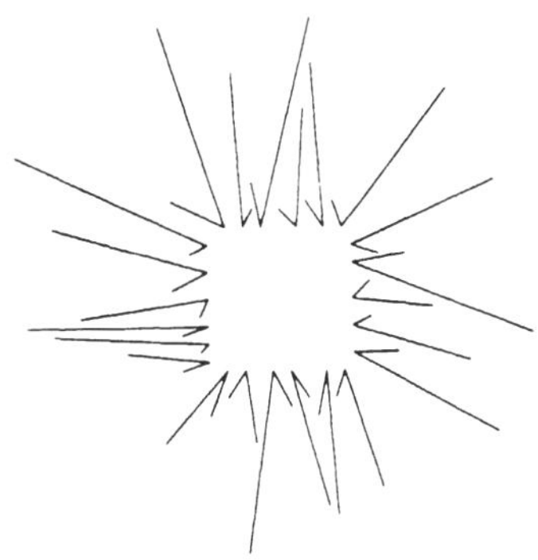

Figure 5

Figure 6

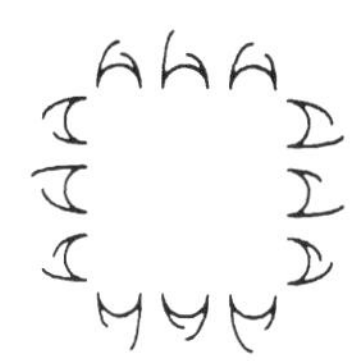

Figure 7

ing the stimulus conditions for IC perception (for more details see Albert, submitted).

In Figure 6 subjects perceive an ICO defined by the endings of the semicircular arcs. In Figure 7 we have added another coterminating arc for each ending of a semicircular arc in Figure 6 in such a way that their tangents agree at the cotermination points. Thus, the strong impression of line endings is preserved. Yet the IC has all but vanished. This seems inconsistent with the line-end contrast theory of IC perception put forward by Frisby and Clatworthy (1975), and with the theory of Grossberg and Mingolla (1985). However, using the genericity principle we can readily understand why Figure 6 produces an ICO and Figure 7 does not.

Analogous results can be obtained with the "neon color spreading" effect (van Tuijl, 1975). If we start with a display that produces neon color spreading using colored lines, and then add to it lines that intersect the original lines at their points of color change, then the neon color spreading is greatly reduced, and the perception of transparency disappears (see Albert and Hoffman, to appear).

While we have stated our theory in terms of the generic viewpoint assumption, analogous arguments can be made using the assumption that the physically independent objects in a scene are placed "randomly" with respect to each other in space. This constraint can explain the effects discussed above even if the illusory surface is seen as being only "infinitesimally" closer to the observer than the inducers when occlusion is perceived. Perhaps both constraints are influencing our perceptions.

In addition, although we have stated the theory in terms of "rules" for image interpretation and ICO perception, we do not claim that it is impossible to perceive an occluding illusory figure in displays for which such a percept in nongeneric. We only claim that, other things being equal, such a percept is much less likely, especially for naive subjects, than it is in displays in which the percept is generic. Similarly, it is possible to perceive "special" viewpoints on Necker cubes as cubes, but such interpretations are rarely made by naive subjects. Genericity is only one among many factors that are weighed by the visual system when it interprets images. It can be violated when other factors, which contradict its prediction for a particular image, are given greater weight by the visual system. We do not believe that constraints such as genericity (or, for example, rigidity in structure from motion) are strict rules of image interpretation. Our view is that these constraints can interact and compete with each other and with other visual cues to determine image interpretation.

Figures 4, 5, 6, and 7 were used in experiments with 25 naive human subjects. They rated the IC's in Figures 4 and 6 as much stronger than those in Figures 5 and 7, respectively. For Figures 4 and 6, 21 subjects said that the black elements were occluded, whereas only 2 said so for Figure 5, and none for Figure 7.

Many researchers have pointed out that outlines of pac-men (or other blob inducers) fail to generate IC's. Using genericity we can understand this outcome as follows: The short line segments that follow the potential IC cannot be seen as being partially occluded by an illusory surface because if they are viewed as being part of a larger blob-like element, then it is highly improbable that just a very thin edge of that blob would be visible (see also Kellman and Shipley, 1991). On the other hand, if they are viewed simply as line segments, then if any of them were at a different depth from the illusory edge which they appear to lay on (or next to) in the image, it would imply that our viewpoint on the scene was highly improbable. Thus, the short line segments must be at the same depth as the illusory edge, possibly interpreted as a highlight, or a surface irregularity, or as something attached to the side of the surface. Now those short segments also terminate at common points with the circular arcs, so by rule 2 the circular arcs must also be at the same depth as the short line segments at their points of intersection. Therefore, the potential inducers cannot appear to be occluded by an illusory surface.

Kanizsa (1974) has argued that "closure" can explain the perception of IC's with line-end inducers. Supporters of this theory might claim that the effects seen in our displays could be explained in this way (since the curves in Figures 5 and 7 are, at least, closed "on the side of the potential IC"). However, we believe genericity to be a more satisfactory explanation, since it is a valid ecological constraint. It also predicts certain perceived depth relations which closure cannot (see Albert, submitted).

Mathematical Formalization

Koenderink (1990) has proposed a theory of object recognition based on the the idea of generic versus accidental views. In his theory the ambient space of possible viewpoints on a scene is divided into "cells". The cell which contains a particular viewpoint is the largest connected region of the ambient space within which all viewpoints give rise to topologically equivalent images. Intuitively two images are topologically equivalent if the junctions among the image curves (excluding L junctions) have the same qualitative structure. The "cell walls" in this theory define surfaces in space. When an observer crosses a cell wall the qualitative structure of the image changes.

We believe that not only topological structure, but also first order differentiable structure is perceptually important. This entails that corresponding image curves have corresponding tangent discontinuities (i.e., transversality is taken into account). We make the following hypothesis: The visual system prefers not to interpret images in a way that places its viewpoint on a scene within a "cell wall" with regard to first order differentiable structure. The justification for this hypothesis is that if a viewpoint on a scene is chosen "at random", then the probability of ending up in a cell wall is zero. So if the features defining the cells and cell walls are perceptually salient, the visual system can use this probabilistic information in selecting interpretations for images. Nakayama and Shimojo (1990) have made a similar proposal, but they considered only topological structure. They applied their idea to a particular display in a way that is very similar to the style of analysis used in this paper.

Parallelism

Witkin and Tenenbaum (1983) proposed a framework for theories of perception based on the idea of "non-accidental" *3-D* relations. For example, if an image contains a group of parallel curves, then Binford's theory justifies the inference that they are parallel in space. Now, Witkin and Tenenbaum claim that it would be a highly improbable coincidence that they are all parallel to one another, unless they all arouse from a single "cause" or process. And this explains why the visual system is, in a sense, "correct" to group such curves together.

Lowe (1985) used the ideas of Witkin and Tenenbaum to construct a computer vision system. When Lowe's system saw two parallel lines in an image which could plausibly represent edges of objects in the scene, and if those lines were relatively close to each other in relation to the overall density of line segments at that scale in the image, then the system inferred that those lines were opposing edges of a single 3-dimensional object in the world. That is, Lowe instantiated Witkin and Tenenbaum's idea that the two lines arose from a single cause, to the inference that they represented edges of a single object.

Rock (1983) has pointed out that human subjects group parallel curves together to form the boundaries of regions more readily than they do non-parallel curves. For example, in Figure 8a most people see the black regions as figure and the white regions as ground, whereas the reverse is true for Figure 8b.

Now, consider Figure 9. This display has been discussed by many researchers going back to Kanizsa (1955). Note that 1) this display contains more black area than Figure 2, 2) there are equal amounts of the contour of the blobs along the potential IC in both displays, and 3) the length of the IC to be interpolated is the same. However, in spite of this, human subjects perceive an IC only weakly, if at all, in Figure 9, and a strong IC in Figure 2. Kanizsa claimed this as strong supporting evidence for his theory of IC's based on Gestalt ideas. He believed that the visual system creates an illusory surface in Figure 2, for example, so that the pac-men can be amodally completed to disks, which are "good", symmetrical forms. On the other hand, in Figure 9 the crosses are already quite symmetrical, and amodally extending them behind

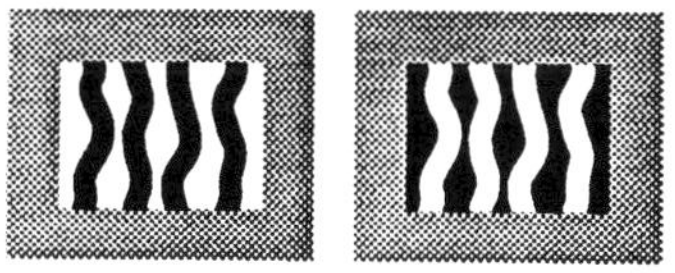

a b

Figure 8

Figure 9

Figure 10

a potential illusory surface would destroy their symmetry. However, it was subsequently shown by Kanizsa and others that strong IC's occurred in displays in which the inducers could not possibly be amodally completed to entities possessing figural "goodness" in the Gestalt sense.

So, the theoretical position researchers found themselves in was the following: Potential amodal completion of inducers into good, symmetrical forms was not a major factor in *causing* IC's to occur. But, if the shape of the inducers was already good and symmetrical, then this *inhibited* the emergence of IC's.

However, consider Figure 10. In this display the inducers are just as symmetrical as in Figure 9, and there is much less alignment of the physically present edges, a factor that is known to have a considerable impact on IC strength (Rock and Anson, 1979). Yet, for most subjects the IC is stronger in Figure 10 than in Figure 9.

We would like to suggest that the major factor inhibiting the emergence of an IC in Figure 9 is *parallelism*. We claim that IC's should be weakened in displays with blob inducers if part of the contour of a blob is reasonably close to, approximately parallel to, and opposite the part of the blob's contour that is along the potential IC (e.g., Figure 11a). The theoretical ideas and psychophysical demonstrations presented above support this claim in the following way. When an IC occurs in a display with blob inducers, the part of the contour of a blob that is contiguous with the illusory surface must be interpreted by the visual system as being "owned" by the illusory surface, and the remainder of the contour interpreted as owned by the blob. Now, if the part of the contour that was meant to be owned by the illusory surface is parallel to and opposite a part of the contour that was meant to be owned by the blob, then the IC should be weakened if the visual system is biased towards interpreting parallel and opposite contours as both belonging to the same object.

We asked 25 naive human subjects to say whether Figure 11a or 11b had a stronger IC. The contours of the blobs in the two displays differ only in edges that are not along and do not intersect the potential IC. However, parts of the contours of the blobs in Figure 11a are parallel to and opposite the parts of the contours that lie along the potential IC. In our experiments 22 out of the 25 subjects said that the IC was stronger in Figure 11b than in Figure 11a. The same 25 subjects were also asked to rank order Figures 12a, b and c in terms of IC strength (from strongest to weakest). Here 20 out of the 25 subjects ordered them as 12c, 12b, 12a, 4 subjects ordered them as 12c, 12a, 12b, and one subject ordered them as 12a, 12b, 12c. Note that Figure 12c has half as much black area as Figure 12a, and that there are 6 possible orderings of the three displays.

Summary and Conclusion

We have explored the hypothesis that the visual system applies the principle of genericity to the whole collection of contours that are perceived in an image. This includes contours that are given by real contrast edges, as well as illusory ones. In addition, we have shown that parallelism strongly influences IC perception, and that displays which had previously been thought to confirm the importance of symmetry might best be understood in terms of the influence of parallelism on the perceived "ownership" of contours.

What is the overall significance of the principle of genericity for understanding IC perception? Of course, it cannot predict the exact strength and perceptual quality of the IC's seen by observers in arbitrary displays. However, we feel it does provide important constraints for a more comprehensive theory.

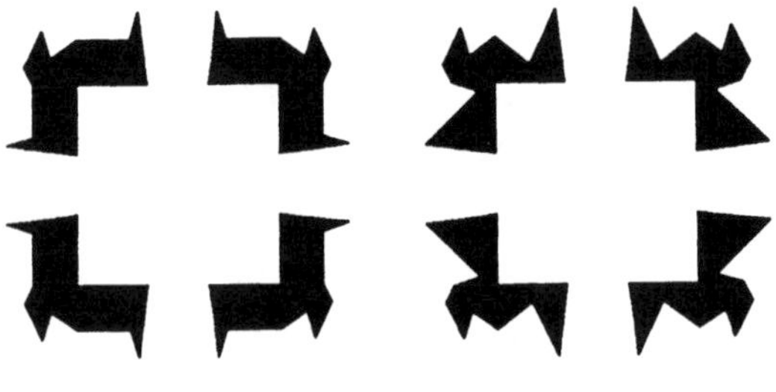

a b

Figure 11

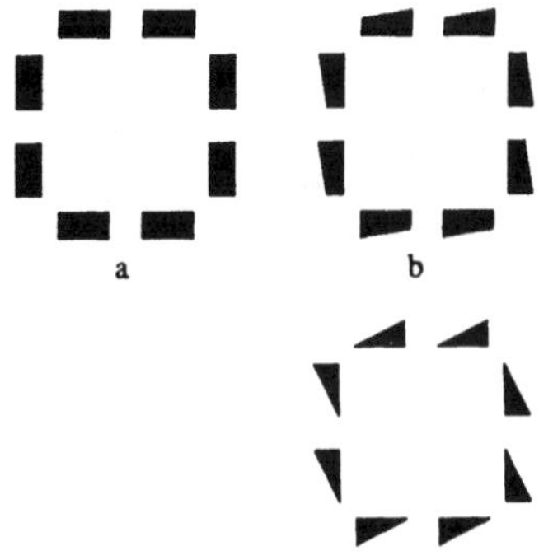

a b c

Figure 12

Acknowledgments

Many thanks to Don Hoffman for numerous discussions about the ideas in this paper. Thanks also to Mike Braunstein, Jack Beusmans, and David Honig.

References

Albert, M. K. submitted. General position assumptions, parallelism, and illusory contours. *Perception*

Albert, M. K.; Hoffman, D. D. to appear. Generic visions: General position assumptions in visual perception. *Scientific American*

Becker, M. F.; Knopp, J. 1978. Processing of visual illusion in the frequency and spatial domains. *Perception and Psychophysics* **23** 521-526

Binford, T. O. 1981. Inferring surfaces from images. *Artificial Intelligence* **17** 205-244

Brady, M.; Grimson, B. 1981. The perception of subjective surfaces. M.I.T. AI memo no. 666

Brigner, W. L.; Gallagher, M. B. 1974. Subjective contour: Apparent depth or simultaneous brightness contrast? *Perceptual and Motor Skills* **38** 1047-1053

Coren, S. 1972. Subjective contours and apparent depth. *Psychological Review* **79** (4) 359-367

Frisby, J. P.; Clatworthy, J. L. 1975. Illusory contours: curious cases of simultaneous brightness contrast? *Perception* **4** 349-357

Ginsburg, A. P. 1975. Is the illusory triangle physical or imaginary? *Nature* **257** 219-220

Gregory, R. 1972. Cognitive contours. *Nature* **238** 51-52

Grossberg, S.; Mingolla, E. 1985. Neural dynamics of form perception: boundary completion, illusory figures, and neon color spreading. *Psychological Review* **92** 173-211

Guillemin, V.; Pollack, A. 1974. *Differential Topology.* (Englewood Cliffs: Prentice-Hall)

Hoffman, D. D.; Richards, W. A. 1984. Parts of recognition. *Cognition* **18** 65-96

Kanizsa, G. 1955. Marzini quasi-percettivi in campi con stimolazione omogenea. *Rivista di Psicologia* **49** 7-30

Kanizsa, G. 1974. Contours without gradients or cognitive contours?
Italian Journal of Psychology **1** 93-112

Kellman, P. J.; Shipley, T. F. 1991. A theory of visual interpolation in object perception. *Cognitive Psychology* **23** 141-221

Kennedy, J. M. 1978. Illusory contours not due to completion. *Perception* **7** 187-189

Koenderink, J. 1990. *Solid Shape.* (Cambridge: MIT Press)

Lowe, D. G. 1985. *Perceptual Organization and Visual Recognition.* (Boston: Kluwer)

Marr, D. 1982. *Vision.* (New York: W H Freeman and Co.)

Nakayama, K.; Shimojo, S. 1990. Towards a neural understanding of visual surface representation. *Cold Spring Harbor Symposium on Quantitative Biology* **55**

Rock, I. 1983. *The Logic of Perception.* (Cambridge: MIT Press)

Rock, I.; Anson, R. 1979. Illusory contours as the solution to a problem. *Perception* **8** 665-681

van Tuijl, H. F. J. M. 1975. A new visual illusion: Neonlike color spreading and complementary color induction between subjective contours. *Acta Psychologica* **39** 441-445

Witkin, A. P.; Tenenbaum, J. M. 1983. On the role of structure in vision. In *Human and Machine Vision* Eds J Beck, B Hope, A Rosenfeld (New York: Academic Press) pp 481-543

Perceiving the Size of Trees Via Their Form

Geoffrey P. Bingham
Department of Psychology and Cognitive Science Program
Indiana University
Bloomington, IN 47405
gbingham@ucs.indiana.edu

Abstract

Physical constraints on growth produce continuous variations in the shape of biological objects that correspond to their sizes. We investigated whether two such properties of tree form can be visually discriminated and used to evaluate the height of trees. Observers judged simulated tree silhouettes of constant image size. Comparison was made to judgments of real trees in natural viewing conditions. Tree form was shown to confer an absolute metric on ground texture gradients. Eyeheight information was also shown to be ineffective as an alternative source of absolute scale.

Introduction

The problem of size perception arises because the size of the image projected from an object varies with the distance of the object from the observer. Image size, by itself, provides no information about object size. The traditional solutions to this problem are size-distance invariance theory and familiar size.

In size-distance invariance theory, the inverse relation between image size and object distance is used to derive perceived object size, assuming that information about distance is available (Gogel, 1977; Holway & Boring, 1941; Kilpatrick & Ittelson, 1953). This confounds the problems of size and distance perception. Because distance perception is itself a difficult problem, an independent approach to size perception would be advantageous. Familiar size does not presume information about distance. For this reason, familiar size is usually included among hypothetical sources of information about distance (Epstein, 1961; Gogel, 1977; Gibson, 1950; Hartmen & Harker, 1957;).

The familiar size solution is simply that the observer knows the size of certain identifiable objects that have highly stable and definite sizes. Familiar size reduces size perception to form perception because object recognition is achieved by identifying characteristic forms. Familiar size is usually considered with respect to man made objects, like playing cards, matchbooks, and watches, because the relevant forms are distinct and the sizes are well restricted. Application to biological objects is more difficult because the sizes for a given type of object are less restricted and the relevant forms are more complex and subject to continuous variations. Can observers use continuous variations in form to perceive variations in size? If so, the generalization would make the familiar size solution very powerful. However, generalization depends, in part, on discriminative abilities in form perception.

A second consideration is associated with a requirement that sizes be restricted and highly stable. Such regularity and predictability is produced by constraints impinging on the formation of the objects in question. The sizes and forms of biological objects are constrained by physical and biological laws. The study of such laws comprises the subject matter of functional morphology and allometry[1] (Calder, 1984; Hildebrand, Bramble, Liem & Wake, 1985; McMahon, 1984; McMahon & Bonner, 1983; Peters, 1983; Thompson, 1961). D'Arcy Thompson (1961) has described organic form as a "diagram of forces" and, following observations of Galileo,

[1] Similar considerations are found in the study of scale models in engineering where object form and materials must be distorted or altered in small scale models to preserve structural integrity and function for purposes of testing (Baker, Westine & Dodge, 1973).

has noted that organic forms alter in the face of scale changes to preserve the integrity of structure and function. The forms must change because various linear or geometric dimensions in an object scale differently to relevant forces.

For instance, as discussed by Galileo, the strength of a bone required to support its weight is proportional to the square of its diameter while the weight to be supported is proportional to the cube of its length. As the bone increases in size, the diameter must increase faster than the length for the strength be adequate to support the weight. Bigger bones must be relatively thicker. Such changes in form are especially prominent in biological objects because their materials remain invariant over scale changes wrought by growth. This is true in particular of the forms assumed by vegetation.

Observers can certainly distinguish a stalk of grass from a tree. The forms are fairly distinct. Is the same true of small versus large trees? In this instance, the size can vary continuously from a couple of feet to a couple of hundred feet. Do specific continuous variations in tree form accompany such variations in size? If so, can observers distinguish such continuous variations in form and use such information to evaluate size?

For a number of reasons, perceiving the size of trees provides a good test case for a reduction of size perception to form perception by virtue of physical constraints on form. First, trees are extremely common in the visual environment and they span the greater part of the range of sizes directly relevant to human activity. Their presence could be used to determine the size of neighboring objects including human artifacts (e.g. buildings) and terrain features (e.g. rock outcrops). Second, their frequency of appearance in the surround means that observers will be familiar with them. Third, tree morphology has been studied extensively. The scaling laws that determine changes in form accompanying changes in size with growth have been described (Borchert & Honda, 1984; Fisher & Honda, 1979a, b; Honda, Tomlinson & Fisher, 1981; McMahon & Bonner, 1983; McMahon & Kronauer, 1976; Turrell, 1961). Fourth, the same scaling laws apply to most other forms of terrestrial vegetation and some apply as well to aspects of the form and structure of vertebrates (McMahon, 1984; McMahon & Bonner, 1983). Fifth, the relevant forms are complex and the variations in form are sufficiently subtle to provide a good test of the ability of the visual system to detect subtle variation in complex forms and employ it as information about size.

Two scaling laws are known to determine characteristic properties of tree form that vary with tree height. First, successful mechanical support is achieved in trees by preserving elastic similarity (McMahon, 1975; McMahon & Kronauer, 1976). The diameter of a branch or tree trunk scales with the remaining length along the branch or trunk to its tip as follows:

$$\text{Diameter} = (\text{Height})^{1.5}.$$

This is consistent with an empirically derived relation which also predicts maximum heights for given climate zones (Kira, 1980). Both relations predict that the ratio of the diameter of the trunk to the height of a tree is specific to the actual height of the tree. (This ratio also applies to any point along a branch using the diameter at that point and the remaining length to the tip.) Because the H/D ratio is well preserved in tree images, the relation determines optical information for tree height. Using the Kira relation for temperate zone trees:

$$\text{Actual Height}=131.23 - 3.28(H/D).$$

Second, the number of terminal branches in a tree scales with the size of the tree (Borchert & Honda, 1984; Turrell, 1961). To an approximation, a tree covers the surface of its branching volume with leaves of constant size to collect light. Branches are required in constant proportion to the leaves. An exponential branching process is constrained by the hydrodynamics of the neutrient distribution producing conformity to a surface law (Borchert & Honda, 1984; Honda, Tomlinson & Fisher, 1981). This has been confirmed (Kira, 1980; Turrell, 1961) and predicts:

$$\text{Number of branches}=a(\text{Height})^2.$$

Thus, the number of branches, a property well preserved in images, also provides information about tree height.

Judging isolated silhouettes and real trees

Can observers use forms generated by such scaling relations to judge tree size? Using the two scaling relations, we produced tree silhouettes of constant image height in 7 different architectures (Halle, Oldeman & Tomlinson, 1978; Honda,

1971; Tomlinson, 1983)[2]. 24 observers first judged the height of 16 real trees observed on the IU campus at distances preserving constant image heights equivalent to our simulated images. Actual heights ranged from 10 ft-90 ft. O's next judged heights of simulated trees viewed as silhouettes (parallel projection) with no background structure.

O's judging real trees were instructed to judge height in feet by glancing rapidly at the specified tree and writing a quick, "off the cuff" assessment of the height. Each judgment was made in a period of about 2-3 seconds. Before making these judgments, participants were shown a short (26 ft) and a tall (64 ft) lighting pole and were told the heights. A regression showing mean judgments (with standard error bars) against actual heights appears in Figure 1. The rather surprising accuracy given the rapidity of judgments is reflected by a mean slope of .9 and intercept near 0. r^2 for the individual judgments was .81.

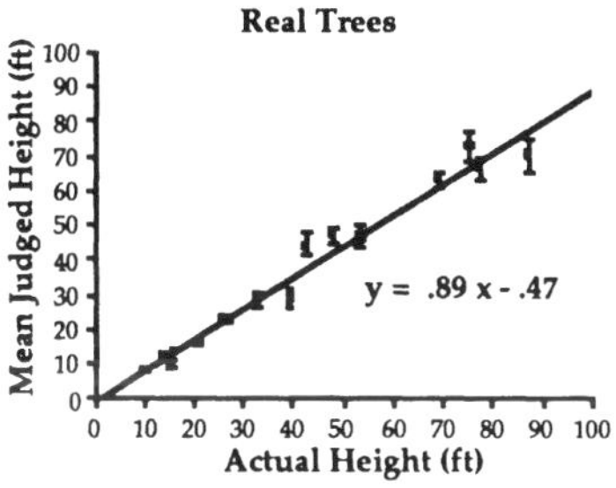

Figure 1.

O's also judged height in feet for the simulated tree silhouettes. O's were given packets in which each page contained a single tree image. 56 trees varying in height and architecture were arranged in 2 random orders. O's flipped through the packets writing their judgments in order and then were allowed to go back to adjust their judgments after having studied the whole set of tree images. The mean height judgments with standard error bars appear in Figure 2 compared to actual modeled heights for each of the 7 architectures.

[2] Rolf Borchert provided us with the program described in Borchert & Honda (1984) which simulated branching as determined by the hydrodynamics. M. Stassen, E. Gutjahr, and I incorporated routines to compute tree diameters and to draw trees, ground texture, and cylinders in perspective.

On average, judgments were monotonically increasing with increasing actual height. Rank orderings of mean height judgments were computed simultaneously across all 56 trees as were orderings according to actual modeled height, number of branches, and the D/H ratio. The ordinal relations for trees across all 7 architectures were accurately reproduced in judgments with the exception of one of the architectures. O's were also asked to rate the naturalness of the tree images. O's rated the architecture with poor ordinal results as the least natural or realistic. Overall, however, the images were rated as realistic.

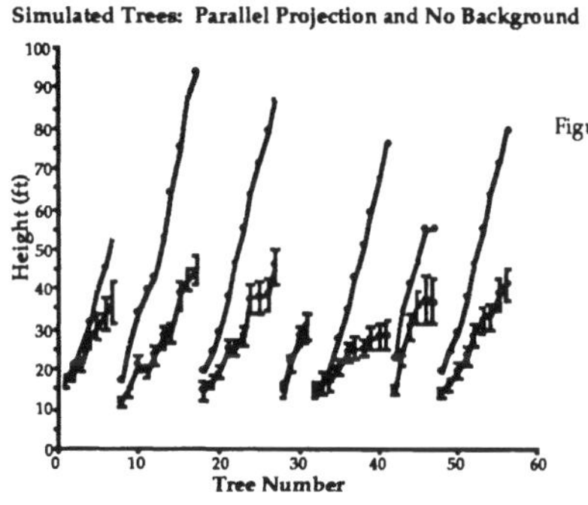

Figure 2.

The scaling relation between actual height and the H/D ratio predicted by the Kira relation was linear. The relation for modeled and judged heights was linear in both cases. When H/D and number of branches were regressed simultaneously on modeled heights, both H/D and number of branches were significant, $p<.001$, with almost equal beta weights of opposite sign. (Overall $r^2=.902$.) In the same regression performed on height judgments, only H/D was significant with a beta weight that dwarfed that for number of branches. (Overall $r^2=.914$.) Thus, judgments seemed to have depended primarily on the H/D ratio. Of course, because the H/D ratio and the number of branches covary to a large extent, the number of branches cannot be irrelevant to either height or size judgments. In pilot studies using simulations that only varied the H/D ratio, not branch number, some O's refused to perform the task because the information was contradictory.

The obvious problem with the simulation results was that, while the judgments were well ordered, the slopes were shallow. The overall slope for actual heights regressed on judged heights was .37. Mean judgments did not exceed 45 ft while modeled heights reached 90 ft. Why should this have been so? One possibility is that the simulation viewing conditions may have been so reduced as to distort the forms and suppress judgments. The tree images were all produced using parallel projection for all sizes and distances. As a result, the images were all extremely flat and, for nearer trees, distorted. Also, if we wished to compare simulation results to the judgments of real trees in more natural viewing conditions, then the lack of a ground texture gradient may have been significant. O's may have had some difficulty in resolving successive increments in either the H/D ratio or the number of branches. Judgments of individual O's tended to exhibit random local reversals or flattening of the judgment curves. Location in a ground texture gradient may enable O's to overcome this difficulty.

Conferring an absolute metric on the field

For the next study, we sought to make simulations more comparable to natural viewing conditions. We used polar projection and placed the trees in the context of a ground texture gradient. A sample image appears in Figure 3.

Use of a ground texture gradient introduced another interesting question. Like motion parallax, texture gradients provide information about relative distances, but not about absolute distances. Might the trees confer an absolute scale on the ordinal (or interval) field set up by the gradient? We investigated whether the trees can be used, in the context of a ground texture gradient, to scale the absolute size of other objects with Platonic forms. We placed 6 cylinders at various locations within the gradient. Cylinder size was varied to preserve image size.

This manipulation also allowed us to control for another hypothesized effect of ground texture information. Use of information associated with eyeheight has been hypothesized to confer an absolute metric on both parallax and texture gradients (Lee, 1980; Mark, 1987; Warren & Wang, 1987). On a flat ground plane, the image of the horizon has been shown to cut across the images of all objects in the field of view at a height corresponding to the height of the point of observation. If this is the source of any observed improvements in the accuracy of judgments made of simulations with ground texture gradients, then similar results should be obtainable for cylinders viewed without trees. We investigated this possiblity by asking a separate group of observers to judge the size of cylinders without viewing or judging trees.

Finally, we also manipulated the viewing of real trees to make it better comparable to the original simulations. Ground texture information was reduced by having O's view trees through a tube with an aperture of visual angle slightly larger than that of the trees. Although the ground extending from the O to the tree was occluded, some ground texture remained visable immediately around a given tree. The results for 10 O's appear in Figure 4. Restricted viewing dropped the slope from .9 to .8. A multiple regression performed on the combined data from restricted and unrestricted viewing with vectors for actual height, viewing condition, and the interaction was significant, $p<.001$, $F(3,372)=504.6$, $r^2=.80$. Actual height was significant, $p<.001$, partial $F=1186.4$. Viewing was not significant, but the interaction was significant, $p<.05$, partial $F=4.06$. Thus, restricted viewing resulted in a change in slope, but no change in intercept. Mean judgments did not exceed 60 ft.

Using the same procedure as before, 17 O's judged simulated trees in the context of a ground texture gradient. Only 6 architectures were used excluding that poorly rated in the previous study.

The mean height judgments with standard error bars appear in Figure 5 compared to actual modeled heights for each of the 6 architectures. Overall mean slope was .48, greater than for

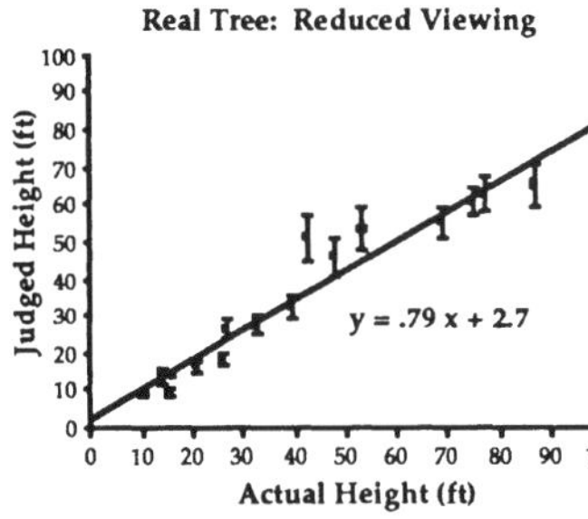

Figure 4.

isolated silhouettes, but still less that for reduced viewing of real trees. However, as shown in Figure 5, the result of a linear fit is misleading. Mean judgments were linear and close to actual values for heights up to about 40-50 ft at which point judgment curves appear to hit a ceiling. The source of this effect remains to be determined. The good fit between mean judgments and actual modeled heights for trees below 40 ft reveals an absence of a 'contraction effect'. The ceiling reached after 40 ft may reflect difficulty in resolving subsequent increases in diameter or branch number.

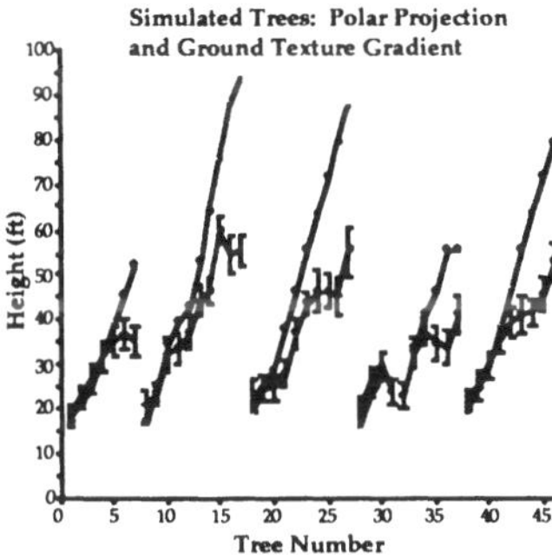

Figure 5.

When asked to judge the height of the cylinders, O's who had not seen the trees produced judgments that were ordinally correct but highly variable in absolute value as shown in Figure 6. Mean slope was .88. Mean judgments were high. Random variability was large. In contrast, O's who had first seen the trees appearing in the context of the cylinders produced judgments that were systematic and much more accurate as shown

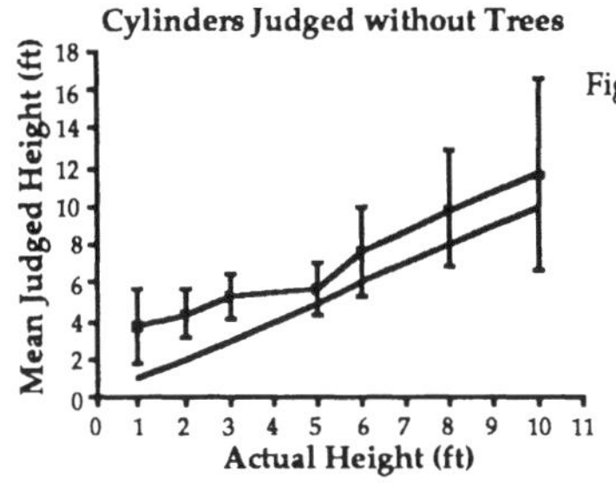

Figure 6.

in Figure 7. Mean slope was .57, close to the slope for the tree judgments. Random variability was low in comparison.

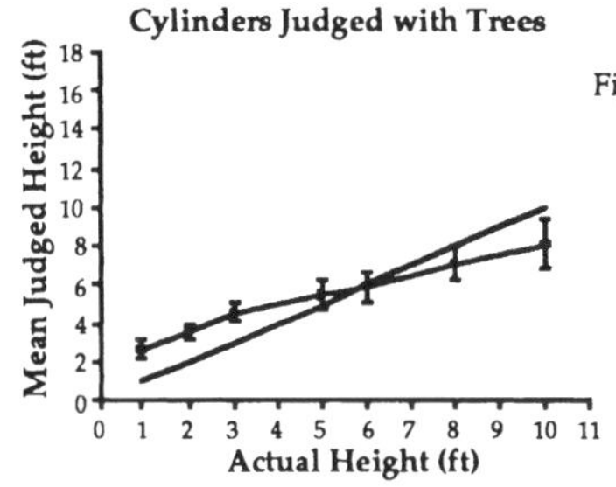

Figure 7.

Conclusions

We have a new solution to the old problem of size perception. The ability to discriminate subtle variations of this complex bioligical form was sufficient to enable people to use the information to judge scale. The information conferred an absolute metric on a ground texture gradient. In contrast, eyeheight information was ineffective.

Acknowledgments.

This research was supported by NSF Grant BNS-9020590. Michael Stassen masterfully programmed the simulations and cheerfully guided observers about in the winter cold. We wish to express our gratitude to Rolf Borchert for sending us his program and to Brayton Wilson as well as Rolf for consultation.

Send correspondence to: Geoffrey P. Bingham, Department of Psychology, Indiana University, Bloomington, IN 47405.

References

Baker, W.E., Westine, P.S. & Dodge, F.T. (1973). *Similarity Methods in Engineering Dynamics: Theory and Practice of Scale Modeling.* Rochelle Park, N.J.: Hayden Books.

Borchert, R. & Honda, H. (1984). Control of development in the bifurcating branch system of tabebuia rosea: A computer simulation. *Bot.Gaz., 145,* 184-195.

Calder, W.A. (1984). *Size, Function, and Life History.* Cambridge, MA: Harvard University Press.

Epstein, W. (1961). The known-size-apparent-distance hypothesis. *AJP, 74,* 333-346.

Fisher, J.B. & Honda, H. (1979a). Branch geometry and effective leaf area: A study of terminalia-branching pattern. I. Theoretical trees. *Am. J.Bot., 66,* 633-644.

Fisher, J.B. & Honda, H. (1979b). Branch geometry and effective leaf area: A study of terminalia-branching pattern. II. Survey of real trees. *Am. J.Bot., 66,* 645-655.

Gibson, J.J. (1950). *The Perception of the Visual World.* Westport, CT: Greenwood Press.

Gogel, W.C. (1977). The metric of visual space. In W. Epstein (Ed.), *Stability and Constancy in Visual Perception.* New York: Wiley.

Hallé, F., Oldeman, R.A.A. & Tomlinson, P.B. (1978). *Tropical Trees and Forests: An architectural Analysis.* Berlin: Springer.

Hartmen, B.D. & Harker, G.S. (1957). The retinal size of a familiar object as a determiner of apparent distance. *Psych. Monog., 71, #442.*

Hildebrand, M., Bramble, D.M., Liem, K.F. & Wake, D.B. (1985). *Functional Vertebrate Morphology.* Cambridge, MA: Harvard University Press.

Holway, A.H. & Boring, E.G. (1941). Determinants of apparent visual size with distance variants, *AJP, 54,* 21-37.

Honda, H. (1971). Description of the form of trees by the parameters of the tree-like body: Effects of the branching angle and the branch length on the shape of the tree-like body. *J.Theo.Biol., 31,* 331-338.

Honda, H., Tomlinson, P.B. & Fisher, J.B. (1981). Computer simulation of branch interaction and regulation by unequal flow rates in botanical trees. *Am. J. Bot., 68,* 569-585.

Kilpatrick, F.P. & Ittelson, W.H. (1953). The size-distance invariance hypothesis. *Psych. Rev., 60,* 223-231.

Lee, D.N. (1980). The optic flow field: the foundation of vision. *Phil.Trans.R. Soc.Lon.B, 290,* 169-179.

McMahon, T.A. (1975). The mechanical design of trees. *Sci. Am., 223,* 97-102.

McMahon, T.A. (1984). *Muscles, Reflexes, and Locomotion.* Princeton, NJ: Princeton University Press.

McMahon, T.A. & Bonner, J.T. (1983). *On Size and Life.* New York: Scientific American Books.

McMahon, T.A. & Kronauer, R.E. (1976). Tree structures: Deducing the principle of mechanical design. *J.Theo.Biol., 59,* 443-466.

Mark, L.S. (1987). Eyeheight-scaled information about affordances: A study of sitting and stair climbing. *JEP: HPP, 13,* 683-703.

Peters, R.H. (1983). *The Ecological Implications of Body Size.* Cambridge University Press.

Thompson, D'A. (1961). *On Growth and Form.* Cambridge: Cambridge University Press.

Tomlinson, P.B. (1983). Tree architecture. *Am.Sci., 71,* 141-149.

Turrell, F.M. (1961). Growth of the photosynthetic area of citrus. *Bot. Gaz., 122,* 284-298.

Warren, W.H. & Wang, S. (1987). Visual guidance of walking through apertures: Body-scaled information for affordances. *JEP:HPP, 13,* 371-383.

Toward the origins of dyslexia

Roderick I. Nicolson and Angela J. Fawcett
Department of Psychology
University of Sheffield
Sheffield S10 2TN, England
email: R.NICOLSON@UK.AC.SHEFFIELD.PRIMEA

Abstract

A series of experiments will be reported comparing the performance of groups of 11 year old and 15 year old children with dyslexia with groups of normal children matched for chronological age (CA) and reading age (RA) respectively. Experiments testing gross motor skill demonstrated that both groups of children with dyslexia showed significant deficits in balance when required to undertake a further task at the same time, whereas the control groups were not affected. It was concluded that the control children balanced automatically whereas the children with dyslexia did not. Further experiments indicated that working memory performance was not disordered, in that deficits in memory span were paralleled by deficits in speed of articulation. Tests of information processing speed led to an interesting dissociation. Simple reaction performance was indistinguishable from that of the CA controls. By contrast, on the simplest possible choice reaction, both groups of children with dyslexia were slowed to the level of their RA controls. It was concluded that the locus of the speed deficit lay within the decision-making process. Further experiments demonstrating deficits in sensory thresholds and abnormal evoked potentials will also be reported. We conclude that an automatisation deficit is consistent with most of the known problems of children with dyslexia.

Introduction

Developmental dyslexia is conventionally defined as "a disorder in children who, despite conventional classroom experience, fail to attain the language skills of reading, writing and spelling commensurate with their intellectual abilities" (World Federation of Neurology, 1968). A typical estimate of the prevalence of dyslexia in Western school populations is 5% (Badian, 1984; Jorm et al., 1986), with roughly four times as many boys as girls being diagnosed. It has been assumed that the problems of children with dyslexia derive from impairment of some skill or cognitive component largely specific to the reading process, and the consensus view (e.g., Stanovich, 1988) is still that the deficits are attributable to some disorder of phonological processing.

Interestingly, however, dyslexia has caught the imagination of researchers from several disciplines, and there is now a wealth of inter-disciplinary information about dyslexia (though inconsistencies of diagnosis bedevil meta-studies of the literature). Frustratingly, different perspectives on dyslexia have led to quite different hypotheses. Neuroanatomical studies by Galaburda and his colleagues have identified both "a uniform absence of left-right asymmetry in the language area and focal dysgenesis referrable to midgestation ... possibly having widespread cytoarchitectonic and connectional repercussions. ... Both types of changes in the male brains are associated with increased numbers of neurons and connections and qualitatively different patterns of cellular architecture and connections" (Galaburda, Rosen & Sherman, 1989, p383). Genetic studies (e.g., Smith et al., 1983) have led to the conclusion that there is a strong genetic component. Studies mapping gross electrical activity in the brain have also uncovered anomalies in processing (e.g., Duffy et al., 1980; Hynd et al., 1990).

At present, there is little or no link between the genetic, the anatomical, the neurological and the cognitive approaches to dyslexia. The 'Holy Grail' of dyslexia research is surely to establish such a link. Not only would such a link between brain and mind prove a breakthrough in dyslexia research, it seems likely that the insights gained would provide a rich source of ideas for modelling normal cognition, bequeathing a research agenda stretching into the next century. This quest has proved the inspiration for our past five years' research, and in this paper we wish to summarise the early results, report recent findings, and outline future research directions.

Our broad research strategy was first to map out the full range of cognitive problems shown by children with dyslexia, and next to attempt to find the lowest common denominator of these problems by attempting to design simpler and simpler tests until

The research reported here was supported by grant F.118P; S893133 from the Leverhulme Trust to the University of Sheffield.

we eventually arrived at a situation where test A showed no deficit whereas test B, which involved addition of a minimal further component to A, revealed a deficit.

Overall Design of the Studies

It is clearly valuable to identify whether children with dyslexia perform significantly worse than their age-matched controls, but one of the key discriminants between theories is a test of performance of children with dyslexia against reading age controls, since a significant impairment compared with reading age controls is indicative of developmental disorder rather than just a developmental lag (cf. Bryant & Goswami, 1986). Since the specific nature of dyslexic children's deficits may also change with age, it is important to examine the effects of age separately. These considerations suggest an experiment with at least six groups of subjects: two groups of children with dyslexia of different mean ages; two groups of normal children matched to the children with dyslexia on chronological age; and two groups of normal children matched to the children with dyslexia on reading age.

Three separate issues are of interest in the statistical analyses for each experiment. First, whether there are any between-group differences at all. This involves a design which treats all the six groups within one factor, irrespective of age and presence/absence of dyslexia. We refer to this as the 'Overall Analysis'. A lack of a significant effect here would suggest that the variable under investigation was unaffected by either age or dyslexia. Second, it is important to identify whether children with dyslexia perform worse than their age-matched controls. This design has the two level factor age and the two level factor presence/absence of dyslexia. We refer to this as 'CA & Dyslexia'. A main effect of age would indicate a developmental trend in the variable in question, while a main effect of dyslexia would suggest a reliable difference between dyslexic and control subjects of equivalent age. Such a difference may, however, be attributable either to a fundamental difference, or to a developmental lag. Deciding between these requires a third analysis, one involving a comparison with reading age controls; this also has two factors, namely a two level factor reading age, and a two level factor presence/absence of dyslexia. We refer to this as 'RA & Dyslexia'. A negative effect of dyslexia on this analysis would indicate that dyslexic subjects are performing more poorly than younger children of equivalent reading age, and would argue against a developmental lag interpretation.

Note that the latter two analyses are based on only four of the experimental groups. Fortunately, as described in the next section we were able to select two groups of children with dyslexia who were sufficiently similar in IQ to allow a single control group to be used both as RA control for the older children with dyslexia and CA control for the younger children with dyslexia, thus leading to a total of only five groups.

The subject panel

Five groups of subjects participated in the initial studies. The groups were; 12 children with dyslexia around 15 years old; 11 children with dyslexia around 11 years old; a group of 12 normal children matched to the older children with dyslexia for age and full IQ; a group of 11 normal children of similar IQ to the two dyslexic groups, matched for chronological age with the younger children with dyslexia and for reading age with the older children with dyslexia; and a fifth group of 10 normal children around 8 years old matched for reading age and full IQ with the younger children with dyslexia. All the children with dyslexia had been diagnosed as dyslexic between the ages of 7 and 10, based on discrepancies of at least 18 months between chronological and reading age. Their IQ levels fell in the normal to superior range on the Wechsler Intelligence Scale for Children (Wechsler, 1976) and they had no known neurological deficit or primary emotional difficulty. For several children with dyslexia IQ and/or reading age deficit had changed since diagnosis, and a criterion of at least one year deficit in reading age compared with chronological age at the time of the experiment was adopted (this led to the exclusion of 5 children from a larger original pool). The children with dyslexia were recruited via the local dyslexia associations, and the normal controls were recruited from local schools. Recently we have recruited a group of 8 year old children with dyslexia and a group of 6 year old RA controls, and hope to report the combined results for all seven groups. The following analyses, however, are based only on our original five groups.

Study 1. Working Memory and Dyslexia

We first report briefly a set of experiments exploring the relationship between phonological processing, working memory and articulation rate. Recent research has demonstrated that in addition to their phonological deficits, children with dyslexia suffer impairments in working memory performance (Jorm, 1983; Snowling et al., 1986; Gathercole & Baddeley, 1990). In principle either a phonological deficit or a working memory deficit could underlie both sets of symptoms. We undertook a series of experiments with the above five groups of children designed to

Figure 1. Memory Span as a function of Articulation Rate

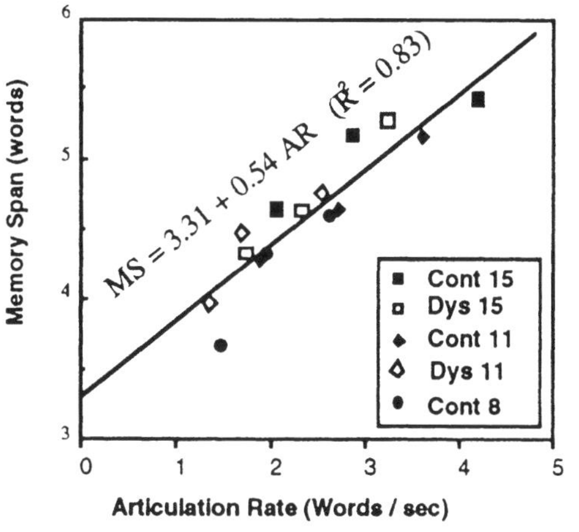

tease out which factors underlay the deficits (Nicolson, Fawcett & Baddeley, 1992).

In summary, the performance of the older children with dyslexia across the range of tasks was slightly but not significantly worse than their CA controls and indistinguishable from that of their RA controls. By contrast, the performance of the younger children with dyslexia was significantly worse than that of their CA controls on phonological discrimination, articulation rate, and nonword repetition, and was significantly worse even that of their RA controls on repetition of longer nonwords. On tests of memory span, all five groups showed the normal phonological similarity effect and the normal word length effect. When memory span was regressed as a function of articulation rate, there was no evidence of impaired slope or intercept for the children with dyslexia (see figure 1).

We concluded that the 11 year old children with dyslexia show residual problems on phonological processing, especially for tasks involving unfamiliar stimuli, but that by their mid-teens children with dyslexia have largely overcome these problems. The major remaining problem for the children with dyslexia appeared to be a continuing lack of fluency in articulation, a factor which is sufficient to account for the slight deficits on memory span. Furthermore, we argued that the articulation rate deficit provides a parsimonious explanation of the range of deficits shown by the younger children with dyslexia. We concluded that neither impaired phonological skills nor impaired working memory is sufficient in itself to explain the deficits, but that some deeper explanation must be sought.

Study 2. Motor Balance and Dyslexia

This set of experiments was designed to identify whether children with dyslexia showed any deficits in a skill as far removed as possible from reading, namely gross motor balance. Clearly any balance deficits would suggest that the phonological deficit hypothesis was insufficiently broad to account for the range of problems suffered by children with dyslexia. An initial study is reported in Nicolson & Fawcett (1990), with subsequent studies using the current subject groups reported in Fawcett & Nicolson (1991). In brief, we established that children with dyslexia balanced just as well as their controls under 'just balance' conditions, but that the children with dyslexia showed a significant balance deficit (ie. they wobbled more) when required to balance while carrying out a further task (even as simple a task as a selective choice reaction, in which they had to say 'Yes' or press a button on hearing a low tone, but to make no response if a high tone was presented).

We interpreted these results as evidence that dyslexic children's balance was not fully automated, unlike that of their controls, and consequently formulated our 'Dyslexic Automatisation Deficit' (DAD) hypothesis, which states that children with dyslexia suffer from extreme difficulties in fully automatising skills (whether cognitive or motor). In addition to DAD we formulated the 'Conscious Compensation' hypothesis which states that in normal circumstances children with dyslexia are able to mask their lack of automatisation by concentrating harder on the task (conscious compensation), and thus that the deficits will show up primarily under adverse conditions (such as in dual tasks; when high speed is required, as in reading; or in general when the child is stressed or tired). The DAD hypothesis has an inherent plausibility, and accounts for a range of findings about children with dyslexia such as quicker tiring, greater distractibility, problems in shoe lace tying, and the like (see Augur, 1985). However, it was by no means the only possible interpretation of the results, with one key issue being whether the dual task deficits were actually attributable to some general problem with attention sharing rather than a problem specific to balance.

We decided to investigate this issue further by running a further experiment in which we compared the balance performance of our subjects under normal conditions and when they were blindfolded. The latter condition is not a dual task condition but the subjects are prevented from using the normal visual cues to assist with balance. Consequently an attention-sharing deficit theory predicts equivalent performance to that of the controls, whereas DAD (and any motor skill deficit hypothesis) predicts impaired performance

Figure 2. Blindfold Balance

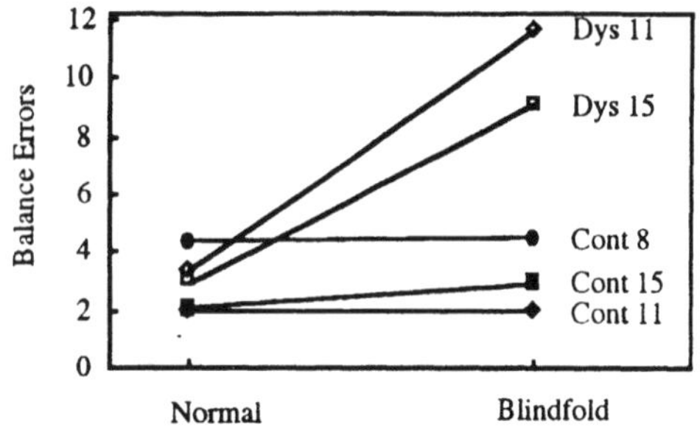

relative to the controls when blindfolded. The results were clearcut, following exactly the same pattern as we obtained in our earlier dual task balance experiments (see Fig. 2).

It may be seen that the children with dyslexia balanced as well as their controls in the normal balance condition, but, unlike the controls, the dyslexic children's balance was impaired in the blindfold balance condition. Statistical analyses of the blindfold balance deficit (the increase in errors when blindfolded) showed that both groups of children with dyslexia were significantly more impaired than all three control groups (p<.01).

Study 3. Speed of Processing and Dyslexia.

In an effort to trace the dyslexic deficit back to its source, we decided to investigate speed of information processing, using a variety of reaction time tasks in the hope that at some point we would find a cut off where tasks of lesser complexity would show no deficit, whereas more complex tasks would result in a deficit. We tried a variety of simpler and simpler tasks — 'Coding', lexical access, choice reaction, and finally simple reaction to a tone in an attempt to find normal performance. Deficits were found all the way down to 2-choice reactions, and so we present the data for the two simplest conceivable tasks, a simple reaction and a selective choice reaction task (SCRT). The study is presented in detail in Nicolson & Fawcett (1992) and so a summary should suffice here.

In both tasks, subjects sat with a single button in their preferred hand, and their task was to press it as quickly as possible whenever they heard a low tone. In the simple reaction task, no other tone was ever presented, but in the SCRT task, there was an equal probability of a high tone being presented. The subject had to ignore the high tone, thereby make a selective choice reaction to the low tone. These tasks were introduced by Donders well over a century ago. His rationale was that the only difference between the

Figure 3. Simple Reactions and Selective Choice Reactions

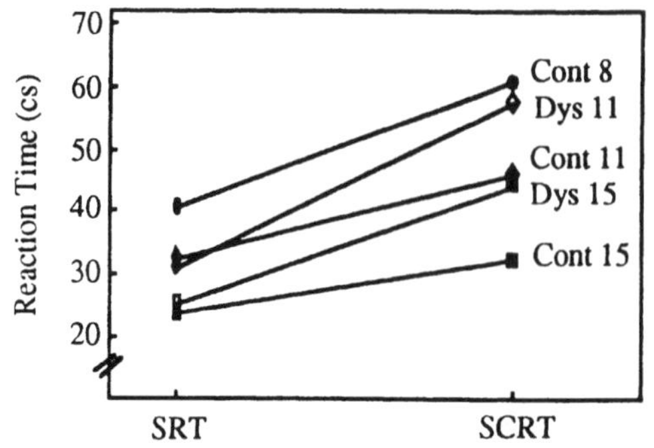

tasks was the need to classify the stimulus before responding in the SCRT trials, and he argued that subtracting the simple reaction time (SRT) from the SCRT time gave an estimate of 'stimulus classification' time. The experiment was computer-controlled, and 100 stimuli were presented at an average rate of 1 per 2 seconds. The results are shown in figure 3, plotted on the same graph to facilitate comparison.

It is clear that, although the subjects with dyslexia performed at the same level as their same-age controls on the simple reactions, they were slowed down more by the need to make an SCRT. Analysis of the simple reactions indicated that both groups of older subjects were significantly faster than all three groups of younger subjects.

Analysis of the SRT's in terms of age and dyslexia indicated a highly significant effect of age (p<.0001), whereas there was no effect whatsoever of dyslexia. Analysis of the SCRTs indicated a significant effect of both age and dyslexia at the .01 level. Pairwise comparisons indicated that performance of the older controls was significantly faster than for the other four groups.

A third analysis of variance was conducted on the SRT's, omitting the oldest controls and comparing the two dyslexic groups with their reading age controls. The main effects of both age and dyslexia were highly significant at the .001 level. Analysis of the SCRTs indicated a significant effect of age (p<.01) but no effect of dyslexia.

To summarise the RT results: for the simple reaction task both groups of dyslexic subjects performed at the appropriate level for their age, and significantly faster than their RA controls. However, for the SCRT condition the dyslexic subjects slipped back, to the extent that they were significantly slower than their CA controls and equivalent to their RA controls. The same pattern of results applied for both mean and median latency. Furthermore, the dyslexic children were, if anything, less accurate than their age-matched controls, showing that the SCRT results cannot be attributed to some speed-accuracy trade-off effect.

A number of possible explanations can be offered for the slower decision problem — presumably the problem must lie either in the time taken to analyse the stimuli, the time taken for the 'central executive' to 'notice' that the stimulus has been classified, the time taken to determine the appropriate motor response, or the time taken to 'load' it ready to despatch the neural impulses to the finger. Further research would be needed to distinguish between these possibilities, but our guess is that the problem is not a perceptual one — the time taken to classify the input — since this is not sufficiently general to explain the range of deficits suffered. Consequently it seems most likely that the problem lies somewhere within the central processing system. We return to this issue in the overall discussion of the findings in the three studies.

Discussion

Simplifying greatly, there are three major groups of theory for the origins of dyslexia: the phonological deficit hypothesis, the visual deficit hypothesis, and the automatisation deficit hypothesis (henceforth DPD, DVD and DAD respectively). The most recent formulation of the DVD hypothesis is that there is a deficit in the transient visual system, caused by neuroanatomical abnormalities in the magnocellular pathways to the lateral geniculate nucleus (Livingstone et al., 1991). The DVD hypothesis is unable to account for the established phonological deficits or for the balance deficits reported here, and so can only be a partial explanation. Phonological processing is of course a skill built up via continual practice throughout childhood. It would therefore be expected under DAD that phonological deficits would arise, and indeed, the predictions of DAD appear to be indistinguishable from those of DPD in the area of phonological processing, except that DAD also predicts reduced speed of articulation. One might interpret, therefore, the DPD hypothesis as a specialisation of DAD to the phonological domain. It may, however, be significant that although automaticity for phonological skills develops in a similar fashion to visual automaticity, at least in the early stages it tends to be less resistant to interference from multi-channel input (Mullinex, Sawusch & Garrison, 1992). The two theories differ in that DAD predicts deficits outside the phonological domain (indeed, for any skilled activity where automatisation deficits cannot be masked by conscious compensation). This was the reason why we selected balance as a task to investigate in the first place — it was a sharp test which discriminated between DPD and DAD.

Unlike DPD, the DAD hypothesis provides a natural explanation of both the dual task balance deficits and the blindfold balance deficits. It is also able to explain the working memory deficit naturally in terms of the reduced (less fluent) articulation rate. The remaining issue from the studies reported here is whether DAD is able to give a principled account of the reason for the dichotomy between SRTs and SCRTs. Here the argument is less convincing. Why, for instance, is there no deficit in SRTs, but a deficit in SCRT and other more complex choice reactions? Presumably the reason lies in the need to make a decision in the latter cases, rather than merely execute a pre-determined response. The slowed cognitive decision is quite consistent with the framework of automatisation — for the same amount of practice the decision is less automatic — but unfortunately for DAD, it can offer no principled reason why automatisation should be difficult only for decision-based tasks. Once one is attempting to explore central processes, the DAD hypothesis offers too coarse grain an analysis.

One speculation, given the involvement of the cerebellum not only in motor skill but also in automatisation (Ito, 1984) and in the development of cognitive skill and language (Leiner, Leiner & Dow, 1989), is that some cerebellar abnormality is mediating the range of deficits shown. Given the complexity of the brain circuitry involved, however, it seems likely that abnormalities in almost any component of the cerebro-cerebellar loop might lead to qualitatively similar symptoms.

Conclusions

To summarise our research findings, we established that a phonological deficit alone was not capable of accounting for the range of deficits found. There was strong evidence of an automatisation deficit on motor balance, and the automatisation deficit hypothesis appeared capable of explaining not only the balance deficits, but also the phonological deficit and the established problems of working memory. However, the automatisation deficits are best seen also as a symptom of some deeper underlying cause. The lack of a deficit for simple reactions, taken together with the appearance of a deficit in even the simplest choice reactions, suggest that the most likely cause is some problem within the central brain processes. One speculation is that these deficits derive from abnormalities within the cerebro-cerebellar neural pathways.

One must be cautious in generalising from the results from two small groups of children with dyslexia, but, if our findings are replicated on a wide range of children with dyslexia, we believe that this analysis sets new and exciting agenda for dyslexia research for the next decade. The research agenda are clear: inter-disciplinary collaboration towards a clearly identified target that promises to disperse some of the mists which presently shroud the relationship between mind, brain and behaviour.

Acknowledgments. We acknowledge gratefully the support of the participants in this study, the continuing support and advice of Prof. T.R. Miles, and the participation of Dr. Alan Baddeley in the overall design of the study, and in particular his contribution to the analyses of working memory. The potential role of the cerebellum was suggested to us by Dr. Paul Dean.

References

Augur, J. 1985. Guidelines for teachers, parents and learners. In M. Snowling (ed). *Children's written language difficulties.* Windsor: NFER Nelson.

Badian, N.A. 1984. Reading disability in an epidemiological context: Incidence and environmental correlates. *Journal of Learning Disabilities* 17: 129-136.

Bryant, P., and Goswami, U., 1986. Strengths and weaknesses of the reading level design. *Psychological Bulletin* 100: 101-103.

Duffy, F.H.; Denckla, M.B.; Bartels, P.H., and Sandini, G., 1980. Dyslexia: Regional differences in brain electrical activity by topographic mapping. *Annals of Neurology* 7: 412-420.

Fawcett, A.J., and Nicolson, R.I., 1991. Automatisation Deficits in Balance for Children with Dyslexia. Technical Report, LRG 12/90, Dept. of Psychology, University of Sheffield.

Galaburda, A.M.; Rosen, G.D., and Sherman, G.F., 1989. The neural origin of developmental dyslexia: Implications for medicine, neurology and cognition. In A.M. Galaburda (ed.) *From Reading to Neurons.* Cambridge, MA: MIT Press.

Hynd, G.W.; Semrud-Clikeman, M.; Lorys, R.R.; Novey, E.S., and Eliopulos, D., 1990. Brain morphology in developmental dyslexia and attention deficit disorder / hyperactivity. *Archives of Neurology* 47: 919-926.

Ito, M. 1984. *The cerebellum and motor control.* New York: Raven Press.

Jorm, A.F. 1983. Specific reading retardation and working memory: a review. *British Journal of Psychology* 74: 311-342.

Jorm, A.F.; Share, D.L.; McLean, R., and Matthews, D., 1986. Cognitive factors at school entry predictive of specific reading retardation and general reading backwardness: A research note. *Journal of Child Psychology and Psychiatry and Allied Disciplines* 27: 45-54.

Leiner, H.C.; Leiner, A.L., & Dow, R.S., (1989). Reappraising the cerebellum: what does the hindbrain contribute to the forebrain? *Behavioural Neuroscience* 103: 998-1008.

Livingstone, M.S.; Rosen, G.D.; Drislane, F.W., & Galaburda, A.M., 1991. Physiological and anatomical evidence for a magnocellular deficit in developmental dyslexia. *Proceedings of the National Academy of Sciences of the USA* 88: 7943-7947.

Mullinex, J.W.; Sawusch, J.R., and Garrison, L.F., 1992. Automaticity and the detection of speech. *Memory and Cognition* 20: 40-50.

Nicolson, R.I. and Fawcett, A.J., 1990. Automaticity: a new framework for dyslexia research. *Cognition* 30: 159-182.

Nicolson, R.I. and Fawcett, A.J., 1992. Reaction Times and Dyslexia. *Quarterly Journal of Psychology.* Forthcoming.

Nicolson, R.I.; Fawcett, A.J., and Baddeley, A.D., 1991. Working Memory and Dyslexia. Technical Report, LRG 3/91, Dept. of Psychology, University of Sheffield.

Smith, S.D.; Kimberling, W.J.; Pennington, B.F., and Lubs, H.A., 1983. Specific reading disability: identification of an inherited form through linkage analysis. *Science* 219: 1345-1347.

Snowling, M.J.; Goulandris, N.; Bowlby, M., and Howell, P., 1986. Segmentation and speech perception in relation to reading skill: a developmental analysis. *Journal of Experimental Child Psychology* 41: 487-507.

Stanovich, K.E. 1988. The right and wrong places to look for the cognitive locus of reading disability. *Annals of Dyslexia* 38: 154-177.

Wechsler, D. 1976. Wechsler Intelligence Scale for Children (WISCR). Slough, England: NFER Nelson.

World Federation of Neurologists, 1968. *Report of research group on dyslexia and world illiteracy.* Dallas: WFN.

A Memory Architecture for Case-Based Argumentation*

Eric Shafto
shafto@nwu.edu

Ray Bareiss
bareiss@ils.nwu.edu

Lawrence Birnbaum
birnbaum@ils.nwu.edu

The Institute for the Learning Sciences
Northwestern University
1890 Maple Avenue
Evanston, IL 60201

Abstract

This paper describes a memory organization that supports intelligent memory-based argumentation. Our goal is to build a system that can argue opposite sides of an issue by retrieving stories that support or oppose it. Rather than attempting to determine how a story relates to a point on the fly, we explicitly represent the points that the stories support or oppose, as well as how they support or oppose those points. We have developed a hierarchy of story point types; associated with each type is a set of *rhetorical templates*, which describe the ways that a story could support or oppose a point of that type. Each template consists of a series of assertion types on which the argument depends. This enables the program to attack intelligently the foundations of the point it is trying to refute. Our approach is being developed within the context of the ILS Story Archive, a large multimedia case base which includes stories from a wide variety of domains.

Introduction

Argumentation is fundamentally a memory process. When we are presented with a point in an argument, we generally find that examples or facts that support or refute that point come to mind automatically and without conscious effort[McGuire *et al.*, 1981]. Clearly, complex inference is involved in determining whether some aspect of a given story can be used to strengthen or weaken a given argument. However, it is not at all clear that the bulk of this inference must take place in real time. For any issue about which a person has an opinion, there are sure to be a large number of anecdotes, experiences, and memories of previous arguments that are permanently associated with that issue.[1] Our model is based on the idea that pre-existing links between points in an argument and stories that support or oppose those points play a key role in argumentation.

Our approach is intended to model a small but important part of the human ability to find stories that support or refute points. Obviously, the ability to hear a new story and determine that it supports a point, the ability to learn new points, and to re-evaluate a story with regard to other points is important to argumentation, and to understanding in general. These are interesting and difficult problems. However, we believe that a system that had to rely on arguing from first principles in this way would bog down immediately in an intractable memory search process. We wish to show here that much interesting behavior can be obtained through a system that relies on a corpus of pre-analyzed stories that are linked to the points they represent. The core of our theory thus lies in the taxonomy of points and link types embodied in the story point/rhetorical template hierarchy.

Our goal is to build a system that can argue for or against a point by recalling appropriate stories, and explaining how they support or oppose that point. This work is being carried out within the context of the ILS Story Archive Project. The Archive is a large multimedia case base of stories, including news footage, speeches, lectures, and interviews with experts. The overall goal of the system is to enable users to easily gain access to stories relevant to their interests, and in particular, to facilitate the retrieval of follow-up stories relevant to whatever story the user has just viewed. The underlying metaphor is that of a normal human conversation, in which the interaction flows naturally and coherently from one story to the next[Schank, 1977, Bareiss *et al.*, 1991, Ferguson *et al.*, in press]. At their best, arguments are a particularly informative kind of conversational

*This research was supported in part by the Defense Advanced Research Projects Agency, monitored by the Office of Naval Research under contracts N00014-90-J-4117 and N00014-91-J-4092, and by the Office of Naval Research under contract N00014-89-J-1987. The Institute for the Learning Sciences was established in 1989 with the support of Andersen Consulting. The Institute receives additional support from Ameritech, an Institute Partner, and from IBM.

[1] We use the term "stories" to refer to all of these types of information in memory.

interaction. Our goal, then, is to extend the memory organization and retrieval strategies of the Archive to enable this kind of interaction.

Suppose that the user has just viewed the story "Brown on Violence," in which activist H. Rap Brown argues that violence is a justified means of achieving racial equality:

> I say violence *is* necessary. Violence is part of America's culture. It is as American as cherry pie ... America taught the black man to be violent. We will use that violence, if necessary ... we will be free by any means necessary.[2]

Roughly speaking, Brown's main argument here is that the ends justify the means: Because racial equality is a desirable goal, violence is justified to the extent that it helps to achieve this goal.[3] An argument of this sort can be opposed in at least four ways: By arguing that the means are unacceptable for some other reason, by arguing that they will not achieve the goal, by arguing that there are alternative means of achieving the goal, or by arguing that the goal itself is not desirable. Corresponding to each of these, the Archive contains at least one story, as follows:

- In "Birmingham Riots," Martin Luther King declares that violence is unacceptable because it is counter to Christian ethics.
- In "Race and Riots," Arthur Ashe argues that violence will not achieve racial equality, stating that it has never succeeded in the past.
- In "I Have a Dream," Martin Luther King claims that an alternative plan, non-violence, will achieve the goal of racial equality.
- Finally, in "Segregation Forever," George Wallace states his view that racial equality is not a proper goal for American society.

Our goal is to create a program that can oppose the points represented in a story by recalling appropriate stories, as illustrated above, and explaining their relevance. This requires, first, a representation of story points and how they can be supported or opposed; second, a retrieval method that can find stories that support or oppose a particular point; and, third, a method of generating bridging explanations between stories.

Representation

"The ends justify the means," as argued by H. Rap Brown, is exactly the sort of generic argument structure that our representation seeks to capture, in the form of *rhetorical templates* (cf *e.g.* [Alvarado *et al.*, 1986, Birnbaum, 1982, Flowers *et al.*, 1982]). The two templates that make up Brown's argument are that the plan achieves a desirable goal, and that no other plan achieves that goal. Our representation of "Brown on Violence" includes instances of these templates, among others. It is of course apparent that there is more to Brown's argument than this, and some of it is quite subtle. However, such a representation of the point is sufficient to enable the retrieval of a number of relevant stories, as we have shown.

Much work has been done on the problem of representing and indexing stories for retrieval (see, *e.g.*, [Schank and Osgood, 1990, Kass, 1990, Bareiss and Slator, 1992, Ferguson *et al.*, in press]. The earlier work on this problem generally tried to create detailed representations of the entities, actions, and relations that comprised a story. This approaches the ideal of a general-purpose representation — sufficient to infer the story's point, in whatever context one needs to understand the story. All too often, however, the points of stories are not readily emergent from this kind of representation, either to a human or a program. For example, it is difficult to determine computationally that some story is a case in which goal conflict was resolved by turn-taking, when the representation consists entirely of actions, agents, goals, plans, intentions, etc.

Our approach is to back off from this general-purpose ideal, and explicitly represent the points that the stories support or oppose (see also [Bareiss *et al.*, 1991]). Instead of creating detailed representations of the stories, we represent only the points of the stories, as perceived by human indexers. Furthermore, we do not build complex representations of the points themselves. Instead, we represent how the story supports or opposes the point. For example, one can argue that a person is good by arguing that they do good things, that they try to do good things, that they have good beliefs, that they possess a valuable skill, and so on.

For a point of a given type, there is a set of these rhetorical templates — ways a point of that type could be supported or opposed. We have developed a hierarchy of point types that is very well-behaved with regard to the rhetorical templates that are applicable to each type. That is, a template usable for some story-point type is usable by all the children of that type, and, in general, the same templates do not appear in different branches of the story-point hierarchy. A simplified portion of this hierarchy is shown in Figure 1.

The root of this hierarchy is *Any Assertion.* All story point types are specializations of this basic type. When an indexer categorizes a point as being of this very abstract type, all they are saying about it is that some statement was made. Nonetheless, there are a variety of weak, but commonly-used, templates that can be used to support or refute a point of this type. Assertion, authority, logical proof and appeals to common sense are all examples of rhetorical templates for supporting or opposing virtually any point. As we specialize the assertion, we gain more specific (and, in

[2] All of the examples in this paper are actual stories from the Archive.

[3] One subtlety missed by this representation is that violence needs some justification—that it is considered bad by default.

general, more powerful) templates. For example, if the assertion is of the form *Something is good,* we can use the templates *Best available*, or *Fulfills function.*

However, when the *Something* in *Something is good* is specialized to be a plan, a resource, a goal, an agent, etc., the set of templates becomes extremely rich. For example, to support a point of type *Plan is good,* we can use any of the templates we have listed so far. In addition, we can use *Plan succeeds*, *Plan has positive side effects*, *Plan is low-risk*, *Agent has nothing to lose by executing plan,* and so on. This level of specificity seems to be a natural one for categorizing rhetorical templates. As we get more specific than this, we begin to talk about ways to support individual plans, or to show that certain people or types of people are good or bad. At that level, we would expect an explosion of categories, most of which would contain very little in the way of new, more-specific templates (see, *e.g.*, [Rosch, 1988]).

Each instantiated rhetorical template is a sequence of assertions that are necessary to the story's ability to support or oppose the main point. They are represented in the same way as points, and this is what allows the program to undercut an argument. To rebut a point, the Archive can retrieve a story that opposes the main point directly, or it can retrieve stories that oppose the points that were used to support the main point. For example, as shown in Figure 1, the point type *Plan is good* can be supported by the template *Achieves good end.* This is elaborated as two statements (actually represented as clauses): *The plan meets some goal,* and *that goal is good.* As we saw in the Introduction, the program can oppose the *Plan is good* point by finding stories that oppose either of these two supporting points.

A partial representation in these terms for the H. Rap Brown story is shown in Figure 2. The point being indexed is that violence is a good plan for achieving racial equality. Since this is an assertion of the type, *Plan is good,* we must use one of the templates available to this type. There are two templates that are used to support this contention. The first is that violence achieves a good goal. The instantiated conjunct of clauses that supports that point can be seen in the figure. These clauses, as well as the point of the story itself, are used as queries to find opposing stories.

- Any Assertion
  - A is good
    - +Consistent with good theme
    - +Only available
    - -Does not fulfill function
    - ...
  - A is good: A is a plan.
    - +Best available
      - *There are no plans B, such that B achieves the goal, and B is better than A.*
    - +Achieves good end
      - *The plan achieves a goal, and the goal is good.*
    - +Low risk
    - +Agent has nothing to lose
    - +Has succeeded in past
    - -Risky
    - -Costly
    - ...
  - Person A is good
    - +Does good thing
      - *The person does something, and that something is good.*
    - ...
  - Goal A is good
    - +Subgoal to greater end
    - ...
  - Theme A is good
    - +Engenders good goals
    - ...
  - Resource A is good
    - ...
  - A promotes B
  - A is like B
  - A requires B
  - A is supposed to B
  - A achieves B
  - A knows B
  - A risks B
  - ...

Figure 1: **A portion of the story-point hierarchy.** Pluses and minuses represent supporting and opposing templates, respectively. The italicized sentences are actually represented by a conjunctive list of assertion types, which are instantiated with fillers from the point.

Retrieval

Given this representation, a very simple mechanism suffices for story retrieval. Any story point has pointers to all the stories that support or oppose it. When a story is needed to oppose a point, all stories listed as opposing it are retrieved.

In our first example, the probes *Violence is good*, *Violence achieves racial equality*, *No plan but violence achieves racial equality*, and *Racial equality is a good goal* are each used to find stories that oppose the idea

"Brown on violence"
Speaker: H. Rap Brown

Supports: "Violence is a good plan"
    Type: Plan is good.
    Plan = violence
    Agent = American-Blacks
    Goal = Racial-Equality
    Template: No alternative.
        Relies on:
        There is no plan that achieves racial equality except violence.
    Template: Achieves good goal.
        Relies on:
        Violence achieves Racial equality.
        Racial equality is a good goal.

"King: Birmingham Riots"
Speaker: Martin Luther King

Opposes: "Violence is a good plan"
    Type: Plan is good
    Plan = violence
    Agent = American-Blacks
    Goal = Racial-Equality
    Template: Inconsistent with good theme.
        Relies on:
        Violence conflicts with Christianity.
        Christianity is a good theme.
        ...

Figure 2: **A sample representation.** Each of the templates represents one index. There are other points that these stories support that are not shown here. The italicized sentences are actually represented by conjunctive lists of clauses.

that violence is a good plan[4]. Any story that opposes one of those points weakens Brown's argument in favor of violence.

However, this leaves us with a problem. Each of these retrieved stories may have many points. Sometimes, the point for which the story was retrieved is only a small part of the story as a whole. A user of the program might not understand why the story is being shown. This means that it is not sufficient to simply retrieve an appropriate story and present it to the user. Research on hypermedia presentations (*e.g.* [Oren *et al.*, 1990]) has demonstrated that users often have difficulty determining the relevance and intended meaning of a story. It is essential that any system that is intended to argue be able to explain why it is telling a given story — what its relevance is to the conversation.

Bridging

An important aspect of the work on the Story Archive is focused on the building of these "bridging" explanations. Viewing the Archive as a case-based reasoning program, bridging can be seen as a form of case adaptation. Instead of modifying the cases themselves[5], we are modifying the user's perceptions and understanding. This modification takes the form of a short text explaining how the story fits in the given context before the story is shown.

From the viewpoint of the arguing system being described here, bridging is essential to the argument's coherence. It is part of the job of an arguer to explain how the evidence presented is connected to the point being argued. Otherwise, the story presentation will seem cryptic and incoherent. The retrieval shown in the Introduction is a good example of this. The first story, "Brown on violence," makes many points. In addition to advocating violence to achieve civil rights, Brown comments on American culture, and threatens other black leaders. The first story retrieved as a counterargument, "Birmingham Riots," is fairly long, and is not entirely about violence as a plan for achieving civil rights. It is also about courage, morality, withstanding violence, brotherhood and so on. The stories are clearly related, and a viewer might understand the transition, but the intended relationship between the stories is much clearer when the second story is preceded by an explanation such as the following:

[4]Since the part of the Archive that is most thoroughly indexed is the Civil Rights area, the majority of stories that are found will be from the domain of the US Civil Rights struggle in the 1960's. When the Archive has grown, we might expect stories about Gandhi, Sitting Bull and Spartacus to be retrieved as well: Whether the program should prefer more similar or more distant analogies is an ongoing research question.

[5]Due to the fact that all our materials are digitized video images, we have little ability to modify the cases.

> In the story you have just viewed, H. Rap Brown makes the point that violence is a good plan because it achieves a good end, racial equality. In the story you are about to see, Martin Luther King makes the point that violence is a bad plan because violence is counter to Christianity, which is a good theme.

Before the story, "Segregation Forever," the bridge would be:

> In the story you have just viewed, H. Rap Brown makes the point that violence is a good plan because it achieves a good end, racial equality. In the story you are about to see, George Wallace makes the point that racial equality is a bad goal, undermining H. Rap Brown's argument.

These bridges are assembled at run time, out of pieces of text associated with the point types, objects and templates. Taking the last bridge as an example, we have associated with the story point type *Plan is good* the string, "$speaker makes the point that $plan is a good plan to achieve $goal, because $instantiated-template." This is known as a half-bridge. Half-bridges are combined according to information associated with the type of opposition being used to oppose the point — in this case, undermining. The "boilerplate" for undermining is, "In the story you just viewed, $half-bridge-1. In the story you are about to see, $half-bridge-2, undermining $speaker1's argument." We have not yet built a large set of bridging strategies, but the problem appears to be quite tractable.

Conclusion

Our story-point type hierarchy currently contains about 25 point types, although we expect this number to grow dramatically. Some of these types have as many as ten rhetorical templates, and many have only one. We are now in the process of indexing new stories on paper to determine which types and templates need to be added to our hierarchy. We have been encouraged by the fact that we have been able to index many new stories without needing to add many new items to our hierarchy.

We recognize, of course, that a system such as ours necessarily misses many of the more subtle points of a story. For example, in the story "Birmingham Riots," King argues that violence harms the aggressor more than the victim, and that it is better to be beaten and have a clear conscience than to beat someone else and be guilty. There are many more interesting conclusions that are suggested in King's speech that we lack the mechanisms to deal with. Given its limitations, however, the model exhibits rather sophisticated behavior with a minimum of knowledge-intensive programming.

One trend at ILS (as opposed, say, to the CYC project[Lenat and Guha, 1990]) has been to move away from large, inference-intensive AI reasoning systems, with their detailed and complex representations. The emphasis in projects like this one is on building AI systems that require a less fine-grained knowledge representation. We believe this is essential if we are to meet our goals of scalability. Our goal is not to build a system that functions very well on a small set of interesting examples. Rather, we are trying to create a system that can exhibit interesting behavior over a broad range of cases, drawn from a wide variety of domains.

References

[Alvarado *et al.*, 1986] S.J. Alvarado, M.G. Dyer, and M. Flowers. Editorial comprehension in oped through argument units. In *Proc. AAAI-86*, pages 250–256, Philadelphia, PA, August 1986. AAAI.

[Bareiss and Slator, 1992] Ray Bareiss and Brian M. Slator. From protos to orca: Reflections on a unified approach to knowledge representation, categorization, and learning. Technical Report #20, Institute for the Learning Sciences, Northwestern University Evanston, IL, January 1992.

[Bareiss *et al.*, 1991] Ray Bareiss, William Ferguson, and Andy Fano. The story archive: A memory for case-based tutoring. In *Proceedings of the 1991 DARPA Case-Based Reasoning Workshop*. DARPA, 1991.

[Birnbaum, 1982] Lawrence Birnbaum. Argument molecules: A functional representation of argument structure. In *Proceedings, AAAI 1982*, pages 63–65, Philadelphia, PA, August 1982. AAAI.

[Ferguson *et al.*, in press] William Ferguson, Ray Bareiss, Lawrence Birnbaum, and Richard Osgood. Ask systems: An approach to the realization of story-based teachers. In *The Journal of the Learning Sciences*, in press.

[Flowers *et al.*, 1982] Margot Flowers, Rod McGuire, and Lawrence Birnbaum. Adversary arguments and the logic of personal attacks. In Wendy G. Lehnert and Martin H. Ringle, editors, *Strategies for Natural Language Processing*. Lawrence Erlbaum Associates, Morristown, NJ, 1982.

[Kass, 1990] Alex M. Kass. Developing creative hypotheses by adapting explanations. Technical Report #6, The Institute for the Learning Sciences, Northwestern University Evanston, IL, November 1990.

[Lenat and Guha, 1990] D. Lenat and R. V. Guha. *Building Large Knowledge-based Systems*. Addison-Wesley, Reading, PA, 1990.

[McGuire *et al.*, 1981] R. McGuire, Lawrence Birnbaum, and M. Flowers. Opportunistic processing in arguments. In *Proceedings of the Seventh IJCAI, Vancouver, B.C.*, pages 58–60, 1981.

[Oren *et al.*, 1990] T. Oren, G. Salomon, and K. Kreitman. Guides: Characterizing the interface. In B. Laurel, editor, *The Art of Human-Computer Interface Design.* Addison-Wesley, Reading, PA, 1990.

[Rosch, 1988] Eleanor Rosch. Principles of categorization. In A. Collins and E. Smith, editors, *Readings in cognitive Science: a perspective from Psychology and Artificial Intelligence*, pages 312–322. Morgan Kaufman, San Mateo, CA, 1988.

[Schank and Osgood, 1990] Roger Schank and Richard Osgood. A content theory of memory indexing. Technical Report # 2, The Institute for the Learning Sciences, Northwestern University Evanston, IL, March 1990.

[Schank, 1977] Roger C. Schank. Rules and topics in conversation. *Cognitive Science*, 1:421–441, 1977.

Constructive Similarity Assessment: Using Stored Cases to Define New Situations

David B. Leake
Department of Computer Science
Indiana University
Bloomington, IN 47405
leake@cs.indiana.edu

Abstract

A fundamental issue in case-based reasoning is similarity assessment: determining similarities and differences between new and retrieved cases. Many methods have been developed for comparing input case descriptions to the cases already in memory. However, the success of such methods depends on the input case description being sufficiently complete to reflect the important features of the new situation, which is not assured. In case-based explanation of anomalous events during story understanding, the anomaly arises *because* the current situation is incompletely understood; consequently, similarity assessment based on matches between known current features and old cases is likely to fail because of gaps in the current case's description.

Our solution to the problem of gaps in a new case's description is an approach that we call *constructive similarity assessment*. Constructive similarity assessment treats similarity assessment not as a simple comparison between fixed new and old cases, but as a process for deciding which types of features should be investigated in the new situation and, if the features are borne out by other knowledge, added to the description of the current case. Constructive similarity assessment does not merely compare new cases to old: using prior cases as its guide, it dynamically carves augmented descriptions of new cases out of memory.

Introduction

Case-based reasoning (CBR) systems facilitate processing of new cases by retrieving stored information about similar prior episodes, and adapting solutions from the prior episodes to fit the new situation (for a selection of current CBR approaches, see (Bareiss, 1991)). A fundamental issue in applying the CBR process is similarity assessment: how to judge the similarity between new cases and those retrieved from memory. The decisions of whether a retrieved case applies, and of where to adapt it if it fails to apply, depend on similarity judgements; consequently, similarity criteria have been the subject of considerable study. Many approaches have resulted (see (Bareiss & King, 1989) for a sampling), but they share a common property: they compare some subset of the features provided by the input case to features of cases stored in memory.

When input case descriptions contain all the information that is relevant to assessing the applicability of the new case, comparing features in the input case description to the features of old cases works well. However, for the task of case-based explanation construction during story understanding, the input cases presented to the understanding system will seldom provide sufficient information for feature comparisons to determine the relevance of prior cases. Consequently, case-based explanation requires not just comparing a static new case description to stored cases, but elaborating and expanding the new case's incomplete description.

Elaborating the new case requires seeking additional information about the current situation, either by inference from existing system knowledge or by investigation in the world. For example, a detective who knows nothing about person X and is informed of X's death cannot hope to find an appropriate explanation by trying to remember the most similar previous episodes of death—the new case does not yet include sufficient information. Likewise, a story understander facing an anomalous situation is unlikely to begin with explicit knowledge of the important factors to consider during similarity assessment: the central problem for explanation is not matching fixed sets of features, but building up what the new case really is. Thus for both detective and story understander, the information provided by explicit inputs is likely to be too sparse for feature matching to be reliable.

It might appear that the problem of incomplete input case descriptions to a CBR component could be solved by preprocessing the new case. If the input case can be elaborated to select important features and fill in missing information, the case-based reasoning process can apply traditional similarity assessment procedures to the resulting case description. However, a circularity vitiates this method: the preprocessing phase of identifying the needed features would need to analyze the current situation in order to select the features to fill in, and that analysis of the current situation is the entire problem that the system originally needed to solve.

We propose an alternative approach: using stored cases themselves to guide case elaboration during the similarity assessment process. The goal of a case-based explanation system is to form a coherent view of the new situation; previously-explained cases can suggest features relevant to the new explanation, even if those features were omitted from the original case description. In this view, what is important is not the match between old and new features *per se*, but the ability of the old case to guide formation of a coherent view of the new situation. Here similarity assessment goes beyond comparison to become the constructive process of hypothesizing, from prior cases in memory, important features beyond the case description of the current situation, and attempting to build up a new candidate case description including those features. We call this approach *constructive similarity assessment.*

Constructive similarity assessment differs from traditional methods in two fundamental ways. First, rather than treating the features of an input case as fixed by the input case description, it treats the input description as a starting point for further elaboration. Second, the types of features of the new case that are elaborated depend on *suggestions* from the old case, but considerable adaptation may be required to fit new circumstances. Thus unlike straightforward feature matching, which compares features of a static new case description with prior cases, constructive similarity assessment uses the current contents of case memory as starting point for *deriving* features to consider as part of the new situation.

By using information in case memory to guide elaboration of an input case, constructive similarity assessment allows case-based reasoning systems to deal more flexibly and effectively with incomplete input cases and to better guide their search for additional information. The following sections expand on the process, showing how it extends the capabilities of a case-based reasoner to deal with poorly-defined or incomplete input cases, and sketching how the process has been investigated in the context of case-based explanation of anomalous events.

The Case Description Problem

In certain domains, input cases routinely provide all the information that needs to be considered during similarity assessment. For example, the standard input to a planner or problem-solver is a set of goals and constraints. Descriptions of those goals and constraints provide the essential information that CBR systems such as CHEF (Hammond, 1989) and JULIA (Kolodner, 1987) need in order to judge the similarity of prior cases stored in their memories. Likewise, in legal domains, input cases are routinely described in legal briefs that include all relevant features of the situation under consideration; they provide all the information that needs to be considered by CBR systems such as HYPO (Ashley & Rissland, 1987) and GREBE (Branting & Porter, 1991) as those systems identify similar cases. In such domains, for which input cases are guaranteed to include sufficient relevant features, traditional similarity assessment—comparison of the new case's features with features of a stored case—is appropriate.

However, input information is not always complete. For example, a lawyer taking on a case will probably not be content with the information initially provided, and will need to seek additional information. Likewise, it is crucial for story understanders that explain anomalous situations to build up the relevant details of the anomalous new case.

Stories are incomplete, and in principle, any of the many possible inferences from the text of a story could be relevant to explaining an anomaly, but forming all those connections is an overwhelming task (Rieger, 1975). For example, suppose an understanding system attempts to explain the breakdown of a car during a routine shopping trip. The features of the situation relevant to the breakdown's explanation may not be included in the story at all—even if the breakdown is caused by a ruptured hose, the story is unlikely to identify the hose's weakness before the fact.

Thus when explaining anomalies, adequate descriptions of input cases are hard to generate because no case description can include all the features of a real-world situation, and it may be impossible to identify *a priori* which factors of the stated situation to include in the description of the new case. In the example of the car breakdown, *a priori* schemes might suggest that the case should include features known to be directly related to the engine, such as engine noises, while omitting other features, such as which groceries were purchased. However, as the following mechanic's anecdote shows, unexpected aspects of cases may be important to their analysis:

> An elderly customer told a mechanic that her car's starting depended on the type of ice cream she bought: whenever she bought peppermint the car refused to start, but whenever she bought vanilla it started perfectly. The mechanic knew that peppermint could not affect the car's starting, but humored her with a test drive to buy ice cream. She parked at the store, went inside and bought a pint of vanilla; the car started perfectly. She drove around the block a few times, parked and bought peppermint. When she came out, the engine would not start. (Porter, 1973, pp. 253-254)

In this example, the flavor *was* implicated in the explanation: vanilla was sufficiently popular to be prepackaged, while peppermint was hand packed. Hand packing caused the purchase to take a few minutes more, allowing fuel from the carburetor to percolate into the engine and flood it. Because of the wide range of possible features, identifying important features of a new case is a difficult problem for case-based explanation.

Using Experience to Decide Case Features

Although it is impossible for a case-based explainer to determine important case features *a priori*, the case-based explanation framework suggests a method for deciding which features to consider adding to an initial case description. Because a case-based explainer attempts to use suggestions from prior explanations to explain the current case, it is natural to look for the additional features that are suggested by comparisons between the new case description and previous explanatory cases. If relevant features are not already present in the input description of the new case, the system can determine whether they apply by pursuing their connections to prior knowledge; if they apply, or if they suggest adapted features that apply, it can add the derived features to its description of the new case. Thus constructive similarity assessment uses retrieved cases to suggest which paths to pursue when elaborationing a current case, and the result is not just an evaluation that a prior case is relevant, but a new picture of the input case. This process is summarized in figure 1.

Constructive similarity assessment models our intuitive picture of the process that detectives, doctors or mechanics apply to deal with sketchy initial information. Rather than simply comparing old and new cases, they use old cases to guide their development of a picture of the new situation. For example, a detective hearing of the death of a millionaire might immediately be reminded of the previous murder of a millionaire by his heirs. If the only available information is that a millionaire died, the match between old and new cases is quite partial. However, the initial comparison shows that it is possible the old case is relevant, and suggests possible features to try to establish. If the previous millionaire was killed by hostile heirs, the detective looks for another feature in the current situation: did the dead millionaire have heirs who disliked him?

Features omitted from the input case may be built up in two ways: either by seeking external information, as a detective might do by interviewing the millionaire's acquaintances, or by inference based on the information already in the system's memory. In either case, the guidance of the prior case makes it possible to focus processing. If the decision of what to include in the description of a new case were made entirely before retrieval, the system would have no way to choose important features from the countless aspects of the situation that could theoretically be relevant. However, when the decision of what to include is made in light of a prior case, consideration is constrained to look at a much smaller set of features, and those features are guaranteed to give useful information: their presence or absence helps confirm or disconfirm the applicability of the prior case that the system is attempting to apply.

Literal Matching vs. Matching Adapted Features

Even after a retrieved case has suggested the features to consider in an input, constructive similarity assessment does not reduce to simple feature matching with the elaborated input case. Because the appropriateness of the retrieved case is determined entirely by its ability to suggest a coherent view of the new situation, much more abstract similarity relationships may apply. For example, if the anecdote connecting peppermint ice cream to starting problems were retrieved from memory to be applied to a new case, relevance of that story to a new situation would depend on whether the

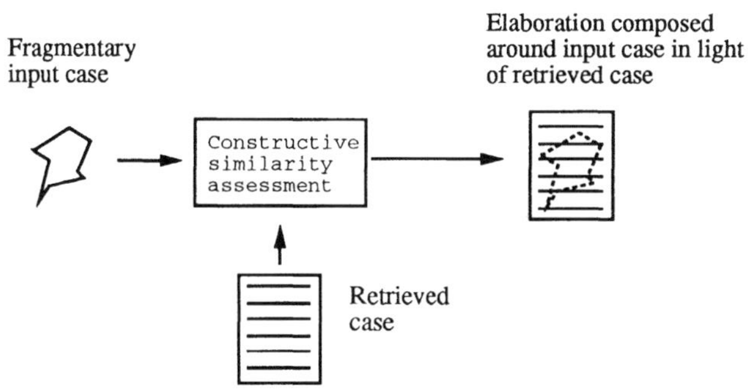

Figure 1: How constructive similarity assessment elaborates new cases based on experience.

new situation involved something *causally equivalent* to the ice cream purchases: delays correlated with an engine problem.

Thus unlike traditional similarity assessment methods, constructive similarity assessment does not require that the same predicates apply in order to consider two situations similar; what must match is the more abstract causal structure. Accordingly, similarity assessment need not actually evaluate the match between individual features of the old or new cases. What is important is whether, with guidance from the retrieved case, the features of the new case can be adapted into a coherent picture of the current situation. Because this method makes similarity assessment depend strongly on the ability to hypothesize elaborations of new cases and to judge the reasonableness of possible elaborations given current knowledge, its results depend on the contents of memory. In addition, the results depend on adaptation strategies for fitting an old case to a new situation: constructive similarity assessment treats a retrieved case as appropriate to the new situation, regardless of how dissimilar its features might be, if an adaptation of the retrieved case provides a suitable elaboration of the new situation.

Programs for Constructive Similarity Assessment

As a more concrete illustration, we consider the constructive similarity assessment process investigated in the case-based explanation framework of SWALE (Kass, 1986; Leake & Owens, 1986; Schank & Leake, 1989) and of ACCEPTER, a system which began as the case evaluation component of SWALE. (Leake, 1992). Both SWALE and ACCEPTER are story understanding systems that use case-based reasoning to explain anomalous events in news stories. The primary example of SWALE is the story of Swale, who was a superstar 3-year-old racehorse in peak shape, decisively winning major victories, who collapsed and died without warning a few days after winning the Belmont stakes. Anomalies processed by ACCEPTER also include the death of basketball star Len Bias the day after being first choice in the basketball draft, the explosion of the space shuttle Challenger, and the news that the American warship Vincennes shot down a civilian airliner.

For each of SWALE's and ACCEPTER's stories, inputs to the system are highly incomplete: They correspond to the information contained in newspaper headlines. In general, there will be many explanations in memory for an event such as a sudden death, and any of those explanations would match the few supplied features of the input case equally well. Both as a component of SWALE, and as a stand-alone system, the job of ACCEPTER is to guide constructive similarity assessment: it uses knowledge of likely events to evaluate elaborations of the input case in light of experience, to guide elaboration of input case descriptions based on prior cases.

Unlike most CBR approaches, the case-based explanation model does not treat matches between given features in old and new cases as necessarily important to using the new case: what is important is simply whether consideration of the old case gives rise to a (possibly quite different) coherent scenario for the new situation. For example, one of the explanations that SWALE retrieves for Swale's death is the explanation for the death of the rock star Janis Joplin: death from an overdose of recreational drugs taken to escape stress. Few features of this explanation match at a literal level, but the suggestion of drugs leads to a plausible alternative more directly connected to horses: that Swale might have died from an overdose of performance-enhancing drugs. Another retrieved explanation is the death of the runner Jim Fixx because jogging overtaxed a hereditary heart defect, which again fails to match at a literal level (horses are not joggers, and the input case provides no information about a heart defect). Nevertheless, that explanation leads to a reasonable hypothesis: Swale's racing overtaxed a heart defect to lead to his death.

The features implicated in these generated explanations (death from performance-enhancing drugs, or death from racing with a heart defect) are not part of the input case presented to SWALE, which include only Swale's age, the information about SWALE's Belmont victory and the fact that he died. Consequently, the resultant explanations might not receive high rankings by traditional similarity assessment. However, each explanation can be connected to known information about Swale to form a larger new case with which they share important features. For example, the use of performance-enhancing drugs fits stereotypes for horse racing, supporting that Swale's trainer might have given him a drug overdose; a heart defect follows from inbreeding, which might be associated with purebred animals. The two candidate explanations suggest that these links be pursued to form a more complete picture of the input case, either confirming or refuting those hypotheses. By following such links and considering their ramifications, constructive similarity assessment incrementally builds up a richer picture of a sketchy input case.

Relationship to Other Perspectives

Constructive similarity assessment is particularly relevant to three questions in CBR: what to include in a case representation, how changing circumstances affect criteria for similarity assessment, and how case features should be compared.

What to include in a case representation: Rather than requiring a new case to be completely specified in advance, constructive similarity assessment considers how a chosen candidate case can suggest inference of features that may not be included in the input case description. Owens (88) also considers how a case library can guide interpretation of unanalyzed situations, but from a different perspective: That work concentrates on how the contents of memory can guide selection of features to derive from an input case in order to discriminate between possibly-relevant stored cases.

How changing circumstances affect similarity assessment: Previous CBR research has proposed models of similarity assessment that are dynamic with respect to changing system goals for applying retrieved cases; in those models current goals determine which features to consider important in a given situation (Ashley & Rissland, 1987; Kolodner, 1989; Leake, 1991). The constructive similarity assessment process is dynamic in a different way: It elaborates new situations according to the current contents of memory—specific cases, general background knowledge and specific beliefs—and dynamically alters how it will understand a given new case, based on the current contents of memory.[1]

Comparison of case features: The constructive similarity assessment process we propose allows old cases to be applied to new situations not only if they match the facts of new situations, as in most similarity assessment criteria, but if their features are related on a much more abstract level. Both GREBE (Branting & Porter, 1991) and PROTOS (Bareiss, 1989) also go beyond requiring literal feature matches, using explanation of more abstract relevance of features to decide similarity (e.g., PROTOS can match the legs of one chair to the pedestal of another, because both serve as a seat support). Constructive similarity assessment, however, uses this process to suggest directions to investigate in order to suggest properties of the current case that may not be present in the initial representation of the new case. For ACCEPTER, old cases are applicable if adaptations of their features yield a picture of the new situation that fits other knowledge in memory. For example, the explanation of a recreational drug overdose is considered relevant because it can be adapted to suggest a *new* picture of the death that makes sense in light of stereotypes for horse racing: that Swale was drugged by his trainer.

Conclusion

Similarity assessment processes traditionally assume that all relevant information about a new case is available at the time of case retrieval,

[1]In this respect, the flavor of this model is very similar to that of Schank's dynamic memory theory (Schank, 1982).

so that similarity assessment is simply matching of given features. We have shown that this assumption does not hold in real-world explanation; nor will *a priori* methods work for elaborating the important features to consider in a new case. Consequently, a more flexible means of building new cases is needed. The solution we present is constructive similarity assessment. Rather than treating cases as having a fixed set of predetermined features, it treats them as starting points for further elaboration based on suggestions of the retrieved case; rather than evaluating only the match between the new situation and a fixed set of features in the retrieved case, it takes features in the retrieved case only as a starting point, allowing them to be adapted as long as the result is a plausible scenario given other system knowledge. This method guides processing of poorly-understood situations by combining sketchy input information with suggestions from memory to go beyond simple comparison and produce a sharper view of the events being understood.

Acknowledgment

I would like to thank Chris Riesbeck for helpful comments and suggestions leading to this paper.

Reference

Ashley, K. & Rissland, E. (1987). Compare and contrast, a test of expertise. In *Proceedings of the Sixth Annual National Conference on Artificial Intelligence*, pp. 273–284 Palo Alto. AAAI, Morgan Kaufmann, Inc.

Bareiss, R. (1989). *Exemplar-Based Knowledge Acquisition: A Unified Approach to Concept Representation, Classification, and Learning.* Academic Press, Inc., San Diego.

Bareiss, R. (Ed.)., Bareiss (1991). *Proceedings of the Case-Based Reasoning Workshop*, Palo Alto. DARPA, Morgan Kaufmann, Inc.

Bareiss, R. & King, J. (1989). Similarity assessment in case-based reasoning. In Hammond, K. (Ed.), *Proceedings of the Case-Based Reasoning Workshop*, pp. 67–71 San Mateo. DARPA, Morgan Kaufmann, Inc.

Branting, K. & Porter, B. (1991). Rules and precedents as complementary warrants. In *Proceedings of the Ninth National Conference on Artificial Intelligence*, pp. 3–9 Anaheim, CA. AAAI.

Hammond, K. (1989). *Case-Based Planning: Viewing Planning as a Memory Task.* Academic Press, San Diego.

Kass, A. (1986). Modifying explanations to understand stories. In *Proceedings of the Eighth Annual Conference of the Cognitive Science Society* Amherst, MA. Cognitive Science Society.

Kolodner, J. (1987). Extending problem solving capabilities through case-based inference. In *Proceedings of the Fourth International Workshop on Machine Learning* Irvine, CA. Machine Learning, Morgan Kaufmann.

Kolodner, J. (1989). Selecting the best case for a case-based reasoner. In *Proceedings of the Eleventh Annual Conference of the Cognitive Science Society*, pp. 155–162 Ann Arbor, MI. Cognitive Science Society.

Leake, D. (1991). ACCEPTER: a program for dynamic similarity assessment in case-based explanation. In Bareiss, R. (Ed.), *Proceedings of the Case-Based Reasoning Workshop*, pp. 51–62 San Mateo. DARPA, Morgan Kaufmann, Inc.

Leake, D. (1992). *Evaluating Explanations: A Content Theory.* Lawrence Erlbaum Associates, Hillsdale, NJ.

Leake, D. & Owens, C. (1986). Organizing memory for explanation. In *Proceedings of the Eighth Annual Conference of the Cognitive Science Society*, pp. 710–715 Amherst, MA. Cognitive Science Society.

Owens, C. (1988). Domain-independent prototype cases for planning. In Kolodner, J. (Ed.), *Proceedings of a Workshop on Case-Based Reasoning*, pp. 302–311 Palo Alto. DARPA, Morgan Kaufmann, Inc.

Porter, J. (Ed.). (1973). *The Family Car.* Time-Life Books, New York.

Rieger, C. (1975). Conceptual memory and inference. In *Conceptual Information Processing.* North-Holland, Amsterdam.

Schank, R. (1982). *Dynamic Memory: A Theory of Learning in Computers and People.* Cambridge University Press, Cambridge, England.

Schank, R. & Leake, D. (1989). Creativity and learning in a case-based explainer. *Artificial Intelligence*, *40*(1-3), 353–385. Also in Carbonell, J., editor, *Machine Learning: Paradigms and Methods*, MIT Press, Cambridge, MA, 1990.

Generic Teleological Mechanisms and their Use in Case Adaptation

Eleni Stroulia and Ashok K. Goel *
College of Computing
Georgia Institute of Technology
Atlanta, GA 30332-0280
eleni@cc.gatech.edu, goel@cc.gatech.edu

Abstract

In experience-based (or case-based) reasoning, new problems are solved by retrieving and adapting the solutions to similar problems encountered in the past. An important issue in experience-based reasoning is to identify different types of knowledge and reasoning useful for different classes of case-adaptation tasks. In this paper, we examine a class of non-routine case-adaptation tasks that involve patterned insertions of new elements in old solutions. We describe a model-based method for solving this task in the context of the design of physical devices. The method uses knowledge of generic teleological mechanisms (GTMs) such as cascading. Old designs are adapted to meet new functional specifications by accessing and instantiating the appropriate GTM. The Kritik2 system evaluates the computational feasibility and sufficiency of this method for design adaptation.

Overview

In experience-based (or case-based) reasoning, new problems are solved by retrieving and adapting solutions of similar problems encountered in the past. Once a new solution is created, it can be stored in memory for potential reuse in future. Much of previous work on modeling experience-based reasoning uses simple modification operators and rules for "tweaking" the solution in the retrieved case (Alterman 1988; Ashley & Rissland 1988; Hammond 1989; Kolodner & Simpson 1989). These methods are often sufficient for routine case-adaptation where the needed modifications involve changing the parameter of an element in the old solution or substituting one solution element by a similar one. Many adaptation tasks, however, appear to require modifications that go beyond parameter changes or component substitutions. Case adaptation in innovative design, for example, often involves insertion of new components in old designs.

Reasoning about insertions of new components in a design structure can be very complex. This is because the insertion of a new component can potentially have a non-local impact on the functionality of the design. A general, computationally feasible and cognitively plausible model for this type of reasoning is not yet known. Nevertheless, it seems reasonable to assume that (human) designers use additional knowledge to constrain their reasoning about design modifications, and thus manage the complexity of the task. An important and open research issue in modeling experience-based reasoning, then, is to identify additional types of knowledge useful for different classes of case-adaptation tasks and to develop process models of their usage.

Informal observations of designers have led us to hypothesize that (i) one class of case-adaptation tasks is characterized by *insertion of specific patterns of components* into the structure of the design retrieved from the case memory, and (ii) the insertion of these patterns is based on knowledge of *generic teleological mechanisms*. Examples of generic teleological mechanisms (GTMs) in design include cascading, feedback, and feedforward. These mechanisms are "teleological" in that they result in specific functions. The mechanism of feedback, for example, takes information about the deviation of the output of a system from its desired output, feeds it back into an input to the system, and this results in the specific function of reducing the deviation in the output. Also, these mechanisms are "generic" in that they are device independent.

The feedback mechanism, for example, is independent of the specific device in which it might be instantiated, and in principle can be instantiated in any control system. The instantiation of such a mechanism in the context of a particular device leads to a patterned insertion of components in the structure of the system. The instantiation of the feedback mechanism in a system, for example, may result in the insertion of components that can measure the deviation of the system's output from its desired output, compo-

*This work has been supported by a research grant from the Office of Naval Research, contract N00014-92-J-1234, a graduate fellowship from IBM, a research gift from NCR, and equipment donated by IBM and Symbolics.

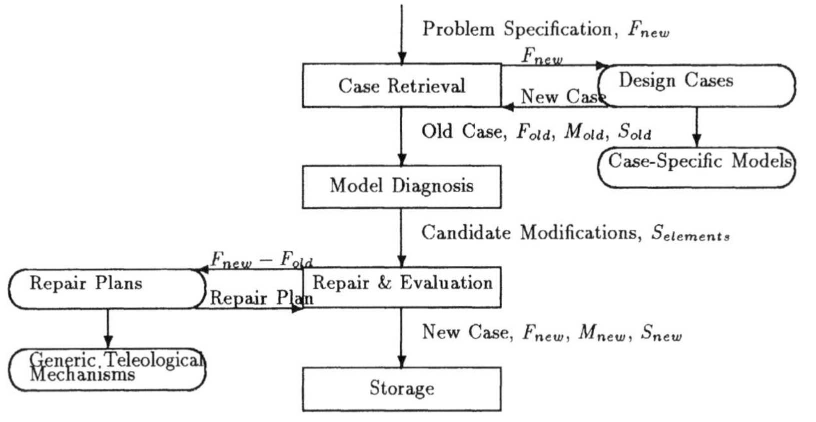

Figure 1: Kritik2's Process Model

nents that can relay this information to an input to the system, and components that can control the system input accordingly. This hypothesis about case adaptation raises a number of issues concerning the representation, indexing, access, and use of GTMs. The Kritik2 project investigates these issues in the context of designing physical devices such as simple electrical circuits and heat exchangers.

Process Model

Figure 1 depicts Kritik2's process model for experience-based design. Kritik2 takes as input the specification of a function desired of a new device F_{new}. In the case-retrieval step, it uses F_{new} as a probe into a functionally indexed case memory and retrieves the closest matching case. Each design case in the case memory contains a pointer to the corresponding device model M_{old} that specifies how the structure of the known device S_{old} delivers its functions F_{old}. In the diagnosis step, Kritik2 uses M_{old} to generate candidate modifications to S_{old} so as to achieve F_{new}. In the repair step, it uses the difference between F_{new} and F_{old} as a probe into a memory of repair plans and retrieves the applicable plans. The candidate modifications generated by the diagnosis step are used as a secondary index to discriminate among the applicable plans. Some repair plans contain pointers to GTMs. If selected, such a repair plan instantiates the corresponding GTM in the context of M_{old} and synthesizes the instantiated GTM with S_{old} to produce a candidate design for achieving F_{new}.

Case-Specific Device Models

Let us consider as an example the task of designing a Nitric Acid cooler (NAC_{new}) to reduce the temperature of some quantity of Nitric Acid from some initial temperature $T1$ to some final temperature $T2_{new}$. Let us also suppose that the case-retrieval task returns the design and model of a Nitric Acid cooler (NAC_{old}) which reduces the temperature of the same quantity of Nitric Acid from temperature $T1$ to temperature $T2_{old}$, where $T1 - T2_{new} >> T1 - T2_{old}$. Clearly, the desired function of cooling Nitric Acid from $T1$ to $T2_{new}$ is similar to but different from the delivered function of cooling Nitric Acid from $T1$ to $T2_{old}$. The difference between the two functional specifications, which we will denote as $F_{new} - F_{old}$, lies in the range by which the Nitric Acid is cooled.

The structure of NAC_{old} is shown in figure 2(a). It consists of a pump that pumps cold water into the device, a pipe through which hot Nitric Acid flows in the device, and a heat-exchange chamber which contains the cold water pumped into the device and includes the Nitric Acid pipe. The model for NAC_{old} specifies how the device works, i.e., how its structure delivers its function of cooling Nitric Acid from $T1_{old}$ to $T2_{old}$. The functioning of this device can be informally described as follows: Hot Nitric Acid flows through a pipe, a part of which is enclosed in a heat-exchange chamber. The chamber contains cold water that is pumped into the device by a water pump. Inside this chamber, heat is transferred from the hot Nitric Acid to the cold water. As a result of this transfer of heat, the temperature of out-flowing Nitric Acid is lower than the temperature of in-flowing Nitric Acid; the temperature of cold water increases correspondingly.

Kritik2 explicitly represents the functions, the structure and the *internal causal behaviors* of the device, where the internal causal behaviors specify how the device structure delivers its functions (Goel 1991). Its behavioral representation language generalizes the *functional representation scheme* (Sembugamoorthy & Chandrasekaran 1986) and grounds it in *component-substance ontology* of physical devices (Bylander & Chandrasekaran 1985). The internal causal behaviors in this language are represented as partially ordered sequences of states and state-

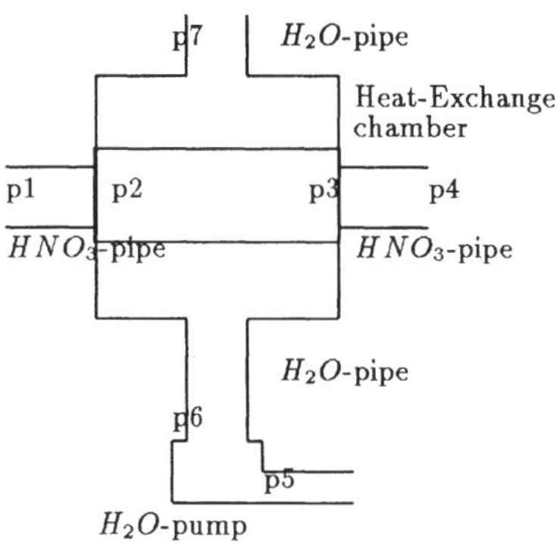

Figure 2(a): Structure of NAC_{old}

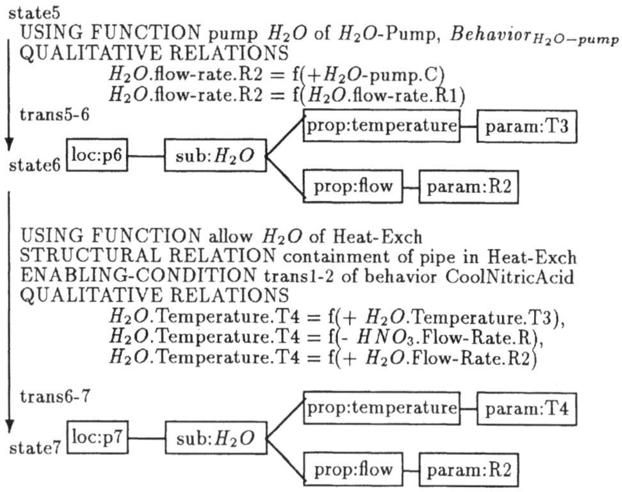

Figure 2(b): Behavior HeatWater of the old HNO_3 cooler,NAC_{old}

transitions. A state is a partial description of some substance of the device at some particular point in the device structure. A state transition describes how the parameters of a substance change as the substance flows from one point to another.

Figure 2(b) shows a fragment of an internal behavior of the Nitric Acid cooler called behavior `HeatWater`. This behavior describes the flow of the water through the device. Initially, at `state5`, water has a flow-rate R1, and temperature T3. In `state6`, the water flow-rate has increased to R2 due to the functionality of the water pump. After flowing through the heat-exchange chamber the water temperature is T4. The transition `trans6-7` is due to the multiple functions of the heat-exchange chamber, which allows the flow of the water and also allows the heat flow between hot Nitric Acid and cold water. This transition occurs simultaneously with `trans2-3` in the behavior `CoolNitricAcid` of the device, where behavior `CoolNitricAcid` specifies the state transitions of the Nitric Acid. Each transition is annotated with enabling conditions that need to be true in order for the transition to occur and with qualitative equations that relate the state changes. (Goel 1991) provides a more detailed description of case-specific device models.

Generic Teleological Mechanisms

Kritik2 posits a memory of GTMs such as cascading, feedback, and feedforward. The representation of a GTM encapsulates (i) knowledge about the difference between the functions of a known design and a desired design that the mechanism can help to reduce, and (ii) knowledge about modifications to the internal causal behaviors of the known design that are necessary in order to reduce this difference. A GTM thus associates a type of functional difference with a type of behavioral modification, with the former acting as an index to the latter.

Figure 3 illustrates Kritik2's representation of the cascading mechanism. Figure 3(a) specifies that the cascading mechanism is applicable when the known design (Design1) changes the value of some substance property from $val11$ to $val21$ by some known internal behavior B1, the desired design (Design2) changes the value of the same substance property from $val12$ to $val22$ by some Behavior B2, and $|val22 - val12|$ is many times $|val21 - val11|$. Figure 3(b) illustrates Kritik2's representation of the modifications necessary to reduce the functional difference. It specifies that Behavior B2 might be achieved by replicating Behavior B1 as many times as needed. Since, in general, $|val22 - val12|$ might not be a multiple of $|val21-val11|$, Behavior B2 also includes the possibility of forming a new goal to reduce the functional difference left after replicating B1. Note that the behavioral model of the cascading mechanism is indexed by the functional difference it can reduce.

Given a specific type of functional difference between the desired design and the retrieved one, Kritik2 uses the functional difference to access the applicable GTM. For example, if the difference between the desired function and the delivered function is that the delivered function alters some substance property by some "small amount" and

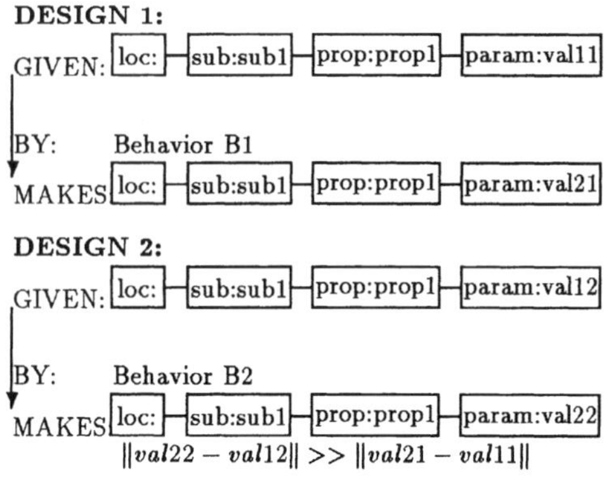

Figure 3(a): Functional Difference that Cascading reduces

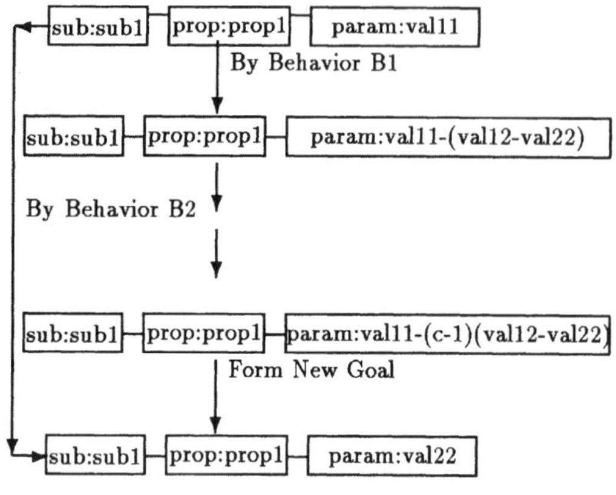

Figure 3(b): Behavior Modification that Cascading suggests

the desired function is the alteration of this property by a "large amount", then this functional difference can be used to access the GTM for cascading. Once accessed, the GTM can be applied to the internal causal behaviors of the known design.

Case Adaptation

Let us now consider how knowledge of the cascading mechanism helps in the task of adapting the structure of NAC_{old}, which reduces the temperature of the same quantity of Nitric Acid from temperature $T1$ to temperature $T2_{old}$, to design NAC_{new} which can cool the same quantity of Nitric Acid from $T1$ to $T2_{new}$, where $T1 - T2_{new} >> T1 - T2_{old}$.

Diagnosis : First, the diagnosis task identifies the set of structural elements that influence the substance properties that need to be changed, and the set of the specific behavioral state-transitions in which each element plays a role. In the above example the output of the diagnosis task is $S_{elements} = \{$ water-pump.C $\}$, where C is the capacity of the water pump. If $S_{elements}$ contains more than one element, then they can be ordered heuristically. (Stroulia, Shankar, Goel & Penberthy 1992) provide a more detailed description of the diagnosis step.

Repair : Next, the repair task instantiates the cascading mechanism in the context of the model of the known device. The repair plans in Kritik2 specify compiled sequences of operations that need to be performed for repairing a design, given a specific type of difference desired in its function. Kritik2 uses the difference $F_{new} - F_{old}$ as a probe into the memory of repair plans and retrieves the applicable plans. The candidate modifications $S_{elements}$ are used as a secondary index to discriminate among the applicable plans. Some repair plans contain pointers to GTMs. These plans also contain procedural knowledge of how to synthesize the behavior of a GTM with the model of the known device. If a repair plan of this type is selected, Kritik2 retrieves the appropriate GTM, instantiates it in the context of the model of old device, and synthesizes the behavior of the GTM with the model.

For our Nitric Acid cooler example, Kritik2 uses the type of functional difference between the desired and the retrieved designs as an index into the repair plan memory, and selects the repair plan called the *structure-replication plan*. This plan contains a pointer to the cascading mechanism shown in Figure 3. It synthesizes the behavior of the cascading mechanism with the device model of NAC_{old} in two steps: (i) behavior revision and (ii) structure revision. First the behavior $B_{waterpump}$ is replicated in the internal behavior `HeatWater` shown in Figure 2(b) because the water pump was the structural element identified by the diagnosis task.

More specifically, since the water pump plays a role in `trans5-6` of behavior `HeatWater`, this transition is replicated to obtain the modified behavior `HeatWater` shown in figure 4(a). The changes in the values of state variables caused by this are propagated forward throughout the behavior. Since the changed values affect another behavior in the device model, namely, the behavior `CoolNitricAcid`, the values are propagated in this behavior as well.

Once the behavior revision is completed, the structure is revised. Since each state-transition explicitly specifies the structural elements which are responsible for the transition, the behavioral modifications are directly translated into structural modifications. The structure of the resulting design, with multiple water pumps, is shown in figure 4(b). Kritik2 now evaluates the candidate design by qualitatively simulating the case-specific device model. If the simulation reveals inconsistencies between the desired functions of the device and its output behaviors, redesign is needed.

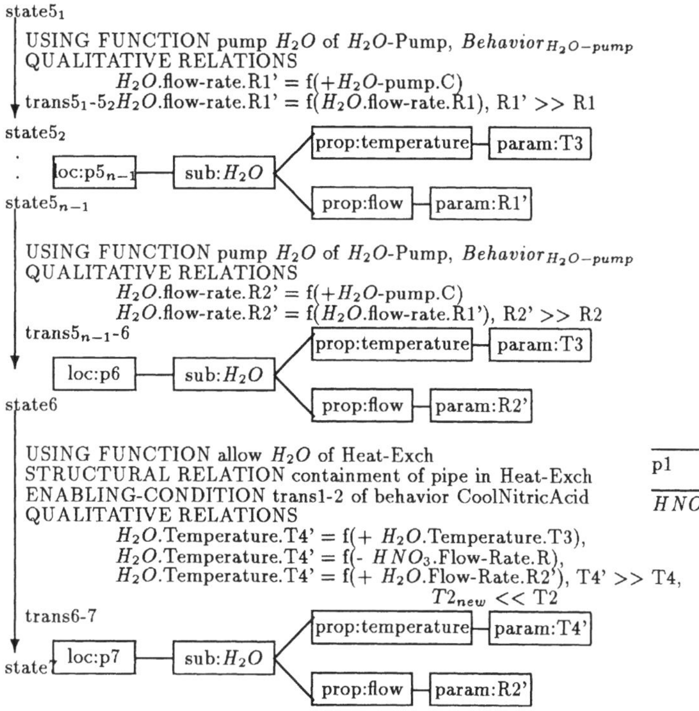

Figure 4(a): Revised Behavior HeatWater of the new HNO_3 cooler, NAC_{new}

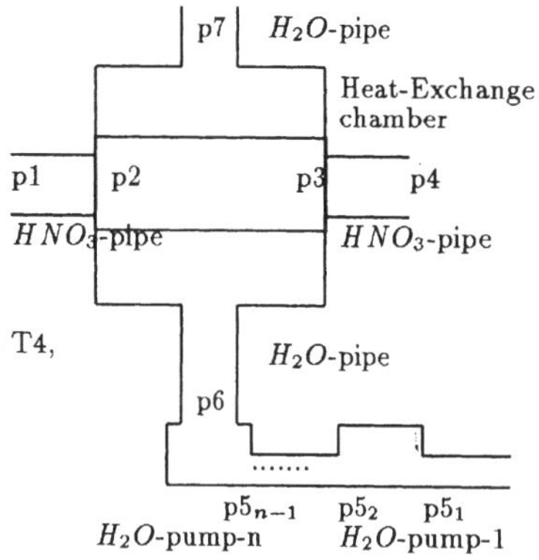

Figure 4(b): Revised Structure of the new HNO_3 cooler, NAC_{new}

Evaluation and Analysis

Kritik2 provides a computational testbed for conducting controlled experiments with GTMs and their use in case adaptation. The case memory in Kritik2 contains designs of four types of physical devices: simple heat exchangers of the type described above, electrical circuits such as the circuit in a household flashlight, electromagnetic devices such as the household buzzer, and complex angular momentum controllers such as those aboard the Hubble Space Telescope. This indicates that its component-substance ontology and behavioral representation language are not limited to any specific device domain. Kritik2 demonstrates the sufficiency of the scheme for representing, indexing, accessing, and using the cascading mechanism in two of these four domains: replication of pumps in heat exchangers and replication of batteries and resistors in electrical circuits. Again, this indicates that the method of using GTMs for case adaptation is not limited to any specific device domain. However, the present implementation of Kritik2 contains only the GTM for cascading. We are presently adding more mechanisms to our library of generic teleological mechanisms and evaluating them in more domains.

Ablation experiments with Kritik2, in which specific types of knowledge and methods of reasoning in the system are "lesioned" (Cohen & Howe 1988), indicate that its process model for use of GTMs in case adaptation is quite flexible. In general more than one repair plan in the plan memory may be applicable for a given case-adaptation task. For instance, the *component-replacement* plan is another repair plan applicable in our Nitric Acid cooler example. If selected, this plan probes the memory of components for a water-pump with higher capacity. If such a water-pump is available, the execution of this plan results in a simple substitution of one component (water-pump with a low capacity) by another (water-pump with a higher capacity). In that case, Kritik2 behaves like most previous case-based systems, in that it uses component replacement to tweak the retrieved design. If, however, a water-pump with a higher capacity is not available in the component memory, the component-replacement plan would fail. In this case, Kritik2 resorts to the use of the structure-replication plan which instantiates the cascading mechanism.

Additional ablation experiments with Kritik2 indicate a different kind of flexibility pertaining to the diagnosis task in the process model. Although the process model includes a diagnosis task, the diagnosis step actually is optional. The method of instantiating GTMs, however, results in poorer designs if the diagnosis task is not performed. In the Nitric Acid cooler example, for instance, we found that instantiating the cascading mechanism results in the replication of the water pump if the diagnosis task is performed, and in the replication of the entire heat exchanger if the diagnosis task is not performed. The former design is more parsimonious and hence better than the latter one. This leads us to conclude that while the quality of the solution appears to improve when diagnosis task is performed, instantiating GTMs appears to be a useful strategy for case adaptation whether or not the diagnosis task is performed.

Related and Further Research

Experience-based reasoning is a model of human decision making and problem solving (Riesbeck & Schank 1989). Previous work on experience-based reasoning has investigated the use of modification operators and rules for routine case adaptation in which the needed modifications involve changing the parameter of an element in the old solution or substituting one solution element by a similar one (Alterman 1988; Ashley & Rissland 1988; Hammond 1989; Kolodner & Simpson 1989). Exploration of more robust methods of case adaptation based on derivational traces (Carbonell 1986) and causal models (Goel 1991; Koton 1988; Sycara & Navinchandra 1989) also has been largely limited to relatively routine case adaptation.

Our research on generic teleological mechanisms builds on earlier research on model-based case adaptation. The Kritik2 system provides a model of how knowledge of GTMs might complement knowledge of case-specific device models and help designers to reason about patterned insertions of new components in old designs.

Darden has proposed that scientific theories can be viewed as devices and theory revision can be viewed as a design-adaptation task (Darden 1990). In recent personal communication (Darden 1991), she has further conjectured that Kritik2's use of GTMs for design adaptation might provide a basis for modeling the formation of early theories of heredity and genetics. If this is correct, it would indicate that the use of GTMs is a very general domain-independent method of case adaptation. Our current work on GTMs involves modeling how designers learn GTMs from specific design cases and use this knowledge in analogical reasoning across different domains (Bhatta & Goel 1992).

Acknowledgements We wish to thank Sambasiva Bhatta for many discussions on the subject and comments on earlier drafts of this paper.

References

Alterman, R. 1988. Adaptive Planning. *Cognitive Science*, 12:393-422.

Ashley, K. and Rissland, E. 1988. A Case-Based Approach to Modeling Legal Expertise. *IEEE Expert*, Summer 1988.

Bhatta, S. and Goel, A. 1992. Discovery of Principles and Processes from Design Experience. To appear in *Procs. of Workshop on Machine Discovery*, ML-92.

Bylander, T. and Chandrasekaran, B. 1985. Understanding Behavior Using Consolidation. *Proc. Ninth International Joint Conference on Artificial Intelligence*, 450-454.

Carbonell, J. 1986. Derivational Analogy: A Theory of Reconstructive Problem Solving and Expertise Acquisition. *Machine Learning: An Artificial Intelligence Approach, Volume II*, R. Michalski, J. Carbonell and T. Mitchell (editors). San Mateo, CA: Morgan Kauffman.

Cohen, P. and Howe, A. 1988. How Evaluation Guides Research. *AI Magazine*, 9(4):35-43, Winter 1988.

Darden, L. 1990. Finding and Fixing Faults in Scientific Theories. *Computational Models of Discovery and Theory Formation*, J. Shrager and P. Langley (editors). Hillsdale, NJ: Erlbaum.

Darden, L. 1991. Personal Communication.

Goel, A. A Model-Based Approach to Case Adaptation. *Proc. Thirteenth Annual Conference of the Cognitive Science Society*, Chicago, August 7-10, 1991, pp. 143-148.

Hammond, K. 1989. *Case-based Planning: Viewing Planning as a Memory Task*, Boston, MA: Academic Press.

Kolodner, J.L. and Simpson, R. 1989. The MEDIATOR: Analysis of an Early Case-Based Reasoner. *Cognitive Science*, 13:507-550.

Koton, P. 1988. Combining Causal and Case-Based Reasoning. *Proc. Tenth Annual Conference of the Cognitive Science Society*.

Riesbeck, C. and Schank, R. 1989. *Inside Case-based Reasoning*. Hillsdale, NJ: Erlbaum.

Sembugamoorthy, V. and Chandrasekaran, B. 1986. Functional Representation of Devices and Compilation of Diagnostic Problem-Solving Systems. In *Experience, Memory and Reasoning*, J. Kolodner and C. Riesbeck: (editors), Hillsdale, NJ: Lawrence Erlbaum, pp. 47-73.

Stroulia, E., Shankar, M., Goel, A. and Penberthy, L. 1992. A Model-Based Approach to Blame Assignment in Design. To appear in *Proc. Second International Conference on AI in Design*, Pittsburg, June 1992.

Sycara, K. and Navinchandra D. 1989. A Process Model of Case-Based Design. *Proc. Eleventh Cognitive Science Society Conference*.

Representing Cases as Knowledge Sources that Apply Local Similarity Metrics

David B. Skalak
Department of Computer Science
University of Massachusetts
Amherst, MA 01003
skalak@cs.umass.edu

Abstract [1]

A model of case-based reasoning is presented that relies on a procedural representation for cases. In an implementation of this model, cases are represented as knowledge sources in a blackboard architecture. Case knowledge sources define local neighborhoods of similarity and are triggered if a problem case falls within a neighborhood. This form of "local indexing" is a viable alternative where global similarity metrics are unavailable. Other features of this approach include the potential for fine-grained scheduling of case retrieval, a uniform representation for cases and other knowledge sources in hybrid systems that incorporate case-based reasoning and other reasoning methods, and a straightforward way to represent the actions generated by cases. This model of case-based reasoning has been implemented in a prototype system ("Broadway") that selects from a case base automobiles that meet a car buyer's requirements most closely and explains its selections.

Introduction

This paper addresses two fundamental problems of case-based reasoning (CBR): case representation and case similarity. Its central point is that a procedural, locally-indexed representation for cases provides several benefits. We use the term "locally indexed" to refer to a case retrieval technique that uses similarity metrics that are applicable only within a neighborhood of a case in the space of cases. Local indexing is distinguished from a global indexing method that relies on a single function to assess similarity throughout the case base. Metrics that are locally defined can be viewed as an attempt to approximate *piecewise* an ideal — but often difficult to construct — function that measures the similarity of a problem situation to any case, where each piece of the local metric is applicable only in a suitable area of the case.

The intuition behind local indexing is that each case is in the best position to map the topography of the case space in a neighborhood of itself. What then counts as similar depends on where the case is located in case space. Informally, the general perspective of this research is to impose problem-solving responsibility on the cases themselves by including in the case representation knowledge that is usually external to cases, including similarity metrics. In addition, this case-centric perspective regards cases as active entities, rather than as responsive to external procedures.

Seminal work in CBR has exploited the notion that how one assesses the similarity of a stored case depends on the problem situation, e.g., [Ashley, 1990], [Bareiss, 1989], [Kolodner, 1983], [Sycara, 1987]. The model we describe tries to exploit the complementary idea that cases of a particular sort possess features — independent of a problem situation — that will help determine how similarity will be assessed for cases of that variety. Informally, regardless of whether you're looking to buy a Cadillac or a Miata, if on the used car lot you encounter a pickup truck with huge tires whose body is six feet off the

[1]This research was supported in part by the National Science Foundation, contract IRI-890841, the Air Force Office of Sponsored Research under contract 90-0359, the Office of Naval Research under a University Research Initiative Grant, contract N00014-87-K-0238, and a grant from GTE Laboratories, Waltham, MA.

ground, your assessment will be governed by whether you want a vehicle with a six-foot high cab. Following this intuition, we try to build into each case several ways of assessing similarity that are useful for a case of that variety.

Briefly, the model of retrieval presented here regards a case as a procedural entity that is activated when a problem situation falls within a local neighborhood of the case. We claim that the following benefits flow from this approach:

• A procedural case representation yields a consistent knowledge representation for hybrid architectures that combine CBR with other reasoning methods.

• Local indexing is a useful alternative in situations where no available similarity metric can be applied uniformly across cases.

• Fine-grained scheduling for case retrieval is facilitated and permits focused control of problem-solving.

• Multiple perspectives on a case can be easily represented through local similarity functions.

Implementation

Our implementation of the model uses a blackboard architecture in which knowledge sources respond to changes on a global blackboard. The Broadway prototype uses GBB v.2.0, a toolkit for developing high-performance blackboard applications [Blackboard Technology Group, 1991]. The GBB Agenda Shell enables a user to define knowledge sources that may be triggered, checked for the fulfillment of their respective preconditions, and, if fulfilled, instantiated in knowledge source activation records with some execution rating. Knowledge source activations are then placed on an agenda of activations pending execution.

A correspondence between cases and blackboard knowledge sources can be exploited. A [case/knowledge source] is [similar/activated] when its [index/ precondition function] is [triggered/ satisfied], and therefore the [case's suggested action/ knowledge source function] should be executed. The current implementation fuses these corresponding aspects: cases are represented as knowledge sources. The precondition of a case knowledge source is a similarity predicate. If a problem situation satisfies this locally defined similarity predicate, the case knowledge source is activated. This incorporation of case similarity into knowledge source precondition functions yields one immediate by-product. The greater the similarity of a problem situation to a case, the higher the execution rating returned by the precondition of the knowledge source corresponding to the case, and the earlier the case's action is executed. The most similar cases execute their constituent actions first.

The Domain

The Broadway prototype addresses the almost century-old quandary "Which car should I buy?". For many people this question appears to invite anecdotal case-based reasoning. As in many common-sense areas, no strong domain theory is available to resolve the competing constraints involved in automobile purchase. However, this domain does not present the complexity of such classical blackboard applications as speech understanding [Erman et al., 1980] or sonar signal interpretation [Nii et al., 1982]. For our purposes the automobile domain is an interim vehicle for investigating the utility of procedural case representation and associated ideas of similarity, indexing and control.

Description of Control Flow

The skeletal flow of control for Broadway is given in general terms in Figure 1.

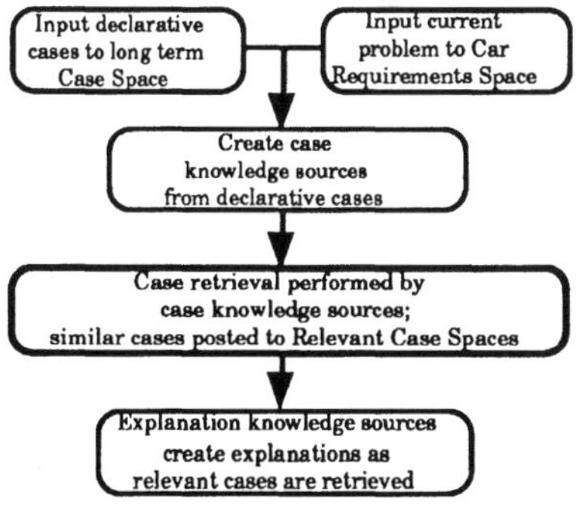

Figure 1: Flow of control in Broadway.

When the system is initialized, declarative, frame-based car case representations are loaded onto a blackboard space that is Broadway's long term case memory. Broadway's current case base consists of 93 cases that represent automobiles of a particular make and model, e.g., an Eagle Talon. Domain engineering has been aided by the 1991 annual automobile issue of Consumer Reports magazine [Consumer Reports, 1991]. Each declarative case then is used as the basis for creating several case knowledge sources that are activated when a similar problem case is posted to a problem space. See Figure 2 for a simplified example,

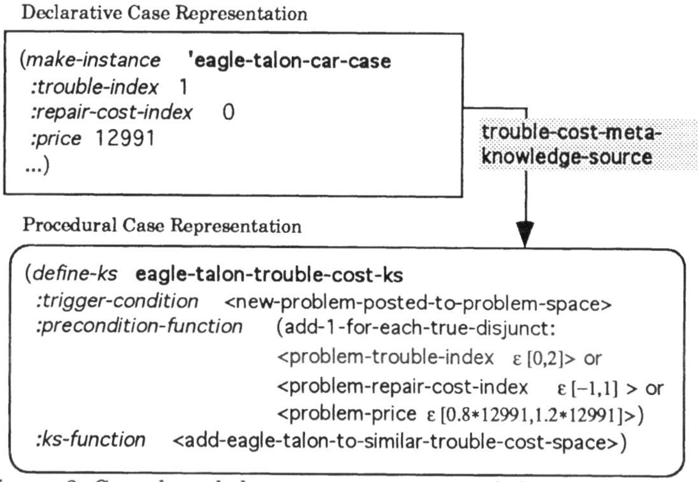

Figure 2: Case knowledge sources are created from case-frame representations by meta-knowledge-sources. The scale for the trouble-index and repair-cost-index is -2 (much worse than average) to 2 (much better than average).

which prescribes an interval of similarity for each feature and then adds the number of problem features that fall within those intervals to yield an execution rating.

These procedural representations of cases assume the usual CBR tasks of retrieving similar cases and manipulating them. Next, the user inputs a problem case, which is a specification of the features that the user desires in a car, presented in the same form as the declarative case frames. Cars that partially or completely meet the user's specifications are then retrieved through the activation of car case knowledge sources whose preconditions recognize the similarity of that case to the problem requirements. In Broadway, the action taken by a case upon activation is to post the corresponding declarative case to a relevant case space on the blackboard. This posting in turn triggers explanation knowledge sources, which we do not describe further here except to say that stereotypical explanation patterns are applied to explain to the user why these cars are appropriate recommendations. In general, the action taken by each case knowledge source depends on the application, for example, to suggest a repair to a plan case (e.g., CHEF [Hammond, 1989]) or to supply an argument fragment (e.g., HYPO [Ashley, 1990], CABARET [Rissland and Skalak, 1991]).

Procedural Cases

Procedural representations are created by several meta-knowledge-sources, which encode the knowledge to create a case knowledge source for a given perspective on a case. For example (Fig. 2), Broadway applies a meta-knowledge-source that uses a modified Manhattan metric to create a similarity neighborhood based on several case features related to the perspective of economy: purchase price, repair frequency and repair cost. Each neighborhood is local partly because the size of the interval of similarity for each case feature depends on its value in that case. To take a simple example, the interval surrounding the price of an expensive car is larger than that around the price of a cheap car. We have also experimented with metrics that are local in that they are applicable only to certain types of cases, e.g., to sports cars. Perspectives are reflected in metrics and include any means of evaluating cases that corresponds to a particular way of assessing similarity [Ashley, 1989], [Kolodner, 1989] including, e.g., dimensions [Rissland, Valcarce and Ashley, 1984], [Ashley, 1990], signatures of feature values [Samuel, 1967], or the reasoning or explanation captured in a case [Barletta and Mark, 1988], [Branting, 1991]. To assess case similarity, Broadway currently uses two perspectives: a perspective based on figures for reliability and cost provided by Consumer Reports and a perspective based on comparison of the signatures of features of cases, resulting in 186 case knowledge sources.

Some Advantages and Shortcomings of this Representation

Local Neighborhoods and Indexing. Local similarity neighborhoods provide the framework to tailor the measurement of similarity to each case and to the region of the case space where the case resides. To take an example from a classical AI program, imagine a case-based system that plays checkers and stores board positions in a case library. [Samuel, 1967] applied polynomial evaluation functions to evaluate checkerboard positions, and noted improved performance when the game was divided into opening, middle-game, and end-game phases, with a different set of evaluation function coefficients used for each phase. As many as six game phases were used. In a CBR system that used similarity metrics based on these evaluation functions, the appropriate metric would be determined by the game phase, reflected in the location in case space of the cases under consideration.

The primary advantage of localized metrics stems from the practical and theoretical difficulties of capturing in a global approach to similarity the nuances of case similarity across all cases in the case base. Case-based retrieval mechanisms generally rely on a system-wide method or metric to compute case similarity. As a practical matter, it may be difficult to reflect in a single metric or global evaluation function the important differences among all cases and account for interactions between related features, but still avoid implicit comparison of features that are incommensurate. Also, entirely different means of assessing similarity may be required for different types of cases. We speculate that since local metrics need only work in a neighborhood of a case, spurious feature interactions or inappropriate comparisons may more likely be avoided. Since CBR is often useful in poorly understood, "weak theory" domains, a globally applicable similarity function may be hard to come by, as it implicitly would reflect strong knowledge about the domain that holds across features, cases, perspectives and contexts.

A second advantage to local metrics is the facility with which multiple views of a case can be captured (see, e.g., [Rissland, Valcarce and Ashley, 1984], [Ashley, 1989], [Kolodner, 1989]). Different perspectives are reflected in distinct similarity metrics that capture the varying importance to be accorded features when reasoning from diverse vantage points.

Additionally, exceptional cases may have unusual features that are known in advance and should be considered if relevant to one's specifications, but that are hard to incorporate into a global calculus of similarity. These exceptional features can be captured in a similarity metric local to the case.

A minor benefit of local similarity metrics may be in the observance of the software engineering principles of modularization and data encapsulation. Unusual or salient aspects of a case that are important to determining similarity can usefully be encoded in a metric local to the case.

Scheduling Granularity. Case retrieval may be scheduled at a fine level of granularity in this model. Case-based retrieval has sometimes been modeled and implemented as a monolithic action, "Search the case base for relevant cases and return them." See, e.g., [Rissland, Kolodner and Waltz, 1989] for a description of the classical control flow of CBR. A retrieval mechanism that is both large-grained and uninterruptible will potentially consume computing resources that may be applied more efficiently than to additional search of a large case base [Veloso and Carbonell, 1991]. For example, the cases initially retrieved may suggest a modification of the current case probe [Owens, 1989]. However, the current implementation of the model in Broadway does not schedule case knowledge source preconditions, and so does not exhibit this benefit. An extension to the blackboard control shell would be required to realize this advantage.

Consistent Hybrid Representation. In a hybrid system cases may be represented consistently with knowledge sources from other reasoning paradigms. A uniform representation supports the use of CBR as a component in a hybrid architecture where cases and other sources of expertise respond uniformly and cooperatively to progress and failure in problem-solving.

Shortcomings in the Implementation of the Model. A primary disadvantage of this implementation is that identifying a similarity rating with a knowledge source execution rating reduces a complex assessment to an information-losing numeric scale. Previous research on analogy and on case retrieval has cast doubt that similarity can be captured so simply, e.g., [Ashley, 1990], [Carbonell, 1986], [Falkenhainer et al., 1989], [Gentner, 1983], [Holyoak and Thagard, 1989]. On the other hand, on a serial computer, cases must be individually retrieved in some order, which implicitly ranks cases ordinally.

Secondly, it is not at all clear that every case should be proceduralized. Creating knowledge sources for rarely referenced cases or perspectives incurs computational overhead without apparent benefit. This problem will have to be addressed if case bases are to be scaled up to realistic levels, possibly consisting of tens or hundreds of thousands of cases [Schank, 1991]. Reserving procedural representation only for prototypical cases may present one way to deal with knowledge source proliferation.

Related Work

This project benefits from a long history of thought about the relative benefits of representing knowledge declaratively or procedurally, including [Anderson, 1983] (ACT), [Bobrow and Winograd, 1977] (frame-driven dialog), [Minsky, 1975] (procedural attachment), and [Schank and Abelson, 1977] (scripts). A more dynamic approach to cases was inspired by PANDEMONIUM [Selfridge, 1959]. Several systems have also used a blackboard architecture to combine CBR with other reasoning methods, but all have used a declarative representation for cases (FIRST [Daube and Hayes-Roth, 1988], PROLEXS [Oskamp et al., 1989] and ABISS [Rissland, et al. 1991].) The memory-based approach of [Stanfill and Waltz, 1986], which represents both rules and stored experience within the MBR paradigm, suggested the search for a uniform knowledge representation for hybrid systems with a CBR component.

Summary

Our approach tests the fit of cases and knowledge sources, reflecting an alternative model of case-based retrieval. This procedural, locally-indexed approach is characterized by similarity metrics that are local to cases, fine scheduling granularity for case retrieval, case-generated actions that are incorporated within cases themselves, and the uniform representation of knowledge in hybrid systems.

Acknowledgments

I thank Edwina Rissland, who also suggested the Consumer Reports automobile domain, Dan Corkill, Jamie Callan, Claire Cardie, Jody Daniels, Kevin Gallagher, Ellen Riloff, Zack Rubinstein, Dan Suthers and Richard Weiss. Thanks also to Blackboard Technology Group, Inc., Amherst, MA, for providing GBB v.2.0.

References

Anderson, J. R. (1983). *The Architecture of Cognition*. Cambridge, MA: Harvard.

Ashley, K. D. (1989). *Toward a Computational Theory of Arguing with Precedents: Accommodating Multiple Interpretations of Cases*. Vancouver, BC: ACM, 93-102.

Ashley, K. D. (1990). *Modelling Legal Argument: Reasoning with Cases and Hypotheticals*. Cambridge, MA: M.I.T.

Bareiss, E. R. (1989). *Exemplar-Based Knowledge Acquisition*. Academic Press.

Barletta, R. and Mark, W. (1988). Explanation-Based Indexing of Cases. *Proceedings, Case-Based Reasoning Workshop*. Clearwater Beach, FL: Morgan Kaufmann, 50-60.

Blackboard Technology Group (1991). *GBB Reference Manual, Version 2.0*. Amherst, MA: Blackboard Technology Group, Inc.

Bobrow, D. G. and Winograd, T. (1977). An Overview of KRL, a Knowledge

Representation Language. *Cognitive Science*, 1, 3-46.

Branting, L. K. (1991). *Integrating Rules and Precedents for Classification and Explanation: Automating Legal Analysis*. Ph.D. Thesis, AI Laboratory, Univ. Texas.

Carbonell, J. G. (1986). Derivational Analogy: A Theory of Reconstructive Problem Solving and Expertise Acquisition. In R. S. Michalski, J. G. Carbonell & T. M. Mitchell (Ed.), *Machine Learning: An Artificial Intelligence Approach*, 371-392. Los Altos, CA: Morgan Kaufmann.

Consumer Reports. (1991). *Annual Auto Issue, The 1991 Cars*, 56(4), 206-294.

Daube, F. and Hayes-Roth, B. (1988). FIRST: A Case-Based Redesign System in the BB1 Blackboard Architecture. *Proceedings, Case-Based Reasoning Workshop*. AAAI-88. St. Paul, MN.

Erman, L. D., Hayes-Roth, F., Lesser, V. R. and Reddy, D. R. (1980). The HEARSAY-II Speech Understanding System: Integrating Knowledge to Resolve Uncertainty. *Computing Surveys*, 12.

Falkenhainer, B., Forbus, K. and Gentner, D. (1989). The Structure-Mapping Engine: Algorithm and Examples. *Artificial Intelligence*, 41, 1-63.

Gentner, D. (1983). Structure-Mapping: A Theoretical Framework for Analogy. *Cognitive Science*, 7, 155-170.

Hammond, K. J. (1989). *Case-Based Planning: Viewing Planning as a Memory Task*. Boston, MA: Academic Press.

Holyoak, K. J. and Thagard, P. (1989). Analogical Mapping by Constraint Satisfaction. *Cognitive Science*, 13, 295-355.

Kolodner, J. L. (1983). Maintaining Organization in a Dynamic Long-Term Memory. *Cognitive Science*, 7(4), 243-280.

Kolodner, J. L. (1989). Judging Which is the "Best" Case for a Case-Based Reasoner. *Proceedings, CBR Workshop*. Pensacola Beach, FL: Morgan Kaufmann, 77-81.

Minsky, M. (1975). A Framework for the Representation of Knowledge. In P. Winston (Ed.), *The Psychology of Computer Vision* New York: McGraw-Hill.

Nii, H. P., Feigenbaum, E. A., Anton, J. J. and Rockmore, A. J. (1982). Signal-to-Symbol Transformation: HASP/SIAP Case Study. *AI Magazine*, 3(2), 23-35.

Oskamp, A., Walker, R. F., Schrickx, J. A. and Berg, P.H.v.d. (1989). PROLEXS, Divide and Rule: A Legal Application. *Proceedings, ICAIL-89*, Boston, MA: ACM.

Owens, C. (1989). Integrating Feature Extraction and Memory Search. *Proceedings, The 11th Annual Conference of the Cognitive Science Society*. Ann Arbor, MI: Lawrence Erlbaum.

Rissland, E. L., Basu, C., Daniels, J. J., McCarthy, J., Rubinstein, Z. B. and Skalak, D. B. (1991). A Blackboard-based Architecture for CBR: An Initial Report. *Proceedings, CBR Workshop*. Washington, DC: Morgan Kaufmann, 77-92.

Rissland, E. L., Kolodner, J. and Waltz, D. (1989). Case-Based Reasoning, Introduction. *Proceedings, CBR Workshop*. Pensacola Beach, FL: Morgan Kaufmann.

Rissland, E. L. and Skalak, D. B. (1991). CABARET: Rule Interpretation in a Hybrid Architecture. *International Journal of Man-Machine Studies*, 34, 839-887.

Rissland, E. L., Valcarce, E. M. and Ashley, K. D. (1984). Explaining and Arguing with Examples. *Proceedings, AAAI-84*, Austin, TX. AAAI.

Samuel, A. L. (1967). Some Studies in Machine Learning using the Game of Checkers II — Recent Progress. *IBM J. Research and Development*, 11, 601-617.

Schank, R. (1991). Where's the AI? *AI Magazine*, 12(4), 38-49.

Schank, R. and Abelson, R. (1977). *Scripts, Plans, Goals, and Understanding*. Hillsdale, NJ: Lawrence Erlbaum.

Selfridge, O. G. (1959). Pandemonium: A Paradigm for Learning. *Proceedings of the Symposium on the Mechanization of Thought Processes*, 511-529.

Stanfill, C. and Waltz, D. (1986). Memory-Based Reasoning. *Communications of the ACM*, 12(12), 1213-28.

Sycara, K. P. (1987). *Resolving Adversarial Conflicts: An Approach Integrating Case-Based and Analytic Methods*. Ph.D. Thesis, School of Information and Computer Science, Georgia Institute of Technology.

Veloso, M. M. and Carbonell, J. G. (1991). Variable-Precision Case Retrieval in Analogical Problem Solving. *Proceedings, Case-Based Reasoning Workshop, May 1991*. Morgan Kaufmann, San Mateo, CA.

Multicases: A Case-Based Representation for Procedural Knowledge*

Roland J. Zito-Wolf and Richard Alterman
Computer Science Department - Center for Complex Systems
Brandeis University, Waltham MA 02254
rjz@cs.brandeis.edu; alterman@cs.brandeis.edu

Abstract

This paper focuses on the representation of procedures in a case-based reasoner. It proposes a new method, the *multicase*, where several examples are merged without generalization into a single structure. The first part of the paper describes multicases as they are being implemented within the FLOABN project (Alterman, Zito-Wolf, and Carpenter 1991) and discusses some properties of multicases, including simplicity of use, ease of transfer between episodes, and better management of case detail. The second part presents a quantitative analysis of storage, indexing and decision costs based on a decision-tree model of procedures. This model shows that multicases have significantly reduced storage and decision costs compared to two other representation schemes.

Introduction

A currently popular reasoning paradigm for AI systems is case-based reasoning (CBR: Rissland and Ashley 1986; Stanfill and Waltz 1986; Alterman 1988, Kolodner 1983). CBR proposes to de-emphasize reasoning from general principles in favor of a more memory-intensive approach. The representation of large numbers of cases is crucial to practical applications of CBR. However, to date case representation alternatives have not been explored in a systematic way.

This paper examines representations for procedures in a case-based reasoner. Two basic organizations have been proposed: *individual cases* and *microcases*. Individual cases (e.g., Kolodner 1983; Lebowitz 1983; Hammond 1990; McCartney 1990) equate the unit of knowledge presentation, the *example*, with the unit of retrieval from memory, the *case*. Under this method, case retrieval returns a single complete example episode for the target task. Microcases (e.g., Stanfill and Waltz 1986; Langley and Allen 1990; Goodman 1991) convert each example into multiple cases, one for each step of the episode. Procedure retrieval occurs incrementally, one retrieval per procedure step. Hybrids of these methods have also been proposed (e.g., Redmond 1990; Robinson and Kolodner 1991).

Although many representations have been proposed, no serious analysis of these alternatives has been attempted. Intuitively, individual cases suffer from redundancy and fragmentation of knowledge, while microcases suffer from increased retrieval effort due to the expanded number of cases.

This paper advocates a new representational method, the *multicase*, in which related examples are merged into a single structure. This paper will take a first step toward quantifying the consequences of different case representation and indexing methods. Multicase organization reduces both storage and retrieval costs, facilitates transfer, and helps manage the accumulation of case detail.

The Multicase

A multicase representation for a procedure is similar to a decision tree, in that it describes sequences of actions and decisions that achieve a given result. A multicase for making photocopies is shown in Figure 1. We define a *decision point* (DP) as any point in a plan requiring selection among alternatives: choices of action, expectations about events, or determination of values for step parameters. Each decision point contains the knowledge relevant to making a single decision, represented in terms of cases. For example, the branch to the **get copy card** step (marked **A** in the figure) would be associated with example episodes in which that action was appropriate. To make a decision at a DP, the features of the current situation are matched against the cases stored with each alternative. (For simplicity, the cases associated with each DP are not shown.)

Contrast this to the individual-case representation, where each (distinct) photocopying episode would be stored in memory as a separate case. To represent all the possibilities captured in Figure 1, one would have to store a case for every possible path through the multicase, or at least a high proportion of them (e.g., McCartney 1990; c.f. Hammond 1990). For individual cases, procedure execution basically consists of a single

*This work was supported in part by the Defense Advanced Research Projects Agency, administered by the U.S. Air Force Office of Scientific Research under contract #F49620-88-C-0058.

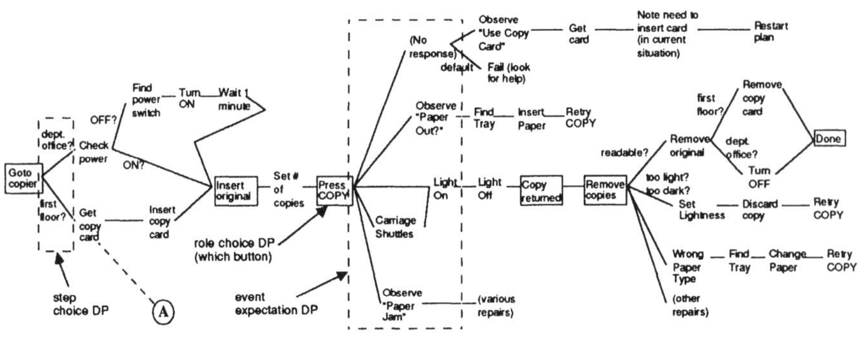

Figure 1: A Multicase for Photocopying

up-front decision among alternative cases (Figure 2a).

One popular variant of the individual-case method includes systems such as CYRUS (Kolodner 1983) and IPP (Lebowitz 1983) in which cases are organized via hierarchies of generalizations (MOPS, Schank 1982) as shown in Figure 3. A major disadvantage of this method is that it entails redundant index hierarchies having large storage requirements (which we will quantify later). Multicases reduce the need for generalization hierarchies by *conditionalizing* parts of the procedure – e.g., its steps and step parameters. A multicase corresponds to a generalized procedure, but **it is not a generalization** in the usual sense, as all of the details of its constituent examples are retained. While procedures containing branches are not new (e.g., Turner 1989, Robinson and Kolodner 1991) the substitution of conditionalization for generalization is novel.

The microcase approach breaks up the example multicase into many subcases, one (at least) for each step. The subcases will contain sufficient information to identify when each step should be executed, so that the structure of the photocopying procedure is encoded implicitly in the applicability conditions for the steps. Using microcases, procedure execution is a sequence of decisions, where each decision must decide among all the known steps (Figure 2b).

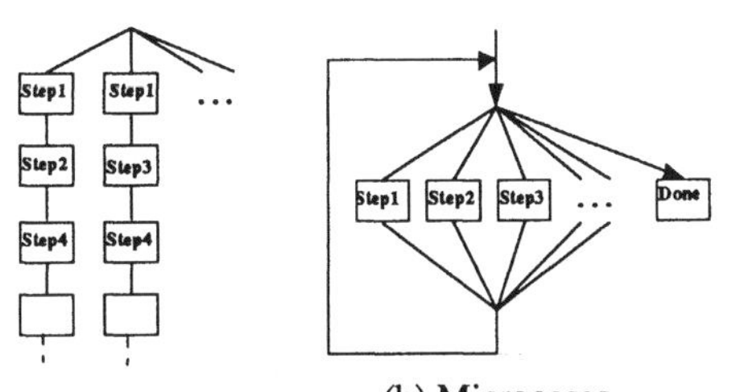

(a) Individual Cases (b) Microcases

Figure 2: Procedure Representations

A Multicase Implementation

FLOABN (Alterman Zito-Wolf, and Carpenter 1991) is a project exploring the acquisition of plans for using everyday mechanical and electronic devices, such as photocopiers and telephones. FLOABN acquires and revises plans via adaptation and through the interpretation of instructions and messages read or received during interaction with a simulated environment. The core of FLOABN is an adaptive planner SCAVENGER (Zito-Wolf and Alterman 1991) which uses a multicase plan representation. The multicase provides SCAVENGER with background knowledge for operationalizing instructions or messages received during execution.

The multicase is well-suited to this type of application. It is simple to execute as it is in a directly procedural form. Incorporating knowledge from new episodes is simplified because the multicase is stored in a compatible (i.e., episode-like) form. New episodes are first *assimilated* to an appropriate multicase through adaptation (Alterman 1988), after which the multicase is *elaborated* by adding DP's to account for any elements or decisions of the episode not already present in the multicase. An interesting feature of this domain is that over time the multicase becomes customized to the agent's normal situations of action. For photocopying, the plan becomes habituated to the photocopiers the agent normally uses.

The algorithm SCAVENGER uses to act and to learn

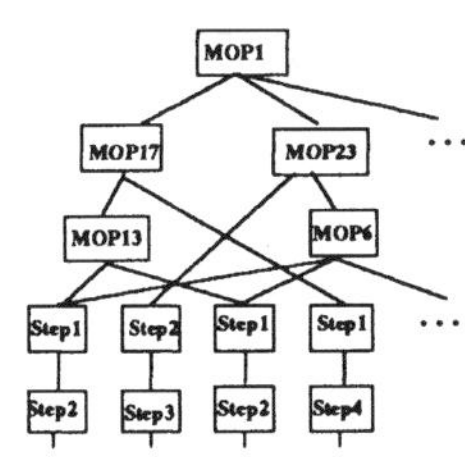

Figure 3: Individual Cases Plus MOPs

```
Given: a multicase-base and a goal
1. Select a multicase appropriate to the goal
2. Execute the next step specified by multicase
  a. if no steps remain to execute, then return(success)
  b. if type(current step(s))=EVENT then wait for one of the events to occur
     if excessive time passes (based on previous experience at this step), adapt or return(event failure).
  c. else    i. if multiple step alternatives exist
                then select one whose context best matches the current situation
             ii. Check preconditions (if missing then adapt, subgoal, or fail)
             iii. Execute step
             iv. Check post-conditions (either adapt, subgoal, or fail)
3. Check for unexpected events: if found then
  a. if adaptation limit exceeded then return(fail: situation too unfamiliar)
  b. if event is receipt of instructions relevant to current plan
     then interpret instructions (Alterman et al. 1991)
4. Whenever adding a step or alternative to the plan:
  a. Add decision point
  b. Find a way to discriminate from other alternatives at that point
  c. Call this the relevant context
```

Figure 4: Pseudo-Code for Multicase-Based Activity with Learning

using a multicase is summarized in Figure 4. The system begins with a skeleton procedure such as an agent might acquire by having the task explained to it or seeing it performed. The boxed steps in Figure 1 indicate this initial copying procedure. Each additional detail arises from some specific experience. Some experiences add new paths (e.g. running out of paper), some add detail to existing paths (e.g. observing lighting and movement as copies are made), and some modify existing steps or decision criteria (e.g., learning where to look for a power switch). Most paths through the multicase reflect contributions from several experiences.

Case Representation Comparisons

Storage Requirements While memory is becoming increasingly plentiful, it remains a finite resource. Individual cases have significant redundancy, as many steps will recur across cases. Worse, as the case-base fills up with variant episodes, retrieval becomes more expensive. Our analysis will show that multicases have the least storage requirements of the methods discussed.

Retrieval Cost The primary execution-time cost in case-base reasoning is in finding and processing the relevant examples. Retrieval cost is influenced by the number of cases searched and the complexity of matching them to the current situation.

Multicases, and to a lesser degree microcases, reduce the cost of each retrieval though partitioning of the procedure and the case-base, as will be shown later. In addition, the cost is distributed throughout procedure execution rather than incurred prior to execution.

These problems can *not* be solved simply though indexing, for two reasons. Indexing may not always be practical, due to, for example, the type of data involved. Second, our analysis points out that indexes involve costs of their own, and that not all indexes are created equal. Indexes require space on the same order as the case-base; hence individual cases have the largest index requirements. Moreover, schemes involving redundant index hierarchies (e.g., MOPs) have indexes much *larger* than the case base itself.

Level of Detail How much detail should be stored when a case is acquired is a perennial CBR dilemma. It is desirable to make case memory as detailed as possible, since the system cannot know in advance which features will prove important and which not. On the other hand, irrelevant details complicate matching, proliferate cases, and increase retrieval cost, so that one desires to store only "relevant" details.

Multicases permit a different approach to the management of detail, by allowing it to be acquired incrementally through the overlay of old episodes with newer ones. The learner can postpone the detail decision: since a multicase is always growing, there is no rush to recall any one case in complete detail.

Transfer When an exact match to a given situation is not available, relevant cases (i.e., partial matches) can usually still be found. Knowledge is in effect distributed over the relevant cases, and the reasoner needs to be able to select and transfer the relevant items of knowledge to the new situation.

Multicase methods facilitate transfer between episodes because a path through the multicase may be formed by plan modifications contributed by several different experiences. For the photocopier multicase, one episode might add the steps for using the ON switch, another those to fill the paper tray when it runs out, and a third the steps needed for doing reductions. Since the relevant information for performing a task is stored together, the need for patching together that information at execution time is largely avoided. Contrast this to individual cases, where transfer of knowledge across cases requires additional mechanisms such as abstraction of specific modifications into repair schema (e.g., Hammond 1990).

Microcases facilitate transfer also, by allowing any known step to be executed at any point in a procedure.

While this may be an advantage in domains that are described by a few underlying rules, such as a Tower-of-Hanoi type problem, it has the disadvantage (at least in domains like commonsense procedures) that each step selection must differentiate between every known step rather than the smaller set of steps known to be relevant in some past example.

Another measure for transfer is the ease of access to the underlying procedure. A multicase is simple to access as it is a single unit. In the other representations, procedures are represented in a distributed form, and must be reconstructed during execution.

Analysis of Storage and Decision Cost

This part of the paper presents a more formal comparison of the multicase, individual cases and microcases focusing on storage and decision cost. The approach is to first formally define a model of a procedure and then show how it would be encoded into cases using each representation method. Two assumptions of this model are that each retrieval returns exactly one case, and that irrelevant features are already "factored out".

Procedure Model

Let the procedure to be acquired be a complete binary decision tree T of uniform depth n. Each node $i \in T$ contains a *step* (or chunk of steps) to be performed plus a *decision* selecting the next node to be executed. T contains $|T| = 2^n - 1$ steps and $2^{n-1} - 1$ decisions (those in the leaves are ignored). Each procedure execution *episode* will consist of n steps and decisions along some path in T from root to leaf.

Let the input to the decision at a node i be the set of binary features F_i. Then $F = \bigcup_{i \in T} F_i$ is the set of all features referenced by the procedure. We assume there exists some upper bound $f = \max_{i \in T} |F_i|$ on the number of features tested by any specific decision, and that f is small compared to $|F|$. We have $n - 1 \leq |F| \leq (2^{n-1} - 1)f$ since there must be at least $n - 1$ features for all the paths in T to be distinguishable, while the upper bound applies in the case that all the F_i are disjoint. A reasonable estimate for $|F|$ should be proportional to f, and should allow for each feature to be referenced in some significant percentage of paths through the tree. We therefore estimate $|F|$ as $(n-1)f$, which models for example a procedure containing $n - 1$ distinct decisions in a fixed order.

Representation of Procedures

Case-based reasoning for procedure execution is the example-based selection of a sequence of steps to achieve a given goal. Each occasion for selection is a *problem* P_i, the process of searching through the case-base to solve a problem is a *retrieval*, and and the number of steps retrieved per problem is the *problem size* S_P. The solution to each problem will be encoded in memory as some set of *cases* C_P. The case is the unit of memory storage and retrieval. Each case pairs a problem solution with a conjunction of features for which it applies. Since it has been stipulated that a given decision references at most f features, at most 2^f cases will be required to represent a decision, one for each possible conjunction of the features and their negations.

We will first consider a linear search model of case retrieval[1], in which the *retrieval effort per problem* E_P is proportional to the number of feature tests made. E_P is the product of the number of cases to be searched through and the number of features to be tested per case. (We assume for simplicity that the cases can be used as retrieved, meaning that we do not attempt to account for adaptation costs.) Letting $|P|$ be the number of problems per episode, the *total retrieval effort* per episode $E = E_P|P|$.

The input parameters of the model are the depth n of the procedure tree T, the maximum number of features f referenced by a decision, and the total number of features $|F|$ referenced by the procedure. $|P|$ and S_P will vary with the specific representation. The outputs of the model are estimates for $|C|$, E_P and E.

Individual cases store each episode of (i.e., path through) T as a case, so $S_P = n$, the number of steps in an episode. Since the entire mapping from situation features to step sequence is performed in one retrieval, $|P| = 1$, with 2^{n-1} potential outcomes. The number of cases can be estimated from the total number of features referenced, yielding $|C| = 2^{(n-1)f}$.

Microcases represent a procedure as a set of independent decisions, making procedure execution a series of case retrievals, one per step. To encode T as microcases we make each selection of a step a separate problem. Then $S_P = 1$, and $|C_P| = 2^f$ cases per problem. There are $|P| = n$ problems per episode, but $2^n - 1$ problems to be encoded to represent the entire procedure, giving $|C| = 2^f(2^n - 1)$. Retrieval effort per problem is $E_P = (|F| + n)2^{n+f}$. Note that n additional features are added to distinguish the 2^{n-1} potential "current positions" within the represented procedure.

Multicases allow us to represent a procedure as a *sequence* of context relative decisions. The retrieval effort is divided up according to separate decisions, and the branching structure of the plan is expressed explicitly as part of the multicase rather than implicitly as extra features referenced by the cases. We have $S_P = 1$, $|P| = n - 1$ problems per episode, $|C_P| = 2^f$ with $2^{n-1} - 1$ problems overall, for a total[2] of $|C| = (2^f + 1)(2^n - 1)$.

Because we focus on only one decision at a time, the number of cases that must be searched through and features needing to be consulted at any given decision point are greatly reduced. For the multicase, only f features need be consulted per decision, and only 2^f cases need

[1] A model for indexed retrieval will introduced shortly.

[2] Since we are counting case nodes, the leaves of the tree add another 2^{n-1} steps.

	Individual cases		Microcases		Multicases	
Item	formula	example[1]	formula	ex.	formula	ex.
Total cases $\|C\|$	2^F	256	$O(2^{n+f})$	124	$O(2^{n+f-1})$	75
CB size[2]	$n2^F$	1280	$O(2^{n+f})$	124	$O(2^{n+f-1})$	75
Effort/problem E_P (unindexed)[3]	$F2^F$	2048	$O(F2^{n+f})$	1612	$f2^f$	8
Effort/problem E_P (indexed)	F	8	$O(n+f)$	7	f	2
Effort/episode E (indexed)	F	8	$O(n^2+F)$	35	F	8
Index size	2^{F+1}	512	$O(2^{n+f+1})$	256	$O(2^{n+f})$	128
Effort/problem E_P ($\|C\|$ fixed)	$\log_2 \|C\|$		$\log_2 \|C\|$		$\max(\log_2 \|C\| - n, f)$	

Notes: 1. Example figures are for a complete binary tree with $n = 5$, $f = 2$, $F = (n-1)f = 8$
2. Total case-base size is product of total decisions and case/decision.
3. Effort/problem is product of features/problem and number of cases per problem C_P.

Table 1: Storage and Retrieval Cost Summary

be examined; the rest of the cases are only relevant to *other decisions*. Thus $E_P = f \cdot 2^f$.

Indexing

Because case-retrieval via linear search involves effort exponential in case-base size, most CBR systems use some form of *indexing*[3] for faster retrieval. An index can be treated as a boolean discrimination network which tests just enough features to discriminate all the cases. Assuming that the network is balanced, the decision cost is proportional to the depth of the index, which is the log base 2 of the number of cases entering into a given decision: $O(nf)$ for individual cases, $O(n+f)$ for microcases, and $O(f)$ for multicases. This simple index model provides a lower bound on access costs; more complex retrieval processes – e.g., inexact matching, choosing among multiple retrieved cases, or features that interact in other than boolean combinations – will have larger decision costs.

The above model of indexing applies to an *optimal* index, that is, one which is (a) balanced, and (b) provides a single path to each case. If the first assumption is violated, the average access cost will be increased. In the second case, storage cost will be significantly increased. For example, a fully redundant generalization hierarchy (such as proposed by MOPs-based systems) involves $F! > (F/3)^F$ nodes[4], a quantity which is much larger than 2^F for any practical value of F.

There are a number of reasons why optimal indexing of the cases may not be practical in all situations: because the features to be indexed must be identifiable in advance for the index to be constructed; because the feature values must be enumerable (for example, features containing variables, features derived through calculations, and continuous features will in general not meet this requirement); and because efficient indexes require significant effort to update as new episodes are acquired. To the extent that indexing falls short, some degree of actual search through cases is required, and the numbers for unindexed decision cost apply.

It might be argued that the above formulas overestimate index costs since no CBR system will ever have available to it more than, say, 2^{15} cases, so that in practice index size is bounded. Let us explore this assumption. First, the relative costs of the indexing schemes are unaffected. Second, for a given $|C|$, individual-case and microcases will have decision effort $E_P = \log_2 |C|$, while multicases will have on the average $E_P = \log_2(|C|/2^n) = \log_2 |C| - n$, a significant improvement. Third, our original estimates for $|C|$ can now be used to estimate the coverage provided per additional example. The larger the space of possible cases, the harder it is to acquire a representative sample of the entire procedure T. Viewed this way, multicases yield the most "knowledge" per example.

Results

The results of this analysis are summarized in Table 1. The formulas derived here are for a complete case-base resulting after all of the possible situation configurations and hence procedure sequences have been observed. Although in general one's case-base is never complete, presumably it must contain a significant percentage of the relevant cases in order to perform adequately, so that the formulas given are expected to be of the correct order of magnitude. The table includes illustrative values for a complete decision tree with $n = 5, f = 2$.

Case-Base Storage. The three alternatives use different amounts of case-base storage. A multicase representation requires the least cases – $O(2^{n+f-1})$ – of the three methods; microcases require 2 times as much, and individual cases require about 2^{nf} times as much.

Effort. Two alternatives were evaluated. If complete indexing is not possible, multicases offer much better

[3]It is important to note that "indexing" is used in the CBR literature in at least three distinct senses: as a *performance* method, to accelerate access to a desired case or cases; as an *organizing* method, for grouping cases observed to have similar features, typically in the service of making generalizations (e.g., CYRUS and IPP); and as a *knowledge-encoding* method, for defining sets of cases with related content. We model the first, we do not model the second, and we assume the third can be encoded as additional features.

[4]Both IPP and CYRUS provide mechanisms for trimming away useless generalizations, but do not evaluate their effectiveness.

per-problem and per-episode retrieval cost than microcases or individual cases ($O(2^f)$ vs. $O(2^{n+f})$ and $O(2^{nf})$ respectively). If complete indexing *is* possible, multicases and individual cases have the same per-episode retrieval cost of $(n-1)f$, but the multicase reduces *per-step* retrieval cost by a factor of $O(n)$. Microcases are the most costly alternative, with $O(n+f)$ retrieval cost per problem and $O(n^2)$ per episode. This is a significant difference when executing a procedure under temporal constraints, where it is desirable to minimize computation per step as well as overall. If all the decision effort is lumped into one large computation, it may become an execution bottleneck.

Index Size. Multicases require less index space than the other two methods, though the difference between multicases and microcases is not large. A more significant advantage is that, if f is small compared to n (e.g, $f \leq 4$) multicases may permit one to avoid indexing entirely, saving not only space but the effort to construct and update them.

Concluding Remarks

A few points deserve drawing out. Representational choices in CBR systems *do have* a significant effect on performance and resource requirements as case-bases increase in size. The multicase embodies two key ideas: exploiting the underlying structure of the problem domain to partition case retrieval into a number of smaller, cheaper retrievals, and keeping the representation as "concrete" (episode-like) as possible (cf. McCartney 1990).

Our formal results depend on our assumption that $f \ll F$ – for procedures, that (a) most choices depend on only a few of the available features, and (b) most steps are only relevant at certain points in a process. Given these, partitioning is very effective. Not all procedures have this property; for example, the Tower-of-Hanoi problem can be described using just one rule which is applied at every step. Issues for future work include the impact of representation on the process of case acquisition, and the complexity and consequences of approximate matching during case retrieval.

The three CBR methods described here can be ordered by increasing constraint on the sequencing of steps. At one extreme, microcases allow any step to be chosen at any time; at the other extreme, individual cases fix entire step sequences. Microcases occupy an intermediate position. Though we have focused on procedure representation, this distinction may be useful in other areas. Consider a case-based design system in the domain of electronic amplifiers. The individual-case approach would correspond to a library of off-the-shelf designs. Microcases would correspond to general design rules for building amplifiers out of smaller functional units. Multicases would correspond to a library of designs plus knowledge of how to adapt them to suit various requirements. In a sense, each multicase would represent the "procedure" for customizing a particular design.

In summary, this paper makes several contributions to the analysis and evaluation of case-based reasoners. It defines an abstract model of CBR to which a variety of architectures can be fit and compared, and defines four criteria on which to evaluate such systems – case-base size, indexed and unindexed retrieval effort, and index size. It provides one of the first detailed complexity analyses of case representation and organization, and uses it to contrast several schemes appearing in the literature.

Acknowledgements

The first author thanks Marc Goodman for sharing his expertise in CBR systems in general and indexing in particular.

References

[1] Richard Alterman. Adaptive planning. *Cognitive Science*, 12:393–421, 1988.

[2] Richard Alterman, Roland Zito-Wolf, and Tamitha Carpenter. Interaction, comprehension, and instruction usage. *Journal of the Learning Sciences*, 1(4):361–398, 1991.

[3] Marc Goodman. A case-based, inductive architecture for natural language processing. In *AAAI Spring Symposium on Machine Learning of Natural Language and Ontology*, 1991.

[4] Kristian J. Hammond. Case-based planning: A framework for planning from experience. *Cognitive Science*, 14:385–443, 1990.

[5] Janet L. Kolodner. Reconstructive memory: A computer model. *Cognitive Science*, 7:281–328, 1983.

[6] Pat Langley and John A. Allen. Learning, memory, and search in planning. In *Proceedings of the Thirteenth Annual Conference of the Cognitive Science Society*, pages 364–369, Chicago, Illinois, 1991.

[7] Michael Lebowitz. Generalization from natural language text. *Cognitive Science*, 7:1–40, 1983.

[8] Robert McCartney. Reasoning directly from cases in a case-based planner. In *Proceedings of the Twelfth Annual Conference of the Cognitive Science Society*, pages 101–108, Hillsdale, NJ, 1990. Lawrence Erlbaum Associates.

[9] Michael Redmond. Distributed cases for case-based reasoning: Facilitating use of multiple cases. In *Proceedings of the Eighth National Conference on Artificial Intelligence*, pages 304–309, 1990.

[10] Edwina Rissland and Kenneth Ashley. Hypotheticals as heuristic device. In *Proceedings of the Fifth National Conference on Artificial Intelligence*, 1986.

[11] Stephen Robinson and Janet Kolodner. Indexing cases for planning and acting in dynamic environments: Exploiting hierarchical goal structures. In *Proceedings of the Thirteenth Annual Conference of the Cognitive Science Society*, pages 882–886, 1991.

[12] Roger Schank. *Dynamic Memory: A Theory of Reminding and Learning In Computers and People*. Cambridge University Press, Cambridge, 1982.

[13] Craig Stanfill and David Waltz. Toward memory-based reasoning. *Communications of the ACM*, 29(12):1213–1239, 1986.

[14] Roy M. Turner. A schema-based model of adaptive problem solving. Technical Report GIT-ICS-89/42, Georgia Institute of Technology, 1989.

[15] Roland Zito-Wolf and Richard Alterman. Ad-hoc fail-safe plan learning. In *Proceedings of the Twelfth Annual Conference of the Cognitive Science Society*, pages 908–914. Lawrence Erlbaum Associates, 1990.

Locally-to-Globally Consistent Processing in Similarity

Robert L. Goldstone
Indiana University
rgoldsto@ucs.indiana.edu

Abstract

SIAM, a model of structural similarity, is presented. SIAM, along with models of analogical reasoning, predicts that the relative similarity of different scenes will vary as a function of processing time. SIAM's prediction is empirically tested by having subjects make speeded judgements about whether two scenes have the same objects. The similarity of two scenes with different objects is measured by the percentage of trials on which the scenes are called the same. Consistent with SIAM's prediction, similarity becomes increasingly influenced by the global consistency of feature matches with time. Early on, feature matches are most influential if they belong to similar objects. Later on, feature matches are most influential if they place objects in alignment in a manner that is consistent with other strong object alignments. The similarity of two scenes, rather than being a single fixed quantity, varies systematically with the time spent on the comparison.

Introduction

The similarity of two things is not simply a relation between the two things; it is a relation between the two things and the comparison-maker. Similarity assessments must be constructed by a process that compares the items in question. Sometimes the process is straightforward. The similarity of cigars and cigarettes is easily determined. Determining the more abstract similarity between cigarettes and time bombs (Ortony, Vondruska, Foss, & Jones, 1985) seems to take a longer time. The fact that similarity develops along a time course suggests that similarity does not immediately impinge upon our perceptual system. Instead, perceptual and cognitive processes actively build a conception of similarity.

The time course of similarity assessments provides a useful tool for investigating the comparison process. If we dispatch with the assumption that similarity is "out there" in the objective world, then the question of "How does similarity develop?" becomes crucial. One method for understanding how comparisons are made is temporal analysis.

Dynamic Models of Similarity

General models of similarity have not often addressed temporal aspects of processing (Carroll & Wish, 1974; Tversky, 1977). These models do not consider similarity to be a dynamically evolving quantity. Instead, their equations for similarity give single "endpoint" estimates. However, specific process models have been developed for some specialized tasks. For example, similarity has often been measured by the time elapsed, or the errors made, when subjects determine if two displays are different. The assumption made is that the longer it takes to respond that the displays are different, or the more times that different displays are erroneously thought to be the same, the more similar the displays are. Specific processing mechanisms have been hypothesized to account for how this speeded same/different task is executed (for a review, see Farrell, 1985). The speeded same/different task will be used to measure similarity in the experiment to be reported. The speeded same/different task cannot replace subjective ratings as a method for investigating similarity, but it does provide a converging measure that is relatively immune to experimenter demands and subjects' high-level reasoning strategies.

Recently, a general model of similarity has been developed called SIAM that also hypothesizes a dynamic time course for comparisons (Goldstone 1991; Goldstone & Medin, in press). According to SIAM (Similarity as Interactive Activation and Mapping), when structured scenes are compared, the parts of one scene are aligned, or placed in correspondence, with the parts of the other scene. Emerging correspondences influence each other as processing continues. With sufficient time, the strongest correspondences will be those that are consistent with other correspondences. Similarity is determined by a process of interactive activation between feature and object correspondences. The degree to which features from two scenes are placed in correspondence depends on how strongly their objects are placed in correspondence. Reciprocally, how strongly two objects are placed in correspondence depends on the correspondence strength of their features.

The details of SIAM are discussed elsewhere (Goldstone, 1991). Essentially, SIAM's network architecture is composed of nodes that excite and inhibit each other. Nodes represent hypotheses that two entities correspond to one another in two scenes. For the present purposes, two types of nodes are important: feature-to-feature nodes, and object-to-object nodes. Each feature-to-feature node represents an hypothesis that two features correspond to each other. One feature-to-feature node is assigned to every possible pair of alignable features. As the activation of a feature-to-feature node increases, the two features referenced by the node will be placed in stronger correspondence. Object-to-object nodes represent hypotheses that two objects correspond to each other.

Network activity starts by features being placed in correspondence according to their physical similarity. After this occurs, SIAM begins to place objects into correspondence that are consistent with the feature correspondences. As objects begin to be put in correspondence, activation is fed back down to the feature (mis)matches that are consistent with the object alignments. In this way, object matches influence activation of feature matches and feature matches influence the activation of object matches concurrently.

Activation spreads in SIAM by two principles: 1) nodes that are consistent with one another send excitatory activation to each other and 2) nodes that are inconsistent inhibit one another. Nodes are inconsistent if they produce many-to-one mappings, and are consistent otherwise. Processing in SIAM starts with a description of the scenes to be compared. Scenes are described in terms of objects that contain feature slots that are filled with particular feature values. On each "slice" of time (cycle), activation spreads between nodes. Nodes that are highly active are weighted heavily in the similarity assessment.

SIAM shares architectural commonalities with McClelland and Rumelhart's (1981) interactive activation model of word perception and Marr and Poggio's model of depth perception (1979), and is highly related to the SME (Falkenhainer, Gentner, and Forbus, 1990) and ACME (Holyoak and Thagard, 1989) models of analogical reasoning. In ACME, SME, and Marr and Poggio's model, there are pressures against developing many-to-one mappings, and pressures in favor of developing mutually consistent mappings. The models of McClelland and Rumelhart, Holyoak and Thagard, and Marr and Poggio are all examples of what Marr (1982) calls "cooperative algorithms." Cooperative algorithms create globally consistent mappings by local interactions between units. SME also moves from locally determined mappings to globally consistent mappings with more processing. As we will see, SIAM incorporates a similar local-to-global processing principle.

A Behavioral Prediction of SIAM

In SIAM, object correspondences depend on feature and object correspondences[1]. SIAM

[1] In the full version of SIAM, object correspondences also depend on role

initially begins to place objects in correspondence on the basis of their featural overlap; the more featural commonalities two objects have, the more strongly they will be placed in correspondence. However, the strength of an object correspondence is also influenced by its consistency with other object correspondences. If two objects from one scene correspond to a single object in another scene, then the two correspondences are inconsistent and will decrease each others' strength. SIAM, like ACME and SME, predicts that object correspondences will become increasingly influenced by other object correspondences with time, as activation spreads between nodes .

One prediction of this temporal processing is that feature matches that are inconsistent with the set of globally consistent correspondences should tend to influence similarity less with time. Globally consistent feature matches should become more influential with time. A set of mappings between objects is globally consistent if it a) yields only one-to-one mappings, and b) maximizes the number of matching features that belong to corresponding entities. Even though object A from scene 1 may be most similar to object B from scene 2, these objects may not be a part of globally consistent set of mappings. In particular, if other objects from scene 1 are also fairly similar to B, and other objects from scene 2 are fairly similar to A, and if we only allow one-to-one correspondences, then placing A in correspondence with B may not maximize the number of feature matches between aligned objects.

In SIAM, object correspondences will first be based on feature matches, the only information available. Objects that are featurally similar will begin to be placed in correspondence. With time, object correspondences will be inhibited by inconsistent object correspondences, and excited by consistent object correspondences. By these interactions, object correspondences that are consistent with many other object correspondences become stronger. In turn, the feature matches that belong to these globally consistent correspondences will receive more weight. In this manner, the global consistency of feature matches comes to influence similarity more with increased processing time.

correspondences that serve to align objects that play similar roles in their scenes.

Experimental Support for a Local-to-global[2] Processing Shift

To test the influence of processing time on globally consistent and inconsistent feature matches, subjects are shown pairs of scenes; sample scenes are shown in Figure 1. Each scene contains two butterflies, and each butterfly contains four features. Subjects must decide whether the two scenes contain the same butterflies within a specified deadline. A symbolic representation is shown below each of the butterflies in Figure 1. For example, the target scene is composed of butterflies "AAAA" and "BBBB," where the letters refer to different values along the four dimensions {body shading, head type, tail type, wing shading}. The butterfly "XABA" has feature matches on the second and fourth dimensions with "AAAA", and a feature match on the third dimension with "BBBB."

First, consider trials in which the target scene is compared to the base scene. Both of the butterflies in the target scene have more matching features in common with the base scene's left butterfly than right butterfly. The base butterfly "BABA" has two matches in common with both of the target scene's butterflies. Thus, if we only consider the locally preferred mappings, we would map both target butterflies onto the top butterfly of the base. However, if the global consistency of object mappings is maintained, then this many-to-one mapping is not permitted. The best globally consistent mapping is to map the left butterflies onto each other, and the right butterflies onto

[2] The term "local-to-global," as used here, is only distantly related to previous researchers' (e. g. Navon, 1977) claim for a processing shift from global (holistic) to detailed/analytic similarities. The current claim concerns the increasing importance of globally consistent features matches on similarity.

each other. In short, "BBBB" corresponds to "BABA" if we only consider local feature matches, but "BBBB" corresponds to "XXXB" if we consider the influence that object correspondences can have one another.

The target scene is also compared to two derivatives of the base scene. Each derivative differs from the base scene by only a single feature. For Figure 1A, one of the local feature matches is removed, leaving all of the globally consistent matches intact. For Figure 1B, one of the globally consistent matches is removed, and all of the local matches are preserved. The empirical questions of primary interest is "Is the target scene more similar to the scene in Figure 1A or 1B, and does the relative similarity of the scenes depend upon the processing time allowed?"

Thirty-three undergraduates were presented with 608 displays each. On half of the trials, two copies of the target scene were displayed. On these trials, the subjects' correct response was "same." On the other half of the trials, displays consisted of the target scene and one of the other three scenes shown in Figure 1. Butterfly position, dimension order, dimension values, scene location, and display type were all randomized. Displays were presented on Macintosh SE30 computers.

The subjects' task was to press a key with one hand if the butterflies of one scene were the same as the butterflies of the other scene, and to press a key with the other hand if the two scene's

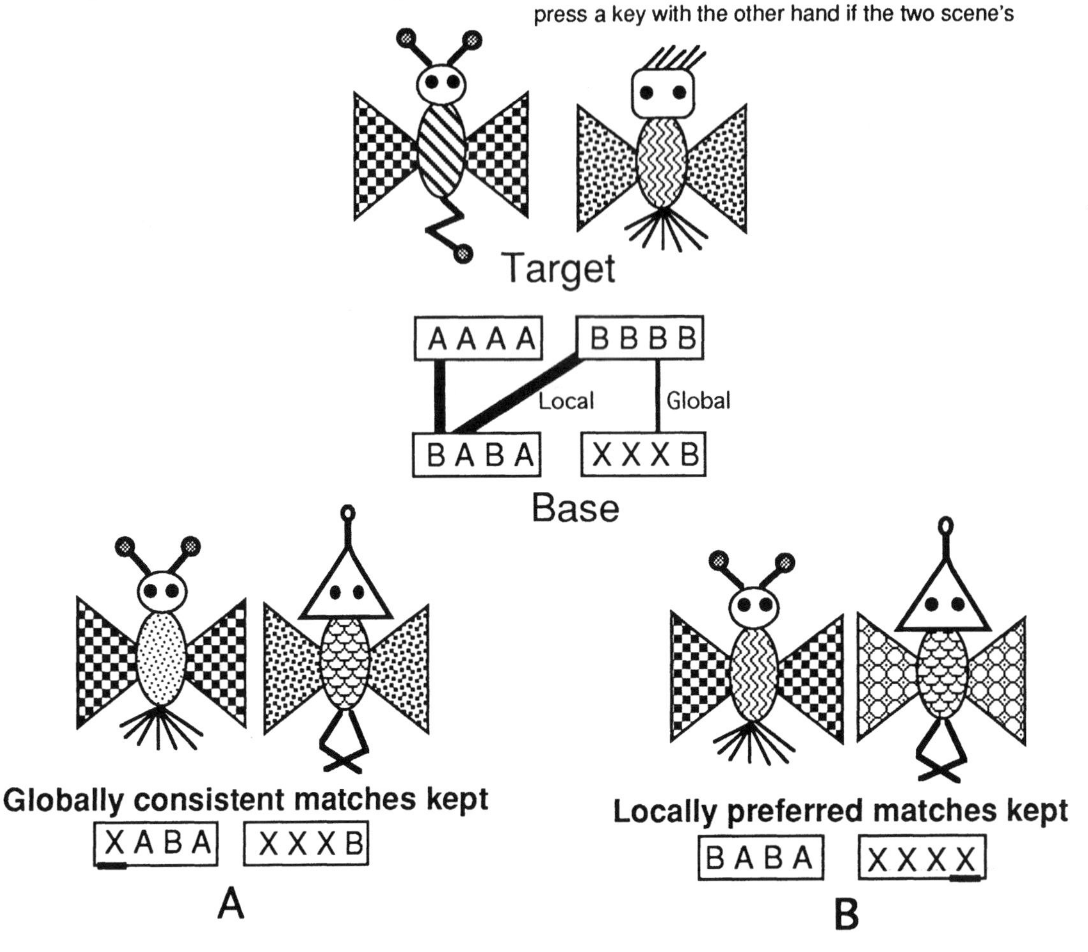

Figure 1. Sample scenes used in the experiment.

butterflies were different. It was stressed to subjects that the same butterflies did not have to be in the same positions in their scene in order to respond "same." The experiment was divided into 19 blocks. On each block, subjects were given a "very fast," "fast," or "fairly slow" deadline (1, 1.84, and 2.68 sec respectively). If a subject did not respond before the deadline passed, the message "OVERTIME" appeared on the screen.

The significant (F (4, 288) = 3.94, mse = 8, p<.05) cross-over interaction between deadline and type of display is shown in Figure 2. If subjects are forced to respond within a short deadline, the display that preserves the locally preferred match is more often incorrectly responded to as "same" than the display that preserves the globally consistent match. The opposite effect is found when subjects are given longer to respond. The four mean error rates of particular interest are: slow-deadline/global-matches-kept = 5%, slow-deadline/local-matches-kept = 3%, fast-deadline/global-matches-kept = 18%, and fast-deadline/local-matches-kept = 21%. A planned comparison of these four data shows a significant interaction between deadline and type of scene on error rate (F (1,288) = 3.46, mse=6.8, p<.05). The overall times to correctly respond "Different" to the different displays are not significantly different (base = 1.147 sec, global match kept = 1.137 sec, local match kept = 1.135 sec).

Implications

If similarity is measured by the percentage of trials that scenes with different butterflies are incorrectly judged to be the same, then the obtained results are consistent with SIAM's prediction. More incorrect "same" judgments are found for short deadlines when local matches are preserved. More incorrect "same" judgments are found for the longest deadline when global matches are preserved. This is consistent with SIAM's dynamic account of similarity. The influence of one object-to-object mapping on another takes time to develop, and until it is developed, object-to-object mappings will be largely determined by feature-to-feature matches. Locally consistent matches are more important than globally consistent matches for similarity early in processing (fast deadline). Later in processing, globally consistent matches gain in importance relative to local matches. At first, both butterflies of the target are mapped onto one butterfly of the other scene, but with time the influence of one mapping redirects the other mapping.

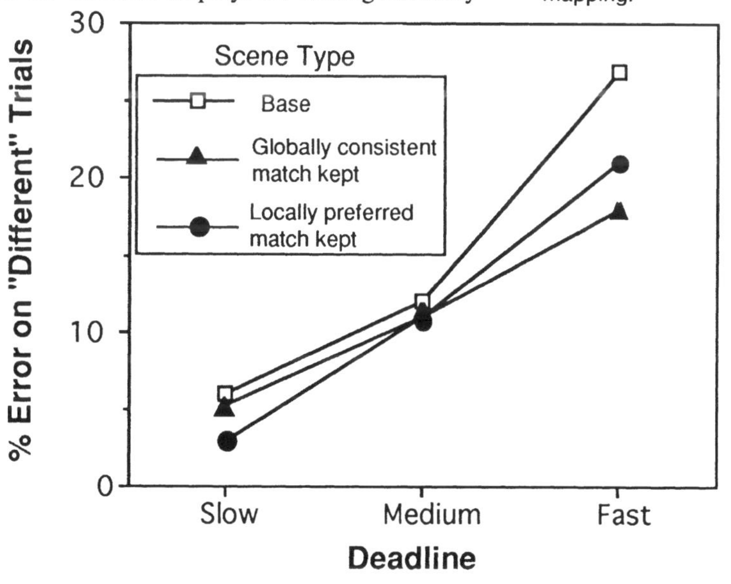

Figure 2. Results showing an interaction between deadline and type of scene.

The experiment indicates that featural similarity cannot completely predict object correspondences. Objects will tend to be aligned if they share many features, however object alignment also depends on the similarity of other objects pairs. Butterfly BBBB from the target scene of Figure 1 is most similar to butterfly BABA of the base scene, but it is placed in proper alignment with butterfly XXXB. BBBB corresponds to XXXB and not BABA because BABA is also similar to the target scene's other butterfly, AAAB. By aligning BBBB with XXXB and BABA with AAAB, the number of matching features between consistently (one-to-one mapping) aligned objects is maximized. With increased processing time, SIAM and subjects both seem to base object correspondences more on global consistency than on the local similarity of objects.

The experiment supports a notion of similarity as constructed over time. In fact, the results are problematic for any model that hypothesizes that two entities have a single process-independent similarity value. We cannot assign single estimates for the similarity of the target scene and Figure 1A, and the target scene and Figure 1B, because at different times each is more similar than the other. Figure 1A is more similar to the target scene on slow deadlines, but Figure 1B is more similar to the target scene on fast deadlines. The similarity of two entities seems to depend on the particular mechanisms of the comparison process. In the current case, comparisons seem to involve a process in which locally determined correspondences give way to globally consistent ones. More generally, the outcome of a comparison seems to depend not just on the things compared, but also on the process that is doing the comparing.

Acknowledgements

This paper has benefitted greatly by many useful suggestions by Douglas Medin, Dedre Gentner, Ed Smith, Keith Smith, and Keith Holyoak. The research was supported by NSF Grant BNS-87-20301 awarded to Dedre Gentner and Douglas Medin.

References

Carroll, J. D., & Wish, M. (1974). Models and methods for three-way multidimensional scaling. In D. H. Krantz, R. C. Atkinson, R. D. Luce, & P. Suppes (Eds.) Contemporary developments in mathematical psychology (Vol. 2, pp. 57-105). San Francisco:Freeman.

Falkenhainer, B., Forbus, K.D., & Gentner, D. (1990). The structure-mapping engine: Algorithm and examples. Artificial Intelligence, 41, 1-63.

Farell, B. (1985). "Same"- "Different" judgments: A review of current controversies in perceptual comparisons. Psychological Bulletin, 98, 419-456.

Goldstone, R. L. (1991). Similarity, Interactive Activation, and Mapping. Unpublished doctoral dissertation, University of Michigan.

Goldstone, R.L., & Medin, D.L. (forthcoming). Interactive Activation, Similarity, and Mapping. in K. Holyoak and J. Barnden (Eds.) Advances in Connectionist and Neural Computation Theory, Vol. 2: Connectionist Approaches to Analogy, Metaphor, and Case-Based Reasoning. Ablex

Holyoak, K. J., & Thagard, P. (1989). Analogical mapping by constrain satisfaction. Cognitive Science, 13, 295-355.

Marr, D. (1982). Vision. San Francisco: Freeman.

Marr, D., and Poggio, T. (1979). A computational theory of human stereo vision. Proceedings of the Royal Society of London, 204, 301-328.

McClelland, J. L., & Rumelhart, D.E. (1981). An interactive activation model of context effects in letter perception: Part 1. An account of basic findings. Psychological Review, 88, 375-407.

Navon, D. (1977). Forest before trees: The precedence of global features in visual perception. Cognitive Psychology, 9, 353-383.

Ortony, A., Vondruska, R. J., Foss, M. A., and Jones, L. E. (1985). Salience, similes, and the asymmetry of similarity. Journal of Memory and Language, 24, 569-594.

Tversky, A. (1977). Features of similarity. Psychological Review, 84, 327-352.

Goal-Directed Processes in Similarity Judgement

Hiroaki Suzuki
Human Communication Research Centre
University of Edinburgh
2 Buccluech Place
Edinburgh EH8 9LW Scotland
hsuzuki@cogsci.ed.ac.uk

Hitoshi Ohnishi
Department of Systems Science
Tokyo Institute of Technology
4259 Nagatsuda-Cho Midori-Ku
Yokohama, Japan
ohnishi@sys.titech.ac.jp

Kazuo Shigemasu
Department of Education
Tokyo Institute of Technology
Ohokayama Meguro-Ku
Tokyo, Japan
kshigema@nc.titech.ac.jp

Abstract

This study explored the effects of a goal and subject's knowledge in similarity judgements. We hypothesized that the process of computing similarity consist of two phases: the processes of explanation and feature comparison. When a goal is salient and the knowledge required to achieve it is available, people compute similarity by explaining the goal in terms of a given state by using domain knowledge. Thus, in this case, rated similarity should be a function of the distance between the goal and the state. When the explanation fails, the judgements should instead to be based on the feature comparison. Expert, novice, and naive subjects were asked to solve the Tower of Hanoi puzzle. The subjects were required to judge the similarity between the goal and various states of the puzzle. The results showed that their judgements differed, depending on their expertise. While experts' ratings were best characterized by the number of operators necessary to transform a given state to the goal, those of naive subjects were completely based on the number of shared features. The second experiment revealed that the experts' judgements of similarity are not be due to learned contiguity through practice.

Introduction

People's flexible and intelligent behavior is enabled by their ability to use appropriate and relevant past experiences. Since knowledge required to deal with a situation is likely to have application in similar situations, similarity is often a good measure of the relevance and the appropriateness of such experiences. People retrieve the knowledge of a past experience which is similar to the present situation, and use specific facts or procedures stored in the knowledge to understand, explain, and learn from, the situation.

In this sense, judgements of similarity are ubiquitous in every kind of cognition. Actually, categorization has been considered as processes in which people compute similarities between an instance and its prototype (Smith, 1990). Problem solving may also be characterized as a process by which similar experiences are retrieved and modified (Hammond, 1990). Transfer, one of the central issues in the study of learning, can be thought of as a matter of which types of similarities learners should, or are apt to, attend to — deep ones (school of formal discipline) or shallow ones (Thorndykian). Whichever similarities learners may attend to, it is central to the issue of transfer.

If judgements of similarity are involved in these various kinds of cognition, such judgements cannot be stable. In fact, similarity changes dynamically across situations. For example, studies on novice-expert differences have shown that experts' judgements of similarity of one problem to another are fundamentally different from novices' (Chi, Feltovich, & Glaser, 1981). Smith (1990) suggested that young children's preference for global similarity is gradually replaced by dimensional similarity. Moreover, people's judgements of similarity are highly context sensitive. Tversky and Gati (1982) showed that rated similarity between two countries varied depending on the type of stimulus sets in which they were included. In addition, Goldstone, Medin, & Gentner (1991) revealed that people are sensitive to relational similarities when relations are more salient, and are sensitive to attributional similarities when attributes are salient. In Pylyshyn's terms, similarity judgements are *cognitively penetrable* (Pylyshyn, 1984).

Although the studies described above consistently show that the selection of feature changes across situations, current theories of similarity do not take it into account. For example, the "contrast model" (Tversky, 1977) does not provide any constraints on which features are selected and which features should be considered salient. This problem becomes more serious when one notices that the "frame problem" (McCarthy & Hayes, 1969) also matters in similarity judgement. Since there are potentially infinite number of features, it is impossible to collect all of the features of an object by asking people to list them. In addition, people create idiosyncratic features so that a man who gets drunk sometimes "jumps into a swimming pool with all his clothes on" (Murphy & Medin, 1985).

The second problem is that most of the experiments in the previous studies were conducted in rather "neutral" settings where "disturbing" factors such as subjects' prior knowledge, or goals spontaneously generated by subjects, were carefully removed. Although this approach might be suitable for the investigation of object-level similarity, the results cannot easily be extended to problem-solving and learning. This is because goals and knowledge play critical roles in these activities (Glaser, 1984; Resnick, 1990).

Therefore, a model of similarity should be developed that can deal with the above-mentioned problems. In problem-solving and learning, people have to focus on important features: What determines the importance of each feature? Features contributing to the achievement of the goal are those which should be judged to be important. However, it is not sufficient simply to have the goal, because it is quite often the case that features satisfying the goal are not readily accessible. Thus, the next question is: What determines the degree to which a feature contributes to the goal. It is knowledge of the domain which determines the degree of the contribution to the goal. If features satisfy the triggering condition of knowledge which contributes to achieving the goal, these features should be considered to be contributing to the goal.

The above analysis leads to the idea that the process of computing similarity is *explanation*, because people *explain* the goal, using given features and the domain knowledge (Mitchell, Keller, & Kedar-Cabelli, 1986). For example, suppose that you are looking for an ashtray, and that there is a paper cup, a juice can which has not been opened yet, and a cookie can with some cookies in. The goal, knowledge, and given features of each object in this case are as follows:

Goal:		
ashtray(X)	←	heat-proof(X)
	∧	open-concavity(X)
	∧	unimportant-in(X)
Knowledge		
heat-proof(X)	←	can(X)
unimportant-in(X)	←	empty(X)
empty(X)	←	mv-content(X,¬X)
open-concavity	←	can(X)
heat-proof(X)	←	can(X)
heat-proof(X)	←	in(water, X)
⋮	⋮	⋮
Features		
can(juice-can)		
can(cookie-can)		
white(paper-cup)		
delicious(cookie)		
⋮		

The three conditions of the goal are satisfied by transforming the features of the cookie-can, using the domain knowledge of the shape and material. As a result, recognition of similarity between the cookie can and an ashtray is obtained. It is important to note that irrelevant features such as the color and taste of objects are not picked up for the computation of similarity in this process.

It is often the case that there is a difference of complexity between one explanation and another. For example, one can make the paper cup an ashtray by pouring juice into the cup. However, the required explanation is more complex in this case than in the previous case. Generally, an item whose features require more transformation is judged to be less similar. This would be a source of a degree of similarity. Complexity of the explanation may partly be affected by what kinds of knowledge people have. If one has well-organized and readily accessible knowledge, the derived explanation is likely to be much simpler. We may therefore expect to find a difference between experts and novices in similarity judgement.

What happens if the explanation fails? There are two cases when the explanation fails: a case where no explicit goal is concerned, and that where required knowledge is not accessible. In this case, the only available information is features of items to be compared with. As a result, judgement of similarity is carried out on the basis of features.

Thus, we hypothesize that processes of computing similarity consist of two subprocesses. In the first phase, people judge the degree of similarity by explanation. A degree of similarity in this phase is defined as the number of operators that is required to explain one item in terms of the goal. When the goal is not obvious, or relevant operators are not easily accessible, the second subprocess follows. In this phase, people's judgements are based on features shared with or specific to items, as is modeled by the contrast model.

There are several advantages to this model. First, as is obvious, the model can be applied to problem-solving and learning where goal and knowledge play dominant roles. Second, the model can provide an account of feature selection. Since, in theory, one can create an infinite number of attributes which characterize given items, it is crucial for models of similarity to select relevant features. In this model, these are resolved by relevance of features to the goal. Thus, f in the contrast model is no longer defined adhoc in this model.

In order to explore the effects of a goal and knowledge on similarity judgement, we conducted a series of experiments using the Tower of Hanoi puzzle. Subjects were asked to rate similarities of a given state to the goal where all disks were placed at the right peg. The reason why we chose this puzzle is that it is easy to specify the goal, the knowledge (operators), as well as features.

According to our hypothesis, whether subjects know the rule of the puzzle determines whether judgements are based on explanation or feature comparison. If subjects know the rules of the puzzle, judged similarity should be a function of the distance, that is, the number of operators required to transform a given state to the goal. On the other hand, if subjects do not know the rule, the judgements should be carried on the basis of the number of shared features. In addition, the model assumes that the accessibility of the operators also affects similarity judgements. It is likely that novices' judgements are based on the number of shared features, since those who have just been taught the rules would find it more difficult to access an appropriate operator than experts. Therefore, it is predicted that while experts' judgements should be best characterized by the number of operators, novices' ones should be based on both the number of operators and on the number of shared feature.

Experiment 1

Method

Subjects Subjects were 21 Tokyo Institute of Technology graduate and undergraduate students. They were randomly assigned to one of the three conditions: expert, novice and control. None of the subjects in the novice or control conditions had any prior experience with the Tower of Hanoi puzzle.

Procedure Subjects in the expert condition first read instructions that described the goal, available operators, and constraints of the Tower of Hanoi puzzle. Then they proceeded to the training session. In this session, they were given the puzzles with varying initial states and required to solve them within 15 seconds. The goal was fixed so that all disks were placed on the rightmost peg. After subjects could solve them successfully, they proceeded to the next session: pre-judgement session.

In this session, subjects were given a twenty-six page booklet. On each page, one of 26 states of the three-disk Tower of Hanoi puzzle was printed, paired with the goal where all disks were placed on the rightmost peg. Subjects were asked to rate how similar the states were, and to circle "7" if the pictures were very similar, "1" if they were least similar, and other numbers for the intermediary degrees of similarity. Subjects were instructed to respond as quickly as possible. After practice, subjects were given another booklet which consisted of three blocks of 26 pairs. Thus, subjects were required to compare the 78 pairs in the same way as they had done in the previous session.

The procedure for the novice condition was basically same, except that there was no training session. Thus, they read the instructions of the puzzle, then proceeded to the pre-judgement session, and finally rated the similarities of each 78 pair. The control group performed the pre-judgement session and the similarity judgements only. Thus, they had no idea that the presented stimulus was the puzzle. The orders of the stimulus presentation in the pre-judgement and final session were randomized across subjects.

Results and Discussion

In order to examine the effects of the number of shared features and operators, we calculated Spearman's rank order correlation coefficients between the rated similarity and the number of operators and the number of shared features. The number of operators was defined as the distance between a given state and the goal in the problem space of the puzzle. The number of shared features was calculated by adding attributional and relational similarities. The degree of attributional similarity was the number of disks on the target (rightmost) peg. That of relational similarity was the number of *on*-relations. For

Table 1: Spearman's rank order correlation coefficients of rated similarity and the number of features and distance

	No. Shared Feature	No. Distance
Expert	0.328**	−0.534**
Novice	0.425**	−0.150*
Control	0.624**	−0.034

Note: * shows $p < .05$, ** $p < .01$
r_s between the distance and the feature is 0.038.

example, suppose a state where the largest disk is located on the leftmost peg and the other disks are on the rightmost peg. In this case, the degree of attributional similarity is 2 because two disks are on the rightmost peg, and that of relational similarity is 1 because the smallest disk is *on* the medium disk.

The results are shown in Table 1. The similarity ratings in the expert condition are greatly affected by the number of operators, although the number of shared features also affects the ratings. In contrast, ratings in the control condition are based solely on the number of shared features. The more features two states have in common, the more they are judged to be similar. The performance of the novice condition is in between the expert and control conditions. Although the effects of operators are observed, the judgements are oriented mainly by the number of shared features. Additionally, we performed separated ANOVAs, taking the distance and the conditions as the independent variables, and the rated similarity as the dependent variable. All the ANOVAs show significant interaction (for one shared feature, $F(6, 240) = 3.38(p < 0.01)$; for two shared features, $F(2, 120) = 3.80(p < 0.05)$; for three shared features, $F(4, 159) = 6.41(p < 0.01)$).

Our hypothesis is supported by the results that the number of operators affects the rating greatly for the expert condition, moderately for the novice condition, and hardly for the control condition. The differences between the control and the other conditions suggest that the recognition of the goal causes subjects to compute the similarity by explanation. The difference between the expert and novice conditions suggests that the accessibility of the appropriate operators determines whether the judgements are carried out by the explanation or the feature comparison. The reason why the effect of the number of shared features was found even in the expert condition may be due to the fact that the experts carried out the judgements by feature comparison when the required explanation was very complicated.

Although we concluded that differences between the novice and expert are attributed to the difference in the accessibility of operators, there might be an alternative interpretation. Since subjects in the expert condition have a lot of opportunities to observe a sequence of solutions in the training session, they may recognize that states closer to the goal are always contiguous to the goal. Therefore, judgements of the experts might reflect the contiguity rather than the accessibility of operators.

Experiment 2

In order to examine whether the expert's performance reflects the learned contiguity, we conducted another experiment where comparisons were made between two expert groups.

It is well known that there are several different strategies to solve the Tower of Hanoi puzzle (Simon, 1975). One of the strategies, called the "perceptual strategy", can be described as follows: To construct the tower of the disks on the target peg, the largest disk must be placed on the target peg first, then the next largest, and so on. This strategy does not always specify the appropriate operator at each state, because some moves directed by the strategy violate the constraints of the puzzle. However, this strategy gives the better understanding of the subgoal structure for solving the puzzle. Another strategy is called "move-pattern strategy." This strategy can be described as follows: On odd-numbered moves, move the smallest disk; On even-numbered moves, move the next-smallest disk that is exposed; The smallest disk is always moved from the left to the right to the center to the left peg, and so on. As is obvious from the above description, this strategy is quite the opposite of the perceptual strategy. This strategy always specifies the appropriate operators. However, this strategy is rather mechanical or rote, in a sense that people do not have to recognize the subgoal structure at all. What is necessary for the strategy is only to keep track of the parity of the move and the cycling direction for the smallest disk (Simon, 1975).

What happens if the two strategies are used in the similarity judgement? Since a subject using the perceptual strategy understands the subgoal structure, he or she is likely to give good, but not exact, estimates of the number of operators to achieve the goal. For example, when the first subgoal of the strategy has not been achieved yet, he or she may judge the given state to be less similar to the goal. When the second subgoal has been achieved, the state may be judged to be very similar to the goal. On the other hand, a subject who uses the move-pattern strategy may have difficulties in estimating the distance. If a subject tries to explain the given state, he has to move the disks mentally and count the number of operators to be applied. Since it places a substantial burden upon working memory, it is likely that the subject would give up the explanation and shift to feature comparison.

It is important to note that the solution paths are identical between the two conditions. Thus, the "contiguity" hypothesis predicts that there is no difference between the two, because subjects in both conditions observed approximately the same number of the sequence of states in the practice session. On the other hand, our model predicts differences between the two. Subjects who use the perceptual strategy should be more sensitive to the number of operators required to achieve the goal, because they are more likely to recognize the subgoal structure which provides a good basis for the estimation of the distance. By contrast, a subject who uses the move-pattern strategy is likely to judge the similarity on the basis of feature comparison. That is because mentally executing this strategy places a substantial burden upon working memory.

Since the pilot study revealed that the differences were very subtle, we made several changes in order for the experiment to be sensitive to possible differences. First, the 7 points scale of the rating in Experiment 1 was replaced with a 10 points scale. In addition, the five-disk Tower of Hanoi was used for rating, so as to avoid the "ceiling effect."

Method

Subjects Twelve undergraduate students were randomly assigned to the subgoal or rote condition. None of them had experienced with the Tower of Hanoi puzzle prior to the experiment.

Procedure Subjects in both conditions first read instructions which described the rules of the puzzle. Then they were given a description of the strategies to be learned: Subjects in the subgoal condition read the description of the perceptual strategy; Subjects in the rote condition read the description of the move-pattern strategy. Then subjects were asked to understand the procedure. When they did not understand it, an experimenter taught them the strategy according to the instruction. After reading it, they were required to solve the three-disk puzzle, using the taught strategies. The initial state of the practice was fixed so that all the disks were placed on the leftmost peg. If subjects solved the puzzle within ten seconds without mistakes, they were allowed to proceed to the next session. After the practice for rating, subjects were given a nine-page booklet, and asked to judge the similarity of a state to the goal. The goal and one of the nine states were printed on each page. Subjects were asked to judge the similarity of the pairs, as quickly as possible. The nine states that were used for the comparison were selected to approximately balance the number of features and the distance from the goal.

Results and Discussion

It took 116 seconds for subjects in the subgoal condition and 126 for those in the rote condition to solve the five-disk puzzle. This suggests that there is no difference in efficiency of the strategy use in both conditions.

However, Spearman's rank order correlation coefficients between the rated similarity and the number of operators show that there exist differences between the two groups of subjects (for the subgoal condition, $r_s = -0.479 (p < 0.01)$; for the rote condition, $r_s = -0.219 (p = 0.11)$.

These results indicate that the differences observed in the experiment 1 could not be attributed to mere recognition of contiguity. The difference is due to the understanding of the subgoal structure which provides a good basis for estimating the number of operators necessary to achieve the goal.

General Discussion

Similarity must be sensitive to goals, since it is involved in various kinds of human activities in which goals play privileged roles. The experiments presented here clearly show that the judgements of similarity are affected by the goal and knowledge of operators. When the goal is salient, people's judgements of similarity are carried out by *explanation*, sensitive to the number of operators required to transform given states to the goal. On the other hand, judgements come to be based on the number of shared features when there is no explicit goal, or when relevant knowledge to achieve the goal is not readily accessible.

By incorporating goal and knowledge into the model of similarity, we can provide adequate accounts for several phenomena found in people's judgement of similarity. First, our model has direct relevance to the "surface-structural" argument in studies of similarity. Gentner & Landers (1985) found that while people retrieved superficially similar stories in a memory recall task, they tended to choose structurally similar ones in tasks which required the rating of the soundness of analogy between stories. The shift from superficial to structural similarity can be attributed to the fact that there is no explicit goal in the recall task, whereas the goals and the solutions are salient in the rating of analogical soundness. As we suggested before, the recognition of the goal leads people to compute similarity by explanation. In this case, the subjects were sensitive to the goal-subgoal hierarchy which corresponds to the "structure" of the task. This would be the reason why subjects' ratings were based on the structure of the stories in the soundness rating.

More evidence of the "surface-structural" distinction comes from studies on expert-novice differences. It is well known that whereas experts attend to structural aspects of problems, novices attend to superficial ones. These results can be explained by the accessibility of knowledge. In the categorization task in Chi et al's experiments (Chi, Feltovich, & Glaser, 1981), novices seemed to know that they should attend to structural similarity among problems because they had been taught an elementary physics. However, the lack of the appropriate knowledge of physics which related one problem to another caused them to compute the similarity via on the number of shared features.

Our model has much in common with the MAC/FAC model (Gentner & Forbus, 1991). The MAC/FAC model consists of two stages. While in the MAC stage, computationally cheap matchers act on content vectors of items in LTM, structural examinations are made in order to compute "deep" similarity in the FAC stage. The MAC and FAC stages correspond to the feature comparison and the explanation, respectively. This suggests that theories which aim at modeling the processes of similarity judgements in problem-solving contexts should have two subprocesses to compute deep as well as shallow similarities.

However, there are several differences between the two. First, the role of the goal is not explicitly mentioned in the MAC/FAC model. Although the MAC/FAC might be able to explain the effects of goal by modifying content vectors, an initial set of features has to be changed. In contrast, our model explains the effects not by modifying the description of objects, but by adding the goal and the operators. Second, although the accessibility of appropriate operators determines whether the judgements are carried out by explanation or feature comparison, there seem no mechanisms in the MAC/FAC model to explain the effects.

Acknowledgements

This research was partly supported by the British Council. We thank Dedre Gentner, Jon Oberlander, Keith Stenning and Sam Stern for their helpful comments.

References

Chi, M. T. H., Foltovich, P. J., and Glaser, R. 1981 Categorization and representation of physics problems by experts and novices. *Cognitive Science* **5**:121–152.

Gentner, D. and Forbus, K. D. 1991 MAC/FAC: A model of similarity-based retrieval. In Proceedings of the Thirteenth Annual Conference of the Cognitive Science Society, August, 1991.

Gentner, D. and Landers, R. 1985 Analogical reminding: A good match is hard to find. *Proceedings of the International Conference on Systems, Man, and Cybernetics.* Tucson, AZ.

Glaser, R. 1984 Education and thinking: The role of knowledge. *American Psychologist* **39**:93–104.

Goldstone, R. L., Medin, D. L. and Gentner, D. 1991 Relational Similarity and the Nonindependence of Features in Similarity Judgments. *Cognitive Psychology* **23**:222–262.

Hammond, K. J. 1990 Case-based planning: A framework for planning from experience. *Cognitive Science* **14**:385–443.

McCarthy, J. and Hayes, P. J. 1969 Some philosophical problems from the standpoint of artificial intelligence. *Machine Intelligence* **4**:463–502.

Mitchell, T. M., Keller, R., and Kedar-Cabelli, S. 1986 Explanation-Based Generalization: A Unifying View. *Machine Learning* **1**:47–80.

Pylyshyn, Z. W. 1984 *Computation and Cognition: Toward a Foundation for Cognitive Science.* Cambridge, MA.: MIT Press.

Resnick, L. B. ed. 1990 *Knowing, Learning, and Instruction.* Hillsdale, NJ.: Erlbaum.

Smith, E. E. 1990 Categorization and induction, In M. I. Posner ed. *Foundations of Cognitive Science.* Cambridge, MA.:MIT Press.

Smith, L. B. 1989 From global similarities to kinds of similarities: The construction of dimensions in development. In S. Vosniadou. and A. Ortony eds. *Similarity and Analogical Reasoning.* Cambridge, MA.: Cambridge University Press.

Simon, H. A. 1975 The functional equivalence of problem solving skills. *Cognitive Psychology* **7**:268–288.

Tversky, A. 1977 Features of Similarity. *Psychological Review* **84**:327–352.

Tversky, A. and Gati, I. 1978 Studies of similarity. In E. Rosch and B. Lloyd eds. *Cognition and Categorization.* Hilldale, NJ: Lawrence Erlbaum Associates.

Correlated properties in artifact and natural kind concepts

Ken McRae
Department of Psychology
Meliora Hall
University of Rochester
Rochester, NY 14627
kenm@psych.rochester.edu

Abstract[1]

Property intercorrelations are viewed as central to the representation and processing of real-world object concepts. In contrast, prior research into real-world object concepts has incorporated the assumption that properties are independent and additive. In two studies, the role of correlated properties was explored. Property norms had been collected for 190 natural kinds and artifacts. In Experiment 1, property intercorrelations influenced performance in a property verification task. In Experiment 2, concept similarity, as measured by overlap of *independent* properties, predicted short interval priming latency for artifacts. In contrast, concept similarity, as measured by overlap of *correlated* property pairs, predicted short interval priming for natural kinds. The influence of property intercorrelations was stronger for natural kinds because they tended to contain a higher proportion of correlated properties. It was concluded that people encode knowledge about independent and correlated properties of real-world objects. Presently, a Hopfield network is being implemented to explore implications of allowing a system to encode property intercorrelations. Finally, results suggest that semantic relatedness can be defined in terms of property overlap between concepts.

Property-based representations of concepts were used to investigate aspects of the semantic representation of simple real-world objects, such as DOG and COUCH. Models based on conceptual primitives or properties have been used in a number of areas of cognitive science, including vision (Biederman, 1987), lexical representation (e.g., McClelland & Rumelhart, 1981), and concepts and categorization (e.g., Gluck & Bower, 1988; Hintzman, 1986; Kruschke, 1992; Rosch & Mervis, 1975; Malt & Smith, 1984).

In previous investigations of real-world object concepts, the simplifying assumption has been made that concepts are composed of independent and additive properties (e.g., Rosch & Mervis, 1975). In direct contrast, a number of people have claimed that property co-occurrences are an important organizing principle for objects in the world (e.g., Boyd, 1984; Keil, 1989; Malt & Smith, 1984; Medin, 1989; Rosch et al., 1976). In fact, it has been claimed that concepts are organized around clusters of intercorrelated properties. For example, the concept BIRD is assumed to be organized around a cluster of intercorrelated properties that includes: HAS WINGS, HAS FEATHERS, FLIES, and HAS A BEAK.

Evidence for encoding and use of property co-occurrence information has been found in **artificial** concept experiments by Medin and Schwanenflugel (1981), Medin et al. (1982), and Younger and Cohen (1983). Due to these findings, recent computational models of categorization have incorporated mechanisms capable of simulating effects of correlated properties (e.g., Gluck, Bower, & Hee, 1989; Hintzman, 1986; Kruschke, 1992; Nosofsky, 1986). However, only one study has investigated the influence of correlated properties in **real-world** concepts. Malt and Smith (1984) found **no** evidence to support the hypothesis that correlated properties predict typicality judgements above and beyond the predictive power of independent properties. Therefore, at this point in time, despite the fact that property intercorrelations are viewed as central to concepts and categorization, no empirical evidence exists to suggest that they are encoded in conceptual representations of real-world objects. The purpose of this paper was to explore potential effects of property intercorrelations in semantic tasks involving real-world object concepts.

[1] This research was partially supported by an FCAR (Quebec) graduate fellowship and an NSERC (Canada) postdoctoral fellowship.

Target property: HUNTED BY PEOPLE

Concept	DEER		DUCK	
	Correlated Property	**Shared variance (%)**	**Correlated Property**	**Shared variance (%)**
	1) is herbivorous	71.9	1) an animal	22.0
	2) has antlers	68.3	2) lives in water	14.9
	3) lives in the woods	49.2	3) migrates	12.9
	4) lives in the wild	43.3	4) swims	10.9
	5) a mammal	23.5		
	6) an animal	22.0		
	7) is brown	14.8		
	8) has hooves	10.2		
	9) has 4 legs	10.0		
	10) has fur	6.8		
	11) has legs	6.6		

strength (DEER) = 71.9+68.3+49.2+43.3+23.5+22.0+14.8+10.2+10.0+6.8+6.6 = 326.6
strength (DUCK) = 22.0+14.9+12.9+10.9 = 60.7

Note: shared variance refers to percentage of shared variance between HUNTED BY PEOPLE and the listed property.

Figure 1 - Computing strength of intercorrelation

Semantic property norms were collected for 19 exemplars from 10 categories, 190 in total. There were 4 natural kind categories: BIRDS, MAMMALS, FRUITS, and VEGETABLES; and 6 artifact categories: CLOTHING, FURNITURE, KITCHEN ITEMS, TOOLS, VEHICLES, and WEAPONS. Three hundred subjects were divided into 10 groups. Each group was asked to produce lists of properties for 19 of the 190 exemplars. The exemplars and property lists were used to create stimuli and representations for the following experiments.

Experiment 1 - Property Verification

I tested the hypothesis that intercorrelational strength among properties within a concept influences performance in a property verification task. In this task, a concept name (e.g., DEER) was presented, followed by a property name (e.g., HAS ANTLERS). A subject was asked to indicate whether or not the property was reasonably true of the object to which the concept name referred. For example, the correct response was "yes" to DEER-HAS ANTLERS, but "no" to HORSE-HAS SCALES.

To select stimuli, a target property was paired with two concepts: one in which the correlations between the target property and other properties within it were as **strong** as possible, and a second concept in which the correlations between the target property and other properties within it were as **weak** as possible. It was hypothesized that a property that was strongly intercorrelated with other properties within a concept would be more easily verified than a property that was weakly intercorrelated.

Method

Subjects. Twenty McGill university undergraduates were paid $2 each to participate.

Materials. The stimuli were 37 pairs of concepts, each sharing a target property. The intercorrelations between the target property and other properties in one concept were as strong as possible, but were as weak as possible in the matched concept. For example, as shown in Figure 1, HUNTED BY PEOPLE is more strongly intercorrelated with properties of DEER than properties of DUCK. The stimuli were chosen using the following algorithm. Properties that were part of fewer than three concepts were removed, then proportion of shared variance was computed between each property pair. Proportion of shared variance was computed between properties represented as 190-element vectors of production frequencies (1 element per concept). Strength of intercorrelation of a property within a concept was measured as the sum of the proportion of shared variance between the target property and all properties within the concept with which it was significantly correlated. This computation is illustrated in Figure 1.

Mean intercorrelational strength differed significantly between groups (strong: 175; weak: 20). Because a critical property appeared in both groups, all variables associated with it, such as reading time, were equated on an item by item basis. The groups were also equated for production frequency of a target property, concept familiarity, and number of properties produced for a concept. Two lists were constructed so that a subject saw each target property only once. Further methodological detail can be found in McRae (1991).

Procedure. Each trial proceeded as follows: a 1500 ms intertrial interval; an asterisk in the centre of the screen for 500 ms; a blank screen for 100 ms; the concept for 400 ms; and the property until the subject responded. Subjects were instructed to press the "yes" key as quickly and accurately as possible if the property was reasonably true of the concept, or press the "no" key otherwise.

Results

One-tailed paired samples t-tests were used to determine the effect of intercorrelational strength on decision latency and (square root) errors. Subjects were faster to judge that a property was part of a concept if there were strong intercorrelations (820 ms) than if there were weak or no intercorrelations (912 ms), $t(19) = 4.22$, $p < .0003$ by subjects, $t(36) = 4.23$, $p < .0002$ by items. Furthermore, fewer errors were committed for items in the strong condition (4.3%) than in the weak condition (13.0%), $t(19) = 4.34$, $p < .0003$ by subjects, $t(36) = 3.24$, $p < .002$ by items.

Discussion

A clear effect of property intercorrelations was found on the ease with which properties were verified. Property verification necessarily involves computation of an exemplar concept and computation of the meaning of a property. The effect of property intercorrelations was attributed to computing the meaning of the target property, rather than computing the exemplar concept. Given that production frequency is assumed to indicate the strength with which a property is activated when a concept name is read, there should be no difference between groups due to computing exemplar concepts because production frequency was equated across groups. In contrast, if computing the meaning of a property is viewed as initiating a pattern completion process, its effect on verification is apparent. If co-occurrence information is encoded in memory, then when a property's representation is computed, properties correlated with it become activated. Because the item pairs were constructed so that the target property was more highly intercorrelated with the properties of one concept, the pattern for that concept would have been completed to a greater extent by the target property, resulting in an easier verification decision. For example, when the meaning of HUNTED BY PEOPLE was computed, it more strongly activated properties associated with wild mammals that have antlers and live in the woods, than properties associated with birds. In summary, it can be concluded that knowledge about independent and correlated properties is encoded in memory for the meaning of words.

Experiment 2 - Short Interval Semantic Priming

If property intercorrelations are encoded in semantic memory, then concepts must be represented in terms of independent and correlated properties. Therefore, similarity between concepts may depend on shared correlated property pairs, as well as independent properties. I investigated this possibility with a short interval semantic priming task. Each target appeared with a similar prime (e.g., EAGLE-HAWK) and a dissimilar prime (e.g., SANDALS-HAWK). Short interval semantic priming should be a sensitive measure of concept similarity for two main reasons. First, a number of studies strongly suggest that short interval priming is sensitive to effects associated with computing the meaning of the prime and target (e.g., de Groot, 1984; den Heyer, Briand, & Dannenbring, 1983; Neely, 1977). Second, because each target appears with a similar and a dissimilar prime, it serves as its own control. Therefore, variance due to the decision, as well as variance due to nuisance variables such as target length and frequency, can be factored out in the analyses. It was hypothesized that the magnitude of priming effects can be predicted by concept similarity in terms of independent and correlated properties.

Method

Subjects. Forty-eight McGill university undergraduates were paid $3 each to participate.

Materials. Ninety similar prime-target pairs served as stimuli, 9 similar prime-target pairs from each of the 10 categories (e.g., TRUCK-VAN, EAGLE-HAWK). In every case, the prime and target had been normed by different subjects. In each of the 4 tasks, a subject made a semantic decision to the target, the decision being: "is it animate?", "is it an object?", "is it made by humans?", or "Does it grow?". Two lists were constructed so that no subject saw a target twice.

Further methodological detail can be found in McRae (1991).

Procedure. Each subject performed all 4 tasks, which were blocked. Each trial proceeded as follows: a 1500 ms intertrial interval; an asterisk in the centre of the screen for 250 ms; a blank screen for 250 ms; the prime for 200 ms; a mask, &&&&&&&&&, for 50 ms; and the target until the subject responded. Subjects were instructed to silently read the prime, then respond as quickly and accurately as possible to the target. Examples were given to portray what was meant by: an animate thing (something that is alive, a cockroach is but a keyboard is not); an object (something that is tangible or concrete, a pair of scissors is but the sky is not); something made by humans (something that is manufactured by people, a razor is but a butterfly is not); and something that grows (something that grows on its own, a spider grows but a door does not).

Results

The frequency of errors for each task was less than 4% and were not further analyzed. Two items, EMU and STARLING, were removed because it was apparent from this and other studies that their referents were unknown to many subjects.

Subjects were faster to make decisions to targets preceded by similar primes in 3 of the 4 tasks: "is it animate?" (63 ms priming effect), $F(1, 188) = 13.10$, $p < .01$ by subjects, and $F(1, 96) = 12.46$, $p < .01$ by items; "is it an object?" (50 ms), $F(1, 188) = 9.98$, $p < .01$ by subjects, and $F(1, 96) = 15.70$, $p < .01$ by items; and "is it made by humans?" (53 ms), $F(1, 188) = 6.93$, $p < .01$ by subjects, and $F(1, 96) = 8.82$, $p < .01$ by items. The priming effect in the "does it grow?" task was not significant (23 ms), $F(1, 188) = 3.40$, $p > .5$ by subjects, and $F(1, 96) = 1.66$, $p > .1$ by items.

Regression analyses. In the independent properties representation, each concept was represented as a list of individual properties weighted by production frequency. Properties that were produced for an exemplar by a minimum of one-sixth of the subjects were included, resulting in 1,242 properties to describe the 190 exemplars. Concept similarity was computed by measuring the cosine of the angle between concept vectors in 1,242-dimensional property space. To construct a correlated properties representation, the set of 1,242 properties was first reduced to the 240 that were part of three or more concepts. Proportion of shared variance between each pair of properties was computed; it measured the tendency for two properties to occur together in the world (as described by the norms). The correlated properties representation contained one to ten vectors for each pair of significantly correlated properties ($p < .01$), contingent upon amount of shared variance. The number of units assigned to each property pair is shown in parentheses: (10) $R^2 > .964$; (9) $R^2 > .864$; (8) $R^2 > .764$; (7) $R^2 > .664$; (6) $R^2 > .564$; (5) $R^2 > .464$; (4) $R^2 > .364$; (3) $R^2 > .264$; (2) $R^2 > .164$; (1) $R^2 > .064$.

With this algorithm, properties that participated in a greater number of correlated property pairs received greater representation, as did properties that participated in highly correlated pairs. A concept was represented as a pattern across the resulting 2,630 vectors, including 1,190 unique property pairs. A value was given to each unit on the basis of a three-part function. If a concept had neither property from the correlated property pair, the unit was set to 0. If a concept possessed both properties, the unit was set to the sum of the production frequencies. If a concept contained one of the properties, then it violated the correlation and the unit's value was the negated production frequency of the property that was possessed by the concept. Therefore, the more strongly a concept possessed one property of the pair without the other, the greater the violation. For example, the correlated property pair, FLIES-HAS FEATHERS, shared 42.5% of their variance. The pair was represented by 4 units. According to the norms, a CARROT neither FLIES nor HAS FEATHERS (activation = 0); an EAGLE both FLIES and HAS FEATHERS (activation = 13 + 16 = +29), and an OSTRICH HAS FEATHERS, but never FLIES (activation = -22). Concept similarity was measured by the cosine of the angle between concept vectors in the 2,630-dimensional space defined by these pairs.

In the regression analyses, prime-target concept similarity was used to predict mean response latency for similar prime-target pairs (LAMP-CHANDELIER) after mean response latency for dissimilar prime-target pairs (GOOSE-CHANDELIER) had been entered. Similarity for dissimilar prime-target pairs was excluded from the regression analyses because it carried little information. Regression analyses were conducted separately for the complete set of 88 stimulus pairs, the 54 artifact pairs, and the 34 natural kind pairs. Similarity, as measured by overlap of independent properties, was used to predict amount of priming. Because any existing theory assumes that independent properties are encoded in semantic memory, the burden of proof was placed on demonstrating an effect of correlated properties; the predictive ability of correlated properties was tested with concept similarity in terms of independent properties already forced into the regression equation. The independent properties predicted a significant proportion of variance in priming effects for the 88 prime-target pairs ($R^2 = .125$, $p < .001$), but the correlated properties did not ($R^2 = .004$, $F < 1$). More importantly, overlap of independent properties accounted for a significant proportion of variance of priming effects for artifact prime-target pairs ($R^2 = .150$, $p < .005$), but not for natural kinds ($R^2 = .040$, $p > .2$). Conversely, for natural kinds ($R^2 = .211$, $p < .009$), but not for artifacts ($R^2 = .003$, $F < 1$), overlap of property intercorrelations accounted for a significant proportion of residual variance of priming effects.

Discussion

A clear influence of property intercorrelations on concept similarity was found. Artifact similarity is captured by independent properties, and natural kind similarity by the correlations among them. These results are consistent with those found by McRae (1991) in a same/different category decision task.

Predictions for natural kinds were more dependent on property intercorrelations for two reasons. A greater proportion of natural kind property pairs (10.9%) than artifact pairs (5.8%) were significantly correlated ($p < .01$). Natural kinds also had much richer representations over property pairs. On average, they possessed a significantly greater number of correlated property pairs than did artifacts (83 versus 36). In contrast, there was no difference in terms of individual properties (natural kinds: 17; artifacts: 15). Furthermore, on average, for the stimuli of Experiment 2, natural kind prime-target pairs shared a significantly greater number of correlated property pairs (55) than did artifact prime-target pairs (22).

Why did successful prediction by correlated properties preclude prediction by independent properties? Concept similarity in terms of independent properties increases monotonically with number of shared properties. In contrast, it was possible for an additional shared property to **decrease** concept similarity in terms of correlated property pairs. If $concept_a$ possessed both properties of a correlated pair, then $concept_b$ was more similar to $concept_a$ if it possessed neither property than if it possessed one property of the pair (i.e., violated the correlation). Predictions based on the independent properties representation may have been affected to the extent that these cases were present. Across the 88 similar prime-target pairs used in Experiment 2, the mean number of cases where one concept possessed both properties of a pair and the other possessed only one was significantly greater for natural kinds (43) than for artifacts (17). Therefore, concept similarity in terms of independent properties failed to make reliable predictions about natural kinds because of its insensitivity to shared and violated correlated property pairs.

In summary, because natural kinds possess a rich structure of property intercorrelations, similarity among natural kind concepts is largely determined by overlap of correlated property pairs. In contrast, similarity among artifacts is largely determined by independent properties; because few property intercorrelations exist for artifacts, they carry little weight in determining concept similarity. These results provide support for an hypothesis in the philosophical literature concerning differences between natural kinds and artifacts. Boyd (1984) has argued that natural kinds are natural precisely because they are structured around sets of "contingently clustered" properties. That is, Boyd has claimed that the network of causally interconnected properties is much richer for natural kinds than for artifacts, confirmed empirically in the analyses reported above. Similarly, Keil (1989) has claimed that natural kinds have an "essence", which is the underlying source of a dense network of property intercorrelations. In contrast, artifact properties have weaker causal interconnections because they cohere around intended function of the creator.

Semantic relatedness. Experiment 2 is also important from the perspective of investigating the source of short interval semantic priming effects. To date, these effects have been attributed to "semantic relatedness", but the source of semantic relatedness has not been clearly defined. For example, in a recent study by Hodgson (1991), he compared priming effects for six types of semantic relatedness. He found that no type of relatedness produced consistently greater priming effects. I have shown that short interval semantic priming effects can be predicted by concept similarity as measured by independent properties and correlated property pairs. In other words, semantic relatedness can be defined, at least in part, by property-based concept similarity.

Conclusions

Property intercorrelations are viewed as an important element of human knowledge. Although it has been found that property intercorrelations affect performance in artificial concept studies, and recent models of categorization have incorporated mechanisms that can account for these effects, an effect of correlated properties on semantic tasks involving real-world object concepts had not been demonstrated. The experiments reported here provide strong evidence that property intercorrelations are encoded in conceptual representations of real-world objects and affect performance on semantic tasks. Therefore, any reasonable model of the representation and use of concepts must incorporate a mechanism by which co-occurrences among properties are encoded. Currently, Virginia de Sa and I are exploring implications of using a Hopfield (1982) network to learn and process property-based conceptual representations.

References

Anderson, J. R. (1991). The adaptive nature of human categorization. *Psychological Review*, **98**, 409-429.

Biederman, I. (1987). Recognition-by-components: A theory of human image understanding. *Psychological Review*, **94**, 115-147.

Boyd, R. (1984). Natural kinds, homeostasis, and the limits of essentialism. Paper presented at Cornell University.

de Groot, A. M. B. (1984). Primed lexical decision: Combined effects of the proportion of related prime-target pairs and the stimulus-onset asynchrony of prime and target. *The Quarterly Journal of Experimental Psychology*, **36A**, 253-280.

den Heyer, K., Briand, K., & Dannenbring, G. L. (1983). Strategic factors in a lexical-decision task: Evidence for automatic and attention-driven processes. *Memory & Cognition*, **11**, 374-381.

Gluck, M. A., Bower, G. H., & Hee, M. R. (1989). A configural-cue network model of animal and human associative learning. In *Proceedings of the Eleventh annual Conference of the Cognitive Science Society*, Ann Arbor, Michigan. August 16-19, 1989 (pp. 323-332). Hillsdale, NJ: Erlbaum.

Hintzman, D. L. (1986). "Schema abstraction" in a multiple-trace memory model. *Psychological Review*, **93**, 411-428.

Hodgson, J. M. (1991). Informational constraints on pre-lexical priming. *Language and Cognitive Processes*, **6**, 169-264.

Hopfield, J. (1982). Neural networks and physical systems with emergent collective computational abilities. *Proceedings of the National Academy of Science, USA*, **79**, 2254-2258.

Keil, F. C. (1989). *Concepts, kinds, and cognitive Development.* Cambridge, MA: MIT Press.

Kruschke, J. K. (1992). ALCOVE: An exemplar-based connectionist model of category learning. *Psychological Review*, **99**, 22-44.

Malt, B. C., & Smith, E. E. (1984). Correlated properties in natural categories. *Journal of Verbal Learning and Verbal Behavior*, **23**, 250-269.

McClelland, J. L., & Rumelhart, D. E. (1981). An interactive activation model of context effects in letter perception: Part 1. An account of basic findings. *Psychological Review*, **88**, 375-407.

McRae, K. (1991). *Independent and correlated properties in artifact and natural kind concepts.* Unpublished doctoral dissertation, McGill University, Montreal.

Medin, D. L. (1989). Concepts and conceptual structure. *American Psychologist*, **44**, 1469-1481.

Medin, D. L., Altom, M. W., Edelson, S. M., & Freko, D. (1982). Correlated symptoms and simulated medical classification. *Journal of Experimental Psychology: Learning, Memory, and Cognition*, **8**, 37-50.

Medin, D. L., & Schwanenflugel, P. J. (1981). Linear separability in classification learning. *Journal of Experimental Psychology: Learning, Memory, and Memory*, **7**, 355-368.

Murphy, G. L., & Medin, D. L. (1985). The role of theories in conceptual coherence. *Psychological Review*, **92**, 289-316.

Neely, J. H. (1977). Semantic priming and retrieval from lexical memory: Roles of inhibitionless spreading activation and limited-capacity attention. *Journal of Experimental Psychology: General*, **106**, 22-254.

Nosofsky, R. M. (1986). Attention, similarity, and the identification-categorization relationship. *Journal of Experimental Psychology: General*, **115**, 39-57.

Palmer, S. E. (1978). Fundamental aspects of cognitive representation. In E. Rosch and B. B. Lloyd (Eds). *Cognition and categorization* (pp. 259-303). Hillsdale, NJ: Lawrence Erlbaum.

Rosch, E. (1975). Cognitive representations of semantic categories. *Journal of Experimental Psychology: General*, **104**, 192-233.

Rosch, E., & Mervis, C. B. (1975). Family resemblances: Studies in the internal structure of categories. *Cognitive Psychology*, **7**, 573-605.

Younger, B. A., & Cohen, L. B. (1983). Infant perception of correlations among attributes. *Child Development*, **54**, 858-867.

Extending the Domain of a Feature-based Model of Property Induction

Steven Sloman[1]
Department. of Psychology
University of Michigan
sloman@psych.stanford.edu

Edward Wisniewski
Psychology Department
Northwestern University
Evanston, IL 60208
edw@nwu.edu

Abstract

A connectionist model of argument strength, which applies to arguments involving natural categories and unfamiliar predicates, was proposed by Sloman (1991). The model applies to arguments such as *robins have sesamoid bones, therefore hawks have sesamoid bones.* The model is based on the hypothesis that argument strength is related to the proportion of the conclusion category's features that are shared by the premise categories. The model assumes a two-stage process in which premises are first encoded by connecting the features of premise categories to the predicate. Conclusions are then tested by examining the degree of activation of the predicate upon presentation of the features of the conclusion category. The current work extends the domain of the model to arguments with familiar predicates which are nonexplainable in the sense that the relation between the category and predicate of each statement is difficult to explain. We report an experiment which demonstrates that both of the phenomena observed with single-premise specific arguments involving unfamiliar predicates are also observed using nonexplainable predicates. We also show that the feature-based model can fit quantitatively subjects' judgments of the strength of arguments with familiar but nonexplainable predicates.

Introduction

One of the most striking capacities of the human mind is the ease with which it can generate new beliefs from old ones. One form of this capacity is property-induction: The ability to express degrees of belief that one category of things exhibits some property given that other categories do. This ability can be expressed as a judgment of the strength of an argument in which the premises specify the relevant old beliefs and the conclusion specifies the newly hypothesized category-property relation. An example of such an argument is

i. Robins secrete uric acid crystals.
<u>Penguins secrete uric acid crystals.</u>
Therefore, Hawks secrete uric acid crystals.

How do people transmit belief from the premises to the conclusion of such an argument and what kind of systematicities in human judgment can we expect as a result of this process?

As an alternative to a model proposed by Osherson et al. (1990), Sloman (1991) proposed a simple connectionist network to model the subjective strength of a restricted class of arguments. Each argument consisted of a set of propositions, with all but one taken as statements of fact (premises). Subjects judged the validity of the remaining proposition (the conclusion) in light of the premises. Each proposition consisted of a one-place predicate (e.g., "secretes uric acid crystals") and a natural-kind object-category (e.g., "robins") to which it applied. Within an argument, all propositions shared a single predicate; only the category differed. The task was further constrained by allowing only predicates that were unfamiliar to subjects (such as "secretes uric acid crystals"). Unfamiliar predicates were used because they severely limit subjects' ability to reason about them. This allows theorists to focus on the transmission of belief amongst the categories of an argument, ignoring the role of the predicate. As described in Sloman (1991), the model was able to account for a host of qualitative phenomena involving arguments with unfamiliar predicates and showed good quantitative fits to subjects' ratings of argument strength.

The current work aims to extend the domain of the model to a class of arguments involving familiar

[1]Steven Sloman is now at the Department of Cognitive and Linguistic Sciences, Box 1978, Brown University, Providence, RI 02912

predicates. Limiting ourselves to single-premise arguments that are *specific* (a superordinate that properly includes one category also includes the other), we show that when subjects cannot explain the relation between the categories and the familiar predicate of an argument, they behave in the same way as they do with unfamiliar predicates. Before describing our evidence for this extension of the model's domain, we briefly identify the class of phenomena that the model was designed to account for, and describe the model itself.

Argument Strength Phenomena

Psychologists have identified about a dozen phenomena or general tendencies concerning the subjective strength of arguments involving unfamiliar predicates (cf. Osherson et al., 1990; Rips, 1975; Sloman, 1991). One example is the diversity phenomenon: People prefer arguments whose premises are less similar. To illustrate, people tend to believe that argument i. above is stronger than an argument with more similar premises like "Robins have X, Sparrows have X, therefore Hawks have X." (Because all predicates are unfamiliar, they can be referred to generically as predicate X.) In the course of describing our feature-based model and its extension to familiar predicates below, we outline four other phenomena: feature exclusion, nonmonotonicity, similarity, and asymmetry.

Feature Coverage

The model is based on the hypothesis that the strength of an argument is directly related to the proportion of the conclusion category's features or attributes that it shares with the premise categories -- the extent to which the features of the premise categories *cover* those of the conclusion category. The key assumptions are that all categories can be represented as a list of features, and that these features can be obtained from subjects. Roughly, an argument is strong to the extent that the features of the conclusion category are spanned by the features of the premise categories.

Feature-Based Induction

By representing categories as feature sets, we are able to distribute the representation of a category over a set of variables or units, where each unit represents a particular feature. One advantage of such a representational scheme is that any learning involving a feature of one category will automatically generalize to other categories sharing that feature. To model the

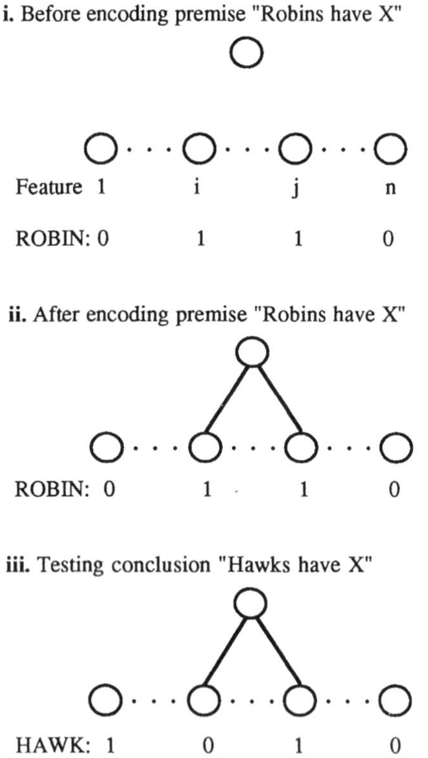

Figure 1. Illustration of the feature-based model for the argument Robins have X, therefore Hawks have X.

transmission of belief from one category to another, we take advantage of this automatic generalization property. Our model consists of a network of n input units, which are used to represent categories consisting of n features each, and a single output unit, which is used to represent the unfamiliar predicate X. In brief, the model posits that premise categories are first encoded as a vector of weights by connecting the units representing the features of the premise categories to the predicate unit. The network is then presented with the conclusion category. The process is illustrated in Figure 1. The value of the output unit upon presentation of the conclusion category is the model of argument strength. Strength is proportional to the extent to which the features of the conclusion category have been connected to the predicate unit by virtue of the encoding of the premises.

We now describe the two stages of the model in more detail. First, premises are encoded by connecting the features of their categories to the predicate unit. The connection at time t from feature i to the predicate unit ($w_{i,t}$) is updated for each premise using the following delta rule:

$$w_{i,t+1} = w_{i,t} + [1 - w_{i,t}][1 - a_x(P)]f_i(P),$$

in which $a_x(P)$ is the activation of the predicate unit upon presentation of premise P and $f_i(P)$ is the value of feature i of the premise's category. The coefficient $[1 - w_{i,t}]$ is used in place of the usual learning-rate parameter to keep each weight between 0 and 1.

In the second stage, the conclusion is tested by presenting its category to the input units and observing the activation of the predicate unit. The activation rule is

$$a_x(C/P_1,...,P_j) = \frac{W(P_1,...,P_j) \cdot F(C)}{|F(C)|^2}$$

which reads: The activation of unit X upon presentation of category C given that premises P_1 to P_j have been encoded equals the dot (or inner) product of the weight vector encoding the premises with the C feature vector, all divided by the squared length of the C vector. When C is a conclusion category, this activation value is the model of argument strength.

The weight vector is a non-linearly derived representation of the premises. The projection of the weight vector onto the conclusion category vector is therefore a representation of the projection of the premise categories onto the conclusion category. It corresponds to the features that the conclusion category has in common with the combined premise categories. Geometrically, the model proposes that argument strength is equal to the ratio of the length of this projection to the length of the conclusion category vector. This is the sense in which argument strength is hypothesized to be proportional to the coverage of the conclusion category's features by the premise categories.

Results using Unfamiliar Predicates

Making use of a simple model of the similarity between categories, the feature-based model can be shown to account for 11 of 12 argument strength phenomena (Sloman, 1991). For example, it accounts for the diversity phenomenon above because more diverse premises tend to cover the feature space better than less diverse ones. Another example is a phenomenon that acts as a boundary condition on diversity, feature exclusion. If a premise category shares few features with the conclusion category, it provides little additional coverage and therefore does not contribute to argument strength even if it is dissimilar to other premises.

The one phenomenon not accounted for by the model is called nonmonotonicity. Sometimes, adding a premise can reduce argument strength. For example, introductory psychology students prefer, on average, the argument "Flies have X, therefore Bees have X" to the argument "Flies have X, Aardvarks have X, therefore Bees have X." One interpretation of this phenomenon is that feature consistency is important; perhaps features that appear in one premise but are inconsistent with other premises are given less weight in the feature-matching process. This idea could be implemented in the feature-based model in several ways. A particularly simple way would entail introducing weight decay. If weights are reduced every time that they are updated, then the representations of features appearing in all but the last premise will have lower values if those features do not re-appear in later premises. In the current example, the strength of both arguments depends primarily on the overlap of the features of flies and bees because aardvarks and bees have so few common features. Because flies and aardvarks also have very few features in common, the weights corresponding to the representation of flies would decay in the second argument and therefore be lower than in the first. The reduced values of the flies' representation would lead us to expect the first argument to be stronger. None of the results that we report below would be affected by this generalization of the model because weight decay would have no effect on the model of single-premise arguments.

The model has been tested quantitatively by correlating its predictions to ratings of argument strength provided by subjects. To obtain the predicted strength of an argument from the feature-based model, the model must be given a featural description of each category appearing in the argument. Such featural descriptions were obtained from feature ratings for a set of mammals collected by Tony Wilkie (cf. Osherson et al., 1991). Varying a single parameter (a cutoff which determined a threshold below which feature ratings were set to 0), correlations of 0.96, 0.97, 0.59, 0.83, and 0.77 were obtained on five different data sets, respectively.

Extending the Model to Familiar Predicates

We define a "nonexplainable" predicate as one which is familiar but for which subjects cannot explain the relation between category and predicate. A nonexplainable argument is one containing nonexplainable predicates. We ran an experiment to test our hypothesis that nonexplainable arguments will be treated in the same way as arguments involving unfamiliar predicates. Subjects rated the strength of arguments with familiar predicates and, afterward, tried to explain the relations among the various categories and predicates. We evaluate our hypothesis in two ways. First, we expect that we should observe the same phenomena with nonexplainable arguments as we do with those using unfamiliar predicates. Our use of single-premise, specific arguments limits us to two such phenomena, similarity and asymmetry. The similarity phenomenon states that arguments tend to be stronger the greater the judged similarity between the premise and conclusion categories. We therefore test for this phenomenon by examining correlations between argument strength and similarity judgments. The asymmetry phenomenon states that the strength of arguments can be changed by reversing the premise and conclusion categories. We evaluate asymmetry by testing the feature-based model's ability to predict differences between the judged strengths of a set of arguments and their reversed counterparts. Finally, we expect the model to make predictions consistent with subjects' strength ratings for nonexplainable arguments. We test this prediction by examining correlations between the feature-based model's predicted argument strengths and subjects' judgments. We compare the correlations we obtain for explainable versus nonexplainable arguments.

Experimental Procedure

We constructed 16 arguments which we expected to be explainable and, using the same categories, another 16 which we expected to be nonexplainable. By exchanging the premise and conclusion of each argument, we obtained a total of 32 arguments of each kind. An argument was deemed explainable if it seemed that subjects would base their judgments on only a small set of features. For example, we believed that the argument

Collies are susceptible to heat stroke.
Siamese cats are susceptible to heat stroke.

would suggest features like "have fur" while

Wolves sometimes attack their mates.
German shepherds sometimes attack their mates.

would suggest features like "can be fierce." Examples of nonexplainable arguments include

Collies hate salted peanuts.
Siamese cats hate salted peanuts.

and

Wolves have dark tongues.
German shepherds have dark tongues.

Each of two groups of 12 University of Michigan students from Introductory Psychology courses were tested on different sets of 8 explainable and 8 nonexplainable arguments. Two other groups of 12 students were tested on corresponding arguments with premise and conclusion statements reversed. Each subject first rated the likelihood of each of the 16 conclusions on an integral scale from 0 to 10. We refer to these estimates as prior likelihoods. Next, they rated the likelihood of the conclusion given the premise on the same scale. The wording of the likelihood question can be inferred from the following example: "Collies hate salted peanuts. How likely do you think it is that siamese cats also hate salted peanuts?" We refer to these estimates as conditional likelihoods. Next, they were asked to briefly explain each premise and conclusion. They were given some example explanations and were encouraged to provide explanations that were sensible though they need not be true. Subjects were also told that if no possible explanation came to mind, they could skip that statement. They also provided a confidence rating of the validity of their explanations but we will not report these data. Finally, they rated the similarity (from 1 to 7) of each premise category to its corresponding conclusion category.

Results

To verify our assessment of explainability, we counted the number of explanations provided for each statement of each argument (out of a possible 24). For each argument, we averaged the number of explanations given for the premise and conclusion. All arguments which had an average of greater than 18 explanations were labelled "explainable" and all others were labelled "nonexplainable". On this basis, 8 of the arguments that we had expected to be nonexplainable were categorized as explainable and 2 explainable arguments were relabelled as nonexplainable. We thereby ended up with 38 explainable arguments and 26 nonexplainable ones.

Similarity. We found evidence for the similarity phenomenon for both explainable and nonexplainable arguments. Because we were interested in the role of similarity in the transfer of belief from premise to conclusion (the conditional likelihood), without the influence of any spurious correlation between similarity and prior likelihood, we looked at the part correlation between i. similarity judgments and ii. conditional likelihoods with priors partialed out. These correlations were significant for both explainable ($r = .40$, $p < .001$) and nonexplainable arguments ($r = .43$, $p < .001$). We conclude that the similarity phenomenon does indeed hold for nonexplainable arguments and in fact holds for explainable ones as well.

Asymmetry. The feature-based model predicts that reversing premise and conclusion categories will lead to an argument of more or less strength depending on the relative richness or magnitude of the representations of the two categories. The richness of a representation refers to the extent of featural information that is known about a category. Richness would tend to increase with a category's familiarity and complexity. To see why the model predicts these asymmetries, consider its activation rule. The model of the arguments P therefore C and its reversed counterpart C therefore P have identical numerators ($F(P) \cdot F(C)$; cf. Sloman, 1991), but different denominators. The denominators are the magnitudes of the conclusion categories. Therefore, the model predicts that the strength of the argument with the lower magnitude conclusion category will be greater. For example, people often judge "tigers have X, therefore buffaloes have X" to be stronger than its reversal because, according to Osherson et al.'s (1991) feature ratings, the buffaloes vector has a smaller magnitude than the tigers one. Furthermore, the degree of asymmetry should be directly related to the size of the difference between the magnitudes of the two categories.

To test this prediction, we calculated the magnitude of each category using the feature ratings. Based on these magnitudes, we determined whether an argument or its reversal should be stronger. To measure the actual strength of an argument, we used the mean difference between its conditional and prior likelihood judgments. Each strength measure was weighted by the difference between the magnitudes of that argument's categories. This weight reflects the degree of expected asymmetry. A 2 x 2 analysis of variance with one between-argument factor (the explainability of the argument -- explainable or not) and one within-argument factor (predicted asymmetry -- the argument predicted to be stronger or its reversal) revealed a statistically reliable main effect for the predicted asymmetry, $F(1,30) = 4.42$, $p < .05$. No significant main effect for explainability or for the interaction was observed (both F's < 1). Apparently, the model was able to successfully predict not only the direction of the asymmetry for nonexplainable arguments, but for explainable ones as well.

Fit of the model. The feature-based model was fit to the data using the equations and feature ratings described above. Because of the feature-rating method used, ratings tended to overestimate the value of nonsalient features (cf. Sloman, 1991). We therefore varied a cutoff which determined a threshold below which feature ratings were set to 0. The cutoff was varied in small discrete increments. The model's predictions were generated using the cutoff that maximized the correlation between the predictions and the data. The data consisted of the mean difference, for each argument, between each subject's conditional and prior likelihood estimates. Because these means represent a combination of judgments by subjects for whom the argument was explainable and those for whom it was nonexplainable, we do not expect these correlations to be extremely high. The relatively small number of times that subjects failed to provide any explanation prevented us from obtaining reliable likelihood estimates for each argument using only those cases. Nevertheless, the maximum correlation (taken over cutoffs) for nonexplainable arguments was 0.66 ($p < .001$). Notice that this correlation is greater than that obtained between argument strength and similarity ratings. The maximum correlation for explainable arguments was much less (0.36; $p < .05$). The difference between the two correlations was marginally significant ($z = 1.57$; $p = .06$). We conclude that these quantitative tests provide some support for the feature-based model as an account of subjects' judgments of the strength of nonexplainable arguments.

Conclusion

A simple model of property induction, alike in many respects to connectionist models of concept-learning, is consistent with a variety of phenomena in a domain of confirmation -- people's willingness to assert properties of natural-kind categories. Our experiment supports our contention that the domain is larger than previously shown. It includes not only arguments with unfamiliar predicates, but those with familiar but nonexplainable predicates as well.

References

Osherson, D., Smith, E. E., Wilkie, O., Lopez, A., and Shafir, E. 1990. Category-based induction. *Psychological Review* 97:185-200.

Osherson, D., Stern, J., Wilkie, O., Stob, M., and Smith, E. E. 1991. Default probability. *Cognitive Science*, 15: 251-269.

Rips, L. 1975. Inductive judgments about natural categories. *Journal of Verbal Learning and Verbal Behavior*, 14:665-681.

Sloman, S. A. 1991. Feature-based induction. Tech Report No. 40, Cognitive Science and Machine Intelligence Laboratory, Univ. of Michigan.

An Instantiation Model of Category Typicality and Instability

Evan Heit
Department of Psychology
University of Michigan
330 Packard Road
Ann Arbor, MI 48104
Evan.Heit@um.cc.umich.edu

Lawrence W. Barsalou
Department of Psychology
University of Chicago
Chicago, IL 60637
L-Barsalou@uchicago.edu

Abstract

According to the instantiation principle, when we make a judgment about a relatively superordinate category, we follow a two-step process. First, we instantiate the category into one or more subordinates. Second, we make a judgment based on the subordinates. Instantiation theory applied to typicality judgments makes the following predictions. When subjects judge the typicality of a category A with respect to category B, their mean typicality judgment should equal the weighted mean typicality (with respect to B) of subordinate categories of A. Furthermore, typicality judgments for category A will be unstable (i.e., have a high standard deviation) to the extent that A has a large number of diverse subordinates. The instantiation principle was implemented in a computer simulation, which used production frequencies and typicality ratings for subordinates to predict ratings for superordinate-level categories. In two experiments, subjects judged the typicalities of various animal and food categories. The instantiation model successfully predicted the means and standard deviations for the observed distributions of responses for these categories. Extensions and other applications of the instantiation principle are also briefly discussed.

Introduction

Vertical Category Structure and Instantiation

Semantic categories may be described and related to each other in terms of their taxonomic, or vertical, structure (Rosch, 1978). Some categories, such as mammal and beverage, are relatively superordinate; they are at a high level of abstraction. Other categories, such as dog and milk, are subordinate; they are more specific than their respective superordinates.[1] The rationale behind the instantiation principle is that it is hard to think about relatively superordinate categories. Members of superordinate categories often do not share many features and or even look like each other. A superordinate category may be quite abstract or ambiguous. On the other hand, reasoning about more subordinate categories ought to be easier because members of these categories are more alike. The instantiation principle says that when we make a judgment about a relatively superordinate category (e.g., mammal), we perform this task by first instantiating the superordinate into one or more subordinate categories (e.g., dog and human), then second, making judgments about the subordinates.

There is some prior research bearing on instantiation. Contextual information has been shown to lead people to perform instantiation during reading comprehension. For example, after reading the sentence "The fruit was made into wine," the word "grape" will serve as a better retrieval cue for this sentence than the word "fruit" (Anderson, Pichert, Goetz, Schallert, & Stevens, 1976). The instantiation principle has also been successfully applied to reasoning. Osherson, Stern, Wilkie, Stob, and Smith (1991) proposed that when people evaluate the strength of an inductive argument such as

[1]The terms subordinate and superordinate are used here solely to describe taxonomic position *relative to each other*. These terms are not intended to describe taxonomic position relative to basic-level categories (as in Rosch, 1978).

"Canines have sesamoid bones; therefore mammals have sesamoid bones," they first instantiate the superordinate categories, canine and mammal. Osherson, et al. tested this proposal by finding subjects' subordinates of various superordinate categories. In the present example, they found that the argument's strength could be predicted from the similarity between instantiations of canine (e.g., dog and wolf) and instantiations of mammal (e.g., dog and human).

Typicality Structure and Instability

In addition, it has been found universally that categories have a typicality structure: some category members are judged to be better members than others. The typicality variable is an excellent predictor of how people perform other categorization tasks, such as the time to classify exemplars and the ease of learning category members. Interestingly, typicality structure does not remain fixed across contexts, between people, or even within a single person at different times (Barsalou, 1987). Categories may differ not only in their mean typicalities but also in terms of the standard deviations, reflecting their instabilities. One goal of this study is to investigate whether and how different categories differ systematically in their degree of instability.

Instantiation Model of Typicality Judgments

Following is a simple version of the instantiation principle, applied to typicality judgments. When a person judges the typicality of a category A with respect to category B, they do two things. First, they produce one subordinate of A, subA. Second, they judge the typicality of subA with respect to B. So, someone judging the typicality of mammal with respect to animal, might instead judge the typicality of dog with respect to animal.

More complex variants of this model are possible, with additional assumptions about how or how many instantiations are produced. Also, this model does not make any claims about how the typicality of subA is judged; it might involve the same instantiation process, or it might involve a simpler, non-recursive process. This model only specifies the relation between typicality judgments on A and typicality judgments on subordinates of A. Nonetheless, this simple instantiation model is worth testing, to see what it reveals about category structure.

What does the instantiation principle predict about typicality judgments? In general, people's responses when they judge the typicality of A with respect to B should resemble people's responses when they judge the typicality of instantiations of A with respect to the same B. If mammals are typical animals, then particular mammals, such as dogs and humans, should also be typical animals. Imagine that a group of subjects makes typicality judgments for A and for instantiations of A. (A set of instantiations of A may be obtained by asking another group of subjects to each name a single subcategory of A.) The two distributions of typicality judgments, for A and for its instantiations, should be alike. In particular, the respective means and standard deviations of the two distributions should be the same.

Some superordinate categories may be particularly ambiguous, leading people to make highly variable, unstable judgments using the different possible instantiations. According to the instantiation principle, instability of typicality judgments has a few sources. First, the typicality of A will be more unstable as the typicality ratings for the subordinates of A are more unstable. However, even if A has fairly stable subordinates, a second source of instability may be the diversity of its subordinates. If it has two subordinates that are very different from each other, then typicality judgments about A would show a lot of variation. In general, the instantiation principle predicts that typicality judgments about categories with more subordinates will be more unstable, unless these subordinates are all alike in terms of how typical they are.

Experiment 1

Testing the instantiation principle for typicality judgments was accomplished with two groups of subjects. The Production group produced subordinates for a set of relatively superordinate categories (e.g., mammal, reptile). The Rating group made typicality judgments about the superordinate and subordinate categories, always with respect to the category "animal."

Method

Subjects. The Production and Rating groups each consisted of 20 Stanford University undergraduates, recruited in dormitories. The Production subjects were all run before the Rating subjects.

Stimuli. For the Production group, the stimuli were 7 superordinate categories: amphibian, bird, fish, insect, mammal, microorganism, and reptile. The Production subjects produced 63 different subordinate categories in response to these superordinates. For the Rating group, the stimuli were 63 subcategories produced by the first group, plus the 7 superordinate

Table 1. Data for Mammal Category, Experiment 1.

Instantiation	Prod. Freq.	Typicality with Respect to Animal
Human	5	10, 8, 10, 10, 1, 5, 10, 1, 10, 7, 10, 10, 9, 6, 10, 8, 2, 9, 6, 2
Bear	3	10, 9, 10, 10, 6, 10, 10, 9, 10, 9, 4, 10, 7, 8, 10, 8, 9, 10, 10, 10
Kangaroo	3	8, 8, 10, 9, 4, 10, 10, 5, 2, 7, 4, 7, 7, 6, 5, 6, 5, 7, 8, 8
Whale	2	8, 7, 10, 7, 6, 10, 10, 1, 5, 4, 8, 10, 6, 8, 5, 6, 10, 5, 9, 4
Ape	1	10, 8, 10, 10, 7, 10, 10, 9, 10, 9, 8, 10, 7, 6, 8, 6, 9, 10, 7, 8
Cat	1	10, 8, 10, 9, 6, 10, 8, 10, 8, 9, 1, 6, 10, 7, 10, 10, 10, 10, 10
Cow	1	10, 9, 10, 9, 6, 10, 10, 7, 10, 8, 10, 9, 8, 9, 9, 9, 10, 10, 10, 10
Dog	1	10, 9, 10, 10, 10, 10, 10, 9, 10, 9, 10, 8, 8, 10, 9, 10, 10, 10, 10, 10
Elephant	1	10, 10, 10, 10, 5, 10, 10, 8, 10, 9, 7, 9, 7, 9, 7, 7, 10, 10, 10, 10
Horse	1	10, 9, 10, 10, 7, 10, 10, 10, 10, 8, 10, 9, 7, 10, 10, 10, 10, 10, 10, 10
Platypus	1	4, 6, 10, 6, 4, 10, 10, 3, 7, 7, 5, 8, 5, 4, 5, 8, 5, 7, 10, 3

Superordinate	Typicality with Respect to Animal
Mammal	9, 6, 10, 10, 1, 5, 10, 10, 10, 9, 10, 10, 9, 7, 10, 10, 9, 10, 10, 5

categories. Each subcategory appeared once as a Rating stimulus, regardless of the number of mentions by Production subjects.

Procedure. Production subjects were instructed to produce one subordinate for each of the 7 categories. Rating subjects were instructed to rate the 7 superordinate categories and the 63 subordinates on typicality with respect to "animal," using a 1-10 scale, for which higher numbers meant greater typicality. Sample questions would be "How typical is a mammal for the category animal?" and "How typical is a dog for the category animal?"

The Instantiation Model

The predicted distributions of typicality ratings for the superordinate categories relied on two sources of data: the production frequencies from the Production group and the subordinate typicality ratings from the Rating group. The predictions for each superordinate category were created by a computer simulation consisting of an iterated two-step process. First, a subordinate was chosen, according to the production frequencies for that superordinate. For example, 5 of 20 subjects produced "human" as an instantiation of "mammal," so "human" was used 25% of the time on this step, when predicting the distribution of responses for "mammal." Second, a typicality rating for the subordinate was chosen from the 20 typicality judgments that the Rating subjects made for this item. In this case, one of the 20 ratings that people made for "human" was chosen. This simulation did not operate stochastically, rather it exhaustively produced a distribution of 400 ratings for each predicted distribution of superordinate category responses. For each of the 20 productions of subordinates, each of its 20 corresponding typicality ratings were added to the predicted distribution. Thus, for each superordinate category, the predicted distribution represented 400 simulated subjects, with each simulated subject assumed to have made one production of a subordinate then one typicality rating on this subordinate. The observed distributions for the 7 superordinate categories were simply the 20 ratings made for each superordinate by the Rating subjects.

Table 1 illustrates how this model was applied to the data collected for the superordinate category "mammal." The 20 Production subjects responded with 11 distinct subordinates of this category, with production frequencies as shown in the table. Then the 20 Rating subjects judged the typicality of each subordinate, as well as the category "mammal," with respect to "animal." The 20 typicality ratings for

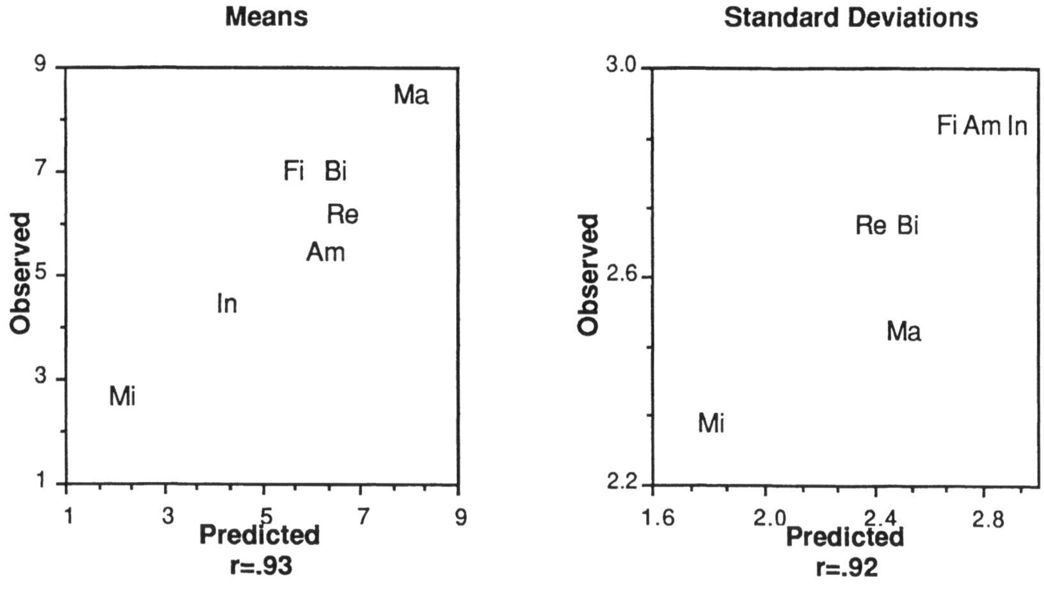

Figure 1. Statistical comparisons of observed and predicted distributions, Experiment 1.

each category are shown in the table. The observed distribution for "mammal" was simply the 20 typicality ratings for this item. The predicted distribution contained 400 ratings: each rating for "human" taken 5 times, each rating for "bear" and "kangaroo" taken 3 times, each rating for "whale" taken twice, and each rating for the other subordinates taken once.

Results

Of primary concern is whether the 7 observed distributions of typicality ratings for the superordinate categories correspond to the 7 predicted distributions, obtained by application of instantiation theory. The observed and predicted distributions are described in Figure 1, in terms of mean and standard deviation. (Each data point is shown as the first two letters of the superordinate category, e.g., Ma for mammal.) These categories did indeed differ in terms of their means and standard deviations. The correspondence between the observations and the predictions is excellent; $r=.93$ for the means and $r=.92$ for the standard deviations, one-tailed $p<.01$ in both cases. The predicted standard deviations generally underestimate the observed standard deviations, suggesting that additional sources contributed to instability of the ratings on the superordinates.

Experiment 2

To provide generality, this experiment was a replication of Experiment 1, with typicality judgments with respect to a different category (food) and more subjects.

Method

Subjects. The Production group consisted of 40 University of Michigan undergraduates, recruited in public places in Ann Arbor. The Rating group consisted of 40 Michigan undergraduates, who participated as part of a course requirement.

Stimuli. For the Production group, the stimuli were 9 superordinate categories of food: beverages, dairy products, desserts, fish, fruits, meats, poultry, seasonings, and vegetables. For the Rating group, the stimuli were the 9 superordinate categories plus 88 subordinates produced by the first group.

Procedure. The procedure was like Experiment 1, except Rating group subjects rated the categories on a scale from 1 to 9, in terms of typicality with respect to "food."

Results

Again, the primary analysis was to compare the observed and predicted typicality distributions for the superordinate categories. As in Experiment 1, the predicted distribution for each of the 9 categories was obtained by exhaustively simulating the 1600 possible combinations of subordinate productions (from the first group) and subordinate typicality ratings (from the second group).

The observed and predicted distributions for the superordinate categories are described in Figure 2, in terms of mean and standard deviation. The correspondence is excellent for the means, $r=.89$, $p<.01$. For the standard deviations, $r=.64$, $p<.05$ one-tailed. As in Experiment 1, the predictions of category instability are good but not quite as good as the predictions of mean typicality.

General Discussion

Evaluation of the Instantiation Model

These experiments demonstrate a constancy of categorical structure: judgments about a category tend to resemble judgments about its instantiations. Most strikingly, the mean typicality judgment of a superordinate was quite close to the weighted mean typicality judgments for its subordinates. This particular result is consistent with the simple instantiation model presented here, in which a subject making a judgment about an superordinate category relies on a judgment about one subordinate. However, the relatively less impressive predictions of category instability (standard deviations) suggest that extensions must be made to the simple model.

One extension would be relax the assumption that people evaluate exactly one instantiation for each category. Another source of stability--or instability--could be that subjects produce different numbers of subordinates for different categories, due to knowledge differences. If a subject's judgment for a category is based on the mean value for several instantiations rather than based on a single instantiation, then the judgment will be more stable, just as the standard error of a mean decreases as the sample size increases. And if some subjects find it difficult to produce even one subordinate of some category (e.g., amphibian), then they might respond randomly, also affecting the standard deviation of responses.

Another direction for extending this model is to further apply the instantiation principle to explain the typicality judgment process. The model, as presented, assumes that to judge how typical mammals are of the category "animal," a subject compares instantiations of "mammal" to the category "animal." This comparison process itself might

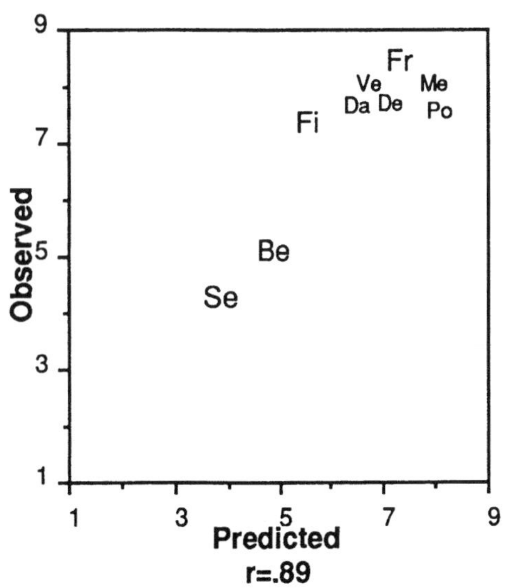

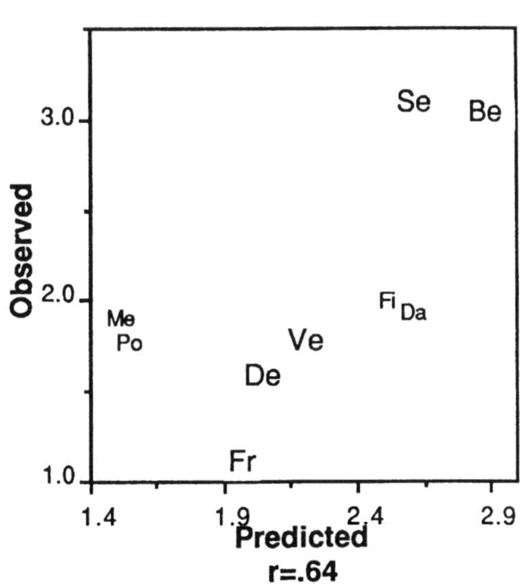

Figure 2. Statistical comparisons of observed and predicted distributions, Experiment 2.

depend on further instantiations, such as instantiating "animal" into its subordinates. The Osherson, et al. (1991) model successfully applied such a scheme, assuming that instantiation takes place for every superordinate category considered in a judgment.

A third future direction for extending this model would be to allow the basis of judging typicality to vary with context. In particular, Barsalou (1987) has proposed that when people make a judgment about a category member, they do not consider every piece of featural information. Instead, certain features may only be activated in certain contexts. For example, people might not usually consider cost when they judge the typicality of various foods, but the costs of the foods could indeed influence judgments if costs were made more relevant or salient.

While the preliminary tests of the simple version of the instantiation model were encouraging, the clear directions for extending this approach suggest that further experiments and more advanced models will be needed, as well as comparison to alternate approaches.

Applications of the Instantiation Principle

More generally, the instantiation principle sheds light on other results about instability of category structure. Barsalou (1987) identified three kinds of instability. Both within-subjects instability and between-subjects instability may be explained in terms of different instantiations being produced at different times. Recent experiences and differences in knowledge are likely to contribute to these differences in instantiations. Context-dependent instability (such as categories having different structures when they are considered from different perspectives or in different environments) can be partly explained in terms of different instantiations for different contexts. However, the typicality judgments on the instantiations could also reflect different knowledge about the different contexts, even if the same instantiations are made for both contexts. For example, someone might judge fish to be a typical "ocean animal" but an atypical "desert animal." For both judgments, "fish" might be instantiated the same way, say as "shark," but sharks would then be judged as typical for ocean animals and atypical for desert animals, based on knowledge about oceans and deserts.

Much more generally, the instantiation principle's successful applications to comprehension, inductive reasoning, and typicality judgment suggest further applications to other cognitive abilities. For example, to make a decision about whether to eat at restaurant X or restaurant Y, someone might compare specific meals that they have had at each restaurant (cf., Kahneman & Miller, 1986). This person would consistently choose restaurant X over restaurant Y if the meals at X are better overall than meals at Y, and the particular meals do not vary much. However, the decision-making process would be more unstable to the extent that the particular instances of restaurant X meals and restaurant Y meals were more variable.

Acknowledgments

We are grateful to Douglas Medin, David Thau, and Edward Wisniewski for comments on this paper, and to Rachael Jeffries and William Krooss for help in collecting the data.

References

Anderson, R. C., Pichert, J. W., Goetz, E. T., Schallert, D. L., Stevens, K. V., & Trollip, S. R. (1976). Instantiation of general terms. *Journal of Verbal Learning and Verbal Behavior, 15*, 667-679.

Barsalou, L. W. (1987). The instability of graded structure: Implications for the nature of concepts. In U. Neisser (Ed.), *Concepts and conceptual development: Ecological and intellectual factors in categorization.* Cambridge: Cambridge University Press.

Kahneman, D., & Miller, D. T. (1986). Norm theory: Comparing reality to its alternatives. *Psychological Review, 93*, 136-153.

Osherson, D. N., Stern, J., Wilkie, O., Stob, M., & Smith, E. E. (1991). Default probability. *Cognitive Science, 15*, 251-269.

Rosch, E. (1978). Principles of categorization. In E. Rosch & B. B. Lloyd (Eds.), *Cognition and Categorization.* New York: Wiley.

Inhibition and Brain Computation

Steven L. Small*† and Gerhard H. Fromm*

Cognitive Modelling Laboratory
Department of Neurology* and Program in Intelligent Systems†
University of Pittsburgh
325 Scaife Hall, Pittsburgh, PA 15261
sls+@cs.cmu.edu

Abstract

The synapse plays a fundamental role in the computations performed by the brain. The excitatory or inhibitory nature of a synapse represents a (simplified) characterization of both the synapse itself and the computational role it plays in the larger circuit. Much speculation concerns the functional importance of excitation and inhibition in the physiology of the cerebral cortex. The current study uses neural network (connectionist) models to ask whether or not the relative proportion of inhibition (i.e., inhibitory synapses) and excitation (i.e., excitatory synapses) in the brain affects the development of its neural networks? The results are affirmative: An artificial neural network, designed to perform a particular task involving winner-take-all output nodes, is sensitive to the initial configuration of positive (excitatory) and negative (inhibitory) connections (synapses), such that it learns considerably faster when started with 60-75% inhibitory connections than when it includes a greater or lesser proportion than this. Implications of this result for neuroanatomy and neurophysiology are discussed.

Introduction

The brain computes through a distributed network of discrete neural elements whose pattern of connections gives rise to particular types of computations. Many morphological and functional features of these objects contribute to their ability to compute, with the synapse playing a fundamental role [Shepherd and Koch, 1990]. The excitatory or inhibitory nature of a synapse represents a (simplified) characterization of both the synapse itself and the computational role it plays in the larger circuit. The functional importance of excitation and inhibition in the brain is the subject of significant speculation [Fromm, 1992], which has led to assertions about the importance of inhibition in the physiology of the brain, particularly within the cerebral cortex.

Acknowledgment: The support of the NIH-NIDCD under grant number DC00054-02 is gratefully acknowledged.

In this paper, we use a neural network (connectionist) model to examine the question: Does the relative proportion of inhibition (i.e., inhibitory synapses) and excitation (i.e., excitatory synapses) in the brain affect the computational efficiency of its neural networks? Two parallel issues devolve from this question, one involving the development of computational circuits, and the other concerned with the operation of already learned circuits.

We investigate these questions in the context of the cortical visual system, particularly the results of Mishkin and his colleagues [Mishkin, et al., 1983] on macaque visual processing. When required to perform the dual task of visual object recognition and spatial localization, the macaque uses two separate visual systems to perform the two tasks, a temporal "what" system and a parietal "where" system [Desimone, et al., 1985; Mishkin, et al., 1983]. Rueckl, Cave, and Kosslyn [1989] have constructed a computer model of this system, which was used as a testbed for the present study of inhibition and excitation.

Two hypotheses motivated the current study: (1) The development of the visual system (to perform the object recognition and spatial localization task) takes place faster when the initial neural network contains a predominance of inhibitory synapses; and (2) Fully developed neural networks of the (two pathways of the) visual system operate more accurately when containing predominantly inhibitory synapses. We tested these two hypotheses by teaching numerous initial configurations of the Rueckl model (with different fractions of excitatory and inhibitory synapses) to perform the visual task.

This model represents one of a class of neural network (or connectionist) models currently under investigation by researchers from diverse disciplines. Such models attempt formally to understand biological neuronal networks, at the levels of both individual neuronal processing (e.g., dendritic computations, synaptic behavior) and of large assemblies of neuronal processing (e.g., cerebellar cortex) [Sejnowski, et al.,

1988; Sun, et al., 1988]. In addition, neural network models of high level cognitive processing, in such areas as vision [Feldman, 1989] and language [McClelland and Rumelhart, 1986; Seidenberg and McClelland, 1989], are reshaping accepted notions in information processing psychology. While these models differ greatly in the formal specifications of their neuronal units, and in the particular manner in which the units are connected to perform computations, they share similar underlying principles of organization.

The overall goal of the experimental method is to use a theoretical analysis to make suggestions for empirical scientists. As a necessity of the approach [Churchland and Sejnowski, 1987], we address computational questions about the brain at a high level of abstraction. Thus, we will not be able to show how much inhibition is actually used in a particular area of the brain or for a particular neurological task. Nonetheless, we do provide some suggestions about how the overall balance of inhibitory and excitatory synapses might make a difference in the computations that are possible. While this will not answer the morphological questions, it may help motivate research to establish a better correlation between anatomical and physiological results.

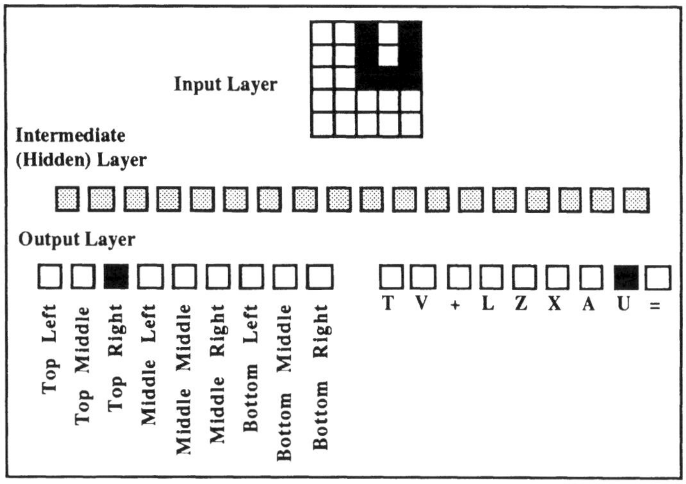

Figure 1: Network Representation of Visual Pattern Recognition

Methods

As with most connectionist models, and with all models at this level of analysis, many simplifying assumptions are made about neurons and synapses [Sejnowski, et al., 1988]. The model presented here uses a very simple model of a neuron, and an even simpler model of a synapse. These simple models are shown in Table 1, and are representative of the strategies common in connectionist modelling. (See our previous discussion of this in [Small, 1991].) A parallel distributed processing (PDP) approach [Rumelhart and McClelland, 1986], employing a layered feed forward (i.e., non-recurrent) network with one hidden layer of computational units in addition to the input and output layers, formed the basis of the implementation.

Such networks are able to learn by example to perform some task (i.e., associating a number of input patterns with desired output patterns) by adjusting their connection strengths according to particular error minimization rules. The back propagation learning algorithm [Rumelhart, et al., 1985] represents a useful learning strategy. One criterion for the success of a PDP model using back propagation is the number of presentations of each training example required to teach the model to perform the desired task. The models discussed in this paper use this criterion; a network that learns a task quicker (i.e., with fewer presentations of each training instance) is considered to be superior to one that learns the task slower. Of course, these numbers are subject to statistical interpretation, and the concepts of faster and slower must conform to standard criteria of significance.

As noted, the experiments presented here were conducted with the visual system model of Rueckl and his colleagues [1989]. The model performs the classification of two-dimensional visual images into two categories, one representing what object was shown and the other representing where on the input grid the image appeared.

CNS Concept	Model Analogue	Nature	Description
Neuron	Unit	Abstraction	Associated values and functions
Synaptic strength	Connection weight	Value	Real number
Axon firing rate	Unit potential	Value	Real number
Synapse	Unit input	Value	Weighted unit potential
Inhibition	Negative weight	Value	Negative real number
Excitation	Positive weight	Value	Positive real number
Depolarization	Potential function	Function	Adjusted sum of inputs
Threshold	Bias	Value	Real number

Table 1: Computer Model Correlates of Neurobiological Concepts

This dual task of visual object recognition and spatial localization has a neurobiological basis in the temporal "what" system and parietal "where" systems identified in the macaque monkey

[Desimone, et al., 1985; Mishkin, et al., 1983]. Rueckl et al showed that a computational neural network learned the two tasks much faster if instead of a single process, the network were subdivided into two parallel network processes, one for object recognition and the other to perform the spatial localization.

The specific experimental hypotheses, based on the general hypotheses discussed above, are that the feed forward layered connectionist network designed by Rueckl et al [1989] both (1) learns faster when started with predominantly negative weights than with random weights or with predominantly positive weights; and (2) performs more accurately when containing predominantly negative weights. While the specific hypotheses concern mathematical networks, we suggest that the principles of operation apply to biological networks and their synapses.

This model consists of a feed forward network of units containing three layers: (1) an input layer, (2) an output layer, and (3) an intermediate (hidden) layer. These three layers are illustrated graphically in Figure 1. The input layer consists of a linear representation of a two dimensional visual pattern. The 5 x 5 input grid of Figure 1, representing the letter "U", is actually represented as the linear vector of binary digits, with a "1" representing a pixel in the pattern and a "0" one that is not in the pattern. The output layer represents *what* pattern was presented and *where* in the two dimensional input space the pattern was presented.

The output layer of Figure 1 illustrates these two sublayers of representation: One sublayer (the right hand part of the output layer in the Figure) contains the information on what input object was presented — in this case the letter "U" — and the other sublayer (the left hand part of the output layer in the Figure) contains the information as to where in the input grid that object appeared — in this case, in the top right position of the input grid. The network representation of the output layer uses binary digits, with a "1" for the correct identification and location, and a "0" otherwise.

The units of one layer are fully connected to those of the next layer, and the connections can be either positive (excitatory) or negative (inhibitory). The units within a particular layer are not connected. The network starts with random connection strengths among the units of the adjacent layers. It is then repeatedly presented with nine different input patterns in all nine possible positions, along with the desired output values, indicating what pattern was presented where. The example input pattern and correct output value of Figure 1 are illustrative. The model then uses the back propagation algorithm [Rumelhart et al, 1985] to change the connection strengths (weights) of the network in a way that minimizes the overall network error. Ultimately the network learns to classify all the input patterns.

In the original model of Rueckl et al [1989], learning was significantly faster when the network was split into separate "what" and "where" systems than when the task was attempted by a single undivided network. The current study used the split network for all trials.

In order to test the hypothesis about the relative importance of inhibitory versus excitatory connections in brain computations, another network parameter was varied, namely, the percentage of initial network weights with positive values. Recall that positive connection weights represent excitatory synapses, and negative weights represent inhibitory synapses.

A pseudo-random number generator was used to generate two values, a real number between 0 and +2, and an integer sign (either -1 or +1). Thirteen experiments were conducted: For each connection in the network, the probability of it receiving an initial positive weight was 0% in one trial, 6.25% the next trial, and 12.5%, 18.75%, 25%, 31.25%, 37.5%, 43.75%, 50%, 56.25%, 62.5%, 68.75%, and 75% in the twelve additional trials. The learning algorithm was constructed to present input/output pairs (training instances) repeatedly until either (a) the sum squared error of the network dipped below 4.0; or (b) the total number of presentations of the entire corpus of training instances (one epoch = 9 images x 9 positions = 81 individual training instances) reached 200. These numbers were chosen following several pilot experiments that showed that a network error of about 4.0 represented good performance. The limit to 200 epochs was a practical decision motivated by limitations in computational resources.

For this project, the basic model was reimplemented using the DYSNET simulator. Specific choices regarding potential functions, learning parameters, error measure, and weight updating function are shown in Table 2. Note that these choices may or may not reflect those of the original model by Rueckl et al [1989] and are practically, but not theoretically, important.

Results

The results of these experiments are summarized in

Attribute	Value	Reference
Network Structure	Feed Forward	[Rumelhart and McClelland, 1986]
Hidden Layers	One	[Rumelhart and McClelland, 1986]
Layer Widths	25 x 18 x 18 units	[Rueckl, et al., 1989]
Substructures	Splitting	[Rueckl, et al., 1989]
Potential Function	Logistic Function	[Rumelhart, et al., 1985]
Learning Algorithm	Generalized Delta Rule	[Rumelhart, et al., 1985]
Weight Updating	QuickProp Algorithm	[Fahlman, 1988]
Error Measure	Sum Squared Error	[Rumelhart, et al., 1985]
Learning Rate	0.5 + Unit Fan In	[Fahlman, 1988]

Table 2: Computational Features of the Model

Table 3. These data reflect the average of one hundred individual learning experiments at each fraction of initial positive weights.

Note that both the average network error and the average number of epochs vary with the fraction of positive initial random weights, reaching a nadir between 25% and 37.5%, but increasing as the fraction of initial positive weights decreases below or increases above this level. The statistical results compare the number of epochs required to learn the task at each starting configuration (i.e., percentage of initial positive weights) with the minimum number required when 25% of the initial weights are positive. The Student t-test using a two tailed distribution was used for this comparison. When adjusted for multiple comparisons, it still shows a significant effect: A starting configuration reflecting a preponderance of negative weights (within a specific range) leads to faster network convergence than with a preponderance of positive weights.

Fraction Positive Weights	Mean Network Error	Mean Number Epochs	Probability (number of epochs)	Paired T Value
0.125	5.290	149.51	5.649	$p < 0.0001$
0.250	4.183	111.93	***	***
0.375	4.095	117.49	0.869	$p = 0.3824$
0.500	4.276	150.87	6.236	$p < 0.0001$
0.625	5.128	181.10	13.129	$p < 0.0001$
0.750	7.181	197.18	17.977	$p < 0.0001$

Table 3: Experimental Results

Figure 2 shows a graphic illustration of one portion of the initial configuration when the fraction of initial positive weights was set to 25% of all connection weights. In the Figure, white squares represent positive values and black squares represent negative values. The area of the box represents the real number value (in this example, the largest box encodes an absolute value of 2.0). The Figure illustrates the connection strengths between each of the eighteen hidden units and the fifth unit of the output layer.

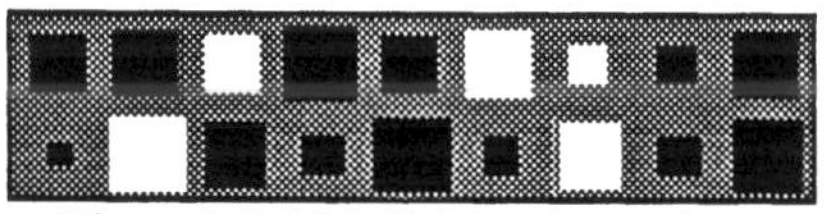

Figure 2: Initial Weights from Hidden Layer to Fifth Unit of Output Layer

Figures 3 and 4 illustrate graphically the results of the thirteen experiments. (Note that the standard errors of the means, which are not shown in the Figures, are extremely small). Figure 3 shows the average minimum network error achieved in 100 separate learning trials, when the starting configuration included random connection strengths in which the percentage of positive and negative values was varied. The abscissa of this graph measures the fraction of positive initial weights and the ordinate measures the sum squared error of the network. Note that the average network error reaches a minimum with a starting configuration of 37.5% excitatory weights, and increasing or decreasing this percentage sharply increases the total error (0% excitation is not shown: the error exceeds 8).

Figure 4 shows the average number of trials required to reach either a network error of 4.0 or a total of 200 trials. Numbers close to 200 therefore represent failure to converge in 200 trials. As in Figure 3, these data were accumulated from 100 separate learning trials, with a varied initial percentage of positive and negative connection weights. The abscissa of this graph measures the fraction of positive initial weights and the ordinate measures the number of trials. The minimum number of epochs required to learn the task occurs at a starting configuration of 31.25% inhibition, with alterations in this percentage significantly impairing learning. Analysis of the final network demonstrated a linear correlation between the inhibition fraction of the initial network (before learning) and that of the completely trained network.

Discussion

The present study demonstrates that an artificial neural network, designed to perform a particular task, is sensitive to the initial configuration of positive (excitatory) and negative (inhibitory) connections (synapses). The particular network examined uses winner-take-all output representations, and learning is considerably faster when the structure of the network includes 60 - 75% inhibitory connections than when it includes a greater or lesser proportion than this. While there are many intuitive analyses of the importance of inhibition for brain computations, both at the level of individual neurons as well as at the level of the

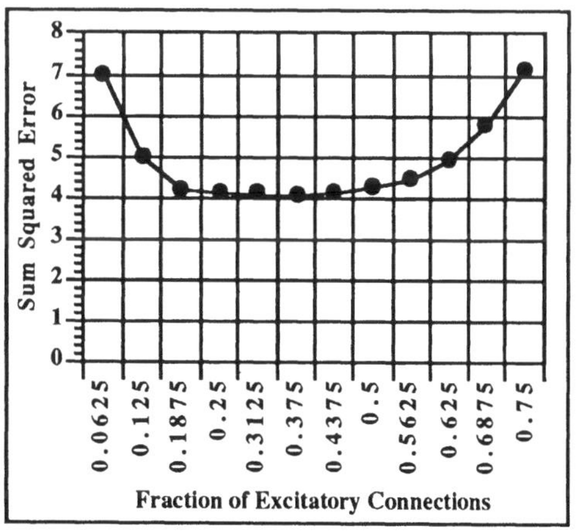

Figure 3: Error vs. Weights

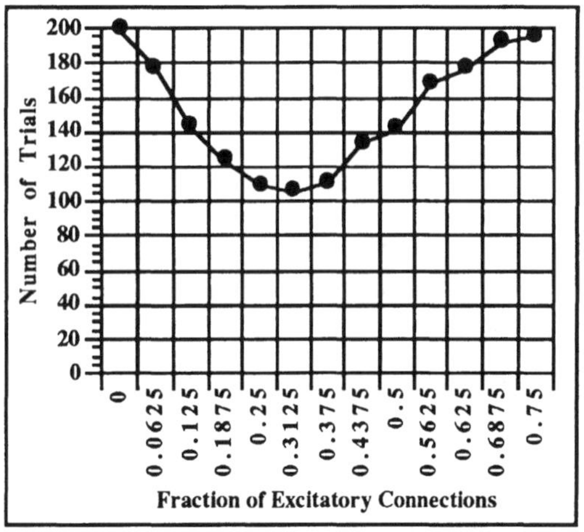

Figure 4: Trials to Converge vs. Weights

organism as a whole, there has not been a similar suggestion from computational simulation.

A complex set of events must occur in a neuron for it to initiate an electrical signal, and this depends largely on the architecture of the individual neuron and the nature of the chemical signals it receives. Each afferent signal (neuronal input) can be viewed as having either excitatory (facilitatory) or inhibitory effects on the development of an action potential. The signals constituting these inputs manifest a variety of interesting temporal and spatial organizations as well as complex local interactions [Koch, et al., 1983], all of which contribute to their ultimate computational conclusion — whether or not to initiate an action potential.

While the relative proportion of inhibitory and excitatory synapses in the central nervous system is not known, attempts have been made to quantify these proportions by a variety of methods. Immunocytochemical analyses have led to the view that gamma-amino butyric acid (GABA) is the most prevalent inhibitory neurotransmitter of the central nervous system [Kandel and Schwartz, 1985]. Smith [1989] even suggests that GABA is the most widely used neurotransmitter of any kind in the CNS, with over 40% of all synapses using GABA.

Support for this idea comes from studies of cortical interneurons, which suggest that most are GABAergic. These studies, summarized by Jones and Hendry [1986], use three different techniques in arriving at the conclusion: (a) [^{3}H]GABA uptake; (b) immunoreactivity for GABA; or (c) immunoreactivity for glutamic acid decarboxylase (GAD).

In the prestriate visual system of the macaque, the lateral geniculate nucleus contains significant immunoreactivity for glutamic acid decarboxylase (GAD), an enzyme required for GABA synthesis [Shaw and Cynader, 1986]. In the optic tectum of the frog, nearly one third all tectal cells are immunoreactive for GABA. In the striate cortex of the macaque monkey, layers 2, 3, 4A, and 4C contain large concentrations of GABA receptors [Shaw and Cynader, 1986].

Physiological study has led to the notion of orientation selectivity as a fundamental organizing principle of the visual cortex [Hubel and Wiesel, 1962]. The computational implementation of orientation selectivity requires that a bar of excitation be surrounded by a massive ring of inhibition, in order to eliminate ambiguity in the perception of an edge in the desired orientation. This computational constraint suggests that a large number of synapses act principally in an inhibitory manner.

Pharmacological evidence to support this postulate comes from studies of the selective GABA antagonist bicuculline. Application of bicuculline to orientation selective nerve cells in the cortex of the cat abolishes their response to the correctly oriented bar of light [Sillito, 1986].

Investigation of the ultrastructural (anatomical) differences between two types of synapses has led to different results. Type I (round asymmetric or RA) synapses, which are frequently excitatory, have asymmetrical densification of their pre- and post-synaptic membranes, and are associated with round synaptic vesicles. Type II (flat symmetric or FS) synapses, frequently inhibitory, have symmetrical densification and have flattened or pleomorphic vesicles [Gray, 1959; Shepherd and Koch, 1990].

Beaulieu and Colonnier [1985] studied the cat's visual cortex using these methods and concluded that about 84% of the synapses are of the RA type (usually excitatory) and 16% of the FS type (usually inhibitory). Two main questions remain in interpreting this data (and other data like it): (1) Do these RA synapses contain primarily an excitatory neurotransmitter, an inhibitory neurotransmitter, or both? (2) What is the relationship between the number of inhibitory synapses and the magnitude of their computational effects? Ultimately, we need to know the extent to which anatomical and physiological information bears on the computational issues and vice versa.

There is an apparent discrepancy between anatomical and physiological data regarding neuronal processing in the primary visual cortex of the cat. The anatomical results are particularly difficult to interpret, since knowledge of the number of synapses with a particular morphological structure does not necessarily indicate how these synapses are used computationally in the actual physiological setting. Combinations of excitatory and inhibitory synapses in complex topographical arrangements lead to intricate local circuit behaviors that may not correlate in any simple way with their absolute numbers. For example, a single inhibitory synapse, appropriately placed, can negate multiple excitatory stimuli.

Conclusion

While the integration of ultrastructural, physiological and computational data may require the development of new techniques, the goal of doing so may have important consequences for the understanding of structure/function relationships. Using computer modelling techniques and abstract representations of neurons and synapses, the present study suggests a preeminent role for inhibition in the computational organization of the brain.

In the brain, local circuit organization of inhibitory synapses, regardless of their absolute numbers, can have a controlling effect. When these are located closer to the soma than the excitatory synapses, they can (under certain circumstances) totally negate the excitatory effects [Koch, et al., 1983; Shepherd and Brayton, 1987; Shepherd and Koch, 1990]. Whether or not this computational effect bears on the situation in the visual cortex is not clear. However, the computational and physiological data suggest that the apparent preponderance of (typically excitatory) RA synapses in this area does not correlate with a preponderance of overall excitatory activity there.

The present study was initiated in response to speculation about the importance of inhibition in the physiological function of the human brain [Fromm, 1992]. The modelling results demonstrate a highly significant role of inhibition in particular artificial neural networks (containing sparsely coded output representations) and support the concept of inhibition as a basic computational feature of the brain.

References

Beaulieu, C. and M. Colonnier (1985): A Laminar Analysis of the Number of Round-Asymmetrical and Flat-Symmetrical Synapses on Spines, Dendritic Trunks, and Cell Bodies in Area 17 of the Cat, *J Comp Neurol*, 231:180-189.

Churchland, P. S. and T. J. Sejnowski (1987): Neural Representation and Neural Computation, The Johns Hopkins University.

Desimone, R., S. J. Schein, J. Moran and L. G. Ungerleider (1985): Contour, Color, and Shape Analysis Beyond the Striate Cortex, *Vision Res*, 25:441-452.

Fahlman, S. E. (1988): An Empirical Study of Learning Speed in Back-Propagation Networks, Carnegie Mellon University.

Feldman, J. A. (1989): Neural Representation and Neural Computation, in *Neural Connections, Mental Computation.*

Fromm, G. H. (1992): Neurophysiological Speculations on Zen Enlightenment, *J Mind Behav*, 13(2):163-168.

Gray, E. G. (1959): Axo-somatic and Axo-dendritic Synapses of the Cerebral Cortex: An Electron Microscope Study, *J Anat*, 93:420-433.

Hubel, D. H. and T. N. Wiesel (1962): Receptive Fields, Binocular Interaction, and Functional Architecture of Monkey Striate Cortex, *J Physiol (London)*, 160:106-154.

Jones, E. G. and S. H. Hendry (1986): Co-Localization of GABA and Neuropeptides in Neocortical Neurons, *TINS*, 9:71-76.

Kandel, E. R. and J. H. Schwartz (ed.) (1985): *Principles of Neural Science (Second Edition).*

Koch, C., T. Poggio and V. Torre (1983): Nonlinear Interactions in a Dendritic Tree: Localization, Timing, and Role in Information Processing, *Proc Natl Acad Sci USA*, 80:2799-2802.

McClelland, J. L. and D. E. Rumelhart (1986): *Parallel Distributed Processing: Explorations in the Microstructure of Cognition: Volume 2: Psychological and Biological Models.*

Mishkin, M., L. G. Ungerleider and K. Macko A. (1983): Object Vision and Spatial Vision: Two Cortical Pathways, *Trend Neurosci*, 6:414-417.

Rueckl, J. G., K. R. Cave and S. M. Kosslyn (1989): Why are "What" and "Where" Processed by Separate Cortical Visual Systems? A Computational Investigation, *J Cog Neurosci*, 1:171-186.

Rumelhart, D. E., G. E. Hinton and R. J. Williams (1985): Learning Internal Representations by Error Propagation, University of California San Diego.

Rumelhart, D. E. and J. L. McClelland (1986): *Parallel Distributed Processing: Explorations in the Microstructure of Cognition: Volume 1: Foundations.*

Seidenberg, M. S. and J. L. McClelland (1989): A Distributed, Developmental Model of Word Recognition and Naming, *Psych Rev*, 96:523-568.

Sejnowski, T., C. Koch and P. Churchland (1988): Computational Neuroscience, *Science*, 241:1299-1306.

Shaw, C. and M. Cynader (1986): Laminar Distribution of Receptors in Monkey (Macaca Fascicularis) Geniculostriate System, *J Comp Neurol*, 248:301-312.

Shepherd, G. M. and R. K. Brayton (1987): Logic Operations are Properties of Computer-Simulated Interactions between Excitable Dendritic Spines, *Neuroscience*, 21:151-166.

Shepherd, G. M. and C. Koch (1990): Introduction to Synaptic Circuits, in *Synaptic Organization of the Brain.*

Sillito, A. M., Functional Considerations of the Operation of GABAergic Inhibitory Processes in the Visual Cortex, in *Cerebral Cortex, Volume 2: Functional Properties of Cortical Cells.*

Small, S. L. (1991): Focal and Diffuse Lesions of Cognitive Models, *Proceedings of the Thirteenth Annual Meeting of the Cognitive Science Society.*

Smith, C. U. M. (1989): *Elements of Molecular Biology.*

Sun, R., E. Marder and D. Waltz (1988): The Modelling of Lobster Stomatogastric Neural Networks, Brandeis University.

Relearning after Damage in Connectionist Networks: Implications for Patient Rehabilitation*

David C. Plaut
Department of Psychology
Carnegie Mellon University
Pittsburgh, PA 15213–3890
plaut+@cmu.edu

Abstract

Connectionist modeling is applied to issues in cognitive rehabilitation, concerning the degree and speed of recovery through retraining, the extent of generalization to untreated items, and how treated items are selected to maximize this generalization. A network previously used to model impairments in mapping orthography to semantics is retrained after damage. The degree of relearning and generalization varies considerably for different lesion locations, and has interesting implications for understanding the nature and variability of recovery in patients. In a second simulation, retraining on words whose semantics are atypical of their category yields more generalization than retraining on more prototypical words, suggesting a surprising strategy for selecting items in patient therapy to maximize recovery.

Introduction

Cognitive neuropsychology aims to extend our understanding of normal cognitive mechanisms by studying their pattern of breakdown following brain damage in neurological patients. An underlying motivation for many researchers is that a more detailed analysis of the normal mechanism, and the way it is impaired in particular patients, should lead to the design of more effective therapy to remediate these impairments (Howard & Hatfield, 1987). Significant progress has been made in analyzing cognitive mechanisms and their impairments in terms of "box-and-arrow" information-processing diagrams, particularly in the domain of written language (Coltheart et al., 1980, 1987; Patterson et al., 1985). However, relatively few remediation studies have been based directly on these cognitive analyses, and while these few have been fairly successful, the specific contribution of the analysis is often unclear (for examples and general discussion, see Byng, 1988; Caramazza, 1989; Seron & Deloche, 1989; Wilson & Patterson, 1990). In large part the limited usefulness of box-and-arrow diagrams in this regard may stem from the general lack of attention paid to specifying the actual representations and computations that perform a task (Seidenberg, 1988).

Recently, a number of researchers employing connectionist models have attempted to go beyond the box-and-arrow approach by demonstrating that a fully-specified implementation of the normal process, when damaged, actually behaves like patients with analogous brain damage (e.g. Farah & McClelland, 1991; Hinton & Shallice, 1991; Mozer & Behrmann, 1990; Patterson et al., 1990; Plaut, 1991; Plaut & Shallice, 1991a, 1991b, 1992). This paper attempts to extend connectionist modeling in neuropsychology to address issues in cognitive rehabilitation. These issues concern degree and speed of recovery through retraining, the extent of generalization to untreated items, and how treated items can be selected to maximize this generalization.

The domain of investigation is impaired word reading, known as "acquired dyslexia." First, studies on remediation in acquired dyslexia based on cognitive models of normal reading are summarized, focusing on a study by Coltheart & Byng (1989) that attempted to reestablish the mapping between written words (orthography) and their meanings (semantics). A set of simulation experiments are presented in which a network, previously used to model impaired reading for meaning (Hinton & Shallice, 1991), is retrained after different lesions in which a proportion of the connections between groups of units are removed. The amount of recovery and generalization depends on the location of the lesion in the network and has interesting implications for understanding the effects seen in patients. The paper concludes with a second simulation demonstrating that retrain-

*I'd like to thank Marlene Behrmann and Geoff Hinton for their help with the research described in this paper. All of the simulations were run on a Silicon Graphics Iris-4D/240S using the Xerion simulator developed by Tony Plate. This research is supported by grant 87-2-36 from the Alfred P. Sloan Foundation, grant T89-01245-016 from the Pew Charitable Trusts, and grant ASC-9109215 from the National Science Foundation. Plaut (1992) presents an abstract of this work.

ing on words whose semantics are atypical of their category yields more generalization than retraining on more prototypical words, suggesting a surprising strategy for selecting items in patient therapy to maximize recovery.

Remediation of reading for meaning

Coltheart & Byng (1989) undertook a remediation study with an acquired dyslexic, EE, a 40-year-old left-handed postal worker who suffered left temporal-parietal damage from a fall. On the basis of a number of preliminary tests administered about 6 months later, they determined that EE had a specific deficit in deriving semantics from orthography. To improve the patient's word reading ability, Coltheart & Byng designed a study involving words containing the spelling pattern -OUGH (e.g. THROUGH, COUGH, BOUGH), which have highly irregular pronunciations and, thus, are difficult to read without semantics. EE was retrained on 12 of 24 such words, in which he studied the written words augmented with mnemonic pictures for their meaning (e.g. a picture of a tree drawn on the word BOUGH). Prior to therapy, four of the treated words were read correctly; after therapy, all 12 were read correctly. In addition, the *untreated* words also improved, from one correct prior to therapy, to seven correct after therapy. Thus, the improvement in the untreated set (6 words) was 75% as large as the improvement in the set that was actually treated (8 words). This generalization to untreated words is surprising because a word and its meaning are arbitrarily related—there is no intuitive reason why relearning the meanings of some words should help reestablish performance on other words with unrelated meanings.

In a second study, EE was given the 485 highest frequency words for oral reading. The 54 words he misread were divided in half randomly into treated and untreated sets. EE again learned to read the treated words by studying cards of the written words augmented with mnemonics for their meanings. As a result, his reading performance on the treated words improved from 44% to 100% correct. Once again, the untreated words also improved, from 44% to 85% correct (73% generalization). This improvement was not due to "spontaneous recovery" nor to other non-specific effects because performance on the words was stable both before therapy and after therapy.

Thus, in at least one patient, retraining the mapping from orthography to semantics for some words can generalize to other words. However, it should be noted that such improvement and generalization does not always occur. Some patients learn

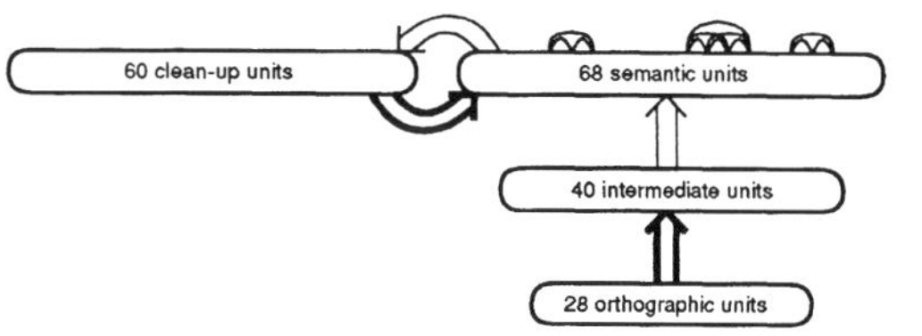

Figure 1: The network used by Hinton & Shallice (1991). Arrows in bold represent sets of connections that are lesioned in the present study.

the treated items but show no generalization to untreated items. Others show generalization within but not between modalities. Still others may have difficulty learning the treated items themselves. For instance, Behrmann (1987) found no generalization from treated to untreated homophonic word pairs (e.g. RIGHT and WRITE) in the writing of acquired dysgraphic CCM, although the writing of 75 irregular words did improve significantly. Scott & Byng (1988) found that retraining the reading of homophonic word pairs of an acquired dyslexic, JB, generalized to reading untreated pairs but not to his *writing* of either treated or untreated pairs.

Why some patients improve while others do not is not at all clear. An explanation of the effects seen in patient therapy in this domain should account not only for the occurrence of generalization in some patients and conditions, but also for its absence in others. Connectionist networks are proving useful in understanding the nature of impaired word reading—can they provide insight into the nature and variability of its recovery?

Modeling impaired reading for meaning

Hinton & Shallice (1991) have put forward a connectionist account of the process of accessing semantics from orthography, and the pattern of errors this process exhibits under damage. Based on previous work by Hinton & Sejnowski (1986), they trained a recurrent back-propagation network to map from the orthography of 40 three- or four-letter words to a simplified representation of their semantics, described in terms of 68 predetermined semantic features. The architecture of the network they used, shown in Figure 1, has two main pathways: (1) a "direct" pathway, from 28 orthographic units to 68 semantic units via 40 intermediate units, and (2) a "clean-up" pathway, from the semantic units to 40 clean-up units and back to the semantic units. The direct pathway generates initial semantic activity from visual (orthographic) input, while the clean-up pathway iteratively refines this initial activity into the exact correct semantics of the word.

After training the network, Hinton & Shallice systematically lesioned it by removing proportions of units or connections, or by adding noise to the weights. They found that the damaged network occasionally settled into a pattern of semantic activity that satisfied the response criteria for a word other than the one presented. These errors were more often semantically and/or visually similar to presented stimuli than would be expected by chance. While the network showed a greater tendency to produce visual errors (e.g. CAT ⇒ "cot") with lesions near the input layer and semantic errors (e.g. CAT ⇒ "dog") with lesions near the output layer, both types of error occurred for almost all sites of damage. This pattern of errors is similar to that of patients with deep dyslexia (Coltheart et al., 1980).

More recently, Plaut & Shallice (1991a, 1991b) have extended these initial findings in two ways. First, they established the generality of the co-occurrence of semantic, visual, and mixed visual-and-semantic errors by showing that it does not depend on peculiar characteristics of the network architecture, the learning procedure, or the way responses are generated from semantic activity. Second, they extended the approach to account for many of the remaining characteristics of deep dyslexia, including the effects of concreteness/imageability and their interaction with visual errors, the occurrence of visual-then-semantic errors, greater confidence in visual as compared with semantic errors, relatively preserved lexical decision with impaired naming, and the existence of different subvarieties of deep dyslexia.

The replication of the diverse set of symptoms of deep dyslexia through unitary lesions of the network strongly suggests that the underlying computational principles of the network capture important aspects of the process of mapping orthography to semantics in humans. Extending this claim further, we would expect relearning in the lesioned network to show similar effects to those observed in rehabilitation studies with analogous neurological patients. The following experiments test this claim.

Experiments in relearning after damage

A version of the Hinton & Shallice network was trained without momentum until it could read all 40 words perfectly (see Plaut, 1991, for details). The effects of lesions near orthography (orthography ⇒ intermediate connections) were compared with those of lesions within semantics (clean-up ⇒ semantics connections). For each of these two sets of connections, a severity of lesion was selected which lowered correct performance to near 20% (30% of orthography ⇒ intermediate connections, and 50% of clean-up ⇒ semantics connections).

For a given instance of a lesion, the responses to the 40 words were categorized as correct or incorrect. A response was considered correct if the proximity (i.e. normalized dot-product) of the semantics generated by the network was within 0.8 of the correct semantics of the presented word, and the proximity of the next best word was at least 0.05 further. Half of the correct words and half of the incorrect words were randomly selected and placed in the "treated" set; the remaining words were placed in the "untreated" set. Thus, both the treated and untreated sets always contained 20 words and were balanced for correct performance.

The lesioned network was then retrained for 50 sweeps on the treated words only. Performance was measured at each sweep during relearning separately for the treated and untreated word sets, in terms of the average percentage of words read correctly using the response criteria. The two sets were then exchanged and the retraining was repeated, starting from the same initial (lesioned) set of weights. Finally, the weights were again reinitialized and the lesioned network was retrained on all 40 words.

Figure 2 presents the retraining results for both locations of lesion, averaged over all 20 lesion instances and over exchanges of the treated and untreated word sets. First consider lesions within semantics (left of the figure). The treated words are quickly relearned by the network, with performance improving from near 20% to over 90% correct in under 20 sweeps through the word set. In addition, there is considerable generalization from the treated to untreated word sets (mean generalization 0.61, $t(39) = 28.1$, $p < .001$). Correct performance on the untreated words improves from 20% to 68% even though these words are never presented to the lesioned network. In fact, relearning on all of the words is quite dramatic, with performance recovering completely after 50 sweeps. These results replicate earlier findings on relearning and generalization in connectionist networks after corrupting weights with noise (Hinton & Plaut, 1987; Hinton & Sejnowski, 1986).

In contrast, retraining after lesions near orthography results in a quite different pattern of performance (see the right of Figure 2) . Relearning the treated words proceeds more slowly, with over 40 sweeps required to raise performance above 90%. Relearning all 40 words is even slower and more erratic. More importantly, there is no evidence of generalization to the untreated words—if anything, average correct performance on these words shows a trend towards getting slightly worse (mean generalization: -0.024,

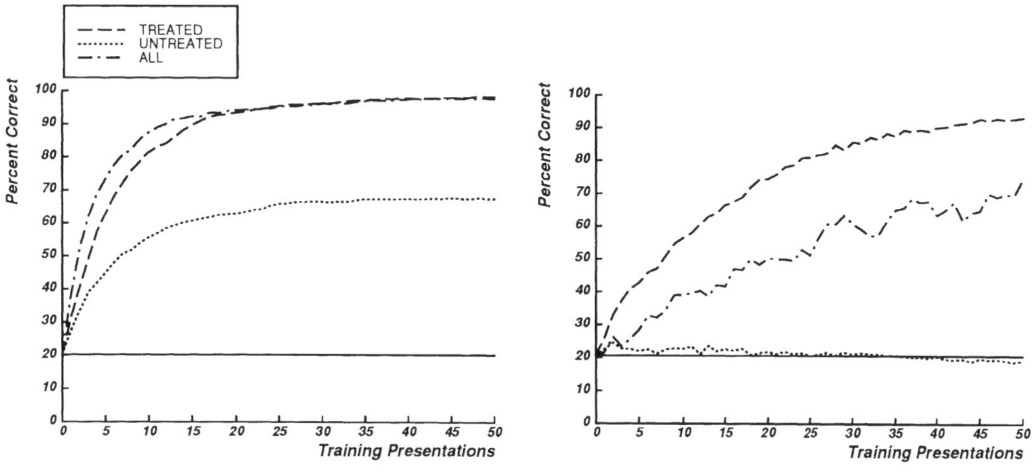

Figure 2: Retraining performance after clean-up ⇒ semantics lesions (left) and orthography ⇒ intermediate lesions (right). The solid horizontal lines represent the levels of performance at the onset of retraining.

$t(39) = 1.17$, $p = .25$).

Why does retraining after lesions within semantics yield rapid relearning and considerable generalization while retraining after lesions near orthography produces much worse relearning and no generalization? The degree of relearning and generalization depends on the consistency of the weight changes (i.e. directions of movement in "weight space") that would be optimal for individual words. While this is typically described in terms of the degree of overlap in the distributed representations of words, it depends more precisely on the consistency or structure in the mapping from input to output. Viewed as an abstract task, there is no systematic structure in mapping orthographic strings onto their semantics—input similarity is unrelated to output similarity. However, when instantiated in a network, the task is broken down by the learning procedure into a number of separate transformations involving intermediate representations carried out by different parts of the network. These transformations constitute "subtasks" that may differ considerably in their degree of structure. For example, the subtask of the clean-up ⇒ semantics connections is to refine the initial semantic activity generated by the direct pathway into the exact correct semantics of the presented word. Since semantically similar words require similar clean-up, this subtask is highly structured. In contrast, the subtask of the orthographic ⇒ intermediate connections is to generate intermediate layer representations that are as semantically organized as possible from visually organized inputs. Since semantic similarity is unrelated to visual similarity, there is no structure in this subtask. However, to the extent that the orthographic ⇒ intermediate connections succeed in generating semantically organized representations, the subtask of the intermediate ⇒ semantics connections becomes (semantically) structured. Consistent with this interpretation, relearning after lesions to these connections yields moderate but significant generalization (24%; see Plaut, 1991). Thus, the effectiveness of relearning after a lesion to a set of connections reflects the degree to which the mapping those connections carry out is structured.

As described above, studies of cognitive rehabilitation of acquired dyslexics in the domain of reading for meaning have demonstrated considerable relearning of treated items and (often) improvement on untreated but related items. Relearning after lesions to a network that operates in the same domain results in similar qualitative effects for lesions within semantics but not for lesions near orthography. Thus, at a general level, the cause of rapid relearning and generalization in the network—distributed representations and structure in subtasks—may provide an explanation for the nature of recovery, and lack of recovery, in these patients.

A specific hypothesis that comes out of the relearning simulations relates to the systematic differences observed in the degree of relearning and generalization as a function of lesion location. The simulations predict that a patient with a functional impairment close to or within semantics should show considerable generalization, while one with an impairment close to orthography should show little or none. Conversely, the degree of generalization observed in a patient can be used to predict the fine-grained location of their functional impairment *within* the mapping from orthography to semantics.

Designing therapy to maximize generalization

Ideally, we would like to use our understanding of the impairment in a particular patient to lead to the design of a rehabilitation strategy that maximizes recovery. The previous simulation clarifies the conditions under which retraining yields generalization. Under these conditions, how can items be selected for retraining so as to maximize this generalization? A critical variable in semantic representation is prototypicality—how close a concept is to the central tendency of its category (Rosch, 1975). The question is, is it better to retrain on prototypical or non-prototypical words?

Unfortunately, the limited size and complexity of the original training set precludes a reasonable comparison. Accordingly, a second simulation study was carried out, analogous to the first except that it involved 100 "words" whose orthographic and semantic representations were artificially generated. First, a single semantic "prototype" was created by randomly setting each of 50 semantic features to be present with probability $p = 0.2$. Two sets of 50 word meanings were generated from this prototype using different levels of random distortion (Chauvin, 1988). A "prototypical" set consisted of small distortions of the prototype (each feature of a word had a probability $d = 0.1$ of being randomly regenerated with $p = 0.2$). A "nonprototypical" set consisted of large distortions ($d = 0.5$). Orthography was represented as random patterns of activity ($p = 0.2$) over 20 input features. Using the same architecture and learning procedure as in the first study, a network was trained to generate the appropriate semantic features from each orthographic pattern. We investigated relearning after lesions to the intermediate ⇒ semantics connections because they yielded only moderate generalization. Seventy instances of lesions of severity 0.25 reduced overall correct performance to 35.6% on average.

After each lesion, words were divided into prototypical and non-prototypical groups as described above, and then one group was further divided in half (balanced for correct performance). One of these halves formed the treated set, while the other formed one untreated set, and the words of the opposite type formed a second untreated set. The network was then retrained for 50 presentations of the treated set. Figure 3 presents the average generalization (i.e. ratio of untreated to treated improvement in correct performance, using a simple best-match criterion) from prototypical and non-prototypical treated sets to prototypical and non-prototypical untreated sets. Overall, retraining on non-prototypical words produces more generalization than retraining on prototypical words ($F(1,69) = 337.4$, $p < .001$). The figure shows that this effect is due primarily to the fact that retraining on prototypical words significantly *reduces* performance on untreated non-prototypical words.

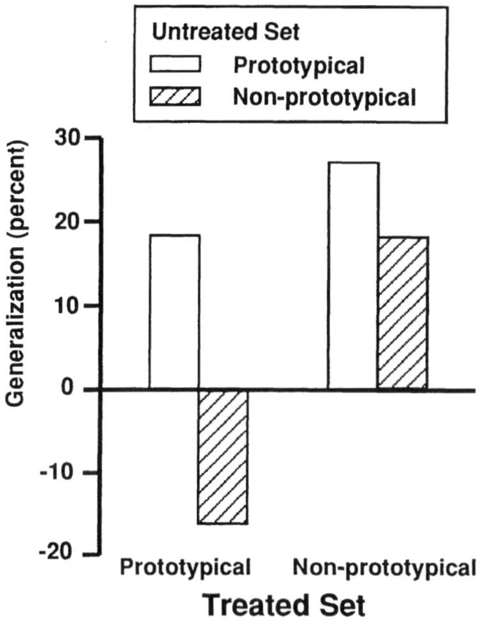

Figure 3: Generalization from prototypical or non-prototypical treated sets, to prototypical or non-prototypical untreated sets.

We can understand this effect by analogy with a set of randomly distributed points, where each point represents the effects of training on a particular word. The average of the outliers (non-prototypical words) may well-approximate the central points (prototypical words), but the average of the central points is still quite far from the outliers.

Summary

Theoretical analyses of cognitive impairments following brain damage should lead to the design of more effective strategies for rehabilitation. Simulations in this paper extend the relevance of connectionist modeling in neuropsychology to address issues in rehabilitation.

Attempts at cognitive rehabilitation of the mapping between orthography and semantics in patients have resulted in considerable improvement in performance on treated words, as well as significant generalization to untreated but related words, although the degree of recovery can vary considerably. The degree of relearning and generalization after damage in a network that performs the analogous task depends considerably on the location of lesion. These differences can be understood in terms of the amount of structure in the subtasks performed by parts of the network. The differences also provide a possible explanation for the variability in recovery observed in patients, and generate hypotheses about the specific location of their underlying functional impairment.

A potential benefit of connectionist modeling in neuropsychological rehabilitation is that it provides a framework for investigating the relative effectiveness of alternative rehabilitation strategies. A second simulation found that retraining on less prototypical words produced more generalization that retraining on more prototypical words, suggesting a surprising strategy for selecting items in patient therapy to maximize recovery.

Overall, the results demonstrate that investigations of relearning after damage in connectionist networks can provide an account of the general nature of relearning and generalization in patients and can generate interesting hypotheses about the design of effective patient therapy.

References

Behrmann, M. (1987). The rites of righting writing: Homophone remediation in acquired dysgraphia. *Cognitive Neuropsychology*, 4(3):365–384.

Byng, S. (1988). Sentence processing deficits: Theory and therapy. *Cognitive Neuropsychology*, 5(6):629–676.

Caramazza, A. (1989). Cognitive neuropsychology and rehabilitation: An unfulfilled promise? In Seron, X. & Deloche, G., editors, *Cognitive Approaches in Neuropsychological Rehabilitation*, chapter 12, pages 383–398. Lawrence Erlbaum Associates, Hillsdale, NJ.

Chauvin, Y. (1988). *Symbol Acquisition in Humans and Neural (PDP) Networks*. PhD thesis, University of California, San Diego.

Coltheart, M. & Byng, S. (1989). A treatment for surface dyslexia. In Seron, X. & Deloche, G., editors, *Cognitive Approaches in Neuropsychological Rehabilitation*, chapter 5, pages 159–174. Lawrence Erlbaum Associates, Hillsdale, NJ.

Coltheart, M., Patterson, K. E., & Marshall, J. C. (1980). *Deep Dyslexia*. Routledge, London.

Coltheart, M., Sartori, G., & Job, R. (1987). *The Cognitive Neuropsychology of Language*. Lawrence Erlbaum Associates, Hillsdale, NJ.

Farah, M. J. & McClelland, J. L. (1991). A computational model of semantic memory impairment: Modality-specificity and emergent category-specificity. *Journal of Experimental Psychology: General*, 120(4):339–357.

Hinton, G. E. & Plaut, D. C. (1987). Using fast weights to deblur old memories. In *Proceedings of the 9th Annual Conference of the Cognitive Science Society*, pages 177–186, Seattle, WA.

Hinton, G. E. & Sejnowski, T. J. (1986). Learning and relearning in Boltzmann Machines. In Rumelhart, D. E., McClelland, J. L., & the PDP research group, editors, *Parallel Distributed Processing: Explorations in the Microstructure of Cognition. Volume 1: Foundations*, chapter 7, pages 282–317. MIT Press, Cambridge, MA.

Hinton, G. E. & Shallice, T. (1991). Lesioning an attractor network: Investigations of acquired dyslexia. *Psychological Review*, 98(1):74–95.

Howard, D. & Hatfield, F. M. (1987). *Aphasia therapy*. Lawrence Erlbaum Associates, Hillsdale, NJ.

Mozer, M. C. & Behrmann, M. (1990). On the interaction of selective attention and lexical knowledge: A connectionist account of neglect dyslexia. *Journal of Cognitive Neuroscience*, 2(2):96–123.

Patterson, K. E., Coltheart, M., & Marshall, J. C. (1985). *Surface Dyslexia*. Lawrence Erlbaum Associates, Hillsdale, NJ.

Patterson, K. E., Seidenberg, M. S., & McClelland, J. L. (1990). Connections and disconnections: Acquired dyslexia in a computational model of reading processes. In Morris, R. G. M., editor, *Parallel Distributed Processing: Implications for Psychology and Neuroscience*. Oxford University Press, London.

Plaut, D. C. (1991). *Connectionist Neuropsychology: The Breakdown and Recovery of Behavior in Lesioned Attractor Networks*. PhD thesis, School of Computer Science, Carnegie Mellon University. Available as Technical Report CMU-CS-91-185.

Plaut, D. C. (1992). Rehabilitating reading for meaning: Experiments in relearning after damage in connectionist networks. *Journal of Clinical and Experimental Neuropsychology*, 14(1):49.

Plaut, D. C. & Shallice, T. (1991a). Deep dyslexia: A case study of connectionist neuropsychology. Technical Report CRG-TR-91-3, Connectionist Research Group, Department of Computer Science, University of Toronto, Toronto, Ontario, Canada. Submitted to *Cognitive Neuropsychology*.

Plaut, D. C. & Shallice, T. (1991b). Effects of abstractness in a connectionist model of deep dyslexia. In *Proceedings of the 13th Annual Conference of the Cognitive Science Society*, pages 73–78, Chicago, IL.

Plaut, D. C. & Shallice, T. (1992). Perseverative and semantic influences on visual object naming errors in optic aphasia: A connectionist account. Technical Report PDP-CNS-92-1, Series on Parallel Distributed Processing and Cognitive Neuroscience, Department of Psychology, Carnegie Mellon University, Pittsburgh, PA. Submitted to *Journal of Cognitive Neuroscience*.

Rosch, E. (1975). Cognitive representations of semantic categories. *Journal of Experimental Psychology: General*, 104:192–233.

Scott, C. & Byng, S. (1988). Computer assisted remediation of a homophone comprehension disorder in surface dyslexia. *Aphasiology*.

Seidenberg, M. (1988). Cognitive neuropsychology and language: The state of the art. *Cognitive Neuropsychology*, 5(4):403–426.

Seron, X. & Deloche, G. (1989). *Cognitive Approaches in Neuropsychological Rehabilitation*. Lawrence Erlbaum Associates, Hillsdale, NJ.

Wilson, B. & Patterson, K. E. (1990). Rehabilitation for cognitive impairment: Does cognitive psychology apply? *Applied Cognitive Psychology*, 4:247–260.

Modelling Paraphasias in Normal and Aphasic Speech

Trevor A. Harley & Siobhan B. G. MacAndrew
Department of Psychology
University of Warwick
Coventry CV4 7AL
England
EMAIL: psrds@csv.warwick.ac.uk

Abstract

We model word substitution errors made by normal and aphasic speakers with an interactive activation model of lexicalization. This comprises a three-layer architecture of semantic, lexical, and phonological units. We test four hypotheses about the origin of aphasic word substitutions: that they result from pathological decay, loss of within-level inhibitory connections, increased initial random noise, or reduced flow of activation from the semantic to the lexical level. We conclude that a version of the final hypothesis best explains the aphasic data, but with random fluctuations in connection strength rather than a uniform decrement. This model accounts for aspects of recovery in aphasia, and frequency and imageability effects in paraphasias. Pathological lexical access is related to transient lexical access difficulties in normal speakers to provide an account of normal word substitution errors. We argue that similar constraints operate in each case. This model predicts imageability and frequency effects which are verified by analysis of our normal speech error data.

Introduction

Paraphasias are the erroneous substitution of one word for another in speech. They occur as errors in normal speech (e.g. "warm" -> *cold*), and as word substitutions which arise in the acquired speech deficit of jargon aphasia (see Butterworth, 1985). The speech of these patients is copious but characterized by gross word finding difficulties. We outline how aphasic paraphasias might be explicable within the context of a model of normal speech production. We "lesion" a connectionist model of lexicalization so that it produces paraphasias similar to those of aphasic speakers. Finally, we argue that similar mechanisms are involved in the production of errors in normal speech.

The model of lexical access upon which our simulation is based is close to that of Harley (1990) and Stemberger (1985), and uses an interactive activation architecture (McClelland & Rumelhart, 1981). It accounts for much that is known about normal speech production. It shares some features with that of Dell (1986), but differs importantly in that we use a different type of semantic representation. This enables us to explore semantic word substitutions in a more plausible way. Our model also has intra-level inhibitory connections that have the computational consequence of increasing the effects of within-level competition, and inter-level inhibitory connections that speed up processing by more quickly suppressing inappropriate competitors. These are motivated by findings of inhibitory priming in lexical access in a naming task (Wheeldon, 1989), and phonological blocking in the tip-of-the-tongue state (Jones, 1989).

Units in our model are organized into semantic, lexical, and phonological levels. There is converging evidence in the literature for two stages in speech production, with semantic representations first mapped into abstract lexical forms (our lexical level), followed by the retrieval of phonological forms (e.g. Levelt et al., 1991a). Motivated by speech error data such as phonological facilitation and lexical bias (Harley, 1984; Stemberger, 1985), our model postulates interaction between these stages. There is currently debate about the extent to which these stages are modular (Levelt et al., 1991b). However, interactive models can be shown to be consistent with naming data which at first sight support the modular hypothesis (Dell & O'Seaghdha, 1991).

Each unit in the model is connected to every unit in the following layer. Appropriate between-level connections (such as the lexical unit **dog** to the phonological unit /d/) are excitatory, whereas inappropriate connections (such as **dog** to /k/) are inhibitory. Units within the lexical and phonological levels are completely inter-connected by inhibitory connections. There are feedback connections between the phonological and lexical levels. As usual the net input net_u to a unit u is the sum of the products of all inputs a_j from j units with the weights of the appropriate connections w_{ju}, $\Sigma a_j w_{ju}$. In each cycle the change in activation of unit u is given by the equations:

This research was supported by a grant from the UK Tri-Council Initiative on Cognitive Science/HCI, No. SPG 9018232.

$$\Delta a_u = (\max - a_u)net_u - decay(a_u - rest)$$

if $net_u > 0$; otherwise,

$$\Delta a_u = (a_u - \min)net_u - decay(a_u - rest)$$

Max and *min* are the maximum and minimum levels of activation, and *rest* a resting level of activation dependent for lexical units upon the frequency of those words and close to zero for other units. Separate parameters control the decay of activation at each level. The degree to which the resting levels of lexical units vary around a mean of zero is determined by the value of a parameter *freqgain* whereby:

$$rest = freqgain * (\log_e(\text{item frequency}) - \text{mean} \log_e(\text{item frequency}))$$

Units also possess a variable amount of normally distributed random noise at the beginning of each processing epoch. The standard deviation of this distribution is determined by a parameter. Our simulated lexicon contains 70 lexical units, which receive input from 26 semantic feature units and send output to 21 phonological units for each of five positions in a serial order phonological output frame. For simplicity, each input feature is a simple on-off binary unit, a semantic representation similar to that of Hinton and Shallice (1991).

The model's architecture is shown in Figure 1. Figure 2 illustrates the normal time course of lexicalization of the word **cow**, the target in all subsequent examples. It shows the activation level of units plotted against processing cycle (time) when the semantic features corresponding to the semantic representation of **cow** are activated. For illustrative purposes, the activations of a semantically and phonologically competing lexical neighbour (**calf**) and of an unrelated control word (**dart**) are also shown, as are the activation values of the target's initial phoneme (/k/) and a non-target phoneme (/d/). Such simulations produce an accurate account of normal lexicalization. The model can also account for findings such as data on the time course of lexicalization in picture naming and facilitation in speech error data.

Lesioning the network

We wish to show that under certain conditions paraphasias are produced by the model when it is nevertheless given the target semantic input. Although it is our goal to produce all types of error, at present we are concentrating upon failures of lexical access. This is a vital first step towards explaining a further phenomenon of jargon aphasia, the production of neologisms (non-words). If there is a clear competitor to the lexical target, a word substitution is likely to occur. If there is either no competitor or a number of equally activated competitors, then the conditions for the generation of neologisms have been met. Finally, we contend that pathological paraphasias are very similar in key respects to normal paraphasias.

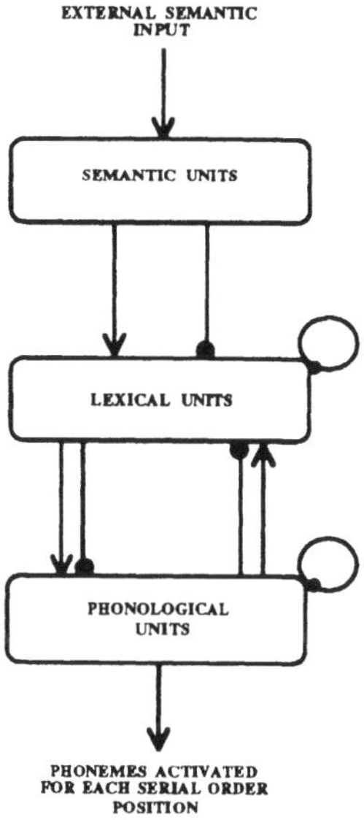

Fig. 1. General architecture of our lexicalization model, with excitatory connections shown by an arrow, inhibitory connections by a filled circle.

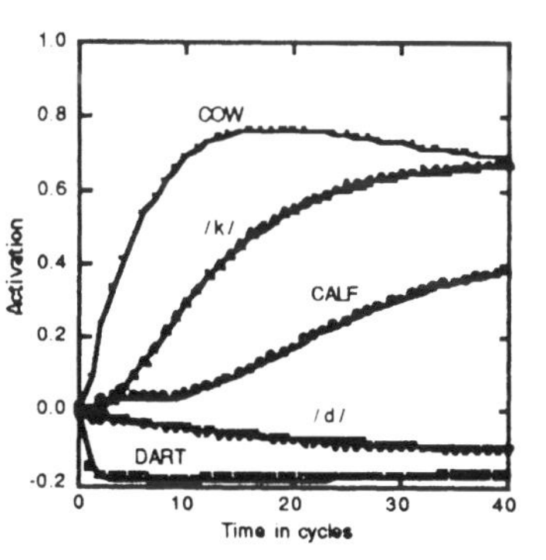

Fig. 2. The time course of a simulation of the normal lexicalization of the word "cow".

To produce substitutions we must somehow disrupt the flow of activation from the input semantic units to the output phonological units. We mimic lesioning by manipulating the parameters of our model. These include both connection strengths, and the control parameters of the network (e.g. the rate of decay of activation, the amount of random noise, the effect of lexical frequency, and the time external semantic input is received by the semantic units). Rather than trying all possibilities, we specifically test predictions derived from the aphasia literature on the origin of paraphasias.

Increase in the Rate of Decay. Martin and Saffran (1991) describe a deep dysphasic patient NC whose speech output includes a large number of paraphasias and neologisms. NC also has a severely restricted phonological short-term memory. They argue that his symptoms arise from a pathological increase in the decay rate of the target lexical nodes. This increases the probability that phonologically or semantically related lexical words will replace the target. If this hypothesis is correct then increasing the parameter controlling the rate of decay of the lexical units in our simulation should increase the probability of paraphasias occurring. In particular, the activation level of the target should decrease as the rate of lexical decay increases, while those of its competitors should increase. The results of these simulations are shown in Figure 3. Although increasing lexical decay causes the activation of the target to fall, it is still considerably higher than those of its close competitors, and remains well above its resting level, even at exceptionally high levels of decay (0.99). Furthermore, with increasing decay, the activation levels of the competitors level off at a low value.

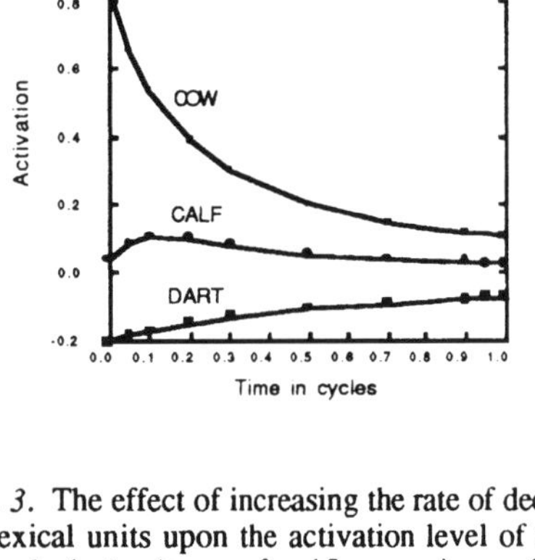

Fig. 3. The effect of increasing the rate of decay of lexical units upon the activation level of the target lexical unit **cow** after 15 processing cycles.

The effect of combining pathological decay with curtailing the time semantic units receive external activation that can in turn be passed on to lexical units is shown in Figure 4. Here the time external input is given to the semantic units is reduced to only 3 cycles. We do not think this is a plausible account of the generation of paraphasias and neologisms for two reasons. First, it requires two simultaneous deficits. Second, though the activation level of the target unit is reduced to near zero, so are those of its competitors. We propose therefore that a pathological increase in lexical decay is unlikely to produce paraphasias.

The Loss of Intra-level Inhibitory Connections. Harley (1990) proposed that the paragrammatisms often associated with neologistic jargon result from excessive blending of syntactic fragments as a consequence of the loss of within-level inhibitory connections. Can a similar mechanism also account for the presence of word substitutions? If so, then decreasing the value of the parameter that controls the degree of intra-lexical inhibition, *gammall*, should decrease the activation of the target unit and increase those of its competitors. Figure 5 shows the effect upon the target activation value for different levels of *gammall*.

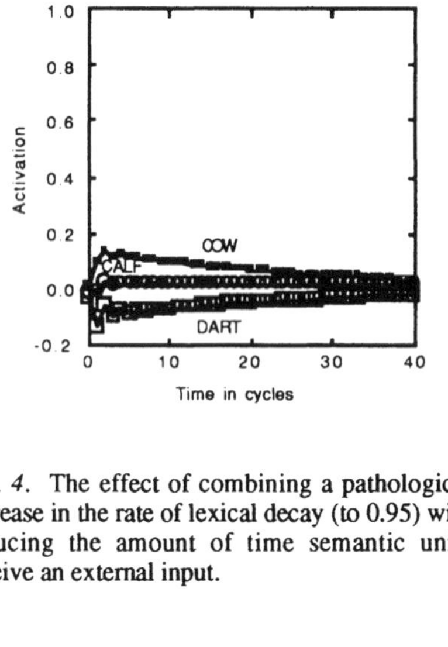

Fig. 4. The effect of combining a pathological increase in the rate of lexical decay (to 0.95) with reducing the amount of time semantic units receive an external input.

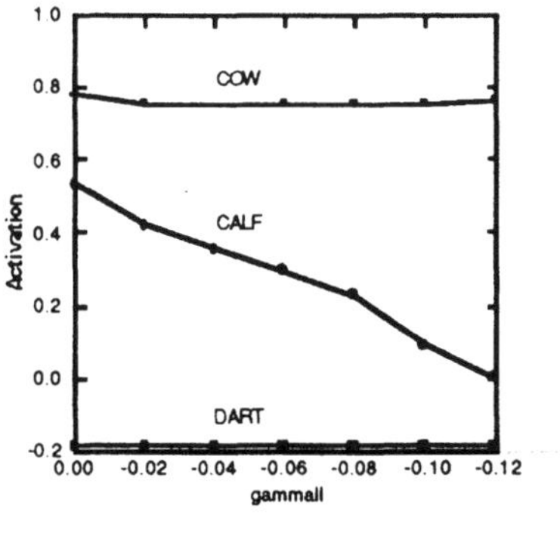

Fig. 5. The effect of decreasing the amount of intra-lexical inhibition upon the activation level of the target unit **cow** after 15 cycles.

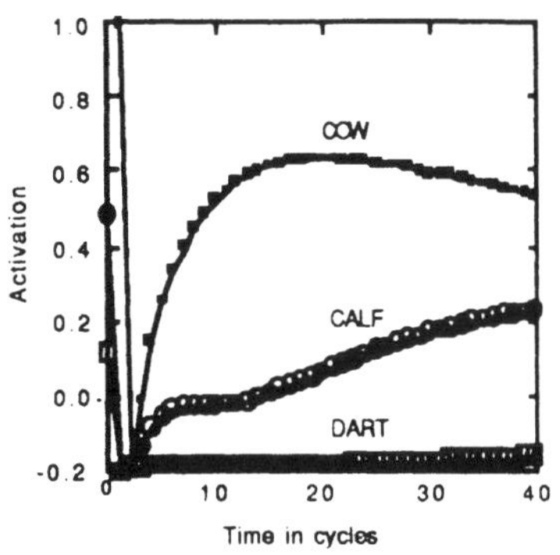

Fig. 6. The time course of activation for a number of competing units with high initial random noise.

Although decreasing the strength of these connections does increase the activation levels of the competitors, it also slightly increases the activation of the target. Even after 25 cycles the target unit is still the most activated, although semantic competitors also have high levels of activation. Such a mechanism then is unlikely to be able to account for the jargon data.

Increased Initial Random Noise. Another possibility is that increasing the amount of initial random noise will increase the probability of paraphasias occurring. Simulations were run with exceptionally high levels of initial lexical noise. Figure 6 shows the effect of increasing the lexical noise level a hundred fold. The target unit quickly recovers and then progresses normally. We can rule out high initial random noise as a causal factor in jargon paraphasias.

Weak Lexical Activation. Our final hypothesis is based on Miller and Ellis (1987). They argue that the impairments found in their patient RD can be explained by difficulty in activating lexical units in the speech output lexicon. They propose that the flow of activation from the semantic level to the lexical level in neologistic jargon aphasics is reduced to a trickle. As units at the lexical level have received insufficient activation, they cannot in turn properly activate the target phonemes. Other phonemes, which have high activation levels due to random noise, are usually accessed in preference.

If this hypothesis is correct then it should be possible to generate substitutions by reducing the value of the parameter that governs the rate of spread of activation between the semantic and lexical levels, *alphasl*. The results are shown in Figure 7. Manipulating *alphasl* does not behave exactly as predicted by the weak lexical activation hypothesis. Although a decrease in *alphasl* does decrease the activation of the target, over part of the range the activation levels of the semantically competing items decrease even more rapidly until very low levels of *alphasl* are reached. Hence the weak lexical activation hypothesis makes an additional prediction: if lower levels of *alphasl* are reflected in increasingly severe symptoms, more severe cases of jargon aphasia should show a lower level of semantic paraphasias relative to other types of word substitutions. We know of little data that address this issue, although Kertesz and Benson (1970) provide evidence from the evolution of aphasia that supports this prediction. They show that during recovery, there is a general progression from neologistic jargon to semantic jargon and then to circumlocutory anomic speech.

At a very low level of *alphasl* (0.0001) the activation level of the **cow** lexical unit has reached 0.662 by only the hundredth processing cycle, and is clearly distinguished from other lexical candidates. This suggests that if jargon aphasics had sufficient time, they would eventually retrieve the correct target. Clearly this is not the case, as lexicalization attempts do not improve over time and do not converge upon the target (Miller & Ellis, 1987). It is necessary to make the further assumption that the semantic units are unable to send activation to the lexical units for more than a fixed time. (Note that this time is not pathologically low, as in Figure 4.) This is consistent with data on the time course of lexicalization in picture naming (Levelt et al., 1991a). Even then, if semantic units send activation to lexical units for only 10 processing cycles, **cow** still reaches an activation level of 0.25 after 100 processing cycles, and the associated /k/ phoneme reaches a level of 0.55.

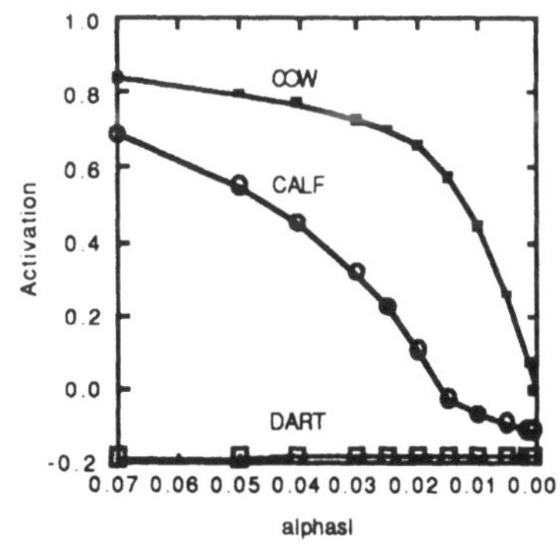

Fig. 7. The effect of reducing the strengths of the excitatory connections between the semantic and lexical levels, *alphasl*.

This hypothesis further predicts an interaction with frequency such that reducing *alphasl* has more effect upon low frequency items. More frequent words should be more robust because they have a higher resting level. They should hence be relatively well preserved at low levels of semantic-lexical facilitation compared with low frequency items. With aphasics, a major determinant of success for a patient attempting to produce a word is its frequency (Ellis, Miller & Sin, 1983). We compute the ratio of the activation level of a lexical unit at a high level of semantic-lexical facilitation (*alphasl* = 0.03) to its activation level at a low level of facilitation (*alphasl* = 0.005) after 15 processing cycles. We call this ratio the *sensitivity ratio* for a particular item as it reflects a lexical unit's sensitivity to different levels of *alphasl*. High frequency items should have lower sensitivity ratios than low frequency items. The sensitivity ratio was computed across a range of lexical frequencies. Figure 8 shows the result of these simulations. Although other factors are clearly operating, frequency does behave as predicted by the weak lexical activation hypothesis. Inspection suggests that the main origin of the residuals in a regression of frequency onto sensitivity ratio is the number of the semantic units that are "on" for any particular lexical item. That is, the effect of lowering the strength of the semantic-lexical connections is moderated not only by frequency, but also by the richness of the underlying semantic representation for each item. We take this to be reflected in the *imageability* of words, in the same way as Plaut and Shallice (1991). Further simulations teased out the differing contributions of lexical frequency and imageability. Two types of simulations were run with artificial lexical items. In the first, the effect of varying the frequency of the target lexical units was investigated while the semantic representation was held constant. In the second, the number of "on" semantic units in the input was varied while the frequency was held constant. In both cases near linear relationships are found between the sensitivity ratio and pure frequency and pure imageability. This further predicts that high imageability words should also be preferentially preserved in jargon aphasia independent of frequency. Again, we know of no data that directly address this issue, though deep dyslexics perform better on more imageable words (Coltheart, 1980).

Although this gives a more satisfying distribution of lexical activations than the other accounts, it still fails to satisfy the criterion that, on some occasions, the activation of competitors should be above that of the target. To achieve this, it is necessary to introduce some random variation into the weakening of the semantic-to-lexical connections. Hence the excitatory semantic-to-lexical connections were randomly lesioned. This was achieved by adding an amount of normally distributed random noise to each connection. The severity of lesioning is mimicked by increasing the standard deviation of the noise distribution. Lexical units then behave as desired (Figure 9). Random lesioning of *alphasl* affects the target lexical unit such that the greater the severity of the lesioning, the lower the probability of the target unit being highly activated. Further, the greater the lesioning, the higher the probability of other lexical units being highly activated. Of course, because this manipulation of *alphasl* is random, actual results vary from trial to trial, and this variation increases as the amount of lesioning increases.

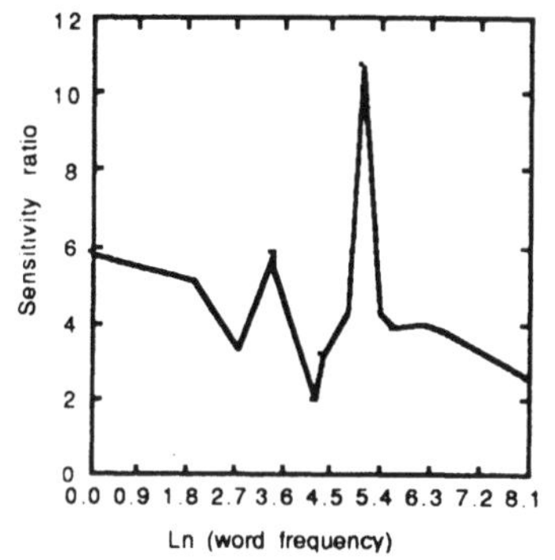

Fig. 8. The sensitivity ratio plotted against $\log_e$ (lexical item frequency).

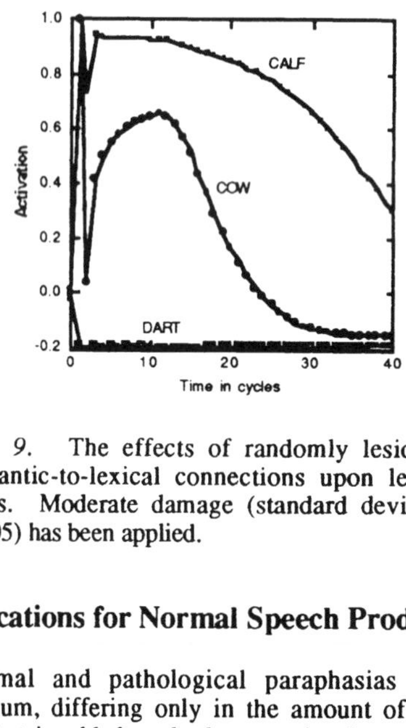

Fig. 9. The effects of randomly lesioning semantic-to-lexical connections upon lexical units. Moderate damage (standard deviation =0.05) has been applied.

Implications for Normal Speech Production

If normal and pathological paraphasias form a continuum, differing only in the amount of random noise that is added to the between-level connections, then they should share many characteristics. In particular, we can make two predictions about frequency and imageability. Those words that are most robust under noisy conditions are going to be the more frequent and imageable words in the language. Hence the words upon which errors occur should be of lower frequency than average, because these are just those items that are particularly susceptible to disruption. We also predict that when normal speakers make a spontaneous word substitution, the target word should be replaced by one more frequent and imageable. It is possible to test these predictions against our corpus of 5468 naturally occurring speech errors. We looked at completed content word substitutions where there was either a semantic or phonological relationship between the target and error words. Both semantic ($t[798] = 4.01$, $p < 0.001$, all results two-tailed) and phonological ($t[448] = 3.62$, $p < 0.001$,) targets were significantly lower in frequency than control words in the corpus. Semantic word substitutions resulted in more imageable words replacing the target ($t[201] = 2.42$, $p < 0.02$), although there was no effect for pure phonological cases ($t[25] = 0.57$, $p > 0.5$). This final result perhaps says no more than that semantic and phonological word substitutions arise at different loci, and that the latter are less affected by semantic constraints. It is a reminder that our model only adequately addresses semantic substitutions. Finally, it has been argued that jargon paraphasias and normal tip-of-the-tongue states share many properties (Miller & Ellis, 1987). Our current simulations suggest that they arise from weakened lexical-to-phonological connections.

Conclusions

The model described here has two important limitations. First, there are no phonotactic constraints: any string of phonemes is permissible. A related problem is that the slot-and-filler mechanism used to implement the serial ordering of phonemes is primitive and inconsistent with the connectionist, non-explicitly rule-based foundations of the model. However, as Miller and Ellis (1987) point out, because phoneme substitutions are random and within-word phoneme exchanges occur no more than would be expected by chance, for RD at least it is not necessary to postulate an additional phoneme ordering mechanism deficit. Another limitation of our work so far is that it is limited to monosyllabic, morphologically simple content words. Nevertheless, lesioning this model by adding noise to the semantic-to-lexical connections can account for a number of important characteristics of jargon aphasic speech.

We would like to conclude by pointing out that connectionist explanations of this type are not inconsistent with earlier hypotheses concerning the origins of jargon, but explain what is happening at a lower level of explanation. Earlier models point to a failure of lexical access; we hypothesize how that failure occurs.

References

Butterworth, B. (1985). Jargon aphasia: Processes and strategies. In S. Newman & R. Epstein eds., *Current perspectives in dysphasia.* Edinburgh: Churchill Livingstone.

Coltheart, M. (1980). Deep dyslexia: A review of the syndrome. In M. Coltheart, K. Patterson, & J. Marshall eds., *Deep dyslexia.* London: Routledge & Kegan Paul.

Dell, G. (1986). A spreading-activation theory of retrieval in sentence production. *Psychological Review* 93:283-321.

Dell, G., & O'Seaghdha, P. (1991). Mediated and convergent lexical priming in language production: A comment on Levelt *et al.* (1991). *Psychological Review* 98:604-614.

Ellis, A., Miller, D., and Sin, G. (1983). Wernicke's aphasia and normal language processing: A case study in cognitive neuropsychology. *Cognition* 15:111-144.

Harley, T. (1984). A critique of top-down independent levels models of speech production: Evidence from non-plan-internal speech production. *Cognitive Science* 8:191-219.

Harley, T. (1990). Paragrammatisms: Syntactic disturbance or failure of control? *Cognition* 34:85-91.

Hinton, G., & Shallice, T. (1991). Lesioning an attractor network: Investigations of acquired dyslexia. *Psychological Review* 98:74-95.

Jones, G. (1989). Back to Woodworth: Role of interlopers in the tip-of-the-tongue phenomenon. *Memory and Cognition* 17:69-76.

Kertesz, A., & Benson, D. (1970). Neologistic jargon: A clinicopathological study. *Cortex* 6:362-386.

Levelt, W., Schriefers, H., Vorberg, D., Meyer, A., Pechmann, T., & Havinga, J. (1991a). The time course of lexical access in speech production: A study of picture naming. *Psychological Review* 98:122-142.

Levelt, W., Schriefers, H., Vorberg, D., Meyer, A., Pechmann, T., & Havinga, J. (1991b). Normal and deviant lexical processing: A reply to Dell and O'Seaghdha. *Psychological Review* 98:615-618.

McClelland, J., & Rumelhart, D. (1981). An interactive activation model of context effects in letter perception: Part 1. An account of the basic findings. *Psychological Review* 88:375-407.

Martin, N., & Saffran, E. (1991). A connectionist account of deep dysphasia: Evidence from a single case study. Forthcoming.

Miller, D., & Ellis, A. (1987). Speech and writing errors in "neologistic jargonaphasia": A lexical activation hypothesis. In M. Coltheart, G. Sartori, and R. Job eds. *The cognitive neuropsychology of language.* London: Erlbaum.

Plaut, D., & Shallice, T. (1991). Effects of word abstractness in a connectionist model of deep dyslexia. In Proceedings of the 13th Annual Conference of the Cognitive Science Society, 73-78. Chicago, IL.

Stemberger, J. P. (1985). An interactive activation model of language production. In A. W. Ellis ed. *Progress in the psychology of language* (Vol. 1). London: Erlbaum.

Wheeldon, L. (1989). Competition between semantically related words in speech production. Paper presented to the Experimental Psychology Society, London.

Linguistic Permeability of Unilateral Neglect: Evidence from American Sign Language

David P. Corina, Ph.D.
U.S.C. Program in Neuroscience
HNB 18C University Park, Los Angeles CA 90089-2520
corina@gizmo.usc.edu

Mark Kritchevsky, Md.
U.C.S.D. Dept. of Neurosciences
San Diego CA, 90036

Ursula Bellugi, Ed.D.
L.L.C.S. The Salk Institute, La Jolla CA 90037

Abstract

Unilateral visual neglect is considered primarily an attentional deficit in which a patient fails to report or orient to novel or meaningful stimuli presented contralateral to a hemispheric lesion (Heilman et al. 1985). A recent resurgence of interest in attentional disorders has led to more thorough investigations of patients exhibiting neglect and associated disorders. These studies have begun to illuminate specific components which underlie attentional deficits, and further serve to explicate interactions between attentional mechanisms and other cognitive processes such as lexical and semantic knowledge. The present paper adds to this growing literature and presents a case study of a deaf user of American Sign Language who evidences severe unilateral left neglect following a right cerebral infarct. Surprisingly, his ability to identify visually presented linguistic signs is unaffected by the left neglect, even when the signs fall in his contralesional visual field. In contrast the identification of non-linguistic objects presented to the contralesional visual field is greatly impaired. This novel and important finding has implications for our understanding of the domain specificity of attentional disorders and adds new insights into the interactions and penetrability of neglect in the face of linguistic knowledge. These results are discussed in relation to the computation model of neglect proposed by Mozer and Behrmann (1990).

American Sign Language

American Sign Language (ASL), is a manual gestural system passed down from one generation of deaf people to the next. It has evolved into an autonomous language with its own internal linguistic mechanisms for relating visual form with meaning. These linguistic mechanisms are not derived from English or any spoken language, but rather are deeply rooted in the visual modality. One of the most significant and distinguishing aspects of sign language structure is the unique role of space. Spatial contrasts and spatial manipulations figure structurally at all linguistic levels; phonological, morphological and syntactic. For example, in the syntactic domain nominals introduced into the discourse are assigned arbitrary reference points in a horizontal plane of signing space; signs with pronominal function are directed toward these points and verb signs obligatorily move between such points in specifying grammatical relations. Thus a grammatical function served in many spoken languages by case marking or by linear ordering of words is fulfilled in ASL by spatial mechanisms (Klima and Bellugi, 1979). The existence of a language in which linguistic forms are developed and communicated through visual spatial devices provides a unique opportunity to examine interactions between the neural systems underlying language and spatial cognition.

Visual Spatial Neglect

Unilateral neglect manifests as a disorder in which patients appear unaware of, or fail to respond to stimulation occurring contralateral to the damaged hemisphere. Behaviorally these patients may not orient to tactile or visual stimulation in the contralesional visual field. In constructional tasks, they may only draw one half of a figure, or write only on one side of a page (Heilman et al. 1985). Another common characteristic of neglect, especially in its later stages, is extinction. A patient who can detect a single contralesional stimulus may fail to report that stimulus when a second stimulus appears simultaneously in the ipsilesional space. Extinction may manifest in tactile, auditory and visual modalities. Left neglect associated with right hemisphere damage tends to be more common that right neglect associated with left hemisphere damage.

A priori neglect could disrupt ASL production at two levels. In a patient with neglect for left space, there might be omission of the left half of signs, or failure to use the left part of articulatory space. Similarly, neglect could impair the ability to comprehend ASL, either because of failure to process part of individual signs or failure to attend to signs communicated in the neglected space. Below we report the case study of a deaf signer, patient J.H., who following right hemisphere damage exhibits severe visual spatial neglect. We systematically explore the extent to which J.H.'s hemi-neglect interferes with sign language and visual object processing.

Background & Medical Report

J.H. is a 61 year old right handed congenitally deaf male who suffered a right hemisphere CVA 9/78. The large right hemisphere stroke involved central portions of the frontal, parietal and temporal lobes as well as associated deep white matter and basal ganglia structures. Born to normally hearing parents, J.H. attended a residential school for the deaf when he was 5. He is a fluent signer and was an active member of the Deaf community prior to his CVA. We tested J.H. several times, beginning in 1988 and the most recently in 1991. Thus our data reflect a stable rather than transient condition.

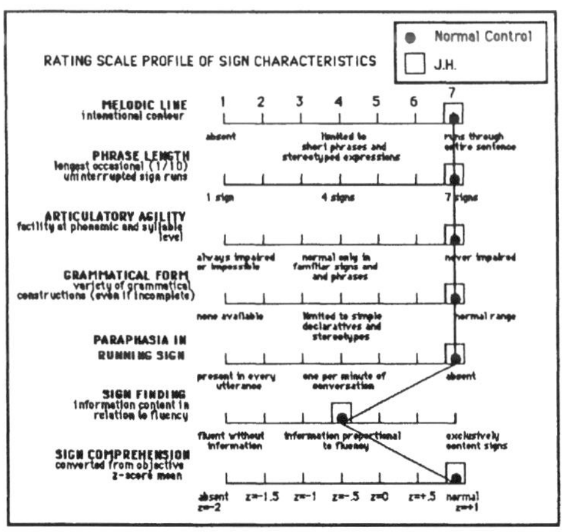

Figure 1

Spared Linguistic Capacity. We administered several standard aphasia tests including the Salk Institute adaptation of the Boston Diagnostic Aphasia Examination (BDAE) (Goodglass & Kaplan, 1972) . The test results indicate that J.H. was not at all aphasic for sign language. Figure 1 illustrates the language profile for the Salk Institute adaptation of the BDAE. The profile for J.H. is compared to a non-brain damaged elderly control. As is evident from the figure, J.H. shows no evidence of core language impairment in either production or comprehension. In production his language is fluent and grammatically complex. He shows no signs of articulatory difficulty, and no semantic or formational paraphasias.

Non-language Visual Spatial Abilities. In contrast to his well preserved linguistic abilities, J.H.'s performance on a variety of non-language visual spatial tests reveal frank disruptions. J.H. showed profound impairments in line orientation judgments (Benton et al. 1977), form perception (Delis et al. 1986), and visuoconstructive tasks.

Left Visual Field Neglect

in a Right Lesioned Signer

BDAE Drawings from Copy

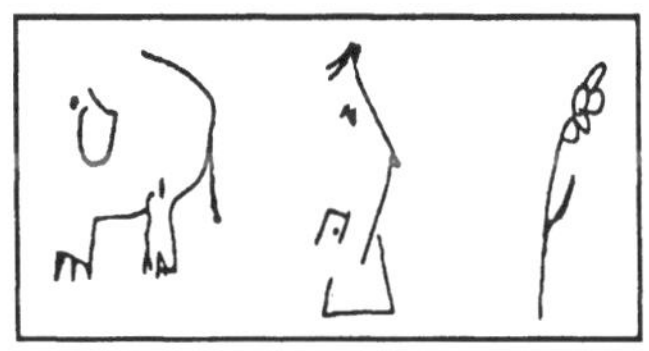

elephant house flower

BDAE Drawing from Command

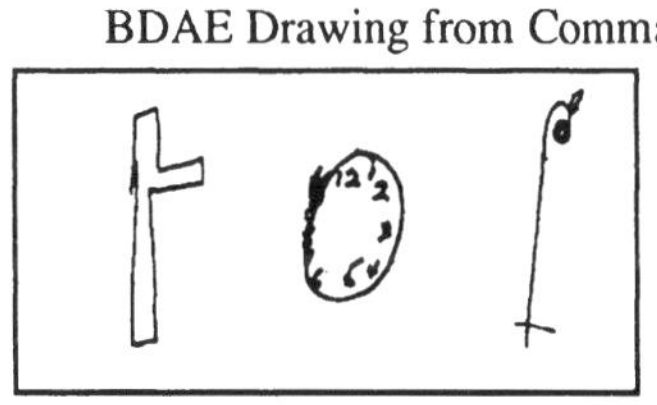

cross clock flower

Figure 2

Severe Visual-Spatial Neglect. One persistent finding in our testing of J.H.'s visual spatial abilities was visual spatial neglect. Figure 2 (top half) illustrates J.H.'s performance on the Albert's (1973) neglect test. In this test the subject is given a page with a random distribution of short lines, and is asked to cross each of these lines. J.H. systematically fails to cross lines appearing in the upper left half of this figure. Figure 2 (middle and bottom half) shows drawing from the parietal lobe battery of the BDAE. All of J.H.'s drawing evidence omissions of left sides of these figures.

Summary. The test reported above are consistent with a profile of left neglect following right hemisphere lesion. What is most surprising however, is that despite this severe and persistent neglect, our preliminary findings found no evidence for comprehension or productive sign language deficits. We chose to extend these preliminary findings and explore in a systematic fashion the effects of left neglect on visual sign and object processing using a visual half field procedure.

Sign Neglect Test.

The structure of ASL provides a unique opportunitiy to investigate visual-linguistic processing in patients with neglect. Several sign pairs can be identified that are minimal pairs (or near minimal pairs) which are distinguished solely on the presence or absence of a single hand. For example the one handed sign FATHER in many contexts is signed with a handshape in which all fingers are upright and spread (an "open-5" handshape) with the palm facing away from the signer, the thumb touching the forehead. A common form of the sign DEER is identical except both the dominant and non-dominant hand simultaneously touch the forehead. In face to face signing, under normal viewing conditions, the receiver focuses attention about the lower half of the signers' face. Thus while observing the sign DEER, one hand will fall into the right visual field while the other hand will fall into the viewer's left visual field. Given this scenario one could well imagine how neglect of one half of visual space could severely disrupt comprehension of signing. In particular under some conditions, neglect may render the sign DEER (a two handed sign) to be misinterpreted as FATHER (a one handed sign). Another example is the pair COMPARE vs. MIRROR Illustrated in figure 3.

A second way that sign language can be used to investigate issues of neglect capitalizes upon a discourse convention in which the signer chooses to articulate each half of a semantic pair with the opposite hand. For example in signing the ASL equivalent of "apples and peaches are my favorite fruits", one may choose to sign APPLE with the dominant hand and PEACH with the non-dominant hand. Thus for the viewer one of the signs will fall into the right visual field and the other will fall into the left visual field. A signers with visual neglect may interpret the sentence APPLE PEACH MY-FAVORITE (apples and peaches are my favorite fruit) as the semantically plausible "apples are my favorite fruit", neglecting the sign PEACH which falls into the contralesional visual field.

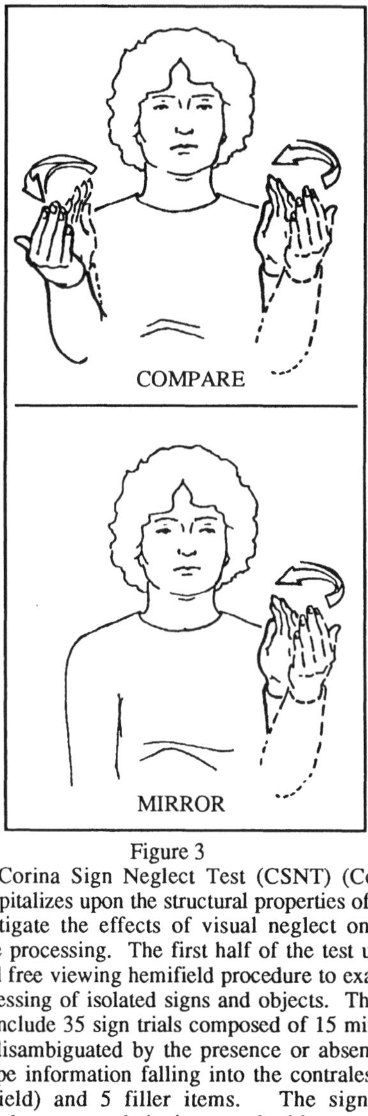

Figure 3

The Corina Sign Neglect Test (CSNT) (Corina, 1991) capitalizes upon the structural properties of ASL to investigate the effects of visual neglect on sign language processing. The first half of the test uses a modified free viewing hemifield procedure to examine the processing of isolated signs and objects. The test stimuli include 35 sign trials composed of 15 minimal pairs (disambiguated by the presence or absence of handshape information falling into the contralesional visual field) and 5 filler items. The signs are articulated at a normal signing speed with an average duration of 1.25 seconds. In addition to the sign processing condition the test includes a non-linguistic control condition. A variety of common objects (e.g. a camera, hair brush, telephone etc.) are presented to the left and right visual field for identification. To make the presentations of objects comparable to the sign condition, the objects are moved into place by a "signer" in locations comparable to the articulatory space used for signing. The object test includes 20 bilateral trials and 10 single object trials. Duration of object trials averaged 2 seconds. All stimuli trials were videotaped for presentation. The second half of the CSNT investigates comprehension of signs in simple sentences which utilize the alternating dominant/non-dominant hand discourse convention. This test includes 18 sentences, half of the sentences are alternating hand format (e.g. APPLES[dominant hand], PEACHES[non-dominant hand] MY-FAVORITE FRUIT) in the remaining nine sentences only one noun is mentioned (e.g. SNAKE DANGEROUS CAN) (snakes can be dangerous). Following each sentences a comprehension question was asked, "What's my favorite fruit?". In this way we can probe to see if the subject had processed the entire sentence or alternatively only attended to "half" of the message.

Methodology. The patient was seated 190.5 cm from a screen where the test videotape was projected. A 63.5 cm. X 63.5 cm. video image was projected onto a screen through the use of a Sharp XG-1100 video projector with back projection capabilities. The image of the signer was slightly larger than life size, all testing was done in a darkened room. Visual angle for object and sign presentations ranged from 3.1^{o} to 5.1^{o} of visual angle during testing. During the first condition, J.H. was told that he would see some signs and some objects, and he was to simply report (i.e. name) what he had seen. During testing the subject's eyes were videotaped for later eye-gaze analysis. In the second condition, the subject was told that he would see some sentences followed by a question, and he was to answer each question as best possible.

Note that the signer who was videotaped on the test tape was right handed. Thus in face to face signing a right handed signer's dominant hand falls largely into the right visual field. However, the patient tested evidences left visual field neglect. To make the right handed signer appear to be left handed, we used the back projection feature of the Sharp XG-1100 video projector, which in essences "flips" the projected image. Thus for the signing experiments disambiguating handshape information falls in to subject left visual field. In fact ,for completeness, the hemifield portion of the test was administered in both "left hand and right hand dominant" orientations, permitting important comparisons which are detailed below.

Results. We first report data from the most demanding left hand dominant condition, whereby the disambiguating hand (i.e. the non-dominant hand) falls in J.H.'s neglected visual field. On the sign identification trials out of a possible 35 signs, J.H. identified 33 correctly or 94%. In contrast, on the object recognition test out of 30 objects to be identified J.H. identified only 15/30 or 50%. An analysis of the object errors revealed that J.H. consistently failed to report the left visual field object during bilateral trials. The impaired object identification performance stands in marked contrast to the excellent identification of two handed signs.

In the mirror reverse condition, numerically the results were nearly identical. For signs J.H. correctly identified 33/35 signs (94%). On the object trials J.H. was correct only for 17/30 trials (57%) trial. Once again he consistently missed bilateral object trials, failing to report the object presented to the left visual field. What is surprising is that these were the same objects he correctly identified in the first condition. For example in the first condition, a camera was presented in the right visual field, simultaneously a package of cigarettes was shown in the left visual field. In this condition J.H. reports only the camera, and fails to report the cigarettes. In the second orientation condition where the entire video image is mirror reversed, the cigarettes appear in the right visual field, and the camera in the left visual field. Now J.H. reports only the cigarettes. However for sign identification flipping the video image has no effect whatsoever on J.H. ability to report both one and two handed signs.

In the sentential condition of the CSNT, where sign sentences are presented with an alternating hand format, J.H. also showed excellent appreciation of information falling into the otherwise neglected visual field. Out of 18 sentences, he correctly comprehended 15/18 sentences, missing one non-alternating sentence and two alternating hand sentences. Thus for sentences where complete comprehension required processing sign information in both the right and left visual fields, J.H. correctly reported 7/9 sentences.

Discussion

The CSNT test was used to compare recognition of signs and objects presented in the left and right visual fields in a patient with neglect. J.H.'s performance indicates that he is accurate in identifying signs whose composition require processing of sign information from the left (neglected) visual field. This is particularly striking as the double handed stimuli chosen have a plausible one handed interpretation. In contrast, J.H. showed significant impairment in identifying objects presented in the neglected visual field. This was most evident in bilateral presentation trials, where J.H. consistently reported only the items in the left visual field. This latter finding is compatible with the description of visual extinction, whereby a subject may report single objects in both ipsilesional and contralesional visual fields, but under simultaneous stimulation fails to report or extinguishes visual information presented to the contralesional visual field (Heilman et al. 1985).

The results reported here extends the phenomena of extinction in a new and exciting way. Despite consistent extinction of objects, J.H. shows excellent comprehension of bimanual signs, (signs in which the two hand simultaneously articulate in the right and left visual fields). This finding at first blush suggest that visual linguistic information is treated qualitatively different from visual object information. However recent computational models of neglect suggest a different conception of this problem.

Mozer and Behrmann (1990) discuss a computational model MORSEL which simulates several neglect findings. Crucially MORSEL demonstrates how a single lesion to the connections that help draw attention to an object in the models visual field can result in performance which appears to implicate higher level impairment, for example neglect-dyslexic errors.

Neglect dyslexics may ignore the left side of an open book, the beginning words on a line of text, or the beginning letters of a single word. However Behrmann et al. 1990 have shown that the ability of a neglect dyslexic patient to select the left most of two words is influenced by the relation between the words. When their patient was shown pairs of semantically unrelated three letter words separated by a space e.g. SUN_FLY and asked to read both words the left word was reported on only 12% of the trials; however when the two words could be joined to form a compound word, e.g. COW_BOY the left word was read 28% of trials (Behrmann 1990; Mozer and Behrmann 1990). From this finding Behrmann concluded that operation of attention to select among stimuli interacts with higher order stimulus properties. The most crucial aspect of MORSEL's simulations of neglect dyslexia owes to the presence of the "pullout" network with its semantic-lexical units. The presence of higher order lexical knowledge i.e. the overt representations of items corresponding to "cow", "boy" and "cowboy" permits the lesioned network to recognize the visual input "boy" as "cowboy" some small percentage of the time.

The case of J.H. poses a challenge to the Mozer and Behrmann model as 94% of bilateral trials are reported as correct two handed signs. To the extent that a MORSEL-like model is an approximation of the mechanism underlying neglect, we are forced to consider the crucial stimulus properties which underlie the representations of signs, objects and words. Importantly the representational properties must correctly lead to the quantitative differences in the percentages of whole form retrievals. One property which may go far in explaining these results is the notion of neighborhood competition (Glushko 1979). We note that for bilateral signs, there are few similar phonological forms and thus a sparsely populated neighborhood. Partial activation of a bilateral sign is sufficient to fill out a unique entry. In the case of orthographic compound forms the neighborhoods are more densely populated (consider the possible neighborhood for the partial activation of "cow_boy"; cowboy, cowgirl, cowslip, highboy, batboy, waterboy...). In this case the partial activation of this orthographic form will not uniquely map to a single form. The result is a lower percentage of reliable unique compound response (but statistically higher than for a totally unrelated word pair "sun_fly"). Objects on the other hand typically do not participate in compound membership and thus do not benefit from higher order "compound" representation.

In summary the case of J.H. who demonstrates an unusual ability to process visually presented signs but not objects adds new insights into the interactions and penetrability of neglect in the face of linguistic knowledge. When considered in relation to a current computational model of neglect, we are forced to make explicit representational differences underlying sign, word and object perception. The existence of a language in which linguistic forms are developed and communicated through visual spatial devices provides a unique opportunity to examine interactions between the neural systems underlying language and spatial cognition.

Acknowledgments

This work was supported in part by the Laboratory for Language and Cognitive Studies at the Salk Institute, NIH grants DC00146, DC00201, HD12349, and HD266022, NSF grant BNS86-09085, and Andrus Gerontology Center grant NIA 5T32AG 00037. I would like to thank J.H. for his participation, and also Lucinda O'Grady-Batch, Freda Norman, Don Baer, Karen Emmorey, and Karen vonHoek. for their assistance.

References

Albert, M. 1973. A simple test of visual neglect. *Neurology*, 23:685-664.

Behrmann, M., Moscovitch, M., Black, S.E., & Mozer, M.C. 1990. Perceptual and conceptual mechanisms in neglect dyslexia: Two contrasting case studies. *Brain* 113:1163-183.

Benton A.L., Varney, N.R., deS Hamsher, K. 1977. Test of judgment of line orientation. Iowa City, IA: Benton Laboratory of Neuropsychology, University of Iowa.

Corina, D (1991). Corina Sign Neglect Test. m.s. The Salk Institute, La Jolla CA.

Delis, D., Robertson, L.C., Effron, R. 1986. Hemispheric Specialization of Memory for Visual Hierarchical Stimuli.. *Neuropsychologia*, 24 :265-214.

Goodglass, H., & Kaplan, E. 1972. Revised edition, 1983. *The Assessment of Aphasia and Related Disorders*. Philadelphia:Lea and Febiger.

Glushko, R.J. 1979. The organization and activation of orthographic knowledge in reading aloud. *J.E.P.: Human Perception and Performance*,5:674-691.

Heilman, K. M., Watson, R.T., & Valenstein, E. 1985 Neglect and related disorders. In K.M. Heilman & E. Valenstein (Eds.) *Clinical Neuropsychology, 2nd ed.* New York: Oxford University Press.

Klima, E. & Bellugi, U.1979 *The Signs of Language*. Cambridge: Harvard University Press.

Mozer, M.C. & Behrmann, M. 1990 On the Interaction of Selective Attention and Lexical Knowledge: A Connectionist Account of Neglect Dyslexia. *Journal of Cognitive Neuroscience*. 2, 2:96-123.

Hippocampal-System Function in Stimulus Representation and Generalization: A Computational Theory

Mark A. Gluck **Catherine E. Myers**

Center for Molecular and Behavioral Neuroscience
Rutgers University
197 University Ave., Newark, NJ 07102
gluck@pavlov.rutgers.edu *myers@pavlov.rutgers.edu*

Abstract

We propose a computational theory of hippocampal-system function in mediating stimulus representation in associative learning. A connectionist model based on this theory is described here, in which the hippocampal system develops new and adaptive stimulus representations which are predictive, distributed, and compressed; other cortical and cerebellar modules are presumed to use these hippocampal representations to recode their own stimulus representations. This computational theory can been seen as an extension and/or refinement of several prior characterizations of hippocampal function, including theories of chunking, stimulus selection, cue-configuration, and contextual coding. The theory does not address temporal aspects of hippocampal function. Simulations of the intact and lesioned model provide an account of data on diverse effects of hippocampal-region lesions, including simple discrimination learning, sensory preconditioning, reversal training, latent inhibition, contextual shifts, and configural learning. Potential implications of this theory for understanding human declarative memory, temporal processing, and neural mechanisms are briefly discussed.

Introduction

The hippocampus and adjacent cortical regions in the medial temporal lobe have long been implicated in learning and memory via lesion data in both humans (Scoville & Millner, 1957; Squire, 1987) and animals (Mishkin, 1982; Squire and Zola-Morgan, 1983). While there is general agreement that this region plays an essential role in many aspects of learning and memory, there is little consensus as to the precise specification of this role. One approach has been to seek to define the class of learning and memory tasks which require an intact hippocampal region. Squire (1987) emphasizes the critical role of this brain region for the formation of explicit declarative memories in humans. Studies of lower (non-primate) mammals have focussed on place-learning and spatial navigation as tasks which require an intact hippocampal region (Morris, et al., 1982; O'Keefe and Nadel, 1978; McNaughton & Nadel, 1990).

Another approach to functional theories of hippocampal-region processing has been to characterize an underlying information-processing role for the hippocampal region and then seek to derive a wider range of task-specific deficits. Two broad classes of hippocampal-based deficits have been characterized: those dealing with temporal processing and those concerned with modifying stimulus representations. We are concerned here solely with the latter, representational, properties of hippocampal-region function. Thus, in the analysis to follow, we will consider only trial-level properties of associative learning.

Several different representational theories of hippocampal-region function have been proposed. Wickelgren (1979) suggested that the hippocampus participates in a process whereby the component features within a stimulus pattern are recognized as co-occurring elements and thus come to be treated as a unitary whole or "chunk." Others have viewed the hippocampal-region as an attentional control mechanism which alters stimulus selection through a process of inhibiting orienting or attentional responses to irrelevant cues (Douglas & Pribram, 1966; Schmajuk & Moore, 1985). More recently, Wickelgren's "chunking" idea has been extended and elaborated by Sutherland and Rudy (1989) who proposed that the hippocampus provides the neural basis for the acquisition and storage of configural events, while other brain systems store only direct cue-outcome associations. A related characterization of hippocampal-function suggests that it is best understood as providing a "contextual tag" for associative learning (Nadel & Willner, 1989; Winocur, et al., 1987; Hirsch, 1974). Eichenbaum and colleagues have also emphasized the representational role of hippocampal function, particularly in the flexible use of conjunctive associations in novel situations (Eichenbaum & Buckingham, 1991).

These characterizations of hippocampal-lesion deficits can all qualitatively describe subsets of the empirical data. What is lacking, however, is a clear mechanistic interpretation of hippocampal-region function which can be formally and rigorously tested against a broad range of empirical data. Computational models of abstract connectionist networks offer one possible approach for exploring candidate functional roles of the hippocampal system. Models developed at this abstract level do not directly yield a physiological understanding of circuit-level processing in the hippocampal region; nevertheless, these models may suggest how effects of hippocampal-region lesions might emerge from an underlying processing function which is localized in the hippocampal region. By developing a connectionist theory of hippocampal processing in conditioning, we seek, in the work to be described here, to address the question: Is there a simple underlying *computational* function which can derive the representational processes subserved by the hippocampal system in associative learning?

Cortical-Hippocampal Model

We begin with the single key idea that the representational function of the hippocampal system can be approximated by a simple network architecture, called a predictive autoencoder. This type of network develops novel and flexible representations with three key properties: they are **distributed**, **predictive** (of future sensory inputs), and **compressed** (i.e., reduced in size by compressing statistical redundancies). In our model, the hippocampal module is conceptualized as a predictive autoencoding network, which learns sensory-sensory mappings through a narrow hidden layer, as shown in Figure 1 (c.f., Hinton, 1989). As a result, the narrow hidden layer is forced to discover a representation which compresses regularities and irrelevant stimuli while allocating more resources to predictive stimuli. Other learning "modules" such as cerebellar and cerebral cortices (one such module is shown in Figure 1) are restricted to learn simple associations via procedures such as the Rescorla-Wagner rule (Rescorla & Wagner, 1972). If, however, a linear recombination of the hippocampal module's

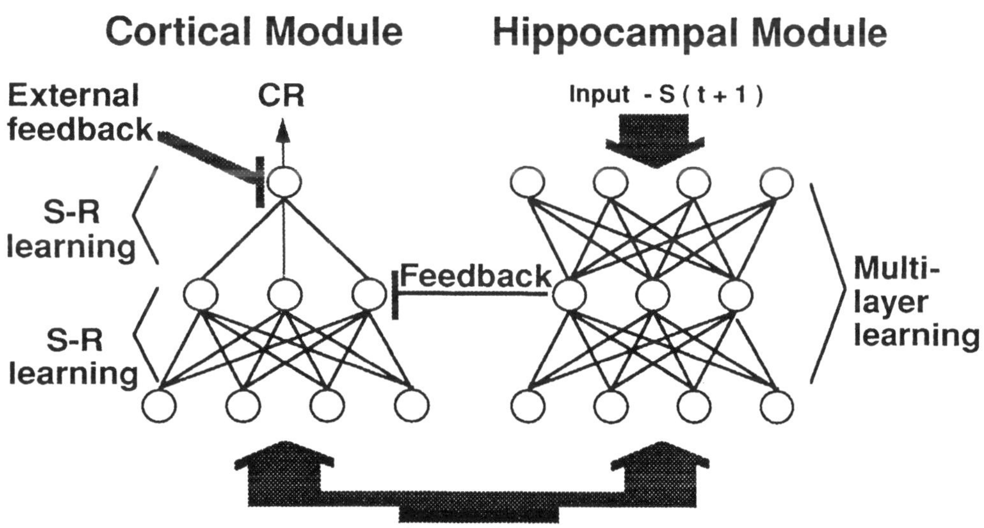

Figure 1. *The intact cortico-hippocampal model. Learning in the hippocampal module (on right) mediates the development of novel stimulus representations in cerebral and cerebellar modules (one shown on left). The hippocampal module has the capacity for multi-layer learning, which results in a novel recoding (or re-representation) of its stimulus inputs. The cortical module is restricted to using only (single-layer) S-R learning, e.g., the LMS rule of Widrow & Hoff (1960). Hippocampal-lesion experiments are modeled by removing the hippocampal module. This results in the bottom layer of the cortical module remaining fixed (e.g., with non-modifiable weights). The upper layer of the cortical module, however, can still be trained to learn based on the fixed recoding of the cortical inputs which occurs in the cortical bottom layer. Learning without the hippocampal module is thus limited to discriminations which can be solved without learning a new stimulus representation.*

hidden layer is provided as feedback to the cortical module hidden layer, the cortical module can use its simple learning rules to map from sensory inputs to this representation and from the representation to outputs, allowing it to learn complex associations. This is our connectionist conceptualization of normal, intact cortico-hippocampal interaction in associative learning.

Within this framework, a hippocampal lesion is characterized by removing the hippocampal module. This has the effect of eliminating the hippocampal feedback which the cortical module would otherwise use to construct hidden layer representations. When the hippocampal-system module is so lesioned, the lower layer of cortical module weights remains fixed, and the system learns associative stimulus-response (S-R) relationships based on this fixed encoding (representation) of the stimulus inputs. In comparison, the intact system learns the same S-R relationships, but does so based on a flexible and distributed re-coding of the stimulus inputs which reflect both predictive S-R relationships as well as sensory-sensory correlations. For example, simple discrimination learning (A+/B-) is largely unaffected or even facilitated after hippocampal lesion (Schmaltz & Theios, 1972; Eichenbaum, et al., 1988). As shown in Figure 2A, both the intact and the lesioned network models can solve a simple discrimination task. Furthermore, the lesioned network shows some facilitation. This occurs because the initial representation is sufficient to learn the task, whereas in the intact model, this initially sufficient representation is altered by hippocampal influence, retarding learning.

Applications to Data

We turn now to examining the behavior of the cortico-hippocampal model in several other key paradigms. These simulations will illustrate how the model instantiates or refines aspects of four prior

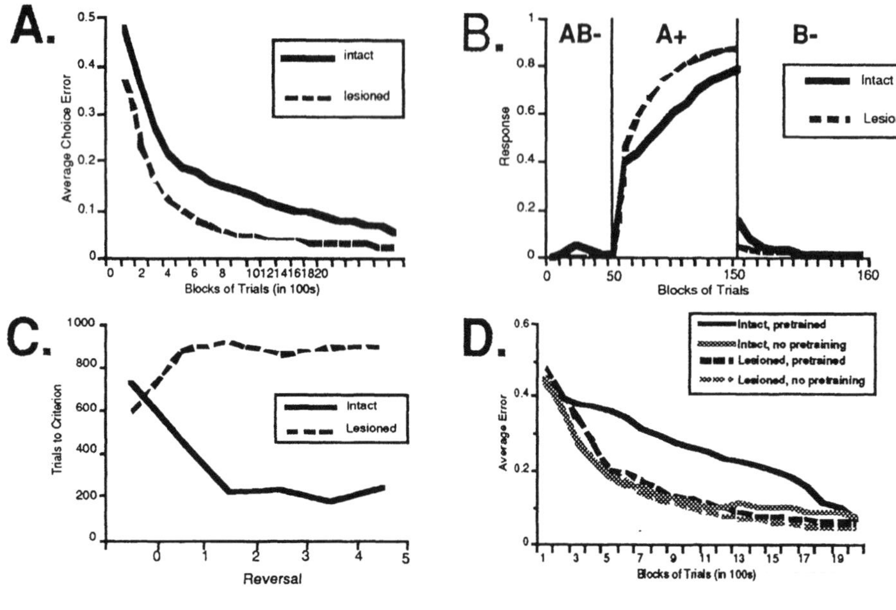

Figure 2. *Simulations of intact and lesioned cortico-hippocampal model.* **(A) Simple Discrimination***: Training to A+/B-. Lesioned model learns somewhat faster as intact model must first learn a new representation in the hippocampal module and then transfer this representation to cortical module.* **(B) Sensory Preconditioning***: Pre-exposure to a non-reinforced AB- compound followed by training to A+. Intact and lesioned systems are similar through first two phases, but only intact system shows transfer of response when tested with B in third phase.* **(C) Multiple Reversals***: Training on A+/B-, followed by reversal A-/B+, then A+/B-, etc. Intact system shows a progressive decrease in the number of trials required to learn each discrimination; lesioned system has difficulty with all but the first discrimination.* **(D) Latent Inhibition***: Non-reinforced A- training followed by A+ training. Intact but not lesioned system impaired on A+ learning compared with control condition of pre-training to another cue (e.g., C+).*

information-processing theories of hippocampal-system function: chunking, stimulus-selection, cue-configuration and contextual labeling.

Chunking

As suggested by Wickelgren (1979), the model incorporates a "chunking" mechanism through the sensory compression which occurs in the hippocampal-system module to static and co-occurring (most clearly redundant) features, including the context. This is seen in **sensory preconditioning**, where an animal is first pre-exposed to an unreinforced AB stimulus compound, and then given A+ training. In a final training phase, the animal shows partial transfer of the learned response to stimulus cue B. Port & Patterson (1984) demonstrated that the hippocampal-region is necessary for sensory preconditioning. Figure 2B shows that our model is consistent with this result. The intact system builds an internal recoding during pre-exposure to AB- training which "chunks" A and B together. This chunked representation persists during the A+ training. Therefore, some of the A+ association transfers to B through a process akin to an acquired form of stimulus generalization.

Stimulus Selection

Theories of stimulus selection in classical conditioning can be differentiated into two classes: those based on a modulation of the reinforcing value of the unconditioned stimulus (e.g., Rescorla & Wagner, 1972) and those which presume an attentional or salience modulation of sensory inputs (e.g., Pearce & Hall, 1980; Mackintosh, 1975). Our hippocampal-system model incorporates both forms of stimulus selection. Reinforcement modulation is localized in the cerebellar and cerebral cortices, while sensory modulation localized in the hippocampal region. This mapping is consistent with results indicating that behaviors which are uniquely explained by sensory modulation (e.g., reversal facilitation, latent inhibition) show the clearest deficits after hippocampal lesion. In comparison, stimulus selection behaviors which can be uniquely explained by reinforcement modulation (e.g., conditioned inhibition) show no hippocampal deficit. Phenomena which can be explained by both mechanisms (and hence are assumed in our theory to be multiply determined across several brain regions) have resulted in inconclusive or partial deficits (e.g., blocking and overshadowing). We focus now on the two phenomena which we expect to show clearest hippocampal dependence.

In **reversal learning**, an animal is first trained on a simple A+/B- discrimination. This is then followed by reversal training on A-/B+. These two discriminations are then repeatedly reversed. Normal intact animals shown a progressive facilitation in learning the new discriminations; in contrast, HL animals show an impairment (Berger & Orr, 1983). As illustrated in Figure 2C, the intact cortico-hippocampal system shows a progressive decrease in the number of trials required to learn each reversal – reflecting the fact that the hippocampal-module's distributed stimulus recoding devotes more and more resources to the relevant cues. The lesioned system, with no such mechanism for stimulus selection, must first "unlearn" previous discriminations before starting afresh to learn each new reversal.

Latent inhibition, first described by Lubow (1973), is an especially important "marker" of hippocampal processing because the hippocampal-damaged animals show *increased* responding in a transfer task. When animals are first given A- (unreinforced) trials, and then switched to A+ (reinforced) training, their acquisition of A+ is impaired relative to animals with no A- pre-training. Error-correcting models such as the Rescorla-Wagner (1972) model and error backpropagation networks (Rumelhart, et al. 1986) fail to predict latent inhibition, because they expect no learning during the (errorless) A- training phase. In our model, however, sensory-sensory learning does take place during A- pre-training. The hippocampal module learns to "chunk" A together with the context because neither is predictive of the US or of any other significant event. Later, in A+ training, the representation of A must be "de-chunked" from the context before learning can occur (Figure 2D). Solomon and Moore (1975) showed that latent inhibition is absent in HL animals, and Figure 2D shows that it is absent in the lesioned system as well.

Cue-Configuration

Sutherland and Rudy's (1989) cue-configuration theory proposes that the hippocampus is necessary for the acquisition and retention of configural associations. Our cortico-hippocampal model implies a similar hippocampal involvement in configural learning: configural tasks will typically entail a stimulus recoding necessitating an intact hippocampus. Simulation results (not shown here) demonstrate a lesioned system deficit for **configural tasks** such as negative patterning (A+/B+/AB-).

Contextual Labeling

Our model can also be viewed as an instantiation of theories suggesting a key role for the hippocampus in developing a "contextual tag" for stimulus-response associations (Hirsh, 1974; Nadel & Willner, 1989, Winocur, et al., 1987). Hippocampal-lesioned animals are often shown to have difficulty **encoding context**. Given training with A+ in one context, normal intact animals show a decreased response when tested with A in a new context; HL animals show no such decrease (Penick & Solomon, 1991). Likewise (simulations not shown), the intact cortico-hippocampal model shows a decreased response if contextual cues are changed; the lesioned model shows no such deficit. In the intact model, the contextual cues which co-occur with A become part of that stimulus's representation; when A is shown in a different context, the representation is less strongly activated than usual, and the response strength drops.

Summary and Discussion

The model we have presented here shows how a specific network architecture can form compressed, predictive, and distributed representations of stimuli, which are made available to other learning systems (such as the cerebellum and cerebral cortex). This model incorporates and refines aspects of many prior, qualitative information-processing theories of hippocampal function, including chunking (Wickelgren, 1979), stimulus selection Rescorla & Wagner, 1972; Pearce & Hall, 1980; Mackintosh, 1975), contextual labeling (Hirsh, 1974; Nadel & Willner, 1989) and cue-configuration (Sutherland & Rudy, 1989). Our theory also relates to other behaviors sensitive to hippocampal damage. For example, several task-specific theories of hippocampal-region function have noted the impairment of HL animals in spatial navigation (O'Keefe & Nadel, 1978). In our connectionist cortico-hippocampal model, place learning could be another kind of representational learning; the hippocampus would be responsible for mapping from a partial view of an environment into a full representation of a place. Linear autoassociator models of the hippocampus (e.g., McNaughton, 1989; Rolls, 1990) have previously been proposed. A predictive autoencoder, used here as a conceptualization of the hippocampal-system, generalizes the properties of a linear autoassociator.

In its current form, the model does not address hippocampal mediation of temporal and sequential processing, functional roles implied by the failure of hippocampal-damaged animals at conditioning with long ISI delays or trace conditioning (Moyer, et al. 1990). These additional temporal roles may either be interpreted as requiring refinements of the same mechanisms proposed here, or they may be localized within different brains structures in the medial temporal lobe. Future efforts will be needed to better understand the interaction between temporal and representational processing in the hippocampal-region and the precise neurobiological locus (or loci) and physiological characteristics of these two functions.

Acknowledgments

This work was supported by a Young Investigator Program award from the Office of Naval Research to MAG. For their helpful suggestions and critical feedback, we are indebted to Gyorgy Buzsaki, Ian Creese, Howard Eichenbaum, Paul Glauthier, Bruce McNaughton, Lynn Nadel, Jerry Rudy, Larry Squire, and Richard Thompson.

References

Berger, T., and Orr, W. 1983. Hippocampectomy disrupts discrimination reversal learning of the rabbit nictitating membrane response. *Behavioral Brain Research* 8:49-68.

Buzsaki, G., 1989. Polysynaptic long-term potentiation: A physiological role of the perforant path – CA3/CA1 pyramidal cell synapse. *Brain Research* 45:192-195.

Cohen, N. and Squire, L., 1980. Preserved learning and retention of pattern analysing skill in amnesia: Dissociation of knowing how and knowing that. *Science* 210:207-209.

Douglas, R. and Pribram, K., 1966. Learning and limbic lesions. *Neuropsychologia* 4:192-220.

Eichenbaum, H. and Buckingham, J., 1991. Studies on hippocampal processing: Experiment, theory and model. In *Memory: Organization and locus of change*, ed. J. McGaugh, Oxford University Press, Oxford (in press).

Eichenbaum, H.; Fagan, A.; Mathews, P.; and Cohen, N., 1988. Hippocampal system dysfunction and odor discrimination learning in rats: Impairment or facilitation depending on representational demands. *Behavioral Neuroscience* 102:331-339.

Gluck, M. and Bower, G., 1988. Evaluating an adaptive network model of human learning. *Journal of Memory and Language* 27:166-195.

Grastyan, E.; Lissak, K.; Madarasz, I.; Donhoffer, H., 1959. Hippocampal electrical activity during the development of conditioned reflexes. *Electroencephalography and Clinical Neurophysiology* 11:409-430.

Hinton, G., 1989. Connectionist learning procedures. *Artificial Intelligence* 40:185-234.

Hirsh, R., 1974. The hippocampus, conditional operations and cognition. *Physiological Psychology* 8:175-182.

Lubow, R., 1973. Latent inhibition. *Psychological Bulletin* 79:398-407.

Lynch, G., and Granger, R., 1991. Serial steps in memory processing: Possible clues from studies of plasticity in the olfactory-hippocampal circuit. In, *Olfaction as a model system for computational neuroscience*, ed. J. Davis, MIT Press, Cambridge MA.

Mackintosh, N., 1975. A theory of attention: Variations in the associability of stimuli with reinforcement. *Psychological Review* 82:276-298.

McNaughton, B., 1989. Neuronal mechanisms for spatial computation and information storage. In, *Neural connections, mental computations*, ed. R. Harnish, MIT Press, Cambridge MA.

McNaughton, B.; Chen, L.; and Markus, E., 1991. "Dead reckoning", landmark learning and the sense of direction: A neurophysiological and computational hypothesis. *Journal of Cognitive Neuroscience* 3:190-202.

McNaughton, B. and Nadel, L., 1990. Hebb-Marr networks and the neurobiological representation of action in space. In, *Neuroscience and Connectionist Theory*, ed. D. Rumelhart, Lawrence Erlbaum Associates, Hillsdale NJ.

Mishkin, M., 1982. Memory in monkeys severely impaired by combined but not separate removal of the amygdala and hippocampus. *Nature* 273:297-298.

Morris, R.; Garrud, P.; Rawlins, J.; and O'Keefe, J., 1982. Place navigation impaired in rats with hippocampal lesions. *Nature* 297:681-683.

Moyer, J.; Deyo, R.; and Disterhoff, J., 1990. Hippocampectomy disrupts trace eye-blink conditioning in rabbits. *Behavioral Neuroscience* 104:243-252.

Nadel, L. and Willner, J., 1989. Context and conditioning: A place for space. *Physiological Psychology* 8:218-228.

O'Keefe, J. and Nadel, L., 1978. *The Hippocampus as a Cognitive Map*, Claredon University Press, Oxford UK.

O'Keefe, J.; Nadel, L.; Keightly, S.; and Kill, D., 1975. Fornix lesions selectively abolish place learning in the rat. *Experimental Neurology* 48:152-166.

Pearce, J. and Hall, G., 1980. A model for Pavlovian learning: Variations in the effectiveness of conditioned by not of unconditioned stimuli. *Psychological Review* 87:532-552.

Penick, S. and Solomon, P., 1991. Hippocampus, context and conditioning. *Behavioral Neuroscience* 105:611-617.

Port, R. and Patterson, M., 1984. Fimbrial lesions and sensory preconditioning. *Behavioral Neuroscience* 98:584-589.

Rescorla, R. and Wagner, A., 1972. A theory of Pavlovian conditioning: Variations in the effectiveness of reinforcement and nonreinforcement. In, *Classical Conditioning II: Current Research and Theory*, ed. W. Prokasy, New York.

Rolls, E., 1990. Theoretical and neurophysiological analysis of the functions of the primate hippocampus in memory. *Cold Spring Harbor Symposia on Quantitative Biology* 55:995-1006.

Rumelhart, D.; Hinton, G.; Williams, R., 1986. Learning internal representations by error propagation. In, *Parallel Distributed Processing - Explorations in the Microstructure of Cognition*, vol. 1, eds. D. Rumelhart & J. McClelland, MIT Press, Cambridge MA.

Schmajuk, N. and Moore, J., 1985. Real-time attentional models for classical conditioning and the hippocampus. *Physiological Psychology* 13:278-290.

Schmaltz, L. and Theios, J., 1972. Acquisition and extinction of a classically conditioned response in hippocampectomized rabbits (Oryctolagus cuniculus). *Journal of Comparative and Physiological Psychology* 79:328-333.

Scoville, W. and Millner, B., 1957. Loss of recent memory after bilateral hippocampal lesions. *Journal of Neurology, Neurosurgery and Psychiatry* 20:11-21.

Solomon, P. and Moore, J., 1975. Latent inhibition and stimulus generalization of the classically conditioned nictitating membrane response in rabbits (Oryctolagus cuniculus) following dorsal hippocampal ablation. *Journal of Comparative and Physiological Psychology* 202:1192-1203.

Squire, L., 1987. *Memory and Brain*, Oxford University Press, New York.

Squire, L. and Zola-Morgan, S., 1983. The neurology of memory: Qualitative assessment of retrograde amnesia in two groups of amnesic patients. *Journal of Neuroscience* 9:828-839.

Sutherland, R. and Rudy, J., 1989. Configural association theory: The role of the hippocampal formation in learning, memory and amnesia. *Psychobiology* 17:129-144.

Wickelgren, W., 1979. Chunking and consolidation: A theoretical synthesis of semantic networks, configuring in conditioning, S-R versus cognitive learning, normal forgetting, the amnesic syndrome, and the hippocampal arousal system. *Psychological Review* 86:44-60.

Winocur, G.; Rawlins, J.; and Gray, J., 1987. The hippocampus and conditioning to contextual cues. *Behavioral Neuroscience* 101:617-625.

Learning Distributed Representations for Syllables

Michael Gasser
Departments of Computer Science and Linguistics
Indiana University
Bloomington, IN 47405
gasser@cs.indiana.edu

Abstract

This paper presents a connectionist model of how representations for syllables might be learned from sequences of phones. A simple recurrent network is trained to distinguish a set of words in an artificial language, which are presented to it as sequences of phonetic feature vectors. The distributed syllable representations that are learned as a side-effect of this task are used as input to other networks. It is shown that these representations encode syllable structure in a way which permits the regeneration of the phone sequences (for production) as well as systematic phonological operations on the representations.

Linguistic Structure and Distributed Representation

If the language sciences agree on one thing, it is the hierarchical nature of language. The importance of hierarchical, structured representations is now generally recognized for the phonological pole, where syllables and metrical units now play a major role (see, e.g., Frazier (1987) and Goldsmith (1990)), as well as for the syntactic/semantic pole of language and language processing. The major reason for believing in structured representations is the significance of structure-sensitive operations in language processing. A semantic inference rule may need to know where the subject of a clause is; a morphological reduplication rule may need to know where the coda (final consonant(s)) of a syllable is.

Traditional symbolic representations are based crucially on the simple notion of **concatenation** (van Gelder, 1990). A syllable representation, for example, is a (bracketed) string of concatenated phones. Recent connectionist work offers as an alternative to this widely accepted approach **distributed** representations, for which it is generally impossible to isolate which elements of the representation denote which of the lower-level units comprising the structure being represented.

What good are distributed representations? They certainly are harder to interpret directly, at least by external "users" of the system that creates them. And at first blush it seems cumbersome, if not impossible, to implement structure-sensitive operations on them, operations which present no particular difficulty for symbolic representations (Fodor & Pylyshyn, 1988). Clearly distributed representations would be useless for most purposes if they were not amenable to such operations. Recently, however, it has been shown that it is possible to arrive at a set of connection weights which implements structure-sensitive operations on distributed representations. Where the representations arise on hidden layers through training, the operations on them are also implemented through training (Chalmers, 1990). Where the representations arise as a result of the application of a set of primitive operations analogous to the filling of roles in symbolic models, the operations on them can be implemented more directly (Legendre, Miyata, & Smolensky, 1991).

There are three reasons to prefer distributed over symbolic representations for structured objects such as syllables and sentences.

1. Distributed representations do not necessarily increase in size as the complexity of the represented object increases. In the case of some types of representations, for example, those described in this paper, representations for objects of the same type are of fixed width (Pollack, 1990). This seems more important for syntax/semantics than for phonology, where there is apparently no recursive embedding, but in a learning context, it is a desirable feature for phonological representations too since a system cannot be expected to know beforehand how complex the representations will need to be and therefore how much memory to allot to them.
2. Complex transformations can be performed on distributed representations in a single parallel step, rather than through a series of symbolic `conses`, `cars`, and `cdrs` (Legendre et al., 1991).
3. There are relatively simple algorithms for **learning** the structure in distributed representations (Elman, 1990; Pollack, 1990).

Most work concerned with distributed representations for structured objects has examined syntax or semantics. It remains to be shown whether it is possible to learn distributed syllable representations which embody the structure required for phonological operations of various sorts. This is in part what this study seeks to establish.

Linguistic Structure and Time

Language takes place in time: input to hearers and output from speakers is sequential. If linguistic knowledge is organized hierarchically, at least part of what hearers do in perceiving language must consist in taking in sequences of elements at one level and classifying them as belonging to a single unit at a higher level. Something temporal is turned into something static. In this sense a syllable is a static **summary** of a temporal sequence of phones. Speakers in turn carry out the reverse process: they turn static representations into temporal sequences. Given a syllable representation, they must unpack it into its component onset (initial consonant(s)) and rime (remaining segments). The sorts of syllable representations we seek should be accessible via the categorization that takes place during perception and should be expandable into their component elements during production.

The temporal nature of language is related intimately to the issue of short-term memory. The process by which a sequence of elements at one level is recognized as a single element at a higher level requires access to more than just a single element at a time; a context is necessary. The production of a sequence of elements, given a higher-level summary representation as input, requires as a context some representation of what has already been produced.

One approach to short-term memory is to give a system access to a buffer of some fixed width. This has several drawbacks, in particular the problem of how the system is to know beforehand how wide the buffer should be (Port, 1990). An alternative is an approach that permits a system to develop its own short-term memory. This is possible in connectionist networks with recurrent connections (Elman, 1990; Jordan, 1986; Port, 1990). It is this method that is utilized in the study described here.

The Learner's Task

Language acquisition begins with perception, so we expect the representations for syllables and other prosodic units to result from perceptual processes. There are several possibilities for how this might happen, though the most reasonable is probably some combination.

1. The hearer/learner may be learning phonology for its own sake, that is, either simply looking for regularity in the input, or looking for evidence that would allow the setting of some innate parameters (Dresher & Kaye, 1990).
2. The hearer/learner may be attempting to map perceptual features onto representations of articulatory gestures, as in various versions of the motor theory of perception (Liberman & Mattingly, 1986).
3. The hearer/learner may learn prosodic representations as a side-effect of word recognition.

It is the third possibility that is pursued here. The idea that phonology emerges as the child learns to recognize and produce words is an appealing idea, and an old one. It is based on the notion that phonology is not just arbitrary patterning, but rather a phenomenon with functions for the language processing system: to facilitate word recognition and to organize word production. According to the third view in the list above, the child acquires phonological representations in the context of using them.

Consider the relationship between the acquisition of word recognition and the acquisition of syllable structure. In learning to distinguish an initial subset of the words in the target language, a learner is provided with relatively direct information about the distinctiveness among a sizable subset of the possible syllables in the language. Because the syllables are contrastive units, the learner is forced to distinguish them in order to distinguish the words. The question addressed here is whether the word recognition task suffices to develop representations which support phonological operations.

A human learner/hearer is presented with unsegmented, continuous input. The task of the system studied here is a considerably simpler one: its input consists of sequences of phones, each in the form of a phonetic feature vector. The phones appear one at a time, and the internal state of the system on the previous time step provides the necessary context for recognition. The system's initial task is simply to assign sequences of phones (representing words in the language) to lexical categories. As a side effect of performing this task, it develops internal representations for various subsequences making up the words, in particular for the syllables in the language. These subsequence representations can then in turn be investigated by treating them as inputs to components with other tasks. Two further tasks are dealt with here: the transformation of a static sequence representation into the sequence of phones it represents (the production task), and the systematic mapping of one sequence representation onto another. In both cases, what is of interest is whether the sequence representations permit **generalizations** to be made. That is, trained on a subset of the sequence representations, does the system respond to others on which it was not trained?

The Approach

The networks used in the study described here are **simple recurrent networks** of the types first investigated by Jordan (1986) and Elman (1990). They consist of feedforward networks supplemented with recurrent connections from the hidden and/or output layers and are trained using the familiar back-propagation learning algorithm (Rumelhart, Hinton, & Williams, 1986). Figure 1 shows the architectures of the networks used for the recognition and production tasks in the experiments described below. Earlier experiments indicated the superiority of these particular architectures over other variants of simple recurrent networks for these tasks.

The recognition network is presented with a sequence of phones, one at a time, each phone consisting of a vector of phonetic features. Among the features is **sonority**, which tends to correlate with proximity to the nucleus of a syllable. Each sequence ends with a boundary symbol, represented by an input pattern consisting entirely of zeros. The network is trained to auto-associate the input phone pattern (that is, simply to copy it to a set of output units), and to categorize the input sequence as belonging to one of a set of morphemes in the language. The auto-association task, while not directly related to word recognition, has the effect of forcing the network to distinguish the phones making up the sequences. The network is provided with targets for both the auto-association and recognition tasks. The lexical target remains constant throughout the presentation of the sequence. Via recurrent connections the network also has access to a copy of its hidden layer on the previous time step. A distributed syllable representation, to be used as input to other networks, is obtained by presenting the sequence of phones making up the syllable followed by a boundary symbol and saving the pattern which appears on the hidden layer at the end of this sequence.

The production network takes a distributed syllable representation (from the recognition network) as input. This remains constant throughout the production of the sequence. The network is trained to output, one at a time, the phones making up the sequence followed by a boundary pattern following the sequence. Each phone takes the form of a feature vector, identical to the pattern used as input to the recognition network. Targets are provided for each of the output phones. The production network has recurrent connections on both the hidden and output layers. The output pattern is added to a decayed version of the previous sequence of outputs and sent to the network as part of its input (on the STATE layer).

Experiments

Stimuli

Stimuli for the experiment consisted of phones and phone sequences in an artificial language. Phones were represented by vectors of 11 phonetic features. Possible syllables in the language are characterized as follows:

onset → {0,p,f,m,t,s,n,k,x}

nucleus → {i,e,a,o,u}

rime → {0,n,s}.

Thus there were 135 possible syllables in all.

Procedures

Each experiment began with the training of a recognition network to categorize a set of words in the artificial language. Each word consisted of two legal syllables in the language, and the set of words was generated by randomly combining pairs of syllables, with the restriction that no identical pairs were included. Once the recognition network had been trained on the words, representations for each of the 135 syllables in the language, consisting of hidden layer patterns following the presentation of the syllable sequences, were extracted from the network. These syllable representations were then used as inputs to other networks.

Experiment 1

First 100 two-syllable words were generated. This resulted in a set which contained 104 of the 135 possible syllables in the language. Next the recognition network was trained to identify the phone sequences representing the words. Previous experiments have shown that word recognition training on a relatively large set is more effective if the words are introduced gradually to the network rather than all at once, an idea inspired by the regimen used by Plunkett & Marchman (1991) to train a network to learn English past tense forms. Three new words were introduced to the training set each time the mean square error per pattern for the current training set dropped below 1.0. Training continued for 600 repetitions of the training set (43,048 words), by which time all 100 words had been introduced to the training set.

Performance on word recognition at this point was far from impressive. Only 17 of the 100 words were correctly identified at the point where the final word boundary was presented. Still it was felt that in attempting to learn to distinguish the words, the network might have developed distinct representations for the syllable sequences that made them up. Representations for all 135 possible syllables were set aside by presenting the network with the phone sequences and then saving the final pattern on the hidden layer. The hidden layer of the recognition network, and hence the width of the distributed syllable representations, was 25 units.

Next these syllable representations were used as inputs to a production network. 20% of the syllables were randomly selected to be set aside for testing the network for generalization. These included sequences which had been parts of the words in the original recognition training set and others which were not included in the set. The production network was trained to output each syllable sequence followed by a boundary symbol. Training continued for 110 repetitions of all patterns, at which point the network made errors on 7 of the 384 segments making up the training syllables. Errors were made on 7 of the 95 segments in the test sequences. Only one of these segments was one which did not lead to a legal syllable in the language.

These results indicate that the recognition network is able to generalize about syllable structure on words containing a subset of the possible syllables and that the distributed representations developed during training can be used for production as well. The fact that the errors made are reason-

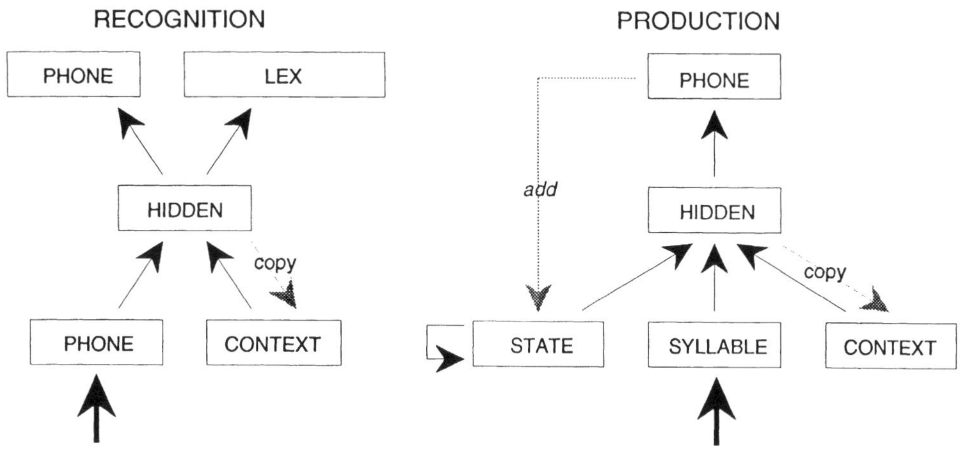

Figure 1: Network Architectures

able ones indicates that the representations are encoding syllable structure in a systematic way.

Next the trained recognition network was presented a representative set of 142 bogus syllables, sequences which did not conform to the language the network had been trained on. These included sequences with phones not among the phoneme inventory of the language (e.g., *b* and *d*), sequences with illegal codas (e.g., *fap*), sequences with long nuclei (e.g., *mua*), sequences with cluster onsets, and sequences with no nuclei. The hidden-layer representations for each of these sequences were saved and presented to the trained production network. The output of the production network was then examined to determine whether the networks would in effect correct the representations. The production network responded to 97 of the 142 sequences (68%) by replacing the original sequence with a legal syllable in the language. Typical responses included the following: *kn* → *ken*, *kfe* → *ke*, *xou* → *xu*, *pik* → *pi*, *zan* → *nan*.

These results are further evidence that the recognition and production networks have learned about the structure of syllables in the language. They also indicate that the representations are robust.

Experiment 2

Finally, the syllable representations from the recognition network were used as inputs to simple feedforward networks which were designed to determine whether the representations could be used for phonological transformations. Each feedforward network took as input a syllable representation and yielded as output the syllable representation that resulted when applying a particular rule to the input syllable. Three rules (and three networks) were used: a rule which replaced the vowel in a syllable with *u*, a rule which made the coda of the syllable *-s*, and a rule which replaced the onset of the syllable with the fricative in the same place of articulation as the onset of the original syllable (or by *s* if there was no onset).

Each network was trained on 80% of the syllables until there were no errors, then tested on the remaining 20%. Training required about 25 repetitions of all of the patterns. The network's response was taken to be that syllable (of the 135 possible) whose distributed representation was closest (in Euclidian distance) to the network's output pattern. For each rule, over 95% of the test syllables were generated correctly. In all cases errors resulted in syllables which satisfied the basic constraint imposed by the rule in question (*u* nucleus, *s* coda, fricative onset).

These results indicate that the syllable representations learned by the recognition network encode syllable structure in a way which makes it accessible to the sorts of operations which are common in the phonological systems of natural languages.

Discussion

The experiments reported on here demonstrate that simple recurrent networks can be trained to develop representations of syllables which encode information about structure in a distributed form. These representations present a viable alternative to traditional concatenative types of representations. Like their symbolic counterparts, the distributed representations can be unpacked into the sequences they represent and can be transformed in systematic ways. Unlike their symbolic counterparts, the distributed syllable representations are learned; are of fixed width; and permit parallel, single-step operations.

There are at least two other connectionist approaches to the acquisition of syllables. Goldsmith & Larson (1990) model the syllabification of words in a variety of languages using a constraint satisfaction network in which units represent segments in the word and activations represent the "derived sonority" of the segments, an indication of their role in the syllabic structure of the word. Two simple parameters characterize syllabification in each language. The model provides an elegant account of a range of phenomena, but it is not clear what it has to do with processing since what is modeled is abstract, atemporal derivation. It is also not specified how a language learner might have access to the derived sonorities needed to learn the parameters.

More in the spirit of the present approach is an experiment by Corina (1991), in which a simple recurrent network was presented with sequences of phonetic segments from a database of spoken English utterances. Trained simply to predict the next segment, the network showed clear evidence of having discovered the statistical regularities that characterize the structure of the English syllable. That is, its output predictions corresponded closely to the actual probabilities of particular segment classes in particular positions. This is evidence that a network can also learn about syllable structure from training on an unsupervised task. It remains to be seen whether the hidden layer patterns from Corina's network are suited for recognition and production or whether there is anything to be gained by combining the supervised recognition and unsupervised prediction tasks.[1]

How might the syllable representations learned in the network fit into to a more complete model of word recognition and production? I noted above that the recognition network was not especially successful in learning to distinguish the 100 words it was trained on. As the number of words to be recognized increases to more plausible ranges, we can expect very serious degradation in this capacity, though increasing the hidden layer size would offset the degradation to some extent. Yet the problem might go away in a hierarchically organized system with simple recurrent networks operating with different units as inputs. Word recognition might then be a process of assigning sequences of syllables and/or larger metrical units to word or morpheme units. Thus the syllable representations learned in the network described here would provide the input to a syllable-level network. See Gasser (1991) for more on this proposal.

From the perspective of its plausibility as a model of phonological acquisition, the present model has a number of inadequacies and gaps. First, I have only scratched the surface in terms of what might be required of such a model. How, for example, might this approach account for learning how to assign stress to novel words (in a language which does this in a non-arbitrary way)? Recently, Gupta & Touretzky (1991) have shown that perceptrons can learn to assign stress to syllable sequences from 19 natural languages (apparently encompassing the range of possible stress systems). The present approach would attempt to achieve this in the context of the hierarchical architecture referred to above, by training a sequential network which takes distributed syllable representations (one at a time) as input to recognize words involving one or more metrical units (sequences of stressed and unstressed syllables). The hope would be that distributed representations for these units, and eventually for the entire words, would arise, and that these would provide the input to the word production process, where stress assignment takes place. While considerably more involved than the approach of Gupta & Touretzky (1991), this would respect the sequential nature of language and maintain the relationship between word recognition and phonological learning.

A further weakness of the framework in its current state involves the learning of production. While the learning of syllable representations as a side-effect of the process of word (or morpheme) recognition seems reasonable, the learning of the reverse process is another matter. The network trained on the production task was provided with targets for each output phone, a degree of supervision that clearly does not correspond to anything in the experience of the human language learner. For now it may be best to view this task as nothing more than an existence proof that the representations can be unpacked for production or alternatively a technique for analyzing the distributed representations, which, unlike their symbolic counterparts, are not directly interpretable. Of course, the issue of how children learn to produce, as well as perceive, linguistic forms, when they are not provided with targets, is one facing any approach to language acquisition.

Finally, the present approach presupposes some mechanism for segmentation, first, at the level of the phones that are the inputs to the recognition process, and second, at the level of the syllables (or words) themselves. Again, segmentation is a problem for all sorts of acquisition models. Recently Doutriaux & Zipser (1990) have had some success in training simple recurrent networks to discover segments in speech. Thus this seems to be a problem that can be approached within the framework outlined in this paper.

[1] In some preliminary experiments, I have not found better performance on word recognition from networks which are also expected to predict their next sequence.

References

Chalmers, D. (1990). Syntactic transformations on distributed representations. *Connection Science, 2*, 53–62.

Corina, D. P. (1991). *Towards an Understanding of the Syllable: Evidence from Linguistic, Psychological, and Connectionist Investigations of Syllable Structure.* Ph.D. thesis, University of California, San Diego.

Doutriaux, A. & Zipser, D. (1990). Unsupervised discovery of speech segments using recurrent networks. In Touretzky, D., Elman, J., Sejnowski, T., & Hinton, G. (Eds.), *Proceedings of the 1990 Connectionist Models Summer School*, pp. 303–309. Morgan Kaufmann, San Mateo, CA.

Dresher, B. E. & Kaye, J. D. (1990). A computational learning model for metrical phonology. *Cognition, 34*, 137–195.

Elman, J. (1990). Finding structure in time. *Cognitive Science, 14*, 179–211.

Fodor, J. & Pylyshyn, Z. (1988). Connectionism and cognitive architecture: a critical analysis. *Cognition, 28*, 3–71.

Frazier, L. (1987). Structure in auditory word recognition. *Cognition, 25*, 157–187.

Gasser, M. (1991). Sequence comparison and simple recurrent networks. *Center for Research in Language Newsletter.*

Goldsmith, J. & Larson, G. (1990). Local modeling and syllabification. In Deaton, K., Noske, M., & Ziolkowski, M. (Eds.), *Papers from the 26th Annual Regional Meeting of the Chicago Linguistics Society: Parasession on the Syllable in Phonetics and Phonology.* Chicago Linguistics Society.

Goldsmith, J. (1990). *Autosegmental and Metrical Phonology.* Basil Blackwell, Cambridge, MA.

Gupta, P. & Touretzky, D. S. (1991). Connectionist networks and linguistic theory: investigations of stress systems in language.. Unpublished report, Carnegie-Mellon University.

Jordan, M. (1986). Attractor dynamics and parallelism in a connectionist sequential machine. In *Proceedings of the Eighth Annual Conference of the Cognitive Science Society*, pp. 531–546 Hillsdale, New Jersey. Lawrence Erlbaum Associates.

Legendre, G., Miyata, Y., & Smolensky, P. (1991). Distributed recursive structure processing. In Lippmann, R. P., Moody, J. E., & Touretzky, D. S. (Eds.), *Advances in Neural Information Processing Systems 3*, pp. 591–597. Morgan Kaufmann, San Mateo, CA.

Liberman, A. M. & Mattingly, I. G. (1986). The motor theory of speech revised. *Cognition, 21*, 1–36.

Plunkett, K. & Marchman, V. (1991). U-shaped learning and frequency effects in a multilayered perceptron: implications for child language acquisition. *Cognition, 38*, 1–60.

Pollack, J. B. (1990). Recursive distributed representations. *Artificial Intelligence, 46*, 77–105.

Port, R. (1990). Representation and recognition of temporal patterns. *Connection Science, 2*, 151–176.

Rumelhart, D. E., Hinton, G., & Williams, R. (1986). Learning internal representations by error propagation. In Rumelhart, D. E. & McClelland, J. L. (Eds.), *Parallel Distributed Processing*, Vol. 1, pp. 318–364. MIT Press, Cambridge, MA.

van Gelder, T. (1990). Compositionality: a connectionist variation on a classical theme. *Cognitive Science, 14*, 355–384.

Finding Linguistic Structure with Recurrent Neural Networks

Nick Chater
University of Edinburgh
Departments of Psychology &
7, George Square
Edinburgh EH8 9JZ
nicholas@uk.ac.ed.cogsci

Peter Conkey
University of Edinburgh
Artificial Intelligence
80 South Bridge
Edinburgh EH1 1HN
pbc@uk.ac.stir.cs

Abstract

Simple recurrent networks have been used extensively in modelling of learning various aspects of linguistic structure. We discuss how such networks can be trained, and empirically compare two training algorithms, Elman's "copyback" regime and back-propagation through time, on simple tasks. Although these studies reveal that the copyback architecture has only a limited ability to pay attention to past input, other work has shown that this scheme can learn interesting linguistic structure in small grammars. In particular, the hidden unit activations cluster together to reveal linguistically interesting categories. We explore various ways in which this clustering of hidden units can be performed, and find that a wide variety of different measures produce similar results and appear to be implicit in the statisticsof the sequences learnt. This perspective suggests a number of avenues for further research.

Introduction

Simple recurrent neural networks (SRNs) developed by Jordan (1986) and Elman (1988) provide a powerful tool with which to model the learning of many aspects of linguistic structure (for example, Elman 1990, 1991; Shillcock, Levy & Chater 1991) and there has been some exploration of their computational properties (Chater 1989; Cleermans, Servan-Schrieber & McClelland 1989; Servan-Schrieber, Cleeremans & McClelland 1991). The presence of recurrent connections allows past activation to influence current output, which means that output can respond sequential structure in the input. The extent to which such networks can be taught to learn interesting sequential structure depends on the learning algorithm employed. A natural approach is to apply the back-propagation training algorithm which has proved so successful in training non-recurrent feedforward networks to learn interesting static input-output patterns.

The structure of the paper is as follows. First we discuss a number of ways in which the back-propagation algorithm can be adapted to train recurrent networks to learn sequences, concentrating on two options, Elman's (1990) "copyback" scheme, and back-propagation through time (Rumelhart, Hinton & Williams 1986). We note that there are theoretical reasons to suppose that the copy-back regime will learn less well, and this conclusion is borne out in our simulations. However, the copy-back approach is computationally inexpensive and has provided impressive results in a number of language processing tasks. We investigate the scope of this method further, and follow Elman in investigating the nature of the hidden unit representations developed for a network which learns to predict the next element in sequences generated by a simple grammar. A number of very different measures over the hidden units are found to generate very similar syntactic/semantic clustering, and these clusters are also implicit in the statistics of the sequences learnt. This suggests that network performance can usefully be analysed in terms of the statistical structure of the input sequences, and that the applicability of SRNs to real natural language data can be assessed by analysing relevant aspects of its statistical structure

Training SRNs

Backpropagation cannot directly be applied to SRNs since the algorithm applies only to feedforward networks. For these, back-propagation performs gradient descent in the sum-squared error over the finite number of input-output pairs in the training set. To apply back-propagation to recurrent networks, the SRN must be "unfolded" into a feedforward network which is then trained in the conventional way. The most popular method of training SRNs involves unfolding the network by providing an additional input - the context units - which corresponds to the previous values of the hidden units (Elman 1990) (Figure 1a, 1c). The context units are dependent on the previous inputs, among which is the previous value of the context units. Hence the behaviour of the network is influenced not just by the current input but by the sequence of past inputs. While activation is propagated forwards through the network from arbitrarily far back in time, error is only propagated back to the context units.

An alternative approach is to unfold the network through several time steps (Rumelhart, Hinton & Williams 1986) so that each weight has several "virtual incarnations" and to back-propagate through the resulting network (Figure 1a, 1c). The overall weight change is simply the sum of the changes recommended for each incarnation. This "back-propagation through time" can in principle be back-propagated through the entire training history of the network (Rohwer personal communication) but is typically implemented by unfolding through a small number (here we shall use 5) time steps. The copy-back scheme can be viewed as a special case of back-propagation through time, in which the back-propagation of error stops at the first copy of the hidden units - the context units.

The more the network is unfolded, the better the approximation of the feedforward network to the underlying recurrent network, and the better the network learns to respond to sequential material. The minimal unfolding embodied in copy-back scheme should therefore produce the poorest learning, although it has the considerable advantage of being the cheapest computationally. This is borne out below in the comparitive studies below.

A natural assumption would be that the number of steps back that error is propagated will precisely fix the number of steps "back" the network can learn about. If this were true, the copy-back scheme would only be able to respond to the current and previous input, and would thus not be able to learn any interesting sequences. However, as long as the relevant temporally distant information "percolates through", even in some degraded form, to a point in the unfolded network to which error is propagated, the weights forward of that point can be adjusted to utilize that information successfully. Hence, the last point in the network to which error is propagated forms a "bottleneck", at which temporally more distant information must pass if the network is to be able to learn to respond to it (see Figure 2) (Chater 1989).

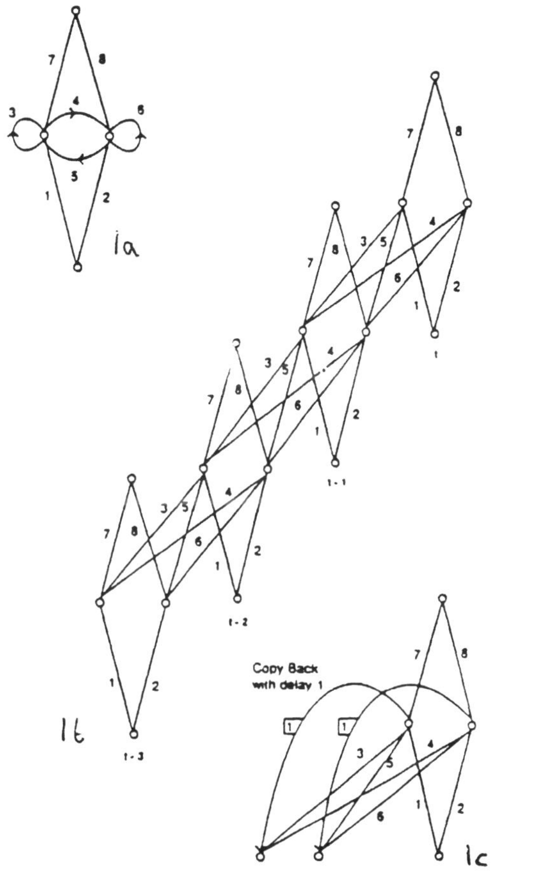

Figure 1 *Unfolding a recurrent neural network (1a), for back-propagation through time (1b) and copy-back training (1c).*

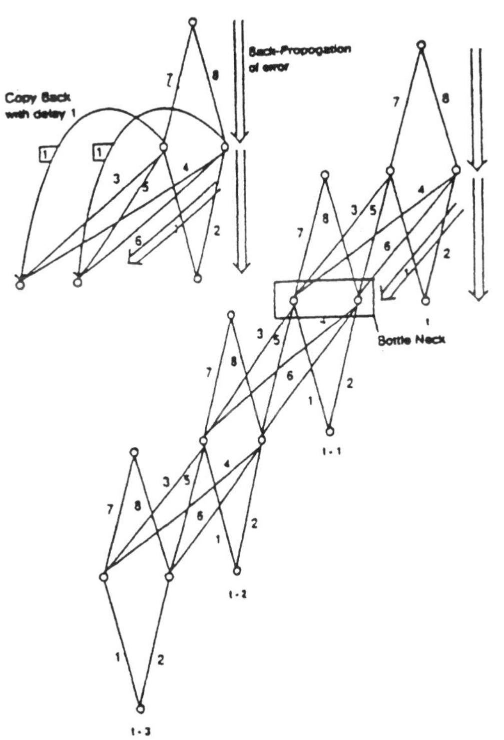

Figure 2 *Back-propagation in the copy-back training regime.*

In many tasks, temporally distant information beyond the bottleneck, is relevant to predicting intervening material. This means that the network encodes that information in its hidden units, to predict that more local information, which forces this information through the bottleneck. Hence the SRN is able to learn to respond to this distant information successfully. However, in a task in which temporally distant information is not correlated with intervening material, such as the task of learning to be delay line, reported below, learning with the copy-back scheme should be poor.

While our primary interest in this first set of simulations was comparing the performance of copy-back and back-propagation through time, a secondary interest was in the effect of using or not using context units in back-propagation through time. If the network is unfolded several time-steps (5 in the simulation we report), the contribution of the context units at the bottleneck to the final output may be very small, and the large number of intervening layers may make it difficult to learn to respond to this input, even if it is informative. From a theoretical point of view, not using context units is attractive, since the network can then be viewed as learning a fixed input-out put set (or a sample from a fixed distribution), and hence the proof that back-propagation performance gradient descent is valid. For most problems, the presence or absence of context units seems to have little effect on performance, and we discuss this briefly below.

Copy-Back and Backpropagation Through Time

We report simulations on two very simple tasks using binary sequences, discrete XOR and learning to the a delay line.

Discrete XOR

Consider a binary sequence in which two out of three bits are generated at random, and the third is the XOR of the previous two. The task is to attempt to predict the next value in the sequence. This task is difficult, since only every third bit can is in principle predictable. Optimal performance is to correctly predict these bits, and to output 0.5 otherwise.

Architecture	Hidden Units	Average squared errors Posn 1	Posn 2	Posn3
copy back	4	0.278 ± 0.003	0.273 ± 0.003	0.16 ± 0.02
copy back	7	0.282 ± 0.002	0.288 ± 0.002	0.11 ± 0.01
copy back	10	0.283 ± 0.001	0.289 ± 0.002	0.10 ± 0.01
unfolded with contexts	2	0.267 ± 0.004	0.271 ± 0.002	0.20 ± 0.02
unfolded with contexts	3	0.276 ± 0.004	0.280 ± 0.003	0.16 ± 0.03
unfolded with contexts	4	0.275 ± 0.004	0.280 ± 0.004	0.18 ± 0.03
unfolded with contexts	5	0.278 ± 0.004	0.283 ± 0.003	0.12 ± 0.02
five unfolds no contexts	2	0.268 ± 0.004	0.270 ± 0.004	0.19 ± 0.02
five unfolds no contexts	3	0.282 ± 0.003	0.285 ± 0.003	0.12 ± 0.01
five unfolds no contexts	4	0.281 ± 0.003	0.286 ± 0.003	0.11 ± 0.02
five unfolds no contexts	5	0.289 ± 0.001	0.287 ± 0.001	0.089 ± 0.004

Table 1 : Performance on the discrete XOR task with 50 epochs of training

Copy-back and back-propagation schemes (both with and without context) were trained on XOR (Table 1). The results were averaged over 50 trials, with 50 training epochs over 3000 input-output pairs with learning rate 0.1 and momentum 0.9. For back-propagation through time, the net was unfolded 5 time steps. The weights were initialised randomly between -5 and 5. If the weight starts are smaller than this, "copyback" learning is slow, perhaps because for the copy-back regime, perhaps because for small inputs the sigmoid activation function is nearly linear, and hence unable to compute XOR . Notice that the standard deviations of the errors obtained are small throughout.

For a network of a given size, performance is far better with back-propagation through time than using the copy-back scheme, which require far more hidden units to attain comparable results. This pattern is consistently obtained in a comprehensive range of simulations (Conkey 1991).

Turning to our second concern, performance using back-propagation through time is not significantly different with or without context, despite the fact that context could in principle have provided very useful information. This is because the no context network may not be able to determine from just 5 time steps which bits are predictable and which are not. If the last five bits were

...0 1 1 1 0

then the third and fifth bits are both the XOR of their predecessors. In this case it is not in principle possible for these unfolded nets without context units to know whether the next bit is the result of an XOR or is random. Since this ambiguity occurs in almost 60% of cases, the ability to use past context to disambiguate (effectively storing a regular "pulse" indicating which bits are predictable) would be advantageous. However, it does not appear to be possible to learn to utilize this information in practice.

Learning to be a delay line.

The analysis of the copy-back learning algorithm above suggested that it should be poor at learning to respond to temporally distant input, unless the temporally distant information has been used in intermediate predictions. This suggests that while in many interesting problems (such as that of learning a grammar with some recursive structure, detailed in Elman (1991)) the net can respond to temporally distant information, this will extremely difficult if the nature of the distant dependency is independent of the intervening material. The simplest such task is learning to be a delay line - to reproduce a random binary input stream delayed by several time steps.

A recurrent network with n+1 hidden units can act as a delay line of n, given appropriate weights. One intuitively attractive solution is for the hidden units to act as buffers for the input so that one unit has output at time t of i(t-1), another i(t-2) and so on back to i(t-n). Figure 3 illustrates weights that would implement this solution for a delay line of 1 and a network with two hidden units.

Table 2 shows a typical sample of results. While back-propagation through time is able to learn the delay line task quite well (and with only n+1 hidden units for small delays n), the copy-back scheme can only learn to respond to small delays with relatively large numbers of hidden units. This is explicable in terms of the theoretical discussion above - the more hidden units the more likelihood that relevant information will by chance percolate through the network and thus that the network will be able to learn to use this information. Learning performance is also very much less consistent with the copy-back regime.

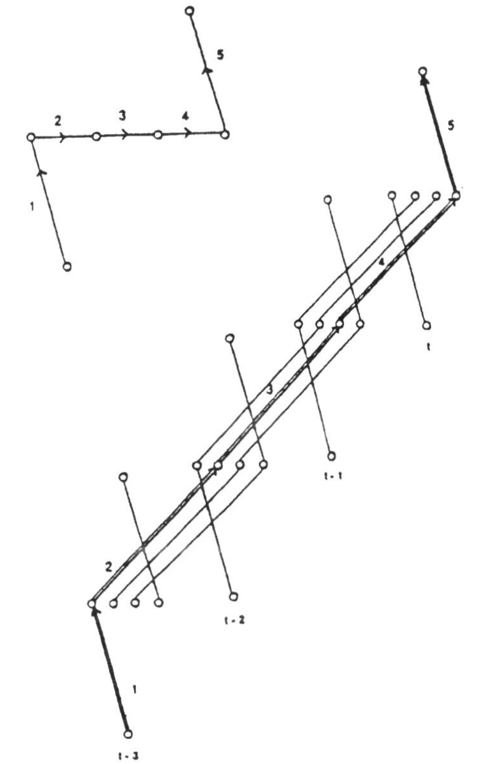

Figure 3 *Implementing a delay of n time steps with n+1 hidden units.*

The inability of the copy-back scheme to learn to respond to long time delays contrasts with good performance reported predicting dependencies in small scale language tasks where the intervening material is relevant (Elman 1991).

The results of these experiments bear out the theoretical analysis that back-propagation through time leads to better learning than the copy-back scheme. However, back-propagation through time is computationally more expensive, and the copy-back scheme may be able to learn many interesting tasks. One particularly intriguing result is that the averaged hidden unit patterns appear to encode the syntactic/semantic categories for a toy grammar (Elman; 1990) . The studies reported below repeat, extend and analyse this result, and argue that such a clustering is to be expected given the statistics of the sequences leant.

Architecture	Delay	Hidden units	Learning rate	Average squared error	Passes
copy back	1	2	0.05	0.22	< 100
copy back	1	2	0.10	0.21	< 100
copy back	1	3	0.05	0.20	< 100
copy back	1	3	0.10	0.16	< 100
copy back	1	4	0.05	0.13	< 100
copy back	1	4	0.10	0.097	< 100
unfolded	1	2	0.10	0.001	3
copy back	2	3	0.05	0.254	< 100
copy back	2	3	0.10	0.256	< 100
copy back	2	4	0.05	0.262	< 100
copy back	2	4	0.10	0.260	< 100
copy back	2	7	0.05	0.278	< 100
copy back	2	7	0.10	0.264	< 100
copy back	2	10	0.05	0.160	< 100
copy back	2	10	0.10	0.239	< 100
unfolded	2	3	0.10	0.031	3

Table 2 : Learning to be a delay line

Incidently Recognising Linguistic Structure

Elman used to copy-back regime to train a net to predict the next item in a continuous text sequence, generated by a simple grammar (borrowed from Elman 1988; 1990). Whereas Elman represented each "word" by a random bit vector, used a completely localist representation, thus using 29 input units to represent the 29 words. As in Elman's simulations there is no explicit marker for the end of a sentence. 150 hidden units and 150 corresponding context units were used.

Conditional probabilities for the next word, given the sentence so far were calculated from the data set. The RMS errors relative to this benchmarkwere 0.2 per patternin both cases, whereas Elman obtained 0.05. This difference may be a result of our choice of a localist input representation. We then followed Elman in cluster analysing the hidden unit activation evoked on presentation of each word. These were averaged to give a single 150 element vector for each of the 29 words.

The results from a typical net (Figure 4) do not give as good a clustering as that obtained by Elman. There is poor separation of nouns and verbs, and some confusion between different classes of each. Clustering on the basis of current input may not be the best measure, since the hidden unit values must encode previous input relevant to prediction, not just the current word. We also clustered hidden unit states averaged by the entirey Hence a more attractive alternative is to average hidden unit patterns together on the basis of the word predicted.

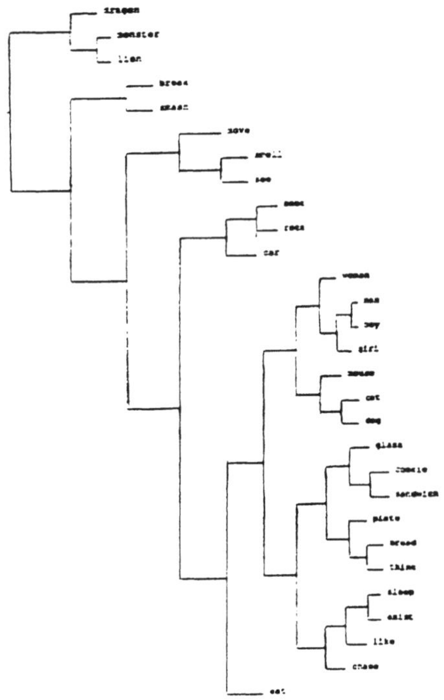

Figure 4 *Clustering by current word*

This measure, which completely cross-classifies the data with respect to the original measure, does indeed produce much better clusters, shown in Figure 5. Using this measure the clusters obtained well reflect the underlying syntactic categories of the grammar, with, for examples, nouns being separated from verbs, and different kinds of nouns being very well segrated and verbs segregated somewhat less precisely.

A further possibility is to cluster not the hidden unit pattern associated with an incoming word, but the change in hidden unit representation brought about by that word. Again a good clustering is obtained (Figure 6), comparable in quality with that obtained by clustering with respect to the target word.

It seems that a variety of measures of hidden unit values produce clusters corresponding to linguistically interesting categories. Is there an "optimal" clustering of this data set to which all of these measures are approximating? Not really - each of the measures considered above correspond to statistics of the data sets, which can be directly measured. For example, Elman's original measure of averaging hidden units on the basis

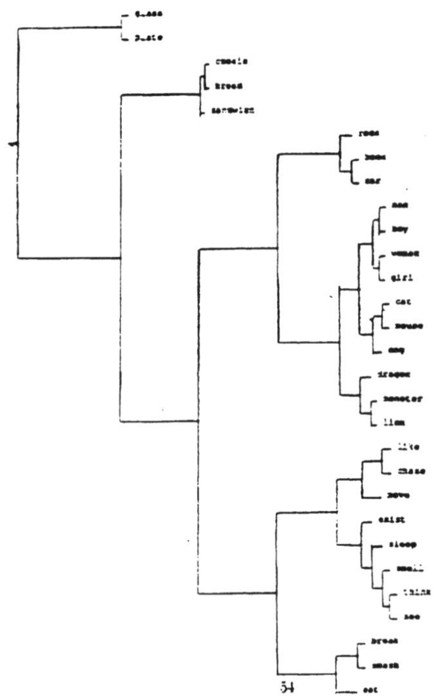

Figure 5 *Clustering by predicted word*

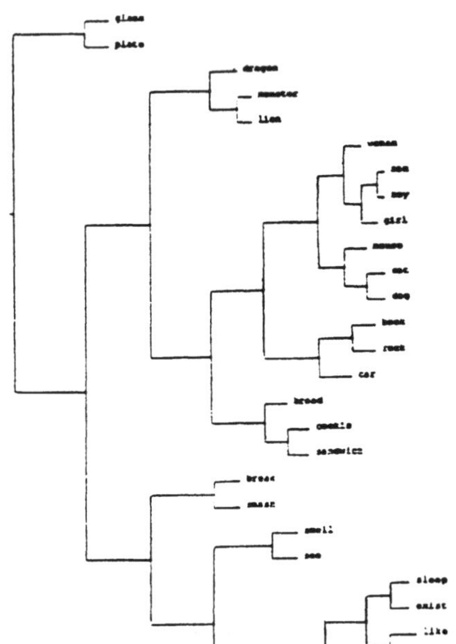

Figure 6 *Clustering by change of hidden unit pattern*

of the past word corresponds to grouping words by the conditional probabilities of successive words. We measured this quantity directly, and then cluster analysed (Figure 7) to produce very similar ressults to those obtained from the network (Figure 4) - this means that the network is successfully sampling the relevant statistic. Similar results can be obtained by comparing the two other measures with statistical analogues (clustering words on the basis of the conditional probabilities of the preceding words, and the change in conditional probabilities expected after a word is input,

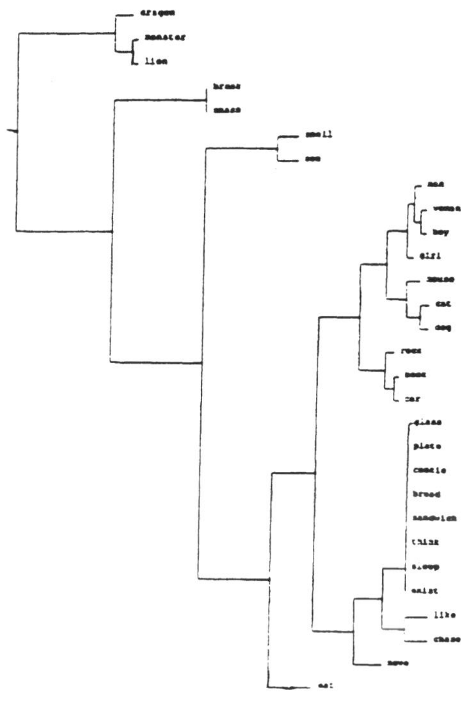

Figure 7 *Conditional probabilities clustered by preceding word*

respectively). Since the copy-back scheme is sampling these statistics successfully, there seems to be no room for improvement using back-propagation through time, and thus we predict that the clusters from back-propagation through time will produce similar results. The limitation on performance is the structure of the data rather than nature of the network used.

These results suggests that the hidden unit patterns that recurrent neural networks develop can be viewed as reflecting quite directly the statistical structure of the sequences learnt. Furthermore, particular statistical measures of hidden unit activation may closely correspond to a related statistic of the sequence itself.

Conclusion

The first set of simulations reported confirmed the theoretically motivated expectation that the back-propagation through time is superior to (less expensive) copy-back training for learning sequential structure. Experiments with large copy-back networks suggest that the hidden unit representation is successfully sampling statistics of the underlying sequential material, and we predict that back-propagation through time should, therefore, produce very similar clusters. Of course, if the underlying grammar, and hence the relevant statistics, are more complex, then back-propagation through time may then be able to sample these statistics better.

This suggests three interesting avenues for further research: 1) to investigate further the relationship between statistical analysis of the hiddenunit representations and direct analysis of the original data set, both using the copy-back and back-propagation through time regimes; 2) to explore real natural language data directly by cluster analysing using simple statistics to explore what peformance can be expected from a neural network model; 3) to investigate if statistics which are revealing of linguistic structure can be implemented more directly in a network, so that a full-size network can be built which is able to handle real natural language data. These last two avenues have recently been explored by Finch & Chater (1991) with encouraging results.

Chater, N. (1989) "Learning to Respond to Structures in Time" (Technical Report, Research Initiative in Pattern Recognition, St Andrews Road, Malvern, Worcs. RIPRREP/1000/62/89.

Cleeremans, A., Servans-Schreiber, D. & McClelland, J. L. (1989) Finite state automata and simple recurrent networks. Neural Computation, 1. 372-381.

Conkey, P. (1991) Sequence Prediction Using Re current Neural Networks. MSc Thesis, Department of Artificial Intelligence, University of Edinburgh.

Elman, J. L. (1988) Finding structure in time. Technical Report, CRL TR 8801, Centre for Research in Language, UCSD.

Elman, J. L. (1990) Finding structure in time. *Cognitive Science*, 14:179-211.

Elman, J. L. (1991) Distributed Representations, Simple Recurrent Neural Networks, and Grammatical Structure. Machine Learning, 7, 195-225.

Finch , S. & Chater, N. (1991) A Hybrid Approach to the Automatic Learning. *AISB Quarterly*, 78, 16-24.

Jordan, M. (1986) Serial order: a parallel distributed approach. Institute for Cognitive Science Report, 8604, University of California, San Diego.

Rumelhart, D. E., Hinton, G. E. & Williams, R. J. (1986) Learning internal representations by error propagation. In D. E. Rumelhart & J. L. McClelland (Eds) *Parallel Distributed Processing* Vol 1, 318-362, Cambridge, Mass: MIT Press.

Servans-Schreiber, D., Cleeremans, A. & McClelland, J. L. (1991) Graded state machines: The representation of temporal contingencies in simple recurrent networks. *Machine Learning*. , 7, 161-193.

Shillcock, R., Levy, J. & Chater, N. (1991) "A connectionist model of word recognition in continuous speech" Proceedings of the Cognitive Science Society of America Annual Conference, 340-345.

A phonologically motivated input representation for the modelling of auditory word perception in continuous speech

Richard Shillcock
Centre for Cognitive Science
University of Edinburgh
2 Buccleuch Place
Edinburgh
rcs@cogsci.ed.ac.uk
rcs@cogsci.ed.ac.uk@nsfnet-relay.ac.uk

Geoff Lindsey
Dept. of Linguistics
University of Edinburgh
AFB, George Square
Edinburgh
geoff@ed.ling.ac.uk

Joe Levy
HCRC
Univ. of Edinburgh
2 Buccleuch Place
Edinburgh
joe@cogsci.ed.ac.uk

Nick Chater
Dept of Psychology
Univ. of Edinburgh
7 George Square
Edinburgh
nicholas@cogsci.ed.ac.uk

Abstract

Representational choices are crucial to the success of connectionist modelling. Most previous models of auditory word perception in continuous speech have relied upon a traditional Chomsky-Halle style inventory of features; many have also postulated a localist phonemic level of representation mediating a featural and a lexical level. A different immediate representation of the speech input is proposed, motivated by current developments in phonological theory, namely Government Phonology. The proposed input representation consists of nine elements with physical correlates. A model of speech perception employing this input representation is described. Successive bundles of elements arrive across time at the input. Each is mapped, by means of recurrent connections, onto a window representing the current bundle and a context consisting of three such bundles either side of the current bundle. Simulations demonstrate the viability of the proposed input representation. A simulation of the compensation for coarticulation effect (Elman and McClelland, 1989) demonstrates an interpretation which does not involve top-down interaction between lexical and lower levels. The model described is envisaged as part of a wider model of language processing incorporating semantic and orthographic levels of representation, with no local lexical entries.[1]

Introduction

Psychologists wishing to model spoken language perception have typically assumed that the physical speech signal may be translated into a featural level of representation, which then maps onto a phonemic level, which, in turn, supplies activation to a lexical level. In the TRACE model (McClelland & Elman 1986) these three levels of representation are instantiated in an interactive-activation architecture, in which patterns of activation percolate up and down between adjacent levels. In models of speech production, similar assumptions are made. In the Plaut and Shallice (*in press*) model of pronunciation in deep dyslexics, patterns of activity at the semantic level are mapped onto position-specific phonemic representations. There have been some departures, however, from explicit, local phonemic representations. In the Seidenberg and McClelland (1989) model of word naming, orthographic patterns are mapped onto Wickelphones consisting of triples of consecutive features. In the model of word recognition proposed by Norris (1990), bundles of distinctive features arrive across time at the 11 input nodes and activation is mapped onto an output layer consisting of local representations of words.

In some of these models, the phonological representations employed, although internally consistent, are simplifications of what might be thought adequate by phonologists. In others, while the input representations are relatively sophisticated, they do not reflect phonologists' more recent disenchantment with the phoneme and with bundles of features organized strictly linearly. From a different perspective, many researchers in automatic speech recognition (ASR) have become disillusioned with the notion that ASR is best attempted by recognizing SPE-style features (Chomsky & Halle, 1968) in the physical signal and subsequently parsing them into phonemes and words.

Although the models cited above have achieved considerable success in capturing many of the qualitative aspects of language processing, the enterprise should be substantially improved by the use of a speech input representation which, first, is consonant with current phonological theory and, second, has a consistent relationship with the physical speech signal. Such an input representation will more

[1] This research was carried out under E.S.R.C. grant number R000 23 3649.

adequately reflect the structure of the real-world problem of speech recognition. Below we present such an input representation – an alternative to the orthodox SPE-style framework – and describe its instantiation in a recurrent network. We then review its success as a psychological model.

Phonological motivation

The input representation described in Table 1 is based on recent work in Government Phonology (Kaye, Lowenstamm & Vergnaud, 1985, 1990). The speech signal is decomposed into nine elements, defined briefly as follows.

A: oral cavity openness; alone, the vowel quality of *palm*.
I: palatality; alone, the vowel quality of *see*.
U: labiality; alone, the vowel quality of *boot*.
?: occlusion; abruptness; alone, glottal stop.
h: aperiodic energy; alone, [h].
N: nasality.
R: apicality/coronality/coronal formant locus.
@: velarity/centrality.
H: voicelessness.

The elements are represented principally in a binary way. In four cases a value of 0.5 is used; this is a representational compromise which reflects the notion of "government" within the phonological theory. The lefthand column gives Machine-Readable Phonetic Alphabet (MRPA) equivalents for short vowels and syllable-initial consonants; elements may be subtracted from the definitions in Table 1 to represent segments in other environments (*e.g.* /k/ would lack the element **h** when unreleased as in *act*). In Table 1, the initial glides in *yet* and *wet* have the same element representations as

The input representation

	elements								
segment	?	h	U	N	R	@	H	I	A
p (pat)	1	1	1	0	0	0	1	0	0
t (tap)	1	1	0	0	1	0	1	0	0
k (cat)	1	1	0	0	0	1	1	0	0
b (bat)	1	1	1	0	0	0	0	0	0
d (dot)	1	1	0	0	1	0	0	0	0
g (got)	1	1	0	0	0	1	0	0	0
m (mill)	1	0	1	1	0	0	0	0	0
n (nil)	1	0	0	1	1	0	0	0	0
ng (sing)	1	0	0	1	0	1	0	0	0
f (fit)	0	1	1	0	0	0	1	0	0
th (thin)	0	.5	0	0	1	0	1	0	0
s (sin)	0	1	0	0	1	0	1	0	0
sh (shin)	0	1	0	0	1	0	1	1	0
zh (measure)	0	1	0	0	1	0	0	1	0
h (hat)	0	1	0	0	0	0	0	0	0
v (vat)	0	1	1	0	0	0	0	0	0
dh (that)	0	.5	0	0	1	0	0	0	0
z (zen)	0	1	0	0	1	0	0	0	0
l (lip)	1	0	0	0	1	0	0	0	0
r (rip)	0	0	0	0	1	0	0	0	0
y (yell)	0	0	0	0	0	0	0	1	0
w (well)	0	0	1	0	0	0	0	0	0
i (bin)	0	0	0	0	0	0	0	1	0
e (den)	0	0	0	0	0	0	0	1	.5
a (ban)	0	0	0	0	0	0	0	.5	1
o (don)	0	0	1	0	0	0	0	0	1
uh (bud)	0	0	0	0	0	1	0	0	1
u (wood)	0	0	1	0	0	0	0	0	0
@ (about)	0	0	0	0	0	1	0	0	0
A (see text)	0	0	0	0	0	0	0	0	1

Table 1. Definition of short vowels and syllable-initial consonants of standard Southern British English in terms of elements. The special element A does not appear singly in isolation (see Table 2, overpage).

segment	expansion
ch (chair)	t sh
jh (journey)	d zh
@@ (bird)	@ @
ou (bode)	@ u
oi (boy)	o i
oo (bored)	o o
ii (bee)	i i
uu (boot)	u u
ei (bade)	e i
ai (bide)	A i
au (loud)	A u
u@ (poor)	u @
i@ (beer)	i @
e@ (pair)	e @
aa (bard)	A A

Table 2. Expansions of the 2 affricates and the 13 diphthongs and long monopthongs of standard Southern British English.

the vowels of *pit* and *put*, respectively; the differences are attributed to location in syllable structure and are not explicitly encoded in the model described below. Affricates, diphthongs and long monopthongs are decomposed into two consecutive segments, as shown in Table 2.

The model

The goal is a comprehensive model of speech processing which will allow the detailed modelling of psycholinguistic data, going beyond the essentially qualitative results achievable with models such as TRACE. The earliest point at which to begin the psychological modelling of speech processing is with the transduction of the physical signal into some common currency of activation. A noisy acoustic signal is converted into a representation which has psychological significance. Sensitivity to a window of context is essential if this conversion is to be reliable. Some parts of the signal will be captured more securely than others, either because they are inherently more distinctive or because of context. The input representation detailed above is seen as the most appropriate description of the input at this point.

The approach taken below is defined, first, by the fact that, although the elements have a relationship with the physical speech signal, there is no discrete set of acoustic entities which might be employed in a mapping from an element level of representation. Still less is there available a corpus of natural speech transcribed in acoustic terms. The second constraint on the modelling is the principle of eschewing intermediate levels of linguistic representation, phonemes in particular. In this approach, the aim is to develop a single, psychologically realistic level of representation of the speech stream and to map this directly to a level at which the semantics of individual words is expressed. (This approach postpones segmentation and binding issues.)

In accordance with these constraints, the model described below captures the context-dependence of the initial transduction of the speech signal by auto-associating the patterns of elements as they arrive at successive time-slices. We envisage a subsequent mapping onto semantic representations; only the initial auto-association is described here, however. It is predicted that certain processing, typically ascribed to higher (lexical, morphological) levels of description, will be more or less weakly foreshadowed at this low level. (Norris (*in press*) demonstrates that, in principle, connectionist models which learn are likely to encode aspects of higher-level generalizations at lower levels of representation, giving rise to behaviour which looks like traditional "top-down" interaction.) The model we describe, once trained, accepts transcribed stimulus materials from experiments on spoken word recognition and outputs quantitative data on the facility with which each segment may be identified and represented.

Previously some of the present authors have suggested that some aspects of auditory word perception can be captured in a mapping between a featural level of representation and a phonemic level, in a recurrent network (Shillcock, Levy and Chater, 1991; Levy, Shillcock and Chater, 1991). This was an attempt to model speech processing in a comprehensive, full-scale way which postponed incorporating any local representations of lexical entries. The model was motivated by the Seidenberg and McClelland (1989) model of word naming, in which psychologically interesting behaviour falls out of a simple mapping between orthography and phonology, and in which a "lexical entry" is a distributed entity. Our goal was to remain within the auditory modality and to assess the extent to which behaviour previously attributed to higher-level (morphological and lexical) representations could be captured at the lower, featural and phonemic levels. This model illustrated Norris's observation about lower levels of representation embodying aspects of higher-level generalizations. Thus, for instance, the model correctly predicted five out of the six phoneme restorations reported by Elman and McClelland (1989) in their demonstration of compensation for coarticulation being triggered by a restored phoneme. The model was able to achieve the same sort of restoration on the basis of a learned features-to-phonemes mapping, with no explicit representations of words.

The new input representation described above allows us to continue with the investigation of a mapping between observable and necessary levels of representation (orthography, phonology, semantics) while avoiding local representations, at intermediate levels, of less defensible categories (phonemes, local lexical entries). Accordingly, the earlier model of

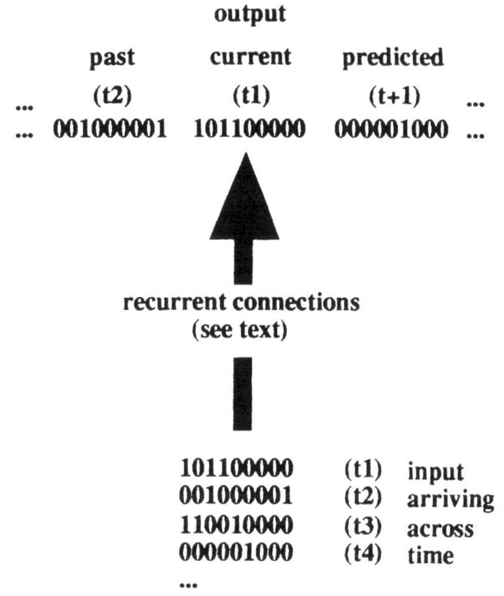

Figure 1. Input across time is mapped onto a more stable representation.

speech processing is superseded by that shown in Fig. 1.

The mapping currently implemented between the two representations is a "cut-down" version of "back-propagation through time", unfolding the network once rather than many times (Servans-Schreiber, Cleeremans & McClelland 1989; Chater 1989; Chater & Conkey, *submitted*) and thus sacrificing some of the ability reliably to pick up long distance dependencies, in exchange for speed of training. This "copyback" structure was introduced by Elman (1988, 1990) and Norris (1988). It will be superseded, in the present model, by a full "back-propagation through time" algorithm (Rumelhart, Hinton & Williams, 1986), in which a recurrent network is unfolded into many copies arranged in a feed-forward architecture, with standard back-propagation being applied to the result.

There are nine input nodes corresponding to the nine elements described above. There are thirty hidden units. The output nodes are grouped into seven sets of nine representing the bundles of elements at particular time-steps. The exact choice of the number of time-steps represented in the output is arbitrary, to an extent, and at this stage is motivated by the desire to study the effects of phonotactic constraints. (In practice, most of the observation of the behaviour of the model has involved recording the activation pattern for the current time-step.) A window which stretches over several segments allows right-context effects to be studied. In reality, the effects due to coarticulation with the immediately adjacent time-step are the most important, although vowel-to-vowel effects require a window of several time-steps. Note that including the surrounding feature-bundles in the output layer forces the network to learn the context in which any current feature-bundle occurs.

Although current simulations have used the idealization of one segment, or bundle of elements, to each time-step, it is possible to relax this aspect of the model by, for instance, centring each bundle of elements around three consecutive time-steps. Elements may then be allowed to spread into adjacent time-slices occupied predominantly by the adjacent bundle of elements. Thus, the nasalization of the vowel in *don* may be represented by the element **N** being present in one or all of the time-slices occupied by the elements corresponding to the vowel. This move would also facilitate the vexed issue of the representation of diphthongs and long monopthongs.

Training the network

The current version of the model employs the "copyback" structure described above. In the simulations reported below, a learning rate of 0.1 was employed; momentum was not used. The network was trained until it began to show signs of overfitting – training that resulted in a decrease of error for the training set but led to increasing error for a separate test set was disregarded. This required between 500 and 600 epochs. To encourage the network to employ context, noise was added to the input; there was an 11% probability that any element in the input would have its value changed to or from 0. The noise was generated on-line and was different for every epoch of training. The learning phase of the simulations was quite computationally intensive, using 30-40 CPU hours on a variety of SUN SPARC-based machines, using a customized version of the Rumelhart and McClelland (1988) simulation package.

The initial, limited training data was derived from some 3490 words worth of spoken discourse, taken largely from the LUND Corpus (Svartik & Quirk 1980). This corpus consists of a word level transcription and includes filled pauses, false starts and corrections. This was converted automatically to an idealized segment-level transcription which was, in turn, converted to the nine elements described above. The training set was made up of 9097 segments and a test set of 3285 segments was used to test for overfitting. No attempt was made to impose phonological reduction or coarticulation.

Modelling psycholinguistic data

The success of the model in representing a particular bundle of elements in the output level was measured by calculating the sum of squares of the error associated with that bundle when compared with each of the possible input patterns. This allowed us to determine, for instance, whether a particular output resembled the expected pattern for /s/ or /sh/. Initial simulations illustrate the potential of the model and of the wider approach.

Sensitivity to context

The model is sensitive to segmental context. Auto-associated patterns of elements in "current" position are more accurate, in terms of sums of squares of the error, when the model is given input from transcribed normal discourse than when the same bundles of elements are presented in random order. The model relies on previous context to identify the current bundle of elements; when this context is aberrant, it hinders correct recognition.

Human listeners employ context both before and after the segment in question. Training with noise forced the network to rely on both "left" and "right" context. The scores for the bundles of elements in "past" positions for normal and abnormal discourse indicated that the model is sensitive to right context in recognizing bundles of elements, and was misled by abnormal right context.

Phoneme restoration and compensation for coarticulation

Listeners' perception of degraded individual speech sounds in words is often restored (Warren 1970), particularly when the intended phoneme and the replacing sound (*e.g.* white noise, a click, silence) are similar, and when replacement occurs after the uniqueness point of the word.

This effect is often not compelling, however, and there are inherent difficulties in interpreting the effect. An important exception to this latter problem is the demonstration of compensation for coarticulation reported by Elman and McClelland (1989). This particular experiment is of crucial theoretical interest because of the claim that it demonstrates top-down lexical influences on lower-level phonemic representations. It is the strongest experimental evidence for phoneme restoration, and for top-down influences on perception in general. Most of the six words employed in that study (*Christmas, copious, ridiculous, foolish, English, Spanish*) end with suffixes, suggesting that the phoneme restoration reported for the final segment might emerge from a model which only encoded low-level statistical generalizations about spoken English. Suffixes are frequent sequences of segments and such a model might simply encode the knowledge that the sequences corresponding to *-ish* and *-ous* are more likely in some contexts than in others, without having anything like an adequate representation of morphological categories.

The model was given transcriptions of the words listed above, in neutral left-contexts, with the final segment replaced in each case by an identical hybrid segment intermediate between /s/ and /sh/. (/s/ and /sh/ differ only on the palatality, I, element; this was replaced by 0.5 to create the intermediate segment.) The scores assigned to the critical segment in "current" position were recorded and the sum of squares of the error calculated for the bundles of elements corresponding to /s/ and /sh/ respectively. The model has an overall preference for the /s/ interpretation, reflecting the relative preponderance of /s/ over /sh/ in the training corpus. Crucially, however, when the difference between the two sums of squares, for /s/ and /sh/ for each word, is calculated the model exhibits precisely the pattern of restoration found in human subjects. In Fig. 2. "preference for /s/" is the difference between the two sums of squares. It would therefore be possible for a categorical perception criterion to be placed on the /s/–/sh/ continuum so as to ensure that appropriate phoneme restoration occurs for each word.

This simulation suggests that there is no need to posit top-down interaction, as traditionally conceived, to explain Elman and McClelland's demonstration of phoneme restoration.

Conclusions

The proposed input representation, based on the elements of Government Phonology, is a viable alternative to ones comprising SPE-style features, for the connectionist modelling of speech processing. The input representation used gave simulation results closer to the human data than did the previous input based on SPE-style features. Simulation of the compensation for coarticulation effect suggests that this effect, which was previously interpreted as a top-down lexical effect, may be the result of learning simple statistical generalizations within speech input. Apparent "higher-level" generalizations are apparent, to differing degrees, at very low levels. A more accurate view of what is possible at such an early stage limits what it is necessary to explain at putative higher levels.

Increasing the training corpus from the current 3490 word tokens (905 word types) will improve the performance of the model. Introducing into the training corpus plausible levels of coarticulation (by means of element spreading) and phonological reduction (by means of element and segment elimination) will improve the performance by adapting the context of any particular element or bundle of elements in a predictable way.

Many psycholinguistic phenomena which have been taken to involve access to specific representations of spoken words may be explained in terms of the low-level statistical structure of the speech input, as encoded in connectionist models of the process. There

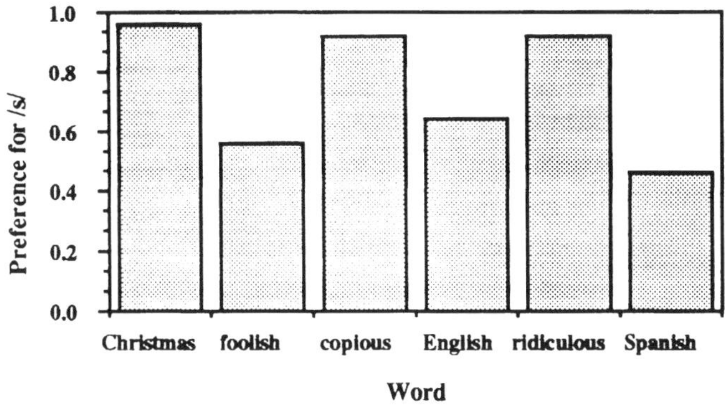

Figure 2. Preference for the bundle of elements corresponding to /s/ is greatest in the appropriate words.

is a methodological imperative within psycholinguistic research to allow "higher level" interpretation of data only when low level explanations can be ruled out.

References

Chater, N. 1989. Learning to respond to structures in time. Technical Report RIPRREP/1000/62/89, RSRE, Malvern, Worcs.

Chater, N, & Conkey, P. (*submitted*). Finding linguistic structure with recurrent networks.

Chomsky, N. & Halle, M. (1968). *The Sound Pattern of English.* Harper & Row, New York.

Elman, J. L. 1988. Finding structure in time. Technical Report, CRL TR 8801, Centre for Research in Language, UCSD.

Elman, J. L. 1990. Finding structure in time. *Cognitive Science*, 14:179-211.

Jakobson, R., G. Fant and M. Halle 1952. *Preliminaries to Speech Analysis.* Technical Report 13, M.I.T. Acoustics Laboratory, Mit Press.

Kaye, J. D., Lowenstamm, J. & Vergnaud, J. -R. (1985) The internal structure of phonological elements: a theory of charm and government. *Phonology Yearbook* 2, 305-328.

Kaye, J.D., Lowenstamm, J. & Vergnaud, J.-R. (1990) Constituent structure and phonological government. *Phonology* 7.

Levy, J., Shillcock, R.C. & Chater, N. (1991). Connectionist modelling of phonotactic constraints in word recognition. Proceedings of the *IJCNN*, Singapore, 1991.

McClelland, J. L. & Elman J. L. 1986. Interactive processes in speech perception: the TRACE model. In D. E. Rumelhart & J. L. McClelland eds. *Parallel Distributed Processing*, Vol. 2., 58-121, Cambridge, Mass: MIT Press.

McClelland, J. L. & Rumelhart, D. E. 1988. *Explorations in Parallel Distributed Processing: Models, Programs and Exercises.* Cambridge, Mass: MIT Press.

Norris, D. G. 1990. A dynamic-net model of human speech recognition. In (G. Altmann, ed.) *Cognitive Models of Speech Processing: Psycholinguistic and cognitive perspectives*, MIT Press.

Norris, D. G. (*in press*). Bottom-up connectionist models of "interaction". To appear in (G. Altmann & R. Shillcock, eds.) *Cognitive Models of Speech Processing: 2nd Sperlonga workshop.*

Plaut, D.C. & Shallice, T. (1991). Deep dyslexia: A case study of connectionist neuropsychology. *Ms.*

Seidenberg, M. S. & McClelland, J. L. 1989. A distributed, developmental model of word recognition and naming. *Psychological Review*, 96, 523-568.

Servans-Schreiber, D., Cleeremans, A. & McClelland, J. L. 1989. Learning sequential structure in simple recurrent networks in D. Touretsky ed. *Advances in Neural Information Processing Systems*, Vol 1, Morgan Kaufman, Palo Alto, 643-653.

Shillcock, R.C., Levy, J., & Chater, N. (1991). A connectionist model of auditory word recognition in continuous speech. *Proceedings of the Cognitive Science Society Conference*, pp. 340-345, Chicago, 1991.

Svartvik, J., & Quirk, R. 1980. *A Corpus of English Conversation.* Lund: Gleerup.

Warren, R. M. 1970. Perceptual restoration of missing speech sounds. *Science*, 167, 392-393.

A PDP Approach to Processing Center-Embedded Sentences

Jill Weckerly
Jeffrey L. Elman
Center for Research in Language
University of California, San Diego
weckerly@crl.ucsd.edu; elman@crl.ucsd.edu

Abstract

Recent PDP models have been shown to have great promise in contributing to the understanding of the mechanisms which subserve language processing. In this paper we address the specific question of how multiply embedded sentences might be processed. It has been shown experimentally that comprehension of center-embedded structures is poor relative to right-branching structures. It also has been demonstrated that this effect can be attenuated, such that the presence of semantically constrained lexical items in center-embedded sentences improves processing performance. This raises two questions:

(1) What is it about the processing mechanism that makes center-embedded sentences relatively difficult?

(2) How are the effects of semantic bias accounted for?

Following an approach outlined in Elman (1990, 1991), we train a simple recurrent network in a prediction task on various syntactic structures, including center-embedded and right-branching sentences. As the results show, the behavior of the network closely resembles the pattern of experimental data, both in yielding superior performance in right-branching structures (compared with center-embeddings), and in processing center-embeddings better when they involve semantically constrained lexical items. This suggests that the recurrent network may provide insight into the locus of similar effects in humans.

The Problem

It has been known for many years that not all embedded sentences are processed equally easily by listeners. Over a variety of measures, the comprehension and general processing of center-embedded structures has been found to be worse than that of right-branching sentences (Blaubergs & Braine, 1974; Blumenthal, 1966; Blumenthal & Boakes, 1967; Cairns, 1970; Fodor & Garrett, 1967; Larkin & Burns, 1977; Marks, 1968; Miller & Isard, 1964; Schlesinger, 1968). Thus, in

(1)(a) The woman saw the boy that heard the man that left. (RB)

(b) The man the boy the woman saw heard left. (CE)

Sentence (1a), which involves a right-branching structure (RB), is more readily processed than sentence (1b), which involves a center-embedding (CE).

The are various reasons why these two classes of sentences might differ with regard to intelligibility. These include (a) adherence to canonical word order, (b) difficulty of subject-verb matching in the matrix and embedded clauses, (c) distance between subjects and verbs, and (d) consistency of role assignments for nouns in both main and subordinate clauses. While canonical SV-O word order in (1a) is maintained through the matrix clause, in (1b), word order diverges considerably. The processor is faced with three adjacent nouns followed by three adjacent verbs. In (1a), the processor must be able to match the verb encountered in the first relative clause, *heard* with the previous noun, *boy*. This means the processor must "store" some notion of this noun until *heard* is reached. Once past the verb, it goes on to repeat the same action with the next relative clause.

In sentences such as (1b), these resolutions are not made as easily. The processor is required to simultaneously keep track of three nouns before it reaches the first verb, *saw*. It then must determine all the subject-verb-object relationships from representations of items occurring very early in the sentences. As each verb is encountered, the noun that serves as its subject was encountered progressively further back in the sentence, making the distance between the last verb and its subject considerable.

The difficulties of a subject-verb-object match in structures such as (1b) tax the storage capacity of the processing mechanism by requiring the simultaneous activation of a number of items and over a great distance. This is in contrast with sentences such as (1a) where the storage of information must span over one intervening item at most and whose S-V-O relationships can be determined one at a time. A difficulty that both sentences in

(1) share is that nouns serving as subject in one clause instantiate the object role in another. This shift in perspective is more difficult in (1b) because the nouns are adjacent, thereby increasing the chance of confusion.

A second finding of note is that, despite the problems posed by CE sentences, their comprehensibility may be improved in the presence of semantic constraints. Compare the following in (2)

(2)(a) The man the woman the boy saw heard left.

(b) The claim the horse he entered in the race at the last minute was a ringer was absolutely false.

In (2b), the three subjects nouns create strong—and different—semantic expectations about possible verbs and objects. This semantic information might be expected to help the hearer more quickly resolve the various subject and object lineups and as such aids in processing (Bever, 1970; King and Just, 1991; Schlesinger, 1966; Stolz, 1967). The verbs in (2a), on the other hand, provide no such help. All three nouns might plausibly be the subject of all three verbs.

We believe that such phenomena may provide valuable clues as to the nature of the processing mechanisms which subserves language. Our goal in this work has been to test a particular model of language processing (the simple recurrent network) in order to see whether it might provide an explanation for these effects.

The Model

Elman (1990, 1991a) showed that simple recurrent networks (SRNs) were able to develop internally structured representations which provide the basis for abstract, productive, and systematic behavior as required in syntactic processing. Such networks (shown in Figure 1) were able to use cooccurrence statistics to develop representations which captured type/token distinctions, lexical category distinctions, and aspects of grammatical structure.

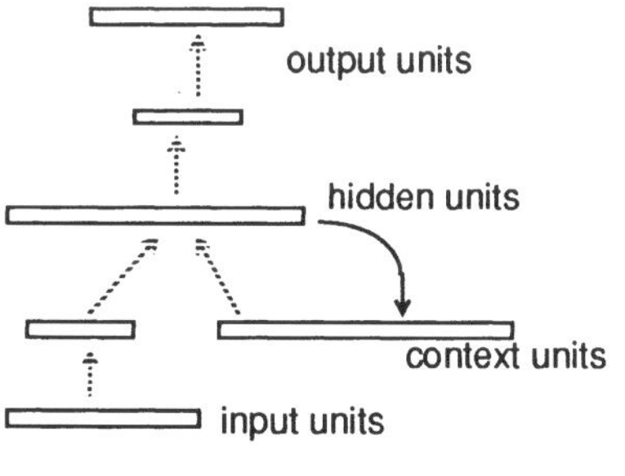

The SRN model of language processing suggests that the processing differences between center-embedded and right-branching structures arise as a basic consequence of the processing mechanism itself, rather than from limitations in a memory system which is separate from (although used in) language processing. The state machine metaphor embodied in a PDP model delineates a more plausible role for the capacity of memory based on the nature of the representations used in the processing of a sentence. The PDP processor in the time course of a sentence creates a representation which integrates previous context with present input and can be thought of as a state vector that reflects the processor's current position in the sentence. As this vector is continually passed through a "squashing" function, it has only finite precision. Finite precision and degradation over time are also qualities of the human processor.

Using a PDP network, it will be shown how a processor that employs this type of representation degrades, and hence, is finitely precise in such a way that mimics the pattern of processing center-embedded and right-branching sentences by human processors. As the state vector cannot hold information for an infinitely long period of time, it is suggested that the representations used by the human processor are captured best by the state vector metaphor and are similarly limited.

As it has been experimentally demonstrated that the processing of center-embedded sentences is aided by the use of verbs that are semantically constrained, the simulation results of the processing of these structures will show that the architecture and mode of representation in the network support this behavior as well. It can best be understood by considering what types of information are available to the processor and how they are stored. Different types of information interact in a PDP model and influence the output such that it is the product of multiple constraints. The fact that a model such as this uses information other than "purely" syntactic information is nothing new; many theories posit the interaction of this sort as a cornerstone to processing and comprehension. What a PDP model suggests is that with regard to the processing of embedded sentences, the representations used are ones where information present in the various levels of embedding is simultaneously visible and allowed to interact either to facilitate or encumber processing. Unlike a stack-device metaphor for storage, representation in a state vector is not encapsulated and unavailable. The simulations will demonstrate how a PDP model with these properties produces behavior similar to that of the human processor and how viewing processing and representation in this way accounts for behavioral patterns in a straightforward way.

Simulations

For the purposes of the simulation, a small vocabulary was created consisting of 26 words: 10 nouns, 14 verbs, complementizer "that", and an end-of-sentence marker, ".". Since one of the behaviors of interest is processing sentences with semantic bias, some notion of meaning must be represented. The network can never be semantically grounded in the sense that it knows what words mean; semantic relatedness is captured in the co-occurrence restrictions of the verbs. Classes of nouns serving as subjects and objects fall into classes of humans (NH), animals (NA), documents (DOC), and inanimate objects (INOBJ). The semantic structure of the artificial language is shown below:

VERB	POSSIBLE SUBJECTS	POSSIBLE OBJECTS
walk	NH, NA	—
live	NH, NA	—
write	NH	DOC
send	NH	DOC
love	NH	NH, NA
kick	NH	NH, NA
bite	NA	NH, NA
chase	NA	NH, NA
see	NH, NA	NH, NA, DOC, INOBJ
hear	NH, NA	NH, NA
advise	NH	NH
thank	NH	NH
own	NH	NA
tame	NH	NA

Table 1

The preceding words were constituents of an artificial grammar that generated both simple and complex sentences. Sentence types were produced with the basic pattern of NOUN-VERB-NOUN with verbs equifrequent and every instance of noun able to serve as head of an object- or subject-relative. In this way many different sentence types were generated including center-embedded/object-relative, right-branching, and subject-relative constructions. Sample sentences are shown in (3).

(3)(a) Wizard that advises dorothy tames lion.
(b) Dog that dorothy loves bites witch.
(c) Tiger chases lion that hears dorothy that kicks witch that sees slippers.
(d) Tinman thanks wizard.

Simple sentences were produced by restricting noun phrases to simple nouns and thus followed the strict NOUN-VERB-NOUN pattern. In every case, semantic restrictions were observed. All subject and object relatives were constructed with the appropriate verbs. A subject relative for animal nouns was instantiated only by those verbs for object- relatives which an animal subject is possible. For object relatives, only verbs which specified the head noun as possible object type were used to fill out that relative clause construction. Hence, sentences of the form shown in (4) did not occur.

(4)(a) *Wizard that bites dorothy tames lion.
(b) *Dog that dorothy advises bites witch.

Using the back propagation learning (Rumelhart, Hinton & Williams, 1986), an SRN of the form shown in Figure 1 was trained in a prediction task on data sets varying in composition. The network was presented with sentences, a word at a time. Each word was represented with a 26-bit vector in which a single bit was turned on. As a result, input representation contained no explicit information about the semantic or grammatical characteristics of lexical items. This information had to be learned by the network based on cooccurrence facts.

An incremental training strategy was used, based on the results reported in Elman (1991b), which indicated that the successful induction of hierarchical grammatical structures requires incremental learning. Accordingly, the network was trained on an initial data set of simple (monoclausal) sentences; over time, the percentage of complex sentences was increased until a final ratio of 75% complex/25% simple was achieved. The network was trained on a total of 40,000 sentences, each of which was presented 10 times.

Results

Network performance was evaluated by seeing how closely the network predictions approximated the (empirically derived) likelihood of occurrence of possible next words, given the prior sentence context; optimal performance would be achieved if the network learned the conditional probability distributions. We measured this by computing the mean cosine of the angle between the output activation vectors and the empirical likelihood vectors based on the final training data set. By the end of training, the network was good at predicting the following word in a variety of sentence structures as well as predicting the semantically appropriate verbs and objects for subjects and verbs respectively. The average cosine between the two sets of vectors was 0.8784. Perfect performance would have been 1.0; i.e., the vectors would have been parallel).

Test 1: center-embedded and right-branching sentences

The network was tested on subsets of center-embedded and right-branching sentences. Performance was evaluated on 192 novel sentences, each containing two levels of embedding as shown in (5).

(5)(a) Tinman hears tiger that sees witch that tames lion. (RB)
(b) Witch that tiger that tinman hears sees tames lion. (CE)

Both the likelihood and network output vectors were computed from these 192 test sentences. A four-bit vector that gave the distribution of outputs and likelihoods for each of the categories NOUN, VERB, THAT, and S (end of sentence) was calculated. The mean cosine of center-embedded structures was 0.7137. The mean cosine for right-branching constructions was 0.8484. We can conclude from this that, given the prediction task, the network is more successful at right-branching structures than center-embedded ones.

Discussion

If we equate the network's error as measured by the cosine of the output and likelihood vector with a general processing difficulty then we have results that closely model the human data. It is not the case that center-embedded sentences are impossible; they are simply more difficult relative to other constructions. Another aspect of the model's behavior is reminiscent of the human data: the network's performance decreases drastically at three embeddings which is the limit to comprehension reported in the literature as well.

The network's performance can be understood if we consider the way it represents grammatical structure at the hidden layer. As each new word is presented, the hidden units receive input from both the current word and the previous hidden unit state. Thus, a given word's internal representation always reflects the prior context. Among other things, this context indicates where, in the space of possible grammatical sentence trajectories, the processor is; this context also indicates what may be expected before a sentence-final state is reached. For example, if the current word is *hears*, and the previous hidden unit state reflects the network having seen *dog*, for instance, then the internal state will be such that the network will not predict the end of the sentence. The grammatical structure it has inferred demands that the object of the verb be present before a final state can be achieved.

The representation of relative position in a sentence makes certain demands on the processor regardless of the structure being processed. However, processing structure type also makes its own demands on the representational capacity of the processor. The difference in processing of center-embedded vs. right-branching sentences very much depends on the amount of information that must be stored for further processing in the sentence. As each THAT clause is introduced in center-embedded sentences, the information about the head noun as well as its position relative to the matrix sentence must be represented and stored until the verb of its clause is found. Consider (6):

(6) Dog that dorothy that tiger chases loves bites witch.

After it has seen *tiger*, the network must "remember": (1) that it has seen three nouns, two animals, one human; (2) the fact that the human noun came between the two animals; (3) that the verbs that "go with" these nouns will be of a certain class; and (4) that it must find three verbs in order for the sentence to complete. This places heavy demands on a processor whose actions are executed via a state vector representation.

In right-branching constructions, the representational demands are not as extreme. Consider 7.

(7) Tiger chases dorothy that loves dog that bites witch.

The initial noun that the network encounters is followed immediately by a verb. After seeing this verb, the network can forget about the initial noun because its verb has been found. For the verb, the network need only store information about an appropriate object in generating its predictions. As it encounters the object of the matrix sentence, the processor expects that the sentence be resolved or that the previous noun be the head of another relative clause. In the case of right-branching structures, the processor need only keep information about one noun after encountering the relative pronoun. Thus, there is less information to be stored and over a much shorter distance.

This disparity is clear in the behavior of the network. With right-branching constructions the state vector need only contain representations of two previous words as well as the general position in the sentence. No level of embedding need be stored, because no resolution crucially depends on it. In contrast, with center-embedded sentences, the state vector must reflect sentence position *and* current level of embedding within the sentence. Furthermore, it must also keep information about the previously introduced nouns without having the verbs to advance it into the next state. This "state of suspension" imposes a significant tax on the representational capacity of the hidden unit layer, and approaches its limits of precision.

Researchers have often cited the limitations of working memory to explain certain processing biases of the human parsing mechanism, and specifically, to explain the difficulty in processing center-embedded sentences. In that view, working memory is seen as distinct from the mechanism which contains the grammatical information. The current account provides a somewhat different way of thinking about the asymmetry in processing center-embedded vs. right-branching structures. The account also appeals to the notion of representational storage capacity. However, the representational limitations are seen as intrinsic to the grammatical processor itself, rather than arising from a separate working system. If we view the process of sentence parsing/comprehension as movement from one state to another as in a connectionist network, then memory limitations are not an arbitrary number, but due to the nature of representations in human memory in sentence processing. This capacity specifies that a state like representation can only hold so much information over a certain distance. A reduction in the amount or information or in the distance to be stored would facilitate processing, as in right-branching structures.

Test 2: Semantically biased and unbiased structures

The other major finding of interest to us was the fact that not all center-embedded sentences are equally difficult to process. We thus proceeded to test the network on different types of center-embedded sentences. Two sets of center-embedded sentences were created: one set of 192 sentences with semantically biased verbs, and another set of 192 sentences with semantically unbiased verbs. Bias in this case means there is some information in the verb that uniquely links it with either its subject or object or both. For instance, in sentence (8a) each verb encountered can only be resolved with one noun as subject whereas in (8b), any subject is compatible with any verb.

(8)(a) Dog that dorothy that bear bites tames chases tiger.
(b) Dog that dorothy that bear sees hears walks.

The network's outputs in response to the two sets of 192 sentences were collected. Likelihood vectors were calculated based on the two sets of center-embedded sentences combined. Comparisons were made between biased and unbiased sentences with one embedding, and then with two levels of embedding. The results were as predicted. For sentences containing one level of embedding, the mean cosine between the activation and likelihood vectors for unbiased sentences was 0.5719; the mean cosine for the sentences with semantic biases was 0.6311. For sentences with two levels of embedding, the overall performance decreased but the same basic pattern remained. In the unbiased condition the mean cosine was 0.5385 and in the biased condition it was 0.5719[1]. It can be concluded that semantic information which uniquely linked a subject with its verb in center-embedded sentences aided the network.

Discussion

We see again that the network's performance parallels that of human listeners. The network benefits from the semantic constraints associated with words in order to represent embedded structures more clearly. The semantic information provided by the verb helps in two ways. First, it helps the network pinpoint the noun which serves as its subject by incompatibility of the other nouns in storage. Second, because this resolution can be made with higher probability, it puts the network in a more precise state of expectation for the next word. As the network goes through the sentence, word by word, there will be various points where it must be able to link words often separated a great distance with their conceptual dependencies.

As soon as it encounters the first verb it must be able to determine which noun is its subject and which noun served as its object. The resources of the network are heavily taxed at this point, because it has information about three nouns that it must keep active at some level. As it encounters the first verb, it is able to make a relatively easy match. As there no words intervening between the last mentioned noun and the first verb, this will be the easiest subject-verb resolution the network has to make. It is at this point where the network might be aided by the nature of the verb. If the verb provides some information by virtue of its co-occurrence restrictions, and to a lesser extent its argument structure, the next subject-verb-object resolution might be greatly aided. That is, if the first verb encounters is compatible with only the last mentioned noun, the resolution can be made quickly and puts the network in a state of awaiting the next word with stronger expectations.

When the next verb is encountered, the network is forced to make another subject-verb resolution. If the nature of this second verb is such that it is compatible only with the intermediary noun and not the first or last mentioned nouns, then the network will benefit greatly from this information. The noun instantiating the subject role of this verb will be determined with less chance of confusion with the other two nouns. The network has heavy representational demands at this point because the potential subjects of the current verb have occurred quite a long time ago and chance of confusion and intermixing of information are high. Thus, information that would clearly delineate a match will be used by the network, and again, will put the network in a more precise state of readiness for the next word.

As it encounters the last verb, the network will be aided by stronger expectations about this verb. Also if this verb is compatible with only the first noun, then the network will be able to match it with its noun, which at this point has occurred many words previous. Semantically biased words aids in putting the processor in a more precise state of readiness. A precise state in this case means that in predicting the word, the activations for incompatible verbs are lower and the activations for appropriate verbs are higher. For example, consider (9)

(9) Dog that tinman that bear chases tames bites witch.

After the network has seen the verb *chases* it must predict a verb that is compatible with a human subject. In

[1] To a large extent, the low values here are only an artifact of the measure used. The likelihood vectors are calculated specific to a data set. The test data set only contains one structure of the many that the network has mastered, and therefore skews the likelihood vectors in a way that makes the network's performance appear low.

general, the network is pretty good at this. It strongly activates human compatible verbs that humans can do compared to the very low activations that are present for bits corresponding to verbs which require animal subjects. Where one can see the effect of type of verb is in this pattern of activation. When the network is presented with a verb that has definite subject specifications relative to the other words, the activations for appropriate verb for the next word are higher and the activations for verbs that are incompatible, and would constitute mistakes, are lower.

If we consider the experimental human data in processing these same type of sentences, we can compare the network's decrease in error and activation of appropriate expectations with a general measure of comprehension. The pattern is basically the same. With the inclusion of semantic cues, error goes down and appropriate activation increases. This would find its correlate in better comprehension in human subjects.

We do not claim that poor comprehension in human subjects is solely due to imprecise predictions, and of course we recognize that the prediction task captures only a small part of natural language processing. Although not the only factor in sentence comprehension, there is evidence that comprehension is in part driven by the ability to anticipate (e.g., Grosjean, 198; Marslen-Wilson & Tyler, 1980). The present findings illustrate general processing characteristics of our PDP model, and we believe similar behaviors would be observed in a comprehension task as well.

The network, its architecture, and its representations suggest similar properties in the human processing mechanism. That the network uses semantic information in what would be considered syntactic parsing, is suggestive of a parallel, interactive system. Additionally, the nature of the interaction of semantic constraints points to a system that allows the simultaneous availability of all types of pertinent information up to that point in the sentence. In other words, information is also available not only across semantic and syntactic modules, insofar as they exist, but also across levels of embedding.

The network represents what it has seen in a sentence by a state vector. Within this vector, the network has "stored" information about properties of nouns and verbs as well as the number of embeddings. Contrary to a stack-like mechanism, information is simultaneously available from all levels. There is no encapsulation of information. Upon encountering a verb, the fact that the network has information about all the previous nouns from different levels of embedding and co-occurrence restrictions of the verb facilitates in the processing of that word and subsequent words. It is suggested that the human processing mechanism has the same properties in order for there to be better comprehension with semantically biased verbs. Any model which designates a traditional stack as its primary storage device is hard pressed to account for the processing difference observed in the experimental data.

References

Bever, T. (1970a). The cognitive basis for linguistic structure. In J.R. Hayes (Ed.), *Cognition and the development of language*. New York: Wiley.

Blaubergs, M.S. & Braine, M.D.S. (1974). Short-term memory limitations on decoding self-embedded sentences. *Journal of Experimental Psychology*, **102**, No.4, 745-748.

Blumenthal, A. (1966). Observations with self-embedded sentences. *Psychonomic Science*, **6**,453-454.

Blumenthal, A.L. & Boakes, R. (1967) Prompted recall of sentences. *Journal of Verbal Learning and Verbal Behavior*, **6**, 674-676.

Elman, J.L. (1990). Finding structure in time. *Cognitive Science*, **14**, 179-211.

Elman, J.L. (1991a). Distributed representations, simple recurrent networks, and grammatical structure. *Machine Learning*, **7**, 195-225.

Elman, J.L. (1991b). Incremental learning, or the importance of starting small. Technical Report 9101, Center for Research in Language, University of California, San Diego.

Foss, D.J. & Cairns, H.S. (1970). Some effects of memory limitation upon sentence comprehension and recall. *Journal of Verbal Learning and Verbal Behavior*, **9**, 541-547.

Grosjean, F. (1980). Spoken word recognition processes and the gating paradigm. *Perception & Psychophysics*, **28**, 267-283.

King, J & Just, M.A. (1991). Individual differences in syntactic processing: the role of working memory. *Journal of Memory and Language*, 30, 580-602.

Larkin, W. & Burns, D. (1977). Sentence comprehension and memory for embedded structure. *Memory and Cognition*, **5**, 17-22.

Marks, L.E. (1968). Scaling of grammaticalness of self-embedded English sentences. *Journal of Verbal Learning and Verbal Behavior*, **7**, 965-967.

Marslen-Wilson, W., & Tyler, L.K. (1980). The temporal structure of spoken language understanding. *Cognition*, **8**, 1-71.

Miller, G. & Isard, S. (1964). Free recall of self-embedded English sentences. *Information and Control*, **7**, 292-303.

Schlesinger, I.M. (1968). *Sentence structure and the reading process*. The Hague: Mouton.

Stolz, W.S. (1967). A study of the ability to decode grammatically novel sentences. *Journal of Verbal Learning and Verbal Behavior*, **6**, 867-873.

Forced Simple Recurrent Neural Networks and Grammatical Inference

Arun Maskara
New Jersey Institute of Technology
Department of Computer and Information Sciences
University Heights, Newark, NJ 07102
arun@hertz.njit.edu

Andrew Noetzel
The William Paterson College
Department of Computer Science
Wayne, NJ 07470

Abstract

A simple recurrent neural network (SRN) introduced by Elman [1990] can be trained to infer a regular grammar from the positive examples of symbol sequences generated by the grammar. The network is trained, through the back-propagation of error, to predict the next symbol in each sequence, as the symbols are presented successively as inputs to the network. The modes of prediction failure of the SRN architecture are investigated. The SRN's internal encoding of the context (the previous symbols of the sequence) is found to be insufficiently developed when a particular aspect of context is not required for the immediate prediction at some point in the input sequence, but is required later. It is shown that this mode of failure can be avoided by using the auto-associative recurrent network (AARN). The AARN architecture contains additional output units, which are trained to show the current input and the current context.

The effect of the size of the training set for grammatical inference is also considered. The SRN has been shown to be effective when trained on an infinite (very large) set of positive examples [Servan-Schreiber *et al.*, 1991]. When a finite (small) set of positive training data is used, the SRN architectures demonstrate a lack of generalization capability. This problem is solved through a new training algorithm that uses both positive and negative examples of the sequences. Simulation results show that when there is restriction on the number of nodes in the hidden layers, the AARN succeeds in the cases where the SRN fails.

Introduction

The problem of inferring a regular grammar from examples has often been studied. An overview of traditional algorithms can be found in [Angluin and Smith, 1983]. Following the development of the back-propagation algorithm [Rumelhart and McClelland, 1986], various recurrent neural architectures using back-propagation have been shown to have some capability for grammatical inference when trained from examples [Servan-Schreiber *et al.*, 1991, Pollack, 1991, Giles *et al.*, 1990].

The idea of training recurrent neural networks with back-propagation was first introduced by Jordan [1986]. In the *recurrent* neural network, the symbols of a sequence are presented sequentially as network inputs. Also, the output of a higher level layer during one time interval is fed back as an input to a lower level layer at the next interval.

Two training schemes are associated with two different paradigms for grammatical inference. In the *classification* paradigm, it is assumed that the desired output is not known until the end of the sequence. Training this case requires back-propagation of error in time [Rumelhart and McClelland, 1986]. It is hard to train a recurrent neural network using this paradigm since the network makes many decisions before it is influenced by the results of those decisions. The classification paradigm has been studied by Giles *et al.* [1990] and Pollack [1991].

The other paradigm is the *predictive* paradigm. In this case, it is assumed that the desired output at each time interval is known. To use this paradigm in the problem of grammatical inference, the network is trained to predict the next input in any sequence. An on-line training algorithm can be used to back-propagate error at the end of each interval. The simple recurrent network (SRN) introduced by Elman [1990] has been shown to behave as a finite state automata (FSA) when trained in the predictive paradigm [Servan-Schreiber *et al.*, 1991].

The current work is based on the the predictive paradigm. It is shown that when the SRN predicts incorrectly it is because it has failed to encode the aspects of the context (the previous symbols of the sequence) that are necessary to predict the next input. We show that this problem can be solved by modifying the SRN to include features of an auto-associative network. The result is the auto-associative recurrent network

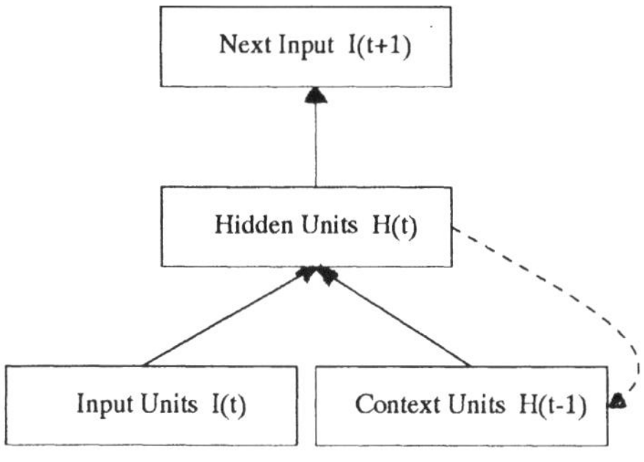

Figure 1: Simple Recurrent Network

(AARN). When trained to predict the next input, the AARN generally performs better than the SRN.

However, when only a finite set of positive examples are used for training, all networks of the SRN type exhibit a limited ability to generalize. Their generalization capability is improved when a finite set of positive and negative examples are used for training.

The Simple Recurrent Network

A diagram of the simple recurrent network (SRN) introduced by Elman [1990] is shown in Figure 1. In this model, the state of the hidden units is copied into the context unit at the end of each interval in the temporal sequence. At time t the context unit holds the state of the hidden units at $t-1$. And at t, the network is trained to predict the input at $t+1$. In order to produce an output corresponding to the next symbol of any sequence, the SRN creates an internal representation, or *encoding*, of the previous symbols of the sequence. The symbols preceding any input symbol will be called the *context* for that input.

The input and output layers use a local representation for the symbols of the sequence: each symbol is represented by a single node. The back-propagation algorithm [Rumelhart and McClelland, 1986] is used to train the network from the positive examples of the sequences. For a particular context, a positive error is back-propagated for the next symbol, and a negative error is back-propagated for all the other symbols. After completion of training, when a symbol sequence is presented at the input, the the network generates output node values that represent likelihoods for the next input symbol.

If a particular context is always followed by one particular symbol, then in that context a positive error is back-propagated only for that symbol, and a negative error for all other symbols. In an ideal case the network should learn to generate a high likelihood (1.0) for that symbol and low likelihood (0.0) for all other symbols. When different instances of a particular context are followed by different symbols, then during training the successor symbols will sometimes back-propagate a positive error and at other times a negative error. The number of times each error is back-propagated will depend on the frequency of the symbol. Therefore the value of likelihood shown by a symbol in a context will depend on the frequency of occurrence of the symbol in that particular context. If in the training data p different symbols follow a particular context with equal frequency, then after training each of those p symbol should ideally generate a likelihood of $1/p$.

The network is said to *accept* a sequence if each symbol in the sequence is predicted; and a symbol is said to be predicted if its associated output unit shows a value of greater than the positive threshold τ on the preceding symbol. The value of τ depends on the number of different symbols which can follow a context. If at most two symbols can follow a context then ideally a threshold of 0.5 should be used. Since the momentum is used in training, in case of two successor symbols a nominal value of 0.3 is used for the threshold τ [Servan-Schreiber *et al.*, 1991]. The sequence is said to be *rejected* if the network is unable to predict any symbol.

The process of encoding context in the hidden layer can be conceptualized in the following way. Early in training, each symbol forms its own distinct code in the hidden layer. Since the output of the hidden layer at $t-1$ is used as the context input at t, the encoded context gradually begins to show a representation of the previous symbol. Then the hidden layer can begin to encode combinations of the previous and current symbols. With this combination as the context, the hidden layer begins to encode the relevant aspects of three consecutive symbols. Eventually, it encodes the relevant aspects of the entire sequence.

The Auto-Associative Recurrent Network

Sometimes the SRN is unable to encode the context required to predict the next input. During training, at time t the network encodes in the hidden layer the information necessary to predict the next input. The context at t is the output of hidden layer at $t-1$. If a particular aspect of any context is not encoded at t, it will not be propagated to times greater than t. The network fails to predict properly if at time t a particular aspect of the context is not encoded (since it is not relevant to predict the input at $t+1$), but that aspect of context is required for prediction at a later time.

An auto-associative network is one in which the output is trained to be the same as (or similar to) the input. Adding the auto-associative feature to the SRN results in auto-associative recurrent network (AARN). The AARN succeeds in the cases where the SRN fails. The AARN has output units that show the current context and the current input, as well as the predicted next input. A diagram of the AARN is shown in Figure 2. At the beginning of interval t, the previous activation pattern in the hidden units (the encoded context) is copied into the context unit. The network is trained through back-propagation to show the current input and the current context that is active in the context unit, as well as to predict the next input.

The idea of auto-association in recurrent networks was initially used by Pollack [1990], in a model called the recursive auto-associative memory (RAAM). Gharamani and Allen [1991] showed that the AARN performs better than the SRN for the XOR problem. The AARN will encode aspects of the context that are not required to predict the next symbol, but are required at a later time. The encoding of the context is more efficient in the AARN because the hidden layer is forced to represent simultaneously the past (the current context) and the present (the current input).

The formation of internal code in the AARN begins same as in the SRN. Each symbol initially becomes associated with a single internal representation. But in the SRN, two symbols that result in a common prediction could form similar internal representations. And the AARN architecture guarantees that each symbol will have a unique representation, since the hidden layer must learn to represent the current symbol. In the SRN, only the context necessary to predict the next input is gradually encoded. But since the AARN forces the hidden layer to show the previous context, the entire sequence is always gradually encoded.

We have simulated the training of the SRN from examples in which two different contexts must have different hidden layer activations, and yet must make the same prediction. In terms of finite state automata (FSA), this is the same as the requirement that the minimal FSA has two different states that have same set of successor symbols. Let u be the number of states in the minimal FSA from which the training sequences were generated. (We will call the FSA used to generate the training set the 'desired' FSA.) For many examples of FSA, the SRN with $O(\lceil \log_2 u \rceil)$ units in the hidden layer was unable to correctly encode the desired FSA. An AARN with $O(\lceil \log_2 u \rceil)$ units succeeded in the cases where the SRN failed.

Training with an Infinite Data Set

We have performed a number of experiments where the SRN fails but the AARN succeeds. Here we discuss the results of two such experiments. For these experiments the training sequences were generated randomly from the desired FSA. If the desired FSA has n edges leading out of a state, then each edge was assigned a probability $1/n$ of being selected.

In both experiments, the common weights of both SRN and AARN were initialized with the same random values for each trial. During training, the same random sequences were generated for both the cases. A training trial consisted of 60,000 randomly generated sequences. We performed 15 different training runs and results were evaluated at the end of each run. After completion of the training the performance of the network was evaluated with threshold $\tau = 0.3$ (since each state of the FSA has at the most two possible successors).

The performance of the network was evaluated by two types of randomly generated sequences. The first type are those sequences that were randomly generated from the FSA. These are positive examples of the language of the FSA. We will call these *random positive sequences*. The second type are random sequences of valid symbols. Each was started with the symbol S, and then valid symbols other than S were picked randomly until the symbol E was encountered. The special symbols S and E are used to indicate the beginning and the end of each sequence. We will call these *random sequences*. Most of the random sequences are the negative examples of the language of the desired FSA.

For the experiments discussed next, the performance was evaluated on 60,000 sequences: 10,000 are random positive sequences and 50,000 are random sequences. The network was declared a failure if it was unable to correctly classify even one of the 60,000 test sequences.

The first experiment consisted of training the networks to recognize the sequences generated by the FSA shown in Figure 3. In this FSA, state 7 is the final state. For each of the 15 runs the SRN

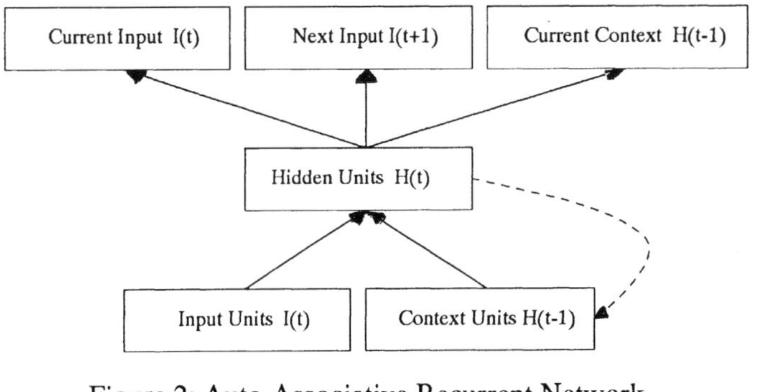

Figure 2: Auto-Associative Recurrent Network

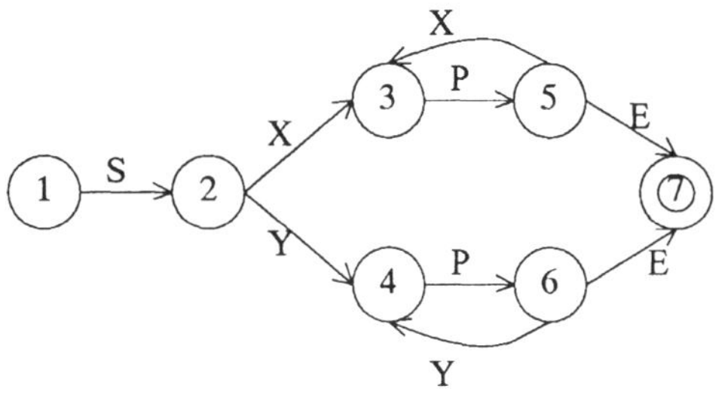

Figure 3: The FSA for Experiment 1

with three units in the hidden layer was never able to correctly classify the entire test set. By examining the hidden layer encodings, we found that at the end of the each training run the SRN had generated similar encodings for the contexts SX and SY.

The AARN with three units in the hidden layer was able to correctly encode the desired FSA in 9 out of 15 runs. After investigating, it was found that in each of the six failure, the context SX and SY had different encoding, the reason of failure seems to be lack of generalization.

The second experiment consisted of training the networks to recognize the sequences generated by the FSA shown in Figure 4. In this FSA, state 10 is the final state. An SRN with four units in the hidden layer was trained in on the language of the FSA in 15 training runs. At the end of each training run, the SRN correctly predicted the next symbols of all the random positive sequences, but failed for few of the random negative sequences. By examining the hidden layer code of the SRN it was found that the network had similar hidden layer representations for the contexts $SPXP$ and $SPPXP$.

The failures occurred because the number of P's preceding X is not required to predict the next symbol. This aspect of the context was therefore not encoded. After the context $SPXP$ and $SPPXP$, the SRN predicted a value above the threshold τ for the symbols X and P. In other words, the SRN learned to recognize the sequences of the regular expression $((P|PP)X)^+$, instead of the expression $((PX)^+|(PPX)^+)$.

The AARN with four units in the hidden layer was able to correctly encode the desired FSA in 10 out of 15 runs. Again the reason of failures in five runs seems to be lack of generalization.

In the related set of experiments we have run some simulations to show the necessity of forcing SRN to show both the current input and the current context. The SRN which was only forced to show the current input succeed for the FSA used shown in Figure 3, where the SRN always failed. The SRN and the SRN with current input failed for the FSA shown in Figure 4, but the SRN with current context succeeded in this case. When a FSA that combines the FSA's shown in Figures 3 and 4 was used, the SRN, the SRN with only current input, and the SRN with only current context were unable to encode that FSA. However, the AARN was able to succeed in that case. More detail of the experiment can be found in [Maskara and Noetzel, 1992].

Training with a Finite Data Set

In the problem of grammatical inference, we cannot assume that the FSA is given to us and the sequences are generated randomly from the FSA. Rather, a small set of training data is provided, and the problem is to find the rules (FSA) which will accept the entire class of language implied by the training data. Depending on the application and the inferring algorithm the training could be carried out by a finite set of positive examples, or a finite set of positive and negative examples.

In the next section we will show that when a finite set of positive data is used for training, the SRN encounters some additional problems. However, we show that these problems can be solved by using positive and negative training examples.

Using Only Positive Examples

After training, the SRN learns to predict the likelihood of the next symbol for each context. If the training sequence is randomly generated from the desired FSA, and each successive symbol is picked up with equal probability, then the SRN learns to generate equal likelihood values for each possible next symbol.

If a finite set of positive data is used for train-

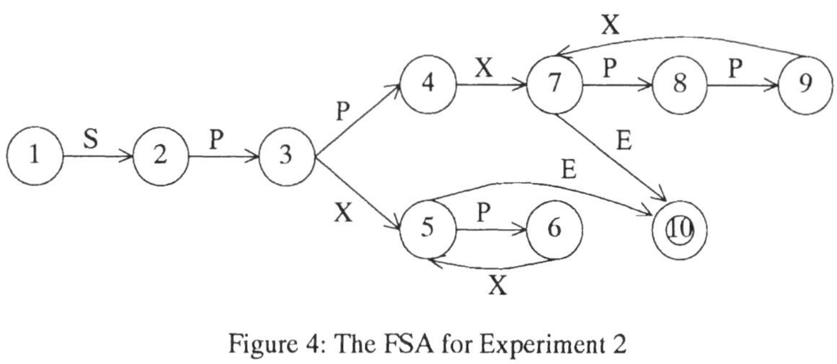

Figure 4: The FSA for Experiment 2

ing, then the value of the symbol predicted by the SRN will depend on the frequency of the occurrence of the symbol in the particular context with in the training set. For example, suppose in a data set with five sequences, the context SX is followed by E in one sequence, and by P in the rest. In this case, after training the SRN should learn to predict E with a value of 0.2, and P with 0.8. This creates a serious problem in determining the value of threshold τ, which is important in deciding whether a sequence is accepted or rejected.

To illustrate this problem, the SRN was trained to learn the regular expression XP^*, from the following six sequences: SXE, $SXPE$, $SXPPE$, $SXPPPE$, $SXPPPPE$, and $SXPPPPPE$. After 5,000 training trials for the context SX, the SRN learned to predict E with a value of 0.19, and P with value 0.82. As the training continued the problem worsened. After 20,000 trails, for the context SX it predicted E with a value 0.15, and X with a value 0.83. The situation did not improve with further training.

The AARN, when trained with the finite set of positive examples, behaved in a similar way. We have found that the problem of dependence on the frequencies of cases in the training set can be avoided by using both positive and negative examples.

Using Positive and Negative Examples

We consider a training algorithm that uses both positive and negative examples. This has two main advantages. The first is that the threshold is not dependent on the frequency of the symbols, so a predetermined value can be used. The second advantage is that the training can be halted as soon as the network is able to correctly classify all the given examples. This will help the network to avoid overgeneralization. If the training is continued after the correct classification of the examples, the network continue to encode the entirety of each sequence in the training set, which will decrease its capability for generalization.

In the training algorithm, the error is back-propagated only if the example is incorrectly classified. A positive example is correctly classified if each context results in the correct prediction of the next symbol. That is, for all contexts, the output for next symbol has a value greater than the positive threshold τ. If a positive sequence is incorrectly classified, a positive error is back-propagated for the instances of the context for which the next symbol prediction failed. If the AARN is used, it is also trained to show the current input and the current context throughout the sequence. This will allow the context to be propagated even when the next symbol is correctly predicted.

A negative example is correctly classified if at some point in the sequence the network does not predict a following symbol. A prediction failure in a negative example is indicated by an output cell value that is less than a threshold (called the negative threshold ρ) that is generally less than the threshold τ used to indicate prediction in positive examples. For the negative examples, the location of the classification error is unknown, so a negative error is back-propagated for all contexts. Along with the back-propagation of a negative error for the next symbol, the AARN is also trained to show the current context and the current input. The training is stopped as soon as the network is able to correctly classify all the positive and negative examples.

The results of two experiments show the advantage of using a finite set of positive and negative examples. During training, a positive threshold of $\tau = 0.7$ and a negative threshold of $\rho = 0.3$ was used. After training, a threshold equal to 0.5 was used to check the validity of the sequences. The number of sequences used to measure the performance will be mentioned in the results.

In the first experiment, the SRN was trained to recognize the sequences generated by the regular expression XP^*. These four positive examples were again used: SXE, $SXPE$, $SXPPE$, $SXPPPPE$. In addition, five negative examples were used: SE, SPE, $SXXE$, $SXPPPXE$, $SXPXE$. The SRN was trained by picking a sequence randomly, and back-propagating the error if the sequence was incorrectly classified. This process was repeated until all the nine sequences were correctly classified. After the completion of the training, the performance was evaluated by using 100 random positive sequences, and 1,000 random sequences.

A network was said to generalize correctly, if it correctly classified all of the randomly generated sequences. A SRN with two hidden units was trained for 15 different runs. For each of the runs, the SRN correctly classified all of the randomly generated sequences. The average number of training trials required were 1,800. For the same experiment the AARN succeeded with an average of 1,427 training trials.

The second experiment was trained the networks to recognize the sequences generated from the automata shown in Figure 3. The following positive and negative examples were used.

Positive Examples: $SXPE$, $SXPXPE$, $SXPXPXPE$, $SYPE$, $SYPYPE$, $SYPYPYPE$.
Negative Examples: SE, SP, SXX, SXY, SXE, $SXPY$, $SXPP$, $SXPXX$, $SXPXY$, $SXPXE$, SYX, SYY, SYE, $SYPX$, $SYPP$, $SYPYX$, $SYPYY$, $SYPYE$.

The SRN with three hidden units was trained for 60,000 random presentation (trials) of the above sequences. After 15 different training runs, the SRN never correctly classified all of the training sequences. An AARN with three hidden units was trained for 15 different runs. After each the AARN correctly classified all the training sequences. To check the performance of the network, 10,000 random positive sequences, and 50,000 random sequences were generated. The AARN was able to correctly classify all the random generated sequences in 11 out of 15 runs. In each of the remaining four runs, it failed for one particular long sequence. On the average the training lasted for 12,424 trials.

A *cutoff* point is defined as a position in a negative example at which the prediction of next symbol should be below the threshold ρ. The cutoff point is determined by the associated set of positive examples. For example, the sequence $SXPXX$ has a cutoff point after the context $SXPX$, that is, the symbol X should not be predicted after the context. Since the context $SXPX$ is used in the positive example, each of the next symbol should have a value greater than τ. The only symbol which can have a value less than ρ is the symbol X after context $SXPX$.

Each negative examples in the experiment described above has exactly one cutoff point. But if the examples with more than one cutoff points are used in the training, the generalization will deteriorate. For example, the sequence $SXXPE$ has three cutoff point, the symbol X after the context SX, the symbol P after the context SXX, and the symbol E after the context $SXXP$. During training, the network is required to learn just one of the cutoff point, since the output value below the threshold ρ for any of the next symbol will be taken as correct negative classification. During training, the algorithm can pick any cutoff point, but the maximal generalization will be done in the case when the symbol X after the context SX is picked as the cutoff point. We are still working to develop a training algorithm for an SRN type network which will always find the optimal cutoff point such that the generalization is at its maximum.

Related Work

Pollack [1991] and Giles *et al.* [1990] have developed neural network architectures that learn from positive and negative examples. They used a classification paradigm in which the error is backpropagated at the end of each example. In the classification paradigm, a network does not have a trap state. An error followed by a long sequence of correct examples will be accepted by such a network [Pollack, 1991]. However, the predictive paradigm has a trap state, so that the network stops accepting the input as soon an error is detected.

Conclusions

The AARN architecture can be used in any application where the SRN has been used. We show that there are cases for which the SRN fails to encode the FSA, but the AARN with same number of units in the hidden layer succeeds. The training algorithm used by Servan-Schreiber *et al.* [1991], is suitable for training the network from randomly generated sequences from the FSA. For most applications, the FSA is not known in advance. Hence only a finite set training data may be available. We show that when a finite set of positive examples is used, the network will not perform well if a predetermined threshold value is used for classification. This problem can be solved by using a finite set of positive and negative examples. The results of using both positive and negative examples are encouraging, but the performance of the network deteriorates if the negative examples have more than one cutoff points. Work is in progress to remove this deficiency.

Acknowledgement

The author Arun Maskara is a Ph.D. Candidate at the Polytechnic University, Brooklyn, New York. This work has been done as part of his Ph.D. dissertation.

References

[Angluin and Smith, 1983] Dana Angluin and Carl H. Smith. Inductive inference: Theory and methods. *Computing Survey*, 15(3):237–269, 1983.

[Elman, 1990] Jeffrey L. Elman. Finding structure in time. *Cognitive Science*, 14:179–211, 1990.

[Ghahramani and Allen, 1991] Zoubin Ghahramani and Robert B. Allen. Temporal processing with connectionist networks. In *Proceedings of the International Joint Conference on Neural Networks*, pages 541–546. Lawrence Erlbaum, 1991.

[Giles *et al.*, 1990] C. Lee Giles, G. Z. Sun, H. H. Chen, Y. C. Lee, and D. Chen. Higher order recurrent networks & grammatical inference. In *Advance in Neural Information Processing Systems 2*, pages 380–387. Morgan Kaufmann, 1990.

[Jordan, 1986] Michael I. Jordan. Attractors dynamics and parallelism in a connectionist sequential machine. In *Proceedings of the 8th Annual Conference of the Cognitive Science Society*, 1986.

[Maskara and Noetzel, 1992] Arun Maskara and Andrew Noetzel. Forced learning in simple recurrent neural networks. In *Proceedings of the Fifth Conference on Neural Networks and Parallel Distributed processing*. Indiana-Purdue University, Fort Wayne, Indiana, 1992.

[Pollack, 1990] Jordan B. Pollack. Recursive distributed representation. *Artificial Intelligence*, 46:77–105, 1990.

[Pollack, 1991] Jordan B. Pollack. The induction of dynamical recognizer. *Machine Learning*, 7(2/3):227–252, 1991.

[Rumelhart and McClelland, 1986] David E. Rumelhart and James L. McClelland, editors. *Parallel Distributed Processing*. MIT press, 1986.

[Servan-Schreiber *et al.*, 1991] David Servan-Schreiber, Axel Cleeremans, and James L. McClelland. Graded state machine: The representation of temporal contingencies in simple recurrent networks. *Machine Learning*, 7(2/3):161–193, 1991.

Decomposition of Temporal Sequences

Judith Avrahami[1] and Yaakov Kareev
The Goldie Rotman Center for Cognitive Science and Education
School of Education
The Hebrew University of Jerusalem
Jerusalem, 91905, Israel
Email:Bauky@vml.huji.ac.il

Abstract

This paper deals with the decomposition of temporal sequences and the emergence of events. The problematic nature of various definitions of events is first reviewed and an hypothesis - the cut hypothesis - is proposed. The cut hypothesis states that a sequence of stimuli is cut out to become a cognitive entity if it is repeatedly experienced in different contexts. The hypothesis can thus explain the emergence of events on the basis of former experience. Two experiments were conducted to compare the predictions of the cut hypothesis to the predictions of two other explanations, explanation by association and explanation by changes along the sequence of stimuli. The first experiment showed that subjects better recognized a certain sequence after seeing it repeated as a whole than after seeing it as a part of another repeating sequence. The second experiment demonstrated that after experiencing a certain repeating sequence subjects would hardly consider dividing in its midst even though that point was a point of maximal change, as evidenced by divisions made by control subjects who did not experience the repeating sequence.

This paper deals with the decomposition of temporal sequences of stimuli into parts, into entities. We look for the basis of decomposition, and ask whether it is objective, independent of the person who performs the decompositon or subjective, depending on that person's former experience.

The question can be phrased much more simply as: "What is an event?" Events are generally treated as basic, even atomic, entities of which all occurances are composed. The error is evident, however, when considering the hierarchical nature of events (J. J. Gibson, 1979). In an earlier work (Avrahami & Kareev, 1990), we studied how easy it was to describe the components of an occurance and then immediately divide the components into sub-components. Even children (6th and 8th graders) easily divided and subdivided the sequences, which shows that the components are not 'atoms'.

We do not claim that atomic units of perception do not exist: it is, of course, these atomic units which make up the whole perceived sequence (Treisman & Gormican, 1988). The question is how sequences of atomic units turn into entities which are considered as events.

Some say that an event includes both transformation and invariance (E. J. Gibson & Spelke, 1983; J. J. Gibson, 1979). But this does not clarify how the begining and the end of an event are determined.

Quine (1985) proposes that events should be reified by their spatial and temporal characteristics, leaving the determination of their boundaries to "what concerns us". Yet, for something to concern us it has to be known as a 'something' first.

[1] This work was funded by a doctoral fellowship to the first author from the Israel Foundations Trustees.

Davidson (1969), dealing with the individuation of events, says that events are often related to changes in a substance and are often thought to be uniquely determined by their time and place. Yet, he claims, neither substance nor time and place can uniquely determine an event because one can think of different events which occur in the same place, at the same time and on the same substance. Davidson proposes to individuate events by their causes and their effects. However, since we cannot point at the boundaries between the causes and the events (causes and effects being events too) this definition does not answer the question of how temporal sequences are decomposed into events.

Some, indeed, tie events to objects and define an event as an action which is connected to some object. E. Gibson (1969), for instance, suggests that a ball rolling over the floor is an example of an event and that everything connected to this rolling is one event. Still, it is easy so see that the question of decomposition into objects is not solved either: If a ball is an object, is a pinball machine one object or several? and if two balls are shot in the machine together are they participating in one event or two? and so forth.

Following J. J. Gibson, who said that in every event there is invariance, we may be able to define an event by that which does not change in it: e.g. the agent, the object, the location. The boundaries of an event would then be the points where maximal changes occur. Of the same approach is the explanation of events as the products of the perception of dynamic geometric patterns (McCabe, 1986) with the beginings and ends determined by seams in the perceived patterns of energy (Michaels & Carello, 1981).

However, even a cursory inspection of the components mentioned by subjects in various studies shows that such changes may occur both within and between events. To take two examples: in the birthday party mentioned by Nelson, Fivush, Hudson & Lucariello (1983), in which "You cook a cake and eat it", it is clear that there is no less change within the "cooking" of the cake (materials, utensils, the baking oven) than there is between cooking and eating. In the restaurant script, mentioned by Schank and Abelson (1977), which is composed of entering, being seated, ordering, eating and paying, it is obvious that eating itself may include several courses, hence change of plates and their contents, change of cutlery, etc. It is therefore possible that though maximal changes may serve as breakpoints in sequences of stimuli not experienced before, experience may dictate other organizations which ignore those changes.

The explanation we offer for the decomposition of sequences into events rests on an hypothesis which was forwarded in a previous work, the "cut hypothesis" (Avrahami & Kareev, 1990). According to this hypothesis a sub-sequence of stimuli is cut out of a general sequence to become a cognitive entity for someone, if that person has experienced the sub-sequence many times, with the sub-sequences preceding and following it differing on the various occasions; the boundaries of the emergent entity are the boundaries of the repeating sub-sequence. The cut hypothesis can explain decompositions both of space and of time. The hypothesis resembles James' words: "What is associated now with one thing and now with another tends to become dissociated from either, and grow into an object of abstract contemplation by the mind" (James, 1890, p. 506). In other words, a cognitive entity is individuated from the general sequence by virtue of its having appeared in different contexts. To illustrate negatively: it is known that people often fail to describe or define the eyebrows of even close relatives, whose faces they know very well. This cannot be explained by claiming that they have seen the whole faces more often than they did the eyebrows but it can be explained by the cut hypothesis: Since the eyebrows have always been repeated on the same background (the face) they were not individuated into a separate entity, while the face, which was repeated on different occasions on a different background, has become such an entity.

The cut hypothesis is closely related to the theory of associationism which also explains the emergence of cognitive entities as a result of repeating experience with certain stimuli. But according to associationism the entities are created not by cutting a sequence into sub-sequences but by lumping basic entities together: repetition will cause lumping irrespective of what precedes and what follows the sub-sequences.

Two experiments were conducted: One to juxtapose the cut hypothesis with the associationistic explanation and the other to juxtapose the cut hypothesis. with the claim that the boundaries of events are determined by changes.

The materials for the experiments were video films. The films were constructed from an inventory of 50 basic units, each lasting three seconds. The basic units were extracted from nine different cartoons. Cartoons were used because they generally consist of short "takes" with abrupt cuts between them. Each basic unit depicted a specific action but care was taken that none would convey a sense of a beginning or an end of an episode (e.g., no one crashed in any). No two basic units were adjacent in the original cartoons. All the cartoons were about the same two characters, and in each basic unit either one of them or both were performing an action (See Appendix for a description of some of the units).

In the first experiment subjects watched a film which contained an arbitrary sequence of basic units that was repeated several times among various other basic units which appeared only once.

Both according to the cut hypothesis and according to the laws of association the repeating sequence was expected to become an entity - an event. But according to the associationistic explanation the event would be created by association of its basic units, hence parts of the event would be remembered no less than the whole; in contrast, according to the cut hypothesis the event is carved out from the whole film by virtue of its having different preceding and following sequences, therefore the whole event would be better remembered than its parts.

In the second experiment, two groups of subjects were asked to suggest points at which a test film, made of randomly ordered basic units, could best be divided into parts. The control group saw just the test film. The experimental group saw first a training film which contained a repeating sequence extracted from the test film; it was chosen in such a way that the most popular breakpoint (among the control subjects) was at its midst. The experimental group was then asked to divide the same test film as the control group.

Lacking any former experience with the film, control subjects were expected to choose points of maximal change for their divisions. The question was whether the experimental subjects would partition the film at the same points - supporting the claim that decomposition is based on maximal changes or, having experienced those points within a repeationg sequence, they would ignore the point of maximal change - thus supporting the cut hypothesis.

Experiment 1

In Experiment 1 we compared the recognition of a sequence of basic units which appeared either as a whole repeating sequence within one film or as a part of a longer repeating sequence in another film.

Method

Design. Four films were used - films I, II, III and IV - with repeating sequences of length three, seven, four and eight, respectively. The sequence presented at the recognition task was either the whole repeating sequence (for the shorter ones) or the middle part of the repeating sequence (for the longer ones). Thus the length of the test sequence was either three (for films I and II) or four (for films III and IV).

Subjects. 48 college students participated in the experiment as paid volunteers, 12 with each film.

Materials. The Films: Out of the total inventory of basic units seven were randomly chosen for the repeating sequence of film II and another eight for film IV. Other units have been randomly chosen to serve as fillers or noise which surrounded the repeating sequence. The repeating sequences of films I and III were the three and four middle units of the seven and eight of films II and IV. The repeating sequences occured five times in the whole film. Films I and II were 58 units long and films III and IV were 68 units long. To achieve equal length films I and III had 20 more noise units than films II and IV.

The Tests: The tests consisted of three test sequences, all of the same length: three for films I and II, and four for films III and IV. Subjects were asked whether each test sequence appeared in the film they had seen before. The first question, which was the one of interest, was either about the whole repeating sequence (in films I and III) or about the middle part of the repeating sequence (in films II and IV). Note that the questions were identical for films I and II and for films III and IV.

Procedure. Subjects first watched the film and then performed the recognition task in which they had to mark down whether or not each test sequence had appeared in the film.

Results and Discussion

The number of subjects (out of 12) answering the critical question correctly was 11, 8, 11 and 7 for films I, II, III and IV, respectively. The proportion of correct answers was, then, .92 for the sequence when it had been experienced as a whole and .63 for the same sequence when experienced as a part. A two-way analysis of variance with Part/Whole and Film as its factors revealed a significant effect only of Part/Whole ($\underline{F}(1,44) = 6.04$, $\underline{p} = .018$).

The results show that though watching one of two films of the same length and tested for recognition of the same sequence, which appeared the same number of times in both films, subjects who experienced the sequence as a whole repeating sequence recognized it more easily than subjects who experienced it as a part of a longer repeating sequence. Note that the advantage of the whole repeating sequences is particularly impressive since the sequences of films I and III were embedded in much more noise than the sequences of films II and IV. The results thereby support the claim that a sequence which is repeated in different contexts is cut out to become a whole, a cognitive entity, and is remembered as such. In contrast, when the same sequence always appears as part of a longer sequence, hence always in the same immediate context, it is less likely to become a well recognized entity.

Experiment 2

In Experiment 2 we tested whether a breakpoint within a sequence of stimuli, chosen by subjects who watched the sequence for the first time, would be ignored by other subjects who first experienced repeated presentations of a sub-sequence surrounding this point.

Method

Design. Two groups of subjects, a control and an experimental group, participated in this experiment. The control-group subjects watched one of two test films made of randomly chosen basic units and were asked to suggest where they would divide it. Then two training films were created on the basis of the answers of the control group: the most popular dividing point of each test film was identified and the sub-sequence surrounding it was used as the repeating sequence in the training films. The experimental subjects, having first watched one of the training films, were also asked to divide one of the test films.

Subjects. 34 college students participated in the experiment as paid volunteers: 16 in the control group and 18 in the experimental group.

Materials. The materials were video films made of the same basic units as the films of Experiment 1.

Test film I was of length 15, and the sequence taken out of it was of length three. This sequence repeated five times in a training film of total length of 35. The corresponding numbers for film II were 20, 4, and 45.

Procedure. Subjects in the control group were told that they were going to see a film made up of short units, and should suggest where the film could best be divided into parts. They watched the film twice: Once, to get familiar with it and a second time to say aloud where they would divide it. The experimenter marked down subjects' answers. Subjects in the experimental group were told they were going to see a film made of short units which they should just watch. Having watched this training film they performed the same task as the control subjects.

Results and Discussion

Though the test films were made of randomly combined unrelated units, it turned out that some division points were more popular than others. The most popular breakpoint in test film I was chosen by six out of the eight control subjects who watched it, and the most popular one in test film II was chosen by seven out of eight control subjects. In both cases the proportion of subjects choosing the point was significantly above chance.

As for the experimental subjects, who first saw the corresponding training film, only two out of nine chose the popular point of film I and two out of nine chose the popular point of film II. A two-way analysis of variance revealed a significant difference between the two groups ($F(1,30) = 16.15$, $p < .001$). As it turned out, the total number of division points provided by the control subjects was higher than that of the experimental subjects. To compensate for that, and employ a more conservative estimate of the difference between the two groups, each subject received a score of 1 divided by her/his total number of division points if they divided the test film at the critical point and 0 otherwise. The difference between the two groups remained significant ($F(1,30) = 10.08$, $p = .003$).

The results show that repeated experiencing of a sequence of basic units embbeded in different others makes them into one entity irrespective of significant changes which occur within that sequence.

Discussion

The experiments demonstrate that the repetition of a sequence of stimuli in various contexts is the basis for the emergence of events. This emergence is best described as cutting the whole repeating sequence out of its context rather than associating its basic units into a whole. The experiments explore the cut hypothesis only in situations of perfect repetition. It is obvious that in reality no sequences of stimuli recur identically. Further research is needed to establish the degree of similarity required for one sequence to be considered a recurrence of another. See Kareev & Avrahami (1990) for a study of a related question.

Appendix

Descriptions of some of the basic units used in the films:

A character throws a dart.

A character adds pepper to a cauldron.

A character hides explosives under a bridge.

A character runs.

A character aims a gun.

The two characters meet.

A character rolls an iron ball off a cliff.

References

Avrahami, J., and Kareev, Y. 1990. *Decomposition*, Working paper no. 33, The Goldie Rotman Center for Cognitive Science and Education, The Hebrew University, Jerusalem.

Kareev, Y., and Avrahami, J. 1990. Is there a default similarity distance for categories? In *Proceedings of the Twelfth Annual Conference of the Cognitive Science Society*, 125-132. Cambridge, MA: Erlbaum.

Davidson, D. 1969. The individuation of events. In N. Resche and D. Reidel (Eds.) *Essays in the honor of Carl G. Hempel*, pp. 216-234, Dordrecht-Holland: D. Reidel Publishing Co. Also appeared in *Essays on action and events*, 1989, Oxford: Clarendon Press.

Gibson, E.J. 1969. *Principles of perceptual learning and development*. New York: Appleton-Century-Crofts.

Gibson, E.J. and Spelke, E.S. 1983. The development of perception. In P.H. Mussen (Ed.), J.H. Flavell and E.M. Markman (Vol. Eds.). *Handbook of child psychology,* Vol. 3. New York: Wiley.

Gibson, J.J. 1979. *The ecological approach to visual perception*. Boston: Houghton Mifflin Company.

James W. 1890. *The principles of psychology*. London: MacMillan.

McCabe, V. 1986. Introduction: Event Cognitiion and the conditions of existence. In V. McCabe and G.J. Balzano (Eds.) *Event cognition: An ecological perspective*. Hillsdale, NJ: Erlbaum.

Michaels, C.F. and Carello, C. 1981. *Direct perception*. Murray Hill, NJ: Prentice-Hall.

Nelson, K., Fivush, R., Hudson, J., and Lucariello, J. 1983. Scripts and the development of memory. *Contributions to Human Development,* 9, 52-70.

Quine, W.V.O. 1985. Events and reification. In E. LePore, and B.P. McLaughlin, (Eds.) *Actions and events: Perspectives on the philosophy of Donald Davidson*. Oxford: Blackwell.

Schank, R.C., and Abelson, R.P. 1977. *Scripts, plans, goals and understanding*. Hillsdale, NJ: Erlbaum.

Treisman, A. & Gormican, S. 1988. Feature analysis in early vision: Evidence from search asymmetries. *Psychological Review,* 95, 15-48.

The Role of Correlational Structure in Learning Event Categories

Alan W. Kersten
Dorrit O. Billman
School of Psychology
Georgia Institute of Technology
Atlanta, GA 30332-0170
email: billman@pravda.gatech.edu

Abstract

How do people learn categories of simple, transitive events? We claim that people attempt to recover from input the predictive structure that is the basis of 'good', inferentially rich categories. Prior work with object categories found facilitation in learning a component relation (e.g. feathers covary with beak) when that correlation was embedded in a system of other, mutually relevant correlations. Little research has investigated event categories, but researchers have suggested that verb meanings (hence perhaps event categories) might be organized quite differently from noun meanings (and object categories). Thus it is far from clear whether the learning biases or procedures found for object categories will also appear for event learning. Two experiments investigated the effects of systematic correlational structure on learning the regularities comprising a set of event categories. Both found the same pattern of facilitation from correlational coherence as found earlier with object categories. We briefly discuss relations to 1) other constraints on concept learning that focus on the organization of the whole system of concepts and 2) learning paradigms that produce competition, not facilitation, between correlated cues.

Event Category Learning

What makes some event categories harder or easier to learn than others? By events we mean simple, "verb-sized" interactions between agents and patients. Categories are generalizations across multiple such events. In our experiments subjects view animations of simple, transitive events in an unsupervised learning paradigm and we assess what regularities they learn. Our work has two broad motivations: 1) to identify what makes some systems of categories natural and coherent but others arbitrary and ad hoc and 2) to assess whether proposals developed for object categories apply to event categories.

We propose that 'natural', coherent categories 1) support useful predictions about new instances and 2) facilitate the learning of attribute relations within a category and relations to contrast categories. We look at ease of learning, particularly in unsupervised tasks, as an important index of category 'goodness'.

Our experiments ask how the organization of a system of categories affects learning components of that system. We believe that understanding coherence or ease of learning requires considering a system of categories, not each category in isolation. Investigating system-wide, structural constraints on learning has been a small but visible component of current research. Keil's (1979) concept of predicability, Markman's (1989) mutual exclusivity constraint, and perhaps the suggestions about the role of background theory in categorization (Murphy and Medin, 1985) are examples of work in this area.

We believe that the notion of correlational structure is critical for understanding category coherence. Rosch (1978) claimed that a good category captures rich correlational structure in the world and hence is inferentially rich; knowing something is a bird allows you to predict many of its properties. We extend Rosch's claim about structure to two learning principles. First, learners are biased to seek out categories with rich correlational structure so that learning any component correlation (say between feathers and flying) is facilitated if other attribute values also correlate (singing and having a beak). Thus we predict facilitation among correlated cues (Billman and Heit, 1988), not competition as in several other learning models (Rescorla & Wagner, 1972; Gluck & Bower, 1988). Second, learning benefits if correlations among the same attributes are present consistently across a system of contrast categories (values of body covering and locomotion correlate for FISH as well as BIRD).

The current experiments test the first principle, asking if learning a correlation between two attributes is facilitated in a system where these attributes also covary with others. We compare learning the correlation in this context to learning the same correlation in differently organized systems. These experiments paralleled prior work on object categories which found support for this principle (Billman & Jeong, 1989; Billman & Knutson, 1990). To investigate events we had to identify attributes relevant to event categories; we did this by reference to work on perceptual properties of simple events (Michotte, 1946/1963) and to work identifying aspects of event meaning that are reliably identified in syntax and verb meaning.

Although event categories are widely acknowledged as important, concept research has overwhelmingly focused on object categories. Given the paucity of research on event categories, it was unclear whether the results for event categories would parallel those for objects. The closest research comes from study of verbs, not event categories per se. Gentner (1981) has hypothesized that relational concepts such as verbs have little correlational structure. Huttenlocher and Lui (1979) further claim that verbs, unlike nouns, possess a matrix-like organization, with little correlation between elements of meaning such as direction, instrument, intent, and manner. If real-world event categories indeed have little correlational structure, category learners would most likely not have a bias toward learning novel event categories with such structure. However, should correlational structure facilitate the learning of event as well as object categories, this would suggest that the presence of a coherent system of correlations plays a very pervasive role in learning. In addition, a commonality between object and event categorization would be identified, encouraging comparison between analyses of object categories (and nouns) and event categories (and verbs).

Experiment 1

Given the view that categories capture correlational structure, we investigated the ease of learning component correlations. The first experiment tested whether a correlation is more easily learned in isolation or when the attributes participating in that correlation also participate in other correlations. We predicted that correlations would be easier to learn in the presence of other correlations, due to a bias of category learners to seek out such correlational structure.

Method

Subjects. Thirty-six undergraduates at the Georgia Institute of Technology volunteered as subjects for class credit.

Stimuli.

Learning Phase. Every event consisted of the actions of a square character, the *agent*, and a circular character, the *patient*, on a varying background, the *environment*. The characters left their starting locations when the subject pressed the mouse button, with the agent always moving towards the patient. When the agent reached the patient, the *state change*, or change in appearance of the patient, took place. The patient then always moved away from the agent.

Each event varied on 7 attributes, each with 3 possible values. Three attributes specified static properties of the objects or their environment. Four specified dynamic properties which combined to produce collisions, chases etc. with different outcomes for the patient. The attributes were: (1) the agent color: red, green, or blue; (2) the patient color: purple, brown, or yellow; (3) the nature of the environment: a fine grain, squiggly lines, or small ovals; (4) the state change: blowing up, shrinking, or flashing; (5) the path of the agent after the state change: movement toward the patient, away from the patient, or remaining at the place were it met the patient; (6) the path of the patient before the state change; and (7) the agent's manner of motion: smooth, direct motion; oscillation perpendicular to the direction of motion; or surging forward in bursts.

For each subject, at least two of these attributes were correlated, such that the value of one attribute could be predicted given the value of the other. The correlation between these two attributes was designated the *target rule*. This was the only correlation present for subjects in the *isolating condition*. For subjects in the *structured condition*, these two attributes were correlated with two other attributes chosen at random. Three different target rules were used and we compared the ease of learning each target rule in the structured versus isolating conditions. Figure 1 shows an example of the correlations present in the structured and isolating conditions.

Test Phase. Each subject's knowledge of his or her target rule was tested. This was done by collecting ratings of events with correctly or incorrectly matched values of the attributes in the target rule. To test knowledge of the target rule in isolation, it was necessary to hide any attributes which were correlated with the attributes in that rule. If this were not done and one attribute had an incorrect value in the structured condition, multiple rules

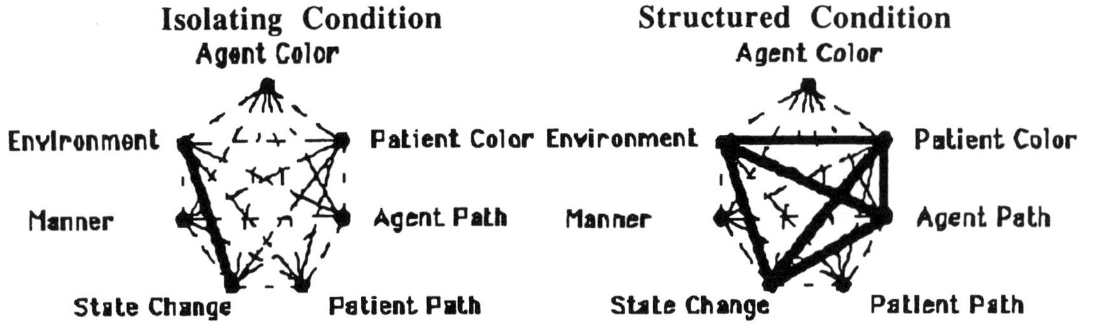

Figure 1: Example schemas for two subjects in Experiment 1. Dark lines indicate correlations. Target rule between environment and state change is present in both conditions. For structured condition, these two attributes are correlated with two other attributes.

besides the target would also be incorrect. Thus, two correlated attributes were hidden from view on each test event for structured condition subjects. Two random attributes were hidden for each test of the target rule for isolating condition subjects. To make the test phase minimally instructive, two filler 'rules' were also tested. For structured condition subjects, these were two correlations which were present in the learning phase. For isolating condition subjects, these rules had not been present, so these subjects could not have had knowledge of the correct pairings on the filler rules. The target rule and each filler rulewere each tested 18 times. On half of these trials, the values of the attributes in the rule were correctly matched, while on half they were matched differently than in the learning phase.

Procedure. The learning phase consisted of 120 animated events. Subjects were instructed that they would be seeing events on another planet, and were to learn the kinds of events which took place on this planet. Each event was initiated by the subject. There were five breaks during learning. After the learning phase, subjects were told that their knowledge of events would be assessed. Six correct, instruction displays familiarized subjects with how attributes would be hidden from view. Subjects were then told to rate each test display, based on the available attributes, for how good an example it was of possible events on planet Daysee. Subjects could repeat each of the 54 events they were to rate up to 3 times.

Design. The design consisted of two factors. One of these was correlational structure. The two levels of this factor were structured and isolating. The other factor was the target rule. The three correlations comprising this factor were between patient color and agent path, state change and environment, and manner of motion and patient path. The dependent measure was rating accuracy on the 18 events testing the target rule.

Results

Accuracy scores were derived from rating scores by finding the difference between each subject's rating and the correct rating and then subtracting this from 2. Thus, a perfect rating for an event was awarded a score of 2, while the most incorrect rating received a -2.

Structured condition subjects had higher accuracy scores, averaging .89, compared to the average isolating condition score of .31. The correlation between state change and environment was rated most accurately, followed next by patient color and agent path, and finally by patient color and manner of motion (see Figure 2).

Due to significant heterogeneity of variance, a Brown-Forsythe test was performed instead of an ANOVA. This test revealed a significant effect of correlational structure, with an $F(1,15)$ of 8.82 ($p < .01$). The effect of rule was also found to be significant, producing an $F(2,15)$ of 15.83 ($p < .001$). The interaction was insignificant, with an $F(2,15)$ of 0.56.

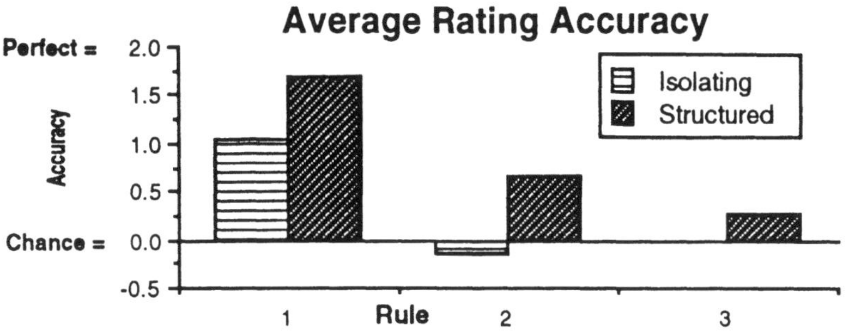

Figure 2: Experiment 1 accuracy scores for the three rules. Rule 1: state change and environment. Rule 2: patient color and agent path. Rule 3: patient path and agent manner of motion.

Experiment 2

The results of Experiment 1 supported the notion that correlational structure, in the sense of multiple related correlations, facilitates the learning of individual correlations. However, there is an alternative interpretation of the data. Since four of the seven attributes were correlated for the structured condition, only three attributes varied randomly. In contrast, five of the seven attributes varied randomly for the isolating condition. Thus, the effects attributed to correlational structure in Experiment 1 could instead have been a result of differences in the amount of randomness in the two conditions.

Experiment 2 was an attempt to replicate Experiment 1 while controlling for the amount of randomness. Randomness was conceptualized as the number of possible events allowed within the specification of correlational structure for a given subject. For example, in the isolating condition of Experiment 1, the five randomly varying attributes were free to form any combination with one another and with the two correlated attributes, allowing 3^6 possible combinations. The structured condition only allowed 3^4 combinations (from A1-A2-A3-A4 all covarying with each other). In Experiment 2, the number of possible combinations in the comparison, or crossed, condition was reduced to 3^4 by introducing two correlations which were independent of the target rule (A1-A2, A3-A4, and A5-A6 pair-wise correlations). Thus, the number of possible events was equalized across conditions, while the crossed condition was still low in systematic, correlational structure.

Method

Subjects. Thirty undergraduates at the Georgia Institute of Technology volunteered as subjects for class credit.

Stimuli.

Learning Phase. The learning phase differed from that of Experiment 1 in that subjects in the crossed condition saw correlations between three pairs of attributes. No attribute was correlated with more than one other attribute for these subjects. Every structured condition subject was presented with correlations among the same four attributes: agent path, environment, manner of motion, and state change (see Figure 3).

Test Phase. The test phase differed from that of Experiment 1 in that crossed condition subjects were tested on each of the three rules present in the learning phase 18 times. One attribute from each pair which was not being tested was hidden for each test phase event for these subjects. The test phase procedure was identical to that of Experiment 1 for structured subjects.

Procedure. Same as Experiment 1.

Design. The design of Experiment 2 differed from that of Experiment 1 in the rule factor. The three target rules for this experiment were between agent path and environment, agent path and manner of motion, and manner of motion and environment.

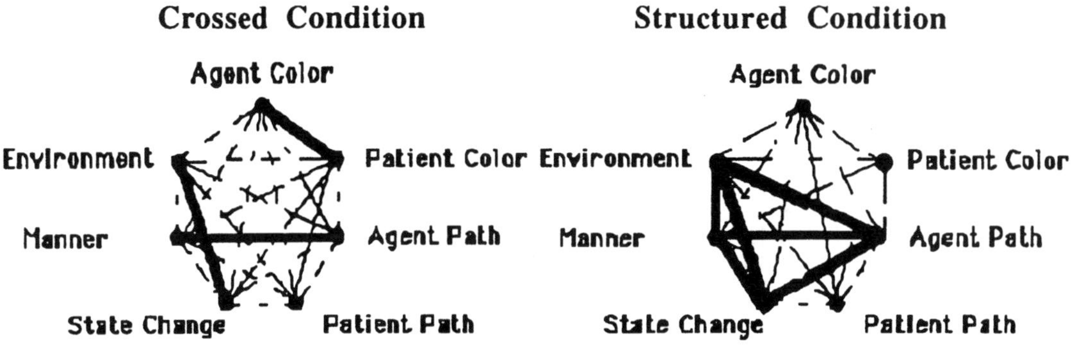

Figure 3: Schemas for two conditions in Experiment 2. Dark lines indicate correlations. Target rule between manner and agent path is present in both conditions. For structured condition, these two attributes are correlated with two other attributes.

Results

Structured condition subjects again scored higher in rating accuracy. These subjects averaged 1.08, compared to .29 for crossed subjects. The correlation between agent path and manner of motion was easiest to learn, followed next by agent path and environment, and finally by manner of motion and environment (see Figure 4). An ANOVA revealed significant effects of correlational structure ($F = 8.18$, $p < .01$) and rule ($F = 3.49$, $p < .05$). The interaction was not significant ($F = .03$).

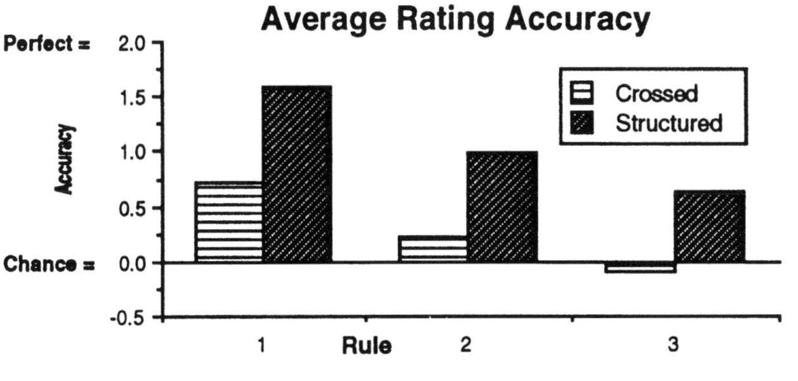

Figure 4: Experiment 2 accuracy scores for the three rules. Rule 1: agent path and agent manner of motion. Rule 2: agent path and environment. Rule 3: environment and agent manner of motion.

Discussion

Our experiments found that a systematic set of correlations determining event categories facilitated learning of a component relation. Experiment 1 showed that a given relation between two event attributes was more likely to be learned when that relation was part of a system of interpredictive attributes (Structured Condition) than in a 'simpler' system where the target rule was the only rule in the system (Isolating Condition). Experiment 2 showed facilitation of learning the target rule embedded in a more structured system over a system where there were a) the same number of possible events but b) differently organized correlations.

This research investigated correlational structure of events. We believe that other types of relations, most notably causal ones, are also important and are part of peoples' representations of event categories. However, we have focused on the 'data-driven' aspect of correlation; the learner can observe correlations in data, but can only infer cause. Clearly, the two are related. In particular, we believe that a learner biased to recover clusters of mutually relevant, correlated attributes will be finding just those correlations likely to reflect a common underlying cause. Thus, we investigated how properties of the data affect data-driven learning, but anticipate that the biases or principles investigated are just those that usefully interface with theory-driven processes.

Our findings with event categories parallel our earlier findings with object categories and suggest that the study of events can be a means of investigating the domain generality of principles initially proposed for object categories. Our findings also suggest that verb meanings may have more correlational structure than has been noted, if one takes the correspondence between verbs and events and between nouns and objects seriously. For example, verbs such as "gallop" and "read" suggest strong predictions about other components of the described event: gallop will probably have a horse for its agent; "read" will have an animate (and literate) agent and something written as patient. Event concepts and verb meaning may interact in interesting ways.

In both experiments we did not just ask if it is easier to learn about a category when there were many category predictors than when there were a few, but whether an individual, identical component pattern is learned faster. This identifies one important way in which the organization of a category system as a whole impacts learning its components. Further, the finding of facilitation among components identifies a useful bias for unsupervised learning and suggests a quite different view than that of competitive cue models from highly supervised tasks (e.g. classical conditioning).

References

Billman, D.O. & Heit, E. 1988. Observational learning from internal feedback: A simulation of an adaptive learning method. *Cognitive Science* 12:58-69.

Billman, D., & Jeong, A. 1989. Systematic correlations facilitate learning component rules in spontaneous category formation. Paper presented at the Psychonomics Society, Atlanta, November, 1989.

Billman, D., & Knutson, J. 1990. Systematic versus orthogonal sets of correlations benefit unsupervised concept learning. Paper presented at the Psychonomics Society, New Orleans, November, 1990.

Gentner, D. 1981. Some interesting differences between verbs and nouns. *Cognition and Brain Theory* 4;161-178.

Gluck, M.A., & Bower, G.H. 1988. Evaluating an adaptive network model of human learning. *Journal of Memory and Language* 27:166-195.

Huttenlocher, J., & Lui, F. 1979. The semantic organization of some simple nouns and verbs. *Journal of Verbal Learning and Verbal Behavior* 18:141-162.

Keil, F.C. 1979. *Semantic and Conceptual Development: An Ontological Perspective.* Cambridge, Mass.: Harvard University Press.

Markman, E.M. 1989. *Categorization and naming in children: Problems of induction.* Cambridge, Mass.: MIT Press, Bradford Books.

Michotte, A. 1963. *The perception of causality.* (T.R. Miles & Elaine Miles, Trans.). New York: Basic Books (Original work published 1946).

Murphy, G.L. & Medin, D.L. 1985. The role of theories in conceptual coherence. *Psychological Review* 92:289-316.

Rescorla, R.A., & Wagner, A.R. 1972. A theory of Pavlovian conditioning: Variations in the effectiveness of reinforcement and non-reinforcement. In A.H. Black, & W.F. Prokasy eds., *Classical conditioning: II. Current research and theory.* New York: Appleton-Century-Crofts.

Rosch, E. 1978. Principles of categorization. In E. Rosch, & B.B. Lloyd, *Cognition and categorization.* Hillsdale, NJ: Erlbaum.

DEVELOPMENT of SCHEMATA DURING EVENT PARSING:

Neisser's Perceptual Cycle as a Recurrent Connectionist Network

Catherine Hanson
Department of Psychology
Temple University
Phildelphia, PA 19122

Stephen José Hanson †
Learning Systems Department
SIEMENS Research
Princeton, NJ 08540

Abstract

The present work combines both process level descriptions and learned knowledge structures in a simple recurrent connectionist network to model human parsing judgements of two videotaped event sequences. The network accomodates the complex event boundary judgement time-series and provides insight into the activation and development of schemata and their role during encoding.

Perceiving and Encoding Events

Day to day experience is characterized, remembered, and communicated as a series of events. We think about *driving to work*, we remember *having an argument* with our spouse, and we tell a friend about our plans to *attend the theatre* next Saturday. Abreviated phrases such as *driving to work* act as a type of shorthand notation for describing complex action sequences. Thus, our abilty to communicate sucessfully with others using such labels as *driving to work* reflects a certain level of familiarity with the referenced activities that we share or presume to share with our intended audience.

How common is our knowledge about common events? Empirical work suggests that there is considerable consensus concerning the constituent actions of familiar events (Bower, Black, & Turner, 1979). Bower, et al. found that subjects showed considerable agreement about the composition of common events (e.g., *going to a restaurant*), many responses being offered by more than 70% of their subjects and very few being unique. Considerable agreement about event boundaries extends to online measures of parsing as well (e.g., Newtson, 1973; Hanson & Hirst, 1989) suggesting that familiarity with events may provide the basis for understanding and encoding new information.

Neisser's Perceptual Cycle

Neisser (1976) has suggested that perception is a cyclical activity in which: (1) memory in the form of schemata guides the exploration of the environment, (2) exploration yields samples of available information, and (3) data collected from the exploration process modifies the prevailing schema. By focusing on the interaction of perception and memory, Neisser's "perceptual cycle" model offers a particularly fertile context for studying the processsing of event information. However, because this is a processing model, rather than a model of knowledge representation, little emphasis is placed on the structure of schematized knowledge. Thus, it is not clear how *turning ignition* might be related to *driving home* or even what role the decomposition of events might play in generating the expectations purportedly used to guide sampling of available information. Germane to this issue is another that arises in relation to the proposed modification process. How does the prevailing schema change in response to the sampling process? In particular, what is the basis for the

† Also a member of the Cognitive Science Laboratory, Princeton University, 221 Nassau Street, Princeton, NJ 08540

similarity between the ongoing situation and the schemata that are subsequently activated?

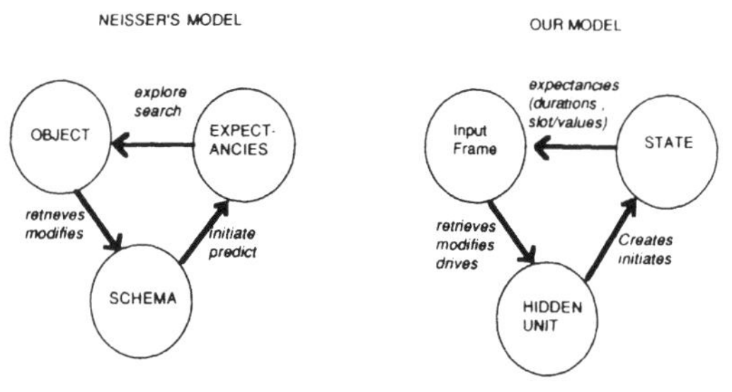

Figure 1: Neisser's Perceptual Cycle and Recurrent Net

The Problem with Scripts

Perhaps the best known attempt to address the kinds of questions raised here has been made by Schank (1982) within an artificial intelligence framework. Schank's approach to the parsing problem is essentially a taxonomic one in which relatively abstract knowledge structures (i.e., MOPs and TOPs) are posited to emerge from relatively specific action sequences (i.e., scripts). He suggests that comprehension emerges from a "reminding" process in which we "pass through old memories while processing a new input" (p.25).

"Reminding" is posited to occur when an online event activates an appropriate knowledge structure as a function of the similarity between the two. Thus, "reminding" is a process not unlike that posited in exemplar based categorization models (e.g., Medin & Schaffer, 1978) or the myriad of "nearest neighbor" algorithms posited to account for pattern recognition performance (Dasarathy, 1990). But, defining similarity remains as much a problem for Schank as for others wrestling with categorization issues.

Regrettably, similarity is invoked again when questions about structure development are raised. Structures at high levels in the hierarchy are posited to function as prototypes and to be abstracted from lower order structures. According to Schank (1982), these new high level structures develop "where the essential similarities between different experiences are recorded" (p. 81).

In addition to an inherent vagueness about the mechanism underlying the retrieval and development of knowledge structures, another problem with Schank's (1982) approach is its failure to deal with the temporal character of event knowledge in any straightforward way. Events persist for a given duration. Moreover, not only do different events persist for different durations, but the same event may last for different periods of time as a function of any number of factors such as the age of the actor, the experience of the actor, the time of day when the event takes place, the location of the event, and so on. An event not only persists for a given duration but derives its meaning from the context in which it occurs, that is, the events that precede and follow it. In itself, *ordering* means very little and *ordering* after *eating* makes little sense. It is only when *ordering* occurs in its rightful place among *sitting*, *eating*, and *leaving* that any real understanding can occur.

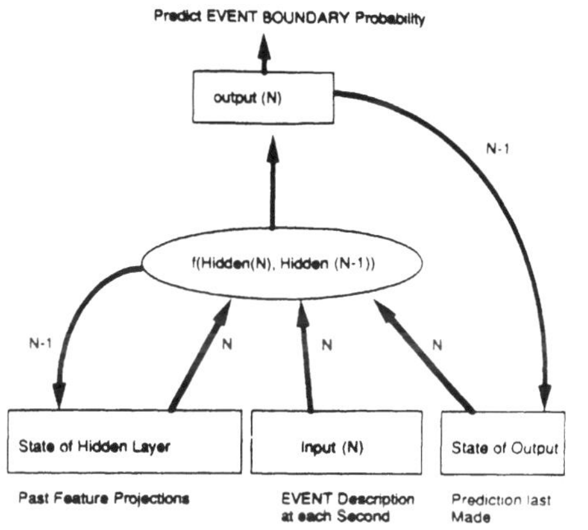

Figure 2: The Present Recurrent Net

Event Parsing

One way to avoid some of the difficulties that arise when a script structure is implemented is to model data derived from a task that creates context in terms of meaningful sequences of actions. That is, the nature of organizing schemata can be abstracted from human judgements concerning event boundaries for everyday situations. Data from a study by Hanson and Hirst (1989) provide such information. Briefly, subjects in this study were asked to watch videotapes of common event sequences. One videotape showed two people playing a game of Monopoly and the other showed a woman in a restaurant who drinks coffee and reads a newspaper.

TAPE	AGENT	ACTION	TYPEofVERB	OBJECT	MOVEMENT
game	mark	puts	transitive	money	yes
restaurant	pam	puts	transitive	money	yes

Figure 3: Input Encoding Example

Subjects watched the videotapes under various orientations and pressed a response button whenever they believed a new event was beginning. In the present study, we used reponses made when subjects had been oriented toward "small" events while viewing the tapes. This orientation produced the greatest number of perceived event boundaries and therefore a rich data set for use in training and transfer simulations.

Recurrent Nets and the Perceptual Cycle

A connectionist simulation provides an opportunity to examine how prior experience affects the parsing of actions into events. Recurrent networks, for example, inherently resemble Neisser's perceputal cycle [1] (See Figure 1). A recurrent net provides feedback information from hidden layers or from outputs creating information from either past actions at various moments in time or from past judgements about the presence or absence of an event change. For the net, an input frame consisting of a set of features and an arbtrary unit of time represents an object and moment of time in the world. The hidden layer of the network, which is driven by the input, also retrieves a learned category (Neisser's schema) which causes some moments in time to have a certain similarity to others (based on features). The feedback to the hidden layer creates a state (Neisser's expectancies) that influences in a top-down fashion judgements about the similarity of the present moment to an active schema retrieved via the hidden layer by the input frame.

A second reason for using a connectionist simulation is the opportunity it affords to examine the "black box" between input and output. By analyzing the hidden units of the network we hoped to gain some understanding about the kind of information needed to represent events and additionally, to learn how memory about events changes with experience. Thus, we hoped to be able to shed some light on how event knowledge is: (a) acquired, (b) represented, and (c) used to guide parsing. Our approach was a direct one; we examined how the representation of event knowledge changes with experience and observed the net's ability to transfer its knowledge about events to related and unrelated action sequences.

1. Rumelhart first suggested this connection between the perceptual cycle and reccurrent nets.

Network Structure and Training

A simple recurrent network used sources of information including features of events from the present moment in time, past event-moment features and past predictions of an event change (see Figure 2). It is known that simple recurrent networks (Elman, 1988; Rumelhart, Hinton & Williams, 1986) can represent *at least* a finite state machine (Servan-Schreiber, Cleremans & McClelland, 1988; Watrous & Kuhn, 1991; Giles et al., 1991) and thus are good candidates for encoding temporal event sequences. The present recurrent network received feedback from hidden layers and outputs delayed by one time step. Inasmuch as these activation values were combined over time they potentially can represent a complete sequence from the start of the event parsing.

Input Encoding. As stated before, two kinds of videotaped action sequences were used as data, one involving two people playing a Monopoly game and another involving two people in a restaurant sequence. Each tape was transcribed to the resolution of one *second.* Five variables were chosen to represent each second of the event sequences. These variables included AGENT, ACTION, OBJECT, TYPE of VERB (transitive or intransitive) and MOVEMENT (whether any movement occurred in that second). In Figure 3 are examples of a single second transcribed for each kind of videotape: Hereafter, this attribute-value structure will be referred to as the frame-second. The combined information from both tapes included a total of 4 AGENTS, 33 ACTIONS and 43 OBJECTS. Sixty percent of ACTIONS and 9% of OBJECTS overlapped between the two event sequences. The network was provided a binary representation (17 bits) of this input frame.

Training. The network's task was to learn to map the current frame-second, any past frame-second, and the past event change probability to the next event change probability. Event change probability was computed from the number of subjects (out of 20) who judged that an event boundary had occurred (by pressing a response button) during that frame-second. Shown in Figure 4 are the event change probablities[2] for the Monopoly game. On the x-axis are the 420 frame seconds corresponding to the transcribed features. The y-axis shows the relative frequency of button presses at that second.

The network was trained by 1st-order gradient descent ("back-prop in time") to produce the event change time series. Due to the noise present in the time series other methods such as line-minimization or conjugate gradient methods (e.g. BFGS optimization) fared poorly in terms of speed of convergence and reliability to the same solution as a function of starting point. Simple 1st-order back-prop, converged quickly and reliably to the same solution in spite of the target noise.

Standard Models. The event change probabilities were not modelled well as an ARIMA (Box-Jenkins) time-series suggesting few periodicities were present in the time-series independent of the frame-seconds. A standard multiple regression accounted for less than 5% (Pearson r correlation of .07) of the data variance suggesting that the mapping was significantly nonlinear.

Learning of Game and Restaurant Tapes

Using split halfs of the Game and Restaurant tapes the recurrent network was able to acount for over 45%-50% of the data variance with Pearson r correlations of .75 for the Game and .68 for the Restaurant tape. The difference appears to be related to the differences in length of the tapes (Game-420 seconds, Restaurant-287 seconds) and the higher diversity of actions in the Game event sequence.

2. Subsequent figures of event change probabilities show a smoothed (4-second window) version of this data in order to make visual comparisons only. All prediction values from network training are for the raw data shown in Figure 4.

Transfer. The Game tape learning was then transferred to the Restaurant tape[3]. Shown in Figure 5 in the left panel is the result of training on the Game tape. The dashed line represents the event-change data and the solid line is the second-by-second prediction of the reccurrent network. Using all the data for the Game tape boosted the variance accounted for to 80% (r=.9). Transfer to the Restaurant tape, was significant (40%, r=.65) in spite of large attribute-value differences at each second of each tape. Hidden unit sensitivity was explored for 5 to 30 hidden units. Variance accounted for on transfer went up slowly, reaching asymptote near 15-20 hidden units.

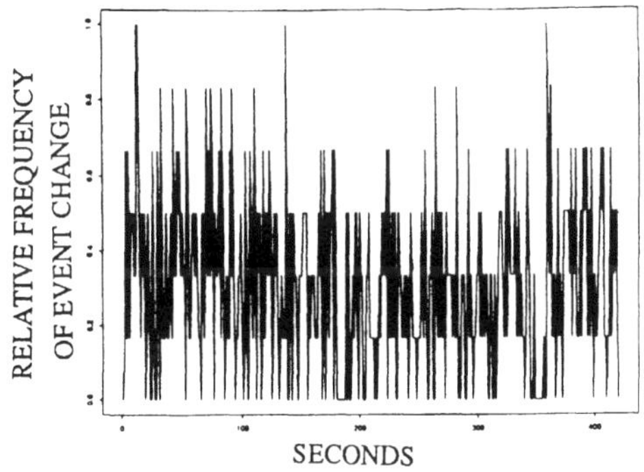

Figure 4: Event Change Judgements over Time (seconds)

Internal Representation

Hidden unit patterns were analyzed (Hanson & Burr, 1990) over each second in order to determine the similarity of frame-seconds that the reccurrent net discovered to make the event change predictions. A hierarchical cluster analysis (Centroid, and Farthest Neighbor agreement) was performed on the hidden unit activations over the 420 seconds and over the 287 seconds. Very regular dendrograms were produced and an examination of successive differences over the merge history indicated 10-15 clusters to be present.

Insofar as clusters represent groups of frame-seconds that are similar from the reccurrent net's point of

3. The Restaurant tape was also tranferred to the Game tape, but with less success, probably because the sample size of the restaurant tape was about 75% that of the Game tape.

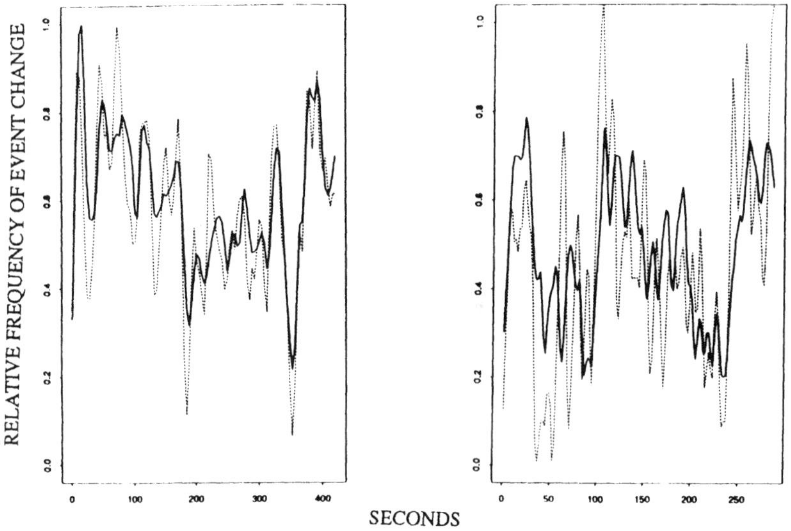

Figure 5: Transfer from the Game to the Restaurant Event Sequence

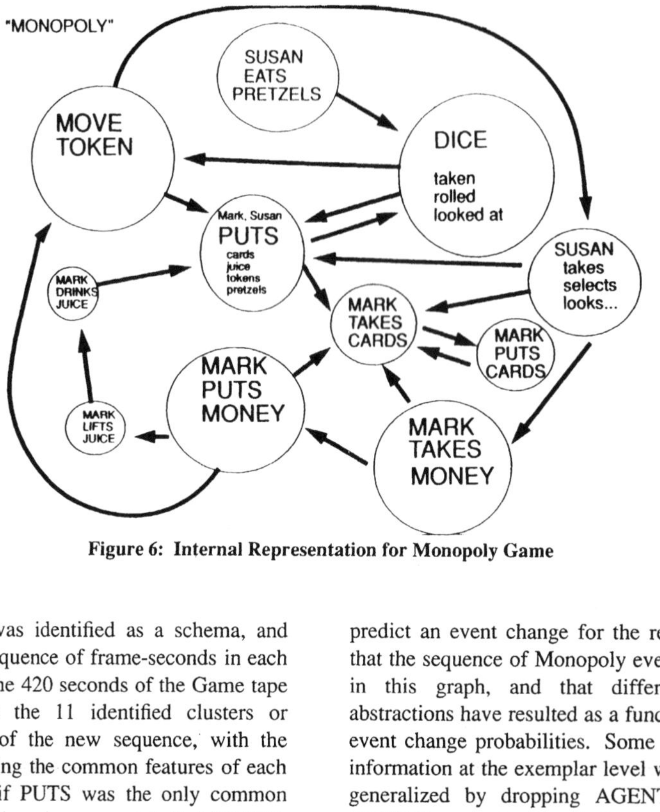

Figure 6: Internal Representation for Monopoly Game

view, each cluster was identified as a schema, and used to relabel the sequence of frame-seconds in each tape. For example, the 420 seconds of the Game tape was relabelled with the 11 identified clusters or schemata. A graph of the new sequence, with the schemata labelled using the common features of each frame second (e.g., if PUTS was the only common feature in the cluster the schema was labelled "PUTS", if MARK TAKES MONEY was common to every frame-second then the schema was labelled "MARK TAKES MONEY"), is plotted in Figure 6.

The size of each ball is based on the relative frequency of the schema in the tape, and the arrows represent the state transition (a state transition would predict an event change for the recurrent net). Note that the sequence of Monopoly events is representated in this graph, and that different schema level abstractions have resulted as a function of learning the event change probabilities. Some schemata represent information at the exemplar level whereas others have generalized by dropping AGENT or ACTION or OBJECT or subsets of these variables. Finally, notwithstanding the differences in frame-second content (especially in terms of objects) and poorer prediction performance, a representation was extracted that did correspond to action sequences consistent with the actual events in the Restaurant sequence.

The Nature of Event Perception

We conclude by providing answers to the questions about schemata posed earlier in this paper. Based on the computation of the recurrent net and the internal representations of schemata extracted, several aspects of event perception might be clarified.

On what basis are schemata activated? Several factors determine whether a schema is activated or not. One, similarity (in this case dot-product to the hidden layer) of a schema to a present frame-second (in terms of attribute and value presence) can activate and retrieve new schemata in a bottom-up fashion. Two, past schemata will resist bottom-up input at a given frame-second and will tend to block the activation of new schemata. The more specific a schema is (in terms of attributes and values) the less likely transitions to a new schema will occur. Three, each schema has been associated with an expected duration. The duration of a schema can be determined by clamping the plan vector with a given attribute value and starting input values at ambiguous values (.5) and counting the number of seconds passed before the output approaches a value between .75 and 1.0. All 11 schemata for the Game tape were clocked in this way. If a schema is expected to continue for a long time, inconsistent input data will be ignored until a sufficient number of instances appear. Some schemata occur frequently having brief durations while others occur rarely but at longer durations. In fact, there was a significant negative correlation (-.52) between schema duration and frequency.

What role do schemata play during encoding? The active participation of schemata help to select input and maintain resistance to change. Within the context of the recurrent nets, schemata create expectations about the level of abstraction that will appear in the input frame and the specific content that should be found. Finally, once a schema is activated there is an expectation about its duration (due to the feedback) and a search for confirmation continues until the schema terminates.

How do schemata develop? The frequency of events and their duration within the frame-seconds determine how schemata develop and what properties they will possess. As stated above, there is an inverse relation between duration of schemata and their frequency in the Game tape. High-frequency, short-duration schemata tend to be more asbtract or general, whereas low-frequency, long-duration schema tend to be more specific or exemplar based. Examination of the schemata as they develop during learning indicate that they tend to evolve from specific exemplar based clusters into more abstract based clusters by accepting increasingly more diverse input over time.

References

Bower, G.H., Black, J.B., & Turner, T.J. (1979). Scripts in memory for text. *Cognitive Psychology*, 11, 177-220.

Dasarthy, B. (1990). NN Pattern Classification Techniques, Los Alamitos, CA: IEEE Computer Society Press.

Elman, J.L. (1988). *Finding structure in time*. CRL Technical Report 8801. Center for Research in Language, UCSD.

Giles, L., Miller, C. B., Chen D. Chen, H. H. Sun G. Z., Lee, Y.C. (1992). Learning and Extracting Finite State Automata with Second-Order Recurrent Neural Networks, *Neural Computation*, (in press).

Hanson, C. & Hirst, W. (1989). On the representation of events: A study of orientation, recall, and recognition. *Journal of Experimental Psychology: General*, 118, pp. 124-150.

Hanson, S.J. & Burr, D. J. (1990). What Connectionist Models Learn: Learning and Representation in Connectionist Networks. *Behavioral and Brain Sciences*, 13, 3 pp. 477-518.

Medin, D.L., & Schaffer, M.M. (1978). Context theory of classification learning. *Psychological Review*, 85, 207-238.

Neisser, U. (1976). *Cognition and reality: Principles and implications of cognitive psychology*. San Francisco: W.H. Freeman.

Newtson, D. (1973). Attribution and the unit of perception of ongoing behavior. *Journal of Personality and Social Psychology*, 28, 28-38.

Rumelhart, D., Hinton, G. & Williams, R. (1986). Learning internal representations by error propagation. In D.E. Rumelhart and J.L. McClelland (Eds.), *Parallel Distributed Processing I: Foundations*. Cambridge, Mass: MIT Press.

Schank, R.C. (1982). *Dynamic Memory: A theory of reminding and learning in computers and people*. Cambridge: Cambridge University Press

Servan-Schreiber D., Cleeremans, A. & McClelland, J. (1988). *Encoding sequential structure in simple recurrent networks*. CMU Technical Report CS-88-183.

Watrous, R. & Kuhn G. (1992). Induction of Finite -State Languages Using Second -Order Recurrent Networks, *Neural Computation*, (in press).

Skill as the Fit Between Performer Resources and Task Demands: A Perspective from Software Use and Learning

Catherine A. Ashworth
Institute of Cognitive Science
Campus Box 345
University of Colorado
Boulder, CO 80309-0345
ashworth@clipr.colorado.edu

Abstract

This paper goes beyond the routine vs. adaptive expertise distinction seen most recently in Holyoak (1991) by offering a framework which locates skill in the fit between performer resources and task demands. Empirical support for this framework is derived from a review of the literature about "real world" software learning and usage.[1]

Introduction

When we speak of skill or expertise in some endeavor or domain we are typically referring to both depth and breadth of knowledge and performance. In addition, we usually expect an expert to perform more quickly, accurately, and with less effort than a novice. Despite this consensus, some researchers have asserted the existence of more than one type of expertise. The goal of this paper is to sketch a framework that embraces both routine and adaptive expertise, explains how each is acquired, and sketches the knowledge base required for each.

Routine and Adaptive Expertise

A recent treatment of this position (Holyoak, 1991) presents a list of established properties of expertise with challenging findings for each. Holyoak concluded that there are two types of expertise, "routine" and "adaptive." The canonical view of expertise, well modeled by production system architectures such as ACT* (Anderson, 1987) and Soar (Newell, 1990), is termed "routine." In contrast, the contrary findings are collected under the rubric "adaptive." In brief, routine expertise is rapid and accurate on highly stereotyped tasks but its inflexibility and sparse meaning precludes transfer to novel tasks or creativity. Adaptive expertise can also be swift, but is characterized primarily by knowledge breadth and the invention of new procedures.

Holyoak's list contains six important and contentious features of skill: 1) constant improvement with practice, 2) automaticity of responses, 3) domain specificity, 4) ease of task solution, 5) superior domain memory, and 6) forward search. Exceptions to these generalizations about expertise include: 1) improvement plateaux (Ericsson & Krampe, 1991), 2) successful remapping of conditions and actions (Singley & Anderson, 1989; Allard & Starkes, 1991), 3) transfer (Singley & Anderson, 1989; Dorner & Scholkopf, 1991), 4) more effortful or elaborate solution (Scardamalia & Bereiter, 1991; Jeffries, Turner, Polson, & Atwood, 1981), 5) poorer memory for some information (Adelson, 1984; Patel & Groen, 1991), and 6) flexible or breadth-first search (Jeffries, Turner, Polson, & Atwood, 1981; Dorner & Scholkopf, 1991). A review of the literature involved in the disagreement on these six issues supports the following characterization of two families of tasks -- routine and adaptive -- which differ in duration, variety, and knowledge requirements.

Routine tasks are fairly well constrained. The givens-solution organization of these tasks renders them fairly brief in duration, residing in Newell's (1990) cognitive timeband. This is due to a lack of extended reasoning such as constraint elaboration, coherence checking among subplans, et cetera.[2] In addition, there is very little task variety. Consider a

This work was conducted while the author was being supported by grant MDA 903-89-K-0025 from the Army Research Institute

[1] A much more thorough treatment of these and other related topics can be found in the author's predissertation paper. Please contact.

[2] From another perspective, tasks which require strategy, internal coherence checks, or elaboration of constraints cannot be represented in a straightforward set of <condition> <action> links, which can be rapidly executed.

typical task such as solving algebra problems. A certain type of problem (perhaps a dilution problem) is always properly solved with a certain method. Because this ability is rooted in recognition of problem types (beneath varying surface features) and the application of specified solutions, improvement means increasing speed and accuracy in the application of these schemas. That is, schema construction during practice essentially eliminates search during problem solving on these tasks. Skilled performance on this type of task gives the impression of extreme ease, automaticity, forward reasoning, constant (inevitable) improvement, and recall of details (for there is no higher structure).

Adaptive expertise may occur on tasks which are marked by longer duration, fundamental problem variety, and greater knowledge requirements. These tasks reside in Newell's (1990) rational time band. Example tasks are essay writing, computer programming, or medical diagnosis. Because these tasks are complex and varied, successful solution requires strong metacognition such as a detailed knowledge of plans and organizational requirements. These tasks can be regarded as problem solving tasks on which search and planning cannot be severely reduced. Therefore these tasks cannot be successfully completed through the application of so-called automatic responses. This family of tasks provides many of the exceptional results which challenge the canonical description of expertise. On these tasks, skilled performance is marked by improvement discontinuities, great effort, lack of automaticity, varied reasoning and search direction, and memory for more abstract information than for specifics.

In short, one might map the distinction between routine and adaptive expertise onto the distinction between well-defined and ill-defined problems or between closed and open tasks. Although the degree of external constraint on a task is important, this paper asserts that the sort of expertise observed is not determined solely by this factor. Many and most tasks lie between the two extremes of the routine-adaptive continuum and can be accomplished with either approach. On these tasks the performer is a major determinant of the nature of the performance. That is, the choices and style of a performer is another factor, crossed with external constraints, that must be considered. The following review of a portion of the human-computer interaction literature strongly supports this assertion.

Software Usage and Learning

The human-computer interaction (HCI) literature contains intriguing findings which challenge current thinking about skill. From a skill perspective, one of the most provocative findings in HCI is found in studies of software learning and usage. Specifically, long term users of a piece of software (application) frequently know very few commands (Carroll & Rosson, 1987). For many users this is the result of a conscious choice. Rare additions to the small repertoire are driven by tasks that cannot be accomplished with known methods. Also, knowledge of commands may be noncumulative such that a given repertoire may not include all commands of a smaller repertoire. This pattern challenges the notion that skill necessarily and almost artifactually increases through practice in an orderly manner.

The decoupling of knowledge breadth and experience or a noncumulative progression of command acquisition has been documented in various applications. Experienced library database searchers employ a small proportion of available, useful commands (Fenichel, 1981). Bank clerks also use a small portion of available information retrieval commands, often using a sequence of two or three known commands to avoid relearning a more direct function (Kraut, Hanson, & Farber, 1983). Even computer professionals with broad technical and software experience press known functions and software into service to avoid learning (Nielsen, Mack, Bergendorff, & Grischkowsky, 1986).

Experienced users with small command repertoires are also the rule in operating system usage. The number of unique operating system commands issued by users at a large university computing site was uncorrelated with the length of the computing tradition of their various disciplines (Anstey, 1988). A study of electronic mail (e-mail) with experimental tasks revealed that long term ("regular") users issued significantly fewer commands than did system managers or consultants (Akin & Rao, 1985). Interestingly, each of these user groups employed unique commands. This noncumulativity of commands was similarly observed in an operating system command logging study which also found most users had small command sets (Draper, 1984).

Not only do most users have small, idiosyncratic command sets, they also resist complex or highly specific commands. Experienced database searchers use very simple query structures which do not exploit the system's power (Fenichel, 1981). E-mail commands which were unique to the "regular" users were more general than those unique to the system managers (Akin & Rao, 1985). In addition, the use of powerful, tailorable editing functions such as macros can be as strongly correlated to years of programming experience as to experience with the editor (Rosson, 1984). Even when self tailoring of software and workstations is aggressively taught and supported, heavy users of the software are remarkably conservative (MacLean, Carter, Lovstrand, & Moran, 1990). In summary, long term users of software typically know only a handful of general commands.

There are two basic arguments offered by those who use a small command set and routinize their work. One is that their goal is to accomplish a bit of work such as drawing a graph or reading mail from a colleague, and not to learn a piece of software. The software is strictly regarded as a tool. Following from this perspective is the strategy of attempting to achieve maximum coverage from a minimal investment, which is the other argument often made. Most users with a small command set are quite satisfied with that repertoire and feel prepared to deal with anything that might arise (Anstey, 1988). Individual users acquire a new command or function only when a specific task forces them to, which produces the observed noncumulativity of command repertoire across users.

It is not clear that users who maintain a small command set suffer. In fact, they avoid an entire family of problems for it turns out that increasing skill breadth with an application may yield low returns or even handicaps and is effortful and error prone. A lack of performance advantage from a larger command repertoire was obtained in the e-mail study: In the end, those users who employed more commands were no speedier nor more accurate (Akin & Rao, 1985). From one perspective, those users who employed fewer unique commands were more efficient. Spreadsheet based experiments reveal that learning or performance costs can accompany knowing more than one method for a task (Olson & Nilsen, 1987; Ashworth, 1992) which is highly likely in a moderate or larger command set. Learning and performance costs are observed in error rates and increased planning times. One must learn an almost overwhelming amount of information of many different sorts to profit from the work (Olson & Polson, 1992; Doane, Pellegrino, & Klatzky, 1990).

Moreover, learning new functions requires venturing into poorly understood, error prone territory. Errors during exploration are particularly problematic because 1) they are difficult to identify and 2) it is not clear how to recover from an action which itself was poorly understood. In fact, a user has no guarantee that the target method will work, will be appropriate, or will save time. In addition, in a domain such as software in which the action to initiate a command is fairly arbitrary, there may be severe memory problems associated with using a large command set. Thus, acquiring enough knowledge to be adaptive is very effortful, risky, may entail costs, and may not yield benefits.

In summary, most users purposely (and perhaps wisely) maintain a small command repertoire. These users operationalize incoming tasks so that they can be accomplished with known methods, minimizing their knowledge overhead. Although this description characterizes the majority of users an interesting exception is a minority group of users, often called "wizards," who are usually system managers or informal troubleshooters. These users possess an extensive repertoire of immediately usable commands and also a large pool of commands which can be quickly reconstructed or derived. These wizards demonstrate that a given task a user's style can be routine or adaptive. Interestingly, wizards can make more errors and access on-line help or manuals more often than do other users (Draper, 1984; but see Vaubel & Gettys, 1990). These findings represent another (apparent) conflict in the skill literature for we expect experts to make fewer errors.

A small command repertoire and a low frequency of errors both derive from an approach which routinizes and avoids additional learning. The user who avoids learning new commands or functions and instead assimilates tasks to a small well known set of functions avoids three things: 1) the error prone process of identifying and employing a new command, 2) the effort of learning and remembering a new command which may well have to be relearned for its next use, and 3) selection among known commands. However, this strategy may also reduce the ease with which new functions can be acquired (should the user select this option). The contrasting approach, knowledge seeking, produces the opposite pattern of an extensive command repertoire and a high rate of errors and on-line help accesses. These users bear the memory, learning, and performance liabilities of a large repertoire, but can do more things and may be better able to use new commands. The software learning and usage literature supports the assertion that a given task, in and of itself may not require routine or adaptive expertise. Instead, for many tasks, a performer can elect one style or the other.

A Conceptual Framework

The combination of these results from the HCI literature and the distinction between routine and adaptive expertise (discussed above) supports the following framework embracing both routine and adaptive expertise. The central tenet of this framework is that the performer determines whether she will accomplish a given task in a routine or adaptive manner. This decision is made in reference to the performer's task operationalization which incorporates the personal goal to seek or avoid additional knowledge. Although the HCI literature emphasizes the influence of the performer, expertise is located in the fit between two factors: 1) constraints inherent in the task (such as grainsize) or the situation (such as time pressure) and 2) the performer's resources. Concepts important in this argument include: performer resources, personal learning goals, and task operationalization

Real world tasks typically have an external definition based on the final goal, such as "produce a manuscript in APA format." Although such definitions may be constrained by specific process instructions or time pressure, they often contain many choices. It is this task freedom which permits the influence of performer choices. For instance, the manuscript preparer decides how and when to achieve the required margins. Because the task definition is rarely completely specified for the performer, it is detailed by the performer. The performer's operationalization of the task is developed in reference to two strongly related criteria: 1) achieving personal goals and 2) achieving a fit between performer resources and task plan demands.

Personal learning goals are best understood by example. For example, in addition to satisfying the manuscript requirements, the preparer may be pursuing personal goals including: completing the task using currently known methods only, using the opportunity to learn how to do related (potential) tasks, or using the task simply to explore the software. These personal goals fall into two general styles: 1) learning avoidance and 2) knowledge seeking. The performer considers these personal learning goals when operationalizing the task. The plan generated to accomplish the task therefore describes the particulars of what the performer is willing to do. These particulars include the knowledge that the performer has committed to bringing to the task. The relationship of this knowledge to the performer's current resources determines the nature of the fit, routine or adaptive. Although the performer also chooses the degree of fit achieved, the focus here is on the manner in which it is achieved.

When a task is acquired there may appear to be a gap between performer resources and task demands. To bridge this gap, the performer can change: 1) performer resources or 2) task demands. These two ways of bridging the gap are tied to personal learning goals just discussed. When a task seems to require unknown methods, the performer may acquire that knowledge (increase user resources) and complete the task. In this case, the personal goal is of the knowledge or understanding seeking variety. Another approach is to accomplish the task while avoiding resource expansion. This occurs when the performer introduces the personal goal of accomplishing the task with known methods. In accordance with this personal goal, the performer creates an operational definition of the task that can be achieved with current resources. This assimilates the task to known methods and avoids learning. Of course, there are many mixtures of these two extremes.

In summary, a performer has personal learning goals which shape the task demands to those that the performer is willing to accomplish. In this way, the performer selects the degree to which, and manner in which, performer resources will fit task demands. Additionally, each task episode can shape the performer's resources for the next. As discussed below, both approaches accomplish the task, both have costs and benefits, and neither is inherently superior or more skilled than the other.

Discussion

Current skill theories which espouse the canonical view of expertise are rooted primarily in accounting for the power law of practice. In fact, investigating how these theories model the power law is a useful way to become acquainted with them for it exposes most of their mechanisms and processes. And, the power law of practice summarizes several core beliefs about skill including: constant improvement, increased accuracy, specificity, and automaticity. The performance of one who has practiced a fairly closed task of brief duration is well modeled by these theories. Example tasks include: transcription typing, naming state capitals, recalling the contents of a chess board, or sorting and solving physics problems. All of these tasks are fairly constrained by their timeband, their lack of variety, and the extremely tight guidance of mature problem schemas.

According to the power law, improvement is the reduction of problem solving processes in favor of proceduralized processes. That is, skill is regarded as a strengthening or sharpening process. Tasks which are not well captured in current theories are tasks on which improved performance or increased skill requires an effortful broadening of available procedures forced by a diet of task variety. A review of the literature cited by Holyoak (1991) makes it clear that the rogue results which he gathers together under the umbrella of adaptive expertise come from these sorts of tasks. Examples include social problem solving, writing about one's summer vacation, designing software, or controlling a complex computer simulation.

It is not clear how a theory of skill based on the power law and its entailments could explain adaptive skill. Conceptually, performance gains accrue in two loci. First, single procedures must become speeded. Second, the knowledge of pre- and post-conditions for every procedure must be represented and processed in a way that supports rapid creation of novel chains. A constraint based approach which does not model skill through the creation of large precompiled rules appears more useful (Kitajima & Polson, 1992; Mannes & Kintsch, 1989).

Although task constraints can be strong, there is another way for a performance to appear either routine or adaptive. On many tasks, the performer has the option of either assimilating or accommodating the task. The software usage literature convincingly demonstrates that performers determine where their tasks fall along a continuum from extremely routine to extremely novel. The majority of users assimilate incoming tasks to known methods. From the perspective of efficient tool use (and controlling surprises during tool use) this is a reasonable approach. There is also an identifiable group of users who accommodate incoming tasks, taking the opportunity to learn new software functions and facts. This approach also has costs and benefits. Thus, the performer is also an important factor in the nature of the performance. This paper has sketched a framework which explicitly acknowledges and explores the consequences of the performer's choice.

This framework goes beyond the distinction between routine and adaptive expertise which focuses solely on task constraints by identifying the learner or performer as another important factor in the sort of performance observed. The relation of these two factors, the degree of external task constraints and the performer's approach (internal performer constraints), is the location of skill. That is, skill is located in the performance of a task which is the product of task and performer constraints.

This perspective has several entailments. First, in general, a given task does not inherently require or call forth one sort of expertise or the other. (See above for a discussion of the exception to this statement: tasks at the ends of the continuum.) Second, even within one domain a performer may not use solely a routine or an adaptive approach. For example, a performer may assimilate one task to current knowledge but be willing to learn a little something for another. Routine and adaptive expertise also coexist in an individual in that it is not clear how adaptive expertise could develop without the support of routine expertise.[3] Consistent practice in these different styles, routine or adaptive, engenders knowledge representations which differ in content and size. Because there are costs and benefits to both ways of accomplishing a task it is inappropriate to label one superior to the other

This framework explains some recently mentioned, apparently anomalous, findings in the area of skill acquisition and skilled performance. For instance, we have discovered how expert performance can be lengthier, more effortful, and more errorful than that of less skilled performers. In addition, we have employed the same simple ideas to understand how a performer can engage in an activity for thousands of tasks and not appear to improve in expertise. This framework is currently being elaborated and expanded to account for other variations in experimental results.

[3]In contrast, it is entirely possible for a performer who typically routinizes not to possess the knowledge required to support a more adaptive, flexible approach.

Acknowledgements

The author expresses gratitude to Gene Gollin for championing the "system perspective" and to Marita Franzke and Peter Polson for useful discussions of these ideas. Appreciation to Marita Franzke, Evelyn Ferstl, Susan Davies, Cathleen Wharton, Julia Moravcsik, Adrienne Lee, and Abby Harrison who provided guidance, encouragement, and proofreading.

References

Adelson, B. (1984). When novices surpass experts: the difficulty of a task may increase with expertise. *Journal of Experimental Psychology: Learning, Memory, and Cognition, 10*, 483-495.

Akin, O., & Rao, D.R. (1985). Efficient computer-user interface in electronic mail systems. *International Journal of Man-Machine Studies, 25*, 557-572.

Allard, F. & Starks, J.L. (1991). Motor skill experts in sports, dance, and other domains. In K.A. Ericsson & J. Smith (Eds.), *Toward a General Theory of Expertise: Prospects and Limits.* Cambridge: Cambridge University Press.

Anderson, J.R. (1987). Skill acquisition: Compilation of weak-method problem solutions. *Psychological Review, 94*, 192-210.

Anstey, P. (1988). How much is enough? A study of user command repertoires. In D.M. Jones & R. Winder (Eds.), *People and Computers IV, Proceedings of the Fourth Conference of the British Computer Society*. Cambridge: Cambridge University Press.

Ashworth, C.A. (1992). Specialized methods do not increase efficiency. In *CHI '92 Poster and Short Talks Proceedings of Human Factors in Computing Systems*. New York: ACM.

Carroll, J.M. & Rosson, M.B. (1987). Paradox of the active user. In J.M. Carroll (Ed.), *Interfacing Thought: Cognitive aspects of human-computer interaction*. Cambridge, MA: MIT Press.

Doane, S.M., Pellegrino, J.W., & Klatzky, R.L. (1990).Expertise in a computer operating system: conceptualization and performance. *Human-Computer Interaction, 5*, 267-304.

Dorner, D & Scholkopf, J. (1991). Controlling complex systems or expertise as "grandmother's know-how". In K.A. Ericsson & J. Smith (Eds.), *Toward a General Theory of Expertise: Prospects and Limits*. Cambridge: Cambridge University Press.

Draper, S.W. (1984). The nature of expertise in UNIX. In B. Shackel, (Ed.), *INTERACT '84 - IFIP Conference on Human-Computer Interaction*. Amsterdam: Elsevier-Science.

Ericsson, K. A. & Krampe, R. (1991) The role of deliberate practice in the acquisition of expert performance. Technical Report of The Institute of Cognitive Science, Univerisity of Colorado, Boulder, CO.

Fenichel, C.H. (1981). Online searching: measures that discriminate among users with different types of experiences. *Journal of the American Society for Information Science, ?*, 23-32.

Holyoak, K.J. (1991). Symbolic connectionism: toward third-generation theories of expertise. In K.A. Ericsson & J. Smith (Eds.), *Toward a General Theory of Expertise: Prospects and Limits*. Cambridge: Cambridge University Press.

Jeffries, R., Turner, T., Polson, P., & Atwood, M. (1981). Processes involved in designing software. In J.R. Anderson (Ed.), *Cognitive Skills and Their Acquisition*. New Jersey: Hillsdale.

Kitajima, M. & Polson, P.G. (1992). A computational model of skilled use of graphical user interfaces. In *CHI '92 Proceedings of Human Factors in Computing Systems*. New York: ACM.

Kraut, R. E., Hanson, S.J., & Farber, J.M. (1983). Command use and interface design. In *CHI '83 Proceedings of Human Factors in Computing Systems*. New York: ACM.

MacLean, A., Carter, K., Lovstrand, L. & Moran, T. (1990). User-tailorable systems: pressing the issue with buttons. In *CHI '90 Proceedings of Human Factors in Computing Systems*. New York: ACM.

Mannes, S.M. & Kintsch, W. (1989). Planning routine computing tasks: Understanding what to do. ICS Tech Report 89-09. Boulder, Colorado: Institute of Cognitive Science, University of Colorado.

Newell, A. (1990). *Unified Theories of Cognition*. Cambridge, MA: Harvard University Press.

Nielsen, J., Mack, R.L., Bergendorff, K.H., & Grischkowsky. (1986). Integrate softare usage in the professional work environment: Evidence from questionnaires and interviews. In *CHI '86 Proceedings of Human Factors in Computing Systems*. New York: ACM.

Olson, J.R. & Polson, P. G. (1992). The cognitive cost paradigm. Manuscript in preparation.

Olson, J.R., & Nilsen, E. (1987). Analysis of the cognition involved in spreadsheet software. *Human-Computer Interaction, 5*, 221-265.

Patel, V.L. & Groen, G.G. (1991). The general and specific nature of medical expertise: a critical look. In K.A. Ericsson & J. Smith (Eds.), *Toward a General Theory of Expertise: Prospects and Limits*. Cambridge: Cambridge University Press.

Rossen (1984). The role of experience in editing. In B. Shackel, (Ed.), *INTERACT '84 - IFIP Conference on Human-Computer Interaction*. Amsterdam: Elsevier-Science.

Scardamalia, M. & Bereiter, C. (1991). Literate expertise. In K.A. Ericsson & J. Smith (Eds.), *Toward a General Theory of Expertise: Prospects and Limits*. Cambridge: Cambridge University Press.

Singley, M.K. & Anderson, J.R. (1989). *The Transfer of Cognitive of Skill*. Cambridge, MA: Cambridge University Press

Vaubel, K.P. & Gettys, C.F. (1990). Inferring user expertise for adaptive interfaces. *Human-Computer Interaction, 5*, 95-117.

The Nature of Expertise in Anagram Solution[1]

Laura R. Novick & Nathalie Coté
Dept. of Psychology & Human Dev.
Box 512 Peabody
Vanderbilt University
Nashville, TN 37203
novicklr@vuctrvax.bitnet
cotenc@vuctrvax.bitnet

Abstract

Second-generation theories of expertise have stressed the knowledge differences between experts and novices and have used the serial architecture of the production system as a model for both expert and novice problem solving. Recently, Holyoak (1991) has proposed a third generation of theories based on the idea of expertise-related differences in the processing of solution constraints. According to this view, the problem solving of experts, in contrast to that of novices, often is better characterized as a process of satisfying multiple solution constraints in parallel than as a process of serially testing and rejecting hypotheses. We provide data from three experiments that are consistent with this hypothesis for the domain of anagram solution.

Background

How does the problem solving behavior of experts differ from that of novices? According to Holyoak (1991), the first generation of research on expertise was based on Newell and Simon's (1972) theory that experts are distinguished by their superior ability to employ general heuristic search methods. Subsequently, research in domains such as chess and physics discovered that heuristic search was a weak method actually used more often by novices than by experts. This research spawned a second generation of theories that focussed on expertise-related differences in knowledge. According to the new theories, what distinguished experts from novices was the larger size, superior organization, and greater accessibility of their knowledge base within the domain of expertise. Regardless of level of expertise, however, problem solving was conceptualized as a serial process based on the architecture of the production system.

Recently, Holyoak (1991) has challenged the second-generation view of expertise. In addition, he has proposed a third generation of research that would be focussed on processing differences rather than knowledge differences. According to this new view, theories of expert performance would be based on the parallel architecture of connectionism (e.g., Rumelhart, McClelland, & the PDP Research Group, 1986). The hypothesis is that problem solving in experts, in contrast to that of novices, often is better characterized as a process of attempting to satisfy multiple task constraints in parallel than as a process of serially testing and rejecting hypotheses.

A complete understanding of expertise most likely will require reference to both knowledge and processing components of performance. In this paper, we highlight the issue of processing differences. Studying processing differences in knowledge-rich domains such as chess and physics is difficult, however, because solvers at different levels of expertise also will differ in terms of their knowledge bases. Given this potential confound, we chose anagram solution as the domain of study, because expertise-related differences in knowledge could be minimized.

Anagram Solution

Anagram solution involves unscrambling a string of letters into an English word (e.g., "iasyd" becomes "daisy"). Clearly, anyone literate in English has the knowledge necessary to solve most anagrams (e.g., knowledge of spelling

[1] The research reported in this paper was supported by a grant from the University Research Council of Vanderbilt University.

constraints). To further minimize any impact of knowledge differences in our research, we used only five-letter anagrams of common words (such as the example just given), and our subjects were college students at a highly selective university.

A second important criterion for selecting a domain is that there is some *a priori* reason for expecting experts to be more likely than novices to engage in parallel processing of solution constraints. Anagram solution meets this criterion as well. An intriguing phenomenon is that sometimes the answer to an anagram seems to "pop out" very quickly without any conscious awareness of a solution attempt. Anecdotal reports from self-proclaimed experts suggest that they solve many five-letter anagrams (e.g., "erjko", "dnsuo", "rcwdo", "iasyd") in less than 1.5-2 seconds. In contrast, novices do not often report pop-out solutions.

The hypothesis that experts attempt to solve anagrams by trying to satisfy in parallel the multiple, often conflicting, constraints on the rearranged order of the letters suggests a potential mechanism for the occurrence of the pop-out phenomenon. Moreover, experts' intuitions concerning pop-out solutions contrast with the conclusions of the experimental literature (which presumably is heavily weighted by data from non-expert solvers), which characterizes anagram solution as a deliberate, serial process of testing and rejecting hypotheses.

Although we believe that serial processing of solution constraints characterizes much of anagram solution (even by experts), we do not believe that it provides a complete account of expert behavior. Consistent with research in other domains (e.g., chess), we suspect that anagram experts often unscramble anagrams by a parallel rather than a serial process.

Serial and Parallel Models of Anagram Solution

How would serial and parallel models of anagram solution differ? Numerous answers to this question are possible. Our goal here is simply to provide a brief description of what each type of model could look like, not to construct detailed simulations of such models. To understand the experimental predictions we make later, it will help to have in mind a concrete example of each type of solution process.

Mendelsohn (1976) has proposed a serial, hypothesis-testing model of anagram solution. The first phase of solution involves forming hypotheses about the correct letter order based on the judged likelihood of each possible bigram (two-letter combination). These hypotheses are formed in decreasing order of the bigrams' frequencies in the language. For example, given the anagram "dnsuo," the first three hypotheses to be tested concerning the initial bigram of the solution would be, in order, "un", "so", "do" (Mayzner & Tresselt, 1965). The second phase of solution involves testing each hypothesis by retrieving from memory words that match the hypothesized partial reorganization of the anagram. As each hypothesis fails to match a word, the next most probable one is tested until a solution is found. An alternative serial model of anagram solution might propose that the second phase involves rearranging the remaining (i.e., non-initial) letters of the anagram and attempting to find a match between the candidate solutions and entries in one's mental lexicon.

Now consider a parallel model based on a connectionist architecture. In one such model (Novick, in progress), the (symbolic) processing units correspond to hypotheses about possible combinations of letters and positions (e.g., D in position 1, denoted here by D1, for the anagram "dnsuo"). Excitatory and inhibitory links (denoted by "<+>" and "<≠>", respectively) between the letter/position units embody constraints on the rearrangement of the letters. Inhibitory links instantiate the constraints that each letter can occupy only a single position (e.g., D1<≠>D2) and each position can contain only a single letter (e.g., D1<≠>N1). Excitatory links enforce English spelling rules by favoring bigrams that are more common in the language (e.g., U1<+>N2 would be greater than U1<+>S2). Constraints on individual letter/position units also can be modeled. Such constraints might include a bias to begin words with consonants, to put vowels in the middle of a word, and to keep the letters in the same positions in which they occur in the anagram. Mayzner and Tresselt (1958) provided experimental evidence for the last constraint. The letter/position units accumulate activation over time, as a function of the other units to which they are connected and the weights on the connections, until a steady state is reached. To a first approximation, a solution is achieved if the five units with the highest activations form an English word.

The models just described represent extremes along a continuum of degree of serialism versus parallelism. Hofstadter (1983) has proposed a model of anagram solution that incorporates both types of processing. In this model, letters float in a "cytoplasm" looking for other letters with which to form clusters of increasing size until a word is created. Although the progression from isolated letter to bigram to syllable to word occurs serially, the clustering of units at different levels happens in parallel. Different clusterings are explored to different depths, depending on such constraints as rules of spelling, and this also occurs in parallel. As structures coalesce, the model gradually makes a transition from primarily parallel to primarily serial processing.

Experiments 1a and 1b

Our first goal was to provide scientific evidence for the existence of very fast solutions (i.e., the pop-out phenomenon) and for the association of pop-out with expertise. In addition, for the design of Experiment 2, we needed to identify words that experts were more likely to solve quickly than novices. We will refer to such stimuli as discriminating anagrams.

Method

Subjects. The subjects in Experiment 1a were 17 psychology graduate students and 2 undergraduates who were selected to represent the full range of self-reported ability on a 1 (awful) to 9 (excellent) scale. In the experimental session, subjects completed an objective test of anagram-solving ability (the Scrambled Words Test) in which they were given 10 min to solve 20 difficult five-letter anagrams selected from Arnold and Lee (1973). The subjects in Experiment 1b were 20 undergraduates who were preselected based on their Scrambled Words Test scores: eight high solvers had scores of 12 or more, four intermediate solvers had scores between 7 and 11 inclusive, and eight low solvers had scores of 6 or less.

Materials, design, and procedure. Subjects in Experiment 1a solved 110 core anagrams. An additional 40 filler anagrams that required only one letter move for solution (e.g., "pkoer") were interspersed throughout this list to ensure that all subjects would have some success at the task. Subjects in Experiment 1b solved 120 core anagrams. Some of the anagrams were the same as those used in Experiment 1a. Others were new scramblings of the old words because the earlier results did not enable us to identify a sufficiently large set of discriminating anagrams. Such anagrams are difficult to identify, because numerous anagram problems can be constructed for a word, and the different anagrams are not equivalent in difficulty (e.g., "rcwdo" results in more fast solutions than does "dwcor"). The anagrams were divided into two blocks. Subjects also completed a block of 10 practice anagrams. All anagrams were printed in lower-case letters.

The anagrams were presented one at a time on a computer screen for a maximum of 10 sec each. Subjects pressed a button on a response box as soon as they solved the anagram, and then they reported their solution out loud to the experimenter. The computer recorded the solution time. Feedback was given concerning the correct solution for each trial.

Subjects in both experiments also completed the Concealed Words Test from the Kit of Reference Tests for Cognitive Factors (French, Ekstrom, & Price, 1963). This is a speeded test in which subjects have to identify words that have been partially erased (see Figure 1). French, Ekstrom, and Price indicate that performance on this test is correlated with anagram solution. More important for our purposes, successful performance on the Concealed Words Test would seem to require parallel processing, because each partial letter is ambiguous in isolation, and in fact McClelland and Rumelhart (1981) used their computer model's ability to "read" such items as supporting evidence for the use of parallel processing in the identification of letters.

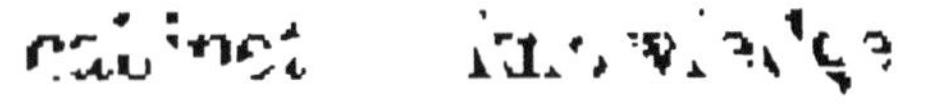

Figure 1. Two examples of the type of item that appears on the Concealed Words Test. The solutions are "cabinet" and "knowledge," respectively.

Results

Because the literature provides no guidance in defining fast solutions, any criterion is somewhat arbitrary. We chose 2 sec as our cut-off time for a fast solution, because below about 2 sec experts get clear intuitions of the solution popping out, whereas above about 2 sec experts get clear intuitions of using a serial strategy. To control for the fact that by definition better anagram solvers solve more anagrams than poorer solvers overall, our measure of proficiency at solving anagrams quickly was the percent of each subject's solutions that occurred in under 2 sec. We will refer to this measure as %rapid.

Not surprisingly, accuracy on our relatively unspeeded test of anagram expertise (the Scrambled Words Test) was highly correlated with accuracy on the core anagrams in our speeded experimental task: $r = .79, p < .01$, and $r = .54$, $p < .05$, for Experiments 1a and 1b, respectively. More importantly, expertise was highly correlated with *speed* of solution of the experimental items, as defined by the %rapid measure: $r = .71$, $p < .01$, and $r = .46$, $p < .05$, respectively, for Experiments 1a and 1b. Although the correlation between expertise and fast solutions does not illuminate directly the issue of strategy use, the data from the Concealed Words Test provide evidence consistent with the hypothesis that experts are more likely to use a parallel solution strategy than are novices. In both experiments, accuracy on this test was highly correlated with the %rapid measure: $r = .61$ and $r = .57$ for Experiments 1a and 1b, respectively, both $p < .01$.

Experiment 2

Gathering experimental evidence for parallel versus serial processing is difficult, because typically it is possible to construct models of the two types that mimic each other. Nevertheless, it is important to distinguish the two models (Townsend, 1990). The purpose of Experiment 2 was to begin to explore the consequences of the hypothesized greater frequency of parallel processing of solution constraints among experts than novices. We used a methodology described by Townsend in which subjects' processing was interrupted prior to completion. After any fixed amount of time, more partial information will be available to those engaged in parallel rather than serial processing, because in the former case multiple pieces of information are processed simultaneously, whereas in the latter case processing proceeds sequentially. For example, assume that processing of a five-letter anagram is interrupted after N ms. With parallel processing, one has N ms of partial information on each letter, including information on the various constraints concerning the locations of the letters and bigrams. With serial processing, however, one has either N ms of partial information on one letter or $N/5$ ms of partial information on each letter (or something between these two extremes; in any case, less information is available than with parallel processing). Information about the constraints on the letter positions would be sparser, because that information cannot be accessed until all of the letters have been encoded.

Method

Solvability judgment task. The subject's task was to judge whether a string of letters could be unscrambled to form an English word (e.g., "dnsuo" forms "sound," but "rusyb" does not form a word no matter how the letters are rearranged). The stimuli were presented at one of three display durations: short, intermediate, or long. Even at the relatively long durations, however, the time allotted for processing was so brief that most subjects were not expected to be able to solve the anagrams. Thus subjects were forced to make their solvability judgments based on partial information. Because the unsolvable items were very similar to the anagrams, considerable information about the possible positions of *all* letters would be needed to distinguish the two types of stimuli. If subjects have considerable information on only a few of the letters, or very little information on all letters, their performance should be near chance level.

Our hypothesis of greater parallel processing among experts than novices leads to the prediction that high solvers will be more accurate than low solvers at distinguishing solvable from unsolvable items. At the shortest display durations, we expect that the low solvers' performance will be at or near chance. In contrast, high solvers may be above chance even at the short durations.

Materials. Based on the results of Experiments 1a and 1b, we chose 30 anagrams that discriminated high and low solvers in terms of solution time, with pop-out being more likely for the high solvers. The items were selected to meet the following criteria: (a) the correlation between solution time and expertise was at least as extreme as -.24 (M = -.39, ranging from -.24 to -.67), (b) there was a gap in the solution time distribution of at least 200 ms at or before 2 sec (M = 476 ms, ranging from 204-1313 ms), and (c) the difficulty of solution was moderate, defined as 20-65% of subjects solving the anagram at or below the lower limit of the break in the solution time distribution (see (b) above; henceforth referred to as the rapid solution cut-off time; M = 41%). In addition to the 30 experimental anagrams, there were 45 training and 15 warm-up anagrams.

Each anagram was matched to a "nonanagram" (i.e., an item that could not be unscrambled to form an English word; e.g., "clnai"), which was constructed as follows: First, a word was selected that began with the same letter as the anagram. Then, one letter of that word was replaced by another letter (vowel for vowel or consonant for consonant) such that the resulting set of five letters could not be rearranged to form a word. The letter that was substituted met the restriction that its frequency of occurrence in English differed from that of the replaced letter by a ratio of no more than 2:1 (see Pratt, 1942). Finally, the letters of the nonword were scrambled such that the absolute difference between the summed bigram frequency (SBF) of the nonanagram and that of its matched anagram was no more than 25 points (M = 10.38). The SBF of "clnai," for example, is 48, which is obtained by summing the frequency of CL in positions 1 and 2, LN in positions 2 and 3, etc. (see Mayzner & Tresselt, 1965).

Subjects, design, and procedure. The subjects were 30 undergraduates who were preselected based on their Scrambled Words Test scores: 15 high solvers had scores of 12 or more, and 15 low solvers had scores of 6 or less. We crossed the two levels of expertise with three levels of display duration. Because the rapid solution cut-off times differed for the 30 experimental anagrams (ranging from 1318-1999 ms), the short, intermediate, and long display durations were defined as percentages of the cut-off times: 45% (M = 803 ms), 70% (M = 1249 ms), and 85% (M = 1517 ms), respectively. Three stimulus lists were constructed so that each subject would see a given experimental item at only a single display duration, but across lists each item would appear at each of the three durations. The nonanagram display durations were yoked to their matched anagram times.

A deadline procedure was used to force subjects to make their solvability judgments based on partial information. The stimulus was displayed for a predetermined duration. Coincident with the offset of the stimulus, a beep sounded indicating that subjects were to respond. If subjects did not respond within 250 ms, another beep sounded to indicate the end of the response period. Responses were recorded up to 125 ms after the second beep. Feedback on each trial included the response time for that trial, the average response time, and the solution (or "not a word" for the nonanagrams). Subjects completed 90 training trials prior to the experimental items. Responses to only 5% of the items were lost due to subjects failing to respond within the 375 ms deadline.

Results

Our primary measure of performance was d', a sensitivity measure from signal detection theory. Applying the theory to our task, we assume that each stimulus yields an impression of solvability at some value along a "sensory" continuum. Then, d' is defined as the distance between the means of the distributions for the anagrams and the nonanagrams (in standard score units). The more sensitive subjects are to the solvability of the items, the farther apart their two distributions and the higher their d' scores. A d' of 0 indicates chance performance.

A d' score was computed for each display duration for each subject. An ANOVA on these data indicated that high solvers were more sensitive to item solvability than were low solvers, $F(1, 28) = 18.65, p < .001$. In addition, sensitivity increased as display duration increased, $F(2, 56) = 2.88, p < .07$. There was no interaction between level of expertise and display duration, $F(2, 56) < 1$. The mean d' scores are shown in Table 1. For the high solvers, all of the means were reliably above chance. In contrast, low solvers' performance was reliably above chance only at the longest (85%) duration. In further support of the hypothesis that good performance on the solvability judgment task involves parallel processing, mean d' scores (collapsed across display duration) were positively correlated with Concealed Words Test scores, $r = .43, p < .05$.

	Sensitivity (d')			
Exposure	High Solvers		Low Solvers	
Duration	M	SD	M	SD
45%	0.79	0.84	0.20	0.55
70%	1.17	0.73	0.31	0.84
85%	1.34	0.68	0.43	0.64
mean	1.10	0.57	0.31	0.42

Table 1. Sensitivity to Solvability (d') at Each Display Duration for High and Low Solvers.

We also analyzed subjects' criteria for choosing a response. "Sensory" values above and below the criterion lead to responses of "solvable" and "not solvable," respectively. We used C as a measure of the location of the response criterion, because it is independent of d' (Snodgrass & Corwin, 1988). Mathematically, C is the distance (on the sensory continuum) of the criterion from the intersection of the anagram and nonanagram distributions. Positive C scores indicate a strict criterion (bias to respond "not solvable"), and negative scores indicate a lenient criterion (bias to respond "solvable"). C = 0 means that the subject is unbiased. An ANOVA indicated a marginally reliable effect of expertise, $F(1, 28) = 3.49, p < .08$, with high solvers setting a stricter criterion than low solvers (see Table 2). There also was a main effect of display duration, $F(2, 56) = 3.79, p < .03$, indicating that response criteria became stricter as display duration increased. It makes sense that subjects would require more evidence before deciding that an item is solvable if they already have been working for a relatively long time without getting an answer. There was no interaction between level of expertise and display duration, $F(2, 56) < 1$.

Exposure Duration	Response Criterion (C) High Solvers M	SD	Low Solvers M	SD
45%	0.00	0.60	-0.13	0.64
70%	0.39	0.50	0.02	0.39
85%	0.36	0.43	0.08	0.58
mean	0.25	0.38	-0.01	0.40

Table 2. Response Criterion (C) at Each Display Duration for High and Low Solvers.

Discussion

The research reported here has been conducted within the framework of the third generation of expertise theories (Holyoak, 1991), which proposes that expert performance is based on parallel processing of solution constraints, in contrast to the serial processing of novices. Our data on anagram solution are consistent with this hypothesis. First, we provided evidence of a pop-out phenomenon of very fast anagram solution (within 2 sec) that is highly correlated with expertise. Second, when subjects had to judge whether letter strings were solvable based on incomplete processing of those strings, performance was directly related to expertise. In fact, at the shorter display durations, only the more expert solvers performed reliably above chance. Finally, both the frequency of pop-out solutions and the ability to discriminate solvable and unsolvable stimuli based on partial information were highly correlated with scores on the Concealed Words Test, which prior research would suggest requires parallel processing for good performance.

In sum, the research reported here has provided evidence concerning fundamental processing differences as a function of expertise. Clearly, any attempts to facilitate the acquisition of expertise must be based on a solid understanding of what is to be acquired. Although anagram solution probably is not a skill at which most people wish to become expert, it is an excellent domain for testing the hypothesis under consideration. Once the nature of processing differences is understood in simpler domains (such as anagram solution), the work can be extended to more knowledge-intensive and "messier" domains, such as physics problem solving or medical decision making.

References

Arnold, H., & Lee, B. (1973). *Jumble #7*. NY: The Berkeley Publishing Group.

French, J. W., Ekstrom, R. B., & Price, L. A. (1963). *Kit of reference tests for cognitive factors*. Princeton, NJ: Educational Testing Service.

Hofstadter, D. R. (1983). The architecture of Jumbo. *Proceedings of the International Machine Learning Workshop* (pp. 161-170). Monticello, IL.

Holyoak, K. J. (1991). Symbolic connectionism: Toward third-generation theories of expertise. In K. A. Ericsson & J. Smith (Eds.), *Toward a general theory of expertise: Prospects and limits* (pp. 301-335). NY: Cambridge University Press.

Mayzner, M. S., & Tresselt, M. E. (1958). Anagram solution times: A function of letter order and word frequency. *Journal of Experimental Psychology, 56,* 376-379.

Mayzner, M. S., & Tresselt, M. E. (1965). Tables of single-letter and digram frequency counts for various word-length and letter-position combinations. *Psychonomic Monograph Supplement, 1,* 13-32.

McClelland, J. L., & Rumelhart, D. E. (1981). An interactive activation model of context effects in letter perception: Part 1. An account of basic findings. *Psychological Review, 88,* 375-407.

Mendelsohn, G. A. (1976). An hypothesis approach to the solution of anagrams. *Memory & Cognition, 4,* 637-642.

Newell, A., & Simon, H. A. (1972). *Human problem solving*. Englewood Cliffs, NJ: Prentice-Hall.

Novick, L. R. (in progress). [Alfred: An implementation of parallel constraint satisfaction as a model of anagram solution by experts].

Pratt, F. (1942). *Secret and urgent: The story of codes and ciphers*. Garden City, NY: Blue Ribbon Books.

Rumelhart, D. E., McClelland, J. L., & the PDP Research Group. (1986). *Parallel distributed processing: Explorations in the microstructure of cognition* (Vol. 1). Cambridge, MA: MIT Press.

Snodgrass, J. G., & Corwin, J. (1988). Pragmatics of measuring recognition memory: Applications to dementia and amnesia. *Journal of Experimental Psychology: General, 117,* 34-50.

Townsend, J. T. (1990). Serial vs. parallel processing: Sometimes they look like Tweedledum and Tweedledee but they can (and should) be distinguished. *Psychological Science, 1,* 46-54.

Allocation of Effort to Risky Decisions

Kip Smith
Paul E. Johnson
Department of Information and Decision Sciences
Carlson School of Management
University of Minnesota
Minneapolis, MN 55455
kip@vx.acs.umn.edu

Abstract

This research investigates expertise at decision making under risk and the allocation of cognitive effort as risky decisions are made[1]. We conceptualize risk within a space defined by decision variables that managers monitor in their environment. We present a representation of the risk space that captures how foreign exchange traders understand risk in spot currency markets. Results from an experiment with professional traders as subjects show that the risk space explains whether and when traders make decisions to buy, sell, and hold spot positions in foreign currencies. An index of cognitive effort is presented that can be used to predict subjects' level of confidence in their assessments of market behavior. Effort is relatively high when conditions are likely to trigger uncertainty. Effort is relatively low when markets act as expected.

Introduction

We propose that the decision making behavior of managers can be explained in terms of a risk space defined by the decision variables they monitor in their environment. Events are traced and evaluated within the risk space relative to thresholds for acceptable risk (Senders, 1966; Moray, 1986). The thresholds for risk parse the risk space into decision regions. Each region is associated with a specific variety of risk. A representation of the risk space, the phase plane (Phatak & Bekey, 1969), is presented. We propose that the time-trace of decision variables through the phase plane explains the actions managers take to manage risk.

[1] This work was funded by Doctoral Dissertation Fellowships to the lead author from the Graduate School of the University of Minnesota and the Carlson School of Management.

The domain of study is foreign exchange (currency) trading. Subjects in our experiment are professional traders employed by a large regional bank to buy and sell foreign currencies for the purpose of generating profits. They routinely manage the risks associated with exposure to the volatile international currency markets.

Foreign exchange markets are an amalgam of the composite activity of traders around the world. As forums for action by a large group of goal-seeking agents, foreign exchange markets act with a quixotic intentionality that traders continually attempt to assess to infer its future behavior. The assessments or expectations largely drive traders' "pro-active" decision making behavior.

We propose that assessing market intentionality requires and receives deliberate, effortful thought. We hypothesize that associating a continuous measure of cognitive effort with overt decision behavior should reveal that effort is high when the intentionality of the market is hard to fathom and is low when the market behaves in accord with expectations.

In this paper we present one episode from an experimental session to illustrate how the phase plane representation of the trader's risk space explains whether and when subjects implement decisions to manage the risks associated with foreign exchange trading. The scenario also illustrates that cognitive effort varies with and can be used to predict the level of fit between expectations and actual market behavior.

Risk in Foreign Exchange Trading

Traders take "positions" in foreign currencies by buying and selling dollars in exchange for a foreign currency. A position may be "long" dollars or "short" dollars. A trader takes a long dollar position by buying dollars and selling an equivalent amount of foreign currency. Long dollar positions profit when the dollar appreciates against (buys more of) the foreign currency. Taking a short dollar position involves selling dollars and buying a foreign currency. Short dollar positions profit when the value of the dollar declines.

Because foreign currencies are negotiable only in specialized international markets, holding a foreign currency entails assuming the risk of losing equity while owning a non-negotiable asset. For traders, the level of risk is determined solely by the magnitude of the potential loss (Zaheer, 1991). In contrast to the conceptualization of risk presented by behavioral decision theory (e.g., Fischhoff, Slovic, Lichtenstein, Read, & Combs, 1978), probability does not figure into traders' thinking about risk (Zaheer, 1991). Risk increases with the size of a position.

The sources of risk in foreign exchange are (1) changes in the price at which currencies can be bought and sold, (2) the rate at which a price changes, and (3) uncertainty about future market behavior. The actual value of a price has little consequence; it is the change in price that determines profit or loss, and hence, risk.

The direction of price movement is calculated relative to a reference point called the "all-in rate". The all-in rate is the price at which a trader would break even on the position she has taken. This reference point is dynamic. It changes as she makes trades. For instance, taking a profit on a portion of a long position reduces the price at which she would break even on the remaining position. Thus, the reference point against which risk is judged is a function not only of the environment (the market) but also of her trading activity. A trader can manipulate this dimension of risk by taking action in her environment.

The effect of a dynamic reference point is dramatic. Taking a profit makes it easier to continue to take profits. Conversely, taking a loss makes it more difficult to turn a profit.

The second dimension of risk in currency trading is the time rate at which prices change. Currency markets are volatile and have a knack for turning suddenly against a position. Market movements can rapidly outstrip traders' ability to attend and respond. The trader who fails to read a significant headline or who gets caught in a contrary market stands to lose a large sum of money in a very few minutes. This dimension of risk is purely environmental and independent of a trader's actions.

The third dimension of risk is the trader's uncertainty of the markets. We conceptualize a two-step process to assess this uncertainty. The first step is the attribution of intentionality to the market. For example, we can imagine a trader's mental soliloquy (Dennett, 1983): "Strikes are crippling the German economy. I think the market believes the dollar will rise against the mark". The second step in this process is an evaluation of confidence in this attribution of intentionality: "I'm sure that's what they'll do". Confidence about future market behavior reduces the risk of taking a long dollar position against the mark.

There are three varieties of risk in currency markets: (1) paper losses, (2) diminishing gains, and (3) neglected profits. For long positions, the risk of a paper loss appears whenever the price at which a currency can be exchanged falls below the all-in rate. The risk of diminishing gains is opportunity lost that appears whenever the sign (+ or -) of the time rate of price change devalues the position. The risk of neglected profit appears whenever profits are so large that the trader cannot afford not to take them: "You never go broke taking profits".

Different cues signal the three varieties of risk. For long positions, prices below the all-in rate signal the risk of paper losses. Falling prices signal the risk of diminishing gains. Prices well above the all-in rate signal the risk of neglected profit. Traders attend to four quantitative decision variables to assess these risks: the all-in rate, position size, the direction of and the rate of price movements.

Risk Space

We conceptualize the direction and rate of price change as defining a risk space that explains whether and when traders allocate effort. Figure 1 introduces an observer construct called the "phase plane". This representation of risk space was originally designed to serve as a graphic aid to assist agents performing the task of monitoring time-varying phenomena (Phatak & Bekey, 1969; Moray, 1986). The axes of the phase plane are simple qualitative functions of the four quantitative decision variables that traders monitor.

The horizontal axis of the phase plane is defined by the difference between the current price, p_n, and

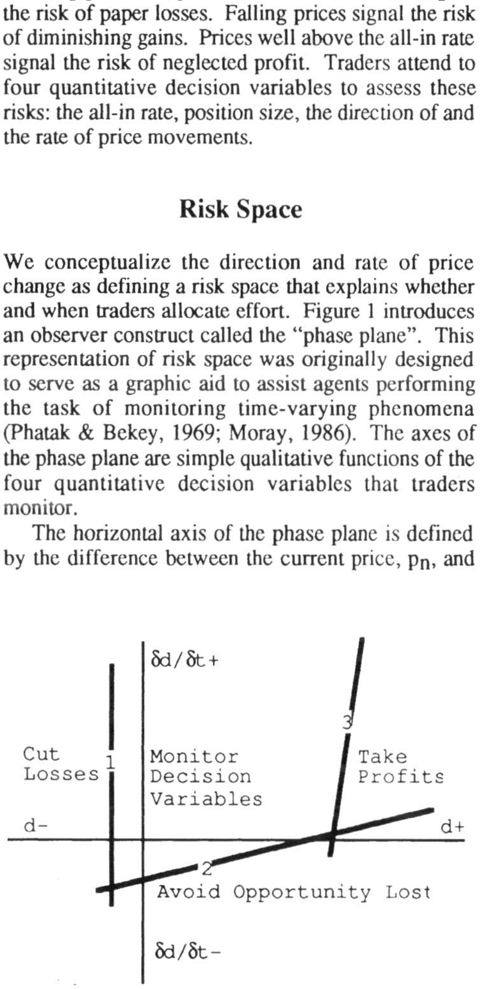

Figure 1 Phase plane representation of risk space in foreign exchange trading

the trader's all-in rate, p_a, and is scaled by the size of her position, S: $d = (p_n - p_a)*S$. As shown in Figure 1, positions profit with positive values of d. This axis quantifies the profit (loss) that would be made by squaring (closing) the position at the current market price. On the job, traders keep track of this dimension of risk with qualitative "ball park" estimates that they make precise when time permits.

The vertical axis of the phase plane is defined by a qualitative calculus that assesses the time rate of price change: $\delta d/\delta t = (p_n - p_{n-1})/(t_n - t_{n-1})$. This axis accounts for the second dimension of risk in the markets: its volatility and seemingly contrary intent. The vertical axis is scaled by heuristics about how fast a price is likely to turn. It captures the risk associated with market behavior that can readily outstrip the trader's ability to attend and respond.

Thresholds for acceptable risk parse the risk space into homogeneous regions associated with specific varieties of risk. Three constraints on a trader's behavior guide the location of these thresholds. Constraint 1 is "cut losses". Taking large losses is bad business. The threshold for cutting losses parallels the vertical axis at some small value of paper losses. Market events that drive a position to the left of this threshold signal the risk of paper losses. Constraint 2 is "avoid opportunity lost". Traders trim profitable long positions when they consistently lose value at an unacceptable rate, $\delta d/\delta t-$. The threshold for opportunity lost is sub-parallel to the horizontal axis. Market events that drive a position below this threshold signal the risk of diminishing gains. Constraint 3, "take the money and run", is a variation on avoiding opportunity lost. Traders generally lock-in a proportion of their paper profits when d+ reaches heady proportions. The threshold for reaping a large gain is sub-parallel to the vertical axis at some large value of paper gain. Market events to the right of this threshold signal the risk of neglected profits.

The three thresholds bound a decision region associated with the action of monitoring information. Events that fall in this region trigger the default action: don't trade, hold the position, and focus attention on market events. If a trader is responsible for more than one position, events that place one currency in this region are likely to lead her to shift attention to another currency.

Experiment

The decision making task studied in the experiment is trading in the spot (delivery in three days) foreign exchange markets. Three traders from the foreign exchange department of a large customer-oriented Twin Cities bank agreed to serve as subjects. Subjects have four to twelve years experience at the trading desk and are responsible for their bank's exposure to foreign currencies. Their task in the experiment is to generate "profits" by trading a pair of currencies simultaneously as a market scenario unfolds. The illusory nature of these profits is likely to be the most serious constraint to the generality of the experimental results.

Eight trading scenarios were constructed. Each scenario presents a realistic but hypothetical global socioeconomic setting based on historical precedent. During each scenario, market events - news and price changes - unfold as they do on the traders' job. The platform for experimentation that simulates their work environment is Treasury Risk Manager (© 1987–92 Chisholm Roth & Co Ltd), a commercial PC-based training package that emulates the Reuters on-line foreign exchange dealing screen. While Treasury Risk Manager simulates spot, forward, and money markets, this research focuses on spot markets where traders routinely encounter time pressure. The scenarios were presented to subjects individually in random order over two to four days.

Measures and apparatus

We gather two types of data to ascertain how traders think about the risks associated with foreign exchange. Videotape recordings and concurrent verbal reports (protocols) reveal overt decision making behavior. This behavior includes shifting the focus of attention among several information displays, answering telephone requests to quote prices (quoting), contacting other traders to ask them to quote prices (dealing), and executing trades. The salient measure of overt behavior is profit or loss.

Spectral analysis of heart-rate variability reveals a covert aspect of risky decision making: the level of cognitive effort (Kahneman, 1973; Vicente, Thornton, & Moray,1987). The amplitude of the 0.10 Hz component of the power density spectrum of the heart-beat interval time series is inversely proportional to the on-going level of cognitive effort (Mulder, 1980). The amplitude of the power spectrum at a given frequency is directly proportional to variance - amount of information - in the time series at that frequency (Bingham, Godfrey, & Tukey, 1967). These complementary lines of evidence test the explanatory power of the phase plane construct. We calculate an index of cognitive effort by subtracting the total estimate of variance between 0.07 and 0.14 Hz from an arbitrary large number.

We assume that an increase in the index of effort in the absence of physical exertion or other sources of stress reflects an increase in cognitive effort. While these conditions are met in our experiments, spectral analysis of cardiac arrhythmia is currently a

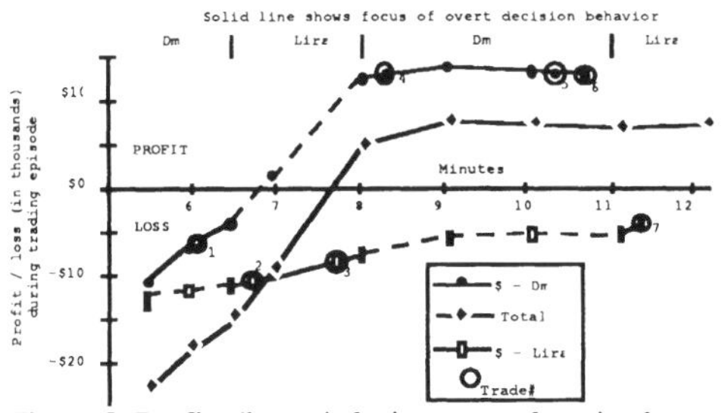

Figure 2 Profits (losses) during example episode

controversial index of cognitive effort (Hancock, Meshkati, & Robertson, 1985).

Results

We present an episode from one experimental session to illustrate our understanding of how professional foreign exchange traders allocate effort as they make risky decisions. There are typically three or four such episodes in each session. In this scenario the subject is responsible for positions in both the German mark (Dm) and the Italian lira (L); she has "inherited" two losing positions from a cohort: long $8 (million) against marks and long $7 against lira. The dollar-lira position starts further out of the money than the dollar-mark position. Figure 2 tracks her profits and losses during the episode. Trades are numbered and highlighted with open circles.

Figure 3 traces the index of cognitive effort during the episode. Estimates of the relative amount of effort the subject expends during 40 second intervals are calculated every 20 seconds and plotted midway through the interval. We analyze the graph for trends in the level of effort, not for single point anomalies. Trades are again highlighted with open circles. Text below the effort trace highlights market events that appear to drive the subject's decision making behavior. Text near the trace summarizes the actions she takes.

We begin with a review of the effort and profit data. We then offer an explanation of her decision making behavior by recasting the decision variables in the phase plane.

Effort and decision making

In the five minutes prior to this episode, the subject reads an "overnight report" like that she gets every morning at work. This summary of global socioeconomic conditions and events suggests that the market expects the dollar to be relatively strong against the lira and weak against the mark. The

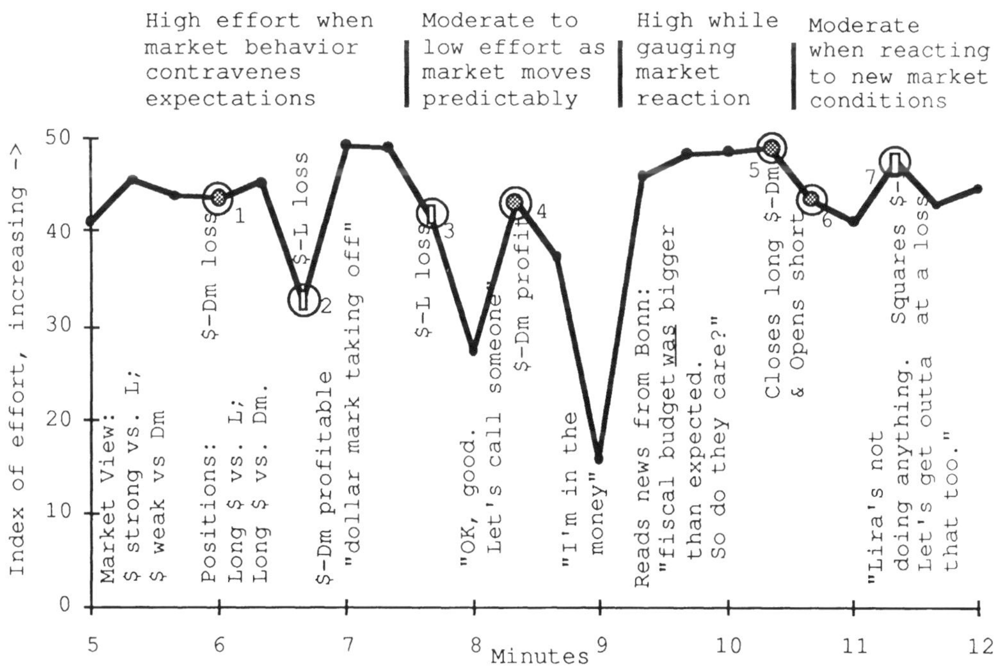

Figure 3 Time trace of cognitive effort during example episode

prevailing wisdom leads her to be more concerned with her dollar-mark position than her dollar-lira position.

The subject faces the risk of paper losses in both positions. Her first act is to trim her long dollar-mark position, taking a small loss in the process. At six minutes into the session, the dollar unexpectedly rallies against the mark while remaining relatively weak against the lira. This price movement prompts the following line of protocol:

```
6:14 Dollar mark kinda rocketing
     here.  What was the news?
```

As shown in Figure 3, the subject responds with a relatively high level of effort. Effort remains high even after the dollar-mark position turns profitable.

By 7:30 (minutes), it becomes clear that the dollar is rallying:

```
7:26 Dollar marks taking off
```

Her level of effort moderates as she takes stock of her positions and the effects of her actions. She has slashed the unprofitable dollar-lira position and held most of the profitable dollar-mark position.

```
7:59 So now I'm long 1 against
     lira ... Plus 1 (L) plus 5
     (Dm) OK good ... Lets call
     someone.
```

The index of effort dips precipitously when the markets provoke the following assessment:

```
9:04 I'm in the money.
```

The effort data suggest the trader may be taking a short mental break when a news item from Germany grabs her attention:

9:19 Bonn - Germany's Fiscal Budget was in surplus by D1.23 bln in September, compared with a surplus of D1.04 in the previous month, Finance Ministry said.

```
9.23 fiscal budget was bigger
     than expected
9:30 so do they care?
9:50 It was kinda expected to be
     there so
9:55 well its (Dm) picking up a
     little well [Deals]
```

The headline is a scheduled piece of news discussed in the overnight report. The budget figure comes in slightly above market expectations. This news may

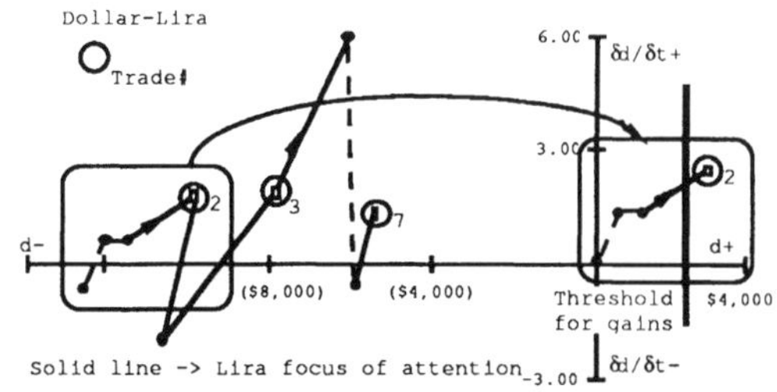

Figures 4 Subject's position in dollar-lira recast in the risk space. Trading behavior is triggered when position crosses threshold for gains.

be bearish for the dollar. The subject faces a quandary. Is the market going to move on the news against her profitable dollar-mark position? Her response to the headline is a high level of effort as she focuses on her dealing screen to track the rate of movement in the mark. When she sees that the mark is "picking up" - moving against the dollar - she calls out to take profits. She executes a pair of trades that yield substantial profits. Effort moderates. It is relatively easy to take profits in predictable markets, even when they are predictably contrary.

The risk space

Figures 4 and 5 transforms the profit data of Figure 2 into the phase plane representation of the risk space to trace the time-course of the two positions through the episode. In these figures, the horizontal axis reflects (paper) profits and losses. It intersects the vertical axis at the all-in rate. The vertical axis is a measure of the time rate of price change. Trades are highlighted with open circles. When a position is the current focus of attention, it is shown with a solid line.

The risk space for dollar-lira, Figure 4, illustrates decisions prompted by a position crossing the threshold for gains. The risk space for dollar-marks, Figure 5, illustrates decisions prompted by a position

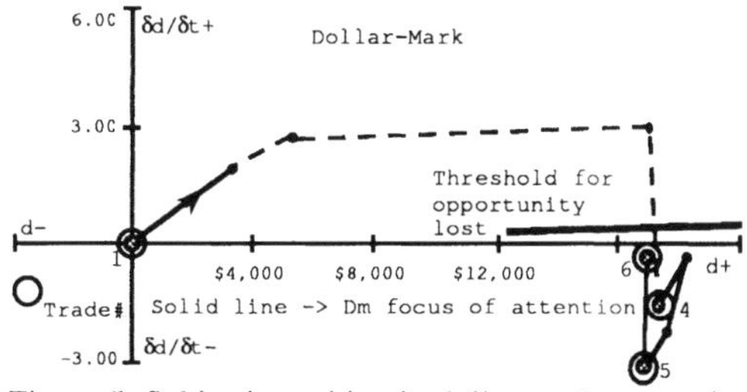

Figure 5 Subject's position in dollar-marks recast in the risk space. Trading behavior is triggered when position crosses threshold of opportunity lost.

crossing the threshold for opportunity lost.

The episode begins as the subject comes to work and inherits two positions from a cohort. Both positions plot in the decision region where the appropriate action is to cut losses; this is what she does.

The timing for cutting losses in the dollar-lira can be predicted by assuming she re-sets her all-in rate to the price at which she inherits the position. As shown in Figure 4, re-setting the all-in rate effectively translates the position to the intersection of axes. She cuts her losses as if she were taking gains wherever the position crosses a threshold for gains near +$2500. Later, when queried about this type of behavior, the subject said "I really started at zero since this was an overnight position". She manages foregone losses as if she were capturing gains.

This episode provides no evidence for a threshold for gains in the dollar-mark. Rather, she allows the dollar-mark to appreciate as she cuts her losses in the dollar-lira. Managing foregone losses is more pressing than taking certain profits.

She shifts attention to the dollar-mark when the rate of change of the dollar-mark price turns negative. This turn from rising to falling prices is reflected in the risk space by an abrupt drop below the horizontal axis. The subject takes profits as the position closes in on the horizontal axis, that is, just before the rate of change can be predicted to turn from positive to negative. She acts just before it appears the mark will reach its peak. Her threshold for acceptable risk of opportunity lost lies above the horizontal axis.

An implication of the risk space representation is that decision variables in the trader's environment define coherent regions of within the risk space. Each decision region is associated with a set of appropriate actions. Thresholds of acceptable risk parse the space and isolate actions. We suggest that key components of expertise at risky decision making are tacit knowledge of (1) thresholds for risk and (2) the dimensions of risk that place current values of decision variables within a decision region.

Conclusions

Two points emerge from the example episode. First, experienced risk managers monitoring time-varying signals act when those signals approach or cross thresholds for acceptable risk (Senders, 1966; Moray, 1986). This result suggests that one component of expertise at risky decision making is judgment that compares signals generated by the environment to thresholds of acceptable risk.

The second point is that cognitive effort appears to vary in response to the level of confidence the manager has in her understanding of the environment. The index of cognitive effort may provide insight into decision making about uncertainty.

References

Bingham, C.; Godfrey, M.D.; and Tukey, J.W., 1967. Modern techniques of power spectrum estimation. *IEEE Transactions on Audio and Electroaccoustics*, AU-15, No. 2, p. 56-66.

Dennett, D.C., 1983. Intentional systems in cognitive ethology: The "Panglossian paradigm" defended. *The Behavioral and Brain Sciences*, 6, 343-390

Fischhoff, B.; Slovic, P.; Lichtenstein, S.; Read, S.; and Combs, B., 1978. How safe is safe enough? A psychometric study of attitudes towards technological risks and benefits: *Policy Sciences*, 9, 127-152.

Hancock, P.A., Meshkati, N., & Robertson, M.M., 1985. Physiological reflections of mental workload. *Aviation, Space, and Environmental Medicine*, 1110-1114.

Kahneman, D., 1973. *Attention and Effort.* Englewood Cliffs, NJ: Prentice-Hall.

Moray, N., 1986. Monitoring behavior and supervisory control. In K.R. Boff, L. Kaufman, and J.P. Thomas (Eds.) *Handbook of Perception and Human Performance*, 2, 40. New York: Wiley.

Mulder, G., 1980. The heart of mental effort. Studies in the cardiovascular psychophysiology of mental work. Ph.D. diss., University of Groningen, Groningen, Netherlands.

Phatak, A.V.; and Bekey, G.A., 1969. Decision processes in the adaptive behavior of human controllers. *IEEE Transactions of Systems, Science and Cybernetics*, SSC-5, 339-352.

Senders, J.W.,1964. The human operator as a monitor and controller of multidegree of freedom systems. *IEEE Transactions on Human Factors in Electronics*, HFE-5, 2-5.

Vicente, K.J.; Thornton, D.C.; and Moray, N., 1987. Spectral analysis of sinus arrhythmia: A measure of mental effort. *Human Factors*, 19, 2, 171-182.

Zaheer, S.A., 1991. Organizational context and risk-taking in a global environment: A study of foreign exchange trading rooms in the U.S. and Japan. Ph.D. diss., Sloan School of Management, Massachusetts Institute of Technology.

A Constraint Satisfaction Model of Cognitive Dissonance Phenomena

Thomas R. Shultz
Department of Psychology
McGill University
1205 Penfield Avenue
Montréal, Québec, Canada H3A 1B1
shultz@psych.mcgill.ca

Mark R. Lepper
Department of Psychology
Stanford University
Jordan Hall, Building 420
Stanford, CA 94305-2130
lepper@psych.stanford.edu

Abstract

A constraint satisfaction network model simulated cognitive dissonance data from the insufficient justification and free choice paradigms. The networks captured the psychological regularities in both paradigms. In the case of free choice, the model fit the human data better than did cognitive dissonance theory.

Cognitive Dissonance

Cognitive dissonance theory (Festinger, 1957) has been a pillar of social psychology for some 30 years. The theory holds that dissonance is a psychological state of tension which people are motivated to reduce. Two cognitions are dissonant when, considered by themselves, one of them follows from the obverse of the other. The amount of dissonance is a function of the ratio of dissonant to consonant relations, with each relation weighted by its importance. Dissonance can be reduced by decreasing the number and/or the importance of the dissonant relations, or by increasing the number and/or the importance of consonant relations. How dissonance gets reduced depends on the resistance to change of the relevant cognitions, with less resistant cognitions being more likely to change. Resistance derives from the extent to which change would produce new dissonance, the degree to which the cognition is anchored in reality, and the difficulty of changing those aspects of reality.

Festinger (1957) used dissonance theory to account for a number of existing psychological phenomena, including the evaluation of choices, attitude change following attitude-relevant actions, and responses to the disconfirmation of beliefs. It has since been successfully applied in a wide variety of both predictive and postdictive contexts.

Consonance Model

In this paper, we present a computational model of cognitive dissonance. The model is based on the idea that dissonance reduction is a constraint satisfaction problem. Such problems are solved by the simultaneous satisfaction of many soft constraints which can vary in their relative importance. In this framework, beliefs are represented as units in a network and implications among the beliefs are represented as connections among the units. The units can be variously active and the connections (weights) can vary in strength. Hopfield (1982, 1984) has worked out the mathematics for solving such constraint satisfaction problems in parallel networks.

Hopfield networks are capable of simulating a variety of psychological phenomena, including belief revision, explanation, schema completion, analogical reasoning, and content-addressable memories (Holyoak & Thagard, 1989; Rumelhart, Smolensky, McClelland, & Hinton, 1986; Thagard, 1989). Unless used to model memory, these networks are generally considered ephemeral in the sense that they are created on line to deal with some particular task, although the creative process is not usually modeled. Hopfield networks function by reducing energy (equivalently, maximizing goodness) subject to the constraints supplied by the connections and any external input. Our Consonance Model for reducing cognitive dissonance is a Hopfield network lacking some of the parameters of other Hopfield networks and introducing some special parameters of its own.

Maximizing the consonance (goodness) of any pair of connected units depends on the sign of the connection between them. Assume an activation range of 0 to 1. If connected by a positive weight, both units should be active in order to maximize consonance. With a negative weight, consonance is maximized when both units are not active, that is, when both are inactive or only one is active. Activations change over time cycles so as to satisfy weight constraints and maximize consonance.

More formally, the consonance contributed by a particular unit i is

$$consonance_i = \sum_j w_{ij} a_i a_j \qquad (1)$$

where w_{ij} is the weight between units i and j, a_i is the activation of unit i, and a_j is the activation of unit j.

The overall consonance in the network is the sum of the values given by (1) over all units in the network

$$consonance_o = \sum_i \sum_j w_{ij} a_i a_j \quad (2)$$

Activation spreads over time cycles by two simple update rules:

$$a_i(t+1) = a_i(t) + net_i (ceiling - a_i(t)) \text{ when } net_i >= 0 \quad (3)$$

$$a_i(t+1) = a_i(t) + net_i (a_i(t) - floor) \text{ when } net_i < 0 \quad (4)$$

where $a_i(t+1)$ is the activation of unit i at time t + 1, $a_i(t)$ is the activation of unit i at time t, ceiling is the maximal level of activation, floor is the minimal activation, and net_i is the net input to unit i, defined as

$$net_i = resist_i (\sum_j w_{ij} a_j) \quad (5)$$

The parameter $resist_i$ is a measure of the resistance of unit i to having its activation changed. The larger the value of the resistance multiplier, the less the resistance to change. The default values for floor and ceiling are 0 and 1, respectively.

At each time cycle, n units are randomly selected and updated according to rules (3) and (4). By default, n is the number of units in the network.

A few additional parameters concerning the construction of the networks are described later in the context of particular simulations.

Simulations

With more than 1000 published entries in the cognitive dissonance literature, there is considerable choice in deciding what to simulate. Here we present two of our current simulations, one representing each of two of the major paradigms in dissonance theory: insufficient justification and free choice.

Insufficient Justification

The insufficient justification paradigm deals with situations in which subjects engage in some counter-attitudinal action with rather little justification. Dissonance theory predicts that the less the justification for the behavior, the greater the dissonance and, at least when it is difficult to retract one's action, the more people will be motivated to change their attitudes so as to provide additional justification for their action.

Several different types of experiments have been developed to test these insufficient justification predictions (e.g., Aronson & Carlsmith, 1963; Aronson & Mills, 1959; Festinger & Carlsmith, 1959). In the present paper we simulate one of the best studied and most robust of these.

In one of the seminal studies within this paradigm, nursery school children were forbidden to play with a desirable toy under either mild or severe threat (Aronson & Carlsmith, 1963). Both of these threats were sufficient to prevent the children from playing with the desirable toy during a play period in which the experimenter was absent from the room. In subsequent ratings, the children derogated the forbidden toy more under mild threat than severe threat. The theoretical explanation is that the children committed themselves to the dissonant behavior of not playing with the desirable toy. Since dissonance increases with the fewer cognitions that support the behavior, there was more dissonance in the mild threat condition than in the severe threat condition. Because the counter-attitudinal behavior could not be retracted, dissonance was reduced by derogating the forbidden toy. The greater the dissonance, the greater the derogation.

Alternative explanations of these findings included the notion that severe threat focused more attention on the toy or made it seem more desirable and the idea that the experimenter was more likeable or more credible in the mild threat condition. To rule out such alternatives, Freedman (1965) added surveillance conditions to the experiment in which the experimenter stayed in the room while the child played. In the surveillance conditions, the same threats were used but temptation, and thus dissonance, was lowered by the experimenter's continued presence. Actual play with the previously forbidden toy five weeks later indicated greater derogation in the mild than in the severe conditions only when there was no surveillance, thus supporting the dissonance explanation against the alternatives.

Our simulation focused on the Freedman (1965) experiment. The constraint satisfaction network for the non-surveillance conditions of this simulation is presented in Figure 1. Because unit activations have a floor of 0, two units are used to encode each dimension of interest: *toy evaluation*, *threat*, and *play with toy*. In each pair of units, the unit coded + represents the positive end of the dimension and the unit coded - represents the negative end of the dimension. The units in each pair are connected by a negative weight so that only one of them is active at a time. In these network diagrams, negative weights are symbolized by dashed lines and positive weights by solid lines. Each pair of units is surrounded by an ellipse to convey idea that they refer to opposite ends of the same dimension.

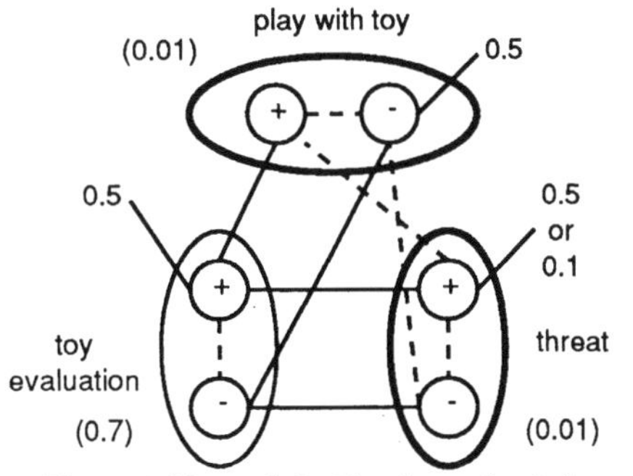

Figure 1. Network for Freedman simulation, non-surveillance condition.

Connections across different dimensions (ellipses) reflect assumed psychological implications among the beliefs. For simplification, we connect positive units only to other positive units and negative units only to other negative units across dimensions. For the Freedman simulation, there were positive connections between toy evaluation and play (the better liked the toy, the more it would be played with), positive connections between toy evaluation and threat (the better liked the toy, the more threat would be required to prevent play), and negative connections between play and threat (the bigger the threat, the less the toy would be played with).

Resistance of units to activation change is portrayed by the thickness of the ellipse. Resistance values for a particular boundary thickness are presented in parentheses in Figure 1: 0.70 for toy evaluation (low resistance) vs. 0.01 for the other two beliefs (high resistance). These resistance values are based on the assumption that, whereas play and threat are relatively fixed, evaluation of the toy should be allowed to vary. In a more complete model, resistance might be implemented by constraining connections to many other beliefs. For simplification, this can be accomplished with an explicit resistance parameter.

Initial activations provided to units are indicated in Figure 1 by pointers coming from outside the units. The toy is given a moderately positive evaluation (0.5) to reflect its desirability, play is given a moderately negative (-0.5) evaluation because it was not done, and the amount of threat is either 0.5 or 0.1 to represent the two severity conditions.

A cap parameter, when set to a high negative proportion, prevents activations from growing to the ceiling of 1.0. Our default setting for cap is -0.8. Mathematically, cap is the value of the connection between each unit and itself, w_{ii}.[1] Hopfield (1982, 1984) had assumed that such self-connections are 0. Allowing self-connections to be other than 0 produces additional spurious states in the neighborhood of a desired attractor, thus increasing the variability of solutions (Hertz, Krogh, & Palmer, 1991). We use cap to enforce the psychologically realistic assumption that the events in most dissonance experiments are not of major importance to the subjects. Therefore, activations should not reach maximal values.

The wrange parameter represents the range of positive weights below 1 and negative weights above -1. We employ a default value of 0.2 for wrange. The weights are not identical across networks, but rather are mainly positive or mainly negative within this specified range. Again, the purpose is to introduce some degree of psychological realism. Such variation is not necessary to qualitatively capture the predicted dissonance phenomena. This randomization of weights violates the symmetry assumed by Hopfield (1982, 1984), in that $w_{ij} <> w_{ji}$. He reported that violations of the symmetry assumption increased memory errors and instability in network solutions. Such results may correspond to psychological variation.

Use of the cap and wrange parameters effectively nullifies the mathematical guarantee that these nets will maximize consonance. It is our view that psychological plausibility should outweigh guaranteed maxima in the context of simulating human data.

The rand% parameter was defined by default as wrange/2. It represents a random percentage added to or subtracted from the initial values of activations, resistances, and caps. This too was for psychological realism; presumably not everyone has precisely the same parameter values.

For the surveillance condition, there was no connection between toy evaluation and play, represented by weights of 0. No matter how much you like toy, you won't be tempted to play with it as long as the experimenter is present. The impact of both threats was scaled up by a multiplier in the spirit of update rule (3): new_threat = old_threat + (0.5 * (1 - old_threat)). This made the value of threat 0.75 in the severe/surveillance condition and 0.55 in the mild/surveillance condition. This reflects the idea that surveillance enhances the value of both threats, but in accordance with the way that activations change.

As a simulation begins, activations of units are updated in a random, asynchronous fashion. On each time cycle, n units are randomly selected and updated using rules (3) and (4). By default, n is the number of units in the network, 6 in this simulation. Updating

[1]Thanks to Denis Mareschal for this suggestion.

continued for 20 cycles because asymptotes were reached well within that period. We ran 10 networks in each condition.

Mean evaluation of the toy after cycle 20 is shown in Figure 2. This was computed as the difference between activation of the positive unit and the negative unit. As in Freedman (1965), there was an interaction between surveillance and severity of threat, $F(1, 36) = 169.02$, $p < .001$. There was more derogation in the mild than in the severe condition, but this effect was much larger without surveillance.

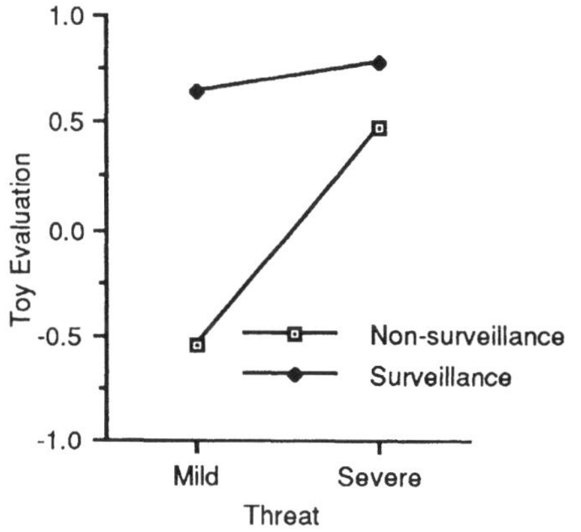

Figure 2. Results for Freedman simulation.

Free Choice

Another major paradigm in cognitive dissonance concerns free choice. Choosing between alternatives creates cognitive dissonance due to the fact that the chosen alternative is never perfect and the rejected alternative often has desirable aspects which are foregone when a final choice is made. Dissonance can be reduced either by making the chosen object more desirable or by making the rejected object less desirable. Thus, dissonance reduction further separates the alternative choices in desirability. The magnitude of dissonance is greater the closer the alternatives are in desirability, and hence the more difficult the choice between them is, before the choice is made.

The classic free choice experiment asked female university students to rate eight small appliances (Brehm, 1956). They were then given a difficult choice, between two objects that they had rated high, or an easy choice, between one object they had rated high and one they had rated low. Then the objects were rated again. Degree of separation was measured by subtracting the second rating from the first rating for each object. Although the dissonance theory prediction was for greater separation in the difficult choice condition than in the easy choice condition, most of the actual separation obtained was due to a relatively large decrease in the value of the rejected alternative in the difficult choice condition.

The network for simulating the Brehm experiment is portrayed in Figure 3. There were pairs of units to represent each of the three critical dimensions: chosen alternative, rejected alternative, and decision. There were positive weights between chosen and decision, and negative weights between rejected and decision. The initial activations were 0.5 for chosen, 0.4 for rejected difficult, 0.1 for rejected easy, and 0.7 for decision. There was high resistance for the decision and low resistance for evaluation of the two alternatives, with the default values of 0.01 and 0.7, respectively. Other parameter settings were the same as in the Freedman simulation.

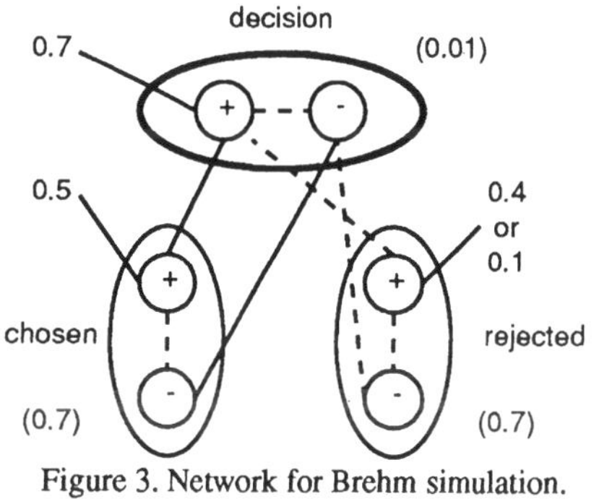

Figure 3. Network for Brehm simulation.

The mean difference scores (re-evaluation - initial evaluation) are plotted in Figure 4. Each evaluation was computed as the difference in activation between the positive and negative units. Evaluation of the chosen object increased and evaluation of the rejected object decreased in both conditions. The amount of change was greater in the difficult condition, as predicted by dissonance theory, $F(1, 18) = 57.70$, $p < .001$. Notice that most of the change in the difficult condition is due to a decrease in evaluation of the rejected alternative. This outcome fits Brehm's (1956) results more precisely than does dissonance theory, which predicts only a larger separation of the alternatives following a difficult choice than following an easy choice.

Discussion

The simulation results matched the psychological findings and, in the case of free choice, provided even better coverage of the psychological data than did dissonance theory. In the free choice simulation, the locus of most of the action was in the re-evaluation of the rejected alternative in the difficult condition. This was indeed what Brehm (1956) found, although he did not comment on the discrepancy from strict dissonance theory predictions.

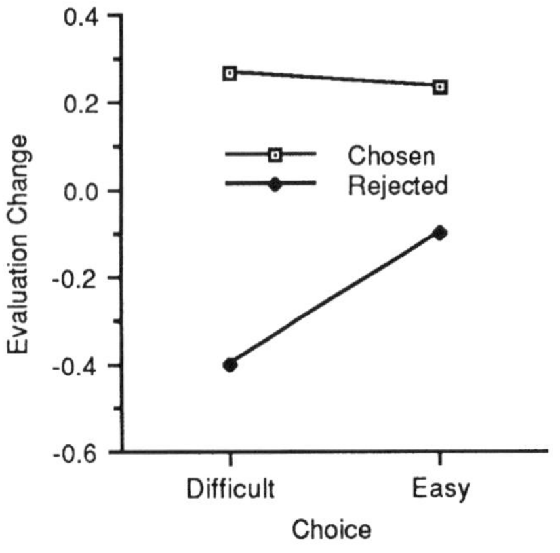

Figure 4. Results for Brehm simulation.

We have found similar instances of the constraint satisfaction model fitting psychological data from other dissonance experiments better than dissonance theory does. These superior fits derive from the capacity of constraint satisfaction models to deal with variables other than those unique to dissonance theory and the increased precision that is inherent to computational formulations.

The present simulations were conducted with a minimum of parameter adjustment. Network weights were positive, negative, or zero; resistance was high or low; and initial levels of activation were either high or low. Additional experimentation revealed that these effects were robust against parameter variation, and that the default parameter settings were applicable to a variety of other dissonance simulations.

The present simulations began with some units having initial, non-zero activation values. More conventionally, constraint satisfaction programs start all units at zero activation and provide some units with external inputs. Activations then gradually build up from zero as a function of both external input and internal network input. This conventional scheme did not seem appropriate for cognitive dissonance phenomena because it yielded results indicating a gradual increase in consonance, but no dissonance. To ensure that the networks modeled dissonance, we initialized some unit activations in conformity with procedures in the psychological experiments.

Although connection weight values can be learned for constraint satisfaction models (e.g., Anderson & Mozer, 1981), there was no such learning in the present simulations. This reflects the fact that the typical dissonance experiment is not an occasion for learning. Instead, acculturated, experienced subjects enter a situation in which they commit themselves to some behavior under the influence of a few salient, experimentally engineered cognitions. These cognitions, the behavioral commitment, and existing knowledge act as constraints on the subject's subsequent re-evaluations. Thus, the typical dissonance experiment capitalizes on past learning, but does not involve much in the way of new learning. The principal thing a subject in a dissonance experiment might learn is how he or she feels about something.

Indeed, there is a sense in which cognitive dissonance phenomena are antithetical to learning phenomena. In both contexts, subjects behave in a way that is less than ideal. In some cases, subjects are able to learn to change their behavior to improve their payoff. But that avenue is closed in dissonance experiments by the fact that subjects remain committed to their behavior. Reduction of cognitive dissonance by re-evaluation is an exercise in coping with behavior that cannot be undone.

Cognitive dissonance phenomena have traditionally been considered as distinct from less counter-intuitive psychological phenomena. But since constraint satisfaction models also account for a wide variety of other phenomena, there is considerable scope for novel theoretical unification.

Cognitive dissonance theory is but one of a number of theories in social psychology emphasizing that people try to achieve consistency among cognitions (Abelson, Aronson, McGuire, Newcombe, Rosenberg, & Tannenbaum, 1968; Abelson & Rosenberg, 1958; Heider, 1958). Although these consistency theories have enjoyed considerable success as verbal formulations, the underlying reasoning mechanisms for establishing consistency have not been precisely specified. It may be that connectionist constraint satisfaction models could serve as a general modeling technique and explanatory device in these areas (cf. Holyoak & Spellman, 1991).

Acknowledgements

This research was supported by a grant to the first author from the Social Sciences and Humanities Research Council of Canada and a grant to the second author from the U. S. National Institute of Mental Health.

References

Abelson, R. P., Aronson, E., McGuire, W. J., Newcombe, T. M., Rosenberg, M. J., and Tannenbaum, P. H. eds. 1968. *Theories of cognitive consistency: A sourcebook*. Chicago: Rand McNally.

Abelson, R. P., and Rosenberg, M. J. 1958. Symbolic psycho-logic: A model of attitudinal cognition. *Behavioral Science* 3:1-13.

Anderson, J. A., and Mozer, M. C. 1981. Categorization and selective neurons. In G. E. Hinton and J. A. Anderson eds., *Parallel models of associative memory*, pp. 213-236. Hillsdale, NJ: Erlbaum.

Aronson, E., and Carlsmith, J. M. 1963. Effect of severity of threat on the devaluation of forbidden behavior. *Journal of Abnormal and Social Psychology* 66:584-588.

Aronson, E., and Mills, J. 1959. The effect of severity of initiation on liking for a group. *Journal of Abnormal and Social Psychology* 59:177-181.

Brehm, J. W. 1956. Post-decision changes in the desirability of choice alternatives. *Journal of Abnormal and Social Psychology* 52:384-389.

Festinger, L. 1957. *A theory of cognitive dissonance*. Evanston, IL: Row, Peterson.

Festinger, L., and Carlsmith, J. M. 1959. Cognitive consequences of forced compliance. *Journal of Abnormal and Social Psychology* 58:203-210.

Freedman, J. L. 1965. Long-term behavioral effects of cognitive dissonance. *Journal of Experimental Social Psychology* 1:145-155.

Heider, F. 1958. *The psychology of interpersonal relations*. New York: Wiley.

Hertz, J., Krogh, A., and Palmer, R. G. 1991. *Introduction to the theory of neural computation*. Reading, MA: Addison-Wesley.

Holyoak, K. J., and Spellman, B. A. 1991. If Saddam is Hitler then who is George Bush? Analogical mapping between systems of social roles. Forthcoming.

Holyoak, K. J., and Thagard, P. 1989. Analogical mapping by constraint satisfaction. *Cognitive Science* 13:295-355.

Hopfield, J. J. 1982. Neural networks and physical systems with emergent collective computational abilities. *Proceedings of the National Academy of Sciences, USA* 79:2554-2558.

Hopfield, J. J. 1984. Neurons with graded responses have collective computational properties like those of two-state neurons. *Proceedings of the National Academy of Sciences, USA* 81:3008-3092.

Rumelhart, D. E., Smolensky, P., McClelland, J. L., and Hinton, G. 1986. Schemata and sequential thought processes in PDP models. In D. E. Rumelhart and J. L. McClelland eds., *Parallel distributed processing: Explorations in the microstructure of cognition*, Vol. 2, pp. 7-57. Cambridge, MA: MIT Press.

Thagard, P. 1989. Explanatory coherence. *Behavioral and Brain Sciences* 12:435-502.

A Rational Theory of Cognitive Strategy Selection and Change

Quanfeng Wu and John R. Anderson
Department of Psychology
Carnegie Mellon University
Pittsburgh, PA 15213
wu@psy.cmu.edu, anderson@psy.cmu.edu

Abstract

This paper presents a rational theory of cognitive strategy selection and change in which the cognitive agent in consideration is proposed to be adaptive in choosing the "best" or optimal strategy from a set of strategies available to be employed. The optimal strategy is assumed to maximize the difference between the expected utility of the goal which the selected strategy would lead to and the computational cost associated with achieving this goal. We considered an example of strategy selection and change in computer programming and interpreted the results from a set of experimental studies we had conducted in this domain in the light of this rational framework. We also substantiated our theoretical claims by developing a computer simulation of this example. The simulation was implemented in ACT-R, a cognitive model constrained by rational analysis as well as by experimental data.

Introduction

Cognitive research on strategy selection and change has flourished in the last three decades. Wason (1960) studied the hypothesis-testing behavior of human subjects on a concept attainment task and found that the dominant strategy used by subjects on the task was a kind of "biased" confirmation strategy which had been supposed to be "irrational" according to Popper's (1959) philosophy of science. Wason's research has since promoted a large body of controversy and triggered a series of ensuring studies (Tukey, 1986). Recently, from the perspective of information theory, Hoenkamp (1989) has suggested that "the biased strategy is not necessarily a bad one; moreover, it reflects a healthy propensity of subjects to optimize the expected information on each trial" (p.651). Early in 1960s, Bruner and his colleagues were also among the first to investigate how subjects chose among various strategies in concept attainment (Bruner, Goodnow, & Austin, 1962). Contrary to what Wason's research seemed to reveal, however, they arrived at the conclusion that "in a formal sense it may be said that the subjects in this experiment were seeking to maximize the expected utility of their decisions and in this way to regulate the risk involved in their problem-solving behavior" (p.124). Consequently, according to Hoenkamp and Bruner et al., subjects' performance on concept attainment tasks can in fact be characterized as "rational" or "adaptive" in terms of information or utility maximization.

Recently, there has also been research on strategy selection and change in other cognitive domains—e.g., in memory retrieval (Reder, 1987), in arithmetic (Siegler & Jenkins, 1989), and in programming (Wu & Anderson, 1991). Though conducted in different cognitive domains, these related studies have so far yielded convergent evidence suggesting that subjects are quite adaptive in their strategy selection in that they are highly sensitive to problem types and that they choose appropriate strategies accordingly. Reder & Ritter (1992) have further demonstrated that there is a general tendency of shifting from a computing-on-site strategy to a retrieving-from-memory strategy as subjects practice more and more on problems of a certain type. Thus, taken together, this line of research shows that subjects' strategy selection behavior is not only very sensitive to environmental cues (i.e., problem types) but also strongly influenced by their experience of learning and practicing.

The present paper consists of three parts: First, we attempt to outline a rational theory of strategy selection and change. As mentioned above, the notion of rational strategy choices has existed in the cognitive psychology literature for some time; our objective in this regard is simply to articulate this notion in a more formal manner and to put it in a more general framework of studying human cognition—namely, the Rational Analysis (RA) perspective. Second, we illustrate this rational theoretical framework with an example of strategy selection and change in computer programming. Third, we put forward a simulation model for these results to substantiate some of the theoretical claims we make.

A Rational Framework of Strategy Selection and Change

RA is a new theoretical framework for understanding human cognitive behavior (Anderson, 1990). A fundamental assumption underlying this approach is the *Principle of Rationality* which basically claims that the human cognitive system is adaptive to its informational environment. It is worth noting that the "rationality" here does not mean optimization without any limitation; it merely implies that an individual can achieve an optimal solution only within his or her cognitive or computational constraints. Since the difference between optimization -without-limitation and optimization-with-limitation is somewhat subtle, this issue has in fact caused considerable controversy in evolutionary theory, in economics, as well as in psychology (Simon, 1983).

The idea of human rationality in decision making rooted in economics; in fact, the expected utility model of strategy selection proposed by Bruner et al. (1962) came directly from economics. The traditional notion of rationality in economics, nevertheless, was that there were not cognitive or computational limits on people for decision making. This was an unrealistic assumption of human rationality. In opposing this traditional view, Simon has long been propounding the notion of bounded rationality and arguing that people only adopt satisficing solutions but never optimal ones. Considered that the RA approach only assumes optimization within computational constraints, RA and Simon's notion of bounded rationality can in fact be reconciled (for more detailed discussion, see Anderson, 1990; pp. 246-250).

In Anderson (1990), much attention has been paid to developing RA theories of memory, categorization, causal inference, and problem solving. Recently, Anderson and Kushmerick have also endeavored to develop an RA theory of strategy selection (Ch. 5; Anderson, in preparation). Basically, strategy selection can be conceptualized as choosing among a set of branches in a conceptual tree with each branch leading to a certain goal. For each branch, there would be a probability of success P_i, an utility value for its goal U_i, and a computational cost C_i associated with. Under this characterization, a rational choice among all available strategies would amount to choosing the branch which satisfies the following:

$$\mathrm{Max}\{P_iU_i - C_i \mid i \text{ goes through all available strategies}\}. \quad (1)$$

Note that the set consisting of all available strategies may be dynamic; that is, some new strategies may only become available during the course of problem solving. Thus, (1) is the basic tenet of our rational theory of strategy selection and change presented in a formal fashion; it directly corresponds to the underlying concept of the cost-benefit analysis approach in economics (Varian, 1987). In cognitive psychology, Russo & Dosher (1983) and Payne, Bettman, & Johnson (1988) have similarly argued that human strategy selection involves not only maximizing expected utility but minimizing cognitive effort as well.

Relating to the memory structure of human cognition (e.g., see Anderson, 1983), there are basically two factors involved in the computational cost term C_i in (1): the cost of retrieving from long-term memory (LTM) and the cost of calculation. In other words, to choose and apply a certain strategy, there would be a cost associated with retrieving that strategy from LTM as well as a cost of calculating the details and actions of that strategy. As these two factors are contributing to the same term, we would expect that a tradeoff between these two factors may sometimes be involved in strategy selection and change. We would further expect that the more one practices with a certain strategy, the lower the retrieving cost for that strategy would be, and consequently the more often that strategy would be selected. In fact, this is the shift from calculating-on-site to retrieving-from-memory which Reder & Ritter (1992) observed in their experiments. On the other hand, if an individual is only naive with a certain strategy, we can conceive that in such a case the calculation cost would be smaller than the retrieval cost. In other words, what is already in the individual's current consciousness may influence his or her strategy selection behavior in subsequent situations; in consequence, it can be expected that there be some lateral transfer effect occurring from solving earlier problems to solving later problems for such an individual.

Iterative Strategy Selection and Change in Programming

We had conducted a set of experimental studies on how PASCAL programmers would choose and change iterative strategies in their programming. Since the quantitative results had been published elsewhere (Wu & Anderson, 1991), here we only show some major points of these results. However, we will analyze an episode of strategy selection and an episode of strategy change in some detail since these episodes are very illustrative of our theoretical claims of rational strategy selection and change and that they had not been reported previously.

There are two indefinite looping constructs in PASCAL—namely, the **while...do** and the **repeat...until** constructs. For convenience, these two constructs or strategies will hereafter be referred to as **W**- and **R**-constructs or strategies. To implement any kind of looping program, either construct alone would suffice; nonetheless, in certain cases using the **W**-construct would produce a more concise and well-structured program than using the **R**-construct, and in other cases it is just the opposite. The general principle for choosing between the **W**- and the **R**-constructs is to use the **W**-construct for looping programs where the looping body may not be executed at all and to use the **R**-construct for cases where the looping body must be executed at least once. According to this principle, we can classify looping problems either as **W**-problems, for which it is easier and more natural to use the **W**-construct, and **R**-problems for which it is better to use the **R**-construct. An example of **W**-problem and an example of **R**-problem, which we actually used in our experiments, along with their modal PASCAL solutions are illustrated in Figure 1.

The first of our experiments was to investigate how programmers would choose between the two looping strategies on different types of problems. The subjects involved in the experiment were recruited from CMU; their programming experience ranged from having just finished an introductory programming course in PASCAL to highly skillful (e.g., having more than ten years of programming experience). The results from the experiment turned out to be that there was a minority of the subjects (about 20%) who idiosyncratically used only one type of looping construct over all the problems tested in the experiment while the majority (the rest 80%) did vary their choices of looping strategies on different problems. For these 80% subjects, Figure 2 (a) shows the pattern of their choices on the two types of looping problems. As the figure shows, these subjects were in fact very sensitive to problem types and quite adaptive in choosing appropriate looping strategies; a one-way ANOVA performed on the data for the **W**-strategy in programming revealed that the effects due to problem types were significant. In another experiment, we tried to see whether subjects' performance would deteriorate if they were forced to use an unnatural strategy, i.e., to use the **W**-construct on **R**-problems or *vice versa*. Figure 2 (b) shows the major results from this experiment; statistical analyses revealed that the effect due to experimental manipulations and its interaction with problem types were both significant. Therefore,these results clearly indicated that when the subjects were forced to use a non-perferred strategy their performance in terms of programming time did suffer. As to be shown, this performance deterioration can be accounted for by the higher computational cost associated with the non-perferred iterative strategy.

A W-problem:

Copy a file into another file until a record with its name field as 'END' is reached, but excluding this record.

Its modal solution:
(only main portion shown):

```
• • • • • • • • •
while File1^.Name <> 'END' do
  begin
    Read(File1, TempRecord);
    Write(File2, TempRecord)
  end;
• • • • • • • • •
```

A R-problem:

Copy a file into another file until a record with its name field as 'END' is reached, including this record as well.

Its modal solution:
(only main portion shown):

```
• • • • • • • • •
repeat
    Read(File1, TempRecord);
    Write(File2, TempRecord)
until TempRecord.Name = 'END';
• • • • • • • • •
```

Figure 1. Examples of W-problem and R-problem and their modal PASCAL solutions.

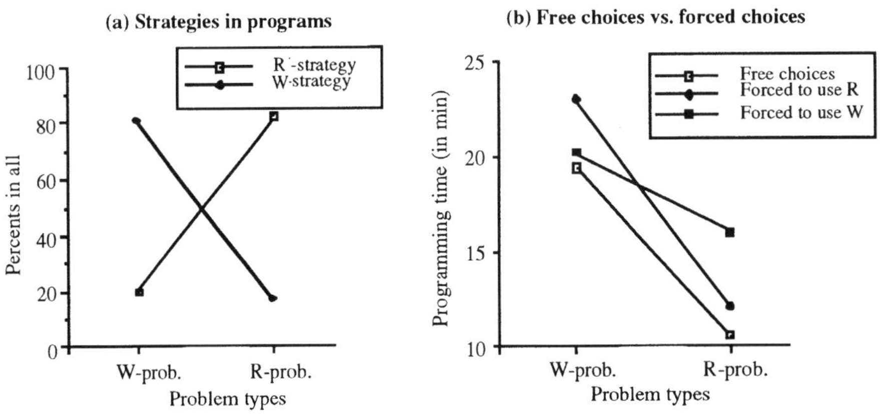

Figure 2. Major results from experiments on iterative strategy selection and change.

Among the 80% adaptive subjects, certainly some were highly adaptive and others were only moderately adaptive (for more details, see Wu & Anderson, 1991). As discussed before, we can expect transfer from solving earlier problems to solving later problems if the subject was only moderately adaptive; furthermore, if the effect was of a negative type, this would amount to the well-known Einstellung effect in problem solving (Luchins & Luchins, 1959). We collected verbal protocols in our experiments, and some programming episodes showed that this type of negative lateral transfer effect did occur in the experiments. Figure 3 presents an example of such transfer effect. The subject involved in this example solved the **W**-problem shown in Figure 1 (a) first; his solution to the problem, however, was not well-structured since he used the **R**-construct instead and an inside-loop **if...then** statement which would be unnecessary otherwise. Simply judging from this solution, we could infer that the subject was a novice programmer. The problem subsequent to the **W**-problem was the **R**-problem shown in Figure 1 (b). When the subject came to this **R**-problem, he (mentally) retrieved his solution to the preceding **W**-problem and only made a minor change of it to fit the new problem. Although on this **R**-problem the subject did use the **R**-construct, his solution was obviously not concise compared to the modal solution shown in Figure 1 (b). This example clearly showed that the subject was negatively transferring what he had done before to what he had to do subsequently, and this negative transfer can easily be interpreted as a result of the subject's attempt to reduce his computational cost, or in other words, to spare his cognitive effort of devising a new but better solution.

For our current example of iterative strategy selection and change, since both the **W**- and the **R**-strategies would lead to working programs, we could suppose that the success probabilities for both strategies are the same, i.e., both equal to one. On the other hand, as the two strategies produce solutions of different styles and of different execution efficiencies, we would assume that they had different utilities. Specifically, we would assume that the strategy leading to more concise, more well-structured, and more efficient programs has a higher utility than the other strategy on a particular type of problems. Do subjects indeed evaluate the utility of their selected strategy? and how would they do this? For those highly-adaptive subjects, they seemed to adopt the preferred strategy at the very beginning of their programming in most cases, and their evaluation of utility seemed to be realized subconsciously without any manifestation in their verbal protocols. On the other hand, for those moderately-adaptive subjects, they usually performed this evaluation very explicitly, and their verbal protocols would have corresponding episodes for the evaluation process. Moreover, in some cases there were strategy changes occurring in such subjects' courses of programming. Figure 4 shows an example of strategy change which reveals the utility evaluation process. The problem was again the **W**-problem shown in Figure 1 (a). The subject first tried to use the **R**-construct; however, having almost completed her first solution, she paused for a while and found that the **if...then** statement would be redundant if using the **W**-construct instead. Upon the reflection, the subject modified her solution to be a **W**-constructed one, and her verbal protocols in fact revealed that she was evaluating the utility of her first choice during the pause and before making the change.

Modeling Strategy Selection and Change

As pointed out in Anderson (1990), the RA approach should not be pursued independently from other existing practices in cognitive science; rather, there should be an intimate interplay among building RA theories, computer modeling, and experimentation. The recent research by Anderson and his group has vigorously employed multiple practices and involved strong interactions among them (see Anderson, in preparation). We have so far proposed a rational theory of strategy selection and change and interpreted some of the results from our experiments on iterative strategy selection and change in PASCAL programming in terms of this rational

The subject's solution to the W-problem:

```
• • • • • • • • •
  repeat
    Read(File1, T);
    if T.Name <> 'END' then
      Write(File2, T);
  until T.Name = 'END';
• • • • • • • • •
```

The subject's solution to the R-problem:

```
• • • • • • • • •
  repeat
    Read(File1, T);
    if T.Name <> 'END' then
      Write(File2, T);
  until T.Name = 'END';
  Write(File2, T);
• • • • • • • • •
```

Figure 3. An example of lateral transfer effect in strategy selection.

The subject's 1st solution to the W-problem:

```
• • • • • • • • •
Read(File1, Predata);
if Predata.Name <> 'END' then
    repeat
        Write(File2, T);
        Read(File1, T)
    until Predata.Name = 'END';
• • • • • • • • •
```

The subject's 2nd solution to it:

```
• • • • • • • • •
Read(File1, Predata);
while Predata.Name <> 'END' do
    begin
        Write(File2, T);
        Read(File1, T)
    end;
• • • • • • • • •
```

Figure 4. An example of strategy change involving utility evaluation.

theory. In what follows we will present a computer simulation we developed for our example of iterative strategy selection and change; we hope that the simulation will exemplify some of our theoretical claims in a more concrete way.

Our simulation was accomplished in the ACT-R (Adaptive Character of Thought—Rationality) cognitive model. The ACT-R model maintains the same basic architecture as its predecessor—the ACT* model (Anderson, 1983), having a declarative LTM with data chunks as its knowledge elements, a procedural LTM with productions as its knowledge elements, and a dynamic WM. Compared to ACT*, however, the ACT-R model has some new features derived from RA. One such feature which is particularly relevant to the present rational theory of strategy selection and change is that each production in ACT-R has an associated probability of succeeding its goal actions and a cost of retrieving and applying it. Another particularly relevant feature is related to the conflict resolution mechanism embedded in ACT-R. Basically, each time when a production is matched or instantiated, the conflict resolution mechanism estimates the utility of its goal and the computational cost for executing the production, evaluates the formula expressed in (1), and chooses the optimal one among all the productions instantiated at that point of time. Based on these new features of ACT-R, Anderson & Kushmerick have successfully simulated subjects' route-choosing behavior on a computer-based navigation task (Anderson, in preparation).

For simulating iterative strategy selection and change, we considered the two looping constructs as two high-level productions; these productions can also be conceived as plans with details to be filled in (for the notion of plan in programming, see Soloway et al., 1988). Under this consideration, for those idiosyncratic subjects who used only one strategy on all problems, it was reasonable to suppose that they had only developed one such production—i.e., either P1 or P2 shown in Figure 5—or that they had developed both productions, but one was highly practiced while the other seldom practiced. In the latter case, it was conceivable that the retrieval cost associated with the frequently-used production was overwhelmingly lower than the cost for the other production; as a consequence, these subjects also displayed the type of idiosyncratic behavior observed in our experiments.

As to the adaptive subjects, since there was a difference manifested in their protocols between those who were highly-adaptive and those who were only moderately-adaptive, we simulated their strategy selection behavior differently with different productions. In ACT-R, it is a premise that only declarative knowledge coming into WM can be verbalized; the mere execution of a production is procedural and not verbalizable without much reflection. Thus, to simulate the strategy selection behavior of those highly adaptive subjects we used two adaptive productions, i.e., P1' and P2' shown in Figure 5, for which the evaluation of their utilities is an integrated part of themselves and is automatically performed by the conflict resolution mechanism. On the other hand, for the moderately-adaptive subjects, since their verbal protocols revealed an explicit process of utility evaluation and of conflict resolution, it seemed most suitable to use the non-adaptive productions P1 and P2 together with the assistance of the WM mechanism to model their performance. Specifically, we assumed that these subjects had developed both P1 and P2 associated with comparable costs of retrieval. Consequently, since either production could fire, the simulation would indiscriminately choose any one looping construct and then proceeded along with it. During the next several steps of simulation while the elements in WM are accumulating, some further productions would evaluate whether there was any redundant statement (i.e., the same statement in more than one place) in the constructed program and then, depending on the evaluation, either proceed to finish the selected course or to make a change. This was basically the type of behavior we observed in the example presented in Figure 4.

Conclusion

In this paper we have presented a rational perspective for understanding human cognitive behavior of strategy selection and change; we have

Non-adaptive looping-construct selection productions

PRODUCTION P1:

IF the program involves an indefinite loop;
THEN choose the **while...do** construct.

PRODUCTION P2:

IF the program involves an indefinite loop;
THEN choose the **repeat...until** construct.

Adaptive looping-construct selection productions

PRODUCTION P1':

IF the program involves an indefinite loop;
AND the loop may be executed zero times;
THEN choose the **while...do** construct.

PRODUCTION P2':

IF the program involves an indefinite loop;
AND the loop will at least be executed once;
THEN choose the **repeat...until** construct.

Figure 5. High-level productions for simulating looping-construct selection.

also illustrated this theoretical perspective with a concrete example of strategy selection and change in computer programming and substantiated our theoretical claims in terms of a computer simulation for this example. To summarize, we would conclude from this work the following points:

1. Several related studies conducted within different cognitive domains, including our own one on iterative programming, have provided convergent evidence indicating that subjects are highly adaptive in their strategy selection; that is, they choose their strategies appropriately in response to environmental information.

2. Subjects' behavior of strategy selection and change can be better understood within a rational framework which proposes that in making strategical choices subjects usually attempt to optimize the difference between the utility (e.g., information gain, execution efficiency, or design elegance) of the expected goal and the cognitive or computational cost associated with attaining the goal; this framework is integrative in that it incorporates within it such earlier models as proposed by Bruner et al. and by Hoenkamp.

3. In the course of choosing a strategy, subjects' evaluation of the expected utility of the chosen strategy may be performed implicitly (subconsciously) without any manifestation in their verbal protocols or explicitly (consciously) with corresponding episodes in their verbal protocols.

4. In the ACT-R cognitive model, cognitive strategies can be modeled as high-level productions. Specifically, strategy selection with implicit evaluation of expected utility can be modeled as instantiation of specialized productions, whereas strategy selection with explicit evaluation relies on some evaluation of working memory elements.

References

Anderson, J. R. 1983. *The Architecture of Cognition*. Cambridge, MA: Harvard University Press.

Anderson, J. R. 1990. *The Adaptive Character of Thought*. Hillsdale, NJ: Erlbaum.

Anderson, J. R. In Preparation. *Rules of the Mind*. Hillsdale, NJ: Erlbaum.

Bruner, J. S.; Goodnow, J. J.; & Austin, G. A. 1962. *A Study of Thinking*. New York: Science Editions, Inc.

Hoenkamp, E. 1989. "Confirmation bias" in rule discovery and the principle of maximum entropy. *The Proceedings of the Eleventh Annual Conference of the Cognitive Science Society*. 651-658. Hillsdale, NJ: Erlbaum.

Payne, J. W.; Bettman, J. R.; and Johnson, E. J.1988. Adaptive strategy selection in decision making. *Journal of Experimental Psychology: Learning, Memory, and Cognition*, 14: 534-552.

Popper, K. R. 1959. *The Logic of Scientific Discovery* (2nd ed.). New York: Basic Books.

Reder, L. R. 1987. Strategy selection in question answering. *Cognitive Psychology*, 19: 90-138.

Reder, L. R.; and Ritter, F. 1992. The effect of feature frequency on feeling of knowing and strategy selection for arithmetic problems. *Memory and Cognition*, Forthcoming.

Russo, J. E.; and Dosher, B. A. 1983. Strategies for multi-attribute binary choice. *Journal of Experimental Psychology: Learning, Memory, and Cognition*, 9: 676-696.

Siegler, R. S.; and Jenkins, E. A. 1989. *How Children Discover New Strategies?* Hillsdale, NJ: Erlbaum.

Simon, H. A. 1983. *Reasons in Human Affairs*. Stanford, CA: Standford University Press.

Soloway, E.; Adelson, B.; and Ehrlich, K. 1988. Knowledge and processes in the comprehension of computer programs. In M. T. H. Chi; R. Glaser; and M. J. Farr (Eds.), *The Nature of Expertise*. Hillsdale, NJ: Erlbaum.

Tukey, D. D. 1986. A philosophical and empirical analysis of subjects' modes of inquiry in Wason's 2-4-6 task. *The Quarterly Journal of Experimental Psychology*, 38A: 5-33.

Varian, H. R. 1987. *Intermediate Microeconomics: A Modern Approach*. New York: Norton.

Wason, P. C. 1960. On the failure to eliminate hypotheses in a conceptual task. *The Quarterly Journal of Experimental Psychology*, 12: 129-140.

Wu, Q.; and Anderson, J. R. 1991. Strategy selections and changes in PASCAL programming. In J. Koenemann-Belliveau, T. G. Moher; and S. P. Robertson (Eds.), *Empirical Studies of Programmers: Fourth Workshop*. pp. 227-238. Norword, NJ: Ablex.

Simultaneous Question Comprehension and Answer Retrieval

Scott P. Robertson, Jonathan D. Ullman, Anmol Mehta

Psychology Department
Rutgers University - Busch Campus
New Brunswick, NJ 08903

Abstract

A model is described for question comprehension in which parsing, memory activation, identification and application of retrieval heuristics, and answer formulation are highly interactive processes operating in parallel. The model contrasts significantly with serial models in the literature, although it is more in line with parallel models of sentence comprehension. Two experiments are described in support of the parallel view of question answering. In one, differential reading times for different question types were shown to be present only when subjects intended to answer the questions they were reading. In another, reading times for words in questions increased and answering times decreased when a unique answer could be identified early in the questions. The results suggest that source node activation and answer retrieval begin during parsing. Both symbolic and connectionist approaches to modeling question answering are potentially influenced by this perspective.

Question Answering

Question answering is a process that has interested researchers in several disciplines within cognitive science, especially cognitive psychology (Graesser & Franklin, 1990; Graesser, Robertson, & Anderson, 1981; Singer, 1984a, 1984b, 1986; Robertson & Weber, 1990), artificial intelligence (Dyer, 1983; Lehnert, 1977, 1978), philosophy of language (Belnap & Steel, 1976), and PDP modeling (Miikkulainen & Dyer, 1990). Question answering is also an important applied problem in query-directed information retrieval systems and in the context of education (Schank, 1986).

Question answering is interesting because it involves question-specific retrieval operations over complex mental representations. Researchers in this area have concentrated mainly on the relation between question types and retrieval heuristics, or on the heuristics themselves. Largely as a simplifying assumption, they have considered question answering to be independent of the language comprehension processes involved in question parsing or the language generation processes involved in answer production. In this paper we take issue with this view of the independence of question parsing, answer retrieval, and answer production. Following from Robertson & Weber (1990) we argue that retrieval and parsing, at least, occur simultaneously and may interact.

The main components of question answering are parsing to produce a conceptual representation of linguistic input, source node activation based on the conceptual representation, identification of retrieval heuristics appropriate for the identified question type, application of retrieval heuristics to identify or generate answer candidates, pruning of answer candidates based on pragmatic, appropriateness, and other criteria to isolate a single answer, and production of the answer in linguistic form. The most explicit models of question answering in the literature--Dyer (1983), Graesser & Franklin (1990), Lehnert (1978), and Singer (1986)--treat these as stages in a serial process. Indeed, this is the easiest thing to do since there are many dependencies among these processes, and most of the dependencies move from parsing toward production. For example, in some cases the question category can only be uniquely identified after it is determined whether the question presupposition is a motivated action or part of an unmotivated causal sequence.

The serial bias rests on many assumptions that may not be valid, however. In particular, serial models make assumptions about the need to pursue an answer in a single category or according to a unique retrieval rule. For example, in answering a question that begins with "Why did John...," serial models would be unable to begin because it is unclear at this point whether retrieval heuristics should be applied to search goal structures (as in "Why did John go to the store?") or causal chains (as in "Why did John fall down?," if he didn't do it on purpose!). If we imagine, however, that both retrieval processes could begin, activating all causal consequents involving "John" and all goals that "John" had, then we no longer need to assume an ordered relationship between parsing and retrieval. Instead, we are faced with describing how independent, but simultaneous, processes might interact and share resources.

The TSUNAMI Model

As an alternative framework for thinking about question answering, we are developing a model called TSUNAMI, for "Theory of Simultaneous UNderstanding Answering and Memory Interaction." At this point the model is offered as a broad architecture for supporting highly interactive application of the mechanisms already identified by question answering researchers. It remains to be seen how the nature of these mechanisms will change, and what new mechanisms might be necessary, when implemented in the TSUNAMI framework.

The TSUNAMI model, depicted in Figure 1, utilizes two working memory components. One memory stores question candidates and the other stores answer candidates. These working memory components act like "blackboard" data bases in that items stored there may be inspected and altered by several processes operating at once (Erman & Lesser, 1980). The question candidate memory and answer candidate memory are the only knowledge structures that take output from processes in the model (processes are indicated by ovals). The influences of processes on these memories might be to add propositions, update proposition contents, or delete propositions. The behaviors of processes that use the data in the question and answer candidate memories, in turn, are influenced by the contents of those memories.

Parsing and Matching

Processing begins when the *parser* starts receiving input from a question. the parser utilizes grammatical, case, and pragmatic information in semantic memory to produce various propositional representations which are then stored in the question candidate buffer. Multiple arrows from the parser into the question candidate buffer suggest that the parser can produce many candidates, often only partially specified propositions, in response to the input at any given time. As words come into the parser, prior candidate structures may be updated or disconfirmed and deleted by the parser.

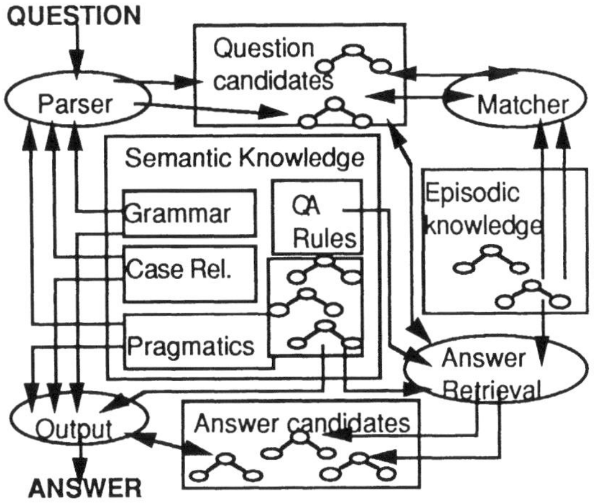

Figure 1. The TSUNAMI model of question answering.

As soon as there is any information in the question candidate buffer, a *matcher* begins comparing question candidate structures with activated information in episodic memory; specifically, the subset of episodic memory that is considered relevant to the question (e.g. a story just read, a set of memories foregrounded in the conversational context, etc.). We assume that the matching process occurs in parallel for several candidates (as indicated by multiple arrows from both the question buffer and episodic memory into the matcher) but that its speed is affected by the number of candidates and the number of matches. If the matcher finds a proposition in episodic memory that corresponds to a question candidate, then this is identified as a likely source node for answer retrieval processes. Influences of the matcher on question candidates are indicated by an arrow from the matcher into the question candidate memory. Matches in episodic memory raise the activation level of this information and make it more available to future analysis by any processes that utilize episodic memory (in this way the model is like Anderson's 1983 ACT* model). When a match is found for a question candidate, other candidates become less likely interpretations of the input. We assume that partial propositions in the question candidate memory can be matched to complete propositions in memory, thereby updating the question candidate list to include *expectations*. Predicted question candidates can influence answer retrieval as the retrieval process, described below, is blind to the status or origin of propositions in the question buffer.

Answer Retrieval

As soon as there is information in the question candidate buffer that might suggest a question or question type, the *answer retrieval* process can begin. This mechanism examines question candidates, attempts to determine question categories appropriate to the various candidates, and begins applying retrieval heuristics in episodic memory. The answer retrieval mechanism utilizes question answering rules stored in semantic memory and other, relevant general knowledge. This process operates simultaneously with the parser and the matcher, but is dependent on the contents of the question candidate memory and the activity in episodic memory (which is influenced by the matcher). The answer retrieval mechanism operates to highlight relevant portions of episodic memory. If it is guessed that a question is about a goal, for example, then the answer retrieval mechanism may activate goal hierarchies in the episodic trace. On the other hand, if the mechanism is expecting a causal antecedent question, then causal sequences may become more active. This activity will affect the behavior of the matcher.

The outputs of the answer retrieval process are propositions that are candidate answers to questions that the system finds consistent with the input (and memory) and any given time. Potential answers produced by the answer retrieval process are stored in the *answer candidate buffer* as propositions. These propositions may also be partially specified, and are subject to subsequent modification and deletion by the operation of the answer retrieval process. For example, if a question candidate that spawned a retrieval process is later disconfirmed, then the answer candidate built by that process will be deleted by the answer retrieval process.

Output

The answer candidates are examined by an *output preparation* process which also utilizes grammatical, case, pragmatic, and relevant general knowledge in semantic memory. This process can influence the answer candidate set. For example, if pragmatic concerns dictate that an answer is inappropriate, then the output preparation process will delete it from the answer buffer. Finally, when one candidate remains in the answer buffer and all of the question has been input to the parser, the final answer is formulated. It is reasonable to assume that the output preparation mechanism will not commit to a final interpretation until all of the input has been processed since a final phrase on a question can change its focus, and hence the appropriate answer, tremendously.

Experiment on Comprehension Instructions

The TSUNAMI model posits that parsing, matching, and retrieval processes share resources. Effects of retrieval processes on parsing can be seen in increased reading times for the words of a question. Such increases would be due to the increased workload resulting from simultaneous processes sharing resources. In Robertson & Weber (1990), we showed that knowledge of the question type during reading of a question (when the question word was at the beginning of a question) increased word-by-word reading times but decreased answer retrieval time when compared to conditions in which the question type was not known (when the question word was at the end of a question). Increased reading times suggested parallel retrieval and parsing. Decreased answering times reinforced this interpretation by showing that the answer was "closer" and that the workload effect was related to the answer retrieval process.

In that study we also observed increased answering times for time questions ("When did...") relative to reason questions ("Why did..."). Reason questions were answered 116ms faster than time questions. We have observed this discrepancy in two subsequent extensions of that study (299ms in one case and 259ms in the other), and it was the only reliable effect in another study on presentation speed of questions (455ms). In short, the reason-time discrepancy is a highly reliable effect related in some way to differences in the retrieval processes for these two types of questions.

In this experiment we exploited the reason-time discrepancy and sought to find it during reading. Also, we asked if a reason-time effect would be present in reading times only when subjects were reading with the intention of answering a question. If subjects were not intending to answer a question, then the answer retrieval mechanism would be inactive and the reason-time discrepancy should not be apparent.

Method

Subjects. Twenty-six subjects participated in this study for credit in Introductory Psychology.

Materials. Forty-eight short (5-7 line) stories were written. In each story a character went to some location. A reason question and a time question were prepared for each story. The reason questions read "Why did <ACTOR> go to the <LOCATION>?," whereas the time questions read "When did <ACTOR> go to the <LOCATION>?"

Design and Procedure. There were three instruction conditions in the experiment: *no-story paraphrase, story paraphrase,* and *story answer.* In the *no story paraphrase* condition subjects read questions (self-paced one word at a time) and were told to come up with a paraphrase. When they had reached the last word of the question they were to press the response key "when the meaning of the question was understood." They then wrote down their paraphrase. After this they pressed the response key and saw a computer generated question and were asked to judge if it "meant the same thing" as the question. The latter task was intended to reinforce the paraphrase instruction. Subjects worked through eight reason questions and eight time questions randomly intermixed in a block. In the *story paraphrase* condition subjects read a paragraph-long story which was then followed by a question. The question was presented in the same manner as the *no story paraphrase* condition and subjects were instructed to come up with a paraphrase in the same way. Each question was followed by a "means the same thing" judgement and there were again eight reason and eight time questions randomly intermixed in a block. Finally, in the *story answer* condition the subjects received stories followed by questions as in the *story paraphrase* condition, but this time their instruction was to come up with answers to the questions and press the response key "when an answer comes to mind." Similarly, they were asked to judge if the second question "has the same answer" as the first. Stories were randomly assigned to conditions and rotated through the conditions across subjects. Instruction block orders were counterbalanced.

Table 1
Mean reading time (ms/syllable) for all but the last word of reason and time questions read under three comprehension instructions: no-story paraphrase (NSP), story paraphrase (SP), and story answer (SA).

	INSTR			
QUEST	NSP	SP	SA	Mean
Reason	390	367	332	363
Time	401	370	371	381
Mean	395	368	351	372

Results and Discussion

Table 1 shows the mean reading times per syllable for all of the words of the question except the last. Reading time for the last word includes time for memory retrieval and answer/paraphrase formulation, and it is not of interest for studying processes occurring during parsing. When subjects were paraphrasing, the question type did not affect their reading times. When they were answering, however, the time questions took longer to read than the reason questions. An interaction between comprehension instruction and question type confirmed this interpretation, $F(2,50)=3.24$, $p<.05$. The results support the hypothesis that question-related retrieval processes are being activated during reading.

One explanation for the reason-time discrepancy is that the set of possible answers to a reason question is a subset of the set of possible answers to a similar time question. A goal is always an acceptable answer to a time question ("John went to the store WHEN he wanted some milk") whereas a time is a bad answer to a goal question ("John went to the store BECAUSE it was Saturday") except in highly specific circumstances (e.g. if John worked at the store on weekends in our example). Hence, in the TSUNAMI framework when a time question is being processed the question buffer contains more possible question interpretations relative to a reason question, the matcher would activate more nodes, and the retriever would generate more answer candidates. This would slow the overall operating time of the parser as seen in this experiment.

Experiment on Number of Unique Answers

In this experiment we concentrated on the role that the matcher and answer retriever play in the TSUNAMI model. In the model, the matcher tries to find antecedents in episodic memory for propositions in the question buffer. This process raises the activation level of the antecedents making them more likely to serve as source nodes for the retrieval process. The fewer memory items there are that are consistent with the input at any given point, the further the answer retriever can go.

We manipulated the contents of episodic memory in such a way that sometimes the source node from which retrieval processes would begin could be identified early in parsing by virtue of a unique actor. Subjects read stories in which an action was performed at four different times for four different reasons. In each story one actor performed the action on three occasions while the other performed the action on one occasion. In the following story, for example, Mary is the unique actor and John is a multiple actor:

John went to the store to buy bread on Monday.
John went to the store to buy milk on Tuesday.
John went to the store to buy cheese on Wednesday.
Mary went to the store to buy eggs on Thursday.

Now consider questions like the following:

q1. Why did John drive to the store on Tuesday?
q2. Why did Mary drive to the store on Thursday?

In answering q1 it is impossible to identify the exact source node in memory that corresponds to the question presupposition until the end of the question. In q2, however, it is possible to identify the unique source node in memory as early at the subject, Mary. In a question answering architecture with parallelism, like TSUNAMI, answer retrieval heuristics should begin earlier when reading q2 than q1. If simultaneous parsing and retrieval compete for resources as we have argued, then increased reading times should be observed for the words in q2 relative to q1 as the parser slows down. Additionally, if the increased workload is due to retrieval processes, then the answering time at the end of the questions should be faster for q2 than q1. In strict serial models, in contrast, source node activation and application of retrieval heuristics would be delayed until after question parsing and no reading time differences should be apparent (if anything, spreading activation theory for antecedent concepts predicts longer reading times for q1 over q2, Anderson 1976).

Method

Subjects. Twenty-eight Rutgers undergraduates participated for credit in Introductory Psychology.

Materials. Sixteen four-sentence stories like the one above were constructed for the experiment. Each story consisted of four instances of the same action performed at four different times for four different purposes. In each story there were two characters. When the stories were presented, one character was associated with three actions while the second was associated with a single action. For each story, one action was chosen as the "query action," about which a question would be asked. The actor associated with the query action was varied across subjects so that for some subjects the query action was performed by the unique character and for other subjects the query action was performed by the character who did several things.

Design and Procedure. Each subject read the sixteen stories and answered two questions about each one. The entire text of each story was presented on a computer screen and subjects spent as long as they liked reading it. When they were finished they pressed a response key. At this time a prompt appeared on the screen. Each subsequent keypress revealed a word of the question, and the words appeared side-by-side in their normal positions. Subjects were instructed that on the last word of the questions they should press the response key "as soon as an answer comes to mind." Reason questions were in the form "Why did NOUN1 VERB PREP1 DET NOUN2 PREP2 TIME?" Time questions were of the form "When did NOUN1 VERB PREP1 DET NOUN2 AUX VERB2 NOUN3?" The reading times for each word were recorded.

Subjects were first asked a reason or time question about each story. Each subject was asked reason questions about eight stories and time questions about eight stories. For each subject, half of the questions were about the action performed uniquely by one actor (unique action) and half were about one of the three actions performed by the other actor (non-unique action). The question type condition (reason/time) and action uniqueness condition (unique/non-unique) were crossed. Across subjects, the stories were rotated through the conditions.

Since it was possible to answer the time and reason questions without paying attention to the actors in the story, each reason/time question was followed by a "who" question about each story. The antecedent for the who question was chosen randomly between the unique action and one of the non-unique actions. The who-question guarantied that subjects would pay close attention to the actors. Reading times were not collected for these questions.

Table 2
Mean reading time (ms/word) averaged across the five common words

Question	Actor		
	Unique	NonUnique	Mean
Reason	483	463	473
Time	507	465	486
Mean	495	464	479

Results and Discussion

Table 2 shows the mean reading times averaged across NOUN1, VERB, PREP, DET, and NOUN2 for the reason questions and time questions in the unique and non-unique actor conditions. As predicted, the time to read the questions was greater in the unique action condition relative to the non-unique action condition, $F(1,27)=7.50$, $p<.05$. There was no effect of question type and no interaction.

Our second prediction, that answering time would be faster in the unique action condition relative to the non-unique action condition, was also confirmed. The mean answer times were 2449ms vs 3554ms in the unique vs. non-unique conditions respectively, $F(1,27)=19.11$, $p<.001$. In contrast to other experiments in our lab, reason questions were answered more slowly overall than time questions (3256ms vs 2746ms), $F(1,27)=9.39$, $p<.05$. There was no interaction.

The results support the hypothesis that if a source node can be identified early during parsing, retrieval heuristics can be identified and applied. The simultaneous operation of the parser and answer retriever slows both, but pays off in the end with a faster answer.

Final Comments

We have proposed a new architecture for question answering that has many parallel components and presented empirical evidence in support of it. A parallel view of question answering would bring this important aspect of language processing into line with current thinking on parallel processes in sentence parsing (Gorrell, 1989; McClelland & Kawamoto, 1986; Miikkulainen & Dyer, 1991; St. John & McClelland, 1990; Waltz & Pollack, 1985). Of course, the experiments support the general idea of parallelism, not the specifics of the TSUNAMI model. However, the model is general enough to incorporate many specific instantiations. As it develops it will be interesting to see how, or if, changes will be necessary in the retrieval heuristics proposed by researchers working within a serial paradigm. More than likely a new class of problems will arise having to do with conflict resolution among competing question interpretations and answer possibilities in the face of partial input.

Recently there has been considerable progress on connectionist models of sentence parsing (McClelland & Kawamoto, 1986; Miikkulainen & Dyer, 1991), and PDP models will inevitably begin to approach the problem of question answering. In this paradigm too it will be necessary to face the issue of whether the output of a parsing network should be the input to a question answering network, or whether these processes are more closely intertwined. Our results suggest the latter approach.

References

Anderson, J.R. (1976). *Language, memory, and thought*. Hillsdale, N.J.: Erlbaum.

Anderson, J.R. (1983). *The architecture of cognition*. Cambridge, MA: Harvard University Press.

Belnap, N.D., & Steel, T.S. (1976). *The logic of questions and answers*. New Haven: Yale University Press.

Dyer, M.G. (1983). *In-depth understanding: A computer model of integrated processing for narrative comprehension*. Cambridge, MA: MIT Press.

Erman, L.D., & Lesser, U.R. (1980). The HEARSAY-II speech understanding system: A tutorial. In W. Lea (Ed.), *Trends in speech recognition*. Englewood Cliffs, NJ: Prentice-Hall.

Gorrell, P. (1989). Establishing the loci of serial and parallel effects in syntactic processing. *Psycholinguistic Research, 18*, 61-74.

Graesser, A.C., & Franklin, S.P. (1990). QUEST: A cognitive model of question answering. *Discourse Processes, 13*, 279-304.

Graesser, A.C., Robertson, S.P., & Anderson, P.A. (1981). Incorporating inferences in narrative representations: A study of how and why. *Cognitive Psychology, 13*, 1-26.

Lehnert, W.G. (1977). Human and computational question answering. *Cognitive Science, 1*, 47-73.

Lehnert, W.G. (1978). *The process of question answering*. Hillsdale, NJ: Erlbaum.

McClelland, J.L., & Kawamoto, A.H. (1989). Mechanisms of sentence processing: Assigning roles to constituents. In J. McClelland & D. Rumelhart, (Eds.), *Parallel distributed processing: Explorations in the microstructure of cognition. Volume 2: Psychological and biological models*. Cambridge, MA: MIT Press.

Miikkulainen, R., & Dyer, M.G. (1991). Natural language processing with modular PDP networks and distributed lexicon. *Cognitive Science, 15*, 343-400.

Robertson, S.P., & Weber, K. (1990). Parallel processes during question answering. *Proceedings of the 12th Meeting of the Cognitive Science Society*. Hillsdale, NJ: Erlbaum.

Schank, R.C. (1986). *Explanation patterns.: Understanding mechanically and creatively*. Hillsdale, NJ: Erlbaum.

Singer, M. (1984a). Mental processes of question answering. In A.C. Graesser & J.B. Black (Eds.), *The psychology of questions*. Hillsdale, NJ: Erlbaum.

Singer, M. (1984b). Toward a model of question answering: Yes-no questions. *Journal of Experimental Psychology: Learning, Memory, and Cognition, 10*, 285-297.

Singer, M. (1986). Answering wh- questions about sentences and text. *Journal of Memory and Language, 25*, 238-254.

St. John, M.F., & McClelland, J.L. (1990). Learning and applying contextual constraints in sentence comprehension. *Artificial Intelligence, 46*, 217-258.

Waltz, D.L., & Pollack, J.B. (1985). Massively parallel parsing: A strongly interactive model of natural language interpretation. *Cognitive Science, 9*, 51-74.

IMPLICIT ARGUMENT INFERENCES IN ON-LINE COMPREHENSION[1]

Gail A. Mauner and **Michael K. Tanenhaus**
phooph@psych.rochester.edu mtan@psych.rochester.edu
Department of Psychology
University of Rochester
Rochester, NY 14627

Greg N. Carlson
carlson@psych.rochester.edu
Department of Foreign Languages, Literature and Linguistics
University of Rochester
Rochester, NY 14627

Abstract

While people are capable of constructing a variety of inferences during text processing, recent work on inferences suggests that only a restricted number of inferences are constructed on-line. We investigated whether implicit semantic information associated with the arguments of verbs is automatically encoded. Short passives such as "The ship was sunk" are intuitively understood as containing an implicit Agent, e.g. that someone is responsible for the ship's sinking. To investigate whether implicit Agents are encoded automatically, short passives were compared to intransitive sentences with the same propositional content. The experimental logic used depended on a specific property of rationale clauses such as "to collect an insurance settlement"; namely, that the contextual element associated with the understood subject of the rationale clause must be capable of volitional action. If people encode an implicit Agent while processing short passives, then they should be able to associate it with the understood subject of a rationale clause. No such association should be possible with intransitives. In two experiments, intransitives elicited longer reading times and were judged to be less felicitous than short passives at the earliest point possible in the the rationale clause. Short passives were judged fully felicitous and their reading times did not differ from control sentences with explicit agents.

[1] This work was supported by NIH grant HD27206.

Introduction

While a reader's representation of a text or discourse clearly contains a mixture of information that is explicitly represented in the text and information that is inferred, it remains unclear what types of inferences are typically drawn during immediate or "on-line" comprehension. Research in the 1970's that relied mostly on memory paradigms led to the conclusion that a rich variety of inferences are encoded when a reader constructs an interpretation of a text. However, recent work using on-line paradigms has suggested that inferencing may be much more restricted and limited (see McKoon and Ratcliff, 1991). In fact, McKoon and Ratcliff argue for a minimal inferencing model in which the only inferences that are routinely drawn are those that are easily accessible (e.g. based on associations) and those that are required for local coherence. This last type might include inferences triggered by syntactic or semantic information that is implicitly represented in sentences.

Many sentences in English contain implicit syntactic and semantic information. For example, many verbs can occur with optional arguments. Often these arguments are not expressed in a sentence when their content is provided by information in the context. For example, the prepositional phrase that expresses the recipient of the donation in sentence (1) is likely to be omitted when the recipient is provided by the context, as in (2). In (2), there is an implicit "Recipient" argument that may function anaphorically to integrate the sentence with the context (Carlson and Tanenhaus, 1988).

(1) John donated five dollars to the United Fund.

(2) The United Fund asked John for a contribution.
John donated five dollars.

In addition, there may be other implicit arguments that are not anaphoric, but which may also be encoded as part of the normal understanding of a sentence (Fillmore, 1986). One likely candidate is the English passive construction. Full passives contain an explicit "by-phrase" which introduces either an "Agent" or an instrument, as in (3). However, short passives such as (4) do not specify an Agent. Nonetheless, we have the strong intuition that the interpretation of the sentence includes an understood or implicit Agent. These intuitions can be highlighted by comparing a short passive to a sentence with the same explicit propositional content, as in (5). In contrast to (4), sentence (5) does not seem to imply that someone was responsible for the sinking of the ship.

(3) The ship was sunk by the captain.

(4) The ship was sunk.

(5) The ship sank.

This paper presents two experiments that were conducted to determine whether the hypothesized implicit Agent associated with short passives is, in fact, encoded during sentence processing.

How one should go about finding evidence for the encoding of non-anaphoric implicit arguments is not immediately clear. Studies investigating empty categories and anaphoric implicit arguments have often looked at whether the referential expression primes its antecedent. However, the one study that has used a priming methodology to investigate implicit Agents did not find reliable priming effects (MacDonald, 1989). However, this result is not surprising because priming depends upon the anaphoric properties of the implicit Agent. To the extent that implicit Agents are not, as we have argued, anaphoric, then one should not expect to see priming to a potential Agent that has been introduced earlier in the discourse.

The logic we used depends on a specific property of rationale clauses. Rationale clauses are adverbial infinitive modifiers that carry a connotation of purpose (see Jones, 1991 for extensive discussion). Like all infinitive clauses, rationale clauses have an understood subject. Although some infinitive clauses allow their subjects to be interpreted arbitrarily, the understood subject of many others must be associated with a noun in the preceding context. Rationale clauses are of this second type. Moreover, rationale clauses require the contextual element that is associated with its understood subject to be capable of volitional action (e.g. to be agentive), as the sentences in (6) illustrate.

(6) a. John_i hit the man_j [e_i to stop him_j]

b. *The bat_i hit the man_j [e_i to stop him_j]

Thus, the logic for the following studies is this: if people encode an implicit Agent as part of their representation of a short passive, then this implicit argument, being agentive, should provide a contextually appropriate antecedent for the understood subject of a rationale clause. In contrast, if a rationale clause is preceded by an intransitive, there should be no contextually appropriate antecedent available for interpretation. As a result, comprehension difficulties should be encountered. The sentences in (7) appear to confirm this hypothesis; (7a) appears to be felicitous, whereas (7b) is not.

(7) a. The ship was sunk to collect a settlement from the insurance company.

b. The ship sank to collect a settlement from the insurance company.

These intuitions suggest that interpretation of a short passive includes the representation of an implicit agent. However, intuitions alone do not provide evidence about when in the time course of comprehension the implicit Agent is encoded; and whether encoding an implicit Agent requires making a resource-demanding inference. The experiments that we report were designed to answer these questions.

Experiment 1.

Experiment 1 was designed to experimentally establish the contrast between short passives and intransitives presented in (7). In order to do so, we used a word-by-word self-paced reading task in which subjects also pressed a button if the sentence stopped making sense at any point. This "stop-making-sense" task has proved useful in studying the time course of processing of sentences with other types of infinitive clauses (Boland, Tanenhaus, and Garnsey, 1990) and sentences with filler-gap relationships e.g. Tanenhaus, Garnsey & Boland, 1990).

Method

Stimuli consisted of 20 sentence pairs formed from a passivized transitive verb and its intransitive correlate followed by a rationale clause, like the pair given in (7). Within each stimulus pair, the same inanimate subject and rationale clause were used. Both presentation lists contained 10 passive-initial sentences and 10 intransitive-initial sentences randomly interspersed among 36 distractor sentences, 11% of which did not make sense. Items from the two experimental conditions were counterbalanced for length and condition across the two lists. The critical region of each sentence consisted of the transitive or intransitive verb and the first four words of the rationale clause, which always consisted of the infinitive marker "to" followed by a verb, a determiner, and a noun. A 4-word post-critical region was included to avoid contaminating critical region latencies with a "wrap-up" effect (end of sentence increases in response times). No item was longer than a single line. Prior to completing 10 practice trials, 26 native English-speaking undergraduates from the University of Rochester were given examples of sentences that did not make sense and explanations of why they did not make sense. In both practice and experimental trials, after presentation of a trial number, subjects controlled the word-by-word presentation rate of sentences with a button press, as each sentence accumulated across the screen of a video monitor. Subjects continued pressing a "Yes" button as long as a sentence made sense. When a sentence ceased to make sense, subjects pressed a "No" button which then ended the current trial and initiated a new one.

RESULTS AND DISCUSSION

We collected two types of data, the percentage of "no" judgments and the amount of time it took subjects to make "yes" judgments (latency data). Recall that when a "no" response is made at a given word position, the trial ends. As a result, simple frequencies of "no" responses at each word position are dependent on whether any "no" responses were made at previous word positions. Thus, to minimize the dependence of values at later positions on earlier ones, judgment data were transformed into percentages adjusted to reflect the remaining number of opportunities to respond "no" at each word position. These percentages of remaining "no" responses were then entered into analyses of variance.

Judgments. The cumulative percentages of "no" judgments in Figure 1 and analyses of variance on the percentages of remaining "no's" reveal no differences between passives and intransitives at the first and second word positions and, as expected, intransitives elicited a significantly greater proportion of "no" responses than passives at the verb, $F1(1,24)=6.63$, $p<.02$; $F2(1,18)=9.27$, $p<.01$, in both subject and items analyses.

Latencies. Latencies for "yes" judgments provide information about processing differences that may not be reflected in judgment data. In word-by-word reading, subjects frequently encounter locally incoherent points that resolve coherently in a word or two. Consequently, even when a sentence stops making sense, subjects may wait for a word or two before pressing the "no" button. It is assumed that when subjects encounter a point of local incoherence but still respond "yes", that they must be encountering some processing difficulty and that this difficulty will be reflected as an increase in response times. The response latencies for "yes" judgments shown in Figure 2 revealed that intransitives elicited longer response times than short passives at the verb $F1(1,24)=6.12$, $p<.03$; $F2(1,18)=1.35$, $p>.2$, and at the determiner $F1(1,24)=7.01$, $p<.02$, $F2(1,18)=2.85$, $p>.1$, but that the effect was significant only in the subjects analysis.

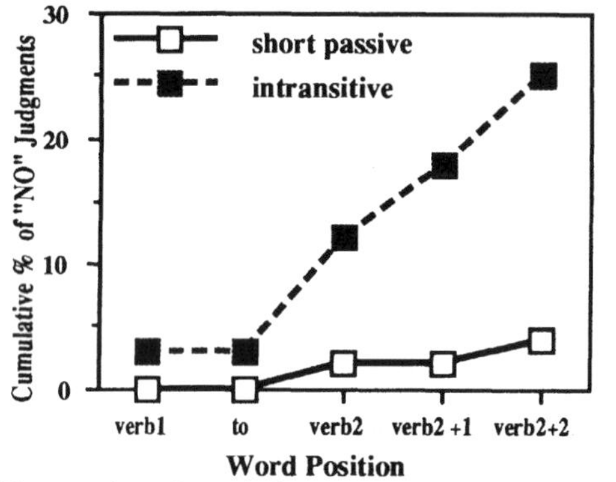

Figure 1. Cumulative percentages of "NO" judgments to short passives and intransitives by word position.

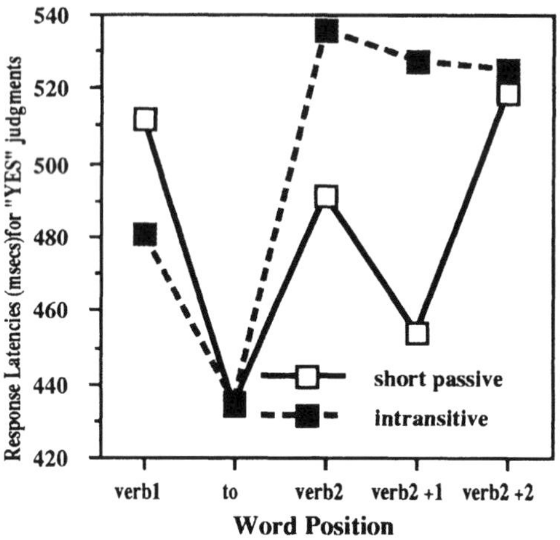

Figure 2. Latencies (msecs) for "yes" judgments for short passives and intransitives by word position.

The reader may wonder why clear effects in the subject analyses do not approach significance in the item analyses for the latency data. The reason is discussed in detail in Boland et al (1990). Subjects who are reading quickly tend to respond "no" a word or two later than subjects who are reading more slowly. However, their "no" responses are often extremely fast, indicating that they probably detected the oddity on the preceding word. As a result, fast subjects contribute relatively more data than slow subjects to the intransitive conditions beginning at the verb in the rationale clause, and this varies somewhat by item. If item means are adjusted (using a conservative data replacement method) to estimate what the means would have been given equal contributions from all subjects, then the item statistics show the same pattern as the subject statistics.

Discussion. While the results of Experiment 1 clearly establish a difference between short passives and intransitives, they do not indicate whether or not encoding the implicit Agent for the short passives required a resource-demanding inference. In order to answer this question, it is necessary to compare the short passives to constructions that provide an explicit Agent. Experiment 2 was conducted for this purpose.

Experiment 2.

Two control conditions, an active declarative and a full passive, were added to the short passive/ intransitive manipulation of Experiment 1. Unlike the short passive, both control conditions contained an explicit Agent. Full passives are structurally similar to short passives, but their Agent by-phrases can introduce unintended infelicities (Mauner, 1991). Actives, while dissimilar structurally, have the same explicit arguments as full passives and are fully felicitous. Both control conditions were included because, a priori, it was not clear which would be the more appropriate control. A set of sample materials is given in (8).

(8) a. One of the owners sank the ship to collect a settlement from the insurance company. **active**
b. The ship was sunk by one of its owners to collect a settlement from the insurance company. **full passive**
c. The ship was sunk to collect a settlement from the insurance company. **short passive**
d. The ship sank to collect a settlement from the insurance company. **intransitive**

Because both control conditions have explicit agents they should both be judged felicitous. Short passives should pattern with active and passive controls if they too contain an Agent in their representation. Judgment data alone may not provide clear evidence about when readers encode an implicit Agent in short passives. An Agent could be encoded at the first verb if the semantic information associated with the verb is accessed and interpreted when that verb is recognized. Alternatively, an Agent may be inferred only after the verb in the rationale clause is encountered. On this view, an inference that creates an agent is easier following a short passive than following an intransitive, but it is the properties of the rationale clause that drive the inference. The latencies for subjects' "yes" decisions, can be used to decide between these two alternatives. This is because subjects might require time to make an inference but still say “yes”. If an inference is required to create an Agent following the short passive, then we should expect longer latencies early in the rationale clause for short passives as compared to actives and full passive controls. If, however, the Agent was already encoded, then latencies for short passives should pattern with full passives and actives.

Method

The stimuli for Experiment 2 consisted of 20 sets of four sentences, each containing a short passive/ intransitive pair from Experiment 1 and their active and full passive correlates, as illustrated in (8). Although some control sentences extended onto a second display line, the critical region plus at least one additional word were always displayed on the first line. Distractor items were modified to include the display and structural characteristics of the new items. Items in each set of materials were counterbalanced for length and condition across four presentation lists and within each list, the 20 experimental items were interspersed among 36 distractor sentences, 11% of which were constructed to not make sense. Forty native English-speaking undergraduates from the University of Rochester participated in this experiment following the same procedure as Experiment 1.

RESULTS AND DISCUSSION

Judgment and latency data were collected and treated in the same manner as in Experiment 1. However because the first word of the critical region differed in terms of category across the conditions, this word position was not included in the latency analyses.

Judgments. The adjusted percentages of "no" responses in Figure 3 and the analyses of variance performed on the percentages of remaining “no”responses reveal no differences across conditions until the verb in the rationale clause, at which point "no" responses rise sharply for intransitives. At the verb, intransitives elicited more "no" responses than short passives $F1(1,36)=11.92$, $p<.001$, $F2(1,16)=12.70$, $p<.003$; actives $F1(1,36)=7.05$, $p<.02$, $F2(1,16)=12.89$, $p<.003$ and full passives $F1(1,36)=8.75$, $p<.006$, $F2(1,16)=12.90$, $p>.003$, when either subjects or items were random. These differences continued to be significant at later word positions with probability levels of $p=.05$ or less.

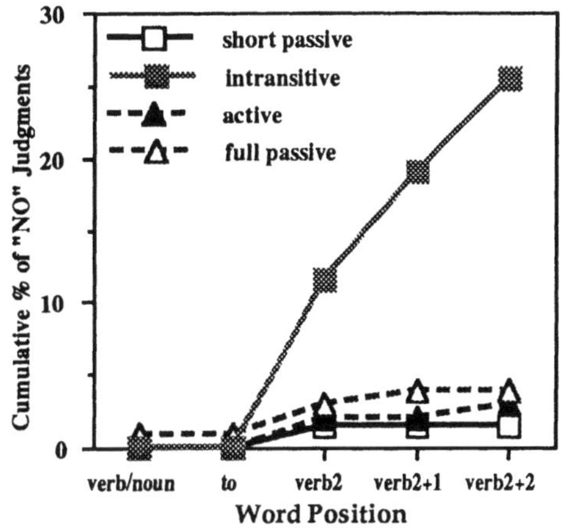

Figure 3. Cumulative percentages of "No" judgments to short passives, full passives, actives and intransitives by word position.

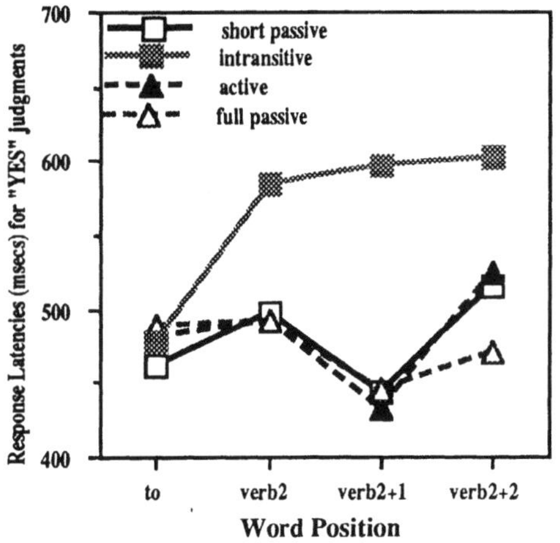

Figure 4. Latencies (msecs) for "Yes" judgments to short passives, full passive, actives and intransitive by word position.

Latencies. As the adjusted percentages of "No" response and cumulative percentages in Figure 4 show, intransitives elicited longer response times than the short passives $F1(1,33)=6.68$, $p<.01$, $F2(1,16)=.86$, $p<.3$; or active $F1(1,33)=5.58$, $p=.02$, $F2(1,16)=3.85$, $p<.07$, or full passive controls $F1(1,33)=5.31$, $p<.03$, $F2(1,16)= 2.90$, $p>.12$. These differences were weakly significant at the verb and clearly significant ($p<.02$) at following word positions. Moreover, short passives did not elicit longer response times than either control condition early in the rationale clause. Except at the noun position, there were no differences among these three conditions. At the noun, mean latencies were found to differ significantly ($p<=.05$) when subjects but not items were random. Further investigation revealed that while short passives and actives did not differ from each other, both elicited significantly longer latencies than full passives at the noun.

Discussion. In Experiment 2, short passives patterned with full passive and active controls in both judgment and latency data. Unlike the intransitives, which began to show felicity effects at the verb, short passives were judged to be as felicitous as full passive and active controls throughout the critical region. This suggests that the implicit Agent is indeed encoded as part of subjects' representations of short passives. Also in contrast to intransitives, short passives did not elicit longer response latencies than control conditions at any word position in the critical region. This suggests first, that no costly inference is required to encode implicit Agents, and second, that the processes that are involved in determining the grammatical relationships that exist between a verb and its explicit arguments are also likely to be involved in the encoding of implicit Agents.

General Discussion

These experiments demonstrated that rationale clauses that are preceded by short passives are no more difficult to process than rationale clauses that are preceded by clauses that introduce an explicit agent (full passives or active sentences with transitive verbs). In contrast, rationale clauses that are proceeded by a clause with an intransitive verb (with a subject that does not have agentive properties) are difficult to comprehend. Taken together, these results demonstrate that the encoding of implicit Agents in short passives takes place rapidly and that it does not involve a resource-demanding inference. While subjects could have inferred the presence of an Agent in the intransitive cases, it appears that either they did not, or that this inference was not constructed automatically. These results are consistent with McKoon and Ratcliff's (1991) suggestion that inferences are constructed automatically only if they are required to satisfy demands for local coherence. This minimalist position might seem to be too restrictive, at first. However, if one takes into account the range of implicit arguments and relations, as well as contextually dependent expressions in natural language, then the minimalist and constructivist positions are not as far apart as they might appear, although they differ in spirit. One can think of the implicit information in sentences as structural triggers that indicate what aspects of context, real-world knowledge, etc., are likely to be relevant to interpreting a sentence in context. In addition, some of the information that is part of the understanding of a sentence may not be specific enough to be revealed by many of the methodologies that are currently most popular. The work we have presented here is a small step towards exploring the role that implicit information that is part of the syntactic and semantic structure of a sentence plays in guiding comprehension processes.

ACKNOWLEDGMENTS

We would like to thank Rich Campbell for many helpful discussions at the beginning of this project.

REFERENCES

Boland, J.E.; Tanenhaus, M.K; and Garnsey, S.M., 1990. Evidence for the Immediate Use of Verb Control Information in Sentence Processing. *Journal of Memory and Language* 29:413-432.

Carlson, G.N. and Tanenhaus, M.K. 1988. Thematic Roles and Language Comprehension. In W. Wilkins ed., *Thematic Relations*. New York: Academic Press.

Fillmore, C. J. 1986. Pragmatically Controlled Zero Anaphora. Proceedings of the Twelth Annual Meeting of the Berkeley Linguistics Society, 97-107.

Jones, C. 1991. *Purpose Clauses*. Dordrecht: Kluwer Academic Publishers.

MacDonald, M. 1989. Priming Effects from Gaps to Antecedents. *Language and Cognitive Processes*, 4(1):35-56.

McKoon, G. and Ratcliff, R. Inference during Reading. Forthocoming.

Mauner, G. Syntactic Context and the Interpretation of VP Anaphors. Proceedings of the Twenty-Second Annual Meeting of the New England Linguistics Society. Forthcoming.

Tanenhaus, M. K., Garnsey, S.M., and Boland, J.E. 1991. Combinatory Lexical Information and Language Comprehension. In G. Altman ed. *Cognitive Models of Speech Processing: Psycholinguistic and Computational Perspectives*. Cambridge, MA: MIT Press.

Another Context Effect in Sentence Processing: Implications for the Principle of Referential Support

Michael J. Spivey-Knowlton
Department of Psychology
University of Rochester
Rochester, NY 14627
spivey@psych.rochester.edu

Abstract

A major goal of psycholinguistics is to determine what sources of information are used immediately in language comprehension, and what sources come into play at later stages. Prepositional phrase attach-ment ambiguities were used in a self-paced reading task to compare contexts that contained one or two possible referents for the verb phrase (VP) in the target sentence. With one set of sentences, a VP-attachment preference was observed in the 2-VP-referent context, but not in the 1-VP-referent context. With another set of sentences, no effect of context was observed. This result falls outside of the scope of the principle of referential support (Altmann & Steedman, 1988) as currently formulated. It suggests that a similar but more broadly-based theory is required.

Introduction

The strong claim for modularity in language processing (Fodor, 1983) has inspired a wealth of research occasionally supporting the existence of an informationally encapsulated syntactic processor (e.g. Ferreira & Clifton, 1986; Frazier, Clifton & Randall, 1983; Rayner, Carlson & Frazier, 1983) and occasionally questioning it (e.g. McDonald, 1992; Spivey-Knowlton, Trueswell & Tanenhaus, 1992; Taraban & McClelland, 1988 Trueswell & Tanenhaus, 1991, this volume; Trueswell, Tanenhaus & Garnsey, 1989). See Altmann (1989) for a review of some of the work on this issue.

The vast majority of this research examines subjects' reading times[1] in sentences that contain temporary syntactic ambiguities. If a subject makes an incorrect syntactic commitment at the ambiguity, reading time will be slow when she encounters a region that resolves that ambiguity.[2]

A frequently studied type of syntactic ambiguity is the prepositional phrase (PP) attachment ambiguity. Consider the sentence, "*Johnny attacked the cat with the rubber mouse.*" The PP "*with the rubber mouse*" can be syntactically attached either to the verb phrase (VP) "*attacked*" or to the noun phrase (NP) "*the cat*". The former attachment corresponds to an interpretation in which the attacking was done with an instrument called a rubber mouse. The latter attachment would mean that the cat was somehow distinguishable by its association with a rubber mouse -- perhaps it was playing with it.

Some studies have shown that, upon encountering the ambiguously-attaching preposition, readers tend to prefer the VP-attachment (Altmann, 1986; Frazier, 1978; Rayner, et al., 1983). That is, when the sentence is more plausibly an NP-attachment (i.e., "*Johnny attacked the cat with the short hair.*"), readers find themselves "garden-pathed" when they reach the disambiguating region ("*short hair*").

The predominance of the VP-attachment bias has been interpreted as evidence for the minimal attachment principle (Frazier, 1978, 1987). This explanation rests on the fact that the constituent structure of the NP-attached sentence contains more nodes and more depth of branching than the constituent structure of the VP-attached version. It is claimed that, regardless of context, the sentence processor will initially commit to the structurally less complex attachment.

In response to this, Altmann & Steedman (1988) proposed what is now generally known as the Referential Theory of syntactic disambiguation. Their principle of referential support (derived from Crain & Steedman's (1985) principle of referential success) is based on the assumptions that a definite

[1] Reading times are measured either by a self-paced reading task in which the subject presses a button to present each successive word or phrase in the sentence, or by monitoring eye-movements and fixation durations while the subject reads text on a computer screen.

[2] This is called a "garden-path". It is assumed that the reader must then reinterpret the syntactic and thematic relations of the sentence, which takes additional processing time.

NP (e.g.,"*the cat*") presupposes the existence of a unique referent in the discourse model, and when a definite NP is encountered, the reader attempts to link it to the appropriate entity in her discourse model. When the syntactically ambiguous preposition "*with*" is encountered, both attachments are considered in parallel and their presuppositions weighed. The VP-attachment, or simple NP analysis, "*attacked the cat with (instrument)*" maintains the previous presupposition of a unique referent. However, the NP-attachment, or complex NP analysis, "*attacked the cat with (attribute)*" presupposes more than one cat in the discourse model, to one of which a distinguishing reference is being made. Thus when the context contains zero cats[3] or one cat, the VP-attachment upholds the correct presupposition. On the other hand, when the context contains two or more cats, only the NP-attachment upholds the correct presupposition. Hence, the reader's attachment preference is modulated by referential pragmatics, not by syntactic complexity.

To experimentally test this theory, Altmann & Steedman (1988) manipulated discourse contexts so that they had either one or two referents for the definite NP preceding the ambiguous PP ("*the cat*" in "*Johnny attacked the cat with...*"). For example, one version of a context would contain two possible referents for the definite NP: a cat that has short hair and a cat that has long hair, while the other version would contain one such referent: a cat that has short hair and a dog that has long hair. It was demonstrated that, while a 1-NP-Referent context did not change the VP-attachment bias in the target sentence, a 2-NP-Referent context produced a clear preference for NP-attachment in the target sentence (Altmann & Steedman, 1988; Spivey-Knowlton, 1991).

However, Altmann & Steedman's referential explanation of their findings, though more explicit than broader accounts, may be too narrowly focused. A less restrictive, albeit more vague, explanation of this result is simply that they have set up a minimal pair of entities in context between which only an NP-attachment can discriminate. Their contexts systematically begin by introducing a character carrying an instrument or tool ("*A burglar broke into a bank carrying some dynamite.*"), his intention is stated ("*He planned to blow open a safe.*"), and then the one or two NP referents are introduced ("*He saw a safe which had a new lock and a strongbox/safe which had an old lock.*"). Aside from setting up two referents, such a context may also set up a kind of conceptual uncertainty in which the reader anticipates that, since the burglar planned to blow open a safe, he's going to have to pick one to blow open first. The reader will expect to be told *which safe* the burglar decided to blow up. Assuming that the subject begins the target sentence with the specific goal of discriminating between the entities of this suspended minimal pair, immediate effects of an NP-attachment expectation should be observed. This idea that discourse may produce expectations for greater specificity in particular aspects of upcoming information makes unnecessary the distinction between presuppositions of the simple and complex definite NP.[4] This proposal also makes the prediction that setting up a minimal pair of *events* should increase the preference for VP-attachment.[5] Extending the logic above, a context that introduces two possible, but yet-to-occur, events should create an expectation that they will be distinguished in the description of subsequent related events. Attaching a PP to the VP is a common way to convey more detail of the event.

The Experiment

Recent findings by Taraban & McClelland (1988) have also warranted a theory of syntactic disambiguation that is very different from the single encapsulated rule (Minimal Attachment) proposed by Frazier and colleagues. Taraban & McClelland (1988) argue that, in conjunction with certain prepositions, some verbs may produce strong expectations for particular thematic roles that would be violated by attaching the PP to the verb. The result would be an indirect preference for NP-attachment, even in the absence of a context. Unlike Minimal Attachment, however, this hypothesis does not explicitly exclude simultaneous influences from context. To experimentally support this hypothesis, Taraban & McClelland (1988) demonstrated that the thematic role biases of several verb - noun - preposition combinations, as indicated by sentence completion and rating tasks, accurately predicted attachment biases in self-paced reading. They constructed a small

[3] In a context that contains no cats at all [or in the absence of context, as in Rayner et al. (1983)] it is assumed that the reader will, upon reading the definite NP, "create" a single unique referent in her discourse model, thus biasing her toward the simple NP analysis.

[4] In fact, some preliminary findings (Spivey-Knowlton & Sedivy, in preparation) indicate that, in the absence of context, Altmann & Steedman's stimuli have a significant, though smaller, VP-attachment bias *even when the NP is indefinite*. As indefinite NPs carry no presuppositions, Referential Theory predicts *no attachment preference* for such sentences.

[5] Because it has been argued that tense of the verb can be treated as referential (Webber, 1988), it is conceivable that the VP referents in context may act as individuals to one of which the VP is referring. This may be the basis upon which a reformulation of the principle of referential support may account for a VP-attachment bias due to 2 VP referents in context.

set of novel sentences to compare with those of Rayner et al. (1983), and showed that the attachment preferences were nearly equal and opposite.

With the opposition of these two groups of sentences in mind, this experiment compared contexts that contained one NP referent and one VP referent with contexts that contained one NP referent and *two VP referents*. According to discourse-driven structural expectations, the latter context should bias the reader toward a VP-attachment. To examine effects of these contexts on sentences that have context-free VP-attachment biases and those that have context-free NP-attachment biases, stimuli from Rayner, et al. (1983) and Taraban & McClelland (1988) were used as target sentences. Because Taraban & McClelland's stimuli were specifically constructed to have thematic role expectations that bias the reader toward an NP-attachment, and Rayner et al. stimuli were not constructed with such thematic factors in mind, it was possible that the contexts might have different effects on the two groups of stimuli. Nonetheless, evidence for an increase in preference for VP-attachment that is due to having two VP referents in context would, at the very least, require an extension of the principle of referential support.

Method

Subjects. Twenty-four undergraduates of the University of California, Santa Cruz participated in the experiment for course credit. All subjects were native English speakers.

Stimuli and Design. Sixteen context pairs were constructed to accompany sixteen target sentence pairs. Eight of the target pairs were taken from Rayner, et al. (1983) and eight were taken from Taraban & McClelland (1988). Referent order was counterbalanced across stimuli to avoid a referent recency effect (Clifton & Ferreira, 1989). Sixteen filler stimuli were constructed with contexts and targets that were superficially similar to the experimental stimuli. To minimize the difference between the two contexts, the 1-VP-referent context was formed by substituting the second candidate referent with an alternate verb. To avoid lexical priming effects, the structurally ambiguous preposition in a target sentence was not included in the corresponding context paragraph. See Figure 1.

Procedure. Contexts were presented in full on an 80-column computer screen by an IBM PC. Target sentences were subsequently presented in a non-cumulative word-by-word self-paced moving window fashion (Just, Carpenter & Woolley, 1982). Order of

CONTEXT:
One day on the subway, a kid got on carrying a weapon in each hand. He almost hit someone by swinging a whip and pretended to *threaten/hit* someone else using a baseball bat. Then he started to approach the people sitting next to me. There was a girl who had a wart and a boy who had a scar.
TARGET:
That kid hit the girl with a *wart/whip* before he got off the subway.

Figure 1. The 2 VP referent context contains *hit* and *hit*, while the 1 VP referent context contains *hit* and *threaten*. The NP attached target had *wart*, and the VP attached version had *whip*. This target sentence was taken from Rayner et al. (1983).

experimental and filler trials was randomized with the first 6 trials being fillers, to give the subject practice. An open-ended comprehension question followed each trial. Reading times were recorded for the disambiguating noun (*wart* or *whip*) and for the next four words.

Results

A repeated-measures analysis of variance (ANOVA) was computed for the answers to the comprehension questions. A main effect of Target Attachment was observed such that NP attachments produced more errors (9.4%) than did VP attachments (1.6%): $F1(1,16)=10.99$, $p<.005$; $F2(1,8)=11.67$, $p<.005$. This off-line result is consistent with the minimal attachment principle (Frazier, 1978, 1987).

Data for trials in which the subject gave an incorrect answer to the comprehension question (5.5%) were excluded from the reading time analysis. Trials in which reading time was greater than 2.5 standard deviations from the mean for a given word position (2.1%) were also excluded.

A repeated-measures ANOVA collapsing across recorded word positions revealed a hint of a three-way interaction of Sentence Source (Rayner et al. / Taraban & McClelland) X Context Bias (1-VP-Referent / 2-VP-Referents) X Target Attachment (NP-attached / VP-attached). It appeared that an increase in the VP attachment bias due to 2 VP referents in context was evident in Rayner et al.'s sentences but not in Taraban & McClelland's sentences. However, planned comparisons between means showed no significant differences in reading time.

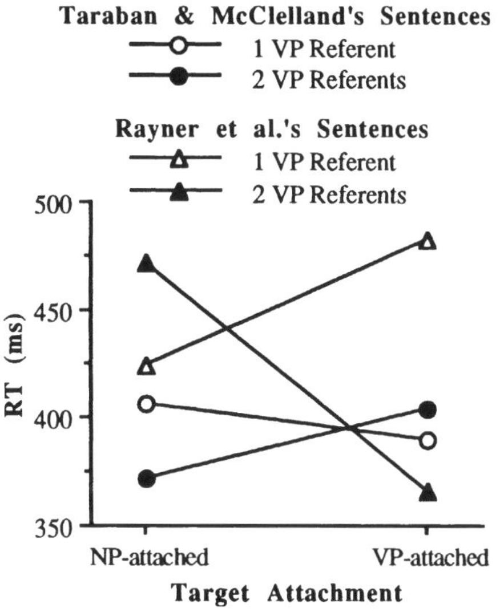

Figure 2. The effect of a second VP referent in context on reading times of the disambiguating noun in NP-attached and VP-attached sentences. Only small differences are seen with Taraban & McClelland's sentences. But with Rayner et al.'s sentences, a robust interaction between context and attachment is observed. This difference in context effect between the stimulus groups is evidenced in a significant three-way interaction (see text).

To test for an immediate effect of context, I conducted an analysis of reading times for the disambiguating noun. This showed a more robust interaction of Sentence Source X Context Bias X Target Attachment: $F1(1,16)=6.62$, $p=.02$; $F2(1,8)=4.34$, $p=.07$ (see Figure 2). Planned comparison protected t-tests indicated that: 1) with Rayner et al.'s sentences, a significant VP attachment bias (NP-attached vs. VP-attached) was observed in the 2 VP referent context ($p<.01$) but no significant attachment bias was observed in the 1-VP-referent context, and 2) Rayner et al.'s VP-attached targets were read faster when preceded by a 2-VP-Referent context than when they were preceded by a 1-VP-Referent context ($p<.01$).

Discussion

The nature of the differential effects between stimulus groups complicates interpretation of the data. I initially expected that the effect of the second VP referent would be most visible in Taraban & McClelland's stimuli because, in the absence of context, they have an NP-attachment bias. Curiously, context had no effect on the reading times of these sentences. In contrast, reading times for Rayner et al.'s sentences, that have a VP-attachment bias outside of context, were modulated by the number of VP referents in context.

A possible, though post hoc, interpretation of this interaction between context effect and sentence source derives from a difference in thematic relations that characterize the VP-attached versions of the two groups of sentences. In six of the eight sentences from Rayner et al., the VP-attachment entails the thematic tole type, *instrument*. Thus, the corresponding contexts contained two like events that were distinguished by their instruments (see Figure 1). Conversely, of the several thematic roles that characterize the VP-attachments of Taraban & McClelland's stimuli, only one sentence involved an instrument role. It is possible that, in context, minimal pairs of yet-to-occur events that are distinguishable by their *instruments*, rather than by location, manner or time, are more effective in producing a conceptual expectation for greater specification of a related VP. Whether one accepts this post hoc interpretation of the differences observed between sentence sources or not, the fact remains that an increase in VP-attachment preference is observed in Rayner et al.'s stimuli due solely to introducing a second VP referent in context.

As stated before, a contextual influence from having two yet-to-occur events in the discourse does not falsify the principle of referential support. By expanding its notions of definite reference and pragmatic presupposition to allow verb phrases to act in much the same way as noun phrases, Referential Theory might be able to account for this effect. Such expansion, however, is just a smaller version (though evolving in the same direction) of exactly what I am proposing. Gradually expanding a highly-restricted theory each time an effect is observed that it previously ruled out is, arguably, a too narrowly focused approach to the problem that may prevent us from considering other fruitful interpretations.

Alternatively, a much broader theory may better characterize the solution space with which we have to work. The less restrictive theory of discourse-driven structural expectations readily predicts appropriate effects from 2-NP-referent contexts and 2-VP-referent contexts. This theory posits that when an explicit conceptual uncertainty (such as a minimal pair of entities or events) is introduced into context, a reader may develop an expectation for conceptual disambiguation that would select, on pragmatic grounds, between alternatives of a subsequent ambiguity. If the reader encounters a syntactic ambiguity, of which only one alternative promises to dismabiguate the conceptual uncertainty, that syntactic alternative will be the preferred structural assignment.

As an example of when modification of the VP would be preferred on grounds of conceptual uncertainty, consider a context in which there is a little boy, Jimmy, and he is having dinner with the Queen of England. Jimmy's mother has instilled in him an overwhelming fear of the consequences of him using the wrong utensil at the wrong time in the presence of royalty. As desert comes around, he breaks into a cold sweat because he can't remember whether to eat his cake using the fork farthest left from the plate or to eat it using the fork above the plate. (Here, we have the two possible events between which a distinction is crucial, at least to Jimmy and, perhaps, the Queen.) If we later read, "*Finally, Jimmy ate some cake...*", we clearly expect to be told the important discriminating information, not which cake he ate, but which fork he used.

Each of these minimal pair contexts no doubt has a different degree of influence on attachment preference. The principle of referential support, however, has no obvious way of accounting for this variability. The degree of mutual exclusivity between two entities or events is likely, according to the theory of discourse-driven structural expectations, to modulate the strength of the context effect.[6] For an extreme example, in a 2-NP-Referent context that contains two guns and someone intends to pull the trigger on someone else, a case of one gun having bullets and the other having blanks would produce a stronger expectation for entity (or referent) distinction (hence, modification of the NP "*the gun*") than would a case where one gun has .32 caliber bullets and the other has .38 caliber bullets. "*John picked up two guns; one that had bullets and one that had blanks. He pointed them both at me. Then, he pulled the trigger of the gun with...*"

In light of very recent evidence for numerous simultaneous constraints on syntactic ambiguity resolution, such as availability of syntactic alternatives, frequency of co-occurrence, semantic plausibility, NP definiteness, referential and temporal contexts, noun animacy, and verb subcategorization information (Burgess & Tanenhaus, in preparation; Hindle & Rooth, 1990; McDonald, 1992; Pearlmutter & McDonald, 1992; Spivey-Knowlton & Sedivy, in preparation; Spivey-Knowlton et al., 1992; Trueswell & Tanenhaus, 1991, this volume; Trueswell, et al., 1989; Trueswell, Tanenhaus & Kello, in preparation), it would be imprudent to argue that the principle of referential support is completely wrong, or, for that matter, that the minimal attachment principle is completely wrong. In fact, with respect to these two opposing theories, results from experiments manipulating definiteness of the NP that precedes the PP suggest that Minimal Attachment's predictions fail precisely where Referential Theory's succeed, and vice versa (Spivey-Knowlton & Sedivy, in preparation). (However, the effects predicted by Minimal Attachment can also be accounted for by other factors.[7]) We find that, contrary to Referential Theory and in accordance with Minimal Attachment, NP-attached sentences with *indefinite NPs* elicit significant garden-paths. However, contrary to Minimal Attachment and in accordance with Referential Theory, NP-attached sentences with *definite NPs* elicit significantly larger garden-paths than those with indefinite NPs.

Given this reliable effect of NP definiteness (the backbone of Referential Theory), I do not wish to subsume the principle of referential support as an epiphenomenon of discourse-driven structural expectations. The two theories, however, are by no means completely separable. Finding and testing those places where the theories clearly diverge (e.g., NP definiteness, continuous degree of constraint) will be a challenge for the future.

This examination of discourse-driven structural expectations is in its initial stages and needs a great deal more refinement. Future work will require independent measures of the specificity of the expectation set, such as sentence completion and rating tasks. The resulting normed range of strongly-constraining to weakly-constraining contexts must then map appropriately onto degree of context effect in on-line reading time results. Moreover, demonstrating context effects with target sentences that contain *indefinite NPs* would be compelling evidence for a theory that is broader than the principle of referential support. The present result is a first step in finding the applicability of one such "broader theory" and developing ways to test it. This line of work has eventual implications for the issue of modularity in language processing and for the general goal of finding what sources of information are important in sentence processing at any stage, early or late.

[6] The greater effect from events that are distinguished by instrument roles, mentioned above, may be due to their mutual exclusivity being more salient than that of events that are distinguished by location, manner or time roles.

[7] Minimal Attachment's predictions for PP-ambiguities will often coincide with other locally-determined accounts of attachment preference, such as thematic role expectations (Taraban & McClelland, 1988) and frequency of co-occurrence (Hindle & Rooth, 1990).

References

Altmann, G. (1986). Reference and the resolution of local syntactic ambiguity: The effect of context during human sentence processing. Unpublished Ph.D. dissertation, University of Edinburgh.

Altmann, G. (1989). Parsing and interpretation: An introduction. *Lang. & Cog. Proc.*, 4, 1-19.

Altmann, G., & Steedman, M. (1988). Interaction with context during human sentence processing. *Cognition*, 30, 191-238.

Burgess, C. & Tanenhaus, M. (in preparation). The interaction of semantic and parafoveal information in syntactic ambiguity resolution.

Clifton, C. & Ferreira, F. (1989). Ambiguity in context. *Lang. & Cog. Proc.*, 4, 77-104.

Crain, S. & Steedman, M. (1985). On not being led up the garden path: The use of context by the psychological parser. In Dowty, Kartunnen, & Zwicky (Eds.) *Natural language parsing*. Cambridge: Cambridge U. Press.

Ferreira, F. & Clifton, C. (1986). The independence of syntactic processing. *JML*, 25, 348-368.

Fodor, J. (1983). *The modularity of mind: An essay on faculty psychology*. Cambridge: MIT Press.

Frazier, L. (1978). On comprehending sentences: syntactic parsing strategies. Unpublished Ph.D. dissertation, University of Connecticut.

Frazier, L. (1987). Theories of syntactic processing. In Garfield (Ed.) *Modularity in knowledge representation and natural language processing*. Cambridge: MIT Press.

Frazier, L., Clifton, C. & Randall, J. (1983). Filling gaps: Decision principles and structure in sentence comprehension. *Cognition*, 13, 187-222.

Hindle, A. & Rooth, M. (1990). Structural ambiguity and lexical relations. In *Proceedings of the 28th Annual Meeting of the Association for Computational Linguistics.*

Just, M., Carpenter, P. & Woolley, J. (1982). Paradigms and processes in reading comprehension. *JEP:General*, 111, 228-238.

McDonald, M. (1992). Probabilistic constraints and syntactic ambiguity resolution. Manuscript submitted for publication.

Pearlmutter, N. & McDonald, M. (1992). Garden-paths in "simple" sentences. Presented at the 5th Annual CUNY Sentence Processing Conference.

Rayner, K., Carlson, M. & Frazier, L. (1983). The interaction of syntax and semantics during sentence processing: Eye movements in the analysis of semantically biased sentences. *JVLVB*, 22, 358-374.

Spivey-Knowlton, M. (1991). Discourse-driven expectations in the on-line resolution of syntactic ambiguity. Unpublished B.A. thesis, UC Santa Cruz.

Spivey-Knowlton, M. & Sedivy, J. (in preparation). The effect of noun phrase definiteness on parsing preferences: An interactive constraints account.

Spivey-Knowlton, M., Trueswell, J. & Tanenhaus, M. (1992). Context effects in parsing reduced relative clauses. To appear in *Canadian Journal of Psychology: Special Issue.*

Taraban, R. & McClelland, J. (1988). Constituent attachment and thematic role expectations, JML, 27, 597-632.

Trueswell, J. & Tanenhaus, M. (1991). Tense, temporal context and syntactic ambiguity resolution. *Lang. & Cog. Proc.*, 6, 303-338.

Trueswell, J. & Tanenhaus, M. (this volume). Consulting temporal context in sentence comprehension: Evidence from the monitoring of eye movements in reading. Presented at the 14th Annual Conference of the Cognitive Science Society. Bloomington, IN.

Trueswell, J., Tanenhaus, M. & Garnsey, S. (1989). Semantic influences on parsing: Use of thematic role information in syntactic disambiguation. Presented at the 2nd Annual CUNY Sentence Processing Conference.

Trueswell, J., Tanenhaus, M. & Kello, C. (in preparation). Evidence of the immediate use of verb subcategorization information in sentence processing.

Webber, B. (1988). Tense as a discourse anaphor. *Computational Linguistics*, 14, 61-73.

Acknowledgements: Part of this work was supported by a Sproull Graduate Fellowship from the University of Rochester. I would like to thank Bill Farrar, Ray Gibbs, Ken McRae, Mark Steedman, Mike Tanenhaus, and John Trueswell for helpful comments and discussions. The experiment described was part of a B.A. honors thesis in Psychology at UC Santa Cruz.

Consulting temporal context during sentence comprehension: Evidence from the monitoring of eye movements in reading

John C. Trueswell and Michael K. Tanenhaus
trueswel@psych.rochester.edu mtan@psych.rochester.edu

Department of Psychology
University of Rochester
Rochester, NY 14627

Abstract

An important aspect of language processing is the comprehender's ability to determine temporal relations between an event denoted by a verb and events already established in the discourse. This often requires the tense of a verb to be evaluated in relation to specific temporal discourse properties. We investigate the time course of this process by examining how the temporal properties of a discourse influence the initial processing of temporarily ambiguous reduced relative clauses. Much of the empirical work on the reading of reduced relative clauses has revealed that readers experience a large mis-analysis effect (or 'garden-path') in reduced relatives like "The student spotted by the proctor received a warning" because the reader has initially interpreted the verb "spotted" as a past tense verb in a main clause. Recent results from an eye movement study are provided which indicate that this mis-analysis of relative clauses can be eliminated when the temporal constraints of a discourse do not easily permit a main clause past tense interpretation. Such a finding strongly suggests that readers process tense in relation to the temporal properties of the discourse, and that constraints from these properties can rapidly influence processing at a structural level.

Introduction

Understanding a discourse requires the listener or reader to develop and maintain a representation of the events and entities under discussion. In addition, many linguistic expressions can only be interpreted by making reference to information in this discourse representation. It is this contextually dependent property of language which motivates incremental models of sentence processing. Such models propose that, as a sentence unfolds in time, readers and listeners make rapid commitments to interpretations by incrementally evaluating linguistic input in relation to some portion of the discourse (Crain & Steedman, 1985; Altmann & Steedman, 1988; Ni and Crain, 1990). In this way, sentence processing is made sensitive to the constraints provided by various aspects of the discourse. Most of the empirical evidence supporting this approach focuses on the contextual dependencies of definite noun phrases. Such work provides evidence suggesting that the reference to individuals, sets and their properties via noun phrase expressions influences sentence processing at a structural level. We report here complementary research which examines whether similar processes apply to the establishment of, and reference to, temporal and event discourse entities.

Within computational linguistics, there is an increasing interest in the topic of tense and aspect and its relation to discourse processing problems (Webber, 1988). In particular, researchers wish to characterize, at a processing level, how to interpret and represent complex events in a changing discourse. Two issues of critical importance are the problem of encoding new event descriptions (typically denoted by tense-marked verb phrases), and the problem of manipulating references to events already established in the discourse. A number of researchers have pointed out that the tense of a verb can only be interpreted in relation to the temporal information associated with the event structure of the discourse (Dowty, 1986; Hinrichs, 1986; Partee, 1984; Steedman, 1982; Webber, 1988). In fact, Webber (1988) has recently argued that tense, by definition, should be considered a discourse anaphor, similar to that of a definite noun phrase.

Most models of tense interpretation take as their starting point the work of Hans Reichenbach (1947),who proposed that interpretation of tense requires three separate semantic entities: time of speech (S), time of event (E), and time of reference (R). Time of speech, S, is the time at which the utterance occurs (usually thought of as "now"). The time of event, E, is the time at which an event (usually denoted by a verb) takes place. The time of reference, represents the temporal perspective from which the described event is viewed. When these ideas are incorporated into a model of discourse, S and R are typically considered variables which change during the progression of the discourse.

Without going into the details of how S and R are established for a current discourse segment (see Trueswell and Tanenhaus, 1991), consider the following sentence pair, in which the first sentence establishes a particular relation between S and R.

1a. Tomorrow, Ms. Brown will announce she is running for president. (R > S).

b. She ran for president in the last election as well.

The notation (R>S) indicates that sentence (1a) establishes a time of reference for the current discourse segment which is ahead of the current time of speech ("now"). Consider the past tense verb "ran" in 1b. A past tense verb requires the opposite relation between R and S from that established in 1a (ie., it requires time of reference precede time of speech (R<S)). Such a tense shift has serious consequences to the discourse organization. For, not only does a new time of reference need to be established, but, most likely, a new discourse segment as well (Grosz & Sidner, 1986).

Webber (1988) has proposed that discourse processing often proceeds along the "path of least resistance" (minimal inference). Readers and listeners will avoid costly discourse processing, such as segmentation, unless explicitly marked (as in 1b). Natural questions that arise from this processing perspective are: a) at what point in processing do listeners and readers consider tense in relation to temporal properties of the discourse, and b) can discourse constraints related to such considerations influence processing at a structural level.

To explore these questions, we made use of the well-known temporary ambiguity associated with restrictive reduced relative clauses. Ambiguous phrases such as "The student spotted..." can be continued either as a main clause with a past tense form of the verb (e.g., "The student spotted the proctor and went back to work") or a relative clause with a participial form of the verb (e.g., "The student spotted by the proctor received a warning"). Crucially, a past tense in a main clause and a participle in a relative clause require different temporal and event information to be present in the discourse. A past tense verb in a main clause requires time of reference to precede time of speech. However, participial verbs in relative clauses make no restrictions on R and S, because these phrases only refer to events already established in the discourse.

Much of the empirical work on the reading of reduced relative clauses has revealed that readers experience a large mis-analysis effect (or 'garden-path') in such sentences because the reader has initially interpreted the verb as a past tense verb in a main clause (Ferreira and Clifton, 1986; Trueswell et al, 1992). We explored whether this mis-analysis of relative clauses can be eliminated when the temporal constraints of a discourse do not easily permit a main clause past tense interpretation. Such a finding would demonstrate that readers process tense in relation to the temporal properties of the discourse, and that this temporal information can rapidly influence syntactic processing.

Method

Twenty University of Rochester students participated in the study. Eye movements of each subject were recorded using a Stanford Research Institute Dual Purkinje Eyetracker. The eyetracker transmitted information concerning horizontal and vertical eye position angle of the subject's right eye to a Macintosh II computer equipped with an analog to digital conversion board. Eye position was determined by sampling every millisecond both the horizontal and vertical eye angle and blink signals from the eyetracker. The position and duration of each fixation was computed and stored to disk. Stimuli were displayed on a high resolution RGB monitor, with the subject's eyes approximately 64 cm from the screen. The visual angle of each character was slightly greater than 12 minutes of arc, allowing for one character resolution from the eyetracker position signals. For each subject, the eyetracker was aligned and the signal was calibrated to the screen coordinates. Each trial consisted of the presentation of a three to five sentence paragraph. The subject read the sentences silently and then pressed a button to signal that he or she was finished. On about a third of the trials, a yes/no comprehension question appeared on the screen prior to the line trace test. Subjects were given feedback as to whether their button response answer was correct. (See Trueswell et al, 1992, for a more complete description of a similar procedure.)

An example target paragraph is shown in Table 1. The 16 target paragraphs were embedded in 44 distractor paragraphs containing various constructions, and various temporal relations. See Trueswell & Tanenhaus (1991) for a complete list and description of the target stimuli.

Table 1: Example of target stimuli.

Context
Past
Several students were sitting together taking an exam yesterday. A proctor came up and noticed one of the students cheating.
Future
Several students will be sitting together taking an exam tomorrow. A proctor will come up and notice one of the students cheating.

Target
Reduced:
The student spotted by the proctor received/will receive a warning.
Unreduced:
The student who was spotted by the proctor received/will receive a warning.

Predictions

The past context establishes the appropriate temporal parameters for a new past event to be introduced into the discourse, whereas the future context does not. Therefore, a fragment such as "The student spotted..." should be interpreted as part of a main clause in the past context and as part of a reduced relative clause in the future context. This prediction was confirmed in a sentence completion study (Trueswell & Tanenhaus, 1991). The present study was conducted to determine whether these temporal constraints would affect online processing.

Results

Total Reading times

Each target sentence was divided into three scoring regions: the initial noun phrase ("The student"), the relative clause ("spotted by the proctor"), and the verb phrase region ("will receive" or "received a"). Figure 1a presents mean total reading times for each region (the total amount of time spent within a region, including rereads of a region). To better see any effect of ambiguity, Figure 1b shows the difference in total reading times between ambiguous reduced relative clauses and unambiguous unreduced relative clauses for both future and past contexts. Positive differences reflect increased difficulty in processing a region when the relative clause did not contain the "who was" (ie., when it was reduced).

The temporal information in the discourse context clearly mediated the magnitude of the garden-path for sentences with reduced relative clauses. In past contexts, the (ambiguous) reduced relative clauses took longer to read than the (unambiguous) unreduced relative clauses. Such a result indicates that readers initially interpreted the first verb as a past tense verb in a main clause. The subsequent "by-phrase" signalled that the segment was in fact a relative clause, requiring a reanalysis. However, in the future contexts, reduced relatives took only slightly longer to read than the unreduced relatives.

A) Total reading times

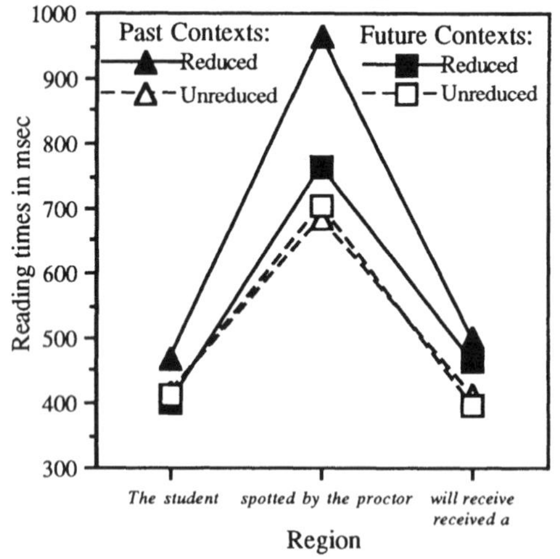

B) Total reading time differences (Reduced minus Unreduced)

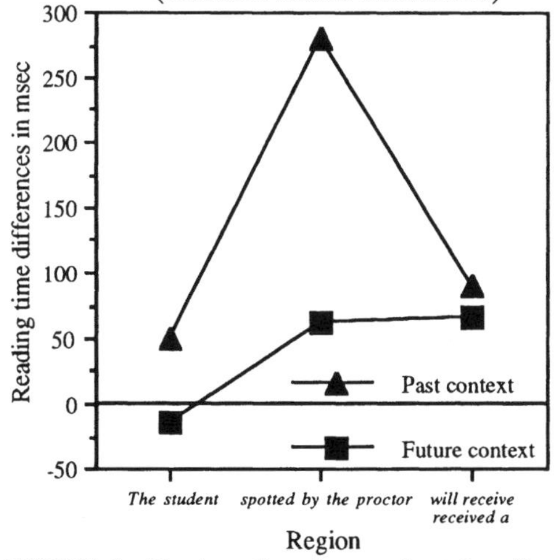

FIGURE 1: Total reading times and total reading time differences.

An analysis of variance was conducted on the data at each scoring region. At the initial noun phrase region, there were no reliable effects or interactions. At the relative clause region, however, there was a clear interaction between the type of clause (reduced and unreduced) and the type of context (future or past), $F_1(1,16)=10.87$, $MSe=22102$, $p<0.01$; $F_2(1,12)=4.88$, $MSe=31098$, $p<0.05$. In past contexts (triangles), subjects spent more time reading the reduced relative clauses as compared to unreduced relatives ($F_1(1,16)=48.11$, $MSe=16312$, $p<0.01$; $F_2(1,12)=10.40$, $MSe=57405$, $p<0.01$). In future contexts (squares), subjects took only slightly longer to read the reduced relatives as compared to unreduced relatives, and the difference was not reliable, $F_1(1,16)=1.64$, $F_2(1,12)=1.46$). Finally, at the verb phrase region, there was an effect of relative clause type, $F_1(1,16)=24.14$, $MSe=5759$, $p<0.01$; $F_2(1,12)=5.07$, $MSe=17015$, $p<0.05$.

First and Second pass reading times

Total reading times do not differentiate between initial processing and secondary processing (re-reads). To investigate better the time course of these effects, total reading times were separated into first pass and second pass reading times. First pass reading times were obtained by summing the durations of all left-to-right fixations in a region plus any regressions made to other points within that region. When the reader made an eye movement out of a region (either a regressive eye movement to a prior region or a forward movement to a following region), first pass reading was considered complete for that region. Second pass reading times include all re-readings of a region. It is commonly believed that first pass reading times reflect the processes associated with deriving an initial interpretation of a region whereas large second pass reading times reflect reanalyses caused by arriving at an incorrect initial interpretation.

The difference in reading times between the reduced and unreduced relative clauses for first and second pass readings are shown in Figures 2a and 2b respectively. Both the first and second pass data show the same pattern as the total reading times. There were large differences in reading times between the reduced and unreduced relative clauses in the past contexts, and only small unreliable differences in the future contexts. Moreover, all of the effects and interactions at the relative clause region that were significant in the total reading times were also reliable in the first pass ($p<0.05$). Such a result indicates that temporal information influenced ambiguity resolution early on in processing.

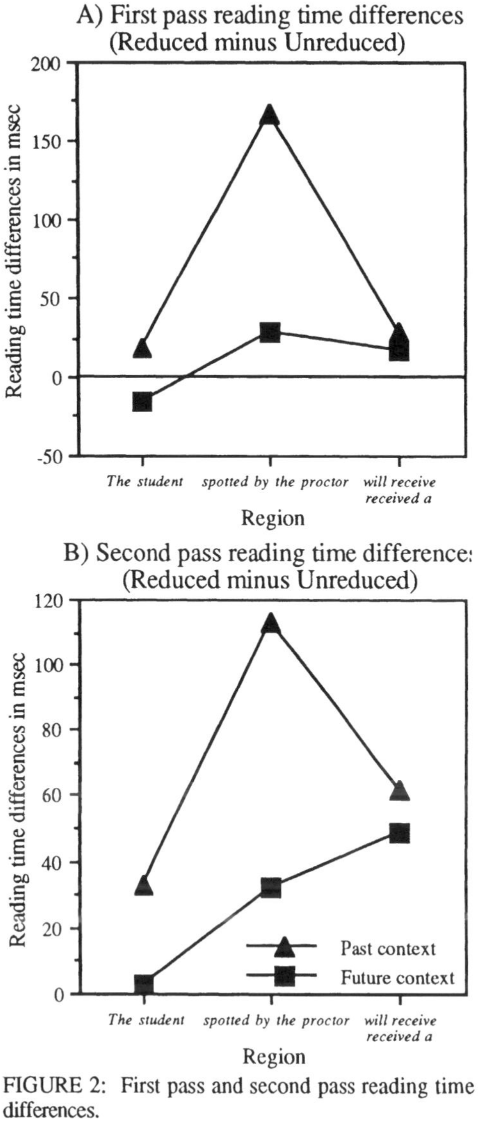

FIGURE 2: First pass and second pass reading time differences.

Dividing up the relative clause region

Finally, additional analyses were conducted in which the relative clause scoring region was divided into two smaller regions ("spotted by" and "the proctor"). Such a division provides a breakdown of relative clause processing into early and late processing of the clause. Figure 3a and 3b present reading time differences between reduced and unreduced relative clauses for both total and first pass reading times. In the first half of the relative clause, reduced relatives are more difficult than unreduced, regardless of temporal context, in both the total reading times ($F_1(1,16)=21.92$, $MSe=13459$, $p<0.01$; $F_2(1,12)=8.52$, $MSe=24855$, $p<0.05$) and the first pass reading times ($F_1(1,16)=5.13$, $MSe=5217$, $p<0.05$; $F_2(1,12)=3.68$, $MSe=24990$, $p<0.1$). In the second half of the relative clause, an interaction between clause type and context type occurs in the total reading times ($F_1(1,16)=15.16$, $MSe=7128$, $p<0.01$; $F_2(1,12)=5.59$, $MSe=11582$, $p<0.05$) and the first pass reading times ($F_1(1,16)=5.11$, $MSe=4109$, $p<0.05$; $F_2(1,12)=3.68$, $MSe=4491$, $p<0.1$). Thus, in both first pass and total reading times, we see a reduction in processing difficulty in the second half of reduced relatives when the temporal context is in the future. Such results suggest that readers do have some early processing difficulties with reduced relative clauses in future contexts, but that the temporal constraints of the discourse rapidly eliminate these problems.

General Discussion

In this paper, recent work on tense and discourse was combined with basic incremental assumptions about sentence processing to predict that the processing of reduced relative clauses would be affected by the tense of the preceding discourse. This prediction was confirmed. Readers only showed evidence of a consistent misanalysis of reduced relative clauses when the context permitted the introduction of a new past event (past contexts). Conversely, readers showed little evidence of a misanalysis when context made the introduction of a new past event difficult (future contexts). Such results indicate that readers rapidly consult the temporal context of the discourse to interpret the tense of a verb.

The only indication of a misanalysis in the future contexts comes from small elevations found in the first half of the relative clause. These elevations may be due to conflicting local lexical and syntactic constraints supporting a main clause interpretation. It is likely that the comprehension system is sensitive to the fact that most sentences beginning with a noun phrase followed by a verb are main clause active sentences (Bever, 1970). Moreover, the semantic information of the head noun phrase supports a main clause interpretation (e.g., "student" is an animate object and a likely Agent of "spotted") (Trueswell, Tanenhaus & Garnsey, 1992). It is possible that discourse constraints cannot immediately "veto" these cues, but rather the temporal constraints from the discourse can only combine rapidly with these other constraints to arrive quickly at a relative clause interpretation (e.g., via constraint satisfaction). In addition, some of this small effect may be due to some localized complexity

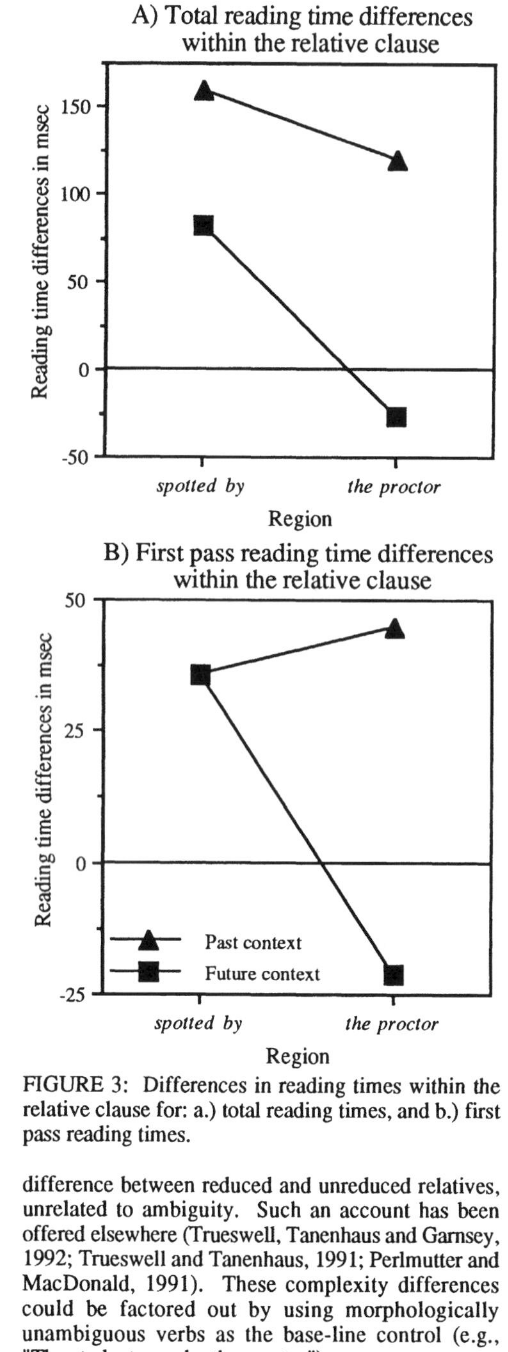

FIGURE 3: Differences in reading times within the relative clause for: a.) total reading times, and b.) first pass reading times.

difference between reduced and unreduced relatives, unrelated to ambiguity. Such an account has been offered elsewhere (Trueswell, Tanenhaus and Garnsey, 1992; Trueswell and Tanenhaus, 1991; Perlmutter and MacDonald, 1991). These complexity differences could be factored out by using morphologically unambiguous verbs as the base-line control (e.g., "The student seen by the proctor").

The present results, and those of Trueswell and Tanenhaus (1991), demonstrate that readers establish temporal relations between an event being introduced in a sentence and events already established in the context by making use of tense and temporal parameters such as time of reference and time of speech. An important question for future research will be to determine the extent to which other types of temporal information in the discourse model influence immediate sentence processing. It is possible that only temporal parameters such as time of reference and time of speech are consulted when a tense marked verb is encountered. Alternatively, readers may attempt more detailed updating in which the event denoted by a verb is related to other relevant events in the discourse. If this is the case, real-world knowledge about determining the temporal and causal relations between events might also be consulted during immediate sentence processing. A likely candidate would be information about the plausible antecedents and consequences of an event (cf. Moens & Steedman, 1988).

Acknowledgements

This work was supported by NIH grant # HD 27206 and the Center for Visual Science NIH grant # EY 01319.

References

Altmann, G. & Steedman, M. (1988). Interaction with context during human sentence processing. *Cognition, 30*, 191-238.

Bever, T. G. (1970). The cognitive basis for linguistic structures. In J. R. Hayes (ed.) *Cognition and the development of language*. New York: John Wiley and Sons.

Crain, S. & Steedman, M. (1985). On not being led up the garden path: The use of context by the psychological parser. In D. Dowty, L. Karttunen and A. Zwicky (eds), *Natural language parsing: Psychological, computational and theoretical perspectives*. Cambridge: Cambridge University Press.

Dowty, D. (1986). The effects of aspectual class on the temporal structure of discourse: Semantics and pragmatics. *Linguistics and Philosophy, 9*, 37 - 62.

Ferreira, F. & Clifton,C. (1986). The independence of syntactic processing. *Jrnl. of Mem. and Lang., 25*, 348 - 368.

Grosz, B. & Sidner, C. (1986). Attention, intention, and the structure of discourse. *Comp. Ling.,12(3)*, 175-204.

Hinrichs, E. (1986). Temporal anaphora in discourses of English. *Linguistics and Philosophy, 9*, 63 - 82.

Johnson-Laird, P.N. (1983). *Mental models: Towards a cognitive science of language, inference and consciousness*. Cambridge: Cambridge University Press.

Moens, M. & Steedman, M. (1988). Temporal ontology and temporal reference. *Comp. Linguistics, 14*, 15 - 28.

Ni, W. & Crain, S. (1990). How to resolve structural ambiguities. *Proceedings to NELS, 20, Vol 2*, 414 - 427.

Partee, B. (1984). Nominal and temporal anaphora. *Linguistics and Philosophy, 7*, 287 - 324.

Perlmutter, N. & MacDonald, M. (1991). Effects of pragmatic cues on syntactic ambiguity processing. Presented at the *4th Annual CUNY Sentence Processing Conference*, Rochester, NY.

Reichenbach, H. (1947). *Elements of Logic*. New York: Free Press.

Steedman, M. (1982). Reference to past time. In R. Jarvella and W. Klein (eds), *Speech, place and action*. New York: John Wiley and Sons.

Trueswell, J. C., Tanenhaus, M. K., & Kello, C. (1992). Verb-specific constraints in sentence processing: Separating effects of lexical preference from garden-paths. Manuscript submitted for publication.

Trueswell, J. C., Tanenhaus, M. K., & Garnsey, S. M. (1992). Semantic influences on syntactic ambiguity resolution. Manuscript submitted for publication.

Trueswell, J. C. & Tanenhaus, M. K. (1991). Tense, temporal context, and syntactic ambiguity resolution. *Language and Cognitive Processes, 6(4)*, 303 - 338.

Webber, B. (1988). Tense as a discourse anaphor. *Comp. Linguistics, 14*, 61 - 73.

Plausibility and Syntactic Ambiguity Resolution

Neal J. Pearlmutter
Dept. of Brain & Cognitive Sciences, E10-109
Massachusetts Institute of Technology
Cambridge, MA 02139
neal@psyche.mit.edu

Maryellen C. MacDonald
Hedco Neurosciences Building
University of Southern California
Los Angeles, CA 90089-2520
mcm@gizmo.usc.edu

Abstract

Different theories of human syntactic parsing make conflicting claims concerning the role of non-syntactic information (e.g. semantics, real world knowledge) on on-line parsing. We address this debate by examining the effect of plausibility of thematic role assignments on the processing of syntactic ambiguities. In a self-paced reading experiment, ambiguous condition reading times were longer than unambiguous condition times at the point of syntactic disambiguation only when plausibility cues had supported the incorrect interpretation. Off-line measures of plausibility also predicted reading time effects in regression analyses. These results indicate that plausibility information may influence thematic role assignment and the initial interpretation of a syntactic ambiguity, and they argue against parsing models in which the syntactic component is blind to plausibility information.

Introduction

An important issue in psycholinguistic research concerns the type of information that the sentence processor or *parser* uses in deciding which interpretation(s) to pursue for a syntactic ambiguity. The most prominent position on this issue has been that the parser consults only syntactic information during the initial parse and is blind to all non-syntactic information (Frazier, 1987; Rayner, Carlson & Frazier, 1983; Ferreira & Clifton, 1986). This position is accompanied by two other claims: 1) that the parser can pursue only one interpretation at a time, and 2) that the parser must make its choice as soon as the ambiguity is encountered. A parser with these features could make many costly mistakes, and in order to increase the chances of making correct choices, the parser could rely on parsing strategies. The best known of these is Minimal Attachment (Frazier, 1987): The parser chooses the "simplest" interpretation, building only the syntactic (phrase structure) representation with the fewest nodes. On this view, semantic and/or discourse information may affect the parse only after the parser has begun building a structure; the semantic information aids in reanalysis when the parser's first analysis fails (Rayner et al., 1983; Ferreira & Clifton, 1986).

In this paper, we argue that the parser is not as limited in the resources available to guide decisions about ambiguities as Frazier's "garden path" model suggests. We present evidence that some non-syntactic information can guide the parser in the initial interpretation of syntactic ambiguities. Specifically, we claim that lexical or discourse information affects decisions about *thematic roles* that are assigned to noun phrases in a sentence, and that these role assignments can in turn affect the choices the parser makes when confronted with a syntactic ambiguity.

A thematic role is an abstract semantic relation between a verb and one of its noun phrase (NP) arguments; these roles include the *Agent* of the action described by the verb, the *Theme* (sometimes called *Patient*) of the action, the *Instrument* of the action, and others. For example, the verb *cut* has two different argument structures; it may appear with the three arguments Agent, Theme, and Instrument, as in *Mary cut the rope with a knife*, and it may also omit the Instrument role, as in *Mary cut the rope*. Tanenhaus and Carlson (1989) have hypothesized that when a verb is encountered in the input, its alternative argument structures are activated in parallel, and tentative thematic role assignments are attempted for the NPs that have been encountered to that point. Possible assignments are evaluated by comparing information in the noun's lexical entry with the verb's argument structure. For example, Agents must be animate, and so an inanimate NP will not be assigned an Agent thematic role. If one argument structure produces compatible assignments for the NPs encountered to that point and other structures do not, then this argument structure will be adopted by the parser. The parser will then adopt a phrase structure representation of the input that is compatible with the chosen argument structure.

Trueswell, Tanenhaus & Garnsey (1992) tested these claims in a study measuring reading times in sentences containing "main verb/reduced relative" ambiguities, in which it is temporarily ambiguous whether a verb in the sentence is the main verb (e.g., *examined* in *The defendant examined the document*) or is modifying a noun in a reduced relative clause (as in *The defendant examined by the lawyer was nervous.*). All critical sentences in the study were resolved with the reduced relative interpretation; the key manipulation was the animacy of the subject noun, as in the pair *The [defendant/evidence] examined by the lawyer turned out to be unimportant.* The noun *defendant* is animate and so could be the Agent of *examined* (permitting the simple main verb structure, in which the defendant examined something), but *evidence* is inanimate and thus cannot receive an Agent thematic role, thereby eliminating the simple main verb interpretation as a candidate for this ambiguous sentence. In the animate (*defendant*) condition, which permits the incorrect main verb interpretation, reading times were longer at the disambiguation (*by the...*) compared to unambiguous controls. In the inanimate *evidence* condition, however, there was no ambiguity effect at the disambiguation, suggesting that subjects had used animacy information early in the parse to guide thematic role assignment (assigning the Theme role to *evidence*), which in turn guided them to the correct interpretation of the ambiguity.

The Trueswell et al. results suggest that thematic role assignments, rather than strategies like Minimal Attachment, guide the parser from the outset. If so, an obvious question is how the parser evaluates the possible thematic role assignments as it is constructing a syntactic representation of the input. The Trueswell et al. data point to the importance of animacy, but other data suggest that some non-syntactic information (e.g., real world knowledge such as who is likely to send or receive flowers, Rayner et al., 1983) does not influence early parsing. One reason for this difference might be that animacy can be represented in nouns' lexical entries, independent of context: *evidence* is inherently inanimate and the Agent role for *any* verb must be assigned to an animate noun. Perhaps the feature ±Animate can influence thematic role assignment, but real world plausibility information, such as the likelihood of someone sending flowers, cannot. However, the Rayner et al. (1983) stimuli differ from the Trueswell et al. (1992) stimuli not only in the real world *vs.* lexical (animacy) nature of the information, but also in the timing of the constraints and the ambiguity. The animacy constraints arrive at the ambiguous verb (e.g. the conjunction of *evidence* and *examined*), but Rayner et al.'s real world constraints appear well after the ambiguity is introduced and so might arrive too late to influence the initial parse. If we are to examine further the role of non-syntactic information in parsing, these factors must be unconfounded so that a plausibility cue is available at the point of ambiguity.

The study below manipulates features of the subject NP in a main verb/reduced relative ambiguity, as in Trueswell et al. (1992), but instead of a lexical feature like ±Animate, we manipulate the *plausibility* of a thematic role assignment within animate NPs. For the verb *applauded*, for example, *musician* is more plausible as a Theme than as an Agent, and *audience* is a more plausible Agent than Theme. NPs that are better Themes than Agents should be implausible in the main verb interpretation and will therefore promote the reduced relative interpretation, whereas NPs that are better Agents than Themes will promote the main verb interpretation. When the disambiguation reveals that the reduced relative interpretation is correct, readers should have difficulty (evidenced by slowed reading times) only when a Poor Theme had promoted the incorrect interpretation; Good Theme sentences should not be problematic. This pattern of results would indicate that real world plausibility information can influence thematic role assignment and parsing and would be difficult for Frazier's model to explain. If plausibility does not guide the parser, however, this would be evidence that only basic lexical information like animacy can influence thematic role assignment, but that plausibility information cannot.

Method

Ratings. To develop a set of stimuli manipulating plausibility, preliminary ratings were conducted on a large pool of NP-verb combinations such as *prisoner captured, student taught, audience applauded, musician applauded,* etc. We obtained ratings of the NP as both a subject (Agent role assigned, main verb reading) and as an object (Theme role assigned, relative clause reading) of a particular verb. We also obtained a comparative preference rating between these two readings. Because raters were directly comparing the two readings, both readings were conveyed with sentence fragments, as in *The audience applauded the...* and *The audience that was applauded by the...*, for the subject and object rating respectively. Each fragment was rated on a scale from 1=good to 7=bad; subjects were instructed to base their ratings on whether the fragment made sense and described something that could happen in real life. The comparative rating was also performed on a 7-point scale, with 1 corresponding to a very strong subject preference and 7 a very strong object preference. Four rating lists were prepared so that subjects would see a given verb in only one pair of fragments. Forty-five subjects participated in the ratings.

Materials. Using the pool of rated fragments, 24 stimulus items were constructed. Each item was constructed from two pairs of rated fragments that had contained the same ambiguous verb. The NP in one fragment pair had received a very low comparative rating (such that the NP was strongly preferred as an Agent, and thus was a Poor Theme, e.g. *The audience applauded the...* was preferred to *The audience that was applauded by the...*), and the NP in the other pair had received a very high comparative rating (such that this NP was strongly preferred as a Theme, e.g. *musician applauded*). Introductory phrases and completions were constructed for each item, as were unambiguous controls, so that each stimulus item could appear in four versions by crossing two independent variables: Ambiguity (ambiguous (A) *vs.* unambiguous (U)) and Role Plausibility (Good Theme (GT) *vs.* Poor Theme (PT)) as shown in Table 1. Sixty-one filler items with a variety of syntactic structures were also prepared, as were yes/no comprehension questions for all items.

Subjects. Fifty-two MIT students who had not participated in the ratings pretest were paid for their participation. All were native speakers of English.

Procedure. Subjects read the sentences on a CRT in a Moving Window display (Just et al., 1982) in which they pressed a key to see each word of the sentence. A trial began with a display of dashes indicating all nonspace characters of the sentence. When the subject pressed a computer terminal key, the first word appeared, replacing its dashes. With each successive keypress, the visible word reverted to dashes and the next word appeared. Pressing the key after the last word removed the sentence and displayed a comprehension question. The subject responded by pressing a key labeled "Yes" or "No" and did not receive feedback on accuracy. Subjects saw five practice items, followed by the experimental and filler items in pseudo-randomized order; at least one filler preceded any experimental item.

Results

Reading time. Reading times at each word position were trimmed for each subject, replacing data points more than 3 SD over the relevant mean with the 3 SD cutoff value, affecting less than 3% of the data. The experimental sentences were divided into four regions for the reading time analysis, as shown in Figure 1. Region 1 consisted of the Good or Poor Theme NP. Region 2 contained the ambiguous verb, as well as *that was/were* in the unambiguous conditions. Region 3 consisted of the by-phrase, and Region 4 contained the first two words of the main verb phrase (the disambiguation). The remainder of the sentence was analyzed separately. Region 4 is considered the disambiguation because the by-phrase in Region 3 does not necessarily disambiguate the sentence. The preposition "by" is ambiguous between an agentive phrase (which would disambiguate) and a locative (e.g., *by the sea*, which would not disambiguate if the preceding verb had an intransitive interpretation, as some stimuli did have; see MacDonald, 1992; Pritchett, 1989). Only the last word of Region 3 suggests the correct interpretation, and a locative interpretation is never ruled out.

Table 1--Example sentence set

The producer said that the live broadcast went smoothly, and	
Good Theme (GT)	
Ambig.	the musician applauded by the host enjoyed the show immensely from start to finish.
Unambig.	the musician that was applauded by the host enjoyed the show immensely from start to finish.
Poor Theme (PT)	
Ambig.	the audience applauded by the host enjoyed the show immensely from start to finish.
Unambig.	the audience that was applauded by the host enjoyed the show immensely from start to finish.

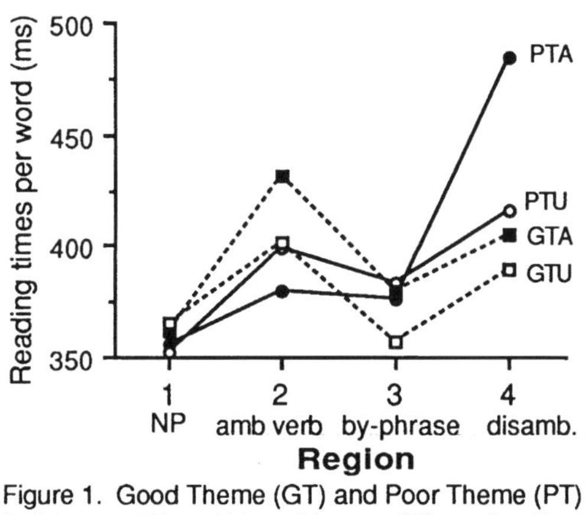

Figure 1. Good Theme (GT) and Poor Theme (PT) Ambiguous (A) and Unambiguous (U) reading times

An omnibus ANOVA revealed the predicted interaction of Role, Ambiguity and Region [$F_1(3,153) = 5.71, p = .001, F_2(3,69) = 2.87, p < .05$], justifying examination of effects at each Region separately.

As expected, there were no significant effects in Region 1, before the ambiguity was introduced. In Region 2, however, reading times were identical in the two unambiguous conditions (401 ms and 399 ms for the Good Theme (GT) and Poor Theme (PT) conditions, respectively), but there was an effect of Role on the Ambiguous items. The PT Ambiguous reading times were relatively fast (380 ms), while GT Ambiguous times were relatively slow (432 ms), a significant difference [$t_1(51) = -2.58, p < .05$; $t_2(23) = -3.14, p < .005$]. This result replicates MacDonald's (1992) finding that reading times in an ambiguous region are faster when the available information promotes a simple interpretation (here, the PT condition, which is plausible in the simple main verb interpretation), than when information promotes a more complex interpretation (the GT condition here, which is more plausible in the reduced relative structure than in the main verb structure). In other words, the plausibility information appears to be influencing reading times (and parsing) at the earliest possible point, and thus does not appear to be limited to later backtracking.

Analyses in Region 3, containing the by-phrase, revealed a Role x Ambiguity interaction [$F_1(1,51) = 4.28, p < .05$; $F_2(1,23) = 4.15, p = .05$]. The source of this interaction was that the GT Unambiguous condition (358 ms) was at least marginally faster than the other three conditions (all 377-384 ms), while these latter three did not differ from one another.

The analysis for Region 4, containing the disambiguation, also revealed a Role x Ambiguity interaction [$F_1(1,51) = 4.68, p < .05$; $F_2(1,23) = 2.22, p = .15$]. Further analysis revealed that in the PT conditions, reading times were significantly slower in the ambiguous sentences (485 ms) than in the unambiguous controls (416 ms) [$t_1(51) = 2.81, p < .01$; $t_2(23) = 2.58, p < .05$], but there was no effect of ambiguity in the GT conditions (405 *vs.* 390 ms in ambiguous and unambiguous conditions, *t's* < 1). In other words, subjects had no difficulty when plausibility favored a Theme assignment for the first NP, but they were slow when a Poor Theme had promoted the incorrect main verb interpretation. In addition, GT Unambiguous reading times were marginally faster than PT Unambiguous reading times [$t_1(51) = 1.90, p < .10$; $t_2(23) = 2.10, p < .05$], indicating an effect of plausibility even in the absence of ambiguity.

Analyses of the remaining words showed small effects of Role, but neither the Ambiguity factor nor its interaction with Role was robust. There was also no reliable effect of any factor on comprehension question accuracy.

In sum, the reading time data show a clear difference between the ambiguity effects for the Good and Poor Theme conditions. In the PT condition, containing no helpful plausibility information, reading times at the ambiguous verb were faster than in the unambiguous condition, but they increased substantially over unambiguous times at the disambiguation. Subjects appear to parse these sentences as the garden path model predicts, initially interpreting the ambiguous verb as the main verb of the sentence, and then showing surprise at the disambiguation. By contrast, the GT Ambiguous condition produced slower reading times at the ambiguous verb, suggesting that subjects are already considering the (correct) relative clause interpretation at this point. Later, at the disambiguation, subjects show no significant increase in reading times over the unambiguous condition. This pattern suggests that either subjects are using the plausibility information to choose an interpretation well before they reach the disambiguation, or, if Minimal Attachment is acting first and the plausibility information is affecting backtracking later, then this sequence of events must all be taking place very early, during the reading of the ambiguous verb in the GT condition, because subjects seem to have completely settled on the correct interpretation by the disambiguation. While it is conceivable that misanalysis and reanalysis are both occurring during the reading of a single word, the simpler explanation is that the plausibility information is guiding thematic role assignment and parsing from the outset. Additional evidence is provided by the relationship between the reading time and rating data, which we consider next.

Ratings. The reading data demonstrated that the sentences in the GT conditions did not show the same ambiguity effects as those in the PT conditions; the ratings will investigate whether the plausibility constraints operate along a continuum. The ambiguous and unambiguous reading times and ambiguous-unambiguous difference scores in the four regions were correlated with the object ratings we obtained in our initial norming procedures. The object ratings were those in which subjects rated the goodness of the NP as a direct object in unreduced relative clause fragments such as *The audience that was applauded by the....* The object ratings therefore reflect the plausibility of the eventual disambiguation of the ambiguity in the experimental items.

Because the stimuli were selected with the aid of comparative ratings (in which subjects indicated which interpretation was more plausible), it is important to examine the range of object ratings at each level of Role to be sure that these ratings are not merely recreating the dichotomous Role variable. There was a small range of object ratings in the Good Theme condition (1.07-2.27) and a wide range of ratings within the Poor Theme condition (1.95-5.27). The distributions of ratings in the two conditions thus overlap, with some Good Theme items having poorer ratings than some Poor Theme items. Given this range, correlations between ratings and reading times should be informative. It should also be useful to examine the correlations within each level of Role, though only the strongest relationships could be identified, because the range of ratings and statistical power are reduced when only half of the items enter into the analysis.

The rating/reading time correlations for the entire stimulus set are shown in Table 2. As expected, correlations for Region 1 (before the ambiguity was introduced) were all non-significant, but in Region 2 (the ambiguous verb), the negative correlation with ambiguous reading times was marginally significant, indicating that ambiguous verbs following better Theme NPs tended to be read more slowly than when the NP was a less plausible Theme. In other words, subjects read the ambiguous verb more slowly when plausibility information provided some reason to expect a complex relative clause structure, and they read more quickly when plausibility information promoted the simple main verb interpretation. As mentioned above, this effect is important, as it suggests that the plausibility of alternate interpretations is affecting the *initial* interpretation of the sentence. Moreover, the correlation was present within the items in the Good Theme condition [$r = -.36, p < .10$]. Thus the best of the Good Theme items were read more slowly at the ambiguity than were slightly poorer Good Theme items.

In Region 3 (the by-phrase), ambiguous reading times were unrelated to the ratings, but unambiguous reading times decreased as the ratings improved. These results show that within a few words after encountering the Region 2 verb, the language processor is sensitive to the anomaly of assigning a Theme role to a poor Theme in an unambiguous relative clause. These effects on unambiguous reading times were also present, though smaller, within the Poor Theme items considered separately [$r = .26, p < .25$] and within the Good Theme items [$r = .38, p < .10$].

Table 2--Rating/Reading time correlations

	Region			
	1 NP	2 amb vrb	3 by-phr.	4 disamb
Amb	-0.10	-0.27*	-0.05	0.49***
Un	-0.22	0.07	0.34**	0.28**
A-U	0.07	-0.23	-0.29**	0.26*

A-U = Amb-Unamb Difference score. Ratings are 1=good to 7=bad; positive correlations therefore indicate that better ratings produce faster reading times or smaller A-U differences.

*$p < .10$. **$p < .05$. ***$p = .0005$.

In Region 4 (the disambiguation), both the ambiguous and unambiguous reading times correlated well with the ratings, though the effect was much stronger for the ambiguous items, suggesting that plausibility accounts for some effects in all sentences regardless of ambiguity, but that it has an additional effect on the ambiguous items. The correlation between the ratings and the difference scores indicates that as the initial NP becomes a better Theme, the ambiguity effect decreases at the point of disambiguation: Subjects expect the reduced relative structure when the plausibility information favors it, and not when plausibility promotes a main verb structure.

In sum, the correlations confirm and extend the major effects in the reading times: As the Region 1 NPs become poorer Themes and so promote the incorrect simple interpretation, reading times on the ambiguous verb in Region 2 become faster, and reading times in the disambiguating Region 4 become slower. As mentioned above, it is difficult to ascribe these effects to backtracking and revision; it appears that the early plausibility information supporting a relative clause encourages subjects to adopt that structure well before the definitive disambiguation is reached. Moreover, the fact that meaningful correlations can be found within the Good and Poor Theme items considered separately demonstrates the graded nature of the plausibility cues investigated here. It is not only dichotomous lexical features like ±Animate that influence syntactic parsing: Our graded plausibility effects also have a rapid influence on ambiguity resolution.

Discussion

The results of the present experiments are consistent with the Trueswell et al. (1992) findings but contrast with the Rayner et al. (1983) results that suggested that plausibility information could not help to determine the course of first-pass parsing. The differences across studies suggest that sufficiently strong plausibility cues present at the onset of an ambiguity can aid in initial syntactic parsing, as in the present experiments, but that weaker and late-occurring plausibility cues in the Rayner et al. study could not influence early processing, so that subjects still had difficulty in those sentences at the disambiguation. Our work in progress pursues these issues in two ways: 1) manipulation of the information between the ambiguity and the disambiguation (which was always a by-phrase here, cf. MacDonald, 1992; Rayner et al., 1983), and 2) examination of the role of working memory capacity (Just & Carpenter, 1992) on the ability to compute the plausibility information that guides thematic role assignment and ambiguity resolution.

Our results to date suggest that probabilistic plausibility cues do not aid parsing in an all-or-nothing manner, but rather that the ease with which the parse is handled is directly related to the strength of the plausibility cue. These findings do not lend support to a two-stage model like the garden path model (Frazier, 1987), in which plausibility information can influence only the second stage of analysis (backtracking), unless, as we have noted, the first and second stages can take place during the processing of a single word, a modification that substantially weakens such models. By contrast, our results are easily explained by the class of models discussed by Tanenhaus and Carlson (1989), in which the syntactic parser builds its syntactic interpretation based on the best-supported of the available verb argument structures. Trueswell et al. (1992) showed that the "best-supported" argument structure could be influenced by relevant lexical semantic information such as noun animacy; the current data indicate that plausibility information arising from the combination of noun and verb meanings can also influence this determination.

Our results are also compatible with Altmann & Steedman's (1988) "weakly interactive" model of parsing in which the parser proposes multiple structural analyses in parallel, and pragmatic factors determine which interpretation will be pursued by the parser. What is not yet clear from any of this research is the level of representation that is mediating the plausibility effects: It may be the discourse representation of the sentence, incorporating real world knowledge, or it may be that activation of a rich lexical/semantic representation of nouns and verbs like *audience* and *applauded* is sufficient to guide thematic role assignments. An additional possibility is that multiple sources of information constrain thematic role assignments, but over different time courses, so that rapidly computed information (e.g. animacy) has an earlier effect than more subtle information such as the plausibility information investigated here. Additional study of this issue, especially that focusing on the graded nature of the cues, will have important implications for theories of both syntactic parsing and lexical representation.

Acknowledgments

Much of this research was completed while M. MacDonald was at MIT, where this research was supported by BRSG 2 S07 RR07047-23 from the National Institutes of Health.

References

Altmann, G. & Steedman, M. 1988. Interaction with context during human sentence processing. *Cognition, 30*, 191-238.

Ferreira, F., & Clifton, C. 1986. The independence of syntactic processing. *Journal of Memory and Language, 25*, 348-368.

Frazier, L. 1987. Sentence processing: A tutorial review. In M. Coltheart (Ed.), *Attention and performance XII: The psychology of reading.* Hillsdale, NJ: Erlbaum. 559-586.

Just, M.A. & Carpenter, P.A. 1992. A capacity theory of comprehension: Individual differences in working memory. *Psychological Review, 99*, 122-149.

Just, M. A. , Carpenter, P. A., & Woolley, J. D. 1982. Paradigms and processes in reading comprehension. *Journal of Exp. Psychology: General, 3*, 228-238.

MacDonald, M. C. 1992. *Probabilistic constraints and syntactic ambiguity resolution.* Submitted.

Pritchett, B. L. 1988. Garden path phenomena and the grammatical basis of language processing. *Language, 64*, 539-576.

Rayner, K., Carlson, M., & Frazier, L. 1983. The interaction of syntax and semantics during sentence processing. *Journal of Verbal Learning and Verbal Behavior, 22*, 358-374.

Tanenhaus, M. K., & Carlson, G. 1989. Lexical structure and language comprehension. In W. Marslen-Wilson, (Ed.), *Lexical Representation and Process.* Cambridge, MA: MIT Press.

Trueswell, J. C., Tanenhaus, M. K., & Garnsey, S. M. 1992. *Semantic influences on parsing: Use of thematic role information in syntactic disambiguation.* Submitted.

The Time Course of Metaphor Comprehension

Phillip Wolff and Dedre Gentner
Department of Psychology
Northwestern University
2029 Sheridan Road, Evanston, IL 60208
wolff@ils.nwu.edu, gentner@ils.nwu.edu

Abstract

This research investigates the process by which people understand metaphors. We apply processing distinctions from computational models of analogy to derive predictions for psychological theories of metaphor. We distinguish two classes of theories: those that begin with a *matching* process (e.g Gentner & Clement, 1988; Ortony, 1979) and those that begin with a *mapping* process (e.g. Glucksberg and Keysar, 1990). In matching theories, processing begins with a comparison of the two terms of the metaphor. In mapping theories, processing begins by deriving an abstraction from the base (or *vehicle*) term, which is then mapped to the target (or *topic*).

In three experiments, we recorded subjects' time to interpret metaphors. The metaphors were preceded by either the base term, the target term, or nothing. The rationale was as follows. First, interpretations should be faster with advanced terms than without, simply because of advanced encoding. The important prediction is that if the initial process is mapping from the base, then seeing the base in advanced should be more facilitative than seeing the target in advanced. Matching models predict no difference in interpretation time between base and target priming. The results generally supported matching-first models, although support for mapping-first models was found with highly conventional metaphors.

Introduction[1]

How are metaphors understood? For instance, on hearing *A surgeon is a butcher*, we apprehend the meaning of the metaphor to be something like "a surgeon is someone who cuts flesh." How do we derive this interpretation? Most current psychological theories of metaphor describe the process of comprehension in only very general terms. In this research we apply processing distinctions made in computational models of analogy to specify and test broad classes of processing models in psychology.

[1]This work was supported by NSF grant BNS 87-20301 and ONR grant N00014-89-J1272 awarded to the second author.

Approaches to Processing

From computational models of analogy we can derive three general classes of processing algorithms, which can be described as *matching, mapping*, and *matching-then-mapping* (Gentner, 1989).(See Figure 1.) In *matching* models, the commonalities of two representations are recognized by aligning their parts and structures (Winston, 1982). We might interpret the metaphor *A surgeon is a butcher*, for instance, by noting that both cut flesh. Ortony's (1979) salience imbalance theory of metaphors is primarily a matching theory. In *mapping* theories, processing begins by accessing or creating from the base term a higher-order category or abstraction which is then used to attribute properties to the target term (e.g., Burstein, 1983; Carbonell, 1983; Greiner, 1988; Kedar-Cabelli, 1985). Glucksberg and Keysar (1990) have recently proposed a theory of metaphor that fits into this processing framework. According to their model, a metaphor like *My job is a jail* is understood by accessing or deriving an abstraction or category associated with the *base* term, or *vehicle, jail*; this abstraction (e.g., confining institution) is then applied to the *target* term, or *topic, job*. We will refer to this particular instantiation of the mapping perspective as *Category-mapping*.

The third kind of processing can be termed *Matching-then-mapping* (Falkenhainer, Forbus, & Gentner, 1989; Hofstadter, Mitchell & French, 1987; Holyoak & Thagard, 1990). For example, in the

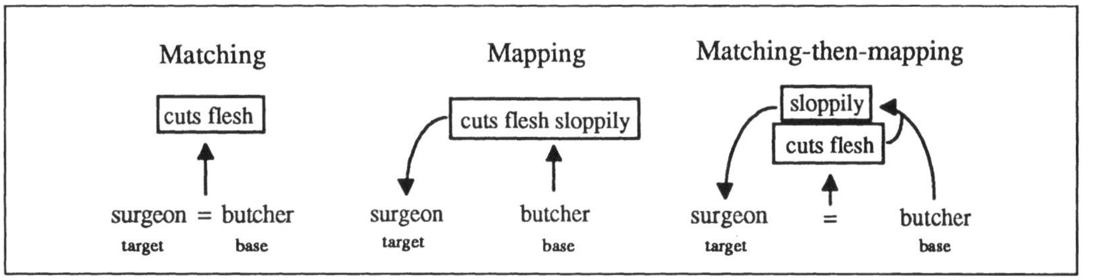

Figure 1. Models of metaphor comprehension

Structure-mapping Engine (SME), processing begins by matching the representations of the base and target. Once a global alignment is discovered, predicates may be carried over from the base to the target using the systematicity criterion: given a common system of interconnected predicates, predicates that belong to that system in the base but are not yet present in the target are mapped to the target as *candidate inferences* (Falkenhainer, Forbus, & Gentner, 1989). Returning to the metaphor *A surgeon is a butcher*, a Match-then-map process asserts that people first find the commonality that both surgeons and butchers cut flesh; then further properties belonging to this common 'cutting-flesh' system are mapped from butchers to surgeons: e.g., that butchers cut sloppily.

Since the research here primarily addresses the initial stages of processing, we will lump match-then-map models with the simple matching models and call them collectively *match-first models*; these will then be contrasted with *map-first* models.

Strengths of the Processing Accounts

Match-first models capture the intuition that metaphor shares with similarity a focus on commonalities (e.g., Tversky, 1977). A further advantage of the match-first models is their ability to deal with the problem of property selection when the same base is compared with different targets. For example, consider the metaphors *The surgeon is a butcher* and *The general is a butcher*. Though they have the same base, they convey quite different meanings: the first suggests a clumsy surgeon, the second a ruthless and efficient general. This property selection problem is not as easily handled by map-first models, which must account for how *butcher* gives rise to two different abstractions in these two contexts (Map-first models can be augmented with the assumption that people try abstractions sequentially, until one fits the target, though this explanation seems cumbersome at best.) Finally, the match-first view can predict further inference as part of a secondary mapping stage.

The map-first perspective, as exemplified by Glucksberg and Keysar's (1992) Category-mapping theory, has its appealing aspects as well. First, it captures the intuition that there should be an intimate relation between metaphor and categorization. Second, the map-first view explains why metaphors are often directional. Just as the class-inclusion statement "A surgeon is a doctor." cannot be reversed to make "A doctor is a surgeon.", neither can the metaphorical statement "A vacation is a doctor." be reversed to make "A doctor is a vacation." The map-first view also offers an intuitive explanation for our ability to understand metaphors that convey new inferences about the target. On hearing "The waiter is a skyscraper" we understand that the waiter is tall even through "tallness" is not necessarily a feature present in our prior representation of waiters. This importing of new features in to the target is a problem for simple matching models.

Testing the Models

The match-first and map-first models make different predictions for the time course of processing of metaphors. According to the Category-mapping model (a map-first model), processing begins with the base term, from which a category must be derived to apply to the target term. In contrast, according to match-first models, processing begins with a comparison of the two terms. This suggests a way to test these theories. If processing begins with the base, as is implied by a category-mapping process, then metaphoric processing should be facilitated if people see the base in advance of the metaphor. More specifically, there should be greater facilitation when the base is given in advance than when the target is given in advance. (Some facilitation is expected in either case, under any model, simply by virtue of permitting a head start in encoding.)

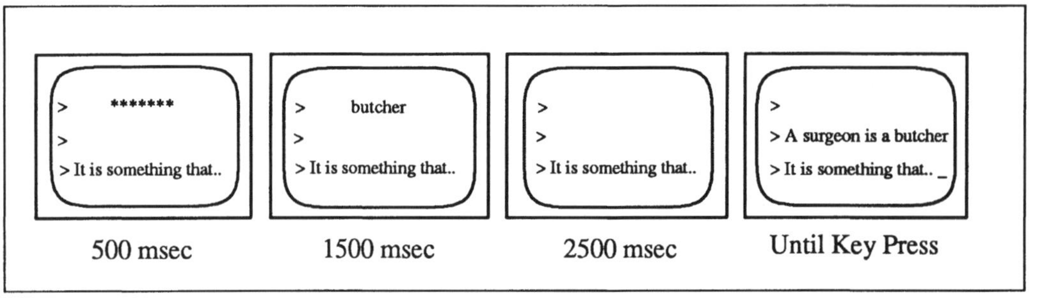

Figure 2: General order and timing of presentations in Experiments 1, 2 and 3.

In contrast, if metaphor interpretation begins with a matching process, there should be no differential advantage for seeing the base vs. the target prior to seeing the metaphor.

Based on this logic, we carried out three experiments to compare the adequacy of these models as psychological accounts of metaphor interpretation. In all three experiments, we recorded subjects' time to interpret metaphors. The metaphors could be primed by an immediately prior presentation of the base term, the target term, both terms, or nothing (no prime). The key question is whether there is more facilitation for base primes than for target primes.

Experiment 1

Method

Subjects. The subjects were 24 Northwestern University students.

Materials. Twenty-four metaphors were drawn from the literature. Prior to the study, in order to establish the preferred direction of the metaphors, we asked 40 subjects to order each pair of terms to make their preferred metaphor. In this way we were able to independently determine the most natural base and target.

Procedure. Subjects were run on individual computers in groups of 2 to 5 subjects at a time. Subjects were shown a series of metaphors, randomly ordered, and instructed to begin typing in an interpretation to each metaphor as soon as they had it well formulated. (These instructions were given to forestall subjects' adopting a strategy of starting to type immediately and then pausing to develop an interpretation.) Prior to seeing each metaphor, subjects were shown either the target, the base, or a blank line. Figure 2 shows the order in which information appeared. Subjects were instructed to use the preceding words to get a head start on their interpretation. Interpretation-time was recorded from the moment the metaphor appeared on the screen until the subject pressed the first key for his or her interpretation. The time to type each interpretation was also recorded.

Design. All prime types (i.e. TARGETs, BASEs, or BLANKs) were presented to each subject and counterbalanced across all metaphors.

Results and Discussion

The mean interpretation times are shown in Table 1. (All analyses throughout this paper are based on subjects' mean responses.) Contrary to the predictions of the Category-mapping model, there was no advantage for BASE primes over TARGET primes; indeed, there was a nonsignificant difference in the reverse direction. A one-way repeated-measures analysis of variance showed a significant effect of Prime type, $F(2,46) = 7.72$, $p < 0.001$. Pairwise Bonferroni tests indicated that at the 0.05 level, interpretation times were faster for BASEs than for BLANKs, $t(23) = 3.25$, and also for TARGETs than for BLANKs, $t(23)=3.16$. Interpretation times for TARGETs and BASEs, however, did not differ significantly, $t(23)=0.63$.

These results provide no evidence for the category-mapping prediction that processing begins with the base. The results are consistent with the match-first models (although only by default). However, one concern in interpreting these results is that the method may not have been fair to the Category-mapping model, since subjects were not told whether the prime was a base or a target. One might argue that the assumptions of the Category-mapping model are that subjects treat the base term differently from the target term. On this reasoning, it is to be expected that subjects would be unable to begin

TABLE 1

Mean Times to Begin an Interpretation from Experiment 1

Prime Type	BASE	TARGET	BLANK
Means	4951	4828	5648
Standard Error	996	1090	1475

TABLE 2

Mean Times to Begin an Interpretation from Experiment 2

Prime Type	BOTH	BASE	TARGET	BLANK
Means	3048	4086	4420	4829
Standard Error	1576	1474	1658	1436

interpretation until they know that they have the base term. To address this possibility, in Experiment 2 the prime's role (i.e. whether it was the target or base) was made explicitly clear. This was done to encourage the fullest possible use of the primes. One other change was to add a fourth condition in which both primes were shown together. This amounts to simply showing the whole metaphor from the start. This condition tested the prediction of the match-first views that having both terms at the outset should be faster than having only one (since interpretation can begin only when both terms are present). The Category-mapping model makes no strong predictions as to whether seeing both primes should lead to faster interpretation times than seeing the base. Thus, finding an advantage for both terms over one would not distinguish match-first from map-first positions, but *failure* to find such an advantage would count against match-first models.

Experiment 2

Method

Subjects. The subjects were 40 Northwestern University students.

Materials. The metaphors were the same as those used in Experiment 1.

Procedure. The procedure was the same as that used in Experiment 1 with two exceptions. First, the roles played by the primes were made explicit by putting them into sentence frames. So, if *butcher* was used as a base prime, subjects saw *A something is a butcher.* Similarly, if *surgeon* was used as a target prime, subjects saw *A surgeon is a something.* In the BLANK condition, subjects saw *A something is a something,* and in the newly added BOTH condition, subjects saw the complete metaphor: e.g. *A surgeon is a butcher.*

Design. As in Experiment 1, all prime types (i.e. TARGET, BASE, BLANK, BOTH) were presented to each subject and counterbalanced across all metaphors.

Results and Discussion

The mean interpretation times are shown in Table 2. Again contrary to the predictions of the Category-mapping model, there was no significant advantage for BASE primes over TARGET primes, although this time the difference was in the predicted direction. A one-way repeated measures analysis of variance indicated an overall significant effect of Prime type, $F(3,114) = 22.24$, $p < 0.001$. Pairwise contrasts using the Bonferroni t statistic at the .05 level indicated that interpretation times for BASEs (i.e., for metaphors given BASE primes) were faster than for BLANKs, $t(40) = 3.14$. In addition, BOTHs were faster than BASEs, $t(39) = 5.05$; and BOTHs were faster than TARGETs, $t(39) = 5.945$. No significant difference was found between TARGETs and BLANKs, $t(40) = 1.65$. Coming to the key result, no significant difference was found between BASEs and TARGETs, $t(40) = 1.79$.

Thus, as in Experiment 1, no support was found for the Category-mapping model. The match-first account fared better: as predicted, seeing BOTH primes resulted in faster interpretation times than seeing either BASEs or TARGETs alone. While this is not inconsistent with category mapping, it is a central prediction of the matching accounts.

However, although the Category-mapping model has received no strong support, there are some patterns that deserve consideration. First, although there is no significant BASE advantage over TARGETs, the direction of the means is consistent with the predictions of Category mapping. In addition, BASEs, but not TARGETs, were found to show an advantage over BLANKs. To examine the data more closely, we plotted cumulative curves of reaction times, as shown in

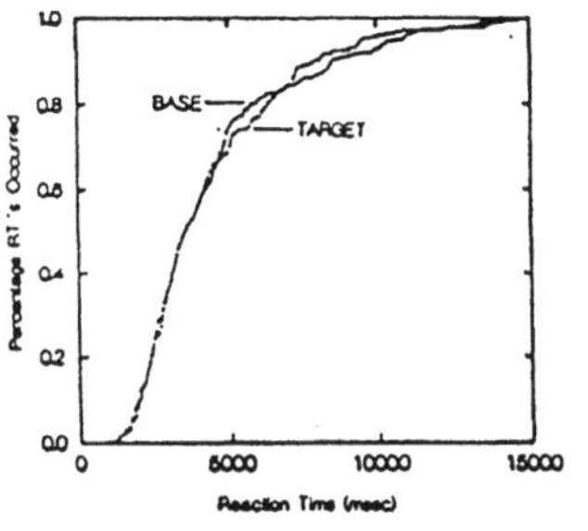

Figure 3: Exp. 2 Cumulative Percentage of Interpretation Times for Targets and Bases.

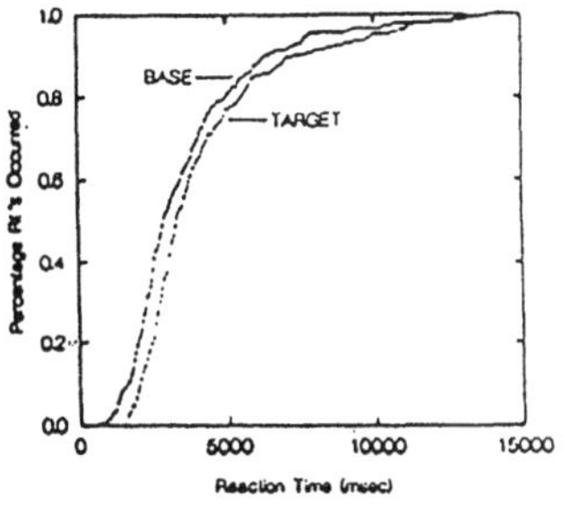

Figure 4: Exp. 4 Cumulative Percentage of Interpretation Times for Targets and Bases.

Figure 3. That is, we plotted the cumulative percentage of responses that had occurred by each duration from the beginning (i.e., the point when the metaphor appeared). If BASEs have an early advantage over TARGETs, we should see an initial difference between the cumulative curves (that is, there should be more BASE responses than TARGET responses at short durations). In fact, the two curves lie virtually on top of one another in the initial part of the distribution. The (nonsignificant) advantage of BASEs over TARGETs appears only at longer durations [2].

In summary, even when a prime's role was explicitly marked, no evidence was found for a category mapping. So far, the results are consistent with a match-first processing account, and inconsistent with a category-mapping account. In Experiment 3 we considered another factor that might affect whether people use a category-mapping model, namely, the conventionality of the metaphors. In the first two experiments the metaphors used were for most people relatively novel. Possibly this degree of novelty contributed to the apparent superiority of the match-first model over the map-first model in accounting for the patterns. Perhaps the category-mapping process is most probable when people are given metaphors whose bases have pre-stored 'stock' metaphorical interpretations. This would be reasonable, because with stock metaphors people could draw on relatively unambiguous existing abstractions. In Experiment 3 we tested this possibility. We applied the same basic priming procedure as in Experiment 2 to metaphors with conventional bases: e.g., *That waiter is a skyscraper* and *That plane is a dinosaur.*

[2] Conceivably, this pattern of BASE advantage that appears only for interpretation times of longer duration could suggest a Match-then-map model in which priming of the base term facilitates a later mapping and inference process.

Experiment 3

Method

Subjects. The subjects were 40 Northwestern University students.

Materials. Twenty metaphors were constructed with bases that possessed a stock metaphorical meaning.

Procedure. The procedure was the same as that used in Experiment 2.

Design. As in Experiment 3, all prime types (i.e. TARGETs, BASEs, BLANKs, and BOTHs) were presented to each subject and counterbalanced across all metaphors.

Results and Discussion

In Experiment 3, there was a major shift in the pattern of results: BASE primes led to significantly faster responding than TARGET primes, as predicted by the category-mapping model. A one-way repeated measures analysis of variance indicated a significant effect of Prime type, F (3,117) = 23.4, $p < 0.001$. Pairwise contrasts at the .01 level using the Bonferroni t statistic indicated that BASEs (M=3577) were faster than TARGETs (M=4218), $t(39) = 3.08$. In addition, BOTHs (M= 2770) were faster than BASEs, $t(38) = 3.74$; BOTHs were faster than TARGETs, $t(38) = 6.620$; and BASEs were faster than BLANKs (M= 4632), $t(39) = 4.05$. No significant difference was found between TARGETs and BLANKs, $t(39) = 1.62$.

This effect is consistent with the predictions of the Category-mapping model. This conclusion is supported by an inspection of the TARGET and BASE cumulative graphs in Figure 4, which show the BASE distribution clearly precedes the TARGET distribution. That is, in Experiment 3, subjects who saw BASE primes had a higher percentage of early responses than subjects who saw TARGET primes. This suggests that category-first models may apply well to metaphors whose bases have stock metaphorical meanings, but not to novel metaphors.

As in Experiment 2, BOTHs were faster than BASEs and TARGETs, as is compatible with both the Category-mapping and the Match-first. Also as in Experiment 2, BASEs were faster than BLANKs, while TARGETs were not.

Summary

The results of these three experiments can be summarized by the following conclusions: 1) Interpretation of novel metaphors begins with matching, and 2) Interpretation of stock metaphors may begin with mapping. The first conclusion was supported by the failure to find an interpretation advantage for metaphors preceded by bases in the first two experiments (which used relatively novel metaphors). The second conclusion was supported by the results of Experiment 3, in which a clear advantage was observed for metaphors following base primes when the metaphoric bases possessed stock metaphoric meanings.

This suggests implications for psychological theories of metaphor as follows. Match-first theories, such as those of Ortony and Falkenhainer, Forbus & Gentner, best reflect how people understand metaphors that are relatively novel. But match-first theories, notably Glucksberg and Keysar's recent proposal, best represent how people understand metaphors with relatively conventional 'stock' base meanings.

The findings of these experiments connect well with recent work on conventional metaphors in natural language understanding. Both Martin (1991) and Gibbs (in press) have noted that any proper treatment of metaphor must not ignore the influence of preexisting structures in long-term memory. The findings of these experiments affirm the importance of these preexisting structures by demonstrating how they might have an effect on processing.

These findings also highlight a more general consideration. Most theorists would agree that the target and base of a metaphor either belong to some common category or share some common attributes. However, a question remains as to whether it is the common category that determines which features two things share, or whether it is shared features that determine a common category. We suggest that in the case of relatively novel pairings, a common category is derived from a Matching/Mapping process. That is, a process of matching (and subsequent mapping) leads to a common system which can serve as a common category. In time, this category may come to have conventional status, in which case the time course of subsequent metaphors is changed.

References

Burstein, M. (1983). Concept formation by incremental analogical reasoning and debugging. In R. Michalski, J. Carbonell, & T. Mitchell (Eds.), *Machine Learning* (Vol. 2, pp.351-369). Los Altos, CA: Morgan Kaufman.

Carbonell, J.G. (1983). Derivational Analogy: A theory of reconstructive problem solving and expertise acquisition. In R. Michalski, J. Carbonell, & T. Mitchell (Eds.), *Machine Learning* (Vol. 2, 371-392). Los Altos, CA: Morgan Kaufman.

Falkenhainer, B., Forbus, K. D., and Gentner, D. (1989/90). The structure-mapping engine: Algorithm and examples. *Artificial Intelligence*, 41, 1-63.

Gibbs, R. W. (in press). Categorization and metaphor understanding. *Psychological Review*.

Gibbs, R. W. (1990). Psycholinguistics studies on the conceptual biases of idiomaticity, *Cognitive Linguistics*, 1, 207-246.

Gentner, D., Falkenhainer, and Skorstad. (1988). *Viewing metaphor as analogy*. In D.H. Herman (ed.), Analogical Reasoning (pp. 171-177), New York: Kluwer Academic Publishers.

Glucksberg, S., Keysar, B. (1990). Understanding metaphorical comparisons: Beyond Similarity. *Psychological Review*, 97, 3-18.

Greiner, R. (1988). Abstraction-based analogical inference. In D.H. Herman (ed.), *Analogical Reasoning* (pp. 147-170), New York: Kluwer Academic Publishers.

Hofstadter, D. R., Mitchell, M., & French, R. M. (1987). *Fluid concepts and creative analogies: A theory and its computer implementation* (Tech. Rep. No. 87-1). Ann Arbor, MI: University of Michigan, Fluid Analogies Research Group.

Kedar-Cabelli, S. (1985). Purpose-directed analogy. *Proceedings of the Seventh Annual Conference of the Cognitive Science Society*. (pp. 150-159), Irvine, California.

Martin, J. H. (1991). *Computational theory of metaphor*. San Diego: Academic Press.

Ortony, A. (1979). Beyond literal similarity. *Psychological Review*, 66, 161-180.

Tversky, A. (1977). Features of similarity. *Psychological Review*, 84, 327-352.

Winston, P.H. (1980). Learning and reasoning by analogy. *Communications of the ACM*, 23, 689-703.

Is the future always ahead? Evidence for system-mappings in understanding space-time metaphors.

Dedre Gentner and **Mutsumi Imai**
Department of Psychology
Northwestern University
2029 Sheridan Road Evanston, IL, 60208
gentner@aristotle.ils.nwu.edu, imai@aristotle.ils.nwu.edu

Abstract

Languages often use spatial terms to talk about time. FRONT-BACK spatial terms are the terms most often imported from SPACE to TIME cross-linguistically. However, in English there are two different metaphorical mapping systems assigning FRONT - BACK to events in time. This research examines the psychological reality of the two mapping systems: specifically, we ask whether subjects construct global domain-mappings between SPACE and TIME when comprehending sentences such as "Graduation lies before her" and "His birthday comes before Christmas."

Two experiments were conducted to test the above question. In both experiments, subjects' comprehension time was slowed down when temporal relations were presented across the two different metaphorical systems inconsistently. This suggests that people had to pay a substantial remapping cost when the mapping system was switched from one to the other. The existence of domain mappings in on-line processing further suggests that the two SPACE/TIME metaphorical mapping systems are psychologically real.

Introduction[1]

We often talk about time in terms of space. It has been pointed out that the correspondence between the two domains is orderly and systematic (e.g. Bennett, 1975; Clark, 1973; Lehrer, 1990; Traugott, 1978). Table 1 shows some SPACE-TIME correspondence in English (taken from Lehrer, 1990). Language in the TIME domain is roughly divided into three aspects: tense, sequencing, and aspect (Traugott, 1978). Our concern here is with sequencing, the system whereby events are temporally ordered with respect to each other and to the speaker. There are some universal properties in importing language about SPACE to describe TIME (Clark, 1973; Traugott, 1978). First, since TIME is usually conceived as an uni-dimensional property, the spatial terms that are borrowed are uni-dimensional terms (e.g., *front/back, up/down*) rather than terms that capture two or three dimensions (e.g. *narrow/wide, shallow/deep*). Second, to capture the sequential order of events, the time-line has to be directional. Thus ordered terms such as *front/back* and *before/after* are used, rather than symmetric terms such as *right/left*. Overall, spatial terms referring to *FRONT/BACK* relations are the ones most widely borrowed into the TIME domain cross-linguistically.

Two systems for sequencing events

In the English language, it has been pointed out that there are actually two SPACE-->TIME metaphoric systems. The first system can be termed the EGO-MOVING metaphor, where EGO or the observer's context progresses along the time-line towards the future. The second system is the TIME-MOVING metaphor. In this metaphor, a time-line is conceived of as a river or conveyor belt on which events are moving from the

Table 1. SPACE-TIME correspondence in Language.

SPACE	TIME
at the corner	*at* noon
from here *to* there	*from* two o'clock *to* four o'clock
through the tunnel	*through* the night
He stood *before* the house	it happened *before* evening
He was running *ahead of* me	He arrived *ahead of* me

[1] This research was supported by ONR grant N00014-89-J1272 and NSF grant BNS 87-20301 awarded to the first author.

FUTURE to the PAST (See Fig. 1). These two systems lead to different assignments of FRONT/BACK terms to a time-line (Clark, 1973; Fillmore, 1971; Lakoff & Johnson, 1980; Traugott, 1978). For example, in the EGO-MOVING system, FRONT is assigned to the *future (later)* event (e.g. " The war is *behind* us." or "His whole future is *before* him."). In the TIME-MOVING system, in contrast, FRONT is assigned to a *past (earlier)* event. (e.g. "I will see you *before* 4 o'clock" or " The reception *is after* the talk.")

Our question here is whether these two metaphoric systems -- the EGO-MOVING and the TIME-MOVING systems -- are psychologically real metaphors: that is, are they processed as metaphoric mappings in real time. This question is part of a larger psychological issue of metaphoric processing, namely, whether conceptual metaphors are psychologically processed as generative domain mappings (Gentner & Gentner, 1983; Gentner & Boronat, 1991; Gibbs & O'Brien, 1990; Lakoff & Johnson, 1980).

Conceptual metaphors and the domain-mapping hypothesis

Lakoff and his colleagues have pointed out the presence of large-scale systems of conventional conceptual metaphors: language from one domain that is habitually used in other domains (Lakoff & Johnson, 1980; Kövecses, 1986). These metaphors can often be characterized as originating in one or more abstract schemas in the base (or source) domain: e.g. ANGER IS HEAT/FIRE, ARGUMENT IS WAR. Here are some examples of conventional expressions reflecting the ANGER IS FIRE schema (from Kövecses, 1986):

Those are *inflammatory* remarks.
She was *doing a slow burn*.
He was *breathing fire*.
Your insincere apology just *added fuel to the fire*.

Such linguistic patterns suggest that many conceptual domains can be described and organized systematically in terms of more tangible and familiar domains. However, the question remains as to whether such metaphoric systems are psychologically processed as domain-mappings. That is, do people actually comprehend these domain metaphors by carrying out analogical mappings from a base domain to a target domain? Alternatively, the metaphoric meaning could be stored as an additional meaning sense of the base term. In this case, a word like "inflammatory" would have (at least) two word senses associated with it: 'causes fire' and 'causes/promotes anger'. There would be no global domain-mapping, only a series of local polysemies. In this case such conceptual systems are informative about the history of language, but not about current processing.

Some evidence for the domain-mapping hypothesis has been found in research studying how people make inferences from science analogies (Gentner & Gentner, 1983) and in comprehending and imaging idiomatic phrases (Gibbs & O'Brien, 1990). However, these studies did not bear on on-line comprehension processes.

What would it mean for conceptual metaphors to be processed as domain mappings? In structure-mapping theory, the knowledge representations of the two domains are structurally aligned and further relations (*candidate inferences*) connected to the common system of relations are mapped from the base to the target (Gentner, 1983, 1989; Falkenhainer, Forbus & Gentner, 1989; see also Burstein, 1988). Gentner and Boronat (1991) applied this framework to extended discourse metaphors. They reasoned that when a series of metaphors is presented from a single cohesive schema, people should be able to integrate each local metaphor into a global mapping from one conceptual system to the other. In this case, processing should be fluent. In contrast, when metaphors from different domains are juxtaposed, the hearer must shift from one base-target mapping to another, and consequently processing should be disrupted.

Gentner and Boronat devised a paradigm in which a series of conceptual metaphors from a single coherent source domain is presented in a connected text. This should establish a global mapping which serves as a setting for the final test sentence. In the Consistent mapping condition, the same metaphor is maintained throughout; in the Inconsistent mapping condition, the metaphor shifts between the initial passage and the final sentence. Using this technique, Gentner & Boronat (1991) found that subject's reading time for the final sentence was slowed down following a shift from one metaphor to another. This *cost of remapping* whe the underlying metaphor shifted suggests that the metaphors were processed as domain mappings.

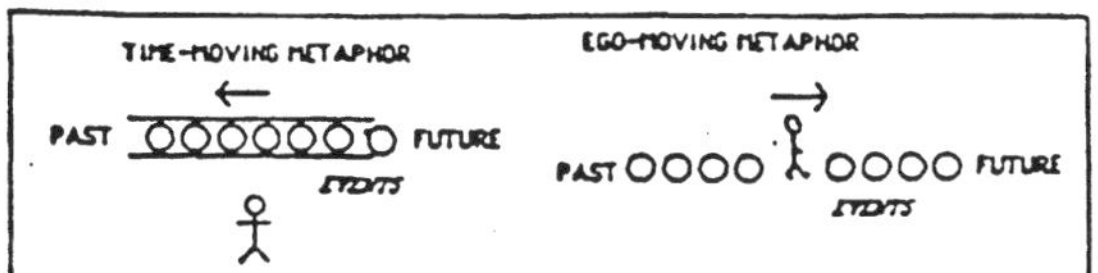

Fig 1. TIME-MOVING and EGO-MOVING Metaphors

Are SPACE/TIME metaphors processed as domain mappings?

The central question here is whether the two metaphoric systems -- the EGO-MOVING and the TIME-MOVING systems -- are processed as domain mappings in real time. There is reason to doubt this. First, Gentner and Boronat obtained evidence for domain mappings only when conceptual metaphors were relatively novel; tests using highly conventional metaphors (such as 'get this topic across') did not reveal a significant cost for remapping. The SPACE/TIME metaphors are highly conventional and frequently used in everyday language; indeed, the expressions reflecting the two metaphoric systems are almost invisible; people rarely notice that there are two different SPACE-TIME mapping systems in their everyday language. This might suggest that there will be no domain-mapping effect. It could be that the two mapping metaphors were alive in the history of language, but now are only stored as alternate word-senses of the spatial terms.

A second reason we might not expect to see a mapping effect is that the contrast between metaphors here is quite subtle. In the materials used by Gentner and Boronat, two metaphors from different base domains (e.g. HEAT, DANGEROUS ANIMAL) were applied to the same target (ANGER). In the present case, however, we have two conceptual systems from the *same* base domain, SPACE, to the same target domain, TIME. For this reason, we will call these mappings *system mappings* rather than *domain-mappings*. Evidence for distinct global system-mappings here would be particularly interesting, since it would suggest considerable representational specificity in the on-line mapping process.

The two metaphoric systems both serve to sequence events in a time-line, yet produce different temporal orders. Therefore, the cost of shifting from one system to the other may be substantial. Indeed, there are cases in which shifting from one system to the other produces opposite temporal order for the same words: e.g., *before, ahead* and *behind.* Compare the following two sentences:

(1) Christmas is six days *ahead of* New Year's Day.
(2) The holiday season is just *ahead of* us.

Let us denote the event first mentioned in each sentence E_1 and the second event E_2 in Fig.2. The two time lines below show how E_1 and E_2 are placed in the time-line. The relative PAST and FUTURE of both E_1 and E_2 is reversed between (1') and (2'), even though the same term *ahead of* is used to describe both temporal relations.

To test the system mapping hypothesis, we employed a variation of the paradigm used by Gentner & Boronat (1991). Because the linguistic expressions reflecting the two mapping systems are highly conventional, we were concerned that merely having subjects read might not be fully able to capture the phenomena. Because of the high familiarity of the expressions, subjects might mistakenly think they "understood" what the sentence means without deeply processing the sequential relations between events described in the sentence. To be sure that subjects fully comprehended the sentences, we employed a paradigm that requires deeper processing than merely reading text.

Figure 3 shows how the experimental materials were presented. Sentences were presented one at a time on the top of the CRT screen, with a time line below. The event mentioned in the second place (E_2 in the notation given earlier) was located on the time-line. Subjects pressed one of two keys to indicate whether the first-mentioned event (E_1) was located PAST or FUTURE of E_2 in the time-line (see Figure 3). Subjects' accuracy and response time were recorded.

The general method was very similar to the Gentner & Boronat study. A test sentence describing a temporal relation between E_1 and E_2 was preceded by Setting sentences. In the Consistent mapping condition, the Setting sentences and the Test sentence used the same metaphoric system -- either EGO-MOVING or TIME-MOVING. In the Inconsistent mapping condition, the Setting sentences utilized a different mapping system from that in the Test. According to the domain-mapping

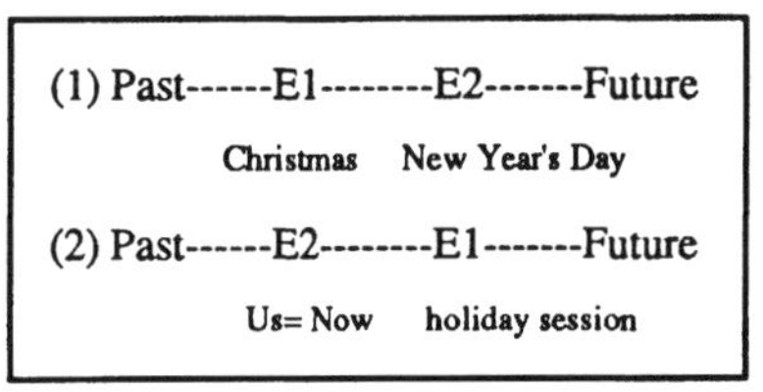

Figure 2: Sequencing 2 events in (1) and (2)

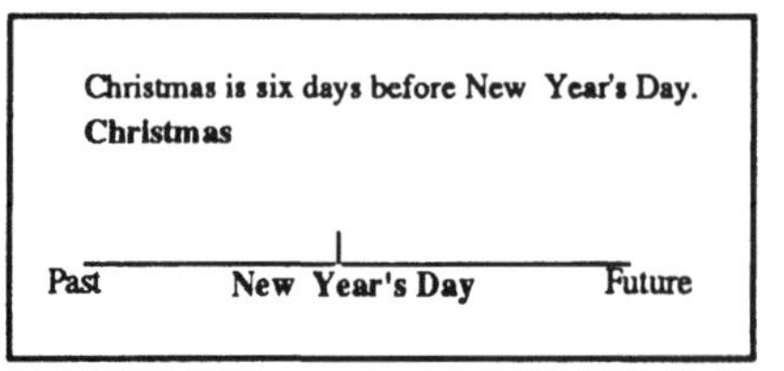

Figure 3. An example of material in Ex 1& 2

hypothesis, processing should be faster in the Consistent mapping condition than in the Inconsistent mapping condition. This is because in the Consistent condition, subjects can continue to build on the same systematic mapping as they progress from the Setting sentences to the Test sentence, but in the Inconsistent condition, to understand the Test sentence subjects must discard their existing mapping and set up another.

The alternative possibility would be that people process temporal relations at a local, purely lexical level and the systematic mapping will not take place. In this case, we should find no difference where a series of temporal relations are provided systematically in a single system or haphazardly from the two different systems. Two experiments have been conducted in order to examine this question.

Experiment 1

Method

Subjects: Subjects were 112 Northwestern University students who received course credit for their participation.

Materials: The materials consisted of two SETTING sets and two Test sets. One Setting set consisted of 15 sentences from the EGO-MOVING mapping system; the other consisted of 15 sentences from the TIME-MOVING mapping system. Likewise, one Test set consisted of five sentences from the EGO-MOVING system; the other consisted of 5 sentences from the TIME-MOVING system.

Design: The design was a 2 (Metaphor Type) X 2 (Consistency) between-subject design. There were four between-subject conditions, consisting of the four possible combinations of Setting set and Test set. Condition 1: TIME-MOVING SETTING --TIME-MOVING TEST; Condition 2: EGO-MOVING SETTING--TIME-MOVING TEST; Condition 3: EGO-MOVING SETTING -- EGO-MOVING TEST; and Condition 4: TIME-MOVING SETTING--EGO-MOVING TEST.

Procedure: Subjects saw a sentence and a diagram on the CRT screen as depicted in Fig. 2. They were instructed to respond by pressing one of two keys to indicate whether the first event (E_1) in the sentence takes place in the PAST or FUTURE relative to the second event (E_2). The second event (E_2) was located on a time line as shown in Fig. 2. Subjects saw five blocks of three Setting sentences. After each such block a Test sentence was presented. The organization of Setting sentences within blocks and the presentation order of the Test sentences were randomized. Subsequently, in the two Consistent Conditions (Conditions 1 & 3), the subjects saw all the 20 sentences from a single mapping system (either EGO or TIME-MOVING system). In the two Inconsistent conditions, the mapping system was switched in every fourth sentence.

Results and Discussion

As predicted, subjects in the consistent conditions responded faster [Mean=4228.0] than those in the inconsistent conditions [Mean=4798.7]. A 2 X 2 ANOVA confirmed a significant effect of Consistency, $F(1, 108)=5.074$, $p<.05$. Error responses were excluded from the analyses. The overall accuracy rate was 93.08% and was evenly distributed across the four conditions. There was no main effect of Metaphor type nor interaction between Consistency and Metaphor Type.

The fact that subjects were faster to make inferences when the test sentences continued the same metaphor as the setting sentences is consistent with the system mapping hypothesis. This pattern suggests that people understand these metaphors via a systematic mapping of metaphors, so that processing further metaphors belonging to the same system is facilitated relative to shifting to metaphors belonging to a different system.

However, because of our randomization procedures, we were concerned about local effects that could have inflated the effect for Consistency. The most important of these was repeated words. If a Test sentence was preceded by a Setting sentence using the same spatio-temporal term (e.g., *before-before,*), local lexical associations could clearly affect the results through same-word priming. In our design, the probability of such same-word repetition was low but non-zero. We were also concerned about the possibility of response bias in cases when the same response occurred in the last setting sentence and in the test sentence (e.g., PAST/PAST).

Experiment 2 was designed to control those local factors strictly. In Experiment 2, the Test sentences utilized only three terms: *ahead, before* and *behind.* All of these are common to both the Space/Time mapping systems. In order to separate the global mapping effects from possible local effects, we manipulated the Setting sentences just prior to the Test sentences. In both the Consistent mapping and Inconsistent mapping conditions, the Test sentences were preceded equally often by Setting sentences of the following three types: the Same term

(Context Word) (e.g. *before* in Setting and *before* in Test); the Opposite term (e.g. *after-before*); or a Neutral term (e.g., *preceding-before*). If the advantage for the Consistent conditions obtained in Experiment 1 was merely due to local lexical priming and response bias effects, no overall advantage should be found for the Consistent mapping conditions in Experiment 2. Rather, the difference between the Consistent and Inconsistent mappings should be observed only when the preceding Setting sentence uses the same term as the Test.

Experiment 2

Method

Subjects: The subjects were 72 students at Northwestern University who received course credit for their participation.

Design and Materials: Each subject received one of three sets of materials. Each of the three sets consisted of 12 blocks of 3 Setting sentences and a Test sentence. Therefore, the subjects saw total of 48 sentences, 36 Setting and 12 Test . Of the 12 Test sentences, 6 contained the temporal relations reflecting the EGO-MOVING metaphor, and the other 6 reflected the TIME-MOVING metaphor. Of the six blocks in each metaphor type, 3 blocks were in the Consistent Condition (i.e., the Setting and Test sentences were in the same mapping system). The other 3 blocks were in the Inconsistent Condition. Further, the Setting sentences appearing prior to a Test sentence included one of the following Context Word Types: 1) the Same word; 2) the Opposite word; 3) the Neutral word. Thus, each of the 12 blocks reflected each combination of 2 Metaphors types, 2 Consistency in mapping and 3 Context word types. As mentioned earlier, we restricted ourselves to *ahead, before* and *behind* as test relational terms in the Test sentences. Thus, we have 3 Test Word types in the two mapping metaphors, yielding 6 types of item sets consisting of 6 Test sentences. We divided the six item sets into 3 Assignment Groups on the contingency that each Group received both types of mapping metaphors in the different Test Word. Thus, the design of the experiment was 3 (Assignment Group) X 2 (Metaphor) X 2 (Consistency) X 3 (Context Word).

Procedure: The method of stimulus presentation and response was the same as in Experiment 1. The 12 blocks reflecting the 12 within subject conditions was totally randomized. Each subject saw the same Test sentence only once but each Test sentence was assigned to the Consistent and Inconsistent conditions in each Context Word Type an equal number of times across subjects.

Results and Discussion

As predicted, responses were faster in the Consistent than in the Inconsistent conditions, as shown in Table 2. However, before considering the reaction-time results in detail, we should discuss the accuracy rate. Overall accuracy rate was high and evenly distributed across conditions except when the word in the Test was *behind* in the TIME-MOVING mapping metaphor. The overall accuracy rate without *behind-Time Moving* item set was 95.27 % and no significant difference was obtained between the Consistent and Inconsistent conditions. In contrast, in the *behind*-TIME-MOVING Metaphor item set, the difference in accuracy between the Consistent and Inconsistent conditions was almost significant (81.9% in Consistent vs. 68.1% in Inconsistent), t=2.00, p=0.0568. Also, there was a speed-accuracy trade-off for this item set. Therefore, we analyzed the reaction time both including and excluding the assignment group in which this item set was contained (Group 3).

A 3 (Group) X 2 (Consistency) X 2 (Metaphor type) X 3 (Context Word type) mixed-measures ANOVA was conducted. We will report only the effects involving Consistency and Context Word Type.[2]

In the analysis including Group 3, the effect for Consistency was only marginally significant, $F(1,69)=3.743$, $p=0.057$. However, there was a significant interaction between Consistency and Assignment Group and Consistency X Metaphor due to the speed-accuracy trade-off observed in the *behind-Time Moving* item set in Group 3. A similar analysis excluding Group 3 revealed a significant effect of Consistency, $F(1, 46)=12.714$, $p<0.01$. No significant Consistency X Group or Consistency X Metaphor effects were found. No significant main effect for Context Word or interaction effect for Consistency X Context Word was found in the analyses including or excluding Group 3. No significant overall Consistency X Context Word

[2] There were significant main effects for Group and Metaphor. However, the effects were due to the length or other properties of the items. There were a few other high-way interactions significant at $p<.05$. However, they seemed all to be due to particular item properties and small number of data points.

Table 2. Means for the Consistent and Inconsistent conditions at three levels of Context Word Type

	Consistent	Inconsistent
Same	4345.67	4998.04
Opposite	4549.57	4659.24
Neutral	4597.17	4650.23

interaction was found, $F(2,69)=0.034$, either. Means for the Consistent and Inconsistent Conditions at the three levels of Contextword are given below in Table 2. The pattern of means suggests that the subjects were slower in the Inconsistent Condition than the Consistent Condition at all levels of Context Word.

In addition, we conducted an ANOVA excluding the Same Context Word condition to examine if there is still a main effect for Consistency when the Context word was Neutral or Opposite. The Consistency effect was significant, $F=5.452$ (1,46), $p<.05$.

The results of Experiment 2 replicated the results obtained in Experiment 1. Further, the effect of Consistency was obtained when the possible local factors were strictly controlled by the manipulation of the Context Word. This strongly indicates that the global system mappings take place when people make an inference about temporal relations.

General Discussion and Conclusion

The results obtained from the two experiments are evidence for system mappings in SPACE-->TIME. The two metaphoric mapping systems discussed in this paper are very natural and are rarely noticed in everyday language. Yet our experiments showed that when people make inferences about temporal relations in text, they process more fluently if the sequence of metaphors belongs to the same global mapping system. The results here suggest that when a current mapping system is shifted to another system during the global mapping, people have to redo the mapping between the two domains (i.e. SPACE and TIME) with a cost in on-line processing. This in turn suggests that the two SPACE-TIME metaphoric mapping systems are psychologically real and not a mere historic relic.

Acknowledgements

We thank Phillip Wolff for the program used in the experiments and also, along with Arthur Markman, for many helpful discussions.

References

Bennett, D. C. (1975). *Spatial and temporal uses of English prepositions: an essay in stratificational semantics*. London: Longman Group.

Burstein, M. (1988). Incremental learning from multiple analogies. In A. Prieditis (Ed.), *Analogica*. London: Pitman.

Clark, H.H. (1973). Space, time semantics, and the child. In T. E. Moore (Ed.), *Cognitive development and the acquisition of language*. New York: Academic Press.

Falkenhainer, B., Forbus, K. D., & Gentner, D. (1989). The structure-mapping engine: Algorithm and examples. *Artificial Intelligence* 41:1-63.

Fillmore, C. J. (1971). *The Santa Cruz lectures on deixis*. Bloomington, IN: Indiana University Linguistic Club.

Gentner, D. (1983). Structure-mapping: A theoretical framework for analogy. *Cognitive Science* 7:155-170.

Gentner, D. (1989). The mechanisms of analogical learning. In S. Vosinadou & A. Ortony (Eds.), *Similarity and analogical reasoning*. New York: Cambridge University Press.

Gentner, D., & Boronat, C. (1991) *Metaphors are (sometimes) processed as domain mappings*. Paper presented at the symposium on Metaphor and Conceptual Change, Meeting of the Cognitive Science Society, Chicago, IL.

Gentner, D., & Gentner. D. (1983). Flowing waters or teeming crowds: Mental models of electricity. In D. Gentner & A. L. Stevens (Eds.), *Mental models*. Hillsdale, NJ: Erlbaum.

Gibbs, R., & O'Brien, J. (1990). Idioms and mental imagery: The metaphorical motivation for idiomatic meaning. *Cognition* 36:35-68.

Lakoff, G., & Johnson, M. (1980). *Metaphors we live by*. Chicago: University of Chicago Press.

Kövecses, Z. (1986). *Metaphors of anger, pride, and love*. Amsterdam: John Benjamins Publishing.

Lehrer, A. (1990). Polysemy, conventionality, and the structure of the lexicon. *Cognitive Linguistics* 1: 207-246.

Traugott, E. C. (1978). On the expression of spatio-temporal relations in language. In J. H. Greenberg (Ed.), *Universals of human language: Vol. 3. Word structure*. Stanford, CA: Stanford University Press.

Indirect Analogical Mapping

John E. Hummel and Keith J. Holyoak
Department of Psychology
University of California, Los Angeles, CA 90024-1563

Abstract[1]

An Indirect Analogical Mapping Model (IMM) is proposed and preliminary tests are described. Most extant models of analogical mapping enumerate explicit units to represent all possible correspondences between elements in the source and target analogs. IMM is designed to conform to more reasonable assumptions about the representation of propositions in human memory. It computes analogical mappings indirectly -- as a form of guided retrieval -- and without the use of explicit mapping units. IMM's behavior is shown to meet each of Holyoak and Thagard's (1989) computational constraints on analogical mapping. For their constraint of pragmatic centrality, IMM yields more intuitive mappings than does Holyoak and Thagard's model.

Introduction

The central function of analogical thinking is to aid in creating coherent, structured representations of important novel situations. By finding a mapping -- that is, a set of correspondences -- between a known situation (the *source* analog) and a novel one (the *target* analog), the structure of the source can be used as a kind of blueprint for building a representation of the target. While analogy involves a number of component processes, the mapping process is pivotal because the correspondences it establishes constrain the inferences that can be generated about the target.

This paper presents our preliminary investigations into an Indirect Analogical Mapping Model (IMM). The primary goal of this effort is to develop an algorithm for analogical mapping consistent with reasonable assumptions about the representation of propositional information in human memory. Extant models of analogical mapping typically posit explicit processing units for all possible correspondences between the elements of the source and target analogs (e.g., Falkenhainer, Forbus & Gentner, 1989; Holyoak & Thagard, 1989). There are a number of serious problems associated with such enumeration of mapping units (Hofstadter & Mitchell, in press). Although it can be argued that mapping units are a notational convenience rather than a literal claim about the nature of mental representations, it is unclear how the critical processes posited by such models (e.g., parallel constraint satisfaction) would operate under more natural representational assumptions. A related difficulty with explicit mapping units is that they exist strictly for the purpose of analogical mapping and have no obvious usefulness for other cognitive processes. The primary goal of IMM is to simulate analogical mapping within an architecture more consistent with realistic assumptions about the representation of propositions in memory.

[1]This research was supported by Contract MDA 903-89-K-0179 from the Army Research Institute, and by NSF Grant DIR-9024251 to the UCLA Cognitive Science Research Program.

Theoretical Motivation

Representation of Propositions. The central problem in representing propositions involves encoding their internal structure. Representing a proposition entails creating a set of bindings between the arguments of the proposition and the case roles they fill. For example, to distinguish the representation of (chase Arnold Bill) from (chase Bill Arnold), Arnold must be bound to the agent role of "chase" in the first proposition and to the patient role in the second. A basic tenet of our approach is that active representation of propositional information (i.e., in working memory) and its long-term storage require different solutions to this binding problem.

Let us first consider the problem of case role-argument binding in an active representation. It is possible to imagine a representation for propositions in which dedicated units (or patterns of activation) represent each role-argument binding. For example, units could be created *de novo* each time a proposition enters working memory, or -- as proposed by Smolensky (1990) -- bindings could be represented by explicitly calculating a tensor product of the activation vectors representing the individual case roles and arguments (Halford, Wilson, Guo, Gayler, Wiles & Stewart, in press). In both these cases, the bindings are *static* because they are represented by units dedicated to specific conjunctions of elements.

This approach to the representation of attribute conjunctions suffers numerous limitations (cf. Hummel

& Biederman, 1992). The most serious is that by coding conjunctions of case roles and arguments, static binding units cannot represent the individual case roles (predicates) and arguments (objects); hence, the natural similarity structure of the predicates and objects is lost. For example, if separate static units represent (a) Arnold as the agent of chasing, (b) Arnold as the patient of chasing, (c) Bill as the agent, and (d) Bill as the patient, then the proposition (chase Arnold Bill), represented by units a and c, would be no more similar to (chase Bill Arnold), represented by b and d, than it is to (says My-doctor Caffeine-makes-me-nervous). And although this example assumed a localist representation, the underlying problem cannot be solved simply by postulating a more distributed representation. In Smolensky's (1987) tensor product representation, which uses distributed representations, the representation of a given object bound to one case role will not necessarily overlap *at all* with the representation of the identical object bound to a different case role.

An alternative to static binding is *dynamic binding*, in which units representing case roles are *temporarily* bound to units representing the arguments of those roles. Following Shastri and Ajjanagadde (1990) and others, IMM represents dynamic case role-argument bindings as synchronized firing of units representing the bound elements. For example, (chase Arnold Bill) is represented by units for the agent role of "chase" firing in synchrony with units for Arnold, while units for the patient role of "chase" fire in synchrony with units for Bill. Naturally, the agent/Arnold set must fire out of synchrony with the patient/Bill set.

Dynamic binding permits a small set of units to be reused in an unlimited number of specific bindings. The capacity to reuse units allows the representation of case roles and objects to be completely independent of one another. The theoretical and practical advantages of this independence are vast, but the most important is that it preserves similarity across different bindings. For example, all propositions in which Bill serves as an argument will be similar by virtue of their sharing the units that represent Bill; likewise, all propositions involving the predicate "chase" will employ the same "chase" units. As such, the independence afforded by dynamic binding permits essentially complete isomorphism between the meaning of a proposition and its representation: the representation of two propositions will overlap exactly to the extent that their meanings overlap. Some practical advantages of this isomorphism will become clear when IMM's operation is described.

Although dynamic binding affords critical benefits in the active representation of propositions, it is of course completely impractical as a solution to the storage of role-argument bindings in long-term memory. In long-term memory, bindings must be represented in a static form (e.g., as "synaptic" strengths) that can remain dormant until the proposition is reactivated. Importantly, the long-term representation must be capable of reinstating the original dynamic bindings of arguments to case roles when it is reactivated[2]. To this end, IMM encodes propositions into its long-term memory as connections from units representing objects and predicates to semantically empty units called *sub-proposition* (SP) units. A proposition is retrieved from long-term memory by activating the SP units that encode it. When an SP unit fires, it activates and synchronizes the object and predicate units to which it is connected. Separate SPs within a proposition remain out of synchrony (desynchronized) with one another. Together, a proposition's SPs reconstruct the synchronized firing of predicate and object units that represents the structured semantic content of the proposition.

Computational Constraints on Mapping. The computational theory underlying IMM as a model of analogical mapping is borrowed from Holyoak and Thagard's (1989) Analogical Constraint Mapping Engine (ACME). ACME posits three broad classes of constraints on natural correspondences between the elements of analogs. (1) The structural constraint of *isomorphism* has two components: (a) *structural consistency* implies that if a particular source and target element correspond in one context, they should do so in all others; (b) *one-to-one mapping* implies that each element should have a unique correspondent in the other analog. (2) *Semantic similarity* implies that elements with some prior semantic similarity (e.g., by virtue of joint membership in a taxonomic category) should tend to map to each other. (3) *Pragmatic centrality* implies that a mapping should give preference to elements that are deemed especially important to goal attainment, and should maintain correspondences that can be presumed on the basis of prior knowledge.

The Indirect Mapping Model

ACME implements the mapping constraints directly, via parallel constraint satisfaction on explicit mapping units of the type described previously. Our goal is to achieve analogical mapping according to these constraints, but to do so indirectly -- i.e., without directly implementing the constraints as connections among mapping units. Rather, IMM treats analogical mapping as a form of guided retrieval: propositions in a source analog drive the activation of propositions in a target analog. This process is mediated by a set of predicate units that are shared by the propositions in

[2]This is true by definition; any long-term representation that could not reproduce the active representation of a binding would not encode that binding in any meaningful way.

both analogs. When a proposition in the source becomes active, its SPs create a synchronized pattern of firing across the predicate units. This pattern then activates the proposition(s) in the target analog to which it most closely matches. The resulting match (i.e., mapping) is then learned by updating modifiable connections between units across the analogs. The asymptotic strengths of these connections are interpreted as the model's preferred mappings.

The implementation described here was designed to test IMM's basic capacity for this type of mapping. To unconfound the properties of the architecture from the properties of any specific units of which it might be composed, we have made a number of strong simplifying assumptions that idealize IMM's operation. These assumptions will be relaxed in future implementations.

Architecture

Figure 1 illustrates IMM's basic architecture using the following analogs:

Source	Target
(chase Arnold Bill)	(eat fox goose)
(chase Bill Charles)	(eat goose corn)

IMM is composed of three types of units: predicate units, object units, and sub-proposition (SP) units. Predicate units represent the semantic content of predicates in a distributed fashion. For example, the predicate "chase" is represented by one pattern of activity over these units, and the predicate "pursue" would be represented by a different but overlapping pattern. (These patterns are not detailed in the figure.) Similarly, objects such as Arnold and Bill share some predicates (e.g., both are human and male) and differ on others. The similarity between two objects or two predicates is defined by their degree of overlap on the predicate units. The precise content of these representations is less important for our current purposes than the architecture in which they reside.

Propositions are encoded into IMM's long-term memory by symmetrical, excitatory connections from predicate and object units to SP units. Each SP permanently encodes a binding of one object to some number of single-place predicates and to *one role* of one multi-place predicate. For example, (chase Arnold Bill) is represented by two sub-propositions. The first encodes Arnold as the agent of chasing, and is denoted chase(Arnold _). Chase(Arnold _) has excitatory links to (a) the object unit for Arnold, (b) each single-place predicate unit that describes Arnold (e.g., person, male, etc.), and (c) the units for agent role of the two-place predicate "chase". The second SP, chase(_ Bill), has excitatory links to Bill, Bill's single-place predicates, and the patient role of "chase". Predicate units do not directly communicate; their only connections are to SP units. Predicate and object units are temporally yoked to SP units -- i.e., they fire only when they receive excitatory inputs from SPs. Therefore, when an SP fires, all predicate and object units to which it is connected also fire.

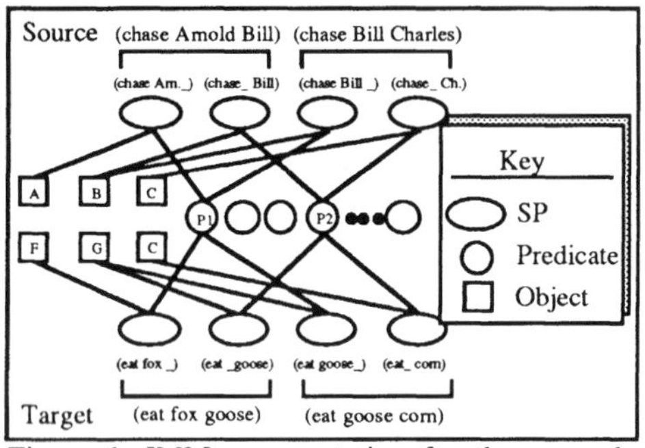

Figure 1. IMM representation for the example above. Not shown: Modifiable connections; connections between SPs within an analog; full SP-predicate connectivity.

Because predicate and object units are yoked to SPs, it is critical that separate SPs fire out of synchrony with one another. If chase(Arnold _) and chase(_ Bill) fired in synchrony, all their predicates and objects would also fire in synchrony, and it would be impossible to tell who was chasing whom. Therefore, SPs in the same proposition are assumed to share links that *desynchronize* their outputs. In the current implementation, all SPs within an analog are forcibly desynchronized (i.e., they are forced to fire one at a time).

Modifiable connections (of initial strength zero) exist between SP units across analogs and between object units across analogs; their function is described below.

The IMM Algorithm

The state of the network is updated in discrete cycles. The following sequence of operations is performed on each cycle:

1) One SP in the source analog fires; its output is set to 1.0 and propagated to the predicates, objects, and target SPs to which it is connected.
2) The SPs in the target update their activations (A_i) based on their excitatory inputs (E_i) and their lateral inhibitory inputs from one another. Lateral inhibition is implemented by the equation:

$$A_i = E_i^3 / \Sigma_j E_j^3.$$

3) The object units in the target update their activations based on their excitatory inputs from

the target SPs and their lateral inhibitory inputs from one another. Lateral inhibition is implemented by the above equation.

4) The SPs in the target recalculate their activations based on their excitatory inputs from the predicates, source SPs, and target objects, and their lateral inhibitory inputs from one another.

5) SPs in the target update their connections to SPs in the source, and objects in the target update their connections to objects in the source by the Hebbian rule

$$\Delta W_{ij} = A_i A_j,$$

where W_{ij} is the connection weight from source element j to target element i. Reflecting the one-to-one mapping constraint, connections to a target unit (both SP and object) and from a source unit are constrained to add to 1.0. This constraint is enforced by normalizing the modifiable connections at the end of each cycle by the ratio:

$$W_{ij} = W_{ij}/(W_{ij} + \Sigma_k W_{kj} + \Sigma_l W_{il}), i = k, l = j.$$

This normalization resets each connection according to its weight and the weights of all other connections to the same target SP and from the same source SP.

Simulations

Six tests of IMM are reported here, five based on small examples designed to test specific capacities of the model, and one based on a larger example. All tests were run ten times. The modifiable (SP-to-SP and object-to-object) connection weights were initialized to zero at the beginning of each run. Each run consisted of 20 iterations through the source analog, and each iteration consisted of one cycle (as described above) for each SP in the source analog. The firing order of the SPs was randomized at the beginning of each iteration. Mapping results are reported below in terms of the mean modifiable connection weights (object-to-object and, in one case, SP-to-SP) developed across analogs over the ten runs.

Test 1 was based on the analogs depicted in Figures 1 and 2. In this example, the predicates and objects are assumed to have no semantic overlap across the analogs, so the mapping must be solved purely on the basis of structural isomorphism. The most natural solution maps Bill to goose because they share the structural property of appearing in both the second place of the first proposition and the first place of the second. Because of the one-to-one mapping constraint, Arnold should then map to fox, and Charles to corn.

A detailed illustration of IMM's operation on the first cycle of this test is given in Figure 2. (1) The SP chase(Arnold _) fires and sends activation to the object unit for Arnold and the predicate P1 (shaded cells in Figure 2). P1 is a structural predicate indicating that its argument (Arnold) appears in the first place of some multi-place predicate, i.e., that its argument is the subject of some proposition. (2) The target SPs eat(fox _) and eat(goose _) each receive an excitatory input of 1.0 from P1. Since this is the first cycle through the source, all SP-to-SP connections and object-to-object connections are zero. After lateral inhibition, both target SPs' activations are 0.5. (3) The target objects fox and goose each receive inputs of 0.5 from their respective SPs, and after lateral inhibition, their activations are 0.5. (4) The target SPs recalculate their activations, again settling on 0.5 each. (5) SP-to-SP and object-to-object connections are updated. The weights from chase(Arnold _) to both eat(fox _) and eat(goose _) become 0.5, and the weights from Arnold to both fox and goose become 0.5.

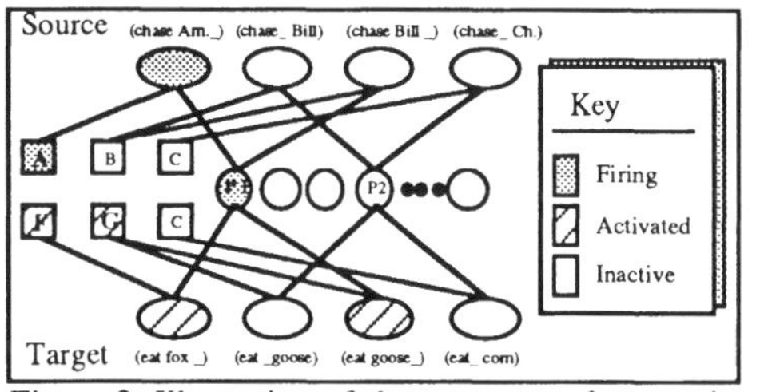

Figure 2. Illustration of the sequence of events in one cycle with the example above.

On the second, third and fourth cycles, these steps are repeated for chase(_ Bill), chase(Bill _) and chase(_ Charles), respectively. The second and third cycles are the most critical for finding the mapping from Bill to goose. On the second, Bill updates its connections both to goose and corn; on the third, it updates it connections to goose and fox. Over repeated iterations, Bill updates its connections to goose twice as often as it does to fox or corn, and -- due to the normalization of connection weights -- the Bill-goose connection eventually overpowers all other connections involving either Bill or goose. This reduction of other connections involving goose allows Arnold and Charles to map less strongly to goose, and more strongly to fox and corn, respectively. IMM successfully mapped these analogs on the basis of their structure alone. Over 10 runs, the mean object-to-object connection weights (after 20 iterations per run) were: Arnold --> fox = 0.99; Bill --> goose = 0.98; Charles --> corn = 0.99. All other object-to-object connections were zero.

Our second example tests IMM's sensitivity to the structure of information within propositions. It is based on the analogs:

Source	Target
(bite dog man)	(pet boy cat)
	(sit-on cat boy).

The predicates bite, pet, and sit-on are assumed to have no semantic overlap. The objects dog and cat share the single-place predicate unit for "animal", and the objects boy and man share the predicate "human". Both target propositions share four predicate units with the source proposition (p1, p2, animal, and human). However, (bite dog man) should map to (sit-on cat boy) because, in each case, an animal appears in the agent role, and a human in the patient role. In (pet boy cat), the human and animal are bound to opposite roles. Thus, this example constitutes a test of semantic similarity in which successful mapping depends on sensitivity to structure. Run with this example, IMM unambiguously mapped man to boy and dog to cat: dog --> cat = 0.99; man --> boy = 0.99; all other object-to-object connections were zero. As indicated by the the SP-to-SP connections, IMM also correctly mapped (bite dog man) to (sit-on cat boy), and correctly mapped corresponding roles within those propositions:

	bite(dog__)	bite(__ man)
sit-on(cat __)	0.66	0.00
sit-on(__ boy)	0.00	0.66
pet(boy __)	0.04	0.27
pet(__ cat)	0.27	0.04

Holyoak and Thagard's (1989) ACME model encounters difficulty with the constraint of pragmatic centrality. It does not respond appropriately to source elements that are marked as "important" (Spellman & Holyoak, in preparation; Hummel, Burns & Holyoak, in press). Therefore, IMM's treatment of important elements is a particularly critical test. Tests 3 - 5 examine the effect of object importance on mapping. Importance is implemented in IMM by allowing SPs containing important objects to fire more often than SPs containing objects not given extra importance. This convention is based on the assumptions that (1) firing rate reflects the activation of a unit, and (2) more important elements are more active than elements not given special importance. In these simulations, important SPs were allowed to fire twice (rather than only once) on each iteration through the source analog.

Tests 3 - 5 were based on the following analogs:

Source	Target
(chase coyote roadrunner)	(chase pig rabbit)
(eat Popeye spinach)	(eat rabbit carrot).

On every test with this example, IMM correctly mapped coyote exclusively to pig (connection weight 0.99) and spinach exclusively to carrot (connection weight 0.99). The interesting question concerns the degree to which Popeye vs. roadrunner will map to rabbit based on which (Popeye or roadrunner) is deemed "important". With neither given importance (Test 3), IMM mapped both equally to the rabbit (connection weights were 0.49), reflecting the ambiguity of the mapping. With special importance given to roadrunner (Test 4), IMM mapped roadrunner to rabbit more strongly than Popeye to rabbit (0.65 vs. 0.34). Similarly, with special importance given to Popeye (Test 5), it mapped Popeye to rabbit more strongly than it did roadrunner (0.66 vs. 0.33). Thus, IMM was able to adjudicate between ambiguous mappings on the basis of the relative importance of an element. In contrast, ACME produces less clear mappings for these simple examples (Hummel et. al., in press).

How does IMM differ from ACME so that the former succeeds on these simple tests of pragmatic centrality? In ACME, the success of a particular mapping depends upon the activity of the corresponding mapping unit relative to its competitors. An element (object or predicate) is marked as important by increasing the activities of all units representing mappings involving it. The increased activity associated with an important element's mapping units has the effect of increasing the tendency for those mappings to dominate other mappings. As such, the important element tends to map more to *everything*, rather than selectively mapping more to those other elements with which it already matches well.

By contrast, consider how an element in the source analog (SE) establishes a mapping with a target element (TE) in IMM. Each time an SE fires, its tendency to map to a specific TE is a function of (1) how closely the pattern of which the SE is a part matches the pattern(s) of which the TE is a part (as determined by the number of predicate units they share) and (2) how often and how strongly the SE has mapped to that TE in the past (as captured in the modifiable object-to-object and SP-to-SP connections). Like ACME, IMM implements increased importance as increased activity. In IMM, increased activity results in an increased firing rate. But note that an SE's *tendency* to map to any given TE, as defined by (1), has nothing to do with how often either unit fires; rather it is strictly a function of how well they match when they do fire. Therefore, increasing an SE's firing rate simply increases the *number of opportunities* that the SE has to map to those TEs for which it already has a preference. Each time an SE maps to a TE, they strengthen the connection between them at the expense of their other connections. Thus, a greater firing rate (i.e., more importance) means more opportunities for an SE to monopolize its preferred TE's connections.

The first five examples were designed to test specific capacities of the IMM architecture, and were deliberately kept small. The sixth test was designed to reveal IMM's capacity to deal with larger analogies. Test 6 is based on the "radiation to lightbulb" problem from Holyoak and Thagard (1989, Table 3). Space limitations prohibit full elaboration of the analogy, but it can be summarized as follows: The source analog states that there is a lightbulb with a broken filament that can be fused back together by a laser beam. The laser can generate either strong or weak beams. The strong beam would break the glass bulb surrounding the filament, but a single weak beam is too weak by itself to fuse the filament. The goal is to fuse the filament without breaking the glass bulb. The target analog states that there is a tumor surrounded by healthy tissue, and there is a radiation machine that can destroy the tumor. The radiation machine can generate either strong or weak rays. A strong ray would damage the healthy tissue surrounding the tumor, but the weak ray is too weak to destroy the tumor by itself. The goal is to destroy the tumor without damaging the healthy tissue. The intuitively correct mapping between these analogs generates the following object correspondences: laser --> radiation machine; strong laser beam --> strong rays; weak laser beam --> weak rays; tumor --> filament; glass bulb --> healthy tissue. IMM discovered all the correct mappings (mean modifiable connection strengths corresponding to correct mappings were all greater than 0.97) and did not discover any incorrect mappings (mean modifiable connection strengths corresponding to incorrect mappings were all zero).

Discussion

The initial simulations reported here, although run with a highly idealized version of IMM, have yielded encouraging results. IMM clearly demonstrates sensitivity to all the mapping constraints postulated by ACME: isomorphism, semantic similarity, and pragmatic centrality. It also scaled well to the larger analogy on which it was tested. Importantly, this behavior emerges from an architecture exploiting deliberately general principles for the representation of propositional information.

One strength of the IMM representation that we have not yet discussed is its capacity to scale with larger knowledge bases. Each proposition is encoded by a small number of SP units (typically three or fewer, depending on the number of argument places in the proposition). Therefore, the number of SP units required to represent an analogy grows linearly with the size of the analogs, and the number of modifiable connections between SPs across analogs grows linearly with the product of the number of propositions in the source and target.

The modifiable weights on object-object and SP-SP connections allow a relatively stable representation of the mapping between source and target elements to emerge. These modifiable connections are analogy-specific, making it possible for the system to learn contextually constrained correspondences between analogs without necessarily altering the structure of semantic memory. For example, the fact that a tumor maps to a filament in the context of the radiation/lightbulb analogy need not imply that these two concepts should now be closely related in semantic memory. At the same time, the asymptotic weights on the modifiable connections may provide inputs to post-mapping mechanisms that support the generation of analogical inferences about the target, as well as induction of relational generalizations based on the mapping between the source and target analogs.

It remains to be seen how IMM will perform with more realistic processing assumptions. The current implementation works largely because the sequence of events is globally and tightly controlled. If IMM proves highly sensitive to imperfections in the timing of events, it could be difficult to make it work with locally-controlled mechanisms for dynamic binding (i.e., for maintaining synchrony). Nonetheless, the indirect approach to analogical mapping seems sufficiently promising as to merit further exploration.

References

Falkenhainer, B., Forbus, K. D., & Gentner, D. (1989). The structure-mapping engine: Algorithm and examples. *Artificial Intelligence, 41*, 1-63.

Halford, G. S., Wilson, W. H., Guo, J., Gayler, R. W., & Wiles, J., & Stewart, J. E. M. (in press). Connectionist implications for processsing capacity limitations in analogies. In K. J. Holyoak & J. A. Barnden (Eds.), *Advances in connectionist and neural computation theory, Vol. 2: Analogical connections.* Norwood, NJ: Ablex.

Hofstadter, D. & Mitchell, (in press). An overview of the Copycat project. In K. J. Holyoak & J. A. Barnden (Eds.), *Advances in connectionist and neural computation theory, Vol. 2: Analogical connections.* Norwood, NJ: Ablex.

Holyoak, K. J., & Thagard, P. (1989). Analogical mapping by constraint satisfaction. *Cognitive Science, 13*, 295-355.

Hummel, J. E., & Biederman, I. (1992). Dynamic binding in a neural network for shape recognition. *Psychological Review*, in press.

Hummel, J. E., Burns, B. & Holyoak, K. J. (in press). Analogical mapping by dynamic binding: Preliminary investigations. In K. J. Holyoak & J. A. Barnden (Eds.), *Advances in connectionist and neural computation theory, Vol. 2: Analogical connections.* Norwood, NJ: Ablex.

Shastri, L., & Ajjanagadde, V. (1990). From simple associations to systematic reasoning: A connectionist representation of rules, variables and dynamic bindings. Technical Report MS-CIS-90-05, Computer and Information Science Department, University of Pennsylvania.

Smolensky, P. (1990). Tensor product variable binding and the representation of symbolic structures in connectionist systems. *Artificial Intelligence, 46*, 159-216.

Spellman, B. A., & Holyoak, K. J. (in preparation). Pragmatics in analogical mapping. Research in progress, Department of Psychology, University of California, Los Angeles.

Visual Analogical Mapping

Paul Thagard, David Gochfeld, and Susan Hardy
Cognitive Science Laboratory
Princeton University
221 Nassau St., Princeton, NJ 08542[1]

Abstract

This paper describes some results of research aimed at understanding the structures and processes required for understanding analogical thinking that involves images and diagrams. We will describe VAMP.1 and VAMP.2, two programs for visual analogical mapping. VAMP.1 uses a knowledge representation scheme proposed by Janice Glasgow that captures spatial information using nested three-dimensional arrays. VAMP.2 overcomes some limitations of VAMP.1 by replacing the array representation with a scheme inspired by Minsky's Society of Mind and connectionism.

Introduction

Part of analogical thinking involves finding correspondences between structures that represent analogous problems. Various computational models of how mapping between analogs can be conducted have been proposed (SME: Falkenhainer, Forbus, and Gentner 1989; Gentner 1983; ACME: Holyoak and Thagard 1989).[2] Like the vast majority of AI programs, analogy programs such as SME and ACME represent analogs propositionally rather than visually.

But many analogies have a strong visual component. Consider the Duncker tumor problem that has been widely used in psychological experiments (Gick and Holyoak 1980, 1983). Subjects are told to try to figure out how to use an x-ray machine to destroy a tumor inside a patient without damaging the patient's flesh. The solution is to use a number of x-ray sources producing rays of diminished intensity that converge on the tumor and destroy it. Subjects are aided in coming up with this solution if they are told of a general whose strategy for attacking a fortress involved dispersing his army and having them converge on the fortress from different directions. Although all the information necessary can be represented propositionally, it is natural to produce a diagram or mental picture that shows the army and the rays converging on the tumor and the fortress from different directions. Gick (1985) and Beveridge and Parkins (1987) found that the use of diagrams improved subjects problem solving effectiveness on this problem.

But it is difficult to model the visual aspect of analogical reasoning using the knowledge representation techniques that have been most common in AI. Ideally, visual representations should serve to make mapping between analogs much easier than propositional representations, which require considerable work to place appropriate predicates and arguments in correspondence. If, for example, we had a visual representation of the Duncker problem, we could map the tumor problem to the fortress problem by simply superimposing an image of the one onto the other and identify by inspection the objects that correspond to each other, such as the tumor and fortress. We cannot expect the visual representation to do all the work of analogical mapping, since many predicates such as **cause** will not lend themselves to visual representation, but visual representation should help greatly with aspects of the problems that are easily pictured. Finke (1989) provides a convenient summary of the large body of psychological experimentation that supports the contention that human thinking involves an important visual component.[3]

[1] This research was supported by contract MDA903-89-K-0179 from the Basic Research Office of the U.S. Army Research Institute for the Behavioral and Social Sciences. Paul Thagard's current address is: Department of Philosophy, University of Waterloo, Waterloo, Ontario, Canada, N2L 3G1. Email: prthagard@logos.waterloo.edu.

[2] Mapping is also implicit in computational models of case-based reasoning (e.g. Riesbeck and Schank 1989). See also Mitchell and Hofstadter (1990).

[3] Finke and others distinguish between visual information (how things look) and spatial information (how things relate to each other) but I shall include both of these under the heading "visual."

This paper describes some results of research aimed at understanding the structures and processes required for understanding analogical thinking that involves images and diagrams. We will describe VAMP.1 and VAMP.2, two programs for visual analogical mapping. VAMP.1 uses a knowledge representation scheme proposed by Janice Glasgow that captures spatial information using nested three-dimensional arrays. VAMP.2 overcomes some limitations of VAMP.1 by replacing the array representation with a scheme inspired by Minsky's Society of Mind and connectionism.

VAMP.1

VAMP.1 (Visual Analogical Mapping Program) is based on the knowledge representation scheme for computational imagery that Glasgow and her colleagues have been developing (Glasgow 1990; Glasgow and Papadias in press; Papadias and Glasgow 1991). In the earlier computational model of Kosslyn (1980), quasi-pictorial images were represented by a configuration of points in a matrix; an image is displayed by selectively filling in cells of the matrix. An image, then, is construed as a two-dimensional array, with each entry like a pixel that is either on or off. Glasgow's scheme is more complex in two key respects. First, it takes images to be inherently three-dimensional, although two-dimensional projects can also be handled as a special case. Greater dimensionality obviously makes possible representation of more complex images such as those required for mental rotation. Second, the entries in the three-dimensional arrays can be encoded hierarchically, in that each entry is represented symbolically by a entry that can have a subimage. For example, a house can be represented by the array shown in Figure 1, with each symbolic entry such as "window" providing a pointer to another array. In sum, Glasgow's representational scheme takes images to be three-dimensional symbolic hierarchical arrays. Numerous important visual operations can be defined on Glasgow's arrays, including constructing symbolic arrays from propositional representations, comparing images using array information, and moving and rotating images.[4]

We have developed a Common LISP implementation of parts of Glasgow's scheme and extended it to produce VAMP.1, a visual analogical mapping program. Given two arrays, VAMP.1 can do simple analogical mapping, putting the elements of the two arrays in correspondence with each other. VAMP.1 first checks to see if the arrays are the same size. If not, it scales them up to the size of the least common multiple of their sizes. For example, to compare a 4x4x4 array and a 6x6x6 array, VAMP.1 converts both arrays to 12x12x12 arrays. When both arrays are equal in size, VAMP.1 superimposes them and gives a list of all parts which are in corresponding cells.

roof	roof	roof
window		
	door	

Figure 1: Array representation of a house.

As described in Thagard and Hardy (1992), VAMP.1 has been used to model the use by John Dalton (1808) of an analogy between the structure of the atmosphere involving molecules and a pile of shot. It is natural to construct a mental image of a pile of cannon balls with one ball nesting on four below which nest on nine below, and then transform this into a picture of the atmosphere consisting of atoms surrounded by heat similarly nesting. The representation of the pile of balls is not just the various slices shown, but the whole array which encapsulates a very large amount of spatial information. This encapsulation makes creating a visual analog trivial: all we have to do to produce a representation of the structure of the atmosphere is to replace each entry of BALL with an entry of ATOM. The hierarchical nature of the representation scheme is important because it allows us to substitute a complex of atom and heat, as Dalton recommended, rather than just atom.

Glasgow's knowledge representation scheme is very useful in suggesting how visual/spatial information can be stored and used. But it has some clear limitations. Arrays are too "boxy" to capture more complex spatial arrangements than *left, right, above, below*: a cannon ball sits above four others at roughly 60 degree angles, not directly above. Also not naturally represented in Glasgow's scheme are relations of containment. In the tumor problem, for example, the patient's flesh contains the tumor: there are not distinct objects of flesh filling all the adjacent boxes. From the perspective of processing, the Glasgow scheme has advantages in making the appropriate maps readily identifiable when the arrays coincide,

4 Other computational models of visual thinking have been developed by Funt (1980), Shrager (1990), and Chandrasekaran and Narayanan (1990).

but does not suggest how partial maps might be found. In addition, using the array structures seems potentially inefficient, since they will contain various empty cells and have to be worked with in monolithic fashion. Accordingly, we have tried to retain some of the advantages of Glasgow's scheme while producing more flexible mappings.

VAMP.2

According to Marvin Minsky's provocative "Society of Mind" theory, each mind is made up of many small processes he calls *agents*. Minsky says (1986, p. 17): "Each mental agent by itself can only do some simple thing that needs no mind or thought at all. Yet when we join these agents in societies - in certain very special ways - this leads to true intelligence." We propose to reconceptualize Glasgow's scheme by imagining that corresponding to each box in the 3-D array there is a simple agent that can communicate with other agents representing other boxes. Each agent knows what other agents are adjacent to it in various directions. The agents can process information in parallel to provide answers to simpler questions. For example, if you want to know what is above the door in a visual representation, you can query all agents until you find one that has the door, then have that agent ask the agent above it what it has. This corresponds to simply looking at the door and then looking up above it.

Once you have a set of agents each of which has knowledge of the adjacent agents, you no longer need the array structure at all. The same information captured by the boxes in the 3-D array can be captured more locally by what the individual agents know about themselves and the adjacent agents. Moreover, much more flexible spatial structures can be used than simply left, right, above, and below as in the array: an agent can know that there is an agent above it and to the left at a particular angle. Agents can also possess another important kind of spatial information: what agents contain them or are contained by them.

For visual analogical mapping, each analog can be represented by a set of agents, and the computational problem is to put agents from different sets in communication with each other in such a way that the appropriate correspondences are found. For example, the agent for **tumor** in the Dunker problem must be put in contact with the agent for **fortress** in the other problem. Think of two competing baseball teams whose members shout at each other to find the players in corresponding positions; after an initial noisy display, the shortstops on each team will find each other, and so on. This example shows that the mapping may not be simple, since more than one pitcher on each team may correspond to more than one pitcher on the other.

VAMP.2 is a program that implements this kind of analogical mapping. It is written in the Common LISP Object System, for it is natural to encode society-of-mind ideas using object-oriented programming. Each thing in an analogy is represented by an agent, implemented as a CLOS object. For mapping purposes, we want to avoid the complexity of having every agent in one analog try to correspond to every agent in the other analog, so visual similarity used to screen for agents of mutual relevance: two agents only begin a relationship if the things they represent have similar appearance or containment relations. But once such a relationship is established between agents S1 from the source analog and T1 from the target analog, the agents adjacent to S1 can be put in correspondence with agents adjacent to T1 even if they have different shapes and spatial relations. We want, for example, to have S2 which is to the right of S1 establish a connection with T2 which is to the right of T1. Matters obviously become much trickier when the target contains more than one agent that is similar in shape to S1 so that we cannot tell right away which one should correspond to it.

To solve this problem, we have used connectionist techniques of parallel constraint satisfaction that worked well in earlier models of analogical mapping and retrieval, ACME and ARCS (Holyoak and Thagard 1989; Thagard, Holyoak, Nelson and Gochfeld 1990). ACME and ARCS showed that analogs can be retrieved from memory and mapped by satisfying a combination of semantic, structural, and pragmatic constraints. These constraints are represented in a connectionist network of units with excitatory and inhibitory links, and a simple settling process selects out what correspondences best satisfy the constraints. VAMP.2 uses constraints specific to visual representations that can however be viewed as special cases of constraints in the more general programs. We want to encourage mappings between things of similar appearance, encourage mappings between things with similar adjacencies and containment relations, and discourage one-many and many-one mappings.

For each pair of agents who establish a relationship for appearance or containment relations, VAMP.2 creates a mapping unit that represents the plausibility of their being in correspondence: we will write the mapping unit that pairs S1 and T1 as S1=T1. Mapping units form packages that tend to go together. If S1 is adjacent to S2, and T1 is adjacent to T2 in the same way, then the unit S2=T2 will be formed. We want the mappings S1=T1 and S2=T2 to go together, so a symmetric excitatory link is established between these two units. Similarly, if S1 contains S2 and T1 contains T2, then we want the mappings S1=S2 and T1=T2 to encourage each other, so an excitatory link is established between those units. A special unit that is always active is used to encourage mappings between agents representing things that are visually similar. VAMP.2 is given a verbal description of things and uses this to compute visual similarity, or visual similarity is specified by the programmer.[5] If S1 and T1 are visually similar, then an excitatory link is established between S1=T1 and the special unit. To discourage mappings that are not one-to-one, an inhibitory link will be created between S1=T1 and any units S*=T1 and S1=T* representing other ways of mapping S1 and T1. In VAMP.2, all links are symmetric. Once these networks are created, a simple connectionist settling algorithm is used to adjust the activation of units in parallel until they all settle and the winning and losing units are apparent. The appendix contains a precise description of the algorithms used by VAMP.2.

Now let us look at a simple example of VAMP.2 in operation. Holyoak and Koh (1987) did experiments using the Duncker tumor problem with another problem that is more isomorphic to it than the fortress problem. The filament problem requires finding a way to use a laser to fuse a broken filament inside a glass bulb without breaking the glass. Our representation of the two problems is portrayed in figure 2. The tumor is contained in flesh which is contained in a hospital room along with an x-ray source and the rays, which are to the left of the the patient. To provide a greater challenge for VAMP.2, the representation of the other problem has the laser and beam to the right of the the filament and glass, which are contained in a laboratory. VAMP.2 is given the information that there is some visual similarity between the beam and ray and between the laser and x-ray source.[6] It therefore creates the units BEAM=RAY and LASER=SOURCE. Similarity in containment relations leads to creation of units LAB=ROOM, GLASS=FLESH, and several others. Figure 3 shows all the units created by VAMP.2 along with their inhibitory links. After 44 cycles of updating, the units all achieve stable activations and the appropriate mappings, BEAM=RAY, LASER=SOURCE, FILAMENT=TUMOR,

[5] Ideally, the program would make this sort of judgment itself on the basis of a pictorial representation.

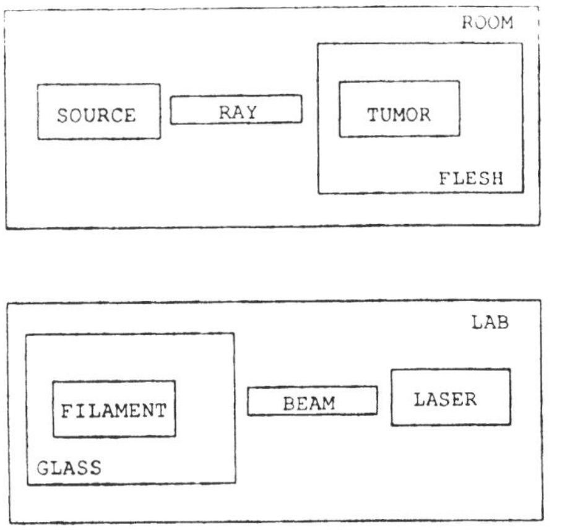

Figure 2. Diagrammatic representation of tumor and filament problems.

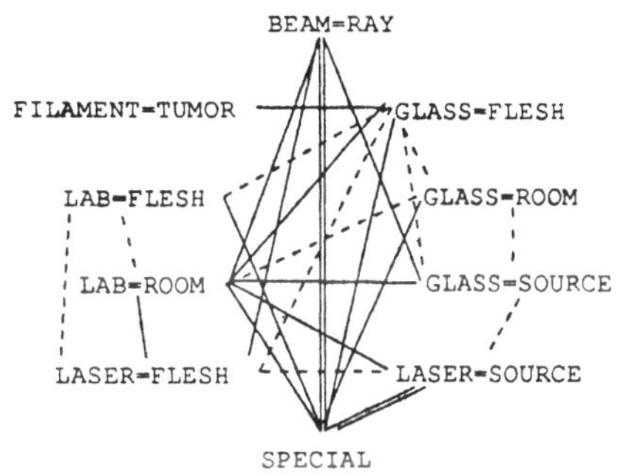

Figure 3. Network created to map tumor and filament representations.

Links to the special unit based on visual similarity are shown with double solid lines. Other links to the special unit based on adjacency and containment relations are shown by single lines, as are links between other units based on adjacency and containment relations. The dotted lines indicate inhibitory links that serve to discourage mappings that are not one-to-one.

LAB=ROOM, and GLASS=FLESH all settle with positive activation while the other four units all settle with negative activation. To compare VAMP.2's operation on this problem with our previous mapping program, ACME, we propositionally represented the two problems at the same level of detail using predicate calculus. ACME produces far more units to calculate the mapping than VAMP.2 (115 versus 9) and has difficulty recovering from the mismatch that is suggested by having the tumor and filament on opposite sides of the ray and beam.

VAMP.2 has also been run successfully on several other examples: the tumor/fortress analogy, the atom/solar system analogy, and Dalton's 3-D analogy between molecules and piles of shot. The atom/solar system example shows that VAMP.2, like ACME but unlike SME, can perform one-many mappings when it is appropriate to do so. Given representations of a hydrogen atom with one electron and a solar system with several planets, VAMP.2 correctly maps the electron to each of the planets. The Dalton analogy is much trickier for VAMP.2 than for VAMP.1, where array structure makes finding the correspondences between atoms and balls very easy. Nevertheless, despite the much greater number of possible correspondences between atoms and balls that VAMP.2 must deal with, it manages to sort out appropriate mappings using 107 units. In contrast, when ACME is given a long non-visual encoding of the analogs using representations such as

(LEFT-OF (BALL2 BALL3)) and
(ABOVE-LEFT (BALL1 BALL2)),

it creates 3500 units, more than our SPARCstation 2 could handle.

VAMP.2 is by no means the final word on visual analogical mapping. While it has a much more flexible scheme for knowledge representation than VAMP.1, it still is limited in how well it can represent such visually complex matters as how rays converge at a point. Moreover, it does not address the crucial question of visually representing dynamic information of the sort that might be found in a movie-like mental image of rays shooting out and converging. The atom/solar system analogy can most effectively be conveyed by imagining electrons and planets in moving orbits. Finally, VAMP.2 performs mapping by visual representations alone, ignoring many cues that might be provided by proposition-based mapping. A powerful integrated mapping scheme could be built by having VAMP.2 work in concert with a program like ACME, with each program passing partial results back and forth, getting the most out of the different kinds of representation available. Our new system CARE already integrates analogical mapping with retrieval and rule based reasoning, and it should be possible to fit VAMP.2 into it gracefully (Nelson, Thagard, and Hardy, in press). VAMP.2 is already capable of modeling some of what is involved in transferring a solution to a source problem into one for the target problem: given a description of the filament problem that includes its solution with convergent beams, it maps part of the solution back to the target tumor problem.

[6] If this information is omitted, VAMP.2 still creates the units BEAM=RAY and LASER=SOURCE because of similar containment relations and number of adjacencies.

Many analogies in ordinary life and in science have a substantial visual component. We have shown that it is possible to start to model visual aspects of analogy without having to simulate the entire human perceptual system. While structured array representations have many attractive features, visual analogical mapping of complex examples requires a more flexible representation such as that inspired by Minsky's Society of Mind theory. The price of this flexibility is that additional mechanisms of parallel constraint satisfaction are needed to accomplish the mapping.

Appendix: VAMP.2 Algorithms

A. Map visually similar things.

For each thing S in the source image, and any thing T in the target image such that

a) T is the same type of thing as S,

b) T has been declared to be visually similar to S, or,

c) both S and T contain something,

create a mapping unit "S=T" and add this to M, the list of mapping units. Make an excitatory link between this unit and the visual special unit.

B. Map adjacencies of previously mapped things.

Copy M into N, and then repeat the following steps until there are no mapping units left in N:

1) Let N1 be the first unit in N.

2) Let S be the source thing that is mapped in N1, and T be the target thing mapped in N1.

3) Let A(S) be the list of things adjacent to S, and A(T) be the things adjacent to T.

4) For each thing AS1 in A(S), and any thing AT1 in A(T) which is the same direction from T as AS1 is from S, create a mapping unit "AS1=AT1", if it does not already exist.

5) Add AS1=AT1 to N and M, and create an excitatory link between it and unit N1.

6) Remove N1 from N, and repeat the above steps until N is empty.

C. Map contents of previously mapped things.

Again copy M into N and repeat the following step:

1) Let N1 be the first unit in N.

2) Let S be the source thing that is mapped in N1, and let T be the target thing that is mapped in N1.

3) Let C(S) be the list of things directly contained in S, and C(T) be the things directly contained in T.

4) For each thing CS1 in C(S), and any thing CT1 in C(T) that has the same number of adjacencies as does CS1, create a mapping unit "CS1=CT1", if it does not already exist.

5) Add CS1=CT1 to N and M. Create an excitatory link between this and N1.

6) Remove N1 from N, and repeat the above steps until N is empty.

D. Inhibit multiple mappings by creating inhibitory links.

E. Update activation of units until network settles.

The algorithms for D and E are the same as in ACME; see Holyoak and Thagard (1989), p. 314.

References

Beveridge, M., and Parkins, E. (1987). Visual representation in analogical problem solving. *Memory and Cognition, 15:* 230-237.

Chandrasekaran, B., and Narayanan, N. (1990). Integrating imagery and visual representations. *Proceedings of the 12th Annual Conference of the Cognitive Science Society.* Hillsdale, NJ: Erlbaum, 670-677.

Dalton, J. (1808). *A new system of chemical philosophy.* London: Bickerstaff.

Falkenhainer, B., Forbus, K., and Gentner, D. (1989) The structure-mapping engine: Algorithms and examples. *Artificial Intelligence*, 41, 1-63.

Finke, R. (1989). *Principles of mental imagery.* Cambridge, MA: MIT Press.

Funt, B., (1980). Problem solving with diagrammatic representations. *Artificial Intelligence, 13,* 201-230.

Gentner, D. (1983). Structure-mapping: A theoretical framework for analogy. *Cognitive Science, 7,* 155-170.

Gick, M. (1985). The effect of a diagram retrieval cue on spontaneous analogical transfer. *Canadian Journal of Psychology, 39:* 460-466.

Gick, M., and Holyoak, K., (1980). Analogical problem solving. *Cognitive Psychology, 12,* 306- 355.

Gick, M., and Holyoak, K. (1983). Schema induction and analogical transfer. *Cognitive Psychology, 15,* 1-38.

Glasgow, J. (1990). Imagery and classification. *Proceedings of the 1st ASIS SIG/CR Classification Research Workshop.* Toronto.

Glasgow, J., and Papadias, D. (in press). Computational imagery. *Cognitive Science.*

Holyoak, K. J., and Koh, K. (1987). Surface and structural similarity in analogical transfer. *Memory and Cognition, 15,* 332-340.

Holyoak, K. and Thagard, P. (1989) Analogical mapping by constraint satisfaction. *Cognitive Science, 13,* 295-355.

Kosslyn, S. (1980). *Image and mind.* Cambridge: Harvard University Press.

Minsky, M. (1986). *The society of mind.* New York, Simon and Schuster.

Mitchell, M., and Hofstadter, D. (1990). The right concept at the right time: How concepts emerge as relevant in response to context-dependent pressures. *Proceedings of the Twelfth Annual Conference of the Cognitive Science Society*, Hillsdale, NJ: Erlbaum, 174-181.

Nelson, G., Thagard, P., and Hardy, S., (in press). Integrating analogies with rules and explanations. In J. Barnden and K. Holyoak (Eds.), *Advances in Connectionism*, vol. 2, *Analogical Connections*, Norwood, NJ: Ablex.

Papadias, D., and Glasgow, J., (1991). A knowledge representation scheme for computational imagery. *Proceedings of the Thirteenth Annual Conference of the Cognitive Science Society*, Hillsdale, NJ: Erlbaum, 49-54.

Riesbeck, C., and Schank, R. (1989). *Inside case-based reasoning.* Hillsdale, NJ: Erlbaum.

Thagard, P., Holyoak, K., Nelson, G., and Gochfeld, D. (1990). Analog retrieval by constraint satisfaction. *Artificial Intelligence, 46,* 259-310.

Thagard, P., and Hardy, S. (1992). Visual thinking in the development of Dalton's atomic thinking. *Proceedings of the Ninth Canadian Conference on Artificial Intelligence*, Vancouver.

Probing the Emergent Behavior of Tabletop, an Architecture Uniting High-level Perception with Analogy-making

Douglas R. Hofstadter and Robert M. French
Center for Research on Concepts and Cognition • Indiana University
510 North Fess Street • Bloomington, Indiana 47408

Abstract

Tabletop is a computer model of analogy-making that has a nondeterministic parallel architecture. It is based on the premise that analogy-making is a by-product of high-level perception, and it operates in a restricted version of an everyday domain: that of place-settings on a table. The domain's simplicity helps clarify the tight link between perception and analogy-making. In each problem, a table configuration is given; the user, hypothetically seated at the table, points at some object. The program responds by doing "the same thing", as determined from the opposite side of the table. Being nondeterministic, Tabletop acts differently when run repeatedly on any problem. Thus to understand how diverse pressures affect the program, one must compile statistics of many runs on many problems. Tabletop was tested on several families of interrelated problems, and a performance landscape was built up, representing its "likes" and "dislikes". Through qualitative comparisons of this landscape with human preferences, one can assess the psychological realism of Tabletop's "taste".

Cognitive generality in a microdomain

Tabletop, a computer model of analogy-making, is based on the premise that analogy-making and perception are inseparably intertwined. The nondeterministic and parallel program builds representations of situations and makes analogies in a familiar but restricted domain: place-settings on a table. Though simple, the domain brings out general issues in high-level perception and analogy-making.

Imagine Henry and Eliza facing each other across a table. Henry touches an object and says, "Do this!" Eliza must respond by touching some object. The program plays the role of Eliza, with *selection* playing the role of touching. One obvious possibility would be to "touch" *literally the same* object. This option is always open, no matter what the configuration and no matter what Henry touches. Often, though, there are aspects of the situation—pressures—that make the literal-sameness option less appealing than touching some other object. The possibility of any number of pressures coexisting, and their often subtle interactions, lend the domain considerable complexity and depth.

Suppose both individuals have coffeecups before them. Most people would perceive the cups as *counterparts*. Thus, if Henry touches *his* cup, it would seem more natural for Eliza to touch *her* cup than to reach across the table to touch his. But now suppose the "counterparthood" is weakened by changing Eliza's cup to a glass (Fig. 6). Here, although the two objects remain counterparts in terms of *position*, their *categories* no longer match exactly.

However, as "cup" and "glass" are closely related categories, there remain reasons—pressures—for Eliza to see her glass and Henry's cup as counterparts. Chances are good that Eliza will rank touching her glass higher than touching his cup. Of course, if the category mismatch is further increased—give Eliza a fork, not a glass—the sense of counterparthood will be so diminished that Eliza may revert to the literal-sameness option (touching Henry's cup).

Other pressures that might influence Eliza's choice include: the arrangement, category memberships, and orientations of objects, etc. Often, perceptual groupings ("chunks"), whose plausibility depends on the physical and conceptual proximity of the items involved, play a role in determining what items Eliza is prone to see as counterparts.

Obviously, not all possible groupings can be considered by a person or by a program—after all, with just a dozen objects on the table, hundreds of potential ways of grouping them exist; moreover, if smaller chunks are allowed to be members of larger chunks (a key feature of human perception, which routinely builds up such hierarchical representations), the number is even higher. Of course, not only efficiency but cognitive plausibility militates strongly against a computer model in which brute-force strategies of any sort play any role. Thus a crucial design philosophy of the Tabletop program is that it does not routinely invoke all possible pressures in each situation; rather, it lets a limited number of context-dependent pressures *emerge* as each situation is perceptually processed.

Parallel emergent perceptual processes

The central challenge of the Tabletop project is to model the simultaneous existence and interaction of multiple pressures in a human mind perceiving (*i.e.*, building representations of) a complex situation. We stress perception rather than analogy-making, since our philosophy is that *analogies emerge automatically as a by-product of high-level perception* (see [Chalmers, French, & Hofstadter 91]). This idea is at the crux of Tabletop; we contend that the program should be judged not only on the accuracy with which it mimics human performance in its narrow domain, but also on its general principles, intended to apply to any domain, irrespective of size. (In fact, Tabletop's forerunner Copycat uses a similar architecture in a different microdomain. See [Mitchell 90] and [Hofstadter & Mitchell 91].)

In Tabletop, "high-level perception" means the concurrent carrying-out of the following tasks:

- initial labeling of objects in terms of basic categories;
- further labeling, on higher levels of abstraction, of already-labeled table objects;

- hierarchical building-up of *groups* (tentative perceptual chunks) on the basis of:
  - physical proximity of component items (*i.e.*, table objects or already-built groups);
  - conceptual proximity of component items;
  - structural similarity of component subgroups;
- building-up of *correspondences* (links between two items establishing them tentatively as each other's counterparts) on the basis of:
  - corresponding physical positions of the two items;
  - conceptual proximity of the two items;
  - structural similarity of the items (if they are groups);
- assignment of a time-varying *salience* to each perceived item (object, group, or correspondence);
- assignment of a time-varying *strength* to each perceived correspondence;
- competition among rival perceptual structures, giving rise to a pruning of weaker structures.

All these tasks are carried out in parallel. A key idea of the architecture is that each type of task is implemented as a sequence of small, independent *micro-actions*. Building a group, for instance, involves an escalating series of "microtests" that check the physical and conceptual distances between prospective group members. If any such test fails, the potential group is aborted; if it succeeds, the way is clear for further tests; if all requisite hurdles are cleared, the group gets built by a specific micro-action.

To carry out all these tasks requires many micro-actions, which are interleaved *at random*. For instance, a microtest checking out the attractiveness of a potential group on Eliza's side might run, followed at random by a micro-action that proposes attaching an abstract label to some object on Henry's side, followed by another microtest that checks out some other aspect of Eliza's potential group, followed by a micro-action that tests some aspect of a proposed correspondence elsewhere, etc. In sum, many different sorts of small things happen, one after another, at different places on the table. Through such interleaving of scattered local micro-actions, *large-scale perceptual structures gradually emerge in parallel in all areas of the table*. Because all these processes have mutual influences, the perceptual structures that they build tend to form conceptually coherent sets.

Tabletop's parallelism thus lies at the *task* rather than the *micro-action* level. The degree of effective parallelism is determined by the grain of the break-up of tasks into micro-actions. The finer the grain, the more evenly will emerge the different perceptual structures. A *totally unbiased* selection of micro-actions would result in all large-scale tasks getting carried out, on average, at the same speed—a completely "fair" sharing of attention over the table. However, such perceptual fairness is far from Tabletop's strategy; rather, Tabletop *accelerates* avenues of exploration that offer promise while *retarding* ones that appear uninteresting. For instance, Tabletop is not equally likely to inspect all objects on the table; at any given moment, probabilities bias its choice of what to look at. Metaphorically speaking, certain objects and areas of the table are perceptually "hot" while others are "cool", and these biases are dynamic: they change as new perceptions are made.

Dynamic biasing is realized by assigning each micro-action an *urgency*—effectively its probability of being chosen. Urgencies are assigned according to the perceived promise of a given micro-action. For instance, a proposed micro-action involving a salient object would get a higher urgency than one involving an object of low salience (all other things being equal). Many factors are taken into account in urgency assignment: location on the table, types of items involved, strengths and/or saliences of objects or correspondences involved, etc. Since micro-actions having high urgencies tend to be chosen swiftly, sequences of logically related high-urgency micro-actions will tend to be accelerated, while sequences of low-urgency micro-actions will tend to be slowed down. In this way, different perceptual structures emerge naturally at different speeds, depending on the program's best *a priori* estimate of their significance.

Since the relative saliences of table items are critical in determining what makes a region of the table "hot", one might ask, "What makes an item salient?" Many factors are involved, including: location relative to the touched object; conceptual proximity to the touched object; being a group (as opposed to a mere object); size of group; physical position in a group; prior perception of other objects in the same category; having a counterpart or not; etc. Obviously, some of these attributes vary over time, so that saliences also change, which implies that various areas of the table become stronger or weaker probabilistic foci of the program's attention. Not only saliences of objects but also strengths of correspondences play a role in determining an area's (probabilistic) perceptual attractiveness, and a similar list of factors is taken into account in the computation of each correspondence's (dynamically varying) strength.

Any group can be disbanded and any correspondence taken down. Often the reason for dismantling a perceptual structure is the discovery of another structure of comparable or greater strength. Thus Tabletop's perceptual process is a rough-and-tumble contest among conflicting interpretations (often just fragmentary), the outcome of which, in the end, is hopefully a strong set of mutually-reinforcing perceptual structures.

Structure value is a dynamically varying number that represents the total strength of all currently existing perceptual structures. This number can be considered a measure of how well the program has so far done in "making sense" of the scene before it; at the end of a run, structure value can serve as a "quality measure" of the answer produced by the program. An important pressure on Tabletop is to maximize structure value; counterbalancing this, however, is a competing pressure—time pressure—that pushes for the program to finish within a reasonable amount of time.

As perceptual structures emerge around the table, *mappings* also emerge. Indeed, a mapping is just one type of perceptual structure: a family of one or more mutually compatible (often mutually reinforcing) correspondences. A mapping, needless to say, is an analogy. The basic premise of Tabletop, then, is that *analogy-making is a high-level by-product of perception*. In other words, analogies

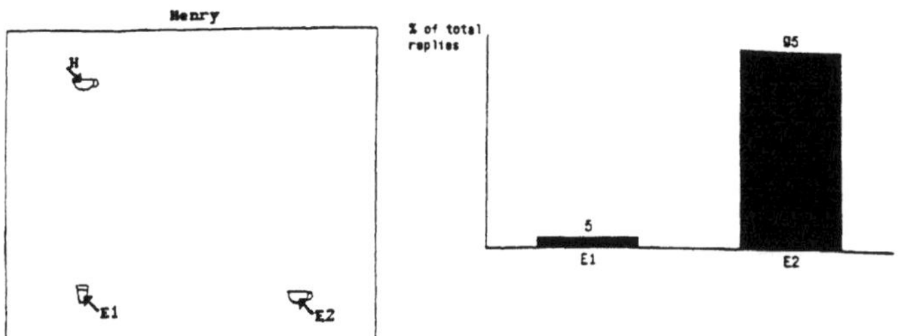

Eliza touches	Av. structure value	Av. no. of micro-actions
E1	58	52
E2	89	50

Figure 1. In Figs. 1 through 5, representing the *Surround* family, H touches his cup. Though the literal-sameness answer (*i.e.*, touching H's cup) is always possible, the main rivalry is between E's glass and cup. The variants explore combinations of pressures by surrounding E's glass and H's cup with various sets of objects. Fig. 1, the "base case", has no surrounding objects; here, just two pressures contribute to the decision: category membership and physical position. The former favors the cup (category *identity* is better than category *proximity*). What about the latter? People are more likely to seek a corner object's counterpart in the diagonally opposite corner than in the mirror-image corner, so such a bias was built into Tabletop. Therefore, position pressure also favors the cup. Overall, then, the pressure in favor of E's cup is very strong; indeed, Tabletop chooses her glass only 5% of the time.

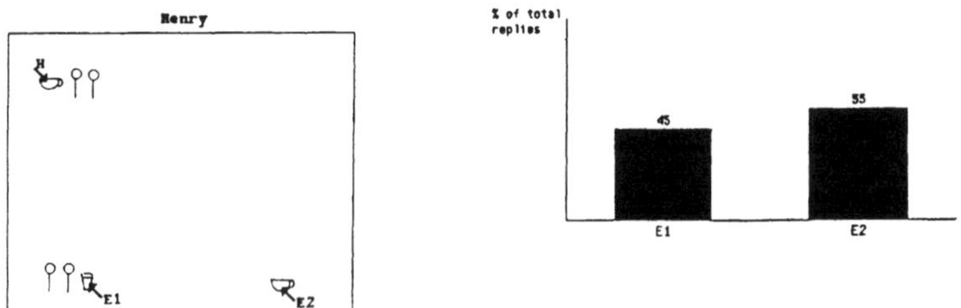

Eliza touches	Av. structure value	Av. no. of micro-actions
E1	186	64
E2	107	56

Figure 2. When humans look at this setup, they effortlessly perceive two groups — one containing H's cup, the other containing E's glass. (Of course other groupings are possible, but virtually never come to mind.) Tabletop is similarly inclined; it sees a group consisting of H's cup and two spoons (the spoon pair is likely to be seen as a *subgroup*), and a group consisting of E's glass and two spoons (also likely to be seen as a subgroup). Not only do these groups map onto each other as wholes, but their subgroups (if seen) map strongly onto each other, thus pushing the structure value up and increasing the pressure for mapping H's cup onto E's glass. Indeed, Tabletop now touches the glass 45% of the time. As might be expected, the average structure value when it does so is significantly higher than when it touches her cup. This is a case where highest frequency and best structure disagree.

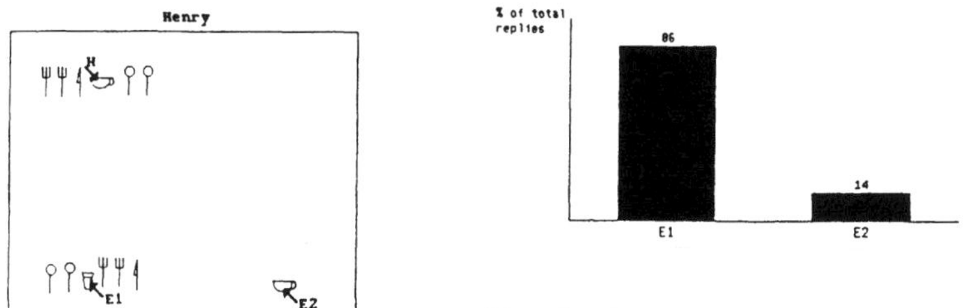

Eliza touches	Av. structure value	Av. no. of micro-actions
E1	248	101
E2	146	94

Figure 3. The groups around H's cup and E's glass are very similar: they have the same number of objects and their subgroups are identical. Humans see the mappings as very strong and see E's cup as a loner, thus feel much pressure to touch her glass. Not only does the program do this far more often than touch her cup, but the structure value for the former averages far higher than for the latter.

The strong mappings push so hard for touching E's glass that one might wonder what would *ever* induce Tabletop to pick E's cup. Two factors are involved. One is, Tabletop sometimes simply fails to build those mappings. On such runs, the pressures do not so greatly favor her glass. More rarely, Tabletop may build the mappings but simply choose (stochastically) to ignore them and touch E's cup. Though this may seem irrational, people often act similarly. In a survey, subjects were asked to draw all relevant correspondences in this setup. Some, after drawing a line linking the two spoon-groups, another linking the fork-groups, and a third linking the knives, *ignored* all these lines and chose E's cup. In this light, the "anomalous" 14% of runs in which Tabletop touches the lone cup seem justified.

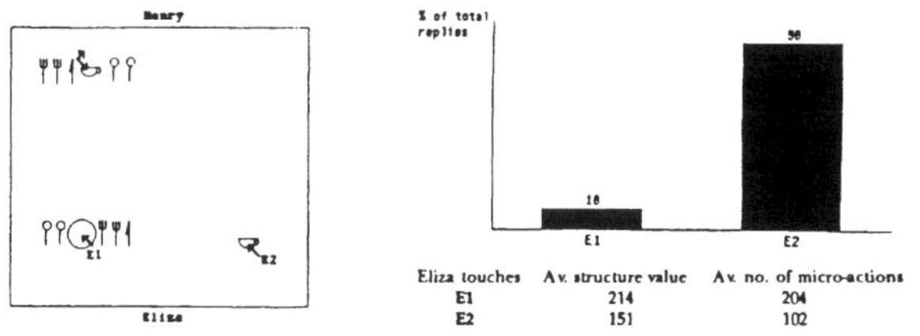

Figure 4. E's glass has been replaced by a plate, conceptually remote from the touched object. This should shift the pressures back to favoring the isolated cup. Indeed, Tabletop now touches E's cup 90% of the time, and her plate just 10%. Still, the structure value associated with the plate-answer remains over 40% higher than that for E's cup.

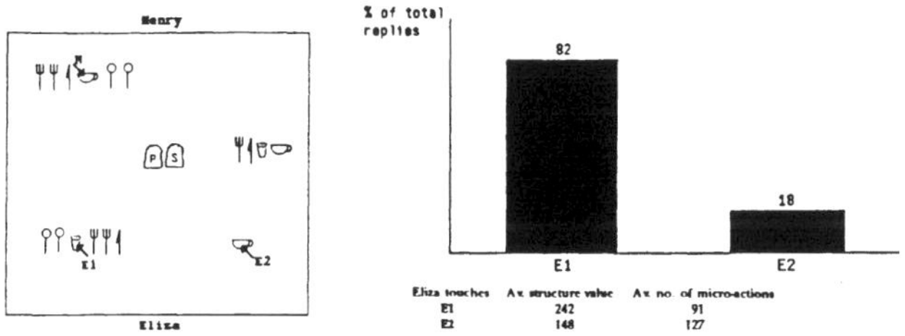

Figure 5. Distractors have been added to Fig. 3, more than doubling the number of potential object-correspondences. But if Tabletop's focusing mechanisms operate well, this should have little effect on the amount of processing and the distribution of answers. Indeed, there is little contrast between these results and Fig. 3. The average run length is almost exactly the same as in Fig. 3, which had no distractions at all. Thus Tabletop essentially ignores objects in unlikely locations on the table, focusing its attention primarily on *a priori* preferred regions. (However, when no objects are in *a priori* preferred regions, Tabletop does examine *a priori* unlikely regions.)

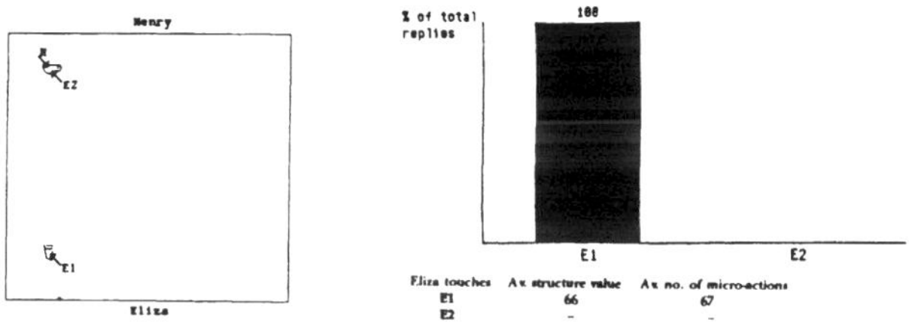

Figure 6. Figs. 6 through 10 represent the *Blockage* family. The cup and glass facing each other are not identical, but almost so: the Slipnet nodes "cup" and "glass" are very close. Also the glass is in a favorable position with respect to the cup. There is thus much pressure to choose the glass, and Tabletop always does so here. In variants, the pressures for touching H's cup are increased by creating correspondences that "usurp" E's glass. In this, the base case, there is no attempt at blockage.

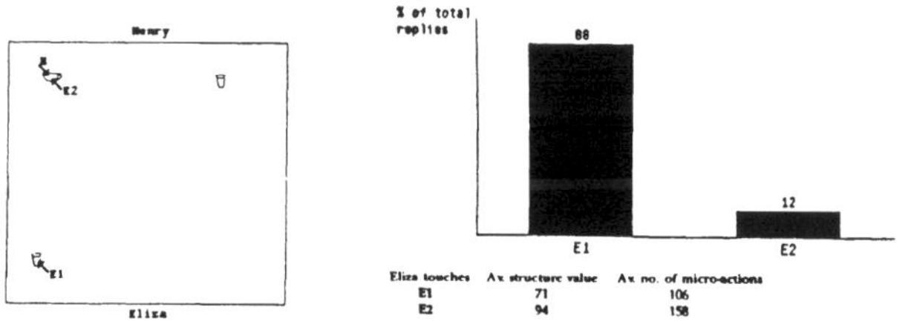

Figure 7. A glass has been added; being on H's side, it is most unlikely to be touched. Unlike the additions in the *Surround* family, this addition creates no new group. One might thus expect that this addition, like the distractions in Fig. 5, would have little effect on Tabletop's answers or resources expended. But another effect — the distant glasses' *identicality* — gives rise to a pressure to build a *correspondence* between them. When this is done, the glasses are seen as part of a single, albeit weak, structure, which exerts a *blockage* effect. (Correspondences and groups are both perceptual chunks and are similar in many ways; the former, however, tend to be weaker since their constituents, usually being far apart on the table, are not tightly bound together.)

Many subjects (40% [French 1992]) saw the glasses as counterparts. When this happens, since E's glass cannot be the counterpart *both* of H's glass *and* of his cup, just one answer remains: the literal-sameness answer, H's cup. Tabletop occasionally (12% of the time) sees the glasses as counterparts and touches H's cup. Note that the structure value of this answer is 30% higher than for E's glass, though the latter is chosen far more often. In addition, runs on which Tabletop chooses H's cup average roughly 50% longer than for E's glass. Once again, this is not surprising: answers involving deeper perception should take longer to find than those with less.

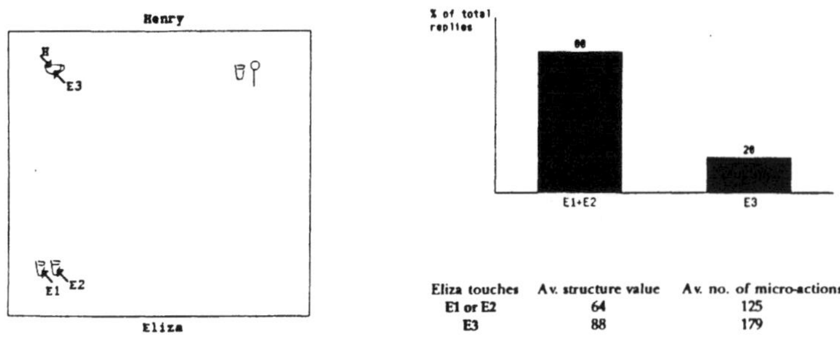

Eliza touches	Av. structure value	Av. no. of micro-actions
E1 or E2	64	125
E3	88	179

Figure 8. Two objects have been added, strongly suggesting groups. Tabletop almost always builds the group on E's side, since the two glasses are not just neighbors but identical objects. The group on H's side has less appeal, since "spoon" and "glass" are distant Slipnet nodes. Still, on many runs, both groups get built. When, in addition, a diagonal correspondence between them is built, despite its weakness, it "usurps" both glasses on E's side, forcing Tabletop to go for H's cup.

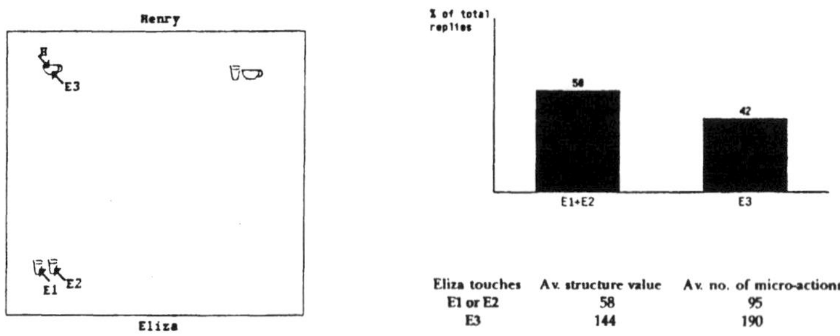

Eliza touches	Av. structure value	Av. no. of micro-actions
E1 or E2	58	95
E3	144	190

Figure 9. The group on H's side now has more appeal, as "cup" and "glass" are far closer in the Slipnet than "spoon" and "glass". Often the two objects are seen as physically *and* conceptually close. This makes for a stronger group, which in turn makes for a stronger diagonal correspondence, leading Tabletop to choose H's cup more than twice as often as in Fig. 8.

Sometimes E's group is built but not mapped to anything as a unit; in such runs, the touched cup tends to be mapped onto one of E's glasses. There is pressure to map her other glass onto H's glass (diagonally opposite identical objects make strong counterparts). But there is also counterpressure: to map E's two glasses, which have been grouped and are thus a conceptual unit, onto unrelated objects would be to disrespect their unity. Yet Tabletop does this occasionally, in which case the structure value suffers markedly. When Tabletop goes for H's cup, the structure is much better than when it chooses one of E's glasses. Also note that Tabletop takes significantly longer to build the structure that gives rise to the better answer.

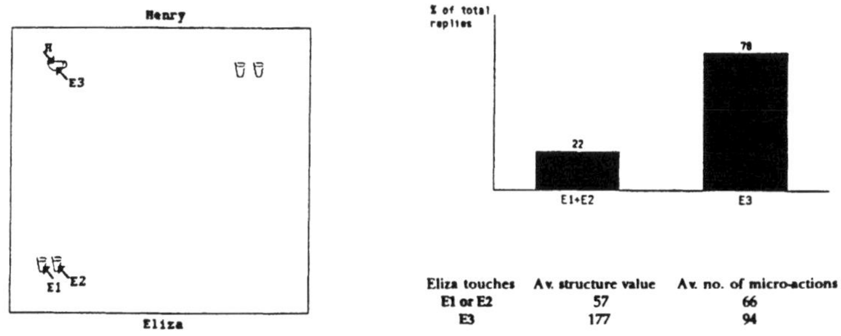

Eliza touches	Av. structure value	Av. no. of micro-actions
E1 or E2	57	66
E3	177	94

Figure 10. A turning point in the *Blockage* family: Tabletop chooses H's cup over half the time. The reason is simple. Both glass–glass groups are very strong, as is the correspondence between them—strong enough, it turns out, to make Tabletop very reluctant to break it by mapping H's cup onto either of E's glasses. Tabletop thus picks H's cup 78% of the time. (Human subjects chose H's cup 66% of the time [French & Hofstadter 1991].) As one might expect, the average structure value for this answer is better than when Tabletop chooses one of E's glasses. Also as usual, it tends to take Tabletop longer (by about 40%) to get the answer having better structure.

represent the highest (most abstract) level of perception. It would thus not be exaggerated to describe Tabletop as a model of high-level vision. Of course, the raw input to Tabletop must be thought of as being the output of a prior module that carries out perceptual processing at a lower (and more modality-specific) level. Tabletop is not a model of all of vision, but of vision's high end—the end that interfaces with concepts at various levels of abstraction.

Many aspects of Tabletop's architecture can barely be hinted at herein. ([French 92] gives a much fuller presentation.) In particular, *conceptual proximity* is implemented in the *Slipnet*, a network in which each node represents a concept (strictly, the *core* of a concept), and links to other nodes establish a metric defining conceptual distances. Each node (conceptual core) has a dynamically varying "halo" (the *full* concept)—a diffuse region centered on it and probabilistically including nearby nodes. For instance, the degree to which the node "glass" is, at any moment, included in the halo of "cup" represents the current likelihood of those two concepts to be "equated" (the likelihood that their non-identity will be "forgiven") in the act of considering whether a particular cup and glass deserve being deemed counterparts (at least tentatively).

Tabletop's overall "personality"

[French & Hofstadter 91] presented Tabletop's architecture along with a few runs on three problems. Though of interest, this afforded only a limited perspective on the program's behavior. Because of its stochastic nature, Tabletop follows different pathways on different runs, and thus often comes up with different answers on different runs. Therefore, to get a feel for the program's overall behavior, one must run it not only on many different problems, but *many times on each given problem*. Only thus can one gain a clear perspective on how different combinations of pressures "pull" the program. Since the heart of the model is its ability to handle multiple interacting pressures, this is a key test.

We have probed Tabletop's "personality" by running it many times on a great variety of configurations. Inevitably, once any problem was devised, several close variants would spring to mind in which the altered pressures would alter the appeal, to humans, of various answers. By testing Tabletop on such tightly interrelated *families* of problems, we learned how it responds to diverse combinations of pressures.

In the figures we sample two families, each represented by five problems. In each problem, the table is shown on the left, with the object Henry touched indicated by an arrow with an "H". Possible responses by Eliza are indicated by arrows labeled "E1", "E2", etc. On the right, a bar graph is shown; each bar represents the frequency of one answer. All problems were run 50 times. On each run, a monitor recorded the *answer*, the final *structure value*, and the *run-length* (total number of micro-actions). Below the graph is a table giving, for each answer, the average final structure value and the average run-length for all runs yielding that answer.

Of particular interest are cases where the highest-frequency answer is not the answer having the highest final structure value. Such cases, rather than reflecting a defect of the architecture, reflect an inevitable fact about high-level perception: deep perceptions are often hard to discover; it is easy to be distracted by routes having more surface appeal. Thus Tabletop often prefers "shallow" answers, provided they have at least a modicum of plausibility, over "deep" ones (where depth is measured by structure value). It is, however, a virtue of Tabletop's parallel stochastic architecture that, by allowing simultaneous exploration at different rates along rival routes showing different degrees of promise, it is not always seduced by surface glitter, and can on occasion come up with deeper visions.

By exploring several families of "Do this!" problems, each family having many members, we built up a "performance landscape" of the program—a surface in the abstract multidimensional space of all Tabletop problems, where each dimension roughly corresponds to a given pressure. The "ridges" in this landscape represent critical combinations of pressures where the program switches from one preference to another (e.g., Fig. 10 in the Blockage family). Likewise, "peaks" and "valleys" correspond to clear and stable "likes" and "dislikes" on the program's part. By making qualitative comparisons of the locations of Tabletop's ridges, valleys, and peaks with our own personal preferences, as well as with statistics summarizing the preferences of experimental subjects, we were able to assess the psychological realism of Tabletop's "taste". (Experimental results can be found in [French 92].)

Analogy-making as high-level perception

From our point of view, the Tabletop program did a creditable job, on a qualitative level, of simulating the taste of a typical human playing the role of Eliza. (Readers can look at the bar graphs and decide for themselves whether they agree.) Despite this success, we reiterate our contention that the program is not to be judged primarily on this basis, but rather on its overall architecture, in which analogy-making falls out as a natural by-product of high-level perception, a cognitive activity that is realized by parallel processes guided by dynamically evolving pressures that emerge in response to the situation being faced.

References

[Chalmers, D., French, R., & Hofstadter, D. 91] "Perception, High-level Representation, and Analogy-making: A Critique of Artificial Intelligence Methodology". Technical Report, Center for Research on Concepts & Cognition, Indiana U., Bloomington, 1991. To appear in *Journal of Experimental and Theoretical Artificial Intelligence*, 1992.

[French, R. 92] "Tabletop: An Emergent, Stochastic Model of Analogy-Making". Ph.D. diss.,. Computer Science & Engineering Dept., U. of Michigan, Ann Arbor, 1992.

[French, R. & Hofstadter, D. 91] "Tabletop: An Emergent, Stochastic Model of Analogy-Making". In *Proceedings, 13th Annual Cognitive Science Society Conference*, pp. 708–713. Hillsdale, NJ: LEA, 1991.

[Hofstadter, D. & Mitchell, M. 91] "The Copycat Project: A Model of Fluid Concepts and Analogy-Making". Tech. Rpt., Ctr. for Resch. on Concepts & Cog'n, Indiana U., Blgtn, 1991. To appear in Barnden, J. & Holyoak, K. (eds.) *Advances in Conn'ist & Neural Computation Theory. Vol. 2: Analogical Connections*. Hillsdale, NJ: LEA, 1992.

[Mitchell, M. 90] "Copycat: A Computer Model of High-level Perception and Conceptual Slippage in Analogy-making". Ph.D. diss., Comp. & Communic'n Sci's Dept., U. Mich., Ann Arbor, 1990.

Concept Learning and Flexible Weighting

David W. Aha
Applied Physics Laboratory
The Johns Hopkins University
Laurel, MD 20723
aha@cs.jhu.edu

Robert L. Goldstone
Department of Psychology
Indiana University
Bloomington, IN 47405
rgoldsto@ucs.indiana.edu

Abstract

We previously introduced an exemplar model, named GCM-ISW, that exploits a highly flexible weighting scheme. Our simulations showed that it records faster learning rates and higher asymptotic accuracies on several artificial categorization tasks than models with more limited abilities to warp input spaces. This paper extends our previous work; it describes experimental results that suggest human subjects also invoke such highly flexible schemes. In particular, our model provides significantly better fits than models with less flexibility, and we hypothesize that humans selectively weight attributes depending on an item's location in the input space.

We need more flexible models of concept learning

Many theories of human concept learning posit that concepts are represented by prototypes (Reed, 1972) or exemplars (Medin & Schaffer, 1978). Prototype models represent concepts by the "best example" or "central tendency" of the concept.[1] A new item belongs in a category C if it is relatively similar to C's prototype. Prototype models are relatively inflexible; they discard a great deal of information that people use during concept learning (e.g., the number of exemplars in a concept (Homa & Cultice, 1984), the variability of features (Fried & Holyoak, 1984), correlations between features (Medin *et al.*, 1982), and the particular exemplars used (Whittlesea, 1987)).

Exemplar models instead represent concepts by their individual exemplars; a new item is assigned to a category C if it is relatively similar to C's known exemplars. Exemplar representations are far more flexible than prototype representations since they retain sensitivity to all of the information listed above. This flexibility often translates to increased categorization accuracy. For example, unlike prototype models, humans and exemplar models can learn some non-linearly separable categories as easily as linearly separable categories (Medin & Schwanenflugel, 1981). This capability is not limited to flat learning architectures; several researchers capture this flexibility in radial basis networks (e.g., Kruschke, 1992; Hurwitz, 1991).

While existing exemplar models are more flexible than prototype models, they are still not sufficiently flexible. We argue that people represent categories not only with category exemplars, but also with a set of specific weights associated with each exemplar's (or set of exemplars) attributes. The subject experiments described in Section 2 suggest that the weight given to an attribute depends on its exemplar's "neighborhood" in psychological space, where exemplars are assumed to be describable by their attributes' values. Our claim is that concepts are not represented simply by a set of attribute weights. Rather, an attribute's importance in similarity calculations depends on its *context* – the other attributes that are true for a particular exemplar. For example, the relative importance of the "date of next deadline" attribute for predicting membership in the "will work this weekend" category varies depending on the "upcoming computer downtime" attribute's value (e.g., when a deadline exists for the middle of the following week, one might be more likely to work during the preceding weekend when it is known that the computers will not be functioning on the days immediately preceding the deadline). Moreover, people can learn the importance of an attribute in concept-learning situations even when they have little guidance for assigning attribute weight settings. Since

[1] Other summary information may also be stored by more advanced prototype models; our concerns primarily target problems with "pure" prototype models. More accurately, we are interested in supporting the learning behavior displayed by the advanced exemplar models described in Section 3 regardless of the models' representation for categories (Barsalou, 1989).

people have almost no background information on the artificial stimuli used in the experiments described in Section 2, and since the exemplars in those experiments exhibit somewhat arbitrary regularities, we can be confident that our subjects are actively attending to the stimuli's regularities rather than applying knowledge that they previously acquired.

The work presented here has several precursors. Medin and Schaffer (1978) developed an exemplar model for representing concepts that was subsequently generalized by Nosofsky (1984; 1986). In turn, Aha and McNulty (1989) created a learning algorithm for Nosofsky's model and extended its selective attention mechanism to be a function of the target concept. We further augmented this learning model to include exemplar-specific weights; each exemplar in each concept was given its own set of attribute weights (Aha & Goldstone, 1990). This new model, named GCM-ISW, achieved faster learning rates and higher asymptotic performance than other models on artificial categorization tasks whose concepts were best modeled by using context-sensitive settings for attribute weights.

In Section 2, we extend our previous work by showing that human subjects are highly flexible in that they can selectively weight an attribute differently depending on the region of the instance space in which it is located. In Section 3, we show that GCM-ISW can fit these subjects' predictions better than two concept-learning systems with less flexible weighting schemes for warping the instance space. Like humans, GCM-ISW can allow the importance of attributes to be a function of its region of instance space.

Experiments on weighting attributes

An experiment was conducted to determine whether human subjects can learn categories that require attributes to be weighted differently for different category exemplars. That is, this experiment investigates whether subjects are constrained to weight attributes equally regardless of their context. This experiment also investigates whether subjects subsequently generalize their categories according to the attribute weights that they have learned. First, the subjects learn to distinguish category A from category B exemplars until they can accurately classify a set of *training* exemplars. The subjects are then given a set of *test* exemplars to classify. We can indirectly ascertain the weights that subjects assigned to the attributes by observing how these test exemplars were classified.

During training, 40 undergraduate subjects were told to categorize picture items, corresponding to exemplars, into category A or category B. These pictures varied along two dimensions: size of square and position of line in square. Each dimension is defined over eight evenly-spaced values. Square size varied from 2.0 cm to 7.5 cm. Position of line in square varied from the far left side to the far right side.

Subjects were presented with twelve training items, where half belonged to each category. The particular items shown to the subjects in Experiments 1 and 2 are shown in Figure 1. The first matrix shows the two groups of items in Experiment 1. Each cell in this matrix represents a possible stimulus item. For example, the bottom-leftmost cell represents the item *very small square with line on the far left side of the square.* The twelve items that were shown in the training stage were labeled A or B according to their category. The cluster of items in the top-right of the first matrix is characterized by relatively large squares with lines relatively far to the right. The other cluster has relatively small squares with lines further to the left. Line position was the more important dimension for distinguishing category A from category B items for the first cluster; items with the value six for line position belonged in category B whereas items with the value seven belonged in category A. Conversely, size was the more important dimension for the other cluster of items; items with a value of seven on the size dimension were exemplars of category B, while items with a value of six belonged in category A.

During training, after the twelve items' ordering was randomized, they were subsequently presented to the subjects on a Macintosh SE. For each item, the subject pressed A or B to indicate their category prediction. Subjects were told whether their classification was correct immediately after their response. Training continued until the subject performed four error-free classifications of the complete set of training items.

During testing, all 64 possible combinations of line position and square size were displayed to subjects in a random order. For each item, subjects indicated whether they believed the item belonged in category A or B. Only twelve of these items were previously shown to the subjects; the remaining 52 were novel items, and their placement in category A or B represent generalizations of these categories.

The results from the test stage of Experiment 1 are displayed in Figure 2. The number in each cell indicates the percentage of times that subjects placed the item into category B during testing. The percentages indicate fairly good retention of the items that were presented during training and widespread generaliza-

Experiment #1:

	1	2	3	4	5	6	7	8
1	.	.	.	.	.	.	.	.
2	.	.	.	.	.	B	Z	.
3	.	.	.	.	.	B	A	.
4	.	.	.	.	.	B	A	.
5	.	.	.	.	.	Y	A	.
6	.	A	A	A	X	.	.	.
7	.	W	B	B	B	.	.	.
8	.	.	.	.	.	.	.	.

Experiment #2:

	1	2	3	4	5	6	7	8
1	.	.	.	.	.	.	.	.
2	.	.	.	B	B	B	Z	.
3	.	.	.	Y	A	A	A	.
4	.	A	X	.	.	.	.	.
5	.	A	B	.	.	.	.	.
6	.	A	B	.	.	.	.	.
7	.	W	B	.	.	.	.	.
8	.	.	.	.	.	.	.	.

Figure 1: Training sets and critical test items for the two experiments. The horizontal and vertical axes denote (increasingly right) line positions and (decreasing) square size dimensions respectively. The categories of the training items are shown as A and B. The four critical test items per experiment are marked with one of $\{W, X, Y, Z\}$.

Average of the Subjects' Predictions:

	1	2	3	4	5	6	7	8
1	50	45	45	40	50	65	10	30
2	65	60	80	80	85	90	20	25
3	55	50	55	60	70	85	0	20
4	40	55	45	40	65	85	0	15
5	20	20	15	25	30	70	0	20
6	10	0	5	10	10	5	5	15
7	85	90	90	95	90	85	75	80
8	55	60	65	60	55	60	55	60

GCM-ISW's Predictions:

	1	2	3	4	5	6	7	8
1	67	74	77	78	78	76	34	31
2	50	63	72	76	78	78	33	30
3	31	43	58	69	74	74	29	28
4	20	26	37	52	64	69	26	24
5	17	19	23	32	46	57	23	21
6	21	21	22	27	37	48	26	25
7	74	75	77	78	78	74	53	61
8	78	78	78	78	76	72	42	49

Figure 2: The subjects' averaged predictions and GCM-ISW's probabilistic guess that the test items in Experiment 1 belong to category B.

tion of the training knowledge to the new items.

Particular test items of interest to us are labeled by the letters W, X, Y, and Z in Figure 1. These items were not presented during training. Their pattern of classification seems to confirm that subjects generalized their categories by differentially weighting the attributes for different items. For example, item W was categorized as an exemplar of category B by 90% of the subjects in Experiment 1, although it is as close to category A items as it is to category B items. Similarly, item X in Experiment 1 was categorized as a member of category A by 90% of the subjects. These results indicate that subjects strongly weight the size dimension in these categorizations. It is as if the subjects are *stretching* the vertical axis in this area of the space, so that the A and B items become separated by a greater psychological distance. However, the entire vertical axis is not stretched. Instead, it is selectively stretched in this single region of the space (i.e., the lower-left). Similarly, the horizontal axis is selectively stretched in the upper-right region; item Y was categorized as a B by 70% of the subjects while item Z was categorized as an A by 80% of the subjects, indicating that subjects considered line position to be more important than square size for categorizing items in this region. In summary, subjects generalized their concepts on the basis of the square size dimension for one cluster and on the basis of the line position dimension for the other cluster.

Experiment 2 replicates Experiment 1 with a relocation of the training items. One possible explanation of Experiment 1's results is that, perceptually, there was a bigger difference between size six and size seven squares than there is between size two and three squares and/or a relatively large perceptual difference between lines in positions six and seven. If this were true, then our generalization results could be explained without requiring that subjects learned to selectively weight dimensions in particular regions of the space. Experiment 2's results refute this possible explanation; if one assumed that there is a large perceptual difference between size six and size seven squares, then precisely the wrong prediction would be

Average of the Subject's Predictions:

	1	2	3	4	5	6	7	8
1	65	50	65	55	55	60	65	60
2	70	75	85	90	90	90	90	80
3	15	10	15	15	10	10	0	10
4	15	5	85	35	15	15	5	25
5	20	5	90	60	35	40	50	35
6	20	0	85	70	50	60	50	60
7	30	15	90	80	75	75	60	60
8	30	10	70	55	55	50	55	50

GCM-ISW's Predictions:

	1	2	3	4	5	6	7	8
1	47	39	69	75	77	78	78	78
2	59	50	72	77	77	77	76	74
3	24	25	46	36	27	23	21	21
4	22	22	58	49	36	28	23	21
5	25	26	68	65	55	42	32	25
6	28	30	75	74	70	61	48	36
7	30	31	76	77	76	72	65	54
8	29	31	74	77	77	76	74	68

Figure 3: The subjects' averaged predictions and GCM-ISW's probabilistic guess that the test items in Experiment 2 belong to category B.

GCM-SW's Predictions:

	1	2	3	4	5	6	7	8
1	55	64	71	74	72	67	59	51
2	44	53	64	73	69	63	47	39
3	33	38	49	51	38	31	25	27
4	28	27	43	47	39	32	27	25
5	31	33	59	57	50	41	34	30
6	37	40	66	67	62	53	45	38
7	48	59	76	74	70	63	56	48
8	58	66	74	76	74	69	63	57

Figure 4: GCM-SW's probabilistic guess that the test items in Experiment 2 belong to category B.

made for Experiment 2, where item W is now placed in category A based on its line position. More specifically, 85% of the subjects categorized both items W and Y as members of category A, whereas items X and Z were predicted to belong to category B by 85% and 90% of the subjects respectively. The results for Experiment 2 are summarized in Figure 3.

Protocols were also obtained from the subjects. In Experiment 1, the modal protocol, given by 15 out of the 20 subjects, can be expressed by the following subject's statement:

> I looked at the size of the square. If it was big, then I looked at where the bar was. If it was a little further to the right, then I put it in A. Otherwise, I put it in B. If the square was small, I looked carefully at its size. A squares were slightly bigger than B squares.

This protocol reveals a two-step process whereby a subject (1) determines the region in which an item belongs and (2) focuses on the particular dimension that is important for that region.

Simulations on weighting attributes

Three exemplar-based process models were evaluated in simulations for their ability to fit the subjects' responses. These models, GCM-NW, GCM-SW, and GCM-ISW, were previously described in (Aha & Goldstone, 1990), are all derived from Nosofsky's (1986) Generalized Context Model (GCM), and differ only in how they weight attribute dimensions. The least flexible, GCM-NW, weights all attributes equally. GCM-SW instead uses a single set of attributes and, in keeping with Nosofsky's *attention-optimization hypothesis*, tunes attribute weights so as to optimize categorization performance. Finally, GCM-ISW is an extension of GCM-SW that maintains a separate set of attribute weight settings with each stored exemplar.

These models process training items incrementally and, for each item x, compute an estimate of the probability that x is a member of each category C as follows:

$$\text{Probability}(x \in C) = \frac{\sum_{y \in S_C} \text{Similarity}(x, y)}{\sum_{y \in S} \text{Similarity}(x, y)},$$

where S_C is category C's stored exemplars and S is the set of all stored exemplars. Similarity is defined as:

$$\text{Similarity}(x, y) = e^{-c\text{Distance}(x,y)},$$

where

$$\text{Distance}(x, y) = \sqrt{\sum_i f(i, x, y) \times (x_i - y_i)^2},$$

and where i ranges over the set of attributes used to describe the exemplars, parameter c's setting (fixed at 10 in our experiments) determines the slope of the exponential decay, and function f determines the normalized weight for attribute i (i.e., $\sum_i f(i,x,y) = 1$ and $\forall i\{0 \leq f(i,x,y) \leq 1\}$).

Function f is a constant function for GCM-NW. For GCM-SW, $f(i,x,y) = w_i$, which is an estimate of the conditional probability that two exemplars will be in the same category given that they have high similarity and highly similar values for attribute dimension i. Weights are initially equal and their settings are updated after each training item is presented via a strategy akin to the delta rule (Rumelhart, McClelland, & the PDP Research Group, 1986).[2] Finally, GCM-ISW's function f combines the category-specific weight settings learned by GCM-SW with a separate set of weight settings stored with exemplar y. Exemplar-specific weight settings are updated in the same manner as category-specific weights except that they are only updated for similarity computations involving their exemplar. More specifically, $f(i,x,y)$ interpolates between the category-specific weight for attribute i and y's exemplar-specific weight for i. This value is more similar to the exemplar-specific setting when $|x_i - y_i|$ is small and more similar to the category-specific setting when this difference is high.

GCM-NW, GCM-SW, and GCM-ISW have three, four, and six free parameters respectively. Informal manual searches were used to find values for these parameters that allowed the models to perform well: 10 for c, which determines the slope of the exponential decay defining similarity; 1 for the GCM's concept bias parameters; 0.01 for GCM-SW's and GCM-ISW's learning rate parameter for updating category-specific weights; 0.1 for GCM-ISW's similar parameter for exemplar-specific weights; and 0.5 for GCM-ISW's parameter for combining exemplar- and category-specific weights in function f. GCM-ISW's additional parameters certainly contributed to its superior performance. However, alternative values for the other models' parameters would not affect their relative behavior because the concepts were equally probable during training and different slopes would still not allow GCM-NW and GCM-SW to locally warp the instance space.

These models were trained and tested in the same way as the subjects except that their items were represented as two-dimensional vectors and they yield estimates of the *probability* that items are members of category B rather than a binary categorization prediction.

[2] Briefly, the magnitudes of weight changes are a decreasing function of Similarity(x,y) and an exponentially decreasing function of $|x_i - y_i|$. Weight settings are increased when x and y are in the same category and otherwise are decreased.

The testing results for GCM-ISW in Experiment 1 are summarized earlier in Figure 2 alongside the subjects' average predictions. Fisher's method for converting correlations (r) to Z-scores was used to evaluate the fits of each model to the subject data. The correlation between GCM-ISW's results and the averaged subject data for the 64 test items was 0.81 and 0.85 for Experiments 1 and 2 respectively. GCM-SW's was 0.66 for both experiments and GCM-NW's was 0.65 and 0.68. GCM-ISW's results correlated significantly better with the subject data from the first experiment than did GCM-NW ($Z = 2.75, p < 0.01$) and GCM-SW ($Z = 2.61, p < 0.01$). This is also true for Experiment 2's results (i.e., ($Z = 3.62, p < 0.0005$) and ($Z = 3.36, p < 0.002$) respectively). For example, visual inspections help to confirm that GCM-SW's predictions for Experiment 2, shown in Figure 4, are not as similar to the subjects' predictions as are GCM-ISW's, as shown in Figure 3.

GCM-ISW's correlations with the subjects' averaged responses for the four critical test items were significantly better than GCM-SW's and GCM-NW's for both experiments (i.e., $Z(1) = 1.92, p < 0.1; Z(1) = 2.53, p < 0.025$ and $Z(1) = 3.05, p < 0.0025; Z(1) = 2.28, p < 0.025$ respectively). More specifically, GCM-ISW's correlations for these two sets of four test items were 0.97 and 0.95 respectively. GCM-SW's respective correlations were 0.17 and -0.84 while GCM-NW's were -0.42 for both experiments. GCM-ISW's categorization predictions matched the predictions made by the majority of subjects on all eight critical test items, whereas GCM-SW agreed on only two and GCM-NW on only four.

In summary, GCM-ISW provides a better fit to the subject data than do the other models. Its combination of category-specific and exemplar-specific attribute weights captures the context sensitivity of attribute importance in these experiments. Thus, these results support our claim that *a psychologically plausible learning algorithm's selective attention processes must be a context-dependent function*; a simple strategy of using one weight per attribute will not necessarily provide optimal fits to subject data.

Discussion

Many other exemplar models of human concept formation can "locally" stretch the input space. For example, Nosofsky, Clark, and Shin (1989) described a model that associates a weight with each value of each dimension. However, this strategy is less flexible than GCM-ISW's; it constrains items sharing an attribute's value to also share its weight setting. Medin and Edelson (1988) proposed a process model similar to GCM-ISW that uses exemplar-specific attribute weights to account for subjects' context-specific sensitivity to base rate information during categorization tasks. However, their model does not ensure that exemplar-specific weights are used only in a local region of the instance space; they may be used to help classify dissimilar items. This constraint should always be applied to models with localized weighting schemes. Medin and Shoben (1988) investigated an *exemplar-directed* attribute-weighting scheme that distinguishes between directions along numeric-valued attribute dimensions. We plan to evaluate an extension of GCM-ISW that incorporates this increased flexibility. We also plan to study models with *region-specific* weighting schemes, in which a region's weights are abstracted so as to specify the relative importance of attributes for similarity decisions within a small region of the instance space. Such models blur the distinction between rule- and exemplar-based models since they use both exemplars and rule-like abstractions derived from them to guide categorization decisions. Furthermore, our model will vary the degree to which abstraction is performed in a region-specific manner, thus increasing its flexibility to represent complex concepts.

Several other researchers have also advocated that psychologically plausible process models should categorize items in a context-sensitive manner (e.g., Barsalou & Medin, 1986; Tversky, 1977). We believe that many future models will incorporate a context-sensitive categorization capability and that they will continue to fit subject data significantly better than models that do not support this flexibility.

References

Aha, D. W., & Goldstone, R. L. (1990). Learning attribute relevance in context in instance-based learning algorithms. In *Proceedings of the Twelfth Annual Conference of the Cognitive Science Society* (pp. 141–148). Cambridge, MA: Lawrence Erlbaum.

Aha, D. W., & McNulty, D. (1989). Learning relative attribute weights for independent, instance-based concept descriptions. In *Proceedings of the Eleventh Annual Conference of the Cognitive Science Society* (pp. 530–537). Ann Arbor, MI: Lawrence Erlbaum.

Barsalou, L. W., & Medin, D. L. (1986). Concepts: Static definitions or context-dependent representations? *Cahiers de Psychologie Cognitive*, *6*, 187–202.

Barsalou, L. W. (1989). On the indistinguishability of exemplar memory and abstraction in category representation. In T. K. Srull & R. S. Wyer (Eds.), *Advances in social cognition*. Hillsdale, NJ: Lawrence Erlbaum.

Fried, L. S., & Holyoak, K. J. (1984). Induction of category distributions: A framework for classification learning. *Journal of Experimental Psychology: Learning, Memory, and Cognition*, *10*, 234–257.

Homa, D., & Cultice, J. (1984). Role of feedback, category size, and stimulus distortion on acquisition and utilization of ill-defined categories. *Journal of Experimental Psychology: Learming, Memory and Cognition*, *8*, 37–50.

Hurwitz, J. B. (1991). Learning rule-based and probabilistic categories in a hidden pattern-unit network model. Unpublished manuscript. Harvard University, Department of Psychology, Cambridge, MA.

Kruschke, J. K. (1992). ALCOVE: An exemplar-based connectionist model of category learning. *Psychological Review*, *99*, 22-44.

Medin, D. L., Altom, M. W., Edelson, S. M., & Freko, D. (1982). Correlated symptoms and simulated medical classification. *Journal of Experimental Psychology: Learning, Memory, and Cognition*, *8*, 37–50.

Medin, D. L., & Edelson, S. M. (1988). Problem structure and the use of base-rate information from experience. *Journal of Experimental Psychology: General*, *117*, 68–85.

Medin, D. L., & Schaffer, M. M. (1978). Context theory of classification learning. *Psychological Review*, *85*, 207–238.

Medin, D. L., & Schwanenflugel, P. J. (1981). Linear separability in classification learning. *Journal of Experimental Psychology: Human Learning and Memory*, *7*, 355–368.

Medin, D. L., & Shoben, E. J. (1988). Context and structure in conceptual combination. *Cognitive Psychology*, *20*, 158–190.

Nosofsky, R. M. (1984). Choice, similarity, and the context theory of classification. *Journal of Experimental Psychology: Learning, Memory, and Cognition*, *10*, 104–114.

Nosofsky, R. M. (1986). Attention, similarity, and the identification-categorization relationship. *Journal of Experimental Psychology: General*, *15*, 39–57.

Nosofsky, R. M., Clark, S. E., & Shin, H. J. (1989). Rules and exemplars in categorization, identification, and recognition. *Journal of Experimental Psychology: Learning, Memory, and Cognition*, *15*, 282–304.

Reed, S. K. (1972). Pattern recognition and categorization. *Cognitive Psychology*, *3*, 382–407.

Rumelhart D. E., McClelland, J. L., & The PDP Research Group (Eds.), (1986). *Parallel distributed processing: Explorations in the microstructure of cognition* (Vol. 1). Cambridge, MA: MIT Press.

Tversky, A. (1977). Features of similarity. *Psychological Review*, *84*, 327–352.

Whittlesea, B. W. A. (1987). Preservation of specific experiences in the representation of general knowledge. *Journal of Experimemtal Psychology: Learning, Memory, and Cognition*, *13*, 3–17.

Adaptation of Cue-Specific Learning Rates in Network Models of Human Category Learning

Mark A. Gluck **Paul T. Glauthier**
Center for Molecular and Behavioral Neuroscience
Rutgers University

Richard S. Sutton
GTE Laboratories Incorporated

Abstract

Recent engineering considerations have prompted an improvement to the least mean squares (LMS) learning rule for training one-layer adaptive networks; incorporating a dynamically modifiable learning rate for each associative weight accellerates overall learning and provides a mechanism for adjusting the salience of individual cues (Sutton, 1992a,b). Prior research has established that the standard LMS rule can characterize aspects of animal learning (Rescorla & Wagner, 1972) and human category learning (Gluck & Bower, 1988a,b). We illustrate here how this enhanced LMS rule is analogous to adding a cue-salience or attentional component to the psychological model, giving the network model a means for discriminating between relevant and irrelevant cues. We then demonstrate the effectiveness of this enhanced LMS rule for modeling human performance in two non-stationary learning tasks for which the standard LMS network model fails to adequately account for the data (Hurwitz, 1990; Gluck, Glauthier, & Sutton, in preparation).

Introduction

In earlier papers, we have explored a simple adaptive network as a model of human learning (Gluck & Bower, 1988a,b; Gluck, Bower, & Hee, 1989; Gluck, 1991). This network model is based on Rescorla & Wagner's (1972) description of classical conditioning; the learning rule is the same as the least mean squares (LMS) learning rule for training one-layer networks (proposed by Widrow & Hoff, 1960), where the goal of learning is to minimize the discrepancy between the expected and the actual outcome.

There are, however, many different engineering algorithms for minimizing classification error; the LMS rule is only the simplest of them. Recent engineering considerations have prompted an improvement to the least mean squares (LMS) learning rule for training one-layer adaptive networks; incorporating a dynamically modifiable learning rate for each associative weight accellerates overall learning and provides a mechanism for adjusting the salience of individual cues (Sutton, 1992a,b). From a psychological perspective, this enhanced LMS rule is analogous to adding a cue-salience or attentional component to the psychological model, giving the network model a means for discriminating between relevant and irrelevant cues. Thus, it is similar to psychological ideas of learning cue-specific saliences, associabilities, and attentional parameters (Pearce & Hall, 1980; Mackintosh, 1975; Frey & Sears, 1978). We call the class of learning methods that dynamically adjust cue-specific learning rates *dynamic-learning-rate* (DLR) methods.

Dynamic-Learning-Rate Methods

Dynamic-learning-rate (DLR) methods are meta-learning algorithms for adapting step-size parameters (i.e. learning rates) during a base-level learning process, which in this paper is the Rescorla-Wagner (1972) or LMS rule (Widrow & Hoff, 1960). The step-size parameters are incrementally adjusted by a gradient descent process to optimize convergence and tracking performance. Such methods have been of interest within the neural network community as a way of speeding the relatively slow convergence of learning methods such as back-propagation (e.g., Jacobs, 1988; Silva & Almeida, 1990; Lee & Lippman, 1990; Sutton, 1986; Barto & Sutton, 1981; Tollenaere, 1990) and have also been proposed as relevant to a key problem in machine learning: finding good individualized learning rates to speed and direct learning (Sutton, 1992a). Recently, Sutton (1992b) has argued that these dynamic-learning-rate

Correspondence should be addressed to Mark A. Gluck, Center for Molecular and Behavioral Neuroscience, Rutgers University, 197 University Avenue, Newark, NJ 07102. E-Mail to gluck@pavlov.rutgers.edu

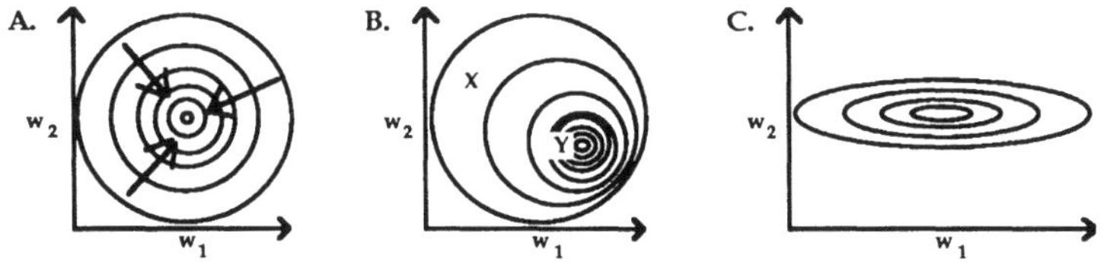

Figure 1. **A.** Expected squared error surface in the weight space of a network with two weights of equal diagnosticity (relevance). The two axes shown represent the possible values of weights 1 and 2, respectively. The third axis (shown as a contour plot) represents the expected squared error, and the global minimum of the surface specifies the ideal weight solution. The LMS network learns by changing the weights in the direction of steepest descent. **B.** A surface with different slopes in different places. In the standard model, the largest changes are made to those weights where the surface drops most sharply, as in part Y. A better strategy would be to take larger steps in the shallow, gently curving part X to traverse it efficiently, and smaller steps in part Y to prevent instability and oscillation in the network. **C.** A ravine: a surface with different slopes in different directions. Here, the network should take larger steps along the horizontal axis where the slope is gentle (along the ravine) and smaller steps along the vertical axis where it is steep (across the ravine) (After Sutton, 1986).

methods may improve classical engineering methods for estimation such as least-squares methods and the Kalman filter. The idea behind these DLR methods is a generalization of Kesten's (1958) method for accellerating stochastic approximation. Consider one of the base-level modifiable parameters--one of the weights in a connectionist network, for example--and how it changes over time. If the weight changes are all in the same direction--e.g., all increases--this signifies that the step-size parameter is too small. The weight could reach its asymptotic value faster if it took larger steps. On the other hand, if the weight changes are in opposite directions--e.g., first up and then down--this signifies that the step-size parameter is too large. For example, opposite-signed weight changes will occur when the weight is overshooting its optimal value. The basic idea behind current DLR methods is to adjust the step size according to the correlation between successive weight changes, with the goal of obtaining zero correlation. Jacobs (1988) proposed correlating the current weight change with a recency-weighted average of previous weight changes. This update rule was written $\bar{\Delta}(t-1)\Delta(t)$ and was called the Delta-Bar-Delta algorithm. The extension of this method to the incremental case is called the Incremental Delta-Bar-Delta (IDBD) method (Sutton, 1992a). This is the method we use to form the extended psychological model explored in this paper.

DLR methods have advantages for both static problems, in which the correct solution does not change, and for non-static problems, in which the correct solution does change over time and must continually be tracked.

In static problems, DLR methods help overcome well-known limitations of *steepest descent* methods such as LMS and backpropagation. In the weight space of a network (i.e. the space formed by assigning each connection weight its own dimension), the expected squared error forms a surface. The minimum of this surface is the point at which the error is smallest; this identifies the ideal asymptotic weights for a particular learning task. In the standard model, a step is taken in weight space at each trial in the direction in which performance is expected to improve most rapidly (Figure 1A). Steepest descent methods are well known to perform poorly for surfaces with different slopes in different places (Figure 1B) and for those containing *ravines*--places which curve more sharply in some directions than others (Figure 1C). In both of these cases, following the direction of steepest descent does *not* take you directly to the minimum. Jacobs (1988) and others have shown that DLR methods can significantly increase the speed of convergence on static problems.

Another advantage of DLR methods is on non-static "tracking tasks", in which the correct solution is not fixed, but continues to change. For example, suppose a subject is faced with a sequence of categorization tasks. Even if the correct solution differs from task to task, the same subset of cues may always be relevant. If cue-relevance can be learned on the early tasks, learning performance on later tasks can be greatly improved. Advantages of this sort have been shown in an engineering context for DLR methods (Sutton, 1992a,b).

The DLR Model

In this section we present the specifics of the standard LMS model and of its extension with a DLR method. We will refer to the extended model as *the DLR network model*. In the standard LMS model, the network operates in a training environment in which feedback (the US or the correct classification) is given after each stimulus pattern. At each time step, or trial, t, the learner receives a set of inputs, $x_1(t)$, $x_2(t)$, ... , $x_n(t)$, computes its output, $y(t)$, and compares this to the desired output, $\lambda(t)$. In the standard model, the w_i 's change on each trial according to:

$$w_i\,(t{+}1) = w_i\,(t) + \alpha\,\delta\,(t)\,x_i\,(t)\,,$$

$$\text{where } \delta\,(t) = \lambda\,(t) - y\,(t), \text{ and } y\,(t) = \sum_{i=1}^{n} w_i\,(t)\,x_i\,(t)$$

The learning rate, α , is a positive constant (on the order of .01 in most simulations) that determines how much all the weights change when the output differs from the training signal.

With the DLR algorithm, however, there is a different learning rate, α_i , for each input, x_i , and these change according to a meta-learning process. The base-level learning rule is:

$$w_i\,(t{+}1) = w_i\,(t) + \alpha_i\,(t{+}1)\,\delta\,(t)\,x_i\,(t)$$

(The α_i are indexed by $t + 1$ rather than t to indicate that their update, by a process descibed below, occurs *before* the w_i update.) To insure that the learning rates remain positive, they are expressed and stored in the form: $\alpha_i\,(t) = e^{\beta_i\,(t)}$. The IDBD algorithm we used updates the β_i by:

$$\beta_i\,(t+1) = \beta_i\,(t) + \frac{\theta\,\delta(t)\,x_i\,(t)\,h_i\,(t)}{\sqrt{\alpha_i\,(t)}}$$

where θ is a positive constant, the *meta-learning rate*, and h is an additional per-input memory variable initialized at zero and updated by:

$$h_i\,(t+1) = h_i\,(t)\left[1 - \alpha_i\,(t+1)\,x_i^2\,(t)\right]^+ + \alpha_i\,(t{+}1)\,\delta(t)\,x_i\,(t)$$

where $[x]^+$ is x , if $x > 0$, else 0. The first term in the above equation is a decay term; the product $\alpha_i\,(t{+}1)\,x_i^2\,(t)$ is normally zero or a positive fraction and this causes a decay of h_i towards zero. The second term increments h_i by the previous error. The memory, h_i, is thus a decaying trace of the cumulative sum of recent errors (Sutton, 1992).

Results

To assess the capabilities of the extended DLR model, we look at model fits to data from both an XOR classification task (Hurwitz, 1990) and a new reversal experiment (Gluck, Glauthier, & Sutton, in preparation). Both of these experiments involved relevant and irrelevant dimensions. The Gluck et al study is a non-stationary task in the sense that the correct response to the stimuli changes over time. Non-stationary tasks are especially appropriate here because they test the DLR model's ability to learn biases during early learning and then use these biases to improve later learning.

XOR Experiment (Hurwitz, 1990)

Hurwitz (1990) describes an experiment in which subjects learned to classify words from a new language into one of two categories. The design involved 16 patterns, each defined on 4 binary dimensions. The assignment of patterns to categories A and B was determined by two relevant dimensions, related to the categories by the XOR rule: the patterns 11•• and 22•• were assigned to category A, and 12•• and 21•• to category B; the irrelevant dimensions are indicated by bullets (•). The trials were broken into four divisions of 80 trials each with only a subset of the patterns presented during each division (Figure 2). Each subset was designed so that both the first and second pairs of dimensions could produce XOR relationships to the categories. However, combinations of the subsets preserved this relationship only for the relevant pair of dimensions. Subjects who used the XOR relationship defined on the irrelevant pair of dimensions in making their categorization responses would therefore show reduced performance when a new subset was introduced. Generalization to new patterns in a new subset could only occur if the relevant features had been discovered during training on previous subsets.

A.

Set	Patterns A	Patterns B	XOR Classification ●●--	XOR Classification --●●
1	●●○○ ○○●●	●○●○ ○●○●	A	A
2	●●●● ○○○○	●○○● ○●●○	A	A
3	●●●○ ○○○●	●○○○ ○●●●	A	B
4	●●○● ○○●○	●○●● ○●○○	A	B

B.

Block	Pattern Set 1	2	3	4
1	10	10	0	0
2	5	5	10	0
3	0	0	5	15
4	5	5	5	5

Figure 2. Hurwitz' XOR experimental design. **A.** Patterns were divided into 4 subsets of 4 patterns each so that within each subset both the first and second pairs of dimensions could produce XOR relationships to the categories. Combinations of subsets preserved this relationship only for the first pair of dimensions. **B.** Subsets were presented with varying frequencies in each block of 80 trials.

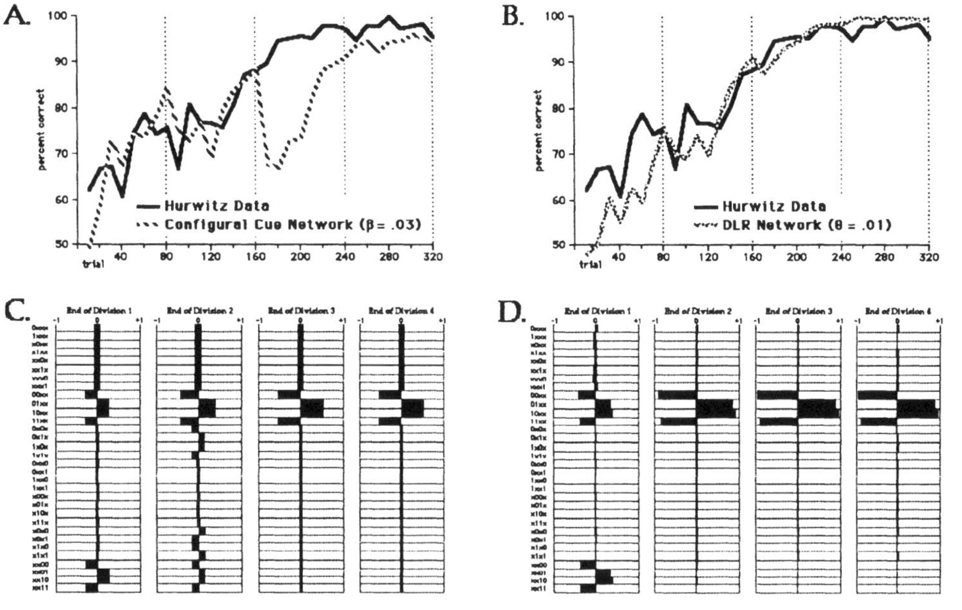

Figure 3. Fits to performance on Hurwitz' XOR experiment. **A** and **B** show average percent correct (y axis) throughout training in blocks of 10 trials each (x axis). The vertical lines mark the divisions where a new subset of patterns was introduced. **A** Standard Configural-Cue LMS Network (β=.03). **B.** DLR Network (θ=.01). **C** and **D** show network weights at the end of each division. Each of the 32 rows represents the connection weight for a feature node (both component and configural) indicated at the left. The middle of a row represents a weight of 0, the left represents -1, and the right represents +1. **C.** Standard Configural-Cue LMS Network Model which reaches a stable solution by the end of Division 3. **D.** DLR Network Model which stabilizes by the end of Division 2.

The results of Hurwitz's experiment suggest that many subjects were able to distinguish the relevant dimensions from the irrelevant ones. Subjects' acquisition curve was relatively smooth, showing a downward trend only at the beginning of the second trial division. In the third division, their curve remained smooth, despite the fact that the exemplar distribution was changing again. By the fourth division, they were at virtually perfect performance.

The model fits from Hurwitz's simulations show that the LMS configural-cue network (Gluck & Bower, 1988b; Gluck, Bower, & Hee, 1989) and exemplar models are unable to account for the subjects' ability to generalize to the new patterns introduced at the beginning of the third trial division. As shown in Figure 3A, performance of the standard configural-cue LMS model drops significantly at this point, whereas the subjects' performance continued to improve.

Hurwitz argued that the configural-cue LMS model has no mechanism to allow it to differentiate relevant from irrelevant cues. Thus, it does not predict the solvers' generalization to new patterns in the third division of trials.

The extended model presented here, however, does have the ability to differentiate between relevant and irrelevant cues. As with the standard network model, the DLR network does not use hidden layers or backpropagation. The individual, dynamic learning rates on each connection allow the network to differentiate relevant from irrelevant features. As shown in Figure 3B, performance of the configural-cue DLR model continues to improve at the beginning of the third trial division, despite the fact that the exemplar distribution changes again. This is consistent with the empirical data and suggests how the model successfully distinguishes relevant from irrelevant dimensions.

To illuminate the difference between the standard configural-cue model and the DLR model, we compared their solutions at the end of each trial division. The major difference between the models' solutions was apparent at the end of Division 2. Whereas the standard model still attributed significant weight to the local XOR relationships defined in pattern subsets 2 and 3 (Figure 3C), the DLR model reduced weights on all irrelevant dimensions to virtually zero (Figure 3D). Only the XOR solution that remained consistent across subsets 1 and 2 was reflected in the weights. For the remainder of the trials, this solution remained essentially unchanged.

A New Reversal Experiment

The DLR model is designed to acquire an appropriate bias from previous learning experience. An ideal testing ground is a series of problems requiring non-identical solutions but similar biases. Reversal experiments are a good example of this kind of problem. In such experiments, subjects learn to classify stimuli into one of two mutually exclusive categories. At some point in learning, typically after some criterion has been reached, the contingencies are reversed. For a deterministic classification task, this means that all stimulus exemplars that had previously belonged to category A now belong to category B, and vice versa. For a probabilistic task, the probability that an exemplar belongs to category A and the probability that it belongs to category B are interchanged. When an experiment involves frequent reversals and irrelevant cues, a bias for the relevant cue(s) must be generated for optimal performance. A subject who has identified the relevant cue(s) will be able to recover from a reversal much more quickly than one who has not.

Our experiment involved four binary cues, one of which was the relevant dimension. The binary cues for the relevant dimension determined the category assignment of exemplars with a probability of 0.9. The cues for each of the three irrelevant dimensions were assigned to each category equally often. The contingencies were reversed 11 times over the course of the experiment; the cue that had been diagnostic of category A became diagnostic of category B, and vice versa. The reversals occurred more frequently as the experiment progressed. Of the 420 trials presented, reversals occurred after trials 80, 140, 200, 240, 280, 300, 320, 340, 360, 380, and 400. Recovery from the later reversals was possible only if the relevant dimension was discovered earlier in training. Otherwise, performance would drop after each reversal and never get much above chance.

Predictions. Prior to examining any human behavioral data, we can compare the engineering value of the standard LMS network model with that of the DLR network model for this particular task. For each model, we can ask: what is the best it can do? Figure 4A shows the performance of each model with the best value of its free parameter.

The standard model predicts that with frequent contingency reversals, ideal weight values will not be reached. Regardless of the size of the learning rate, weights on all connections will begin to change to compensate for the sudden increase in error. Thus, the weights on the irrelevant dimensions will begin to acquire non-zero values in response to local characteristics of the trial ordering. These non-zero values take the model farther from the solution, and keep the performance close to chance until they are again brought down to zero. This type of solution can be characterized as "local" or "unbiased". No matter how much previous learning has suggested that the weights on the irrelevant dimensions be kept at zero, weights for all present features still change.

The DLR model predicts that the ideal weights for a block of trials between reversals will be approached more and more rapidly as the experiment progresses. With each reversal, the bias toward the relevant dimension will become stronger. In other words, the learning rates on the connections from the node(s) for the relevant dimension will become larger, while all other learning rates will drop toward zero. When a reversal occurs, the model is "biased" toward

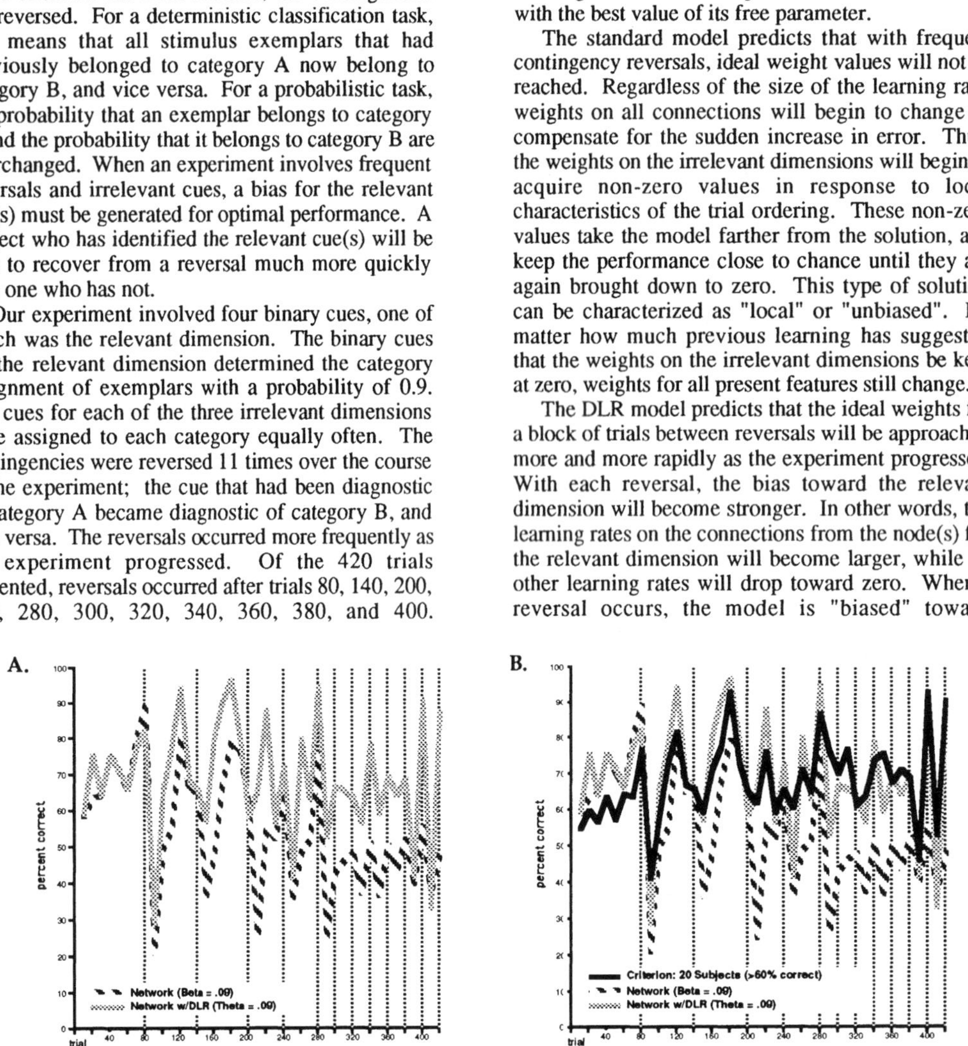

Figure 4. **A.** Optimal parameters for the standard and DLR network models. The graph shows average percent correct (y axis) throughout training in blocks of 10 trials each (x axis). The vertical lines mark the divisions where a reversal occurred. **B.** Fits to performance on Reversal Experiment. Subjects who met criteria were able to recover from reversals late in training. Whereas the standard model exhibited only chance performance, the DLR network was able to form a selective bias to the relevant dimension and recover from the reversals.

changing only the weights on the relevant dimension. Weights on the irrelevant dimensions stay close to zero, allowing the model to improve its performance more quickly and efficiently.

Results. Subjects showed a gradual increase in performance over the first block of 80 trials and a sharp drop in performance after the first reversal. It took subjects an average of 35 trials to recover from this first reversal. However, the drop in performance on subsequent reversals was considerably less, even though the reversals occurred more and more frequently. Recovery rates were likewise improved.

The data we obtained suggest that many subjects were able to identify the relevant dimension and use this bias to help them recover performance after later reversals. The last 7 reversals, occuring every 20 trials, were of particular interest because of their frequency. Whereas the standard LMS network is unable to account for the solvers' performance on these later reversals, the DLR model fits the data well (Figure 4B).

Conclusion

The new psychological model presented here emerges from engineering considerations; it comes from a search for a better way to minimize the expected error (Sutton, 1992a,b). Giving each input its own dynamic learning rate is analogous to adding a salience or attentional component to the learning mechanism. Learning is significantly accellerated on static learning tasks. On series of related tasks with common relevant cues and on selected non-stationary tasks, the extended model shows superior learning capacity and provides a better account of human learning behavior. We have shown that for two such tasks, in which the standard LMS configural-cue model (Gluck & Bower, 1988b) fails to account for the behavioral data, the extended DLR model succeeds far better.

References

Barto, A.G. & Sutton, R.S. (1981) Adaptation of learning rate parameters. Appendix C of *Goal Seeking Components for Adaptive Intelligence: An initial assessment*. Air Force Wright Aeronautical Laboratories/Avionics Tech. Report.

Frey, P.W. & Sears, R.J. (1978) Model of conditioning incorporating the Rescorla-Wagner associative axiom, a dynamic attention process, and a catastrophe rule. *Psychological Review, 85*, 321-340.

Gluck, M.A. & Bower, G.H. (1988a) From conditioning to category learning: an adaptive network model. *Journal of Experimental Psychology: General, 117*, 227-247.

Gluck, M.A. & Bower, G.H. (1988b) Evaluating an adaptive network model of human learning. *Journal of Memory and Lang., 27*, 166-195.

Gluck, M.A., Bower, G.H., & Hee, M. (1989) A configural-cue network model of animal and human associative learning. *Proceedings of the 11th Annual Conference of the Cognitive Science Society*, Ann Arbor, MI.

Gluck, M.A. (1991) Stimulus generalization and representation in adaptive network models of category learning. *Psychological Science, 2(1)*, 50-55.

Gluck, M.A., Glauthier, P.T., & Sutton, R.S. (in preparation) A Theory of dynamic variations in stimulus associability.

Jacobs, R.A. (1988) Increased rates of convergence through learning rate adaptation. *Neural Networks 1*, 295-307.

Hurwitz, J.B. (1990) A hidden-pattern unit network model of category learning. Ph.D. thesis, Harvard Psychology Dept.

Kesten, H. (1958) Accelerated stochastic approximation. *Annals of Mathematical Statistics 29*, 41-59.

Lee, Y. & Lippmann, R.P. (1990) Practical characteristics of neural network and conventional pattern classifiers on artificial and speech problems. In *Advances in Neural Information Processing Systems 2*, D.S. Touretzky, Ed., 168-177.

Mackintosh, N.J. (1975) A theory of attention: Variations in the associability of stimuli with reinforcement. *Psychological Review, 82*, 276-298.

Pearce, J.M. & Hall, A.G. (1980) A model for Pavlovian conditioning: Variation in the effectiveness of conditioned but not of unconditioned stimuli. *Psychological Review, 87*, 532-552.

Rescorla, R. & Wagner, A. (1972) A theory of Pavlovian conditioning: Variations in the effectiveness of reinforcement and non-reinforcement. In A. Black & W. Prokasy (Eds.) *Classical Conditioning II: Current Research and Theory*. New York, Appleton-Century-Crofts.

Sutton, R.S. (1982) A theory of salience change dependent on the relationship between discrepancies on successive trials on which the stimulus is present. Unpublished working paper.

Sutton, R.S. (1986) Two problems with backpropagation and other steepest-descent learning procedures for networks. *Proceedingd of the Eighth Annual Conference of the Cognitive Science Society*, 823-831.

Sutton, R.S. (1992a) Adapting bias by gradient descent: An incremental version of Delta-Bar-Delta. AAAI-92.

Sutton, R.S. (1992b) Gain adaptation beats least squares?. *Proceedings of the Seventh Yale Workshop on Adaptive and Learning Systems.*

Tollenaere, T. (1990) Super SAB: Fast adaptive back propagation with good scaling properties. *Neural Networks 3*, 561-574.

Widrow, B. & Hoff,, M. (1960) Adaptive switching circuits. *Inst. of Radio Engineers, WESCON Convention Record, 4: 96-194.*

Abstractional and Associative Processes in Concept Learning: A Simulation of pigeon data.

Helena Matute
Departamento Psicología Básica
Universidad de Deusto
Apartado 1; 48080 Bilbao
Spain
Email: matute@deusto.es

Eugenio Alberdi
Computing Science Department
University of Aberdeen.-King's College
Old Aberdeen, AB9 2UB
Scotland, U.K.
Email: eugenio@csd.abdn.ac.uk

Abstract

Symbolic and associative theories have been claimed to be able to account for concept learning from examples. Given that there seems to be enough empirical evidence supporting both claims, we have tried to integrate associative and symbolic formulations into a single computational model that abstracts information from empirical data at the same time that it takes into account the strength with which each hypothesis is associated with reward. The model is tested in a simulation of pigeon data in a fuzzy concept learning task, where only a few abstractions are stored in representation of all the training patterns and strengthed or weakened depending on their predictive value.

Introduction

In concept learning from examples, subjects are required to incrementally be able to describe the relevant characteristics of a concept and to correctly classify new instances as either members or nonmembers of the category. For example, after seeing many instances of tall and short people, a child may come up with a useful --though fuzzy-- description of a tall person. Similar experiments have been conducted with pigeons (Pearce, 1988; 1989), and in some sense, the task to which the pigeons were exposed was more complex than that in our example, since the tall and short instances presented to the pigeons were not single objects (like the tall or short people in our example), but rather, each exemplar was a tall or short group of bars.

In general, it can be said that if subjects are exposed to several exemplars of a concept (S+), and several negative exemplars (S-), they will eventually be able to discriminate the relevant characteristics of the concept and to respond differently to members and nonmembers of the category. This result has been widely observed, and reported either as concept or category learning, in both animal and human cognitive research, or as discrimination learning in conditioning experiments conducted mostly with animals (Estes, 1985; Medin & Schaffer, 1978; Pearce,1989).

At the theoretical level, however, there is no general agreement about the internal processes involved, or even as to whether a single learning process can account for results obtained in concept learning and conditioning experiments. Research in the animal and human traditions has been conducted separately for many years and many would view animal learning as a purely associative mechanism which is much simpler than human cognitive learning (see Catania, 1985). On the other hand, several models have been recently proposed in the attempt to offer a unified view of conditioning and category learning, assuming that the same type of learning takes place when a human is learning concepts from examples, as when an animal is

This research was conducted in the context of a joint research project of Deusto University and Labein laboratories (Bilbao) and was carried out while the second author was at Labein supported by a grant from the Engineering School at Bilbao. The research was also partially supported by the Trade and Industry Department of the Basque Government.

learning to respond to positive instances and to avoid responding to negative ones. But while some explore the role of associative learning processess, traditionally studied in conditioning research, when accounting for concept learning tasks (Gluck & Bower, 1988; Pearce, 1989), others prefer to emphasize the symbolic components of conditioning and concept learning, arguing that they can not be reduced to associative learning (Holland et al., 1986; Holyoak, Koh & Nisbett, 1989; Waldmann & Holyoak, 1990).

The dispute is not new. Associative (Spence 1936) and hypothesis testing (Krechevsky 1932) theories have been proposed to account for discrimination learning in animals; studies of human conditioning have been plagued by the controversy between purely associative mechanisms versus awareness, hypothesis testing and symbolic representation of the contingencies (see Boakes, 1989; Davey, 1987). Similarly, two main families of theories of concept learning have been traditionally distinguished. The first theory was the associative account proposed initially by Hull in 1920 and further developed in greater depth by Hull (1943) and Spence (1936). This theory postulated the existence of similar associative mechanisms in animal and human learning, but it was abandoned by most psychologists since the publication of the book by Bruner, Goodnow and Austin (1956), who viewed concept learning in terms of hypothesis testing and emphasized the symbolic aspects of learning and representation. Although Hull had conducted experiments which favored an associative interpretation of concept learning, experiments conducted thereafter by Levine (1975) and others, supported the symbolic account of Bruner, Goodnow and Austin who viewed subjects as hypotheses generators and testers. More recent research has shown that this is not a complete view either, and that most experiments in the hypothesis testing tradition were using well-defined concepts with an all-or-none structure which can be defined by necessary and sufficient conditions, and that this does not correspond to natural concepts which are usually fuzzy, ill-defined and with a graded structure of more typical and less typical exemplars (Rosch, 1978). On the other hand, learning of natural concepts has also been reported in animals (see Herrnstein 1984 for a review) and associative theories have been claimed, once again, to account for the learning of ill-defined concepts (Gluck & Bower, 1988; Pearce, 1989).

Given that enough empirical evidence seems to support each of the above claims, we have tried to integrate associative and symbolic formulations into a single model, called IKASLE[1], which we have implemented in LISP code and tested successfully in simulations of human concept learning during problem solving (Alberdi & Matute, 1991). In order to explore the generality of our model, here we present a simulation of pigeon behavior during concept learning as reported by Pearce (1988), where pigeons were exposed to a fuzzy discrimination task of compound stimuli with overlapping features.

Pearce's Data

Pearce (1988; 1989) reported several experiments conducted with pigeons that were exposed to a series of compound stimuli which were exemplars of the "tall" and "short" categories. Each stimulus was composed of three colored bars against a blue background. In the short category, the mean height of each bar was 3 units (+-2) and the sum of the heights of the three bars was 9 units. For instance, the pattern 3-5-1 is an example of this category (the numbers refer to the heights of each of the three bars). In the tall category, the mean height of each bar was 5 units (+-2) and the sum of their heights was 15 units. An example of this category is the stimulus 7-3-5. There were 36 compound stimuli; 18 exemplars of each category.

In the first experiment (Pearce, 1988), pigeons were randomly allocated to two groups. For group "Category", the short patterns were consistently reinforced in an autoshaping paradigm whereas the tall patterns were never reinforced. For group "Random", half of the tall and half of the short patterns were followed by food. As expected, subjects in group "Category" learned to discriminate between both types of patterns whereas subjects in group "Random" did not show a discriminative behavior.

In a subsequent test phase, Pearce presented new stimuli which were not used during acquisition. The

[1]IKASLE means learner in Basque and stands for "Incremental, Knowledge-independent, Associative and Symbolic Learning from Examples".

stimuli 3-3-3 and 5-5-5 represented the respective means of the short and tall categories. In the test task, however, pigeons in group "Category" showed a greater excitation toward the 1-1-1 than to the 3-3-3 stimulus, and a greater inhibition to the 7-7-7 than to the 5-5-5 stimulus. This "shift of the peak" (Hanson, 1959) was replicated in similar concept learning experiments conducted thereafter (Pearce, 1989) and interpreted in terms of the interaction between the excitatory and inhibitory associative gradients (Spence, 1936) that generalize to similar stimuli from the exemplars stored during the learning phase.

The Elements of the Association

Accepting an associative view of concept learning requires, as Pearce noted, specifying the elements of the association. Several alternative explanations for the above results were discussed by Pearce (1888; 1989), including the association of reinforcement with a single feature (area of blue background) which is constant for all members of a category, and the formation of a prototype (as central tendency of a category), both of which did not seem to be supported by the experimental data. In Pearce's view, two alternative explanations could account for the data: associations appear to be formed, "either between reinforcement and individual *elements* of the patterns, or between reinforcement and separate *configurations* of the elements that represent the different training patterns" (1989, p. 405, italics added). Finally, Pearce argues for an exemplar view of concept learning (Medin & Schaffer, 1978), suggesting that the pigeons remember each pattern and its significance (Pearce, 1988; 1989).

Although it is difficult to empirically determine the elements of the associations, in the simulation described below, we show that, in principle, an associative-symbolic approach, as implemented in IKASLE, is also able to account for the data. By "symbolic" representation we do not necessarily mean that pigeons share the "human" ability to encode relational descriptions (see Pearce, 1988 for data suggesting that pigeons are different from humans in this respect), but rather, that associations can be formed between reinforcement and *abstractions* of the training patterns, instead of between reinforcement and representations of the individual training patterns. In our view, the advantage of this approach is that it allows a more economical treatment of memory when many training patterns are used (Herrnstein 1984), and of generalization gradients, since only a few abstractions are stored and weighted in representation of all the training patterns. Below we present a brief outline of IKASLE (see Alberdi & Matute, 1991 for more details) and its results in the simulation of the above data.

IKASLE

IKASLE is an associative-symbolic computational model of learning which preserves a symbolic representation of events and hypotheses as postulated by cognitive theories while at the same time, in order to strengthen or weaken the alternative hypotheses that the system is forming while learning, it makes use of the associative capabilities demonstrated in animals and humans.

The information provided by positive and negative stimuli --or exemplars-- is summarized in two sets of hypotheses (positive and negative). Hypotheses are abstractions from empirical data and are formed through a generalization process (Michalski, 1983).

Old hypotheses are not abandoned upon the creation of a new one. Instead, the process of hypothesis testing is made more flexible and adaptive by taking into account the predictiveness of each hypothesis, or, in other words, the strength with which each hypothesis becomes associated with reward. IKASLE deals with the assignment and revision of associative strengths implementing the Shanks and Dickinson (1987) adaptation of the Rescorla-Wagner (1972) model of conditioning. Hypotheses are reinforced if they correctly predict the outcome of future trials and lose strength if their prediction is incorrect.

In this way the system is able to cope with inconsistent or noisy data, --or imperfect correlations between events-- as well as with concept drift or changes over time (Schlimmer & Granger, 1986). Whereas acquired concepts are mutable and flexible during the earlier phases of learning, resistance to extinction is increased as the hypotheses generated acquire enough associative strength. Thus, dealing with concept drift will be easier for concepts not yet fully acquired than for concepts already established through enough empirical data supporting them --and hence, with high associative strength. On the other hand, complete lack of correlation between events (such as the treatment of group "Random") can not lead to discriminative behavior since the descriptions which are being formed can never get enough strength.

One of the collateral effects that we have obtained this way has been the simulation of typicality effects: given that the system generates and tests several hypotheses, and stores each of them with a different associative strength, a hierarchical structure of possible descriptions of the concept is formed, and therefore, typical exemplars of the category are predicted with high accuracy by the best descriptions, whereas atypical exemplars are only predicted with low strength by weak hypotheses.

Results of the Simulation

In our attempt to test IKASLE in quite different learning tasks, the program was not modified for this simulation and the way in which IKASLE learned in this experiment was identical to the way it learned concepts in other domains, such as the game of "mus" (a card game similar to poker; Alberdi & Matute, 1991).

Table 1 clarifies the learning process of IKASLE

S	H+	H-
(1-3-5)+	(1 3 5) .24	
(4-7-4)-		(4 7 4) .24
(5-1-3)+	((1-5)(1-3)(3-5)) .24	
(2-2-5)+	→ .42; .58	
(4-4-7)-		(4 (4-7) (4-7)) .24
(4-3-2)+	((1-5)(1-3)(2-5)) .24	
(7-3-5)-		((4-7)(3-7)(4-7)) .24
(4-2-3)+		
.	.	.
	(1 3 5) .24	(4 7 4) .24
	((1-5)(1-3)(3-5)) .86	(4 (4-7) (4-7)) .24
	((1-5)(1-3)(2-5)) .42	((4-7)(3-7)(4-7)) .88
	((1-5)(1-4)(2-5)) .56	((3-7)(3-7)(4-7)) .56
	((1-5)(1-5)(2-5)) .42	((3-7)(3-7)(3-7)) .56
	((1-5)(1-5)(1-5)) .58	

Table 1: Summary of the behavior of IKASLE when simulating a hypothetical subject in Pearce's experiment. The upper panel shows some of the stimuli to which the subject is exposed and the process of hypothesis generation and weighting. The lower panel shows the descriptions acquired by this subject at the end of the training phase, along with their associative strengths.

during the simulation of Pearce's experiment. First, IKASLE did not store every single training exemplar. Rather, it abstracted regular characteristics from positive and negative exemplars, thus forming a set of hypotheses to describe both the short and tall --positive and negative-- categories at the same time that it kept a record of empirical validity of each hypotesis through its associative strength. In the absence of more information, the first positive and negative hypotheses are formed by the first positive and negative instances respectively. If a new instance in the next trial is not covered by the current hypotheses, previous feature intervals in the descriptions are generalized so as to cover this new exemplar. This abstraction summarizes information from empirical data. If it correctly predicts the outcome of future trials, its associative strength is augmented. If it incorrectly covers a given exemplar (or predicts incorrectly the outcome of a given trial), its strength is reduced.

The results obtained by group "Category" in the Test phase of our simulation are shown in Figure 1. The overall probability of responding was greater in Pearce's experiment since the pigeons responded in an autoshaping paradigm and we did not attempt a search for the best parameters, but rather, we simply replicated the values used by Shanks & Dickinson (1988) for changes in the strength of the associations. Although with a lower response rate,

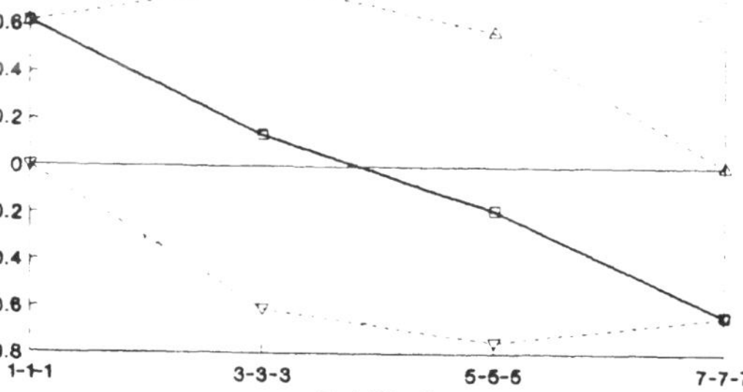

Figure 1: Probability of responding to the test stimuli depending on the difference between the maximum excitatory and inhibitory strength with which the pattern is predicted by both positive and negative descriptions.

the results that we obtained in the simulation showed the same tendency of those of the pigeons reported by Pearce. The shift of the peak is also shown here for group "Category", with a greater probability of responding to the 1-1-1 than to the 3-3-3 stimulus and a greater inhibition to the 7-7-7 than to the 5-5-5 stimulus. Group Random did not show any clear pattern of discrimination.

Peak shift effects ocurred in group "Category" as a consecuence of the interactions between the excitatory and inhibitory strengths of the hypotheses that the system formed abstracting the common feature values for each category. Although not all the possible configurations of the stimuli had the same strength (see Table 1), in general, it can be said that, for the short category, values between 1 and 5 were possible for each of the 3 bars, whereas for the tall category, the possible range of values for each bar was between 3 and 7. Thus, test stimulus 3-3-3 was receiving both excitatory and inhibitory strengths and so, responding to it was less probable than responding to the 1-1-1 pattern which was possible only in the short category. Similarly, for inhibitory or negative stimuli, test stimulus 7-7-7 was clearly a tall --negative-- pattern whereas responding to the 5-5-5 test stimulus was dependent on the interaction of inhibitory and excitatory strengths of both the short and tall categories. The greater inhibitory and excitatory response rates are shown with the tallest and shortest stimuli respectively.

Discussion

Through the paper we have shown a way in which associative and symbolic formulations of conditioning and category learning can be integrated. Hypotheses about the best description of the concept are formed from individual training patterns, and are associated with either reinforcement or its abscence, depending on their ability to predict the outcome of future trials. The advantage of this approach is that it can account for quite different types of data, such as human hypothesis testing in problem solving, pigeon discrimination learning between categories with overlapping features, and typicality effects.

Certainly, we do not claim to account for all of the complexity of concept learning. Selective attention, a greater representational potential, and similarity are just some important factors that should be dealt with.

For instance, for the simulation of the experiment presented above, the introduction of a similarity measurement (Medin & Schaffer, 1978) and a generalization rule for excitatory and inhibitory gradients has not been needed, and IKASLE has been tested in its original version without modification. However, we do not mean to imply that a similarity measurement is not necessary in other tasks. In general, the inclusion of a similarity measurement should permit the system to respond to a new pattern that is not predicted by any of the hypotheses that the system has already generated and tested, as well as to simulate typical conditioning experiments where generalization gradients are reported after training with just one single instance (e.g., Pearce 1989 experiment 2). Note that this approach is not incompatible with the formation and weighting of abstractional descriptions. Given that organisms are exposed to hundreds of training objects and that many of them are certainly very similar to each other, it is probably more adaptive to compute the similarity of a new pattern to a few abstractional representations summarizing information from previous training patterns than to compute the similarity of this new instance to the representations of all the training patterns (Pearce, 1989) or to a random subset of all the individual representations (Medin & Schaffer, 1978).

In this view, and assuming that learning and responding are better understood as separate processes (Miller & Matzel, 1988), subjects can respond to a new pattern not covered by current hypotheses, by means of its similarity to previously stored and weighted descriptions. The information provided by this new instance is then incorporated into previous knowledge, modifying the already existing hypotheses. If, on the other hand, the test patterns are already predicted by the current descriptions, only associative strengths are modified depending on the accuracy of the prediction.

Acknowledgments

Thanks are due to Santi Rementeria and Anselmo DelMoral for their help and suggestions at many stages of the research.

References

Alberdi,E. & Matute,H. (1991): Aprendizaje a partir de ejemplos en un contexto ruidoso de resolución de problemas. [Learning from examples in a noisy context of problem solving]. *Actas del IV Congreso de la Asociación Española para la Inteligencia Artificial.* Madrid: AEPIA.

Boakes,R.A. (1989): How one might find evidence for conditioning in adult humans. In T.Archer & L.G.Nilsson (Eds.): *Aversion, Avoidance and Anxiety.* Hillsdale,N.J.: Erlbaum.

Bruner,J.S., Goodnow,J.J. & Austin,G.A. (1956): *A Study of Thinking.* New York: Wiley.

Catania,A.C. (1985): The two Psychologies of learning: Blind alleys and nonsense syllables. In S.Koch & D.E.Leary (Eds.): *A Century of Psychology as Science.* New York: McGraw-Hill.

Davey,G. (Ed) (1987): *Cognitive Processes and Pavlovian Conditioning in Humans.* Chichester: Wiley.

Estes, W.K. (1985): Some common aspects of models for learning and memory in lower animals and men. In L.G.Nilsson & T.Archer (Eds.): *Perspectives on Learning and Memory.* Hillsdale, N.J.: Erlbaum.

Gluck,M.A. & Bower,G.H. (1988): From conditioning to category learning: An adaptive network model. *Journal of Experimental Psychology: General, 117,* 227-247.

Hanson,H.M. (1959): Effects of discrimination training on stimulus generalization. *Journal of Experimental Psychology, 58,* 321-334.

Herrnstein,R.J. (1984): Objects, categories, and discriminative stimuli. In H.T.Roitblat, T.G.Bever, & H.S.Terrace (Eds.): *Animal Cognition.* Hillsdale,N.J.: Erlbaum.

Holland,J.H., Holyoak,K.J., Nisbett,R.E. & Thagard,P.R. (1986): *Induction: Processes of Inference, Learning and Discovery.* Cambridge, MA: MIT press.

Holyoak,K.J., Koh,K. & Nisbett,R.E. (1989): A theory of conditioning: Inductive learning within rule-based default hierarchies. *Psychological Review, 96,* 315-340.

Hull,C.L. (1920): Quantitative aspects of the evolution of concepts. *Psychological Monographs, 28,* (whole No. 123).

Hull,C.L. (1943): *Principles of Behavior.* New York: Appleton.

Krechevsky,I. (1932): "Hypotheses" in rats. *Psychological Review, 39,* 516-532.

Levine,M (1975): *A Cognitive Theory of Learning.* Hillsdale,N.J.: Erlbaum.

Medin,D.L. & Schaffer,M.M. (1978): Context theory of classification learning. *Psychological Review, 85,* 207-238.

Michalski,R.S. (1983): A theory and methodology of inductive learning. In *Machine Learning: An Artificial Intelligence Approach, vol.1.* Los Altos,CA: Morgan Kaufmann.

Miller,R.R. & Matzel,L.D. (1988): The comparator hypothesis: A response rule for the expression of associations. In G.H. Bower (Ed): *The Psychology of Learning and Motivation, vol.22.* New York: Academic.

Pearce,J.M. (1988): Stimulus generalization and the acquisition of categories by pigeons. In Weiskrantz (Ed.): *Thought without language.* Oxford: Clarendon.

Pearce,J.M. (1989): The acquisition of an artificial category by pigeons. *Quarterly Journal of Experimental Psychology, 41B,* 381-406.

Rescorla,R.A. & Wagner,A.R. (1972): A theory of Pavlovian conditioning: Variations in the effectiveness of reinforcement and nonreinforcement. In A.H. Black & W.F. Prokasy (Eds.): *Classical Conditioning II: Current Research and Theory.* New York: Appleton.

Rosch,E. (1978): The principles of categorization. In E.Rosch & B.B. Lloyd (Eds.): *Cognition and Categorization.* Hillsdale,N.J.: Erlbaum.

Schlimmer,J.C & Granger,R.H. (1986): Beyond incremental processing: Tracking concept drift. In *Proceedings of the National Conference on Artificial Intelligence.* Los Altos,CA: Morgan Kaufmann.

Shanks,D.R. & Dickinson,A. (1987): Associative accounts of causality judgment. In G.H. Bower (Ed.): *The Psychology of Learning and Motivation, vol.21.* New York: Academic.

Shanks,D.R. & Dickinson,A. (1988): The role of selective attribution in causality judgment. In D.J.Hilton (Ed.): *Contemporary Science and Natural Explanation.* Sussex: Harvester.

Spence,K.W. (1936): The nature of discrimination learning in animals. *Psychological Review, 43,* 427-449.

Waldmann,M.R. & Holyoak,K.J. (1990): Can causal induction be reduced to associative learning?. In *Proceedings of the Twelfth Annual Conference of the Cognitive Science Society.* Hillsdale,N.J.: Erlbaum.

Multivariable Function Learning: Applications of the Adaptive Regression Model to Intuitive Physics

Paul C. Price and David E. Meyer
Department of Psychology
University of Michigan
330 Packard Road
Ann Arbor, MI 48104-2994
paul price@um.cc.umich.edu
david meyer@um.cc.umich.edu

Kyunghee Koh
Department of Psychology
University of Rochester
Rochester, NY 14627
koh@cvs.rochester.edu

Abstract

We investigated multivariable function learning--the acquisition of quantitative mappings between multiple continuous stimulus dimensions and a single continuous response dimension. Our subjects learned to predict amounts of time that a ball takes to roll down inclined planes varying in length and angle of inclination. Performance with respect to the length of the plane was quite good, even very early in learning. On the other hand, performance with respect to the angle of the plane was systematically biased early in learning, but eventually became quite good. An extention of Koh and Meyer's (1991) adaptive regression model accounts well for the results. Implications for the study of intuitive physics more generally are discussed.

Introduction

Our research concerns function learning--the acquisition of quantitative mappings between continuous stimulus and response dimensions. In tennis, for example, your distance from the net is a continuous stimulus dimension, and the force with which you should hit the tennis ball is a continuous response dimension. Furthermore, there is a function that relates your distance from the net to the appropriate amount of force with which you should hit the ball. You must learn this function if you are to become a competent tennis player.

Of course, in reality optimal response magnitudes are often functions of more than one stimulus dimension. The amount of force with which you should hit a tennis ball depends on more than just your distance from the net. It is also, for example, a function of the velocity with which the ball is approaching you. The present article, therefore, is about multivariable function learning--the acquisition of quantitative mappings between multiple continuous stimulus dimensions and a continuous response dimension.

Before presenting our results on multivariable function learning, we will review some recent work by Koh and Meyer (1989, 1991) on single-variable function learning. The work by Koh and Meyer--particularly their adaptive regression model of function learning--is highly relevant here because it forms the basis for our current research on multivariable function learning.

The Work of Koh and Meyer (1991)

Koh and Meyer (1991) studied the learning of a motor response to a single dimension of a perceptual stimulus. On each trial of their experiments, subjects were presented with a stimulus consisting of two vertical lines separated by some horizontal distance. Given this distance, subjects had to make two successive finger taps such that the amount of time between the taps equalled a correct response duration, which was a predetermined function of the stimulus length. There were two types of stimulus-response pairs: practice and test. At the end of each trial with a practice stimulus-response pair, subjects received auditory feedback consisting of two beeps, with the amount of time between the beeps equalling the correct response duration for the current stimulus. Subjects were also informed about whether their response durations had been too long or too short, and given a point score between 0 and 100, depending on how close their actual response duration had come to the correct response duration. Subjects received no feedback after trials with test stimulus-response pairs.

In each of three separate experiments performed by Koh and Meyer (1991), the correct response duration was related to the stimulus length by one of three strictly monotone mappings: a power function with a positive exponent less than one, a logarithmic function, or a linear function with a positive intercept. Koh and Meyer (1991) found that regardless of which function had to be learned, subjects' initial responses appeared to be a power function of the stimulus length. This resulted in the power function of Experiment 1 being learned quickly and accurately, whereas the logaritmic and linear functions of Experiments 2 and 3 were learned only after considerable practice.

This power-function bias is explained by Koh and Meyer (1991) with an adaptive regression model of function learning. According to the adaptive regression model, the magnitudes of the stimulus lengths and feedback about correct response durations are transformed logarithmically and stored--along with some noise--in a procedural memory. After each learning trial, a polynomial regression of the transformed responses onto the transformed stimuli is performed. Subsequent responses are then chosen on the basis of parameters derived from this regression.

Under the adaptive regression model, the regression function is initially constrained to be linear in log-log coordinates at the start of learning. This constraint allows people to learn power functions quickly and accurately, because such functions are linear in log-log coordinates. Because other (non-power) functions are not linear in log-log coordinates, the initial linearity constraint imposed by the adaptive regression model implies that they would be learned less easily, consistent with Koh and Meyer's (1991) results. As practice progresses, however, the adaptive regression model assumes that the initial linearity constraint is gradually relaxed, eventually allowing various non-power functions to be learned accurately, just as Koh and Meyer also found.

Mathematically, the adaptive regression model is embodied in the four equations shown below.

(1) $$\ln R = \sum_{j=0}^{k} a_j (\ln S)^j.$$

(2) $$L = \lambda L_1 + (1 - \lambda) L_2.$$

(3) $$L_1 = \sum_{i=1}^{n} \left\{ \left[\sum_{j=0}^{k} a_j (\ln S_i)^j \right] - \ln R_i \right\}^2.$$

(4) $$L_2 = \int_{S_S}^{S_L} \left[\sum_{j=2}^{k} j (j-1) a_j x^{j-2} \right]^2 dx.$$

Here responses are chosen according to Equation 1, where R is the current response duration to be produced and S is the current stimulus length. The coefficients of Equation 1, the a_js, are estimated through the regression process so as to minimize the quantity L of Equation 2. As Equation 2 shows, L is a weighted combination of components L_1 and L_2, whose values appear in Equations 3 and 4. L_1 is simply the sum of squared deviations between predicted response durations and stored feedback about values of correct response durations, and L_2 represents the degree of curvature of the fitted function.

Including L_2 as part of minimizing L biases the computed coefficients of the polynomial's nonlinear terms to have small absolute values whenever the weight parameter λ (which takes a value between 0 and 1) is much less than 1. Consequently, the regression algorithm will tend initially to yield a linear function in log-log coordinates. Nevertheless, if λ is significantly greater than zero, placing some weight on L_1, the adaptive regression model can overcome its initial tendencies. As more and more stimulus-response pairs are experienced, the sum of squared deviations in L_1 will increase, eventually overshadowing the curvature constraint L_2, which remains essentially constant. With an appropriate value of λ, therefore, the model can closely mimic the rate at which people learn non-power functions. (See Koh & Meyer, 1991, for more details).

The Inclined-Plane Experiment

To study multivariable function learning and to extend the adaptive regression model, we had subjects learn to predict the amount of time that a ball takes to roll from the top to the bottom of an inclined plane, based on the length of the plane and its angle of inclination. Our stimuli were presented on a video display, and consisted of line segments varying in length and angle of inclination. Responses consisted of two key taps such that the amount of time between the taps constituted a prediction about the motion time of a ball on the displayed inclined plane. This methodology is advantageous because it is procedurally very similar to the work of Koh and Meyer (1991), and conceptually very similar to prior work with the inclined-plane task (e.g., Anderson, 1983; Bjorkman, 1965).

Method

Subjects. Three University of Michigan students participated in the experiment. They were paid a base wage of $5.00 per 75-minute session, plus a performance bonus described below.

Design. Subjects learned the function

(5) $$T = k\, [L / \sin A]^{1/2},$$

Stimulus Angle	Stimulus Length				
	11.96	18.47	28.47	43.97	67.83
80.00	300	372	463	575	714
39.66	372	463	575	714	887
24.43	463	575	714	887	1102
15.54	575	714	887	1102	1370
10.00	714	887	1102	1370	1701

Table 1. *Correct-Response Duration (ms) as a Function of Stimulus Length (mm) and Stimulus Angle (degrees from horizontal).*

where T is the time that a ball takes to roll from the top to the bottom of an inclined plane, k is a constant of gravitation, L is the length of the plane, and A is its angle of inclination. The stimuli were line segments anchored in the lower left corner of a Hewlett-Packard 1437a graphics display and extending diagonally upward to the right. Each of five stimulus lengths was combined with each of five stimulus angles to form 25 unique stimuli. The constant k was chosen so that the correct response times ranged from 300 to 1701 ms (see Table 1).

The subjects participated in one 75-minute experimental session per day over five consecutive days. Each session began with four warm-up trials; responses to the warm-up trials were highly variable and were therefore not analyzed. The remainder of each session was divided into 30 blocks of 25 trials each. Each of the 25 stimuli was presented once per block, in random order.

Procedure. At the start of each trial, one of the 25 stimuli was presented on the display screen. The subject responded by tapping the slash ("/") key twice so that the amount of time between the two taps was his or her prediction about the amount of time that a ball would take to roll down the displayed inclined plane. After responding, subjects received three forms of correct-response feedback. At 300 ms after the second tap, the first of two short beeps (15 ms, 1000 Hz tones) occurred; a second beep followed. The time between the two beeps was the correct response time. The stimulus display was cleared at the onset of the second beep, and the subject was presented with additional information about the accuracy of his or her response. One of three messages--"LONG," "SHORT," or "PERFECT"--was presented, depending on whether the response was greater than, less than, or equal to the correct response duration. Accompanying this message was a numerical point score, ranging from 0 to 100, which indicated how close the subject's response duration had been to the correct duration. This point score was calculated according to the equation $P = 100 - .2 \mid T_C - T_S \mid$, where P is the number of points, T_C is the correct response duration, and T_S is the duration produced by the subject. If P happened to be negative, a score of zero points was awarded. The message and point score were visible for 700 ms and were followed by a 500 ms intertrial interval.

After each trial block, the subject was presented with his or her point total for the block and cumulative point total for the session. Ultimately, the cumulative point total was converted to a bonus payment of $.05 per 1000 points earned. In general, this resulted in a bonus of between $2.00 and $2.50 per session.

Results

Tables 2a and 2b show subjects' mean response durations for each inclined plane during sessions 1 and 5, respectively. To make the patterns in these tables clearer, Figure 1 shows subjects' mean response durations for sessions 1 and 5 versus the log of the stimulus length, averaged across the five stimulus angles. Response durations have been transformed

Stimulus Angle	Stimulus Length				
	11.96	18.47	28.47	43.97	67.83
80.00	324	451	498	557	597
39.66	434	567	630	734	794
24.43	534	653	773	879	915
15.54	618	833	945	1024	1132
10.00	741	1032	1171	1328	1475

Table 2a. *Subjects' Mean Response Duration (ms) as a Function of Stimulus Length (mm) and Stimulus Angle (degrees from horizontal). Session 1.*

Stimulus Angle	Stimulus Length				
	11.96	18.47	28.47	43.97	67.83
80.00	317	395	483	578	651
39.66	383	463	596	730	870
24.43	464	577	743	865	1010
15.54	557	731	893	1010	1196
10.00	744	948	1175	1277	1596

Table 2b. *Subjects' Mean Response Duration (ms) as a Function of Stimulus Length (mm) and Stimulus Angle (degrees from horizontal). Session 5.*

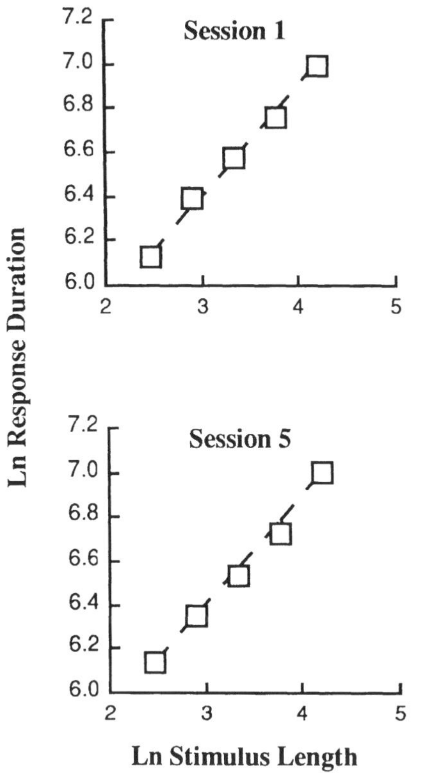

Figure 1. *Log Response Duration versus Log Stimulus Length, Averaged over Subjects and Stimulus Angles for Sessions 1 and 5, respectively.*

logarithmically before averaging the data and making the graphs, so as to produce a linear correct-response function (the dashed lines in Figure 1).

Note that even during session 1, subjects' mean response durations appear to be a nearly log-linear function of stimulus length; that is, they learned this aspect of the inclined-plane task very quickly and accurately. To confirm this, a log-polynomial regression of mean response durations onto stimulus length was performed separately for each subject, session, and angle. Both linear and quadratic coefficients were obtained. The mean value of the linear coefficients was .48, which is not significantly different from the optimal linear coefficient of .50 [$t(2) = 1.54$, $p > .05$]. (Note: The optimal linear coefficient equals 0.50 because of the square-root exponent in Equation 5.) The mean value of the quadratic coefficients was -.00003, which is not significantly different from the optimal quadratic coefficient of zero [$t(2) = .03$, $p > .05$]. This implies that there was essentially no bias in subjects' responses with respect to length even during session 1.

The linear coefficients, or slopes, were then treated as the dependent variable in an ANOVA with session number and stimulus angle as fixed factors and subjects as a random factor. These coefficients did not change significantly across sessions [$F(4,8) = .72$, $p > .05$], nor did they differ significantly across stimulus angles [$F(4,8) = .71$, $p > .05$]. In other words, no significant learning took place with respect to stimulus length after session 1, and performance with respect to stimulus length did not depend on the angle of the stimulus.

Figure 2 shows subjects' mean log response durations for sessions 1 and 5 versus the log of the reciprocal of the sine of the stimulus angle. Again, this produces a linear correct-response function (the dashed lines in Figure 2).

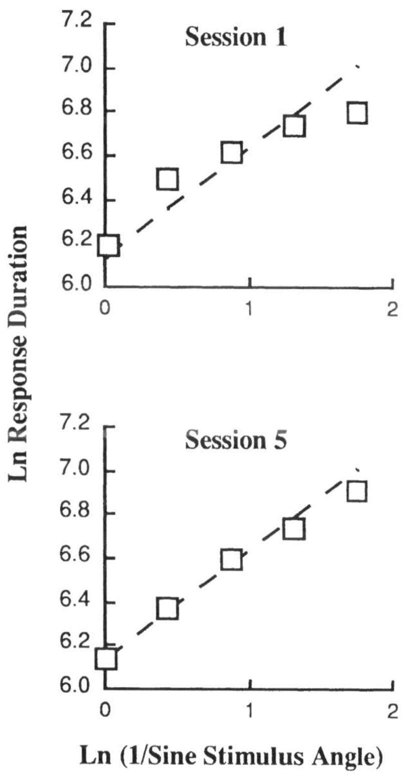

Figure 2. *Log Response Duration versus the Log of the Reciprocal of the Sine of the Stimulus Angle, Averaged over Subjects and Stimulus Lengths for Sessions 1 and 5, respectively.*

Note that during session 1, there is a pronounced curvature in the plotted response times. To confirm this, a log-polynomial regression of mean response duration onto the reciprocal of the sine of the angle was performed for each subject, session, and stimulus length. Both linear and quadratic coefficients were obtained. The mean value of the linear coefficients was .34, which again is not significantly different from the optimal linear coefficient of .50 [$t(2) = 1.80$, $p > .05$]. However, when the data were analyzed separately for each subject, we found that each subject produced linear coefficients that were significantly less than the optimal linear coefficient of .50 [$t(4) = 15.54$, $p < .05$; $t(4) = 12.46$, $p < .05$; $t(4) = 13.16$, $p < .05$].

The mean value of the quadratic coefficients was -.013, which is not significantly different from zero [$t(2) = 1.64$, $p >.05$]. However, it is three orders of magnitude larger than the mean quadratic coefficient for length, and it is easily perceived as the downward curvature in the top panel of Figure 2. Moreover, analyzing these data separately for each subject reveals that two out of the three subjects produced significantly negative mean quadratic coefficients [$t(4) = 13.26$, $p < .05$; $t(4) = 8.26$, $p < .05$], while the third produced a non-significantly negative one [$t(4) = .00005$, $p > .05$].

An ANOVA analogous to the one described earlier showed that the linear coefficients increased significantly across sessions [$F(4,8) = 5.15$, $p < .05$], and the quadratic coefficients decreased somewhat across sessions [$F(4,8) = 2.34$, $p < .15$]. Although the latter change in the quadratic coefficients was not significant at conventional levels, it was in the direction predicted by the adaptive regression model (see below). In other words, subjects' initial response biases with respect to angle did decrease with practice. Neither the linear nor quadratic coefficients differed across stimulus lengths [$F(4,8) = 1.46$, $p > .05$; $F(4,8) = .37$, $p > .05$].

Extending the Adaptive Regression Model

The adaptive regression model--as originally formulated by Koh and Meyer (1991)--quite readily explains performance with respect to stimulus length in the present experiment. According to this model, the relationship between the logarithmically transformed stimulus length and response duration is assumed to be linear. Because the relationship between the length of an inclined plane and a ball's rolling time is in fact linear in log-log coordinates, the model predicts that it should be learned quite rapidly. This is, of course, exactly what was found in the present experiment (Figure 1).

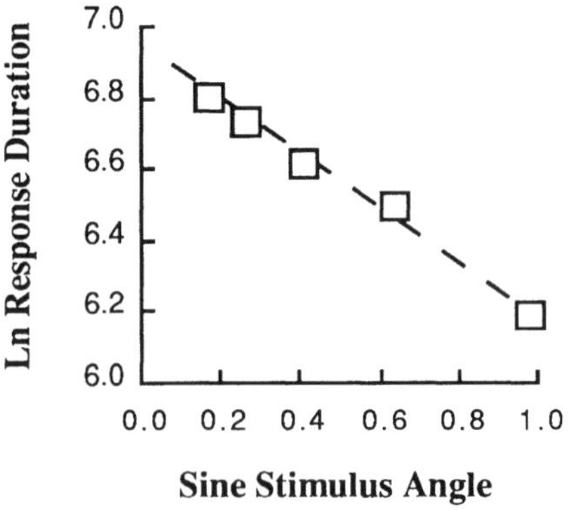

Figure 3. *Log Response Duration versus the Sine of the Stimulus Angle, Averaged over Subjects for Session 1.*

The adaptive regression model can also be extended to account for performance with respect to stimulus angle. Our extended version of the model initially assumes that the relationship between the transformed stimulus angle and logarithmically transformed response duration is linear. In the case of angle, however, the transformation performed is not logarithmic. Instead, the angle is transformed by taking its sine. That subjects actually do assume an initial linear relationship between the log of the response duration and the sine of the angle can be seen in Figure 3. This figure shows subjects' mean log response durations versus the sine of the stimulus angle; the relationship is clearly linear ($r^2 = .994$).

Our extended adaptive regression model assumes further that the transformed response duration is an additive combination of the two transformed stimulus magnitudes: ln L and sin A. Early performance in the inclined-plane task, therefore, is characterized by the equation

(6) $\ln R = a + b \ln L + c \sin A$,

where R is the subject's response duration, L is the length of the inclined plane stimulus, and A is its angle of inclination. Equation 6 accounts for 99.4% of the variance in subjects' mean response durations during session 1. Finally, to explain how subjects' performance with respect to stimulus angle improves over sessions, our extended adaptive regression model gradually adds non-zero, higher-order polynomial terms for sine A into Equation 6 as practice progresses, using the same sort of relaxation process posited in Equations 1 through 4.

Discussion

It is especially interesting to consider the implications of these results for the domain of intuitive or "naive" physics. A number of studies have demonstrated that people have faulty initial intuitions about the behavior of objects in a variety of simple physical situations (e.g., Bjorkman, 1965; McCloskey, 1983). For example, people may not realize that the mass of an object is irrelevant to the amount of time that it takes to fall to the ground, and therefore may predict that a heavier object should fall faster than a lighter object.

Such faulty intuitions are usually attributed to a lack of knowledge about the laws of physics. However, the present results suggest that people may have faulty initial intuitions about many such physical situations for a very different reason. Namely, the nature of the transformations that people perform on stimulus variables, as well as their preferred modes of psychologically combining those variables, may lead to biased expectations and predictions, as in the present experiment. If so, then even the most highly educated physicists would exhibit the same biases as naive subjects. Perhaps this is a possibility worth exploring in future experiments.

References

Anderson, N.H. (1983). Intuitive physics: Understanding and learning of physical relations. In J. Tighe & B.E. Shepp (Eds.), *Perception, cognition, and development* (pp. 231-265). Hillsdale, NJ: Erlbaum.

Bjorkman, M. (1965). Studies in predictive behavior. *Scandinavian Journal of Psychology, 6,* 129-156.

Koh, K., & Meyer, D.E. (1989). Induction of continuous stimulus-response relations. In G.M. Olson & E.E. Smith (Eds.), *Proceedings of the 11th annual conference of the Cognitive Science Society* (pp. 333-340). Hillsdale, NJ: Erlbaum.

Koh, K., & Meyer, D.E. (1991). Function learning: Induction of continuous stimulus-response relations. *Journal of Experimental Psychology: Learning, Memory, & Cognition, 17,* 811-836.

McCloskey, M. (1983). Intuitive physics. *Scientific American, 248,* 122-130.

Memory for Multiplication Facts

Richard Dallaway*
School of Cognitive & Computing Sciences
University of Sussex
Brighton BN1 9QH, UK
richardd@cogs.susx.ac.uk

Abstract

It takes approximately one second for an adult to respond to the problem "7 × 8". The results of that second are well documented, and there are a number of competing theories attempting to explain the phenomena [Campbell & Graham 1985; Ashcroft 1987; Siegler 1988]. However, there are few fully articulated models available to test specific assumption [McCloskey, Harley, & Sokol 1991]. This paper presents a connectionist account of mental multiplication which models adult reaction time and error patterns. The phenomenon is viewed as spreading activation between stimulus digits and target products, and is implemented by a multilayered network augmented with a version of the "cascade" equations [McClelland 1979]. Simulations are performed to mimic Campbell & Graham's [1985] experiments measuring adults' memory for single-digit multiplication. A surprisingly small number assumptions are needed to replicate the results found in the psychological literature—fewer than some (less explicit) theories presuppose.

Phenomena

When asked to recall answers to two digit multiplication problems "as quickly and accurately as possible" [Campbell & Graham 1985], both children and adults exhibit well documented patterns of behaviour. In general, response times (RTs) increase across the multiplication tables: problems in the nine times table tend to take longer to answer than problems in the two times table. However, this "problem size effect" has plenty of exceptions (e.g., the five times table is much faster than its position would suggest—see figure 1). In addition, "tie" problems (2 × 2, 3 × 3 etc.) are recalled relatively quickly. Campbell & Graham [1985] found that adults under mild time pressure make errors at the rate of 7.65 per cent, and 92.6 per cent of those errors fall into the following five categories (after McCloskey et al. [1991]):

- Operand errors, for which the erroneous product is correct for a problem that shares a digit (operand) with the presented problem (e.g., 6 × 4 = 36, because the problem shares 4 with 9 × 4 = 36).
- Close operand errors, a subclass of operand errors, where the erroneous product is also close in magnitude to the correct product. That is, for the problem $a \times b$, the error will often be correct for the problem $(a \pm 2) \times b$ or $a \times (b \pm 2)$ (e.g., 6 × 4 = 28). This phenomenon is referred to as the "operand distance effect".
- Frequent product errors, where the error is one of the five products 12, 16, 18, 24 or 36.
- Table errors, where the erroneous product is the correct answer to some problem in the range 2 × 2 to 9 × 9, but the problem does not share any digits with the presented problem (e.g., 6 × 4=15).
- Operation errors, where the error to $a \times b$ is correct for $a + b$.

Despite being drilled on the multiplication tables at school, children and adults make these systematic slips in recall. The problem is to produce a model which has correctly learnt the multiplication tables, yet can make slips when recalling answers. Given the observations on the types of erroneous responses, and the RT for correct responses, what assumptions must be made to account for these phenomena? The model presented here suggests that the initial skew in the frequency and order of presentation of multiplication facts [Campbell 1987, p. 118] is one of the important factors.

*Thanks to Harry Barrow & David Young. Funded by the SERC in conjunction with Integral Solutions Ltd. Simulations were performed using a modified version of the McClelland & Rumelhart [1988] *bp* program, and POPLOG POP-11.

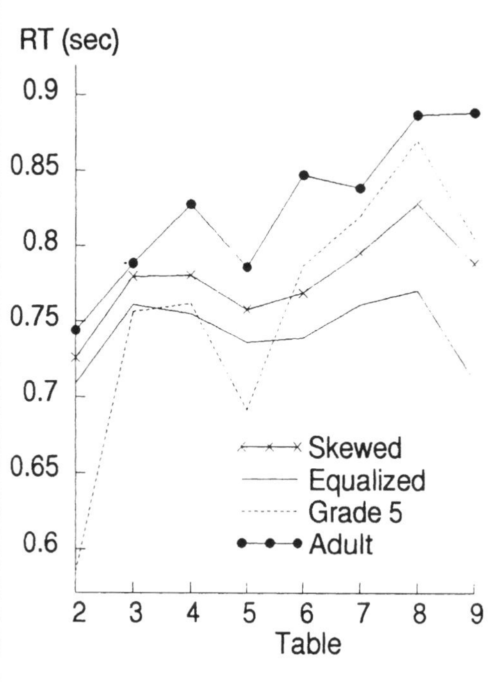

Figure 1: Plot of mean correct RT per multiplication table collapsed over operand order for mean RT of: 60 adults [Campbell & Graham 1985, app. A]; 26 children in grade 5, RT scaled down from a range of 1.19–2.97 seconds to fit graph [ibid. app. B]; 20 networks trained on skewed frequencies; and, the same 20 networks after continued training on uniform frequencies. The RT for both networks has been scaled by the same amount.

Architecture of the model

The structure of the network is shown in figure 2. This architecture has evolved in a number of stages since it was first used as a subnetwork in a sequential network for long addition (and later long multiplication). Initially the output layer was divided into "tens" and "units", and by adding a simple RT measure it was found that the network produced a prominent dip in the RT curve for the five times table. This effect was increased by training sequentially through the tables, but the network did not produce the kinds of mistakes reported by Campbell & Graham [1985]. Changing the output layer to a representation of products, and using a coarse encoding of the input digits produced more realistic errors.

The current network is trained on all the problems 2×2 through 9×9 in a random order using backpropagation. The two digits that comprise a problem are coarse encoded on the two sets of eight input units, with the activation decaying exponentially from the presented digit (e.g., when encoding "5", the input vector would contain 1.0 for the five unit, and 0.5 for the four and six units, and so on). For tie problems, an additional tie bit is set to 1.0. Without this, the tie problems were consistently among the slowest problems. The tie bit can be thought of as reflecting the perceptual distinctiveness of tie problems. Activation flows through a hidden layer of ten units to the output layer. There is one output unit per product type plus a "don't know" unit. The network is trained to activate one output unit per problem (a one-of-N encoding).

During training the presentation frequency of each pattern is linearly skewed in favour of the smaller problems (relative frequency of 1.0 for 2×2 to 0.1 for 9×9, based on correct product). Although small problems do occur more frequently in textbooks, there is no reason to believe this skew continues into adulthood [McCloskey et al. 1991, p. 328]. Hence, after training to an error criterion (total sum squared, TSS) of 0.05 on the skewed training set (taking approximately 8 000 epochs), the network is trained for a further 20 000 epochs with equal frequencies (reaching a mean TSS of 0.005). At the end of training both the "skewed" networks and "equalized" networks correctly solve all problems. An initial worry was that the skew would lead the networks into a local minima from which the task could not be completed. To avoid this possibility, a low learning rate of 0.01 was used during training (momentum was 0.9).

The skew was produced by storing the relative frequency (between zero and one) of a problem alongside the problem in the training set. When a problem was presented to the network, the weight error derivative was multiplied by the relative frequency value for that pattern. (This can be thought of as providing each input pattern with a different learning rate.) This method allowed accurate control over the presentation frequencies, without duplicating entries in the training set.

The "cascade" activation equation [McClelland & Rumelhart 1988, p. 153] is used to simulate the spread of activation in the network. Each unit's activation is allowed to build up over time:

$$\text{net}_i(t) = k \sum_j w_{ij} a_j(t) + (1 - k)\ \text{net}_i(t-1),$$

where k is the cascade rate which determines the rate with which activation builds up, w is the weight matrix, and $a_j(t)$ is the activation of unit j at time t. For the simulations described here, $k = 0.05$. The net_i is passed through a logistic squashing function to produce the activation value, a_i. The response values are taken to be the normalized activation values (the sum of the output layer activity is 1.0).

McClelland & Rumelhart [1988] point out that the asymptotic activation of units under the cascade equation is the same as that reached after a standard feed-forward pass. Hence, the network is trained without the cascade equation (or with $k = 1$, if you prefer), and then the equation is switched on to monitor the network's behaviour during recall.

At the start of cascade processing the initial state of the network is the state that results from processing an all-zeros input pattern. This gives a common starting point for all problems. The network is trained to activate the "don't know" unit for an all-zeros input. Figure 3 is a time plot of output activation using the cascade equations.

Simulations

Method

On each trial (presentation of a problem) the network randomly selects a threshold between 0.4 and 0.9. Processing then starts from the all-zeros ("don't know") state, and proceeds until a product unit exceeds the threshold. The RT (number of cascade steps) is recorded for a correct response, and erroneous responses are classified into the five categories itemized above. The network is presented with each of the 64 problems 50 times, and the mean correct RT is recorded. This is repeated with 20 different networks (different initial random weights).

Given enough time (usually no more than 50 cascade steps), the network will produce the correct response for all 64 problems. For example, figure 3 shows the response of a network to the problem 3×8. After the "don't know" unit has decayed, the unit representing 27 becomes active until the network settles into the correct state, 24. This is a demonstration of the operand distance effect, but there is slight activation of other products: $3 \times 7 = 21$, $2 \times 8 = 16$, $4 \times 8 = 32$, $3 \times 3 = 9$, and $2 \times 7 = 14$.

With a high threshold the networks will reliably produce the correct response to a problem. However, early in processing erroneous products are active (e.g., 27 in figure 3), and with a low threshold these errors are reported. Note that this is rather different to previous connectionist (Brain-state-in-a-box, BSB) model of mental arithmetic [Viscuso 1989; Anderson, Spoehr, & Bennett 1991]. The full details of the BSB model have not yet been published, and only small scale simulations have been performed. In essence, the model is an auto-associator that settles into attractor states representing the answer to the presented problem. However, this means that the model, as presented, simply lacks the ability to correctly answer some problems, or fails to respond at all. This runs against the notion that slips are one-off run-time errors, rather than permanent disabilities. McCloskey et al. [1991] comment that the Viscuso, Anderson, & Spoehr [1989] "proposal has several limitations and cannot be considered a well-articulated model", but add that "the [neural net] approach probably merits further exploration" [p. 395].

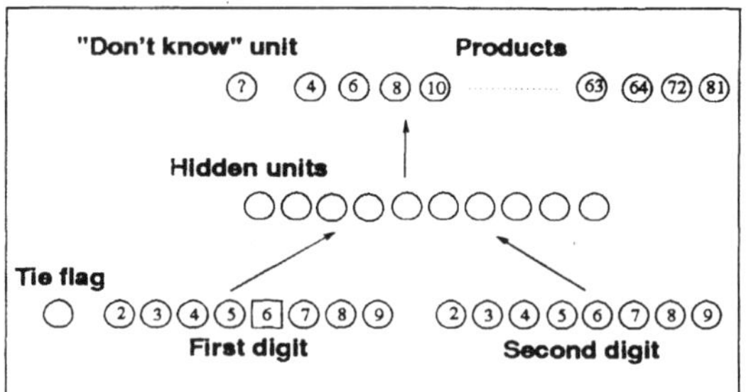

Figure 2

Results

The mean RTs plotted in figure 1 show some of the basic features of the problem size effect. For the skewed networks the RT correlates $r = 0.36$ ($p = 0.0018$) with adult RT [Campbell & Graham 1985]. This falls to $r = 0.19$ ($p = 0.063$) after substantial training on the equalized patterns. Note that the RTs have reduced and flattened out for the equalized network, which is just what is expected after continued practice [Campbell & Graham 1985, p. 349]. The obvious feature of the RT plot is the drop in RT for the nine times table. Children in grades 3 to 5 respond faster to $9\times$ than $8\times$ problems [Campbell & Graham 1985], but this levels out for adults.

The inclusion of a ties unit is necessary to ensure that ties are among the fastest problems. Implicit in this is an assumption that there is something perceptual about ties which results in a flagged encoding—perhaps the effect of being taught the notion of "same" and "different". The RTs of 6 (out of 8) of the tie problems were below the mean RT for their table, increasing to 7 ties for the equalized networks (6×6 remaining above the mean for the six times table).

Table 1 shows the error percentages of the net-

	Networks		Adults
	Skewed	Equalized	
Operand errors	90.04	86.51	79.1
Close operand errors	78.98	73.75	76.8*
Frequent product errors†	27.76	23.68	30.6
Table errors	9.74	13.49	13.5
Operation error	3.98	3.22	1.7*
Error frequency	14.10	18.64	7.65

* Approximate percentage.

†Percentage of operand errors.

Table 1: Percentage breakdown of errors. Figures are mean values from twenty different networks, and mean values from sixty adult subjects [Campbell & Graham 1985, app. A]. Note that the model has not been trained on addition facts, so the frequency of operation errors is coincidental.

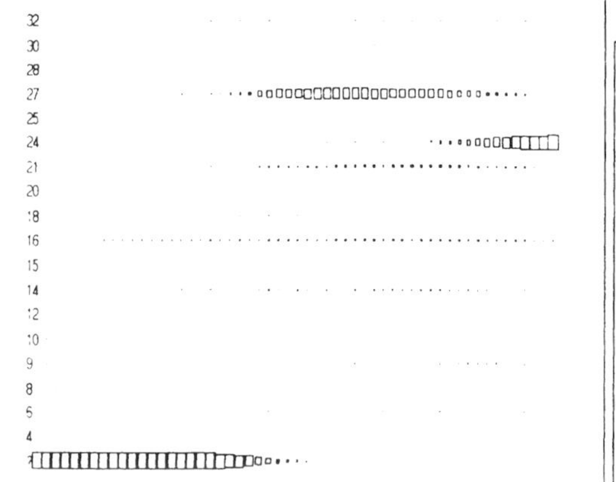

Figure 3: Response of the output units over 40 time steps for the problem 3 × 8. Output units representing products over 32 are not shown on this graph.

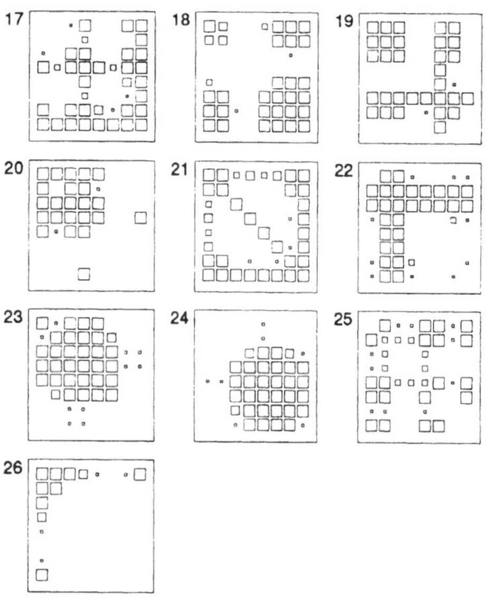

Figure 4: Hidden unit activation for one network. Each large rectangle represents one hidden unit. Within each rectangle, the size of the smaller rectangles represents the activation of the hidden unit to a particular problem. Each large square mimics the multiplication table (top-left for 2 × 2, and bottom-right for 9 × 9). For example, unit 22 responds to problems in the three and four times tables.

works compared to those of adults. Both sets of networks have error distributions that are similar to that of adults, and there is little difference between the skewed and equalized networks.

It should be noted that human subjects sometimes respond with a number that is not a correct product for any of the problems 2 × 2 to 9 × 9 (e.g., 2 × 3 = 5). The current network cannot produce non-table errors. However, Campbell & Graham [1985] report that only 7.4 per cent of errors are of this kind. (An account of non-table errors might begin by augmenting the network with a tens and units read-out layer.)

A further point of interest is the correlation between problem error rate and correct RT. Campbell [1987, p. 110] reports a correlation of 0.93 for adults. For the skewed and equalized networks $r = 0.74$ and $r = 0.76$ respectively. It is not obvious that any model would necessarily predict that slower problems produce more errors.

Analysis

RT depends on the net input to a unit, and this can be increased by having some large (or many small) weights. Although there is no easy way to determine why certain weights develop, some of the factors involved can be described.

The presentation frequency of a problem and product should have a strong effect on the weights: those problems seen more often should develop larger weights. Simulations with networks trained only on patterns with equal presentation frequencies have demonstrated that the frequency of presentation is important. Typically these networks produce high-frequency products as errors, and have poor correlations to human RT.

Frequency does not explain why the five times table should be faster than the four times table. "Product uniqueness" may explain why: none of the products in the five times table occur outside the context of five (unlike the two times table, where the products 12, 16 and 18 occur in other tables). Hence, the error signals for the fives products are not diluted through differing hidden representation for different problems. The same is true of the seven times table, but for a lower presentation frequency.

The nine times table has the largest range of all the tables. This seems to give the table a RT advantage because many hidden units respond to the nine times table: the nine times table is the third "most encoded" problem (typically five hidden units respond to the nine times table; seven for the two times tables; six to the three times). This is because the hidden units respond to a range of input problems. For example, figure 4 shows that unit 26 responds to small products; unit 23 to medium products; and unit 24 responds to larger products.

The hidden units' preference for responding to bands of inputs explains the mechanism behind the operand distance effect. Hidden units' activities change smoothly during the course of processing, but at differing rates. This change affects groups of related products, and due to the overlap in encoding (e.g., between unit 23 and 24 in figure 3), some hidden units may force incorrect products to exceed threshold.

Discussion

Apart from the training frequency skew, the other main assumption of the model is the coarse coding of the input pattern. The importance of this assumption has been demonstrated by simulations using a one-of-N input encoding (the same encoding that was used for the outputs). The results of those simulations produced comparable RT correlations, but poor error distributions. The assumption is that the coarse encoding is due to general knowledge of number (perhaps from counting).

This study has focused on mean adult performance on the problems 2×2 to 9×9 because these problems have detailed published results. There are persistent statements in the literature that zero and one times tables are governed by procedural rules (e.g., Campbell & Graham [1985, p. 341]; Miller, Permutter, & Keating [1984, p. 51]; Stazyk, Ashcraft, & Hamann [1982, p. 334]). The motivation for this seems to stem from the fact that it is easy to produce answers for the zero and one tables. Initial experiments with the architecture confirm what is expected of backpropagation when the zero and one times tables are included in the training set. The zero table is by far the fastest and least prone to error, followed by the one times table. This is consistent with RTs reported by Miller et al. [1984]. On this basis there is no reason to assume that there is a separate mechanism for the zero and one tables.

Some of the assumptions posed by other models may be accounted for by differences in methodology. For example, in models that assume direct links (no hidden units) between stimulus digits and target products there must be additional information for the model to be capable of producing the correct answer. There must be either: different (token) answer nodes for each problem (e.g., multiple copies of the "12" node for 2×6 and 3×4 as used by Ashcroft [1987]); or input nodes representing whole problems (e.g. a "3×4" input node as in Campbell & Graham [1985]); or both [Siegler 1988].

However, other assumptions were not found to be needed in this model. For example, there was no need for explicitly learning incorrect associations, as suggested by both Siegler [1988] and Campbell & Graham [1985]. Nor was there need for connections between product units (Campbell & Graham [1985] and Ashcroft [1987]), nor connections from general "magnitude" units as used by Campbell & Graham [1985]. These models have been criticised by McCloskey et al. [1991] for not specifying the rationale for these additional connections.

Of course, there are a number of shortcomings to

the model presented here. There is no empirical evidence to suggest that adults are exposed to a skew in the frequency of multiplication problems, and this was modelled by further training the skewed networks on equal frequency problems. Although the RTs for the equalized networks diverge from the adult RTs, they retain the basic features of the problem size effect and error distributions. One conclusion that can be drawn from this is that it is quite possible for the effect of training on skewed problems to continue to be felt even after a significant period of training on non-skewed problems.

As it stands the model makes no attempt to account for a number of important aspects of arithmetic. Future directions for this work could focus on: modelling single digit addition; the role of back-up (counting) procedures; error priming; and the model's position in long (multi-digit) arithmetic procedures.

The backpropagation cascade model presented here has detailed the spread of activation, response selection, training regime and minimal assumptions needed to replicate results on adult performance. This has been done in the context of attempting to mimic the experiments performed by Campbell & Graham [1985], and hence the results are of a statistical nature. The explicitness of the model is one of its strong points, and as McCloskey et al. [1991] point out it is now time to "shift from a demonstration of the framework's basic merit to the hammering out of detailed, fully elaborated models" [p. 394]. This has been an attempt to do just that.

References

Anderson, J. A., Spoehr, K. T., & Bennett, D. J. [1991]. A study of numerical perversity: Teaching arithmetic to a neural network. Technical report 91–3, Department of Cognitive and Linguistic Sciences, Brown University. To appear in Levine, D. S. and Aparicio, M. (eds) *Neural Networks for Knowledge Representation and Inference*. Lawrence Erlbaum Associates, Hillsdate, NJ.

Ashcroft, M. H. [1987]. Children's knowledge of simple arithmetic: A developmental model and simulation. In Bisanz, J., Brainerd, C. J., & Kail, R., eds., *Formal Methods in Developmental Psychology*, chapter 9, pp. 302–338. Springer-Verlag, New York.

Campbell, J. I. D. [1987]. The role of associative interference in learning and retrieving arithmetic facts. In Sloboda, J. A., & Rogers, D., eds., *Cognitive Processes in Mathematics*, pp. 107–122. Clarendon Press, Oxford.

Campbell, J. I. D., & Graham, D. J. [1985]. Mental multiplication skill: Structure, process, and acquisition. *Canadian Journal of Psychology*, *39*(2), 338–366.

McClelland, J. L. [1979]. On the time relations of mental processes: An examination of systems of processes in cascade. *Psychological Review*, *86*(4), 287–330.

McClelland, J. L., & Rumelhart, D. E. [1988]. *Explorations in Parallel Distributed Processing: A Handbook of Models, Programs, and Exercises*. MIT Press, Cambridge, MA.

McCloskey, M., Harley, W., & Sokol, S. M. [1991]. Models of arithmetic fact retrieval: An evaluation in light of findings from normal and brain-damaged subjects. *Journal of Experimental Psychology: Learning, Memory, and Cognition*, *17*(3), 377–397.

Miller, K., Permutter, M., & Keating, D. [1984]. Cognitive arithmetic: Comparison of operations. *Journal of Experimental Psychology*, *10*(1), 46–60.

Siegler, R. S. [1988]. Strategy choice procedures and the development of multiplication skill. *Journal of Experimental Psychology: General*, *117*(3), 258–275.

Stazyk, E. H., Ashcraft, M. H., & Hamann, M. S. [1982]. A network approach to mental multiplication. *Journal of Experimental Psychology: Learning, Memory, and Cognition*, *8*(4), 320–335.

Viscuso, S. R. [1989]. *Memory for Arithmetic Facts: A Perspective Gained from Two Methodologies*. Ph.D. thesis, Department of Psychology, Brown University, Providence, RI.

Viscuso, S. R., Anderson, J. A., & Spoehr, K. T. [1989]. Representing simple arithmetic in neural networks. In Tiberghien, G., ed., *Advances in Cognitive Science*, Vol. 2. Ellis Horwood, Chichester, UK.

Calculating Salience of Knowledge

Lisa F. Rau
GE Research and Development Center
Schenectady, NY 12301 USA
and
Computer Science Department
University of Exeter, Exeter, UK

Abstract

As information systems continue to grow in size and scope, advances in data management become more and more on the critical path for usability of these systems. This paper reports on the implementation and applicability of an important function - that of calculating the conceptual *salience* of knowledge or data in a knowledge base or database.

Salience is calculated with a method based on Tversky's formulation of salience as composed of two factors: intensity and discriminability. The salience computation has been implemented and tested on a database and is independent of the particular knowledge area.

Introduction

This paper reports on a theory and implementation of the cognitive notion of conceptual *salience*, a concept not typically modelled from a computational perspective. While the concept of *salience of knowledge* has a clear intuitive meaning, this work aims to formalize the notion and provide a computational mechanism for its determination.

Salience of knowledge is, intuitively, the prominence or conspicuousness of knowledge. It is important from a practical perspective because salient knowledge is typically buried along with insignificant knowledge in a large database system. Potentially important facts and relationships are represented in the same way as unimportant information. Discovering what is salient *adds* knowledge of hitherto unknown relationships that can, in turn, be used to reason with and increase the utility of the data represented. Moreover, salient knowledge should be more *accessible* and more *relevant* to a given task, and should be chosen preferentially over more obscure facts.

Salience is a characteristic of knowledge that is important for case-based reasoning [Seifert, 1989], analogical reasoning [Gentner, 1983], knowledge discovery in databases [Piatetsky-Shapiro and Frawley, 1991], understanding metaphor [Makoto *et al.*, 1990] and information retrieval in general. Knowing what characteristics of knowledge are salient allow processes to only deal with those characteristics. Salience, however, is not typically encoded at the time the knowledge is entered, and in fact, salience can only be computed with respect to the other knowledge in the system. The salient portions of a database are in effect only the highlights.

This paper reports on a system that automatically extracts information from text, stores it in a database and discovers salient features. Although the methods are applicable to an arbitrary database or knowledge base, the origins of this database from a real-world source are suggestive of their use in automatically computing the salience of general or specific knowledge.

This work is one portion of a larger project to provide computational methods for automatically deriving what a system knows, for example, in terms of its *breadth of knowledge* [Rau, 1992]. The other "meta-properties" of knowledge are computed based on similar computational foundation as the salience computation.

Overview of Methodology

This paper reports on the theory, implementation and testing of a method for computing what, in an arbitrary knowledge or database, is salient. This section describes both the intuition behind the formalization, and the formalization itself. Following sections describe the database used in demonstrating the implementation, and detail the results.

Tversky [Tversky, 1977] hypothesized that salience is composed of two factors: (1) the intensity of an aspect of a concept (the amount of information), which we denote $\mathcal{I}$, and (2) discriminability, denoted $\mathcal{D}$, corresponding to how well an aspect of a concept distinguishes that concept from related concepts. Iwayama [Makoto *et al.*, 1990] proposed a method, using information theory, to compute and combine these two factors. The method uses a probabilistic model of conceptual categorization. With this representation, the intensity of a concept is equal to its information theoretic redundancy (the inverse of the Shannon [Shannon and Weaver, 1949] entropy); a measure of the amount of uncertainty present in the frequency distribution of values. The discriminability is the ratio of this intensity to the sum of related concepts' intensities. The exact computation is detailed below.

This computational method has been extended to uncover salient combinations of features in a database. This is accomplished by taking all binary combinations of database fields, and computing the salience of the fillers of those fields with respect to related fillers.

Probabilistic Model of Conceptual Categorization

Underlying the calculation of salience is a probabilistic model of conceptual categorization [Smith and Medin, 1988]. Given a database composed of fields $\mathcal{F}$ that contain fillers $\mathcal{F}_i$, we treat the fields as *concepts* or *conceptual categories* and the fillers as *features* or *aspects* of those concepts. Then the frequency of occurrence of each filler in the database approximates the frequency of occurrence of features of a conceptual category. Note that this assumption limits the salience computation to what is salient with respect to the area of expertise in a database.

For example, suppose a database field of **sex-of-person** contained two fillers with the following frequencies of occurrences:

`sex-of-person = {(male, .5) (female, .5)}`

We denote these probabilities p_i. The category of **sex-of-person** is assumed to be composed of and defined by two features, male and female, each of which occurs with equal probability. On the other hand, a database from a medical office that deals exclusively with pregnancy would have a different defining notion of what sex the patients were. This is important as what is salient to a particular database is necessarily dependent on the particular context and bias of that database.

Amount of Information

Taking the database field to be $\mathcal{F}$ and the individual fillers to be the $\mathcal{F}_i$, the amount of information is the normalized inverse of the well-known Shannon [Shannon and Weaver, 1949] measure of entropy $E(F_i)$:

$$E(F_i) = -\sum_{j=1}^{m} p_{i,j} \log_2 p_{i,j}$$

This entropy measure is adjusted to reflect the total number of distinct fillers (m) by dividing by a normalization factor $\log_2 m$ to obtain relative entropy $e(F_i)$:

$$e_i = \begin{cases} 0 & \text{if } m = 1 \\ \frac{E(F_i)}{\log_2 m} & \text{otherwise} \end{cases}$$

The amount of information varies inversely with the relative entropy, so we define the amount of information $\mathcal{I}$ to be $I = 1 - e(F_i)$. The amount of information is zero when all values are equiprobable, and is one when all values are the same, i.e., when there is only one value for the field. Intuitively, the amount of information measures the variability in frequency of occurrence among different fillers of a field. If all the fillers occur with roughly equal frequencies, than no one filler "stands out" from the rest, hence this component of the salience is low.

Example

To make the calculation concrete, we calculate an example from the domain of experimentation; incidents involving terrorism in Latin America. We calculate the amount of information of **location-of-kidnapping**. Note that we could also calculate the amount of information of **location-of-incident-types** in general (a category that includes other incidents such as **bombing** and **attack**. However the computation of salience takes *slices* of the database, looking at the distribution of fillers with respect to a particular value. We denote the distribution of fillers with respect to a particular value $\mathcal{S}_i$, as opposed to $\mathcal{F}_i$. This is discussed in the next section.

The field of **location** has nine possible fillers, appearing below. To compute the discriminability of **kidnapping** with respect to the location of the incident, we first generate the frequency list of this slice of the database; what proportion of the **kidnappings** occurred with respect to each of the nine **locations**; there were 119 total. This yields:

location-of-kidnapping

	p	-plogp		p	-plogp
Colombia	.50	.50	Venezuala	.01	.07
ElSalvador	.21	.47	Peru	.04	.19
Guatemale	.19	.46	Ecuador	.02	.11
Chile	.02	.11	Brazil	.02	.11
Panama	.01	.07			

The most frequent value is **COLOMBIA**. Under the formula given above for amount of information we can compute: $\mathcal{I}(location - of - kidnapping) = 1 - \frac{2.08}{log_2 9} = 1 - \frac{2.08}{3.17} = 1 - .66 = .34$

Discriminability

In order to compute the measure of discriminability $\mathcal{D}$, there must be a notion of what the concept is to be differentiated with respect to. In database terms, it is possible to look at the variation in the distribution of fillers of a given field with respect to a different field. For example, a database that contained people's occupations and education levels could determine how differentiating a given occupation is with respect to education level, or how good a differentiator or discriminator a certain education level is with respect to occupation. The relation is not symmetric because the measure of discriminability incorporates the amount of information of related concepts; this measure is different for occupation than for education level. We call the distribution of fillers of a field with respect to a different field a *slice* of the database.

The discriminability is calculated by taking the ratio of the amount of information of a slice to the sum of all the amount of information of slices of related concepts. Only related concepts that have the same most frequently occurring filler contribute to this sum.

Continuing with the example, to compute the discriminability, we look at the similar concepts to **kidnapping** that have the same most frequent value of **location**, in this case, **COLOMBIA**. This entails computing the frequency distribution of each slice of incident-type with respect to location. There is only one similar concept that has this same most frequent value of location; **BOMBING** and its amount of information is .36. Hence the discriminability is:

$$.34/(.36 + .34) = .34/.70 = .48$$

The ratio of the amount of information of location-of-kidnapping to location-of-incident is the discriminability, in this case, .48. This value ranges from nearly zero to one.

This value approaches zero when the denominator is large, which corresponds to when there are many similar concepts with this most frequent value. This value is one when no other similar concepts have this most frequent value, in which case the numerator and denominator are the same.

Salience

Finally, the salience is obtained by multiplying the two terms, $\mathcal{I}$ and $\mathcal{D}$. In the example of **location-of-kidnapping**, the salience is simply:

$$.34 \times .48 = .16$$

This reflects the contribution of two factors; the amount to which the filler and field combination discriminates among similar concepts, and the inherent *amount of information* that that filler has with respect to its field. If it is no more common than any other filler, the amount of information is very low or zero, thus diluting this filler's salience. Conversely, if it is the only filler that field has, the amount of information conveyed by that filler is a maximum.

The limiting cases cannot be determined by more simple counting measures. For example, it can be easily determined (1) when there is only a unique value for any given database slice and (2) when the most frequent value does not occur in any other similar concept. However, the combination of (1) and (2) is extremely unlikely, and for (2) the information theoretic redundancy (amount of information) is equal to the salience, as the discriminability is equal to one (numerator and denominator are equal). In this case, the salience is equal to the amount of information inherent in that data slice. Salience, therefore, combines discriminability with information content. Features that are highest in both will be the most salient. In practice, fillers that occur with very low frequencies (once or twice) tend to give high salience results, however they are typically errors or anomalies. The demonstration performed here excluded fillers that occurred less than twice. Examples are given in the next section.

Implementation

In order to determine what in the database is salient, the system examines all possible slices and computes the relative salience of each slice with respect to various fillers. After inverting a database with n distinct fillers, (an $\mathcal{O}(n)$ operation), it is computationally tractable to compute intersections of pairs of fillers (an $\mathcal{O}(n^2)$ operation). Holding one field constant, the system looks at the distribution of fillers with respect to the constant field, and its related fields. This allows the system to make relative comparisons between, say, the pattern of **instrument** used among different **countries**. In this case, the particular filler of the **country** field is kept constant, and the distribution of values of the instruments within that country are examined.

For every pair of fields, the amount of information, salience and discriminability are calculated and classified according to the following categories:

Discriminability of One: A discriminability of one indicates that the numerator (the amount of information) and the denominator (the amount of information of the related concepts that have the same most likely values) are equal. This means that no other related concept has the same most likely property as the numerator concept. In this case, the salience is equal to the amount of information. If there are many roughly equally frequently occurring fillers, the salience is low.

Amount of Information of One: An amount of information value equal to one indicates that there is only *one value* for this value with respect to this field. In this case, the discriminability and the salience are identical.

Salience of One: If all three values are one, then there is *only one value* for this combination of value with respect to field, and *no other related field has the same most likely property*. Intuitively, that singular value distinguishes among related slices. If the amount of information is one and the discriminability is zero, which implies a zero salience, it is the case that the slice is empty, see below under **Identical**.

Amount of Information of Zero: An amount of information of zero means that all values occur with equal values, that is the frequency of occurrence of all the values is identical. In this case, we define the discriminability to be NIL, and the salience is at a minimum of zero as well. A salience of zero may also indicate that the product of amount of information with the discriminability is very small; so a check for a null discriminability is necessary to distinguish this case.

Ordinary: If none of the above conditions hold, there is at least one other related concept that has the same most likely value. This is the typical case, and here the salience and discriminability are just as defined.

In the cases where more than one value occurs with the same maximum frequency, the amount of information, salience and discriminability are calculated for each of these most likely values.

Demonstration

This section describes the database used to demonstrate the methods just described. The database used contained almost 2,000 database records, each of which has 24 fields of information. The highest accuracy records were manually created to be used to test the accuracy of automated methods of data extraction [Krupka *et al.*, 1991], and it was these manually created records that were used in this demonstration. The fields that contains strings of natural language were made canonical (and conceptual) by running them through the same natural language program that generates the entire templates. These records were created from texts reporting on terrorist activities in Latin America, and we have natural language text processing programs described elsewhere [Jacobs and Rau, 1990; Jacobs and Rau, 1993] capable of generating these records with close-to-human accuracy. Using news stories as a source suggests that this work has the potential to operate on arbitrary and general knowledge, as well as specific databases. Figure 1 shows a sample message and template from this set.

Results of Demonstration

This section describes the major results of the demonstration. 22,320 salience measurements were computed by looking at all slices of the database that contained over two members. From these, the top scoring results are reprinted here. The most salient slices that contained null values are not included in this summary, although the salience of slices that contained null values was computed. The correlations with missing information can be useful, but they convey less information than the other associations.

In what follows, the `Magnitude` is the number of times this combination occurred in the database,

DEV-MUC4-0351

BOGOTA, 18 AUG 89 (EFE) -- [TEXT] SENATOR LUIS CARLOS GALAN, LIBERAL PARTY PRESIDENTIAL HOPEFUL, WAS SHOT THIS EVENING WHEN HE WAS ABOUT TO GIVE A SPEECH AT MAIN SQUARE OF SOACHA, 15 KM SOUTH OF BOGOTA, IT WAS CONFIRMED BY POLICE AND HEALTH AUTHORITIES.

ACCORDING TO THE FIRST REPORTS, AT LEAST ONE MAN FIRED ON THE SENA- TOR FROM AMONG THOSE GATHERED. THE SENATOR IS CURRENTLY AT THE EMERGENCY ROOM OF A HOSPITAL IN BOSA, CLOSE TO SOACHA. TWO OTHER PERSONS WERE WOUNDED DURING THE ATTACK.

0. MESSAGE: ID	DEV-MUC3-0351
1. MESSAGE: TEMPLATE	1
2. INCIDENT: DATE	18 AUG 89
3. INCIDENT: LOCATION	COLOMBIA: SOACHA (CITY)
4. INCIDENT: TYPE	ATTACK
5. INCIDENT: STAGE OF EXECUTION	ACCOMPLISHED
6. INCIDENT: INSTRUMENT ID	-
7. INCIDENT: INSTRUMENT TYPE	GUN: "-"
8. PERP: INCIDENT CATEGORY	TERRORIST ACT
9. PERP: INDIVIDUAL ID	"AT LEAST ONE MAN" / "ONE MAN"
10. PERP: ORGANIZATION ID	-
11. PERP: ORGANIZATION CONFIDENCE	-
12. PHYS TGT: ID	*
13. PHYS TGT: TYPE	*
14. PHYS TGT: NUMBER	*
15. PHYS TGT: FOREIGN NATION	*
16. PHYS TGT: EFFECT OF INCIDENT	*
17. PHYS TGT: TOTAL NUMBER	*
18. HUM TGT: NAME	"LUIS CARLOS GALAN"
19. HUM TGT: DESCRIPTION	"LIBERAL PARTY PRESIDENTIAL HOPEFUL" / "SENATOR": "LUIS CARLOS GALAN" "TWO OTHER PERSONS"
20. HUM TGT: TYPE	GOVERNMENT OFFICIAL: "LUIS CARLOS GALAN" CIVILIAN: "TWO OTHER PERSONS"
21. HUM TGT: NUMBER	1: "LUIS CARLOS GALAN" 2: "TWO OTHER PERSONS"
22. HUM TGT: FOREIGN NATION	-
23. HUM TGT: EFFECT OF INCIDENT	INJURY: "LUIS CARLOS GALAN" INJURY: "TWO OTHER PERSONS"
24. HUM TGT: TOTAL NUMBER	-

Figure 1: Example Text and Data Extracted

the `Salience` is the actual numerical salience of the result. Recall that the slices compute frequency distributions of database fields with respect to a particular filler, so that that `Filler-2` comes from a different database field than the `Field-1`. The `Field-1` and `Filler-2` define the slice of the database, where `Filler-2` comes from `Field-2`. The `Most Likely Value` is the most frequently occurring filler in this slice; it is one of the fillers of the `Field-1`. For example, the first result indicates that the salience of `human-effect-of-accomplished` is .531, and that the most likely human effect when the event is accomplished is `DEATH`. Recall in the earlier example, we calculated the salience of `location-of-kidnapping` where the most likely value was also `COLOMBIA`. All combinations that appeared over 50 times are shown and with a salience of over .2 are shown.

Analysis

It is always a difficult problem to evaluate automated discovery systems - the human cannot determine what discoveries were not found, and there are no general methods of judging the inherent goodness of any given discovery. However it is safe to say that the relationships categorized as "salient" by this method indeed serve to discriminate among related concepts, and are prominent in terms of relative frequency of co-occurrence when compared to other similar data slices.

Field-1	Filler-2	Field-2	Most-Likely-Value	Magnitude	Salience
HUMAN-EFFECT	ACCOMPLISHED	STATE	DEATH	487	.5310
PHYSICAL-EFFECT	ACCOMPLISHED	STATE	SOME-DAMAGE	212	.5110
LOCATION	TERRORIST-ACT	CATEGORY	COLOMBIA	335	.4710
PERPETRATOR-ORG	PERU	LOCATION	SHINING-PATH	53	.4460
INSTRUMENT-TYPE	ATTACK	TYPE	GUN	123	.4090
INSTRUMENT	BOMBING	TYPE	BOMB	155	.4070
PERPETRATOR-ORG	EL-SALVADOR	LOCATION	FMLN	155	.3910
LOCATION	ACCOMPLISHED	STATE	EL-SALVADOR	424	.3450
INSTRUMENT	BOMBING	TYPE	BOMB	108	.3440
LOCATION	SHINING-PATH	PERPETRATOR-ORG	PERU	53	.3330
HUMAN-TYPE	ACCOMPLISHED	STATE	CIVILIAN	502	.3240
PERPETRATOR-ORG	TERRORIST-ACT	CATEGORY	FMLN	158	.3200
INSTRUMENT-TYPE	TERRORIST-ACT	CATEGORY	BOMB	136	.3140
HUMAN-EFFECT	SOME-DAMAGE	PHYSICAL-EFFECT	INJURY	61	.3060
PHYSICAL-EFFECT	TERRORIST-ACT	CAT	SOME-DAMAGE	92	.2980
HUMAN-EFFECT	BOMBING	TYPE	INJURY	86	.2970
INSTRUMENT	TERRORIST-ACT	CATEGORY	BOMB	96	.2720
CATEGORY	ACCOMPLISHED	STATE	TERRORIST-ACT	790	.2140
INSTRUMENT	ACCOMPLISHED	STATE	BOMB	148	.2020

Figure 2: Sample of Results of Salience Computation

Uses

The salience result can be used in a variety of different ways. Some of the high-salience data reflect logical associations between slots, such as that when a **PHYSical-TARGET** suffers **NO DAMAGE**, any **HUMAN-EFFECT** is likely to be **NO INJURY OR DEATH**. Another example of these logical associations is the relationship between certain perpetrator organizations (**PERP-ORG**, for example **SHINING PATH**) and the location **PERU** where these organizations reside. Detecting such slot inter-dependencies is critical in order to correctly apply any future machine learning methods that assume independence.

Another use of these results is to aid in the determination of which questions to ask to effectively differentiate an event. For example, suppose an analyst is interested in discriminating **TERRORIST ACT**s from **STATE-SPONSORED VIOLENCE**. The most effective slot to know is that which is most salient with respect to the slot one wishes to differentiate upon. This gives the analyst guidance as to which information is most likely to differentiate one from the other. For example, in this case, the **PERPetrator-ORG** differentiates these two types of events very well. The salience results also allow a system analyst to make predictions. For example when a bombing of an **ENERGY** structure is encountered, the above results lend credence to the hypothesis that it was **DESTROYED** in **EL SALVADOR** and that it was a **TERRORIST ACT**. This prediction is justified because these fillers occur more frequently than any other, and discriminate between other types of structures that are **PHYSical-TARGETs**.

Related Work

This computation of salience builds upon the original formulation of Tversky [Tversky, 1977] and the implementation outlined in Iwayama, et. al. [Makoto *et al.*, 1990]. In particular, this paper expands the applicability to an arbitrary database by abandoning the distinction between concepts and features of concepts. We examine all combinations of fillers and fields exhaustively. That is, Iwayama assumes there are categories such as **fruit**, with members such as **apples, lemons**, and that these category members have certain *attributes* or *features* such as **color** and **shape**. Here, we propose one database field (say **incident-type**, which has the fillers such as **kidnapping** and **bombing**) to be a category and examine the salience with respect to another database field (say **location**), putting the **location** field in place of the attribute or role. We also compute the salience of the reverse situation, examining all locations with respect to the incident-types that occur with those locations. While it may seem as if fruit somehow "makes a better category" than color, and color makes a better "role" than fruit, in fact these distinctions are artificial. It is possible to compute the salience of **color-of-apple** just as it is possible to compute the salience of **fruit-of-red**. In the one case, apples are fillers of fruit categories that have color roles. In the other case, red is a filler of color categories that have fruit roles. This blurring of the distinction between categories and roles enables the determination of what is most salient in an arbitrary database. Finally, we have shown the utility of the measure by running it on a real database where the frequency, values are empirically determined.

A great deal of research has addressed the problem of what a system might know or believe [Halpern, 1986; Vardi, 1988]. The work described here contributes to that body of research by adding a new metric that is calculated from what is known, the salience of knowledge. This work is related to recent work in the area of knowledge discovery in databases [Piatetsky-Shapiro and Frawley, 1991] that attempts to learn new knowledge from the structure and content of databases. However, the particular problem of computing salience of knowledge has not been directly addressed in this new research area.

Limitations and Future Directions

The primary area for future work is in the application of the techniques described here to improve the efficiency and accuracy of real programs. Some possibilities are to focus search on salient items, relax an information request for case-based reasoning or information retrieval along salient dimensions, filter out salient discoveries from the output of a machine learning program, and focus reasoning processes on salient characteristics of a problem domain.

One theoretical issue still to be investigated here is the effect of context on the set of "related items" used in this salience computation. As has been shown by Ortony [Ortony *et al.*, 1985], the features of concepts and concepts themselves judged as similar (related) is heavily influenced by context. One artifact of this implementation has to do with individual styles of creating the answer key from which that data was obtained. Each participant created 100 templates, and some had particular ways of indicating certain events that other sites did not. This makes the peculiarities of a given individual's template filling style appear salient. This artifact can be an advantage in that the methods described here can detect such pecularities to improve the consistency of any database where data is entered manually by a variety of individuals.

Conclusions

This paper began with an analysis of the notion of conceptual salience. A specific computation was detailed for automatically determining the conceptual salience of a knowledge or database. The computation combines the amount of information with the discriminability to produce a numerical score. This calculation was validated by computing the salience of combinations of database fillers on a 1,900 record database.

This work is important not only for the methods and computations described here, but for investigating new questions we would like large knowledge based system to be able to answer - questions such as "what do you know that is important?" and "what stands out?". Looking at areas traditionally reserved for the purely cognitive realm, such as meta-questions of knowledge scope and extent, offers a new perspective from which to develop computational answers.

References

[Gentner, 1983] D. Gentner. Structure-mapping: A theoretical framework for analogy. *Cognitive Science*, 7(2), 1983.

[Halpern, 1986] J. Y. Halpern, editor. *Proceedings of the First Conference on Theoretical Aspects of Reasoning About Knowledge.* Morgan Maufmann, Los Altos, CA, March 1986.

[Jacobs and Rau, 1990] P. S. Jacobs and L. F. Rau. The GE NLToolset: A software foundation for intelligent text processing. In *Proceedings of the Thirteenth International Conference on Computational Linguistics*, volume 3, pages 373–377, Helsinki, Finland, 1990.

[Jacobs and Rau, 1993] P. S. Jacobs and L. F. Rau. Innovations in text interpretation. *Artificial Intelligence (Special Issue on Natural Language Processing)*, 48, To Appear 1993.

[Krupka *et al.*, 1991] G. R. Krupka, P. S. Jacobs, L. F. Rau, and L. Iwańska. Description of the GE NLToolset system as used for MUC-3. In *Proceedings of the Third Message Understanding Conference (Muc-3)*, San Mateo, Ca, May 1991. Morgan Kaufmann Publishers.

[Makoto *et al.*, 1990] I. Makoto, T. Takenobu, and T. Hozumi. A method of calculating the measure of salience in understanding metaphors. In *Proceedings of the American Association for Artificial Intelligence*, pages 298–303, Los Altos, CA, August 1990. AAAI, Morgan Kaufmann, Inc.

[Ortony *et al.*, 1985] A. Ortony, R. J. Vondruska, M. A. Foss, and L. E. Jones. Salience, similes, and the asymmetry of similarity. *Journal of Memory and Language*, 24:569–594, 1985.

[Piatetsky-Shapiro and Frawley, 1991] G. Piatetsky-Shapiro and W. J. Frawley. *Knowledge Discovery in Databases.* MIT Press, Cambridge, MA, 1991.

[Rau, 1992] L. F. Rau. Calculating breadth of knowledge. In *Proceedings of the Fourteenth Annual Conference of the Cognitive Science Society*, Hillsdale, NJ, July 1992. Cognitive Science Society, Lawrence Erlbaum Associates.

[Seifert, 1989] C. Seifert. Analogy and case-based retrieval. In *Proceedings of the Darpa Case-Based Reasoning Workshop*, pages 125–129, San Mateo, May 1989. Darpa, Morgan Kaufmann Publishers.

[Shannon and Weaver, 1949] C. E. Shannon and W. Weaver. *The Mathematical Theory of Communication.* University of Illinois Press, Urbana, Il, 1949.

[Smith and Medin, 1988] E. E. Smith and D. L. Medin. *Categories and Concepts.* Harvard University Press, Cambridge, Ma, 1988.

[Tversky, 1977] A. Tversky. Features of similarity. *Psychological Review*, 84(4):327–352, 1977.

[Vardi, 1988] M. Y. Vardi, editor. *Proceedings of the Second Conference on Theoretical Aspects of Reasoning About Knowledge.* Morgan Maufmann, Los Altos, CA, March 1988.

The Interaction of Memory and Explicit Concepts in Learning*

Susan L. Epstein
Department of Computer Science
Hunter College and The Graduate School of The City University of New York
695 Park Avenue, New York, NY 10021
sehhc@cunyvm.cuny.edu

Abstract

The extent to which concepts, memory, and planning are necessary to the simulation of intelligent behavior is a fundamental philosophical issue in AI. An active and productive segment of the research community has taken the position that multiple low-level agents, properly organized, can account for high-level behavior. The empirical research relevant to this debate with fully operational systems has thus far been primarily on mobile robots that do simple tasks. This paper recounts experiments with Hoyle, a system in a cerebral, rather than a physical, domain. The program learns to perform well and quickly, often outpacing its human creators at two-person, perfect information board games. Hoyle demonstrates that a surprising amount of intelligent behavior can be treated as if it were situation-determined, that often planning is unnecessary, and that the memory required to support this learning is minimal. The contribution of this paper is its demonstration of how explicit, rather than implicit, concept representation strengthens a reactive system that learns, and reduces its reliance on memory.

Introduction

This paper is about the interaction among explicit concept representation, memory requirements, and the ability to learn. *Learning*, in this context, is defined as the transformation of subsequent behavior by previous experience. Learning during problem solving may manifest itself as a change in the speed with which one solves a problem, as a change in the path one takes to a solution, or as a change in the solution at which one arrives.

Although there is general agreement that an intelligent artifact learns, there is less certainty about what is required to learn. Clearly, by the definition of learning, experience is necessary. Most would argue that memory is also necessary for learning, although a machine that rewired itself to incorporate new knowledge, rather than recorded it in some "softer" manner, would meet the criterion. The necessity for concepts, reasoning, and planning in learning, however, has recently come under careful scrutiny by proponents of reactive systems.

The thesis of this paper is that reactive, hierarchical systems can minimize deliberation, but that both memory and explicitly represented concepts are necessary if a program is to learn to perform intelligently. The discussion focuses on a domain previously cited as inhospitable for a reactive system: two-person, perfect information board games (Kirsh, 1991). The paper demonstrates how reactive systems that learn to play games can have unreasonable memory requirements, and discusses concepts and their role in cerebral tasks. Empirical evidence shows how explicit concept representation can reduce memory requirements and improve performance while preserving the essential features of a reactive system: refusal to plan, reluctance to search, and reliance on low-level responses to achieve high-level goals.

The Control-Concept Controversy

One of the lessons of empirical AI is that general state space search heuristics are weak methods, and that power requires domain specialization. A program may have the appropriate knowledge prespecified or may learn it (Laird, Rosenbloom, & Newell, 1987; Minton, 1988; Mitchell et al., 1989). When search and learning are not enough, many systems *plan*, i.e., reason about possible actions and their outcomes before committing to them.

Biology, however, offers many examples of seemingly intelligent and planned behavior that can be explained as prespecified, i.e., "hard-wired." Ants transporting food cooperatively or young birds avoiding precipices, it is said, do not reason about hunger or danger, although their behavior simulates a creature that does. Some researchers have extrapolated from this to suggest that, in the simulation of intelligence, planning, goals, and representation are unnecessary; that when behavior is cast as reaction to environmental stimulus, only appropriate control is required. Such

* This work was supported in part by NSF 9001936 and PSC-CUNY 668287.

programs are called *reactive systems*.

Brooks has provided the following "representation-free" description of a reactive system: "Low-level simple activities can instill the Creature with reactions to dangerous or important changes in its environment.... By having multiple parallel activities, and by removing the idea of a central representation, there is less chance that any given change in the class of properties enjoyed by the world can cause total collapse of the system.... Each layer of control can be thought of as having its own implicit purpose.... The purpose of the Creature is implicit in its higher-level purposes, goals, or layers." (Brooks, 1991)

Reactive systems are built from small components called *agents*. Each agent has a simple task to accomplish, for example, looking, feeling a force, or moving forward. Each agent "decides" what to do by processing input sensory data. The agent's reaction is its output. The entire program performs as a collection of competing behaviors to which an observer may impute motives and goals where none are ever explicitly represented, i.e., reactive systems do not *deliberate* (plan from concepts).

The coordination of these agents to effect such control is non-trivial. A *layer* is a subsystem of agents that produces an activity, i.e., pursues some implicit purpose. Experiments indicate that a hierarchical *subsumption architecture* that coordinates its agents in layers is the key to proper control for a reactive system (Brooks, 1991; Connell, 1990). One Brooks robot, for example, has a layer to avoid obstacles, another to wander, and one to explore.

The simulation of intelligence in reactive systems is purely a control issue, their proponents claim, without any concern for representation or focus of attention. A few robotic reactive systems have been able to learn their own control strategy (Maes & Brooks, 1990; Mahadevan & Connell, 1991).

Preliminary successes with robots have been predicted to scale up to any task because "there need be no explicit representation of either the world or the intentions of the system to generate intelligent behaviors for a Creature" (Brooks, 1991). Kirsh, however, claims that Brooks has worked only on *situation-determined behavior*, i.e., problems where an egocentric perception of the "indicators that matter" is sufficient to determine the appropriate course of action. (Kirsh, 1991). He characterizes *cerebral tasks*, the kinds of tasks on which he believes a reactive system would fail: tasks that involve other independent agents, that require planning, that require an objective viewpoint, that require problem solving. Between them they pose the *control-concept controversy:* Should a program learn explicit concepts that generalize experience, as in the traditional AI paradigms, or should it learn control for a reactive system? The remainder of this paper explores that issue in a domain Kirsh predicts as too difficult for a reactive system: game-playing.

Reactive Playing

An obvious reactive system to play a specific game perfectly would construct one agent for each possible game state, an agent that would output the perfect move whenever it sensed a match with its state description. Challenging games, however, would require far too many such agents. Thus, this ideal reactive system must somehow be supplemented with knowledge. The four reactive programs described below are goal-free; all they do is sense patterns and respond to them.

Henri demonstrates how pure pattern recognition can be insufficient for learning even a simple game in a noise-free environment (Painter, 1992). For several different games on a three-by-three board, Henri learns values for three-symbol (X's, O's, and blanks) patterns and applies those values to each of the eight possible three-position lines on the board. Henri learns, for example, that in tic-tac-toe the pattern "X-X-blank" is more valuable than the pattern "blank-X-blank." Values are calculated by a primitive kind of reinforcement learning based on contest outcome. On its turn, Henri evaluates each possible legal move, and selects one with the highest pattern score. Against a programmed expert, after training in 200 tic-tac-toe contests, Henri still loses 15% of the time, because of inaccuracies in the pattern values. It is unclear how long Henri would take to learn to play perfect tic-tac-toe, or if it ever would.

N-N/Tree shows how pattern recognition plus search can still fail to learn a simple game in a realistic environment. This program uses temporal differences to learn weights for a neural net that accepts nine-position pattern input for games on a three-by-three board (Flax et al., 1990). N-N/Tree is also permitted a 3-ply search. It plays against a programmed expert that may err as often as 5% of the time. After 1000 tic-tac-toe training contests, approximately 9000 training examples, N-N/Tree still loses 8% of its contests.

Dooze suggests that learning only control, while adequate, may require more memory than a machine can offer. Dooze is a classifier system that learns to play games on a three-by-three board (Esfahany, 1992). Learning is the introduction and deletion of decision-making rules, called *classifiers*, at the end of each contest. Each classifier has the form "when the board matches the following pattern, move to position i." A pattern describes each of the nine positions as an X, an O, a blank, or a "don't care" symbol. After 63 contests, on average, Dooze learns to play apparently perfect tic-tac-toe. Its better performance may be attributable, however, to its larger memory requirements. There are $9 \cdot 4^8$ possible Dooze classifiers for tic-tac-toe. The program must maintain a set of 150 of them, about 15%, to learn to play expertly. Many of the learned classifiers are quite restrictive, i.e., entail patterns that would apply to very few game states. For five men's morris, a relatively simple game with 10 markers and 16 positions, 15% of the possible classifiers would be about 2^{31} rules.

Morph highlights a possible learning tradeoff between memory size and number of training experiences required to learn. It learns patterns while playing chess against a competent commercial program (Levinson & Snyder, 1991). Morph is characterized as a search-free and purely syntactic game player, i.e., one that reacts only to patterns, without planning or reasoning. A Morph pattern is a labeled graph that describes how selected markers and positions on the board relate to each other. Such a pattern is more sophisticated and less specific than the ones used by Dooze and N-N/Tree, and often applicable to more game states. Given an appropriate, hand-crafted pattern language, Morph's methods can be applied to any game. On tic-tac-toe, a Morph-like program learned to play perfectly after approximately 250 contests and learned approximately 50 patterns (Levinson, 1991). The difference in learning rate and storage requirements between this program and Dooze suggests a trade off between memory size and number of training experiences required to learn.

Careful analysis reveals that each of these "representation-free" programs actually incorporates concepts, generalizations about game playing understood by the programmer and incorporated into the code. Henri only uses knowledge about lines and how positions lie on them in two-dimensional space; its performance is also the weakest. N-N/Tree uses knowledge about the minimax algorithm for search control and how to apply it three-ply deep. Dooze's learning algorithms value the winning move highly, value every move the expert model makes, and recognize that good positions for X are good for O when the markers are interchanged. This is a hefty dose of primitive game-playing commonsense. Dooze's don't-care symbols also support abstractions, such as "If X holds the center, …." Morph's pattern language embeds ideas like threat and defense in both the pattern learner and in memory. In summary, although reactive game players are possible, they rely on hidden knowledge to achieve acceptable performance, and probably have some tradeoff between memory size and learning speed.

Concepts and their Representation

A *concept* is defined here as some recognized set of regularities detected in some observed world. *Regularity* means repeated occurrence and/or consistency of use. In this context, a concept includes not only the necessary and sufficient descriptions called definitions, but also defaults, associations, and expectations. Thus a concept may incorporate error, bias, and inconsistency (Wierzbicka, 1985). From an AI perspective, a concept is generalized domain knowledge, a description of what has been encountered. Although specific examples may be remembered, a concept is not a set of instances but a summary of experience.

If a machine is constructed to meet a goal, either implicit or explicit concept representation is necessary. A cherry pitter, for example, implicitly references the concept of a cherry as a small, round object containing an even smaller object which can be extracted when pressure is appropriately applied. Although the architecture of a sufficiently elaborate machine, like a robot, may obscure its concepts, they are present implicitly, in circuitry and mechanical devices. Any program claimed "representation-free" is characterized here as a program with implicit concept representation. In contrast, explicit concept representation offers several benefits to a machine that learns: organization of knowledge, focus of attention, and ability to discard experience. Thus explicit concept representation reduces the need for induction and deduction, as it flexibly makes regularities immediately accessible.

People find it convenient to expect regularity in the world, and they have many devices to represent the regularities they detect. Four kinds of *concepts*, ways that people generalize about regularities in their experience, are identified here. *Compiled* regularities, like how to ride a bicycle, abbreviate a reliable response to specific situations. Compiled knowledge is experienced as reactive behavior; it has lost the detail, rationale, and instructions that once accompanied it. When the lost information is needed, reconstruction often requires observation from fresh experience. *Categories* are sets of objects with common features. For example, a chair is a category and every chair, physical or hypothetical, is an *instance* of the category, with specifically noted values for some of its features. *Scripts* are regularities about what is expected of an experience and those who participate in it (Schank & Abelson, 1977). For example, any visit to a restaurant is a walk through a script, a partially ordered set of expectations for everyone's behavior there. *Meta-principles* are regularities applicable to many different kinds of experience, ones to fall back upon when more detailed knowledge fails. Examples of meta-principles include efficiency, safety, and propriety. The instantiation of a meta-principle for a particular domain results in a *principle,* behavioral guidance that may curtail search. The application of efficiency, safety, and propriety to driving a car, for example, would result in directives to drive rapidly, to drive carefully, and to obey the driving laws, respectively. Note the evident conflict among these principles.

People behave appropriately and learn quickly in part because they retrieve and apply these regularities, or concepts, continually and effectively. There is ample psychological and anthropological evidence that concepts are both learned and culturally determined, and that people prefer them to logical reasoning for any but the simplest examples (D'Andrade, 1991). In any culture, those judged experts are those who give more modal responses, i.e., agree most with commonly held regularities (D'Andrade, 1990). Thus *an expert learns compiled knowledge, categories, scripts, and principles, and knows when and how to apply them.* Given those regularities, learning and problem solving with them may not be trivial, but it should be easier and require less memory.

The Power of Concepts and Memory

Hoyle is a learning program that now equals or outperforms its human mentors at more than a dozen two-person, perfect information board games. The complexity of the program prevents a full technical description in this abbreviated space; interested readers are referred to (Epstein, 1992a, 1992b) for additional detail. Hoyle explicitly represents, integrates, and exploits each of the four kinds of concepts in its memory, learning, and behavior. There is a script for game playing that provides predefined, uniform, procedural direction, so that the system performs as if it were accustomed to playing games. There is a category representation for games and another for *useful knowledge* (knowledge that is possibly relevant and probably correct) that may be acquired during play. Hoyle's compiled knowledge resides in its *Learner,* as pre-specified, uniform, game-independent heuristic procedures to compute and selectively store useful knowledge. Finally, Hoyle's *Advisors* are principles, implemented as heuristic agents and layered in a subsumption architecture. They accept current knowledge and make comments on moves they favor or oppose.

Given a game, the Learner initiates a series of tournaments against an *expert model* (Kirsh's "other agent") that is only observed, never queried. Whenever it is Hoyle's turn to move, the Advisors comment based upon the current state of their cerebral reality: the game state, the legal moves, and any useful knowledge about the game already acquired. Move selection is a simple arithmetic calculation, part ordering and part voting, that mediates among the Advisors' disagreeing comments. After contests and after tournaments, the Learner's algorithms compute and record useful knowledge.

Hoyle is a reactive system for a cerebral task. The Hoyle cycle is pause-sense-react, where "pause" cedes control to the expert model, "sense" is the collection of current information, and "react" is the collective response of the agents to their input. Each Advisor is an agent, a low-level intelligence that does not plan, that merely senses the input data and responds to it with output signals. The control mechanism is based upon a hierarchical subsumption architecture. *Each move choice is a rapid and simplistic mathematical computation, a reaction without search or deliberation.*

Hoyle meets the postulated reactive system criteria as follows. The low-level simple activities are its quick reactions to short-term possibilities of success and failure. The multiple parallel activities are its Advisors, each of which processes sensory data independently. When Hoyle's world changes, with a new game to play or new opposition to play against, the program is robust and degrades gracefully. The implicit purpose of each Advisor is to forward, in its own particular way, the meta-principle it instantiates. Hoyle can play legally with any subset of its Advisors. The purpose implicit in its higher-level layers is to learn to play perfectly, but there is no explicit representation, in Hoyle's game-playing algorithm or in its control mechanism, of intention, belief, plan, goal, subgoal, win, loss, or draw.

Figure 1 shows performance curves for several 20-contest tic-tac-toe tournaments. The bottom line (#1) is a reasonable lower bound for performance; it shows how a program lost all but one contest in 20 when it made random legal moves against a programmed expert. The top line in Figure 1, for absolute expertise, is a reasonable upper bound; it shows how a program that made perfect moves achieved repeated draws against a programmed expert. Once a program learns to play perfectly, its performance curve should parallel that for absolute expertise indefinitely.

Hoyle's useful knowledge is a compendium of the regularities expert game players look for and exploit. A *significant state* is an inevitable win or loss when both participants play expertly. Such a state is a deduced, compiled regularity computed by the Learner at the end of a contest and stored in memory. The regularity captured by a significant state is that *every* time it occurs the outcome when two experts play is inevitable, not that it is an abstraction of a game state. A significant state may either be treated as a concept (used in computation) or treated as a reflex action (turned toward or avoided). Retrieval of significant states is from a hash table, and is assumed to require no search. Besides significant states, useful knowledge includes selected contest histories, moves experts have made that may have served them well, whether or not it is an advantage to go first, the length of the average contest, data gathered on the relevance and reliability of individual Advisors, and relevant *forks,* game-independent concepts whose instantiation with the current game state can provide powerful offensive and defensive advice (Epstein, 1990). The memory requirement

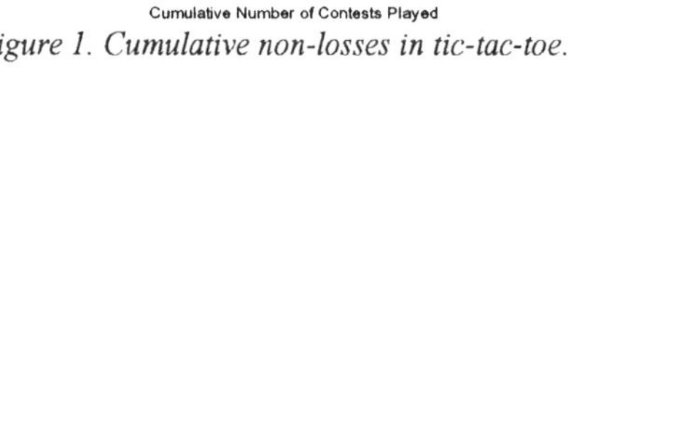

Figure 1. Cumulative non-losses in tic-tac-toe.

for learning a game is essentially a function of the number of significant states and expert moves.

At present Hoyle plays correctly (has the rules for) 24 games gathered from almost as many cultures (Bell, 1969; Zaslavsky, 1982). Hoyle's task is to learn to play each game expertly. Although none of the games is as difficult as checkers or chess, they incorporate a variety of challenges: boards of varying shapes and sizes, stages where the rules change, cycles, and very large search spaces. One of the games it learns to play expertly is Qubic, which has more than a billion game states and is generally acknowledged to lie on the border between simple games and the difficult ones.

The performance of the full version of Hoyle at tic-tac-toe against a programmed expert is shown as curve #4 in Figure 1. In 11 contests, Hoyle learned to play perfectly, and stored an average of 3 significant states and 4 expert moves. Compare this with Dooze's 63 contests and 150 rules, and the Morph-like 250 contests and 50 patterns.

Are Hoyle's concepts or its memory responsible for its ability? Elimination of all concepts from Hoyle would deprive it of its game-playing algorithm, and make it unable to play at all. Less radically, if all the Advisors and all learning were removed, Hoyle would make random moves and would play no better than curve #1 in Figure 1. An interesting reactive version of Hoyle with severe concept restrictions learns responses to game states but uses them only in one way. This *concept-poor* version of Hoyle is a flawless imitator; the Advisors react to the input knowledge but do not perform simple calculations from past experience. This version restricts learned useful knowledge to detailed recollection of the contests it has played and of significant states as reflexes, not as concepts; that way it is permitted only performance repetition, rather than simple computation, with its knowledge. The concept-poor version recognizes previously encountered certain wins and losses, imitates moves the expert made in the identical situation, and tries to avoid reproducing its own failing moves in an identical situation. The performance of this concept-poor version against a programmed expert is shown as curve #2 in Figure 1. The partially-disabled Advisors immediately learn and recommend the successful openings of the human participant, but find the play later in a contest more difficult. In this tournament Hoyle loaded up its memory while it very gradually learned to avoid losing moves, as if it were building the obvious one-agent-per-state reactive system. Although theoretically Hoyle could learn about all possible states by backing up such experience, Figure 1 shows that this memory-greedy process is also slow.

A *memory-free* but concept-dependent version of Hoyle is analogous to an intelligent participant that played every contest at the same game as if it were the first. To explore whether the Advisors that apply useful knowledge really need memory to learn to play expertly, Hoyle played a tournament against a human expert (#3 in Figure 1), this time only with those Advisors omitted from the concept-poor version and without memory. Now the program could rely only on its concepts. After a few contests, a human opponent unaware that Hoyle lacked memory tried a simple strategy for X that defeated the program, one the memory-free version could not learn to avoid. On a hunch the person repeated the same strategy and immediately observed that Hoyle did not learn from its mistakes. (This accounts for the step-like pattern of #3; Hoyle played perfectly in the alternate contests.) Thus the program without memory plays reasonably intelligent contests, but performs unintelligently in a tournament situation. Learning requires memory, and without memory Hoyle never would develop expertise. In all of the games, except the very easiest, it has been repeatedly observed that *Hoyle's power derives from this synergy between memory and concepts.*

Thus far, Hoyle has learned to play as well or better than each of its 14 game-specific external experts, without planning and with only minimal search. When Hoyle has had difficulty learning a new game, its useful knowledge has been very gradually extended to include new concepts and the low-level agents to apply them. This gradual debugging process is much like Brooks' robot-control layering: "so far, so good" (Brooks, 1991).

Conclusions

In a domain that is not situation-determined, Hoyle is a successful, reactive, hierarchical system that retains only a small fraction of what it experiences. The program pays an interesting price for its reactivity, however: it must rely on concepts to learn to perform intelligently. Hoyle offers evidence that learning cerebral tasks demands more explicit concepts than Brooks would like, and far fewer than Kirsh would assume. Hoyle may not resolve the control-concept controversy, but it should certainly influence our attitude on the significance of low-level agents in high-level tasks.

Four other reactive game playing programs have been shown here to employ concepts implicitly. Their pattern generalizations, however, are tailored to a single set of board-specific algorithms, and their memory requirements grow dramatically with the number of positions on the board. Hoyle outperforms these programs, this paper has argued, because it explicitly represents and exploits its concepts.

Hoyle's concepts organize the way it remembers experience, focus its attention on what is important to learn, force it to apply its experience, and permit it to discard experience that is judged unlikely to be useful. As a result it learns with smaller memory requirements and applies its compact useful knowledge more flexibly. Although Hoyle is reactive, the full version of the program incorporates and remembers concepts: knowledge about the regularities that people learn, prefer, and exploit when playing games, and how people use those regularities. When the program is partially disabled and the results observed, it is clear that the synergy between memory and concept application provides the program with its power.

Hoyle's ability to learn with only 15 relatively simple Advisors suggests that *more high-level behavior is available through low-level reactive processes than one might initially suspect.* As the games become more difficult, new concepts are necessary to support performance. *Learning high-level behavior efficiently with a limited memory requires concepts.* After a recent improvement that provided symmetry discovery, Hoyle learned faster and required less memory. *Low-level sensory data can offer an immediate improvement in high-level processing.*

For the time being, several tasks have been relegated to the human system designer: the framework of the categories for game definition and useful knowledge, the correct identification of the culturally determined meta-principles (characterized as "commonsense" but by no means trivial), the instantiation of the meta-principles to construct low-level agents, the assignment of Advisors to tiers based upon knowledge about relations among meta-principles, the specification of which Advisors access which concepts, and the description of how they apply that knowledge. This author believes that all of these can eventually be automated. Work continues on the specified sequence of games; for the moment search during play is limited to two-ply and there is no planning.

Hoyle's results demonstrate for at least one broad cerebral task, game playing, that *a reactive system without memory is impractical, and that reliance only on extensive, detailed memory is brittle and often impossible.* This paper has shown how concepts can structure resource-efficient memory, provide flexibility, and regularize knowledge to support performance. Will a reactive program ever, then, have to search and plan and believe? Hoyle's answer is not yet, perhaps not explicitly, and far less than we ever expected.

Acknowledgments

The author thanks Jack Gelfand, Alice Greenwood, Cullen Shaeffer, and Rick Shweder for their insightful comments and suggestions.

References

Bell, R.C. 1969. *Board and Table Games from Many Civilizations.* London: Oxford University.

Brooks, R.A. 1991. Intelligence without Representation. *Artificial Intelligence* 47: 139-160.

Connell, J. 1990. *Minimalist Mobile Robotics.* New York: Academic Press.

D'Andrade, R.G. 1990. Some Propositions about the Relations between Culture and Human Cognition. In *Cultural Psychology*, ed. J.W. Stigler, R.A. Shweder & G. Herdt. Cambridge: Cambridge University Press.

D'Andrade, R.G. 1991. Culturally Based Reasoning. In *Cognition and Social Worlds*, ed. A. Gellatly and D. Rogers. Oxford: Clarendon Press.

Epstein, S.L. 1990. Learning Plans for Competitive Domains. Proc. 7th International Conference on Machine Learning, 190-197. Morgan Kaufmann.

Epstein, S.L. 1992a. Hard Questions about Easy Tasks. In *Computational Learning Theory and Natural Learning Systems: Constraints and Prospects*, Cambridge, MA: MIT Press. Forthcoming.

Epstein, S.L. 1992b. Prior Knowledge Strengthens Learning to Control Search in Weak Theory Domains. *International Journal of Intelligent Systems.* Forthcoming.

Esfahany, K. 1992. A Pattern Classifier that Learns to Play Games. In preparation.

Flax, M.G., Gelfand, J.J., Lane, S.H. & Handelman, D.A. 1990. Integrating Neural Network and Tree Search Approaches to Produce an Auto-Supervised System that Learns to Play Games. In Proceedings of the 1992 International Joint Conference on Neural Networks, Beijing. Forthcoming.

Kirsh, D. 1991. Today, the Earwig, Tomorrow Man? *Artificial Intelligence* 47: 161-184.

Laird, J. E., Rosenbloom, P. S. and Newell, A. 1987. SOAR: An Architecture for General Intelligence. *Artificial Intelligence* 33: 1-64.

Levinson, R. 1991. Personal communication.

Levinson, R. & Snyder, R. 1991. Adaptive Pattern-Oriented Chess. In Proceedings of the Eighth International Machine Learning Workshop, 85-89. San Mateo, CA: Morgan Kaufmann.

Maes, P. and Brooks, R.A. 1990. Learning to Coordinate Behaviors. In Proceedings of the Eighth National Conference on AI, 796-802. AAAI Press.

Mahadevan, S. and Connell, J. 1991. Scaling Reinforcement Learning to Robotics by Exploiting the Subsumption Architecture. In Proceedings of the Eighth International Machine Learning Workshop, 328-332. San Mateo: Morgan Kaufmann.

Minton, S. 1988. *Learning Search Control Knowledge.* Boston: Kluwer Academic.

Mitchell, T., et al. 1990. Theo: A Framework for Self-Improving Systems. In *Architectures for Intelligence*, ed. K. Vanlehn. Boston: Erlbaum.

Painter, J. 1992. Pattern Recognition for Decision Making in a Competitive Environment. Master's thesis, Dept. of Computer Science, Hunter College. In preparation.

Schank, R. and Abelson, R. 1977. *Scripts, Plans, Goals, and Understanding: An Inquiry into Human Knowledge Structures.* Hillsdale, NJ: Erlbaum.

Wierzbicka, A. 1985. *Lexicography and Conceptual Analysis.* Ann Arbor, MI: Karoma Publishers.

Zaslavsky, C. 1982. *Tic Tac Toe and Other Three-in-a-Row Games.* Crowell.

REMIND: Integrating Language Understanding and Episodic Memory Retrieval in a Connectionist Network*

Trent E. Lange
Artificial Intelligence Lab, Department of Computer Science
University of California, Los Angeles

Charles M. Wharton
Department of Psychology
University of California, Los Angeles

Abstract

Most AI simulations have modeled memory retrieval separately from language understanding, even though both activities seem to use many of the same processes. This paper describes REMIND, a structured spreading-activation model of integrated text comprehension and episodic reminding. In REMIND, activation is spread through a semantic network that performs dynamic inferencing and disambiguation to infer a conceptual representation of an input cue. Because stored episodes are associated with the concepts used to understand them, the spreading-activation process also activates any memory episodes that share features or knowledge structures with the cue. After a conceptual representation is formed of the cue, the episode in the network with the highest activation is recalled from memory. Since the inferences made from a cue often include actors' plans and goals only implied in its text, REMIND is able to get abstract remindings that would not be possible without an integrated understanding and retrieval model.

Introduction

The most parsimonious account of comprehension and reminding is that they "amount to different views of the same mechanism" (Schank, 1982). Consider:

There were sightings of Great Whites off Newport, but Jeff wasn't concerned. The surfer was eaten by the fish. They found his board with a big chunk cut out.

When reading this passage, we may think of other stories of people being eaten by sharks, or, more abstractly, of others who knowingly ventured into mortal danger and died (e.g., skiers buried under avalanches). Why? In order to comprehend stories, a reader must find memory structures that provide inferences such as the goals and plans of story characters and the characteristic features of events and locations. Thus, while understanding a text, we may be reminded of analogous past episodes because they were understood with (and remembered with) the same knowledge structures.

In spite of the interweaving of comprehension and memory, AI simulations of memory have usually modeled reminding separately from language understanding. While this makes accounts of the phenomena more manageable, it is undeniable that real-world retrieval results from comprehension processes. Further, how an elaborated interpretation is constructed from a text will influence what is retrieved from memory. Consider:

John put the pot inside the dishwasher because the police were coming. **(Hiding Pot)**.

First appears John is cleaning a cooking pot, but later it seems he was hiding marijuana from the police to avoid being arrested. **Hiding Pot** might remind a person of superficially-similar stories involving police and marijuana. Or it might lead to more abstract remindings of hiding something to avoid punishment, such as *Billy put the Playboy under his bed so his mother wouldn't see it and spank him.* **(Dirty Magazine)**.

To retrieve episodes only related by similar plans and goals, a model must be able to infer them in the first place. As **Hiding Pot** shows, such interpretations often require both the ability to make multiple inferences and resolve ambiguities. Only with such language understanding capabilities can a retrieval model go directly from input texts to remindings of episodes that are analogous in terms of inferred plans, goals, and abstract relationships. Thus, a model that integrates the process by which a retrieval cue is understood with the process by which it is used to recall information can make an important contribution to the understanding of episodic reminding. In this paper we describe REMIND (Retrieval from Episodic Memory through INferencing and Disambiguation), a structured connectionist model that integrates language understanding and memory retrieval.

Previous Memory Retrieval Models

Memory retrieval has generally been explored in isolation from the process of language understanding. In case-based reasoning (CBR) models (cf. Hammond, 1989; Riesbeck & Schank, 1989), memory access is performed by recognition of meaningful *index patterns* that allow retrieval of episodes (or cases) most likely to aid their current task. CBR models are therefore generally models of expert reasoning within a given domain, rather than models of general human reminding. Whereas expert memory retrieval may be satisfactorily modeled by only retrieving cases matching expected indices within the domain of interest, general reminding

* This research was supported by NSF grant DIR-9024251, Army Research Institute contract MDA 903-89-K-0179, and by a grant from the Keck Foundation.

seems to be substantially "messier", being affected not only by the sort of useful abstract indices used in CBR models, but also by superficial semantic similarities that often lead to quite inexpert remindings. Further, the problem of selecting and recognizing appropriate indices becomes substantially more difficult when reading ambiguous texts outside of limited expert domains.

General, non-expert reminding has been modeled in systems such as MAC/FAC (Gentner & Forbus, 1991) and ARCS (Thagard et al., 1990). These systems model retrieval without using specific indexing methods. Instead they retrieve episodes whose representations share superficial similarities with cues, with varying degrees of preference towards episodes that are also analogically similar. However, unlike CBR models, these systems do not specify how they construct the representation of cues from a source input or text, and so cannot explain how inferences and comprehension affect reminding.

Previous Language Understanding Models

In REMIND, the understanding mechanism constructs an interpretation from its input that not only serves as the model's representation of the meaning of the text, but also serves as an elaborated cue for episodic memory retrieval. Symbolic, rule-based systems have had some success performing the inferencing necessary for this, but have substantial difficulties with ambiguous texts.

Distributed connectionist models can be trained to perform disambiguation and understand script-based stories (c.f. Miikkulainen & Dyer, 1991; St. John, in press). However, it is unclear whether they can be scaled up to handle language that requires the inference of causal relationships between events for completely novel stories. This requires chains of *dynamic inferences* over simple known rules, with each inference resulting in a potentially novel intermediate state. Other distributed connectionist models are able to partially handle this problem by explicitly encoding variables and rules in the network (c.f. Touretzky & Hinton, 1988). Unfortunately, these models are *serial at the knowledge level*, i.e. they can only select and fire one rule at a time, a serious drawback for language understanding, in which multiple alternative interpretations must often be explored in parallel (Lange & Dyer, 1989).

Marker-passing models (c.f. Riesbeck & Martin, 1986; Norvig, 1989) solve many of these problems by spreading symbolic markers across semantic networks in which concepts are represented by individual nodes. Because of this, they can perform dynamic inferencing and pursue multiple candidate interpretations of input in parallel as markers propagate across different parts of the network. A drawback of marker-passing models is that they must use a separate serial path evaluator to select the best interpretation path among the often large number of alternative paths generated. This is a particularly serious problem for ambiguous text in which many alternative paths must be evaluated (Lange, 1992).

Structured connectionist networks (c.f. Waltz & Pollack, 1985) are well-suited to disambiguation because it is achieved automatically as related concepts provide graded activation evidence and feedback to one another in a form of constraint relaxation. Like marker-passing models, they have the potential to pursue multiple candidate interpretations in parallel, since each interpretation is represented by activation in different local areas of the network. A number of researchers have recently shown how structured connectionist models can perform dynamic inferencing (c.f. Ajjanagadde & Shastri, 1989; Barnden, 1990). However, most of these new models no longer perform disambiguation. An exception is ROBIN (Lange & Dyer, 1989; Lange, 1992), a structured spreading-activation model that propagates *signature* activation patterns to perform dynamic inferencing and generate multiple possible interpretations of an input in parallel. At the same time, ROBIN uses the network's evidential constraint satisfaction to perform lexical disambiguation and select the contextually most plausible interpretation. This makes ROBIN a promising start for an integrated understanding and memory retrieval model.

Language Understanding in REMIND

REMIND is a structured spreading-activation model that integrates aspects of the language understanding and memory retrieval problems. REMIND is an extension of ROBIN, whose capabilities allow it to perform the high-level inferencing and disambiguation necessary to build interpretations of syntactically-parsed input for short texts such as **Hiding Pot** and **Dirty Magazine**. These interpretations are added to the network to represent the model's long-term memory episodes.

In REMIND, memory retrieval is a natural side-effect of the spreading-activation understanding process. The knowledge structures used to understand an input cue activate similar episodes that were understood and stored in the network earlier. For example, **Dirty Magazine** becomes active when **Hiding Pot** is being understood. An episode is retrieved from memory when there are enough similarities between it and a cue's interpretation to make it the most highly-active episode. Because inferencing and retrieval occur within a single spreading-activation network, both processes strongly interact and affect each other, as appears to be the case in human memory. In the following section, we give an overview of the language understanding portion of the model.

Knowledge Given To REMIND

As with ROBIN, REMIND uses structured networks of simple connectionist units to encode semantic networks of frames and rules representing world knowledge, such as the scripts, plans, and goals (Schank, 1982) necessary for understanding stories in a limited domain. Its knowledge base is hand-built, as in most structured models. However, it is given no information about specific episodes that the network is used to understand.

The knowledge given to REMIND is used to *construct* the actual structure of the network before any processing begins. As with other structured connectionist models, nodes in the network represent each frame or role. Rela-

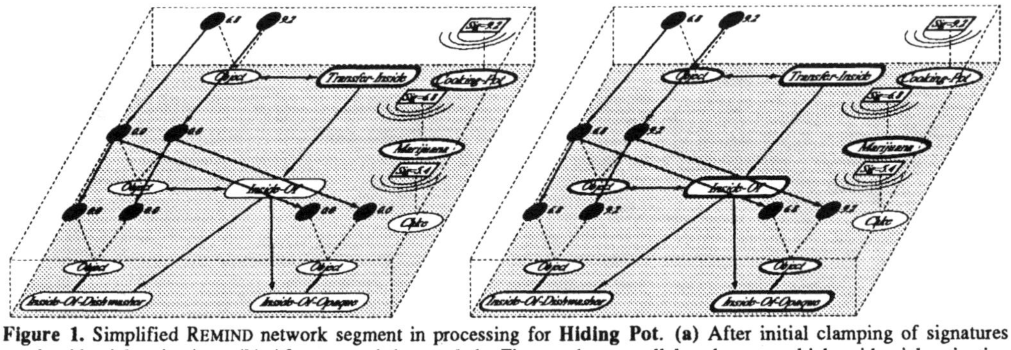

Figure 1. Simplified REMIND network segment in processing for **Hiding Pot**. **(a)** After initial clamping of signatures and evidential activation. **(b)** After network has settled. Figures show parallel paths over which evidential activation (bottom plane) and signature activation (top plane) spread. Signature units (outlined rectangles) and binding units (solid black circles) are in the top plane. Thickness of concept nodes (ovals) represents their evidential activation. Shown are two binding units per role; actual network has enough to hold meanings of network's most ambiguous word.

tions between concepts are represented by weighted connections between the nodes. Activation on concept nodes is *evidential*, corresponding to the amount of evidence available for them in the current context. However, as described earlier, simply representing the amount of evidence available for concepts is insufficient for language understanding. Solving the variable binding problem requires a means for *identifying* the concept dynamically bound to a role. Furthermore, the network's structure must allow these bindings to propagate across the network to dynamically instantiate inference paths and form an elaborated representation of the input.

Dynamic Inferencing With Signatures

Variable bindings are handled in REMIND by network structure holding *signatures* — activation patterns that uniquely identify the concept bound to a role (Lange & Dyer, 1989). Every concept has a set of *signature units* that output its signature, a constant activation pattern different from all other signatures. A dynamic binding exists when a role's *binding units* hold an activation pattern matching the bound concept's signature.

Figure 1a shows a simplified portion of REMIND after **Hiding Pot** has been input. The nodes in the lower layer of the network form a normal semantic network whose weighted connections represent world knowledge. The knowledge represented here is that: (a) transferring an object inside of another (Transfer-Inside) results in it being inside it (Inside-Of), and (b) that two possible concept refinements (or reasons) for it being inside are (1) because it is inside of a dishwasher (Inside-Of-Dishwasher), which will lead to further inferences about it being cleaned, or (2) because it is inside of an opaque object (Inside-Of-Opaque), which will lead to inferences about it being hidden.

Signature activations for variable binding and inferencing are held by the black binding units in the top plane of Figure 1a. In this simplified example, signatures are arbitrary scalar activation values. Here Marijuana is signified by 6.8, Cooking-Pot by 9.2, and Cake by 5.4. As shown, unit-weighted connections between binding units allow signatures to be propagated to other roles defined by general knowledge rules. For example, there are connections from the binding units of Transfer-Inside's Object to the respective binding units of Inside-Of's Object, since the object transferred inside is always the object that ends up inside.

To represent *John put the pot inside the dishwasher*, Transfer-Inside is clamped to a high level (dark oval in Figure 1a). Since Transfer-Inside's Object is either Marijuana or a Cooking-Pot, its Object's binding units is clamped to their signatures' activations, 6.8 and 9.2, respectively[1]. Similarly, one of the binding units of its Actor role is clamped to John's signature and of its Location role clamped to Dishwasher's signature.

Once the activations of the initial signature bindings and conceptual nodes are clamped, both types of activation spread through the network. Figure 1b shows the result after the network has settled from the inputs of Figure 1a and the rest of **Hiding Pot**. The signature activations representing the bindings have propagated along paths of corresponding binding units, so that the network has inferred that the Cooking-Pot or Marijuana is Inside-Of the Dishwasher. This is shown by the fact that their signatures are on the appropriate binding units. As can be seen, the propagation of signatures has also instantiated two different candidate interpretation paths. One path goes through Inside-Of-Dishwasher and continues through other cleaning frames such as $Dishwasher-Cleaning and Clean. Another path goes through Inside-Of-Opaque and continues through frames representing the object being blocked from sight (Block-See), the goal of avoiding detection (Avoid-Detection), and so on. Figure 2 shows a partial overview of the rest of the network.

[1] Other bindings can be presented by simply clamping the binding units to the activations of different signatures. REMIND does not currently address the problem of deciding upon the original syntactic bindings, e.g. that the *"pot"* is bound to Transfer-Inside's Object role.

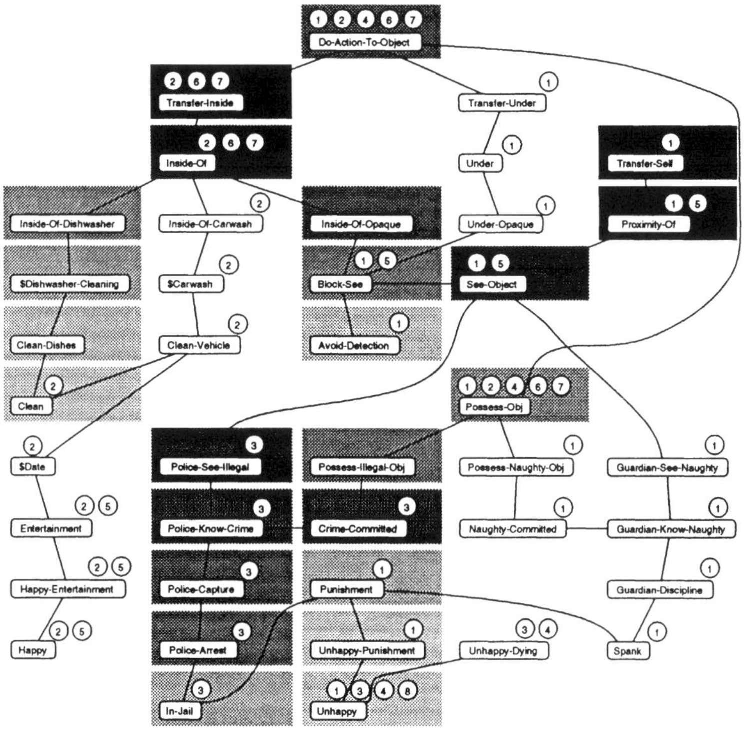

Figure 2. Overview of part of the network after activation has settled in processing **Hiding Pot**. Gray boxes represent level of evidential activation on the frames (darker = higher activation). Circles above frames indicate a long-term instance of that frame in an episode. Episodes understood and stored here: 1: **Dirty Magazine**. 2: *Fred put his car inside the car wash before his date with Wilma* (**Car Wash**). 3: *Jane shot Mark with a Colt-45. He died.* 4: *Betty wanted to smoke a cigarette, so she put it on top of the stove and lit it.* 5: *The pleasure boat followed the whales to watch them.* 6: *Barney put the flower in the pot, and then watered it.* 7: *Mike was hungry. He ate some fish.* 8: *Suzie loved George, but he died. Then Bill proposed to her. She became sad.*

This view includes the other frames instantiated by the propagation of signatures from Figure 1a and the clamped input for the remainder of **Hiding Pot** starting from Transfer-Self (*the police were coming*).

At the same time as signatures propagate to perform inferencing, activation spreads and accumulates along the bottom layer of conceptual nodes to disambiguate between those inferences. Initially the Inside-Of-Dishwasher path receives the most evidential activation because of feedback between it and its strong stereotypical connections to Cooking-Pot and Dishwasher. However, activation feedback between Inside-Of-Opaque and inferences from the police coming (Transfer-Self...Block-See) and the Police-Capture frames causes Inside-Of-Opaque to end up with more activation than Inside-Of-Dishwasher and Marijuana to end up with more activation than Cooking-Pot.

The network's final interpretation of **Hiding Pot** includes the most highly-activated path of frames in Figure 2 and their signature bindings. This interpretation includes the inferences that (a) Marijuana was inside of an opaque dishwasher (Inside-Of-Opaque) and has been blocked from sight (Block-See), (b) John possesses illegal marijuana (Possess-Illegal-Obj), and (c) John is in danger of being arrested by the police (Police-Arrest). Note that alternative interpretation paths retain activation for future possible reinterpretation, since REMIND uses a form of inhibition that normalizes activations rather than driving losers to zero. See Lange & Dyer (1989) and Lange (1992) for further details on how the network performs such inferencing and disambiguation for **Hiding Pot** and other inputs.

Memory Retrieval

In REMIND, memory retrieval occurs automatically as a side-effect of the spreading-activation understanding process. Representations of previously-understood episodes are connected directly to the semantic network that understood them in the first place. This direct form of "indexing" causes episodes that share many conceptual similarities with the cue to become active as REMIND interprets it. The most active (and hence most similar) episode gets chosen as the retrieved episode.

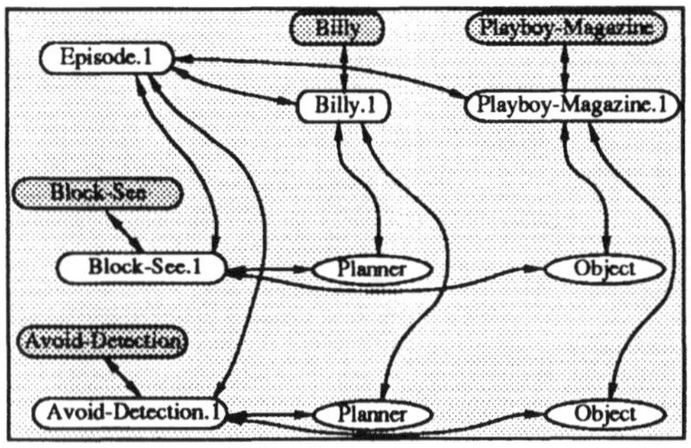

Figure 3. Part of Episode.1. Gray units are pre-existing concepts. White units are new episode units.

Network Encoding of Episodes

Whereas the general world knowledge used to initially build REMIND's networks is hand-coded, REMIND is not given any information about the particular episodes it processes and stores in long-term memory. The representations used for these target episodes are created entirely by REMIND's spreading-activation understanding process. Input for each episode's text is presented to the network, which infers an interpretation by the spread of signature and evidential activation. Next, units and connections are added to store the episode's entire resulting interpretation in the network. Thus, each episode's representation includes all aspects of its interpretation, from its disambiguated surface features (such as the actors and objects in the story) to the plans and goals that the network inferred that the actors were using.

As an example, consider how **Dirty Magazine** (*Billy put the Playboy under his bed so his mother wouldn't see it and spank him)* is processed and stored in the network as a memory episode. First, input for its phrases is clamped and an interpretation inferred, as described for **Hiding Pot**. As in **Hiding Pot**, the network infers that somebody is hiding something (Avoid-Detection) and that it is blocked from sight (Block-See). Here, however, the inferred signatures show that it is Billy hiding a Playboy-Magazine rather than John hiding Marijuana. Several other knowledge structures involved in **Hiding Pot** (e.g. Proximity-Of, Possess-Obj, Punishment) are also activated by **Dirty Magazine.** However, there are a number of differences, e.g. frames of the Guardian-Discipline structure are part of **Dirty Magazine**'s interpretation, but the Police-Capture frames are not. The rest of the frames activated as part of **Dirty Magazine**'s interpretation are shown by nodes that have a circled "1" above them in Figure 2. Other circled numbers represent elements of other stored episodes' interpretations.

To encode an episode after interpreting it, units are added to the network (by hand) for each of its interpretation's elements. Figure 3 shows a simplified part of the network's evidential layer after **Dirty Magazine** (Episode.1)'s interpretation has been added to the network. As can be seen, nodes have been added to represent the particular instances of Billy and Playboy-Magazine appearing in Episode.1. These nodes, Billy.1 and Playboy-Magazine.1, are connected with bidirectional connections to their respective frames. They are also connected to a node representing the entire episode (Episode.1). In addition, the fact that the network inferred that Billy had the goal of hiding the Playboy and that he caused it to be blocked from sight is encoded by nodes added for Avoid-Detection.1 and Block-See.1. Finally, each instance in the episode is connected to units representing their roles (e.g. the Planner and Object unit for Avoid-Detection.1), which are in turn connected to the concepts bound to them (e.g. Avoid-Detection.1's Planner is connected to Billy.1, and its Object is connected to Playboy-Magazine.1). All of the above connections have unit weight, with the exception of the connections from the episode units (e.g. Episode.1) to their elements, which have a small weight (0.05). The rest of the interpretation of each episode (e.g. the remaining parts of Episode.1 in Figure 2) is encoded similarly with units and connections that represent all of its instantiated frames and elements.

The Retrieval Process

With episodes understood and stored within the network, retrieval is performed simply by presenting an input cue to the network to be understood. Because the instance units representing episodes are connected directly to the normal evidential units, they become activated by the spread of signature and evidential activation. The more similarities an episode shares with the inferred interpretation of a cue, the more of its instances become active and the more activation its episode unit receives.

Figure 4 shows activations of the eight episodes from Figure 2 during understanding of **Hiding Pot**. Episode.6 (*Barney put the flower in the pot, and then watered it*) initially becomes highly active because it shares a number of surface features with **Hiding Pot** — e.g. both involve a Transfer-Inside, both have humans, and Planting-Pot is activated from the word *pot*. Similarly, Episode.2 and other episodes having varying degrees of shared features become active. However, as time goes on, the hiding and punishment frames are inferred and become active. Because of this, Episode.1 (**Dirty Magazine**)'s activation climbs and eventually wins, because it shares the most surface *and* abstract features of any episode with **Hiding Pot**'s interpretation (see Figure 2). **Dirty Magazine** is therefore retrieved as the episode most similar to **Hiding Pot.**

An example of how strongly the inferencing and disambiguation of the model affects retrieval is shown in Figure 5, which shows activations after presentation of input for *John put the pot inside the dishwasher because company was coming* (**Dinner Party**). Note that although this cue differs from **Hiding Pot** by only a single word (*company* instead of *police*), the interpretation REMIND reaches is completely different (i.e. that he was trying to clean a cooking pot to prepare for a dinner party). This causes a different episode to be re-

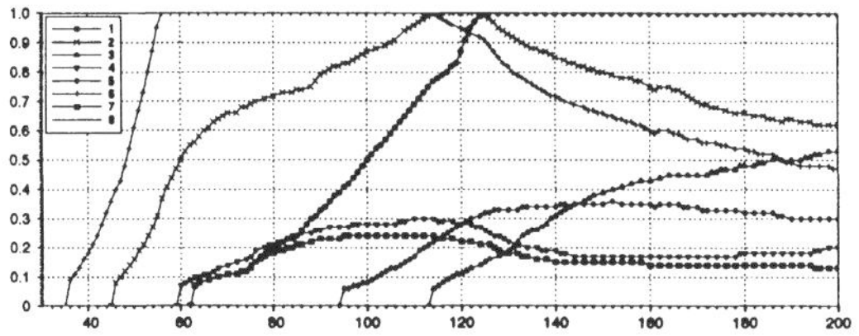

Figure 4. Episode unit activations for **Hiding Pot**.

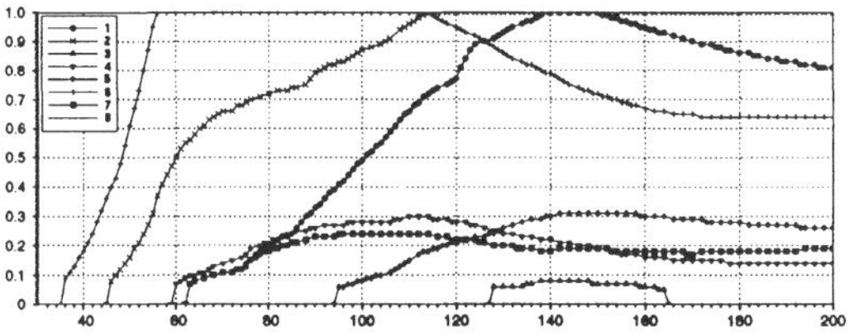

Figure 5. Episode activations for **Dinner Party**.

called, Episode.2, which shares the goals of cleaning something in preparation for an entertainment event.

As appears to be the case in human reminding, REMIND often retrieves superficially similar episodes to a cue even when a better analogy exists. For example, when REMIND has an episode in memory explicitly about smoking marijuana (such as *Cheech put the grass inside the bong because Chong was coming*), it retrieves it for **Hiding Pot** even though **Dinner Party** is a better analogy. In cases like this, REMIND often retrieves a superficially similar episode even though it has a different goal structure than the cue's interpretation and a better analogy exists in memory.

Discussion

Theoretically, REMIND lies somewhere between case-based reasoning models and general analogical retrieval models such as ARCS and MAC/FAC. Like ARCS and MAC/FAC, REMIND is meant to be a psychologically-plausible model of general human reminding, and therefore takes into account the prevalence of superficial feature similarities in remindings. However, we believe that many of the types of high-level planning and thematic knowledge structures used as indices in case-based reasoning systems also have an important effect on reminding. REMIND is thus partially an attempt to bridge the gap between case-based and analogical retrieval models. As it turns out, this gap is naturally bridged when the same spreading-activation mechanism is used to both understand cues and retrieve episodes from memory. Using the same mechanism for both processes causes retrieval to be affected by all levels that a text was understood with. This is the case in REMIND, in which the understanding mechanism is given the superficial features and actions of a text and attempts to explain them by inferring the plans and goals being used — causing memory episodes to be activated by both. This seems to give a more psychologically-plausible form of reminding than previous models, because the episodes it retrieves have varying degrees of superficial and abstract similarities to the cue, as seems to be the case in human reminding (Wharton et al., 1992). However, significant improvements in the network's language understanding abilities (see discussion in Lange, 1992) will have to be made before it can retrieve episodes of the complexity that some CBR models can.

A final aspect to note about REMIND is how its language understanding and retrieval processes come full circle. The episode retrieved depends crucially on the interpretation of the cue from the spreading-activation network's inferences. Once an episode is retrieved, it in turn primes the activation of the evidential spreading-activation network, perhaps leading to a different disambiguation and therefore interpretation of the next cue.

References

Ajjanagadde, V., and Shastri, L. 1989. Efficient Inference with Multi-Place Predicates and Variables in a Connectionist System. In Proceedings of the 11th Annual Conference of the Cognitive Science Society.

Barnden, J. 1990. The Power of Some Unusual Connectionist Data-Structuring Techniques. In J. A. Barnden and J. B. Pollack eds. *Advances in Connectionist and Neural Computation Theory*, Ablex.

Gentner, D., and Forbus, K. D. 1991. MAC/FAC: A Model of Similarity-Based Retrieval. *Proceedings of the 13th Annual Conference of the Cognitive Science Society.*

Hammond, K. 1989. *Case-Based Planning*. Academic Press.

Lange, T., and Dyer, M. 1989. High-Level Inferencing in a Connectionist Network. *Connection Science* 1:181-217.

Lange, T. 1992. Lexical and Pragmatic Disambiguation and Reinterpretation in Connectionist Networks. *International Journal of Man-Machine Studies* 36:191-220.

Miikkulainen, R., and Dyer, M. G. 1991. Natural Language Processing with Modular PDP Networks and Distributed Lexicon. *Cognitive Science* 15: 343-399.

Norvig, P. 1989. Marker Passing as a Weak Method for Text Inferencing. *Cognitive Science* 13: 569-620.

Riesbeck, C. K., and Schank, R. C. 1989. *Inside Case-Based Reasoning*. Hillsdale, NJ: Lawrence Erlbaum.

Riesbeck, C. and Martin, C. 1986. Direct Memory Access Parsing. In Kolodner and Riesbeck eds. *Experience, Memory, and Reasoning*, 209-226. Lawrence Erlbaum.

Schank, R. C. 1982. *Dynamic Memory*. New York: Cambridge University Press.

St. John, M. in press. The Story Gestalt: A Model of Knowledge Intensive Processes in Text Comprehension. *Cognitive Science.*

Thagard, P., Holyoak, K. J., Nelson, G., and Gochfeld, D. 1990. Analog Retrieval by Constraint Satisfaction. *Artificial Intelligence* 46:259-310.

Touretzky, D. and Hinton, G. 1988. A Distributed Connectionist Production System.*Cognitive Science* 12:423-466

Waltz, D., and Pollack, J. 1985. Massively Parallel Parsing: A Strongly Interactive Model of Natural Language Interpretation. *Cognitive Science* 9: 51-74.

Wharton, C., Holyoak, K., Downing, P., Lange, T., and Wickens, T. 1992. The Story with Reminding: Memory Retrieval is Influenced by Analogical Similarity. In Proceedings of the 14th Annual Conference of the Cognitive Science Society. Hillsdale, NJ: Lawrence Erlbaum.

A Model of the Role of Expertise in Analog Retrieval

L. Karl Branting
Department of Computer Science
University of Wyoming
Laramie, Wyoming 82071-3682
karl@eolus.uwyo.edu

Abstract

This paper presents a model of the use of expert knowledge to improve accuracy of analog retrieval. This model, match refinement by structural difference links (MRSDL), is based upon the assumption that expertise in domains requiring analogical reasoning consists in part of knowledge of the structural similarities and differences between some pairs of the source analogs. In an empirical evaluation on four data sets, MRSDL consistently retrieved the most similar or nearly most similar source analog. Achieving comparable accuracy on these data sets with a two-stage retrieval technique such as MAC/FAC would require exhaustive matching with more than half of the source analogs. The evaluation also showed that parallel competitive matching is often substantially faster than exhaustive matching or MRSDL.

Similarity in Analogical Reasoning

The terms "reasoning by analogy" and "case-based reasoning" subsume a variety of different problem-solving and learning activities. Common to all these activities, however, is attributing conclusions to a new situation based on its relevant similarity to some previous situation to which the same conclusions applied.

There is a consensus among researchers in analogical reasoning that *structural consistency* is a central component of similarity for the purposes of analogical reasoning (Winston, 1980; Gentner, 1983; Falkenhainer et al., 1989; Holyoak and Thagard, 1989). Two analogs are structurally consistent if objects in the two analogs can be placed into correspondence so that relations also correspond. This correspondence is generally modeled as a mapping from the objects in one analog (the *source*) to those in another (the *target*).

A number of factors have been identified that may influence the process of constructing a mapping from a source to a target analog. Holyoak *et al.* (Holyoak and Thagard, 1989) have stressed the role of *semantic similarity*, preference for mappings that put semantically similar objects and relations into correspondence, and *pragmatic centrality*, preference for mappings that are directly related to problem solving goals, in constraining the mapping process. Genter has emphasized *systematicity*, preference for mappings between "higher-order" relations, *i.e.*, those that take propositions as arguments, over first-order relations, *i.e.*, those that take objects as arguments. Other research, *e.g.*, (Faries and Reiser, 1990) and (Branting and Porter, 1991), has studied the effect of *elaboration* of the target analog, that is, inferring facts not explicit in the target analog. Finally, (Branting, 1991) and (Branting and Porter, 1991) illustrated use of general domain theory to reformulate a problem in a manner that can lead to improved structural consistency with its most similar analog. Following (Holyoak and Thagard, 1989), semantic similarity, pragmatic centrality, systematicity, target elaboration, and problem reformulation will be collectively referred to as *constraints* on the mapping process.

Methods for Analog Retrieval

The task of analog retrieval is to determine the potential source analog in memory that shares the greatest structural consistency with a target analog, or *probe*, under a given set of mapping constraints. The simplest approach to analog retrieval is exhaustive matching between a target analog and all potential source analogs in memory. However, exhaustive matching is psychologically implausible and computationally intractable for large knowledge bases.

Implemented alternatives to exhaustive matching include ARCS (analog retrieval by constraint satisfaction) (Thagard et al., 1990) and MAC/FAC (many are called but few are chosen) (Gentner and Forbus, 1991). Given a target probe, ARCS first finds a set of candidate source analogs that "in some degree" share semantic similarities with the probe. For each candidate analog, ARCS constructs a constraint network. A connectionist relaxation algorithm is then used to settle into a state that indicates the relative correspondence of the various stored structures to the probe under the given constraints. MAC/FAC is also two-stage model. A computationally inexpensive measure of surface similarity is used to retrieve an initial set of candidates. Exhaustive matching is then used to determine which of the candidates is structurally most similar to the probe.

ARCS and MAC/FAC both successfully account for the widely observed phenomenon that surface (*i.e.*, semantic) similarity is a stronger predictor of memory access in novices than structural consistency (Ross, 1989; Gentner, 1989) (although structural consistency is also a predictor of retrieval (Wharton et al., 1991)). There is reason to question, however, whether these approaches to retrieval are equally successful at modeling analog retrieval by experts. There is empirical evidence that experts are better than novices at using structurally similar analogs and are less prone to use analogs with misleading surface similarities (Novick, 1988). The hallmark of expertise in many fields is the ability to find the structurally most similar analog irrespective of surface differences. In legal reasoning, for example, the legal precedent most relevant to a given case may have very different facts. Skillful attorneys are adept at finding such precedents.

Modeling analogical retrieval in experts therefore requires showing how the most similar (or nearly most similar) source analog can be found without exhaustive search of memory. The difficulty of two-stage retrieval methods such as ARCS and MAC/FAC is in determining the size of the set of initial candidates. If the initial candidate set is too small, then the most similar analog may not be found. If the initial candidate set is too large, then exhaustive search of the candidate set will not be significantly less expensive than searching the entire library of analogs. When surface similarity is unreliable, a sufficiently poor choice of candidate set size can conceivably lead to the worst of both worlds: exhaustive search of a significant portion of the analog library that nevertheless fails to retrieve the most relevant analog. Improving upon the two-stage retrieval models requires showing how expert knowledge can improve retrieval accuracy.

One form of knowledge that experts can be expected to have and novices lack is knowledge of the structural similarities and differences between at least some pairs of the analogs in memory. Suppose, for example, that a law student is asked to analyze a hypothetical *H1*, and the student recalls a superficially similar precedent *P1*. Suppose that the student is then told that the controlling precedent is instead *P2* because of the greater structural similarities between *H1* and *P2*. To profit from this lesson, the student must understand the structural differences both between *H1* and *P1* and between *H1* and *P2* in order to appreciate that the former are greater than the latter. Perforce, the student must also understand the structural differences between *P1* and *P2* that led to the differences in their degree of structural similarity to *H1*. If on a later occasion the student encounters hypothetical *H2* that is superficially similar but structurally dissimilar to *P1*, the student can use knowledge of the structural differences between *P1* and *P2* to recover from the spurious match to *P1* and find the structurally most similar precedent *P2*.

The next section describes an algorithm that uses preexisting knowledge of structural differences among source analogs to recover from spurious matches.

Match Refinement by Structural Difference Links

In match refinement by structural difference links (MRSDL), an initial candidate is selected on the basis of surface semantic similarity. Precomputed information on the structural relations among analogs is then used to refine the match. Specifically, if the structural differences between an analog A_{cur} and a probe P have been determined, difference links containing precomputed information about the structural differences between A_{cur} and alternative analogs $A_1 \ldots A_n$ can be used to estimate inexpensively the similarity between P and each A_i. The idea behind this approach is that A_i is a better match to P than A_{cur} to the extent that A_{cur} differs from A_i and P in the same way. However, to the extent that A_i has additional differences from P, the match between A_i and P is worse. The most promising A_i is the case for which the differences with A_{cur} shared by A_i and P are greatest and the additional differences between A_i and P are least.

Consider a simple example involving the following brief narratives, represented in figure 1:

- **Probe.** John gave flowers to Mary because he likes her.
- **Analog-1.** Jimmy likes Billy because Billy gave him a snake.
- **Analog-2.** Bob gave flowers to Sally because he likes Sally's mother, Jane.

The highest degree of surface semantic similarity is between the probe and Analog-1: the probe and Analog-1 have identical relations, whereas Analog-2 has a relation, *mother*, not found in the probe. An initial retrieval based on surface semantic similarity would therefore favor Analog-1. However, Analog-1 differs structurally from the probe. The mapping that maximizes the structural congruence between the probe and Analog-1, Analog-1⇒Probe, is the following:

Billy→John
Jimmy→Mary
like2→like1

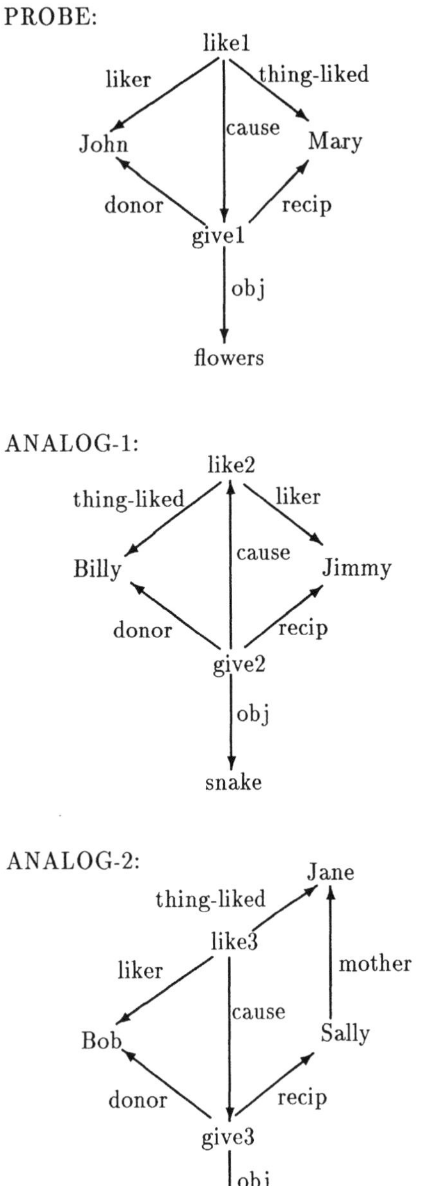

Figure 1: A probe and two analogs.

give2→give1
snake→flowers

Under this mapping, the following propositions in Analog-1 have no corresponding propositions in the probe:

(thing-liked like2 Billy)
(liker like2 Jimmy)
(cause give2 like2)

These unmatched propositions constitute a *difference* denoted Analog-1 – Dom(Analog-1⇒Probe) (where Dom(Analog-1⇒Probe) is the set of propositions having an image under Analog-1⇒Probe).

Assume that the following structural information concerning Analog-1 and Analog-2 has been precomputed:

- Analog-1⇒Analog-2, the mapping that maximizes the structural congruence between Analog-1 and Analog-2, and Dom(Analog-1⇒Analog-2), the Analog-1 propositions that have an image in Analog-2 under Analog-1⇒Analog-2.
- Analog-2⇒Analog-1, the mapping that maximizes the structural congruence between Analog-2 and Analog-1, and Dom(Analog-2⇒Analog-1), the Analog-2 propositions that have an image in Analog-1 under Analog-2⇒Analog-1.

Using this information, the number of propositions of Analog-2 that would have no image in the probe under the best mapping from Analog-2 to the probe can be estimated by the sum of the number of Analog-1 propositions having an image in Analog-2 but no image in the probe, *i.e.,* |Dom(Analog-1⇒Analog-2) – Dom(Analog-1⇒Probe)|, which in this case is zero, and the number of Analog-2 propositions that have no image in P under the composition of Analog-2⇒Analog-1 and Analog-1⇒Probe (*i.e.,* |Analog-2 – Dom(Analog-2⇒Analog-1 o Analog-1⇒Probe)|. In this case, Analog-2 – Dom(Analog-2⇒Analog-1 o Analog-1 ⇒ Probe) is the following:

(thing-liked like3 Jane)
(mother Sally Jane)

This is fewer than the three propositions in Analog-1 – Dom(Analog-1⇒Probe), so Analog-2 is a closer match to the probe than Analog-1.

The full algorithm for match refinement is as follows:

Given:

- P, a probe (*i.e.,* target analog)
- A_{cur}, the source analog that is currently the best match to P
- A_{cur} – Dom(A_{cur}⇒P), the propositions of A_{cur} that have no image in P under A_{cur}⇒P, the best mapping from A_{cur} to P

- Precomputed difference links between A_{cur} and cases $A_1 \ldots A_n$ containing the following information for each A_i:

  $A_{cur} \Rightarrow A_i$, the best mapping from A_{cur} to A_i, and Dom($A_{cur} \Rightarrow A_i$), the propositions of A_{cur} that have an image in A_i under $A_{cur} \Rightarrow A_i$

  $A_i \Rightarrow A_{cur}$, the best mapping from A_i to A_{cur}, and Dom($A_i \Rightarrow A_{cur}$), the propositions of A_i that have an image in A_{cur} under $A_i \Rightarrow A_{cur}$

Do:

1. Select the A_i for which $|A_i - \text{Dom}(A_i \Rightarrow P)|/|A_i|$, the proportion of propositions unmatched under the best mapping from A_i to P, is estimated to be least, where $|A_i - \text{Dom}(A_i \Rightarrow P)|$ is estimated by the cardinality of the following set:

   $\text{Dom}(A_{cur} \Rightarrow A_i) - \text{Dom}(A_{cur} \Rightarrow P) \bigcup A_i - \text{Dom}(A_i \Rightarrow A_{cur} \circ A_{cur} \Rightarrow P)$

2. Calculate the actual value of $A_i - \text{Dom}(A_i \Rightarrow P)$
3. If $|A_{cur} - \text{Dom}(A_{cur} \Rightarrow P)|/|A_{cur}| < |A_i - \text{Dom}(A_i \Rightarrow P)|/|A_i|$, then P matches A_{cur} better than any of the A_i's, so return A_{cur}. Otherwise, call the procedure again with A_i as the current best match.

As illustrated in (Branting, 1991), the composition of two best-mappings may fail to be itself a best mapping. Under these circumstances the algorithm may either over- or underestimate the true degree of structural difference between an analog A_i and a probe. As a result, difference-link refinement is a heuristic procedure.[1]

Comparison of MRSDL with Other Retrieval Techniques

To determine whether MRSDL represents an effective model of the use of expert knowledge in analog retrieval, a comparative evaluation was performed in which MRSDL was compared to three other retrieval techniques. The first alternative retrieval technique was exhaustive matching. The second technique was Best-First Incremental Matching (BFIM) (Branting, 1991). BFIM consists of best-first search of the space of partial mappings between each analog in memory and the probe. BFIM resembles ARCS in that it is a form of parallel matching involving competition among analogs (although in BFIM this competition doesn't consist of inhibition between competing match hypotheses, but merely of directing computational resources to the most promising match). The third technique was surface semantic retrieval. Degree of surface match was determined by the proportion of relations occurring in an analog that also occurred in the probe.[2]

These techniques were compared on four sets of analogs. The first two consisted of 100 fables and 26 plays (25 Shakespearean plays and West Side Story), generously provided by Paul Thagard, consisting of approximately 21 propositions per play and 55 propositions per play. The remaining two sets of analogs, taken from the worker's compensation law knowledge base of GREBE (Branting, 1991), consisted of 11 precedents of employment activities (averaging 29 propositions per case) and 10 precedents of near-miss noninstances of employment activities (averaging 30 propositions per case).

In each retrieval trial, the fables and plays were randomly divided into 5 or 3 (respectively) approximately equal partitions. Each analog of each partition was then used as a probe with the cases of the remaining partitions as analogs. Thus, each retrieval of each fable was tested using 80 other fables as analogs, and retrieval of each play was tested using 17 or 18 other plays as analogs. A set of 21 worker's compensation hypotheticals (averaging 89 propositions per case) were used to test retrieval of the instance and noninstance precedents of employment activities.

Before MRSDL could be run on each collection of analogs, some set of difference links had to be installed among them. The behavior of MRSDL depends heavily upon the configuration of difference links among analogs (Branting, 1991). For example, if there is no sequence of difference links connecting an initial surface match with the closest analog, then clearly no series of match refinements can retrieve the closest analog.

In this experiment, no effort was made to achieve an optimal configuration. Instead, a configuration of analogs connected by difference links was incrementally built up in a manner consistent with the scenario presented at the end of section 2: Each configuration was initialized with a single randomly selected analog. The remaining analogs were added in random order. For each new analog A, a difference link was installed between A and the superficially most similar analog, $SS(A)$. Exhaustive search was then used to determine the analog structurally most similar analog, $Ex(A)$. If $Ex(A)$ and

[1] To compensate for the possible inaccuracy of the estimate of the degree of structural difference between an analog A_i and a probe, the implementation of MRSDL described below modifies step 1 of the algorithm by selecting not only the analog A_i for which the estimated structural difference is least, but also all other analogs $A_j \ldots A_k$ whose estimated structural differences are within .05 of those of A_i. The actual closest structural match to the probe among $A_i, A_j \ldots A_k$ is then determined in step 2.

[2] Weighting the relations by their relative abundance in analogs was not found to increase retrieval accuracy.

Data Set	MRSDL		
	% exact	% close	comparisons
fables (80)	51.0	81.3	9.0
plays (17-18)	73.1	94.2	5.3
EA+ (11)	76.2	78.6	3.1
EA− (10)	78.6	92.9	2.5

Table 1: The proportion of MRSDL retrievals that were identical to the best match as determined by exhaustive match, the proportion of retrievals that returned an analog whose degree of match was within 5% of the closest analog, and the average number of structural comparisons required in each of the data sets. "EA+" and "EA−" represent instances and near-miss noninstances of employment activities, respectively.

$SS(A)$ were distinct, then difference links were installed between $Ex(A)$ and $SS(A)$, and between A and $Ex(A)$. This approach was chosen because the number of difference links required is linear in the number of analogs and because the approach is consistent with a plausible scenario for acquiring knowledge of structural relations among analogs. A distinct configuration of difference links was constructed for each set of partitions used as source analogs.

In each of the retrieval approaches (except surface semantic retrieval) structure matching was performed by the best-first algorithm described in (Branting, 1991) running in greedy mode. To isolate the task of finding the structurally most similar analog from the contribution of various mapping constraints and to expedite the trials, the algorithm was run with information concerning semantic similarity among relations and case elaboration rules removed. Degree of structural similarity was measured by the proportion of propositions in the source analog that have an image in the target under the mapping that maximizes structural congruence.

Table one sets forth the performance of MRSDL averaged across four trials. The first column sets forth the proportion of MRSDL retrievals for each data set that were identical to the best match as determined by exhaustive search.[3] There are often several analogs having an almost identical degree of match with a probe. The second column sets forth the proportion of MRSDL retrievals that returned an analog whose degree of match was within 5% of that of the closest analog found by exhaustive search. The third column contains the average number of structural comparisons required for each

[3]A separate comparison with BFIM was unnecessary because BFIM always finds the same match as exhaustive search

Data Set	Surface Similarity		
	$AV_{p \epsilon Probes}$ *Min-exact*$_p$	$MAX_{p \epsilon Probes}$ *Min-exact*$_p$	$MAX_{p \epsilon Probes}$ *Min-dl*$_p$
fables	14.9	71.3	68.0
plays	4.3	12.5	8.5
EA+	3.3	9.0	9.0
EA−	3.1	6.0	6.0

Table 2: Min-exact$_p$ is the minimum number of candidates that must be retrieved by surface similarity to insure that the analog closest to probe p is in the candidate set. Min-dl$_p$ is the smallest candidate set size guaranteed to contain an analog whose degree of match is at least as great as the degree of match of the analog returned by MRSDL.

MRSDL retrieval.

The first two columns of table two contain information concerning *Min-exact*$_p$, the minimum number of candidates that must be retrieved by surface similarity to insure that the analog closest to probe p is in the candidate set.[4] The first column sets forth the average of *Min-exact*$_p$ for all probes p in each data set. This represents the average number of candidates that would be necessary for two-stage retrieval if one somehow knew *Min-exact*$_p$ for every probe p. The second column sets forth the maximum of *Min-exact*$_p$ for all probes p. This represents the smallest candidate set size that would guarantee for all probes that the candidate set would contain the best analog. The last column represents the maximum of *Min-dl*$_p$,[5] the smallest candidate set size guaranteed for all probes p to contain an analog whose degree of match at least as great as the degree of match of the analog returned by MRSDL.

Table three sets forth the average retrieval time in seconds of user CPU time for exhaustive search, BFIM, and MRSDL.

Discussion

Table one shows that MRSDL performs reasonably well, although not infallibly. In the fable and employment activity noninstance data sets MRSDL was over 90% accurate in retrieving analogs that were within 5% of the optimal match. Table two illustrates the shortcomings of two-stage retrieval. Although the average value of *Min-exact*$_p$ was comparable to the average number of structural comparisons performed by MRSDL, each data set con-

[4]If $Ex(p)$ is the analog found by exhaustive search of a given analog set with probe p and $SS(p, n)$ is the set of n closest surface matches to p, then *Min-exact*$_p$ = $\min\{n | Ex(p) \in SS(p, n)\}$.

[5]If $Dl(p)$ is the analog found by MRSDL, *Min-dl*$_p$ = $\min\{n |\ SS(p, n)$ contains some analog that matches p at least as well as $Dl(p)\}$.

Data Set	Exh.	BFIM	MRSDL
fables	10.7	2.9	5.0
plays	5.8	4.7	6.3
EA+	4.1	2.6	1.7
EA−	2.9	1.4	1.1

Table 3: Average retrieval times (in seconds of user CPU time) for exhaustive search, BFIM, and MRSDL.

tains some probe p for which $Min\text{-}exact_p$ is at least half the size of the data set. Thus, on these data sets at least, no two-stage retrieval scheme can simultaneous insure correctness and search less than half of the analogs in memory. The last column of table two illustrates that the smallest initial candidate set size guaranteed to equal the accuracy of MRSDL is at least half the size of the analog library.

Table three illustrates that MRSDL is usually substantially faster than exhaustive matching. The surprising exception was in the plays data set where MRSDL was actually slower than exhaustive matching. BFIM was consistently faster than exhaustive matching, more than three times as fast in the fable data set. Surprisingly, BFIM was also faster than MRSDL in two of the data sets. Parallel competitive matching has been criticized on grounds of psychological implausibility (Gentner and Forbus, 1991), but these data suggest that this retrieval technique can be relatively efficient.

Conclusion

This paper has presented a model of the use of expert knowledge to improve the accuracy of analog retrieval. This model, match refinement by structural difference links (MRSDL), is based upon the assumption that expertise in domains requiring analogical reasoning consists in part of knowledge of the structural similarities and differences between some pairs of the source analogs. In an empirical evaluation on four data sets, MRSDL generally found the most similar or nearly most similar source analog. Achieving comparable accuracy on these data sets with a two-stage retrieval technique such as MAC/FAC would require exhaustive matching with more than half of the source analogs.

References

Branting, L. K. (1991). *Integrating Rules and Precedents for Classification and Explanation: Automating Legal Analysis.* PhD thesis, University of Texas at Austin.

Branting, L. K. and Porter, B. W. (1991). Rules and precedents as complementary warrants. In *Proceedings of Ninth National Conference on Artificial Intelligence*, Anaheim. AAAI Press/MIT Press.

Falkenhainer, B., Forbus, K., and Gentner, D. (1989). The structure-mapping engine: Algorithm and examples. *Artificial Intelligence Journal*, 41(1).

Faries, J. M. and Reiser, B. J. (1990). Terrorists and spoiled children: Retrieval of analogies for political arguments. To appear in *Proceedings of the AAAI Spring Symposium on Case-Based Reasoning*, Palo Alto, California, March 27-29.

Gentner, D. (1983). Structure mapping: A theoretical framework for analogy. *Cognitive Science*, 7(2):155–170.

Gentner, D. (1989). Finding the needle: Accessing and reasoning from prior cases. In *Proceedings of the Second DARPA Case-Based Reasoning Workshop.* Morgan Kaufmann.

Gentner, D. and Forbus, K. (1991). MAC/FAC: A model of similiarity-based retrieval. In *Thirteenth Annual Conference of the Cognitive Science Society*, pages 504–509.

Holyoak, K. and Thagard, P. (1989). Analogical mapping by constraint satisfaction. *Cognitive Science*, 13(3).

Novick, L. (1988). Analogical transfer, problem similarity, and expertise. *Journal of Experimental Psychology*, 14:510–520.

Ross, B. (1989). Some psychological results on case-based reasoning. In *Proceedings of the Second DARPA Case-Based Reasoning Workshop.* Morgan Kaufmann.

Thagard, P., Holyoak, K., Nelson, G., and Gochfeld, D. (1990). Analog retrieval by constraint satisfaction. Technical Report CSL-Report 41, Princeton University.

Wharton, C., Holyoak, K., Downing, P., Lange, T., and Wickens, T. (1991). Retrieval competition in memory for analogies. In *Thirteenth Annual Conference of the Cognitive Science Society*, pages 528–533.

Winston, P. H. (1980). Learning and reasoning by analogy. *Communications of the ACM*, 23(12).

The Story with Reminding: Memory Retrieval is Influenced by Analogical Similarity

Charles M. Wharton[1], Keith J. Holyoak[1], Paul E. Downing[1], Trent E. Lange[2], and Thomas D. Wickens[1]
[1]Dept. of Psychology and [2]Dept. of Computer Science, University of California, Los Angeles, CA 90024

Abstract*

AI models of reminding (ARCS, MAC/FAC) that predict that memory access is influenced by analogical similarity are tested. In Experiment 1, subjects initially studied a set of 12 target stories. Later, subjects read 10 other cue stories and were asked to write down the stories they were reminded of from the first set. Cue stories were associated with either an analogous and disanalogous target (competition condition), an analogous target (singleton condition), or a disanalogous target (singleton condition). An effect of analogical similarity was found only in the competition condition. Experiment 2 used the same design but targets and cues were simple subject-verb-object sentences. Cue sentences shared similar nouns and verbs with target sentences. Materials were constructed such that associated nouns either consistently mapped or cross-mapped between cues and targets. Consistent-mapped sentences were recalled more than cross-mapped sentences in both conditions. Issues for future research are addressed.

Introduction

One of the central issues in analogical reasoning theory concerns the degree analogical similarity affects reminding (e.g., McDougal, Hammond, & Seifert, 1991). This issue is important because what one is reminded of in problem-solving affects further action. If one is reminded of a analogous problem in which a similar goal was solved, the plan that achieved that goal can be used to solve the current problem. However, if one is reminded of past situations on the basis of superficial resemblances, little of the knowledge associated with that reminding can be used to achieve the current goal. This question remains unanswered. In the present experiments, we explore the retrieval conditions that are necessary to demonstrate analogical reminding.

*Preparation of this article was supported by Contract MDA 903-89-K-0179 from the Army Research Institute and a Keck Foundation grant and NSF Grant DIR-9024251 to the UCLA Cognitive Science Research Program. We thank Mary Jo Ratterman and Dedre Gentner for assistance with Experiment 1.

Computer Models of Reminding

The extent to which reminding theories assert that memory retrieval is influenced by analogical similarity appears to be partially a function of the amount of domain expertise assumed about the reasoner. Some case-based reasoning (CBR) models implicitly represent reasoning done within a domain of expertise (see review in Reisbeck & Schank, 1989). As such, memory access in these systems is determined mostly by the plan or goal similarity between the current problem and cases in memory. Case retrieval will be only minimally based on surface similarity. In contrast, models of more general reminding such as ARCS (Thagard et al., 1990) and MAC/FAC (Gentner & Forbus, 1991) or assert that memory retrieval is influenced by surface (i.e., cue/target lexical overlap) as well as analogical similarity.

In ARCS theory, reminding is governed by three types of constraints: direct semantic similarity of concepts, isomorphism (consistent mapping of predicates and arguments), and pragmatic centrality (problem-solver's goals) Retrieval first proceeds with a search of all symbolic representations in memory to find targets that overlap semantically with propositions in the cue. Second, a connectionist "mapping network" is formed to represent competing potential mappings that are created between the cue and semantically-related episodes in memory. The connections of this mapping network form pressures on reminding. Excitatory and inhibitory connections embody the three constraint types. Finally, a connectionist process of this mapping network produces retrieval of the episode(s) in memory that best satisfy these constraints.

MAC/FAC theory is similar to ARCS in that retrieval is based on semantic and structural overlap. Computationally, retrieval is a two-step process. In the first stage (MAC), the episode in long-term memory that has the most surface commonalities with the probe story is retrieved. Stories are represented in MAC as vectors, where each element represents a word. Similarity is computed by taking the dot product between the probe and each story in memory. Any other stories whose dot product is within 10% of the best match are also retrieved. The second stage (FAC) computes how well each retrieved first stage story matches the cue, based on common relational structure and object descriptions. Stories are represented in FAC in predicate calculus form. The episode with the highest match, along with any story within 10% of the best match, is retrieved.

The predictions of ARCS and MAC/FAC seem relatively simple: Both semantic and relational similarity should influence reminding. However, actually demonstrating the affect of relational similarity empirically has proven difficult.

Empirical Studies of Reminding

Analogical reminding has been directly examined by Ratterman and Gentner (1987). Here, subjects initially read a number of stories. Two weeks later subjects read more stories and were asked to write down any stories that they were reminded of from the previous session. Ratterman and Gentner varied object attribute similarity and higher-order relational similarity in their materials. Crossing these these two types of similarity resulted in four different cue/target similarity matches, (a) shared object attributes and higher-order relations (literal similarity), (b) shared object attributes only (mere appearance), (c) shared higher order relations only (true analogy), and (d) no shared object attributes or higher-order relations (false analogy). No cue story was ever matched to more than one target story. The only reliable differences found were advantages for the literal-similarity and mere-appearance conditions relative to the true-analogy and false-analogy conditions (recall proportions were 0.56, 0.53, 0.12, and 0.09, respectively). Ratterman and Gentner concluded that reminding is primarily, though not exclusively, influenced by object or "surface" similarity (see also Gentner & Forbus, 1991; Gentner & Landers, 1985; Seifert et al., 1986).

Why might the role of analogical similarity in reminding be so difficult to demonstrate? An answer can be found by examining computational models of analogical and case-based retrieval. Most such reminding models attempt to retrieve at least the single *best* or most similar target that can be accessed semantically for a given cue. The most similar target in memory is usually considered to be one that which is both analogous to the cue *and* shares significant semantic overlap with it. If there are no analogous targets that are semantically related to the cue, then a semantically-related non-analogous target will be

Table 1. Example of One Materials Set (Lexically-Associated Cue and Targets) from Wharton et al. (1991)

Consistent Target
Having just been fired from a high level job, he decided to go to his church for counseling. *The pastor calmed the businessman.*

Inconsistent Target
The church was having trouble approaching local corporations for contributions to the shelter. *The executive soothed the priest.*

Target-Cue Sentence
The rabbi reassured the chairman.

considered to be the best match, and therefore be retrieved. Accordingly, most contemporary models of analogical and case-based reminding would predict only small effects of cue/target analogical similarity in experiments in which each cue is semantically associated to a *single* target (a *singleton design*). This is because a single related target is likely to be retrieved regardless of whether it is analogous or not. This, in fact, is what previous studies of analogical reminding, all of which have used singleton retrieval designs, have generally found, .

In Wharton et al. (1991), we demonstrated an effect of analogical similarity on reminding by using a design in which each cue was semantically associated with *both* a consistently- and inconsistently-mapped target in memory (a *competition design*). An example of our materials and design is shown in Table 1. Targets were short texts built around a single sentence whose roles either mapped consistently with those of the cue (e.g., *rabbi/pastor*, *chairman/businessman*), or mapped inconsistently with those of the cue (e.g., *rabbi/executive*, *chairman/priest*).

A schematic of our design is shown in Figure 1. Subjects saw items in both competition (left boxes) and singleton conditions (right boxes). Wharton et al. found an overall effect of analogical similarity on reminding (in the form of consistent object-level mappings). However, the effect of analogical similarity was much larger in the competition condition than in the singleton condition.

Figure 1. Design of Wharton et al. (1991)

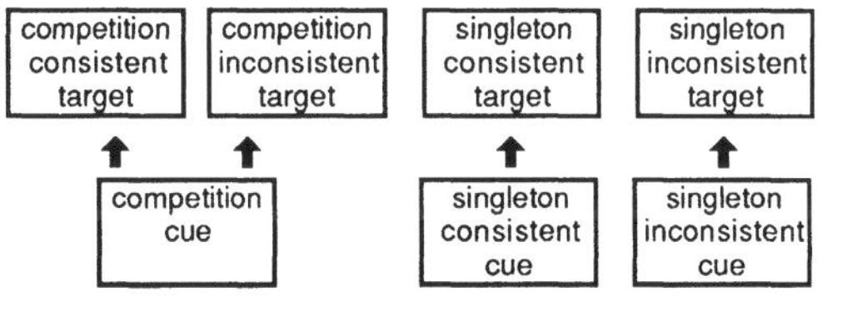

Table 2. Example of a Materials Set Used in Experiment 1 (after Seifert et al., 1986).

Set A, Theme 1: Ernie was really encouraged about his interview for a security guard at the new factory in town. He thought he was saved. Ernie went to the shopping mall, hunted for a dark blue security guard uniform, and bought several. The next day he received a phone call from the factory personnel office about the security guard position. Ernie was dismayed that he had wasted money. He didn't have a job.

Set A, Theme 2: Dan wasn't working and he was very concerned because he had very little left in his bank account. Several days later he had lunch with the chairman about becoming a broker. Dan thought he had made a good impression when he gave his resume to the investment partnership. Dan went to the department store and tried on some suits, and got a few. He felt that he was moving up again.

Set B, Theme 1: Ronnie thought she had it made because she thought she had done well in the audition for a musician. Ronnie went to the music showroom, played some electronic keyboards, and then purchased one. Later she got a message from the guitarist about her playing keyboards. She wasn't in a band. Ronnie was dejected that she had run up her credit card.

Set B, Theme 2: Pam was worried she that she had blown her savings. She was between jobs. Pam was really excited about her tryout as a dancer for a new musical. That evening she met the director about the dancer role. Pam got over to some stores, searched for, and bought some leotards. She believed her troubles were over.

Experiment 1

The view we took in Wharton et al. (1991)—that our results with varied object (or role)-level mappings generalized to mean that analogy influences reminding—could be disputed. Although our materials varied the consistency of object-level mappings, we never manipulated the consistency of higher-order relational mappings. As can be observed with the materials set in Table 1, *reassured* consistently maps to both *calmed* and *soothed.* That is, there is no difference in higher-order relational consistency between the consistent and inconsistent targets. Thus, our previous experiments did not constitute a complete test of the effect of analogical similarity on reminding.

We constructed new set of stories in which cue/target relational consistency has been varied. An example of a story set is shown in Table 2. Here, the same underlying set of propositions are used to construct two different story lines, *counting your chickens before they're hatched* (theme 1), and *finding desperately-needed employment* (theme 2). No content words are used in more than one story and the characters and objects used in set A that map to those in set B are not closely related semantically (e.g., stock broker and dancer). Thus, the paired stories represent analogous rather than literally similar instantiations of the same common theme.

In the present study, subjects rated 12 stories of this type for imageability. Later, subjects read 10 more stories and were asked to write down any of the rated stories of which they were reminded. We used the same design as in Wharton et al. (1991). The event-related cue and target stories were either instantiations of the same theme (analogous) or of different themes (disanalogous). We predicted that analogous stories would be retrieved more than disanalogous stories, and that this difference would be greater in the competition than in the singleton condition.

Method

Materials. Materials consisted of 14 sets of 4 stories (see Table 2). Some stories were derived from Ratterman and Gentner (1987) and Seifert et al. (1986). Within each set of stories, we constructed one core set of propositions. Two different unique story plots were created for each story set by rearranging the sequence of propositions. To avoid having the surface order of propositions covary with analogical similarity, event sequences shared between thematically similar stories were changed as much as was possible without altering the underlying story plot. Each story within a set was written about a different set of actors such as roommates, countries, or siblings. No content words or proper names were used in more than one story across the entire set of materials.

In order to determine if, when reading our materials, people are sensitive to the factors we manipulated, 20 undergraduates attending the University of California, Los Angeles (UCLA), completed questionnaires designed to assess the perceived similarity of the cue and target stories. Subjects were asked to rate "...how similar are the scenes being described" on a 6-point Likert scale (range: 1, completely dissimilar to 6, completely similar). Analogous story pairs were rated more similar than disanalogous story pairs (4.70 vs. 3.40), $F(1, 19) = 34.86$, $p < .0001$; disanalogous story pairs were rated more similar than unrelated story pairs (2.00), $F(1, 19) = 67.37$ $p < .0001$.

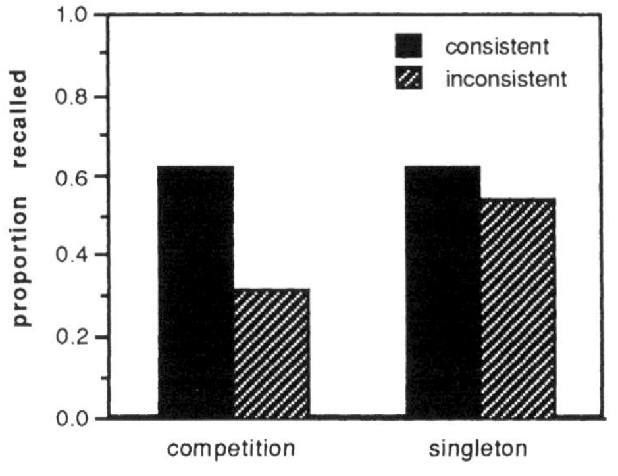

Figure 2. Proportion of remindings for each condition of Experiment 1.

Two supersets of materials were created by assigning the stories in each of story sets to one of two groups. The two stories from each story set assigned to each superset contained dissimilar themes. An equal number of cues and targets came from each superset. In order to have all stories appear in all conditions, we created 14 separate configurations of cue and target stories. Each configuration of materials was administered to three subjects.

Target and cue stories were made into separate booklets. A different random order of targets and cues was created for each subject. Each booklet of target stories contained 12 stories (2 stories each in the competition analogy, competition disanalogy, singleton analogy, singleton disanalogy conditions, and 4 stories not related to any cue). Each booklet of cue stories contained 10 stories (2 stories each in the competition, singleton analogy, singleton disanalogy conditions, and 4 stories not related to any target).

Procedure. Subjects were 42 UCLA undergraduates who participated either for pay or in order to meet requirements of one of several psychology survey courses.

During the encoding portion of the experiment, subjects rated the target stories on a 6-point Likert scale for imageability. Subjects were not informed that they would later try to recall these stories. Subjects were given a 5 min distractor task, after which they were asked to read the cue stories and write down anything that they were reminded of from the stories that they had previously rated.

Data Analysis. Reminding protocols were scored with respect to which story was accessed in response to a given cue. For each separate attempt at recalling a story (i.e., each attempt to report what the subject considered a single passage), credit for access was given to whichever passage had content words recalled; if content words from two passages were included, credit was given to the passage that contributed the most content words. If subjects wrote down content words from two passages in *separate* retrieval attempts in response to a single cue, access credit was given for each. Because synonym substitutions could be confused with interchanges of hyponyms across paired passages, a criterion of literal recall had to be used in scoring content words.

In order to be able to generalize our findings beyond the specific materials we created, the conservative min F' formulation of analysis of variance was calculated for all tests of mean differences (Clark, 1973). One story from each of the 28 story themes was used in every condition. Thus, there were 28 observations per condition in the item ANOVAs.

Results and Discussion

The proportion of target story types of which subjects were reminded by cue stories is shown in Figure 2. Subjects recalled more analogous (consistent) stories than disanalogous (inconsistent) stories, min F' (1, 54) = 6.29, $p < .05$. The interaction between competition and analogical similarity was not reliable, min F' (1, 54) = 2.39, $p > .10$; however, both the subject and item ANOVAs were significant, $p < .05$ (F (1, 28) = 4.38, $F(1, 27) = 5.27$). As predicted, there was more access of analogous stories than disanalogous stories within the competition condition, min F' (1, 54) = 6.76, $p < .05$, but not within the singleton condition, min $F' < 1$.

These findings strongly support the claims of general reminding models such as ARCS and MAC/FAC that analogical similarity influences memory. As in Wharton et al. (1991), an effect of analogical similarity was only obtained with a competition design. This would seem to indicate that the effect of higher-order cue/target relational consistency on reminding is not very strong. Thus, our results also imply that direct semantic similarity of individual concepts (objects and predicates) dominates reminding for novice subjects.

Experiment 2

In Experiment 1, we demonstrated an effect of analogical similarity with story materials. Wharton et al. (1991) showed a similar effect with target passages consisting of several sentences that described a single scene (see Table 1). Thus, in the present Experiment 1 and Wharton et al. (1991), target materials consisted of multi-sentence scene or story descriptions. It is likely that the text contexts encouraged subjects to make inferences that augmented their text representations and therefore increased the chance that analogous targets would recalled more than disanalogous targets.

The passages in Table 1 can be used to illustrate how sentential context might make readers add additional knowledge to their text representations. For the consistent target in Table 1, it is natural to infer from the surround statement, *Having just been fired from a high level job...*, that the businessman went to the pastor about a personal problem, and that the priest helped the businessman to deal with this problem. For the inconsistent target, the surround statement, *The church was having trouble approaching local corporations for contributions...* might lead to the inference that the priest went to the executive about a financial problem, and that the executive helped the priest to deal with it. The cue sentence, *The rabbi reassured the chairman*, although not itself embedded in a text context, seems more likely to elicit inferences that parallel those generated when reading the consistent target, *The pastor calmed the businessman*, than those generated when reading the inconsistent target, *The executive soothed the priest....* Thus, additional inferences triggered by the surrounding text context should increase semantic and structural overlap between subjects' representations of the cue sentence and of the consistent target, relative to that of the cue with the inconsistent target.

It is possible that the inferences that readers create from multi-sentence scene descriptions are necessary for analogy and consistency effects in reminding. This view contrasts with that of ARCS and MAC/FAC, which predict that targets with consistent object mappings will be recalled more than targets with inconsistent object mappings, even in the absence of differential inferences. In order to test these contrasting predictions, single sentence targets *without* any surrounding text context (e.g., *The pastor calmed the businessman*) were presented to subjects during initial encoding.

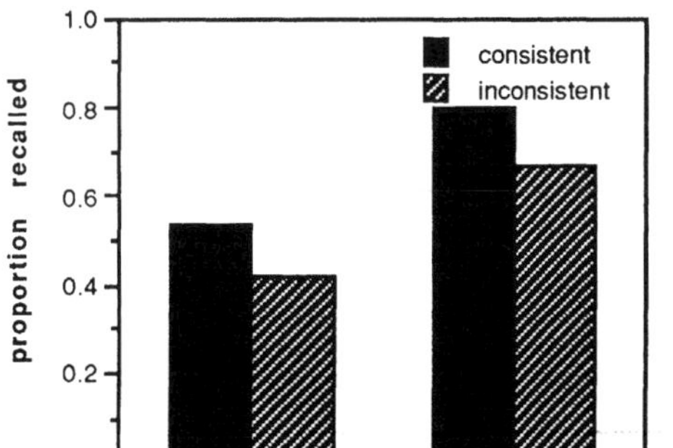

Figure 3. Proportion of remindings for each condition of Experiment 2.

Method

Materials consisted of 24 sets of sentences. Each set consisted of two target passages and one cue sentence. Both target sentences were related to the set's cue sentence. The matched target and cue sentences shared two sets of associated nouns (e.g., pastor, priest, rabbi; businessman, executive, chairman) and a single set of associated verbs (e.g., calm, soothe, reassure). The nouns and verbs within a set were chosen so that the nouns would jointly make sense in either the object or subject position. Within each passage pair, verbs were randomly assigned to pairs of target nouns, after which each target sentence was randomly assigned to one of the two passages. In order to avoid confounding cue/target consistency with surface order of the noun hyponyms, an equal number of active and passive cue sentences and target sentences were constructed in each condition. Random assignment was used to decide whether cue and target sentences would be active or passive and which target passage would be analogously cued.

The methodology for similarity ratings was virtually identical to that Experiment 1. Subjects were 96 UCLA undergraduates. Consistent pairs were rated more similar than inconsistent pairs (4.10 vs. 3.12), min F' (1, 56) = 19.76, $p < .001$. Inconsistent pairs were rated more similar than unrelated pairs (2.07), min F' (1, 51) = 19.11, $p < .001$.

There were 24 different configurations of cue and targets sentences. Each configuration was administered to three subjects. The design of target and cue booklets was identical to that in Experiment 1.

Subjects were 72 UCLA undergraduates who participated either for pay or in order to meet requirements of an introductory psychology course. We used the same design and procedure as in Experiment 1 except that subjects made three separate ratings passes, respectively, for plausibility, meaningfulness, and imageability. Subjects were given 20 s to read and rate each sentence during each pass through their target sentence booklets.

All 48 target sentences appeared in all conditions. Consequently, we treated each target as an observation for the item ANOVAs (i.e., n = 48 observations). We used Experiment 1's scoring rules.

Results and Discussion

The proportion of target sentences retrieved is shown in Figure 3. Although the difference is relatively small, access was more likely with structurally consistent targets than with structurally inconsistent targets, min F' (1, 93) = 5.76, $p < .05$. There was less access of each target type in the competitor condition than in the singleton condition, min F' (1, 92) = 33.39, $p < .001$.

That consistent SVO targets were recalled more than inconsistent SVO targets is an important finding for theories of analogical reminding. Target passages were not seen with a surrounding context which could have biased subjects' interpretations of target sentences and so accounted for our previous results. This finding implies that even minimal configural differences between episodes in memory will influence memory retrieval, as claimed by ARCS and MAC/FAC.

Future Directions

There are several general areas in which investigation would contribute much to theories of analogical reminding. One possible weakness with our studies concerns the short time delay that we have always used, 5 min. CBR systems generally model retrieval of episodes long after they have been experienced. Thus, the effect of time delay on reminding with story material is a critical area in need of investigation.

Perhaps the most pressing area for future research on case-based reminding is on the differences between expert and novice remindings. Such work is necessary to resolve the claims of some CBR models relative to those of general reminding systems such as ARCS and MAC/FAC.

Finally, future work needs to address the role of comprehension and inference processes in reminding. ARCS, which uses only taxonomic relations between individual concepts as retrieval paths, lacks any capacity to infer, for example, that the overall theme of a passage is "retaliation". As a result, the model is unable to use such implicit abstractions to guide retrieval by indexing episodes with similar abstract themes. Accordingly, ARCS might have difficulty modeling Experiment 1. We have addressed this problem with REMIND (Lange & Wharton, 1992), a structured connectionist model that integrates text inferencing and episodic reminding. Future experimental studies should address the questions of *why* structural and analogical similarity affect remindings — do syntactic structural isomorphisms play a role in retrieval in and of themselves, or can the entire effect of analogical similarity be explained by inferred similarities produced in the course of understanding the theme of a story?

References

Clark, H. H. 1973. The Language-as-Fixed-Effect Fallacy: A Critique of Language Statistics in Psychological Research. *Journal of Verbal Learning and Verbal Behavior* 12: 335-359.

Gentner, D., and Forbus, K. D., 1991. MAC/FAC: A Model of Similarity-Based Retrieval. In Proceedings of the Thirteenth Annual Conference of the Cognitive Science Society, 504-509. Hillsdale, NJ: Lawrence Erlbaum.

Gentner, D., and Landers, R., 1985. Analogical Reminding: A Good Match is Hard to Find. In Proceedings of the International Conference on Systems, Man and Cybernetics, 607-613. Tucson, AZ.

Lange, T., and Wharton, C., 1992. REMIND: Integrating Language Understanding and Episodic Memory Retrieval in a Connectionist Network. In Proceedings of the Fourteenth Annual Conference of the Cognitive Science Society. Hillsdale, NJ: Lawrence Erlbaum.

McDougal, T., Hammond, K., and Seifert, C., 1991. A Functional Perspective on Reminding. In Proceedings of the Thirteenth Annual Conference of the Cognitive Science Society, 510-521. Hillsdale, NJ: Lawrence Erlbaum.

Ratterman, M. J., and Gentner, D., 1987. Analogy and Similarity: Determinants of Accessibility and Inferential Soundness. In Proceedings of the Ninth Annual Meeting of the Cognitive Science Society, 22-34. Hillsdale, NJ: Lawrence Erlbaum.

Reisbeck, C. K., and Schank, R. C., 1989. *Inside Case-Based Reasoning*. Hillsdale, NJ: Lawrence Erlbaum.

Seifert, C. M., McKoon, G., Abelson, R. P., and Ratcliff, R. 1986. Memory Connections Between Thematically Similar Episodes. *Journal of Experimental Psychology: Human Learning and Memory* 12: 220-231.

Thagard, P. Holyoak, K. J. Nelson, G. and Gochfeld, D., 1990. Analog Retrieval by Constraint Satisfaction. *Artificial Intelligence* 46: 259-310.

Wharton, C. M., Holyoak, K. J., Downing, P. E., Lange, T. E., and Wickens, T. D., 1991. Retrieval Competition in Memory for Analogies. In Proceedings of the Thirteenth Annual Conference of the Cognitive Science Society, 528-533. Hillsdale, NJ: Lawrence Erlbaum.

Abductive Explanation of Emotions

Paul O'Rorke*
Phone & Fax: (714) 854-2894
E-Mail: ororke@ics.uci.edu
Department of Information and Computer Science
University of California, Irvine
Irvine, CA 92717

Andrew Ortony†
Phone: (708) 491-3500
E-Mail: ortony@ils.nwu.edu
Institute for the Learning Sciences
Northwestern University
1890 Maple Avenue
Evanston, IL 60201

Abstract

Emotions and cognition are inextricably intertwined. Feelings influence thoughts and actions which in turn give rise to new emotional reactions. We claim that people infer emotional states in others using common-sense psychological theories of the interactions between emotions, cognition, and action. We have developed a situation calculus theory of emotion elicitation representating knowledge underlying common-sense causal reasoning involving emotions. We show how the theory can be used to construct explanations of emotional states. The method for constructing explanations is based on the notion of abduction. This method has been implemented in a computer program called AMAL. The results of computational experiments using AMAL to construct explanations of examples based on cases taken from a diary study of emotions indicate that the abductive approach to explanatory reasoning about emotions offers significant advantages. We found that the majority of the diary study examples cannot be explained using deduction alone, but they can be explained by making abductive inferences. The inferences provide useful information relevant to emotional states.

Introduction

Explaining people's actions often requires reasoning about emotions. This is because experiences give rise to emotional states which in turn make some actions more likely than others. For example, if someone strikes another person, we may explain the aggression as being a result of anger. As well as reasoning about actions in terms of emotional states, we can reason about emotional states themselves. Explaining emotional states requires reasoning about the cognitive antecedents of emotions. In the right context, we might reason that a person was angry because he or she had been insulted. This paper focuses on explanations of this kind.

We present a computational model of the construction of explanations of emotions. The model is comprised of two main components. The first component is a situation calculus theory of emotion elicitation. The second component is a method for constructing explanations. The representation of emotion eliciting conditions is inspired by a theory of the cognitive structure of emotions proposed by Ortony, Clore, and Collins (1988). In addition to codifying a set of general rules of emotion elicitation inspired by this theory, we have also codified a large collection of cases based on diary study data. We have implemented a computer program that constructs explanations of emotions arising in these scenarios. The program constructs explanations based on a first order logical abduction method.

Abductive Explanation

Peirce used the term abduction as a name for a particular form of explanatory hypothesis generation (Peirce, 1931–1958). His description was basically:

The surprising fact C is observed;
But if A were true,
C would be a matter of course,
hence there is reason to suspect that A is true.

Since Peirce's original formulation, many variants of this form of reasoning have come to be known as abduction. Examples of abduction methods proposed in AI research include abductive approaches to diagnosis (Peng & Reggia, 1990) and natural language comprehension (Hobbs, Stickel, Martin, & Edwards, 1988). We focus on a logical view of abduction advocated by Poole (e.g., Poole, Goebel, & Aleliunas, 1987). In this approach, observations O are explained given some background knowledge expressed as a logical theory T by find-

*Supported in part by National Science Foundation Grant Number IRI-8813048.

†Supported in part by National Science Foundation Grant Number IRI-8812699.

Table 1: Elicitation Conditions for 20 Emotion Types

$joy(P, F, S)$	$\leftarrow$	$wants(P, F, S) \wedge holds(F, S)$.
$distress(P, F, S)$	$\leftarrow$	$wants(P, \overline{F}, S) \wedge holds(F, S)$.
$happy\_for(P_1, P_2, F, S)$	$\leftarrow$	$joy(P_1, joy(P_2, F, S_0), S)$.
$sorry\_for(P_1, P_2, F, S)$	$\leftarrow$	$distress(P_1, distress(P_2, F, S_0), S)$.
$resents(P_1, P_2, F, S)$	$\leftarrow$	$distress(P_1, joy(P_2, F, S_0), S)$.
$gloats(P_1, P_2, F, S)$	$\leftarrow$	$joy(P_1, distress(P_2, F, S_0), S)$.
$hopes(P, F, S)$	$\leftarrow$	$wants(P, F, S) \wedge anticipates(P, F, S)$.
$fears(P, F, S)$	$\leftarrow$	$wants(P, \overline{F}, S) \wedge anticipates(P, F, S)$.
$satisfied(P, F, S)$	$\leftarrow$	$precedes(S_0, S) \wedge hopes(P, F, S_0) \wedge holds(F, S)$.
$fears\_confirmed(P, F, S)$	$\leftarrow$	$precedes(S_0, S) \wedge fears(P, F, S_0) \wedge holds(F, S)$.
$relieved(P, \overline{F}, S)$	$\leftarrow$	$precedes(S_0, S) \wedge fears(P, F, S_0) \wedge holds(\overline{F}, S)$.
$disappointed(P, \overline{F}, S)$	$\leftarrow$	$precedes(S_0, S) \wedge hopes(P, F, S_0) \wedge holds(\overline{F}, S)$.
$proud(P, A, S)$	$\leftarrow$	$agent(A, P) \wedge holds(did(A), S) \wedge praiseworthy(A)$.
$self\_reproach(P, A, S)$	$\leftarrow$	$agent(A, P) \wedge holds(did(A), S) \wedge blameworthy(A)$.
$admire(P_1, P_2, A, S)$	$\leftarrow$	$agent(A, P_2) \wedge holds(did(A), S) \wedge praiseworthy(A)$.
$reproach(P_1, P_2, A, S)$	$\leftarrow$	$agent(A, P_2) \wedge holds(did(A), S) \wedge blameworthy(A)$.
$grateful(P_1, P_2, A, S_1)$	$\leftarrow$	$agent(A, P_2) \wedge holds(did(A), S_1) \wedge precedes(S_0, S_1) \wedge$
		$causes(A, F, S_0) \wedge praiseworthy(A) \wedge wants(P_1, F, S_1) \wedge holds(F, S_1)$.
$angry\_at(P_1, P_2, A, S_1)$	$\leftarrow$	$agent(A, P_2) \wedge holds(did(A), S_1) \wedge precedes(S_0, S_1) \wedge$
		$causes(A, F, S_0) \wedge blameworthy(A) \wedge wants(P_1, \overline{F}, S_1) \wedge holds(F, S_1)$.
$gratified(P, A, S_1)$	$\leftarrow$	$agent(A, P_2) \wedge holds(did(A), S_1) \wedge precedes(S_0, S_1) \wedge$
		$causes(A, F, S_0) \wedge wants(P, F, S_1) \wedge holds(F, S_1) \wedge praiseworthy(A)$.
$remorseful(P, A, S_1)$	$\leftarrow$	$agent(A, P_2) \wedge holds(did(A), S_1) \wedge precedes(S_0, S_1) \wedge$
		$causes(A, F, S_0) \wedge wants(P, \overline{F}, S_1) \wedge holds(F, S_1) \wedge blameworthy(A)$.

ing some hypotheses H such that

$$H \wedge T \vdash O.$$

In other words, if the hypotheses are assumed, the observation follows by way of general laws and other facts given in the background knowledge.

We construct explanations using an abduction engine based on an early approach to mechanizing abduction described in (Pople, 1973). The method is implemented in a PROLOG meta-interpreter called AMAL. It takes as input a collection of PROLOG clauses encoding theories. One theory represents background knowledge, another captures the facts of the case at hand. An observation to be explained is given as a query. AMAL is also given an operationality criterion and an assumability criterion. The output includes an explanation of the given observation, possibly including some assumptions that must be made in order to complete the explanation.

In general, many explanations are possible and it is important to constrain the search to avoid large numbers of implausible hypotheses and explanations. In early experiments, we found that the abduction engine conjectured large numbers of implausible causal relationships. This problem was solved by disallowing assumptions of the following forms:

$$preconditions(A, F)$$
$$causes(A, F, S)$$

In other words, the abduction engine was not allowed to assume that an arbitrary fluent might be a precondition for an action, nor was it allowed to assume unprovable cause-effect relationships between actions and fluents.

Emotion Elicitation

Our first order logical theory of emotion elicitation contains rules covering eliciting conditions of twenty emotion types (see Table 1). In addition, we have coded variants of a number of them, details of which have been omitted due to space constraints. (See O'Rorke & Ortony, 1992 for a presentation of the full theory.)

The theory draws upon knowledge representation work on situation calculus (McCarthy, 1968) and conceptual dependency (Schank, 1972). It includes axioms that support causal reasoning about actions and other events that can lead to emotional reactions.

For example, the first law below mediates positive and negative effects of actions. The second law states that a precondition of a physical transfer from one location to another is that one must first be at the initial location. The remaining laws state the effects of a physical transfer.

$$holds(F, do(A, S)) \leftarrow causes(A, F, S) \wedge poss(A, S).$$
$$poss(ptrans(P, To, From, T), S)$$
$$\leftarrow holds(at(T, From), S).$$
$$causes(ptrans(P, To, From, T), at(T, To), S)$$
$$causes(ptrans(P, To, From, T), \overline{at(T, From)}, S).$$

Emotion types are represented as fluents and their eliciting conditions are encoded in rules. As examples, consider the rules for the emotion types *fear* and *relief*, shown in Table 1. The *fear* rule captures the idea that people may experience fear if they want an anticipated fluent not to hold. Relief may be experienced when the negation of a feared fluent holds. Fear usually occurs before the fluent holds. Note that, although many examples of fear involve expectations, we use the predicate *anticipates* in an effort to suggest the notion of "entertaining the prospect of" a state of affairs. The purpose of this is to avoid suggesting that hoped-for and feared events necessarily have a high subjective probability.

Explaining Emotions

In this section, we use an example to illustrate the abductive construction of explanations involving emotions. The example is based on data taken from a diary study of emotions. Most of the subjects who participated in the study were sophomores at the University of Illinois at Champaign-Urbana. They were asked to describe emotional experiences that occurred within the previous 24 hours. They typed answers to a computerized questionaire containing questions about which emotion they felt, the event giving rise to the emotion, the people involved, the goals affected, and so on. Over 1000 descriptions of emotion episodes were collected, compiled, and recorded on magnetic media. We have encoded over 100 of these examples using our situation calculus representation language. The following case provides examples of *relief* and *fear*.

Mary wanted to go to sleep.
Karen returned.
T.C. finally left her place.
Mary was relieved.

The case is encoded as shown in Table 2. The case fact says that Mary wants sleep. The query asks why Mary is relieved that T.C. is not at her home in the situation that results after T.C.'s departure. T.C.'s departure occurred in the situation resulting from Karen's return. (Note the abbreviations for the relevant situations at the bottom of the Table.)

The explanation shown in Table 2 was constructed automatically by AMAL. The program works by backward chaining on observations to be explained. It tries to reduce the observation to known facts by invoking general laws (e.g., causal laws of situation calculus and laws of emotion elicitation). In this case, the eliciting condition for relief is invoked in order to explain Mary's relief. This generates new questions that must be answered, and so on. The resulting explanation (shown in Table 2) states that Mary is relieved that T.C. is no longer at her home. The explanation assumes that Mary fears T.C.'s presence in her home because she wants T.C. not to be in her home but she anticipates that he will be there. A deeper explanation connecting this desire and anticipation to Mary's desire for restful sleep should be possible. For example, the presence of T.C. might interfere with Mary's sleep. The explanation of his abcence does not include the possibility that he may have been driven away by Karen's return. But it does serve to illustrate the use of causal laws to infer negative fluents relevant to emotional reactions. In this case, since T.C. moved from Mary's home to another location, it can be inferred that he is no longer at Mary's home.

Discussion

Like the example of relief and fear, the majority of the cases in the diary study data require assumptions. The kinds of assumptions needed include missing preconditions, goals, prospects, and judgements. In the example, the assumption that T.C. was at Mary's home in the initial situation helped explain why he was there after Karen came home. This in turn was a precondition for T.C.'s leaving Mary's home. The example also required an assumption that Mary wanted T.C. to go somewhere else in order to explain Mary's fear that T.C. would be at her home. Assumptions about other's goals also occur in explaining emotions that involve the "fortunes of others." Abductive assumptions about other mental states include assumptions about whether agents anticipate events. In the example of *relief*, it was necessary to assume that Mary anticipated T.C.'s continued (unwelcome) presence in her home. Assumptions about judgements of blameworthiness and praiseworthiness are important in explaining a number of emotions not present in the example.

The explanation constructed in the example, and

Table 2: Explanations of Relief and Fear

```
Case Facts
    wants(mary, sleep(mary), _)
Query
    why(relieved(mary, not at(tc, home(mary)), s2))
Explanation
    relieved(mary, not at(tc, home(mary)), s2)
        precedes(s1, s2)
        fears(mary, at(tc, home(mary)), s1)
            wants(mary, not at(tc, home(mary)), s1)
            anticipates(mary, at(tc, home(mary)), s1)
        holds(not at(tc, home(mary)), s2)
            causes(ptrans(tc, _29887, home(mary), tc), not at(tc, home(mary)), s1)
            poss(ptrans(tc, _29887, home(mary), tc), s1)
                holds(at(tc, home(mary)), s1)
                    not causes(ptrans(karen, home(mary)), not at(tc, home(mary)), s0)
                    holds(at(tc, home(mary)), s0)
                    poss(ptrans(karen, home(mary)), s0)
Abbreviations
    s1=do(ptrans(karen, home(mary)), s0)
    s2=do(ptrans(tc, _591, home(mary), tc), s1)
```

many other explanations (see O'Rorke & Ortony, 1992), could not have been constructed by the abduction engine without its abductive inference capability, given the background knowledge and codifications of the cases provided with the observations to be explained. Given the same information, a purely deductive PROLOG-style interpreter would fail to find an explanation. Admittedly, the knowledge base could conceivably be extended so that some assumptions could be eliminated and replaced by deductive inferences. For example, if knowledge of ethics and standards of behavior could be provided, the number of assumptions in explanations requiring judgements of blameworthiness and praiseworthiness could be reduced. But it is not likely that all relevant preconditions, desires, prospects, and judgements can be provided in advance.

Related and future work

We give a complete description of the situation calculus of emotion elicitation in (O'Rorke & Ortony, 1992). That paper also contains additional examples and details of the mechanism used to generate explanations.

A previous study formalizing commonsense reasoning about emotions is summarized in (Sanders, 1989). This work takes a deductive approach, using a deontic logic of emotions. The logic focuses on a cluster of emotions involving evaluations of actions — including what we have called admiration, reproach, remorse, and anger. The evaluation of actions is ethical, and involves reasoning about obligation, prohibition, and permission. The logic was used to solve problems involving actions associated with ownership and possession of property (e.g., giving, lending, buying, and stealing) by proving theorems. For example, the fact that Jack will be angry was proved given that he went to the supermarket, parked his car in a legal parking place, and when he came out, it was gone. It is not clear whether the theorems were proved automatically or by hand so questions of complexity of inference and control of search in the deontic logic remain unanswered. We have argued that abduction offers advantages over deduction alone when applied to the task of constructing explanations involving emotions. And our situation calculus of emotion elicitation is more comprehensive than the deontic logic for emotions in that it covers more emotion types. But our approach could benefit from Sanders' treatment of ethical evaluations. We hope to undertake a detailed comparison and integration of the best parts of the two approaches in future work.

The present work focuses on explaining emotions in terms of eliciting situations. But while situations give rise to emotional reactions, emotions in turn give rise to goals and actions that change the state of the world. Applications such as plan recognition will require a theory specifying causal connections between emotions and subsequent actions. For a brief description of a system for recognizing plans involving emotions, see Cain, O'Rorke, and Ortony (1989). This paper also describes how explanation-based learning techniques can be used to learn to recognize such plans. For a fuller discussion of reasoning about emotion-induced actions, see Elliott and Ortony (1992).

In Ortony, Clore, and Foss (1987) about 270 English words are identified as referring to genuine emotions from an initial pool of 600 words that frequently appear in the emotion research literature. In another study, 130 of these emotion words were distributed among 22 emotion types. Some emotion words map to several different types, e.g., "upset" is compatible with distress, anger, or shame. Many words map to the same type. Encoding the relationship between the affective lexicon and the emotion types is an important topic for future research aimed at automatically processing natural language text involving emotions.

Conclusion

We have developed a theory of the cognitive antecedents of emotions and an abductive method for explaining emotional states. We sketched a computer program, an abduction engine implemented in a program called AMAL, that uses the theory of emotion elicitation to construct explanations of emotions. We presented an explanation of an example based on a case taken from a diary study of emotions.

The most important advantage of our approach to explanatory reasoning about emotions is that abduction allows us to construct explanations by generating hypotheses that fill gaps in the knowledge associated with cases where deduction fails. In most cases, emotional states cannot be explained deductively because they do not follow logically from the given facts. The abduction engine explains the emotions involved in these cases by making assumptions including valuable inferences about mental states such as desires, expectations, and the emotions of others.

Acknowledgments

We thank the reviewers for suggestions that improved the paper. Terry Turner kindly provided diary study data. Steven Morris, Tim Cain, Tony Wieser, David Aha, Patrick Murphy, Stephanie Sage, Clark Elliott, and Milton Epstein participated in various stages of this work.

References

Cain, T., O'Rorke, P., & Ortony, A. (1989). Learning to recognize plans involving affect. In A. M. Segre (Ed.), *Proceedings of the Sixth International Workshop on Machine Learning* (pp. 209–211). Ithaca, NY: Morgan Kaufmann.

Elliott, C., & Ortony, A. (1992). *Point of view: Modeling the emotions of others.* Manuscript submitted for publication.

Hobbs, J. R., Stickel, M., Martin, P., & Edwards, D. (1988). Interpretation as abduction. *Proceedings of the Twenty Sixth Annual Meeting of the Association for Computational Linguistics* (pp. 95–103). Buffalo, NY: The Association for Computational Linguistics.

McCarthy, J. (1968). Programs with common sense. In M. Minsky (Ed.), *Semantic Information Processing* (pp. 403–418). Cambridge, MA: MIT Press.

O'Rorke, P., & Ortony, A. (1992). *Explaining emotions* (Technical Report 92-22). Submitted for publication. Irvine: University of California, Department of Information and Computer Science.

Ortony, A., Clore, G. L., & Collins, A. (1988). *The cognitive structure of emotions.* New York: Cambridge University Press.

Ortony, A., Clore, G. L., & Foss, M. A. (1987). The referential structure of the affective lexicon. *Cognitive Science,11*(3), 361–384.

Peirce, C. S. S. (1931–1958). *Collected papers of Charles Sanders Peirce (1839–1914).* Cambridge, MA: Harvard University Press.

Peng, Y., & Reggia, J. A. (1990). *Abductive inference models for diagnostic problem solving.* New York: Springer-Verlag.

Poole, D. L., Goebel, R., & Aleliunas, R. (1987). Theorist: A logical reasoning system for defaults and diagnosis. In N. Cercone, & G. McCalla (Eds.), *The Knowledge Frontier: Essays in the Representation of Knowledge.* New York: Springer-Verlag.

Pople, H. E. (1973). On the mechanization of abductive logic. *Proceedings of the Third International Joint Conference on Artificial Intelligence* (pp. 147–152). Stanford, CA: Morgan Kaufmann.

Sanders, K. E. (1989). A logic for emotions: a basis for reasoning about commonsense psychological knowledge. In E. Smith (Ed.), *Proceedings of the Annual Meeting of the Cognitive Science Society* (pp. 357–363). Ann Arbor, MI: Lawrence Erlbaum and Associates.

Schank, R. C. (1972). Conceptual dependency: A theory of natural language-understanding. *Cognitive Psychology,3*(4), 552–631.

Assessing Explanatory Coherence: A New Method for Integrating Verbal Data with Models of On-line Belief Revision

Patricia Schank and Michael Ranney
Mathematics, Science, and Technology Division; Graduate School of Education
University of California; Berkeley, CA 94720
ranney@cogsci.berkeley.edu & schank@garnet.berkeley.edu

Abstract[1]

In an earlier study, we modeled subjects' beliefs in textually embedded propositions with ECHO, a computational system for simulating explanatory evaluations (Schank & Ranney, 1991). We both presumed and found that subjects' representations of the texts were not completely captured by the (*a priori*) representations generated and encoded into ECHO; extraneous knowledge likely contributed to subjects' biases toward certain hypotheses. This study builds on previous work via two questions: First, how well can ECHO predict subjects' belief evaluations when *a priori* representations are not used? To assess this, we asked subjects to predict (and explain, with alternatives) an endpoint pendular-release trajectory, while collecting believability ratings for their on-line beliefs; subjects' protocols were then "blindly" encoded and simulated with ECHO, and their ratings were compared to ECHO's resulting activations. Second, how similar are different coders' encodings of the same reasoning episode? To assess intercoder agreement, we examined the fit between ECHO's activations for coders' encodings of the same protocols. We found that intercoder correlations were acceptable, and ECHO predicted subjects' ratings well—almost as well as those from the more diminutive, constrained situations modeled by Schank and Ranney (1991).

Introduction

People often differentially evaluate the plausibility of similar or even identical beliefs when reasoning or arguing about a situation. How do people decide what description of the world is most plausible? Thagard (1989) and others characterize the plausibility of a belief as generally increasing with its increasing simplicity (e.g., fewer necessary cohypotheses), increasing breadth (i.e., more coverage of observation), and decreasing competition with alternate (especially entrenched) beliefs (cf. Johnson & Smith, 1991). These principles play important roles in evaluations of the quality of an explanation (Schank & Ranney, 1991; Read & Marcus-Newhall, 1991).

Science is a rigorous interpretive system that we overlay on our experiences to understand and use them. Although individuals' beliefs may not fit precisely into a scientific framework (i.e., their beliefs may conflict with established scientific hypotheses), people often hold their beliefs as long as they help explain many of their experiences. For example, students learning physics tend to hold strong intuitive beliefs about the physical world that tend to resist revision (e.g., Ranney 1987/1988; Hartley, Byard, & Mallen, 1991). Ranney and Thagard (1988) characterize belief revision as the result of seeking explanatory coherence between theories and observations. The Theory of Explanatory Coherence (TEC) is intended to account for a variety of explanatory evaluations. This theory has been implemented in a computational model called ECHO, based on the claim that beliefs and data are related explanatory entities, and evaluating their plausibility is an interactive, principled, coherence-seeking process (Thagard, 1989; Ranney, in press).[2]

We describe here an empirical study that extends our previous research (Schank & Ranney, 1991) by focusing on two questions: First, how well can ECHO predictively model how strongly individuals believe the assertions they make in the course of an explanation or argument? Second, how subjective is the ECHO encoding process? To further assess ECHO's predictive ability, we asked subjects to predict the path a bob follows when released from the endpoint of a pendular swing, to explain their own (and others') predictions, and to rate the strength of the beliefs used. Their verbal protocols were then encoded and simulated with ECHO, and its activations compared to the subjects' ratings. To assess intercoder reliability, multiple coders encoded the protocols,

1 Preparation of this article was supported by the Spencer Foundation, the Evelyn Lois Corey Fund, and a Faculty Research Grant from the University of California, Berkeley.

2 ECHO's "theoretical/systemic" coherence differs from (and is generally orthogonal to) standard notions of "linguistic" coherence (Ranney, Schank, & Ritter, 1992). In ECHO, coherence is seen from the perspective of competing theories, where the dynamic tension represented as explicitly conflicting theories *reduces* the overall coherence of a system of propositions (compared to a single-theory network). In contrast, textual/discourse coherence is generally viewed as increasing with more explicit relations among various entities and assertions in a text, and less reliance on implicit background knowledge for making inferences (such as anaphoras). (E.g., the textual stimuli used in Schank & Ranney, 1991 were designed to be low in systemic coherence and high in linguistic coherence).

and we examined differences among ECHO's eventual activations with respect to the coders' encodings of identical protocols.

TEC and its ECHO Model

In TEC, coherence involves relations among two or more propositions that may "hold together" or "resist holding together." (We use "proposition" for something proposed, e.g., a piece of evidence or hypothesis, such as "gravity pulls objects down"—in contrast to a concept, such as "gravity.") For current purposes, the following principles establish the local pairwise relations among cohering and incohering propositions (nb. these principles are, selectively, from Schank & Ranney, 1991, Thagard, 1992, and Ranney & Thagard, 1988): (1) Coherence and incoherence are symmetric relations. (2) Hypotheses that together explain a proposition cohere with each other and with the explained proposition. (3) Simplicity: The plausibility of a belief is inversely related to the number of cohypotheses it needs to explain a proposition. (4) Data Priority: Results of observations, such as evidence and acknowledged facts, have a degree of acceptability on their own. (5) Contradictory hypotheses incohere. (6) The acceptability of a proposition depends on its coherence within the system of propositions in which it is embedded. A proposition's acceptability increases as it coheres more with other acceptable propositions and *incoheres* more with *unacceptable* propositions. (In ECHO, a proposition's acceptability is measured by its activation value, ranging from -1, complete rejection, to 1, complete acceptance.) (7) The overall coherence of a network of propositions depends on the local pairwise cohering of its propositions.

Schank and Ranney (1991) and Read and Marcus-Newhall (1991) show that these principles play important roles in explanations. They found that subjects prefer explanations that account for more data, are simpler, and involve hypotheses that can be further explained. Subjects' evaluations of explanations are also changed by the availability of competing explanations.

ECHO uses a connectionist architecture in which each node represents a proposition. Hypothesis evaluation is treated as the satisfaction of many constraints, determined from the explanatory relations and from a few parameters that provide degrees of freedom. Given declared input propositions and relations between them, node activations are updated using a simple settling scheme. For more complete descriptions of ECHO's algorithms, see Thagard (1989 & 1992), Schank and Ranney (1991), and Ranney and Thagard (1988).

Why Physics, and Why ECHO?

This study focuses on questions regarding (a) ECHO's ability to predictively model individual subjects' evaluations of their beliefs about physical motion (encoded from verbal protocols), and (b) intercoder reliability regarding ECHO. We chose to model beliefs about motion since studies have shown that individuals tend to persistently hold naive beliefs about the natural world that sometimes conflict with scientific explanations (e.g., Brewer & Chinn, 1991; Ranney, 1987/1988).

ECHO has been used (mostly *ex post facto*) to model changes in subjects' beliefs about physical motion (Ranney & Thagard, 1988), scientific and juror reasoning (Thagard, 1989, etc.) and social situations (e.g., Read & Marcus-Newhall, 1991), and to examine and foster students' scientific reasoning skills (e.g., Carlock, 1990; Ranney, in press). We used ECHO to model subjects' beliefs in textually embedded propositions, and found that subjects sometimes entertained competing hypotheses as nonexclusive and seemed to presume an implicit backing (i.e., other evidence or beliefs) that supported certain hypotheses (Schank & Ranney, 1991). Despite attempts to decontextualize texts, our subjects (not surprisingly) brought extraneous knowledge to bear when reasoning about the texts' statements. Consequently, the subjects' representations of the texts were not completely captured by the representation encoded into ECHO, and this unrepresented extraneous knowledge likely contributed to their relative biases. We coarsely modeled this tendency for subjects to presume backing behind "superordinate" hypotheses in ECHO by assigning them a fraction of *data priority* (usually reserved for evidence). Still, the study raised the question: Can ECHO predict subjects' beliefs as well or better if they make their implicit backings (coarsely modeled by Schank & Ranney, 1991) explicit?

Other computational models. Several models of explanation evaluation and belief change are compatible with TEC and ECHO. Ranney (in press) points out that TEC does not explicitly account for memorial capacity and processing limitations, inspiring Bar-On's (1991) theory of local coherence within *views*, an attempt to account for attentional and short term memory effects via limited capacities. Bar-On argues that localist connectionist models provide more appropriate levels of abstraction (than distributed models) for simulating locally coherent views. Also similar to ECHO is HEIDER, Gabrys' (1989) simulation, which seeks consistency (coherence) within its world view in the face of new information.

ECHO may initially seem less compatible with other computational models. For example, Ram and Leake (1991) argue that people best learn to accept explanations when they come with needed information, and present a goal-based computational model that focuses on finding "useful" (vs. "valid") explanations by incorporating the goals into explanatory evaluations. Okada and Klahr (1991) code subjects' naive, complex, idiosyncratic, beliefs (garnered from transcribed protocols) as a hypothesis space, but they view belief revision as a search through this space of beliefs (vs. parallel constraint satisfaction, as in ECHO). However, both of these models highlight *goal-* or *utility-based* reasoning, which ECHO does not attempt to model. These models are more comparable to MOTIV-ECHO (Thagard, 1992), a program that allows ECHO's inferences to be biased by goals.

Models of text and discourse analysis. Our previous methodology was limited in that the extraneous knowledge subjects brought to bear when reasoning about assertions in the texts were not completely captured by the representation encoded into ECHO (Schank & Ranney, 1991). Hence, we asked subjects in the current study to *explain* their beliefs and predictions, and encoded their protocols as ECHO belief networks.

Other researchers represent mental interpretations of text and discourse by systems of interrelated propositions (e.g., Trabasso, van den Broek, & Suh, 1989; Givon, 1991). Kintsch's work (in press, cf. 1988) supports the use of higher-level (e.g., causal or explanatory) relations between belief propositions as a powerful level of abstraction. He describes text comprehension as the construction of representations consisting of primary concepts and higher level propositions, in a network with associative relations at the conceptual and propositional levels and causal (explanatory) relations at the proposition level. Including both types of links in the representation enabled more accurate predictions of subjects' immediate recall (r=.76), but causal links alone explained most of the variation (r=.61).

Method

In this study, we use ECHO encodings, produced from a qualitative analysis of the subjects' protocols, for a more ecologically valid and comprehensive test of the ECHO model and encoding schemes (cf. Ranney and Thagard, 1988). Design decisions followed the desire to represent subjects' explanations and believability ratings as completely and accurately as possible, and to assess intercoder reliability for ECHO networks. We used a novel combination of convergent methods (cf. Ranney 1987/1988) in that we collected (a) drawn trajectory predictions, (b) quantitative believability ratings (to avoid subjective bias about the strength of subjects' beliefs), and (c) verbal protocols (with evaluative comments and ratings edited out), used as the basis for the ECHO encodings.

Subjects and Procedure

Ten subjects, four men and six women, were chosen from the University of California (Berkeley) student population, from responses to an advertisement. The subjects had various backgrounds, but little or (usually) no formal physics background. During the 30-60 minute sessions, subjects were asked to make predictions about pendular–release situations, and to rate the believability of the hypotheses and evidence they verbalized as they reasoned about the task. The interviewer recorded subjects' beliefs on paper (in real-time, as they completed an utterance) using the subjects' terminology. (Subjects were later given feedback on the situations' outcome, and then again asked to rate the believability of the same, noted, propositions.) Audiotaped protocols were collected and transcribed for encoding and intercoder reliability analyses.

Tasks. Subjects were first shown an animated pendular–release situation (from Ranney, 1987/1988) in which the swinging motion of the pendulum–bob was frozen in time at the endpoint of its swing. Subjects were asked to imagine that the pendulum string broke at the extreme of the swing and/or the bob was released, and to predict (draw and explain) the subsequent trajectory of the bob. They were then shown, serially, five alternative, commonly predicted, paths (generated by subjects in Ranney, 1987/1988, and in a recent pilot study), and asked to explain why each path may or may not be correct.

As a subject reasoned out loud about the plausibility of the paths, the interviewer noted the subject's assertions. After the subject finished reasoning about the endpoint-release situation, the interviewer read back to the subject the list of beliefs she had noted. Subjects were then asked to rate (on a scale from 1, "completely unbelievable," to 9, "completely believable") how strongly they believed the propositions they had verbalized, and to rate how strongly they believed in the alternative paths (now displayed in parallel).[3]

Encoding. Subjects' stated believability ratings were edited out of copies of the transcribed protocols, as were evaluative statements that qualitatively revealed the strength of their beliefs. The edited protocols were then encoded into ECHO–style input by one to four, variously experienced, "blind" coders. (I.e., coders segmented and categorized subjects' assertions into beliefs, evidence, explanations, and contradictions.) Encodings of only the first (prediction) part of the first (pendular–release) task are reported here (see footnote 3). This portion generally accounted for over half of the transcribed sessions.

Coders 1 (who coded all of the protocols at least once) and 2 had experience encoding previous protocols (e.g., from Ranney, 1987/1988, and the pilot study), but coders 3 and 4 had virtually none. Coders discussed encoding principles at length, and incorporated agreed–upon princi-

[3] Data from what follows are not analyzed here: After elaborating on their predictions on the endpoint pendular-release task, subjects read and made predictions about an isomorphic playground swing situation. Subjects were asked to predict the path that a child leaving the seat at the endpoint of a playground swing would follow, and to rate both their beliefs (again, noted "on-line" by the interviewer) and their predicted path.

Following their predictions in the pendular–release and playground situations, subjects were allowed to change their predictions, then given trajectory feedback: The pendular-release task's simulation was repeated and subjects were asked (1) if they chose to modify or draw a new path, and (2) to rate their beliefs (read back by the interviewer), including the alternative paths (displayed in parallel). Subjects were then shown the *actual* path (i.e., a dynamic feedback simulation of both the swing and the subsequent vertical trajectory after release; Ranney, 1987/1988). They were then asked, with hindsight, (a) to try to explain why the (usually surprising; Ranney & Thagard, 1988) feedback was correct, and (b) to again rate the strength of their beliefs. Similarly, on the isomorphic playground task, subjects could change their prediction and re-rate their beliefs before being given (verbal) feedback. After feedback (indicating a vertical, post-release trajectory), they were asked to give a hindsight explanation of the vertical trajectory, and to re-rate their beliefs.

ples into an "explanation encoding guide" that included a list of syntactic and grammatical cues that tend to indicate explanation structures (e.g., verbs that signal beliefs; conjunctions and verbs that signal explanations; negations, conjunctions, and verbs of conflict that signal contradictions;[4] see Table 1 for an example encoding).

ECHO simulations of the encodings were run with the parameter settings used by Schank and Ranney (1991), which were also in the midrange of those used by Ranney and Thagard (1988).[5] Each subject's belief ratings (i.e., of assertions encoded as ECHO propositions for which the subject reported a rating) were then compared with the ECHO activations from the simulations. Comparisons between ECHO's activations and subjects' ratings (just prior to the time of feedback) were made for each subject–coder pair and for each coder overall. For each subject, ECHO activations among coders were also compared.

By having coders unfamiliar with the subjects' (excised) believability ratings and qualitative evaluations encode the protocols, we could use the results of ECHO simulations on these encodings to assess how well ECHO predicts *a priori* the subjects' beliefs. Thus, we compared ECHO's activations for the various encodings with subjects' evaluations (i.e., believability ratings). We assessed intercoder agreement by comparing different encodings and simulation results of the same protocols.

Table 1. Example protocol and its encoding.

Convention:	**Indicates:**
underlines	explanations
italics	contradiction/competition
[bracket]	beliefs/data propositions

Protocol: "I've changed my mind. (no coding; monitoring statement.). I like P1[the arch path]. I think P1[the bob will fly out in an arch curve] since H1[its got motion right] and H2[up], and H3[gravity is pulling it down]. *But* H1[it has motion right] so I guess P2[the diagonal path] is possible. NP3[It won't drop straight down], though, because H1[it has motion right]. Hm. I remember E1[jumping off a swing and flying out in an arch], though, so I think P1[its going to fly up and out like it does in that arch path you have there]. Yeah, I really think P2[the diagonal path] won't happen (no coding; evaluative statement.). "

Encoding:

P1 The arch path	P2 contradicts P1 (implicit)
H1 Bob has motion right	NP3 Not straight down path
H2 Bob has motion up	H1 explains NP3
H3 Gravity is pulling down	E1 I fly out in an arch when I jump off a swing
H1, H2, and H3 explain P1	data E1
P2 Diagonal path	
H1 explains P2	E1 explains P1

Results

Figure 1 shows subjects' average ratings (prior to feedback) for each path. We computed correlations between ECHO's activations and subjects' ratings for each subject–coder pair (see Figure 2), and for each coder. Analyses of variance of the subjects' ratings were performed; we also computed correlations between ECHO's activations among the various coders for each protocol. Analyses were computed both (a) over all beliefs, and (b) for the path propositions only.

Protocol and Encoding Analyses

Encoded portions of the protocols averaged about 600 words in length. On average, about 23 propositions, 2 data (e.g., observations or remembered experiences), 13 explanations (3 of which were considered "implicit" by coders), and 21 contradictions (17 of which were considered implicit, including 15 essential contradictions between the competing release paths) were encoded per protocol. As expected, subjects did not make all of their explanations and contradictions explicit, even with prompts for elaboration (cf. Grice, 1975, on the conversational *maxim of quantity*, which predicts that people typically avoid being overly informative). The interviewer captured on-line (and hence had subjects rate) about 60% of the belief propositions that were later encoded by coders, so ratings for 60% of the encoded propositions were available for comparison with ECHO activations. Ratings for most of the subjects' key central beliefs, and all alternative trajectories, were collected. Analyses of the encodings also revealed that beliefs *not* rated played more peripheral explanatory roles, compared to their rated counterparts. (E.g., compared to rated beliefs, unrated beliefs were about twice as likely to not be part of any explicit explanation of a path, and about *thrice* as likely to not be part of *any* explicit explanation. Further, rated beliefs were about twice as likely as their unrated counterparts to be within two explanatory links from any path-prediction, and almost 50% more likely to *directly* explain a path-prediction. These differences were significant at p<.01.)

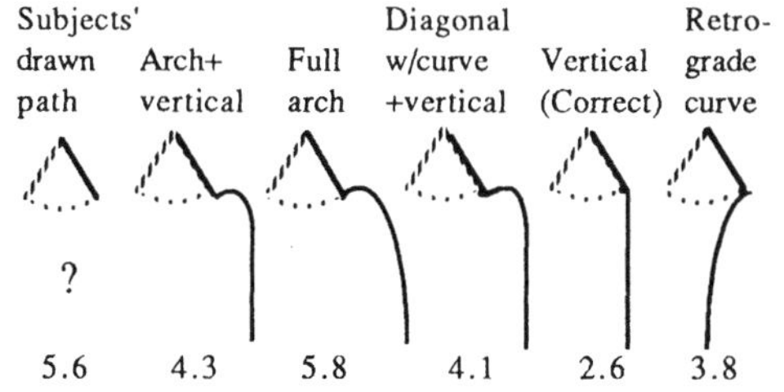

Figure 1. Mean path believability ratings (1-9 scale; prior to feedback). (Nb. Four subjects drew unique paths not among the 5 alternatives. Mean ratings for drawn paths were about 5.6 both for unique paths and for non–unique paths.)

[4] We encoded the alternative pendular–release trajectories in this study as contradictory, based on the assumption that subjects believe one unique trajectory exists.

[5] The parameter values used were: decay = .04, excitation = .03, inhibition = -.06, and data excitation = .055.

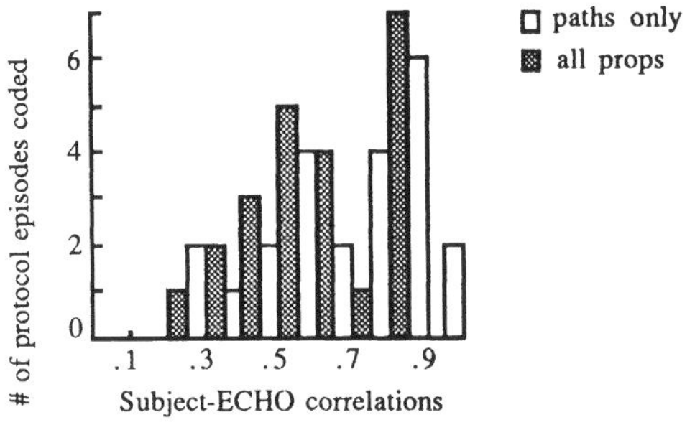

Figure 2. Distribution of correlations between ECHO's activations and subjects' ratings on *all* rated propositions (grey) and on endpoint-release paths only (white).

Prediction of Subjects' Ratings

ECHO simulations were run with the implicit explanations and contradictions, added by the coders, included. Final ECHO activations, over all rated beliefs, were positively correlated (at p<.05) with subjects' ratings in 14 of the 23 simulations. Correlations were as high as r >.80 on seven simulations, and as low as r=.25 on one (see Figure 2).[6] Overall, there was a significant positive correlation (r =.56) between ECHO's activations and the subjects' ratings (r=.65 for most highly correlated coder).

Correlations between ECHO activations and the subjects' ratings for the path-propositions alone mirrored those computed over all beliefs, with correlations as high as r= .99 and as low as r=.23. The overall correlation for paths was nonsignificantly higher (at r=.61) than the correlation over all beliefs (r=.56). The overall correlations were also slightly higher for most of the coders (r=.70 for the most highly correlated coder).[7] In general, results from coders who coded the most protocols yielded higher correlations between coding-based ECHO activations and subjects' ratings.

Intercoder Agreement

Over all beliefs, and for the paths alone, ECHO's final activations were significantly correlated with subjects' ratings for coders 1 and 2 (r >.57; p<.0001). These coders were the most experienced, and encoded more of the protocols (and thus had more data to correlate); the correlations between ECHO's final activations and subjects' ratings generally increased with encoding experience.

Comparing ECHO networks to assess intercoder agreement is difficult; tractable methods of comparing such network topologies have not, to our knowledge, been developed. Therefore, we used approximating measures to gauge intercoder agreement. We analyzed the propositions common among coders; pairs of propositions were judged to be the same if both their wording and source locations in the protocol were virtually the same. On average, about 60% of encoded propositions were judged to be the same between pairs of coders. For these, the overall intercoder correlation (i.e., between ECHO activations for common propositions) was significant (r=.49; p<.001).

Analysis of variance results for the subjects' ratings over all beliefs (with ECHO activations as a covariate) later indicated that ECHO's activations explain about 28% of the variance in the subjects' ratings, while individuals account for about 8% of the variance in the ratings (both p<.001). As one would hope, the ANOVAs also indicate that *none* of the variance is accounted for by the coder, so systematic coder effects are negligible. Similar results are found when only the path propositions are considered, except that the type of path considered also accounts for a significant amount of variance.

Discussion and Conclusions

Schank and Ranney (1991) raised the question of whether ECHO can predict subjects' beliefs as well as or better than they had observed (r >.7) if subjects make their "implied backings" explicit. However, we found slightly lower ECHO-subject correlations (about r =.6).[8] This might suggest that ECHO does not predict subjects' beliefs better (or perhaps even as well as) when they make their implicit backings explicit. There are several reasons to resist this conclusion. First, of the ten individuals' ratings in this study, six were predicted with a correlation of r >.80 (seven, when only the paths were considered); the data also suggest that the ECHO-subject correlation tends to increase when only the (more central) path beliefs are examined. Second, combined with other variables, ECHO helped account for about 40% of the variation in subjects' ratings. Third, the texts used in our prior study reflected topologies defined *a priori*, which likely constrained subjects' representations of the situation. (I.e., subjects in Schank & Ranney, 1991, were not encouraged to elaborate on their beliefs and bring other knowledge into their representations, as they were encouraged to do here). For these (and other content– and context–dependent) reasons, the task of modeling the subjects' beliefs in Schank and Ranney (1991) was, in essence, of smaller scale. Fourth, the ECHO networks generated here were, by salient measures (e.g., the number of propositions, the number of links), about two to over 20 times larger and much less explicit than the networks in the our prior study. This extra complexity may have caused difficulties for subjects who, unlike ECHO, have limited attention

[6] Earlier versions of this article, due to a single coding error, spuriously reported two negatively correlated simulations.

[7] As predicted by Ranney and Thagard's (1988) simulation, feedback on the vertical *pendular–release* trajectory influenced subjects' believability ratings for the (uncoded) isomorphic *playground* situation: Belief in predicted, non–vertical release paths for the playground swing went from strong belief (mean = 7.1) prior to pendular feedback, to slight disbelief (mean = 4.0) after such feedback.

[8] Recall Kintsch's (in press) aforementioned r=.61 between his activations and his subjects' propositional recall.

and memory (Ranney, in press). We plan to incorporate such limitations in future modeling efforts.

In sum, ECHO predicted subjects' ratings fairly well, and the overall intercoder agreement was acceptable (r=.49; so, the coders' simulations currently correlate better with the subjects' data than with each other). Still, the model did not fit the data as well as one might have expected, based on the more diminutive, constrained, theoretical conflicts modeled by Schank and Ranney (1991). But the correlation between ECHO's activations and subjects' believability ratings seems to increase with encoding experience, suggesting that the encoding process is successful and refinable. Further, this study's method is novel, so these nascent attempts to establish intercoder reliability will, no doubt, improve. However, the moderate intercoder correlation also suggests that, for modeling purposes, a better approach may be to have subjects encode their own representations directly into ECHO—with a user–friendly interface (e.g., Carlock 1990). Thus, we are now incorporating ECHO into a computer-based learning environment in which subjects can directly encode their own representations. We plan to use the environment ("the reasoner's workbench;" Ranney, in press) to teach coherent argumentation, and to encourage students to think about consistency and coherence as metrics of reasoning and rationality.

Acknowledgements

We thank Peter Pirolli, Paul Thagard, Margaret Recker, Marsha Lovett, Christopher Hoadley, Joshua Paley, Edouard Lagache, Michelle Million, two reviewers, and the Reasoning Group for their help and comments.

References

Bar-On, E. (1991, May). *Mental capacity and locally-coherent views: Towards a unifying theory*. Draft. Department of Science Education, Technion, Israel.

Brewer, W., & Chinn, C. (1991). Entrenched beliefs, inconsistent information, and knowledge change. In L. Birnbaum (Ed.), *Proceedings of the International Conference on the Learning Sciences*, 67-73. Charlottesville, VA: Association for the Advancement of Computing in Education (AACE).

Carlock, M. (1990). *Learning coherence through MacECHO IFE: An analysis of coherence.* Second Year Project Report, Univ. of California, Berkeley, Graduate School of Education.

Gabrys, G. (1989). HEIDER: A simulation of attitude consistency and attitude change. In S. Ohlsson (Ed.), *Aspects of cognitive conflict and cognitive change.* Univ. of Pittsburgh, Technical Report, KUL-89-04.

Givon, T. (1991, August). *Coherence: Toward a cognitive model.* Paper presented at the First Annual Meeting of the Society for Text and Discourse.

Grice, H. P. (1975). Logic and conversation. In P. Cole and J. L. Morgan (Eds.). *Syntax and Semantics*, vol 3: Speech Acts. New York, NY: Seminar Press.

Hartley, J. R., Byard, M. J., & Mallen, C. L. (1991). Qualitative modeling and conceptual change in science students. In L. Birnbaum (Ed.), *Proceedings of the International Conference on the Learning Sciences*, 222-230. Charlottesville, VA: AACE.

Johnson, T. & Smith, J. (1991). A framework for opportunistic abductive strategies. *Proceedings of the Thirteenth Annual Conference of the Cognitive Science Society*, 760-764. Hillsdale, NJ: Erlbaum.

Kintsch, W. (1988). The role of knowledge in discourse comprehension: A construction-integration model. *Psychological Review, 95*, 163-182.

Kintsch, W. (in press). How readers construct situation models for stories: The role of syntactic cues and causal inferences. To appear in A. F. Healy, S. Kosslyn, & R. M. Shiffrin (Eds.), *Essays in honor of William K. Estes*, vol. 2. Hillsdale, NJ: Erlbaum.

Okada, T., & Klahr, D. (1991). Searching an hypothesis space when reasoning about buoyant forces: The effect of feedback. *Proceedings of the Thirteenth Annual Conference of the Cognitive Science Society*, 842-846. Hillsdale, NJ: Erlbaum.

Ram, A. & Leake, D. (1991). Evaluation of explanatory hypotheses. *Proceedings of the Thirteenth Annual Conference of the Cognitive Science Society*, 867-871. Hillsdale, NJ: Erlbaum.

Ranney, M. (1988). Changing naive conceptions of motion (Doctoral dissertation, University of Pittsburgh, Learning Research and Development Center, 1987). *Dissertation Abstracts International, 49*, 1975B.

Ranney, M. (in press). Explorations in explanatory coherence. To appear in E. Bar-On, B. Eylon, and Z. Schertz (Eds.) *Designing intelligent learning environments: From cognitive analysis to computer implementation.* Norwood, NJ: Ablex.

Ranney, M., Schank, P., & Ritter, C. (1992, January). *Studies of explanatory coherence using text, discourse, and verbal protocols.* Paper presented at the Third Annual Winter Text Conference, Jackson, Wyoming.

Ranney, M. & Thagard, P. (1988). Explanatory coherence and belief revision in naive physics. *Proceedings of the Tenth Annual Conference of the Cognitive Science Society*, 426-432. Hillsdale, NJ: Erlbaum.

Read, S. J. & Marcus-Newhall, A. (1991). *The role of explanatory coherence in the construction of social explanations.* Draft. Univ. of Southern California.

Schank, P., & Ranney, M. (1991). An empirical investigation of the psychological fidelity of ECHO: Modeling an experimental study of explanatory coherence. *Proceedings of the Thirteenth Annual Conference of the Cognitive Science Society*, 892-897. Hillsdale, NJ: Erlbaum.

Thagard, P. (1989). Explanatory coherence. *Behavioral and Brain Sciences, 12*, 435-502.

Thagard, P. (1992). *Conceptual Revolutions.* Princeton, NJ: Princeton University Press.

Trabasso, T., van den Broek, P., & Suh, S. (1989). Logical necessity and transitivity of causal relations in stories. *Discourse Processes, 12*, 1-25.

Educating Migraine Patients Through On-Line Generation of Medical Explanations

Johanna Moore and Stellan Ohlsson
Intelligent Systems Laboratory and
the Learning Research and Development Center,
University of Pittsburgh, Pittsburgh,
PA 15260, U. S. A.
jmoore@speedy.cs.pitt.edu
stellan@vms.cis.pitt.edu

Abstract

Computer support for learning in technical domains such as medicine requires an intelligent interface between the non-expert and the technical knowledge base. We describe a general method for constructing such interfaces and demonstrate its applicability for patient education. The employment of this technology in a medical clinic poses problems which are linguistic, psychological, and socio-cultural, rather than technological, in nature.[1]

Real-Life Learning

People learn in many situations which are not classified as either schooling or training. Applying for a visa at a foreign consulate, appearing in court as plaintiff, defendent, or juror, placing an order with a travel agent, and visiting a medical clinic are examples of situations which confront the non-expert with technical knowledge that does not fit squarely into any traditional school subject. To participate successfully in such situations, the non-expert often needs to acquire some understanding of the relevant technical knowledge.

In this paper we focus on learning in the medical clinic. There is evidence that patients with more knowledge about their disease and their therapy get well faster because they comply more accurately and more conscientiously with the physician's prescriptions (Eraker, Kirscht, & Becker, 1984). The distinction between abortive and prophylactic treatments of migraine provides an illustration of the relation between knowledge and cure. Migraine patients who experience symptoms that consistently precede a migraine attack (e. g., visual disturbances) are typically given drugs that abort the attack. Because abortive drugs are ineffective if taken after the onset of an attack, patients without such warning signals are instead given prophylactic drugs that have to be taken on a regular schedule. When a patient complains that a prophylactic drug is ineffective, questioning might reveal that he or she stopped taking the drug when the headaches stopped; the headaches then

[1] The preparation of this report was supported by grant No. 1 R01 LM05299-01 from the National Library of Medicine, National Institute of Health. The opinions expressed are not necessarily those of the sponsoring agency and no endorsement should be inferred.

returned. Incorrect or incomplete understanding can jeopardize the therapy.

Several factors limit how much physicians can engage in patient education. (a) Doctor's time, particularly the time of specialists, is already a bottleneck in the health care system. (b) Doctors are not trained to communicate with people who do not share their expertise. (c) The doctor's interest and professional pride is typically invested in diagnosing the disease and finding a cure, not in explaining the same seemingly simple matters over and over again many times a day.

One solution to this dilemma is to use Cognitive Science technology to provide the patient with access to the doctor's knowledge without taking up the doctor's time. In collaboration with Bruce Buchanan and Diana Forsythe at the Intelligent Systems Laboratory and Gordon Banks at the Presbyterian University Hospital we are building a computer system which can generate answers and explanations, in English and on-line, in response to questions from patients (Buchanan, Moore, Forsythe, Banks, & Ohlsson, 1992). We are focussing on migraine for a variety of reasons, including the fact that it is a frequent and often disabling disease (Stewart et al, 1992). We first describe the technology we are using and then some non-technological problems that arise in its employment.

On-Line Medical Explanations

Unlike other A. I. systems in the medical domain, the purpose of our system is neither to automate diagnosis nor to train medical students, but to educate patients. The intended system responds to patient questions with answers and explanations which are generated on-line and adapted to the individual patient. We describe the knowledge base, explanation module, and query analyzer of our current prototype, and indicate where they fall short of our goals. The prototype was implemented by Claudia Tapia (Tapia, 1991).

The knowledge base. The topics a migraine patient might want or need information about include (a) the physiology of migraine, (b) potential triggers for headaches, (c) the accompanying symptoms, (d) possible treatments, and (e) their side effects. Our goal is to encode a significant proportion of the medical profession's extensive knowledge about these topics (Raskin, 1988). The current knowledge base contains approximately 400 concepts referring to types and properties of headaches, symptoms, treatments, and drugs. No causal knowledge has as yet been encoded. The knowledge base is not an expert system; there is no inference engine for diagnosis or therapy planning. It is an open question whether on-line medical reasoning will ultimately be needed or whether we can encode everything we might want to explain to the patient in the knowledge base. The knowledge base is implemented in Loom (MacGregor, 1988).

The explanation module. Previous research on the generation of natural-language explanations indicates that an informative explanation cannot be generated from a knowledge base by translating internal code (procedures, rules, or schemas) into English (Buchanan & Shortliffe, 1984; Moore & Swartout, 1988), but requires a dedicated problem solver, called a *text planner*. The text planner described by Moore (1989), Moore and Paris (1989), and Moore and Swartout (1989) operates according to means-ends analysis: Post a goal, activate operators that can achieve that goal, and post the subgoals required by those operators; recurse until all posted operators are primitive. However, the goals and operators are not interpreted as physical situations and motor actions as in typical problem solvers and planners (see, e. g., Wilkins, 1988). The goals are *discourse goals*, i. e., effects that a speaker might want his or her utterance to have on a hearer. Examples are to make the hearer *believe* some proposition and to *persuade* the hearer to perform some action. A different type of discourse goal is to establish some rhetorical relation, e. g., that an assertion P is *evidence* for some other assertion Q or that descriptions P and Q *differ* with respect to some attribute A. The operators encode rhetorical strategies by which discourse goals can be accomplished. For example, to make the hearer understand the differences between two objects, first describe what they have in common and then list their differences. Syntactically, an operator consists of an effect--a goal--and a conjunction of subgoals, the satisfaction of which is sufficient to achieve that effect. The application of an operator is guided by *constraints*, i. e., tests on the knowledge base (or on some other knowledge source; see below). The primitive operators are individual speech acts, e. g., to assert or to ask[2].

The details of this method for text planning have been published in Moore (1989), Moore and Paris (1989), and Moore and Swartout (1989). It was originally implemented in the context of an expert system for programming style (Moore, 1989), but we have successfully transferred it to the migraine domain. Our prototype contains approximately 35 operators, although this number is expected to grow; more than 75 operators were needed to produce satisfactory performance in the programming domain. The current system generates answers to three types of questions with only one or two seconds' delay. When asked to compare prophylactic and abortive treatments (i. e., given the request "COMPARE migraine prophylactic treatment and migraine abortive treatment"), the system generates the following text (Tapia, 1991):

"Migraine prophylactic treatment and migraine abortive treatment are migraine pharmacological treatments. Migraine prophylactic treatment is used to prevent migraine while migraine abortive treatment is used to abort migraine. Migraine prophylactic treatment requires you to take a drug daily whereas migraine prophylactic treatment requires you to take a drug at the immediate onset of headaches. Migraine prophylactic treatment is suitable for frequent or severe headaches while migraine abortive treatment is suitable for infrequent or non-severe headaches."

As the example shows, the text planner needs to be fine tuned to make the text more idiomatic, but this is a low priority at this time. To produce this text, the system posts the goal to make the patient know the contrast between the two treatments. This goal is achieved through an operator that posts the two subgoals to inform the patient about a superordinate concept of which both treatments are instances (i. e., both are pharmacological treatments) and to make the hearer know their contrasting attributes. The latter subgoal in turn activates an operator which posts the subgoals to inform the patient about each individual difference (i. e., the different purposes, treatment protocols, and indications).

To adapt a text to an individual patient, the system needs a user (patient) model. We do not anticipate implementing a runnable user model, or even an overlay model, but will settle for a *global description* (Ohlsson, in press). Relevant global descriptors include age, current therapy, educational background, health state (other diseases, fitness, pregnancy, etc.), past treatment attempts, and gender. Our system will access the user model in the same way as the knowledge base:

[2]The speech acts are only primitive relative to the text planner. Each speech act generates a complex description of the desired utterance which is passed to the FUF language generator (Elhadad & Robin, 1992) for translation into English.

through constraints on the operators. For example, an operator to inform the patient that pregnancy is a counterindicator for drug X might include a constraint that the patient is female. We anticipate that the patient model will consist mainly of information gathered by the physician and to a lesser extent of information gathered during the patient-system interaction. The current system does not have a user model.

The query analyzer. The top-level goal in a text plan derives from the user's question. The current system has a parser which can accept three types of user requests typed in from the keyboard: (a) describe X, (b) describe property Y of X, and (c) compare X and Y, where X and Y can be either treatments or drugs. Due to the difficulties of parsing open ended keyboard input and the need for robust performance in the medical clinic, the finished system will use other technologies for accepting user queries. The user will be able to select questions from a menu. In addition, if he or she wants the system to clarify its response, the user can highlight the problematic portion of the text and receive a context-sensitive menu of possible follow-up questions. See Moore and Swartout (1990) for a description of such an interface.

Problems of Employment

We envision placing our completed system in a neurology clinic where migraine patients can interact with it as desired. Interaction with the system is not intended to *replace* visits with the physician, but to help the patient make better use of the limited time with the physician. However, the employment of advanced technologies in real life situations is a non-trivial endeavor. As we have argued elsewhere (Ohlsson, 1991), most of the problems involved in the design and use of instructional technologies are not technological. We have so far identified three groups of non-technological problems which we need to address with respect to the migraine tutor.

Linguistic problems. Although we have a technology for generating English text on-line, this technology does not tell us which text we ought to generate. To produce idiomatic, comprehensible, and non-redundant text, our system must be sensitive to at least some of the factors that shape people's utterances. One such factor is that people adapt their formulations to what has already been said in previous parts of the dialogue. For example, an object like a drug can be referred to as "it", if it has been recently mentioned, but not otherwise. As a more complex example, if a patient first asks a doctor about the side effects of Inderal (a migraine drug) and then later asks him or her to describe Elavil (an alternative drug), the doctor is likely to respond to the second request by contrasting Elavil with Inderal. Our system can adapt its text plans to what has been said earlier by including tests on the stored dialogue history among the constraints associated with the operators. However, we do not yet know *how much* of the previous dialogue to take into account. We are currently collecting data on human dialogues in order to categorize such backward references and to determine how far back into a dialogue they extend.

Psychological problems. Students frequently distort the content of science instruction by incorporating what they are taught into their prior misconceptions about the relevant topic (Confrey, 1990). Similarly, migraine patients are likely to have prior beliefs about human physiology and medicine that affect what they will or can learn from our system (Arnaudin & Mintzes, 1985; Furnham, 1988, Chap. 5). Oversimplified causal reasoning is one potential source of difficulty (Einhorn & Hogart, 1986). Migraine attacks are triggered probabilistically by a wide range of factors (chocolate, red wine, stress, etc.) and a patient can be sensitive to more than one factor.

Furthermore, these factors can be additive, so that red wine *and* chocolate taken together might trigger an attack even when either factor by itself would not. Correct understanding of these facts might be hindered by a common tendency to think in terms of single, deterministic causes (Konold, 1989). For example, one physician reported the case of a patient who discovered that red wine triggered her migraines, but who later concluded that she was mistaken on the basis of *a single instance* of a glass of wine that did not cause a headache. We are currently planning a series of studies of people's conception of medical causality and its effect on learning from medical explanations.

Socio-cultural problems. People who work in a medical clinic typically perceive themselves as engaged in the rational enterprise of fixing the ailments of the visiting patients. However, a clinic is also a social system with its own mores and customs. If our migraine tutor is to make a constructive contribution to the life of the clinic, it has to be designed with this system in mind. For example, it is not *a priori* obvious which types of information patients typically request of physicians, nor which kinds of explanations physicians usually give to patients. We began this project with the notion that patients are always interested in the physiological mechanism of their disease and that docors spend at least some of their time explaining disease mechanisms, but we no longer believe this. Patients ask mainly for instrumental information, e. g., information about headache triggers, and doctors report to us that they rarely volunteer information about physiological mechanisms. Our colleague Diana Forsythe at the Intelligent Systems Laboratory is currently conducting ethnographic research in four medical settings in order to study these and related issues (Forsythe, 1992).

Conclusions

Our belief in the viability and general applicability of our approach to online generation of explanations is considerably strengthened by the fact that it could be transferred from the domain of programming style to the rather different domain of migraine treatments. Thus, we now have a general technology for providing a nonexpert with an intelligent interface to expert knowledge, as long as that knowledge is encoded in a computer knowledge base. The implications of such a technology obviously reach beyond our immediate objective of improving patient education.

However, the employment of this or any other instructional technology in situations in real-life situations is a difficult enterprise. The problems of employment are not themselves technological in nature. Linguistic conventions, the psychology of learning, and the structure of the social system in which the technology is to be employed must all be considered.

As our society becomes more knowledge-driven, non-experts will increasingly find that they must acquire at least a rudimentary familiarity with some expert knowledge base in order to stay in control of their own lives. People will more often be learning in real life situations which are not conceptualized as instructional. We believe that the field of applied Cognitive Science would benefit from studies of how to design instructional systems for a variety of such situations.

References

Arnaudin, M. W., & Mintzes, J. J. (1985). Students' alternative conceptions of the human circulatory system: A cross-age study. *Science Education, 69*, 721-733.

Buchanan, B., Moore, J., Forsythe, D., Banks, G., & Ohlsson, S. (1992, February). *Explanation in the clini-*

cal setting (Technical Report). Pittsburgh, PA: University of Pittsburgh.

Buchanan, B., & Shortcliffe, E., (Eds.), (1984). *Rule-based expert systems*. Reading, MA: Addison-Wesley.

Confrey, J. (1990). A review of the research on student conceptions in mathematics, science, and programming. In C. Cazden, (Ed.), *Review of Research in Education* (pp. 3-56). Washington, DC: American Educational Research Association.

Elhadad, M., & Robin, J. (1992). Controlling content realization with functional unification grammars. In R. Dale, E. Hovy, D. Rosver, & O. Stock, (Eds.), *Proceedings of the Sixth Workshop on Natural Language Generation*. Berlin, Germany: Springer-Verlag.

Einhorn, H. J., & Hogarth, R. M. (1986). Judging probable cause. *Psychological Bulletin, 99*, 3-19.

Eraker, S. A., Kirscht, J. P., & Becker, M. H. (1984). Understanding and improving patient compliance. *Annals of Internal Medicine, 100*, 258-268.

Forsythe, D. E. (1992, February). *Using ethnographic studies to build a working system: A patient education system in migraine* (Technical Report). Pittsburgh, PA: University of Pittsburgh.

Furnham, A. (1988). *Lay theories: Everyday understanding of problems in the social sciences*. Oxford, UK: Pergamon.

Konold, C. (1989). Informal conceptions of probability. *Cognition and Instruction, 6*, 59-98.

MacGregor, R. (1988). A deductive pattern matcher. In *Proceedings of the Seventh National Conference on Artificial Intelligence*, August, St Paul, Minnesota.

Moore, J. D. (1989). A reactive approach to explanation in expert and advice-giving systems. Ph. D. dissertation. University of California at Los Angeles, Los Angeles, CA.

Moore J. D., & Paris, C. L. (1989). Planning text for advisory dialogues. *Proceedings of the 27th Annual Meeting of the Association for Computational Linguistics* (pp. 203-211), 26-29 June, Vancouver, British Columbia, Canada.

Moore, J. D., & Swartout, W. R. (1988). *Explanation in expert systems: A survey* (Technical Report No. ISI/RR-88-228). Marina del Rey, CA: Information Sciences Institute.

Moore, J. D., & Swartout, W. R. (1989). A reactive approach to explanation. *Proceedings of the 11th International Joint Conference on Artificial Intelligence* (pp. 1504-1510) August 20-25, Detroit, Michigan, USA.

Moore, J. D., & Swartout, William R. (1990). Pointing: A way toward explanation dialogue. *Proceedings of the National Conference on Artificial Intelligence* (pp. 457-464), July 29 - August 3, Boston, MA.

Ohlsson, S. (in press). The impact of cognitive theory on the practice of courseware authoring. In R. Lewis, (Ed.), *Authoring environments for computer-based courseware*. Berlin, Germany: Springer-Verlag.

Ohlsson, S. (1991). System hacking meets learning theory: Reflections on the goals and standards of research in artificial intelligence and education. *Journal of Artificial Intelligence in Education, 2*, 5-18.

Raskin, N. H. (1988). *Headache* (2nd ed). New York, NY: Churchill Livingstone.

Reason, J. (1990). *Human error*. Cambridge, MA: Cambridge University Press.

Stewart, W., Lipton, R., Celentano, D., & Reed, M. (1992). Prevalence of migraine headache in the United States. *Journal of the American Medical Association, 267*, 64-69.

Tapia, C. P. (1991, September). *MEMO: A migraine therapy explanation module* (Technical Report). Pittsburgh, PA: University of Pittsburgh.

Wilkins, D. E. (1988). *Practical planning: Extending the classical AI planning paradigm*. San Mateo, CA: Kauffmann.

Validating COGNITIO by Simulating a Student Learning to Program in Smalltalk

Yam San CHEE and Taizan CHAN
Department of Information Systems and Computer Science
National University of Singapore
Lower Kent Ridge Road
Singapore 0511
cheeys@nusdiscs.bitnet
chantz@nusdiscs.bitnet

Abstract

We describe COGNITIO, a computational theory of learning and cognition, and provide evidence of its psychological validity by comparing the protocols of a student learning to program in Smalltalk against a COGNITIO-based computer simulation of the same. COGNITIO is a production system cognitive architecture that accounts parsimoniously for human learning based on three learning mechanisms: *schema formation*, *episodic memory*, and *knowledge compilation*. The results of simulation support the validity of COGNITIO as a computational theory of learning and cognition. We also draw some implications of COGNITIO for the teaching of complex problem solving skills.

Introduction

Existing computational theories of cognition appear unable to account for the complexities of human learning and cognition in a comprehensive and integrated manner (Chan, Chee, & Lim, 1992). The major extant architectures seem incapable of providing us with a suitable framework for modeling students learning to program in Smalltalk.

Although Anderson's ACT* and PUPS theories (Anderson, 1983; Anderson, 1989) are able to account, in detail, for skill acquisition, they are unable to describe other aspects of learning such as *assimilation* of new declarative domain knowledge, reliance on prior problem solving *episodes*, and the *formation* of memory *schemata*. Similarly, SOAR (Laird, Rosenbloom, & Newell, 1986; Laird, Newell, & Rosenbloom, 1987) is able to account for skill acquisition but shares the same deficiencies of ACT*.

Other computational theories such as Schank's MOPs and TOPs (Schank, 1982) and Kolodner's E-MOPs (Kolodner, 1983; Kolodner, 1987), on the other hand, account for cognition only from a restricted point of view, namely, that of case-based reasoning (see Slade (1991) for a discussion of the case-based reasoning paradigm). Case-based reasoning and problem solving based on episodic memory are important elements of human cognition that ACT* and SOAR do not account for. Unfortunately, case-based reasoning, on its own, is unable to account for the finer problem solving and learning behavior that ACT* and SOAR offer through their production system architecture and their knowledge compilation (Neves & Anderson, 1981; Anderson, 1987) and chunking mechanism (Laird, Rosenbloom, & Newell, 1986) respectively.

Theories such as Induction (Holland, Holyoak, Nisbett, & Thagard, 1987) and Repair Theory (Brown & Vanlehn, 1980) are also problematic from the perspective of accounting for complex problem solving behavior. The theory of Induction accounts for learning in a problem solving domain through activation: a rule which has been successfully applied will be more highly activated than another which has been unsuccessfully applied. Consequently, when a similar problem arises the next time, the more highly activated rule will be selected first. The theory of Induction seems to describe cognition at a level which is too low for modeling complex problem solving behavior. Furthermore, it omits the role of schematic knowledge in problem solving, an important indication of increasing expertise in a domain (Chi, Feltovich, & Glaser, 1981; Rumelhart, 1980; Rumelhart & Norman, 1978). Repair theory, on the other hand, has been most successfully applied to the study of subtraction. However, it is difficult to extend it to the modeling of programming behavior. While modeling two or three column subtraction can be achieved using a very small number of operators, the programming process entails a much more complex sequence of planning and reasoning steps. Unlike subtraction, the range of possible impasses is virtually unbounded in programming. In addition, Repair Theory also excludes the phenomenon of schematic memory organization. Consequently,

Repair Theory is ill-suited to studying programming behavior.

In response to the perceived deficiencies of existing cognitive theories, we have formulated COGNITIO as an integrated theory of learning and cognition. COGNITIO is particularly well-suited to accounting for learning and cognition by incorporating the formation of schematic knowledge, the use of episodic memory, and the compilation of knowledge.

COGNITIO: An Extended Theory of Cognition

COGNITIO is essentially an extension to the ACT* theory. The architecture of cognition embodied in COGNITIO is given in Fig. 1.

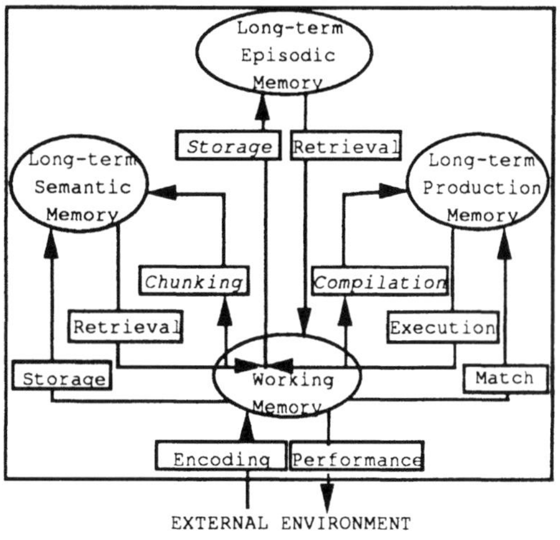

Fig.1 The Cognitive Architecture *COGNITIO*

COGNITIO is a production system theory of cognition. It contains a *working memory* and three separate long-term memories – *semantic memory*, *episodic memory*, and *production memory*. Working memory is a short-term, limited capacity memory which is the activated portion of the semantic memory. Episodic memory contains prior problem solving episodes which are represented as plan-trees (Chan, Chee, & Lim, 1992). When the conditions in a production rule contained in the long-term production memory match the state of the working memory, the rule is fired, and the actions (mental and possibly physical) are performed. Cognitive behavior is the result of a series of production matchings and firings.

More importantly, COGNITIO is also a theory of *learning*. COGNITIO postulates that the long-term memories are transformed *as a result of production firings*. That is, learning in a domain occurs only if that domain knowledge is used in solving problems of the domain. These long-term memory transformations are evidenced by increasing competence in a domain. The transformations occur through three learning mechanisms: *schema formation*, *episode storage/retrieval*, and *knowledge compilation*.

(i) *Schema Formation* (or Declarative Chunking). Related memory elements from semantic memory which are often accessed together in solving problems are organized together into higher-level memory structures – *schemata* – which can be accessed later as individual units in working memory. Schema formation reduces the demand on working memory and enables a person to view a complex concept or problem in an appropriate context. The degree of coherence and understanding achieved by virtue of a newly acquired schema will depend on how well elaborated it is. In addition, a schema that has become a unit by itself can become part of a larger schema as the learner continues with his learning. Such schema formation corresponds to the ability to build a better mental model of a problem at hand since more information is made available, in a coherent form, at the time a person reasons about the problem.

(ii) *Episode Storage/Retrieval*. A person can rely on his prior experience to guide the solution to a new but similar problem. Each problem-solving experience is considered an *episode* and is stored in the long-term *episodic memory*. When a new problem bears some resemblance to a problem that has been solved previously, a person could be reminded of that previous problem-solving episode. He would then rely on that episode for some of the steps performed or decisions made previously, instead of attempting to solve the current problem anew. An episode is retrieved when an episode's goal and conditions match the current goal and current working memory content.

(iii) *Knowledge Compilation*. Knowledge compilation accounts for increasing fluency in skill acquisition and ultimately to a high degree of automaticity. The two submechanisms of knowledge compilation are *proceduralization* and *composition*. Proceduralization creates a new production by removing or modifying conditions that require access to long-term semantic memory so that the semantic knowledge is built into the new production itself. Composition creates new productions by composing two productions that are fired in sequence in achieving related goals into another production that, when fired, will have the same effect as the two original productions fired in sequence.

In summary, COGNITIO is a computational theory of learning and cognition. It improves upon earlier models of cognition by incorporating the role of episodic memory in a production-based architecture and by specifically including a mechanism for schema formation and accommodating the role of schemata in domain knowledge assimilation.

Validating COGNITIO

This section describes a COGNITIO-based simulation of a student's learning and problem-solving behavior. The first subsection describes the nature of the protocols of a student learning to program in Smalltalk. These protocols provide the psychological data for testing the validity of COGNITIO. The second subsection describes the simulation itself. The third subsection summarizes the simulation results.

Protocol Collection

We performed a simulation of a student learning to program in Smalltalk based on the protocols collected from a first-year computer science undergraduate at this university. The student had only studied Pascal programming for one semester. Both video and audio protocols were collected of the entire Smalltalk study session (total duration of approximately six hours). The student read aloud the instruction and verbalized his thoughts while trying to understand the instructions and while solving problems. The instructions explained Smalltalk concepts and the programming interface. They also required the student to solve problems periodically by being engaged in a financial game. An instructor was present to provide assistance whenever necessary. The video recordings captured the interaction between the student and the Smalltalk programming environment. The audio recordings captured the verbalizations of the student and the dialogs between the student and the instructor.

The Simulation

This section describes the simulation of the opening segment of the protocols. A schematic version of the protocols is contained in the Appendix. The first subsection describes our approach to simulating the learning embodied in the protocols while the second subsection describes the simulation in detail.

Process of simulation. A computer system based on the COGNITIO architecture and theory was implemented to simulate the student's protocols. The process of simulation runs directly parallel to the unfolding learning behavior. Consequently, at points where the student read instruction, knowledge corresponding to the instruction was encoded (in propositional form) and entered into the system. Similarly, at points where a problem had to be solved, the system was given the corresponding problem solving goal to achieve. Thus, a detailed, step-by-step simulation of the learning process was performed. Such an approach is significant in that knowledge acquired at time *t1* affects not only behavior at time *t2* (later) but also how new knowledge entered into the system at time *t2* is interpreted and integrated into the system memory. Consequently, assimilation of new declarative knowledge is directly dependent on what is already known.

The simulation results. At the commencement of the simulation, the system only contains a few productions that represent the weak problem solving method of *divide-and-conquer*. These productions, like other productions that are used for modeling problem-solving behavior, have actions that would normally generate subgoals to achieve a given goal. For example, one weak-method production encodes the following rule:

> **If** a goal is to achieve a certain function and
> some operation can achieve that function and
> that operation has a number of steps
> **then** set subgoals to perform those steps.

The long-term semantic memory also contains the propositional form of the instruction that was read by the student before he attempted the first problem, namely, to determine the balance in the account object `MyAccount`. Like the student (in steps 4 to 8), the system solves the problem in the three steps given, namely, by typing the expression "`MyAccount` **`queryBalance`**", highlighting the expression, and selecting "print it" from the operate menu. The steps were generated as three subgoals in the simulation. After solving this problem, no schema is formed and no productions are compiled because the level of activation of semantic memory elements involved in solving the problem does not exceed a prespecified threshold. However, the system's episodic memory has been registered with the steps the system took to evaluate the expression "`MyAccount` **`queryBalance`**".

The preceding episode guides the solution to the second problem of determining the interest rate of `MyAccount`. The system, like the student (in steps 10 to 12), is able to type out the expression that was thought to be needed, to highlight it, and to select "print it". This behavior is best explained as behavior that results from invoking the knowledge stored from the first problem solving episode. It cannot be explained in terms of knowledge compilation because the student is, at this point, still verbalizing the need to highlight the expression (*"so we select the whole thing"*) and no production was compiled after the first problem solving episode.

The system, like the subject (in step 10), however, fails to type the object `MyAccount` in the expression evaluated. There are two reasons for this. First, the previous episode only encoded what steps needed to be performed to achieve the desired function, namely, to type the expression, highlight it, and select "print it". The episode did not encode the knowledge about the structure of a message expression – that a message expression consists of an object followed by a message. Secondly, there was a lack of explicit specification of the object in the instructions but strong association of **interestRate** to the function of determining interest rate. As a result, the expression to be evaluated is associated with only the message **interestRate** in the working memory leading to the incorrect behavior observed. However, when the cause of the error was pointed out by the instructor (missing receiver object, MyAccount, in the message expression), the student (in step 14) was able to make the appropriate correction (steps 15-17).

When the system (and the student in step 18) is given the goal to start up the trading place, it (like the student) is able to code the expression correctly (steps 19-21). Furthermore, from the verbalizations of the student (*"TradingPlace is the object"* and *"**startUp** is the message"*), it can be seen that a schema encoding the structure of a message expression is being formed. The system simulates this in corresponding fashion by creating a *message expression* schema that comprises an object slot, a message slot, and a constraint slot specifying that the receiver object must precede the message selector.

As the simulation proceeds, the message expression schema is enhanced with a slot indicating that a Smalltalk object is returned from the evaluation of a message expression (steps 23-25). Evidence of this becomes apparent later (step 31) by the way the student interprets the nested message expressions in step 30. Furthermore, a production that proceduralizes the steps to evaluate a message expression is also formed. The omission of verbalization and the speed with which the expression was evaluated (that is, highlighting and selecting "print it" in step 23) are supportive of the proceduralization having occurred.

As the student progressed with his learning, he acquired more schemata, more experience from prior problem-solving episodes, and more productions formed in production memory. For example, after about an hour and a half after the protocol extract shown in the Appendix, the student had also formed an appropriate *method* schema that resulted from coding several expression series, examining and understanding methods in detail, and coding a method. Supporting evidence for this schema formation is provided when the student was trying to understand the following given method:

```
buyShareOfName: aString
 SharesPortfolio add:(TradingPlace
           buyShareOfName:aString
           usingAccount: self)
```

in which the global variable SharesPortfolio is accessed in the method. The student elaborated his *method* schema in the following verbalization: *"SharesPortfolio being a global variable, we can do this"*. This is because the method schema specifies the parts of a method as well as the constraint that only instance and class variables are accessible in the method of the receiver. Thus, upon encountering Sharesportfolio which is neither the instance or class variable of an account object, the student is trying to interpret the method in a manner which is consistent with his current schema.

Evidence for problem solving based on episodic memory is provided at a point in time after the student coded the above method. The student remembered that he had previously used another message which drew cash directly from the account to get a higher discount: *"remember there is **usingCashOf**"*, (after locating the actual message name) *"so, if I were to rewrite my method...of the...to **usingCashOf withdraw** amount, wouldn't that be better?"*. As a result, he changed the given method to the one that gave a higher discount. This behavior can be simulated by inserting an additional condition in the working memory that a maximum discount is required so that the episode involving the use of the nested message:

```
"TradingPlace buyShareOfName:aShareName
         usingCashOf:(MyAccount
              withdraw:anAmount)"
```

is retrieved. Without the learning mechanism of episodic memory, the working memory must first be augmented with the various facts and relationships between the messages **withdraw:** and **buyShareOfName:usingCashOf:**. A series of reasoning steps must then be performed to construct the appropriate nested message. However, the protocol of the student did not reveal the occurrence of such reasoning steps. In addition, knowledge compilation cannot be used to explain the behavior because the student was unable to reproduce the whole nested expression readily. Instead, he was reminded of the episode, but still had to rely on the text for the exact form of the message expression required.

There was also ample evidence of knowledge compilation. However, most of the rules compiled were interface-related; for example, copying a method from one class to another through the system browser. Since the knowledge compilation mechanism operates the same way as in ACT*, we do not elaborate on this further.

Simulation Conclusions

The simulation results described above demonstrate that COGNITIO provides a faithful computational account of cognition and learning. In particular:

(i) The basic production system architecture can be used to simulate cognition, and especially learning. This is consistent with the ability of ACT* (Anderson, Farrell, & Sauers, 1984) and SOAR (Lewis, et al, 1990) to simulate human problem solving behavior.

(ii) The three learning mechanisms, namely, schema formation, storage/retrieval of episodes, and knowledge compilation, as embodied in COGNITIO are all essential components of any account of human learning. Taken individually, each mechanism may account for one aspect of learning but not another. By integrating the three learning mechanisms in a parsimonious way, COGNITIO is able to account for learning in a more integrated and powerful way.

Implications of COGNITIO

The simulation highlights some implications for teaching Smalltalk and other skills in general:

(i) It is important that a student be given ample practice and exposure to concepts (for example, the structure of message expressions) before he has to proceed further in a course of instruction. This is to ensure that appropriate knowledge has been schematized so that it will leave more working memory capacity available for new concepts to be acquired.

(ii) Instruction for teaching a skill, such as Smalltalk programming, should be designed with the aim of equipping the student with the various schemata of the essential concepts of the skill. For example, in Smalltalk, some relevant schemata are the message expression schema, the method schema, the class schema, and so on. The observation is that a schema can aid the student in understanding related concepts that will be introduced later. For example, a schema for message expressions is essential for understanding more complicated concepts such as nested message expressions or methods.

(iii) Parts of a knowledge or skill can be acquired through episodic learning but other parts might be better taught explicitly. This is illustrated by the earliest part of the simulation when the student had no difficulty performing the steps to evaluate an expression but forgot to include the object in the message expression. Thus, a "semantic" relationship, such as that between an object and a message, is best taught explicitly; failure to do so entails the risk of the relationship being ignored by a student.

Conclusions

This paper has described a computational theory of learning and cognition, as embodied in COGNITIO. The theory is able to account for the schematic organization of memory, the role of episodic memory in learning, and skill acquisition in cognition. Evidence of the validity of COGNITIO was demonstrated by its application to the simulation of a student learning to program in Smalltalk. The simulation also highlights some implications for teaching problem-solving skills such as Smalltalk programming.

Appendix. A schematic protocol of a student learning to program in Smalltalk is shown below. The student's actual verbalizations are shown in italics in square brackets.

1. Student learns that to achieve something in Smalltalk, he must send a message to some object.
2. Student learns that the basic format to do so is "object message-name".

[Problem Solving Episode 1: Steps 3-8]

3. Student is asked to determine the balance in `MyAccount` by sending the message **`queryBalance`** to the object `MyAccount` in the Workspace window (already opened).
4. The steps given are: (a) type "`MyAccount` **`queryBalance`**", (b) highlight the expression, and (c) select "print it" from the operate menu.
5. Student types the expression.
6. Student highlights the expression.
7. Instructor teaches how and where to select "print it"
8. Expression is evaluated – 5000 (which is the current balance in `MyAccount`) is shown beside the evaluated expression.

[Problem Solving Episode 2: Steps 9-12]

9. Student is asked to determine the interest rate offered in `MyAccount` by sending it the message `interestRate`.
10. Student thinks aloud [*"**interestRate**"...*] as he types "**`interestRate`**"
11. Student thinks aloud [*"so we select the whole thing"...*] as he highlights "**`interestRate`**"
12. Student selects "print it".
13. Error message appears (because the system treats **`interestRate`** as an object according to the syntax of Smalltalk, and such an object does not exist).
14. Instructor points out the mistake that the object in the expression has been omitted.

[Problem Solving Episode 3: Steps 15-17]

15. Student types "`MyAccount`" before "**`interestRate`**".
16. Student thinks aloud [*"**interestRate** is not object, **interestRate** is message"*].
17. Student evaluates the expression – 0.04, which is the interest rate, is shown.

[Problem Solving Episode 4: Steps 18-21]

18. Student is asked to start a `TradingPlace` running by sending it a message, **`startUp`**.
19. [*"`TradingPlace` is the object"*] types "`TradingPlace`".
20. [*"**startUp** is the message"*] types "**`startUp`**".
21. Student evaluates the expression by highlighting it and selecting "print it" (a window representing a trading place appears).

[Problem Solving Episode 5: Steps 22-25]

22. Student is asked to buy shares by sending the message **buyShare** to `TradingPlace`.
23. Student types "`TradingPlace`", and then "**buyShare**"; he then evaluates it – `anUserShare` is shown. [In Smalltalk, an object is always returned as a result of executing a message expression. If a user selects "print it" as the command for executing an expression, the object will be printed. If an object is not a literal, its class will be printed.]
24. Student tries to determine what is `anUserShare`.
25. Instructor points out that it is an indication that a share object is returned.
[Problem Solving Episode 6: Steps 26-27]
26. Student is asked to buy shares by evaluating the expression

```
"TradingPlace buyShareOfName:'Emtex'
              usingAccount:MyAccount".
```

27. Student types "`TradingPlace`", then types "**buyShareOfName**:`'Emtex'`" and then types "**usingAccount**:`MyAccount`" and then evaluates it – `anUserShare` is shown.
28. Student is told that the he has lost access to the shares bought because they were not stored away somewhere.
29. Student is told that the shares to be bought later can be stored in the set object `SharesPortfolio` and that a share can be added into the set by sending it the message "**add**:`aShare`".
30. Student is asked to evaluate one of the messages in order to buy shares and add it into the `SharesPortfolio`.

```
(i) SharesPortfolio add:TradingPlace
                        buyShare
(ii) SharesPortfolio add:
(TradingPlace buyShareOfName:shareName
              usingAccount:anAccount)
(iii) SharesPortfolio add:
      (TradingPlace
           buyShareOfName:shareName
           usingCashOf:
             (anAccount withdraw:amount))
```

31. Student interprets the above messages (i) and (ii) as *["**add:** is the message...message a share...the argument being sent in is...when is returned is `anUserShare` and this is an object so goes into **add:** and this **add:** goes into `SharesPortfolio` which is the set object"]* and then (iii) as *["**withdraw:** is another message whereby it allows you to say how much to withdraw ...and same thing...and account withdraw will return you an object..."]*

References

Anderson, J. R. 1983. *The Architecture of Cognition.* Cambridge, MA: Harvard University Press.

Anderson, J.R. 1987. Skill acquisition: Compilation of weak-method problem solutions. *Psychological Review*, 94(2): 192-210.

Anderson, J.R. 1989. A theory of the origins of human knowledge. *Artificial Intelligence*, 40: 313-351.

Anderson, J. R., Farrell, R., and Sauers, R. 1984. Learning to program in Lisp. *Cognitive Science*, 8: 87-129.

Brown, J. S., and Vanlehn, K. 1980. Repair theory: A generative theory of bugs in procedural skills. *Cognitive Science*, 4: 379-426.

Chan, T., Chee, Y. S., & Lim, E. L. 1992. COGNITIO: An Extended Computational Theory of Cognition. *Proceedings of ITS'92*. New York, NY: Springer-Verlag. Forthcoming.

Chi, M. T. H., Feltovich, P., and Glaser, R. 1981. Categorization and representation of physics problems by experts and novices. *Cognitive Science*, 5: 121-152.

Holland, J. H., Holyoak, K. J., Nisbett, R. E., and Thagard, P. R. 1987. *Induction: Processes of Inference, Learning, and Discovery*. Cambridge, MA: MIT Press.

Kolodner, J. L. 1983. Towards an understanding of the role of experience in the evolution from novice to expert. *International Journal of Man-Machine Studies*, 19: 497-518.

Kolodner, J. L. 1987. Extending problem solving capabilities through case-based inference. In *Proceedings of Forth Annual International Learning Workshop*. Los Altos, CA: Morgan-Kaufmann.

Laird, J. E., Rosenbloom, P. S., and Newell, A. 1986. Chunking in Soar: The anatomy of a general learning mechanism. *Machine Learning*, 1: 11-46.

Laird, J. E., Newell, A., and Rosenbloom, P. S. 1987. Soar: An architecture for general intelligence. *Artificial Intelligence*, 33: 1-64.

Lewis, R. L., Huffman, S. B., John, B. E., Laird, J. E., Lehman, J. F., Newell, A., Rosenbloom, P. S., Simon, T., and Tessler, S. G. 1990. Soar as a unified theory of cognition. In *Proceedings of the 12th Annual Conference of the Cognitive Science Society*. Hillsdale, NJ: Lawrence Erlbaum.

Neves, D. M., and Anderson, J. R. 1981. Knowledge compilation: Mechanisms for the automatization of cognitive skills. In J.R. Anderson (ed.), *Cognitive Skills and Their Acquisition*. Hillsdale, NJ: Lawrence Erlbaum.

Rumelhart, D. E. 1980. Schemata: The building blocks of cognition. In R. J. Spiro, B. C. Bruce, and W. F. Brewer (eds.), *Theoretical Issues in Reading Comprehension*. Hillsdale, NJ: Lawrence Erlbaum.

Rumelhart, D.E., and Norman, D. A. 1978. Accretion, tuning, and restructuring: Three modes of learning. In J. W. Cotton and R. L. Klatzky (eds.), *Semantic Factors in Cognition*. Hillsdale, NJ: Lawrence Erlbaum.

Schank, R. C. 1982. *Dynamic Memory – A Theory of Reminding and Learning in Computers and People*. Cambridge: Cambridge University Press.

Slade, S. 1991. Case-based reasoning: A research paradigm. *AI Magazine*, 12(1): 42-55.

Using Theory Revision to Model Students and Acquire Stereotypical Errors *

Paul T. Baffes and Raymond J. Mooney
Department of Computer Sciences,
University of Texas at Austin,
Austin, Texas 78712
baffes@cs.utexas.edu, mooney@cs.utexas.edu

Abstract

Student modeling has been identified as an important component to the long term development of Intelligent Computer-Aided Instruction (ICAI) systems. Two basic approaches have evolved to model student misconceptions. One uses a static, predefined library of user bugs which contains the misconceptions modeled by the system. The other uses induction to learn student misconceptions from scratch. Here, we present a third approach that uses a machine learning technique called *theory revision*. Using theory revision allows the system to automatically construct a bug library for use in modeling while retaining the flexibility to address novel errors.

1 Introduction

One of the most important components of an Intelligent Computer-Aided Instruction (ICAI) system is the student model (Wenger, 1987). Some researchers have argued (Carbonell, 1970; Laubsch, 1975) that the effectiveness of an ICAI system depends heavily upon its student modeling component. Without the flexibility to model novel student errors, ICAI systems will not progress much beyond today's electronic page turners with canned responses tuned to the average student.

Over the last two decades, several techniques for student modeling have been developed. One method, called *overlay* modeling (Carr and Goldstein, 1977), assumes a student's knowledge is always a subset of the correct domain knowledge. While simple to implement, this method is incapable of capturing misconceptions, or *bugs*, that represent faulty student knowledge.

To capture such misconceptions, other researchers (Brown and Burton, 1978; Burton, 1982; Brown and VanLehn, 1980; Sleeman and Smith, 1981) have focused on the use of bug libraries. In these approaches, models are built by matching student behavior against a catalog of bugs. Typically, such catalogs are either difficult to construct or fail to cover a wide enough range of behaviors.

A third method of student modeling attempts to model student misconceptions without overlays or a bug library (Langley et al., 1984; Ohlsson and Langley, 1985). Here, *induction* is used to construct a student model from examples of student behavior. While this provides more flexibility, in general accurate induction requires a large number of such examples. Moreover, this approach cannot take advantage of likely misconceptions which could be preprogrammed.

Here we present a new algorithm for student modeling called ASSERT (Acquiring Stereotypical Student Errors using Revision of Theories). There are two main contributions of our algorithm. First, ASSERT demonstrates a new method for constructing student models using a machine learning technique called *theory revision*. Theory revision allows ASSERT to build models more accurately with fewer examples of student behavior. Theory revision also enables ASSERT to utilize a complete or partial bug library. Second, ASSERT provides a new method for automatically constructing and extending a bug library by combining multiple student models into a *stereotypical student model*. Thus ASSERT can can create and use a bug library, while retaining the flexibility to address novel student errors.

2 Overview of Theory Revision

Theory revision algorithms modify existing rule bases to make them consistent with a given set of examples. Unlike induction algorithms which receive only examples as input, theory revision systems expect both the examples and a set of rules (theory). Typically, theory revision algorithms are used under the assumption that the rules are partially correct but not yet completely defined to cover all examples. The theory is successively refined, by specialization and generalization, until it is consistent with the examples. Most theory revision systems attempt to change the input rules as little as possible to accom-

*This research was supported by the NASA Graduate Student Researchers Program under grant number NGT-50732, the National Science Foundation under grant IRI-9102926, the NASA Ames Research Center under grant NCC 2-629, and a grant from the Texas Advanced Research Program.

modate the examples.

Unfortunately we do not have the space to describe theory revision in detail. However, note that theory revision systems have been implemented using a variety of techniques, including both logic and connectionist frameworks. Here we use the EITHER system, which revises theories expressed in an extended propositional logic. In EITHER, theories consist of rules written as Horn clauses and examples represented as vectors of observable features. EITHER uses abductive and inductive reasoning to affect six types of changes. Antecedents can be specialized, generalized, added or deleted, and rules may be added or removed from the theory. For a more detailed overview of theory revision and the EITHER algorithm, see (Ourston and Mooney, 1990).

3 The ASSERT Algorithm

3.1 Student Modeling as Theory Revision

Our description of the ASSERT algorithm begins with the observation that student modeling can be viewed as theory revision. Due to the restrictions imposed by EITHER, it is assumed that the tutoring task is a categorization problem (multiple concept lesson). While categorization problems have not been a major focus of ICAI modeling efforts, concept lessons have a well understood pedagogy (Dick and Carey, 1990) and are common CAI applications. Furthermore, as Gilmore and Self (Gilmore and Self, 1988) have pointed out, machine learning has been successfully applied in categorization domains making it natural to explore its potential in concept tutorials. Other tasks, specifically procedural ones, in general cannot be represented using propositional Horn clauses. It is important to point out, however, that the basic technique of using theory revision for student modeling is not limited to categorization domains since other theory revision algorithms may use different underlying representations.

Given this assumption, the correct knowledge for the task can be represented in a straightforward manner using Horn clause rules. Each concept is represented as the head of a Horn clause, and the components of that concept make up the predicates that form the body of the clause. Disjunctive concepts are represented using multiple clauses. Figure 1 shows part of a theory for animal classification.

The full theory classifies examples as one of twelve different animals. The rules form a hierarchy where the consequents of some rules are referenced as antecedents in others. Disjunctive concepts (*e.g.* "`mammal`") are represented by multiple rules. Examples are classified by the theory via rule chaining. For instance, the following example

```
mammal    ←  birth=live
mammal    ←  feed-young=milk
ungulate  ←  mammal & ruminate
giraffe   ←  ungulate & neck=long &
             pattern=spots
```

Figure 1: Animal classification rules.

```
(birth=live & ruminate & neck=long &
 feed-young=milk & pattern=spots)
```

is classified by the rules in Figure 1 as a giraffe. Either `birth=live` or `feed-young=milk` suffices to prove `mammal` which, when combined with `ruminate` proves `ungulate`. The rest of the facts combine for the final categorization as a giraffe.

3.2 Constructing Student Models

Modeling faulty student knowledge now becomes a matter of modifying the theory to make it consistent with responses generated by the student. In other words, the *correct* theory modeling a *perfect* student is altered using theory revision to match actual student behavior. This is contrary to the typical use of theory revision, but in principle there is no difference. We are simply reversing the notion of "goodness": instead of fixing incorrect theories, we use theory revision to *introduce* faults to model the incorrect knowledge of the student.

As an example, consider again the giraffe example above. If the feature `pattern=spots` were absent and the student still classified the example as a giraffe, there would be an inconsistency between the rules and the student's observed behavior. EITHER would modify the rules to account for the discrepancy by removing the `pattern=spots` antecedent from the giraffe rule, as long as this change remained consistent with other classifications made by the student.

ASSERT begins with a correct theory of perfect student behavior and a list of examples as categorized by a particular student. These student categorizations could be collected in any number of ways including a multiple choice test where the student classifies examples represented as lists of features (such tests are common in the instructional design of CAI systems). Here we assume that a multiple choice test can be constructed from a pool of examples and the results given to EITHER as examples of a particular student's behavior. EITHER then changes the correct input theory to match the student's erroneous classifications.

3.3 Building the Stereotypical Model

For many tutoring domains, it is possible to outline typical misconceptions that a student might exhibit. This is one of the justifications for the bug library approach described earlier. Assuming that such errors will be common, it makes sense to collect several student models and note the commonalities that exist across students. Using these commonalities one can form a representative student model which we call a *stereotypical student model*. Construction of the stereotypical student model proceeds in four phases as follows.

Phase 1: *Collection of student models.* First, several student models must be generated from the same input theory using the process already described.

Phase 2: *Sorting of rule changes.* Next, all changes from all the student models are grouped by the rule altered and the *type* of change made. As mentioned earlier, there are six types of changes or *deviations* that EITHER can make to a rule. Each deviation may consist of multiple *component* changes to the rule. The result of this sorting is a list of proposed deviations to each rule, grouped by type. The size of the group equals the number of different student models that proposed a deviation to the given rule.

Phase 3 *Thresholding.* Each group of deviations associated with a rule is discarded if the size of that group does not exceed a desired *threshold*. This ensures that only those changes which are common to multiple students are incorporated into the stereotypical student model. The threshold can be modified to make the system more or less conservative about what deviations are considered stereotypical.

Phase 4: *Extraction of common changes.* After thresholding, all the deviations within a group represent a particular type of change to the rule. However, since these changes come from different student models, they will not necessarily be the same. To pull out only what is common among all the deviations, ASSERT uses the *common component extraction* algorithm shown in Figure 2. This algorithm measures commonality using two metrics: (1) the number of student models that contain the component and (2) the size of the component, where larger components represent more specific changes. Large frequent changes are preferred. The algorithm is iterative; as each common component is selected, that component is removed from all the deviations of the group before selecting the next component.

To illustrate the steps for constructing a stereotypical student model, refer again to Figure 1. Assume that three student models have been generated, and all have proposed changes to the last rule of the theory as shown in Figure 3. Assume further that the value of the threshold is 2. Since there are two different types of changes, two groups of deviations will be formed. The first, for adding rules, will contain all the `giraffe` rules from each student model. The second, for generalizing the mammal rule, will contain only the `mammal` rule from student model 2. This second group will be thrown out during thresholding since only one deviation is in the group and the threshold is set at 2.

1. Compare all components of all deviations.
2. For each component-component comparison, find the common subcomponent.
3. Store each subcomponent with a count of the number of *different* student models in which it was present. Call this count "N".
4. Select the "best" subcomponent based on the formula "L*N" where "L" is the length of the subcomponent. Add the subcomponent to the common deviations to be returned.
5. Remove all components from all deviations that are subsumed by the "best" subcomponent of step 4.
6. Repeat steps 1-5 until there are no common subcomponents (*i.e.*, step 2 produces the empty set). Return the subcomponents collected in step 4.

Figure 2: Common component extraction algorithm.

Student Model 1: 2 rules added
    `giraffe` ← `foot-type=hoof` & `ungulate`
    `giraffe` ← `color=tawny`

Student Model 2: 1 rules added, 1 rule changed
    `giraffe` ← `color=tawny` & `ungulate`
    `mammal` ← `birth=live or egg`

Student Model 3: 2 rules added
    `giraffe` ← `foot-type=hoof` & `ungulate`
    `giraffe` ← `color=tawny` & `ruminate`

Figure 3: Example student models.

This leaves the three student models proposing added rules. Table 1 shows how each of the components of these deviations is measured by the common component extraction algorithm. While `color=tawny` and `ungulate` appear the most frequently ($N = 3$), the conjunct `foot-type=hoof & ungulate` has a larger product ($L * N = 4$) and is thus selected first.

Next, all of the rules that are subsumed by `foot-type=hoof & ungulate`, are removed from the student models. This leaves the second rule from model 1, the first rule from model 2, and the second rule from model 3. The only remaining common element is `color=tawny` which is extracted as the second component of the stereotypical model. This last extraction covers the rest of the remaining

Subconjunct	*L*	*N*	*L* ∗ *N*
`foot-type=hoof`	1	2	2
`ungulate`	1	3	3
`color=tawny`	1	3	3
`foot-type=hoof & ungulate`	2	2	4

Table 1: Subconjunct comparison table.

rules. The final stereotypical model is

giraffe ← **foot-type=hoof** & **ungulate**
giraffe ← color=tawny

3.4 Using the Stereotypical Model

Once the stereotypical student model has been generated, it can be used directly as a bug library. However there are two different ways of incorporating its information into the modeling process. One method would be to modify the search mechanisms employed by EITHER to prefer bugs in the stereotypical model over the normal theory revision process. This would mirror the traditional use of bug libraries; bugs would be tried singly or combined in groups to predict student behavior. If no bug combination produced an accurate model, the normal theory revision process would be invoked.

A second method for incorporating the stereotypical model relies on the fact that theory revision is input/output compatible. Specifically, the input to theory revision (a theory) is identical in form to the output (a revised theory). Thus the bugs stored in the stereotypical student model can be used by simply incorporating them into the theory used to model subsequent students. Due to its simplicity, this was the approach taken here for our initial test of ASSERT.

Of course, it is unlikely that any one student will exhibit exactly the bugs of the stereotypical student model. The result is that the theory revision algorithm may be forced to repair bugs just introduced. On the other hand, it is rare to find a student who has no misconceptions in common with the average bugs. On the average, it was hoped that revising a stereotypical set of rules would be superior to revising a correct theory.

4 Empirical Results

4.1 Experimental Design

Two hypotheses formed the basis of our testing methodology. First, we expected theory revision to be more accurate at student modeling than inductive modelers due to the extra information available in the input rules. Second, we expected revising stereotypical theories to be more effective than revising correct theories since common student errors are part of the stereotypical model.

For the preliminary experiments presented here, we chose to work with artificial data for the animal classification domain referenced earlier (see Figure 1). We are currently planning experiments using actual student data collected with a CAI system for more realistic testing. As an initial domain, the animal classification rules represent a rich enough task to test our hypotheses on a variety of potential student misconceptions.

Our tests were run from a pool of 180 examples randomly generated using the correct animal classification rules (15 examples for each of the 12 categories). Artificial students were generated by making modifications to the correct theory. As each student theory was formed, it was used to relabel the 180 examples to simulate the behavior of that student. These relabeled examples act as "answers" the student would generate to the 180 "multiple choice questions."

Modifications made to the correct theory to create students were of two types. One set of modifications was predefined, with a given probability of occurrence. These simulated common errors that occurred in the student population. We used four common deviations, each with a 0.75 probability of occurrence. Two of these deleted antecedents from rules, one added an antecedent, and one changed an antecedent. To simulate individual student differences, each student theory was further subjected to random antecedent modifications with a probability of 0.10.

ASSERT was tested against both normal theory revision and induction using a two-phased approach. The first phase was used to build a stereotypical model for the second phase as follows:

1. First, 20 artificial students were created using the methods described above.
2. For each student, all 180 examples were relabeled using the student's buggy theory.
3. From these 20 students, 20 student models were generated using EITHER on all 180 relabeled examples.
4. A stereotypical student model was then built from the 20 student models using the algorithm from section 3.3 with a threshold of 10 (*i.e.*, half the students had to exhibit a bug for it to be considered "common").

Three of the four common predefined bugs ended up in the stereotypical student model. The fourth was more difficult for EITHER to generate, since it required a deletion of an antecedent followed by an addition of a different antecedent. This fourth bug ended up as two different rules in the stereotypical student model. All four resulting deviations were

Initial Rules	*modeling time (sec.)*	*test set accuracy*
stereotypical	72	96%
correct	124	84%
stereotypical, no revision	n/a	67%
correct, no revision	n/a	61%
none (induction)	12	52%

Table 2: Effect of initial rules on modeling.

applied to the correct rules to form a stereotypical student theory.

For the second phase, additional artificial students were generated to test EITHER using various initial theories. A series of experiments were run starting EITHER with (1) the correct animal rules, (2) the stereotypical student rules from phase 1 above, or (3) no initial theory. With no initial theory, EITHER defaults to an inductive learning process which uses the ID3 (Quinlan, 1986) algorithm. This phase ran as follows:

1. First, 10 new artificial students were generated using the same techniques used in phase 1. For each, the 180 examples were relabeled using the student's buggy theory.
2. 50 examples were randomly chosen from the 180 relabeled by the student as training examples. Each new student was modeled using EITHER with the same 50 examples and one of the three initial theories described above.
3. The other 130 examples were reserved for testing the accuracy of each student model as follows. Recall that the output of EITHER is a revised theory representing the student model. This theory was used to label each of the 130 test examples. These labels were compared to those generated using the student's buggy theory from step 1 to compute a percentage accuracy.

Table 2 compares the average accuracy and modeling times of EITHER started with each of the three different initial theories. For comparison purposes, we also measured the accuracy of both the correct and stereotypical theories. Statistical significance was measured using a Student t-test for paired difference of means at the 0.05 level of confidence (*i.e.*, 95% certainty that the differences were not due to random chance). All the differences shown in table 2 are statistically significant.

4.2 Discussion of Results

Both of our hypotheses were borne out by the results presented in table 2. First, it is apparent that theory revision is superior to induction in terms of accuracy. This is not surprising since induction must model *correct* as well as buggy student behavior, whereas theory revision need only alter correct rules to capture the misconceptions. The difference is even more pronounced when theory revision proceeds from the stereotypical model. Induction simply has more work to do.

Second, our results show that theory revision models students faster and more accurately when given an initial rule base that approximates typical student errors. Since all the students were generated using the same criteria, providing EITHER with the stereotypical rules effectively gives it a head start over the correct theory.

Running the correct and stereotypical theories without revision also produced interesting results. Both outperformed induction, further illustrating the disadvantage of trying to model students from scratch using a small number of examples. It is also apparent that revision is essential to effective modeling, even if the initial rules model buggy student behavior. Without revision, modeling novel student errors is simply not possible since a static library of bugs will not contain the needed information.

5 Related Work

There are two systems directly related to the work described here. Both make use of machine learning techniques to dynamically model student behavior.

Sleeman describes an extension to his PIXIE system, which models arithmetic errors, called INFER* (Sleeman et al., 1990) INFER* starts with a library of known bugs and induces rules to fill gaps between *one* student's solution and the correct rules known to the system. INFER* also relies heavily upon domain-dependent heuristics for controlling its search, and it is not clear that these techniques can be used for domains other than arithmetic. Furthermore, INFER* does not make any attempt to generalize across students in an effort to extend its library of bugs.

Like INFER*, the theory revision techniques used by ASSERT can be biased with specific heuristics and known bugs, but will operate effectively without them. Furthermore, ASSERT contains an algorithm for extracting common elements from multiple student models and thus can automatically extend its library of buggy rules.

Langley et. al. (Langley et al., 1984; Ohlsson and Langley, 1985) describe the ACM system which uses induction to generate a production system model of an individual student from a problem space of operators describing the domain. While ACM is a domain independent algorithm, Langley et. al. make the assumption that student errors are only the result of *correct* actions taken in an incorrect *context*. This prohibits ACM from modeling illegal actions. Also, each run of ACM starts with no knowledge of when operators should be applied, forcing it to spend time modeling both correct and buggy student control knowledge that could be preprogrammed. Finally, there is no facility within ACM for building in typical student bugs nor for using the output of one run to aid subsequent modeling efforts.

6 Future Work

There are two chief disadvantages to the current ASSERT system. First, we have not tested ASSERT on real student data. Our current efforts are focused on obtaining data to run such tests. Second, as discussed above (section 3.4), our simple method of incorporating all bugs from the stereotypical model should be replaced with a revised theory revision algorithm that is biased towards preferring the bugs. Common misconceptions would be tried before more general purpose revisions so that bugs would only be considered for students who actually exhibit problems. Finally, ASSERT could also be extended to use a first-order theory revision algorithm (Richards and Mooney, 1991). This might enable ASSERT to model relational and procedural problem domains.

7 Conclusions

This paper has described a new algorithm for student modeling called ASSERT. ASSERT uses theory revision to dynamically construct student models. Multiple student models are combined to automatically construct a bug library of stereotypical student errors. Theory revision has been shown to be more effective than static bug library approaches as well as inductive modeling techniques. Revising a rule base of stereotypical student errors allows ASSERT to build and refine a bug library while retaining the flexibility to address novel misconceptions.

8 Acknowledgments

The authors would like to thank Stellan Ohlsson for his detailed insights and comments which have helped clarify this presentation.

References

Brown, J. S. and Burton, R. R. (1978). Diagnostic models for procedural bugs in basic mathematical skills. *Cognitive Science*, 2:155–192.

Brown, J. S. and VanLehn, K. (1980). Repair theory: a generative theory of bugs in procedural skills. *Cognitive Science*, 4:379–426.

Burton, R. R. (1982). Diagnosing bugs in a simple procedural skill. In Sleeman, D. H. and Brown, J. S., editors, *Intelligent Tutoring Systems*. London: Academic Press.

Carbonell, J. R. (1970). Mixed-initiative man-computer instructional dialogues. Technical Report BBN Report No. 1971, Cambridge, MA: Bolt Beranek and Newman, Inc.

Carr, B. and Goldstein, I. (1977). Overlays: a theory of modeling for computer-aided instruction. Technical Report A. I. Memo 406, Cambridge, MA: MIT.

Dick, W. and Carey, L. (1990). *The systematic design of instruction*. Glenview, IL: Scott, Foresman/Little, Brown Higher Education. Third edition.

Gilmore, D. and Self, J. (1988). The application of machine learning to intelligent tutoring systems. In Self, J., editor, *Artificial Intelligence and Human Learning*, chapter 11. New York, NY: Chapman and Hall.

Langley, P., Ohlsson, S., and Sage, S. (1984). A machine learning approach to student modeling. Technical Report CMU-RI-TR-84-7, Pittsburgh, PA.: Carnegie-Mellon University.

Laubsch, J. H. (1975). Some thoughts about representing knowledge in instructional systems. In *Proceedings of the Fourth International Joint conference on Artificial intelligence*, pages 122–125.

Ohlsson, S. and Langley, P. (1985). Identifying solution paths in cognitive diagnosis. Technical Report CMU-RI-TR-85-2, Pittsburgh, PA.: Carnegie-Mellon University.

Ourston, D. and Mooney, R. (1990). Changing the rules: a comprehensive approach to theory refinement. In *Proceedings of the Eighth National Conference on Artificial Intelligence*, pages 815–820. Detroit, MI.

Quinlan, J. R. (1986). Induction of decision trees. *Machine Learning*, 1(1):81–106.

Richards, B. and Mooney, R. (1991). First-order theory revision. In *Proceedings of the Eighth International Workshop on Machine Learning*, pages 447–451. Evanston, IL.

Sleeman, D., Hirsh, H., Ellery, I., and Kim, I. (1990). Extending domain theories: two case studies in student modeling. *Machine Learning*, 5:11–37.

Sleeman, D. H. and Smith, M. J. (1981). Modelling students' problem solving. *Artificial Intelligence*, 16:171–187.

Wenger, E. (1987). *Artificial Intelligence and Tutoring Systems*. Los Altos, CA: Morgan Kaufmann.

Knowledge Tracing in the ACT Programming Tutor[1]

Albert T. Corbett and John R. Anderson
Psychology Department
Carnegie Mellon University
Pittsburgh, PA 15213
corbett @psy.cmu.edu
anderson @psy.cmu.edu

Abstract

The ACT Programming Tutor provides assistance to students as they write short computer programs. The tutor is constructed around a set of several hundred programming rules that allows the tutor to solve exercises step-by-step along with the student. This paper evaluates the tutor's student modeling procedure that is used to guide remediation. This procedure, termed *knowledge tracing*, employs an overlay of the tutor's programming rules. In knowledge tracing, the tutor maintains an estimate of the probability that the student has learned each of the rules. The probability for a rule is updated at each opportunity to apply the rule, based on the student's performance. The predictive validity of the modeling procedure for tutor performance accuracy and posttest performance accuracy is assessed. Individual differences in learning parameters and cognitive rules are discussed, along with possible improvements in the modeling procedure.

The ACT Programming Tutor

This paper reports an assessment of student modeling in the ACT Programming Tutor (APT). APT is a practice environment for students learning to program in Lisp, Pascal or Prolog (Anderson, et. al., in press). The tutor presents exercises that require students to write short programs and provides assistance as the students code their solutions. This report focuses on the initial seven sections of the Lisp curriculum. Table 1 displays example exercises from the first and last of these sections. The first section introduces two basic data types, atoms (symbols) and lists (groupings of symbols), and three functions that extract information from a list. By the seventh section, students are learning to define new functions that employ these three extractor functions in combination with three other functions that construct new lists.

In working with the tutor, students enter exercise solutions top-down with an interface that is similar to a structure editor. The first exercise requires two coding cycles: first the student enters *car* (either by typing or through menu selection) then types the literal list '(c d e). The second example requires eight coding cycles. The student enters *defun* (to define a new function), then codes the new function name, declares two variables (two cycles) and codes the body of the function (four cycles). The last exercise also requires three additional interface manipulation cycles in which unneeded editor nodes are deleted. The tutor monitors student performance on a cycle-by-cyle basis and provides immediate feedback to keep the student on a correct solution path. If the student makes a mistake, the tutor notifies the student, and allows the student to try again. The tutor does not volunteer any verbal feedback on errors, but the student can request help at each step.

We have been using such tutors to teach programming courses over the past eight years and the Lisp and Prolog modules are currently used to teach a self-paced introductory course. Overall, the tutors have proven effective. Students using the tutor generally work through exercises more quickly and perform as well or better on posttests (Anderson & Reiser, 1985; Corbett & Anderson, 1990, 1991). Despite this general effectiveness, however, some students flounder. As a result, we incorporated a student modeling and remediation mechanism into the tutor. Recently we have begun evaluating the validity of this mechanism.

1 This research was supported by the Office of Naval Research, grant N00014-91-J-1597.

Section 1 Example Exercise

Write a Lisp function call that returns c from the list (c d e).

Answer: (car '(c d e))

Section 7 Example Exercise

Define a function named replace-first that takes two arguments. Assume the second argument will be a list. The function replaces the first element of the list with the first argument. For example,

```
(replace-first 'rose '(tulip daisy iris))
      returns (rose daisy iris).
```

Answer:
```
(defun replace-first (itm lis)
      (cons itm (cdr lis)))
```

Table 1. Two Lisp Exercises drawn from the initial and final sections of the curriculum under review.

Student Modeling

APT is constructed around a set of several hundred production rules for writing programs, called the *ideal student model*, which allows the tutor to solve the exercises step-by-step along with the student. The tutor attempts to match the student's action at each step to an applicable rule in the ideal model in a process we call *model tracing*. The ideal student model also serves as an overlay model of the individual student's knowledge state (Goldstein, 1982). As the student works, the tutor maintains an estimate of the probability that the student has learned each rule in the ideal model. At each opportunity to apply a rule, the probability that the student knows the rule is updated contingent on the accuracy of the student's action. This process, which we call *knowledge tracing*, serves as the basis for remediation in the tutor. A small set of coding rules is introduced in each section of the curriculum and after the student completes a minimal set of required exercises, the tutor continues presenting remedial exercises in the section until the student has "mastered" each rule in the set. Mastery is defined in the tutor as a learning probability of at least 0.95.

The knowledge tracing mechanism passed a minimal validity test when it was first introduced. Posttest scores were higher when the remediation mechanism was in operation (Anderson, Conrad, & Corbett, 1989). Recently, we completed a more detailed assessment of knowledge tracing on the basis of both tutor and posttest data (Corbett & Anderson, 1992). While knowledge tracing is intended to infer a student's knowledge state for the purpose of guiding practice, the underlying cognitive and learning models can be used to predict a student's accuracy in completing tutor exercises. At each goal (step) in solving the exercises, we can estimate the probability of a correct response given the student's history. To assess the model, we compared actual and predicted accuracy across subjects at each goal. Knowledge tracing performed moderately well in this validity check. Across the 158 coding steps in the required exercises, actual and predicted accuracy were reliably correlated, $r = 0.47$.

In principle, the final production rule learning probabilities should predict posttest performance just as they predict tutor accuracy. However, the final probabilities are tightly distributed between 0.95 and 1.0 after knowledge tracing, so there is little potential to test this hypothesis. A related prediction is that if mastery learning is successful, posttest performance should not correlate with the number of mistakes students make in achieving mastery. The student modeling mechanism did not do as well by this criterion. The number of exercises required to reach criterion was reliably correlated with posttest performance, $r = -0.52$. The more exercises students completed in reaching criterion, the worse they did on the quiz.

Revising the Models

While the model predicted tutor performance reasonably well, there were systematic deviations in the fit that could be traced to deficiencies in both the underlying mathematical and cognitive models. As described below, the mathematical model employs four parameters. The values employed by the tutor, which were estimated from prior tutor data and held constant across production rules in the ideal model, resulted in a substantial underprediction of the variability in accuracy across goals in the lesson. We refit the data after the fact, allowing the four parameter values to vary across productions. The best fitting estimates yielded a substantially better fit, r =

0.85. In this fit, predicted and actual accuracy still deviated substantially for some rules, suggesting that the ideal model requires revision. In this paper we report an evaluation of a revised model.

The Study

The study assesses the internal and external validity of the APT knowledge tracing mechanism over the first seven sections of the Lisp curriculum.

The Students

Twenty five students worked through the curriculum in the course of completing a class in introductory programming. This was the first college level exposure to programming for all students and the first exposure to Lisp.

Procedure

Students worked through the exercises at their own pace. In each section of the tutor curriculum students read about Lisp in a text, completed a set of required exercises that covers all the rules introduced in the section, then completed remedial exercises as needed. Remedial exercises are selected by the tutor to bring all productions introduced in the section to a minimum learning probability of 0.95. At the end of the first lesson the students completed a quiz.

The Curriculum

In the seven sections of the curriculum under investigation, students are introduced to (1) two data types, atoms (symbols) and lists (hierarchical groupings of symbols), (2) function calls (operations) and (3) function definitions. The first section in the curriculum introduces simple lists (flat lists of atoms) and three basic extractor functions, *car*, *cdr* and *reverse*, that return components of or a transformation of a list. The second section introduces three basic functions that form new lists, *append*, *cons* and *list*. In the third and fourth sections students learn to apply the same six functions to hierarchically nested lists. In the fifth section, students are introduced to extractor algorithms - nested function calls applying multiple extractor functions to lists. In the sixth section, students learn to define new functions that perform such extractor algorithms. In the final section students define functions that employ both extractor and constructor functions. A minimum of forty exercises is required to complete this curriculum.

The Cognitive Model

The cognitive model consists of 35 productions rules. Three rules govern the coding of a single extractor function, either, *car*, *cdr* and *reverse* with a simple list in section 1. A single rule governs the transfer of all these extractors to nested lists in section 2. Five additional rules govern the coding of these three functions in more complex algorithms in section 5. The model distinguishes five additional extraction contexts in the last two sections on function definitions. A single rule governs the coding of any extractor or extractor algorithm in each of these five contexts.

The three constructor functions are employed in three contexts across sections 2, 4 and 7 - simple lists, nested lists and functional arguments. Given students' difficulty with constructor functions, the tutor makes no assumptions concerning the generalization of constructor knowledge. Thus, nine rules are employed to model the three rules in the three contexts. Twelve additional rules model the coding of data structures, the elements of function definitions, notably variable declaration and usage, and some editor manipulations.

This model incorporates three revisions stemming from the prior assessment: (1) separate rules for coding functions with flat and with nested lists, (2) the modeling of extractor algorithms with five functionally defined rules rather than two syntactically defined rules and (3) two distinct rules for declaring the first variable and subsequent variables in a function definition.

Knowledge Tracing

Knowledge tracing in the tutor assumes a simple two-state learning model with no forgetting. Each rules is either in a learned or an unlearned state. A rule can make the transition from the unlearned to the learned state at each opportunity to apply the rule, but rules cannot make the transition in the opposite direction. The goal in knowledge tracing is to estimate the probability that each rule is in the learned state. After each step in problem solving, the tutor updates this learning probability estimate for the applicable production rule, based on the student's

P_0	the probability that a rule is in the learned state before the rule is employed in problem solving for the first time (i.e., from reading)
P_T	the probability that a rule will make the transition from the unlearned to the learned state following an opportunity to apply the rule
PG	the probability a student will guess correctly if the rule is in the unlearned state
PS	the probability a student will slip and make an error when the applicable rule is in the learned state

Table 2. The four parameters employed in knowledge tracing.

action. The Bayesian computational procedure is a variation of one described by Atkinson (1972). It employs two learning parameters and two performance parameters, as displayed in Table 2. These parameter values are estimated empirically from the previous study and vary freely across the thirty-five rules in the tutor. See Corbett and Anderson (1992) for complete details on estimating learning probabilities and performance accuracy.

Results

Students completed an average of 23 remedial exercises in addition to the 40 required exercises in working through the curriculum. The number of remedial exercises ranged from 1 to 49. The revised cognitive model and parameter estimates improved the fit of the student model to the tutor accuracy data. A correlation of 0.71 was obtained between empirical and predicted values across the 203 goals in the required exercises. However, a moderate correlation persisted between number of tutor exercises required to reach criterion and posttest performance, $r = -0.44$. We again refit the data after the fact, and the best fitting parameter estimates yielded a better fit to the tutor data, $r = 0.90$. The final learning probabilities in this fit are still tightly distributed and do not correlate reliably with posttest performance.

There are a variety of reasons that the posttest performance may correlate with number of errors required to reach criterion in the tutor. First, the quiz involves transfer to a different coding environment. We might expect such transfer to correlate with the amount of practice required to reach criterion in the tutor. Second, students may be preparing differentially for the posttest, since the assessment draws on data from a course. Again, study habits may correlate with learning rate in the tutor. As a result, some correlation of number of tutor exercises and posttest performance might be expected even if the student model is essentially valid. Nevertheless, we plan to explore alternatives that do reflect on the validity of the student model: individual differences in parameter estimates and the nature of cognitive rules.

Individual Differences in Parameter Estimates. While the four parameter estimates vary across productions, they are held constant across subjects. Consequently, we may be underestimating the learning probabilities for students who are performing well and overestimating for students making more errors. As a result, students who are doing few remedial exercises would nevertheless tend to be overlearning while students doing many remedial exercises would tend to be underlearning. This pattern would result in a negative correlation of posttest performance with the number of errors in reaching criterion in the tutor. To assess the magnitude of individual differences, we divided the students into two groups based on posttest accuracy and generated best fitting parameter estimates for their tutor performance. The average estimates for the two learning parameters, P_0 and P_T, for twelve production rule categories are displayed in Table 3. As can be seen, the mean estimates for the two parameters are roughly 30% and 10% higher respectively, for students who performed well on the quiz. This suggests that we may obtain better fits by individualizing parameter estimates. An immediate goal is to investigate whether applying an individualized multiplicative constant to the group parameter estimates improves the performance of the model.

Individual Differences in Cognitive Rules. A second possibility is that different students acquire different cognitive rules. While we can track a student's ability to manipulate symbols in specific contexts, we cannot directly track the student's understanding of those manipulations. Knowledge tracing may insure that students are

Production Rule Category	High Test Accuracy		Low Test Accuracy	
	P_0	P_T	P_0	P_T
Extractors - Flat Lists	0.78	1.00	0.68	0.97
Constructors - Flat Lists	0.73	0.73*	0.76	0.93*
Literal Data Structures	0.45	0.48	0.06	0.72
Extractors - Nested Lists	0.87	1.00	0.68	0.69
Constructors - Nested Lists	0.68	0.37	0.35	0.37
Extractor Algorithms	0.69	0.58	0.70	0.28
Defun/Function Name	0.96	0.00*	0.93	0.00*
Variables	0.32	0.53	0.24	0.40
Extractor Algorithms Function Body	0.58	0.10	0.70	0.24
Extractor Algorithms Novel Contexts	0.63	0.31	0.25	0.09
Constructor Function Body	0.31	0.35	0.03	0.22
Interface	0.56	0.91	0.51	0.81
Mean	0.63	0.53	0.49	0.48

*One rule in this set is excluded in computing pT. p0 = 1 for that rule, so pT is inestimable

Table 3. Best fitting estimates for P_0 (the probability a production is in learned state initially) and P_T (the probability of a transition to the learned state) for twelve production rule categories

learning rules that enable them to complete exercises, but they may not be the rules assumed in the ideal student model. For example, one of the earliest stumbling blocks in learning Lisp is understanding the hierarchical structure of lists. In section 1, however, students are introduced to extractor functions and constructor functions with flat, non-hierarchical lists. As a result, students can apply everyday knowledge of lists to learn the extractor functions without fully grasping the structure of lists. Constructor functions are a second stumbling block in learning Lisp. To master constructors, students must understand the structure of lists and grasp the relationship between the arguments to the constructor function and structure of the resulting list. However, when constructor functions are introduced in the tutor curriculum with flat lists, students can learn rules based on the structure of the arguments alone that do not generalize to later sections.

The parameter estimates in Table 3 suggest that this may be happening. The two learning parameter estimates are quite similar across the two groups for the extractor function rules in section 1 and the constructor functions in section 2. However, when these functions are employed with more complex data structures and in more complex algorithms in later sections (rows 4, 5, 10 and 11 of Table 3) the learning parameter estimates are generally higher for the high posttest accuracy group. Students read text at the beginning of each section, so these parameter estimates may partly reflect differential initial comprehension of each section. However, such differences would be expected if some students are learning rules that generalize more readily to later contexts.

A final result is also consistent with this possibility. We generated best fitting parameter estimates for just the required exercises that every student completes at the beginning of a section. These estimates fit the required exercise data quite well, r = 0.91. We employed these parameter estimates to generate production learning probabilities on the basis of just the required exercises, as if students had not completed the remedial exercises. The mean probability estimates obtained from the required exercises correlated reliably with posttest accuracy, r = 0.43. This pattern would be expected if the number of opportunities required to master a rule is inversely related to the probability of acquiring an optimal understanding. Some suboptimal rules may capitalize on accidental characteristics of the tutor environment that do not transfer to the posttest environment. Other suboptimal rules may transfer in principle, but may in fact be retained less well. We might be able to model this possibility by decreasing the transition probability P_T with practice. However, it also suggests that students may benefit from explanatory feedback in the context of correct actions as well as in the context of errors.

Conclusion

On balance, the tutor's knowledge tracing procedure performed fairly well in this assessment. The model accounted for about 50% of the variance in students' performance with the tutor, and about 80% when best fitting parameter estimates were derived. On the other hand, students' learning rate (the number of errors students made in satisfying the model's mastery criterion) correlated negatively with posttest performance. While the student's final knowledge state should predict posttest performance, learning rate in reaching that state should not. We plan to explore possibilities for accomodating individual differences in learning rates by adjusting the model's learning parameters and the underlying cognitive rule set.

The correlation of tutor learning rate and posttest performance in this study may also reflect the fact that the tutor allows students to practice one programming skill, code generation, while the posttest environment allows the students to exercise other skills, e.g., debugging. It should be possible to generalize the knowledge tracing mechanism to other skills, however. Knowledge tracing does not depend on a unique solution path for each exercise, nor on immediate feedback, although these characteristics simplify the task. Rather, what is required is a cogntive model of the task consisting of rules that map onto observable behavior. As a result it should be possible to trace debugging and other skills.

References

Anderson, J.R., Conrad, F.G. and Corbett, A.T., 1989. Skill acquisition and the Lisp Tutor. *Cognitive Science* 13: 467-505.

Anderson, J.R., Corbett, A.T., Fincham, J.M., Hoffman, D. and Pelletier, R., in press. General principles for an intelligent tutoring architecture. In V. Shute an W. Regian (eds.) *Cognitive approaches to automated instruction*. Hillsdale, NJ: Erlbaum.

Anderson, J.R. and Reiser, B.J., 1985. The Lisp Tutor. *Byte, 10*, (4), 159-175.

Atkinson, R.C., 1972. Optimizing the learning of a second-language vocabulary. *Journal of Experimental Psychology*, 96, 124-129.

Corbett, A.T. and Anderson, J.R., 1990. The effect of feedback control on learning to program with the Lisp Tutor. In The Proceedings of the Twelfth Annual Conference of the Cognitive Science Society. Hillsdale, NJ: Erlbaum.

Corbett, A.T. and Anderson, J.R., 1991. Feedback control and learning to program with the CMU Lisp Tutor. Paper presented at the Annual Meeting of the American Educational Research Association.

Corbett, A.T. and Anderson, J.R., 1992. Student modeling and mastery learning in a computer-based programming tutor. In The Proceedings of the Second International Conference on Intelligent Tutoring Systems. New York: Springer-Verlag.

Goldstein, I.P., 1982. The genetic graph: A representation for the evolution of procedural knowledge. In D. Sleeman and J.S.Brown (eds.) *Intelligent tutoring systems*. New York: Academic.

Integrating Case Presentation with Simulation-Based Learning-by-Doing

Robin Burke and Alex Kass[1]

Institute for the Learning Sciences
Northwestern University, 1890 Maple Avenue
Evanston, IL 60657

Abstract

In this paper we argue that the key to teaching someone to perform a complex task is to interleave instruction and practice in a way that exploits the synergism between the two effectively. Furthermore, we argue that computer simulations provide a particularly promising environment in which to achieve this interleaving. We will illustrate our argument by describing a simulation-based system we are building to train people to perform complex social tasks, such as selling consulting services. In particular, we will focus on the system's ability to present real-world cases at the moment that they are relevant to the student's simulated activities. In doing so, we hope to contribute both to the construction of useful teaching systems and to the theory of case-based reasoning, particularly in case retrieval.

Linking Practice and Instruction

In order to perform a complex skill effectively, a student needs to understand the abstract principles at work in the skill domain and must also learn how those principles apply in practice. When instruction and practice are combined appropriately, the student's actions in the practice environment are guided by instruction, and the abstract principles described by the instructor are motivated, operationalized, and made memorable by the student's experiences in the practice environment.

Computers can provide a vehicle for integrating instruction and practice through "Intelligent Learning-By-Doing Environments" (ILDE's). An ILDE provides two things for the user,

1. an interactive task environment, and
2. a suite of teaching modules.

The task environment puts the student into an active learning role, allowing the student to practice the target skill. The teaching modules monitor the student's interaction with the task environment. They treat the student as a traditional craftsman would treat an apprentice, providing coaching, modeling, and scaffolding for the student during the practice sessions. (See (Collins, Brown and Newman, 1989) and (Lave and Wenger, 1991).)

We are building several ILDE's, the largest and most sophisticated of which, GuSS-Sales (GUided Social Simulation-Sales)[2], is an ILDE we constructed to teach consultants how to sell consulting services. The program integrates simulated client interaction that actively engages the student with explicit discussion of real-world consulting cases and principles.

[1]This work is supported in part by the Defense Advanced Research Projects Agency, monitored by the Air Force Office of Scientific Research under contract F49620-88-C-0058 and the Office of Naval Research under contract N00014-90-J-4117, by ONR under contract N00014-J-1987, and by the AFOSR under contract AFOSR-89-0493. The Institute for the Learning Sciences was established in 1989 with the support of Andersen Consulting, part of The Arthur Andersen Worldwide Organization.

[2]Some of the other ILDE's we are building teach geography through the task of taking car trips, social studies through broadcast journalism, and second languages through simulated conversations. They are described in (Kass and Guralnick, 1991) and (Ohmaye, 1992). The first incarnation of GuSS-Sales was called ESS (Engagement Simulation System). It is discussed in more detail in (Blevis and Kass, 1991) and (Kass and Blevis, 1991).

The social simulation architecture we have developed attempts to accomplish for a social environment what the flight simulator accomplishes for the physical environment of the cockpit. This architecture is particularly appropriate for teaching the complex interpersonal skills required in domains such as diplomacy, negotiation, and business. For instance, inexperienced business consultants need to learn the delicate skills required to successfully interact with client organizations. Such skills include

- discovering the official and unofficial structures of client organizations,
- dealing effectively with different personality types within those organizations, and extracting from people the information necessary to make recommendations, and
- making appropriate recommendations in a style that is convincing but not threatening.

The GuSS-Sales task environment includes many of the same obstacles and resources that a consultant would find in the real business world. Communication with other agents can be performed through face-to-face conversation, telephone calls, memos and written reports. Scanned in drawings and photos are used to show what the clients and their offices look like. The student can turn to charts, reports, higher-ups, and friendly coworkers for advice and assistance. Prospective clients, fellow consultants, and sellers from competing firms act on the basis of their own goals, beliefs, expectations, and attitudes. See (Kass and Blevis, 1991) and (Kass, et al., in prep.) for a detailed discussion of the design of the task environment.

The GuSS-Sales task environment attempts to provide the student consultant with an experience that is realistic, and is as memorable, in its own way, as the flight simulator experience for the aspiring pilot. The teaching modules in GuSS-Sales play a role analogous to that of an expert pilot watching over the actions of a student who is using a flight simulator. The modules can perform several different kinds of intervention. They can coax the student to try actions that might be useful, discuss an abstract concept currently at work in the simulation, or tell a relevant story that illuminates the current situation.

We are working on modules for each of these different types of intervention. In the remainder this paper, we will focus on the storytelling module, which is called SPIEL (Story-Producer for Interactive Environmental Learning). SPIEL monitors the simulation and presents cases from its library when they are relevant to the student's situation. We believe that case presentation (*i.e.* telling relevant stories) in the context of a simulation is a *particularly* useful way to link the experience with the simulation to general principles because stories describe principles *as they apply* in action. The following example illustrates the role played by SPIEL's stories in a typical session with GuSS-Sales.

GuSS-Sales and SPIEL in action

Each GuSS-Sales scenario begins with the student in his/her own office, where they can receive assignments in the same way a real consultant would. One representative scenario is as follows:

- The student gets a memo from her boss: there is a potential sales opportunity at a client that the boss is unable to follow up on.
- The student goes for an initial sales call and meets Bill Bell, the CEO of a department store chain.
- Bill expresses doubt about the value of consultants and asks for some concrete information about possible solutions to his problem. The student manages to allay his fears, side-step questions of detail and gather some needed information.
- At the end of the interview the student proposes that they begin by performing a "High Spot Review." (a standard procedure).
- Bill replies that a high spot review is a lot more than what he wanted.
- The student defends the review.
- Bill, unconvinced, becomes irritated, and begins to wonder whether consultants are worth the trouble.

This is a good time to give some feedback to the student. She knows that the High Spot Review is the best way to proceed, but she has lost sight on the fact that pleasing the client is more important that defending every detail of the prospective job. One of the 171 stories currently stored in SPIEL's library of stories is relevant to this situation:

> **"Review Present Procedures"**
>
> You told Mr. Bell about your proposed approach. He disagreed. You continued to argue for your approach. Unfortunately, he is now angry with you. Here is a story about a similar situation in which the salesperson used a different method and was successful.
>
> We were working for a two-billion-dollar world-wide distribution company. We had sold the job and were presenting our work program, and when you present a work program, you have step one, step two, step three. Step one is always "Review present procedures," in [our] charts. Step two is a design or action step.
>
> So I was explaining this to the group using Japanese graphics on the board. I said the first thing we're going to do is review present procedures. [The senior managing director], said "No!" in Japanese. Of course, I asked him "Why not?" He said "We will decide what the future operation of the plant should look like, then we will use that vision, that new operation...we will use the system to enforce that operation." So I immediately took the transparencies and crossed off all the Japanese words and told the manager who was with me, the Japanese manager, to write down: "determine how to operate plant in the future." And he did that. Then we went on. Of course when we got on the job, part of the work to determine future operations was to review the present procedures.
>
> Your plan of arguing with the client didn't work well. In the future, you might consider agreeing when the client proposes an approach.

Paragraphs 2 and 3 above are verbatim presentations of a story from SPIEL's memory. SPIEL precedes each story with introductory paragraph called a **bridge** and summarizes each with a **coda** paragraph. The bridge explicitly connects the story context of the student's activity; the coda brings the student back to the events of the simulation by suggesting some possible actions. This example illustrates the synergistic interaction between the simulation and the explicit instruction and the simulation. Without the story provide an explanation for Bill's reaction, the student might be confused about how to interpret the events in the simulation. Without the active engagement provided by the simulation, however, the student might lack the motivation and context to read, understand and remember the story.

Storage and Retrieval

How does SPIEL manage to find relevant stories from its library to tell to the student? Tutorial storytelling is a case-retrieval task (Kolodner and Jona, 1991). Like case-based problem-solving systems, such as CHEF (Hammond, 1986), SPIEL must locate knowledge structures in its memory. CHEF indexed its recipes using features of the cooking goals that they could achieve. SPIEL's goal is to tell its stories in instructive ways. SPIEL's indexing system is therefore based on a theory of educational storytelling. This theory has three parts:

1. A representation language for expressing the indices attached to each story,
2. A theory of storytelling purposes,
3. A set of storytelling strategies which map an index and a storytelling purpose to a set of opportunity-recognition rules.

Since the stories must be recalled quickly as the student interacts with the task environment, SPIEL does as much preprocessing as possible at storage-time.

Storage Time:

1. Indices are attached to each story to be included in the database.
2. Each of SPIEL's 6 storytelling strategies examines each story index. If the strategy is applicable to the index, the strategy will generate an opportunity-recognition rule for that index, along with a bridge and coda template.
3. An optimized opportunity-recognition rule set is generated, which improves the speed of matching at retrieval time by eliminating redundant matching of identical clauses across the rule set.

Retrieval Time:

1. Opportunity-recognition rules are matched against the state of the simulation.
2. When a story is successfully retrieved, natural language text is generated that integrates the story into the student's current context.

Indices

A SPIEL index labels a story in terms of one of the points of the story. SPIEL labels each story in its library multiple times because a story can have multiple points. For instance, "Review present procedures" can be used to make a point about the need to match presentation content to the buyer's beliefs, the disparity between what the consultant says and what he does, the pitfalls of using standardized presentation materials, or the differences between Japanese business practices and those in America. A given story will have one SPIEL index attached to it for each point that the story can convey. Since SPIEL does not yet incorporate any natural-language understanding, the indices are generated manually.

The structure of SPIEL's indices is derived from the Universal Indexing Frame (Schank, et al. 1990). Since SPIEL is intended to teach planful activity, its indices center around plans, goals and expectations. Each index contrasts some component of the story as it actually occurred, with that component viewed from the perspective of a character in the story. There are five types of perspectives used in SPIEL's indices:

Expected: Some aspect of the story turns out differently that some character expected.

Perceived: There is a discrepancy between what some character perceived and what actually happens in the story.

Ideal: The actual events in the story vary from an ideal, usually that of the story-teller.

Feared: The story contrasts a character's fears with actual events in the story.

Wanted: The story contrasts a character's desires against the actual events in the story.

The interpretation of "Review present procedures" that gets activated in the example focuses on the salesperson's method for dealing with the client's objection. Since the consulting firm puts great store by their problem-solving methodology, a consultant should, ideally, be able to defend that methodology when it is challenged. In this story, the consultant does not defend the firm's methodology.

Perspective type: Ideal

Ideal component: Consultant employs defend-methodology plan to achieve sell-work goal.

Actual component: Consultant employs agree-with-client plan to achieve sell-work goal.

Story-Telling Strategies

For any given index SPIEL may have several applicable story-telling strategies. A story-telling strategy represents the class of situations in which a story of a particular type is likely to be worth telling. For instance, the story-telling strategy that brought up the above example is as follows:

Strategy 1: Demonstrate alternative plan: Tell a story about a successful plan to achieve a particular goal when the student has executed a different plan and failed to achieve the goal.

The same story can also be used to *explain* the actions of someone in the simulation whose actions might otherwise be mysterious. For instance, suppose the student were collaborating with a more senior consultant on a presentation. If the other consultant knew that a normal step in the process was objectionable to the client he might omit that step from the presentation. The student would probably be confused be confused by this, but. The "Review present procedures" story can explain the partner's actions, and help the student generalize this experience. The SPIEL strategy responsible for bringing the story up in this situation is as follows:

Strategy 2: Explain other's plan: Tell about a successful plan that the student may not know about when someone has just executed a similar plan.

SPIEL has a total of six storytelling strategies. The remaining four are as follows:

Strategy 3: Reinforce plan: Tell a story about a successful plan to achieve a particular goal when the student has just executed a similar plan.

Strategy 4: Warn about plan: Tell a story about an unsuccessful plan when the student has begun executing a similar plan.

Strategy 5: Demonstrate alternative result: Tell a story about the result of a particular course of action when the student has just executed a similar course of action but experienced different results. (similar to 3 and 4 above, except that the contrast need not involve success and failure).

Strategy 6: Warn about perspective: Tell a story about someone's unrealized expectation (or perception, fear, ideal, or desire), when a student appears to have that same expectation.

These strategies correspond to goals a tutor would have when trying to teach someone how to plan and engage in an activity. **Explain other's plan** in particular applies only to a social domain, in which there are other individuals whose plans may need to be explained. When the tutor's goal is not to teach how to plan a course of action, but to teach something else, such as design, the same basic storage and retrieval algorithm would apply, but different indices and different storytelling strategies would be required.

CreANIMate (Edelson, 1991), for example, tells stories about animals in the course of a tutorial dialog centered around a design task: putting together an imaginary animal. CreANIMate's stories are about animals, not about students who have tried to design animals, so its stories relate to the product of the student's design activity, not the design activity itself. An analogous task in the selling domain might be looking at a contract the student has negotiated and retrieving stories about other contracts, based purely on the features of the document itself. Additional storytelling strategies would be needed for SPIEL to tell stories in this mode.

SPIEL uses its storytelling strategies at storage-time to precompute a set of all situations in which a given story would be relevant as well as to precompute the template used to produce the bridge and coda.

SPIEL's strategies have three parts:

- **Applicability test:** The applicability test determines whether the index is appropriate for this strategy.

  Example: Strategy 1 is obviously only applicable to stories making a point about successful plans. Its applicability test involves checking the ACTUAL-RESULT slot of the index to see if it contains a successful outcome.

- **Recognition-condition generator:** The recognition-condition generator uses the index and the system's domain knowledge to generate specific opportunity-recognition conditions for the index.

  Example: Strategy 1 prepares SPIEL to tell "Review present procedures" in situations where the student is pursuing a goal similar to please-the-client, but is pursuing a plan that is *not* similar to agree-with-client-about-procedure. The result is an opportunity-recognition rule that looks like this:

TELL "Review present procedures" AS A
  Demonstrate-alternative-plan story WHEN
    the student has been talking to a client
    whose opinion affects a sale.
    AND the student told the client about a
      proposed method or approach.
    AND the client reacted negatively or
      present an alternative.
    AND the student argued for the
      proposal and/or against alternative.
    AND the client reacted negatively to
      the student's action.

- **Natural language templates:** The natural language templates are used at retrieval time to generate explanations for the student, indicating why the retriever believes that the story may be relevant. Each storytelling strategy employs a different set of bridge and coda templates.

Conclusion

We believe that our work on GuSS-Sales represents a contribution on two fronts: to the practical technology of education and training, and to the important theoretical problem of case retrieval. The GuSS architecture consists of an interactive task environment and a suite of expert teaching modules that run concurrently. That combination is educationally powerful because instruction and practice are interleaved, each complimenting the other. The overall result is therefore much stronger than either practice or instruction alone. Since the teaching modules are able to monitor and interrupt student's practice activities, they can deliver their instructional messages in the most timely possible fashion. The system is thus able to help the student combine information sources. The teaching modules can help the student generalize the lesson of a concrete experience, and the practice environment can help the student experience the principles that the teaching modules describe.

Stories about real-world cases provide a form of instruction that is particularly well-suited to teaching complex skills in the context of a simulation. They allow the student to tie principles to action, and they bridge the gap between the student's experiences within the simulation and the broader set of real-world challenges they will face. The main question we have addressed in designing SPIEL is how a system can notice when it has a story that is relevant. This requires two theories. The first is a theory of how a story can be represented in a way that is useful for retrieval. This theory is implemented in SPIEL's indexing scheme. The second is a theory of the set of purposes that telling a story can serve. This is the theory behind SPIEL's storytelling strategies. By combining an index for a particular story with a story-telling strategy, SPIEL can determine when and how to tell each of its stories.

Acknowledgments

For extensive advice and assistance with this research, including much of the implementation and design work, we are indebted to Eli Blevis, Chip Cleary, Marita Decker, Lucian Hughes, Tom Murray, Scott Rose, Michelle Saunders, Roger Schank, Chris Sorensen, and Mary Williamson.

References

Blevis, E., and Kass, A. 1991. Teaching by Means of Social Simulation. In *Proceedings of the International Conference on the Learning Sciences*. Association for the Advancement of Computing in Education, Evanston.

Burke, R., and Ohmaye, E. 1990. Case-Based Environments for Teaching. In *Proceedings of the AAAI Symposium on Knowledge-Based Environments for Learning*. Palo Alto.

Collins, A.; Brown, J. S.; and Newman, S. E., 1989. Cognitive Apprenticeship: Teaching the Crafts of Reading, Writing, and Mathematics. In Resnick, L. B. (Ed.), *Knowing, learning and instruction: Essays in honor of Robert Glaser*. Hillsdale, NJ: Lawrence Erlbaum Associates.

Edelson, D. C., 1991. Why Do Cheetahs Run Fast? Responsive Questioning in a Case-Based Teaching System. In *Proceedings of the International Conference on the Learning Sciences*. Association for the Advancement of Computing in Education, Evanston.

Hammond, K. J., 1986. Case-based Planning: An Integrated Theory of Planning, Learning and Memory. Ph.D. Thesis, Yale University.

Kass, A., and Blevis, E. 1991. Learning Through Experience: An Intelligent Learning-by-Doing Environment for Business Consultants. In *Proceedings of the Intelligent Computer-Aided Training Conference*. NASA.

Kass, A.; Blevis, E.; Burke, R.; and Williamson, M., (in preparation). *Guided Social Simulation: An Architecture for Teaching*. Technical Report, Institute for the Learning Sciences.

Kass, A., and Guralnick, D. 1991. Environments for Incidental Learning: Taking Road Trips Instead of Memorizing State Capitals. In *Proceedings of the International Conference on the Learning Sciences*. Association for the Advancement of Computing in Education, Evanston, IL.

Kolodner, J. L., and Jona, M. Y. 1991. *Case-based Reasoning: An overview*. Technical Report #15, Institute for the Learning Sciences.

Lave, J., and Wenger, E. 1991. *Situated Learning: Legitimate peripheral participation*. New York: Cambridge University Press.

Ohmaye, E., 1992. *Role Playing and Social Simulation*, Ph.D. Thesis, Dept. of EECS, Northwestern University.

Orr, J., 1989. Sharing knowledge, celebrating identity: War stories and community memory among service technicians. In D. S. Middleton and D. Edwards (eds.), *Collective remembering: Memory in society*. Beverly Hills, CA: Sage Publications.

Schank, R. C. et al., 1990. *A Content Theory of Memory Indexing*. Technical Report #2, Institute for the Learning Sciences.

Diagnosis can help in intelligent tutoring

Roderick I. Nicolson
Department of Psychology
University of Sheffield
Sheffield S10 2TN, England
email: R.NICOLSON@UK.AC.SHEFFIELD.PRIMEA

Abstract

Recently there has been controversy about whether Intelligent Tutoring Systems are, even potentially, more effective than standard CAL programs, that is, whether it is educationally more valuable to attempt to identify the cause of user's mistakes rather than merely explain the correct method. This issue was addressed by comparative testing of two versions of the SUMIT Intelligent Tutoring Assistant for arithmetic using a diagnostic version, which diagnosed errors and gave appropriate messages, and a 'CAL' version was identical in all respects except that it made no diagnoses and therefore gave standard error messages indicating the correct method. In a comparative study of the two versions, a class of 9 year old children were first divided into two matched groups on the basis of a pencil and paper pre-test, then both groups had two 30 minute individual sessions with the appropriate version of SUMIT, and then performance was assessed on a subsequent pencil and paper post-test. Both groups improved significantly in their performance from pre-test to post-test, but the diagnostic group showed significantly greater reductions in the number of bugs. It is concluded that diagnostic remediation can be more effective than non-diagnostic approaches.

Introduction

Traditional Computer Aided Learning (CAL) programs have been criticised on the grounds that they do not *understand* the domain for which they were devised, and so they cannot give the adaptive help expected of a human teacher. This critique proved the stimulus for the creation of Intelligent Tutoring Systems (ITSs) which did understand their domain sufficiently to provide the same adaptive quality of guidance and instruction as a human teacher. Intelligent Tutoring Systems have made impressive progress in the intervening years, making contributions not only to pedagogical theory but also allowing empirical tests of theories of learning (see Anderson et al., 1990, for a recent review). However, the educational credibility of the ITS approach has recently been called into question by Sleeman et al. (1989), who were evaluating the effectiveness of remediation by human tutors in the domain of linear algebra problems. In a series of studies Sleeman and his colleagues compared the effectiveness of 'model-based remediation' (in which the tutor identified the type of error made, and explained why it was wrong), with 'reteaching' in which the tutor ignored the type of error made and merely explained the correct procedure. Both procedures were effective (as compared with a control group who received no remediation), but they were equally effective, leading the researchers to conclude that *"when initial instruction and remediation are primarily rule-based and procedural, remedial reteaching appears to be as effective as model-based remediation. From this it follows that 'classical' CAI would be as effective as an ITS"*. (1989, p563).

Recently I have developed the SUMIT system which is intended to function as an 'Intelligent Tutoring Assistant' for early school arithmetic (see Nicolson, 1990 for a full description of the design issues and studies of its effectiveness). SUMIT provides an ideal opportunity to assess the added value of diagnosis in tutoring in that diagnostic feedback is normally available, but can be 'turned off' if required by setting the appropriate flag. Both versions are otherwise identical, with the non-diagnostic version (henceforth SUMIT-ND) giving support in terms of the correct way to answer the problem, and the diagnostic version (henceforth SUMIT-D) giving not only that support but also a brief diagnosis of why the user's answer was wrong. The design of the study is therefore straightforward. We took a class of 9 year old children, gave them a pencil and paper pre-test on subtraction sums selected to investigate a range of potential problems, ranked them in order of score, split the class into two matched groups via this ranking, gave group 1 two sessions of individual practice with SUMIT-D, and group 2 two sessions with SUMIT-ND, then gave them a pencil and paper post-test equivalent to the pre-test, and compared the resulting gains in score and understanding. Before describing the study in detail, it is valuable to provide some more information on the SUMIT system.

The SUMIT Intelligent Tutoring Assistant

SUMIT was inspired by Brown and Burton's seminal work (1978) on diagnosis of the reasons underlying arithmetic errors, which led to the creation of the DEBUGGY system for bug diagnosis. In many ways, their research program was exemplary cognitive science, starting with identification of an important theoretical issue, collecting a large corpus of human performance data relating to that issue (children's substraction errors in this case), then constructing an offline diagnostic system intended to infer from the errors manifested which procedures were not fully understood, thus moving from performance assessment to competence assessment. The approach proved very fruitful, to the extent that most subsequent ITSs incorporated a 'bug catalogue' as part of their diagnostic armoury, and also in providing a rich source of data and ideas for important theoretical developments such as VanLehn's Sierra theory of procedural learning (e.g., 1990). But noone actually constructed a working, fully interactive, ITS for school arithmetic! SUMIT was the result of a longstanding 'spare time' project, conducted jointly with Margaret Nicolson, an experienced teacher of middle school arithmetic, to do just that.

The development program followed an 'evolutionary' strategy. Extensive knowledge engineering studies were undertaken over a period spanning three years in which first a detailed analysis of the traditional methods of teaching arithmetic was performed (based on three classroom studies). These studies were intended to identify areas of strength and weakness in the traditional approach, thus allowing the program to be targetted on relief of the weaknesses of traditional teaching, rather than duplication of the strengths. In particular, we identified the ability to give immediate feedback as critical, together with the ability to generate sums at a difficulty level appropriate for the child. These two capabilities would essentially allow a child to get on with practice at sums without the need for continual checking by the teacher. By contrast, the ability to explain why the methods used were the appropriate ones seemed much better suited to the traditional classroom demonstrations, where the teacher was able to use a range of techniques, adapted to his/her preferred teaching style and to the capabilites of the children, to explain the basis of the procedures. The analysis led us to undertake the construction of an 'Intelligent Tutoring Assistant' (ITA), less ambitious than an ITS, aimed at providing adaptive, generative practice at the procedural skills, with support for *which* procedure to use, and *how* to do it, but not for *why* the procedure should be used. The ITA was aimed to assist, rather than replace, the teacher.

Figure 1. The traditional stages in pencil and paper addition

Stage 1	Stage 2	Stage 3	Stage 4
3 7	3 7	3 7	3 7
5 8 +	5 8 +	5 8 +	5 8 +
?	5	? 5	9 5
	?	1	1

The analyses also identified the target skills required, a teaching strategy for imparting them, and a range of teacher preferences important for smooth incorporation of the ITA within the traditional teaching methods. A non-diagnostic program (SUMS) was then developed and tested extensively in the school setting. This investigation led to the introduction of further teacher support facilities, but its main function was to collect automatically a large corpus of arithmetic mistakes made in free use of the program for each of the four operations — addition, subtraction, multiplication and division. Extensive hand analyses were carried out on the corpora, leading to the identification of the error types (including their incidence), and, following analysis of how to automatically diagnose the major bugs, we were then able to 'bolt on' an online diagnostic capability, thus creating the SUMIT prototypes. Further details are provided in Nicolson (1990).

Using the SUMIT system

The following description shows how SUMIT is able to give a reasonably faithful replication of the traditional approach to arithmetic. Figure 1 demonstrates the traditional stages in pencil and paper arithmetic. The sum is written down on paper and the computation is carried out in stages as shown below — from units through carries to tens. The question mark is, of course, imaginary and it is included here to indicate which stage is involved. A clear difference between this written arithmetic and mental arithmetic is that it occurs step by step and, most important, intermediate steps are explicitly entered. Completion of the sum is often accompanied by muttered self-instructions somewhat like seven add eight is ... fifteen, so write down the 5 (stage 1) and carry the 1 to the tens (stage 2). Now three add five is ... eight, add the one carried, that's nine, so put the 9 in the tens (stage 3)'. Exactly the same procedure is used by SUMIT, with the child required to complete all five stages in the appropriate order, and if no mistakes are made, the procedure is essentially identical. Following successful completion of a sum, SUMIT generates a further sum at the appropriate difficulty

Figure 2. The non-diagnostic adaptive help available in the SUMS program

{The user has nearly completed the sum, but is unsure how to complete the addition of the tens column, and so presses H. The right hand side illustrates the help given in such a situation.}

Sum	SUMS Help
3 7	You are adding the tens:
5 8 +	that is: **3 + 5 + 1** carry
[H] 5	*press RETURN to continue*[1]
1	The total is **9**
	so put the **9** in the box
	press RETURN to continue

level and so on. The advantage of CAL becomes apparent if a mistake is made. Since the appropriate answer is always known for each stage in completing the sum, any error is noted immediately, and the user is warned of the error and required to try again. In the original SUMS program, adaptive help was available either on demand or following three errors on a given sum, but this only explained the correct method for continuing the sum, and made no effort to diagnose what the user's misconception might have been. On the basis of the extended studies of performance on SUMS, SUMIT is able to diagnose up to 20 different bugs for each of the four arithmetic operations. This allows an immediate diagnosis of the likely cause of any error. For instance, if the user typed in '8' instead of '9' at stage 3, the program decides that the most likely bug is 'failure to add in carry' and is therefore able to offer the suggestion 'Remember to add in the carry 1'. Adaptive standard help is again available on demand or after three helps (see figure 2).

In both diagnostic and non-diagnostic versions, an error results in a warning tone, and the user is not allowed to proceed until the correct answer has been entered. Non-diagnostic help is normally given automatically following three errors on a sum. Diagnostic help following an error of typing in '8' in the above situation would involve the short message "Remember to add in the 1 you've carried from the units". In the non-diagnostic form, following an error only the warning tone is presented, followed by the message "Bad luck, please try again".

In view of the greater complexity of subtraction, and in recognition of its special status in the ITS literature, for the investigation of diagnostic versus standard feedback we decided to investigate the effects of diagnostic support on subtraction skills.

[1]Initially only the first part of the message is displayed. Pressing the Return key adds in the next part, and so on.

Figure 3. A subtraction sum which involves 'borrowing'

Stage 1	Stage 2	Stage 3
	?	3
4 5	4 5	4 ?5
2 7 -	2 7 -	2 7 -
?	-	-

Stage 4	Stage 5	Stage 6
3	3	3
4 15	4 15	4 15
2 7 -	2 7 -	2 7 -
?	? 8	1 8

The standard 'decomposition' approach to teaching subtraction is shown in Fig. 3. Note the complexity of the procedures involved.

The self-instructions for this sum might go as follows: *"5 take 7 won't go, so put a dash in the box (stage 1) and try to take help from the tens column (stage 2). 4 take 1 leaves 3, so cross out the 4 and put 3 (stage 3). Next take the ten help we were given [borrowed] and give it to the units — that makes 15 (stage 4). We can now take the 7 from our 15, that makes 8, so put the 8 in the box (stage 5) and go to the tens column. 3 take 2 is 1, so put the 1 in the box (stage 6)".*[2]

The commonest subtraction bug (1S) occurred at stage 2, where rather than subtracting 1 from the tens, the child got confused and performed the subtraction on the tens column (thereby yielding 2 in this case). Bug 1 appears to occur only on the computer, and the most generic subtraction bug (2S) [smaller from larger] occurs at stage 1, where the child enters 2 for 5-7. This is a beginner's error, symptomatic of difficulty in knowing how to cope with a negative outcome.

The bugs diagnosed by SUMIT-D, and their incidence in the initial corpus are shown in Table 1. Note that the use of —> in the example indicates that the user entered the digits in the order shown. For instance, for bug 1S, the sequence for answering 83-24 was - {correct}, then 6, {the error, reflecting subtraction of the two entries in the tens column (8-2) rather than subtraction of the borrowed 1 from the tens column, leading to the answer of 7}. It is much easier to follow this exposition if the sum is laid out as shown in figure 1!

[2]The 'decomposition' procedure for subtraction is now preferred to the older 'equal additions' method which would add 10 to both top and bottom (ie turning the 5 of 45 into 15, and turning the 2 of 27 into 3) on the grounds that for decomposition the manipulation is only on one number, and can easily be shown to be valid by means of Dienes' blocks etc.

Table 1. Subtraction Bugs Diagnosed by SUMIT-D

Bug	Description	Example
1S	Subtract current column in mid-borrow	83-24 —> -6 etc.
2S	smaller from larger	3-6=3
3S	Put 1 in before decrementing column	83-24 -> -1 etc.
4S	miss out stage in initial borrow	83-24 -> 7 etc.
5S	Don't decompose 10 in borrowing over 0	803-24 —> --71x etc where x<>9
6S	0-n=n (specialis'n of 2S)	0-7=7
7S	0-n=0	0-7=0
8S	use non-decremented minuend	583-124—>-7196
9S	Lose place in mid-borrow	83-24 —> -79
10S	'Add 10' bug	83-24 —> -710
11S	response perseveration (repeat prev press)	803-24—>- - -
12S	subtract 1 'for luck' from last column	83-22=51
13S	Missed out step	eg. 10-7=3
14S	0-1=0 when borrowing across 0	803-24 —> -0
AS	arithmetic error	13-6=8
US	*unclassified (non-borrow)*	
UbS	*unclassified (in borrow)*	

Experiment. Diagnostic help vs non-diagnostic help using SUMIT

Two groups of 9 year old schoolchildren from the same class were selected, individually matched on performance on a pre-test. Both groups then experienced two 30 minute individual sessions of SUMIT, one group using SUMIT-D and the other group using SUMIT_ND (with the standard feedback and help facility). Children used the program individually, with two children at a time taken out of their normal arithmetic lesson. The experimenter was Chris Harrop, a third year undergraduate student, who had chosen to undertake the work as part of his final year undergraduate dissertation in Psychology. The experimenter's role was to ensure that the appropriate version of SUMIT was selected, to check that each child started the session at an appropriate level of difficulty, and to provide general encouragement. He gave no direct instructional support. In the first computer session each child started at the baseline level, and sums were automatically generated at levels of increasing difficulty until mistakes started to emerge, at which stage the program generated sums of the appropriate difficulty subsequently. In the second session the child was encouraged to start at a level one below that reached in the first session. Finally, performance on a pencil and paper post-test equivalent to the pre-test was measured. For each test the written answers were scored, and any error made was assigned to one of the bug categories (see Table 1). The total bug count was determined by including the unclassified bugs (which correspond to a bug not included in the diagnostic help) but not the arithmetic bugs. Comparison of pre-test and post-test scores and bugs for the two groups should reveal whether diagnostic help really does help or not..

Results

Results for the pre-test and post-test scores are shown in Figure 4a and those for bugs in Figure 4b. It may be seen that, as expected, both groups improved as a result of the sessions with SUMIT, and that the diagnostic group improved somewhat more in overall score, and markedly more in terms of the overall bugs. An analysis of variance on the scores indicated a significant main effect of time-of-test ($p<.01$) but no significant main effect of group, and no significant interaction. In terms of the effectiveness of the learning induced, the non-diagnostic group's mean score improvement was 0.30 sd units [based on the original standard deviation of scores of both groups together, cf. Bloom (1984)], well below that of the diagnostic group (0.75). An analysis of variance on the bugs data (omitting children who obtained pre-test scores of 29 or 30 out of 30) indicated a significant main effect of time-of-test ($p<.05$), no significant effect of group, but a significant interaction between group and time-of-test ($p<.05$), indicating that the diagnostic group eliminated their bugs significantly more effectively than the standard group. The individual results are displayed in Figure 5. Comparing the histograms for the two groups, it is clear that the major effects are attributable to those children who were initially performing badly. For the diagnostic group, there are large improvements (see especially OT who improved from 5/30 to 30/30), whereas this improvement was less consistent for the non-diagnostic group.

Discussion

It remains to consider the wider significance of these results. First it is important to stress that the results relate only to two groups of children in one school on one task, and that the results are attributable to only a few of these children. Next, the major improvement is attributable to the SUMIT program itself, and the further improvement due to the diagnostic element is of only secondary importance.

Fig. 4a. Scores for the two groups

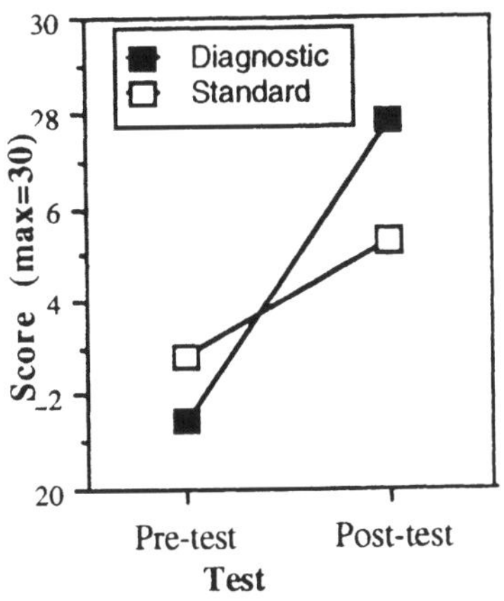

Fig. 4b. Bugs for the two groups

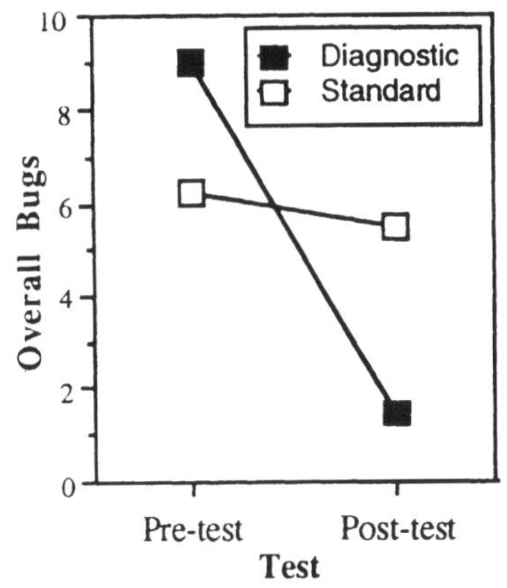

Figure 5. Individual Scores at pre-test and post-test

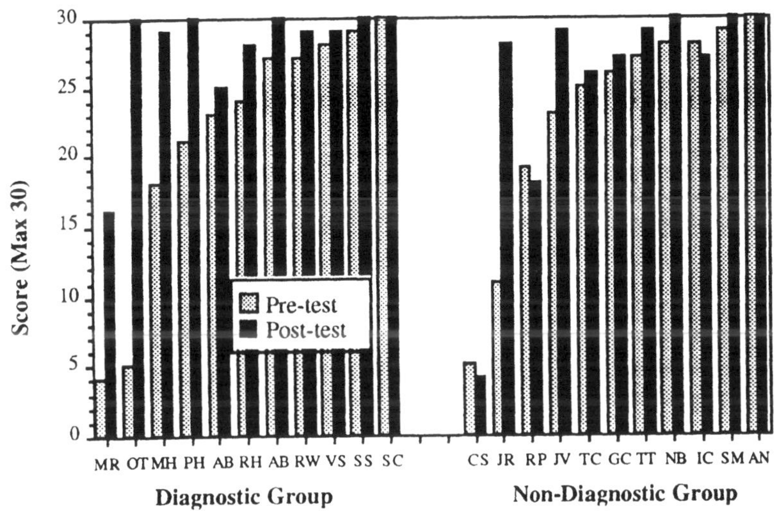

Further studies are needed to assess the reliability and the generality of these results. If the results are representative, one must consider why this study obtained a differential effect, unlike the three studies on highschool algebra reported by Sleeman et al. (1989):

(i) our differential test was more sensitive, in that diagnosis was the only factor differing between the two conditions, whereas for Sleeman et al. human tutoring was involved, which may have increased the variability of the effects

(ii) the arithmetic diagnosis was explicitly linked to corpora of data collected in previous studies, and thus likely to be well-tuned to the types of mistakes made.

(iii) the arithmetic diagnostic messages were very brief and to the point, whereas the 'model based remediation' used in the algebra studies was a lengthy process. Clearly, the latter would reduce the time available for 'reteaching'.

(iv) the arithmetic diagnosis appeared particularly valuable for the weaker children, leading to large improvements in their performance. It may be that for children with more advanced understanding many errors are careless slips, and all that is required is some indication that they have made a mistake, and they can easily identify for themselves the cause of their mistake. For these children diagnostic-based remediation is not necessary. For the weaker children, who have only a shaky knowledge of the procedures, it may be that non-diagnostic error information may cause them to invent a 'patch' (Brown & VanLehn, 1980) to their procedures, which, if faulty, will be difficult to eradicate and cause lasting confusion.

Of course, a large number of other possible reasons may be advanced for the differences. As Sleeman et al. (1989) conclude, more research is needed to identify those situations in which diagnosis-based teaching is more effective. We conclude that although SUMIT is effective in helping children learn the rules of arithmetic with or without diagnostic help, SUMIT's diagnostic help facility does indeed confer a further advantage in terms of the elimination of bugs, especially for those children who are weaker at arithmetic.

Acknowledgments. It is a pleasure to acknowledge the contributions of many of my students over the years, and, most recently, Chris Harrop, who carried out the empirical work reported here. My thanks also to Lydgate Middle School, Sheffield, and in particular to its head, Geoffrey Hall.

References

Anderson, J.R.; Boyle, C.F; Corbett, A.T., & Lewis, M.W. 1990. Cognitive Modeling and Intelligent Tutoring. *Artificial Intelligence* 42: 7-49.

Bloom, B.S. 1984. The two sigma effect: the search for methods of group instruction as effective as one-to-one tutoring. *Educational Researcher* 13: 3-16.

Brown, J.S. & Burton, R.R. 1978. Diagnostic models for procedural bugs in basic mathematical skills. *Cognitive Science* 2: 155-192.

Brown, J.S. & VanLehn, K. 1980. Repair Theory: A generative theory of bugs in procedural skills. *Cognitive Science* 4: 379-426.

Nicolson, R.I. 1990. Design and Evaluation of the SUMIT Intelligent Teaching Assistant for Arithmetic. *Interactive Learning Environments* 1: 265-287.

Sleeman, D.; Kelly, A.E.; Martinak, R.; Ward, R.D., & Moore, J.L. 1989. Studies of diagnosis and remediation with high school algebra students. *Cognitive Science* 13: 551-568.

VanLehn, K. 1990. *Mind Bugs: the origins of procedural misconceptions.* Cambridge, MA: MIT Press.

"The Proper Treatment of Cognition"

Tim van Gelder
Department of Philosophy, Indiana University
Bloomington IN 47405
(812) 855 7088 tgelder@ucs.indiana.edu

Abstract

The apparent contradiction between Smolensky's claim that connectionism is presenting a dynamical conception of the nature of cognition as an alternative to the traditional symbolic conception, and Giunti's recent elaboration of computational systems are special cases of dynamical systems can be resolved by adopting a framework in which (a) cognitive systems are dynamical systems, (b) cognition is state-space evolution in dynamical systems, and (c) differences between major research paradigms in cognitive science are differences in the kind of dynamical systems thought most appropriate for modeling some aspect of cognition, and in the kinds of concepts, tools and techniques used to understand systems of that kind.

If cognition consists of those internal, knowledge-based processes which underlie sophisticated human or animal behavior, then the primary question that cognitive scientists address is: what kind of processes are these?[1] A wide range of answers have been proposed, varying with the particular cognitive domain (vision, language processing, etc) under consideration and the level of abstraction at which the answer is framed. It is now becoming increasingly apparent, however, that most if not all such answers can be subsumed under one very general empirical hypothesis: cognition is state-space evolution in dynamical systems. This hypothesis follows naturally from two key insights, discussed below. The first is Smolensky's realization that a dynamics-based conception of cognition provides a deep alternative to traditional computational approaches. The second, paradoxically, is Marco Giunti's demonstration that traditional computational systems are special cases of dynamical systems. The apparent contradiction is resolved by seeing cognitive systems and models as drawn from a wide range of possible kinds of dynamical systems. The deep contrast is not between computational systems on one hand and dynamical systems on the other; it is between *kinds* of dynamical system, and corresponding kinds of concepts, tools and techniques for analyzing them.

1. The Proper Treatment of Connectionism

In his widely-read and influential article The Proper Treatment of Connectionism (PTC) (1988), Smolensky's aim was to articulate the connectionist approach to cognitive science, and to contrast it with the traditional "symbolic" approach. Since the latter approach has been described in detail in many places (e.g., Pylyshyn 1984), I will not elaborate on it here; suffice to say that, for current purposes, it can be summarized as the view that cognition is essentially computation: (something like) the rule-governed manipulation of symbolic representations with "conceptual" level semantics. Smolensky discussed many points of contrast between the symbolic and the connectionist approaches, but of particular concern here is the general account of the nature of cognition itself that he claimed to find embodied in connectionist

[1] This characterization of cognitive science is not intended to exclude the detailed study of actual human or animal performance. As Chomsky for one pointed out, often the most appropriate first stage in the study of cognition is to gain an adequate description of the performance itself. This characterization is also not intended to beg any questions about the extent to which those processes underlying sophisticated performance need to be knowledge-based.

work. A synthesized version of that account is summarized in the following claims:

PTC Dynamical Cognition Hypothesis

(1) Connectionist networks are high-dimensional, continuous and non-linear dynamical systems consisting of networks of interconnected units.
(2) Cognitive systems are connectionist networks.
(3) Cognition is state-space evolution within connectionist networks.
(4) The most appropriate tools for the study of cognition are dynamical modelling and dynamical systems theory.

It turns out that this cluster of claims has been largely ignored in subsequent discussion; for example, almost no mention of these themes is made in the interdisciplinary peer commentary that accompanied PTC in *Behavioral and Brain Sciences*. This is somewhat surprising, since Smolensky is here articulating, apparently for the first time, a deep and exciting new description of the nature of cognition, one very different from the dominant symbolic conception.

The PTC approach can be assessed from at least two directions: as a description of connectionism and its conceptual innovations, on one hand, and as an hypothesis concerning the nature of cognition on the other. As an account of connectionism it is in some ways misleading. It is probably true nowadays that most connectionist networks are high-dimensional, continuous and non-linear, but of course there have been and still are strains of connectionist work that reject properties such as continuity, or non-linearity. More importantly, only a relatively small portion of connectionist researchers bring genuinely dynamical methods to bear in their descriptions of network functioning or cognitive processes, at least in any extensive or systematic way. Indeed, most connectionist researchers seem to shy away from dynamical methods even though the networks they set up are in fact dynamical systems defined by differential or difference equations. There are at least two fairly standard strategies for doing this. One is to observe the behavior of the system only over very few time steps - often, as few as one or two, in standard feed-forward backpropagation networks. The other strategy is to focus attention at any given time only on restricted portions of the state-space - e.g., on the possible activity patterns over the hidden units, or those over the output units. Indeed, it is standard practice to *shift* attention from one portion of the state-space to another at each time step, as when one observes how an input pattern is transformed into a pattern over the first hidden layer, and so forth. Most connectionists critically depend on manoeuvres such as these, but they both represent ways of avoiding thinking of cognitive processing as general state-space evolution, and consequently enable and encourage connectionists to use analytical techniques quite different from those standard in dynamical modeling and dynamical systems theory. In short, the PTC perspective may be true to most connectionist work in some respects, but in others describes only a small portion of it. It is probably best regarded as at least partly normative: as describing, in other words, what might be thought of as the "most interesting" connectionist approaches.[2]

As an account of a novel conception of the nature of cognition the PTC perspective is also somewhat misleading. A fundamental component of PTC is a shift to a dynamics-based conception of cognition. Smolensky implies that this is a distinctively connectionist contribution. However, there are increasing numbers of researchers adopting dynamical approaches to the study of various aspects of cognition without being connectionists (see, e.g., van Geert 1991, Townsend 1989, Skarda & Freeman 1987). These researchers deploy dynamical systems theory in constructing models of cognitive processes, even though those models do not come in the form of networks of interconnected processing units. They believe that cognitive systems are dynamical systems, and that cognition is state-space evolution; they see the importance of properties such as continuity and non-linearity as much as any connectionist. From their point of view, connectionist networks are *just one way* to implement genuinely dynamical approaches.

The upshot of these points is that the dynamical conception of cognition Smolensky articulates in PTC (a) accurately characterizes only part of connectionist work, but (b) is held in common with various other non-connectionist strands of research. Together, these suggest that it would be wrong to tie the exciting idea that cognition is a dynamical phenomenon too closely to connectionism in particular. There is a

2 For a brief discussion of the use of dynamical explanatory methods in connectionist psychological modelling, see van Gelder 1991.

different and more natural conceptual boundary to be drawn. It does not identify the dynamical conception of cognition with connectionism, but rather uses the dynamical conception as the central commitment tying together a diverse group of researchers which *includes* some connectionists. We can thus think of "dynamicists" as those researchers committed to something like the following very general claims:

General Dynamical Cognition Hypothesis:

(a) Cognitive systems are [non-computational; see below] dynamical systems.

(b) Cognition is state-space evolution within such dynamical systems.

(c) The most appropriate tools for the study of cognition are dynamical modelling and dynamical systems theory.

It is important to see that this hypothesis has two sides. One specifies (in very abstract terms) the nature of cognitive processes. The other is methodological: it recommends certain kinds of tools as most appropriate for the detailed investigation of cognition. As will become more clear below, these two sides are complementary.

2. Computational Systems as Dynamical Systems

In a recent PhD dissertation Marco Giunti has exhaustively elaborated the thesis that computational systems, including those deployed in the mainstream symbolic approach to the study of cognition, are special cases of dynamical systems (see Giunti 1991). Here I will only illustrate his position with a coarse description of a paradigm example of computational systems, the Turing Machine, as a dynamical system. The overall state S of a Turing Machine at time t is fully specified when we know the contents of every cell on the tape, the current head state, and the current location of the head. Since the tape is unbounded in both directions, a useful way to represent this overall state is S(t) = ...aaaaqaaaa.... where each "a" designates the contents of a cell of the tape, q is the head state, and the head is positioned over the cell immediately to the right of q. The evolution equation F for the Turing Machine is a general specification of behavior of the machine, i.e., a specification of what state the machine will go to at time t+1 depending on the state it is in at time t, as depicted in the following schema:

t	F	t+1
....aaaaqaaaa....	====>	....aaaaqaaaa....

Each state transition in a Turing Machine involves three elementary changes: writing in the current cell, changing head state, and moving the head either right or left. The exact nature of the state transition depends on the contents of the current cell and the current head state, in a way that is specified in the machine table. Thus the machine table really is the evolution equation, though encoded in a somewhat unusual form. The general form of an evolution equation for a discrete system is S(t+1) = F(S(t)). In this case the equation is a tedious conditional easily reconstructible from the machine table. The table below, for example, gives the evolution equation for Minsky's seven state, four symbol universal Turing Machine (Minsky 1967).

Of course, there is nothing distinctive about Turing

S(t+1) = F(S(t)) =	....aaa1a_aaa..... if S(t) = ...aaaa1Yaaa.....	....aaa7a1aaa.... if S(t) = ...aaaa41aaa....
	....aaa1a_aaa..... if S(t) = ...aaaa1_aaa.....	....aaa4a1aaa.... if S(t) = ...aaaa4Aaaa....
	....aaa2a1aaa..... if S(t) = ...aaaa11aaa.....	....aaaaY5aaa.... if S(t) = ...aaaa5Yaaa....
	....aaa1a1aaa.... if S(t) = ...aaaa1Aaaa....	....aaa3aYaaa.... if S(t) = ...aaaa5_aaa....
	....aaa1a_aaa.... if S(t) = ...aaaa2Yaaa....	....aaaaA5aaa.... if S(t) = ...aaaa51aaa....
	....aaaaY2aaa.... if S(t) = ...aaaa2_aaa....	....aaaa15aaa.... if S(t) = ...aaaa5Aaaa....
	....aaaaA2aaa.... if S(t) = ...aaaa21aaa....	....aaaaY6aaa.... if S(t) = ...aaaa6Yaaa....
	....aaaaY6aaa.... if S(t) = ...aaaa2Aaaa....	....aaa3aAaaa.... if S(t) = ...aaaa6_aaa....
	....aaa3aYaaa.... if S(t) = ...aaaa3Yaaa....	....aaaaA6aaa.... if S(t) = ...aaaa61aaa....
(halt)	....aaaa3_aaa.... if S(t) = ...aaaa3_aaa....	....aaaa16aaa.... if S(t) = ...aaaa6Aaaa....
	....aaa3aAaaa.... if S(t) = ...aaaa31aaa....	....aaaa_7aaa.... if S(t) = ...aaaa7Yaaa....
	....aaa4a1aaa.... if S(t) = ...aaaa3Aaaa....	....aaaaY6aaa.... if S(t) = ...aaaa7_aaa....
	....aaa4aYaaa.... if S(t) = ...aaaa4Yaaa....	....aaaa17aaa.... if S(t) = ...aaaa71aaa....
	....aaaaY5aaa.... if S(t) = ...aaaa4_aaa....	....aaaa_2aaa.... if S(t) = ...aaaa7Aaaa....

Machines in this regard, although their relative familiarity makes it particularly easy to illustrate the point. From this perspective, computation - a particular sequence of symbol manipulations within a computational system - turns out to be a matter of state-space evolution within the particular kind of discrete state space offered by a digital computer. (Indeed, we might say that computation is a matter of *touring* the state-space.) Consequently, when the symbolic approach to cognition construes cognitive processes as computational processes, it also is construing them as state-space evolution within (computational) dynamical systems.

3. The Space of Cognitive Systems

The fact that computational systems can be described as dynamical systems has important implications for the discussion in the first section. The PTC dynamical cognition hypothesis and its more general counterpart were both intended as presenting *alternatives* to the symbolic conception of cognition. The deep difference between the symbolic approach and the dynamical alternatives cannot, however, be a contrast between symbol manipulation on one hand and state-space evolution in dynamical systems on the other, for the former is a special case of the latter. Rather, *the significant differences must lie in the kind of dynamical system employed, and the kinds of concepts and tools one uses in describing these systems.*

Occasionally, snippets of the official rhetoric of the computational approach to cognition has been dynamical in flavor; recall, for example, Newell & Simon's definition of a *physical symbol system* as "a machine that produces through time an evolving collection of symbol structures"[3]. Typically, however, the tools, techniques and concepts of dynamical modelling and dynamical systems theory are completely absent from standard discussions of computational systems. Why is this, if computational systems are special cases of dynamical systems? The answer is that certain ways of thinking about the behavior of systems lend themselves most naturally to certain kinds of systems. Various deeply different ways of understanding behavior can be applied to dynamical systems, but are most effective when applied to systems of particular kinds. Further, it is in the nature of standard computational systems, as dynamical systems, to encourage algorithmic rather than dynamical ways of thinking.

Consider the Turing machine again. This kind of computational device originated as Turing's own formalization of the process of elementary arithmetical calculations using pencil and paper. It is from this humble origin that the extremely simple nature of basic Turing machine operations derive. Consequently, Turing Machines as dynamical systems are fundamentally:

(1) Discrete. Each transition change takes place at a distinct point in "time".

(2) Digital. State transitions involve a jump from one unambiguously identifiable state to another. The symbols which can appear in the cells, the head states, and the head positions, are all digital in character.

(3) Deterministic. From each state there is only one next state to which the machine can change.

(4) Low interdependency of state variables. The Turing Machine system contains an unbounded number of state variables, but, in general, the change in any given state variable depends on only a very small number of these. For example, each cell on the tape corresponds to a distinct state variable. Will the value of that variable change in a given state variable? That depends on the values of only two other variables - i.e., on head position, and head state.

(5) Local. Each state transition involves changes in only three of the unbounded variables. The outcome of each transition is another point very "close" in state space.

It is, of course, no accident that the Turing Machine exhibits this particular combination of features. Basically, they make it possible to think of the behavior of the machine as the following of an *algorithm*. The low interdependency of state variables and local nature of state transitions enable one to *ignore* most of the state of the machine at any given time; after all, any change depends on only two variables and affects at most three. These features, in other words, encourage one to think of processing steps not as transitions from one total state of the system to another, but rather as localized alterations in particular

3 Newell & Simon (1981) p.40. In Human Problem Solving (1972; pp.11-12) they maintain that "the explanations of cognitive science are not in principle different from the explanations of any other science which is concerned with the dynamical behavior of some system.".

variables. They encourage thinking of the behavior of the system not in the geometrical sense of how the system is moving through its state space, but in the mechanical or syntactic sense of how particular constituents are being manipulated. Consider then the effect of adding the other three major features - discreteness in time, and digital and deterministic state transitions. Combined, these have the effect that each of these highly local alterations can be specified by its own simple rule. The total behavior of the machine is then a sequence of elementary rule-governed steps. By careful ordering of these steps, the desired overall effect is achieved - i.e., the behavior of the machine is specifiable by an algorithm. That is, it is in the very nature of a Turing machine, as a dynamical system, that all its basic state transitions can be *micro-managed* by the designer of the machine. The whole idea is to enable the designer to achieve controlled complexity in the overall behavior of the machine by orderly sequencing of carefully defined elementary operations. Complexity of global behavior is supposed to flow from simplicity and order at the base.

My claim, then, is that fundamental features of the Turing Machine as a dynamical system directly facilitate thinking about its behavior in algorithmic terms - and algorithmic modes of analysis are deeply different from dynamical ones. Generalizing, I ambitiously assert that what is true here of Turing Machines holds true of computational systems more generally. Computational systems (von Neumann machines, LISP machines, production systems, etc) form a natural class by virtue of sharing certain fundamental characteristics that enable us to most effectively describe their behavior in basically algorithmic terms. Conversely, the kind of connectionist systems that Smolensky had in mind when he formulated the PTC conception of cognition share certain other fundamental characteristics which render dynamical techniques fundamentally appropriate in their analysis.

Waxing metaphorically, we can think of particular dynamical systems as falling into a vast space of possible kinds of dynamical systems. The axes of this space are the fundamental properties that such systems can have - properties such as continuity vs discreteness, degree of interdependence of state variables, linearity vs non-linearity, and so on. Typical computational systems possess a certain characteristic set of properties and so "cluster" in one region of the space of possible systems. The symbolic approach to cognition can then be seen as the empirical hypothesis that real cognitive systems are dynamical systems which also fall into that particular region of the space - somewhere relatively "close" to Turing Machines. PTC, by contrast, focuses on systems that are "connectionist" - typically high-dimensional, continuous and non-linear - and is committed to the empirical hypothesis that real cognitive systems belong in this corner. The more general dynamical conception proposed at the end of Section 1 can then be seen as the suggestion that the PTC perspective circumscribes the corner into which cognitive systems fall a little too narrowly, though even the wider area embraces only systems which demand dynamical methods in their analysis. Most generally of all, each of these perspectives shares the fundamental assumption that real cognitive systems are located *somewhere* in this space of possibilities. This is equivalent to the the broad empirical hypothesis that cognition is state-space evolution in dynamical systems.

If this is right, it poses at least three major questions for further research. First, what really are the key dimensions of the space of possible dynamical systems? What are the deep properties which make for fundamental differences among kinds of dynamical systems? To stretch the spatial metaphor to its limits, what *principal components* can we abstract from the kinds of dynamical systems we already know about? A variety of important issues have already figured in the discussion so far (i.e., continuity, non-linearity, degree of interdependence of state variables, number of state variables, digital, deterministic), but there are also many other relatively obvious candidates (e.g., systems might have numerical vs arbitrary symbolic state variables, or be time-invariant, homogeneous, reversible, or chaotic), and no doubt a variety of not-so-obvious ones as well. From the perspective being advanced here, properly understanding the possible forms that cognitive processes might take, and the relationships between different research programs in cognitive science, presupposes clearly understanding the most basic kinds of properties that dynamical systems can have.

Second, what are the natural clusters within this space of possibilities? This is really the question: what are the natural kinds of dynamical systems (if any), based on their deep properties? It seems plausible that, for example, classical computational systems and perhaps connectionist systems (or various sub-categories of them) cohere into identifiable types, on the basis of which more specific hypotheses concerning the nature of cognitive processes can be framed. But are there other candidates as well? Perhaps certain types of analog computers, or the kinds of systems deployed by the non-connectionist dynamicists mentioned above, are equally candidates.

Third, what arguments can be formulated for supposing that real cognitive systems belong to a given kind? Mainstream computational cognitive science can be understood as making a bet - underwritten by some respectable arguments - that real cognitive systems belong in their corner of the space of possible dynamical systems, and hence that computational systems will provide the best models, and computational methods will provide the best analyses. From the point of view of others, such as dynamicists, the computational corner looks more like a ghetto, a particularly narrow and confining nook which people stay in not out of choice but because of unfortunate historical contingencies. Their bet is that real cognitive systems are to be found in relatively remote regions inhabited by systems which demand genuinely dynamical techniques if they are to be properly understood. The general arguments in favor of this position are yet to be worked out in detail, but at least one intuition is worth mentioning at this stage: if computational systems are attractive cognitive models for agents conceived as *abstract reasoners*, certain kinds of non-computational systems appear more deeply suited for models of agents conceived as *situated actors*.

References

Giunti, M. 1991. *Computers, Dynamical Systems, Phenomena, and the Mind.* PhD. Dissertation, Department of History and Philosophy of Science, Indiana University.

Minsky, M. 1967. *Computation: Finite and Infinite Machines.* Englewood Cliffs, NJ: Prentice-Hall.

Newell, A., & Simon, H. 1981. Computer science as an empirical enquiry. in Haugeland, J. ed. *Mind Design.* Montgomery VT: Bradford Books.

Newell, A. & Simon, H. 1972. *Human Problem Solving.* Englewood Cliffs, NJ: Prentice-Hall.

Pylyshyn, Z. 1984. *Computation and Cognition.* Cambridge MA: MIT Press.

Skarda, C.A. & Freeman, W.J. 1987. Brain makes chaos to make sense of the world. *Behavioral and Brain Sciences* 10:161-195.

Smolensky, P. 1988. On the proper treatment of connectionism. *Behavioral and Brain Sciences* 11:1-74.

Townsend, J.T. & Busemeyer, J.R. 1989. Approach-Avoidance: Return to Dynamic Decision Behavior. in Hizuko Izawa ed. *Current Issues in Cognitive Processes.* Hillsdale NJ: L.Erlbaum Associates.

van Geert, P. 1991. A dynamic systems model of cognitive and language growth. *Psychological Review* 98:3-53.

van Gelder, T.J. 1991. Connectionism and Dynamical Explanation. *Proceedings of the 13th Annual Conference of the Cognitive Science Society.* Hillsdale N.J.: L.Erlbaum Associates; 499-503.

Are Computational Explanations Vacuous?

Vinod Goel
Institute of Cognitive Science
University of California, Berkeley
goel@cogsci.berkeley.edu

Abstract

There is a certain worry about computational information processing explanations which occasionally arises. It takes the following general form: The informational content of computational systems is not genuine. It is ascribed to the system by an external observer. But if this is the case, why can't it be ascribed to any system? And if it can be ascribed to any system, then surely it is a vacuous notion for explanatory purposes. I respond to this worry by arguing that not every system can be accurately described as a computational information processing system.

Introduction

Computers play a very different role in cognitive science than they do in other disciplines, such as meteorology, city planning, or physics. No one claims that traffic patterns, thunder storms, and galaxies *are* computational systems. The claim is simply that many aspects of their "behavior" — in fact any aspect to which we can give an algorithmic description — can be simulated on a computer (given sufficient resources).

The cognitive science claim is of a very different nature. We do not simply claim that cognitive processes can be simulated on computational systems, but rather that cognitive processes *are* computational processes. The interesting versions of the cognitive claim (Fodor, 1975; Fodor, 1981; Fodor, 1987; Johnson-Laird, 1983; Newell, 1980; Pylyshyn, 1984) make it clear that this is not hand waving or a metaphorical way of speaking. It is meant to be taken very literally.[1]

Thus, cognitive science has a very special claim on the notion of computation not shared by other disciplines. Computation is not simply a modeling tool for us. It lies at the center of our theoretical/explanatory apparatus. Our theories of cognition quantify over the notion of *computational information processing* and use this notion in the explanation of cognitive behavior.

There is, however, a worrisome problem with the notion of computational information processing which is often raised. On most accounts, computational information processing is *as-if*. It is a matter of ascription. But — so the objection goes — anything can be described *as-if* it is doing computational information processing. If anything can be described *as-if* it is doing computational information processing, then to explain cognition as computational information processing is not to advance a substantive thesis.

A number of researchers have noted the problem and have been worried by it to different degrees (Chomsky, 1980; Cummins, 1989; Dietrich, 1990; Fodor, 1975, p.74, footnote 15; Searle, 1984; Searle, 1990). However, no satisfactory response has been forthcoming.

In this paper I would like to respond to the form of this problem raised by Searle (1990). I will argue that, from the fact that computational information processing is *as-if*, and thus ascribed to a system, it does not follow that it can be ascribed to *any* system. In fact, there are some very stringent constraints that systems have to meet before they can be described as doing computational information processing. Both my discussion of the problem, and my response, shall be restricted to "classical" computational systems and explanations (i.e., the Language of Thought and Physical Symbol Systems type accounts).[2]

The Vacuousness Objection

Searle's (1990) specific objection takes the following form: Computation is defined syntactically. But syntax is not intrinsic to the physics. It is *assigned* to the physics by an outside observer. In fact it can be assigned to any physical system. This is disastrous because we want "to know how the brain works." It is no help to be told that "the brain is a digital computer in the sense in which the stomach, liver, heart, solar system, and the state of Kansas are all digital computers." We want to know what fact about brains makes them digital computers. "It does not answer that question to be told, yes, brains are

[1] The cognitivist claim is not the very strong claim that computation is both necessary and sufficient for cognition. It is the more interesting claim that computation is necessary for cognition.

[2] A very brief discussion of connectionist information processing claims can be found in Goel (1991).

digital computers because everything is a digital computer" (Searle, 1990, p.26).

The logical structure of Searle's argument is the following:

P1) Computation is defined syntactically.
P2) Syntax is not intrinsic to the physics.
P3) Syntax can be assigned to any system.
Therefore: Any system can be described as a computational system.

I will argue that premise P1 is false, and thus the conclusion does not follow.

I think that Searle's intuition about syntax (premise P2) is correct. The syntax of external symbol systems is not intrinsic to the physics. Perhaps the way to construe syntax is as *an arbitrary property of the world that we use to individuate elements to which we will assign a semantical interpretation.* But the notion of syntax, while necessary, is not sufficient for the notion of computation that we use in classical cognitive science (contrary to P1). We also need notions of causation and interpretability. And if there is more to computation than syntax, then from the facts that syntax is ascribed to a system (P2), and that it can be assigned to any system (P3), nothing interesting about computation follows.

The burden of my response will be to argue that there is more to computation than syntax — specifically, causation and interpretability — and these notions in turn place stringent restrictions on the assignment of syntax to physical systems for purposes of describing them as computational systems.

Structure of Classical Computational Explanations[3]

I have argued elsewhere (Goel, 1991) that what cognitive science wants/needs from computer science is a notion of information processing, where information processing requires (i) quantification over the *content* of states of the system, and (ii) a *causal implication* of that content in the behavior of the system. Such a notion of information processing is derived from our folk psychology and seems to be the the one desired by an number of writers (Dretske, 1989; Fodor, 1975; Fodor, 1987; Newell, 1980; Newell & Simon, 1981; Pylyshyn, 1984). I will call any notion of information processing which satisfies these two criteria, a notion of *cognitive information processing.*

Such a notion of cognitive information processing does not, on most accounts, satisfy the requirements of "respectable scientific explanations". It uses mental or intentional predicates, which themselves require explanation. What we need is a mechanistic account of cognitive information processing which cashes out the intentional predicates. It is for this that we turn to computer science.

As it turns out, computer science can not currently deliver such an account (Goel, 1991; Searle, 1980; Searle, 1984). It can, however, deliver an epistemic counterpart to it in the notion of *computational information processing.* The notion of computational information processing which we get from computer science involves (i) the systematic individuation of physical states into computational states, (ii) the assignment of content to those states, and (iii) the systematic recoverability of computational states and contents at each step in the trajectory of the system over time.

Minimally, such assignment and interpretability involves the following:

A) One needs to be able to (i) assign (at the initial state of the system, **t=0**) a subset of the physical states of the system to equivalence classes of physical states (i.e computational states); (ii) correlate a subset of the computational states with reference-classes; and (iii) once the assignments and correlations have been set up, one must be able to look at the physical states of the system and systematically recover the computational states and reference-classes. To recover computational states means, minimally, that it is possible to identify equivalence-classes of physical states and "read off" their content. In certain cases this content will be an address of another computational state or device. To recover reference-classes means, minimally, to trace through the pointers to the actual computational state or device being referred to.

B) One must be able to (i) maintain the assignment and interpretation of the system as it evolves through time; i.e., given any instantaneous description of the system one should be able to recover the computational states, the reference-classes (as above), and a pointer to the next instantaneous description in the sequence; (ii) given a temporal sequence of instantaneous descriptions, it must be the case that some set of the computational states of the instantaneous description at **t** *cause* the computational states and/or device activations at instantaneous description **t+1**, and do so by virtue of the very property which gained them membership into that equivalence class of states; and (iii) the computational story one tells of the system must parallel the causal story.

We can consider these necessary criteria for a notion of computational information processing. Any system which can satisfy these criteria may be called a CIP system.

The relationship between cognitive and computational information processing is the following: In the

[3]Parts of this and the following sections are adopted from Goel (1991).

case of cognitive information processing there is an ontic fact of the matter as to the content of a mental state, independent of assignability and interpretability. But since there can't be such a fact without genuine reference/content, the systematic assignability and interpretability of computational information processing gives us an epistemic fact, or at least that's the intuition. Similarly, in the cognitive case, we have the content of mental states causally implicated in behavior, but again there can be no such ontic fact without genuine reference/content. But again, being able to trace through the evolution of the system (by maintaining the assignability and interpretability of the instantaneous descriptions), and discovering a parallelism between the causal and logical levels in the computational case, gives us an epistemic fact; or again, that's the intuition. In going from cognitive information processing to computational information processing we are in effect trading in some ontology for epistemology, a move that is not without precedent.

To summarize, the form that I am suggesting that classical computational explanations take is depicted in Figure 1. We have a notion of cognitive information processing, derived from folk psychology, that we appeal to in explaining cognitive behavior. However, it contains mental predicates which need to be cashed out. We turn to classical computational mechanisms for this purpose. However, these mechanisms cannot directly satisfy the criteria of cognitive information processing. They can, however, give us a related notion of computational information processing, which is underwritten by a reasonably well-understood mechanism. So the strategy is to map the notion of cognitive information processing onto the notion of computational information processing and to explain the latter notion with a classical computational mechanism.

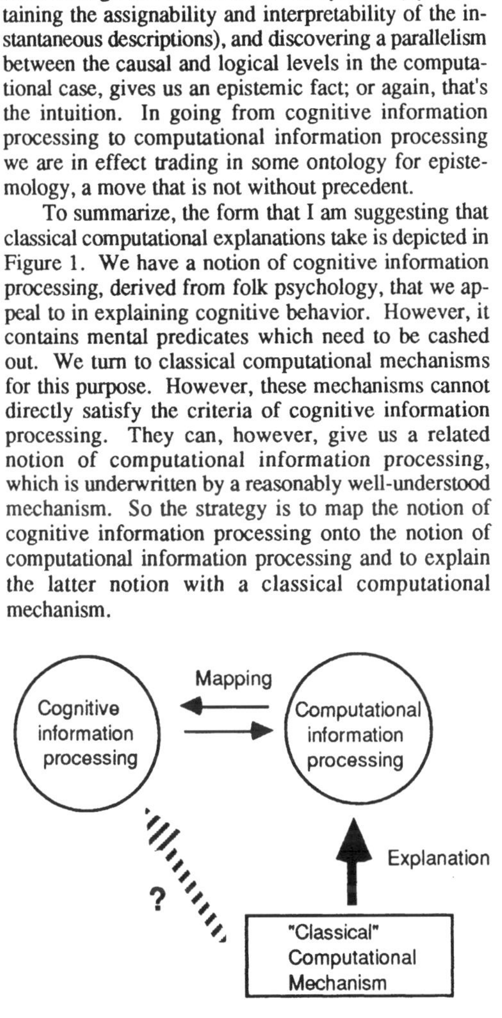

Figure 1. Structure of classical computational explanation. See text.

This, of course, leaves us with some deep questions about what, if anything, is gained by accounting for computational information processing, when our real interest is cognitive information processing. However, this question will not be pursued here. The focus of this paper is to show that the notion of computational information processing we get is not vacuous. Its relationship to cognitive information processing will be considered elsewhere.

Constraints on CIP Systems

Satisfying the criteria for computational information processing places rather severe constraints on any system. In fact, it is the case that only a system which meets the following constraints can be interpreted as a CIP system:

1) Equivalence classes of physical states of the system must be specified in terms of some function of causally efficacious characteristics such as shape or size.
2) These equivalence classes of physical states must be disjoint.
3) Membership of physical states in equivalence classes must be effectively differentiable, where differentiability is ultimately limited by physical possibilities.
4) Each state in the trajectory of the system must be "causally connected in the right way". While the specification of "causally connected in the right way" is obviously problematic, the intuition is something like the following: Certain physical states in the instantaneous description at t_n must have a direct causal connection to certain physical states in instantaneous descriptions at t_{n-1} and t_{n+1}. The connection must be such that certain physical states at t_{n-1} cause or bring about certain physical states at t_n, which in turn bring about certain states at t_{n+1}, and so on. Furthermore the transformation of the computational state CS_n at t_n into CS_{n+1} at t_{n+1} must be *realized as* the causal transformation of physical state PS_n at t_n into PS_{n+1} at t_{n+1}, where PS_n at t_n and PS_{n+1} at t_{n+1} are a subset of physical states of the system which are to be mapped onto computational states.
5) The correlation of equivalence classes of physical states with contents and/or reference-classes — within each instantaneous description of the trajectory — must be unambiguous in the sense that each member of an equivalence class of physical states must pick out the same, single, content and/or reference-class.
6) The membership of entities in reference-classes must be effectively differentiable.
7) The transformation of the system from one instantaneous description to the next instantaneous

description must be such that the above six criteria are preserved.

These are necessary constraints on CIP systems and may be called CIP constraints. It is worth noting that some are constraints on the individuation of syntactic elements, while others are constraints on semantic interpretation. If any of these constraints are violated, then some criteria on computational information processing will also be violated. For example:

- If equivalence classes of physical states are not specified in terms of some causally efficacious property, then B(ii) will be violated.
- If the individuation of equivalence classes of physical states is not disjoint, there will be not be a fact of the matter as to which computational state some physical state belongs to, thus thwarting the assignment of computational states to physical states. This would be a violation of A & B(i).
- If the individuation of computational states of the system is not effectively differentiable, then — whether they are disjoint or not — no procedure will be able to effectively make the assignment of physical states to computational states. For example, if the individuation of computational states is dense, then in the assignment of physical states to computational states, there will always be two computational states such that one cannot be ruled out as not belonging to a given physical state. This would also violate A and B(i).
- If the correlation of computational states with reference-classes is ambiguous, then there will be no fact of the matter as to the referent of any given computational state, and the systematic interpretability of the system will be impaired. This would violate A(iii) and B(i).
- If membership in reference-classes is not effectively differentiable, then no effective procedure will be able to specify which object any given computational state refers to. For example, if the reference classes are densely ordered, then in the assignment of objects to classes, there will be two classes for any object O, such that it is not effectively possible to say that O does not belong to one. This would violate A(ii, iii) & B(i)
- If the causal constraint is violated, we will not get an isomorphism between the physical and computational story and violate B(ii, iii). Furthermore, we will get the absurd results that time-slice sequences of arbitrary, unconnected patterns (e.g. the conjunction of the physical states consisting of craters on the moon at t1, the meteor shower on Neptune at t2, the food on my plate at t3, the traffic pattern on the Bay Bridge at t4, etc.) qualify as computational systems.
- If at any instantaneous description of the system, any of the above constraints are violated, then at that point some constraint on computational information processing will be violated.

Not Every System is a CIP System

The final step in the argument is to show that not every physical system is a CIP system, and that there is indeed a fact of the matter as to whether some system is, or is not, a CIP system. Given the nature of CIP constraints, determination of CIP systems can be made at just the syntactic level, or the syntactic and semantic levels. Both situations are discussed below.

Syntactic Individuation

Let's take a particular dynamical system — for example, the solar system — and ask whether it is a CIP system. If we accept the physical/causal story given by Newtonian mechanics — which recognizes things like planets, gravitational force, the shape of orbits, etc. — and use it to individuate the states and transformations of the systems (which are mapped onto computational states and transformations), our question becomes something like, "do the orbits of the planets around the sun constitute a CIP system?". I think one can unproblematically say they do not. For one thing, the instantaneous descriptions of the system will be densely ordered and thus violate the effective differentiability constraints.

Of course, it is possible to take the solar system and individuate components and relations in such a way that the CIP constraints are met. For example, a colleague suggested the following individuation: "we can divide up the orbit into quadrants, assign them numbers, think of them as states, and observe that each is followed by the next with law-like regularity."[4] While this is logically coherent, the point is that there is nothing in our physics (i.e., our science of the solar system) that requires, necessitates, or sanctions such an individuation. There are two reasons why such individuations are not generally sanctioned. First, they do not pick out higher-level regularities which deepen our understanding of the system. (If they did pick out such regularities, we would incorporate them into our scientific story.) Second, they may not even coincide with our scientific story. For example, where a planet is located in a quadrant at time t_i does not matter for this particular individuation, but it may matter very much to the physical/causal story. It may be the case that particular locations in the quadrant are associated with varying degrees and types of causal interactions with other heavenly bodies. If this is the case, this

[4]Kirk Ludwig

individuation does not coincide with our physics and can be dismissed on that basis.

Semantic Individuation

Can we make the same claims about the semantic constraints? Given an arbitrary dynamical system, can there be a fact of the matter as to whether it does, or does not, satisfy the semantic constraints on CIP systems? If one chooses not to interpret the system semantically, clearly there can be no such fact. The question will never arise. However, the important point is that, if one does choose to interpret the system, then *relative to a specific individuation of states and transformations* (ie. a particular syntactic individuation) *and a specific semantic interpretation,* there is a matter of fact as to whether the system is a CIP system or not. To get this matter of fact, one proceeds as follows:

(i) Decide on the system and phenomenon you are interested in and the level at which it occurs.
(ii) Understand the system/phenomenon on its own physical/causal terms; i.e., explicate the structure and dynamics of the system which are causally relevant in the production of the phenomenon under investigation.
(iii) Use the physical/causal structure to individuate equivalence classes of physical states and transformations which are to be assigned to computational states and transformations (i.e., the syntactic interpretation).
(iv) Specify the program the system is supposed to be running (i.e., the semantic interpretation) and again use the causal structure and dynamics of the system to interpret the computational states and transformations.
(v) Ask whether this individuation and interpretation meets the constraints on CIP systems.

The system under investigation may or may not meet the CIP constraints. It may fail in the first instance because the causal structure and dynamics of the system result in an individuation of (computational) states and transformations which do not meet the syntactic constraints. It may fail in the second instance because — since reference is correlated with causation — the causal network of the system may not support the interpretation of computational states and transformations required by the program which the system is supposed to be running (i.e., the semantic constraints).

Is our stomach — as a processor of food — a CIP system with respect to a certain individuation and interpretation of computational states and transformations? It is an empirical question. There is no *a priori* answer independent of the causal structure and dynamics of the system and a specific semantic interpretation. One needs to proceed as above and discover the answer. Is our brain a CIP system under the relevant individuation and interpretation of computational states and transformations? That is, do the structure and dynamics of the brain which are causally relevant in the production of mental life satisfy the CIP constraints? Maybe they do; maybe they don't. It is, as cognitive science claims, an empirical question.

Since the facts about CIP systems are relative to some individuation and interpretation of computational states and transformations, they need not be unique facts. A system may turn out to be a CIP system with respect to several individuations and interpretations. But there is no reason to believe that it will turn out to be a CIP system with respect to every individuation and interpretation because the CIP constraints tie the individuation and interpretation into the physical/causal structure of the system.

Conclusion

If it is indeed the case that (i) we appeal to computational systems for a notion of computational information processing, (ii) only CIP systems can satisfy the criteria on computational information processing, and (iii) not every system is a CIP system, then from the (correct) premise that syntax is not intrinsic to the physics, it does not follow that the notion of computation as used by cognitive science is vacuous.

Indeed, to say the brain is a computer is to make a very substantial empirical claim. What cognitive science is doing by appealing to computation — and claiming it is a necessary condition for cognition — is putting forward the empirical hypotheses that the mechanism that underwrites computational information processing is the very same mechanism which underwrites cognitive information processing. This mechanism is a dynamical system that satisfies the CIP constraints. Thus the cognitive system on this view is accurately described as a CIP system. This claim is not vacuous, nor harbors an homunculus. It may of course be false, but that is a separate question which can be determined only by empirical enquiry.

Acknowledgements

The author is indebted to John R. Searle and Brian C. Smith for both inspiration and assistance in the course of developing this argument. They are of course not responsible for its shortcomings. This work has been supported by a Gale Fellowship, a Canada Mortgage and Housing Corporation

Fellowship, and a research internship at System Sciences Lab at Xerox PARC and CSLI at Stanford University.

References

Chomsky, N. (1980). Rules and Representations. *Behavioral and Brain Sciences, 3*, 1-61.

Cummins, R. (1989). *Meaning and Mental Representation.* Cambridge, Massachusetts: The MIT Press.

Dietrich, E. (1990). Computationalism. *Social Epistemology, 4* (2), 135-154.

Dretske, F. (1989). Putting Information to Work. In P. P. Hanson (Eds.), *Vancouver Studies in Cognitive Science*. Vancouver, Canada: University of British Columbia Press.

Fodor, J. A. (1975). *The Language of Thought.* Cambridge, Massachusetts: Harvard University Press.

Fodor, J. A. (1981). Methodological Solipsism Considered as a Research Strategy for Cognitive Psychology. In J. Haugeland (Eds.), *Mind Design.* Cambridge Mass.: MIT Press.

Fodor, J. A. (1987). *Psychosemantics: The Problem of Meaning in the Philosophy of Mind.* Cambridge, Massachusetts: The MIT Press.

Goel, V. (1991). Notationality and the Information Processing Mind. *Minds and Machines, 1* (2), 129-165.

Johnson-Laird, P. N. (1983). *Mental Models: Towards a Cognitive Science of Language, Inference, and Consciousness.* Cambridge, Mass.: Harvard University Press.

Newell, A. (1980). Physical Symbol Systems. *Cognitive Science, 4*, 135-183.

Newell, A., & Simon, H. A. (1981). Computer Science as Empirical Inquiry: Symbols and Search. In J. Haugeland (Eds.), *Mind Design.* Cambridge, Mass.: MIT Press.

Pylyshyn, Z. W. (1984). *Computation and Cognition: Toward a Foundation for Cognitive Science.* Cambridge, Massachusetts: A Bradford Book, The MIT Press.

Searle, J. R. (1980). Minds, Brains and Programs. *The Behavioral and Brain Sciences, 3*, 417-457.

Searle, J. R. (1984). *Minds, Brains and Science.* Cambridge, Mass.: Harvard University Press.

Searle, J. R. (1990). Is the Brain a Digital Computer? In *Sixty-fourth Annual Pacific Division Meeting for the American Philosophical Association.* Los Angeles, CA, March 30, 1990.

Taking connectionism seriously:
the vague promise of subsymbolism and an alternative.

Paul F.M.J.Verschure
AI lab, Institute for Informatics, University of Zürich,
Winterthurerstrasse 190, CH- 8057 Zürich,
Switzerland.
e-mail: verschur@ifi.unizh.ch

Abstract

Connectionism is drawing much attention as a new paradigm for cognitive science. An important objective of connectionism has become the definition of a subsymbolic bridge between the mind and the brain.

By analyzing an important example of this subsymbolic approach, NETtalk, I will show that this type of connectionism does not fulfil its promises and is applying new techniques in a symbolic approach.

It is shown that connectionist models can only become part of such a new approach when they are embedded in an alternative conceptual framework where the emphasis is not placed upon what knowledge a system must posses to be able to accomplish a task but on how a system can develop this knowledge through its interaction with the environment.

Introduction

Connectionism has been gaining much attention in cognitive science. On of the reasons is that problems of the traditional cognitivistic approach, like the need for noise and fault tolerance and the capability to generalize, are solvable with connectionist, brain-like, techniques.

This proposal makes the problem of complete reduction (PCR) (Haugeland, 1978), or of how a symbolic description of cognition can be reduced to a non-symbolic one, again highly relevant.

In the traditional cognitivistic view cognition is seen as formal symbol manipulation. The basic steps of this approach can be defined as: "1, Characterize the situation in terms of identifiable objects with well defined properties. 2, Find general rules that apply to situations in terms of those objects and properties. 3, Apply the rules to the situation of concern, drawing conclusions about what should be done." (Winograd and Flores, 1986, p.15).

The physical symbol system hypothesis (Newell, 1980) can be taken as the most influential formulation of this approach. The hypothesis states that a physical symbol system (PSS) constitutes the necessary and sufficient conditions for general intelligence. A PSS consists of a set of actions and is embedded in a world that consists of discrete states; objects and their relations. Moreover, a PSS has a "body of knowledge" that specifies the relations between the events in the world and the actions of the system, we can also refer to this body of knowledge as a world model built up with symbolic representations. The actions of the system, either in the world, or internal inferences, are organized around the goals of the system according to the principle of rationality: roughly a system will use its knowledge to reach its goals. An important implication of this conceptualization of cognition is that it can (and must) be modelled at the abstract level of symbol manipulation. The specifics of the implementation are, therefore, of no importance. PCR is no longer an issue since the non symbolic level of brain dynamics is not taken to be very relevant in explaining cognition.

The hypothesis of physical symbol systems is often seen as the only plausible model for general intelligence which has no serious competitors (e.g. Pylyshyn, 1989). Despite this claim this paradigm also confronts some serious problems. One of these problems is the symbol grounding problem (Harnad, 1990), or the question of how symbols acquire their meaning. In the cognitivistic tradition the meaning of symbols is taken as given (Newell, 1981), which implies that cognitivism has to resort to a nativistic position: that the "body of knowledge" is just present from the start on. Moreover, one has to assume that the system possesses very reliable transduction functions that allow the coupling between events and objects in the world and their internal symbolic representation. These assumption have been criticized on several grounds. For instance, the genome does not have the coding capacity to represent this body of knowledge (Edelman, 1987), or it still needs to be explained how during evolution this "body of knowledge" could have been acquired (Piaget in Piatelli-Palmarini, 1980). Moreover, practical applications developed within this paradigm, for instance robot control architectures, have not been very successful (see Malcolm et al., 1989, for an overview).

A related issue is the frame of reference problem (FOR) (Clancey, 1992) which conceptualizes the relation between the designer, the observer, and the system. The designer of a system develops this system out of his/her domain ontology (i.e. a categorization of the task domain into events, objects, and relations). The consequence of this is that the knowledge on which the system is based is grounded in the experience of the designer and that this domain ontology is static.

An alternative position towards explaining cognition can be found in traditional connectionism (e.g. Rosenblatt, 1958). Here the hypothesis of formal symbol manipulation was rejected in favour of theories that take the dynamics of the brain into account. The appropriate tool here was not logic but statistics. It is assumed that by interacting in its environment an organism, which does not possess prior knowledge of this environment, develops preferences for specific responses to certain stimuli. The evolving associations between stimuli and responses are directly related to the development of distinct connection patterns in its nervous system. The classical example of this approach is the perceptron proposed by Rosenblatt (1958). Also in this case PCR is dissolved since the intentional level of symbol manipulation is not taken to be relevant in explaining behavior.

When we compare the solutions of PCR of both approaches they have two contradictory positions. While cognitivism emphasizes the importance of a formal symbol manipulating mind traditional connectionism underlines the importance of the dynamical brain. This contrast can be seen as a mind-brain dilemma (Verschure, 1992). Subsymbolic connectionism has an alternative position towards this dilemma.

Smolensky (1988) tried to define a theoretical framework for connectionism where he assumes that cognition, as described within classical cognitivism, is an *emergent* property of the interaction of a large number of units which are subsymbolic. His proposal is based on developments in the present main stream of connectionist research (e.g. Rumelhart and McClelland, 1986).

Smolensky assumes that in a connectionist model symbols are encoded by the 'complex patterns of activity over many units. Each unit participates in many such patterns ... The interactions between individual units are simple, but these units do not have conceptual semantics: they are subconceptual' (Smolensky , 1988, p. 6).

The subsymbolic description of cognition at the level of units is supposed to be, in principle, reducible to brain processes. The limited knowledge we have of the brain is here seen as the only barrier we have to take to complete this subsymbolic reduction of cognition.

Subsymbolic connectionism offers a new perspective on the relation between the mind and the brain. It assumes that both levels can be joined up by specifying "bridging principles" between the cognitivistic symbol manipulating mind and the dynamic brain. If this approach can show how PCR can be solved without rejecting one of the levels of description involved it can indeed be taken as progress.

To evaluate this claim of subsymbolic connectionism I will first analyze its paradigmatic example, NETtalk. This analysis will show that subsymbolic connectionism does not fulfil its promise to solve the mind-brain dilemma, but still constitutes, in essence, a symbolic approach. Next I will sketch an alternative framework which does allow a solution to this dilemma. Central to this alternative position is that in order to understand cognition the focus should not be on a predefined "body of knowledge", but on how this can be acquired through the system-environment interaction.

NETtalk: the example of subsymbolic reduction

NETtalk, the famous 'parallel network that learns to read aloud' by Sejnowski and Rosenberg (1986, 1987) is put forward by Smolensky, and others, as *the* example of subsymbolic reduction.

With NETtalk Sejnowski and Rosenberg have successfully built a model that could pronounce English words. Although they acknowledge the differences between the architecture of NETtalk and the brain they assume that NETtalk can teach us how information (in this case letter to phoneme mappings) is represented in 'large populations of neurons'.

The input layer of NETtalk consists of 7 identical groups of 29 units each. The letters of the alphabet plus 3 extra features representing word boundary and punctuation are coded in every group by a special unit. The hidden layer of NETtalk has no pre-assigned interpretation but is necessary to accomplish the mapping between the input- and the output layer. Every unit of the output layer represents one of 23 articulatory features or one of 3 features representing stress and syllable boundaries. The network learns, by means of back propagation, to associate the letter coded for by the active unit of the fourth group of the input layer with a specific set of articulatory features represented by a specific pattern of active output units. The other 6 groups of the input layer provide a context. The coupled activation patterns of the input- and output layer are determined by the designers of the system.

NETtalk is able to learn to correctly pronounce 95% of the presented words after training with 50000 words. It could correctly generalize to new cases in 78% of the test words.

Sejnowski and Rosenberg next tried to determine the features coded by the hidden units of NETtalk by clustering input patterns that lead to the same activation patterns of these elements. This cluster analysis of NETtalk showed that the activity

	Vowels:	Consonants:
Tensed	9	0
Medium	8	0
High	**6**	**1**
Central 1	**5**	**1**
Front 1	**5**	**1**
Front 2	**5**	**2**
Central 2	4	0
Low	4	0
Back 1	2	0
Back 2	2	0

	Vowels:	Consonants:
Voiced	**1**	**21**
Unvoiced	**1**	**12**
Fricative	0	9
Palatal	0	8
Velar	**1**	**8**
Labial	0	7
Stop	0	7
Affricative	0	6
Alveolar	0	6
Nasal	0	6
Dental	0	5
Liquid	0	4
Glide	**1**	**3**
Glottal	0	1

Table 1: the frequency of occurrence of the articulatory features in coding vowels and consonants.

patterns of the hidden units could be understood as separating two main features: vowels and consonants. These results where considered to be an important proof of the power of subsymbolic computing: the emergence of a 'symbolic' separation of the letter to phoneme mapping in vowels and consonants.

A closer analysis of the letter to phoneme mapping the network has to learn shows, however, that the patterns presented to the network can beforehand be separated into two global categories: vowels and consonants. To illustrate this in Table 1 the 24 articulatory features represented by the units of the output layer are shown with their frequency of being involved in coding a vowel or a consonant. Articulatory features that are used to code both vowels and consonants are printed in bold face.

Table 1 shows that the features that are used to code about 95% of the vowels only code about 5% of the consonants and vice versa. Only 8 of the 24 features show an overlap and are used for coding vowels *and* consonants. Notice, however, that this overlap is always rather limited. For instance the feature "Unvoiced" is used 12 times in encoding a consonant and only once in encoding a vowel. Because every input letter is related to *a number* of articulatory features it can unambiguously be coded as a vowel or a consonant. Only one of the 51 symbols learned is completely defined by features related to the opposite class (the letter c as pronounced in logic is completely defined by articulatory features which mostly code vowels). See Verschure (1992) for an elaborate analysis.

NETtalk is put forward as a clear example of a model possessing subsymbolic representations. In this analysis it is shown, however, that the subsymbolic reduction given by NETtalk of the pronunciation of English words, expressed in the separation of vowels and consonants, is put in by the designers of the system. The vowels are always translated to a set of articulatory features of which we know beforehand that they distinguish vowels from consonants. Therefore, it is not surprising that NETtalk learns to discriminate them from the category of patterns coding consonants. The trick of subsymbolic reduction seems to lie in the transformation from the symbols (in this case articulatory features) to the actual activation patterns that NETtalk learns. This transformation, which conserves symbolic regularities (a vowel-consonant distinction), is made by the designers: Sejnowski and Rosenberg and not by NETtalk. Therefore, the claim that NETtalk started out *without* 'considerable "innate" knowledge in the form of input and output representations that were chosen by the experimenters' (Sejnowski & Rosenberg, 1987, p.158) does not relate to the reality behind the model.

The analysis of NETtalk suggests that subsymbolic reduction seems to boil down to a circularity consisting of the following steps: 1, The designer of the system defines basic symbolic properties in which a certain task can be described (in NETtalk articulatory features and characters); the knowledge the system must have to accomplish the task is defined. 2, These properties get translated to regularities of activation patterns presented to a connectionist model (in the case of NETtalk this is expressed in which letter should be pronounced with which set of pronunciation features). 3, The connectionist model learns to separate the patterns on their differences and groups them together on their regularities. These separations and groupings get expressed in the dynamics of the network, for instance in the activation of the hidden layer or in a specific distribution of the weights. 4, The regularities expressed in the dynamics of the network, which are completely determined by the regularities put in by the designers of the system, are symbolically interpreted by the designer (in the case of NETtalk as a vowel/consonant distinction). Steps 1 to 3 show a strong similarity to the ones of the cognitivistic approach listed earlier. It can be shown (Verschure, 1992) that this hypothesis concerning the circularity of subsymbolic reduction can easily be generalized to other connectionist models which have 'emergent' properties and models that rely on completely distributed representations.

Subsymbolism seems to be based on a misconception of the epistemological status of the representations of the model (Verschure, 1990). Knowledge that is put in by the designers, that relates to their domain ontology (the symbolic categorization of the domain consisting of characters and phonetic features), is erroneously interpreted as an emergent property of the model. This provides another example of the seriousness of the FOR problem.

The solution of the mind-brain dilemma that subsymbolic connectionism offers remains a vague promise.

From symbols to dynamics

The analysis of NETtalk showed that subsymbolic connectionism can be seen as a new methodology in a well known theoretical framework: cognitivism. The initial ambition of connectionism to form an alternative paradigm for cognitive science is not fulfilled. It seems useful to reevaluate the role that connectionism can play in cognitive science.

To reassess the ambition of connectionism it is useful to first evaluate the nature of connectionist models. Connectionist models are dynamical structures with a brain-like flavor, but they can also be applied to model other phenomena like the immune system or auto catalysis (e.g. Farmer, 1990). This implies that these models are neutral to any interpretation and cannot by themselves constitute a new paradigm.

In defining an alternative conceptual framework the FOR problem can be taken as a starting point. To understand behavior it is important not to confuse the different perspectives involved. If the design of a system is based on an external domain ontology (from the designer or observer) and its behavior is interpreted as if it were related to the experience of the artefact we are suddenly confronted with the symbol grounding problem. Because, it is not recognized that the representations of the system are founded in this external domain ontology. In this respect the symbol grounding problem can be seen as an artifact of a symbolic approach which ignores the FOR problem.

It is obvious that intelligence is related to knowledge. The point is, however, that this knowledge should from the start on be grounded in the experience of the system and not in that of the designer or observer. Moreover, symbolic descriptions of behavioral regularities can be taken as being part of an observer ontology. But there is no reason to automatically assume that the behavior of the system is produced by internal symbolic processes that mirror these regularities.

Given the above mentioned problems there is no reason to subscribe immediately to the assumptions made by cognitivism. In our own work, which relates to the emerging field of "New AI" (Brooks, 1991), a different set of assumptions is made. The first assumption is that cognition can only be modelled using autonomous agents (see also Brooks, 1991): systems that have realistic sensors and effectors with which they interact with the world. Next, these systems do not assume highly reliable transduction functions that take care of the perception of, for instance, a letter, but they span the whole domain from sensing to acting. This allows the development of representations that are grounded in the experience of the system. The behavior of the system is not separable from its environment. It is the result of the ongoing interaction between the two and not a distinct property of one of these elements (e.g. Ashby, 1960) Furthermore, a different set of assumptions about the world is made: First, the real world is constantly changing, only partially knowable, and only partially predictable. Therefore there cannot be a predefined body of knowledge that approximates the properties of the real world (see also Agre & Chapman, 1988; Suchman, 1987). Second, the world does not consist of a collection of events. The notion event is completely connected to the interaction between a system and the world. This last point will be further dealt with in the discussion section.

While cognitivism assumes that there is a "body of knowledge" to be able to explain behavior and postpones the question of learning (Haugeland, 1985) this proposal takes the opposite strategy. The central theme is how a system can acquire knowledge from its interaction with the world: how does adaptation take place and what are its prerequisites. Moreover, all processes, internal and external, are in principle dynamic. The observed behavior can, however, be described in symbolic terms.

Starting with the assumptions outlined above we have developed a design methodology for autonomous agents, distributed adaptive control (DAC) (Pfeifer & Verschure, 1992; Verschure et al., 1992) which is based on a model for classical conditioning (Verschure & Coolen, 1991). The basic properties of the system are related to a value scheme, which is taken to be defined by the genetic setup of the system (Edelman, 1987). The value scheme defines the properties of the sensors and effectors and some initial sense-act relations (reflexes). The value scheme allows a coarse adaptation to the environment, for instance, when there is a collision to the left turn to the right. The system is also equiped with a more sophisticated sensor: an inverse range finder which represents, in essence, time to contact. The states of this more sophisticated sensor are gradually integrated into the basic reflexes of the system due to the system environment interaction. This integration process, which is based on a Hebbian learning mechanism, will lead to a fine tuned adaptation to the specifics of the environment. In Figure 1 the set up of the control architecture is depicted. The three sensors project their state onto specific neural fields.

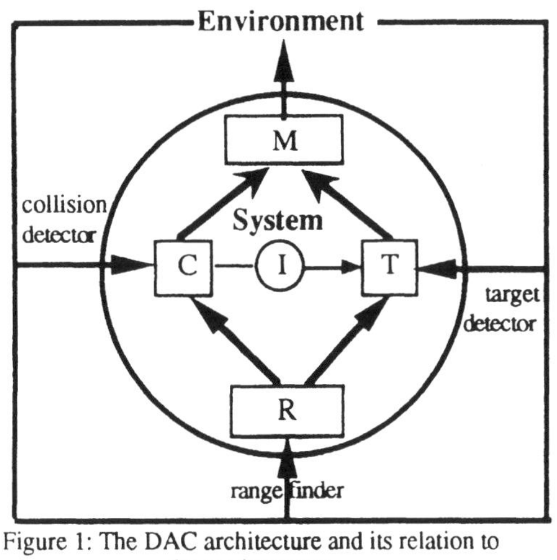

Figure 1: The DAC architecture and its relation to the environment.

Activation of units in the fields that relate to the collision detector (C) and the target detector (T) will automatically trigger an action, avoid or approach respectively. The basic reflexes can be described as: "collision left -> turn right" and symmetric for the other side, and "target left -> turn left" and symmetric for the other side. The default action of the system is to move forward. These actions are represented by motor units in field M. The connections between C, T and M are prewired. These connections implement the reflexes by connecting the related sensing and acting components. Since the change of the connections between C and T and the range finder field (R) is based on a Hebbian learning rule any state in R which occurs congruently with an action will be associated with the activation in C and/or T that triggered this action. Over time specific prototypical states of R will develop that will trigger specific actions. Next to these four fields a special inhibitory unit, I, is defined that regulates the interaction between avoid and approach actions: activation in C will inhibit the output from T.

We showed (Verschure et al., 1992) that a system based on these properties can develop emergent behaviors like wall following in an environment where targets are placed behind holes in walls. This regularity leads the system to associate being parallel to a wall with approach actions. Over time this behavior was generalized to any situation were the system was next to a wall. It follows a wall wiggling along switching between approach and avoidance actions.

This wall following behavior can be described in symbolic terms like a strategy or rule which is based on the representation of a wall and the action to follow it. The properties of the control architecture, however, indicate that such a rule is not present in the system. This behavior will only emerge when a specific regularity is present in the system-environment interaction (e.g. targets behind holes in walls). This indicates that although behavioral regularities might be based on special internal regularities of a system this does not have to be the case. Moreover, this emergent behavior is only present from the point of view of the observer. The system can only act on its immediate sensory states, while wall following behavior is displayed over several time steps consisting of many actions. This behavior that for an observer looks very structured can only be explained when it is decomposed into the actions that constitute it. This decomposition, however, shows that the system is acting like it always would, whether it is following walls or doing something else, that for an observer might look not that well organized.

Discussion

The analysis of subsymbolic connectionism has shown that it is in fact applying new techniques in a well known conceptual framework: cognitivism. Therefore, it does not provide a new perspective on the mind brain dilemma. It was argued that to assess the role connectionism could play in cognitive science it is of importance to find an alternative conceptual scheme in which it can be applied. The reason for this is not to find a justification for doing connectionism, but to address the mind-brain dilemma.This alternative framework can be found in the developing field of "New AI". The contrast between the two approaches now becomes that assumptions of cognitivism, which lead to the symbol grounding problem, become central research issues. One of this issues is, for instance, what is the role and nature of knowledge in adaptive behavior.

In doing this it becomes clear that the issue of emergence should also be viewed from the perspective of FOR. Emergent behavior then relates to an observer who specifies a specific time and or spatial frame in which "interesting" behavior is displayed by the system. This emergent behavior is not a property of the system but of the interaction with the environment. The chunk of action that an observer can call wall following is related to a set of actions that become a connected whole in the frame of reference of the observer. To explain this behavior it should be viewed from the perspective of the system. Which in the case of the presented example means that what is wall following from the observer perspective can only be explained from the system's perspective as a sequence of approach or avoid actions given the immediate sensory and the internal states.

This perspective on behavior gives a different status to notions that are taken for granted in the symbolic paradigm. For instance, the latter assumes that the world consists of objects and events which are somehow mirrored by the internal representations of the system. In our case we see that the notion of event and object is defined from the perspective of the system where an event always relates to actions. For instance, initially an action can only be triggered by one of the basic reflexes defined by the value scheme. Due to the learning mechanism this can be transferred to range finder states. What will now become a situation in which a specific action will be triggered cannot be predicted but depends on the specifics of the system environment interaction. Only from the perspective of the learning history of the system the notion event can be defined.

An important issue is how this proposal will scale up to the phenomena traditionally studied in cognitive science like reasoning and language. The central question is, however, whether we should see this issue as a conflict between two approaches. From the perspective of FOR we can see that the accounts offered by traditional approaches can be taken as observer characterizations of behavioral regularities. Which would mean that it is possible to describe some parts of behavior, like language, in terms of discrete elements that we call symbols. From the systems perspective linguistic behavior is still behavior built up out of many actions.

The mind brain dilemma can be addressed from the presented perspective. Supposedly conflicting paradigms in fact provide a different perspective on the phenomenon of behavior. With this we can overcome the isolated position of the study of the mind as a special science and focus on the initial ambition behind cognitive science to develop a fruitful interaction between the behavioral and the neurosciences.

Acknowledgements

The research reported in this paper was partly sponsored by grant 21-30269.90 of the Swiss National Science Foundation to Rolf Pfeifer. The author thanks Rolf Pfeifer and Thomas Wehrle for valuable discussions.

References

Agre, P.E.; and Chapman, D. 1987. Pengi: An implementation of a theory of activity. *AAAI-87*, Seattle, WA: 268-272.

Ashby, W. R. 1960. *Design for a brain: The origin of adaptive behavior.* New York: Wiley.

Brooks, R.A. 1991. Intelligence without reason. *IJCAI-91, Proceedings of the twelfth international conference on artificial intelligence, vol 1:* 569-595.

Clancey, W.J. 1992 The frame of reference problem in the design of intelligent machines. In: K.V.Lehn Ed.: *Architectures for intelligence. Proc. 22nd Carnegie Symposium on Cognition.* pp. 357-423, Hillsdale, N.J.: Erlbaum.

Edelman, G.M. 1987. *Neural Darwinism: The theory of neuronal group selection.* New York: Basic Books.

Farmer, J.D.1990. A Rosetta stone for connectionism, *Physica D, 42*: 153-187.

Harnad, S. 1990. The symbol grounding problem. *Physica D, 42*: 335-346.

Haugeland, J. 1978, The Nature and Plausibility of Cognitivism, *The Behavioral and Brain Sciences, 2*: 215-260.

Haugeland, J.(1985). *Artificial Intelligence: The very idea.* Cambridge Ma.: MIT.

Malcolm,C., Smithers, T.,& Hallam, J. (1989). An emerging paradigm in robot architecture. In: T.Kanade, F.C.A. Groen, & L.O.Herzberger (Eds.): *Intelligent Autonomous Systems 2*, Amsterdam: Elsevier.

Newell, A. 1980. Physical symbol systems. *Cognitive Science, 4*: 135-183.

Newell, A. 1981. The knowledge level. *AI magazine, 2*: 1-20.

Piatelli-Palmarini, M., Ed. 1980. *Language and Learning. The debate between Jean Piaget and Noam Chomsky.* Cambridge Ma.: Harvard University Press.

Pylyshyn, Z. 1989. Computing in cognitive science. In: M.I.Posner Ed. *Foundations of cognitive science.* pp. 51-91. Cambridge Ma.: MIT.

Rosenblatt, F.1958 The Perceptron: a probabilistic model for information storage in the brain. *Psychological Review, 65*: 386-408.

Rumelhart, D.E., McClelland, J.L. and the PDP research group. 1986. *Parallel Distributed Processing; explorations in the microstructure of cognition, volume 1: foundations.* Cambridge: MIT press.

Sejnowski, T. & Rosenberg, C. 1986. NETtalk: a Parallel network that learns to read aloud. The Johns Hopkins University Electrical Engineering and Computer Science Technical Report JHU/EECS-86/01.

Sejnowski, T. & Rosenberg, C.1987 Parallel Networks that Learn to Pronounce English Text. *Complex Systems, 1*: 145-168.

Smolensky, P. 1988 On the Proper Treatment of Connectionism.*Behavioral and Brain Sciences, 11*: 1-73.

Suchman, L.A. 1987. *Plans and situated actions.* Cambridge University Press.

Verschure, P.F.M.J.1990 Smolensky's Theory of Mind, *Behavioral and Brain Sciences, 13*: 407.

Verschure, P.F.M.J. 1992 Connectionist Explanation: taking positions in the mind-brain dilemma. *submitted.*

Verschure, P.F.M.J., & Coolen, A.C.C. 1991 Adaptive Fields: Distributed representations of classically conditioned associations. *Network, 2:* 189-206.

Verschure, P.F.M.J, Kröse, B.J.A., & Pfeifer, R. 1992. Distributed Adaptive Control: The self-organization of structured behavior. *Robotics and Autonomous Agents*, In Press.

Winograd, T., & Flores, F. (1986). *Understanding Computers and Cognition: A new foundation for design.* Reading Ma.: Addison Wesley.

Compositionality and Systematicity in Connectionist Language Learning

Robert F. Hadley
School of Computing Science
Simon Fraser University
Burnaby, B.C. V5A 1S6
hadley@cs.sfu.ca

Abstract

In a now famous paper, Fodor and Pylyshyn (1988) argue that connectionist networks, *as they are commonly constructed and trained*, are incapable of displaying certain crucial characteristics of human thought and language. These include the capacity to employ *compositionally structured representations* and to exhibit *systematicity* in thought and language production. Since the appearance of Fodor and Pylyshyn's paper, an number of connectionists have produced what seem to be *counter-examples* to the Fodor-Pylyshyn thesis. The present work examines two of these apparent counter-examples; one is due to Elman and the other to St. John and McClelland. It is argued that although Elman's and St. John & McClelland's networks discover a *degree* of compositionality, and display a degree of systematic behaviour, the degrees involved are substantially less than that found in humans, and (consequently) are less than what Fodor & Pylyshyn require (or presumably would require if the question were put to them).

1. Introduction

In a now famous paper, Fodor and Pylyshyn (1988) argue that connectionist networks, *as they are commonly constructed and trained*, are incapable of displaying certain crucial characteristics of human thought and language. These include the capacity to employ *compositionally structured representations* and to exhibit *systematicity* in thought and language production.[1] Since the appearance of Fodor and Pylyshyn's paper, an number of connectionists have produced what seem to be *counter-examples* to the Fodor-Pylyshyn thesis. In the present work I examine two of these apparent counter-examples; one is due to Elman (1990), the other to St. John and McClelland (1990). I have chosen these works because, on the face of it, both constitute *strong* counterexamples, and because both are directly concerned with language acquisition, which is a focal point of my discussion here. In (Hadley, 1992) I examine four other recent, apparent counter-examples (due to Pollack, Smolensky, Small, and Chalmers.) As will emerge, I argue that although Elman's and St. John & McClelland's networks discover a *degree* of compositionality, and display a degree of systematic behaviour, the degrees involved are substantially less than that found in humans, and (consequently) are less than what Fodor & Pylyshyn require (or presumably would require if the question were put to them).

2. Compositionality and Systematicity

In this section I examine two experiments which establish, to varying degrees, that connectionist networks (hereafter, c-nets) can discover the compositionality implicit in a training corpus of *sentences.* When describing the results of these learning experiments, researchers commonly argue from the fact that a network can correctly process *novel* sentences (not contained in the training corpus) to the conclusion that the network has indeed induced a compositional structure, and as a consequence is able to exhibit a degree of *systematicity*. As we examine the c-net experiments described below, it will be useful to distinguish different degrees of systematicity, according to the *degree of novelty* of sentences which a c-net is able to recognize (given the c-net's training regime). I shall distinguish three degrees of systematicity. No doubt, it would be possible to make even finer distinctions, but for our purposes the following should suffice. The degrees of systematicity are: weak, quasi, and strong.

1) *Weak Systematicity.* Networks exhibiting weak systematicity can perform at least the following kind of generalization: Suppose that a training corpus is "representative" in the sense that every *word* (noun, verb, etc.) that occurs in some sentence of the corpus also occurs (at some point in the training corpus) in every permissible syntactic position. Thus, although the training corpus omits some sentences permitted by the target grammar, any network trained on this corpus will have been trained to recognize every word in every syntactic position that the word will occupy in the set of novel test sentences which are used to demonstrate the network's generalization capacity. Assuming that this set of novel sentences contains only sentences which are syntactically isomorphic to sentences in the

[1] 'Compositionality' here presupposes that representations have a combinatorial syntax and semantics, whereas 'systematicity' refers to the systematic relationships which *result* when such combinatorially structured representations are employed.

training corpus, and that no new vocabulary is present, we shall say that a c-net exhibits at least *weak systematicity* if it is capable of successfully processing (by recognizing or interpreting) novel test sentences, once the c-net has been trained on a corpus of sentences which are *representative* in the sense described above. I describe such c-nets as (at least) weakly systematic in order to reflect the fact that their generalization capacity has only been tested upon sentences which are *weakly novel* with respect to the training corpus.

2) *Quasi-Systematicity.* We shall say that a system exhibits only quasi-systematicity if (a) the system *can* exhibit weak systematicity, (b) the system successfully processes novel sentences containing embedded sentences, such that both the larger containing sentence and the embedded sentence are (respectively) structurally isomorphic to various sentences in the training corpus, (c) for each successfully processed novel sentence containing a word in an embedded sentence (e.g., 'Bob knows that Mary saw *Tom*') there exists some *simple* sentence in the training corpus which contains that same word in the same syntactic position as it occurs within the embbeded sentence (e.g., 'Jane saw *Tom*'). A system would be *merely* quasi-systematic if 'Tom' needed to occur (in the training corpus) in the *object position* of a *simple sentence*, before the system could correctly process embedded occurrences of 'Tom' in object position. Analogous remarks apply to subject position, verb position, etc.

3) *Strong Systematicity.* We shall describe a system as strongly systematic if (i) it *can* exhibit weak systematicity, (ii) it can correctly process *simple* novel sentences containing words in positions where they *do not appear* in the training corpus (i.e., the word within the novel sentence does not appear *in that same syntactic position* within any *simple or embedded* sentence in the training corpus). Note that a system which has not been trained on embedded clauses may still exhibit strong systematicity, because neither condition (i) or (ii) requires that embedded sentences be present.

Having now distinguished three degrees of systematicity, I should emphasize that although these degrees are directly related to issues of learnability, their primary relevance to the Fodor-Pylyshyn controversy stems from the fact that degrees of novelty are at issue. That novelty is the central issue is underscored by the fact that Elman and others (cf. Hadley, 1992) base their claims to have undermined the Fodor-Pylyshyn thesis upon evidence that particular c-nets can process sentences which are novel with respect to training sets. Since these authors (and St. John and McClelland) take the ability to process novel input as evidence for generalization and systematicity, it seems fair turnabout and epistemically sensible to suppose that the ability to process various degrees of novelty should be taken as indicative of various degrees of systematicity having been induced. Moreover, quite apart from the Fodor-Pylyshyn controversy, I contend that the trifold distinction introduced here illuminates important differences between the respective abilities of humans vs. existing c-nets to process novel *kinds* of input. To establish this thesis, I shall first argue that humans exhibit the strongest of my three forms of systematicity.

To begin with, there is good reason to believe that even young children, who have not yet reached the stage of producing multi-word utterances, are frequently able to obey *simple* imperative sentences which contain words in syntactic positions where the child has never encountered the word before.

It is well known, for example, that in the few weeks which precede a child's first multi-word utterances, a "spurt" occurs in a child's acquisition of *nominals* (both common and proper nouns), and that during this period children are able *rapidly* to acquire the use of nominals by means of "what's that" games (cf. (Ingram, 1989; Dromi, 1987)). Once they have acquired nominals in this fashion, children are soon thereafter (i.e., within minutes) able to comprehend these words in sentences they encounter. This fact is established by Katz, Baker, and Macnamara (1974) who also present a strong case that the ability of young children to distinguish proper nouns from common nouns is much more a function of a child's *prior* ability to distinguish re-identifiable individuals from classes of objects than it is a function of some capacity to distinguish words which are syntactically preceded by an *article* from those which are not.

Moreover, as children begin to produce simple, multi-word utterances, they will often produce semantically reasonable, *albeit non-grammatical* combinations of the words they have acquired in previous contexts. In fact, children do this sufficiently often that some psycholinguists posit the existence of a "child grammar" (Ingram, 1989). Now, whether or not we accept the existence of a child grammar, the fact that children are able to recombine words in patterns that are not present in their training corpus strongly suggests that (at least in the early weeks of multi-word utterance) children have a much greater grasp of the *semantic* content of particular words than they do of their syntactic roles (in adult grammars). (The results of Katz, Baker, and Macnamara also reinforce this conclusion.) Furthermore, and more to the point, the ability of children to *sensibly* recombine words in patterns they have not been trained to produce clearly demonstrates that children are not nearly as dependent upon syntactic context as systems which are only weakly (or quasi) systematic.

As we consider somewhat older children, who have acquired a rudimentary syntax (but not necessarily the use of prepositional phrases or relative clauses), it becomes transparently clear that humans can learn to use nominals long before they have encountered them in all possible positions. For example, a child visiting a zoo with her parents may hear her mother exclaim, "Susie, look at the otter". Susie may reply, "What's an otter?" The mother, pointing, replies "Here, this is an otter". If Susie is adept at language, she may learn the (approximate) meaning of 'otter' rapidly, by this ostensive means, and may soon utter, "Look, Mommy, this otter is chasing the other one". Although the child has never encountered the word in *subject* position, she is able to use it in that position once its meaning has been surmised. Of course, most children will require a few repetitions before an ostensively introduced word enters long term memory, but these repetitions need not present the word in all legal positions. With adults, new words may enter the vocabulary even more rapidly, as when one surmises a word's meaning during the course of conversation, or when listening to a brief exchange during a meeting. Once a word's meaning has been surmised, most adults can use it freely in embedded sentences and simple sentences, although they may only have heard the word used in a single syntactic position.

We turn now to consider connectionist systems which, *prima facie*, challenge Fodor and Pylyshyn's view on the limitations of c-nets vis-a-vis compositionality and systematicity. In considering these systems we should bear in mind that Fodor and Pylyshyn are concerned with the kind of full-fledged compositionality and systematicity that human thought and language exhibit.

2.1 St. John and McClelland

St. John and McClelland (1990) present a connectionist model which learns to assign "semantic representations" to English sentences which are presented as input. Although the details of their model are somewhat complex, the overall gist is that, via backpropagation, the network is trained to produce a correct semantic representation of the situation *described* by each input sentence. Situations (or events) described by input sentences consist of relationships, and the objects involved in those relationships. Input sentences are fed into the network in presegmented constituents. As each constituent is processed, an inspection is made to see whether the network has output the desired, complete representation of the target situation. Backpropagation is performed after each such inspection. Because it is usually not possible to predict the *entire* target representation on the basis of isolated sentence constituents, the network is forced to learn associations between individual constituents and particular objects or relations in the target situation.

Sentences which serve as input constitute a highly simplified version of English, in that all articles are deleted and only singular nouns are present. However, certain prepositional phrases are permitted. Each target semantic representation consists of an *ordered* series of role/filler pairs. Roles are *agent, action, patient,* etc., and fillers are "concepts" (*my* scare quotes) corresponding to individual nouns and verbs. Thus, each semantic representation is a *structured, concatenated* sequence of pairs. By itself, this aspect of the model would seem to undermine any potential the model might possess for deposing Fodor and Pylyshyn's thesis that human thought requires structured, internal representations. For the experimental design presupposes the existence of such representations (at the point where backpropagation is employed). Moreover, the ability to form such representations *presupposes* that the learner has *already discovered* a compositional, systematic method of representing situations. Thus, from Fodor and Pylyshyn's standpoint, the model's design concedes one of their major contentions. However, the question still remains whether the model acquires knowledge of the compositional, systematic nature of its *input* sentences. St. John and McClelland (hereafter, St.J&Mc) clearly claim that it does (p. 250, 1990), and it is this claim we now consider.

As mentioned, St.J&Mc's training corpus includes sentences containing prepositional phrases. Unfortunately, when testing their network for the acquisition of compositional knowledge (which is manifested as *systematicity*) St.J.&Mc used simpler training corpora, which lacked prepositional phrases. Two experiments were conducted to test for systematicity of behaviour – one syntactic, the other semantic. Tests for syntactic systematicity involved only 10 objects and 10 reversible actions. Each object (action) uniquely corresponds to a particular noun (verb) in the training corpus. Both active and passive verb forms were permitted, and each input sentence had the general form: [noun verb-form noun]. Given that both active and passive forms are possible, a total of 2000 sentences are possible. All 2000 sentences were generated. Of these, 1750 comprised the training corpus, and the remaining 250 were set aside for later testing. Although St.J&Mc do not explicitly say so, their remarks elsewhere (p. 243, 1990) suggest that these 250 sentences were randomly selected. Assuming they were, it is highly probable that the remaining 1750 sentences contained occurrences of every word in every legal syntactic position. (Otherwise, 80% of the 250 test sentences would have to contain the *same* particular noun or verb in the same syntactic position. Given that there are 10 nouns and 10 verbs, this is extremely unlikely.[2]) Moreover, St.J&Mc give *no indication* that the training corpus (of 1750 sentences) *does not* include every possible word in every possible position. On the available evidence, therefore, it is reasonable to believe that the training corpus does present every word in every possible position. This conclusion is reinforced by St.J&Mc's remark that "What makes this a generalization task is that some of the sentences were set aside and not trained: *some agents were never paired with certain objects*" (my emphasis). The fact that sentences in the *test* corpus describe novel agent-object combinations does present convincing evidence of generalization, but does not suggest that anything stronger than weak systematicity and compositionality were tested for. To be sure, the network does display some degree of systematicity. The network assigns the correct semantic representation to 97% of the novel 250 sentences. However, given the above considerations, it seems entirely likely that the network displays only *weak syntactic* systematicity.

The test for *semantic* generalization is analogous, in relevant respects, to the one just described. The se-

[2] Note that 10% of the 2000 original sentences contain a given noun or verb in a given position. So, if a given noun or verb does not occur in a given position within the 1750 training sentences, then 200 of the 250 test sentences must contain that given word in the given position.

mantic test involved a set of 400 possible sentences, of which 350 were used for training and the remaining 50 were used for testing. St.J&Mc explicitly note that the 50 test sentences were randomly selected from the set of 400. As before, the set of 400 sentences exhausts the space of possible sentences. Now, since the 50 test sentences were randomly chosen, it is extremely probable (by analogy with the reasoning given in the previous footnote) that each word occurred in a syntactic position within the *test* corpus that it also occupied within the training corpus. Thus, it is virtually certain that the test for *semantic* generalization established only weak systematicity. Certainly, we are given no reason to suppose otherwise. Also, it is clear that St.J&Mc's model was not even intended to display the kind of strongly systematic behaviour and rapid integration of semantic knowledge which our example of the child at the zoo illustrates (involving the word 'otter').

It should be acknowledged, however, that despite the weaknesses mentioned above, the network we have considered yields some impressive results, including the ability to learn "to disambiguate ambiguous words; instantiate vague words; assign thematic roles; and immediately adjust its interpretation as each constitutent is processed" (p. 220, 1990). Even the ability to demonstrate weak systematicity is no small feat. However, it should be remembered that humans appear to exhibit a much stronger form of systematicity than this.

2.2 Elman

We turn now to the work of Elman (1989, 1990) on connectionist learning of syntactic structure. Elman contends that "the sensitivity to context which is characteristic of many connectionist models, and which is built-in to the architecture of the networks used here, does not preclude the ability to capture generalizations which are at a higher level of abstraction." In addition, Elman clearly opposes his results (and those of others, including St.J&Mc) to the conclusions advanced by Fodor and Pylyshyn (1988), and to Fodor's (1976) Language of Thought thesis. Yet, while it is clear that Elman's networks do generalize and acquire a degree of systematicity, it is by no means clear that they display the degree of systematicity that humans exhibit. Moreover, since Elman's research does not address issues of *semantic* systematicity and compositionality, it is unclear whether this work actually threatens Fodor's views on the Language of Thought. After all, we saw that St.J&Mc were able to train their network to discover *semantic* compositionality only when they assumed the prior existence of a concatenative, structured set of internal representations. However, let us consider Elman's results in some detail.

Elman (1989, 1990) describes two experiments, both employing recurrent networks with a context layer feeding back into the hidden layer. The training procedure for both networks is essentially the same. Simplified English sentences (articles are absent) are fed into the network one word at a time, and backpropagation is used in a (*prima facie*) attempt to train the network to *predict* the next word it will receive as input. However, since a large training corpus is employed (10,000 sentences in each experiment), the network cannot learn to predict the next input word, but does learn (in essence) to predict the *syntactic category* of the following word. The first of the two experiments is designed, in fact, to demonstrate that the network does indeed develop a set of syntactic categories which correspond to the traditional grammatical categories. Cluster analysis on the network's hidden-layer activation values reveals that the network acquires *approximately* traditional categories, as well as (approximate) subcategories corresponding to animate noun, inanimate noun, transitive verb, etc.[3] The syntactic corpus for this experiment consists entirely of simple 2 and 3 word sentences. Both singular and plural nouns are included, and the network does learn to detect number agreement.

The second experiment is designed to test whether a somewhat more complex recurrent network can discover syntactic structure. In this experiment the training corpus includes relative clauses, embedded to a maximum depth of two (judging by examples provided). Now, although the acquisition of *approximate* syntactic categories in the first experiment seems to indicate that a *degree* of systematicity has been discovered, only in this latter experiment is a test for systematicity explicitly performed. We therefore concentrate our attention upon the latter experiment.[4]

The training regime for the second experiment consisted of four phases, the first of which presented the network with a continuous stream of 10,000 sentences, containing no relative clauses. The three remaining phases each built upon the preceding phases, and involved increasingly high percentages of relative clauses. This controlled, graduated exposure to relative clauses raises questions about the psychological plausibility of the design, which I shall explore in section 3. However, our present concern is with systematicity. Given that the initial training phase involved 10,000 sentences, comprised only of 8 common nouns, 2 proper nouns, and 12 verbs, we have good reason to suppose that the *initial* training corpus presented every word in every syntactically legal position.[5] Assuming this is so, it is clear that the first phase could induce only weak systematicity. Moreover, there is no reason to suspect that any stronger form of systematicity is established by the later training phases, since every possible arrangement

[3]These categories only approximate traditional categories because (for example) the representations developed for subject and object tokens of the same noun are not identical, though they do cluster together.

[4]Also, it is quite clear that the training corpus for the first experiment presented every word in every syntactically legal position. This can readily be established on the basis of the number of nouns and verbs available. It follows that the first experiment establishes only weak systematicity at best.

[5]Note that even if we assumed that every verb optionally takes a direct object, the total number of possible *simple* sentences is: [10 nouns × 12 verbs × 10 nouns = 1200] plus [10 nouns × 12 verbs = 120].

of nouns and verbs that could occur as the complement of a relative clause appears to have been present within simple sentences in the first training corpus.

In passing, it is worth noting that Elman does not say whether his test corpus included greater *depths* of embedding than were present in the training corpus. This is unfortunate, since the ability to generalize to greater depths is an important component of human thought.

3. Plausibility of Training Regimes

In the preceding pages I have occasionally commented upon psychologically problematic aspects of certain of the training regimes invloved. Although none of the authors considered here make strong claims for the psychological plausibility of their methods, it is important to consider whether the results obtained actually *require* learning conditions which are truly implausible. For, even *competence* models of cognitive behaviour (as well as performance models) are normally expected to preserve (or at least approximate) extensional relationships between an agent's *real* input and *real* output. If a particular c-net training regime *requires* the existence of input copora or external error feedback which simply do not occur in human conditions, then serious doubts arise as to whether the c-net model can even provide insight into human cognition. This is especially true when there appears to be no way to *modify* the c-nets involved such that more realistic sets of input and output can be accommodated.

In what follows I examine aspects of the work of St.J&Mc (1990) and Elman (1989, 1990) which *prima facie* (at least) involve seriously unrealistic assumptions about certain learning and/or biological conditions involved in human language acquisition.

3.1 St. John and McClelland (1990)

Recall that the training regime of St.J&Mc presupposes that the learner has *already* apprehended, at the time a given input *sentence* is processed, the particular external state of affairs that the sentence describes.

The learner apprehends this state of affairs by having a *structured*, sequentially ordered representation of this state of affairs in mind. I have already remarked that these structured representations resemble, in spirit at least, those of Fodor's Language of Thought. However, our present concern is with a different problem, *viz.*, is it legitimate to assume that among all the various states of affairs perceptually available to the agent at the time the sentence is presented, the agent's attention is drawn to the particular state of affairs described by the sentence?

St.J&Mc briefly address the above difficulty when they say (p. 249, 1990) "The problem of discovering which event in the world a sentence describes when multiple events are present would be handled in a similar way, though we have not modelled it. Again, the aspects of the world that the sentence actually describes would be discovered gradually over repeated trials, while those aspects that spuriously co-occur with these described aspects would wash out". However, there may be a serious problem with St.J&Mc's suggestion. For, given their experimental design, if spurious states of affairs were frequently presented to the backpropagation algorithm as the *intended* target state of affairs, the *number of iterations* required to wash out the spurious information may well be utterly implausible. Even without spurious information, the network requires over 300,000 iterations *before* it begins to master sentences in passive voice. Were a substantial percentage of spurious states of affairs to be presented, the complexity of the learning task would certainly increase, and we have no reason to suppose the number of iterations involved would fall within anything resembling a plausible range. Even the existing figure of roughly 300,000 raises doubts. These doubts would not be so unsettling if we had reason to believe that alternative architectures would dramatically decrease the iterations involved, but the authors present no arguments to that effect. Moreover, we must bear in mind that the learning task has already been dramatically oversimplified by (a) the absence of articles in the input copora and (b) the fact that the agent's internal representation of the target state of affairs contains a *marker* indicating whether the input sentence is in active or passive voice. It is very difficult to see how a *perception* of an external state of the world could yield an indication as to whether the given *sentence* was active or passive. Also, it seems implausible that the agent would represent voice information *before* the active-passive distinction had been at least partially mastered. Note that *that* distinction is discoverable in the relationships between the input sentences and the internally represented states of affairs, not in the latter representations alone.

Another difficulty concerns the way in which St.J&Mc employ the backpropagation algorithm. During the training procedure, as each constituent of a sentence is processed in turn, the resulting pattern on the output layer is compared to the *entire target* state of affairs. Differences are noted, and backpropagation of error is employed *after each constituent is processed* (although weight changes are accumulated and adjusted after each 60 trials). Now the lack of a biological correlate for the standard backpropagation algorithm is a well known problem, but the defense is commonly made that there may exist some unknown biological process whose effects are roughly analogous to those of this algorithm. This defense is reminiscent of the kind of hand-waving that some connectionist find lamentable in classical AI. However, even if this hand-waving response is accepted, and even if we accept a suggestion of Smolensky that connectionist processing occurs at a more abstract level than the neural level, still there must be *some* biological process which is presumed to support the more abstract process which is supposed to (roughly) correspond to the backpropagation algorithm. Moreover, this biological process would presumably occur each time the backpropagation algorithm is executed in the training regime, and this biological process requires time. Given the complexity of backpropagation, it is difficult to believe that a biological process supporting the algorithm's abstract analogue could occur during the interval between the uttered constituents in a sentence. In light of this, serious doubts arise as to the legitimacy of invoking the backpropagation algorithm *each time* a sentence constituent is heard.[6] At best, the burden of proof rests upon St.J&Mc to show that this application of the backpropagation algorithm has even a *rough* physiological basis.

3.2 Elman (1989, 1990)

Like St.J&Mc, Elman employs backpropagation, but he does attempt a justification for doing so. Recall that, in an effort to teach his networks the syntactic categories of lexical items, Elman trains the network, via backpropagation, to *attempt* to predict the *next* word in a successive stream of words. In defense of this "error-feedback" strategy, Elman remarks that "it does seem to be the case that much of what listeners do involves anticipation of future input". Presumably, Elman takes this as evidence that listeners are constantly attempting to predict the next word they hear. This strikes me as a dubious extrapolation needing empirical support. However, a more serious objection is that Elman's invocation of backpropagation after each word is processed is subject to the same criticism as St.J&Mc's usage. It is difficult to believe that a biological process supporting anything analogous to backpropagation could occur between succeeding words in an utterance.

Another difficulty with Elman's approach is that, when relative clauses are involved, training occurs in 4 distinct phases. Phase 1 presents the network with a concatenated string of 10,000 *simple* grammatical sentences (no relative clauses are included). This string of 10,000 sentences is presented to the network 5 times over. Now, not by the wildest stretch of the imagination is this a psychologically plausible regime. Normally, a child would encounter many breaks even during a series of 20 sentences. During some of these breaks the child may hear sentence fragments, or even simple names. Almost certainly, the child would be exposed to a substantial percentage of unfinished and ungrammatical sentences. The question naturally arises, would Elman's networks be able to induce systematic regularities under these conditions? Not likely, but if not, what are the real implications of this research?

Returning to the succeeding phases of Elman's regime, phase 2 modifies phase 1 by having 25% of the 10,000 sentences contain relative clauses. Phase 3 contains 50% relative clause sentences, and phase 4 contains 75% relative clauses. Clearly, this training regime is highly contrived. Children are not exposed to anything like this artificial partitioning of the input copora.

In fairness, I should note that the artificiality of Elman's training regime is certainly not unique to his work, and he would no doubt readily concede its artifice. Somewhat analogous remarks would apply to St.J&Mc (who also train their c-nets with implausibly long strings of sentences). It is doubtful whether these authors would attempt serious defenses of the size and presentation of their input copora. However, in the absence of such defense we must ask whether these networks could discover even the moderate degree of compositionality they do discover if they were subjected to the erratic, mixed, and often ungrammatical input that humans receive.

[6]Note that even if weights are modified only after every N invocations of the algorithm, the strategy described requires equally as many invocations just to enable the information to be gathered for later weight modification.

4. Summary

In the foregoing I have examined c-net experiments which arguably establish that c-nets can be trained to discover compositionality and exhibit systematicity. In neither of the cases examined does there appear to be any reason to suppose that the c-nets involved exhibit anything stronger than weak systematicity. I have also argued that humans exhibit a much stronger form of systematicity than these c-nets, and thus there is no reason to suppose that the results of Elman and St.J&Mc defeat the Fodor-Pylyshyn thesis. Moreover, I have argued that the experiments considered here involve seriously unrealistic training regimes , and this in turn casts doubt upon the cognitive significance of the experiments. I do not suggest that these experiments are uninteresting; it may be that they will ultimately illuminate an important aspect of the overall puzzle. However, as it stands, it is difficult to see what the cognitive implications of these experiments are.

References

Dromi, E. (1987) *Early Lexical Development*, Cambridge University Press, Cambridge, UK.

Elman, J.L. (1989) "Representation and Structure in Connectionist Models", CRL Technical Report 8903, University of California at San Diego.

Elman, J.L. (1990) "Finding Structure in Time", *Cognitive Science*, Vol. 14, pp. 179-212.

Fodor, J.A. (1976) *The Language of Thought*, Harvard University Press, Cambridge, Mass.

Fodor, J.A. & Pylyshyn, Z.W. (1988) "Connectionism and Cognitive Architecture: A Critical Analysis", *Cognition*, Vol. 28, pp. 3-71.

Hadley, R.F. (1990) "Connectionism, Rule Following, and Symbolic Manipulation", Proceedings of AAAI-90, MIT Press, Boston.

Hadley, R.F. (1992) "Connectionism and Systematicity in Connectionist Language Learning", Technical Report, CSS-IS TR 92-03, Centre for Systems Science, Simon Fraser University.

Ingram, D. (1989) *First Language Acquisition*, Cambridge University Press, Cambridge, UK.

Katz, N., Baker, E. & Macnamara, J. (1974) "What's in a Name? A Study of How Children Learn Common and Proper Names", *Child Development*, Vol. 45, pp. 469-473.

St. John, M.F. & McClelland, J.L. (1990) "Learning and Applying Contextual Constraints in Sentence Comprehension", *Artificial Intelligence*, Vol. 46, pp. 217-257.

The (Non)Necessity of Recursion in Natural Language Processing

Morten H. Christiansen*
Centre for Cognitive Science
University of Edinburgh
2 Buccleuch Place
Edinburgh EH8 9LW
Scotland, UK
morten@uk.ac.ed.cogsci

Abstract

The *prima facie* unbounded nature of natural language, contrasted with the finite character of our memory and computational resources, is often taken to warrant a *recursive* language processing mechanism. The widely held distinction between an idealized infinite grammatical competence and the actual finite natural language performance provides further support for a recursive processor. In this paper, I argue that it is only necessary to postulate a recursive language mechanism insofar as the competence/performance distinction is upheld. However, I provide reasons for eschewing the latter and suggest that only data regarding observable linguistic behaviour ought to be used when modelling the human language mechanism. A connectionist model of language processing—the simple recurrent network proposed by Elman—is discussed as an example of a non-recursive alternative and I conclude that the computational power of such models promises to be sufficient to account for natural language behaviour.

Introduction

Is it necessary to postulate a recursive language mechanism in order to account for the apparently unbounded complexity and diversity of natural language (NL) behaviour, given the finite nature of the memory and computational resources that underly the human production of this behaviour? What seems to be needed in the first place is a mechanism which is able to generate, as well as parse, an infinite number of NL expressions using only finite means. Obviously, such a mechanism has to be of considerable computational power and, indeed, recursion provides a very elegant way of achieving this property. Consequently, recursion has been an intrinsic part of most accounts of NL behaviour—perhaps due to the essentially recursive character of most linguistic theories of grammar.[1]

It is often noted that the *prima facie* existence of recursion in NL behaviour poses serious problems for connectionist approaches to NL processing (e.g., Fodor & Pylyshyn, 1988) since recursion—*qua* computational mechanism—is defined as being essentially symbolic. However, the existence of recursion in NL presupposes that the grammars of linguistic theory correspond to *real* mental structures, rather than mere structural *descriptions* of NL *per se*. Yet, there are no *a priori* reasons for assuming that the structure of the observable public language necessarily must dictate the form of our internal representations (van Gelder, 1990b). Still, many linguists and psychologists (e.g., Chomsky, 1986; Frazier & Fodor, 1978; Kimball, 1973; Pickering & Chater, 1992; Pulman, 1986) take grammars as corresponding to in-the-head representations that are manipulated by computational processes. But, since human NL behaviour is limited under normal circumstances, a distinction is typically made between the bounded observable performance and an infinite competence inherent in the internal grammar.

In what follows, I start off by arguing from a methodological perspective that the alleged distinction between linguistic competence and actual NL performance must be rejected if linguistic theories are to encompass representational claims regarding the human NL mechanism. Then, I drive a wedge between the (quasi-) recursive nature of NL, as described in most current linguistic theories, and the actual NL processing mechanism. In particular, I suggest that recursion is a *conceptual artifact* of the competence/performance distinction (C/PD), instead of a necessary characteristic of the underlying computational mechanism.[2] In this light, the prob-

*This research was made possible through award No. V910048 from the Danish Research Academy.

[1] For example, in GB (e.g., Chomsky, 1981) the underlying principles of $\overline{X}$-theory are recursive, as are the ID-rules of GPSG (Gazdar *et al.*, 1985).

[2] I will therefore not discuss connectionist models of NL processing that merely simulate—or mirror—symbolic recursion. An example of such models is provided by McClelland & Kawamoto (1986) who apply

lem facing connectionist models of NL processing is *not* whether they can implement some kind of recursive mechanism, but whether they will be able to account for the (limited) ***recursive structure*** found in NL behaviour purely in terms of non-symbolic computation. I therefore consider a connectionist model—designed by Elman (1990, 1991)—which exhibits recursive behaviour without implementing a symbolic recursion mechanism, and conclude that such connectionist models provide a psychologically appealing way of modelling NL processing.

The Competence/Performance Distinction

In most—if not all—linguistic theories of NL, recursion is unbounded. However, since the main source of data of modern linguistics implies intuitive grammaticality judgements (e.g., Horrocks, 1987), the fact has to be explained that the greater the length and complexity of utterances, the less sure people are of their respective judgements. To explain this phenomena, a distinction between an idealized infinite linguistic *competence* and a limited NL *performance* is made. The performance of a particular individual is limited by memory limitations, attention span, lack of concentration, etc. (e.g., Fodor & Pylyshyn, 1988; Horrocks, 1987).

This methodological separation of the infinite linguistic competence of a recursive grammar from the limited performance of observable NL behaviour has been strongly advocated by Chomsky:

> One common fallacy is to assume that if some experimental result provides counterevidence to a theory of processing that includes a grammatical theory T and parsing procedure P (say, a procedure that assumes that operations are serial and additive, in that each operation adds a fixed "cost"), then it is T that is challenged and must be changed. The conclusion is particularly unreasonable in the light of the fact that in general there is independent (so-called "linguistic") evidence in support of T while there is no reason at all to believe that P is true. (Chomsky, 1981: p. 283)

The main methodological implication of this position, which I will refer to as the ***strong*** C/PD, is that it leads to what I call the ***'Chomskian paradox'***. On the one hand, the strong C/PD makes T immune to all empirical falsification, since any falsifying evidence can always be dismissed as a consequence of a false P. However, on the other hand, all grammatical theories rely on grammaticality judgements that (indirectly via processing) display our knowledge of language. Consequently, it seems paradoxical that only certain kinds of empirical material is acccepted—i.e., grammaticality judgements—whereas other kinds are dismissed on what appears to be relatively arbitrary grounds. Thus, the strong C/PD provides its proponents with a protective belt that surrounds their grammatical theories and makes them empirically impenetrable to psycholinguistic counterevidence.

In contrast, a more moderate position, which I will refer to as the ***weak*** C/PD, contends that although linguistic competence is supposed to be infinite, the underlying grammar must support an empirically appropriate performance. This is done by explicitly allowing performance—or processing—considerations to constrain the grammar. Pickering & Chater (1992) have suggested that such constraints must be built into the representations underlying the grammatical theory, forcing a closer relation to the processing theory. This ensures that the relation between the theory of grammatical competence (Chomsky's T) and the processing assumptions (Chomsky's P) is no longer arbitrary, resulting in an opening for empirical testing. Nevertheless, inasmuch as T and P are still functionally independent of each other, the option is always open for referring any falsifying empirical data questioning T to problems regarding the independent P, i.e., to performance errors.

To compare the methodological differences between models of NL processing that adopt, respectively, the strong or the weak C/PD, it is illustrative to conceptualize the models as rule-based production systems. In such a system, the grammar would correspond to a knowledge base consisting of a set of declarative rules, each corresponding to a rule in the grammar. The system has a working memory (WM) in which intermediate processing results are stored. The content of the WM is changed by the system through the application of the rules in its knowledge base. A rule can be applied when its right-hand side matches the current content of the WM (or an appropriate part of it).[3] For example, if the content of the WM consists of the two words, say, the_{Det} and dog_N the system would be able to apply a rule such as NP $\rightarrow$ Det N changing the content of WM to, say, $[_{NP}\ \text{the}_{Det}\ \text{dog}_N]$.

Within this framework, the grammar of a particular linguistic theory corresponds to the system's knowledge base. The system can therefore be said to have an infinite linguistic competence in virtue of its independent knowledge base, whereas its per-

the standard way of implementing symbolic recursion in Von Neumann architectures—i.e., using a push-down stack and multiple subroutines—in their modelling of the human sentence processing mechanisms. This kind of connectionist solution to the problem of recursion in NL behaviour is orthogonal to the subject of this paper, since it merely implies a non-symbolic *implementation* of a symbolic model.

[3]Although it is assumed here that we are dealing with a bottom-up parser, no significant changes would have to be made were we to parse top-down instead.

formance through processing is constrained by WM limitations. This is in direct correspondence with the strong C/PD, since the grammar is completely separated from processing. Models adhering to the weak C/PD would similarly have an independent, declarative knowledge base corresponding of the grammar, but in addition they would also have an extra knowledge base consisting of what we might coin *linguistic meta-knowledge*. This knowledge consists of various performance motivated parsing heuristics that provide context-dependent constraints on the *application* of grammatical rules—such as, for example, the *'minimal attachment principle'* (Frazier & Fodor, 1978). Thus, the performance of the model is constrained not only by limitations on WM but also by linguistic meta-knowledge.

From the production system analogy it can be seen that proponents of both the strong and the weak C/PD stipulate grammars that are functionally independent from processing. As a consequence, empirical evidence that appears to falsify a particular grammar can always be rejected as a result of processing constraints—either construed as limitations on WM (strong C/PD) or as a combination of WM limitations and false linguistic meta-knowledge (weak C/PD). In short, as long as the C/PD—weak or strong—is upheld, potentially falsifying evidence can always be explained away by referring to performance errors. This is methodologically unsound insofar as linguists want to claim that their grammars have representational reality. By evoking the distinction between grammatical competence and observable NL behaviour, thus disallowing negative empirical testing, they cannot hope to find other than speculative support for their theories. In other words, if linguistic theory is to warrant representational claims, then the C/PD will have to be abandoned.[4]

In contrast, a connectionist perspective on NL promises to eschew the C/PD, since it is not possible to isolate a network's representations from its processing. The relation between the "grammar", which has been acquired through training, and the processing is as direct as it can be (van Gelder, 1990b). Instead of being a set of *passive* representations of declarative rules waiting to be manipulated by a central executive, a connectionist grammar is distributed over the network's memory as an *ability to process language* (Port & van Gelder, 1991). In this connection, it is important to notice that although networks are generally "tailored" to fit the linguistic data, this does not simply imply that a network's failure to fit the data is passed onto the processing mechanism alone. Rather, when you tweak a network to fit a particular set of linguistic data, you are not only changing how it will *process* the data, but also what it will be able to *learn*. That is, any architectural modifications will lead to a change in the overall constraints on a network, forcing it to adapt differently to the contingencies inherent in the data and, consequently, to the acquisition of a different grammar. Thus, since the representation of the grammar is an inseparable and *active* part of a network's processing, it is impossible to separate a connectionist model's competence from its performance.

However, this leaves open the question of what kind of performance data should be accepted. For the purpose of empirical tests of NL mechanisms we need to distinguish between 'real' performance data as exhibited in normal NL behaviour and examples of abnormal or *'pathological'* performance such as 'slips-of-the-tongue', blending errors, etc. It might be objected that by proposing such a distinction I am letting the C/PD in by the back door. Yet, this is not the case, since we can plausibly assume that the language processor is an informationally encapsulated, modular system and that pathological performance is due to factors outside the language module.

In this way, what counts as valid data is not dependent on an abstract, idealized notion of linguistic competence but on observable NL behaviour under statistically 'normal' circumstances. Consequently, we should be able to filter out the pathological performance data from a language corpora simply by using 'weak' statistical methods. For example, Finch & Chater (1992) applied simple bigram statistics to the analysis of a *noisy* corpus consisting of 40,000,000 English words and were able to find phrasal categories defined over similarly derived approximate syntactic categories. It seems very likely that such a method could be extended to a clausal level in order to filter out pathological performance data. Thus, having suggested what qualifies as empirical evidence with respect to models of NL behaviour, I will discuss below whether such data warrant a recursive processing mechanism.

Recursion and Natural Language Behaviour

The history of the relationship between grammar and language mechanism dates back to Chomsky's (1957) demonstration that language can, in principle, be characterized by a set of generative rules.[5] In addition, he argued that NL cannot be accounted for by a finite state automaton, because the latter can only produce regular languages. This class of

[4] By this I do not mean that the present linguistic theories are without explanatory value. On the contrary, I am perfectly happy to accept that these theories might warrant certain *indirect* claims with respect to the language mechanism, insofar as they provide means for describing empirical NL behaviour.

[5] For a detailed historical overview, see Pickering & Chater (1992).

languages—although able to capture *left-* and *right-*embedded recursive structures—cannot represent *centre*-embedded expressions.[6] For linguistic theories adhering to the C/PD (weak or strong), such a restriction on the power of the finite-state grammars prevents them from being accepted as characterizations of the idealized linguistic competence. On this view, NL must be at least context-free, if not weakly context-sensitive (cf. Horrocks, 1987). However, having eschewed the C/PD, the question is how much processing power is needed in order to account for observable NL behaviour. Do we need to postulate a NL mechanism with the full computational power of a recursive context-free grammar?

Before answering this question, it is worth having a look at some examples of different kinds of recursive NL expressions. Since the crucial distinction between regular and other richer languages is that the former cannot produce expressions involving unbounded centre-embedding, we will look at such sentences first. As the following three examples show, the difficulty of processing a centre-embedded sentence increases with the depth of embedding:

(1) The boy the girl saw fell.
(2) The boy the girl the cat bit saw fell.
(3) The boy the girl the cat the dog chased bit saw fell.

The difficulty of understanding such centre-embedded sentences has been the subject of much debate (e.g., Frazier & Fodor, 1978; Kimball, 1973; Pulman, 1986; Reich, 1969; Wanner, 1980). Proponents of the C/PD have explained the difficulty in terms of performance limitations. For example, in order to account for the problems of parsing recursively centre-embedded sentences, both Kimball's (1973) parser and Frazier's & Fodor's (1978) 'Sausage Machine' parser apply a performance-justified notion of a viewing 'window' (or look-ahead). The window, which signifies memory span, has a length of about six words and is shifted continuously through a sentence. Problems with centre-embedded sentences are due to the parser not being able to attach syntactic structure to the sentences because the verb belonging to the first NP is outside the scope of the window. However, this solution is problematic in itself (cf. Wanner, 1980) since triply centre-embedded sentences with only six words do exist and are just as difficult to understand as longer sentences of similar kind; e.g.,

(4) Boys girls cats bite see fall.

[6] I will adopt the standard notion of these three kinds of embedded recursion. In case X is a non-terminal symbol, and α and β are strings of terminal and non-terminal symbols, we have left-embedding when $X \Rightarrow X\beta$ (i.e., there is a derivation from X to $X\beta$), a centre-embedding when $X \Rightarrow \alpha X\beta$, and a right-embedding when $X \Rightarrow \alpha X$.

A plausible way out of this problem due to Reich (1969) is to argue that centre-embedded sentences, such as (1)–(4), are ungrammatical. Pulman (1986) has opposed this move by contending that with increased computational resources (e.g., pen and paper) or practice, performance on centre-embedded sentences generally increases, whereas this is not the case for ungrammatical strings. However, when we abandon the C/PD, the distinction between 'grammatical' and 'ungrammatical' becomes less important, since we seek to account for performance data as exhibited by typical NL behaviour, rather than abstract grammatical competence. Thus, the difficulty encountered when parsing centre-embedded sentences suggests that NL models need to display the same problems when confronted with this kind of recursive expressions. Still, this solution leaves left- and right-recursion to be dealt with. That these structures cannot be easily dismissed, but seem to be relatively ubiquitous in NL, can be seen from the following examples involving such phenomena as multiple prenominal genetives (5), right-embedded relative clauses (6), multiple embeddings of sentential complements (7), and PP modifications of NPs (8):

(5) [[[Bob's] uncle's] mother's] cat]...
(6) [This is [the cat that ate [the mouse that bit [the dog that barked]]]].
(7) [Bob thought [that he heard [that Carl said [that Ira was sick]]]].
(8) ...the house [on [[the hill [with the trees]][at [the lake [with the ducks]]]]].

Furthermore, *prima facie* there seems to be no immediate limits to the length of such sentences.

Even though (5)–(8) are describable in terms of left- or right-recursion, it has been argued—with support from, e.g., intonational evidence (Reich, 1969)—that these expressions are not recursive but *iterative* (Ejerhed 1982; Pulman, 1986). In case these structures are iterative, rather than recursive, then it is possible to account for NL solely in terms of a finite state automaton (FSA). Strong support for this claim comes from Ejerhed (1982) who demonstrated that it is possible for a FSA, comprising a non-recursive context-free grammar, to capture the empirical data from Swedish (provided that unbounded dependencies are dealt with semantically). This demonstration is significant because Swedish is normally assumed to require the power of context-sensitive languages (e.g., cf. Horrocks, 1987). Thus, we have strong reasons for believing that a non-recursive FSA provides sufficient computational power to account for NL performance without needing to postulate a functionally independent infinite competence. As we shall see below, certain kinds of connectionist models, that is, simple recurrent networks, have the ability to mimic FSAs in a psychologically interesting way.

A Connectionist Account of "Recursive" Natural Language Behaviour

We have found that it is not necessary to invoke the C/PD in order to account for NL processing. Moreover, we have seen that a non-recursive FSA has sufficient computational power to function as a NL processing mechanism. So, the remaining question is whether a connectionist model can mobilize such power—or whether we have to give in to Fodor's & Pylyshyn's (1988) negative claims concerning connectionist NL processing. In the rest of this paper, I provide arguments to the effect that a particular kind of connectionist model—the simple recurrent network (SRN) (Elman, 1990, 1991)—promises to have sufficient power to capture NL behaviour.

An SRN is a connectionist feed-forward network that has an extra set of hidden, so-called *'context'* units (Elman, 1990, 1991). At time t, the hidden unit activation is copied over into the context units. Via recurrent links, the activation over the context units is fed back (as part of the input) to the hidden units at time $t + 1$. In this way, the presence of recurrent links, together with the context units, allows past activation to influence the current output, thus enabling the network to encode temporal sequences. The latter is typically encoded in terms of a prediction task in which the SRN is trained to predict the next item in a sequence (e.g., the next word in a sentence).

Simulation results obtained by Servan-Schreiber, Cleeremans & McClelland (1991) show that an SRN is able to mimic an FSA in a quite unique way. Instead of encoding the discrete finite representations corresponding to particular inputs, as in a traditional FSA, the network encodes an *association* between a given input and the appropriate prediction of the next output state. This allows the network to capture long-distance dependencies by *shading* its internal representations; that is, by picking up subtle statistical contingencies. In other words, the network learns to respond to temporally distant information by encoding contextually relevant cues in a condensed form in the recurrent links. In addition, SRNs appear to have *functional compositionality* (van Gelder, 1990a) insofar as they are able to process functionally compound representations in a way that is sensitive to their constituent structure. These results demonstrate that SRNs have sufficient power to develop representations that possess the rich internal structure that is necessary for the explanation of systematic NL behaviour.

In a particularly interesting simulation, Elman (1991) demonstrated that a SRN, though inherently sensitive to context, can learn the abstract and general grammatical structure implicit in a language corpus. The network was trained on four different corpora, each of which had an increasing number of relative clauses, and which together totalled 10,000 sentences with a length between 2 and 16 words. The distributed representations developed by the network through training were analyzed in terms of the trajectories through state space over time. More specifically, the trajectories correspond to the internal representations evoked at the hidden unit layer, as the network processed a given sentence (Elman, 1991). Elman's analysis showed that the network was able to capture *agreement* between subject nouns and verbs. The network also developed *verb argument structure*; that is, the network learned to behave in an appropriate manner according to whether it encountered intransitive, transitive, or optionally transitive verbs.

From the viewpoint of the present paper, the most interesting result of this simulation was that the network developed a differentiated capacity with respect to the processing of complex sentences with *recursive* structure. For example, the network was able to process the following *centre-embedded* sentence involving long-distance agreement dependencies:

(9) Boys who girls who dogs chase see hear.

Trajectory-analysis of similar sentences evinced that successive embedded clauses are represented in the same way as the first embedded clause, but slightly displaced in state space. This systematic displacement of recursive clauses in state space enabled the network to keep track of the depth of recursion, while at the same time acknowledging structural similarities between the recursive clauses. However, the network's performance on recursive sentences was limited. An interesting fact about the network's degrading recursive performance was that sentences involving centre-embedded recursion were more badly affected than sentences involving right-embedding. This is consonant with our earlier psycholinguistic observations regarding the parsing of recursive structures.

Pace Fodor & Pylyshyn (1988), connectionist models are suitable for the modelling of NL processing. Indeed, as we have seen, the SRN is particularly interesting from a psycholinguistic perspective in that it appears to exhibit the same behaviour as humans when confronted with complex, recursive sentences (Elman, 1991)—without reverting to *explicitly* programmed limitations on memory. The simulations conducted by Servan-Schreiber, Cleeremans & McClelland (1991) have shown that an SRN is able to mimic a graded FSA—but, more importantly, the SRN *learns* how to behave as if it was an FSA *with a limited stack*, enabling it to deal with centre-embedded sentences with a limited depth of nesting. In contrast, performance orientated symbolic approaches to NL processing (typically also based on FSAs) need to build in such limitations *explicitly*; for example, as limitations on the number of iterations in a regular grammar (Reich, 1969), or as a procedure that clears a stack-like memory structure when a certain threshold is met (Pulman, 1986). Hence, connectionists models provide an appealing non-symbolic account of recursion in linguistic descriptions, while respecting actual psycholinguistic constraints on human NL processing. Crucially, performance aspects do not have to be programmed explicitly outside a connectionist model—they, so to speak, "fall" out in a natural way as a side-effect of the processing of recursive sentences.

Conclusion

In this paper, I have argued that recursion in NL is best construed as a descriptive phenomenon, rather than a basic processing mechanism. In addition, I have questioned the psychological plausibility of the C/PD and have advocated the incorporation of psychological constraints into the NL processing mechanism. It is possible for a connectionist model to account for recursion in NL insofar as the notion of infinite competence is dropped and replaced with a psychologically constrained processing ability. However, such a model must be able to explain empirical data from NL behaviour as an interaction between processing abilities and limitations inherent in the model itself. Recursion in NL is therefore only a problem insofar as linguistic theories are viewed as having explanatory adequacy, and insofar as the notion of an infinite competence is maintained. Work within the connectionist paradigm indicates that descriptive recursion can be accounted for in a way which follows empirical constraints on NL behaviour. However, it is too early to say whether connectionism in the long run will be able to account for the full complexity of human NL behaviour. However, at least presently, connectionism provides a promising framework for non-symbolic NL research.

Acknowledgements

Many thanks to the members of the Foundations of Cognitive Science workshop at the Centre for Cognitive Science, especially Nick Chater and Martin Pickering, for comments and suggestions regarding earlier drafts of this paper. Thanks are also due to Elisabet Engdahl and Ewan Klein for commenting on the penultimate draft.

References

Chomsky, N. (1981). *Lectures on Government and Binding*. Dordrecht: Forris Publications.

Chomsky, N. (1986). *Knowledge of Language*. New York: Praeger

Ejerhed, E. (1982). The Processing of Unbounded Dependencies in Swedish. In E. Engdahl & E. Ejerhed (eds.), *Readings on Unbounded Dependencies in Scandinavian Languages*. Stockholm: Almqvist & Wiksell International.

Elman, J. L. (1990). Finding Structure in Time. *Cognitive Science*, **14**, 179-211.

Elman, J. L. (1991). Distributed Representations, Simple Recurrent Networks, and Grammatical Structure. *Machine Learning*, **7**, 195-225.

Finch, S. & Chater, N. (1992). Bootstrapping Syntactic Categories by Unsupervised Learning. Forthcoming in Proceedings of the 14th Annual Conference of the Cognitive Science Society.

Fodor, J. A. & Pylyshyn, Z. W. (1988). Connectionism and Cognitive Architecture: A Critical Analysis. *Cognition*, **28**, 3-71.

Frazier, L. & Fodor, J.D. (1978). The Sausage Machine: A New Two Stage Parsing Model. *Cognition*, **6**, 291-325.

Gazdar, G., Klein, E., Pullum, G. & Sag, I. (1985). *Generalized Phrase Structure Grammar*. Oxford: Basil Blackwell.

Horrocks, G. (1987). *Generative Grammar*. London: Longman.

Kimball, J. (1973). Seven Principles of Surface Structure Parsing in Natural Language. *Cognition*, **2**, 15–47.

McClelland, J. & Kawamoto, A.H. (1986). Mechanisms of Sentence Processing: Assigning Roles to Constituents of Sentences. Chapter 19 in J. McClelland & D. Rumelhart (eds.), *Parallel Distributed Processing*, Volume 2. Cambridge, Mass.: MIT Press.

Pickering, M. & Chater, N. (1992). Processing Constraints on Grammar. Ms.

Port, R. & van Gelder, T. (1991). Representing Aspects of Language. In Proceedings of the 13th Meeting of the Cognitive Science Society, 487–492, Chicago, Illinois: Cognitive Science Society.

Pulman, S.G. (1986). Grammars, Parsers, and Memory Limitations. *Language and Cognitive Processes*, **2**, 197–225.

Reich, P. (1969). The Finiteness of Natural Language. *Language*, **45**, 831–843.

Servan-Schreiber, D., Cleeremans, A. & McClelland, J. (1991). Graded State Machines: The Representation of Temporal Contingencies in Simple Recurrent Networks. *Machine Learning*, **7**, 161–193.

van Gelder, T. (1990a). Compositionality: A Connectionist Variation on a Classical Theme. *Cognitive Science*, **14**, 355–384.

van Gelder, T. (1990b). Connectionism and Language Processing. In G. Dorffner (ed.) *Konnektionismus in Artificial Intelligence und Kognitionsforschung*. Berlin: Springer-Verlag.

Wanner, E. (1980). The ATN and the Sausage Machine: Which One is Baloney? *Cognition*, **8**, 209–225.

POSTERS

Encoding and Retrieval Processes: Separate Issues in Problem Solving

Lea T. Adams
Department of Psychology
Illinois State University
Normal, IL 61761
LTADAMS@ILSTU.BITNET

Abstract

Studies investigating the facilitation of spontaneous access during problem solving by manipulating encoding processes suggest that similar processing at acquisition and test (i.e., problem-oriented processing) enhances spontaneous access (Adams et al., 1988; Lockhart et al., 1988). Bowden (1985) argues that access difficulty is due to problem solving time (i.e., retrieval) constraints rather than acquisition processes. Ross et al. (1989) have challenged Bowden by suggesting that an increase in retrieval time allows subjects to "catch on" to the the experimental procedure. This study investigates this claim and also attempts to separate acquisition and retrieval factors by crossing problem solving time (40, 80, 120 sec) with acquisition processing factors (problem-oriented, fact-oriented, and mixed orientation). The mixed condition includes problem-oriented and fact-oriented as a within subjects variable. Results show an increase in performance from 40 sec to 80 sec, but no added benefit beyond 80 sec. Problem-oriented processing facilitates spontaneous access. The critical evaluation is that of the mixed condition. Performance in the mixed condition also shows a faciliation of spontaneous access for those acquisition materials that involve problem-oriented processing, but not fact-oriented processing, suggesting that one form of encoding facilitates later access.

Introduction

The ability of people to problem solve simply requires the access and application of previously acquired information to a new or unique situation. However, studies in human problem solving have consistently demonstrated a lack of spontaneous access abilities. Spontaneous access in problem solving means that people spontaneously, on their own, and with no external hints, access or retrieve the information necessary to solve a problem. In studies of access, psychologists have generally used the following procedure. First, information, necessary to solve problems that will later be attempted, is presented to subjects in an incidental learning task (acquisition information). Some of the subjects are informed as to the information's relevance to the later problems (informed subjects) and others are not (uninformed subjects). Then subjects are given a set of problems to solve. The number of problems solved is the measure of access. The results show that uninformed and baseline subjects perform equally, solving few, if any, of the problems while informed subjects solve a high percentage of the problems. In other words, humans do not make effective use of potentially relevant information during problem solving unless direction to do so is provided (Gick & Holyoak, 1980; Perfetto, Bransford, & Franks, 1983; and Weisberg, Dicamillo, & Phillips, 1978).

Perfetto et al. (1983) developed a very obvious set of solution sentences to 12 insight problems adapted by Gardner. An example of the insight problems used is "One night my uncle was reading an exciting book when his wife turned out the light. Even though the room was pitch dark, he continued to read. How could he do that?" The corresponding solution sentence was "A blind person can read braille in the dark." This study was intended to address the issue of access (i.e., do subjects access the acquisition materials during problem solving) versus application (i.e., do subjects access the acquisition materials but reject them as being relevant to the problem). It was believed that, because the solution sentences were so obvious, if subjects accessed them at all application was insured. Using the past methodology, Perfetto et al. (1983) presented the acquisition information to informed and uninformed subjects and compared uninformed subjects' subsequent problem solving performance to the performance of informed and baseline subjects. Uninformed subjects performed no better than baseline subjects and informed subjects out performed all other groups. In other words, spontaneous access of the relevant information had still not been demonstrated in the laboratory.

A current research focus of spontanteous access during problem solving is in the area of facilitation during retrieval (Ross, 1984; 1987; Bowden, 1985). Bowden (1985) stated that the lack of laboratory demonstrated spontaneous access was due to the experimental methodology. He argued that humans usually are not under the time constraints to solve problems imposed by typical problem solving studies. These time constraints do not allow a full search in a problem space and, therefore, do not allow the demonstration of access that is naturally occuring in humans. He maintained that spontaneous access was not as critical an issue as most problem solving researchers believe. He demonstrated this by conducting a study using Perfetto et al.'s (1983) materials and general methodology. The critical difference involved the amount of time subjects were given to solve the problems (i.e, 120 seconds per problem as opposed to the usual 40 seconds per problem). With the increased time for problem solving, uninformed subjects performed as well as the informed subjects and both groups were superior to baseline performance.

More recently, Ross, Ryan, and Tenpenny (1989) attempted to replicate Bowden's (1985) study and found that the results did not replicate if a different order of problems was used. Ross et al. (1989) applied a simple mathematical model to the results that attributed the differences between the two studies to Bowden's uninformed subjects "catching on" to the relevance after solving some problems. Ross et al. (1989) maintained that subjects in Bowden's study were presented easier problems first and were able to generate a solution on their own. Once subjects generated a solution they realized they had recently heard similar information during the acquisition phase. This realization let subjects "catch on" to the procedure used in the study. Therefore, it was concluded, that extra problem solving time does not lead to spontaneous access unless subjects become "reminded" of the acquisition material and subsequently focus on it. Furthermore, this reminding is governed by the similarity between generated answers and the sentences presented during acquisition.

Other researchers have focused on encoding processes, as opposed to retrieval processes, as a way of facilitating spontaneous access. Adams, Kasserman, Yearwood, Perfetto, Bransford, and Franks (1988) developed a set of follow-up studies based on the Transfer Appropriate Processing model of memory retrieval. Simply put, Transfer Appropriate Processing suggests that memory access/retrieval is enhanced if the processing used at retrieval of information is similar to the processing used during encoding of that information. Adams et al. (1988) developed a series of materials based on the materials used in the Perfetto et al. (1983) studies. However, instead of using factually stated sentences for the acquisition information, as in the earlier studies, the acquisition information was cast into a problem oriented format. It was believed that if the acquisition sentences first presented an ambiguity, followed by a clarifier or solution, that processes necessary to comprehend the acquisition sentence would be similar to the processes used during problem solving. Therefore, the problem-oriented form of the fact-oriented sentence, "A blind person can read braille in the dark" became "It is possible to read in the dark; if you are reading braille." By using the problem-oriented form of acquisition materials, spontaneous access was demonstrated in the laboratory. That is, uninformed subjects performed at the same level as informed subjects. Other researchers, working in parallel and using a very similar approach, demonstrated spontaneous accesss as well (Lockhart, Lamon, & Gick, 1988).

Though researchers have focused on retrieval processes and encoding processes as facilitators of spontaneous access, little research exists that investigates the relationships between these processes. A recent study (Adams, 1992) attempted to identify the more critical process (i.e., encoding or retrieval) by crossing the two variables, acquisition sentence form and retrieval time. Four experimental conditions were investigated: 1) fact-oriented acquisition sentences with a 40-sec solution time, 2) fact-oriented, 120-sec, 3) problem-oriented, 40-sec, and 4) problem-oriented, 120-sec. It was hypothesized that, because problem-oriented acquisition sentences already facilitated spontaneous access, increasing retrieval time would not notably improve problem solving performance because subjects in problem-oriented conditions are already reminded of the acquisition materials. Catching on should not be an issue for those in the problem oriented conditions. Therefore, when the four experimental groups are compared, an interaction between acquisition sentence form and retrieval time would be expected. Instead, main effects for retrieval time and acquisition sentence form were found. Results indicated that while an increase in problem solving time increases spontaneous access, problem-oriented acquisition processing enhanced access well beyond the benefits attributed to increasing problem solving time, suggesting that these are additive, independent effects. Those results did not support a simple "catching on" explanation of the spontaneous access during problem solving associated with an increase of problem solving time.

Instead, new questions were generated. If acquisition sentences are encoded as a list upon which, after catching on, one could focus a solution search, then no difference in performance would be expected between problem-oriented and fact-oriented conditions. If the list can be recognized, found, and searched in 2 minutes, why would one list, once found, be better searched than another? Is one list more easily found than another? If similarity between a subject generated response and the acquisition material triggers the finding of a memory list (as suggested by Ross et al, 1989), fact-oriented acquisition material would seem to be more similar in form to subjects' answers and, therefore, be more easily found. This is contrary to the study's results.

Another possible explanation to findings of the earlier research is that the acquisition sentences are not simply stored as lists, but are somehow integrated into exisiting conceptual frameworks or stored as miscellaneous information grouped according to other unique, identifying characteristics, such as meaning ambiguity or encoding processes. Indeed, this encoding strategy was proposed by Adams et al. (1988) as well as Lockhart et al. (1988). If the information is not encoded as a list, the catching on explanation of performance facilitation with an increase of problem solving time is also inappropriate.

The objective of the present study was to provide for a more rigorous replicaton of the above study (Adams, 1992) and to incorporate an extension that investigated the encoding issues raised by the that study's results. To that end, subjects were exposed to either fact-oriented or problem-oriented acquisition materials and were given either 40, 80, or 120 seconds to solve each problem in a susbsequent problem solving task. The addition of the 80 second condition allowed for the determination of a linear relationship, orthogonal and additive, between the acquisition sentence form and the retrieval time constraints. Additionally, baseline and informed conditions were included in the study for a more complete design and provide for more rigorous analyses of problem solving performance.

In order to investigate the encoding issues discussed earlier, a new condition was incorporated into the methodology. In this condition, subjects were presented a list of acquisition sentences that contained both fact-oriented and problem-oriented sentences (mixed orientation condition). The rationale for this condition is as follows. If the acquisition sentences are encoded as a single list and subjects simply find the list and search it, spontaneous access for all problems, regardless of the form of the associated acquisition sentence will be facilitated. On the other hand, if the sentences are encoded and stored differentially according to unique characteristics or qualities of the sentences, it is expected that the fact-oriented and the problem-oriented sentences would not be stored together due to the form differences. Because previous research demonstrates a facilitation of spontaneous access for problem-oriented acquisition sentences (Adams, et al., 1988, Lockhart, Lamon, & Gick, 1988), better performance for those problems that had problem-oriented acquisition sentences would be expected.

Methods

Subjects

Subjects were 420 undergraduate psychology students from the Illinois State University subject pool.

Procedure

Subjects were randomly assigned to one of 3 experimental conditions: the fact-oriented condition, the problem-oriented condition, and the mixed-orientation condition. Within these conditions, subjects were either allowed 40, 80, or 120 seconds per problem during the problem solving task. Baseline groups were also run, where subjects attempted to solve the problems, without exposure to the acquisition sentences, with 40, 80, or 120 seconds per problem. Additionally, informed groups (subjects are told of the relationship between the acquisition sentences and the problems prior to problem solving) mirroring the experimental conditions were run. Subjects were tested in groups of 8 to 12 subjects at a time.

Prior to participation, subjects were told that they would be asked to do a series of unrelated tasks to help finish several in-progress experiments. This cover story was presented so the subjects would not automatically assume that the acquisition sentences were related to the subsequent problem solving task.

During the acquisition phase, subjects, excluding those in the baseline conditions, listened to a taped presentation of the acquisition sentences. The first and last sentence in the presentation were filler sentences, taking fact-oriented form in the fact-oriented and mixed-orientation conditions and a problem-oriented form in the problem-oriented conditions. After each sentence, during a 20-sec pause, the subjects rated the sentence on general truthfullness using a 1 to 5 scale (1 = true only in a specific instance to 5 = always true).

There was a 4 minute interval between the acquisition phase and the problem solving task. The subjects were then given a booklet containing three filler problems followed by 10 insight problems. The 10 insight problems were presented in one of seven random orders.

The subjects were told not to open their booklets until instructed to do so. They were then told that they had 40/80/120 seconds to solve each problem and that the experimenter would tell them when to go on to the next problem. They were also told not to work ahead or go back to an earlier problem. Subjects were instructed to write an answer for every problem, even if they did not believe it was correct. At this point, subjects in the informed conditions were also told of the relationship between the acquisition sentences and the problems. Upon completion of the problem solving task, subjects were asked to fill out a questionnaire that asked if subjects noticed the relationship between the acquisition sentences and the problems and if they were familiar with any of the problems prior to participating in the study. Data from subjects familiar with 2 or more insight problems was discarded.

Results and Discussion

An overall 4(Acquisition Sentence Type) X 3(Problem Solving Time) X 2(Informed Status) ANOVA was performed on the number of correcly solved target problems. Significant main effects for Acquisition Sentence Type, Problem Solving Time, and Informed Status were found, $F(3,419) = 29.6$, $p = .0001$, $F(2,419) = 13.7$, $p = .0001$, and $F(1,419) = 45.4$, $p = .0001$. No significant interactions were found. Though several different types of analyses were performed, the analyses that directly address issues of spontaneous access are those related to the performance of uninformed subjects. The overall problem solving performance of uninformed subjects is presented in Table 1.

A 4(Acquisition Sentence Type) X 3(Problem Solving Time) ANOVA performed on the number of correctly solved problems for those subject in the Uninformed conditions maintained significant main effects for both factors, $F(3,236) = 18.1$, $p = .0001$, $F(2,236) = 7.0$, $p = .001$, respectively. No significant interactions were found. Using Dunn's multiple comparison procedure, the fact-oriented and problem-oriented conditions performed significantly better than baseline for all problem solving times,

Table 1
Number of Problems Solved in Relation to Acquisition Sentence Form, Problem Solving Time for Uninformed Subjects

Acquisition Sentence	Problem solving time (in sec)		
	40	80	120
Baseline	1.5	2.5	2.5
Fact-oriented	3.8	4.2	3.8
Problem-oriented	4.2	5.1	5.5
Mixed-condition	3.2	5.5	4.8

$\underline{d}(3,236) = 0.53$, $\underline{p} < .01$, $\underline{d}(3,236) = 0.55$, $\underline{p} < .01$, and $\underline{d}(3,236) = 0.65$, $\underline{p} < .01$, for the 40, 80, and 120 sec conditions, respectively. Subjects in the problem-oriented conditions solved significantly more problems than those in the fact-oriented conditions if allotted 80 or 120 sec for problem solving. There was no significant difference in performance between the fact-oriented and problem-oriented groups in the 40 sec condition.

Similar comparisons between the different problem solving times for the baseline, fact-oriented, and problem-oriented groups suggest that subjects solve fewer problem in the 40 sec condition than the 80 or 120 sec conditions. No significant difference in performance was found between the 80 and 120 sec conditions except for the fact-oriented groups. Subjects in the fact-oriented groups actually performed better in the 80 sec condition than in the 120 sec condition. These results replicate previous research in that an increase (up to 80 sec) in problem solving time and problem-oriented acquisition sentences both seem to facilitate spontaneous access.

An investigation of performance in the mixed-orientation condition showed no facilitation of spontaneous access for acquisition sentences that were presented in a fact-oriented form even though they were presented with problem-oriented acquisition sentences. When the proportion of problems solved by subjects in the problem-oriented condition (0.42, 0.51, 0.54 for the 40, 80, and 120 sec conditions, respectively) is compared with the proportion of problems solved that were associated with problem-oriented acquisition sentences in the mixed condition (0.36, 0.58, and 0.50.), no difference in performance is found. Furthermore, when making the same comparison between performance in the fact-oriented conditions (0.37, 0.41. 0.38, for the 40, 80, and 120 sec conditions, respectively) and the fact-oriented materials in the mixed-orientation conditions (0.28, 0.51, and 0.45), no differences are found. If the acquisition material was simply stored as a list during presentation, once subjects solved a few problems and "caught on" to the experimental manipulation, facilitation for fact-oriented and problem-oriented problems would be expected. However, this was not the case. This suggests that the acquisition sentences are encoded and stored differentially and that one form of encoding facilitates later access of that information. Furthermore, a catching on explanation associated with an increase of problem-solving time would be inappropriate if the information is indeed encoded separately.

References

Adams, L. T. 1992. Enhancing spontaneous access of relevant information during problem solving: Strategic processing vs. time constraints. Paper presented at the XXV International Congress of Psychology, Brussels, Belgium.

Adams, L. T., Kasserman, J. E., Yearwood, A. A., Perfetto, G. A., Bransford, J. D., Franks, J. J. 1988. Memory access; The effects of fact-oriented versus problem-oriented acquisition. Memory & Cognition, 16(2):167-175.

Duncker, K. 1945. On problem-solving. Psychological Monographs, 58(270).

Gick, M. L., & Holyoak, K. J. 1980. Analogical problem solving. Cognitive Psychology, 12:306-355.

Lockhart, R. S., Lamon, M., & Gick, M. L. 1988. Conceptual transfer in simple insight problems. Memory & Cognition, 16(1):36-44.

Perfetto, G. A., Bransford, J. D., & Franks, J. J. 1983. Constraints on access in a problem solving context. Memory & Cognition, 11:24-31.

Ross, B. H. 1984. Remindings and their effects in learning a cognitive skill. Cognitive Psychology, 16:371-416.

Ross, B. H. 1987. This is like that: The use of earlier problems and the separation of similarity effects. Journal of Experimental Psychology: Learning, Memory, & Cognition, 13:629-639.

Ross, B. H., Ryan, W. J., & Tenpenny, P. L. 1989. The access of relevant information for solving problems. Memory & Cognition, 17(5):639-651.

Weisberg, R., Dicamillo, M., & Phillips, D. 1978. Tranferring old associations to new situations: A nonautomatic process. Journal of Verbal Learning & Verbal Behavior, 17:219-228.

A Grounded Mental Model of Physical Systems: A Modular Connectionist Architecture

Amit Almor
Department of Cognitive and Linguistic Sciences
Brown University
almor@drew.cog.brown.edu

Abstract

Some basic characteristics of subjects' use of mental models of physical systems are discussed. Many representations for physical knowledge suggested so far, including qualitative-reasoning-based models, do not account for these experimental findings. This paper presents a connectionist architecture which suggests an explanation of these experimental results. Two simulation experiments are described which demonstrate how mental models of physical systems may evolve and why grounding symbols used by a mental model to a quantitative representation is necessary.

Mental Models of Physical Systems

Recent studies of physical knowledge acquisition have focused on the way a mental model of a physical system can be created from a set of elementary pieces of knowledge about the physical world. In this context, mental model means a structured representation of knowledge about a specific system. Norman (1983) observes the following facts. (1) Mental models evolve through interaction with the system they model. (2) Mental models are used to facilitate the interaction between the subject and the physical system, and are not accurate descriptions of the physical system. (3) Mental models are runnable, i.e. subjects can run their mental models and predict a particular future state of the system. (4) People are notoriously bad in running mental models through a large number of stages or for a long time. Also, people are often hesitant about the validity of their mental-model-based judgements. All these characteristics relate to the performatory aspect of mental models, that is to the actual behavior of subjects in experiments in which, presumably, they use their mental models. Most research in this domain has focused on the form of knowledge representation which gives rise to these behavioral patterns.

Qualitative Reasoning Theory

The qualitative reasoning theory (Weld & deKleer, 1990) evolved out of research into mental models of physical systems. Often, when subjects apply their physical knowledge, their behavior and reports are incompatible with any theoretical law of physics. Thus, an alternate, simpler "qualitative" physics theory has been formalized. A qualitative-reasoning based mental model is a list of qualitative equations describing the physical system. The qualitative equation is an expression describing the interaction between coarse-valued variables. Special qualitative arithmetic is defined to operate on these qualitative values. Expressed this way, some physical concepts e.g. "flow" can be expressed by specifying their interactions with other concepts such as height of liquid columns (deKleer & Brown, 1990).

Difficulties. Critics of the symbolic paradigm however, claim that a qualitative-reasoning-based mental model, is not a satisfactory model for any cognitive process since it gives rise to the symbol grounding problem (Harnad, 1990). Symbols cannot be arbitrary forms which are assigned meanings independently of the cognitive model. Rather, their form must be causally determined in a bottom up manner.

A further difficulty with symbolic knowledge representation is its artificial distinction between competence and performance. The theory of qualitative reasoning does not account for how mental models evolve through interaction, why mental models are runnable, and why subjects are so bad in running them over many stages. The symbolic framework excludes these confounds from any discussion about the knowledge representation form. An alternative framework, under which both competence and performance confounds will be explained by the postulated knowledge representation form should be preferred on the grounds of parsimony.

This paper presents a modular connectionist architecture for mental models of physical systems which allows the transition from quantitative to qualitative knowledge, and which avoids the problems described above. The architecture generates symbols which are assigned "real-world meanings" as a natural and necessary quality of the processes by which they evolve. Relations between the generated symbols constitute an alternative to the symbolic notion of compositional structure (Fodor and Pylyshyn, 1988).

The distinction between competence and performance is eliminated by using a connectionist knowledge representation form. (1) Due to the connectionist training process, the representation gradually evolves through interaction with the environment. (2) By imprinting the behavior of the physical system in a connectionist network, a model can be "rerun" later in order to make predictions on the system's future state. (3) The statistical nature of the knowledge representation built makes it hard to run the model for many stages or for a long period as errors propagate and accumulate quickly.

I further describe two simulation experiments with the architecture, which lead to a number of interesting observations. In the current model, adequate symbol generation is possible only if the system has reached some level of familiarity with the real environment. Once this level of familiarity is attained, improving the system's knowledge of the environment is faster using the generated symbols than by increasing the system's familiarity with the environment. The question arises as to the computational status of symbols. First, "grounding" the symbols is no longer a mere philosophical requirement. Rather, it is a computational requirement in order for symbols to be functional. Second, the role of symbols might be conceived as efficient knowledge modifiers rather than arbitrary shapes used as building blocks for some compositional structure.

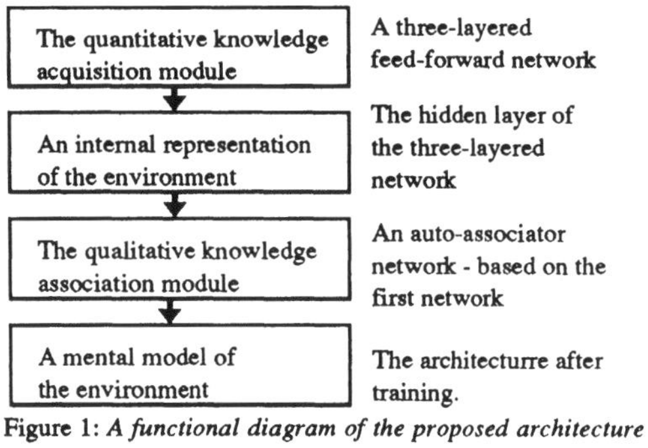

Figure 1: *A functional diagram of the proposed architecture*

The Modular Architecture: From Quantitative to Qualitative

The proposed architecture consists of two inter-related modules. The first interacts with the environment to construct a non-symbolic mental model of it. The second uses the internal analog representations built by the first module and associates qualitative symbols with these representations (see Figure 1). The entire model is then able to make qualitative statements and predictions about the state of the environment, given any qualitative specification of an initial scenario.

The Quantitative Module

The first module is a feed-forward three-layered network which is exposed to a representation of the environment (The exact form will be discussed in the next section). The input consists of a representation of the state of the environment at time *t*. The expected output is a representation of the state of the environment at time *t+1* (See Figure 2). The module is trained using the error back-propagation rule (Rumelhart, Hinton, & Williams, 1986).

There are two important facts regarding the coupling of this module with the environment. First, even though (technically speaking,) back-propagation is a supervised training scheme, in this case, with the environment supplying both the input and the correct output, the training is teacher-less and thus psychologically plausible. Second, since the forms of the input and the output of this network are identical, the network should be viewed as a recurrent network with one input/output layer and one hidden layer. For computational simplicity, the network is trained as a three-layers feed-forward network.

The motivation for using this particular architecture for the first module is twofold. First, a three layered feed-forward network trained by error back-propagation is capable of learning complex interactions in the environment. Second, it allows for the generation of an internal representation of the environment over the hidden layer which already encompasses some information about what the next state of the environment will be. This internal representation is then available for further processing.

The Qualitative Module

The second module auto-associates verbal labels and qualitative values with activation patterns over the hidden layer of the first model. As demonstrated in the next section, the labels and qualitative values do not have to correspond to explicit representations in the input for the first module. This module consists of a recurrent network trained using the Widrow-Hoff (1960) learning rule. An expansion of the "Brain State in a Box" (BSB) algorithm (Anderson, Silverstein, Ritz and

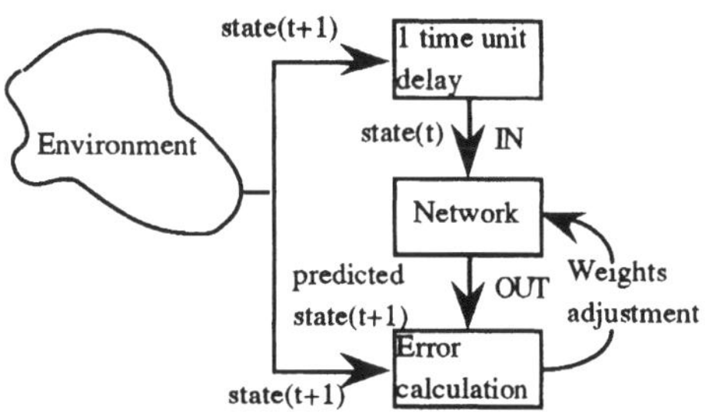

Figure 2: *A functional diagram of the first, quantitative module.*

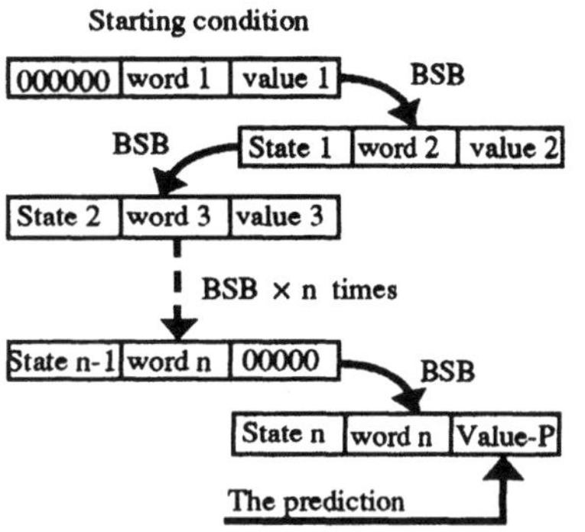

Figure 3: *A diagram of the full activation cycle of the proposed architecture. The first component of the state vector is used as a "state" descriptor or a memory cell. The second and third components have the word and a qualitative value representations. The flow prediction is eventually extracted from the last 40 units of the last state vector.*

Randal, 1976) is used as the network activation scheme. The proposed architecture employs the basins of attraction of the BSB model to achieve qualitative linguistic judgements. (See Hopfield, 1982; Anderson, Silverstein, Ritz and Randal 1976; and Golden, 1986; for a formal analysis of the effect of basins of attraction).

The manner in which the entire architecture functions is similar to a finite-state-automaton where the "internal representation" component of the activation vector functions as the "state." A word, and possibly a qualitative value, is the input which allows transition from the current state to the next state using the BSB dynamics (see Figure 3). The following sections describe a low-scale implementation of the architecture for modelling the generation of mental models of physical systems.

Simulation Experiments

The architecture was used to construct a mental model of liquid flow between reservoirs (See Figure 4). Liquid flow was chosen because: (1) it is familiar, subjects can make good qualitative predictions about its basic behavior; and (2) though fairly simple, liquid flow presents the difficulties mentioned above concerning people's representation of its behavior.

If the mental model simulation generates a concept which corresponds to the physical measure of flow from the sensory information, without starting with an explicit representation of flow, it exemplifies how physical concepts might emerge and how symbols (the symbol for "flow" in this case) may be grounded. In addition, if a training process leads to the generation of a system of qualitative relationships describing the physical system being modeled, it demonstrates how a qualitative-reasoning based mental model could arise.

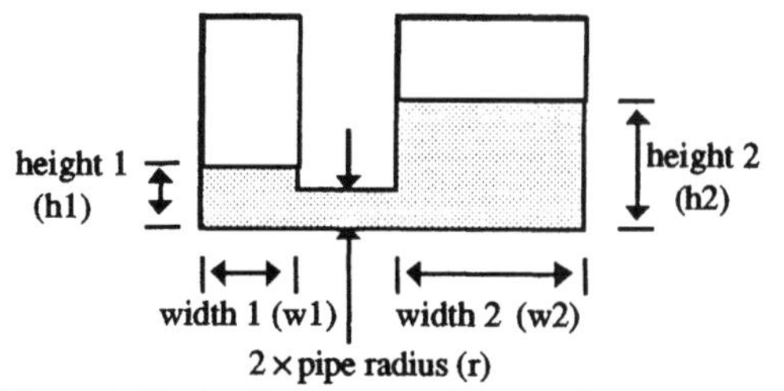

Figure 4: *The liquid flow system being modeled. The real, continuous time dimension across which the process of flow occurs is divided into digital units at which the state of the physical system is sampled.*
Q_t-The flow-rate at time t, is given by Torriceli's law:

$$Q_t = 2\pi r^2 \sqrt{2g|h1_t - h2_t|}$$

where h1 and h2 are the heights of the liquid column in the two reservoirs and r is the radius of the pipe connecting them. Therefore, the heights of the liquid level after a single time unit will be:

if h1>h2:	*if h1<h2:*	*if h1=h2:*
$h1_{t+1}=h1_t-Q_t/\pi w1$	$h1_{t+1}=h1_t+Q_t/\pi w1$	$h1_{t+1}=h1_t$
$h2_{t+1}=h2_t+Q_t/\pi w2$	$h2_{t+1}=h2_t-Q_t/\pi w2$	$h2_{t+1}=h2_t$

where w1 and w2 are the widths of the two reservoirs.

Input and Output Representation and Training Order

The first module consisted of a three layered network with 200 input units, 80 hidden layer units and 80 output units. The second module consisted of 160 fully connected units (See Figure 5).

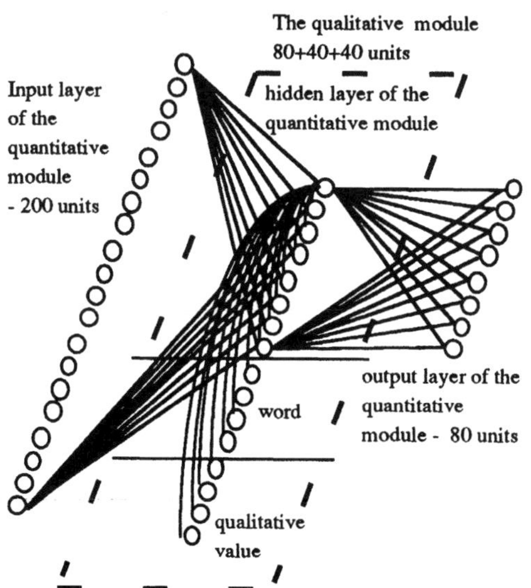

Figure 5: *The proposed architecture*

The input to the first module consisted of quantitative representations of the heights of the water columns in the two reservoirs, the width of the two reservoirs and the width of the pipe connecting them. Each measure was represented in an area of 40 units in which a sliding bar of five units indicated the value. For example, a water height of 50 in the first reservoir is represented by:

00000000000000000000000000000000000 11111

The total dimensionality of the input layer was therefore 5x40=200 units. For reasons of simplicity and computational feasibility, the output layer was only 80 units long and had only the two heights of the two water columns. It should be noted that bar code representations have some biological appeal since they resemble brain maps that have been described in different areas of the brains of many species. It is therefore plausible to assume bar-code representations as a general quantitative representation form.

Given any quantitative initial condition, this module predicts the quantitative condition at the next time unit. The training phase included exposing this module to 2000 different flow scenarios; each scenario begins by choosing random values for the five input measures and then uses Toricelli's law to calculate the states of the system in the following time units until equilibrium is achieved.

The input to the second module consisted of the internal representation of the physical system's state generated over the hidden layer of the first, quantitative, module, plus an arbitrary representation for a word and a qualitative value. Since the word and the qualitative value are arbitrary symbols (grounded by training this module), they were represented by arbitrary representations that maximize orthogonality. The hidden layer of the first model was 80 units long. The word and qualitative value were each represented by 40 units. Therefore, this module consisted of 160 fully connected units.

The words represented were: "HEIGHT-1", "HEIGHT-2", "WIDTH-1", "WIDTH-2", "WIDTH-PIPE" and "FLOW". The qualitative values represented for all the words but "FLOW", were "HIGH", "MEDIUM" and "LOW". For the word "FLOW", the value field was segmented into two parts, the first representing the qualitative strength of the flow: "HIGH", "MEDIUM" or "LOW" and the second representing the direction of the flow: "FROM-1-TO-2" or "FROM-2-TO-1" or "NONE". For flow value of "NONE", the magnitude field was ignored.

Training the second module started by presenting each one of the 2000 initial states used to train the first module to the first module. Then, the internal representation generated over the hidden layer of the first module was auto-associated with each of the six words and with the qualitative value corresponding to the quantitative initial condition. The quantitative range associated with each qualitative value was chosen so that all qualitative values would occur with equal frequency.

Running the Qualitative Mental Model

In order for the system to make a prediction, the qualitative initial condition must be specified. As mentioned earlier, the process works much like a finite state automaton (Figure 3). The process is described below:

1) Zero the internal representation component of the second, qualitative module.
2) Load the first word and qualitative value of the initial condition into the appropriate areas.
3) Run the BSB activation scheme until saturation.
4) Leave the internal representation area as is (since it functions like the state in a FSA) and load the next word and qualitative value specifying the initial conditions.
5) Repeat 3-5 until all the qualitative initial measures have been specified.
6) Zero the qualitative value.
7) Load the representation for the word "FLOW" to the word area.
8) Run the BSB activation scheme until saturation.
9) Take the prediction for the flow direction and magnitude from the qualitative value area (More precisely, the closest qualitative value to whatever is taken out of the qualitative value area in terms of vector cosine).

The evaluation of a prediction takes into account the fact that when an initial state is specified qualitatively, more than one qualitative prediction can be correct. Based on the probability of each prediction given any initial condition, the following evaluation scheme is used:

1) Test the network for predictions for all possible initial condition.
2) Categorize each prediction as either correct (and most probable), second best choice, third best choice or direction error.
3) Assign a grade to the overall performance of the model by:
   Grade = 4 * (%correct - %direction error)
   - 2 * %third best errors
   - 1 * %second best errors.

The grade is mostly affected by the percentage of correct predictions versus the percentage of direction errors.

Experiment 1: The Importance of Grounding Symbols

The first experiment tested the importance of the "grounding knowledge" to the overall performance of the system. While the amount of training put into the second module (the symbol associator) was kept constant, different types of quantitative internal representations were used.

1) In order to test the importance of the internal representation component in the dynamics of the symbol associator, random internal representations were used rather then real activation patterns over the hidden layer of the trained first module. The weights connecting the units of the first module were assigned randomly, yielding random activation patterns over the hidden layer. The same 2000 exemplars were used for training the second module. If this network is trainable, then the real knowledge of the specific system must have no importance; the model would work due to there being an internal representation component that "anchors" the symbols arbitrarily, regardless of their contents. Such a result would suggest that this model does not offer any advantage over symbolic architecture because it is indifferent to the meaning assigned to the symbols.

2) Even if the random internal representations proves unsuccessful, it is still possible that the system can function with the real internal representations due to the general structure present in these representations rather than their particular contents. That is, the presence of structure might be sufficient to "anchor" the symbols. In order to rule out this possibility, another version of the first module was used. As before, the first module was not trained but was assigned weights in a structured pattern yielding structured and systematic activation patterns over the hidden layer which were still non related to the real flow system[1]. Again, if this network is trainable, then the real knowledge of the specific system has no importance.

3) Finally, it needs to be shown that given the appropriate training, the system can work. To clarify the difference between the previous cases and the case of the real training, I used the two arbitrary weights setups (the random and the structured) as initial weights for the first module, and trained it using one introduction of each of the 2000 flow scenarios. The training of the qualitative module started only after the quantitative knowledge was generated. This training was done by associating the appropriate words and values with the 2000 initial states. Each of the associations was used 10 times on average during the training.

4) To further explore the importance of the qualitative grounding knowledge, the previous cases were replicated with the exception that the first module was further trained by using the same training set once more before starting the training of the second module.

Over all there were two control systems in which there was no grounding knowledge, two systems in which there was a certain amount of grounding knowledge, and two other systems in which there was more grounding knowledge. The training of the qualitative module was identical for the six systems. The qualitative performance of each system was then evaluated by the scheme described above and assigned a grade.

Results of Experiment 1: The results are shown in Table 1. Although in no cases were the predictions made by the system perfect, the results still suggest the following points: (1) Real grounding knowledge is necessary for better qualitative performance. (2) Arbitrary structure of the internal representations is not sufficient for qualitative knowledge generation. (3) Better grounding knowledge consistently yields better qualitative results. (4) Symbols (such as the word "FLOW" and its associated qualitative values) can serve to generate novel concepts from internal representations of "sensory inputs." The relation between the novel concept and the concepts associated with the "sensory input" is the alternative this framework offers to the notion of "compositionality" in the symbolic framework.

Initial weights	Control groups	Single quantitative training cycle	Two quantitative training cycles
random	-105	-3	2
structured	-105	-29	8

Table 1: *Results of experiment 1. The untrained control groups did the worst. Two quantitative training cycles improved the performance meaning that "stronger grounding", improved the overall results.*

Experiment 2: The Importance of Symbols for Teaching

This experiment examines how the system's predictions can be improved. One method is to further train the first module; i.e., let the system "watch" more flow scenarios. An alternative method would be to retrain the second, qualitative module if it failed to make the correct prediction about scenarios which were part of its training set; i.e., "tell" the system more about how flow behaves qualitatively. In this method, the first module is not retrained. The last method I consider is to only correct the qualitative errors the system does in the evaluation test. This is much like the manner in which a teacher would qualitatively test a student on novel situations and correct his/her errors.

I also wanted to inspect the effect of "grounding knowledge" on the ability to improve the predictions made by the system, by making qualitative corrections. Twelve systems were compared in this experiment. Four were the systems from the previous experiment which were grounded to the real physical system. For each of these four systems, the two qualitative

[1] The weights were setup according to a Gausian formula to ensure that systematic changes in the input values would yield systematic changes in the activation patterns over the hidden layer.

correcting procedures were applied separately. The twelve networks were evaluated as before.

		Initial weights	Single qualitative training cycle	qualitative correction with the training set	qualitative correction with the complete evaluation
Single quantitative training cycle	random	-3	-105	-105
	structured	-29	-116	-116
Two quantitative training cycles	random	2	7	7
	structured	8	7	16

Table 2: *Results of experiment 2. Both methods of qualitative correction worsened the overall performance of the less grounded networks but slightly improved the overall performance of the more grounded networks.*

Results of Experiment 2: The results are shown in Table 2. There are two interesting observations: (1) qualitative correction does not improve performance for the less grounded systems. On the contrary, it reduces total performance[2]. Grounding is not only necessary for overall performance but also for making qualitative corrections. (2) In most cases, with some degree of grounding established, the more efficient qualitative correction methods enable further learning beyond that achieved by quantitative retraining.

General Discussion

The paper sketches a general connectionist architecture that remedies the symbol grounding problem without giving up the notion of symbols. The generation of the novel concept of flow, by associating symbols to an internal representation of simpler interacting factors, demonstrates how a compositional or hierarchical conceptual structure might evolve. The connectionist modelling techniques eliminate the artificial distinction between competence and performance that prevails in much of the research on mental models. The proposed architecture gives a unified account for both the form of the knowledge representation, and for the empirical evidence about how subjects perform tasks using this knowledge.

The generation of a mental model of a the liquid flow physical system demonstrates the two essential aspects of the symbols suggested in this paper. On the one hand, symbols need to be grounded to real-world meanings for the model to work. On the other hand, once this grounding condition is satisfied, using symbols has some evident advantages. These results suggest a wider interpretation of the symbol grounding problem. Not only do symbols need to be grounded to explain real-world meaning assignment, but their grounding is a necessary computational condition. The grounding is what causally determines the compositionality (Fodor and Pylyshyn, 1986) of the symbols. Obviously, this assumption is valid only in the framework sketched in this paper. The computational necessity for grounding, however, may contribute to the failure of the symbolic architecture to meet Turing's (1950) vision.

[2] Because the correction process is used only for wrong predictions, it can cause the network to forget predictions that it previously got right, therefore decreasing the overall grade.

References

Anderson, J. A.; Silverstein, J. W.; Ritz, S. A.; and Randal, J. S., 1976. Distinctive Features, Categorical Perception, and Probability Learning: Some Applications of a Neural Model. *Psychological Review* 84: 413-451.

Fodor, J. A., and Pylyshyn, Z. W., 1988. Connectionism and Cognitive Architecture: A Critical Analysis. In: *Connections and Symbols.* Pinker, S., and Mehler, J. eds. 3-72. Cambridge, MA: The MIT Press.

Golden, R. M., 1986. The 'Brain-State-in-a-Box' Neural Model is a Gradient Descent Algorithm. *Journal of Mathematical Psychology* 30:73-80.

Harnad, S., 1990. The Symbol Grounding Problem. *Physica D.* 42:335-346.

Hopfield, J. J., 1982. Neural Networks and Physical Systems with Emergent Collective Computational Abilities. Proceedings of the National Academy of Sciences 79:2554-2558.

Norman, D. A., 1983. Some Observations on Mental Models. In: *Mental Models.* Gentner, D. & Stevens, A.S. eds. Lawrence Erlbaum Associates, Publishers.

Rumelhart, D. E., Hinton, G. E., and Williams, R. J., 1986. Learning Internal Representations by Error Propagation. In: *PDP: Explorations in the MicroStructure of Cognition,* Vol I.:318-362. Cambridge, MA: The MIT Press.

Searle, J. R., 1980. Minds, Brains, and Programs. *The Behavioral and Brain Sciences* 3:417-424.

Turing, A. M., 1950. Computing Machinery and Intelligence. *Mind* 59:433-460.

Weld, D.S., and deKleer, J. eds. 1990. *Readings in Qualitative Reasoning about Physical Systems.* Morgan Kaufman Publishers, INC.

Widrow, B., and Hoff, M. E., 1960. Adaptive Switching Circuits. 1960 IRE WESCON Convention Record, New York: IRE. 96-104.

Self-Organization of Auditory Motion Detectors

Sven E. Anderson*
Department of Linguistics
Indiana University
Bloomington, Indiana 47405
sven@cs.indiana.edu

Abstract

This work addresses the question of how neural networks self-organize to recognize familiar sequential patterns. A neural network model with mild constraints on its initial architecture learns to encode the direction of spectral motion as auditory stimuli excite the units in a tonotopically arranged input layer like that found after peripheral processing by the cochlea. The network consists of a series of inhibitory clusters with excitatory interconnections that self-organize as streams of stimuli excite the clusters over time. Self-organization is achieved by application of the learning heuristics developed by Marshall (1990) for the self-organization of excitatory and inhibitory pathways in visual motion detection. These heuristics are implemented through linear thresholding equations for unit activation having faster-than-linear inhibitory response. Synaptic weights are learned throughout processing according to the competitive algorithm explored in Malsburg (1973).

The Perception of Spectral Motion

The processing of sequential stimuli is an essential component of auditory and visual perception in many animals. Recent efforts have resulted in learning algorithms that can be used to encode sequential patterns within autoassociative (Reiss & Taylor, 1991; Metzger & Lehmann, 1990; Elman, 1990) and supervised paradigms (Wang & Arbib, 1990; Fōldiák, 1991). We believe that these approaches can be successfully extended to the self-organization of sequential pattern detectors through the integration of a hierarchy of network layers, each of which is sensitive to particular attributes of a sequential input stream. This report details an implementation of the first module of a system for building representations of sequential auditory patterns that are statistically salient in an animal's environment. When exposed to an environment consisting of frequency sweeps, sound bursts, and constant frequency components this module learns to detect direction of motion from a 1-dimensional tonotopic input array.

Neural patterns of response to auditory stimuli travel from the basilar membrane to the auditory cortex via the cochlear nucleus, inferior colliculus, and medial geniculate. There is little doubt that higher and higher centers of auditory processing respond to auditory stimuli of increasing complexity and duration (Pickles, 1988). Stimuli with changing frequency are important to many species for communication, navigation, and target tracking. For example, within the mustached bat, sensitivity to frequency modulated tones (FM) has been found at the cochlear nucleus (Suga, 1990). Auditory cortex apparently contains large numbers of units that respond best to species specific calls (Aitkin, 1990) which are usually temporally and spectrally complex. Whitfield and Evans (1965) discovered that the majority of a sample of 104 cells of auditory cortex responded only to frequency modulation in a particular direction. The effect of rate of frequency modulation on cell response was minimal though perceptible for some cells. We thus propose that an important first task of the auditory pathway is the rate-invariant determination of direction of motion across the spectrum for non-constant stimuli.

The motion detection model presented here converts an inherently temporal pattern into a spatial code (unit activity). This may be useful if further processing is to isolate sequential patterns using spatial learning mechanisms like competitive learning or the delta rule. Since the non-stationary aspect of signals is more important to speech that steady frequency components, direct representation of frequency change emphasizes functionally relevant aspects of acoustic signals. Finally, the predictive aspect of motion detection should permit sequence tracking to be robust in the complex acoustic environment faced by most animals.

Auditory Motion Layer

Preprocessing

For testing on actual auditory stimuli, input to the model approximates response characteristics of the auditory nerve. These characteristics could be modeled using a model like that studied in

*Supported by the Indiana University Graduate School and the Armed Forces Communications and Electronics Association. This work was also supported by ONR grant N00014-91-J1261 to Robert Port.

(Delgutte, 1982), but are merely simulated here. The important property of the preprocessor is that it consists of an array of linear bandpass filters, each followed by adaptation that leads to rapid ON-type response and then decay to a much lower value. Consequently, the input stimuli used in these simulations consists of sequences of binary-valued patterns sweeping across the input field as though ON-type responses had been filtered through a cutoff-threshold.

Model and Learning

The topology and learning heuristics of the present model are adapted from those presented in (Marshall, 1990) for the processing of visual motion and velocity information. Marshall's model employed the shunting equations studied by Grossberg (1973) and the competitive learning equations outlined in (Carpenter & Grossberg, 1987). In early simulations it was found that the shunting equations proposed by Marshall contain a number of strong linearities that require careful numerical integration and are therefore computationally expensive. These equations were revised in order to make possible the eventual simulation of much larger networks necessary to speech processing over the entire audible spectrum. We found that the essential features of Marshall's model are retained in the current formulation.

The motion detection layer (Figure 1) is a tonotopic layer of inhibitory clusters, the units of which are connected to the units of all other inhibitory clusters in the same layer by excitatory connections having fixed delay. The clusters themselves are on-center off-surround anatomies that emphasize the activation of the unit with greatest activation. Each input line connects to all units in a single cluster corresponding to the receptive field represented by the input line, thus preserving the tonotopic arrangement of the input units.

Initially the lateral excitatory connections between units of the motion detection layer are randomly connection. Over time these connections organize themselves to represent the spatio-temporal correlations present in the input environment. Learning is proportional to the degree to which bottom-up input to a unit coincides with input from units in other clusters. A unit that receives both bottom-up and lateral excitation tends to suppress other units in its cluster and, as a result, learns more strongly than other units in its cluster. Competition between units ensures that the units of each cluster respond to different input patterns.

Unit Equations. The essential attributes of units in the inhibitory clusters are determined by the necessity that all units in a cluster activate in the presence of bottom-up input, and the opposing requirement that combined lateral and bottom-up excitation cause winner-take-all behavior. Moreover, the selection of the most strongly activated unit in a cluster must be rapid, or intermediate activation values will corrupt learning. One can distill

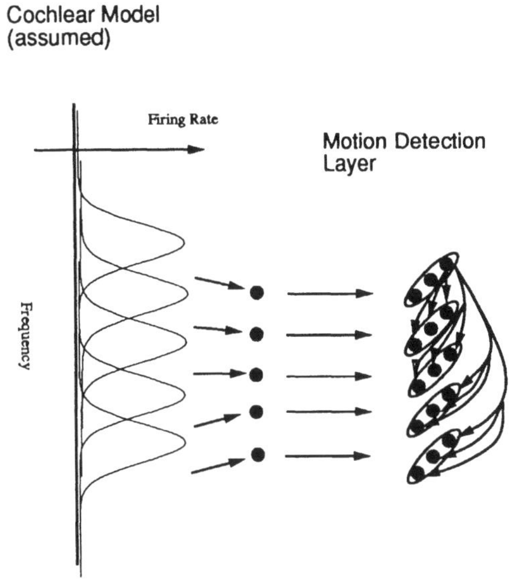

Figure 1: Layer of units responsible for detection of motion. Inhibitory clusters are enclosed by an ellipse. All excitatory connections from a single unit of one cluster are shown.

the essential behavior of Marshall's motion detection model to the following four points:

1. The network must be stable.
2. When input to a unit falls to zero the unit's activation must rapidly decay to zero.
3. At low activation values units in a cluster can be simultaneously active.
4. At high activations the unit having greatest activation rapidly saturates while simultaneously suppressing other units in a cluster.

All units in the motion detection layer obey the equation

$$(1)\quad x_j(t+1) = x_j(t)[1 - \gamma\tau\Delta t] + \tau\Delta t\, f(I_j + x_j(t) + \sum_i^N w_{ij}^+ x_i(t-k) - \sum_i^N w_{ij}^-(1 + x_i(t))^2)$$

where f is the linear threshold function

$$f(z) = \begin{cases} 0 & z \leq 0 \\ z & 0 < z < 1 \\ 1 & z \geq 1 \end{cases}$$

The unit activation function is the Euler approximation to the corresponding differential equation, and discretization is controlled by the value of Δt. The parameters τ and γ are the time constant of a unit and its decay rate, respectively. These parameters are the same for all units. The w_{ij}^+ are excitatory synaptic weights from unit i to j, and the

w_{ij}^- are their inhibitory counterparts. Inhibitory weights were all set to one value as described in the next section. For all simulations reported below the value of the delay along excitatory connections was $k = 10$. Bottom-up connections had a delay of one time step. The use of the linear threshold function ensures boundedness, whereas the faster-than-linear inhibition satisfies conditions 3 and 4 above. At high activation values, the winning unit quickly saturates and suppresses other units, whereas at low activation values all units in a cluster remain active for considerably longer.

Ignoring the nonlinearity f, one can solve for the equilibrium solution of (1).

$$x_j = \frac{I_j + \sum_i^N w_{ij}^+ x_i(t-k) - \sum_i^N w_{ij}^- x_i^2(t)}{(\gamma - 1)}$$

When $I_j = 0$ for all j, the network settles to an equilibrium value of $\vec{x} = 0$. If the lateral excitatory connections are ignored, and the matrix of inhibitory connections is symmetric, then the network converges to its equilibrium (Cohen & Grossberg, 1983). Unless properly chosen, non-zero lateral excitatory connections will introduce positive feedback that can cause all units in the network to permanently saturate. In practice this does not occur because when a connection from one unit to another is large, the corresponding recurrent connection is very small.

Network Initialization. Initially we set all of the inhibitory weights within clusters to $\frac{-1}{N_c}$, where N_c is the number of units per cluster. Excitatory weights between all units outside a cluster favor local connections and were set to

$$w_{ij}^+ = (1.0 + r)e^{(\mu \| Z_j - Z_i \|)}$$

where r is a random variable drawn from a uniform distribution on $[-0.3, 0.3]$. The variable Z_i is the location of the ith unit in the array of units and corresponds to the index of that cluster within the entire layer, thus the third cluster has $Z_i = 3$ for all units i in the third cluster. At present inhibitory connections are not shaped by learning.

Learning Equations. Learning of excitatory weights is Hebbian, and follows Malsburg (1973) in requiring that the sum of all excitatory weights to a unit remain constant over time. Weight normalization implements competition between incoming signals that heavily favors connections between simultaneously active units.

$$\tilde{w}_{ij}^+ = w_{ij}^+ + \epsilon x_i x_j^2$$

$$w_{ij}^+ = E \frac{\tilde{w}_{ij}^+}{\sum_i^N \tilde{w}_{ij}}$$

The network learns on every time cycle. Over time the synaptic weights encode the spatio-temporal correlations that occur when delayed lateral excitation is strongly correlated with bottom-up activation from input to the motion detection layer. Because shorter connections are initially stronger, units are more likely to encode local transition information.

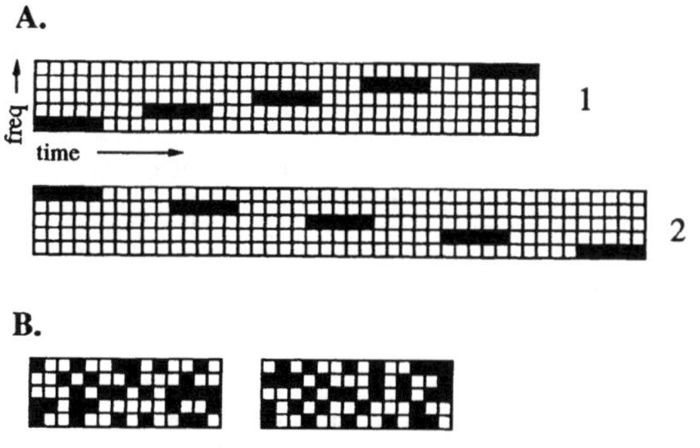

Figure 2: Artificially produced stimuli used to examine self-organization of motion detection. **A.** Frequency modulated "up" and "down" sweeps at two different rates (1/8 and 1/10). **B.** Bursts consisting of random input for limited durations.

Simulations

FM Sweep Stimuli. The self-organizing properties of the model were studied in conditions corresponding to ideal realizations of input from cochlear preprocessing. A network consisting of 10 input units and 10 clusters of 3 motion detection units was exposed to FM sweeps beginning at all 10 of the units (i.e., all ten different frequencies) for 10,000 time steps. Monotonically increasing and decreasing sweeps occurred at 3 different rates (1 frequency step per 8, 9, and 10 time steps). The input for each frequency simulated ON-type cell response by remaining on for 5 time steps and then falling to 0. Spectral representations of some of these stimuli are shown in Figure 2. The inclusion of stimuli that begin at all frequencies enhances learning of units along the edges of the motion detection layer by reinforcing delayed connections from tonotopically near neighbors. If stimuli begin only at the edge of a detection layer, distant connections are most relevant to detection at the other edge and motion is not disambiguated as well for those units. All stimuli were separated by periods of zero input to permit previous activations to decay. In the absence of input most activations decayed after about 3 iterations.

The values of parameters used in all simulations are listed below.

Description	Parameter	Value
discretization	Δt	0.2
unit time const.	τ	3.3
unit decay const.	γ	1.1
sum of weights	E	1.3
lateral delay	k	10
learning rate	ϵ	0.07

As would be expected, those networks exposed to stimuli at a single rate (not shown) develop the most discriminative code. The presence of stimuli at other rates blurs the temporal correlations arriving at successive units in the motion detection layer, but motion detection is quite robust across the three different rates. The output of input and some of the output units is shown in Figure 3 for upward and downward sweeps at the fastest and slowest rates. For the sake of clarity only a subset of the motion detection outputs are shown, although the units not shown here responded similarly. In the figure the 10 input units are shown at the bottom of the graph, while the last 4 clusters of units are grouped and drawn above maintaining their tonotopic relationship. The first 2 sequences exhibit upward FM sweeps, whereas the latter 2 exhibit downward sweeps. Dashed lines have been placed at activation values of 0.7 to permit comparison of unit activities. Consider downward sweep first. Note that units 28-30, the first units to fire for downward sweeps, are stimulated only by bottom-up activation and therefore remain moderately active but do not show winner-take-all behavior. Later, unit 27 of the next cluster of units (25-27) and then unit 20 of (19-21) receive both bottom-up and time-delayed lateral activation and thus go supra-threshold, consistently encoding direction of motion for downward sweeps at all rates. In like manner, unit 25 encodes direction of motion for upward sweeps. Disjoint subsets of units in each cluster learn to encode the two possible directions, though the cluster (22-24) does not respond well to downward sweeps. Finally, note that the response to upward sweeps is both greater in value and longer in duration. This occurs because later firing units receive input from a larger set of coherent motion detection cells already responding to direction of motion.

Bursts. It is extremely important that motion detection learning be robust despite the introduction of noise and constant frequency components, since both types of stimuli are well represented in natural environments. We did not examine constant frequency stimuli, since these involve self-excitation of one cluster and therefore produce no correlation between bottom-up excitation caused by spectral motion and lateral excitation patterns. However, the effect of bursts like those shown in Figure 2, which produce spurious correlations, were simulated. When noise bursts of duration 5, 7, and 10 (random input) were added to the FM task outlined above, the motion detection layer still reliably encoded direction of motion at all rates.

Discussion

There is an important relationship between stimulus duration and the constant k that determines the duration of transmission delay. If stimulus duration approaches the value of k, distant units may be simultaneously active, leading to ambiguity in the direction of motion. This can cause incorrect learning or, worse, the development of weights that cause some units of the network to permanently saturate. Thus, stimulus duration must be sufficient to permit competition between units in a cluster, but must not be so great as to cause too many simultaneously active units in the motion detection layer.

It is instructive to compare this formulation of feature processing with that which is implied by bottom-up time-delay systems. If one were to assume that motion sensitivity of the sort advocated in this report were founded on bottom-up time delays, the motion detection layer would necessarily have to map dissimilar inputs to the same output (See Figure 4.) This problem arises because as a time-delayed pattern sweeps across L2, its manifestations at different points in time are entirely unrelated. In Figure 4 the same input pattern at two successive points in time is labelled $P(t-1)$ and $P(t)$. As Rumelhart and McClelland (1986) note, solutions to this problem can be found by incorporating a hidden layer of units. Unfortunately, in this case a hidden layer of units leads to a very abstract, non-tonotopic code for motion that is not easily learned without some form of supervision. These problems are overcome in a very simple manner if bottom-up time delays are replaced by lateral delays that permit the learning of spatiotemporal correlations.

This report shows how the shunting equations used by Marshall (1990) can be reformulated and combined with a different learning rule to endow a network layer with the ability to encode direction of motion. The motion detectors arise as chains of active units in response to statistical regularities that would occur over a 1-dimensional tonotopic array of units with receptive fields limited to a small band of frequencies. Members of the chains of motion detectors that arise for more rapid spectral patterns continue to encode direction of motion, although the code becomes spatially and temporally sparse. From the standpoint of local computational constraints, detection of auditory spectral motion provides a means for discriminating two patterns that may well excite the same group of neurons on the basis of direction of motion. Output from the motion detection network can then be interpreted by networks that learn spatial patterns, leading to more general sequential pattern recognition.

References

Aitkin, L. 1990. *Information Processing in Mammalian Auditory and Tactile Systems.* New York: Wiley.

Carpenter, G. and Grossberg, S. 1987. A massively parallel architecture for a self-organizing neu-

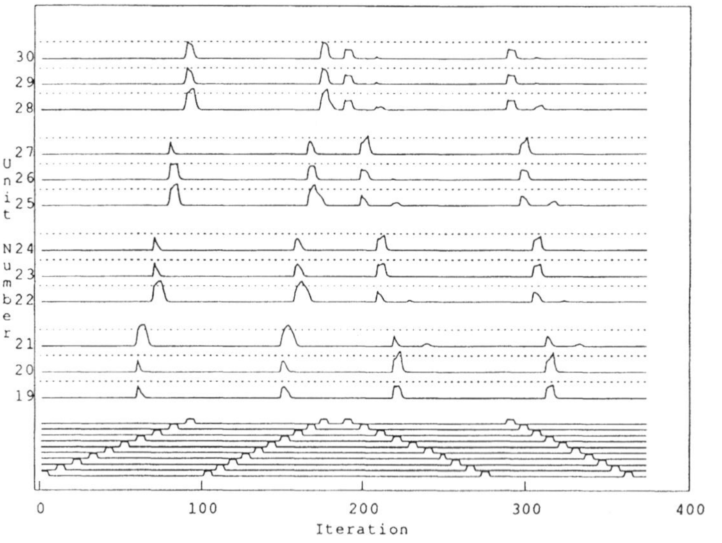

Figure 3: Unit activations over time during presentation of 4 FM sweep stimuli. The unit number of each unit in the motion detection layer is listed at the left of the graph. The first two stimuli are progressively faster upward sweeps at the rates 1/10 and 1/8; the last three stimuli show response to downward sweeps in the same order. Dashed lines indicate an activation of 0.7. Responses below this value are considered sub-threshold. Note that for each cluster, units that win the competition for one upward sweep also win for all other upward sweeps. (Input stimuli have been scaled for purposes of illustration.)

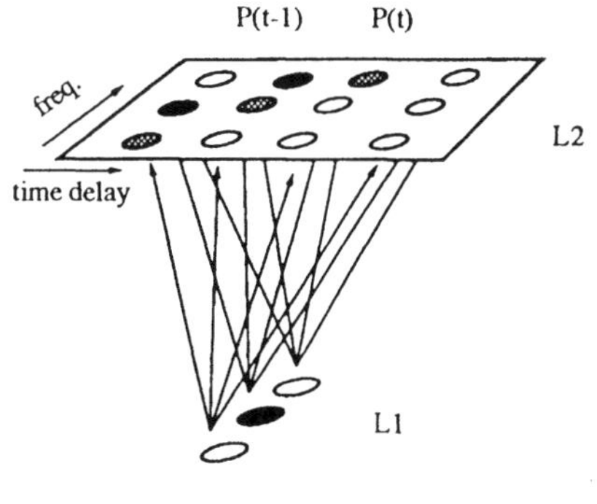

Figure 4: Pattern of activations as a pattern sweeps across L1 and leaves a time-delay trace across layer L2. Activations shown in L2 are the superimposition of two discrete time steps; the activations arising from motion of a single pattern are shown in two shades to indicate that they are not simultaneous.

ral pattern recognition machine. *Computer Vision, Graphics and Image Processing*, 37:54–115.

Cohen, M. and Grossberg, S. 1983. Absolute stability of global pattern formation and parallel memory storage by competitive neural networks. *IEEE Transactions on Systems, Man, and Cybernetics*, SMC-13:813–825.

Delgutte, B. 1982. Some correlates of phonetic distinctions at the level of the auditory nerve. In Carlson, R. and Granstrom, B., editors, *The Representation of Speech in the Peripheral Auditory System*, pages 131–149. Elsevier Biomedical Press.

Delgutte, B. 1986. Analysis of french stop consonants using a model of the peripheral auditory system. In Perkell, J. and Klatt, D., editors, *Invariance and Variability in Speech Processes*. Hillsdale, New Jersey: Erlbaum Associates.

Elman, J. 1990. Finding structure in time. *Cognitive Science*, 14:179–211.

Földiák, P. 1991. Learning invariance from transformation sequences. *Neural Computation*, 3:194–200.

Grossberg, S. 1973. Contour enhancement, short term memory, and constancies in reverberating neural networks. *Studies in Applied Mathematics*, 52:217–257.

Kohonen, T. 1984. *Self-Organization and Associative Memory*. New York: Springer-Verlag.

Malsburg, C. 1973. Self-organization of orientation sensitive cells in the striate cortex. *Kybernetik*, 14:85–100.

Marshall, J. 1990. Self-organizing neural networks for perception of visual motion. *Neural Networks*, 3:45–74.

Metzger, Y. and Lehmann, D. 1990. Learning temporal sequences by local synaptic changes. *Network*, 1:169–188.

Pickles, J. O. 1988. *An Introduction to The Physiology of Hearing*. New York: Academic Press. Second Edition.

Reiss, M. and Taylor, J. 1991. Storing temporal sequences. *Neural Networks*, 4:773–787.

Rumelhart, D. and McClelland, J. 1986. *Parallel Distributed Processing: Explorations in the Microstructure of cognition*, volume 1. Cambridge, Massachusetts: MIT Press.

Suga, N. 1990. Cortical computational maps for auditory imaging. *Neural Networks*, 3:3–21.

Wang, D. L. and Arbib, M. 1990. Complex temporal sequence learning based on short-term memory. *Proceedings of the IEEE*, 78:1536–1542.

Whitfield, I. C. and Evans, E. F. 1965. Responses of auditory cortical neurons to stimuli of changing frequency. *Journal of Neurophysiology*, 28:655–672.

Simple+Robust = Pragmatic : A Natural Language Query Processing Model for Card-type Databases

Seigou Arita   Hideo Shimazu   Yosuke Takashima

C&C Information Technology Research Laboratories
NEC Corporation
1-1,Miyazaki 4-chome, Miyamae-ku,Kawasaki,Kanagawa 216 Japan
arita%joke.cl.nec.co.jp@uunet.uu.net

Abstract

Real users' queries to databases written in their natural language tend to be extra-grammatical, erroneous and, sometimes just a sequence of keywords. Since most conventional natural language interfaces are *seminatural*, they cannot treat such real queries very well. This paper proposes a new *natural* language query interpretation model, named SIMPLA. Because the model has a keyword-based parsing mechanism, it is very robust to cope with extra-grammatical sentences. The strong keyword-based parsing capability is very dependent upon its target database's being a "card"-type. SIMPLA provides several operators to define peripheral knowledge, regarding the target database. Such peripheral knowledge is stored *virtually* in parts of the target "card"-type database. Since the target database with the peripheral knowledge remains "card"-type, SIMPLA does not decrease its robust natural language processing capability, while it embodies the ability to respond to questions concerning peripheral questions.

1. Introduction

As the number of commercial databases increases, the expectations of ordinary people, with regard to *pragmatic* natural language (NL) interface to databases, are increasing. Though many research efforts on user interfaces, including menu systems and friendly command language, are running, they are still *unnatural* to ordinary people.

Many research efforts on NL interfaces have also been implemented for more than 20 years (Winograd, 1977)(Simmons, 1970)(Hendrix et al., 1978)(Tennant, 1981). Some of them, like INTELLECT (Shneiderman, 1987), have been used in real business applications. However, they still have only *seminatural* language processing capabilities. Most conventional NL interface systems cannot treat queries that are written really *naturally*. Real users' queries to databases, written in their natural language, tend to be extra-grammatical, erroneous and, sometimes just a sequence of keywords. A few research efforts (Carbonell et al., 1984) have been struggling with extra-grammaticality. However, they have not yet succeeded in coping with such real queries.

This paper proposes a new *natural* language query interpretation model. Because the model has a keyword-based parsing mechanism, it is very robust to cope with extra-grammatical sentences. The proposed natural language interface model, **SIMPLA**(SIMPle Language Analyzer), has the following characteristics:

- Parsing is keyword-based:
  SIMPLA extracts only keywords from an input sentence, and generates its interpretation from the extracted keywords. Therefore, the parsing is very robust to extra-grammatical expressions. Even a sequence of keywords, as an input sentence, can be correctly interpreted.

- Operators are provided to define peripheral knowledge and to put it into a target "card"-type database virtually:
  SIMPLA can retrieve appropriate data from its target "card"-type database, as the response to a natural language query. NL interface must respond to queries which are not just a direct data retrieval of the target database, but are concerning questions. To interpret and reply

to such extended questions, the NL interface must hold peripheral knowledge for the target database. In SIMPLA, several operators are provided to define such knowledge. However, the strong capability of SIMPLA's keyword-based parsing is very dependent upon its target database's being "card"-type. So, SIMPLA holds such knowledge as if it were placed in a part of the target "card"-type database. It looks like the target databases are extended, but the extension is *virtual*. The authors call the extended database a **virtual database**. Since the form of the target database remains "card"-type, SIMPLA does not decrease its robust natural language processing capability, while SIMPLA embodies the ability to respond to peripheral questions.

Section 2 shows an example that provides the concept regarding how SIMPLA processes a query. Section 3 describes notions of label base and vocabulary space. Section 4 illustrates the algorithm for interpreting queries. Section 5 explains knowledge representation for domain-oriented thesaurus through the virtual extension of target databases.

2. Basic SIMPLA Idea

SIMPLA's interpretation mechanism is novel. SIMPLA analyzes queries using much simpler grammar than that involved in conventional linguistic approaches.

First, look at the process for interpreting the following sample query towards a sample database "world handbook", shown in Figure 1. It provides the concept regarding how SIMPLA interprets queries (Arita et al, 1991).

Nation	Area	Population	Capital	Language
Korea	99016	42380	Seoul	Korean
Canada	9976139	26250	Ottawa	English,French
Japan	377801	123120	Tokyo	Japanese

Figure 1: World handbook

S1: "Where is the capital of Canada?"

When S1 is given to SIMPLA, it extracts only keywords from the sentence. Here, the keywords are "capital" and "Canada". "Capital" is an attribute name. "Canada" is a value of the "nation" attribute in a record. SIMPLA constructs the meaning of this sentence, using only these bits of information. SIMPLA searches the target database for the records whose "nation" attribute has the value, "Canada", then, gets the value of the "capital" attribute for the extracted records. In this case, a record, which includes the attribute value "Canada", is the second record. A value for the attribute name "capital" for the second record is "Ottawa". So, "Ottawa" is an output.

Like the above example, most of those queries fall into the SELECT-FROM-WHERE type queries in SQL. SIMPLA regards all the queries as SELECT-FROM-WHERE instructions.

Thus, the process has been implemented in a keyword based manner. The above process is tolerant to extra-grammatical queries. The algorithm is far more easily implemented than, the ones using conventional linguistic approaches.

SIMPLA's actual mechanism is more complex. It includes the notion of a virtual database. The virtual database is a virtually extended target database with thesaurus. First, SIMPLA analyzes an input query as a query targeted to the virtual database, and generates its internal representation. Then, it translates the representation to the real query command to the actual database management system.

3. Label Base And Vocabulary Space

SIMPLA regards a query as a sequence of *descriptions on a relationship between an attribute name and its value*. For instance, the S1 sentence, "Where is the capital of Canada" are regarded as a sequence of descriptions: (1) "capital" is an attribute name with no attribute value assigned in the query. (2) "Canada" is equal to the value of "nation". On this semantic, which pairs of an attribute name and its value are existing in a target database form the most basic information. In SIMPLA, such information is prepared in **label base** and **vocabulary space**.

Label base is a set of **basic pairs** of binary relation:

```
basic_pair(Attribute, Value).
```

Here, "Attribute" and "Value" indicate an attribute name and an attribute value in a target database, respectively. Vocabulary space is a set of **vocabulary pairs** of binary relation :

```
vocabulary_pair(VirtualAttribute,
VirtualValue).
```

Here, "VirtualAttribute" and "VirtualValue" indicate an attribute name and an attribute value in a virtual database, respectively. Correspondences between basic pairs in label base and vocabulary pairs in vocabulary space are given in **anchor**. For example, label base, vocabulary space and anchor of the "world handbook" in Figure 1 are shown in Figure 2 and in Figure 4, respectively.

Virtual database is an image of the database, which is seen by SIMPLA, with vocabulary space in place of label base (Figure 5). SIMPLA's parser refers to vocabulary space in order to determine that an input word is either an attribute name or an attribute value. An attribute name and an attribute value respectively appear as a left component and a right component in a vocabulary pair in vocabulary space. This is the formal definition for an attribute name and an attribute value in SIMPLA:

```
attribute_name(AttributeName)
:- vocabulary_pair(AttributeName, _).

attribute_value(AttributeValue)
:- vocabulary_pair(_,AttributeValue).
```

For example, with vocabulary space in Figure 4 in place of label base in Figure 2, the target database seems to have "east/west" and "population density" attributes.
Because, the `attribute_name('east/west')` and `attribute_name('population density')` hold in that definition. So, the database appears to have more attributes than really existing ones ,as "virtual world handbook", shown in Figure 3.

4. SIMPLA's Structure

SIMPLA's structure is shown in Figure 6. First, Parser translates the user's query into **virtual form**. Second, Realizer translates the virtual form to **real form**. Finally, Generator generates a raw retrieval form from the given real form , which is executable by the target database system.
Both virtual form and real form are sequences of **units**. Unit shows the relation between an attribute name and value.

Parser

Parser translates a query to a virtual form, with reference to vocabulary space, as follows: (1). To extract a sequence of relational words, attribute names and attribute values from the query through reference to vocabulary space. (2). To apply simple

(nation,Korea) (nation,Japan) (nation,Canada)
(area,99016) (area,9976139) (area,377801)
(population,42380) (population,26250)
(population,123120) (capital,Seoul) (capital,Ottawa)
(capital,Tokyo) (language,Korean) (language,English)
(language,French) (language,Japanese)

Figure 2: Label base

Nation	East/west	Area	Population	Pop.density	Capital	Language
Korea	East	99016	42380	0.428	Seoul	Korean
Canada	West	9976139	26250	0.003	Ottawa	English,French
Japan	East	377801	123120	0.326	Tokyo	Japanese

Figure 3: Virtual world handbook

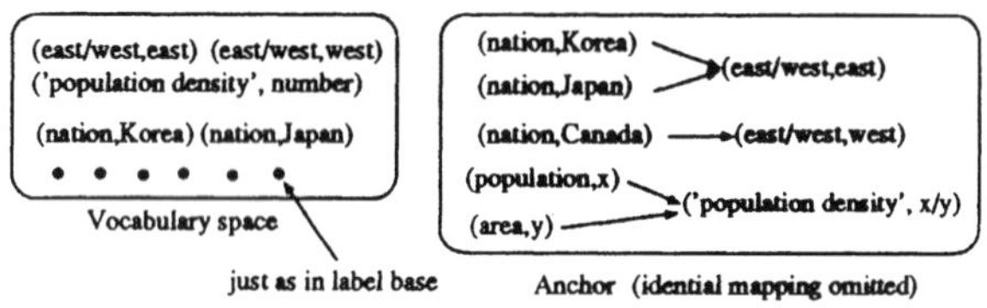

Figure 4: Vocabulary space and anchor

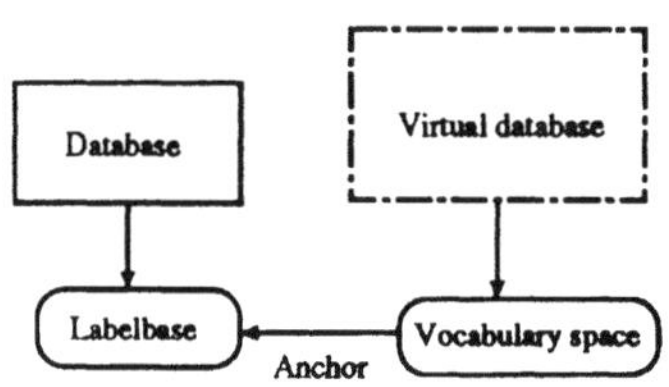

Figure 5: Virtualization by vocabulary space and anchor

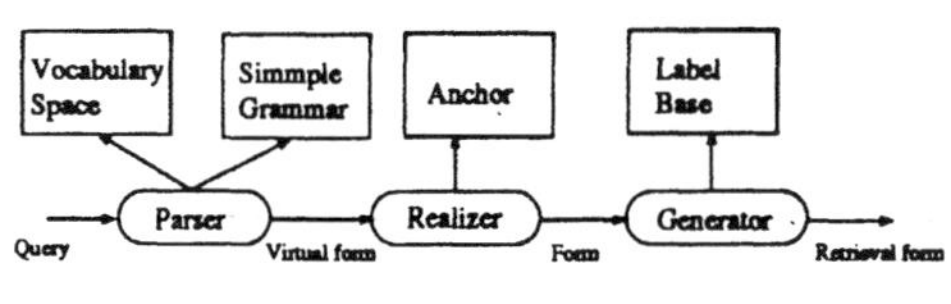

Figure 6: SIMPLA's structure

```
parse([]) --> [].
parse([Unit | Units]) --> gen_unit(Unit), parse(Units).

gen_unit(unit(Type,Name,Value)) -->
    [attribute_name(Name),attribute_value(Value),relational(Type,R)],
    {vocabulary_pair(Name,Value)}.   (1)
gen_unit(unit(Type,Name,Value)) -->
    [attribute_value(Value),relational(Type,R),attribute_name(Name)],
    {vocabulary_pair(Name,Value)}.   (2)
gen_unit(unit(eq,Name,Value))    -->
    [attribute_name(Name),attribute_value(Value)],
    {vocabulary_pair(Name,Value)}.   (3)
gen_unit(unit(eq,Name,Value))    -->
    [attribute_value(Value),attribute_name(Name)],
    {vocabulary_pair(Name,Value)}.   (4)
gen_unit(unit(eq,Name,Value))    -->
    [attribute_value(Value)],
    {vocabulary_pair(Name,Value)}.   (5)
gen_unit(unit(eq,Name,_))    -->
    [attribute_name(Name)],
    {vocabulary_pair(Name,_)}.       (6)
```

Figure 7: SIMPLA's simple grammar

grammar to the sequence in order to obtain the virtual form. **Relational word** describes a relation between an attribute name and its vale.

```
relational(gtr, ''greater than'').
relational(sml, ''smaller than'').
...
```

The simple grammar is a set of rules used to transform a sequence of relational words, attribute names and attribute values into a virtual form. Some of these rules are shown in Figure 7, in DCG style.

For instance, gen_unit (1) means: if attribute name `Name`, attribute value `Value` and relational word `R` for category `Type` appear at the top of the sequence and if `(Name, Value)` appears in the vocabulary space as a vocabulary pair, then generate `unit(Type, Name, Value)`. The rest of the sequence are processed in the same way. The number of applied rules are equal to the number of generated units. Parser's output is virtual form, which is a sequence of those units.

For instance, look at the process of parsing the following query:

S2:"List languages in the East."

First, the query is filtered with vocabulary space into a sequence:

```
[attribute_name(language),
attribute_value(east)].
```

Second, the simple grammar is applied to the sequence. The gen_unit (6) matches the top of the sequence:

```
gen_unit(virtual_unit(eq, language, _))
← [attribute_name(language)]
```

and vocabulary pair (language, _) belongs to the vocabulary space [1]. So, unit `unit(eq, language, _)` is generated. Similarly, by gen_unit (5) is applied to the rest of the sequence `[attribute_value(east)]`, unit `unit(eq, east/west, east)` is generated. Thus, Parser outputs virtual form:

```
[unit(eq, language, _), unit(eq,
east/west, east)].
```

Realizer

Since a virtual form may include "virtual" attribute names or values (as "east/west" in the above example), virtual form must be realized to real form by Realizer. Realization of virtual form to real form is accomplished by realizing each unit in virtual form. The unit realization is driven by anchor, which is a map between basic pairs in label base and vocabulary pairs in vocabulary space. For instance, the virtual form [unit(eq, language, _), unit(eq, east/west, east)] is realized to a real form [unit(eq, language, _), unit(eq, nation, korea), unit(eq, nation, japan)], because anchor in Figure 4 holds a mapping

[1] "_" indicates an uninstantiated variable, just as in Prolog

Anchor: (nation, Korea), (nation, Japan)
→ (east/west, East)

Generator

Finally, Generator translates real form into a retrieval form, which is executable by the target database system. Unit in real form generates Select-clause, if its attribute value remains an uninstantiated variable. Other units in real form generate Where-clause. For instance, real form [unit(eq, language, _), unit(eq, nation, korea), unit(eq, nation, japan)] is translated into the retrieval form "SELECT language FROM world_handbook WHERE nation = "Korea" OR nation = "Japan" " , if represented in SQL.

SIMPLA doesn't completely interpret all queries correctly. However, with emphasis on practical use, SIMPLA's target is to interpret most practical queries immediately and robustly, avoiding make the model too naive and too large regarding the cost for processing very complicated and unusual queries.

5. Thesaurus

This section describes how SIMPLA virtualizes a database. Virtualizing a database is just generating vocabulary space and anchor associated to the database. SIMPLA provides several operators to define a schema for a virtual database, as the extension of an original database. These operators are:

Name index operator
`name_index(Attribute, AttributeName)`. To name an attribute **`Attribute`** in a database as **`AttributeName`**.

Value index operator
`value_index(AttributeName, Value, ValueName)`. To name a value `Value` for an attribute **`AttributeName`** as a **`ValueName`**.

Grouping operator
`grouping(A, {V1,...,Vn}, NewA, NewV)`. To register a set of some attribute values {`V1`, ..., `Vn`} for an attribute name `A` as a new attribute value **`NewV`** for a (new) attribute name **`NewA`**.

Compound operator
`compound(A, {A1, ..., An}, ψ)`. To define a new attribute `A` using already defined attributes `A1`, ..., `An` in a database, where ψ is an expression that defines `A`.

Using a name index operator and a value index operator, the natural language interface designer can assign natural language fragments to each corresponding bit of data in the database. By using a grouping operator, the designer can align the database attributes into a hierarchy. By a compound operator, the designer can form associations for the attributes. For example, regarding the "World handbook" database(Figure 1), the operators in Figure 8 can be implemented, so that the virtual database "Virtual world handbook"(Figure 3) is obtained.

```
name_index(Any, Any).                                              (10)
value_index(AttributeName, AnyValue, AnyValue).                    (11)
group(nation, {korea, japan}, 'east/west', east).                  (12)
group(nation, {canada}, 'east/west', west).                        (13)
compound('population density', {population, density}, /).         (14)
```

Figure 8: The "virtual world handbook" definition

Virtualizing Database by Vocabulary Space

Here, virtualizing operators, defined in the previous section, are implemented as operations on vocabulary space.

First, label base is constructed using name index operator and value index operator. In the "World handbook" case, operators (10) and (11) have been applied, in Fig8. Because they specify nothing special, label base consists of all pairs of attribute names and attribute values. That is, the label base for the "World handbook" is as shown in Figure 2. An initial state of vocabulary space is equal to label base, when the anchor is trivial.

Anchor: vocabulary_pair(A, V)
→ basic_pair(A, V)

The grouping operator and the compound operator modify vocabulary space and anchor.

The grouping operator **`grouping(A, {V1,...,Vn}, NewA, NewV)`** operates on vocabulary space and anchor, as follows: (1). To add vocabulary pair (**`NewA`**, **`NewV`**) to vocabulary space VS. (2). To define the image for anchor *Anchor* versus the above vocabulary pair , as follows:

Anchor: vocabulary_pair(NewA, NewV)
→ basic_pair(A, $\{V_1, \ldots, V_n\}$*)*

(The right-hand side means that attribute name A takes any one of $V_1, \ldots, V_n$ as a value).

In "World handbook" case, operator (12) **grouping(nation, {korea, japan}, east/west, east)** is applied in Figure 8. It adds vocabulary pair (east/west, east) to the vocabulary space VS. The anchor image for this pair is as follows:

Anchor: vocabulary_pair(east/west, east)
→ basic_pair(nation, {korea, japan})

The compound operator **compound(A, {A1, ..., An}, ψ)** operates on vocabulary space and anchor as follows: (1). To add vocabulary pair (**A, X**) having the first argument **A** as an attribute name and a variable **X** as an attribute value to vocabulary space VS. (2). To define the image for anchor *Anchor* versus the above vocabulary pair , by the rest of the arguments as follows:

Anchor: vocabulary_pair(A, X)
→ basic_pair($\psi(A_1, \ldots, A_n)$, X)

In "World handbook" case, operator (14) **compound('population density', {population, density}, /)** is applied in Figure 8. It adds vocabulary pair ("population density", X) to vocabulary space VS. The anchor image for this pair is as follows:

Anchor: vocabulary_pair('population density', X)
→ basic_pair(population/density, X)

After processing all operators displayed in Figure 8, SIMPLA gets vocabulary space and anchor, as illustrated in Figure 4.

Conclusion

In this paper, the authors described SIMPLA, a new natural language query processing model, and its implementation. SIMPLA has a very simple parsing mechanism augmented with the notion of a virtual database. SIMPLA can interpret extragrammatical sentences as well as ordinary natural language sentences. It even accepts just a sequence of keywords, as an input sentence. Therefore, SIMPLA can be placed among conventional natural language processing model, command language interpreter, and keyword-based information retrieval model.

Of course, the linguistic capability for SIMPLA is not as strong as conventional NL systems, which accept predicted *seminatural* language descriptions. However, in the real applications, SIMPLA's hybrid interpretation ability is more *pragmatic* for conventional natural language processing model, command language interpreter, and keyword-based information retrieval model. Users can generate queries in multi-style. If a user can not generate a natural language query, which SIMPLA can interpret correctly, he/she just makes a sequence of keywords, instead. SIMPLA may accept the sequence appropriately.

The SIMPLA's response time is faster than that for a conventional NL interface, because SIMPLA's processing work load is far lighter, compared with the fully parsing approach. According to early experimental results, the response time for a query to a database, whose size is more than 2500 records, was less than two seconds, implemented on Quintus-Prolog, on Sparc-station-1.

Giving up an attempt to process very complicated and unusual queries, SIMPLA has gained robust ability to interpret most of simple queries immediately and correctly. Now, to interpret even those unusual queries efficiently, the authors are trying to combine Case-based method with this approach (Shimazu et al, 1991)(Shimazu et al, 1992).

References

ARITA, S.; SHIMAZU, H.; and TAKASHIMA, Y. 1991. Simple Natural Language Interface Model, In Proc. of the 5th Annual Conference of JSAI.

CARBONELL, J.G.; and HAYES, P.J. 1984. Recovery strategies for parsing extragrammtical language, Technical Report CMU-CS-84-107, Dept. of Computer Science, CMU.

HENDRIX, G.G.; SACERDOTI, E.D.; SAGALOWICZ, D., and SLOCUM, J., 1978. Developing a Natural Language Interface to Complex Data, In ACM Trans. on Database Systems.

SHIMAZU,H.; ARITA,S.; and TAKASHIMA,Y. 1992. Design Tool Combining Keyword Analyzer and Case-based Parser for Developing Natural Language Database Interfaces, Proc. of COLING-92.

SHIMAZU,H.; and TAKASHIMA,Y. 1991. Acquiring Knowledge for Natural Language Interpretation Based On Corpus Analysis, Proc. of IJCAI-91 Natural Language Learning Workshop.

SHNEIDERMAN, B. 1987. Designing the User Interface, Addison-Wesley Pub.

SIMMONS, R.F. 1970. Natural Language Question-Answering Systems, CACM, Vol. 13, Jan.

TENNANT, H. 1981. Natural Language Processing, Petrocelli Books.

WINOGRAD, T. 1977. Understanding natural Language, Academic Press.

Integrating Reactivity, Goals, and Emotion in a Broad Agent

Joseph Bates and **A. Bryan Loyall** and **W. Scott Reilly**
School of Computer Science
Carnegie Mellon University
Pittsburgh, PA 15213

Abstract

Researchers studying autonomous agents are increasingly examining the problem of integrating multiple capabilities into single agents. The Oz project is developing technology for dramatic, interactive, simulated worlds. One requirement of such worlds is the presence of broad, though perhaps shallow, agents. To support our needs, we are developing an agent architecture, called Tok, that displays reactivity, goal-directed behavior, and emotion, along with other capabilities.

Integrating the components of Tok into a coherent whole raises issues of how the parts interact, and seems to place constraints on the nature of each component. Here we describe briefly the integration issues we have encountered in building a particular Tok agent (Lyotard the cat), note their impact on the architecture, and suggest that modeling emotion, in particular, may constrain the design of integrated agent architectures.

Broad Agents

The Oz project [Bates, 1992] at Carnegie Mellon is developing technology for artistically interesting, highly interactive, simulated worlds. We want to give users the experience of living in (not merely watching) dramatically rich worlds that include moderately competent, emotional agents.

An Oz world has four primary components. There is a simulated physical environment, a set of automated agents which help populate the world, a user interface to allow one or more people to participate in the world [Kantrowitz and Bates, 1992], and a two-player adversary search planner concerned with the long term structure of the user's experience [Bates, 1990]. Oz shares some goals with traditional story generation systems [Meehan, 1976, Lebowitz, 1985], but adds the significant requirement of rich interactivity.

One of the keys to an artistically engaging experience is for the user to be able to "suspend disbelief". That is, the user must be able to imagine that the world portrayed is real, without being jarred out of this belief by the world's behavior. The automated agents, in particular, mustn't be blatantly unreal. Thus, part of our effort is aimed at producing agents with a broad set of capabilities, including goal-directed reactive behavior, emotional state and behavior, and some natural language abilities. For our purpose, each of these capacities may be as shallow as necessary to allow us to build broad, integrated agents [Bates *et al.*, 1991].

Oz worlds are far simpler than the real world, but they must retain sufficient complexity to serve as interesting artistic vehicles. The complexity level is somewhat higher, but not exceptionally higher, than typical AI micro-worlds. Despite these simplifications, we find that our agents must deal with imprecise and erroneous perceptions, with the need to respond rapidly, and with a general inability to fully model the agent-rich world they inhabit. We suspect that some of our experience with broad agents in Oz may transfer to other domains, such as social, real-world robots.

Building broad agents is a little studied area. Much work has been done on building reactive systems [Brooks, 1987, Georgeff *et al.*, 1987, Firby, 1989, Simmons, 1991], natural language systems, and even emotion systems [Dyer, 1983, Ortony *et al.*, 1988, Mueller, 1990]. There is growing interest in integrating action and learning (see [Laird, 1991]) and some very interesting work on broader integration [Vere and Bickmore, 1990, Newell, 1990]. However, we are aware of no other efforts to integrate the particularly wide range of capabilities needed in the Oz domain. Here we present our efforts, focusing on integration mechanisms and their impact on components of the architecture.

Tok and Lyotard

In analyzing our task domain, we concluded that the capabilities needed in our initial agents are perception, reactivity, goal-directed behavior, emotion, social behavior, natural language analysis, and natural language generation. Our agent architecture, Tok, partially (but not fully) partitions these tasks into several communicating components. Low-level perception is handled by the Sensory Routines and the Integrated Sense Model. Reactivity and goal-directed behavior are handled by Hap [Loyall and Bates, 1991]. Emotion

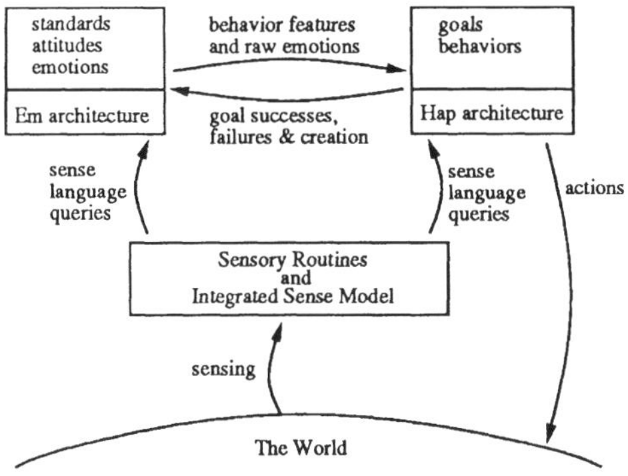

Figure 1: Tok Architecture

Emotions	Behaviors		Features
hope†	wanting to be pet	purring	curious
fear	chasing ball/*creatures*	arch back	content
happy	wanting to go out/in	hiss	aggressive
sad	*pouncing on creatures*	swat	ignoring
pride	wanting to eat	bite	friendly
shame	getting object	run away	*proud*
admiration	(using human)	have fun	energetic
reproach	searching for thing	eating	
gratification	cleaning self	*crazy hour*	
remorse	playing with ball	rubbing	
gratitude	playing with mouse	licking	
anger	*carrying mouse*	watching	
love	hiding (anger/fear)	sit in sun	
hate	pushing things around		

†*italicized items were not included in final implementation*

Table 1: Original Lyotard Task

and social relationships are the domain of Em [Reilly and Bates, 1992]. Language analysis and generation are performed by Gump and Glinda [Kantrowitz, 1990, Kantrowitz and Bates, 1992]. Figure 1 shows how these components, excluding Glinda and Gump, are connected to form Tok.

In the remainder of this section we describe the components of Tok in just enough detail to allow discussion of the integration issues. For a more complete description see [Bates *et al.*, 1992]. We illustrate the description using examples from an existing Tok agent, a simulated house cat named "Lyotard", which exercises most of the capabilities of the architecture.

Table 1 lists the emotions and behaviors from our original informal design document for Lyotard. Our goal in developing Lyotard was to build a creature that could believably pass for a cat in an Oz micro-world. The emotions shown are those naturally available in the current version of Em, though in the end we did not use all of them. The behaviors were developed over several hours by the cat owners in our group. The behavioral features are used to modify the style of particular behaviors. They are usually derived from Lyotard's emotional state, though they also can be directly adjusted by behaviors.

The Simulated World and Perception

The Oz physical world is an object-oriented simulation. Agents sense the world via sense data objects which propagate from the item sensed through the world to the agents. Each sense datum describes the thing sensed as a collection of property/value pairs. Unique names are not used to identify objects; agents must derive identity from other properties. Sense data can be transformed as they travel. For example speech behind a closed door can be muffled. In general, the sense data available to an agent can be incomplete, incorrect, or absent. Agents perform actions by invoking methods on appropriate sets of objects. These methods may alter the world, propagate sense data, and succeed or fail.

Each Tok agent runs by executing a three step loop: sense, think, act. During each sense phase a snapshot of the perceivable world is sensed and the data is recorded in the sensory routines. These snapshots are time-stamped and retained. An attempt is then made to merge them into the Integrated Sense Model (ISM), which maintains the agent's best guess about the physical structure of the whole world. The continuously updated information in the sensory routines and the longer term, approximate model in the ISM are routinely queried when choosing actions or updating the emotional state of Lyotard.

Action (Hap)

Hap is Tok's goal-directed, reactive action engine [Loyall and Bates, 1991]. It continuously chooses the agent's next action based on perception, current goals, emotional state, behavioral features and other aspects of internal state. Goals in Hap contain an atomic name and a set of parameters which are instantiated when the goal becomes active, for example `(goto <object>)`. Goals do not characterize world states to accomplish, and Hap does no explicit planning. Instead, sets of actions (which we nonetheless call "plans") are chosen from an unchanging plan library which may contain one or more plans for each goal. These plans are either ordered or unordered collections of subgoals and actions which can be used to accomplish the invoking goal. Multiple plans can be written for a given goal, distingiushed in part by a testable precondition. If a plan fails, Hap will attempt any alternate plans for the given goal, and thus perform a kind of backtracking search in the real world.

Hap stores all active goals and plans in a hierarchical structure called the active plan tree (APT). There are various annotations in the APT to support reactivity and the management of multiple goals. Two important annotations are *context conditions* and *success tests*. Both of these are arbitrary testable expressions over the perceived state of the world and other aspects of internal state. Success tests may be associated with selected goals in the APT. When a success test is true, its associated goal is deemed to have been accomplished and thus no longer needs to be pursued. This can happen before the goal is attempted in which case it is skipped or can happen during execution of the goal in which case it is aborted.

Similarly, context conditions may be associated with APT plans. When a context condition becomes false, its associated plan is deemed no longer applicable in the current state of the world. That plan fails and is removed from the tree along with any executing subgoals. The parent goal then choses a new plan or fails.

Hap executes by first modifying the APT based on changes in the world. Goals whose success test is true and plans whose context condition is false are removed along with any subordinate subgoals or plans. Next one of the leaf goals is chosen. If the chosen goal is a primitive action, it is executed. Otherwise, the plan library is indexed and the plan arbiter chooses a plan for this goal from among those whose preconditions are true. The plan arbiter will not choose plans which have already failed, and prefers more specific plans over less specific ones. At this point the execution loop repeats.

Emotion (Em)

Em models selected emotional and social aspects of the agent. It is based on ideas of Ortony et al. [Ortony *et al.*, 1988]. Like that work, Em develops emotions from a cognitive base: external events are compared with goals, actions are compared with standards, and objects are compared with attitudes. Most of Em's possible emotions are shown in Table 1. We describe them very briefly here, but see [Reilly and Bates, 1992] for details.

Happiness and sadness occur when the agent's goals succeed or fail. Hope and fear occur when Em believes that there is some chance of a goal succeeding or failing.

Pride, shame, reproach, and admiration arise when an action is either approved or disapproved. These judgments are made according to the agent's *standards*, which represent moral beliefs and personal standards of performance. Pride and shame occur when the agent itself performs the action; admiration and reproach develop in response to others' actions.

Some emotions are combinations of other emotions. Lyotard doesn't like to be touched when he's in the wrong mood and doing so will cause him sadness and reproach. These give rise to the composite emotion of anger at whomever pet him. Similarly, gratitude is a composite of happiness and admiration, remorse is sadness and shame, and gratification is happiness and pride.

Finally, love and hate arise from noticing objects toward which the agent has positive or negative *attitudes*. In Lyotard we use attitudes to model the human-cat social relationship. Lyotard initially dislikes the user, a negative attitude, and this attitude varies as the user does things to make Lyotard angry or grateful. As this attitude changes, so may his resulting love or hate emotions.

Emotions fade with time, but attitudes and standards are fairly stable. An agent will feel love when close to someone liked. This fades if the other agent leaves, but the attitude toward that agent remains relatively stable.

Component Integration

The arcs in Figure 1 denote the communication paths between the parts of Tok. The main interactions are Hap and Em querying perceptual information, Hap querying both the emotional state and behavioral features derived from the emotional state, and Em receiving notification of the creation, failure, and success of goals. Here we discuss only the communication between Hap and Em.

Hap's Communication with Em

As Hap runs, its active plan tree changes. These changes include goal creation and goal removal due to success or failure. As these events occur, Hap informs Em of what goals were affected, how they were affected, and the degree of importance that the agent builder associated with each of the goals. Em then uses this information to generate many of its emotions, as described above.

Behavioral Features

When we began the design of Lyotard, we expected that Hap would directly query Em's state. However, we found that Hap's decisions were often based on complex tests of the emotional state, and that the same tests arose repeatedly. Further, we sometimes wanted to produce behavior as if Em held a certain emotion which in fact was absent. It became clear that Lyotard's emotion-related behavior depended on an abstraction of the emotional state.

The abstraction, called "behavioral features", consists of a set of named features that modulate the activity of Hap. Features are adjusted by Hap or Em to control how Hap achieves its goals. Em adjusts the features to express emotional influences on behavior. It continuously evaluates a set of functions that determine certain features based on the agent's emotional state. Hap modifies the features when it wants to force a style of action. For example, it may decide to act friendly to help achieve a goal, even if the agent isn't feeling especially friendly.

Features may influence several aspects of Hap's execution. They may trigger demons that create new top-level goals. They may occur in the preconditions, success tests, and context conditions of plans, and so influence how Hap chooses to achieve its goals. Finally, they may affect the precise style in which an action is performed.

Table 1 shows Lyotard's behavioral features. The "aggressive" feature, for example, arises when Lyotard is either angry or mildly afraid (which might be considered bravado). This feature may affect Hap by giving rise to a new goal, such as bite-human, by influencing the choice of plan for a goal, such as nipping instead of meowing to attract attention, or by modifying the style of an action, such as swatting a toy mouse with unusual emphasis.

We have no structured set of features, and know of no source that suggests one. Besides those in Lyotard, we have seen the following suggested: curious, belligerent, persistent, depressed, patient [Carbonell, 1979]; timid, reckless, quiet, arrogant [Hovy, 1988].

The feature mechanism, while very *ad hoc*, appears to provide a useful degree of abstraction in the interface between emotion and behavior. It is not merely a

```
Lyotard:
  (*go-to "closet")
  (*lookaround "closet")
  (*lookaround "closet")
  (*jump-on "plant")
  (*lookaround "plant")
  (*walk-along "plant")
  (*nibble "plant")
  (*walk-along "plant")
  (*jump-off "plant")
  (*go-to "bedroom")
  (*go-to "sunroom")
  (*go-to "spare room")
  (*jump-on "chair")
  (*sit-down)
  (*lick "lyotard")
  (*lick "lyotard")
Player:
  (*go-to "spare room")
L: (*jump-off "chair")
  (*run-to "sunroom")
P: (*go-to "sunroom")
L: (*lookaround nervously)
P: (*pet "lyotard")
L: (*bite "player")
  (*run-to "diningroom")
P: (*go-to "spare room")
L: (*lookaround nervously)
  (*go-to "sunroom")
  (*pounce-on "superball")
  (*lookat "superball")
  (*nudge "superball")
  (*pounce-on "superball")
  (*pounce-on "superball")
  (*go-to "diningroom")
  (*go-to "kitchen")
  (*meow)
P: (*go-to "sunroom")
L: (*meow)
P: (*go-to "diningroom")
L: (*wait)
P: (*take "glass jar")
L: (*go-to "diningroom")
P: (*go-to "kitchen")
L: (*jump-on "table")
  (*jump-off "table")
  (*go-to "kitchen")
  (*meow)
P: (*pour "jar" in "bowl")
L: (*eat "sardine")
  (*eat "sardine")
  (*eat "sardine")
  (*eat "sardine")
  (*eat "sardine")
P: (*pet "lyotard")
L: (*close-eyes lazily)
P: (*take "lyotard")
L: (*close-eyes lazily)
```

Figure 2: Section of an interaction with Lyotard

mechanism to vary Tok's behavior and thereby possibly increase the appearance of richness. Rather, it is an initial solution to the integration problem of driving behavior from both goals and emotion.

Results of Integration in Lyotard

We discuss here the beginning portion of a fragment of behavior that Lyotard has exhibited. The complete fragment is given in figure 2, and a correspondingly complete discussion can be found in [Bates *et al.*, 1992]. The purpose of the trace is not to show the breadth of Lyotard's capabilities, which are better indicated by Table 1, but to demonstrate the integration of Tok and Em.

As the trace begins, Lyotard is engaged in exploration behavior in an attempt to satisfy a goal to amuse himself. (The explore behavior was not in the original Lyotard design presented in Table 1, but was added to Lyotard at a later stage). This behavior leads Lyotard to look around the room, jump on a potted plant, nibble the plant, etc.

After sufficient exploration, Lyotard's goal is satisfied. This success is passed on to Em which makes Lyotard mildly happy. The happy emotion leads to the "content" feature being set. Hap then notices this feature being active and decides to pursue a behavior to find a comfortable place to sit, again to satisfy the high-level amusement goal. This behavior consists of going to a bedroom, jumping onto a chair, sitting down, and licking himself for a while.

At this point, a human user whom Lyotard dislikes walks into the room. The dislike attitude, part of the human-cat social relationship in Em, gives rise to an emotion of mild hate toward the user. Further, Em notices that some of Lyotard's goals, such as not-being-hurt, are threatened by the disliked user's proximity. This prospect of a goal failure generates fear in Lyotard. The fear and hate combine to generate a strong "aggressive" feature and diminish the previous "content" feature. In this case, Hap also has access to the fear emotion itself to determine why Lyotard is feeling aggressive. All this combines in Hap to give rise to an avoid-harm goal and its subsidiary escape/run-away behavior that leads Lyotard to jump off the chair and run out of the room. (Space restrictions forbid us from continuing our discussion beyond this point, but see [Bates *et al.*, 1992].)

The Oz system is written in Common Lisp and CLOS. Of the 50,000 lines of code that comprise Oz, the Tok architecture is roughly 7500 lines. Lyotard is an additional 2000 lines of code. On an HP Snake (55 MIPS), each Tok agent takes roughly two seconds for processing between acts. (Most of this time is spent sensing, which suggests that even in the interactive fiction domain it may be desirable to use task specific selective perception.)

Discussion of Tok and Related Work

Developing Tok forced us to consider several issues which may be of general interest: the requirements emotion may place on reactive architectures, using goals without building world models, producing coherent overall behavior from independent particular behaviors, and modeling personality and its influence on behavior.

Emotion, Explicit Goals, and World Models

Some researchers have argued in recent years that representing goals explicitly in agents presents serious obstacles to the production of robust, reactive behavior [Brooks, 1987, Agre and Chapman, 1990]. Others, of course, disagree with this view and feel that goals are necessary to organize action (for instance, see many of the papers in [Laird, 1991] and the varied work on Soar [Newell, 1990]).

It is essential that Oz agents be reactive, so we have been sympathetic to the reactivity arguments made by Brooks, Agre, and others. However, it is also necessary for our agents to at least appear clearly to have goals and for them to exhibit emotion in response to events affecting those goals. This latter requirement, in particular, seemed unsolvable to us without explicitly representing goals within the agent.

Once we accepted the importance of reactivity and grounding in sensory inputs, which was forced upon us by facing our task squarely, it was not difficult to develop an architecture that represented goals explicitly while retaining reactivity. We were not forced to adopt the view of operators as pre/post condition pairs and goals as predicates on world states. It was not even necessary to view goals as testable expressions. Rather, we could view them simply as internalized tokens (perhaps with arguments) that guide Hap in choosing appropriate behaviors. Those behaviors may in turn contain other tokens, and so on.

Thus, we suggest that robust, reactive behavior is not diminished by the presence of explicit goals in an agent, but by the attempt to model the agent's choice of action as a planning process over characterizations of the world. Our view of goals allows us to avoid many of the unpleasant consequences of trying to model the world, while preserving the strengths of goals as a mechanism for organizing action. (Though we note that it may well be possible to combine these views in "plan-and-compile" architectures [Mitchell, 1991, Mitchell, 1990], of which Soar is a particularly rich example [Laird and Rosenbloom, 1990].)

Mixing Independent Behaviors

As we have used the word, a behavior is a cluster of related goals and plans that produces some recognizable, internally coherent pattern of action. A behavior is often represented by a single high-level goal.

We initially developed the notion of a behavior to allow us to specify Lyotard. We needed some concise way to represent the major components of Lyotard's action, and attaching suggestive names to a set of high-level goals seemed helpful.

The goals were implemented independently, resulting in a set of independent behaviors. Each of these behaviors is composed of a set of Hap plans and subgoals, with appropriate context conditions and success tests.

The context conditions and success tests were developed to make each behavior robust in the face of changes in the world, be they unexpected failures or serendipitous success. We expected these surprises to be due to external events performed by other agents or unforeseen complexities in the physical nature of the world. However, it has turned out that the agent's own actions, performed by other interleaved behaviors, are one of the main causes of unexpected changes. The context conditions and success tests allow these independent behaviors to mix together fairly well, without much explicit design effort to consider the interactions. Thus, adding reactivity to goal-directed behavior seems to help support the production of coherent, robust overall behavior from independently executing particular behaviors.

Modeling Personality

Tok must support construction of a variety of agents by the artists building Oz worlds. A key facet of this support is allowing different personalities to be modeled without requiring that every agent be built from scratch. The behavioral feature mechanism appears to provide a simple means to help achieve this.

One can build behaviors to respond to a standardized set of behavioral features, in ways consistent with the names of the features. For example, the aggressive feature is uniformly used to produce aggressive behavior. With the feature to behavior mapping thus fixed, facets of a personality can be determined by the mapping from emotion to features.

The standard fear emotion, for instance, might lead to any of a number of features, such as fright, or flight, or even frozen, depending on the artist's choice of the emotion to feature mapping. Each of these features would cause the previously constructed behaviors to react appropriately. Another example might be an agent where goal failures were seen as learning experiences and used to enable the proud feature. This approach takes advantage of the feature mechanism's role as an abstract interface between emotion and action.

There are several components internal to Hap that may also help model personality. An agent that overestimated the likelihood of goal success or failure might be an optimist or pessimist. One that consistently overrated the importance of goals would tend to have extremes of hope, fear, happiness, and sadness. Making an agent's success tests too easily satisfied would produce sloppiness or incompetence, while making its context conditions too difficult to maintain would produce a kind of perfectionism.

Conclusion

We have described Tok, an architecture that integrates mechanisms for reactivity, goals, and emotion. Several mechanisms, including behavioral features, success tests, and context conditions, support the integration. Lyotard, a particular agent, has been built in Tok and exhibits signs of success in integration.

While Tok maintains various kinds of memory, including perceptual memory, a richer learning mechanism is conspicuously absent from the architecture. There are two reasons for this. First, Oz worlds exist for only a few hours, perhaps too short a time for interesting learning to occur. Thus, the integration issues discussed here seem more important for our application. Second, the integration of learning with action is widely studied, and we want to build on this substantial effort rather than compete with it. As a result, we may be failing to see essential constraints that could guide us to a better architecture. To help judge this possibility, one of our colleagues is implementing Lyotard in the Soar architecture.

We are engaged in several efforts to extend Tok. First, Gump and Glinda, our natural language components, are attached to Tok only as independent Lisp modules invokable from Hap rules. It would be best if they were expressed as complex behaviors written directly in Hap. We have increasingly observed similarities in the mechanisms of Hap and Glinda, and are exploring the possibilities of merging them fully.

Second, since the Oz physical world is itself a simulation, it would be conceptually straight-forward to embed a (possibly imprecise) copy inside Tok for use as an envisionment engine. This might allow Tok, for instance, to consider possible re-orderings of steps in behaviors, and to make other decisions based on a modicum of foresight.

Finally, we have built and are continuing to build several realtime, multi-agent, animated, interactive Oz worlds. This is imposing hard timing constraints and genuine parallelism on Tok, and has caused substantial changes to the implementation and smaller changes to the architecture.

It has been suggested to us that it may be impossible to build broad, shallow agents. Perhaps breadth can only arise when each component is itself modeled sufficiently deeply. In contrast to the case with broad, deep agents (such as people), we have no *a priori* proof of the existence of broad, shallow agents. However, at least in the Oz domain, where sustained suspension of disbelief is the criteria for success, we suspect that broad, shallow agents may be possible. Tok is an experimental effort to judge the issue.

Acknowledgments

This research was supported in part by Fujitsu Laboratories, Ltd. We thank Phoebe Sengers, Peter Weyhrauch, and Mark Kantrowitz for their assistance, and the reviewers for helping us clarify our presentation.

References

[Agre and Chapman, 1990] Philip E. Agre and David Chapman. What are plans for? In *Robotics and Autonomous Systems*. Elsevier Science Publishers, 1990.

[Bates *et al.*, 1991] Joseph Bates, A. Bryan Loyall, and W. Scott Reilly. Broad agents. In *Proceedings of AAAI Spring Symposium on Integrated Intelligent Architectures*, Stanford, CA, March 1991. Available in *SIGART Bulletin*, Volume 2, Number 4, August 1991, pp. 38-40.

[Bates *et al.*, 1992] Joseph Bates, A. Bryan Loyall, and W. Scott Reilly. An architecture for action, emotion, and social behavior. Technical Report CMU-CS-92-144, School of Computer Science, Carnegie Mellon University, Pittsburgh, PA, May 1992. Submitted to the Fourth European Workshop on Modeling Autonomous Agents in a Multi-Agent World, S.Martino al Cimino, Italy.

[Bates, 1990] Joseph Bates. Computational drama in Oz. In *Working Notes of the AAAI-90 Workshop on Interactive Fiction and Synthetic Realities*, Boston, MA, July 1990.

[Bates, 1992] Joseph Bates. Virtual reality, art, and entertainment. *PRESENCE: Teleoperators and Virtual Environments*, 1(1):133–138, 1992.

[Brooks, 1987] Rodney Brooks. Intelligence without representation. In *Proceedings of the Workshop on the Foundations of Artificial Intelligence*, June 1987.

[Carbonell, 1979] Jaime Carbonell. Computer models of human personality traits. Technical Report CMU-CS-79-154, School of Computer Science, Carnegie Mellon University, Pittsburgh, PA, November 1979.

[Dyer, 1983] Michael Dyer. *In-Depth Understanding*. The MIT Press, Cambridge, MA, 1983.

[Firby, 1989] James R. Firby. *Adaptive Execution in Complex Dynamic Worlds*. PhD thesis, Department of Computer Science, Yale University, 1989.

[Georgeff *et al.*, 1987] Michael P. Georgeff, Amy L. Lansky, and Marcel J. Schoppers. Reasoning and planning in dynamic domains: An experiment with a mobile robot. Technical Report 380, Artificial Intelligence Center, SRI International, Menlo Park, CA, 1987.

[Hovy, 1988] Eduard Hovy. *Generating Natural Language under Pragmatic Constraints*. Lawrence Erlbaum Associates, Hillsdale, NJ, 1988.

[Kantrowitz and Bates, 1992] Mark Kantrowitz and Joseph Bates. Integrated natural language generation systems. In R. Dale, E. Hovy, D. Rosner, and O. Stock, editors, *Aspects of Automated Natural Language Generation*, volume 587 of *Lecture Notes in Artificial Intelligence*, pages 13–28. Springer-Verlag, 1992.

[Kantrowitz, 1990] Mark Kantrowitz. Glinda: Natural language text generation in the Oz interactive fiction project. Technical Report CMU-CS-90-158, School of Computer Science, Carnegie Mellon University, Pittsburgh, PA, 1990.

[Laird and Rosenbloom, 1990] John E. Laird and Paul S. Rosenbloom. Integrating planning, execution, and learning in soar for external environments. In *Proceedings of the Eighth National Conference on Artificial Intelligence*, pages 1022–1029. AAAI Press, 1990.

[Laird, 1991] John Laird, editor. *Proceedings of AAAI Spring Symposium on Integrated Intelligent Architectures*, March 1991. Available in *SIGART Bulletin*, Volume 2, Number 4, August 1991.

[Lebowitz, 1985] Michael Lebowitz. Story-telling as planning and learning. *Poetics*, 14:483–502, 1985.

[Loyall and Bates, 1991] A. Bryan Loyall and Joseph Bates. Hap: A reactive, adaptive architecture for agents. Technical Report CMU-CS-91-147, School of Computer Science, Carnegie Mellon University, Pittsburgh, PA, June 1991.

[Meehan, 1976] James Meehan. *The Metanovel: Writing Stories by Computer*. PhD thesis, Computer Science Department, Yale University, 1976. Research Report #74.

[Mitchell, 1990] Tom M. Mitchell. Becoming increasingly reactive. In *Proceedings of the Eighth National Converence on Artificial Intelligence*, pages 1051 – 1058. AAAI Press, 1990.

[Mitchell, 1991] Tom M. Mitchell. Plan-then-compile architectures. In *Proceedings of AAAI Spring Symposium on Integrated Intelligent Architectures*, Stanford, CA, March 1991. Available in *SIGART Bulletin*, Volume 2, Number 4, August 1991, pp. 136-139.

[Mueller, 1990] Erik T. Mueller. *Daydreaming in Humans and Machines*. Ablex Publishing Corporation, 1990.

[Newell, 1990] Allen Newell. *Unified Theories of Cognition*. Harvard University Press, Cambridge, MA, 1990.

[Ortony *et al.*, 1988] A. Ortony, G. Clore, and A. Collins. *The Cognitive Structure of Emotions*. Cambridge University Press, 1988.

[Reilly and Bates, 1992] W. Scott Reilly and Joseph Bates. Building emotional agents. Technical Report CMU-CS-92-143, School of Computer Science, Carnegie Mellon University, Pittsburgh, PA, May 1992.

[Simmons, 1991] Reid Simmons. Concurrent planning and execution for a walking robot. In *Proceedings of the IEEE International Conference on Robotics and Automation*, Sacramento, CA, 1991.

[Vere and Bickmore, 1990] S. Vere and T. Bickmore. A basic agent. *Computational Intelligence*, 6:41–60, 1990.

Dedal: Using Domain Concepts to Index Engineering Design Information

Catherine Baudin*
Jody Gevins**
Artificial Intelligence Research Branch
NASA Ames Research Center, M/S 269-2
Moffett Field, CA 94305
<baudin,gevins>@ptolemy.arc.nasa.gov

Vinod Baya
Ade Mabogunje
Center For Design Research
Stanford University
<baya,ade>@sunrise.stanford.edu

Abstract

The goal of Dedal is to facilitate the reuse of engineering design experience by providing an intelligent guide for browsing multimedia design documents.

Based on protocol analysis of design activities, we defined a language to describe the content and the form of technical documents for mechanical design. We use this language to index pages of an Electronic Design Notebook which contains text and graphics material, meeting reports and transcripts of conversations among designers. Index and query representations combine elements of the design language with concepts from a model of the designed artifact. The information retrieval mechanism uses heuristic knowledge from the artifact model to help engineers formulate questions, guide the search for relevant information and refine the existing set of indices. Dedal is a compromise between domain-independent argumentation-based systems and pure model-based systems which assume a complete formalization of all design documents.

We describe an experiment where a subject must design a new shock absorber by modifying a similar design. Toward this end, we indexed on-line documents and videotaped material associated with a shock absorber designed at Stanford's Department of Mechanical Engineering. The subject uses Dedal to access these multimedia documents while solving the problem.

* Employed by RECOM Inc.

** Employed by Sterling Software.

Introduction

Dedal is part of the Design Reuse Assistant project whose goal is to facilitate the reuse of previous design experience for designing mechanical devices.

In an attempt to capture design information, researchers have investigated ways of acquiring both formal design knowledge (Baudin et al 90) (Gruber & Russell 90) and informal design information using media such as videotapes (Stults 88), audiotapes and text and graphic documentation aids (Lakin et al 89). While complete models of an artifact's design are difficult to acquire, canned-text design information or videotapes of meetings are easy to capture but difficult to retrieve by systems that have no representation of the information *content* (Blair & Maron 85).

Based on protocol analysis of a designer's information seeking behavior, we identified a language to describe the *content* and the *form* (text, table, equation, etc.) of design records such as meeting summaries, pages of an electronic design notebook, technical reports and transcripts of conversations among an expert designer and a novice. The language incorporates elements of a model of the artifact being designed with a vocabulary of topics usually covered by design documents.

Dedal is a system that uses this language to: (1) enable the description of the design record content, (2) help engineers formulate questions, and (3) select appropriate records in answer to a question. We tested the ability of an engineer to formulate queries using this language and the ability of Dedal to retrieve records related to the user's questions. Section 2 describes a pilot study to identify the design language. Section 3 presents an overview of Dedal. Section 4 describes empirical results of the usefulness of this indexing scheme by two engineers using Dedal while modifying the design of an automobile shock absorber.

Language to Describe Design Records

Design records associated with an Engineering device cover prescribed features such as requirements, structure and behavior of the final artifact and elements of design history such as decisions, alternatives considered and rationale for design choices. This information has different *levels of detail* ranging from detailed descriptions of a part to global views of an assembly and can take the form of text, graphics, tables, photos or equations.

Our first task was to define a language to describe these aspects of design documents by observing the information-seeking behavior of a designer solving a problem.

We conducted a pilot study at Stanford's Center for Design Research and NASA Ames to identify the classes of information engineers need to access when they *redesign* an existing mechanical device in response to a requirement change. A redesign task requires an understanding of the original design and triggers questions that cover most aspects of its characteristics and history. For this study we chose a variation of an automobile shock absorber (or damper) developed at Stanford's Department of Mechanical Engineering for Ford Motor Corporation by three designers over a seven month period. The design was documented in the designers' personal notebooks and in three technical reports written at different stages of the design process.

- *Pilot study:* A mechanical designer (*subject*) adapted a damper previously designed for a compact car to a vehicle for multiple terrains. While solving the redesign problem the subject was encouraged to think aloud and ask questions of one of the damper designers (*expert*) who could either answer the questions or act as a quick index into the documentation. We videotaped and transcribed this six-hour session.
- *Questions extraction:* We extracted 80 questions from the verbal protocol that the subject asked during the session.
- *Contextualization:* Each question was reformulated to incorporate contextual elements, such as the subject of the question, that were implied but not explicitly stated in the question.
- *Identification of design topics:* We identified categories that encompassed the extracted questions. These classes are an extension of the classification performed by researchers at Oregon State University (Kuffner & Ullman 90) to find out what kind of information designers are interested in.

We duplicated this study with a NASA designer and found that *both* the questions asked by the designer and the concepts addressed in the design documents could be expressed by combining an element from the topic list presented in Figure 1 with elements from a model of the artifact being designed. The meaning of the vocabulary extracted from our study (Figure 1) is described in a separate paper [Baya et al 92]. The concepts of the topic, level-of-detail and media lists are likely to be addressed in *any* mechanical design document. If instead of *design* our main task was *diagnosis* or *manufacturing*, chances are that this vocabulary would have to be changed to take into account the important features of these domains. The topic, levels of detail, and media lists are generic *task-dependent* concepts whereas the subject-class list refers to concepts that depend on a *particular design*.

Accordingly, any information in a design record can be described by several *indexing patterns* of the form: **information-about** topic T **regarding** subject S **with** level of detail L **using** medium M. T, L and M are selected from the topic, level-of-detail and media lists, respectively, whereas S is one of the following types: feature, component, assembly, requirement. In addition, each indexing pattern contains a pointer to the record and segment corresponding to the starting location of the information (e.g.,. document name and page number or video counter).

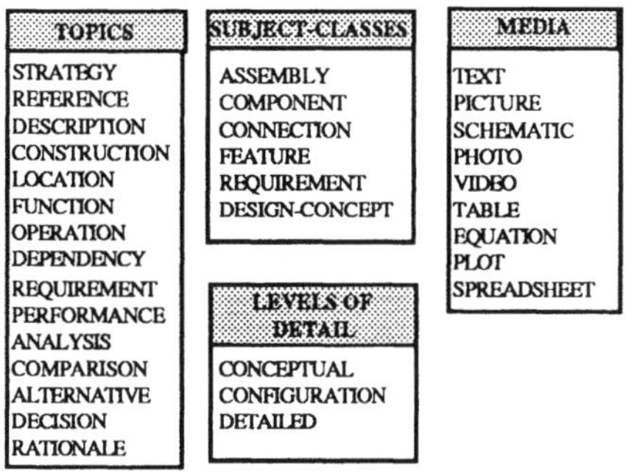

Figure 1: Task and Design Dependent Elements of Dedal's Language

For instance, if:*"the solenoid induces a magnetic field generating a force which pushes the lever..."* is a segment of information in the document Report-3344 , it can be described in Dedal by the following indexing patterns:

<information-about topic: **operation**,
regarding subject: **solenoid**
with level-of-detail: **conceptual**,
using medium: **text**,
in record: Report-3344, segment: 12>

<information-about topic: **dependency**,
regarding subject: **magnetic-field of solenoid**,
with level-of-detail: **conceptual**,
using medium **text**,
in record: Report-3344, segment 12>.

Questions in Dedal mirror the structure of the indexing patterns. The format of a question is: **ask-about** *Topic* **regarding** *subject* with preferred medium or level of detail. For instance "where is the solenoid?" can be formulated as:
(q1) <ask-about topic: **location**, regarding subject: **solenoid** with preferred-medium:**schema**>

Interacting with Dedal

Engineers at Stanford's Center for Design Research are using an Electronic Design Notebook™[1] (EDN) to capture information such as technical reports, meeting summaries and design notes usually recorded in the designer's paper notebooks. Dedal is an interface to these electronic records. It includes two main components: (1) an indexing component to describe the records, and (2) a retrieval component. These two components interact with a knowledge-base of indexing patterns describing the records.

Indexing

Indexing with Dedal includes two phases:

In the first phase, a knowledge engineer sits with an expert familiar with the artifact described in the design records. Together they define the device dependent concepts associated with the design and the relations among these concepts. In our tests with indexing the automobile shock absorber we started by focusing on the structure of the prototype produced at the end of the first conceptual design phase. The structure of a device is usually easier to identify than other aspects, such as the geometry of the parts or the description of its behavior. We also included the quantities associated with the attributes of the device, dependency links among these quantities, the main decision points and alternative considered. These concepts and links constitute what we call the device related concept model (DRC model).

In the second phase, an "indexer" describes each record. During this content description phase, the following questions are triggered: What element of the DRC model is the subject of this information segment (subject selection)? What is being said about this subject (topic selection)? At what level of detail? What is the medium used to convey the information? The information can be attached to different levels of the subject hierarchy. For instance, the content of a segment of text detailing the interactions between the parts of a mechanism M can be described by several indexing patterns or "summarized" by saying that this text is about the topic *description* of the subject *mechanism M*.

Querying

The query module consists of:
- A **question formulation component** in the form of a graphic interface displaying the DRC

---
[1] The Electronic Design Notebook is a trademark of performing graphics inc.

components of the mechanical assembly and the generic task dependent vocabulary.

• A **retrieval component** which takes as input a question from the user and matches it to the set of indexing patterns, returning a set of references. The retrieved references are grouped into *answer-sets* of no more than five references. The retrieval component uses a set of heuristics to loosen the match when either no indexing patterns exactly correspond to the current question or the user instructs Dedal to retrieve another answer-set. The retrieval strategy of Dedal can be summarized as follows: given a question of the form <topic, subject, preferred medium, preferred level-of-detail>.
1. Find candidate indexing patterns.
2. Order the indexing patterns retrieved by preferred medium and level of detail.
3. Get feedback from the user on the relevance of the answers. If a relevant reference is retrieved using a heuristic, the question asked by the system is turned into a new index.

Dedal currently uses fourteen **retrieval heuristics** to find related answers to a question. For instance: segments described by concepts like *<decision for lever material>* and *<alternative for lever material>* are likely to be located in nearby regions of the documentation. Another heuristic is "look-for-superpart": if *A* is a part of a mechanism B, information about: *<location of A >* is likely to be found in a picture describing mechanism *B*. Other heuristics reason about the dependencies among the quantities describing the device. For instance: if a question is about *rationale* for the value of quantity Q1, and Q1 depends on Q2, then look for a segment of information describing the *rationale* for the value of Q2.

If the question **q1** (page 3) about the **location** of the **solenoid** is posed to Dedal and if no indexing pattern matches this description, Dedal will activate the retrieval heuristics. In this example, given that *solenoid* is a part of the *force generating mechanism*, the application of the "look-for-superpart" heuristic will retrieve the pattern Ip1 which reference a record showing a picture of this mechanism:

Ip1: <topic: **description**
subject: **force-generation-mechanism**
level-of-detail: configuration
medium: schema
reference: (record:damper-transcribed-data-subject-one, segment: 2)>

The reference associated with this pattern is the transcription of a conversation between one of the original damper designers and an engineer asking questions about this design. The reference has two parts, the record name and an EDN page where the relevant information segment is located. If a reference in an answer set is on-line, the user can select it and go directly to the page using the hypertext facility associated with the EDN environment.

• **The index refinement component**: Each time Dedal uses the inference mechanism to find a reference, the user has the option to validate the retrieved reference. Dedal then creates a new index: The topic and the subjects associated with this new index are from the current question, the reference of this new index is the validated reference. The next time the same query is asked to the system, this new index will be retrieved as an exact match in the fist answer set. In our example, If the indexing pattern Ip1 is validated by the user, Dedal will create a new pattern Ip2:

Ip2: <topic: **location**
subject: **solenoid**
level-of-detail: configuration,
medium : schema,
reference:(record:damper-transcribed-data-subject-one,, segment: 2)>

Next time the same question is asked, Dedal will retrieve Ip2 as an exact match without calling the retrieval heuristics. In our current version the new descriptions created by the system are stored in a file associated with a particular user and redesign session.

Empirical Results

We conducted an experiment to evaluate the ability of an engineer to describe design records using our language, the ability of a user to formulate questions to the system, and the retrieval performance of Dedal. Among the questions extracted from the damper redesign protocol, we selected 47 questions whose answers were available in the design records. We asked the mechanical engineer (the subject) who was involved in the redesign pilot study to formulate and submit these questions to Dedal. The goal for Dedal was to reproduce the "intelligent retrieval behavior of the expert" during the redesign study. The answers retrieved by the system were then rated by the subject and by the expert. We considered two classes of ratings: relevant and irrelevant.

Indexing the Damper Design Records:
Most damper design documents were captured in the Electronic Design Notebook, except for a videotape showing the operation and testing of the original prototype. One of the experts who designed the damper indexed the design records. The expert defined a "first draft" of a damper model and then refined it while indexing the design records by adding a new subject when no suitable concept could be found in the DRC model. After each indexing session, the expert and a knowledge engineer reviewed the new concepts added during the session and decided how to incorporate them in the damper model.

An average of 4 minutes was needed to index an EDN page of the damper requirement report. The time was less for the appendix pages which were mainly articles and manufacturing catalogue information. Indexing the videotape transcription of the conversation among designers required about 2 minutes per EDN page. It should be noted that the information in the records was familiar to the expert.

Ability to formulate questions in Dedal:
Of the 47 questions asked, only 7 had to be reformulated because the user was unsure of which subject to select. This was the case for the extracted question "What did they (the designers) do to limit the heat dissipation coming from the solenoid?" This was first translated to "What is the isolation of the solenoid?" but isolation was not in the model. The user then switched to "What is the material of the solenoid shaft?" and the system was then able to retrieve a set of references.

Retrieval Performance
The subject asked 47 questions. Dedal returned a relevant answer in the first answer set for 37 of those questions. For 6 questions the subject had to ask for more information to find a good answer, and for 4 questions the expert could find a better answer in the design records than the ones retrieved by Dedal. Table 1 summarizes these results.

	subject	expert
relevant-1st-answer-set	37	36
relevant-other-answer-sets	6	7
relevant-answer-not-retrieved	4	4
rated questions	47	47
%-of-question-answered	.91	.91

Table 1: Retrieval Performance

The percent of relevant answers in each answer set was estimated to be 71% by the expert and only 62% by the subject for the same questions and the same references retrieved. Reasons why irrelevant references were retrieved include: incorrect question formulation, bad information description, overgeneralization of the retrieval heuristics. In some cases irrelevant references referred to relevant information previously retrieved and therefore were no anymore interesting to the user.

Although the numbers reported in Table 1 are the same for the subject and the expert, it is interesting to note that in many cases the expert and the subject disagreed on which references were relevant. The main source for this difference in appreciation is related to the "context" dependence attached to the notion of *relevant* answers (Graesser 85). This means that the relevance of the answers retrieved depends on other contextual elements than the ones included in the question formulation. Additional contextual elements that could come into play are: the previous questions asked and the problem goal (in the context of a redesign problem, high level goals could be identified).

Conclusion

Our system uses a model of the artifact being designed and of the design process to describe the content and the form of design information and to formulate queries to the system. More experimentation is needed to evaluate Dedal. However, our preliminary results tend to show that: (1) The analysis of the designer's information-seeking behavior is a relevant way of identifying a vocabulary to describe design documents, (2) using these descriptions to index design records lead to good information retrieval performance, and (3) a user has been able to use our language to formulate queries about the original damper design in the context of a redesign task.

This type of conceptual indexing is performed interactively with a human. This raises questions about *who* should index the records, what his level of expertise should be, and *when* indexing should take place. It is interesting to note that indexing time can be broken into three time segments: a time T1 to understand the information in the record, T2 to select the proper concepts to describe the information and T3 to create the index in Dedal. If indexing is performed by a user familiar with the records and shortly after they have been generated T1 will be minimized. If indexing is performed by a knowledge engineer familiar with modeling techniques T2 might be shortened. We are currently investigating these questions

We are considering two ways to alleviate the index acquisition burden. The first is to integrate the indexing phase with the design process to help designers *generate* the design documentation as in (Russell 89). The second is to investigate further the use of incremental *question-based* indexing techniques (Mabogunje 90) where the questions asked by a designer are used to create new indices. This is facilitated by our assumption that the query language and the information description language can use the same vocabulary and have similar representations.

Acknowledgements

We are grateful to Fred Lakin for his support of the EDN system. Thanks to Larry Leifer and to the other members of the GCDK group for their feedback and support on this project.

References

Baudin, C.; Sivard, C.; Zweben, M., 1990. Recovering rationale for design changes: A knowledge-based approach. In Proceedings of IEEE, Los Angeles, CA.

Baya, V.; Gevins, J.; C. Baudin C.; Mabogunje, A.;, Leifer, L., 1992. An Experimental Study of Design Information Reuse. In Proceedings of the 4th International Conference on Design Theory and Methodology.

Blair, D. C.; Maron, M. E., 1985. An Evaluation of Retrieval Effectiveness for a Full-Test Document-Retrieval System. Communications of the ACM Volume 28, Number 3.

Conklin, J.; Begeman, M.L., 1988. gIBIS: A Hypertext Tool for Exploratory Policy Discussion. ACM-0-89791-282-9/88/0140.

Graesser, A.; Black, J., 1985. The Psychology of Questions. Lawrence Elrlbaum associates.

Gruber, T.; Russell, D., 1990. Design knowledge and design rationale: A framework for representation, capture, and use. KSL internal report.

Kuffner, T. Ullman, D., 1990. The information requests of mechanical design engineers. In Proceedings of Design Theory and Methodology. DPRG 90-1.

Lakin F.; Wambaugh J.; Leifer L.; Cannon D.; and Sivard C., 1989. The Electronic Design Notebook: Performing Medium and Processing Medium. THE VISUAL COMPUTER, *International Journal of Computer Graphics* 5: 214-226.

Mabogunje, A. 1990. A conceptual framework for the development of a question based design methodology. Center for Design Research Technical Report (19900209).

Russell, D.M., 1988. IDE: The Instructional Design Environment, in: J.Psotka, J., D. Massey, & S. Mutter (Eds), *Intelligent Tutoring Systems: Lessons Learned*. Hillsdale, NJ: L. Erlbaum Associates.

Stults, R., 1988. Experimental Uses of Videotapes to Support Design Activities. Xerox PARC Technical Report SS1-89-19.

No Logic? No Problem! Using A Covariation Analysis On A Deductive Task

John B. Best
Department of Psychology
Eastern Illinois University Charleston, IL 61920
cfjbb@ux1.cts.eiu.edu

Abstract

Subjects were presented with previously played Mastermind games in the form of "Mastermind problems". Although each problem was formally deducible, and in some cases, overdetermined, subjects nevertheless usually failed to make more than a third of the potential deductions. A Bayesian model that treated the task as one of "probabilistic reasoning" rather than "logical deduction" accounted well for the performance of the lower performing subjects. It is argued that at least some of the reasoning failures seen on hypothesis evaluation tasks such as this one are produced in part by the solver's replacement of a "deduction" representation with a "probabilistic reasoning" representation.

Introduction

Studies using hypothesis evaluation paradigms suggest that the "confirmation bias" seen in the 2-4-6 task (Wason, 1960) results in part from subjects' inability to generate the hypotheses. That is, when the appropriate disconfirmatory hypotheses are already generated for them, then subjects who are instructed to do so can recognize disconfirmatory hypotheses, suggesting that the necessary logical operators are intact (Farris & Revlin, 1989). But there is at least one important proviso to this finding. In hypothesis evaluation paradigms (Farris & Revlin, 1989), or in rule discovery tasks in which the subjects are "debiased" by instructions (Gorman & Gorman, 1984), the context in which the materials are presented to the subjects almost always involves "logical deductions or scientific thinking". a contextual effect that seems very likely to be reflected in the subjects' representation of the task. This implies that the ability to generate or recognize disconfirmatory response may be a product of both the presence of the disconfirmatory hypotheses, and the "right" elements in the solver's representation. Whether people can routinely engage in disconfirmatory analyses when the context, and therefore perhaps the person's representation, are not so explicitly presented as "logical deduction" is the issue motivating this paper. I argue that, in such contexts, some failures on formally deductive tasks are produced in part by the subject's replacement of the concept of "logical necessity" with a concept of "probabilistic reasoning". What follows is some evidence to support this claim, as well as a model that duplicates some of the effects of "probabilistic reasoning" on a purely formal deductive task.

Method

Materials and Procedure

Four previously played Mastermind games of moderate complexity were presented as "Mastermind problems" in this study. Thus, in this form of Mastermind, subjects did not generate their own hypotheses. Rather, their task was to evaluate the hypotheses and feedback that had been produced in an effort to deduce the code. In each of the four problems, the set of hypotheses and feedback that were displayed contained information that was necessary and sufficient to permit the deduction of that problem's code.

Each of the 48 subjects was run individually. The experimenter explained the rules of Mastermind. also stating that in this form of the task, the subjects would not be generating their own hypotheses. The problems were presented in a way that simulated actual Mastermind play. That is, subjects saw each hypothesis and its associated feedback individually. Following the presentation of each hypothesis, the subject was asked to indicate the extent of his or her deductions on a response form. Subjects indicated two principal types of deductions: The subject marked an "assignment" when he or she was convinced that a particular letter was definitely a code member. The subject also indicated the purported location of the assigned letter. "Exclusions" were marked by the subject when he or she was positive that a particular letter was definitely not a code member. All previously presented hypotheses from that problem remained on view until after the problem's penultimate hypothesis was presented. The presentation order of the four problems was completely counterbalanced.

Results

The data were scored by counting the number of accurate assignments and exclusions marked by the subjects following the nth row of each of the four problems. Following the nth row of each problem, enough information was present to permit four assignments and two exclusions. One point was awarded for each such deduction. Mean performance across the four problems was 8.6 (maximum score = 24). Even though the specific assignments and exclusions were all logically deducible following the presentation of the nth hypothesis and feedback of each problem, the likelihood of the subject's correctly deducing assignments varied significantly both within and across problems. Of the 16 assignments in the four problems only 4 assignments were made accurately by a majority of the subjects. Five assignments were made correctly by 40-49% of the subjects. Three assignments were made correctly by 30-39% of the subjects, and four of the assignments were made correctly by only 10-29% of the subjects.

Moreover, it appears that the likelihood of a particular letter's being correctly assigned to a specific position was influenced by the number of times that the letter appeared in the problem at the same position. Specifically, in cases in which a letter appeared several times at the same location, and in which black feedback was given, subjects were likely to conclude that the letter must be correctly placed at the position where it appeared most of the time.

For example in Problem 2, 69% of the subjects correctly deduced that B must be placed at position 4. B appeared three times in the problem, each time at the same location (ratio of total appearances/different positions = 3/1), and was accompanied by a total of seven black feedback pins. But substantially fewer subjects (23%) correctly deduced that E must be placed at position 3 (ratio of total appearances/different positions = 3/3, or 1/1). This analysis suggests that the subjects were basing their deductions on the degree of covariation between a letter's placement and the occurrence of black feedback. Table 1 shows the results of similar computations carried for this and the other three problems used in the study, and it confirms the findings suggested by the initial problem. When a particular letter appeared in the array frequently and remained more or less stationary, the subjects were likely to conclude that this covariation enabled a necessary logical connection. But when a letter appeared infrequently, or appeared frequently in different locations, then subjects were less likely to deduce its assignment .

Table 1

Likelihood of Correct Assignment as a function of Appearance/Position Ratio

Correct Assignments Percentage Deducing	Total Appearances	Positions	Ratio
51% (or more), N=4	4.25	1.75	2.43
40-49%, N=5	3.40	2.40	1.42
30-39%, N=3	3.33	2.67	1.25
10-29%, N=4	2.00	2.00	1.00

Modeling the Deductions of Low-Performance Subjects

Subjects who accept the experimenter's depiction of the task may indeed represent the four problems as involving logical deduction. Such a representation would likely include logical operators whose function is to take various inputs from the problem array and produce deductions. While such operators may not always succeed in producing valid deductions, the situations in which they fail are presumably describable in terms of memory, or other exogenous demands on the cognitive system.

However. there are certainly other plausible ways of representing the problems, and these other forms of representation may predict outcomes that are more consistent with the observed findings than are the predictions of the "logical" model. For example, subjects who treat the logical problems as analogous to everyday problems in the real world may view covariation as a useful heuristic in making assignments. One way of conceptualizing the reasoner's task is to consider the reasoner as using the evidence that has accrued (the black feedback) in an effort to assess its effects on the likelihood of a particular hypothesis (namely, that a specific letter is assignable at a specific location) being true. Using a standard form of the Bayesian equation to represent this state of affairs we have:

$$p\,[L(P)/F = \frac{p\,[F/L(P)\ p\,[L(P)]}{p\,[F/L(P)\ p\,[L(P)] + p[F/{\sim}L(P)\ p[{\sim}L(P)]}$$

where p [L(P)/F] represents the probability of a particular letter (L) to position (P) assignment being true given that a certain feedback pattern (F) has been observed; P [F/L(P)] represents the likelihood of observing a certain pattern of feedback given that a letter to position assignment is true, and p [L(P)] represents the prior probability of any specific letter to position assignment being true. p [F/~L(P)] represents the probability of the feedback pattern being observed given that the letter to position assignment is not true, and p [~L(P)] represents the prior probability that the letter to position assignment is not true. Computing some of the equation's terms is straightforward: Given that any of the six available letters can be assigned to, any specific position, p[L(P)] can be estimated at .17, and p [~L(P)] = 1 - p [L(P)]. The estimation of p [F/L(P)] involves computing for any specific letter to position assignment (as in letter A in position 1) the proportion of all codes (of which there are 360) that would generate this particular feedback pattern through this hypothesis, if A were indeed correctly located at position 1. That is, of the 60 codes in which A is correctly located at position 1, what proportion would be followed by the specific feedback "1 Black, 1 White" if this hypothesis had been played? The same logic is used to estimate p [F/~L(P)]. That is, of the 300 codes that do not have A correctly located at position 1, what proportion would be followed by the specific feedback "1 Black, 1 White" if the hypothesis A B C D had been played?

To run the Bayesian model, each of the six letters was initialized as a four place vector with prior probabilities of .17 in each of the four slots, thus creating a 6 X 4 matrix. The probabilities for each letter to position assignment were updated after each hypothesis, treating the previous hypothesis's posterior likelihood as the current hypothesis's prior likelihood. A normalization procedure was applied after each updating cycle for any row vector whose probabilities exceeded unity.

Testing the model's predictions involved splitting the sample of subjects into two subgroups based on their overall performance. The logic here is that subjects who have access to logical operators, and who are motivated enough to use them, will be unlikely to rely on the covariation analysis to assign letters, and thus will be likely to correctly assign letters whose logical status is clear regardless of how such letters look to the covariation analysis. On the other hand, subjects who do not have access to such operators, or who are not motivated enough to engage in the fairly effortful analysis required to use them should be likely to rely on covariation analysis which the Bayesian model should pick up. The subjects were divided into two groups. High performance subjects (N = 25, M = 12.6) were those who scored 9 or better on the four problems, while low performance subjects (N =23, M = 3.7) were those whose score ranged from 0 to 8. Expected frequencies of letter to position assignments were computed for each letter in each problem by multiplying the elements of the letter's row vector by the number of subjects in each subgroup who actually made assignments of that letter. Eight chi-squares (4 problems X 2 subgroups) were used to evaluate goodness of fit between expected and observed frequencies of letter to position assignments. For all four problems in the high performance group, the chi-squares were significant at $p < .001$, indicating poor goodness of fit. But for all four problems in the low performance group, the chi-squares failed to reach statistical significance ($p > .05$), suggesting a reasonable conformity between the model's predicted assignments and those that the low performance subjects actually made. Moreover, the deviations from the model's predictions made by the high performance subjects were always in the direction indicated by a logical analysis rather than by a covariation analysis.

Discussion

These findings suggest that the deductive performance of subjects who do particularly poorly on this task is the result of an over-reliance on a covariation analysis, and this covariation analysis can be modeled effectively using principles of Bayes' theory. This is not to say that the lower-performing subjects are engaged in a sophisticated Bayesian analysis. Rather, such subjects seem to be engaged in a type of probability estimation, and these estimations seem to be capturable in a Bayesian model.

In addition, the findings are suggestive that the use of the covariation analysis is driven by the subject's representation of the problem as an example of "real life" reasoning, rather than as a problem in formal logic. That is, although few of the subjects in the study had studied logic formally, as college students they were well aware that logic is a formal discipline, and the topic of university coursework. And from what they know of all academic disciplines, such subject matter is approached with considerably more rigor and intensity than is the "corresponding" subject matter in day-to-day life. Thus for example, I may mix the ingredients in a cake recipe with considerably less precision than I would use to mix the ingredients in the chemistry lab, knowing that the cake will probably turn out regardless. When applied to the current situation, the typical subject may be well aware that to be "logical" might mean exercising greater precision in the reasoning process, including being more demanding of evidential standards, being more skeptical, being more alert to discrepancies of appearance and reality, and so on. Presented as it is in this context, that is, as a game, Mastermind might be not necessarily invite the more rigorous approach characteristic of subjects in studies of "logical deduction".

One of the issues in the literature on hypothesis evaluation concerns the ability of humans to recognize disconfirmatory hypotheses when such hypotheses have been generated for them. Researchers typically find that people are good at discerning disconfirmatory hypotheses in this situation. To the extent that the Mastermind problems used in this study can be seen as analogs to the reasoning vignettes used by Farris and Revlin (1989), these findings suggest that there are situations in which simply having the relevant hypotheses available does not mean that the subjects can engage in the modus tollens like reasoning of necessary logical operations. As we saw in this study, in an absolute sense the subjects did not make all the necessary logical operations that were available. Finally, these findings may be seen as an instantiation of the rational analysis approach (Anderson, 1990) that has proved useful in the areas of memory and categorization. As applied to reasoning, such an analysis suggests that, given that some individuals understand formal reasoning as equivalent to model of causation that might be derived from daily experiences, their deductions within that context are orderly and plausible.

References

Anderson , J. R. (1990). *The adaptive character of thought.* Hillsdale, NJ: Erlbaum.

Farris, H. H., & Revlin, R. (1989). Sensible reasoning in tasks: Rule discovery and hypothesis evaluation. *Memory & Cognition, 17,* 221-232.

Gorman, M. E., & Gorman, M. E. (1984). Comparison of disconfirmatory confirmatory and control strategies on Wason's 2-4-6 task. *Quarterly Journal of Experimental Psychology, 36A,* 629-648.

Wason, P. C. (1960). On the failure to eliminate hypotheses in a conceptual task. *Quarterly Journal of Experimental Psychology, 12,* 129-140.

Projected Meaning, Grounded Meaning and Intrinsic Meaning

C. Franklin Boyle
CDEC
Carnegie Mellon University
Pittsburgh, PA 15213
fb0m@andrew.cmu.edu

Abstract

It is proposed that the fundamental difference between representations whose constituent symbols have intrinsic meaning (e.g. mental representations) and those whose symbols have meanings we consider "projected" (e.g. computational representations) is causal. More specifically, this distinction depends on differences in *how* physical change is brought about, or what we call "causal mechanisms". These mechanisms serve to physically ground our intuitive notions about syntax and semantics.

Introduction

One of the defining characteristics of mind is that the contents of mental states, i.e. concepts and conceptual relationships, are intrinsically referential; they refer to things in the world without requiring external agency to realize this capacity — their meanings are intrinsic. Moreover, it seems likely that this quality of mind is responsible for the kind of understanding we experience when reading or when listening to spoken language.

It has been argued that formal symbols like those instantiated in digital computers do not have intrinsic meaning; that is, formal symbol manipulation is not sufficient for semantics (Searle, 1980; 1990). Any meaning such symbols purportedly have is projected onto them by us (Harnad, 1990). Yet it is often pointed out that "at bottom" everything is just syntax or, as Haugeland (1989) cogently observes, "meanings do not exert mechanical forces". Unless we are willing to believe in the existence of some sort of non-physical "mindstuff", we either have to agree with the major tenets of this latter, functionalist view or accept the burden of proving that there is some fundamental physical difference between mental states and computational states that might explain differences in their referential capacities.

Of course, not all the potentially supporting evidence for functionalism has been gathered and may not be for quite some time; computer systems are far from being robust enough to match the functional complexity of mind and, therefore, to test the "multiple instantiations" hypothesis (Thagard, 1986) which asserts that it is only the causal *relationships* between mental states that are physically relevant to mind, not particular substrates or architectures. There are plenty of reasons for believing that this hypothesis will never be proved: the system complexity required to test it, the operationally ill-defined nature of Turing-type tests, the "other minds" problem, etc. If, however, such causal associations are the only physical requirement on which having a mind depends (whether or not this can actually be demonstrated), it would mean that semantics is very likely the product of a system's functioning, implying that computational systems could understand the way we do. Searle's counter to this position is an intuitive argument about the nature of our understanding and a claim that brains, not computers, have the right "causal property" to produce intentional states (Searle, 1980). Though he says nothing further about this causal property, we might infer that he does not believe it to be just the set of causal relationships between mental states, since that would render it indistinguishable from functionalism.

Instead of waiting for the set of causal relationships between mental states to be instantiated computationally (if, in fact, that is possible) or relying on intuitive arguments about understanding, our strategy here is to advance Searle's position by 1) introducing a set of *causal mechanisms* which determine the kinds of physical changes that can occur when physical objects interact, and 2) arguing that differences in these causal mechanisms give rise to different ways representing entities mean. Since Searle does not explain what "causal property" is, we take the liberty of equating it with causal mechanism and then identify different causal mechanisms as causal properties of different kinds of information processing systems. This provides us with a physically-principled basis for arguing that digital computers (indeed, all pattern-matching systems, as we will see) differ from brains in how their respective representing entities mean. This latter difference strongly suggests that understanding, insofar as it depends on how representing entities mean, will *always* be different for computers and brains; that having a set of causal relations between states is not equivalent to having a particular causal mechanism that enables those relations, i.e. that transforms one state into the next. In other words, there are special physical processes that are the basis for mental representations having intrinsic meaning.

Three Types of Meaning

We begin by listing three ways representing entities can mean, that is, three ways they come to be about the things they purportedly represent:

> Projected Meaning — representing entities have meaning by virtue of our projecting it onto them. The association between representation and referent is arbitrary; that is, inputs are encoded by us or by procedures we construct.
>
> Grounded Meaning — representing entities have meaning by virtue of their being grounded in the analog projections of sensory stimuli -- the relationship between stimulus and internal representation is non-arbitrary (Harnad 1990).
>
> Intrinsic Meaning — representing entities have meaning by virtue of their being both grounded in analog projections *and* causal by the same kind of structure-preserving process that underlies their grounding (Boyle 1991).

We will argue that differences in these are due to differences in a specific aspect of physical object interactions that has so far been overlooked as being fundamental to our understanding of so-called information-processing systems — *how* physical objects are causal (Boyle, 1991), to be distinguished from *what* changes they cause. Our claim is that this is the only *principled* criterion for distinguishing the above ways representing entities can mean, not correlation or form similarity which are customarily used to reason about the nature of meaning in representational systems.

In what follows, we first discuss correlation and form and their shortcomings with respect to determining how symbols mean. We then investigate the causal aspects of representations and explain how the above types of meaning depend on what we identify as "causal mechanisms".

Correlation

As the designers of cognitive models, *we* determine what meanings the constituent symbols and symbolic expressions have simply by designating their referents. We then proceed to make these designations (which we store in our heads) consistent with the effects those symbols and symbolic expressions have on system behavior via procedures that associate the symbols and symbolic expressions with actions. Such actions are "grounded" in the interaction of system and environment — behavioral grounding — so that the meanings of the symbols behaviorally correlate with what we initially intended them to be about.

Clearly, the meanings of the symbols and symbolic expressions *before* we integrate them into a functioning system are *projected*, just as any object in the world could be interpreted as representing something else. But does making this pre-implementational, projected meaning consistent with system behavior change it from projected to intrinsic? In other words, is consistency based on correlation sufficient for semantics? According to functionalism it should be; if a computational system's behavior is indistinguishable from our own, then from the multiple-instantiations hypothesis, so are its "computational" states indistinguishable from our mental states to the extent that the relevant features are causal relations between mental states. Thus, its constituent symbols and symbolic expressions must mean in the same way the contents of our mental states mean, which we consider to be intrinsic. This sort of reasoning seems to be invoked in the so-called "systems reply" to Searle's Chinese Room argument — if the system is behaviorally indistinguishable from a native Chinese speaker, then it must understand the input (words) it processes in a manner similar to the way we understand language.

There are, however, two issues which suggest that this hypothesis about meaning in such systems, based as it is on correlation, is not empirically testable. The first is a practical one; because verification depends on behavior, if we fail to actually build such a system, we may be unable to determine if our failure was due to the omission of certain internal state relationships (e.g., state X causes state Y) or because computer systems lack some physical property that prevents us from successfully implementing all such relationships. The second issue is a reminder of the limitations inherent in making inferences about the nature of a system's internal characteristics based on its behavior. *Any* system whose internal representation of the world affects its behavior requires some (presumably high) degree of consistency between its symbols and their referents (what Haugeland (1989) sees as a strong constraint on the number of possible interpretations of its symbols) if that system is to behave in a manner we would call rational. But since this should be the case for any coherent, representing system, whether brain or machine, it implies that behavior-based correlation is not adequate for distinguishing between systems with intrinsic meaning, like brains, and those whose meanings may only be projected, like computational models of cognition.

Functionalists might respond by pointing out that completeness is also necessary; that a system will have intrinsic meaning only when it is as behaviorally robust as the brain. But why would a more complete system, which differs from one that is less complete only in the number of state relationships it instantiates, necessarily be more consistent except to the extent that there is simply more of it to be consistent? The only plausible answer, one which avoids the implication that the symbols in *any* consistent program, no matter how simple, have intrinsic meaning, is that cognitive properties might emerge when system complexity (in terms of the number of rules, for instance) is increased beyond some threshold. But until we determine what might cause this sort of emergence, if indeed it could actually happen, we must depend on consistency.

Thus, it seems clear that using behavioral criteria to explain how representing entities mean leaves too many questions unanswered. Since we believe there is something quite specific that gives rise to intrinsic meaning and since we reject explanations based on any sort of non-physical Cartesian mind-stuff or emergence, the only alternative at this point is to look for *physical* differences that depend on the structural characteristics of representing entities, independent of the particular medium, and on how these physically affect system behavior. For example, what are the medium-independent physical differences between mental representations of trees and their computational counterparts?

There appear to be two kinds of physical differences. The first is associated with the similarity between a representing entity's form and that of its referent. The second is based on *how* symbols in various systems bring about change; *how* they are causal. As we noted above, with respect to meaning, the latter is fundamental.

Form

If the physical forms of symbols are unlike those of their referents, which is the case for formal symbols in computers, then how can they represent what it is they are purportedly about unless we say they do, that is, unless their meanings are projected? On the other hand, if the structures of symbol and referent are similar or nearly so, can we say the meanings of such symbols are intrinsic? In the Chinese Room, for example, it could be argued that understanding is very different from our own because there are no forms accompanying the structurally-arbitrary input symbols that are isomorphic to the forms of the referents of those symbols. *We* acquire this kind of "form information" visually and associate it with words in our language, presumably to understand them, so should not computers require the same to understand language? Perhaps, but using form similarity to determine whether the meaning of a representation is intrinsic or not, and, hence, whether a computer's understanding is like our own, is problematic for two reasons.

First, similarity between two shapes or structures is a matter of degree, whereas meaning is *either* intrinsic *or* it is not. Otherwise, we might end up with a representation in which the meanings of some symbols, or even parts of symbols, are intrinsic while others are not, implying that somehow specific objects in the world give rise to different types of meaning, or that the system understands different objects differently, both of which seem highly unlikely. A second problem with form similarity concerns the issue of what physically makes a particular representation similar in form to its referent, and to whom. At first glance, the answer to the first part seems obvious. After all, the bitmap of a tree is clearly similar in structure to its referent. However, this may only be a similarity *to us;* digital computers probably do not "see" it that way. For them it is just another pattern to be matched, no different than any other bitmap or arbitrary combination of symbols because it is only the presence of a matcher which "fits" the pattern that is relevant to the pattern's effect on system behavior, not its *particular* form, i.e., not its *appearance*. Hence, such structures might be characterized as "intrinsically meaningless" to digital computers because they are not causal according to appearance. This will become clearer after we introduce causality as the basis for distinguishing different types of meaning.

Form, therefore, is really a criterion for distinguishing different kinds of representations (at least for us) such as extrinsic (e.g. propositional) and intrinsic (e.g. iconic) representations (Palmer, 1978), not meanings. That is, form has to do with how a representation encodes what it represents rather than how it means.

Causal Criteria

Having argued that behavior-based symbol-referent correlation and form similarity are not adequate for distinguishing how representing entities mean, we now turn to causality, but causality considered in a non-standard way. Typically, causality is expressed in terms of *why* something happens (cause) or *what* happens (effect). These are combined to form cause-and-effect pairs which associate a particular entity, a symbolic expression for example, with the effects it brings about — a highly functional characterization of physical change akin to "if-then" rules. The physical processes which actually produce the effects get buried in the structureless connection between the antecedent and consequent of such forms. In other words, there is no sense in which the associative link conveys *how* the cause actually brings about the effect, only that it does.

Here, however, we consider causality *deterministically*. That is, we determine *how* particular effects could be brought about when physical objects interact. The different ways effects are physically brought about we refer to as "causal mechanisms".

Three Causal Mechanisms

There are only three causal mechanisms for bringing about change in physical interactions: *nomologically-determined change, pattern matching* and *structure-preserving superposition* (Boyle, 1991). Each mechanism depends on a particular aspect of physical objects that is responsible for the resulting changes. These are *measured attributes, form* and *appearance*, respectively. Though physical objects have only two *physical* aspects — measured attributes and extended structure — we describe the latter as form or as appearance depending on whether the causal mechanism is pattern matching or structure-preserving superposition, respectively.[1]

1). *Nomologically-determined change* is the causal mechanism that underlies most physical interactions. Exemplified by what is customarily described in the literature as "billiard ball collisions", the effects of such interactions are determined according to nomological relationships between measured attributes (e.g. momentum) of the colliding objects. When two billiard balls collide, the outcome of the interaction is determined by the law of conservation of momentum along with constraint relationships that depend on structural aspects of the particular situation, such as the angle of closest approach. Thus, the changes that result from an interaction depend only on the *values of measured attributes* of the colliding objects, which determine the magnitude and direction of the forces that bring about those changes.

Informationally (i.e. if measured attributes are taken to represent), the changes are not indicative of the particular *objects* which interacted, only of the values of their measured attributes. Certainly initial measured-attribute values of the particular objects will cause specific value changes, but these are situation-specific, not object-specific — they do not identify particular objects — since a) there are, in essence, an infinite number of configurations for two objects to be in when they collide and, therefore, an infinite number of different values to describe them, and b) this is true for *any* two objects. Analog computers are exemplary of systems that utilize nomologically-determined interactions informationally; specific measured attributes of their component parts are taken to represent quantities in mathematical and physical models of different phenomena and the interactions of these parts are engineered to produce value changes which correspond to value changes of the represented quantities in the models.

2). *Pattern matching* is the *physical* process underlying many biomolecular interactions, such as enzyme catalysis, as well as computational changes in digital computers. Unlike nomologically-determined change, pattern matching physically depends on the *forms* of interacting objects because a successful pattern match can only occur if the pattern and matcher structurally "fit". The values of measured attributes of pattern and matcher are not relevant to the pattern matching process except insofar as they physically enable it to happen. That is, structure fitting involves forces like any other physical interaction. Indeed, if the interacting structures, such as a key and a door lock, do not fit, there is no set of measured-attribute values that could lead to an outcome which would have been produced if they had.

Thus, pattern matching depends on the structural forms of interacting objects. The actual change caused by this kind of interaction, however, is "simple" (Pattee, 1986) in that it does not embody or transmit structural features of the pattern, and, in fact, is generally structureless — e.g. the switching of a computer circuit voltage from "high" to "low" as the output of an electronic comparator. Because the pattern is matched *as a whole*, we say that its *form* causes the change. Informationally, the particular pattern is relevant only to the extent that there is a matcher which fits it. This is the reason symbols in formal symbol systems can have any form as long as they admit of a consistent interpretation.

3). Like pattern matching, the third causal mechanism, *structure-preserving superposition*, or

[1] Because of space limitations, these claims about the existence of only three causal mechanisms and two physical aspects of objects will have to remain unsubstantiated, though we do consider the latter to be self-evident. Objects also have functional and various relational aspects (e.g. part/whole), but these are not physical aspects.

SPS, depends on extended structure, but in a very different way. Whereas pattern matching is based on the existence of two structures which fit, that is, on the forms of *both* pattern and matcher, SPS actually causes a change that is the *transmission* of a pattern, like a stone imprinting its surface structure in a piece of soft clay, so that the *effect* is a structural formation of the specific features of the pattern's extended structure; that is, its *appearance* rather than form. Informationally, the structure of the input is transmitted to the system receiving it, in contrast to pattern-matching systems whose constituent matchers recognize input patterns, but do not transmit them. SPS is "automatic" in that, as a physical process, it can create new structures simply by physically superimposing structures.

To reiterate, the above three causal mechanisms are the only ways physical objects cause change; there are no other ways that one physical object can affect another except by one (or both) of its only two *physical* aspects: measured attributes and extended structure. Insofar as physical objects can be taken to represent, these causal mechanisms explain how what we tend to call information affects the behavior of information processing systems. Thus, they serve as a critical link between information and the physical world. But only in the cases of nomologically-determined change and SPS are the representing entities actually changed. In pattern matching, extended structure is used only to control nomological changes, such as voltage switching.

Causal Mechanisms and Types of Meaning

It was suggested above that form isomorphism between representation and referent is not sufficient for intrinsic meaning; just because a symbol looks *to us* like it represents does not mean it is not arbitrary to the system within which it is embedded. We are not talking here about the kind of arbitrariness that would result from our designating a tree bitmap to represent a cow, for example, but, rather, the arbitrariness that arises when the form of a symbol does not matter to the change it produces, which is the case for pattern matching systems.

In pattern matching systems (which include all artificial information processing systems except analog computers) the matcher and pattern physically fit, so that the forms of symbols and, hence, the structural similarity of symbol and referent, does not matter. *Any* form can be used to trigger a particular effect because form is used strictly for control. For example, the information about tree structure could be encoded as a bitmap (iconic representation) or a textual description (propositional representation), but in both cases matchers that fit the representing forms have to be present in order for that representation to be causal, i.e. for it to affect the system's functioning. The result of a match is a structureless change which *triggers* the next informationally-relevant physical change, such as the execution of a subroutine. Thus, for all pattern matching systems, which includes digital computers as well as current connectionist systems (Boyle, 1991), the meanings of representing entities are *projected* because the physical process of pattern matching eliminates any presumed functional significance from their forms, regardless of how they encode what they purportedly represent. Patterns in such systems seduce us into believing they are inherently meaningful because, in fact, they are to us. But they are not inherently meaningful to the systems precisely because matchers physically fit them, i.e. their appearances are not relevant to their functioning. Only *output* has no matcher, so *its* appearance *does* matter; that is, the appearance of the output determines the interaction of the system with its environment. But "inside" the system there are no such criteria for constraining structure. In effect, structure fitting renders the internal behavior "mindless".

In contrast to the "arbitrary" encodings of referent structures in strict pattern matching systems, arbitrary whether encoded by us or procedures we write for accepting input, *symbol grounding* involves what Harnad (1990) calls the "analog re-presentation" and "analog reduction" of sensory stimuli which generate perceptual category representations that are *not* arbitrary with respect to their referents. Based on his description, we take these analog processes to be examples of SPS. The resulting iconic structures that form perceptual categories are then associated with abstract symbols. The meanings of these symbols are not exactly projected because the relationship between symbol and referent is *physically* grounded in the sensory input; that is, SPS enables the extended structure of the input signal to directly create perceptual representations by transmitting them (in pattern matching systems, the input is not transmitted but *encoded* through a set of matchers).

However, if SPS is involved only in the *formation* of symbols, then we claim that their meaning is not intrinsic because subsequent to their formation, their extended structures are matched. Their grounding may be "fixed", but if they are not causal through SPS, then their appearances are no longer relevant and, therefore, meaningful to the system -- that is, they become causal through pattern matching, in which case they are like symbols in any pattern-matching system. In other words, SPS grounds the structural relationship between symbol and referent, but from then on the symbols behave as *form*al patterns. There is nothing about their particular structures that is necessary for the specific changes they bring about. Only the presence of identically structured matchers is important. We could have done as much by encoding them ourselves because once the symbols become patterns to be matched, any groundedness they had is superfluous to their effects on system functioning. Nevertheless, we will refer to their meaning as grounded, which, in essence, is projected meaning with a non-arbitrary form relationship between symbol and referent.

According to the present thesis, only if these initially grounded symbols are subsequently causal through SPS would their meanings be intrinsic. To be meaningful to a system, they must cause changes which actually embody their structural features, not be "collapsed" into a formless outcome. Thus we believe SPS to be the physical basis for semantics and, hence, the causal mechanism underlying cognition, which is partly supported by evidence from sensory perception, in the form of retinotopic mappings on the primary visual cortex, for example. Furthermore, as Churchland (1989) notes, "there are many other cortical areas, less well understood as to exactly what they map, but whose topographical representation of distant structure is plain." Thus, it is SPS which we offer here as the fundamental difference between symbols in computers and brains; that the latter are semantic while the former are only syntactic. SPS, we believe, is Searle's hypothesized "causal property".

Summary and Conclusions

Intrinsic meaning is identified here with representations that have a causal capacity to effect physical change through structure-preserving superposition or SPS. Without this causal mechanism underlying physical change in a representing system, any meaning associated with the representation is projected onto it by us. Physical systems exhibiting this latter property are pattern-matching systems, such as digital computers (pattern matching is *their* underlying causal property). Grounded meaning is exhibited by symbols in systems which have a structure-preserving relationship between internal representations and their referents, but whose subsequent effects on system behavior are enabled through pattern matching — their meaning is projected, though they are rooted in a non-arbitrary form relationship with their referents.

In summary, we have tried to show that there is a plausible physical explanation for the apparent differences in meaning and understanding possessed by computers and brains that begins to forge a connection between our intuitions about mind and the physical world. It is based on a previously unexplored analysis of causality; *how* cause effects change. Its implications are that purely syntactic structures are those which are causal through pattern matching, while semantic structures are those which are causal through SPS and, hence, are those whose meanings are intrinsic. This further implies that pattern matching systems may never be able to instantiate the set of causal relationships between mental states and, therefore, may not be capable of simulating mind because the physical process of pattern matching is fundamentally different from SPS. We speculate that this shortcoming will likely manifest itself as a learning deficit.

References

Boyle, C. F. 1991. On the Physical Limitations of Pattern Matching. *Journal of Experimental and Theoretical Artificial Intelligence* 3:191-218.

Churchland, P.M. 1989. *A Neurocomputational Perspective*. Boston, Mass.: MIT Press.

Harnad, S. 1990. The Symbol Grounding Problem, *Physica D* 42:335-346.

Haugeland, J. 1989. Artificial Intelligence and the Western Mind. In J.R. Brink and C.R. Haden (eds), *The Computer and the Brain: Perspectives on Human and Artificial Intelligence*. Elsevier.

Palmer, S. E. 1978. Fundamental Aspects of Cognitive Representation. In E. Rosch and B. Lloyd (eds), *Cognition and Categorization*. Hillsdale, NJ: Lawrence Earlbaum.

Pattee, H.H., 1986. Universal Principles of Language and Measurement Functions. In J.L. Casti and A. Karlqvist (eds), *Complexity, Language and Life: Mathematical Approaches*. New York: Springer-Verlag

Searle, J. 1980. Minds, Brains and Programs, *Behavioral and Brain Sciences* 3:417-457

Searle, J. 1990. Is the Brain's Mind a Computer Program? *Scientific American* 262(1):26-31

Thagard, P. 1986. Parallel Computation and the Mind-Body Problem. *Cognitive Science* 10:301-318

Seeing is Believing: Why Vision Needs Semantics

Matthew Brand, Lawrence Birnbaum and **Paul Cooper**
Northwestern University
The Institute for the Learning Sciences
Evanston, Illinois
brand@ils.nwu.edu

Abstract

Knowledge about the functional properties of the world constrains and informs perception. For example, looking at a table, chair, a building or a sculpture, we are able to resolve occluded attachments because we know that in order to stand, an object's center of gravity must lie within its footprint. When when we see a floating wheel in the interior of a vehicle, we know that it is probably the means by which the driver communicates steering information to the chassis. Movable handles imply input to machines; fixed handles imply an upside and a downside to any object they grace.

We are constructing a machine-understanding machine with which to explore the usefulness of semantics in perception. This system will investigate simple mechanical devices such as gear trains, simultaneously building a representation of the structures and functions of parts, and using that representation to guide and disambiguate perception. In this paper we discuss how this work has led to an understanding of perception in which a semantics of structure and function play a central role in guiding even the lowest level perceptual actions.

Vision is Cognition

We distinguish ***visual understanding*** from ***visual recognition*** by the central questions that drive the two activities. For recognition, the question is "What is out there?" For understanding, the question is "What is happening/can happen in this scene?" or more specifically, "How can I interact fruitfully with the scene?"

Humans see and understand the world in terms of its affordances [Gibson 66], which signal the potential for function and for interaction. To see and act purposefully, robots must likewise be designed with a capacity for the visual understanding of the affordances of their worlds [Brand & Birnbaum 92].

Visual understanding is, firstly, explaining the scene with regard to the goals and causal knowledge of the viewer, and secondly, explaining the image with regard to the scene[1]—what is known as image understanding[Birnbaum et al. 92]. An explanation should capture the why and how of a scene: the causal relations between objects, the sources of motion and stability, and the potential uses of these causal properties for the viewer. For example, in the reduction engine pictured in figure 1, two gears are causally related to each other in that they will transmit (and reverse) rotational motion. The handle is causally related to the viewer in that it affords the viewer an opportunity to inject motion into the situation. A visual understanding of figure 1 will include the assessment, "This is a device which, when powered from one handle, causes the opposite wheel to rotate at a much higher torque and slower speed."

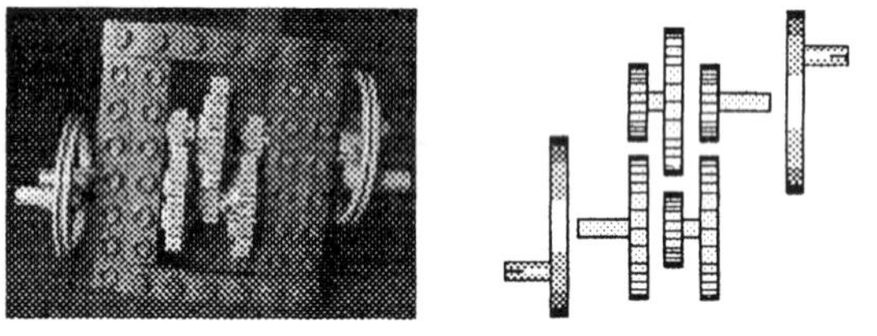

Figure 1: A reduction engine as it appears to the camera, and an exploded schematic of its drivetrain. To understand it, we piece together a coherent explanation of the what, why, and how of this drivetrain, using clues from the image, from knowledge of function and structure, and from new views procured by perceptual acts.

Our prescription for visual understanding dispenses with the conventional notion of an order of processing (e.g. [Marr 82, Barrow & Tanenbaum 78][2]). Instead of a visual front end which outputs image segmentation descriptions for a back end to use, understanding is a matter of negotiation between the constraints and hypotheses of a generative semantics of function and the activity of low-level visual routines. The semantics guide the activity of the visual hardware through queries: in the course of building an explanatory model

[1] I.e. what is traditionally called computer vision: grouping high-contrast gradients into edges, finding flow boundaries, matching to models, etc. [DARPA 92]

[2] This filter-then-analyze approach has has previously been called into question by, among, others, Tanenbaum himself; see [Witkin & Tanenbaum 83]

from visual reports of clues to function, gaps and inconsistencies in the model to are used to general queries for the visual routines, which cue various perceptual acts that result in new reports.[3] The visual processes answer queries by testing for features and tracking invariants in the scene that have functional significance, such as tracking parallel lines (generated by the edges of a rod) to find the end of the rod. We are in the process of constructing such a system and have analyzed several image sequences in the manner we suggest. This papers describes aspects of a number of these analyses.

Why and how are functional properties detectable in an image? Scenes are structured, and the causally 'loaded' regions of a scene tend to be where parts interface: where they are joined or where there are contrasts in motion. This means that the parts of the scene where change is most likely and most significant—the parts of interest to a robot—often have characteristic manifestations at predictable locations in the image. For example, meshed gears produce adjacent regions of optical flow with opposite curl. At the junction itself, the flow will converge, then diverge. Statically, a gear meshing introduces characteristic textures into the image because of the toothing, and this manifests itself as a local peak of a high-frequency component in the image. This is what robots would look for if they were made to fix car transmissions.

Understanding just a small part of a picture—even a single component of a structure—immediately yields a rich set of expectations about neighboring part boundaries, structural concomitants, typical axes of motion, and so forth, and these in turn have characteristic manifestations in the image. This is because most things in our visual experience have the *quality of design*: their construction reflects a host of functional constraints. Even the simplest functional constraint—resistance to the pull of gravity—profoundly influences design and appearance, and generates for us many expectations that guide visual cognition. This is equally true within and outside the realm of man-made objects: The world is pervaded by function.

In this paper we present the beginnings of a generative functional semantics for vision, with enough detail to account for example scenes ranging in complexity from sticks and strings to common machines.

The Importance of Being Connected

It is generally understood that the causal properties of the scene are usually mediated by physical connections between the parts it contains. Understanding an object or scene requires visually tracing through the causally most "loaded" connections between subparts.[4] Toward this end, we have been developing a catalog of connection types, in which each connection is indexed along with a description of function, typical structural correlates, and characteristic visual manifestations. We now have a rich catalog of mechanical connections ranging from E-clasp fasteners to gear meshings to hub-axle interfaces—nearly 40 connection types at time of writing. The descriptions of function and structural correlates provide great leverage in visual search, generating hypotheses about neighboring parts, as-yet-unperceived assemblies, and the relative locations of parts.

Knowledge about connections provides a reasoner with a special and highly useful set of expectations about the world. In order to use this knowledge, we also need to have good theories of how and why parts are put together, and of the capabilities of our vision system to extract useful features and invariants from the image. This requires a large rule base which expresses the principles of rational design, and which describes—in terms of the visual routines—the perceptible artifacts of design. Design semantics tell us a good deal about what kinds of image processing we need. This is true of both abstract and specific constraints. At the abstract level, for instance, we have a constraint such as the following:

> A drivetrain assembly has function if it transduces, regulates or switches motion. In visual terms, this function is manifuest in the following rule: *A patch of the scene is explained if it connects to two patches of differing motion (transduction); if it connects to just one patch of motion but appears to have significant mass (regulation); or if over time, its position relative to connecting patches changes so that their optical flow is no longer related (connection and disconnection)*

Similarly, at the specific level, we have a rule such as the following:

> Most axis- and rail-mounted machine parts have some symmetry with respect to their axes of motion so to reduce vibration (and simplify manufacture). This includes gears, carriages, and pistons. In visual terms, this functional constraint implies the following rule: *For most moving parts, there is a way to orient the camera so that motion of the part causes a minimal change in its visual profile.*

We have developed a set of such rules sufficient to produce explanations for the objects pictured in the paper. This knowledge combines with an explicit (if somewhat simplified) theory of the image-processing and camera-orienting subsystems to make predictions about which visual routines (e.g. [Ullman 84]) to engage and what misclassifications they can make about features in the image. To describe the vision subsystem, we identify the assumptions and strategies built into its camera-orienting and feature-extracting processes, and then produce characterizations of when and how various low-level routines will produce spurious reports:

- A change of perspective usually suffices to distinguish adjacency from occlusion. *A report of non-*

[3] This is similar in spirit to work in text-understanding by [Ram 89]

[4] Indeed, when we ask colleagues to look at the objects and pictures in this paper, we see them visually trace out the "functional drivetrain" of an object.

adjacency from the visual system is reliable; reports of adjacency can be mistaken if the parts are close.

- A gradual dip and then recovery in the frequency of the strongest signal component taken along a line through the image implies a periodic texture mapped onto a curved object, such as a gear. [Bajcsy & Lieberman 76] *If the camera is not oriented in the plane of a gear before using the visual toothed-wheel detector, there may be spurious negative reports.*

In sum, not only do we need knowledge of the causal structure of the world, but we need knowledge of how that causal structure is revealed (and sometimes mistaken) by perceptual actions. One kind of knowledge tells us what is missing or wrong in our explanation of a scene; the other kind tells us how to find missing information in the image, or where to find mistaken interpretations in the explanation.

Examples

Figure 2: Views of a tensegrity object standing up, on its side, and from above. Different views lead to different explanations.

Tensegrity

A tensegrity [Fuller 75] is a rigid structure of rods and cables. The simplest possible construction, consisting of three rods and nine cables is pictured in figure 2. None of the rigid elements touch each other, yet the whole structure stands. People find tensegrity constructions fascinating because a very basic assumption of visual explanations fails to apply, namely that a rigid object is decomposable into substructures that support each other [Birnbaum et al. 92]. The only means of connection in the tensegrity is tethering; there is no support and only the illusion of suspension. In fact, gravity plays no role in its stability.

It is, however, the illusion of suspension that allows the tensegrity to be explained. A first view of the structure will reveal a large part (a rod) which looks as if it should be falling. To explain its stability, one scans up the rod, looking for an attachment which prevents it from falling in the direction that it leans. Near the top, a cable from another rod prevents this collapse. However, this does not explain why the rod doesn't pitch in a direction perpendicular to the cable, and a further scan reveals a nearly perpendicular cable which partially fulfills this function: it keeps the rod from falling "outwards." To explain why it does not fall inward, we look for a third fixating element, and find a third cable attached to the endpoint that has a small vector component contrary to inward motion. The rod endpoint is now considered stable, as all motions are restricted (some, apparently, by gravity). The rod as a whole, however, may not be. Thus, to explain why the rod doesn't slide out from underneath itself, a similar set of scans discovers three more tethers. Now, the rod is provisionally considered stable. Yet all the cables need to be explained, and this leads to similar explanations of the other two rods. At the end, every part has been assigned a function, and every force appears to have been countered.

However, this results in a circular explanation, where each part is, ultimately, held up by itself. In order to "ground" this explanation, we must invoke the principle of symmetry. Symmetry is a design stratagem for canceling out all forces. It is necessary to know about symmetry, and how to look for some kinds of symmetry in an image, if one is to explain why static objects stand up. Symmetry is the most common form of balance, which is often the ultimate explanation of stability.

Symmetry is also a way of resolving explanatory loops. For the tensegrity object we propose a three-fold rotational symmetry around a vertical axis, and orient the camera above the structure. The endpoints are used to estimate where the symmetrical axis is, and once the camera is collinear with this axis, a visual routine processes the image to find evidence of rotational symmetry. Finding symmetry completes the explanation.

Reduction Engine

A reduction engine works on the principle that a small gear connected to a large gear will reduce speed and increase torque. To explain such a machine, the input and output must be found, the drivetrain must be traced, and the parts that serve to frame and stabilize the object must be identified. The order of discovery of all these assemblies is not important—finding any one or part of any one produces many functional clues about where and how to look for other parts.

For example, finding a protuberance from the face of a wheel (an ellipse in the image) is a good indicator of an axis or handle. An ellipsis-finding Hough transform will tell us where to expect the axis. If the protruberance is off-center, then it is a handle, which indicates that the part is an input or output to the machine.

In the reduction engine, a wall lies directly behind the wheel, so the axis is invisible. However, it is reasonable to expect that the axis is fixed in place by the frame, so the wall is hypothesized to be part of the frame, and the axis is hypothesized to pass through it. Scanning along the line of the hypothesized axis brings a toothed-texture into center view, which can be verified as a gear with the appropriate curvature for the axis. The axis is now provisionally explained.

To explain the gear, it must mesh with at least one other gear (or a chain). [Brand & Birnbaum 92] describes a system for scanning a camera across a train of meshed gears, reporting when a bounding wall has been hit or no more gears have been found. When this finds a meshing gear, the first gear is explained. Explaining this second gear requires looking for an axis to carry along the motion to another part, since no other connecting gears can be found. The axis is almost entirely hidden, so the same strategy that verified the first axis is used. At this point, the operations just described repeat to explain the remainder of the mechanism.

Other Examples

Even without the ability to move the camera to scan for new information, functional expectations will resolve ambiguities in the interpretation of a scene. [Halabe 92] has implemented a program with a modest semantics of attachment and stability that will "reattach" legs of Tinkertoy constructions that have been "severed" by occlusion.

Figure 3: A house of cards and a toy horse. Stability constraints make it possible to reason about occlusions such as the horse's hind leg and the obscured card.

Similar analyses have been done with various houses of cards (e.g. figure 3), in which connections mediate support or friction, but there is no attachment.

A Functional Analysis of Vision

We believe that the use of functional semantics in the design of vision systems and visual primitives applies to the whole range of systems that process images and/or perform visual reasoning. Whether designing a hardwired animat, or compiling the knowledge to be used in a mechanical reasoning system, a functional analysis will outline the kinds of features that are needed for choosing actions [Brand 91], the kinds of image-predicates that are necessary to support those features, and the kinds of ambiguities that the system will face.

Semantic constraints are pervasive in the world, thus we need a functional semantics for every kind of scene. We outline below the main functional relationships inherent in different kinds of scenes, to show the basic building blocks of visual semantics for different tasks. We identify three general types of scenes, briefly sketch the fundamental questions that drive explanation in each, outline the basic principles used in these explanations, and describe how they may be detected in an image.

Static Objects

- Will it fall apart? (How is the motion of all parts constrained?) The means of static constraint are blockage (support and containment are special cases), attachment, and tethering (again, suspension is a special case). Blockage is detectable though adjacency in the image, especially in the horizontal plane, where one part is supporting another. Attachment can be inferred from off-horizontal adjacency, partial containment, and characteristic attachment artifacts such as screw and rivet heads. Tethering can be inferred from adjacency to the end of a long thin object such as a cable.
- Will it fall on me? (How is the center of gravity placed within the footprint? Or how is the object affixed to something heavier?) Typical means of standing include spread feet (or narrow tops), counter-balancing, and anchoring. Spread feet are often visible as protuberance on the ground plane diverging from the object-image center. Narrowing can be calculated, for example, as a gross geometric predicate, or by looking for finer detail higher up in the image (e.g. a greater proportion of high-frequency components). Counter-balancing and symmetry are profoundly difficult to find in an image; we are compiling a host of methods, including looking for anomalously thick or long projections to diagnose counter-balancing. Anchoring often requires projections into the ground plane, often accompanied by bumps in the plane (e.g. tree roots).
- What can it hold up? (What devices of support, attachment, etc., does it have that are not used in its own skeletal integrity?) This is often a matter of identifying objects which afford support or attachment but do not participate in the explanation of the object's stability. Unused high horizontal surfaces (tabletops), hook shapes or vertical points (coatstands), and regions of concavity (bowls) are good indicators of overall function.

Objects with internal motion (Machines)

- How is motion constrained and channelled[5]? In machines, the means of constraining motion always leave a dimension or two of freedom. This is principally achieved by partial containment (eyes, hubs, sockets, etc.) in the man-made world, and by flexion in the natural world. This is a difficult problem for us, since most of a containment device is obscured from view. At present, we plan to simply infer containment devices from the limited motions of parts. There is some potential in developing a library of visual signatures for containment devices, much as the screw-head is a signature for a largely invisible part.

[5]This is very similar to the question asked of static objects. In fact, we had analyzed several machines before realizing that static objects are a special case, in which *all* motion is restricted.

- Why are all these parts moving? (How is motion communicated? What kinds of connections are there?) The principal means for communication of motion are attachment and friction. Communication produces characteristic patterns of flow in adjacent regions. Optical flow algorithms may only suffice to reveal regions of varying motion, requiring other visual processes to close in on and resolve details of how motion is communicated.
- What kind of motion does this produce? Classification of motion into rotation, translation, lifting, swinging, hammering, etc., provides a useful index to function, and often suggests a likely mechanism. For example, repeated translational motion along a line almost always requires an associated rotational motion.
- How do I connect with it? (What is the interface to the rest of the world?) This is similar to the use question asked of static objects. There is a fairly limited range of control devices which specifically interface to the human hand, and which have characteristic shapes: handles, buttons, dials, and steering wheels, for example. These will have to be resolved by local searches in the image for characteristic shapes.

Terrains

- Where are the animate objects? (What's moving and what are our relative positions in the food chain?) This is largely a matter of noticing independent translational motion in the image sequence. Visually, we look for small regions of depth change, as well as texture anomalies.
- Where can I pass or flee? (What part of the terrain is navigable for an agent with legs or wheels like mine?) The most important constraint for land navigation is continuity of ground plane, followed by smoothness. Another important affordance for navigation are things that can be climbed. For this reason, it is useful to look for low-frequency texture on objects that rise out of the ground plane, for example a tree with rough bark. One special case–stairs–adds the constraint that the vertical texture have a single strong frequency component.
- Where can I take shelter or hide? (What part of the terrain has limited accessibility and/or limited visibility?) The key to this function is identifying places in the world where vision itself doesn't work very well. One hides in caves or overhangs, which are bounded regions of relative darkness and low contrast, or one hides in underbrush: areas of omnidirectional high frequency image noise.

Vision Requires Outlook

Vision has long suffered the notion that an artificial visual cortex will be a "front end" for an intelligent system that itself is not necessarily visually sophisticated. A consequence of this view is that much talent and energy has been invested in trying to find an appropriate form of output for vision systems. Once an output representation has been invented, there is the usual struggle of finding a robust algorithm to map images to reasonable (literally) outputs. This has typically resulted in recognition systems, which match the image to a database of models via reverse optics transformations (e.g. [Horn 86]). We have learned from this work that no single algorithm or image transform ever works more than perhaps 80% of the time[6].

Recently, some researchers have given attention to the *use* of visual processing, that is, what happens in the "back end" (e.g. [Ballard 89]). This has led to a reformulation of vision in which processing is specifically aimed at quickly extracting the features that are most decisive for the immediate pursuit of a goal. In the "active vision" paradigm, the back end is reciprocally considerate of the front end, reorienting the camera to procure ever better input for the feature detectors. This is typical in visual navigation systems, which extract surprisingly few topographic features from the image, and then make strikingly good use of them. This is a significant development because it incorporates the notions of (1) functionally derived features and (2) focus of attention, both deployed according to an analysis of the dynamics of the task.

Recognition vision builds a model of the scene by explaining the image in terms of the physics of light and the configuration of the scene. It incorporates analyses of optical physics and of shape, which give it a mathematical, nonfunctional slant. Active vision work tends to be miserly in its representation, but tries to participate directly in the causality of the scene. It incorporates analyses of the task and of visual invariance across motion, which give it a decided functional slant and, significantly, a good measure of robustness[7].

What is missing in vision, though hinted at by active vision, is a functional analysis of the *world*—of what is being looked at. The purpose of vision is not to describe the image in terms of segmentation candidates, but to explain the scene in terms of what we believe about the world. The primary visual belief that humans enjoy is the dictum that "form follows function." The world that we see is one of design, everywhere imbued with function, and interesting mainly because we have to interact with it.

The questions we ask of our eyes are: "Will it fall on me?" "Will it support my weight?" "Where can I pass?" "What does it do?" These functional questions lead straight to structural questions: "Does the center of gravity lie outside the footprint?" "What are the load-bearing lines?" "Where is the ground plane navigable by foot?" "How does its motion relate to a human activity?" The structural questions in turn lead to questions posed of the world (of the image or of an image stream): "Where above the ground plane is the visual centroid?" "How thick is the train of connect-

[6] Minsky, personal communication

[7] Active vision aims to reduce uncertainty through tracking; thus the importance of invariance across motion.

ed substructures that rises from the ground plane to carry my weight?" "Where is the illumination gradient smooth or striped (steps)?" "Where is a handle-shaped object and the drivetrain that it moves?"

One might object to our emphasis on questions such as, "Why is this part here?" and, "How do these things relate?" when humans seem able to answer, "What is out there?" so effortlessly. Humans have prodigious visual memories, and equally uncanny powers of recognition. However, it is not recognition we are trying to explain; it is the original cognition. Given the amount of work this takes, it is not surprising that we are equipped with a caching mechanism which uses the memory of the first cognition to speed perception of the same object later on.

Related Work

Recent work in the understanding of diagrams indicates that researchers have found it useful to employ a simple semantics in conjunction with a simulated visual search for "regions of interest" in the diagram. [Narayan & Chandrasekaran 91] give an example of a straight flat line that is a shared boundary between two objects, which consequently have the potential to slide against each other. [Forbus et al. 87] provide a model for the qualitative analysis of rigid body interactions, given a qualitative description of the scene. Both are primarily post-visual paradigms, whereas we intend for our semantic analyses to interactively guide and disambiguate visual processes. It is also worth noting that most kinematic analyses of scenes, whether qualitative, diagrammatic, or truly visual, use a semantics of *motion*. In contrast, we are interested firstly in a semantics of *function*; and only secondly in its manifestation as motions, shapes and textures.

The work of the Vision and Modeling Group at the MIT Media Lab is also of note because, in trying to model the objects in the scene in terms of bent and deformed superquadrics [Pentland 90], they are also, in a sense, explaining the scene. This interesting approach differs from ours in that it is functionally neutral; such explanations tell how the scene could be made from simple lumps of clay that are deformed and combined to produce complex shapes. No hypotheses about causal relationships and function are present in these explanations, nor does such knowledge guide explaining at the level of image-processing either. However, their work has interesting possibilities because the models, once constructed, are imbued (via simulation) with a causality which includes rigid and elastic body dynamics, mass, and gravity. This could be used to provide feedback to an image-to-model constructor, by telling it whether or not the model is stable and static, or unbalanced and lacking in structural integrity.

Acknowledgements

Thanks to Ken Forbus, Dan Halabe, Bruce Krulwich, Peter Prokopowicz, and Louise Pryor for many useful discussions. This work was supported in part by the Defense Advanced Research Projects Agency, monitored by the Office of Naval Research under contract N00014-91-J-4092, and by the National Science Foundation, under grant number IRI9110482. The Institute for the Learning Sciences was established in 1989 with the support of Andersen Consulting, part of The Arthur Andersen Worldwide Organization. The Institute receives additional support from Ameritech, an Institute Partner, and from IBM.

References

[Bajcsy & Lieberman 76] R. Bajcsy and L. Lieberman. Texter gradients as a depth cue. In *Computer Graphics and Image Processing* **5**, 1976.

[Ballard 89] Dana H. Ballard. Reference Frames for Animate Vision. *Proceedings of AAAI-89*, 1989.

[Barrow & Tanenbaum 78] Recovering Instricsic Scene Characteristics from Images. *Computer Vision Systems*, A.R. Hanson & E.M. Riseman (eds.), New York: Academic Press. 1978.

[Birnbaum et al. 92] Lawrence Birnbaum, Paul Cooper, and Matthew Brand. *Every Picture Tells a Story.* Forthcoming Technical Report, Northwestern University, The Institute for the Learning Sciences. 1992.

[Brand 91] Matthew Brand. Incorporating Resource Analyses into an Action System. 12th *Proceedings of the Cognitive Science Society*, 1991.

[Brand & Birnbaum 92] Matthew Brand and Lawrence Birnbaum. Perception as a Matter of Design. In *Proceedings the AAAI Spring Symposium on Selective Perception*, 1992.

[DARPA 92] *Proceedings of the 1992 DARPA Image Understanding Workshop.* San Mateo, CA: Morgan Kaufmann Publishers, Inc. 1992.

[Forbus et al. 87] Ken Forbus, Paul Nielsen, & Boi Faltings. Qualitative Kinematics: a framework. *Proceedings of the* 10th *IJCAI*, 1987.

[Fuller 75] R. Buckminster Fuller. *Synergetics.* New York: MacMillan Publishing Co., Inc. 1975.

[Gibson 66] J.J. Gibson. *The Sensense Considered as Perceptual Systems.* Botson: Houghton Mifflin, 1966.

[Halabe 92] Daniel Halabe. *A Leg to Stand On.* Unpublished manuscript, Northwestern University, The Institute for the Learning Sciences, 1992.

[Horn 86] Berthold Klaus Paul Horn. *Robot Vision.* Cambridge, MA: The MIT Press, 1987.

[Marr 82] David Marr. *Vision.* New York: W.H. Freeman and Company, 1982.

[Narayan & Chandrasekaran 91] N. Hari Narayan and B. Chandrasekaran. Reasoning Visually about Spatial Interactions. *Proceedings of the IJCAI*, 1991.

[Pentland 90] Alex Pentland. *THINGWORLD 2.0.* Vision and Modeling Group, The Media Lab, MIT. 1990.

[Ram 89] Ashwin Ram. *Question-driven understanding.* Dissertation, Department of Computer Science, Yale University, 1989.

[Ullman 84] Shimon Ullman. Visual Routines. *Cognition* **18**, 1984.

[Witkin & Tanenbaum 83] Andrew Witkin & Jay Tanenbaum. On the Role of Structure in Vision. In *Human and Machine Vision*, Beck, Hope, & Rosenfeld, (*eds.*), 1983.

A Case-Based Approach to Problem Formulation

L. Karl Branting
Department of Computer Science
University of Wyoming
Laramie, Wyoming 82071-3682
karl@eolus.uwyo.edu

Abstract

In domains requiring complex relational representations, simply expressing a new problem may be a complex, error-prone, and time-consuming task. This paper presents an approach to problem formulation, termed *case-based formulation* (CBF), that uses previous cases as a model and a guide for expressing new cases. By expressing new problems in terms of old, CBF can potentially increase the speed and accuracy of problem formulation, reduce the computational expense of retrieval, and determine the relevant similarities and differences between a new case and and the most similar old cases as a side-effect of expressing the new case. Three forms of CBF can be distinguished by the extent to which the retrieval and adaptation of previous cases are automated and the extent to which the facts of multiple cases can be combined. An initial implementation of one form of CBF is described and its ability to use previous cases to increase the efficiency and accuracy of new-case formalization is illustrated with a complex relational case.

The Task of Problem Formulation

Problem formulation, the expression of a problem in a representation amenable to manipulation by a computer, is an essential step in every form of automated problem solving. In systems that use featural representations of cases, problem formulation is typically quite straightforward. In MYCIN, for example, a new case is described by specifying values for parameters appearing in subgoals during a consultation. Similarly, a new case is represented in Protos as a vector of feature-value pairs. Problem formulation is easy in such systems because they use featural representations that have been engineered to represent only those aspects of cases known *a priori* to be relevant to the single specific task for which the systems were designed.

However, there is a growing recognition that featural representations are inadequate for a wide variety of applications. General-purpose knowledge bases intended to support a variety of different tasks clearly cannot use featural representation languages capable of expressing only case attributes relevant to a single task. Instead, such knowledge bases require relational representations that are capable of expressing a wide variety of relations among domain entities (Lenat and Feigenbaum, 1991; Porter et al., 1988). Moreover, featural representations

> cannot be used in areas requiring essentially relational knowledge representations. These areas include temporal reasoning, scheduling, planning, qualitative reasoning, natural language and spatial reasoning. Problems also occur in other areas involving arbitrarily complex structural relationships such as prediction of protein folding and DNA gene mapping (Muggleton, 1991).

However, relational representations are typically much more complex than featural representations because relational information implicit in the latter is made explicit in the former. This complexity can drastically complicate the process of formulating new cases. For example, in the context of a qualitative simulation program such as QSIM, a case consists of a set of qualitative differential equations specifying the structure of a physical system (Kuipers, 1989). Qualitative differential equations are represented in a relational language capable of expressing qualitative constraints among domain variables. The price of the expressive power of this representation is that creating and debugging qualitative differential equations is a complex, lengthy, and error-prone process (Farquhar et al., 1990).

The practical consequences of the difficulty of problem formulation in relational representation languages are illustrated by GREBE, a legal reasoning system that integrated case-based with rule-based reasoning (Branting and Porter, 1991; Branting, 1991a). GREBE used a representation language in which arbitrary causal, temporal, and intensional relations could be stated explicitly. This representation contributed significantly to GREBE's performance: in a preliminary evaluation, GREBE's analysis of 18 worker's compensation hypotheticals was found to compare well with analyses by law students (Branting and Porter, 1991). However, the expressiveness of this representation came at a cost. Representations of GREBE's test cases consisted on average of 89 propositions, each of which had to be entered by hand. Entering cases of this size was a lengthy process—typically several orders of magnitude longer than GREBE's run-time—and the resulting representation often required considerable debugging. Moreover, it was the experience of the legal reasoning group at the University of Texas that different knowledge enterers often chose to represent identical facts differently, creating the danger of inconsistent analyses of equivalent cases. Limitations in problem formulation, rather than problem solving, prevented GREBE from being usable in any practical setting (Branting, 1991a).

This paper proposes an approach to problem formulation in which previous cases are used as a model and a guide for expressing new cases. The use of existing information as a model for expressing new information makes it possible to "use what we know to help us process what we receive" (Schank, 1982). I refer to this approach to problem formulation as *case-based formulation* (CBF).

The Elements of Case-Based Formulation

The fundamental assumption underlying CBF is that new situations can be efficiently formalized using previous cases as models. CBF exploits the phenomenon that useful knowledge seldom consists of isolated facts, but instead tends to consist of collections of related facts. A simple example is a frame. The object/slot/value triples constituting a frame can be viewed as a collection of propositions that are related because they all concern the same object. In the context of CBF, a case can be any fact-collection/abstract-description pair[1]. The fact collection is referred to as the *facts of the case*, and the abstract description is referred to as the *consequent* of the case.

Although various approaches to case-based formulation are possible, all share the following basic steps:

1. **Retrieval.** Fetch an appropriate previous case from memory. Let F be the facts of the previous case.
2. **Substitution.** Substitute the names of the entities to which the new facts apply for the entities in F.
3. **Adaptation.** Add any necessary and delete any superfluous facts from the resulting set of new facts.
4. **Match Refinement.** If additions or deletions make the facts of some other case more similar to the new facts than the current case, refine the match by fetching the more similar case and go to step 2.
5. Return the description of the new case together with a record of all substitutions, additions, and deletions, since these constitute the relevant similarities and differences between the new facts and the previous case.

Three different forms of case-based formalization can be distinguished by the extent to which the steps of retrieval, substitution, and match refinement are automated and the extent to which the facts of multiple cases can be combined.

Copy and Edit

The simplest form of case-based formulation is the *copy and edit* approach to knowledge entry used extensively in the development of the Cyc knowledge base (Guha and Lenat, 1990). In the copy and edit methodology, a frame is added to a knowledge base by finding a similar frame, copying it, and modifying the copy.

The copy and edit methodology can speed knowledge entry and tends to enforce representational consistency. However, it doesn't address the potentially difficult task of retrieving the most appropriate frame, and it requires that the correspondence between entities in the new and old frames be determined manually. Moreover, it provides no mechanism for reuse of groupings of related knowledge larger than individual frames.

Single-Case CBF

The shortcomings of copy and edit can be addressed by automating the retrieval and substitution steps. I will refer to case-based formulation in which retrieval and substitution are automated and in which only a single case at a time can be used as a model for the new facts as *single-case CBF*. Single-case CBF begins when the user asserts some small number of facts. These facts are used by the system as a cue or memory probe to retrieve the facts of the most similar case. The system determines the correspondence between the new entities and the entities in the case that leads to the best match and makes the appropriate substitutions. If the system detects that the addition or deletion of facts makes some other case more closely match the new facts, the more similar case is automatically substituted for the current case.

Multiple-Case CBF

A new collection of facts is often best represented as a combination of several cases. In the domain of law, for example, a new case may match portions of several different precedents more closely

[1]This use of the term "case" is somewhat broader than in traditional CBR usage, where the term usually refers either to reusable plans (Hammond, 1986) or designs (Goel et al., 1991; Sycara and Navinchandra, 1991) or to *exemplars* (Porter et al., 1990), distinguished points in an instance space.

than it matches all the facts of any single precedent (Branting, 1991b). Similarly, designs are often best modeled as combinations of portions of multiple previous designs (Goel et al., 1991; Sycara and Navinchandra, 1991). Moreover, a single fact in a new case may itself be the consequent of some other case. I refer to the extension of single-case CBF to permit multiple cases to be combined as *multiple-case CBF*.

An Implementation of Single-Case CBF

CBF1 is an initial implementation of single-case CBF that provides a set of utilities for creating, viewing, and manipulating relational representations. CBF1 has been tested with a small knowledge-base of vehicles and with GREBE's knowledge base of worker's compensation precedents and hypotheticals.

The algorithm of CBF1 is as follows:

GIVEN:

- A partial description D consisting of a collection of propositions $(Pred_1\ A_{11} \ldots A_{1m}) \ldots (Pred_n\ A_{n1} \ldots A_{nk})$
- Optionally, a goal $(Pred_g C_1 \ldots C_p)$.

DO:

1. **Retrieval.** Fetch the case, C, whose facts, $F = (Pred_1\ B_{11} \ldots B_{1m}) \ldots (Pred_i\ B_{i1} \ldots B_{ik})$, most closely match D using structural congruence (Winston, 1980; Branting, 1991a; Holyoak and Thagard, 1989) as a similarity metric (limiting the search to cases whose consequents have the same predicate, $Pred_g$, as the goal, if a goal has been specified). F will be the model for the new description.
2. **Substitution.** Let M be the structurally most consistent mapping from entities in F to entities in D, that is, the mapping that maximizes the number of corresponding relations. Let D be the result of replacing each entity A_{ij} in F with $M(A_{ij})$, creating a new variable name if M is not defined for A_{ij}.
3. **Adaptation.** Add any necessary and delete any superfluous facts from D.
4. **Match Refinement.** If additions or deletions make the facts of some other case C' more similar to the D than F, let F equal the facts of C' and go to step 2.
5. Return D, along with all substitutions, additions, and deletions, since these constitute the relevant similarities and differences between D and F, the facts of the most similar case C.

The behavior of CBF1 is illustrated by the following example in which CBF1 was used to represent a case from GREBE's domain.

One of the cases used to compare the performance of GREBE to that of law students in GREBE's evaluation (Branting, 1991a) concerned Stanley, the head of a surveying crew at a large construction site. Stanley performed some of his duties—making architectural charts—at home during hours he set himself. One day, after doing some work at home, Stanley was injured in an accident while driving to the construction site.

The manually-constructed representation of Stanley's case used in the evaluation of GREBE consisted of 51 tuples. This representation took a number of hours to construct, and it is likely that a different knowledge-enterer (or even the same knowledge enterer on a different occasion) would have represented the case somewhat differently. With CBF1, however, entry of the case is relatively simple and the resulting representation is consistent with the conventions of the cases that have already been entered.

Representing Stanley's case using CBF1 begins with the assertion of the basic facts that Stanley was employed by Tower Construction Company to direct a surveying crew:

```
(employee Stanley-employment
          Stanley)
(employer Stanley-employment
          tower-construction-company)
(had-duties Stanley-employment
            directing-surveying-crew)
```

CBF1 uses a user-specifiable retrieval technique to determine the cases that most closely match these facts.[2] The system retrieves three candidate cases, each an instance of an employment-related activity: the Vaughn case, the Brown case, and the Prototypical Work Case. For each of these cases, CBF1 displays:

- The mapping from the entities in the case to entities in the new case that leads to the best match.
- The facts of the case that are unmatched in the new case. These facts constitute default conclusions about the new case under the assumption that the case is used as a model.
- The proportion of facts of the case that are matched in the new case.

For example, the best mapping from the entities in the Prototypical Work Case includes the following:

```
typical-employee ⇒ Stanley
typical-employer ⇒ Tower-Construction-Co.
```

[2]Three techniques for retrieving cases represented relationally were empirically compared in (Branting, 1991a). For convenience, the simplest of these algorithms, retrieval by best-first incremental matching (RBIM), is used in this example. Various alternative approaches to retrieval of relationally represented cases are discussed in (Branting, 1990).

typical-work-activity ⇒ directing-surveying-crew

...

Fourteen defaults are associated with the match to the Prototypical Work Case, including the assumptions that Stanley had some work hours and received a salary, that Stanley's being at the construction site was a prerequisite for Stanley's directing the surveying crew, and that Tower Construction Company had some goal that was achieved by Stanley's directing the surveying crew. Variable names are gensymed for entities in the case for which no corresponding entities exist in the new case. For example, the default that Stanley's being at the construction site was a prerequisite for Stanley's directing the surveying crew is represented by the tuple:

```
(prerequisite-for being-at-place.1635
                  directing-the surveying-crew)
```

All of the defaults of the Prototypical Work Case are true of Stanley's case, whereas each of the other cases has defaults that are not true of Stanley's case. As a result, the user selects the Prototypical Work Case as the initial model.

The adaptation step consists of the user entering the distinguishing facts of Stanley's case that are not true in the Prototypical Work Case. Such facts include that Stanley had the additional duty of making architectural charts, that he performed this duty at home, that he set his own hours for making the charts, and that he traveled from home to the construction site:

```
(had-duties Stanley-employment
            making-architectural-charts)
(activity-occurring-there Stanley-home
                          making-architectural-charts)
(prerequisite-for Stanley-being-at-home
                  making-architectural-charts)
(determined-by Stanley-work-hours
               Stanley)
(destination traveling-to-the-construction-site
             construction-site)
(source traveling-to-the-construction-site
        Stanley-home)
```

CBF1 permits the user to specify a goal to constrain the matching process. In Stanley's case, we can specify the goal of determining whether Stanley's traveling to the construction site is an employment-related activity (this determines whether Stanley is entitled to worker's compensation for his injuries). This goal will constrain matches only to cases of employment-related activities. Moreover, the mappings between each such case and Stanley's case will be constrained to pair the employment-related activity in the case with Stanley's traveling to the construction site, and the employment relation in the case with Stanley's employment.

After the user asserts the additional facts and the goal, the system performs a match refinement step in which the cases that most closely match the description under construction are retrieved. The closest match is with the Meyer case, which involved a real estate broker who was injured while traveling between his home, where he performed part of his job duties, and his office. This match provides an additional 11 defaults (*e.g.*, that Stanley was the driver in traveling to the construction site). Two of these defaults are not true in Stanley's case (Meyer had an additional job duty). When the user deletes these two tuples, the representation of Stanley's case is complete.

The representation of Stanley's case produced by CBF1 is more concise than the manual representation—32 propositions as opposed to 51 propositions—because facts irrelevant to the goal (*i.e.*, facts not contributing to the match with the controlling precedent) were omitted. Of these 32 propositions, only 9 tuples—28%—had to be explicitly asserted. The other tuples were obtained as defaults from matches with the Prototypical Work Case and the Meyer case. CFB1 reduces the time required to represent Stanley's case from hours to minutes and insures that the resulting representation is consistent with previous cases. Moreover, no additional retrieval or matching is necessary to analyze Stanley's case. This is because the most relevant precedent, Meyer, has been found and the relevant similarities and differences between Meyer and Stanley's case determined by the process of formulating the case.

Integrating Problem Formulation with Problem Solving

CBF is an application of case-based reasoning to the task of problem formulation. In domains for which problem formulation is complex enough to impede system use, CBF can be the first part of a two-step process: (1) case-based formulation of the problem, followed by (2) applying the appropriate problem-solving method to the problem thus formulated.

However, there are many tasks, such as such as precedent-based legal reasoning and case-based heuristic classification, for which problem solving consists at least in part of determining the relevant similarities and differences between a new case and the most similar past cases. CBF can perform these tasks, in part or entirely, as a side-effect of problem formulation. Problem solving in GREBE, for example, consists of (1) determining the mapping from the most similar precedents of the concept at issue to the facts of a new case, and (2) using this information to construct one or more explanation structures. As discussed in the previous section, CBF performs the first of these steps in the very process of formulating the facts of a new case. Thus, for tasks amenable to case-based reasoning, CBF can integrate problem formulation with problem solving.

This integration is desirable because a rigid separation between problem formulation, retrieval, and case comparison can exacerbate the difficulty and computational expense of each of these steps. The previous section illustrated how interleaving case retrieval and comparison with problem formulation can improve the accuracy and efficiency of the latter. The converse relation holds as well: interleaving these steps makes case retrieval and comparison more tractable. The combinatorics of matching relational cases makes the computational expense of finding the structurally most consistent case in memory steeply increase with the complexity of the probe[3]. By using a small set of initial facts as a probe and then incrementally refining the initial match as facts are added or deleted, CBF can avoid the computational expense of using a complete case description as a probe.

Range of Applicability of CBF

CBF as Knowledge Acquisition.

Problem formulation is a form of *knowledge acquisition*, the process of extracting knowledge from non-computer sources and encoding that knowledge in a form that is usable by a computer for problem solving. CBF is not restricted to problem formulation, but is applicable to acquisition of any type of knowledge organized around collections of related facts that can be manipulated as wholes. Viewed as a knowledge-acquisition technique, CBF has the virtue of being interactive, of automatically insuring consistency with existing collections of related facts, and of potentially improving, rather than degrading, as the knowledge base expands and a larger set of models for new cases becomes available.

Multiple-Case CBF

Extending the applicability of CBF to domains in which problems are best described as compositions of multiple previous cases will require implementing multiple-case CBF. Two mechanisms are required for multiple-case CBF. First, combining cases at the same level of abstraction requires the ability to partition the description under construction, apply single-case CBF to the partitions, and combine the results. Second, combining cases at different levels of abstraction requires the ability to view a single fact in a description under construction as the consequent of a collection of facts at a lower level of abstraction. For example, a swamp boat could be represented by using a motor boat as a model at a high level of abstraction (*e.g.*, a motor boat consists of a hull, a rudder, an engine, etc.) but, at a lower level of abstraction, using an airplane engine as a model of the swamp boat's engine. Current research is directed toward developing these two mechanisms.

Limitations

The effectiveness of any case-based reasoning system depends upon the existence of a library of cases relevant to the task addressed by the system. As an application of case-based reasoning to the task of problem formulation, CBF depends on the existence of a library of cases that can serve as suitable models for the problems that the system will encounter. If there is little similarity between new problems and past problems, CBF can provide little assistance.

Although CBF's use of past cases as models can reduce the danger of inconsistent representations of patterns of case facts, the current implementation of CBF nevertheless presupposes a consistent vocabulary of representational primitives. For example, if the initial description of Stanley's Case had been

(had-job Stanley Stanley-employment)

CBF1 would have failed to find any relevant past case because had-job is not part of the vocabulary in which the past cases were described. CBF1 would be improved by some mechanism for detecting possible inconsistent uses of primitives while at the same time permitting new primitives to be added when necessary.

A second limitation of CBF1 is its rudimentary *knowledge presentation* (Musen, 1988), *i.e.*, the conceptual model presented to the user. CBF1's knowledge presentation consists simply of the tuples that constitute the facts of a case in GREBE's representation idiom. An iconic presentation or a subset of English would greatly improve interaction with CBF1.

Conclusion

Relational knowledge representations are necessary for general-purpose knowledge bases intended for multiple tasks and for any of a wide variety of individual tasks. However, the price of the increased expressiveness of relational representations is that they make the task of expressing new cases correspondingly more complex. In domains involving sufficiently complex cases, simply expressing the facts of a problem in a relational representation language can itself be a complex, error-prone, and time-consuming task. This paper has presented an approach to problem formulation that uses previous cases as a model and a guide for expressing new cases.

[3]While various approaches to retrieval of relationally represented cases have been proposed, *e.g.*, MAC/FAC (Gentner and Forbus, 1991), ARCS (Thagard et al., 1990), and MRSDL (Branting, 1992), no approach has been shown both to guarantee a high level of accuracy and to cost significantly less than exhaustive matching.

CBF has a number of potential benefits. As illustrated by the example of Stanley's case, CBF can reduce the time necessary to pose new problems because modifying an existing representation is often much simpler than creating a new representation *ab initio*. CBF can reduce the danger of representational inconsistency by reusing conventions for representing particular patterns of facts rather than requiring them to be recreated in every new case. Moreover, when new cases are expressed in terms of old, the relevant similarities and differences between new and old cases are determined *a fortiori* by the very process of formulating each new case. CBF can also simplify case retrieval. By using a small set of facts as a probe and then incrementally refining the initial match as facts are added or deleted, CBF can avoid the computational expense of using a complete case description as a probe.

Finally, psychological plausibility argues for CBF over a rigid division between problem formulation and problem solving. Previous experience is not merely the yardstick by which new experiences are measured, but is the very medium in which they are expressed.

References

Branting, L. K. (1990). Techniques for retrieval of structured cases. *Working Notes of the AAAI Spring Symposium on Case-Based Reasoning*, Palo Alto, California.

Branting, L. K. (1991a). *Integrating Rules and Precedents for Classification and Explanation: Automating Legal Analysis*. PhD thesis, University of Texas at Austin.

Branting, L. K. (1991b). Representing and reusing explanations of legal precedents. In *Proceedings of the Third International Conference on Artificial Intelligence and Law*, pages 145–154, Oxford, England.

Branting, L. K. (1992). A model of the role of expertise in analog retrieval. In *Proceedings of the Fourteenth Annual Conference of the Cognitive Science Society*, Bloomington, Indiana.

Branting, L. K. and Porter, B. W. (1991). Rules and precedents as complementary warrants. In *Proceedings of Ninth National Conference on Artificial Intelligence*, Anaheim. AAAI Press/MIT Press.

Farquhar, A., Kuipers, B., Rickel, J., and Throop, D. (1990). QSIM: The program and its use. Technical report, Artificial Intelligence Laboratory, Department of Computer Sciences, University of Texas at Austin.

Gentner, D. and Forbus, K. (1991). MAC/FAC: A model of similiarity-based retrieval. In *Thirteenth Annual Conference of the Cognitive Science Society*, pages 504–509.

Goel, A., Kolodner, J., Pearce, M., and Billington, R. (1991). Towards a case-based tool for aiding conceptual design problem solving. In *Proceedings of the Third DARPA Case-Based Reasoning Workshop*, pages 109–120. Morgan Kaufmann.

Guha, R. V. and Lenat, D. B. (1990). Cyc: a midterm report. *AI Magazine*, 11(3):32–59.

Hammond, K. (1986). *Case-Based Planning: An Integrated Theory of Planning, Learning, and Memory*. PhD thesis, Yale University.

Holyoak, K. and Thagard, P. (1989). Analogical mapping by constraint satisfaction. *Cognitive Science*, 13(3).

Kuipers, B. (1989). Qualitative reasoning: modeling and simulation with incomplete knowledge. *Automatica*, 25:571–585.

Lenat, D. B. and Feigenbaum, E. A. (1991). On the thresholds of knowledge. *Artificial Intelligence Journal*, 47(1–3).

Muggleton, S. (1991). Inductive logic programming. *New Generation Computing*, 8.

Musen, M. A. (1988). *Generation of Model-Based Knowledge-Acquisition Tools for Clinical-Trial Advice Systems*. PhD thesis, Stanford University.

Porter, B., Lester, J., Murray, K., Pittman, K., Souther, A., and Acker, L. (1988). AI research in the context of a multifunctional knowledge base project. Technical Report AI-TR8-88, Artificial Intelligence Laboratory, Department of Computer Sciences, University of Texas at Austin.

Porter, B. W., Bareiss, E. R., and Holte, R. C. (1990). Concept learning and heuristic classification in weak-theory domains. *Artificial Intelligence Journal*, 45(1–2).

Schank, R. C. (1982). *Dynamic Memory, A Theory of Reminding and Learning in Computers and People*. Cambridge University Press, Cambridge, England.

Sycara, K. and Navinchandra, D. (1991). Influences: A thematic abstraction for creative use of multiple cases. In *Proceedings of the Third DARPA Case-Based Reasoning Workshop*, pages 133–144. Morgan Kaufmann.

Thagard, P., Holyoak, K., Nelson, G., and Gochfeld, D. (1990). Analog retrieval by constraint satisfaction. Technical Report CSL-Report 41, Princeton University.

Winston, P. H. (1980). Learning and reasoning by analogy. *Communications of the ACM*, 23(12).

Orthographic and Semantic Similarity in Auditory Rhyme Decisions

Curt Burgess
Department of Psychology
University of California-Riverside
Riverside, CA 92521
Michael K. Tanenhaus
Nancy Marks
Department of Psychology
University of Rochester
Rochester, NY 14627

Abstract[1]

Seidenberg and Tanenhaus (1979) demonstrated that orthographic information is obligatorily activated during auditory word recognition by showing that rhyme decisions to orthographically similar rhymes *pie-tie* were quicker than rhyme decisions to orthographically dissimilar rhymes *rye-tie*. This effect could be due to the fact that orthographic and phonological codes are closely inter-related in lexical memory and the two dimensions are highly correlated. However, it could also be a example of a more general similarity bias in making rhyme decisions, in which subjects cannot ignore irrelevant information from other dimensions. We explored this later possibility by having subjects make rhyme decisions to words that vary in orthographic similarity and also to words that vary in semantic similarity (*good-kind, cruel-kind*). This possibility is ruled out in two experiments in which we fail to find an interference effect with semantically related trials, while replicating the basic orthographic interference and facilitation results.

[1]This research was supported by National Institute of Child Health and Human Development Grant HD 18694 awarded to M. K. Tanenhaus.

Introduction

There have been a number of studies demonstrating that orthographic information is activated during auditory word recognition. One of the clearest demonstrations was originally reported by Seidenberg and Tanenhaus (1979). They found that that rhyme decisions were faster to orthographically similar rhymes such as *pie-tie* than to orthographically dissimilar rhymes such as *rye-tie*. One explanation for these results is that orthographic and phonological codes become closely inter-related in the process of learning to read when the mapping of orthographic to phonological codes occurs. As a consequence of this learning, both phonological and orthographic information are activated during auditory word recognition. During the rhyme decision, the orthographic information causes a Stroop-like effect in the decision process. Subjects seem unable to use an optimal strategy that would rely solely on the phonological codes that are required for the rhyme decision.

There is an alternative explanation which does not assume any special correspondence between these two lexical codes. Kahnemann (personal communication, 1985) suggested that these orthographic effects are an example of a more general phenomenon that he refers to as cognitive Stroop effects. These effects arise when subjects cannot ignore information from an irrelevant dimension. On this view, when multiple aspects of a representation are activated, it becomes difficult to ignore irrelevant information. As a result, the orthographic interference obtained with auditory rhyme decisions is not a product of the linkage between phonological and orthographic codes, but would be due to a general similarity bias with *Yes* decisions to words being facilitated when they are similar along any dimension. As a result, rhyme decisions would be quicker to *pie-tie* than to *rye-tie* since *pie-tie* are similar along more dimensions than are *rye-tie*. Seidenberg and Tanenhaus (1979) also showed that rhyme decisions to orthographically similar non-rhymes like *touch-couch* were slower than to orthographically dissimilar non-rhymes like *dutch-couch*. Since a *No* response is required for a non-rhyme, the orthographic similarity results in a slower *No* decision. Thus, the general similarity bias explanation can neatly account for the orthographic effect in rhyming without positing any special linkage between phonology and orthography.

The proposal that a general similarity bias might produce an artifactual pattern of results corresponding to orthographic interference is important to consider given that there have been a number of task differences in the lexical-semantic priming literature that have been attributed to post-lexical bias. For example, when subjects read a sentence and then have to name a target that is highly related to the sentence, responses to targets are facilitated. However, if the subject response is a lexical decision, responses to the targets are inhibited (Fischler & Bloom, 1979; Stanovich & West, 1983). Presumably, the facilitation reflects a passive automatic spread of activation between highly related or predictable concepts and naming is sensitive to this effect. Likewise, responses made by lexical decision include this automatic component but also include a more strategic aspect. Seidenberg (1985) suggests that the lexical decision creates a situation similar to that of a Stroop task. The target words are highly related to the sentential context. When the expectation or bias for a particular concept is violated, the lexical decision proves to be sensitive to this relatedness aspect of the task and the incongruity inhibits the lexical decision response. In a similar fashion, when there is incongruent orthography between the stimuli that do rhyme, the *overall* similarity between the rhyming stimuli decreases. Thus, in experiments such as these, the phonological similarity crucial to the rhyme decision and the orthographic incongruity could result in an inhibitory component in the decision task. If the similarity bias hypothesis alone is sufficient to account for the effects of orthography on rhyming, it should also be difficult to make a *No* response with a rhyme decision to semantically related non-rhymes such as *good-kind*.

In experiment 1, we sought to replicate the Seidenberg and Tanenhaus (1979) result and also to include a condition to test the similarity bias hypothesis. We tested the similarity bias hypothesis by including synonyms and antonyms. If a similarity strategy is used by subjects, we would expect longer *No* response times for the synonyms compared to the antonyms.

Experiment 1

Stimuli

Monosyllabic stimuli consisted of primes that varied in their phonological and orthographic similarity to the target. In the rhyme condition, targets were preceded by an orthographically similar or dissimilar prime, for example, *plate-gate* or *freight-gate*. In the non-rhyme condition, targets were preceded by an orthographically similar or dissimilar prime like *touch-couch* or *dutch-couch*. The additional non-rhyme condition involved the use of synonyms like *kind-good* and antonyms like *kind-cruel*. There were twelve of each of these kinds of trials in each of two lists. Twelve orthographically similar and twelve orthographically dissimilar rhymes were included as filler trials.

Procedure

Thirty-two native English speaking students participated. Primes and targets were presented on one channel of a two-headed stereo tape recorder (Sony TC-270). A brief 1000 Hz trigger tone was placed on the other channel, precisely at the onset of the second word. The trigger tone, which was inaudible to the subject, was connected to a silent solid-state voice relay. An Apple IIe equipped with a Digitry CTS system was used to time the duration from the onset of the second item to the subject's response. Subjects listened to the prime words binaurally through stereo headphones.

Subjects received 10 practice trials and 94 experimental trials. Subjects were instructed to attend to the two words presented over the headphones, and quickly decide if the items rhymed or not, and press either the YES or NO button.

Results

There was an interaction between rhyme decision (yes-no) and orthographic match (match or no-match), $F(1, 30) = 11.82$, $p < .001$. Subjects were slower by 42 msec to make rhyme decisions to orthographically dissimilar rhymes (741 msec) than to orthographically similar rhymes (699 msec), $F(1, 30) = 7.47$, $p < .02$. However, decisions to orthographically similar non-rhymes (861 msec) were slower by 57 msec than the orthographically dissimilar non-rhymes (804 msec), $F(1, 30) = 9.21$, $p < .005$. There was no difference in rhyme decision latencies between the synonym (727 msec) and antonym (746 msec) word pairs, $F(1, 30) = 1.45$, $p = .238$. A parallel effect was obtained in the error analysis.

Discussion

The orthographic interference which was obtained for non-rhymes with a rhyme-decision task replicated Seidenberg and Tanenhaus (1979). Even in an auditory task, orthographic differences affected rhyme decisions. If a similarity strategy had been used by subjects we would have expected longer *No* response times for the synonyms compared to the antonyms, however, this did not occur.

Experiment 2

It may be that comparing synonyms and antonyms was not a strong enough test of the similarity hypothesis. It is possible that semantic similarity along any dimension[2] (i.e., synonyms or antonyms) produces interference in a rhyme decision.

In experiment 2, we manipulated semantic similarity for non-rhymes by using synonym and antonym pairs, but we rotated these trials such that a semantically unrelated control condition was included. For example, *good-kind* and *cruel-kind* can now be compared to the completely unrelated *fast-kind.* This should provide a better baseline for an interference effect rather than comparing two differently related conditions.

We manipulated orthographic similarity for non-rhymes (like in experiment 1) and also rotated these items to provide for a completely unrelated condition as well. In this case, *touch-couch* and *dutch-couch* can be compared to *leaf-couch.* If a similarity strategy is being used in the rhyme decision, we would not expect an interaction between the three levels of the semantic condition and the three levels of the orthography condition, since semantically related trials would show interference when compared to the semantically unrelated trials. Filler trials were included, resulting in an equal number of rhyme and nonrhyme decisions. Forty-eight native English speaking students participated. The procedure and task was identical to experiment 1.

[2] We want to acknowledge that there are other stimulus dimensions in which one could use to manipulate similarity bias. However, the nature of the 'yes/no' rhyme decision task and the similarity/dissimilarity of the synonyms and antonyms led us to believe that this particular semantic manipulation would be particularly sensitive to a possible response bias.

Results

An interaction obtained with semantic nonrhymes and orthographic nonrhymes, $F(2, 94) = 20.34$, $p < .001$. There was no difference in reaction time between the three levels of the semantic condition (synonyms, 698 msec; antonyms, 701 msec; unrelated, 703 msec), $F(2, 94) < 1$. However, orthographically similar nonrhymes (848 msec) were responded to more slowly than the dissimilar nonrhymes (765 msec) which were, in turn, responded to even more slowly than their unrelated control (723 msec), $F(2, 94) = 36.58$, $p < .001$. A parallel effect was obtained in the error analysis.

Discussion

The results of this experiment are consistent with those of experiment 1, which did not use the completely unrelated control condition. The interaction between the semantic condition and the orthographic condition suggests that the subjects utilize the orthographic information in making rhyme decisions and do not allow some more abstract similarity metric to influence the decision.

Conclusions

We have replicated the finding that orthographically similar words that rhyme can be detected more quickly than orthographically dissimilar rhymes. These results are consistent with a model in which multiple codes become activated in parallel, even if initially such activation is not useful for the task. Such activation would seem to be an automatic effect since subjects did not adopt a strategy where they used only the phonological information in the rhyme decision. Had a subject been selectively able to use phonological information, no interference would have been obtained with the orthographically dissimilar trials, and there should also have been no difference between the orthographically similar and dissimilar non-rhyme conditions.

This effect appears to be specific to the orthographic code, in that we can rule out a general similarity strategy or response bias, since we failed to find semantic interference with the rhyme decision in both experiments. While there are clearly a number of cognitive Stroop phenomena that can be explained with a similarity heuristic, the orthographic interference obtained with auditory rhyme decisions appears to be a product of the linkage between phonological and orthographic lexical codes, and is not due to a general similarity bias with *Yes* decisions to words being facilitated when they are similar along any dimension.

Acknowledgments. We want to thank Robert R. Peterson, Catherine H. Decker and an anonymous reviewer for their helpful comments and criticism. Send correspondence to Curt Burgess, Psychology Department, University of California-Riverside, Riverside, CA 92521. [email: curt@prodigal.psych.rochester.edu]

References

Fischler, I., & Bloom, P. (1979). Automatic and attentional processes in the effects of sentence contexts on word recognition. *Journal of Verbal Learning and Verbal Behavior, 18*, 1-20.

Kahnemann, D. (1985). Personal communication to Mike Tanenhaus at the Cognitive Science Meeting at Irvine, CA.

Seidenberg, M. S. (1985). The time course of information activation and utilization in visual word recognition. In D. Besner, T. G. Waller, & G. E. MacKinnon (Eds.), *Reading research: Advances in theory and practice, Vol. 5.* New York: Academic Press.

Seidenberg, M. S., & Tanenhaus, M. K. (1979). Orthographic effects on rhyme monitoring. *Journal of Experimental Psychology: Human Learning and Memory, 5*, 546-551.

Stanovich, K. E. & West, R. (1983). On priming by a sentence context. *Journal of Experimental Psychology: General, 112*, 1-36.

Analogy and Representation: Support for the Copycat Model

Bruce D. Burns and Maureen E. Schreiner
Department of Psychology
University of California, Los Angeles
Los Angeles, CA 90024, USA
burns@cognet.ucla.edu
missy@cognet.ucla.edu

Abstract[1]

We report two experiments which assessed the psychological validity of the Copycat framework for analogy, which proposes that analogy is a process of creating a representation. Experiment 1 presented subjects with two letter string analogies: "If **abc** is changed to **abd** how would **kji** be changed in the same way?", and the same statement but with **mrrjjj** as the string to be changed. Each subject attempted to solve both analogies and order of presentation was varied. The predictions of Copycat very closely matched the performance of human subjects on the first analogy people solved. However, the second analogy task showed substantial asymmetrical transfer effects that the model does not directly predict. Substantially greater transfer was observed from the **mrrjjj** analogy, for which it is hard to produce a highly structured representation, to the easier to represent **kji** analogy, than vice-versa. In Experiment 2 the first part of the statement of the problem was "If **aabbcc** is changed to **aabbcd**...". In this case **kji** becomes harder to represent than **mrrjjj**. As predicted, this version yielded more transfer from **kji** to **mrrjjj** than the reverse. In both experiments transfer was asymmetrically with greater transfer from less structured to more structured problems than the reverse. Overall the study supported Copycat's contention that representation is a vital component for understanding analogical processors.

Copycat (Hofstadter, 1985; Hofstadter & Mitchell, 1988; Mitchell, 1990; Mitchell & Hofstadter, 1989, 1990) is a program that creates analogical inferences in a simple domain, a micro world consisting of the 26 letters of the alphabet and associated concepts. The program solves problems such as, given that **abc** changes to **abd**, what does **srqp** change to analogically? Or how would **mrrjjj** change? (Note that the string to be changed will be presented in **bold-face**, while strings that are possible answers will be in *italics*) Copycat differs from most other models of analogy (e.g., Holyoak & Thagard, 1989; Falkenhainer, Forbus, & Gentner, 1989) in that other models take an existing representation of a situation and build an analogy from that representation. Copycat instead proposes that building the representation interacts crucially with the mapping of the analogy. In Copycat, representations are not just something from which analogies are derived; rather, the construction of a representation is viewed as the most basic analogical process. Consequently, analogy is viewed as a much more general process than most other models of analogy envisage.

Copycat limits its domain to letter strings because analogy is viewed by Copycat's authors as too complicated to be initially studied in a complex domain (Hofstadter, 1985). For Copycat, representing an analogy involves forming relevant, coherent structures at an appropriate level of abstraction. For example, a link may be formed between the letters *a* and *b* because *b* is the alphabetic successor to *a*. Or a link may be formed between the letters *a* and *m* because they are both the left-most letters in their strings. Forming structure can also require deciding which aspects are to be taken literally and which are allowed to "slip" to related concepts. For example, *c* is the predecessor of *d*, but under appropriate conditions a representation may be built that allows the concept of "predecessor" to slip such that the "predecessor" relationship is mapped to the opposite "successor" relation. In Copycat these slippages, like all aspects of building the representation, occur as the result of competition between different interpretations. What these interpretations may be is not limited by the initial representation as it is in the other analogy models mentioned above. Letter string analogies incorporate all of the problems Hofstadter and

[1]This research was supported by Contract MDA 903-89-K-0179 from the Army Research Institute.

Mitchell consider important: they have structure, they may lead to "slippage", and they often have competing interpretations. Because letter string analogies are a relatively simple domain, they make it possible to isolate the important aspects of analogy without having to incorporate many domain-specific mechanisms.

However, because Copycat solves a different domain of problems than that of most other computational models of analogical processes, it is very difficult to directly test Copycat against other models of analogy. It is in its underlying approach that Copycat differs from these other models, rather than in predicting specific different results. It is also unclear how best to test Copycat as a psychological model of analogical reasoning. Copycat is not designed to model human behavior at a fine grain, although Mitchell (1990) intends the Copycat internal architecture and external behavior to provide a plausible model of human analogy making. Another factor that makes it difficult to test Copycat as a model is that clearly many aspects of human experience are excluded from its micro world (e.g., people generally know the alphabet better forwards than backwards). In this study we make some initial attempts to experimentally test implications of the Copycat approach.

It would increase the psychological plausibility of the Copycat model if it could produce similar answers to letter-string analogy problems to those people generate. Copycat is intended as a model of a single individual, but it is impractical to ask the same person to solve one of these problems a thousand times and compare the frequency of responses to that generated by a thousand runs of the Copycat program. Therefore we instead gave analogy problems to a number of people, with the assumption that their responses would reflect the biases of a single individual over time. Mitchell (1990) gave analogy problems to people to solve and found that Copycat could produce the most common answers people advanced, but it was unable to generate the full range of answers that people did. However, even human data for responses that Copycat was able to generate had a clearly different frequency distribution for those responses than that Copycat produced. This suggest that Copycat may be limited as a model of human behavior, reflecting either limitations of the current instantiation of Copycat (as Mitchell acknowledges) or the model's general approach to analogy making.

Rather than simply comparing the frequency of Copycat's responses to human data, we sought to evaluate Copycat's claim that the interaction of representation and mapping is critical to analogical reasoning. We were unable to test this claim directly, but instead sought to test a prediction that appears to be consistent with this claim. If representation is critical, then particular transfer effects may be predicted when subjects try to solve successive analogy problems. (The Copycat program has never attempted to model transfer; however, there appears to be no conceptual problems with doing so) In Experiment 1 we presented subjects with two analogies: 'Suppose that the letter string **abc** was changed to **abd**; how would you change the letter string **kji** in the same way?'; and the same analogy but with **mrrjjj** as the string to be changed. These two problems were chosen from among the many that Copycat has attempted because they both had at least two reasonably frequent answers, suggesting that they had enough inherent ambiguity of interpretation to be affected by attempts to solve preceding problems. Copycat generated three main answers to the **kji** problem: *kjh*, *kjj*, and *lji*. Each of these answers reflect a different representation of the problem. To produce *kjh* Copycat builds a representation in which the **kji** string is a left-to-right sequence of letters that precede each other in the alphabet, while **abc** is a string in which the letters are alphabetic successors. The opposing nature of these structures causes the concept of changing the right-most letter (derived from "**abc** changes to **abd**") to its successor to "slip" to "change the right-most letter to its predecessor". The string *lji* results from a similar representation but instead of letting "change predecessor" slip to "change successor", the idea of "change right-most" slips to "change left-most". In contrast, the answer *kjj* can be produced with relatively little structure, by directly transferring the idea "change to successor of left-most" with no direct influence from the string to which it is being applied. Little structure can be effectively built however to represent the **mrrjjj** analogy problem. The most common answers Copycat produces are *mrrjjk* and *mrrkkk*, both of which are the result of transferring the unmodified rule of "change right-most element to successor", with the only aspect producing differences being the question of what constitutes the "right-most element". More structure is built for *mrrkkk* than for *mrrjjk*, as the answer *mrrkkk* requires the grouping of *jjj* into a single element that then maps to the element *c*. However, Copycat gives both of the common **mrrjjj** solutions a much higher *temperature* (temperature is Copycat's measure of the amount of structure it builds for an analogy, with low temperature indicating large amounts of structure are built) than it calculates for the *kjh* and *lji* answers for the **kji** analogy.

To examine transfer effects, approximately half the subjects received **kji** first and half received **mrrjjj** first; then each subject received the other analogy. It would be unsurprising to find some transfer effects, as the idea of changing the right-most letter to the successor is applicable to both analogies. But if subjects are trying to build comprehensive representations of these problems (as Copycat seems to suggest they would) then there should be an asymmetry of transfer effects. It should be relatively difficult to transfer **kji** answers to **mrrjjj** because it is expected that subjects will build a relatively highly structured representation of **kji**, a representation that will then be a poor fit to **mrrjjj** if subjects try to make use of the whole representation they have built. However, little structure can be built for **mrrjjj**, allowing relatively unencumbered transfer of the idea "change to successor of right-most letter".

Experiment 1

Subjects were from the introductory psychology subject pool at the University of California, Los Angeles. A total of 140 subjects attempted the analogies either during or after participating in another experiment. Sixty-six subjects did the **kji** analogy first and 74 did **mrrjjj** first. Subjects had to decide on one best response to each analogy and were given as long as they wanted to complete the task. However, they could not return to the first analogy after completing it, or start the second without completing the first.

	kjh	kjj	ijl	lji	ijk	Other
first	24	14	5	15	0	8
second	9	43	3	9	2	8

Table 1. Frequencies of each answer to the **kji** analogy for subjects receiving **kji** first or second.

Table 1 reports the frequency of each response to **kji** that occurred more than twice for subjects solving this analogy first or second[2]. Let us first consider the results for subjects solving the problem first. Copycat generated the most common answers that people produced, but subjects form a number of answers that Copycat is not reported to have produced, although

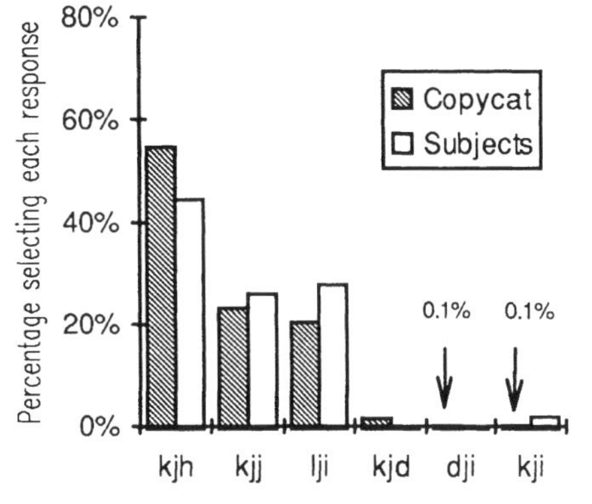

Figure 1. A comparison of the frequency of each response to the **kji** analogy when solved first for Copycat and human subjects (as a percentage of the subset of responses that Copycat is capable of generating).

	mrrjjk	mrrkkk	mrriii	mrrjjjj	jjjmrr	mrsjjk	mrsjjj	Other
first	11	34	0	1	3	5	1	10
second	5	11	4	3	1	3	3	36

Table 2. Frequencies of each answer to the **mrrjjj** analogy for subjects receiving **mrrjjj** first or second.

none have a high frequency. As discussed above, Mitchell (1990) also found this limitation of Copycat, and explanations she offers account for many of the non-Copycat answers. In order to test Copycat on its own terms we used only the subset of answers that Copycat was also capable of generating and computed the frequency of each answer as a percentage of the frequency of all answers in this subset. These percentages are presented in Figure 1, together with the frequency with which Copycat generated each answer over a thousand runs as reported by Mitchell (1990). Copycat's frequencies are very close to those of the subjects. The hypothesis that the distribution of results is the same for Copycat and for people cannot be rejected (using all cells with an expected value greater than 1.0, $X^2(3) = 2.65$, *n.s.*).

Table 2 reports the frequency of all responses to **mrrjjj** which more than two subjects generated[3]. This

[2] Other answers were: *kln, kji, jig, kli, hij, kij, lkj, jkl, kjd, kji, qji, kjk, ijk, klj, kjp.*

[3] Other answers were: *jms, mrt, mrrkkkk, mrtjjk, mmrrkk, mrrjjjwwww, mrrk, mrrjji, mrrfff, rjmjrj, msskkk, mtrjjj, mjjjrr, jjrrmj, rrmjjj, mrrjjjkkkk,*

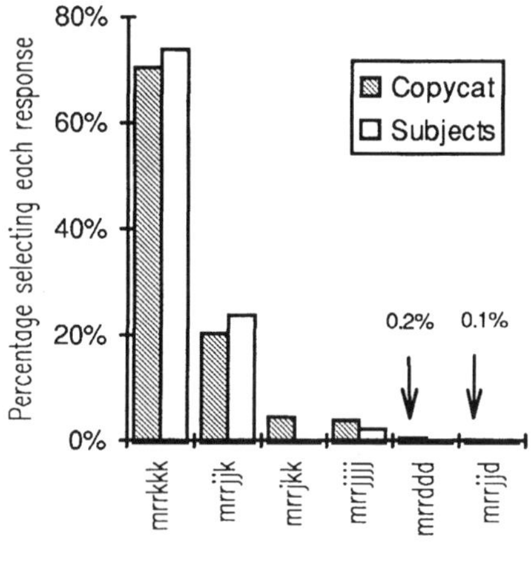

Figure 2. A comparison of the frequency of each response to the **mrrjjj** analogy when solved first for Copycat and human subjects (as a percentage of the subset of responses that Copycat is capable of generating).

data follows a similar pattern to the **kji** results. The main difference is that a greater variety of answers are produced and fewer subjects give the more common answers. This reflects the intuition that **mrrjjj** is a harder analogy for which to produce a systematic answer. Figure 2 presents the comparison of Copycat's results with those of the human subjects, again limiting the comparison to answers that Copycat had produced. Again the hypothesis that frequencies are the same for Copycat and people cannot be rejected (using all cells with an expected value greater than 1.0, $X^2(4) = 2.79$, *n.s.*). These data suggest that the underlying ideas on which Copycat is based have some validity, given that within its limitations Copycat produces results remarkably close to those for human subjects. It is unclear why Mitchell (1990) was less successful in matching Copycat's results to human performance, as Mitchell does not give enough details of her methodology to know if there were major procedural differences with the present study. One difference is that she used fewer subjects (32 for **mrrjjj** and 10 for **kji**).

Comparing the frequencies of responses to the same problem when done second instead of first shows strong order effects. As Table 1 indicates, for the **kji** problem the most common answer, *kjh*, becomes significantly less common ($X^2(1)= 11.34$, $p<.05$) when the problem is solved second. The answer *kjj* becomes the most common answer and is significantly more frequently generated than when **kji** is solved first ($X^2(1)= 19.67$, $p<.05$). For the **mrrjjj** analogy the two most common answers (*mrrkkk* and *mrrjjj*) collectively are generated less often when **mrrjjj** is solved second, $X^2(1)= 30.76$, $p<.05$. (While *mrrkkk* is generated relatively less often than *mrrjjk* when **mrrjjj** is solved second, this difference is not significant, $X^2(1)= 2.19$, *n.s.*) The hypothesized asymmetry of transfer was found. Of the 34 subjects who produced *mrrkkk* when doing **mrrjjj** first, 23 answered *kjj* to their second analogy. However, of the 14 subjects answering *kjj* when they were given **kji** first, only two answered *mrrkkk*. This asymmetry is significant ($X^2(1) = 11.32$, $p<.05$). Similarly, of the 11 subjects answering *mrrjjk*, 10 subsequently answered *kjj*, but of the 14 answering *kjj* only three give *mrrjjk* as their answer to **mrrjjj**, again demonstrating a significant asymmetry of transfer ($X^2(1)= 11.91, p<.05$). The greater variety of answers generated when **mrrjjj** is solved second is also consistent with the claim underlying the asymmetry hypothesis, that it will be more difficult to systematically transfer a highly structured representation than a less well structured one. If subjects try to fit their relatively highly structured representations formed for the **kji** analogy, they may end up distorting their representation of the **mrrjjj** analogy in unusual ways. An example of this was provided by subjects who generated *mrriii*, an answer never produced except when **mrrjjj** is solved second. It appears that subjects may be transferring the idea of taking the predecessor of the last element from their answer for **kji** (three of the four subjects who did this had answered *kjh* to the **kji** analogy). In contrast, there is relatively little difficulty in transferring the unstructured representations of the **mrrjjj** analogy to the **kji** analogy, so there is no increase in the generation of unusual answers when **kji** is solved second. If this interpretation is correct then it is not simply the idea of "change last element" that is transferred between the two problems, but the whole representation of the problem. While Copycat does

---

mrruuu, mrrjjd, mssjjj, mrrkk, mrrsss, mrrjkk, mrrhhh, jmmrrr, mrrggg, mqqjjj, nrjjjj, mrsjjj, mrrttt, jjjmrs, mrrnnnn, jjjrrm, mrrzbd, mrrjjj, mrrmmmm, mrrlll, mjjrr, morij, mrrzzzz, mrsjkl. These responses also illustrate a limitation of Copycat that people do not seem to share, namely, the same element cannot have two interpretations in Copycat. For example, *mrrkkkk* violates this constraint because the *kkkk* element is both the numerical and alphabetic successor to *jjj*.

not directly deal with transfer between analogies, these results are consistent with its general approach of emphasizing analogy making as an interaction between representation and mapping.

Experiment 2

In Experiment 2 we constructed analogies that should exhibit a reversal of the asymmetry of transfer found in Experiment 1. This was accomplished by making **kji** the analogy with less structure to represent, so that it should transfer to a **mrrjjj** analogy; and **mrrjjj** the analogy for which more structure can be built, so that it should show less transfer to **kji**. To create problems to test this hypothesis we replaced the initial part of the analogy "**abc** was changed to **abd**" with "**aabbcc** was changed to **aabbcd**". The same two strings then had to be changed: **kji** or **mrrjjj**. The representation of **kji** is now much more difficult, because there is no simple map of its three elements to the six elements of **aabbcd**. Even greater difficulties arise if subjects try to incorporate the idea of "splitting" the last element (as **cd** can be seen as the splitting of the **cc** element) into their representation. In contrast, the **mrrjjj** analogy now has a straightforward representation available: simply map each letter in **aabbcd** to each in **mrrjjj** and change the rightmost element to its successor. The major complication in representing **mrrjjj** in Experiment 1 was trying to group six elements so as to map to three, but this difficulty is now eliminated. Neither of these analogies have been given to Copycat, though there is little doubt that it would be able to produce some plausible answer to even these poor analogies, just as people proved able to do so. The aim of this experiment was to test the hypothesis that an asymmetry of transfer would be found, consistent with Experiment 1. The nature of the asymmetry is hypothesized to be that subjects solving the **kji** analogy (for which little structure may be formed) first would display more transfer to the **mrrjjj** analogy (for which relatively more structure can be formed) than subjects solving **mrrjjj** first would display transfer to the **kji** analogy.

Subjects were 104 students from the introductory psychology subject pool at the University of California, Los Angeles who participated in this experiment as a filler task within other experiments (primarily a memory experiment). Sixty subjects solved **kji** first and 44 solved **mrrjjj** first.

The frequency of responses for the subjects to the **kji** analogy for all answers that occurred more than twice are presented in Table 3[4]. The difficulty of this analogy is reflected in the greater variety of answers that are constructed relative to the **kji** analogy in Experiment 1, especially when **kji** is solved second. The difficulty of representing this analogy is apparent from the large number of subjects who added letters to the string in attempting to map three elements to six.

	kjh	kjj	lji	kkjjih	kji	kkjjij	iijjkl	kjih	kjk	Other
first	14	17	5	6	3	0	3	4	0	8
second	3	14	2	1	1	4	2	1	3	13

Table 3. Frequencies of each answer to the **kji** analogy for subjects receiving **kji** first and second.

	mrrjjk	mrrjji	mrrjjg	mrrjjh	mrrjk	mrrjj	mrrjk	Other
first	23	0	2	2	1	1	2	13
second	27	5	1	2	2	3	1	19

Table 4. Frequencies of each answer to the **mrrjjj** analogy for subjects receiving **mrrjjj** first and second.

The frequencies of responses for the subjects to the **mrrjjj** analogy for all answers that occurred more than twice are presented in Table 4[5]. The **mrrjjj** analogy does not show a clear order effect. Both before and after solving the **kji** analogy, about half of the subjects appear to have settled on the simplest representation: changing the rightmost letter to its successor and therefore producing *mrrjjk*.

Examination of the transfer effects supported the asymmetry hypothesis. Of the 17 subjects who produced *kjj* as their answer to **kji** when doing that analogy first, 14 subsequently generated *mrrjjk* as their answer to **mrrjjj**. However, of the 23 subjects who produced *mrrjjk*, only 10 generated *kjj* as their answer to **mrrjjj**. This difference in proportions was significant ($X^2(1)= 6.16, p<.05$). Similarly, five of the fourteen subjects who produced *kjh* as their answer to **kji** subsequently generated *mrrjjk* as their response to **mrrjjj**. But only one of the 23 subjects who produced *mrrjjk* subsequently generated *kjh* ($X^2(1)= 6.30, p<.05$). Thus transfer tended to be from **kji** to **mrrjjj**

[4] Other responses recorded were: *jih, kjil, kkjjil, ijl, lkj, kjid, kjf, kkjjl, kki, iijjk, kkjjim, lkji, lkji, lkjjii.*

[5] Other responses recorded were *mmmmrrrrjjkk, mrsjjj, jjmmrr, mrrjjd, mrrjjh, rrjjmj, mrrjjq, jjrrmn, mrrjjo, mrrjjp, jjjmrs, mrrjjn, mrrjd, mrrjkl, mrrjkk, mrrjjt, mrrjjj, jjjmr, mrrjcj, nrrjjj, mrr, mrrjjm.*

rather than the reverse, which is the opposite direction to that found in Experiment 1. However, this reversal is consistent with previous results if we accept that in Experiment 2 it was more difficult to build a highly structured representation of **kji** than for **mrrjjj**.

Conclusions

The general approach of Copycat appears to be supported by our results. Subjects who appeared to adhere to the limitations of Copycat's micro world actually produced very similar behavior to Copycat when solving their first analogy problem. The asymmetrical transfer results indicated that the apparent ease of transferring a representation predicts the amount of transfer observed, which is consistent with Copycat and its emphasis on building representations as the crucial part of making an analogy. However, we have not definitively tested Copycat or rejected alternative models of analogy making (e.g., Holyoak & Thagard, 1989; Falkenhainer, Forbus, & Gentner, 1989) which may be able to handle the transfer effects we found if appropriately setup. But it may not be possible to construct definitive tests given that currently Copycat solves very different problems from those that other analogy making models attempt and Copycat represents a different approach to the whole question.

In ongoing work we are trying to further investigate the psychological plausibility of Copycat and its approach. In particular, we are examining how subjects represent these types of problems by using the generation and transfer methodologies we describe here and by having subjects rate the quality of a set of possible answers.

Acknowledgments

We thank Keith Holyoak for comments on an earlier draft of this paper.

References

Falkenhainer, B., Forbus, K.D., and Gentner, D. 1989. The structure-mapping engine: Algorithm and examples, *Artificial Intelligence* 41:1-63.

Hofstadter, D. R. 1985. Analogies and roles in human and machine thinking. In *Metamagical themas* (pp. 547-603). New York: Basic Books.

Hofstadter, D. R., and Mitchell, M. 1988. Conceptual slippage and analogy-making: A report on the copycat project. In Proceedings of the Tenth Annual Conference of the Cognitive Science Society, 601-607. Hillsdale, NJ: Lawrence Erlbaum Associates.

Holyoak, K.J., and Thagard, P. 1989. Analogical mapping by constraint satisfaction. *Cognitive Science* 15:295-355.

Mitchell, M. 1990. Copycat: A computer model of high-level perception and conceptual slippage. Ph.D. diss., Dept. of Computer and Communication Sciences, University of Michigan, Ann Arbor.

Mitchell, M., and Hofstadter, D. R. (1989). The role of computational temperature in a computer model of concepts and analogy-making. Proceedings of the Eleventh Annual Conference of the Cognitive Science Society, 765-772. Hillsdale, NJ: Lawrence Erlbaum Associates.

Mitchell, M., and Hofstadter, D. R. 1990. The right concept at the right time: How concepts emerge as relevant in response to context-dependent pressures. Proceedings of the Twelfth Annual Conference of the Cognitive Science Society, 174-181. Hillsdale, NJ: Lawrence Erlbaum Associates.

Using Cognitive Biases to Guide Feature Set Selection

Claire Cardie*
Department of Computer Science
University of Massachusetts
Amherst, MA 01003
cardie@cs.umass.edu

Abstract

Although learning is a cognitive task, machine learning algorithms, in general, fail to take advantage of existing psychological limitations. In this paper, we use a learning task from the field of natural language processing and examine three well-known cognitive biases for human information processing: 1) the tendency to rely on the most recent information, 2) the heightened accessibility of the subject of a sentence, and 3) short term memory limitations. In a series of experiments, we modify a baseline instance representation in response to these limitations and show that the overall performance of the learning algorithm improves as increasingly more cognitive biases and limitations are explicitly incorporated into the instance representation.

Introduction

Inductive concept acquisition has always been of primary interest for researchers in the field of machine learning. In this task, a system typically learns one or more concepts by analyzing a set of examples (and possibly counterexamples) of the concepts. In fact, a number of systems for the acquisition of concepts now exist (e.g., ID3 (Quinlan, 1979), ARCH (Winston, 1975), COBWEB (Fisher, 1987), UNIMEM (Lebowitz, 1987)).Independently, psychologists, psycholinguists, and cognitive scientists have examined the effects of numerous psychological limitations on human information processing. However, despite the fact that concept learning is a basic cognitive task, most machine learning systems for concept formation fail to exploit these limitations and make no attempt to model human concept learning.

*This research was supported by the Office of Naval Research, under a University Research Initiative Grant, Contract No. N00014-86-K-0764, NSF Presidential Young Investigators Award NSFIST-8351863 awarded to Wendy Lehnert, and the Advanced Research Projects Agency of the Department of Defense monitored by the Air Force Office of Scientific Research under Contract No. F49620-88-C-0058.

In this paper, we show that the explicit encoding of known cognitive biases into the training instance representation can improve the performance of the learning algorithm for cognitively-based learning tasks. More specifically, we use a well-known concept acquisition system and focus on a single learning task from the field of natural language processing (NLP). After training the system using a baseline instance representation, we modify the representation in response to three cognitive biases: 1) the tendency to rely on the most recent information, 2) the heightened accessibility of the subject of a sentence, and 3) short term memory limitations. Each modification explicitly incorporates one or more cognitive biases into the feature set. In a series of experiments, we compare each of the modified instance representations to the baseline.

Finding the Antecedents of Relative Pronouns

Although the use of cognitive biases to guide feature set selection is a domain-independent technique, we will use a learning task from NLP to illustrate the performance of the technique throughout the paper. Our task for the machine learning system is the following: Given a sentence with the relative pronoun"who," learn to recognize the phrase or phrases that represent the relative pronoun's antecedent. (Note: This paper focuses only on the technique with respect to machine learning issues. For a detailed discussion of the viability of this approach for the disambiguation of relative pronouns from the NLP perspective, see (Cardie, 1992a) and (Cardie, 1992b)). Finding the antecedents of relative pronouns is a crucial task for natural language systems because the antecedent must made available to the subsequent clause where it implicitly fills the *actor* or *object* roles.[1] Consider the following example:

Igor shook hands with *the skater* **who** beat him in the race.

[1]In practice, the antecedent of "who" sometimes fills semantic roles other than the actor or object.

A correct semantic interpretation of this sentence should include the fact that "the skater" is the actor of "beat" even though the phrase does not appear in the embedded clause. Only after the natural language system associates "the skater" with "who" can it make this inference. Locating the antecedent of "who" may initially appear to be an easy problem because the antecedent often immediately precedes the word "who." Unfortunately, this is not always the case as shown in S1 and S2 of Figure 1. Even when the antecedent does immediately precede the relative pronoun, it does not appear in a consistent syntactic constituent. In S3, for example, the antecedent is the subject of the preceding clause; in S4, it is the direct object; in S5, it is the object of a preposition. Furthermore, the antecedent of "who" may contain more than one phrase. In S6, for example, the antecedent is a conjunction of three phrases and in S7, either "our sponsors" or its appositive "Gatorade and GE" is a semantically valid antecedent. Occasionally, there is no apparent antecedent at all (e.g., S8).

S1. *The woman* from Philadelphia who played soccer was my sister.
S2. I spoke to *the man* in the black shirt and green hat over in the far corner of the room who demanded to meet the skiers.
S3. *The skater* who won the medal was from Japan.
S4. I saw *the skater* who won the medal.
S5. Igor ate dinner with *the skater* who won the medal.
S6. I'd like to thank *Nike, Reebok, and Adidas,* who provided the uniforms.
S7. I'd like to thank *our sponsors, Gatorade and GE,* who provide financial support.
S8. We wondered who would win the race.

Figure 1: Antecedents of "who"

Despite these ambiguities, we will describe how a machine learning system can learn to locate the antecedent of "who" given a description of the clause that precedes it. In effect, we are teaching the system to recognize the "relative pronoun antecedent" concept. More importantly, we will show that performance of the learning system improves as the instance description explicitly encodes increasingly more cognitive limitations and cognitive biases.

COBWEB and the Representation of Training Instances

For our experiments we chose COBWEB (Fisher, 1987) – a well-known concept formation system that is one of a relatively small number of concept acquisition systems designed to model some aspects of human concept learning.[2] Given a set of training instances, COBWEB discovers a classification scheme that covers the instances. Instead of forming concepts at a single level of abstraction, however, COBWEB organizes instances into a classification hierarchy where leaves represent instances and internal nodes represent concepts that increase in generality as they approach the root of the tree. In addition, COBWEB's construction of the hierarchy is cognitively economical in that new objects are incrementally added to the hierarchy as they arrive. To evaluate the concepts it creates, COBWEB employs the *category utility* metric (Gluck, & Corter, 1985) – a measure developed in psychological studies of basic level categories.

COBWEB takes as input a set of training instances described as a list of attribute-value pairs. Because the antecedent of a relative pronoun usually appears as one or more phrases in the clause preceding "who," the attribute-value pairs in each training case represent the constituents that precede "who." At first glance, it may seem that only syntactic information needs to be encoded. However, finding the antecedent of a relative pronoun actually requires the assimilation of syntactic and semantic knowledge. For this reason, each *constituent attribute-value pair* takes the following form:

- The *attribute* describes the syntactic class and position of the phrase.

- The *value* provides its semantic classification.

Consider, for example, the sentences in Figure 2. In the training instance for S1, we represent "the man" with the attribute-value pair *(s human)* because it is the subject of the sentence and the noun "man" is human. We represent "from Oklahoma" with the pair *(s-pp1 location)* because it is the first prepositional phrase that follows the subject and "Oklahoma" is a location. All noun phrases are described by one of seven general semantic features: human, proper-name, location, entity, physical-target, organization, and weapon.[3] When clauses contain conjunctions and appositives, each phrase in the construct is labelled separately. In S2, for example, the real direct object of "thank" is the conjunction "Nike and Reebok." However, in our instance representation, "Nike" is tagged as the direct object *(do)* and "Reebok" as the

[2]The COBWEB/3 system was provided by Kevin Thompson, NASA Ames Research Center.

[3]These features are specific to the domain from which the training instances were extracted. A different set would most likely be required for nouns in a different domain.

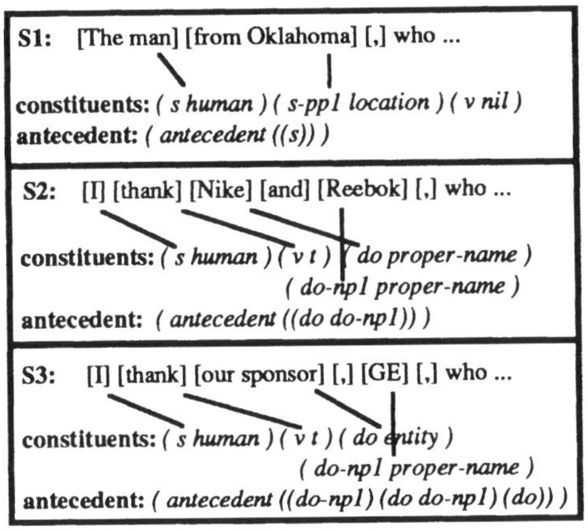

Figure 2: Training Cases

first noun phrase that follows the direct object *(do-np1)*.[4] For verb phrases, we currently note only the presence or absence of a verb using the values *t* and *nil*, respectively.

In addition to the constituent attribute-value pairs in every training instance, we include information about the correct antecedent in the form of an *antecedent attribute-value* pair.[5] (This feature is labelled "antecedent:" in the examples of Figure 2.) The value of the antecedent attribute is a list of the constituent attributes that represent the location of the antecedent or *(none)* if there is no apparent antecedent. In S1, for example, the antecedent of "who" is "the man." Because this phrase appeared as the subject of the previous clause, the value of the antecedent attribute is *(s)*. Sometimes, however, the antecedent is actually a conjunction of constituents. In these cases, we represent the antecedent as a list of the constituent attributes associated with each element of the conjunction. Look, for example, at sentence S2. Because "who" refers to the conjunction "Nike and Reebok," the antecedent is described as *(do do-np1)*. S3 shows yet another variation of the antecedent attribute-value pair. In this example, an appositive creates three semantically equivalent antecedents, all of which become part of the antecedent feature: 1) "GE" — *(do-np1)*, 2) "our sponsor"— *(do)*, and 3) "our sponsor, GE"— *(do do-np1)*.

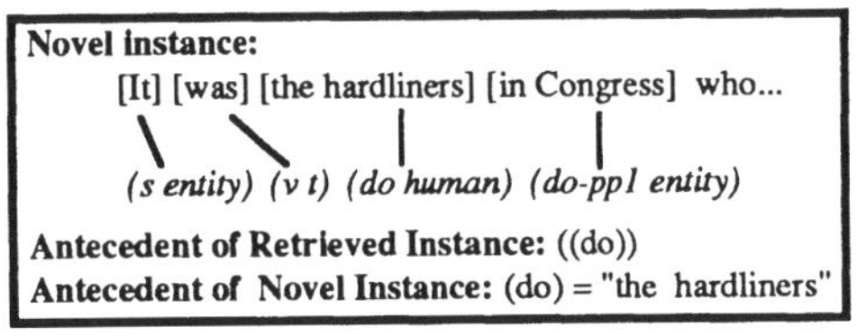

Figure 3: Instance Retrieval

Training instances are generated automatically from unrestricted texts by the UMass/MUC-3 NLP system (Lehnert et al., 1991) as a side effect of parsing. Only the antecedent must be specified by a human supervisor via a menu-driven interface that displays the antecedent options. After training, we use the resulting COBWEB classification hierarchy of relative pronoun disambiguation decisions to predict the antecedent of "who" in new contexts. Given a new instance to classify, COBWEB retrieves from the hierarchy the most specific concept that adequately describes the instance. Then, the antecedent of the retrieved concept guides selection of the antecedent for the novel case. Given the test instance in Figure 3, for example, COBWEB retrieves an instance that specifies the direct object (*do*) as the location of the antecedent. Therefore, we choose the contents of the *do* constituent — "the hardliners" — as the antecedent in the novel case. Sometimes, however, COBWEB retrieves a concept that lists more than one option as the antecedent. In these cases, we choose the option that appeared most often in the underlying instance(s) and whose constituents overlap with those in the current context. (For a description of a better, but more complicated adaptation heuristic, see (Cardie, 1992b).)

The Baseline Experiments

We tested this baseline instance representation by extracting all examples of "who" from 3 sets of 50 texts from the MUC-3 corpus[6]. In each of 3

---

[4]In a separate paper (Cardie, 1992a), we explain this representational decision in more detail. In general, NLP systems do not reliably handle complex conjunctions and appositives. They can, however, accurately locate lower level phrases like individual noun phrases, verbs, and prepositional phrases. As a result, we let the the machine learning system recognize conjunctions and appositives and allow the NLP system that generates the training instances to ignore these tasks.

[5]COBWEB is designed to perform unsupervised learning. However, many applications of COBWEB, including our own, encode pseudo-supervisory information, i.e., class information, as part of the instance representation (see (Fisher, 1987)).

[6]The MUC-3 corpus consists of 1500 texts (e.g., newspaper articles, TV news reports, radio broadcasts) containing information about Latin American terrorism and was developed for use in the Third

experiments, 2 sets were used for training and the third reserved for testing. The results are shown in Figure 4 and indicate that COBWEB finds the correct antecedent of "who" an average of 59% of the time when using the baseline instance representation. In the next two sections, we modify this baseline representation in response to three cognitive biases and show the results of these modifications on COBWEB's performance.

Exp #	Training Sets (# instances)	Test Set (# instances)	Baseline Rep
1	set1 + set2 (170)	set3 (71)	63%
2	set2 + set3 (159)	set1 (82)	47%
3	set1 + set3 (153)	set2 (88)	66%

Figure 4: Baseline Results (% correct)

Incorporating the Recency Bias

In processing language, people consistently show a bias towards the use of the most recent information (e.g., (Kimball, 1973), (Frazier, & Fodor, 1978), (Gibson, 1990)). In particular, the mechanisms people use for finding the antecedents of pronouns and missing subjects have been investigated in a series of recent experiments (see (Nicol, 1988)). The results show that in locating antecedents during language processing, people consider all noun phrases preceding the pronoun starting with the most recent noun phrase and working backwards to the most distant noun phrase.

We translate this recency bias into representational changes for the training instances in two ways. First, we label the constituent attribute-value pairs with respect to the relative pronoun. This establishes a right-to-left labelling rather than the left-to-right labelling of the baseline. In Figure 5, for example, "in Congress" receives the attribute *pp1* because it is a prepositional phrase one position to the left of "who." Similarly, "the hardliners" receives the attribute *np2* because it is a noun phrase two positions to the left of "who." Notice, however, that the subject of the sentence retains its original *s* attribute. We based this decision on studies that indicate that the subject of a sentence remains highly accessible even at the end of a sentence (e.g., (Gernsbacher, Hargreaves, & Beeman, 1989)). Consider the following sentences: 1) "it was a message from *the hardliners* in Congress, who..." and

Message Understanding System Evaluation and Message Understanding Conference (Sundheim,1991).

Sentence:
[It] [was] [the hardliners] [in Congress] who...
Baseline Representation:
(s entity) (v t) (do human) (do-pp1 entity) (antecedent ((do)))
Right-to-Left Labelling:
(s entity) (v t) (np2 human) (pp1 entity) (antecedent ((np2)))
Duplicate Information:
(s entity) (v t) (do human) (do-pp1 entity) (most-recent entity) (part-of-speech prep-phrase) (antecedent ((do))

Figure 5: Incorporating the Recency Bias

2) "it was from *the hardliners* in Congress who ...". The right-to-left labelling tags the antecedents in each sentence with the same attribute (i.e., *pp2*), indicating the similarity of the examples with respect to the location of the relative pronoun antecedent. In the baseline representation, however, the antecedents retain distinct attributes – *do-pp1* and *v-pp1*, respectively.

Alternatively, given the baseline instance representation, we can incorporate the recency bias by including more than one attribute-value pair for the most recent information. Figure 5 also shows this second representational change. The most recent constituent ("in Congress") is represented three times[7]: 1) as a constituent attribute-value pair – *(do-pp1 entity)*, 2) as the most recent constituent – *(most-recent entity)*, and 3) via its part of speech – *(part-of-speech prep-phrase)*. In this representation, we also allow the antecedent attribute-value pair to refer to the more general *most-recent* constituent rather than the equivalent, but more specific, constituent attribute-value pair. If, for example, the antecedent in Figure 5 had been *do-pp1*, it would become *most-recent* in the new representation.

The results of experiments that use each of these representations separately and in a combined form are shown in Figure 6. In this table, the MR1 representation used the right-to-left labelling, the MR2 representation included extra information about the most recent constituent, and the MR1+MR2 representation combined both the right-to-left labelling and the duplicate information formats. In general, it is clear that incorporating the recency bias into the instance representation improves performance. On average, the right-to-left labelling

[7]We used all information about the most recent constituent readily available from the parser.

Exp #	Training Sets (# of instances)	Test Set (# of instances)	Baseline Rep	MR1: R-to-L Labelling	MR2: Duplicate Info	MR1 + MR2
1	set1 + set2	set3	63%	75%	83%	84%
2	set2 + set3	set1	47%	62%	65%	73%
3	set1 + set3	set2	66%	66%	71%	74%

Figure 6: Experiments Using the Recency Bias (% correct)

increased the percentage of correctly identified antecedents from 59% to 68% while including extra information for the most recent constituent increased the percentage correct to 73%. The best results, however, occurred using the combined representation, where the percentage correct increased to an average of 77%.

Incorporating the Short Term Memory Bias

Psychological studies have determined that people can keep at most seven plus or minus two facts in short term memory (Miller, 1956). More recently, Daneman and Carpenter ((Daneman, & Carpenter, 1980), (Daneman, & Carpenter, 1983)) show that working memory capacity affects a subject's ability to find the referents of pronouns over varying distances. Also, King and Just (King, & Just, 1991) show that differences in working memory capacity can cause differences in the reading time and the comprehension of certain classes of relative clauses. Moreover, it has been hypothesized that language learning in humans is successful precisely because limits on information processing capacities allow children to ignore much of the linguistic data they receive (see (Newport, 1990)).

COBWEB, however, clearly does not make use of short term memory (STM) limitations either in its learning algorithm or in its attribute-value instance representation. Each training and test instance has to be normalized with respect to all attributes across the training instances.[8] In the baseline representation, this normalization resulted in instances of 35 attribute-value pairs as compared to an average of 5 attribute-value pairs in the original, unnormalized instances. The short term memory bias implies that not all of the 35 features should be retained for the task of finding relative pronoun antecedents. In an attempt to incorporate this limitation, we ran a series of experiments using instances with successively fewer features. We let n = 1,2,3,4,5,7,9,15,20, and 50, and included in the training and test instances only those features that occurred at least n times in the original, unnormalized instances of the training set. As n increases, the number of attributes per instance decreases.

When this STM cutoff was applied to the baseline representation, the performance of the learning algorithm gradually declined as n increased. The percentage correct declined from 63% to 37%, from 47% to 19%, and from 66% to 41% for experiments 1, 2, and 3, respectively. Although the STM cutoff did not improve performance when applied to the MR1 training sets that use the right-to-left labelling, the decline in percentage correct was not nearly so drastic. For experiment 1 (originally 75% hit rate), the percentage correct never dropped below 69%. For experiment 2, results ranged from 62% (with no cutoff) to 49%; and in experiment 3 (originally 66% correct), results ranged from 51% to 67% correct.

Figure 7 shows the results of the STM cutoff for the instance representations of MR2 (extra information for most recent phrase) and MR1+2 (right-to-left labelling and extra information for most recent phrase). In these experiments, the STM bias actually improved COBWEB's performance. In the MR2 experiments, the original hit rate for experiment 1 increased from 83% (37 attributes / instance) to 87% at n = 7 (16 attributes / instance). In experiment 2, the percentage correct moved from 65% (34 attributes / instance) in the original representation to 74% at n = 9 (13 attributes / instance). In experiment 3, the percentage correct increased from 71% (36 attributes / instance) in the original representation to 76% at n = 2 (25 attributes / instance). Similar results occurred for the MR1+2 instance representation. There were increases from 84% (25 attributes / instance) to 87% (17 attributes / instance, n = 4) and from 74% (29 attributes / instance) to 76% (10 attributes / instance) for experiments 1 and 3, respectively. Performance for experiment 2, however, declined.

[8]Our fixed feature set includes every attribute that appears in the training set. To create a training instance, we generate a unique value for any missing attribute, i.e., for any attribute that is irrelevant for the instance.

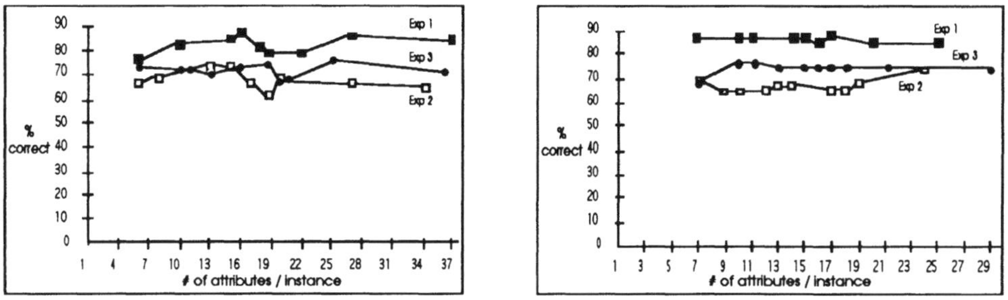

(a) Applying STM Bias to MR2 **(b) Applying STM Bias to MR1+2**

Figure 7: Experiments Using the STM Bias

Conclusions

Based on the preliminary experiments presented in the last three sections, we conclude that explicit incorporation of cognitive biases into the instance representation can greatly improve learning algorithm performance. In addition, although the technique was tested on only one task from NLP, the use of cognitive biases to guide feature set selection is a domain-independent technique that can be applied to any cognitively-based learning task. It is clear, however, that further experimentation is required to explore the effects of additional cognitive limitations, to determine the biases that work well together, and to find the correct parameters for those biases. Finally, further research is required before we can use cognitive biases to automate, rather than guide, feature set selection.

References

Cardie, C. (1992a). Corpus-Based Acquisition of Relative Pronoun Disambiguation Heuristics. *Proceedings, 30th Annual Meeting of the Association for Computational Linguists.* University of Delaware, Newark.

Cardie, C. (1992b). Learning to Disambiguate Relative Pronouns. *Proceedings, Ninth National Conference on Artificial Intelligence.* San Jose, CA.

Daneman, M., & Carpenter, P. A. (1980). Individual differences in working memory and reading. *Journal of Verbal Learning and Verbal Behavior, 19,* 450-466.

Daneman, M., & Carpenter, P. A. (1983). Individual differences in integrating information between and within sentences. *Journal of Experimental Psychology: Learning, Memory, and Cognition, 9,* 561-584.

Fisher, D. H. (1987). Knowledge Acquisition Via Incremental Conceptual Clustering. *Machine Learning,2,* 139-172.

Frazier, L., & Fodor, J. D. (1978). The sausage machine: A new two-stage parsing model. *Cognition, 6,* 291-325.

Gernsbacher, M. A., Hargreaves, D. J., & Beeman, M. (1989). Building and accessing clausal representations: The advantage of first mention versus the advantage of clause recency. *Journal of Memory and Language, 28,* 735-755.

Gibson, E. (1990). Recency preferences and garden-path effects. *Proceedings, Twelfth Annual Conference of the Cognitive Science Society.* Cambridge, MA.

Gluck, M. A., & Corter, J. E. (1985). Information, uncertainty, and the utility of categories. *Proceedings, Seventh Annual Conference of the Cognitive Science Society.* Lawrence Erlbaum Associates.

Kimball, J. (1973). Seven principles of surface structure parsing in natural language. *Cognition, 2,* 15-47.

King, J., & Just, M. A. (1991). Individual differences in syntactic processing: the role of working memory. *Journal of Memory and Language, 30,* 580-602.

Lebowitz, M. (1987). Experiments with Incremental Concept Formation: UNIMEM. *Machine Learning, 2,* 103-138.

Lehnert, W., Cardie, C., Fisher, D., Riloff, E., & Williams, R. (1991). University of Massachusetts: Description of the CIRCUS System as Used for MUC-3. *Proceedings, Third Message Understanding Conference (MUC-3).* San Diego, CA. Morgan Kaufmann Publishers.

Miller, G. A. (1956). The magical number seven, plus or minus two: Some limits on our capacity for processing information. *Psycol. Review, 63(1).*

Newport, E. (1990). Maturational Constraints on Language Learning. *Cognitive Science, 14,* 11-28.

Nicol, J. (1988). *Coreference processing during sentence comprehension.* Ph. D. Thesis. Massachusetts Institute of Technology.

Quinlan, J. R. (1979). Discovering Rules from Large Collections of Examples: A Case Study. In D. Michie (Ed.), *Expert Systems in the Microelectronics Age.* Edinburgh: Edinburgh University Press.

Sundheim, B. M. (1991). Overview of the Third Message Understanding Evaluation and Conference. *Proceedings, Third Message Understanding Conference (MUC-3).* San Diego, CA. Morgan Kaufmann Publishers.

Winston, P. H. (1975). Learning Structural Descriptions from Examples. In P. H. Winston (Ed.), *The Psychology of Computer Vision.* New York: Mc Graw-Hill.

The Interaction of Principles and Examples in Instructions

Richard Catrambone and Ronald M. Wachman
Georgia Institute of Technology
School of Psychology
Atlanta, GA 30332-0170
rc7@prism.gatech.edu gt4375a@prism.gatech.edu

Abstract

Learners often have difficulty following instructions written at a general enough level to apply to many different cases. Presence and type of example (example either matched the first task, did not match the first task, or was not present) and presence of a principle (that provided a rationale for part of a procedure) were manipulated in a set of instructions for computer text editing in order to examine whether initial performance and later transfer could be improved. The results suggest that a principle can aid initial learning from general instructions if no example is given or the example does not match the first task. The principle could help users disambiguate the instructions by providing a rationale for potentially misunderstood actions. However, if the example matches the first task, then the presence of a principle seems to slow initial performance, perhaps because the learner tries to compare and integrate the example and the principle. On later training tasks, however, a principle improves performance. These results suggest that the features of instructions that aid initial performance and those that aid later performance are different and careful research on how to integrate these features is important.

Introduction

People frequently have difficulty following instructions (e.g., Reed & Bolstad, 1991; Wright, 1981). One reason for this difficulty is that the procedures described in the instructions are ambiguous or abstract at certain points. These points are often places where options in the procedure exist. If the procedure is described too concretely at these points, that is, a particular choice for that point is described, the learner might not understand that other choices are possible and thus, fail to generalize when confronted with new tasks (Catrambone, 1990). Ideally, the learner needs to understand, or instructions need to convey, the necessary generality. However, if the learner is new to the domain then he or she will have difficulty comprehending the generality and determining how to instantiate the procedure for initial tasks. This is one place where a principle or example could be useful.

A principle provides the learner a way to generate for him- or herself an explanation of the ambiguity (Mitchell, Keller, & Kedar-Cabelli, 1986). This presumably would help the learner apply the instructions to the initial tasks as well as later tasks.

An example can help the learner instantiate the instructions. If the example matches the task the learner needs to perform, it is likely the learner will perform the task successfully. However, if the initial task and the example do not match, the learner typically has difficulty applying the example successfully (Catrambone & Holyoak, 1990; Ross, 1987), perhaps because the learner is unable to distinguish superficial features of the example from ones that are relevant for carrying out the procedure.

The current paper examines the effects of examples and principles that accompany general instructions for the task of deleting text using a word processor. Learners were given instructions that were general enough to cover all deletion situations. With the word processor used in the experiment, deletion is done by placing the cursor at the beginning of the to-be-deleted text, selecting the "delete" option from a menu, highlighting the text, and pressing Enter. To highlight the text the user presses the final character of that text. The instructions say to "Type the character at the end of the text you want to delete, typing it over and over until the text is highlighted." When the user specifies the target, the computer then highlights the text up to that character. The target character can be pressed several times if it occurs more than once in the to-be-deleted text.

The notion of multiple target specification is bewildering to new users who do not understand the idea of target searches (Catrambone, 1990). It is at this point that a principle or example is helpful. An explanation of what the computer is doing when the target character is pressed could help the learner understand why it is sometimes necessary to press the final character more than once (see Table 1). Alternatively, an example showing a to-be-deleted word such as "telephone" that requires its final letter to be typed more than once would help the learner to explain to him- or herself why multiple presses of the target is sometimes necessary.

Another factor to consider is the nature of the initial task the learner attempts. If the first task subjects faced was to delete a word that required only a single keypress (i.e., its final letter did not occur earlier in the word), then the learner could be confused by the instruction that says to type the end character over and over. This in fact was the initial task in this experiment. Two factors were manipulated in order to examine their effects on this potential confusion. The first was whether the instructions were accompanied by a principle that explained why multiple presses of the target letter are sometimes necessary. This is predicted to help the learner determine more quickly why the final letter of the first to-be-deleted word needs to be pressed only once. The other factor was the type of example that accompanied the instructions. If the example is a word whose final letter occurs earlier in the word (such as "telephone"), this could help the learner understand why the final letter sometimes has to be pressed more the once; however, this example does not match the first task and thus, could confuse the learner. Conversely, if the example is a word whose final letter does not occur earlier in the word (such as "airplane"), prior research suggests the example would help the learner with the first task since the example and the task mesh. However, there remains the potential for confusion since the example is in some sense at odds with the instructions. In this situation, the learner would be likely to pay more attention to the example since it meshes with the first task and pay less attention to the instruction (LeFevre & Dixon, 1986). Finally, if no example is given then a principle again becomes important in helping the learner to understand the instructions.

The purpose of this study was to explore the relationship between general instructions, principles, examples, and performance on initial and later tasks.

Prior Work Examining Elaborations and Examples in Learning Procedures

One difficulty that faces new users of a set of instructions is understanding what is really going on when they execute a series of steps. Prior work has suggested that a principle or explanation, at some level, of what the system "really" does, even if that explanation is only an approximation, would help learners understand those steps more rapidly (Kieras & Bovair, 1984) and even apply them to novel situations more effectively (Gentner & Gentner, 1983).

In a similar vein, other studies suggest that background knowledge or elaborations may help initial performance (Barsalou & Hale, 1992). For example, Reder, Charney, and Morgan (1986) found that subjects learning various DOS commands were more successful if the instructions contained elaborations about the commands instead of primarily syntax information. Smith and Goodman (1984) found that subjects who received elaborations of instructions for building circuits that included information about the structure or function of the circuits more accurately built the circuits and showed superior transfer when building new circuits.

Despite the demonstrated value of elaborations and mental models on learning, people prefer to learn from examples (Chi, Bassok, Lewis, Reimann, & Glaser, 1989; LeFevre & Dixon, 1986; Pirolli & Anderson, 1985). One reason examples are often preferred might be that they provide a concrete guide to behavior. The learner typically can visually compare the example to the procedure as well as the current task and decide how to make changes appropriate to the current task. However, one well-established difficulty is that learners often have trouble adapting examples to novel problems (Catrambone & Holyoak, 1990; Reed, Dempster, & Ettinger, 1985).

In the current study it is predicted that learners will do the initial deletion task most successfully if the example matches the task or if the instructions contain a principle. It is unclear, based on prior work, what the nature of the interaction between example type and principle will be.

Experiment

Method

Subjects. Subjects were 61 students at the Georgia Institute of Technology who participated for course credit. Subjects, as indicated on a questionnaire, had computer experience confined to a Macintosh whose interface is considerably different than the interface for the word processor used in this experiment.

Procedure. Subjects performed word processing tasks on an IBM PS/2 Model 80 computer (this will be referred to as the "task" computer). In addition, a second PS/2 80 was used to present instructions to subjects (this will be referred to as the "instruction" computer).

Subjects were first shown several features of the task computer and the word processing program. The task computer screen displayed an "empty" document. Subjects were shown how to move the cursor around the screen with the arrow keys. Subjects were then asked to type a paragraph to allow them to get comfortable with the keyboard.

Next, subjects were shown how to use the instruction computer to read the instructions on how to do various tasks. Instructions consisted of a series of screens of information. The instructions included procedures for retrieving and exiting documents and inserting text into a document. These instructions were the same for all subjects. In addition, the instructions included the procedure for deleting text. Table 1 contains the part of the instructions for deleting text that varied from group to group.

There were six groups in the experiment defined by the presence or absence of the principle and the type of example (matching first task, not matching, no example). The deletion instructions were identical for all subjects except for Screen 7 (see Table 1). "Principle" subjects received an explanation of the searching the computer does for a target character. "Example" subjects received an example of a word being deleted. The word required either a single keypress (matching first task) or multiple keypresses (not matching first task). If a subject received both a principle and an example, the principle preceded the example. The type of instructions subjects received was confounded with length.

Presentation of Instructions. The instructions could be viewed on the instruction computer one screen at a time. The contents of a screen became visible when the subject held down the space bar. When the space bar was not held down, an outline of the instructions appeared on the display. The outline consisted of rows of dashes where each row corresponded to a screen. Each row that represented the first screen of information for a particular topic (such as retrieving a document) consisted of the title of the topic rather than dashes. This allowed subjects to keep visual track of where they were in the instructions. In addition, one row in the outline was always at a higher intensity than the others. This row corresponded to the screen that would appear if the space bar was pressed. Subjects could go forwards or backwards through the instruction screens by pressing the Next Page key or the Previous Page key. Subjects' movements through the instructions were automatically recorded.

Training and Test Phases. After learning how to read instructions on the instruction computer, subjects were shown the first document on which they were to work. The document was marked-up to show the changes that were to be made. Items to be deleted were underlined in red ink. The name of the document was printed in the upper left-hand corner since the name was needed in order to retrieve the document.

Prior to doing a task (such as retrieving a document) in the training phase for the first time, a subject read the instructions for that task all the way through before attempting to do the task. This was done in order to make sure subjects saw all the steps for the procedure at least once and would be less tempted to guess about how to do a step later. Subjects were told that once they were done reading a section, they could not look back at it while they attempted to do the task.

Once subjects began a task, if they did not know what to do at a particular point or made a mistake from which they could not recover, they had to re-read that section of the instructions and then redo the task. Once subjects successfully completed a task, they did not have to read that section of the instructions again unless they later made a mistake from which they could not recover. The time to do a deletion task was defined as the moment the function key was pressed (that opened the menu containing the delete option) until the Enter key was pressed, causing the appropriate text to disappear from the screen (including time spent redoing the task if the subject made a mistake).

The experiment was broken into a training phase and a test phase. Training tasks included deleting a total of six words, six sentences, and three paragraphs. The test tasks consisted of deleting words, sentences, and paragraphs as well as other entities, such as garbage letters in the middle of a word, that required subjects to apply the procedure to unfamiliar units.

The first three documents (constituting the training phase) each required the following tasks in order: 1) retrieve the document, 2) delete a word, 3) insert a phrase, 4) delete a sentence, 5) insert a phrase, 6) delete a paragraph, 7) insert a phrase, 8) delete a word, 9) insert a phrase, 10) delete a sentence, and 11) exit the document.

The phrase insertions were always seven words long. The word deletions during the training phase never involved a word whose last letter also occurred earlier in the word. Thus, these word deletions required only a single specification of the target letter. Similarly, all sentences ended in a period and no sentence contained any internal periods. The insertion and document retrieval and exiting tasks were included to make the tasks somewhat realistic.

During the test phase subjects performed only deletion tasks. Each of the three test phase documents contained five deletion tasks, two involving words, two involving sentences, and one involving a paragraph. The first deletion task in the test phase was to delete the word "mysterious." The second task was to delete a sentence that had internal periods (and thus a period had to be pressed three times in order to completely highlight the sentence). The third task was to delete a paragraph in which the last character, a period, occurred only once in the paragraph. Other test phase deletion tasks differed from the word, sentence, and paragraph deletion tasks in the training phase in various ways. First, some tasks began in the middle of words, sentences, and paragraphs rather than at the beginning as was the case in the training phase. Second, some tasks did not include the end of some obvious unit (e.g., deleting the first few words of a sentence without deleting the rest of the sentence). Third, some tasks, other than paragraph deletions, required multiple keypresses.

Results

Given that a prior study (Catrambone, 1990) demonstrated a long start-up time for general instructions, but good transfer to novel tasks, the result of most interest is time to do the first deletion task (deleting a word). The times varied from 31 to approximately 80 seconds (see Table 2). A two-way analysis of variance showed no main effect of either principle ($F(1,55)=.23$, $p=.63$) or example ($F(2,55)=.02$, $p=.98$). However, the interaction was significant, $F(2,55)=3.31$, $p=.04$. The three fastest groups were the groups with the matching example and no principle, the group with the principle and no example, and the group with the mismatching example and the principle.

Table 1: Deletion Instructions

Screen 7:
Type the character at the end of the text you want to delete, typing it over and over until the text is highlighted.

Principle:
Each time you type the character, the computer "searches" in a forward direction, starting from the point at which the cursor is located, until the computer finds the character. When the computer finds the character, it highlights all the text it searched through on the way to finding the character.

Example:
Matched initial task: For example, if the word you wished to delete was airplane then you would type the letter e.

Did not match initial task: For example, if the word you wished to delete was telephone then you would type the letter e three times.

Table 2: Time to Perform Deletion Tasks (seconds)

	Example Matches First Task		Example Does Not Match First Task		No Example	
Deletion Task	Principle (n=10)	No Prin (n=10)	Principle (n=11)	No Prin (n=10)	Principle (n=11)	No Prin (n=9)
1st (Delete Word)	75.9	31.0	42.6	60.6	31.0	79.5
2nd (Delete Sentence)	242.3	112.4	180.4	171.6	103.7	159.6
3rd (Delete Paragraph)	13.3	18.1	13.5	14.4	31.2	26.1
Remainder of Training Phase Tasks	8.9	10.7	9.7	10.1	9.4	11.7
1st Word Requiring Multiple Keypresses	8.7	16.5	9.1	10.2	9.8	13.0
1st Sentence Requiring Multiple Keypresses	10.4	10.9	12.1	12.6	12.2	12.3
1st Paragraph Requiring a Single Keypress	10.2	10.4	9.5	16.0	9.7	10.5
Remainder of Test Phase Tasks	10.0	10.6	11.8	11.2	11.6	11.4

The pattern of results in Table 2 for the first deletion task suggests the following interpretation. If no example is present or the example does not match the current task, then having a principle appears to help the learner apply the procedure. If the example matches the first task and there is no principle, the learner also performs well. However, if the example matches the first task and the principle is present, the learner is slowed, perhaps because the learner spends time trying to reconcile the discrepancy between a principle that explains why multiple presses are needed with an example that only requires a single press. If no principle is given and either no example or a mismatching example is given, performance is also slowed.

The second deletion task, deleting a sentence, caused problems for subjects, primarily because they had difficulty realizing that a period could be used as a target character to specify the range to be highlighted. There was no effect due to the principle, example, or their interaction. This is not entirely surprising given that the major difficulty, realizing that a period is a legitimate target character, does not appear to be benefited in any obvious way by the principle or examples used here.

The third deletion task, deleting a paragraph, showed a trend favoring subjects who received the principle ($F(1,55)=3.60$, $p=.06$). Performance time for all groups dropped considerably from the sentence deletion time (see Table 2). This is reasonable given that subjects had learned from the prior task that a period can be used as a target. The only new feature here is that the target needs to be pressed more than once. The first two deletion tasks, deleting a word and a sentence, only required a single keypress. Perhaps subjects developed an expectation of only having to type the target once. However, principle subjects could have possessed the necessary understanding to realize more quickly why a single keypress was not sufficient to highlight the entire paragraph. There was also a trend towards subjects without an example taking longer, but this was primarily due to two outliers.

Performance on the rest of the training tasks showed a trend favoring subjects who received the principle ($F(1,55)=3.39$, $p=.07$). This suggests some benefit of a principle beyond the performance of the initial tasks. Perhaps the principle provides an additional pathway for helping subjects recall or reconstruct the details of how to specify a target character for both single and multiple specification cases. The example may help only on initial cases that match it.

The test phase tasks involved novel features such as deleting a word using multiple keypresses or deleting only part of a word or sentence. Performance on the first test phase task, deleting a word that required multiple specifications of the target, favored subjects who received the principle ($F(1,55)=8.14$, $p=.006$). It is surprising that subjects who received the word deletion example that involved multiple keypresses did not show superior performance on this task. However, this result is consistent with the training phase result showing no benefit of an example beyond the initial task. Perhaps an example is accessible for initial tasks similar to the example whereas a principle is accessible for, and therefore applied to, many tasks.

Performance on the next task, the first sentence deletion task that involved multiple specifications of the target, did not show performance differences as a function of either manipulation. This makes sense since subjects had just completed a task requiring multiple specifications.

Performance on the next task, the first paragraph deletion task in which the target character had to be specified only once, showed a trend favoring principle subjects ($F(1,56)=3.60$, $p=.06$). This probably occurred because by this time subjects were used to specifying a target multiple times for a paragraph and some subjects pressed the target multiple times before realizing they did not have to do so. Subjects who received the principle were less likely to make this error since they presumably were more aware of what the specification did. Nevertheless, this is a surprising result given that subjects had already done 17 deletions prior to this task.

There were no differences in performance for the remainder of the test phase tasks as a function of the manipulations. Subjects had, by this time, been exposed to all the oddities they would encounter later and presumably were able to handle them (principle: $F(1,55)=.09$; example: $F(2,55)=1.34$, $p=.27$; interaction: $F(2,55)=.48$).

Discussion

The results indicate that a principle is useful in helping learners to follow general instructions initially, particularly if no example or a mismatching example is present. An example is useful on the initial task if it matches that task. Interestingly, if the example matches the initial task, then a principle seems to get in the way of performance of the initial task. These findings suggest that instructions could be written at a general level and still be relatively easy to use initially if certain elaborations or principles are provided. Other research has suggested that instructions can be written in detail for specific tasks that aid initial performance and allow generalization to novel tasks (Catrambone, 1990, Experiment 2).

The decision to write general instructions with principles or to write specific instructions that promote generalization could be a function of other factors. For example, if the user will be performing a limited set of tasks, then specific instructions for each task is the best approach. However, if it is likely that the user will eventually have to do novel and perhaps unforeseen (by the instruction writer) tasks, then general instructions with principles is the best approach. Clearly, these issues need to be tested in additional experiments.

Experiments that manipulate the ambiguity of general instructions should show differential effects of including principles. It may be the case that well-written general instructions do not benefit from the inclusion of principles and examples. In fact, users might perform best with a well-written minimalist set of instructions (Carroll, Smith-Kerker, Ford & Mazur-Rimetz, 1987-88).

The general instructions used in this experiment were probably not optimal. Screen 7 of the instructions (Table 1) said to type a target character over and over until the text was highlighted. The instructions could have been better worded (e.g., "type the character one or more times...") with little loss of generality. If this improved wording had been used, it is not clear that the inclusion of the principle would have had the same impact as it did in the current study. In any case, the effect of providing a principle needs to be examined for other tasks and with instructions involving varying degrees of generality before strong conclusions can be drawn about the effects of principles on the comprehension and application of instructions.

It would also be useful in future work to build a model to explain how a principle or an example can aid in the comprehension and carrying-out of general instructions. One approach is suggested by Kieras and Bovair (1984) who argue that mental model information is information that maps on to the requirements of a procedure, thus allowing a learner to infer a procedure. Thus, in the current study, if a subject could not remember part of the procedure or had difficulty determining how to apply it, the presence of the principle helped them reconstruct the necessary steps. Another possibility is that the principle helps disambiguate instructions. An explanation of why a certain step is needed could help point the user towards the correct interpretation of the instructions. Thus, a principle should be more useful as the instructions are more general.

References

Barsalou, L.W., & Hale, C.R. (1992, February). *Roles of explanation in the acquisition of symptom-fault rules during troubleshooting.* Presentation made at the Army Research Institute Contractors' Meeting, Winter Park, Florida.

Carroll, J.M., Smith-Kerker, P.L., Ford, J.R., & Mazur-Rimetz, S.A. (1987-88). The minimal manual. *Human-Computer Interaction, 3*, 123-153.

Catrambone, R. (1990). Specific versus general procedures in instructions. *Human-Computer Interaction, 5(1)*, 49-93.

Catrambone, R., & Holyoak, K.J. (1990). Learning subgoals and methods for solving probability problems. *Memory & Cognition, 18(6)*, 593-603.

Chi, M.T.H., Bassok, M., Lewis, R., Reimann, P., & Glaser, R. (1989). Self explanations: How students study and use examples in learning to solve problems. *Cognitive Science, 13*, 145-182.

Gentner, D., & Gentner, D.R. (1983). Flowing waters or teeming crowds: Mental models of electricity. In D. Gentner & A.L. Stevens (Eds.), *Mental models.* Hillsdale, NJ: Erlbaum.

Kieras, D.E., & Bovair, S. (1984). The role of a mental model in learning to operate a device. *Cognitive Science, 8*, 191-219.

LeFevre, J., & Dixon, P. (1986). Do written instructions need examples? *Cognition and Instruction, 3(1)*, 1-30.

Mitchell, T.M, Keller, R.M., & Kedar-Cabelli, S.T. (1986) Explanation-based generalization: A unifying view. *Machine Learning, Volume 1*, 47-80.

Pirolli, P.L., & Anderson, J.R. (1985). The role of learning from examples in the acquisition of recursive programming skill. *Canadian Journal of Psychology, 39*, 240-272.

Reder, L.M., Charney, D.H., & Morgan, K.I. (1986). The role of elaborations in learning a skill from an instructional text. *Memory & Cognition, 14(1)*, 64-78.

Reed, S.K., & Bolstad, C.A. (1991). Use of examples and procedures in problem solving. *Journal of Experimental Psychology: Learning, Memory, and Cognition, 17(4)*, 753-766.

Reed, S.K., Dempster, A., & Ettinger, M. (1985). Usefulness of analogous solutions for solving algebra word problems. *Journal of Experimental Psychology: Learning, Memory, and Cognition, 11*, 106-125.

Ross, B. (1987). This is like that: The use of earlier problems and the separation of similarity effects. *Journal of Experimental Psychology: Learning, Memory, and Cognition, 13*, 629-639.

Smith, E., & Goodman, L. (1984). Understanding written instructions: The role of an explanatory schema. *Cognition and Instruction, 1*, 359-396.

Wright, P. (1981). "The instructions clearly state...." Can't people read? *Applied Ergonomics, 12*, 131-141.

Strategies for Contributing to Collaborative Arguments[1]

Violetta Cavalli-Sforza
sforza@cs.pitt.edu

Alan M. Lesgold
al+@pitt.edu

Arlene W. Weiner
arlene@vms.pitt.edu

LRDC, Univ. of Pittsburgh
Pittsburgh, PA 15260

Abstract

The Argumentation Project at LRDC aims to support students in knowledge building by means of collaborative argumentation. A component of this project is a system for helping students generate arguments in a dialogical situation. Empirical research suggests that students generally have difficulty generating arguments for different positions on an issue and may resort to giving arguments that are insincere or irrelevant. Our system will assist the arguer by constraining him to respond relevantly and consistently to the actions of other arguers, suggesting appropriate ways to respond. This assistance will be provided by strategies derived from conversational maxims and "good conduct" rules for collaborative argumentation. We describe a prototype system that uses these strategies to simulate both sides of a dialogical argument.

Introduction

Argumentation is an important part of many intellectual activities, spanning situations as diverse as justifying a policy decision, proposing a new scientific theory, and determining whom to vote for in an election. Yet, secondary education has little impact on the ability to carry out a reasoned argument. [Perkins, 1985] found that students' ability to generate arguments for and against a position on an issue is generally disappointing and does not significantly improve between the first year of high-school and graduate school. Students' shortcomings in the area of reasoned discourse may be attributed, in part, to the scarce opportunities for this activity provided by secondary education. Where opportunities do exist – commonly in the form of essay writing, informal logic courses, and debate teams – they may promote a narrow view of argument. Studying argumentation only in the context of essay writing may cause students to confuse argument structure with essay structure [Lesgold, 1989]. Informal logic courses focus on identifying fallacies in individual argument steps, while ignoring the larger context in which the arguments are embedded and how arguments are generated. The debate format restricts the interaction between arguers and encourages an adversarial view of argument.

While recognizing the value of traditional uses of argument and methods of teaching argumentation, we are interested in fostering a different view and use of argument. Collaborative argumentation seeks to achieve a clear articulation of each arguer's perspective on an issue and of the beliefs and reasoning underlying it. Participants in collaborative arguments are not trying to win by any means available, as in some adversarial arguments; nor are they attempting to persuade, as do advertisements or propaganda. Collaborative argumentation occurs in some tutoring interactions [Cavalli-Sforza & Moore, 1992], in group design or decision-making, and in high-productivity team work. Collaborative argumentation helps to uncover both the common ground of different viewpoints and the areas of fundamental disagreement, suggesting starting points for building consensus and areas in need of further investigation.

Our research group is building a computer environment to support students in practicing collaborative argumentation. We envision the system as providing a "blackboard" on which two (or more) students record their argument in a graphical language that makes explicit the structure of the argument as it develops. Other environments have focused on graphical tools and languages for representing arguments [Smolensky *et al.*, 1987, Conklin & Begeman, 1988, Streitz, Hanneman, & Thuring, 1989]. While such tools will be an important part of the environment, our system will also assist students by suggesting and critiquing ways of contributing to an argument based on overall argument structure, and by providing on-line knowledge of different types of argument steps. In selected domains, it will be able to argue with a user and make content suggestions in an argument between users.

Although arguing in any domain requires knowledge of its content and its argumentation practices, some argumentation knowledge is shared across domains. In this paper, we describe the domain-independent strategies we use to guide the generation of collaborative arguments. We have only started building the practice environment, but we have used these strategies in a program that simulates an argument between two persons. The proposed strategies capture an intuitive understanding of the dynamics of a collaborative argument; therefore we think a computer coach can use them to

[1]This research has been funded by grants from the James S. McDonnell and the Andrew W. Mellon foundations. We thank Kevin Ashley, John Connelly, and Johanna Moore for helpful comments.

guide an argument between students. Pedagogically, our approach differs from traditional ones in that we are concerned with helping students contribute to ongoing arguments as well as criticize isolated inferences. Our model of argument generation differs from [Reichman-Adar, 1984] in using prescriptive and explicit strategies. Unlike [Flowers, McGuire, & Birnbaum, 1982], we focus on non-adversarial arguments; we also base our strategies on general principles of argumentation rather than on abstract configurations of support and attack links.

Overview and Justification

In designing an environment for practicing collaborative argumentation, we were influenced by experimental results on high-school and first year college students' handling of argument analysis and generation [Lesgold, 1989, Lesgold *et al.*, 1990]. In generating arguments, some subjects did not think it necessary to support a claim with reasons, and many found it difficult to generate reasonable arguments for positions other than their own (as in [Perkins, 1985]). It was also common for subjects to respond to pieces of an argument separately, not considering how they might be related to each other; [Voss *et al.*, 1983] observed similar piecemeal behavior among novice problem-solvers in ill-structured domains. Finally, subjects tended to discard an entire argument if they could fault any part of it, thus removing the need for further analysis. These results suggest that students have difficulty handling the complexity of arguments and could benefit from an environment which reifies the structure of an extended argument and suggests ways of responding to and generating arguments.

We address students' weaknesses by providing several types of assistance [Cavalli-Sforza, 1991]. The student and the system interact through a visual representation of the *argument graph* [Flowers, McGuire, & Birnbaum, 1982], which records the structure of a developing argument in terms of claims, supporting reasons, responses, and the results of other types of *argument actions* an arguer can take. The system keeps a taxonomy of possible actions and suitable response *tactics*. For example, tactics for responding to a *microargument* (support) for a claim include: 1) counterarguing the claim by arguing for an opposing claim, 2) attacking or requesting further elaboration of some aspect of the support, 3) or conceding the claim. Some tactics may be carried out in several ways. Both the action taxonomy and the argument graph are based on an augmentation of Toulmin's model [Toulmin, 1958] of argument, described briefly below.

In suggesting ways of responding to an action, the system also uses a set of *strategies* based on principles of cooperative conversation, standard practices of sound and effective argumention, and rules embodying the spirit of collaborative arguments. In collaborative arguments, the goal is not winning but critical exploration. Participants must be willing to put themselves at risk by articulating their reasoning, they must examine critically others' arguments, and they must respond to criticisms of their own arguments [Keith, Weiner, & Lesgold, 1990]. Some of our strategies rely on the changing level of support received by (pro)positions advanced by each arguer during the argument. This measure of support both shows the effect of applying the strategies and justifies their use.

Strategies for Contributing to Collaborative Arguments

Strategies for contributing to a collaborative argument are applied when one arguer completes his contribution and it becomes the other arguer's turn to respond. Deciding what action to take requires the following steps:

1. Determine *where* to address one's response. This yields candidate action *classes* with specific *targets*, e.g., "attack support 1" or "counterargue claim 1". An action class such as *attack support* subsumes other action classes, e.g., *attack premises* and *attack reasoning*, each of which might be instantiated in different ways using the arguer's beliefs and knowledge.
2. Eliminate poor candidates based on the action class and target. This may require considering the components of the target in greater detail. For example, a support is a complex structure relating multiple propositions (e.g., Microargument 1 in the example below, Figure 1(a)). One way of attacking support 1 is to attack any of the propositions it relies on, but attacking GROUNDS 3 is a poor choice because, presumably, this proposition is shared factual knowledge.
3. Try to instantiate the candidate actions using the arguer's beliefs, knowledge of the other arguer and of different types of reasoning.
4. Eliminate undesirable instantiations of actions.
5. Select the best of the remaining actions.

Steps 3-5 are applicable only to actions that construct microarguments or rebuttals (as opposed to requests for further support or concessions). Different strategies are used at different stages in the above process. We think of strategies used in step 1 as determining *where* to respond; strategies used in steps 2, 4, and 5 as determining *how* to respond.

Determining Where to Respond

Recency Preference. In a collaborative argument, each move is normally a response to the most recent question or challenge. This expectation of locally relevant responses is related to discourse focus and is also implicit in the model described by [Reichman-Adar, 1984]. Both the recency preference and the response tactics maintain continuity in the argument and encourage relevant responses.

Good Conduct Rules. Arguers may change the topic or shift attention to another part of the argument when they are at a loss for how to defend their position. In collaborative arguments these shifts are undesirable. Three rules of *good conduct* require locally relevant responses before changing topic. In the following rules, the term "challenge" is used to mean either an attack on an argument, or a request to provide further support or clarification for a part of that argument. The expression "live support for statement P" means that at least one of the arguments provided in support of P has not been dismissed. A "direct counterargument" against a statement P is an argument in support of P', where P' and P are mutually exclusive.

IF Arguer A *directly counterargues* a statement P by Arguer B AND P has no live support
THEN Arguer B must support P or challenge A's argument or abandon P

IF Arguer A *attacks an argument* in support of a statement P by Arguer B AND P has no other live support
THEN Arguer B must resupport P or challenge A's attack or abandon P

IF Arguer A *requests support or clarification* for a statement P by Arguer B AND P has no live support
THEN Arguer B must support or clarify P or abandon P

In keeping with the spirit of critical investigation, good conduct rules also constrain the challenger.

IF Arguer B responded to Arguer A's challenge AND did not abandon P
THEN Arguer A must challenge B's response OR concede P

Level of Support Heuristic. The support a microargument brings to a claim is as strong as its weakest component. Therefore, in choosing among the possible responses, an arguer should concentrate on supporting his own weak positions or weakening the other arguer's stronger positions.

Determining How to Respond

Response tactics and strategies for determining where to respond suggest a set of candidate actions specified in terms of the general type of action (e.g., support, counterargue, concede) and the target proposition or support relation. The following sets of strategies select a response through processes of elimination, instantiation, and voting.[2]

Eliminating Classes of Responses. The following strategies, implemented as a set of elimination rules, discard response classes that are inappropriate based on the target's content.

- *Avoid actions inconsistent with one's beliefs.* An arguer should avoid attacking a position he believes.
- *Avoid responses inconsistent with one's past actions in the argument.* If an arguer uses a proposition held by another arguer to support his own position, without necessarily believing it himself, he should not attack that proposition at a later time since this will invalidate his own argument.
- *Avoid wasteful actions.* An arguer should avoid actions that impact only propositions that have already been abandoned or conceded.

The first rule partially implements Grice's Maxim of Quality[Grice, 1975], the others are common sense planning advice.[3]

Eliminating Instantiations of Responses. To further weed out undesirable responses from the remaining candidates, the arguer must consider different ways in which he could instantiate them based on his knowledge and beliefs (step 3). The following rules are then used to eliminate potential responses that are undesirable based on the content of the propositions they use.

- *Avoid actions inconsistent with one's beliefs.* An arguer's action should not rely on propositions he believes to be false. This rule also implements Grice's Maxim of Quality.
- *Avoid responses inconsistent with one's past actions in the argument.* An arguer should not take actions that rely on the truth of propositions he attacked earlier in the argument, unless he later conceded them.
- *Avoid "irrational" responses.* An arguer should not directly counterargue a claim unless he can attack all existing live support for the claim since, if he cannot find fault with all justifications, he ought to accept the claim.
- *Avoid dangerous responses.* An arguer should avoid responses that use propositions on which an attack can be anticipated from knowledge of the other arguer, unless he is prepared to meet the challenge.

Selecting the Best Response. If more than one response survives the elimination phase, each of the following heuristics votes for responses meeting its criteria. The response with the most votes is selected.

- *Speak to audience.* An argument that appeals to beliefs of the other arguer is generally more persuasive.
- *Many-in-one.* An action is more effective if it helps the arguer in more than one way.
- *Prefer stronger arguments.* Depending on the domain, some microargument types may be preferred to others.

[2] Some of the elimination strategies should be relaxed if the arguer is playing devil's advocate, is constructing an indirect argument, and in some other situations.

[3] Really the first submaxim: "do not say what you believe to be false"; the second submaxim, "do not say that for which you do not have adequate evidence" is handled by requiring that claims be supported.

An Example

An example will show how the system, simulating two arguers, uses the strategies described above to guide the generation of a bilateral argument. Consider the following argument, whose structure is depicted by the argument graph of Figure 1(a).[4]

[1] ANN: Israel has an obligation to return the West Bank (WB) to the Palestinians. It was their land and it's unfair for Israel to keep it.
[2] BOB: But it's highly undesirable for Israel to return the WB. It would be a threat to its security.
[3] ANN: On the contrary, returning the WB would improve Israel's security.
[4] BOB: How? Returning the WB would make it easier for the Palestinians to launch an attack on Israel.
[5] ANN: Yes, but if the Palestinians had the WB, they wouldn't want to attack Israel.

The graph shows the arguers' microarguments and their interrelations, filling in implicit assumptions (in italics). The analysis of *support* (S) relations uses Toulmin's model [Toulmin, 1958]. Claims are supported by other situation-specific propositions, the *grounds* (g), and a more general principle, the *warrant* (w). A warrant may hold in general but admit of exceptions. The presence of exceptions is captured in the graph by a *rebuttal* (R) relation between a proposition and a support relation. Propositions may also be in an *opposition* relation, either *directly*, if the propositions are contradictory, or *indirectly*, if they denote opposition but on the basis of different modalities (criteria).

Bob can respond to Ann's first microargument [1] in a number of ways, for example, by arguing that "Israel does not have an obligation to return the WB because .." (a direct counterargument to claim 1), or by attacking some of Ann's grounds. Note that he cannot attack the proposition that "Israel holds the WB" since Bob presumably believes this too. In [2] he takes another option: he *indirectly counterargues* Ann's claim [1], which is based on a criterion of fairness, by appealing to a criterion of desirability. Figure 1(b) shows that, after Bob's response, Bob and Ann's claims each have some level of support (LOS), less than if unopposed but more than if directly opposed.[5]

The kind of shift embodied in indirect counterarguments would be inadmissible in a more adversarial situation, but is an integral part of exploring multiple perspectives on an issue. It constrains permissible responses for Ann. For example, if Ann resupports her claim, the two arguers will be arguing past each other. Rather, she will need to respond directly to his argument, argue that the desirability criterion is not relevant to the discussion, or consider the relative weight of the two criteria.[6] In [3], Ann directly counterargues Bob's claim with a parallel microargument: appealing to the same motivational principle (the warrant), she argues to the opposite conclusion from contradictory grounds. This is a good move for Ann because it completely "neutralizes" the support for Bob's claim as shown in Figure 1(c). Ann's move is preferred over other actions by the *many-in-one critic*, because her grounds also attack Bob's grounds, and by the *speak to audience* critic, because it shares a proposition with Bob's support. Ann is able to counter Bob's argument directly because she can attack his grounds (and will do so in [5]); i.e., the *avoid irrational responses* critic does not apply.

The resulting configuration of parallel argument and counterargument arises frequently in controversial issues.[7] It constrains Bob's response options. He can resupport his claim [2] differently. He can shift criteria again, returning his attention to the initial argument in [1] or introducing yet another perspective. He can also try to find some fault with Ann's counterargument, although clearly he should not attack the warrant, on which his own argument depends. None of these actions, however, would address the underlying source of the disagreement, i.e., their contradictory grounds. Bob chooses to focus on this disagreement by supporting his own grounds while simultaneously attacking Ann's. As shown in Figure 1(d), Bob's action strengthens his line of argument at Ann's expense. His grounds for claim 2 overpower Ann's contradictory grounds, allowing claim 2 to "win out" over Ann's opposing claim 3.

Bob's action in [4] also activates the rule of Good Conduct for direct counterargument. Ann must respond by challenging Bob's argument, by supporting her grounds for claim 3, or by abandoning them. With her response in [5], she both supports her position and rebuts Bob's argument, and she does so using a shared belief. This has the effect of reversing the LOS situation: now Ann's claims 3 and 5 have support and Bob's claims 2 and 4 do not. Bob's eventual response in [6] will be subject to Good Conduct rules in two ways: 1) since Ann has responded to his challenge by supporting her grounds for claim 3, he must now address that re-

[4] The simulator's output and input, including the arguers' beliefs, are represented in a frame language.

[5] In the interface, the evolving argument graph will be depicted in a manner similar to Figure 1(a), and display the LOS attached to each proposition directly. Figures 1(b) through 1(e) are used here to illustrate different stages of the argument. Currently, we compute the LOS using a weighted sum scheme that takes into account rebuttals and oppositions; we are also evaluating alternative schemes (e.g., ECHO [Thagard, 1989]).

[6] We haven't yet addressed these types of response, really meta-arguments, but we will do so in the near future. We could use an additional Good Conduct Rule to post an obligation to consider the relative weight of the two criteria later in the argument.

[7] The similar "standoff" pattern analyzed by [Flowers, McGuire, & Birnbaum, 1982] differs from the present situation in two respects: 1) in a standoff, the arguer's and the opponent's grounds are not contradictory, and 2) in the present situation, the intent is not to dismiss the issue of security as a line of argument.

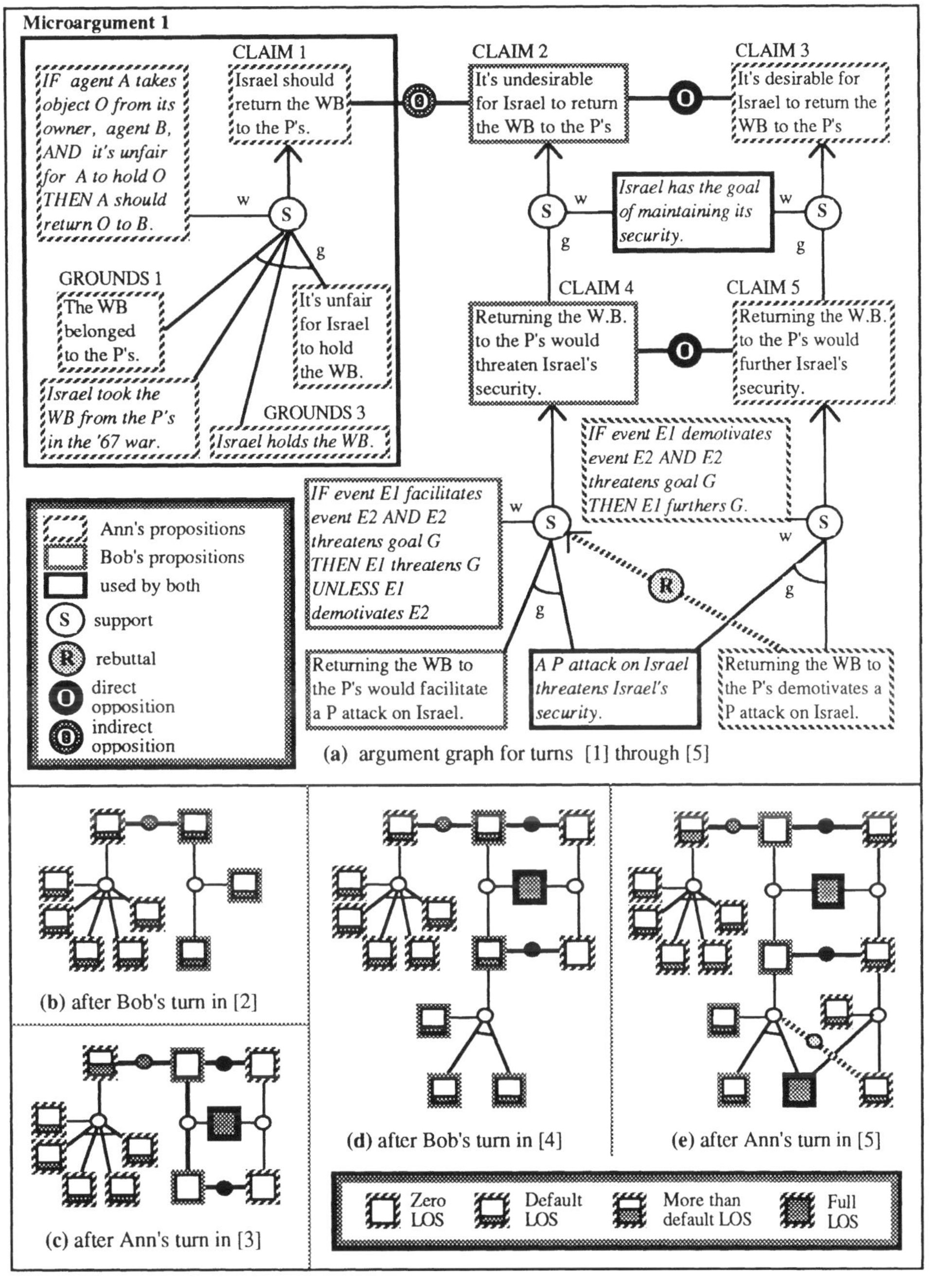

Figure 1. A sample argument with changes in Level of Support as the argument develops.

sponse; 2) since Ann has also attacked his only argument for claim 4 by providing a rebutting exception, he must now respond to that attack. Jointly, these rules suggest that he challenge Ann's support for claim 5, the intuitively desirable action. If he cannot address Ann's response, he must abandon claim 4 and examine other ways of continuing the argument. The structure of the argument to date and the LOS for propositions in the argument graph can be used as a basis for suggesting alternative continuations.

Summary and Future Work

We have described a design for a practice environment to help students argue collaboratively and explore arguments in support of different viewpoints on an issue. A unique feature of our design is a set of tactics and strategies for guiding users' contributions to the argument. Combined with the reification of the developing argument structure, a measure of support for statements advanced by the arguers, and on-line knowledge of different types of argument steps, these strategies will help students handle the complexity of extended arguments.

Previous work on argumentation environments has focused on tools for recording arguments (e.g., [Conklin & Begeman, 1988]) or composing argumentative texts (e.g., [Streitz, Hanneman, & Thuring, 1989]), with no assistance provided for argument construction or analysis. Models of argument generation have emphasized discourse phenomena [Reichman-Adar, 1984]), or they have focused on different kinds of arguments (e.g., legal arguments [Ashley, 1991], informal adversarial arguments [Flowers, McGuire, & Birnbaum, 1982], arguments in editorials [Alvarado, 1990]). Because our objective is to coach collaborative argumentation between students, the knowledge we use to guide argument generation is explicit, prescriptive, and embodies the properties of collaborative arguments. We want our tactics and strategies to be teachable, so they embody a few intuitive principles of argumentation and cooperative discourse.

We have used our strategies in a system that simulates the interaction between two arguers. We intend to test them with students by having an experimenter guide the interaction between two arguers as the system would. We are also developing an interactive environment through which students can argue with each other, allowing us to test these and other forms of argumentation assistance. Finally, recognizing that argumentation relies on domain specific as well as domain-independent knowledge, we are working on incorporating domain-dependent argumentation strategies as part of the assistance available to the student.

References

Alvarado, S.J. 1990. *Understanding Editorial Text: A Computer Model of Argument Comprehension.* Boston, MA: Kluwer.

Ashley, K. 1991. Reasoning with cases and hypotheticals in HYPO. *International Journal of Man-Machine Studies* 34:753-796.

Cavalli-Sforza, V. 1991. An Environment for Tutoring Argumentation Skills. In Working Notes of the AAAI '91 Spring Symposium Series: Argumentation and Belief, 71-84. Stanford, CA.

Cavalli-Sforza, V.; and Moore, J. 1992. Collaborating on Arguments and Explanations. In Working Notes of the AAAI '92 Spring Symposium Series: Producing Cooperative Explanations, 61-68. Stanford, CA.

Conklin, J.; and Begeman, M.L. 1988. gIBIS: A hypertext tool for argumentation. *ACM Transactions on Office Information Systems* 6:303-331.

Flowers, M.; McGuire, R.; and Birnbaum, L. 1982. Adversary arguments and the logic of personal attacks. In Lehnert, W.G., and Ringle, M.H. eds. *Strategies for Natural Language Processing*, 275-297. Hillsdale, NJ: LEA.

Grice, H.P. 1975. Logic and Conversation. In Cole, P., and Morgan, J.L. eds. *Syntax and Semantics* 3. New York, NY: Academic Press.

Keith, W.K.; Weiner, A.W.; and Lesgold, A.M. 1990. Toward computer-supported instruction of argumentation. In Proceedings of the Second International Conference on Argumentation, 1144-1153. Amsterdam, The Netherlands: SICSAT.

Lesgold, A.M. 1989. Skills of Argumentation in School Subjects: Year 1, A Report to the James S. McDonnell Foundation, LRDC, Univ. of Pittsburgh.

Lesgold, A. *et al.* 1990. Skills of Argumentation in School Subjects: Year 2, A Report to the James S. McDonnell Foundation, LRDC, Univ. of Pittsburgh.

Perkins, D. N. 1985. Postprimary Education Has Little Impact on Informal Reasoning. *Journal of Educational Psychology* 77(5):562-571.

Reichman-Adar, R. 1984. Extended Person-Machine Interface. *Artificial Intelligence* 22:157-218.

Smolensky, P.; Fox, B.; King, R.; and Lewis, C. 1987. Computer-Aided Reasoned Discourse, or, How to Argue with a Computer, Technical Report, CU-CS-358-87, Dept. of Computer Science, Univ. of Colorado.

Streitz, N. A.; Hannemann, J.; and Thuring M. 1989. From Ideas and Arguments to Hyperdocuments: Travelling through Activity Spaces. In Proceedings of HyperText '89, 343-364.

Toulmin, S. 1958. *The Uses of Argument.* New York, NY: Cambridge University Press.

Thagard, P. 1989. Explanatory Coherence. *Behavioral and Brain Sciences* 12:435-502.

Voss, J. F. *et al.* 1983. Problem Solving Skill in the Social Sciences. In Bower, G.H. ed. *The psychology of learning and motivation: Advances in research theory* 17. New York, NY: Academic Press.

The Zoo Keeper's Paradox
A Decision Theoretic Analysis of Inconsistent Commonsense Beliefs

Kwok Hung Chan
Dept. of Mathematical Sciences
Memphis State University
Memphis, TN 38152
U.S.A.
chand@hermes.msci.memst.edu

Abstract

Default reasoning is a mode of commonsense reasoning which lets us jump to plausible conclusions when there is no contrary information. A crucial operation of default reasoning systems is the checking and maintaining of consistency. However, it has been argued that default reasoning is inconsistent: Any rational agent will believe that it has some false beliefs. By doing so, the agent guarantees itself an inconsistent belief set (Israel, 1980). Perlis (1986) develops Israel's argument into an argument for the inconsistency of recollective Socratic default reasoning systems. The Zoo Keeper's Paradox has been offered as a concrete example to demonstrate the inconsistency of commonsense beliefs.

In this paper, we show that Israel and Perlis' arguments are not well founded. A rational agent only needs to believe that some of its beliefs are possibly or probably false. This requirement does not imply that the beliefs of rational agents are necessarily inconsistent. Decision theory is used to show that concrete examples of seemingly inconsistent beliefs, such as the Zoo Keeper's Paradox, can be rational as well as consistent. These examples show that analyses of commonsense beliefs can be very misleading when utility is ignored. We also examine the justifications of the exploratory and incredulous approaches in default reasoning, decision theoretic considerations favor the exploratory approach.

Default Reasoning

The goal of artificial intelligence is to build electronic agents which can use knowledge to solve problems. A large part of what we know is commonsense knowledge consisting of general laws/rules which are almost always true, with a few exceptions (Reiter, 1980).

Example 1:

(R1) We can start a car by turning the key while stepping slowly on the gas pedal.

R1 is a general rule which is almost always true. There are exceptions to the rule, such as, "the gas tank is empty" and "the battery is low." However, usually we do not check to make sure that everything is normal. We just assume *by default* that the car is in working condition, *unless* there is information to the contrary. Say, if we notice that the ignition switch is lying on the floor with a loose wire, then we conclude that the car is out of order and it will not start. Such reasoning is not deductive, and has been called *default reasoning* in the literature.

Example 1 illustrates the *non-monotonicity* of default reasoning: A sentence A which is derivable from a theory T may not be derivable from a superset of T. Because of this property, formalizations of default reasoning have been called *non-monotonic logics*. In this paper we will use "non-monotonic logics" as a general term covering all formalizations of default reasoning with the property of non-monotonicity.[1]

The Consistency of Default Reasoning

A number of authors have worried about the integrity and consistency of commonsense/default reasoning. Israel (1980) claims that non-monotonic logics are not well motivated, because they rest on the confusion of proof-theoretic with epistemological issues. Israel also suggests that commonsense beliefs are very often inconsistent. Since most non-monotonic logics perform

[1] Ginsberg (1987b) contains original papers of major works before 1987. Besnard (1989) is a more recent introduction to non-monotonic logics.

a consistency test before making a default assumption, they would be paralyzed by inconsistent beliefs.

Perlis (1986) develops Israel's argument for the inconsistency of commonsense beliefs into an argument for the inconsistency of non-monotonic logics under some natural conditions. According to Perlis, ideal thinkers capable of appropriate commonsense reasoning must be able to reflect on their past errors. They must be aware of the fallibility of their use of defaults (Socratic) and able to recall what default assumptions they have made (recollective). However, recollective Socratic reasoning is inconsistent. Perlis also presents the Zoo Keeper's Paradox as a concrete example that illustrates the inconsistency of commonsense beliefs. The performance of the major formalizations, namely, Circumscription (McCarthy, 1980), Non-monotonic Logic (McDermott & Doyle, 1980) and Default Logic (Reiter, 1980), is compromised: they do not produce the intuitively correct commonsense default conclusions in cases like the Zoo Keeper's Paradox.

In this section, we will consider briefly the general arguments for the inconsistency of commonsense beliefs and default reasoning. A detailed analysis is presented in Chan (1992).

The Goal of Non-monotonic Logics

Israel's (1980) major complaint about non-monotonic logics is that the motivation behind non-monotonic logics is based on a confusion of proof-theoretic with epistemological issues. This has been misinterpreted by some authors as an issue of terminology: Logic is, by its very definition, monotonic, and the notion of "non-monotonic logic" is a contradiction in terms (Ginsberg, 1987a). Such misinterpretation misses the point of Israel's argument as well as the chance to show that Israel is mistaken.

Default reasoning makes a default assumption *A* only if there is there is no information to the effect that *A* is false. This requirement is implemented in non-monotonic logics as a consistency check. Before *A* is concluded by default, the system checks to see if *A* is consistent with the set of current beliefs. *A* is also required to be consistent with the justifications of default assumptions made previously. Hence, a default assumption will remain consistent with subsequent default beliefs. This consistency requirement is interpreted by Israel as follows:

> [To make an assumption that *A* is to believe that *A* is] both compatible with everything that a given agent believes at a given time and *remains so* when the agent's belief set undergoes certain kinds of changes *under the pressure of* both *new information* and further thought, and where those changes are the result of rational epistemic policies (Israel, 1980).

Based on this understanding Israel takes non-monotonic logics to be formal systems for *general belief fixation.* He then argues that there is no logic of belief fixation and scientific procedures are the only methods for belief fixation. Because there are no logics of belief fixation, non-monotonic logics can never achieve their goal.

Even if Israel is correct in claiming that there is no logic of belief fixation, his criticism of non-monotonic logics is not justified. This is because he has misinterpreted the consistency requirement and the aim of non-monotonic logics. Non-monotonic logics do not aim to be general logics for belief fixation. A default assumption *A* is *not* required to remain consistent when we add *new information.* Actually, the non-monotonic nature of default reasoning requires that *A* should be deleted when it is not consistent with new information! The correct intuitive understanding of the consistency requirement is: A new default assumption *A* should be consistent with current beliefs and should not falsify the justifications of default assumptions *previously* made. Hence, a default assumption is only guaranteed to remain consistent with *subsequent default beliefs.* Non-monotonic logics are not logics for making general hypotheses. That is the job of scientific procedures. Non-monotonic logics have a rather moderate aim. They are logics for the proper extensions of beliefs by, and *only by,* default assumptions supported by default rules such as "Typically P's are Q's."

The Consistency of Commonsense Beliefs

Israel also argues that the consistency requirement cannot be met in practice, because commonsense beliefs are mostly inconsistent. Any rational agent will believe that it has some false beliefs. By doing so, the agent guarantees itself an inconsistent belief set; there is no possible interpretation under which all of its beliefs are true (Israel, 1980).

If being rational requires our having an inconsistent set of beliefs, this notion of rationality is too strong, and should be replaced by a weaker notion. To be rational an agent does *not* need to believe that it *actually* has some false beliefs. It only needs to believe that some of its beliefs are *possibly or probably* false. Such belief sets may be consistent. Hence commonsense beliefs of a rational agent are not necessarily inconsistent.

Recollective Socratic Agents

Perlis (1986) develops Israel's argument for the inconsistency of commonsense beliefs into an argument for the inconsistency of default reasoning under some natural conditions.

According to Perlis, default reasoning consists of a sequence of steps involving, in its most general form, oracles, jumps, and fixes. Since consistency check is only semidecidable, we need to appeal to an *oracle* to tell us that a given default assumption is consistent with the current beliefs. Because default reasoning *jumps* to conclusions, it is error-prone and *fixes* are necessary to preserve (or re-establish) consistency. For rational agents to be capable of appropriate commonsense reasoning, they must be able to reflect on their past errors, and indeed, on their potential future errors. They must be aware of the fallibility of their use of defaults (Socratic) and able to recall what default assumptions they have made (recollective). Perlis (1986) shows that recollective Socratic agents are inconsistent.

As in the case of Israel's argument, Perlis' definition of Socratic thinkers is too strong. A rational agent does *not* need to believe each of its default assumptions and *simultaneously* believes that some of its default beliefs are in fact false. A rational agent only needs the weaker belief that some of its default beliefs are *possibly* false. Such recollective *weakly Socratic* agents are not necessarily inconsistent.

The Zoo Keeper's Paradox

In additional to a general argument for the inconsistency of rational agents, Perlis also offers the Zoo Keeper's Paradox as a concrete example of inconsistent commonsense beliefs.

Example 2: (the Zoo Keeper's Paradox)
Bob works as a zoo keeper and keeps a written record of the animals there. Ten American bare eagles have been recorded by Bob as in good health (and so able to fly). One day Bob receives a message from a laboratory saying that blood samples from the eagles show that some eagles in the zoo are infected by virus (and as a result cannot fly). However, the laboratory has mixed up the blood samples, so we cannot tell which eagle is infected. Bob still believes that each individual eagle at the zoo can fly, that he is highly unwilling to leave any of their cage doors open, and that he is also unwilling to call any one of them to the attention of the zoo veterinarian. Yet, he is also very concerned at the verterinarian's failure to arrive for work at the usual hour, because he also believes that some (unspecified) eagles in the zoo are sick (and cannot fly).[2]

Are Bob's beliefs consistent? What conclusions should a default reasoning system make? We may formalize the hard facts in this example as follows:

[2] This is a modified version of the Zoo Keeper's Paradox in Perlis, 1986. A similar paradox is the Lottery Paradox discussed in McDermott (1982), Shoham (1987) & Poole (1991).

$$\text{eagle}(e_1) \wedge \cdots \wedge \text{eagle}(e_{10}) \quad \text{(C1)}$$

$$\text{sick}(e_1) \vee \cdots \vee \text{sick}(e_{10}) \quad \text{(C2)}$$

$$\text{eagle}(x) \wedge \text{sick}(x) \rightarrow \neg\, \text{fly}(x) \quad \text{(C3)}$$

We have the following default rule:

Eagles typically fly. (D)

Let us consider what default assumptions we should rationally make. For each eagle we would like to conclude that it can fly by default, because we do not have specific information about any individual eagle that it cannot fly. Applying this reasoning we conclude that the first nine eagles (in some arbitrary order) fly.

$$\text{fly}(e_1) \quad \text{(A1)}$$

$$\vdots$$

$$\text{fly}(e_9) \quad \text{(A9)}$$

There are three possibilities regarding the last eagle e_{10}:

1. Since we have no evidence to single out e_{10} from the rest, we may apply the same reasoning and conclude by default that e_{10} can fly. However, the addition of this last default conclusion results in a set of inconsistent beliefs. According to Perlis this is what Bob believes.
2. We may deduce from C1–C3 and A1–A9 that e_{10} does not fly. Since there are ten ways to pick this last eagle, there are ten possible extensions, each as good as the other. An *exploratory* system (Reiter, 1980) would pick an arbitrary extension.
3. An *incredulous* system (McCarthy, 1980; McDermott & Doyle, 1980) would consider as default conclusions only those shared by all extensions:
   Nine of the eagles fly. (G1)
   One of the eagles does not fly. (G2)

In the rest of this section we will consider the consistency of Bob's beliefs. The exploratory and incredulous approaches will be examined in the next section.

Are Bob's beliefs inconsistent? First of all, how do we know what Bob's beliefs are? Perlis proposes a behavioral criterion of *use-belief*.

Definition 1: (use-belief)
An agent believes a proposition p if it trusts and uses p in planning and acting, "as if it were true." The agent should be willing to recognize p as theorems and ignore the possibility that p may be false. If the agent also does something that is appropriate only if p is false, then the agent only believes that it is highly probable that the proposition is true (Perlis, 1986).

This definition tries to identify an agent's beliefs by its actions. However, actions are not determined only by beliefs. The rationality of an action also depends on its utility. In what follows, we will apply decision theory (Savage, 1972) to find out if Bob's belief/behavior is rational.

Suppose Bob believes with probability p that an eagle e is sick. Being a zoo keeper, Bob is responsible for keeping e healthy as well as keeping e in the zoo. Let us represent the utility of different possibilities as follows:

Event	escape	stay	dead	sick	healthy
Utility	0	1	0	v	u

Table 1. Utility of events for Bob

We will consider the decision trees for closing/opening the cage door in two scenarios.

Scenario 1: Closing the cage door does not make the condition of a sick eagle worse.

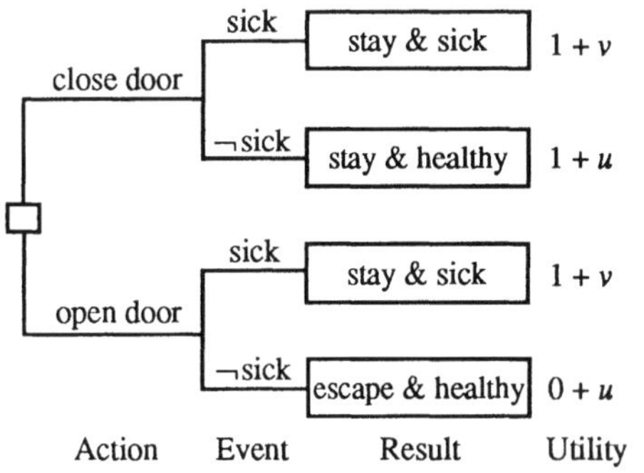

Figure 1. Decision tree for Scenario 1

The expected utility of closing the cage door is greater than the expected utility of opening the cage door by $(1-p)$ $(= p(1+v) + (1-p)(1+u) - p(1+v) - (1-p)u)$. As long as Bob is not absolutely certain that e is sick ($p < 1$), he better keeps the door closed. If $p = 1$, then it makes no difference if the door is closed or open. Hence, in Scenario 1 Bob should close the cage door no matter whether he believes that e is sick or not. Bob is probably in this situation in Example 2. This shows the possibility of interpreting Bob's behavior as rational without attributing an inconsistent set of beliefs to him. What he does is rational and is consistent with his belief that one of the eagle is sick.

Scenario 2: A sick eagle will die if the cage door is closed.

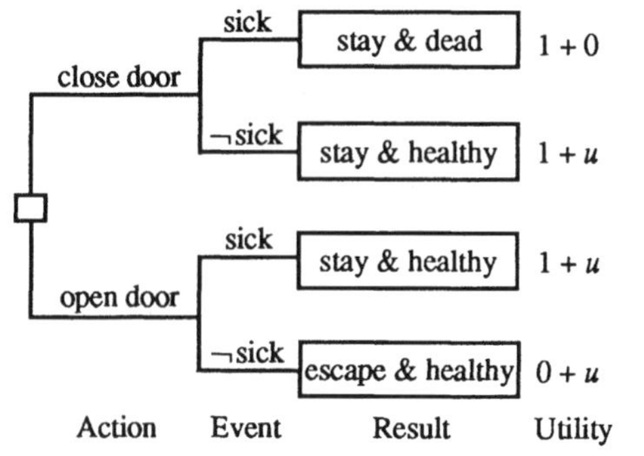

Figure 2. Decision tree for Scenario 2

The expected utility of closing the cage door is greater than the expected utility of opening the cage door by $(1-p) - pu$ $(= p + (1-p)(1+u) - p(1+u) - (1-p)u)$. The values of $EU(\text{close}) - EU(\text{open})$ for some representative values of p and u are shown in Table 2. In this scenario Bob has to make a choice between opening or closing the cage door. What he should do depends on p as well as u. If keeping the eagle in the zoo is as important as keeping the eagle healthy $(u \approx 1)$, then Bob should keep the cage door closed if he believes that probably the eagle is healthy, but he should open the cage door if he believes that probably the eagle is sick. However, if the eagle is an endangered species and it is very important to keep it healthy $(u \gg 1)$, he should open the cage door even if he believes that probably the eagles are not sick $(p \leq 0.5)$. Because the penalty for mistake is so high, it is rational for Bob to consider an unlikely proposition (sick(e)) to be true by default.

Scenario 2 is a special case in which two goals compete for an action. Keeping the cage door closed achieves the goal of keeping the eagle in the zoo, but violates the goal of keeping the eagle healthy. According to the policy of minimizing expected loss, it is rational to perform the action appropriate to the unlikely event if the penalty of overlooking the event is too

EU(close) – *EU*(open)		*u*							
		.01	.11	.5	1	5	9	10	20
p	.1	.899	.889	.85	.8	.4	0	- 0.1	- 1.1
	.5	.495	.445	.25	0	- 2	- 4	- 4.5	- 9.5
	.9	.091	0	- 0.35	- 0.8	- 4.4	- 8	- 8.9	- 17.9

Table 2. The value of *EU*(close) – *EU*(open) for different *p* and *u*.

great. In other scenarios, the required remedy/preventive action for an unlikely event may not compete with the normal action appropriate to the more likely event. For example, the probability p of having a car collision is low. However, the penalty of not wearing a seat belt is very high if a car collision does occur. Fortunately, the preventive measure of wearing a seat belt can be performed simultaneously with other actions appropriate for the much more likely event of no car collision. In such cases we usually entertain two belief sets, belief set S1 is consistent with the occurrence of a normal event E, whereas belief set S2 is consistent with the unlikely possibility of $\neg E$. Actions appropriate to S1 are performed if they do not compete with remedy or preventive actions appropriate to $\neg E$.

The Zoo Keeper's Paradox illustrates that it is highly misleading to consider rational behavior without taking utility into account. Given a belief set, rational behavior is determined by utility/penalty. It is rational to make a default assumption only if the penalty for making a mistake is (very) low. If the penalty is great enough, even an unlikely proposition should be considered to be true by default: preventive measures or remedies are implemented as if the unlikely event will occur or has occurred.

Exploratory vs. Incredulous Approaches

Can we apply decision theory in the context of non-monotonic logics? Although probability and utility are *not* considered *within* non-monotonic logics, the *practice* of default reasoning is justified by decision theoretic considerations. Under the normal operating conditions of default reasoning, each default rule has a high probability of being true, it is desirable to draw the default conclusions and the penalty for drawing a false conclusion is low. In this section we will consider the decision theoretic justifications of the exploratory and incredulous approaches in default reasoning when these normal operating conditions are satisfied.

Let us take a detour and consider Bob's beliefs before he received the message from the laboratory. At that time Bob was presumably justified by the default rule D to conclude that all ten eagles can fly. Suppose that the actual world is a *maximally typical world* in which all individuals not known to be atypical are indeed typical, then all ten eagles can fly. Of course, the actual world is not maximally typical. However, because exceptions are rare, *most of our default conclusions are true*. The successful rate depends on how typical the world is. Although we may make mistakes from time to time, that is acceptable, because the penalty for such mistakes are low and in the long run true default assumptions out number mistaken default assumptions.

Now, consider again the original version of the Zoo Keeper's Paradox in which Bob knows that at least one of the eagles is sick. Since consistency check is an essential step in the normal operation of non-monotonic logics, the inconsistent belief set acknowledging ten healthy eagles cannot be tolerated. There are ten different consistent (maximal) extensions of the core beliefs. An incredulous non-monotonic logic does not commit itself to any one of the competing extensions. Only default conclusions shared by all consistent maximal extensions are made. Such shared conclusions are true in all maximally typical world with one sick eagle, and we can appeal to the same statistical justification for this conservative approach.

Should we commit ourselves to any one of the ten extensions? If we have some empirical evidence that makes one of eagle, say e_1, the prime suspect, then we should prefer an extension in which e_2 to e_{10} can fly. Otherwise, all ten extensions are equally justified and we have no reason to prefer one rather than another.

Suppose Bob knows that exactly one of the eagles is sick and he uses an exploratory non-monotonic logic to pick one of the ten extensions. Do we have any statistical justification for such a practice? There are ten extensions, so the chance of getting *all ten* default conclusions right is only 10%. This is a low percentage. However, let us compute the expected number of correct conclusions. Let p_i be the probability that extension i is correct and N_i be the number of correct conclusions if extension i is correct. The expected number of correct conclusions is

$$\sum_{i=1}^{10} N_i \times p_i = 10 \times 0.1 + 9 \times 8 \times 0.1 = 8.2.$$

The expected number of correct conclusions is summarized in Table 3.

Suppose Bob knows that at least one eagle is sick. Using an exploratory default logic, he would pick an extension with only one eagle being sick. If the world is a maximally typical world, then only one eagle is sick and the expected number of correct default conclusions is the same as the previous case (8.2). However, if each eagle has a 50% chance of being sick, then the expected number of correct default conclusions is reduced to only 4.6. In general if the default rule in question is very strong with a very high percentage of typical members,

No. of eagles known to be sick	1	2	3	4	5	6	7	8	9
Expected no. of correct conclusions	8.2	6.8	5.8	5.2	5	5.2	5.8	6.8	8.2

Table 3. Expected number of correct default conclusions

then the expected number of correct default conclusions would be very close to 8.2. Although there is no empirical reason to prefer one extension to another, *any one* of the extensions would serve just as well. Using an exploratory default logic, we can always backtrack and try another extension when we find out later that we have picked the wrong one.

From this example, we can see that an exploratory default logic may be justified even when we do not have any empirical evidence to prefer one extension over another. Moreover, under the normal operating conditions of non-monotonic logics, each default rule has a high probability of being true and incurs a very low penalty for false conclusions. Using an exploratory system we can make more default assumptions without incurring heavy penalty. In the long run the advantages of making more correct assumptions will outweigh the small penalty incurred by occasional false conclusions. On the other hand, we will miss the chance to make many useful default assumptions if we follow the incredulous approach.

In special cases where the normal operating conditions of default reasoning are not satisfied, neither the exploratory nor the incredulous approach would work as such. We need a more powerful mode of reasoning which can entertain competing belief sets and act on the basis of both sets. It is interesting to see how we can extend current non-monotonic logics or develop new systems to handle defaults with heavy penalty.

Conclusions

In this paper, we have proposed that a rational agent only needs to believe that some of its beliefs are possibly or probably false. This requirement does not imply that the beliefs of rational agents are necessarily inconsistent. Decision theory is used here to show that concrete examples of seemingly inconsistent beliefs can be rational as well as consistent. Such examples show that analysis of commonsense beliefs can be very misleading when utility is ignored. Justifications of the exploratory and incredulous approaches in default reasoning are examined and decision theoretical considerations favor the exploratory approach.

The use of decision theoretic analysis in default reasoning solves some old issues but also presents some new challenges. In particular, there are two types of default assumptions: (i) Some propositions are assumed to be true by default because they are probable. (ii) Other propositions are assumed to be true by default because it is too risky to assume that they are false. Default assumptions of the second type are numerous in practical applications and we need to extend existing systems or develop new systems to incorporate this overlooked type of default reasoning.

References

Besnard, P. 1989. *An Introduction to Default Logic*. Berlin: Springer-Verlag.

Chan, K. H. 1992. The Consistency of Commonsense and Default Reasoning. Unpublished.

Ginsberg, M. L. 1987a. Introduction to Readings in Nonmonotonic Reasoning. In Ginsberg, M. L. Ed. *Readings in Nonmonotonic Reasoning*, 1–23. Los Altos, CA: Morgan Kaufmann.

Ginsberg, M. L. Ed.. 1987b. *Readings in Nonmonotonic Reasoning*. Los Altos, CA: Morgan Kaufmann.

Israel, D. J. 1980. What's Wrong with Non-monotonic Logic? In *Proceedings of the First Annual National Conference on Artificial Intelligence*, 99–101. Los Altos, CA: Morgan Kaufmann.

McCarthy, J. 1980. Circumscription—A Form of Non-Monotonic Reasoning. *Artificial Intelligence*, 13:27–39.

McDermott, D. V. 1982. Nonmonotonic Logic II: Nonmonotonic Modal Theories. *JACM*, 29(1):33–57.

McDermott, D. & Doyle, J. 1980. Non-Monotonic Logic I. *Artificial Intelligence*, 13:41–72.

Perlis, D. 1986. On the Consistency of Commonsense Reasoning. *Computational Intelligence*, 2:180–190.

Poole, D. 1991. The Effect of Knowledge on Belief: Conditioning, Specificity and the Lottery Paradox in Default Reasoning. *Artificial Intelligence*, 49:281–307.

Reiter, R. 1980. A Logic for Default Reasoning. *Artificial Intelligence* , 13:81–132.

Savage, L. J. 1972. *The Foundations of Statistics, 2nd Ed.*. New York: Dover.

Shoham, Y. 1987. A Semantical Approach to Nonmontonic Logics. In *Proceedings of the Tenth International Joint Conference on Artificial Intelligence*, 388–392. Los Altos, CA: Morgan Kaufmann.

A Connectionist Architecture for Sequential Decision Learning

Yves Chauvin*
Psychology Department
Stanford University
Stanford, CA 94305
chauvin@psych.stanford.edu

Abstract

A connectionist architecture and learning algorithm for sequential decision learning are presented. The architecture provides representations for probabilities and utilities. The learning algorithm provides a mechanism to learn from long-term rewards/utilities while observing information available locally in time. The mechanism is based on gradient ascent on the current estimate of the long-term reward in the weight space defined by a "policy" network. The learning principle can be seen as a generalization of previous methods proposed to implement "policy iteration" mechanisms with connectionist networks. The algorithm is simulated for an "agent" moving in an environment described as a simple one-dimensional random walk. Results show the agent discovers optimal moving strategies in simple cases and learns how to avoid short-term suboptimal rewards in order to maximize long-term rewards in more complex cases.

Introduction

Learning from Long-Term Rewards

If we imagine an agent (machine, animal or human) making decisions and acting in an environment, how can long-term payoffs received from the environment influence present decisions? Similar questions have been raised in different disciplines, including human and animal psychology, machine learning, engineering, robotics and economics (e.g., Bellman, 1957; Sutton & Barto, 1987; Samuel, 1959; Slovic, Lichtenstein & Fishoff, 1988; Watking, 1989). A number of mathematical and numerical tools in the decision sciences have been proposed to answer these questions. In particular, the theory of *dynamic programming* (Bellman, 1957) was specifically developed as an optimization method to solve sequential decision problems.

Recently, Barto, Sutton, and Watkins (In Press) integrated methods from dynamic programming and parameter estimation methods to construct a framework for sequential decision learning. In the same spirit, this paper integrates the classical decision theory framework with learning principles from connectionist theories. Whereas Barto et al. mainly suggested connectionist networks could be used as parametric models to compute an evaluation function, this paper mainly proposes the use of connectionist networks to compute action policies. First, "connectionist representations" for probabilities and utilities in static environments are presented. The framework is then extended to dynamic environments and compared to previous formalisms (e.g., Sutton 1990). Simulations of a simple random walk then show how an agent can maximize the long-term expected utility computed over the decision period.

*also with Net-ID, Inc., Menlo Park, CA.

A Connectionist Framework for Decision Making

The connectionist decision making framework suggested in Chauvin (1991) introduces "connectionist representations" of probabilities and utilities. The final layers of a connectionist network are composed of sets of *e-units*, *p-units* and *u-units*. The *e-units* are exponential units with activations $e_i = e^{\beta s_i}$ where s_i is the input to the *e-units* and β a sensitivity parameter. The *p-units* compute probabilities using the Boltzman distribution:

$$\pi_i = \frac{e_i}{\sum_j^n e_j} = \frac{e^{\beta s_i}}{\sum_j e^{\beta s_j}} \tag{1}$$

Utilities (considered as monotonic functions of rewards) are represented as "utility weights" u_{ij} from the *p-units* to the linear *u-units*. This set of weights can be given *a priori*, observed from the environment, or estimated during learning. The set of *u-units* then computes an expected utility:

$$U_i = \sum_j^n u_{ij}\pi_j \tag{2}$$

where i is an index taken over a given set of categories.

In this framework, learning consists in maximizing the expected utility computed by the *u-units*. Backpropagation principles can be used to compute the gradient of expected utility with respect to the inputs to the *e-units*:

$$\frac{\partial U_t}{\partial s_j} = \beta \pi_j (u_{tj} - U_t) \quad (3)$$

where t represents a target category given an input pattern. Note there is no "error" in this formulation: The algorithm directly maximizes an expected utility. Also note that the *p-units* compute "decision beliefs" π_j representing a decision behavior rather than estimated probabilities of future environmental events. These probabilities represent how the network should classify input patterns to maximize expected utilities. With simple environments, it is possible to show the algorithm converges to optimal behavior. In particular, Bayesian optimality (pure decisions) is obtained when the environment is stochastic. It is also possible to show that, using the Widrow-Hoff procedure, the network parameters (decision and utility weights) can be adapted "on-line" after each observation of the environmental response (Chauvin, 1991).

Sequential Decision Learning

Markovian Decision Hypotheses

Markovian decision problems are defined in terms of a finite set of states X and state transition probabilities p_{xy}. At each time step k, an agent makes a decision a among a set of permissible actions A_x function of the current state x. Depending on the chosen action, the environment will switch from state x to $y \in Y(x,a)$ with probability $p_{xy}(a)$. The set of permissible actions for each state x can be characterized by a probability distribution of actions π_{xa} called a *policy P*. For each state transition, the agent receives a reward/utility (or incurs a cost) u_{xy}.

The goal of the agent is to maximize the long-term expected utility V_i^P from state $x(0) = i$:

$$V_i^P = E^P[\sum_{t=0}^{\infty} \gamma^t u_t | x(0) = i] \quad (4)$$

where $u_t = u_{x_t y_t}$ is the utility received at time t by moving from state x_t to y_t and where γ is a discount factor. The term V_i^P is called the evaluation function of state i given the policy P. For the rest of this paper, we assume that the environment has absorbing states with 0 utilities and set γ to 1.

In the most general case, the learning environment is supposed to be stochastic: $0 < p_{xy}(a) < 1$. For interesting optimized functions, the agent's optimal policy can be shown to be deterministic: $\pi_{xa} \in \{0,1\}$. Dynamic programming approaches to sequential decision problems generally consider *a priori* that the agent's actions have to be deterministic and provide mechanisms to maximize V over a set of finite policies. Barto et al. (In Press) and others consider cases where no model of the environment is known *a priori*. Using parameter adaptation methods, they assume the agent's actions can be stochastic and can be continuously adapted as the agent *learns* about its environment.

In this paper, the stochasticity of the agent is assumed *a priori* and is an essential property of the learning method. The environment itself is for now considered as deterministic: There is a one-to-one mapping between actions a and resulting states y. We will see that the prior assumption of agent stochasticity allows us to derive an interesting gradient ascent method on the current estimation of the evaluation function in a "policy" network.

Proposed Formalism

We now extend the "static" connectionist representations for probabilities and utilities to dynamic environments by introducing time and delayed rewards. The agent's total expected utility is now a function of the decision behavior over the complete decision period. Suppose the agent is in state x, for a given policy P, the immediate expected utility can be written as:

$$U_x^P = \sum_a u_{xa} \pi_{xa} \quad (5)$$

where the action probabilities π_{xa} characterize the policy P. From the current state x, suppose the agent can reach a state y by taking action $a \in A$, the long-term expected utility from state x can then be written as:

$$V_x^P = U_x^P + \sum_{a/y} V_y^P \pi_{xa} \quad (6)$$

For a fixed policy P, that is for a fixed set of action probabilities π_{xa}, we could compute V_x^P by "backing up" the state evaluation function one step from V_y^P.

Suppose that at stage k, only an estimation $\widehat{V}_y^P(k)$ of V_y^P is available, we can then compute the estimation $\widehat{V}_x^P(k+1)$ of V_x^P using:

$$\widehat{V}_x^P(k+1) = \sum_{a/y} [u_{xa} + \widehat{V}_y^P(k)]\pi_{xa} \quad (7)$$

This process is similar to *value iteration* in dynamic programming. For a given policy P, with a small number of assumptions about the environment and about the order of computations, value iteration will converge to the value V_x^P for each state x.

Equations 7 can be written as:

$$\widehat{V}_x^P(k+1) = \sum_a \widehat{v}_{xa}(k) \pi_{xa} \quad (8)$$

with $\widehat{v}_{xa}(k) = u_{xa} + \widehat{V}_y^P(k)$. This equation has the same form as Equation 2. (Note, however, that the indices have different interpretations.) The set of action probabilities π_{xa} can be computed using a set of *e-units* and *p-units* and from a given connectionist representation of each state x. The resulting network can then be called the *policy network*. At each time step k, the estimate of the long-term expected utility V_x^P can be estimated with one linear *u-unit* where the utility weights u_{ij} of Equation 2 now become $\hat{v}_{xa} = u_{xa} + \hat{V}_y^P$.

Our goal is to find the optimal policy P^* which maximizes the long term expected utility $V_x^{P^*} = V_x^*$:

$$V_x^* = Max_{\pi_{xa}} \sum_{a/y} (u_{xa} + V_y^*)\pi_{xa} \tag{9}$$

The idea is then to maximize V_x^P by gradient ascent on the current estimate of the evaluation function with respect to the parameters of the policy network. From Equations 3 and 8, for each time step, we can obtain the gradient of $\hat{V}_x^P(k)$ with respect to the inputs s_a of the exponential *e-units*. Simplifying the notation for clarity, we obtain:

$$\begin{aligned} \frac{\partial V_x^P}{\partial s_a} &= \beta\pi_{xa}(v_{xa} - V_x) \\ &= \beta\pi_{xa}(u_{xa} + V_y^P - V_x^P) \end{aligned} \tag{10}$$

We can now imagine various methods to organize the computations of the evaluation function and of the corresponding policy. A possible *on-line* method is the following. At each time step, an action a is chosen in function of the current action probability distribution implemented by the policy network. From the resulting state y, the current estimation $\hat{V}_y^P(k)$ and the state transitions utilities u_{xa} are used to compute $\hat{V}_x^P(k)$ using Equation 7. The weights of the policy network are then changed by gradient ascent on the current estimate of the evaluation function. Equation 10 computes this gradient with respect to the inputs of the *e-units*. Back-propagation techniques can be used to propagate this gradient further in the policy network as a function of the chosen architecture.

With this organization, the policy is adapted *on-line*, in function of the estimation of the state evaluation function at each time step. Furthermore, the state evaluation updating schedule is itself a function of the current policy. Exploitation by gradient ascent is therefore simultaneous with exploration, determined by the set of action probabilities. The balance of exploitation and exploration may be obtained by tuning the various model parameters and by modifications of the organizations of the computations.

Various connectionist network architectures can be used to compute the action probabilities π_{xa} from the set of possible states. The simplest network consists in having one unit per state and direct connections between states and *e-units*. Such a network architecture can be called "exhaustive" since there is one parameter per state-action pair. Such an exhaustive network is used in the simulations below and is shown in Figure 2. For an exhaustive network, the weight update obtained by gradient ascent can be derived from Equation 10:

$$\Delta w_{ax} = \alpha\beta\pi_{xa}(u_{xa} + V_y^P - V_x^P) \tag{11}$$

where α is a learning rate. Of course, it might be more interesting to provide state representations and network architectures which are specifically adapted to the environment and to the application. In particular, layers of hidden units might be used to discover compact internal state representations that would be generated by the learning algorithm.

Decision Learning and Parameter Estimation

Equation 7 can be seen as a standard backward dynamic programming technique. It is also related to what Sutton (1988) calls the *Temporal Difference* method. Barto et al. (In Press) point out the relationships between Temporal Difference methods and dynamic programming in more complex situations. They also suggest how an evaluation function V could be estimated using connectionist networks. By contrast, in the framework proposed above, V is actually a table look-up whereas the policy P is implemented with a connectionist network. Of course, we can imagine combinations of evaluation networks and policy networks and various techniques to integrate the connectionist computations of the evaluation functions and corresponding policies.

Barto, Bradtke and Singh (1991) use a variety of algorithms to estimate evaluation functions, also inspired from dynamic programming procedures. In their examples, policies are implemented using a Boltzman distribution:

$$p_{xa} = \frac{e^{\hat{V}_y(k)/T}}{\sum_{z \in Y} e^{\hat{V}_z(k)/T}} \tag{12}$$

where T acts as a "computational" temperature. In their simulations, the temperature is annealed as a function of the learning performance. The agent's behavior becomes deterministic over the complete set of states simultaneously as the temperature decreases. In the framework proposed above, the level of determinism depends on the weights of the policy network. These weights are updated by gradient ascent on the current estimation of the total expected utility. If from a given state x, the current estimates are identical for all permissible states y, the gradient is null and actions remain equally random for that state. If for a given state, the long-term expected utilities are well differentiated over the set of admissible actions, the gradient descent approach will make the behavior's agent deterministic for that particular state. The agent should learn how and where it should take deterministic decisions, or whether it should keep exploring or not, as a function of the value of this gradient. If reaching a goal provides a high reward for the agent, the agent's behavior will quickly become deterministic when getting close to the goal. The "amount" of determinism will then "move backward" from the goal.

Sutton (1990) also suggests a Boltzman distribution

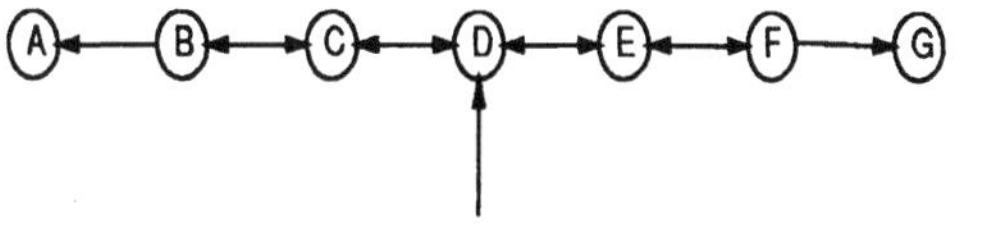

Figure 1: Environment used in the simulations. The starting state is D. The goal of the agent is to reach G.

for action probabilities:

$$p_{xa} = \frac{e^{w_a x}}{\sum_{i \in A} e^{w_{ix}}} \quad (13)$$

Furthermore, after each action taken by the agent, the distribution parameters are changed according to:

$$\Delta w_{ax}(t+1) = \alpha(u_{xa} + V_y - V_x) \quad (14)$$

Note the similarities between Equation 14 and Equation 11. The difference in the two weight update equations resides in the action probability π_{xa}. But in Sutton, actions are chosen according to the multinomial probability distribution parametrized by the p_{xa}. Therefore, if we imagine the evaluation function is updated only after a large number of actions have been sampled according to this probability distribution, the weight update becomes $\Delta w_{ax} = \alpha' f_{xa}(u_{xa} + V_y - V_x)$ where f_{xa} represents a frequency of actions and is an unbiased estimate of the "propensity" of action π_{xa}. Therefore, Sutton's weight update equation can be seen as a stochastic form of a gradient ascent on the estimation of the evaluation function in an exhaustive network.

Simulations

Random Walk Environment

The environment used in the simulations is inspired from the random walk process introduced by Sutton (1988). It consists of seven possible states, as shown in Figure 1. The agent's initial position is state D. In each state, the agent has to make a decision about the direction of the following move, left or right. When the agent moves, it might receive a payoff which depends on the current state and on the moving decision. These payoffs may be negative (e.g., they may represent an amount of energy being spent for the move) or positive (e.g., they may represent a received amount of food). These payoffs may then be represented in a two-dimensional *utility* matrix. When the agent reaches the absorbing states A or G, it is put back to the initial state D. The goal of the agent is to maximize the total utility received from the initial state to the goal state G. In the simulations below, we look at the agent's learning behavior as a function of given arbitrary utility matrices.

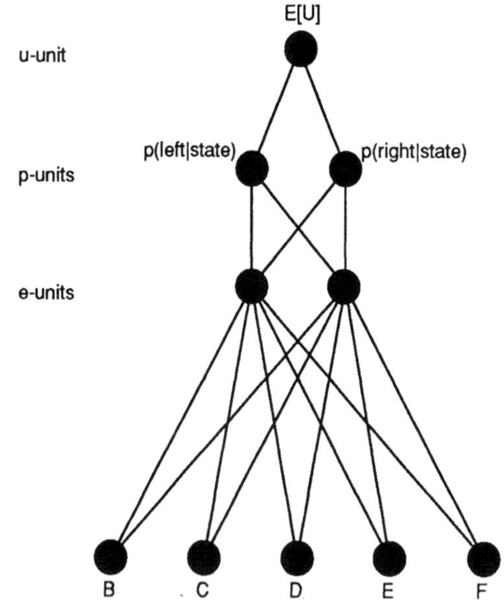

Figure 2: Policy network. The architecture is "exhaustive": there is one parameter per state-action pair.

Policy Network Architecture

The policy network architecture is exhaustive (Figure 2) and can be seen as composed of two subnetworks. The first subnetwork computes move decision probabilities $\pi_{xl} = p(left|state)$ and $\pi_{xr} = p(right|state)$ from each possible state, where states are represented by single binary input units. The second decision subnetwork computes long-term expected utility from decision probabilities and utilities. During learning, the decision weights between states and *e-units* are changed by gradient ascent on the long-term expected utility using the algorithm described above. In this framework, we suppose the agent stores present and estimated future utilities in memory. Although an evaluation network could compute these utilities as needed (Barto et al., In Press), we simply assume for now that learning operates through utilities perfectly retrieved from memory by the agent.

Results

Case 1: Learning from Long-Term Rewards

The approach is first illustrated with the utility matrix shown in Table 1. A simulation *run* is defined as a new set of initial decision weights, representing a new agent. A simulation *trial* is defined as a sequence of moves from the initial state D to the goal G. The network performance can then be judged by the num-

State	B	C	D	E	F
Left move	0	0	0	0	0
Right move	0	0	0	0	10

Table 1: Cost/utility matrix for case 1.

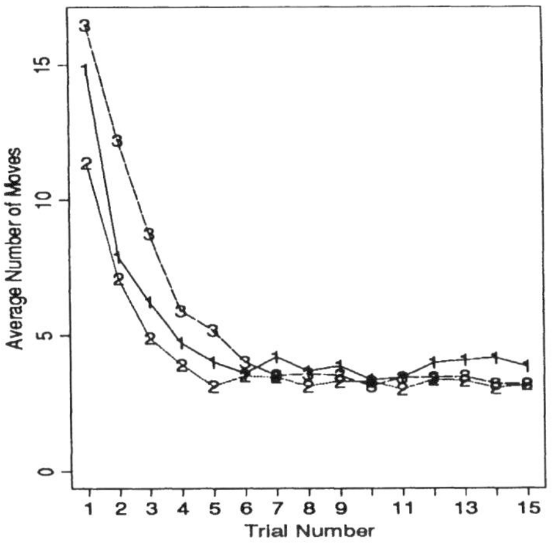

Figure 3: Average number of moves to reach the goal as a function of trial number in cases 1, 2, and 3.

ber of moves (averaged over a given number of runs) it takes for the agent to reach the absorbing state G as a function of trial number. If the agent first reaches absorbing state A, it is put back to the initial state D, incrementing the move counter by 1. Figure 3 shows the agent's performance up to 15 trials averaged over 50 runs. Simulations show the agent always reaches optimal performance (3 steps to the right from state D to G). Absolute performance characteristics obviously depend on the learning parameters (such as the learning rate).

As the network learns how to estimate the correct long-term expected utilities, it also learns how to estimate optimal decision probabilities. This process works "backward in time". At first, the agent reaches goal G from initial state D by chance. In doing so, it observes current payoffs, updates future utility estimation, and adjusts its behavior through the learning mechanism. Eventually, the agent's behavior in state F then converges to optimal behavior. The total expected utility subsequently converges to the optimal value V_F^*. When in state E, the agent reaches state F by chance: the long-term utilities and policy will then be adapted using the estimations obtained for state F. The learning mechanism back-propagates the gradient of long-term expected utility through updated utility weights to modify the decision weights. The utility received from state F to G will make the agent's behavior more deterministic in state F, then backward from F to D. In some sense, both evaluation function *and* policy's determinism are "backed up" from the goal to the initial state.

Of course, the agent's behavior resulting from the implementation of this process might not be as sequential as it sounds. As explained above, evaluation and policies are changed as a function of the current state and of the current decision move, which are stochastic and depend on the organization of the computations. Similarly to on-line policy iteration methods, the agent does not wait to reach the goal to update action probabilities. The agent just looks one step ahead to adjust its "propensities" of action for the current state. For the given utility matrix, because the agent eventually visits the goal and because no other reward may modify the adaptation of behavior, it should always reach optimal behavior.

State	B	C	D	E	F
Left move	-2	-2	-2	-2	-2
Right move	-2	-2	-2	-2	10

Table 2: Cost/utility matrix for case 2.

State	B	C	D	E	F
Left move	-1	-1	+1	-1	-1
Right move	-1	-1	-1	-1	10

Table 3: Cost/utility matrix for case 3.

Case 2: Learning from Long-Term Rewards with Moving Costs

In the second set of simulations, the agent obtains a +10 utility when it reaches the absorbing goal G and a -2 utility when moving from any state to a neighboring state. The corresponding utility matrix is shown in Table 2. Figure 3 shows the agent's learning performance up to 15 trials averaged over 50 runs. With this new utility matrix, the agent reaches optimal performance faster than with the utility matrix used in case 1. This result might not be intuitive since the state expected utilities have now become smaller in reason of the -2 costs "spent" between step moves. The reason for this result is actually that there is now a differential expected utility between going left and going right. For example, the optimal long-term expected utilities V_E^* and V_F^* from states E and F are respectively 8 and 10. This differential expected utility creates a differential utility gradient between each move, forcing the agent to become deterministic earlier and to learn more rapidly how to move in the correct direction.

Case 3: Avoiding Suboptimal Immediate Rewards

One of the motivations for studying sequential decision making and for using dynamic programming methodologies is to avoid suboptimal short-term decisions which may prevent future optimal decisions. The utility matrix shown in Table 3 illustrates this situation. In this case, the agent gets an immediate reward by moving left from the initial state. However, the late reward from state F to G should force the agent to ignore the immediate reward on the left, to move right and to receive the late reward at the goal. When the agent learns about the environment, it will probably be attracted to the immediate reward at first. But by exploration, the agent should learn about the delayed reward and should adjust its behavior over time to ignore the early reward. The short-term reward from D to C should simply delay learning of the optimal strategy. Figure 3 shows the network learning performance for case 3. The learning curve reflects the predicted behavior. At first, it takes more steps to reach the goal because the short-term reward leads the agent in the wrong direction. During early learning, the agent actually learns how to move to the left. However, after sufficient learning, for the given utility matrix, the agent always learns how to avoid short-term rewards and to move directly to the goal. In general, the exact behavior learned by the agent will depend on the balance between exploration and exploitation, which in turn will depend on the model parameters and on the organization of the computations.

Conclusion

A sequential decision learning formalism is proposed which integrates elements of standard decision theory with connectionist principles. In statistical pattern recognition, standard procedures may first estimate model parameters to estimate class probabilities. Costs may then be invoked to compute minimal risk classification. In dynamic environments, dynamic programming techniques, such as *policy iteration* may generated successive evaluate decision strategies and long-term expected rewards until optimal decision behavior is obtained. The present approach directly updates the parameters of a policy network by gradient ascent of the current estimate of the long-term expected utility. The formalism may be seen as a generalization of some of the policy adaptation methods proposed by Sutton (1990) and Barto et al. (1991).

The learning procedure was simulated and tested in a simple environment. In various cases, the procedure was actually shown to generate interesting and intelligent looking learning dynamics. There are many ways the proposed formalism could now be integrated with other dynamic programming concepts or combined with other parameter estimation methods. Of course, it remains to be seen if these learning principles may be powerful enough to generate intelligent decision behavior in more complex environments. But the formalism can be seen as a generalization of previously proposed mechanisms and the gradient ascent approach appears to be conceptually satisfying and promising.

References

Barto, A. G., Bradtke, S. J., & Singh, S. P. (1991). *Real-time learning and control using asynchronous dynamic programming.* Technical Report 91-57, Computer and Information Science, University of Massachusetts, Amherst.

Barto, A. G., Sutton, R. S., & Watkins, C. J. C. H. (In Press). Learning and sequential decision making. In M. Gabriel & J. W. Moore (Eds.), *Learning and computational neuroscience.* Cambridge, MA: MIT Press.

Bellman, R. E. (1957). *Dynamic Programming.* Princeton, NJ: Princeton University Press.

Chauvin, Y. (1991). Decision making connectionist networks. In *Proceedings of the 13th Cognitive Science Society conference, Chicago, IL.* Hillsdale, NJ: Lawrence Erlbaum.

Samuel, A. L. (1959). Some studies in machine learning using the game of checkers. In E. A. Feigenbaum & J. Feldman (Eds.), *Computers and thoughts.* New York: McGraw-Hill.

Slovic, P., Lichtenstein, S., & Fischhoff, B. (1988). Decision making. In R. C. Atkinson, R. J. Hernstein, G. Lindsey, & R. D. Luce (Eds.), *Steven's handbook of experimental psychology.* New York: Wiley. 2nd. Edition

Sutton, R. S. (1988). Learning to predict by the method of temporal differences. *Machine Learning, 3*, 9–44.

Sutton, R. S. (1990). Integrated architectures for learning, planning, and reacting based on approximating dynamic programming. In *Proceedings of the Seventh International Conference on Machine Learning.* San Mateo, CA: Morgan Kaufmann.

Sutton, R. S. & Barto, A. G. (1987). A temporal-difference model of classical conditioning. In *Proceedings of the 9th Cognitive Science Society conference, Seattle, WA.* Hillsdale, NJ: Lawrence Erlbaum.

Watkins, C. J. C. H. (1989). *Learning from delayed rewards.* PhD thesis, Cambridge University, Cambridge, England.

Early Warnings of Plan Failure, False Positives and Envelopes: Experiments and a Model

Paul R. Cohen, Robert St. Amant, David M. Hart
Experimental Knowledge Systems Laboratory
Department of Computer Science
University of Massachusetts, Amherst

cohen @cs.umass.edu 413 545 3638

Abstract

We analyze a tradeoff between early warnings of plan failures and false positives. In general, a decision rule that provides earlier warnings will also produce more false positives. Slack time envelopes are decision rules that warn of plan failures in our Phoenix system. Until now, they have been constructed according to ad hoc criteria. In this paper we show that good performance under different criteria can be achieved by slack time envelopes throughout the course of a plan, even though envelopes are very simple decision rules. We also develop a probabilistic model of plan progress, from which we derive an algorithm for constructing slack time envelopes that achieve desired tradeoffs between early warnings and false positives.

1 Introduction

Underlying the judgment that a plan will not succeed is a fundamental tradeoff between the cost of an incorrect decision and the cost of evidence that might improve the decision. For concreteness, let's say a plan succeeds if a vehicle arrives at its destination by a deadline, and fails otherwise. At any point in a plan we can correctly or incorrectly predict that the plan will succeed or fail. If we predict early in the plan that it will fail, and it eventually fails, then we have a *hit*, but if the plan eventually succeeds we have a *false positive*. False positives might be expensive if they lead to replanning. In general, the false positive rate decreases over time (e.g., very few predictions made immediately before the deadline will be false positives) but the reduction in false positives must be balanced against the cost of waiting to detect failures. Ideally, we want to accurately predict failures as early as possible; in practice, we can have accuracy or early warnings but not both.

The false positive rate for a decision rule that at time t predicts failure will generally decrease as t increases. We analyze this tradeoff in several ways. First, we describe a very simple decision rule, called a *slack time envelope*, that we have used for years in the Phoenix planner (Sections 2 and 3). Then, using empirical data from Phoenix, we evaluate the false positive rate for envelopes and show that envelopes can maintain good performance throughout a plan (Section 4). An infinite number of slack time envelopes can be constructed for any plan, and the analysis in Section 4 depends on "good" envelopes constructed by hand. To be generally useful, envelopes should be constructed automatically. This requires a formal model of the tradeoff between when a failure is predicted (earlier is better) and the false positive rate of the prediction (Section 5). Finally we show how the conditional probability of a plan failure given the state of the plan can be used to construct "warning" envelopes.

2 Slack Time Envelopes

Imagine a plan that requires a vehicle to drive 10 km in 10 minutes. Figure 1 shows progress for three possible paths that the vehicle might follow, labeled A, B and C. Case A is successful: the vehicle makes rapid progress until time 3, then slows down from time 3 to time 4, then makes rapid progress until time 8, when it completes the plan. Case B is unsuccessful: progress is slow until time 4, and slower after that; and the required distance is not covered by the deadline.

The solid, heavy line is a slack time envelope for this problem. Our Phoenix planner (Cohen et al., 1989; Hart, Anderson, & Cohen, 1990) constructs such an envelope for every plan and checks at each time interval to see whether the progress of a plan is within the envelope. Case A remains within the envelope until completion; case B violates the envelope at time 6.

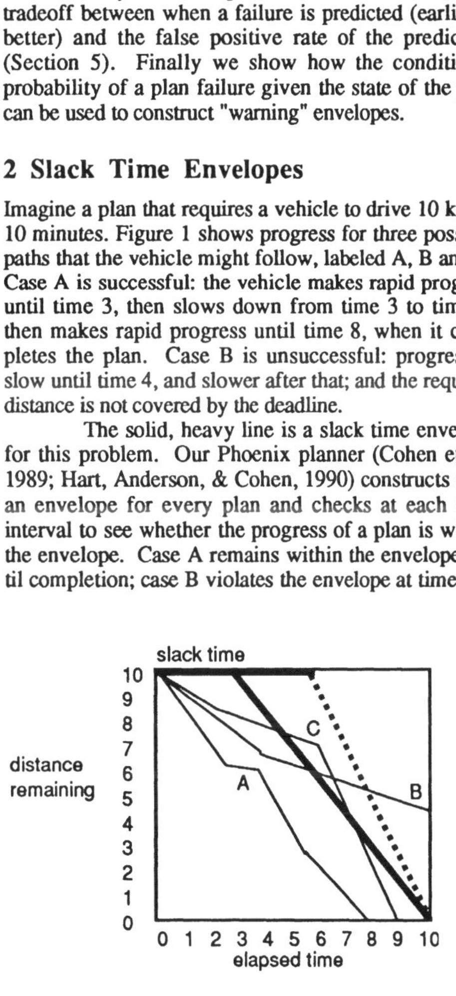

Figure 1. Illustration of envelopes

When an envelope violation occurs, the Phoenix planner modifies or completely replaces its plan. It should not wait until the deadline has expired to begin, but should start replanning as soon as it is reasonably sure that the plan will fail. Clearly, envelopes can provide early warning of plan failure; for example, in case B, the envelope warned at time 6 that the plan would fail. The problem is that progress might pick up after an envelope violation, as shown in case C. At time 5 the envelope is violated, but by time 8, the plan is back within the envelope. If in this case the Phoenix planner abandoned its plan at time 5, it would have incurred needless replanning costs. Case C is a false positive as we defined it earlier: a plan predicted to fail that actually will succeed. Note that a different envelope, shown by the heavy dotted line, will avoid this problem. Unfortunately, it doesn't detect the true failure of case B until time 8, two minutes after the previous envelope. This illustrates the tradeoff between early warnings and false positives. (This and other concepts in the paper derive from signal detection theory, e.g., (McNicol, 1972; Coombs, Dawes, & Tversky, 1970).)

Slack time envelopes get their name from the period of no progress that they permit at the beginning of a plan. The Phoenix planner adds slack time to envelopes so that plans will have an opportunity to progress before they are abandoned for lack of progress. Until recently, this was all the justification for envelopes we could offer. In the following sections, however, we show why the simple linear form of envelopes achieves high performance, and how to select a value of slack time.

3 The Data Set

One way to evaluate slack time envelopes is to generate hundreds of plans, monitor their execution at regular intervals, and, at each interval, use an envelope to predict success or failure. We generated 1139 travel plans, or *paths*, for vehicles in our Phoenix simulation. Phoenix is based on a machine-readable map of Yellowstone National Park that includes roads, obstacles, a variety of elevations and ground covers, and other terrain features. The Phoenix planner fights simulated forest fires in this environment by surrounding the fires with fireline built by bulldozers. Envelopes in Phoenix monitor fire spread rate, fireline digging, and progress in different bulldozer tasks. The focus of this paper, however, is a simpler problem: getting from one point on the map to another by a deadline. To generate our data set, we repeatedly selected pairs of points 70 km apart as the crow flies, and asked the Phoenix planner to construct a path between each pair. Then we simulated the traversal of each path, monitoring it every 1000 simulated seconds. At each monitoring step we estimated the distance remaining to the destination. Because of obstacles, terrain, and so on, the distribution of remaining distances at a given monitoring interval

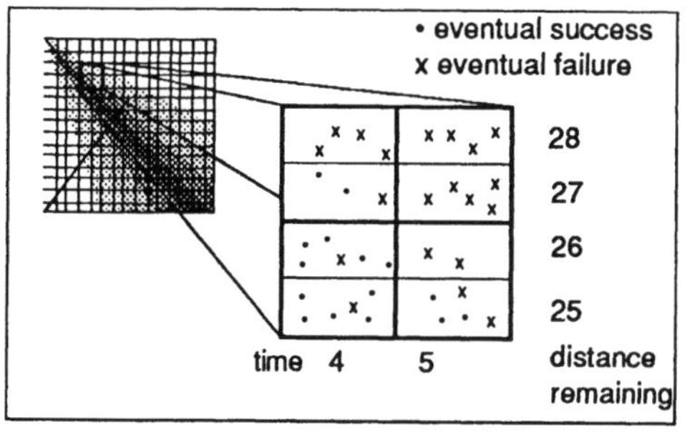

Figure 2. How we generated distributions of DR for successes and failures at each time interval.

was considerable (including many greater than 70 km). For example, after 5000 seconds, the mean remaining distance was about 54 km with a range of 13.6 to 79.1 km. We generated, executed and monitored 1139 paths in this manner.

3.1 Distributions of Eventual Successes and Failures Before the Deadline

We chose a deadline of 15,000 seconds to divide the paths into two groups: paths that reached their goals by the deadline were called *successes*, and those that did not were called *failures*. Of 1139 paths, 654 succeeded and 485 failed. We looked at each path 15 times, once every 1000 seconds, and recorded an estimate of the number of "distance units" remaining to the goal. For a variety of reasons, a distance unit is 2 km, so the distance remaining to the goal, abbreviated DR, is 35 at the beginning of the plan and zero for successful paths at the end of the plan. Henceforth, we use "time x" as shorthand for "x thousand seconds elapsed." For example, in Figure 2, at time 4, all the paths with DR = 28 are failures; at time 5, all the paths with $26 \leq DR \leq 28$ are failures; but at time 5, DR = 25, three paths are successes and two are failures.

We plotted frequency polygons for DR for successes and failures at each of the 15 time intervals. Figure 3 shows the distribution of successes and failures at time 5. Note that most failures still have a long way to travel at time 5: the bulk of the distribution lies to the right of DR = 30 (the mean DR for failures at time 3 is 33). The distribution of successes, however, is made up of paths with relatively short remaining distances to the goal (mean DR = 17).

3.2 Empirical Hit Rates and False Positive Rates for DR Thresholds

Let's predict that a path will fail to reach its goal by its deadline if, at time 5, the remaining distance to the goal is 30 or more, that is, the *threshold* $DR \geq 30$. The dark shaded area in Figure 3 represents false positive

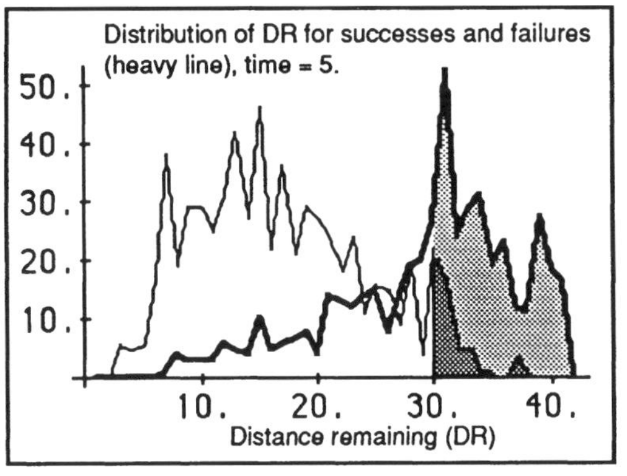

Figure 3. Frequency polygons for DR at time 5.

errors, paths we predict will fail but that eventually succeed. Of the 654 paths that eventually succeed, 37 lie in the dark shaded area; the probability of a false positive is therefore 37/654 = 0.056. The light shaded area represents hits, paths that are predicted to fail and that actually do fail. The probability of a hit is 261/485 = 0.538. The ratio of the probabilities is 9.60.

At time 5, the threshold DR ≥ 30 seems pretty good because the ratio of hit probability to false positive probability is high, but we cannot say it is the *best* threshold unless we know the relative values of hits and false positives.

This is just part of the analysis of our data set. In particular, we haven't shown our analyses of success and failure distributions at other times; nor hit and false positive probabilities for different DR thresholds at different times; nor success and failure distributions for stricter or more lenient deadlines. We can summarize these analyses as follows: At later time intervals, the success distribution is increasingly right skewed, with most of its mass around low values of DR. At intermediate time intervals (e.g., time = 7) the failure distribution is roughly uniform. Later, it is right skewed like the success distribution, but with more mass in its tail than the success distribution. Shifting the DR threshold to the right decreases both hits and false positives, though false positives decrease faster (as in Fig. 5, only more so at later time intervals). These patterns hold for stricter and more lenient deadlines; the main effect of a stricter deadline is to reduce the number of successes. The following evaluations of envelopes are based on the 15,000 second deadline illustrated above because it produces a nearly even split between successes (654, total) and failures (485, total).

4. Evaluation of Slack Time Envelopes

Slack time envelopes are decision rules for predicting whether paths will succeed or fail. As illustrated in Figure 1, if the path is within the boundary of an envelope at a particular time, then we predict success, otherwise we predict failure. Each point on an envelope boundary specifies a DR threshold for a particular time, and so has an associated hit rate and false positive rate. In this section we use slack time envelopes to predict whether paths in our data set, discussed above, will succeed or fail. We evaluate the predictive performance of slack time envelopes according to this criterion: An envelope should provide performance approaching optimal throughout a plan.

This depends of course on our definition of optimal. Consider a decision rule based on distance remaining (dr) to the goal at time t:

If $l(dr, t) > \beta(t)$, *then predict plan failure*

$$\text{where } l(dr, t) = \frac{\Pr(dr \mid plan\ fails, t)}{\Pr(dr \mid plan\ succeeds, t)}$$

Intuitively, we have an observation of dr at time t, and we must decide whether this observation has been produced by an eventual success or failure. We base our decision on whether the likelihood is greater than the threshold $\beta(t)$. A basic result from signal detection theory is that the utility of this decision is maximized if

$$\beta(t) = \frac{\Pr(plan\ succeeds, t)}{\Pr(plan\ fails, t)} (Payoff(t))$$

where

$$Payoff(t) = \frac{Val(correct\ rej, t) + Cost(false\ pos, t)}{Val(hit, t) + Cost(miss, t)}$$

In the simplest case, $\beta(t)$ is constant over the course of a plan. A more realistic assessment requires analysis of the terms in $\beta(t)$. The first term, the prior probabilities, decreases with time, as plans begin to succeed. The second term *Payoff(t)* determines the relative importance of hits, false positives, correct rejections, and misses. The value of correctly predicting a plan failure decreases over time; early warnings are worth more. At the same time the cost of a false positive increases over time; if we are going to unnecessarily abandon a plan, it is better to do so early in the plan than later when we have invested a lot of time in the plan. It is more difficult to assess the value of a correct rejection and the cost of a miss, but if we assume they are constant relative to the other parameters, then the value of the second term in *Payoff(t)* increases over time. We consider the cases in which *Payoff(t)* is constant and also in which *Payoff(t)* increases linearly with time.

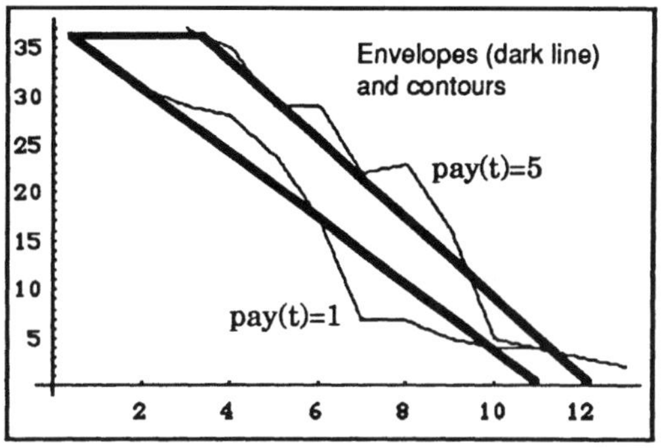

Figure 4. Hand-constructed slack time envelopes superimposed on constant *Payoff(t)* contours.

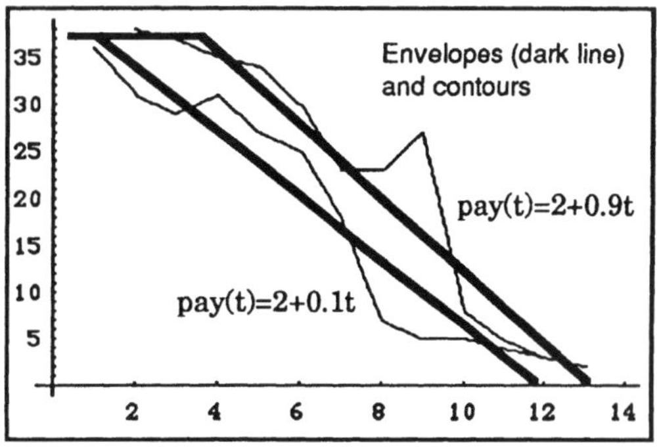

Figure 5. Slack time envelopes on linear *Payoff(t)* contours.

4.1. Comparing Slack Time Envelopes with Empirical Utility Contours

Because envelopes are just straight lines, it is unclear whether they can satisfy the optimal performance criterion. In particular, for constant or linear payoff functions, the DR threshold required to maintain a constant ratio of hit probability to false positive probability might not change linearly over time. To find out, we calculated utility ***contours*** from the empirical data for different *Payoff(t)*, as shown in Figures 4 and 5. A contour represents a fixed *Payoff(t)* function; each point on a contour is the DR threshold (y axis) that is required at a particular time (x axis) to ensure that the utility of the decision is maximized. In Figure 4 we let *Payoff(t)* be constant at 1 and 5, and in Figure 5 we let *Payoff(t)* vary as a function of time.

An important characteristic of these contours is that they require high DR thresholds for the first few time intervals, but then gradually smaller thresholds for later time intervals. Utility contours are roughly linear, which suggests that a slack-time envelope, fit to one of these contours, ought to provide performance approaching optimal, given our payoff function. For our data set, at least, Figures 4 and 5 tells us that an envelope can be constructed to satisfy our performance criterion. We will formalize this result in the next section.

5. Constructing Slack Time Envelopes: How Much Slack?

Our focus now turns to the task of constructing slack time envelopes. We assume that the end points of the envelope are the distance to the goal and the deadline, so the only parameter is how much slack time to allow. Next we present a model that predicts utility for different values of slack time. The model also predicts the ***early warning premium*** for values of slack time. Early warning premiums accrue when, by constructing a tight envelope with little slack time, we detect failures earlier than we would with a looser envelope. Empirically, early warning premiums come at the expense of false positives. We assess a cost for each hit proportional to the time interval in which it is detected; this places a premium on early hits. We assess a constant cost for each false positive. This is described further in the following sections.

5.1. A Probabilistic Model of Progress

If we know the distributions of distance remaining (DR) for successes at each time interval (e.g., those in Figure 3) then we can predict the false positive rate for a given DR threshold. A simple model of the distribution of DR begins with the assumption that in each time interval a vehicle can progress at its maximum rate c with probability p, or makes no progress at all with probability q = 1 - p. Then the distribution of progress is binomial, as shown in Table 1: the probability of having made r units progress by time n is just the binomial probability $\binom{n}{r}p^r q^{(n-r)}$.

For example, the probability of one unit progress by time 4 is $4pq^3$ because there are four ways to achieve this result, each with probability pq^3: we could make no progress until time 3 (with probability q^3) and then progress at the maximum rate for one time unit (total probability, pq^3). Or we could make one

Progress \ Time	1	2	3	4	5
0c	q	q^2	q^3	q^4	q^5
1c	p	pq	$3pq^2$	$4pq^3$	$5pq^4$
2c		p^2	$3p^2q$	$6p^2q^2$	$10p^2q^3$
3c			p^3	$4p^3q$	$10p^3q^2$
4c				p^4	$5p^4q$
5c					p^5

Table 1. Progress in each time interval follows a binomial distribution.

unit of progress by time 3 (with probability $3pq^2$) and then make no progress for the remaining time unit (total probability $3pq^3$). The sum of these options is $4pq^3$.

The expected progress after N time units is cNp and the variance is cNpq. If p = q = .5 then the distributions of progress in each time interval are symmetric. Otherwise the mass of the distribution at time N tends toward cN (if p > q) or zero (if q > p). Important characteristics of this model are that progress is linear and variance changes linearly with time.

5.2 Utility Contours Using the Model

This model explains the shape of utility contours and slack time envelopes, and it predicts the probability of false positives for a given envelope. Let us elaborate the model a little: Our goal is to travel some distance D_g by a deadline time T_g. At any time t, we can assess the progress that has been made, D(t), and the progress that remains to be made, DR(t); and the time remaining, $TR = T_g - t$. A success is defined as $D(t) \geq Dg$ and $t \leq Tg$. The conditional probability of a success given DR(t), Dg and Tg is:

$$\mathbf{Pr}(success \mid DR(t)) = \sum_{r=DR(t)}^{TR} \binom{TR}{r} p^r q^{(TR-r)}$$

A similar equation holds for the conditional probability of a failure. If *Payoff(t)* is for example constant, this means means that the ratio of these conditional probabilities must be constant as well. Now imagine that we have DR(t) distance remaining at time t and we extrapolate forward TR time units to the deadline. At this point we have a binomial distribution with N = TR, divided into a portion below the DR=0 line (the successes, those cases that have arrived by the deadline) and a portion above the line (the failures.) The ratio of the areas of the two portions gives us the ratio of the conditional probabilities. If we want to find at each time the distance for which this ratio is constant, we plot a constant z-score for distributions with N ranging from T_g to 0.

Figure 6 shows contours for constant *Payoff(t)*. Contours for comparable linear *Payoff(t)* are very similar, with identical slack times, but more pronounced curve. To generate the figure we assumed D_g = 25, T_g = 50, p = .5, and c = 1, and applied the above analysis to get conditional probabilities of success and failure for every value of t.

Imagine that a vehicle has made 10 units of progress at time 25, that is, DR(25) = 15, illustrated by the large dot near the center of Figure 6. Because this dot lies on the contour labelled *Payoff(t)* = 5, we know that Pr(failure | DR(25) = 15) / Pr(success | DR(25) =15) = 5. If the vehicle makes no progress for another

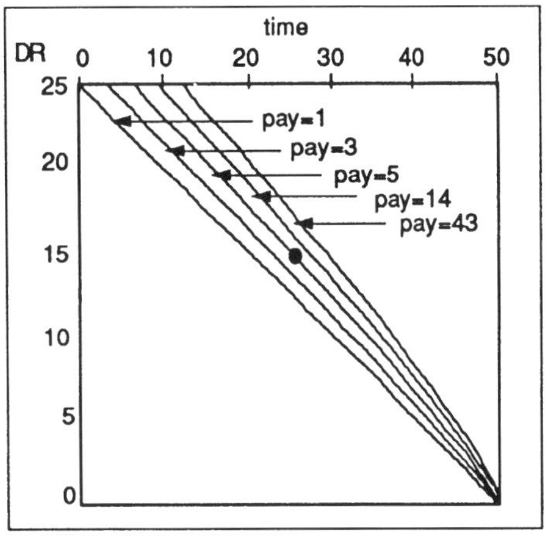

Figure 6. Contours of constant payoff from each point in the space.

five time units, then the dot would lie to the right of the contour labeled *Payoff(t)* = 43, so the probability ratio is much higher.

These contours vary as $\sqrt{t}$. At the scale on which we monitor, linear envelopes provide a good approximation of the contours, as long as the envelope boundaries have the right slope, that is, if they are constructed with the right amount of slack time. Note, too, that Figure 6 justifies the use of slack time in envelopes: The contours associated with high payoffs (and thus high ratio of hit probability to false positive probability) allow a period of no progress at the beginning of the plan.

5.3 Setting Slack Time

A slack time envelope is just a pair of lines, one representing the period in which no progress is required—the slack time—and another connecting the end of the first to the deadline, as shown in Figures 1 and 6. Slack time is the only parameter in slack time envelopes, but we must still show how to set it.

We desire a balance of false positives against early warning premiums. We have not yet derived from our binomial model a closed-form expression for the expected number of false positives and early warnings, but we have an algorithm that produces these expectations for a given value of slack time, if we assume that $D_g = .5\ T_g$:

For each possible value of DR, dr_i:

a. calculate t_e, the time at which the envelope boundary will be crossed, given dr_i; for example, in Figure 1, when $dr_i = 5$ and t>8, the solid envelope boundary is crossed, so for $dr_i = 5$, $t_e = 8$.

b. use the binomial model to calculate p_e, the probability of reaching t_e; for example if $dr_i = 3$ and $t_e = 5$, Table 1 tells us that $p_e = 10p^2q^3$.
c. use the model to find the probability of a false positive, $p_{fp} = Pr(success \mid DR(t_e) = dr_i)$.
d. $p_e \times p_{fp}$ is the probability of a false positive for this value of dr_i
e. $p_e \times (T_g - t_e)$ is the expected *early warning premium* for this value of dr_i.

$T_g - t_e$ is the time that remains before the deadline at the envelope boundary at dr_i; this is why $T_g - t_e$ is called the early warning premium. The expected early warning premium for a value of dr_i is just $T_g - t_e$ times the probability of crossing the envelope boundary. The *mean expected early warning premium* is the mean over all values of dr_i of $p_e(T_g - t_e)$. We expect it to have higher values for lower slack times, because the envelope boundaries for low slack times are further from the deadline. The *mean probability of false positives* is obtained by summing $p_e\, p_{fp}$ for all values of dr_i and dividing by the number of these values. We expect it to rise, also, as slack time decreases, as suggested by the contours in Figure 6.

With a table of values for the mean probability of false positives and the mean expected early warning premium, and utilities for early warning and false positives, we can make a rational decision about slack time.

6 Conclusion

Although we rely heavily on slack time envelopes in the Phoenix planner, we have always constructed them by heuristic criteria, and we did not know how to evaluate their performance. In this paper we showed that high performance can be achieved by hand-constructed slack time envelopes, and we presented a probabilistic model of progress, from which we derived a method for automatically constructing slack time envelopes that balance the benefits of early warnings against the costs of false positives.

Other work has been done in this area, e.g., (Miller, 1989) constructs an execution monitoring profile of acceptable ranges of sensor values for a mobile robot (this profile is also called an "envelope"). If, during plan execution, a sensor value exceeds the envelope boundaries, a reflex is triggered to adjust the robot's behavior in such a way that the sensor readings return to the acceptable range. (Sanborn and Hendler, 1988) have used monitoring and projection in a simulated robot that tries to cross a busy street. The robot has a basic street-crossing plan, but monitors oncoming traffic and predicts possible collision points which trigger reactive avoidance actions. Our contribution has been to cast the problem in probabilistic terms and to develop a framework for evaluation. We are currently extending our work to other models of progress and different, more complex domains. A technical report covering this work in more detail is in preparation.

Acknowledgements

This research is supported by DARPA under contract #F49620-89-C-00113, by AFOSR under the Intelligent Real-time Problem Solving Initiative, contract #AFOSR-91-0067, and by ONR under a University Research Initiative grant, ONR #N00014-86-K-0764, and by Texas Instruments Corporation. We wish to thank Eric Hansen and Cynthia Loiselle for many thoughtful comments on drafts of this paper. The US Government is authorized to reproduce and distribute reprints for governmental purposes notwithstanding any copyright notation hereon.

References

Cohen, P.R.; Greenberg, M.L.; Hart, D.M.; and Howe, A.E., 1989. Trial by Fire: Understanding the Design Requirements for Agents in Complex Environments. *AI Magazine*, 10(3):32-48.

Coombs, C.; Dawes, R.; and Tversky, A., 1970. *Mathematical Psychology: An Elementary Introduction*. Ch. 6. The Theory of Signal Detectability. Prentice Hall.

Hart, D.M.; Anderson, S.D.; and Cohen, P.R., 1990. Envelopes as a Vehicle for Improving the Efficiency of Plan Execution. In *Proceedings of the Workshop on Innovative Approaches to Planning, Scheduling, and Control*. K. Sycara (Ed.) San Mateo, CA.: Morgan-Kaufmann, Inc. Pp. 71 – 76.

McNicol, D., 1972. *A Primer of Signal Detection Theory*. George Allen and Unwin, Ltd.

Miller, D.P., 1989. Execution Monitoring for a Mobile Robot System. In SPIE Vol. 1196 *Intelligent Control and Adaptive Systems*. Pp. 36-43.

Sanborn, J.C.; and Hendler, J.A., 1988. Dynamic Reaction: Controlling Behavior in Dynamic Domains. *International Journal of Artificial Intelligence in Engineering*, 3(2).

Syllable Priming and Lexical Representations: Evidence from Experiments and Simulations

David P. Corina Ph.D.
U.S.C. Program in Neuroscience
HNB 18C, University Park, Los Angeles CA 90089-2520
corina@gizmo.usc.edu

Abstract.

This paper explores the composition of syllable structure in lexical representations. Data from auditory lexical decision experiments are presented which demonstrate that syllable structure is represented in the mental lexicon and that the effects of syllable structure are separable from shared segmental overlap. The data also indicate that syllable representations correspond to a *surface* syllable rather than an abstract underlying syllable posited by some linguistic theories. These findings raise questions concerning the origin of syllable structure in lexical representations. A connectionist simulation utilizing the TIMIT data base shows that syllable-like structure may be induced from exposure to phonetic input. Taken together these results suggest that knowledge of surface syllable structure is actively used in understanding language and this knowledge may derive from a speaker's experience with language.

Introduction

An important enterprise in psycholinguistic research is determination of the lexical properties which underlie our knowledge of language. Early work in this area has identified the importance of semantic relatedness in the organization of the mental lexicon (Neely, 1977). Less well understood is whether phonological properties factor in this organization. Studies indicate that when a subject is asked to determine whether the second member of a word pair is a well formed English word, reaction times are significantly shorter if the preceding word shares some phonological similarity (e.g. MAKE/BAKE vs. RUN/BAKE) (Meyer, Schaneveldt, & Ruddy, 1974; Hillinger, 1980; Jakimik, Cole & Rudnicky, 1985; Slowiaczek, Nusbaum, & Pisoni, 1987; and Emmorey, 1987.) However, the exact locus of these priming effects remain unclear. Two competing factors are implicated in phonological form based priming: First, the amount of phonological overlap (e.g. word pairs like BLAND/BLACK where three segments overlap are more likely to prime than BLEED/BLACK where only two segments overlap). Second, the structure of the overlap (e.g. words pairs like BA.LOON/SA.LOON, where final syllables overlap, are more likely to prime than BREA.KING/SMI.LING, where the final syllables are phonetically different). Unfortunately in past studies these two factors, amount of phonological overlap and the structure of the overlap, have been confounded. The present study was designed to disentangle the effects of shared segmental and syllabic overlap by directly pitting segments and syllables against one another while holding constant the absolute number of shared phonemes.

Experiment 1.
Syllable vs. Segment Priming

To explore the effects of shared syllabic and segmental overlap, we compare priming effects in two groups of words; word pairs which share syllabic overlap (either an initial syllable; PAM.PER/PAM.PHLET, or a final syllable; DU.RESS/CA.RESS) and words which share only segmental overlap (e.g. STACK/STAB and BLIS.TER/BLIZ.ZARD). In all cases phonological overlap is approximately three phonemes.

Comparing magnitude of priming for these groups of words permits us to systematically factor out priming effects arising from the quantity of segmental overlap from priming arising from the structure (i.e., syllable structure) of this overlap. Specifically, if it is simply the amount of shared segmental overlap which determines phonological form-based priming, then monosyllabic words such as STACK/STAB should show greater priming than the bisyllabic words which share a syllable (PAM.PER/PAM.PHLET & DU.RESS/CA.RESS). Note that in monosyllabic words, approximately 3/4 of the segments are identical whereas in the bisyllabic words only approximately 1/2 of their total segments overlap. If on the other hand it is the structure of the overlap which is important in phonological priming, we expect greater priming for bisyllabic words which share an initial or final syllable relative to monosyllabic words which do not share syllabic overlap. Importantly, in each case, the amount of segmental overlap is approximately three phonemes. Finally in this last comparison there is a possible source of confound. Specifically, if greater priming is found for the bisyllabic words relative to the monosyllabic words, we cannot be sure the whether observed priming is uniquely attributable to the shared syllables or rather some independent property of bisyllabic words, such as sheer acoustic duration. That is, since bisyllabic words are acoustically longer they might show more robust priming than the shorter monosyllabic words. To control for this confound, we include bisyllabic words which share only a consonant cluster and a vowel but not an entire syllable (e.g. BLIS.TER/BLIZ.ZARD). The inclusion of these word forms will permit examination of the acoustic duration factor in the present experiment. To summarize, this experiment is designed to determine whether the amount of segmental overlap or the structure of the overlap (i.e. syllable structure) is critical in phonological form-based priming. We evaluate phonological priming using an auditory lexical decision paradigm.

Method. Eighteen college students listened to auditorily presented word pairs with an I.S.I. of 100 msec. in an acoustically controlled room. Subjects were to decide the lexical status of the second word of each pair. Subjects pressed one of two computer keys to indicate their choice. Reaction time to respond measured from the end of the second word constituted the dependent variable. For each condition stimuli lists consisted of 18 related-pair trials, 18 unrelated-pair trials and 18 filler items. To determine if priming effects are present reaction times to related-pair trials (e g. STACK/STAB) are compared to unrelated-pair trials (e.g. TRIM/STAB).

Shared Syllables		Examples
Initial	35.7*	PAM.PER/PAM.PHLET
Final	60.0*	DU.RESS/CA.RESS
Shared Phonemes		Examples
Monosyllabic	59.1*	STACK/STAB
Bisyllabic	- 6.1	BLIZ.ZARD/BLIS.TER

Priming in msec. (* $p < .01$)
Table 1

Results. The results presented in Table 1 indicate significant priming for words which shared syllable overlap (both initial and final). In addition, segment priming was found only for monosyllabic words, bisyllabic words which shared an initial consonant cluster and a vowel (but not an entire syllable) did not show priming. The results are consistent with a spreading activation model of lexical access in which both segments and syllables are overtly represented. Moreover the data suggest that higher order syllable representations permit propagation of activation over time. Note for example, we find significant priming for bisyllabic words which share an initial syllable (PAM.PER/PAM.PHLET) but not for bisyllabic words which overlap only in segments (BLIS.TER/BLIZ.ZARD). This result indicates that the syllable has a status independent of the absolute number of shared phonemes. Two important findings emerge from this study: 1) The results support a model of lexical representation in which syllable structure is overtly represented in the lexicon. 2) The results reveal that the effects of shared syllable overlap are separable from shared segmental overlap. However questions remain as to the exact nature of these syllable representations.

Experiment 2. Underlying vs. Surface Syllable Priming

We may make a distinction between surface syllabification and underlying syllabification. Consider for example the word "pony". In slow careful speech speakers syllabify this word as /po.ni/, reflecting perhaps the underlying syllable boundaries. However in fast, everyday speech, we might represent the syllabification as /pon.ni/ where the /n/ appears to be a member of both first and last syllable. This parse is a reflection of surface level syllable structure. Recent work in linguistic theory has greatly elaborated the differences between underlying syllable structure and surface syllable structure. Two differences which factor the construction of the stimuli used in the present experiment include: 1) scope of syllabification and 2) the degree of internal constituency. These differences provide a basis for determining whether the syllable priming observed in Experiment 1 is a reflection of surface or underlying syllable representations.

The scope of syllabification differs for underlying and surface syllables. It has been argued that the domain of syllabification for underlying syllables is limited, whereas surface syllabification is considered exhaustive. The "smaller" underlying syllable serves as a constraint which interacts with word formation processes (Borowsky, 1989). A second difference concerns the presence or absence of internal syllable constituency. Internal constituency refers to hypothetical sub-units which comprise the syllable, these include the onset, nucleus, rime, and coda. Their is considerable evidence from speech production and speech planning for the importance of these constituents in surface syllables (Fromkin 1971, Stemberger 1985; Yaniv, Meyer, Gordon, Huff, & Sevald, 1990). However there is far less evidence for syllable constituency in *underlying* syllable representations (Clements & Keyser 1983).

With these two differences in mind we can construct stimuli sets which contrast surface and underlying syllable structures. Table 2 illustrates underlying and surface representations of the word pairs FAKE/FATE and FAKE/BAKE. In monosyllabic words with long vowels, final segments are extrametrical at the level of underlying syllable structure (Meyers, 1987; Borowsky, 1989). In the diagrams, long vowels are represented as a double occurrence of /a/ in keeping with current phonological theory. Note that in the underlying representations, only the initial consonants and the vowel fall under the scope of the syllable, final consonants are excluded from this domain on the basis of extrametricality (hence marked "ex." in the diagram).

Underlying:

```
σ          σ
/\         /\
fa a  k    f a a  t
       \           \
        ex.         ex.
```

Underlying:

```
σ          σ
/\         /\
fa a  k    b a a  k
       \           \
        ex.         ex.
```

Surface:

```
σ          σ
/\         /\
O R        O R
|  |\      |  |\
|  N C     |  N C
|  /\ |    |  /\ |
f  a a k   f  a a t
```

Surface:

```
σ          σ
/\         /\
O R        O R
|  |\      |  |\
|  N C     |  N C
|  /\ |    |  /\ |
f  a a k   b  a a k
```

Table 2

Extrametricality expresses the general tendency for domain-peripheral elements to be skipped over by rules sensitive to metrical structure (Hayes, 1982). While originally conceived to aid in the description of stress systems, more recently extrametricality has been shown to interact with the realization of segmental content at the syllable level. As stated above the domain of underlying syllables structure is limited, extrametricality is one formal device for expressing limitations on underlying syllable structure. Given these underlying representations, the words FATE and FAKE will share the underlying syllable /faa/, whereas the words FAKE and BAKE differ in their initial underlying syllables (e.g. /faa/ and /baa/ respectively).

In contrast, in the surface representations all phonemes are exhaustively syllabified and internal constituency is represented. (e.g. onset (O), rime (R) nucleus (N) and coda (C). While neither word pair shares and entire syllable, the word pair FAKE/BAKE, do share the internal rime constituent /aak/, contrariwise the pairs FATE/FAKE lack this relationship.

We predict that if underlying syllabic structure is being primed then the word pairs sharing an entire underlying syllable (e.g., FAKE/FATE) should show a priming effect greater than that of the words which only share partial segmental overlap (e.g., FAKE/BAKE). On the other hand, if surface level syllabification is contributing to priming, we predict that the words FAKE and BAKE may show greater priming due to their shared surface syllable constituency. In summary, comparing priming in words sets which share these characteristics (shared underlying syllables versus shared surface syllable constituents) we may assess whether syllable priming observed in Experiment 1 owes to the activation of surface level or underlying syllable representations.

Method. The data set consisted of ten monosyllabic target words. Three different primes were constructed for each target word. In one case the primes shared underlying syllabic structure (e.g., FATE/ FAKE) in another case, the primes were rhyming pairs sharing surface syllable constituents (e.g., BAKE/FAKE). These two classes of relatedness were compared to phonologically unrelated primes (e.g., WIPE/FAKE). Method and subjects were the same as in Experiment 1.

Underlying Syllable	28 (n.s.)	FATE/FAKE
Surface Syllable	54*	BAKE/FAKE

Priming in msec. (*$p < .02$)

Table 3

Results. The results shown in table 3 reveal that words which share a surface syllable constituent (i.e. rime) showed significant priming effects, while words which shared underlying syllabic structure did not. These findings suggest that the locus of syllabic priming observed in lexical decision experiments derives from surface rather than an abstract underlying syllabic structure.

Discussion These two experiments argue for a model of lexical representation in which surface syllable structure is overtly represented. These findings raise questions concerning the origin of syllable structure in lexical representations. This issue was explored in a simulation which examined whether syllable like representations could be derived from surface level phonetic input.

Simulation 1.
Induction of Syllable Structure.

The present simulation examines whether syllable-like structure is derivable from naturalistic, phonetically transcribed speech. The simulation uses as input a large data base constructed for studies of automatic speech recognition and uses a neural network to predict structural regularities in the data base. The simulation is highly successful in illustrating the extraction of syllable-like structure from a natural language corpus.

Data The data for the simulation is a subset of the TIMIT data base (Zue, Seneef, & Glass, 1990). The data used in the simulation is derived from the phonetic transcription of the TIMIT sentences provided with the data base. The coding scheme identifies 62 distinct speech sounds and includes demarcations of pauses and ends of sentences. A subset of entire data base was used for the simulation. Ten sentences from 77 randomly chosen male speakers were used, yielding a total of 770 sentences. The phonetic transcription of these sentences was concatenated and arranged sequentially, one phonetic label to a line. All sentence boundary information and pauses were removed from the data set. This yielded a total data set of 27,689 phonetic labels. Importantly, the input data was continuous, no information about word or syllable boundary information is represented in the data set.

Method. A sequentially recurrent network was used in a prediction task as outlined in Elman (1989a). In this case, a sequential network's task is to take successive phonemes from the input and to predict the subsequent phonemes on the output layer. After each phoneme was input, the output was compared with the actual next phoneme, and the back propagation of error learning algorithm (Rumelhart, Hinton, & Williams, 1986) was used to adjust the weights. Localist encodings of the 59 phonetic labels were presented in order with no breaks between words or syllables.

The network (shown in figure 1) consisted of 299 nodes configured to accept 59 inputs and 59 outputs Input was fed to an intermediate layer of 30 units, which in turn was passed to a recurrent layer with 90 units. This was fed to another layer of 30 units and finally back out to the 59 unit output layer. The network was trained through 15 passes through the corpus yielding a total summed squared error of .9278.

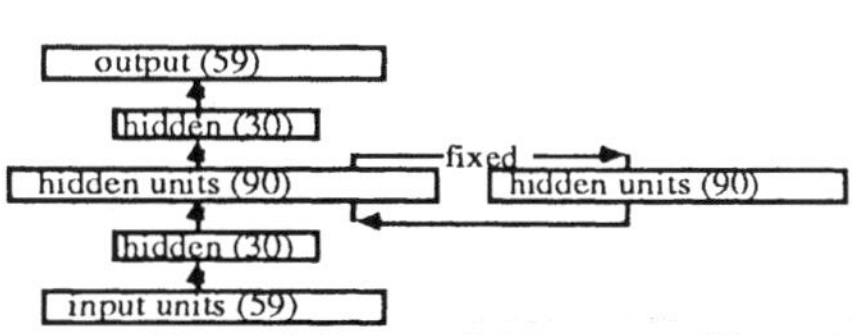

Figure 1. Sequential Recurrent Network

As discussed in Elman (1989), the prediction task is non-deterministic, and short of memorizing the sequences, the network cannot succeed in exact predictions. It is only by virtue of inherent underlying regularities in the corpus that such predictions can be made. Thus, the prediction task provides an avenue by which to discover the regularities of the data set. . The question of interest here is the extent to which this structure may correspond to canonical syllable structure.

Results. To examine the network's success in discovering syllable structure, the extant structure of the input was compared to that predicted by the network. To make this comparison, it was necessary to determine from the *input* data an averaged representation of syllable structure. Recall that the input to the network

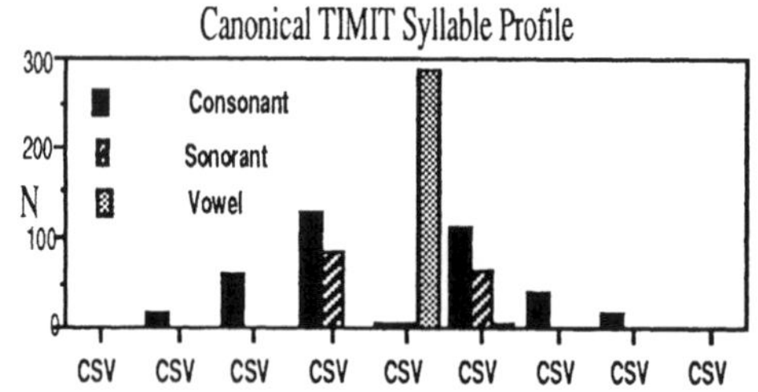

Figure 2a. TIMIT Syllable Profile

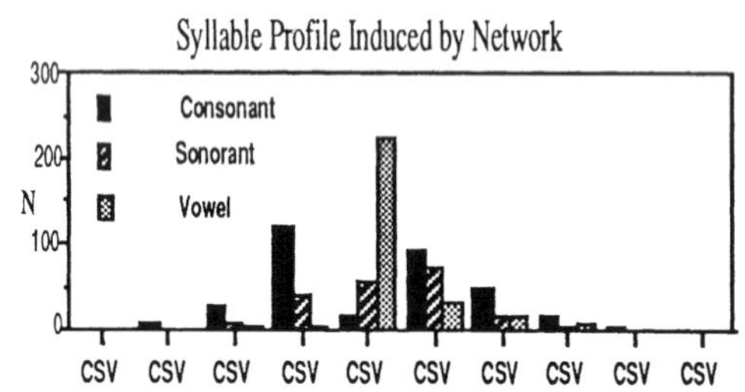

Figure 2b. Network output.

had no demarcation of syllable boundaries. Sampling from the beginning and end of the data base, 300 syllables were identified by the author by reference to the target words. Each phonetic transcription symbol was recoded as belonging to one of three classes of segments: consonants, sonorants, or vowels. The resulting syllables, ranged in length from one to nine segments in length. Next, these 300 syllables were "averaged" to yield the data base's canonical syllable profile. This consisted of determining, for each syllable, the nucleus (typically the vowel) and aligning all 300 syllables according to their nuclei. Finally, counting the number of segment types to the left and right of the nuclei relative to the three classes of segments, provides a "histogram" of canonical syllable preference for the TIMIT data base.

Figure 2a illustrates the canonical TIMIT syllable profile. The graph shows the number of times a particular segment type was found in a position relative to the nucleus. The graph illustrates the preference for English syllables to have a central vocalic nucleus, and the immediate left of the nucleus to be either a consonant or a sonorant. As one moves further from this position, typically only consonants are found. A near reverse profile is found to the right of the nucleus. The diminishing frequency of complex onsets and codas simply corresponds to the fact that CVC words are more common than CCSVSCC words. The profile depicted in figure 2a provides reference template for examining the results of the networks output.

To determine whether the *network* had discovered syllable-like patterns of regularity, averaged output patterns were compared to the canonical syllable template determined from the input data.

The phonetic labels of the output sequences were translated into corresponding consonant, vowel, and sonorant labels. The lengths of the input syllables served as guides to parse the output data. This provided a conservative method for determining output groupings. These sequence groupings were "averaged" in an identical fashion to the input sequences. Specifically, the most sonorant segment of a sequence was considered the nucleus of the grouping. Aligning all 300 output groupings by nuclei, we tally the frequency with which a consonant, sonorant or vowel was associated with a given position. The results are shown in Figure 2b. The graph shows a striking similarity to the graph of the input data. As one moves immediately to the right or left of the nucleus position, one finds that consonants and sonorants are favored, in a proportion which looks very similar to that observed of the input data. As one moves outward from this position, a greater percentage of consonants are found. The resulting "template" is broader than the input template, due to the conservative method of determining nucleus position in the parsed output groupings. In addition, we find vowels in positions beyond those of the input data, again an artifact of the conservative scoring method. Taken together, these facts suggest that an even tighter syllable template could be produced from a less conservative method of analysis. However based on this analysis, it appears that the network, through prediction, is able to extract regularities which bear remarkable similarity to syllable structure regularities observed in the input data. It is important to emphasize that the input data was continuous, with no demarcation of syllable or word boundaries yet the inherent regularities of segment position was sufficient for a network to induce syllable-like structure from exposure to positive instances of data.

Summary. We have examined the ability of a network to extract regularity from a phonetically coded English language data base. The output data revealed that the network was able to extract predictable structure which corresponded, in a striking manner, to syllable structure. The simulation provides extremely strong evidence for the ability to extract syllable structure from positive instances of phonetically labeled data in a natural language data base.

Discussion

Data from two experiments was presented which revealed that syllable structure is used in lexical access. Importantly the data support a model of lexical representation in which facts about surface syllable structure, rather than an abstract underlying syllable structure, is represented. This finding raises questions concerning the origin of this knowledge. A connectionist simulation was presented which demonstrated that a simple learning mechanism, when exposed to a natural language data base, was successful in uncovering syllable-like information. This simulation in part mirrors the child's experience with language. Specifically, we observe that the structure is implicit in the corpus and also in the input to the child. The child's task, like that of the network, is to discover these inherent regularities. However for the model, the input data, while continuous, is nevertheless segmented into phonetic labels. This segmentation gives the model an obvious head start which is not available to the child. Recent work using speech spectrograms as input to a recurrent connectionist network has demonstrated some success in segmentation of speech. (Doutriaux & Zipser 1990; see also Elman 1989b). We may conjecture that knowledge induced from inherent regularities in language provides one basis of organization for the mental lexicon and this knowledge serves in the recognition of words. Taken together these results suggest that the knowledge of syllable structure that a speaker actively utilizes in understanding language may be derived from that speaker's experience with the language. These results have important implications for models of lexical representation.

Acknowledgments

This work was supported in part by Andrus Gerontology Center Grant NIA 5T32AG 00037. I would like to thank Jeff Elman, Liz Bates and Mark Seidenberg for helpful comments on various versions of this work. Alan Petersen and Kim Daugherty also provided valuable input.

References

Borowsky, T. 1989. Structure preservation and the syllable coda in English. *Natural Language and Linguistic Theory*, 7(2):145-166.

Doutriaux, A & Zipser, D. 1990. Unsupervised Discovery of Speech Segments Using Recurrent Networks. In *Connectionist Models, Proceedings of the 1990 Summer School*; D. Touretzky, J.L. Elman, T.J. Sejnowski, and G.E. Hinton Eds. Morgan Kaufmann, San Mateo, CA.

Elman, J. L. 1989a. Representation and structure in connectionist models. CRL Technical Report, 8903, Center for Research on Language, Univ. of California, San Diego.

Elman, J. L. 1989b. Connectionist Approaches to Acoustic/Phonetic Processing. In William Marslen-Wilson (ed.) *Lexical representation and process* MIT Press/Bradford Books, Cambridge MA.

Emmorey, K. D. 1987. Morphological structure and parsing in the lexicon. Ph.D. diss., Univ. of California, Los Angeles, CA.

Hayes, B. 1982. Extrametricality and English Stress. *Linguistic Inquiry*, 13(2):227-276.

Hillinger, M. L. 1980. Priming effects with phonemically similar words: The encoding-bias hypothesis reconsidered. *Memory and Cognition*, 8(2):115-123.

Jakimik, J., Cole, R. A., & Rudinicky, A. I. 1985. Sound and spelling in spoken word recognition. *Memory and Language*, 24:165-178.

Meyer, D. E., Schvanevldt, R .W., & Ruddy, M. G. 1974. Function of graphemic and phonemic codes in visual word recognition. *Memory and Cognition*, 2:309-321.

Meyers, S. 1987. Vowel Shortening in English. *Natural Language and Linguistic Theory*, 5(4):485-518.

Neely, J. H. 1977. Semantic priming and retrieval from lexical memory: Roles of inhibition less spreading activation and limited-capacity attention. *Journal of Experimental Psychology: General*, 106:226-254.

Rumelhart, D. E., Hinton, G., & Williams, R. J. 1986. Learning internal representations by error propagation. In D. E. Rumelhart, J. L. McClelland, & the PDP Research Group (Eds.), *Parallel distributed processing: Explorations in the microstructure of Cognition. Volume 1: Foundations.* Cambridge, MA: MIT Press/Bradford Books.

Slowiaczek, L., Nusbaum, H. C., & Pisoni, D. 1987. Phonological priming in auditory word recognition. *Journal of Experimental Psychology: Learning Memory and Cognition*, 13(1):64-75

Slowiaczek, L., & Pisoni, D. 1986. Effects of phonological similarity in auditory word recognition. *Memory and Cognition*, 14:230-237.

Yaniv, I., Meyer, D., Gordon, P., Huff, C. A., & Sevald, C. A. 1990. Vowel similarity, connectionist models, and syllable structure in motor programming of speech. *Journal of Memory and Language*, 29:1-26.

Zue, V., Seneef, S., & Glass, J. 1990. Speech database at MIT: TIMIT and Beyond. *Speech Communication*, 9:351-356.

An empirically based computationally tractable dialogue model

Nils Dahlbäck and Arne Jönsson
Department of Computer and Information Science
Linköping University, S- 581 83 LINKÖPING, SWEDEN
nda@ida.liu.se, arj@ida.liu.se

Abstract

We describe an empirically based approach to the computational management of dialogues. It is based on an explicit theoretically motivated position regarding the status of computational models, where it is claimed that computational models of discourse can *only* be about computers' processing of language. The dialogue model is based on an extensive analysis of collected dialogues from various application domains. Issues concerning computational tractability has also been decisive for its development. It is concluded that a simple dialogue grammar based model is sufficient for the management of dialogues with natural language interfaces. We also describe the grammar used by the dialogue manager for a Natural Language interface for a database system.

Introduction

Most, if not all, work on dialogues in present- day computational linguistics do not make explicit to which extent the models and theories developed should be seen as theories about the processing of dialogue by computers or people or both. Though never explicitly stated, the underlying assumption seem to be that the theories are to be general theories of discourse for all kinds of agents and situations. There are, however, a number of reasons for assuming that the cognitive architecture of present day computers and people are sufficiently different to make it necessary to clarify to which extent a computational theory of discourse (or any other cognitive phenomenon, for that matter) is primarily to be seen as a psychological account or an account of computer's processing of discourse. This is not only true for those that are critical to the computational theory of mind, but also for the defenders of that view (cf. Pylyshyn, 1984). It is thus, in a sense, an uncontroversial position. But what is perhaps less so, is the consequences that we claim of necessity follows from it.

As far as the internal, or representational, aspect is concerned, we want to claim that procedural computational accounts of the process of discourse using concepts from present day computer technology cannot be seen as a psychological account. "Two programs can be thought of as strongly equivalent or as different realizations of the same algorithm or the same cognitive process if they can be represented by the same program in some theoretically specified virtual machine" Pylyshyn (1984, p 91). A consequence of this is that "*any* notion of equivalence stronger than weak equivalence[1] must presuppose an underlying functional architecture, or at least some aspects of such an architecture." (ibid , p 92) "Typical, commercial computers, however, are likely to have a far different functional architecture from that of the brain; hence, we would expect that, in constructing a computational model, the mental architecture must first be emulated (that is, itself modelled) before the mental algorithm can be implemented" (ibid., p 96).

There are some obvious consequences that follows from this. The most important is that most, if not all, present day theories in computational linguistics are about computer's processing of language, and nothing else. Why then, is this important? Because we know that language use is situation dependent. Content and form differs depending on the situation in which occurs (e.g. Levinson, 1981, 1983), but also depending on the perceived qualities of the interlocutors; language directed to children is different from language directed to grown-ups (Phillips, 1973, Snow, 1972), as is the case with talking to foreigners, brain-injured people, and people that do not know who John Lennon was. The ability to modify the language to the perceived needs of the speaker seem to be present already at the age of four (Shatz & Gelman, 1973).

One simple but important consequence of the position outlined above is therefore that goals of research on dialogue in computational linguistics such as "Getting computers to talk like you and me" (Reichman, 1985), or developing interfaces that will "allow the user to forget that he is questioning a machine" (Gal, 1988), are not only difficult to reach. They are misconceived. We always adapt to the qualities of our dialogue partner, and there is every reason to believe that NLI-users will adapt to the fact that they are interacting with a computer.

Another important consequence is that the lan-

This research was financed by the Swedish National Board for Technical Development (STU) and the Swedish Council for Research in the Humanities and Social Sciences (HSFR). The authors names are in alphabetical order.

[1] I.e. realizing the same input-output function (N.D. & A.J.)

guage samples used for providing the empirical ground of the computational theories should come from relevant application domains for such software technology. We are therefore advocating a sub-language approach (Grishman & Kittredge, 1986) to studies of dialogue in computational linguistics, where language samples used to develop, motivate or illustrate computational theories are taken from the relevant application domains.

Finally, since the functional architecture of man and machine are different, psychological realism on the representational level is of no interest here. We therefore argue for a position of 'representational agnosticism' (Dahlbäck, 1991b).

Previous Empirical Studies

An increasing number of researchers have acted on positions similar to the one outlined above (though not necessarily with similar explicit theoretical commitments), and there is accumulating evidence in support of the theoretical assumptions presented here. One important source of information has been the use of so-called Wizard-of-Oz investigations. (For reviews, see Dahlbäck 1991b, Jönsson & Dahlbäck, 1988, and Gilbert & Fraser, 1991). A number of linguistic differences between the language used when communicating with a computer and characterizations of human dialogues have been observed: The syntactic variation is limited (Reilly, 1987). The use of pronouns is rare (Guindon, 1988 and Dahlbäck & Jönsson, 1989, Kennedy et al. 1988) and the antecedent of a pronoun is mostly found in the immediate linguistic context (Dahlbäck & Jönsson, 1989). So-called 'ill-formed input' is very frequent (Grosz, 1977, Guindon et al., 1986). A limited vocabulary seem to be sufficient for communication in restricted domains (Malhotra, 1975). In our own work we have found that indirect speech acts are rare, lack of cue phrases, abrupt dropping of topics (which creates problems for plane-based models), frequent use of domain-specific conceptual relations and, most important for our present purposes, a dialogue structure which differs from the one often found in human dialogues.

Dialogue Management for Natural Language Interfaces

Managing the dialogue in an NLI can be performed in various ways. There are today two competing approaches to dialogue management. One is the plan based approach, i.e. to reason about the user's goals and intentions using plans describing the actions which may possibly be carried out in different situations (c.f Cohen & Perrault, 1979; Allen & Perrault 1980, Litman, 1985, Carberry, 1990). The other approach is to model speech act information in a dialogue grammar.

The plan based approach is mostly used in search for a general computational model of discourse. This is a more comprehensive goal than dialogue management for natural language interfaces. (For a survey of plan based approaches see Carberry, 1990.)

Central to the plan based approach is the recognition by the listeners of the speakers goals, where goals are modelled using plans. There exists, however, today no efficient plan recognition algorithm for general "STRIPS"-like planners. Attempts have been made by adding restrictions to plans to get them more tractable. Chapman (1987) was the first to present a plan generator that could be theoretically analysed. He presented a planning algorithm that subsumed most previous planners, for instance STRIPS. Chapman showed that, under certain conditions, planning is undecidable. Bäckström & Klein (1991) showed that it is not possible to construct a polynomial-time planning algorithm for a more restricted class of problems, the SAS-PU[2] class. Furthermore, the SAS-PU class is probably too restricted for practical use in natural language processing. However, it should be noted that recent results, (Bylander, 1991) regarding the problems to be solved by polynomial planners might be a bit more optimistic. Moreover, both Bylander and Bäckström & Klein state that a careful examination of the problem might provide a polynomial planner for some problem classes, but there seems to be no single domain-independent planning algorithm.

Vilain (1990) presents a parser that can recognize plans in polynomial time using Earley's algorithm (Earley, 1970). The plan formalism used by Vilain is developed by Kautz (Kautz, 1991). Kautz developed a plan recognition formalism for recognizing plans whose types appear in an event hierarchy. Thus, he uses more restricted plans than those proposed by Allen, Cohen & Perrault, where new plans can be recognized by chaining together the preconditions and effects of other plans. Kautz maintains this restriction because otherwise ".. it would lead to massive increase in the size of the search space, since an infinite number of plans could be constructed by chaining on preconditions and effects." (Kautz, 1991, p. 72). It seems therefore that plan recognition for natural language dialogue is exponential if it is based on the STRIPS-formalism.

Another reason for our doubt concerning the use of plan recognition for dialogue management in certain natural language interface applications is that in many situations it is overkill: the interaction between a human and a computer using written language through a terminal does not include all the many difficult phenomena that arrive in human-human interaction, c.f. the previous section. Furthermore, it is difficult to correctly describe the different goals and intentions that can be carried out

2. SAS is a simplified version of the action structures (Sandewall & Rönnquist, 1986) where the simplification reduces the parallelism that is modelled in the action structures and is thus similar in expressiveness to that of regular planners like STRIPS. P stands for post-unique which means that one action achieves only one effect in the world; U means that it is Unary, i.e. every operator has only one effect in the world.

in a dialogue situation (Guindon 1988).

The other approach when building a dialogue manager that can efficiently handle a limited set of dialogue features is to identify adjacency-pairs (Schegloff & Sacks, 1973) and to use a dialogue grammar (e.g. Reichman, 1985 , Polanyi & Scha, 1984, Frohlich & Luff, 1990, and Bilange, 1991). This approach has been criticised for not adequately describing a naturally occurring discourse (see for instance Levinson, 1983). However, for a restricted sublanguage, such as natural language communication with computers, we believe that this can be a very efficient way of managing the dialogue (cf. Levinson, 1981, p 114).

Our work differs, however, from previous proposed dialogue grammars. Reichman and Polanyi & Scha try to manage discourse in general. Thus, they need rules to cover a wide variety of phenomena that seldom occur in interface interactions. Frohlich & Luff also present a rich grammar, basing their menu-based natural language interface grammar on studies of human-human conversations. Problems with this approach is pointed out in Dahlbäck & Jönsson (1989).

Bilange designed his system for oral communication which suggests a number of interesting differences compared to typed dialogue; for instance, his need for elaboration as the third part of an adjacency-pair, i.e. he demonstrates that the structure negotiation-reaction-elaboration is very common in oral dialogue. Stubbs' (1983) model for human dialogues also includes a third confirmatory move. This pattern seems not to occur in written human-computer communication (Dahlbäck, 1991a, b).

As for dialogue grammars, one might ask whether they are also complex, requiring exponential algorithms for parsing? The reply is that if a dialogue grammar can be written using a context-free grammar, then there are well-known polynomial-time algorithms. The question then arises as to whether it is possible to write a context-free grammar for the dialogues that we are interested in?

The Empirical Study

The dialogue model is based on the analysis of a number of dialogues collected by the means of Wizard of Oz NLI-simulations[3]. We have used five different background systems, varying not only the content domain, but also the 'intelligence' of the systems, and the number and types of tasks possible to perform by the user. The most detailed analysis has been conducted on a corpus of 21 dialogues.

We have used two database systems. PUB is a library DB in use at our department. C-line is a simulated DB containing information about the computer science curriculum at Linköping University. In the HiFi-system the user can order HiFi-equipment after having queried a (simulated) DB containing information about the available equipment. The Travel system simulates an automated travel agency offering charter holidays to Greek islands. These systems differs from the two above in two respects; the system is more 'cognitively' advanced, and there are more actions that can be performed by the user, i.e. not only asking for information but also order something. The Wine system is a simulated advisory system, capable of suggesting suitable wines for different dishes, if necessary within a specific price range. (The experimental settings are described in more detail in Jönsson & Dahlbäck, 1988, Dahlbäck & Jönsson 1989, and Dahlbäck, 1991a, b)

The total number of dialogues is 21; PUB: 4, HiFi: 5, C-line: 5, Wine: 4, Travel: 3. The total number of utterances is 1055, where we count each turn by user or system as one utterance. This gives us an average of 50 utterances/dialogue. The longest are in the travel domain, where the average dialogue is 92 utterances long, and the shortest are the PUB dialogues with an average of 25 utterances. Apart from the dialogues analysed here, we have collected more than 60 others, using four other real or simulated background systems. Dahlbäck (1991b) describes some of these in more detail. A current project has collected another set of 60 dialogues, some of which are described below.

Analysis and Results

The dialogue structure is analysed using only two basic types of moves, initiatives (I) and responses (R). The definition of the categories is only based on local information. If the move is seen as introducing a goal it is scored as an initiative, if it is a goal-satisfying move, it is scored as a response. One important reason for this is that the categories are domain independent. We can therefore compare dialogues from different domains. Another advantage is that the categories are (fairly) simple to define and identify, making it possible to code the dialogues with sufficient inter-rater reliability (97%).

Discourse management moves such as ***Welcome to WingHolidays. What can we do for you?, Can I help you with anything more?*** and ***Bye*** etc. are all scored as initiatives. We subcategorize them as DO (discourse opening), DC (discourse continuation), and DE (discourse ending), to make it possible to exclude them from some of the analysis presented below. (Responses to these kind of initiatives are optional in the model).

Since we only used local information when ascribing a category to a move, we can get a measure of the structural complexity of the dialogues by analysing them using a simple dialogue tree model called LINDA.

[3]. The model is implemented as a module for the Swedish NLI developed in the LINLIN-project. Ahrenberg, Jönsson and Dahlbäck (1990) gives an overview of the project. Dahlbäck (1989, 1991a, 1991b), Dahlbäck and Jönsson (1989), Jönsson (1990), and Jönsson and Dahlbäck (1988) presents other aspects of the empirical issues. Further aspects of the implemented system can be found in Jönsson (1991a and 1991b)

(For LINköping DiAlogue, see Dahlbäck, 1991a, b for a detailed description) The model only accepts units consisting of an initiative followed by a response or embeddings of such units in higher IR-units, e.g. (I R), or successive and recursive embeddings such as (I (I R) R), (I (I R) (I R) R), or (I (I (I R) R) R) etc. All moves must belong to some discourse segment, and no segments with the structure (I I R) or (I R R) are allowed.

We find that 92% or more of the dialogues fit this structure, see Figure 1. Furthermore, the use of recursive embeddings is limited, as seen in the high number of adjacency pairs in the dialogues.

	LINDA model fit	Adjacency pairs
PUB	100%	75%
C-line	98%	96%
HiFi	99%	98%
Travel	99%	88%
Wines	92%	78%

Figure 1: LINDA model fit.

This does not mean that the dialogues consist of a sequence of isolated questions and answers, as there is frequent use of anaphoric expressions. In fact 49% of the *initiatives* contain some kind of anaphoric expression (Dahlbäck & Jönsson, 1989). What the figures show is rather that in spite of being clear cases of connected discourse, these dialogues have a much simpler structural complexity than most other genres. It thus seems as if most man-machine dialogues in natural language, even when no restrictions on the users' way of expressing themselves, lack most of the complexity found in other types of discourse. Our corpus is admittedly of a limited size, but it covers some of the most typical possible applications for NLI technology, and, apart from the advisory type of system, is not tied to one particular topic domain. Taken together, this gives us confidence in believing that the results have some generalizability

We have also found (Dahlbäck, 1991b) that the LINDA-structure can be used to direct the search of antecedents to anaphors. It is thus not only possible to describe the dialogues using the IR tree structure, but this structure can then be used to guide further processing of the dialogue.

The LINLIN Dialogue Manager

We have developed a dialogue manager based on the LINDA-model and in this section the dialogue grammar will be presented. However, there are some notions from the LINLIN-system that needs to be presented before we can present the dialogue grammar.

We refer to the constituents of a dialogue by the name of *dialogue objects*. The communication is hierarchically structured using three different categories of dialogue objects. There are various proposals as to the number of levels needed and they differ mainly on the modelling of complex units that consist of sequences of discourse segments, but do not comprise the whole dialogue. For instance the system developed by Polanyi & Scha (1984) uses five different levels to hierarchically structure a dialogue and LOKI (Wachtel, 1986) and SUNDIAL Bilange (1991) uses four.

The feature characterizing the intermediate level is that of having a common topic, i.e. an object whose properties are discussed over a sequence of exchanges. When analysing our dialogues we found no certain criteria concerning how to divide a dialogue into a set of exchanges. In fact, a sequence of segments may hang together in a number of different ways; e.g. by being about one object for which different properties are at issue. But it may also be the other way around, so that the same property is topical, while different objects are talked about. (This is discussed and illustrated in more detail in Ahrenberg, Jönsson and Dahlbäck (1990))

In our model the instances of dialogue objects form a dialogue tree which represent the dialogue as it develops in the interaction. The root category is called Dialogue (D), the intermediate category Initiative-Response (IR), and the smallest unit, the move.

An utterance can consist of more than one move and is thus regarded as a sequence of moves. A move object contains information about a move. They are categorized according to type of illocutionary act and topic. Some typical move types are: Question (Q), Assertion and declaration of intent (AS), Answer (A) and Directive (DI). Topic describes which knowledge source to consult — the background system, i.e. solving a task (T), the ongoing discourse (D) or the organisation of the background system (S). For brevity when we refer to a move with its associated topic, the move type is subscribed with topic, e.g. Q_T.

Following the LINDA-model, the only intermediate level consists of recursively embedded IR-units. The initiative can come from the system or the user. A typical IR-unit in a question-answer data base application is a task related question followed by a successful answer Q_T/A_T. Other typical IR-units are: Q_S/A_S for information about the system, Q_T/AS_S when the requested information is not in the data base, Q_D/A_D for questions about the ongoing dialogue, e.g. requests for clarification.

A Dialogue Grammar for the Cars Database

The dialogue manager is implemented for yet another dialogue domain; an existing INGRES-database containing information on used cars. To customize the dialogue manager to the new application, we ran a new set of Wizard of Oz-experiments. The number of dialogues is five and the average number of utterances per dialogue is 32.

The structural analysis has been carried out according to the principles described above. On the level of a move we have only identified two different illocutionary types: Question (Q) and Answer (A). The module responsible for translating the syntactic form of an utterance to these categories is called the instantiator (Ahrenberg, 1988). The instantiator will identify the illocutionary type of an utterance. So, for instance, the instantiator will interpret the utterance *Show data for Mercedes* as a request for information and it will thus categorize it as a question, although it's syntactic form is directive. The instantiator will not be considered further in this paper, a similar module for syntactic and semantic analysis is used by for instance Litman (1985, p 15) and Carberry (1990, p 75).

The resulting grammar is context free. It is very simple and consists merely of sequences of task-related questions followed by answers Q_T/A_T or in some cases an embedded reparation sequence Q_D/A_D, initiated by the system, see Figure 2.

$$D ::= IR^+$$
$$IR ::= Q_T/A_T \mid Q_x/A_S$$
$$Q_T/A_T ::= Q_T\,(Q_D/A_D)^*\,(A_T)$$
$$Q_D/A_D ::= Q_D\,(A_D)$$
$$Q_x/A_S ::= Q_T\,A_S \mid Q_S\,A_S \mid Q_D\,A_S$$

Figure 2. A dialogue grammar for the Cars application[4]

The grammar is not to be regarded as providing an accurate description of every database information retrieval application, but it will accurately describe the dialogue used by five different experimental subjects interacting with such a system in the domain of used cars.

The grammar presented here only shows two of the attributes of our dialogue objects. In fact, we use a number of descriptors with attributes describing for instance focused objects (Ahrenberg, Jönsson & Dahlbäck, 1990, and Jönsson, 1991a), but this does not affect the type of grammar. Furthermore, it is a simplified version of the grammar that is to be used in the Cars application. Acknowledgement phrases like *Wait...Searching* occur in all our dialogues, but they only serve to indicate that the user utterance is received. Thus they are omitted as they pose no interesting problems.

Summary

We have described a computational model of dialogue management for human-computer dialogues in natural language. The development is based on a sublanguage approach, on the belief that it is necessarily to distinguish between computational models for efficient processing of natural language and simulations of human processing of natural language, on the concern with computational tractability and empirical validity. The essential characteristics of the model is the use of a simple context-free dialogue grammar generating a dialogue structure of sequential and recursively embedded initiative-response (IR) units. It is not to be seen as a psychologically realistic cognitive model, but as a model that will successfully emulate human linguistic behaviour in the situations for which it is intended to be used, i.e. natural language interfaces.

4. The * is the closure operator meaning zero or more instances and the + is the positive closure denoting one or more instances. Parenthesis denote optionality and vertical bars denote disjunction.

Acknowledgements

This work on dialogue management in the LINLIN project is much inspired by the work that we have done in collaboration with Lars Ahrenberg, Åke Thurée and Mats Wirén.

References

Ahrenberg, Lars (1988) An Object-Oriented Dialogue System for Swedish, *Nordic Journal of Linguistics*, Vol. 11, Nos 1-2, pp 3-16

Ahrenberg, Lars, Arne Jönsson & Nils Dahlbäck (1990) Discourse Representation and Discourse Management for Natural Language Interfaces, *Proceedings of the Second Nordic Conference on Text Comprehension in Man and Machine*, Täby, Stockholm.

Allen, James. F. & C. Raymond Perrault (1980) Analyzing Intention in Utterances, *Artificial Intelligence*, 15, pp 143-178.

Bilange, Eric (1991) A Task Independent Oral Dialogue Model, *Proceedings of the Fifth Conference of the European Chapter of the Association for Computational Linguistics*, Berlin, 1991.

Brown, G & Yule, G. (1983) *Discourse Analysis*.- Cambridge: Cambridge University Press.

Bylander, Tom (1991) Complexity Results for Planning, *Proceedings of the Twelfth International Joint Conference on Artificial Intelligence*, Sydney, Australia

Bäckström, Christer & Inger Klein (1991) Parallel Non-Binary Planning in Polynomial Time, *Proceedings of the Twelfth International Joint Conference on Artificial Intelligence*, Sydney, Australia.

Carberry, Sandra (1990) *Plan Recognition in Natural Language Dialogue*, MIT Press, Cambridge, MA.

Chapman, David, (1987) Planning for conjunctive goals, *Artificial Intelligence*, **32**, pp 333-377.

Cohen, Philip. R. & C. Raymond Perrault (1979) Elements of a Plan-Based Theory of Speech Acts, *Cognitive Science*, **3**, pp 177-212.

Dahlbäck, Nils (1989) A Symbol is Not a Symbol. *Proceedings of the 11th Joint Conference on Artificial Intelligence (IJCAI'89)*Detroit, MIchigan, USA.

Dahlbäck, Nils (1991a) Empirical Analysis of a Discourse Model for Natural Language Interfaces, *Proceedings of the Thirteenth Annual Meeting of The Cognitive Science Society,* Chicago, Illinois.

Dahlbäck, Nils (1991b) *Representations of Discourse. Cognitive and Computational Aspects.* Ph.D. Thesis, Linköping University. Sweden.

Dahlbäck, Nils & Arne Jönsson (1989) Empirical Studies of Discourse Representations for Natural Language Interfaces, *Proceedings of the Fourth Conference of the European Chapter of the ACL,* Manchester.

Earley, Jay (1970) An Efficient Context-free Parsing Algorithm, *Communications of the ACM,* Vol. 13, No 2.

Frohlich, David & Paul Luff (1990) Applying the Technology of Conversation to the Technology for Conversation, In Luff, Gilbert & Frohlich (Eds) *Computers and Conversation,* Academic Press.

Fraser, N & Gilbert, N.S. (1991) Simulating Speech Systems. *Computer Speech and Language,***5**, pp 81-99.

Gal, A.(1988) *Cooperative Responses in Deductive Databases* Ph.D. Thesis. University of Maryland, College Park.

Grishman, R. & Kittredge, Ralph (Eds.) 1986. *Analysing language in restricted domains.* Lawrence Erlbaum.

Grosz, Barbara J. (1977) The representation and use of focus in dialogue understanding. Unpublished Ph.D. thesis. University of California, Berkeley.

Guindon, Raymonde (1991) Users Request Help from Advisory Systems With Simple and Restricted Language: Effects of Real-Time Constraints and Limited Shared Context, *Human Computer Interaction,* Vol. 6, pp 47-55.

Guindon, Raymonde (1988) A Multidisciplinary Perspective on Dialogue Structure in User-Advisory Dialogues, In Guindon (Ed) *Cognitive Science and Its Applications For Human-Computer Interaction,* Lawrence Erlbaum.

Guindon, R. Sladky, P., Brunner, H. & Conner, J. (1986) The structure of user-advisor dialogues: Is there method in their madness? *Proceedings of the 24th Conference of the Association for Computational Linguistics.*

Jönsson, Arne (1991a) A Dialogue Manager Using Initiative-Response Units and Distributed Control, *Proceedings of the Fifth Conference of the European Chapter of the Association for Computational Linguistics,* Berlin.

Jönsson, Arne (1991b) A Natural Language Shell and Tools for Customizing the Dialogue in Natural Language Interfaces. Internal Report, LiTH-IDA-R-91-10.

Jönsson, Arne (1990) Application-Dependent Discourse Management for Natural Language Interfaces: An Empirical Investigation, *Papers from the Seventh Scandinavian Conference of Computational Linguistics*, Reykjavik, Iceland.

Jönsson, Arne & Nils Dahlbäck (1988) Talking to a Computer is not Like Talking to Your Best Friend. *Proceedings of The first Scandinavian Conference on Artificial Intelligence*, Tromsø, Norway.

Kautz, Henry A. (1991) A Formal Theory of Plan Recognition and its Implementation, In: James F. Allen, Henry A. Kautz, Richard N. Pelavin & Josh D. Tenenberg (Eds.), *Reasoning About Plans*, pp. 69-125, Morgan Kaufmann.

Levinson, Stephen C. (1981) Some Pre-Observations on the Modelling of Dialogue, *Discourse Processes,* No 4, pp 93-116.

Levinson, Stephen C. (1983) *Pragmatics.* Cambridge University Press.

Litman, Diane J. (1985) Plan Recognition and Discourse Analysis: An Integrated Approach for Understanding Dialogues, PhD Thesis, TR 170, University of Rochester, Rochester, NY

Malhotra, A.(1975) Knowledge-based English Language Systems for Management Support: An Analysis of Requirements. *Proceedings IJCAI'75*

Perrault, C. Raymond & James F. Allen (1980) A Plan-Based Analysis of Indirect Speech Acts, *American Journal of Computational Linguistics*, Vol. 6, No 3-4.

Philips, J.R. (1973) Syntax and vocabulary of mothers' speech to young children: Age and sex comparisons. *Child development* **44** 182-185.

Polanyi, Livia & Remko Scha (1984) A Syntactic Approach to Discourse Semantics, *Proceedings of COLING'84,* Stanford.

Pylyshyn, Z. (1984) *Computation and Cognition* Cambridge, MA: The MIT Press.

Reichman, Rachel (1985) *Getting Computers to Talk Like You and Me,* MIT Press, Cambridge, MA.

Reilly, Ronan G. (1987) Ill-formedness and miscommunication in person-machine dialogue, *Information and software technology,* Vol. 29, No 2.

Sandewall, Erik & Ralph Rönnquist (1986) A Representation of Action Structures, *Proceedings of the Fifth National Conference on Artificial Intelligence*, Philadelphia.

Schegloff, Emanuel, A. & Harvey Sacks (1973) Opening up closings, *Semiotica,* **7**, pp 289-327.

Shatz, M. & Gelman, R. The development of communication skills: Modifications in the speech of young children as a function of listener. *Monographs of the Society for research in child development* **38**, No 152.

Snow, C. (1972) Mothers' speech to children learning language.*Child Development,* **4**, 1-22.

Stubbs, M (1983) *Discourse Analysis* Oxford: Blackwell.

Vilain, Marc (1990) Getting Serious about Parsing Plans: a Grammatical Analysis of Plan Recognition, *Proceedings of the AAAI,* Boston.

Wachtel, Tom (1986) Pragmatic sensitivity in NL interfaces and the structure of conversation. *Proceedings of the 11th International Conference of Computational Linguistics*, University of Bonn, pp. 35-42.

Learning Context-free Grammars: Capabilities and Limitations of a Recurrent Neural Network with an External Stack Memory

Sreerupa Das
Department of Computer Science
University of Colorado
Boulder, CO 80309
rupa@cs.colorado.edu

C. Lee Giles
NEC Research Institute
4 Independence Way
Princeton, NJ 08540
giles@research.nec.nj.com

Guo-Zheng Sun
Institute for Advanced Computer Studies
University of Maryland
College Park, MD 20742
sun@sunext.umiacs.umd.edu

Abstract

This work describes an approach for inferring Deterministic Context-free (DCF) Grammars in a Connectionist paradigm using a *Recurrent Neural Network Pushdown Automaton (NNPDA)*. The NNPDA consists of a recurrent neural network connected to an external stack memory through a common error function. We show that the NNPDA is able to learn the dynamics of an underlying pushdown automaton from examples of grammatical and non-grammatical strings. Not only does the network learn the state transitions in the automaton, it also learns the actions required to control the stack. In order to use continuous optimization methods, we develop an analog stack which reverts to a discrete stack by quantization of all activations, after the network has learned the transition rules and stack actions. We further show an enhancement of the network's learning capabilities by providing hints. In addition, an initial comparative study of simulations with *first, second* and *third* order recurrent networks has shown that the increased degree of freedom in a higher order networks improve generalization bu not necessarily learning speed.

Introduction

Considerable interest has been shown in language inference using neural networks. (For more traditional approaches to inference of grammars see [Miclet 90].) Recurrent networks in particular, with various training algorithms, have proved successful in learning regular languages, the simplest in the Chomsky hierarchy. Work by [Elman 90], [Giles 90], [Mozer 90], [Pollack 91], [Servan-Schreiber 91], [Watrous 92], and [Williams 89] have demonstrated that the recurrent nature of these networks is able to capture the dynamics of the underlying computation automaton. [Giles 92a] and [Watrous 92] have used higher order (higher dimensional weights) recurrent neural networks with no hidden layer and showed that such models are capable of learning state machines and appear to be at least as powerful as any multilayer network. Using a heuristic clustering method, [Giles 92a] showed that finite state automata could be extracted from the neural networks both during and after training. [Giles 92b] successfully demonstrated a method for learning an *unknown* grammar.

This work is concerned with inference of DCF grammars - moving up the Chomsky hierarchy. This recurrent neural network model, previously described by [Sun 90] and [Giles 90], has an external stack memory integrated through a hybrid error function, hence making it powerful enough to learn DCF grammars. Previous work by [Williams 89] showed that, given both the training set and action information of the read/write head of a Turing Machine, a recurrent network is capable of learning the finite state machine part of the Turing Machine that recognizes the training set. The model described here learns both the stack control (*push*ing and *pop*ing of the stack) and the state transitions of the underlying finite state automaton of the pushdown automaton. This is performed by extracting information only from the training data. The learning capabilities of the inferred Pushdown Automaton is enhanced by providing more information, *hints*, about the training strings. For other work on the use of recurrent neural networks for DCF inference, see [Allen 90] and [Pollack 90].

The stack is *external* and *continuous*. The reason for using an external stack, as opposed to an internal one, [Pollack 90], is that the external stack requires lesser resources for training. The continuous part permits the use of a continuous optimization method, in our case gradient-descent. We present a brief description of the model, discuss the dynamics of the stack action and give simulation results of learning performance.

Neural Network Pushdown Automaton (NNPDA)

The network consists of a set of fully recurrent neurons, called *State Neurons* which represent the states and permit classification and training of the NNPDA. One of the state neurons is designated as the *Output Neuron*. The *State Neurons* get input (at every time step) from three sources, namely, from their own re-

a	.4
b	.5
c	.8
..	..

Table 1: *Left column indicates the content of the stack; Right column indicates the quantity of each alphabet on stack. Top of the stack is a.*

c	.6
a	.4
b	.5
c	.8
..	..

Table 2: *After pushing* 0.6 *of* c *onto stack shown in Table 1.*

a	.1
b	.5
c	.8
..	..

Table 3: *After poping* 0.9 *from the stack in Table 2.*

current connections, from the *Input Neurons* and from the *Read Neurons*. The *Input Neurons* register external inputs to the system. These external inputs consist of sequences of characters of strings fed in one character at a time. The *Read Neurons* keep track of the symbol(s) on top of the stack. One non-recurrent neuron, called the *Action Neuron* indicates the stack action (push, pop or no-op) at any instance. The continuous valued activation of this neuron is used to perform analog actions (namely push and pop) on the stack. The architecture of the Neural Network is shown in *Figure 1*.

Many appropriate error functions could be devised. The one we chose to train the network consists of two error functions: one for legal strings and the other for illegal strings. For legal strings we require 1). the NNPDA must reach a final state and 2). the stack must be empty. This criterion can be reached by minimizing the error:

$$Error = 1/2[(1 - S_o(l))^2 + L(l)^2] \quad (1)$$

where $S_o(l)$ is the activation of an $OutputNeuron$ with its target value for legal strings as 1.0 and $L(l)$ is the stack length, all after a string of length l has been presented as input a character at a time. For illegal strings, the error function is modified as:

$$Error = S_o(l) - L(l) \quad if\ (S_o(l) - L(l)) > 0.0 \quad (2)$$

otherwise $Error = 0.0$. Equation (2) reflects the criterion that, for an illegal pattern we require either the final state $S_o(l) = 0.0$ or the stack length $L(l)$ to be greater than 1.0.

Stack Control

The analog stack is external to the network and is manipulated by the action neuron with continuous activation values. Since the activation of the action neuron is continuous valued, the pushing and popping is also continuous. Associated with each element on the stack is an analog value. An example of the stack would be the one shown in *Table 1*. It has 0.4 of *a* stacked over 0.5 of *b* and so on. Operations on the stack are determined by the activation of *Action Neuron*, S_a. The value of S_a is allowed to vary between +1 and −1. The operations will be described as follows:

PUSH: If the activation of *ActionNeuron*, S_a is significantly positive the action taken is *push*. In our simulations we performed *push* when the magnitude of $S_a > 0.1$. In case of *push* the current input is pushed on the stack and its value is determined by the magnitude of the activation of *ActionNeuron*. Therefore, for the stack shown in *Table 1*, $S_a = 0.6$ and the current input is *c*, then, after the operation, the stack would appear as shown in *Table 2*.

POP: If activation of *ActionNeuron* is sufficiently negative, the action taken is *pop*. In this case, quantities stored on the stack are removed up to a depth denoted by the magnitude of S_a. Therefore, for the stack in *Table 2* and $S_a = -0.9$, after the *pop* operation stack would appear as shown in *Table 3*. For our simulations we performed *pop* if $S_a < -0.1$.

READING from the stack: At every time step (or with processing of every element of the input string), the information on *top of the stack* has to be updated every time an action is taken. This is done as follows. All the elements on the top of the stack up to a *depth* of 1.0 (i.e., all the symbols whose quantities add up to 1.0 from the top) are considered. Then their individual quantities on the stack are used as the corresponding activations of the *Read Neurons* in the next time step. For example, the *Read* information of the stack shown in *Table 3* would be $R_a = 0.1; R_b = 0.5, R_c = 0.4$ if we consider only three input symbols. It should be noted that our goal is to train the network to take the correct actions, and as training proceeds all magnitudes of S_a should approach 1 or 0. Hence, the quantities of symbol pushed and popped on the stack would also approach 1. Thus, after training, a specific *reading* of the stack should contain only *one* symbol and the performance of the analog stack should approximate that of a discrete one.

NO OPERATION: If the magnitude of S_a is significantly small, no operation is taken. For our simulations we performed a *no-operation* if $-0.1 < S_a < 0.1$.

Training of the NNPDA

The activation of *State Neurons* (and *Action Neuron*) may be written as

$$S(t+1) = F(S(t), I(t), R(t); W) \quad (3)$$

where I is the activation of the *Input Neurons* and R is the activation of the *Read Neuron* and W is the weight matrix of the network. We use a localized representation for Input and Read symbols (thus, a symbol is uniquely represented by a vector which has only one 1 and all other elements 0). We now describe the different forms equation (3) take for different orders of the *State, Read and Input Neurons.*

For *First Order*, let V(t) represent a concatenation of vectors $I(t)$, $R(t)$ and $S(t)$, i.e., $V(t) = I(t) \oplus R(t) \oplus S(t)$. Then equation (3) becomes

$$S_i(t+1) = g(\sum W_{ij} V_j(t)) \tag{4}$$

For *Second Order*, let V(t) represent concatenation of vectors $I(t)$ and $R(t)$, i.e., $V(t) = I(t) \oplus R(t)$. Equation (3) becomes

$$S_i(t+1) = g(\sum\sum W_{ijk} S_j V_k(t)) \tag{5}$$

For *Third Order* equation (3) becomes

$$S_i(t+1) = g(\sum\sum\sum W_{ijkl} S_j(t) I_k(t) R_l(t)) \tag{6}$$

where $g(x) = 1/(1+exp(-x))$.

At the end of each input sequence of alphabets $a_0, a_1, a_2, ...a_{l-1}$, a distinct symbol called the end-marker is presented to the network. The activation of the *Output Neuron* at this point is compared with the *Target*. The end symbol is useful because there may be more than one final state and we want to accept a string whenever the string reaches *some* final state. The end symbol facilitates computation by effectively constructing an extra hidden layer. Adjusting the weights connected to the end symbol neuron (since the input has a local representation, only one input neuron turns on to represent a symbol) corresponds to the training of a *super*-final state.

There are two *coupled* functions that the network needs to learn in the process of training: the state transition function and the stack manipulation function. During training, input sequences are presented one at a time and activations are allowed to propagate until the end of the string is reached. Once the end is reached the *Target* is matched with the *Output Neuron* and weights are updated in accordance with the learning rule. The learning rule used in the NNPDA is a significantly enhanced extension to *Real Time Recurrent Learning* [Williams 89].

For the First-order network, using the objective function defined by equation (1) and (2) in a gradient-descent weight update expression $\Delta W_{ij} = -\eta \partial Error/\partial W_{ij}$, the weight update rule becomes

$$\Delta W_{ij} = \begin{cases} \eta((Target - S_o(l))\partial S_o(l)/\partial W_{ij} - \\ \quad L(l)\partial L(l)/\partial W_{ij}) \quad \textit{for equation 1} \\ -\eta(\partial S_o(l)/\partial W_{ij} - \partial L(l)/\partial W_{ij}) \\ \quad \textit{for equation 2} \end{cases} \tag{7}$$

where η is the learning rate. Then, $\partial S_o(l)/\partial W_{ij}$ can be calculated from the following recurrence relation by setting $\partial S_m(0)/\partial W_{ij} = 0.0$.

$$\partial S_m(t+1)/\partial W_{ij} = g'(\delta_{mi} V_j(t) + \sum W_{mn} \partial S_n(t)/\partial W_{ij} + \sum W_{mn} \partial R_n(t)/\partial W_{ij}) \tag{8}$$

where $\delta_{mi} = 1\ if\ m = i$, $g' = d(g(x))/dx$.

How do we obtain $\partial R(t)/\partial W_{ij}$? Since the current stack reading depends on its entire history, no simple recurrence relation can be found. However, the following approximation appears valid. It may be noted that we are able to differentiate R only because the stack is continuous. Also, after the network has been trained sufficiently and action values are large (> 0.5), each reading may not contain much information of the past. We obtain an approximate value of $\partial R(t)/\partial W_{ij}$ as follows:

$$\partial R(t)/\partial W_{ij} = (\partial R(t)/\partial S_a(t))(\partial S_a(t)/\partial W_{ij})$$

where $S_a(t)$ is the activation of the *Action Neuron.*

During *push* and *pop*, any incremental (or decremental) change of ΔS_a in S_a would cause an increase (or decrease) of R in the top of the stack with the same amount. Therefore,

$$\partial R_i/\partial S_a = 1$$

if R_i corresponds to the symbol on top of the stack. Also, since the total reading length (equal to 1) is fixed, any incremental (or decremental) change of ΔS_a in S_a would also cause a decrease (or increase) of R in the bottom of the stack. Hence,

$$\partial R_i/\partial S_a = -1$$

if R_i corresponds to the symbol at the bottom of the stack. It may be noted that, these are only first order approximations with the assumption that the network has been trained sufficiently so that actions are large in magnitude (close to 1.0).

Therefore $\partial R_m(t)/\partial W_{ij}$ may be approximated as:

$$\partial R_m(t)/\partial W_{ij} \approx (\delta_{mr_1} - \delta_{mr_2}) \partial S_a(t)/\partial W_{ij}) \tag{9}$$

where r_1 and r_2 are the indices of the symbols on top and bottom of the stack respectively, and $\delta_{mr_i} = 1$ if $m = r_i$. Having defined $\partial R(t)/\partial W_{ij}$ and assuming all partial derivatives at $time = 0$ to be 0, $\partial S_m(l)/\partial W_{ij}$ can be evaluated, where l is the length of the input string being processed.

Since the stack length $L(t)$ may be recursively evaluated by

$$L(t+1) = L(t) + S_a(t) \tag{10}$$

the second partial derivative, $\partial L(l)/\partial W_{ij}$, in equation (7) may be expressed as

$$\partial L(t+1)/\partial W_{ij} = \partial L(t)/\partial W_{ij} + \partial S_a(t)/\partial W_{ij} \tag{11}$$

For an initial condition let $\partial L(0)/\partial W_{ij} = 0.0$, then $\partial L(l)/\partial W_{ij}$ can be evaluated by the above recursion. Therefore, by imposing the "on-line" learning algorithm, the derivatives of the weights are propagated forward using the recursive formula and the final correction ΔW_{ij} is made at the end, after one whole input string has been presented. The learning rules for second and third order networks are exactly the same in nature but vary in the type of interconnections or the W matrix.

To determine the time complexity of the learning algorithm, let S and I be respectively the number of fully-connected recurrent and input neurons and l the length of the input string. Then the number of operations required per time step is of the order $l * (S+1)^2 * (S+R) * (I+S+R)$ for a first-order recurrent network (primarily dominated by the computation of the partial derivatives in equation (8)) The 1 in $S+1$ takes into account the action neuron. Similarly a second and a third order network require respectively $l * S^2 * (S+1)^2 * (I+R)^2$ and $l * S^2 * (S+1)^2 * I^2 * R^2$. Note that for large S, the complexity goes as $O(S^4)$.

Learning with Hints

Our training sets contained both positive and negative strings. One problem with training on incorrect strings is that, once a character in the string is reached that forces the string to a *reject state*, no further information is gained by processing the rest of the string. For example, if we are training the network on language $a^n b^n$ and we come across a string that begins with *aaaaba...*, no matter what follows the last *a* in the string, it is unnecessary to parse and train the network on rest of the string any further. In order to incorporate this idea we have introduced the concept of a *Dead State*.

During training, we assumed that there is a *teacher* or an *oracle* who has some knowledge of the grammar and is able to identify the points on the strings (of negative examples) that takes the strings to a *reject state.* When such a point is reached in the input string, further processing of the string is stopped and the network is trained so that one designated *State Neuron* called the *Dead State Neuron* is "on". To accommodate the idea of a *Dead State* in the learning rule, the following change is made: if the network is being trained on illegal strings that end up in a *Dead State* then the length $L(l)$ in the error function in equation (1) is ignored and simply becomes $Error = 1/2(Target - S_o(l))^2$. Since such strings have an illegal sequence, they cannot be a prefix to any legal string. Therefore at this point we do not care about the length of the stack.

For strings that are either legal or illegal but do not go to a dead state (an example of such a string would be a prefix of a legal strings, that ends prematurely); the objective function remains the same as described earlier in equation (1) and equation (2). *Hints* in this form made learning faster, helped in learning of exact pushdown automata and made better generalizations. For certain languages, these *hints* actually made learning possible. There are methods for inserting hints (rules) directly into recurrent neural networks [Omlin 92]; it would be interesting to see the effect of using these methods in training a NNPDA.

Simulations

The training data consisted of sequence of strings generated in *alphabetical order* from the input alphabet set. *Incremental, real-time learning* was used to train the NNPDA. In other words, the length of the strings in the training set was increased in steps, gradually as the network learned the smaller ones. At the beginning of each run the weights were initialized with a set of random values chosen between [-1.0, 1.0]. Training began with the shortest possible strings (of length *one*).

Once the network learned to recognize the strings in the current training set, longer strings (of length one more than the longest string in the current set) were added to the training set. Longer strings were added when either of the two criteria was satisfied: (1) a threshold number of epochs were completed, (2) network learned to recognize all strings in the training set before completing the threshold number of epochs. Epochs here imply one pass over the training set. A training set was considered to be successfully learned when all the strings in the set were recognized correctly. In general, for every language trained, this threshold was varied until the performance (in terms of total number of epochs needed for training) could not be further increased. For most simulations, the threshold for the number of epochs ranged between 20 and 40.

If the correct stack actions are learned by the NNPDA, then adding longer strings would not increase the error. This was used to estimate an upper bound for the maximum length of training strings to be used. The maximum length of the strings required for training was usually limited to *ten*. For simple languages like $a^n b^n$, training strings of length up to *six* were sufficient to train the NNPDA. For a particular length, since the number of positive strings was much smaller than the number of possible negative strings, a positive string of the same length was placed every third string in the training set. Thus, a small set of positive strings were repeated many times in the training set. Once the network was trained, the actions and states were quantized so as to *extract* a perfect pushdown automaton. This *extracted pushdown automaton* can recognize strings of arbitrary length. For a discussion of this extraction method, see [Sun 90] and [Giles 90] and, more recently, for finite state automata [Giles 92a].

The same simulation criteria and initial conditions described above were used for training NNPDA of various orders. A comparative performance of the networks of *first, second* and *third* orders in terms of number of iterations required, generalization capability and number of neurons are shown in *Tables 4, 5 and 6.* The values in the tables were typical ones obtained in our simulations; changing the initial conditions resulted in values of similar orders of magnitude. These tables show statistics for the *minimal machines* learned.

Conclusions

A neural network pushdown automaton (NNPDA) was constructed by connecting a recurrent neural network state controller to an external stack memory through a joint error function. This NNPDA was shown to be capable of learning a range of small, but interesting, deterministic context-free (DCF) grammars. A *continuous* external stack was constructed that permitted the successful use of continuous optimization methods (gradient-descent). The NNPDA learned to make efficient use of this stack. When it was trained on regular languages, e.g. (*single parity*, where the odd or even occurrence of a single symbol is checked for acceptance), the network learns the state transitions without making use of the stack. However, a language like parity could have been learned using a stack, that is, it could have used the stack by *pushing* a symbol on every odd occurrence of a character and *popping* the stack on every even occurrence. But the NNPDA error function apparently allows the network to selectively avoid using the stack when the language can be learned without it.

Simulations varying the order of the recurrent network showed that, in general, the higher the order of the net, the easier it was to learn grammars. (For some grammars, higher order proved to be a necessity for successful training!) However, it proved possible to learn a simple DCF Language such as the *parenthesis matching* grammar by using only *first-order* networks. We also observed that the stack was able to learn to change its stack actions. For example, in learning the language $a^n b^n c b^m a^m$, the stack had to learn to push a's and push b's when it saw an a and then reverse that process. Third order networks do not necessarily perform much better than second order networks. One possible explanation is that in the higher order networks the increase in the degrees of freedom slows down convergence. Of course the network has only learned small DCF grammars; larger grammars should be much more difficult. However, the NNPDA was able to learn how to efficiently control and use *an external stack* while at the same time learning its neural network state machine controller.

Acknowledgement

The authors would like to acknowledge helpful and useful discussions with M. Goudreau, C. Miller, M. Mozer, C. Omlin, H. Siegelmann, P. Smolensky and D. Touretzky.

References

[Allen 90] Allen, R.B., 1990. Connectionist Language Users. *Connection Science* 2(4): p 279.

[Elman 90] Elman, J.L., 1990. Finding Structure in Time. *Cognitive Science* 14:p. 179.

[Giles 90] Giles, C.L.; Sun, G.Z.; Chen, H.H.; Lee, Y.C.; Chen, D., 1990. Higher Order Recurrent Networks & Grammatical Inference. *Advances in Neural Information Systems 2*, D.S. Touretzky (ed), Morgan Kaufmann, San Mateo, Ca:p. 380.

[Giles 92a] Giles, C.L.; Miller, C.B.; Chen, D.; Chen, H.H.; Sun, G.Z.; Lee, Y.C., 1992. Learning and Extracting Finite State Automata with Second-Order Recurrent Neural Networks. *Neural Computation* 4(3):p. 393.

[Giles 92b] Giles, C.L.; Miller, C.B.; Chen, D.; Sun, G.Z.; Chen, H.H.; Lee, Y.C., 1992. Extracting and Learning an *Unknown* Grammar with Recurrent Neural Networks. *Advances in Neural Information Systems 4*, J.E. Moody, S.J. Hanson, R.P. Lippmann (eds), Morgan Kaufmann, San Mateo, Ca.

[Miclet 90] Miclet, L., 1990. Grammatical Inference, *Syntactic and Structural Pattern Recognition; Theory and Applications*, H. Bunke and A. Sanfeliu (eds), World Scientific, Singapore, Ch 9.

[Mozer 90] Mozer, M.C.; Bachrach, J., 1990. Discovering the Structure of a Reactive Environment by Exploration. *Neural Computation* 2(4):p. 447.

[Omlin 92] Omlin, C.W.; Giles, C.L., 1992. Training Second-Order Recurrent Neural Networks Using Hints. Proceedings of the Ninth International Conference on Machine Learning, D. Sleeman and P. Edwards (eds). Morgan Kaufmann, San Mateo, Ca.

[Pollack 90] Pollack, J.B., 1990. Recursive Distributed Representations. *J. of Artificial Intelligence* 46:p. 77.

[Pollack 91] Pollack, J. B. 1991. The Induction of Dynamical Recognizers. *Machine Learning* 7:p. 227.

[Servan-Schreiber 91] Servan-Schreiber, D.; Cleeremans, A.; McClelland, J.L., 1991. Graded State Machine: The Representation of Temporal Contingencies in Simple Recurrent Networks. *Machine Learning* 7:p. 161.

[Sun 90] Sun, G Z.; Chen, H.H.; Giles, C.L.; Lee, Y.C.; Chen, D., 1990. Neural Networks with External Memory Stack that Learn Context-Free Grammars from Examples. Proceedings of the Conference on Information Science and Systems, Vol. II: p. 649. Princeton University, Princeton, NJ: Conference on Information Science and Systems, Inc.

[Watrous 92] Watrous, R.L.; Kuhn, G.M., 1992. Induction of Finite-State Languages Using Second-Order Recurrent Networks, *Neural Computation* 4(3).

[Williams 89] Williams, R.J.; Zipser, D., 1989. A Learning Algorithm for Continually Running Fully Recurrent Neural Networks. *Neural Computation* 1(2):p. 270.

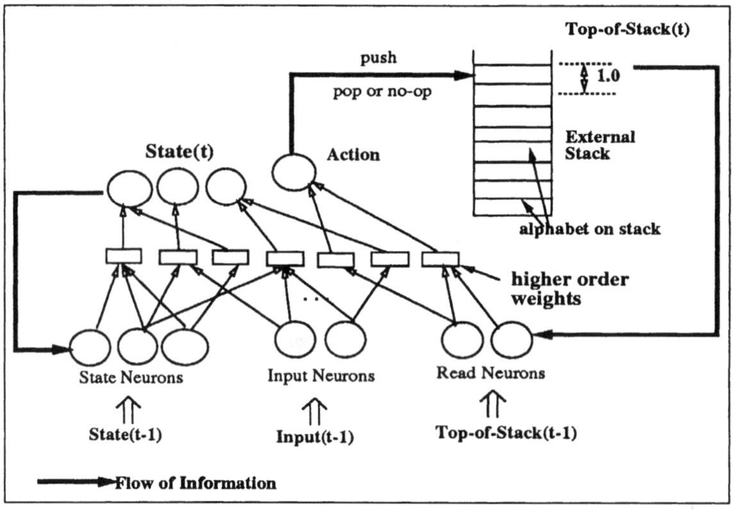

Figure 1: *The figure shows the architecture of a third-order NNPDA. Each weight relates the product of Input(t-1), State(t-1) and Top-of-Stack information to the State(t). Depending on the activation of the Action Neuron, stack action (namely, push, pop or nooperation) is taken and the Top-of-Stack (i.e. value of Read Neurons) is updated.*

Order	parenthesis		$a^n b^n$		$a^n b^n cb^m a^m$		$a^{n+m} b^n c^m$	
of NN	hints	w/o hints	hints	w/o hints	hints	w/o hints	hints	w/o hints
1st	50–100	***	300–500	***	***	***	***	***
2nd	50–80	80–100	150–300	300	500	***	200-250	***
3rd	50–80	50–80	150–250	150–250	150	***	150–250	***

Table 4: *Iterations required by first, second and third order networks to learn various languages with and without hints and under same initial conditions, namely, same initial learning rate, same initial value of state neurons, same random number and same input set ("****" in the table implies that the simulation did not converge).*

Order	parenthesis		$a^n b^n$		$a^n b^n cb^m a^m$		$a^{n+m} b^n c^m$	
of NN	hints	w/o hints	hints	w/o hints	hints	w/o hints	hints	w/o hints
1st	0.0	***	8.9	***	***	***	***	***
2nd	0.0	3.07	0.0	2.67	5.56	***	0.0	***
3rd	0.0	0.0	0.0	1.03	3.98	***	0.0	***

Table 5: *Generalization (in % error on all possible strings up to length 15, starting from length 1, that is, with 65534 strings).*

Order	parenthesis		$a^n b^n$		$a^n b^n cb^m a^m$		$a^{n+m} b^n c^m$	
of NN	hints	w/o hints	hints	w/o hints	hints	w/o hints	hints	w/o hints
1st	3+1	***	3+1	***	***	***	***	***
2nd	1+1	2	1+1	3	1+1	***	1+2	***
3rd	1+1	2	1+1	2	1+1	***	1+1	***

Table 6: *Minimal number of State Neurons required to learn the languages in various orders (for the simulations with hints one neuron was required explicitly for dead state and hence the "+1"s).*

The Role of Expertise in the Development of Display-Based Problem Solving Strategies

Simon P. Davies
Department of Psychology,
Nottingham University,
University Park, Nottingham, NG7 2RD, UK
spd@uk.ac.nottingham.psych

Abstract

This paper reports two experiments which explore the relationship between working memory and the development of expertise. Consideration is given to the role played by external memory sources and display-based problem solving in computer programming tasks. Evidence is presented which suggests that expertise in programming is dependent upon the development of strategies for effectively utilising external displays. In this context, it appears that novices rely extensively upon working memory to generate as much of a solution as possible before transferring it to an external source. In contrast, experts make extensive use of an external display as an information repository. These results are discussed in terms of a framework which emphasises the role of display-based problem solving and its contribution to strategy development.

Introduction

A pervasive finding of recent research into the cognitive aspects of programming is that code is not generated in a linear fashion - i.e., in a strict first-to-last order (Davies, 1991; Rist, 1989). Rather, many deviations are made from linear development, where programmers leave gaps in the emerging program to be filled in later. Green et al (1987) have proposed a model to account for this finding. Their Parsing/Gnisrap model introduces a working memory component into the analysis of coding behaviour which forces the model to use an external medium (eg the VDU screen) when program fragments are completed or when working memory is overloaded. This means that programmers will frequently need to refer back to generated fragments in order to recreate the original plan structure of the program which may have only been partially implemented in code. The parsing element of the model describes this process, while gnisrap (the reverse of parsing) describes the generative process.

Davies has looked at the nature of the nonlinearities in program generation for programmers of different skill levels. One finding to emerge from this work was that experts perform a greater number of between-plan jumps than novices and that novices tend to perform more within-plan jumps - that is, adopt a linear generation strategy. This might seem anomalous, since if we assume that the re-parsing of a generated output involves some cognitive cost, then one might expect the development of programming skill to be partly dependent upon a programmer's ability to generate as much of the program internally before writing it to an external source, thus reducing the need to re-parse. However, the opposite appears to be the case. The results of Davies (1991) suggest that skilled programmers make much use of external memory sources (i.e., a VDU screen) while novices tend to rely upon the use of internal memory to develop as much of the solution as possible before transferring it to external memory.

One question that arises is the extent to which expertise in programming and other complex skills can be explained by recourse to an extended working memory model as opposed to a model which places emphasis upon the role of externalised memory structures and display-based comprehension? The following experiments attempt to address this issue directly. The first experiment considers the role of working memory in the determination of strategy for novice and expert programmers. The second experiment looks at the effects upon certain error forms of restricting the kinds of manipulations programmers can make within an environment.

Experimental Studies

In the first experiment, subjects carried out a simple articulatory suppression task while engaged in a program generation activity. If working memory limitations cause programmers to make use of an external medium, as suggested by Green et al, then the act of loading working memory through a concurrent task should give rise to an increase in nonlinearities, since subjects would have to engage more fully in the parsing/gnisrap cycle in order to make use of the external display.

The second experiment looks at the way in which restricting the use of an external medium affects performance. Here, if programmers are not able to correct already generated code at later stages in the coding process, then this should have some effect upon their performance. In this experiment, subjects created a program using a full-screen editor that provided no opportunity for the revision of existing text. The use of such an editor clearly places a significant load upon a subjects working memory capacity since they will be required to internally generate as much of the program as possible before externalising it. By placing emphasis upon the use of working memory it should be possible to induce error prone behaviour which parallels that evident when working memory is loaded in other ways, for instance via articulatory suppression.

We might expect a detriment in expert performance when the device used to create the program is restricted in such a way as to make retrospective changes impossible. This is based upon the assumption that experts make greater use of external sources to record partial code fragments which are then later elaborated. Conversely, it has been suggested that novices will tend to rely more upon generating as much of the program internally before writing it to an external source. It is clear that these strategic differences will be supported to a greater or a lesser extent by the device used to create the program. Hence, for expert programmers, it might be suggested that restricting the device will cause them to revert to a novice strategy, since they will then be unable to use the external display in the normal way.

Establishing support for this hypothesis would have a number of implications. Firstly, it would suggest that the development of expertise may not be based simply upon the acquisition of knowledge about a given domain. If this were the case, we would expect experts to perform better than novices regardless of the constraints imposed by the task environment. Secondly, it would indicate that increased working memory availability does not necessarily lead to better performance. Moreover, if working memory availability is correlated with expertise, then experts should perform better that novices in situations where they must rely upon internal sources. If this is not the case, then we might question the status of working memory in theories dealing with the development of complex skills. An alternative explanation is that experts have developed particular strategies for dealing with task complexity that involve close interaction with external information repositories in order to record partial solution fragments as they are generated. If novices have failed to develop such strategies, then it is unlikely that their performance would be affected significantly by restricting the task environment.

This analysis can be extended by classifying the errors in the programs generated by subjects. A scheme devised by Gilmore and Green (1988) suggests four main categories of error:

1 - Surface level errors caused mainly by typing and syntactic slips: (e.g. confusion between < and >, missing or misplaced quotes etc).

2 - Control-Flow errors: (e.g. missing or spurious else statements, split loops etc).

3 - Plan-Structure errors: Including, guard test on wrong variable, update wrong variable etc.

4 - Interaction errors: A class of errors occurring at the point where structures of different types interact: (e.g. a missing 'Read' in the main loop, initialisations within the main loop).

Clearly some of these errors will be knowledge-based (specifically, plan-structure errors) while others will be dependent upon working memory limitations. For example, both control-flow and interaction errors, since they depend upon establishing referential links and dependencies between code structures, are likely to be affected by working memory constraints. In terms of the first experiment, we might expect both control-flow and interaction errors to predominate in novice solutions where working memory availability is reduced. In the case of experts, it is argued that the interactions between code structures will be evaluated in the context of an external memory source. That is, by re-parsing existing code fragments in order to reconcile them with the code the programmer is currently working on. Thus, that the act of loading working memory should not affect the occurrence of these types of error.

In the case of the second experiment, we would expect the converse. If experts are not able to use the external display in the manner predicted, then it might be hypothesised that interaction and control-flow errors will predominate in the condition where use of the device is restricted. It might also be predicted that this experimental manipulation will not affect the occurrence of plan-structure errors since these are hypothesised to be knowledge-based rather than strategy-based.

Experiment 1. Effects of articulatory suppression on strategy and errors

Method

Subjects: Twenty subjects participated in this experiment. One group of ten subjects consisted of professional programmers. All the subjects in this group used Pascal on a daily basis and all had substantial training in the use of this language. Members of this group were classified as experts. A second group consisted of second year undergraduate students, all of whom had been formally instructed in Pascal syntax and language use during the first year of their course. Members of this group were classified as novices.

Procedure and Design: Subjects were asked to carry out a simple articulatory suppression task which involved repeating a string of five random digits. At the same time, subjects were requested to generate a simple pascal program that could read a series of input values, calculate a running total, output an average value and stop given a specific terminating condition. This specification was derived from Johnson and Soloway (1985) and was chosen because it has formed the basis of evident for the novice group. In the case of the expert group the same comparison proved not to be significant.

Error classification analysis: In the case of experts, there is a fairly even distribution of error types across the two experimental conditions. Indeed, further statistical analysis revealed no significant differences between error types both within and between conditions (multiple t-tests). In the case of the novice group, the distribution of error types is less straightforward. In the non-suppression condition, novices produced a significantly greater number of plan errors in comparison to the other categories (t-test). Moreover, the only significant difference between the novice and experts groups in this condition was the number of plan errors produced by the novice group (t-test). In the second condition, the distribution of errors across classification types for expert subjects was again fairly even. No significant differences between any of the error classifications were evident. For the novice group, significantly more control-flow and interaction errors were evident in comparison to the other two error classifications (t-test). Moreover, for the novice group,the number of plan errors occurring in the second condition was significantly less than in the first condition (t-test).

Discussion: This experiment shows that expert performance in programming tasks is not significantly affected by articulatory suppression. Hence, for experts the number of errors produced is not significantly different comparing the suppression condition to the non-suppression condition. Moreover, it appears that strategy is similarly unaffected. Hence, the prevalence of between-plan jumps in the non-suppression condition for the expert group is not diminished in the suppression condition. Similarly, the occurrence of within-plan jumps does not differ significantly in the two experimental conditions.

Conversely,the novice group produced significantly more errors in the suppression condition when compared to the non-suppression condition. In addition, the nature of the coding strategy that they adopt is also affected. In particular, it appears that novice programmers revert from a linear generation strategy characterised by the prevalence of within-plan jumps, to a strategy more characteristic of experts. That is, to a strategy which reflects a greater number of between-plan jumps.

Earlier it was stated that expert programmers appear to rely much more extensively than novices upon the use of external sources to record partial code fragments and that the act of loading working memory or of otherwise reducing its availability would not affect this process. It was suggested that experts will tend engage in very closely linked cycles of planning, subsequent code generation and evaluation. Since it is posited that this process relies little upon the programmer's working memory capacity it is reasonable to expect that articulatory suppression would not affect the nature of performance in the context of this task. The results of this experiment provide support for this view. Further support for this view is evident in the error data. In the non-suppression condition, novice subjects are clearly more error prone. This finding is not unexpected. However, in the suppression condition, the error rate for the expert group changes little from this base line whereas the novice error rate more than doubles. This may indicate that when working memory is loaded novices must externalise information and that this constitutes a strategy which they find unnatural, thus leading to an increased error rate.

A more detailed analysis of these errors reveals a change in the nature of errors for novice subjects between the two experimental conditions. In the non-suppression condition, the novice group make a greater number of plan errors, suggesting knowledge-based difficulties. Conversely, in the suppression condition a greater proportion of control-flow and interaction errors are evident. In terms of the present analysis, the preponderance of control-flow and interaction errors may reflect problems keeping track of the interdependencies between elements in the emerging program. When working memory availability is reduced it appears that novices experience some difficulty with these interdependencies. Unlike experts, it appears that novices cannot use the external display as an aid to memory to its full extent.

An alternative explanation for these findings is that experts simply have an extended working memory capacity. Such an account would presumably have no difficulty predicting the results of the experiment reported above. In order to assess the cogency of this alternative explanation, the second experiment reported in this paper adopts a different approach for exploring the relationship between working memory and the development of programming skill. In particular, if experts, for whatever reason, are able to extend their effective working memory capacity or increase its availability in other ways then restricting the task environment should not significantly affect their performance.

Experiment 2. Effects of restricting the task environment

The second experiment is complementary to the first. Whereas the first experiment attempted to reduce the subjects' available working memory capacity, this experiment has been designed to encourage subjects to rely upon working memory. Hence, if experts have an extended working memory capacity they should demonstrate performance equitable to that displayed in the first experiment. Moreover, if the extended capacity notion is correct, then experts should out perform novices even in the situation where the task environment is severely restricted as in this second experiment.

many empirical studies. Hence, the resulting programs could easily analysed for errors and plan structures.

Subjects were allowed to study the specification for 5 mins. and were then asked to generate a program corresponding to this specification while engaged in the concurrent suppression task. The subjects were given 15 mins. to complete this task, typing their solutions onto a familiar text editor. Subjects' keystrokes were recorded for further analysis. This analysis provided an indication of the temporal sequence in which programs were generated. Three independent raters were asked to analyse all the resulting program transcripts for the presence of common plan structures (Soloway and Ehrlich, 1984) and for errors (using the classification described above). Within and between-plan jumps were defined as follows: Within-plan jumps were classified as movements between a particular line of the program text to another line which formed part of the same plan structure. Between-plan jumps were defined as movements from the current line to lines within different plan structures (see Davies, 1991). These protocols applied only to situations where the jump was followed by an editing action. The experiment was a two-factor design, with the following independent variables: 1. Articulatory suppression/No suppression and 2. Level of expertise (Novice/Expert).There were two dependent variables: 1. The number of Between/Within-plan jumps and 2. Errors remaining in the final program.

Results

Plan-jumps: Figure 1 shows the number of within and between-plan jumps performed by novice and expert programmers in the two experimental conditions. Analysis revealed main effects of suppression ($F_{1,72}$ = 8.47, p<0.01) and expertise ($F_{1,72}$ = 12.56, p<0.01) on jump-type and a more complex interaction between suppression and expertise ($F_{1,54}$ = 4.73, p<0.05). A number of post-hoc comparisons were carried out using the Newman-Keules test with an adopted significance level of p<0.01. This procedure indicated that experts produced significantly more between plan jumps than novices in the non-suppression condition. Conversely, novices produced a greater number of within plan-jumps in this condition. In the case of the suppression condition, there were no significant differences.

Errors: Figure 2 shows the total mean number of errors remaining in the programs on task completion for novice and expert subjects in the two experimental conditions. Analysis revealed a main effect of expertise ($F_{1,36}$ = 9.37, p<0.01) and suppression ($F_{1,36}$ = 4.54, p<0.05) and an interaction between these two factors ($F_{1,36}$ = 15.89, p<0.01). Once again a number of post-hoc comparisons were carried out using the Newman-Keules test with an adopted significance level of p<0.01. This indicated a significant difference in error rates in the both experimental conditions when comparing the novice and expert groups. In addition, a significant difference between error rates across conditions was

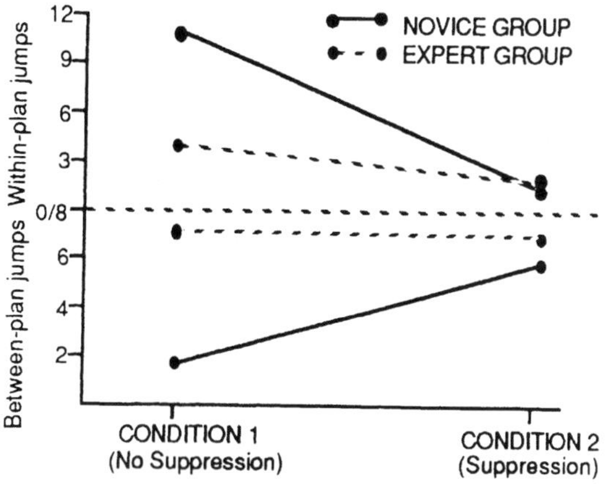

Figure 1 Within and Between-Plan jumps by novices and experts during the first experiment

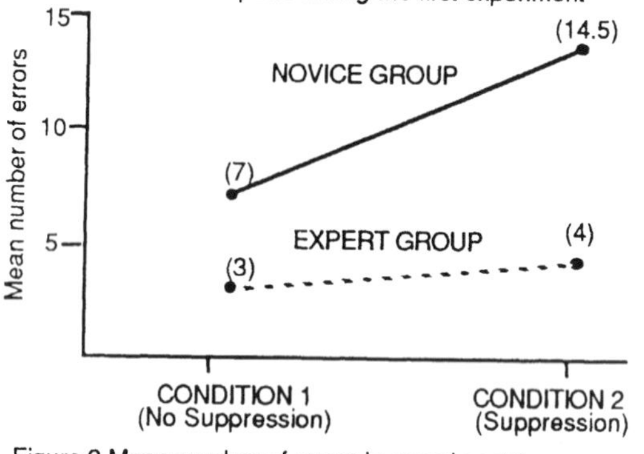

Figure 2 Mean number of errors in experiment 1 for novice and expert subjects

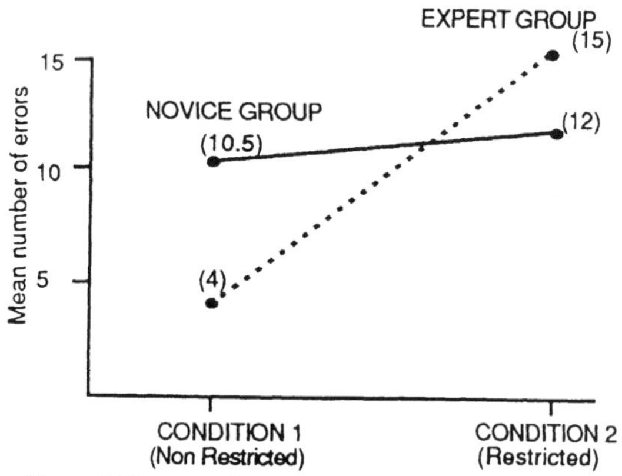

Figure 3 Mean number of errors in experiment 2 for novice and expert subjects

Method

***Subjects, Procedure and Design*:** The same subjects took part in this experiment, with the order of participation randomised. Subjects were asked to produce a program corresponding to a brief specification which involved processing simple bank transactions. Here, the nature of the task environment formed the basis for the two experimental conditions. In one condition, subjects used a familiar full-screen text editor. In the second condition subjects used a modified version of the same editor,which allowed only restricted cursor movement. That is, from the top of the screen to the bottom. and only between adjacent lines. Once a subject had generated a line and pressed the return key, they were unable to then return to that line to perform other editing operations. The editor did however allow edits to the current line being generated. Subjects were informed about the basic modifications to the editor. and were asked to attempt to generate a program from the specification and were asked to check each line of their program before pressing the return key, in order to determine whether they were satisfied with their response. 15 mins were allowed for this task. This experiment was a two-factor design with the following independent variables:Environment (restricted/ unrestricted) and Level of expertise (Novice/Expert).

Results

***Errors*:** The results of this experiment are shown in figure 3. These data were analysed using a two-way analysis of variance with the following factors; Environment (restricted or unrestricted) and Level of expertise (Novice/Expert) This analysis revealed a main effect of Environment ($F_{1,36} = 5.74$, $p<0.05$), a main effect of Level of expertise ($F_{1,36} = 4.21$, $p<0.05$) and an interaction between these two factors ($F_{1,36} = 9.76$, $p<0.01$). Post-hoc comparisons were carried out using the Newman-Keules test with a significance level of $p<0.01$. This analysis revealed a significant difference between the number of errors produced by novices and experts in condition 1.

***Error classification*:** The resulting program transcripts were analysed according to the classification scheme described previously. In the case of experts, analysis revealed no significant differences between error types within this condition (t-tests). In the case of the novice group, the distribution of error types in the first condition suggests a greater proportion of plan errors in comparison to the other categories (t-test). In the second condition, the distribution of errors across classification types for expert subjects was more complicated. This showed a greater proportion of control-flow and interaction errors compared to the other classifications (t-tests). In addition, experts produced significantly more control-flow and interaction errors in comparison to the first condition.

***Discussion*:** These results provide a striking demonstration of the effects of restricting a task environment. We have argued that experts rely to a great extent upon using the external display to record fragments of code that are then further elaborated at subsequent points during the generation process. This led to the hypothesis that if programmers were unable to return to previously generated fragments then they would be forced into a situation where they would have to rely extensively upon working memory. However, it appears that while novices are seemingly unaffected by changes to the task environment, experts not only perform worse than novices but also produce the kinds of errors that are indicative of an inability to internally construct links and dependencies between code structures. These results reveal that experts produce more errors than novices in the restricted task environment. Moreover, experts produce a significantly greater number of control-flow and interaction errors in this second condition.

It was suggested previously that the first experiment that the results might be interpreted as indicating that experts have an extended working memory capacity. However, if this is the case then the results of this second experiment would appear to be rather anomalous. If we assume that experts have an extended working memory capacity in comparison to novices, then we might expect that situations which cause experts to rely upon working memory would not give rise to such an extensive decriment in performance. Moreover, in terms of this view there appears to be no reasonable explanation as to why experts produce many more control-flow and interaction errors in comparison to novices.

A more cogent explanation for these findings might simply involve suggesting that experts rely upon external sources and are not able to efficiently revert to a strategy that demands extensive reliance upon working memory. This would account for both sets of experimental findings. In the first experiment a reduction in working memory availability did not affect expert performance. This could clearly be accounted for in two ways. On the one hand, it could be argued that experts simply have an extended working memory capacity. Conversely, we might claim that experts rely extensively upon external sources and find it difficult to adopt other alternative strategies. However, the second experiment appears to suggest that the first of these explanations is incorrect. In particular, if experts have an extended working memory capacity then we would expect them to perform better than novices in situations where a reliance upon working memory is necessitated. This appears not to be the case.

Another finding relating to this data was that in the restricted environment condition the expert group produced fewer surface and plan errors. An explanation for this may be that, in the restricted environment condition, the normally automatic aspects of programming skill are disrupted. This may lead the programmer to attend to the knowledge-based components of programming skill leading to a reduction in surface and plan-based errors. There is evidence in the literature which suggests that so called 'skill' and 'knowledge-based' errors are to some extent disassociable (Reason, 1979).

Conclusions

These experiments have a number of implications. Firstly, it appears that experts rely upon external sources to record code fragments as these are generated and then return later, in terms of the temporal sequence of program generation, to further elaborate these fragments. It has been suggested that a major determinant of expertise in programming may be related to the adoption or the development of strategies that facilitate the efficient use of external sources. The externalisation of information clearly has a high cost in terms of the reparsing or recomprehension of generated code that is implied. Hence, it might seem counter intuitive to suggest that problem solvers will tend to rely upon this kind of strategy rather than upon a strategy which involves the more extensive use of working memory. However, this explanation is consonant with existing work which has implicated display-based recognition skills in theoretical analyses of complex problem solving (Larkin, 1989). The contribution of these analyses has been important, but they have neglected to consider the relationship between display use and expertise and the consequent effect that this may have upon the nature of problem solving strategies.

The work reported here poses implications for the way in which we might attempt to explain the occurrence and distribution of error types. In particular, it is clear that a certain classes of error can be attributed to working memory limitations and that such errors are not distributed at random. In terms of the error classification employed here, it appears that interaction and control flow errors predominate in situations where working memory availability is reduced. Previous work (Anderson, 1989) suggests that errors arising from working memory failures will occur at random. However, the results of the studies presented here suggest that working memory related errors may have a more systematic distribution, and that the type of errors one might expect to occur may to some extent be predictable.

It also appears that the nature of display-based problem solving in programming may be highly dependent upon features of the programming language considered. Green (1991) suggests that some programming languages are "viscous" in that they are highly resistant to local change. Hence, adding a line to a Basic program may involve renumbering lines such that the correct control flow is maintained. In terms of the present analysis, less viscous languages will provide better support for the kind of incremental problem-solving processes that are proposed.

Such language features are important in the present context, since they will clearly affect the incremental nature of code generation and comprehension/ recomprehension. This analysis extends existing work by suggesting ways in which language features and strategy may interact with features of the task environment to give rise to particular forms of behaviour. Such effects would not be taken into account by display-based views, since the salience of particular features of the display remains undifferentiated. Moreover, existing accounts of display-based problem solving give no consideration to the effects of the kinds of information manipulation that are possible in different display spaces.

Summary

While this paper has indicated the importance of display-based performance in programming, it has also suggested two primary limitations of this general approach. Firstly, existing accounts of display-based problem solving ignore the apparent relationship between expertise and the development of strategies for utilising display-based information. Secondly, such accounts fail to consider the possibility that different forms of display-based information will be differentially salient in the context of a given task. Further developments of display-based accounts of problem solving will need to address these issues if they are to provide a coherent description of human performance in the context of complex tasks.

References

Anderson, J. R. 1989. The analogical origins of errors in problem solving. In D. Klahr and K. Kotovsky (Eds.), *Complex information processing: The impact of Herbert A. Simon*. LEA, Hillsdale, NJ.

Davies, S. P. 1991. The role of notation and knowledge representation in the determination of programming strategy: A framework for integrating models of programming behaviour. *Cognitive Science*, **15**, 547 - 572.

Gilmore, D. J. and Green, T. R. G. 1988. Programming plans and programming expertise. *Quarterly Journal of Experimental Psychology*, **40A** (3), 423 - 442.

Green, T. R. G. 1991. Describing information artefacts with cognitive dimensions and structure maps. In D. Diaper and N. Hammond, Eds., People & Computers 6, Cambridge University Press.

Green, T. R. G. Bellamy, R. K. E. and Parker, J. M. 1987. Parsing and gnisrap: a model of device use, Proc. INTERACT'87, H. J. Bullinger and B. Shackel (Eds.), Elsevier Science Publishers.

Johnson, W. L. and Soloway, E. 1985. PROUST: Knowledge-based program understanding. *IEEE Trans. on Software Engineering*, **SE-11**, 3, 423 - 442.

Larkin, J. H., 1989. Display-based problem solving. In D. Klahr and K. Kotovsky, Eds, *Complex Information Processing; The impact of Herbert A. Simon.*

Rist, R. S. 1989. Schema creation in programming. *Cognitive Science*, **13**, 389 - 414.

Soloway, E. and Ehrlich, K. 1984. Empirical studies of programming knowledge. *IEEE Trans. SE*, **SE - 10** (5), 595 -609.

Taxonomies and Part-Whole Hierarchies in the Acquisition of Word Meaning – A Connectionist Model

Georg Dorffner

Dept. of Medical Cybernetics and Artificial Intelligence, University of Vienna, and Austrian Research Institute for Artificial Intelligence[1]
Schottengasse 3, A-1010 Vienna, Austria
georg@ai.univie.ac.at

Abstract

The aim of this paper is to introduce a simple connectionist model for the acquisition of word meaning, and to demonstrate how this model can be enhanced based on empirical observations about language learning in children. The main sources are observations by Markman (1989, 1990) about constraints children place on word meaning, and Nelson (1988), as well as Benelli (1988), about the role of language in the acquisition of concept taxonomies. The model enhancements based on these observations, and those authors' conclusions, are mainly built on well-known neural mechanisms such as resonance, reset and recruitment, as first introduced in the adaptive resonance theory (ART) models by Grossberg (1976). This way the strength of connectionist models in plausibly modeling detailed aspects of natural language is underlined.

A Simple Model of Word Meaning

The connectionist model introduced in Dorffner (in press) is designed to demonstrate the abilities of a self-organizing system to acquire the meaning of simple words. Virtually no knowledge about the world is included a priori, mainly general pre-wired architecture. It concentrates on some important basic aspects while (necessarily) leaving out many details. The core ideas of the original model are the following.

- Words are primarily symbols in their referential sense (see, for instance, Dorffner 1992b). Learning the meaning of words in a first approach therefore means learning to identify symbols and their function.
- A basic task for every cognitive agent is to categorize environmental situations based on rich sensory stimuli and thus form concepts "about the world".
- Words as symbols in their simplest form refer to concepts. As a result, interpreting a symbol means building an internal link to one of the concepts in the above sense.
- Concepts are mental states clearly separable from other states. As such they are independent from each other, but can grow associative links so as to establish relations among each other. As a result, any conceptual structure such as hierarchies, usually attributed to conceptual schemata, is not reflected in the model architecture beforehand. It is viewed as being either "in the eye of the observer" of the model, or at best localizable after learning as associative traces between conceptual states.

In summary, meaning becomes defined with respect to the subjective experiences of the individual agent, leading to a constructivist core theory of word semantics. According to these ideas, the model consists of two sensory inputs, two components for concept formation based on categorization, and a set of layers for building the *referential links* (Fig. 1). For the sake of simplicity the two parts of the model (including the two different inputs) are kept separate—one being used for perceiving and clearly identifying the words (or so-called *external embodiments* of the symbols), the other for forming concepts about the perceived environment other than the elements of language. In the implementation, primitive acoustic input (stationary speech signals) was used for the former, simple visual input for the latter. In essence, both parts work the same way. Recognizing words is done by categorizing acoustic stimuli, the same way forming concepts is done by categorizing visual stimuli,

[1] *The Austrian Research Institute for Artificial Intelligence is supported by the Austrian Federal Ministry of Science and Research*

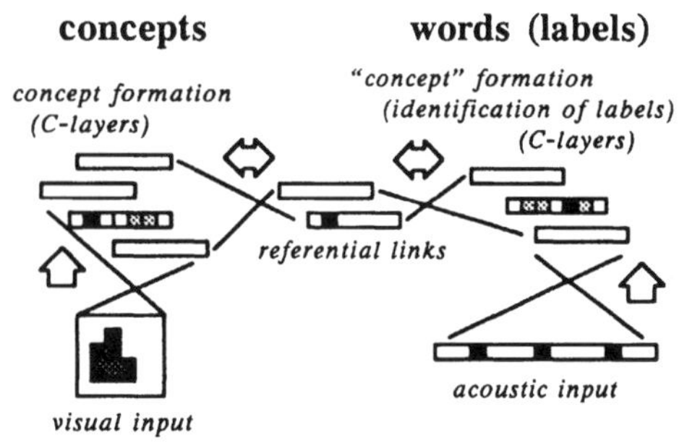

Fig. 1: The outline of the model

although for the latter internal activations can also have a major influence.

Concept Formation

The component for concept formation consists of a pool of so-called *C-layers*, which are connected to other C-layers and to the input via full associative feed-forward connections. Those connections are adapted with a "soft" competitive learning mechanism (Dorffner 1992a) that gradually compresses initially distributed patterns. Through learning it develops states we call *conceptual* or *identifiable states*. In the implementation they are basically defined as states where the winner of a C-layer is considerably larger than the average of the others. For this, the interactive activation rule (McClelland & Rumelhart, 1981) and negative weights between competing units, which are also adapted by the learning rule, are used.

It should be noted that concept formation based on, for instance, visual input would have to incorporate the great complexities of any natural visual system if it wanted to remain plausible in all respects as compared to humans. For the aspects this model is focused on such complexities would go far beyond the available computational resources. Therefore it is assumed that the patterns in the input are the result of preprocessing mechanisms, such as feature detectors or other transformations. Categorization as discovering invariants in stimuli classes starts after this preprocessing has happened. It is assumed that appropriate preprocessing can always transform real sensory stimuli into patterns showing similarity structures that can be processed by neural networks of the introduced kind.

Activation spreads to and within a pool of C-layers in two phases. First, any activation pattern originating outside the layer pool can activate patterns in the C-layers, plus the C-layers can cross-activate each other, all depending on the weight matrices shaped through learning. Secondly, competition sets in, both within layers and *between* layers. By introducing such an interconnected pool of C-layers the model is able to develop more than one conceptual state given one sensory stimulus. Examples would be concepts on different taxonomic levels. Furthermore, conceptual states develop either cleanly bottom-up (based on sensory input alone), or influenced by any other conceptual state to any degree. Competition in the second phase of a cycle ensures that always one (or at best, a few) conceptual state remains active. The underlying assumption is that a cognitive agent at one point in time can only concentrate on one conceptual perspective. For instance, a dog is either seen as a *poodle*, a *pet*, or an *animal*, but hardly as all three at the same time. It is, however, unclear as to which mechanism should select the most appropriate concept to win in competition—one drawback that has lead to the extensions described below.

An additonal part of concept formation based on sensory input is an automatic *focus of attention*—in this case a window whose size and location are adaptive. It can automatically be centered on any part of the original input in order to cut out (or enhance) this part. This window is then used as the input to the C-layers. By changing the size, the system can focus on parts of an object, building conceptual states based on those. Again, it is rather unclear when and how a shift of window position and size should be triggered.

Referential Links

For the implementation of referential links between identifiable states, a layer of specialized units—called *SY-layer*—with a winner-take-all (WTA) characteristics was introduced. This layer learns to identify co-occuring concepts and link them via a link unit. Learning of symbolic reference can be divided into the following phases. In the *fuzzy phase* no link unit has learned to respond to an identifiable state (clear concept). In the *identification phase* two identifiable states occur at the same time in different parts of the model. The winning link unit initiates WTA and the weights are strengthened. If this happens often enough, in the *recognition phase* one identifiable state is sufficient to let a link unit win and associate the corresponding concept. A special value of that unit—called the "symbol status"—must be above threshold to distinguish this case from the fuzzy phase. The links are built by a special "one-to-one rule" that lets weights grow for co-occuring identifiable states on both sides of the SY-layer. In other words, labels (words) are only linked to concepts that consistently occur at the same time, neither less nor more often.

The important properties of referential links implemented this way are as follows. First, not just any stimulus can be linked to a concept, it has to be clearly identified first. This means that referential links are built relatively late, i.e. after a considerable number of concepts have been learned. This corresponds to observations on early child language (e.g. Aitchison 1987), where consistent names for object classes are learned well after sounds are recognized and reproduced. Secondly, through WTA it is ensured that similarity structures on one side of the model do not map onto similarity structures on the other one. This corresponds to the arbitrariness of symbols by which their form is not related to their meaning (see, e.g., Lyons 1976 – ch.2, Dorffner 1992b, for more details).

As SY-layers are connected to C-layers they too can influence concept formation. For this, however, two modi operandi of referential links have to be distinguished. The modus described above presupposes equal treatment of the two model parts. Put differently, words (labels) are identified at the same time concepts are, whereafter the two get linked. This is plausible for the acquisition of the very first words, when neither the words, nor their potential referential power are known. Later, however, words are identified as referring to something even if no concept is activated at the same time. Words, in order to influence concept formation, should thus be permitted to grow referential links from the label side only ("directed link"). Therefore, the model can be switched to a mode where novel link units can be activated by identifiable states on the acoustic part. Examples for concepts influenced by language this way are superordinates such as 'furniture' or 'vehicle.' This, too, leads us to the model extensions described in the next section.

Problems with the Simple Approach

In this model, assigning word meaning becomes the problem of developing an appropriate referential link between the identifiable state corresponding to an external symbol embodiment (i.e. word or label) and the concept the symbol should refer to. Not surprisingly, this is not as trivial as described above. Words do not simply name categories of objects in a unique way. Among others, complexities arise as

- there are words on different taxonomic levels, such as 'poodle,' 'dog,' 'pet,' or 'animal.' This means that for one given object a large set of words would be applicable, depending on context and the intent of the reference.
- there are words for parts of objects. When the object is presented, its parts are too, which leads to the problem as to which of these the word should refer to.
- there are words for overall properties of an object, such as size or color. The same problems arise as with parts of an object.

The one-to-one learning rule briefly described above already captures some of the complexities in that it builds consistent links only for consistent pairs of co-occuring conceptual states. Words on a higher taxonomic level (say 'animal') than a given concept (say *dog*) will not develop such a link, as it co-occurs with other concepts as well. Words on a lower lever (say 'poodle'), on the other hand, co-occur with only a few instances of *dog*. However, such a mechanism is not sufficient. Complexities like the ones described cannot simply be explained by using mere statistical correlations. As a result, extensions to the simple approach need to be made. In the spirit of the introduced model such extensions should preferably not require too complex an architectural implementation, while the model's behavior should remain at least psychologically plausible.

Constraints Children Place on Word Meaning

One source for the model extensions we have chosen are proposals put forward by Markman (1989, 1990). Markman has observed that children must face very similar problems during the learning of word meanings as the ones identified above for the model. Time is too short and the possible combinations of concepts and words are too large to let statistical correlations decide alone upon how to construct the mappings between the two. She suggests that there must be some constraints children implicitly apply when learning words, and derives the following, empirically supported, assumptions.

(a) the **whole object assumption**:

When faced with one or more objects, each consisting of several parts, children obviously assume that "a novel label is likely to refer to the whole object and not to its parts, substance or other properties" (Markman 1990, p.59).

(b) the **taxonomic assumption**:

"This assumption states that labels refer to objects of the same kind rather than to objects that are thematically related" (Markman 1990, p.59). In other words, children obviously assume that the concept a label refers to is based on categories built on similarities and not on thematic relations. For instance, 'dog' refers to a class of objects that look and behave similarly, and not to a class of objects in a certain thematic context (e.g. objects being petted by the mother).

(c) the **mutual exclusivity assumption**:

This assumption states that usually only one label can be attached to each concept. Thus it "helps children override the whole object assumption, thereby enabling them to acquire terms other than object labels" (Markman 1990, p.66). In experiments it could be shown that children faced with an object—for which they do not have a word yet—and a label tend to take this label for the whole (assumption (a)). Children who know a word already tend to take the new word as standing for a salient part of the object (e.g. a receiver of a telephone – see Markman 1990), due to assumption (c).

These three assumptions together, among other principles, permit children to efficiently learn the meanings of words (or better, nouns). As it turns out, all three assumptions can be nicely transferred to the connectionist model for word meaning introduced earlier.

Assumption (a) can be introduced by assuming that the model starts with a large focus of attention.[2] In other words, each label is first attached to concepts corresponding to whole objects rather than parts. Assumption (b) has already implicitly been built into the model by letting concepts be based on similarities.

According to assumption (c), novel identifiable states corresponding to labels should not be linked to previously labeled concepts. This is reminiscent of principles realized in *adaptive resonance theory (ART)* models (Grossberg 1976). There, in a kind of competitive learning, the goal is to prevent overgeneralization of categories to patterns that just happen to activate the same winner "by accident." It is achieved by letting each category learn a prototype and comparing this prototype with the current pattern. If the two patterns are similar enough ("resonance") the category is maintained. If they are not ("mismatch") the winner is reset and a new unit gets a chance of becoming active.

The same mechanism can be used to implement the mutual exclusivity assumption. After training, a conceptual state can activate the corresponding link unit representing its label. Now consider the case that in such a situation a novel link is about to be built, i.e. a novel link unit is active together with the same concept. This situation can be detected by letting the concept activate its learned label (link unit) and comparing the two in a kind of resonance or mismatch. In fact, comparison simply comes down to checking whether the two link units are identical. If they are (resonance), learning can continue. If they are not (mismatch) then the concept is supressed (reset), and another concept can be linked, provided the same cycle now leads to resonance.

If no other conceptual state can be activated given the current situation, mismatch can further trigger a shift in focus. Remember that it was stated earlier that principles have to be found to automatically set the size and center of the focus of attention. The presence of a word, mismatch and the absence of another concpetual state can now be introduced as one mechanism to trigger a shift. It can either be a positional one onto another object, or one in size onto one of the salient parts of the object. Although this has only partially been implemented (mainly the positional shift), the technical realization of the shift itself appears straightforward. If the shift is onto one of the object's part, the exact same behavior can be realized in the model that was roughly observed with children—namely that the model first tends to link words with whole object concepts, and later, only if a label has already been attached, with parts of objects.

The Origins of Taxonomies

With the inclusion of the three assumptions it is still not clear how the model should deal with sub- and superordinate concepts and their labels. They are not a matter of focus. If concepts on several taxonomic levels were activated when

[2] In this discussion we assume that objects do not overlap, thus bypassing the problem of how to focus on one object in a scene.

presenting sensory input, it would still be a problem as to which link should be built, aside from what can be handled by the one-to-one rule (see above). We need to look at further observations.

It is generally acknowledged that in human categorization there exists something called a 'basic level' (Rosch 1978, et al.), where concepts are formed most naturally and easily. For sensory-based concepts this level can be identified as the one where categories are found mainly due to natural similarities. Objects are named on the basic level (e.g. 'dog' or 'cat') much more frequently than on any other taxonomic level (e.g. 'poodle' or 'animal') and children learn such words much earlier than others (see, e.g., Aitchison 1987). Further empirical studies by Nelson (1988) and Benelli (1988) suggest that for learning concepts that are not on the basic level, such as superordinates, language itself plays a major role. They even go as far as suggesting that "superordinate terms are defined *in the language* and not in the world" (Nelson, 1988, p.4, emphasis by author).

Nelson (1988) further distinguishes two cases. First, she introduces superordinates as so-called "slot filler" categories. With this she means categories of objects (which might be in different basic-level categories) that are seen as belonging together through a specific context, such as food one usually eats for breakfast. Secondly, Nelson describes real taxonomic categories which she claims are defined through language and can only be learned linguistically.

This observations can, at least partially, also be reflected in the introduced model. The conceptualization mechanism based on similarities very nicely shows basic-level behavior, in that some categories are learned naturally without further reinforcement or influence ("bottom-up"). For superordinate categories at least two cases can be distinguished.

(a) They are categories of patterns that have too little similarities to be naturally thrown into one class.
(b) They are categories of patterns whose "similarities" are mainly defined through context, that is through invariant activations in other model components.

Case (b) corresponds to the slot-filler case, case (a) to real taxonomic concepts. In both cases further influence is necessary in order to cause a conceptual state to develop. This further influence, according to Nelson and Benelli, is language, that is, the labeling of those categories by words. In the model this can be implemented by conceptual states that are induced by a referential link. In other words, novel link units that cannot be mapped to any existing concept are permitted to recruit uncommitted units in a C-layer. This too is similar to the ART-model where units can be recruited to stand for novel classes. This recruitment can be supported by existing similarities, however small they may be. A learning rule—which is a slight variation of the soft competitive learning described above—can adapt weights so as to grow strong links between the invariants that are there and the concept unit. Thus after some training in many cases the concept can be activated bottom-up without first presenting a label.

A Simulation Run

The results of a simple simulation run of the major model components should make clear the basic functions of the extended model. The eight patterns in Fig. 2, chosen rather arbitrarily, were used as visual input. Eight different acoustic signals (call them 'a' to 'h') were used as labels ("words"). The model started in modus one and began to categorize both types of inputs. For this all eight visual, as well as the eight acoustic patterns were presented, each around 50 times. The visual patterns were presented in random order. The acoustic patterns were presented so as to correspond to one of the visual patterns' labels. For instance, label 'a' was used together with visual pattern 1 or 2, label 'e' for patterns 1 through 4, and so on (among the possibilities the choice was again random).

On the visual side, three basic level categories were developed, grouping patterns 1&2, 3&4, and 5&6 (the fourth grouping did not happen to be discovered in this particular run). All eight acoustic patterns lead to different identifiable states. As the first four labels were used to name the four expected basic level concepts (1&2,

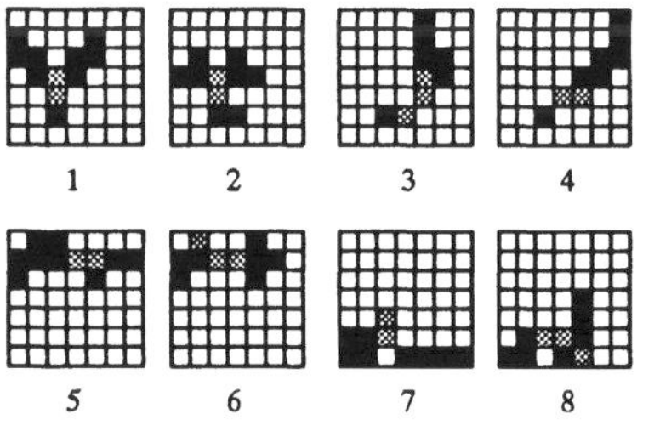

Fig. 2: Patterns used in the simulation

3&4, 5&6, and 7&8), three of them (the first three) were indeed learned perfectly. The one-to-one learning rule prevented any of the other labels to be associated with these concepts.

Then the model was switched to the modus of directed links. Now, whenever the new label 'e' was used to name the patterns 1 through 4 (superordinate concept), the novel link unit was compared with the perviously trained one, which lead to mismatch and reset of the conceptual state. No other concept could be activated. Shift of focus was not considered in this run. Therefore, after mismatch a new unit in one of the C-layers was recruited. Weight adaptation lead to the strengthening of the little similarities among patterns 1 through 4 (such as unit 3 in row 6), so that the new concept could also be activated bottom-up after some time. The model had thus learned a superordinate concept induced by naming. The previously learned basic level labels remained unchanged.

In another run, shift in focus of attention was considered, as well. Before recruiting a new unit, mismatch first triggered a shift. So labels for the most salient parts of each pattern (e.g. 'f' for the upper part in patterns 1 and 2) were learned, but only if a label for the whole object had been attached before. Otherwise the label was associated with the basic level concept, roughly mirroring Markman's observation on child language.

Conclusion

In this paper we have presented a psychologically plausible model for many aspects in the acquisition of word meaning. Many properties of the model can directly be compared to rough observations about early language learning in children; the model shows basic level conceptualization, builds links between labels and concepts after both have been identified clearly, and mirrors assumptions by Markman (1989, 1990) and Nelson (1988). It has been demonstrated that for achieving this mostly well-known neural mechanisms have been used. In the case of the extensions described in this paper, principles from the ART-model (Grossberg 1976) were transferred into this model.

Of course, many limits and weaknesses remain. Although many oversimplifications have been made, some complexities of the approach could nevertheless be demonstrated. Thus the strength of connectionist models to achieve a detailed model behavior has been underlined. Therefore the model is seen not only as a model with its psychological value, but also as a fruitful step toward alternative natural language understanding systems. Only the level of words—and there mainly nouns, and some adjectives—has been captured. It is believed that this is the primary level to be thoroughly understood if one wants to approach the phenomenon of natural language.

References

Aitchison J. 1987. *Words in the Mind*, an Introduction to the Mental Lexicon, Oxford: Basil Blackwell.

Benelli B. 1988. On the Linguistic Origin of Superordinate Categorization, *Hum.Dev.*, 31:20–27.

Dorffner G. 1992a. "Winner-take-more"—A mechanism for soft competitive learning, Technical Report 92–12, Austrian Research Institute for Artificial Intelligence.

Dorffner G. 1992b. On redefining symbols and reuniting connectionism with cognitively plausible symbol manipulation, Technical Report 92–13, Austrian Res.Inst.f.Artificial Intelligence.

Dorffner G. in press. A Step Toward Sub-Symbolic Language Models without Linguistic Representations, in Reilly R., Sharkey N. eds., *Connectionist Approaches to Natural Language Processing*, Hove: Lawrence Erlbaum, forthcoming.

Grossberg S. 1976. Adaptive pattern classification and universal recoding, I: Parallel development and coding of neural feature detectors, *Biological Cybernetics* 21:145–159.

Lyons J. 1977. *Semantics, Vol. I & II*, Cambridge: Cambridge University Press.

Markman E.M. 1990. Constraints Children Place on Word Meanings, *Cognitive Science* 14: 57–77.

Markman E.M. 1987: How children constrain the possible meanings of words, in Neisser U. ed., *Concepts and Conceptual Development*, Cambridge: Cambridge University Press.

McClelland J.L., Rumelhart D.E. 1981. An Interactive Activation Model of Context Effects in Letter Perception: Part 1. An Account of Basic Findings, *Psych.Review* 88: 375–407.

Nelson K. 1988. Where Do Taxonomic Categories Come from?, *Hum.Dev.* 31: 3–10.

Rosch E. 1978. Principles of Categorization, in E.Rosch, B.B.Lloyd eds., *Cognition and Categorization*, Hillsdale: Lawrence Erlbaum.

Point of View: Modeling the Emotions of Others*

Clark Elliott
The Institute for the Learning Sciences
Northwestern University
1890 Maple Avenue
Evanston, Illinois, 60201
and
The Institute for Applied Artificial Intelligence
DePaul University, 243 South Wabash Ave.
Chicago, Illinois, 60604
email: elliott@ils.nwu.edu

Andrew Ortony
The Institute for the Learning Sciences
Northwestern University
email: ortony@ils.nwu.edu

Abstract

When people reason about the behavior of others they often find that their predictions and explanations involve attributing emotions to those about whom they are reasoning. In this paper we discuss the internal models and representations we have used to make machine reasoning of this kind possible. In doing so, we briefly sketch a simulated-world program called the *Affective Reasoner*. Elsewhere, we have discussed the Affective Reasoner's mechanisms for generating emotions in response to situations that impinge on an agent's concerns, for generating actions in response to emotions, and for reasoning about emotion episodes from cases [Elliott, 1992]. Here we give details about how agents in the Affective Reasoner model each other's point of view for both the purpose of reasoning about one another's emotion-based actions, and for "having" emotions about the fortunes (good or bad) of others (i.e., feeling sorry for someone, feeling happy for them, resenting their good fortune, or gloating over their bad fortune). To do this, agents maintain *Concerns-of-Others* representations (COOs) to establish points of view for other agents, and use cases to reason about those agents' expressions of emotions.

Introduction

The Affective Reasoner is a program that reasons about the emotions of agents in a simulated multi-agent world. Agents in this world are given a simple emotional life, consisting of twenty-four emotion classes and approximately 1200 different expressions of these emotions. They are given idiosyncratic personalities which allow them to have different interpretations of identical situations in their world and different response tendencies for identical emotional states. In addition to "having" and "expressing" their own emotions, agents also reason about, and have emotions in response to the emotions of other agents.

To date the Affective Reasoner has been used primarily to simulate the interpersonal interactions of taxi drivers with one another and with their passengers in an instantiation called TaxiWorld [Elliott and Ortony, 1992]. However, experiments can be and have been conducted in other simple areas of interpersonal relations including the one used in a story about a rookie quarterback discussed later in this paper. In fact, all the Affective Reasoner requires is a domain in which emotions can arise, provided the domain can be represented using a discrete-event simulator.

We consider reasoning about emotions to have both a strong-theory component, from emotion-inducing situations to the emotions they engender, and a weak-theory component, from emotions to actions. To represent the different characteristics of individual agents we accordingly break our rudimentary personality representations into two parts. The first of these, the *interpretive personality*, is used for determining whether some event, act, or object is of concern with respect to the goals, standards, or preferences (hereafter GSPs) of some agent. Rule-based, strong-theory reasoning is used to tie interpretations of situations to emotional states. The second part of each personality representation is the *manifestative personality* component which is used for determining how an agent will act, or "feel," in response to these emotional states. This component contains a set of *temperament traits* which are dynamically activated so as to tune an agent's action tendency profile. Case-based, weak-theory reasoning is used to reason back from observed actions to an agent's emotional states. After a brief introduction to these two constructs, this paper focuses on how agents form, and maintain, internal representations of the personalities of other agents.

To understand how the *interpretive personality* works, suppose a basketball player on a team misses a shot at the buzzer in a game so that the team loses by one point. One observing agent might be unhappy that his team had

*Preparation of this article was supported in part by National Science Foundation grant IRI-8812699, and in part by Andersen Consulting through Northwestern University's Institute for the Learning Sciences.

lost the game, whereas another observing agent, being a fan of the victorious team, might be happy that her team had won. This is an example of how the same situation (the missed shot at the buzzer) can be interpreted either as one of having achieved, or of having failed to achieve, a goal. There are other alternative or additional interpretations that might be made. For instance, an observing agent might have admiration for the player for making a heroic attempt to win the game, an example of mapping the act into the perception of a principle being upheld, or an observer might simply be enthralled by the beauty of the move to the basket, an example of mapping the move into the perception of an appealing object, and so on.[1]

One important aspect of the Affective Reasoner's representation of agents' *interpretive personalities* is that it treats them as modular data structures, representing them as inheritance hierarchies of frames.[2] Any leaf node frame (i.e., one of the frames used to match and interpret events, acts, and objects in the world), and its inheritance path, may be combined with any other leaf node frame, and its inheritance path, when forming an *interpretive personality* representation. As a consequence, multiple and even conflicting interpretations of a situation may arise for some agent. This not a limitation of our system, but rather a requirement of emotion representation. For example, we may see a woman sad over the death of her favorite aunt, and yet relieved because she knows that her inheritance will be her financial salvation. So, the same situation can give rise to conflicting emotions which, in turn, could even give rise to similar actions (people can cry from sadness or joy). To sum up, the interpretive component of personality representations in the Affective Reasoner exploits the fact that individuals have different goals, standards, and preferences (GSPs) by allowing these different concerns to be the basis for interpreting emotion-inducing situations.

We turn now to a brief account of the *manifestative personality* component. Suppose that some agent is feeling proud about some admirable act she has performed. If she is a quiet type, she may simply manifest this emotion through a quiet *somatic* response (e.g., a feeling of general well-being). If she is verbally inclined, she may express her pride through *verbal communication* with another agent (e.g., telling someone about how proud she is). If she tends to be manipulative she may manifest her pride by attempting to *modulate the emotions of others* (e.g., seeking to have them admire her by calling attention to her praiseworthy act). Agents in the Affective Reasoner can have many of these different temperament traits active at the same time. Together these give the agent its idiosyncratic *manifestative personality.*

[1]See [Ortony *et al.*, 1988] for a full treatment of the emotion eliciting condition theory.

[2]Actually this is an over-simplification. The matching of frames against situations in the world also involves pattern-matching variables, a specialized unification algorithm, procedural attachment to the slots, and so forth.

There is at best only a loose mapping from emotional states to particular actions; few actions are unambiguously indicative of the emotions that initiated them. For example, people can *smile* because they are happy, because they are gloating over the misfortune of an adversary, or because they are pretending not to be afraid. In addition people can express the same emotion in many different ways (e.g., frowning because they are angry and want it to be known, or smiling because they are angry and do not want it to be known). Complicating this is the fact that individuals have different emotion expression styles, and are affected by moods as well. Because of such ambiguities we have chosen a case-based approach to reasoning in this portion of the emotion domain.

People model one another's points of view. This enables them to both explain and predict the responses of others to situations, and to have emotions regarding the fortunes of those others. In the Affective Reasoner these points of view are captured in the two-part rudimentary personality representation just discussed. To model such points of view, agents can maintain internal models of both the concern structures of other agents (i.e., their *interpretive personalities*), and their response action structures (i.e., their *manifestative personalities*). A model of the former allows the agent to make inferences about emotions other agents are likely to have in certain situations. A model of the latter allows the agent to make explanatory inferences about antecedent emotions when seeing other agents acting in a certain way. These two aspects of capturing and maintaining knowledge about other agents are discussed in the next two sections.

Representing the Concerns of Others

For an agent to understand how another agent is likely to construe a situation, he or she must see that situation from the other agent's point of view. Because in the Affective Reasoner an agent's interpretations of the world are derived from its GSP database it follows that an observing agent must also have some internal representation of the observed agent's GSPs. This knowledge is captured in data structures known as *Concerns-of-Other* (COO) databases. They are, essentially, imperfect copies of other agents' GSPs, and represent their concerns as modeled by an observing agent.

Thus, in addition to the GSP database representing an agent's own concerns, a COO database can be maintained for each other agent the observing agent is modeling. Using the same machinery that causes emotions to be generated by the system for some agent when a situation is filtered through that agent's *interpretive personality* (i.e., its own GSP database) that agent may instead filter the situation through the internally modeled *interpretive personality* of the observed agent (i.e., its COO representation for that other agent) to see the situation from the other agent's (supposed) point of view.

A perfect COO representation, of course, would be an exact duplicate of the observed agent's GSP database, and would always lead to the same interpretations that the observed agent has. However, because as discussed above, GSPs are built out of interpretation modules (i.e., frames), partial COOs can be created incrementally. Even though they are imperfect, these partial representations are useful because they allow the observing agent to interpret at least some situations correctly. For example, Harry might know that Sarah is a passionate Cubs fan, and that if the Cubs lose she will be upset, and yet not know more about her. Still, if the Cubs do lose, and Sarah is jumping up and down, then Harry probably knows why.

Because the Affective Reasoner was developed as a general research platform, several options are available with respect to the establishment of COO databases. They may be constructed at start-up time as part of the initial composition of agents, or they may be learned as the simulation proceeds and as agents come into contact with one another. In the former case a number of difficulties are avoided, such as having to work out the details of when agents are permitted to observe each other. In the latter case, many useful knowledge acquisition issues can be studied. For example, if the system is to be used to store knowledge about interesting agents and to study emotions that arise when they interact, then the domain-analysis investment required for setting up the COO learning process will have little return. On the other hand, if one is studying user-modeling from an emotion perspective, such a component could be very useful.

Collecting Construal Frames for COOs

When the Affective Reasoner is set up so that agents learn about one another's concerns through interactions, COOs are built up incrementally by locating and incorporating construal frames that seem to explain another agent's emotional states in response to observed situations. For example, when Harry sees that Sarah is always unhappy when the Cubs lose he might infer that Sarah construes some aspect of this situation as blocking one or more of her goals. Harry might then try to determine exactly which goals are involved: is Sarah a Cubs fan, or has she just been betting on them to win? In the following algorithm, which describes this process, we assume that the observing agent has already discovered the emotion(s) present in the other agent.[3] The observing agent now attempts to explain that emotion in terms of the eliciting situation, and possible construals of that situation. To do this the observing agent first consults its COO for the observed agent, and then, if necessary, a set of databases containing alternate construal frames (see **Defaults**). Here is the algorithm for incrementally building COO representations:

1. Locate the *Concerns-of-Other* representation (COO) for the observed agent. If one does not exist or it does not contain an interpretation for this *type* of eliciting situation[4] then go to 5,
2. Filter the situation through the COO, producing an emotion. If this emotion is the same as the emotion that actually was present in the observed agent then the COO has probably given a correct interpretation of the eliciting situation so go to 8,
3. Because the interpretation produced by the COO is incorrect (i.e., the emotion based on the COO's interpretation of the eliciting situation does not match the emotion known to be present in the agent whose concerns it is supposed to represent) the construal frame used to make the interpretation should not be part of the concern structure for the agent. Remove it from the COO,
4. Mark the construal frame ineligible for this agent. Eligible frames are those frames that can produce interpretations for this *type* of situation. Ineligible frames are previously eligible frames which have been found to produce incorrect interpretations,
5. Search through the global (or default) database for the next eligible interpretation of this situation,
6. Evaluate the situation using the new interpretation as a filter. If the resultant emotion is not the same then go to 4,
7. Add the construal frame to the COO,
8. Generate an explanation based on the current construal frame.

Once a COO has been established for some other agent it can be used for two purposes. First, it is now possible for an observing agent to have emotions based on its perceptions of the fortunes of the second agent. In the Affective Reasoner this may come about if the agents are in one of the following three (possibly only unidirectional) relationships: *friendship, animosity*, and *empathetic unit.*[5] For example, if the observing agent knows that a second agent is a Cubs fan, then if they are *friends* the observing agent can feel sorry for the second agent when the Cubs lose. On the other hand, if they are *adversaries* then the observing agent can gloat when the Cubs lose. Lastly, should the bond between the two agents be so strong in some situation that the observing agent temporarily takes some of the second agent's concerns on as its own then an *empathetic unit* has been formed. The observing agent will temporarily suspend its own GSP database, using its COO for the second agent to generate direct emotions instead. Note that even in this case the observing agent might actually be wrong about the import of a particular situation for the observed agent, since

[3]Obviously, to make inferences about *why* an agent is in some emotional state we must first know what that state is. Because of space limitations our approach to this will only be discussed briefly in a later section.

[4]All eliciting situations are typed. Construal frames which interpret them have the same type.

[5]These three relationships have a very specific meaning here. *Friendship* means that an agent will tend to have similarly valenced emotions in response to the emotions of another agent. *Animosity* means that the emotions will tend to be oppositely valenced. *Empathetic unit* means that the particular situation is seen "through the eyes" of the other agent, so that the emotions are experienced as the observing agent's own.

it is the observing agent's *representation* of the observed agent's concerns that is being used to generate emotions, not the actual GSPs of the observed agent. The second use of COOs is that once they are established it is possible to explain, and sometimes predict, the emotional responses of other agents based on the eliciting condition rules, as in the previously discussed case of Sarah the Cubs fan, which opened this section.

Satellite COOs

Agents in the Affective Reasoner may be set up to do more than model the simple, direct concerns of agents whom they observe. The need for more complex internal models is illustrated by the following story, on which one of our simulation runs was based.

> A rookie quarterback is, as usual, sitting on the bench during a football game. His brother and a woman friend of his brother are in the stands. Suddenly the starting quarterback goes down with a knee injury. The woman smiles because she is happy for her friend who's brother will now be placed in the game.

When reasoning about the emotions that arise in this situation we must consider the following sets of concerns and relationships:

1. The actual concerns of the rookie quarterback, i.e., his GSPs. Implied in the story is that he will be pleased about achieving a *getting-to-play* goal.
2. The supposed concerns of the rookie quarterback as represented by his brother (i.e., the COO representing the brother's beliefs about the GSPs of the rookie quarterback).
3. The relationship between the rookie quarterback and his brother. Specifically the *friendship* relationship, or even an *empathetic unit* relationship.
4. The *friendship* relationship between the woman and the brother.
5. The supposed concerns of the brother as represented by the woman. This must include, recursively, her *supposed* supposed concerns of the brother for the quarterback as well, and the supposed empathetic relationship between the brother and the quarterback. In other words, the woman must have a belief that the brother will believe that the rookie quarterback will be happy about the starting quarterback's injury. Furthermore, she must believe that the relationship between the brother and the rookie quarterback is such that a positive outcome for the rookie quarterback maps to a positive outcome for the brother.

Because the story gives no clues as to the emotional states of either the quarterback or his brother it should be obvious that neither the actual concerns of the quarterback nor those of his brother are necessary for understanding the episode. To make this clear, consider the following possible continuation to the story:

> ...But the smile quickly fades when the brother says, "Oh no, I *told* him he shouldn't have drunk that case of beer at lunch."

Clearly the woman's beliefs leading to emotional states and action expressions of those states are not dependent upon all of the actual facts. Similarly, even if her understanding of the facts is correct, this still does not mean that her emotions have to be in line with them. Consider the following alternate continuation to the story,

> Instead of playing the rookie quarterback, however, the coach puts in a third-string quarterback. The woman, who unbeknownst to the brother had consumed a case of beer with the rookie quarterback at lunch and was sworn to secrecy about it, is relieved. The brother, however, feels terrible for the rookie quarterback because he will not get to play. Consequently the brother is very unhappy. The woman is sorry to see him in this state.[6]

In this case the woman *knows* that the brother's beliefs are incorrect, and she does not share them, but she still is capable of having emotions based on the brother's fortunes, which in turn are based on those incorrect beliefs.

It can be seen then, that for observing agents to represent the fortunes of another agent, to have emotions regarding those fortunes, and to interpret their actions with regard to those fortunes, the observing agents must not only be able to represent the concerns of the observed agents, but sometimes must also be able to model the observed agents' own representations of the concerns of those important to them. In the Affective Reasoner we capture such knowledge in a second-level set of COOs, called *satellite COOs*. These are used in conjunction with a set of supposed relationships between the observed agents and those whom the satellite COOs are intended to represent. For example, if Harry believes that Sarah is in a relationships with both Joan and Eva, then Harry will maintain a COO for Sarah and satellite COOs for Joan and Eva as seen through Sarah's eyes.

The three distinct *interpretive personality* representations used by the system are all structurally and functionally the same. It does not matter whether the representation is to be used as a system-level GSP or as an agent-level COO or satellite COO. The emotion machinery that is applied to GSPs for the generation of direct emotions may also be applied to COOs used for the generation of the fortunes-of-others emotions and to satellite COOs used for representing an agent's beliefs about another agent's beliefs.

[6]Situations in which the feelings of the other agent are not in accord with the known facts, and yet where the observing agent responds only to those feelings, are in fact not very common. In general this is because if the observing agent knows that the observed agent will soon find out the facts, the observing agent is much less likely to base his or her emotions on the temporary happiness or unhappiness of the other agent. When this occurs it adds an element of secrecy, or of quirky twists of fate (e.g., someone dying before they find out) which in itself almost always complicates the situation and the resulting emotions. There seems to be some difference between the negatively and positively valenced emotions with respect to this as well. One is more likely to be sad that a friend is temporarily unhappy because she misunderstands a situation that will ultimately make her happy, than one is to be happy that a friend is temporarily happy because she misunderstands a situation that will soon make her sad.

The process for making use of each of these GSP and COO databases is also the same in all cases. The eliciting event, act, or object is filtered through each respective GSP or COO database to produce an interpretation with respect to the antecedents of emotions. In the direct case, the result is emotions that the system generates for the agent. In the once-removed case the result is an interpretation based on *imagining* what it is like for the other agent (possibly incorrectly), which, when combined with a relationship, may yield a fortunes-of-other emotion in the observing agent. In the twice-removed case, when combined with beliefs about relationships, the interpretation may lead to a belief about the emotional state of the other agent, which in turn may also lead to fortunes-of-others emotions.

Defaults

In some cases little may be known about another agent. Nonetheless, one may feel sorry for a stranger, and one certainly may wish to explain the actions of strangers. Thus we must give agents a mechanism by which they may still reason about the emotions of other agents, even if nothing specific is known about those agents.

Because, for the purpose of generating emotions in the Affective Reasoner, one GSP database is as good as another, and because even the component construal frames may be mixed at will, we may use a system of defaults for reasoning under uncertainty. Two of these are rather obvious. The first is a system-wide default GSP which corresponds to the knowledge source one might consult in addressing such questions as *How might a typical agent interpret this situation?* The interpretations produced by this default database are useful when producing explanations such as *When someone is hit they get mad* and *Losing money increases distress.* The next obvious default GSP is an agent's own GSP database, which corresponds to the knowledge source one consult's when asking *How would I interpret and react to this situation?* The resulting emotional states can then be projected onto the other agent.

In addition to these two defaults, observing agents in the Affective Reasoner might also make use of an extant COO for some third agent, provided that one exists and that it is consistent with what has already been observed about the new modeled agent. As long as this COO is suitable it remains in use. When the COOs diverge (i.e., when one of the construal frames in the existing COO is found to be incorrect for the new modeled agent) then a copy of the existing COO is made and the offending construal frame is removed. This becomes the current representation of the COO for the new agent. The use of COOs in this manner corresponds roughly to reasoning that because *Agent A* seems just like *Agent B,* then assume they are alike in all ways until learning differently.

Nor are we restricted to using only one COO when searching for an explanation. For agents then, the order of precedence is as follows: (1) search through the COO for the other agent to look for an interpretation of some eliciting situation; if there is none, or it is found to be in error, then (2) search through a COO for some other agent that appears to be similar to this one, if one exists; next, (3) search through the system default GSP database to see *how a typical agent* would interpret the situation; failing in this, then (4) search through one's own GSP database to see *how I might interpret* this situation, and lastly (5) search through the global shared database of construal frames for all possible interpretations of the situation.

Representing the Action Tendencies of Others

In the Affective Reasoner observing agents attempt to make sense of the way observed agents respond to situations that arise by using emotion-specific knowledge to limit the search space. Something happens, this gives rise to an emotion, or set of emotions, and these in turn give rise to emotion-induced reactions. So far in this paper we have discussed representations that agents keep of others' concerns. This knowledge may be used in two ways. First, reasoning backwards it may be used abductively to explain how an agent sees the world. Second, once established for an agent, it may be used deductively to predict what emotions that agent may have in response to future situations. By contrast, in this section we discuss an alternate source of knowledge which allows agents to observe features of a situation and some agent's response to that situation (i.e., an emotion *episode*), and reason back to an emotion category using past cases. This roughly corresponds to lines of thought such as *Is the agent smiling? He is probably happy. Is he shaking his head? He may be reproachful.*

With respect to such reasoning the Affective Reasoner may be run in three major modes, two of which will be discussed here. In the first mode agents make use of a heuristic classification component based on the *Protos* program developed by Bareiss [Bareiss, 1989]. Using their own set of cases drawn from past experience, agents make determinations about which emotion is present based on the features in an eliciting situation and on the observed agent's responses to that situation. In this mode agents are free to ask questions of the "teacher" (in this case the user) to acquire knowledge about the relationship of features to emotion categories. This knowledge includes such relationships as *highly correlated, mutually exclusive*, and so forth. For example, in this mode an agent may decide that another agent who is shouting is expressing anger, because the present case reminds the agent of a previous case of anger that had that feature. Should there be additional features present that the agent does not understand, or if the classification is made incorrectly, then the agent, through Protos, asks for domain knowledge from the teacher. Suppose for example that in addition to shouting, the observed agent was also represented as shaking its fist in the air. In this case the observing agent might ask for an explanation of how this feature relates to anger, and would then either update the present exemplar, and possibly the set of remindings that lead to its selection, or create an entirely new exemplar if the new one is sufficiently different from the old one. In the second mode Protos is used only to make classifications for agents using an existing case base. No knowledge is acquired and the case base remains static. This mode is useful for running simulations without input from the researcher, where the case bases have reached a certain level of maturity. As with COOs, case bases may be established at start-up time as part of the initial composition of agents, they may be acquired entirely as part of the current simulation, or they may be established at start-up time and then enhanced as part of the current simulation.[7]

In using this scheme for capturing weak-theory knowledge about the features of emotion episodes one might wish to store cases for *each* other agent, just as one maintains distinct COO representations for each other agent. Such knowledge would be equivalent to knowing that *Tom shakes his fist when he gets angry,* and *Harry always wrings his hands when he gloats.* We have not taken this approach in the Affective Reasoner. Instead, each agent maintains a single case base. This is the equivalent of having *seen a case of anger before where the agent was shaking his fist.* However, since the name of the agent is counted among the features of an episode which may be recorded by Protos, it is nonetheless still possible to capture some agent-specific knowledge about the expression of emotions through actions.

Conclusion

We have described components of a system, the Affective Reasoner, in which a number of simulated agents, each with their own "personality" interact. These agents respond to situations in their world in emotionlike ways, but they also respond to what they take to be the emotions of others. We conclude with a brief discussion of some caveats pertaining to this work, particularly with respect to relation between the "emotions" of our simulated agents and emotions as we know and experience them as human beings.

We do not claim that our simulated agents "have" or "feel" emotions. Such a claim would be uninterpretable at best and nonsense at worst. Human emotions are comprised of interacting cognitive components, behavioral components, physiological components, and phenomenal components. The Affective Reasoner only seeks to model (aspects) of the cognitive and behavioral components, and is best considered as an attempt to generate the ingredients required to reason about emotions rather than as an attempt to produce emotions. When we humans speak of having or feeling emotions, we are implicitly focusing on the phenomenal and, perhaps therefore by necessity, the physiological components. Human emotions are not cold cognitions leading to detached behaviors, they are hot cognitions integrated with (sometimes dysfunctional) behaviors. It is their physiological and phenomenal qualities that give them their special "feel," and we neither attempted to, nor even would know how to begin to model these aspects of emotions. Rather, we concentrated on the more manageable aspects, namely the cognitive and behavioral ones. Human emotions are not randomly related to how people perceive their world. There is some order, and the Affective Reasoner seeks to capture some of that order by embodying a strong theory of the relation between construed situations and emotions. Similarly, whereas the linkage between human emotions and actions may be somewhat weak, still, the relation is not arbitrary. The Affective Reasoner incorporates a weak theory of emotion-to-action relations in an effort that we claim to be little more than a first step. Thus, the affective Reasoner could be viewed as a system that attributes emotions to its agents by reasoning, rather than as a system in which emotions simply arise in agents. In this paper, we have focussed on this reasoning process at one level of embedding in that we have described what emotions the system attributes to its agents when they are interpreting the situated behaviors of others as being emotion-induced.

Trying to build systems that understand anything at all about emotions is a not easy. In order to prevent the system from becoming unmanageable, consideration of many important aspects of emotions and emotion-related behavior had to be postponed. The most obvious of these is the omission of considerations of emotion intensity which, in future efforts, is likely to be handled using qualitative reasoning techniques. In fact, as experiencers and observers of human emotions we frequently use intensity-relevant inferences to predict and explain behaviors. For example, given knowledge about particular individuals and their emotional "styles" we can infer whether or not they will react with intense or mild emotions in a particular types of situations. We can make similar inferences about emotion-induced behaviors because we know that generally speaking, mild emotions do not give rise to extreme behaviors, and so on. We hope to address these and other limitations in future work on this project.

[7] In the third mode, classification of episodes is bypassed entirely, and observing agents are simply informed directly by the system what emotion(s) the observed agent was experiencing.

References

[Bareiss, 1989] Ray Bareiss. *Exemplar-Based Knowledge Acquisition, A Unified Approach to Concept Representation, Classification, and Learning.* Academic Press, Inc., 1989.

[Elliott and Ortony, 1992] Clark Elliott and Andrew Ortony. The affective reasoner: Modeling emotions in a multi-agent system. Submitted to: Fourth European Workshop on Modeling Autonomous Agents and Multi-Agent Worlds, April 1992.

[Elliott, 1992] Clark Elliott. *The Affective Reasoner: A Process Model of Emotions in a Multi-agent System.* PhD thesis, Northwestern University, May 1992. Also forthcoming technical report from The Institute for the Learning Sciences.

[Ortony *et al.*, 1988] Andrew Ortony, Gerald L. Clore, and Allan Collins. *The Cognitive Structure of Emotions.* Cambridge University Press, 1988.

Using Stories to Enhance and Simplify Computer Simulations for Teaching

Richard G. Feifer
Thomas R. Hinrichs
The Institute for the Learning Sciences
Northwestern University
1890 Maple Ave
Evanston, IL 60201
feifer@ils.nwu.edu
hinrichs@ils.nwu.edu

Abstract

Computer-based simulations are a valuable teaching tool because they permit a learner to explore a phenomenon on his own and to learn from his mistakes. Two factors, however, limit the use of computer simulations in teaching: good simulations are hard to build and learners can flounder with just a simulation.

We have built HeRMiT[1], a case-based tutor that integrates a simulation with a library of videotaped stories. The stories make up for any lack of depth or fidelity in the simulation by facilitating the generalization and application of underlying principles.

Introduction

The best way to learn how to do something is to try to do it and learn by your mistakes. Computer simulations allow you to make mistakes in situations that would normally be dangerous or expensive (i.e., flying a plane or disarming a bomb). Two factors, however, limit the use of computer simulations in teaching:

1. good simulations are hard to build;
2. learners can flounder with just a simulation.

Building a good simulation in any but the most trivial domains is difficult. Allowing the user a wide range of actions and simulating the results of any combination of those actions requires a complete model of the domain.

Even if we could build a simulation with sufficient fidelity, it is difficult to learn from mistakes without some guidance. Failure only shows you what not to do. It is often not obvious why you failed, in what range of circumstances the actions you took would lead to failure and what you should have done instead.

One solution to the floundering problem is the addition of a computer-based coach to watch over the learner's shoulder and advise (Burton & Brown, 1979; Goldstein, 1979). The coach, however, also requires a good model of the domain, plus a model of the learner. Even if we could model a learner's misunderstandings in a simulation, we still have the problem of generating a dialog to ameliorate those misunderstandings.

We have built HeRMiT (Human Resource Management Tutor), a case-based tutor (Feifer & Soclof, 1991; Schank, 1991) that teaches without true fidelity in a simulation, without a learner model, and without computer generated responses by adding human stories (Bell & Feifer, 1992). Instead of the computer generating instruction, good story tellers, experts in the domain, tell their stories on video tape. These stories are then indexed to the kinds of failure for which they are relevant.

These stories make teaching through simulations more practical in three ways:

1. the simulation need only provide a context and motivation for the story, the stories make up for any lack of depth or fidelity in the simulation;
2. it is easier to index failures than to model the learner sufficiently to provide intelligent coaching;
3. it is easier to show a video than to generate instruction, and more compelling to the learner;

Supplementing Simulations

In a case-based tutor we begin with a simulation of the task we want the learner to accomplish. The

[1]HeRMiT was built in cooperation with the Professional Education Division of Andersen Consulting.

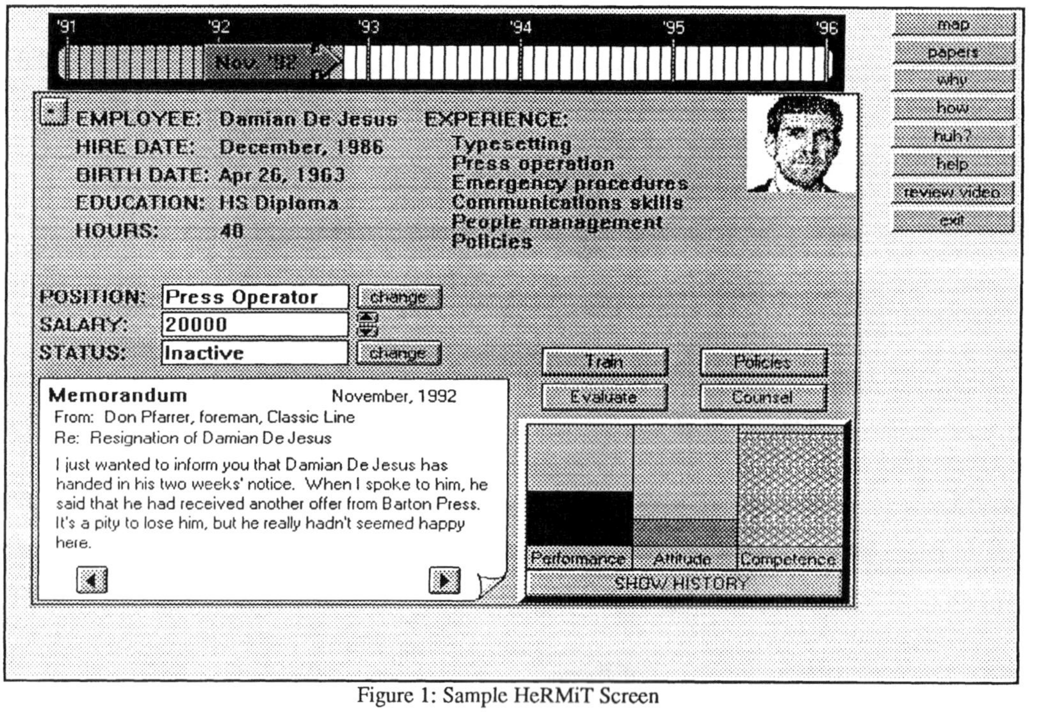

Figure 1: Sample HeRMiT Screen

interface to the simulation provides the learner with a range of actions that would be possible in the situation. The simulation provides immediate feedback: what might happen in the real world if the learner took the same actions?

HeRMiT's goal is to teach the basic issues involved in managing human resources, and to convince the learner that there is a connection between the manner in which a company manages its human resources and the company's bottom line. To accomplish this goal the learner is asked to manage the human resource function of a simulated company (figure 1).

The *Position, Salary, Training, Evaluate, Counsel,* and *Status* buttons allow the learner to take actions that would be available to a Human Resource Manager in a company. The bar charts in the lower right indicate the results of any of those actions. In addition there are meters on the main screen that reflect the company's overall productivity and morale as a result of actions taken on individual employees.

Simulating Human Resource Management

To be a useful teaching device, the HeRMiT simulation had to be broad enough to illustrate basic principles, such as the 'Peter Principle' (promoting someone to their own level of incompetence) and the 'Hawthorne Effect' (employees tend to be more motivated when management pays attention to them).

It was also critical to capture individual differences among employees. Human resource management is not formulaic — there are no cookbook solutions to human problems. To emphasize the importance of paying attention to the individual, we represent employees as differing in their levels of dedication, aptitude, ambition, experience, and education. These factors determine how quickly a new employee learns his job, how effectively he performs, and how rapidly his expectations of salary and position will grow. Accounting for individual differences is especially critical when making hiring, promotion, and salary decisions.

The simulation also had to be deep enough to emphasize the nature of fundamental tradeoffs in personnel management. For example, paying high salaries may help employee morale, but it will eat into corporate profits. Likewise, promoting people when they are ready may be best for the individual, but it may also lead to top-heavy management and large payroll costs. HeRMiT forces the student to balance these considerations.

Despite these complexities, the simulation is not particularly sophisticated. We were able to restrict the range of phenomena it covers (the breadth of the simulation) by restricting the variety of its inputs and outputs. The inputs to the simulation are the discrete actions that the student can take: changing an employee's salary or position, training the employee, evaluating, counseling, firing and hiring an employee. The outputs of the simulation are the values of a few parameters: the attitude, competence, and performance of each employee and the overall morale and productivity of the company. These few parameters still permit the simulation to be open ended because the student controls the timing and magnitude of his actions.

Learning From Failure

What happens once the learner's actions have led the simulation to a negative outcome? In Figure 1, for example, the memo in the lower left informs the learner that a valued employee has just resigned. The resignation is not a good thing, indicating that the learner in some way mis-handled this simulated employee. The learner is wondering: 'Why did he quit?' or 'What could I have done to avoid it?' In this particular case the employee quit because she had become overqualified for her current position and the learner did not promote him.

Relying on the Simulation

Ideally, the learner forms a hypothesis for the cause of the negative outcome and tests it in the simulation. Through trial and error the learner can find a correct path. Ideal here refers to both the learner and the simulation. The ideal learner has the motivation and the skills to form and test hypotheses. The ideal simulation provides accurate feedback for the full range and combination of learner actions.

Unfortunately HeRMiT's simulation is not ideal. One of the problems in developing social simulations such as HeRMiT is the availability and precision of theories of human behavior. Existing models of motivation and performance tend to be descriptive and statistical in nature and are often defined in terms of influence systems. To translate this into a generative model we had to specify behaviors more precisely than theory would warrant.

One possible solution to this problem is to use a 'non-deterministic' simulation in which the behavior of an individual obeys a probability distribution. For example, in our domain this would mean that an employee might or might not quit in a given situation. For this approach to be effective, however, the learner would have to generalize this behavior by running the simulation many times. Since we expect the learner to run the simulation only a few times, we felt that the complexity and unpredictability of a non-deterministic simulation would be unnecessary and undesirable.

Our solution was to build a deterministic simulation by translating qualitative influences into cumulative, decaying, and one-shot effects and scaling those effects to demonstrate (and somtimes exaggerate) the human resource management principles. For example, we model the Hawthorne Effect as a short-term boost in attitude after any positive management action (e.g., a raise, promotion, or evaluation) and a gradual, cumulative decline in attitude that begins after the employee has been ignored for too long (where the time of onset and rate of decline depend on characteristics of the individual employee). While this oversimplifies the real phenomenon, more precise theories are unavailable, and this model is sufficient to communicate the ideas we wish to teach.

Using Stories

If the simulation were to stand on its own, its shortcomings would be more readily apparent. However, because the simulation is one component of a case-based tutor, it successfully provides a motivating context for learning the theory of human resource management, a responsive environment for making H.R. decisions, and an opportunity to make mistakes and fail with impunity.

Once a learner has failed, he is thinking about the context of that failure. He is motivated to learn anything that might help to avoid this failure in the future. There is also a good chance that he will store any new information appropriately in his long term memory, such that he will be able to recall it in relevant future contexts. Such failures provide the teaching system an opportunity to tell good stories. Thus we do not correct a learner when he makes a mistake. Rather, we wait for the mistake to lead to a recognized failure and for the learner to attempt to diagnose the failure on his own. Stories in HeRMiT are told by charismatic storytellers, who lived through similar disasters. Each story contains a description of the first warning signs, a dramatic description of how the problem led to some horrible outcome and the lessons learned from the situation.

The learner viewing the screen in Figure 1 should be wondering what he did wrong. While still viewing the personnel folder, he can look for clues and may form a hypothesis. Once he closes the folder, the learner has satisfied the above criteria, and HeRMiT presents a story. In this case the story is about a mid-level manager who was due for a promotion. The company, however, needed her in her present position because of some special skills that she had. They explained the situation to her, and thought that she understood and was being a good sport. Six months later she quit to accept a higher position with a competitor and is currently making life miserable for her original company.

The story provides the learner with at least one more example, this one in the real world, of the type of failure they encountered in the simulation. To use the story the learner must generalize their failure to a class of failures. Instead of thinking about the mistake of "Not promoting someone who has been doing the same job for 4 years," they are encouraged to think about the more general issue of "Failing to meet an employee's expectation of growth within a job."

Since the failure described in a story will not perfectly match the failure that led to the learner's negative outcome, the story does not provide an "answer" to the learner's problem. Thus the learner must derive the general principles in order to adapt the warning signs or solutions indicated in a story to the simulation.

Indexing Stories to Failures

An effective case-based teaching system must tell the right story at the right time. Knowing the right time to tell a story turns out to be easy in HeRMiT because failures are only manifested in a few ways: when the company's productivity hits zero, when overall morale drops below a threshold, or when a good employee quits. The difficulty is in knowing the right story that will best explain the learner's error.

Picking the right story is difficult for two reasons. First, reconstructing the events that led to a failure is beyond the ability of HeRMiT's simulator. Second, failures often have multiple causes in this domain.

Tracing back from a failure to its causes is hard partly because there are no backward links from the outputs of the simulation to the inputs, and partly because the simulation does not automatically record its history of events. Even if it did, it would still be difficult to determine how far back to trace and how best to explain the failure in meaningful terms. For example, if an employee were to quit, is it because his attitude was low or because he was hired and not properly trained for the job?

Therefore, rather than trying to trace back from a failure to its causes, we record the learner's mistakes as they are made and before a failure actually occurs. To do this, HeRMiT maintains a list of typical Human Resource problems for which it has video stories, such as training too little and too late, underpaying, promoting too soon, promoting too far, and evaluating too infrequently. Each type of mistake is recognized by a 'demon' that checks whether its conditions are satisfied after each simulated month. When a demon recognizes a mistake, it adds it to a list of the learner's mistakes that could lead to failure. If a failure later occurs, HeRMiT considers any mistake that could have led to the particular failure to be a possible cause.

Unlike traditional intelligent tutoring systems, there is no learner modelling. HeRMiT models only causality within the domain, not between the learner's cognitions and the domain. We do not try to blame the failure on some belief the learner might have, which would be very difficult. Rather we blame the failure on an action we know the learner took.

The list of possible causes is used to choose an appropriate story to tell. The difficulty here is that there may be many factors that contribute to a failure. For example, one way the simulated company could fail would be if the learner hired the wrong applicant, neglected to train him, paid little attention while the employee's attitude declined, and then tried to correct the mistake by overpaying everybody. In this case, the most recent mistake is probably not the most critical.

Instead of choosing the most recent mistake, we rank the types of mistakes roughly in order of their importance from a Human Resources perspective. HeRMiT then selects the most critical mistake and presents its corresponding video story. The story doesn't necessarily explain 'the' cause of the failure, but suggests how one of the learner's mistakes may have contributed to the failure. This approach of indexing by mistakes is simple but effective for domains with stereotypical classes of mistakes and well defined failure modes.

Conclusion

We have built a tutor that combines two knowledge elements:

First, there is a simplified simulation of the human resource function of a small company. Users have reported that it is fun to use. They enjoy playing with the simulation, trying out different actions, seeing how long they can stay afloat.

Second, there are 40 minutes of indexed stories about actual human resource management disasters. People enjoy listening to the stories. Some people have even sat through all 40 minutes, from beginning to end.

As engaging as each of these elements is, however, neither standing alone would effectively teach the principles of human resource management. It is only by combining them and giving the computer the ability to find the right story for a failure, that we have a cognitively sound tutoring environment.

The first learner test of HeRMiT was conducted in April of 1992 with five subjects. Four subjects had no previous exposure to human resource management. After using HeRMit all learners expressed a new appreciation of the importance of human rescource management to the health of a company and were able to demonstrate an understanding of the basic issues (through written and oral debriefing). At least as important, all subjects and the 30 learners who informally used the program reported that it was engaging.

Acknowledgements

This research was supported in part by the Defense Advanced Research Projects Agency, monitored by the Air Force Office of Scientific Research under contract F49620-88-C-0058 and the Office of Naval Research under contract N00014-90-J-4117, by the Office of Naval Research under contract N00014-J-1987, and by the Air Force Office of Scientific Research under contract AFOSR-89-0493. The Institute for the Learning Sciences was established in 1989 with the support of Andersen Consulting, part of The Arthur Andersen Worldwide Organization. The Institute receives additional support from Ameritech, an Institute Partner, and from IBM.

In addition to the authors, the following people are involved in the development of HeRMiT: Larry Langelier, Donna Fritzsche, Wayne Schneider, Michael Korcuska, Joshua Tsui, Leena Nanda, Cheryl Jindra, Matt Greising, Alan Nowakowski, and Don Jastrebsky. Special thanks to the HeRMiT storytellers: Bill Braemer and Tim Coan.

References

Bell, B. L., & Feifer, R. G. (1992). Intelligent tutoring with dumb software. In *Proceedings of the International Conference on Intelligent Tutoring Systems*, . Montreal:

Burton, R., & Brown, J. S. (1979). An investigation of computer coaching for informal learning activities. *International Journal of Man-Machine Studies*, 11, 51-77.

Feifer, R. G., & Soclof, M. S. (1991). Knowledge-based tutoring systems: Changing the focus from learner modelling to teaching. In *Proceedings of the International Conference on the Learning Sciences*, (pp. 151-157). Northwestern University, Evanston, IL: Association for the Advancement of Computing in Education.

Goldstein, I. P. (1979). The genetic graph: A representation for the evolution of procedural knowledge. *International Journal of Man-Machine Studies*, 11, 51-77.

Schank, R. C. (1991). *Case-based teaching: Four experiences in educational software design* (Technical Report No. 7). The Institute for the Learning Sciences, Northwestern University, Evanston IL.

Bootstrapping Syntactic Categories

Steven Finch
University of Edinburgh, Centre for Cognitive Science, 2, Buccleuch Place, Edinburgh.
steve@cogsci.ed.ac.uk
Nick Chater
University of Edinburgh, Dept. of Psychology, 7, George Square, Edinburgh.
nicholas@cogsci.ed.ac.uk

Abstract

In learning the structure of a new domain, it appears necessary to simultaneously discover an appropriate set of categories and a set of rules defined over them. We show how this *bootstrapping* problem may be solved in the case of learning syntactic categories, without making assumptions about the nature of linguistic rules. Each word is described by a vector of bigram statistics, which describe the distribution of local contexts in which it occurs; cluster analysis with respect to an appropriate similarity metric groups together words with similar distributions of contexts. Using large noisy untagged corpora of English, the resulting clusters are in good agreement with a standard linguistic analysis. A similar method is also applied to classify short sequences of words into phrasal syntactic categories. This statistical approach can be straightforwardly realised in a neural network, which finds syntactically interesting categories from real text, whereas the principal alternative network approach is limited to finding the categories in small artificial grammars. The general strategy, using simple statistics to find interesting categories without assumptions about the nature of the irrelevant rules defined over those categories, may be applicable to other domains.

The Bootstrapping Problem

One reason why learning the structure of a domain without any prior knowledge is so difficult is that both an appropriate set of categories to describe the phenomena and the rules defined in terms of those categories must be learned from scratch. Thus the learner must solve a "bootstrapping" problem: the specification of a set of rules presupposes a set of categories, but the validity of a set of categories can only be assessed in the light of the utility of the set of rules that they support. *Prima facie*, at least, this implies that both rules and categories must somehow be derived together. However, the space of possible of rule/category combinations is so large that it seems unlikely that such an approach will be feasible for learning the structure of any but the simplest domains.

Although the focus here will be natural language, the bootstrapping problem arises in the context of learning about any new domain. For example, in learning some new subject, say elementary physics, learners must somehow acquire both the relevant concepts and the correct rules of inference defined over those. For example, learners must grasp the concepts of momentum, force and so on, as well as how these concepts may be manipulated and interrelated using the formal rules. The bootstrapping problem is acute since these two projects are thoroughly interdependent - understanding the concepts presupposes some understanding of the rules in which they figure, and the statement of the rules presupposes the concepts that they interrelate. In the terminology of the philosophy of science, the development of science requires both new *natural kinds* and new *scientific laws* relating those kinds together. Thus the bootstrapping problem is at the heart of the problem of theory change, both in scientific enquiry and in individual cognitive development.

Rather than attempt to tackle the bootstrapping problem in its full generality, we shall focus on the test case of learning syntax as an illustration of a particular way in which the bootstrapping problem may be overcome. In syntax learning the bootstrapping problem is to learn the set of syntactic categories and the syntactic rules defined over them. Most work on formal models of syntax acquisition does not encounter the bootstrapping problem, since the syntactic category of individual lexical items are taken as given, and the focus is on deriving the set of rules defined over these items (that is, the corpus used in learning is *tagged*). Even given this restriction, of course, the problem of rule induction is very difficult, and there are a number of formal results (Gold 1967; Pinker 1984; Osherson, Stob & Weinstein 1986) which suggest that constraints on possible linguistic rules must be innately specified. We pursue a parallel approach, using an untagged corpus, and tackling the bootstrapping problem directly. We give no prior information to the learner, and attempt to derive both the stock of syntactic categories and the syntactic category of individual words from scratch.

The general strategy that we use is straightforward: we collect very simple statistics from of the data set, in the hope that a similarity measure defined in terms of these statistics will reflect useful underlying categories. We then derive a set of categories on the basis of their similarity with respect to these simple statistics. Despite the simplicity of these statistics in relation to the complexity of the rules of syntax of natural language, redundancy in the data means that the categories generated are close to the categories given by standard linguistic theory. Thus, the bootstrapping problem can be solved by inferring categories directly from simply, readily available statistics, without needing to make assumptions about the nature of the relevant rules.

Once these categories have been found, we can tag the previously untagged corpus, marking each word with its syntactic category, and to attempt to find rules defined over these categories. This can allow us to find a set of higher level phrasal categories defined over categories for words already derived. Thus a hierarchy of categories and rules can be derived by iterating this process. This method also promises to allow the revision of initial categorisation decisions, based on impoverished assumptions concerning the set of rules, in the light of the rules derived (we shall discuss this below). Below, we outline how this approach has been applied to learning aspects of the structure of natural language.

An Algorithm for Bootstrapping Syntactic Categories

In order to illustrate the above suggestions concerning how empirical measures of similarity can be exploited to solve the bootstrapping problem, we now derive a linguistic taxonomy which is remarkably close to the orthodox view of the various species of syntactic category. In order to achieve this, a measure of similarity between words and phrases inspired by the "replacement test" of theoretical linguistics was used.

Empirical Similarity and Numerical Taxonomy

In traditional linguistics, words and phrases are categorised into several standard linguistic categories: nouns, verbs, noun phrases, and so on. One justification for this taxonomy is afforded by a number of "distributional tests", which assume that words and phrases which are distributed similarly should receive similar linguistic categories. Probably the best known test is the "replacement test"(e.g. Radford 1988):

> Does a word or phrase have the same distribution (*i.e. can it be replaced by*) a word or phrase of a known type? If so, then it is a word or phrase of that type.

In traditional linguistics, "distribution" is grounded in linguistic intuitions as to whether a purported sentence is syntactically 'well-formed'. In the present context such intuitions cannot, of course, be presupposed, but the replacement test can be made empirically relevant by operationalising it as follows:

> Statistical Replacement Test
>
> Has the word or phrase been observed to occur in a corpus in similar contexts to another word or phrase? If so, then these should be given similar linguistic categories.

It remains to give formal accounts of what constitutes the "context" in which a word or phrase appears, and to define some measure of "similarity" between two such contexts.

To avoid unnecessary presuppositions about the structure of language, we assume an extremely simple definition of the context of a word - the context is simply the preceding two and following two words. To keep the computations tractable, attention was restricted to context words which were among the 150 most common words observed in the corpus. The context we used can therefore be thought of as four vectors of 150 dimensions, each dimension corresponding to one of the 150 most common words. The value of the vector is then given by the number of times the focal word appeared in the relevant relation (i.e., preceding, following, last but one, next but one).

There were several candidates for this which were quite good at uncovering structure automatically. In the spirit of the statistical replacement test described above, we propose that any reasonable measure of similarity defined to elucidate linguistic distributional similarity should be insensitive to the absolute frequency of ocurrance of any particular word, but should be dependent on the position it is observed to occur at relative to other words. That is, it should satisfy the following criterion:

Replacement Criterion If every occurrence of a word, w, is replaced throughout the whole corpus independently and at random by w' with probability p, and w'' with probability $1-p$, and neither w' nor w'' previously occurred in the corpus, then w' and w'' should have similar contextual distributions according to the chosen similarity metric.

A metric which gives hierarchical structure in accord with linguistic orthodoxy was found to be the Spearman Rank Correlation Coefficient between the vectors of frequencies of context words. Since Rank Correlation between two vectors of ranks is in the range $[-1,1]$, we used an appropriate rescaling of values into the range $[0,1]$.

Since Sokal & Sneath (1963) first introduced techniques of numerical taxonomy to the biological community, hierarchical cluster analysis has found a wide range of applications, especially in the biological and social sciences. We use our distributional similarity metric as the basis for a hierarchical cluster analysis of words, which places words with similar distributions nearby in the hierarchy. Nodes in the resulting taxonomy correspond closely to traditional syntactic categories.

The goal, in the first instance, is to induce a standard syntactic categorisation. Then we analyse short phrases in a similar way to deduce similarities between phrases of various length, and thereby induce facts about the grammar describing them.

Computational Experiments

We have conducted a number of studies deriving syntactic categories from artificial data generated by a phrase structure grammar, and classifying letters and phonemes into linguistically interesting classes using corpora of real text (Finch & Chater 1991). Here we concentrate on the problem of finding syntactic categories in real corpora.

Syntactic categories in natural language

A 40,000,000 word corpus of USENET newsgroup data was stripped of headers, footers and the like. Even before cluster analysis, a list of the ten nearest neighbours of sample words shows that the Rank Correlation metric reveals at least some linguistic structure.

three: three, four, five, six, several, real, black, old, high, local, white.

I: I, we, they, he, she, you, I've, doesn't, don't, I'm, didn't.

south: south, east, west, north, war, public, government, tv, system, dead, school.

Clustering results

The tree structure for the entire set of words analysed, the 1000 most common words in the corpus, is much too large to display in a single diagram. Therefore, an overview of the structure of the tree is given, with labels a node corresponding to the predominant syntactic category of the items dominated by that node. A small number of items have no well defined syntactic category (for example, single letters of the alphabet and words connected with newsgroup administration such as "edu" and "com") and these were rejected from the analysis. Of the remainder, less than 5% are misclassified with respect to the label that we have given to their dominating node. Figure 1 therefore shows that the gross taxonomy of the lexical items is very close to a standard taxonomy of syntactic categories.

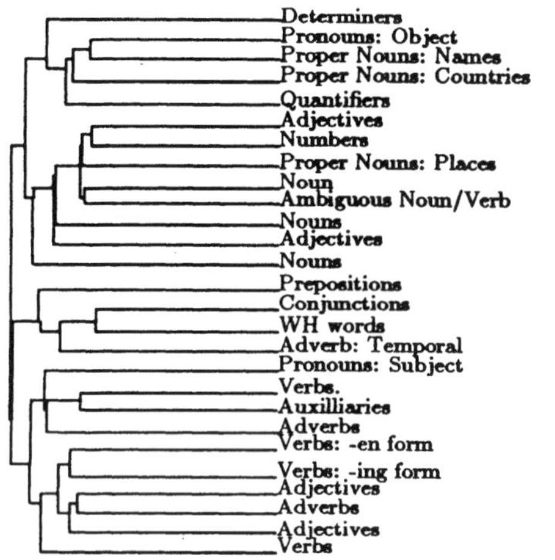

Figure 1

Figure 2(a) shows some of the low-level structure apparent within the whole dendrogram. The left hand dendrogram corresponds to part of the "adverbs" category of Figure 1. Note that some semantic regularities are apparent (really/actually, finally/eventually, thus/therefore, and so on). The other two dendrograms show respectively that low-level semantic features are revealed (being a computer term) and the dendrogram of subject-position pronouns shows a (relatively) orthodox syntactic analysis of pronoun/auxiliary contractions.

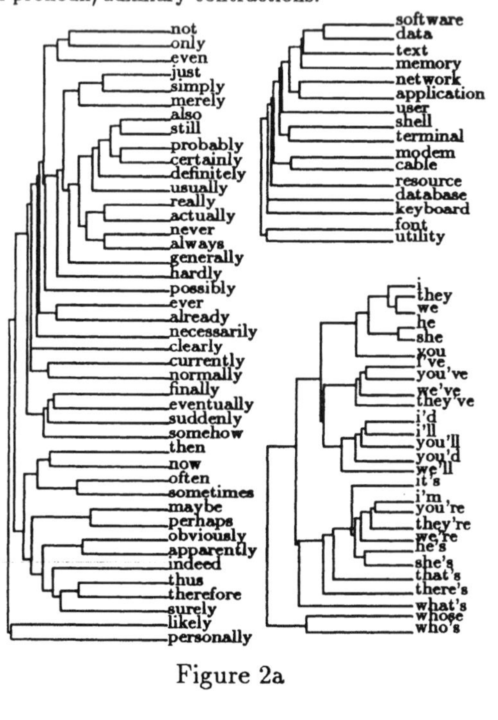

Figure 2a

Figure 2(b) shows low level structure for some adjectives, object position pronouns, countries, and numbers. Again it is clear that there is considerable accord between empirical and syntactic/semantic similarity.

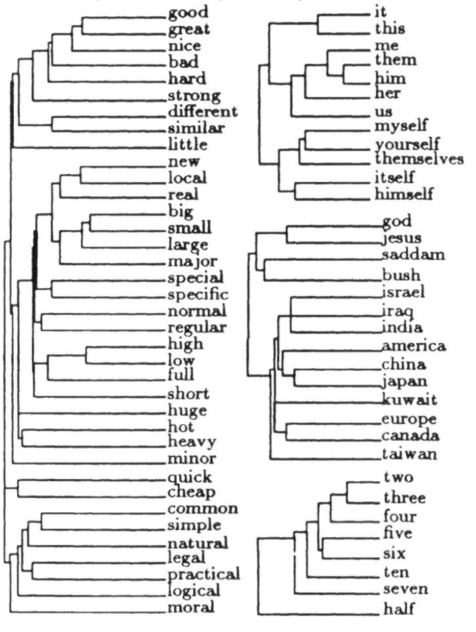

Figure 2b

Sequences and Similarity

After a hierarchical classification of lexical items has been derived, we can use this to classify sequences of categories hierarchically, and the derived similarity metric will again turn out to reveal interesting linguistic structure. This section details the experimental techniques and results of this analysis.

Classification The lexical hierarchy derived above was used to classify each lexical item by cutting the dendrogram at a particular level of dissimilarity, and thereby obtain several disjoint classes of words. Individual words were replaced with a code which corresponded to the class to which they belong, and the corpus was "parsed" accordingly. For instance, the two word sequences "the women", "the file" and "most data" were replaced by the sequence of labels "C30 C16". The principle advantage of this is one of sample size. If 600 words were in C16, and 20 in C30, for example, then the bigram "C30 C16" comprises, in principle, 12,000 word-level bigrams. This means that reliable statistics can be gathered on the "C30 C16" bigram with a much smaller corpus than needed for word-level bigrams. The situation is clearly exponentially worse for trigrams. For instance, the "C30 C16 C16" (Determiner Noun Noun) trigram corresponds to a possible 7,200,000 word-level trigrams. The size of the corpus is a major limitation to how far this unsupervised statistical approach can uncover the structure of language, and classification can be seen as a means of elucidating generalisations from (relatively) small corpora.

Results Rather than present dendrograms as we did for individual words, in order to show that interesting linguistic structure has been captured we instead show some of the "tightest" clusters. That is, the dendrogram is "cut" at a particular level of dissimilarity, and some of the resulting clusters are given as an illustration.

Noun Phrase Det Noun, Det Adjective Noun, Det Noun Noun, Det Verb/Noun, Det Adjective Verb/Noun, Det Inf, Det Verb/Noun Noun, Det Noun Verb/Noun, Det Inf Noun, Det ing Noun, Det PastPpl Noun, Det Det Noun, Det Adjective Noun, Det Adjective Inf, Det Adjective Verb/Noun, Det ing, Det Noun Adjective, Det Place Noun, Det Adjective QuantProNP

Note that the ambiguous category "Verb/Noun", which contains words judged to occur roughly equally frequently as non-finite verbs and nouns, behaves very much like "Noun" when preceded by a determiner. Even words which are typically non-finite verbs are judged similar to nouns when preceded by a determiner.

Verb Phrase Inf ProObj, Inf ProObj Noun, Inf Det Noun, Inf Det Verb/Noun, Inf Det Inf, Verb/Noun Det Noun, Verb/Noun ProObj, Inf ProObj Prep/Adv, Inf QuantNP, Inf QuantProNP, Inf ProObj Adjective, Inf Countries, Inf Noun, Inf Adjective Noun, Inf Noun Noun, Inf PastPpl, PastPpl PastPpl, PastPpl Adjective

Note that when followed by an object position pronoun, or a noun phrase, the ambiguous category "Verb/Noun" now behaves as (appears in the same contexts as) non-finite verbs.

Prepositional Phrase Prep Noun, Prep Det Noun, Prep Adjective Noun, Prep Det Verb/Noun, Prep Inf, Prep Det Inf, Prep Adjective Noun, Prep Verb/Noun, Prep Adjective, Prep QuantProNP, Prep ProObj Noun, Prep Conj&WH Noun, Prep Noun Noun, Prep QuantProNP Noun

Complex Nouns Noun
Noun, Noun, Noun Verb/Noun, Noun Preposition Noun, Noun Conj&WH Noun

Nouns are similarly distributed to compound nouns.

Auxiliaries Auxiliary Adverb, Auxiliary Adverb Adverb, Adverb Auxiliary, Auxiliary, Auxiliary TempAdvb, Auxiliary AdjMod, Auxiliary Adjective

As can be seen, auxiliaries can appear close to adverbs of various sorts, and the resulting phrase is similarly distributed to auxiliaries alone.

Relation to Neural Network Approaches

Outside the statistical tradition, there has been much interest in using neural networks to extract linguistic categories from raw data. In particular, Elman (1990, 1991; see also Chater 1989; Cleeremans, Servans-Schrieber & McClelland 1989) has shown how a recurrent neural network, trained to predict the next element in a sequence of inputs generated by a simple grammar, can develop patterns of hidden units which, when appropriately averaged and cluster analysed reveal underlying syntactic categories.

Elman's approach has a number of limitations. Firstly, it does not readily generalise to handle more realistic grammars, with many grammatical rules and a large lexicon. This is because the prediction tasks rapidly becomes extremely difficult, and learning is extremely inefficient and slow, if it occurs at all. Secondly, the linguistic categories are only implicit within the network, and can only be revealed using cluster analysis. However, cluster analysis on simple bigram statistics of the training corpus provide equally good clusters (Chater & Conkey in submission), so it is not clear how much statistical work the network is doing in uncovering the underlying linguistic categories.

The statistical analysis presented above suggests an alternative neural network approach, in which a network learns through simple Hebbian learning to represent words by their distributional context. Since similar words are assigned similar patterns, the network can find the relevant syntactic categories by performing a cluster analysis of the patterns. An attractive paradigm for unsupervised clustering is due to Kohonen (1982). This implements a variant of k-means clustering, where the k output units (or more exactly their weight vectors) correspond to the k-means which compete to account for portions of the data, to which they are most similar.

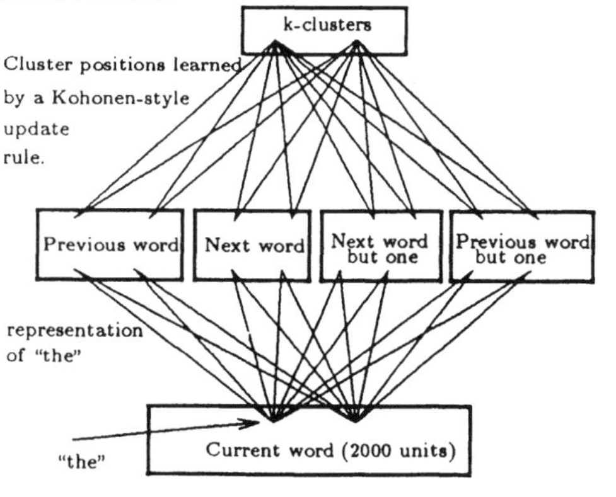

Figure 3

The network shown in Figure 3 corresponds to that used in our simulations with large corpora of real text. We used a similar, scaled down, version in the letter and phoneme level simulations reported below. The lower set of units use a localist representation of the current word (there are 2000 units, each corresponding to a different word under study). The middle set of units are divided into 4 banks, one bank corresponding to each of the four contextual bigram relations considered: last word but one, previous word, next word, next word but one. Only the most common 150 words were considered, and appearances of all other words in these contextual relations were ignored. The first layer of the net was trained with the 40,000,000 word newsgroup corpus simple Hebbian learning, with normalisation. After training, when a "current word" is presented, the middle layer represents the distribution of contexts in which that word occurs. The pattern representing each of the 2000 words are then clustered into 100 groups using a Kohonen network.

Network Simulations

First, a small network was given the task of clustering together letters, which were represented by the distribution of their surrounding context as described above for words. When the network consisted of two cluster nodes, it precisely divided vowels from consonants. The clusters resulting from a small (12,000 phoneme) corpus of phonemically transcribed speech (Svartvik & Quirk 1980), also approximately divided vowels from consonants as shown below.[1]

Vowels: @@ @ uu uh u oo o ng nd ii i e aa a

Consonants: zh z y @r w v th t sh s r p n m l k jh hi h g dh d ch b

In the word-level experiments, Some of the clusters obtained are shown below. In general words in the same cluster tend to have the same syntactic category, although there is sometimes more than one cluster which corresponds to the same syntactic category. Also some clusters appear to correspond to no linguistic category. Some of the clusters are shown below. Notice that one of clusters corresponds not to a single linguistic category, but consists of words which are ambiguous between two linguistic categories, nouns and verbs. In many of the categories there are one or two apparently spurious items, and some of the smaller categories, not shown, do not appear have any coherent linguistic basis. Although the categories are generally in accord with an orthodox syntactic classification, more linguistically perspicuous categories can be found by cutting the dendrogram produced in a full hierarchical cluster analysis at a particular dissimilarity level, to give disjoint clusters (as shown in Figure 1). Hence it may be possible to improve network performance further.

your those this these their the our one's my its his every each another an a

[1] We use the Machine-Readable Phonetic Alphabet.

why whom whether where what though that how because

two three ten six several half four five few fairly very

you've you're who's what's we're wasn't they've they're there's that's suddenly she's knowing it's i'm he's haven't comes being

washington v steve robert president peter mike michael math m john jesus japan iraq india george engineering david dave bell

yourself whatever us themselves them something someone somebody saddam myself me kuwait himself him her forth everyone anything

without within with when via unless under toward on near in if from for during by beyond between before at as among against across about

writing willing watching using turning trying thrown taking supporting showing sending selling seeing running putting printing playing paying passing making looking keeping giving getting flying finished finding doing considering coming changing calling buying behind acting

wanted used tried treated taught taken suggested stopped stated started sold shown seen saw saved responsible reported removed released received published provided produced presented posted played placed paid opposed noticed needed moved met looked led intended included heard found experienced done discussed died designed caught carried assumed associated asked applied allowed added accepted

window warning wall voice unit train track tape table stock statement stack signal screen sample role ring results ram purpose program process performance object months menu market map list link letter image ii frame format form foot flow filter film file faith entry effect dog distribution disks course contents chip box book article animal address addition account

walk wait use try stick sign share send save rid respond refer recognize reach protect pick pass offer occur miss keep judge include ignore hurt handle follow focus fix fill exist drop define count convert continue compile cause bring bother belong beat answer

words women views versions types tools tapes stories states sites responses questions programs products postings parents papers opinions numbers names movies laws ideas functions friends fonts fans experiences examples elements effects documentation discussions computers children cases canada applications advice

update transfer trade test split spell ride return report reply release register record present post plan move log lead force fly figure feed face escape end email die deal copy charge call break benefit attack

wonder wish win trust tell see say respect remember realize prove notice mention know imply imagine hope hear guess forget feel explain expect except doubt determine deny decide claim care blame believe assume ask argue agree

valid tough stupid somewhat slow simple silly separate related practical possible nice negative neat logical less intelligent important hot greater good faster expensive excellent easy correct closer blind better appropriate accurate

Discussion

We have shown how it is possible derive good approximations to the syntactic categories for English, without having a good account of the rules of syntax, by collecting statistics, deriving a similarity metric, and applying hierarchical cluster analysis. Further it was possible to use the lexical level categories derived to find phrasal categories defined over these. The mechanisms for finding lexical categories can be implemented as a neural network, which learns to classify words into syntactically interesting classes.

One feature of the present version of this iterative procedure which is not attractive is that there is no mechanism for correcting inaccuracies in early categories, based on an oversimple model of the rules of the domain, even when a more elaborate model of these rules has been derived. For example, the initial bigram model does not allow for the possibility that there are some surface forms (for example, FIRE) which correspond to more than one underlying lexical representation, with a different lexical category (in this case, NOUN and VERB). This difficulty can be overcome by using the observed context of ocurrance of the ambiguous word to disambiguate it. This can be achieved, as we noted above, by using the analysis of the similarity between phrases.

We hope that the general approach to the bootstrapping problem that we have outlined can be applied to other domains, as well as learning linguistic categories, and other problems involving the analysis of sequential structure. For example, in learning the structure of a visual domain, simple statistics concerning neighbouring values in the image (either grey scale values, or values which are the output of some pre-processing) can be used as basis for constructing statistical models of visually interesting categories. There will, of course, be no easy general solution to the bootstrapping problem - after all, this would be tantemount to a general theory of the processes of cognitive development or scientific enquiry. However, we hope that we have shown that in specific contexts, it is possible to bootstrap successfully using statistical methods.

References

Chater, N. (1989) Learning to respond to Structures in Time. Research Inititiative in Pattern Recognition: Technical Report RIPRREP/1000/62/89. Malvern, UK.

Chater, N. & Conkey, P. (in submission) Finding Linguistic Structure with Recurrent Neural Networks. Submitted to ICANN 1992, Brighton, UK.

Cleeremans, A., Servan-Schrieber, D. & McClelland, J. L. (1989) Finite State Automata and Simple Recurrent Networks. *Neural Computation*, 1, 372-381.

Elman, J. L. (1990) Finding Structure in Time. *Cognitive Science*, 14, 179-211.

Elman, J. L. (1991) Distributed Representations, Simple Recurrent Networks, and Grammatical Structure. *Machine Learning.*

Finch, S. & Chater, N. (1991) A Hybrid Approach to the Automatic Learning of Linguistic Categories. *AISB Quarterly*, 78, 16-24.

Gold, E. M. (1967) Language Identification in the Limit. *Information and Control* 16, 447-474.

Kohonen, T. (1982) Self Organised Formation of Topologically Correct Feature Maps. *Biological Cybernetics*, 43, 59-69.

Osherson, D., Stob, M. & Weinstein, S. (1986) *Systems That Learn: An Introduction to Learning Theory for Cognitive and Computer Scientists.* Cambridge, Mass: MIT Press.

Pinker, S. (1984) *Language Learnability and Language Development.* Cambridge, Mass: Harvard University Press.

Radford, A. (1988) *Transformational Grammar.* 2nd Edition, Cambridge: Cambridge University Press.

Sokal, R. R. & Sneath, P. H. A. (1963) *Principles of Numerical Taxonomy.* San Francisco: W.H. Freeman.

Svartvik, J. & Quirk, R. (1980) *A Corpus of English Conversation.* Lund: LiberLaromedel Lund.

Frequency Effects on Categorization and Recognition

Judy E. Florian
University of Michigan
330 Packard Road
Ann Arbor, MI 48104
(313) 764-0318
e-mail: judy_florian@ub.cc.umich.edu

Abstract

An experiment investigating effects of familiarity (indicated by presentation frequency) on categorization and recognition behavior is presented. Results show frequency influenced performance under speeded response conditions only, producing increased categorization of new, similar items with the frequent item, and differentiation (a decrease in false alarms to these same items) in recognition. These results are evaluated with respect to different versions of an exemplar model of categorization and recognition (Medin & Schaffer, 1978; Nosofsky, Clark & Shinn, 1989). Models that include a mechanism for *differentiation*, or changes in the similarity computation to a familiar example, provided better descriptions of both categorization and recognition behavior than models without this added aspect. The addition of a differentiation mechanism improved fits to categorization data of all three versions of exemplar models considered: the *type* model (in which repetitions do not produce separate memory traces), the *token* model (which posits individual memory traces for each repetition of an item) and the frequency parameter model (which includes frequency weighting as a free parameter).

Introduction

If you live in the Midwest, you are probably familiar with species of birds different from someone living in another geographic location. For example, you see geese and ducks throughout winters and cardinals heralding every spring. A person from Florida, in contrast, would probably have more encounters with seagulls, egrets and pelicans, birds you only see on vacations. An interesting question to ask is, does your *bird* concept differ from a Floridian's because of your familiarity with different birds?

One possibility is that your concept of bird is influenced by your exposure to geese, and you are more likely to think of a goose-like bird when someone mentions the category *bird*. A second possibility is that your experience has led you to differentiate your bird category into the two categories of *geese* and *birds*. This would leave your bird category unaffected, or possibly even less goose-like than the Floridian's. A last possible effect of your experience might be no effect at all. That is, your first encounter with a new bird would affect your concept, but repeated encounters with that animal might not. This would happen if the category structure is more important in people's concept formation than category familiarity (Rosch, Simpson & Miller, 1976).

In this paper, I present an experiment that investigates effects of familiarity (as measured by frequency of encounter) on categorization and recognition, and evaluate these results with respect to different versions of an exemplar model of categorization and recognition. In particular, I will compare models which include a mechanism for differentiation, or changes in the similarity metric to a familiar example, to models without this added aspect.

The modeling framework employed was the context theory of classification (Medin & Schaffer, 1978). According to this theory, people's representations of categories consist of stored memory traces of every category exemplar observed. Categorization decisions are made by comparing an item's summed similarity to members of different categories. Nosofsky (1988, 1991) has extended the exemplar-based classification theory to account for recognition performance. Recognition of a stimulus item is predicted by summing its similarity to all exemplars stored in memory. One variant of exemplar theory is a *type* model, in which repetitions of examples are not stored in memory as additional traces. This kind of model predicts that there are no effects of repetitions, or familiarity on categorization or recognition. A type model can be contrasted with a *token* model, in which every experience produces an additional memory trace. A *differentiation* model was constructed by allowing likelihood of retrieval of a stored example (a function of similarity to the exemplar, in this model) to vary with familiarity of an item. In the differentiation model, similarity to a familiar exemplar is computed separate from similarity to other items in the sense that the dimension weightings that make up similarity are different. For example, if geese are differentiated from your bird category, a duck-like animal might not cause you to retrieve geese, but rather ducks and other birds. Whereas, a person who is not as familiar with geese might retrieve a memory trace of a goose in response to presentation of a duck.

Frequency Effects on Classification and Recognition

Nosofsky (1988, 1991) found exemplar frequency effects on both categorization and recognition behavior. Furthermore, an exemplar model with a free parameter for frequency weighting provided the best fit (over a straight type or token model) to his data (Nosofsky, 1991). Nosofsky (1991) found exemplar frequency had less pronounced effects on recognition when compared with effects on classification. Furthermore, Ratcliff, Clark & Shiffrin (1990) conducted studies in which they increased an item's study time or frequency of presentation relative to other items. Recognition accuracy (d') to the frequently presented items increased with presentation frequency. However, Ratcliff et al. expected to find a greater effect of frequency when lists consisted of frequent and infrequent items than when lists of all frequent and all infrequent items were compared. Frequency improved recognition performance in both cases, and sometimes even more for the pure lists (all frequent vs. all infrequent presentations) than mixed lists. Shiffrin, Ratcliff & Clark (1990) account for these data with a variant of the exemplar model SAM which includes differentiation of exemplars. Differentiation is accomplished in this model by reducing the activation of a "strengthened" target item in memory in response to presentation of an unrelated item.

In the present experiment, the effects of exemplar frequency on categorization and recognition under both speeded and unspeeded response conditions was studied. In previous research we found minimal effects of exemplar frequency when subjects were given as much time as they needed to respond. Therefore, we included two more conditions in which subjects were forced to respond quickly. Decreased response time was expected to produce less accurate recognition performance, or more false alarms to foils similar to the frequent exemplar (Ratcliff, 1978). Categorization models predict a corresponding effect on categorization performance; that is, more categorization of those same items with the frequent exemplar.

Method

Design. This experiment used geometric figures comprised of four dimensions, as stimuli. Two categories were constructed, each consisting of three exemplars. The stimulus dimensions were pattern (striped or shaded), form (oval or rectangular), size (small or big) and orientation (horizontal or vertical).

In the Frequent condition, one exemplar was presented with tripled frequency relative to the other five items. In the Control condition, all exemplars were presented with equal frequency. Both a recognition test and a classification test consisting of new and old items were given. New transfer items were either similar to the frequent exemplar (matching on 3 of the 4 dimensions), or neutral in similarity to the frequent exemplar (mismatching on 2 or 3 dimensions).

Procedure. During the learning phase, a subject categorized the six category items, receiving feedback, until a block of 18 or 24 (for control and frequency conditions, respectively) was completed without any errors. Next, subjects were tested on recognition or classification of seen and unseen items. This test phase was followed by a second learning phase (which ended under the same criterion as the first) and the remaining test. Subjects who were tested under speeded conditions were instructed to respond as quickly as possible. In addition, these subjects were alerted after 1.5 seconds that they were taking too long to respond to an item. (Test order, dimension coding, and category labels were counterbalanced.)

Subjects. Subjects were 113 University of Michigan undergraduates who participated as a course requirement for an introductory psychology class. The subjects were randomly assigned to the four experimental conditions.

Results and Discussion

Figure 1 shows the proportion of categorization responses of new transfer items into the same category as the frequent exemplar. There is an interaction between test conditions (speeded versus unspeeded) and frequency condition on categorization responses; $F_{(1,109)}=15.05$, $p=0.0002$. That is, under speeded conditions, exemplar presentation frequency produced an increase in the categorization of new exemplars with the frequent item (from .63 in the control condition to .70 in the triple frequency condition). This did not occur under unspeeded response conditions (which yielded categorization probabilities of .65 and .60 in the control and triple frequency conditions, respectively).

Figure 1 also shows results of the recognition test. Speeded response conditions had the unexpected effect of increased recognition accuracy in the frequent condition. False alarms to new items which were similar to the frequently presented exemplar decreased with exemplar presentation under speeded conditions. False alarm probabilities were

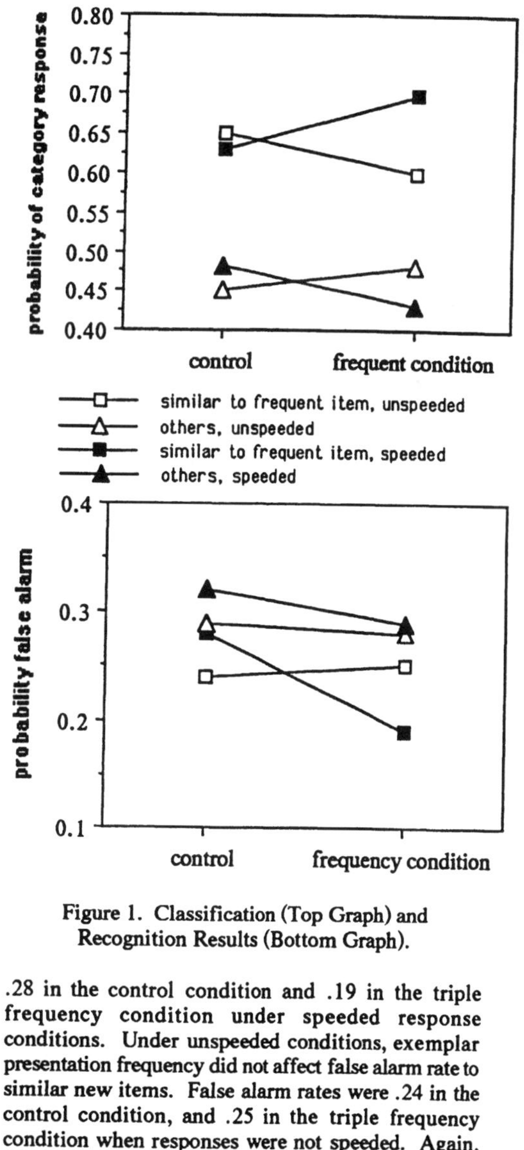

Figure 1. Classification (Top Graph) and Recognition Results (Bottom Graph).

.28 in the control condition and .19 in the triple frequency condition under speeded response conditions. Under unspeeded conditions, exemplar presentation frequency did not affect false alarm rate to similar new items. False alarm rates were .24 in the control condition, and .25 in the triple frequency condition when responses were not speeded. Again, this interaction between exemplar frequency and response conditions was significant; $F_{(1,109)}=11.76$, $p=.0009$.

Modeling of Classification Results. Various versions of the context model of categorization (Medin & Schaffer, 1978; Nosofsky, Clark & Shinn, 1989) were fit to both classification and recognition data. Table 1 contains a summary of the observed data and the results of model fits, including values of G^2 which is a measure of goodness of fit of the model. Decreases in G^2 can be used as a measure of significant improvement of a model with added parameters because it conforms to a χ^2-distribution with degrees of freedom equal to the number of additional parameters. Reductions in G^2 exceeding expectation according to a χ^2-distribution are used here to identify significant improvements in model fits.

The first comparison which can be made is between the fits of the type model, in which exemplar repetitions are not included, and the token model, in which each repetition is included as an additional exemplar in the model. The type model provides a better fit to the data than the token model, which predicts large effects of exemplar frequency. A third model included frequency as a free parameter weighting of the repeated exemplar. This third model, called the *frequency parameter* model, fared no better than the type model, however.

An additional variant (called the differentiation model) of each model (type, token and frequency parameter) was fit to the data by calculating similarity to the frequent item with a separate computation from similarity to other stored items. In the context model, similarity between two items is computed by multiplying similarity measures for each dimension, with a match along any dimension yielding a similarity value of 1 and a mismatch being assigned a value between 0 and 1. (These similarity measures are free parameters in the model.) In the differentiation model, these similarity values of mismatched dimensions are different for the frequently seen item than for other learned items. That is, the differentiation model allows the similarity space around the familiar item to differ from the similarity space of other exemplars. This model can account for differentiation, or decreases in item retrieval in response to a probe with increases in familiarity, if mismatches along dimensions are assigned small similarity values. This would have the effect of drastically reducing similarity between a target and probe item, thereby suppressing memory retrieval of the target item. In this experiment, categorization performance of similar new items decreased with increases in frequency of presentation under unspeeded conditions.

Columns 5-7 of Table 1 show the results of the differentiation model simulations. For all three kinds of models (type, token, and frequency parameter), the addition of exemplar-specific similarity parameters for the frequent exemplar

Table 1.
Summary of Observed and Predicted Category Response Probabilities (top portion) and "Old" Recognition Probabilities (bottom portion) for Test Items by Similarity to Frequent Example.[a]

	Observed	Type Model	Token Model	Frequency Parameter Model	Type Model with Differentiation	Token Model with Differentiation	Frequency Parameter Model with Differentiation	Type Model with Differentiation + Accessibility	Token Model with Differentiation + Accessibility	Frequency Parameter Model + Differentiation + Accessibility
Classification:										
New Exemplars Similar To Frequent Item:										
Unspeeded, Control	.65	.64	.63	.64	.64	.64	.64	.64	.64	.64
Unspeeded, Triple Freq.	.60	.64	.72	.67	.66	.64	.65	.63	.60	.63
Speeded, Control	.63	.64	.63	.64	.64	.64	.64	.64	.64	.64
Speeded, Triple Freq.	.70	.64	.72	.67	.66	.64	.65	.69	.70	.69
Other New Exemplars:										
Unspeeded, Control	.45	.44	.44	.44	.44	.44	.44	.44	.44	.44
Unspeeded, Triple Freq.	.47	.44	.49	.45	.45	.41	.41	.43	.41	.41
Speeded, Control	.48	.44	.44	.44	.44	.44	.44	.44	.44	.44
Speeded, Triple Freq.	.43	.44	.49	.45	.45	.41	.41	.45	.41	.41
G^2 =		176.2	199.5	174.1	161.4[b]	155.8[b]	151.8[b]	158.7[b]	151.9[b]	151.0[b]
Recognition:										
New Exemplars Similar To Frequent Item:										
Unspeeded, Control	.24	.25	.21	.25	.25	.25	.25	.25	.25	.26
Unspeeded, Triple Freq.	.25	.25	.33	.23	.22	.22	.22	.24	.25	.24
Speeded, Control	.28	.25	.21	.25	.25	.25	.25	.25	.25	.26
Speeded, Triple Freq.	.19	.25	.33	.23	.22	.22	.22	.19	.18	.18
Other New Exemplars:										
Unspeeded, Control	.29	.29	.27	.29	.29	.29	.29	.29	.29	.29
Unspeeded, Triple Freq.	.28	.29	.28	.29	.30	.30	.30	.30	.30	.30
Speeded, Control	.32	.29	.27	.29	.29	.29	.29	.29	.29	.29
Speeded, Triple Freq.	.29	.29	.28	.29	.30	.30	.30	.30	.29	.30
G^2 =		57.7	88.1	57.0	55.8	57.6[b]	55.4	53.9	53.8[b]	53.6

[a]Note that model fittings were to performance on individual transfer items, and only a summary of this data is presented here.
[b]This model provides a significantly better fit to the data than the model without added parameters (i.e., type, token or frequency parameter models), $p<.05$.

provided a significant increase in the model's fit to the data. An interesting observation is that the token model which includes similarity parameters specific to the frequent exemplar now provides a better fit than the type model ($G^2 = 155.8$ versus 161.4).

A further variant of each of the three models was fit to the data in an attempt to predict the effect of response conditions. This variant included a weighting (referred to as *accessibility*) of the frequent exemplar under speeded response conditions only, because the data show an effect of frequency under these conditions. The results of these simulations are shown in columns 8-10 of Table 1. With the additional accessibility parameter, the differentiation model predicts the observed interaction between presentation frequency and response conditions. That is, the effect of frequency on categorization is different under speeded and unspeeded conditions, and the accessibility parameter allows prediction of that interaction. However, the additional parameter did not provide a significant quantitative improvement in the model's fit to the data, when compared with the models with exemplar specific similarity parameters.

Modeling of Recognition Results. The frequency parameter model was able to model differentiation of items, or decreases in false alarms to similar, new items, from the frequent exemplar. (See the bottom half of column 4 of Table 1.) The differentiation versions of the type, token and frequency parameter models also predicted differentiation with frequency. This qualitative improvement of the differentiation model is only statistically significant for the token model, however. With the addition of an accessibility parameter (which is a weighting of the frequent exemplar under speeded response conditions only), all three models were able to account for the interaction between response conditions and frequency.

Conclusion

In sum, effects of exemplar frequency were best described by exemplar models having a differentiation mechanism. These models fared better than a type, a token or a model which included frequency as a free parameter. Without a differentiation mechanism, the type and frequency parameter models fared better than a token model, in which every repetition produces an additional memory trace. An interesting observation is that once differentiation is added to the straight token model, its fit became as good as the other two. Therefore, multiple-trace models that store memory traces for each repetition may not be as inadequate as previously found. An additional parameter representing accessibility of a familiar item over the course of responding was necessary to fit the interaction between response conditions (speeded vs. unspeeded) and exemplar frequency observed in the data.

In addition, categorization and recognition tests were fit separately because subjects' strategies on these two tasks diverged. Differentiation, or decreased retrieval, of a frequent item occurred on the recognition test under speeded conditions, but generalization, or increased retrieval, occurred on the test of categorization.

It should be noted that the SAM model of recall and recognition also has a mechanism for differentiation with increased presentation strength (Gillund & Shiffrin, 1984; Shiffrin et al., 1990). In this variant of SAM, the parameter signifying activation of a trace by any stimulus item is allowed to change as trace strength increases in response to such factors as frequency or length of presentation. In addition, repetitions of items are captured in one trace representation in this model. However, the SAM model with differentiation does not capture changes in similarity between items in response to increased frequency. Rather, SAM predicts decreases in overall activation with increases in exemplar strength. In the differentiation model presented here, the parameters comprising similarity varied with exemplar strength, so the similarity space surrounding a strengthened exemplar could change. The model fits suggest that this kind of differentiation is necessary to model classification performance.

In another experiment, the position of exemplar repetitions during the learning phase (throughout or only after learning had occurred) was manipulated. In this experiment, the frequency effects observed above were replicated. That is, categorization of similar items increased in response to presentation frequency, and recognition accuracy was greater for the frequent item. However, when exemplar repetitions occurred after learning, they did not affect categorization or recognition. Therefore, repetition probably has a greater effect on categorization and recognition early in learning (Medin & Bettger, 1991). Therefore, total frequency does not appear to be producing the frequency effects observed. Further research will hopefully determine what leads to frequency effects at different points in learning.

Acknowledgments

This research was supported by National Science Foundation Grant 91-10245. I thank Doug Medin for numerous discussions of this project and Evan Heit for comments on an earlier version of this paper.

References

Gillund, G. & Shiffrin, R.M. (1984). A retrieval model for both recognition and recall. *Psychological Review, 91*, 1-67.

Medin, D.L. & Bettger, J.G. (1991). Sensitivity to changes in base-rate information. *American Journal of Psychology, 104*.

Medin, D.L. & Schaffer, M.M. (1978). A context theory of classification learning. *Psychological Review, 85*, 207-238.

Nosofsky, R.M. (1988). Similarity, frequency and category representations. *Journal of Experimental Psychology" Learning, Memory, and Cognition, 14*, 54-65.

Nosofsky, R.M. (1991). Tests of an exemplar model for relating perceptual classification and recognition memory. *Journal of Experimental Psychology" Human Perception and Performance, 17*, 3-27.

Nosofsky, R.M., Clark, S.E. & Shinn, J.H. (1989). Rules and exemplars in categorization, identification, and recognition. *Journal of Experimental Psychology: Learning, Memory, and Cognition, 15*, 282-304.

Ratcliff, R. (1978). A theory of memory retrieval. *Psychological Review, 85*, 59-108.

Ratcliff, R., Clark, S.E. & Shiffrin, R.M. (1990). List-strength effect: I. Data and discussion. *Journal of Experimental Psychology: Learning, Memory, and Cognition, 16*, 163-178.

Rosch, E., Simpson, C. & Miller, R.S. (1976). Structural bases of typicality effects. *Journal of Experimental Psychology: Human Perception and Performance, 2*, 491-502.

Shiffrin, R.M., Ratcliff, R. & Clark, S.E. (1990). List-strength effect: II. Theoretical mechanisms. *Journal of Experimental Psychology: Learning, Memory and Cognition, 16*, 179-195.

Declarative Learning: Cognition without Primitives

Edmund Furse and Roderick I. Nicolson

Department of Computer Studies
The Polytechnic of Wales
Pontypridd
Mid Glamorgan
CD37 1DL
UK
efurse@uk.ac.pow.genvax

Department of Psychology
The University of Sheffield
Sheffield
S10 2TN
UK
r.nicolson@uk.ac.sheffield.primea

Abstract

Declarative learning by experience is a foundation cognitive capability, and we argue that, over and above the normal processes of declarative learning, the ability for truly novel learning is the critical capability which bootstraps human cognition. Next we assert that none of the established models of machine learning and no established architecture for cognition have adequate declarative learning capabilities, in that all depend for their success on some pre-characterisation of the learning domain in terms of state space or pre-existing primitives geared to the domain. Finally we describe briefly the Contextual Memory System, which was designed explicitly to support all five declarative learning capabilities. The CMS underlies the Maths Understander machine learning system which 'reads' mathematics texts from scratch, assimilating mathematics concepts, and using them not only to check proofs but also to solve problems.

Introduction

The origins of human knowledge have provided a fertile source of speculation over the millenia, and even today the issue is far from resolved. Meno's paradox, the 'learning paradox' derives from the ancient Greek sophists who argued that truly novel learning was impossible in that *"novel knowledge cannot be derived completely from old knowledge, or it would not be new. Yet the transcending part of it cannot be completely new either, for then it could never be understood."* (Boom, 1991, p274). Plato sidestepped this paradox by asserting that all knowledge was innate, but initially dormant, and the ensuing debate between empiricist and nativist positions has echoed down the centuries. Piaget (1952, 1985) tackled the issue by arguing that the process of equilibration, the striving to change cognitive structures to avoid cognitive disequilibrium, was the key to truly novel learning, and went on to argue that his four stages in cognitive development derived from three such qualitative changes in cognition. Piaget's stage theory has, of course, been frequently criticised, but his primary emphasis was on genetic epistemology, the origins of knowledge. Fodor (1980) has revived the Meno paradox, and adopted a nativist position, arguing that Piaget's equilibration concept was not powerful enough to support the acquisition of novel cognitive structures, and the issue remains unresolved in the developmental literature (see Boom, 1991 and Juckes, 1991 for recent analyses). Unfortunately, the debate has focused on development rather than learning, even though it is clearly nonsense to suggest that true learning occurs only at three or four stages in one's lifetime, and in reality children (frequently) and adults (sometimes) are confronted by the need to acquire knowledge in a completely new domain. Adults are able to acquire the concepts of new games, new academic subjects (mathematics, computer programming, statistics, geology, etc.), new social conventions. Children have to start from scratch. They are refuting the learning paradox most of the time in the early years. Consequently, we argue that the learning paradox is at least as great a challenge for cognitive science as it is for developmental psychology.

We use the term declarative learning to refer to all aspects of the learning of declarative knowledge. It has many facets, and in Furse and Nicolson (1991) we stated the necessary competences of a declarative learner, which we termed comprehension, assimilation, utilisation, and accommodation. We now make explicit the fifth requirement, that of supporting truly novel learning.

(i) Comprehension (cf. Encoding). The system must be able to encode novel information so as to cause it to enter working memory. We use the term comprehension to emphasise the need to recode the input into the format used in the declarative memory structures.

(ii) Assimilation (cf. Storage). The system must be able to create a long-term memory entry for information in its working memory. Furthermore, the system must be able to incorporate the new information into its memory structures in such a way as to facilitate the adaptive use of that information in other contexts. It is worth noting the need to store not only the new information but also the context in which it occurs. Tulving (1983) refers to this as the 'cognitive environment'.

(iii) Utilisation (cf. Retrieval). The system must be able to retrieve the stored information given an appropriate cue or an appropriate context. For many theorists, the term retrieval suggests too automatic a process, and terms such as reconstruction (Bartlett, 1932) or ecphory (Tulving, 1983) capture better the complex search and matching processes involved. We adopt the neutral term to indicate the adaptive use of the existing knowledge to satisfy the system's requirements.

(iv) Accommodation. Regardless of how broadly one interprets the above three processes, we believe that they are incomplete for a viable declarative learner. In addition the system must be able to modify the information in the light of subsequent experience so as to improve its adaptivity of use. This includes the making of new links, the strengthening of salient associations, the forgetting of useless associations, and the creation of new features.

(v) Novel Learning. While the above processes provide a reasonable description of much declarative learning, it is possible to envisage a declarative learner which showed all four competences but was unable to undertake truly novel learning. Consequently, as a fifth, stringent criterion for declarative learning, we argue that it is necessary to demonstrate learning 'from scratch' without any pre-characterisation of the space to be learned, or any pre-characterisation of the primitives required for learning in that domain.

Existing Cognitive Science Approaches to Declarative Learning

The problems in acquiring knowledge have recently moved centre stage in cognitive science. Anderson (1990) argued that the origins of human knowledge were one of the major issues for cognitive science, concluding by reduction that from the viewpoint of ACT*/PUPS the weak problem solving principles (by which all subsequent domain-dependent problem solving productions are derived) must themselves be innate. Lenat and Feigenbaum (1991) present as one of the guiding principles of AI the 'Knowledge Principle' — *"If a program is to perform a complex task well, it must know a great deal about the world in which it operates. In the absence of knowledge, all you have left is search and reasoning, and that isn't enough."* Indeed, Lenat argues that this principle mandates a new direction for the main enterprise of AI, and accordingly has devoted the last seven years to his ambitious CYC project for hand-coding a significant subset of human knowledge, thus providing what he hopes will be a knowledge-base sufficiently rich to bootstrap natural language understanding systems, and, ultimately, machine learning systems.

Encouragingly, the three major architectures for cognition — Soar (Laird et al, 1986), ACT* (Anderson, 1983) and connectionist approaches — have a learning mechanism as the cornerstone of their approach to cognition. ACT*/PUPS suggests that declarative knowledge must be acquired initially, then this declarative knowledge is 'proceduralised' by a 'knowledge compilation' process consequent upon successful performance, turning it into a production rule format. The production rules may subsequently be tuned by extended practice. Soar learns primarily by a process of problem-space search, with learning taking place by automatic processes of 'subgoaling' following a failure and by 'chunking'. Connectionist models learn by processes of differential link strengthening procedures based on learning procedures such as gradient descent (eg. Hinton, 1989).

Very surprisingly, summarising an analysis presented in Nicolson and Furse (1992), not one of the above cognitive architectures is able to cope with declarative learning.

Neither of the above symbolic architectures are able to cope with declarative learning, in that they both precharacterise the space in which learning is to take place. Anderson hand-crafts the appropriate declarative knowledge in each of the domains in which he works, and Soar makes the assumption that all learning can be characterised as search through a problem space, an assumption criticised by Boden (1988) as quite untenable.

The ability of connectionist models to perform at all at the symbolic level remains controversial. In brief, we argue that none of these approaches addresses the Learning Paradox.

One might expect that the problem of declarative learning would have been extensively studied in the machine learning literature, but, bafflingly, this is not so. Space precludes a detailed analysis of the machine learning literature here (see Ellman, 1989 for a useful review), and we shall merely note four problems of the established machine learning demonstrations. Thus four major critiques of the machine learning research are that most of the models lack psychological plausibility; that the declarative knowledge being acquired is relatively unstructured and semantically arid; that the learning tasks are too simple to be ecologically valid; and that all the models operate within a closed world in which the knowledge representations are pre-specified.

In summary, declarative learning is of major theoretical and applied importance, yet no established cognitive architecture offers any mechanism for it, and no current machine learning approach offers a principled and psychologically plausible account of the processes involved. Consequently, approaches to modelling human cognition are critically incomplete.

Overcoming the learning paradox- the Contextual Memory System

Let us start by trying to derive an informal requirements analysis for overcoming the learning paradox. Consider a person confronted by some novel event or situation — watching a new game; examining some complex, unfamiliar machine; or trying to understand some unfamiliar branch of mathematics or computer programming language. First impressions are of a mass of detail — pieces and actions; pipes, cables and bolts; or series of symbols. It is not initially a problem of telling the wood from the trees, it is a problem of even telling what the trees are! One can initially identify neither which features are salient nor the overall purpose of the components of the game, machine or language. An expert may help by labelling a few key components — the pieces, the components, or the commands. One stores this information, blindly, without understanding, but it forms the basis around which further information can be accreted, slowly building a better ability to describe the components and the key features. Eventually, over a period of time, one acquires the ability to tell the wood from the trees, understanding not only what the purpose of the components are, but also which features are salient. Several approaches, including ACT*, Soar and connectionist approaches, give a reasonably plausible account of the later learning events, the accommodative processes which tune the existing knowledge to achieve better task performance, but the learning paradox arises in the initial stage — how can we take in new information when we understand neither what is relevant nor what features to look for?

We argue that one plausible approach is to generate as many features as possible for each event, and to attempt to maintain and adapt this population on the basis of subsequent events. Over time, the feature population will 'evolve' into one which fits better into the current evolutionary niche, in that salient features should emerge, and useless ones should die out.

This analysis suggests that one needs

(i) a mechanism for generating features automatically from input without the need for some prior characterisation in terms of domain-specific primitives

(ii) a mechanism for tuning the population of features on the basis of subsequent experience so as to encourage it to adapt to the domain.

This is made more formal in the Contextual Memory System which at the top level can be thought of as providing mechanisms for the storage and retrieval of information, both storage and retrieval being in terms of dynamic features. Information from the external environment is stored within the CMS as items which are indexed by features. The CMS starts with no items and no features. When an item is remembered, feature analysis takes place dynamically to create new features. This process of remembering will use a mixture of old features and new features. The features and items have energy which decays with time, but increases on recall. The links between the features and items have strengths which are adjusted on recall to ensure that the most salient features have the higher strengths. Since features are created dynamically the resulting memory configuration is complex and depends upon the history of memory processes.

When an item is first stored, it will be in terms of the currently high energy features. When subsequently recalled, it may be found with a different set of features, and a process of adjusting the CMS takes place to improve future access of the item. We now turn to how it is possible to generate the features dynamically without them being built in, and how the features are updated through experience.

Building features from the environment

It is possible to build features of new information without using built in primitives, by using built in feature building *mechanisms*, rather than built in features themselves. The essential idea is to break down the input into parts. The parts may not be meaningful to the agent on initial encoding, but through subsequent experience, these parts acquire meaning. We have found in the area of mathematics that the use of positional information is useful for subsequent retrieval of the information, and this also provides a simple model of attention. This process is best illustrated by means of an example from Winston's definition of a cup (Winston et al., 1983). This is normally represented as:

cup(x)
⇔ liftable (x) and stable (x) and open-vessel (x)

In the FEL language (Furse, 1990), it would be represented as:

Definition of cup
x isa cup
iff x isa liftable and x isa stable and x isa open-vessel

In whatever manner "cup" is initially represented, it is not necessary to have already represented the notions of "liftable", "stable" and "open-vessel", or even "and", "⇔" in order to build features of the object. This is most easily demonstrated by the mechanism used by the MU system (Furse, 1992a) and the current implementation of the CMS (Furse and Nicolson, 1991, Furse 1991, Furse 1992b), whereby the input is first parsed into a predicate calculus like representation:

(<=> (cup x)
    (and (liftable x)
        (and (stable x) (open-vessel x))))

We do not claim that the predicate calculus is used by people as an internal representation, this is just used for illustrative purposes. This datastructure can be thought of as a tree, and it is possible to analyse it into a number of features using various feature building mechanisms. The space of feature names is formally defined in BNF by:

<feature-name> ::= <pos><spec><type><term>
<pos> ::= LHS- | RHS- | null
<pos> ::= LHS-<pos> | RHS-<pos>
<spec> ::= IS- | HAS-
<type> ::= FORM- | TERM-

Most of this apparatus is to give a formal notion of the focus of attention, namely what part of the information the agent is attending to when generating the feature. The end of the feature-name, namely the <term> ensures that the feature space is infinite and only determined by its inputs, ie no pre-characterisation. LHS- means one is focusing on the left hand side of the tree, RHS- means the right hand side, and composition takes one further down left or right branches of the tree. HAS- means that the term occurs somewhere within the focus of attention, whilst IS- means that the term occurs exactly at the specificed focus of attention. A form is a canonical representation of a term useful in mathematics whereby the leaves of the tree are replaced by the letters a, b, c, ... to result in a canonical representation. Abstract HAS-forms introduce abstraction whereby parts of the tree are replaced by nodes. For a given term the number of abstract HAS-forms is in general very large.

Some of the mechanisms to generate features are:
Break the tree into single level HAS- forms without any positional information, eg
HAS-FORM-[LIFTABLE_A]
HAS-FORM-[STABLE_A]
Break the tree into single level HAS- forms with positional information, eg
RHS-HAS-FORM-[LIFTABLE_A]
LHS-RHS-IS-FORM-[STABLE_A]
Break the tree into abstract HAS- forms, at whatever level, eg to capture the notion that one has noticed that it is a definition of a cup:
HAS-FORM-[<=>_[CUP_A]_B]
One has noticed that there are three anded expressions on the right:
RHS-HAS-FORM-[AND_A_[AND_B_C]]
One has noticed the liftable and stable notions somewhere:
HAS-FORM-
[AND_[LIFTABLE_A]_[AND_[STABLE_B]_C]]]

Using these mechanisms it is possible to generate scores or often hundreds of different features of the input. This process is described in greater detail in Furse (1992b). Clearly the current implementation needs the use of a tree data structure, but mechanisms could be built to work on a string representation, for example one could break up
"x isa cup iff x isa liftable and x isa stable
and x isa open-vessel"
into features like:
HAS-FORM-"liftable"
START-HAS-FORM-"a isa cup"
HAS-FORM-"a isa liftable and a isa stable and b"
HAS-FORM-BEFORE-"liftable"_"open-vessel"
HAS-FORM-ADJACENT-
"a is liftable"_"a isa stable"
although one might want to allow the form abstraction to range over all substrings and not just the "x" as in the examples above, or not use it at all.

Learning the appropriate features

The processes described above allow the agent to generate a large number of features of the input information. But many of these features will be redundant. It is only through experience that the agent discovers which of the features are relevant for making future retrievals from memory. But it is essential that redundant features are generated at the outset, otherwise we are in danger of being in a closed box. Thus it is in the process of accommodation that the CMS models this learning experience.

In the CMS all features are given an energy value, and the features of highest energy can be thought of as representing the current attentional state. The CMS starts as a tabula rasa with no features, but as it stores information it generates features dynamically from the input as described above. The very first item will of course be represented in terms of brand new features, each of which will be given a default energy value. When subsequent items are stored, feature analysis will generate a mixture of old and new features. In the CMS a mixture of old and new features is stored, with the old features being the ones of highest energy. At the time of initial storage this may not be a very good choice of features for retrieval purposes, but provided that enough were generated and the agent is given useful subsequent experience, the features may subsequently be refined.

Learning in the CMS takes place during retrieval of information from memory. Any given probe gives rise to a number of features, and of course only the old features are used to index the memory. This search is implemented sequentially, but conceptually could be considered a parallel process whereby features activate items in memory that they index. Some items will be indexed by more than one feature, and once their activation energy (technically within the CMS this is known in the retrieval process as a transient energy) rises above a threshold, the item fires and is tested against the probe. When the item is not in the memory this is discovered very fast. If it is present, retrieval will result in the found item which matches the probe, and a number of failures. Learning is then the processes of accommodation whereby the CMS memory structures are adjusted so that in future it is easier to retrieve the found item than the failures.

There are four accommodative processes: storing uncomputed features, increasing the energy of useful features, decreasing the energy of unuseful features, and creating new distinguishing features.

Uncomputed Features. Uncomputed features arise because not all the features that are found in the probe may be stored as indexing the recalled item. The context of recall may be very different from when the item was first stored and other contexts when it was recalled, so that at the time of recall it may not be known whether the item has one of the probe features or not. Features thus have a 3-valued logic: positive, negative or uncomputed; only positive feature links are stored in the CMS. Thus the features which have not been previously computed for the item, and are found to be positive for the item are now given new links to the item.

Useful and Unuseful Features. Features are deemed to be useful in the search if they index the found item, but not the failed items. These features can have their energies increased and also the energy of the link from the feature to the item. Conversely, features are considered unuseful if they index the failed items but not the desired item, and they have their energies decreased.

Creating new distinguishing Features. The final process of learning from the experience of recall is to dynamically create new features to distinguish the found item from the failures. In the CMS this is currently only done when no existing features succeed in doing this distinction. New features can be created by two different approaches. In the first the found item and a failure are analysed to discover differences, from which a feature can be derived. In the second approach both the found item and a failure are broken up into a large number of features, and a feature in the found item set and not the failure is chosen. The CMS currently uses the latter approach.

Through these processes of accommodation, the CMS continually adjusts itself with experience so that the items in memory that are the most important are most easily retrieved because specialised features will have been built and they will have high energy. Conversely items rarely needed take more time to retrieve because they may only have features of low energy, and furthermore several of these features may not be useful in any context other than the original presentation, ie they are redundant. This could be called the "Redundant Feature Hypothesis", and further empirical work is planned to test whether people need to store redundant features.

MU: the Mathematics Understander

While the CMS is intended as a generic architecture for learning by experience, it has to date only been thoroughly investigated in the domain of pure mathematics (Furse, 1992a). The CMS forms the basis of MU, a large computer program which models the reading of mathematics texts by students. MU has competence in all aspects of declarative learning, and is unique both in machine learning and in cognitive architectures in its ability to truly capture declarative learning.

To date, MU has been applied to two branches of pure mathematics, namely Group Theory (using the textbook by Herstein, 1975) and Classical Analysis (using the textbook by Anderson, 1969). The approach adopted involves first rewriting the text by

Definition 2.1.7
G is abelian iff G isa group and $\forall$ a,b a $\in$ G and b $\in$ G ab=ba

....

Problem 2.3.3
Prove (G isa group and $\forall$a,b $\in$ G $(ab)^2 = a^2b^2$)
$\Rightarrow$ G is abelian

Solution:
Suppose G isa group
and $\forall$a,b a $\in$ G and b $\in$ G $\Rightarrow$ $(ab)^2 = a^2b^2$
RTP G isa abelian
RTP G isa group and $\forall$a,b a $\in$ G and b $\in$ G
$\Rightarrow$ ab = ba by definition of abelian
Part 1
RTP G isa group
Follows logically
QED Part 1
Part 2
RTP $\forall$a,b a $\in$ G and b $\in$ G $\Rightarrow$ ab = ba
Suppose a $\in$ G and b $\in$ G
RTP ab = ba
Now $(ab)^2 = a^2b^2$
$\Rightarrow$ (ab)(ab) = (aa)(bb) since x^2 = xx
$\Rightarrow$ a((ba)b) = a((ab)b) since (ab)(cd) = a((bc)d)
$\Rightarrow$ (ba)b = (ab)b since au = aw $\Rightarrow$ u = w
$\Rightarrow$ ba = ab since ua = wa $\Rightarrow$ u = w
QED Part 2

Figure 1. Problem Solving in Group Theory by MU

hand into FEL — a 'formal expression language' (Furse 1990), which captures the essential semantics of the domain — thus forming a machine understandable text for MU to 'read'. The basic requirement in reading the text is to 'comprehend' each input line, and then for each section either to assimilate new knowledge, to check through each step of a proof, or to attempt to solve a problem. An example of successful problem solving is shown in Figure 1 (Definition 2.1.7 for abelian has been read in and assimilated earlier, as has the definition for group).

The key problem in understanding proofs and solving problems is to be able to find the appropriate mathematical result. MU uses the CMS to ensure that it does not suffer from a combinatorial explosion, by focusing its search to only inference rules that are relevant to the current context. Fig.1 shows an example of MU's ability to solve problems in group theory. When MU is trying to reason forwards from the step $(ab)^2 = a^2b^2$, it does a feature search using features such as:

IS-FORM-[=>_A_B]
LHS-IS-OP-=
HAS-FUNCTION-*
HAS-FUNCTION-=
RHS-LHS-IS-OP-*
HAS-FUNCTION-SQUARE

These features have partly already been refined through the experience of checking proofs. The features retrieve a number of relevant inference rules for example $x^2 = xx$, and also other inference rules which do not match the input. The CMS is then adjusted to ensue that the right inference rule can be found more easily in future.

Because of MU's extremely rich knowledge, all of which is learned, MU demonstrates much more complex mathematical understanding than nearly all programs derived from the theorem-proving tradition of AI, a paradigm case of Lenat's Knowledge Principle.

Conclusions

In conclusion, one of the major computational problems facing the human infant, and also the human adult, is how to make sense of new situations. This is surely achieved by dynamic, adaptive learning. We argue, therefore, that a dynamic, declarative learning capability should be the cornerstone of architectures for cognition. The CMS has provided an existence proof that a declarative learner can be constructed for the mathematics domain. We hope that this demonstration will prove the catalyst for the construction of a new generation of cognitive architectures.

References

Anderson, J. 1969. *Real Analysis*. Logos Press.

Anderson, J.R. 1989. A theory of the origins of human knowledge. *Artificial Intelligence*, 40: 313-351.

Boden, M. 1988. *Computer models of mind*. Cambridge: Cambridge University Press.

Boom, J. 1991. Collective development and the learning paradox. *Human Development*, 34: 273-287.

Ellman, T. 1989. Explanation-based learning: a survey of programs and perspectives. *ACM. Comp. Survey*, 21: 163-221.

Furse, E. and Nicolson R.I. 1991. The Contextual Memory System and Learning Mathematics, *AISB Quarterly Autumn/Winter 1991 no.78* (Special Issue on Hybrid Models).

Furse, E., 1990. A Formal Expression Language for Pure Mathematics, Technical Report CS-90-2, Department of Computer Studies, The Polytechnic of Wales.

Furse, E., 1991. Mathematical Expertise and the Contextual Memory System, Technical Report CS-91-6, Department of Computer Studies, The Polytechnic of Wales.

Furse, E., 1992a. The Mathematics Understander. *Proceedings of the International Conference on Artificial Intelligence in Mathematics*, Oxford University Press (in press).

Furse, E., 1992b) The Contextual Memory System: a cognitive architecture for learning without prior knowledge. *Cognitive Systems* (in press).

Herstein, I.N. 1975. *Topics in Algebra*. London, Wiley.

Juckes, T.J. 1991. Equilibration and the learning paradox. *Human Development*, 34: 261-272.

Lenat, D.B. and Feigenbaum, E..A. 1991. On the thresholds of knowledge. *Artificial Intelligence*, 47: 185-250.

Nicolson, R.I. and Furse, E. 1992. Declarative Learning: a Challenge for Cognitive Science. *Submitted*.

Piaget, J. 1952. *The origins of intelligence in children*. NY: International Universities Press.

Piaget, J. 1985. *The equilibration of cognitive structures*. Chicago: University of Chicago Press.

Winston, P.H., Binford, T.O., Katz, B., and Lowry, M. 1983. Learning physical descriptions from functional definitions, examples and precedents. In Proceedings of the American Association of Artificial Intelligence, (AAAI83), 433-439, Morgan Kaufmann.

Hebbian Learning of Artificial Grammars

Giorgio Ganis
Department of Cognitive Science
9500 Gilman Dr., 0515
UC, San Diego
La Jolla, CA 92093-0515
ganis@cogsci.ucsd.edu

Haline Schendan
Program in Neurosciences, 0608
UC, San Diego
La Jolla, CA 92093
hschenda@igrad1.ucsd.edu

Abstract*

A connectionist model is presented that used a hebbian learning rule to acquire knowledge about an artificial grammar (AG). The validity of the model was evaluated by the simulation of two classic experiments from the AG learning literature. The first experiment showed that human subjects were significantly better at learning to recall a set of strings generated by an AG, than by a random, process. The model shows the same pattern of performance. The second experiment showed that human subjects were able to generalize the knowledge they acquired during AG learning to novel strings generated by the same grammar. The model is also capable of generalization, and the percentage of errors made by human subjects and by the model are qualitatively and quantitatively very similar.

Overall, the model suggests that hebbian learning is a viable candidate for the mechanism by which human subjects become sensitive to the regularities present in AG's. From the perspective of computational neuroscience, the implications of the model for implicit learning theory , as well as what the model may suggest about the relationship between implicit and explicit memory, are discussed.

Introduction

An artificial grammar (AG) is produced according to a "Markovian process in which a transition from a state S_i to any state S_j produces a letter" (Reber, 1967). Experimental psychology has extensively studied the learning of AG's (Reber, 1989). The data reveal several things. I) Subjects can learn and use the grammatical structure of the strings to facilitate the learning itself. This has been shown by comparing the learning time for grammatical and random letter strings. II) Subjects can generalize the knowledge they have acquired to novel strings. This was found by using a discrimination task in which previously unseen strings were classified as grammatical or non-grammatical. III) Depending upon the instructions, subjects have differing degrees of awareness of the nature of the grammar and of the underlying rules. Nevertheless, both the processes involved in AG learning and the resultant knowledge structure remain largely unknown. The purpose of our model was to shed light upon both of these aspects of AG learning.

There are several ways in which AG knowledge may be represented. I) Subjects could represent the AG by explicit rules that correspond to the Markovian process. II) Subjects may discover the correlation between symbols, combinations of symbols, and their string positions, as well as the frequency of occurrence of such correlations. III) According to the "distributive position" (Vokey and Brooks, 1991), the knowledge that subjects acquire is composed of a set of specific instances. These instances can be used to categorize new stimuli by means of a similarity evaluation process. IV) In contrast to III, the "abstractive position" concludes that subjects learn by means of a nonconscious abstraction system without retention of specific instances (Reber, 1989). According to this position, both learning and knowledge are implicit.

While there is evidence for all of these positions, this may not be problematic because there may be multiple learning mechanisms which thus result in multiple forms of knowledge representation. Additionally, these positions may not necessarily be mutually exclusive. For example, the distributive position bears some resemblance to exemplar models of perceptual categorization and and the abstractive position resembles prototype models (Posner and Keele, 1968). Parallel distributed processing (PDP) models have been proposed which result in representations that encompass both exemplars and prototypes (Rumelhart and McClelland, 1986). PDP models are an attempt to use what is known about the brain to constrain psychological explanation. With regard to the kind of sequence learning that may be involved in AG learning, Cleeremans and McClelland (1991) used a simple recurrent PDP model to explain how human subjects could encode the temporal context of complex sequential material. They suggest that "sequence learning may be based solely on associative learning processes." The model presented here suggests that

* Authors names are in alphabetical order. Research supported by the McDonnell-Pew Center for Cognitive Neuroscience in San Diego.

associative learning, as may be involved in AG learning, may take place according to a hebbian mechanism.

Though not a PDP model, the competitive chunking model of Servan- Schreiber and Anderson (1990), one of the few computational models of AG learning, is also of interest because it is based on the notion of hierarchical chunking. Chunks within a level compete with each other, while those between levels do not. The main shortcoming of the model, however, is that it suggests little about neuronal mechanisms, failing to explain how the mind/brain system can accomplish AG learning by chunking.

One way in which chunks may be formed is through a process that causes multiple representations to be connected together. The mechanism proposed by our model is that of hebbian modification of synaptic weights between neurons in a network. The Hebb rule is a neurobiologically plausible mechanism of learning (for review, cf. Brown et al, 1990). It states that when cell A excites cell B "some growth or metabolic" change will take place in both cells such that there is an increase in the ability of cell A to fire cell B: (Hebb, 1949). The main points of hebbian plasticity are that (1) it is activity between two neurons which determines plasticity, (2) the level of activity in cell B must be high enough to generate action potentials, and (3) the change in efficacy is specific to the connection between cell A and B. In other words, correlated firing in two neurons enables an increase in efficacy of the connection between them.

One of the hallmarks of hebbian plasticity is that it is a mechanism whereby correlations in the environment are the parameters which determine learning and memory. Correlations between environmental stimuli are a ubiquitous characteristic of both the structure of the world and the temporal sequencing of events. The PDP model presented in this paper incorporates the Hebb rule since it is the appropriate computational function for picking up the correlations present in AG strings. In this paper, we focus upon the computations and products of neuronal plasticity which enable a rapprochement between the various theories of AG learning. The model is tested on the seminal learning experiments conducted by Reber (1967). Additionally, our model may have general implications for how the brain acquires and structures its knowledge of the world.

The model

The model is a feedforward net composed of three layers, as shown in Figure 1. The input layer is a 6X8 two dimensional array in which the rows correspond to the set of symbols that may appear in a string plus a *blank* symbol, and the columns correspond to the position of the symbol within each string. Each unit projects forward onto several units in the next layer. The dimensions of layers 2 and 3 are 10X8. Each unit in layers 2 and 3 receives projections from all units in the previous layer corresponding to its position and to positions immediately adjacent to it.

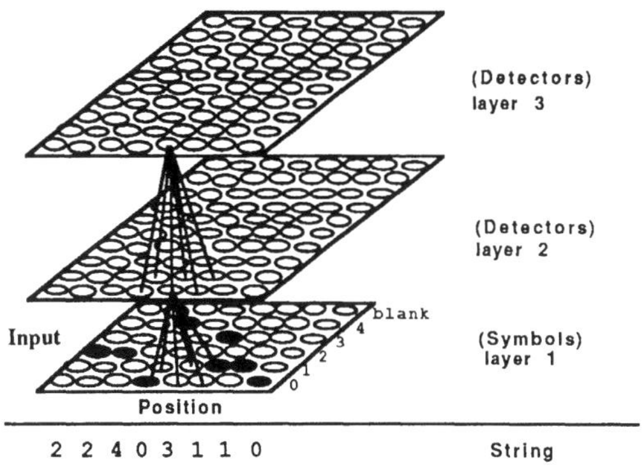

Figure 1. Architecture of the network. Only a few connections are depicted for simplicity. The filled ellipses represent the units in the input layer that are activated by the string -bottom.

The training process can be divided into stages. First, a string was presented and the activation of each unit was computed in a feedforward fashion by taking a weighted sum of the incoming inputs. The formula used was:

(1) $Output_i = ß \sum_j w_{ij} Input_j$, where ß is a scaling factor.

In order to prevent the activations from becoming too high, they were clipped at top and bottom. Then, for each position, the most active unit was selected and its weights updated according to a hebbian rule:

(2) $w = Input_j \, Output_i \, Delta$, where Delta is the learning rate.

After being updated, the weights were normalized in such a way that the sum of their absolute value remained constant. Such a normalization can be justified in neurobiological terms by assuming that weights have a metabolic cost; there is a limit to the amount of connection weighting that a neuron may support. This process is repeated for every string and the changes in the weights are cumulative. The value of Delta was selected following pilot simulations and analysis of the developed weights and activations. Delta of 0.2 was chosen because this value resulted in more local weight representations, while allowing more units to develop than did higher values. Lower values of Delta resulted in more distributed weight representations.

Experiment 1

The first experiment simulated experiment 1 of Reber (1967). In this experiment there were two types of 6-8 letter strings,generated either by a Markovian (grammatical) or a random process. Fig. 2 shows the

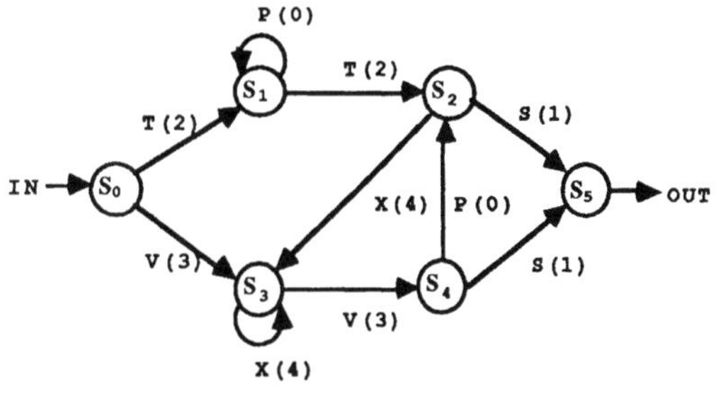

Figure 2. The artificial grammar used both in Reber (1967) and in our experiments.

Markovian process that was used both in our experiments and in Reber (1967). Subjects were trained on recall of 4 letter strings. For each of 7 sets of strings, training continued until a criterion per set was reached. Grammatical strings were learned faster than random, as reflected in the mean number of recall errors per set (ME). Supposedly, the acquisition of knowledge about the AG made it easier to recall, and/ or learn to recall, the grammatical strings. Analogously, the model was trained on either grammatical or random strings, and the training list was divided into sets of 4 strings. The network learned by updating its weights following each presentation of a string. To evaluate performance of the net, it was necessary to select a criterion that was analogous to ME. Average mean activation (avgMA) per set, which is inversely related to ME, was chosen because it incorporates the values of the weights that develop during training. AvgMA was calculated for both grammatical and random strings for 4 subjects.

Results

The graphs of the model's data for both layers 1 (Fig. 3a) and 2 (Fig. 3b) approximate Reber (1967) well. For both layers 1 and 2, the graph for grammatical strings rises with each set, while that for the random strings remains relatively flat. Performance improved with training set and grammatical strings were easier to learn than random. A two-way ANOVA was performed. For layer 1, there was a significant increase in avgMA across sets for both grammatical and random strings [($F = 72.78$), $p<0.001$]. String type, grammatical versus random, was also significant [($F = 225.37$), $p<0.001$]. Additionally, there was an interaction effect of string type with set number [($F = 29.33$), $p<0.001$]. For layer 2 also, there was a significant increase across training sets for both grammatical and random strings [($F = 70.18$), $p<0.001$]. String type , grammatical versus random, was also significant [($F = 306.95$), $p<0.001$]. Additionally, there was an interaction effect of string type with set number [($F = 22.63$), $p<0.001$]. A Tukey test for the avgMA for

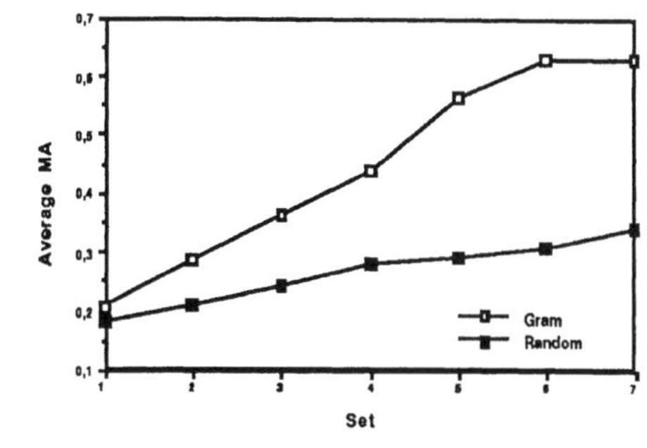

Figure 3a. Results of Experiment 1 for layer 1.

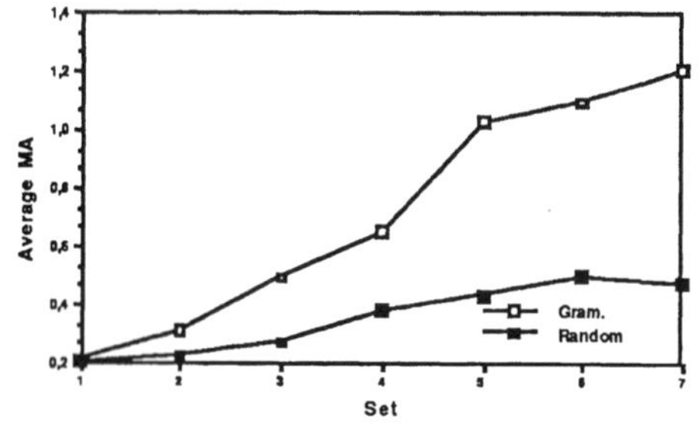

Figure 3b. Results of Experiment 1 for layer 2.

sets 1 and 2 was not significant for both layers 1 and 2. This accounts for the interaction. The most striking difference between the experimental and simulation graphs corresponds to the steep change that Reber found between sets 1 and 2 for both string types. This is missing from both simulation graphs.

Discussion

The absence of a steep rising phase during training on sets 1 and 2 is consistent with Reber's explanation of this finding. He attributed this rise to "a rather complicated warm-up effect" (Reber, 1967). This learning is unrelated to the stimuli. Rather, it is a kind of procedural learning of the training task, irrespective of the stimuli. It is of interest that procedural learning is also a kind of implicit learning (Squire, 1987). Thus there may be at least two forms of implicit learning active that are involved in AG learning. One that is stimulus specific and one that is task general. It is only the former which interested Reber and which the model was designed to emulate.

Experiment 2

Experiment 2 of Reber (1967) was also simulated. Here, subjects were trained with 20 grammatical letter strings until a criterion level of performance was reached. At test, subjects classified novel strings as grammatical or ungrammatical. These grammaticality judgments were about 79% correct (Reber, 1967).

While the simulation training procedure was identical to that used in experiment 1, the training strings differed as per Reber (1967). 20, as opposed to 28, grammatical strings were learned, and the range of string lengths was broader, 3-8 symbols long. Each sequence of 20 grammatical strings was presented to the net 5 times, for a total of 100 training trials. The testing phase and the construction of grammatical, ungrammatical, and random test strings were identical to that of Reber (1967). The trained net was presented with all 44 test strings twice. 5 subjects were simulated by using different initial weights. In experiment 1, avgMA was used as the dependent variable because it was a rough index of learning; in experiment 2, fine grammaticality judgments were required, based upon subtle irregularities present in the strings. We therefore used a more sensitive criterion for grammaticality judgment. The criterion was such that, after being trained with grammatical strings, the net classified such strings as grammatical. At the end of training, the net was tested with its training strings. Each string generated different activations in the various units. For each one of the eight possible spatial positions, the minimum activation obtained over the whole set of strings was taken as criterion for that spatial position. A string was judged as grammatical by a layer if and only if the activations generated by that string were above criterion for all the eight possible spatial positions. If there was a conflict between the judgments made by the two layers, the choice was made at random. It is obvious by construction that all the training strings were judged grammatical.

Results

The frequencies of errors made by the network matched those of Reber (1967), as shown in Table 1a and 1b, respectively. The type of errors made by the network on ungrammatical items was analyzed. As in Reber (1967), strings with multiple errors had a significantly higher detection rate than any of the others, $[X^2(1) = 6.21, p<0.01]$, the detection rate of strings with a single error in the last position was higher than that of strings with an error in the middle, $[X^2(1) = 6.42, p<0.01]$, and the detection rate of strings with a single error in the first position was higher than that of strings with an error in the last position, $[X^2(1) = 6.94, p<0.01]$. The analysis of the weight matrices indicated that n-gram detectors had developed. N varied from 1-3 symbols for units in layer 1; this means that the units could be selective for a single symbol, bigram, or trigram. An average of 40% of the units in layer 1 became selective for a single n-gram. 5% of the units became selective for two n-grams. The remaining units had very small weights, non-selective for any particular n-gram. An example of a single symbol detector is a unit which was highly activated only by a "4" in the second position. An example of a single bigram detector is a unit which became selective only for the bigram "44" in positions 3 and 4. Examples of trigram detectors are a unit selective for the trigram "200" in positions 1, 2, and 3, and a unit selective for the trigram "430" in positions 4, 5, and 6. Units in layer 2 were always highly activated by multiple symbols, bigrams, trigrams, and/ or higher n-grams.

Discussion

The consistency of the model with Reber's data support the hypothesis that the learning principle of the model is analogous to that used by human beings. Assuming this, then the structure of the weight matrices developed according to this learning principle may reveal something of the nature of mental/brain representation of AG knowledge. While Reber (1967) found that errors in the last position were most consistently detected, our model was most consistent at detecting errors in the first position. This discrepancy between Reber (1967) and the model arose because there was less variability that position than in any other position. Specifically, while the last symbol in a string was always a '1' and the first symbol either a '2' or '3', the position of the final symbol could appear in positions 3-8, whereas the initial symbol could only appear in position 1. Thus, considering both variability in symbol type and position, symbols in the first position were most determined. Additionally, units in the first and last positions received input from only 2 columns of units in the previous layer, whereas all other units received input from 3 columns, and the constraint on maximum total weight yielded higher weights per connection between units for those receiving 2 columns of input. Since symbols at the end of a string could be represented by a unit in positions 3-8, strings with the final symbol in either of positions 3-7 would have lower connection strengths than for symbols in either the first or 8th position. Thus the end symbols were disadvantaged, relative to the those in the first position, in terms of the strength of weights that may project to them. This further contributed to the superior error detection for symbols in the first position.

Judged		Gr	Ungr
	Gr	79%	21%
	Ungr	22%	78%

Table 1a. Frequencies of errors obtained in Reber (1967).

		Gr	Ungr
Judged	Gr	73%	27%
	Ungr	23%	77%

Table 1b. Frequencies of errors obtained in the model.

The analysis of the weight matrices confirmed that pooling knowledge present in both layers 1 and 2 is a good strategy. Both layers developed units that preferred certain n-grams to others. However, there was a broad range of n-gram specificity in each layer. Some units were specific to only one n-gram, while others responded well to many. In general, this specificity reflected two characteristics of the AG used. One characteristic is the frequency of occurrence of each n-gram. More frequent n-grams had a higher probability to develop responsive units, and those units were also more likely to respond only to that n-gram. The second characteristic is number of different symbols or n-grams which may occur per position, or set of positions, respectively. The more n-grams, the less specificity there was for any particular n-gram at that position.

One problem with the representation of n-grams within layers 1 and 2 needs to be emphasized. Because of the two characteristics of the AG mentioned above, the broad tuning of some units tended to make such units more highly activated by some ungrammatical n-grams. These ungrammatical n-grams were mixtures of the multiple, grammatical n-grams that the unit had come to prefer. For example, a unit that preferred legal n-grams "444" and "002" in positions 2, 3, and 4 could combine these in several ways that result in a strong activation for ungrammatical n-grams, such as "404." This property of broadly tuned units indicates that for AG learning, distributed representation can increase error rates. Indeed, several of the errors made in grammaticality judgments were the results of this. Had there been more unit specificity, the accuracy of the model might have been greater. However, since the model performed about as well as Reber's subjects, humans may also use broadly tuned units to code AG information. Thus the successes and pitfalls of hebbian learning mirror those of implicit learning mechanisms in humans.

It should also be mentioned that layer 2 units were more broadly tuned than layer 1 units. This follows naturally from the fact that layer 2 made use of the units in layer 1, many of which themselves were broadly tuned. That the combined information from both layers yielded greater accuracy is also of relevance to theories about brain processing. According to neuroscience, it is the more advanced the levels of processing are the ones that most strongly influence overt behavior. Assuming the model captures the most relevant aspects of the neurobiology of learning, then the simulation results argue against this idea. It is perhaps more realistic to say that the level of processing than is most relevant to the overt response depends upon the characteristics of the task. In the case of AG learning, each layer contributes different factors to the grammaticality judgments. Layer 2 has units which prefer longer strings than any layer 1 unit. For strings, or n-grams, less than six symbols long, there may be a unit in layer 1 that prefers it to any other string. Thus, the most important function of layer 2 is to represent longer strings (than layer 1). The most important function of layer 1, however, may be to be specific to only one n-gram. This may counteract the effects of the broadness of layer 2 units. Thus it is a combination of narrow and broad tuning in conjunction with distributed representation that constitutes the useful structure of AG knowledge.

General Discussion

In this paper, a PDP model was presented that used a hebbian learning rule to acquire AG knowledge. To evaluate the validity of the model, two experiments from the classic paper by Reber (1967) were selected. It was found that the model simulated the Reber results well. The increase in avgMA of the net paralleled the decrease in ME that humans exhibit. Novel stimuli that display the same invariants as learned stimuli elicit similar responses both in the model and in human subjects. Thus both are capable of generalization. This was demonstrated by the testing phase of experiment 2. Thus, in general, a hebbian learning mechanism is capable of computing the correlations between aspects of a stimulus and is able to use that knowledge to generalize and make decisions. These are basic cognitive abilities.

The model may have important implications extending beyond the domain of AG learning. While the model may be presumed to reveal characteristics only of implicit learning, our results are consistent with the hypothesis that implicit learning is carried out in the brain by means of a hebbian mechanism. Perhaps, the representations formed consist of units distributed across areas of cortex. with both narrow and broad tuning for correlations in the stimuli.

This characterization of the structure of knowledge resulting from implicit learning may be used to address issues from the implicit/explicit memory literature, such as whether or not there is a relationship between implicit and explicit stores and, if so, what the nature of the interaction may be. Specifically, a grammaticality judgment must be verbalized. Thus implicit knowledge must be accessible to verbalization, or explicit memory, systems. The simulations of Reber's experiments suggest how implicit knowledge may be used by the explicit system.

Once the model has been trained on a set of AG strings, within it is stored the level of activation that corresponded to acceptable sequences. This information exists as the weights connecting units in the model and, perhaps, as the relative efficacy of synapses in the brain of a human subject. In this type of neural representation, units coding grammatical n-grams have higher weights for symbols in correct positions within a string. Thus when a novel string is presented, mainly grammatical n-gram units will be activated. These grammatical n-gram units will produce higher activation over the net than when ungrammatical strings are presented. In the ungrammatical case, fewer of these highly weighted n-gram units are activated, resulting in relatively lower net activation for such strings.

Through training, systems involved in generating a response may learn what level of activation tends to be elicited by the training strings. Any novel string that elicits a similar level of net activation, its minimum determined by a threshold, may tend to be treated in the same way by the response system. This may be related to perceptual categorization (Posner and Keele, 1968).

That this may be a consequence of AG learning is supported by the Reber et al (1980) finding that the earlier explicit instructions are presented to a subject during training, the better will be performance when tested with a well formedness task. Well formedness was substituted for grammaticality by Reber in subsequent experiments (Reber, 1989). The implications of this study for the results of the model are twofold. One, the function of the explicit instructions may be to provide the learning system with feedback. This may enhance the ability of the system to categorize the stimuli as belonging to one kind. Additionally, this may enable a response system to better differentiate between the levels of activations that correspond to the category of training stimuli. For example, in terms of the simulation, the threshold, or range of thresholds, may be set more precisely.

The second implication of the Reber et al (1980) results for the model is that the explicit system is capable of interaction with implicit knowledge systems. The nature of this interaction probably depends upon the specific task demands. However, it is not clear in what ways they interact. The explicit system may not interact with the kind of learning system modelled here, but rather with a system involved in generating responses. However, since in the Reber experiments, training did not involve grammaticality feedback, the human learning system can acquire the ability to make accurate well formedness judgments without being specifically taught to do so. Rather, human subjects can use the knowledge gleaned from exemplar presentation to influence such judgments. Our model supports this view. The hebbian learning mechanism permits knowledge of correlations within exemplars to be stored in a form that may be used by an explicit, verbalization, or response system to successfully complete tasks like grammaticality or well formedness judgments. The model suggests also that the learning system may function largely in isolation from one involved in determining the overt behavioral response. However, that human subjects are able to use explicit information under certain conditions to facilitate their performance, and because the results of our simple model did not conform perfectly to the human data suggests that under most conditions, the implicit and explicit knowledge systems may interact. This includes the possibility that the structure of the implicit knowledge system may be modified by the explicit system. In terms of the model, it must be remembered that the model is restricted to coding the stimuli in isolation from any other functions that are performed by a real neural net. Therefore, it is possible that systems outside that modelled here are responsible for the effect of explicit instructions found by Reber, et al (1980). Adding such systems to the model may allow it to explain how implicit and explicit systems interact. The model may best be described as simulating both how knowledge is stored implicitly and the form of such knowledge, while also being suggestive of how implicit knowledge may be used by a verbalizable knowledge system.

Acknowledgments

We thank Prof. Marta Kutas and Prof. Mark. St John for helpful comments and discussion.

References

Brown, T.H., Kairiss, E.W., and Keenan, C.L. 1990. Hebbian Synapses: Biophysical Mechanisms and Algorithms. *Ann. Rev. of Neurosci.* 13:475-511, .

Cleeremans, A. and McClelland, J.L. 1991. Learning the Structure of Event Sequences. *J. of Exp. Psych., Gen.* 120(3):235-253.

Hebb, D.O. 1949. *The Organization of Behavior.* Wiley Press, New York.

Posner, M.I. and Keele, S.W. 1968. On the Genesis of Abstract Ideas. *J. of Exp. Psych.*, 77(3):353-363.

Reber, A.S. 1967. Implicit Learning of Artificial Grammars. *J. of Verb. Learn., Verb. Behav.*, 77:317-327.

Reber, A.S., Kassin, S.M., Lewis, S., & Cantor, G.W. 1980. On the relationship between implicit and explicit modes in the learning of a complex rule structure. *J. of Exp. Psych., Hum. Learn.,Mem.* 6:492-502.

Reber, A.S. 1989. Implicit Learning and Tacit Knowledge. *J. of Exp. Psych.: Gen.* 118(3):219-235.

Rumelhart, D. E. and McClelland, J. L. 1986. Parallel distributed processing : explorations in the microstructure of cognition. Cambridge, MA: MIT Press.

Servan-Schreiber, E. and Anderson, J.R. 1990. Learning Artificial Grammars With Competitive Chunking. *J. of Exp. Psych.: Learn., Mem., Cog. 16(4):592-608.*

Squire, L.R. *Memory and Brain*, Oxford Univ. Press, New York, (1987).

Vokey, J.R. and Brooks, L.R. 1992. The Salience of Item Knowledge in Learning Artificial Grammars. *J. of Exp. Psych.: Learn., Mem., Cog.* Forthcoming.

Comparison of Well-Structured & Ill-Structured Task Environments and Problem Spaces

Vinod Goel
Institute of Cognitive Science
University of California, Berkeley
goel@cogsci.berkeley.edu

Abstract

Many of our results in the problem-solving literature are from puzzle-game domains. Intuitively, most of us feel that there are differences between puzzle problems and open-ended, real-world problems. There has been some attempt to capture these differences in the vocabulary of "ill-structured" and "well-structured" problems. However, there seem to be no empirical studies directed at this distinction. This paper examines and compares the task environments and problem spaces of a prototypical well-structured problem (cryptarithmetic) with the task environments and problem spaces of a class of prototypical ill-structured problems (design problems). Results indicate substantive differences, both in the task environments and the problem spaces.

Introduction

At the core of any theory of cognition, there will need to be a robust model of reasoning and problem solving. Over the years we have made considerable progress in developing such models (Chandrasekaran, 1983; Duncker, 1945; Ernst & Newell, 1969; Fikes & Nilsson, 1971; Greeno, 1978b; Kleinmuntz, 1966; Newell, 1980; Newell & Simon, 1972; Sacerdoti, 1980; Simon, 1978; Simon, 1983). However, much of this work has been done in the domain of puzzle-type problems, such as cryptarithmetic and the tower of hanoi.

While the investigation of puzzle domains has resulted in very important and significant results, most people share the intuition that there are important differences between solving a cryptarithmetic puzzle and, say, writing a novel or designing a bridge. It is not *a priori* clear that the results from the former will generalize to the latter domains. In fact, there are reasons to believe the contrary.

Reitman (1964), in a seminal paper, argued for a classification of problem types based on the distribution of information in the problem vector. We generally try to capture this distinction in the vocabulary of "well-structured" and "ill-structured" problems. Puzzle games are said to be well structured because the start states, goal states, evaluation functions, and transformation functions are well specified. For example, in cryptarithmetic, the start state is completely specified, as is the goal state. The transformation function, which is also specified, is restricted to two operations: replace a letter with a digit between 0 and 9, and add. Tasks such as writing a novel and designing a bridge are considered ill defined because the start state is incompletely specified, the goal state is specified to an even lesser extent, and the transformation function is completely unspecified.

The distinction is not, however, universally accepted (Simon, 1973), and there seem to be no empirical studies directed at it. In this paper I would like to argue that there is a substantive difference between ill-structured and well-structured problems. I would like to point out that there are a number of crucial differences in the task environments of ill-structured and well-structured problems, and present data indicating corresponding differences in the structure of problem spaces. This paper is a brief summary of work presented in full elsewhere (Goel, 1991; Goel & Pirolli, in press).

The general strategy is to examine and compare prototypical cases of well-structured problems and prototypical cases of ill-structured problems. Cryptarithmetic will be used as an example of a well-structured category, while various forms of design problem solving will be used as examples of the ill-structured category. It may seem odd to restrict the discussion in this fashion, but the strategy has a number of advantages. The design and puzzle game distinction is finer grained, and thus internally more homogeneous. This internal homogeneity will sharpen and highlight any differences across the two categories.

Comparison of Task Environments

There are a number of differences in the structure of the task environments of ill-structured and well-structured problems, in addition to the differences in the distribution of information in the problem vector noted by Reitman (1964). Some which are specific to design and cryptarithmetic task environments are briefly discussed below.

One very important — but little-noted — difference has to do with the nature of the constraints in the two cases. In cryptarithmetic, as in all puzzles and games, the constraints are logical or constitutive of the task. That is, if one violates a constraint or rule, one is simply not playing that game. For example, if we are playing chess, and I move my rook diagonally across the board, I am simply not playing chess.

However, the constraints we encounter in most nongame situations are of a very different character. Some of these constraints are nomological; many of them are social, economic, cultural, etc. I will encompass the latter category under the predicate "intentional". While there is much to be said about this category, what is important for our purposes is that these constraints are not definitional or constitutive of the task. On the contrary, they are negotiable. For example, if you go to an architect and ask him to build you a new house, and he convinces you to renovate your existing house instead, or to live in a tree in the local park, it seems odd to say that he is not playing the game of design.

Nomological constraints are constraints dictated by natural law. So, for example, if a beam is to support a downward thrust of x psi, it must exert an upward thrust of equal or greater amount. These constraints, while never negotiable, are also not definitional or constitutive of the task. They, in fact, vastly underdetermine design solutions.

Another difference between design and cryptarithmetic problems is one of size and complexity. Cryptarithmetic problems take on the order of minutes to hours to complete. Design problems typically take on the order of days to months to complete.

There are also differences with respect to the lines of decomposition and the interconnectivity of parts. In both cases, the problems decompose into smaller problems. However, in cryptarithmetic, the lines of decomposition are determined by the logical structure of the problem. (So, for example, each row is treated as a component or module.) In design, on the other hand, lines of decomposition are determined by the physical structure of the world, practice within the community, and personal preference.

In terms of the interconnectivity of parts, one finds logical interconnections in cryptarithmetic (i.e., there is always the possibility that any row will sum to greater than 9 and affect the next row). Thus the subject has no choice or selectivity in attending to interconnections. Interconnections in design problems are contingent. This gives the designer considerable latitude in determining which ones to attend and which ones to ignore.

It is also the case that in design problems, as in most nongame situations, there are no right or wrong answers, though there are certainly better and worse answers (Rittel & Webber, 1974). In cryptarithmetic, as in most puzzle games, there are right and wrong answers, and clear ways of recognizing when they have been reached.

In design, as in many real-world tasks, there are consequential costs associated with errors. Resources and lives are often at stake. In cryptarithmetic, as in most games, errors may cause some embarrassment to the subject, but that is about the extent of the "damage."

Lastly, in design problems, as in many real-world situations, there is no immediate feedback from the world. Hence, it must be simulated, or self-generated. This requires considerable resource allocation for modeling and performance predicting. In cryptarithmetic there is genuine feedback after every operator application. It is, however, local feedback, and the final solution needs to satisfy global constraints.

This list is meant to be neither unique nor exhaustive. It is meant to indicate that there are a number of *substantive differences* in the task environments of at least some well-structured problems (cryptarithmetic) and some ill-structured problems (design problems). Given the logic of information processing theory, such differences should have psychological consequences at the level of the problem space. The balance of the paper describes a study which explores and articulates some of the differences in design and cryptarithmetic problem spaces.

Methodology and Database

The results presented here are based on single subject protocol studies (Ericsson & Simon, 1984). A total of sixteen protocols, twelve from design situations, and four from puzzle-game situations, were examined and compared. The design protocols were gathered from expert designers from the disciplines of architecture, mechanical engineering, and instructional design. The four puzzle protocols were from the domains of cryptarithmetic and the Moore-Anderson Tasks.[1] They were extracted from Newell and Simon (1972). The methods of collection and analysis of the data are described below. The results of the analysis of three of the design protocols — one from each discipline — and two cryptarithmetic protocols, are presented below.

Design Protocols

Subjects, Tasks, and Procedure: As noted above, the design protocols were collected from professional designers from the disciplines of

[1] The Moore-Anderson Task is a string transformation task isomorphic to theorem proving in the propositional calculus.

architecture, mechanical engineering, and instructional design. The architectural task called for the design of a self-help automated post office for the UC-Berkeley campus. The mechanical engineering task required the design of an "automated postal teller machine" for the above post office. The instructional design task was unrelated. It involved the design of a self-contained instructional package to teach lay people a reasonably complicated computational environment.

The procedures for collecting the protocols were the same in each case. Each subject was given a one-page design brief, and any related documents, and asked to specify a solution to the problem, to the degree of specificity allowed by time and resource constraints. They were allowed to use any external drawing aids/tools that they desired. All chose to use paper, pencil and/or pen.

The durations of the sessions varied from two to three hours. The experimenter was present to answer any questions relating to the experiment, and otherwise assume the role of the client. Subjects were encouraged to ask clarification questions as the need arose. The experimenter answered all questions, but at no time initiated the conversation.

Subjects were asked to "talk aloud" as they solved the problem. They were cautioned against trying to explain what they were doing. Rather, they were asked to vocalize whatever was "passing through their minds" at that time. Most of the subjects did not have much difficulty in doing this. Where subjects did lapse into periods of silence, the experimenter prompted them by asking "what are you thinking now?".

The sessions were videotaped. The tapes, along with the written and drawn material, constituted the data.

Coding Scheme: The protocols were transcribed, cross-referenced with the written material, and coded. The coding involved breaking the protocols into individual statements representing single "thoughts" or ideas. Content cues, syntactic cues, and pauses were used to effect this individuation. This resulted in very fine-grained units with a mean duration of eight seconds and a mean length of fifteen words. Each statement was coded for the operator applied (e.g. add, delete, justify, etc.), the content to which the operator was applied, the mode of output (verbal or written), and the source of knowledge (design brief, experimenter, self, inference).

These statements were then aggregated into modules and submodules, which are episodes organized around artifact components. For example, for the architect subjects, the modules were components like site, building, and services. The site submodules were components like circulation, landscaping, and site illumination. The building submodules included such things as doors, roof, and mail storage. The modules were then further aggregated into design-phase levels. The design-phase level coded for several things, the most important being design development phases such as problem structuring, preliminary design, refinement, and detail design. These categories were further coded for the aspect of design development attended to (e.g. people, purposes, behavior, function, and structure).

This resulted in a reasonably complex three-level, hierarchical scheme, similar in spirit, but not detail, to the one employed by Ullman, Dietterich, and Stauffer (1988). It is fully detailed elsewhere (Goel, 1991; Goel & Pirolli, in press).

Cryptarithmetic Protocols

Subjects, Tasks, and Procedures: The two cryptarithmetic protocols were gathered from published sources (Newell & Simon, 1972, Appendices 6.1, 7.1), recoded, and compared with the design protocols. These particular ones were chosen on the basis of their duration. The subjects for these studies were undergraduate students. The procedure of collecting the protocols was similar in relevant aspects to the one described above.

The task for both subjects (NS6.1 and NS7.1) was the following problem:

```
  DONALD          D=5
+ GERALD
 ________
  ROBERT
```

Each letter stands for a digit. The digits encoded as DONALD and GERALD add up to the digits encoded as ROBERT. The task is to transform the letters into the appropriate digits. The clue given is that D=5.

Coding Scheme: The protocols were re-coded with a modified subset of the scheme devised for the design protocols. Three changes were required. First, it was found that while one could differentiate between problem structuring and problem solving, it was not possible to further differentiate problem solving into preliminary, refinement, and detail phases. Second, the aspect of design development category was not applicable. Third, information about mode of output was not available.

Comparison of Problem Spaces

This section discusses some of the characteristics of design problem spaces and notes how they differ from cryptarithmetic problem spaces. The results are presented in more detail elsewhere (Goel, 1991; Goel & Pirolli, in press).

Stopping Rules and Evaluation Functions: The stopping rules and evaluation functions in design problem spaces are determined by the designer rather than the structure of the problem. The decisions are based on personal preference and experience, professional standards and practice, and client expectations. The personalized nature of the stopping rules and evaluation functions can be explained by appealing to three factors in the task environment. First, there is not enough information in the problem statement to make these decisions. Second, there are no right and wrong terminating states. Third, there are few, if any, logical constraints.

In cryptarithmetic, the stopping rule is explicitly supplied and evaluation functions, at least locally, are determined by the structure of the problem. The issue of personal preference just does not enter the picture.

Memory Retrieval & Inferences: On a related front, a very small percentage of statements in design protocols (1.3%) is generated by overt deductive inferences. Most seem to be the result of memory retrieval and modification and/or nondemonstrative inference. The cryptarithmetic problem spaces had a much higher percentage of statements generated by demonstrative inference (41%).

This large difference in deductive reasoning is what one would expect, given the structures of the task environments. Deductive systems require logical constraints. As already noted, the constraints on cryptarithmetic are logical, while the constraints on nongame tasks, like design, are nonlogical.

Direction of Transformation Function: In well-structured domains, the transformation function maps the start state onto the goal state. In the design problem spaces, it was noted that the subjects would stop and turn things around. That is, they would try to manipulate the problem constraints and client expectations so as to change the start state to one which better fits their knowledge, experience, and expertise. One might call this phenomenon "reversing the direction of the transformation function," because the subject has prior knowledge of some goal state and is trying to transform the problem parameters to fit that goal state.

Again, the reason that this can occur is that the problem is incompletely specified, and the constraints are nonlogical, therefore manipulable. It cannot, and does not, occur in cryptarithmetic because of the logical nature of the constraints. Any attempt by the subject to change the parameters would be viewed as an inability or lack of desire to participate in the assigned task.

Solution Decomposition: Another interesting difference across the two problem spaces has to do with solution decomposition. There are two interesting findings. First, design problems are decomposed into many more modules than cryptarithmetic modules, and second, the density of interconnections between modules is higher in the cryptarithmetic case than the design case.

For example, subject S-A, working on the architectural task of designing a post office, decomposed the solution into 34 modules corresponding to structural and functional components such as roof, door, location of equipment, flow of traffic, etc. Given 34 modules, 1,122 interconnections are logically possible. The subject actually made only 7.4% (83.03) of these connections.

In cryptarithmetic, on the other hand, the problems were decomposed into 6 modules (corresponding to the six columns). But while the actual number of modules were fewer, the density of interconnections between modules was considerably greater. Subject NS6.1, for example, made 20% (6) of the logically possible connections.

The denser interconnectivity of the cryptarithmetic modules is what one might expect, given that they are intended to be multiple constraint satisfaction problems, and all the constraints are logical (so must be attended to). This is perhaps why such problems can have so few components and still be challenging. The reason design problems can have so many components and still be tractable is that the interconnections are contingent rather than logical. This gives the designer considerable flexibility in determining which one to attend to and which ones to ignore.

Development of Solution: Yet another interesting difference has to do with the incremental development of solutions in design problem spaces. One of the most robust findings in the literature of design problem solving is that, as design solutions are generated, they are retained, massaged and nurtured to completion (Kant, 1985; Ullman et al., 1988). They are not easily discarded.

A number of aspects of the design task environment favour such a strategy. First, there is the obvious fact that the problems are large, and given the sequential nature of human information processing, cannot be completed in a single cycle. Second, since there are few logical constraints to be violated, and no right or wrong answers, there is little reason to give up on a partial solution to start again from scratch. Third, incremental development is compatible with the "least-commitment" control strategy used by designers (see below).

In contrast, traversal of cryptarithmetic problem spaces have an all-or-nothing character about them. Most paths searched are wrong and independent of the correct path(s). Thus there is no sense of building up to a solution. Once a path is searched, and it turns out not to be on the solution path, the subject is no better off than before the search began. He must start again on another path.

Control Structure: There are also a number of interesting differences with respect to control strategies in the two cases. The design subjects used a control strategy, not unlike the "least-commitment" control strategy identified by Stefik (1981). The basic feature of this strategy is that, when working on a particular module, it does not require the designer to complete that module before beginning another. Instead, one has the option of putting any module on "hold" to attend to other related (or even unrelated) modules, and returning to the first at a later time. This embedding can go several levels deep.

The control structure of the design subjects is naturally analyzed into three hierarchical levels: movement from module to module, movement from submodule to submodule, and movement internal to submodules. The first two levels are task-specific; that is, the modules and submodules vary from task to task. The third level, however, generalized across all three design tasks. The control structure within any level is repetitive, cyclical, and flexible. One effect of this repetition and reiteration is that most modules and submodules are considered in more than one context.

The cryptarithmetic strategy was interestingly different in some respects. While one could trace a cyclical, repetitive control structure, as in the design case, most of the problem solving occurred internal to modules/episodes. There was little carryover from previous visits to a module/episode. In fact, Newell and Simon (1972), in their original analysis of these protocols, claimed that in returning to a former state, the subject is in fact returning to a previous knowledge state with respect to the problem. If the subject goes down the wrong path and returns to the previous state, all that he knows is that the path just explored does not lead to the goal state. He does not have an enriched understanding of the state he is returning to. The complete control structures of a design and cryptarithmetic subject are traced out in Goel (1991).

Making & Propagating Commitments: While the least-commitment control strategy allows design subjects to keep options open, the solution must ultimately be brought to closure. This requires that one make and propagate commitments through the problem space. In the cryptarithmetic protocols, while commitments are certainly made, they are propagated only until a local evaluation function accepts or rejects them.

The last aspects of design and cryptarithmetic problem spaces that I would like to discuss have to do with the phases of solution development.

The development of a design solution has several distinct phases. Four of these phases are: problem structuring, preliminary design, refinement, and detailing. These phases differ with respect to the type of information dealt with, the degree of commitment to generated ideas, the level of detail attended to, and the number and types of transformations engaged in.

Problem structuring is the process of retrieving information from long-term memory and external memory and using it to construct the problem space; i.e., to specify start states, goal states, operators, and evaluation functions. Problem structuring relies heavily on the client and design brief as a source of information, considers information at a high level of abstraction, makes few commitments to decisions, and involves a high percentage of add and propose operators.

Preliminary design is a classical case of creative, ill-structured problem solving. It is a phase where alternatives are generated and explored. Alternative solutions are not, however, fully developed when generated. They emerge through incremental transformations of a few kernel ideas. These kernel ideas are images, fragments of solutions, etc. to *other* problems which the designer has encountered at some point in his life experience. Since these "solutions" are solutions to other problems which are being mapped onto the current problem, they are, not surprisingly, always out of context or in some way inappropriate and need to be modified to constitute solutions to the present problem.

This generation and exploration of alternatives is facilitated by the abstract nature of information being considered, a low degree of commitment to generated ideas, the coarseness of detail, and a large number of lateral transformations. A lateral transformation is one in which movement is from one idea to a slightly different idea, rather than a more detailed version of the same idea. These transformations are necessary for the widening of the problem space and the exploration and development of kernel ideas.

The *refinement* and *detailing phases* are more constrained and structured (though still very different from puzzle games). They are phases where commitments are made to a particular solution and propagated through the problem space. They are characterized by the concrete nature of information being considered, a high degree of commitment to generated ideas, attention to detail, and a large number of vertical transformations. A vertical transformation is one in which movement is from one idea to a more detailed version of the same idea. It results in a deepening of the problem space.

While these phases of design development may seem trivially obvious, they are rendered interesting by the fact that cryptarithmetic problem spaces cannot be individuated into similar phases. As already noted, in such game problems one gets more of the same activity. Either one is on a path which will abruptly lead to the solution, or one is not. There is no sense in which one builds up to a solution.

Conclusion

I have presented arguments and data to suggest that there are interesting differences in the task environments of (at least some) well-structured and (at least some) ill-structured problems, and that these lead to some nontrivial differences in ill-structured and well-structured problem spaces.

The reader may have noted, however, that the comparison of problem spaces is not carried out at the level of states and operators, which is the level at which problem spaces are generally defined. It is conducted at a much more abstract level. The fact of the matter is, if one compares the problem spaces at the level of states and operators, it is difficult to differentiate the two problem spaces. It is only when one abstracts away from the low-level details — the sequence of states and operators — that the differences emerge. However, the fact that they emerge at this more abstract level does not make them any less real or interesting. On the contrary, generalizations at this level may serve to fill the theoretical gap that some argue exists in information processing theory between implementations of specific problem spaces and the general notion of an information processing system (Chandrasekaran, 1983; Goel & Pirolli, 1989; Greeno, 1978a).

Acknowledgements

The author is indebted to Peter Pirolli, Susan Newman, and Mimi Recker for helpful discussions and comments. This work has been supported by a Gale Fellowship, a Canada Mortgage and Housing Corporation Fellowship, a research internship at System Sciences Lab at Xerox PARC and CSLI at Stanford University, and an Office of Naval Research Cognitive Science Program grant (# N00014-88-K-0233 to Peter Pirolli).

References

Chandrasekaran, B. (1983). Towards a Taxonomy of Problem Solving Types. *AI Magazine, winter/spring*, 9-17.

Duncker, K. (1945). *On Problem Solving*. Westport, Connecticut: Greenwood Press.

Ericsson, K. A., & Simon, H. A. (1984). *Protocol Analysis: Verbal Reports as Data*. Cambridge, Massachusetts: The MIT Press.

Ernst, G. W., & Newell, A. (1969). *GPS: A Case Study in Generality and Problem Solving*. N.Y.: Academic Press.

Fikes, R. E., & Nilsson, N. J. (1971). Strips: A New Approach to the Application of Theorem Proving to Problem Solving. *Artificial Intelligence, 2*, 189-208.

Goel, V. (1991) *Sketches of Thought: A Study of the Role of Sketching in Design Problem Solving and its Implications for the Computational Theory of Mind*. Ph.D. Dissertation, University of California, Berkeley.

Goel, V., & Pirolli, P. (1989). Motivating the Notion of Generic Design within Information Processing Theory: The Design Problem Space. *AI Magazine, 10* (1), 18-36.

Goel, V., & Pirolli, P. (in press). The Structure of Design Problem Spaces. *Cognitive Science.*

Greeno, J. G. (1978a). Natures of Problem-Solving Abilities. In W. K. Estes (Eds.), *Handbook of Learning and Cognitive Processes, Volume 5: Human Information Processing*. Hillsdale, N.J.: Lawrence Erlbaum Associates.

Greeno, J. G. (1978b). A Study of Problem Solving. In R. Glaser (Eds.), *Advances in Instructional Psychology, Vol. 1*. Hillsdale, N.J.: Lawrence Erlbaum Associates.

Kant, E. (1985). Understanding and Automating Algorithm Design. *IEEE Transactions on Software Engineering, 11*, 1361-1374.

Kleinmuntz, B. (Ed.). (1966). *Problem Solving: Research, Method, and Theory*. NY: John Wiley.

Newell, A. (1980). Reasoning, Problem Solving, and Decision Processes: The Problem Space as a Fundamental Category. In R. S. Nickerson (Eds.), *Attention and Performance VIII*. Hillsdale, N.J.: Lawrence Erlbaum.

Newell, A., & Simon, H. A. (1972). *Human Problem Solving*. Englewood Cliffs, N.J.: Prentice-Hall.

Reitman, W. R. (1964). Heuristic Decision Procedures, Open Constraints, and the Structure of Ill-Defined Problems. In M. W. Shelly & G. L. Bryan (Eds.), *Human Judgements and Optimality*. N.Y.: John Wiley and Sons.

Rittel, H. W. J., & Webber, M. M. (1974). Dilemmas in a General Theory of Planning. *DMG-DRS Journal, 8* (1), 31-39.

Sacerdoti, E. D. (1980). Problem Solving Tactics. *AI Magazine, 2* (1), 7-15.

Simon, H. A. (1973). The Structure of Ill-Structured Problems. *Artificial Intelligence, 4*, 181-201.

Simon, H. A. (1978). Information-Processing Theory of Human Problem Solving. In W. K. Estes (Eds.), *Handbook of Learning and Cognitive Processes, Vol.V*. (pp. 271-295). Hillsdale, N.J.: Lawrence Erlbaum Associates.

Simon, H. A. (1983). Search and Reasoning in Problem Solving. *Artificial Intelligence, 21*, 7-29.

Stefik, M. (1981). Planning and Meta-Planning (Molgen: Part 2). *Artificial Intelligence, 16*, 141-170.

Ullman, D. G.,Dietterich, T. G., & Stauffer, L. A. (1988). *A Model of the Mechanical Design Process Based on Empirical Data* (Tech. Report No. DPRG-88-1). Dept. of M.E., Oregon State University.

Prediction Performance as a Function of the Representation Language in Concept Formation Systems*

Mirsad Hadzikadic

Department of Computer Science
University of North Carolina
Charlotte, NC 28223
mirsad@unccvax.uncc.edu

Abstract

Existing concept formation systems employ diverse representation formalisms, ranging from logical to probabilistic, to describe acquired concepts. Those systems are usually evaluated in terms of their prediction performance and/or psychological validity. The evaluation studies, however, fail to take into account the underlying concept representation as one of the parameters that influence the system performance. So, whatever the outcome, the performance is bound to be interpreted as 'representation-specific.' This paper evaluates the performance of INC2, an incremental concept formation system, relative to the language used for representing concepts. The study includes the whole continuum, from logical to probabilistic representation. The results demonstrate the correctness of our assumption that performance does depend on the chosen concept representation language.

Introduction

Concepts lie at the core of human thought, perception, speech, and action. Consequently, the issue of *concept formation* represents an important research problem of interest to researchers from diverse disciplines, including psychology, philosophy, linguistics, and artificial intelligence. The section on concept formation partially summarizes past work in the above disciplines.

One of the far-reaching decisions to be made by every investigator/system designer is the language(s) for representing concepts and instances. The representation language defines not only how *easily* a concept can be learned, but, more importantly, *what* kind of concept can be acquired. Also, it seems plausible that the same representation cannot be equally well suited for different tasks in different application domains under different circumstances. Therefore, the goal of this paper is to *evaluate the relationship between performance and representation language in concept formation systems.* The 'Concept Representation' section provides a brief overview of different representation formalisms, while the following section explains the specifics of two evaluation methods, i.e., prediction accuracy and psychological validity.

The experimental tool used in this process is INC2 (Hadzikadic and Elia, 1991; Hadzikadic and Yun, 1989), an incremental, similarity-based concept formation system. The INC2's architecture, briefly explained in the 'Representation Continuum' section, allows us to easily modify its representation language both statically and dynamically in order to understand a potential correlation between performance and representation.

The remaining sections of the paper summarize the results of our analysis with respect to both prediction performance and psychological evaluation.

Concept Formation

Concept formation refers to the incremental process of constructing a hierarchy of concept descriptions (categories) which characterize objects in a given domain. A system which can accomplish this task can be used both as an aid in organizing and summarizing complex data and as a retrieval system which can predict properties of previously unseen objects. Such a system will be useful in domains where knowledge is incomplete or classifications and/or human experts do not exist.

Most existing *concept formation* systems use hill-climbing methods to find suboptimal clusterings of objects to be characterized. Six existing systems which share all of the above features are COBWEB (Fisher, 1987), CLASSIT (Gennari, Langley, and Fisher, 1989), UNIMEM (Lebowitz, 1987), CYRUS (Kolodner, 1984), WITT (Hanson and Bauer, 1989), and INC2 (Hadzikadic and Elia, 1991).

Researchers from disciplines other than computer science, e.g., psychology, philosophy, and linguistics, have been very active in this area as well. For example, Wittgenstein's research (1953) is associated with the ideas of *family resemblance.* Family resemblance introduces the idea that members of a category may be related to one another without all members having any properties in common that define that category.

Brown (1958) begins the study of what will later

*This work was supported by the grants from the College of Engineering and the Office of Academic Affairs, UNCC.

become known as basic-level categories. *Basic-level* categorization places the cognitively basic categories in the 'middle' of a general-to-specific hierarchy. Generalization and specialization, then, proceed upward and downward, respectively, from the basic level.

Finally, Rosch and her collaborators (1976) suggest that thought in general is organized in terms of *prototypes* ('best' examples) and basic-level structures. Their work establishes research paradigms in cognitive psychology for demonstrating family resemblance and basic-level categorization.

Concept Representation

The system that established the field of conceptual clustering, CLUSTER/2 (Michalski and Stepp, 1983) used a logic-based representation to represent both instances and concepts. The concepts were represented as conjunctions of necessary and sufficient features (logic expressions). The membership in a class was defined as all or none, depending on whether the instances possessed the required features or not.

In contrast, many researchers (as indicated in the previous section) have suggested that some instances are better examples of the concept than others, and that instances of the concept are distributed all over the space defined by the concept features. The best example (prototype) is the center of that space, with 'good' examples gravitating toward the center, while the 'bad' ones lie at the concept's periphery. Clearly, a logic-based representation, in its original form, cannot capture such distributional information. *Probabilistic* concept representations (Smith and Medin, 1981), however, handle this problem easily by associating a probability (weight) with each feature of a concept definition. This weight is usually implemented as the conditional probability $p(f \mid C)$ of the feature f's presence, given category C. In literature, it is often referred to as *category validity* of the feature. The retrieval and prediction, using probabilistic concepts, are usually based on the comparison between the sum of the feature weights and a given threshold (Smith and Medin, 1981). Both COBWEB and INC2 systems are based on a hierarchical probabilistic representation of concepts, where the hierarchical structure eliminates the weakness of simple probabilistic representations, namely their inability to capture non-linear correlations among features.

Probabilistic representations are more general than the logic-based ones in a sense that the former can simulate the latter by dropping all features with the category probability of less than 1.0. In addition, it is easy to imagine a continuum of probabilistic representations which differ in the value of their feature drop threshold. The drop threshold will range from 0.0 (initial probabilistic representation) to 1.0 (logic representation).

Performance Tasks

The choice of the drop threshold (and ultimately the representation) may influence the performance of the system. Prediction and psychological validity are the two performance tasks most frequently used in concept formation systems.

Prediction refers to the process of drawing inferences in regard to the category membership of previously unseen instances. It assumes existence of two key components: (1) a set of concepts known to the system, and (2) a domain-independent heuristic which indicates the likelihood of each concept being the target category. Concept formation systems usually rely on heuristics developed in psychology to guide the classification process. For example, INC2 utilizes the *contrast model* (Tversky, 1977) to compute the similarity between two objects/concepts and *family resemblance* (Wittgenstein, 1953) to decide whether to place an object into the category or not. On the other hand, COBWEB makes use of *category utility* (Gluck and Corter, 1985) to find the optimal clustering at each level of the hierarchy.

Psychological validity, on the other hand, emphasizes the importance of psychological findings (human subject studies) and measures the extent of their overlap with the results of the concept-formation systems. These findings include *typicality, basic level categories, and intra- and inter-category similarity.* More often than not, concept formation systems rely on their heuristic evaluation function (category validity in COBWEB; contrast model and family resemblance in INC2) to demonstrate 'human-like' performance as a side effect.

Representation Continuum

The experimental tool used in this evaluation study is INC2, an incremental concept formation system which builds a hierarchy of concept descriptions. The leaves of the hierarchy are objects (singleton concepts). The root of the hierarchy has associated with it a description which is a summary of the descriptions of all objects seen to date by the system.

In addition to features and hierarchical pointers, each concept description contains an estimate of its cohesiveness, given in the form of *family resemblance* (Wittgenstein, 1953). Family resemblance is defined as the average similarity between all possible pairs of objects in a given category. The similarity function used by INC2 represents a variation of the contrast model (Tversky, 1977), which defines the similarity between an object and a category as a linear combination of both common and distinctive features. As a result, INC2 implements a hill-climbing strategy which encourages advancement toward the maximal improvement of the hierarchy as measured by the increase in the family resemblance of the host concept.

INC2 uses a probabilistic representation to store concept descriptions. A description of each concept C is defined as a set of features f (attribute-value pairs). Each feature has a conditional probability $p(f \mid C)$ associated with it. Thus, representing the color feature of red apples would take the form (*color red* 0.75). The 0.75 means that members of this category are red 75% of the time. Since members of a given concept may reside in distinct portions of the hierarchy, the adopted representation formalism is referred to as a *distributed probabilistic concept hierarchy.*

The only threshold introduced in INC2 is a drop threshold. This threshold allows for concept descriptions to be either probabilistic or logical. It can be set anywhere between 0.0 and 1.0, and means that any feature with the conditional probability below this threshold should be dropped[1] from the concept description. The value of 1.0 for this threshold would yield a logical concept description. It is easy to imagine systems with different values for the drop threshold, e.g., 0.75 (each instance should have at least 3/4 of the features in common with other instances of the category), or 0.5 (at least 1/2 common features).

The drop threshold is static in nature, i.e., the same value is used at every level of the hierarchy and for all instances, no matter what their time of arrival or path of incorporation happens to be. However, the nature of classification calls for a dynamically adjusted threshold rather than a fixed one. For example, all features are important at the top level of the hierarchy, no matter how low their probabilities might be, due to the diversity of objects in the domain as well as the potential noise in object descriptions. Therefore, the drop threshold should be set close to 0.0. At the lower levels of the hierarchy, however, certain patterns have been detected, resulting in high conditional probabilities for 'relevant' features and low probabilities for the ones not significantly present in those patterns. Since all categories at the lower levels have few members, all the features found in their descriptions will have relatively high conditional probabilities. To avoid the interference of irrelevant features with the retrieval process, the drop threshold should be set close to 1.0. The intermediate categories will, then, require the drop threshold somewhere between 0.0 and 1.0, depending on the level of the hierarchy (the lower the level, the higher the drop threshold).

In order to accommodate this type of reasoning, INC2 relies on family resemblance to provide an estimate of the drop threshold value. Family resemblance is naturally set close to 0.0 at the root (summarizing the whole universe) and to 1.0 at the leaves. Consequently, INC2 automatically sets the drop threshold to the value of the family resemblance of the parent category during both classification and retrieval. That value increases with the object traversing the hierarchy downward. INC2, therefore, performs a context-sensitive classification/retrieval due to its adaptive behavior that changes from level to level of the hierarchy. In that process, INC2 uses different representations to describe objects/categories at different levels of the hierarchy, possibly moving from the probabilistic representation (drop threshold = 0.0) at the top level to the logical one (drop threshold = 1.0) at the leaves.

The idea of a dynamically adjusted drop threshold, coupled with the fact that features are only dropped temporarily (until the changing environment will have brought them back into the foreground of the system's attention), effectively emulates the idea of tracking *concept drift* (i.e., adapting to concepts that change over time) as advanced by Schlimmer and Granger (1986).

Prediction Performance Evaluation

At this point, the reader should have a sufficient understanding of INC2's representation formalism to appreciate the context in which the probabilistic-vs-logical-representation experiment has been carried out. We will briefly describe, next, the domain of clinical audiology in which the experiment took place, and then the experiment itself.

The audiology domain consists of 200 cases, 58 features, and 24 ideal categories[2]. The distribution of cases across the categories varies from one to 48 per category. Half of the categories are represented by only one or two cases. Such a distribution certainly makes learning almost impossible for those categories that are under-represented. The cases include noise in the form of incorrect and/or missing features. On average, each case has only 11 features with known values.

The probabilistic-vs-logical-representation experiment involved four different sizes of the training set (20, 50, 100, and 150) and six different values for the drop threshold (variable, 0.0, 0.25, 0.5, 0.75, and 1.0). The size of the test set was kept constant at all times (45 objects -- 22.5% of the total object set). Figure 1 summarizes the percentage of correct responses, averaged over five runs with randomly chosen objects, for all of the above cases.

[1] This happens only temporarily since new object acquisitions may bring that feature back into the concept description.

[2] Provided by Prof. Jergen from the Baylor College of Medicine and Bruce Porter of the University of Texas at Austin.

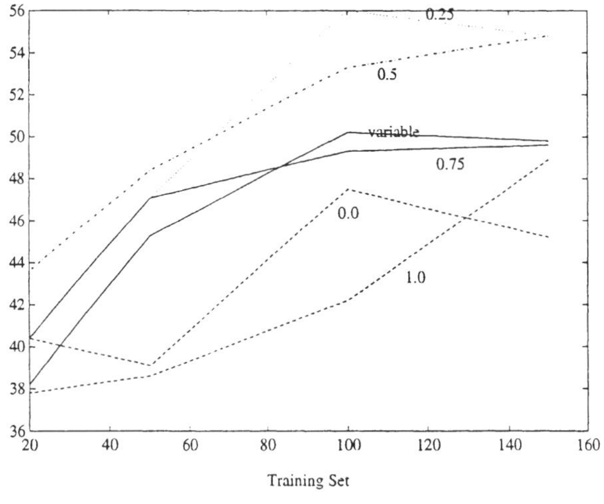

Figure 1: Prediction performance for both variable and fixed drop thresholds.

In the case of the set with a low number (20) of input objects, the variable drop threshold was outperformed by all the fixed-version values except for 1.0. The picture, however, changed for larger sets (e.g., 50, 100, and 150). The variable threshold clearly outperformed both probabilistic and logical representations, while scoring comparably to the 0.75 case. At the same time, the probabilistic representation consistently demonstrated better performance than the logical one, though not decisively so.

Unexpected results, however, came from the strong performance of the 0.25 and 0.5 cases, which clearly proved to be the best choice in our experiments. The 0.5 performed better than the 0.25 in the experiments with a low number of training objects (actually, even the 0.75 case was as good as the 0.25 under those conditions), while the 0.25 demonstrated its strength in the cases with a large number of input objects. These results seemed to indicate that neither storing all features nor 'forgetting' those that do not hold for all instances of the concept maximizes the performance of the system or provides a clear advantage over one another.

In addition, the results demonstrated the need for 'forgetting' those features that were irrelevant for the category membership. It remained unclear, however, how to 'recognize' them. Forgetting the features that do not hold for at least a half of the concept instances proved to be beneficial for the low number of training instances. An increased number of training objects provided some new evidence about the importance of certain features, and the drop threshold had to be lowered in order to improve the system performance. This evidence is in line with the reasoning behind the variable drop threshold, which adopts higher values for the nodes closer to the leaves (summarizing but a few input cases) and lower values for the nodes closer to the root (those that accumulate higher levels of experience).

Psychological Evaluation

In addition to its prediction performance, INC2 has been evaluated in terms of the psychological validity of its results. There are three issues of special interest here: typicality, basic level categories, and intra-category similarity vs. inter-category dissimilarity.

Due to the uneven distribution of instances, two classes (*cochlear age* and *cochlear unknown*) accounted for 70% of all retrievals. In order to evaluate the quality of retrieved objects in this domain, we decided to closely examine the objects from one of those classes, *cochlear age*. First, we calculated the average similarity of each object with all other members of the category. The similarity ranged from 0.0 to 0.527. The objects with the similarity greater than or equal to 0.5 were considered to be 'good' examples of the category. Then, we reviewed the list of often-retrieved objects and noticed that over 60% of them were among the examples regarded as 'good.' This finding was consistent with the prototype theory.

In addition, we reviewed all objects retrieved at least once, and for each such object calculated its average similarity. As expected, the frequency of retrieval was roughly proportional to the average similarity of the object. Consequently, we can conclude that the INC2-generated hierarchies demonstrate typicality effects similar to those generated by human subjects.

Due to the strategy adopted in its concept formation algorithm (place an object into the category if it increases the family resemblance of the category), INC2 always incorporates the object at its basic level. While traversing the hierarchy, and before it will have reached the basic level, the object encounters more and more familiar objects and categories, i.e., the ones it has more features in common with than with any previously encountered object/category. That will stop at the basic level, however, since the remaining objects/categories will begin having more and more differing features due to their increased specialization within the hierarchy. It is important to notice that objects may have their basic level at different levels of the hierarchy (depending on the order of objects and local context), thus leading to the notion of a *distributed basic level.*

Finally, the issue of intra-category similarity vs. inter-category dissimilarity is addressed implicitly in INC2, again through its algorithm. Namely, the system will place an object into the category which maximizes the increase in the category's family resemblance (compactness). Consequently, the category that receives the object will pull its instances somewhat closer to its imaginative center, thus positioning itself away from other 'gravitation points' in the instance/category space. This process will automatically reduce the force (similarity) between the category and the surrounding concepts.

Summary

This paper has evaluated the relationship between performance and adopted category/object representation. We varied the representation from probabilistic to logical, and compared their corresponding performance on the prediction task. An alternative approach, variable representation, was evaluated as well. It was characterized by the constant switching among different representation schemas according to the value of the compactness of the categories stored at different levels of the hierarchy. The variable-threshold approach worked consistently better than either the probabilistic or the logical representation. It did not, however, match the success of the fixed, middle-of-the-road-valued drop threshold.

This last observation represents our research agenda. We will continue to search for the ways to automatically set the optimal value for the variable drop threshold. In addition, we will extensively evaluate the system in terms of the cost/accuracy trade-off as it moves from probabilistic to logical representation.

References

Brown, R. (1958). How Shall a Thing be Called? *Psychological Review,* **65,** 14-21.

Fisher, D. H. (1987). Knowledge Acquisition Via Incremental Conceptual Clustering. In *Machine Learning,* **2,** 2, 139-172.

Gennari, J. H., Langley, P., and Fisher, D. H. (1989). Models of Incremental Concept Formation. In *Artificial Intelligence,* **4,** 1-3, 11-61.

Gluck, M. A. and Corter, J. E. (1985). Information, Uncertainty, and the Utility of Categories. *Proceedings of the Seventh Annual Conference of the Cognitive Science Society,* 283-287, Irvine, CA, Lawrence Erlbaum.

Hadzikadic, M. and Yun, D. Y. Y. (1989). Concept Formation by Incremental Conceptual Clustering. *Proceedings of the Eleventh International Joint Conference on Artificial Intelligence,* 831-836, Detroit, MI.

Hadzikadic, M. and Elia, P. (1991). Context-Sensitive, Distributed, Variable-Representation Category Formation. *Proceedings of the Thirteenth Annual Meeting of the Cognitive Science Society,* 269-274, Chicago Illinois.

Hanson, S. J. and Bauer, M. (1989). Conceptual Clustering, Categorization, and Polymorphy. In *Machine Learning,* **3,** 4, 343-372.

Kolodner, J. L. (1984). Retrieval and Organizational Strategies in Conceptual Memory: A Computer Model. Lawrence Erlbaum Associates, Publishers, London.

Lebowitz, M. (1987). Experiments with Incremental Concept Formation: UNIMEM. In *Machine Learning,* **2,** no. 2, 103-138.

Michalski, R. S., and Stepp, R. E. (1983). Learning From Observation: Conceptual Clustering. In *Machine Learning: An Artificial Intelligence Approach,* R. S. Michalski, J. G. Carbonell, and T. M. Mitchell (Eds.), Morgan Kaufmann Publishes, Inc., Los Altos, CA.

Rosch, E., Mervis, C., Gray, W., Johnson, D., and Boyes-Braem, P. (1976). Basic Objects in Natural Categories. In *Cognitive Psychology,* **18,** 382-439.

Schlimmer, J. C. and Granger, R. H., Jr. (1986). Beyond Incremental Processing: Tracking Concept Drift. *Proceedings of the Fifth National Conference on Artificial Intelligence,* 502-507, Philadelphia, PA.

Smith, E. E. and Medin, D. L. (1981). Categories and Concepts. Harvard University Press, Cambridge, MA.

Tversky, A. (1977). Features of Similarity. *Psychological review,* **84,** 327-352

Wittgenstein, L. (1953). Philosophical Investigations. MacMillan, New York.

Transitions Between Modes of Inquiry in a Rule Discovery Task

Naftali Halberstadt[1] and Yaakov Kareev
Goldie Rotman Center for Cognitive Science and Education
School of Education
The Hebrew University of Jerusalem
Jerusalem 91905, Israel.
Email: BAUHN@VM1.HUJI.AC.IL

Abstract

Studies of rule discovery behavior employ one of two research paradigms: In the reception paradigm the item evaluated on each trial is provided by the researcher; in the selection/generation paradigm the item to be evaluated is selected or generated by the subject. The prevalence of both paradigms and their correspondence to well established modes of scientific inquiry led us to the hypothesis that if given the choice, subjects would employ both modes of inquiry. To test this hypothesis 27 adults and 27 8th graders solved three rule discovery problems in a computer environment which allowed free transitions between item reception and generation. Almost all the adults and roughly half the children employed both modes of inquiry on at least one problem, with adults much likelier to generate items. The use of a method of inquiry came in blocks with generation tending to follow reception. An inverse relationship was found between item generation and the proportion of positive instances supplied by the environment. Within both age groups, consistent individual differences were found regarding inquiry style. These results shed new light on inquiry behavior and demonstrate the desirability of letting subjects freely choose between differing modes of inquiry.

Theories which attempt to explain behavior in rule discovery tasks usually adopt one of two orientations regarding the process: Data driven or theory driven. This differentiation has also been characterized as "bottom-up" or "top-down."

Most empirical studies of the inquiry process, have implicitly adopted one of these two approaches, in that they employed one of two research paradigms: The reception paradigm, or the selection paradigm.

Under the reception paradigm the stimuli to be classified are presented by the experimenter in a pre-arranged order, giving the experimenter complete control over the stimuli encountered by the subject. Under the selection paradigm the subject is presented with the entire array of stimuli at the outset and then freely chooses stimuli in order to discover the concept. Under the closely related generation paradigm the subject creates or generates the instances to be tested in the search for the concept or rule. The advantage of this mode of inquiry is that it is relatively easy to infer the subject's thought processes given the item chosen and the current hypothesis.

The modern study of rule-discovery and hypothesis testing started with Bruner, Goodnow and Austin (1956) and continued to generate much interest and activity throughout the years (for recent studies and theoretical developments see Evans (1989), Klahr and Dunbar (1988), Klayman and Ha (1987, 1989). Two findings regarding rule discovery behavior are particularly relevant to us. First, solving such tasks subjects exhibit a

---

[1] This paper is based on data collected for a doctoral dissertation submitted to the Hebrew University of Jerusalem by the first author. The work was supported in part by grants from the Israel Foundations Trustees and The NCJW Research Institute for Innovation in Education.

strong confirmation bias (Wason, 1960) which as been found very difficult to modify (Mynatt, Doherty and Tweney, 1978; Wason and Johnson-Laird, 1972). More recently Evans (1989) offered an alternative perspectives on the phenomenon while Klayman and Ha (1987) view the search for positive instances of a target category as part of an efficient heuristic. Second, Klahr and Dunbar (1988) employing a generation paradigm in a scientific inquiry task found that their subjects could be characterized as being either "theorists" or "experimenters," depending upon their preferred mode of inquiry.

A number of studies have been carried out to assess developmental trends in scientific abilities. Moshman (1979) investigated the development of formal hypothesis testing ability in a selection paradigm. Even amongst Moshman's oldest sample (college students) only a third regularly acknowledged the use of falsification. Moshman implicates an inability to distinguish between hypotheses (theory) and evidence (data) in explaining this lack of experimental sophistication. This is precisely the conclusion drawn by Kuhn and her associates (Kuhn, Amsel and O'Loughlin, 1988; Kuhn, 1989), in explaining the differences between children and adults as scientists. A recent study by Sodian, Zaitchik, and Carey (1991) indicates that the age differences reported above might apply only in a task which demands the generation of a critical hypotheses.

A recurrent finding in studies of rule discovery is the less than optimal performance by subjects. Perhaps this finding reflects the effects of employing an unsuitable strategy at a particular stage of the inquiry process. That is, it is possible that subjects were forced to employ an inappropriate mode of inquiry either because they were forced to continue with a less efficient method beyond the point at which they would have employed a more powerful one, or because they were provided with a powerful method when not yet ready to extract the maximum benefit from it.

It follows that a more successful paradigm for the study of inquiry would be to provide subjects with an environment in which they have the opportunity to choose between data-driven and theory-driven modes of inquiry - between observation alone and active experimentation. One may then observe whether one of the methods is preferred over the other at some or all stages of the inquiry process. Will any characteristic patterns of transition between modes of inquiry emerge? Will we find a more efficient use of experimentation? If groups of subjects differ in their respective use of either or both modes of inquiry how do they differ? How, if at all, would the incidence of positive instances in the inquiry environment affect subjects' choice of method? Finally, will there be any developmental trends in the use of the two modes of inquiry?

Method

Subjects. The subjects were 27 college students and 27 8th graders.

Procedure and Materials. Each subject solved three problems. Items were created through combinations of six binary dimensions. The subject's task was to distinguish between members and non-members of a target group.

A computer-based environment allowed the subject to freely choose between the reception or generation mode of inquiry. On each trial the subject could request a ready-made item or construct one (by choosing the desired values from a menu). In either mode, the subject had to state his/her belief regarding the categorical membership of the item and received feedback on the accuracy of that belief. In addition, the subject could choose a "Test Me" option on any trial (referred to as the Exit Sequence) in which s/he had to correctly classify ten consecutive test items in order to prove that s/he had learned the rule. An error during the Exit Sequence returned the subject to the main menu.

Design. The experiment had a three-way factorial design with factors being percent of positive instances supplied by the environment (10%, 25%, or 50%), number of Critical Values (1,2 or 3), and Problem Isomorph (one involved a personnel director evaluating job applicants, another involved trees which grew or did not grow in a particular type of soil, and a third involved a collection of geometric shapes). Each subject completed three problems, organized according to a Greco-Latin Square design.

Results and Discussion

Each step in the subjects' protocol was classified as one of four behavior types: "Get" (a request for ready made item), a "Positive Build" or a "Negative Build" (the generation of a test item by the subject whereby "Positive" and "Negative" refer to whether the subject believed the item built to be a positive or negative instance of the target group), or an "Exit" (a request to enter the Exit Sequence).

A preliminary analysis established that problem content did not significantly affect any one of nine basic measures. Consequently, content does not serve as a factor in the analyses.

Global Measures of Inquiry Behavior

To characterize the overall nature of the inquiry behavior we tabulated the frequency of the four major behavior types. The results, including a comparison between age groups, are presented in Table 1.

Adults used the generative mode more often than did children. They used it on 29.1% of the trials (5.5% for the children), employed it at least once in 79% of the problems (35% for children) and 25 out of 27 adult subjects used it on at least one problem (14 out of 27 children). Children relied more on the reception strategy, entered into (and thus, failed out of) the Exit Sequence more often, and took longer to solve problems.

Transitions Between Modes of Inquiry

It was hypothesized that when given the freedom to choose between modes of inquiry, subjects would begin by opting to view ready made exemplars and only later progress to a strategy in which they would generate test items. The mean trial numbers of the four behavior types appear in Table 2.

	8th Graders		Adults			
Measure	N of Trials	% of Trials	N of Trials	% of Trials	t value	p
GET	15.98	62.6	9.18	55.2	3.19	.002
POSITIVE BUILD	1.17	4.6	3.59	21.6	-4.95	<.001
NEGATIVE BUILD	0.23	0.9	1.25	7.5	-3.76	<.001
EXIT	8.15	31.9	2.61	15.7	3.94	<.001
TRIALS	25.53	100.0	16.63	100.0	3.11	.002

Table 1: Means and t-Values for Behavior Types by Age

Behavior	8th graders [a]	Adults
GET	8.04	6.54
POSITIVE BUILD	11.37	8.97
NEGATIVE BUILD	12.89	10.52
EXIT	14.60	14.32

Table 2: Mean Trial Number for Four Categories of Inquiry Behavior

[a] 8th grade statistics reflect only those protocols in which the subject performed at least one "Get" step and one "Build" step.

The mean trial numbers were in the predicted direction both for the adults and for the 8th graders who used the generative procedure. Moreover, the results indicate that within generation the more common positive test precedes the less frequently used negative test.

We then checked whether or not the four behaviors come in clusters (i.e., if each inquiry behavior was most likely to have been preceded by a behavior of the same type). As it turned out for both age groups, every behavior type was most likely to have been preceded by a like behavior. This indicates that beyond the tendency to first observe and later experiment, subjects progressed through the process in an orderly fashion in which once a stage of inquiry was entered they tended to stay in it for a while before going onto the next stage.

Our next question concerned the effect of mistakes on the progression through the inquiry process. We compared the number of times that a correctly identified generation and an incorrectly identified generation preceded a request for seeing a ready-made item (out of all generations). The comparison revealed that for both age groups there were significantly more instances in which a request for a ready-made item was preceded by an incorrectly identified generation ($\underline{t}(159) = 1.87$ $\underline{p} < .05$). Thus, it may be seen that the orderly nature of the inquiry process was preserved in a regressive as well as progressive manner, when subjects encountered difficulties in their investigations.

Environmental Mediation: The Availability of Positive Instances

People engaged in a rule discovery task exhibit a strong confirmation bias, manifested in the tendency to generate positive, rather than negative test items. Our results (see Table 1) show a similar trend, with positive tests outnumbering negative ones by a ratio of 3:1 among adults and 5:1 among the 8th graders.

As it turned out, the tendency to prefer positive over negative test items was mediated by the environment. An analysis of the effect of the percent of positive instances in the environment on the total incidence of positive tests revealed that the number of positive tests increased as the percentage of positive instances in the environment decreased ($\underline{F}(2,154) = 4.76$, $\underline{p} = .01$). Furthermore, the analysis revealed a significant interaction between age and percent of positive instances ($\underline{F}(2,154) = 3.17$, $\underline{p} < .05$), with the above trend observed only amongst the adult sample.

Characteristics of Experimentation

Experimentation calls for the controlled manipulation of variables in order to study their effects. To gain an insight into the nature of experimentation in the present task, we recorded the number of values changed in each generated item relative to the item just preceding it. We then checked which independent variables were related to the number of changes. It was found that fewer values were changed when: 1- the previous item belonged to the target group than when it did not belong to it (1.84 vs. 2.66 changes; $\underline{F}(1,433) = 39.03$, $\underline{p} <.001$); and 2- the previous trial already involved a generation rather than a reception or a failed attempt at the Exit Sequence (2.04 changes following a Positive Build, 2.19 changes following a Negative Build and 2.82 following a Get or Exit; $\underline{F}(2,433) = 18.61$, $\underline{p} <.001$).

Interestingly, in this analysis age was not found to be a critical factor. Apparently on the occasions that children did construct an item, the strategies they adopted were similar to those employed by the adult builders.

Individual Differences

For a final analysis, we analyzed the profiles of subjects' behavior in search of consistent individual differences in the process of inquiry. In view of the strong effect of age, we carried out the analyses separately in each age group.

For the adults, the factor analysis revealed two main factors which included 22 of the 27 subjects (10 subjects in Factor I, 12 in Factor II). Adult Factor I subjects experimented earlier as well as more frequently than did their Adult Factor II counterparts. In contrast, Adult Factor II subjects were characterized by a more frequent use of the reception mode of inquiry. Factor I subjects were also faster to solve the problem (14.7 trials per problem vs. 18.8 for the others).

Among the 8th graders there was one prominent factor including 17 of the 27 children. These children were quite similar to Factor II adults, exhibiting strong preference for the reception rather than the generation mode of inquiry. Two additional types of behavior were observed amongst the children, consisting of 4 and 5 subjects, respectively. Both types were characterized by a tendency to follow a correct response with the initiation of the Exit Sequence. 8th graders included in Factor III solved their problems in fewer steps than did the 8th graders in Factors I or II.

In summary, our results demonstrate a complex but coherent inter-relationship between the reception and generation paradigms. This inter-relationship is marked by a clear progression in the transition from one to the other. In addition to consistent individual differences in the use of these strategies, their use is related to age and the availability of positive instances in the environment. We propose that any future study of rule discovery should provide subjects with the ability to use both the reception and the generation modes, and to freely switch between them. Such a paradigm, and the study of variables which affect transitions between modes of inquiry, would greatly enhance the future study of rule discovery behavior.

References

Bruner, J., Goodnow, J. and Austin, G. 1956. *A study of thinking*. N.Y.: Wiley.

Evans, J.St.B.T. 1989. *Bias in human reasoning: Causes and consequences.* London: Erlbaum.

Klahr, D. and Dunbar, K. 1988. Dual space search during scientific reasoning. *Cognitive Science* 12:1-48.

Klayman, J. and Ha, Y. 1987. Confirmation, disconfirmation and information in hypothesis testing. *Psychological Review* 94:211-228.

Klayman, J. and Ha, Y. 1989. Hypothesis testing in rule discovery: Strategy, structure and content. *Journal of Experimental Psychology: Learning, Memory and Cognition* 15:596-604.

Kuhn, D., Amsel, E. and O'Loughlin, M. 1988. *The development of scientific thinking skills.* Orlando, Fla.: Academic Press.

Kuhn, D. 1989. Children and adults as intuitive scientists. *Psychological Review* 96:674-689.

Moshman, D. 1979. Development of formal hypothesis-testing ability. *Developmental Psychology* 15:104-112.

Mynatt, C.R., Doherty, M.E. and Tweney, R.D. 1978. Consequences of confirmation and disconfirmation in a simulated research environment. *Quarterly Journal of Experimental*

Psychology 30:395-406.

Sodian, B., Zaitchik, D. and Carey, S. 1991. Young children's differentiation of hypothetical beliefs from evidence. *Child Development* 62:753-766.

Wason, P.C. 1960. On the failure to eliminate hypotheses in a conceptual task. *Quarterly Journal of Experimental Psychology* 12:129-140.

Wason, P.C. and Johnson-Laird, P.N. 1972. *Psychology of reasoning: Structure and content.* Harvard University Press.

Are Rules a Thing of the Past? The Acquisition of Verbal Morphology by an Attractor Network.

James Hoeffner
Department of Psychology
Carnegie Mellon University
Pittsburgh, PA 15213
jh6s@andrew.cmu.edu

Abstract

This paper investigates the ability of a connectionist attractor network to learn a system analogous to part of the system of English verbal morphology. The model learned to produce phonological representations of stems and inflected forms in response to semantic inputs. The model was able to resolve several outstanding problems. It displayed all three stages of the characteristic U-shaped pattern of acquisition of the English past tense (early correct performance, a period of overgeneralizations and other errors, and eventual mastery). The network is also able to simulate direct access (the ability to create an inflected form directly from a semantic representation without having to first access an intermediate base form). The model was easily able to resolve homophonic verbs (such as *ring* and *wring*). In addition, the network was able to apply the past tense, third person *-s* and progressive *-ing* suffixes productively to novel forms and to display sensitivity to the subregularities that mark families of irregular past tense forms. The network also simulates the frequency by regularity interaction that has been found in reaction time studies of human subjects and provides a possible explanation for some hypothesized universal constraints upon morphological operations.

Introduction

In recent years the status of rules in cognitive science has become an issue of heated debate. Many cognitive scientists believe that explicit rules are necessary to explain human behavior (Pinker & Prince, 1988; Lachter & Bever, 1988). Others have challenged this reliance on rules and symbolic systems (Rumelhart & McClelland, 1986; MacWhinney & Leinbach, 1991).

At the center of this debate has been the study of inflectional morphology; the acquisition of the past tense has been of particular importance. Researchers have found that the acquisition of the English past tense involves three distinct stages (Kuczaj, 1977, 1978; MacWhinney, 1978). In the first stage, children correctly produce a small number of both regular and irregular forms. In the second stage they sometimes "overregularize" irregular forms, producing errors such as *goed* and *ated*. Particularly striking is the fact that children will sometimes overregularize irregular forms that they had previously produced correctly. In the third stage, which is only reached gradually over a period of years, the children exhibit total mastery of both regular and irregular forms with very few errors of any kind.

The existence of both overregularizations and a U-shaped curve have been considered strong evidence for the belief that language learning involves organizing linguistic knowledge into a system of rules and exceptions to those rules. Rumelhart & McClelland (1986) challenged this belief by creating a connectionist network which, they claimed, simulated the process of learning the English past tense without making use of any explicit rules. Several authors (Pinker & Prince, 1988; Lachter & Bever, 1988) criticized the Rumelhart and McClelland account on a variety of grounds. These criticisms can be grouped into several clusters: 1) Problems with the phonological representation, 2) Criticism of the training regimen, 3) Problems stemming from the lack of a semantic representation in the model, 4) The failure to incorporate constraints on possible forms, 5) The failure to simulate the differential effects of frequency on regular and irregular forms. The criticisms raised in 1 and 2 have been addressed with some success by later connectionist models (MacWhinney & Leinbach, 1991; Plunkett & Marchman, 1991). This paper addresses the criticisms in 3, 4 and 5.

Problems Stemming from the Lack of a Semantic Representation

All previous connectionist models of past tense acquisition, with the exception of Cottrell & Plunkett (1991), have been phonology to phonology models. They took the verb stem as their input representation and converted it to the past tense (or in the MacWhinney & Leinbach model, to a variety of inflected forms). Because of this, these models were unable to display some elementary properties of natural languages. One of these properties is homophony. Homophones are pairs of words that sound identical but have different meanings. English and other languages have many homophones, for example, there are the verbs *ring* (a bell), *wring* (your hands) and *ring* (form a circle around). The phonology to phonology networks cannot learn to inflect verbs that have homophonic stems but different past tenses (*ring-rang* vs. *wring-wrung* or *ring-ringed*).

Another related problem is that of direct access. The phonology to phonology models take a base form and transform it into an inflected form. People seem to display an ability similar to this. But as MacWhinney and Leinbach point out, "...intuition, theory (MacWhinney, 1978) and experimentation (Stemberger & MacWhinney, 1978) all suggest that we can also access derived forms directly. In other words, we learn that *ran* means *running in the past* and use this knowledge to access *ran* directly without starting off at *run*." Phonology to phonology models cannot simulate this process of direct access.

There are at least two ways to address these problems within a connectionist framework. One way is to add a semantic representation to the input of a phonology to phonology network as MacWhinney and Leinbach did in a small subsidiary model. In principle, this approach can resolve the homophony problems but it does not address the direct access problem. Also, in practice, the net result of adding semantic representations to the input of a phonology to phonology network is to reduce the overall performance of the model, since the network must devote considerable resources to learning the largely arbitrary associations between semantic and phonological representations.

The approach taken in this paper is a different one. The network is trained to form associations between semantic and phonological representations. In this type of model, direct access is the basic process and the ability to inflect a novel form is a secondary ability parasitic on the direct access process. By taking this route, we may be able to solve both the homophony and direct access problems simultaneously and in a principled way.

Another criticism that can be addressed by using a semantics to phonology network is the question of "double-marked" forms like *ated* or *wented*. In the Rumelhart & McClelland model, these errors are produced by blending the specific *go->went* mapping with the regular stem->stem+ed mapping. But Pinker & Prince claimed that the fact that children occasionally produce errors like *wenting* and *ating* even when the progressive is fully regular shows that these errors are caused by feeding the irregular past into the regular suffixation process.

In MacWhinney & Leinbach's model, they were able to produce double-marked errors simply by feeding the past tense (e.g., *ate*) into the network as if it were the stem. In the present model we take a different approach, one based on the nature of the semantic relationships among verbs (Bybee, 1985). In our model, all the members of a particular verbal paradigm share an identical "core" semantic representation. The semantic representations of two members of the same paradigm (e.g., *jump* and *jumped*) are distinguished only by the relatively small number of units which code for inflections. Because of the proximity of the semantic representations of the members of a verbal paradigm it is possible for the network to produce phonological blends of various kinds including the combination of an irregular past form like *went* with a regular suffix like the progressive *-ing*.

Constraints on Possible Forms

Pinker & Prince raised an intriguing criticism of the Rumelhart & McClelland model. There seem to be logically possible inflectional devices that are never used in any language. For example, no language inflects a verb for tense by transposing all the phonemes in the word (making *tih* the past tense of *hit* or *pals* the past tense of *slap*). Another ubiquitous feature of natural language is preservation of the stem in inflected forms. This is true not only of regular forms but of most irregulars as well. In most irregulars the majority of the phonetic material in the stem is preserved in the inflected form (such as *sing-sang* , *give-gave*, etc.). Pinker & Prince claimed that Rumelhart & McClelland's model was insensitive to such universal constraints on possible morphological operations and therefore was invalid as a model of how people actually learn and represent language.

We believe that there is a strong bias against certain types of morphological operations and biases in favor of other types of operations (for example, affixation of a stem or base form). At least some of these biases may result from the nature of semantic level relationships among different verb forms. For example, in our model, the phenomena of preservation of the stem in inflected forms can be explained in the following way. The semantic representations of the members of an inflectional paradigm are very close. All things being equal, connectionist models prefer mappings where the similarity structure of the output set reflects the similarity structure of the input set. This built in bias means that the network will find it much more difficult to output *hit* as the present tense and *tih* as the past then it would be to learn a verbal paradigm where much of the stem was preserved in every inflected form.

Differential Frequency Effects

Prasada, Pinker & Snyder (1990) tested subjects' reaction times in a past tense generation experiment. They found that higher frequency irregulars were produced faster than low frequency irregulars but the same frequency effect was not found for regular past tense forms. They claim that this evidence supports the qualitative distinction between regular and irregular inflected forms. The regulars are stored in an associative memory device that is sensitive to frequency and similarity but the regularly inflected forms are not. Only the stems need to be stored, since there is a second mechanism, an affixation procedure that can take any stem and append the *-ed* suffix to it.

Seidenberg and Bruck (1990) found a similar frequency by regularity interaction in the production of past tense forms, but gave the results a different interpretation, one that is consistent with the behavior of a single connectionist model that learns to produce both regular and irregular verbs. With the attractor network presented in this paper, we can simulate some of the effects of frequency and regularity on reaction time and show that the interaction need not be taken as evidence for a dual mechanism account.

The Model

The network presented in this paper is an attractor network (Hinton & Sejnowski, 1986; Hopfield, 1984). Attractor networks have recurrent connections; these enable the network to develop stable resonant or attractor states. There are at least three advantages to using an attractor network.

1. Learning arbitrary associations: The mapping from semantics to phonology is difficult because the two representations are arbitrarily related. Feed-forward networks need large weights and prolonged training to learn such a mapping. But, as Plaut & Shallice (1991) point out: "They [attractor networks] are also more effective at learning arbitrary associations because the reapplication of unit non-linearities can magnify initially small state differences into quite large ones. "
2. Flexibility: We can interrogate the knowledge stored in the network by presenting it with pieces of information and allowing the network to fill in the rest of the learned pattern. This characteristic of attractor networks will enable us to probe the network's ability to use its knowledge of morphology productively in ways that could not be done with feed-forward or simple recurrent networks.
3. Reaction time measure: We can use the network's settling time as an analog of reaction time in psychological experimentation and simulate the effects of frequency and regularity on subjects' reaction times.

Network Architecture: The network contains 185 units arranged in three layers: the semantic, hidden and phonological layers. The semantic layer contains 96 units, the hidden layer has 68 units and the phonological layer has 21. The semantic layer has bi-directional connections to the hidden layer and the hidden layer has bi-directional connections to the phonological layer. All three layers are fully intra-connected (each unit in a layer is connected to every other unit in that particular layer).

Semantic representation: In the semantic layer, 84 of the 96 units are used to represent the "core" meaning of each verbal paradigm. 12 more units are dedicated to representing inflectional markings. The semantic representation of a verb consists of a randomly generated pattern over the 84 "core" units plus the 12 inflectional units. The semantic representations of the members of an inflectional paradigm (e.g., *jump, jumped, jumps and jumping*) will have an identical "core" representation. The only difference will be in the 12 units that code the inflections. Different verbs (e.g., *jump* and *walk*) will have "core" semantic representations that must differ by at least 24 of the 84 units. This ensures that the semantic distance between different verbs will be greater than the semantic distance between two members of the same paradigm.

Phonological representation: The phonological representation was based on an artificial language devised by Plunkett & Marchman (1991). In this language each phoneme is uniquely represented by a 6 bit vector. Each word is made up of three phonemes. The words can have a CVC(hit), CCV(try) or VCC(ask) syllabic structure. There are also three units in the phonological layer that code for the inflectional suffixes.

Training corpus and schedule: The corpus consists of 800 verb forms (200 paradigms). Each paradigm consists of an unmarked "stem" and three inflected forms: the past tense, the third person singular, and the progressive. The 3s and progressive forms are totally regular, the only irregulars are past tense forms. There are four types of past tense forms in the artificial language. Regular verbs, Vowel Change(VC) verbs, No Change(NC) verbs, and Arbitrary verbs. The past of the Regular verbs is the stem + ed. The past tenses of the Vowel Change verbs are formed by altering features of the vowel. The past tense of the No Change verbs is identical to the stem. And the past tense of the arbitrary verbs bear no systematic phonological relationship to the stem. These four types correspond to the predominant types of stem to past mappings found in natural languages. For example, in English, *go-went* is an Arbitrary, *put-put* is a No Change, and *give-gave* is a Vowel Change. There are 20 irregular pasts in the training set, 2 arbitraries, 6 no change and 12 vowel change. The vowel change verbs are in two clusters, in VC_1 all the stems have the form the C+ing and their pasts have the form C+ang. The VC_2 verbs are conjugated in the following way: C+aif ->C+If.

Both the regular and irregular verbs are organized into high, medium and low frequency groups. For the irregulars there are 8 high, 6 medium and 6 low frequency verbs. For the regulars, there are 40 high, 100 medium and 40 low frequency verbs.

The most frequent inflected forms are presented 9 times per epoch. The least frequent are presented only once. For a given paradigm, all three inflected forms have the same frequency. In all paradigms, the stem is presented twice as often as any of the inflected forms. For the entire corpus, the regular to irregular *type* frequency ratio is 9:1, but the regular to irregular *token* frequency ratio is less than 3:1 because of the higher frequencies of the irregular verbs.

Training schedule: The network is first presented with 30 paradigms, 20 regular and 10 irregular. The network is trained for two epochs on this subset. Then the corpus is expanded incrementally with 4 new paradigms added each epoch up till epoch 45. At no point is there a sudden influx of regular verbs or a rapid change in the relative proportions of regular and irregular verbs.

Learning Algorithm: The network was trained with the Contrastive Hebbian algorithm (Peterson & Anderson, 1987).

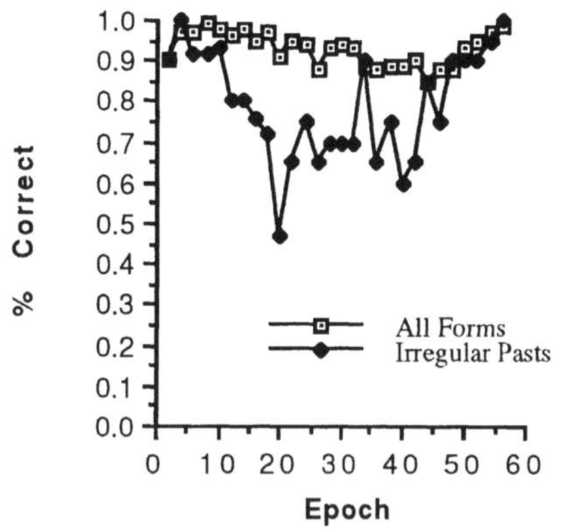

Figure 1: Percent correct on all forms and irregular past tense forms only.

Results

Overall Performance: The network had no difficulty learning the task. Initially, performance was very strong, reaching the high 90's by the 4th epoch. And even though the network was absorbing 16 new forms per epoch, its performance never dipped below 85% correct. Across all 50 epochs, the network averaged 93% correct. At the end of training (epoch 58) the network was producing the 800 verb forms with 98% accuracy. (See figure 1 and Table 1).

U-shaped curve: The network displayed all three stages of the U-shaped developmental pattern. From the beginning of training till epoch 10, the network averaged 93% correct on irregular pasts. During this period, it did not make any overregularization errors. At epoch 12, the network began to overregularize the past tense suffix and its performance on irregular pasts fell to 73%. The network continued to overregularize the irregular pasts throughout most of the training period. By epoch 44, the number of overregularization errors began to diminish as the network learned how to inhibit the suffix when producing an irregular past. By the end of training, the network achieved perfect performance on the irregular pasts, thereby completing the third phase of the U-shaped curve.

In Table 2, all the overregularization errors made by the network are listed. The network made all three possible types of overregularization errors, Base+ed (goed), Past+ed (wented) and No Change+ed (hitted). The network displayed several of the characteristics of a *micro*-U-shaped developmental curve (Plunkett & Marchman, 1991). There was never an across the board overregularization of all or even most of the irregulars. Instead, some verbs were overregularized frequently and others not at all. In addition, overregularization was highly variable for many verbs. For example, No Change Verb #3 was produced correctly till epoch 14,overregularized on epoch 16, was correct on 18 and 20, incorrect again on epoch 22, correct again on 24 and 26 and overregularized once more on epoch 28.

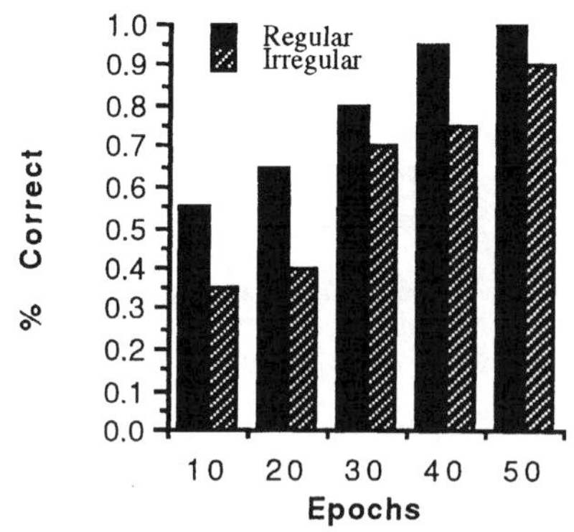

Figure 2: Past tense inflection of novel forms with regular and irregular phonology.

Homophones: There were two pairs of homophones in the training set. Both homophonic pairs were mastered by the network.

Double-Marked Errors: Most of the network's overregularizations were Base+ed(goed) or No Change+ed(hitted) but the network did produce several "double-marked" Past+ed(wented) errors. It also produced a small number of errors similar to "wenting", by combining the irregular past with the 3s or -ing suffix.

Productivity and sensitivity to subregularities: Figure 2 shows the model's ability to generalize its knowledge and apply the three inflectional suffixes correctly to novel forms. The network was presented with a corpus of 40 novel stems. Half were phonologically similar to the irregular verbs the network had been trained on. The other 20 novel verbs did not resemble the irregulars. The network was tested in the following way: the semantic layer inflectional units are "soft-clamped". The novel stem is also "soft-clamped" to the phonological layer. The 3 suffix units in the phonological layer and the rest of the network are left free. The task of the network as it settles is to properly activate the suffix units, thereby inflecting the novel stem. This task is analogous to probing a human subject's ability by presenting him with a novel stem like "glorp" and asking what the past tense of it would be.

The network was able apply all three suffixes productively. Figure 2 shows that the network was able to reach perfect performance in applying the past tense suffix to the phonologically regular stems by epoch 50. This compares very favorably with the generalization performance of Cottrell & Plunkett's(1991) semantics to phonology network, which was was unable to achieve greater than 55% correct generalization to novel stems.

The network also showed sensitivity to the subregularities that mark families of irregulars. The irregular phonology novel pasts were consistently less likely to be suffixed than the regular phonology novel pasts. This shows that the network has learned the phonological regularities that mark irregular families (such as the fact that all the No Change verbs end in t/d) and can use this information to block the regular suffixation process.

The network was also able to apply the -s and -ing suffixes productively as well (see Figure 3). In fact, because they are totally regular, the network learned to use these suffixes more quickly than the it did for the past tense (80-85% correct after only 20 epochs). Also, for these suffixes, there was no consistent effect of irregular phonology. The network was able to inflect novel irregulars as easily as novel regulars.

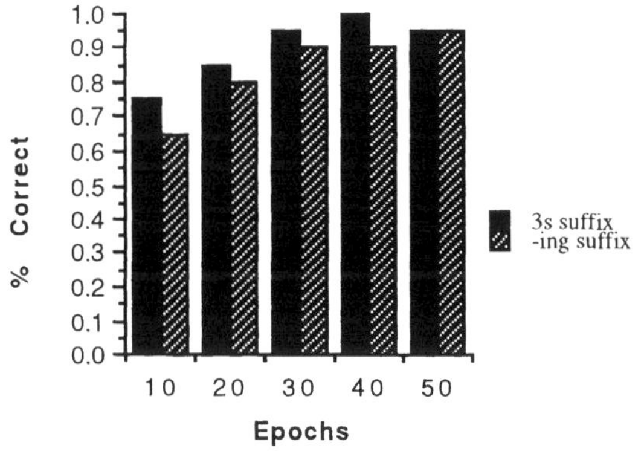

Figure 3: Inflection of novel forms with -s and -ing suffixes.

Frequency Effects: We can use the network's settling time as an analog of reaction time in psychological experimentation. In reaction time experiments the subject is presented with a base form (walk) and told to generate the past tense (walked) as quickly as possible. We can simulate this effect by "soft-clamping" a stem to the phonology layer at the same time that we clamp the full semantic pattern for the verb's past tense. The network has to convert the soft clamped stem to the proper past tense form.

Figure 4 shows that the network also displays the frequency by regularity interaction that has been found in studies of human subjects. (Only correct responses were used, this removed ~2% of the responses). For the irregulars there is a pronounced frequency effect, but for the regular verbs, there is very little effect of frequency.

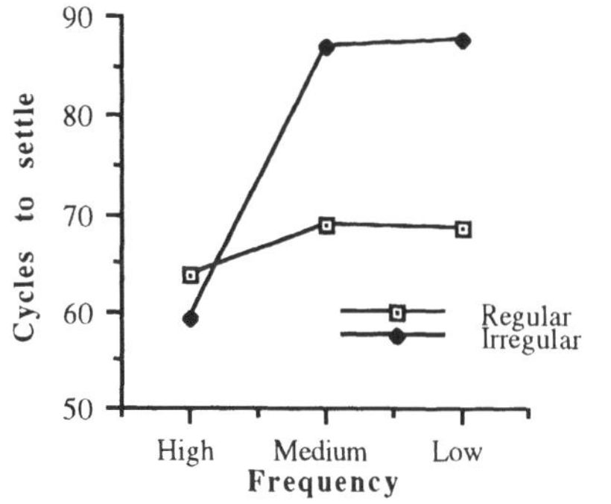

Figure 4: Frequency by regularity interaction

Epoch	2	4	6	8	10	12	14	16	18	20	22	24	26	28	30	32	34	36	38	40	42	44	46	48	50	52	54	56	58
All forms	90	97	97	99	98	96	98	95	97	91	95	94	88	93	94	93	89	88	89	89	90	85	88	88	93	94	94	96	98
Stems	100	100	100	100	98	100	100	96	99	96	98	94	90	98	96	97	92	92	91	89	89	90	90	92	96	96	96	97	97
Past tense	63	89	89	96	97	92	93	90	93	85	88	90	85	90	91	86	87	85	82	81	84	85	88	87	89	93	94	94	98
3s	97	100	100	100	100	97	100	98	99	90	98	96	88	93	94	97	92	89	88	92	92	80	88	88	93	94	95	97	98
-ing	97	100	95	100	98	97	100	95	98	94	96	95	89	92	93	93	86	87	94	89	92	85	87	86	95	94	95	97	98
Irregular past	90	100	92	92	93	80	80	76	72	47	65	75	65	70	70	70	90	65	75	60	65	85	75	90	90	90	95	90	100

Table 1: Percent correct by epoch.

Epoch	12	14	16	18	20	22	24	26	28	30	32	34	36	38	40	42	44	46	48	Totals
Base+ed	1	2	2	1		3	1	3	3	2	1		1	1	3	3	1			28
NoChange+ed	1	1	1	2	2	3	2	2	2	1	2		3	1		2	1	2	1	29
Past+ed					2	1	2						1							6
Totals	2	3	3	3	4	7	5	5	5	3	3	0	5	2	3	5	2	2	1	63

Table 2: Distribution and types of overregularization errors.

Discussion

Our network was able to show the most salient aspects of English past tense acquisition. The network passed through all three stages of the U-shaped developmental curve and did so without the need for a sharp input discontinuity. The model demonstrated the two behaviors that were thought to be paradigmatic of rule acquisition and use (overgeneralizations and *micro* U-shaped development). By using a semantics to phonology attractor network, we were able to solve the direct access, double-marking and homophony problems. We were also able to shed light on the basis for certain universal morphological biases such as preservation of the stem in inflected forms and the constraint against mirror image morphological mappings. Furthermore, we were able to replicate the frequency by regularity interaction that has been found in studies of human subjects.

In the future, this network can be used to study many other aspects of morphological acquisition and processing such as the reasons for the differential acquisition of various inflectional morphemes, the differences between derivational and inflectional morphology and the effects of lesions on morphological processing.

Acknowledgements

I'd like to thank Brian MacWhinney, Jay McClelland, Randy O'Reilly and Priti Shah for their aid and advice.

References

Bybee, J.L., 1985. *Morphology: A study of the relation between meaning and form.* Philadelphia: Benjamins.

Cottrell, G. W., and Plunkett, K. , 1991. Learning the Past Tense in a Recurrent Network: Acquiring the Mapping from Meaning to Sounds. In Proceedings of the Thirteenth Annual Meeting of the Cognitive Science Society.

Hinton, G.E., and Sejnowski, T.J., 1986. Learning and Relearning in Boltzmann machines. In D.E. Rumelhart, J.L. McClelland and the PDP Research Group (Eds.)., *Parallel Distributed Processing: Explorations in the Microstructure of Cognition.* (Vol. 2). Cambridge, MA: Bradford Books.

Hopfield, J. J., 1984. Neurons with Graded Responses have Collective Computational Properties like those of Two State Neurons. In Proceedings of the National Academy of Science, U.S.A., 81, 3088-3092.

Kuczaj, S. A., 1977. The Acquisition of Regular and Irregular Past Tense Forms. *Journal of Verbal Learning and Verbal Behavior*, 16, 589-600.

Kuczaj, S. A., 1978. Children's Judgments of Grammatical and Ungrammatical Past Tense Forms. *Child Development*, 49, 319-326.

Lachter, J., and Bever, T.G., 1987. The Relation Between Linguistic Structure and Associative Theories of Language Learning-A Constructive Critique of Some Connectionist Learning Models.*Cognition*, 28,195-247.

MacWhinney, B., 1978. The Acquisition of Morphophonology. *Monographs of the Society for Research in Child Development* , 174, vol. 43, nos.1-2.

MacWhinney, B. and Leinbach, J., 1991. Implementations are not Conceptualizations: Revising the Verb Learning Model. *Cognition*, 39.

Peterson, C. and Anderson, J.R. ,1987. A Mean Field Theory Learning Algorithm for Neural Networks. *Complex Systems* 1, 995-1019.

Pinker, S., and Prince, A., 1988. On Language and Connectionism: Analysis of a Parallel Distributed Processing Model of Language Acquisition. *Cognition*, 28, 73-193.

Plaut, D., and Shallice, T., 1991. Deep Dyslexia: A Case Study of Connectionist Neuropsychology. Unpublished manuscript.

Plunkett, K., and Marchman, V., 1991. U-shaped Learning and Frequency Effects in a Multilayered Perceptron: Implications for Child Language Acquisition. *Cognition*, 38, 43-102.

Prasada, S., Pinker, S., and Snyder, W., 1990. Some Evidence that Irregular Forms are Retrieved from Memory but Regular Forms are Rule Generated. Paper presented at the 31st annual meeting of the Psychonomic society (November: New Orleans).

Rumelhart, D. E., and McClelland, J.L., 1986. On Learning the Past Tense of English Verbs. In D.E. Rumelhart, J.L. McClelland and the PDP Research Group (Eds.)., *Parallel Distributed Processing: Explorations in the microstructure of cognition.* (Vol. 2). Cambridge, MA: Bradford Books.

Seidenberg, M.S., and Bruck, M., 1990. Consistency Effects in the Generation of Past Tense Morphology. Paper presented at the 31st meeting of the Psychonomic Society (November: New Orleans).

Stemberger, J., and MacWhinney, B., 1985. Frequency and the Lexical Storage of Regularly Inflected Forms. *Memory and Cognition*, 14, 17-26.

Memory and Discredited Information: Can You Forget I Ever Said That?*

Hollyn M. Johnson
University of Michigan
330 Packard Rd., Rm. 220B
Ann Arbor, MI 48104
hollyn.johnson@um.cc.umich.edu

Colleen M. Seifert
University of Michigan
330 Packard Rd., Rm. 216
Ann Arbor, MI 48104
seifert@um.cc.umich.edu

Abstract

Previous research has found that when information stored in memory is discredited, it can still influence later inferences one makes. This has previously been considered as an editing problem, where one has inferences based on the information prestored in memory before the discrediting, and one cannot successfully trace out and alter those inferences. However, in the course of comprehending an account, one can potentially make inferences *after* a discrediting, which may also show influence from the discredited information. In this experiment, subjects read a series of reports about a fire investigation, and their opportunity to make inferences before a correction appeared in the series was manipulated. Subjects received a correction statement either directly following the information it was to discredit, or with several statements intervening. The results show that subjects who received the correction directly after the information it corrected made as many inferences based on the discredited information as subjects who received the correction later (and thus could presumably make many more inferences before the correction occurred). This suggests that discredited information can influence inferences made after a correction, as well as those made before. Several hypotheses accounting for this effect are proposed.

When information stored in memory is shown to be false or unfounded, ideally one would want to diminish or eliminate its effects on future reasoning and understanding processes. Some previous studies on text comprehension (Wilkes & Leatherbarrow, 1988) and jurors' use of inadmissible evidence (Carretta & Moreland, 1983) have presented subjects with instructions to disregard previously presented information. The results show that subjects remember that instruction when queried about it directly. However, they still show influence from the discredited information when asked to make judgments or inferences (further conclusions not directly presented), relative to subjects who were never exposed to that information. Other studies providing instructions to disregard previous information have also found influence from the discredited information on judgments of personality attributes like friendliness or kindness (Wyer & Budesheim, 1987) and of success in social tasks (Ross, Lepper, & Hubbard, 1975). To understand why information that is discredited still influences inferences, one must look at how and when such inferences are generated.

Research on inferences in text comprehension has proposed that inferences can differ on two dimensions which could be helpful in understanding how discredited information influences inferences. The first, how spontaneously inferences are made, distinguishes between *on-line* inferences (made automatically in the course of comprehension) and *requested* inferences (not normally made during comprehension, but can easily be made when one is asked a question). This is similar to the distinction Hastie and Park (1986) make between on-line judgments (immediate and automatic evaluation) and memory-based judgments (unanticipated, and made only after retrieving the original information from memory). The second distinction about inferences is that they may also differ in their direction: *Forward* inferences occur as predictions or expectations about what will appear next in the text, whereas *backwards* inferences are those that link current information to preceding information, often to provide text coherence and causal connections. Studies have found that backwards inferences occur on-line for anaphoric reference (Corbett & Chang, 1983; Dell, McKoon, & Ratcliff, 1983), text coherence (Keenan, Baillet, & Brown, 1984; McKoon & Ratcliff, 1986), and for

* This research was supported by the Office of Naval Research under contract N00014-91-J-1128 to the University of Michigan.

establishing causal relations (Graesser & Clark, 1985). In contrast, forward inferences tend to occur on-line only for very stereotypical or scriptlike texts (McKoon & Ratcliff, 1986; Seifert, Robertson, & Black, 1985). Kintsch (1988) has also proposed that random inferences, based on common associates of text material, are generated during comprehension. Considering the spontaneity and direction of inferences, and their timing relative to a disregard instruction, leads to different predictions about how discredited information may be used in drawing inferences.

One cause for the continued influence of discredited information may be problems in editing one's memory. That is, when one has already made inferences based on the information, and then encounters an instruction to disregard the information, one has both the original information and the inferences based on it in memory. One may successfully discredit the original information, but may not be able to trace out all the inferences that information supported and properly discount them as well. Thus, the inferences can remain in memory, and one can retrieve them when asked about them later. This could occur if, before the discrediting is introduced, one made either on-line forwards or random inferences, or on-line backwards inferences linking concepts to the information. This latter situation would require a "window" between original presentation of the information and the disregard instruction, during which intervening concepts would be linked back to the information. Finally, one may have generated and stored inferences based on the information during the course of any judgment or inference tasks before discrediting occurs.

On the other hand, if little opportunity for context-relevant from the information is provided before discrediting, then the influence the discredited information has on later inferences would be mainly due to the retrieval and use of the discredited information to generate new inferences. Such an effect may seem counter-intuitive, given findings that subjects do recall that the information is invalid when asked about it directly (Wilkes & Leatherbarrow, 1988; Carretta & Moreland, 1983). In the cases where an inference is made when questioned even after discrediting, or when on-line backwards inferences link later information to the discredited information, the comprehension process must "jump over" the correction notice, illicitly retrieving and using the discredited information.

Little previous research has looked at whether continued influence from discredited information is due to problems in editing prestored inferences, or whether it also can involve illicit retrieval and use of the discredited information itself. Some social psychology research suggests pre-discrediting inferences are difficult to edit (Anderson, Lepper, & Ross, 1980; Anderson, New, & Speer, 1985; Hastie & Park, 1986; Wyer & Budesheim, 1987). Wilkes and Leatherbarrow (1988) demonstrated the influence effect in a text comprehension experiment, but remained neutral as to its cause. In their original experiment, subjects read a series of reports on a fire investigation, with an original statement that some volatile materials (cans of paint and pressurized gas cylinders) were stored in a closet. Several messages later, a correction occurred, stating that the closet was empty and thus did not contain volatile materials. Subjects who received this statement to disregard the information about the closet's contents still reported inferences consistent with or mentioning the volatile materials more often than did control subjects, who never received the information about possible storage of volatile materials. This occurred whether the correction directly repeated the information that one was to disregard (direct edit condition) or whether it was just indirectly referred to (indirect edit condition).

The experiment reported here uses a modified set of materials taken from Wilkes and Leatherbarrow (1988), where subjects are presented with a series of reports, one to a page, which they read through at their own pace. In the delayed correction condition, which replicates the direct edit condition in the original Wilkes and Leatherbarrow (1988) paper, subjects read about volatile materials stored in a closet early in the report series, and received a correction five messages later. In the no-mention control group, subjects hear no mention of any volatile materials stored in the closet. This replicates the control group in the original experiment, and is consistent with control conditions used in many belief persistence studies (e.g., Anderson, Lepper, & Ross, 1980; Ross, Lepper, & Hubbard, 1975). Finally, in the immediate correction group, no information intervenes between presentation of the information about the volatile materials and the correction, so subjects have little opportunity to make inferences based on the volatile materials, before hearing that those materials do not exist. In this immediate correction condition, the information about the volatile materials would still be in working memory (along with any random inferences constructed from common associates of the message propositions, and remaining after integration with surrounding context (Kintsch, 1988)) when the discrediting occurred. This should make it easier for subjects to determine that the first and the second, discrediting message about the closet share reference. Other research has proposed that proximity facilitates establishing coreference (Cirilo, 1981), and that establishing coreference is important for detecting contradictions (Epstein, Glenberg, & Bradley, 1984). Subjects could then potentially resolve the contradiction in working memory, and use that representation for further interpretation of subsequent information. Also, for all conditions, the messages prior to the correction are written so as to limit opportunities for forwards and backwards inferences linking the information to other content. Thus, in the immediate correction condition, one might have only those random inferences that one could generate based on the volatile materials message and that happened to be context-relevant, and so survived an integration process, whereas in the delayed correction condition, one would have much more opportunity for inferencing (both random and more strategic, bridging inferences) before the discrediting occurred.

If the influence from discredited information occurs because subjects have difficulty editing inferences made before the disregard instruction, or correction, one would expect more influence from the discredited information to be evident in the delayed correction group. Subjects would have a window within which they could make additional, strategic, causal and coherence-maintaining inferences prior to the correction, which they might not be able to successfully track down and alter when the correction occurs. In the immediate correction condition, subjects will not have the opportunity to make as many context-relevant inferences before the correction occurs. If the problem lies in editing pre-stored inferences, one would not expect as much influence from the discredited information for the immediate correction condition, since fewer inferences could be formed. However, if making illicit post-correction retrievals leads to continued influence of discredited information, then one would expect influence from the discredited information in the immediate correction group, and in the delayed correction group, compared to the no-mention control. Thus, the immediate and delayed correction conditions, together with the control condition having no discrediting, will determine whether influence of discredited information occurs just due to problems in editing pre-correction inferences, or can also occur due to illicit post-correction retrieval and use of the discredited information.

Further, to test whether the effect depends on the wording of the correction message, two versions of the message are used. One version presents the correction directly but in a complex clause, similar to the version used by Wilkes and Leatherbarrow (1988). A second version presents the correction as a direct assertion that no volatile materials were stored in the closet, rather than embedding the reference to the materials in a subordinate clause.

Method

Subjects. Sixty-four University of Michigan undergraduates participated in a single session lasting approximately 50 minutes. They received course credit in an introductory psychology class for participating. Subjects were run in groups of 8 to 10.

Materials. The materials were modified versions of a series of reports used by Wilkes and Leatherbarrow (1988), describing the investigation of a warehouse fire. The series consisted of 13 individual messages, each 2-4 sentences long. The messages were combined into a booklet, with one message per page. The critical messages concerned the contents of a storage closet on the premises. For the no-mention control group, the fifth message in the series stated that this closet was empty, and this information was not controverted later. For the two correction groups, the fifth message stated that the closet contained cans of oil paint and pressurized gas cylinders. Then, for the immediate correction group, Message 6 stated that the previous message regarding the closet's contents was incorrect and that the closet was empty. For the delayed correction group, this statement appeared as Message 12. Half the subjects in both the delayed and the immediate correction groups received a complexly worded correction message, similar to that in Wilkes and Leatherbarrow (1988); the other half received a more direct wording (see Table 1).

Two memory tests were also prepared: a free recall summary of the reports' contents and a questionnaire adapted from Wilkes and Leatherbarrow (1988), including ten questions on facts directly presented in the messages, ten other questions requiring the subjects to make inferences about the event, and two final questions assessing whether subjects were aware of any correction or contradiction in the series (sample questions are shown in Table 2). All questions appeared in the same order for each subject, with all fact questions appearing before any inference questions to prevent the latter from introducing biases, and the two contradiction questions appearing at the end.

Direct message version:
10:40 a.m. A second message received from Police Investigator Lucas regarding the investigation into the fire. It stated that there were no cans of paint or gas cylinders in the closet that had reportedly contained them; the closet had actually been empty before the fire.

Complex message version:
10:40 a.m. A second message received from Police Investigator Lucas regarding the investigation into the fire. It stated that the closet reportedly containing cans of paint and gas cylinders had actually been empty before the fire.

Table 1: *Style of Corrections*

What was the possible cause of the toxic fumes?
What could have caused the explosions?
Why do you think the fire was particularly intense?
For what reason might an insurance claim be refused?

Table 2: *Sample Inference Questions from the Memory Questionnaire*

Procedure. Each subject received a booklet of reports and was instructed to read through it at his or her own pace, but not to go back and reread any of the messages. Subjects were also told that they would be asked to recall the information later. When individual subjects had finished reading, they were given the free recall test. Then all subjects did an unrelated distractor task for 10 minutes. After this time had elapsed, subjects received the memory questionnaire and were instructed to answer each question based on their understanding of the reports.

Results

A complete description of the results is presented in Johnson and Seifert (1992). A coder (blind to the experimental conditions) scored the responses to the inference questions as consistent with either a "negligence" theme or a "supplies" theme. The negligence theme encompassed responses that were consistent with believing that the warehouse contained carelessly stored volatile materials, as would be reasonable if the information about the volatile materials had not been discredited. References to the presence of gas cylinders and paint, carelessness, or the closet itself without indications that it was empty were coded with this theme. The supplies theme was coded if the responses presented a reasonable inference about the fire that was not included in the previous categories, such as references to stored stationery at the warehouse or the structure of the building. One would expect control subjects to make inferences consistent with this theme, because they received no information about the volatile materials.

To assess specific influences of the information one was to disregard, all uncontroverted references to paint and gas cylinders in either memory test were counted and analyzed in a 3 x 2 analysis of variance, with Group (no-mention, delayed correction, immediate correction) and Message (complex or direct) as factors. The mean number of references to the stored volatile materials for each cell is shown in Table 3. The results showed a main effect of group, $F(2, 59) = 10.21$, $p < .0001$, with both the correction groups showing more influence than the no-mention control group. The main effect of message showed a trend towards significance, $F(1, 59) = 9.87$, $p < .09$, with the direct version resulting in fewer references to the stored volatile materials. However, the interaction between group and message was not significant, indicating that the same overall pattern was found in both message conditions. A post-hoc comparison, collapsed over message, showed a significant difference between the control group, which never heard about the stored volatile materials, and the two groups that did, $t(62) = 4.11$, $p < .0001$.

To determine whether the groups differed in the number of inferences consistent with the different possible themes, further 3 x 2 analyses of variance were done. The inference categories were considered separately because the scores are not statistically independent of one another. The mean number of inferences per subject for the negligence and supplies themes by group are shown in Table 3. The main effects of group showed a significant difference on both the "negligence" and the "supplies" themes; $F(2, 59) = 16.99$, $p < .0001$ for negligence; $F(2, 59) = 10.31$, $p < .0001$ for supplies. Planned comparisons revealed significant differences between the correction groups and the control group, with the control group making significantly more inferences consistent with the supplies theme than the correction groups did. They also made significantly fewer inferences consistent with the negligence theme, relative to the correction groups; $t(62) = 4.31$, $p < .0001$ for supplies; $t(62) = 5.52$, $p < .0001$ for negligence. There were no significant differences in either the number of responses consistent with the arson theme, or in number of questions left blank. No other effects or interactions were significant in analyses of the inference variables.

A subject was scored as noticing the correction if it was referred to accurately in either of the memory tests. Both the correction groups showed high levels of recall of the correction, with 100% of the delayed correction group and 90.9% of the immediate correction group recalling it. Analyses of the inference categories, omitting subjects who did not recall the correction, showed the same patterns of significance as reported above.

The free recall summaries were scored for component idea units, using an adaptation of procedures described in Kintsch (1974). Only

Groups	Delayed	Immediate	Control
Supplies theme	3.3	2.9	5.4
Negligence theme	3.5	4.0	1.3
References to volatile materials	2.7	3.5	1.0

Table 3: *Number of Inferences Consistent with Story Themes, by Group*

messages common to all three conditions were scored (i.e., no correction information was included in this measure), and a unit was scored as recalled if the subject reproduced a recognizable portion of its content. Fact questions from the questionnaire were also scored for accurate content. There were no group differences in summary recall or in fact recall ($F < 1$). However, there were main effects of message for both variables: $F(1,59) = 4.87$, $p < .03$ for summary recall, and $F(1,59) = 10.81$, $p < .002$ for fact recall, with subjects receiving the direct message showing somewhat poorer recall of the rest of the passage. Complex message subjects recalled 14.3 summary units whereas direct message subjects recalled 12.2 units; the number of facts correctly recalled was 9.2 and 8.3 for complex and direct message subjects, respectively.

Discussion

The results show that the two correction groups made more negligence inferences based on the volatile materials, and more direct references to those materials, than did the control group. Further, there was no difference between the delayed and immediate correction groups on either of these measures. Thus, when subjects saw the incorrect information, whether corrected early or late in the sequence, they showed influence from it, relative to the control group. This replicates Wilkes and Leatherbarrow (1988), and is consistent with work on belief perseverance (Ross, Lepper, & Hubbard, 1975). Here, the effect occurred even in the immediate correction group, where subjects had little opportunity to make, and therefore little need to edit, inferences. These effects cannot be accounted for by other failures of memory: none of the groups differed in recall of the reports, and over 90% of those in the correction groups recalled the correction. Thus, subjects in both correction groups had the raw materials available to be able to make correct inferences, yet they did not use it with optimal success. Additionally, the fact that the immediate correction group used the discredited information provides some evidence that influence from discredited information can also occur due to illicit post-correction retrieval and use of the discredited information, as well as due to problems in editing prestored inferences (as in the delayed correction condition).

The results of the message manipulation generally support the interpretation advanced here, with some limitations. There were no main effects of message for the inference variables, so type of message did not lead to significant differences in the number of negligence theme inferences or direct references to the stored volatile materials, which suggests that the effect does not entirely depend on correction style. However, as the subjects receiving the direct message also showed significantly fewer free recall units and facts recalled, further work on message effects is needed.

Because the experiment does not directly assess whether subjects in the immediate correction condition made any inferences before the correction occurred, the result could still be due to problems editing prestored inferences; however, there are several reasons why this interpretation is not plausible. First, due to the fact that the two correction groups did not differ in the number of inferences, it seems unlikely that random inferences generated before the discrediting message account for the effect. One would have to argue that all or most of the context-relevant inferences could be generated by this process in both groups. Kintsch (1988) distinguishes between this method of inference generation and says that often one must make additional, more strategic bridging inferences to augment this process and come up with a coherent text. Because the delayed correction condition allows for both processes to occur before the discrediting, one might expect it to show a higher number of prestored (and unedited) inferences, but this was not the case.

Second, the messages limited the number of backward inferences that could be made upon encountering the volatile materials information because they did not present any characteristics of the fire that could be linked with the volatile materials via backward inferences once one heard about them. The characteristics of the fire, which the stored volatile materials could potentially explain, were all mentioned *after* the correction for the immediate correction group. Also, because the earlier messages just mentioned the existence of a fire, it is unlikely that subjects would make a lot of forward inferences. Van den Broek (1990) argues that forwards inferences are more likely to be made when constrained by necessary and sufficient causal conditions; in this case, the mention of the fire gives few clues to its specific characteristics, and so one might not expect many predictions until more information comes in. Lastly, the questionnaire only asked about characteristics of the fire mentioned after the correction, and so would be more likely to catch backwards, bridging inferences subject may have made, which should favor the delayed correction group, where these could be made before the discrediting occurred.

Thus, overall, the results suggest that discredited information can influence inferences due to processes occurring *after* a disregard instruction or correction occurs. One explanation is that subjects may make illicit backward inferences following the correction. That is, as subjects read post-correction statements, they may make connecting inferences to make the text coherent. This may involve the discredited information simply because it fills the need to find causal antecedents. A second explanation for the effect may be that simply mentioning the stored volatile materials acts to make that information more available in memory. Subjects may not make inferences involving the discredited information on-line after the correction, but may instead retrieve any available information at the time of question and use it in further inferencing.

The possibility that subjects make illicit backward inferences to maintain the account's coherence raises some interesting issues. Making backwards inferences involving the discredited information in order to establish causal connections and preserve coherence in an account presents a serious problem for comprehension accuracy. The results here suggest that asserting information results in its propagation through later inferences despite direct, immediate correction. Thus, in all understanding contexts, such as those involving reports of news events, discrediting alone appears to be an insufficient method for removing the traces and influence of incorrect information. Further studies may ascertain what factors lead to illicit post-correction retrieval, when it occurs, and whether some forms of correction might overcome the persistent influence of discredited information.

References

Anderson, C. A., Lepper, M. R., & Ross, L. (1980). Perseverance of social theories: The role of explanation in the persistence of discredited information. *Journal of Personality and Social Psychology, 39*(1-6), 1037-1049.

Anderson, C. A., New, B. L., & Speer, J. R. (1985). Argument availability as a mediator of social theory perseverance. *Social Cognition, 3*(3), 235-249.

Carretta, T. R., & Moreland, R. L. (1983). The direct and indirect effects of inadmissible evidence. *Journal of Applied Social Psychology, 13*(4), 291-309.

Cirilo, R. K. (1981). Referential coherence and text structure in story comprehension. *Journal of Verbal Learning and Verbal Behavior, 20*(3), 358-367.

Corbett, A. T., & Chang. F. R. (1983). Pronoun disambiguation: Accessing potential antecedents. *Memory and Cognition, 11*, 283-294.

Dell, G. S., McKoon, G., & Ratcliff, R. (1983). The activation of antecedent information during the processing of anaphoric reference in reading. *Journal of Verbal Learning and Verbal Behavior, 22*, 121-132.

Epstein, W., Glenberg, A. M., & Bradley, M. M. (1984). Coactivation and comprehension: Contribution of text variables to the illusion of knowing. *Memory & Cognition, 12*, 355-360.

Graesser, A. C., & Clark, L. F. (1985). *Structures and procedures of implicit knowledge.* Norwood, NJ: Abex.

Hastie, R., & Park, B. (1986). The relationship between memory and judgment depends on whether the judgment task is memory-based or on-line. *Psychological Review, 93*(3), 258-268.

Johnson, H. M., & Seifert, C. M. (1992). Memory and discrediting information. Unpublished manuscript.

Keenan, F. M., Baillet, S. D., & Brown, P. (1984). The effects of causal cohesion on comprehension and memory. *Journal of Verbal Learning and Verbal Behavior, 23*, 115-126.

Kintsch, W. (1974). *The representation of meaning in memory.* Hillsdale, NJ.: Erlbaum.

Kintsch, W. (1988). The role of knowledge in discourse comprehension: A construction-integration model. *Psychological Review, 95*(2), 162-182.

McKoon, G., & Ratcliff, R. (1986). Inferences about predictable events. *Journal of Experimental Psychology: Learning, Memory, and Cognition, 12*, 82-91.

Ross, L., Lepper, M. R., & Hubbard, M. (1975). Perseverance in self-perception and social perception: Biased attributional processes in the debriefing paradigm. *Journal of Personality and Social Psychology, 32*(5), 880-892.

Seifert, C. M., Robertson, S. P., & Black, J. B. (1985). Types of inferences generated during reading. *Journal of Memory and Language, 24*, 405-422.

van den Broek, P. (1990). Causal inferences and the comprehension of narrative texts. In Graesser, A. C., & Bower, G. (Eds.), *The Psychology of Learning and Motivation (Vol. 25): Inferences and Text Comprehension.* San Diego: Academic Press, Inc.

Wilkes, A. L., & Leatherbarrow, M. (1988). Editing episodic memory following the identification of error. *Quarterly Journal of Experimental Psychology, 40A*, 361-387.

Wyer, R. S., & Budesheim, T. L. (1987). Person memory and judgments: The impact of information one is told to disregard. *Journal of Personality and Social Psychology, 53*, 14-29.

A Fine-Grained Model of Skill Acquisition: Fitting Cascade to Individual Subjects

Randolph M. Jones & Kurt VanLehn
Learning Research and Development Center and
Computer Science Department
University of Pittsburgh
Pittsburgh, PA 15260
JONES@CS.PITT.EDU VANLEHN@CS.PITT.EDU

Abstract

The Cascade model of cognitive skill acquisition was developed to integrate a number of AI techniques and to account for psychological results on the *self-explanation effect*. In previous work, we compared Cascade's behavior to aggregate data collected from the protocols of 9 subjects in a self-explanation study. Here, we report the results of a fine-grained analysis, in which we matched Cascade's behavior to the individual protocols of each of the subjects. Our analyses demonstrate empirically that Cascade is a good model of subject behavior at the level of goals and inferences. It covers about 75% of the subjects' example-studying behavior and 60% to 90% of their problem-solving behavior. In addition, this research forced us to develop general feasible methods for matching a simulation to large protocols (approximately 3000 pages total). Finally, the analyses point out some weaknesses in the Cascade system and provide us with direction for future analyses of the model and data.

Introduction

Cascade is an integrated model of cognitive skill acquisition. It incorporates a number of methods from artificial intelligence, and was designed with attention to robust psychological findings. Elsewhere (VanLehn, Jones, & Chi, 1991, 1992), we have demonstrated that Cascade's mechanisms interact to account for the main qualitative findings involved in the self-explanation effect (Bielaczyc & Recker, 1991; Chi, Bassok, Lewis, Reimann, & Glaser, 1989; Chi, de Leeuw, Chiu, & LaVancher, 1991; Fergusson-Hessler & de Jong, 1990; Pirolli & Bielaczyc, 1989). In that research, we compared Cascade's behavior to aggregate data taken from the protocols of the 9 subjects in Chi et al.'s (1989) study.

In this paper, we refine the evaluation of Cascade by matching its behavior to the individual protocols of Chi et al.'s subjects. This research is similar to Newell and Simon's (1972) classic study on human problem solving, in that both attempt to determine how closely an AI program can simulate the protocols of individual subjects. However, there are three important differences between our study and Newell and Simon's. First, our task domain is physics, which is arguably much richer than the task domains they studied. Second, a considerable amount of learning occurs in the Chi et al. protocols. Third, the Chi et al. data consist of 252 protocols, each averaging 12 pages in length, so it would be infeasible to analyze them with problem-behavior graphs.

This work makes two important contributions. First, it provides evidence that Cascade's model of the subjects in Chi et al.'s study is quite accurate, even at the level of individual rules and goals. Second, it demonstrates a practical method for large scale comparisons of a simulation system to protocol data. We begin with an overview of the Cascade system. This is followed by a description of our paradigm for matching Cascade to the protocols and a brief discussion of our results. The paper concludes by describing implications of the results on future research with Cascade.

The Cascade system

Cascade is an AI system that integrates multiple strategies for problem solving and learning. Although the system has been applied to elementary probability and naive physics, the current analysis involves the domain of Newtonian physics, because this is the domain studied by the subjects. Due to space restrictions, we can only provide a summary of the system here. A detailed treatment can be found elsewhere (VanLehn & Jones, in press-a; VanLehn, Jones, & Chi, 1992).

Problem solving mechanisms

Cascade's overall control structure is based on a backward-chaining theorem prover (similar to Prolog), but it distinguishes between *explaining examples* and *solving problems*. Problems are presented as a set of literals describing a physical situation and a list of quantities for which Cascade must find values. An example is a problem along with a solution that consists of a sequence of lines describing partial results that lead up to the answer to the problem. To explain an example, Cascade explains (proves) each line. Whereas in problem solving Cascade must find a value, V, for a sought quantity, Q, in explaining an example line Cascade must prove why Q has a given value, V. Explanation is simpler than ordinary problem solving because the provided values help control search. People rarely explain every detail of every line, so Cascade can also *accept* that the current quantity has the stated value, instead of explaining it.

As Cascade explains an example, it stores a trace of its explanation (useful for solving subsequent problems), so more explanation leads to a larger stored derivation.

During problem solving, Cascade attempts to use its rule-based knowledge to find a value for a sought quantity. If this fails, the system tries to use a form of *transformational analogy* (Carbonell, 1983). That is, the system retrieves an example that is similar to the current problem and looks for a line in the example that mentions the sought quantity. If possible, it uses such a line to determine a value for the quantity. As Carbonell also found, this type of reasoning often leads to incorrect results. However, it is a strategy that subjects exhibit quite often.

Learning mechanisms

Cascade also includes two learning mechanisms for improving its problem-solving behavior. First, as we have mentioned, Cascade stores a trace of its solution as it explains examples. Many of the problems are analogous to one or more of the examples. Therefore, when the system works on a problem, it first attempts to create an analogical mapping between the problem and any similar examples. Then, when Cascade needs to solve a particular goal, it checks whether an analogous goal appeared in the example. If so, the system determines how it solved the goal in the example and uses that action in the current problem. In general, this mechanism leads to less search because it implies a better ordering of the rules in memory. This is a symbol-level learning mechanism called *analogical search control* (Jones, in press).

The second mechanism learns at the knowledge level (Dietterich, 1986), and is called *explanation-based learning of correctness* (VanLehn, Ball, & Kowalski, 1990). Cascade's knowledge base contains a number of overly general rules that are not used for general problem solving. However, when the system reaches an impasse on a problem or example and decides that the impasse is due to missing knowledge, it can use the overly general rules to patch its knowledge and introduce new standard rules. This method of learning is similar to knowledge-level learning methods proposed by Schank (1986), Lewis (1988), Anderson (1990), and others.

Fitting to individual subjects

For each subject, we set Cascade's parameters in order to approximate the subject's initial knowledge and example-explaining behavior. We then ran Cascade on the examples and problems that the subject worked on, collected data from the run, and analyzed them in several ways in order to determine Cascade's empirical accuracy. First we describe the parameter fitting and then the results of our analyses.

Initializing Cascade's parameters

Cascade's model of the subjects includes two parameters: the subjects' knowledge just before they explain the examples, and the subjects' decisions about which pieces of the examples to accept without explanation. Each paramater will be discussed in turn.

The subjects acquired their initial knowledge by reading the first several chapters of the textbook and from their earlier studies of physics and mathematics. Because we have no access to their learning history, nor a detailed test of their initial knowledge, we must guess their initial knowledge. Cascade's initial knowledge base for each subject was a subset of a fixed "rule library." The rule library consisted of 110 rules, including rules from the textbook, common sense rules, rules that are learnable via overly general rules, and 3 buggy physics rules that some subjects appeared to have. One buggy rule applies $F = ma$ to any force and not just a net force. Another asserts that the mass of a body is equal to its weight. The third assumes that the sign of all vector projections is positive.

There is no easy way to determine what a subject's initial knowledge is, but we made the best approximation we could by looking for rule use throughout each subject's entire set of protocols. As we found later, we sometimes made mistakes in selecting the initial knowledge. In these cases, we need to fix the mistakes, rerun the simulations

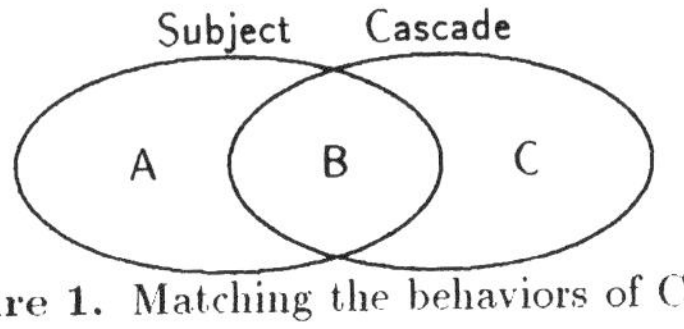

Figure 1. Matching the behaviors of Cascade and a subject

and redo our analyses. However, this will require months of work, so for now, we report the analyses with our imperfect choices of initial knowledge left intact. There were only a few of these cases, so we don't feel that the qualitative nature of our results will change.

The second parameter concerns how deeply the subjects explained the examples. When studying examples, subjects choose to explain some lines but not others. Even when they do explain a line, they may explain it only down to a certain level of detail and decide to take the example's word for the rest. For example, they might explain most of the line, $F_{a_x} = -F_a \cos(30)$, but not bother to explain where the minus sign comes from. Cascade does not model how the subjects decide which lines to explain and how deeply to explain them, so it must be told explicitly which sections to explain. Therefore, whenever Cascade is about to explain the the proposition, $Q = V$, it first checks to see if the literal `accept(Q,V)` is in the example's description. If the system finds such a literal, it merely accepts that Q's value is V without attempting to explain it.

We added `accept` propositions to Cascade by inspecting the subject's example protocols. If the subject merely read a line and said nothing else about it, then we entered an accept literal for the whole line. If the subject omitted discussion of a detail in a line, then we only accepted that detail, allowing Cascade to explain other goals involved in the line. In this fashion, the protocol data completely determined which lines and parts of lines Cascade explained.

The fit between Cascade and individual subjects

We are interested in two types of comparisons between the model and subject data. Suppose the diagram in Figure 1 represents the behaviors of a particular subject and Cascade's model of that subject. Region A represents subject behavior that Cascade failed to match. Region B represents the behavior that Cascade and the subject have in common. Region C represents Cascade behavior that the subject did not exhibit. The two comparisons we want are the ratio of region B behavior to Cascade's total behavior (regions B and C) and to the subject's total behavior (regions A and B). Table 1 shows the specific analyses conducted and their types.

Table 1. Analyses of Cascade's simulation of individual subjects.

1. How many of Cascade's example studying inferences were also made by the subject (BC vs. B)?
2. How many of the subject's example studying inferences were also made by Cascade (AB vs. B)?
3. How many of Cascade's problem solving inferences were also made by the subject (BC vs. B)?
4. How many of the subject's problem solving inferences were also made by Cascade (AB vs. B)?
5. Do the search control decisions made by the subject match those made by Cascade (AB vs. BC)?

In order to carry out these five analyses, we needed a way to quantify behaviors, which implies choosing a unit of analysis. This was not hard for matching region B to Cascade's behavior, because Cascade's behavior is well defined and explicit. For analyses 1 and 3, we used goals as the unit of analysis. After running Cascade, we classified each of its goals depending on the type of action Cascade took at that point. When explaining examples (analysis 1), these actions included deductively explaining the goal, accepting the goal without attempting to explain it, and encountering an impasse and learning a new rule. For problem solving (analysis 3), the actions included regular rule-based problem solving, regular use of transformational analogy, forced use of transformational analogy (for cases where the subject used transformational analogy but Cascade's normal control structure would have used regular problem solving), and encountering an impasse and learning a new rule. After classifying Cascade's behaviors, we determined what the subject's behavior was at each goal. We used the same classifications for subject behavior, but included the possibility of having an impasse and *not* learning a new rule (because the impasse was never resolved).

It was not as easy to determine a unit of analysis for matching region B to the subjects' behavior, so we used a variety of units, depending on the type of analysis being conducted. For analysis 2, we extended an earlier encoding of inferences made by subjects while explaining examples (Chi & VanLehn, 1991) and compared those to the inferences made by Cascade. For analysis 4, we coded a sample of the protocols at the level of Cascade-like goals. These goals are at the same grain size as

Cascade's goals, so the comparison is direct.

Because we are more interested in Cascade's simulation of the subjects' acquisition of physics rules than in its simulation of the chronology of their reasoning, analyses 1–4 ignored the order in which Cascade and the subject made inferences. Both Cascade's behavior and the subject's behavior were reduced to sets of inferences. We simply calculated the intersections and differences between the sets, just as shown in Figure 1. However, we cannot entirely ignore the chronology of inferences, because an earlier study indicated that analogical search control affects the location of impasses, which in turn determines what can be learned during problem solving (VanLehn & Jones, in press-a). Therefore, we used analysis 5 to determine whether subjects' rule choices during problem solving could be predicted by analogical search control.

Results of the analyses

Unfortunately, there is not enough space here to present the simulation runs and analyses in detail, so we will present a general summary and conclusions from the analyses. The details are presented elsewhere (VanLehn & Jones, in press-b).

Results on example explaining. We found that 95% of the example-explaining behavior generated by Cascade was matched by the subjects' behavior (analysis 1). This is not surprising because most of Cascade's example-studying behavior is determined by the parameter settings.

In analysis 2, we found that Cascade successfully accounted for 63% of the 227 explanation episodes in the subjects' example-studying protocols. Of the unmatched explanations, 61 were concerned with cognitive skills that we are not interested in modeling, such as algebraic equation solving. That left only 23 explanations (10% of the 227 total explanations) that Cascade should have been able to model. These fell into two groups: incorrect explanations (14 cases) and general comments (9 cases). The incorrect explanations indicate that Cascade needs more buggy rules than it currently has. In particular, many of the missing rules contained misconceptions about the relationship between acceleration and motion. The general comments indicate that the subjects have an ability to break out of Cascade's strict backward-chaining control structure and do plan recognition or mental modeling. These are certainly interesting and important cognitive skills, but we were surprised that they were used so rarely in this study. When we began developing Cascade, we expected plan recognition to be the most important kind of explanation. This analysis indicates that it occurs rarely and probably has little influence on subsequent problem solving. Overall, Cascade fails to model only 23 (14%) of the 166 explanation episodes that are relevant to the task domain, and we are encouraged by this result.

Results on problem solving. In analysis 3, we found that 97% of the 3947 goals generated by Cascade during problem solving were handled in the same way by the subjects. Of the 118 episodes that weren't matched by the subjects, most (98) involved transformational analogy. We were surprised by the prevalence of transformational analogy during problem solving, although it was certainly due in part to the fact that 12 of the 21 problems in the study were isomorphic (or nearly so) to one of the three examples.

Cascade's model of transformational analogy is too simple to describe adequately all the ways that transformational analogy was used by the subjects. A large number of the 98 cases occurred when subjects used a force diagram from an example to aid in drawing the force diagram for a problem. Cascade currently represents force diagrams in its standard equation-based representation, whereas the subjects were almost certainly using some type of visual representation. This partially explains why Cascade's transformational analogy fails in these cases.

Analysis 4 was quite time consuming, so we were only able to examine a small sample. Of the 225 total problem-solving protocols, we selected 4 that we thought were representative of the variety of approaches used by the subjects. Two protocols were from "good" problem solvers who got correct answers and two were from "poor" problems solvers who got incorrect answers. In addition, each of the pairs included a protocol that used mostly transformational analogy and a protocol that used mostly regular rule-based problem solving. This sample is clearly much too small, but it is a start. In the four protocols, we counted 151 total goals or inferences, excluding trivial arithmetic and algebraic goals. We found 15 cases in this analysis that the current implementation of Cascade failed to account for, so 90% of the subjects' problem-solving behavior is matched by the Cascade model. After several years of experience with these protocols, we feel intuitively that this figure is too high, and that a larger sample might yield a match that could be as low as 60%.

Results on search control. The first four analyses concentrated on matching the knowledge content of Cascade and the subjects without paying attention to *when* the knowledge is used. The order of inferences is determined by Cascade's control structure (backwards chaining) and its mechanism for choosing which rule to try first for achieving a goal (analogical search control, or if no analogical advice is available, then a default ordering of rules). As part of analysis 4, we fit the 151 subject goals to a backward-chaining control structure. Only 3 goals could not be fit, indicating that subjects occasionally make opportunistic inferences about the current situation that are not directly relevant to the current goal.

In order to evaluate Cascade's policy for choosing rules to apply to the current goal, we matched its choices for all 3947 goals to the choices of the subjects, and they agreed in 97.7% of the cases. In short, Cascade's simple control regime turned out to be a fairly good predictor of the order in which subjects make inferences.

Discussion

One contribution of this work is that it demonstrates a method for comparing large-scale AI simulations with protocols. Our general method consists of comparing the amount of shared behavior between the simulation and the subjects to the total simulation behavior and the total subject behavior. The unit of analysis for matching simulation behavior is straightforward, because Cascade's behavior is explicit for each goal it considers when explaining examples or solving problems. For matching subject behavior, we used two separate measures. In analysis 2 (explaining examples), we coded the subject protocols at the level of individual physics or math explanations, and compared the inferences with Cascade's. In analysis 4 (solving problems), we undertook a much more ambitious method, coding the protocols at the level of Cascade-like goals. This analysis allowed us to match the subjects' behavior to Cascade goal by goal, noting the locations where Cascade's model diverged from the subjects' behavior. Although rather time-consuming, our success with this type of encoding encourages us to continue the analysis with a larger sample of protocols.

The second contribution of this research is an empirical evaluation of Cascade's ability to model the behavior of individual subjects at a fine grain size. We discovered that Cascade can explain most of the subjects' example-studying and problem-solving behavior with its three major performance mechanisms: deduction, simple acceptance of example statements, and transformational analogy. Analyses 1–4 indicate that these three processes cover about 75% of the example studying behavior and 60–90% of the problem solving behavior. In addition, the behavior they do not cover mostly involves mathematical manipulations or other types of cognition that are outside the domain of study.

Finally, analysis 5 demonstrates that Cascade rather accurately models the subjects' overall control structure and local control choices. We were pleasantly surprised by this result, because we did not concentrate on these aspects during the system's development.

To put these results in perspective, we look at two other attempts to match cognitive models to individual subjects. Newell and Simon (1972) used GPS to match 80% of an indidividual subjects' behavior on cryptarithmetic problems. VanLehn's (1991) model for strategy discovery accounted for 96% of the behavior of a subject solving the "tower of Hanoi" problem. It is important to note that both of these studies involved modeling the behavior of a single subject. We used Cascade to model the behavior of several individuals, which is almost guaranteed to reduce the model's overall accuracy. With this in mind, Cascade's account of human behavior compares well with the older models.

Perhaps the most important benefit of this research is that it has shown us where some of Cascade's weaknesses are, and it has pointed out some more aspects of the data that should also be analyzed. For example, we found that Cascade's simple model of transformational analogy is inadequate. Subjects were quite clever at forming useful analogies with the examples, and especially their force diagrams. In addition, we were surprised to find that there were so few clear-cut cases of impasse-driven learning in the protocols. During analyses 1 and 3, we found that subjects only showed signs of impasses at 18 of the 44 times that Cascade encountered an impasse and used explanation-based learning of correctness to get out of it. Our initial hypothesis is that these events arise either from ingenious use of transformational analogy by the subjects, or they were actual impasses that were simply not verbalized in the protocols. Our future analyses will concentrate on these learning aspects and should tell us exactly why there were so few clear cases of learning.

Acknowledgements

This research was supported in part by contract N00014-88-K-0086 from the Cognitive Science division and contract N00014-86-K-0678 from the Information Sciences division of the Office of Naval Research.

References

Anderson, J. R. (1990). *Adaptive control of thought.* Hillsdale, NJ: Lawrence Erlbaum.

Bielaczyc, K., & Recker, M. M. (1991). Learning to learn: The implications of strategy instruction in computer programming. In L. Birnbaum (Ed.), *The International Conference on the Learning Sciences.* Charlottesville, VA: Association for the Advancement of Computing in Education.

Carbonell, J. G. (1983). Learning by analogy: Formulating and generalizing plans from past experience. In R. S. Michalski, J. G. Carbonell, & T. M. Mitchell (Eds.), *Machine learning: An artificial intelligence approach.* Los Altos, CA: Morgan Kaufmann.

Chi, M. T. H., Bassok, M., Lewis, M. W.,

Reimann, P., & Glaser, R. (1989).

Self-explanations: How students study and use examples in learning to solve problems. *Cognitive Science, 13,* 145–182.

Chi, M. T. H., de Leeuw, N., Chiu, M., & LaVancher, C. (1991). *The use of self-explanations as a learning tool.* Manuscript submitted for publication.

Chi, M. T. H., & VanLehn, K. (1991). The content of self-explanations. *Journal of the Learning Sciences, 1,* 69–106.

Dietterich, T. G. (1986). Learning at the knowledge level. *Machine Learning, 1,* 287–316.

Ferguson-Hesler, M. G. M. & De Jong, T. (1990). Studying physics texts: Differences in study processes between good and poor solvers. *Cognition and Instruction, 7,* 41–54.

Jones, R. M. (in press). Problem solving via analogical retrieval and analogical search control. In S. Chipman & A. Meyrowitz (Eds.), *Machine learning: Induction, analogy, and discovery.* Boston: Kluwer Academic.

Lewis, C. (1988). Why and how to learn why: Analysis-based generalization of procedures. *Cognitive Science, 12,* 211–256.

Newell, A., & Simon, H. (1972). *Human problem solving.* Englewood Cliffs, NJ: Prentice-Hall.

Pirolli, P., & Bielaczyc, K. (1989). Empirical analyses of self-explanation and transfer in learning to program. *Proceedings of the Eleventh Annual Conference of the Cognitive Science Society.* Hillsdale, NJ: Lawrence Erlbaum.

Schank, R. C. (1986). *Explanation patterns: Understanding mechanically and creatively.* Hillsdale, NJ: Lawrence Erlbaum.

VanLehn, K. (1991). Rule acquisition events in the discovery of problem solving strategies. *Cognitive Science, 15,* 1–47.

VanLehn, K., Ball, W., & Kowalski, B. (1990). Explanation-based learning of correctness: Towards a model of the self-explanation effect. *Proceedings of the Twelfth Annual Conference of the Cognitive Science Society.* Hillsdale, NJ: Lawrence Erlbaum.

VanLehn, K., & Jones, R. M. (in press-a). Integration of analogical search control and

explanation-based learning of correctness. In S. Minton & P. Langley (Eds.), *Planning, scheduling and learning.* Los Altos, CA: Morgan Kaufmann.

VanLehn, K., & Jones, R. M. (in press-b).

Learning by explaining examples to oneself: A computational model. In S. Chipman & A. Meyrowitz (Eds.), *Cognitive models of complex learning.* Boston: Kluwer Academic.

VanLehn, K., Jones, R. M., & Chi, M. T. H. (1991). Modeling the self-explanation effect with Cascade 3. *Proceedings of the Thirteenth Annual Conference of the Cognitive Science Society.* Hillsdale, NJ: Lawrence Erlbaum.

VanLehn, K., Jones, R. M., & Chi, M. T. H. (1992). A model of the self-explanation effect. *Journal of the Learning Sciences, 2* 1–59.

A Re-examination of Graded Membership in Animal and Artifact Categories

Charles W. Kalish
Department of Psychology
3433 Mason Hall
University of Michigan
Ann Arbor, MI 48109
Charles_Kalish@um.cc.umich.edu

Abstract[1]

Previous studies of gradedness have failed to distinguish between the issues of typicality and category membership. Thus, data which have been taken to demonstrate that membership is a matter of degree may only demonstrate that typicality is graded. The present paper reports the results of two studies that attempt to overcome limitations of past methods. In the first study, subjects were asked to rate both typicality and category membership for the same stimuli as a way of distinguishing the two questions. A second study was based on the notion that there may be no definitive answer to questions about membership in graded categories. Thus, disagreements about membership in all-or-none and graded categories may have different qualities Stimuli included animal and artifact categories as well as animals that had undergone different kinds of transformations. Results from both studies suggest some support for claims that membership in animal and artifact categories is graded.

Graded Categories

Some categories clearly have all-or-none membership (e.g., "even number", "square"). Others obviously admit degrees (e.g., "red": things can be more or less red). For most concepts, though, intuitions are not so clear. Plant, animal, and substance categories (often called "natural kinds") and human artifacts (e.g., vehicles, furniture) have been of particular interest to researchers on concepts (Barr & Caplan, 1987; Keil, 1989; Rosch & Mervis, 1975). Whether or not membership in these categories is graded has important implications. For example, based in part on beliefs about the all-or-none nature of membership, it has been argued that mixed models or essentialist models are most appropriate for natural kinds (Keil, 1989). Yet what is the basis for these beliefs about category gradedness? Do we have good means for determining when membership is all-or-none and when it is a matter of degree?

At one time, there was wide consensus that membership in all categories was a matter of degree. Data from a number of studies showed that some instances of a category were better members than others (e.g., Barr & Caplan; 1987; McCloskey & Glucksberg, 1978; Oden, 1977, Rosch & Mervis, 1975). However, questions about the validity of existing measures of category membership suggest we should re-examine these conclusions.

Many of the studies arguing for graded membership have failed to distinguish between typicality and categorization (Rey, 1983). Studies which asked subjects for typicality ratings (e.g., "How representative/typical/characteristic is a penguin of a bird") were often taken as evidence about category membership (see Lakoff, 1987). Even when methods did not call for explicit typicality ratings (e.g., memory tasks or judgments of sentence appropriateness) there has been no way to determine whether results reflected the gradedness of typicality or the gradedness of category membership. We don't know which studies have been measuring typicality and which categorization.

The failure to distinguish between the two kinds of judgments is problematic because graded typicality may not indicate a gradedness of category membership. For example, 34 will receive a low typicality rating as an "even number" yet is a much a member of the category as is a high typicality even number (e.g, 2;

[1] This research was supported by an NSF Graduate Fellowship.

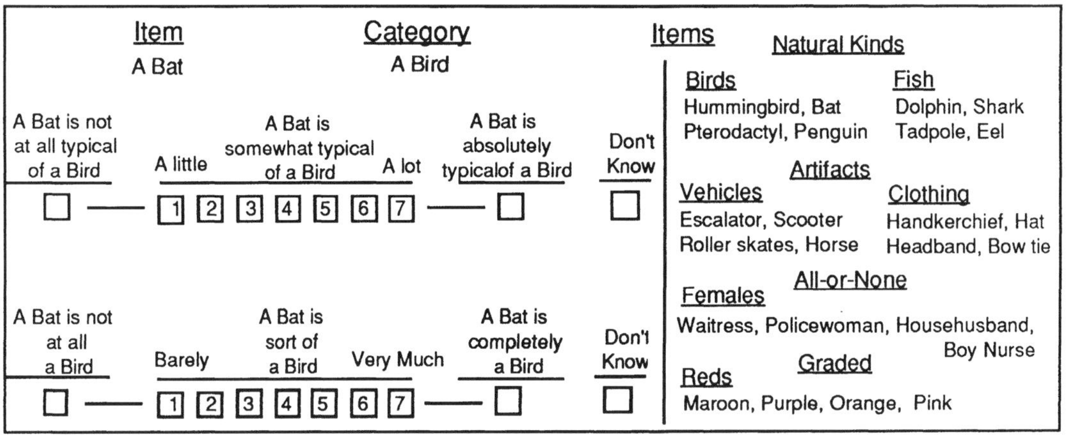

Figure 1: Stimuli for Study 1

Armstrong, Gleitman & Gleitman, 1983). Other studies (Keil, 1989; Rips, 1989) have demonstrated that certain kinds of transformations may affect an animal's typicality but not change the category the animal is assigned to. However, a crucial question not answered by these studies is whether or not the degree of category membership remains unchanged after the transformations. In other words, can category membership be reduced by degrees?

The fact that gradedness of typicality does not imply gradedness of categorization is not, in itself, positive evidence about the nature of category membership. Such evidence must come from studies which specifically address the question of category membership. In the rest of the paper, we present the results of two studies which attempted to asses category membership rather than typicality.

Study I

The purpose of this study was to collect data that reflected judgments of category membership rather than typicality (Rey, 1983; Rips, 1989). In Study 1 subjects were asked to make both typicality and categorization judgments for the same stimuli. It was reasoned that this procedure, along with careful instructions and control items, would tend to distinguish the question of categorization from the question of typicality. These categorization ratings could then be examined for evidence of gradedness. Included in this study were items depicting animals which had undergone certain alterations. These items were included to see if transformation might alter degree of membership.

Methods

Subjects. Subjects were 19 University of Michigan undergraduates participating in experiments as part of the requirements of an introductory psychology course.

Procedure. Subjects were asked to make two judgments about the relationship of an instance (e.g., a robin) to a category (e.g., bird); first its typicality and then its degree of membership. Subjects first read brief instructions describing typicality and categorization, and the differences between the two. They were then presented with 32 computer "screens." One instance and one category were presented on each screen. The first task was to rate how typical the instance was of the category. Following this, subjects were to rate the degree to which the instance was (or was not) a member of the category. A modified Likert scale was used to collect both ratings. The scale included two absolute end points (0 and 8) and a graded scale in the middle (1-7) (see Figure 1). This was done to minimize demand characteristics for either absolute or graded responses. Subjects were also able to indicate "don't know" to either rating.

Stimuli. There were four instances to be rated for each of eight categories. All instances were chosen to have low typicality. Eight instances were to be rated as members of natural kinds. Eight instances were rated as artifacts. There was one control category which was assumed to have all-or-none membership and graded typicality and one control category with graded membership and graded typicality (see Figure 1). Subjects were also asked to rate instances of natural kind categories that had undergone different kinds of transformation. Four items depicted surface transformations (following Keil, 1989; see Appendix 1, item 1). Four other items described more radical, deep transformations (following Rips, 1989; see Appendix 1, item 2). These transformed items were always the last eight screens. Items were otherwise presented in random order.

Results

Figure 2 presents the mean proportion of absolute (0 or 8) answers for a given kind of categorization and typicality judgment. Approximately 80% of the judgments about membership in the category "Female" were absolute. Categorization responses to natural kind and artifact items were compared with responses to "Female" items (the baseline) to test for gradedness. The difference between each subject's mean number of absolute responses to "Female" items and his/her mean response to the test category was computed (Female-target). If the only difference between the test items and "Female" was error variance, this difference should average zero; if the test items had fewer absolute ratings, this difference should be positive. Animal instances were judged as more graded than "Female" instances ($t(18) = 2.3$ $p<.05$). Graded responses were more frequent to transformation items than to "Female" items (surface vs Female, $t(18) = 2.5$ $p<.05$; deep vs Female, $t(18) = 8.5$ $p<.001$). Artifact items also showed high levels of graded responses ($t(18) = 6.6$ $p<.001$). Large differences between items within a kind were also found. Within "Bird," 40% of categorization responses were graded to "Penguin," while less than 10% were graded for "Hummingbird." Analyses of responses to typicality judgments, however, revealed pattern analogous to that found with categorization judgments (see Figure 2). Instances of "Female" were given more absolute typicality ratings than were instances of animals, for example. Only 46% of the typicality judgments for "Female" were absolute, though, compared with 80% of the categorization judgments.

Discussion

Categorization judgments were predominantly absolute for females and largely graded for reds. Comparisons show that membership judgments for artifact categories and natural kind categories showed a significant degree of gradedness compared with controls. However, a similar pattern of results would have been obtained by comparing typicality ratings. Typicality may have been affecting categorization judgments. It is unlikely that subjects were simply using typicality as the basis of both typicality and categorization

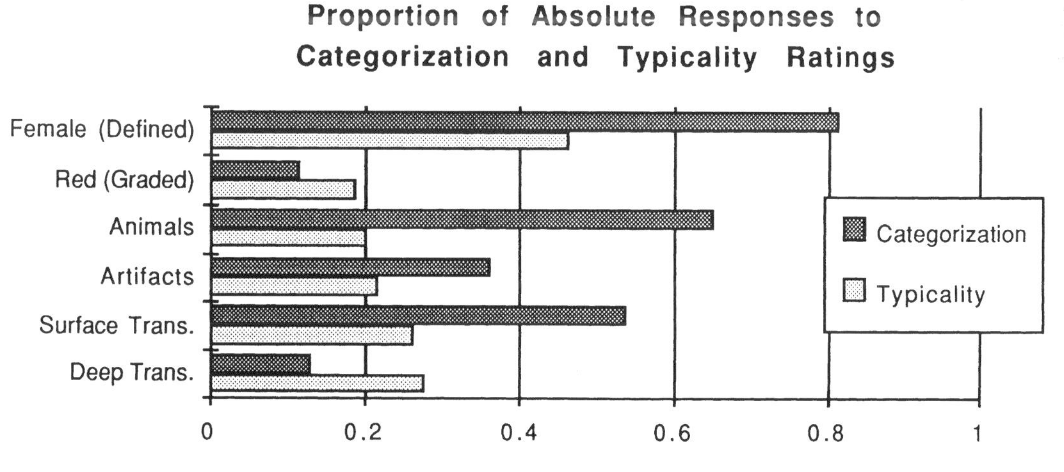

Figure 2: Categorization and Typicality Ratings: Study 1

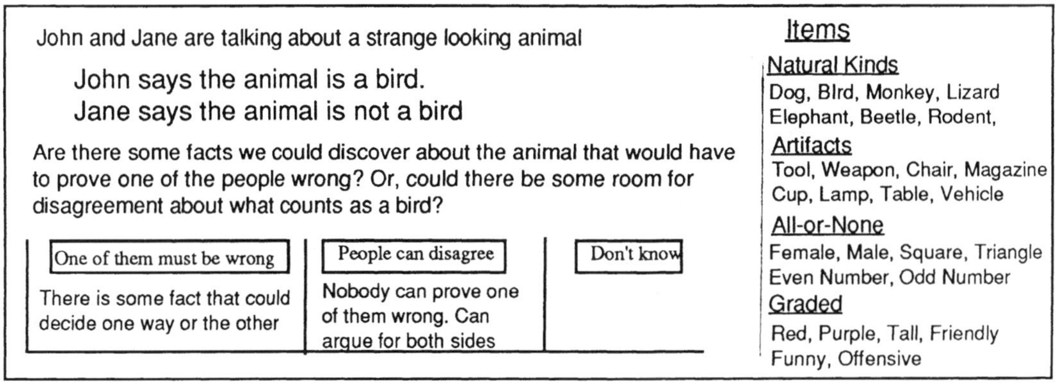

Figure 3: Stimuli for Study 2

judgments because there were significant differences in the percentage of absolute responses to the two kinds of judgments (for some items). However, we cannot conclude with certainty that the categorization ratings truly reflected subjects' beliefs about whether category membership is a matter of degree or not; we might still be measuring the gradedness of typicality[2].

The results of study 1 can be taken as suggestive that subjects view membership in natural kind and artifact categories as a matter of degree. However, such results depend on subjects explicitly distinguishing typicality from categorization. Study 2 reports an attempt to replicate the results of Study 1 using a method that does not require subjects to explicitly interpret questions as asking for categorization or typicality judgments.

Study 2

Whether membership in a category is thought to be absolute or a matter of degree should affect how disagreements about an item's category membership can be resolved. Assume two people disagree about whether a particular instance is or is not a member of a given category. If the category has all-or-none membership, then one of the people must be wrong. Imagine a dispute regarding whether 17341 is a member of the category "prime number." We may not know the answer, but presumably there are some facts we could uncover that would prove the issue one way or the other. If the category admits degrees of membership, on the other hand, then it is possible for there to be no way to demonstrate that one position is incorrect. Since an object may be partly a member of the category, there may be unresolvable disputes about whether it is "enough" of a member to be so identified. Consider a debate about categorizing somebody as friendly or not. We could agree on all the facts of the matter yet not agree on how friendly someone has to be to be called friendly. A similar method was used by Malt (1991; studies 4&5) in a study of people's beliefs about the completeness of their representations of categories. However, that study was not designed to allow conclusions about the gradedness of categories. In the present study, subjects were asked to judge whether certain disagreements might be unresolvable as a way of assessing whether membership in the categories was thought to be all-or-none or a matter of degree. Disputes involving all-or-none categories should be resolvable by factual discoveries.

Methods

Subjects. Subjects were 20 University of Michigan undergraduates participating in experiments as part of the requirements of an introductory psychology course. No subjects in Study 2 had participated in Study 1.

Procedure. Subjects were presented with instructions regarding graded membership and the possibilities of disagreements. After reading

2 We are attempting to replicate the results of Study 1 using different sets of categories where typicality will be matched across item types. All-or-none categories used are,"Even number" and "U.S. Currency."

these instructions they were presented with computer screens describing an encounter with an unusual object and a disagreement about how that object should be categorized. The instances being argued about were described as unusual members of a superordinate class (e.g., John and Jane have come across a strange animal). The disagreement was then presented as to how to categorize the object (e.g., John says the animal is a lizard; Jane says the animal is not a lizard). For each screen, subjects were asked to indicate whether one of the people must be wrong or whether it was possible for people to legitimately disagree (see Figure 3).

Stimuli. Stimuli were 32 disagreements. Four disagreements involved categories with absolute membership; four involved graded categories . Eight disagreements revolved around animals and eight concerned artifacts (see Figure 3). Eight scenes involved a transformed animal (four "surface" transformations and four "deep" transformations). To give subjects an idea of the diversity they should consider, the eight transformation items were presented first. Order of presentation was otherwise random.

Results

Figure 4 presents the proportion of subjects who indicated that the disputants could legitimately disagree. To assess gradedness, responses to target categories were compared with responses to "defined" items. As in Study 1, the difference in each subject's mean number of "can disagree" responses to "defined" and target items was calculated (target-"defined"). If these differences were significantly positive, then this means that more disagreements were accepted for the target category. Subjects were significantly more likely to answer "can disagree" to animal items than they were to "defined" items ($t(19) = 3.3$, $p<.005$). Transformation items also differed from "defined" items ($t(19) = 1.8$, $p<.05$ and $t(19) = 4.4$, $p<.001$ respectively). Disagreements were more often accepted for artifact items than for "defined" items ($t(19) = 6.7$, $p<.001$) . Again there were some differences between items of the same kind; 42% of subjects accepted disagreements for categorizing lizards, while only 20% allowed disagreements for dogs.

Discussion

Results from Study 2 largely support the findings of Study 1. Again, the validity of the measure can be confirmed by the predicted performance on control items. Very few disagreements were accepted for defined items, while disagreements were largely accepted for graded items. Subjects allowed a significant number of disagreements for both natural kind and artifact categories. This suggests that membership in both natural kind and artifact categories may be a matter of degree.

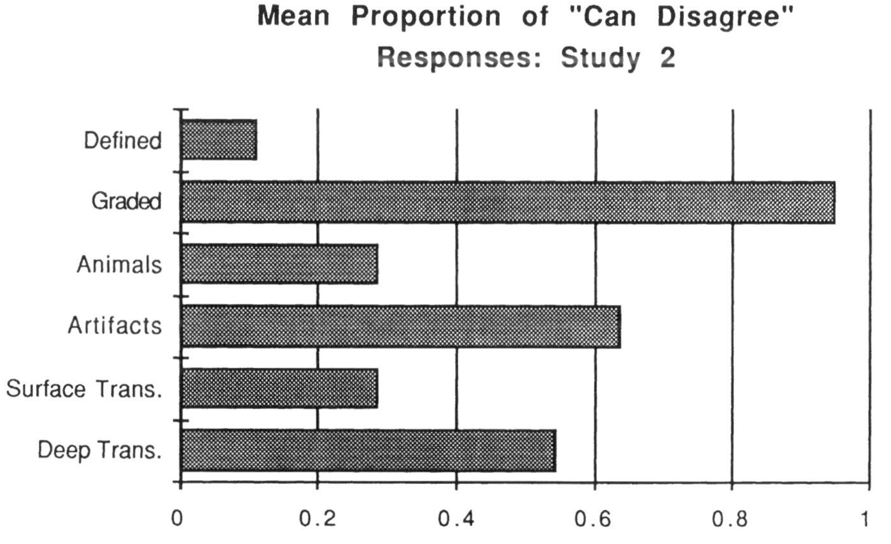

Figure 4: Proportion of judgments that an argument may not be resolvable, Study 2

General Discussion

Our findings generally support earlier conclusions drawn from studies of typicality effects (e.g., Rosch & Mervis, 1975). However, in these studies we have been very careful to distinguish questions of categorization from questions of typicality. In apparent contrast to our findings, Malt (1991) has argued that subjects' believe experts are able to make some definitive categorization decisions. One possibility is that people might be inclined to accept expert opinion even though they believe categories to be graded.[3] On the other hand, gradedness of membership may be the result of having multiple sets of criteria for membership (Lakoff, 1987). Pilot work on Study 2 revealed that many subjects who accepted disagreements about the way animals could be categorized reasoned that the disputants might be using different criteria for category membership. For instance, one might have perceptual, functional and/or biological criteria for categorizing something as a bird. Any one set of criteria, used by one type of expert, may provide an absolute categorization, but multiple sets would allow graded responses. Gradedness could arise because of uncertainty or disagreement about which criteria/expert were relevant. We are currently conducting a study to test this possibility using linguistic hedges (e.g., "biologically speaking") to focus subjects on particular criteria.

Appendix 1

Item 1: A Surface Transformation. Doctors took a horse and did an operation that put black and white stripes all over its body. They cut off its mane and braided its tail. They trained it to stop neighing like a horse, and they trained it to eat wild grass instead of oats and hay. They also trained it to live in the wilds of Africa instead of in a stable. When they were all done, the animal looked just like a zebra. (see Keil, 1989)

Item 2: A Deep Transformation. This fish hatched from an egg laid near the waste pipe of a nuclear power plant which discharged irradiated water. As this fish grew it became long and thin. Its fins never developed. After a while this animal started spending more time breathing air above the surface of the water. Eventually it came to live on land where it slithered around catching and eating small bugs. By the time the animal matured, its head had flattened and it had developed a forked tongue. (see Rips, 1989)

[3]Consider that we accept expert opinion regarding what is a good wine, even though we believe membership in the category "good wine" to be a matter of degree.

References

Armstrong, S. L., Gleitman, L. R., & Gleitman, H. 1983. What some concepts might not be. *Cognition*, 13:263-308.

Barr, R. A., & Caplan, L. J.1987. Category representations and their implications for category structure. *Memory & Cognition*, 15:397-418.

Keil, F. C. 1989. *Concepts, kinds, and cognitive development.* Cambridge: MIT Press.

Lakoff, G. 1987. *Women, fire, and dangerous things: What categories reveal about the mind.* Chicago: University of Chicago Press.

Malt, B. C. 1991. Features and beliefs in the mental representation of categories. *Memory & Cognition* 29:289-315.

McCloskey, M. E.; Glucksberg, S. 1978. Natural categories: Well defined or fuzzy sets? *Memory & Cognition* 6:462-472.

Oden, G. C. 1977. Fuzziness in semantic memory: Choosing exemplars of subjective categories. *Memory & Cognition* 5:198-204.

Rey, G. 1983. Concepts and stereotypes. *Cognition* 15:237-262.

Rips, L. J. 1989. Similarity, typicality, and categorization. In S. Vosniadou & A. Ortony (Eds.), *Similarity and analogical reasoning*, pp. 19-59. New York: Cambridge University Press.

Rosch, E., & Mervis, C. B. 1975. Family resemblances: Studies in the internal structure of categories. *Cognitive Psychology* 7:573-605.

Progressions of Conceptual Models of Cardiovascular Physiology and their Relationship to Expertise

David R. Kaufman, Vimla L. Patel, Sheldon A. Magder
Centre for Medical Education
McGill University
1110 Pine Avenue West, Second Floor
Montreal, Quebec, Canada, H3A 1A3
Email: AX53@MUSICA.MCGILL.CA

Abstract

The application of scientific principles in diverse science domains is widely regarded as a hallmark of expertise. However, in medicine, the role of basic science knowledge is the subject of considerable controversy. In this paper, we present a study that examines students' and experts' understanding of complex biomedical concepts related to cardiovascular physiology. In the experiment, subjects were presented with questions and problems pertaining to *cardiac output*, venous return, and the mechanical properties of the cardiovascular system. The results indicated a progression of conceptual models as a function of expertise, which was evident in predictive accuracy, and the explanation and application of these concepts. The study also documented and characterized the etiology of significant misconceptions that impeded subjects' ability to reason about the cardiovascular and circulatory system. Certain conceptual errors were evident even in the responses of physicians. The scope of application of basic science principles is not as evident in the practice of medicine, as in the applied physical domains. Students and medical practitioners do not experience the same kinds of epistemic challenges to counter their naive intuitions.

Introduction

It is widely recognized that scientific principles play a fundamental role in the organization of conceptual knowledge and procedural knowledge for effective problem solving in diverse science domains and that the use of principled knowledge is a function of expertise (e.g., Chi, Feltovich, & Glaser, 1981). However, in medicine, the role of biomedical or basic science knowledge (e.g., physiology) is a source of considerable controversy. Clinical knowledge and basic science knowledge constitute two distinct bodies of knowledge that are connected only at various discrete points (Patel, Evans, & Groen, 1989). Clinical knowledge is primarily categorical and includes a classificatory scheme for disease entities and associated clinical findings. Basic science knowledge in medicine involves the organization of biomedical models at different levels of abstraction. Basic science knowledge is not easily integrated into clinical contexts and its use frequently does not improve the diagnostic performance of either expert physician or novice medical student (Patel, et al, 1989). It is not clear whether the development of expertise in medicine reflects progressions of increasingly elaborate and refined causal models built around basic science principles.

Empirical studies of many different domains in science indicate that students begin their study of science with strongly held misconceptions of phenomena (Eylon & Linn, 1988). These misconceptions are grounded in experience and are extremely resistant to change, even after instruction. The large majority of science concept learning research has addressed issues in the physical sciences. The relatively few studies in the biomedical sciences have yielded similar results, for example, documenting misconceptions in students' causal understanding of the structure and function of the heart (Feltovich, Spiro, & Coulson, 1989) and in the application of pulmonary concepts in clinical contexts (Patel, Kaufman, & Magder, 1991). These findings underscore a need to characterize students' and physicians' understanding of basic science concepts in different domains of medicine.

Conceptual understanding of physical or biological systems can be characterized in term of progressions of mental models (e.g., Forbus & Gentner, 1986). Mental models refer to the internal models of systems individuals develop from interacting with these systems (Norman, 1983). We can characterize subjects' models and elucidate aspects of subjects' representations that are flawed in terms of the structure and function of a system or in terms of the inferences used to evaluate the behavior.

The purpose of this study is to characterize students' and experts' understanding of concepts related to the mechanical properties of cardiovascular physiology. Specifically, the investigation focuses on the determinants of *cardiac output* (the blood ejected by the heart per unit time), and *venous return* (the blood returning to the heart per unit time). The study addresses individual differences in conceptual understanding and the progression of mental models of the cardiovascular system of subjects at different levels of expertise.

Cardiovascular Physiology

The regulation of cardiac output is a complex abstract topic, which unlike most subject domains in physiology, is lacking in explicit structure-function correspondences. *Cardiac output* is the total amount of blood pumped by the heart per unit time. It is a product of two factors, *heart rate* and *stroke volume.* Heart rate is the number of contractions or heart beats per minute. Stroke volume is the amount of blood ejected by the ventricle during contraction. Stroke volume is determined by three factors: 1) *preload,* which refers to the initial stretch of the cardiac muscle before contraction; 2) *afterload,* which is the tension in the cardiac fibres and is a force in which the heart must pump against; and 3) *contractility,* the functional state of the heart muscle that is defined by the rate and extent of shortening for a given afterload and preload. Preload and contractility are positively associated with stroke volume and therefore cardiac output. Afterload is negatively associated with stroke volume and cardiac output.

Venous return is the amount of blood returning to the heart per unit time. It is a determined primarily by *vascular compliance* and by *venous resistance.* Vascular compliance refers to the ability of a vessel to distend to accommodate more blood volume per unit pressure. Vascular resistance is the opposition to blood flow offered by the vessels and is primarily determined by the radius of the vessel and the viscosity of the blood. *Mean systemic pressure* is a measure that is determined by stressed volume and compliance and is independent of cardiac function. It is the driving pressure for venous return. Stressed volume is the volume that actually stretches the elastic walls of vessels and thus produces pressure in the vasculature.

The circulatory system is a closed system and therefore the blood pumped out by the heart must inevitably return to the heart. Over time, cardiac output has to equal venous return. Cardiac function mainly effects venous return by changing the outflow pressure for the peripheral vasculature. Venous pressures from output are independent from the heart because cardiac volume is small relative to peripheral blood volume.

A causal influence network was generated to represent the set of entities and relations involved in the problem set. This is illustrated in figure 1. This representation is similar to ones used in qualitative simulation of physical systems (e.g., Forbus & Gentner, 1986). The figure represents directional functional dependencies between the variables included in the study. The variables represent quantities, that when changed, can initiate a process that will effect other variables in predictable ways. A variable can exert a positive, negative or neutral influence on another variable. There are relationships that are not explicitly represented and can be deduced from the network. The network, however, does not explicitly reflect temporal relations or enabling conditions that can delimit the circumstances when an influence can be exerted or how to resolve ambiguities from multiple simultaneous influences.

Methods

The subjects consisted of 15 volunteers at several levels of expertise. The subjects included: one student who had completed a degree in biology; students from each of the four years of medical school; two physicians who were completing a cardiology resident training program; an expert physiologist, a cardiologist in private practice, and an academic cardiologist who divides his time between research and hospital practice.

The materials consisted of 49 questions and problems, including (1) 35 basic-level questions about specific factors pertaining to cardiac output, venous return, and pressure-volume and pressure-flow relationships, as well as questions intended to assess the degree to which subjects have integrated coherent models of the circulatory system; and (2) 12 situated problems in which these concepts are to be applied. These include brief clinical and applied physiology problems designed to assess the subjects' ability to recognize the conditions of applicability and use these concepts in context. Subjects were presented with a series of questions and problems on cue cards, one at a time. They were asked to read the question out loud and "talk-aloud", and answer the questions as completely as possible. The subjects were tested one at a time and each session was audio taped and transcribed for analysis. This paper focuses predominantly on subjects' responses to the basic-level questions. A more detailed discussion of the experiment and results can be found in Kaufman, Patel, and Magder (1992).

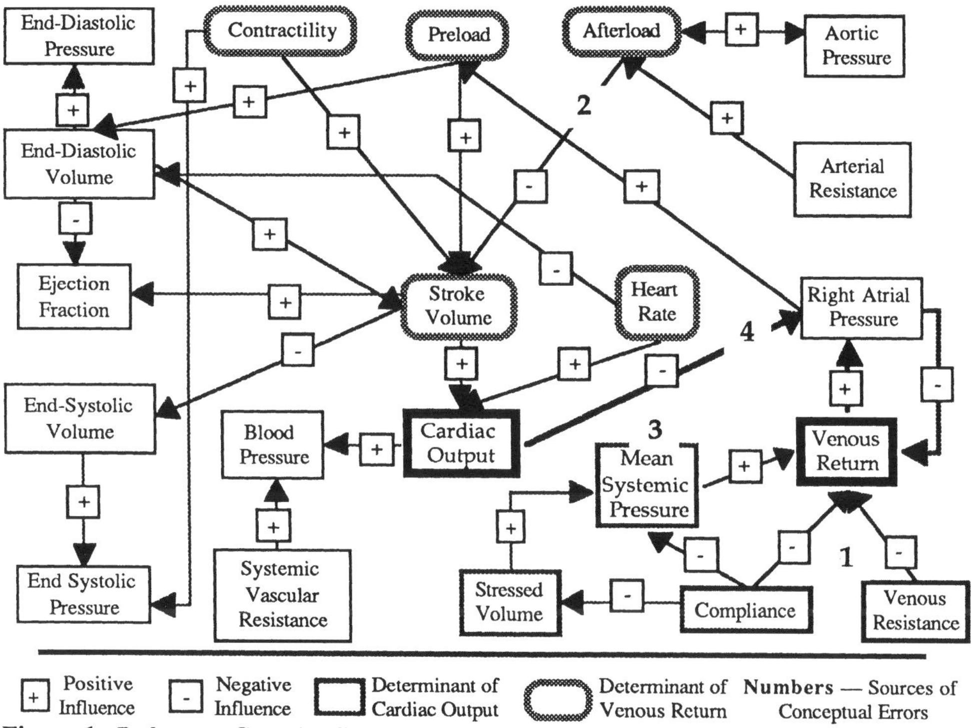

Figure 1: Reference Causal Influence Network of Relationships Between Concepts.

For each problem, a *reference model* response was prepared with the assistance of a consulting expert cardiologist and was used to assess the answers of each subject. Many of the questions required predictions. These took the form of how a particular change in state would affect the value of some measure. For example, a question asked "how does a large increase in *afterload* affect *cardiac output*". The possible responses are *no change*, *increases*, *decreases*. Subjects were required to make predictions in 29 of the questions, for a total of 45 predictions. The probability of randomly generating a correct prediction was approximately 30%. Incorrect responses were categorized according to the difference in direction of prediction. Patterns of subjects responding across questions provided us with information concerning gaps in knowledge and misconceptions. Semantic networks were also used to represent subject's causal explanations for individual questions (Groen & Patel, 1988).

The typology of relations used in semantic networks included in this paper are: **ACT**–engages in an action or process, **CAU**–causality, **COND**–directional conditionality, ***DIR***–direction, **EQUIV**–equivalent in some property, **IDENT**–identity, **LOC**–location, **RSLT**–result of an action, and arrows indicate directionality. A causal influence network, representing subjects' beliefs concerning relationships between variables, was also generated for each subject and then was contrasted with the reference network. This method allows us to characterize aspects of their mental models' of the mechanics of the cardiovascular system.

Results and Discussion

There was a general tendency for an increase in correct predictions with expertise. The medical students predicted a mean of 68.1% (sd=10) of the correct responses. The five more advanced subjects accurately predicted 78.2% (sd=12.5). There were considerable individual differences. The premedical student generated the fewest total correct predictions (38%) and the academic cardiologist (89%) and the physiologist (87%) correctly predicted the highest percentage of responses. Most students tended to have somewhat more difficulty with the venous return questions than cardiac output questions. The expert subjects responded with greater consistency across question types. Surprisingly, a fourth year student and a resident predicted only 51% and 58% of the correct responses, respectively.

Table 1: Percent of Correct Predictions by Subject

	P	1.1	1.2	1.3	2.1	2.2	3.1	3.2	4.1	4.2	R1	R2	Ph	CP	AC	$\bar{X}$	SD
CO	40	73	63	60	67	77	80	70	47	73	53	87	83	77	87	**69.1**	**14.2**
VR	36	64	50	57	64	71	71	64	57	100	64	71	93	71	93	**68.6**	**16.8**
Total	38	71	60	60	67	76	78	69	51	82	58	82	87	76	89	**69.5**	**14.2**

P=Premedical, 1.1=First year medical student subject 1, R=Resident, Ph=Physiologist, EP=Cardiologist Practitioner, AC=Academic Cardiologist. CO=Cardiac Output Predictions, VR=Venous Return Predictions.

Misconceptions

This section will examine two different misconceptions that produce fundamental errors in reasoning. Comprehension of the basic physical principles of hydrodynamics, specifically *pressure-volume* and *pressure-flow* relationships are essential for understanding the flow of blood through the circulatory system. The premedical student (P) exhibits a partial understanding of these principles. He understands that, all other things being equal, *an increase in volume* results in *an increase in pressure.* However, he reverses and extends the relationships to suggest that an increase in pressure implies an increase in volume and an increase in flow. This manifests itself in terms of a fundamental misconception about the nature of pressure-gradients. When a *forward flow* pressure is increased, flow does in fact increase. However, when the pressure is a *back flow* pressure, an increase in pressure results in a decrease in flow because the pressure gradient is narrowed. This is illustrated in a semantic network representation of the subjects' response to a question that asks "what happens when right atrial pressure rises to equal the mean systemic pressure" (Figure 2).

When right atrial pressure rises to equal the mean systemic pressure, the pressure gradient for venous return becomes zero and flow stops. The subject erroneously predicts an increase in flow that propagates throughout the system. The network illustrates that the subject possesses a mental model of the circulatory system and can envision the consequences of the effect of a change in state, however erroneously. The subject demonstrated in many questions that he had an adequate structural representation of the system, but repeatedly made the same kind of error related to pressure gradients.

There are invariably multiple sources of converging knowledge that comprise misconception. In this case, they include, the reversal of a directional relationship (increase in volume leads to an increase in pressure) and failure to differentiate between a driving pressure and a back pressure that opposes flow. This fundamental misconception was not characteristic of any of the other subjects.

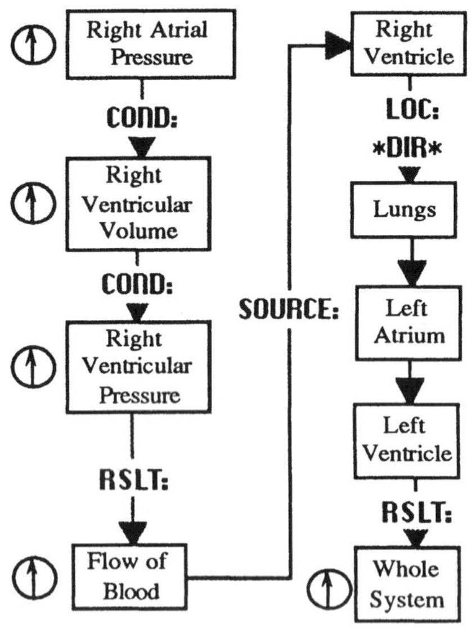

Figure 2. Semantic Network illustrating a Premedical Student's Misconception .

Many misconceptions are grounded in experience and reflect an acceptance of the primacy of experience and intuition over counter-intuitive formal teachings. However, formal learning can also result in the development of significant misconceptions. Resistance is a concept that is well rooted in experience, in the sense that resistance means the slowing down of some process (diSessa, 1983). The most important determinants of venous return are compliance and resistance. Compliance refers to the distensibility of a vessel and its ability to store blood. Venous resistance is primarily a function of the radius of the vessel. An increase in compliance increases the volume storage capacity of the vessel and therefore decreases venous return. Likewise an increase in resistance impedes the flow of blood and slows venous return. It makes sense that an increase in resistance would decrease the diameter of a tube and reduce its compliance. However, they are physiologically independent.

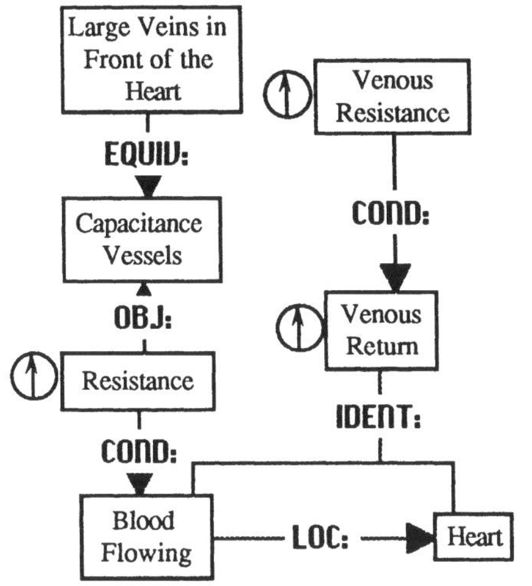

Figure 3. Semantic Network of a Second Year Student's Misconception

A pervasive misconception is the confounding of venous resistance and venous compliance. This was evident in several subjects' protocols, including several of the more senior subjects. This is illustrated in a second year student's (M2.1) response to a question about how "a marked increase in venous resistance affects venous return and cardiac output". The subject reasons that since the large veins are storage vessels, an increase in resistance would diminish storage capacity and increase blood flowing back to the heart. The most significant error here is that the large veins are storage vessels. It is commonly taught that veins are "storers of blood". In fact the large veins are downstream from the capacitance vessels, (the venules are compliance vessels) and are in effect resistance vessels that are critically important in determining blood flow.

Six out of fifteen of the subjects exhibited aspects of this misconception. It is predominantly the more advanced subjects that were most affected by this pattern of thinking. In fact both residents responded to some of the questions in a manner that would indicate that they could not completely disambiguate the effects of compliance from venous resistance. This is in evidence in the response of a resident (R2) to a question concerning the effects of compression of the veins leading to the heart on cardiac output. The subject predicts that this will greatly increase cardiac output. He applies an inappropriate analogy from a common clinical situation whereby the diaphragm is compressing the abdominal structures. This situation is typical of many medical conditions, such as asthma, where you get a sudden increase in respiratory rate and an increased blood flow. However, compressing the veins leading to the heart would dramatically increase resistance and severely reduce venous return.

It should be noted that this misconception concerning the effect of resistance was exclusively related to venous resistance. Each of the subjects, who exhibited this misconception, correctly pointed out that an increase in *arterial resistance* would increase the afterload and therefore reduce cardiac output. There are several bits of erroneous knowledge and beliefs that contribute to this misconception: 1) The belief that venous resistance and compliance are inextricably intertwined; 2) The notion that the large veins are storage vessels, when in fact they are resistance vessels; 3) A malprioritization of factors resulting in a misjudgment concerning the primary effect of resistance; 4) The use of inappropriate clinical analogies.

Mental Models

Causal influence networks were generated for each subject. The correspondences between relations were generated from the subjects' predictions and explanations. The premedical student correctly predicted only 38% of the correct responses. The effect of the pressure-volume misconception is evident in many of the relationships expressed. In general, any of the variables that suggest an increase in tension, resistance, or pressure (e.g., contractility, afterload) is believed to propagate an increase in volume or in flow.

In general, the causal influence networks of the other subjects indicated an increase in conceptual understanding with expertise. The subjects were able to qualitatively derive most behavioral states from changes in quantities to variables. With the exception of the physiologist and the academic cardiologist, each subject demonstrated a partial understanding of the mechanics of the cardiovascular system, exhibiting specific local deficits in their mental models. Figure 1 illustrates the sources of four conceptual errors. The first error relates to the confounding of venous resistance and compliance and was discussed in detail in the last section.

The second conceptual error was evident in the responses of a fourth year student who correctly predicted only 51% of the correct outcomes. It was evident from the subject's explanations that he understood most of the concepts and could apply them in more complex situations. The source of most of the subject's conceptual difficulties is related to the effects of afterload, which is one of the critical determinants of cardiac output. The subject infers that afterload has no effect on stroke volume. Afterload, in fact, decreases stroke volume. The fact that the subject's model is largely coherent, and that he correctly represents the relationship between stroke volume and all other variables, serves to propagate errors throughout the system when a question involves either *afterload, aortic pressure* or *arterial resistance* as causal agents.

The third and fourth sources of conceptual errors were associated with variables related to venous return. The third error reflects a lack of understanding of a primary determinant of venous return, *mean systemic pressure*. The fourth error is related to the functional role of the right atrium as a coupling mechanism relating cardiac output and venous return. Only the academic cardiologist and the physiologist were unaffected by these conceptual errors.

In general, subjects' responses indicated a "cardiocentric" bias, explaining situations in terms of cardiac output factors and excluding venous return factors from consideration. The three experts showed differences in their conceptual understanding. For example, the physiologist could respond with considerable facility to the basic physiology questions and had great difficulty explaining the situated problems. The academic physician was the one subject who could respond to either question type with great facility. The two cardiologists responded very differently to many of the questions. The practitioner correctly predicted only 76% and the academic physician predicted 89% correct. The practitioner tended to focus on a single possible cause, while the academic cardiologist was able to generate several possible alternatives and identify the delimiting factors that could produce different results.

Conclusions

In this study, we examined the conceptual understanding of subjects at several levels of expertise of a rather complex domain, circulatory and cardiovascular physiology. The scope of application of basic science principles is not as evident in the practice of medicine, as in the applied physical domains (e.g., engineering). Students and practitioners cannot experience the same kinds of epistemic challenges to counter their naive intuitions. Consequently, even striking anomalies resulting from fundamental misconceptions can frequently go undetected, and may carry over into clinical practice. Certain conceptual errors are consequences of formal learning. It is important to identify the possible sources of these errors. There is a need to prioritize selected cluster of concepts, and place more effort into the in-depth teaching of these concepts. Medical schools also need to present concepts in diverse contexts and make the relationships between the specific and general aspects, explicit. This entails striking a balance between presenting information in situated contexts, yet allowing the student to derive the appropriate abstractions and generalizations to further develop their conceptual models.

References

Chi, M. T. H.; Feltovich, P.; and Glaser R. 1981 Categorization and Representation of Physics Problems by Experts and Novices. *Cognitive Science*, 5:121-152.

diSessa, A. A. 1983. Phenomenology and the Evolution of Intuition. In D. Gentner and A. L. Stevens (eds.) *Mental Models.* (pp. 15-33), Hillsdale, NJ.: Lawrence Erlbaum Associates.

Eylon, B, and Linn, M. C. 1988. Research Perspectives in Science Education. *Review of Educational Research*, 58:251-301.

Feltovich, P. J., Spiro, R., and Coulson, R. L. 1989. The Nature of Conceptual Understanding in Biomedicine: The Deep Structure of Complex Ideas and The Development of Misconceptions. In V. L. Patel and D. A. Evans (eds.) *Cognitive Science in Medicine.* (pp. 113-172), Cambridge, MA: MIT Press,.

Forbus, K. D. and Gentner, D. 1986. In P. Michalski, J. G. Carbonell, and T. M. Mitchell eds. *Machine Learning: An Artificial Intelligence Approach*, (vol 2), (pp. 311-348), Palo Alto, Ca:: Tioga Press.

Norman, D. A. 1983. Some Observations on Mental Models. In D. Gentner and A. L. Stevens eds.) *Mental Models.* (pp. 7-14), Hillsdale, NJ:: Lawrence Erlbaum Associates.

Kaufman, D. R., Patel, V. L., and Magder, S. A.. 1992. Conceptual Understanding of Circulatory Physiology. Paper presented at the Annual Meeting of the American Educational Research Association, San Francisco, CA. April 20-24, 1992.

Patel, V. L., Kaufman, D. R. and Magder, S. 1991 Causal Reasoning About Complex Physiological Concepts in Cardiovascular Physiology by Medical Students. *International Journal of Science Education.* 13:171-185.

Patel, V. L., Evans, D. A., and Groen, G. J. 1989 Reconciling Basic Science and Clinical Reasoning. *Teaching and Learning in Medicine.* 1:116-121.

Augmenting Qualitative Simulation with Global Filtering

Hyun-Kyung Kim
Qualitative Reasoning Group
Beckman Institute, University of Illinois
405 North Mathews Avenue, Urbana, Illinois 61801
hkim@cs.uiuc.edu

Abstract

Capturing correct changes both locally and globally is crucial to predicting the behavior of physical systems. However, due to the nature of qualitative simulation techniques, they cannot avoid losing some information which is useful for finding precise global behavior. This paper describes how global constraints are represented and manipulated in current simulation systems, using a model of an internal combustion engine. The basic idea of our approach is to automatically generate additional information for maintaining global constraints during simulation so that simulation techniques can filter global behaviors with the sufficient information. This is done by automatically introducing variables and controlling their values to guide correct transitions between the behaviors. We express this idea within the framework of *Qualitative Process* (QP) theory. This technique has been implemented and integrated into an existing qualitative simulation program QPE.

Introduction

Understanding the behaviors of physical systems is an important part of commonsense physics. Given a system description, a qualitative simulator predicts the behavior by finding possible states and the transitions between them.

Conceptually, we view determining possible behavior as a process of filtering illegal states and illegal state transitions (Struss, 1988). Transition filtering should be done at two different levels: *local* and *global* filtering. Local filtering focuses on whether a transition between two states is legal or not, based on the relation between the two without considering the other states. In qualitative simulation, this is determined by the changes of state variables and continuity (this process is called limit analysis). For example, for state variable a and b, if a < b and a is increasing while b is not changing, then next state may be a = b. On the other hand, global filtering concerns finding correct behaviors, i.e, correct sequences of transitions.

Simulation process, whether it is numerical or qualitative, finds behaviors only by local filtering. In spite of the lack of global filtering, numerical simulation can find correct behaviors since it uses precise metric information. However, in the case of qualitative simulation, the localized nature of simulation combined with qualitative description is not sufficient to infer precise global behaviors. Thus, understanding how to automate global filtering and how to integrate it with local filtering is crucial to designing an intelligent reasoning system.

Recent work in global constraints has focused on applying the idea of qualitative theory of dynamic systems to qualitative simulations (Lee & Kuiper, 1988; Struss, 1988). After states and transitions are computed by a qualitative simulator, each state is converted to a point in a phase space[1] and each transition to the segment which connects its predecessor and successor state. Once the behavior of a system is expressed as a trajectory in a phase space, then some geometric constraints are applied to this trajectory for filtering behaviors. The trajectory should be checked in every phase space due to the localized nature of simulations. This approach was useful for some cases, such as the stability of a cycle, even though this approach can be applied only to a limited class of systems. Basically, this method is not adequate for reasoning task where threshold plays an important role (Lee & Kuiper, 1988). Furthermore, understanding this approach is not easy for the people who do not have mathematical background. Obviously, people do not seem to use this phase space for filtering behaviors.

This papers presents an approach to extend qualitative simulation to include global filtering. Instead of checking global behavior after getting locally correct behavior, the global constraints are also filtered during limit analysis. Given the constraints about the behavior after a particular state,

[1] A phase space for a system is a Cartesian product of state variables.

the transitions to the illegal behavior from the state are pruned. Since this filtering is done by limit analysis, our approach can capture the global constraints which are sensitive to the changes of variables. Our approach is illustrated using an implemented model of an internal combustion engine. We describe how subtly different expansion periods with and without combustion in the cylinder are captured.

Problem

Envisioning is a process of deriving all possible behavior of a system given the qualitative descriptions of the system. The behavior is represented by a set of qualitative states and the transitions between them. When these states and transitions are expressed as a graph, the graph is called the envisionment for the system.

Like the state of a finite automaton, each state in an envisionment has the summary of information about its past path (Hopcroft & Ullman, 1979). The next transition from a state is determined only by the constraints between the information in the state. This is the locality nature of simulation. The past behavior needs not be traced for this since the current state has all information about these. This technique can predict accurate behavior on the assumption that each state carries enough information so that local filtering is sufficient to get globally correct behavior. Unlike numerical simulation, which uses precise metric information, qualitative simulation sometimes cannot avoid losing some useful information during suppression of details. We cannot always expect the local filtering to guarantee the global filtering in qualitative simulation. In this section, we illustrate one of the examples using a model of an internal combustion engine.

In an internal combustion engine (Ferguson, 1976), a piston is connected to a crankshaft through a connecting rod (Figure 1). It goes through four phases (i.e., intake, compression, power, and exhaust) to complete one cycle. The rapid heat rise by combustion is translated into pressure which acts on the piston to force it down. Then, by the geometry, positive torque is transmitted to the crankshaft by the connecting rod. The flywheel, which is connected to the crankshaft, also gets positive torque during this power stroke. The heavy flywheel gives back to the crankshaft, during the three other strokes, the surplus energy it took during the power stroke. The interaction of the motions of each part, and pressure changes plays a key role in understanding this system.

In building our model, we have focused on the interaction, ignoring features irrelevant to showing our motivating example, such as intake and exhaust flows. For this, our model is built based

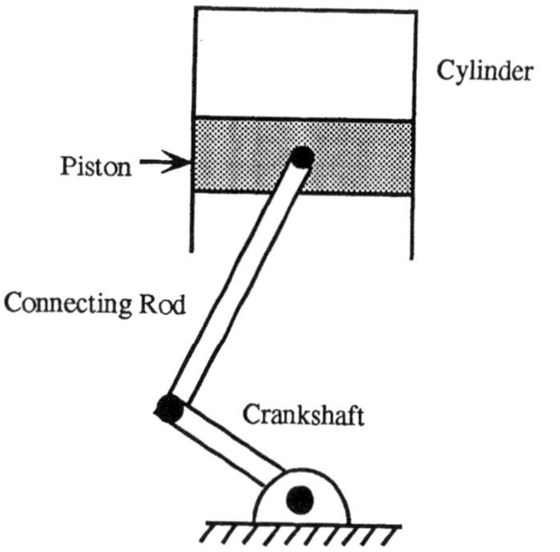

Figure 1: Piston cylinder geometry

on the following assumptions: (1) Working fluid is ideal gas, (2) Fixed mass of working fluid through cycle, and (3) Combustion is modeled as heat addition from external source. In addition to these, friction is considered in motion. With these assumptions, a piston-cylinder repeats the cycle of compression and expansion as the piston moves upward and downward (Figure 2a). When the crankshaft is at *Top dead center (TDC)* and *Bottom dead center (BDC)*, the piston reaches its highest and lowest position, respectively. Each part eventually will stop moving due to friction.

Suppose combustion happens at `TDC`. The expansion period after combustion (i.e., power stroke) is slightly different from the expansion without combustion: each part will not stop during the former while it might stop during the latter. Once the pressure is increased by combustion, say `(pressure ?cylinder)` >= `combustion-pres`, the pressure remains high enough to accelerate the crankshaft even though the pressure is decreasing during following expansion period. The geometry of the engine is designed so that once the pressure reaches `combustion-pres` at `TDC`, the pressure until `BDC` is greater than `nonstop-pres`. The `nonstop-pres` is a pressure sufficient to overcome the friction of each part.

However, it is impossible to distinguish these different behaviors with current qualitative simulators(Figure 2b) since it requires filtering paths, i.e., sequences of states. In other words, correct analysis of the behavior after combustion requires to prevent the transition from the combustion, `(pressure ?cylinder) >= combustion-pres`, to the path ended in stop state, `(pressure ?cylinder)` < `nonstop-pres` during the following

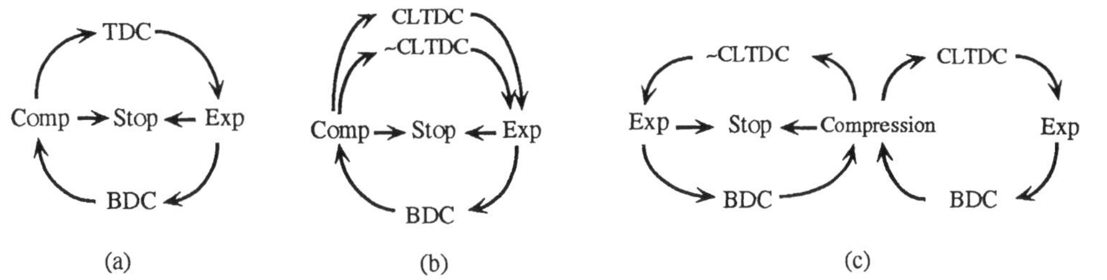

Figure 2: The behavior of a piston-cylinder. These show abstract behaviors rather than actual states in envisionments. While **CLTDC** represents "combustion lasts at TDC", i.e., **(pressure ?cylinder)** >= **combustion-pres**, ~**CLTDC** represents the state **(pressure ?cylinder)** < **combustion-pres**.

expansion, which cannot be done by local filtering. In Figure 2b, the different statuses at TDC are captured while the different expansions are not.

Figure 2c shows the desirable envisionment which captures the accurate behavior of a piston-cylinder. This has more states than Figure 2b and distinguishes the behaviors with and without combustion. Since we do not give any constraint during the compression period after combustion, the behavior during the period are the same as the behavior without combustion.

Modeling Global Constraints

As the example problem shows in the previous section, we need some means to filter next behavior from a particular state. To include this filtering in modeling process, underlying qualitative physics should provide some means to capture behaviors and to give constraints to the behavior. We use *Qualitative Process* (QP) theory (Forbus, 1984) for our basis since it provides the language for conditionalized descriptions: *view* and *process*. View allows us to capture interesting feature, i.e., behavior, by specifying conditions of related individuals. It can also express the constraints in the conditions. On the other hand, process provides the means to control transitions, as described later.

Event

In this section, we introduce a notion of *event* to model the constraints in the behavior after some point. It allows the effect of the state to be explicitly reflected in following paths. Intuitively, it is used to memorize some point, i.e., some special event, and its effects explicitly since they are lost during simulation. This is defined in three parts: *individuals*, *quantity conditions*, and *relations*. Whenever all individuals in *individuals* exist and all conditions in *quantity conditions* are satisfied, the event is active. *relations* contain the constraints about following behavior after the event happens. Figure 3 shows how this is represented

```
Defevent CLTDC ;;; Combustion lasts at TDC
Individuals
 ?pst  :type piston
 ?cyl  :type cylinder
       :conditions (part-of ?cyl ?pst)
 ?crs  :type crankshaft
       :conditions
       (connected ?pst ?crs)
 ?c-g  :type contained-gas
       :form (c-s ?sub GAS ?cyl)
QuantityConditions
 (pressure ?c-g) >=
    (combustion-pres ?pst ?cyl ?crs)
      when (position ?crs) = TDC
Relations
 CLTDC-EXP (pressure ?c-g) >
   (nonstop-pres ?pst ?cyl ?crs)
     when (not ((velocity ?pst) > 0))
```

Figure 3: An example of **defevent**.

in the example of combustion at TDC.

Quantity conditions consists of a set of statements and each statement is expressed by

(*conditions* **when** *configurations or states of objects*)

It says the event occurs if *conditions* are true in some *configurations* or *states of objects* (e.g., the pressure in a cylinder is increasing, decreasing, or not changing). Keyword "when" in quantity conditions is also used to guide the negation. Suppose the quantity conditions are written in the way done in QP theory. (We simply use **combustion-pres** instead of **(combustion-pres ?pst ?cyl ?crs)** as in the previous section.)

```
QuantityConditions
   (pressure ?c-g) >= combustion-pres
   (position ?crs) = TDC
```

This representation implies **CLTDC** does not happen if at least one of these conditions is false. However, what we want to express is whether or not **CLTDC** happens, depending on whether the pressure of a cylinder reaches **combustion-pres** at

TDC. We are not interested in whether or not the pressure reaches the point when the position is not at TDC. The quantity conditions in Figure 3 exactly capture this.

Relations explicitly represent the effects of an event when the event happens. Each statement in the field is expressed by

(*name constrained-behavior*
    when *configurations or states of objects*)

The first part of every statement represents the name of the relation. The remaining parts show the constrained behavior after the event. For instance, in CLTDC, the relation CLTDC-EXP represents the constrained behavior during expansion, i.e., (not ((velocity ?pst) > 0)), after combustion. (When the velocity of the piston is nonnegative, the gas in the cylinder is expanded.)

If the behavior after the event did not happen (i.e., "no event") is also constrained, this is described by a keyword ":neg". Otherwise, as in CLTDC-EXP, it is not specified as a default. Suppose when the pressure at TDC is less than combustion-pres at TDC, the pressure during following expansion cycle is not greater than nonstop-pres. (We simply use nonstop-pres instead of (nonstop-pres ?pst ?cyl ?crs) as in the previous section.) This constraint is then added to relation field as follows:

```
CLTDC-EXP
  (pressure ?c-g) > nonstop-pres
      when (not ((velocity ?pst) > 0))
  :neg (pressure ?c-g) <= nonstop-pres
```

Filtering Behaviors using Event

If we assume underlying simulator finds every correct set of states and local state transitions, what we need for dealing with event is simply to prevent the transitions to the states which lead to the impossible behavior (Figure 4). The remaining part of the envisionment should not be affected by this filtering. For example, in Figure 4, the path from $s1$ is a part of the behavior, even though the transition from s0 to s1 is illegal. Thus it must be protected from the filtering.

Limit analysis is a process to compute every state change by derivative relations of variables and continuity. Thus, if the continuity condition in unwanted transition can be automatically violated, current limit analysis can be used for filtering behavior. We do this by introducing an extra variable whose continuity breaks down in the transitions since limit analysis cannot prevent the transitions with existing variables. Intuitively, this variable is used as the tag which informs afterwards whether the event happened or not at some point. Since each state sometimes cannot include sufficient information to predict the next state, as shown in combustion example, the explicit information about the past is automatically included by specifying an event in that case.

For this, we need some means to express behaviors with an extra variable and to constrain the transitions between them by controlling its values. Views in QP theory can nicely capture the behaviors. Once the behaviors can be identified, filtering some behaviors after a particular state during limit analysis must satisfy the following restrictions: (1) If the behavior is constrained, the transitions from the state should be made only to the paths which describe the behavior. The transitions to others should be prevented (Figure 5a). (2) If the behavior is not constrained, the transitions to every possible path should be made (Figure 5b).

Implementation

In this section, we show how global constraints expressed by defevent are filtered in the framework of QP theory, using the combustion model. A defevent is translated into several views with an extra variable and filtering is done based on the variable.

Generating Views

At first, we need to distinguish whether or not an event happens. This is captured by generating two views for each case with an extra variable. The name of the event and the name prefixed with ~ are used for the names of the two views. For example, two views, i.e., CLTDC and ~CLTDC, are introduced for event CLTDC. The extra variable is set to positive when the event happens and set to nonpositive otherwise. The following shows parts of the views CLTDC and ~CLTDC, respectively. The individuals of both views are same as the individuals of the event CLTDC. An extra variable CLTDC-tag is introduced with different values.

```
CLTDC:
QuantityConditions
 (position ?crs) = TDC
 (pres ?c-g) >= combustion-pres
Relations
 (CLTDC-tag ?pst ?cyl ?c-g) > 0

~CLTDC:
QuantityConditions
 (position ?crs) = TDC
 (pres ?c-g) < combustion-pres
Relations
 (CLTDC-tag ?pst ?cyl ?c-g) <= 0
```

Two different views are also generated for each relation: one for the subsequent behavior after the event and the other for the behavior after no event. The individuals of both are also same as those of the event. The quantity conditions of both include

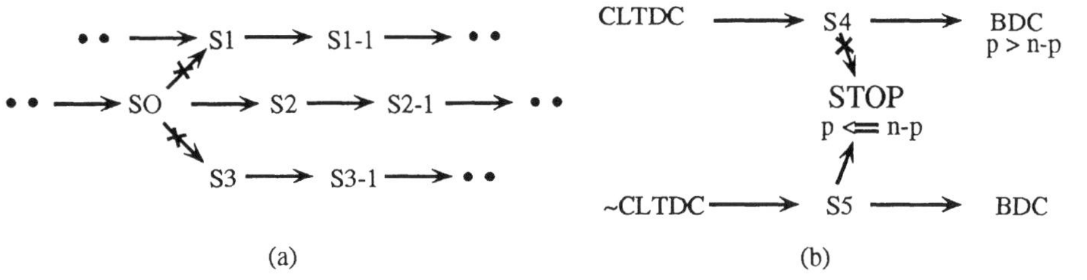

Figure 4: Filtering behaviors. (a) There are three possible paths from s0. Suppose only the path which starts with *s*2 should be selected from *s*0 due to the constraint by a structure description. It requires to prevent the transition to *s*1 and *s*3. (b) The transition from *s*4 to STOP should be prevented. (n-p represents nonstop-pres.)

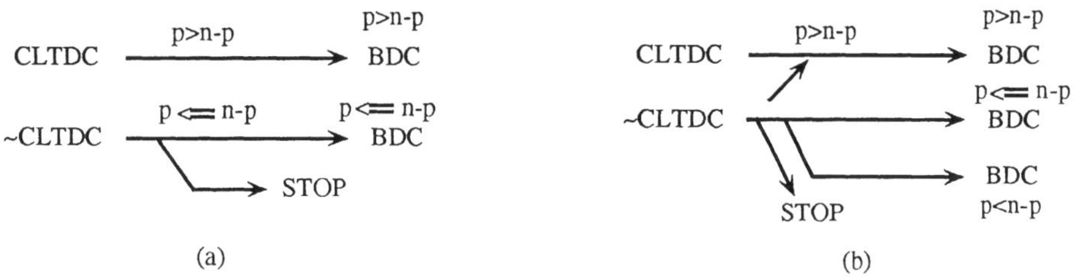

Figure 5: The behavior from TDC to BDC in a piston-cylinder. In (a), expansion after both CLTDC and ~CLTDC are constrained and different from each other. In (b), the expansion after ~CLTDC is not constrained.

the configurations or some states of objects specified after when. In addition to these, the former includes the assumption that the tag is positive while the latter includes the assumption that the tag is non-positive. The relations of the views constrain the behavior and are taken from the relations of the event. The name of the relation and the name prefixed with ~ are used for the names of the views. The followings show parts of view CLTDC-EXP and ~CLTDC-EXP, respectively. Since the behavior during expansion after ~CLTDC is not constrained, the relation field in ~CLTDC-EXP is left empty. If it is also constrained, the constrained behavior will be described in the field.

```
CLTDC-EXP:
QuantityConditions
 (not ((vel ?pst) > 0))
 (CLTDC-tag ?pst ?cyl ?c-g) > 0
Relations
 (pres ?c-g) > nonstop-pres

~CLTDC-EXP:
QuantityConditions
 (not ((vel ?pst) > 0))
 (CLTDC-tag ?pst ?cyl ?c-g) <= 0
Relations
 ()
```

Transition

Once these views are generated, the correct behavior for each case are selected by controlling the continuity of the extra variable during limit analysis.

The first requirement—when the behavior is constrained—is easily solved since the views generated by an event are manipulated to set the tag variable with this in mind. The continuity of the tag variable in the illegal path is violated. On the other hand, the continuity of the variable to the legal path is maintained (Figure 5a).

In case of the second requirement—when the behavior is not constrained, it includes the transition between two states which have different tag values. Note that this case happens only to the behavior after no event since an event is introduced to give constraints to the behavior after the event. If the behavior after no event are not constrained, it implies the transition to the constrained path after the event, since the path is one of the possibilities from no event (Figure 5b).

This transition requires to change the tag from off to on, i.e., from tag $<= 0$ to tag > 0. Thus, we need some means to connect the states which have different values of the tag. There are two approaches to handling this:

Make as a continuous change: In QP theory,

changes are caused only by processes. Thus, legal transition can be done by generating a dummy process which changes the **tag** from **off** to **on**. The process consists of the same fields as the fields of the view for the unconstrained behavior. Thus, it becomes active during the path for the behavior. Its influence field describes the change of the tag variable. Then limit analysis finds the correct transition by checking the continuity of every state variable, including the tag.

Make as a discontinuous change: Even though we assume qualitative physics deals with only continuous changes, reasoning about discontinuous change is important to explain many phenomena. Basically, a discrete change is made by specifying the add-list and delete-list. The transition is made from the state which implies delete-list to the state which implies add-list (Kim, 1992). Thus, we can directly make the transition, even though the continuity of the tag is violated, as far as the transition implies the legal changes of other variables.

In qualitative reasoning, avoiding unnecessary distinctions is important, since this reduces complexity and gives more abstract analysis. Thus it is important to make a distinction only when the effects of an event result in qualitatively different behaviors. Unless the effects clearly make differences, such as the compression period in CLTDC, we do not consider whether or not the event happens. In other words, the extra variable for tag is not used. CLTDC-tag, for instance, is not considered during the compression period. Since the transition between the state with tag and the state without tag does not violate the continuity due to the tag variable, no extra work is done to connect the influenced behavior to the subsequent uninfluenced behavior. For instance, the different paths during expansion are connected to one path for compression in CLTDC (Figure 2c) without extra manipulation. Continuity checking between the state variables is enough.

Though multiple events are defined, they can be handled without any difficulty since they are manipulated by independent additional variables. Thus, there is no interference between the events.

Using this approach, QPE (an implementation of QP theory) produces the envisionment for several examples including an internal combustion engine (Figure 2c), a spring block, and a neon bulb.

Discussion

In this paper, we have shown how simulation technique combined with qualitative information fails to capture global behaviors through qualitative simulation. A model of an internal combustion engine has been used to illustrate this. We extend current simulation methods by providing the means to constrain subsequent behavior after some event. We describe how the constraints are filtered by current limit analysis. Our approach is more intuitive and simpler than previous approaches. Our idea has been implemented and tested on several examples, including combustion in an internal combustion engine.

We hope to eventually analyze a real system. This will require tightly integrating dynamics and kinematics. Our approach can provide the means to link geometric constraints with possible motions. In an internal combustion engine, for instance, the interaction between the behavior during power stroke, and the geometry of a piston-cylinder could be captured by this technique. Our technique is one step towards that final goal, by providing more accurate prediction of behaviors.

Acknowledgements: I would like to thank Ken Forbus for his guidance. This research was supported by the Office of Naval Research, Contract No. N00014-85-K-0225.

References

Falkenhainer, B., and Forbus, K., "Setting up Large-Scale Qualitative Models," Proceedings of AAAI-88, August, 1988

Ferguson, C., *Internal Combustion Engines*, John Wiley & Sons, 1976

Forbus, K., "Qualitative Process Theory," *Artificial Intelligence*, **24**, 1984

Forbus, K., "Qualitative Physics: Past, Present, and Future" in Shrobe, H. (Eds.), *Exploring Artificial Intelligence*. Morgan Kaufmann Publishers, Inc. 1988

Hopcroft,J. and Ullman, J., *Introduction to Automata Theory, Languages, and Computation*, Addison-Wesley Publishing Co., 1979

Kim, H., "Qualitative Reasoning about Discontinuous Changes" submitted to PRICAI-92.

Kuiper, B., "Qualitative Simulation," *Artificial Intelligence*, **29**, 1986

Lee, W. and Kuiper, B., "Non-Intersection of Trajectories in Qualitative Phase Space: A Global Constraint for Qualitative Simulation," Proceedings of AAAI-88, August, 1988

Struss. P., "Global Filters for Qualitative Behaviors," Proceedings of AAAI-88, August, 1988

A Production System Model of Cognitive Impairments Following Frontal Lobe Damage

Daniel Y. Kimberg
Department of Psychology
Carnegie Mellon University
Pittsburgh, PA 15213
(412) 268-8117
kimberg@cmu.edu

Martha J. Farah
Department of Psychology
Carnegie Mellon University
Pittsburgh, PA 15213
(412) 268-2789
farah@psy.cmu.edu

Abstract

A computer model is presented which performs four different types of tasks sometimes impaired by frontal damage: the Wisconsin Card Sorting Test, the Stroop task, a motor sequencing task and a context memory task. Patterns of performance typical of frontal-damaged patients are shown to result in each task from the same type of damage to the model, namely the weakening of associations among elements in working memory. The simulation shows how a single underlying type of damage could result in impairments on a variety of seemingly distinct tasks. Furthermore, the hypothesized damage affects the processing components that carry out the task rather than a distinct central executive responsible for coordinating these components.

Introduction

Patients with damage to the frontal cortex have difficulty with a wide range of tasks, from the execution of simple manual sequences (Luria, 1965; Kimura, 1977) to sorting stimuli into abstract categories (Milner, 1963). One of the challenges of explaining frontal function is to account for the diversity of abilities that can be impaired by frontal damage. In the present paper we attempt to capture a commonality among the failures of frontal-damaged patients in a variety of tasks. We present simulations of four different tasks at which frontal-damaged patients, particularly those with dorsolateral frontal damage, have often been found to show deficits (Stuss & Benson, 1983). They are the Stroop Task, the Wisconsin Card Sorting Test, motor sequencing tasks, and memory for context.

This research was supported by an AFOSR graduate fellowship to the first author, ONR grant N00014-91-J1546, NIMH grant R01 MH48274, NINDS career development award K04 NS01405, and a grant from the McDonnell-Pew Program in Cognitive Neuroscience to the second author, and ONR grant N00014-90-J1489 to John R. Anderson.

Stroop Task. In the Stroop task, subjects are shown color names printed in different colored inks and asked either to read the word or to name the color in which the word is printed. Normal subjects show interference when asked to name the colors of stimuli in which the color and word conflict (e.g., the word "blue" in red ink). A similar pattern of interference exists when naming the word, although the differences are much smaller. In general, frontal-damaged patients have been found to be impaired at this task, showing disproportionate interference when naming colors (Perret, 1974; Dunbar & Bub, in preparation).

Wisconsin Card Sorting Test. In the Wisconsin Card Sorting Test (WCST), patients are asked to sort a number of cards that vary according to the shape of the objects represented, the color of those objects, and the number of objects. The piles into which the cards must be sorted vary according to these same attributes, so that there is exactly one pile for each possible color, shape, and number. Initially, one of these attributes is selected as the sorting category, and the subject will be given positive feedback only if they sort the card according to that attribute. Whenever the subject sorts ten consecutive cards correctly, the category changes.

Milner (1963) found that, as compared to patients with lesions elsewhere in the brain, frontal-damaged patients made an unusually high number of perseverative errors, continuing to sort according to the previous category after the category had shifted.

Motor Sequencing. Frontal-damaged patients have also been widely documented as having difficulty with sequencing tasks, especially the sequencing of motor actions. Kolb and Milner (1981) found that among patients with a variety of lesion sites, left and right frontal-damaged patients were the most impaired at imitating sequences of facial movements, and were also impaired at imitating arm movement sequences. Similarly, Kimura (1982) found that left frontal-damaged patients, in comparison to other patient groups, were the most impaired on all forms of oral movements, but especially sequences of oral movements, and that these same patients were also the most impaired on manual sequences. As well, Jason (1985) examined the performance of a variety of patient groups on a manual sequence task, and found left frontal-damaged patients to be the most impaired.

Memory for Context. While not uniformly amnesic, frontal-damaged patients often show deficits at particular memory tasks. Schacter (1987) has applied the term "spatiotemporal context" to the type of memory tasks at which frontal-damaged patients have been shown to show disproportional impairment. Parkin, Leng and Stanhope (1988) report the results of a detailed case study of a frontal-damaged patient. Among their findings, they report source amnesia, that is, impaired memory for the original context of learning, and impaired memory for temporal sequence. Janowsky, Shimamura and Squire (1989) investigated memory for recently learned facts and memory for the source of the facts in a group of frontal-damaged patients. Although the patients were normal in their ability to recall the facts, compared with age-matched control subjects, they frequently attributed the facts to incorrect sources.

These four types of task appear, on the surface at least, to be quite different from one another. Previous attempts to explain these and other frontal impairments have called into play a variety of mechanisms, including error utilization (Konow & Pribram, 1970), executive or supervisory processes (Shallice, 1982; Norman & Shallice, 1986), planning (Duncan, 1986), temporal integration of behavior (Milner, 1982), and inhibitory processes (Diamond, 1989). Here we propose a single underlying impairment that can account for the failures of frontal-damaged patients in all four of these tasks.

A production system model of the effects of frontal lobe damage

In our view, the effect of frontal lobe damage on behavior is to weaken the associations among working memory representations that include representations of goals, stimuli in the environment, and stored declarative knowledge. Thus, we hypothesize that the representation of the goals themselves is unaffected, consistent with the oft-cited observation that frontal-damaged patients can report the correct goal even while performing an inappropriate action (e.g. Konow & Pribram, 1970). We also hypothesize that the stimulus environment is perceived normally, and the full range of possible actions is available, also consistent with clinical observation. Finally, declarative knowledge is available, consistent with the results of memory research on frontal-damaged patients. We hypothesize a functional attenuation of association strengths among these different working memory representations. In effect, the differing degrees of mutual relevance among goals, stimuli and stored knowledge become less discriminable after frontal lobe damage.

We have chosen to implement our model using a simplified subset of the ACT-R framework (Anderson, 1983; 1989; in preparation). It should be noted that the architecture of this system was designed to account for normal cognition in a variety of tasks, drawing upon empirical findings on normal human learning and memory (Anderson, 1983) and upon a "rational analysis" of human cognition (Anderson, 1989). Thus, to the extent that the present model can account for the behavior of frontal-damaged patients, it does so without any ad hoc features designed specifcally for that purpose.

ACT-R is a production system, incorporating as its procedural knowledge a set of IF-THEN rules. These rules specify actions and the conditions under which they should be performed. In addition, it includes a working memory representation in which declarative knowledge is represented. The behavior of this system depends on which productions are selected for execution, according to two mechanisms: *matching* and *conflict resolution*. Matching refers to the process by which it is determined whether or not each production's conditions hold. This is accomplished by comparing the conditions of the rule to the contents of working memory. If more than one production matches the contents of working memory, as is often the case, then the process of conflict resolution is used to select a single production, as only one can be executed at one time. In the model described below, conflict resolution is accomplished by comparing activation levels, with the most active production being executed. There are four different sources that contribute to each production's activation:

Baseline activation is the invariant activation associated with a particular production. The higher the prior probability that the particular production will be applicable, the higher its baseline activation level. Productions with higher probabilities of being applicable, and therefore higher baseline activation, are more likely overall to fire.

Priming activation is additional activation that a particular production receives when it is executed. Priming activation falls off over the next several cycles of the simulation, and reflects the likelihood that a production that has just been executed will be applicable again in the very near future.

Noise activation is also present in the system.

Data activation is the activation added to a production from the working memory elements (WMEs) with which the production matches. The contribution of data activation is the sum of the activations of those WMEs. A WME's activation is calculated from its previous activation, from the previous activations of those WMEs with which it is connected, and from the strengths of those connections. However, in the present model, the previous activation of each WME is held constant. Thus, the activation of a particular

WME depends only on the strengths of its connections with other WMEs. Furthermore, since each production refers only to a subset of the entire working memory representation, only connections between pairs of WMEs both matched by that production will contribute to its activation.

Consider a production that matches both a goal WME (e.g., name the ink color of a stimulus) and some stimulus attribute WME (e.g., the ink color is red). If the goal is strongly associated with that stimulus attribute (as in this case, most likely) then the production will receive a large amount of activation from its data and will be more likely to fire. If the goal is only weakly associated with the stimulus (if the attribute WME in the above example were the lexical identity instead of the ink color) then the production will receive less activation from its data and will be less likely to fire. In this way, the model is more likely to execute a production for which mutually relevant goal and stimulus attributes are present.

Simulations of the four tasks.

In this section we present a simulation of the four tasks, and examine the effect of weakening association strengths among WMEs on the performance of the simulations. We first describe the undamaged models and then the results of damaging the normal system. Note that the same parameters (noise, priming activation, and decay rate) were used in all four simulations.

Stroop task. The Stroop task simulation consists of two productions, *name-color* and *name-word*, corresponding to the two potential responses to each stimulus. The attribute to be named for a given set of trials is set by strengthening the connection between the appropriate attribute WME (e.g. *colorname* when the goal is to name the color) and a WME which maintains information about the task (such as the current stimulus). Each production only matches against the appropriate attribute WME, so that the *name-color* production will only receive activation from the attribute WME *colorname.* The productions also receive activation from the connection between their attributes and the data they match. So the word naming production receives activation from the strong connections between the word attribute WME and word data, while the color naming task receives activation from the weaker but still strong connections between that attribute and the color data. Also, the baseline activation of the *name-word* production is stronger than that of *name-color*, consistent with the more frequent use of word naming in everyday life (see MacLeod, 1991).

At the presentation of a stimulus, both productions are placed in the conflict set, since all stimuli in this simulation have both color name and word name attributes. However, the correct task WME will receive more activation from the strengthened connection to the relevant attribute. And the discriminability will be greater when the task is word naming, since its production has a greater baseline activation.

Wisconsin Card Sorting Test. The WCST simulation consists of six productions: three for sorting and three for utilizing feedback. Each of the three sorting productions sorts by a particular attribute – color, shape, or number. Thus, whenever the production *sort-by-number* fires, the current card is sorted according to its number attribute.

The three feedback productions model how a subject should ideally utilize feedback, by constraining which categories will be sources of activation. After positive feedback, the current category is made the only category eligible to be a source of activation. After negative feedback, the incorrect category is made no longer a source of activation. And if this results in an empty set, the other two possible categories are then made eligible again. The model always implicitly knows which sorting categories are potentially correct, because only those categories are potential sources of activation. This is analogous to how, in the Stroop task, only the correct attribute (color or word name) is a strong source of activation for its corresponding production. However, the WCST uses the eligibility set to change these biases between trials.

Since the feedback mechanism just affects the eligible sources of activation, not whether or not particular WMEs are in working memory, this information does not directly constrain which productions can match. Instead, it biases which categories will be considered, by providing a source of activation for only those sorting productions whose categories are still eligible. For example, the production *sort-by-color* would receive activation from the connection between the *colorsort* WME and the list of possible categories. Since *colorsort* is most strongly associated with the *color* category, this production would be strongest when *color* was still an eligible category.

Initially, the set contains all three categories, so as to be unbiased. At present, the WCST productions do not wait for a number of correct trials before proceeding, but simply shift after a fixed interval.

When a card is presented, all three sorting productions are in the conflict set. The productions whose categories are still eligible receive activation from the connection between their categories and the corresponding task nodes. After a sort, one of the three feedback productions will match. There is one production to handle positive feedback, and two for negative feedback (when the eligibility set is larger than 1 or equal to 1). After feedback, the model attempts to sort the next card.

Motor sequencing task. The simulation of the motor sequencing task is the simplest. The model is presented with a repeating sequence of stimuli, each of which requires a distinct response. The stimuli can be thought of as devices, and the responses as different motor actions, similar to the task used by Kimura (1977). A different production for each potential response matches against both the action to be performed and the device to be acted upon in working memory. Since matches with congruent actions and devices will benefit from strong connections, they will receive more activation from data. There are five possible devices and five corresponding actions. The sequence of action in the simulation is also straightforward: the first stimulus is presented, and all five motor productions are in the conflict set. The correct one receives more data activation and is most likely to fire. Then the second stimulus is presented.

Memory. Memory for context can be modeled using a single production, *name-context*, to name the context of a presented item. Since the same production is used to name either the correct or incorrect context, in this simulation *name-context* competes only with itself. Different *instantiations* of the production, corresponding to the different contexts, are all in the conflict set simultaneously. However, since these instantiations refer to different subsets of working memory, they receive potentially different levels of activation, and can therefore be discriminated.

Context memory is simulated using a different WME for each context in which information is presented. Each context WME is in turn associated with a set of WMEs which represent the features of the environment. The unique subset of features with which a context WME is associated defines that context. Context memory can then be seen as the ability to name the appropriate subset of features through a label for those features. While this is a oversimplification of the notion of context, it preserves the critical requirement that the model must produce some element unique to the original situation in which a test item was presented, namely a label for that particular conjunction of features. Since the acquisition process is not simulated here, only the testing phase, this requires just a single production – one to name the context of the presented item. Each item is strongly associated with the features of the context in which it originally appeared, and weakly with the other features. The production *name-context* matches all possible available contexts, but each instantiation receives data activation from the connections between a particular context's features and the probe item. Thus, the production will be more likely to fire with the correct context, since that instantiation maximizes the amount of data activation it will receive.

To model the task, a stimulus is presented along with a set of five possible contexts. Thus five instantiations of *name-context* are placed in the conflict set, one for each context. Each instantiation matches against the stimulus and against the features of its context. The instantiation in which the stimulus is most closely associated with the features of the context is the one most likely to fire.

Recognition memory is modeled here as a special case of context memory, in which the subject must decide whether or not the stimulus was originally encoded in the experimental context. In the present simulation, this requires an additional production, *fail-to-recognize*, which produces the default behavior of failing to recognize an item. Because *fail-to-recognize* is a default, it will always match on the basis of its own baseline activation. That baseline activation therefore represents a threshold for recognizing an item. When an instantiation of *name-context* exceeds this threshold, the item is in effect recognized. Otherwise, *fail-to-recognize* will fire as a default. Note that *fail-to-recognize* would probably never fire in the case of context judgements, since it is extremely unlikely that multiple contexts, would all fail to reach threshold on the same trial. In this way, one might say that context memory judgements are between two or more real contexts, while recognition memory judgements are in effect between the correct context and a default context.

The simulation was run on all four tasks, using fixed sequences of stimuli. Each simulation was first run normally, then damaged. The simulation was damaged by weakening all of the connections between working memory elements by either 50% or 80%.

Results

Errors under each condition were tabulated as either non-perseverative or perseverative, except in the memory task, for which there was only one production, and thus no possibility for perseveration in the present model. If the incorrectly fired production had been fired on the immediately preceding trial (whether or not correctly then) it was counted a perseveration. Although it is possible that there would be perseveration due to data priming, this source of activation was not included in this simulation. Also note that the gradual decay of priming activation may cause perseverations over intervening steps, although these are here counted as non-perseverative errors.

The results of the tasks are presented in Table 1.

	Normal		50%		80%	
	NP	P	NP	P	NP	P
StroopColor	9	1	21	56	38	245
StroopWord	0	0	2	0	13	67
WCST	0	4	24	58	81	282
Motor	0	0	23	54	164	367
Context	2		81		379	
Recognition	0		1		22	

Table 1: Total non-perseverative (NP) and perseverative (P) errors on each task (1000 trials for each task for each damage level).

Without damage, the model performs all of the tasks at a high level. Although the noise makes errors possible, the high discriminability of the productions in the conflict set makes errors extremely unlikely. With 50% damage, more errors are made on all tasks, and there is a clear bias towards perseverative errors. Finally, with 80% damage, there is a greater proportion of errors, and a greater proportion of those errors are perseverative.

In the Stroop task, the damaged model shows interference from the unattended attribute. Moreover, as in frontal-damaged patients, there is more interference in the color naming condition than in the word reading condition. What is responsible for this pattern of results? In the undamaged model, the discriminability of the word-naming and color-naming productions is high, due to the strong connections between the appropriate attribute WME and the WME which maintains information about the task. When the connections are weakened, however, the activations of the two productions become more similar, and noise activation is therefore more likely to cause the wrong production to be selected. The fact that color naming is more vulnerable to intrusions by word naming than vice versa is explained by the higher baseline activation of word naming, which results from its more frequent use.

In the WCST, the damaged model simulates patient behavior in perseverating sorting categories even after negative feedback. As before, the reason for this can be understood by first considering the functioning of the normal system. Normally, feedback affects the selection of a sorting category by determining which categories remain in the eligible set. This set biases the model towards eligible categories through connections with the possible sorting WMEs. While the damaged model still uses feedback to constrain the eligible categories, the weakened connections reduce the magnitude of this bias. This reduces the discriminability of the different sorting productions to the level where noise activation can sometimes cause an inappropriate production to be the most active. The perseverative character of many of the errors results from priming activation causing recently selected productions to be especially active.

In the motor sequencing task, damage causes the model to associate incorrect actions with devices. Unlike the previous two tasks, in this task frontal-damaged patients show nonperseverative as well as perseverative errors. Both types of errors are also made by the damaged model. Why does the damaged model behave in this way? While normally the correct action can be discriminated on the basis of its greater association with the current device, this connection is weakened by damage. Noise activation will cause incorrect productions to be selected, out of sequence, and priming activation will bias these errors towards perseveration.

In the memory tasks, damage leads to impaired performance at context memory, but not at recognition memory. As modeled here, discrimination of the correct context depends on the connections between the item to be recognized and the features of that context. When these connections are weakened, the presence of noise makes it more likely that a similar but less appropriate context will receive greater activation. Recognition memory is the only case, among the four tasks simulated, which does not require the discrimination of a particular response among close competitors, and thus is not especially harmed by the damage manipulation. Interestingly, the model predicts that frontal-damaged patients will be much better at recognition than at context memory, but not necessarily normal. Again, there is not sufficient published data to address this definitively. In one study (Parkin, Leng, & Stanhope, 1988) both normal and patient groups appeared to be near ceiling, while in another study (Janowsky et al., 1989), there was a non-significant trend in this direction.

Discussion

We believe that there are two general aspects of the effects of frontal damage that have made the underlying nature of frontal lobe function so elusive. First, frontal damage can affect performance on a wide variety of tasks that do not seem, on the surface, to have anything in common. Second, when frontal damage impairs task performance, it does so without impairing patients' knowledge of the task goals, their perception of the relevant stimuli, their ability to execute the individual actions, or their memory for previously learned facts. These two factors have given rise to the idea of the frontal cortex as a single "central executive," which is called into play regardless of the cognitive domain. This central executive is required when the activities of multiple components of the cognitive architecture must be coordinated, but would not necessarily be required for performing simpler tasks. The model described here, however, provides a unified account of four deficits sometimes arising from frontal lobe damage, and does so without postulating damage to a central executive, but rather to the processing components that are used to perform the task.

This explanation also highlights what is common among the tasks failed by frontal-damaged patients. In all of the tasks modeled here, several sources of information compete to guide behavior. One of these sources in particular, connections among internal representations, is critical for differentiating among several close competitors. When these connections are weakened, other sources of activation (e.g., priming and noise) become more important in determining behavior. While the damaged model might be described accurately as unable to make use of errors (as in the WCST), impulsive (failing to inhibit inappropriate responses, as in the Stroop Task), perseverative (as in the motor task), or impaired in the use of spatiotemporal context (as in the memory tasks), a single functional deficit can account for all of these deficits.

The present model might seem to imply that, in conflict with common clinical observation, all frontal-damaged patients should fail all of the tasks described here. However, there could be distinct areas in frontal cortex sharing the same abstract function, namely the maintenance of working memory associations, but differing as to the types of working memory elements represented. One would therefore not expect the same patient to show impairment on exactly these four or any particular set of tasks. This type of organization, of a large cortical area operating according to common information processing mechanisms but subdivided into distinct and dissociable modules according to the content of the information represented, can also be seen in the visual cortex. Numerous areas in the extrastriate visual cortex share common functional mechanisms (e.g., retinotopy, the integration of information from earlier visual areas, and center-surround organization), but differ in the type of visual information represented in these maps (Cowey, 1982).

Are there any tasks at which the model predicts frontal-damaged patients would be unimpaired? In fact, while the underlying deficit that we have hypothesized is quite general, it is restricted to tasks that tax the conflict resolution process, that is, the ability to select among a group of potentially relevant actions. Thus, we would predict normal performance on tasks in which: the potentially relevant actions or responses are narrowed down by a "structured" or highly constrained task or stimulus environment (in the model, few productions matching the active working memory elements); the appropriate responses are highly routinized (in the model, high baseline activation in the appropriate production); or there are pronounced differences in the relevance of the available responses (in the model, pronounced differences in the association strengths among working memory elements matched by appropriate and inappropriate productions). This accords well with the common clinical observation that frontal-damaged patients may do well on relatively structured tasks or very familiar tasks, somewhat independent of difficulty, despite failing dramatically on the types of tasks modeled here.

Acknowledgements. The authors thank John Anderson, Prahlad Gupta, Paul Reber, Bob Stowe, and Don Stuss, for valuable comments.

References

Anderson, J. R. (1983). *The architecture of cognition.* Cambridge, MA: Harvard University Press.

Anderson, J. R. (1990). *The adaptive character of thought.* Hillsdale, NJ: Erlbaum.

Cowey, A. (in press). Cortical visual areas and the neurobiology of higher visual processes. In M. J. Farah and G. Ratcliffe (Eds.), *The Neural Bases of High-Level Vision: Collected Tutorial Essays.* Hillsdale, NJ: Erlbaum.

Diamond, A. (1989). Developmental progression in human infants and infant monkeys, and the neural bases of inhibitory control of reaching. In A. Diamond (Ed.), *The development and neural bases of higher cognitive functions.* New York: NY Academy of Science Press.

Duncan, J. (1986). Disorganisation of behaviour after frontal lobe damage. *Cognitive Neuropsychology, 3,* 271-290.

Janowsky, J. S., Shimamura, A. P., & Squire, L. R. (1989). Source memory impairment in patients with frontal lobe lesions. *Neuropsychologia, 27,* 1043-1056.

Jason, G. W. (1985). Manual sequences learning after focal cortical lesions. *Neuropsychologia, 23,* 483-496.

Kimura, D. (1977). Acquisition of a motor skill after left-hemisphere damage. *Brain, 100,* 527-542.

Kimura, D. (1982). Left-hemisphere control of oral and brachial movements and their relation to communication. *Philosophical Transactions of the Royal Society of London B, 298,* 135-149.

Kolb, B. & Milner, B. (1981). Performance of complex arm and facial movements after focal brain lesions. *Neuropsychologia, 19,* 491503.

Konow, A. & Pribram, K. H. (1970). Error recognition and utilisation produced by injury to the frontal cortex in man. *Neuropsychologia, 8,* 489-491.

L'hermitte, F. (1983). "Utilization behavior" and its relation to lesions of the frontal lobes. *Brain, 106,* 237-255.

Luria, A. R. (1965). Two kinds of motor perseveration in massive injury of the frontal lobes. *Brain, 88,* 1-10.

Luria, A. R. (1966). *Higher cortical functions in man.* London: Tavistock.

Luria, A. R. (1973). *The Working Brain.* New York: Basic Books.

MacLeod, C. M. (1991). Half a century of research on the stroop effect: an integrative review. *Psychological Bulletin, 109,* 163-203.

Milner, B. (1963). Effects of different brain lesions on card sorting. *Archives of Neurology, 9,* 90-100.

Milner, B. (1982). Some cognitive effects of frontal-lobe lesions in man. *Philosophical Transactions of the Royal Society of London B, 298,* 211-226.

Norman, D. A. & Shallice, T. (1986). Attention to action: willed and automatic control of behavior. In R. J. Davidson, G. E. Schwartz, & D. E. Shapiro (Eds.), *Consciousness and Self-Regulation, Vol. 4.* New York: Plenum Press.

Parkin, A. J., Leng, N. R. C., & Stanhope, N. (1988). Memory impairment following ruptured aneurysm of the antreior communicating artery. *Brain and Cognition, 7,* 231-243.

Perret, E. (1974). The left frontal lobe of man and the suppression of habitual responses in verbal categorical behavior. *Neuropsychologia, 12,* 323-330.

Shallice, T. (1982). Specific impairments of planning. *Philosophical Transactions of the Royal Society of London B, 298,* 199-209.

Stuss, D. T. & Benson, D. F. (1983). Frontal lobe lesions and behavior. In A. Kertesz (Ed.), *Localization in Neuropsychology.* New York: Academic Press.

Inference Evaluation in Deductive, Inductive and Analogical Reasoning*

Boicho Nikolov Kokinov

Institute of Mathematics
Bulgarian Academy of Sciences
Bl.8, Acad. G. Bonchev Street
Sofia 1113, BULGARIA
Tel.: (+359) 2-7133818
FAX: (+359) 2-752078
E-mail: banmat@bgearn.bitnet

Abstract

An experiment with a three factorial design is described which tests the impact of 1) the degree of the mapping isomorphism, 2) the differences in the types of reasoning (deduction, induction, and analogy), and 3) the kind of entities changed (objects, attributes, and relations) on the certainty of the inferences made. All the three factors have been found to have significant main effects and a significant interactions between the first factor and all the rest have also been found. Different particular results are discussed. For example, the certainty in the deductive inferences is not significantly different from the one in induction and analogy when there is no one-to-one mapping between the descriptions. Moreover, deduction, induction and analogy have similar behavior in relation to that factor. This is considered as a possible support of the existence of a uniform computational mechanism for evaluation of inferences in all the three kinds of reasoning, a mechanism which is primarily based on the degree of isomorphism.

*This research has been partially supported by the Bulgarian National Science Fund under Contract No I10/91 as well as by the Bulgarian Academy of Sciences (BAS) under project No 1001002.

Motivation

There is a long tradition in studying human reasoning. Unfortunately, this tradition separates different types of reasoning like deduction, induction (generalization), analogy and typically researchers try to develop separate models of different types of reasoning. The reason for this separation is the claimed differences in their properties. It seems to me, however, that the commonalities in their properties are underestimated and not well known. This paper tries to explore some of these commonalities.

A hypothesis has been made that there could exist a uniform computational mechanism which underlies all three kinds of reasoning (Kokinov, 1988). A specific model, called Associative Memory-Based Reasoning (AMBR), has been proposed. This model includes a number of interacting and parallel running subprocesses like retrieval, mapping, transfer, evaluation and learning. The experimental verification of this hypothesis follows two different and complementary directions: simulation experiments (Kokinov & Nikolov, 1989, Kokinov, 1992) and psychological experiments. As far as the psychological aspects are concerned Kokinov (1990) has demonstrated some common properties of the retrieval processes whereas this paper tries to explore the similarities in the evaluation process.

One widespread opinion is that deductive reasoning produces *absolutely certain* results, while generalization and analogy produce at best *only plausible hypotheses.* This claim is rooted in formal logic, passes through mathematics and is widely imposed on the whole education system.

The present work tries to throw light on the following issues about the evaluation process:

1. Are people indeed absolutely certain in deductive inferences in real-world situations? Do they consider generalization and analogy as unreliable sources of new facts? Are there any common phenomena in the evaluation processes of deduction, generalization and analogy?

2. Is human certainty predefined and based solely on the type of reasoning or does it depend on other factors, e.g. the quality of the established mapping? Is this restricted to analogy or is it a common property of all three kinds of reasoning as AMBR predicts?

3. Are objects, properties and relations processed differently by the evaluation process?

Experiment

Certainty is the reasoner's estimation of the plausibility of an inference. There are at least two sources of evaluation in AMBR (Kokinov, 1992): 1) the reasoner's estimation of the established mapping between a known situation (the base or premises) and the target one and 2) his/her domain knowledge. The goodness of mapping depends on the degree of isomorphism reached, including the number of correspondence pairs found, and on the activation level of the so-called correspondence nodes reflecting the context.

In the present experiment the relation between the degree of isomorphism and human certainty is explored. That is why the evaluation process should be isolated. This is achieved by directly presenting to the subjects what they should know (thus ignoring the retrieval process) together with its graph representation (thus helping them to establish the mapping). Moreover, the domain knowledge as well as the context effects, being sources of evaluation, should be blocked. This condition is fulfilled by presenting problems from a domain completely unfamiliar to the subjects.

In addition to the effect of the *degree of isomorphism* (where we may have the following cases: one-to-one mapping, overspecified new situation, or underspecified new situation) on the subjects' certainty about the inferences made, an attempt is made to examine the effects on the evaluation process of two other factors: the *type of reasoning* (deductive, inductive, or analogical one) and the *type of elements* changed from one situation to another (objects, attributes, or relations).

Thus the problems vary in terms of the relationships between the known and the new situations in three ways:

1. in terms of the degree of *isomorphism* between the descriptions: there is a one-to-one correspondence between the situations (0), in the description of the new situation there is something additional which is unspecified in the known situation and thus remains unpaired (+), in the description of the new situation there is something unspecified and thus something in the known situation remains unpaired (-).

2. in terms of the *inference type,* i.e. the relation type between the two situations: the new situation is a particular case of the known one (deduction - D), the known situation is a particular case of the new one (generalization - G), and there is no pre-specified relation between the situations (analogy - A).

3. in terms of the type of *elements* changed in the second description with respect to the first one: objects (o), attributes (a) and relations (r);

Method

Subjects. Subjects have been 297 volunteers — university students and researchers from different academic disciplines: physics, chemistry, biology, geography, medicine, law, philology. (Mathematicians have been deliberately excluded because a preliminary pilot study has demonstrated that they approach the problems formally and judge always deduction as an absolutely certain inference technique and generalization and analogy as absolutely uncertain ones.)

Procedure. Subjects have been tested individually or in small groups of 5–10 people. They have received a list of 9 target problems preceded by the following written instruction:

> "The present experiment is not a test of your capabilities. It is used to explore some hypotheses about the mechanisms of human thinking. Try to answer the questions although you are unfamiliar with the particular problem domain.
>
> Imagine that you are a researcher in the field of *encelorobes* and you know that all of them consist of *mangovine* and *girofine*, mangovine being *ritalic* and girofine being *tanalic*, and also that mangovine *aceifies* girofine. You also know that encelorobes are *corablic*.
>
> mangovine — ritalic
> | aceifies
> girofine — tanalic
> corablic
>
> During your scientific research you encounter the situations listed below."

For each of them the subjects' certainty in the inferences made has been measured on a 7 point scale. Subjects had as much time as they needed.

Material. Each test problem consists of a situation description and a question. The question is about the subject's certainty of the possible transfer of knowledge (inference) from the known situation described in the instruction to the new one. Situations have been constructed within an artificial nonexisting problem domain in order to avoid subjects' preliminary knowledge about the truth of the inferences and to measure the certainty of the inferences as a result of "pure" reasoning. The concepts used in the situation descriptions are fake but sound scientific (at least in Bulgarian) — something like biochemistry — so that the subjects are not aware of the artificiality of the situations, i.e. they think that correct answers to the questions exist.

Because there is a lot of unfamiliar terminology in the test problems, in addition to the textual description of the situations a simple graph representation is presented.

Several examples of test problems (descriptions of target situations and questions) are presented below in their English translation.

1. You discover a new representative of the class of the *encelorobes* which consists of *didorine* and *caronine*. In it, didorine is *ritalic*, caronine is *tanalic* and the didorine *aceifies* the caronine. It is known that didorine and caronine are kinds of *mangovine* and *girofine* respectively. Are the encelorobes of this type also *corablic?* (The combination of factor levels in this example is [objects, deduction, 0]).

   didorine — ritalic
   | aceifies
   caronine — tanalic

2. Exploring the *stericorobes*, you find out that they consist of mangovine and girofine, the mangovine being ritalic, the girofine being tanalic, the mangovine *privilates* the girofine and the girofine

girovates the mangovine. Are the stericorobes *corablic* as well? (The combination of factor levels in this example is [relations, analogy, +]).

mangovine — ritalic
privilates || girovates
girofine — tanalic

3. It is known that the encelorobes are representatives of the class of the *robes.* Robes consist of mangovine and girofine and the mangovine aceifies the girofine. Are the robes corablic as well? (The combination of factor levels in this example is [attributes, generalization, -]).

mangovine
| aceifies
girofine

Design. The experimental design is a 3x3x3 factorial one, with factors *Elements* (objects, relations, attributes), *Inference Type* (deduction, generalization, analogy) and *Isomorphism* (0,+,-). The dependent variable is the subjects' certainty in the possible inference measured on a 7 point scale.

Six different samples of 9 problems out of the full set of 27 problems have been prepared to randomize the possible order effects. Each sample included problems corresponding to each of the levels of *Inference Type* and *Isomorphism* and to some of the levels of *Elements.* Subjects were randomly assigned to one of these 6 versions. As a result each problem has been presented to about 100 people.

Results

As a result of the experiment 2676 observations have been obtained. A 3 x 3 x 3 ANOVA has been conducted in order to explore the results. In short, the main effects of all three factors are significant.

Mean plausibility ratings vary as a function of the degree of *isomorphism*, $F(2, 2649) = 36.50, p < .001$. Mean plausibility is 5.38 in one-to-one mapping situations, 4.72 when the target situations is overspecified, and 4.87 when the target situation is underspecified.

Mean plausibility ratings vary also as a function of the *inference type*, $F(2, 2649) = 35.18, p < .001$. Mean plausibility is 5.31 in deduction, 5.03 in generalization, and 4.62 in analogy.

Mean plausibility ratings vary also as a function of the *element* changed, $F(2, 2649) = 8.69, p < .001$. Mean plausibility is 4.82 when an object is changed, 5.16 when an attribute changed, and 4.98 when a relation changed.

There are also significant interactions between *Isomorphism* and *Element,* $F(4, 2649) = 27.66$, $p < .001$ (Figure 1a), between *Isomorphism* and *Inference Type,* $F(4, 2649) = 2.75, p < .05$ (Figure 1b), as well as between all the three factors, $F(8, 2649) = 4.19$, $p < .001$.

Discussion

A number of conclusions about main effects, interactions and simple effects can be drawn from the analysis.

1) In contrast with AMBR's predictions *Inference Type* does have a significant effect on the certainty evaluation. It remains, however, to be explored what are the exact causes of this phenomenon: whether this is due to differences in the built- in mechanisms performing the separate types of reasoning, to everyday experience in commonsense reasoning, or to reasoning patterns implanted by the education system.[1]

Let us consider the interaction between *Inference Type* and *Isomorphism* (Figure 1b). Although deduction differs significantly from analogy and generalization and dominates them in the case of a one-to-one mapping (0),

[1]or possibly to the material design (e.g. the base and target have more explicitly paired elements in the cases of deduction and generalization than in the case of analogy, so again the level of isomorphism can be the cause of the decreasing subjects' certainty in analogical inference).

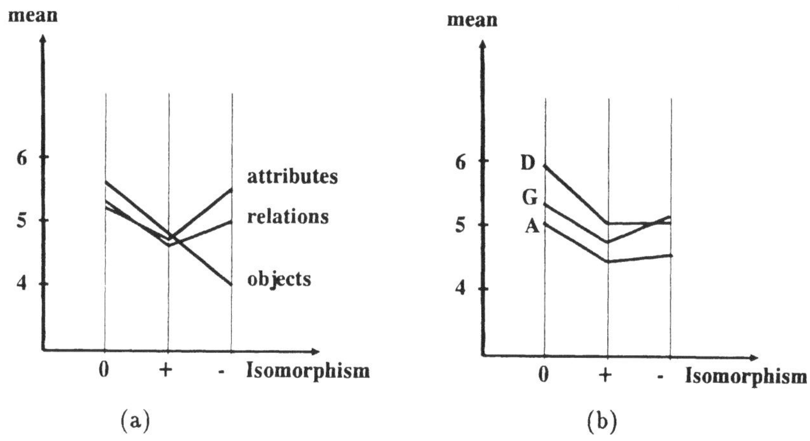

Figure 1: Interactions between: (a) Isomorphism and Elements; (b) Isomorphism and Inference Type

in the other two cases (which are more typical real-world situations) and especially in the case of underspecification of the new situation (–) the results are generally indistinguishable from one another and even sometimes analogy or generalization dominate deduction.

In order to explore the causes of this interaction, let us consider the simple effect of decreasing the certainty level of deductive inference in the non-isomorphic situations. Note that from the point of view of formal logic, if it is known that an object belongs to a class, no more information is needed in order to draw a deductive conclusion with absolutely certainty. As the experiment demonstrates, however, this is simply not true with human reasoning. Moreover, it is clear that each representative of a class has a number of elements (attributes, relations, components) that are specific for it, so there are no logical reasons for questioning the deductive inference solely on the ground that the situation is overspecified (+). According to common sense, the certainty should be at least the same as in the isomorphic case.[2] However, as Figure 1b shows, there is a significant decrease in the subjects' certainty in this situation.

[2]Compare with the "mirror" case, i.e. generalization in the underspecified situation (–), where the subjects' certainty increases.

The high certainty of deduction may be an effect of education. The examples that are most instructive in this respect are given by mathematics, and there only pure isomorphic situations are presented. When, however, subjects are involved in a complex reasoning process in realistic non-isomorphic situations, the reasoning pattern imposed by education may not be triggered and only the built-in mechanisms of evaluation based on isomorphism are activated.

So one possible explanation of the interaction is that the significant differences between deduction, generalization, and analogy in the "0" case are imposed by the learned reasoning patterns, which do not function in the non-isomorphic cases.

Finally, deduction, generalization and analogy have similar behavior in relation to *Isomorphism* (Figure 1b). Moreover, this is true for each of the *Elements* levels: in the case of an object change, the worst case for all inference types is an unspecified object in the new situation, whereas in the case of an attribute or relation change the worst case is the presence of additional attributes or relations. In general, such similarity in behavior is considered to indicate the existence of a uniform computational mechanism for evaluation of inferences in all the three kinds of reasoning.

2) The main effect of Element is due mainly to the "–" and "analogy" cases. The interaction between *Element* and *Isomorphism* is also due mainly to the "–" case. The low certainty measured in the "object" case is probably an artifact due to the material design, because in the case of unspecified objects all their attributes as well as the relations in which they participate can not be specified either, and so in this case the level of isomorphism is additionally decreased.

It is interesting to compare the particular results in the analogical case obtained in this experiment with those of the experiments of Gentner demonstrating the superiority of relations over attributes in the mapping and evaluation process in analogy (Gentner, 1983, 1989, Goldstone et al., 1991). In the case of a one-to-one mapping as well as in the case of overspecification no such priority is demonstrated in the present experiment. In the case of underspecification, however, relations do have a significant priority.

3) Finally, *Isomorphism* has the greatest main effect and in correspondence with AMBR's predictions, the one-to-one mapping gives more certainty than many-to-one or one-to-many mappings. Moreover, *Isomorphism* influences both other factors which should be related to the functioning of a mechanism at the very basic level. This supports the AMBR's hypothesis that the primary mechanism for evaluating inferences is based on the goodness of mapping evaluation.

The results from this complex experiment are very rich and controversial, so I do not consider them as conclusive but rather as a starting point for further experimentation.

Acknowledgements

I am grateful to all participants in the regular seminar of the Bulgarian Society for Cognitive Science for the relevant discussions. I am particularly grateful to Encho Gerganov for his help in data analysis.

References

Gentner, D. 1983. Structure-Mapping: A Theoretical Framework for Analogy, *Cognitive Science* 7:155-170.

Gentner, D. 1989. The Mechanisms of Analogical Learning. In: Vosniadou, S. and Ortony, A. (eds.) *Similarity and Analogical Reasoning.* New York, NY: Cambridge Univ. Press.

Goldstone, R.; Medin, D. and Gentner, D. 1991. Relational Similarity and the Nonindependence of Features in Similarity Judgements. *Cognitive Psychology* 23:222-262.

Kokinov, B. 1988. Associative Memory-Based Reasoning: How to Represent and Retrieve Cases. In: O'Shea, T. and Sgurev, V. (eds.) *Artificial Intelligence III,* Amsterdam: Elsevier.

Kokinov, B. and Nikolov, V. 1989. Associative Memory-based Reasoning: A Computer Simulation. In: Plander, I. (ed.) *Artificial Intelligence and Information-Control Systems of Robots-89,* Amsterdam: North-Holland.

Kokinov, B. 1990. Associative Memory-Based Reasoning: Some Experimental Results. In: *Proceedings of the 12th Annual Conference of the Cognitive Science Society,* Hillsdale, NJ: Lawrence Erlbaum Associates.

Kokinov, B. 1992. A Hybrid Model of Reasoning by Analogy. In: Holyoak, K. and Barnden, J. (eds.) *Analogical Connections, Advances in Connectionist and Neural Computation Theory,* vol. 2, Norwood, NJ: Ablex Publ. Corp. Forthcomming.

Identifying Language from Speech: An Example of High-Level, Statistically-Based Feature Extraction

Stan C. Kwasny
Center for Intelligent Computer Systems
Department of Computer Science
Washington University
St. Louis, Missouri 63130
(314) 935-6160
sck@cs.wustl.edu

Barry L. Kalman
Center for Intelligent Computer Systems
Department of Computer Science
Washington University
St. Louis, Missouri 63130
(314) 935-6160
barry@cs.wustl.edu

Weilan Wu
Center for Intelligent Computer Systems
Department of Computer Science
Washington University
St. Louis, Missouri 63130
(314) 935-6160
wwl@cs.wustl.edu

A. Maynard Engebretson
Central Institute for the Deaf
Department of Computer Science
Washington University
St. Louis, Missouri 63130
(314) 652-3200
ame@cs.wustl.edu

Abstract

We are studying the extraction of high-level features of raw speech that are statistically-based. Given carefully chosen features, we conjecture that extraction can be performed reliably and in real time. As an example of this process, we demonstrate how speech samples can be classified reliably into categories according to what language was spoken.

The success of our method depends critically on the distributional patterns of speech over time. We observe that spoken communication among humans utilizes a myriad of devices to convey messages, including frequency, pitch, sequencing, etc., as well as prosodic and durational properties of the signal. The complexity of interactions among these are difficult to capture in any simplistic model which has necessitated the use of models capable of addressing this complexity, such as hidden Markov models and neural networks. We have chosen to use neural networks for this study.

A neural network is trained from speech samples collected from fluent, bilingual speakers in an anechoic chamber. These samples are classified according to what language is being spoken and randomly grouped into training and testing sets. Training is conducted over a fixed, short interval (segment) of speech, while testing involves applying the network multiple times to segments within a larger, variable-size window. Plurality vote determines the classification. Empirically, the proper size of the window can be chosen to yield virtually 100% classification accuracy for English and French in the tests we have performed.

Introduction

In an international setting, one might overhear parts of conversations in a variety of languages. Given the proper experience, identifying familiar languages can be done easily and accurately. What is it that tells us the identity of a language? How do we know, for example, when the same speaker speaks English or French? Under the right circumstances, people seem to be able to tell immediately, often not from exactly what is being said, but from broad characteristics of the speech.

In fact, it is not necessary that one be competent in French to recognize that people are speaking French. When Arte Johnson speaks English with an accent, then suddenly starts talking in pseudo-German, the audience identifies the language as German, even though he may not use actual German words or phrases. It simply "sounds like German."

Spoken language is perceived on many levels. A variety of judgements about features of speech are constantly being made by a listener. Listeners unconsciously notice many things about speech -- tone of voice, style, pace, gender of the speaker, accent, degree of excitement, who is speaking, etc. These features can be very high level although often not consciously contemplated under ordinary circumstances by the listener. We further observe that spoken communication among humans utilizes a myriad of devices to convey messages, including frequency, pitch, and sequencing, as well as other prosodic and durational properties measurable in the signal. The complexity of interactions among these in the speech signal are impossible to capture in any simplistic model necessitating the use of models such as hidden markov models and neural networks.

While speech understanding research has focused primarily on extracting "meaning" from speech, it is clear that there are many other ways humans process speech. Most of the high-level features mentioned above cannot be tied to any particular, conventional set of phonetic or acoustic features of the speech. Instead, they appear to be related to distributional patterns or statistical aggregates of the speech waveform.

We are investigating the extraction of high-level, statistically-based features from speech. Specifically, in this paper, the task is to determine the language being spoken from samples of raw speech. Bilingual speakers fluent in two languages are recorded and speech samples are separated into training and testing groups. Training attempts to create a network that can reliably determine which language is represented.

We assume that the classification task can be conducted in real time by the model. We further assume that it is only necessary for the model to see very raw speech waveforms, represented as sampled frequency bands over time. We specifically rule out explicit phonetic identification as well as a variety of other intermediate-level structuring that is typically found in speech understanding and recognition systems.

Related Work

There have been several studies that demonstrate the existence of statistically significant differences among spoken languages at the acoustic level (Hanley, et al. (1966); Atkinson (1968)) and also at the level of phonetic features (Denes (1963); Kucera & Monroe (1968)). Abe et al. (1990; 1991) have considered some of the differences in automatically converting a speaker's voice from one language into another. Since these differences are measurable at the low end of the speech chain, then surely it must be possible to exploit those differences to build a model that emulates the human ability to correctly discriminate among languages.

House (1977) proposes a method of language identification which utilizes a language structure component in conjunction with a statistical component. His approach was apparently hindered, at that time, by the lack of sufficient computing power to perform the necessary statistical procedures.

We share some of House's beliefs about the value of statistical procedures in extracting certain high-level features. Being statistically based, the processing will naturally be resistant to noise and tolerant to some variation. We further assume that such processing can be demonstrated in real time. This assumption rules out the existence of a sophisticated language structure component and demands that intermediate levels of processing normally associated with speech understanding be finessed.

Recently, Muthusamy et al. (1990) has followed some of the suggestions made by House in examining this problem for four languages: American English, Japanese, Mandarin Chinese, and Tamil. They recorded six male and six female speakers each speaking 20 utterances in one of the languages. Four waveform and four spectral parameters were extracted and used to segment and label the speech with one of 7 broad phonetic categories with 82.3% accuracy. The segmented speech was then used in a second network designed to classify by language. This proved to be 79.3% accurate in classifying the speech into one of four languages.

Our approach differs from theirs in several respects. We first assume all processing can be conducted in real time. We also wish to finesse the need for intermediate structures as much as possible. We feel there is always some loss of information in mapping the waveform into discrete structures and this loss could have an effect on the success of the classification of the high-level feature.

Data Collection

For our experiments, we collected speech samples from three bilingual speakers: two males and one female. All speakers fluently spoke English and one other language: $male_1$ spoke native French and non-native English; $male_2$ spoke native Japanese and non-native English; and $female_1$ spoke non-native French and native British English. Recordings were made of 12.5 second, randomly chosen samples of each speaker reading the phonetically balanced "rainbow passage" in English and excerpts of spoken passages read from newspaper stories in the other languages. Two different samples were recorded for each language for each speaker. Yielding a total of 12.5 speech samples.

All recordings were made in an anechoic chamber resulting in 16-bit samples at 24kHz. Five Band-Pass filters were used to separate the signal into bands which were low-pass filtered and decimated by a factor of 200. This process is illustrated in Figure 1.

Within the 12 second samples, we selected samples of smaller duration by specifying a start point and a duration and clipping it from the larger sample. This permits numerous overlapping samples to be extracted from each collected sample depending on the size of the sample to be extracted.

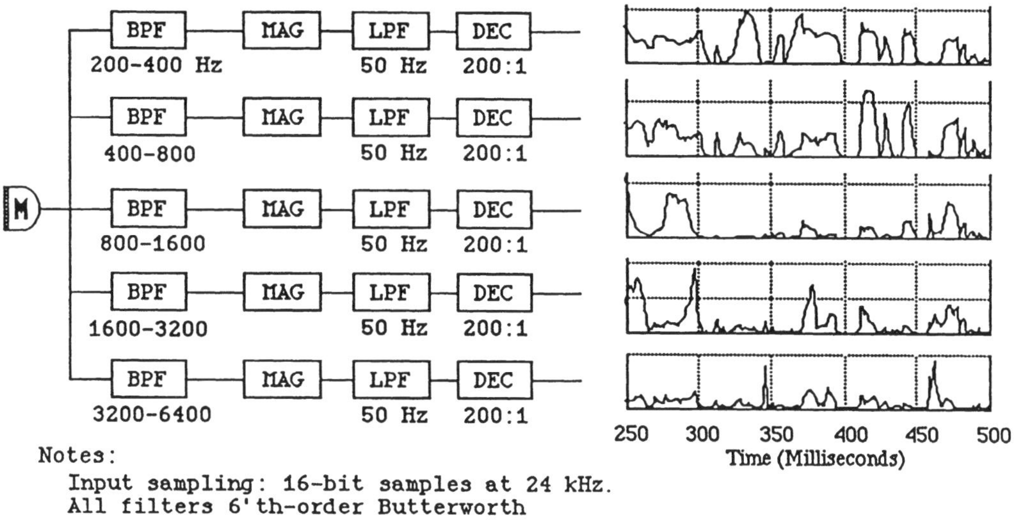

Figure 1: Frontend Processing

System Design

Designing a system for this task requires that proper training is performed and that testing favors correct decisions. We are using a neural network that maps input units representing 750ms. duration of speech to output units representing the range of languages being identified.

The choice of 750ms is based on a compromise between a network that is too small to properly detect the distinctions necessary to identify the language and one that would require enormous computing resources to train. For durations less than 750ms, training patterns contain numerous input similarities which require separation as output dissimilarities. This can be determined by performing boundary pair testing as described in Kalman & Kwasny (1992). Such a situation is unacceptable since it indicates that good training will be extremely difficult to achieve. Durations above 750ms require an enormously large input layer and many network weights to manipulate. While a faster machine or more time could overcome such problems, we felt that this was also unacceptable for us given our current environment.

After choosing the size of the input window, the remainder of the architecture had to be determined. In our preliminary experiment, only two languages, French and English (spoken by $male_1$ and $female_1$), were used. Therefore, the output layer contains only two units, one for French and one for English. The size of the hidden layer is determined by making intelligent guesses. We examine the trainability of the network for all data, training and testing, and find the number of hidden units experimentally where the network maximally accounts for all the data. [Note that this number could also be found through a set of experiments in which training took place with just a randomly determined training set and then tested for generalization among the other patterns, but that method would take much longer.] The final network connects each layer to each layer forward of it, and so there are the standard layered connections as well as connections directly from the input to the output layer. All training was performed using variations on the conjugate gradient method (see Kalman, 1990 and Kalman & Kwasny, 1991).

During network training, its generalization capability is continually being monitored by calculating a confusion matrix for the testing set of patterns and applying a χ^2 test to it. As the χ^2 result continues to increase, training continues. If the test levels off or decreases, adjustments are made in training until the best trained network has been found.

Repetitions	40	60	80	100	120
Threshold	21	31	41	51	61
Duration(secs)	1.725	2.225	2.725	3.225	3.725
Number of Patterns (in each language)	1,729	1,709	1,689	1,669	1,649
English	99.3%	100%	100%	100%	100%
French	92.7%	96.7%	97.3%	99.2%	100%
$Male_2$ (English only)	100%	100%	100%	100%	100%

Figure 2: Performance on Two Speaker/Two Language Task

To evaluate decisions regarding language identification, the short, 750ms segment must be slid across a wider window of speech, creating multiple decisions on which to base the classification. We arbitrarily decided to do so in 25ms intervals. For the two language problem, a simple majority rule is what is used. In effect, the smaller segment result is integrated across the larger time frame. For multiple languages, a plurality vote may be used and may potentially generate "don't know" classifications.

In analyzing the data by bands, the middle (third) band shares much with the adjacent bands. We decided to attempt to train the network from data further reduced by the elimination of band three. We successfully trained the network approximately the same level without including band three. This training is faster since there are fewer weights to adjust and so we used this method of training for all the results reported in the next section.

Results

Our first results were obtained from experiments with two speakers, $male_1$ and $female_1$, each speaking English and French. While this is a very limited task, it represents the technique involved in successfully classifying speech segments for this purpose.

First, the 12.5 second speech samples of the two subjects were divided into training samples and testing samples. Each training sample was processed into 371 overlapping 750ms segments of speech each of which produced 360 numeric values of frequency information across the four bands (90 samples of 4 bands). Training proceeded to settle at 73.7% correct on the test patterns. This trained network was then evaluated on varying durations of windows and performance was measured according to a majority vote. Figure 2 shows the performance while varying the duration from 1.725 seconds to 3.725 seconds. We report figures on all data, both testing and training, to enable us to look at more cases. Clearly, these results would hold rather closely for just the testing patterns. Note that the identification of English examples is total when using the duration of 2.225 seconds, while the French examples require a duration of 3.725 seconds to achieve 100% performance. This level of performance is even more remarkable when we consider that it is based on an evaluation of all 3,298 French and English testing and training patterns.

We then tested the same network with English speech samples from $male_2$. These data are shown in the final row of Figure 2, with perfect performance achievable in a duration of about 1.725 seconds. This illustrates the degree to which the network is capable of generalizing to the speech of subjects for which it has not been trained, in this case $male_2$ whose native language is Japanese.

It is possible to make a theoretical analysis of the tradeoff between achieved performance level on the short segment and the duration of the window necessary for high-level performance (99.5% correct) during testing. Figure 3 shows such a theoretical projection for selected performance levels of the network. For example, if the network performs at the level of 60% correct for the worst category being classified, then assuming independent classificatory decisions (which is not strictly correct, but suitable for this approximation) we use the binomial theorem to yield

$$0.995 \leq \sum_{k=M}^{N} p^k (1-p)^{N-k} \binom{N}{k}$$

Here, N is assumed to be odd to make the calculation simpler, and M is assumed to be $\left\lfloor \frac{N}{2} \right\rfloor$ +2. So, in Figure 3, the initial column determines the probability, p, used in the binomial theorem and N is determined and shown in the second column. The third column can be derived from the second by the formula:

Performance in worst category (percent)	Minumum (odd) N to yield 99.5% performance	Duration of window (seconds)	Normalized χ^2 performance on training set
60	181	5.25	≥ 0.40
65	81	2.75	≥ 0.49
70	51	2.00	≥ 0.55
75	41	1.75	≥ 0.64

Figure 3: Theoretical Projection of Performance

$$0.75 + (N - 1) \times (0.025)$$

since the segment size is 0.75 seconds and the increment for sliding the segment within the window is 0.025 seconds. Further the χ^2 performance when applied to the confusion matrix can be estimated and used in determining when training has reached the proper level to achieve the performance desired.

Conclusions

We have shown how a properly defined neural network is capable of reliably extracting the identification of what language is being spoken from raw speech. In our preliminary study reported here, perfect results were obtained by summing over multiple decisions and using a majority vote to determine a better decision from several individual error-prone ones. In fact, it can be shown that the error decays exponentially as the decision-making window is extended.

In a broader sense, we have illustrated the potential of extracting high-level features from raw speech by a majority decision-making system. The idea of "collecting votes" while sequentially processing input from a source channel is a powerful idea that results in noise tolerant decisions leading to remarkable performance. The majority vote technique exhibited here is a general method for improving the performance of an errorful method to one that is virtually flawless. Successful application of this method requires a task that submits to simple, aggregate classifications of the type demonstrated here and a classification technique that achieves a reasonable level of performance.

Human communication must carry information from one speaker to another by exploiting the characteristics of the channel. The channel of voice communication constrains what is permissible in a natural language utterance and what is not. Each language has developed its own unique system of utilizing the channel of communication to carry messages. While there is considerable overlap from language to language, it is the uniqueness that permits us to determine which language is being spoken and, therefore, which linguistic frame of reference to apply.

Future Work

Ongoing research is investigating how to incorporate voting schemes into the network in natural ways. There is evidence for both spatial and temporal summation in nerve cells found in the brain and we hope to find architectures that better simulate such activity.

A promising approach involves the use of recurrent networks. In preliminary studies, a simple recurrent network was trained to achieve recognition rates competitive with those of non-recurrent ones, but using a much smaller window. Recurrent networks develop a limited memory of past events and can exhibit classification capabilities that consider both immediate inputs and past events.

Experiments have also begun which utilize the data we have collected to train networks for both gender discrimination and speaker discrimination. Here again, the thrust of the work is on reliable identification of high-level features in real time directly from the speech signal. With a small number of speakers, speaker discrimination is proving to be an easy task. This situation is expected to change as data from more speakers is collected. Our voting method is not expected to work quite as well with gender discrimination due to the large degree of overlap in vocal frequency between male and female speakers.

References

Abe, M., and Shikano, K. 1991. Statistical analysis of bilingual speaker's speech for cross-language voice conversion. *Journal of the Acoustic Society of America* 90: 76-82.

Abe, M.; Shikano, K.; and Kuwabara, H. 1990. Cross-language voice conversion. In Proceedings of the International Conference on Acoustics, Speech, and Signal Processing, 345-348.

Atkinson, K. 1968. Language identification from non-segmental cues. *Journal of the Acoustic Society of America* 44, 378(A).

Denes, P.B. 1963. On the statistics of spoken English. *Journal of the Acoustic Society of America* 35, 892-904.

Hanley, T.D.; Snidecor, J.C.; and Ringel, R.L. 1966. Some acoustic difference among languages. *Phonetica* 14, 97-107.

House, A.S., and Neuberg, E.P. 1977. Toward automatic identification of the language of an utterance. I. Preliminary methodological consideration. *Journal of the Acoustic Society of America* 62(3), 708-713.

Kalman, B.L. 1990. Super Linear Learning in Back Propagation Neural Nets. Technical Report WUCS-90-21, Department of Computer Science, Washington University, St. Louis.

Kalman, Barry L., and Stan C. Kwasny. 1991. A superior error function for training neural networks. In Proceedings of the International Joint Conference on Neural Networks, Vol. 2, Seattle, Washington, 49–52.

Kalman, B.L., and S.C. Kwasny 1992. A training strategy for feed-forward neural networks based on input similarities. Submitted for publication.

Kucera, H., and Monroe, G.K. 1968. A comparative quantitative phonology of Russian, Czech, and German. New York: American Elsevier.

Muthusamy, Y.K.; Cole, R.A.; and Gopalakrishnan, M. 1991. A segment-based approach to automatic language identification In Proceedings of the 1991 IEEE International Conference on Acoustics, Speech and Signal Processing, Toronto, Canada.

Toward a Knowledge Representation for Simple Narratives

R. Raymond Lang and **Robert P. Goldman**
Computer Science Department, 301 Stanley Thomas Hall, Tulane University
New Orleans, Louisiana 70118, (504) 865-5840

Abstract

In this paper we report progress on the design of a knowledge representation formalism, based on Allen's temporal logic [Allen, 1984], to be used in a generative model of narratives. Our goal is to develop a model that will simultaneously generate text and meaning representations so that claims about recovery of meaning from text can be assessed. We take as our domain a class of simple stories, based on Grimm's fairy tales. We base our work on story grammars, as they are the only available framework with a declarative representation. We provide the logical foundations for developing a story grammar [Rumelhart, 1975] into a generative model of simple narratives. We have provided definitions to specify the "syntactic" categories of the story grammar and the constraints between constituents.

Introduction

It is not possible to assess the success of natural language processing programs which aim at a 'deep' (e.g., script- or plan-based) understanding of text, absent some criterion of correct understanding. To address this problem, we propose to construct a generative model of a restricted class of narratives. This model must simultaneously generate text and meaning representations so that claims about recovery of meaning from text can be assessed. Furthermore, this model must be specified in such a way that it can be understood and critiqued as a declarative representation, not just as imbedded in a program.

In this paper we report progress on the design of a knowledge representation formalism to be used in a generative model of narratives. For reasons outlined below, we take as our domain a class of simple stories, based on Grimm's fairy tales. We base the formalism on story grammars, as they are the only available framework with a declarative representation.

We wish to emphasize that we are not seeking to model text production. Our model need only capture the correct relationships between the texts and the desired interpretations; although our goal is to develop a model which will map in both directions (i.e. from texts to meanings as well as from meanings to texts). Furthermore, the model need not be *complete* in that factors which are not important to the interpretation problem which interests us can either be left implicit or be omitted from the model.

In designing our KR formalism, we use Allen's temporal logic [Allen, 1984], a language based on first order logic. While we are conscious of criticisms of first order logic as a knowledge representation language (e.g., [McDermott, 1987, Birnbaum, 1991]), we quite simply do not know of an alternative knowledge representation language which has a semantics, an inference calculus (deductive, "scruffy" or otherwise) and is either more expressive or more convenient. We do not wish to be seen as offering an argument on either side of the debate about the use of logic in AI; we are simply choosing the best tool available to us at this time.

Domain: Simple Stories

Much as cooking is often used as a domain for research in planning, simple stories are frequently chosen as a sample domain for testing theories of narrative structure and natural language understanding. Previous work in story generation has followed one of two tracks: (1) procedural and (2) descriptive.

Computer Story Telling

Within the first track, Meehan's TALE-SPIN [1976] is the most well-developed and widely known work. It is fundamentally a simulation of a forest world, producing natural language output describing the interactions of characters pursuing goals such as eating and drinking in a context where duplicity and hostility are possible as well as honesty and friendliness. Among story-telling programs, TALE-SPIN comes closest to what we seek in terms of having access to the meanings (conceptual dependency forms, in this case) from which the natural language text is constructed. However, the model by which the meanings themselves are generated is left implicit; and the relationships among the components of a story are deeply entwined in the procedures which drive the simulation.

1. Story → Setting + Episode
   ⇒ ALLOW (Setting,Episode)
2. Setting → (States)*
   ⇒ AND (State,State,...)
3. Episode → Event + Reaction
   ⇒ INITIATE (Event,Reaction)
4. Event →
   {Episode | Change-of-State | Action | Event + Event}
   ⇒ CAUSE ($Event_1$, $Event_2$) or ALLOW ($Event_1$, $Event_2$)
5. Reaction → Internal Response + Overt Response
   ⇒ MOTIVATE (Internal Response, Overt Response)
6. Internal Response → {Emotion | Desire}
7. Overt Response → {Action | Attempt*}
   ⇒ THEN ($Attempt_1$, $Attempt_2$, ...)
8. Attempt → Plan + Application
   ⇒ MOTIVATE (Plan,Application)
9. Application → (Preaction)* + Action + Consequence
   ⇒ ALLOW (AND(Preaction,Preaction,...),
   {CAUSE | INITIATE | ALLOW}
   (Action,Consequence))
10. Preaction → Subgoal + (Attempt)*
   ⇒ MOTIVATE [Subgoal, THEN (Attempt, ...)]
11. Consequence → {Reaction | Event}

Figure 1: Rumelhart's grammar for stories

Later work in story generation by computer addresses concerns not relevant to our project. Lebowitz's UNIVERSE [1985] is fundamentally a planner in that the output is generated by means of meeting the requirements of a goal. Lebowitz is explicitly interested in author-level goals such as "create suspense" and how these contribute to the content of a story. As such, this work would be more useful if we were trying to automate, say, literary criticism rather than simple, literal understanding. In MINSTREL [Turner and Dyer, 1985], Turner seeks to model the process of human creativity and uses King Arthur-style tales as his domain. Although we are working in a similar domain, we are not making any claims about what a human does when writing a story.

Modeling the Logical Structure of Stories

Work in the modeling of the structure of stories is founded upon seminal works such as Propp's *Morphology of the Folktale* [1968] and Polti's *The Thirty-Six Dramatic Situations* [1921]. Although Propp's perspective is that of an anthropologist and Polti's that of a literary critic, these works (particularly Propp) have been used frequently as a starting point for the development of theories of story structure and as sources for the sorts of categories that pertain to stories.

David Rumelhart [1975] presented a "story grammar" which inspired subsequent work which attempted to express the regular "syntax" of stories by means of grammars. His grammar is reproduced in Figure 1. Rumelhart's grammar, as well as many of the others like it [Bower, 1976, Johnson and Mandler, 1980, Mandler and Johnson, 1977, Stein and Glenn, 1979, Thorndyke, 1977], are context-free grammars. Some, including Rumelhart's, are augmented by "semantic" constraints. The terms *syntax* and *semantics* as used in these grammars may be misleading. The syntactic rules are constraints on event sequences and the semantic constraints restrict the relations between consecutive events.

Interest in story grammars soon lapsed, partially due to an attack on their foundations by Black and Wilensky [1979] (henceforth B&W). They make the case that a story grammar must be an unrestricted rewrite system (Chomsky type 0), but they find only one proposed story grammar [Johnson and Mandler, 1980] that meets this criteria. B&W argue that story grammars fail by trying to capture the idea of what a story is by trying to express the structure of a story text. In particular, they claim there is some purpose in relating a story; and the listener (or reader) of a story has some interest in hearing it. In other words, a story has a "point," i.e. some element that invokes the interest of a reader. The alternative theory of story points given by Wilensky [1982] views the form of a story as a function of its content. Although there is merit to this approach, the points theory fails to account for the regular structure of stories. B&W either ignored or failed to consider results of experiments in story recall supporting the hypothesis that people use story schemas to understand simple stories [Bower, 1976, Mandler and Johnson, 1977, Stein and Glenn, 1979, Thorndyke, 1977].

Part of the reason for the success of B&W's attack was the crude state of techniques for formalizing material like story grammars. In Rumelhart's grammar (as well as those based on this grammar), information about the relationship a story component has to other components is restricted to annotations accompanying the rules. In these grammars, the "syntactic" structure of a portion of a text makes a particular rule applicable, then the relationship of this component to others is gleaned from the annotation to the rule. Unfortunately, the "syntax" given in these grammars doesn't rule out many constructions; while the "semantic" annotations are not formalized rigorously enough.

Since the deficiencies of story grammars are so well documented, the reader is justified in asking why we are resurrecting a formalism laid to rest ten years ago. We believe the idea of a story grammar embodies important insights regarding what there is about a sequence of events such that reporting it constitutes a story. It is these insights we aim to formalize. Among the grammars available to us, Rumelhart's was chosen as a starting point because his use of semantic annotations showed the most promise. Although we use this grammar as a starting point, we also seek to incorporate important intuitions from other grammars, in particular Johnson and Mandler's [1980].

Another reason for the apparent success of B&W's criticisms of story grammars is that their point theory pertains to an different class of narratives than Rumelhart's simple stories. The work of both B&W and that of Rumelhart suffers by lacking a clearly defined domain. It is not possible to say specifically what kinds of stories these authors deal with. B&W's narratives have a more contemporary tone and style than the simple folk tales Rumelhart attempts to describe. In particular, simple folk tales frequently lack antagonists in B&W's sense of an independent agent with his or her own goals in competition with those of the protagonist. Whatever "bad guys" there are in simpler stories are merely animate obstacles that the protagonist must overcome in pursuit of his or her goal. We also find stories in which the main character has no clear goal. These stories are little more than a character's reactions to a sequence of external events. Such stories fall outside B&W's domain, but not outside of ours.

In contrast, Mandler and Johnson [1977, 1980] confine their attention to simple stories limited to a single protagonist in each episode. They explicitly allow stories in which events in one episode lead to another episode in which a different character becomes the protagonist; however, we are presently restricting our work to stories with a single protagonist throughout. Typically, the protagonist in our class of stories will be pursuing an explicit goal which may be given in the setting of the tale or arise as a reaction to an initial event.

We believe the real difficulty with the story grammar approach is that much of the insight is in annotations to the rules, but that the relations used to specify these annotations is vague and fuzzy. Our task, then, is to formalize these intuitions in a rigorous manner. In the next section, we outline the logic we use as a foundation for our work. Following this, we give some examples of how we have used this logic to translate Rumelhart's grammar into a system of logical axioms.

Temporal Logic

We base our formalization upon James Allen's temporal logic [Allen, 1984]. We have chosen this logic because it was specifically designed as a formalism for reasoning about actions in the context of natural language processing, while skirting the complications of modal and higher-order logics.

We have considered some other logics, notably McDermott's temporal logic [McDermott, 1982], and the episodic logic of Schubert and Hwang [1989]. We have chosen Allen's logic for reasons of simplicity. Unlike McDermott's logic, the semantics of Allen's logic is based on a single time line. Since we are primarily concerned with being able to represent sequences of (conceptually) past actions, we see no benefit to branching time semantics to offset the more complicated model. We have been inspired by Schubert and Hwang's work, but find their logic to be heavily biased towards natural language ***understanding***, as opposed to simple representation of action sequences. For example, they introduce mechanisms for referring to existential variables ***outside*** their scopes, and structure the syntax of their logic so as to closely parallel syntactic analyses of natural language sentences. We do not have present use for these complexities, so prefer to skirt them.

The entities in Allen's logic are action and state types, temporal intervals and individual objects and beings. Important relations express the occurrence of events and persistence properties over temporal intervals. E.g., HOLDS(foo(x), t) states that the foo property holds of some entity x over time interval t, and OCCUR(bar(x), t) says that an event of type bar, involving an entity x occurs over time interval t. Other important predicates denote relations between temporal intervals, e.g., BEFORE(t_1, t_2) says that time interval t_1 is before, and does not overlap with t_2. Allen's logic also attempts to give an account of planning and intentional behaviors. Modal aspects of the logic (agents' beliefs) are handled by a quotation method, so the logic as a whole is first-order. We have developed a treatment of intentional action based on Allen's logic, which is reported elsewhere [Goldman and Lang, 1992].

We extend and modify Allen's logic in 4 ways in order to make it more appropriate to the task at hand.

1. We extend the ontology of Allen's logic to include event tokens. This is done as a matter of convenience: our project here is to talk about what it is which makes a series of events a candidate for being recounted as a story. Having to talk about events as pairs of event descriptions and intervals is unnecessarily cumbersome. This extension is captured in the predicate *event(token)*, which is true when *token* denotes a particular instance of *event-class* which occurred during *time-interval*, that is,

   $$\forall x \ \mathsf{event}(x) \rightarrow \mathsf{OCCUR}(\mathsf{eclass}(x), \mathsf{time}(x))$$

   where *eclass* is a function which takes an event token and returns the class of events into which it falls, and *time* is a function over event tokens which returns the time-interval at which the event took place.

2. Allen provides a predicate, IS-GOAL-OF for making statements about an agent's intentions. In his logic, a statement of the form

   IS-GOAL-OF(agent,property,gtime,t)

   denotes the fact that the *agent*, at time t, desires that *property* hold, at time *gtime*.

   It is not clear exactly what this statement says about the intentions of *agent*. Is he or she aware of the extent of *gtime*? The difficulty can be seen most clearly if one wishes to interpret a proposition of the form $\exists gt$ IS-GOAL-OF (agent,property,gt,t). In this case, is there a specific time when the *agent* wants *property* to be true, which we just do not happen to know, although he does? Or does *agent* just want the action to be true at some fairly arbitrary time in the future?

In our domain we are more concerned with the latter case than the former (which arises naturally when one needs to reason about actions under time pressure).

3. Allen characterizes plans as a set of decisions about performing or not performing some set of actions. Unfortunately, Allen's account of goals and plans are only tenuously related. Allen's logic does not allow us to state anything stronger than that x has a goal of g at the same time x pursues plan p. One cannot express the fact that x is pursuing plan p in order to achieve goal g.

   We need this more specific relation for at least two purposes. First, we need to give a definition of plan failure in order to specify sensible action sequences. We see no way to do this absent a relation between the plan and its desired end. We also need to make statements which specify which plans are sensibly applicable to which classes of goals. Finally, for more complex stories, we will have to be able to distinguish between intended effects, and unintended side-effects.

4. Finally, for convenience, we extend the syntax of the logic with the $\oplus$ operator, written in prefix form, to indicate exclusive "one-of" disjunction, following Kautz [1986]. Also, we include quantification over restricted parts of the domain. We follow Schubert and Hwang [1989] in using the notation $\exists x[x : type]P(x)$ as a shorthand for $\exists x : type(x) \wedge P(x)$ and likewise $\forall x[x : type]P(x)$ for $\forall x : type(x) \rightarrow P(x)$

A Logic for Stories

In this section, we outline and give examples of our approach to the two immediate tasks which we face in building our logic: (1) developing semantic categories to match Rumelhart's "syntactic" elements, and (2) giving firm definitions of Rumelhart's "semantic" relations.

As mentioned earlier, we believe the semantic relationships and syntactic categories Rumelhart describes are capable of describing a non-trivial class of simple stories. Our work up to this point has been aimed at eliminating the vagueness and informality in Rumelhart's descriptions of these relationships and categories. For the semantic relationships, we clearly delineate the number and types of arguments and define them in terms of the extensions to Allen's logic outlined above. The categories have been recast as single-place predicates, indicating that the argument, which may be a token, is of the indicated type.

Categories As an example, consider the categories *action* and *change-of-state*. Rumelhart decribes these informally in a way that leaves open the possibility that a token could be both simultaneously. We establish these as mutually exclusive sub-types of a restricted version of Rumelhart's category, *event*. Rumelhart characterizes an *action* as "an activity engaged in by an animate being or a natural force." We use Allen's *ACAUSE* predicate to characterize an *action* as a particular kind of event, namely those which have an animate-being as agent.

$$\forall \text{ x action(x)} \leftrightarrow \exists \text{e event(x)} \wedge \text{eclass(x)} = \text{ACAUSE(agent(x), e)}$$

Otherwise, if an event is a *change-of-state*, this implies that there is no action which causes the event.

The category *reaction* is described by Rumelhart as "the response of a willful being to a prior event." In our characterization of *reaction*, we take Rule 5 as a guide and define it as an event having two components: an internal response and an external response.

$$\forall \text{ x reaction(x)} \rightarrow \text{event(x)} \wedge \text{class(x)} = \text{and(internal-response(x), externalresponse(x))}$$

Internal-response and *external-response* are functions over reactions. The domains of these functions correspond to Rumelhart's syntactic categories *internal response* and *overt response*.

Rumelhart describes the syntactic category *attempt* as "the formulation of a plan and application of that plan for obtaining a desire." It appears in rules 7, 8, and 10. We believe the underlying intuition here is that *attempt* characterizes the action(s) that an agent may take in response to some motivating event. A complication is that *attempt* includes not only the "physical" actions that a character performs, but also the "mental" act of planning those actions.

Clearly, the notion of a plan is central to relating a sequences of actions performed by some character. However, the concept is only vaguely defined by Rumelhart. He is not using the term "plan" in the classical AI sense of a "recipe for action." Rumelhart informally defines *plan* as "the creating of a subgoal which if achieved will accomplish a desired end." When he says "plan" he means more what Pollack [1990] terms a "complex mental attitude".

We move toward a refinement of this interpretation with the following axioms. First, we postulate a two-place predicate *applicable(plan,goal)*, which is true if the *plan*, successfully carried out, will cause the state described by *goal* to hold. Furthermore,

$$\forall \text{ p [p : plan]}, \forall \text{ g [g : goal] applicable(p, g)} \rightarrow \text{goal-owner(g)} = \text{agent(p)}$$

where *goal-owner* and *agent* are functions performing the obvious mappings. In the axiom 1, *animate-being, plan, time-interval*, and *application* are single-position

$$\forall x\ [x : \text{attempt}]\ \text{animate-being}(\text{agent}(x)) \wedge \text{plan}(\text{attempt-plan}(x)) \wedge \text{time-interval}(\text{time}(x)) \wedge \text{application}(\text{attempt-application}(x)) \wedge \text{COMMITTED}(\text{agent}(x), \text{attempt-plan}(x), \text{time}(x)) \wedge \text{applicable}(\text{attempt-plan}(x), \text{attempt-goal}(x)) \quad (1)$$

predicates which are true when the argument is an element of the indicated type. The function *attempt-plan* maps from an *attempt* element to the (abstact) plan embodied in the attempt, whereas the function *attempt-application* returns the concrete actions (considered as a unit) which are the "carrying out" of the plan. The function *attempt-goal* maps from an *attempt* to the state description toward which the attempt is aimed at causing to hold. *COMMITTED(a,p,t)* is Allen's predicate signifying that agent *a* intends to carry out the actions composing plan *p* where time *t* is the time of the *intending*, not of the carrying out of the plan.

Of course, these axioms only constrain what must be true in order for some element to be an *attempt*. Further axioms describe how an attempt is related to the event which leads a character to form the *attempt-goal* in the first place. Other axioms define the relationship of an *attempt* to the *application*, which is "carrying out" phase of an attempt.

One of the most important relationships in this class of stories is that of failed attempts to the surrounding events. Of course, in order to relate a failed attempt to another event, we must first be able to recognize when an attempt has, in fact, failed. Axiom 2 defines failure. There are two cases of plan failure to consider which correspond roughly to a base case and a recursive case. In the base case, a plan fails if every action of the plan has been done and the goal still does not hold. In the recursive case, a plan fails if there is a step which is a subgoal and at some time following the time at which the subgoal was to be achieved, the subgoal still has not been met. This portion of the rule is necessary since it is impossible to talk about a step which is a subgoal having been done if the subgoal was not met.

Relations Rumelhart informally describes six semantic relationships on events. They are: *and, allow, initiate, motivate, cause*, and *then*. With the exception of *allows*, we have accounted for all of these in our logic. We have developed axioms which characterize *initiate, motivate*, and *cause*; *and* and *then* have been dissolved into the logic.

In general, it is difficult to give precise formulations of these relationships without laying groundwork for which we lack the space in this paper, but we will present here some examples of logical axioms we have developed in order to give the reader a sense of the direction in which we are moving. Rumelhart describes the semantic relationship *initiates* as follows: "the relationship between an external exent and the willful reaction of an anthropomorphized being to that event." Using Allen's primitives for relating time intervals, we construct the predicate *starts-before* as follows:

$$\forall\ t_1,t_2\ \text{starts-before}(t_1,t_2) \leftrightarrow \oplus\ [\text{in}(t_2,t_1),\text{before}(t_1,t_2),\ \text{meets}(t_1,t_2),\text{overlaps}(t_1,t_2)]$$

We then use this to make explicit at least part of what must be true when x *initiates* y:

$$\forall\ x,y\ \text{initiates}(x,y) \rightarrow \text{starts-before}(\text{time}(x), \text{time}(y)) \wedge \text{reaction}(y)$$

Rumelhart uses the predicate *motivate* to describe the relationship between "an internal response and the actions resulting from that response." We tighten this by means of a predicate *motivates (mental-event, action)* which is true when the mental-event such as an emotion or the adoption of a goal is the motivation for some external action, which may be a high level action composed of or generated by one or more sub-actions. For example, the internal-response component of a reaction to an event *motivates* the agent's overt-response:

$$\forall\ x\ [x : \text{reaction}]\ \text{motivates}(\text{internal-response}(x), \text{overt-response}(x))$$

Rumelhart defines the semantic relationship *cause* as "the relationship between two events in which the first is the physical cause of the second." Like Allen, we take causality to be a primitive relationship and leave it unanalyzed. Rumelhart's two remaining semantic relationships, *and* and *then* essentially are dissolved into the temporal logic. The former was nothing more than logical conjunction; and the latter was meant to suggest temporal ordering of events, which may be more precisely expressed with Allen's predicates.

Conclusions

In this paper we provide part of the logical foundations for "cashing out" a story grammar like Rumelhart's into a generative model of simple narratives. We have provided examples of definitions which specify the "syntactic" categories of Rumelhart's story grammar and the constraints between constituents (the annotations to the rules).

When this work is completed, we intend to move on to implementing the generative model and using it to create a corpus of stories and their "meanings." A first step will be to generate a set of event sequences from a story grammar (revised to take into account the newly-formalized constraints) and a body of background knowledge.

$$\forall p\ [p : \text{plan}],\ \forall t\ [t : \text{timeinterval}]\ \text{failed}(p,t) \leftrightarrow$$
$$(\forall a,t'\ (\text{TODO}(a,t',p) \rightarrow [\text{OCCUR}(a,t') \wedge \text{BEFORE}(t',t)]) \wedge \neg\text{HOLDS}(\text{goal}(p),t))$$
$$\vee\ \exists\ sg,t'\ (\text{TODO}(\text{achieve}(sg),t',p) \wedge \text{BEFORE}(t',t) \wedge \neg\text{HOLDS}(sg,t')) \quad (2)$$

The most important problem remaining to our formalization project is that of characterizing the relationship of story settings to the corresponding sequences of events. Rumelhart uses the *allows* predicate to express the notion that the states composing the setting set the stage for what follows. That the setting simply not contradict the events which follow is clearly not a strong enough constraint on the setting-body relation. A relevance relationship like this one is difficult to formalize in a framework like the first order logic. We are currently working on characterizing the kinds of facts that are expressed in story settings.

References

[Allen, 1984] James F. Allen. Towards a general theory of action and time. *Artificial Intelligence*, 23:123–154, 1984.

[Birnbaum, 1991] Lawrence Birnbaum. Rigor mortis: a response to Nilsson's "logic and artificial intelligence". *Artificial Intelligence*, 47:57–77, 1991.

[Black and Wilensky, 1979] John B. Black and Robert Wilensky. An evaluation of story grammars. *Cognitive Science*, 3:213–230, 1979.

[Bower, 1976] Gordon H. Bower. Experiments on story understanding and recall. *Quarterly Journal of Experimental Psychology*, 28:511–534, 1976.

[Goldman and Lang, 1992] Robert P. Goldman and R. Raymond Lang. Representing intentional actions in allen's temporal logic. forthcoming, 1992.

[Johnson and Mandler, 1980] Nancy S. Johnson and Jean M. Mandler. A tale of two structures: Underlying and surface forms in stories. *Poetics*, 9:51–86, 1980.

[Kautz and Allen, 1986] Henry Kautz and James Allen. Generalized plan recognition. In *Proceedings of the Fifth National Conference on Artificial Intelligence*, pages 32–38, 1986.

[Lebowitz, 1985] Michael Lebowitz. Story telling and generation. In *Proceedings of the Seventh Annual Conference of the Cognitive Science Society*, pages 100–109, Berkeley, California, 1985.

[Mandler and Johnson, 1977] Jean M. Mandler and Nancy S. Johnson. Remembrance of things parsed: Story structure and recall. *Cognitive Psychology*, 9:111–151, 1977.

[McDermott, 1982] Drew V. McDermott. A temporal logic for reasoning about processes and plans. *Cognitive Science*, 6:101–155, 1982.

[McDermott, 1987] Drew V. McDermott. A critique of pure reason. *Computational Intelligence*, 3:151–160, 1987.

[Meehan, 1976] James R. Meehan. *The Metanovel: Writing Stories by Computer*. PhD thesis, Yale University, 1976.

[Pollack, 1990] Martha E. Pollack. Plans as complex mental attitudes. In Philip R. Cohen, Jerry Morgan, and Martha E. Pollack, editors, *Intentions in Communication*, pages 77–104. MIT Press, 1990.

[Polti, 1921] Georges Polti. *The Thirty-Six Dramatic Situations*. James Knapp Reeve, Franklin, Ohio, 1921.

[Propp, 1968] V. Propp. *Morphology of the Folktale*. University of Texas Press, 1968.

[Rumelhart, 1975] David E. Rumelhart. Notes on a schema for stories. In Daniel G. Bobrow and Allan Collins, editors, *Representation and Understanding: Studies in Cognitive Science*, pages 211–236. Academic Press, Inc., New York, 1975.

[Schubert and Hwang, 1989] Lenhart K. Schubert and Chung Hee Hwang. An episodic knowledge representation for narrative texts. In *First International Conference on Principles of Knowledge Representation and Reasoning*, pages 444–458. Morgan Kaufmann Publishers, Inc., 1989.

[Stein and Glenn, 1979] Nancy L. Stein and Christine G. Glenn. An analysis of story comprehension in elementary school children. In Roy O. Freedle, editor, *New Directions in Discourse Processing*, volume 2, pages 53–120. Ablex Publishng Corporation, Norwood, New Jersey, 1979.

[Thorndyke, 1977] Perry W. Thorndyke. Cognitive structures in comprehension and memory of narrative discourse. *Cognitive Psychology*, 9:77–110, 1977.

[Turner and Dyer, 1985] Scott R. Turner and Michael G. Dyer. Thematic knowledge, episodic memory and analogy in minstrel, a story invention system. In *Proceedings of the Seventh Annual Conference of the Cognitive Science Society*, pages 371–375, Berkeley, California, 1985.

[Wilensky, 1982] Robert Wilensky. Points: A theory of the structure of stories in memory. In W. G. Lehnert and M. H. Ringle, editors, *Strategies for Natural Language Processing*, pages 345–374. Lawrence Erlbaum, Hillsdale, N.J., 1982.

Question Asking During Learning with a Point and Query Interface

Mark C. Langston and Arthur C. Graesser
Department of Psychology
Memphis State University
Memphis, TN 38152
langston@memstvx1.memst.edu
graesserac@memstvx1.memst.edu

Abstract

Educational software would benefit from question asking facilities that are theoretically grounded in psychology, education, and artificial intelligence. Our previous research has investigated the psychological mechanisms of question asking and has developed a computationally tractable model of human question answering. We have recently developed a Point and Query (P&Q) human-computer interface based on this research. With the P&Q software, the student asks a question by simply pointing to a word or picture element and then to a question chosen from a menu of "good" questions associated with the element. This study examined students' question asking over time, using the P&Q software, while learning about woodwind instruments. While learning, the students were expected to solve tasks that required either deep-level causal knowledge or superficial knowledge. The frequency of questions asked with the P&Q interface was approximately 800 times the number of questions asked per student per hour in a classroom. The learning goals directly affected the ordering of questions over time. For example, students did not ask deep-level causal questions unless that knowledge was necessary to achieve the learning goal.

Introduction

Question asking and answering play a crucial role in some learning processes (Collins, 1988; Miyaki & Norman, 1979; Schank, 1986). The number and quality of student questions depends on the student initiative that is required in the learning environment. In a classroom environment, learning is not generally under the control of the student, so student questioning approaches zero (Dillon, 1988; Kerry, 1987). A tutoring environment requires the student to take a more active role in the learning process, and this is reflected in substantially more questions asked by the student. For example, approximately 0.2 questions are asked by a student per hour in a classroom whereas 20 questions per hour are asked in a tutoring session (Person, 1990).

Graesser, Person, and Huber (1992) identified four classes of cognitive mechanisms that underlie human question asking: (1) correction of knowledge deficits, (2) monitoring common ground, (3) social coordination of action, and (4) control of conversation and attention. The number of student questions in the knowledge deficit category is a good measure of student initiative; that is, students who ask a lot of questions in this category are active, inquisitive learners capable of identifying and repairing their own knowledge deficits (Brown, 1988; Brown et al., 1983; Palinscar & Brown, 1984; Pressley et al., 1987). As the learning environment shifts the initiative from teacher to student, one would expect that the number of knowledge deficit questions would increase. This is evident in the questions asked by students during tutoring, where 30% of student questions addressed knowledge deficits (Person, 1990).

As the student takes a more active role in the learning process, the environment must be capable of supporting a large, diverse number of student questions. Therefore, there is a need for question asking facilities in educational software. Unfortunately, most of the existing human-computer interfaces (Tennant, 1987; Williams, 1984; Zloof, 1975) have not had ideal question asking facilities. Each of these interfaces suffers from one or more of the following six problems:

(1) **Questioning time**. With current questioning interfaces, it takes the student a long time to ask a question, often several minutes.

(2) **Ease of use**. Questioning interfaces are often quite complex and require hours for the users to learn how to ask a question.

(3) **Question interpretation**. The computer will sometimes misinterpret the query posed by the student, and respond with the wrong information.

This research was funded by grants awarded to the second author by the Office of Naval Research (N00014-88-K-0110 and N00014-90-J-1492).

(4) **Question answering.** Software designed to answer questions should be grounded in a psychological theory of human question answering.

(5) **System ambiguity.** It is not clear to the user what questions the system can answer.

(6) **System focus.** The knowledge base is not organized around questions and answers to questions.

We have developed a Point and Query (P&Q) interface (Graesser, Langston, & Lang, 1992) that attempts to correct all these problems. First, it is very easy for the student to ask a question with the P&Q interface. With two clicks of a mouse, the student can easily ask a question and receive an answer within two seconds. Second, it is very easy for the student to learn to use the P&Q interface. It takes approximately five minutes to learn how to use the system, if the student is familiar with the use of a computer mouse. Those students unfamiliar with the mouse require an extra few minutes. Third, the P&Q system can quickly and correctly answer a question according to a psychological model of human question answering called QUEST (Graesser & Franklin, 1990). The P&Q system and Schank's ASK TOM system (Schank et al., 1991) are the only systems based on an empirically tested psychological theory of question answering (although ASK TOM has not yet been rigorously tested using human subjects). Fourth, the student has direct feedback on what questions the P&Q system can answer at any point in time because a list of relevant questions is displayed in a menu on the screen. This allows the student to learn what questions are good questions. The menu of relevant questions is contingent on the type of knowledge structure the student is curious about, e.g., goal/plan hierarchy, causal network, taxonomic hierarchy, or spatial information. Fifth, the knowledge base in the P&Q interface is organized around questions and answers to questions.

The QUEST model (Graesser & Franklin, 1990) greatly influenced those questions and answers in the interface. The QUEST model specifies what questions are appropriate for the domain through an analysis of the knowledge structures in the domain to be learned. QUEST specifies which information units from an information source are legal answers to a particular question. Whenever an answer to a question provided little or no new information, the question was not included in the menu of questions.

The P&Q system in this study contained knowledge about woodwind instruments. This domain was chosen because it is knowledge-rich in each of the following types of knowledge, or "viewpoints" (Graesser & Clark, 1985; Souther et al., 1989; Stevens, Collins, & Goldin, 1982):

(1) **Taxonomic knowledge**, which includes taxonomic hierarchies and concept definitions.

(2) **Spatial composition**, incorporating the spatial layout of objects, parts, and features.

(3) **Sensory information**, including visual, auditory, kinesthetic, and other sensory modalities.

(4) **Procedural knowledge**, embodying the actions, plans, and goal structures of agents.

(5) **Causal knowledge**, which captures causal networks and states in technological, biological, and physical systems.

These viewpoints are closely interrelated, with mutual constraints and associative mappings between each of the different types of knowledge.

There has been very little empirical research on patterns of exploring knowledge by asking questions, particularly in the context of knowledge-rich domains. This lack of research prompted the present study. We examined the patterns of student questions while they sampled deep causal knowledge versus comparatively superficial knowledge (i.e., taxonomic, sensory, spatial, and procedural knowledge). We manipulated the goals of the student during the learning process, such that they were to focus either on deep causal or superficial levels of knowledge. The purpose of this study was not to evaluate our software as it affects learning. Instead, we were interested in documenting the subjects' course of exploring knowledge over time, how goals affect this process, and how the subjects' questioning rate compares to other contexts (i.e., classrooms and tutoring).

Methods

Goals of student

Subjects were 32 undergraduate students at Memphis State University. Half of the subjects were expected to acquire deep causal knowledge of woodwind instruments (Design Instrument condition), whereas the other half could rely on superficial knowledge (Assemble Band condition). In the Design Instrument condition, the subjects were told to design a new woodwind instrument that had a low pitch and that was pure in tone. The solution to this task required the student to have a deep causal knowledge of woodwind instruments; the student would have to understand causal relationships between the physical features of a musical instrument and the properties of its sound. Subjects in the Assemble Band condition were instructed to assemble a band with six types of woodwind instruments to play at a New Year's Eve party. The solution to this task does not require a deep causal understanding of woodwind instruments; the task could be completed using only superficial knowledge about what the instruments look like, what they sound like, and what their names are.

Computer software

The computer software consisted of a knowledge base about woodwind instruments in a hypertext environment with a P&Q interface. The computer was a Macintosh microcomputer. The knowledge base consisted of approximately 500 "cards" (screen displays) in the hypertext system. The cards included two "seed" cards: a taxonomy of woodwind instruments and a diagram of air flow through each component of a prototypical instrument. The rest of the cards were answers to possible questions that could be asked by the student.

There were 10 generic question categories that a student could choose from when the student selected an element of information to query. The types of knowledge these questions addressed were: taxonomic hierarchy ('What does X mean?', 'What are the properties of X?', 'What are the types of X?'), definitions ('What does X mean?'), sensory information ('What does X look like?', 'What does X sound like?'), spatial composition ('What does X look like?'), procedural knowledge ('How does a

Figure 1. An example question and answer interaction using the Point and Query Interface.

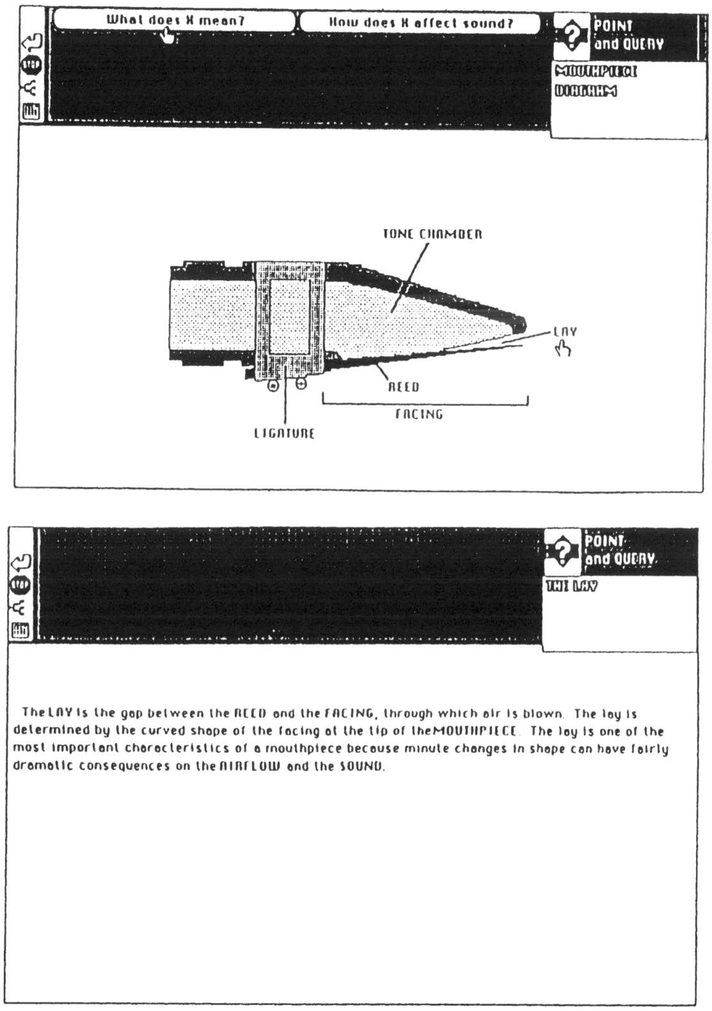

person use/play X?') and causal knowledge ('How does X affect sound?', 'How can a person create X?', 'What causes X?', 'What are the consequences of X?'). These 10 question categories address most relevant questions that could be asked at any given time about particular elements in the knowledge base. Categories that contain instances of the same question type are collapsed and considered one category (e.g., Taxonomic-Definitional, Sensory-Spatial). When the student was presented with a screen of information, there were particular elements that were highlighted. The students pointed to one of the elements they were curious about. A particular subset of the 10 question categories was presented according to (1) the QUEST psychological theory of human question answering (Graesser & Franklin, 1990), (2) the knowledge structures associated with the queried element, (3) the good questions associated with the type of knowledge structure, and (4) whether or not a question has an informative answer. After pointing to a screen element, the student selected a question and received the answer.

For example, in Figure 1, the student was presented with a picture representing a single reed mouthpiece. The student was curious about the LAY of the instrument and pointed to that element. Two questions relevant to the element were presented and the student asked 'What does lay mean?'. Within a second the answer was displayed. This process could then repeat until the students either exhausted the knowledge base, ran out of time, or felt they had acquired enough information to satisfy their learning goals.

Procedure

The subjects were randomly assigned to either the Design Instrument or the Assemble Band condition. The subjects read a three-page packet that described the use of the interface. The experimenter demonstrated the use of the interface to the subject, and allowed the subject approximately one minute to become familiar with the system. At the end of the familiarization phase, the subjects were given their problem solving task and were allowed to interact with the P&Q system for 30 minutes. The computer recorded the elements and questions the subjects pointed to.

Results and Discussion

As a preliminary analysis we computed the mean number of questions asked by the subjects during the 30-minute interaction period. We found that the subjects asked a mean of 75.6 questions per session in the Design Instrument condition and 59.9 questions per session in the Assemble Band condition. Therefore, the rate of student questioning while using the P&Q interface was 135 questions per hour. This is about 7 times the rate of student question asking during normal tutoring (Person, 1990) and 800 times the rate of student questioning in a classroom environment. The high frequency of question asking using this software implies that the P&Q interface has the potential to radically encourage active learning when combined with educational software. However, more research is needed to substantiate this possibility.

The 30-minute interaction period was segregated into three 10-minute time blocks, yielding time block 1, 2, versus 3. We clustered the 10 question categories into four categories that addressed four different types of knowledge: taxonomic-definitional , sensory , procedural, and causal. An analysis of variance was performed on question asking frequencies using a mixed design with three independent variables: condition (Design Instrument versus Assemble Band), time block (1, 2, versus 3), and question type (taxonomic-definitional, sensory, procedural, and causal).

We first analyzed the main effects in the ANOVA. The frequency of questions did not significantly vary as a function of time blocks, with means of 22.8, 23.4, and 21.2 questions in time blocks 1, 2 and 3, respectively. Therefore, the volume of questions was approximately constant across the three 10-minute segments, indicating a constant level of student curiosity and initiative. More questions were asked in the Design Instrument condition than in the Assemble Band condition, $F(1, 28) = 5.00$, $p < .05$. The number of questions per time block significantly differed among the four question types, with means of 8.8, 5.3, 1.1, and 7.3 for taxonomic-definitional, sensory, procedural, and causal knowledge questions, respectively, $F(3, 84) = 27.62$, $p < .05$. This result is not surprising, however, because the baserate frequency of available questions was quite different among the four question types.

There was a significant three-way interaction between condition, time block, and question type, $F(6,168) = 2.89$, $p < .05$. Figure 2 plots the cell means that expose this three-way interaction. The following trends explained the interaction:

Taxonomic-definitional

The frequency of these questions were about equal for the two conditions in time block 1. The frequency of this question type decreased over time in the Design Instrument condition, but remained constant in the Assemble Band condition. This would indicate that the learner must acquire taxonomic and definitional knowledge of a domain during the initial learning phase, regardless of the learner's goals. As learning progresses, the learner asks questions that are more directly related to the learning goal.

Figure 2. Questioning frequencies for different types of knowledge.

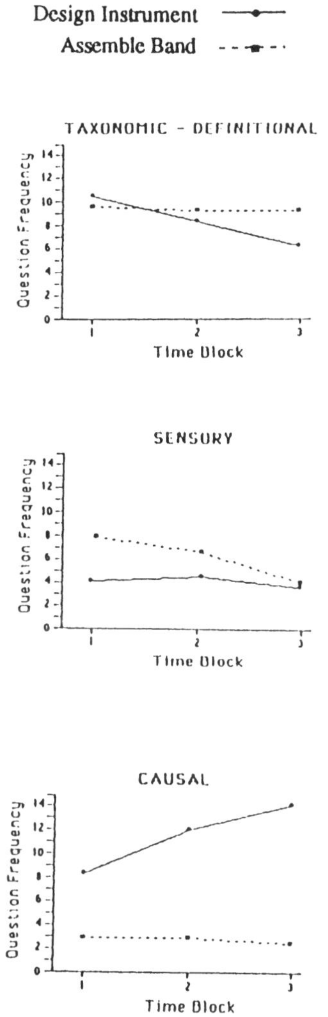

Causal

The frequency of this type of question was extremely high and increased over time in the Design Instrument condition, whereas the frequency was extremely low and constant in the Assemble Band condition. It appears that student initiative in asking causal questions is directly affected by learner goals. Deep level causal knowledge was required to satisfy the task in the Design Instrument condition, and the student was forced to ask many causal questions. Deep level knowledge was not required in the Assemble Band condition, so the subjects did not try to acquire this knowledge.

Sensory

The frequency of sensory questions was low and constant in the Design Instrument condition. The frequency was initially high in the Assemble Band condition but decreased robustly over time. The subjects in the Assemble Band condition wanted to find out what the instruments looked like and sounded like very early in the learning process. This superficial knowledge was necessary for the spatial and aesthetic considerations involved in assembling a band, whereas causal knowledge was unimportant.

Procedural

There was a floor effect for this type of question so it was difficult to decipher trends. Subjects in the Assemble Band condition asked approximately twice as many questions in this category as did subjects in the Design Instrument condition.

Conclusion

This study has documented how a P&Q computer learning environment can stimulate student initiative and questioning. Initiative was measured by the frequency and type of student questions. We found that in a computer environment designed around questioning, students were capable of taking a very active role in the learning process. We have also presented evidence that some student questioning patterns are directly affected by the student's learning goals, whereas other patterns are comparatively impervious to their learning goals. Students are capable of actively monitoring the acquisition of knowledge in a domain, and adjusting this acquisition to satisfy their goals.

The P&Q interface and other similar new interfaces (Schank et al., 1991; Sebrechts & Swartz, 1991) have made it extremely easy for the user to ask questions. It is possible that interfaces like these could have a substantial impact on education to the extent that they rekindle curiosity and good question asking skills. Students can and will take the initiative in the learning process, if given the right environment.

The P&Q interface represents a radical approach to educational software. The only action allowed is question asking, and the student has full initiative in the learning process. As research progresses and the interface evolves, we intend to shift the initiative, allowing for a more realistic mixed-initiative dialogue. Additional research is required to uncover the system's full potential as a learning tool.

References

Brown, A. L. 1988. Motivation to learn and understand: On taking charge of one's own learning. *Cognition and Instruction*, 5:311-321.

Brown, A. L., Bransford, J. D., Ferrara, R. A., & Campione, J. C. 1983. Learning, remembering, and understanding. In J. H. Flavell & E. M. Markman (Eds.), *Handbook of child psychology (4th ed.). Cognitive development, vol.3*. New York: Wiley.

Collins, A. 1988. Different Goals of Inquiry Teaching. *Questioning Exchange*, 2:39-46.

Dillon, J. T. 1988. *Questioning and teaching: A manual of practice*. New York: Teachers College Press.

Graesser, A.C., & Clark, L.C. 1985. *Structures and procedures of implicit knowledge*. Norwood, NJ: Ablex.

Graesser, A.C., & Franklin, S.P. 1990. QUEST: A cognitive model of question answering. *Discourse Processes*, 13:279-303.

Graesser, A.C., Langston, M., & Lang, K.L., 1992. Designing educational software around questioning. *Journal of Artificial Intelligence and Education*, 3:235-243.

Graesser, A.C., Person, N., & Huber, J. 1992. Mechanisms that generate questions. In T. Lauer, E. Peacock, & A. Graesser (Eds.), *Questions and information systems*. Hillsdale, NJ: Erlbaum.

Kerry, T. 1987. Classroom questions in England. *Questioning Exchange*, 1:32-33.

Miyaki, N. & Norman, D.A. 1979. To Ask a Question, One Must Know Enough to Know What is Not Known. *Journal of Verbal Learning and Verbal Behavior*, 18:357-364.

Palinscar, A. S., & Brown, A. L. 1984. Reciprocal teaching of comprehension-fostering and comprehension-monitoring activities. *Cognition and Instruction*, 1:117-175.

Pressley, M., Goodchild, F., Fleet, J., Zajchowski, R., & Evans E. 1987. The challenges of classroom strategy instruction. *The Elementary School Journal*, 89:301-342.

Person, N.K. 1990. The Documentation of Questioning Mechanisms and Types of Questions in Tutoring Protocols. Master's thesis, Dept. of Psychology, Memphis State Univ.

Sebrechts, M.M., & Swartz, M.L. 1991. Question asking as a tool for novice computer skill acquisition. *Proceedings of the International Conference on Computer-Human Interaction*, 293-297.

Schank, R. C. 1986. *Explanations Patterns: Understanding mechanically and creatively*. Hillsdale, NJ: Erlbaum.

Schank, R., Ferguson, W., Birnbaum, L., Barger, J., & Greising, M. 1991. ASK TOM: an experimental interface for video case libraries. In the *Proceedings of the 13th Annual Conference of the Cognitive Science Society*, 570-575, Hillsdale, NJ: Erlbaum.

Souther, A., Acker, L., Lester, J., & Porter, B. 1989. Using view types to generate explanations in intelligent tutoring systems. *Proceedings of the 11th Annual Conference of the Cognitive Science Society*, 123-130, Hillsdale, NJ: Erlbaum.

Stevens, A., Collins, A., & Goldin, S. E. 1982. Misconceptions in students' understanding. In D. Sleeman & J. S. Brown (Eds.), *Intelligent tutoring systems*. New York: Academic Press.

Tennant, H. R. 1987. Menu-based natural language. In S. C. & D. Eckroth (Eds.), *Encyclopedia of artificial intelligence*. New York: John Wiley & Sons.

Williams, M.D. 1984. What makes RABBITT run? *International Journal of Man-Machine Studies*, 21:333-352.

Zloof, M. M. 1975. Query by example. *Proceedings of the National Computer Conference*, 44:431-438. Arlington, VA: AFIPS Press.

TOWARDS A DISTRIBUTED NETWORK LEARNING FRAMEWORK: THEORY AND TECHNOLOGY TO SUPPORT EDUCATIONAL ELECTRONIC LEARNING ENVIRONMENTS

James A. Levin
Michael J. Jacobson
Department of Educational Psychology
University of Illinois at Urbana-Champaign
210 Education Building
1310 So. Sixth Street
Champaign, IL 61820
jim-levin@uiuc.edu

Abstract

Electronic networks are being increasingly used to support a variety of educational activities. Although early research in this area has been promising, there has been less work to date concerning more basic cognitive and theoretical issues associated with the design and use of educational electronic networks. This paper proposes a distributed network learning framework (DNLF) which will be presented through three main aspects: (a) network mediators and the flow of information and knowledge, (b) networks and cognitive theories of learning, and (c) the network-human interface. As an example of an application of the distributed network learning framework, an ongoing research and development project is discussed that involves a cognitively-based educational electronic communication tool, The Message Assistant. In addition to the standard electronic mail features such as creating, sending, and receiving messages, this program includes a user-defined incremental expert system and hypertextual linking functions to assist a network mediator in her or his evaluation, organization, and distribution of network information and knowledge. The distributed network learning framework can function as a flexible--and extendable--set of conceptual views from which to examine and to work with different aspects of dynamically evolving network learning environments.

Credits: This material is based upon work supported by the National Science Foundation under Grant No. TPE-8953392. The Government has certain rights in this material. Any opinions, findings, and conclusions or recommendations expressed in this material are those of the authors and do not necessarily reflect the views of the National Science Foundation.

There has been considerable interest in the development and use of computer network-based learning environments. Early research in this area has identified a number of ways in which successful learning activities may be conducted over electronic networks. For example, elementary students can participate on important scientific research matters such as "acid rain" in collaboration with distant learners and scientific advisors; high school students can use supercomputers, visualization software tools, and network access to scientific advisors to conduct high-quality scientific investigations; and students can conveniently publish their written work to national and even international audiences (e.g., Hunter, 1992). There has been less work, however, on more basic cognitive and theoretical issues associated with the design and use of network learning environments.

The central goal of this paper is to articulate a view of electronic learning environments that we refer to as the *distributed network learning framework*. This cognitively-grounded framework is intended to describe important characteristics of information flow over educational networks and critical cognitive dimensions associated with the use of network-based learning environments. The first portion of this paper presents the major aspects of this cognitively-grounded framework. The second part describes a communications tool we have been developing that implements and supports a number of facets of the framework. The final section of the paper considers research issues for the evaluation of network-based learning activities that are suggested by the learning framework.

The Distributed Network Learning Framework

The *distributed network learning framework* (DNLF) will be describes through three major aspects, each of which considers a different facet of the complex and dynamic network learning environments that are becoming available to students, teachers, and researchers. A central notion of the framework is that of *network mediators and the flow of information and knowledge*. As educational network activities are fundamentally concerned with learning, the second aspect of the framework is *networks and cognitive theories of learning*. The final aspect discusses the *human-network interface*, specifically with respect to the cognitive management of complexities associated with using the current generation of computer networks. Each of these aspects is briefly discussed in the sections below.

Network Mediators and Flow of Information and Knowledge

Network learning: Expected value of information. Fundamental to our view of electronic networks which support learning activities is that there are a variety of mediators--both human and computer-based--at nodes on the network that control the flow of information from an existing location towards a new location where it is currently needed. One general characteristic of electronic networks is that information can flow rapidly through the network. Another aspect of the movement of information on networks involves decisions which are made about the nature and value of the information at network nodes that result in much more gradual movements of stored knowledge. The DNLF is based on the following general principle: *At each node in the network, information appearing at that node is stored locally if the expected value of storing that information is positive.* The expected value of information storage is the expected benefit minus the expected cost. For human mediators, the evaluation of expected value can be quite complex and situation specific, in contrast to computer mediators for which the expected value is typically crudely determined using relatively simple and well-structured criteria or rules.

The expected value depends on an estimate of the probability of needing the information again (i.e., a *prediction* problem). The simplest approach to prediction is to assume that the future will be like the past, and thus to predict that the likelihood of an event occurring in the future will be the same as the occurrence of the event in the past. So, if the computer-based mediator analyzed its log of past needs for a particular kind of information, it could, under this "the future will be like the past" assumption, predict that information needed frequently in the past will be needed again in the future. Mediators may also have other more sophisticated ways of predicting the future, but in the absence of such knowledge, the frequency rule can be used. The value of having stored the information locally is that the mediator does not have to go out on the network to get it when needed, which may entail a monetary cost or a time delay cost or both. Again, the mediator can infer the cost of future access (and the benefit from storing the information) by accessing its log or memory of past costs of accesses. The expected cost may also involve both the direct cost of storage and the increased cost of accessing already stored information. That is, each new piece of information can make the retrieval of previously stored information more difficult. One consequence of higher costs of accessing an increasing body of existing information is that the human mediator could be motivated to *restructure* the stored information to optimize access and reduce storage costs (see below).

As an example of network mediator interactions on an electronic network, consider a *person-to-person* interaction through electronic mail in which one person asks a question to another person on the network. Obviously, the person receiving the question must decide whether to respond, and if so, how much of a response is necessary. For the first individual who asked the question, the answer that is received must be evaluated and a decision made as to what to do with the information. Based on the framework presented above, the probability that the requestor would keep and use the received information would be higher if the expect value of the information was perceived to be high.

This notion of interacting network mediators can also be extended to include human and computer-based mediators. For example, there can be *person-to-computer* interactions over a network, such as when an individual uses a network to access an online database to obtain information on a topic. Perhaps a journal citation and abstract is found on the topic, but the person realizes the paper is old and therefore may contain out-of-date information. In this case, the probability that the human mediator would keep the information would be low. *Computer-based mediator-to-computer-based mediator* network interactions are also possible (although generally much cruder than human mediators), as when a software agent scans an online information source. The expect value of information would be high when a specific search criteria rule is matched, thus causing the software agent to obtain and store the information.

Network learning: Information optimization. There is another aspect of this probabilistic view of how the decision operations of human and computer-based mediators affect the flow of network information. One consequence of the DNLF general principal is that each node in the network attempts to optimize its functioning by storing things that are likely to be used again. Local storage of information thus occurs over time, essentially creating a local database of information. This local storage of information may also be internalized by the mediator, resulting in learning. There are several consequences of this. In a functional learning environment, there would be added value to information that is used repeatedly by a particular node. There would be a tendency to increase the judged probability of needing that information again, and thus the probability would be raised that the mediator at the node would attempt to store or learn the information. This rule also suggests that there would be a *graduate acquisition of expertise*, as over time the local storage of information would become increasingly richer and organized. The optimizing functioning that occurs at a node could be analogous to the learning processes of *accretion, tuning, and restructuring* (Rumelhart & Norman, 1978). Initially, there would be the mere accretion of information (both in terms of computer-based storage and the human memory representations) which would evolve into more differentiated and organized knowledge structures over time. These knowledge structures (e.g., for the computer mediator: rules or hypertextual views; for the human mediator: schema or mental models) would then serve as the basis from which the mediator would evaluate new network-based information and would be *tuned* as new information is locally stored and used at the node. Finally, it may become necessary to *restructure* or develop new organization structures in order to accommodate the ever changing informational flow over the network and the changing educational activities of the learning environment.

Networks and Cognitive Learning Theory

Much of the discussion of the DNLF has so far been at a somewhat abstract level, dealing with general notions of information flow, probabilities of storage, and so forth. However, the *raison d' etre* of an educational network is determined by the context in which it operates, which is to support learning and knowledge dissemination. This second aspect of the DNLF, *networks and cognitive learning theory*, thus provides an important perspective to understanding and structuring productive educational experiences over electronic networks. Due to the scope of this paper, it is not possible to provide a detailed discussion of all the important cognitive theories of learning that are relevant to network instructional activities. This section is therefore a brief and selected overview of a very broad topic.

Given the explicit function of computer networks to link individuals and groups together, there is a natural affinity between the DNLF and general theories of learning that take into account the extended social contexts of learning. We regard the seminal theories of Vygotsky (1978) as being of central importance to understanding the dynamics of network mediated learning activities. One key Vygotskian concept is the *zone of proximal development* (ZPD) which proposes that there is a progressive internalization of knowledge by the learner that occurs through the interaction of the learner and other members of their social environment (e.g., teachers, parents, other students). From this perspective, computer networks can function to extend the ZPD that learners are exposed to by providing a wider and richer set of social contexts for learning activities.

Further insight into the complex context of learning is provided by recent cognitive instructional research that has documented a number of problems learners have with acquiring and transferring complex knowledge (e.g., Gick & Holyoak, 1987; Lave, 1989). As many network-based learning activities involve conceptually demanding content areas or the application of knowledge to real world situations and problems, it is important that cognitive factors associated with learning complex knowledge be considered. Certainly the theories of learning focussing on the social context of learning have much to contribute to this area (e.g., Brown, Collins, & Duguid, 1989; Vygotsky, 1978). Also, there are other contemporary cognitive theories of learning which are relevant to the use of technology-based systems in instruction, such as *generative learning environments* (Cognition and Technology Group at Vanderbilt, 1991) and *cognitive flexibility theory* (Spiro, Feltovich, Jacobson, & Coulson, 1991). These theories of learning focus on different aspects of the learning and instructional processes, such as anchoring instruction in meaningful problem-solving contexts and the structure of the knowledge representations given to the learner. Indeed, the application of multiple theories of learning that each address different facets of the overall learning context may function in a synergistic way that could help inform the design and use of network learning activities to foster better learning of complex subject content.

The Human-network Interface

Given the many technical complexities associated with the current generation of computer networks, it is not surprising that "ease-of-use" has been identified as a significant factor for the successful conduct of instructional activities over electronic networks (e.g., Riel & Levin, 1990). We view this factor broadly to encompass the *human-network interface*, which has two main aspects. First, network "ease-of-*access*" involves logistical problems such as lack of computers, phone lines, or building network wiring pose serious obstacles to the educational use of computers. Ongoing work, such as the establishment of the National Research and Education Network (NREN), will provide critically needed support to help address this problem. Second, network "ease-of-*use*" must be considered, since even with access to the network infrastructure, there are still many computer-human interface factors that must be solved for educators and students using computer networks. There is thus a great need for software tools that are specifically designed to simplify and support the conduct of network-based learning activities.

The Message Assistant and the Distributed Network Learning Framework

The *distributed network learning framework* suggests kinds of software tools that people need to conduct effective network-based learning activities. These tools need to provide specific kinds of functionality that go beyond mere network access or menu driven interfaces. In this portion of the paper, we discuss features of a Hypercard-based interface to educational network resources called *The Message Assistant* (Levin & Jacobson, 1991). This program is intended to assist a network mediator in her or his access, organization, and distribution of network information. At a basic level, this program provides the standard types of electronic mail features, such as creating, sending, receiving, forwarding, and replying. In addition, *The Message Assistant* offers two important additional feature sets that are based on the DNLF: *message preprocessing* and *message organizing*.

Message Preprocessing

Users of national and international electronic networks can easily become inundated with large numbers of electronic messages. This overload has generated the need for electronic message filtering mechanisms (Malone, Grant, & Turbank, 1986). *The Message Assistant* permits the user to have messages preprocessed through a set of user-defined rules with conditions that trigger actions. Messages are initially presented in priority order, with the default priority being reverse chronological order (i.e., the program assigns recent messages a higher priority than older ones). However the user can create rules to influence the priority levels of messages. For example, messages could automatically have their priority level raised moderately by .4 (on a scale of -1 to 1). A different rule could check for messages from a specific group one is only casually interested and lower their priority. The user can easily ignore the message prioritizations since an overview listing of all messages is available and the user can select any messages to read. But we anticipate that when the user is confronted with a large number of new messages to read, the preprocessing and prioritization of messages based on the user's own customized set of rules will prove helpful to the user in deciding which messages to read immediately and which to read later. *The Message Assistant* also allows the user to specify rules for automatically answering, forwarding, and classifying messages.

There are several ways in which these features instantiate aspects of the DNLF. The *network mediator and flow of information and knowledge* aspect holds that probabilistic evaluations of network information are made by a mediator at a particular node. The message rules function as a computer-based mediator that assists the human mediator in making an initial determination of the expected value of the information. The message rules form a user-defined expert system which is incrementally specified and tested over time and thus gradually increases its expertise with use. Such an expert system serves as a computer-mediator that takes over repetitive or low-level evaluations of the expected value of certain types of network information and then automatically initiate actions which filter, route, or store that information. Information that does not match any prespecified rules is passed on to the human mediator to be evaluated, so this kind of "expert system" is robust. If the new information is regarded as being of value, then the human mediator will operate on that information.

The use of rules in *The Message Assistant* also contributes to the *human-network interface* by reducing some of the technical and cognitive load on the human mediator. At a simple level, the prioritization of messages helps the mediator determine what to read first. Also, the rules function as an automatic filtering and routing agent, thus reducing the overall number of messages that the mediator has to explicitly deal with. At a more substantive level, the rules can assist the mediator in organizing the complex knowledge that is contained in a large corpus of messages by automatically creating hypertextual links between messages (see below).

Message Organizing: Hypertextual Links for Knowledge Structuring

As discussed in the first section of the paper, a central assertion of the *distributed network learning framework* is that network learning activities involve the flow of information from locations where the information is stored or available to different nodes on the network based on the learners' information needs. The mere existence of a network infrastructure is a necessary but not a sufficient condition for significant *learning* to occur over educational networks. *The Message Assistant* is being designed to provide message organizing techniques that we hope will prove useful in helping users transform information into usable knowledge while they are participating in learning activities conducted over the network. Message organizing has been implemented both through the user-defined expert system (i.e., message prioritization, see above) and through two types of hypertextual linking mechanisms: *fixed links* and *virtual hypertext links*.

Fixed links. Specific fixed links can be manually created between different messages by the user or automatically by the program when replying or forwarding a message. These "conventional" hypertext links allow the user to organize and access messages in a nonlinear manner.

Virtual conceptual hypertext links. A more powerful feature for knowledge organization involves the ability of the program to create multiple sets of *virtual hypertext links* based on conceptual interrelationships between various messages. We refer to these virtual conceptual hypertext links as "views" of the locally stored messages. Two default views of the messages are "In View" and "Out View" that correspond to received and sent messages. The user may then create new views of the messages that share a common topic, theme, issue, or other conceptual bond. Each view represents one possible set of hypertextual links between a subset of the messages; switching to a different view reconfigures the links between messages. As many messages contain information that deal with several issues, it is possible to assign the same message to several views. In addition, the program allows both the user-defined rules to automatically assign messages to a view and for the mediator to manually assign a message to a view. Overall, this ability to manually and automatically create views and then access the messages from these multiple conceptual perspectives is intended to help the mediator create organized information and lead to the generation of useful knowledge.

Message organizing and the DNLF. As with the preprocessing of messages, there are several ways in which the message organizing functionality of *The Message Assistant* has been guided by the DNLF. The user-defined incremental expert system is very specifically concerned with the *flow of information* into and out of the node. While these rules function to automate the initial decision process concerning the value, local storage, and flow of information in the preprocessing stage, the rules can also serve to record and automatically apply aspects of experience and expertise of the human mediator at a higher conceptual level through the hypertextual message view linkages.

The virtual hypertext links feature helps to instantiate the DNLF goal of *optimization* (i.e., network nodes attempt to optimize their functioning by locally storing potentially useful information). The views function provides the mediator with a mechanism to create a personally meaningful knowledge-base that is also flexible in terms of organizing and accessing the information in the messages. We expect to see a *gradual acquisition of expertise* embedded in the computer-based mediator of *The Message Assistant* in terms of the user-created rules and views and in the human mediator in terms of acquired knowledge. With a functional learning environment, such as using the network for educational activities, there will be information that repeatedly comes to a particular node and thus is judged as having a higher expected value, leading to the local storage of the information and its incorporation into the rule and view structures. Finally, over time the specification of message views and rules would progress from gradual accretion, to the tuning use of a body of accumulated views and rules, to the restructuring of views and rules as the human mediator works with the dynamic and changing learning environment.

There is another aspect of the message organizing features of *The Message Assistant* that is related to the DNLF. The second aspect of the DNLF, ***networks and cognitive learning theory,*** concerns "the bottom line" of a technological learning environment: that students be able to acquire usable information and knowledge as a result of their use of instructional activities. The cognitive demands associated with learning complex knowledge have been receiving increasing attention in the cognitive sciences in terms of theoretical and research work (e.g., Glaser, 1990; Spiro et al., 1988; Vygotsky, 1978). There are several aspects of *The Message Assistant* that attempt to implement various aspects of these current cognitive views of learning. As noted earlier, learning theories concerned with the social and situated contexts of learning are central to the DNLF. *The Message Assistant* is explicitly designed to function in a social context and to be a part of learning activities that involve students in real world problems and issues (e.g., Levin et al., 1987). The preprocessing and message organizing functions of the program are intended to support both collaborative learning activities and a more recent notion of *teleapprenticeships* (Levin et al., 1987). Teleapprenticeships incorporate apprenticeship modes of learning into the network environments through using the network to allow novices and experts to interact in educationally significant ways by overcoming logistical and pragmatic constraints such as work schedules or geographical location. The message preprocessing and organizing features of *The Message Assistant* will prove valuable in supporting teleapprenticeship types of network learning activities and is the one aspect of our current research.

Researching the Distributed Network Learning Framework

In the research to evaluate the DNLF, there are several primary questions we are interested in, such as: What sorts of predictions does the framework make concerning interactional patterns and the large scale flow of network-based information? How can software tools help support the transformation of fluid network information into structured and usable knowledge? What are the implications of the framework for guiding the optimal organization of the network for learning?

Because there are specific probability rules associated with the *flow of information and knowledge* aspect, we plan to construct models of information flow and then to compare those models to specific data on message interactions. *The Message Assistant* is currently being used by network mediators. Data collection routines in the program record a number of facets of the use of the program, particularly with respect to message flow patterns and the specification, evolution, and use of rules and views. Also, an important research agenda concerns factors associated with learning, particularly attempting to ascertain the effectiveness of the learning activities being conducted over the network in terms of helping students acquire knowledge they can use in new ways and in new contexts.

Summary

This paper describes portions of a research project which is developing an electronic mail and knowledge organization program, *The Message Assistant*. This program is intended to promote higher-order learning goals as a part of instructional activities conducted over distributed educational networks. The design of this software is based on a cognitively-oriented view of educational networks that we refer to as the *distributed network learning framework*. The DNLF consists of three main aspects: *network mediators and the flow of information and knowledge*, *networks and cognitive theories of learning*, and *human-network interface*. We regard the framework as a *flexible*--and *extendable*--set of perspectives from which to examine and to work with different aspects of this complex, multifaceted, and dynamically evolving learning technology. While we feel that there are numerous ways in which the *distributed network learning framework* may be applied to educational electronic network research, this paper discusses the framework in terms of the user-defined expert system and hypertextual features of *The Message Assistant* program. The research on models derived from this framework can contribute to the design and use of network-based learning environments that enhance the learning opportunities and outcomes of students.

References

Brown, J. S., Collins, A., & Duguid, P. (1989). Situated cognition and the culture of learning. *Educational Researcher*, **18**(1), 32-42.

Cognition and Technology Group at Vanderbilt. (1992). Technology and the design of generative learning environments. *Educational Technology*, **31**(5), 34-40.

Gick, M. L., & Holyoak, K. J. (1987). The cognitive basis of knowledge transfer. In S. M. Cormier & J. D. Hagman (Eds.), *Transfer of learning: Contemporary research and applications* (pp. 9-46). New York : Academic Press.

Glaser, R. (1990). The reemergence of learning theory within instructional research. *American Psychologist*, **45**(1), 29-39.

Hunter, B. (1992). Linking for learning: Computer-and-communications network support for nationwide innovation in education. *Journal of Science Education and Technology*, **1**(1), 23-34.

Lave, J. (1988). *Cognition in practice: Mind, mathematics and culture in everyday life*. Cambridge: Cambridge University Press.

Levin, J. A., & Jacobson, M. J. (1991). *The Message Assistant: A rule-based electronic mail processor for collaborative electronic learning environments*. Software demonstration given at the Thirteenth Annual Meeting of the Cognitive Science Society. Chicago, IL: University of Chicago.

Levin, J. A., Riel, M., Miyake, N., & Cohen, M. (1987). Education on the electronic frontier: Teleapprentices in globally distributed educational contexts. *Contemporary Educational Psychology*, **12**, 254-260.

Malone, T. W., Grant, K. R., & Turbank, F. A. (1986). *The information lens: An intelligent system for information sharing in organizations* (CISC WP No. 133). Boston, MA: Center for Information Systems Research.

Riel, M. M., & Levin, J. A. (1990). Building electronic communities: Success and failure in computer networking. *Instructional Science*, **19**, 145-169.

Rumelhart, D. E., & Norman, D. A. (1978). Accretion, tuning and restructuring: Three modes of learning. In J. W. Cotton & R. L. Klatzky (Eds.), *Semantic factors in cognition*. Hillsdale: NJ: Lawrence Erlbaum.

Spiro, R. J., Feltovich, P. J., Jacobson, M. J., & Coulson, R. L. (1991). Cognitive flexibility, constructivism, and hypertext: Random access instruction for advanced knowledge acquisition in ill-structured domains.. *Educational Technology*, **11**(5), 24-33.

Vygotsky, L. S. (1978). *Mind in society: The development of higher psychological processes* (M. Cole, V. John-Steiner, S. Scribner, & E. Souberman, Eds.). Cambridge, MA: Harvard University Press.

A Theory of Dynamic Selective Vigilance and Preference Reversal, Based on the Example of New Coke

Daniel S. Levine
Dept. of Mathematics, University of Texas at Arlington, Arlington TX 76019-0408, b344dsl@utarlg.uta.edu

Samuel J. Leven
For a New Social Science, 4681 Leitner Drive West, Coral Springs, FL 33067

Abstract

A neural network theory of preference reversal is presented. This theory includes a model of why New Coke was preferred to Old Coke on taste tests but was unpopular in the market. The model uses competing drive loci representing "excitement" and "security." Context influences which drive wins the competition, hence, which stimulus attributes are attended to. Our network's design, outlined in stages, is based on Grossberg's gated dipole theory. Three sets of dipoles, representing attributes, categories, and drives, are connected by modifiable associative synapses. The network also includes competition among categories and enhancement of attention by mismatch of expectation.

Introduction: Modeling of Irrational Decisions

How rational are we? Tversky and Kahneman (1974, 1981) established that many human decisions do not maximize a measurable utility function. Moreover, deviations from rationality show patterns; for example, decisions among losses are more risk-taking than decisions among gains. Since decision irrationalities are repeatable, they lend themselves to quantitative modeling. Yet models of these effects lag behind models of other cognitive effects, such as pattern classification. Tversky and Kahneman modeled their own data using a non-connectionist theory whereby subjects maximize a nonlinear function of expected gains and losses. However, these authors did not explain how this function arose in the underlying system.

Differential reaction to gains and losses shows that the projected affective value of decisions depends on expectations generated by the current environment. Many neural networks compare current and ongoing values of stimulus or reinforcement variables (Grossberg, 1972; Sutton & Barto, 1981). We use Grossberg's gated dipole theory, to be described below, because it accounts best for stimulus duration effects in conditioning (see Grossberg & Levine, 1987). Grossberg and Gutowski (1987) constructed a gated dipole model for Tversky and Kahneman's data, including the data on gains versus losses. These authors captured the essence of decision under risk despite basing choices on maximizing a single function (affective value). Our model is in the spirit of Grossberg and Gutowski's but adds effects not present in their model: dynamically competing attractions to novel and to familiar stimuli, and competition between drive loci.

Much of Tversky and Kahneman's data involves imagined monetary gains and losses, so their results can be applied to economics. Leven (1987) argues that optimization theory, which dominates economic modeling, is not predictive and must be replaced by theories that include affective factors. Leven and Elsberry (1990) simulate "negotiations" between two neural networks that contain both rational and affective modules. We carry this work further by studying a famous economic example: the failure of New Coke in the market after it had defeated Old Coke in double-blind taste tests. The work of Tversky, Kahneman, and Grossberg readily suggests a qualitative model of the Coke data. Network instantiation of this model, however, led to a complex combination of three sets of gated dipoles representing attributes, categories, and drives; competition among categories and among drives; and associative learning of inter-dipole connections. We first describe the Coke data in detail, then develop our network in stages.

The Coke Data

When the Coca-Cola Company introduced New Coke, it was certain of the flavor's acceptance. Tens of thousands of subjects had undergone highly controlled taste tests. The new flavor had outscored all its competition, including victory over Old Coke by a margin of 2 to 1. Further tests hinted that less than ten percent of Old Coke drinkers would object to the new flavor combined with the old name. As most Americans know, the actual buying situation had very different results. New Coke was so unpopular that the company had to return Old Coke to the market (Oliver, 1986).

Coca-Cola had asked people, "If New Coke were introduced, would you like it?" But the influence of dynamic emotional states means that mental projections of the future are often inaccurate (Holbrook *et al.*, 1985). In the test situation, people based preferences on the direct appeal of taste. In the market, indirect emotional factors, such as memories associated with expected taste, were more important than taste itself. Moreover, buying was different from tests because the Coca-Cola Company was so confident in its research that *Old Coke was unavailable*. The public's reaction against buying New Coke was a *frustrative rebound*. The Coke label created expectation of a particular taste, and of the secure feeling it evoked, which led to frustration when this feeling was absent.

Results of Pierce (1987) support frustration theory. Pierce compared responses to advertisements of old and new versions of Coke by people who had been habitual Coke drinkers and by habitual drinkers of other drinks (such as Pepsi). By a small but significant margin, habitual Coke drinkers were more hostile to products they perceived as New Coke than were non-Coke drinkers.

Frustrative rebound is an example of comparing current with expected or ongoing reinforcement. Just as cessation of a negative reinforcer (*e.g.*, electric shock) is positively reinforcing (provides relief), cessation of a positive reinforcer, or its absence when it is expected, is negatively reinforcing (provides frustration). We will now review how gated dipole networks model both effects.

Background: Gated Dipole Networks

How can a response associated with *offset* of *negative* reinforcement become itself positively reinforcing? To answer this question, Grossberg (1972) introduced the network shown in Fig. 1. The synapses w_1 and w_2 have a chemical transmitter that is depleted with activity. The input J could be shock, for example. The input I is nonspecific arousal to both channels y_1-x_1-x_3 and y_2-x_2-x_4. While shock is on, left channel activity x_1 exceeds right channel activity x_2, leading to net positive activity of the left channel output node x_3. For a short time after shock ceases, both channels receive equal inputs I but the right channel is less depleted of transmitter than the left channel. Hence, x_2 now exceeds x_1, leading to net positive activity of the right channel output node x_4 The active output node excites or inhibits x_5, thus enhancing or suppressing some motor response. The network is called a gated dipole because it has two opposite ("negative" and "positive") channels that "gate" signals based on amounts of transmitter. If the two channels in Fig. 1 are reversed in sign so the channel receiving input is positive, the network explains frustration when positive reinforcement either ceases or is absent when expected.

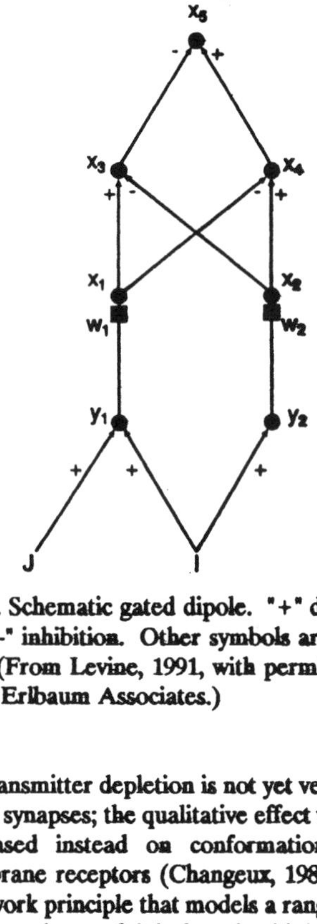

Fig. 1. Schematic gated dipole. "+" denotes excitation, "-" inhibition. Other symbols are explained in text. (From Levine, 1991, with permission of Lawrence Erlbaum Associates.)

Transmitter depletion is not yet verified in many actual synapses; the qualitative effect we model may be based instead on conformation changes at membrane receptors (Changeux, 1981). However, a network principle that models a range of cognitive data can be useful before its biological basis is known, and its later verification is likely even if in a different form than first proposed.

Network Modeling of Coke Data: Combining Sensory and Motivational Dipoles

We build our network for modeling the Coke data in several stages. The network has submodules denoting sensory features, object categories, and drives; simulations are in progress. The rationale for our network architecture can best be shown by starting with simpler networks that model the data partially, then modifying those networks to fit further details of the data.

Gated dipoles instantiate the idea of opponent processing, which applies to vision as well as motivation. For example, there are pairs of opponent colors (*e.g.*, green and red), and each color is transiently perceived after removal of the other. Grossberg (1980) introduced dipoles whose channels consist of "on" and "off" nodes encoding presence or absence of specific stimuli, then joined channel pairs for various stimuli into a *dipole field*. Leven and Levine (1987) discussed how a dipole field could embody competing attractions to previously reinforced stimuli (here, Old Coke) and to novel stimuli (here, New Coke). If drive is high, or reward signals strong, previously reinforced stimuli are favored. If drive is low, novel stimuli are favored. Leven and Levine's first approximation to a model of the Coke data treated testing as a low-motivation state and buying as a high-motivation state, hence explaining preference reversal between the two contexts. They noted an analogy to some monkey data on novelty preference.

Pribram (1961) compared normal rhesus monkeys and those with frontal lobe lesions in a scene with several objects. Successive objects are added to the scene, unobserved by the monkey. Each time a novel object is introduced, a reward (peanut) is placed under the novel object. When the monkey has lifted this object a fixed number of times, the next object is added. Pribram measured the number of errors (liftings of a familiar object) before the monkey first selects the novel object. Frontally lesioned animals are more attracted to novelty than normals so make fewer errors. Fig. 2 shows the dipole field used to model Pribram's data, which was simulated in Levine and Prueitt (1989). The dipole channel pairs in Fig. 2 correspond to an old cue and a novel cue. The nodes $x_{1,5}$ and $x_{2,5}$ represent tendencies to approach given cues. Inhibition between these nodes, and a node $x_{3,5}$ coding some other environmental cue, denote competition between attractions to different cues. The cue with largest $x_{i,5}$ is approached.

The network of Fig. 2 incorporates two competing rules. The on channel corresponding to the novel cue is less depleted than the on channel for the old cue, because that channel has not been active as long. Hence, competition among $x_{i,5}$ nodes favors those corresponding to novel cues, all else equal. But also the reward node is active when the monkey finds the peanut. Each $x_{i,5}$ connects with the reward node via synapses which are strengthened when the corresponding cue is rewarded. Hence, competition also favors $x_{i,5}$'s with strong links to the reward node, all else equal. The frontal lobes are identified with gain from reward nodes to sensory dipoles. If this gain is high, as in a normal monkey, the dipole output $x_{1,5}$ for the previously rewarded cue is the larger. If gain is low, as in a lesioned monkey, the output $x_{2,5}$ for the novel cue is larger.

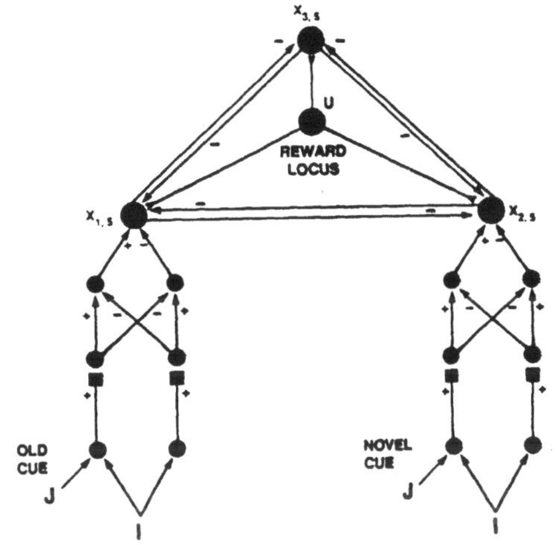

Fig. 2. Dipole field used to simulate novelty data. Semicircles denote modifiable weights. (Adapted from Levine & Prueitt, 1989, with permission of Pergamon Press.)

Leven and Levine (1987) noted that the Coke data could be approximately modeled by the network of Fig. 2, with "New Coke" identified with "novel cue," "Old Coke" with "old cue," "testing" with "frontally damaged," and "buying" with "normal." Of course, most people taking the taste test are not brain-damaged. The analogy is plausible, though, because humans with frontal damage tend to be less goal-directed than normal humans (Fuster, 1989); hence, their day-to-day life is closer to a "play" than to a "serious" situation.

Yet the network of Fig. 2 is inadequate to model the Coke data for at least two reasons. First, in the market, there was not a choice between New Coke and Old Coke as in tests. Hence, relative value attached by buyers to the two drinks must be inferred indirectly from relative preference for New Coke and for non-Coke drinks (*e.g.*, Pepsi). Second, consumers' angry reaction to the change in Coke was not based on taste alone. As one Coca-Cola executive said later, "We were spitting on the American flag and didn't know it." Hence, a realistic model of the Coke data incorporates two competing drives: one for taste, the other for a range of feelings which we label "Security"; these will be added below.

The network of Fig. 2 contains a node that represents a reward signal. We can model frustrative rebound if we replace the reward node by an entire gated dipole representing a specific drive. *Drive representations* have been used in Pavlovian conditioning models (*e.g.*, Grossberg & Levine, 1987). They are based on the theory that conditioning involves learning an association not between a conditioned stimulus (CS) and a specific response or another stimulus, but between a CS and a positive or negative *emotional value*. They are analogs of brain loci (mainly in the hypothalamus) for hunger, thirst, sex, curiosity, *etc*.

In summary, sensory representation dipoles model differences between novel events, whose representations are less depleted of transmitter, and old events, whose representations are more depleted. Motivational (drive) representation dipoles model emotional value attached to changes in received positive or negative reinforcement, such as occur with relief or frustration. Hence, the next stage in our Coke model is to join sensory and motivational dipoles (Fig. 3).

Combined sensory and motivational dipoles may also account for conditioning phenomena such as *unblocking* (Kamin, 1969). *Blocking* has been simulated in neural networks (Grossberg & Levine, 1987; Sutton & Barto, 1981). If a bell, say, is paired with shock and an animal learns a fear response to the bell, then a bell-light combination is paired with the same shock, no fear is learned to the light. The light is *unblocked* when the shock level paired with the bell-light compound is unequal to the level paired with bell alone. Since unblocking involves associating a novel stimulus (light) and a changed affective value, it may be modeled by the network of Fig. 3.

Network Modeling of Coke Data: Categorizations and Multiple Attributes

New Coke elicited strong reactions because of how it was *different* from Old Coke, but also because of how it was *like* Old Coke! The public reacted to a new taste *combined with an old label*. Pierce's (1987) data show that the reaction was stronger when larger positive affect was attached to the old taste. Hence, our modeling problem becomes how to build a network to respond to stimuli that contain both a novel element and a familiar, significant element. For example, Robert Dawes (personal communication) has discussed a network model of seeing the Mona Lisa with a mustache added. We find that picture grotesque because the mustache mismatches expectations produced by the rest of the picture.

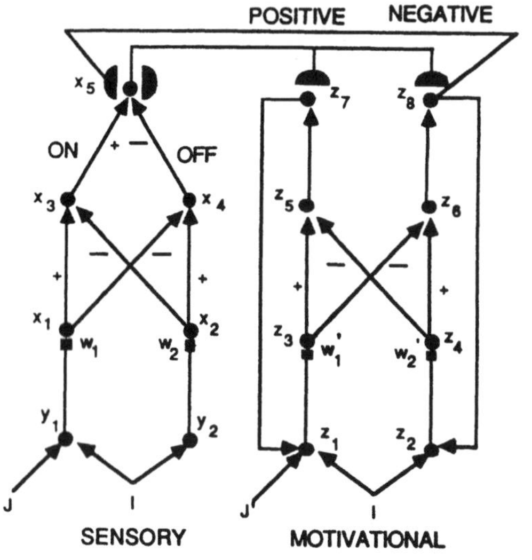

Fig. 3. Combination of sensory and motivational dipoles. Dipole output for a sensory stimulus can be conditioned to positive or negative reinforcement, as shown by modifiable connections at top.

To deal with expectation, we refine our model so that New and Old Coke are no longer single stimuli but vectors of attributes, *each attribute represented by its own gated dipole* (Fig. 4). We use the minimal set of attributes needed: Coke Label; Familiarity; Taste; Pepsi Label. The latter is introduced to model the switch from New Coke to competing cola drinks (lumped together as "Pepsi" for simplicity), or else to avoidance of all soft drinks, when Old Coke was unavailable.

How does this explain the data of Pierce (1987) on habitual Coke drinkers versus habitual Pepsi drinkers? If a network is to represent general cognitive principles, it should also, if possible, account for individual differences. Major behavioral differences can arise from differences in one or a few network parameter values. In Fig. 4, let weights (in both directions) between the on side of the Coke Label attribute dipole and the positive side of the motivational dipole be higher in one copy of the network than in another. Then the first network models a (generic) habitual Coke drinkers, whereas the second models a habitual Pepsi drinker. Analogously, let corresponding weights to and from the Pepsi Label attribute dipole be higher in the second network. Because of feedback from drive to sensory loci in the network of Fig. 4, the expected positive affective value from seeing the Coke label is greater in the habitual Coke drinker. Hence, frustrative rebound from mismatching expectations generated by that label is also greater in habitual Coke drinkers.

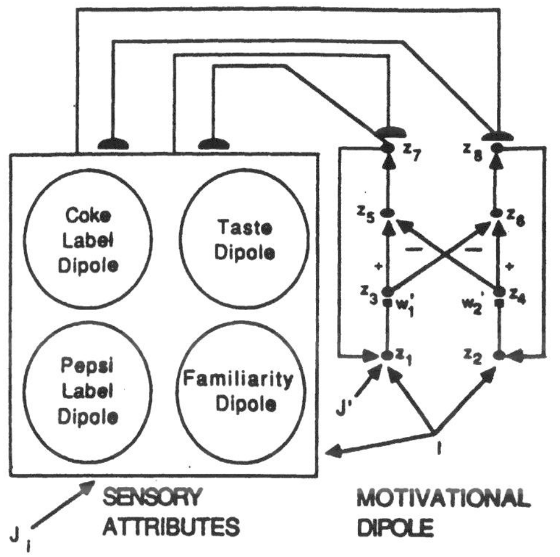

Fig. 4. Extension of network of Fig. 3 to include dipoles for each stimulus attribute, with inputs J_i to each. Circles represent sensory dipoles, not shown in full here. Each sensory dipole has modifiable reciprocal connections with positive and negative motivational channel outputs, z_7 and z_8.

At this point, we must revise our account of the difference between testing and buying. To a first approximation, we have treated testing as a "low motivation" context which thereby disinhibits the attraction to novelty. Yet when we look at attributes, what is actually different between the two contexts is not the *amount* of motivation but rather the *focus* of motivation. During testing, the *intrinsic* (taste-related) attractiveness of the product is important, and the *socially learned* attractiveness of the product much less so. Hence, the Taste attribute (Fig. 4) plays a larger role in categorizations and decisions during testing than does the Familiarity attribute. The Familiarity attribute, by contrast, plays a larger role during buying.

Now we need a theory of context-based attentional switches between attributes. Such a theory includes multiple sensory dipoles *and* multiple motivational dipoles (Fig. 5). Here, two dipoles are labeled "Excitement," *i.e.*, desire for sensory or aesthetic pleasure, and "Security," *i.e.*, desire for a sense of belonging, affiliation, or rootedness in one's society or relationships (*cf.* McClelland, 1961). We posit competition between the positive sides of the Excitement and Security dipoles in Fig. 5, and assume that the "winner" of the competition changes with context (For a history of relevant network models, see Levine, 1991, pp. 133-134). If feedback connections between the Excitement motivational dipole and the Taste attribute dipole, and between the Security motivational dipole and the Familiarity attribute dipole, are much stronger than cross-connections, the winning drive determines which sensory attributes are attended to.

The network we simulate for a future article makes two additions to that of Fig. 5, which are omitted from our figures for space reasons. One addition is category nodes. If habitual Coke drinkers attach positive affect to the *Coke category* as well as the Coke Label attribute, this enhances expectation of positive value from drinking any Coke product, thus increasing frustration when New Coke mismatches that expectation. Our current network includes modifiable feedback between category nodes and attribute nodes, in the manner of ART networks (Carpenter & Grossberg, 1987). Category nodes also connect directly with motivational dipoles. Hence, the affective value of a category can differ from values of the category's exemplars (*e.g.*, one can love humanity and hate people, or vice versa). In ART, the input vector is compared with stored category prototypes, and classified with any prototype that it mismatches to less than a prescribed amount (*vigilance*). Our model posits that vigilance is dynamically feature-selective. If, for example, the current attentional bias favors the Familiarity attribute over the Taste attribute, the network is selectively sensitive to mismatch with the Coke prototype in the Familiarity dimension. The amygdala might be a brain locus for such a bias mechanism (Pribram, 1991).

Our second addition is modulation by mismatch signals of perceived time durations of inputs, which was introduced by Ricart (1992). Ricart identified the modulatory node with the midbrain locus ceruleus, which produces norepinephrine and focuses attention on significant or novel stimuli. This node nonspecifically sharpens perception of both stimuli and reinforcements after mismatch generated by the attribute-category subsystem.

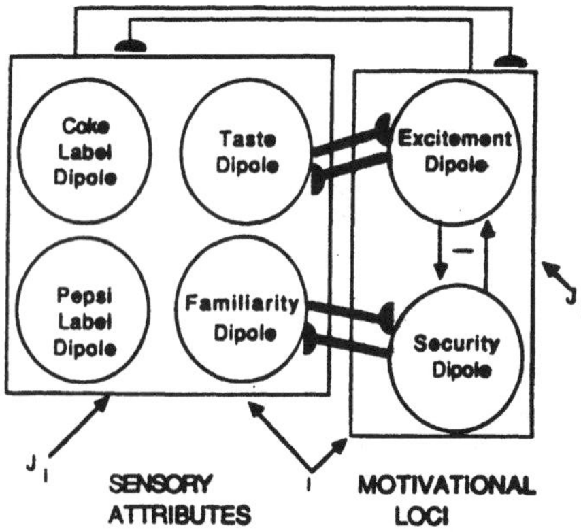

Fig. 5. Extension of network of Fig. 4 to include competing motivational dipoles (shown as circles). Darker lines indicate stronger connections.

References

Carpenter, G., and Grossberg, S. 1987. A Massively Parallel Architecture For a Self-organizing Neural Pattern Recognition Machine. *Computer Vision, Graphics, and Image Processing* 37: 54-115.

Changeux, J.-P. 1981. The Acetylcholine Receptor: an "Allosteric" Membrane Protein. *Harvey Lectures* 75:85-254.

Fuster, J. 1989. *The Prefrontal Cortex*. New York: Raven.

Grossberg, S. 1972. A Neural Theory of Punishment and Avoidance. II. Quantitative Theory. *Mathematical Biosciences* 15, 253-285.

Grossberg, S. 1980. How Does a Brain Build a Cognitive Code? *Psychological Review* 87:1-51.

Grossberg, S. and Gutowski, W. 1987. Neural Dynamics of Decision Making Under Risk: Affective Balance and Cognitive-Emotional Interactions. *Psychological Review* 94:300-318.

Grossberg, S. and Levine, D. 1987. Neural Dynamics of Attentionally Modulated Pavlovian Conditioning: Blocking, Interstimulus Interval, and Secondary Reinforcement. *Applied Optics* 26:5015-5030.

Holbrook, M.; Moore, W.; Dodgen, G.; & Havlena, W. 1985. Nonisomorphism, Shadow Features, and Imputed Preferences. *Marketing Science* 4:215--233.

Kamin, L. 1969. Predictability, Surprise, Attention, and Conditioning. In B. Campbell & R. Church (Eds.), *Punishment and Aversive Behavior*, 279-296. New York: Appleton-Century-Crofts.

Leven, S. 1987. Choice and Neural Process. Ph. D. diss., School of Urban Studies, University of Texas at Arlington.

Leven, S. and Elsberry, W. 1990. Interactions Among Embedded Networks Under Uncertainty. In International Joint Conference on Neural Networks, San Diego, Vol. III, 739-746. Ann Arbor, MI: IEEE Neural Network Council.

Leven, S. and Levine, D. 1987. Effects of Reinforcement on Knowledge Retrieval and Evaluation. *IEEE First International Conference on Neural Networks*, 279-296. San Diego:IEEE/ICNN.

Levine, D. 1991. *Introduction to Neural and Cognitive Modeling*. Hillsdale, NJ: Erlbaum.

Levine, D. and Prueitt, P. 1989. Modeling Some Effects of Frontal Lobe Damage: Novelty and Perseveration. *Neural Networks* 2:103-116.

McClelland, D. 1961. *The Achieving Society*. Princeton, NJ: Van Nostrand.

Oliver, T. 1986. *The Real Coke®, the Real Story*. New York: Random House.

Pierce, W. 1987. Whose Coke Is It? Social Influence in the Marketplace. *Psychological Reports* 60:279-286.

Pribram, K. H. 1961. A Further Experimental Analysis of the Behavioral Deficit that Follows Injury to the Primate Frontal Cortex. *Journal of Experimental Neurology* 3:432-466.

Pribram, K. H. 1991. *Brain and Perception*. Hillsdale, NJ: Erlbaum.

Ricart, R. 1992. Neuromodulatory Mechanisms in Neural Networks and Their Influence on Interstimulus Effects in Pavlovian Conditioning. In D. Levine & S. Leven, eds., *Motivation, Emotion, and Goal Direction in Neural Networks*, 117-166. Hillsdale, NJ: Erlbaum.

Sutton, R. S. and Barto, A. G. 1981. Toward a Modern Theory of Adaptive Networks: Expectation and Prediction. *Psychological Review* 88:135-170.

Tversky, A. and Kahneman, D. 1974. Judgment Under Uncertainty: Heuristics and Biases. *Science* 185:1124-1131.

Tversky, A. and Kahneman, D. 1981. The Framing of Decisions and the Rationality of Choice. *Science* 211:453-458.

Why are Situations Hard?

Michael Lewis

Department of Information Science
University of Pittsburgh
Pittsburgh, PA 15260
ml@icarus.lis.pitt.edu

Abstract

An ecological model of human information processing is introduced which characterizes *intuition* as a state oracle providing information for particular types of situations for which attunements to constraints have been developed. The consequences of this model are examined showing among other things that: for a cognitive task with a fixed problem space difficulty can only be reduced by introducing metaphor, difficulty of translation is minimum for a situationally equivalent metaphor, a situationally equivalent metaphor preserves and reflects extrinsic information about the situation, any situation containing a subcategory isomorphic to a problem situation can be made into a metaphor by supplying instructions, these characteristics can be exploited by an algorithm which chooses a metaphor in such a way that attunements are substituted for problem constraints and instructions are used as an "error term".

Introduction

Despite twenty years of experience and widespread commercial success cognitive principles underlying the effectiveness of direct manipulation and visualization interfaces remain a mystery. Advertising brochures and users glibly describe them as *intuitive*, *direct*, or user-friendly but neither psychologists nor computer scientists can agree on exactly what these terms mean.

Lewis (1991,1992) has proposed an ecological model based on *interactive situations* which operationalizes "directness" and "intuition". Cognition is presumed to operate on situations involving objects in relations rather than propositions about them. The dynamics of these mental situations are governed by attunements to pervasive regularities in our environment such as object constancy. These attunements are presumed to be automatic processes. This "mental model" is animated by initiating actions which are either imagined or perceived. The mental events which follow, unfold in accordance with our attunements to the constraints affecting the situation. By modeling courses of events the mental model makes the resultant states available to cognition. The novelty of this approach to mental models lies in incorporating the decomposition of situations into states of affairs and constraints borrowed from situation theory (Barwise & Perry, 1983) and the characterization of cognitive tasks as search of a problem space (Newell & Simon 1972). This synthesis allows a unified treatment of task difficulty and metaphor as allocation of processing problems. The model and its consequences provide a framework for automating the design of cognitively efficient scientific and problem visualizations and the design and evaluation of graphical user interfaces.

Intuition is presumed to describe the effects of attunements (mental constraints) developed in response to the regularity of certain events in our environment. Attunements are associated with particular *types* of situations and to allow us to imagine/update states of these situations automatically. This process of automatically updating states is referred to as *envisioning*. A related process of *inspection*, makes this state information available for conscious processing. Together, envisioning and inspection form a cycle which acts as a "state oracle" by supplying information about changes in state at essentially no cost.

The model attributes the difficulty of cognitive

This research was supported by
NSF grant IRI-9020603.

Table 1: Problem Isomorphs

	Form	Tower of Hanoi	Monster Globe Move	Monster Globe Change
anchored (ordinal)	a_i	disk size	globe size	monster size
unanchored (nominal)	u_j	disk location	globe location	globe size
Rule 1	$a_i > a_j \wedge u_k.a_i \wedge u_k.a_j \rightarrow \neg \Delta u_l.a_i$	By Attunement	A monster may only pass its largest globe	If monsters hold globes of the same size, only the largest can change
Rule 2	$a_i > a_j \wedge u_k.a_i \wedge u_l.a_j \rightarrow \neg \Delta u_k.a_i$	A larger disk may not be moved ontop of a smaller disk	A monster may not pass its globe to a monster holding a larger globe	A monster may not change its globe to a size held by a larger monster

tasks to two factors: 1) the *intrinsic difficulty* associated with the size and complexity of the problem space and 2) the *extrinsic difficulty* associated with the controlled processing needed to update states and supply constraints in searching that space. Attunement to constraints makes cognitive tasks easier by eliminating some "illegal" events from the problem space. Where problem constraints and attunements perfectly coincide the frame problem is resolved and difficulty is limited to that of searching legal states. This model of cognitive difficulty can be illustrated using the Tower of Hanoi and two of its isomorphs. Subjects find the Monster-Globe problems much more difficult. (Hayes & Simon, 1977), for example, reports differences in average solution times of less than two minutes for the three disk Tower of Hanoi problem, and half an hour for the corresponding Monster-Globe (change) problem. The Monster-Globe (move) problem is of intermediate (14 min). (Kotovsky, Hayes, & Simon, 1985) difficulty. The Monster-Globe change problem is the most difficult because it violates object constancy, a basic attunement which plays a primary role in theories of psychology ranging from cognitive development to perception. Searching its problem space requires the use of limited working memory resources to determine the changes in state resulting from actions because events do not follow environmental constraints to which we are attuned. The Monster-Globe move problem relates states through the movement of objects, to which we are attuned and therefore eliminates the need to use controlled processing and intermediate storage to update states. The problem space made available through these attunements, however, is substantially larger than the official one because we can envision globes being moved among any of the monsters, while the problem rules constrain these movements. Because rule 1 requires information about the initial state of a move and rule 2 requires information about its terminating state, both states and the linking action must be referenced to apply the problem rules. In the Tower of Hanoi rule 1 is subsumed by attunements and violation of rule 2 is determinable by inspection alone, because of the illegal state which results. As a consequence we are mentally constrained to ignore movements of disks from the bottom of stacks (rule 1) and can judge legality by inspecting the terminating state (rule 2) without additional reliance on working memory. This reduces the problem to a controlled search of a space of 50 states and 75 transitions in which each of the 36 prohibited moves are ruled out by inspection for the illegal ''larger on top of smaller'' state at a path of length of 1.

As these examples illustrate, cognition is conceived to be a heterogeneous mixture of automatically updating models and resource consuming rules. A commonsense interpretation of this dichotomy is that cognitive tasks are *direct*, *intuitive*, and easy to the extent that they do not require instructions. The model is analogous to a computer with a limited capacity general purpose processor and a high capacity specialized one. The most efficient program for such a machine will be one which balances the costs of translating data for specialized processing with the savings it offers. This paper examines the consequences of treating human information processing in the same way.

Consequences

The difficulty of interacting with a problem situation is dominated by the rules, f, a user must actively supply. The relation between an interactive situation, S, problem constraints, C, attuned constraints, A, instructed constraints, f, and the problem situation, C∘S, the basis situation, A∘S, and user's situation, f∘A∘S, they define can be expressed as:

$$C \circ S \equiv f \circ A \circ S$$

"The official problem space appears to the user as a situation in which some disallowed events are not imagined (A∘S) but others can only be eliminated by consciously applying rules (f∘A∘S)". The extrinsic difficulty of the problem will depend on the complexity of the rules, f, which must be composed with the attuned constraints, A, to bring the user's constrained situation, f∘A∘S, into agreement with the constrained situation, C∘S, which defines the problem space. Assuming that difficulty measures exist for controlled processing and that attunements are specifiable and indexable by situation-types:

(1) *The difficulty of a cognitive task involving an interactive situation can only be reduced by introducing metaphor.*

This follows from the definitions. The intrinsic difficulty of a task cannot be reduced without altering the task. The only avenue to reducing difficulty is therefore to reduce extrinsic difficulty. The extrinsic difficulty of a task is determined by the constraints in f which must be supplied using controlled processing. The constraints in f are those needed for composition with attunements, A, in order to match C. The attunements, A, are in turn determined by S. The only way to reduce extrinsic difficulty is therefore to introduce a new but equivalent situation S'. Introducing S', however, requires defining a translation, M between S and S'. If the task is incompletely characterized (there may be additional goals or constraints associated with the objects or relations of S) then resource consumption associated with M must be considered as well. Assuming M to require controlled processing, there exists a measure of its difficulty, D(M). The difficulty of a task can therefore be reduced iff there exists an interactive metaphor, M.

M: f∘A∘S → f'∘A'∘S' such that

D(M) + D(f') < D(f) if translation is required

D(f') < D(f) if translation is not required

The three possibilities for the translation between a problem situation and a possible interactive metaphor are shown in the commutative diagrams below:

States of Affairs			Problem Spaces			Interactive Situations		
x_i	$\overset{g}{\rightarrow}$	Gx_i	S	$\overset{h}{\rightarrow}$	HS	x_i	$\overset{g}{\rightarrow}$	Gx_i
r↓		↓r'	C↓		↓f∘A	C∘r ↓		↓f∘A∘r'
x_j	$\overset{g}{\rightarrow}$	Gx_j	S	$\overset{h}{\rightarrow}$	HS	x_j	$\overset{g}{\rightarrow}$	Gx_j

Here x_i are objects in the interactive situation, r and r' are the relations within the two situations, C is the problem constraints in the first situation, A is the attunements to the interactive metaphor, f is the instructed constraints for the interactive metaphor, and G and H are the mappings between the states and the problem spaces respectively of the two situations.

We will show that isomorphism between states of affairs (diagrams 1 and 3) is needed to minimize D(M) the difficulty of translating between situations while isomorphism between problem spaces (diagrams 2 and 3) is needed to satisfy the definition of interactive metaphor. The useful result is that a less constrained state equivalent situation can generally be transformed into a situationally equivalent metaphor by supplying instructions while a problem space equivalent situation cannot.

These distinctions and the role of instructions in creating metaphors can be illustrated in a simple example using the 2-blocks, mouse in the maze, and 2-lights situations shown on the next page. Allowed state transitions are indicated by arrows. The basis situation for the mouse in the maze is an interactive metaphor for the 2 blocks situation because they have isomorphic problem spaces. One mapping is:

F= { mouse-in-left-chamber → white-block-ontop ∧ ¬ black-block-ontop,
mouse-in-middle-chamber → ¬ white-block-ontop ∧ ¬ black-block-ontop
mouse-in-right-chamber → black-block-ontop ∧ ¬ white-block-ontop }

The mouse in the maze situation is not isomorphic to the two blocks problem situation, however, because they are not isomorphic in their constituents <<r_1, black-block,white-block>> vs. <<r_2, mouse, maze>> or <<r_3, mouse, chamber-left, chamber-middle, chamber-right>>.

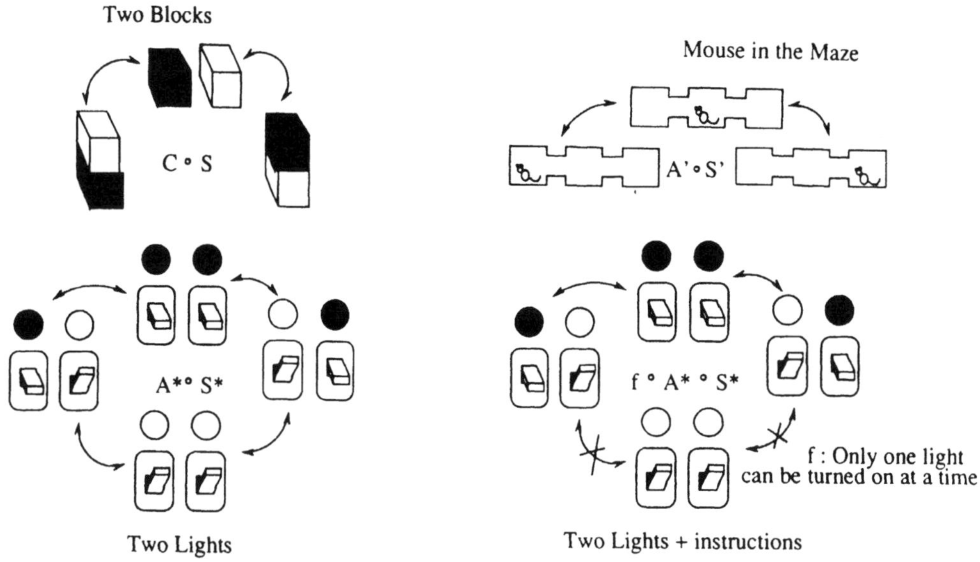

Figure 1: 2-blocks, mouse in the maze, and 2-lights situations

To verify this claim note that there is no mapping of constituents which preserves the equivalence of constraints (e.g., H is not decomposable into constituent mappings).

In its basis situation, states of the 2 lights are isomorphic to those of the 2 blocks but it is not an interactive metaphor for the two blocks situation because the 2-lights problem space includes a state in which both lights are on (or alternately both are off) while the constraints governing the 2-blocks situation prohibit both the black-block and the white block being on-top (of each other) at the same time. Although the representations appear dissimilar. by choosing an appropriate f (both lights cannot be turned on because the circuit has a limited capacity) we produce:

G: C_{blocks} 2-blocks → f ∘ C_{lights} 2-lights

which is now an isomorphic situation. To verify this claim, note that if we assign as constituent mappings:

b: black-block → left-light,
w: white-block → right-light
r: stack → turn-on
then
G= b∘w∘r

As an example of an additional property of equivalent situations consider a new goal in the 2-blocks situation, "move the black block". In the 2-light situation this is translated by G as turn on/off the right-light. In the mouse in the maze situation there is no unique translation under H.

This example highlights properties of the three forms of equivalence which are important for exploiting interactive metaphor.

(2) *The difficulty, D(M) of translating between an interactive situation, S, and its interactive metaphor, S' is minimum if they are isomorphic in states of affairs.*

Assuming situational mental representation, M must translate the objects and relations of S to objects and relations of S' and vice versa. The number of mapping rules in M must therefore be at least as great as |x| the cardinality of objects plus |r| the cardinality of relations in the states of affairs of S. If S and S' are isomorphic in states of affairs then there is a one-to-one correspondence between objects and relations in the two situations and only |x| + |r| rules are required. If S and S' are only problem space isomorphic then M may require up to |S| x (|x| + |r|) translations (one for each object and relation in each state of affairs). To construct such an isomorphism let S' be a situation having one of the numerals 1..|S| as the object in each of its states and identity as its relation. Randomly assign a state of affairs in S to each state of S' and construct a set of constraints, C', to match C. S' is now problem space equivalent but will require a separate set of translations for objects and relations for every state. This is true for the mouse in the maze and the two blocks situation, as well, although only three states were involved. We are presently investigating a less extreme instance of the disruption of translation by state inequivalence. Subjects required to translate between the state equivalent Tic-Tac-TOH and Monster Globe (change) problems have averaged 25.4 translations in a fifteen minute session while only 10 translations are achieved between the state inequivalent Tower of Hanoi and Monster Globe (change) problems.

(3) *If an interactive situation, S, and its metaphor, S', are state equivalent, the metaphor will reflect and preserve extrinsic information about the situation.*

If S and S' are equivalent situations then M is a full, faithful, and representative functor and must reflect and preserve goals or constraints defined over considered relations in S. The translation of the added goal between the 2-blocks and 2-lights problem is an example of this property. For constructing interactive metaphors this means that as long as they remain state equivalent they do not need to be complete.

(4) *Any Situation S' with* $|S'| \geq |S|$, *and containing C∘S as a problem space isomorphic subcategory of A'∘S' can be made an interactive metaphor for S simply by supplying instructions f.*

This is trivially true. In the worst case the constraints in C can simply be transferred to f. Its specialization is more useful.

(a) *Any Situation S' with* $|S'| \geq |S|$, *and containing C∘S as a situation isomorphic subcategory of A'∘S' can be made a state equivalent metaphor for S simply by supplying instructions f.*

This specialization shows that if we reverse the normal process of seeking metaphors in exact graph matches and instead simply look for situations with equivalent states we can generally bring them into agreement and enjoy the full advantages of situational equivalence simply by generating instructions.

(5) *An interactive metaphor f∘A∘S' which minimizes the difficulty of an interactive situation, S, can be found by an algorithm* **A**.

This follows from the existence of a measure D of the difficulty of controlled processing and the indexing of attuned constraints by situation-types. If a metaphor preserving extrinsic constraints is desired, the criterion to be minimized is c=D(f') + D(M), otherwise it is c=D(f') (consequence 1). The algorithm performs an exhaustive search of the taxonomy of situation-types examining every combination of situation-type and each replication of relation-type and object role within combinations of situation-types for each problem space isomorphic subcategory of A∘S'. f, M, and c are determined and the metaphor (f,S') with minimum c is retained. At conclusion, if c' < D(f) the algorithm returns C'∘S', a description of an interactive graphic and f, the constraints defining a set of instructions to be supplied to a user. The description, C'∘S', is a partially determined interactive situation resembling the abstract visual objects used in scientific visualization. Conceptually the algorithm chooses S' so as to shift constraints from f where they require controlled processing to A' where they do not.

(6) *An incremental algorithm,* A_m approximates the results of **A** by re-writing relation-types.

A more tractable algorithm considers only those situations, S', having relations isomorphic to S as candidate metaphors. If c= D(f') + D(M) only state equivalent situations are considered minimizing D(M) by consequence (4). The rewrite method conducts an incremental search for relations which when substituted into S will cause some constraint expression in the new version of C to match a constraint expression in the taxonomy of attunements. When a match is found, the relation is rewritten S → S', C→C', the constraint is marked in C', and f is updated. When no further matches can be found the method terminates returning C'S' a description of the metaphor and its behavior, f, a list of instructions generated from the unmarked constraints, and M the map, S→S'.

Concluding Remarks

In this paper we have examined some consequences of adopting an ecological model human cognition. Although ecological models are commonly believed to be less precise and tractable than their conventional counterparts we have shown that some aspects of cognition, such as intuition, which are difficult to quantify in conventional models can be handled under ecological assumptions. The problem space principle (that difficulty of cognitive tasks is fixed) is habitually made in cognitive science.

Without abandoning this partial truth (conservation of difficulty), we have shown that the extrinsic difficulty of tasks is not fixed and can be systematically manipulated through representation.

References

Barwise, J. & Perry, J. 1983. *Situations and Attitudes*. Cambridge: MIT Press.

Hayes, J.R. & Simon, H.A. 1977. Psychological differences among problem isomorphs. In N.J. Castellan, D.B. Pisoni, & G.R. Potts (Eds.) *Cognitive theory*. Hillsdale,NJ: Erlbaum.

Kotovsky, K., Hayes, J.R. & Simon, H.A. 1985. Why are some problems hard? Evidence from Tower of Hanoi. *Cognitive Psychology*, 17:248-294.

Lewis, M. 1991. Visualization and situations. In Barwise, Gawron, Plotkin, & Tutiya (Eds.) *Situation Theory and Its Applications, Vol 2*. Stanford, CA: CSLI.

Lewis, M. 1992. Situated visualization: Building interfaces from the mind up. *Multimedia Review*,3:23-40.

Newell, A. & Simon, H.A. 1972. *Human Problem Solving*. Inglewood Cliffs, NJ: Prentice-Hall.

Front-end Serial Processing of Complex and Compound Words: The APPLE Model[1]

Gary Libben
Department of Linguistics
University of Calgary
2500 University Drive N.W.
Calgary, Alberta, Canada
T2N 1N4
(403) 220-7315
libben@uncamult

Abstract

Native speaker competence in English includes the ability to produce and recognize morphologically complex words such as *blackboard* and *indestructibility* as well as novel constructions such as *quoteworthiness*. This paper addresses the question: How do subjects 'see into' these complex strings? It presents, as an answer, the Automatic Progressive Parsing and Lexical Excitation (APPLE) model of complex word recognition and demonstrates how the model can provide a natural account of the complex and compound word recognition data in the literature. The APPLE model has as its core a recursive procedure which isolates progressively larger substrings of a complex word and allows for the lexical excitation of constituent morphemes. The model differs from previous accounts of morphological decomposition in that it supports a view of the mental lexicon in which the excitation of lexical entries and the construction of morphological representations is automatic and obligatory.

A fundamental claim of all linguistic approaches to the study of morphology is that words such as UNHAPPINESS, INDESTRUCTIBILITY and BLACKBOARD are composed of smaller units and that these basic units are organized in specific ways to form complex words. This claim is supported by the observation that native speaker competence in a language is characterized by two significant abilities: (1) the ability to understand and produce complex words of the language and (2) the ability to understand and produce novel complex words. (e.g., QUOTEWORTHINESS, COMPUTERIZABILITY, WHITEBOARD).

The question of the relationship between morphological structure and native speaker competence has also been the subject of much investigation in the psycholinguistic literature. It has been assumed that a native speaker's vocabulary is stored in a mental lexicon which is organized to meet the processing demands of access and retrieval speed as well as storage efficiency. One way in which the organization of the mental lexicon could exploit the morphological structure of a language such as English would be to store multimorphemic words in their morphologically decomposed form, thereby greatly reducing the number of entries in the lexicon[2]. This possibility was first suggested by Taft & Forster (1975) who proposed that the morpheme rather than the word is the basic unit of the mental lexicon. In the Taft and Forster model, there is no separate lexical entry for the word UNLUCKY. Rather, the word is decomposed into its morphological constituents during the process of word recognition and is ultimately recognized through the representation of its root morpheme LUCK.

[1] This research was supported by Social Sciences and Humanities Council of Canada Research Grant 410-90-1502 to the author.

[2] Clearly the question of morphological decomposition is language specific. It is extremely unlikely that for agglutinating languages such as Turkish, words could be represented in their full forms. (See Hankamer 1989 for a discussion of this).

In the fifteen years since its first publication, the morphological decomposition hypothesis has been investigated using a variety of experimental paradigms in studies which have addressed both the general issue of decomposition and specific differences that might exist in the representation and processing of particular affix types. However, these research efforts have neither yielded clear support for the morphological decomposition hypothesis nor a clear refutation of it (see Henderson (1985) for a review of this literature).

Despite the lack of empirical consensus, the Taft and Forster model has retained a certain degree of attractiveness because it provides a bold and explicit account of both the organization of the mental lexicon and the process by which multimorphemic words are recognized. The hypothesis has thus far received greatest support from studies which investigate prefixed words such as REVIVE. Although Taft & Forster (1975) claim that decomposition should apply to all affix types, they present evidence solely for prefixed words (as does Taft (1981)). Overall, the data from studies which employ suffixed words have been less supportive of the view that morphological decomposition is achieved through automatic and indiscriminate affix stripping (e.g., Henderson Wallis & Knight 1984; Mandelis & Tharp 1977; Stanners et. al. 1979). Taft (1985) provides an account of this discrepancy by claiming that only prefixes are stripped prelexically, whereas suffixes are stripped by a left-to- right scanning procedure which isolates increasingly larger substrings until a match is found in the mental lexicon. The notion that left-to-right parsing plays an important role in morphological decomposition is also found in Taft & Forster (1975) and Hankamer (1989).

In this paper I argue that left-to-right parsing is a fundamental component of the recognition of all multimorphemic word types. Prefixes hold no special status with respect to morphological decomposition. Rather, left-to-right scanning simply creates the appearance of prefix stripping. I also argue that the prefix stripping hypothesis is only tenable under the questionable assumption that the mental lexicon is restricted to monomorphemic entries. I suggest that a more natural account of the data in the literature is provided under the view that all units of meaning (rather than only the simplest units of meaning) are represented in the mental lexicon.

The APPLE model

As a formalization of this argument, I propose the Automatic Progressive Parsing and Lexical Excitation (APPLE) model of visual word recognition. Below, I provide a description of the details of the model and attempt to show how it provides a natural account of the complex and compound word recognition data in the literature.

The APPLE model contains features of the original Taft & Forster (1975) account of morphological decomposition but begins with a very different view of the purpose and status of prelexical parsing. Early morphological decomposition proposals assumed that the purpose of prelexical parsing is to preprocess multimorphemic strings and thereby simplify access to the lexicon. Taft & Forster's (1975) affix stripper falls into this class of preprocessing procedures because its goal is to identify and remove affixes from a string. The initial assumptions of the APPLE model differ from those of models which see the problem of visual word recognition as a problem of isolating a particular entry in a store of perhaps 100,000 lexical items. The APPLE model assumes that all entries that can be excited are excited- the problem of visual word recognition is not to excite the entries in the first place, but rather to choose between the entries which have been automatically excited. In short, the lexicon is hungry!

Parsing in the APPLE model is essentially goalless. I claim that left-to-right parsing falls out from general properties of the language processing system and is not motivated by a 'desire' to identify any particular type of morpheme (see Cutler, Hawkins & Gilligan (1985) for a

```
Procedure MorphParse

Repeat until all letters of StimulusString are used
Begin
  Add next letter of StimulusString to TargetString;
  Allow lexical excitation of TargetString;
  If TargetString lexical and remainder legal³ then
    Begin
      Make the remainder the new StimulusString;
    Do MorphParse;
    End;
End.
```

Figure 1: *The MorphParse Procedure*

[3]In the face of findings which indicate that illegal nonwords (e.g., FTANG) are rejected more quickly than legal nonwords (e.g., FRANG) it seems reasonable to postulate a well-formedness 'gating' mechanism as the initial component of lexical access. When the inital component of a string excites a lexical entry but the remainder of that string is not legal (e.g., hen-chman), the remainder is not examined further.

1.	F
2.	FO
3.	FOO
4.	FOOT
5.	B
6.	BA
7.	BAL
8.	BALL
9.	FOOTB
10.	FOOTBA
11.	FOOTBAL
12.	FOOTBALL

Table 1: *The operation of MorphParse on the compound FOOTBALL*

discussion of left-right parsing and language universals). The core of the APPLE model is the MorphParse procedure represented in Figure 1. This procedure simply moves automatically across the input string isolating increasingly larger substrings on the left until there are no remaining graphemes on the right. On the way, it generates the exhaustive lexical excitation of all legal substrings in a stimulus.

The key features of the MorphParse procedure are obligatoriness, recursion, the isolation of *initial* substrings, and the independence of parsing and lexical excitation. These features produce morphological parses of compound and complex stimuli which have a particular set of properties. It will be argued below that precisely these properties are required to provide a principled account of the compound and complex word data reported in the literature.

Compound Words

The operation of the MorphParse procedure is represented as a series of derivations such as those provided in Table 1. Note how the excitation of a lexical entry (indicated by an underscored representation) causes MorphParse to proceed in an identical fashion across the remainder of the string. MorphParse 'pops' back to and continues a higher analysis when the graphemes of the current StimulusString have been exhausted. In so doing, it leaves behind a 'path' of lexical excitation (indicated in Table 1 by successive indentations). A property of the APPLE model is that for any multimorphemic string, a left to right morphologically decomposed analysis is available before a whole word analysis.

We may now consider the operation of the APPLE model for novel compound stimuli such as those used in Taft & Forster's (1976) lexical decision study. They found that lexical decision latencies to strings which had real-word initial substrings were longer than latencies to strings which had either no real-word substrings or only final real-word substrings.

Table 2 represents the novel compounds used in the Taft and Forster study and the mean reaction time for each stimulus type. Applying the MorphParse procedure to these stimuli highlights the relationship

	WW (RT=758) DUSTWORTH	WN (RT=765) FOOTMILGE	NW (RT=682) TROWBREAK	NN (RT=677) MOWDFLISK
1.	D	F	T	M
2.	DU	FO	TR	MO
3.	DUS	FOO	TRO	MOW[a]
4.	DUST	FOOT	TROW	MOWD
5.	W	M	TROWB	MOWDF
6.	WO	MI	TROWBR	MOWDFL
7.	WOR	MIL	TROWBRE	MOWDFLI
8.	WORT	MILG	TROWBREA	MOWDFLIS
9.	WORTH	MILGE	TROWBREAK	MOWDFLISK
10.	DUSTW	FOOTM		
11.	DUSTWO	FOOTMI		
12.	DUSTWOR	FOOTMIL		
13.	DUSTWORT	FOOTMILG		
14.	DUSTWORTH	FOOTMILGE		

Note. Taft and Forster's RT data are given above each stimulus type.
[a]The string MOW does not trigger a call to Morphparse because the remainder DFLISK is illegal.

Table 2: *The APPLE Model Analysis of Taft and Forster's (1976) Data.*

between the operation of the model and characteristics of the stimuli. The observed reaction times fall out naturally from the architecture of the APPLE model. If we assume that each iteration of the parsing procedure consumes time which is measurable in a lexical decision task then the model creates a processing cost that increases the greater the number of real-word initial substrings and the closer those substrings are to the beginning of the stimulus.

Affixation

The APPLE model makes no distinction between types of morphemes. Prefixed and suffixed words are treated exactly in the same manner as compounds. Some interesting differences fall out, however, from the fact that prefixes occur at the beginning of strings and suffixes at the end. The APPLE model predicts that although there is no special mechanism to identify prefixes, they will appear to be stripped from their stems, whereas suffixes will not. This can be seen by considering the strings REVIVE and SENDER in Table 3. Each of these strings has a two-character affix and a four-character stem. However, because of the positional differences of the affixes, REVIVE is parsed in 10 steps and SENDER is parsed in 8 steps. Moreover, according to the APPLE model, the difference between the number of steps required to parse prefixed vs. suffixed words increases with the length of the string. This is due to the fact that in any derivation, the number of parsing steps is equal to the sum of the lengths of the TargetStrings (as defined in the MorphParse algorithm). Thus a ten-character stem prefixed by RE would be parsed in 22 steps, whereas a ten-character stem suffixed by ER would be parsed in 14 steps. It seems probable that this characteristic of the model could provide an account of the fact noted at the outset of this paper-- namely that in general morphological decomposition effects have been much more evident in studies which investigate prefixation than in studies which investigate suffixation.

In contrast to the prefixation literature, the suffixation literature presents an unclear, often contradictory, view of whether suffixed words are decomposed in the process of visual word recognition. It has been found that in repetition priming experiments, a suffixed word such as CARING will prime its root constituent CARE (Fowler, Napps & Feldman, 1985; Napps, 1989). The opposite relationship (i.e., one in which CARE primes CARING) has also been found by Murrell & Morton (1974). In a study which employed a frequency mapping paradigm, Burani, Salmaso & Caramazza (1984) found that lexical decision response times to suffixed Italian words is influenced by both the frequency of the root as well as the frequency of the entire string.

The above findings all suggest that at least part of the recognition of suffixed words involves the dissociation of roots and suffixes and may be taken as support for the extension of Taft and Forster's prefix stripping hypothesis to suffixes. On the other hand, the findings in studies which investigate pseudosuffixation seem to argue against such an extension.

As predicted by the Taft and Forster hypothesis, Bergman, Hudson & Eiling (1988), Smith & Sperling (1982), and Lima (1987) found that pseudoprefixed words such as RELISH are more difficult to process than truly prefixed words such as REVIVE . This evidence, which is counter-intuitive and constitutes strong support for the morphological decomposition hypothesis has not been found in studies which investigated the role of pseudosuffixation. Such studies (e.g, Henderson, Wallis & Knight (1984); Mandelis & Tharp (1977); Rossman-Benjamin (1986)) have failed to find differences in processing time between suffixed stimuli such as SENDER and pseudosuffixed stimuli such as SISTER. The absence of suffixation-pseudosuffixation differences constitutes evidence against the view that suffixes are obligatorily stripped from word stems.

I suggest that the contradictory findings referred to above are not contradictory at all, but fall out naturally from the architecture of the APPLE model. Note in Table 3 that although the prefixed words RELISH and REVIVE are both parsed in 10 steps, a lexical decision 'yes' response to REVIVE can be given at Step 6 (the point at which lexical excitation has occurred for both constituents). In the case of RELISH, the 'yes' response can only be given after all 10 steps have been completed.

Turning to the effect of pseudosuffixation, the model again correctly predicts that no difference will be found between the 'yes' latencies to SISTER and SENDER. In both cases the correct response is available at Step 6 of the derivation. Note that in the APPLE model this does not mean that prefixes are stripped but suffixes are not. As has been stated above, the appearance of prefix stripping is simply a consequence of parsing direction.

Finally, the model also provides a natural account of the stem priming effects and the stem frequency effects for suffixed words. Note that in the derivation of SENDER, the units SEND, ER and SENDER are all activated, predicting just the results obtained by Napps (1989) and Burani, Salmaso & Caramazza (1984).

I claim therefore that there never was a contradiction between the pseudosuffixation effects and the stem priming frequency effects. Rather, its appear-

	Stimulus Type			
	Pseudoprefixed	Prefixed	Pseudosuffixed	Suffixed
	RELISH	REVIVE	SISTER	SENDER
1.	R	R	S	S
2.	RE	RE	SI	SE
3.	L	V	SIS	SEN
4.	LI	VI	SIST	SEND
5.	LIS	VIV	SISTE	E
6.	LISH	VIVE	SISTER	ER
7.	REL	REV		SENDE
8.	RELI	REVI		SENDER
9.	RELIS	REVIV		
10.	RELISH	REVIVE		

Table 3: *The APPLE Model Analysis of Affixed Words*

ance resulted from the investigation of different phenomena which turn out not to be two sides of the same coin after all. Under this view, the question of whether suffixes are stripped is quite distinct from the question of whether stems are activated. The investigation of these phenomena requires explicit reference to both the details of the experimental task and the details of the processing model.

Implications of the model

Serial processing in a parallel world

The details of the APPLE model show promise in their ability to provide a unified explanation for a number of seemingly unrelated findings in the visual word recognition literature. An important characteristic of this model is that it is event-driven rather than teleological. It is, however, clearly serial. In my view, the serial nature of this 'front-end' to the word recognition process makes no claims about the nature of the rest of the recognition process or about the preferred nature of human cognition. Rather it seems simply to be a response to the serial nature of morphemic organization. There are currently no parallel models of morphological parsing in the literature and it seems unlikely that the positional effects discussed in this paper could plausibly be accounted for in a parallel model. Nevertheless, there is good reason to suppose that the lexical system which the MorphParse algorithm feeds is characterized by parallel processing. Indeed, an important area of future research in the elaboration of this model concerns the spread of activation resulting from the activation of individual lexical items during the parse. For example, the APPLE model predicts that a 'yes' lexical decision response to a word such as REVIVE is possible as soon as both constituent morphemes have been recognized but that a 'no' response to a novel construction such as RE-DISK requires an exhaustive parse. This effect assumes an automatic spread of activation within the lexicon which is independent from the parsing procedure and is currently being modeled in our laboratory.

The mental lexicon

The reinterpretation of key findings in the word recognition literature in terms of the APPLE model supports a view of the mental lexicon in which the excitation of entries is automatic and obligatory. It points to a view of the lexicon (and of language processing in general) which is radically different from that which guided the work of Taft & Forster (1975; 1976). The role of morphological parsing is not to simplify word recognition by reducing the number of lexical entries which must be activated. Rather, I propose that lexical excitation is essentially cost-free as is the construction of morphological representations. This position is consistent with a view of language processing which has emerged from a number of disparate investigations. It has been shown by Onifer & Swinney (1981), Swinney (1979) and Tannenhaus, Leiman & Seidenberg (1979), that both meanings of a semantically ambiguous word are automatically activated. Tannenhaus, Carlson & Seidenberg (1985) have found similar effects for the processing of sentence ambiguity. These findings support the general view that language processing is characterized by modular multilevel processing in which all possible representations at all linguistic levels (i.e, phonology, morphology, syntax and semantics) are created. These representations may later be acted upon by a set of evaluation processes which unlike the representation- creating processes are not insulated from the effects of context.

References

Bergman, M. W., Hudson, P. T. W., and Eiling, P. A. 1988. How simple complex words can be: Morphological processing and word representations. *Quarterly Journal of Experimental Psychology, 40A*, 41-72.

Burani, C., Salmaso, D., and Caramazza, A. 1984. Morphological structure and lexical access. *Visible Language, 8*(4), 342-352.

Cutler, A., Hawkins, J. A., and Gilligan, G. 1985. The suffixing preference: A processing explanation. *Linguistics 23*, 723-728.

Fowler, C. A., Napps, S. E., and Feldman, L. 1985. Relations among regular and irregular morphologically related words in the lexicon as revealed by repetition priming. *Memory and Cognition, 13*, 241-255.

Hankamer, J. 1989. Morphological parsing and the lexicon. In W. Marslen-Wilson (Ed.), *Lexical representation and process*. Cambridge, Mass: MIT Press.

Henderson, L. 1985. Toward a psychology of morphemes. In A.W. Ellis (ed.), *Progress in the psychology of language, volume 1*. London: Lawrence Erlbaum.

Henderson, L., Wallis, J., and Knight, D. 1984. Morphemic structure and lexical access. In H. Bouma & D. Bouwhuls (Eds.), *Attention and Performance X*. Hillsdale, New Jersey: Lawrence Erlbaum.

Lima, S. 1987. Morphological analysis in sentence reading. *Journal of Memory and Language, 26*, 84-99.

Mandelis, L. and Tharp, D. A. 1977. The processing of affixed words. *Memory and Cognition 5(6)*, 690-695.

Murrell, G. and Morton, J. 1974. Word recognition and morphemic structure. *Journal of Experimental Psychology, 102*, 963-968.

Napps, S. E. 1989. Morphemic relationships in the lexicon: Are they distinct from semantic and formal relationships; *Memory and Cognition, 17(6)*, 729-739.

Onifer, W. and Swinney, D. 1981. Accessing lexical ambiguities during sentence comprehension: effects of frequency and meaning on contexual bias. *Memory and Cognition, 9*, 225-236.

Rossman-Benjamin, T. 1986 . *The psychological reality of morphological structure: evidence from language performance*. Unpublished manuscript. Department of Psychology, University of Pennsylvania.

Smith, P. T., and Sterling, C. M. 1982. Factors affecting the perceived morphological structure of written words. *Journal of Verbal Learning and Verbal Behavior*, 21, 704-721.

Stanners, R.F., Neiser, J. J., Hernon, W. P., and Hall, R. 1979. Memory representation for morphologically related words. *Journal of Verbal Learning and Verbal Behavior, 18*, 399-412.

Swinney, D. 1979. Lexical access during sentence comprehension. (Re)consideration of context effects. *Journal of Verbal Learning and Verbal Behavior, 18*, 645-660.

Taft, M. 1981. Prefix stripping revisited. *Journal of Verbal Learning and Verbal Behavior, 20*, 289-297.

Taft, M. and Forster, K. I. 1975. Lexical storage and retrieval of prefixed words. *Journal of Verbal Learning and Verbal Behavior, 14*, 638-647.

Taft, M. and Forster, K.I. 1976. Lexical storage and retrieval of polymorphemic and polysyllabic words. *Journal of Verbal Learning and Verbal Behavior, 15*, 607-620.

Taft, M. 1981. Prefix stripping revisited. *Journal of Verbal Learning and Verbal Behavior, 20*, 289-297.

Taft, M. 1985. The decoding of words in lexical access: a review of the morphological approach. In D. Besner, T. G. Waller, and G. E. Mackinnon (Eds.), *Reading research: advances in theory and practice, Vol. 5*. Academic Press.

Tannenhaus, M., Carlson, G, and Seidenberg, M. 1985. Do listeners compute linguistic representations: In D. Dowty, L. Karttunen and A.. Zwicky (eds.) *Natural language parsing*, Cambridge, England: Cambridge University Press.

Tannenhaus, M., Leiman, J. and Seidenberg, M. 1979. Evidence for multiple stages in the processing of ambiguous words in syntactic contexts. *Journal of Verbal Learning and Verbal Behavior, 18*, 427-441.

The search image hypothesis in animal behavior: its relevance to analyzing vision at the complexity level *

Dennis R. Lomas
John K. Tsotsos †
Department of Computer Science, University of Toronto
Toronto Ont. M5S 1A4
email: lomas@vis.toronto.edu tsotsos@vis.toronto.edu

Abstract

We show how a concept from animal behavior, the visual search hypothesis, is relevant to complexity considerations in computational vision. In particular we show that this hypothesis is an indication of the validity of the bounded/unbounded visual search distinction proposed by Tsotsos. Specifically we show bounded visual search corresponds to a broad range of naturally occurring, target-driven problems in which attention alters the search behavior of animals.

Introduction

In *Analyzing vision at the complexity level* Tsotsos (1990) develops a method for understanding biological visual search processes, an immeasurably difficult reverse engineering problem. He maintains that, since visual search aside from such things as direct sensing of light on the retina is fundamentally a computational task, any model or theory for human or animal visual search must satisfy computational complexity constraints.[1] This means that algorithms in computationally-based models or theories must compute in reasonable time. In the terminology of algorithmic complexity theory vision algorithms that mimic human or animal visual search need to be ***tractable*** rather than *intractable*.[2] Moreover such models or theories need to accomplish their tasks with the known resources for visual processing in the brain. Furthermore, since general visual search is an intractable problem,[3] such complexity considerations are not just a detail to contend with at implementation but need to inform each stage of model or theory development.[4]

Using this approach he develops (by placing constraints on such things as the type of objects that can be recognized and the number of features that can be used in recognition) a model for human and animal vision that satisfies first order complexity constraints.

Tsotsos begins his analysis by establishing a fundamental dichotomy for visual search problems between ***unbounded visual search*** and ***bounded visual search***. In ***unbounded visual search*** "either the target is explicitly unknown in advance or it is

*This research was conducted with financial support of the National Sciences and Engineering Research Council of Canada and the Information Technology Research Center, a Province of Ontario Center for Excellence.

†Fellow of the Canadian Institute for Advanced Research

[1] For a brief overview of complexity theory see sections 1.3 and 1.4 in *Analyzing vision at the complexity level* (Tsotsos 1990).

[2] A tractable algorithm performing visual search on an image can be viewed as one whose time requirements can be expressed as a polynomial function of the pixels required to represent the image whereas for an intractable algorithm time requirements in the worst case are an exponential function of the pixels required to represent the image.

[3] Tsotsos suggests that general visual search is intractable because it contains as a subprogram an intractable problem: *unbounded visual search* (see below for a definition of *unbounded visual search*).

[4] Tsotsos also uses minimization of cost, a consideration not relevant to this contribution.

Figure 1: The dog and his search image. "When the master orders his dog to retrieve a stick, the dog, ... has a quite specific search image of the stick." Jakob von Uexküll who coined the phrase 'search image.' This figure appeared in an anecdotal article by Uexküll (1934), *A stroll through the worlds of animals and men.*

somehow not used in the execution of the search" while in ***bounded visual search*** "the target is explicitly known in advance in some form that enables explicit bounds to be determined that can be used to limit the search process." The dichotomy arises from a complexity analysis of the two problems:[5] ***unbounded visual search*** is potentially intractable, ***bounded visual search*** is tractable.

Moreover, he suggests "because actual psychological experiments on visual search with known targets report search performance as having linear time complexity and not exponential, the inherent computational nature of the problem strongly suggests that attentional influences play an important role."

The purpose of this contribution is to show that a concept from animal behavior, the ***visual search hypothesis***, introduced some thirty years ago, confirms the validity of dividing visual recognition into the categories of ***unbounded visual search*** and ***bounded visual search***.[6] Experiments and observations of animal behaviorists involved in developing and testing the ***visual search hypothesis*** indicate that ***unbounded visual search*** is an intractable form of visual search. Additionally, these studies suggest ***bounded visual search*** corresponds to a broad range of naturally occurring, target-driven problems in which attention alters the search behavior of animals in the studies.

The search image hypothesis

The ***search image hypothesis*** was first formulated by Tinbergen (1960) making use of the term 'search image' coined by Uexküll (1934). Birds observed by Tinbergen dramatically fail to detect the presence of novel prey even though its dietary appropriateness and abundance warrants predation. Detection of the prey commences only after chance encounters. He suggested these encounters prompt the formation of a search image that subsequently enables the predator to detect the prey. (Pietrewicz and Kamil 1979) Of course such observations fall short of establishing the hypothesis by today's experimental standards. In the last twenty years in the course of designing and performing numerous experiments, animal behaviorists have more precisely formulated the ***search image hypothesis*** and tested it against alternate explanations of predator behavior.

[5] an analysis undertaken in a mathematical, formal setting

[6] See the exchange between Paul R. Kube and John Tsotsos for an interesting discussion of this claim. (See commentary on Tsotsos's ***Analyzing vision at the complexity level*** (Tsotsos 1990) and (Tsotsos 1991)).

In animal behavior literature (for example, see Pietrewicz and Kamil (1979)) the *search image hypothesis* explains behavioral change by postulating a perceptual change in the ability of a predator to detect prey. This perceptual change occurs because the predator has learned to recognize prey (formed a search image) where typically the prey is cryptic (the background and the prey are similar[7]). Thus confronted with prey in a cryptic setting the predator has been able to learn, and to attend selectively to, cues that enable it to distinguish the prey from the background.[8]

The formation of the search image directly influences the predator's ability to see the prey. Moreover, it operates in conjunction with attention mechanisms.

Search image hypothesis: Confirmed in experimentation

In the last twenty years many experiments have confirmed the *search image hypothesis*. In particular the work of Dawkins in 1971 warrants a brief review since it demonstrated for the first time in an experimental setting the validity of that hypothesis and set the stage for subsequent experimental work. Dawkins (1971a) noted that "Tinbergen's basic idea in postulating it (the *search image hypothesis*), namely, that birds become better able to perceive cryptic prey as their experience of it increases, is of great interest. The purpose of this paper is to show that changes in a bird's ability to perceive its cryptic prey do indeed occur, and may be responsible for major changes in its feeding behavior."

Dawkins observed chicks taking grains of rice from backgrounds of stones glued onto hardboard. The grains were either green or orange and could be either the same colour as the background, in which case they were called cryptic, or a different colour from the background, in which case they were called conspicuous. Dawkins found that most chicks took grains more slowly at the beginning of both tests but that the chicks were much less able initially to detect cryptic grains. "It would seem," Dawkins remarked, "that chicks did not take cryptic rice at first because they did not see it. . . ."

The experiments showed the chicks' initial inability to detect cryptic food despite almost certainly looking at it. "...it seems reasonable to suggest," Dawkins argued, "that their subsequent improvement in detection is due to some sort of central perceptual change rather than to more peripheral modifications to vision such as reorientation of the head and eyes."

In a subsequent set of experiments Dawkins (1971b) first presented chicks with grains of one colour and type and then presented them with a choice of grains in another setting, e.g. after cryptic orange grains the chicks are presented with a choice between conspicuous green grain and cryptic orange grain. The experiments thus tested for an attentional mechanism associated with the development of a search image. "...the results," Dawkins notes, "are ...compatible with the idea that chicks become better able to see cryptic grains when they have just been eating other cryptic grains than after eating conspicuous ones. They may temporarily 'shift attention' on cues that enable them to detect such grains."[9]

Dawkins experiments were repeated in 1985 by

[7] I.e. an organism is cryptic if its colour pattern is a random sampling of the background against which predators usually see it. (Lawrence and Allen 1983).

[8] The cryptic prey may be novel or familiar. In the latter case, the formation of a search image signifies "a change in the ability to detect cryptic, familiar prey as a function of recent encounters with that prey." Pietrewicz and Kamil distinguish this case from the former by calling the former where novel,cryptic prey are involved the development of a *specific* search image (Pietrewicz and Kamil 1979).

[9] Like subsequent search-image-hypothesis experiments Dawkins set up the experiments to exclude other explanations of observed behavior such as:

- learning to visit a particular place to find food
- learning to look in a particular type of place to find food
- alteration of the search path to increase the chances of encountering prey
- learning to handle prey more effectively
- preference or avoidance of a prey over others that is independent of the predator's ability to see the different types
- learning of specialized hunting techniques by particular individuals

(taken from Lawrence (1983))

Lawrence (1985) on blackbirds. In reference to experiments testing detection of cryptic prey, corresponding to Dawkins' first set of experiments, Lawrence observes: "the simplest explanation is that the birds failed to see the cryptic prey at first; the alternative (and more unlikely) explanation is that the birds (for some reason) found prey unacceptable only under cryptic conditions. ...The high frequency of background-directed pecks during the first third of the feeding sessions on cryptic prey suggests that initially the birds failed to see the prey." Moreover, he argues that the results of all his experiments "lend support to the idea that wild predators acquire search images as a normal part of their foraging behavior."

Over the past two decades the *search image hypothesis* has similarly been confirmed. These studies include: Murton's (1971) work with wood-pigeons (*The significance of a specific search image in the feeding behavior of the wood-pigeon*); Pietrewicz and Kamil's (1979) study with jays (*Search images and the detection of cryptic prey: An operant approach*); Bond's (1983) experiments with pigeons (*Visual search and selection of natural stimuli in the pigeon*); and Gendron's (1986) work with quail (*Searching for cryptic prey: evidence for optimal search rates and the formation of search images in quail*).[10] An example of recent work on the *search image hypothesis*[11] is Blough's (1989) experiments with pigeons (*Attentional priming and visual search in pigeons*).

The accumulated research confirming the *search image hypothesis* together with the fact that originally it arose from observations in natural settings indicates the robustness of the phenomenon. The findings suggest, as Bond (1983) suggested for his work, "the operation of a robust and pervasive cognitive process, one that may well be characteristic of visual search for cryptic stimuli in other species."

What the search image hypothesis shows

It is evident that ***unbounded visual search*** is a form of visual search prevailing before a predator forms a search image:[12] the predator is searching for prey, prey that is readily available but its presence does not guide the search for food. Moreover, ***bounded visual search*** corresponds to behavioral modifications induced by the formation of a search image, behavioral modifications that are likely accompanied by some form of attention. Thus, the *search image hypothesis* shows the validity both of the unbounded/bounded distinction in visual search and of the suggestion that attentional elements enable ***bounded visual search***. Furthermore, the robustness of the ***search image hypothesis*** and the fact that is has a significant domain – the search for cryptic grain – one that occurs in natural settings, suggests the unbounded/bounded distinction in visual search is a natural distinction at least for some significant aspects of vision.

References

Blough, P. M. 1989. Attentional priming and visual search in pigeons. *Journal of Experimental Pschology: Animal Behavior Process*, pages 358–365.

Bond, A. B. 1983. Visual search and selection of natural stimuli in the pigeon. *Journal of Experimental Psychology: Animal Behavior Process*, pages 292–303.

Dawkins, M. 1971a. Perceptual changes in chicks: Another look at the 'search image' concept. *Animal Behavior*, pages 566–574.

Dawkins, M. 1971b. Shifts in 'attention' in chicks during feeding. *Animal Behavior*, pages 575–582.

[10] After the publication of Gendron's paper Guildford and Dawkins (1987) claimed that the experiments on the *search image hypothesis* to that time had not sufficiently accounted for an alternative hypothesis to explain experimental observations: a decreased search rate to enhance detection of cryptic prey accounts for the observed behavior. However the major prediction of this hypothesis was contradicted in subsequent experiments by Blough (1989).

[11] Blough suggests, "In current terminology, a search image might be described as a representation activated by an exposure sequence." (Blough 1989)

[12] as the result, for example, of chance predation or relaxation of attention to another prey

Gendron, R. P. 1986. Searching for cryptic prey: Evidence for optimal search rates and the formation of search images in quail. *Animal Behavior*, 34:898–912.

Guilford, T. and M.S., D. 1987. Search images not proven: A reappraisal of recent evidence. *Animal Behavior*, pages 1838–1845.

Lawrence, E. S. 1985. Evidence for search image in blackbirds. *Animal Behavior*, pages 929–937.

Lawrence, E. S. and Allen, J. A. 1983. On the term 'search image'. *OIKOS*, pages 313–314.

Murton, R. K. 1971. The significance of a specific search image in the feeding behavior of the wood-pigeon. *Behavior*, pages 10–42.

Pietrewicz, A. T. and Kamil, A. C. 1979. Search images and the detection of cryptic prey: An operant approach. In *Foraging Behavior*, pages 311–330. Garland.

Tinbergen, L. 1960. The natural control of insects in pinewoods. i. factors influencing the intensity of predation by song birds. *Arch. neerl. Zool.*, pages 265–343.

Tsotsos, J. K. 1990. Analysing vision at the complexity level. *Behavioral and Brain Sciences*, 13-3:423–469.

Tsotsos, J. K. 1991. Is complexity theory appropriate for analyzing biological systems? *Behavioral and Brain Sciences*, 13-3:423–469. In continuing commentary on Tsotsos's "Analysing vision at the complexity level".

von Uexküll, J. 1934. A stroll through the world of animals and men: a picture book of invisible worlds. In Schiller, C. H., editor, *Instinctive Behavior*, pages 5–76. International, New York. Book published in 1957.

Learning by Problem Solving versus by Examples: The Benefits of Generating and Receiving Information

Marsha C. Lovett*
Department of Psychology
Carnegie Mellon University
Pittsburgh, PA 15213
lovett@psy.cmu.edu

Abstract

This experiment contrasts learning by solving problems with learning by studying examples, while attempting to control for the elaborations that accompany each solution step. Subjects were given different instructional materials for a set of probability problems. They were either provided with or asked to generate solutions, and they were either provided with or asked to create their own explanations for the solutions. Subjects were then tested on a set of related problems. Subjects in all four conditions exhibited good performance on the near transfer test problems. On the far transfer problems, however, subjects in two cells exhibited stronger performance: those solving and elaborating on their own and those receiving both solutions and elaborations from the experimenter. There also was an indication of a generation effect in the far transfer case, benefiting subjects who generated their own solutions. In addition, subjects' self-explanations on a particular concept were predictive of good performance on the corresponding subtask of the test problems.

Introduction

Solving problems and studying examples are both viable methods for learning to solve problems. Solving practice problems provides subjects with the experience of "doing" and forces them to consider all aspects of the solution. Studying example problems accompanied by their solutions (together, called worked examples) provides students not only with the correct answers but also with some information on how those answers were obtained. So, which method (if either) is better? Experimental results on that question do not yet provide a definitive answer.

Sweller and Cooper (1985) have found that subjects using worked examples required less study time and exhibited better near transfer performance than subjects solving problems. However, their method of learning by examples actually included problem solving half of the time.[1] In addition, many researchers have found that the particular content of worked examples can seriously affect subjects' learning outcomes (e.g. Catrambone, 1991; Pirolli, 1991; Ward & Sweller, 1990). Therefore, the experimental procedure used to implement learning by examples and the content of the examples are important variables to consider when evaluating this method. In the case of learning by problem solving, other variables (e.g. the type and timing of feedback) must be taken into account since they too have been shown to affect subsequent performance (e.g. Lewis & Anderson, 1985; Schooler & Anderson, 1990). Finally, individual differences among subjects can also influence the efficacy of these instructional methods. For example, the relationship between the quality of subjects' self-explanations[2] and subsequent performance has been demonstrated (Chi et al., 1989; Pirolli & Bielaczyc, 1989).

Experiment

The current experiment attempts a preliminary comparison between learning by examples and learning by problem solving, while dealing with many of these issues in a systematic way. Subjects worked on three practice problems in introductory probability. They were either provided with solutions or asked to generate their own, and they were either provided with complete, elaborate explanations of the solutions or asked to

*The author was supported by a NSF Graduate Fellowship. Funding for subjects was provided through MDR-87-51890 granted to Dr. John R. Anderson.

[1] For example, Experiment II contained four pairs of isomorphic practice problems. Subjects in the worked-example group studied a worked example for the first problem in each pair and then solved the second.

[2] Self-explanations are the elaborations generated by subjects, usually while studying an example problem, in which they explain various concepts to themselves.

create their own. This design allowed us to compare the two instructional methods *and* to manipulate the content of worked examples in the same experiment.

We denote these four conditions using the form <x>-<y>, where <x> corresponds to the source of the solutions and <y> corresponds to the source of the elaborations. The conditions can be characterized as follows: *subject-subject* = learning by problem solving and self-explaining; *experimenter-subject* = learning by studying "sparse" examples and self-explaining; *subject-experimenter* = learning by problem solving with explanatory feedback; and *experimenter-experimenter* = learning by studying elaborate worked examples. (See Appendix A for an example of the information provided to subjects in the four conditions.)

In accordance with previous results on the generation effect (e.g. Bobrow & Bower, 1969; Slamecka & Graf, 1978), we expected subjects who generated their own problem information to perform better. For example, generating one's own solution may make it more memorable and so improve subsequent performance. In the case of explanations, however, a generation effect may have to compete with an effect from high quality elaborations. Subjects creating their own elaborations may not perform as well as subjects who receive the experimenter's elaborations, if the latter are of substantially higher quality. Thus, we predicted a generation effect for solutions but not for elaborations.

Method

Subjects. Subjects included 50 undergraduate students at the University of California at Berkeley, who had little or no background in probability theory. All subjects were paid for their participation. Two subjects' data were removed from analysis: one because of inability to learn the material and one due to equipment failure resulting in incomplete data. This left 48 subjects in the experiment, twelve per group. Assignment to these groups was random.

Materials. Before solving the three practice problems, subjects were given a three-page introduction to the necessary concepts in probability. It covered the probability of an event E (defined as the number of outcomes satisfying E divided by the total number of possible outcomes) and the probability of multiple independent events (calculated by multiplying the probabilities of the individual events). The information in the introductory text prepared students to follow only one path to the solution of each problem, even though more than one was possible.

Eleven elementary probability problems were used in this experiment (including four adapted from Ross, 1989). All three practice problems were permutation problems with people choosing objects (e.g. scientists choosing from a pool of computers, see Appendix A). For the eight test problems, half were similar permutation problems (near transfer) and the other half were combination problems (far transfer). Combination problems differ from permutation problems in that the exact order of events does not matter. This difference affects the solution, mainly by changing the calculation of the numerator. In addition, both near and far transfer problems were split according to whether they contained people choosing objects or objects being assigned to people. Ross (1989) first found that these role assignments could affect performance: when an example problem, in which humans choose objects, was followed by role-reversed test problems, performance was worse than when the same roles were maintained between practice and test. (See Appendix B for sample test problems.) The test problems were given in the same order to all subjects -- from "nearest" transfer to "farthest".[3]

Procedure. In this experiment, subjects went through a lesson in probability. They read some introductory text, worked on three practice problems, and then solved eight test problems. All subjects were instructed to provide talk-aloud protocol during the practice and test problems and, when applicable, to read experimenter-provided elaborations out loud. Mistakes made by subjects solving the practice problems were treated as follows: in the subject-subject condition, only the fraction corresponding to the correct step (e.g. 1/11) was provided; and in the subject-experimenter condition, both the fraction and the prepared elaboration for that step were given.[4] Subjects never received coaching or corrections on their own elaborations. In order to equalize subjects' time-on-task for the practice problems, we ensured that all subjects spent approximately three minutes studying/solving each practice problem. Then, for the test problems, subjects were asked to work as quickly and accurately as possible.

[3] Subjects received the test problems in the following order: permutation problems with the same roles as the practice problems (people choosing objects), permutation problems with reversed roles (objects being assigned to people), combination problems with same roles, and finally, combination problems with reversed roles.

[4] Note that the manipulations in this experiment could not be purely executed such that subjects generated everything or nothing according to their condition. For example, 5 of the 12 subjects in the subject-subject cell received at least one correction from the experimenter. Nevertheless, cases like these only made the groups more similar and, hence, made differences between groups harder to find.

Results and Discussion

In analyzing these data, we were mainly interested in finding performance differences that might exist between the four experimental groups. We also wanted to explore two other questions about the data: Did subjects' performance on the test problems vary significantly according to transfer distance (near vs. far) and role correspondence (same vs. reversed) between the practice and test problems? And, in what way did individual differences impact on subjects' performance?

A 2x2x2x2 Mixed ANOVA on subjects' percentage of test problems solved correctly provided an overall analysis of the performance data. The between-subjects factors in this analysis were source of solutions and source of elaborations, and the within-subjects factors were transfer distance and role reversal. (See Table 1 for this breakdown of the data.) With respect to the between-subjects factors, no main effect for source of solution or for source of elaboration was found, but the interaction of these factors was significant ($F_{sol}(1,44) = 1.24$; $F_{elab}(1,44) < 1$; $F_{inter}(1,44) = 5.82, p < .05$; $MSE = 1510$). A post-hoc analysis of these data indicated that the interaction was due to high performance in the subject-subject and experimenter-experimenter groups compared to the other two ($F(1,46) = 6.33, p = .02$; $MSE = 375$).

Since this performance pattern arises in later analyses, it is worthwhile to consider it here. First, the finding that subjects who solve and explain on their own perform well is consistent with the generation effect as described above. However, the high performance of subjects who received all their problem-solving information from the experimenter shows the opposite of a generation effect. It seems that receiving correct solutions *and* high quality explanations benefited these subjects. Indeed, the experimenter-provided elaborations were generally of higher quality than subjects' own elaborations. For example, only five out of the 24 subjects elaborating for themselves verbalized three of the most important concepts found in the experimenter-provided elaborations. (Later, we will present evidence that the subjects with more complete self-explanations *do* perform better.)

	s-s	s-e	e-s	e-e	**all**
near-same	96	92	63	88	84
near-reversed	50	46	50	42	47
near overall	**73**	**69**	**54**	**65**	**66**
far-same	17	0	4	17	9
far-reversed	38	17	4	38	24
far overall	**27**	**8**	**4**	**27**	**17**
all test	50	39	29	46	41

Table 1: Average percentage of test problems correct, by experimental condition and test problem type

This interpretation leads to the prediction that subjects in the experimenter-subject cell should perform poorly because they lack the opportunity to generate solutions and they lack high quality elaborations. In fact, this cell did exhibit the worst performance. The subject-experimenter cell, however, did not perform quite as well as a generation effect and high quality elaborations would predict. One possible reason is the procedural awkwardness involved in this condition; subjects solved a step in the solution and then were asked to read the experimenter's elaboration for that step. Although subjects were forewarned about this procedure, it still might have interfered with their concentration and memory load. In fact, six subjects in the subject-experimenter cell required hints compared to only one in the subject-subject cell ($chi2 = 5.09, p < .05$), even though there is no reason to suspect ability differences existed between these two groups. Receiving hints more often could help explain the performance of the subject-experimenter subjects because each hint made them miss an opportunity to generate a solution step. A second explanation proposes that inconsistent information sources resulted in poor performance in the subject-experimenter and experimenter-subject cells because these subjects had to integrate the experimenter's information with their own. This increased cognitive load may weaken the subjects' problem memories. Unfortunately, the present data cannot tease apart these alternative explanations.

Also, in the 2x2x2x2 ANOVA mentioned above, we found some interesting within-subject effects. (See Table 1) Not surprisingly, subjects performed better on near transfer than far transfer test problems ($F(1,44) = 98.2, p < .001$). In addition, subjects performed better on test problems with the same roles as the practice problems, compared to test problems with reversed roles ($F(1,44) = 6.81, p = .01$). This replicates Ross's (1989) finding and suggests that subjects may be using analogical problem solving in this experiment as well. (See Comparison with Related Work.) The only other significant effect in this analysis was the transfer distance x role reversal interaction ($F(1,44) = 54.8, p < .001$). This interaction might have occurred because of learning during the test phase; in particular, subjects did surprisingly well on the very last test problem.

Near transfer

Since the near transfer and far transfer test problems resulted in different performance, we analyze them separately. The average percentage correct for near transfer problems is presented in Row 3 of Table 1. A 2x2 ANOVA (source of solutions x source of elaborations) of these data did not reveal any significant differences between the cells ($F_{sol}(1,44) = 2.32$, n.s.; all other *F's* < 1). Subjects in all four conditions seemed to do quite well.

Far transfer

The average percentage correct for far transfer problems is presented in Row 6 of Table 1. A 2x2 ANOVA (source of solutions x source of elaborations) on these data revealed no main effects, but it did reveal the same interaction as in the total performance measure ($F(1,44) = 7.94$, $p < .01$, $MSE = 656$). Namely, the subject-subject and experimenter-experimenter conditions performed the best ($F(1,46) = 8.27$, $p < .01$; $MSE = 630$). This pattern of results can be explained in the same way as the pattern for the total performance data. (In fact, the total performance differences are due in most part to these far transfer data, since the near transfer data did not differentiate much between the cells.)

More specific data are available with respect to far transfer performance. Recall that, in the far transfer problems, the numerator's starting value must be calculated differently than in the near transfer problems but that the other solution steps are similar to the near transfer problems. Therefore, looking at subjects' choice of numerator starting value (NSV) provides a sharper measure of far transfer performance. The total number of far transfer problems on which subjects had the correct NSV was: subject-subject 29; subject-experimenter 14; experimenter-subject 8; and experimenter-experimenter 19. A 2x2 ANOVA (source of solutions x source of elaborations) on these data revealed the same interaction found in other analyses ($F(1,44) = 8.14$, $p < .01$; $MSE = 1.73$) as well as a marginal main effect indicating that subjects who solved the practice problems performed better than those provided with solutions ($F(1,44) = 3.08$, $p = .08$; $MSE = 1.73$). Again, these data are indicative of a generation effect and a quality of explanations effect.

Individual data

One of the main features of this experiment is that it contains another experiment within its cells. By analyzing the protocols of the subjects in the experimenter-subject condition, we can look at the effects of individual subjects' self-explanations on subsequent performance. For example, we compared subjects' self-explanations of a particular concept in the practice problems with subsequent performance on corresponding parts of the test problems. We chose NSV for this analysis because it plays an important role in these problems (especially for far transfer), and it happened to be the only important concept that was differentially elaborated upon by subjects.[5] Six subjects self-explained NSV and six subjects did not. Of those who did, four ended up getting NSV correct on the test. Of the subjects who did *not* self-explain NSV, none ended up getting NSV correct on the test. Thus, specific self-explanations had a significant effect on specific performance ($chi2 = 6$, $p < .01$).

These results also provide support for the notion that the quality of explanations can affect performance. Above, subjects who self-explained NSV tended to get NSV correct on the test problems. When subjects received the experimenter's elaborations (which always included NSV), they also exhibited this tendency. The proportions of subjects who got NSV correct in at least one test problem are as follows: subject-experimenter 7/12; experimenter-experimenter 8/12. These two proportions are not significantly different from the proportion of subjects in the experimenter-subject cell (4/6) who explained NSV during practice and got it correct at least once during the test ($chi2 \sim 0$, n.s.). However, these proportions are different from the proportion of subjects in the experimenter-subject cell (0/6) who *did not* explain NSV during practice and then got it correct during the test ($chi2 = 7.5$, $p < .01$). These results suggest that (in three experimental conditions) subjects with an elaboration of NSV during practice tend to get the NSV correct at test, regardless of whether that elaboration was provided by the experimenter or self-explained.

For the subject-subject condition, we also evaluated the effectiveness of self-explaining NSV. Here, the proportion of subjects getting NSV correct on the test problems was higher than in the other conditions, regardless of whether the subjects self-explained about NSV during practice. Specifically, all three of the subjects who self-explained NSV during practice got it right during the test, and eight of the nine subjects who did not self-explain NSV during practice got it right during the test. The latter proportion is much greater than the proportion of subjects in the experimenter-subject cell (0/6) who did not self-explain NSV but still got it right ($chi2 = 11.43$, $p < .001$). This large difference indicates that subjects in the subject-subject condition were learning about NSV, whether or not they self-explained about it. This is an advantage of learning by problem solving that is not otherwise captured in our performance data.

Comparisons with Related Work

Research by Ross (1989) and Catrambone (1991) is especially relevant to this work because it uses the same type of probability problems in the context of learning

[5]Recall that, in the near transfer problems, the numerator is always 1, but in the far transfer problems, subjects must calculate the numerator starting value.

by studying worked examples. Ross (1989) found that human/object roles affected subjects' use of their example problem memories in such a way that reversing the human/object roles between the example and test problems made the test problems more difficult to solve. Likewise, Catrambone (1991) varied the human/object roles between example and test problems and produced a similar result. In the current experiment, we also found subjects' performance to be significantly worse on the role-reversed test problems than on the same-role test problems. Thus, our finding lends further support for Ross's (1989) conclusion that subjects may be using an object-mapping approach that is affected by human/object role correspondences.

Catrambone's (1991) experiment also resembles the current experiment because the content of worked examples was varied. Of Catrambone's four instructional groups, only two are comparable to the current design: the *numerator/denominator-subgoal* group in which subjects received an elaborated description of each step in the solution, including explanations for the choice of numerator and denominator, and the *subgoal* group in which subjects received only a brief description of each step in the solution. Catrambone's numerator/denominator-subgoal condition is virtually equivalent to our experimenter-experimenter condition. His subgoal condition, however, resides somewhere between our experiment-subject and experimenter-experimenter conditions in the amount of information it provides to subjects. Comparing percentages of test problems solved correctly by subjects in these conditions demonstrates substantial consistency between the two experiments. For example, Catrambone's numerator/ denominator-subgoal and subgoal groups averaged 70 and 66 percent correct, respectively, on permutation problems (isomorphic to the example) and approximately 14 and 0 percent correct, respectively, on combination problems. These values are quite similar to those of our quasi-corresponding conditions (experimenter-experimenter and experimenter-subject): 65 and 54 percent correct for the permutation problems and 27 and 4 percent correct for the combination problems. In fact, the ordering of performance (across the two experiments) is fairly consistent with an ordering of information provided to subjects. In the current experiment, however, we embedded this comparison of worked examples with different contents in the larger context of comparing two instructional methods. We have also included the analysis of subjects' protocols within this same experiment. These features allowed us to capture differences between conditions and between individual subjects that might otherwise have been missed.

Conclusions

In this experiment, we found reliable performance differences between subjects generating and receiving different amounts of practice problem information. We found that on far transfer problems, subjects in two groups performed best -- those generating solutions and explanations for the practice problems on their own and those receiving high quality solutions and explanations from the experimenter. In addition, on a particular subtask of the test problems, subjects who solved the practice problems performed better than those who received solutions, regardless of their source of elaborations. These results support the existence of a generation effect which benefits subjects who solve problems on their own in the domain of probability.

We also found evidence that the quality of elaborations during practice can greatly improve subsequent performance. High performance in the experimenter-experimenter cell is one example of this since experimenter-provided elaborations were generally of higher quality than subjects' self-explanations. In addition, we examined the effects of individual subject's self-explanations and found that subjects who verbally elaborated on a particular concept in the practice problems performed better on corresponding steps in the test problems than subjects who did not.

Acknowledgements

I would like to thank Michael Ranney, Peter Pirolli, and the members of the Reasoning and CSM groups at the University of California at Berkeley, School of Education. I would also like to thank my advisor, John Anderson, for all his help and support.

Appendices

Appendix A: Sample practice problem and the information given in different conditions

The supply department at IBM has to make sure that scientists get computers. Today, they have 11 IBM computers and 8 scientists requesting computers. The scientists randomly choose their computer, but do so in alphabetical order. What is the probability that the first 3 scientists alphabetically will get the lowest, second lowest, and third lowest serial numbers, respectively, on their computers?

1/11 is the probability that the first scientist alphabetically will get the computer with the lowest serial number because there is only 1 computer with the lowest serial number and there are 11 computers for the first scientist to choose from.

[...]
To get the overall probability, we multiply the probability of each scientist choosing a particular computer:

ANSWER: 1/11 x 1/10 x 1/9

Note: The subject-subject condition received only the problem statement. The subject-experimenter condition received the problem statement and was asked to read each explanation (in italics) after solving the corresponding step. The experimenter-subject condition received only the problem statement and the answer. The experimenter-experimenter condition received the problem statement and all the elaborations (in italics).

Appendix B: Sample test problems

Near Transfer, Same Roles:
South Side High School has a vocational car mechanics class in which students repair cars. One day there are 12 students and 15 cars requiring repairs. The students randomly choose the cars, but go in order of their grades on the last mechanical exam (highest grade choosing first). What is the probability that the 6 cars in the worst shape are worked on by the 6 students with the highest grades on the last mechanical exam, in order of their grades? (i.e. the highest grade student working on the worst car, etc.)
[Answer: 1/15 x 1/14 x 1/13 x 1/12 x 1/11 x 1/10]

Near Transfer, Reversed Roles:
A group of 8 co-workers went to Pizza Hut to try the Personal Pan Pizzas. When they made their order (8 Personal Pan Pizzas with all the toppings), there were 10 such pizzas being taken out of the oven, one at a time. If Pizza Hut's policy is to serve their pizzas as soon as they come out of the oven and the food server distributes pizzas randomly to the co-workers, what is the probability that the first 5 pizzas to come out of the oven will be given to the 5 most senior co-workers, in order? (i.e. the first pizza going to the most senior, etc.)
[Answer: 1/8 x 1/7 x 1/6 x 1/5 x 1/4]

Far Transfer, Reversed Roles:
In the women's locker room at South Side High School, certain lockers are set aside for the 18 female swimmers on the school team. At the beginning of the season, these lockers are assigned to swimmers, starting with a row of 14 lockers next to the showers. These lockers are assigned at random, starting at the end near the showers and going down the row. What is the probability that the 7 lockers closest to the showers are assigned to the 7 butterfly swimmers?
[Ans.: 7/18 x 6/17 x 5/16 x 4/15 x 3/14 x 2/13 x 1/12]

References

Bobrow, S.A. & Bower, G.H. (1969). Comprehension and recall of sentences. *Journal of Experimental Psychology, 80*, 455-461.

Catrambone, R. (1991). Helping learners acquire subgoals to improve transfer. *Cognitive Science Proceedings*, Chicago, Illinois. Hillsdale, New Jersey: Erlbaum.

Chi, M.T.H., Bassok, M., Lewis, M.W., Riemann, P., & Glaser, R. (1989). Self explanations: How students study and use examples in learning to solve problems. *Cognitive Science, 13*, 145-182.

Lewis, M.W. & Anderson, J.R. (1985). Discrimination of operator schemata in problem solving: Learning from examples. *Cognitive Psychology, 17*, 26-65.

Pirolli, P. (1991). Effects of examples and their explanations in a lesson on recursion: A production system analysis. *Cognition and Instruction, 8,* 207-259.

Pirolli, P. & Bielaczyc, K. (1989). Empirical analyses of self-explanation and transfer in learning to program. *Cognitive Science Proceedings*, Ann Arbor, Mich. Hillsdale, New Jersey: Erlbaum.

Ross, B.H. (1989). Distinguishing types of superficial similarities: Different effects on the access and use of earlier problems. *Journal of Experimental Psychology: Learning, Memory, and Cognition, 15*, 456-468.

Schooler, L.J. & Anderson, J.R. (1990). The disruptive potential of immediate feedback. *Cognitive Science Proceedings*, Cambridge, Mass. Hillsdale, New Jersey: Erlbaum.

Slamecka, N.J. & Graf, P. (1978). The generation effect: Delineation of a phenomenon. Journal of Experimental Psychology: Human Learning and Memory, 4, 592-604.

Sweller, J. & Cooper, G.A. (1985). The use of worked examples as a substitute for problem solving in learning algebra. *Cognition and Instruction, 2*, 59-89.

Ward, M. & Sweller, J. (1990). Structuring effective worked examples. *Cognition and Instruction, 7*, 1-39.

The Phase Tracker of Attention

Erik D. Lumer
Xerox Palo Alto Research Center
Palo Alto, CA 94304
lumer@parc.xerox.com

Abstract

We introduce a new mechanism of selective attention among perceptual groups as part of a computational model of early vision. In this model, selection of objects is a two-stage process: perceptual grouping is first performed in parallel in connectionist networks which dynamically bind together the neural activities triggered in response to related features in the image; secondly, by locking its output on the quasi-peridic bursts of activity associated with a single perceptual group, a dynamic network called the *phase-tracker of attention* produces a temporal filter which retains the selected group for further processing, while rejecting the unattended ones. Simulations show that the network's behavior matches known psychological data that fit in the descriptive framework of object-based theories of visual attention.

Introduction

In most elaborate perceptual systems with limited processing resources, mechanisms that focus the attention on small parts of the sensory inputs are often necessary in order to cope with the complexity of the sensed world. The importance of attention in everyday activity has been a major impetus for its extensive study by psychologists and neurophysiologists (Eriksen & St–James, 1986; Crick, 1984; Duncan, 1984; Treisman & Gelade, 1980). As a result of their work, a number of theories have been developed, that fall in general under either one of two broad classes, known as the location-based and object-based theories of visual attention (Duncan, 1984). The first class stipulates that, at any given moment, attention is entirely allocated to a single convex region of space. In this model, the spatial dimension is the basic cue used to direct attention, which is therefore often compared to a mental spotlight. The psychological evidence that supports location-based theories along with the spotlight metaphor is reminiscent of a variety of experimental paradigms, including response competition (Eriksen & St–James, 1986), spatial precueing, and visual search (Treisman & Gelade, 1980). Location-based theories may be contrasted with object-based theories, which assume that attention can be allocated to one or more perceptual groups, regardless of their spatial locations. Object-based theories describe early perception as a two-stage process: the segmentation of images into distinct perceptual groups is done according to low-level, data-driven mechanisms of perceptual organization that exploit detected properties of proximity, continuity, similarity, or common motion, among others. In contrast with the parallel preattentive stage, a second stage of visual processing, called *focal attention*, is serial and consists in the selection and analysis of a particular perceptual group (Neisser & Becklen, 1975).

The experimental evidence in support of an object-based form of attention is multiple. It indicates in particular that subjects are better able to report two properties of the same object than one from each of two objects that are at the same spatial location (Duncan, 1984). Rock and Gutman (1981) also observed that subjects who were directed to attend to only one of two overlapping and novel figures (say the red figure among a red and green one) showed no recognition of form for the unattended one. Such result is not predicted by standard location-based theories of attention. More recently, Driver and Baylis (1989) showed that in response competition experiments, the grouping of target and distracting elements by common motion can have more influence than their proximity. These results are consistent with the hypothesis that attention can be directed to perceptual groups whose components are not spatially contiguous. Despite these and many other experimental facts, very little has been done to address the computational and neurological issues raised by the existence of an object-based form of attention. This situation contrasts with the flurry of recent work devoted to the modeling of location-based mechanisms of attention (Ahmad, 1991; Mozer, 1988).

The goal of this paper is to map the conceptual framework of object-based theories into a crisp computational model which has better predictive value. Note

This work was partially supported by the Air Force Office of Scientific Research contract No. F49620–90–C—0086 given to B.A. Huberman.

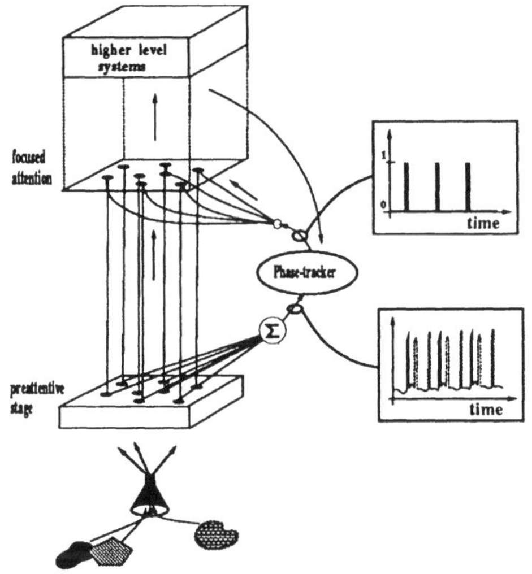

Figure 1 : Functional model of selective attention.

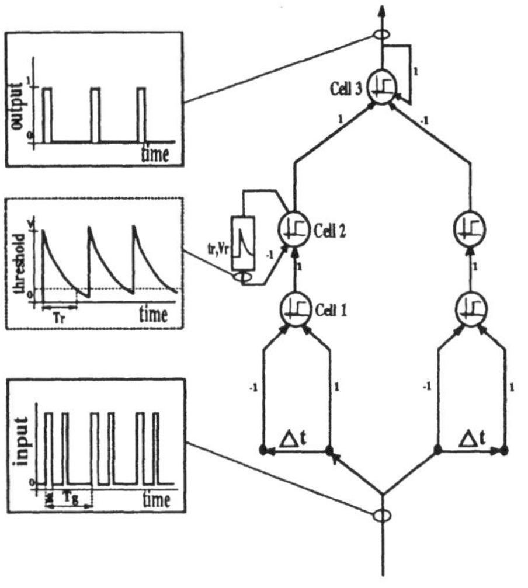

Figure 2 : Connectionist phase-tracker.

that in this framework, the objects that the focal attention can select or discard are defined at a preattentive stage. Thus, one expects mechanisms of attention to be intimately tied to the ones underlying perceptual organization. In what follows, this relationship is unraveled in terms of compatible mechanisms of interactions among neurons. In our model, the organization of visual scenes is based on a biologically inspired mechanism of labeling of perceptual groups (Gray et al., 1989; von der Malsburg, 1981), in which neural assemblies express their membership to a perceptual group by firing simultaneously and in a pseudo-periodic fashion, while being out of synchrony with neurons stimulated by other groups or a background. A number of authors have recently demonstrated the feasibility of perceptual grouping via synchronization of neural activity in large heterogeneous networks (Lumer & Huberman, 1992; Sporns, Tononi, & Edelman, 1991; Baldi & Meir, 1990). We refer to (Lumer 1992) for a description of how this mechanism is implemented in our model. In this paper, we focus more specifically on the issue of internal access to perceptual groups constructed in this way: given *implicit* temporal labels, i.e. the relative phases of neural oscillations distributed across a population of detectors, we still face the problem of how to use them *explicitly* in a mechanism of visual attention that selects among groups for further processing. A solution to this problem is proposed in the next section in the form of a dynamic network called the *phase-tracker of attention*. This system is then tested on examples which mimic the conditions and observed behaviors in a number of psychological experiments. The paper ends with a short discussion about the implications of our work.

The Phase-Tracker

We develop and study below a computational model of a two-staged visual system of the kind described by Neisser (1975) and others.

A coarse schematic representation of the model is given in Figure 1. Let us assume that the segmentation of perceived images is achieved at a preattentive stage via the synchronization of neurons that fire periodically in response to the local properties of a same object. The discussion of how this is actually done is reported elsewhere (Lumer, 1992). In the present context, suffices it to notice that the cumulated activity emerging from the preattentive stage evolves in time as one or several intertwined and periodically bursting signals superimposed over a low level stochastic noise. The noisy activity results from the asynchronous firing of cells stimulated by an incoherent percept or background. Each periodic burst of activity, on the other hand, is associated with a single perceptual group so that its phase can serve as a unique label referencing that group.

In order to make use of such a label to select a desired object, we imagine the following process: the cumulated activity from the preattentive stage is fed into a specialized network, called the *phase-tracker of attention*. This system is capable of locking its output on the periodic burst of activity associated with the target object so that the output equals 1 when the periodic signal is bursting and 0 otherwise. That way, the phase-tracker defines explicitly, that is in terms of its output, the temporal windows during which the neural assemblies representing the target object are firing. If the projections from the preattentive stage onto the input layers of the higher levels of perception are modulated by the output of the phase-tracker, all the non-target objects present in the image will remain undetected beyond the preattentive stage.

A simple implementation of a phase-tracking system is shown on the right hand side of Figure 2. It consists of a hybrid dynamic network exhibiting transient states, delayed propagation and feedforward as well as feedback connections. Each unit in the network connects its total input (i.e. presynaptic) activity, x, with its (postsynaptic) output, y, via a sharp thresholding function that is defined as

$$y = f_\theta(x) \tag{1}$$

where

$$f_\theta(x) = \begin{cases} 1 & if\, x \geq \theta \\ 0 & otherwise. \end{cases} \tag{2}$$

The input to the phase-tracker, $s(t)$, is propagated along a left and a right branch. The two branches act as rising and falling edge detectors, respectively. Let us first take a closer look at how the rising edge detector works. The presynaptic connection to cell 1 (see Figure 2) produces the first order difference of the input signal. When larger than the threshold θ, this difference causes cell 1 to fire. Stated more formally, the output of cell 1, $y_1(t)$, is related to the input signal s(t) by the relation

$$y_1(t) = f_\theta(s(t) - s(t - \Delta t)) \tag{3}$$

where Δt is a positive time delay. With a proper value assigned to Δt, cell 1 will turn on as a result of any sharp increase of its input. It therefore plays the role of a rising edge detector.

The output of cell 1 is fed into cell 2, which possesses a dynamical threshold. The properties of networks of cells with dynamical thresholds have recently been studied by a number of people (Abbot, 1990; Horn & Usher, 1989). In essence, a dynamic threshold is a transient feedback link from the thresholding cell onto itself. It is usually modeled as a leaky integrator which gets charged by the output activity of its cell. Once charged, such threshold inhibits any further firing of its cell for a period determined by the constants of the integrator. In particular, the amplitude of the dynamical threshold of cell 2, $R_2(t)$ evolves in discrete time steps according to

$$R_2(t+1) = V_R.y_2(t) + e^{-1/\tau_R}.R_2(t) \tag{4}$$

where y_2 is the output of cell 2, V_R the gain and τ_R the time constant of the leaky integrator. With the notations of Eq. (1), the output of cell 2 then reads as

$$y_2(t) = f_\theta(y1(t) - R_2(t)). \tag{5}$$

The pulse sent by cell 2 upon detection of a rising input turns on the output of the phase-tracker , that is cell 3. The subsequent inhibition of cell 2 prevents further increases in the input signal from affecting the output of the phase-tracker for a period of time T_R. This refractory period is related to the parameters of cell 2 via

$$T_R = Ceil\left(\tau_R ln\left(\frac{V_R}{1-\theta}\right) + 1\right) \tag{6}$$

where the function *Ceil(.)* rounds its argument up to the nearest integer value and accounts for the discrete nature of the dynamics. Because of the static feedback connection from cell 3 onto itself, its output remains at a high level until a pulse from the falling edge detector resets it to zero. The falling edge detector is very similar to the rising edge detector. By changing the sign of the first order difference computed at the input of the right branch with respect to that used in the left branch, its output will fire in response to any sharp decrease of the input to the phase-tracker.

Consider thus a periodic signal placed at the input of the phase-tracker, which consists of bursts lasting for an interval of time g and separated from each other by regular intervals T_g. The rising edge of the first burst causes the phase-tracker to turn on, a state which is kept by the system until the burst dies off, at which point the falling edge detector emits a pulse and the output of the phase-tracker switches to a low value. The phase-tracker is then inhibited during an interval of time T_R following the rise of the detected burst. It will therefore accurately track the phase of the incoming signal provided that the refractory period is shorter than T_g. Furthermore, other signals added to the input in between two successive bursts of the tracked activity will be ignored by the system as long as they precede or follow a detected burst by an interval of time larger than $T_g - T_R$. This difference defines the *resolution* of the phase-tracker and constrains in part the number of objects which can be separated by the mechanism of attention proposed in this section. This point will be expanded later in the paper.

The regimes just outlined are illustrated on the left side of Figure 2. The lower plot shows the temporal evolution of the incoming signal, *s(t)*. The resulting output of the phase-tracker is represented in the upper plot while the middle one gives the evolution of the dynamical threshold of cell 2. The horizontal line in this figure indicates the threshold value under which the rising edge detector is enabled.

Network Simulations

We have implemented the phase-tracker as part of a connectionist architecture of early perception (Lumer, 1992). In brief, local attributes of 2D-images are detected in parallel by cells organized in a number of feature maps. The segmentation of images via synchronization of activity is done in grouping maps whose outputs are projected in a one to one fashion on the maps which define the first stage of the higher levels of perception. The inputs to the grouping maps can be restricted by a coarse location-based mechanism of attention which works cooperatively with the phase-tracker. More will be said about spatial attention in the final part of the paper. The cumulated output from the grouping maps is fed into the phase-tracker. A simple control mechanism allows the use of top-down information in the selection of objects with specified features: a global detector is associated with each map at the entrance of the higher levels of perception and signals whether the corresponding feature is present in the group currently selected. If this is the case and the feature does not match the description of a target object, the state of the phase-tracker is automatically reset so as to track the next label which is available at its input. To get a rough estimate of the time scales involved in the studied mechanism of selection, we equate one time step in our simulation with 1 msec of real time. The time constants of the grouping cells are set so that the cells fire once every 25 cycles, that is at a frequency of 40Hz. This number is consistent with the observed frequencies of neural oscillations in the primary visual cortex of cats (Gray et al., 1989). The resolution of the phase-tracker, as defined above, is equal to 1 msec. Finally, the visual field in the simulations is an array with 16 by 16 pixels.

The system was tested on a number of examples in which the phase-tracker takes advantage of the temporal separation of perceptual groups that cannot be easily discriminated spatially. Thus, a simple spatial spotlight of attention will fail in these cases. In particular, we have demonstrated the ability of our system to selectively focus on either one of two overlapping figures distinguished from each other by their respective colors. This behavior is in agreement with psychological observations (Rock & Gutman, 1981). Similarly, attention can be restricted to a non-contiguous set of objects animated by a common motion in a setting that mimics, albeit in a caricatural fashion, recent experiments performed by Driver and Baylis (1989). To save space, we will only detail the second example. The image in Figure 3 is composed of five 2x2 objects. The center and two far end elements are animated by a common upward motion while the intermediate objects move downward. The control system is instructed to focus only on the objects moving upward during the first 100 iterations before shifting attention to the other group of objects. Figure 4 displays the input (lower plot) and output (upper plot) of the phase-tracker as a function of time. After a transient period of about 25 iterations during which the temporal labels are formed, the phase-tracker locks on the index to the objects moving upward (their shared label is represented in grey for illustrative purpose only). Attention is released from this group at t=100 msec and redirected towards the objects moving downward after an equivalent time of about 20 msec (the corresponding label for these objects is shown in black). Notice that the locking of the phase-tracker on a label translates into the selection for further processing of the entire group indexed by that particular label.

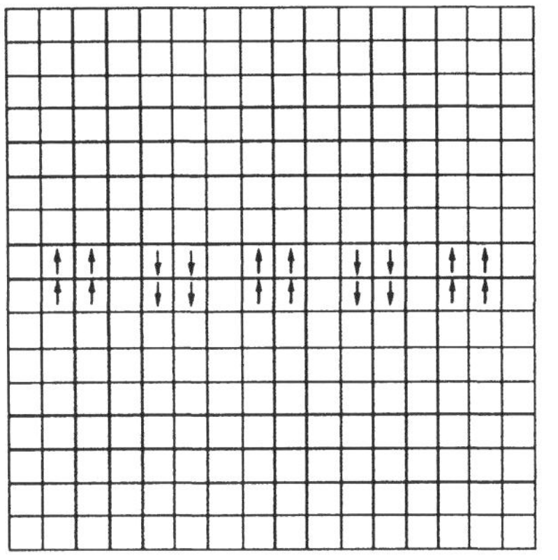

Figure 3 : Image containing three blocks moving upward separated from each other by objects moving downward.

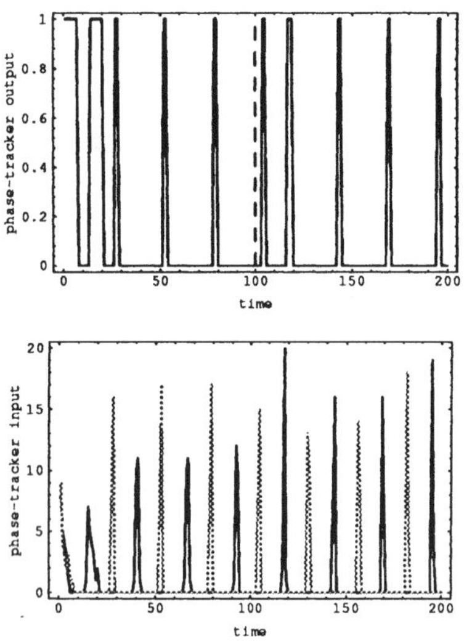

Figure 4 : Tracking of objects with common motion.

Discussion

In this paper, we have presented a non-spatial process of selection among perceptual groups, which overcomes the shortcomings of location–based models of visual attention. The proposed mechanism of selective attention presupposes the segmentation and labelling of perceptual groups via synchronization of the neurons responding to the local properties within a group. Our work is therefore complementary to the flurry of recent reports showing that this type of grouping can be achieved using simple dynamical networks. Indeed, any improvement of image segmentation via dynamic grouping will augment the potential of selection by the phase-tracker. Furthermore, the use of oscillatory dynamics in our model leaves the door open for a better modelling of the cortical tissues in which these regimes have been observed.

The embedding of the phase-tracker in a connectionist architecture of early visual processing reveals its capabilities and limitations. With the parameters used for our simulations, we observed that the bursts of activity, or labels, associated with a single perceptual group have a temporal duration on the order of 3 time steps. Furthermore, the resolution of the phase-tracker, as defined in the second section, is equal to one time steps. We therefore know that each unambiguous label produced by the segmentation system occupies on the time axis about 4 time steps. Since the consecutive firings of grouping cells are spaced by 25 time steps, we conclude that selection cannot operate on more than approximately 6 perceptual groups whose labels are placed at the input of the phase-tracker. This observation places a strong constraint on the interaction between a coarse mechanism of spatial orientation and selection: the former must be tuned so as to limit the number of objects that the dual mechanisms of segmentation and selection operate on at any given moment. To our knowledge, this constitutes the first embodied prediction of a possible interdependence between two modes of attention, i.e. location-based and object-based, that have traditionally been considered as orthogonal. We expect that future work, both experimental and computational, will further elucidate this relation.

Another very interesting observation can be drawn from the fact that the combined mechanisms of segmentation and selection have a maximum "capacity" of about 6 elements. Indeed, in trying to determine how the time required to quantify a collection of n items presented to view was function of n, it was found (Chi & Klahr, 1975) that a striking discontinuity occurs in the region of $n=6\pm1$. This phenomenon, known as subitizing, is characterized by a very rapid apprehension of the number of items below the discontinuity, while the reaction time increases linearly with the number of elements by a much larger increment above the critical point. It is tempting to speculate that a transition from an object-based form of attention to a serial scanning of spatial locations in the display might be related to the observed phenomenon. We also notice that sensory segmentation and selective attention are not the unique attributes of vision. For example, the auditory modality parses complex sound fields into independent streams, each one being associated with a specific external source. This capability is best illustrated in cocktail parties where one is able to distinguish several voices, and selectively attend to one, among a noisy crowd. Since the processes of segmentation of the auditory fields have been modelled as neural oscillators which either synchronize or desynchronize their phases (von der Malsburd & Schneider, 1986), a phase-tracker could likewise be used in this context to implement selective attention.

Last but not least, this paper illustrates the richness of computational mechanisms which can be derived from the use of dynamical networks having transient states. As connectionist models of cognition become larger and develop modularity, the issues of communication and coordination between the heterogenous modules become central. In this context, the connectionist equivalents of communication devices, such as signal multiplexers, clock synchronizers, and phase-locked loops of the kind studied here, are expected to play a fundamental role.

Acknowledgments

I thank Bernardo Huberman, Eric Saund, Roger Shepard and David Rumelhart for helpful discussions as well as Jeff Shrager for mentioning the phenomenon of subitizing.

References

L.F. Abbot. A network of oscillators. *Phys. A: Math. Gen.*, 23:3835–3859, 1990.

S. Ahmad and S. Omohundro. Efficient visual search: a connectionist solution. *In Proc. 13th ann. conf. cog. sci. soc.*, 1991.

P. Baldi and R. Meir. Computing with arrays of coupled oscillators: an application to preattentive texture discrimination. *Neural Comp.*, 2 (4):459–471, 1990.

M.T.C. Chi and D. Klahr. Span and rate of apprehension in children and adults. *Journal of Experimental Child Psych.*, 19:434–439, 1975.

F. Crick. Function of the thalamic retcular complex: The searchlight hypothesis. *Proc. Natl. Acad. Sci. USA*, 81:4586–4590, 1984.

Jon Driver and G.C. Baylis. Movement and visual attention: the spotlight metaphor breaks down. *J. Exp. Psych.: Human Perc. and Perf.*, 15 (3):448–456, 1989.

J. Duncan. Selective attention and the organization of visual information. *J. Exp. Psych.: General*, 113 (4):501–517, 1984.

C.W. Eriksen and J.D. St.James. Visual attention within and around the field of focal attention: a zoom lens model. *Percept. and PsychoPhys.*, 40 (4):225–240, 1986.

C.M. Gray, P. Konig, A.K. Engel, and W. Singer. Oscillatory responses in cat visual cortex exhibit intercolumnar synchronization which reflects global stimulus properties. *Nature*, 338:334–337, 1989.

D. Horn and M. Usher. Neural networks with dynamical thresholds. *Phys. Rev. A*, 40 (2):1036–1044, 1989.

E.D. Lumer. Selective attention to perceptual groups: the phase tracking mechanism. *To appear in Int. Journal of Neural Systems*, 3 (1), 1992.

E.D. Lumer and B.A. Huberman. Binding hierarchies: a basis for dynamic perceptual grouping. *Neural Computation*, 4 (3), 1992.

M. Mozer. A connectionist model of selective attention in visual perception. *In Proc. 10th ann. meet. cog. sci. soc.*, pages 195–201, 1988.

U. Neisser and R. Becklen. Selective looking: Attending to visually specified events. *Cognitive Psychology*, 7:480–494, 1975.

I. Rock and D. Gutman. The effect of inattention on form perception. *J. Exp. Psych.: Human Perc. and Perf.*, 7 (2):275–285, 1981.

O. Sporns, G. Tononi, and G.M.Edelman. Modeling perceptual grouping and figure-ground segregation by means of active reentrant connections. *Proc. Natl. Acad. Sci. USA*, 88:129–133, 1991.

A.M. Teisman and G. Gelade. A feature-integration theory of attention. *Cog. Sci.*, 12:97–126, 1980.

A. Treisman. Focused attention in the perception and retrieval of multidimensional stimuli. *Percep. and Psychophys.*, 22:1–11, 1977.

C. von der Malsburg. The correlation theory of brain function. *Intern. report 81-2*, Dept. of Neurobiology, Max Planck Institute., 1981.

C. von der Malsburg and W. Schneider. A neural cocktail-party processor. *Biol. Cybern.*, 54:29–40, 1986.

An Extension of Rhetorical Structure Theory for the Treatment of Retrieval Dialogues

Elisabeth Maier
Department KOMET
GMD-IPSI
Dolivostr. 15
D-W-6100 Darmstadt
Germany
maier@ipsi.darmstadt.gmd.de

Stefan Sitter
Department COGITO
GMD-IPSI
Dolivostr. 15
D-W-6100 Darmstadt
Germany
sitter@ipsi.darmstadt.gmd.de

Abstract

A unification of a speech-act oriented model for information-seeking dialogues (COR) with a model to describe the structure of monological text units (RST) is presented. This paper focuses on the necessary extensions of RST in order to be applicable for information-seeking dialogues: New relations are to be defined and basic assumptions of RST have to be relaxed. Our approach is verified by interfacing the dialogue component of an intelligent multimedia retrieval system with a component for natural language generation.

1 Introduction and Problem

Approaches to discourse organization especially in the area of computational linguistics are oriented towards the treatment of either monologues or dialogues. Only recently efforts have been reported to develop models which cover both types of discourse (e.g., Sitter & Stein, 1992; Fawcett & Davies, 1992). Our paper is a contribution to this research topic providing a unified model to describe coherence in dialogues in a computational framework.

Background for our work is the development of an intelligent information-retrieval system MERIT (Stein et al., 1992), which makes use of natural language as one of the modalities for system-user interaction. The system integrates the text generation system developed in the KOMET project (Bateman et al., 1991) with an implementation of a speech-act oriented dialogue model called COR (*co*nversational *r*oles).

In this paper we focus on the *theoretical* part of the system integration and discuss the consequences resulting from the application of the monologically-oriented Rhetorical Structure Theory (RST, see Mann & Thompson, 1987) on dialogues. We demonstrate our concepts using a few excerpts from one example dialogue.

2 State of the Art

Since we want to tie the two research strands together - work on dialogue structure and approaches focusing on the treatment of texts - we discuss trends in both fields and also provide a basis for the description of our approach.

2.1 Models of dialogue structure

Many classical systems which are interfaced with natural language systems, like explanation components, database access modules, tutoring systems, etc. (for reviews, see Perrault & Grosz, 1986 and McCoy et al., 1991), lack a model of the information-seeking dialogue and have the simple underlying conception of a dialogue as iteration of adjacent query-answer pairs. More recent work gives an explicit account of the dialogue – i.e., its thematic structure, its relation to an external task, types of failure, etc. (Reichman, 1985; Grosz & Sidner, 1986; Carberry, 1985). Carberry deals with information-seeking problems in a dialogue between the user as the information-seeker and the system as the information-provider. She presupposes that information-seeking takes place in the context of a defined task, and that the user has a plan for his task which (1) can in principle be formulated and (2) has well-defined gaps and misconceptions. The job of the system is to recognize the plan and to assist in plan execution, i.e. provide necessary missing information, inform about hidden obstacles. In case of misconceptions, it also assesses the relevance of the user's questions and uncovers false presuppositions.

Unfortunately, in many realistic situations it seems too restricting an assumption that the information-seeker should be able to verbalize his plan or even have a plan (McAlpine & Ingwersen, 1989; Belkin & Vickery, 1985). If meaningful structures can be construed in information-seeking dialogues in highly vague task settings, this must be done without strong reference to a domain structure.

Winograd and Flores (Winograd & Flores, 1986) give a different account of dialogue structure. The authors argue on philosophical grounds that only on the level of interactional conventions, which specify how to express and negotiate behavioral expectations and commitments, can interactions be formally described. As an example they use two partners', A and B, negotiation of a task which B has to fulfill (*Conversation for Action*, in the following called CfA). The process of negotiation is represented as the traversal of a state-transition network summarizing all possible chainings of A's and B's dialogue actions like 'Request', 'Promise', 'Reject', 'Withdraw', etc., ignoring the contents of the actions. The authors do not treat the computer as a dialogue partner, but as a medium for the structuring of inter-human interaction.

Moore's system (Moore, 1989) for the generation of explanations in the framework of expert systems is able to deal with follow-up questions of users who do not understand parts or all of the system's explanations. Her model differs from the approach proposed by Winograd and Flores insofar as the system plans explanations depending on the user's questions and on the communicative goals which were responsible for the generation of previous system contributions. The system also makes assumptions about the lack of user knowledge and generates clarifications. Unfortunately, the system is restricted to the treatment of a small set of speech acts and to the modeling of a subset of possible interactions in explanatory dialogues. This model therefore can profit from an integration with a more extensive model of human-machine interaction like, e.g., COR - an enterprise we describe in section 3.

2.2 Models of monological discourse

Various approaches for the description of discourse structures emphasize the conventionalized order of discourse segments. Among them are approaches like macrostructures (vanDijk & Kintsch, 1983), grammar-like descriptions for specific genres (Rumelhart, 1975) or schemata (McKeown, 1985; Paris, 1987).

While approaches of this family - of which we consider also the dialogue model CfA a member - describe the sequence of elements in an interaction, they do not give an account of *how* they are related. The recipient of the information has to recognize why the information is presented in the given sequence. A model like RST is able to model this feature of texts, i.e. it makes use of constructs - the relations - which model the semantics of the links between text segments.

RST provides means to represent the structure of *monological texts* hierarchically. In RST, Mann and Thompson defined an open set of relations which are used to describe the semantics of the links between units of texts. Such text units are segments of discourse, the minimal length of which is one proposition. Each pair of text units connected by a relation is again considered to be one unit. A basic assumption of RST is that relations impose an asymmetrical structure on two connected texts units: One unit is more important than the other - it cannot be removed without changing the core meaning of the text. This text unit is called the *nucleus* ($\mathcal{N}$). The segment of text which is of less importance is the *satellite* ($\mathcal{S}$).

The set of relations proposed in RST has been aug-

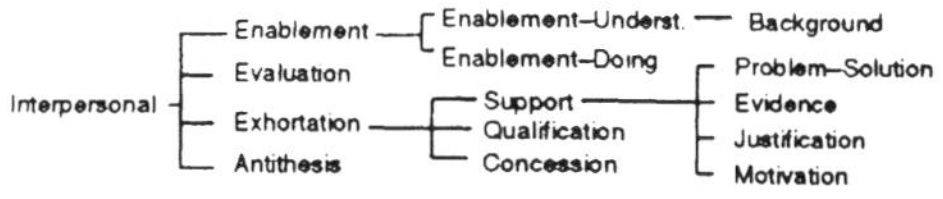

Figure 1: Taxonomy of interpersonal rhetorical relations (taken from Maier & Hovy, 1991).

mented recently by a large number of relations taken from other approaches (e.g. Hobbs, 1990; Sanders, Spooren & Noordman, 1990; Iri, McMillan & Merz, 1990). A classification of these relations (Maier & Hovy, 1991) distinguishes three types: *ideational*, *interpersonal* and *textual* relations. Ideational relations capture links which are concerned with experiential knowledge, while textual relations are used to signal text-internal links, i.e. links which refer to segments of text instead of segments of text-external knowledge. Interpersonal relations are of special importance here since they take the specific features of the discourse participants into account. Among the interpersonal relations are such which affect the ability of the reader (e.g. ENABLEMENT), his willingness to do something (e.g. MOTIVATION), his beliefs (e.g. EVIDENCE), etc. A subtaxonomy for interpersonal relations is given in figure 1 (from Maier & Hovy, 1991). We assume that dialogic discourse implies the need for more relations of this type which leads to the extension of this part of the network. In this paper we show that the application of RST on dialogues leads to an extension of the set of interpersonal relations (section 3.2) and the change of at least one basic assumption of RST (section 3.3).

3 Approach

3.1 The COR Dialogue Model

The COR model describes dialogic information-seeking processes, involving two dialogue partners A and B with the roles of information-seeker and information-provider, respectively (Sitter & Stein, 1992). There are two "ideal" courses of action:

- A formulates his request, B promises to answer and answers the request, A expresses contentment.
- B offers to provide some information (assuming that he has sufficient knowledge about A's information need), A accepts the offer, B provides the information, and A expresses contentment.

These two courses are *expected* in the sense that they match the role expectations which are adequate for the information-seeking situation. In fact, many everyday information-seeking exchanges already follow this expected course, e.g. requesting the time. The state-transition network in figure 2 presents the "expected" courses of action (bold arrows). In more problematic information-seeking situations beyond such short exchanges - e.g., those involving the use of information systems - there are reasons for deviations from the expected course. Information re-

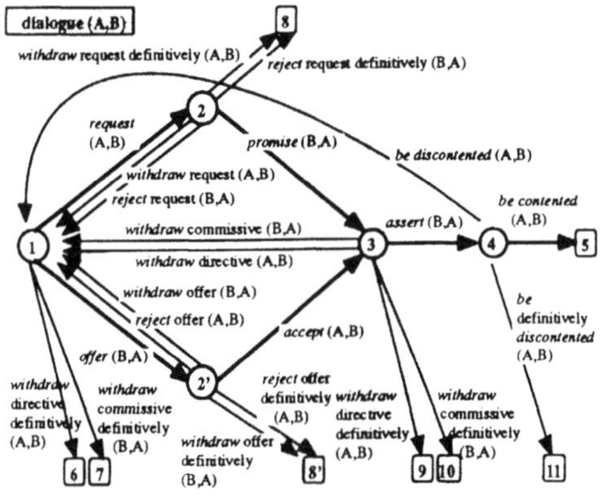

Figure 2: A network representing role expectations in information-seeking dialogues (from Sitter & Stein, 1992). The initial state is < 1 >. The transitions in the network are made by performing the dialogue contributions. The parameters A and B correspond to the speaker and the addressee of the contribution respectively.

quested may not be available; information provided may not satisfy the information-seeker. These examples lead to the introduction of further paths in the network. These additional paths either lead back to the initial state < 1 > in figure 2, if the dialogue is to be continued, or to final states besides state < 5 >. We call these acts *alternative*. Alternative acts reject or withdraw role expectations prevalent in the dialogue situation. E.g., after A's asking a question, B's promise to answer the question is expected, but also B's rejection or A's withdrawal of the question may happen under certain conditions.

The network so far resembles the state-transition network used by Winograd and Flores as an example of their CfA (Winograd & Flores, 1986). However, we interpret the transitions in figure 2 not as atomic *acts* (a request, an answer, etc.), but as possibly extended dialogue *sections* that are subordinated to illocutionary functions (to arrive at a *mutually understood* request, answer, etc.).[1] We will refer to them as dialogue *contributions*.

A contribution can be subdivided into two components. The first has the purpose to express the illocutionary function of the whole section – i.e., instantiate a role expectation or a response to a role expectation (acceptance or rejection). The second delivers some kind of *contextual information* for it, which may become necessary if the first threatens to fail. *Context* here refers to the presence or absence of the conditions which must be fulfilled to render a dialogue contribution successful.[2] E.g, for a question to succeed, there must be some agreement that its answering helps the information-seeker's goals, that the dialogue partner knows the meaning of the words used, etc. These conditions justify behavioral expectations; severe disagreement on them motivates deviation from the expected course of action.

Example 1 shows a 'request' dialogue contribution, consisting of the request proper and contextual information (modified from figure 3 (U = user, S = system)):

Example 1:
U: Tell me about EC-funded projects with enddate after 1992 dealing with 'illocutionary models'. (Request)
'Illocutionary models' is, very roughly, like 'dialogue modeling'. (Contextual information)

Instead of U's supplying contextual information, S might have initiated an embedded dialogue:

Example 2:
U: Tell me about EC-funded projects with enddate after 1992 dealing with 'illocutionary models'. (Request)
S: What do you mean by 'illocutionary models'? (Request for contextual information)
U: It is, very roughly, like 'dialogue modeling'. (Contextual information)

Both variants – U's voluntarily supplying contextual information to his own dialogue contribution (example 1) and S's dialogically exploring the context of U's contribution (example 2) – have the function of increasing the probability of success of the contribution. To cover this similarity, both variants are considered 'request' contributions.

The part of a dialogue contribution addressing the role expectations, labeled as illocutionary types ('Request', 'Reject Request', etc. in figure 2), is called the *nucleus*. The other part addressing the context is called the *satellite*. These terms have been adopted from the analysis of monological texts by means of RST (section 2.2). Often, as in the examples presented throughout this paper, the nucleus is an atomic act not further decomposed by COR. The satellite is either an assertion or an embedded dialogue, or it may be omitted. A dialogue exploring the context of the contribution is itself an information-seeking dialogue and follows the same rules as the top-level retrieval dialogue (see example 2).

Figure 3 shows a dialogue section, mainly elaborating on an initial user request as in example 2. On the left of figure 3, the decomposition of the dialogue is shown. (For more details, see Sitter & Stein, 1992.)

[1] The network, together with further networks for the contributions (Sitter & Stein, 1992), is interpreted as an ATN (augmented transition network).

[2] These conditions include Austin's (Austin, 1962) and Searle's (Searle, 1969) "felicity conditions", which subsume mainly *conditions of the speaker*. However, like Wunderlich (e.g., Wunderlich, 1976), we go beyond these and also include conditions the *addressee* must fulfill and which can be decided about only in the course of the succeeding interaction. Therefore, Wunderlich introduces the term *successfulness* ("Erfolgreichsein") in addition to felicity.

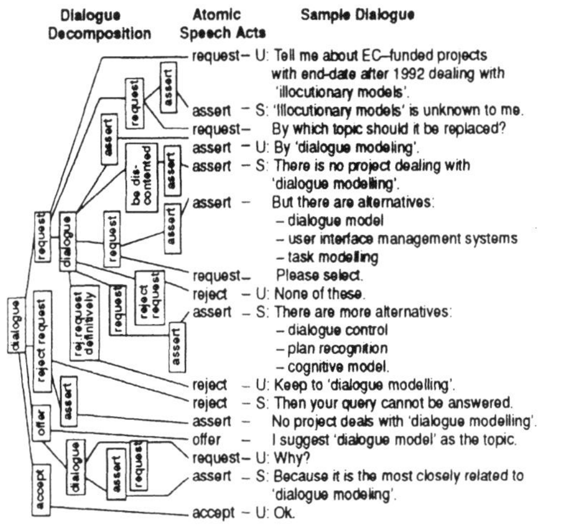

Figure 3: An example MERIT dialogue section negotiating a user request with its decomposition according to COR. MERIT provides information about EC-funded projects.

3.2 Extension of the Set of Interpersonal Relations

Based on Winograd and Flores' approach (Winograd & Flores, 1986), COR is a model of the level of role expectations in an information-seeking dialogue. We abstract from the content of the dialogue (the thematic level) and leave a description of this to a thematic model, upon which we pose only very low demands (e.g., it need not be a plan-based description of the domain). Therefore, we explicitly permit failures and problematic situations in the dialogue; the dialogue model can organize the means to deal with them interactively.

Still underdeveloped is an account of the various types of failures and problematic situations. Possible reasons for deviations should be classified and adequate means for their expression sought.

As mentioned above we assumed that the application of RST for the description of dialogues may lead to the extension of the set of necessary relations, especially of *interpersonal* relations which typically address discourse participants. The analysis of our sample dialogue brought about such an extension. In the following we give two examples from figure 3 for such "new" relations.

Example 3:

S: 'illocutionary models' is unknown to me.
(Contextual information)
By which topic should it be replaced? (Request)

The first proposition gives an explanation for the – otherwise not understandable – reaction of the system. I.e. the system starts a subdialogue in order to repair a potential failure of the whole dialogue. The information requested by the user ('illocutionary models') is not available, which is the reason why the system comes up with the alternative strategy to request a different query from the user. The system also motivates the decision by providing contextual information: It explains why it asks for different / additional information. The nucleus – the central illocution for the whole fragment – is contained in the second proposition, i.e. it is exactly this system request. The semantics of the relation resembles the RST definition for

BACKGROUND:

constraints on $\mathcal{N}$:
$\mathcal{R}$ (the reader) won't comprehend $\mathcal{N}$ sufficiently before reading text of $\mathcal{S}$
constraints on $\mathcal{S}$: none
constraints on the $\mathcal{N}$ + $\mathcal{S}$ combination:
$\mathcal{S}$ increases the ability of $\mathcal{R}$ to comprehend an element in $\mathcal{N}$
the effect:
$\mathcal{R}$'s ability to comprehend $\mathcal{N}$ increases

The new relation has a more restrictive and dialogue-oriented meaning which can be defined as follows:

BACKGROUND-FOR-REQUEST:

constraints on $\mathcal{N}$:
$\mathcal{R}$ won't comprehend the relevance of the 'Request' specified in $\mathcal{N}$ without being given a reason for it
constraints on $\mathcal{S}$:
$\mathcal{S}$ is an atomic 'Assert' speech act
constraints on the $\mathcal{N}$ + $\mathcal{S}$ combination:
$\mathcal{S}$ increases the ability of $\mathcal{R}$ to comprehend why $\mathcal{N}$, i.e. the request, was uttered by specifying the reason for it
the effect:
$\mathcal{R}$'s ability to comprehend the relevance of $\mathcal{N}$ and his willingness to follow the request in $\mathcal{N}$ increase

The definition of BACKGROUND-FOR-REQUEST is more restrictive than BACKGROUND insofar as the satellite explains why a *request* is made in the nucleus. Compared to the definition of BACKGROUND, constraints on the satellite are available refining its semantics. BACKGROUND-FOR-REQUEST therefore can be considered a subtype of BACKGROUND, which is an *interpersonal* relation. By hypothesizing such a new relation we already extend the set of relations concerned with features of the discourse participants. — Another new relation can be found in the following segment of our sample dialogue:

Example 4:

S: But there are alternatives:
- dialogue model
- user interface management systems
- task modeling.
(Background to request for choice)
Please select. (Request for choice)

The satellite provides a list of options the user has to choose from. The nucleus contains the request proper to select an option. The user is not able to carry out the action specified by the request if the satellite is not given. The new definition can be specified as:

BACKGROUND-FOR-CHOICE-REQUEST:

constraints on $\mathcal{N}$: $\mathcal{N}$ is a choice-request. $\mathcal{R}$ won't comprehend the 'Request' without being given a parameter list which determines which actions are possible as follow-up reactions for the request **constraints on $\mathcal{S}$:** $\mathcal{S}$ is an 'Assert' speech act which consists of a list of action parameters **constraints on the $\mathcal{N}$ + $\mathcal{S}$ combination:** $\mathcal{S}$ increases the ability of $\mathcal{R}$ to comprehend $\mathcal{N}$ and to carry out the requested action **the effect:** $\mathcal{R}$'s ability to follow the choice-request in $\mathcal{N}$ increases

We consider this relation another subtype of BACKGROUND refining the definition given above.

The two new relations are interpersonal according to the taxonomy given in figure 1. They are subtypes of BACKGROUND and are therefore directly subordinated. — The set of relations necessary for the description of information-seeking dialogues is being determined by the examination of a large corpus of dialogues.

3.3 Re-examination of the Basic Assumptions of RST

As pointed out in section 2, the application of RST on describing discourse structures in dialogues is likely to lead to extensions or changes in general assumptions of the theory. In this section we give an example for the relaxation of restrictions made by RST. To show this we make use of the contrast between example 4 above and example 5 taken from our sample dialogue:

Example 5:

S: There are more alternatives:
- dialogue control
- plan recognition
- cognitive model.

(Background to request for choice)

Both examples have in common that they serve one request speech act made by the system, demanding input from the user. Example 4 suggests various alternatives from which the user is supposed to choose one. The system then explicitly utters a demand to select from the set of options. In contrast to this the explicit demand is left out in example 5. This means that the nucleus containing the atomic request is missing. This is contradictory to the assumption made by RST that the nuclei of complex textual structures must not be omitted without losing the meaning of the whole textual entity. The possibility to omit nuclei in spans of discourse has not been found in the monological texts examined in work on RST. The motivation for such a phenomenon, therefore, has to be found in the nature of dialogues: In our sample dialogue session example 5 occurs shortly after example 4 so that the user still has the pattern of interaction with the system in mind – it proposes a list of options and then requests a choice. Therefore, the choice-request, i.e. the nucleus, can be inferred by the user and is omitted in the system utterance.

4 Application of the Model in a Computational Environment

The text planner which is integrated in the KOMET text generation system uses rhetorical relations both for the selection of textual content and for text structuring (Hovy et al., 1992). The selection of the most adequate relation at a given state of the generation process is strongly influenced by the *communicative goal* which is to be achieved by means of the text. If the goal requires the description of a physical object (DESCRIBE-OBJECT), relations of the type ELABORATION are to be preferred. This mechanism is used similarly for the automatic construction of dialogue contributions made by the system.

In contrast to the KOMET text generation system where goals are mostly triggered by text-type specific features, the communicative goals for the generation of dialogue contributions are posted by the interaction manager handling the COR model and by user reactions. The illocutions available at each point of the dialogue have the same functions as goals and therefore influence the rhetorical relations employed. To further constrain which relation is going to be used the text planner has to check whether the knowledge to express the relation is available in the pool of knowledge supplied by the retrieval component; e.g., to express a WHOLE-PART relation the candidate concept to which new information has to be related must be a decomposable object with at least one part specified. The choice of relations is additionally influenced by the context (the dialogue history) in order to prevent the presentation of redundant information. After a relation has been determined the content is selected and the dialogue history is incremented by the newly planned discourse segment.

The modules required for the generation of system contributions in the given framework therefore are: (1) a model for interaction (COR); (2) a representation of communicative goals; (3) a representation of rhetorical relations; (4) an incrementally growing dialogue history; (5) knowledge bases and a knowledge pool capturing the output of the retrieval component. (1), (4) and (5) are specific for the production of dialogues and this is where adaptions of the original text planner have been made.

Acknowledgements

We would like to thank Adelheit Stein for inspiring discussions and for valuable comments on a previous draft of this paper.

References

Austin, J. 1962. *How to do things with words.* Oxford: Clarendon Press.

Bateman, J.A.; Maier, E.; Teich, E.; and Wanner, L. 1991. Towards an architecture for situated text generation. In *International Conference on Current Issues in Computational Linguistics*, Penang, Malaysia. Also available as technical report of GMD/Institut für Integrierte Publikations- und Informationssysteme, Darmstadt, West Germany.

Belkin, N.J., and Vickery, A. 1985. *Interaction in Information Systems: A Review of Research from Document Retrieval to Knowledge-Based Systems*, volume 35 of *Library and Information Research Report.* University Press, Cambridge/GB.

Carberry, S. 1985. *Pragmatic modeling in information system interfaces.* PhD thesis, University of Delaware.

Fawcett, R.P., and Davies, B. 1992. Monologue as turn in interactive discourse: Towards an integration of exchange structure and rhetorical structure theory. In *Proceedings of the 6th International Workshop on Natural Language Generation.* Springer.

Grosz, B.J., and Sidner, C.L. 1986. Attention, intentions and the structure of discourse. *Journal of Computational Linguistics*, 12(3).

Hobbs, J.R. 1990. *Literature and Cognition.* CSLI Lecture Notes.

Hovy, E.; Lavid, J.; Maier, E.; Mittal, V.; and Paris, C. 1992. Employing knowledge resources in a new text planner architecture. In *Proceedings of the 6th International Workshop on Natural Language Generation*, April 1992.

Iri, V.; McMillan, D.; and Merz, T. 1990 S-relators. Technical report, University of Zagreb.

Maier, E., and Hovy, E. 1991. A metafunctionally motivated taxonomy for discourse structure relations. In *Proc. of the 3rd European Workshop on Natural Language Generation*, Judenstein, Austria, March 1991.

Mann, W.C, and Thompson, S.A. 1987. Rhetorical structure theory: A theory of text organization. In Livia Polanyi, ed., *The Structure of Discourse.* Ablex Publishing Corporation.

McAlpine, G., and Ingwersen, P. 1989. Integrated information retrieval in a knowledge worker support system. In *Proceedings of SIGIR*, pages 48 - 57, Cambridge/MA, June 1989.

McCoy, K.F.; Moore, J.D.; Suthers, D.; and Swartout, B. eds. 1991. *AAAI-91 Workshop on Comparative Analysis of Explanation Planning Architectures*, Anaheim, July 1991.

McKeown, K.R. 1985. *Text generation - Using discourse strategies and focus constraints to generate natural language text.* Cambridge University Press.

Moore, J.D. 1989. *A Reactive Approach to Explanation in Expert and Advice-Giving Systems.* PhD thesis, University of California, Los Angeles.

Paris, C.L. 1987. Combining discourse strategies to generate descriptions to users along a naive/expert spectrum. In *Proceedings of IJCAI*, pages 626–632.

Perrault, C.R., and Grosz, B.J. 1986. Natural-language interfaces. *Annual Review of Computer Science*, 1.

Reichman, R. 1985 *Getting computers to talk like you and me. Discourse context, focus, and semantics (an ATN model).* Cambridge, MA: Bradford.

Rumelhart, D.E. 1975. Notes on a schema for stories. In Bobrow and Collins, editors, *Language, Thought and Cognition.* Academic Press, Orlando.

Sanders, T.J.M.; Spooren, W.P.M.S.; and Noordman, L.G.M. 1990. Towards a taxonomy of coherence relations. *Discourse Processes.*

Searle, J.R. 1969. *Speech Acts.* Cambridge/GB: University Press.

Sitter, S., and Stein, A. 1992. Modelling the illocutionary aspects of information-seeking dialogues. *Information Processing and Management*, 28(2):165 - 180.

Stein, A.; Thiel, U.; and Tissen, A. 1992. Knowledge based control of visual dialogues. In *Proceedings of AVI '92 (Advanced Visual Interfaces)*, Rome, Italy, May 1992. Singapore: World Scientific Press.

van Dijk, T.A., and Kintsch, W. 1983 *Strategies of Discourse Comprehension.* Academic Press, Inc.

Winograd, T., and Flores, F. 1986. *Understanding computers and cognition.* Ablex Publishing Company.

Wunderlich, D. 1976. *Studien zur Sprechakttheorie.* Suhrkamp.

A Connectionist Solution to the Multiple Instantiation Problem using Temporal Synchrony*

D. R. Mani and **Lokendra Shastri**
Department of Computer and Information Science
University of Pennsylvania
Philadelphia, PA 19104, USA
mani@linc.cis.upenn.edu
shastri@central.cis.upenn.edu

Abstract

Shastri and Ajjanagadde have described a neurally plausible system for knowledge representation and reasoning that can represent systematic knowledge involving n-ary predicates and variables, and perform a broad class of reasoning with extreme efficiency. The system maintains and propagates variable bindings using temporally synchronous—i.e., in-phase — firing of appropriate nodes. This paper extends the reasoning system to incorporate *multiple instantiation of predicates*, so that any predicate can be instantiated up to k times, k being a system parameter. The ability to accommodate multiple instantiations of a predicate allows the system to handle a much broader class of rules, including bounded transitivity and recursion. The time and space requirements increase only by a constant factor, and the extended system can still answer queries in time proportional to the length of the shortest derivation of the query.

Introduction

In (Shastri & Ajjanagadde, 1990a, 1990b and Ajjanagadde & Shastri, 1991), Shastri and Ajjanagadde have described a solution to the variable binding problem (Feldman, 1982, Malsburg, 1986) and shown that the solution leads to the design of a connectionist reasoning system that can represent systematic knowledge involving n-ary predicates (relations) and *variables*, and perform a broad class of reasoning with extreme efficiency. The time taken by the reasoning system to draw an inference is only proportional to the *length* of the chain of inference and is independent of the number of rules and facts encoded by the system. The reasoning system maintains and propagates variable bindings using temporally synchronous—i.e., in-phase—firing of appropriate nodes. The solution to the variable binding problem allows the system to maintain and propagate a large number of bindings *simultaneously* as long as the number of *distinct* entities participating in any given episode of reasoning remains bounded. Reasoning in the proposed system is the transient but systematic flow of *rhythmic* patterns of activation, where each *phase* in the rhythmic pattern corresponds to a distinct *entity* involved in the reasoning process and where variable bindings are represented as the synchronous firing of appropriate argument and entity nodes. A fact behaves as a temporal pattern matcher that becomes 'active' when it detects that the bindings corresponding to it are present in the system's pattern of activity. Finally, rules are interconnection patterns that propagate and transform rhythmic patterns of activity.

Several other researchers have proposed connectionist solutions to the dynamic binding problem using a variety of techniques. These include the use of dynamic connections (Feldman, 1982), parallel constraint satisfaction (Touretzky & Hinton, 1988), position specific encoding (Barnden & Srinivas, 1991), tensor product representations (Dolan & Smolensky, 1989) and signatures (Lange & Dyer, 1989). (Shastri & Ajjanagadde, 1992) compares and contrasts these other approaches with the temporal synchrony approach used in this paper.

The system described in (Shastri & Ajjanagadde, 1990b) has the limitation that any predicate in the reasoner can be instantiated *at most once*.[1] It is not difficult to find examples which suggest that *reflexive* (effortless) reasoning involves dealing with multiple instances of predicates. For example, if we know that Mary is John's spouse, we would not have any difficulty in realizing that John is Mary's spouse. In other words, given *spouse-of(Mary,John)* we can reflexively answer 'yes' to the query *spouse-of(John,Mary)?* Such behavior would require the *spouse-of* predicate to be instantiated *twice*: once with *spouse-of(John,Mary)* and again with *spouse-of(Mary,John)*. As another example, consider the situation in which we know that Mary is older than John's father. If we now hear that John married Mary, we can instantly sense the unusualness of the situation, since Mary is obviously much older than John. But the fact that Mary is older than John has not been explicitly stated. This would suggest that we may have inferred *older-than(Mary,John)* using the facts *older-than(Mary,John's-father)* and *older-than(John's-father,John)*,[2] and the transitive nature of the *older-than* predicate. To model this scenario

*This work was supported by NSF grant IRI 88-05465 and ARO grants DAA29-84-9-0027 and DAAL03-89-C-0031.

[1] An extension for dealing with multiple instantiation using two levels of temporal synchrony was outlined in (Shastri & Ajjanagadde, 1990b). The solution proposed here is distinct from that in (Shastri & Ajjanagadde, 1990b).

[2] *older-than(John's-father,John)* follows from the knowledge that fathers are older than their children.

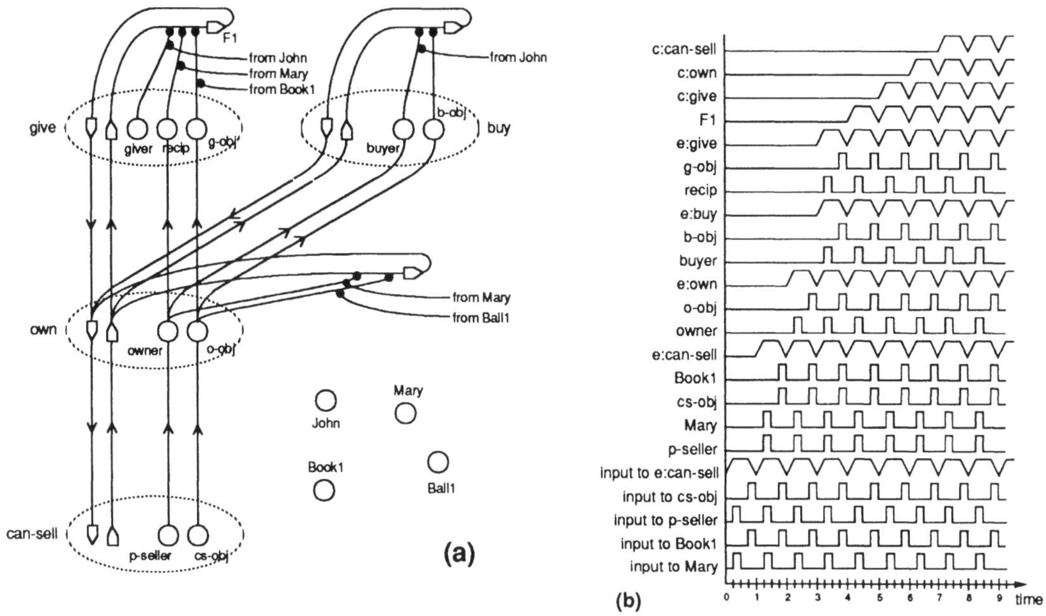

Figure 1: (a) An example encoding of rules and facts. (b) Activation trace for the query *can-sell(Mary,Book1)?*.

in the reasoning system, we would need to simultaneously handle three instantiations of *older-than*. Similarly, we can, without conscious deliberation, infer that John may be jealous of Tom if we know that John loves Mary and Mary loves Tom. Here again, we would need the ability to represent multiple instantiations of the *loves* predicate to capture the situation. Thus, any system which purports to model common-sense, reflexive reasoning should be capable of representing multiple instantiations of predicates.

This paper describes how the basic reasoning system of (Shastri & Ajjanagadde, 1990b) may be extended to incorporate multiple instantiation of predicates. We begin with a brief overview of the rule-based reasoning system, followed by an exposition of multiple instantiation in the reasoning system and its implementation. We will primarily concern ourselves with backward reasoning. Forward reasoning will be considered only briefly towards the end.

The rule-based reasoning system

Fig. 1a illustrates how long-term knowledge is encoded in the rule-based reasoning system. The network encodes the following *rules* and *facts*: i) $\forall x,y,z$ [*give(x,y,z)* $\Rightarrow$ *own(y,z)*], ii) $\forall x,y$ [*buy(x,y)* $\Rightarrow$ *own(x,y)*], iii) $\forall x,y$ [*own(x,y)* $\Rightarrow$ *can-sell(x,y)*], iv) *give(John,Mary,Book1)*, v) *buy(John,x)*, and vi) *own(Mary,Ball1)*.

The encoding makes use of two types of nodes (see Fig. 2): ρ-**btu** nodes (depicted as circles) and τ-**and** nodes (depicted as pentagons). These nodes have the following idealized behavior: On receiving a periodic spike train, a ρ-btu node produces a periodic spike train that is *in-phase* with the driving input. We assume that ρ-btu nodes can respond in this manner as long as the period of oscillation, π, lies in the interval $[\pi_{min}, \pi_{max}]$, where π_{min} and π_{max} are the minimum and maximum periods at which nodes can oscillate. For a discussion of biologically motivated values for these parameters, see (Shastri & Ajjanagadde, 1992).

A τ-and node behaves like a *temporal* AND node, and becomes active on receiving an oscillatory input consisting of a train of pulses *of width comparable to the period of oscillation*. On becoming active, a τ-and node produces an oscillatory pulse train whose period of oscillation and pulse width matches that of the input. A third type of node we use later is the τ-or node which becomes active on receiving *any* activation; its output is a pulse whose width and period equal π_{max}. Fig. 2 summarizes the behavior of the ρ-btu, τ-and and τ-or nodes.

The maximum number of distinct entities that may participate in an episode of reasoning equals $\lfloor \pi/\omega \rfloor$ where π is the period of oscillation and ω is the width of the window of synchronization—nodes firing with a lag or lead of less than $\omega/2$ would be considered to be in synchrony. The encoding also makes use of *inhibitory modifiers*—links that impinge upon and inhibit other links. A pulse propagating along an inhibitory modifier will block a pulse propagating along the link it impinges upon. In Fig. 1a, inhibitory modifiers are shown as links ending in dark blobs.

Each entity in the domain is encoded by a ρ-btu node. An n-ary predicate P is encoded by a pair of τ-and nodes and n ρ-btu nodes, one for each of the n arguments. One of the τ-and nodes is referred to as the *enabler*, *e:P*, and the other as the *collector*, *c:P*. In Fig. 1a, *enablers* point upward while *collectors* point downward. The *enabler e:P* becomes active whenever

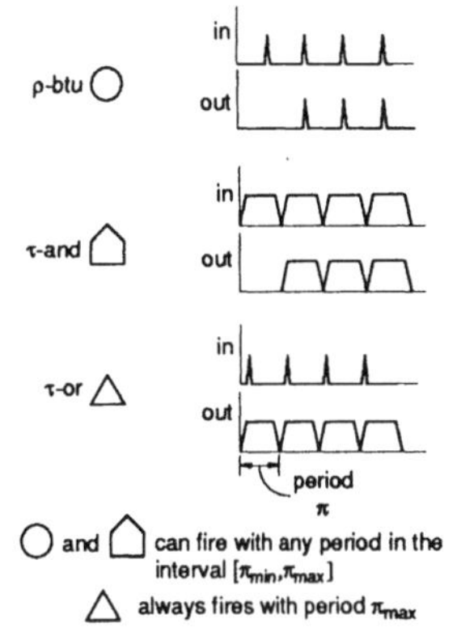

Figure 2: Behavior of the ρ-btu, τ-and and τ-or nodes in the reasoning system.

the system is being queried about P. On the other hand, the system activates the *collector c:P* of a predicate P whenever the system wants to assert that the current dynamic bindings of the arguments of P are consistent with the knowledge encoded in the system. A rule is encoded by connecting the *collector* of the antecedent predicate to the *collector* of the consequent predicate, the *enabler* of the consequent predicate to the *enabler* of the antecedent predicate, and by connecting the arguments of the consequent predicate to the arguments of the antecedent predicate in accordance with the correspondence between these arguments specified in the rule. A fact is encoded using a τ-and node that receives an input from the enabler of the associated predicate. This input is modified by inhibitory modifiers from the argument nodes of the associated predicate. If an argument is bound to an entity in the fact then the modifier from such an argument node is in turn modified by an inhibitory modifier from the appropriate entity node. The output of the τ-and node is connected to the *collector* of the associated predicate (refer to the encoding of the fact *give(John,Mary,Book1)* and *buy(John,x)* in Fig. 1a.)

Inference Process Posing a query to the system involves specifying the query predicate and the argument bindings specified in the query. The query predicate is specified by activating its *enabler* with a pulse train of width and periodicity π. Argument bindings are specified by activating each entity and the argument nodes bound to that entity in a distinct *phase*. Phases are just non-overlapping time intervals within a period of oscillation.

We illustrate the reasoning process with the help of an example. Consider the query *can-sell(Mary, Book1)?* (i.e., Can Mary sell Book1?) The query is posed by i) activating the *enabler e:can-sell* ii) activating *Mary* and *p-seller* in the same phase (say, phase-1), and iii) activating *Book1* and *cs-obj* in some other phase (say, phase-2). As a result of these inputs, *Mary* and *p-seller* fire synchronously in phase-1 of every period of oscillation, while *Book1* and *cs-obj* fire synchronously in phase-2. See Fig. 1b. The activation from the *can-sell* predicate propagates to the *own*, *give* and *buy* predicates via the links encoding the rules. Eventually, as shown in Fig. 1b, *Mary*, *p-seller*, *owner*, *buyer* and *recip* will all be active in phase-1, while *Book1*, *cs-obj*, *o-obj*, *g-obj* and *b-obj* would be active in phase-2. The activation of *e:can-sell* causes the enablers of all other predicates to go active. In effect, the system is asking itself three more queries—*own(Mary,Book1)?*, *give(x,Mary,Book1)?* (i.e., Did *someone* give Mary Book1?), and *buy(Mary,Book1)?*. The τ-and node F1, associated with the fact *give(John,Mary,Book1)* becomes active as a result of the uninterrupted activation it receives from *e:give*, thereby answering *give(x,Mary,Book1)?* affirmatively. The activation from F1 spreads downward to *c:give*, *c:own* and *c:can-sell*. Activation of *c:can-sell* indicates an affirmative answer to the original query *can-sell(Mary,Book1)?*.

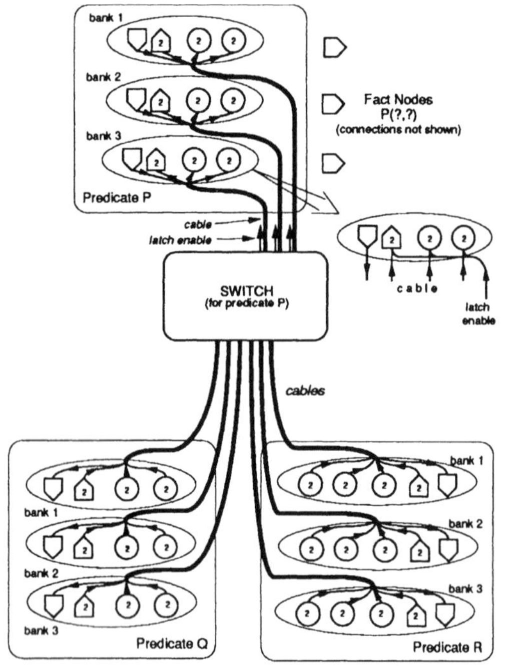

Figure 3: An overview of the multiple instantiation system. P and Q are binary predicates while R is a ternary predicate. The multiple instantiation constant $k = 3$.

Multiple Instantiation in the Reasoning System

Introducing multiple instantiation relies on the assumption that, during an episode of reflexive reasoning, any given predicate need only be instantiated a *bounded* number of times. In (Shastri & Ajjanagadde, 1992), it is argued that a reasonable value for this bound is around three (also see below). We shall refer to this bound as the *multiple instantiation constant*, k.[3]

Representing Predicates Since every predicate must now be capable of representing up to k dynamic instantiations, predicates are represented using k *banks* of units. Each *bank* of an n-ary predicate P consists of τ-and nodes for the *collector* ($c:P$) and *enabler* ($e:P$) along with n ρ-btu nodes representing the arguments of P. *Each* bank is essentially similar to the predicate representation used in (Shastri & Ajjanagadde, 1990b) (see Fig. 1a). Fig. 3 illustrates the structure of predicates in the extended system. Note that the *enabler*, $e:P$, and the argument nodes have a threshold[4] $\theta = 2$.

The Multiple Instantiation Switch Every predicate in the extended system has an associated *multiple instantiation switch*, which we shall refer to as the *switch*. All connections to a predicate are made through its multiple instantiation switch. The switch has k output *cables* (see Fig. 3), each of which connects to one bank of the predicate. A *cable* is a group of wires originating or terminating at a predicate bank; a cable, therefore, has wires from all the units (collector, enabler and argument units) in a bank. Each output cable from the switch is accompanied by a *latch enable* link.

The switch arbitrates input instantiations to its associated predicate and brings about efficient and automatic dynamic allocation of predicate banks by ensuring the following: (1) Fresh predicate instantiations are channeled to the predicate banks only if the predicate can accommodate more instantiations. (2) All inputs that transform to the *same instantiation* are mapped into the *same predicate bank*. Thus, new instantiations selected for representation in the predicate are always unique.

Structure and Operation of the Multiple Instantiation Switch Figures 4a and 4b illustrate the construction of the the multiple instantiation switch. The switch consists of k groups or *ensembles* of units. Each ensemble consists of an *arbitrator bank*, and several *input banks*. The *arbitrator* consists of n ρ-btu nodes representing the arguments of the associated n-ary predicate, $(n-1)$ τ-or nodes and two τ-and nodes for the collector and enabler. Each ρ-btu node except the *first* is associated with a τ-or node, as shown in Fig. 4b. The i-th *arbitrator bank* directly connects with the i-th bank of the predicate. *Input banks* (Fig. 4b) consists of n ρ-btu units representing the arguments of the predicate, and two τ-and nodes representing the collector and enabler of the bank. Each *input bank* also has a τ-or node associated with it. The cable terminating at the *input bank* is an input to the switch; the outputs of the *input bank* connect to the *arbitrator* of the respective ensemble. Corresponding *input banks* across ensembles are interconnected as shown in Figs. 4a and 4b.

Ignoring the associated τ-or nodes, the *input banks* and the *arbitrators* have a structure which exactly mimics the bank structure of the predicate with which the switch is associated. The number of lines in any input cable to the switch is decided by the arity of the predicate originating the cable. The number of lines in the switch output depends on the arity of the predicate associated with the switch.

To start with, only the first ensemble in the switch can respond to incoming activation. Activation in one or more *input banks* of the first ensemble will cause the enabler in the *arbitrator*, $e:Arb$, to become active. All *input banks* with inactive enablers will be inhibited via the τ-or nodes associated with the respective *input banks*. The activation of $e:Arb$ will enable Arb_{arg_1} to pick a phase to fire in. This phase is communicated to all the *input banks*, via the associated τ-or nodes (see Fig. 4b). Each τ-or node checks if the phase selected by Arb_{arg_1} matches the phase of the first argument of its *input bank*. A mismatch shuts off the entire *input bank*. In the meantime, $e:Arb$ would have activated the τ-or node associated with the second argument in the *arbitrator*. This enables Arb_{arg_2} to select a phase from the activation remaining after inhibiting instantiations that did not agree with Arb_{arg_1}. Note that Arb_{arg_2} is enabled by the associated τ-or node *independent* of Arb_{arg_1} and will select a phase to fire in *even if* Arb_{arg_1} *is inactive* (which would be the case if all incoming instantiations have an unbound first argument). The process continues, allowing $Arb_{arg_3}, \ldots, Arb_{arg_n}$ to select phases to fire in. After, Arb_{arg_n} has made its choice, the *latch enable* becomes active and the selected instantiation is transferred to the first predicate bank. A link from the last τ-or node to $e:Arb$ in the second ensemble enables the second ensemble to select a fresh instantiation. Once the second ensemble makes its choice, it enables the third, and so on. The process continues until k instantiations have been channeled to the predicate, after which, any fresh input instantiations are ignored.

Note that if the i-th ensemble ($1 < i \leq k$) is making its choice, it will always select an instantiation which is *different* from those picked by the first $i-1$ ensembles, ensuring that all instantiations channeled to the predicate are unique. A more detailed description of the structure and operation of the switch can be found in (Mani & Shastri, 1992).

Encoding Rules and Facts

Every predicate in the system has k banks of units for representing k instantiations. Further every predicate has an associated multiple instantiation switch which arbitrates the instantiations which will be represented in the predicate. Given this modified underlying framework, we encode rules and long-term facts using an appropriate extension of the scheme detailed in (Shastri & Ajjanagadde, 1990b).

Dynamic instantiation which matches a long-term fact for predicate P could be present in *any one* of its k banks, necessitating a fact-pattern-matcher for *each* of the predicate banks. Thus, any fact of the form $P(C_1, \ldots, C_n)$ will be encoded using k τ-and

[3] For simplicity, we assume that k is the same for all predicates. This need not be the case.

[4] This applies to a predicate in the backward reasoning system. In a forward reasoner, the *collector*, $c:P$, and the argument nodes have a threshold $\theta = 2$.

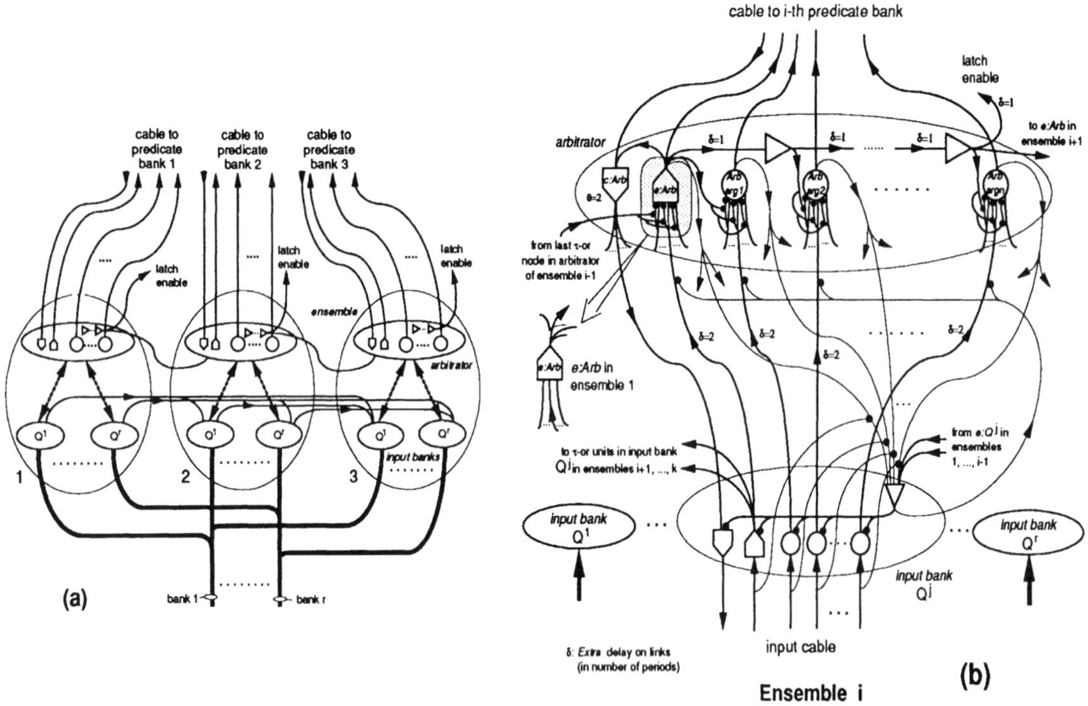

Figure 4: (**a**) Overview of the multiple instantiation switch. The multiple instantiation constant $k = 3$. (**b**) Structure of the i-th ensemble in the switch. Only connections between *input bank* Q^j and the *arbitrator* are shown. Connections between other *input banks* and the *arbitrator* are implied. As indicated, connections to *e:Arb* in the first ensemble are different.

nodes—one for each bank of P (Fig. 3). *Each* τ-and node encodes $P(C_1, \ldots, C_n)$ as described in (Shastri & Ajjanagadde, 1990b) (also see Fig. 1a).

Fig. 3 illustrates rule encoding at a very gross level. Suppose the rule relating P and Q in Fig. 3 is $\forall x,y$ [$P(x,y) \Rightarrow Q(y,x)$]. To encode this rule in a backward reasoning system, each bank of predicate Q is connected to an *input bank* in every ensemble of the switch for P. Thus, the k banks of predicate Q require a total of k^2 *input banks*—k *input banks* in each of the k ensembles of the switch. The *input banks* in the switch for P have a structure identical to the bank structure of predicate P. The cable from a bank of Q connects to the corresponding *input bank* as though the *input bank* itself represented the predicate P. The connection pattern between the bank of Q and the *input bank* is therefore identical to the connection pattern between the actual predicates in the system of (Shastri & Ajjanagadde, 1990b) (see Fig. 1a). Further, activation of the collector in any bank of P is transmitted (via the switch) to the corresponding bank in Q which *originated the instantiation* represented in that bank of P. A detailed description of rule and fact encoding in given in (Mani & Shastri, 1992).

Network Complexity The extended reasoning system requires $O(\mathcal{C} + \mathcal{F} + \mathcal{P})$ nodes and links. $\mathcal{C}$ and $\mathcal{F}$ represent the total number of entities and long-term facts in the system, respectively. $\mathcal{P}$ is the sum of the arities of all predicates in the rule base. The constant of proportionality for the network complexity is proportional to k^2, where k is the multiple instantiation constant. Thus, as in (Shastri & Ajjanagadde, 1990b, 1992), the network complexity is *linear* in the size of the knowledge base although the constant of proportionality is now larger. A similar comment holds for the time complexity: the system can still answer queries in time proportional to the length of the shortest derivation (Shastri & Ajjanagadde, 1990b, 1992); but now, the constant of proportionality is slightly larger, since we also need to consider the time required for the activation to propagate through the switches. Given a predicate P, the worst case propagation time for activation passing through its switch is proportional to k.

Multiple Instantiation in a Forward Reasoning System

To introduce multiple instantiation of predicates in the *forward reasoner*, we structure predicates and their associated switches in a manner similar to the backward reasoning system.[5] Rules with a single predicate in the antecedent can be encoded directly: each bank of the antecedent predicate is connected to *input banks* in every ensemble of the switch for the consequent predicate. For rules like $\forall x,y,z$ [$P(x,y) \wedge Q(y,z) \Rightarrow R(x,y,z)$], which have multiple predicates in the antecedent, we would need to pair each bank of P with all the banks of Q and check if the second argument of P is the same as the first argument of Q. The obvious way to do this requires $O(k^m)$ nodes and links to encode each rule with m predicates in the antecedent.

Typically, we expect m to be around 2, since most predicates in a multiple-predicate antecedent serve to specify constraints on the arguments of a few key predicates. These type-enforcing predicates can be replaced by typed variables, significantly reducing the number of predicates in the antecedent. For example, the rule $\forall x,y$ [*collide(x,y)* $\wedge$ *animate(x)* $\wedge$ *solidobj(y)* $\Rightarrow$ *hurt(x)*] with three predicates in the antecedent is equivalent to $\forall$ *x:animate, y:solidobj* [*collide(x,y)* $\Rightarrow$ *hurt(x)*], which can be directly encoded as rule with a *single* predicate in the antecedent (Mani & Shastri, 1991). Even if this "compression" of the antecedent were not possible, we could introduce dummy predicates and split a rule with several predicates in the antecedent into several rules with just a few predicates in the antecedent, so as to maintain $m \approx 2$.

Maintaining, propagating and using multiple instantiations of a predicate entails a significant cost in space and time. In the context of reflexive reasoning, it is essential that these resources be bounded. This leads us to predict that the value of k is quite small, perhaps on more than 3 (Shastri & Ajjanagadde, 1992). Thus, with $k \approx 3$ and $m \approx 2$, the extra cost of encoding rules in the forward reasoner with multiple instantiation is a factor of about 10 ($\approx 3^2$)— which is about the same as the cost increase for a backward reasoner with multiple instantiation.

Conclusions

Extending the connectionist rule-based reasoning system to accommodate multiple instantiation of predicates enables the system to handle a wider and more powerful set of rules, facts and queries. The extended system can encode and reason with rules that capture symmetry, transitivity and recursion, provided the number of multiple instantiations of a predicate required to draw a conclusion remains bounded. The multiple instantiation reasoning system has been combined with a connectionist type hierarchy (Mani & Shastri, 1991) to provide a more flexible and powerful system. All these features have been successfully included in the simulation system reported in (Mani, 1992).

[5] Except for a few minor functional differences, the multiple instantiation switch associated with a predicate in the forward reasoning system is structurally identical to the switch used in the backward reasoner. See (Mani & Shastri, 1992) for details.

References

(Ajjanagadde & Shastri, 1991) Ajjanagadde, V., and Shastri, L. Rules and Variables in Neural Nets. *Neural Computation* **3**:121–134.

(Barnden & Srinivas, 1991) Barnden, J., and Srinivas, K. Encoding Techniques for Complex Information Structures in Connectionist Systems. *Connection Science* **3**(3):269–315.

(Dolan & Smolensky, 1989) Dolan, C. P., and Smolensky, P. Tensor Product Production System: A Modular Architecture and Representation. *Connection Science* **1**(1):53–68.

(Feldman, 1982) Feldman, J. A. Dynamic Connections in Neural Networks. *Bio-Cybernetics* **46**:27–39.

(Lange & Dyer, 1989) Lange, T. E., and Dyer, M. G. High-level Inferencing in a Connectionist Network. *Connection Science* **1**(2):181–217.

(Malsburg, 1986) von der Malsburg, C. Am I Thinking Assemblies? In G. Palm and A. Aertsen (Eds.). *Brain Theory*. Berlin: Springer-Verlag.

(Mani, 1992) Mani, D. R. Using the Connectionist Rule-Based Reasoning System Simulator. Technical Report, Department of Computer and Information Science, Univ. of Pennsylvania. Forthcoming.

(Mani & Shastri, 1991) Mani, D. R., and Shastri, L. 1991. Combining a Connectionist Type Hierarchy with a Connectionist Rule-Based Reasoner. In *Proceedings of the Thirteenth Annual Conference of the Cognitive Science Society*, 418–423. Hillsdale NJ: Lawrence Erlbaum Associates.

(Mani & Shastri, 1992) Mani, D. R., and Shastri, L. 1992. Multiple Instantiation of Predicates in a Connectionist Rule-Based Reasoner. Technical Report MS-CIS-92-05, Department of Computer and Information Science, Univ. of Pennsylvania.

(Shastri & Ajjanagadde, 1990a) Shastri, L., and Ajjanagadde, V. An optimally efficient limited inference system. In *Proceedings of AAAI-90, the Twelfth National Conference of the American Association of Artificial Intelligence*, 563–570. Cambridge MA: American Association for Artificial Intelligence.

(Shastri & Ajjanagadde, 1990b) Shastri, L., and Ajjanagadde, V. From Simple Associations to Systematic Reasoning: A Connectionist Representation of Rules, Variables and Dynamic Bindings. Technical Report MS-CIS-90-05, Department of Computer and Information Science, Univ. of Pennsylvania.

(Shastri & Ajjanagadde, 1992) Shastri, L., and Ajjanagadde, V. From Simple Associations to Systematic Reasoning: A Connectionist Representation of Rules, Variables and Dynamic Bindings using Temporal Synchrony. *Behavior and Brain Sciences*. Forthcoming.

(Touretzky & Hinton, 1988) Touretzky, D. S, and Hinton, G. E. A Distributed Connectionist Production System. *Cognitive Science* **12**(3):423–466.

Acquiring Rules for Need-Based Actions Aided by Perception and Language

Ganesh Mani and **Leonard Uhr**
Computer Sciences Department
University of Wisconsin—Madison
1210 W. Dayton St., Madison, WI 53706.
{ganesh, uhr}@cs.wisc.edu

Abstract

The CHILDLIKE system is designed to learn about objects, object qualities, relationships among objects, and words that refer to them. Once sufficient visual–linguistic associations have been established, they can be used as foundations for a) further learning involving language alone and b) reasoning about the effect of different actions on perceived objects and relations, and internally sensed need levels. Here, we address the issue of learning efficient rules for action selection. A trial-and-error (or reinforcement) learning algorithm is used to acquire and refine action-related rules. Learning takes place via generation of hypotheses to guide movement through sequences of states, as well as modifications to two entities: the weight associated with each action, which encodes the uncertainty underlying the action, and the potential value (or vector) of each state which encodes the desirability of the state with respect to the current needs. CHILDLIKE is described, and issues relating to the handling of uncertainty, generalization of rules and the role of a short-term memory are also briefly addressed.

Introduction

Perception is crucial to any activity by intelligent agents in an environment. In dynamic environments, perception-mediated reasoning and acting can avoid the problems of planning that goes down blind alleys and expends massive efforts to anticipate situations that never occur. Perceiving the state of the world periodically can also save the embedded agent the trouble of keeping precise track of its moves to infer its position relative to other objects in the environment at each step.

On the performance side, it is instructive to note that humans are able to perceive and recognize scenes containing a few objects within a few hundred milliseconds. Such rapid perception is crucial to adequately fast reaction in an environment. (However, it is rarely necessary for a scene to be fully recognized before an agent reacts. Certain key features or salient objects in a scene may trigger reactions that have been associated with the features or objects by prior learning.)

Rules that facilitate choosing actions without extensive deliberation are important, since planning is time consuming and its utility is limited in dynamic environments. At the same time, it is also important to learn the effect of different actions or operators from experience (e.g., see [Drescher, 1987], [Mason *et al.*, 1989], [Shen, 1989]). Reactive planning (e.g., [Georgeff and Lansky, 1987], [Firby, 1987]) or iterative planning [Kaelbling, 1987] can be cast in a memory-based framework particularly when other tasks such as perception and language are being integrated into the system. (Integrating vision and language is gaining considerable attention in the AI and cognitive science community— e.g., see [Dyer, 1991], [Feldman *et al.*, 1990], [Okada, 1991], [Siskind, 1991].) In this paper we describe how rules can be learned that help the system to react appropriately to its needs, and how generalization of these rules is aided by prior learning of visual and linguistic constructs.

In the next section, we briefly describe the integrated system that we are developing to learn from simple experiences. The subsequent sections focus on the acquisition and refinement of hypotheses (structures of rules) that aid the system in reacting to its internal needs. Prior visual–linguistic associations act as powerful biases for the acquisition and refinement of rules that relate actions to the perceived environmental states and their need-satisfaction potential. The representation as well as generalization of rules is addressed.

Background about the CHILDLIKE System

The CHILDLIKE[1] system [Mani and Uhr, 1991a,b] [Mani, 1992] (in preparation) is a computational information-processing model (implemented in Common Lisp) designed to learn about objects, their qualities, and the words that name and describe them; and, further, to use this knowledge to act towards satisfying its internal needs (e.g., hunger, thirst, sleep, curiosity). Thus the CHILDLIKE system attempts to capture the entire perceive-reason-act-learn loop.

The system is subjected to a series of simple "experiences" from which it attempts to learn. An experience consists of several different types of input— for example, a visual pictorial scene, a short language utterance, an abstracted action, an internal need level.

One component of the system acquires visual–linguistic associations from experience. Initially, tentative associations are formed between words and visual features, and

[1] which stands for Conceptual Hierarchies In Language Development and Learning In a Kiddie Environment.

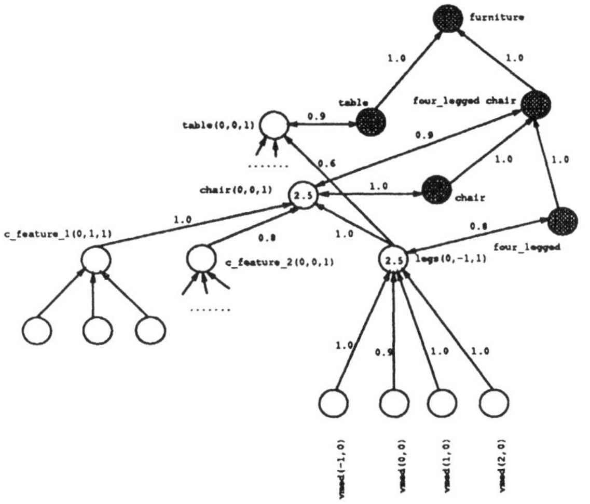

Figure 1: The Network Structure of Hypotheses

groups of these, using learning rules for extraction, aggregation and generation. These associations get strengthened with repeated co-occurrences. Such visual–linguistic associations are refined using ***de-generation*** and generalization mechanisms. For descriptions, see [Mani and Uhr, 1991b] and [Mani, 1992] (in preparation). For an example of the visual–linguistic associations learned by the system in response to inputs such as pictures of chairs and tables, along with words about them, see Figure 1 (only a cross section of the memories is shown). The hatched nodes correspond to structures learned from the linguistic channel; the others to visual features and their compounds. Only some of the highly weighted links are shown. The three numbers alongside a visual feature node represent, respectively, the x and y coordinates of the feature (in the *object*-centered coordinates of its parent node) and its relative size.

This paper concentrates on acquiring knowledge relating to actions and their effects. We also stress how the action-related rules can be generalized and improved using the visual–linguistic associations that have been acquired; integration of these different components is achieved by utilizing a memory-based framework. Mutually grounded representations — that consist of, for example, the visual representation of a fruit, the word that describes it and the action that can be performed on it (eat) to satisfy a certain internal need (hunger) — help in the attempt to span the wide variety of abilities that encompass everyday tasks and reasoning. The current version of the system is a starting point for a realistic architecture and implementation that integrates vision, language and action.

Representation of Rules About Actions

Knowledge for reasoning about actions is usually built into a system ***a priori***, rather than learned. In contrast, CHILDLIKE attempts to learn the effect of various actions (currently, primitive actions are built in, but not

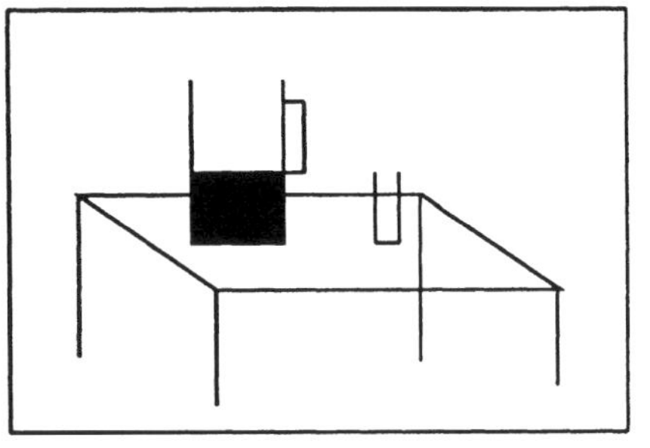

Figure 2: An Example Visual Scene

their effects). An action or an action sequence may impact both the externally perceived entities and the internally sensed need levels. Thus, each state that the system is in can be described in terms of (a subset of) the perceived objects, qualities of objects, relations among them and the internal need levels.

A suitable bias is required to translate the perceived state into an internal state in the plan memories. In CHILDLIKE, the bias used is prior learning from percept–language interactions.[2] Thus, the system translates a visual scene into object names and relations between the objects that are ***known*** to the system.

CHILDLIKE can learn about objects such as `table`, `pitcher`, and `glass`; spatial relations such as `on`, `above`, and `in`; and other relations such as `and`. Such knowledge is acquired by being trained on instances with language strings such as `brown table` and `pitcher on table` along with their corresponding visual scenes. Based on such knowledge, visual scenes (see Figure 2) presented as arrays representing medium- to high-level features are processed by the system to obtain, for example:

```
on(pitcher,table)
on(glass,table)
in(air,glass)
in(juice,pitcher)
yellow juice
.......
```

A hierarchical structure [Uhr, 1978;1987] is used to rapidly imply objects and relations. Needs are introduced by encoding the sensed internal need levels along with the entities perceived from the external world into each state in the long-term plan memories. These memories are graph structures wherein each link (or state transition) connotes an action. Every action has a weight associated with it, encoding the certainty of the action. Apart from the visually perceived information and the internally sensed need levels, each state also encodes its potential to satisfy each need. Currently, four internal needs are modeled: hunger, thirst, rest and curiosity. A state may not necessarily represent all the perceivable information, but simply the features the system is currently attending to.

[2] Note that other candidates may be useful biases. For example, children use perceptual knowledge alone before they acquire any language.

1a. Perceive external world and form a condensed description (VD) (biased by previously learned visual–linguistic associations).

b. Sense internal needs (I) and match the current situation (VD plus I) to states in the long-term action memories.
If there is a match **then**
  Use the resulting state S_c from the long-term memories.
else
  Create a new state S_c and initialize it.

2a. **if** curiosity-need does not dominate **then**
  Follow action-arcs from (S_c) and choose action A_i such that it maximizes the weighted potential $w(A_i)I(S_c) \cdot P(S_j) - \eta e(A_i)$ over all the feasible action–destination pairs from S_c. {$w(A_i)$ is the certainty that action A_i will lead to state S_j, $I(S_c)$ is the numeric need vector at state S_c, $P(S_j)$ is the potential of state S_j, $e(A_i)$ is the energy expended in or cost of executing action A_i and η is a normalizing factor.}
  Execute action A_i.
else
  Execute an action A_i randomly from the action repertoire of S_c.

b. Perceive the new state (S_n). {This is the same as Step 1 above.}
If there exists a link between S_c and S_n labeled with A_i **then**
  Update its frequency-based weight (also update the weights of other links labeled with A_i from S_c).
else
  Form a new link and initialize it.

c. Propagate the potentiality/need-fulfillment information at S_n back to the previous state S_c. Go to 2a.

Figure 3: Algorithm that Creates and Refines the Action Memories

Action selection and refinement of the plan memories takes place using a trial-and-error learning algorithm (the current version used by CHILDLIKE is shown in Figure 3). The algorithm assumes abstracted actions such as `Pick up apple`, `Pour into glass`, and `Drink from glass`. (Future versions of the system will decompose these actions further, into sub-actions and the visual frames that bracket them.)

Rules are acquired implicitly, by updating the memories encoding knowledge about actions after each experience. Initially, all actions (in the set of possible actions associated with each state) are equivalent from the system's point of view, as it starts out with no knowledge about the effect of actions. States also usually have initial potential values of zero— exceptions are need-fulfilling states which have appropriately high potential values (these can be thought of as goal states or states where a reinforcement vector is sensed, changing the need levels). As the values corresponding to the potential of each state to satisfy particular needs get propagated through the learned network, and as the effects of actions are perceived and tabulated using a weight associated with each action (note that an action is represented as a link from one perceptual state to another), the performance of the system improves. The potential $P(S_j)$ of a subsequent state S_j is usually a vector, since there are multiple needs; a dot product with the need vector $I(S_c)$ of the current state S_c is used to reduce it to a single value (see Figure 3).

Figure 4 shows a snapshot of the action memories after a few tens of trials of one experiment. Note that the system starts out with all the weight-like certainty values associated with actions set to 0 and the values of variables

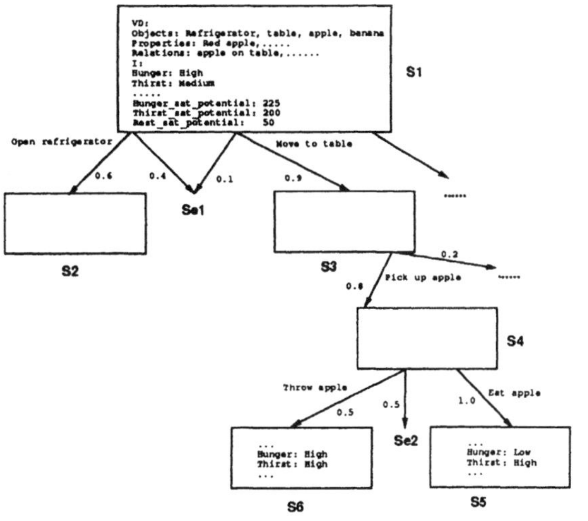

Figure 4: An Example of the Action Memories Acquired by CHILDLIKE

($Se1$ and $Se2$ denote extraneous states which may not be explicitly represented in the system's memories.)

such as `Hunger_sat_potential` (which stands for hunger satisfaction potential) for states like $S1$ set to 0 (or a low initial value). When hunger satisfaction takes place at a state such as state $S5$, the information propagates back (Step 2c in Figure 3); after a number of iterations, states such as $S1$ reflect their true potential. Currently, a simple average of the node's current potential and the potential of its successor (which is known after an action is performed) is used as the new value of the potential. One advantage of this is that a predecessor node's potential value moves towards that of the successor in a smooth trajectory, if there is a preferred action at the predecessor node (which is usually the case after a few trials).

Note that a learned weight such as 0.8 associated with an action such as `Pick up apple` reflects the fact that from S3 (where `apple on table` is perceived), when the action `Pick up apple` is executed, 4 times out of 5 the apple ended up in the hand. This simple approach to handling uncertainty appears to work well on these simple examples.

Notice that the internal need state is combined with the processed visual state in forming a rule for acting (see Figure 3). Approximate rather than perfect matches are usually employed while utilizing these learned rules.

CHILDLIKE's action-selection abilities clearly improve with learning. Table 1 summarizes the results of 20 experiments that involved action sequences of the sort shown in Figure 4. The shortest action sequence was of length 1 and the largest of length 6. Each state had between 1 and 8 possible actions, with an average of about 4. Actual need satisfaction occurred, typically, in two states. One determinant of performance is the number of states the system may have to look at in the course of need satis-

	Before Learning	After Learning
Number of states examined	120	< 12
Probability of success	0.0-0.15	0.75-0.90
Half-life (time units)	22	> 300

Table 1: Performance of the Action Selection Algorithm

faction. (This figure reflects a worst-case scenario. When a small (5) percent of the cases, where the system kept thrashing around without any need-fulfillment, was excluded, an order of magnitude improvement on this measure was noticed in almost all the experiments.) It is important to realize that by choosing wrong actions (such as `Throw apple` or `Topple pitcher`), the system may end up destroying resources which could have helped in need satisfaction. Thus, the probability of success in need-fulfillment is another measure of the system's abilities (for these probability values, the system was allowed only as many action steps as perfect knowledge would require). Another useful metric is ***half-life***: the time units taken to expend half its initial allocation of ***energy***[3]. Learning in these experiments usually involved between 30 and 100 trials (before performance leveled off).

Distinct memories are used to encode the action-related rules and their components; however, they are linked to memories containing encodings of related visual structures and words. Thus a pre-condition representing `on(pitcher,table)` in the action memories as part of a rule is connected to the corresponding visual structures and through them to words.

Discussion

The algorithm shown in Figure 3 is similar to the reinforcement learning algorithms that have been proposed recently (e.g., [Sutton, 1990], [Whitehead and Ballard, 1990]). However, the approach outlined here also has a number of significant differences. First, we attempt to handle uncertainty in the world by keeping track of the reliability of each action as a weight associated with each link that represents an action in the action memories. Second, since the knowledge encoded in the state is based on and linked to other acquired knowledge, it is easier to merge different states in an effort to keep the size of the memories reasonable. Such generalization in the plan memories (based on prior visual–linguistic associations) is one of the crucial mechanisms that stem the combinatorial explosion in the number of states. We are also planning to add a short-term memory component to the action memories to handle situations where action selection may also depend on information perceived at earlier times. We elaborate on some of these issues below.

Generalization, and the Effect of Action Words

Rules which encode actions pertaining to specific objects are initially acquired; as the number of objects experienced by the system grows, mechanisms are needed to compact the memory structures. The generalization mechanisms embodied in CHILDLIKE are designed precisely to do this. The generalization processes implement specific rules from the symbolic inductive learning realm such as the ***turning constants into variable*** rule, ***the closing interval*** rule or the ***climbing generalization tree*** rule [Michalski, 1983]. The generalization tree itself is often learned via language inputs such as `apple is fruit` and `banana is fruit` after visual associations for `apple` and `banana` are learned. When generalizing rules pertaining to actions, the visual and linguistic memories can be exploited. For example, generalization may involve merging states and actions that refer to `apple`, `pear` and `banana`. From these, using the information in the linguistic memories, the system can create states and actions that refer to `fruit`. The weights on the links as well as the need-fulfillment potentials associated with each state are suitably averaged; and further experiences attempt to improve these values, if necessary. An important capability that facilitates such generalization is that actions can be split into their components: a pure action part, objects referred to by the action, and so on. Note that actions described so far are represented using English words for clarity; but the object component of the action (e.g., `apple` in an action represented by `Pick up apple`) is actually an abstract internal symbol that is grounded in terms of both the visual representation and the language equivalent, thus permitting the kind of generalization described above. Using such generalization mechanisms, the system can form rules that refer to classes of objects and their need-satisfying potentialities. An important point to note is that these generalizations are not trivial, since a language input such as `apple is fruit` is not readily interpretable with respect to the plan memories (which typically encode visual features or pointers to them in each state). However, the previously acquired visual–linguistic associations enable going from linguistic descriptions to visual features (and vice versa); this mutual grounding seems to aid each component of the system. States are also merged based on other factors such as visual similarity (e.g., the states involved may refer to `stool` and `chair`, which share some physical attributes), common need-fulfillment (e.g., the states fulfill the `rest` need) and implication of the same action (e.g, the states share the action `Sit on ...`). Generalization keeps the sizes of the acquired memories down to realistic levels. A slight reduction in performance abilities may sometimes be manifested due to generalization. For example, a critical action such as `Peel banana` does not have a counterpart for apple and pear, as experienced by CHILDLIKE. So after generalization, it was found that the `Peel ...` step was skipped over implicitly for all fruits, including a banana. Figure 5 shows the qualitative effect of generalization on CHILDLIKE's action memories; results of experiments with a few hundred examples (or experiences) are described in [Mani, 1992] (in preparation).

Another mechanism that encourages parsimonious representations is the use of *extraneous states*— complete representations of these states are not stored, and they usually encode the results of actions with low weights (or certainty) attached to them. In Figure 4, two examples of

[3] Energy is just a simple function of the inverse of the need levels.

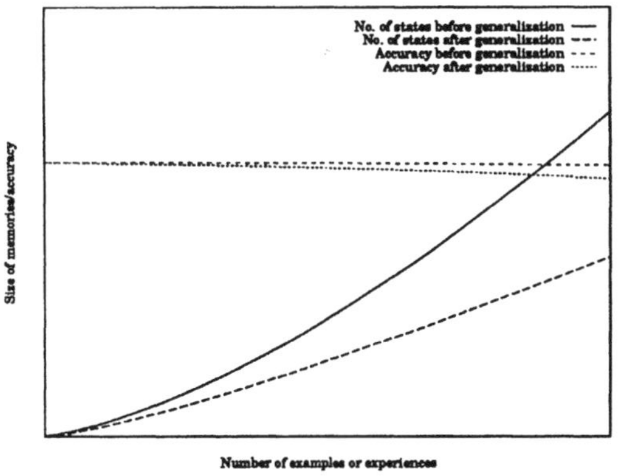

Figure 5: Effect of Generalization on Action Memories

extraneous states (denoted *Se1* and *Se2*) are shown. Extraneous states can be merged without much care; however, keeping a few extraneous states (as opposed to the extremes of storing just one, or all) has the effect of providing weak contexts for reasoning. Preliminary experiments using extraneous states indicate that *perceptual aliasing*[4] may actually turn out to be beneficial in certain cases (for a discussion of the problems usually caused by perceptual aliasing, see [Whitehead and Ballard, 1990]).

Another important issue that is under examination is the learning of links between actions and simple linguistic descriptions of them. Teaching the system words about the various actions is important for two reasons. First, the memories acquired by CHILDLIKE will ground actions in terms of words. It is hoped that this will obviate the building in of actions. As noted earlier, in the current version, actions (but not their effects) are encoded in the system *a priori*. Second, the ability to use words about actions provides an effective way of communicating actions to be executed. Even when CHILDLIKE does not have the necessary motor capability, an action can be effectively achieved by dictating it to another agent that possesses the requisite motor skills. Such dictation can take place using the medium of a language that was acquired by the two agents in similar environments.

Using a Short-term Memory Component

The role of short-term memory in these tasks is obvious—e.g., if the system has to execute a number of "eating steps" each of which reduces hunger gradually, it should continue and finish eating before attending to other pressing needs such as `sleep`. Also, if the system notices the potential of a certain state to quench its `thirst`, while attending to the most pressing `hunger` need, it might be useful to remember the state so that it can come back to it after attending to the `hunger` need. For example, in Figure 4, this translates to going from state *S5* to state *S1* when an action such as `Move to refrigerator (from table)` is not highly implied. Note that this is not an unrealistic scenario even after the apple is eaten, since CHILDLIKE does an approximate match between world states and learned internal plan states. A simple implementation of short-term memory may consist of some selected state information (e.g., that state *S1* has a high `thirst`-satisfying potential) or that a particular action is reliable (i.e., has a very high weight associated with it). Storing and interpreting such information in the short-term memory can be achieved by modifying the action selection algorithm (Step *2a* in Figure 3). Several other approaches to using short-term memory to improve overall action selection are being tested. One is to perform deeper search into the plan memories (e.g., to see whether the target state *S1* above can be reached easily from *S5*)—the rationale behind this is that the current situation may be opportunistic and hence, not fully reflected in the backed-up values of the need-satisfying potentials.

Other Issues

One point that needs to be stressed is that CHILDLIKE's `curiosity` need is slightly different from its other needs (such as `thirst`). The `curiosity` need dominates only when other needs do not, and aids in exploring perceptual states (and actions) that CHILDLIKE may otherwise ignore.

Simple trial-and-error learning appears to be a good mechanism for initial acquisition of knowledge (remember that CHILDLIKE starts out with no *a priori* rules). Once some initial planning knowledge has been acquired, a more deliberative algorithm could be introduced to perform action selection, since the system stores the effect of actions (or at least the effect of reliable actions).[5] Under such a scenario, the system may monitor the state of the world every n steps, where n is a parameter that encodes caution and is dynamically set as a function of the uncertainty expected in the environment and the reaction speeds required; learning an optimal value for this parameter is a good area for future explorations.

Using extraction and aggregation mechanisms, similar to those used in building visual and linguistic structures, to build macro-operators or hierarchies of actions is another issue that merits further exploration.

Conclusions

A survey of conventional planning techniques (for example, see [Allen *et al.*, 1990]) reveals two important related issues. The first is that planning is a hard problem, if only because of the potentially large search space involved. The second is that the environment the agent faces often changes in unpredictable ways, and this uncertainty greatly lowers the utility of planning.

An attractive approach is to learn rules for both perception and planning that can be quickly accessed and applied, choosing actions with very little (or no) deliberation, and constantly checking their effects via perception. This is exactly the design philosophy behind

[4] the phenomenon manifested by a many-to-one mapping from world states to the learner's internal states.

[5] A purely reactive planning system learns the mapping from a set of states $\mathcal{S}$ to a set of actions $\mathcal{A}$ (the $\mathcal{S} \rightarrow \mathcal{A}$ mapping); CHILDLIKE, however, learns a partial $\mathcal{S} \times \mathcal{A} \rightarrow \mathcal{S}$ mapping in addition to the $\mathcal{S} \rightarrow \mathcal{A}$ mapping.

CHILDLIKE, which does not plan in the conventional sense. CHILDLIKE attends to the most pressing need at any point, keeping in short-term memory other pressing needs and possible satisfiers or actions. Moreover, it perceives the world after each step or action.
The current version of the system uses a high-level visual input, but interesting features can be propagated to this layer using a massively parallel, hierarchical structure of processes which starts with real images. Such a recognition-cone based architecture has been successfully employed for rapid recognition of large, digitized TV-frame-like images [Li and Uhr, 1987]. The serial depth of such a system is logarithmic in the size of the sensed input array, and such an architecture appears to roughly mirror the constraints established from neuroanatomical studies of the brain (for a fuller discussion, see [Uhr, 1987]). The CHILDLIKE system is an attempt to build on this parallel-hierarchical framework to handle language inputs and further (as described in this paper) to build and use memory structures that encode learned rules about actions and needs, exploiting the bias provided by similarly learned visual–linguistic associations.

References

[Allen *et al.*, 1990] J. Allen, J. Hendler, and A. Tate. *Readings in Planning*. Morgan Kaufmann, San Mateo, CA, 1990.

[Drescher, 1987] G.L. Drescher. A mechanism for early Piagetian learning. In *Proceedings of AAAI-87*, 1987.

[Dyer, 1991] M.G. Dyer. Symbol grounding in the blobs world with a neural/procedural model. In *Proceedings of the US–Japan Workshop on Language and Vision, CRL, New Mexico State University*, 1991.

[Feldman *et al.*, 1990] J.A. Feldman, G. Lakoff, A. Stolcke, and S.H. Weber. Miniature language acquisition: A touchstone for cognitive science. In *Proceedings of the 12th Annual Conference of the Cognitive Science Society*. Lawrence Erlbaum Associates, 1990.

[Firby, 1987] R.J. Firby. An investigation into reactive planning in complex domains. In *Proceedings of AAAI-87*, 1987.

[Georgeff and Lansky, 1987] M. Georgeff and A. Lansky. Reactive reasoning and planning. In *Proceedings of AAAI-87*, 1987.

[Kaelbling, 1987] L. Kaelbling. An architecture for intelligent reactive systems. In M. Georgeff and A. Lansky, editors, *Reasoning about Actions and Plans*. Morgan-Kaufmann, Los Altos, CA, 1987.

[Li and Uhr, 1987] Z.N. Li and L. Uhr. Pyramid vision using key features to integrate image-driven bottom-up and model-driven top-down processes. *IEEE Transactions on Systems, Man, and Cybernetics*, 17(2), 1987.

[Mani and Uhr, 1991a] G. Mani and L. Uhr. Integrating perception, language handling, learning and planning in the CHILDLIKE system. In *Working Notes of the AAAI Spring Symposium on Integrated Intelligent Architectures*, 1991. Also in the SIGART special issue on *Integrated Cognitive Architectures*, August 1991.

[Mani and Uhr, 1991b] G. Mani and L. Uhr. Perception-mediated learning and reasoning in the CHILDLIKE system. In *Proceedings of the 13th Annual Conference of the Cognitive Science Society*. Lawrence Erlbaum Associates, 1991.

[Mani, 1992] G. Mani. *Learning Language about Objects and Using this Language to Learn Further: The CHILDLIKE System*. PhD thesis, University of Wisconsin, Madison, WI, 1992. In preparation.

[Mason *et al.*, 1989] M.T. Mason, A.D. Christiansen, and T.M. Mitchell. Experiments in robot learning. In *Proceedings of the Sixth International Workshop on Machine Learning*, 1989.

[Michalski, 1983] R.S. Michalski. A theory and methodology of inductive learning. In R.S. Michalski, J.G. Carbonell, and T.M. Mitchell, editors, *Machine Learning: An Artificial Intelligence Approach*. Tioga, Palo Alto, CA, 1983.

[Okada, 1991] N. Okada. Aesopworld: An intellectual and emotional agent. In *Proceedings of the US–Japan Workshop on Language and Vision, CRL, New Mexico State University*, 1991.

[Shen, 1989] W. Shen. *Learning from the Environment Based on Percepts and Actions*. PhD thesis, Carnegie Mellon University, Pittsburgh, PA, 1989.

[Siskind, 1991] J.M. Siskind. Naive physics, event perception, lexical semantics and language acquisition. In *Working Notes of the AAAI Spring Symposium on Machine Learning of Natural Language and Ontology*, 1991.

[Sutton, 1990] R.S. Sutton. Integrated architectures for learning, planning, and reacting based on approximating dynamic programming. In *Proceedings of the Seventh International Conference on Machine Learning*, 1990.

[Uhr, 1978] L. Uhr. 'Recognition Cones' and some test results. In A.R. Hanson and E.M. Riseman, editors, *Computer Vision Systems*. Academic Press, New York, NY, 1978.

[Uhr, 1987] L. Uhr. *Parallel Computer Vision*. Academic Press, Boston, MA, 1987.

[Whitehead and Ballard, 1990] S.D. Whitehead and D.H. Ballard. Active perception and reinforcement learning. *Neural Computation*, 2(4), 1990.

Genetically Generated Neural Networks I: Representational Effects

Leonardo Martí
Center for Adaptive Systems
Boston University
111 Cummington Street
Boston, MA 02215
lmarti@cns.bu.edu

Abstract

This paper studies several applications of genetic algorithms (GAs) within the neural networks field. The system was used to generate neural network circuit architectures. This was accomplished by using the GA to determine the weights in a fully interconnected network. The importance of the internal genetic representation was shown by testing different approaches. The effects in speed of optimization of varying the constraints imposed upon the desired network were also studied. It was observed that relatively loose constraints provided results comparable to a fully constrained system. The type of neural network circuits generated were recurrent competitive fields as described by Grossberg (1982).

Introduction

Genetic Algorithms (GAs) have a lot in common with neural networks. While used in engineering applications, neural networks are noted for their neurobiological foundations. GAs are also based on biological foundations. However, not all known natural genetic functions have been incorporated into GAs. GAs have been used mainly as search and optimization procedures. Natural genetics perform some of these tasks, but more importantly, genetic material contains the "program" for life-building. Although recent discoveries (Ho & Fox, 1988) have changed our view on this "program", it is still undisputed that the genetic material (DNA, RNA) contains enough information to generate an organism.

The genetic system used here consists of a population of organisms or individuals where each member is composed of a gene. A chromosome is composed of a string of alleles. In this particular case, alleles are represented as single bit binary values. An initial population of an arbitrary number of members is created by assigning random values to each allele. Once the population is established, each individual's chromosome is tested against some metric. This can be seen as their "life" performance, generating a probability of reproduction. Once all individuals have been tested (have "lived") the individuals with higher performance will be more likely to procreate and pass on their genes.

This paper uses GAs to search for the parameters that describe a neural network. These parameters will be used to generate such a network and to analyze its behavior when required to perform a specific task. The aim here is not only to use GAs as a parameter search tool, but as a code building tool. Garis (1990), Miller, Todd, & Hegde (1989), and Harp, Samad, & Guha, (1989) have done some work in this direction. The Miller, Todd, & Hegde's system (1989) can't be considered strictly a network building model, but as a network design or configuration system. It basically determines the weight values in a fully interconnected network where the number of nodes is predetermined. Harp, Samad, & Guha (1989) on the other hand, determine the number of nodes and their connectivity in a fairly complete way. Here, the type of network searched for will be of a more biologically based type.

The internal genetic representation is critical to the speed and optimization level of a genetic algorithm. This was shown by testing different approaches to the genetic representation. In order to further accelerate the optimization process, the effects of varying the constraints imposed upon the desired network were also studied.

In this paper, the formation of subsequent generations is based on two genetic operators: mutation and crossover.

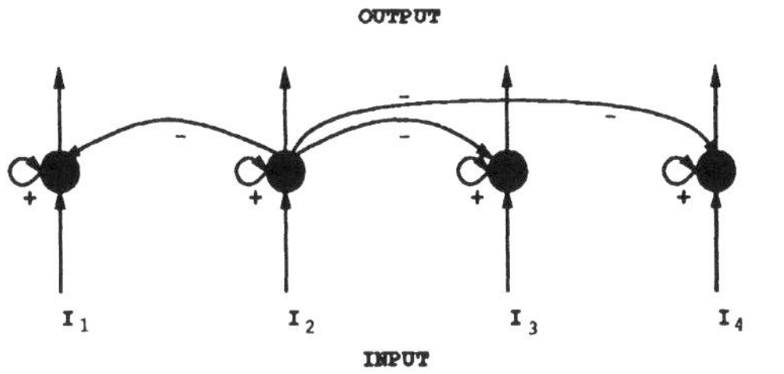

Figure 1: Competitive recurrent circuit. For clarity, only node 2 is shown with all its efferent connections.

Mutation is performed by switching each allele to its complementary value with certain probability. Crossovers are performed by selecting two individuals from the population for reproduction. A crossover point is randomly selected somewhere along the extent of the gene, and two children are generated by switching the genetic material of the two parents after the point chosen. In the simulations carried out here, the probability of mutation was 0.03 per allele. The probability of crossover was 1.00 per chromosome.

In addition, the chromosome of the best individual of each generation was copied unchanged for the next generation. For further reference on genetic operations and implementation details see Goldberg (1989).

System Description

An approach similar to that of Miller, Todd, & Hegde (1989) was used. A network of fixed node size was implemented. The connections between nodes were represented by a 4 by 4 matrix. The GA was used to find which connections should exist and whether these should be inhibitory or excitatory. The activation equation was of the form:

$$\frac{dx_i}{dt} = -Ax_i + (B - x_i)(I_i + \sum f(x_g)) - (x_i + D)\sum f(x_h)$$

where x_i is node i from 1 to 4, A, B and D are constants set at 6.0, 5.0, 5.0 respectively, $f(x)$ is the neuron's feedback equation ($f(x) = x$; if $x > 0$ otherwise $f(x) = 0$), g is the set of excitatory nodes, and h the set of inhibitory nodes. The sets of excitatory and inhibitory nodes are determined by the contents of the genome. For the target circuit, g was the node itself ($f(x_i)$), and h consisted of every other node ($\sum_{h \neq i} f(x_h)$).

The representation of the matrix in the chromosome was implemented by allocating two alleles to each connection. So, locations 1 and 2 specify the type of connectivity between node 1 and itself. Locations 3 and 4 specify the type of connectivity between node 1 and node 2, and so on. The meaning of each pair of alleles is shown in Table 1.

Allele pair	Connection
00	Disconnected
01	Disconnected
10	Inhibitory
11	Excitatory

Table 1: Table for allele representation of connection.

In the current experiment, the exact resulting circuit and its response curve were known a priori, so a measure of the difference from this curve was used as the function to be minimized.

The problem studied with this setup was a network of feedback nodes. The target configuration was a recurrent competitive feedback circuit (Grossberg, 1982), as shown in Figure 1.

Representation Modifications

The setup just described did not converge to an optimal result within a reasonable time (400 generations). An analysis of the schemata (similarity templates) involved in the representation of connections, reveals that it is more likely for the system to change genetic material from one state to a state with lower fitness, rather than to a state with higher fitness. This is due to the allele representation, in conjunction with the metric used, and the manner in which crossover and mutation affects genes.

For example, let's assume that a given connection was initially set as not connected (00) when the optimal setting is excitatory (11). Since the probabilities of mutation are quite low (0.03) compared with the probabilities of crossover (1.0), it is quite unlikely that both alleles will

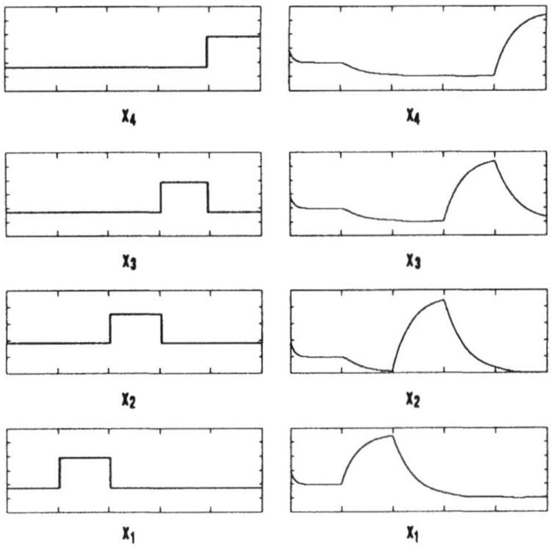

Figure 2: Left: Input sequence. Right: Output of recurrent competitive circuit.

be mutated during the same generation in the same individual, therefore, making crossover the more likely candidate for improving performance. This means that a population member with values of 10 must be combined with another member with values of 01. But a setting of 10 is of lower fitness than the original setting of 00. So the member with the lower fitness is quite unlikely to survive and reproduce, in effect slowing the improvement of genetic material.

In order to avoid this problem, the representation must allow for stepwise improvements in performance through the combination of short length schemata. Crossover should be equally likely to move the schemata to any possible state. This can be achieved in a number of ways. One possible solution would be to use a tri-valued allele, where mutation would be equally likely to switch to any state. This option would avoid the problem of crossover effects on schemata, by not allowing crossover to modify the type of connection used.

The solution chosen still maintains bi-valued alleles, but the meaning of the alleles has been altered. Table 2 shows the table for a connection under the new configuration. Here, it is quite likely that an excitatory connection will eventually move to a disconnected state, and from a disconnected state it is possible to move to either an excitatory or inhibitory state.

A characteristic response curve similar to the one shown in Figure 2 was requested, given the shown inputs. Since this curve does not contain all possible combinations of inputs, the optimal circuit may respond unexpectedly to inputs not tested.

Allele pair	Connection
00	Inhibitory
01	Disconnected
10	Disconnected
11	Excitatory

Table 2: Table for new allele representation of connection.

The resulting network matched exactly that of Figure 1. As desired, the network contains both positive feedback within all the nodes and inhibitory connections to all other nodes. This shows that all possible inputs need not be tested in order to provide sufficient constraints for a unique system. The improvement in fitness across generations is shown in Figure 3. The fitness function used was:

$$O(x) = \frac{1}{1 + \sum(K_t - y_t)}$$

where K_t is the optimal output value at time t, and y_t the actual output from the network at time t.

Constraint Modification

At this point, the fitness function was simplified in order to provide feedback only when the node's activation had settled after each input had changed. The output activations were then compared with two threshold levels, giving

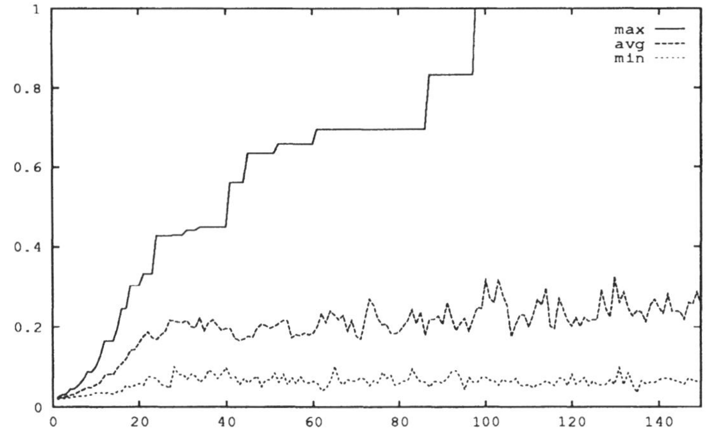

Figure 3: Best, worst, and average population members of the search for a recurrent competitive network over 150 generations and 50 individuals per generation.

three possible states: inactive, inhibited and excited. The discrete result was then compared with a table of the desired network. The disparity from the table was then used as the metric to be optimized. This simpler method of network specification was similarly robust in guiding the GA towards the desired network specification. Since the calculation of the metric is now simpler, the system executed a similar number of generations in less time (about half).

Conclusions

The design and use of neural sub-systems is a complex area that merits further research. In the present study, only small, fix-sized networks were treated. How these networks can increase in size, how they are maintained, modified, and coupled to form more complex systems is an important area that must be investigated to better understand the evolutionary processes.

The present study shows the important interaction between schemata and internal representation. It shows how the variation of the internal representation can modify a GA hard problem (Goldberg, 1989), enabling it to find an optimal network. As a genetic system grows more complex, a methodical testing of the effects of genetic operators is necessary. If novel genetic operations are to prove their usefulness, work of the type performed by De Jong (1975), is required, where a systematic test of the different system parameters is performed. Similarly, other network specification representations should be studied, to observe effects such as the one described here.

The present study also shows how partially constraining a system may be enough to orient the search in the proper direction. It can't be generalized to all problem areas, but it can be used to simplify a genetic algorithm when it becomes too complex.

References

De Jong, K. A. 1975. An Analysis of the Behavior of a Class of Genetic Adaptive Systems. Ph.D. diss., University of Michigan.

de Garis, H. 1990. Genetic Programming. In Proceedings of the Seventh International Conference on Machine Learning, 132-139. San Mateo, Calif.: Morgan Kaufmann.

Goldberg, D. E. 1989. *Genetic Algorithms in Search, Optimization and Machine Learning.* Reading, Mass.: Addison-Wesley.

Grossberg, S. 1982. *Studies of Mind and Brain: Neural Principles of Learning, Perception, Development, Cognition, and Motor Control.* Boston, Mass.: Reidel Press.

Harp, S. A, Samad, T., and Guha, A. 1989. Towards the Genetic Synthesis of Neural Networks. In Proceedings of the Third Conference on Genetic Algorithms. San Mateo, Calif.: Morgan Kaufmann.

Ho, M., and Fox, S. W. 1988. Processes and Metaphors in Evolution. In Ho, M., and Fox, S. W. eds. *Evolutionary Processes and Metaphors* Chichester, England: John Wiley and Sons Ltd.

Miller, G. F., Todd, P. M., and Hegde, S. U. 1989. Designing Neural Networks using Genetic Algorithms. In Proceedings of the Third Conference on Genetic Algorithms. San Mateo, Calif.: Morgan Kaufmann.

Genetically Generated Neural Networks II: Searching for an Optimal Representation

Leonardo Martí
Center for Adaptive Systems
Boston University
111 Cummington Street
Boston, MA 02215
lmarti@cns.bu.edu

Abstract

Genetic Algorithms (GAs) make use of an internal representation of a given system in order to perform optimization functions. The actual structural layout of this representation, called a genome, has a crucial impact on the outcome of the optimization process. The purpose of this paper is to study the effects of different internal representations in a GA, which generates neural networks. A second GA was used to optimize the genome structure. This structure produces an optimized system within a shorter time interval.

Introduction

Though the field of natural genetics is progressing quite rapidly, understanding of the genetic process is still quite incomplete. Even so, knowledge of the natural genetic process has not been completely incorporated into the field of Genetic Algorithms (GAs). To this end, the research reported here tests some new approaches and functions to be used with GAs. In a previous paper (Martí, 1992), and in several other sources (Garis, 1990; Miller, Todd, & Hegde, 1989; Harp, Samad, & Guha, 1989) it has been shown how GAs can be used to generate optimal and novel neural network architectures. Also, it is widely understood how influential the genome representation can be in the success of the genetic search (Davis, 1991; Louis, 1991; Martí, 1992). Here, a genetic algorithm has been used to explore alternative genome representations of another genetic system.

When examined carefully, it becomes clear that natural genetics must possess this functionality in order to provide the flexible evolution that we observe today. Natural genetic evolution is capable of adding new functionality to a species represented by relatively stable genetic material. How new material is added may vary, but one method consists of duplicating a section of existing material. Once this material has been duplicated, variations of it (resulting from other genetic operations such as mutation and crossover) will result in the exhibition of the new functionality. The representation of this new functionality in the chromosome can be based on the representation of the original functionality. But this representation should be able to vary in order to find a more appropriate representation and to survive as stable genetic material.

How these alternative representations are generated and tested is far from being fully understood, but both representation descriptions and actual function descriptions must coexist in the same genetic material. However they do not necessarily reside on the same gene. This paper has assumed that the representation description is more stable than the actual functions. This assumption seems logical if one believes that this representation description must be older genetically than any functions controlled by it, and that older genetic material becomes more stable. An example of this can be seen in the appearance of "homeoboxes" (Gould, 1991) in natural genetic evolution.

In order to carry out the simulation within a reasonable time, a few unbiological simplifications have been made. The representation description was made as simple as possible in order to facilitate its analysis. Even with these simplifications, the system is quite complex to simulate. The representation description is referred to as the "Outer Genetic System" and implemented as a completely separate genetic engine. The function controlled by it is referred to as the "Inner Genetic System". The genome layout of the Inner Genetic System will be optimized by the Outer Genetic System.

GAs can be seen as a Boltzmann-like massively parallel, stochastic gradient descent system, if only mutations are considered. As such, it is uniquely qualified to avoid local minima or sub-optimal results. The separation of genetic material and the genetic engine allow for a certain degree of application independence. It is for these reasons, that GAs are well suited to search the space of possible network topologies. This is a problem space which is too large to be searched exhaustively.

Genetic Environment

A Genetic Algorithm makes use of a string of alleles, called a genome, where it represents the necessary information to describe an individual. This information can be tested for fitness, and the resulting fitness can be compared with that of other individuals. As in natural selection, the individuals with better fitness have a higher probability of reproducing and therefore maintaining their genetic material in subsequent generations. As better organisms appear, and are maintained, the overall fitness of the population also rises.

The genetic engine implemented here makes use of just two genetic operations: mutation and crossover. Mutation consists of randomly selecting an allele and altering its value with certain probability. In all the simulations carried out here, the mutation probability was set unchanged at 0.01 per allele. Crossover is carried out by selecting two genomes and choosing a point where the genomes will be split. At this point, the genomes will be split and recombined with the remaining section of the alternate genome. The crossover probability was also fixed and it was set at 0.85 per chromosome.

In addition, the genome of the best individual of each generation was copied, unchanged, for the next generation.

Inner Genetic System

The Inner Genetic System is the same as the one used in Martí (1992). The purpose of the Inner Genetic System is to generate Neural Networks. Each pair of alleles determines the connectivity among two nodes in a 4 node neural network. Each allele contains a binary value, and when combined determines the connectivity according to the table shown in Table 1. The location of each pair of alleles determines which connection is being specified as a 4x4 connectivity matrix. In the previous paper, this specification remained fixed. For example alleles 1 and 2 determine the connectivity from node 1 to itself, alleles 3 and 4 determine the connectivity from node 1 to node 2 and so on. This representation requires 32 alleles to determine the connectivity of all 4 nodes or 16 connections with 2 alleles per connection.

Allele pair	Connection type
00	Inhibitory
01	Disconnected
10	Disconnected
11	Excitatory

Table 1: Table of allele representation of connections.

In order to calculate the fitness of each individual, a system of equations for the network specified was solved. The target configuration of the network was a competitive feedback circuit (Grossberg, 1982). The system of equations used was:

$$\frac{dx_i}{dt} = -Ax_i + (B - x_i)(I_i + \sum f(x_g)) - (x_i + D)\sum f(x_h)$$

Where x_i is node i, i ranging from 1 to 4, A, B and D are constants set at 6.0, 5.0, 5.0 respectively, $f(x)$ is the neuron's feedback equation ($f(x) = x$; if $x > 0$), g is the set of excitatory nodes, and h the set of inhibitory nodes. The sets of inhibitory and excitatory nodes are determined by the genome. For the target circuit, g was the node itself ($f(x_i)$), and h consisted of every other node ($\sum_{h \neq i} f(x_h)$). The fitness function used was:

$$O(x) = \frac{100}{1 + \sum(K_t - y_t)}$$

where K_t is the optimal output value at time t, and y_t the actual output from the network at time t.

In this Genetic System, representation can be seen as affecting the system in at least two forms. First, the form the table takes, has been shown to heavily influence the outcome of the system (Martí, 1992). The possible values that this table can take are quite few, and can be studied analytically and exhaustively and not treated here.

Another form in which representation plays a role, is in the location of the description of each connection. It is difficult to examine the effect of altering the location of the description of each connection by testing all possible location combinations within the genome. This will be examined with the Outer Genetic System.

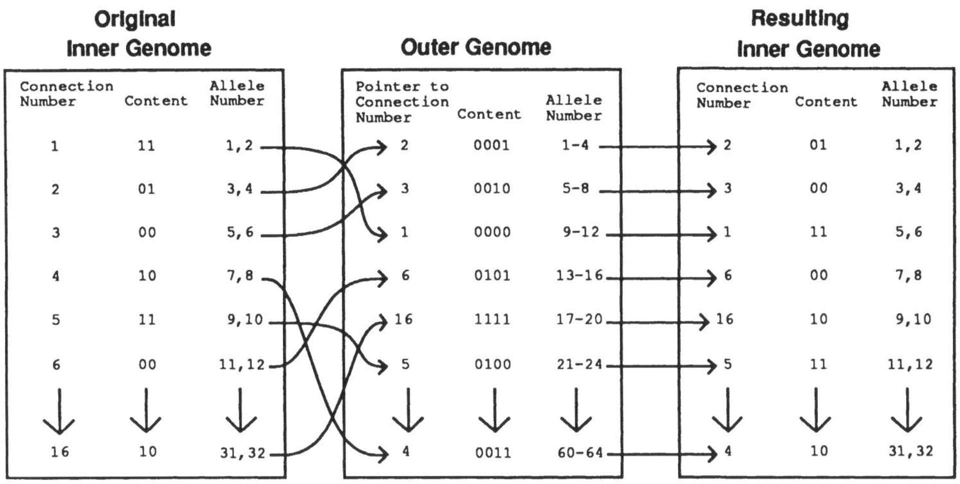

Figure 1: Example of the effect of an outer genome on the location of the connections of an inner genome.

Outer Genetic System

In order to test the effects of placing the description of each connection in different locations, an Outer Genetic System was introduced. This Outer Genetic System can be viewed in two different ways. First, it can be seen as an entirely separate genetic system, where the fitness of each individual depends upon the effectiveness of a certain connection specification placement. The effectiveness of this connection placement is determined by executing an Inner Genetic System, and observing how well it performs.

Another way of viewing this Outer Genetic System is as an additional set of alleles which determines the placement of the description of each connection. According to this view, the location description part of the chromosome (the outer system) varies much slower than and independently of the rest of the genetic material.

In either case, a second and quite independent set of genetic material is needed. The information contained in this genetic material should be able to specify the location of each connection descriptor. Among all the possible manners of determining these locations a relatively simple one was chosen. Basically, the genome was used as a pointer table. A set of four alleles determines the location of each connection descriptor in a binary encoded form. For example, alleles 1 through 4 determine the location where the connection descriptor for the connection from node 1 to itself is to be relocated. Similarly, alleles 5 through 8 determine the location where the connection descriptor from node 1 to node 2 is to be relocated. This representation requires 64 alleles to determine the location of 16 connection descriptors and 4 alleles per connection descriptor. The present system was chosen for its simplicity, and found to be robust enough for the task at hand.

An example of the effect of an outer genome on the location of connections of an inner genome can be seen in Figure 1.

The fitness function for the outer genetic individuals was the fitness of the best individual of the last generation of the Inner Genetic System.

Results

The system was implemented on a Thinking Machines CM2 using 8K processors, and a Sun Sparcstation as the front end. This allowed a population of 90 individuals for each of the genetic systems. Therefore 8100 simultaneous ODE's were solved for each generation.

Figure 2 shows the results of the outer genetic search over 20 generations. Each outer generation consisted of 20 inner generations. Whereas this is not enough to provide us with the optimal population member, partial optimization is sufficient to provide the outer system with the proper direction for the search. This can be seen by the eventual maximization of the best member of the population. But even more importantly, the upward direction of the curve for the average member as shown in Figure 2. This is also seen by comparing a population run of the inner system at an early generation of the outer system with a population

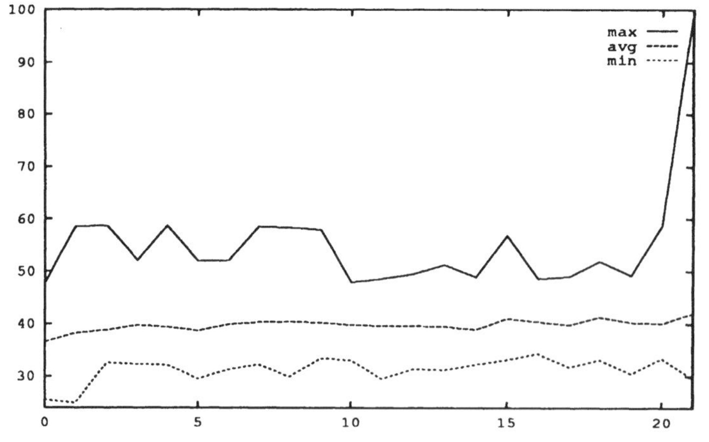

Figure 2: Fitness of best, worst, and average population members of the Outer Genetic System over 20 generations. Fitness of 100 is optimal.

run of the inner system at a later generation. Examples of these two runs can be seen in Figure 3.

Conclusions

The purpose of this paper is not to find "the" optimal representation for a genetic system. It rather intends to show that such an optimization does not have to be done heuristically by the system designer, but that it can be aided by the genetic system itself. And that this leads to an optimized representation.

Variants of the system presented may be suggested. The choice of fitness functions can be changed. The inner system may use a linear function instead of the inverse function used. The outer system may use an average or the best result over many runs for the optimal individual, instead of the last one obtained. As is, the outer system can specify that more than one connection descriptor is located in the same position. By the same token, it can render useless areas of the inner genome. The choice of binary encoding may be modified with the use of gray encoding.

At a more fundamental level, the organization of the genetic material can also be modified. Perhaps a more biologically based approach would be to use a chaotic system to determine the shape of the network or location of each node (Merrill & Port, 1990). Ultimately the two systems can be combined into one genome.

The present paper represents a first step in genetically aided system design and self optimization. Biological evidence for such systems exists from research in natural genetics. As already mentioned, the appearance of homeoboxes represents one of them. Homeoboxes regulate timing and transcription of other genes. Also, the introduction of color vision in primates is believed to have been the result of duplication and later alteration in the representation of specification of retinal cells (Cullis, 1988). At a more basic level, the existence of diploidy in not all but some organisms is yet another indication of a fundamental variation in representation of similar genetic material.

Many more aspects of natural genetics remain to be successfully integrated into artificial genetic system. As more methods from natural genetics are incorporated into GAs, these systems should become more useful and find a wider range of applications.

References

Cullis, C. A. 1988. Control of Variation in Higher Plants. In Ho, M., and Fox, S. W. eds. *Evolutionary Processes and Metaphors* Chichester, England: John Wiley and Sons Ltd.

Davis, L. 1991. Bit-Climbing, Representational Bias, and Test Suite Design. In Proceedings of the Fourth International Conference on Genetic Algorithms, San Mateo, Calif.: Morgan Kaufmann.

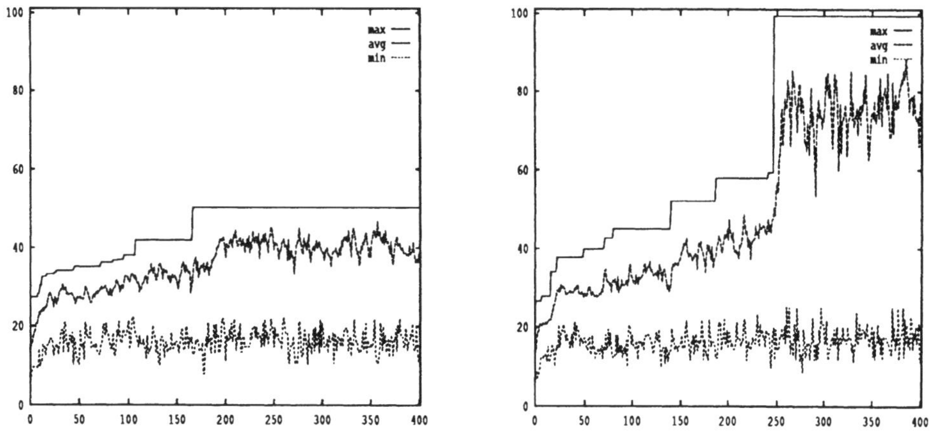

Figure 3: Fitness of best, worst, and average population members of two population runs of the Inner Genetic System over 400 generations. The left run is the result of an early outer individual (generation 2), and the right run is the result of a latter individual (generation 15). Fitness of 100 is optimal.

de Garis, H. 1990. Genetic Programming. In Proceedings of the Seventh International Conference on Machine Learning, 132-139. San Mateo, Calif.: Morgan Kaufmann.

Gould, S. J. 1991. Of Mice and Mosquitoes. *Natural History* July: 12-20.

Grossberg, S. 1982. *Studies of Mind and Brain: Neural Principles of Learning, Perception, Development, Cognition, and Motor Control.* Boston, Mass.: Reidel Press.

Harp, S. A, Samad, T., and Guha, A. 1989. Towards the Genetic Synthesis of Neural Networks. In Proceedings of the Third Conference on Genetic Algorithms. San Mateo, Calif.: Morgan Kaufmann.

Louis, S. J. 1991. Designer Genetic Algorithms: Genetic Algorithms in Structure Design. In Proceedings of the Fourth International Conference on Genetic Algorithms, San Mateo, Calif.: Morgan Kaufmann.

Martí, L. J. 1992. Genetically Generated Neural Networks I: Representational Effects. Forthcoming.

Merrill, J. W. L., Port, Robert F. 1991. Fractally Configured Neural Networks. *Neural Networks* 4: 53-60.

Miller, G. F., Todd, P. M., and Hegde, S. U. 1989. Designing Neural Networks using Genetic Algorithms. In Proceedings of the Third Conference on Genetic Algorithms. San Mateo, Calif.: Morgan Kaufmann.

Using Analogies in Natural Language Generation

Vibhu O. Mittal and **Cécile L. Paris**
USC/Information Sciences Institute
4676 Admiralty Way
Marina del Rey, CA 90292
U.S.A.

Abstract

Any system with explanatory capabilities must be able to generate descriptions of concepts defined in its knowledge base. The use of analogies to highlight selected features in these descriptions can greatly enhance their effectiveness, as analogies are a powerful and compact means of communicating ideas and descriptions. In this paper, we describe a system that can make use of analogies in generating descriptions. We outline the differences between using analogies in problem solving and using them in language generation, and show how the discourse structure kept by our generation system provides knowledge that aids finding an acceptable analogy to express.

Introduction

Good explanation capabilities are crucial for many systems, including expert systems, intelligent tutoring systems and other Knowledge-Based Systems (e.g., to provide on-line documentation). Many such systems can already communicate concepts or processes coherently and effectively to their users. Communication would be enhanced further if a system could employ analogies to help describe concepts and highlight its features. This paper describes a system that can employ analogies in generating descriptions.

Analogies have long been recognized as a valuable tool in problem solving, and it is mainly in this perspective that they have been studied in artificial intelligence – e.g., (Carbonell, 1986; Prieditis, 1988; Falkenhainer *et al.*, 1989); their use in explanation has largely been overlooked. In this paper, we address the problem of finding and using appropriate analogies efficiently to describe concepts. Finding an analogy is computationally expensive: the process essentially involves an attempt to find a consistent mapping between two concepts and their associated relations, sometimes with only a partial and fragmented knowledge about one concept. It is especially expensive if an *exact* mapping is desired, as in the case of most analogical problem solving (APS) systems. However, if *approximate* matches are deemed acceptable, the cost of finding an analogy can be substantially reduced. We illustrate with an example that a generation system does not in fact require the sort of exact matches that most APS systems require.

Our framework is designed around an already existing explanation facility (Moore & Paris, 1989; Moore & Swartout, 1991; Paris, 1991), which is part of the Explainable Expert System (EES) framework (Neches *et al.*, 1985; Swartout *et al.*, 1991). We show how the knowledge about the utterance to be generated and its hierarchical structuring in the form of rhetorical goals can help us identifying and using analogies in the context of explanation generation.

Analogies in Problem Solving vs in Text

Several major differences make the use of analogies in discourse much easier than in problem solving:

First, the features to use to find the analogy are more clearly determined. Unless an APS system builds an explanation tree, there is *a priori* little or no indication of which features are relevant in different situations. Yet, because of the computational cost (since finding an analogy requires searching through a large space of concepts and relations, the fewer the constraints to satisfy, the faster analogies can be found), a minimum number of features should be used in the matching process. Domain-independent theories of feature relevance have been proposed to address this problem: the use of higher-order relations in Gentner's Structure Mapping Engine (SME) (Gentner, 1983; Falkenhainer *et al.*, 1989); the explanation-proof trees by Kedar-Cabelli (1988); the use of abstractions (Greiner, 1988) and justifications of the problem solving (Carbonell, 1986). In discourse generation however, the most important features to match on are given: they are the ones the system is trying to illustrate. Given an appropriate representation of the explanation being constructed, the system can determine the features to highlight or elaborate upon, and use these to find an analogy. Thus, a system designed to employ analogies in generation does not face one of the more difficult problems that APS systems have to address – that of determining the important features on which to match.

[†]The authors can be contacted through electronic mail at: {MITTAL,PARIS}@ISI.EDU. Vibhu Mittal is also in the Computer Science Department of the University of Southern California.

Second, should a system not find a suitable analogy for all the given features, it *should* be able to use analogies for subsets of the original features. Since the determination of these subsets is quite domain specific, most APS systems do not possess a theory of partitioning the features. However, an appropriate representation of the text being constructed contains sufficient information to determine at least one possible subset partitioning for renewed matching attempts in most cases, and thus analogies for a subset of the original features can be used.

Finally and most importantly, expository systems can make use of partial matches, because people seem to have little difficulty in understanding complex and incomplete mappings, as in: *"He is big, and burly, like a bear"*, without trying to resolve how this mapping would affect attributes not mentioned in the sentence. This is *not* possible in APS systems. In addition, if there is a point that is likely to be mis-understood in the analogical mapping, the system can generate an explicit concessionary clause for that point (*"Though A is like B, A is not a ...* "). A system that needs to find analogies for communication purposes therefore has far greater leeway in finding potential matches than an APS system which needs analogues that are consistent in their mapping across many more features.

In summary, then, expository systems can have more guidance as to how to find an analogy and more flexibility in their choice and application of analogies than most APS systems. In the following sections, we briefly describe our generation framework and illustrate our method with simple examples.

The System

Our system is built on an extensive framework for generating explanations for expert systems in natural language using the EES expert system environment (Neches *et al.*, 1985; Moore, 1989; Moore & Swartout, 1991; Paris, 1991; Swartout *et al.*, 1991), the PENMAN Natural Language Generation system (Mann, 1983), and the LOOM knowledge representation language (MacGregor, 1988). A system built using EES can generate descriptions of the terminology it uses, because it has a representation for all its concepts in a terminological knowledge base. The quality of descriptions generated thus depends upon both the detail and the accuracy of the domain knowledge representation. The representation of the terminological knowledge in EES is influenced not only by its operationality in problem solving, but also by the necessity of generating good explanations in the form of descriptions about the concepts themselves. Thus, one of the essential requirements for any system using analogies, a well-detailed domain model, is provided.

To generate grammatically correct English text, the domain model is anchored beneath the Upper-Model, a computational resource for organizing domain knowledge appropriately for linguistic realizations (Bateman *et al.*, 1989). The Upper-Model is a knowledge base of general conceptual categories that provides a domain- and task-independent classification system which supports sophisticated natural language processing.

Given a top level intentional (or discourse) goal to be achieved, text is generated by hierarchical goal decomposition. To ensure the coherence of the text, the text planner makes sure that two subgoal siblings are related with a rhetorical relation, which indicates the functional relationship between any two text spans. Such relations include, for example, `contrast`, `motivation` and `elaboration`. The rhetorical relations used in our system are based on Rhetorical Structure Theory, a theory of coherence (Mann & Thomspson, 1988). As a result of the text planning process, a detailed text plan or Discourse Structure Tree is produced. This text plan contains all the intentional goals and subgoals, as well the rhetorical relations between subgoals. Subgoals in the plan are annotated as either a `NUCLEUS` or a `SATELLITE`, based on their semantic relationship with their parent goal. Nucleic subgoals represent the core or the focus of the information to be presented; satellite subgoals play a supporting role to the nucleus, their relationship being indicated by a rhetorical relation.[1]

Using the Discourse Structure to Find Analogies

There are two ways in which the discourse structure tree (DST) generated by the text planning process can be used to find analogies (which are then incorporated into the DST): (1) to find the features on which to match (2) to find relevant sub-groupings of features to find partial analogies. More specifically, the rhetorical information (RST relations like `elaborate` and `contrast`) and the subgoal specification (`NUCLEUS` vs `SATELLITE`) contained in the DST tree can be used to focus the search for analogies. This section describes the algorithm which makes use of these features.

Our framework uses a slightly modified version of SME to actually find the analogies. Among the differences: during matching, only the features posted are used, and not all the ones that the concept possesses; furthermore, the Upper-Model, as a set of domain independent abstractions that link different knowledge bases together, is used as an additional set of constraints (this enables the system to find analogical concepts that are realized in similar ways in language). Since the system is capable of handling imperfect matches, an efficient but non-optimal algorithm – such as the greedy version of SME (Forbus & Oblinger, 1990) – can be used. It is important to note that the knowledge bases in EES are organized hierarchically, subordinated under the Upper-Model. Although the Upper-Model is by itself too general to work as an effective abstraction hierarchy (as used by Greiner, for instance), it is very useful in categorizing concepts, relations and properties into classes, which can be used to restrict the search. The actual algorithm is presented in Figure 1.

Since the number of features used to constrain the match is relatively small as compared to the number of features employed in APS systems, it is possible to find a relatively

[1]A detailed description on the text planner is out of the scope of this paper. See (Moore, 1989; Moore & Paris, 1989; Moore & Swartout, 1991; Paris, 1991).

Algorithm:

1. Take the text-plan for the text to be generated and scan it to see if there is a concept which is flagged. Concepts are flagged if they are one of the following:
(a) Concepts for which analogies have been used in the past dialogue.[3]
(b) Concepts whose parents have been described using analogies.
(c) Concepts which are *parts* of components that have been described using analogies.
(d) Concepts for which there are no lexical-items.[4]

For any flagged concepts, retrieve the generated analogy and see if it can be extended for this case too. If it cannot, see if a related analogy (that is an analogy from the same domain) can be used in the current text-plan.

2. Traverse the text-plan in a depth-first manner. If there are any CONTRAST or ELABORATION rhetorical relations between two sub-goals posted, attempt to find analogies as follows:
   - **ELABORATION:** Consider the ELABORATION relation and the two sub-goals it relates, as shown in Figure 5, Part (a). The sub-goal on the left-hand side (goal-1) presents some information about a concept, which the sub-goal on the right-hand side (goal-2) elaborates upon. Goal-1 is referred to as the NUCLEUS, and Goal-2 is referred to as the SATELLITE. To generate analogies, the system takes the NUCLEUS sub-tree and gathers all the features that it is supposed to communicate. It also collects all the features expressed by the SATELLITE (to elaborate on the NUCLEUS). The NUCLEUS features are used to find an analogy (necessary features). The SATELLITE features are used as *constraints* that can be relaxed. If an analogy *is* found, it is incorporated (as described below) into the text-plan.
   - **CONTRAST:** The contrast relationship between the two sub-goals indicates that the features expressed by the two sub-goals are contrasting ones. In this case, one can generate analogies for either or both of the sub-goals. However, because of the contrasting nature of the features expressed in the two sub-goals special care must be taken in this case. The two sub-goals are inter-dependent, and an analogy for one of the sub-goals must take into account this dependency. There are two ways in which an analogy may be generated here:
   (a) An analogy can be found by using the features mentioned in one of the sub-goals *and* the negation of the features mentioned in the other sub-goal.
   (b) Two analogies can be found, one for each sub-goal. Using such a compound analogy is allowed only if the two analogies generated belong to the same domain. An example of such a compound analogy is: "Even though he was as big and tough as a *bear*, he was as tender and gentle as a *lamb*". In this case, the analogues bear and lamb are both from the animal domain.

3. For each analogy proposed, evaluate it as follows: If all the features to be presented in the NUCLEUS and the SATELLITE clauses map consistently, incorporate the analogy in the text-plan without further modifications. If, on the other hand, there are any mentioned features that do not map over, generate a CONCESSIVE clause to be appended to the clause containing the analogy ("but... ," "except... ," "although ... ," etc).

Figure 1: Algorithm for incorporating analogies with CONTRAST and ELABORATION relations.

large number of analogues. In such a case, there are two alternatives:

1. The system first checks to see whether other analogues are marked `POSSIBLE` for some other node in the discourse tree. If the current node is related (either a sibling or an ancestor), the system prunes the set of analogues possible for the current node by keeping only those analogues that are compatible with those selected for the related node.
2. If there are no other analogues, the system marks the current set as `POSSIBLE`, and arranges the analogues such that the number of features that match with the base-concept are maximized, while at the same time minimizing the number of dis-similar features.

In case no suitable analogues are found, the system attempts to determine subgroups of features to use for matching partial analogies.

To clarify these two issues (splitting features and selecting from multiple analogues), consider the following example from the domain of local area networks (DEC Manual,1987).

> ... *A communication line can be of three types: simplex, half-duplex and full-duplex ... A simplex line supports data flow in one direction only ... a half-duplex line supports data flow in two directions, but transmissions can only occur in one direction at a time ... a full-duplex line supports data flow in both directions at the same time ...*

[3]Our system retains text plans in order to participate in a dialogue, as described in (Moore & Paris, 1989; Moore, 1989).

[4]Our system cannot currently generate these types of analogies.

It is possible that an attempt to find a suitable analogy for the `communication-line` fails. However, the system can then attempt to determine whether partial analogies can be used to describe the concept. From the DST, the system determines that a communication line is to be described as having three sub-types: simplex, half-duplex or full-duplex. This raises the possibility of finding an analogy for *part* of the original concept. The system tries to find analogies for each of these types, using the features that are most important for each type. The search returns a number of possible analogues for the simplex line: {`electronic-diode, valve, one-way-street, ...`}. This set is marked as `POSSIBLE`, and the system attempts to find analogies for the half-duplex and full-duplex lines. It finds the sets {`two-way-street-with-one-lane-obstructed, ...`} and {`hosepipe, electric-wire, two-way-street`} respectively. The algorithm constrains the system to pick analogies such that related nodes are expressed by related analogies if possible. Since the three nodes are siblings in this case, the system picks a set of related analogies (in this case, sibling concepts subsumed by the concept of `street`). The DST is then modified by the insertion of plan fragments representing these three analogies in the original description. The description generated would be similar to the following, the original description taken from the DEC manual (1987) (p. 2-12):

> ... *A communication line can be of three types: simplex, half-duplex and full-duplex. A simplex line supports data flow in one direction only. This type of signal flow can be compared*

to traffic down a one-way street. A half-duplex line supports data flow in two directions, but transmissions can only occur in one direction at a time. This type of signal flow can be equated to traffic on a two way street with an obstruction in one of the lanes. Two way traffic is possible, but drivers headed in opposite directions must alternate since they have only one lane to use. A full-duplex line supports data flow in both directions at the same time. This is similar to traffic on a two-way street with no obstructions.

As this example demonstrates, the finding of suitable analogies in an explanation framework which maintains a suitable discourse structure can be done quite flexibly, because of the possibility of splitting up the concept and using fewer features (and constraints) with which to find analogies. The following section deals with a short example that illustrates how the algorithm can also make use of the information in the rhetorical relations between clauses to focus the search for analogies.

Focusing the Search with RST Relations

The previous section illustrated how the algorithm could be used to find analogies by splitting up the number of features required to match. There is yet an additional source of information in the discourse structure that can be used: the rhetorical relations that hold between the NUCLEUS and SATELLITE nodes at various levels of the DST. We now illustrate this point with an example, which, for the sake of clarity, only uses a small hypothetical text plan.

Suppose an expert system is asked to select a person for a particular task. The terminological knowledge base of the system will include descriptions of `man`, `person`, etc. The specific domain model will include specific facts about individuals. A portion of the knowledge base together with some specific facts about Mr. Smith is shown in Figure 2. In this case, Mr. Smith is shown as being strong, burly, intelligent, etc. Concepts such as `person` are defined in the Upper-Model (as indicated by the `UM-KB:` prefix in the symbol name). Other domain specific concepts and roles are defined in terms of the Upper-Model to facilitate the generation of appropriate English.

Suppose the system needed to communicate that the assignment needed two features – strength and tactfulness, and that Mr. Smith had these features. It could generate the text shown in Figure 3, Part (a). To generate this text, the system would choose the ELABORATION relation to express these properties of Mr. Smith, as shown in the fragment of the text plan in Figure 5, Part (a).

To enhance this text with an analogy, the system can try to find a concept that matches all the features to be communicated. Failing to do so, the system attempts to find concepts that match partial sets of features, as determined by the features expressed in the NUCLEUS and the SATELLITE, starting with the NUCLEUS, as these are the main features to illustrate, the features of the SATELLITE being considered as desirable but not mandatory. It does find an analogy (the bear), but realizing that the `softspoken` feature is not present in the analogue, generates a CONCESSIVE clause and

```
(defconcept PERSON
  :is (:and UM-KB:Conscious-Being)
  :disjoint-covering
      (UM-KB:Female UM-KB:Male))

(defconcept man
  :is (:and :primitive UM-KB:Male))
  :
(tellm (man Smith))
(tellm (:about Smith
              (:characteristics
               strong burly
               humorous tactful
               gentle softspoken)
              (:UM-KB:age 42)
              (:UM-KB:size big)
              (:profession scientist)
              (:likes caviar)

...))
```

Figure 2: A fragment of the knowledge base in LOOM notation.

inserts it into the text (instead of the ELABORATION). The text generated is shown in Part (b) of Figure 3.

However, based on the context, the system might try to highlight the tactful quality of Mr. Smith. Figure 4, Part (c), shows what text the system generates when it plans the same text using a CONTRAST rhetorical relation so as to highlight the `softspoken` feature. Finally, Figure 4, Part (d), shows the text generated when the system decides to use analogies on both sides of the rhetorical relation. It uses big, strong and not-tactful as features to map on the left-hand side (the NUCLEUS) and softspoken, *not*-big and *not*-strong as the features to match on the right-hand side (the SATELLITE). The system finds many analogies for the left right parts (for example "big and strong as a *rock* ..."), but since the soft-spoken property maps to fewer targets, and the one used, a lamb, is an animal, the other analogue picked is the bear. Figure 5, Part (d), shows the relevant fragments of the skeletal text plan generated in this case.

This point illustrates one of the differences between the way this system uses analogies and the way they are used in APS systems. Here, for instance, the system attempts to find an analogy at the level of the CONTRAST relation. Not finding another concept whose features match the desired characteristics, the system breaks up the features to be matched into smaller subsets (big and strong, soft-spoken or tactful) to see whether an analogy can be found for any of these subsets of features. This is not usually possible in APS systems, since they usually need an analogy which can map *all* the necessary features, and partial matches cannot be used (because of their un-matched attributes).

System: Selected for assignment: John Smith
User: Why?
System: Because the assignment requires both strength and tactfulness. John is both strong and softspoken.

Part (a): No emphasis on `softspoken`. The text is planned using the `ELABORATION` relation.

User: Why?
System: John Smith has been selected because the assignment requires strength and tactfulness. He is big and strong *like a bear*, but he is softspoken.

Part (b): Using an analogy – the `big` and `strong` features matched. A `CONCESSIVE` clause was generated for the `softspoken` feature.

Figure 3: Generating an analogy in the case of an `ELABORATION` relation.

User: Why?
System: John Smith has been selected because the assignment requires strength and tactfulness. Even though he is big and strong, he is very softspoken.

Part (c): Using the `CONTRAST` rhetorical relation to highlight `softspoken`.

User: Why?
System: John Smith has been selected because the assignment requires strength and tactfulness. Even though he is big and strong *like a bear*, he is as softspoken *as a lamb*.

Part (d): Generating analogies for both clauses related by a `CONTRAST` relation, to highlight the `softspoken` feature.

Figure 4: Analogy in the case of an `CONTRAST` relation

Conclusions

Generating effective, understandable descriptions of concepts and terms in the knowledge base is an important requirement for tasks such as explanation, intelligent tutoring and automatic documentation. In this paper, we have described how, by keeping an appropriate representation of the discourse, a generation system can make use of analogies without incurring all of the overhead that APS systems normally do.

The major difference between generating expository analogies for explanation and finding analogies for problem solving lies in two facts: (1) that expository analogies need to be consistent only about a very small region of the feature space and still be understood, and (2) an appropriate discourse representation provides information to partition the set of features to search for partial analogies and to determine which features are mandatory as opposed to desirable in the search for the analogy (features in the `NUCLEUS` as opposed to features in the `SATELLITE`). In generation, then, a system has more information to allow it to find analogies and more flexibility in its use of analogies. Our use of analogies in generating discourse also demonstrates the importance of having task specific criteria for finding and using analogies efficiently and effectively.

We have shown how the relevant features may be extracted from the information available to the text planner. We have also shown that the locality of the information in discourse together with the rhetorical structure can provide information as to the importance of the features to be used in the match process and can thus help focus the search. While there are some unresolved questions about whether the analogies should be generated and integrated with the rest of the text while constructing the text plan, or as our system currently does, by super-imposing the analogy in the text and modifying the text plan, our system demonstrates that it is possible to use analogies in generating discourse in a practical and effective manner.

Acknowledgments

This work was supported in part by the NASA-Ames grant NCC 2-520 and under DARPA contract DABT63-91-C-0025. We are grateful to all members of the EES Project, past and present, for much of the framework's foundation. Our sincere thanks to Profs. Forbus and Gentner for providing us with a version of SME.

References

(Bateman *et al.*, 1989) Bateman, J., Kasper, B., Moore, J., & Whitney, R. (1989). *A general organization of knowledge for natural language processing: The Penman Upper Model.* USC/Information Sciences Institute.

(Carbonell, 1986) Carbonell, J. G. (1986). Derivational Analogy: A Theory of Reconstructive Problem Solving and Expertise Acquisition. In R. Michalski, J. Carbonell, & T. Mitchell (Eds.), *Machine Learning*, volume II chapter 14. Los Altos, CA.: Morgan Kaufmann Publishers.

(DEC, 1987) Network Training Solutions: Introduction to Data Communications – Student Guide. Digital Educational Services, Serial No. EY-6716E-SG-0001.

(Falkenhainer *et al.*, 1989) Falkenhainer, B., Forbus, K. D., & Gentner, D. (1989). The Structure Mapping Engine: Algorithm and Examples. *Artificial Intelligence*, *41*(1), 1–63.

(Forbus & Oblinger, 1990) Forbus, K. D. & Oblinger, D. (1990). Making SME Greedy and Pragmatic. In *Proceedings of the Twelfth Annual Conference of the Cognitive Science Society*, (pp. 61–69)., Boston, MA. Lawrence Erlbaum Associates, Publishers.

(Gentner, 1983) Gentner, D. (1983). Structure Mapping: A Theoretical Framework for Analogy. *Cognitive Science*, *7*(2), 155 – 170.

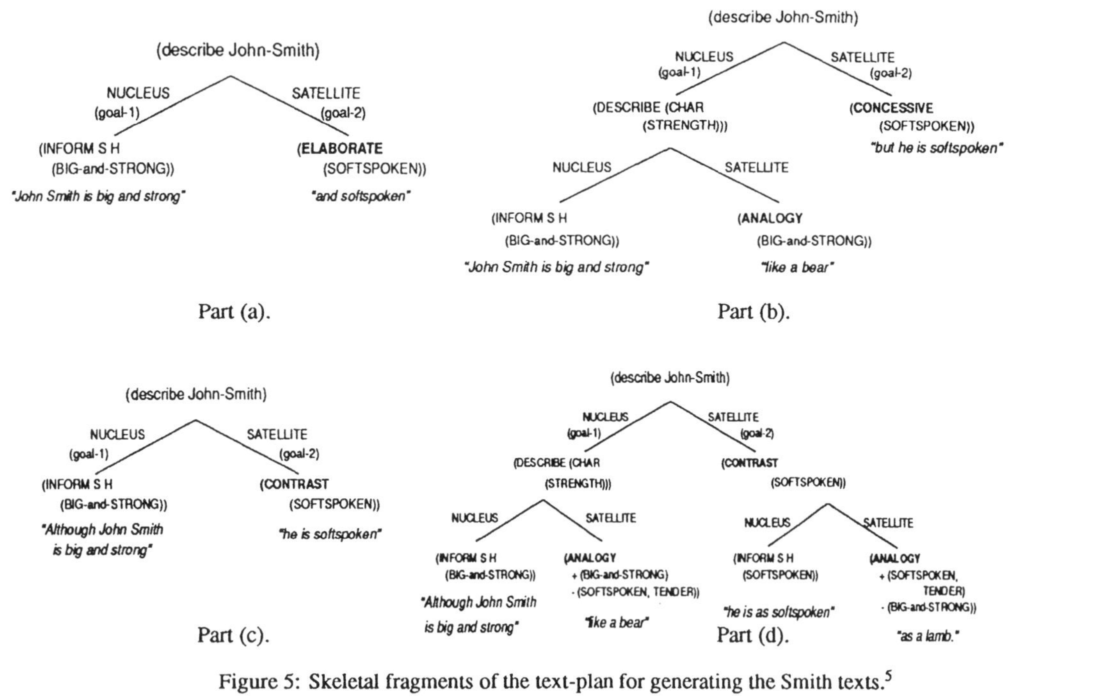

Figure 5: Skeletal fragments of the text-plan for generating the Smith texts.[5]

(Greiner, 1988) Greiner, R. (1988). Learning by Understanding Analogies. In A. Prieditis (Ed.), *Analogica* chapter 1, (pp. 1–31). Morgan Kaufmann Publishers, Inc.

(Helman, 1988) Helman, D. H. (Ed.). (1988). *Analogical Reasoning – Perspectives of Artificial Intelligence, Cognitive Science, and Philosophy*, volume 197 of *Studies in Epistemology, Logic, Methodology, and Philosophy of Science*. Boston: Kluwer Academic Publishers.

(Kedar-Cabelli, 1988) Kedar-Cabelli, S. T. (1988b). *Formulating Concepts and Analogies according to Purpose*. PhD thesis, Rutgers - The State University of New Jersey, NJ.

(MacGregor, 1988) MacGregor, R. (1988). A Deductive Pattern Matcher. In *Proceedings of the 1988 Conference on Artificial Intelligence*, AAAI.

(Mann, 1983) Mann, W. (1983). An Overview of the Penman Text Generation System. Technical Report ISI/RR-83-114, USC/Information Sciences Institute.

(Mann & Thompson, 1988) Mann, W. C. & Thompson, S. A. (1988). Rhetorical Structure Theory: A Theory of Text Organisation. In L. Polanyi (Ed.), *The Structure of Discourse*. Norwood, NJ.: Ablex Publishing Corporation.

(Moore, 1989) Moore, J. D. (1989). *A Reactive Approach to Explanation in Expert and Advice-Giving Systems*. PhD thesis, University of California – Los Angeles.

(Moore & Paris, 1989) Moore, J. D. & Paris, C. L. (1989). Planning text for advisory dialogues. In *Proceedings of the Twenty-Seventh Annual Meeting of the Association for Computational Linguistics*.

(Moore & Swartout, 1991) Moore, J. D. & Swartout, W. R. (1991). A reactive approach to explanation: Taking the user's feedback into account. In C. L. Paris, W. R. Swartout, & W. C. Mann (Eds.), *Natural Language Generation in Artificial Intelligence and Computational Linguistics* (pp. 3–48). Boston: Kluwer Academic Publishers.

(Neches *et al.*, 1985) Neches, R., Swartout, W., & Moore, J. (1985). Enhanced Maintenance and Explanation of Expert Systems through explicit models of their development. *IEEE Transactions on Software Engineering, SE-11*(11).

(Paris, 1991) Paris, C. L. (1991a). Generation and Explanation: Building an explanation facility for the Explainable Expert Systems framework. In C. Paris, W. Swartout, & W. Mann (Eds.), *Natural Language Generation in Artificial Intelligence and Computational Linguistics* (pp. 49 – 81). Boston: Kluwer Academic Publishers.

(Prieditis, 1988) Prieditis, A. (Ed.). (1988). *Analogica*. Los Altos, California: Morgan Kaufmann Publishers.

(Swartout *et al.*, 1991) Swartout, W. R., Paris, C. L., & Moore, J. D. (1991). Design for Explainable Expert Systems. *IEEE Expert*, *6*(3), 58–64.

[5]For examples of the actual text plans, including goals posted, see (Moore, 1989; Moore & Paris, 1989).

Perceiving Size in Events Via Kinematic Form

Michael M. Muchisky and Geoffrey P. Bingham
Department of Psychology and Cogntive Science Program
Indiana University
Bloomington, IN 47405
gbingham@ucs.indiana.edu

Abstract

Traditional solutions to the problem of size perception have confounded size and distance perception. We investigated size perception using information that is independent of distance. As do the shapes of biological objects (Bingham, 1992), the forms of events vary with size. We investigated whether observers were able to use size specific variations in the kinematic forms of events as information about size. Observers judged the size of a ball in displays containing only kinematic information about size. This was accomplished by covarying object distance and actual size to produce equivalent image sizes for all objects and extents in the displays. Simulations were generated using dynamical models for planar events. Motions were confined to a plane parallel to the display screen. Mass density, friction, and elasticity were held constant over changes in size, simulating wooden balls. Observers were able to detect the increasing sizes of the equal image size balls. Mean size judgments exhibited a pattern predicted by a scaling factor in the equation of motion derived using similarity analysis.

Introduction

The textbook solution to the problem of size perception is to use the image size of an object and information about it's distance to derive the actual object size. This requires an ability to determine absolute distance, but most hypothesized sources of distance information specify only relative distance. Size-distance invariance theory confounds the problems of size and distance perception. Given the difficulty of the distance problem, seeking an independent solution to the problem of size perception might be useful.

Functional morphologists and scale engineers have discovered that the shape of objects change to preserve function over changes in scale (Baker, Westine & Dodge, 1973; Calder, 1984; Emori & Schuring, 1977; Hildebrand, Bramble, Liem & Wake, 1985; Thompson, 1961). Scale specific changes in form might provide a source of information about size that is independent of distance. Bingham (1992) has shown that observers are able to discriminate subtle variations in tree form and to use the information to judge the heights of trees. The variations in form were determined by two physically constrained scaling relations. This involved geometric form found in relatively static situations. Consideration of form variations in the context of events leads to forms of motion described via kinematics.

Bingham (1987a; b; 1991) has proposed that kinematic forms may provide visual information about both event identity and scale. In particular, Bingham (1991) suggested that forms in an optical phase space (optical positions vs. optical velocities) might provide visual information enabling observers to identify events. The problem is that spatial metrics are lost in the mapping from event kinematics to optical flows. For example, reference to optical velocities in terms of meters per second is meaningless. Further, the sensory apparatus exhibits strongly nonlinear characteristics in the detection of optical velocities. Ultimately, any scheme for the recovery of metric event velocities is bound to be unrealistic. Nevertheless, experimental evidence shows that observers are able to recognize particular types of events from apprehension of forms of motion. Accordingly, Bingham has argued that metric scaled information cannot be necessary and that ordinal scaling of velocity along continuous extents of optical phase trajectories should be sufficient to develop a taxonomy of qualitative properties used in event recognition. In fact, a level of scaling somewhere between ordinal and interval scaling is most likely.

Watson, Banks, von Hofsten, and Royden (in press) have proposed that visual perception of absolute size and distance might be based on the effects of gravity in constraining event motions and the resulting optical flows. They argue that their analysis is superior to earlier analyses by Chapman (1968) because theirs avoids a dependence on detection of optical accelerations. While the ability to detect optical velocity is well established, the ability to detect optical accelerations is uncertain (Regan,Kaufman, & Lincoln, 1986; Rosenbaum, 1975; Runeson, 1975; Schmerler, 1976; Todd, 1981; see the discussion in Bingham, 1991). In the Watson et al. analysis, distance is specified by a relation requiring detection of optical position, vertical velocity, the location of the focus of expansion, and time. Watson et al. avoid measures of optical accelerations by requiring metric information about both optical velocity at an instant and the time interval from the beginning of fall to the instant when velocity is measured. Both of these measures, however, are problematic. Accuracy in determining metric amounts of time is uncertain. Sufficient accuracy in this case is unlikely. Furthermore, measurement errors in detecting times and velocities would compound (Runeson, 1977). Most problematic, however, is the requirement for detection of metric valued velocities. As discussed above, this is inappropriate. Nevertheless, we should pursue the notion that gravity scales forms of motion in events in a way that can be used by the visual system to determine event scale.

An alternative approach is to expand the analysis from a determination of momentary metric values to spatio-temporally extended qualitative properties of optical trajectories.[1] Similarity methods are appropriate for a determination of the role of gravity in the mapping of trajectory forms into optical phase space. Before describing such an analysis , we describe an investigation as to whether observers can determine object size from motions. The only existing evidence that this might be possible is that produced by Johansson & Jansson (1967) in an unpublished study in which they asked observers to adjust a variable speed film projector.

Experiment 1: Judging in Inches with a Standard

To determine if optical phase space properties provide enough information for observers to accurately judge object sizes we created displays of several different events where all other types of information were eliminated. The image size of the object was the same in all displays. We achieved this by covarying simulated viewing distance and actual object size. If the optical phase space properties did not provide information about size then observers should not have been able to judge the sizes of the objects accurately.

Methods

Participants. Six undergraduates at Indiana University participated in the experiment for credit in an introductory psychology course. All participants had normal or corrected to normal vision.

Displays. Four types of events were used. A ball free falling and bouncing. A ball rolling down an inclined plane. A ball on the end of a string, acting as a pendulum, swinging downward, hitting and knocking over a block. Finally, a stack of four blocks with a ball on top all falling over and coming to rest on a ground surface. Each event was simulated at 5 different scales. In all displays the diameter of the ball was 1 cm on the screen. The simulated actual diameters of the balls ranged from 2.5 cm to 240 cm.

Simulations of planar events were generated using their dynamics and holding gravity, friction, elasticity, and mass density constant. The average frame rate across all of the events was 20 frames per second. Actual frame rates varied from 12 to 30 frames a second depending on the complexity of the displays.

To control for the possibility that observers might use event duration or peak velocity to produce judgments of object size, we manipulated these properties in the pendulum and inclined plane events independent of changes in size. Three levels of event duration (and peak velocity) were created in each of the five sizes of the two events by changing the incline of the surface and the length of the pendulum, for a total of 8 events. If event durations or peak velocities were being used, we expected events of identical sizes but different event durations (or peak velocities) to be judged differently.

Procedure. Before making judgments the observers were given a demonstration to ensure that they understood the situation being modeled. The demonstration involved two rubber playground balls of differing sizes. First, observers were shown how covariation of object size and distance can produce the same image size. The large ball was held approximately 6 feet from the observer while they adjusted the distance of the smaller ball so that it just occluded the larger ball. Second, observers witnessed examples of free fall events where the distances of fall were scaled to the size of the ball (that is, the larger ball was dropped from a greater height).

After this, observers sat facing a computer terminal and were asked to judge the actual diameter of the ball in each event display. They were allowed to view a display as many times as they felt necessary to make their judgment. The judgments were written in inches. Observers judged all 5 sizes of each event in a block of random ordered trials. Each of the 8 events were judged three times in different random ordered blocks. Preceding each of the first set of blocks, observers were shown a standard for that event. They were told the size of the ball in inches. The standard corresponded to the second largest event size. During the latter two trials the standard was only shown if the observer requested it. An experimental session lasted for 75 minutes.

Results and Discussion

As shown in Figure 1, mean judged size increased monotonically but nonlinearly with actual size in all events (save one). A repeated measures analysis of variance (ANOVA) was performed on size judgments with event type, actual size, and trial as factors. The main effects for event type and size were significant, $F(7, 35)=3.88$, $p<.005$ and $F(5,25)=25.36$, $p<.001$, as was their interaction, $F(35,175)=3.30$, $p<.001$. Simple effects tests revealed that within each event type the sizes were judged differently, $p<.001$, in all cases. Simple effects also showed that the judgments did not differ significantly across the 8 event types at the lowest and highest size levels, but that they did at the other size levels. Observers regularly overestimated the smaller sizes and underestimated the larger sizes resulting in a curved pattern of judgments crossing a line representing absolute precision near the value of the standard.

Mean Judged Size vs Actual Size

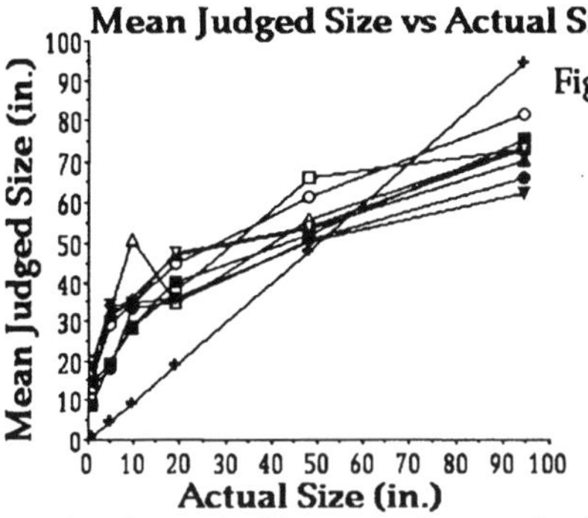

Figure 1.

As shown in Figure 2, the manipulation of event duration for the inclined plane and pendulum events revealed that judgments of size were independent of event durations or peak velocities.

Mean Judged Size vs Period

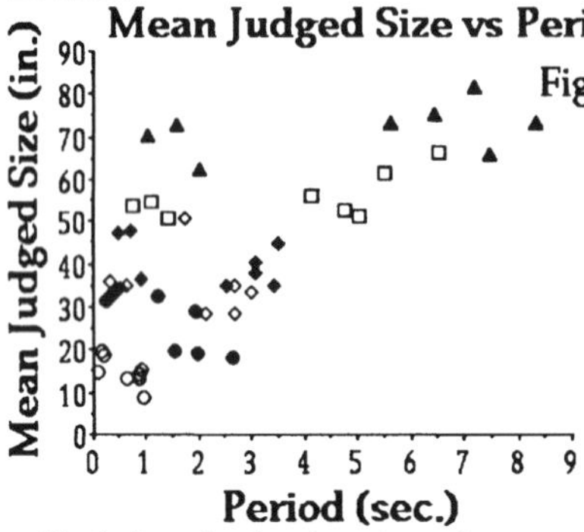

Figure 2.

Variations in the duration of an event within a size level (for instance, the 3 different inclinations of the slope for a given event size) did not prevent observers from correctly grouping objects of the same sizes. Objects of equal size were judged as similar despite different event durations both within a type of event and across the two different types of events. Because peak velocity covaried with event duration, we wished to conclude that event duration and peak velocity were not used to judge object size. However, it was possible that the use of standard trials suppressed the tendency for judgments to reflect event durations. All of event displays were calibrated by the standards at the second largest size level. The next experiment was performed to investigate this possibility.

Experiment 2: Judgments without Numbers or a Standard

To avoid the use of a standard display, we changed the method used by observers to express their judgments. Rather than writing a value in inches, we instructed observers to express ball size using hand span. The observers adjusted the distance between their hands making motions as if they were grasping the ball between their hands. We witnessed several of the observers from the first experiment doing this to estimate size in inches. This measure would avoid the use of numbers requiring an extrinsic system of units and the use of a standard.

Methods

Participants. Five undergraduates at Indiana University participated in the experiment for credit in an introductory psychology course. All participants had normal or corrected to normal vision.

Displays . The same set of displays were used as in Experiment 1 with the exclusion of the largest event size. All other aspects of the displays remained the same.

Procedure. Observers were instructed to place their hands as if they meant to hold the ball shown in the display. The distance between their hands was measured by the experimenter using a standard tape measure. Observers never saw the numbers on the tape. As before, observers were not given feedback on their performance.

No standard displays were shown to the observers. Because judgments across trials in Experiment 1 were not different, we decided to shorten the total session time by collecting only a single judgment of each event display. This was done also to compensate for the extra time required to make the hand span measurements. Observers were first shown the entire set of displays in a random order. No judgments were expressed at this point. Following this, observers viewed and judged the displays in a different random order. An experimental session lasted 1 hr.

Results and Discussion

Once again, mean judgments increased nonlinearly with actual size for all event types. A repeated measures ANOVA was performed on size judgments with event type and actual size as factors. Only the size main effect and it's interaction with event type were significant, $F(4,16)=4.805$, $p<.01$ and $F(28,112)=2.178$, $p<.003$ respectively. In a simple effects test, size was significant for all event types except for the stack and the two largest pendulum events. Alternatively, events were not significantly different from one another at any size levels. Because event types includes variations in event durations, this means that judgments of size were constant when actual sizes were constant despite variations in durations or peak velocities.

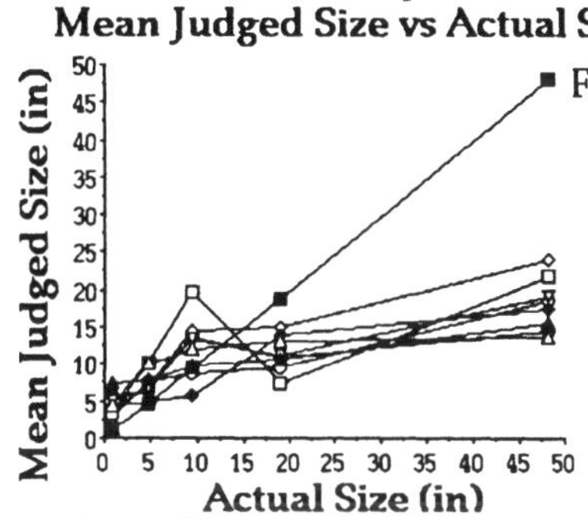

Figure 3.

As in Experiment 1, there was a curved pattern to the mean judgments with overestimation of the smaller sizes and underestimation of the larger sizes. However, in this experiment, judged sizes were lower overall, with less overestimation of smaller sizes and greater underestimation of larger sizes. Mean judgments crossed the line marking absolute precision at 10-15 in actual size. The ceiling on mean judgments was not much above this value which corresponded to the size of the largest ball used in the demonstration preceding the experiment. Such demonstrations are known to produce "transfer effects" affecting the values subsequently used by observers in making magnitude estimations (Poulton, 1989).

The main question addressed in this experiment was whether recalibration via standard trials prevented judgments from following variations in event durations, periods, or peak velocities. The current study was performed without standards. Event durations were manipulated over events of constant size for the inclined plane and pendulum events. The results appear in Figure 4 and should be compared to those from Experiment 1 as shown in Figure 2. The results in the two experiments were essentially the same.

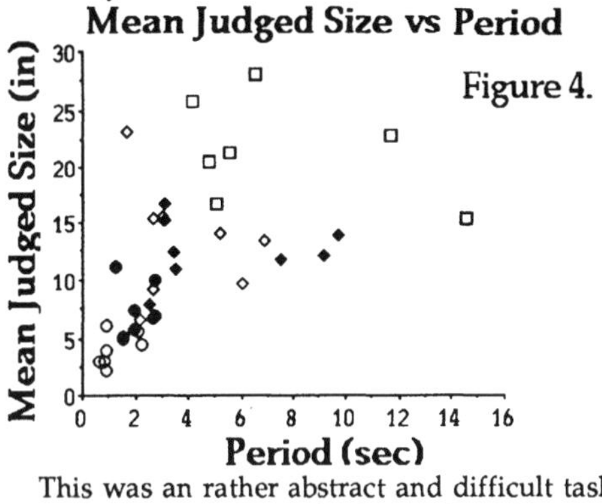

Figure 4.

This was an rather abstract and difficult task for observers who were asked to make judgments of 'actual objects' that appeared as extremely simple line drawings on a small computer screen. Given the complete absence of geometric information about size and the failure of judgments to follow event durations, periods, or peak velocities, the results can be interpreted as preliminary support for the perception of size in events via the form of trajectories as mapped into the optics (of the display).

General Discussion

What qualitative properties of event trajectories might observers have used to judge event scale? We examined the optical phase portraits for all of the events and found a form common to all of the trajectories. All of the trajectories contain segments that were parabolic. In the free fall and bounce, inclined plane, and pendulum events the trajectories were entirely parabolic or nearly so. Because of repeated collisions in the falling stack, only portions of the trajectories were parabolic. The parabolic forms reflect gravitation and (constrained) free fall.

We performed a similarity analysis focusing on free fall to determine whether mapping into the optics distorted the otherwise self-similar forms in a way specific to the scale.[2] Thus the relevant scale transformation was the mapping into optics, performed by dividing all quantities by the viewing distance. By setting the optical height of fall to 1 (that is, the initial height condition in the display), we could perform this transformation by using the actual initial height (ho) to divide instead. The question in performing the similarity analysis was whether, after the transformation had been performed on the equation, the original form of the equation could be recovered in the scale transformed variables (Szücs, 1980). If so, then the scale transformation should be 'benevolent', meaning that the trajectory forms are preserved. If not, then a distortion of the trajectory forms should result directly from the scale transformation. An accessory factor may be isolated which represents the scale specific form of the distortion. Using the equation describing the free fall trajectory in phase space (that is, velocity as a function of position) and ignoring air resistance, we performed the analysis as follows:

Equation of motion for Free Fall

$$\dot{y} = g[2(y-h_o)]^{1/2}$$

Transformation into Optics

$$h_o' = h_o/h_o \qquad y' = y/h_o \qquad \dot{y}' = \dot{y}/h_o$$

Equation in Optical Terms

$$\dot{y}' = g[2(y'-h_o')]^{1/2}\,[h_o]^{1/2}$$

We found that the mapping from event kinematics to optic flows yields a distortion that scales the trajectories by the reciprocal of the initial height to the square root. The square root in the scaling factor accounts for the curvature of the mean judgments. This scaling factor describes the decreasing resolution exhibited in judgments as the object size became larger. Thus, the form of the judgment curves reflected the form of the function determining the information made available via the simple models used to generate these displays.

Acknowledgments

This research was supported by an NSF Grant BNS-9020590 awarded to the second author.
Send all correspondence to: Geoffrey P. Bingham, Department of Psychology, Indiana University, Bloomington, IN 47405.

References

Baker, W.E., Westine, P.S. & Dodge, F.T. (1973). Similarity Methods in Engineering Dynamics: Theory and Practice of Scale Modeling. Rochelle Park, N.J.: Hayden Books.

Bingham, G.P. (1987a). Kinematic form and scaling: Further investigations on the visual perception of lifted weight. Journal of Experimental Psychology: Human Perception and Performance, 13, 155-177.

Bingham, G.P. (1987b). Dynamical systems and event perception: A working paper. Parts I-III. Perception/Action Workshop Review, 2 (1), 4-14.

Bingham, G.P. (1991). The identification problem in visual event perception part I. rate structures in optic flow and the degrees of freedom problem. Cognitive Science Research Reports Series #52, Indiana University, Bloomington, IN.

Bingham, G. P. (1992). Perceiving the size of trees via their form. Proceedings of the 14th Annual Conference of the Cognitive Science Society.

Calder, W.A. (1984). Size, Function, and Life History. Cambridge, MA: Harvard University Press.

Chapman, S. (1968). Catching a baseball. American Journal of Physics, 36, 868-870.

Emori, R.I. & Schuring, D.J. (1977). Scale Models in Engineering: Fundamentals and Applications. New York: Pergamon Press.

Hildebrand, M., Bramble, D.M., Liem, K.F. & Wake, D.B. (1985). Functional Vertebrate Morphology. Cambridge, MA: Harvard University Press.

Johansson, G., & Jansson, G. (1967). The perception of free fall. Unpublished report, Department of Psychology, University of Uppsala, Uppsala, Sweden.

Poulton, E.C. (1989). Bias in Quantitative Judgments. Hillsdale, N.J.: Erlbaum.

Regan, D. M., Kaufman, L., Lincoln, J., (1986). Motion in depth and visual acceleration. In K. R. Boff, L. Kaufman, & J. P. Thomas (Eds.), Handbook of Perception and Performance: Sensory Processes and Perception V1 (pp. 19–1–19-46). New York, NY: Wiley.

Rosenbaum, D. A. (1975). Perception and extrapolation of velocity and acceleration. Journal of Experimental Psychology: Human Perception and Performance, 1, 305- 403.

Runeson, S. (1975). Visual prediction of collision with natural and nonnatural motion functions. Perception & Psychophysics, 18, 261–266.

Runeson, S. (1977). On the possibility of "smart" perceptual mechanisms. Scandinavian Journal of Psychology, 18, 172-179.

Schmerler, J. (1976). The visual perception of accelerated motion. Perception, 5, 167- 185.

Szücs, E. (1980). Similitude and Modelling. Amsterdam: Elsevier.

Thompson, D'A. (1961). On Growth and Form. Cambridge: Cambridge University Press.

Todd, J. T. (1981). Visual information about moving objects. Journal of Experimental Psychology: Human Perception and Performance, 7(4), 795-810.

Watson, J. S., Banks, M. S., von Hofsten, C., and Royden, C. S., (in press). Gravity as a monocular cue for perception of absolute distance and/or absolute size. Perception.

Endnotes

1. Detecting the form of trajectories in optical phase space does not require the detection of optical accelerations. Acceleration can be derived from the slope of a trajectory in phase space, but the rate of change of velocity with respect to position is not the same as acceleration. See Bingham (1991) for discussion.

2. The analysis generalizes directly to the inclined plane and pendulum events

Parallelism in pronoun comprehension

Alexander W. R. Nelson
Human Communication Research Centre
University of Edinburgh
2 Buccleuch Place
Edinburgh EH8 9LW
United Kingdom
sandy@cogsci.ed.ac.uk

Rosemary J. Stevenson
Human Communication Research Centre
Psychology Department
University of Durham
Science Laboratories
South Road
Durham DH1 3LE
United Kingdom
rosemary.stevenson@durham.ac.uk

Keith Stenning
Human Communication Research Centre
University of Edinburgh
2 Buccleuch Place
Edinburgh EH8 9LW
United Kingdom
keith@cogsci.ed.ac.uk

Abstract

The aim of this study was to distinguish between two heuristic strategies proposed to account for the assignment of ambiguous pronouns: a subject assignment strategy and a parallel function strategy. According to the subject assignment strategy a pronoun is assigned to a preceding subject noun phrase, whereas according to the parallel function strategy a pronoun is assigned to a previous noun phrase with the same grammatical function. These two strategies were tested by examining the interpretation of ambiguous subject and non-subject pronouns. There was a strong preference for assigning both types of pronouns to preceding subject noun phrases which supported the subject assignment strategy. However the preference was reduced for non-subject pronouns compared to subject pronouns which we interpreted as evidence for grammatical parallelism. A subsidiary aim of the study was to investigate text-level effects of order-of-mention where a pronoun is assigned to a noun phrase which has been mentioned in the same sequential position. We did not observe any strong effects although we did observe a possible topic assignment strategy where topic-hood depended on order-of-mention. A *post hoc* inspection of the materials revealed possible effects of intra-sentential order-of-mention parallelism. We conclude that a subject assignment strategy, a parallel grammatical function strategy, a topic assignment strategy and a parallel order-of-mention strategy may all constrain the interpretation of ambiguous subject and non-subject pronouns.

Introduction

The comprehension of pronouns is made up of many processes operating at many different levels (syntactic, semantic and nonlinguistic). One such set of processes involves heuristic strategies: mechanical rules of thumb operating over a particular level of representation. Two strategies for pronoun comprehension concern us here: subject assignment and parallel function. The subject assignment strategy (*e.g.*, Broadbent, 1973; Clancy, 1980) proposes that ambiguous pronouns will be assigned to antecedents which function as subjects and the parallel function assignment strategy (Sheldon, 1974) proposes that pronouns will be assigned to antecedents with parallel (identical) grammatical functions.

Sheldon (1974) proposed that parallel function was used in pronoun resolution and noted that (1a) is easier than (1b).

(1) a. Mary hugged John and Betty kissed him.
b. Mary hugged John and he kissed Betty.

Several investigators have studied parallel function in pronoun comprehension and various factors which interact with it (*e.g.*, Caramazza and Gupta, 1979; Cowan, 1980). The general conclusion of such studies is that parallel function does influence the comprehension of ambiguous pronouns. However, Crawley, Stevenson and Kleinman (1990) observed that very few studies of parallel function have used non-subject pronouns and so the results cannot distinguish between a parallel function strategy and a subject assignment strategy because both strategies predict that ambiguous pronouns will be assigned to the subject antecedent. By contrast, the two strategies predict different outcomes when non-subject pronouns are used; the subject assignment strategy predicting subject assignment and the parallel function strategy predicting non-subject assignment. Using three sentence texts, Crawley, Stevenson and Kleinman (1990) found a clear subject assignment bias with non-subject pronouns, from which they concluded that parallel function was not being used.

Unfortunately they only used non-subject pronouns which will only distinguish between subject

[1]The support of the Economic and Social Research Council UK (ESRC) is gratefully acknowledged. The work was part of the ESRC funded Human Communication Research Centre (HCRC)

assignment and parallel function if either strategy is used in isolation. However, it is possible that both strategies are involved in comprehension, so that subject pronouns may show a more marked subject assignment bias than do non-subject pronouns. That is, the subject assignment bias observed by Crawley, Stevenson and Kleinman (1990) for non-subject pronouns might have been attenuated by a concurrent but conflicting parallel function strategy. The main aim of the present paper, therefore, is to test this possibility by investigating both subject and non-subject pronouns.

Stenning (1991) has found effects of order-of-mention in text processing. Given these effects and the possibility that order-of-mention is related to grammatical function because subjects usually come before non-subjects in English, a subsidiary aim was to investigate the existence of order-of-mention in pronoun comprehension. Cowan (1980) investigated order of mention effects in single sentence materials and concluded that there was no effect although there is a slight trend in his data which suggests that such effects might exist. Given Cowan's conclusions and Stenning's findings that order-of-mention operates over sentences, this experiment manipulated order-of-mention over sentences also. This was done by changing the order in which the pronoun's potential antecedents were introduced in the text. Either the order was the same as the order in the target or it was different (the order of antecedents in the target sentence were kept constant).

Crawley, Stevenson and Kleinman's (1990) materials varied in the construction which was used to introduce the potential antecedents. Sometimes they were conjoined subjects and sometimes they were introduced in the subject and predicate of the first sentence. There was some hint in the data that these differences affected the resolution process so the manner of introduction was also included as a design factor.

In summary, this experiment investigated the use of a parallel function assignment strategy in simple three sentence texts, following Crawley, Stevenson and Kleinman (1990). By using both subject and non-subject pronouns a subject assignment strategy could be distinguished from a parallel function assignment strategy. Order of mention effects and the manner of introduction of the potential antecedents were also manipulated.

Experiment

Subjects

The subjects were 192 volunteer students from Durham University.

Example passage
Conjoined antecedents
John and Sammy were playing in the garden. Ellen watched their game with interest. John pushed Sammy and Ellen kicked him.
Antecedents in subject predicate form
John was playing with Sammy in the garden. Ellen watched their game with interest. John pushed Sammy and Ellen kicked him.
Antecedents in separate senteces
John and Ellen were playing in the garden. Sammy watched their game with interest. John pushed Sammy and Ellen kicked him.

Table 1: The same target sentence with three different context sentence pairs. (Same Order conditions. The order in which the two antecedents were introduced into the text was reversed in the Different Order conditions.)

Materials and Design

Each subject read 48 passages consisting of two context sentences followed by a target sentence. Each text described three individuals who were introduced in the two context sentences and repeated in the target sentence. The target sentence was made up of two conjoined clauses: the first mentioned two of the participants and the second mentioned the third participant and a pronoun which referred to one of the two individuals mentioned in the first clause.

The design factors were Text type (conjoined antecedents, subject-predicate antecedents, separate antecedents and target sentence only), Participant order (same order *vs.* different order), and Pronoun position in target (subject *vs.* non-subject). Half the targets contained 3 or 4 words after the *and* and half the targets contained 5 to 7 words after the *and* making a between materials variable of length with two levels. Table 1 shows a reduced text set illustrating three of the levels of the text type variable: conjoined antecedents, subject predicate antecedents and separate antecedents. The target shows the Non-subject pronoun level and the corresponding Subject pronoun level would be "John pushed Sammy and he kicked Ellen". The first two Text conditions are comparable to the materials used by Crawley, Stevenson and Kleinman (1990).

Forty-eight three-sentence texts were constructed (available from the authors on request). Twenty four texts had long targets and twenty four had short targets. Each set of three sentences was used to make 16 versions of itself (four levels of the Text variable, two levels of the Order variable and two levels of the Pronoun variable). These versions were used to produce sixteen lists of materials, four in each of the four text conditions. In each text condition the four lists contained both levels of the Order variable and both levels of the Pronoun variable. Across the four lists, each text occurred in all of its four versions.

Each text was followed by a question derived from the crucial part of the second clause in the target sentence by repeating the second clause with the pronoun replaced by one of the potential antecedents (*e.g.*, Ellen kicked Sammy? or Ellen kicked John?). The question was used to determine the assignment of the pronoun in the preceding text. The number of times each potential antecedent was substituted was balanced across materials.

Procedure

The task was a self-paced reading task, followed by a question about the target sentence. The context sentences of the passages were presented one sentence at a time and the target sentence was presented clause by clause. Subjects were asked to press the space bar as soon as they had read and understood the sentence/clause. When they did so the screen was cleared and the next sentence displayed. Once the final clause had been read, the screen cleared and then the question appeared. After answering the question by pressing one of two keys marked *true* and *false*, subjects were prompted to start the next trial.

The time taken to read the last clause of each target sentence was measured in milliseconds and the answer to each question was recorded. In this paper, we focus primarily on the answers to the questions which were used to infer the assignment of the pronouns.

Results

Comparison over Conjoined,Subject-Predicate,Separate and Single levels of Text: Assignment data Because the Single (target only) level of Text applies to single sentences the Order factor (Same and Different) does not apply, so to make a comparison over Text the Order variable has been collapsed over. The assignment data was prepared for analysis of variance by subtracting the number of non-subject assignments from the number of subject assignments by condition and by subject. This meant that for each subject, each condition was assigned a number between 12 and −12 which represented the assignment bias: a positive number indicated a subject assignment bias and a negative number a non-subject assignment bias.

Figure 1 shows the mean assignment scores for the two types of pronoun in the four text conditions. An analysis of variance was then done on the data using three fixed factors: Text (Conjoined, Subject-Predicate, Separate, Single),

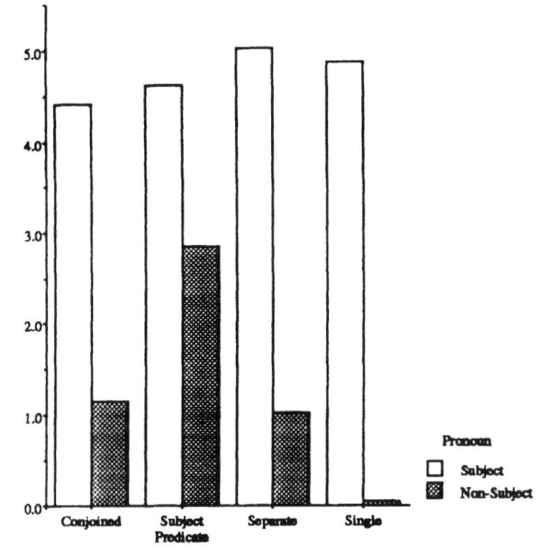

Figure 1: Assignment biases by Text and Pronoun (N = 192).

Length (Short, Long), Pronoun (Subject, Non-Subject). The analysis revealed two reliable effects. Subject pronouns received more assignments to the subject antecedent than to the non-subject antecedent ($minF'(1,83) = 32.32, p < 0.01$, Subject=4.7, SD=4.5 and Non-Subject=1.2, SD=1.2). There was also an interaction between Pronoun and Text, $minF'(3,203) = 2.89, p < 0.05$.

Analysis of simple effects shows that Text has no effect on subject pronouns (F_1 and $F_2 < 1$) but does have a reliable effect on non-subject pronouns ($minF'(3,187) = 4.36, p < 0.01$). Thus subject pronouns are largely unaffected by the preceding sentences whereas the subject assignment bias for non-subject pronouns is affected. The bias is largest when the two potential antecedents are introduced in a subject predicate form, is attenuated if the antecedents are conjoined subjects or subjects introduced in separate sentences, and is smallest for target sentences presented without any context.

A *post hoc* inspection of the materials showed that the structure of the target sentences varied across texts. In most of the targets the structure of the clauses was grammatically parallel, as in the example in Table 1. But in the rest they were not: only the subject pronoun was grammatically parallel to the subject antecedent. The non-subject pronouns and their antecedents occupied different grammatical roles, as in the example in Table 2. These differences could affect the way in which parallel function was operating over the materials because parallel function may operate over strict grammatical categories or over order-of-mention parallelism.

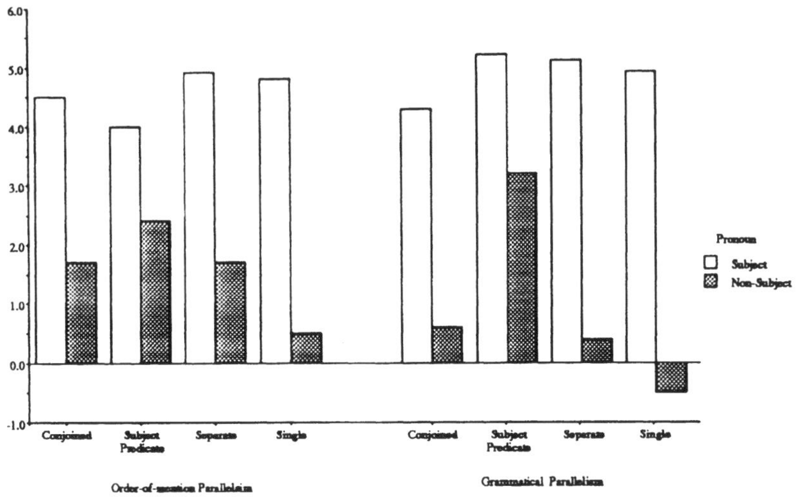

Figure 2: Assignment biases by Text and Pronoun for the Grammatical Parallelism and Order-of-mention Parallelism groups of materials (N = 46 for Order-of-mention parallelism and N = 50 for Grammatical Parallel).

Terry was going to meet Hector at the cricket match.
Their friend Brenda came along to watch as well.
Terry took Hector to the pavilion
and he waved to Brenda.

Table 2: An example of a text with non parallel clauses in the target sentence.

Figure 2 shows the interaction between Text and Pronoun broken down by structure in the target sentence of the materials: those with grammatical parallelism and those with order-of-mention parallelism. Inspection of Figure 2 indicates that the subject bias is reduced in the non-subject pronouns when the targets are grammatically parallel. (We note that Subject-Predicate texts are an exception to this.)

Comparison over Conjoined, Subject-Predicate and Separate levels of Text: Assignment data In order to examine the order effects, the three Text conditions were analysed without the Single Sentence condition. As before, the assignment data was prepared for analysis by subtracting the number of non-subject assignments from the number of subject assignments by condition for each subject. Analysis of variance (with Pronoun, Order, Length and Text as fixed factors) showed that there was one reliable effect of Pronoun where the subject assignment bias was greater for subject pronouns ($minF'(1,163) = 27.3, p < 0.01$, Subject=2.3, SD=2.76, Non-Subject=0.8, SD=2.61). Two interactions approached significance: Order by Text, $F_1(2,141) = 3.95, p < 0.03, F_2(2,69) = 4.33, p < 0.02$ and Pronoun by Text $F_1(2,141) = 3.09, p < 0.05, F_2(2,69) = 5.04, p < 0.01$. Figure 3 and Figure 4 show the means for the two interactions respectively.

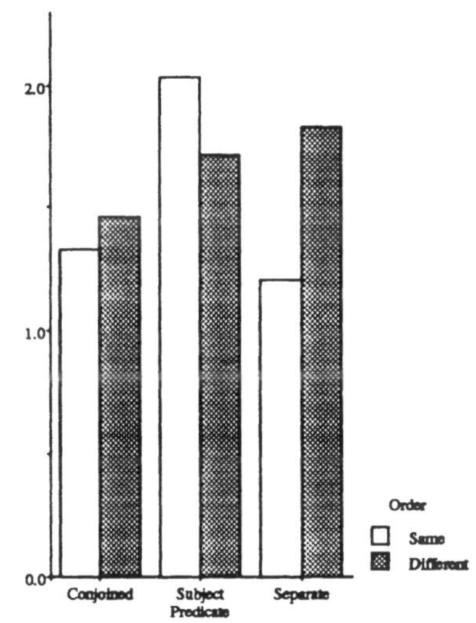

Figure 3: Pronoun assignment bias by Order and Text (N=192).

The interaction between Order and Text appears

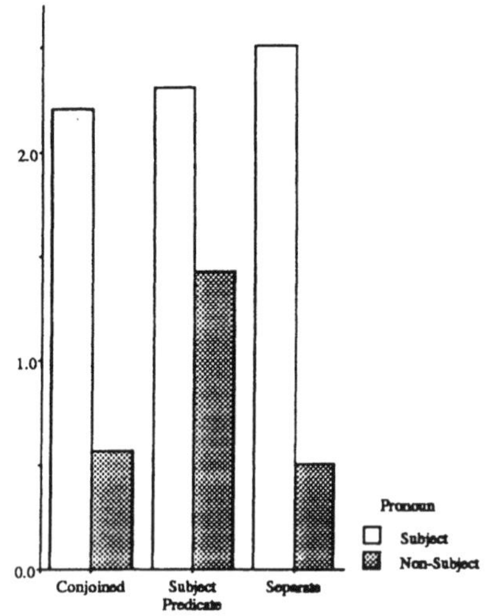

Figure 4: Pronoun assignment bias by Pronoun and Text (N=192).

to be due to the effect of Order in the Separate condition. Analysis of simple effects shows that there is no effect of Order in the Conjoined and Subject-Predicate texts and a marginally significant effect of Order in Separate texts, where the subject assignment bias is greatly increased for non-subject pronouns relative to subject pronouns. ($F_1(1,141) = 6.86, p < 0.01$, $F_2(1,69) = 5.58, p < 0.03$, $minF'(1,171) = 3.08, p > 0.05$) The interaction between Pronoun and Text seems to be caused by the effect of Text on Non-Subject pronouns rather than Subject pronouns, as in the analysis over all the levels of Text. Simple effects analysis shows that at the Subject level of Pronoun there is no effect of Text and a marginally significant effect at the Non-Subject level ($F_1(2,141) = 5.09, p < 0.008$, $F_2(2,69) = 4.49, p < 0.02$, $minF'(2,177) = 2.39, p > 0.05$). Thus, the effect of text type on the assignment of non-subject pronouns is apparent even when the Single Sentence condition is excluded from the analysis.

Reading times The reading time data were not amenable to ANOVA because the cell frequencies were unbalanced and many cells were empty. However, the overall mean reading times for subject and non-subject pronouns when assigned to subject and non-subject antecedents are shown in Table 3. The means are presented separately for parallel order sentences and grammatically parallel sentences as was done for the assignment data in Figure 2. (The pattern of results was similar in the four Text conditions.) Inspection of the Table reveals a clear subject assignment effect: both subject and non-subject pronouns are read most rapidly when the

	Pronoun	
Assignment	Subject	Non-subject
	Parallel Order Sentences	
Subject	2403	2269
Non-subject	2462	2397
	Grammatically Parallel Sentences	
Subject	2314	2338
Non-subject	2545	2475

Table 3: Mean reading times (msec) by Assignment, Pronoun and Parallelism (Order-of-mention and Grammatical).

pronoun is assigned to the subject of the sentence. Further, with the subject pronouns only, the subject assignment effect is much larger in grammatically parallel targets than in parallel order targets.

Discussion

The results show a large subject assignment bias in both subject and non-subject pronouns when preceded by a text, but negligible bias in non-subject pronouns when presented in isolated sentences. The result that both types of pronouns show a subject assignment bias when preceded by a text is consistent with Crawley, Stevenson and Kleinman's (1990) result.

However, in all the Text conditions, the subject assignment bias was reliably greater with subject than with non-subject pronouns. We propose therefore that two strategies are operating: a subject assignment strategy and a grammatical parallelism strategy, and that the consequences of the operation of these two strategies are different for subject and non-subject pronouns. For subject pronouns the two strategies produce the same outcome, hence the marked subject assignment bias. For non-subject pronouns, the two strategies conflict, hence the reduced subject assignment bias.

Textual factors appear to influence the likelihood that a subject assignment will be made with non-subject pronouns, since there was negligible assignment bias when the target sentences were presented in isolation. This presumably reflects the fact that the two strategies have comparable weightings in isolated sentences and so either strategy may be used. However, when the target is preceded by text then other factors, such as topicalisation, may increase the weighting of the subject assignment strategy, since the subject of the sentence is also likely to be the topic of the text. Such an account is compatible with Crawley, Stevenson and Kleinman's (1990) results. Because they used only non-subject pronouns they were unable to distinguish between subject assignment effects and effects due to grammatical parallelism.

In the Separate Text conditions, subject assignment was more marked when the potential antecedents were introduced into the text in a different order from their order in the target sentence (see Figure 3). This may be explained by considering topic effects more closely. Many investigators (*e.g.*, Caramazza & Gupta, 1979) have claimed that the surface subject of a sentence is topicalised and that a new subject in a new sentence introduces a new topic (*e.g.*, Karmiloff-Smith, 1985). Therefore, when the Order is the Same in a Separate text, the second introduced individual (which will be the non-subject antecedent in the target) becomes the most recent topic. This means that if a topic assignment strategy is used to resolve the pronoun, as well as a subject assignment strategy, then the two strategies will disagree (grammatical parallelism has no impact because the interaction between text and order was not modified by a three way interaction with pronoun). However, when the Order is Different the second mentioned individual, which will become the topic, is the subject antecedent in the target. Therefore the two strategies (topic assignment and subject assignment) will agree. The presence or absence of congruence between the two strategies, therefore, may account for the order effect in the Separate Text condition.

This argument does not apply to the other two Text conditions because in those two conditions the individual introduced in the second context sentence is a third individual who is never a potential antecedent of the pronoun. Thus, the effects of the most recent topic that were observed in the Separate condition, are not apparent here and in addition an effect of the initial topic is presumably attenuated.

A subsidiary aim was to investigate text-level order-of-mention parallelism by manipulating Order. However, we have seen that the interaction that was observed between Text and Order is most likely due to topic effect. Thus we have no strong evidence for a text-level order-of-mention parallelism. On the other hand, there does seem to be evidence of order-of-mention parallelism within a single sentence. Figure 2 shows that there is a parallelism effect in both the grammatically parallel targets and the parallel order targets. In the parallel order targets grammatical parallelism cannot be used. Therefore we conclude that the parallel effects in these cases must be due to a parallel order-of-mention strategy. Indeed, the parallelism effect appears to be greater in the grammatically parallel targets, which is what would be expected since the two strategies coincide in these sentences. The reading time data are consistent with these notions. The reading time advantage for subject pronouns is greatest when both order and grammatical parallelism are present. The absence of a similar advantage for non-subject pronouns is presumably due to the precise nature of the processes involved when the outcomes of subject assignment and parallel function conflict.

Conclusion

The results support the existence of a parallel assignment strategy and a subject assignment strategy operating in parallel. There is an effect of text-level ordering which we interpret as a topic effect. There is also an effect of sentence-level ordering which appears to reflect the use of a parallel order-of-mention strategy.

References

Broadbent, D. E. 1973. *In Defence of Empirical Psychology.* London: Methuen.

Caramazza, A. and Gupta, S. 1979. The roles of topicalization, parallel function and verb semantics in the interpretation of pronouns. *Linguistics* 17:497–518.

Clancy, P. M. 1980. Referential choice in English and Japanese narrative discourse. In Chafe, W. L. ed. *The pear stories*, Volume 3: *Advances in discourse processes.* Norwood, N.J.: Ablex.

Cowan, J. R. 1980. The significance of parallel function in the assignment of anaphora. In Kreiman, J. and Ojeda, A. E. eds. *Papers from the parasession on pronouns and anaphora.* Chicago, Illinois: Chicago Linguistics Society.

Crawley, R. A., Stevenson, R. J. and Kleinman, D. 1990. The Use of Heuristic Strategies in the Comprehension of Pronouns. *Journal of Psycholinguisitc Research* 19:245–264.

Karmiloff-Smith, A. 1985. Language and Cognitive Processes from a Developmental Perspective. *Language and Cognitive Processes* 1:61–85.

Stenning, K. 1991. Binding attributes to individuals in terms of sequential information. In Second Winter Text Conference, Jackson's Hole, Wyoming, January, 1991.

Constraints on Models of Recognition and Recall imposed by Data on the Time Course of Retrieval

Peter A. Nobel and Richard M. Shiffrin
Indiana University
Bloomington, IN 47405
pnobel@ucs.indiana.edu
shiffrin@ucs.indiana.edu

Abstract

Reaction time distributions in recognition conditions were compared to those in cued recall to explore the time course of retrieval, to test current models, and to provide constraints for the development of new models (including, to take an example, the class of recurrent neural nets, since they naturally produce reaction time predictions). Two different experimental paradigms were used. Results from a free response procedure showed fundamental differences between the two test modes, both in mean reaction time and the general shape of the distributions. Analysis of data from a signal-to-respond procedure revealed large differences between recognition and recall in the rate of growth of performance. These results suggest the existence of different processes underlying retrieval in recognition and cued recall. One model posits parallel activation of separate memory traces; for recognition, the summed activation is used for a decision, but for recall a search is based on sequential probabilistic choices from the traces. Further constraining models was the observation of nearly identical reaction time distributions for positive and negative responses in recognition, suggesting a single process for recognition decisions for targets and distractors.

Introduction

Neural net and connectionist models have focused more on storage and representations than on retrieval, yet the number of retrieval modes, and the nature of each, is of crucial importance to modelers of memory. In the present research we explore the time course of retrieval in order to ask whether recognition and cued recall are carried out by similar mecha-nisms (in studies that match the response time requirements of the two tasks), to ask whether positive and negative recognition responses are the result of a single process, and to explore the dynamics of memory access. The issue is of con-temporary interest given that many neural net models, particularly of the recurrent variety, provide natural response time predictions.

In a typical long-term memory experiment, subjects are presented during a study phase with a list of items that has to be remembered. In a recall test phase, subjects have to generate the items of the previous studied list in either a random order, i.e. free recall, or a fixed order denoted by the presentation of cues, i.e. cued recall. In a long-term recognition test phase, subjects are presented with words that were either on the list (targets), or that are new (distractors). The subject's task is to identify a word as "old" or "new".

Recognition and recall are improved (for both reaction times and accuracy) by increased study time (see e.g. Ratcliff & Murdock, 1976), decreased list length (see e.g. Roberts, 1972), and lessened delay and/or shortened distractor task between study and test. However, the possibility of different retrieval mechanisms for the two tasks is heightened by several other findings: 1) The use of words having higher natural language frequency increases recall, but decreases recognition (see Hall, 1979). 2) With instructions for maintenance rehearsal, recognition improves (Glenberg & Adams, 1978), but recall is not much affected (see e.g. Dark & Loftus, 1976). 3) Strengthening some list items (by extra study time or extra repetitions) harms the free recall of other items, slightly reduces cued recall of other items, and has no effect on, or even slightly

This research was supported by grant AFOSR90-0215 to the Institute for the Study of Human Capabilities, and by grants NIMH 12717 and AFOSR 870089 to the second author.

helps, the recognition of other items (Ratcliff, Clark, & Shiffrin, 1990).

Memory models

Models that assume the same processes to underlie recall and recognition predict (in their simplest form) the same reaction time distributions, or at least the same shaped distributions, for the two conditions. Models that assume different processes, like the Search of Associative Memory (SAM) model (Gillund & Shiffrin, 1984; Raaijmakers & Shiffrin, 1981) can predict markedly differing distributions.

In SAM, each item is stored in memory as a separate image. The images contain different kinds of information that is rehearsed and coded together in short-term store (Raaijmakers & Shiffrin, 1981). Items are retrieved from long-term store through the weighted strength of association between retrieval cues and stored images. In particular, a given image's activation is determined by the multiplication of the weighted strengths between each cue and that image.

Recognition involves a global familiarity process, in which familiarity results from a single parallel process of activation of all images. Memory is probed with two cues: the context cue and the tested item. The familiarity of the probe is defined as the activation caused by the two probe cues, which is the sum of the activations of all the memory images. This value is compared to a criterion chosen by the subject, and "yes"-responses are made when the familiarity value is higher than this criterion. Such a model predicts sharply peaked response time distributions and similar distributions for 'old' and 'new' responses.

Recall involves an extended serial search, with two phases: sampling and recovery. Again, memory is probed with context and item cues. The probability of sampling a particular image is its activation strength divided by the sum of the activations of all images. After sampling, the information in the image, which is used for the decision and response, must be recovered. The key is that this process continues over and over until a response is found or the subject gives up. Such a model predicts response times spread out over long time periods, and different distributions for correct recalls, intrusions, and the time to 'give up'.

The Composite Holographic Associative Recall Model (CHARM) (Metcalfe Eich, 1982, 1985) is an example of a model that assumes the same retrieval processes underlying recall and recognition. In CHARM, items are represented as feature vectors and are stored in a convolution memory. If pair A - B is presented, the convolution of vectors A and B (A*B) is a vector; it is added to the convolution of A with itself (A*A), and the convolution of B with itself (B*B), and all three are added to the accumulated memory vector for all studied pairs (if not all pairs ever studied).

There is one retrieval process. It operates by correlating the probe vector with the memory vector. In a recall task, the output of this process, a vector itself, is compared to a separate list of words in memory and the response will be the best match above a certain cut-off of activation. In a recognition task, the dot product of the output of the correlation process with the probe is taken, and a positive response is made if the match is above a criterion. Because CHARM treats recognition the same as recall it does not predict differences in the latency distribution for the time to retrieve the trace; any differences would have to be differences in the post-retrieval processes of matching in recognition, or matching in recall.

Numerous memory models share this property that differences in retrieval time distributions for recognition and recall would have to be due to post retrieval operations; e.g. TODAM (Murdock, 1982), Matrix Model (Pike, 1984), MINERVA (Hintzman, 1988), and various connectionist and neural net models (e.g. McClelland & Rumelhart, 1985).

Reaction Times

The literature concerning reaction time (RT) in long-term memory research is mainly restricted to the recognition paradigm. For example, Ratcliff and Murdock (1976) found increasing RT for both hits and correct rejections as a function of output (test) position, decreasing RT as a function of input (study) position, increasing RT when presentation time increases, decreasing RT when the number of presentations increase, and increasing RT as a function of list length.

Some evidence supporting the notion of a sequential search in free recall was collected by Murdock and Okada (1970): Interresponse times increase in a positively accelerated function as recall progresses, interresponse times were shorter the more words were left to recall (for a fixed output position), and at any given output position the interresponse time is a good predictor of the number of words left to recall.

Thus there are data concerning reaction times in recognition and recall tasks separately; there do not seem to be reaction time data when both tasks are given to the same subjects in similar paradigms. In addition, Ratcliff (1978) has argued that testing of models requires closer looks at the reaction time distributions than their central tendencies. He suggests that at a minimum models should account for the shape of reaction time distributions (in particular their skewness), and specify the relationship between speed and accuracy. Ratcliff and Murdock (1976) in fact fit their observed RT distributions with a convolution of exponential and normal distributions. Ratcliff (1978) then fit his model to the parameters of these fitted distributions.

For these reasons a series of studies was designed, using several methodologies to measure reaction times, looking at the effect of several variables in recognition and cued recall conditions, and measuring the entire reaction time distribution.

Experiments

Ten subjects were presented in the study phase with a list of pairs of high frequency words that had to be remembered. The test phase consisted of either single item recognition or cued recall. In the recognition condition, the subject's task was to say whether the test item was on the list , and in the recall condition the subject's task was to recall the other word of the pair. Varied were list length (10 vs. 20 pairs) and presentation time (2 vs. 6 seconds). In order to equate the demand characteristics of the tasks as much as possible, subjects had to press one of two keys when they recognized or recalled, and press the other key if they did not; in the case of recall, a positive response had to be followed by the typing of the word recalled, allowing us to assess accuracy. Two different response procedures were employed. In the free response procedure, subjects were asked to respond as quickly and as accurately as possible after presentation of the test item. This procedure is commonly employed, but suffers from the possibility that subjects might adopt different strategies (e.g. differing biases to respond quickly) in recognition and recall. In the signal-to-respond procedure , which controls for these strategy differences, the subjects were told not to respond until a signal was given (a tone) and then to respond at once (within 300 ms). The delays until the signal ranged in ten steps from 100 ms to 4500 ms.

Results Bearing on Recognition/ Recall Differences

We give representative results because a complete accounting would literally require hundreds of figures. The demonstrated findings hold for the conditions not shown (unless otherwise stipulated). Figure 1 shows the reaction time distributions for correct recognitions of old words (hits) and for correct recalls: The recall distribution has a larger mean, larger variance, larger skewing, and extends over the entire time course of retrieval.

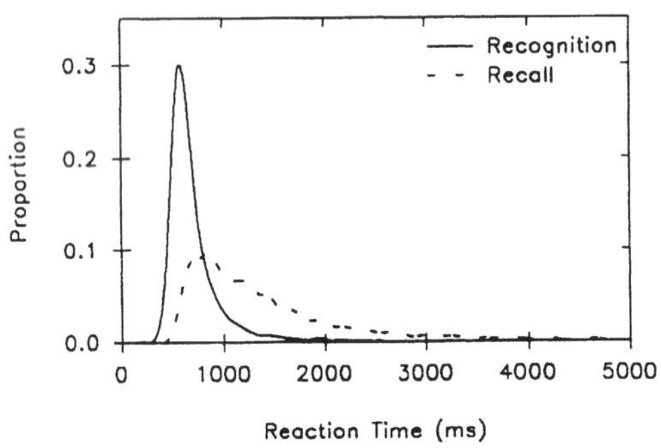

Figure 1. Reaction time distributions for correct recognitions (mean RT=710; st. dev.=299); median=630), and correct recalls (mean RT=1386; st. dev.=769; median=1163).

Figure 2 shows the reaction time distributions for incorrect recognitions of new items (false alarms), and for recalls of list items from other pairs, or, less commonly, non-list items (all termed intrusions): false alarms in recognition have a relatively low mean reaction time and variance, whereas intrusions in recall seem to have an almost uniform distribution.

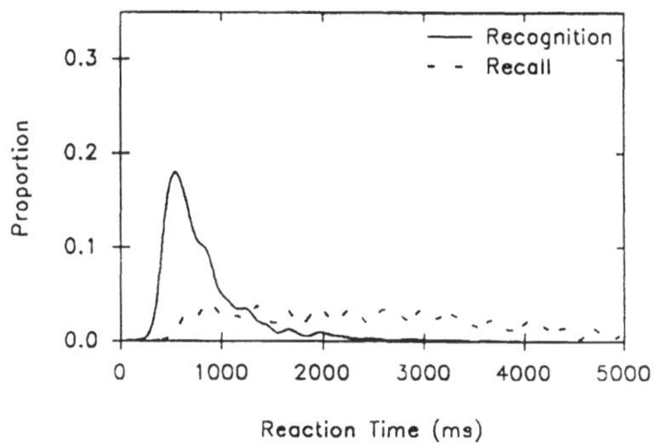

Figure 2. Reaction time distributions for false alarms (mean RT=818; st. dev.=439; median=689), and intrusions (mean RT=2381; st. dev.=1142; median=2320).

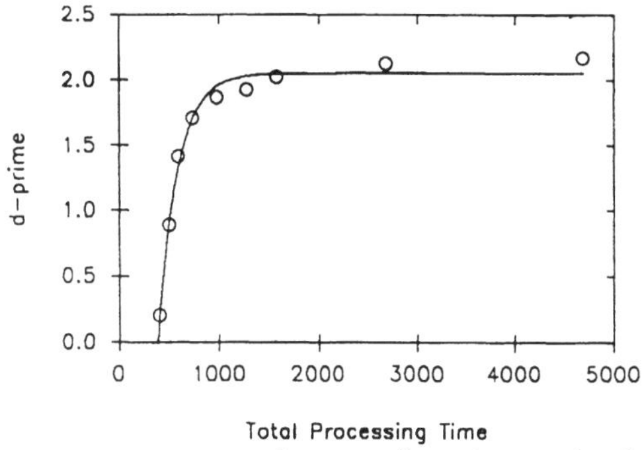

Figure 3. Recognition performance , *d'*, as a function of total processing time in ms (λ=2.06; β=.00503; 1/β=199; δ=373).

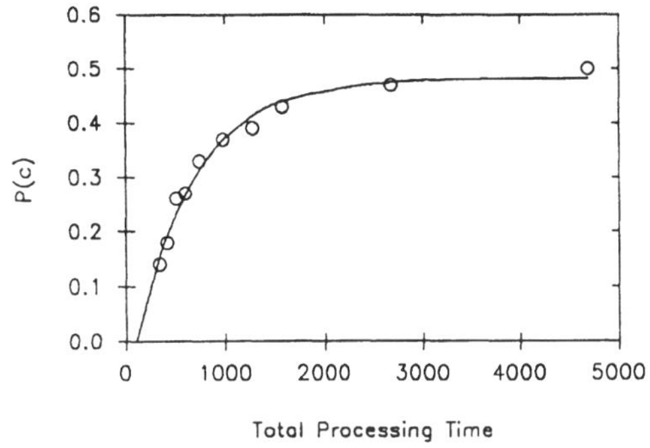

Figure 4. Recall performance, *P(c)*, as a function of total processing time in ms (λ=.48; β=.00161; 1/β=621; δ=46).

The signal-to-respond results can be used to assess the possibility that these large differences might be due to differing strategies or biases in recognition and recall. Presumably the subject will respond with whatever information is available at the time of the signal, whether recognition or recall is being tested. This procedure produces data of a somewhat different sort: The growth of accuracy is measured as a function of the signal delay.

Examination of typical retrieval functions for recognition memory shows an initial period of chance performance, followed by a period of rapid increases in accuracy, and finally, as retrieval time is further increased, accuracy reaches asymptote (see e.g. Dosher, 1984). These functions can be described by an exponential approach to an asymptote with 3 parameters: an asymptotic accuracy parameter that reflects memory information limitations, an intercept (at which point accuracy first rises above chance), and a rate of rise from chance to asymptote. The dynamics of retrieval is summarized by the intercept and the rate parameter. This results in a description of the level of performance, *d'* for recognition and *P(c)* for recall, as a function of total processing time; i.e. delay-of-signal plus response time.

Figures 3 and 4 show performance (observed and predicted) as a function of total processing time for recognition and recall respectively, along with the best fitting exponential functions ($d'(t)$, or $P(c,t)=\lambda(1-\exp[-\beta(t-\delta)]$, for $t-\delta>0$, and 0 elsewhere; in which λ is the asymptote, β the rate, and δ the intercept). It is clear that processes underlying the dynamics of retrieval are quite different: Performance in recognition is characterized by a very fast rate of growth and asymptotic performance is reached fairly quickly, whereas recall performance shows a much more gradual approach to asymptote. These differences are reflected in the parameters of the best fitting functions.

Such results are generally consistent with a two process view of retrieval, such as the SAM model, in which the recall process is spread out in time. The unitary retrieval models would have to posit a difference in post-retrieval mechanisms to explain the recognition-recall differences. For example, in many models noisy information is retrieved from memory (in both recall and recognition). In recall, the process of generating a given word from the noisy information might take a highly variable amount of time, whereas in recognition the time might be relatively fixed (because only a match of the retrieved trace to the test item is needed). In such models it would be necessary to develop a model of post-retrieval response generation that can produce very large response time differences. We are currently carrying out empirical tests contrasting the retrieval time and the post-retrieval time hypotheses, but do not yet have the results.

Results Bearing on Target/ Distractor Differences

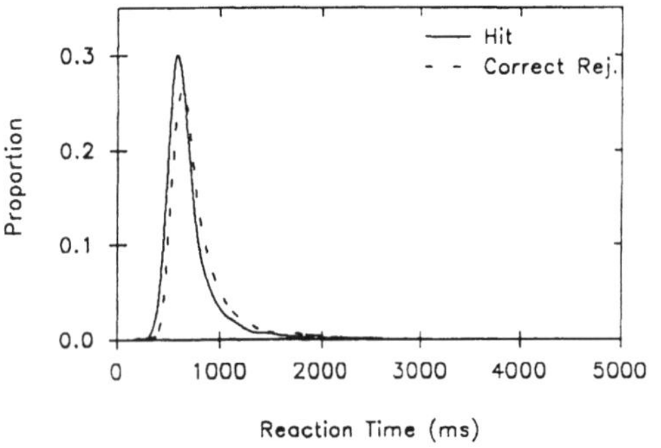

Figure 5. Reaction time distributions for hits (mean RT=710; st. dev.=299; median=630), and correct rejections (mean RT=792; st. dev=334; median=695).

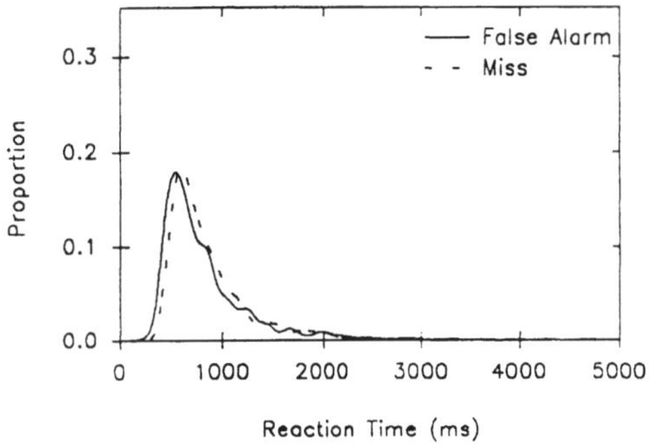

Figure 6. Reaction time distributions for false alarms (mean RT=818; st. dev.=439; median=689) and misses (mean RT=870; st. dev.=458; median=739).

Returning to the free response data, we consider the distributions for positive and negative responses in recognition (Figures 5 and 6). When the responses are correct (hits and correct rejections), the distributions show small differences in both the means and the shape. When these are incorrect (false alarms and misses), the distributions differ slightly in their means, but are identical in shape. Models that use quite different processes for targets and distractors might find such data difficult to predict. On the other hand, careful theoretical work is needed to verify constraints such data place on models. For instance, in the Resonance Model (Ratcliff, 1978) it is assumed that a probe is encoded and then compared in parallel with each item in memory. Each individual comparison is done by a random walk process, and a positive decision is made when any of the parallel comparisons terminates with a match (self-terminating search), and a negative decision is made when all the comparisons terminate with a nonmatch (exhaustive search). With appropriate auxilliary assumptions he was able to show that the model could predict hit and correct rejection distributions that are at least reasonably similar in form.

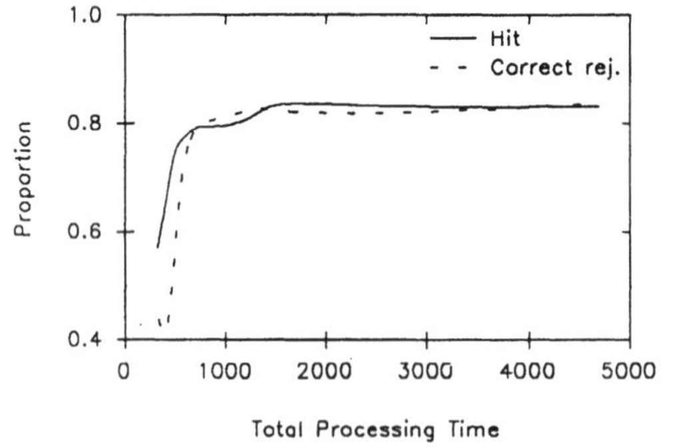

Figure 7. Accuracy growth curves for hits and correct rejections as a function of total processing time (in ms).

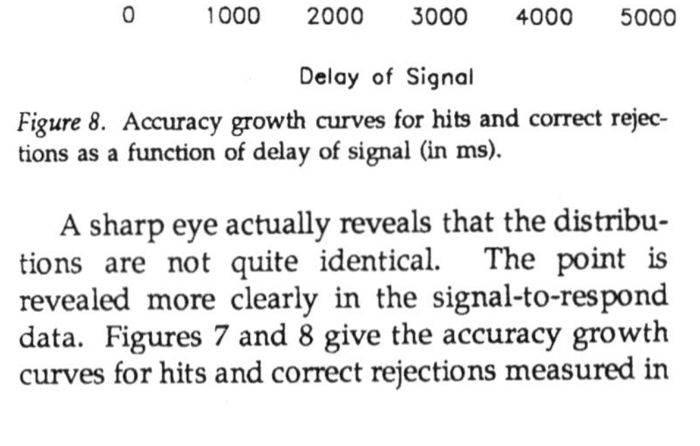

Figure 8. Accuracy growth curves for hits and correct rejections as a function of delay of signal (in ms).

A sharp eye actually reveals that the distributions are not quite identical. The point is revealed more clearly in the signal-to-respond data. Figures 7 and 8 give the accuracy growth curves for hits and correct rejections measured in two ways (in Figure 7, the abscissa includes both the time until the signal and the subsequent time needed to respond). Both methods show that hits start rising sooner than correct rejections, and come together soon thereafter. More research is needed to assess whether this difference is due to a bias to respond 'old' under speed stress, or is due to a real processing difference. Whichever is the case, the remarkable similarities of the target and distractor distributions, and target and distractor signal-to-respond curves, provide strict and informative constraints for models of retrieval.

General Conclusions

We have presented experimental data bearing on the time course of retrieval in both recognition and cued recall, using RT distributions for free response tasks, and accuracy growth curves in signal-to-respond tasks. The large differences between recognition and recall suggest the existence of distinct processes underlying retrieval in the two paradigms. However, we are carrying out further experiments to see whether the differences can be explained in terms of a post-retrieval "clean-up" process in recall (e.g. Metcalfe Eich, 1982). In addition, targets and distractors have nearly identical RT distributions, and fairly similar accuracy growth curves. This suggests a single process for recognition judgments for targets and distractors (such as summation of activation in SAM), and provides general constraints for future model development.

References

Dark, V. J., & Loftus, G. R. (1976). The role of rehearsal in long-term memory performance. *Journal of Verbal Learning and Verbal Behavior, 15,* 479-490.

Dosher, B. A. (1984). Degree of learning and retrieval speed: Study time and multiple exposures. *Journal of Experimental Psychology: Learning, Memory, and Cognition, 10,* 541-574.

Gillund, G., & Shiffrin, R. M. (1984). A retrieval model for both recognition and recall. *Psychological Review, 91,* 1-67.

Glenberg, A., & Adams, F. (1978). Type I rehearsal and recognition. *Journal of Verbal Learning and Verbal Behavior, 17,* 455-463.

Hall, J. F. (1979). Recognition as a function of word frequency. *American Journal of Psychology, 92,* 497-505.

Hintzman, D. L. (1988). Judgments of frequency and recognition memory in a multiple-trace memory model. *Psychological Review, 95,* 528-551.

McClelland, J. L., & Rumelhart, D. E. (1985). Distributed memory and the representation of general and specific information. *Psychological Review, 85,* 159-188.

Metcalfe Eich, J. (1982). A composite holographic associative recall model. *Psychological Review, 89,* 627-661.

Metcalfe Eich, J. (1985). Levels of processing, encoding specificity, elaboration, and CHARM. Psychological Review, 92, 1-38.

Murdock, B. B., Jr. (1982). A theory for the storage and retrieval of item and associative information. *Psychological Review, 89,* 609-626.

Murdock, B. B., Jr., & Okada, R. (1970) Interresponse time in single-trial free recall. *Journal of Experimental Psychology, 86,* 263-267.

Pike, R. (1984). Comparison of convolution and matrix distributed memory systems for associative recall and recognition. *Psychological Review, 91,* 281-294.

Raaijmakers, J. G. W., & Shiffrin, R. M. (1981). Search of associative memory. *Psychological Review, 88,* 93-134.

Ratcliff, R. (1978). A theory of memory retrieval. *Psychological Review, 85,* 59-108.

Ratcliff, R., Clark, S., & Shiffrin, R. M. (1990). The list-strength effect: I. Data and discussion. *Journal of Experimental Psychology: Learning, Memory, and Cognition, 16,* 163-178.

Ratcliff, R., & Murdock, B. B., Jr. (1976). Retrieval processes in recognition memory. *Psychological Review, 83,* 190-214.

Roberts, W. A. (1972). Free recall of word lists varying in length and rate of presentation: A test of total-time hypotheses. *Journal of Experimental Psychology, 92,* 365-372.

Tulving, E., & Watkins, M. J. (1973). Continuity between recall and recognition. *American Journal of Psychology, 86,* 739-748.

A Model of Knowledge-Based Skill Acquisition

Stellan Ohlsson and Ernest Rees
Learning Research and Development Center
University of Pittsburgh
Pittsburgh, PA 15260, U. S. A.
stellan@vms.cis.pitt.edu

Abstract

We hypothesize that two important functions of declarative knowledge in learning is to enable the learner to detect and to correct errors. We describe psychologically plausible mechanisms for both functions. The mechanisms are implemented in a computational model which learns cognitive skills in three different domains, illustrating the cognitive function of abstract principles, concrete facts, and tutoring messages in skill acquisition.[1]

Practice and Knowledge

Practice consists of repeated attempts to solve problems which stretch the learner's competence. The paradox of practice is that the learner is deliberately setting out to solve a problem which he or she knows is beyond his or her current competence. It is far from obvious how this produces learning; and yet, there is no evidence that skills can be acquired without practice.

Beginning with the seminal papers by John R. Anderson and co-workers (e. g., Anderson & Kline, 1979) and by Anzai and Simon (1979), a computational interpretation of learning from practice has been developed over the past fifteen years. It can be summarized in three hypotheses:

(a) *The Weak Method Hypothesis.* When the learner is faced with a problem beyond his or her current competence, he or she uses weak problem solving methods such as analogical inference, forward search, or means-ends analysis to generate task relevant, but possibly inefficient, actions.

(b) *The Memory Storage Hypothesis.* Actions, even inefficient actions, generate information about the task environment, e. g., information about the effects of actions and about the properties of objects. This information is stored in memory.

(c) *The Skill Induction Hypothesis.* One or more learning mechanisms (composition, subgoaling, etc.) revise the current skill on the basis of the information in memory.

Repeated cycles through (a), (b) and (c) result in a domain-specific adaptation of the weak method which can solve the (class of) practice problem(s) efficiently. The three hypotheses can be articulated in different ways to generate a wide variety of specific learning models (see Klahr, Langley, & Neches, 1987).

[1] Preparation of this paper was supported by grant No. N00014-89-J-1681 from the Cognitive Science Program of the Office of Naval Research. The opinions expressed are not necessarily those of the funding agent and no endorsement should be inferred.

The empirical status of this mental bootstrapping theory of practice is still an open question, although its most dubious assumption--that people store large amounts of information in memory while engaged in the capacity demanding process of solving problems--has survived at least one attempt at falsification (Ohlsson, 1991).

The major limitation of the theory is that it depicts procedural knowledge as a closed system: Problem solving skills beget problem solving steps, which in turn beget new problem solving skills. There is no point along this loop at which prior knowledge about the task environment can influence the construction of the new skill. However, humans always learn in the context of prior domain knowledge.

A more complete theory of practice must describe how skill acquisition is influenced by at least three types of knowledge items: abstract principles, concrete facts, and tutoring messages.

Abstract principles are common in mathematics and science. The principle of one-one mapping is a simple example. It plays a crucial role in learning how to count a set of objects (Gelman & Gallistel, 1978). The laws of conservation of mass and energy are examples of principles in science.

Concrete facts are important in both technical domains and in everyday life. The fact that alcohol molecules are characterized by an OH-group is useful when constructing structural formulas in organic chemistry (Solomon, 1988).

Tutoring messages are short verbal instructions, uttered during practice. "Don't borrow unless the minuend digit is smaller than the subtrahend digit," uttered in the context of practice on subtraction with regrouping, is an example. One-on-one tutoring is a very efficient form of instruction (Bloom, 1984).

The purpose of this paper is to describe a computational model which embodies a unified view of the function of abstract principles, concrete facts, and tutoring messages in skill acquisition.

Learning as Error Correction

By necessity, a novice makes many errors while executing a skill; by definition, mastery is characerized by the absence of errors; hence, the gradual increase of competence during practice consists in the successive elimination of errors. Each error provides an opportunity to learn how to avoid similar errors in the future. To make use of such a learning opportunity, the learner must be able to (a) detect that an error as occurred, and (b) compute the appropriate revision of the current skill. We propose that the function of domain knowlege in skill acquisition is precisely to enable the learner to detect and to correct errors.

Learners can detect errors in three different ways: by self-monitoring, by observing the environmental effects, and by being told by others (Reason, 1990, Chap. 6). We focus on the first of these three methods. Learners monitor themselves, we suggest, by testing each new cognitive result (inference or knowledge state) for consistency with prior knowledge about the domain. For example, a Pittsburgh driver who is driving towards the river from the airport on the back roads and who sees a sign saying "route 60 south" recognizes that he or she has made an error, at least if he or she knows that the river is north of the airport. A chemistry student who gets more mass out of an experiment than he or she put in recognizes that an error was made, because this violates the law of conservation of mass. In each instance, detecting the error requires prior knowledge about the domain. During deliberate learning we constantly monitor the situations (problem states) we create for consistency with what we know about the domain and we recognize inconsistencies as errors. The more knowledge, the more powerful the self-monitoring ability.

Learners can correct an error, we suggest, by determining the conditions that produced it and then revising the current skill so as to prevent the relevant action from applying under those conditions. For example, the bewildered Pittsburgh driver might try to figure out which turn was wrong and correct his or her driving accordingly. The identification of the conditions that produced the error will result in a restriction on the relevant action, e. g., "remember not to turn right after exiting the parkway at Clairton."

In summary, according to our theory the learner monitors himself or herself by testing the consistency of each new conclusion or problem state with prior knowledge. Inconsistencies reveal errors which in turn trigger revisions which prevent those errors from occurring in the future. Over the course of practice, the errors are successively eliminated. The new skill has been mastered when no further errors occur.

A Simulation Model

A computational model that instantiates our theory must have (a) a performance component, (b) a representation for prior knowledge, (c) a mechanism for detecting errors, and (d) a mechanism for correcting errors. Our model is called the *Heuristic Searcher* (HS).

Performance component. HS is a vanilla flavored production system language. Rules have a goal and a conjunction of situation features in their left-hand sides and a single problem solving operator in their right-hand sides. Hence, each step in the problem space is controlled by a single production rule. There is no conflict resolution. If more than one rule fires, multiple new knowledge states are created. The system executes best-first search if supplied with an evaluation function by which to rank problem states and either depth-first or breadth-first search otherwise. HS is not an hypothesis about the human cognitive architecture. Our theoretical committment is limited to the two assumptions that cognitive skills are encoded as sets of production rules and that people at least sometimes solve unfamiliar problems through forward search. Both hypotheses are strongly supported by data (Anderson, in press; Newell & Simon, 1972).

Knowledge representation. Prior domain knowledge is encoded in data-structures called *state constraints.* Syntactically, a state constraint is an ordered pair <Cr, Cs>, where Cr and Cs are patterns, i. e., conjunctions of properties similar to the condition sides of production rules. The *relevance pattern* Cr has to match the current knowledge state for the constraint to be relevant and the *satisfaction pattern* Cs has to match for the constraint to be satisfied. States in which Cr match but Cs does not are called *constraint violations*. For example, a fact like "Fifth Avenue is one-way in the easterly direction" would be encoded as "if vehicle X is moving along Fifth Avenue, X had better be going east". Vehicles not moving along Fifth Avenue are not subject to the constraint; a vehicle going east on Fifth Avenue conforms with the constraint; a vehicle going west constitutes a constraint violation. Constraints are not inference rules or operators. They do not generate new conclusions or revise knowledge states. They test whether certain properties are true of the current knowledge state.

Error detection. When the current rule set generates a new knowledge-state, the latter is matched against all state constraints with the same pattern matcher that matches the rule conditions. Constraints in which Cr does *not* match are ignored, as are those in which *both* Cr and Cs match. Neither class of constraints warrant any action on the part of the system. Constraints for which Cr does match but Cs does not are recorded as violated. A constraint violation signals an inconsistency between the system's prior knowledge about the domain and the new outcome generated by the current rule set and it is interpreted as an error. HS assumes that the rule set is at fault.

Error correction. HS assigns blame to the last rule that fired, i. e., the rule that produced the violating knowledge state. A faulty rule is revised in two different ways. First, the relevance pattern by itself is regressed through the rule with a version of the standard goal regression algorithm (Nilsson, 1980) and the *negation* of the result added to the condition side of the rule. This produces a rule that only applies in situations in which the constraint is irrelevant. Second, the entire constraint is regressed through the rule and the result added to the rule condition (without negating it). This produces a rule which only applies in situations in which the constraint is ensured to be satisfied.

Curing a rule from violating one constraint does not garantee that the rule is correct; it might still violate other constraints. Multiple revisions of a rule are common in HS' learning. Because a skill consists of large number of rules, each of which might need multiple revisions, skill acquisition is necessarily gradual.

Three Applications

Three applications of HS have been implemented to date. They are summarized in Table 1. They illustrate that the model can learn from each of the three types of knowledge items specified previously: abstract principles, concrete facts, and tutoring messages.

Learning from abstract principles. Developmental data indicate that children construct the skill of counting a set of objects on the basis of (implicit) knowledge of abstract counting principles (Gelman & Gallistel, 1978). The main supporting phenomena are that children can transfer their counting routines to non-standard counting tasks and that they can evaluate counting performances that they cannot produce (Gelman & Gallistel, 1978; Gelman & Meck, 1986). The counting principles are abstract ideas like the one-one mapping principle. Expressed as a state constraint, this principle becomes "if object X has been assigned object Y, there should not be a third object Z assigned to either X or Y". The HS model learns the correct counting skill if given state constraints

Table 1. Three applications of the HS model.

Problem domain	Type of knowledge given to the model	Skill acquired by the model
Counting	Abstract principles, e. g. the one-one mapping principle.	To count a set of objects (see Ohlsson & Rees, 1991a).
Chemistry	Concrete facts, e. g. that alcohol molecules have OH-groups.	To derive the Lewis structure for a given molecular formula (see Ohlsson, in press-a).
Subtraction	Tutoring messages, e. g., "don't regroup unless the subtrahend is larger than the minuend."	Subtraction with regrouping (see Ohlsson, Ernst, & Rees, in press).

corresponding to the counting principles (Ohlsson & Rees, 1991a) and it can transfer the learned skill to other counting tasks (Ohlsson & Rees, 1991b).

Learning from concrete facts. The skill of constructing a Lewis structure on the basis of the molecular formula is a routine scientific skill taught at the begining of most organic chemistry courses. Textbooks teach this skill by first stating a general but weak procedure and then providing practice problems (e. g., Solomons, 1988). The general procedure is inefficient and must be specialized to particular classes of molecules. We gave HS a version of the general skill and provided it with state constraints expressing facts about three classes of molecules (alcohols, hydrocarbons, and ethers). An example of a fact is that alcohols have an OH-group ("if this is an alcohol molecule, it had better have an OH-group"). The model learned specialized versions of the general procedure for each type of molecule and its learning exhibited the negatively accelerated curve typical of human skill acquisition; see (Ohlsson, in press-a) for a more detailed discussion of these results.

Learning from tutoring messages. Students typically need tutoring to acquire the correct procedures for place value arithmetic. We gave HS an initial rule set which could solve canonical subtraction problems, i. e., problems in which the subtrahend digit is always smaller than the minuend digit in the same column. We then tutored the system through the learning of the regrouping procedure. The state constraints encoded typical tutoring messages, e. g., "don't borrow unless the minuend digit is smaller than the subtrahend digit." The predictions from this simulation experiment contradicted the current wisdom that regrouping is easier to learn than alternative subtraction methods; see Ohlsson (in press-b) and Ohlsson, Ernst & Rees (in press) for a detailed discussion of the results.

Discussion

Skill acquisition always occurs within the context of the learner's prior knowledge about the domain. Models of skill acquisition must explain the interaction between prior knowledge and problem solving experience during practice. We suggest that the cognitive function of domain knowledge is to enable the learner to monitor his or her own performance. The more domain knowledge he or she has, the better he or she can detect and correct errors.

The simulation model we built around this hypothesis learns in three different domains which supports the sufficiency and the generality of the learning mechanism. The model suggests new perspectives on three traditional problems in the theory of procedural learning. First, it predicts negatively accelerated learning curves, because the number of learning opportunities per practice trial will decrease as more and more errors are corrected. Second, it predicts low transfer of training between domains, because generality resides in the declarative knowledge and not in the skill itself. Finally, the model is consistent with the fact that one-on-one tutoring is the most efficient form of instruction, because tutors operate by helping the learner with the two main functions postulated in the model, i. e., to detect and correct errors.

The empirical validity of a complex simulation model is difficult to assess. The derivation of quantitative predictions from a computational model is tricky, because the model's behavior is determined not only by the hypotheses behind it but also by implementation details. Also, different models are seldom applied to the same phenomena, due to differences in the interests of their creators, making comparative evaluations difficult. No strong claims for the empirical validity of HS can be made at this time.

The theory behind HS is similar in spirit to the theory proposed by Schank (1986). According to the latter, people understand events by generating expectations from their current knowledge and they revise their knowledge when their expectations fail. Expectation failures and constraint violations are obviously similar types of events. Schank's theory is focussed on the understanding of other agents' actions rather than on problem solving and it represents knowledge in explanation patterns instead of rules, but the two theories share the hypothesis that learning is a response to an inconsistency between a cognitive outcome and existing knowledge.

This hypothesis might ultimately be undermined by empirical data. However, the problem of the interaction between prior knowledge and experience during practice will not go away. It must be solved before we can claim to fully understand skill acquisition.

References

Anderson, J. R. (in press). *Rules of the mind*. Hillsdale, NJ: Erlbaum.

Anderson, J. R., & Kline, P. J. (1979). A learning system and its psychological implications. *Proceedings of the Sixth International Joint Conference on Artificial Intelligence* (pp. 16-21). Tokyo, Japan.

Anzai, Y., & Simon, H. A. (1979) The theory of learning by doing. *Psychological Review, 86*, 124-140.

Bloom, B. (1984). The 2 sigma problem: The search for methods of group instruction as effective as one-to-one tutoring. *Educational Researcher, 13*, 4-16.

Gelman, R., & Gallistel, C. R. (1978). *The child's understanding of number*. Cambridge, MA: Havard University Press.

Gelman, R., & Meck, E. (1986). The notion of principle: The case of counting. In J. H. Hiebert, (Ed.), *Conceptual and procedural knowledge: The case of mathematics* (pp. 29-57). Hillsdale, NJ: Erlbaum.

Klahr, D., Langley, P., & R. Neches, (Eds.), (1987). *Production system models of learning and development* . Cambridge, MA: MIT Press.

Newell, A., and Simon, H. A. (1972). *Human problem solving*. Englewood Cliffs, NJ: Prentice-Hall.

Nilsson, N. (1980). *Principles of artificial intelligence*. Palo Alto, CA: Tioga.

Ohlsson, S. (1991). Memory for problem solving steps. *Program of the Thirteenth Annual Confernce of the Cognitive Science Society* (pp. 370-375). Hillsdale, NJ: Erlbaum.

Ohlsson, S. (in press-a). The interaction between knowledge and practice in the acquisition of cognitive skills. In A. Meyrowitz & S. Chipman, (Eds.), *Cognitive models of complex learning*. Boston: Kluwer.

Ohlsson, S. (in press-b). Artificial instruction: A method for relating learning theory to instructional design. In P. Winne & M. Jones, (Eds.), *Foundations and frontiers in instructional computing systems*. New York, NY: Springer-Verlag.

Ohlsson, S., Ernst, A. M., & Rees, E. (in press). The cognitive complexity of doing and learning arithmetic. *Journal of Research in Mathematics Eduation*.

Ohlsson, S., & Rees, E. (1991a). The function of conceptual understanding in the learning of arithmetic procedures. *Cognition & Instruction, 8*, 103-179.

Ohlsson, S., & Rees, E. (1991b). Adaptive search through constraint violation. *Journal of Experimental and Theoretical Artificial Intelligence, 3*, 33-42.

Reason, J. (1990). *Human error*. Cambridge, MA: Cambridge University Press.

Schank, R. (1986). *Explanation patterns*. Hillsdale, NJ: Erlbaum.

Solomons, T. W. G. (1988). *Organic chemistry* (4th ed.). New York, NY: Wiley.

Problem-Solving Stereotypes for an Intelligent Assistant

Christopher Owens
Department of Computer Science, The University of Chicago
1100 East 58th Street, Chicago, IL 60637
E-mail: `owens@cs.uchicago.edu`

Abstract

This paper examines the role of case-based reasoning in a *problem-solving assistant* system, which differs from an autonomous problem solver in that it shares the problem-solving task with a human partner. The paper focuses on the criteria driving the system designer's (or the system's) choice of cases, of representation vocabulary, and of indexing terms, and upon how the assumption of a human in the problem-solving loop influences these criteria. It presents these theoretical considerations in the context of work in progress on IOPS, a case-based intelligent assistant for airline irregular operations scheduling.

Introduction

While most work on AI problem-solving has been directed towards the goal of building autonomous systems, capable of reasoning independently from an initial problem description to a successful solution, a growing body of work has begun focusing on the practical and scientific role of *intelligent assistants*: systems that do not solve problems autonomously, that instead enter into a problem-solving partnership with a human user.

This paper examines the role of episodic memory and case-based reasoning in the context of an intelligent assistant system. It focuses on a kind of knowledge that either an autonomous problem solver or an intelligent assistant system might embody: knowledge linking the commonly-occurring threats, opportunities, and failures of the problem-solving domain with appropriate responses to those situations. It explains how the assumption of a cooperative, as opposed to autonomous, problem solver changes the functional constraints on which stereotypical situations to represent, how to represent them, and which predictive features to associate with the stereotypes. It demonstrates how the assumption of a human in the problem-solving loop relaxes certain representational requirements and enables a new kind of feature acquisition, but also how it places additional requirements upon the choice and representation of cases or stereotypes. These theoretical considerations are described against the background of our work in progress on IOPS (Irregular Operations Planning System), an intelligent assistant for the task of airline irregular operations scheduling.

Intelligent assistant systems

The goal of an intelligent assistant systems is to assist the human user in detecting, diagnosing, and analyzing problems and in generating, selecting, and implementing solutions. An assistant might help by performing any or all of the following functions:

- The assistant might perform some specific computation, calculation, or inference at the behest of the human problem solver. A trivial example here is an electronic calculator; a less trivial example is a simulator that lets the user predict the results of some action. Note that the user chooses what calculation to perform and when to perform it; the system responds to specific user requests.
- The assistant might store information that the user would otherwise need to memorize, and provide it to the user at the appropriate time. A trivial example is an on-line reference manual, a more complex example is the "ask-" series of systems [Schank, 1991]. Here, the system provides information in response to the user's request; it is up to the human to interpret the relevance of the information to the current situation.
- The assistant might spontaneously advise the human user. If the system has access to a description of the current situation, the system may detect the applicability of one or more of its stored problem-solving strategies, and suggest it or them to the user.
- The assistant might request additional information from the user, prompted by a need to discriminate among competing strategies to apply to the current situation, or by a need to discriminate among competing hypotheses to explain the origin of the current problem. The behavior of medical diagnostic reasoning systems (e.g. [Shortliffe, 1976]) in suggesting

appropriate laboratory tests is typical of this activity.

- The assistant might perform some of the bookkeeping necessary to help the user carry out a plan. If the user selects a particular abstract, high-level problem solving strategy (e.g. repair a schedule failure by substituting one resource for an unavailable one), the system can fill in some of the details (selecting an appropriate resource to substitute, tracking the state of the old and new resource, etc.)

In a more sophisticated system, combinations of these behaviors are possible. The system might, for example, detect the applicability of several of its problem-solving strategies, request additional information to determine which few are most applicable, partially predict the results of implementing each of the strategies, and present the set of choices to the user. Once the user selects one of the strategies, the system can implement it and update its model of the state of the world.

Underlying each and all of these behaviors is the system's critical need to learn new problem classes, repair strategies, and descriptive features as it interacts with the user and acquires more knowledge about the problem-solving domain. Recent work at Chicago [Hammond, 1992] has described a life cycle of "apprentice" to "assistant" to "advisor" as the system acquires knowledge about the domain, and spends less time asking the human partner questions and more time offering advice and suggestions.

Case-based planning

One mechanism for solving problems is by noticing and exploiting the similarities between the current situation and a **case** — either a specific prior experience or a commonly recurring stereotype — selected from memory. (See, e.g. [Alterman, 1986; Bareiss, 1989; Barletta and Mark, 1988; Hammond, 1989; Kolodner and Simpson, 1989; Schank, 1982] for descriptions of typical case-based problem solvers). For the purposes of this paper, the task of an autonomous case-based planner can be described as:

- **Describe** the current problem in terms of the system's indexing vocabulary,
- **Retrieve** from memory a case whose stored problem description matches the description of the current problem,
- **Analyze** the differences between the current situation and the retrieved situation, and
- **Modify** the solution stored with the old problem description to fit the current situation.
- **Store** the new solution with a description of the current problem, placing particular representational focus on the features that differentiate the current situation from the old case retrieved from memory.

A commonly-articulated argument for the case-based approach is that it is particularly appropriate in situations where a system cannot reasonably proceed by chaining through the problem-solving operators that one would expect to find in a complete descriptive theory of the domain — either because a complete domain theory is unavailable, or because access to it is expensive, where not enough is known about the current situation to precisely determine the applicability of the theory's operators, or where the operator space of the domain theory is so large that the computational costs of searching it are prohibitive. Several of these features characterize the airline irregular operations domain, described below.

When dealing specifically with intelligent assistant systems as opposed to autonomous systems, several more arguments come into play:

- **Case-based advice:** Suggesting relevant cases is an effective mechanism for the system to offer advice to the user. Even if the system lacks sufficient inference capabilities to autonomously derive an appropriate sequence of actions by transforming the case to fit the current situation, it can still assist by presenting the case to the user, and exploiting the user's ability to apply the case.
- **Case-based knowledge acquisition:** The task of case retrieval forms a natural mechanism to control the system's requests for information from the user. The system in effect plays a kind of "20 questions" game with the user, asking for additional descriptive information about the current situation only when that information would play a clear role in discriminating among multiple cases, each of which potentially applies.
- **Case-based feature learning:** The systems retrieval failures (i.e. inability to discriminate between different cases) provides an opportunity to acquire new elements of a descriptive vocabulary from the user. The system can ask, in effect, "How do these two cases differ" and the user can provide, and name, a new descriptive feature.

Case-based knowledge acquisition and case-based feature learning are described further below, in the context of our ongoing work on IOPS, which solves problems in the domain of dynamic schedule repair.

IOPS

Airline irregular operations

The problem solving domain of this research is *airline irregular operations scheduling*. An airline wants to meet anticipated passenger demand over the routes it flies with an efficient allocation of its capital and human resources. To this end, it develops an operations schedule: an assignment of aircraft to scheduled flight operations and scheduled maintenance stops, and of crew to flight legs and rest periods. The schedule is carefully optimized to achieve efficient utilization and distribution of aircraft and crew over the airline's routes.

Unfortunately, schedule disruptions due to weather, traffic congestion, unscheduled equipment maintenance, crew illness, or unanticipated requests for charters or other additional flight operations are inevitable but unpredictable. Because of the massive internal interdependencies inherent in an airline schedule, even a small single-point failure, such as an aircraft temporarily delayed for replacement of a burned-out light bulb, could potentially result in a snowballing sequence of downstream delays, disruptions, and missed connections if actions were not taken to mitigate the consequences of the failure.

To deal with these unexpected events, airlines employ operations controllers: experienced individuals whose job it is to monitor the airline's flight operations and to take steps to minimize passenger delay and inconvenience and cost to the airline. The controllers have access to information about the airline's current and planned operations, and knowledge of current and forecast conditions. Based on the information they receive, they order changes to the airline's operating schedule in an attempt to mitigate the effects of unexpected disruptions.

A content theory

An essential set of decisions in the design of an case-based planner revolves around a **content theory** of the domain: determining what cases ought to be put in memory, what descriptive features ought to be part of the system's representational vocabulary, and how the system ought to extract descriptions of new situations so that those descriptions will be useful in determining the applicability of old cases to new situations. Typical criteria for selecting cases and indices are discussed in [Owens, 1991], [Owens, 1990], and [Birnbaum *et al.*, 1989].

Our initial content theory of failure and repair in the irregular operations domain derives from our observation of several experienced operations controllers over multiple sessions as they detect, diagnose, and solve problems. The result of this analysis has been:

- to categorize, to the extent possible, the different **classes of problems** that the controllers are asked to solve. Examples of such categories include:
  - Unscheduled maintenance delay at a hub airport during peak travel time.
  - Weather-induced bottleneck at a non-hub airport.
  - Traffic congestion restricting outbound flights from a non-hub airport.
- to identify the **primitive operators** the controllers have at their disposal. Examples of primitive operators include:
  - Cancel a flight segment
  - Advance or delay the departure time of a flight segment
  - Add an unscheduled stop to a flight, or skip a scheduled stop.
  - Substitute one aircraft for another
  - Divert a flight to a different destination
  - Ferry an empty aircraft from one airport to another.
- to identify the higher-level **strategies** that the controllers use to solve problems. Higher-level strategies are built from sequences of primitive operators, and they appear to address goals such as:
  - Localize a problem: prevent a disruption at one airport from propagating to the rest of the system, e.g. by rerouting flights around the affected airport.
  - Distribute the impact of a problem, e.g. create small delays across the system to avoid a major bottleneck at one airport.
  - Delay the effects of a problem to increase the chance that an opportunistic solution will present itself, e.g. "borrow" an aircraft from a later flight to cover a shortfall on a current one; cover the later flight by borrowing yet a later aircraft, etc.

The importance of this content theory is that it defines the functional criteria for selecting a problem solver's case library, representation vocabulary, and indexing terms. The cases stored in the system should cover the classes of problems that experienced controllers appear to solve. The representation vocabulary should represent the features necessary to detect the applicability of those cases, and the primitive operators involved in the solutions. The indexing terms should be sufficient to discriminate between the existing cases in determining their applicability to new situations.

While this type of content theory is appropriate to the design of an autonomous problem solver, an additional set of criteria come to bear in the design of an intelligent assistant system.

Some representational problems are easier in a system that can count on a human user to help it diagnose and learn. Difficult tasks of detection and situation assessment, for which no good theory exists, can be deferred to the user. On the other hand, the presence of a human in the problem-solving loop also additional demands upon the system and upon its knowledge representation. Not only must the system's case library and descriptive vocabulary cover the range of expected problems and solutions, but it must do so in such a way as to facilitate communication of partial results and explanations to user, and to enable requests for additional information to be made and understood.

The IOPS system

The knowledge gathered from our observation of schedule controllers is being represented as the case library of IOPS, a case-based intelligent assistant for irregular operations scheduling currently under development.

In addition to the case library, IOPS has access to operating data regarding the current and planned state of an actual airline schedule, including the assignment of aircraft to flight legs, information about passenger loads, connections, and destinations, and the current and planned locations and movement of aircraft.

The function of IOPSis to provide the following kinds of assistance to the human problem solver:

- Given a description of a problem, like "Aircraft 2854 will be out of service for 1 hour for unscheduled maintenance", or "Bad weather is expected to close Denver for 4 hours this evening", use the description, plus all the current operating data, to select potentially applicable repair strategies.
- Given a set of potentially applicable repair strategies, ask the user for information that would discriminate among them. This is not only an opportunity to narrow the current search space, it is also an opportunity, as described below, to acquire new descriptive features about the domain.
- Given a repair strategy, selected either by the search process above or by the user, perform the computation necessary to implement it. If, for example, the strategy is "Delay a flight and use its aircraft to fill in for the temporarily unavailable one", then generate a list of potentially acceptable flights to delay.

Memory and learning for an intelligent assistant

A key role for a case memory in an intelligent assistant system is to direct the interaction between the user and the system. This can effect *short-term* knowledge acquisition, in which the system, over the course of a single problem-solving session, asks the user for help in detecting the presence or absence of abstract properties that it itself cannot detect, and in *long-term* knowledge acquisition, in which the system, through structured interaction with the user acquires new descriptive features for subsequent use in representing and indexing cases. In both short-term and long-term knowledge acquisition, A library of prior cases presents a baseline against which new descriptive information can be acquired from the user.

Short-term knowledge acquisition

Presenting potentially applicable cases to the user along with a request for clarification is a powerful mechanism for managing the interaction between the user and the system. Since, in a cooperative problem-solving context the system cedes some of the feature detection responsibility to its human partner, a mechanism is necessary for the system to request the information it needs. The system cannot simply ask the user "Tell me something about the current situation" — the question is too open-ended.

On the other hand, if the system has a partial description of the current situation, it can use its case library to request additional information. It can accomplish this by retrieving cases that match the current situation, comparing them with each other to identify descriptive features that, if their presence or absence could be determined in the current situation, would discriminate among the potentially-matching cases. The system can then ask the user about the missing features.

For an example from the airline irregular operations scheduling domain, consider a system trying to repair a schedule whose partial description indicates that 5 off-peak flights inbound to a hub airport are reported delayed by over 30 minutes each. The system, searching its library of cases matching these descriptors, finds multiple possible matches. One feature that discriminates between the matches is that some of the prior cases cover situations in which the weather was deteriorating and others covered situations in which the weather was not deteriorating. Although the system lacks sufficient data and inference capability to determine whether or not the weather is deteriorating at the affected airport, it can ask the user and, based upon the user's response, further narrow the search space of relevant cases.

Long-term knowledge acquisition

In contrast to using its case base to acquire knowledge about the current situation as described above, an intelligent assistant system can also use its case base as a mechanism for acquiring new descriptive features about the domain by asking the user. Again, this information can be requested in the context of a failure to find a prior case that satisfactorily matches the current situation. This failure can manifest itself either as:

- The system's inability to find any case that matches the current situation on the basis of the existing description, or
- The system's "best match" case being rejected by the user as inappropriate to the current situation, or
- The system retrieving two or more cases that match the current situation, and the system being unable to identify any features which, if known about the current situation, would discriminate among the cases.

The first manifestation represents a fundamental lack of cases in the system's library; the solution here is not in principle different from the solution adopted by autonomous case-based reasoners: Solve the problem from first principles and store the solution as a case. (The difference being, for an autonomous agent, solving from first principles can include letting the human partner solve the problem.)

The second two manifestations of retrieval failure can be dealt with by requesting the user to identify a new descriptive feature not previously known to the system. The system asks, "As nearly as I can tell, this case exactly matches the current situation, but you don't seem to agree. Please identify a feature that is present in the current situation and absent in the case, or vice versa."

An example of this feature acquisition strategy is taken from our observations of the airline controllers, in which the observer played the role of the intelligent assistant system. The problem being solved is a shortage of baggage cannisters in Toronto:

> Controller: *I'm going to order Vancouver to put some extra empty baggage cannisters onto the next flight to Toronto.*
>
> Observer: *The strategy I know about for solving this problem is to fly in some baggage cannisters from the closest airport having frequent flights to the affected airport and having excess baggage cannisters available. In this case, that might be Detroit or Chicago. Why isn't that appropriate here?*
>
> Controller: *Detroit to Toronto involves crossing a national boundary. Shifting assets across national boundaries involves an additional delay and paperwork expense associated with Customs. In this case, the added time to ship them from Vancouver is not significant relative to these administrative delays and costs.*

At this point, an actual system could have used the interaction to acquire a specific new descriptive feature that was not previously part of the domain theory, that feature being whether or not two stations were in the same country. In future problem-solving sessions, the system could use the new feature to characterize situations and decide among potentially applicable cases.

Naming and detection

Simply because the system has learned about a new feature from the user (like whether or not two cities are in the same country, as above) does not, of course, mean that the system has developed an inference mechanism for detecting whether or not that feature applies in any given situation. While this would likely be an insurmountable problem for an autonomous problem-solver, it is less of a problem for an intelligent assistant because it retains the option of asking the user whether or not the feature describes the current situation.

But this easing of the detection task imposes an additional functional constraint on descriptive features: that they be nameable or describable so as to make communication possible with the user. When the system acquires a new descriptive feature, it must also acquire a name for that feature or a mechanism for describing it so that it can, in the future, ask that user (and other users) whether or not the feature characterizes situations.

Similarly, the system and the user must be able to communicate about the system's cases. It is unreasonable for the system to ask "Do you think `CASE-T0073` is relevant here?", but it is probably also unreasonable to require the system to describe every detail of a case in order to refer to it. This naming and reference problem remains an open one.

Evaluation

Evaluating hybrid systems involving a human-computer partnership is difficult. While there are good objective measures for success in the airline irregular operations domain (typically some function taking into account passenger delay and inconvenience and cost to the airline), IOPS's goal is not to solve problems autonomously. Consequently, the evaluation question is not "Which classes of problems does the system solve?" or "How well (quickly, effectively, cheaply) does it solve these problems?" For the scientific issues discussed in this paper, the evaluation questions are:

- Are the stereotypical situations represented here an appropriate set?
- Are they well represented?
- Are the features used for detection and diagnosis appropriate?

There are several bases on which to evaluate the goodness of the case library and representation/indexing vocabulary:

Are the failure types and repair strategies meaningful? Do experienced controllers recognize them? Can controllers readily answer the question "Is this failure characterization appropriate to the current situation?"

Are the failure types and repair strategies useful to an individual? Can a controller, using the IOPS, develop better solutions than without the system? Or can the controller consider more solutions in a given time, or develop solutions more quickly?

Are the failure types and repair strategies useful across individuals? Can one controller use failure diagnoses and repair strategies developed by observing the behavior of another? Can the failure categories and repair strategies be named or otherwise presented to make this process easier?

Are the failure types and repair strategies an effective mechanism for transferring knowledge? Can novices use the system to obtain results that approximate the results obtained by experts? Can novices learning anything useful about the domain by interacting with the system?

Conclusions

A crucial set of issues in the design of a case-based reasoner revolves around the question of what kind of representation vocabulary should be used to describe the system's cases, and what descriptive features should form the basis for judging the applicability of cases to new situations. The basis for resolving these questions remains one of function: A case-based reasoner whose job it is to repair schedule failures, for example, should describe, categorize and index cases based upon the failures it is able to detect and upon the repairs it is able to perform.

The assumption of a human partner in the problem-solving loop alters these functional criteria in specific ways. While the system may be partially relieved of the burden of detecting every failure itself, it takes on the burden of extracting from the human user an operational description of those failures that the human has detected. While the system may be partially relieved of the burden of ultimately deriving a solution from a retrieved case, it takes on the burden of communicating a partial solution or describing a potentially relevant case. And, if the system is to acquire domain knowledge from the user in the form of new descriptive features, it needs a mechanism for communicating with the user about the success or failure of matches, and for using the failures to prompt the user for new descriptive features.

Another characterization of the functional requirements that derive from a case-based system's interaction with a human partner is to add more tasks to the existing *retrieve*, *debug*, *modify* and *apply* that lie at the core of case-based problem-solving. These new tasks involve asking the user about the presence or absence of a feature in the current situation, asking about the applicability or non-applicability of a case, and asking for a name and sketchy description for a new descriptive feature previously unknown to the system.

Acknowledgments

This work is supported in part by the Air Force office of Scientific Research under contract AFOSR-91-0112, and in part by the Defense Advanced Research Projects Agency and Rome Laboratory under contract F30602-91-C-0028. The author gratefully acknowledges the assistance of United Airlines in providing data and observer access to live operations. Nothing in this paper represents any policy, position, or opinion of United Airlines. John Borse has been an active participant in the data gathering and ongoing system development described in this paper.

References

[Alterman, 1986] R. Alterman. An adaptive planner. In *Proceedings of the Fifth National Conference on Artificial Intelligence*, pages 65–69, Philadelphia, PA, August 1986. AAAI.

[Bareiss, 1989] Ray Bareiss. *Exemplar-Based Knowledge Acquisition*, volume 2 of *Perspectives in Artificial Intelligence*. Academic Press, San Diego, CA, 1989.

[Barletta and Mark, 1988] R. Barletta and W. Mark. Explanation-based indexing of cases. In J. Kolodner, editor, *Proceedings of a Workshop on Case-Based Reasoning*, pages 50–60, Palo Alto, 1988. Defense Advanced Research Projects Agency, Morgan Kaufmann, Inc.

[Birnbaum *et al.*, 1989] Lawrence Birnbaum, Gregg Collins, and Bruce Krulwich. Issues in the justification-based diagnosis of planning failures. In *Proceedings of the Sixth International Workshop on Machine Learning*, pages 194–96. ONR/NSF, 1989.

[Hammond, 1989] Kristian Hammond. *Case-Based Planning: Viewing Planning as a Memory Task*, volume 1 of *Perspectives in Artificial Intelligence*. Academic Press, San Diego, CA, 1989.

[Hammond, 1992] Kristian J. Hammond, editor. *Agency: Planning and acting in a dynamic world*. 1992. In preparation.

[Kolodner and Simpson, 1989] Janet L. Kolodner and Robert L. Simpson. The mediator: Analysis of an early case-based problem. *Cognitive Science Journal*, 1989.

[Owens, 1990] Christopher Owens. *Indexing and retrieving abstract planning knowledge*. PhD thesis, Yale University, 1990.

[Owens, 1991] Christopher Owens. A functional taxonomy of abstract plan failures. In *Proceedings of the Thirteenth Annual Conference of the Cognitive Science Society*, 1991.

[Schank, 1982] R.C. Schank. *Dynamic Memory: A Theory of Reminding and Learning in Computers and People*. Cambridge University Press, 1982.

[Schank, 1991] R. C. et al Schank. Ask (incomplete citation). Technical report, Institute for the Learning Sciences, Northwestern University, 1991.

[Shortliffe, 1976] E.H. Shortliffe. *Computer-based medical consultations: MYCIN*. American Elsevier, New York, 1976.

Communicating Properties Using Salience-Induced Comparisons

T. Pattabhiraman and Nick Cercone
Centre for Systems Science & School of Computing Science
Simon Fraser University
Burnaby, B.C., CANADA V5A 1S6
patta@cs.sfu.ca and nick@cs.sfu.ca

Abstract

A method for generating simple comparison sentences of the form *A is like B* is proposed. The postulated input to the generator consists of the name of an entity A, and a set of descriptors about A in the form of attribute:value pairs. The main source of knowledge that controls decision making is a probabilistic conception of salience of empirically observable properties among concrete objects. We also use a salience heuristic based on the notion of property intrinsicness. The information-theoretic concept of redundancy is used to quantify salience in probabilistic contexts. Salience factors influencing selection decisions are modelled as utilities and costs, and the decision for selecting the best object of comparison is based on the maximization of net expected utility. The method proposed has been implemented in a generation system written in CProlog.

Introduction

In this paper we present a method for generating simple comparison sentences with the aid of the notion of salience. The method has been implemented in a generation system written in CProlog. The postulated input to the generator consists of a set of descriptors about an entity (say, *Mary's cheeks*) in the form of attribute:value pairs (like *colour:red, texture:smooth*, etc). The task we are focussing on consists of describing the entity through a comparison sentence of the form *A is/are like B* (e.g., *Mary's cheeks are like apples*), by means of which the hearer can infer the intended descriptors of A. Based on the format *A is like B*, we will refer to the entity being described (*Mary's cheeks*) as the *A-term* and the chosen example (*apples*) as the *B-term*.

Modelling Property Salience

The approach taken in this work for selecting a good example of an object (dually, concept) possessing a certain property relies mainly upon probabilistic knowledge of the distribution of values for certain essential, empirically observable attributes among concrete (physical) objects considered as objects of comparison. In addition, we employ a salience heuristic based on property intrinsicness.

Concept Representations

The knowledge base includes representations of concrete objects with probabilized value spaces representing the distribution of possible values for attributes. Such representations are assumed to be integral to the *general world knowledge* of the speaker about the concepts.

Formally, a *concept* C is a labelled set of *properties* P_i, and each property P_i is a pair $A_i : \mathcal{V}_i$, where A_i is an *attribute* and $\mathcal{V}_i$ is its *probabilized value space*. $\mathcal{V}_i$ is a set of pairs $V_{ij} : p_{ij}$, with $p_{ij} \in (0,1]$, and $\sum_j p_{ij} = 1$. In our implementation, for example, we have *colour = [yellow:0.25, golden:0.60, green:0.15]* as a property of the *mango* concept. Noting that the probabilities of values in these representations are *conditional* upon the concept possessing them for the respective attributes, we can write, for example: $p(colour = green|mango) = 0.15$.

These representations are a variant of the dimensional approach to probabilistic concept representation, discussed in cognitive psychology literature by [Smith & Medin, 1981]. In our implementation, these representations are embedded in a semantic network. Nominalized versions of property values (like *redness* for *red*) are represented in the network as abstract concepts, with links directed towards their extensions (like *apples* and so on).

Salience Based on Probabilities

When generating comparisons, a concept (say, *apple*) is considered as a candidate example of an input property (say, *colour:red*) if its **most probable** value for the corresponding attribute (*colour*) matches the value (*red*) in the input property. Several objects (all of which are mostly *red*) may present themselves as candidate examples, and the preference is based on a measure of salience of the property. However, salience is not simply *equal to* the probability, since the number of other possible values (say, *green, yellow*) as well as their respective probabilities have the effect of either enhancing or suppressing the prominence of the most probable value in a distribution.

If the most probable value (*red*) has a high probability (*0.85*), and there are very few other possible values (say, only *green* with probability *0.15*), then *red* has high salience in the context of the *colour* distribution. If, on the other hand, even if red was the most probable colour (with probability *0.25*), if other colours were also equally probable (say *green, brown* and *yellow*, each with probability *0.25*), then *red* would not *stand out* in the distribution, and its salience would therefore be very low. Information theory helps us capture these notions of salience precisely, through the concept of **redundancy**, which we use in our work to quantify property salience.

Moreover, when looking for the best example of a red object, not only should the candidate example, say, apple, be mostly red, and highly salient among apples of all colours, but the redness of apples should be more salient than the redness of, say, *strawberries* and other *fruit* which are also mostly red. To model this aspect of salience, for a given property value (say *red*), we consider the redundancy of a concept (say *apples*) in the context of the redundancies of other candidates (say *strawberries*, etc) through the definition of **normalized redundancy**.

Similar information-theoretic measures have been used by [Iwayama, Tokunaga & Tanaka, 1990] in their computational modelling of metaphor comprehension. We adopt their work as a good point of departure to examine the modelling of salience and its role in generating comparisons for describing object properties, and use their example sentence *Mary's cheeks are like apples* to convey our algorithm in this paper. In interpreting the above sentence, their system (called *AMUSE*) calculates the salience of the properties of *apples*, matches the high-salient properties of *apples* with the properties of *cheeks* and infers the properties of *Mary's cheeks* intended by the speaker.

To compute the redundancy of the property of an object, we proceed through the following information-theoretic concepts. Given a discrete probability distribution with n probabilities $p_i \in (0,1]$, with $\sum_{i=1}^{n} p_i = 1$, the **entropy** of the distribution H is given by

$$H = \sum_{i=1}^{n} p_i log_c \frac{1}{p_i} \tag{1}$$

We take c (the base of the logarithm) to be 2 throughout this paper.

The entropy H of a distribution is zero if there is only one possible value (with unit probability). For a given set of n possible values, H is maximum if the values are equiprobable ($p_i = \frac{1}{n}$ for all i), and equals $log_2 n$. H quantifies the 'flatness' or 'dispersion' of a probability distribution, and can be interpreted as measuring the extent to which the prominence of the most probable value in the distribution is *suppressed* in the context of possible values. For instance, for the *colour* property of *mango* represented by *colour = [yellow:0.25, golden:0.60, green:0.15]*, H turns out to be 1.35272. If the colours were equiprobable, H would have been maximum, at 1.58496 (i.e., $log_2 3$). If all mangoes were golden, H would have been 0.

Relative entropy H_{rel} expresses the entropy of a distribution in the unit interval. By normalizing H with respect to maximum entropy for a given set of samples, H_{rel} expresses entropy independently of the sample size. It is defined as

$$H_{rel} = \left(if\ n = 1\ then\ 0\ else\ \frac{H}{log_2 n} \right) \tag{2}$$

For any $n > 1$, $H_{rel} = 1$ if the values are equiprobable, and less if the most probable value is higher in probability, and the number and magnitude of other values, smaller. If only one value is possible (probability = 1), $H_{rel} = 0$. H_{rel} quantifies the extent of suppression, or lack of salience, of the most probable value in a probability distribution. The quantitative complement of H_{rel}, viz., redundancy, therefore measures the degree of salience of the most probable value in a probability distribution:

Redundancy of a distribution is computed by

$$R = 1 - H_{rel} \tag{3}$$

In our system, the *colour* space of *apple* is *[red:0.75, green:0.15, yellow:0.10]*. The redundancy (R) of this distribution is 0.33499. By comparison, the colour of *orange*, with a distribution of *[orange:0.80, yellow:0.10, green:0.10]*, has a (greater) R of 0.41833. As a final example, the colour of *grapefruit* with *[yellow:0.75, pink:0.25]*, has $R = 0.18872$. Even though *red* in *apple* and *yellow* in *grapefruit* occur with the same probability, the former has higher salience (R) in its context of possible values.

Finally, in a set of concepts $C = \{C_1, C_2, \ldots C_n\}$ in which all C_i possess the same most probable value (say *red*) for a given property P (*colour*), the salience of a concept C_k in the context of C due to property P is measured by its **normalized redundancy**, computed as

$$NR(C_k, P) = \frac{R(C_k, P)}{\sum_{i=1}^{n} R(C_i, P)} \quad (4)$$

where $R(C_i, P)$ is the redundancy of the property P of C_i. In a context of fruits which are all mostly red, $NR(apple, colour)$ measures the relative salience of apples, and governs the candidacy of apples as good examples of red fruits. Note also that by the above definition, in a given context C, all the NRs add up to 1, and $NR(C_i, P)$ can be interpreted as the conditional probability $p(C_i|P)$. $p(apple|colour : red)$ is the likelihood of choosing *apple* when looking for a good example of *colour* : *red* among fruits.

Decision Making in Comparison Generation

In formulating the choice of the B-term in the comparison as a decision making problem, we first derive the formal equivalence between information-theoretic redundancy and expected utility.

Redundancy as Expected Utility

Substituting (1) in (2) and (2) in (3), the redundancy R associated with a probability distribution over n possible values ($n > 1$) is:

$$1 - \frac{1}{log_2 n} \cdot \sum_{i=1}^{n} p_i log_2 \frac{1}{p_i}$$

$$= \frac{log_2 n + \sum_{i=1}^{n} p_i log_2 p_i}{log_2 n}$$

$$= \frac{\sum_{i=1}^{n} p_i log_2 n + \sum_{i=1}^{n} p_i log_2 p_i}{log_2 n}$$

$$= \sum_{i=1}^{n} p_i \cdot \left(\frac{log_2(np_i)}{log_2 n} \right) \quad (5)$$

The summation (5) is the familiar form of expected utility, viz., $\sum_{i=1}^{n} p_i u_i$, with $u_i = log_2(np_i)/log_2 n$. The utility u_i is derived from the probability p_i and the size of the value space, n. It is interpreted as the reward associated with the selection of the value with probability p_i. For the special case of $n = 1$, we have $u_i = 1$ and $p_i = 1$. As is evident from the above expression, the reward u_i is maximum for the most probable value, and redundancy measures the expected reward in the context of all possible values. Choosing a concept as an example on the basis of maximum R among competing concepts, and equivalently, on the basis of maximum NR, can hence be modelled as a decision problem of maximizing expected utility.

Generating comparisons describing one property: The algorithm for choosing the best comparison when there is one input property P can now be stated as follows: P in the generator input serves as a source of activation in the knowledge base, which spreads towards concepts (forming a set C) in which P occurs redundantly. For each $C_i \in C$, compute the expected utility $EU(C_i, P) = EU_i = NR(C_i, P)$. Select the concept with maximum EU_i as the best (available) example. In our generator implementation, for the input entity *Mary's cheeks* and the descriptor *[shape:round]*, the following EUs are computed (note that they add up to 1):

C_i	$EU(C_i, [shape : round])$
plum	0.2555
apple	0.3434
grape	0.0908
peach	0.0908
lemon	0.0317
grapefruit	0.1878

The generator outputs *Mary's cheeks are like apples.*

Comparisons Describing Two or More Properties

When two or more properties are intended to be communicated about the input entity (for example, that *Mary's cheeks are red, smooth and round*), each property initiates a search (ideally, in parallel) in the knowledge base for a good example in which the respective property occurs redundantly. The decision criterion for choosing the best example is now one of maximizing *total expected utility* (TEU), the total being the sum of individual EUs from each property.

Table 1 shows a portion of the matrix of EUs computed in our implementation for the input entity *Mary's cheeks* and the descriptors *[colour:red]*, *[texture:smooth]* and *[shape:round]*. The generator outputs *Mary's cheeks are like apples* based on maximum TEU.

The Problem of Zero Credit

The above method is effective as a straightforward extension of the one-property case, and works well when concepts receive non-zero EU for each property to be communicated, as is the case for *apple*

C_i	$EU(C_i, red)$	$EU(C_i, smooth)$	$EU(C_i, round)$	$TEU(C_i)$
apple	0.3333	0.3075	0.3434	0.9842
grape	--	0.4133	0.0908	0.5041
strawberry	0.5284	--	--	0.5284
plum	0.1383	0.0819	0.2555	0.4757
...	...	...	...	...

Table 1: A portion of a matrix of *EU*s

and *plum* in Table 1. Note however in the same matrix, that while *plum* receives an *EU* for each property, *strawberry*, though having no *EU* for *round* and *smooth*, scores a higher *TEU* than *plum* due to sheer intensity of high *EU* for *red*. In this case, *strawberry* isn't exactly a good example of something *red and smooth and round*!

Similarly, when the input contains *[colour:yellow, texture:smooth, shape:round]*, our generator, on the basis of *TEU* alone, would still say *Mary's cheeks are like apples*. Miscommunication results in this case, as the hearer, who also uses knowledge of salience in comprehension, ends up inferring that Mary's cheeks are *red*. If *apple* was not present in the knowledge base, and if the input property values were *red, round* and *smooth*, the generator would have said: *Mary's cheeks are like strawberries*. In this case, there may be either non-communication (nothing at all will be inferred about texture and shape of Mary's cheeks) or anomalous communication (the inferred shape and texture of Mary's cheeks will not be consistent with the general knowledge of the hearer about the shape and texture of cheeks.) In the parlance of connectionism, we may say then that it is not sufficient to have high convergent (*total*) activation: there should also be sufficient activation from *each* source.

While it is tempting to get a quick mathematical fix by defining arbitrary minimum thresholds on the individual *EU*s in the *TEU* criterion, we motivate the solution by considering the comprehensibility of the generated comparisons, and derive *cost* measures (antipodal to *utility*) to use in decision making. Two different cost measures are proposed, corresponding respectively to the problems of zero credit (described above) and unintended properties (described later).

Exception Clauses

It is fairly common in day-to-day speech to come across comparisons in which miscommunication due to zero credit (in the sense discussed above) is averted by generating additional clauses that make explicit the inexactitude of the match: as in *Mary's cheeks are like apples, except...(that) they are yellow*. We call the latter clause an *exception clause*, the property (attribute) described in it (here, *colour*) the *exception property (attribute)*, and the value communicated in it (here, *yellow*) the *exception value*. In simple settings like the ones under consideration, the utility of communicating through comparison drops off rapidly with the number of exceptions that may be sought. One exception is fairly common, as not always do we find *one* best example of all properties we want to communicate about an entity.

We quantify the cost of zero credit by focussing on the generation and acceptability of comparisons with exception clauses. We found it helpful to visualize that the speaker's evaluation of comparisons with exceptions is mediated by an imaginal process in which the B-term object without the exception property is *mentally distorted* into the B-term object with the exception property. For example, in *Mary's cheeks are like apples, except.. they are yellow*, a *red* (salient colour) apple is 'repainted' into a *yellow* one, and offered as an object of comparison describing the intended properties of *Mary's cheeks*. A cost is added to the *TEU* as a negative number, and the less the cost is the better. The cost will be less if it is in some way *easier* to 'distort' a red apple into a yellow apple. We propose the following as a probabilistic measure of cost for the running example:

$$Cost(apple, colour) = \\ p(colour = red|apple) \cdot (1 - p(colour = yellow|apple))$$

The more probable red apples are, and the less probable yellow apples are (in the general knowledge of the speaker), the more difficult will it be to 'mentally distort' a red apple into a yellow one, and the costlier will be the exception clause. In general, given a concept C and an exception attribute P, if the exception value is EV and the most probable value is MPV, then

$$Cost(C, P) = \\ p(P = MPV|C) \cdot (1 - p(P = EV|C))$$

Property Intrinsicness

There seems to be more to the cost of zero credit than probabilistic knowledge as modelled here. For instance, given the input property values of *yellow, round* and *smooth*, compare the acceptability of *Mary's cheeks are like apples.. except, they are yellow*, with what our generator said charmingly oddly in an earlier version: *Mary's cheeks are like bananas, except.. they are round*! Distorting the colour of objects seems easier than distorting the shape of objects as shape is in some sense a more salient attribute than colour. This conception of salience is discussed in cognitive linguistics [Langacker, 1987] under *intrinsicness*. We annotate the properties in the knowledge base with this heuristic measure, giving a higher score to shape than to colour. This is added to the cost of distortion when the distortion entails conception of an impossible object. This cost is zero for *yellow apples* since they do exist in the speaker's knowledge; it is less for *purple apples* than for *round bananas*. Even among impossible objects, some seem more impossible than others!

For every zero credit entry in the TEU matrix we compute such costs. When there is positive EU for an entry, cost is zero, since the property to be communicated is salient in the entry. Finally, when n properties are intended to be communicated, and the TEU matrix is formed in the manner discussed earlier, the probability p_{zc} that any one of the entries is zero is given by $\frac{2^{n-1}-1}{2^n-1}$, analogous to the probability that any one of the inputs of an n-input OR-gate is 0, given that the gate output is 1. The *expected cost* (EC) of an entry in the TEU matrix for the property P of concept C is therefore

$$p_{zc} \cdot Cost(C,P) + (1 - p_{zc}) \cdot 0$$
$$= \left(\frac{2^{n-1}-1}{2^n-1}\right) \cdot Cost(C,P)$$

We can now compute the *total expected cost* (TEC) of a candidate example as the sum of individual ECs for each property. The decision criterion is now one of maximizing

$$TEU - \alpha \cdot TEC$$

where α is a non-negative number. $0 \leq \alpha < 1$ corresponds to *speaker-oriented* generation models, and $\alpha \geq 1$ corresponds to *considerate* or *listener-oriented* generation models. For $\alpha = 1$ and for the values *yellow, smooth and round*, we generated *Mary's cheeks are like grapefruit*, which, though unlikely to be uttered by humans, was the best that our system could do with its modest knowledge base, and without the use of the 'unintended properties' cost, to be described next.

Cost of Unintended Properties

Another problem arises when the list of descriptors in the generator input does not include certain properties in the *generic* representation of the input entity in the knowledge base. For instance, for the input values *yellow* and *smooth*, the generator would say *Mary's cheeks are like bananas*. This is because the input which encodes the speaker's communicative intentions does not include *round* as a descriptor of *Mary's cheeks*. Since both *yellow* and *smooth* receive positive EUs for banana and make its TEU score the highest, the above comparison is generated. The problem arises because the inferred (salient) shape of bananas (B-term) is in conflict with the salient value of the unintended (= not included in the generator input) property in the *general knowledge* about cheeks, viz., *round*. To counter the problem of miscommunication or anomaly due to such interference, we use another cost measure very similar to the one discussed earlier, using both probabilistic knowledge and property intrinsicness. For every property in the general knowledge of the A-term not included in the input, the cost of mismatch with the corresponding salient property of the B-term is computed, and summed up as the total expected cost TEC_{B-A}. The former TEC (for the zero credit problem) can be now relabelled as TEC_{A-B}.

Decision Criterion for Comparison Generation

The final decision criterion we have developed is termed *net expected utility*, computed as

$$NEU = TEU - \alpha_1 \cdot TEC_{A-B} - \alpha_2 \cdot TEC_{B-A}$$

where α_1 and α_2 are non-negative. Although our generation algorithm does not explicitly advert to the concept of *similarity* in its design, its decision criterion NEU has striking formal resemblance to the similarity metrics proposed in the cognitive psychology literature, such as those of [Tversky, 1977] and [Ortony et al., 1985]. NEU is more like Ortony et al's measure, however, since TEU is evaluated with respect to the B-term. With NEU, for the input properties *yellow* and *smooth*, we generate *Mary's cheeks are like lemons* for $\alpha_1 = \alpha_2 = 1$.

General Discussion

The research reported in this paper is integral to our investigation of salience and its role in natural language generation decisions [Pattabhiraman, 1992]. The measure of salience used in this paper is intuitively appealing, while being at the same time mathematically well-grounded. Our method of generating comparisons, while aiming to satisfy cognitive concerns, also relies upon decision theory and information theory for its formal foundations, and thereby lends itself to computational usability in NLG systems, as our implementation has demonstrated. A fresh perspective on the problem of similes emerges when we view it from the speaker's angle. However, several additional factors will have to be treated before presenting our work as a full-fledged descriptive model of human comparison generation, or as a model for computer generators with very large knowledge bases.

First, other determinants of salience like *vividness* and *imageability* ([Osgood & Bock, 1977]) should be explored for their influence on comparison generation. Their influence can be modelled quantitatively in terms of utility functions and incorporated into our decision criterion. A red object may be selected not necessarily because it is redundant in the information-theoretic sense, but because its redness is very vivid. Secondly, preferences for basic-level terms and for concrete objects, which are implicit in our method, should be treated explicitly in the theory. Preference for concrete objects can help generate *indirectly grounded* comparisons like *proud as a peacock*: though peacocks are not quite proud (cf. general discussion in [Ortony et al., 1985]), we suggest that a link exists through *ostentation*: ostentation is a possible symptom of pride, and peacocks when they unfurl their feathers appear ostentatious. The collective noun *an ostentation of peacocks* is another telltale indicator! Such indirect associations may get short-circuited as direct conceptual- or lexical-collocational associations through the process of entrenchment. Thirdly, the initial stage of processing the generator input to form the set of potential objects of comparison can be refined to control selection between literal and metaphorical comparisons.

Finally, while we have assumed tacitly that the property values are independent, covariances among property values and among relations between property values are common in natural concepts [Goldstone, Gentner & Medin, 1989]. The accuracy of decision making increases when knowledge of correlations between properties is represented and used. However, even in small knowledge bases, an immense number of covariances will have to be identified and represented; a decrease in computational efficiency is inevitable with their use. The precise nature of this trade-off and its implications for descriptive models of comparison generation merits further research.

Acknowledgments. Our thanks are due to the anonymous reviewers for their helpful comments and suggestions, and to the CSS and the School of Computing Science at SFU for the use of their facilities.

References

[Goldstone, Gentner & Medin, 1989] Goldstone, R., Gentner, D., and Medin, D. (1989). Relations Relating Relations. In *Proceedings of the 11th Annual Conference of the Cognitive Science Society*, pages 131–138.

[Iwayama, Tokunaga & Tanaka, 1990] Iwayama, M., Tokunaga, T., and Tanaka, H. (1990). A Method for Calculating the Measure of Salience in Understanding Metaphors. In *Proceedings of AAAI'90, Boston*, pages 298–303.

[Langacker, 1987] Langacker, R. (1987). *Foundations of Cognitive Grammar, Volume 1: Theoretical Prerequisites.* Stanford University Press.

[Ortony et al., 1985] Ortony, A., Vondruska, R., Foss, M., and Jones, L. (1985). Salience, Similes and the Asymmetry of Similarity. *Journal of Memory and Language*, 24:569–594.

[Osgood & Bock, 1977] Osgood, C. and Bock, J. (1977). *Salience and Sentencing: Some Production Principles*, In *Sentence Production: Developments in Research and Theory*, Rosenberg, S., Editor, pages 89–140. Erlbaum, Hilldale, N.J.

[Pattabhiraman, 1992] Pattabhiraman, T. (1992). *Aspects of Salience in Natural Language Generation.* PhD Thesis, School of Computing Science, Simon Fraser University. In preparation.

[Smith & Medin, 1981] Smith, E. and Medin, D. (1981). *Categories and Concepts.* Harvard University Press.

[Tversky, 1977] Tversky, A. (1977). Features of similarity. *Psychological Review*, 84(4):327–352.

Dynamic Gating in Vision*

Eric O. Postma, H. Jaap van den Herik and Patrick T. W. Hudson
Department of Computer Science
University of Limburg
P.O. Box 616, 6200 MD Maastricht/The Netherlands
postma@cs.rulimburg.nl

Abstract

Visual attention requires the selection of salient regions and their remapping into a position-invariant format. We propose the *dynamic-gating* model capable of *autonomous* remapping. It combines the localization network of Koch and Ullman (1985) with a modified shifter-circuit network (Anderson & Van Essen, 1987). Autonomous selection and remapping of salient regions result from local gating dynamics and local connectivity, implying that scaling to large problem sizes is straightforward.

1. Introduction

The visual system is equipped with a highly flexible process that enables the allocation of computational resources to a restricted part of the retinal image. When operating independently of eye movements, this process is called *covert attention* (Posner and Presti, 1987). This contribution proposes a neural model of covert attention. In order to allow scaling to large problem sizes (e.g., vision), we impose the implementational restrictions of *local processing* and *local connectivity*. By distributing task load and executing computations in parallel we aim at achieving a performance matching that of the human visual system (Nelson & Bower, 1990). Section 2 discusses behavioral findings on covert attention. In Section 3, dynamic remapping is suggested to play an important role in the visual system. Section 4 presents the dynamic-gating model and Section 5 gives some simulation results. Finally Section 6 evaluates the model.

*IBM is acknowledged for their hardware support under the Joint Study Agreement DAEDALOS (#289651).

2. Covert attention

A common metaphor for covert attention is of a spotlight illuminating part of the retinal image. The processing of stimuli captured by the spotlights' beam is enhanced at the cost of the processing of stimuli lying outside the beam. Behavioral studies revealed that covert attention indeed behaves much like a spotlight. Sagi and Julesz (1986), for instance, found performance on the detection of a test flash to be enhanced when attention was directed on a to-be-identified stimulus in its vicinity. The enhancement was strongest at the stimulus location and dropped gradually at increasing eccentricity. Other findings indicate that the attentional beam can be expanded to cover larger retinal regions (Eriksen, 1990), coincided by an enhancement that is inversely proportional to the size of the attended area. Apparently, then, the computational resources invoked in covert attention are limited causing a trade-off between resolution and viewing angle.

The speed of moving the spotlight to new locations has been reported to exceed the speed of eye movements (even 4 to 5 times, Saarinen & Julesz, 1991). Other findings (Remington & Pierce, 1984; Kwak, Dagenbach, & Egeth, 1991) show that attention jumps in a *time invariant* fashion (i.e., with a speed proportional to distance) suggesting an underlying mechanism reminiscent of the saccades associated with overt attention (Posner, 1980).

Covert attention is here assumed to be based on two processes: *localization* and *identification*. In localization, a conspicuous area is selected very rapidly in order to align the attentional beam with it. In identification, the contents at the attended region are matched against internal object representations. Such a matching requires the computation of an object-centered frame of reference. We will focus on how localization and the construction of an object-centered frame of reference can be modeled given our implementational restrictions of local processing and local connectivity.

3. Dynamic remapping

Detailed visual analysis of complex patterns requires the allocation of large computational resources. It is not feasible to allocate multiple recognizers for all positions in the visual field. Rather a single dedicated recognition module, dynamically linked to a (spatially contiguous) part of the retinal input (cf. the attentional spotlight), is a much more parsimonious solution (cf. Van Essen, Anderson, & Felleman, 1992). Recent neurophysiological findings suggest such a process to operate in concert with visual attention. Duhamel, Colby and Goldberg (1992) found receptive fields of neurons in the parietal cortex to shift in anticipation of eye movements. As a result, the internal representation of the visual scene is remapped to match the retinal pattern after saccade completion. Desimone, Moran and Spitzer (1989) found attentionally modulated receptive-field shifts in inferior temporal cortex. Considering the common functional distinction of a "where" (occipitoparietal) pathway dealing with the task of encoding spatial relations and a "what" (occipitotemporal) pathway involved in the task of object recognition (e.g., Mishkin & Appenzeller, 1987; Goodale & Milner, 1992), a tentative view holds that dynamic remapping accounts for maintaining constancies appropriate for the task at hand.

Construction of perceptual reference frames (e.g., object-centered or viewer-centered) may proceed by appropriate sampling of retinally contiguous spatial (and feature) patterns. Such selective sampling can effectively be realized by dynamic remapping. Two complementary neural models using dynamic remapping have been proposed in the literature. One model deals with the localization of conspicuous patterns, the other accounts for the remapping of patterns into an appropriate reference frame. Below, we briefly discuss these models integrated in the dynamic-gating model presented in Section 4.

The shifter-circuit network

The remapping of patterns has been hypothesized to occur through dynamic routing by Anderson and Van Essen (1987). The shifter-circuit network shown in Figure 1a consists of a hierarchy of three concatenated *shifter circuits* accommodating the routing of a contiguous pattern in the input towards the output. Each shifter circuit shifts incoming patterns to the left or to the right by selectively enabling the transmission lines pointing in one direction while disabling the lines pointing in the other direction. Patterns presented at the bottom of a shifter circuit are, therefore, remapped without distorting the internal pattern structure. Critical for proper functioning of a shifter circuit is the requirement to align enabled (disabled) transmission lines. To fulfil this requirement Anderson and Van Essen (1987) introduced one *shift-control* module for each shifter circuit. A shift-control module contains two neurons, one contacting all transmission lines pointing leftwards and one contacting all transmission lines pointing rightwards. At any time only one of the shift-control neurons is active, effectively disabling signal flow through all the lines it contacts. A shifter-circuit architecture with L layers (i.e., shifter circuits) and k shifting directions ($k = 2$ in Figure 1a) is capable of remapping a subpattern of length i from any position in an input layer of length $i + k^L - 1$ towards the output layer.

The localization network

Koch and Ullman (1985) proposed a hierarchical network that enables localization and detection of conspicuous features. The structure of their model is depicted in Figure 1b. It consists of L layers with k^l links in layer l ($l = 1$ is the top layer, $k = 2$ in Figure 1b). A Winner-Take-All (WTA, Feldman & Ballard, 1982) competition among k adjacent transmission lines (neurons) within a subtree results in the selection of a local maximum input value. After selection, the "winning" value is propagated (remapped) to the next layer upwards. The maximum value is available at the top layer. Figure 1b (bottom) shows the localization of the maximum by a concatenated sequence of "winning" transmission lines (arrows). A feedback network of auxiliary units (each associated with a single upward transmission line) determines the position of the maximum. After the maximum has reached the top layer, auxiliary elements compare the state of their associate transmission line with the state of the one directly above. A unit becomes active if both lines are winners, otherwise it remains inactive. Consequently, activated auxiliary units trace the concatenated winning-line sequence in the reverse direction.

4. The dynamic-gating model

The localization and shifter-circuit networks can be combined to form an autonomous "spotlight" that orients towards conspicuous regions. A salient pattern detected by the localization network can steer

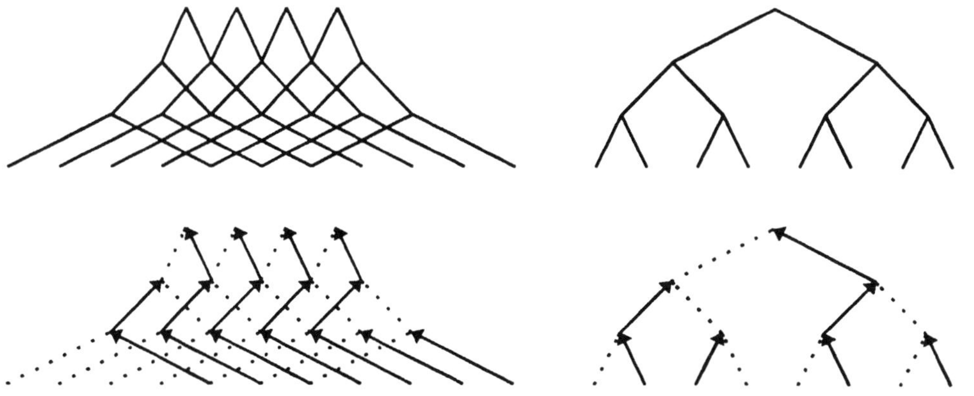

(a) (b)

Figure 1: *(a) The shifter-circuit network proposed by Anderson and Van Essen (1987, top) and its selection/remapping of a pattern of length 4 (bottom). (b) The localization network proposed by Koch and Ullman (1985, top) and its localization of the maximum input value.*

the shifter-circuit network so that the pattern is routed towards its top layer where it can be matched against prototypical object representations. Our implementational constraints of locality of processing and connectivity guide the successive modifications necessary for full integration. This approach manages to retain the neurophysiological plausibility of its component networks. In particular the combination provides a model for dynamic remapping within the occipitotemporal pathway.

Network structure

Although the network structures shown in Figure 1 have similar characteristics they are still incompatible since the magnitude of shifts in localization networks increase (when going upward) whereas in shifter-circuit networks they decrease. To maintain local connectivity, similar functions (i.e., localizing and remapping) must be executed in adjacent areas (e.g., Nelson & Bower, 1990). Therefore we have modified the shifter-circuit network into a structure that matches the localization network by reversing the order of shifter circuits (the *reversed shifter-circuit network*). As shown in Figure 2a and b, the reversed shifter-circuit network matches the structure of the localization network. Although the reversed network employs a larger number of transmission lines in comparison to the original, it achieves the same remapping capacity given an equal number of layers. (Both the standard and reversed network shown in Figure 1a and 2a can remap a pattern of length 4 out of an input of length 11.) Both networks sample the same input, but the localization network samples at a lower resolution (see below).

Gating dynamics

The elements of both networks have a combined gating and competition (selection) function. The gating element (or *gate*, represented by a circle in Figure 2) proposed herein performs both functions simultaneously. It competes with the links in its local neighborhood and, if it wins the WTA-competition, gates a local input value to its output. The dynamics of the attentional spotlight requires considerable and rapid flexibility. Many WTA-schemes react relatively slowly to any change in the input. For this reason we proposed local stochastic gating dynamics based on the neuron model of Little (e.g., Little & Shaw, 1975; Postma, van den Herik, & Hudson, 1992). The intrinsic noise of a gate is exploited to enable it to respond rapidly to changing input. At the same time there is a limit; too much flexibility causes inherent instability (see below).

Horizontal connectivity

The nature of a gate's horizontal interactions (i.e., its local neighborhood) differs for the two networks. In the localization network winners are determined

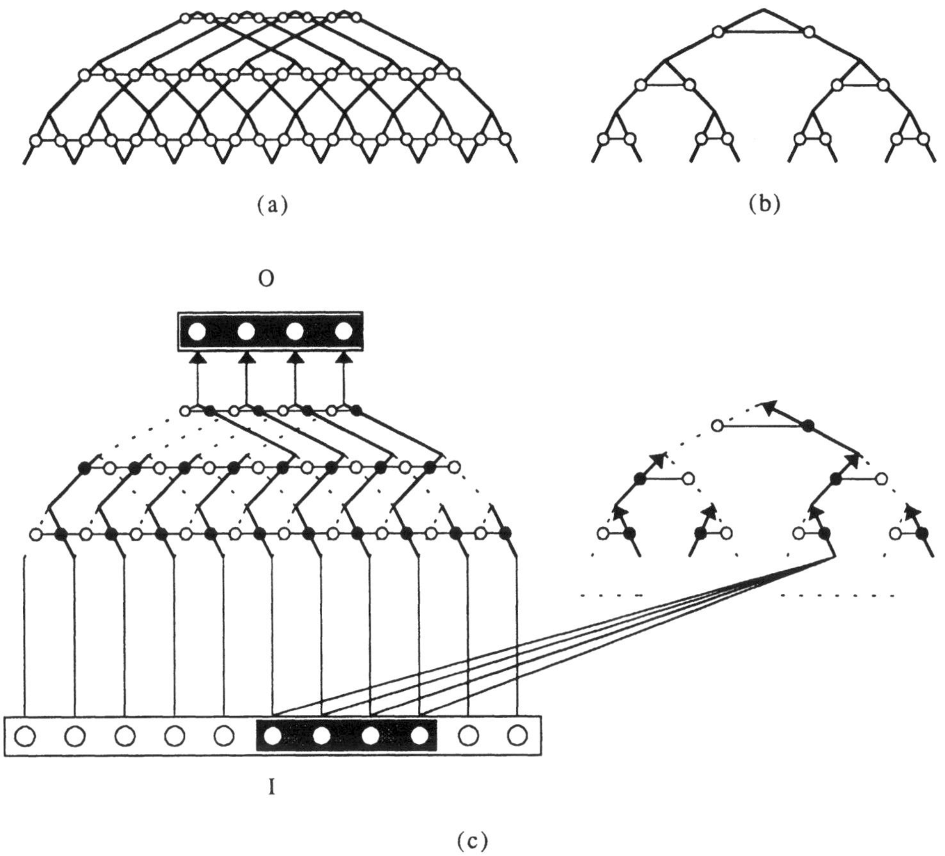

Figure 2: *(a) Reversed shifter-circuit network. (b) Localization network. (c) Integrated dynamic-gating network. The circles represent open (black) or closed (white) gates.*

locally, whereas in the remapping network winners are determined over the length of the output pattern. These opposing characteristics are directly related to the distinct functions of both networks and should be reflected in the definition of the neighborhood of the gates. The horizontal lines in Figure 2 illustrate the appropriate connectivity pattern of both networks. In the localization network, connectivity is restricted to non-overlapping local clusters of 2 gates (for the one-dimensional network and $k = 2$), i.e., each gate is inhibited by a single neighbor (cf. Koch & Ullman, 1985). In the remapping network each gate is inhibited by 2 neighbors, i.e., neighborhoods overlap. In two dimensions, horizontal interactions within a layer form a field of independent clusters in the localization network and a lattice structure in the remapping network. Within each cluster, there is a single open gate. In the lattice, a regular spatial pattern of open gates (each surrounded by a circular neighborhood of closed gates) reflects aligned gating within a single layer. The latter represents a special case of the Ising Lattice of statistical mechanics and exhibits global gating behavior on the basis of local interactions (see Postma, van den Herik, & Hudson, 1992, for a more detailed treatment of these issues).

Integration

In the integrated dynamic-gating model, the auxiliary units *clamp* a small subset of gates within a layer of the remapping network. A single active auxiliary unit effects the proper gating over a large range (of order i) in the lattice. Figure 2c illustrates this: the common input at the bottom is sampled at a course resolution by the localization network and sampled at a high resolution by the remapping network. The winning chunk (the grey box at the bottom of Figure 2c) is found by the localization network in the input pattern I. The sequence of concatenated upward pointing arrows are paired with active auxiliary units (not shown). These units sparsely clamp the appropriate gates in the remapping network so that the contents of the selected pattern is remapped into the output pattern O.

5. Simulations

The integrated dynamic-gating model has been tested by simulations. Here we confine ourselves to illustrating the localization performance of the network. One input in a 16 x 16 two-dimensional input field is assigned a value of 1.0 (target) and the rest randomly distributed values on the interval $[0.0, 0.9]$ (distractors). Figure 3 shows the localization performance (number of localizations per 1000 iterations) for all positions in the input field. The number of target-localizations is specified near the target bar. The three graphs show localization performance for different magnitudes of the intrinsic noise: low noise (left), medium noise (middle), and high noise (right). Although target localization occurs most frequently in all three cases, performance is best at a medium noise level. At low-noise levels, localization tends to "stick" to local maxima (distractors) whereas at high-noise levels it becomes unstable. The optimal (intermediate) noise level combines stability against input noise with vigilance for input change.

6. Evaluation

In conclusion we may state that we have succeeded in formulating a neural model of the attentional spotlight. By obeying the implementational requirements of locality of processing and connectivity, we arrived at an architecture that can be scaled to large visual inputs. As a direct continuation we are currently studying the performance of the dynamic-gating model on a range of covert-attention tasks.

References

Anderson, C.H. and Van Essen, D.C. 1987. Shifter circuits: a computational strategy for dynamic aspects of visual processing. *Proceedings of the National Academy of Sciences USA* 84:6297-6301.

Desimone, R., Moran, J., and Spitzer, H. 1989. Neural mechanisms of attention in extrastriate cortex of monkeys. In Arbib, M.A. and Amari, S. eds. ***Dynamic interactions in neural networks: Models and data***, 169-182. New York: Springer.

Duhamel, J-R., Colby, C.L., and Goldberg, M.E. 1992. The updating of the representation of visual space in parietal cortex by intended eye movements. *Science* 255:90-92.

Eriksen, C.W. 1990. Attentional search of the visual field. In Brogan, D. ed., *Visual Search*, 3-19. London: Taylor & Francis

Feldman, J.A. and Ballard, D.H. 1982. Connectionist models and their properties. *Cognitive Science* 6:205-254.

Goodale, M.A. and Milner, A.D. 1992. Separate visual pathways for perception and action. *Trends in Neurosciences* 15:20-25.

Koch, C. and Ullman, S. 1985. Shifts in selective visual attention: towards the underlying neural circuitry. *Human Neurobiology* 4:219-277.

Kwak, H-W., Dagenbach, D., and Egeth, H. 1991. Further evidence for a time-independent shift of the focus of attention. *Perception & Psychophysics* 49:473-480.

Little, W.A. and Shaw, G.L. 1975. A statistical theory of short and long term memory. *Behavioral Biology* 14:115-133.

Mishkin, M. and Appenzeller, T. 1987. The anatomy of memory. *Scientific American* 256:62-71.

Nelson, M.E. & Bower, J.M. 1990. Brain maps and parallel computers. *Trends in Neurosciences*, 13:403-408.

Posner, M.I. 1980. Orienting of attention. *Quarterly Journal of Experimental Psychology* 32:3-25.

Posner, M.I. and Presti, D.E. 1987. Selective attention and cognitive control. *Trends in Neurosciences* 10:13-17.

Figure 3: *Localization performance for three levels of the intrinsic noise: low (left), medium (middle), and high (right).*

Postma, E.O., van den Herik, H.J., and Hudson, P.T.W. 1992. Dynamic selection through gating lattices. In Proceedings of the International Joint Conference on Neural Networks, Baltimore.

Remington, R. and Pierce, L. 1984. Moving attention: Evidence for time-invariant shifts of visual selective attention. *Perception & Psychophysics* 35:393-399.

Saarinen, J. and Julesz, B. 1991. The speed of attentional shifts in the visual field. *Proceedings of the National Academy of Sciences USA* 88:1812-1814.

Sagi, D. and Julesz, B. (1986). Enhanced detection in the aperture of focal attention during simple detection tasks. ***Nature*** 321:693-695.

Van Essen, D.C. and Anderson, C.H. 1990. Information processing strategies and pathways in the primate retina and visual cortex. In Zornetzer, S.F., Davis, J.L., and Lau, C., eds., *An introduction to neural and electronic networks.* Academic Press, San Diego, CA.

Van Essen, D.C., Anderson, C.H., and Felleman, D.J. 1992. Information processing in the primate visual system: an integrated systems perspective. *Science* 255:419-423.

Understanding Detective Stories

Ian Pratt, Luoping Xu and Ivan Leudar
Departments of Computer Science and Psychology
University of Manchester,
Manchester M13 9PL, UK
email: luoping@cs.man.ac.uk

Abstract

In this paper, we illustrate a general approach to psychological inference by considering its application to a simple detective story. Detective stories provide a fertile ground for the investigation of psychological inference, because their plots so often hinge on the mental states of the characters involved. Although our analysis contains several, logically independent suggestions for how to tackle some of the different problems that arise in understanding the story, one guiding principle underlies our approach: the *re-use thesis*. According to the re-use thesis, certain inferential mechanisms whose primary function has nothing in particular to do with psychological inference can be re-used for psychological inference tasks. In the course of the paper, we present several examples of the re-use thesis in action. Finally, we sketch how these applications of the re-use thesis can contribute to an understanding of our detective story.

1 Introduction

In this paper, we illustrate a general approach to psychological inference by considering its application to a simple detective story. The development of computer programs able to understand simple stories has long been seen as a major challenge for artificial intelligence, and the human being's remarkable ability to infer details left implicit in almost any narrative text, and thereby to forge the story into a coherent explanatory whole must be the envy of any present-day computer reasoner.[1] We focus on detective stories because of their relation to the present authors' wider interest in psychological inference—inference about the beliefs, desires, intentions, and other mental states of agents. The study of psychological inference occupies a central place in cognitive science, not only because of its obvious practical applications, but also because the concepts of *belief, desire, intention* etc. hold centre stage in the philosophy of mind. An account of how these concepts function in thought would thus address a central issue in the philosophy of mind.

Detective stories provide a fertile ground for the investigation of psychological inference, because their plots so often hinge on the mental states of the characters involved. Which of the suspects knew the contents of the heiress's new will? Did the mysterious Australian couple want to kill her for some reason? And why did the art dealer lie about the value of her grandfather's portrait? Answering and—more importantly—raising questions of this kind are essential in understanding a detective story; indeed, it is harder to think of a richer mine of problems connected with psychological inference.

2 A simple example

Consider the following simple detective story.[2]

> Gunner wanted to kill Ridley. He went to see him at his flat shortly before Ridley was due to play a match at the tennis club, and shot him dead. Then he dragged the body over to the door, took the telephone off the hook, and drove to the tennis club. When Ridley failed to turn up for the match, Gunner offered to telephone to see what happened. He went into the telephone booth and dialed. After a few minutes, he dashed out and said that, while he was talking to Ridley, there was a knock at the door, a scuffle of voices, and then shooting, followed by the door closing. The police were called and found Ridley's body. Later, the inspector talked to Gunner again about the telephone call. Then he quietly informed Gun-

*The research reported here was funded by the MRC/SERC/ESRC initiative in cognitive science and human-computer interaction, grant number 8920254

[1]There is a large literature on story-understanding in AI. For a representative example, see (Dyer, 1983).

[2]This plot is taken from one of the stories in a beginners' textbook in German.

ner that, according to the telephone company, Ridley's telephone had been disconnected last week. He then charged Gunner with the murder of Ridley, and Gunner was taken away by two waiting policemen.

The above story illustrates some of the inferences readers of detective stories are called upon to make. The central problem in understanding this story—and no one to whom we showed the story found it difficult to solve—is to work out how the detective inferred that Gunner was the murderer, not just an innocent (telephonic) witness. The pivotal move, of course, is to hit upon the question (from the detective's point of view) of how Gunner knew that Ridley was dead, given that the telephone was disconnected. For known he must have done. And unless he was implicated in the murder, Gunner's behaviour simply does not make sense.[3]

We can get some inkling of just how tricky this understanding process is by noticing that the sentence "Gunner knew that Ridley was dead" appears nowhere in the story; still less is there any reference to *the detective's beliefs as to whether* Gunner knew that Ridley was dead. Readers must therefore infer without prompting that the detective can reason that Gunner must have known that Ridley was dead, and they must then raise the question as to how (from the detective's point of view) Gunner could have come by this knowledge. Then, and only then is it a (relatively) short step to seeing how the detective can infer that Gunner may have shot Ridley himself and faked the telephone call to give himself an alibi.

3 Putting yourself in the detective's shoes

The current literature on psychological reasoning is replete with references to the idea that psychological reasoning may proceed by psychological simulation.[4] The idea is simple enough: if one wants to reason about what some one else will infer or decide to do who believes p, q, r and desires x, y, z, one imagines that p, q, r are true and that x, y, z are desirable, and one sees what one infers or decides to do. One thus uses one's own cognitive state (as it evolves within the imaginary environment) to simulate, and thereby predict, the cognitive state of another person. Well, that is a complicated and certainly controversial story, and we cannot possibly review all of its strengths and weaknesses here. Nor can we discuss the complex problems involved in temporarily suspending certain of one's beliefs. Nevertheless, it is hard to resist the idea that, at least sometimes, one reasons about the thoughts of other people by temporarily entering the make-believe world of those people.

Our detective story is a case in point. Understanding that story really amounts to understanding how the detective worked out that Gunner was the (likely) murderer, a fact of which the reader is informed in the second sentence. And it is plausible that, in trying to understand the detective's inferences, one temporarily suspends that information and imagines knowing what the detective knows: Gunner's rushing out of the telephone booth, the subsequent discovery of Ridley's body, the fact that the telephone was found not to have been working, and so on. Once one is in that cognitive predicament, one's thoughts race on as the detective's must have. The result is the answer to understanding the story.

Certainly, the thesis that psychological reasoning proceeds by psychological simulation can at best constitute a partial account of psychological inference. In particular, simulation (as the term is understood here) is inherently unsuitable for inferring causes from effects ("The *reason* why Gunner faked the telephone call was such-and-such') or for gaining *universal* or *necessary* information ("Gunner *must have* known that Ridley had been shot.") Nevertheless, in favourable situations, simulation is attractive because it holds out the prospect of a certain economy in thought. One needs no special psychological theory to reason about how the detective reasons; one just needs to be able to reason *as* he reasons, and then to observe how that reasoning goes and what it produces. Thus, according to the simulation idea, inferential mechanisms that support one's ability to make inferences and decisions generally can be *reused* in imaginary situations to reason specifically about what inferences and decisions other people will make.

[3]Some people to whom we put this story point out alternative explanations (from the detective's point of view) on which Gunner is not really guilty at all. However, no one we asked failed to understand why the detective might at least strongly suspect Gunner. That is: all those we asked understood the critical issue (for the detective) of how Gunner knew that Ridley had been shot.

[4]Pratt (Pratt, 1989) includes a brief guide to the simulation idea and its manifestations in the literature. Goldman (Goldman, 1989) also analyses the simulation idea (but with a different philosophical orientation); Leudar (Leudar, 1991) discusses the underlying supposition of psychological *similarity* between different persons. Perner (Perner, 1991) provides a comprehensive survey of recent work of the development of psychological concepts in children, in which the simulation idea plays an important role.

4 Understanding Lying

In general, understanding detective stories involves the ability to understand lying and pretence, and why people engage in it. In our story, for example, the detective finally makes sense of Gunner's actions because he sees them as a ploy to fabricate an alibi. How should our story-understanding program work this out?

A useful heuristic in explaining why an agent S performs an action α is to imagine oneself in S's position and see what one can infer about the likely effects of α. If consequences C arise which are, as far as one can judge, positive from S's point of view, then it is sensible to consider the hypothesis that S did α because he intended to bring about C. [5] (Notice that this process is *not* one of simulation.) It goes without saying that that is only part of the story about how one might assign motives: for one thing, S's actions can have unforeseen beneficial consequences; for another, if one is not privy to all of S's beliefs, one cannot unproblematically put oneself in S's position in order to determine the consequences of α by S's lights. Nevertheless, some version of the suggested rule is likely to be a good heuristic for *forming hypotheses* about why S did what he did.

A variation on this heuristic can be used to hypothesize motives for lying. Suppose S tells a lie, P, and one wants to know why. Well, as a special case of the above heuristic, one can consider what S thinks the effects of saying that P are. And one way to solve this problem is to put oneself in the imaginary state of someone hearing the utterance P, but without the special information that S is lying. If, in this imaginary state, one draws a conclusion which S might want one to draw, or reaches a decision which might be favourable to S, then it is sensible to consider the hypothesis that S lied because he wanted one to draw that conclusion or reach that decision. Again, it goes without saying that that is only part of the story about how one might assign motives for lying (or for pretence generally): for one thing, S's utterances have unforeseen consequences; for another, since one is not privy to all of S's beliefs about one's own state of mind, one cannot be sure what S will take one's reaction to his statement to be. Nevertheless, some version of the suggested rule is likely to number among the good heuristics for forming hypotheses about why S lied.

Our detective story is again a case in point. Why, from the detective's point of view, might Gunner have faked the telephone call? Well, if one imagines not knowing that the telephone was disconnected, the natural explanation of Gunner's behaviour is that Gunner heard shooting on the phone because someone shot Ridley while Gunner was on the telephone to Ridley. This would, after all, explain why Ridley's body was found in his flat, and why the telephone was off the hook. Let us call this the *naive explanation*. To be sure, there *are* other explanations of Gunner's actions (there are always other explanations), but the naive explanation seems—without the benefit of knowing that Ridley's telephone was not working—the most plausible.[6] And, of course, on this explanation, Gunner could not have shot Ridley. According to the strategy suggested in this section, then, one can reason as follows about why Gunner lied: if one first imagines not knowing that Gunner was lying about the telephone call, one infers the naive explanation, and concludes that Gunner did not kill Ridley. Now, since Gunner might want one to draw this conclusion, a possible explanation of his actions—we might call it the *alibi-explanation*—is that it was a means to get people to believe he is innocent. Notice how, here again, we are suggesting that a general inference mechanism—the mechanism whereby one determines the consequences of actions—can be *re-used* in psychological inference.

5 Raising questions about beliefs

Let us return to the question of how the detective came to see that Gunner must have known that Ridley had been shot. As mentioned above, this issue is crucial to understanding the story. We proceed via an informal experiment. Imagine learning that someone has telephoned the police to say that there has been a shooting in a house not far from where you live. Imagine in addition that you have no more information about circumstances in which the call was made, or about the identity of the caller. (This is all very improbable, but that does not matter for our purposes.) Suppose now, you are asked to conjecture what might happen next. Presumably, you would answer (roughly) thus: "Armed police will go to the house to see if anything is wrong. If all is quiet, they will knock on the door. If someone answers, they will ask questions about the incident. If, on the other hand, there is no answer, they may try to force their way in. Once inside, they will look for a body or signs of a shooting. If they find no body, they will trace the occupants of the house, ..."

[5] See (Pratt, 1990).

[6] The problem of how, in general, one constructs explanations and decides between competing explanations is not addressed by the present paper. For some discussions of this huge topic, see (Harman, 1986, Lipton, 1991). See also (Leake, 1990, Antaki & Leudar, 1991).

Being able to conjecture the effects of events in the presence of only very partial background information is an indispensable inferential accomplishment. We might call the inferential process involved *scenario-branching*, because of the tree of alternative scenarios that is generated. Do the police find the house deserted? Does anyone answer the door when the police knock? When the police search the house, do they find a body? And so on: many unanswered questions, and correspondingly many branch-points in projecting the possible course of events. But it is absolutely vital not to consider *all* conceivable questions. One should not consider whether the police turned right or left as they left the police station, or whether they drove past a lady with a pram, or (if the knock on the door was answered) whether they were offered a cup of tea, and so on. In short, branching is to be avoided if it is unlikely to yield to interestingly different results; otherwise, projection would be stymied by an infinity of irrelevant possibilities. How people control the branching of possibilities when asked to consider very partially specified scenarios is a question we prefer not to address here. All we claim is that they do.

Now put yourself again in the position of the detective in the story. You have heard Gunner's account of the telephone call, and you know that it cannot be true. You might then wonder what Gunner could possibly have thought he would gain by such a pretence, and to do this, you might simply consider, from Gunner's point of view, what would happen when the police are called in such a situation. Since you are largely in the dark about Gunner's beliefs, there are a number of possibilities. We suggest that, in trying to determine the most interesting of these, you can simply engage in scenario-branching as just described. That is, you can simply consider what happens when the police are called with a report of a shooting, and ask yourself what might transpire. As we have seen, a tree of alternative scenarios will branch out, as the significant unknown factors occur to you. This time, of course, the object of the exercise is not to determine what may happen, but what *Gunner might have thought may happen*. And this changes the way you use the branch-points as the scenario-projection mechanism throws them up. Thus, you ask not whether there will be anyone at home to answer the door when the police call, but whether *Gunner thinks there will*; not whether they police will find a body, but whether *Gunner thinks they will*, and so on.

As we have already observed, the crucial move in understanding the detective story above is to make the unprompted inference (reasoning within the detective's predicament) that Gunner must have known Ridley was dead. The problem here is not so much *verifying* this fact as *thinking of it in the first place*. And, in this section, we have put forward a mechanism for how one might be led to consider this possibility. (The problem of verifying it is not considered in this paper.) Our suggestion is that one can deploy the mechanism for scenario-branching—the mechanism, that is, whereby one conjectures the effects of actions in very partially specified situations. The attractiveness of our suggestion resides in the fact that, according to it, there is no need for a special psychological theory from which we can raise important questions about the beliefs of agents; the ability to conjecture the effects of their actions in unspecified circumstances works, we suggest, well enough.

In each of the previous three sections, we have described how inferential mechanisms whose primary function has nothing in particular to do with psychological inference can be re-used for psychological inference: (i) inferential mechanisms generally can, in favourable cases, be re-used in 'pretend-mode' to reason forwards about the psychological state of another person; (ii) inferential mechanisms for projecting the consequences of actions can be re-used to hypothesize motives of the agents who perform them (a variation on this process involves divining the reasons for lies); and (iii) inferential mechanisms controlling the branching of possibilities when projecting the consequences of actions in incompletely specified situations can be re-used to raise important questions about the beliefs of agents. The idea of the re-use of inferential mechanisms in the keystone in our approach to psychological inference. So much so, it deserves a name:

The re-use thesis: Inferential mechanisms whose primary function has nothing in particular to do with psychological reasoning can be re-used for psychological reasoning tasks.

We stress that, for the purposes of this paper, the re-use thesis is to be considered as a suggestion in AI—that is, as a suggestion for how to design a computer program capable of effective psychological inference. The question of whether *human beings* re-use non-psychological inference mechanisms for psychological reasoning is one we shall not discuss.

6 Putting the bits together

In order to test some of the ideas developed here, a program is currently being written to process some stories requiring psychological inference. In this section, we sketch very briefly how to assemble the foregoing elements into an explanation of the above story.

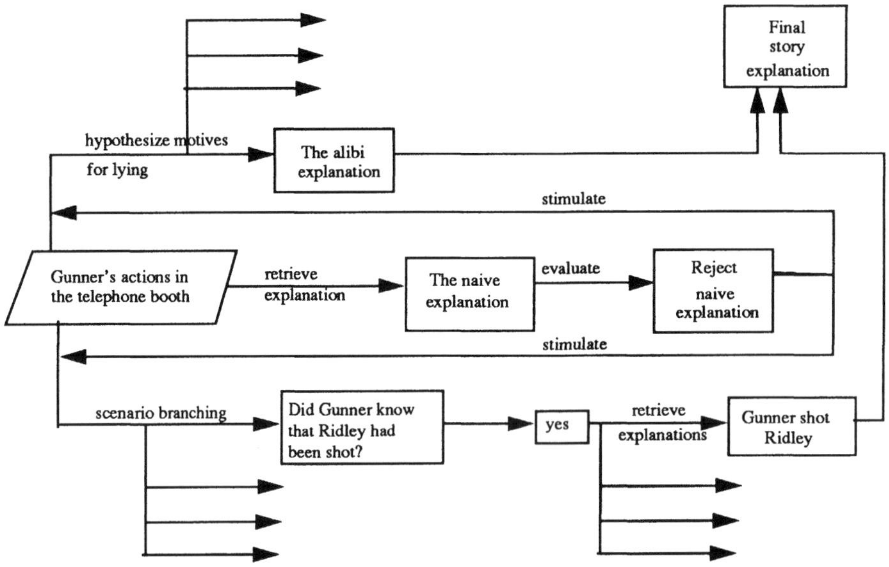

Figure 1: The flow of inference for the detective story.

We certainly do not claim to have a theory of inference general enough to account in an uncontrived way for all of the inferences the story demands of its readers. Our primary concern in this project is the relationship between psychological and non-psychological inference problems, and the present section is included merely to provided an overview of how we envisage our view of that relationship as contributing to the understanding of the above detective story.

The flow of inference is depicted very schematically in fig. 1. The program starts by trying to explain Gunner's assertion that he heard Ridley shot on the telephone. The first explanation the program finds is the *naive explanation*, according to which Gunner really did hear Ridley get shot and wanted to summon help to him. When this explanation is rejected (because it is discovered that the telephone was disconnected) other explanations are considered. At this point, two parallel strands of reasoning are stimulated. The first strand follows the procedure described in section 4 for hypothesizing motives for lying. Here, several alternative explanations are constructed, among them, the so-called *alibi-explanation*, according to which Gunner lied about being on the telephone to Ridley in order to get people to believe he did not kill Ridley. It is important to be clear that, at this stage, the alibi-explanation is only a hypothesis as to why Gunner said what he said—one explanation, that is, among a field of possible contenders. The decisive factor that makes the alibi-explanation (in contrast to its competitors) ultimately acceptable emerges from the second strand of reasoning.

The second strand of reasoning addresses the problem of reconstructing the detective's inference that Gunner must have known all along that Ridley had been shot. The program adopts the procedure described in section 5 for raising important questions about the beliefs of agents, by performing scenario-branching on the proposition that Gunner said that he heard Ridley being shot on the phone. As a result of this process, the question arises as to whether Gunner thinks that Ridley has been shot. Once the program has hit upon this issue, it can confirm that indeed Gunner must have known. The process of confirming this conjecture is somewhat complicated and lies outside the detailed issues discussed in this paper, but, briefly, the program determines that, unless Gunner had good reason to believe that Ridley had been shot, he could expect nothing but negative consequences (including a conviction for wasting police time) to flow from his actions. The next step in this second strand of reasoning is to explain *how* Gunner could have known of Ridley's shooting. A number of possible standard explanations for coming-to-know are considered, many of which can be rejected for reasons that need not detain us here. The one standard explanation for coming to know something that bears fruit is the one on which, in this case, Gunner knew that Ridley had been shot because he actually did the shooting. Having inferred that Gunner shot Ridley, the program deduces that Gunner will have the goal of getting the detective to believe that he did not shoot Ridley (again, the details of this inference are not considered in this paper).

This is where the two strands of reasoning link up. The first strand has hypothesized a number of possible motives for Gunner's actions, among them, alibi-explanation that Gunner wanted people to believe he did not kill Ridley; the second strand has inferred that Gunner did kill Ridley and will therefore have the goal of getting people to believe that he did not. For this reason, the program chooses the alibi-explanation from among its competitors. The result is the inference that Gunner killed Ridley and devised the scene in the telephone booth to fabricate an alibi.

7 Conclusions

In this paper, we have considered a simple detective story with a view to analysing some of the inferential mechanisms needed to understand it. In particular, we have highlighted the need for the reader to reason within a number of imagined cognitive predicaments. Although our analysis contains several, logically independent suggestions for how to tackle some of the different problems that arise, one guiding principle underlies our approach: the re-use thesis. According to the re-use thesis, certain inferential mechanisms whose primary function has nothing in particular to do with psychological reasoning can be re-used for psychological reasoning tasks. We described three examples: (i) inferential mechanisms generally can, in favourable cases, be re-used in 'pretend-mode' to reason forwards about the psychological state of another person; (ii) inferential mechanisms for projecting the consequences of actions can be re-used to hypothesize motives of the agents who perform them (a variation on this process involves divining the reasons for lies); and (iii) inferential mechanisms controlling the branching of possibilities when projecting the consequences of actions in incompletely specified situations can be re-used to raise important questions about the beliefs of agents whose actions are to be explained. For present purposes, we interpret the re-use thesis as a suggestion for AI—that is, as a suggestion for how to construct effective *computer* reasoning systems. All we can claim to have established in the work reported here is a plausible case for its efficacy. The proof of the pudding will be, as ever, in the eating.

8 References

Antaki, C. and Leudar, I. 1991. Explaining in Conversation,*European Journal of Social Psychology*, 22(2):181-194.

Dyer,M.,1983. *In-depth Understanding: A Computer Model of Integrated Processing for narrative comprehension.* Cambridge, MA: MIT Press (1983).

Goldman, A. 1989. Interpretation Psychologized. *Mind and Language*, 4(3).

Harman, G., 1986. *Change in View.* Cambridge, MA: MIT Press.

Leake, D. 1990. Task-Based Criteria for Judging Explanations, In Proceedings of the Twelfth Annual Conference of the Cognitive Science Society, 325-332 Cambridge, MA.

Leudar, I., 1991. Sociogenesis, Coordination and Mutualism. *Journal for the Theory of Social Behaviour*, 21(2):197-220.

Lipton, P. 1991. *Inference to the Best Explanation.* London: Routledge.

Perner, J. 1991. *Understanding the Representational Mind.* Cambridge, MA: MIT Press.

Pratt, I. 1989. Psychological Inference, Constitutive Rationality and Logical Closure, in Vancouver Studies in Cognitive Science, vol.1:366-389. Information, Language and Cognition, Vancouver, BC: University of British Columbia Press.

Pratt, I. 1990. Psychological Simulation and Beyond, in Proceedings of the Twelfth Annual Conference of the Cognitive Science Society, 654-661. Cambridge, MA.

Imagery as Process Representation in Problem Solving*

Yulin Qin **Herbert A. Simon**

Department of Psychology
Carnegie Mellon University
Pittsburgh, PA 15213
qin@psy.cmu.edu has@a.gp.cs.cmu.edu

Abstract

In this paper, we describe the characteristics of imagery phenomena in problem solving, develop a model for the process of forming and observing mental images in problem solving, and check the model against data obtained from subjects. Then, we describe the interaction between imaging and problem solving observed in our experiments, and discuss the use of our model to simulate it. We also discuss the relation between mental models and mental image briefly.

We are interested in exploring the role and characteristics of mental imagery in problem solving and understanding. Our method of inquiry is to observe how subjects understand , with the help of their mental images, the first, kinematic, part of Einstein's 1905 paper: "On the Electrodynamics of Moving Bodies" (Einstein,1905).

The generation, maintenance, inspection, and transformation of images have been well researched in general and at a fine grain size (cf. Kosslyn 1988 for summary). For example, Shepard and Metzler (1971) found that the more one rotates an imaged pattern, the more time is required. Our discussion will focus on more special situations and a rather larger grain size: on complex information processes used in solving a non-trivial problem. In our experiments, subjects used their images to simulate physical processes, and by finding the relation among the related physical quantities, they acquired the information necessary for problem solving -- deriving the appropriate equations.

---

* This research was supported by the Defense Advanced Research Projects Agency, Department of Defense, ARPA Order 3597, monitored by the Air Force Avionics Laboratory under contract F33615-81-k-1539. Reproduction in whole or in part is permitted for any purpose of the United States Government. Approved for public release; distribution unlimited.

Many thanks to Dr. John Anderson and Dr. Jill Larkin for their valuable comments and suggestions.

In Qin and Simon (1992) we reported briefly on the major findings of our project. In Qin and Simon (1990), we discussed in a preliminary way what kinds of information subjects could obtain from their mental images and how they used it in their problem solving. In this paper we will discuss how subjects obtain this information. i.e., the processes subjects use to form, watch (make inferences from) and change their images during their problem solving.

1. Experiment

We will be concerned with the imagery subjects used while deriving the first pair of key equations needed to obtain the Lorentz transformation equations. The reading material states:

> "Let there be given a stationary rigid rod... We now imagine the axis of the rod lying along the axis of x of the stationary system of co-ordinates, and that a uniform motion of parallel translation with velocity v along the axis of x in the direction of increasing x is then imparted to the rod...We imagine further that at the two ends A and B of the rod, clocks are placed which synchronize with the clocks of the stationary system...We imagine further ... Let a ray of light depart from A at the time t_A, let it be reflected at B at the time t_B, and reach A again at the time t'_A. Taking into consideration the principle of the constancy of the velocity of light we find that
>
> $$t_B - t_A = r_{AB} / (c-v) \qquad (1)$$
>
> and
>
> $$t'_A - t_B = r_{AB} / (c+v) \qquad (2)$$
>
> where r_{AB} denotes the length of the moving rod -- measured in the stationary system."

Figure 1 shows the process of light traveling while rod AB is moving relative to the stationary system. Equation (1) can be derived as follows: When light arrives at B at the time t_B, B has moved a distance, $v(t_B - t_A)$, measured in the stationary

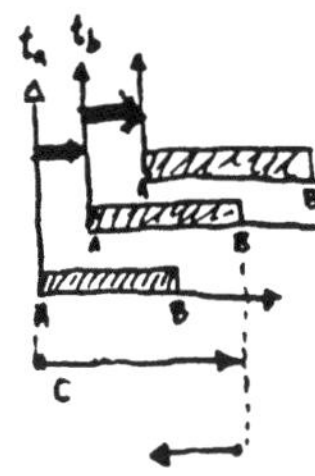

Figure 1

system. So the total distance the light traveled from A to B is $r_{AB} + v(t_B - t_A)$. We assume the constancy of the velocity of light, c, whether it is measured in the stationary system or in the system of the moving rod. Therefore, the total distance the light traveled from A to B is $c(t_B - t_A)$. So we have:

$$c(t_B - t_A) = r_{AB} + v(t_B - t_A).$$

Collecting the terms in $(t_B - t_A)$, we get equation (1). We call this method of derivation the *length method.*

Another method, the *velocity method*, for deriving the equation is to pay attention to velocity instead of length. When light travels from A to B, it travels in the same direction as the moving rod. So, *relative to the moving rod*, still measured in the stationary system, the velocity of the light is c-v, and the distance the light travels is r_{AB}. From the relation: time equals distance divided by velocity, we also get equation (1).

Reference system and *measured system* are key concepts here. Two different reference systems were employed in the two different methods. An interesting phenomenon is that in using the velocity method, but not in using the length method, subjects usually reported the reference system. Equations (1) and (2) contrast with the equations based on Galilean transformations, which assume that the *time* for light to travel, measured in both moving and stationary systems, is the same, and the velocity of light is not a constant.

We obtain information from which we infer what mental images subjects have formed from their protocols, the diagrams they drew, and their gestures.

We have analyzed six subjects' protocols. One subject, with an MS degree in computer science, is a research assistant in psychology at Carnegie Mellon University (CMU). The others are undergraduate or graduate students in electrical engineering or computer science at CMU. None of them were familiar with derivations of the Lorentz equations or could derive them. None were aware that the reading material was from Einstein (1905), until, at the end of the experiment, they were told by the experimenter.

Among these six subjects, S_r and S_m were asked to describe their mental images, draw diagrams, use diagrams; S_b and S_j to describe images, but not to draw diagrams; and S_g and S_s to describe images and draw diagrams, but not to retain the diagrams (The experimenter took each diagram away as soon as it was drawn, and they could not draw diagrams while deriving the equations.) In this paper we will not emphasize differences between the groups, but will try to find some common patterns among them. What the experimental design tells us is that four of six of our subjects could not use diagrams in their problem solving. Allowing some subjects to draw diagrams helps us to clarify their protocol statements about their images. Conversely, protocols can help us to identify the mental images underlying the diagrams they drew, especially in the case of the subjects who were allowed to draw and use diagrams while deriving the equations.

2. The Process of Forming and Watching Mental Images

2.1 Similarities and Individual Differences of Subjects' Mental Images

In his 1905 paper, Einstein invited his readers several times to "imagine" the situations he described. Curiously enough, the published paper contains no diagrams to guide or assist this process. All of our six subjects could form and report their images by means of diagrams and/or protocols. From these we can find the characteristics of their imagery.

Figure 1 is based on S_g's diagram. Figure 2 gives the diagrams drawn by all the other subjects except S_b, who did not draw any diagrams in this experiment. His mental image is inferred by us from his protocol. To make clear the meaning of "triangle image" in S_j's protocol, the experimenter asked S_j to draw the diagram .

To form mental images, subjects needed to add to and clarify the information in the reading material. Without any diagram in the reading material to guide them, there are large individual differences among their images. For example: S_j's "triangle image" has a time dimension (in the horizontal direction), but others do not; Two subjects', S_g and S_s, drew images that had explicit coordinate systems, others did not; S_s' rod and light traveled in a vertical direction, but the others' in a horizontal direction. While traveling from A to B, in most subjects' diagrams, light moved in the same direction as the moving rod, but in S_s' in the opposite direction.

However, underneath the differences, there are some similarities. For example, all the images

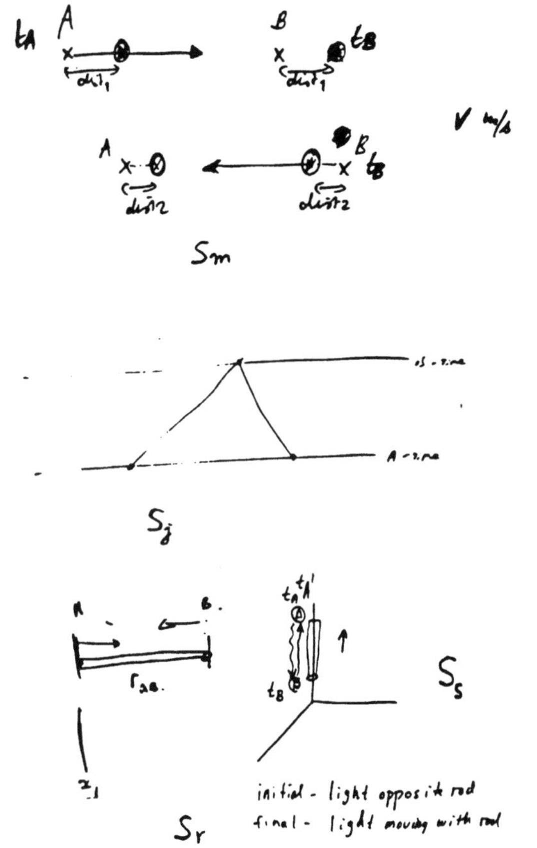

Figure 2

show the process of the light traveling. They represent a process of change. All of them are very simple and neat, containing only information necessary to the subjects' understanding.

2.2 A Model of the Processes of Forming and Watching Mental Images

Based on the data collected in our experiments, we have inferred a model of the processes of forming and watching images, shown in Figure 3.

The boxes in Figure 3 are actions, the circles are data or representations, the solid arrows show the procedure flow. We will discuss them in this section. The dotted arrows in Figure 3 show the information flow, i.e., the influence of problem solving and mental model on forming and watching mental images. We will discuss them in the following two sections.

As shown in Figure 3, the basic process of forming and watching images is as follows:

1. Setting the goal of imaging;
2. Collecting picture information, such as components, structure, and kinematic relations. It consists of these sub-processes:
   a). Choosing type of image: Usually subjects use the default value (running an image of the full situation). However, S_m and S_g also ran a simple partial image to get additional information;
   b). Choosing component: For example, S_j chose the time line as a component in his image;
   c). Assigning attributes and relations: Subjects assign a spatial and a kinematic relation between components (e.g., light traveling from end A to end B of the rod); and assign a given value to an object (e.g.," the length of the moving rod measured in the stationary system is r_{AB}".);
   d). Forming stable image: Integrating all the components (This seems necessary, but no subject mentioned it explicitly);
3. Choosing process and frame, such as moving components, moving process, and reference system, etc. It consists of these sub-processes:
   a). Choosing moving components: The default value is all of the movable components. However, in S_m's simple image, only light is represented as a moving component;

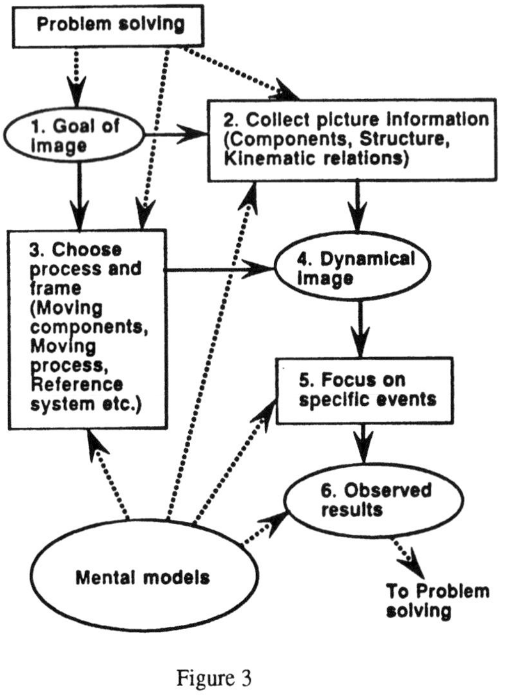

Figure 3

b). Choosing operating method: The choice is based on how clear the dynamic process (e.g., the process of light traveling) is in subjects' minds. When it is clear how to move the moving components, subjects do not report their methods. But when S_m, for example, did not know the path of the light, she tried to determine the path step by step, and then formed her image;

c). Choosing reference system ;

d). Choosing measuring system ;

e). Choosing process, i.e., deciding what process he/she wants to see. For example, S_s used the velocity method to derive the equation. He did not form an image of the whole process of light traveling from A to B and then back to A. Instead, he only imaged light leaving the A end of the rod and the rod moving at the same time;

f). Directing attention: Fixing attention on velocity, or distance, or holding no special focus of attention. For example, a subject using the length method to derive the equation might not pay attention to the velocity of light relative to the moving rod, but instead, focus on the difference between the length of the rod and the distance the light traveled;

4. Forming original dynamical image;
5. Focusing on specific events, forming a snapshot. S_m, for example, reported that when she saw the light arriving at B, she could not see A, but she knew A was moving. It seems possible that, at first, subjects form an image that shows the whole situation, and when trying to measure a particular quantity, they delete some unrelated elements;
6. Getting the result of observing, and giving the explanation of the observed result.

Subjects do not always report all of their steps. A rather complete report, shown below, is S_j's protocol given while he was forming a new image to replace the "triangle image." The number in front of a sentence denotes the (sub-)process to which the sentence corresponds:

(1) I try to figure...(2.b, 2.c) it's the ends of the rod, A is one end of the rod and B is the other end of the rod. (2.b, 2.c) So I guess it [light] is going from one end of the rod to the other and bouncing back to the first end of the rod. (3.c) And I'm trying to figure out if this is all happening in the moving frame of reference as it says RAB goes the length of the moving rod and measured in the stationary system. (3.d) Okay. All right, so this is all measured in the stationary system. (3.e, 4) So it goes from TA to TB it's got, it's going from the beginning of the rod to the end of the rod. (3.f, 5) And it's got a velocity, and the rod is moving at V and the velocity is going and the frame of reference is moving at V. (6) So light impulse then appears to be moving at C minus V, (6) because it's already moving, point A is already moving at V.

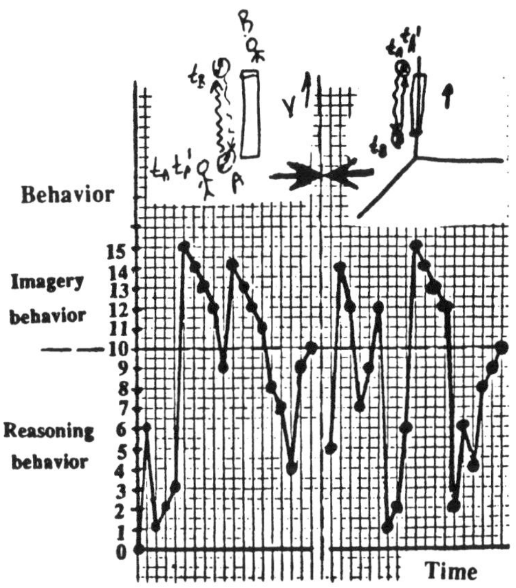

Figure 4

Of the sub-processes that did not appear in this protocol, (2.a) and (3.a) seem to assume their default values; and (2.d) and (3.b) are not usually reported in this kind of situation.

3. The Interaction between Mental Image and Problem Solving

3.1 Subjects' Switches in Attention between Reasoning and Imaging

Figure 4 shows a part of S_s' behavior in the first day of his deriving the equations and the images he formed in this period. The horizontal axis is the time axis, but it only reflects the time order of the behaviors, not time in seconds or minutes. The vertical axis shows his behavior. The meaning of the numbers is as follows (0 to 9 are reasoning behaviors, 11 to 15 are imagery behaviors) :

0. Reading equations;
1. Setting the goal of problem solving;
2. Describing the meaning of tB - tA;
3. Describing the meaning of the equations;
4. The meaning of c - v;
5. The reason for c - v;

6. The meaning of rAB: the length of the rod;
7. The meaning of rAB: the distance the light travels;
8. Basic principle: distance formula t = d/v;
9. Deriving or giving the reason for the equations;
10. Self-monitoring;

11. Qualitative characteristics of the equations;
12. The value of the velocity, by observing the image;
13. The reasoning for the value of velocity;
14. Image, paying attention to the velocity.
15. General image;

As shown in Figure 4, S_s switched his attention between forming and watching images, and reasoning about equations. All of the subjects did so. S_r and S_b's protocols are very short; they only spent 6 and 9 minutes, respectively, in this part of the task, and only switch 2 and 5 times, respectively. Others spent more time on this part and switched more often. For example S_s spent 500 minutes and switched 133 times. Omitting S_m (some of her data were lost), among 5 subjects, the average frequency of switching is once in 4 minutes; the range of frequencies among the subjects is from 0.15 to 0.55 times per minute.

Heavy STM load could be one of the reasons for the attention switching. However, as shown in Figure 4, in the course of this switch process a subject may change his image. So another reason could be the interaction between imaging and problem solving.

3.2 The influence of problem solving on forming and watching images

Images can change the method of problem solving (Qin and Simon, 1990). On the other hand, images are built for the purposes of problem solving, and the problem requirements can change the images.

As shown in Figure 3, our model proposes that problem solving will determine the goal of an image, and influence the processes of collecting picture information and choosing process and frame. We will describe some types of changes in imaging observed in our experiment, and show how we simulate them in our model.

1. Change the image wholly, as S_j does, for example, when the image appears inappropriate to the problem. This can be simulated by changing the choice of components, attributes and relations.

2. Form a new rather simple image to get additional information, as seen in S_m's and S_g's protocols. This can be simulated by choosing a different type of image;

3. Change a local part of an image to clarify a relation. Only after she made this kind of change to show the qualitative relations among unknown quantities, could S_m derive the equations. Such changes can be accomplished by observing the old image to get qualitative results and then matching these results to related quantities;

4. Change the relation among the components of an image and the values of components in the image when a wrong answer is obtained. Figure 4 shows an example from S_S. Such a change can be simulated by altering the mental model and the assignments of attributes and relations;

5. Shift attention from one method to another when difficulties are encountered, e.g. from the velocity to the length method. This is the easiest kind of change, and the one most often used by the subjects. Four of the six subjects changed methods, S_j changing 7 times and S_S 9 times. The method can be changed by altering the focus of attention.

4. Mental Models and Mental Image

Subjects usually gave reasons for their images. For example, S_m reported while forming her image:"Light traveling needs time. So, B is moved..." In Figure 2, section 2, we have seen the influence of subjects' different prior available knowledge on their images.

As seen in S_j's protocol in section 3, when subjects reported an "observed value", they also usually gave their reasons. It seems that the "observed value" is obtained by combining the visual information (based on "seeing" in the "mind's eye"), and reasoning based on prior available knowledge. We call this prior knowledge, which may be evoked by the language of the problem description, the *mental model*. From another viewpoint, imagery in problem solving seems to integrate information already available in the mental model with new information, and thereby reach conclusions that are impossible or difficult to infer without images.

If there is a bug in a mental model, a subject may form a wrong image or "observe" a wrong result from the image. For example, by watching a correct image, the first one shown in Figure 4, S_s derived a wrong value of light velocity, which was opposite to the equations (1) and (2). He reported in his protocol about the situation of light traveling from A to B: "if you look at the bottom of the rod, the speed of light relative to stationary system is in the same direction as the rod so it would simply be the distance divided by C plus V which is the speed relative to the stationary coordinate system." and about from B back to A:"So you have the thing moving at V and the light is being shined back so you have to subtract the two because they are going in different directions." It appears that S_S took Galilean transformations as his mental model of the situation. (For example, in computing the velocity of a boat traveling in the direction of the current as measured from the bank, the velocity is c+v).

5. Discussion and Conclusion

In this paper, we have described the characteristics of imagery in problem solving, have developed a model for the process of forming, watching and changing mental images in problem solving, and have checked the model in a preliminary way. We have also described the interactions between imaging and problem solving that we observed in our experiment and have discussed how our model's behavior can be simulated. We have also discussed the relation between mental models and mental image briefly.

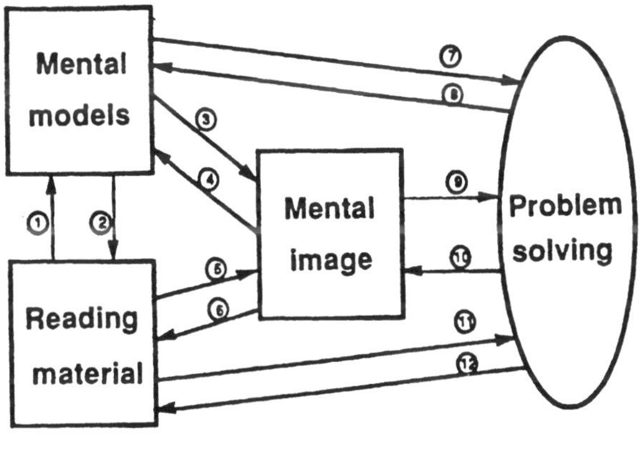

2) : The mental models guide the attention in reading, and influence the process of getting information from the reading material;
3), 5), 10): Mental models, reading material. and the feedback from problem solving are the information sources for image;
7), 9), 11): Mental models, reading material, and image offer information in problem solving;
6), 12) Image and problem solving also influence reading,

Figure 5

On the one hand, the process of forming and watching mental images involves: (1) forming the image (the object to be observed); (2) focussing attention (choosing the quantities to be measured); (3) acquiring the value of the measured quantities. This procedure is similar to people's observing the outside world to get information for problem solving. On the other hand, the result obtained by forming and watching mental images depends on subjects' mental models. Mental imagery seems to involve some combination of the visual information, which is based on "seeing" in the "mind's eye", and reasoning based on given knowledge. These relations among reading material, mental models, mental images and problem solving are depicted in Figure 5.

The consistency in this experiment of our subjects' drawings, gestures, and protocols, and the consistency of these data with the research on the neural basis of mental imagery (Farah, 1985; Kosslyn, 1987) support the usefulness of these four kinds of mutually reinforcing data as sources of information about invisible mental images. A great deal of work remains to be done on the relation between internal and external representations before we can have complete confidence in the generalizations we have drawn from the evidence, and the conditions under which these generalizations hold.

References

Einstein, A., 1905. On the electrodynamics of moving bodies. in A. Einstein, et al. *The Principle of relativity*. Dover Publications, 1923.

Farah, M. J., 1985. The neurological basis of mental imagery: a componential analysis. In S. Pinker, ed. *Visual cognition*. Cambridge MA: MIT press.

Kosslyn, S. M., 1987. Seeing and imagining in the cerebral hemispheres: A computational approach. *Psychological Review*, 94:148-175.

Kosslyn, S. M., 1988. Imagery in learning. In M. S. Gazzaniga, ed., *Perspectives in memory research.* Cambridge, MA: MIT Press.

Qin, Y. and Simon, H. A., 1990. Imagery and problem solving. In Proceedings of the Twelfth Annual Conference of the Cognitive Science Society, 646 - 653. Hillsdale, NJ: LEA publishers.

Qin, Y. and Simon, H. A., 1992. Imagery and mental models in problem solving. In Working Notes of AAAI Spring Symposium on Reasoning with Diagrammatic Representations. Stanford University.

Shepard, R. N. and Metzler, J., 1971. Mental rotation of three-dimensional objects. *Science* 171: 701-703.

What does a System Need to Know to Understand a User's Plans?

Bhavani Raskutti & Ingrid Zukerman
Computer Science Department
Monash University
Clayton, VICTORIA 3168, AUSTRALIA
email: {banu,ingrid}@bruce.cs.monash.edu.au

Abstract

During natural language interactions, it is often the case that a set of statements issued by a speaker/writer can be interpreted in a number of ways by a listener/reader. Sometimes the intended interpretation can not be determined by considering only conversational coherence and relevance of the presented information, and specialized domain knowledge may be necessary in choosing the intended interpretation. In this paper, we identify the points during the inference process where such specialized knowledge can be successfully applied to aid in assessing the likelihood of an interpretation, and present the results of an inference process that uses domain knowledge, in addition to other factors, such as coherence and relevance, to choose the interpretation intended by the speaker. Our mechanism has been developed for use in task-oriented consultation systems. The particular domain that we have chosen for exploration is that of a travel agency.

Introduction

During natural language interactions, it is often the case that a set of statements issued by a speaker/writer can be interpreted in a number of ways by a listener/reader. For instance, there are two possible interpretations of the statements "John was depressed. He bought a rope." One of them is that John has a plan to hang himself. Another is that John has a plan to do some rope skipping. The first one is the interpretation that most people would prefer. This is explained by the fact that in coherent conversations, people expect statements to be linked to a central theme. In addition, the interpretation that the rope is to be used for skipping has no explanation for John wanting to do skipping. Hence, all the information presented cannot be used relevantly. Thus, in this case, the preferred interpretation is arrived at by taking into account conversational coherence and the relevance of the information that was presented.

However, there are situations where the intended interpretation can not be determined by considering these two factors alone, and specialized domain knowledge may be necessary in choosing the correct interpretation. For instance, consider the following (real life) request of a lady to her husband, an hour before they were due to report at the airport for their return flight from overseas, "Can you fill petrol and have the films developed?" The husband inferred that the jobs could be performed in any order and hence filled petrol first. He had neglected to use the domain knowledge that the camera shop needs one hour to process the film, and found out that they could not process the film.

In this paper, we identify the points during the inference process where specialized knowledge can be successfully applied to aid in assessing the likelihood of an interpretation, and present the results from an integrated mechanism that considers multiple interpretations, and uses domain knowledge, in addition to other factors, such as coherence and relevance, to choose the interpretation intended by the speaker. Our mechanism has been developed for use in task-oriented consultation systems. The particular domain that we have chosen for exploration is that of a travel agency.

An interpretation of a user's statements is a set of plans that the user proposes to carry out, and a plan consists of an action with a number of parameters defining the action. For instance, in the travel domain, the proposal to fly from Melbourne to Sydney on December 1st, 1991, is a plan, where *flying* is the action, and the parameters *departure location*, *arrival location* and *departure date* are instantiated.

A number of researchers have considered the various factors that influence the choice of an interpretation. Litman and Allen (1987) take discourse coherence into account to prefer an interpretation. However, since they consider only one interpretation, they cannot cope with situations where a preferred interpretation must be given up in light of new information. For instance, if the above sample statements regarding John are followed by the revelation "Skipping always cheered him up," then the intended interpretation can be arrived at only if the previously discarded interpretation is reinstated. The problem of multiple interpretations has been considered by Carberry (1990) and by Goldman and Charniak (1991). However, the influence of different factors is hidden in the priors chosen for the different hypotheses or contexts. Thus, extensions to their systems to accommodate domain knowledge considerations will not be modular.

In this paper, we present a method for incorporating domain knowledge into a domain independent inference mechanism. In our discussion, we refer to the inference mechanism presented in [Raskutti & Zukerman 1991]. This mechanism addresses the problem of choosing the intended interpretation among multiple possibilities by generating all possible interpretations of a speaker's statements and tagging them with a likelihood measure. The likelihood tag is used to determine the interpretations that have to be maintained during the processing as well as to select the preferred interpretation(s). The likelihood measure used by the mechanism is calculated using Bayesian theory of probability and it is based on two factors: (1) discourse coherence — which favours discourses which are closer to normal patterns of conversation; and (2) relevance of the presented information — which is determined by the *information content* of an interpretation. In this paper, we add another factor, namely domain knowledge, which favours the interpretations that are more likely within a particular domain.

The incorporation of domain knowledge to determine the intended interpretation is based on the following factors: (1) the extent of the usage of a plan for a particular purpose, (2) the feasibility of a single plan and of the overall combination of the plans in an interpretation, and (3) the practicality of a single plan and of the overall combination of the plans in an interpretation. The subsequent sections discuss these factors with reference to the mechanism described below.

The Main Mechanism

In this section, we describe briefly our algorithm for generating and selecting interpretations from a user's utterances. The algorithm may be roughly divided into two inference processes: (1) Direct inference, and (2) Indirect inference. Direct inferences are those that are drawn on the basis of the user's statements, the definition of domain actions and discourse coherence considerations. Indirect inferences are those that are based on domain and world knowledge.

The direct inference stage consists of the inference of interpretations that are based on the definition of the basic actions of the domain and the user's statements. This stage is composed of three parts: (1) the inference of interpretations for each of the user's statements, (2) the inference of discourse relations between the interpretations generated from one statement and the interpretations of the earlier discourse, and (3) the generation and selection of new interpretations based on the first two parts.

The indirect inference stage is used to complete the definition of the plans in the interpretations generated by the direct inference process. During this stage, parameters of a plan are instantiated using sources other than the user's statements, such as plan relationships, and domain and world knowledge. The plan relations we consider are causal relations, such as ENABLE and CONSTRAIN, and temporal relations, such as BEFORE and AFTER. The temporal relations can be either inferred or explicitly stated. Other bases for inference, such as domain and world knowledge, are organized as indirect inference rules.

During both inference stages, a number of factors, such as discourse coherence, relevance of the presented information and domain knowledge, are taken into account to derive the likelihood of each of the interpretations. We have chosen Bayesian probability theory over other numerical methods for reasoning under uncertainty during plan inference (Raskutti & Zukerman 1991). Bayes theory has often been criticized for its computational complexity and the need for independence assumptions during conditional probability calculations. But other numerical methods, such as Dempster-Shafer(D-S) calculus, suffer from the same drawbacks. In addition, they are often explained in terms of probability thus admitting that probability theory is on a much firmer footing than other numerical methods (Goldman & Charniak 1991). For instance, D-S calculations are at least as expensive as probability updating since a limit case of the D-S calculus is the same as point-valued probability theory. Further, D-S calculus requires the availability of a complete set of disjoint hypotheses necessitating extensive independence assumptions.

After each inference stage, the likelihood calculated using Bayes theory is used to choose between competing interpretations by dropping those interpretations with likelihoods lower than a rejection threshold. This rejection threshold is a function of the maximum likelihood of the possible interpretations, so that only those interpretations with a low likelihood relative to the most likely interpretation are dropped. For instance, if there are two interpretations with likelihoods 0.54 and 0.46 (see Section *A Sample Run*), then surely there is no clear winner. Hence, both interpretations are retained and the information in the interpretations is used to query the user intelligently.

Domain Knowledge Considerations

In this section, we discuss the three types of domain knowledge applied by our system, namely extent of usage, feasibility and practicality.

Extent of Usage

The inference of interpretations from one statement issued by a user consists of inferring a set of possible plan schemas which match this statement, and computing the likelihood of each plan-schema. The inference of plan schemas is done by using a STRIPS-like operator library (Fikes & Nilsson 1971) and plan inference rules (Allen & Perrault 1980). Each operator in the operator library represents a domain action, and it is assigned a prior probability indicating the extent of its usage. In the travel domain, the extent of usage of an operator cannot be determined unless some of parameters in the operator are instantiated. Hence, all the operators, such as, FLY and TAKE_TRAIN are assigned an equal prior. However, in other domains, the prior can be used to favour a more commonly used operator. For instance, if a person wants to send a parcel overseas, it is more natural to assume that s/he would be going to the post office to do it, and the priors can be set to reflect this preference.

Extent of usage can also be used during the indirect inference process when indirect inference rules are used to instantiate the parameters of a plan. For instance, in our system, which operates in the travel domain, the transport schedule is organized so that the listing of available transports is in the order of the extent of usage of a particular mode of transport. During the inference process, when the *mode of transport* or the *departure/arrival time* is to be inferred, the first entry in the transport schedule that is in agreement with the earlier inferences is chosen to instantiate the required parameter(s). Thus, the extent of the usage essentially allocates a likelihood of 1 to the most commonly used option. If this option is unacceptable, it can always be altered at the user's request.

Feasibility

Feasibility of a plan indicates whether a plan is achievable. Plans that are not feasible are not executable by any means. For instance, a plan where the departure date is after the arrival date is not feasible. Pollack (1986) also considers plans that are not executable. However, the focus of her research is on understanding the mental processes applied by the speaker in developing these plans, while our focus is on generating an interpretation of a user's request under the assumption that the preferred interpretation should be feasible.

An interpretation is feasible, if all the plans in it are feasible, and it is possible to achieve the plans in the inferred temporal order. For instance, while both the plans to go to Canberra today and Sydney tomorrow are feasible, the interpretation consisting of the two plans is feasible only if the trip to Canberra is before the trip to Sydney. Interpretations that are not feasible are assigned a likelihood of 0, thus eliminating them from the set of likely interpretations. Notice however, that if the user has explicitly requested a plan or a sequence of plans that is not feasible, then the interpretation containing this plan or sequence is retained, and the system must generate clarification queries. The mechanism for generating such queries is the subject of future research.

The feasibility of an interpretation is determined in two stages: (1) by determining the feasibility of all the plans in it, and (2) by determining whether the plans can be performed in the inferred temporal order. The system attempts to determine feasibility at the earliest possible opportunity. To this effect, whenever an inference gives rise to a new instantiation of a parameter, the system checks whether it has enough information to determine feasibility. For instance, if a new inference instantiates the parameter *mode of transport* to be train, and the original plan was to travel from Adelaide to Los Angeles, then clearly this plan, and hence the interpretation containing this plan, are not feasible.

The usage of domain knowledge to determine feasibility is done at the following points: (1) in the direct inference stage, during the inference of discourse relations, e.g., when a user requests for a flight to Hawaii and then asks for a means to get to Sydney, clearly the second request cannot be elaborating on the plan consisting of the flight to Hawaii; (2) in the indirect inference stage, during the inference of a temporal sequence of plans, e.g., the real life situation where the temporal sequence of filling petrol and then going to the camera shop makes the second plan unexecutable; (3) again, in the indirect inference stage, during the inference of parameters using indirect inference rules, e.g., if the *departure time* of a trip is inferred from the user's preference, and there is no transport scheduled at that time, then clearly the plan with the inferred departure time is not feasible.

Practicality

The practicality of a plan is a measure of the ease with which the effects of the plan can be achieved. A plan is practical when it is easily executable. For instance, a plan to service your car at a faraway garage is impractical, though feasible. In the travel domain, a plan to fly overseas from an international airport is practical. The practicality of a plan can be determined at all those points where the feasibility of an interpretation can be assessed. For instance, when both the *origin* and *destination* of a trip are known, then the practicality of the trip can be determined by checking for a direct route between the two places. If a direct means cannot be found, then the likelihood of the interpretation containing the plan is reduced to reflect this fact. However, the reduction is not to the extent that the interpretation would be dropped off, unless it is already an unlikely interpretation. Thus, the system can handle situations where the user intended to achieve an impractical plan by means of a series of sub plans.

The practicality of an interpretation is a measure of the ease with which the sequence of plans in the interpretation can be performed. For instance, the interpretation consisting of three plans, namely, dropping off the car for service, going to work from the garage and picking up the car on the way home, is practical only when the plans are performed in this sequence. A measure of the practicality of an interpretation can be determined only when the plans in the interpretation have been completely detailed. Hence, it is determined

Number of completely defined interpretations is 0

Interpretation 1 with probability 0.6666 **(0.6)** consists of 2 legs:
Fly to *Sydney*
Fly from *Adelaide* to *Hawaii*
departing on 16*th March* 1991 at 11:00 *am*

Interpretation 2 with probability 0.3333 **(0.4)** consists of 2 legs:
Fly from *Adelaide* to *Sydney*
Go to *Hawaii*
departing on 16*th March* 1991 at 11:00 *am*

Figure 1: **Interpretations after Direct Inferences**

Number of completely defined interpretations is 3

Interpretation 1.1 with probability 0.4444 **(0.3478)** consists of 2 legs:
Fly from *Melbourne* to *Sydney*
departing on date < 16*th March* 1991 at time < 10:00 *am*
arriving on date < 16*th March* 1991 at time < 11:00 *am*
Fly from *Adelaide* to *Hawaii*
departing on 16*th March* 1991 at 11:00 *am*

Interpretation 2.1 with probability 0.2222 **(0.2398)** consists of 2 legs:
Fly from *Adelaide* to *Sydney*
departing on date < 16*th March* 1991 at time < 9:30 *am*
arriving on date < 16*th March* 1991 at time < 11:00 *am*
Go from *Sydney* to *Hawaii*
departing on 16*th March* 1991 at 11:00 *am*

Interpretation 2.2 with probability 0.1111 **(0.1449)** consists of 2 legs:
Go from *Melbourne* to *Hawaii*
departing on 16*th March* 1991 at 11:00 *am*
arriving on 16*th March* 1991 at 11:00 *pm*
Fly from *Adelaide* to *Sydney*
departing on date > 16*th March* 1991 at time > 11:00 *pm*
arriving on date > 16*th March* 1991 at time > 12:30 *pm*

Interpretation 1.2 with probability 0.2222 **(0.2173)** consists of 2 legs:
Fly from *Adelaide* to *Hawaii*
departing on 16*th March* 1991 at 11:00 *am*
Fly from *Hawaii* to *Sydney*

Figure 2: **Interpretations during Indirect Inferences**

after all the inferences have been performed in order to enable pruning of the set of likely interpretations.

Our system, operating in the travel domain, determines a measure of practicality of an interpretation as follows: (1) Between an interpretation with a zig-zag route and another one with a straightforward route, the interpretation with the straightforward route is preferred, (2) If there are trips overseas, an interpretation that has unnecessary country crossings is not preferred. We consider country crossings to be unnecessary, when the distances between the countries are relatively large, e.g., between U.S.A. and Australia. On the other hand, in Europe, where the distances between countries are much smaller, an extra trip between countries may be justified.

In addition to the above, other criteria may also be used to determine the practicality of an interpretation. For instance, the duration of travel to a place and the duration of the stay at a place may be taken into account. However, we have not found the necessity to do so in the current implementation.

A Sample Run

Consider the following scenario:
Traveler: "Get me a flight ticket to Sydney.
I am going to fly to Hawaii at
11:00 am the day after tomorrow.
I'll be leaving from Adelaide."

These statements are input to the mechanism as the following three predicates:

(1) FLY (destination = *Sydney*)
(2) FLY (dep_time = *11:00 am*,
dep_date = *day after tomorrow*,

Number of completely defined interpretations is 1

Interpretation 2.1.1 with probability 0.3781 **(0.5400)** consists of 4 legs:
Fly from *Melbourne* to *Adelaide*
departing on date < 16*th March* 1991 at time < 8:00 *am*
arriving on date < 16*th March* 1991 at time < 9:30 *am*
Fly from *Adelaide* to *Sydney*
departing on date < 16*th March* 1991 at time < 9:30 *am*
arriving on date < 16*th March* 1991 at time < 11:00 *am*
Fly from *Sydney* to *Hawaii*
departing on 16*th March* 1991 at 11:00 *am*
arriving on 16*th March* 1991 at 11:00 *pm*
Fly from *Hawaii* to *Melbourne*
departing on date > 16*th March* 1991 at time > 11:00 *pm*
arriving on date > 16*th March* 1991 at time > 11:00 *pm*

Interpretation 1.1.1 with probability 0.6218 **(0.4599)** consists of 4 legs:
Fly from *Melbourne* to *Sydney*
departing on date < 16*th March* 1991 at time < 8:30 *am*
arriving on date < 16*th March* 1991 at time < 9:30 *am*
Fly from *Sydney* to *Adelaide*
departing on date < 16*th March* 1991 at time < 9:30 *am*
arriving on date < 16*th March* 1991 at time < 11:00 *am*
Fly from *Adelaide* to *Hawaii*
departing on 16*th March* 1991 at 11:00 *am*
Fly from *Hawaii* to *Melbourne*

Figure 3: **Interpretations after Completion and Pruning**

destination = *Hawaii*)
(3) LEAVE (origin = *Adelaide*)

The mechanism was run first without domain knowledge, and then run with domain knowledge. The actual output of these runs, with today's date set to the 14th of March 1991, is presented in the Figures 1, 2 and 3. The likelihoods with domain knowledge considerations appear in brackets in boldface, next to the likelihoods without domain knowledge considerations. In this example, all the interpretations generated without using domain knowledge considerations are feasible even when domain knowledge is used. Therefore, the interpretations generated during each run are identical, but the likelihoods of the interpretations are different due to the effect of the practicality considerations during the run when domain knowledge is used. These differences are discussed below.

During the direct inference stage, while inferring discourse relations, the run without domain knowledge prefers the interpretation that the user is flying from Adelaide to Hawaii due to coherence considerations. The run with domain knowledge decreases the likelihood of the first interpretation, since there is no direct means of transport from Adelaide to Hawaii (see Figure 1). During the indirect inference stage, while using the inferred temporal sequence of plans, the likelihoods of interpretations 2.1 and 2.2 are higher with domain knowledge than without, since Hawaii is an overseas location and Melbourne and Sydney are international ports (see Figure 2). During the final stages of both runs, interpretation 2.2 is rejected owing to its low initial probability, while interpretation 1.2 is rejected due to its low information content (Raskutti & Zukerman 1991). The interpretation intended by the user, i.e., interpretation 2.1.1 in Figure 3, is determined to be more practical since the path of its itinerary is less meandering than the path of the other interpretation. Thus, with the use of domain knowledge the interpretation intended by the user is the one with the highest likelihood. However, since the likelihoods of the two interpretations are close there is no clear winner and the user has to be queried to select one of these interpretations. During the generation of a dicriminating query, the information inferred during the plan recognition process can be used to generate a sensible query. Without the use of domain knowledge, the incomplete interpretation that does not represent the user's intent is clearly preferred, and this interpretation needs to be completed by means of an information seeking query that would probably confuse the user. Hence, the application of domain knowledge to aid plan recognition and the subsequent response generation can make sense out of an incoherently phrased request.

Conclusion

In this paper, we have demonstrated the need to use domain knowledge to choose between interpretations. We have described the basis of the application of domain knowledge in our system, and have indicated the points during the inference process where domain knowledge can be successfully applied. We have also presented the results of the application of our ideas in processing a sample request at a travel agency.

Work in Progress

In order to generate cooperative responses, the interpretations generated by the plan inference mechanism have to be analyzed as follows: (1) If there is only one valid interpretation with sufficient relevant information, then it is sent to the planner for planning. If the planner can come up with an itinerary that satisfies the user's needs, then the system has to inform the user about the proposed itinerary; (2) if there are multiple possibilities, as in the case of the example presented above, then the possibilities must be analyzed to determine the points that discriminate between them, so that a clarification question can be generated; finally, (3) if there are no complete interpretations, the user must be queried to complete an interpretation. Presently, we are in the process of incorporating the response generation module into the inference mechanism.

Our mechanism for plan inference has been mainly used in a travel planner where the domain constraints, such as time and location dependencies, are fairly clear. We are currently considering other domains, such as Yellow Pages directory assistance system, to evaluate the portability of our mechanism (Zukerman 1991).

References

Allen, J.F. & Perrault, C.R. (1980), Analyzing Intention in Utterances. In *Artificial Intelligence* 15, pp. 143-178.

Carberry, S. (1990), Incorporating Default Inferences into Plan Recognition. In *Proceedings of the National Conference on Artificial Intelligence*, pp. 471-478.

Fikes, R.E. & Nilsson, N.J. (1971), STRIPS: A New Approach to the Application of Theorem Proving to Problem Solving. In *Artificial Intelligence* 2, pp. 189-208.

Goldman, R. & Charniak, E. (1991), A Probabilistic Model for Plan Recognition. In *Proceedings of the Ninth National Conference on Artificial Intelligence*, pp. 160-165.

Litman, D. & Allen, J.F. (1987), A Plan Recognition Model for Subdialogues in Conversation. In *Cognitive Science* 11, pp. 163-200.

Pollack, M. (1986), A Model of Plan Inference that Distinguishes between the Beliefs of Actors and Observers. In *Proceedings of the 24th Annual Meeting of the Association of Computational Linguistics*, pp. 207-214.

Raskutti, B. & Zukerman, I. (1991), Generation and Selection of Likely Interpretations during Plan Recognition in Task-oriented Consultation Systems. In *User Modeling and User Adapted Interaction — an International Journal* 1(4).

Zukerman, I. (1991), Towards a System for the Generation of Cooperative Responses in Information Seeking Environments. Technical Report, 91/154, Department of Computer Science, Monash University.

Calculating Breadth of Knowledge

Lisa F. Rau
GE Research and Development Center
Schenectady, NY 12301 USA
and
Computer Science Department
University of Exeter, Exeter UK

Abstract

Since the advent of computers, information systems have grown in terms of the quantity of knowledge they deal with. Advances in data management are on the critical path for usability of these systems. This paper reports on a novel approach to an important problem; that of calculating the conceptual *breadth* of knowledge or data in a knowledge base or database. Breadth determination is useful in that ascribing meta-level knowledge of conceptual content can help to predict, for example, the validity of the closed-world assumption or the likelihood of encountering new information of a particular type. The point at which a system determines it is likely to have breadth in a given knowledge area may also serve as the trigger point for calculations that assume relatively complete knowledge in that area. The accurate determination of when a system has complete knowledge in an area is crucial for the accurate application of many AI algorithms.

Introduction

As information systems continue to grow in size and complexity, it becomes more and more critical to develop innovative methods of ascertaining the contents of these information systems. Three features motivate such a construction: (1) someone unfamiliar with the database can pose general questions about the breadth of content of the system (such as "do you know about X?"); (2) the system can now distinguish potentially missing information from information not likely to exist, i.e., the difference between a "no" response and an "I don't know" response and (3) knowing when complete knowledge exists allows for processes that depend on this assumption to operate. For example, the work of Pollack [Pollack, 1986] on plan failure analysis assumes the system has complete knowledge of what the user knows. Knowing when this is assumption is valid is critical in order to correctly apply the methods. The last two motivations address the important problem of knowing when to apply the closed world hypothesis [Reiter, 1978]; that everything not known is *false*.

This paper reports on a theory and implementation of the cognitive notion of conceptual *breadth*, a concept not typically modelled from a computational perspective. While the concept of *breadth of knowledge* has a clear intuitive meaning, this work aims to formalize the notion and provide a computational mechanism for determining breadth. By implementing this computation on an arbitrary knowledge or data base, it is now possible to construct a meta-representation for the scope or limits of what a system knows about. The representation is a "meta-representation" because it is knowledge *about* the knowledge in the system.

This work is one portion of a larger project to provide computational methods for automatically deriving what a system knows, for example, what is salient [Rau, 1992]. Breadth of knowledge is the notion of having a conceptual covering of a subject area. This covering may be shallow or deep. Because virtually all databases necessarily cover an arbitrary knowledge area, it is useful to make a distinction between *absolute* breadth and *relative* breadth. Relative breadth is breadth with respect to the domain of operation of a system. Relative breadth is achieved when a system contains knowledge along all of its possible expected dimensions. For example, consider a database that contains, among other things, information about business occupation for each person in the database. It might be limited to encountering only certain occupations due to the sample of the population in the database, from a socio-economic, regional, age or sex related or functional perspective. For relative breadth, seeing almost all of this subset of occupations is sufficient.

For absolute breadth, the set of occupations encountered should approximate the set of all occupations people have with any regularity. That is, absolute breadth is achieved when the system has seen all the *possible* values a given field can take from the standpoint of an objective measure of human knowledge. Clearly, computing absolute breadth is the more challenging problem.

This paper discusses algorithms for computing absolute and relative breadth with respect to both open class (infinite) categories, and closed class (finite) categories. It details the computational method used to ascertain relative breadth for closed class categories, and provides an experimental validation of the method on a test of a 1,917 record database produced by an NLP system. With these results, users may pose questions such as "What countries of interest does this database cover?"

and obtain, in addition to a list of those countries, the information as to whether these are *all* the countries or just the *countries the system happens to have seen up to that point.* Of additional utility, expert reasoning systems can make likely inferences based on the closed world assumption with this type of analysis.

Overview of Methodology

In this section, I describe the expected functionality of a breadth calculation in general terms, followed by the specific algorithms that implement this functionality. The experimental results are also described.

Functionality

Intuitively, a system has broad knowledge in a given area when it "knows about" (that is, has instances of) virtually all of the possible *manifestations* of knowledge in that area. Translated into database terms, given a database field $\mathcal{F}$ that contains unique fillers (types, not tokens) $(f_1, f_2, \cdots, f_n)$, the system has breadth of knowledge with respect to $\mathcal{F}$ when it has encountered virtually all the f_i, independent of how frequently each of the f_i appears. The assumption here is that a database field filled with a value is equivalent to the database "knowing" that information[1].

This straightforward observation implies that when a system has seen, in a particular field, as many distinct fillers as total numer of fillers, the expectation is that breadth has not been achieved, and that there is a potentially sizable number of unseen fillers not yet encountered. That is, when every filler of a field is unique, it is more likely that the next filler will also be unique. Conversely, when a system has experienced high redundancy in the distinct fillers, and/or has not seen new distinct fillers for a while, the intuition and expectation is that the system has likely seen most if not all the fillers it is likely to encounter, and breadth has been achieved. This scenario is complicated when it is impossible ever to have breadth of knowledge given an infinite number of possible distinct types of fillers. The next section describes how this and the above functionality is achieved.

Implementation

As we have just seen, the key concept for the computation of breadth is the certainty that the system has seen all possible instances of a given concept. To compute this, the system must first make a judgement as to whether the fillers consist of a closed or open class. That is, the system must determine whether the types of fillers are finite or infinite. A closed class is a class of objects with a fixed set of members, and doesn't change. An open class, on the other hand, has no finite number of possible members. For example, the syntactic category of *determiner* is a closed class consisting of a small set of words (such as `a`, `the`, `this`, etc.). The syntactic category of *noun* is an open class, as just about any word can be a noun, and new nouns are created all the time. This qualitative distinction is useful for natural language processing because a system must be able to predict the part of speech for a new, unknown word. It is important for AI systems in general because it gives information as to when it might be correct to apply a closed world assumption.

In summary, in order to compute breadth, the system must be able to perform two functions: (1) differentiate between cases where there are potentially infinitely many unique instance types (open-class fills) and cases where the instances come from a finite set (closed-class fills) and (2) quantify "virtually all" so that the expectation of when breadth is achieved is satisfied. This is done by providing a numerical threshold based on the probability of the given distribution's containing only those distinct types, under the assumption that the distribution of values already seen provides limits on the nature of the distribution in the database.

Determining Class Boundedness

Although the distinction between an open class and a closed class is not always clear [Collins *et al.*, 1975], from the perspective of a database, fillers of a field can always be so classified.

Some databases may have a data dictionary that gives type information for each field of each record. This information does not always map into a correct open/closed class determination, as in the case of person age. Person age in a data dictionary may simply be a number (an open class) whereas in reality it is a closed class of numbers from 0 to 120 or so. However it is quite easy to specify manually, for each field, whether it is closed or open class, so this is an option as well.

The easiest purely automated method for determining whether values of the field or slot are members of a closed class or an open class is to look at the number of distinct values. There are much larger (order of magnitude) numbers of distinct values in open class fields in any data or knowledge base of reasonable size. A reasonable sized database would contain at least thousands of records. Although it is possible to have a closed-class fill that contains thousands of elements (for example in the case of cities of the world), these "borderline" cases are best assumed to be open-class, as it is likely that the set of possible fills is not exhaustive. It is a safe assumption then that closed-class fills, by their nature, are bounded and the number of choices is dwarfed by the number of instances in large samples. Therefore, computing these values for a sample of the data will neatly classify a field as either closed-class or open-class.

Determining Breadth

The next four sections discuss the breadth computation for each of the four cases of relative or absolute breadth with respect to open or closed classes. Recall that relative breadth is breadth of knowledge with respect to a system's domain of operation, as opposed to some objective, absolute measure of human knowledge, termed absolute breadth.

[1]Note that this only applies to knowing the contents of a category, where category is defined by the fields in a database. For example, in the database used here, `country` is a field. More complex concepts, such as "big countries" or "things like countries" can be handled in exactly the same manner as long as a class membership function can be defined.

Relative Breadth, Closed Class: The problem of determining relative breadth for a closed class category is equivalent to the problem of making a determination as to when the system has seen almost all, if not all, of the members of that category. The threshold for this determination can be manipulated according to how certain you would like the system to be in this judgement. Relative breadth of a closed class category translates into the point at which the ratio of the number of distinct fillers seen to the total number of fillers seen (call this `d / n`) is around 1 in $\mathcal{T}$, where $\mathcal{T}$ is the threshold. In the case where the number of distinct fillers is equal to the number of fillers, this ratio is equal to 1. In this case, the system has seen only unique instances and intuitively would expect to see more; it has no reason to believe that it has breadth. In cases where this ratio is small and `n` is large, the system has seen many instances of the finite set of fillers. Taking this distribution to be a good approximation to the expected distribution across the database, the system is unlikely to see any additional distinct fillers. In this case, we can assume the system has breadth. The point at which this cutoff is made depends on how certain you would like the system to be in its breadth determination.

In particular, at any given point in time, the system computes either:

1. Breadth has been achieved. All the most frequently occurring kinds of fillers have been seen, and the only outliers are likely to be anomalous or both extremely infrequent and small in number compared to the total number of distinct fillers.
2. Breadth has not been achieved. The frequency of new fillers is such that it is likely that there are additional fillers that the system has yet to encounter, or there is insufficient data to judge.

We have experimented with more complicated formulations involving information theory and statistics, but this simple heuristic captures the intuition as well as the data. The intuition is that breadth is achieved when you are likely to have seen all the instances there are. The results are described in the next section.

Absolute Breadth, Closed Class: Absolute breadth with a closed class category takes advantage of a generic conceptual hierarchy. This hierarchy was built to support generic text processing, and contains concepts representing the most frequent words of English and appropriate super-categories. Because it has broad coverage of the more frequently used concepts that are used in language, it is a good domain-independent starting point for experiments comparing system knowledge to general human knowledge. Absolute breadth is achieved when over some percentage of the members of the closed-class category, as dictated by the hierarchy, have been seen. The exact percentage depends on what percentage of the knowledge area one assumes needs to be covered before breadth is achieved. Certainly more than 50%, probably more than 75%, and less than 100%. Working with the value of 98%, a simple example is breadth of knowledge of the states in the U.S. Given that there are absolutely and precisely 50 states, the system would have breadth of knowledge of what the states in the U.S. were if it knew of any 49.

Absolute Breadth, Open Class: It is very difficult to determine an absolute measure of breadth for an open class category. For example, consider the problem of ascertaining breadth of knowledge about the different shapes that snowflakes can be; something that at least folk science believes to be infinite. Since it is not possible even for an expert to know all the members of this conceptual category, it is impossible to determine when someone is likely to have breadth in this category, from an absolute perspective.

However if the elements of the open-class can be correctly assigned to superclasses (conceptual parents), the simple percentage of the total possible elements as defined by a generic conceptual hierarchy can give an approximation of when absolute breadth of an open-class category has been achieved. Continuing with the snowflake example, if snowflakes can be mapped into superclasses (one such assignment may be by the shape of the perimeter; hexagonal, round, five-pointed star, etc.), then some breadth is achieved when all the possible shapes have been seen.

There are two obstacles to overcome in order for this approach to succeed. The first is correctly determining the superclass categorization for each individual. In addition to choosing the correct parent from multiple possible parents of each individual, the hierarchy may not contain a parent whose elements correspond to exactly the set of possible individuals. For example, the set of possible shapes of snowflakes in the world may not contain shapes with less than four sides. Second, when the superclass itself is infinite, as the category of "shape" surely is, the superclass must be recursively partitioned into *its* superclass. If no finite partition exists, it is not possible to determine absolute breadth of this open class. These are non-trivial problems requiring some theory of superclasses, still under formulation.

Relative Breadth, Open Class: The solution to this case is similar to the absolute breadth, open class case. Where there are potentially infinitely many unique instances (open-class), it is sufficient to have seen at least one of every *class* of instance of the possible values in that area in order to have achieved breadth. That is, breadth with respect to open class values is calculated by using the conceptual parent of each filler as opposed to the filler itself in the same calculation as for closed class values. The parent of each filler is computed by determining the conceptual category that the unique values belong to. For another example, given a unique person name, we transform that name to the category of **`human`**. The same problems with correct classification still remain. For example, a given person may be subcategorized as a **`military officer`** or **`government official`**, each of these subcategories in turn are **`human`**. Thus determining closed-class "clusters" of the open-class values requires additional mechanics. The methods to perform this clustering accurately are currently under experimentation. After the clusters are determined, they will be treated in the same manner as the truly closed-class values to provide a threshold for the breadth decision.

Experimental Results

This section details some of the results of running the computations just described on a database. The database used contained almost 2,000 database records of information automatically extracted from texts reporting on terrorist activities in Latin America. These records were all created with a natural language text processing program, described elsewhere [Jacobs and Rau, 1990; Jacobs and Rau, 1993; Krupka *et al.*, 1991]. Using news stories as a source suggests that this work has the potential to operate on arbitrary and general knowledge, as well as specific databases. Figure 1 shows a sample message and template from this set.

DEV-MUC4-0351

BOGOTA, 18 AUG 89 (EFE) -- [TEXT] SENATOR LUIS CARLOS GALAN, LIBERAL PARTY PRESIDENTIAL HOPEFUL, WAS SHOT THIS EVENING WHEN HE WAS ABOUT TO GIVE A SPEECH AT MAIN SQUARE OF SOACHA, 15 KM SOUTH OF BOGOTA, IT WAS CONFIRMED BY POLICE AND HEALTH AUTHORITIES.

ACCORDING TO THE FIRST REPORTS, AT LEAST ONE MAN FIRED ON THE SENATOR FROM AMONG THOSE GATHERED. THE SENATOR IS CURRENTLY AT THE EMERGENCY ROOM OF A HOSPITAL IN BOSA, CLOSE TO SOACHA. TWO OTHER PERSONS WERE WOUNDED DURING THE ATTACK.

0. MESSAGE: ID	DEV-MUC3-0351
1. MESSAGE: TEMPLATE	1
2. INCIDENT: DATE	18 AUG 89
3. INCIDENT: LOCATION	COLOMBIA: SOACHA (CITY)
4. INCIDENT: TYPE	ATTACK
5. INCIDENT: STAGE OF EXECUTION	ACCOMPLISHED
6. INCIDENT: INSTRUMENT ID	-
7. INCIDENT: INSTRUMENT TYPE	GUN: "-"
8. PERP: INCIDENT CATEGORY	TERRORIST ACT
9. PERP: INDIVIDUAL ID	"AT LEAST ONE MAN" / "ONE MAN"
10. PERP: ORGANIZATION ID	-
11. PERP: ORGANIZATION CONFIDENCE	-
12. PHYS TGT: ID	*
13. PHYS TGT: TYPE	*
14. PHYS TGT: NUMBER	*
15. PHYS TGT: FOREIGN NATION	*
16. PHYS TGT: EFFECT OF INCIDENT	*
17. PHYS TGT: TOTAL NUMBER	*
18. HUM TGT: NAME	"LUIS CARLOS GALAN"
19. HUM TGT: DESCRIPTION	"LIBERAL PARTY PRESIDENTIAL
HOPEFUL" /	
	"SENATOR": "LUIS CARLOS GALAN"
	"TWO OTHER PERSONS"
20. HUM TGT: TYPE	GOVERNMENT OFFICIAL: "LUIS
CARLOS	
	GALAN"
	CIVILIAN: "TWO OTHER PERSONS"
21. HUM TGT: NUMBER	1: "LUIS CARLOS GALAN"
	2: "TWO OTHER PERSONS"
22. HUM TGT: FOREIGN NATION	-
23. HUM TGT: EFFECT OF INCIDENT	INJURY: "LUIS CARLOS GALAN"
	INJURY: "TWO OTHER PERSONS"
24. HUM TGT: TOTAL NUMBER	-

Figure 1: Example Text and Data Extracted

Closed/Open Class Determination

First, the fields in the records are divided into closed and open classes, according to gross frequency of occurrence of distinct values, as discussed previously. This results in the closed class slots illustrated in Figure 2.

Randomization of Data

After this determination, records are randomly selected from the database and the closed-class fillers are tabulated. Random selection is critical to overcome time-dependent patterns of values typical in any real database.

	Slot Type	Status	Total Distinct Values
3.	LOC OF INCIDENT	CLOSED	(23 distinct values)
4.	INCIDENT TYPE	CLOSED	(6 distinct values)
5.	INCIDENCT STAGE	CLOSED	(3 distinct values
7.	INSTRUMENT TYPE	CLOSED	(19 distinct values)
8.	INCIDENT CATEGORY	CLOSED	(4 distinct values)
11.	PERP ORG-CONF	CLOSED	(7 distinct values)
13.	PHYS TGT TYPE	CLOSED	(17 distinct values)
15.	PHYS TGT NATION	CLOSED	(15 distinct values)
16.	PHYS TGT EFFECT	CLOSED	(8 distinct values)
20.	HUMAN TGT TYPE	CLOSED	(12 distinct values)
22.	HUMAN TGT NATION	CLOSED	(29 distinct values)
23.	HUMAN TGT EFFECT	CLOSED	(11 distinct values)

Figure 2: Closed Class Categories

Threshold Determination

After experimentation, a threshold value $\mathcal{T}$ of 50 was chosen. This was the minimal value that captured almost every distinct type, with any types remaining being anomalous. This number could be larger and still include these same distinct values. Any arbitrary number, even if experimentally determined, is open to suspicions of "picking a number out of a hat". However if breadth is viewed as a binary predicate, then such an arbitrary threshold is required.

An alternative view of breadth is as a continuum, whereby you have more or less breadth, and it monotonically increases the more distinct types encountered. Under this view, the threshold can be a range, and it maps onto the breadth scale linearly. However for ease of exposition, I assume here the binary predicate model. This means that breadth is achieved when $d/n < 1/50$. The number of distinct types seen so far (`d`), the total number of fillers seen to this point (`n`), and the threshold (`50`) are the numbers that estimate the breadth. It is possible to graph `d` *vs.* `n` and see the point at which breadth is achieved, as well as the theoretical projection of where breadth would be achieved if no additional distinct values were seen past the point at which the data ends.

A graphical form of presentation does not capture the frequency with which each distinct value occurs in the sample. In all cases where a cutoff (breadth determination) was made when not all distinct values had been seen, the remaining distinct values were anomalous. Anomalous data is data that occurs only once or twice and is almost always a result of mistyping or other errors. This is to be predicted, in that any values not seen after $\mathcal{T} \times$ `d`, where `d` is the number of distinct values, is likely not to occur with frequency greater than 1 in that amount. For example, suppose as is true in the data presented here, that a breadth determination for the field of `Physical-Target-Type` was made after seeing 1134 instances. In this case, $n = 1134$, $d = 17$ and no additional distinct fill exists in the remainder of this set of data. As such, breadth is still achieved if only truly anomalous data has not been seen. If a significant portion of the data is anomalous, the system would not conclude breadth in the area, as the prediction is that more unseen values would be likely to occur.

3.	LOCATION OF INCIDENT	546 410 133 88 39 32 15 11 7 5 3 2 2 2 2 1 1 1 1 1 1 1 1(no breadth)
4.	INCIDENT TYPE	734 349 145 51 20 7 (0)
5.	INCIDENCT STAGE	1204 60 48 (0)
7.	INSTRUMENT TYPE	550 194 192 145 79 78 50 38 14 12 11 10 8 6 4 3 2 (2 1)
8.	INCIDENT CATEGORY	967 219 130 1 (0)
11.	PERPETRATOR ORG-CONFIDENCE	617 353 148 103 102 80 (2)
13.	PHYSICAL TGT TYPE	580 160 133 121 118 114 77 58 46 42 41 31 23 17 17 1 1 (0)
15.	PHYSICAL TGT FOREIGN NATION	1092 162 25 11 6 5 5 3 2 1 1 1 1 1 (1)
16.	PHYSICAL TGT EFFECT	765 346 162 156 45 12 7 (2)
20.	HUMAN TGT TYPE	1019 382 165 143 99 61 39 37 17 14 13 (1)
22.	HUMAN TGT FOREIGN NATION	1179 49 32 10 10 9 8 6 5 5 4 4 4 4 3 2 2 2 2 2 2 2 2 1 1 1 1 1 1 1 (no breadth)
23.	HUMAN TGT EFFECT	856 457 280 122 65 31 10 10 7 5 1 (0)

Figure 3: Frequency Distribution of Values

Breadth Determination

Figure 3 illustrates the frequency with which each distinct value occurs. For example, there were 734 instances of the first distinct value in the **Type** field, 349 instances of the next distinct value, etc. Numbers in parenthesis indicate those anomalous values that are missed by an early cutoff; `no breadth` indicates that a likely breadth determination cannot be made. Note, for example, how the preponderance of singular values in the **`human-target-foreign-nation`** slot precludes a breadth determination, as the expectation here is towards more such values. When low frequency values contribute significantly towards the total number of distinct values, breadth of knowledge must include even these low frequency values as well. As can be seen from these data, it is possible to answer the question "Do you have breadth?" applied to the fields in a database. Moreover this question can be answered so that in only some of the fields, only a small fraction of the total number of distinct values is unseen; these values are likely to be anomalous at the threshold value of 50. Breadth determinations even more likely to cover 100% of all values can be made at larger thresholds. With these results, a user may get different answers to questions posed to a database where the field referenced had complete breadth of knowledge from one which did not have breadth. For example, all data in which every member of a field participates in some relationship must be qualified as to whether the relationship hold for all-currently-known, as opposed to all-possible.

Probabilistic Analysis

A probabilistic analysis shows that after 50 trials, if only 1 distinct value has surfaced, the chances of seeing a new value next (i.e., on the 51th trial) are:

$$\left(\frac{50}{51}\right)^{51}\left(\frac{1}{51}\right) \simeq .007$$

under the assumption that the *a posteriori* probability equals the *a priori* probability; a good assumption with large n. This only decreases for other values of d, so that if 2 distinct values have occurred in $50 \times 2 = 100$ trials, the likelihood of seeing a pattern consisting of these 2 values in an arbitrary internal distribution, followed by a new value, is .003.

The best case scenario is that all distinct values have been seen. And although taking the pattern of historical values to predict the probability during the next trial assumes a 1/50 chance the next value will be distinct (in the worst case of having seen only one distinct value), there is only a .007 chance that with this weighting we would have seen the pattern just seen, thus lending confidence that any distinct values yet unseen occur with smaller probabilities. This analysis indicates that after this many values, almost all, if not all, distinct values have been seen with good certainty, and any values not seen are likely to be anomalous or extremely infrequent.

The results presented in this section indicate that fairly simple calculations can accurately predict when a database or knowledge base has breadth in closed-class subject areas.

Related Work

A great deal of research has addressed the problem of what a system might know or believe [Halpern, 1986; Vardi, 1988]. This work primarily concerns itself with calculating what a system might be expected to believe, given some set of assertions. The work described here contributes to that body of research by adding a new metric that is calculated from what is known, the breadth of knowledge. This work is related to recent work in the area of knowledge discovery in databases [Piatetsky-Shapiro and Frawley, 1991] that attempts to learn new knowledge from the structure and content of databases. However, the particular problem of computing breadth of knowledge has not been directly addressed in this new research area.

Future Directions

The primary area for future work is in the development and testing of the theory of superclasses to determine breadth in the open-class cases. Another critical step currently on-going is the choice of application area to illustrate the utility of knowing when breadth has been achieved in a real AI/database system. Also of importance is automating the process of threshold determination. This value appears to vary as a function of the raw number of distinct fills as well as the inherent frequency distribution of a given field. Even more accurate determinations may be made if the threshold can be dynamically computed for each slot. Another activity currently on-going is embedding the breadth calculation in a suite of tools under development to construct a meta-profile of the contents of an arbitrary knowledge or database. With this profile, users of an application will be able to get a feeling for the contents of a database that can aid in their judgements of the appropriateness of the database for their information needs, as well as in constructing appropriate and answerable queries. In addition, the system will be able to distinguish negative searches from the result of information requests outside an area of expertise from responses due to closed world assumptions.

Conclusions

This paper began with an analysis of the notion of conceptual breadth, with respect to absolute and relative measures, and with respect to closed class and open class categories. A specific computation was detailed for automatically determining the conceptual breadth of a knowledge or database in the case of computing relative breadth of closed class fields. The point at which breadth is achieved is computed by taking a random sample from the database, and keeping track of the percentage of distinct types to the total number of fillers seen. When this percent falls below a threshold, it is expected that most if not all the distinct types have been seen. This method accurately identifies a point at which breadth is very likely to have been achieved, and a probabilistic analysis supports this expectation.

This work is important not only for the methods and computations described here, but for investigating new questions we would like large knowledge based system to be able to answer - questions such as "what do you know?" and "how complete is your knowledge?". It is critical in determining when to apply the closed-world hypothesis, and when there is not enough information for this assumption to apply. Looking at areas traditionally reserved for the purely cognitive realm, such as meta-questions of knowledge scope and extent, offers a new perspective from which to develop computational answers.

Acknowledgements. The results presented here have benefited from discussions with my colleagues at GE, including Piero Bonissone, Fred Faltin, Paul Jacobs and Barbara Vivier. I am also grateful for the perspectives and suggestions of Noel Sharkey (University of Exeter), and Gregory Piatetsky-Shapiro (GTE Labs). David McDonald also provided comments on a draft of this paper.

References

[Collins *et al.*, 1975] Allan Collins, Eleanor H. Warnock, Nelleke Aiello, and Mark L. Miller. Reasoning from incomplete knowledge. In Daniel G. Bobrow and Allan Collins, editors, *Representation and Understanding*, pages 383–415. Academic Press, Inc., Orlando, FL, 1975.

[Halpern, 1986] J. Y. Halpern, editor. *Proceedings of the First Conference on Theoretical Aspects of Reasoning About Knowledge.* Morgan Maufmann, Los Altos, CA, March 1986.

[Jacobs and Rau, 1990] P. S. Jacobs and L. F. Rau. The GE NLToolset: A software foundation for intelligent text processing. In *Proceedings of the Thirteenth International Conference on Computational Linguistics*, volume 3, pages 373–377, Helsinki, Finland, 1990.

[Jacobs and Rau, 1993] P. S. Jacobs and L. F. Rau. Innovations in text interpretation. *Artificial Intelligence (Special Issue on Natural Language Processing)*, 48, To Appear 1993.

[Krupka *et al.*, 1991] G. R. Krupka, P. S. Jacobs, L. F. Rau, and L. Iwańska. Description of the GE NLToolset system as used for MUC-3. In *Proceedings of the Third Message Understanding Conference (Muc-3)*, San Mateo, Ca, May 1991. Morgan Kaufmann Publishers.

[Piatetsky-Shapiro and Frawley, 1991] G. Piatetsky-Shapiro and W. J. Frawley. *Knowledge Discovery in Databases.* MIT Press, Cambridge, MA, 1991.

[Pollack, 1986] M. Pollack. *Simulating Plan Failure.* PhD thesis, University of Pennsylvania, 1986.

[Rau, 1992] L. F. Rau. Calculating salience of knowledge. In *Proceedings of the Fourteenth Annual Conference of the Cognitive Science Society*, Hillsdale, NJ, July 1992. Cognitive Science Society, Lawrence Erlbaum Associates.

[Reiter, 1978] R. Reiter. On closed world data bases. In H. Gallaire and J. Minker, editors, *Logic and Data Bases*, pages 55–76. Plenum Press, New York, New York, 1978.

[Vardi, 1988] M. Y. Vardi, editor. *Proceedings of the Second Conference on Theoretical Aspects of Reasoning About Knowledge.* Morgan Maufmann, Los Altos, CA, March 1988.

Learning and Problem Solving Under a Memory Load

Paul J. Reber
Department of Psychology
Carnegie Mellon University
Pittsburgh, PA 15217
pr18@andrew.cmu.edu

Kenneth Kotovsky
Department of Psychology
Carnegie Mellon University
Pittsburgh, PA 15217
kotovsky@psy.cmu.edu

Abstract

A problem solving experiment is described where the difficulty Ss experienced in solving a particular puzzle is manipulated using a dual task paradigm. Although Ss show impaired performance solving the puzzle the first time, performance improves considerably on a second trial and Ss are not impaired by a second trial memory load. In spite of the improvement in performance, Ss are unable to report virtually any information about the problem or their solution strategies. A model is presented that describes the pattern of performance across the levels of memory load and across the two trials. The theoretical implications of this model are discussed.

Introduction

Problem solving has traditionally been thought of as search through a problem space (Newell & Simon, 1972). Recent work with "isomorphic" puzzles which have identical underlying problem spaces but different surface features has shown that the working memory demands of the surface features plays an important role in predicting how difficult a particular problem will be (Kotovsky, Hayes & Simon, 1985; Kotovsky & Simon, 1990; Kotovsky & Kushmerick, 1991).

The following experiment demonstrates that the difficulty of a particular puzzle can be manipulated by reducing the working memory available for problem solving using a dual task paradigm. We find that the performance is impaired in proportion to the demands of the secondary task. However, when repeating the problem solving task, the memory load ceases to interfere with problem solving.

The pattern of results indicates that it is a learning process associated with problem solving that is impaired by the secondary task. However, an interesting problem arises in attempting to assess what and how the subjects are learning while problem solving as the subjects are unable to describe what they know about the puzzle or how they solved it.

A reasonably simple mathematical model is presented to attempt to bring together some of the phenomena observed in this experiment. Focusing on weak methods and modeling the learning as an incremental function, the model gives a reasonable first-order account of the structure of the problem solving process for this puzzle.

Method

Seventy-six Carnegie Mellon University undergraduates were asked to solve an isomorph of the "Chinese Ring Puzzle" (Kotovsky & Simon, 1990) called the Balls and Boxes puzzle while performing a secondary task designed to impose a load on working memory. The Balls and Boxes puzzle was presented via computer and is described below. While solving the puzzle, subjects were also asked to listen to a tape containing a stream of letters at a rate of one letter every three seconds with beeps mixed in occasionally. Subjects were divided into four groups, a control group and three different levels of memory load. The three memory load groups were instructed to listen to the letters and remember either the last letter heard (group 1-back), the last two letters heard (group 2-back) or the last three letters heard (group 3-back). Since the tape contains a continuous stream of letters, each time a new letter is heard by the subject, the set of letter(s) to be remembered changes. These groups of subjects were instructed that when they heard a beep they were to write down the first letter of the letter set being remembered, i.e. subjects remembering the last letter wrote this at the beep, subjects remembering two letters wrote the letter before the last letter heard, etc. The control group was instructed to ignore the letters but listen for the beeps and when a beep was heard, write down a random letter. To be sure that the subjects performed the secondary task as well as possible, the importance of listening to the tape carefully was stressed with all subjects and the experimenter monitored subjects performance to be sure they wrote down a letter whenever a beep was heard.

An experimental session consisted of brief instruction on the puzzle and secondary task and two solutions of the puzzle. Between solutions, subjects

---

This research was supported by NIMH training grant 5T32 MH-19102. The authors would like to extend thanks to Gerry Starret who aided greatly in collecting the experimental data; also Chris Schunn and Frank Ritter for insightful comments that facilitated development of the model.

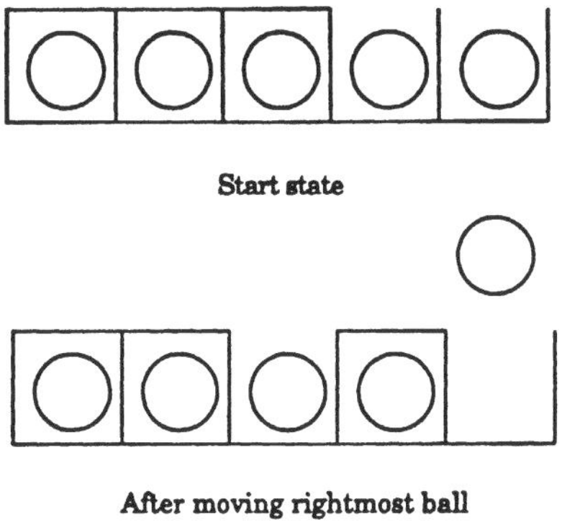

Figure 1: Appearance of the Puzzle to Subjects

were asked to write "how they solved the puzzle, how the puzzle works and especially anything they could say that would help somebody else solve the puzzle." Only two levels of memory load were used on the second trial, the control condition and the two-letter memory load condition.

The Balls and Boxes Puzzle

The Balls and Boxes puzzle is an isomorph of the "Chinese Ring Puzzle" which is a particularly difficult puzzle studied by Kotovsky & Simon (1990). The Balls and Boxes isomorph was designed to be a simpler version of the puzzle where the surface structure was modified so that all legal moves can easily be seen and the operators were "digitized" to reduce the working memory load associated with moving within the problem space. This version of the puzzle was found to be much easier to solve, generally taking 5-10 minutes. The puzzle was presented on a MicroVAX II computer and is shown in Figure 1. This puzzle consists of five balls and five boxes. The object is to get all five balls out of their boxes. A ball may only be moved if its box top is open. As balls are moved in and out of their boxes, the box tops open and close according to a rule that defines the problem. The rule dictates that a ball may be moved (and hence the box top is open) if and only if the ball immediately to its right is in its box and all other balls to the right are out of their boxes. For the subjects, the trick to solving the puzzle is to figure out how to move the balls to get the right boxes to open up so that you can get all the balls out of their boxes.

The problem space, shown in Figure 2, is rather small, containing only 31 possible states, but the starting position was chosen to be 21 moves from the goal (21 moves is the minimum number required to solve the puzzle). Additionally, the problem space is linear, meaning that from any place in the problem space (except the top state) there are exactly two moves that can be made, one leading toward the solution (moving toward lower numbered states in Figure 2) the other leading directly away (moving toward the higher numbered states). Hence, after making a move, the only choices one has are to undo that move or make a new move. If one never retracts or undoes a move when it is not necessary, then one is guaranteed to easily solve the puzzle. However, it is extremely rare for a subject to do this when attempting to solve the puzzle for the first time.

Results

The effect of the memory load the first time the puzzle is solved is clearly seen in Figure 3. As the memory load increases, the number of moves required to solve the puzzle increases. A regression analysis shows this increase is significant (ß=25.82, F=10.82, p<0.01). The results from the second solution of the puzzle are shown in Figure 4. This figure is shown on the same

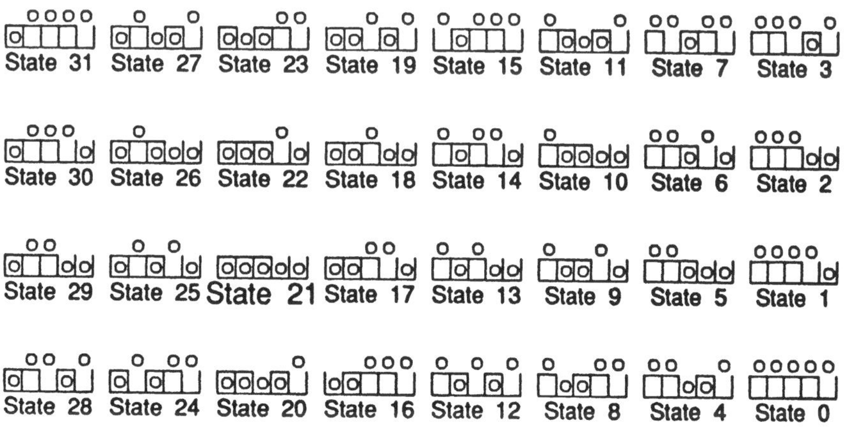

Figure 2: Problem Space of the Balls and Boxes Puzzle

scale as Figure 3. It is clear that there is significant speed up in that the number of moves to solve the puzzle drops dramatically from trial 1 to trial 2 (paired t-test, t=4.57, p<0.001). These results are also broken down by memory load on trial one to demonstrate that there is no difference between the groups in the amount learned on the first trial. It is clear that the memory load did not impair performance on the second trial. This is verified by an ANOVA of trial 2 performance by trial 1 condition and trial 2 condition where neither condition nor the interaction are significant (F=1.5, p>0.22, F=1.85, p>0.17, F=1.76, p>0.33 for the effects of trial 1 condition, trial 2 condition and their interaction respectively).

Examining the progress of a single subject on the puzzle shows that the process of solving the puzzle can be broken down into two phases: *exploratory* and *final path*. The *exploratory* phase contains most of the moves made by the subject and little or no actual progress toward the goal is made during this phase. The *final path* is a sequence of rapid totally error-free moves directly toward the goal. Progress made by subjects over the course of solving the puzzle on the first trial can be seen in Figure 5 (the x-axis is move number and the y-axis measures distance from the goal). The average length of the final path was 18.5 moves and does not vary across groups. The second time the puzzle is solved, 66 of 76 subjects show both exploratory phase and final path behavior while 10 subjects solved the puzzle perfectly the second time (immediate final path). The average length of the final path on the second trial was 19 moves.

The written protocols collected from each subject in between trials were analyzed and rated on a five point scale with 1 indicating that the description had no relevant information about the puzzle; a rating of 2 indicates very little information in the description, i.e. one statement of value; and so on up to 5, indicating that the subject provided a complete description of how to solve the puzzle. The median value of these ratings was 1.5. The average rating did not vary across groups indicating that all groups were equally poor at describing the structure of the puzzle or the strategies that they used to solve it.

Most of the advice subjects gave was very general in nature. The most common statement in the protocols (29 occurrences in 76 protocols) was general advice to remember the combinations or patterns that lead to different boxes opening. In spite of this advice, virtually no subjects described any of these patterns or combinations correctly. Other general advice such as "Be systematic" and "Use trial and error" also occurred frequently (20 times). Subjects did seem to be aware of certain aspects of the problem space: 21 subjects reported that the left ball was hardest, 12 noted that it was important to work from left to right and 10 realized that it was occasionally necessary to put balls back into boxes. No other statement occurred as often as 10 times. Only one subject stated that not undoing the previous move would be a good strategy and two others hinted that not reversing is important. In summary, the protocols were surprisingly uninformative, especially given the fact that most subjects went on to solve the puzzle quite rapidly on the second trial.

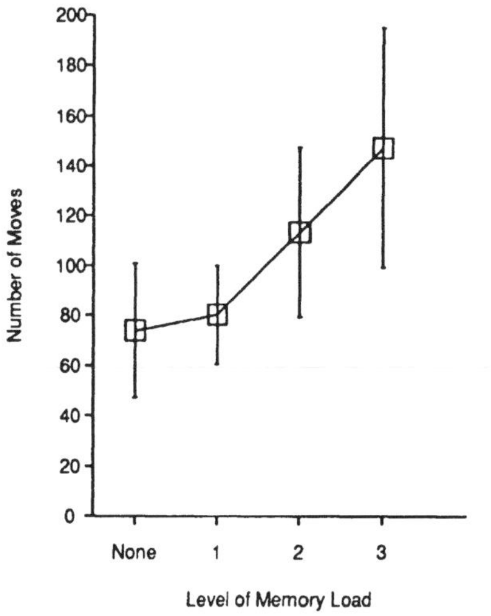

Figure 3: Number of Moves to Solve the Puzzle on Trial 1

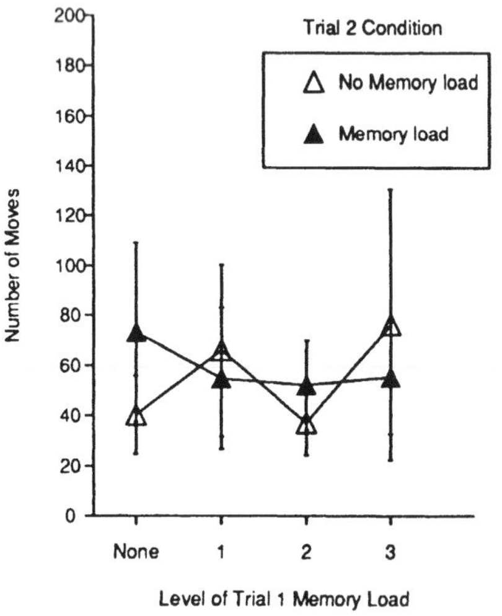

Figure 4: Number of Moves to Solve the Puzzle on Trial 2

The protocols present an interesting contrast to the final path behavior of the subjects. The sudden shift to error-free solution initially seems to suggest insight, yet the protocols suggest that subjects have not settled on a strategy even as simple as consistently avoiding reversals. The fact that most subjects also show exploratory behavior on the second trial also indicates that while subjects know something about the puzzle after the first solution, they don't understand it completely.

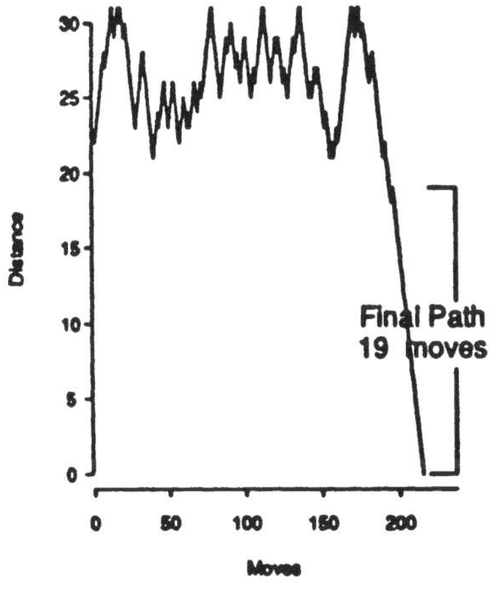

Figure 5: Sample move record

A Model

Although the pattern of results across the two trials is counter-intuitive, a relatively simple model provides a reasonable approximation of the subjects' behavior. The model is based on two basic problem solving strategies: (1) hill-climbing, that is preferring to remove balls from their boxes; and (2) no-reversal, a general preference not to undo the most recent move. Of course, since the problem space is linear, relying solely on the second strategy would lead to optimal solution of the puzzle. Although hill-climbing based on the increasing the number of balls out of their boxes may initially appear useful, reliance on this strategy actually impedes solution of the puzzle. The existence of the barrier areas suggest that at least at the beginning, subjects appear to rely on hill-climbing.

These strategies come into conflict on moves where balls must be replaced to make progress. If there is a decision between two balls is to be replaced (e.g. state 24, Figure 2), then the model will prefer not to make the reversal. However, when the choice is between taking a ball out that was just put in and putting a different ball in, then there is a higher probability of a reversal. An example of this situation would be arriving at state 25 (see Figure 2) from state 24 (the last move was to place the rightmost ball in). The hill climbing strategy indicates taking the rightmost ball out (and returning to state 24) while the no-reversal strategy indicates continuing on to state 26 and putting the secondmost ball from the right back into its box.

The relationship between these two strategies is controlled by a parameter, *bp* (for backup-penalty) describing the relative importance of the no-reversal strategy. The value of the hill-climbing strategy is held constant at +1 for removing a ball and -1 for replacing a ball. To select a move, the values of the two possible moves are calculated. The value of a move is the +1 if a ball is being moved out, -1 if a ball is being replaced. The move that reverses or undoes the previous move is penalized by an additional *-bp*. A small amount of normally distributed noise (mean=0, sd=2) is added to each value. The noise makes the model's behavior more variable and is meant to capture factors that affect subjects' move choice that are not captured by the model. The higher of these values is selected and this move is made. This process repeats until the goal state is reached.

When *bp*=2, these parameters exactly oppose each other in the difficult states described above (choosing between replacing a ball and removing a ball that was replaced on the previous move). When *bp* is less than 2, the model relies more on hill-climbing and when *bp* is greater than 2, the model makes fewer reversals. Since fewer reversals leads to more rapid problem solving, average number of moves to solve the puzzle decreases as *bp* increases.

When *bp* is 2 (or close to 2), the model essentially "wanders" around the problem state. At most states the model prefers moves that avoid reversals, but a number of situations provide problems (generally where two balls must be replaced in succession). These states correspond to the barrier areas observed in the subjects' behavior. The fit between barrier areas can be assessed by tracing subjects' move records with the model and at each point recording whether the model predicts a high, medium or low chance for reversal. This is then compared to the actual rate of reversal for each subject in each of these categories. The mean reversal rates across subjects was 9.7%, 13.6%, and 14.2%, for the states where the model predicted low, medium, and high reversal rates respectively. A repeated measures ANOVA across the reversal rates in each of the three categories for all 76 subjects shows a significant within subject effect ($F=13.6$, $p<0.001$). This indicates that subjects are indeed making reversals at different rates in the three categories and that the hill-climbing nature of the model reflects this strategy in the subject data. From this we can conclude that the *exploratory* behavior of the model is similar to the *exploratory* behavior shown by subjects.

In order to model the data pattern for the experiment across the groups and across trials, the model learns. In the subjects' data, the existence of the *final path* behavior initially appears to suggest insightful behavior, but the stunning lack of information in the protocols strongly suggests otherwise. Correspondingly, the model is constructed using a simple, incremental function to model learning effects. The theoretical implications of this decision are discussed below.

The learning across the two trials is modeled by increasing *bp* incrementally throughout the first trial solution. This has the effect of reducing the model's reliance on hill-climbing over the course of solving the puzzle. Since these strategies compete with each other, the model cannot disambiguate between learning that reduces the perceived value of hill-climbing or increases the perceived value of not making a reversal.

Increasing the value of *bp* increases the likelihood of achieving a final path sequence of moves and solving the puzzle. A higher value of bp speeds problem solution but does not guarantee immediate final path. Hence, carrying the value of bp across from the end of trial one to trial two reduces the number of moves needed to solve the puzzle but does not guarantee immediate final path.

The rate at which *bp* increases also affects the average number of moves it takes the model to solve the puzzle on the first trial. The difference across the memory load conditions can therefore be modeled by having *bp* increase at different rates for the different memory load groups. Since all groups perform with similar efficiency on the second trial, it follows that the value of *bp* carried over to the second trial should be approximately equal for all groups. Accordingly, *bp* increases incrementally, per move, according to the following general formula:

$$(1) \quad bp_t = bp_0 + bp_{max} - ce^{-\alpha t}$$

This formula describes exponential learning from the initial bp_0 toward some maximum value of *bp*, bp_{max}. The overall shape of the learning curve is determined by the quantities *c* and α. The parameter *t*, indexes the number of moves made. Several forms of the learning functions were considered. The function used mirrors the learning function previously used to describe classical conditioning (Rescorla & Wagner, 1972).

The parameters bp_0 and bp_{max} and *c* are the same for all runs of the model. Differences across the groups are captured by using different values of α. The higher memory load groups have smaller values of α, indicating that it takes more moves to reach the same level of *bp*. This implies that the groups are learning at different rates toward the same endpoint.

These parameters were fit to the data pattern with bp_0=1.5, bp_{max}=2.6, *c*=2.0, α=0.0428, 0.0331, 0.0228, 0.0120 for the control, 1-back, 2-back and 3-back groups respectively. These parameter settings start the model out with a reliance on hill-climbing, gradually shifting to relying more on not reversing any moves. The performance of the model on the first and second trial is shown in Figures 7 and 8. The same regression analyses used for the subjects show a significant effect of memory load on the first trial (β=37.4, F=56.9, $p<0.001$) and no effect of either first or second trial memory load on the second trial (F<1 in all cases). The model also shows final path behavior with average final path being 22.9 moves on the first trial and 26.6 moves on the second trial.

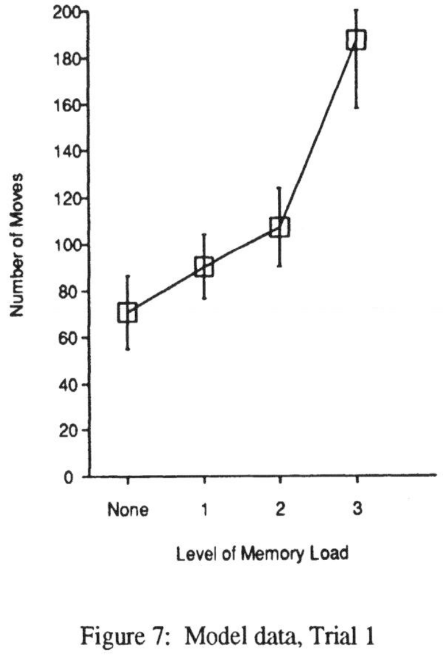

Figure 7: Model data, Trial 1

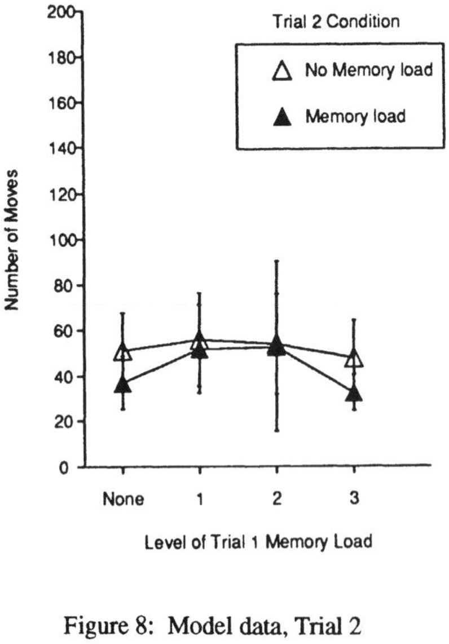

Figure 8: Model data, Trial 2

Discussion

The experiment demonstrated that it is possible to manipulate problem difficulty by reducing available working memory using a dual task paradigm. The fact that the effect of the first trial memory load does not carry over into the second trial and that the second trial memory load does not impair second trial performance indicates that a learning process is impaired on trial one. It follows that subjects from different groups are at essentially the same level of knowledge going into the second trial. The knowledge they have allows them to solve the puzzle much more efficiently on the second trial; but subjects in the high memory load groups on trial one take much longer to reach this level of knowledge.

Interestingly, the knowledge that is acquired about the puzzle is not easy for the subjects to communicate. Other work with this puzzle has similarly found that subjects are unable to describe the puzzle retrospectively and requiring subjects to give a concurrent protocol has also been unenlightening (Reber & Kotovsky, in preparation).

The inability of subjects to report what they are learning or describe the structure of the puzzle after they have demonstrated that they can solve it implies that they are learning to solve the puzzle implicitly or automatically. It follows then that it is critically important to understand when these processes are an integral part of problem solving and how this type of learning is affected by environmental factors such as an additional load on working memory. The research presented here demonstrates that the learning process is impaired by an external memory load, but later application of this knowledge is not impaired by a memory load (witness the lack of an effect of memory load on the second trial).

We also present a model that captures the overall data pattern fairly effectively. The reliance on hill-climbing with a general tendency not to reverse moves generates "search" behavior like the *exploratory* phase found in the subject data. The asymptotic nature of the learning function of the model guarantees that all runs of the model will be at essentially the same "level of knowledge" after the first trial. Changing the learning rate causes the model to take different amounts of time to reach this asymptotic level and hence take different number of moves to solve the puzzle. Thus the model shows differential performance on the first trial based on the learning rate, but performance on the second trial is not affected by changing the learning rate. The match of the model to the subject data further strengthens our claim that the memory load impairs the process of learning to solve the puzzle.

The model also provides a convincing demonstration of how a gradual, incremental learning function can capture the apparently sudden shift in subject behavior from exploratory to final path behavior. The success of the model provides additional support to the idea that subjects may be learning to solve the puzzle by some implicit, incremental function that gradually acquires enough information about the problem space to lead to improved performance.

While the model has been tailored to capture the mathematical structure of the data pattern, it also highlights some important theoretical points. First, the model provides support for the notion that subjects rely heavily on hill-climbing with some attention to not reversing the last move made. Second, the incremental learning function shows how a gradual learning process can result in *final path* behavior. The shape of the learning function further demonstrates how the effect of the memory load can be eliminated across the two trials. Although the model is designed to mathematically describe the data, rather then provide an explicit account of the underlying cognitive processes, it heavily constrains the development of a process model and expands our understanding of the process of solving this puzzle.

Further experimental work is underway that is aimed at uncovering the knowledge acquired on the first trial, together with constraints imposed by the model presented here should allow us to develop a more detailed model of the cognitive processes involved in this and other related tasks. This research represents progress toward a good understanding of the interaction between working memory and learning in problem solving that promises to greatly enhance our overall understanding of problem solving processes.

References

Kotovsky, K., Hayes, J. R. & Simon, H. A. (1985). Why are some problems hard: evidence from the tower of hanoi. *Cognitive Psychology, 17*, 248-294.

Kotovsky, K. & Simon, H. A. (1990). Why are some problems really hard: explorations in the problem space of difficulty. *Cognitive Psychology, 22*, 143-183.

Kotovsky, K. & Kushmerick, N. (1991). Processing constraints and problem difficulty: a model. *Proceedings of the Thirteenth Annual Conference of the Cognitive Science Society*. pp 790-795.

Newell, A. & Simon, H. A. (1972). *Human Problem Solving*. Englewood Cliffs, NJ: Prentice Hall.

Reber, P.J. & Kotovsky K. (in preparation). Implicit and explicit mechanisms of learning and problem solving.

Rescorla, R. A. & Wagner, A. R. (1972). A theory of Pavlovian conditioning: Variations in the effectiveness of reinforcement and nonreinforcement. In A. H. Black & W. F. Prokasy (Eds.), *Classical conditioning II: Current theory and research* (pp 64-99). New York: Appleton-Century-Crofts.

Categorization and stimulus structure

Steven Ritter
Department of Psychology
Carnegie Mellon University
Pittsburgh, Pa. 15213
sritter@psy.cmu.edu

Abstract

Concept discovery experiments have yielded theories that work well for simple, rule-governed categories. They appear less applicable to richly structured natural categories, however. This paper explores the possibility that a complex but structured environment provides more opportunities for learning than the early theories allowed. Specifically, category structure may aid in learning in two ways: correlated attributes may act jointly, rather than individually, and natural structure may allow more efficient cue sampling. An experiment is presented which suggests that each of these advantages may be found for natural categories. The results call into question independent sampling assumptions inherent in many concept learning theories and are consistent with the idea that correlated attributes act jointly. In order to model natural category learning, modifications to existing models are suggested.

Introduction

Thus, commencing our investigation by a careful survey of any one bone by itself, a person who is sufficiently master of the laws of organic structure, may, as it were, reconstruct the whole animal to which that bone belonged.

The quote, written by the naturalist Georges Cuvier in 1812, expresses confidence that the laws of nature so determine the structure of natural kinds that we should be able to deduce the entire form of an animal from only the smallest part. Today, few would argue that natural structure is so strongly determined, but fewer still would argue that the features of a natural object are assembled without regard to those already present. Rather, the laws of nature constrain the co-occurrence of features. Although wings and hollow bones do not necessarily go together, we are likely to find one, having found the other.

Psychologists are only beginning to examine the implications of natural structure for learning. In concept discovery tasks, the issue was usually avoided. A subject might be asked to learn the concept "large red square" or a "group of two items." The dimensions on which concepts varied were selected to be obvious to the subject and convenient for the experimenter to manipulate. There was little concern about whether dimensions were structured or rules defined as they would be in natural kinds. Such experiments taught us quite a bit about learning. For example, they suggested that subjects sampled from a hypothesis space, changing their working hypothesis only when they made an error (e.g., Restle, 1963; Trabasso and Bower, 1968).

Rosch and her colleagues (e.g., Rosch, et al.., 1977) revived interest in the structure of natural kinds. Their work showed that natural categories are not at all like the assemblage of features used to represent a concept in concept discovery experiments. Instead, natural kinds appear to be structured around "family resemblances." Family resemblance categories consist of a large number of highly inter-correlated features. To a large extent, these correlations are not arbitrary; rather, they are the result of natural laws.

This research was supported in part by a graduate student fellowship from the National Science Foundation.

Findings about natural category structure call into question the relevance of concept discovery methods for learning natural kinds because the conditions found in natural categories appear to be precisely those which make concept discovery difficult. Concept discovery experiments turn out to be intractable in all but the simplest versions. If disjunctive rules (e.g., "red or square") are allowed, or if the rule comes from a non-standard space (such as when the correct response depends on a past pattern of responses), subjects may search indefinitely without finding the correct solution (Levine, 1975). If the number of irrelevant attributes is high, subjects will be unable to learn in a reasonable amount of time (Bourne and Haygood, 1959).

Correlations among relevant attributes may help, but only a little. Trabasso and Bower (1968) found that, when stimuli contained features perfectly correlated with each other and with the category label, subjects tended to categorize by either one cue or the other. Only rarely did a subject notice the relationship between both cues and the category label. More importantly, these experiments showed that redundant relevant cues tend to compete for attention so that, although the overall probability of correctly learning increases as the number of relevant dimensions increase, the benefit is only due to having more predictive features. Subjects sample only a small set of features on each trial, so the benefit diminishes as the number of correlated attributes increases.

A central question, then, is how natural categories are learnable at all. Although children may take some time to learn the difference between a dog and a cat, the actual number of exposures to such animals over that time is typically much smaller than the number of exemplars viewed in a concept discovery experiment.

Billman and Heit (1988) present one part of the answer. In their model, each trial provides an opportunity to predict one feature based on another. The prediction uses previously observed contingencies between the predictor value and values of the predicted feature. The choice of predictor and predicted features is governed by salience, which increases for both the predictor and predicted feature when a correct prediction is made. This relationship sets up a feedback loop in which correlations lead to correct predictions, which increase the salience of their constituent features. This results in a greater chance of noticing correlations involving these features, so the cycle continues. Thus, according to this model, the correlation between "has feathers" and "bird" is learned more easily because there is a correlation between "has feathers" and "flies." Billman (1989) provides experimental support for the model.

Correlated attributes can only be part of the answer, however. Natural categories have too many potential correlations to be effectively searched without some guidance. This study aims to provide further evidence for mutual support of correlated attributes and for search guided by structural constraints. The underlying belief is that natural categories may be learnable because their rich structure allows us to make assumptions about which features are likely to go together. While it seems logical to assume that we have developed a categorization system that can take advantage of natural structure, it has been difficult to demonstrate any such advantage (Medin, Wattenmaker and Hampson, 1987; Wattenmaker, et al., 1986; but see Malt and Smith, 1984).

In the current experiment, the stimuli, although artificial, have a rich correlational structure. Redundancy of cues ensures that a stimulus can be unambiguously classified, even if all relevant cues are not present. This is equivalent to the natural situation, where we typically have access to only a few of the many cues we might want to use in identifying something. We make two predictions. First, we expect a well-structured category to be easier to learn than an ill-structured category, even if each provides equivalent information. Second, we expect to replicate Billman's finding that correlated attributes, when embedded in a logical structure, will aid learning beyond the level that would be expected if the cues were acting independently.

Method

Subjects

Subjects were 54 undergraduates taking an introductory psychology class. Participation in the experiment partially fulfilled a course requirement.

Stimuli

Stimuli consisted of pictures of boats, each containing a maximum of five parts: the mainsail, the jib sail, the flag, the rudder and the centerboard. Each of these parts could vary in color and sometimes shape. The body and mast of the boat were identical in all stimuli and were always present (see Figure 1). Any of the variable parts could be missing. The probability that a particular part was present in a stimulus (i.e., the part's availability) is given in Table 1.

Three conditions (control, well-structured and ill-structured) were run. The conditions differed in the composition of stimuli. In the "well-structured" condition, each color corresponded to a shape. For example, all blue flags in this condition were short and wide, while all pink flags were long and thin. In the "ill-structured" condition, the color of a part was independent of its shape. In the control, only color changed.

Stimuli were assigned to two classes, called "gemmer" and "brice." Stimuli could be classified correctly using either color or shape of the parts of the boat. In all conditions, boats could be classified according to the rule: "If the flag is blue, the boat is a gemmer. If the flag is missing and the main sail is blue, the boat is a gemmer. If both the flag and mainsail are missing and the remaining sail is blue, the boat is a gemmer." The rule can be thought of as a hierarchy. The flag is the most important part of

	Flag	Mainsail	Jib	Rudder	Center-board
Availability	65	67	85	67	65
Reliability	100	65	57	35	10

Table 1: Availability and reliability of color for stimuli in the learning phase. In the well-structured group, availabilities and reliabilities for shape are identical to those for color. For shapes in the ill-structured group, values for flag and mainsail apply to the centerboard and rudder (respectively). Availability refers to the percentage of time the part is present. Reliability is the percentage of time the part's color (or shape) predicts the category.

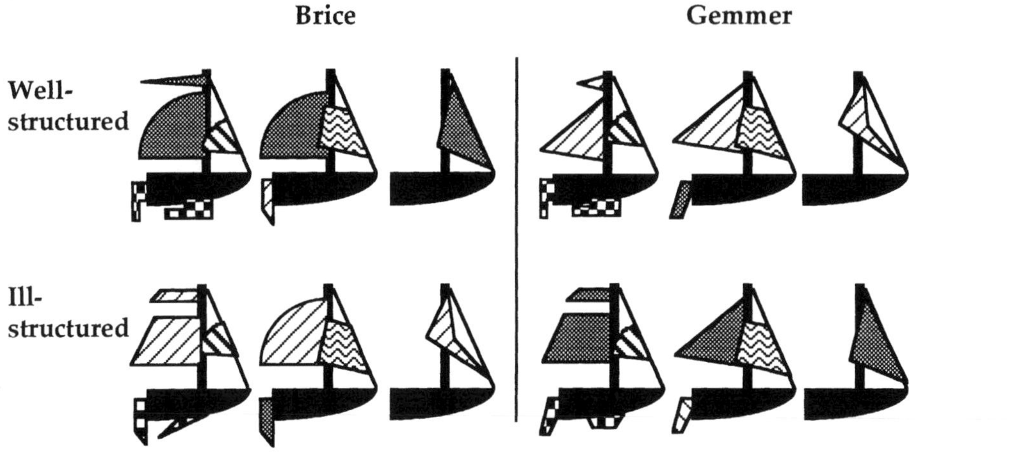

Figure 1: Examples of stimuli used in the experiment. Patterns represent colors used in the experiment. Corresponding boats in A and B have the same coloring. Either set can be categorized by the rule: "If the flag is shaded, it is a brice. If there is no flag and the mainsail is shaded, it is a brice. If there is neither a flag nor a mainsail and the jib is shaded, it is a brice." The ill-structured group can also be classified by a rule of the form: "If the centerboard is triangular, it is a gemmer. If there is no centerboard and the rudder points down, it is a gemmer. If there is no centerboard or rudder and the jib is a concave polygon, it is a gemmer." A similar shape rule applies to the well-structured group.

the boat, followed by the mainsail and the jib sail. In addition, each kind of boat has a color: blue for gemmer and pink for brice. For each stimulus, the topmost part in the hierarchy was guaranteed to be predictive. As a result, flags were always either pink or blue, and, for boats without flags, mainsails were always either pink or blue. In the well-structured and ill-structured conditions, subjects could additionally categorize exemplars by a rule referring to shape, rather than color. In the well-structured condition, the shape hierarchy was the same as the color hierarchy (flag, then mainsail, then jib). In the ill-structured condition, the color hierarchy was flag, then mainsail, then jib; but the shape hierarchy was centerboard, then rudder, then jib. Stimuli in the well-structured and ill-structured conditions were matched such that equivalent stimuli in the two conditions provided both color and shape cues at the same level of the hierarchy. Stimuli in the control condition were matched on the color hierarchy.

Further constraints were put on the stimuli to provide a richer category structure. Four of the five parts had a greater than chance probability of being the color or shape predictive of the category. The reliability of each part of the boat is given in Table 1.

Procedure

The experiment was run on a color Macintosh computer. The experimental session consisted of two stages: training and test. For each trial in the training stage, a stimulus was presented along with the choices "Brice" and "Gemmer." Subjects gave their responses by clicking on the appropriate answer. Feedback followed immediately in the form of "Yes (no) this is a gemmer (brice)." Subjects could study the stimulus along with the feedback and initiate the next trial when they were ready. The training stage continued for 60 trials.

In the test stage, subjects responded to stimuli identical to those in the training stage as well as to stimuli that omitted either the color or shape cues. "Shape-only" stimuli were drawn in gray. "Color-only" stimuli displayed colored circles in place of the studied shapes. Control subjects were not given "shape-only" stimuli. Fourteen stimuli of each kind were presented, for a total of 28 test trials in the control group and 42 in each of the other groups. Test stimuli were constructed so that the color-only, shape-only and color-and-shape sets depended on each level of the hierarchy to the same extent.

Cue Learned	Well-structured	Ill-structured
Color Only	8	2
Shape Only	1	5
Color and Shape	5	2
Neither	4	9

Table 2: Number of subjects learning to respond to each (or both) cues, by group. A subject was considered to be responding to a cue if the subject correctly answered at least 10 of 14 questions providing only that cue.

The procedure for test stimuli was the same as for training stimuli, except that feedback was not given. Instructions for the test stage were not given until the training stage was completed, so subjects were not biased to look at both color and shape. Responses and reaction times were recorded in both stages, but the instructions emphasized correct responding only.

Results

A comparison of the well-structured and ill-structured groups supports our hypothesis about stimulus structure. We may classify subjects as having learned the color cue, the shape cue or both cues, depending on their performance on the testing phase. Subjects correctly answering at least 10 of the 14 color-only questions and less than 10 of the 14 shape-only questions were classified as "color learners." Those correctly answering 10 of 14 shape-only questions and less than 10 color-only questions were classified as "shape learners." Those correctly answering more than 10 in each category were classified as "color and shape learners."[1]

Table 2 shows how subjects from the well-structured and ill-structured groups were

[1] The binomial probability of a subject with no knowledge of either cue falling into either the color-only or shape-only category is .08. The probability of such a subject falling into the "color and shape" category is .01.

classified. Eighty percent of color learners were in the well-structured group. Although the majority of shape learners were in the ill-structured group, this statistic may be misleading. All but one shape learner in the well-structured group also responded to color, thus falling in the "color and shape" group. In total, 13 subjects in the well-structured group and 4 in the ill-structured group learned to respond to color. Six subjects in the well-structured group and 7 in the ill-structured group learned to respond to shape.

There are other indications that the well-structured group found the task somewhat easier. The amount of time taken to learn (in the training stage) was measured by a criterion of trials to second-to-last error (this statistic was more resistant to careless errors than trials to last error). Subjects in the well-structured group reached criterion in an average of 40.0 trials, while those in the ill-structured group took 48.3 trials. An analysis of variance showed the difference to approach significance, $F(1,30)=3.64$, $p<.07$.

Stronger support for the hypothesis comes from the test phase (see Figure 2). A 2 (group) x 3 (learning stimulus order) analysis of variance of percent correct on the color-only questions shows a difference between groups, $F(1,30)=6.17$, $p<.02$. Reaction times tell a similar story. These times were subjected to a log transformation and then analyzed in a 2 (group) x 3 (learning stimulus order) x 14 (question) analysis of variance. The results show a significant difference between groups, $F(1,311) = 10.10$, $P<.01$.

Analyses of shape-only and color-and-shape questions were less conclusive. On color-and-shape questions, the pattern of results (both in percentage correct and reaction time) is the same as for color only, but the differences do not reach significance, $F(1,30)= 0.55$, $p>.46$. Shape-only questions show some advantage for the ill-structured group, but the results do not approach significance, $F(1,30) =1.18$, $p>.28$.

A comparison of the well-structured and control groups speaks to our second prediction, that correlated attributes contribute beyond their individual influences. If correlated attributes support each other, we would expect the well-structured group to be able to answer color-only and color-and-shape questions more easily than

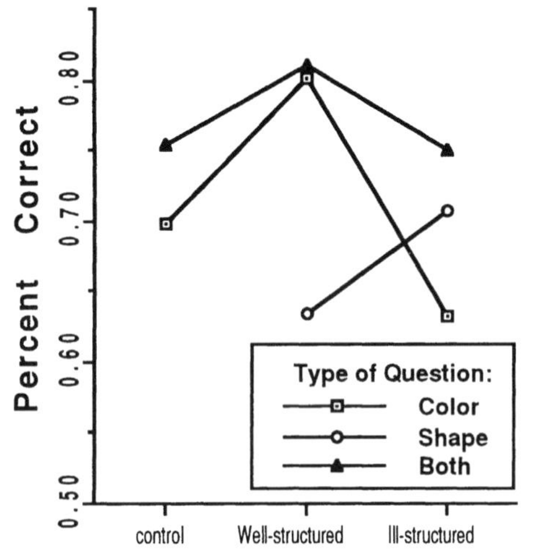

Figure 2: Performance by question type

the control group. This result would obtain if the correlation between shape and color (found in the well-structured stimuli) helped subjects discover the correlation between color and category label. The experimental results are not strong with respect to this hypothesis. Subjects in the well-structured group learned faster than those in the control (by trials to second-to-last error, $F(1,30) = 4.69$, $p<.05$), but this could have been due to the independent influence of shape. Both reaction times and percent correct scores show an advantage for the well-structured group (see Figure 2), but the differences are not statistically significant.

Discussion

The data reflect on two aspects of categorization models. The first is whether correlated attributes contribute individually, as in the Trabasso and Bower model or jointly, as in the Billman and Heit model. While the data show some advantage for the well-structured group over the control, the results do not approach significance. This should not be taken as a failure to replicate Billman and Heit's model, since that model predicts much stronger effects in situations where no feedback is given.

The second issue is whether stimulus structure influences cue sampling. Since the well-structured and ill-structured conditions provide the same number of cues, both the Trabasso and Bower model and the Billman and Heit model predict that there would be no difference between the well-structured and ill-structured groups. If cue-sampling is sensitive to stimulus structure, however, we might expect an advantage for the well-structured condition. In this case, the fact that flag color was relevant might direct us to notice that flag shape is also relevant (in the well-structured condition). The experimental results provide some support for this kind of model. Subjects in the well-structured condition were better able to answer questions about the color cue than subjects in the ill-structured condition, and they discovered the categorization rule faster than other subjects. In addition, these subjects were able to make judgments about color-only stimuli faster than subjects in the ill-structured group. The finding that the advantage for the well-structured group reversed (albeit non-significantly) for the shape-only stimuli is problematic for this explanation, however.

The experiment presented here was designed to examine whether structural properties of complex, real-world objects aid in learning categories. The results suggest that structural properties play a role by influencing the order in which cues are sampled. Clapper and Bower (1991) present a model in which correlated attributes in a category influence the sampling of cues in new exemplars. Pazanni (1991) presents a model in which domain-specific biases affect the order of hypothesis search. The way in which these different influences on search interact is a topic for future research.

Acknowledgments

I thank Brian MacWhinney for his guidance in all stages of this work.

References

Billman, D. (1989). Systems of correlations in rule and category learning: Use of structured input in learning syntactic categories. *Language and Cognitive Processes*, 4, 127-155.

Billman, D. and Heit, E. (1988). Observational learning from internal feedback: A simulation of an adaptive learning method. *Cognitive Science*, 12, 587-625.

Bourne, L. E., Jr. and Haygood, R. C. (1959). The role of stimulus redundancy in concept identification. *Journal of Experimental Psychology*, 58, 232-238.

Clapper, J. and Bower, G. H. (1991) Learning and applying category knowledge in unsupervised domains in G. H. Bower (Ed.), *The Psychology of Learning and Motivation, Volume 27*. (pp. 65-108). New York: Academic Press.

Cuvier, Georges (1812). *Discours préliminaire.* Quoted in Gould, S. J. (1990, June). Everlasting legends. *Natural History*, 12-17.

Levine, M. A. (1975). *A cognitive theory of learning: Research on hypothesis testing.* Hillsdale, NJ: Erlbaum.

Malt, B. C. and Smith, E. E. (1984). Correlated properties in natural categories. *Journal of Verbal Learning and Verbal Behavior*, 23, 250-269.

Medin, D. L., Wattenmaker, W. D. and Hampson (1987). Family resemblance, conceptual cohesiveness and category construction. *Cognitive Psychology*, 19, 242-279.

Pazzani, M. J. Influence of prior knowledge on concept acquisition: Experimental and computational results. *Journal of Experimental Psychology: Learning, Memory and Cognition*, 17, 416-432.

Restle, F. (1962). The selection of strategies in cue learning. *Psychological Review*, 69, 11-19.

Rosch, E., Mervis, C., Gray, W., Johnson, D. and Boyes-Braem, P. (1976), Basic objects in natural categories. *Cognitive Psychology*, 8, 382-439.

Trabasso, T. and Bower, G. H. (1968). Attention in learning: theory and research. New York: Wiley.

Wattenmaker, W. D., Dewey, G. I., Murphy, T. D. and Medin, D. L. (1986). Linear separability and concept learning: Context, relational properties and concept naturalness. *Cognitive Psychology*, 18, 158-194.

"Adaptation" to Displacement Prisms Is Sensorimotor Learning

Jennifer L. Romack, R. Andrew Buss
and
Geoffrey P. Bingham
Department of Psychology
Indiana University
Bloomington, IN 47405
gbingham@ucs.indiana.edu

Abstract

Observers reaching to a target seen through wedge shaped displacement prisms initially reach in the direction of displacement, correcting their reaches over a series of about 12 trials. With subsequent removal of the prisms, observers initially reach to the opposite side of the target, correcting over about 6 trials. This phenomenon has been called "adaptation" because of its similarity to the adaptation of sensory thresholds to prevailing energy levels. We show, however, that this perturbation to visually guided reaching only mimics sensory adaptation initially. Subsequent changes show that this is sensorimotor learning. Error in pointing to targets is the commonly used measure. We measured times for rapid reaches to place a stylus in a target. Participants wearing a prism worked to achieve criterion times previously established with normal, unperturbed vision. Blocks of trials with and without a prism were alternated. Both the number of trials to criterion and the mean times per block of trials decreased over successive blocks in a session, as well as over successive days. By the third day, participants were able to respond rapidly to perturbations. This reflects the acquisition of a new skill that must be similar to that acquired by users of corrective lens.

Introduction

Wedge shaped prisms bend the light projected to the eye so that rays enter the eye at an angle displaced from their original angular location about the point of observation. The direction of displacement is towards the apex of the wedge. The amount of displacement depends on the size of the wedge. Observers reaching towards a target seen through displacement prisms reach, on their first attempt, in the direction of displacement. Over a series of about 12 trials, observers correct their reaches so that eventually they reach directly towards the actual location of the target. With subsequent removal of the prisms, observers initially reach to the opposite side of the target, becoming correct over about 6 trials. This phenomenon has been called "adaptation" because of its similarity to the adaptation of sensory thresholds in response to changing ambient energy levels.

Adaptation of sensory thresholds exhibits time courses invariably described as negatively accelerated exponential curves with an asymmetry in the rate of adaptation depending on the direction of adjustment. For instance, adaptation of visual thresholds to darkness takes 20-30 min while adaptation to bright conditions[1] takes only 2-3 min. Likewise, the curves for gustatory adaptation, adaptation to cutaneous pressure, and adaptation to cutaneous pain exhibit asymmetry depending on direction (Schiff, 1980; Uttal, 1973). Characteristic of sensory adaptation functions is the relative constancy or stability of the relaxation times. Repeated adaptation of visual thresholds to alternating dark or bright conditions does not alter the respective times for adjustment. The magnitudes of these relaxation times reflect the character of the underlying neurophysiological events. The stability of these times reflects the relative simplicity of the underlying dynamics. Only two time scales are involved. A relatively

[1] This is often called "recovery" in recognition of the asymmetry.

fast time scale associated with detection as a threshold is exceeded and a slower time scale corresponding to the adjustment of the threshold level with adaptation.

The effect of displacement prisms is to perturb the perceptuomotor system used, for instance, in reaching. Referring to the response to this perturbation as "adaptation" is to imply that the underlying dynamic is similar to simple dynamic of sensory adaptation. Indeed, the analogy has been made explicitly with the suggestion that a single (correlational) transformation from sensory to motor variables plays the role of the threshold (Dolezal, 1982; Hein & Held, 1962; Held, 1961; 1965; 1968; 1980). The two time scales would be a relatively fast time scale associated with a process of sensorimotor transformation and a slower time scale corresponding to the time for adaptation to the effect of the prism or its removal.

While this holds out the promise of a relatively simple account, we suggest that the commonly successful use of lenses to correct visual dysfunction means that this approach is overly optimistic. The simple account cannot be correct. If it were, the use of corrective lenses would be much less effective and more problematic than it is. Users typically experience difficulties in adjusting to corrective lenses in the first couple of days. Thereafter, however, adjustment to the lenses is almost immediate as is the adjustment to their removal.

The implication is that the time required for adjustment to perturbation by displacement prisms is not constant, but decreases over repeated application of the perturbation. Thus, multiple time scales are involved, certainly more than two. Focusing on paradigmatic reaching with prisms, and starting with relatively fast time scales, there is the time for a single reach. This is on the order of a second. There is the time for adjustment to the prism or its removal. This is on the order of a few minutes. There is the time for change in the period of adjustment. This may be on the order of an hour or two or perhaps, a day or two.

Assuming that his latter change exists as our observations on corrective lenses suggests it must, it would constitute evidence for learning or sensorimotor skill acquisition. Is the adjustment to displacement prisms part of a process of skill acquisition? If so, then we might expect to see improvements in the ability to respond over the course of repeated perturbation on a single day with some retention of skill on a subsequent day and with continued improvements over days leading to an expert's level of skill. How skilled might an expert be? Might he or she be capable of immediate adjustment to prisms of various strengths? This is unlikely given that some period of adjustment is required even for a life long user of corrective lenses when a new prescription is obtained.

We investigated these questions by measuring the time course of rapid reaches to a target performed over alternating application of a displacement prism and its removal, allowing normal vision. We also began to investigate the nature of the potential skill by applying a stronger prism after successive adjustments to a weaker one. Subsequently, we will discuss the new questions that arise with the reconceptualization of this long standing problem in perception/action research.

Methods

Apparatus. All reaches were measured using a two-camera WATSMART system sampling infrared emitting diodes (IREDS) at 100 Hz. IREDS were placed on the dorsal side of the metacarpal-phalangeal joint of the right thumb, on the thumbnail, and around the right eye. The collection period was controlled by an external trigger housed in a launchpad and target. Data collection routines were initiated when a stylus was removed from the launchpad and terminated when the stylus was inserted into a target. Placement of the stylus in the launchpad broke an infrared beam, which set the clock at zero. Removal of the stylus from the launchpad triggered the internal timing mechanism with a maximum delay of 5 ms. Placement of the stylus into the target split a beam which terminated the clock. Movement times were displayed on a CRT at the end of each trial and recorded by the experimenter.

Three pairs of swimming goggles were instrumented to allow measurement of the head and eye position. In all cases, the left eye piece was blackened. The right eye piece was covered with a 9 cm by 4 cm piece of plexiglass which supported three IREDS, placed above, below, and to the right of the eye. Displacement prisms were mounted over the right eye of two of the sets of goggles. Visual displacement was 10° and 15°

to the right, respectively.

Participants. Eight adults, 5 male and 3 female, aged 18–28 years, participated in the experiment. All had good, uncorrected vision and had never worn corrective lens. All were free of motor disabilities. Participants were paid at $5.00/hr.

Procedure. Three experimental sessions were performed on consecutive days at approximately the same time each day. During testing, the participant was seated comfortably. Head movement was unrestricted. The participant's task was to remove a stylus from a launchpad and to place the stylus as rapidly as possible in a target hole by reaching with the right hand. The launchpad was located next to the participant's hip and the target was placed just above the participant's right knee. The target was positioned at a distance reachable by fully extending the arm without moving the shoulder or trunk. The angle of the target was determined by having the subject sight directly down the target hole. The task was performed under four visual conditions: binocular, monocular, monocular with a restricted field of view (clear goggles), and monocular with restricted field of view and displaced vision (prism goggles). The displacement was 10°.

The participant was instructed to move to the target as rapidly and accurately as possible, so as not to collide with the target face at a high speed. The participant was told not to use any targeting strategies other than aiming straight for the target itself. The participant's eyes remained closed except immediately before and during the reach.

The first 2 blocks consisted of 10 trials in each of the binocular and monocular conditions. The remainder of the experiment consisted of alternating blocks of clear goggle and prism trials. The initial clear goggle block consisted of 10 trials which were used to obtain the participant's criterion value. The participant was not aware that a criterion time was being established for use throughout the remainder of the experiment. The criterion value was determined by taking the mean of the participant's movement times for this block (minus the fastest and slowest trials) and adding one standard deviation.

Thereafter, the number of trials for each block varied, depending on the number of trials required for the participant to reach the criterion value during three consecutive trials. Participants were informed that they were trying to achieve reaches at or below criterion times. Alternating blocks of viewing conditions continued until the participant reached the criterion within a maximum of four trials for the prism condition. At this point, an additional round of clear goggles and prism blocks was performed ending with a final block of the clear goggles trials. This was followed by a prism block with a 15° displacement prism.

Results and Discussion

The number of trials and mean movement time for each block of trials were computed for each participant. Based on these results, the data of two participants were removed from the analysis. One male exhibited much slower and more variable times and an exceptionally large number of trials per block compared to the other participants. He was discovered to be left hand dominant. The other was a female who exhibited a distinct lack of motivation while performing the task. Her criterion time was very slow and was found not to be representative of her performance capabilities as revealed in a number of other blocks from baseline and experimental trials. Her data also exhibited an unusually high degree of variability.

Mean movement times and the mean number of trials per block were calculated for the remaining 6 participants. The overall mean criterion time for the 6 participants was 1.1 s (sd=.053 s).

Movement times decreased over trials within blocks. In the first prism blocks on the first day, times started well above criterion and dropped to criterion levels over an average of about 11 trials. In the first clear goggles block following this, times dropped to criterion in about 7 trials. These results were consistent with the standard "adaptation" pattern, including the asymmetric number of trials with and without the prism. However, the amount of decrease in times over trials within blocks itself decreased over successive blocks as did the initial amount of time above criterion at the beginning of blocks. These changes in times revealed that the pattern of results with initial exposure to the prism was merely the first stage in a process of perceptuomotor skill acquisition.

Mean movement time and the mean number of trials per block decreased over successive blocks within a session. In addition, mean movement time and the mean number of trials per block decreased across days. The variability in movement times followed a similar trend showing that participants were performing the task with increasing consistency as well as proficiency.

The mean number of trials per block is shown in Figure 1 for the 3 days (Day 1: filled circles; Day 2: open squares; Day 3: filled triangles). As required by design, the number of trials for the first three baseline blocks (binocular, monocular, and clear goggles 1) remained constant across days at 10 trials. After a slight increase for the first prism block on day 1, the mean number of trials for each subsequent block declined progressively in a nonmonotonic fashion. Mean number of trials to criterion dropped below 4 trials after 14 alternating blocks on day 1. Some participants reached the 4 trial cutoff before others. The number of participants contributing to the means was shown in Figure 1[1].

Mean Number of Trials per Block

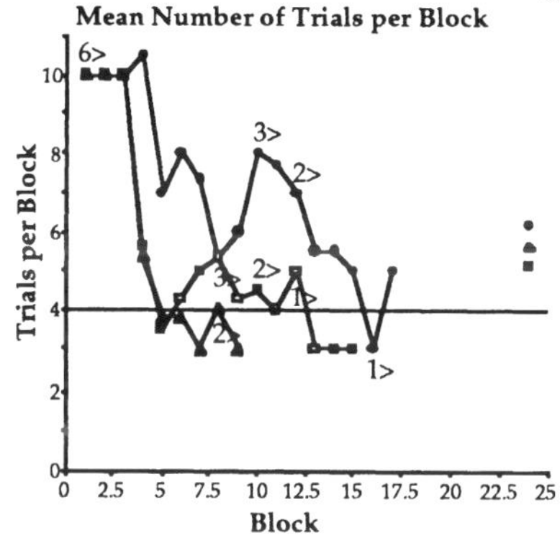

A similar pattern was obtained on day 2, although the mean number of trials in the first prism block was only 5.67, a drop of 4.83 from day 1. This drop also was evident on day 3 in which a mean of 5.33 trials was required to reach criterion in the first prism block. The total number of alternating blocks performed before the cutoff on day 2 was 12, however, the 9-12th blocks were performed by a single participant. For day 3, the number of alternating blocks was 6. The overall mean number of trials to criterion per block dropped over days from 6.5 on day 1 to 4.2 on day 2 to 3.8 on day 3.

As shown in Figure 2, mean movement times for binocular, monocular, and clear goggle 1 blocks were similar to one another on all 3 days. However, the overall mean time for these blocks

Mean Trial Durations per Block

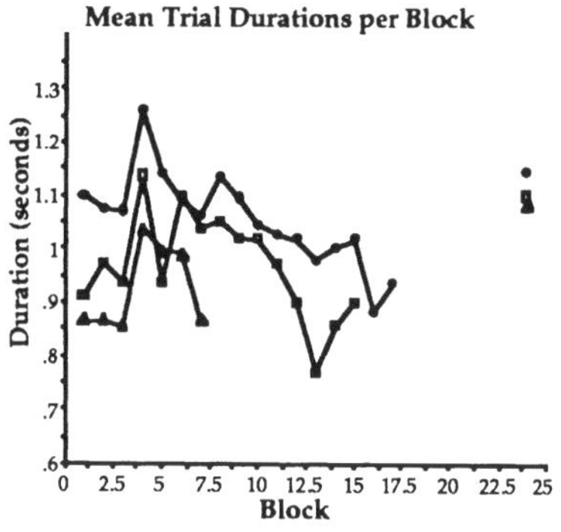

decreased over days from 1.08 s to .94 s to .86 s. This reflected a general improvement in performance across days.

The mean movement time for the first prism block on day 1 was 1.25s (sd=.07s). This was .18 s greater than the mean of the clear goggle 1 block. This increase was exactly the same each day. Prism 1 mean times for days 2 and 3 were 1.12 s and 1.04 s, respectively.

However, subsequent rates of decrease of mean times over blocks were different on different days. We performed linear regressions on the individual times for the first trial of each block from prism 1 on to the last clear goggle block (that is, excluding the 15° prism blocks), regressing block number on times for each day. All 3 regressions were significant, $p<.01$ or better. The slopes represented the decrease in mean time per block. Over days, the slopes increased from -15ms to -47ms to -113ms. When coupled with the mean prism 1 times, these results mean that, on subsequent days, participants started with faster times and proceeded to decrease more rapidly from those times over blocks. However, the change in the rate of decrease occurred only for clear goggle blocks. The result of a linear regression of block number on individual times for the first trials of prism blocks alone was plotted in Figure 3. Almost identical slopes were obtained for all 3 days. The slopes were -69ms, -71ms, and -84ms for days 1-3 respectively. This means that the movement time for the first trial

[1] For instance, 3 participants had reached the 4 trial cutoff by the 9th block overall on the first day, leaving 3 participants contributing to the mean for the 10th block.

of prism blocks decreased at a constant rate of about 70ms per block on all 3 days. The difference in intercept between days 1 and 2 was 220ms (≈3x70ms). Between days 2 and 3, it was 140ms (=2x70ms). This means that about 3 prism blocks worth of progress (or about 1+3=4, 4x2=8 prism/clear goggle alternations) was retained on the second day, while about 3 prism blocks worth of progress beyond day 2 (or about 6 prism/clear goggle alternations) was retained on day 3. The regularity in these numbers surely implies the existence of an underlying learning function.

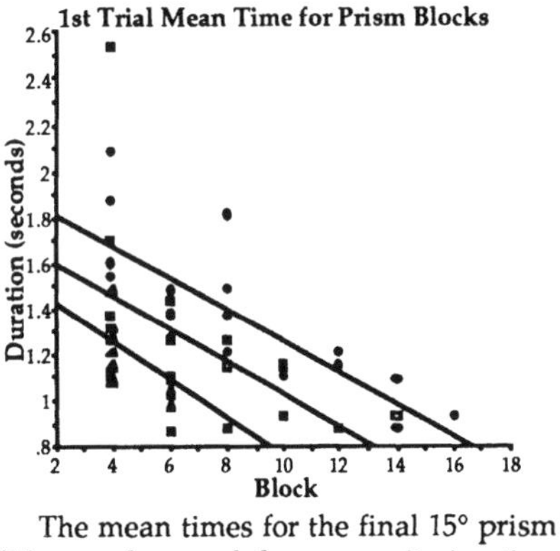

The mean times for the final 15° prism blocks all hovered around the mean criterion time across participants implying that the participants discovered exactly what this time was and tended to optimize to that time. Mean times for 15° prism blocks were less than or equal to those for the first 10° prism blocks on a given day. Because 15° prism trials were performed after participants had re-adjusted to the clear goggle condition, we can infer that skill gained in the context of adjustment to 10° prisms was used to advantage in adjusting to a 15° prism[2]. In the learning literature, this would be called "positive transfer" or "stimulus generalization" (Welch, 1978; 1991).

Finally, the standard deviations for block times were plotted in Figure 4. Variability in movement times decreased over blocks within a day. Further, the rate of decrease from a common initial level of variability increased over days. Linear regressions were performed on standard deviations for blocks starting with prism 1 (and excluding 15° prism blocks), regressing block number on SD's. The slopes increased over days 1-3 from -12ms to -17ms to -31ms, respectively.

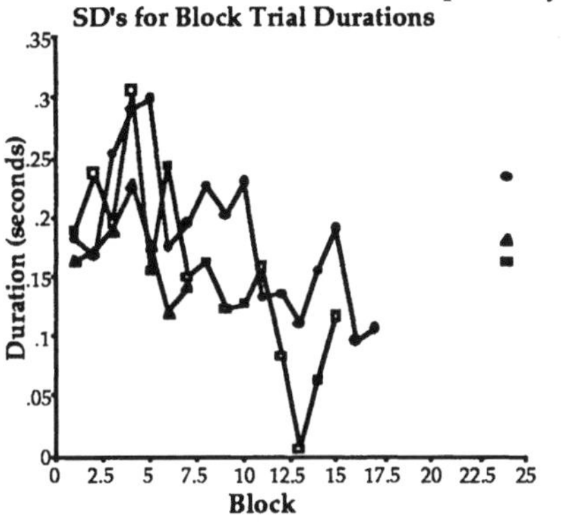

Thus, participants exhibit approximately the same amount of random variability at the beginning of subsequent days. Reduction of this variability preceded more rapidly on successive days. The implication is that participants start with a rough form of the adjustment function on a given day, but proceed with skill to hone in quickly on a stable adjustment function.

Conclusions

We found evidence for processes on multiple time scales including individual reaches performed in about 1 second, adjustment to perturbation over trials within a block occurring over about 1 minute, acquisition of the ability to adjust over a mere couple of trials occurring over 1 hour, and acquisition of the ability to adjust almost immediately on first exposure, which occurred over days. The problem, viewed in the light of this evidence, is to find a common framework enabling us to understand how processes on different time scales can interact and determine one another.

Another, perhaps more familiar way of expressing this question is as follows. We have evidence that adjustments to perturbations by displacement prisms are part of a process of sensorimotor skill acquisition. The questions that arise naturally in this context are: What are the old skills? What are the new skills? How are the new skills related to the old skills?

Our experiments provided us with kinematic data that we can use to begin to address these questions. We will turn to this task in future papers.

[2] Although, participants were not as quick to adjust as they were on the immediately preceding 10° prism block.

Acknowledgments

This work was supported in part by the Institute for the Study of Human Capabilities funded by AFOSR at IU. We are grateful for the assistance of Michael Stassen in programming some of the data analysis routines and for the efforts of Mike Bailey and John Walkie in constructing the triggering apparatus housed in the launchpad and target.

References

Dolezal, H. (1982). *Living in a world transformed: Perceptual and Performatory adaptation to visual distortion*. New York: Academic Press.

Hein, A. & Held, R. (1962). A neural model for labile sensorimotor coordinations. In E. E. Bernard & M. R. Kave (Eds.), *Biological prototypes and synthetic systems* (Vol. 1). New York: Plenum Press.

Held, R. (1961). Exposure-history as a factor in maintaining stability of perception and coordination. *Journal of Nervous and Mental Disease, 132*, 26-32.

Held, R. (1965). Plasticity in sensorimotor systems. *Scientific American, 213* (5), 84-94.

Held, R. (1968). Plasticity in sensorimotor coordinations. In S. J. Freedman (Ed.), *The neuropsychology of spatially oriented behavior.* Homewood, IL: Dorsey Press.

Held, R. (1980). The rediscovery of adaptability in the visual system: Effects of extrinsic and intrinsic chromatic dispersion. In C.S. Harris, (ed.), *Visual coding and adaptability*, (pp. 69-94). NJ : Lawrence Erlbaum Associates.

Schiff,, W. (1980). *Perception: An applied approach*. Boston: Houghton Mifflin Company.

Uttal, W.R. (1973). *The psychobiology of sensory coding*. New York: Harper and Row, Publishers.

Welch, R. B., Bridgeman, B., Anand, S. & Browman, K. (1991). *The acquisition of "dual adaptations" and "adaptation sets"*. A paper presented at the meeting of the Psychonomic Society in San Francisco, CA on November 24th.

Welch, R.B. (1978). *Perceptual modification: Adapting to altered sensory environments*. New York: Academic Press.

An Analysis of How Students Take the Initiative in Keyboard-to-Keyboard Tutorial Dialogues in a Fixed Domain

Gregory A. Sanders, Martha W. Evens, and Gregory D. Hume
Department of Computer Science
Illinois Institute of Technology
10 West 31st Street
Chicago, IL 60616
(312) 567-5153
sanders@iitmax.iit.edu
csevens@minna.iit.edu
ghume@valpo.bitnet

Allen A. Rovick and Joel A. Michael
Department of Physiology
Rush Medical College
Rush Presbyterian-St. Luke's Medical Center
Chicago, IL 60612

Abstract

By student initiatives we mean productions which the student could reasonably expect to modify the course of the tutorial dialogue. Asking a question is one kind of student initiative. This paper describes a system called CircSim-Tutor which we are building, the background of the project, the 28 hour-long tutoring sessions analyzed in this paper, and the analysis done. It compares our work to previous work, gives a classification of the student initiatives found and of the tutor's responses to them, and discusses some examples.

This work was supported by the Cognitive Science Program, Office of Naval Research under Grant No. N00014-89-J-1952, Grant Authority Identification Number NR4422554, to Illinois Institute of Technology. The content does not reflect the position or policy of the government, and no official endorsement should be inferred.

Background of Project

We are building an Intelligent Tutoring System, called CircSim-Tutor, to tutor first-year medical students on how the body maintains a stable blood pressure from minute to minute, compensating for any perturbation of the pressure. The physiological processes involved are an example of a negative-feedback control system. Such control systems also occur in electronic and mechanical systems that have no connection with physiology. CircSim-Tutor communicates with students in English: it both understands and produces English. As background for this effort, we collected and analyzed 28 sessions, each approximately an hour long, with expert human tutors, Physiology professors at Rush Medical College, doing keyboard-to-keyboard tutoring of first-year medical students attending Rush Medical College, who had heard the class lectures about the material. During these sessions, the student was in one room and the tutor in another room, communicating only by typing on the keyboard and reading from the screen of a computer terminal. Each student always knew that the interaction was with a human tutor and also knew the identity of the tutor. In addition, each student also had the tutor as his or her professor in the related physiology course, being taken at the same time as part of his or her first-year medical school coursework at Rush. Thus, the students may have felt the academic or social pressure this would imply. All students were volunteers, recruited from these classes by the tutors, and each was paid a nominal amount for his or her participation. A total of 20 students are represented in these 28 sessions (8 students appear twice).

The two Physiology Professors at Rush who are the tutors in these sessions have taught the related physiology courses to first-year medical students at Rush for many years, and have customarily tutored some students taking these classes face-to-face. Thus, our tutors are highly experienced and expert at teaching the material, both in classroom lecture and in personal tutoring sessions.

Transcripts including timing information were automatically collected. Each sessions was organized as a clinical problem, where a mechanical heart pacemaker suddenly failed, increasing the heart rate (beats/minute) substantially. The student was asked to predict the direction of change, if any, of seven basic cardiovascular parameters, first for the immediate physical effects of the increased heart rate, then for the reflex compensation by the autonomic nervous system to return the blood pressure toward the original value, and finally for the steady state result after this compensation is complete. In addition to making the correct predictions, the tutors want the students to be able to explain why and how each of the changes occurs, and to do so using the "correct" language. It is this concern with language that initially prompted the entire project of building CircSim-Tutor. CircSim-Tutor is a joint project of the Physiology department at Rush Medical College and the Illinois Institute of Technology Computer Science department.

In trying to make CircSim-Tutor handle the discourse phenomena in the sessions with human tutors, we set out to analyze the transcripts of the 28 keyboard-to-keyboard sessions, to identify and categorize each instance where the student took the initiative and to describe how the tutor responded to the initiative. Although the initial purpose of this discourse analysis was to enable the program to respond to such initiatives, we soon became interested in this analysis in its own right.

Related Work

Graesser, Lang, and Horgan (1988) proposed an analytic scheme for questions, covering a corpus of approximately 1000 questions asked by adults in different discourse contexts. They proposed 12 semantic categories for questions.

Verification: Is X true or false?
Disjunctive: Is X or Y the case?
Concept completion: Who? What? When? Where?
Feature specification: What is the value of a variable?
Quantification: How much? How many?
Causal antecedent: What caused some event to occur?
Causal consequence: What happened as a consequence of X occurring?
Goal orientation: Why did an agent do some action?
Enablement: What is needed for an agent to do some action?
Instrumental/procedural: How did the agent perform an action?
Expectational: Why isn't X occurring?
Judgemental: What should an agent do?

We seem to need an added category: questions about ontology or taxonomy.

Graesser et al. also proposed 6 pragmatic categories, intended to be orthogonal to the semantic categories. These categories are: information acquisition, assertions, establishing a context for subsequent discourse, indirect requests for non-verbal behavior, conversation monitoring, and humor. While these may cover the questions we found, it is not clear just where they cover repair questions (e.g., "What did you mean?"), investigated by Fox (1990). Nor is it clear to us just where they cover questions intended to establish the relevance of certain facts or cases to the current discourse focus, so as to enable the dialogue to go forward with necessary shared context.

The following is our classification of the student initiatives. Although primarily semantic or pragmatic (generally, discourse-structure based), some of the categories pick out surface clues that seem to flag a production as an initiative. There are 32 sub-categories in the following table, grouped into 12 major categories. This classification was created from study of the 28 sessions.

Student asks a question.
Straight question about physiology/physics -- about locally current discourse context
Straight question about physiology -- not about locally current context

Student makes a Physiology statement (perhaps incorrect)
Physiology statement -- not in an "answer context"
In response to being asked to make a **corrected** prediction, the student makes some (perhaps accurate) statement of physiology.
I'm not sure if <*stmt*>.
A "complex" statement, hedged by a '?' (not just, e.g., "Up?" or "CC?")
Maybe I should clarify <*previous stmt(s)*>

Student having trouble "seeing" <*X*>
I am having trouble seeing/conceptualizing/grasping ...
I am **still** unclear about <*something just discussed*>
I think I am getting <*X*> mixed up with <*Y*>.
Tutor: "Understand?" Student: "No"

Student requesting Repair (student does not understand)
Student doesn't understand what (or when) the tutor is talking about.
Student not familiar with the physiology lingo, at least in student's opinion.
(Note: This category is to be preferentially picked if it applies.)
The tutor makes a statement of physiology, and the student states he/she does not understand it.
What do I do now?
The student doesn't understand something in the instructions from the tutor.
The tutor got the student confused. (e.g., tutor's mistake)

Student doing Repair (tutor did not understand the student)
Student thinks tutor overlooked or has forgotten something the student typed.

Student asks non-sequitur question OR Student is completely lost
Student asks a non-sequitur question, possibly with backward reference, showing serious misunderstanding or lack of understanding of the material.
Student declares he or she is lost. OR The student doesn't understand a *straightforward* question.
(Note: interesting category for replanning)
The tutor says, "Let me remind you of <*something*>," and the student does not confidently remember.

Student is hedging
"...perhaps...." OR "...??" (OR both) *(category is literal surface strings)*
<*answer*> <*justification for answer*>
Other hedges.

Student not answering a question
Assume *possible* initiative any time we see a **long** pause with no keystrokes.
The student announces reluctance to answer.

Table 1: Classification of Student Initiatives

Student asks an explicitly case-based question
"In one of the cases <*stmt*>. Is that right?" (Note: this is also hedged)
"How is <*fact*> relevant?"

Student makes an explicit backward reference
At <*previous point in session*> we were talking about <*whatever*>.

Request for Confirmation
For example, "So I am correct in my thinking?"

Other initiatives
Administrative questions
Questions specific to the structure of the experiment (e.g., rules of the "game")

Table 1: Classification of Student Initiatives *(continued)*

Our Classification of Student Initiatives and Tutor's Responses

There are two expert tutors represented in the 28 sessions analyzed here. In this paper we discuss the student initiatives from all 28 sessions. In order to present a clearer picture of the tutor's responses, with one less degree of freedom, however, this paper only discusses the tutor's responses from the tutor who did the most sessions (16 of the 28).

Our classification of the student initiatives is in the preceding table. The following is our categorization of the tutor's responses.

Explain or state some material in focus.
Defer handling the initiative: perhaps modifying the tutor's model of the student.
Do repair, stating some material, where the student did not understand the tutor.
Request repair: the tutor doesn't understand what the student means.
Ask student if stuck, or still stuck.
Acknowledge the student's understanding is correct, or state it is not correct.
Replan part or all of the remaining session.
 * perhaps cover material in pieces
 * perhaps make a big backward reference
Give a hint, or perhaps remind student of material already covered in the session.
Ask the student a question. (Socratic tutoring)
State, "you are confusing **X** with **Y**." (Declare a diagnosis)
Invite the student to review his/her thinking with the tutor.

Discussion of Interesting Examples

One of the first things we noticed in the transcripts of the sessions is that the students may use punctuation, if at all, in a personal way, often with minimal relationship to the generally accepted conventions of English punctuation. Thus, punctuation may provide little help in recognizing the mood or clausal structure of sentences. We do not show examples of this. Repeated punctuation (e.g., "???" or "!!") always appeared significant. The students generally capitalize conventionally. Generally, surface clues are what seem to trigger recognition of a student initiative and of its meaning. The Hedging category in the table above has some particularly clear examples of this. It appears the students consistently flag all initiatives in some fashion, so the tutor does not have to notice a departure from the current discourse focus or make similar inferences to recognize initiatives. All examples are given with the original spelling errors, punctuation, capitalization, typographical errors, and so forth.

The following example came at the end of discussing the direct physical effects, before the reflex kicks in. The abbreviations used by the tutor and student in this example are: cc = cardiac contractility, tpr = total peripheral resistance, co = cardiac output, ans = autonomic nervous system, ca = calcium [ions], and i = increase. Note that the student flags the material he wants the tutor to respond to by saying, "I'm not sure if...." Students in our sessions did this sort of thing consistently.

tu - One last question here...
tu - Why did you predict that cc and tpr would be unchanged?

st - Tpr is largely a function of arteriol constriction which takes a while to adjust to co i .
st - Cc changes in response to ans stimulation or ca build up during tachecardia.
st - Im not sure if 120bpm is fast enough to cause that.
tu - Probably not.

The following is another example, starting in the middle of a tutor's production. The only abbreviation is RAP = right atrial pressure.

tu - [. . .] what about the rate at which blood is being removed vfrom the central blood compartmanent?
st - That rate would increase, perhaps increaseing RAP???

In our sessions, the tutors appear to have a well defined picture of what they want the student to demonstrate and what the student should be tutored on if the student does not already know. Interestingly, the mere mention by the student of certain terms not introduced into the session by the tutor is enough to trigger tutoring on the parallels between those parameters and the ones the tutor is using in this session. The parallel in the following example is one of similar values: CVP and RAP are really separate measurements. The abbreviations here are co = cardiac output, RAP = right atrial pressure, and D = decrease.

st - So, when CO I, the central venous pressure will D?
tu - Absolutely correct.
tu - What variable is essentially the same as central venous pressure?
st - RAP.
tu - Right.

Some initiatives are quite brief, their interpretation clear, and the response is fairly obvious.

tu - OK?
st - No

Others are complex. In the following example, SV = stroke volume. The student in this example had previously produced a 209 word response to a question, which the tutor eventually interrupted to tell the student, "you need to be more concise in your answers."

tu - Understand?
st - Not fully.
st - Isn't the amount of filling equivalent to the preload?
st - And doesn't and increased preload invoke Starling's effect?
st - And, most importantly, what is the difference between a length/tension effect (as occurs in Starling's) and the "change in ventricular performance (SV, force,...)" which you say is not related to Starling.

As has been pointed out by research on discourse or dialogue structure, there is always some current focus, often a nested stack of subjects in focus. The preceding example establishes material in local focus, and the tutor responded by tutoring the pieces separately, then returning to the previous course of the session. This question of whether something is in or out of the current focus, seems important in recognizing the intent of student initiatives and in deciding how to handle them. The tutor whose responses are shown in this paper responded to straight questions that were off the current topic in the briefest possible fashion and then simply returned to the previous topic with no surface flagging that the topic was changing back, as if the focus had never changed. For example, an initiative as long and complex as the preceding example got the response, "Yes." On the other hand, questions about the material currently in focus generally got more elaborate treatment. For example, the following initiative took four st/tu pairs of productions to be discussed. It became a significant topic in its own right, even though this question is not part of the "standard" material to be covered in these sessions.

st - Does RAP increase initially with increasing CO and then taper off as CO continues to I?

Agreement Between Raters

All 28 sessions have been independently analyzed by two raters. The first analysis, which created the categorization, picked out 110 initiatives. The second analysis picked out about 210 initiatives, including 108 of the initiatives picked out in the first analysis. We have not yet had the opportunity to do a proper analysis of agreement about the categorization of these initiatives. Of the 110 initiatives picked out in the first analysis, the number per hour-long session ranged from 0 to 11, with a standard deviation of 3.1 and mean of 3.9 per session. The most frequent categories of initiatives in the first analysis were: straight questions about material currently in focus (20 of the 110), and the category, "I am having trouble seeing/conceptualizing/grasping this" (8 of the 110). Four other common categories (each was 7 of the 110) were: straight questions about material not currently in focus, "I'm not sure if *<stmt>*, the student does not understand what/when the tutor is talking about, and the student is not familiar with the physiology lingo.

In one case, it appeared the student had too little grasp of the material to be able to put together a coherent initiative. That student is one of the eight who appear twice, and in the second session the same student, who had learned the material by then, generated six initiatives. This suggests an interesting line for possible future research. It seems to us that the number and depth of initiatives rises as the student's grasp of the material rises, until at some point the student knows the material thoroughly and begins to simply answer questions, with few or no initiatives.

Future Work

We intend to focus on the context in which the initiatives occur. It is clear that how tutors respond depends on the context of the initiative. For us in CircSim-Tutor we thus need to study how to respond. We anticipate that study of the agreement on categorization between different raters could change the description of the categories. After studying agreement on categorization, we expect to have our expert human tutors categorize the initiatives we have identified. Of course, ultimately we want to incorporate what we learn about how tutors understand the students' initiatives and how the tutors respond to them into an enhanced version of the CircSim-Tutor program.

Summary

This work attempts to categorize student initiatives encountered in tutoring a fairly small body of material in depth. It discusses the relationship between our findings and the previous work by Graesser et al. In future work we intend to focus on the context in which student initiatives occur. In applying our result to the design of an approach for CircSim-Tutor to use in responding to student initiatives, we need to understand the way in which the human tutor decides how to respond. The availability of the tutors represented in these sessions for extended discussion should help in trying to understand this.

Selected References

Brennan, S.E., 1991. Conversation with and through Computers. *User Modeling and User-Adapted Interaction* 1:67-86.

Fox, B., 1990. Human Tutorial Dialogue: Final Report. ONR Cognitive Science Program, Contractor/Grantee Meeting.

Galdes, K.K., Smith, P.J., and Smith, J.W., Jr., 1991. Factors Determining When to Interrupt and What to Say: An Empirical Study of the Case-Method Tutoring Approach. In Proceedings of The International Conference on the Learning Sciences, 194-202, Evanston, IL.

Graesser, A.C., Lang, K., and Horgan, D., 1988. A Taxonomy for Question Generation. *Questioning Exchange* 2:3-15.

Graesser, A.C., and Hemphill, D., 1991. Question Answering in the Context of Scientific Mechanisms. *Journal of Memory and Language* 30:186-209.

Grosz, B.J., Pollack, M.E., and Sidner, C.L., 1989. Discourse. In M. Posner, (Ed.), *Foundations of Cognitive Science*, 437-468. Cambridge, MA: MIT Press.

Rovick, A.A., and Michael, J.M., 1992. The Prediction Table: A Tool for Assessing Students' Knowledge. *Adv. Physiol. Ed.* Forthcoming.

Visual Attention and Manipulator Control

Peter A. Sandon
Computer Science Program
Dartmouth College
Hanover, NH 03755
sandon@cs.dartmouth.edu

Abstract

One function of visual attention is as a filter that selects one region of the visual field for enhanced detection and recognition processing. A second function of attention is to provide localization information, which can be used in guiding motor activity. A visual system in which the eyes can be moved requires such localization information to guide eye movements. Furthermore, the control of arm and hand movements for object manipulation is simplified by attentional localization of the hand with respect to a fixation frame centered on the object. This paper describes this role of attention in the visual guidance of simple motor behaviors associated with unskilled object manipulation behaviors.

Introduction

It is often observed that the amount of data contained in an image is too large to be processed completely in the small fraction of a second allowed by many tasks. The obvious solution to this problem is to process only a part of the visual environment according to current task requirements. The animate vision paradigm implements this solution through the use of active control of sensors and task-dependent visual processing (Ballard:ijcai). Animate vision has been proposed as both an approach to designing computer vision systems, and as a model of human visual behavior. While the computational load is reduced when the entire image does not have to be processed, the question of what region of the image to process becomes paramount. Selective visual attention provides the mechanism for answering this question.

The term (selective visual) attention will be used to refer to a specific collection of visual subprocesses which perform the covert selection of retinal regions for further processing. This further processing may involve recognition processing of the selected region, or may involve the use of the corresponding location information for guidance of movements. It is this second aspect that will be the emphasis here, though the recognition processor is involved in this localization function, as will be discussed.

In the remainder of the paper, the mechanisms comprising an attentional visual recognition system are first discussed at a coarse level of detail. This provides a sufficient basis for describing the use of attentional localization in guiding eye and arm movements for object manipulation tasks. In particular, a touching task and a manual tracking task are used to elaborate the concepts, both of which have been implemented in a real-time robotics system to demonstrate the approach.

Attentional visual recognition

A number of computational models of attentional mechanisms have been proposed, including those of (Treisman 1988; Mozer 1988; Cave & Wolfe 1990; Sandon 1990; and Ahmad 1991). While these models differ in a number of details, and in the emphasis they place on various aspects of attentional function, they also share a number of common features which provide a sufficient basis for the current discussion. Thus, the following coarse level description of attentional visual recognition is presented for the benefit of the succeeding discussion.

The visual system consists of three components, a feature processor, an attention processor and a recognition processor (see Figure 1). The feature processor extracts a number of spatially localized features from the image. These features are extracted in parallel over the entire image, and represented retinotopically in feature maps. Though attempts have been made to identify the particular features that are extracted in

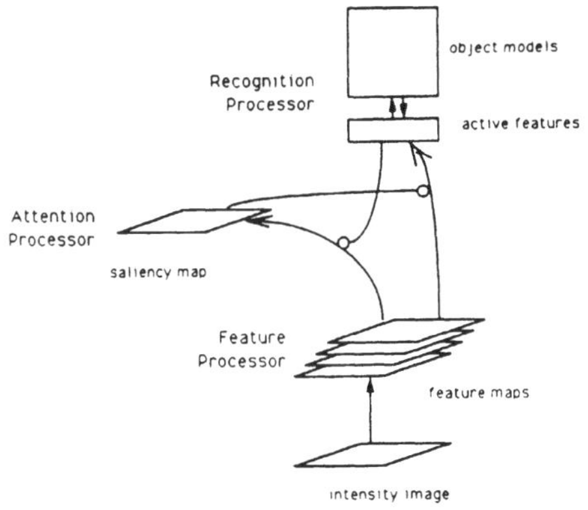

Figure 1 - Schematic drawing of the attentional vision system

human vision, this aspect is not key to the current discussion. Within each feature map, a lateral inhibition network operates on the raw feature activity, to produce contrast enhanced features. The resulting activity in each map is gated to the attention processor to a degree determined by expectancies provided by the recognition processor. Regions of activity in feature maps are also gated to the recognition processor, in this case by localized activity in the attention processor.

The recognition processor has access to a database of object models, which is indexed by feature values. Recognition is performed by having the feature processor pass image feature values to the recognition processor, which are then used to index into the object database. An object is recognized if the feature values are sufficiently close to those defining an object to satisfy some match criteria. Conversely, the recognition processor can use the defining features of an object to modify the gating of the features to the attention processor as mentioned above. We describe this use of the model data for object localization in more detail below.

The attention processor determines the region of the image whose corresponding feature values will be passed up to the recognition processor. The feature map values are combined to provide the input to a saliency map, which represents, in registration with the image, the importance of each image region to the current task. To choose a single region for processing, a selection operator is applied to the activity in the saliency map. This selection operator chooses some region of the image, whose features are then gated to the recognition processor, and whose location can be passed to motor processors.

Given these three component mechanisms, what functions might they implement? In the absence of any task-specific control of the feature map input to the attention processor, the saliency map will be sensitive to all the contrast enhanced features. The resulting saliency activity can be used to implement alerting and orienting behaviors, as well as precategorical image segmentation.

When the recognition processor activates its control of the feature map inputs to the attention processor, according to the features that characterize an object of interest to the current task, the saliency map becomes sensitive only to those specified feature maps. The selection of an active region of the saliency map in this case allows localization of the desired object in retinotopic coordinates. This location information can be used to represent spatial relations among objects, and in particular, can be used to guide motor activity, as we now discuss.

Fixation-based motor control

There has been a great deal of discussion in the literature about the appropriate frame of reference for each different aspect of visuomotor processing. While Marr, for example, emphasized the need for object-centered coordinates in representing visual information for recognition (Marr 1982), others have noted that an egocentric coordinate system would be useful when interacting with objects (Feldman 1985). Ballard argues against the egocentric representation, due to the presumed difficulty of maintaining its currency. Instead, he proposes the use of a coordinate system centered on a particular 'calibration' object (Ballard 1987).

The domain of interest here is visually guided manipulation of objects. Although it is true that the eyes, head and body may all be moving during the execution of such manipulation tasks, even a retinocentric reference frame can be effective for object localization if the spatial relations necessary to the task can be updated in a timely manner. In particular, for a binocular system, the pair of x,y coordinates representing the horizontal and vertical offsets of an object from the center of the image in each eye can be used to compute a location in a three dimensional retinocentric space.

As has been observed elsewhere (Ballard 1989), a reference frame that has particularly desirable properties is the fixation frame, which is centered on the point in space where the two optical axes of a binocular vision system intersect, and is oriented to correspond to the retinal axes and the direction of gaze. The binocular retinocentric frame is the proximal correlate of this distal fixation frame. One version of the projection of the four dimensional binocular retinal coordinates to a three dimensional space is achieved using the horizontal (h) and vertical (v) coordinates of one eye (the dominant eye), and the disparity (d) between the horizontal coordinates in the two eyes. This defines the 3-D retinocentric frame, R, in which locations are expressed as triples of the form (h,v,d). An object at the origin of this coordinate frame is at the fixation point in physical space.

The advantages in representing object location in 3-D retinocentric coordinates are that object locations can be computed quickly and maintained easily, and that the coordinate transformations required for eye movements and arm movements can be easily expressed in terms of this reference frame. The process of localizing objects in each retinal frame is mediated by the attentional mechanisms previously described. For example, to locate a particular object in one image, the recognition processor projects the feature values associated with the object to the feature processor, which differentially gates the corresponding feature maps to the attention processor. The resulting activity in the saliency map reflects the degree of match between the features defining the object and those in any particular region of the image. Selecting the most salient region corresponds to identifying the most likely location of the object in the image.

Given the two retinal locations of an object, equivalently the 3-D retinocentric coordinates, the guidance of eye and hand movements toward the object is relatively straightforward. For eye movements, the motor frame is defined by the gaze angles of the two cameras. Analogous to the 3-D retinocentric frame, the appropriate gaze angle frame, G, for a pair of horizontally offset, fixating eyes, is a 3-D reference frame consisting of the yaw angle (θ) and pitch angle (ϕ) of the dominant eye, and the yaw angle disparity (ψ) between the two eyes.

Eye movements are defined relative to the current gaze, and result in a relative displacement of the retinocentric locations of imaged objects. The kinematic transformation from relative gaze angle, Δg, to relative retinocentric location, Δr, can be approximated by a constant-valued, diagonal Jacobian matrix, J_{GR}:

$$\Delta r = J_{GR} \times \Delta g$$

$$J_{GR} = \begin{bmatrix} C_h & 0 & 0 \\ 0 & C_v & 0 \\ 0 & 0 & C_h \end{bmatrix}$$

where C_h and C_v express the number of pixels per visual angle of the imaging surface in the horizontal and vertical directions, respectively.

This approximation holds when the center of rotation of gaze coincides with the optical center of the lens, and the sensory surface is spherical about this same point. To the extent that these two assumptions are violated, the constant function kinematics will be less accurate, though for small gaze angles and limited depth of field, the accuracy will remain high.

The process of establishing a new fixation point is as follows. If the desired action is to fixate a particular object, the object is first localized in each image as described previously. The locations in the two images are used to compute a location, r, in the retinocentric frame, R. Since the desired location is at the origin in R, the vector -r represents the relative movement in R. This vector is passed to the eye movement control system, which computes the transformation from R to G as:

$$\Delta g = J_{GR}^{-1} \times -r$$

The computed gaze angles are used to direct a saccadic movement of the eyes to the new fixation point.

In the absence of having a particular object specified as the target of fixation, the process remains the same, except that the feature maps are gated to the attention processor according to some default weighting of the individual maps, corresponding to the relative importance of each feature for alerting purposes.

This scheme can be extended to smooth pursuit eye movements by performing an additional filtering step on a sequence of gaze angle values that are obtained by successive executions of the above procedure. To maintain accurate pursuit, a predictive filter such as a proportional-integral-derivative (PID) filter can be used to adjust gaze velocities (Dorf 1986).

Touching and manual tracking

For arm movements, defined with respect to the arm joint coordinate frame, A, analogous computations can be used. Conventionally, control of arm movements is presumed to require a complete model of the arm kinematics in environmental coordinates (Brown & Rimey 1988). Visually guided movements then require that the kinematics of the visual system in environmental coordinates be determined. An alternative approach, that is applicable to the kinds of simple movements considered here, is to express the arm kinematics in the 3-D retinocentric frame. In particular, a representation of the kinematics that is both easy to acquire and to compute with is a local one, where the small change in retinocentric coordinates due to a small change in arm joint positions is used to represent a constant-valued kinematics in that particular region of joint-gaze space (Mel 1989). That is, for a particular joint-gaze configuration, the change in retinocentric coordinates, Δr, for a given change in arm joint positions, Δa, is given by:

$$\Delta r = \hat{J}_{AR} \times \Delta a$$

where $\hat{J}_{AR}$ is the Jacobian evaluated at the particular joint-gaze configuration.

One way to represent the complete kinematics is as a collection of evaluated Jacobian matrices indexed by joint-gaze coordinates in a lookup table. These matrices can be acquired through a calibration procedure prior to use, or through an adaptive process during movement execution. This has advantages for acquisition and for representation of arbitrary relations. Alternately, a representation of the Jacobian terms as low-order functions of joint-gaze space is more efficient and provides better generalization during acquisition when the relations being represented are smooth. The direct kinematic equation above is used for acquisition of the kinematic parameters, while the inverse Jacobian is used for control.

Touching

Perhaps the simplest object manipulation behavior is touching, that is, using arm movements to bring the hand into proximity with some object of interest. Given the previously described attentional mechanism for locating objects in R, and kinematic models for transforming between R and G, and between R and A, the touching task can be accomplished as follows:

TOUCH (object):

r = *Attend*(object)	;locate object in R
$\Delta g = J_{GR}^{-1} \times -r$	;saccade to the object
r = *Attend*(hand)	;locate hand in R
$\Delta a = \hat{J}_{AR}^{-1} \times -r$	;move hand to object

Due to the use of local kinematics, a given move will be inaccurate to the degree that the new joint state is far from the initial one. This approach is appropriate, therefore, when a lack of real-time constraints allows for the use of one or more small compensatory movements to be used to achieve the desired accuracy.

Notice the minimal need for representation of spatial relations in this process. Attention is first used to locate the object of interest. This location information is represented in the state of the selection process, which is transmitted to the eye movement control system. Once the eyes have been moved, the location of the object is implicit in the gaze angles of the eyes, and the attention processor need not maintain that location (which is now out of date in any case). Attention is now used to locate the hand, and the selection process represents the location for the sake of the arm movement control system. There is no need to maintain location information across movements for this simple task, because it can be easily reacquired by repeating the sequence.

Tracking

A relatively simple extension of the touching behavior allows a moving object to be manually tracked. We will use the term pursuit to refer to eye movements that maintain fixation on a moving object, and manual tracking, or simply tracking, to refer to arm movements that maintain proximity of the hand to a moving object. Although the tracking behavior by itself is not one that is commonly executed, it is a necessary component of tasks that require moving objects to be grasped, and a precursor to tasks that require interception of moving objects, such as catching and hitting. More importantly for the present purposes, the tracking behavior demonstrates the use of the attentional mechanism as a shared resource for the concurrent control of the eye and arm motor systems.

The tracking task could be accomplished by simply executing the touching behavior in an iterated loop. However, this yields a sequence of discrete movements for the eyes and the arm, rather than the smooth movements that might be desired. The required modification is straightforward. The attentional processor toggles back and forth to locate first the object, then the hand, as in the touch procedure. The locations that are supplied to the motor control processes are then transformed by a predictive filter. The output of the filter is used to control the gaze and arm joint velocities, such that the object being tracked is maintained at the fixation point, and the hand is maintained close to the object:

TRACK (object):
repeat

r = *Attend*(object)	;locate object in R
$\Delta g = J_{GR}^{-1} \times -r$	;desired gaze change
$\Delta \dot{g} = PID(\Delta g)$	;smooth gaze adjust
r = *Attend*(hand)	;locate hand in R
$\Delta a = \hat{J}_{AR}^{-1} \times -r$	;desired arm change
$\Delta \dot{a} = PID(\Delta a)$	;smooth arm adjust

An implementation of the saccade, pursuit, touching and tracking behaviors just described has been developed for a binocular camera and robotic arm system. The vision system consists of a pair of cameras mounted on a motorized pan-tilt platform, and a Datacube Maxvideo image processing system. The arm is a PUMA 761 six degree-of-freedom arm. A SUN4 workstation runs the control program and mediates communication between the image processing, eye motor control and arm motor control systems.

The features used for defining objects are based on image intensity, edge orientation and edge ratio magnitude. The object of interest is attached to a slowly revolving platform placed within the workspace of the arm. The pursuit behavior has a .4s cycle time, and generates a smooth gaze trajectory that lags the object by up to a degree in each dimension. The tracking behavior has a 1.25s cycle time, and generates discrete arm movements, due to a lack of velocity control in the current arm controller interface. These movements also lag the object movement, and exhibit an appreciable rms error from the expected trajectory, that is four times greater (48mm vs 12mm) in the direction parallel to the line of sight than in the directions perpendicular to the line of sight.

Further details are presented in (Sandon 1992).

Concluding remarks

Although a great deal of consideration has been given to the mechanisms of attention, much less work has addressed the function of attention in everyday visuomotor behavior. This paper describes, and the briefly presented implementation results demonstrate, a computationally simple approach to visual guidance of eyes and arms based on attentional localization and local kinematics. The minimal representation used in the approach has advantages in computational efficiency, both for acquiring and for maintaining a current model of the external world. In addition, minimal representations exhibit advantages in adaptive systems, since the credit assignment problem is reduced (Whitehead & Ballard 1990).

As stated, this approach to object manipulation applies to servo-controlled movements, in which visual feedback is used to repeatedly adjust an eye or arm movement. This is an appropriate model for unskilled behavior, and corresponds to a situation in which the kinematic and dynamic models of the motor systems are not well characterized. While more complete and accurate models are required for modelling skilled movements and for tasks having significant real-time constraints, it seems reasonable to assume that such models are preceded by the approximate ones discussed here. More accurate models are then acquired using the errors that occur while performing these simpler behaviors.

While it may seem intuitive that covert attention should be used to guide overt eye movements, the precise relation between the two systems is not yet clear. On the one hand, Remington found that the enhanced processing associated with attention preceded saccadic eye movements that were initiated by a stimulus onset in the retinal target position (Remington 1980). This provides evidence that attention is being used to guide the eye movement. In addition, there is evidence that one component of saccadic latency is the time needed for attention to disengage prior to localizing a target to be fixated (Fischer & Breitmeyer 1987). However, Remington also found that for eye movements initiated by a central cue indicating the desired direction of movement, attention followed the eye movement to the target position, indicating that saccadic guidance was provided by some other source. As for the guidance of arm movements, there is evidence that eye movements play a part (Ballard, et. al. 1991), but the role of attention is not known.

How does this approach extend to more complex tasks? The introduction of real-time constraints has already been mentioned. These require accurate ballistic movements, which in turn require more accurate kinematic and dynamic models. As previously discussed, these models can be developed during the execution of the simpler behaviors described here. When the task involves the manipulation of additional objects, attention must be shared among the objects to maintain localization information. Furthermore, an explicit short term representation of objects will likely be necessary, in order to maintain continuity of object characteristics, and to predict future object location for guiding the selection process.

Finally, for more complex interactions with objects, in particular, for grasping them, hand movements must be controlled in addition to eye and arm movements. Grasping behaviors require not only localization of an object, but an estimate of object pose. In many cases, scale and major axis orientation information are sufficient for the determination of an appropriate hand configuration for grasping. For more complex objects, detailed pose must be determined. While desirable features for localizing an object are those that do not depend on viewpoint, the features needed to determine pose are those that are viewpoint dependent. In addition, the likely role for attention in detailed pose estimation is in localizing the components of objects to represent the spatial interrelations among parts.

Acknowledgements. This work was done while the author was visiting the University of Rochester. Thanks to Dana Ballard, the Vision Lab research group, and the CVS Visual Attention reading group for many useful discussions. This research was supported by NSF Grant No. IRI-90108999.

References

Ahmad, S. 1991. VISIT: An efficient computational model of human visual attention, Tech. Report, TR-91-049, International Computer Science Institute.

Ballard, D. H. 1989. Reference frames for animate vision. In Proc. of the Eleventh International Joint Conference on Artificial Intelligence, 1635-1641, IJCAI, Inc.

Ballard, D. H. 1987. Eye movements and spatial cognition, Tech. Report, 218, Dept. of Computer Science, Univ. of Rochester.

Ballard, D. H.; Hayhoe, M.; and Li, F. 1991. Hand-eye coordination during sequential tasks. Forthcoming.

Brown, C. M.; and Rimey, R. D. 1988. Coordinates, conversions, and kinematics for the Rochester Robotics Lab, Tech. Report, 259, Dept. of Computer Science, Univ. of Rochester.

Cave K. R.; and Wolfe J. M., 1990. Modeling the role of parallel processing in visual search. *Cognitive Psychology* 22:225-271.

Dorf, R. C. 1986. *Modern Control Systems*, Reading, Mass.: Addison-Wesley.

Feldman, J. A. 1985. Four frames suffice: A provisional model of vision and space. *Behavioral and Brain Sciences* 8:265-289.

Fischer, B.; and Breitmeyer, B., 1987. Mechanisms of visual attention revealed by saccadic eye movements. *Neuropsychologia* 25:73-83.

Marr, D. 1982. *Vision*. San Francisco: Freeman.

Mel, B. W. 1989. MURPHY: A neurally-inspired connectionist approach to learning and performance in vision-based robot motion planning, Tech. Report, CCSR-89-17A, Univ. of Illinois.

Mozer, M. C. 1988. A connectionist model of selective attention in visual perception. In Proc. of the Tenth Conference Cognitive Science Society, 195-201, Hillsdale, NJ.: Lawrence Erlbaum.

Remington, R. W. 1980. Attention and saccadic eye movements *J. Exp. Psych.: HPP* 6:726-744.

Sandon, P. A. 1992. Visually guided touching and manual tracking, Tech. Report, 412, Dept. of Computer Science, Univ. of Rochester.

Sandon, P. A. 1990. Simulating visual attention. *J. Cognitive Neuroscience* 2:213-231.

Treisman, A. 1988. Features and objects: The fourteenth Bartlett memorial lecture. *Quarterly J. Experimental Psychology* 40A:201-237.

Whitehead, S. D.; and Ballard, D. H., 1990. Learning to perceive and act, Tech. Report, 331, Dept. of Computer Science, Univ. of Rochester.

Additive Modular Learning in Preemptrons

Gregory M. Saunders, John F. Kolen, Peter J. Angeline, & Jordan B. Pollack
Laboratory for Artificial Intelligence Research
Computer and Information Science Department
The Ohio State University
Columbus, Ohio 43210
saunders@cis.ohio-state.edu
pollack@cis.ohio-state.edu

Abstract

Cognitive scientists, AI researchers in particular, have long-recognized the enormous benefits of modularity (e.g., Simon, 1969), as well as the need for self-organization (Samuel, 1967) in creating artifacts whose complexity approaches that of human intelligence. And yet these two goals seem almost incompatible, since truly modular systems are usually designed, and systems that truly learn are inherently nonmodular and produce only simple behaviors. Our paper seeks to remedy this shortcoming by developing a new architecture of Additive Adaptive Modules which we instantiate as Addam, a modular agent whose behavioral repertoire evolves as the complexity of the environment is increased.[1]

Introduction

One of the major conundrums of machine learning research, of both the symbolic and neural varieties, is how to produce systems which demonstrate complex cognitive behaviors starting from simple kernels. Simple learning systems, such as feed-forward networks end up with simple behaviors, so are really only theoretical signposts; complex learning systems which start with a large initial software investment, such as explanation-based learning (DeJong and Mooney, 1986; Mitchell, et al., 1986), beg the question of origin. Placing a simple system in a complex environment can work if the environment is non-threatening (Elman, 1988), but often the cost of engineering the environment is greater than that of engineering a working system.

This trade-off between complexity of *specification* and complexity of *environment* has been playing itself out in recent tensions in connectionism between simple systems which do not scale well versus complex (modular) systems whose origins are "not phylogenetically plausible". The current swing to automatic modularization is a response to this tension, but suffers from a lack of distributed control. For example, both Jacobs, Jordan, and Barto (1990) and Nowlan & Hinton (1991) rely on a centralized gating network to select the proper expert module. In the former, the network is given a "task bit" as part of its input, so that the proper expert module is effectively preselected by the input vector. Similarly, the latter permits different inputs to the expert and gating networks, simplifying the modularization process. Furthermore, neither architecture exploits the fact that the outputs of the gating network are continuous; instead, the interactions between modules are encouraged to be binary so that module *i* has no appreciable influence on the output when module *j* is active. In fact, Nowlan & Hinton's work, following Jacobs, et al. (1991) on which it is based, explicitly trains away these interactions.

An alternative approach to modularity is found in the design of autonomous robots, a historically nontrivial control task. Brooks (1986, 1991) offers a task-based subsumptive architecture which has achieved some impressive results. However, since machine learning is not up to the task of evolving these systems, engineers of artificial animals have embedded themselves in the design loop as the learning algorithm, and thus all components of the system, as well as their interactions, must be carefully crafted by the engineer (see, e.g., Connell, 1990).

Research aimed at replacing the engineer in these systems is at an early stage. For example, Maes (1991) proposes an Agent Network Architecture which allows a modular agent to learn to satisfy goals such as "relieve thirst"; however, she presumes detailed high-level modules (such as "pick-up-cup" and "bring-mouth-to-cup"), and her system learns only the connections between these modules. An earlier work that does not presume such an a priori modularization (Maes & Brooks, 1990) allows a six-legged robot to learn to walk, but there are no real modules in the final system. Beer and Gallagher (1991) attack this same problem of robotic mobility, but in a different way. They use a genetic algorithm (GA) that produces a robot which walks well (in simulation), yet they engineered the precise modularity of their system. Lin (1991) similarly presumed a detailed modularization and proceeded to learn each piece.

Thus, the researchers in artificial animals fall into

1. This research has been partially supported by ONR grants N00014-89-J-1200 and N00014-92-J-1195.

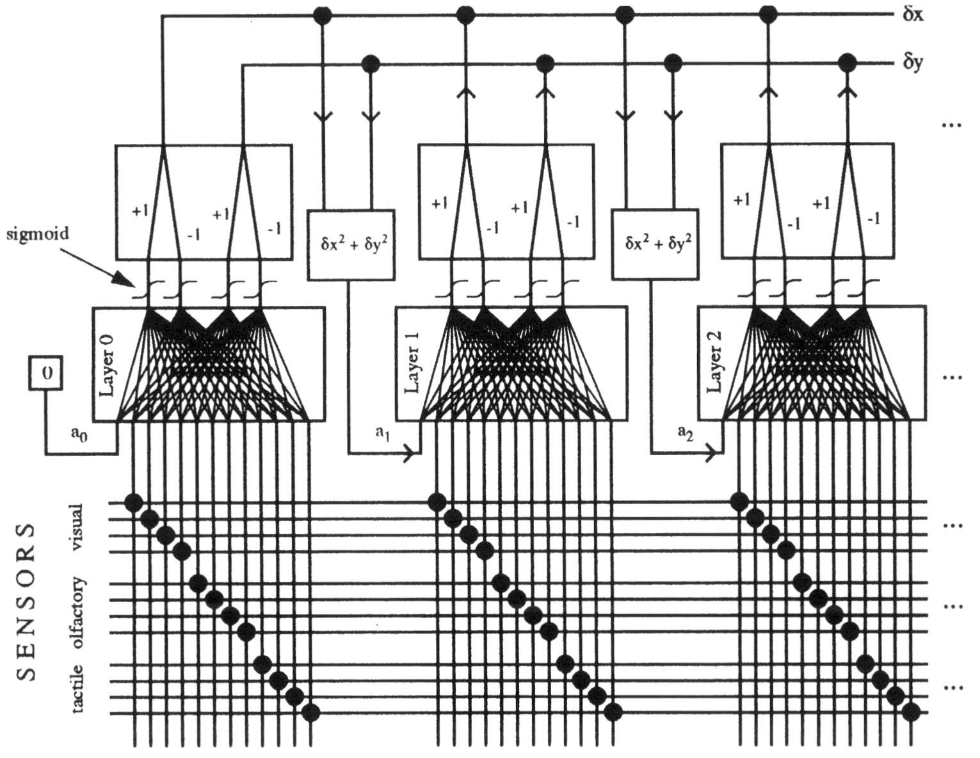

Figure 1: Addam's internal architecture.

the same pitfalls as others who ignore the conundrum of machine learning: either their systems are too complicated to learn, their learning algorithms to simple to scale, or their modularization is arbitrarily indexed to the task.

In this paper, we present a novel approach to modularization, inspired by the work of Brooks, but tempered by the requirements of modular learning. As will be discussed below, our connectionist version of subsumption replaces Brooks' finite state automata (FSAs) with feedforward networks and additional circuitry, combined so that each module in our hierarchy respects the historical prerogatives of those below it, and only asserts its own control when confident.

Additive Adaptive Modules

Our control architecture consists of a set of Additive Adaptive Modules, instantiated as *Addam*, an agent which lives in a world of ice, food, and blocks. To survive in this world, Addam possesses 3 sets of 4 (noisy) sensors distributed in the 4 canonical quadrants of the plane. The first set of sensors is tactile, the second olfactory, and the third visual (implemented as sonar that passes through transparent objects). Unlike other attempts at learning that focus on a single behavior such as walking (Maes & Brooks or Beer & Gallagher, discussed above), we chose to focus on the subsumptive interaction of several behaviors, and hence Addam's actuators are a level of abstraction above leg controllers (similar to Brooks, 1986). Thus Addam is moved by simply specifying δx and δy.

Internally, Addam consists of a set of dynamical systems (instantiated as feedforward connectionist networks) connected as shown above in Figure 1. This architecture is actually quite simple. The 12 input lines are from Addam's sensors; the 2 output lines are fed into actuators which perform the desired movement (δx, δy). Note that we desire $\delta x, \delta y \in (-1, 1)$ so that Addam may move in the positive or negative direction. To keep the outputs in this range, we first tried using the hyperbolic tangent activation function (output range -1 to 1), but this was inadequate because it did not permit 0 as a stable output. We then switched to sigmoids (output range 0 to 1), necessitating the boxes with the fixed -1,+1 connections below. Thus the four outputs of each "Layer i" box represent $+\delta x$, $-\delta x$, $+\delta y$, and $-\delta y$, respectively. This system allows both positive and negative movement, as well as 0 as a stable output for any "Layer i".

Addam's movements are controlled by this system as follows. First, the 12 sensors are sampled and fed into layer 0, which puts its values for δx and δy on the output lines. Layer 1 takes as input these same 12 sensor readings and the sum squared output of layer 0, calculates its values for δx and δy, and adds these to the output lines. Layer 2 works similarly, and the final δx and δy values are translated automatically to motor controls which move Addam the desired amount and direction.

Subsumption in our architecture captures the spirit of Brooks (1986, 1991), where modularity is achieved by a task-based decomposition of complex behavior into a set of simpler behaviors. In his system, layer 0 is obstacle avoidance and layer 1 is wandering. When layer 1 is active, it suppresses the activity of layer 0, and yet obstacles are still avoided because *layer 1 subsumes the obstacle avoidance behavior of layer 0*. Brooks avoids duplicating layer 0 as a subpart of layer 1 by allowing the higher layer random access to the outputs of any of the lower level FSAs. This fact combined with the multiple realizability of layers creates questions regarding Brooks' design methodology of developing a single layer of competence, freezing it, and then building a second layer on top of the first. If layer 0 can be realized equally well by method M_1 or M_2, then under Brooks' methodology we will not know until layer 0 is fixed which methodology's internal modules better facilitate the design of layer 1. Note that Addam does not have this problem with multiple realizability since layer 1 only has access to the *outputs* of layer 0.

In addition to the random access problem, Brooks also permits layer 1 to have unlimited suppression of layer 0's outputs. This works well when a human is engineering the robot, but such unbridled design-space freedom must be limited if we wish to have any chance of evolving the system. Thus Addam's different behavioral layers communicate only in the limited ways shown above.

Instead of being called subsumptive, our architecture is more aptly labeled *preemptive*. The modules are prioritized such that the behaviors associated with the lower levels take precedence over those associated with the higher levels. This is reflected architecturally as well as functionally, so that higher-level modules are trained to relinquish control if a lower-level module is active. For example, suppose that layer 0 behavior is to avoid predators, and layer 1 behavior is to seek out food. In the absence of any threatening agents, layer 0 would remain inactive and layer 1 would move Addam towards food. However if a predator suddenly appeared, layer 0 would usurp control from layer 1 and Addam would flee.

Note that we could have avoided feeding the sum-squared activation line into each module M_i by gating the output of M_i with the sum-squared line. We did not do this because our architecture is more general in that gating can be learned as one of many behaviors by each M_{i}. Our goal was to have each module decide *for itself* whether it should become active – had we used gating, this decision would have been made by M_i's predecessors.

A few more things should be noted about Addam's architecture. First, it has no internal state (or equivalently Addam's entire state is stored external to the agent in the environment, as in Simon, 1969), and thus Addam has no memory. Second, a few of Addam's connections are fixed a priori. (The changeable connections are those in the boxes labelled layer 0, 1, and 2, above.) This minimal structure is the skeleton required for preemption, but it does not assume any prewired behaviors.

Finally, we should point out the similarity of Addam's internal structure to the cascade correlation architecture of Fahlman & Lebiere (1990). There are several important differences, however. First, our system is comprised of several cascaded *modules* instead of cascaded *hidden units*. Second, Fahlman and Lebiere's higher-level hidden units function as higher-level feature detectors and hence must receive input from all the preceding hidden units in the network. This can lead to a severe fan-in problem. Due to the preemptive nature of our architecture, higher-level modules need only know if any lower-level module is active, so they require only a single additional input measuring total activation of the previous modules. Third, Fahlman's system grows more hidden units over time, correlating each to the current error. The nodes of our architecture are fixed throughout training, so that modularity is not achieved by simply adding more units. Finally, there is a difference in training: Fahlman gives his network a single function to learn, whereas our system attempts to learn a series of more and more complex behaviors. (More on this below.)

Training Addam

As mentioned above, Addam's environment consists of three types of objects: ice, food, and blocks. Ice is transparent and odorless, and is hence detectable only by the tactile sensors. Blocks trigger both the tactile and visual sensors, and food emits an odor which diffuses throughout the environment and triggers the olfactory sensors. Addam eats (in one time step) whenever it comes into contact with a piece of food.

Addam's overall goal is to move towards food while avoiding the other obstacles. This makes training problematic – the desired response is a complex behavior indexed over many environmental configurations, and yet we do not wish to restrict the possible solutions by specifying an entire behavioral trajectory for a given situation. Beer & Gallagher (1991) attempted to solve this problem by using GA's, which respond to the agent's overall performance instead of to any particular movement. We take a different approach, namely, we train Addam on *single moves* for a given number of scenarios, defined as one particular environmental configuration. Under this methodology, the *extended moves* which define Addam's behavior emerges from the complex interactions of the adaptive modules and the environment.

Training begins with level 0 competence, defined as

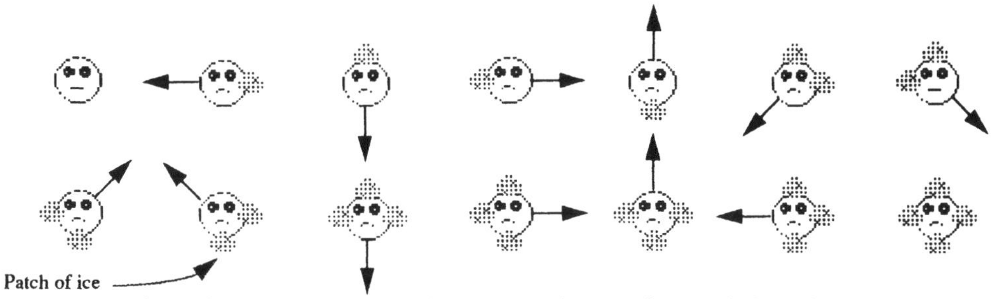

Figure 2: *Training scenarios for level 0 behavior, along with desired responses.*

the ability to avoid ice. The training scenarios are shown below in Figure 2, along with the desired response for each scenario. Module 0 can successfully perform this behavior in about 600 epochs of backpropagation (adjusted so that the fixed +1/-1 connections remain constant), and the connections of this module are then frozen.

We next train Addam on level 1 behavior, defined as the ability to move towards food, *assuming no ice is present*. Once again, training is problematic, because there are a combinatorial number of environmental configurations involving food and ice. We solve this problem as follows. First, we define 14 scenarios as above, but with food replacing ice. This defines a set S of {(SensorValues, MoveToFoodOutput)} pairs. Note that this does not define a value for a_1, the activations of the system prior to module 1. (See Figure 1.) Instead of forcing module 1 to recognize the presence of ice, we assume that module 0 is doing its job, and that when ice is present a_1 will be >> 0. This allows us to define a training set T for level 1 behavior by prepending the extreme values of a_1 to the SensorValues in S, thus doubling the number of configurations instead of having them grow exponentially:

T={ {(0-SensorValues, MoveToFoodOutput)},
{(1-SensorValues, ZeroOutput)} }

Thus layer 1 (which is initially always active) must learn to suppress its activity in cases where it is not appropriate. This training method was motivated by studies on development in which a dynamical system had to learn to suppress its behavior when not appropriate (Thelen, 1990).

After level 1 competence is achieved (about 3500 epochs), a training set for level 2, competence (avoid blocks) is obtained in a similar manner. Note again that this avoids the combinatorial explosion of specifying the many possible combinations of ice, food, and blocks. Level 2 competence is achieved in about 1000 epochs.

Results

Once Addam was trained, we placed it in the complex environment of Figure 3. Its emergent behavior is illustrated in the top half of the figure, where the small dots trace out Addam's path. Each dot is one time step (defined as one application of the trained network to move one step), so the spacing indicates Addam's speed.

Addam begins at (3.5, 1) touching nothing, so its tactile sensors register zero and layer 0 is inactive. The olfactory sensors respond slightly to the weak odor gradient, causing a slight activation of layer 1, disabling the block-avoidance behavior of layer 2. Thus we observe a constant eastward drift, along with random north-south movements due to the noise inherent in the sensors. As Addam approaches the food, the odor gradient increases, the olfactory sensors become more and more active, and layer 1 responds more and more strongly. When the random noise becomes negligible at about (6.5, 1), Addam speeds up, reaches the food, and devours it.

After completing its first meal, Addam detects the faint odor of another piece of nearby food, and once again layer 1 controls its movement. However, at about (9, 5.5) Addam's tactile sensors detect the presence of a piece of ice, activating layer 0, and usurping control from layer 1. In other words, Addam's aversion to cold feet overcomes its zealous hunger, and it moves southeast. After "bouncing off" the ice, the tactile sensors return to zero, and layer 1 regains control, forcing Addam back towards the ice. However this time it hits the ice just a little farther north than the last time, so that when it bounces off again, it has made some net progress towards the food. After several attempts, Addam successfully passes the ice and then moves directly towards the food.

To reach the third piece of food, Addam must navigate down a narrow corridor, demonstrating that its layer 1 behavior can override its layer 2 behavior of avoiding blocks (which would repel it from the corridor entrance). This is shown even more directly in Addam's docking behavior (cf: Lin, 1990) as it eats the fourth piece of food. After finishing the last piece of food, Addam is left near a wall, although it is not in contact with it. Thus both the tactile and olfactory sensors output zero, so both layers 0 and 1 are inactive. This allows Addam's block avoidance behavior to become activated. The visual sensors respond to the open area to the north, so Addam slowly makes its way in that direction. When it reaches the middle of the enclosure, the visual sensors are bal-

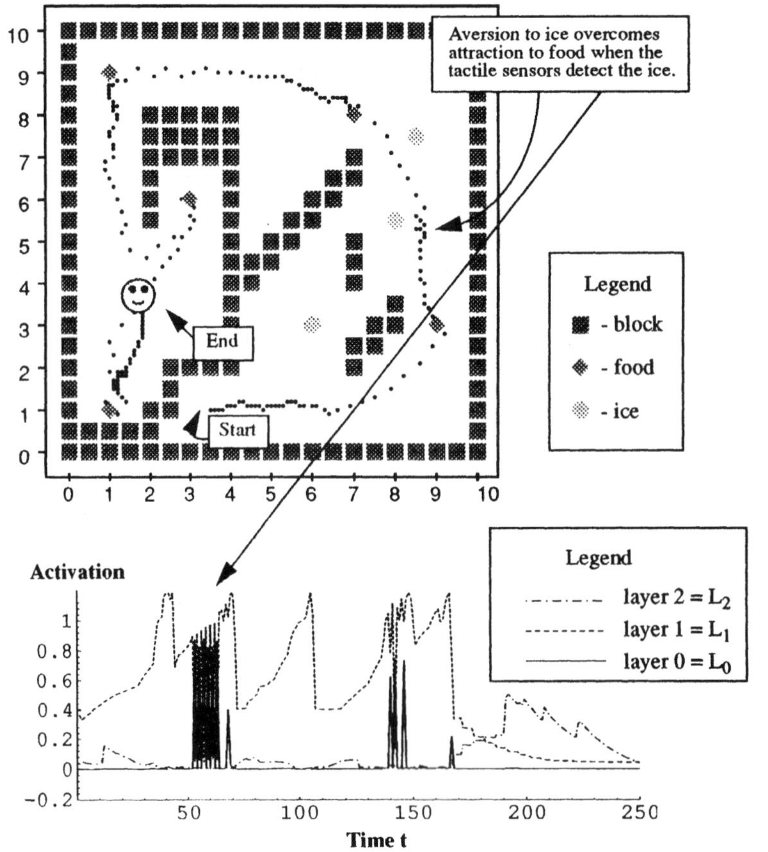

Figure 3: Addam's emergent behavior in a complex environment, with graph showing the activations of layers 0, 1, and 2.

anced and Addam halts (except for small random movements based on the noise in the sensors).

The bottom half of Figure 3 shows the activation of each layer i of the system (where the activation of layer i is $\|(\delta x,\delta y)_i\|$, the norm of layer i's contribution to the output lines). L_0 is generally quiet, but becomes active between time t=52 and t=64 when Addam encounters an ice patch, and shows some slight activity around t=140 and t=168 when Addam's tactile sensors detect blocks. L_1 ("approach food" behavior) is active for most of the session except when preempted by the "avoid ice" behavior of L_0, as between t=52 and t=64. The 5 peaks in L_1's activity correspond to Addam's proximity to the 5 pieces of food as it eat them; when the last piece of food is consumed at t=164, L_1's activity begins to decay as the residual odor disperses. Finally, we see that L_2 ("avoid blocks" behavior) is preempted for almost the entire session. It starts to show activity only at about t=160, when all the food is gone and Addam is away from any ice. The activity of this layer peaks at about t=190, and then decays to 0 as Addam reaches the center of its room and the visual sensors balance.

Remarks

Addam was trained on only 42 simple scenarios, yet it was able to perform well in a complex environment. Unlike other connectionist modular systems, our method of control is distributed – each module decides for itself whether it should exert control in any given situation. Furthermore, there is no gating network which receives a specialized task bit – Addam has three sets of sensors all treated equally and must learn the proper behavior on the basis of these inputs. Finally, instead of limiting activations of the modules to being 0 or 1, we exploited the underlying connectionist nature of our architecture, allowing us to produce interactions between modules more interesting than absolute preemption. For example, the presence of ice overrode Addam's attraction to food, yet Addam's "go-get-it" response to the food had a slight influence on its "runaway" response to the ice. Had preemption been absolute, Addam's attraction to ice and aversion to food would have alternately controlled its movement, with layer 0 exactly countering the effect of layer 1, and Addam would have slowly starved to death as it bounced off the ice indefinitely.

Our system also differed from traditional subsumption. We choose preemption by following Brook's own advice in describing his goal of simplicity: "If you notice that a particular interface is starting to rival in complexity the components it connects, then either the interface needs to be rethought or the decomposition of the system needs redoing" (Brooks, 1986, p. 15). This is a wonderful credo for engineers, but as cognitive scientists, we must generalize it to this: If you notice that your model of a particular aspect of cognition starts to rival in complexity the components of the underlying system, then both the underlying system and the model of the environment need to be reexamined. Learning to adapt to the environment is extremely difficult in Brooks' subsumption architecture, but became possible after switching to a simplified, additive model of modularity.

Our ideas for modular adaptive control are independent of the internal structure of the modules. In fact, the work of Beer & Gallagher or Maes & Brooks is really complementary to ours, for although Addam's modules were instantiated with feedforward networks trained by backpropagation, they could have just as easily been trained by either GA's or correlation algorithms. Moreover, feedforward networks need not have been used either. We could have substituted sequential cascaded networks (Pollack, 1987), endowing Addam with internal state (cf Kirsh, 1991) and allowing even more complex behaviors.

Finally, we note a significant difference in methodology between our work and that of Brooks. In creating his agents, Brooks first performs a *behavioral* decomposition, but in implementing each layer, he performs a *functional* decomposition of the type he himself warns against (Brooks, 1991, p. 146). In training Addam, on the other hand, we first perform a behavioral decomposition, and then let backpropagation decompose each behavior appropriately. This automation significantly lessens the arbitrary nature of behavior-based architectures which has thus far limited the import of Brooks' work to cognitive science.

Acknowledgments

The authors thank David Stucki and Barbara Becker for their critiques of earlier drafts of this paper.

References

Beer, R. D. and Gallagher, J. C. (1991). Evolving dynamical neural networks for adaptive behavior. Technical Report CES-91-17, Case Western Reserve University, Cleveland.

Brooks, R. A. (1986). A robust layered control system for a mobile robot. *IEEE Journal of Robotics and Automation*, 2(1):14–23.

Brooks, R. A. (1991). Intelligence without representations. *Artificial Intelligence*, 47:139–159.

Connell, J. H. (1990). *Minimalist Mobile Robotics: A Colony-style Architecture for an Creature*, Volume 5 of *Perspectives in Artificial Intelligence*. Academic Press, San Diego.

DeJong, G. and Mooney, R. (1986). Explanation-based learning: an alternative view. *Machine Learning*, 1:145-176.

Elman, J. L. (1988). Finding structure in time. Technical Report CRL 8801, University of California, San Diego.

Fahlman, S., and Lebiere, C. (1990). The cascade-correlation learning architecture. Technical Report CMU-CS-90-100, Carnegie Mellon University, Pittsburgh.

Jacobs, R. A., Jordan, M. I., and Barto, A. G. (1990). Task decomposition through competition in a modular connectionist architecture: The what and where vision tasks. *Cognitive Science*, 15:219-250.

Jacobs, R. A., Jordan, M. I., Nowlan, S. J., and Hinton, G. E. (1991). "Adaptive mixtures of local experts", *Neural Computation*, 3(1):79-87.

Kirsh, D. (1991). Today the earwig, tomorrow man? *Artificial Intelligence*, 47:161–184.

Lin, L. J. (1990). Programming robots using reinforcement learning and teaching. In *Proceedings of AAAI-90*, pages 781–786, Menlo-Park, CA. AAAI, MIT Press.

Maes, P. and Brooks, R. A. (1990). Learning to coordinate behaviors. In *Proceedings of the Eighth National Conferences on AI*, pages 769–802. AAAI-90.

Maes, P. (1991). The agent network architecture. In *AAAI Spring Symposium on Integrated Intelligent Architectures*, March.

Mitchell, T., Keller, R., and Kedar-Cebelli, S. (1986). Explanation-based generalization: a unifying view. *Machine Learning*, 1:47-80.

Nowlan, S. J. and Hinton, G. E. (1991). Evaluation of adaptive mixtures of competing experts. In Lippmann, R., Moody, J., and Touretzky, D., editors, *Advances in Neural Information Processing 3*. Morgan Kaufmann, pages 774-780.

Pollack, J. B. (1987). Cascaded back propagation on dynamic connectionist networks. In *Proceedings of the Fourth Annual Cognitive Science Conference*, pages 391-404, Seattle.

Samuel, A. L. (1967). Some studies in machine learning using the game of checkers II – recent progress. *IBM Journal of Research and Development*, pages 601-617.

Simon, H. (1969). *Sciences of the Artificial.* MIT Press.

Thelen, E., Dynamical systems and the generation of individual differences. In Columbo, J. and Fagen, J. W., editors, *Individual Differences in Infancy, Reliability, Stability, and Prediction*, Lawrence Erlbaum Associates, pages 19-43.

MusicSoar: Soar as an Architecture for Music Cognition

Don L. Scarborough, Peter Manolios and Jacqueline A. Jones

Department of Psychology and Department of Computer and Information Science
Brooklyn College of the City University of New York
Brooklyn, NY 11210
dosbc@cunyvm.bitnet, pete@sci.brooklyn.cuny.edu, jajbc@cunyvm.bitnet

Abstract

Newell (1990) argued that the time is ripe for unified theories of cognition that encompass the full scope of cognitive phenomena. Newell and his colleagues (Newell, 1990; Laird, Newell & Rosenbloom, 1987) have proposed Soar as a candidate theory. We are exploring the application of Soar to the domain of music cognition. MusicSoar is a theory of the cognitive processes in music perception. An important feature of MusicSoar is that it attempts to satisfy the real-time constraints of music perception within the Soar framework. If MusicSoar is a plausible model of music cognition, then it indicates that much of a listener's ability is based on a kind of memory-based reasoning involving pattern recognition and fast retrieval of information from memory: Soar's problem-solving methods of creating subgoals are too slow for routine perception, but they are involved in creating the knowledge in long-term memory that then can meet the processing demands of music in real time.

The Soar Architecture

This section is a brief introduction to Soar (Version 5) and is to a great extent based on Newell (1990). Soar is a goal-directed problem-solving cognitive architecture that is built on a parallel production system. Soar displays many of the characteristics of human cognition, and the temporal characteristics of Soar's cognitive behavior are consistent with much of what is known about human cognition.

Goals

A central premise of the Soar theory is that cognition is based on goal-directed problem solving. Soar's problem solving occurs within a *context* that has four predefined *attributes* or *slots*: a *goal*, a *problem space*, a *state*, and an *operator*. Goal-directed cognition begins with the selection of a goal followed by selection of a problem space which delimits the sets of states and of operators that will be considered. Next, a state is selected to represent the current state of the problem. Finally, an operator is selected to change the current state to reach the goal state. The Soar architecture can select a value for one of these context slots on each *decision cycle*.

Decisions about context slot values occur automatically within the Soar architecture if available knowledge is sufficient to guide the decision. Otherwise, an *impasse* occurs and the Soar architecture creates a subgoal which is a new context with the goal of resolving the impasse. This subgoal context then requires selection of a problem space, a state and one or more operators in order to resolve the impasse. Further impasses may occur within this subgoal, leading in turn to additional subgoals.

The initial top-level context is unique and is generally initialized with a predefined goal, problem space and state. Once the top context is initialized, an operator is selected to perform some task. Typically, Soar cannot implement this operator directly which causes an impasse, leading to the creation of a subgoal, as described above, to implement the desired operation.

Long term and working memories

Long term memory consists of productions that contain conditions and actions. If all of the conditions of a production match working memory elements, then the actions of that production fire, adding new information to working memory. Soar differs from conventional production languages, such as OPS5, in many important respects. For example, there is no conflict resolution; all productions that match fire in parallel, adding new and possibly conflicting information to working memory simultaneously. These additions to working memory may make it possible for new productions to fire. This process is called the *elaboration phase* of the decision cycle, and it continues until *quiescence*, when no new productions are triggered by information in working memory. The elaboration phase allows Soar access to all available, relevant knowledge for its decision making. At quiescence, Soar tries to select a value for a context slot. The elaboration phase followed by selection of a context-slot value is the decision cycle.

Working memory holds all the information about currently selected values for context slots. In addition, as long-term memory productions fire, they add to working memory new information consisting of proposed values for context slots, preferences for previously proposed values, and augmentations of information already in working memory. All information in working memory is stored in a network that is linked to context slot values. If the value of a context slot changes, then all information linked to the old slot value is discarded. For example, when a subgoal resolves the impasse that created it, the subgoal context is deleted from working memory along with all information linked to that subgoal. Hence working memory is highly dynamic, allowing elements to disappear when no longer needed.

All I/O is mediated by the top context state. That is, in the Soar theory, perceptual input about the current state of the environment enters the top context state and this information can then be used by Soar in its decision making. Also, motor systems can access output commands that are placed in this top state.

Chunking

All learning in Soar occurs through *chunking*. When Soar finds a solution to an impasse, it creates a chunk, which is a new production that represents the solution to the impasse. The left-hand side of the new production contains the information in working memory that was available when the problem arose and that was used in finding the solution to the problem. The actions of the new production are the results of the problem solving. This production is added to long-term memory, and when Soar finds itself in a similar situation in the future, this production will fire and resolve the problem, thus eliminating the need to solve the problem again.

Mapping Soar to Human Cognition

Newell (1990) argues that the minimum functional neural circuit in the human brain takes about 10 ms to operate. Such a neural circuit can perform a function such as the memory access required to match the left-hand side of a production in long-term memory to conditions in working memory. This matching and firing of productions occurs in parallel in the elaboration phase of the decision cycle. The entire elaboration phase along with the subsequent decision phase takes place automatically in about 100 ms. The implementation of an operator will usually need a sequence of these decision cycles such that even the simplest cognitive tasks will require times on the order of about a sec.

MusicSoar

As noted above, cognition is based on goal-directed problem solving in the Soar theory. To apply Soar to music cognition, we must view listening to music as a form of problem solving. What problem confronts a person listening to music? It seems likely that any intelligent system should attempt to anticipate future events. In MusicSoar, we assume, following Narmour (1991), that the problem in listening to music is to anticipate what is to come, based on music that has already been heard. A listener's knowledge of a specific piece and its style, along with general musical knowledge, provide a basis for expectations of what is to come. If these expectations are accurate and match newly heard events, they become the basis for generating more expectations. On the other hand, expectations may not match what is heard. In our approach, this generates a subgoal to learn new expectations so that, if the same or similar music is heard again, the listener will be better able to anticipate the events that occur.

Listening to Music in MusicSoar

In MusicSoar, as musical events occur, they enter the top state. Musical "problem solving" begins in this top context with the selection of a *listen-to-music* operator. This immediately leads to a subgoal of implementing this operator, and, within this subgoal, MusicSoar creates a new state called *music-working-memory*. It is within this subgoal that MusicSoar listens to the music input and anticipates what will follow. Thus, MusicSoar has two primary states. The top state functions as a passive preattentive sensory input buffer like an "echoic memory," while the music-working-memory represents characteristics of the input that have been attended as well as expectations of what is to come.

Top State Input

Input notes appear in the top state state and disappear after some length of time. We assume that the representation of musical input in this top state is not in terms of waveforms, but rather is a representation of the output of earlier auditory perceptual preprocessing stages. Thus, in MusicSoar, each note in the top state is represented by its pitch and duration, as well as information about its temporal offset from the prior event. Input to the top state is handled by a Lisp function that reads a file containing symbolic representations of musical events. The input function creates an event attribute or augmentation linked to the top state for each new musical event. The value of an event augmentation is information about the event's temporal offset from the previous event as well as information about the pitch and duration of each note in the event. Currently, MusicSoar deals only with music containing a single voice, such as the melody of a folk song without accompaniment.

Music-Working-Memory

Once an intention of listening to music has been selected in the top context (as represented by the choice of the "listen-to-music" operator), MusicSoar creates a listen-to-music subgoal. This subgoal then persists throughout the piece of music. The problem space in this subgoal represents MusicSoar's knowledge of music. The state associated with this first subgoal is called music-working-memory and contains information about top-state musical events that have been attended and processed. The initial representation of top-state events in music-working-memory encodes only some of the simple relational properties such as whether the most recent note is higher, lower or the same in pitch as the previous note, as well as whether the offset of the newest event is the same or longer or shorter than the previous offset. More complex encodings within music-working-memory, such as information about specific pitch intervals, depend upon additional processing. It is also within music-working-memory that anticipations arise. There are only two operators that can be selected within this listen-to-music subgoal context: *attend*, and *learn-expectation.* If a new event occurs in the top state and no operator has been selected, then the attend operator is proposed. Once the attend operator is selected, it is implemented by productions that copy part or all of the musical event information from the top state into the music-working-memory. The productions that implement the attend operator can operate without additional subgoals in a single decision cycle. Additional productions compare the attended event to expectations and, if it matches, the new event is added to a linked list of previously heard events in music-working-memory. Because the attend operator is implemented by productions that fire without requiring decisions about context slots, the attend operator requires a single decision cycle of about 100 ms. Thus, when music conforms to expectations, the attend operator can follow along at about 10 events per second. On the other hand, the learn-expectation operator is proposed whenever an attended event does not match expectations. This operator leads to a subgoal to learn new expectations that match what actually happened.

Expectations

Musical expectations in MusicSoar are stored in its long-term production memory. The left-hand sides of these productions specify particular patterns of events in music-working-memory, while the right-hand sides represent expectations of what should follow the occurrence of these patterns. If the left-hand side of a production matches the previously heard events in music-working-memory, the production fires and adds its expectation to working memory. Generating expectations in this way involves no decision making and thus can occur quickly. Newell (1990) has argued that retrieving information from memory (e.g., matching the conditions of a production to working memory elements) requires about 10 ms. Initially, MusicSoar has only a few default expectations, such as to expect that the next event will have the same properties, e.g. pitch, and offset, as the previous event. As MusicSoar experiences different musical patterns, it learns new productions that are added to long-term production memory.

Learning

Subgoals in Soar arise when it is unclear what to do next, and they result in decisions about what to do. Upon resolving a subgoal, Soar can learn new productions or chunks that store information about how the problem was resolved. The next time that problem arises, the resolution of the problem can be retrieved from long-term production memory without requiring problem solving again. Thus, after experience with a particular problem, Soar can subsequently solve the problem using a form of memory-based reasoning in which the old solution is retrieved from memory rather than solving the problem again from scratch. MusicSoar's expectations arise from learning in the learn-expectation subgoal.

The way in which the learn-expectation subgoal is instantiated depends upon the type of expectation mismatch that occurred and the subgoal problem space that is selected, e.g., metric, rhythmic, melodic, etc. For example, if a new event occurs at a time that is inconsistent with metric expectations, this requires reinterpreting the meter of the music; however, an unexpected event that occurs at a time that is consistent with metric expectations requires learning a rhythmic expectation. Problem solving within the learn-expectation subgoal can occur in several ways. For example, if a note occurs later than expected, MusicSoar can try look-ahead search to see if the expected note appears to lead, in terms of offset, duration and pitch, to the new note. In this case, the expected note may bridge the gap from the previous events to the event just heard. Alternatively, additional musical analysis may reveal new features of the music that has been heard, allowing other productions in long-term memory to fire, and these productions may propose the correct expectation. For example, MusicSoar might look back at previously heard events in working memory to see if some parallel sequence of events has occurred. Finally, the learn-expectation subgoal might use *data chunking* (Rosenbloom, Laird & Newell, 1987) to learn to expect the new event based on the immediately preceding events. In data chunking, some of the previously heard events are learned as a cue that will, in the future, trigger recall of the newly heard event in a sort of paired associate learning. That is, some of the previous events in music-working-memory become the pattern for the left- hand side of a new production that will trigger a new expectation. This lets MusicSoar memorize specific songs. Although MusicSoar can learn expectations corresponding to specific pieces, this is not sufficient. People not only acquire specific knowledge about particular pieces of music, but they also acquire more general schematic knowledge that guides expectations when listening to new pieces of music (Narmour, 1991). For example, experience with Western tonal music leads people to expect particular melodic, harmonic and rhythmic progressions. Thus, a general problem for MusicSoar is to induce schematic musical knowledge and expectations from experience with specific pieces.

The representation of musical input has important effects on learning. People generally perceive events (visual, auditory, tactile, etc.) in terms of the relational properties of the event; e.g., they perceive relative luminance differences in vision rather than absolute luminance levels. In music, the salient perceptual properties involve relative pitch and time differences. To reflect this, the top state representation of musical events is encoded in terms of the pitch and duration of each event relative to previous events. Because MusicSoar learns expectations in terms of these relational properties, this learning will generalize to situations that preserve these relations. Thus, if MusicSoar learns to expect the next note in an arpeggio, this learning will apply to any arpeggio in the same mode (e.g. major or minor) regardless of the key. Thus, learning based on such relational information will generalize directly to transpositions in pitch and time. However, one issue that we have not yet resolved in the data chunking mechanism described above is how much information about previous events should be included in the left-hand side of new productions. Making the left-hand side too specific will prevent the new productions from generalizing to any other situation.

As just indicated, MusicSoar's ability to generalize is influenced by the representation of the music in music-working-memory. A key question for any inductive learning is what knowledge and learning biases exist before the learning actually begins (Dieterich, 1990). For example, we have assumed that human listeners have some understanding of pitch relations without any training. That is, if two sounds differ in frequency, listeners hear the higher frequency as higher in pitch. More complex characterizations of pitch, such as octave relations and perception of consonance may be learned early in life based on auditory stimuli in general (e.g. Terhardt 1991), and thus may be available almost from the very first musical experience. Some primitive temporal knowledge, such as the ability to hear differences in durations, also seems almost certainly innate or at least acquired at a very early age. These considerations determine the design of MusicSoar. We have assumed that, without prior experience, MusicSoar can determine the contour of a melody, i.e., the pattern of ups and downs in pitch. However, we assume that more specific knowledge of pitch relations requires learning about particular intervals. Further assumptions are that MusicSoar can perceive equality of time intervals, and time ratios of two to one and three to one (corresponding to duple and triple meters in music, respectively).

One interesting problem is that Soar has no explicit forgetting mechanism. That is, once a new production is learned, it is never forgotten. However, later learning will create productions that may interfere with older learning. That is, Soar can demonstrate retroactive interference wherein a new production may encode a new and different expectation for a pattern that is similar to the left-hand side patterns of already learned productions. When that pattern occurs, both productions may fire and generate different expectations. One general problem in MusicSoar is how to handle such expectation conflicts.

Knowledge Search

A listener can have both specific and schematic knowledge that is relevant to a particular listening experience. Given this, hearing a particular event sequence may trigger many expectations, some of which may conflict. We have considered and rejected two possibilities for handling such conflicts. First, we might let all expectations be added in parallel to working memory without differentiation. However, this is unacceptable, because if listeners know a particular song, they have specific expectations about what should come next, and they generally will not wander off the track, even though fragments and aspects of the song may have occurred in other previously heard pieces of music. For example, given familiarity with Beethoven's Fifth Symphony, there is no ambiguity about what follows "dit-dit-dit-dah." Thus, we cannot just let all possible expectations based on prior experience have equal status. A second possibility is to use different problem spaces for different pieces of music, e.g. a listen-to-Mozart's-40th and a listen-to-Beethoven's-fifth problem space. The learning that occurred within a problem space would be available only within that problem space. But this also cannot be the right solution, because it is clear that, while listening to Beethoven's Fifth, we do not cut ourselves off from all other musical knowledge. Separate problem spaces would also prevent generalization. What was learned for one piece would be available only in the problem space for that piece. With radically different types of music like Indian ragas and Western tonal music, the expectations and thus the problem spaces are likely quite different, but, given a particular genre or style of music, it seems probable that particular pieces of music within that style are heard within a problem space that is common to that style.

If a listener wants to follow along with Beethoven's Fifth, it must be possible to select quickly the appropriate specific expectations from a much larger set of expectations drawn from general musical knowledge. We are exploring two ways to limit choices in MusicSoar. First, working memory can contain specific elements that can control which expectations are activated. Thus, if music-working-memory contains the information that we are listening to Beethoven's Fifth, then previously learned productions for Beethoven's Fifth can be activated, but not, say, productions specific to Mozart's 40th symphony, because information about the identity of the piece would be included in the left-hand side of the expectation productions for Beethoven's Fifth. This is similar to the idea of setting up different problem spaces for different pieces but is less restrictive. A second related approach is to have production memory contain productions that express preferences for particular expectations. That is, when listening to Beethoven's Fifth, all productions that match a musical fragment may fire, resulting in a rich flood of expectations. However, other productions may also fire that express preferences for those expectations that are linked to MusicSoar's goals. Thus, if the goal is to follow along with Beethoven's Fifth, preferences for expectations specific to Beethoven's Fifth would be activated, keeping MusicSoar's expectations on track.

Temporal Constraints on Processing

Newell has argued that if the human brain implements a Soar-like architecture, a decision to select a value for a context slot (i.e. goal, problem space, state or operator) must take a decision cycle of approximately 100 ms (Newell, 1990). Music unfolds at rates that are not controlled by the listener, and the listener must cope with events as they occur. Thus, we can ask whether MusicSoar, when limited by a 100 ms decision cycle, can meet the real-time demands that are imposed by music. The attend operator in MusicSoar can be selected and implemented in a single decision cycle. This means that MusicSoar can attend to up to 10 events per second if the attend operator is selected repeatedly and without interruption. However, the attend operator for the next input event will be selected only if the current expectations match the most recently attended event. If the generation and selection of expectations were to also require selecting and implementing operators at a rate limited by a 100 ms decision cycle, MusicSoar would only be able to keep up with the slowest music. Thus, expectations must generally arise based on recognition of patterns in music-working-memory that match the left-hand-side of expectation generating productions already available in long-term production memory. As noted above, Newell (1990) argues that it is plausible to assume a memory access time of about 10 ms for production matching. In addition, preferences that select among various expectations must act on music-working-memory directly and cannot require decisions about goals, problem spaces, states or operators. They too can be implemented as quickly as they can be retrieved from production memory.

On the other hand, if expectations do not match the music, MusicSoar enters a learn-expectation subgoal which requires several decision cycles. Given that MusicSoar may be unable to complete this subgoal processing in time to attend to the next event, what happens? One possibility is that the next input event generates an interrupt, forcing MusicSoar to select a new attend operator over any currently selected operator. This would mean that MusicSoar would lose any current "thinking" about the last event. Another possibility is that MusicSoar might continue to think on the basis of the previously attended top state events. But if this occurs, MusicSoar may fall behind, perhaps by several events, which can cause MusicSoar to miss notes, and lead to incomplete processing and representation of the music. Then when MusicSoar completes its thinking and is ready to attend again, what should it attend to?

The temporal processing constraints in MusicSoar may also suggest why one can hear the same piece of music many times and continue to hear new things. A listener may not perceive all aspects of a complex musical event. In MusicSoar, this means that the representation of an event in the top state may be comprehensive, but the attend operator may encode only some of the top state information into working memory. The subset of information that is encoded will then determine the expectations that follow. If a different subset of information is encoded upon later rehearing, then a different set of expectations will arise. Another source of variability in MusicSoar can arise in the learning process. A new event may simultaneously mismatch expectations in several ways, e.g. metric, melodic, harmonic, etc., various musical characteristics which are perceived somewhat independently (Palmer & Krumhansl, 1987). MusicSoar breaks the problem down and tries to learn expectations for these characteristics, each of which is handled within a different subgoal problem space. Because each subgoal usually requires several decision cycles of about 100 ms each, only rarely is there sufficient time to deal with all the characteristics. In a single hearing, MusicSoar can learn only some of the characteristics of the music, and later rehearings then provide opportunities for additional learning. This account suggests why, when a piece is partially learned, a person may be able to imagine hearing it by following the sequence of expectations that are generated. However, any attempt to sing it out loud would be embarrassing as the expectations do not completely specify all the characteristics required for performance.

Hierarchical structure of music

Lerdahl & Jackendoff (1983) argue that perception of music involves creating a subjective hierarchical structure involving meter, rhythm, grouping, tonal movement, etc. Such structures arise in MusicSoar from augmentations added to attended events in the listen-to-music subgoal. The productions that create these augmentations are learned in meter-analysis, grouping-analysis, and tonal-analysis subgoals that arise within the learn-expectation subgoal. The ability of MusicSoar to create such structures depends heavily on past experience, for there is insufficient time for extensive analysis of the music in subgoals. Rather, such structures must generally arise directly from previously learned productions that match the information in working memory.

Composition

One interesting property of the MusicSoar approach is that this system could, with few changes, compose music. That is, because MusicSoar is based on expectations, all that is required is to initialize music-working-memory with a musical fragment. This fragment would then trigger expectations, and these expectations would not be compared to actual inputs but would simply become new values in music-working-memory which are then used to control the generation of still more expectations. The quality of the composition represented by this chain of expectations would be governed by the quality of the learned expectations. However, this perspective brings another point into focus: When there are conflicting expectations, the system must show some indeterminacy. That is, given a set of conflicting expectations, the same one should not be selected on every run or else we have a system that can compose only a few pieces. It is likely that the same indeterminacy should also be true of listening: The same song may be re-heard in different ways.

Summary

There has been relatively little work on building processing theories of music. Soar provides a challenging and exciting framework for such explorations, and, to our knowledge, it has not previously been applied to music cognition. Soar poses several interesting constraints in its application to music cognition. One is that music cognition must be viewed as a problem solving activity. Intuitively, we think it is reasonable to view the problem in music cognition as one of anticipation. A second significant constraint on the application of Soar to music cognition is that Newell has linked the time that is required for a decision cycle in Soar to brain processes that require about 100 ms. This temporal constraint imposes important limitations on how the theory can be applied to a domain such as music and makes it possible to evaluate the Soar theory and its instantiation in MusicSoar in ways that are rarely true of cognitive theories. We think that the Soar framework may generate new insights and questions into the problems that are posed for cognitive theories for domains such as music. MusicSoar is an attempt in this direction.

References

Dieterich, T. G. 1990. Machine Learning. *Annual Review of Computer Science* 4:255-306. Palo Alto, CA: Annual Reviews Inc.

Laird, J., Newell, A., & Rosenbloom, P. 1987. SOAR: An Architecture for General Intelligence. *Artificial Intelligence* 33:1-64.

Lerdahl, F., & Jackendoff, R. 1983. *A Generative Theory of Tonal Music*. Cambridge, MA: MIT Press.

Narmour, Eugene. 1991. The Top-down and Bottom-up Systems of Musical Implication: Building on Meyer's Theory of Emotional Syntax. *Music Perception* 9:1-26.

Newell, Allen. 1990. *Unified Theories of Cognition*. Cambridge, MA: Harvard University Press.

Palmer, C., & Krumhansl, C. 1987. Independent Temporal and Pitch Structures in Perception of Musical Phrases. *Journal of Experimental Psychology: Human Perception and Performance* 13:116-126.

Rosenbloom, P., Laird, J., & Newell, A. 1987. Knowledge Level Learning in Soar. In *Proceedings of AAAI-87: Sixth National Conference on Artificial Intelligence.* Los Altos, CA: Morgan Kaufmann. pp. 499-504.

Terhardt, E. 1991. Music Perception and Sensory Information Acquisition: Relationships and Low-level Analogies. *Music Perception* 8:217-240.

Attention, Memory, and Concepts in Autism.

Haline Schendan
Program in Neurosciences, 0608
University of California at San Diego
La Jolla, CA 92093
hschenda@igrad1.ucsd.edu

Abstract*

In this paper, it is hypothesized that many of the behavioral abnormalities found in autistic persons result from deficits in fundamental cognitive abilities. Memory and attention are the most likely candidates. The memory deficit may be primarily one of retrieval, possibly exacerbated by an encoding deficit. However, both types of memory deficit are probably the result of a primary deficit in attention. This is supported by the observation that the autistic memory deficit resembles that following frontal lobe, rather than mediotemporal lobe, damage. This and other evidence is used to draw a parallel between autism and frontal lobe syndrome. In light of this analogy, how a primary deficit in the fundamental cognitive ability of attention may be responsible for the more secondary autistic deficits in memory and more advanced forms of cognition, such as language acquisition, symbol manipulation, rule extraction, and social interaction, is explored.

Introduction

Several theories have been posited about how the minds of autistic people differ. Some suggest that autistics have an altered theory of mind (Baron-Cohen, 1989). Hobson and Lee (1989) consider the disorder to be primarily one of affect, and still others emphasize the contribution of fundamental cognitive abilities (Gillberg, 1990). This last is the stance argued for here.

It is hypothesized that underlying deficits in attention give rise to a host of other cognitive deficits in autistic persons, such as those in memory, language, and certain thought processes. Because research on basic cognitive abilities in autistic people is relatively sparse, the evidence to be presented is only suggestive. Nevertheless, the ideas put forth in this paper may serve to guide much needed research into the mechanisms which may underlie the variety of behavioral deficits in autistic persons.

* Research funded by the McDonnell-Pew Center for Cognitive Neuroscience in San Diego.

Many cognitive abilities of autistic people differ from those of the general population. There are some areas in which they seem markedly deficient and some in which they are relatively spared, or even unusually proficient. Their apparent social withdrawal not withstanding, the most striking cognitive deficit is in language. It has been estimated that as many as 19% of autistic children of 8-10 years are mute, and 31% speak only some words, though not in conversation (Ricks and Wing, 1975).

Level of language attainment has been correlated with estimates of intelligence (Ricks and Wing, 1975), suggesting a relationship between language and other cognitive abilities. This may best be understood within the framework of the mind as instantiated within a massively parallel and distributed neural system in which there is extensive feed back between systems, the reality of the neocortex. Within the neocortex it is likely that different systems that interact with each other are likely to mutually modify the information processing within each other. The extensive back projections between areas of neocortex indeed support that the effect of a system A, such as attention, feeding forward onto a system B, such as language, may depend upon the nature of the information processing within system B (language). Thus the nature of the neural machinery may enable language systems to affect how attentional systems affect linguistic processing.

Such effects may be most apparent within the context of cognitive abilities which rely heavily on both language and attention. Specifically, symbol manipulation, inference, or deduction, all of which may have been enabled or facilitated by the development of language, may also depend upon systems subserving attention. Autistic persons experience difficulty with tasks requiring symbol manipulation. If their attentional systems are damaged, assuming the symbol manipulation task requires both attention and language, their deficit may most accurately be described as one of attention that creates a deficit in one, the other, or both, language and/or symbol manipulation. This paper focuses upon how deficits in more fundamental cognitive abilities, such as attention, memory, or emotion, may result in language disabilities.

Another reason to shift the focus of research from language to other basic aspects of cognition is that language is not likely to be one of the most basic cognitive abilities (Bates, 1990). This may be argued at least from an evolutionary standpoint. Evolution usually acts by building upon already existing structures (Killackey, 1987), and language ability is one of the most recent to evolve. Thus, it probably depends upon the existence of other cognitive abilities for its function, such as attention, memory, and perception.

Basic Cognitive Deficits

In this regard, memory and attention are likely candidates. Memory is a ubiquitous property of the brain, and attention is known to modify memory. Indeed, both memory and attention seem to be abnormal in autistics. A course of study in which of the interaction of memory, attention, and then language, and the emergent properties to which they give rise, are examined, is more likely to yield a logical and coherent account of the structure of the cognitive deficits in autism. This is in contrast to the approach of addressing the problem exclusively at a high level of cognitive description without consideration of possible underlying, more basic, deficits. However, high level descriptions are necessary to define the problem so that the neural mechanisms which underly them may be discovered. Such descriptions may even be suggestive of the underlying mechanism. Nevertheless, once a high level description of the problem has been put forth to guide research, it may be more fruitful to concentrate on cognitive abilities that are the cornerstones of more complex cognition and build up from there, likely redefining the problem itself. Such an approach is advantageous also because fundamental cognitive abilities should be simpler, relative to higher order cognition.

Memory

Memory and Autism. Paradoxically perhaps, memory is one of the areas of cognition in which autistics seem to be relatively spared, or even especially able. From as early as Kanner's (1943) original article, autistics have been considered to have prodigious rote memory skills, to such an extent that as many as 9.8% of them are categorized as idiot-savants. Nevertheless, a memory deficit could account for at least the memory-dependent cognitive deficits in autism, including their language problems. For example, autistic children tend to use a holistic approach for language acquisition, as well as for other learning situations. This has been proposed to be the reason for their extended echolalia, in which they just tend to repeat back what has been said. (Prizant, 1983). Prizant (1983) suggested that autistics have an impaired semantic memory system, while their episodic memory system, which enables rote memory, remains intact. For example, autistics have trouble segmenting sentences into the meaningful parts which would be stored in a semantic memory system. While current research on memory would modify these proposals, this example illustrates the role that a memory deficit could play in the development of memory dependent cognitive abilities in autistic people.

Further support for the role of memory deficits in the autistic syndrome comes from a number of researchers who have sought to draw a parallel between autism and the amnestic syndrome (Boucher et al, 1976; Heltzer, et al, 1981). This research avenue is supported by the marked learning disabilities of autistic children. After all, memory is the result of the learning process, though other information processing deficits may also be involved. Those who favor a memory explanation of the autistic disorder prefer to attribute the biological abnormality to the mediotemporal lobes. Damage to these structures has been found both in animals and in humans to result in profound anterograde amnesia, as well as some retrograde. This amnesia, however, is restricted to explicit, or declarative, memories (Squire, 1987), accompanied by marked sparing of implicit or procedural memories.

The seminal paper on memory in autistics was that by Boucher and Warrington (1976). Their main finding was that, over a 30 or 60 second interval, autistics were impaired on free recall and recognition, but performed at essentially normal levels on tests of cued recall. In a further series of studies, Boucher (1981) looked into the nature of the memory deficit and sparing in autistics. In general, it was confirmed that autistics do indeed have good cued recall abilities (Boucher, 1981). However, they are impaired at free recall of recent events, and lack any ability at face recognition (Boucher, 1981). Retrieval from memory by autistic persons requires stronger cueing than in normals. Their memory ability, rote learning, like cued recall, is externally cued or predominantly self-cued. It may be this facet which allows rote memory to exist and flourish in the the autistic person, serving well under some, but not all, behavioral conditions. Whereas cues serve well to evoke remembrance in autistics, such people must rely more heavily upon them. Autistic persons apparently are unable to evoke memories internally and spontaneously without distinctive retrieval cues, and will seem to be impaired on tasks requiring free recall. Thus, abnormal retrieval of stored knowledge may be at least partly responsible for both their memory abilities and deficits. With regard to idiot savants, in the absence of a normal memory system, autistics may overuse their intact abilities, such as rote memory, permitting some of them to appear to be mnemonists.

In general, these findings argue against mediotemporal lobe involvement. Mediotemporal lobe amnesia is thought to arise from a deficit in encoding. Such brain injured people cannot consolidate new information, but they have no trouble recalling information acquired prior to their lesion. This suggests that people with mediotemporal lobe lesions have intact retrieval mechanisms. Rather, their deficit is primarily one of encoding. Thus, the autistic memory deficit cannot be solely attributed to an abnormality in mediotemporal lobe function analogous to the amnestic syndrome, though there may be some subtle involvement of this brain region. Nevertheless, autistics do have a memory abnormality, but it may have to be attributed to another cause. The alternative to be explored in this paper is the parallel between frontal lobe syndrome and autism. In addition to the behavioral analogies to be presented, Gedye (1991), in an extensive survey of the literature, has suggested that "the variety of etiologies that cause frontal lobe seizures also accounts for the variety of etiologies traced to autism." Thus the neurology of frontal lobe disorders and autism supports the behavioral parallels.

Frontal Lobe Syndrome and Autism. Memory abnormalities are found in people with lesions of the frontal lobes. This form of memory disorder is qualitatively different from that of organic amnesia. In a well controlled series of studies, Janowsky, et al (1989) found that , in contrast to previous studies showing greater deficits, on most types of memory tasks, frontal lobe patients performed near normal levels, particularly relative to amnesiac and Korsakoff's syndrome patients. They therefore concluded that frontal lobe syndrome does not involve the kind of global amnesia present in mediotemporal patients. Nevertheless, frontal patients did tend to perform at lower levels than the control groups, though this finding did not reach statistical significance. For example, frontal lobe patients were somewhat impaired on free recall. This is congruent with similar findings in autistics, though the evidence suggests that autistics may be more impaired (Boucher et al, 1976; Boucher, 1981). A number of Janowsky, et al's (1989) findings are consistent with those of autistics, including recognition (Ameli et al, 1988) on which both perform well and word fluency (Boucher, 1988) on which both groups are impaired.

This last finding is particularly interesting. Impaired word fluency in frontal lobe patients has been interpreted as a reduction in the fluency and spontaneity of complex behavior (Shimamura, et al, 1991). In contrast, while autistics generated fewer miscellaneous words, Boucher (1988) found that they performed near normal when provided with a category. Perhaps autistics find a category name a better cue than do frontal patients. The autistic people's deficit at free generation in the miscellaneous condition, nevertheless, is consistent with their impaired free recall and their known lack of creativity. Both groups may be less apt to spontaneously search their knowledge base to generate responses. However, when a strong contextual cue is provided, active search may be facilitated, or may be less necessary for performance.

These considerations may suggest that neither autistic persons nor frontal lobe patients have problems consolidating memories of an experience, as do organic amnesiacs. Rather, the problem may be of a different nature: retrieval (Shimamura, et al, 1991). Retrieval from memory involves either implicit, nonconscious, or explicit, conscious, use of stored information. Explicit retrieval is an active process that may require the participation of attention, which may be the mechanism whereby the internal structure of knowledge may be actively searched. Thus, so called explicit memory deficits in frontal patients and autistics may be more fundamentally ones with mechanisms of attention rather than with memory, per se. It is also possible that their retrieval difficulties are exacerbated by a deficit in encoding that results from the interaction of attention with the explicit memory system. Thus there may be a parallel between the dynamics of learning and that of retrieval. This parallel may be that both involve attention. Some form of attention may direct what gets encoded, while the same or a different mechanism of attention aids retrieval.

Attention

Frontal Lobe Syndrome and Autism. With this in mind, the majority of the memory deficits in frontal lobe patients are thought to be associated with their inability to plan and organize their behavior (Mayes, 1988). Learning and memory that require the initiation and maintenance of effortful and organized strategies of encoding and/or retrieval, as well as the ability to switch from one strategy to another pose the greatest problems for frontal lobe patients (Mayes, 1988). This has indeed suggested to some that the frontal lobe syndrome may involve a deficit in sustained attention. *Effortful* and *maintenance* suggest that attention is required, as does *switching ability*. Attention provides the organism with a way to orient, maintain, and shift its awareness to different parts of its knowledge base, to different perceptual systems, and to different aspects of them. Frontal lobe patients may have poor planning ability because they cannot maintain attention. Thus, deficits in mechanisms of attention may disrupt the information processing involved in planning which may lead to an inability to form complex memories, as well as hindering memory retrieval processes.

Attention and Working Memory. These neuropsychological considerations are supported by animal studies of prefrontal working memory, which evolved out of STM research in psychology (Goldman-Rakic, 1989). These studies focus on spatial working memory. Spatial processing involves the parietal lobes, as well as prefrontal, and the parietal lobes have also been implicated in attention.

The relationship between working memory and short term memory (STM) is also important because attention and STM are thought to be closely related. The connection may be that sustained attention is required in order to hold and place information in STM. Frontal patients tend to do poorly on STM tests, such as digit span (Shimamura, 1991). Thus, this may be due to an attentional deficit. It was once commonly thought that memory formation proceeded from STM to long term memory (LTM). Thus, any deficit in STM would also produce a deficit in LTM, or in one's overall knowledge base. Currently memory researchers do not bring STM studies into their theories, nor do most explicitly consider attention. It will therefore be suggested here that frontal lobe patients are abnormal in the way that attention acts on the working memory modules of the frontal lobes. The primary deficit in patients with frontal lobe lesions may be due to the extensive disruption of cortical connectivity necessary for attention to bind cortical modules together both spatially and temporally in a way that is necessary for the performance of complex behaviors, especially those requiring good STM.

Other evidence that supports the frontal lobe parallel are the findings that both populations seem to have a tendency toward perseveration of no longer appropriate responses, as well as in solving problems for which they cannot use a well established routine (Mayes, 1988; Shimamura, et al, 1991). Both of these are strong characteristics of both syndromes. There are several explanations for such behaviors, but the one relevant to this paper is that perseveration may result from inability to shift attention away from a previous problem and well established routines may be required if one cannot shift attention so as to rapidly acquire, or shift to a contextually appropriate, a motor pattern.

Counterevidence. However, there is some evidence that argues against drawing a parallel between frontal lobe syndrome and autism. One problem is that digit span is the part of the WAIS IQ test on which autistics are relatively facile (Lincoln, et al, 1988), arguing against an attentional deficit. However, neurophysiological work suggests that autistic people may indeed have a deficit in their ability to rapidly shift their attention (Courchesne, 1990). There are several possible explanations for this inconsistency. Perhaps, autistics have a compensatory attentional mechanism that allows them to perform near normal on the digit span task. Alternatively, digit span may not require shifting, but rather sustaining, attention, or perhaps digit span does not involve attention to any great degree. However, it is beyond the scope of this paper to determine how to assess the involvement of attention. It is assumed that attention tasks, such as digit span, do assess attention.

Nevertheless, there are some frontal lobe symptoms that are clearly not present in autistics. For example, frontal patients are known to confabulate and have persistent mood changes, including pseudodepression and pseudopsychopathy, autistics may only exhibit extremes of emotion and then only occasionally. Frontal patients also may have problems akin to autotopagnosia, being deficient at behaviors related to egocentric spatial orientation (Mayes, 1988). For example, they cannot point accurately to parts of their bodies as instructed, while having no trouble finding their way around a room via a map. Autistic people are better at spatial tasks than verbal (Lincoln, et al 1988), as evidenced by their performance on the block design and object assembly sections of the WAIS IQ test, whereas frontal patients performed relatively poorly on block design, as well as picture arrangement and digit span (Janowsky, et al, 1989).

Resolution of Discrepancies. However, that there should be significant differences between autistic behavior and that of frontal patients should be expected, even if autism does involve abnormalities in frontal lobe function, After all, the frontal lobe syndrome is usually seen as the result of an extensive cortical lesion in adulthood, whereas autism is thought to be a developmental disorder (Gillberg, 1990). The former results from damage to a mature, normally developed system. The latter may result from the development of abnormal connectivity. Abnormal neuronal connectivity of the frontal lobes, and/or subcortical structures subserving the frontal lobes, may result in symptoms that mimic extensive lesions to frontal cortex in adulthood. In relation to this idea for mediotemporal lobe amnesia, it may also be that some of the differences between autism and organic amnesia are due to the differences that result from a developmental lesion as opposed to one received after maturity. Nevertheless, autism seems to share more cognitive abnormalities in common with frontal lobe syndrome, than with mediotemporal lobe amnesia, though there may in fact be a composite of causes of the autistic disorder, involving both mediotemporal lobe and frontal lobe function. The diversity of lesions classified as frontal, as well as difficulties with diagnosing autism, may then contribute further to discrepancies between frontal lobe syndrome and autism.

Emergence of Higher Order Cognitive Impairment

Assuming that at least some of the symptoms of autism may be attributed to frontal lobe dysfunction, it is relevant to ask whether this could account for any, some, or all of their deficits in higher order cognitive abilities, such as language acquisition, concept formation, categorization, and symbolic representation. Such comparisons are complicated by the fact that in neither autism nor frontal lobe syndrome are the fundamental cognitive deficits characterized, let alone any more complex abilities.

However, one way in which an attentional deficit could result in the complex of cognitive deficits found in these groups may be suggested by a study by Cohen, Ivry, and Keele (1990). They have shown that attention is required for the learning of complex sequences of stimuli. Simpler sequences, in which the previous symbol determines identity of the next symbol, eg. 123123···, may involve the construction of only simple associations between stimuli, but more complex sequences, in which there are multiple, yet constrained, possibilities for the identity of the next symbol, eg. 132312132312···, require the formation of a hierarchical representation. Such less constrained sequences cannot be achieved without attention. Attention may serve to break down the sequence into component parts to which a higher level description may be attached, these higher level descriptions being more determined. This description acts as an additional cue that makes manipulation of the underlying information easier.

These ideas are consistent with the findings of Hermelin (1976) that autistics do not tend to engage in rule extraction to aid them in learning visual sequences. Rather, they learn them by rote. Hermelin's work also suggests that autistics tend to order sequences spatially rather than temporally, the most common strategy in controls. This is consistent with the frontal lobe deficit with temporal order (Shimamura, Janowsky, and Squire, 1991). Additional evidence for a parsing disability comes from work that suggests that autistics may process sensory information centrally in a fundamentally different way (Ornitz 1975, O'Connor, 1975, DeMyer, 1975), resulting in an inability to organize information into modality independent codes. Normally, through the extraction of rules and redundancies, one may arrive at an appropriate integration and interpretation of the components of a perceptual experience. Such codes could be used to reduce information load (Hermelin, 1976), or to facilitate understanding or whatever behavior in which one is engaged. Overall, this research supports the contention that autistics may not be able to use attention to parse an experience in a way necessary to generate such simplifying codes or rules, and is consistent with Prizant's (1983) finding that autistics are impaired at sentence segmentation. Therefore attention plays an essential part in determining the structure of the mind's knowledge. This structure is less a property of the static entity of memory than the result of the dynamic action of attention on the learning experience over space and time. It is with such dynamics that autistics have particular difficulty.

In terms of higher cognitive functions, several studies have shown that autistics have trouble manipulating information, including symbols (Ricks and Wing, 1976). Tager-Flusberg (1985) found that autistic people do not use their acquired cognitive skills in a flexible and appropriate way. While she also found that organization of their semantic knowledge for concrete objects is largely intact, this need not warrant the conclusion that autistics are not deficit in their use of attention at encoding. After all, it takes them an unusually long time to learn these concrete words, and this study did not examine their abilities with abstract words. Prizant's (1983) work on the autistic style of language acquisition suggests that they have a general deficit in using their knowledge and cognitive abilities. While work on concrete versus abstract words are sparse (Hobson & Lee, 1989; Eskes et al, 1990), most researchers consider autistics to be impaired at making abstractions (Ricks & Wing, 1975; Prizant, 1983; Tager-Flusberg, 1985; Hobson & Lee, 1989). Another demonstration of this disability is the tendency of autistic persons to use a holistic rather than an analytic approach (Prizant, 1983). This is true of their language acquisition and of other learning domains. They do not tend to parse their experiences into meaningful, structured components. Rather, autistics form context-bound holistic representations with little meaning, especially for the parts. Inflexible use of any representation formed may indicate that they cannot focus attention on their internal representations efficiently enough to manipulate them, nor can they parse and code representations. Thus, the flexibility of cognition may depend upon attentional capacity.

Conclusions

It has been seen that the autistic syndrome results in a complicated composite of cognitive deficits. It has been argued in this paper that the best way to understand such complexity may be to strip it down to its most fundamental elements. While most researchers have focused on language, this is not likely to be the most basic cognitive ability (Bates, 1990). Thus, it seems reasonable that one should look elsewhere for more basic aspects of cognition. Memory and, particularly, attention are put forth as the most fruitful alternatives. Additionally, comparisons between autistic performance and other patient populations may serve to generate ideas as to the nature of the deficit in either disorder. In terms of cognitive mechanisms, research into autism has tended to be like the deficit itself. Its approach perhaps tends to be holistic and devoid of the foundational meaning that could be provided by an emphasis on cognitive abilities, like attention and memory. Study of the cornerstones of cognition may ground research on autism, contributing to the elucidation of the primary behavioral characteristics which give rise to the more complex behavioral deficits in language and thought seen in the autistic syndrome.

Acknowledgments

Thanks to Dr. Eric Courchesne for his course on Cognitive Development and for helpful comments and encouragement in the preparation of this manuscript.

References

Ameli, R., Courchesne, E., Lincoln, A., Kaufman, A.S., and Grillon, C. 1988. "Visual Memory Processes in High-Functioning Individuals with Autism. *J. of Aut. and Dev. Dis.* 18(4):601-615.

Baron-Cohen, S. 1989. The Autistic Child's Theory of Mind: a Case of Specific Developmental Delay. *J. of Child Psych. and Psychia.*. 30:285-297.

Boucher, J. 1981. Memory for Recent Events in Autistic Children. *J. of Aut. and Dev. Dis.* 11:293-301.

Boucher, J. 1988. Word Fluency in High-Functioning Autistic Children. *J. of Aut. and Dev. Dis.* 18(4): 637-645.

Boucher, J. and Warrington, E.K. 1976. Memory Deficits in Early Infantile Autism. *Brit. J. of Psych.*. 67(1):73-87.

Cohen, A., Ivry, R.I., and Keele, S.W. 1990. Attention and Structure in Sequence Learning. *J. of Exp. Psych.*. 16(1):17-30.

DeMyer, M.K. 1971. Perceptual Limitations in Autistic Children and Their Relation to Social and Intellectual deficits." Rutter, M. In *Infantile Autism: Concepts, Characteristics, and Treatment*, 81-95. Edinburgh, London: Churchill Livingstone.

Eskes, G.A., Bryson, S.E., and McCormick, T.A. 1990. Comprehension of Concrete and Abstract Words in Autistic Children. *J. of Aut. and Dev. Dis.* 20(1) :61-73.

Garretson, H.B., Fein, D., Waterhouse, L. 1990. Sustained Attention in Children with Autism. J. of Aut. and Dev. Dis. 20(1):101-114.

Gedye, A. 1991. Frontal Lobe Seizures in Autism. *Med. Hyp.*, 34:174-182.

Gillberg, C. 1990. Autism and Pervasive Developmental Disorders. *J. of Child Psych. and Psychia.*31(1): 99-119.

Goldman-Rakic, P.S. and Friedman, H.R. 1991. The Circuitry of Working Memory Revealed by Anatomy and Metabolic Imaging. In *Frontal Lobe Function and Dysfunction.* 72-91. New York: Oxford University Press.

Heltzer, B.E. and Griffin, J.L. 1981. Infantile Autism and the Temporal Lobe of the Brain. *J. of Aut. and Dev. Dis.* 11:317-330.

Hermelin, B. 1976. Coding and the Sense Modalities. In *Early Childhood Autism*, 135-168. Pergamon Press.

Hobson, R.P. and Lee, A., 1989. Emotion-Related and Abstract Concepts in Autistic People: Evidence from the British Picture Vocabulary Scale. *J. of Aut. and Dev. Dis.* 19(4):602-623.

Janowsky, J.S., Shimamura, A.P., Kritchevsky, M., Squire, L.R., 1989. Cognitive Impairment Following Frontal Lobe Damage and Its Relevance to Human Amnesia. *Beh. Neurosci.* 103(3):548-560.

Kanner, L. 1943. Autistic Disturbances of Affective Contact. *Nerv. Child.* 2:217-250.

Killackey, H., 1987. Neocortical Evolution.

Lincoln, A.J., Courchesne, E., Kilman, B.A., Elmasian, R., and Allen, M., 1988. A Study of Intellectual Abilities in High-Functioning People with Autism. *J. of Aut. and Dev. Dis.*, 18(4):505-523.

Mayes, A.R. 1988. *Human Organic Memory Disorders.* New York: Cambridge University Press.

O'Connor, N. 1971. Visual perception in autistic childhood. In *Infantile Autism: Concepts, Characteristics, and Treatment*, 69-80.

Ornitz, E.M. 1971. Childhood Autism: A disorder of sensorimotor integration. In *Infantile Autism: Concepts, Characteristics, and Treatment*, 50-68. Edinburgh, London: Churchill Livingstone.

Prizant, B.M., 1983. Language Acquisition and Communicative Behavior in Autism: Toward an Understanding of the "Whole" of it. *J. of Speech and Hear. Dis.* 48:296-307.

Ricks, D.M. and Wing, L., 1975. Language, Communication, and the Use of Symbols in Normal and Autistic Children. J. *of Aut. and Child. Schizo.* 5(3):191-221.

Shimamura, J.P., Janowsky, J.S., and Squire, L.R. 1991. What Is the Role of Frontal Lobe Damage in Memory Disorders? In *Frontal Lobe Function and Dysfunction*, 180-195. New York: Oxford University Press.

Squire, L.R. 1987. *Memory and Brain.* New York: Oxford University Press.

Tager-Flusberg, H., 1985. The Conceptual Basis for Referential Word Meaning in Children with Autism. *Child Dev.* 56: 1167-1178.

Collaborative Mediation of the Setting of Activity

Penelope Sibun **Jeff Shrager**

System Sciences Laboratory
Xerox Palo Alto Research Center

Abstract

Various aspects of task settings, including the actors and the physical environment, interact in complex ways in the construction and selection of action. In this paper, we examine the process of collaborative mediation, that is, how collaborators facilitate activity by making aspects of the setting available or accessible to the principal actor. We investigate collaborative mediation in three activities: verbal descriptions of strongly structured objects, such as one's house; cooperative computer use; and parent-child cooking. In each of these cases, the collaborator's role with respect to the principal actor and the rest of the setting differs, but they are all of similar kind. The collaborator makes available different aspects of the setting (physical setting, goals, tests of success, etc.) as needed at appropriate moments, thus helping to operationalize goals via physical guidance, advice, indication of aspects of the setting to make them accessible or relevant, or the taking of initiative which moves the activity forward more directly. Our analysis elaborates the methods by which agents can mediate one another's construction of the settings in which they find themselves, and so facilitate successful activity. We thus extend and generalize similar analyses and approach a general theory.

Introduction

In this paper we investigate the role of collaboration in activity. Activity is embedded in a setting that includes actors, their goals, the physical environment, and perhaps other aspects. These aspects interact in complex ways in the construction and selection of action. Collaboration is one form of this interaction. Consider the case of a linguist trying to find information about a particular method of linguistic analysis via a computerized database. If the linguist is not an expert user of the database, she will probably be assisted by a research librarian. The librarian's skills contribute significantly to the activity of looking up information. For example, since the librarian knows the structure of the database and what sorts of queries are appropriate (or have worked well in the past), he can guide the linguist's use of the system in a number of ways that are crucial to the success of the whole enterprise.

Authors' addresses are care of Xerox PARC;
3333 Coyote Hill Road; Palo Alto, CA 94304.
Email: Sibun@Xerox.com; Shrager@Xerox.com.

For purposes of analysis, we shall distinguish a *principle actor* who is taking particular actions at particular moments. In addition to the principal actor, we define three further parts of the setting: the domain of discourse; the task setting; and the collaborator set. The *domain of discourse* is approximately the subject of activity (e.g., the topic being discussed, the global goals of the activity). The *task setting* includes the physical and historical aspects relevant to the activity (e.g., the tools at one's disposal, things that have already been said). The *collaborator set* includes other intentional actors that participate in the task. When this set contains a single member, we will refer simply to the *collaborator*. In the above example, from one point of view, our hypothetical linguist is the principal actor; the databases system and its interface constitute the task setting; the domain of linguistic analysis and the linguist's local requirements constitute a likely domain of discourse; and the research librarian is the collaborator. It is important to note that this set of analytic categorizations will vary over the course of activity, and with different points of view. So, for instance, if the participants have a conversation about the database interface itself, then the domain of discourse is no longer linguistic analysis, but perhaps the windows on the screen and their functions. Similarly, we could reanalyze the setting, taking the research librarian as principal actor and the linguist as collaborator.

It is clear that the aspects of the setting are not entirely separate; indeed, it is central to the present project that these do not have separate existences, but rather that they co-construct one another. In a previous study, Agre and Shrager (1990) examined the fine tuning of the complementarity of the principal actor and the domain of discourse and task setting, involving an office worker and a copier. Aspects of the worker's physical and (presumably) mental activity evolved with respect to the rhythms of the copier to produce efficient joint activity. In this paper, we are particularly concerned with the way in which the collaborator set "mediates" the task setting; that is, how collaborators make aspects of the setting available or accessible to the principal actor. We shall use the term *collaborative mediation* for this process.

We have investigated collaborative mediation in three activities: verbal descriptions of strongly structured objects, such as one's house (Sibun, 1991; Sibun, 1992); cooperative computer use, such as that described in the above example; and parent-child cooking (as studied by Shrager & Callanan, 1991).

Three Cases of Collaborative Mediation

The three cases in which we shall examine collaborative mediation lie along a dimension of the role of collaboration in activity: from the relatively passive role of interlocutor in a description activity, through the more active role of assistant in an information access activity, through the very proactive role of parent in a parent-child cooking activity. In each of these very different situations we will identify ways in which collaborators make aspects of the setting available to the principal actor, or make them relevant to the moment.

Description Production

Consider describing your home to another person. The particulars of your home, especially the physical layout, are clearly relevant to what you will say, but what the listener knows, how he or she interacts with you, and other aspects of your shared knowledge are also relevant to the structure of the description. The general form of such descriptions involves the principal actor constructing for an audience text that reflects the structure of the house. The domain of discourse is the house and the collaborator is the person requesting the description. In this case, the task setting is largely irrelevant for the present analysis; its most interesting feature is a tape recorder. We chose this relatively non-interactive version of conversation instead of, say, task-oriented dialogue (e.g., Grosz & Sidner, 1986) because the latter is evidently co-constructed. While the interlocutor of a description is a relatively passive collaborator, all of the examples of house description that we have collected show evidence of participation by the collaborator, even when he or she is trying not to take part in a dialogue.

We show two examples in which collaborative mediation takes place. The text fragments are drawn from descriptions given by people who had spent significant time in a particular house. Each was answering the question: "Can you describe for me the layout of [this] house?" (See Sibun, 1991, for more details and complete transcripts.) In the italicized portions of the fragment in Figure 1, the principal actor explicitly indicates that the form of his description has been affected by his knowledge that the collaborator lives in the house, and presumably is familiar with it. In the fragment in Figure 2, the principal actor not only expects, but insists on input from the collaborator in accomplishing the task.

In these examples, what is being mediated by the collaborator is the principal actor's access to aspects of the domain of discourse, or his interpretation of the relevance of these aspects. That is, in these cases, the seemingly passive listener is actually highly relevant to the speaker, and is particularly relevant in helping the speaker decide, in the first example, what information to give, and in the second example, how to envision the house.

Human-Assisted Information Access

As part of a project to provide computational assistance in information access (approximately, database search), we studied human-assisted information access. Xerox PARC researchers were solicited for help in the study. The first author was familiar with the database system and acted as an assistant. Researchers produced their own goals for the search task. For the present analysis, the searcher will be considered the principal actor, and the assistant will be considered the collaborator. The task setting includes the structure and the content of the databases, the layout of the interface (e.g., where the windows are located on the screen, when buttons are available and what actions they would invoke), and, peripherally, the physical setting (an office) in which the searcher and assistant work.

The clearest examples of mediation in this domain arise from the differential skill and knowledge of the principal actor and collaborator. In the fragment in Figure 3, the assistant helps the searcher translate his search desires into actions in the database interface. The researcher is interested in querying a database of Xerox information to find out how many researchers there are at PARC. The assistant, because she knows both the content of the database and the types of queries

....then there's—there's kind of a big central....room-thing
I mean like when you come in
....this seems very strange telling you this Penni
[...]
....and there's an outside entrance to the—to the basement
....and I guess that's how I would describe the layout of your house
although if I were probably describing it to anyone else
I might have given a little more size information

Figure 1: A description fragment in which the speaker explicitly indicates that the form of his description has been affected by his knowledge that the collaborator lives in the house.

Claire:	my room is....a little longer than wider but it looks pretty square it has two windows....sort of at this one corner that is pretty much diagonally opposed to the door which goes outside to the little hallroom before the bathroom and Ann's room the bathroom is fairly uh....not square *I don't know, does this—the bathroom stick out?*
Penni:	*no!*
Claire:	*it doesn't?*
Penni:	*I don't think you're supposed—*
Claire:	*it doesn't?*
Penni:	*—to ask me though!*
Claire:	*well, I don't—ok—anyway* *so there's the bathroom that has one adjacent wall to mine....*

Figure 2: A description fragment in which the principal actor insists on input from the collaborator.

that would be successful, points out that this query is unlikely to be usefully answered, and suggests instead a query that would search for documents that mention a particular topic. They settle on [linguistics].[1] Notice that in this case the assistant helped the searcher structure his expectations of the search facility by indicating certain capacities that it does not have. Although the searcher had some initial idea of what he was interested in, the goals of the search are jointly developed by virtue of the assistant knowing for what sorts of questions the database is relevant, and how the system will respond to different sorts of queries.

In the fragment in Figure 3 the domain of discourse is initially the domain of research at PARC, but shifts to focus on the searcher's goals and questions and how they related to the contents of the database, as the assistant and searcher negotiate what precisely should be done. The shifting around of the domain of discourse is a common feature of collaborative mediation. The domain of discourse often shifts to the system and the interface, since the searcher is unfamiliar with both, and requires explanations from the assistant.

The fragment in Figure 4 exemplifies this sort of mediation. The searcher enters into this portion of the task wishing to find out "which laboratories (at PARC) do research on [linguistics]"(Figure 4). We enter this fragment after searcher and assistant have discovered one document that seems particularly relevant, and then want to search through that document for the section on research on [linguistics]. Unfortunately, the facility for searching within a document looks in the selected document for *any* of the terms from the original query (including "on" and "do"), thus leading to an enormous amount of useless search. More importantly, in the context of a document that is already relevant to the query, searching once more for the query terms is not usually helpful in locating interesting parts of the document (unless the selected document is very heterogeneous).

We see a progression through domains of discourse; moving from discussion of the topic of

[1]This is not the actual topic. It has been changed for reasons of privacy and clarity. In the protocol fragments we have noted this change by enclosing the modified text in [brackets].

Assistant:	why don't you ask another question [...] another PARC-related question [...]
Searcher:	another PARC-related question
Assistant:	ahhm something that—now that you've seen a little bit— if it's a more general question I think we'll have better luck
Searcher:	okay I okay I got one [...] "how many researchers at PARC?"
Assistant:	know what? I bet that's gonna be— have a similar thing that it's going to mostly hit on "PARC"— I guess it'll hit on "PARC" and "researchers" it may or may not come up with....a number
Searcher:	okay so you want you think—therefore I should try something else?
Assistant:	yeahhh.... [...] if you ask about um like [linguistics] at PARC or....um....
Searcher:	mmhm by content rather than structure
Assistant:	yeah more more content than particular facts
Searcher:	okay....um....okay...."which laboratories do research [in linguistics]"
Assistant:	okay

Figure 3: An assistance fragment in which the assistant helps the searcher translate his search desires into action in the database interface.

	[The searcher and assistant have located a number of documents relevant to the query: "which laboratories do research [in linguistics]?"]
Searcher:	I would guess that "Research Overview" is probably gonna contain something [They select the Research Overview document for detailed examination.]
Assistant:	okay well hit on "Find Key" and see which key it's looking for... [The Find Key button looks for any of the query keywords in the selected document and highlights them in turn.]
Assistant:	"research"....well that's that's good you can now—
Searcher:	where did you see that....oh I see
Assistant:	it lights it up
Searcher:	okay
Assistant:	um so I think if you hit "Find Key" again it will [The searcher continues to press the Find Key button.]
Assistant:	....yes....keep going through with "research" so that....
Searcher:	"research centers"...."Palo Alto Research Center" [...] well here we finally found "[linguistics]" in the title [...] *once I got into this thing probably I'd want to search for the word "[linguistics]"* *I wouldn't want to search for "research"* *or I mean—once we know the context of this document is "research"*
Assistant:	right
Searcher:	and then and I wanna narrow on—narrow in on "[linguistics]"

Figure 4: An assistance fragment exemplifying shifts the in domain of discourse.

desired information, through how to implement the search, discussion of the resulting documents, methods of searching through chosen documents for relevant parts, and finally to discussion of problems with the method (italicized portion). This progression is facilitated by, and in part driven by, the collaborator (the assistant).

Parent-Child Cooking

Child development has long been understood as collaborative. Indeed, it is the origin of at least one central thread of research on activity theory (cf. Vygotsky, 1978; Wertsch, Minick, & Arns, 1984). More recently Rogoff (1990) has characterized the interactionist approach to child development by way of an apprenticeship metaphor. Apprenticeship is inherently collaborative. It is therefore particularly interesting to see how collaborative mediation is deployed in a developmental setting. One version of mediation in development has been advanced by Bruner (1983) and by Wood (1980). They describe the processes of "scaffolding" in which an adult (the collaborator) gives over portions of a task that are doable by the child (the principal actor in this analysis), and structures the setting so that the child can accomplish those aspects and move on to more complete skill. This is similar to the way in which the database search assistant guides the researcher in using the database interface. We shall see that there are other more subtle mediating processes at work in the parent-child setting.

Shrager and Callanan (1991) studied parent-child dyads engaged in baking raisin bran muffins in the family kitchen. Significant changes in the collaborative structure of the activity were observed, and the naturally-occurring "active language" taking place in the setting was examined to identify the various roles that are played by language in such settings of activity. Five functions of active language were identified: object and action labeling; sequencing of expectations (procedure organization); task structuring articulations; explication of non-obvious aspects (e.g., goals and causes) and focusing on relevant aspects; and interaction facilitating articulations. For the purposes of the present paper we are concerned mainly with the function identified as "focusing on relevant aspects" (of the activity).

A number of examples of such focusing can be seen in the collaborative cooking data. It is generally the case in these studies that the parent and child are jointly focussed on a particular object, say, the measuring spoons that are being used to add baking soda to the mixture. The parent has a number of methods by which he or she can obtain, check, and manipulate the child's focus, such as waving the spoons in front of the child to grab his or her attention, and taking the child's hand and touching it to the spoons.[2] These are often (though not always) accompanied by verbalizations that include explanations of what is being pointed out or accomplished. In these ways the parent (the collaborator in the present analysis) is emphasizing certain aspects of the physical setting (part of the task setting), and in so doing is facilitating the joint activity. (Some version of baking activity would go on without these interactional resources, but it might not be very easily understood as collaborative—the child most likely not being very closely engaged with baking.)

In this case, the parent and child are very closely engaged, and almost all aspects of the setting are made available to the child by way of mediating activities of the parent such as naming, explanation, and indication (making aspects of the setting accessible or relevant). (Examples of these are given in Shrager and Callanan, 1991.) The parent's proactive guidance is largely responsible for keeping the task on track, and for enabling it as a collaborative activity.

[2]Transcribed examples of these activities appear in Shrager and Callanan, 1991.

Discussion

We have examined collaborative mediation in three cases that lie along a dimension of the role of collaboration in co-activity, from the relatively passive role of interlocutor in the description activity, through the more active role of assistant, to the very proactive role of parent as assistant, guide, and tutor in baking. In each of these very different situations we are able to characterize ways in which a collaborator makes aspects of the task domain or setting available (or relevant to the moment) for the principal actor. The case of parent-child cooking is perhaps the most obvious; here it is necessary for the parent to do a great deal of explicit mediation of the setting by way of naming, explanation, indication, etc. The parent's proactive guidance is largely responsible for maintaining the directionality of the task, and for enabling it as a collaborative activity. In the case of human-assisted information access, the searcher's access to a significant part of the setting can only be accomplished through the assistant, and both participants act to take advantage of the mediating role of the assistant in this function. The assistant's role enabled the searcher to refine his desires into the specific operations both necessary and sufficient for implementation of the search activity.[3] The case of giving a description is in many ways the most subtle example of mediation. The collaborating person, even in playing the relatively passive role of audience, is still clearly a part of the setting, as evidenced by the appeals of the principal actor to the collaborator.

The collaborator's role with respect to the principal actor and the rest of the setting is similar from one activity to next: actions of the collaborator shape and enable actions of the principal actor, thus facilitating the overall activity. More specifically, the collaborator makes available different aspects of the setting (physical setting, goals, tests of success, etc.) as needed at appropriate moments. This mediation helps to operationalize the princi-

[3]Note once again that our analysis from the point of view of principle actor and collaborator(s) is merely an analytic stance; this analysis can be carried out from any chosen point of view, or from no individual point of view at all. It is rather more complicated to speak about it, though, in the case where there is no individual point of view.

pal actor's goals via physical guidance, advice, indication (making aspects of the setting accessible or relevant), or the taking of initiative to move the activity forward. Our analysis of the functions of collaboration in the construction of activities elaborates the methods by which agents can mediate one another's construction of settings. We thus extend and generalize similar analyses (e.g., Agre & Shrager, 1990; Vygotsky, 1978; Wertsch, Minick, & Arns, 1984) and approach a general theory.

There are many other, more subtle, cases of co-construction of these domains. The most interesting is, perhaps, the construction of which parts of the task setting (which might include people) shall be nominated as "collaborators" and thus, by our definition, become a part of the collaborator set. Thus nominated, a collaborator can take part in the processes of mediation that we have identified. It is interesting to ask what capacities an agent must have in order to be elected to collaborator status. Latour (1988) has analyzed the social role of a mechanical "door-closer"—the hydraulic and spring device that pulls a door closed after one has walked through it. This simple device is hardly a collaborator despite its social role. Rather than mediating the setting for an actor, the door-closer is simply changing the structure of the setting by absolving one of the requirement of pulling the door closed after oneself. Some computational systems exhibit collaboration in a simple sense. A number of systems that attempt to provide mixed-initiative advice to users of computer systems (e.g., Shrager & Finin, 1982), and so-called "learning apprentice" systems attempt to learn the common procedures used by users and then to propose operations in the form of advice when later similar contexts arise (e.g., Mitchell, Mabadevan, & Steinberg, 1990).

The present analysis sheds some light on additional capacities that may be required of such computational agents if they are to become fully-fledged collaborators. A crucial facility for such collaborative systems will be the negotiation of domains of discourse. Such negotiation seems to require the maintenance of joint attention, which may be maintained either by linguistic communications (as in the cases of description and human-assisted information access), or by a number of physical means (as in the case of parent-child cooking).

References

Agre, P. & Shrager, J. (1990). Routine evolution as the microgenetic basis of skill acquisition. Proceedings of the Annual Conference of the Cognitive Science Society. Hillsdale, NJ: Lawrence Erlbaum Associates. 694-701.

Bruner, J. (1983). Child's Talk. New York: W. W. Norton.

Grosz, B. & Sidner, C. (1986). Attention, intentions, and the structure of discourse. Computational Linguistics, (12)3, 175-204.

Latour, B. [Jim Johnston] (1988). Mixing humans and nonhumans together: The sociology of a door-closer. Social Problems. 35(3), 298-310.

Mitchell, T. M., Mabadevan, S. & Steinberg, L. I. (1990). A learning apprentice for VLSI design. In Y. Kodratoff & R. Michalski (Eds.), Machine Learning, Volume III. San Mateo, CA: Morgan Kaufmann. 271-301.

Rogoff, B. (1990). Apprenticeship in Thinking. Oxford University Press.

Shrager, J. & Callanan, M. (1991). Active language in the collaborative development of cooking skill. Proceedings of the Annual Conference of the Cognitive Science Society. Hillsdale, NJ: Lawrence Erlbaum Associates. 394-399.

Shrager, J. & Finin, T. (1982). An expert system that volunteers advice. Proceedings of the National Conference of the American Association for Artificial Intelligence, 339-340.

Sibun, P. (1992). Generating text without trees. Computational Intelligence: Special Issue on Natural Language Generation, 8(1), 102-122.

Sibun, P. (1991). Locally Organized Text Generation. COINS Technical Report 91-73, Department of Computer and Information Science, University of Massachusetts. Also, Xerox Palo Alto Research Center report no. SSL-91-21/P91-00159.

Vygotsky, L. S. (1978). Mind in Society: The Development of Higher Psychological Processes. Harvard University Press.

Wertsch, J. V., Minick, N., & Arns, F. J. (1984). The creation of context in joint problem-solving. In B. Rogoff & J. Lave (Eds.), Everyday Cognition. Harvard University Press. 151-171.

Wood, D. J. (1980). Teaching the young child: Some relationships between social interaction, language, and thought. In D. R. Olson (Ed.), The Social Foundations of Language and Thought. New York: Norton. 280-296.

Incremental Reminding: the Case-based Elaboration and Interpretation of Complex Problem Situations

Brian M. Slator and Ray Bareiss

The Institute for the Learning Sciences
Northwestern University, Evanston, IL 60201
{slator,bareiss}@ils.nwu.edu

Abstract

When solving a complex problem, gathering relevant information to understand the situation and imposing appropriate interpretations on that information are critical to problem solving success. These two tasks are especially difficult in weak-theory domains -- domains in which knowledge is incomplete, uncertain, and contradictory. In such domains, experts may rely on experience for all aspects of problem solving. We have developed a case-based approach to problem elaboration and interpretation in such domains.

An experience-based problem-solver should be able to incrementally acquire information and, in the course of that acquisition, be reminded of multiple cases in order to present multiple viewpoints to problems that present multiple faults. We are addressing issues of 1) elaboration and interpretation of complex problem situations; 2) multiple interpretations; and 3) the role of categories as the foci of reasoning in the context of the Organizational Change Advisor (ORCA). Its model of incremental reminding is a plausible mechanism for this sort of expert problem solving behavior, and one that works well in weak theory domains. Because there is an implicit cost associated with retrieving a complex case, ORCA implements a retrieval time similarity function that requires both general expectations and specific situational relevance be considered before a story is told to the user; this increases the chances that a retrieved case will be useful.

1. Introduction

A "weak-theory" domain, such as business or law, is characterized by a lack of reliable general principles: knowledge is incomplete, uncertain, and even contradictory (Porter, Bareiss, and Holte, 1990). Expert problem solving in such contexts often involves much more than routinely gathering data and generating answers; it involves interpreting a complex problem situation in, perhaps, many ways. Experts answer questions and tell stories; they explore alternative hypotheses and implement intermediate solutions; they also gather further data and revise their assessments, and then incrementally produce new and improved solutions to partially solved problems.

Case-based reasoning (CBR), has been proposed as an effective means of elaborating and interpreting a complex situation (e.g.. Simpson, 1985). In this context, elaboration means acquiring the features necessary to form an interpretation, and interpretation amounts to categorizing a situation as an instance of a known problem. A case serves as a specific model[1] for interpreting a situation and tells the problem-solver which features to attend to (out of a potentially huge range of possibilities) and how important their presence (or absence) might be.

Most implementations of CBR have implicitly assumed that the use of a single retrieved case is sufficient to solve a problem. If another case needs to be retrieved, it is because the current one proved to be inappropriate. This is similar to the single fault assumption in diagnosis (see, for example, MEDIATOR: Simpson, 1985; Kolodner et al., 1985; CHEF: Hammond, 1989).[2] However, there are many complex problem situations that would seem to demand multiple interpretations where, because of the presence of multiple faults, a single retrieval will not do. To account for these situations, we make a weaker assumption: the non-interacting problem assumption,

This research was supported in part by the Defense Advanced Research Projects Agency, monitored by the Air Force Office of Scientific Research under contract F49620-88-C-0058 and the Office of Naval Research under contract N00014-90-J-4117, by the Office of Naval Research under contract N00014-89-J-1987, and by the Air Force Office of Scientific Research under contract AFOSR-89-0493. The Institute for the Learning Sciences was established in 1989 with the support of Andersen Consulting, part of The Arthur Andersen Worldwide Organization. The Institute receives additional support from Ameritech, an Institute Partner, and from IBM.

[1]We use the term model in the sense of Weiss (Casnet; 1978) or Nii et al. (Sonar interpretation; 1982) rather than in the qualitative reasoning sense that the term often now connotes in AI.

[2]Ashley and Rissland (HYPO; 1987) do not attempt to select the single most closely matching case but rather to retrieve all cases which match (or nearly match) the current case on any relevant underlying dimensions; Redmond (CELIA; 1990) does not in principle, but does in published examples; Hinrichs and Kolodner (1991) retrieves a number of cases, then these are decomposed and pieces of several might be employed in a synthesized solution.

i.e., that multiple faults do not interact in such a way that the diagnostically significant features of any problem are masked.

This paper describes a system which implements a multiple-retrieval CBR approach to problem interpretation and elaboration based on iteratively developing a picture of the problem situation. Rather than entering a script-based initial data-gathering dialog to create a featural description of the problem situation, we retrieve a case that will form a solid beginning, and then successively elaborate the problem description through comparison with a sequence of cases retrieved in response to the results of previous comparisons. In other words, a picture of the problem situation is constructed incrementally by comparing and contrasting it with multiple stored cases; and as more stored cases are retrieved, more is learned about the problem, and the interpretations become more "on point".

A typical way to tackle multiple retrieval case-based reasoning is through "difference link" refinement. In these systems (e.g. Protos: Bareiss 1989; Julia: Hinrichs and Kolodner, 1991, and MEDIATOR: Simpson, 1985; Kolodner et al., 1985), this requires a memory in which cases are relatively indexed by significant differences, in order to cache the results of multiple retrievals during problem solving, to provide a shortcut in a similar future episode. With difference links, acquired cases are finely distinguished from each other and problem solving is accomplished by categorization, which is achieved by traversing difference links until an acceptably matching case is located.

However, there are special problems posed when the "weak theory" domain is extremely broad and complex, and the number of distinct cases is large. In this event, the sheer combinatorial magnitude of creating difference links is prohibitive, since evaluation of all the pairwise possibilities simply cannot be accomplished. For domains of this type another strategy must be implemented. Rather than traverse difference links, which are unlikely to exist if they have not been extensively pre-enumerated (because it is unlikely that a similar enough situation was previously encountered), we make a series of retrievals based on the featural differences of the problem situation and the retrieved cases.

2. The Problem/Task Domain

We are studying these issues in the context of organizational change consulting. An "organizational change" consultant is typically contacted by a company when a significant event has occurred that is outside of the company's expertise. The consultant is hired to assess the state of the company, to diagnose its problems, and then to recommend and, typically, to implement changes in the company's structure or way of doing business. This is an extremely complex problem, and in a weak-theory domain such as organizational change, there is no substitute for experience; even though past experiences may not exactly mirror a client's situation, they provide useful analogies illustrating solutions tried in previous situations and how well they have worked.

A knowledge-based system to assist change management consultants should: 1) encourage systematic exploration of a complex problem by asking relevant, context-sensitive questions; 2) suggest necessary information to acquire from the client; 3) propose hypotheses about the client's problems when evidence of relevance is uncovered; 4) present analogies to relevant past cases from the consulting firm's corporate memory; 5) offer assistance in solving the client's problems by providing actions and outcomes associated with past cases.

3 ORCA: Organizational Change Advisor

The Organizational Change Advisor (ORCA; Bareiss and Slator, 1992) is a consulting aid and advice giving program that interacts with a consultant to gather information about a client, and then tells stories that are found to be analogous to the client's situation. The goal of the ORCA system is to build a picture of a complex situation and then to interpret it by categorizing it in several different ways. To do this, ORCA gathers information about a client's situation by posing questions to the consultant. The answers to these questions build up a description in memory which, in turn, reminds ORCA of previous, similar cases, whose features provide expectations that translate to additional questions to ask. This elaboration leads to several different retrievals of previous cases, and these are used to build a description of the client's situation through a series of "follow-up discussions" that incrementally contribute to the further elaboration and building of the client case.

ORCA implements an algorithm for case-based interpretation to retrieve seemingly appropriate cases on the basis of weak remindings. To do this, ORCA operates over a memory of stories and domain elements connected to each other with "reminding" links (where reminding is a heuristic association between domain element such that finding one increases the likelihood of finding the other: i.e. the traditional notion of predictive indexing). ORCA employs the strategy of looking for the important features of the case in order to elaborate the problem situation. If they are found, the new situation can be interpreted as an instance of a known type of experience.

As the user answers questions about the relevance of domain elements to the client's situation, ORCA uses the answers to manage a queue of possibly relevant stories and, when sufficiently reminded of a particular story, shows it to the user as a case related in some way to the client's case. As new information is gathered, either from user input or as a consequence of the user revising their assessment of the client's case, the reminding network considers other types of problems and proposes further stories to the user. In this way, reminding produces expectations which are used to form queries, and it is the success or failure of these expectations that enables interpretation and efficient categorization.

3.1 Representation

ORCA's cases, gathered both from interviewing expert consultants and from searching professional journals, are intended to help the user consider realistic problem solving alternatives and to familiarize the user with the cases in memory most closely related to their client's situation. Unlike many case-based reasoners, however, cases are stories to be presented for use by a human user rather than fully represented entities for use in autonomous problem solving. The sole representation of a case is indexical; that is, the only case features accessible to the system are those used in making retrieval decisions. The bulk of the case, including its problem solving advice is stored as a block of text (or, in some cases, video) that is opaque to the system.

ORCA's memory contains both descriptive features of business situations and inferable abstract problem descriptions. The features are drawn from a vocabulary of descriptors of business situations developed by the consulting firm. These include direct observables, such as "a change in senior personnel has taken place" and reasonably straight-forward inferences, such as "friction exists between organizational units." The abstract problem descriptions include a set of concepts taken from the consulting firm's methodology and another set of common sense descriptions borrowed from the study of conventional, proverbial, wisdom. These abstract descriptions provide an explicit way of organizing the problem-describing features, and a way of interpreting the cases from the firm's corporate memory which are indexed by those features. ORCA represents categories extensionally as sets of retained cases. Rather than being exemplars of a single abstract problem type, ORCA's cases contain features relating to several different abstract problem types. In this way ORCA encodes the corporate reality that, to make a medical analogy, every patient typically has many diseases.

The abstract problem descriptions are based on an indexing strategy that encodes a notion of common sense societal wisdom (Owens 1990). For example, everyone knows that problems arise when an employee is made to report to more than a single boss: conflicting orders are given, time is wasted negotiating priorities, and in the worst cases, the bosses are dissatisfied and the employee is frustrated. Situations like this are so well-known that, over the centuries, society has developed and preserved a shorthand system of aphorisms to describe them. In a case like this, an impartial witness might observe, "No man can serve two masters" or, more picturesquely, "A pig with two masters will starve." In ORCA, proverbial expressions of this sort provide a framework of categories for organizing domain features and interpreting cases.

The basic index in ORCA's memory is the reminding link. ORCA also makes use of censor links to suppress remindings, but has no difference links (see Bareiss, 1989 for further description of these link types). The confirmatory link is a new index type that has been introduced to link abstract problem categories with features which tend to be confirmatory and hence, in general, worth pursuing when reminded of the category. During problem solving, determining the presence of a confirmatory feature in the client's situation suggests the relevance of presenting category instances to the user. In other words, confirmatory features are reasonable, general, things to ask about when reminded of a category.[3]

3.2 Elaboration and Interpretation

The role of a case in elaborating and interpreting a new situation is that the case serves as a specific model for acquiring features and imposing an interpretation. Under this scheme, an initial, possibly weak, reminding leads to an hypothesis which causes a retrieval. The retrieved case is used as a model for interpretation, providing expectations of additional features the case should possess. When the expectations are met, this provides an interpretation (i.e. a classification of the problem situation), and in any event, because cases are multiply classified, new remindings are produced which yield new hypotheses as possible interpretations to be confirmed.

ORCA has a two-level hierarchy of models: MOPs representing abstract problem types and cases representing particular experience. The typical flow of reasoning is that the system is reminded of an abstract problem type, asks questions to confirm its relevance, then retrieves an appropriate exemplar. No particular confirmatory feature is necessary to confirm the relevance of a category; however, one or more must be present. This method is in place because of the relatively high cost of working through a case; therefore, we require confirmation in terms of the norms of the category and in terms of superficial similarity, to assure the case is worth discussing.

Cases have prototypicality ratings with respect to abstract problem types. These qualitative ratings --- strong, medium, and weak --- partially order the cases as exemplars of the corresponding abstract problem types. When presented with a client situation, ORCA tries 1) to systematically acquire a picture of the situation by asking relevant questions, 2) to classify the situation repeatedly, as newly acquired information suggests relevant abstract problem types, and 3) to exemplify those problem types to the user by presenting similar past cases. The primary difficulty is that every situation that ORCA encounters will embody a multiplicity of

[3]These features are the "norms" of the category (Schank, 1982; Kolodner, 1984; Riesbeck and Schank, 1989).

problems. As a consequence, problem solving involves reasoning from a multiplicity of partially matching cases. ORCA is faced with solving a classification problem in which the cases are quite complex and the diagnostic assessment tools are weak, subjective, inconsistent, and inconclusive.

Classification is an iterative process of forming and revising a set of active hypotheses as the user answers ORCA's questions. At the beginning of an ORCA session, the user is asked to identify one or more significant business problems in the client situation, chosen from a pre-enumerated list of problems (i.e. merger or restructuring.) These are connected via reminding links to memory elements representing features and abstract problem descriptions. Each element of which ORCA is reminded is placed on a "best-first" agenda, and questions to confirm each, ordered by importance, are added to a question agenda. At each step in the problem solving process, ORCA asks the question at the head of the agenda.

Answering a question places its associated feature into a confirmation set, activates the associated remindings, and may cause all of ORCA's agendas to be re-evaluated. Generally, such a re-evaluation occurs as a result of a reminding link that associates the newly acquired feature with one or more abstract problem types. For example, if the user confirms that "the company's R&D staff is small," ORCA will be reminded of an abstract problem characterized as "he who looks not ahead looks behind." This reminding will cause ORCA to seek confirming features for the abstract problem, and the questions "Does the client neglect R&D in its planning and budgets?" and "Does the organization use outmoded equipment?" will be placed at the front of the question agenda.

As the features and proverbs are confirmed, the cases indexed by these elements are placed on their own agenda. Cases that come to the front of the agenda may be presented to the user when a heuristic estimation of their relevance exceeds a dynamically computed threshold of reminding strength. Prototypicality ratings suggest the relevance of cases to abstract problem types, but because of the extreme diversity of instances of abstract problem types, they are not the sole determinant of which case is presented. When the problem situation is interpreted to a point where ORCA is reminded of a category, cases for consideration are selected on the joint basis of prototypicality and featural match to the client's situation. No matter how prototypical a case is, however, it will not be presented unless there is some degree of featural match; for example, a story about a bank, and another about a taco stand, might both fall into the category exemplified by the proverb "he who looks not ahead looks behind." -- but a financial institution with the same problem will almost certainly profit less from hearing about the taco stand than from hearing the story of the bank. This two-level system imposes requirements to suppress distant analogies that may not be of great utility to a consultant.

After a case is chosen and its story is told, a follow-up dialog is entered in which the user is asked to elaborate the client's situation by comparing it to the various features of the story. (Some of these elements are already known to be true of the client, of course, since the story was shown on this basis to begin with.) The follow-up dialog amounts to listing the important descriptive elements associated with the story and asking which of these are reasonably associated with the client case. This comparison affords the user the opportunity to think about the most relevant features of their client in the specific context of a previous case. The presentation of past cases aids the user in determining the likelihood of various problems, provides advice by analogy, and provides tangible contexts for acquiring additional information about the client by prompting the user to make explicit comparisons. As discussed earlier, features acquired through these elaborative dialogs remind ORCA of additional possible interpretations of the situation.

4. An Example of ORCA in Action

The objective of an ORCA consultation is to create a case representing the client's situation. This is an iterative picture that develops over days and weeks, as the consultant is able to gather information about the client. At first, the consultant is only expected to answer general or surface-level questions, and often the proper answer to a question is "I don't know." Over time, by following up on ORCA's suggestions, the consultant will find the answers to specific questions to provide a detailed picture of the client's organization and its problems. Imagine that a staff consultant is called into his manager's office, and she tells him only the following basic information:

> *A manufacturing plant renowned for its innovation has encountered difficulty in maintaining its culture since competitive pressure forced the introduction of a "just-in-time" manufacturing system. The plant originally was organized around semi-autonomous production teams, who had been previously allowed to manage their own work at their own individual rates.*
>
> *Though some problems stem from the new system, most arise from the style of the new plant manager. Under pressure from corporate headquarters to use a system designed by an outside consulting firm, he severely limited employee involvement in the process. Employees received neither involvement in the design of the just-in-time system, nor any explanation of the cost problem facing the plant.*

The consultant begins his task by consulting ORCA; he enters the client's name and chooses one or more "change drivers" from a menu of eight (e.g., high-level problem types such as merger/acquisition, business relocation, restructuring/reorganization, competitive threats, strategic planning, political crisis, and so on). In this instance, the consultant enters the client's name and two change drivers: reorganization/restructuring and competitive threats.

Change drivers are initial entry points into ORCA's memory that are directly connected to some of its memory elements. Choosing a change driver reminds ORCA of both surface features and abstract problem types, and causing confirming questions to be placed, best first, on the question agenda. ORCA then poses each of these to the user in turn. The combination of the two change drivers, "reorganization/restructuring" and "competitive threats", reminds ORCA of a number of abstract problem types. The strongest remindings are to problem types exemplified by proverb-23: "Who looks not before finds himself behind.", proverb-17: "He that cannot adapt is obsolete", and proverb-21: "You've got to spend money to make money". ORCA puts questions on the question agenda to attempt to confirm these problems.

The first question ORCA asks is "Has the industry experienced major technological improvements in recent years?" The user answers "Yes" to this question (which confirms feature-1138). This question was suggested by feature-27: "The organization exists in a highly competitive field" which was itself confirmed by the change driver "Competitive Threats". One of ORCA's initial hypotheses, articulated by proverb-17, "He that cannot adapt is obsolete", is confirmed by this answer because of a confirmatory link between feature-1138 and proverb-17. As a consequence of the interaction thus far, ORCA chooses a story to tell. The story is judged sufficiently relevant because it has surface similarity to the problem situation (through feature-27 and feature-1138), as well as abstract similarity (through proverb-17). ORCA tells the following story:

Kodak Cuts Management - Lower Managers Take Control (abridged)

At Eastman Kodak Co.'s apparatus division, which makes parts for Kodak printers, copiers, and film processors, the management team was cut by 30% and layers slashed from seven to three. But some first-line supervisors, who suddenly had to set goals and strategy for their products rather than just carry out orders, "couldn't make it" despite a training course, says Frank Zaffino, the division's vice president and general manager. "They were used to being star technicians, not communicators or leaders." [...]

After a story has been told, ORCA engages the user in a follow-up dialog in order to elaborate the client situation by assessing the degree to which the particular features of the story are relevant to the client. In this instance the story is relevant in some respects, but the cause of the problem is off point: rather than losing autonomy, as in the client's case, the workers at Kodak have difficulty handling increased autonomy. In the follow-up dialog, the features are presented in order of increasing relevance to the story, in the expectation that the most productive contrast can be found in this way.

Through a follow-up dialog, it becomes apparent the two situations differ in significant ways. In particular, ORCA asks, "In order to achieve its vision, does leadership consider downsizing, severe restructuring, or a reallocation of resources imperative?" This feature is weakly relevant to the Kodak story, but the user rates it highly relevant to his client. Then ORCA asks "Is a new vision, image or style planned?" This feature is also weakly relevant to the Kodak story, but the user rates it as moderately relevant to the client. Then ORCA asks, "Is the organization unwilling to risk implementing new systems, processes, or projects? This feature is rated as moderately relevant to the Kodak story but, crucially, the user sees their client in exactly the opposite way and rates this feature as having a negative correlation to the client's problem: the client is more than willing to risk implementing new systems, in fact this seems to be a big part of the problem. Based on the newly acquired information gathered as a consequence of the follow-up interaction, ORCA indicates that another story is immediately available. This story is only moderately prototypical of proverb-17: "He that cannot adapt is obsolete", but it is strongly linked to proverb-23: "Who looks not before finds himself behind.", which is the hypothesis that ORCA has chosen to explore next .

New Zealand Government Struggles with Change (abridged)

The Inland Revenue Department in New Zealand had an interesting situation in that they had some very elderly technology which was not really servicing their needs at all well. They did a major information plan to decide on the technology of the future and ended up with a plan which would involve them over a four year time period basically reinstalling every single system in the place [...]

This story goes on to relate the difficulties a government agency has trying to install a new computer system over the objections of the workforce. The story is more on point than the Kodak story, because it describes how this resistance was overcome through better informing the workforce about planned changes.

The ORCA consultation continues in this way, from story to story, as long as the user believes it to be worthwhile to answer questions and view cases. At the end of the session, the user is given a report that includes a summary of all sessions to date including a list of the features and abstract categories that are confirmed for the client, a list of questions to ask the client (i.e., questions that the user could not answer during the current session), synopses of all cases presented, and other relevant information. Armed with this report, the user is able to gather further information and to use that information to answer more of ORCA's questions in future sessions.

ORCA is able to assimilate the client's case into memory because there is a sense in which the consultant is telling the client's story by answering the questions ORCA poses. At the end of the consulting engagement, because of the answers given, the new client case is indexed in exactly the same way as the pre-existing cases in ORCA's memory. The final task of the consultant, at the end of the consulting engagement, is to provide a narrative account of the engagement to associate with the case built for the client. Then the new case can be added to ORCA's permanent knowledge base and become a part of the corporate memory that ORCA will use to reason about future clients.

5. Conclusion

Case-based reasoning provides an effective means of reasoning about a complex problem situation because it relies on recalling actual past experiences as models for elaborating and interpreting complex problem situations. Incremental elaboration and interpretation is necessary for case-based reasoning in complex, weak-theory domains because there is no "correct answer" assumption -- the problem situations are complex and information becomes available only in the course of problem solving.

The development of ORCA has been motivated by three theoretical issues that we are exploring computationally. The first issue concerns the nature of effective strategies for incremental elaboration and interpretation of complex problem situations. The second is multiple interpretations of a problem situation, perhaps based on multiple, partial views. The third is the role of categories as the foci of reasoning, i.e., the model that a problem solver is reminded of a category and confirms the reminding by retrieving a similar exemplar.

However, interpreting a complex problem situation, albeit through multiple retrievals, incremental refinements, elaboration, and the telling of relevant stories, is only a part of the problem solving process. The harder part, and the more interesting component of modeling expert behavior, is to follow through and find a way to automatically coalesce the advice from multiple cases into a coherent, unified plan. This remains as one of the major open problems waiting to be solved. In the meantime, ORCA provides a useful tool for cooperative human-computer problem solving.

6. Acknowledgments

ORCA is being built at the Institute for the Learning Sciences by Mike Engber, Kerim Fidel, Andy Gordon, Jim McNaught, Leena Nanda, Tamar Offer-Yehoshua, Josh Tsui, and John Welch in collaboration with Beth Beyer, Jeff Davies, Marita Decker, Dan Halabe, and Maureen Zabloudil of Andersen Consulting. Roger Schank and Chris Riesbeck have contributed valuable ideas throughout the process; Tom Hinrichs helped out by commenting on an earlier draft.

7. References

Ashley, K.D. and Rissland E.L. (1987). Compare and Contrast, a test of expertise. *Proceedings of AAAI-87*

Bareiss, R. (1989). *Exemplar-based Knowledge Acquisition: a Unified Approach to Concept Representation, Classification, and Learning*, San Diego: Academic Press

Bareiss, R, and Slator, B.M. (1992). From Protos to ORCA: reflections on a unified approach to knowledge representation, categorization, and learning. (ILS Technical Report #20).

Hammond, K.J. (1989). *Case-based Planning: Viewing Planning as a Memory Task.* San Diego: Academic Press

Hinrichs, T.R. and Kolodner, J.L. (1991). The roles of adaptation in case-based design. *Proceedings of AAAI-91.*

Kolodner, J.L., (1984) *Retrieval and Organizational Strategies in Conceptual Memory.* Hillsdale, NJ.: Erlbaum,

Kolodner, J.L., Simpson, R.L. and Sycara-Cyranski, K. (1985). A Process Model of Case-Based Reasoning in Problem Solving. *Proceedings of IJCAI-85*

Nii, H.P., Feigenbaum, E.A., Anton, J.J., Rockmore, A.J. (1982). Signal-to-Symbol Transformation: {HASP/SIAP} Case Study, *The AI Magazine*, pp. 23--35, Spring.

Owens, C. (1990) Indexing and Retrieving Abstract Planning Knowledge. PhD thesis, Yale University, Department of Computer Science.

Porter, B., Bareiss, R. and Holte, R. (1990) Concept Learning and Heuristic Classification in Weak-Theory Domains. *AI Journal*, 45, 229-263.

Redmond, M. (1990). Distributed Cases for Case-Based Reasoning; Facilitating Use of Multiple Cases. *Proceedings of AAAI-90.*

Riesbeck, C.K. and Schank, R.C. (1989). *Inside Case-Based Reasoning*, NJ: Lawrence Erlbaum Assoc.

Schank, R.C. (1982). *Dynamic Memory: A Theory of Learning in Computers and People*, Cambridge UK: Cambridge University Press.

Simpson, R.L. (1985). *A Computer Model of Case-Based Reasoning in Problem Solving: An Investigation in the Domain of Dispute Mediation.* PhD Dissertation, Georgia Institute of Technology.

Weiss, S.M., Casimir, A.K., Amarel, S. (1978). A Model-Based Method for Computer-Aided Medical Decision-Making, *Artificial Intelligence*, 11, pp. 145--172

Integrating Causal Learning Rules with Backpropagation in PDS Networks

Ronald A. Sumida
Artificial Intelligence Laboratory
Computer Science Department
University of California
Los Angeles, CA, 90024
sumida@cs.ucla.edu

Abstract

This paper presents a method for training PDP networks that, unlike backpropagation, does not require excessive amounts of training data or massive amounts of training time to generate appropriate generalizations. The method that we present uses general conceptual knowledge about cause-and-effect relationships within a single training instance to constrain the number of possible generalizations. We describe how this approach has been previously implemented in rule-based systems and we present a method for implementing the rules within the framework of Parallel Distributed Semantic (PDS) Networks, which use multiple PDP networks structured in the form of a semantic network. Integrating rules about causality with backprop in PDS Networks retains the advantages of PDP, while avoiding the problems of enormous numbers of training instances and excessive amounts of training time.

Introduction

Parallel Distributed Processing (PDP) models [Rumelhart and McClelland, 1986] have demonstrated many desirable characteristics for learning and generalization. Specifically, PDP models: (1) use a simple learning mechanism such as backpropagation that merely modifies link weight values for each training pattern, (2) automatically generalize as a result of the learning process by averaging the training patterns so that the network can correctly respond to new inputs, (3) use the generalizations that are created to perform pattern completion from partial or noisy inputs, (4) naturally account for interference effects since similar concepts share similar representations and (5) are robust against noise and damage to the network.

Backpropagation performs extremely well and provides the advantages listed above when the size of the training set and the number of input features is relatively small. However, there are serious problems with applying backprop to store the large amounts of knowledge needed for higher-level cognitive tasks, such as natural language. Since backprop learns by similarity-based generalization, it must be shown enough training instances so that all relevant features are correlated while all other features are not. As the number of input features increases with the complexity of the problem domain, the number of training examples needed becomes enormous. For example, suppose that we would like to teach the network about the concept of strength, given a number of instances of people successfully and unsuccessfully lifting a heavy object[1]. Two such instances are presented below:

> John, a tall man with short brown hair and a dark complexion successfully lifts a heavy object in the dining room.
>
> Mary, a small child with long blond hair and a light complexion is unsuccessful at lifting the same heavy object in the living room.

Every feature that is present when John lifts the object and every feature that is absent when Mary fails to lift it are possible causes for the different result. Thus, from the data given above, it is possible to conclude that age, height, hair length, hair color, complexion, location or even the person's name are responsible for John's success at lifting

[1]This example is based on one provided in [Pazzani, 1988].

the object and Mary's failure. In order for the network to correctly learn that strength is correlated with age and not with any of the other features, examples of people with all different types of heights, hair lengths, hair colors and complexions in all types of environments must be presented to the network. Assuming that there are 3 values for height (short, medium and tall), 2 for hair length (short and long), 4 for hair color (blond, brown, black, red), 2 for complexion (light and dark), 10 different names, and 10 different locations, 4800 (= 3x2x4x2x10x10) different training instances are needed for the network to only correlate age with strength. If we were to scale the system up further and add more features such as eye color or more values to the features listed above (e.g. adding grey to the hair color list), the number of necessary training instances could easily exceed 100,000. As the number of training instances grows to such unreasonable numbers, the training time becomes so enormous that it is practically impossible to train the network.

One possible solution is to train the network on only a subset of the patterns. The problem with this approach is that the network will find illusory correlations and form improper or very complicated generalizations. For example, if the network is not shown that a specific conjunction of features is not relevant, it may conclude that short brown hair and a dark complexion are correlated with strength.

This paper presents a method for training PDP networks that does not require excessive amounts of training data to generate appropriate generalizations. The method that we present uses information about cause-and-effect relationships within a single training instance to constrain the number of possible generalizations. The following section shows how this method has previously been implemented in rule-based systems. This is followed by a discussion that indicates how the rules are implemented within the PDS Network [Sumida and Dyer, 1989] framework.

Symbolic Approaches to Learning and Generalization

Considerable work has been done on symbolic, rule-based approaches to learning and generalization, examples of which include [Mitchell *et al.*, 1988, Lebowitz, 1990, Pazzani, 1988]. Previous learning systems vary in the amount of prior knowledge that they apply during the learning process. They range from similarity-based learning (SBL) systems, which operate in much the same way (and share the same problems) as backprop and which assume no prior knowledge, to explanation-based generalization (EBL) systems which assume enough prior knowledge so that the system can construct an entire explanation for why an event occured. Our focus here is on the intermediate position between the two extremes, where the only prior knowledge that we assume are rules about causality. [Pazzani, 1988] refers to this form of generalization as theory-driven learning (TDL) and discusses TDL (along with SBL and EBL) in the program OCCAM. In OCCAM, three types of causal generalization rules are used: (1) *exceptionless*, which apply when similar actions yield the same result, (2) *dispositional*, which apply when similar actions yield differing results and focus on the differing features of the actor or object of the action to explain the differing result, and (3) *historical*, which (like the dispositional rules) apply when similar actions yield differing results, but assume that the events that precede the action explain the differing result. In this paper, our focus will be devoted to discussing how dispositional generalization rules are employed.

The first step in employing a dispositional generalization rule is to create a generalized event that represents all the shared features of the set of events with the same result. The generalized event is then matched with the dispositional generalization rule to determine what role of the cause is responsible for the differing effect. If the same feature of that role occurs in two events that have different results, then it could not be responsible for the differing result so it is no longer considered. Of the remaining features, the one that has been most successful at accounting for different results in previous similar situations (i.e., previous situations involving the same act and role) is selected. If there is no reason to prefer one feature over another, then one is selected randomly. The feature that is hypothesized to be responsible is then added to the generalized event. As new inputs are presented, the hypothesis will either be confirmed, in which case a confidence measure associated with the hypothesis will be increased, or it will be refuted, in which case a new feature will need to hypothesized as responsible for the different result.

As an example of the procedure described above, consider the example from the first section. When the events of John successfully lifting the heavy object and Mary failing to lift it are presented, the system applies the following dispositional generalization rule: *If similar actions performed on an object have different results, and they are performed by different actors, the differing features of the actor are responsible for the different results.* The rule indicates that one of the features of the actor is responsible for the difference in their ability to lift the weight. At this point, there is no reason to prefer one attribute over another so a feature is selected at random, for example, hair color (in this case brown). An example of a blond-haired person lifting the weight is then presented, and the system notices that the prediction that brown hair is responsible has been contradicted. The confidence measure associated with the hair color hypothesis is very low, so the system rejects the hypothesis and selects another feature at random as being responsible for the difference. As other features are selected and refuted, the system will quickly select age as being responsible. As further events are encountered that substantiate this hypothesis, its confidence measure grows and the strength *disposition* is created. The disposition is associated with the action (PROPEL) and the role (ACTOR). Thus, when a similar set of events is encountered in the future, such John being able to remove a tightly attached lid from a jar and Mary not being able to, the system will prefer the strength dispositional attribute to features such as hair color and eye color.

Integrating Backpropagation with Causal Learning Rules

In order to implement the above rules with PDP, we first need to represent the information provided in the training instances. In previous papers, [Sumida and Dyer, 1989, Sumida and Dyer, 1991, Sumida, 1991] we showed how Parallel Distibuted Semantic (PDS) Networks store high-level knowledge. The following section describes how PDS Networks store information using only backprop. This is followed by a description of how causal learning rules are integrated with backprop to avoid the problems with traditional PDP approaches.

PDS Networks

The PDS Network approach is to store all knowledge over multiple PDP networks using backprop, with each network representing a class of concepts and with related networks connected in the general manner of a semantic net. For example, the network shown in Figure 1 encodes acts that involve an application of force (PROPEL) and has roles for the actor, object and result. The network is connected to other PDP networks, such as HUMAN, PHYS-OBJ and OUTCOME, that store information about humans, physical objects, and outcomes of events. Each network functions as a type of *encoder net*, where: (1) the input and output layers have the same number of units and are presented with exactly the same pattern, (2) the weights of the network are modified so that the input pattern will recreate itself as output, and (3) the resulting hidden unit pattern represents a *reduced description* of the input. In the networks that we use, a single set of units is used for both the input and output layers. The net can thus be viewed as an encoder with the output layer folded back onto the input layer and with two sets of connections: one from the single input/output layer to the hidden layer, and one from the hidden layer back to the i/o layer. In Figure 1 for example, the actor, object, and result role-groups collectively constitute the input/output layer, and the PROPEL ensemble constitutes the hidden layer.

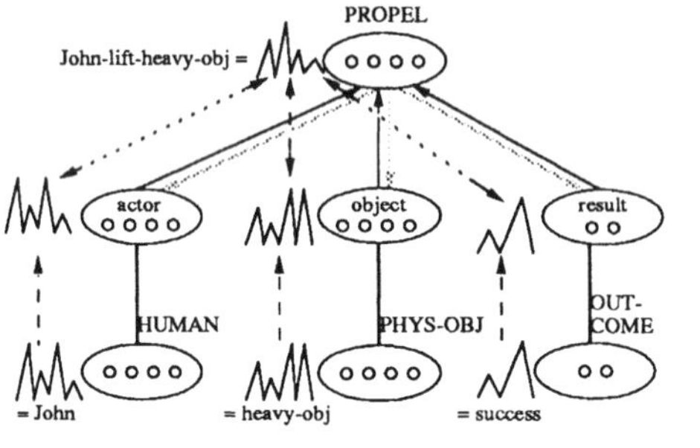

Figure 1: The network that stores information about acts involving an application of force, in this case John-lift-heavy-object. The black arrows represent links from the input layer to the hidden layer and the grey arrows indicate links from the hidden layer to the output layer. The thick lines represent links between networks that propagate a pattern without changing it.

Knowledge is stored in a network by teaching it to encode the items in its training set. For each item, the patterns that represent the features of the item are presented to the input role groups, and the weights are modified using backpropagation so that the patterns recreate themselves as output. For example, in Figure 1, the pattern for John successfully lifting the heavy object is presented to the PROPEL network by propagating the John pattern[2] from the HUMAN network to the actor role, the pattern for the heavy object from the PHYS-OBJ network to the object role, and the success pattern from the OUTCOME network to the result role. The PROPEL network is then trained on this pattern by modifying the weights between the input/output role groups and the PROPEL hidden units so that the John-lift-heavy-object pattern recreates itself as output. The network automatically generalizes since the hidden units: (1) become sensitive to common features of the training patterns and (2) classify a new concept that was not seen during training based on its similarity to familiar concepts.

Implementing Causal Learning Rules in PDS Networks

In order to implement causal learning rules within the PDS Network framework, we need to: (1) accomplish the same result as backprop, that is, modifying link weights so that a hidden unit becomes responsible for recognizing significant correlations in the input as in [Hinton, 1986], but (2) use a a theory-driven learning algorithm rather than one based upon SBL. We therefore need to implement the equivalent of the structure/hypothesis building and rule matching operations by using weight modifications within the network. The idea is to modify the weights so that a particular hidden unit represents the current hypothesis and correlates the pattern for the significant feature (i.e., the feature that is hypothesized to be responsible for the different result) with the pattern for the result. For example, the hypothesis that age is responsible for John's ability to lift the object and for Mary's inability to do so is represented by a hidden unit that is assigned to correlate the pattern for age with the pattern for success (Figure 2).

The following steps are used to apply a dispositional generalization rule and to generate an appropriate hypothesis in PDS Networks. Recall that the first step in symbolic systems is to create a generalized event that contains the shared features of those events with the same result. The

[2] The John pattern represents a reduced description of John's features, such as age, height, hair-color, etc. The procedure for obtaining the John pattern is the same as that described here for the John-lift-heavy-object pattern.

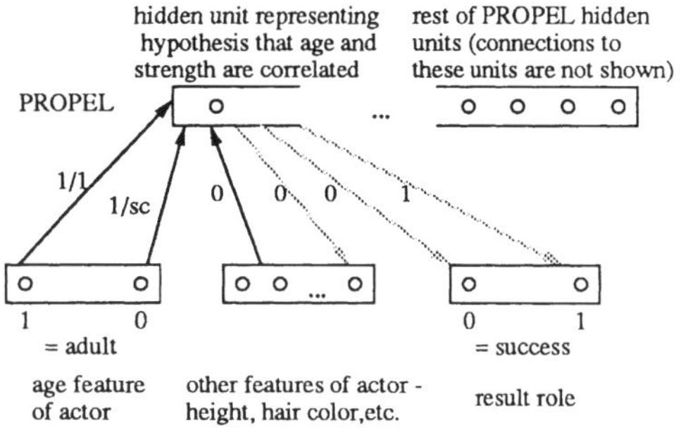

Figure 2: The connections between the hidden unit (that represents the hypothesis that the age feature is responsible for the difference in ability to lift a heavy object) and the actor and result role groups. The dark arrows indicate the connection *to* the hidden unit and the grey arrows indicate the connection *from* the hidden unit. The numbers to the left of each arrow indicate the weight (sc = small constant). All connections other than the ones to the age and result units have a weight value of 0.

equivalent step is accomplished in PDS Networks by clamping the pattern for success or failure (depending on the result of the new event) over the result units and letting the network settle into a stable configuration. Since the network's knowledge of previous events is stored using backprop, the resultant pattern represents the shared features of events with the same result. The next step in symbolic systems is to match the generalized event with the dispositional generalization rule to determine what role is responsible for the different result. This step is accomplished using a mechanism similar to a Propagation Filter [Sumida and Dyer, 1991, Sumida, 1991].

Propagation Filters use the pattern over a selector group of units to determine which of a number of filter groups to enable. Each filter group: (1) gates the connection from a group of source units to a group of destination units, (2) is sensitive to a particular pattern over the selector, and (3) allows the pattern over the source to be propagated to the destination when the particular pattern occurs over the selector. We apply a mechanism similar to Propagation Filters since the pattern for the generalized event acts as a selector. However, rather than have the selector open up a particular group of units, it merely indicates which role group is potentially responsible for the different result. For example, in the PROPEL network, a Propagation Filter with the PROPEL units as its selector indicates that the actor role may be responsible for the different outcome.

The third step in TDL is to build a hypothesis that selects a specific feature from the role group that is responsible for the different result. The equivalent PDS Network operation is to allocate a hidden unit to correlate the feature with the result. The hidden unit is allocated by the following procedure: First, the system finds a free hidden unit. Concepts are stored in PDS networks by training an individual network so that patterns recreate themselves as output. The network only uses as many hidden units as is necessary for learning the training data (i.e., if there are additional hidden units, then backprop leaves them unused). A free hidden unit is chosen from among the unused units. Note the resemblance between the algorithm we are using and a destructive learning algorithm. In a destructive learning algorithm, the system starts with a large number of hidden units and progressively deletes the ones that aren't used. Instead of deleting the unused hidden units, we are using them to represent a hypothesis for which feature is responsible for the different result.

If the connection to the selected hidden unit is from a unit that represents an *irrelevant* feature, its weight is set to 0. Thus, the hidden unit will be unaffected by the values of the irrelevant features. If a unit represents a *relevant* feature, then we would like to have that unit send the hidden unit a value of 1[3]. Thus, we set the weight from the feature to the hidden unit to be 1/(component of the pattern that represents the feature). If the component of the pattern includes a 0, then the weight is set to 1/(a very small constant) since 1/0 is undefined. For example, if age is a relevant feature for success in lifting the heavy object and adult is represented by the pattern "1 0", then the weight from the first unit of age to the hidden unit is set to 1/1 or 1, and the weight from the second unit of age is set to 1/(a very small constant) (see the weights on the links in Figure 2 for an example). The thresold for the hidden unit is set equal to the number of units that represent relevant features. For example, since the pattern for adult is represented over 2 units, the threshold is set to 2. The hidden unit only responds when its activation value is near threshold, not when it is too far above or below. This assures that the unit will only be active when the proper pattern for the relevant feature occurs. Thus, when the "1 0" pattern is encountered over the age units, the hidden unit will be turned on. We now need to have the hidden unit correlate the relevant features with the result. Thus, when the hidden unit is active, we need it to cause the pattern for the correct result to occur. We set the weight from the hidden unit to each result unit to be the value that is expected for that unit. For example, if we expect the result to be success, and success is represented by the pattern "0 1", then the weight to the first success unit is 0 and the weight to the second result unit is 1 (again see Figure 2).

To illustrate the above procedure, consider again the example from the first section. When we show the system the patterns for John successfully lifting the object and Mary failing to lift it, the system notices that the dispositional generalization rule from the second section (in our discussion of symbolic approaches) is appropriate since the pattern for the object role is the same in both events, while the patterns for the actor and the result are different. The mechanism similar to a Propagation Filter suggests that the actor role is responsible for the different result. Since there is no reason to prefer one feature over another, hair color is selected at random. The system now selects one of the free hidden units to represent the hypothesis that hair color is responsible for the different result. Since all other features besides hair color are hypothesized to be irrelevant, the weights from the units representing all features besides hair color are set to 0 (Figure 3). The pattern for brown hair is "1 1 0", so the weight from the first hair color unit to the hidden unit is set to 1/1, the weight from the second hair color unit is set to 1/1, and the weight from the third unit is set to 1/(a very small constant). The threshold for the hidden unit is set to 3, since there are three units for the relevant feature of hair color. The pattern for success is "0 1", so to correlate brown hair with success at lifting the weight, the weight from the hidden unit to the first result unit is set to 0, and the weight to the second hidden unit is set to 1.

An example of a blond-haired person lifting the weight is then presented, which contradicts our hypothesis so another feature is hypothesized to be responsible for the different result and the weights to the hidden unit are changed so that the new feature is correlated with success. As with the symbolic approach, after a small number of examples, the system refutes the hypotheses involving

[3] For the sake of simplicity, we choose the value 1. In reality we can choose a different constant and merely adjust the threshold appropriately.

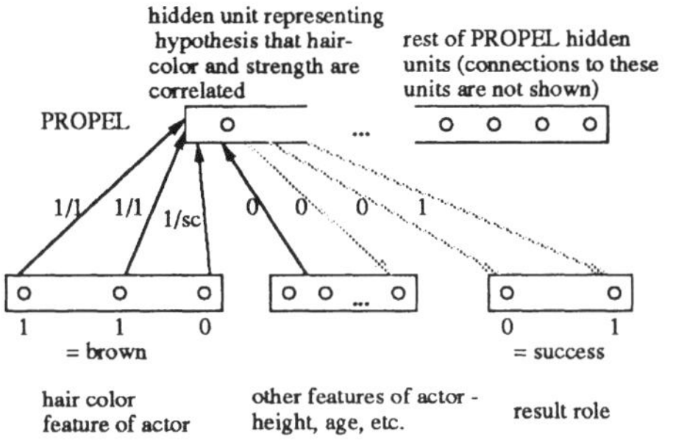

Figure 3: The connections between the hidden unit (that represents the hypothesis that hair color is responsible for the difference in the ability to lift the heavy object) and the age and result units.

irrelevant features and realizes that age is the important feature. The network configuration shown in Figure 2 is therefore the hypothesis that the system builds.

Related Work

Recent work has been done in combining explanation-based learning with neural network approaches, for instance [Shavlik and Towell, 1989, Katz, 1989]. Some of this recent work is similar to the work presented in this paper. For example, some of the rules discussed in [Shavlik and Towell, 1989] bear a resemblance to the rules that we have presented here. However, these systems have not yet provided a neural network framework for representing the type of high-level knowledge that is needed for generating complex explanation chains. We believe that PDS Networks will provide the framework that is necessary for representing complicated knowledge structures. Thus, we have chosen to integrate rules about causality with backpropagation in the PDS Network framework.

Conclusions

In this paper, we have presented a method for training PDP networks that integrates knowledge about cause-and-effect relationships with backpropagation. This approach has a number of important advantages over PDP systems that only use backpropagation: (1) It does not require enormous amounts of training data since rules about causality within a single training instance are used to constrain the number of possible generalizations. In contrast, similarity-based generalization methods such as backpropagation need to compare enormous numbers of instances to determine which features are relevant in forming a generalization. (2) Training time is dramatically decreased because far fewer training instances are examined. (3) The integration of causal learning rules with backprop is implemented in the framework of PDS Networks, so the high-level knowledge necessary for natural language processing can be represented, the advantages of PDP are retained, and the problems encountered in training PDP networks are avoided.

References

G. E. Hinton. 1986. Learning Distributed Representations of Concepts. In *Proceedings of the Eighth Annual Conference of the Cognitive Science Society*, Amherst, MA.

B. F. Katz. 1989. Integrated Learning in a Neural Network. In *Proceedings of the Sixth International Workshop on Machine Learning*, Ithaca, NY.

M. Lebowitz. 1990. Utility of Similarity-Based Learning in a World Needing Explanation. In *Machine Learning, Vol. 3*, Morgan Kaufmann, Los Altos, CA.

Mitchell, T., Kedar-Cabelli, S. and Keller, R. 1988. Explanation-based Learning: A Unifying View. *Machine Learning*, Vol. 1(1).

M. J. Pazzani. 1988. Learning Causal Relationships: An Integration of Empirical and Explanation-Based Learning Methods. UCLA Artificial Intelligence Laboratory Technical Report UCLA-AI-88-10 (PhD Dissertation).

D. E. Rumelhart and J. L. McClelland. 1986. *Parallel Distributed Processing*, Volume 1. MIT Press, Cambridge, Massachusetts.

J. W. Shavlik and G. G. Towell. 1989. Combining Explanation-Based Learning and Artificial Neural Networks. In *Proceedings of the Sixth International Workshop on Machine Learning*, Ithaca, NY.

R. A. Sumida and M. G. Dyer. 1989. Storing and Generalizing Multiple Instances while Maintaining Knowledge-Level Parallelism. In *Proceedings of the Eleventh International Joint Conference on Artificial Intelligence*, Detroit, MI.

R. A. Sumida. 1991. Dynamic Inferencing in Parallel Distributed Semantic Networks. In *Proceedings of the Thirteenth Annual Conference of the Cognitive Science Society*, Chicago, IL.

R. A. Sumida and M. G. Dyer. 1991. Propagation Filters in PDS Networks for Sequencing and Ambiguity Resolution. In *Advances in Neural Information Processing 4 (NIPS-91)*, Denver, CO.

Fuzzy Evidential Logic: A Model of Causality for Commonsense Reasoning

Ron Sun
Honeywell SSDC
Minneapolis, MN 55418

Abstract

This paper proposes a fuzzy evidential model for commonsense causal reasoning. After an analysis of the advantages and limitations of existing accounts of causality, a generalized rule-based model FEL (*Fuzzy Evidential Logic*) is proposed that takes into account the inexactness and the cumulative evidentiality of commonsense reasoning. It corresponds naturally to a neural (connectionist) network. Detailed analyses are performed regarding how the model handles commonsense causal reasoning.

Shoham's Causal Theory

The issue of causality has recently received a lot of attentions from various perspectives (cf. Shoham 1987, Iwasaki & Simon 1986, de Kleer & Brown 1986, Pearl 1988, etc.). The issue has wide ranging impact on areas such as learning, control, and recognition. However, most of these logic based models are aimed for modeling truth functional aspects of causal knowledge, and they tend to ignore some important characteristics of commonsense causal reasoning, for example, gradeness of concepts, inexact causal connections, evidentiality of causal rules, etc., while probabilistically motivated models are mainly concerned with the probabilistic aspect of causal events, and they are more computationally complex and oftentime have only marginal cognitive plausibility in terms of mechanisms involved. Connectionism provides a new and different kind of models that might be of help in accounting for causality in commonsense reasoning; these models entertain a number of interesting properties that other models lack (for example, massive parallelism, generalization, fault/noise tolerence, and adaptability; see Waltz & Feldman 1986, Sun & Waltz 1991) and present a new perspective of reasoning as a complex process in a dynamic system; it will be worthwhile to look into the question of how such models can deal with the issue of causality.

Let us look into Shoham's account of causality (Shoham 1987), which is undoubtedly one of the most notable accounts of causality with rule-based formalisms. His temporal modal logic formalism has a close resemblance to Horn clause logic, and therefore is very suitable for use in rule-based systems. According to Shoham's Causal Theory (CT), causes are primary conditions which, together with other conditions, will bring about the effect. These "other" conditions are somewhat secondary. In reasoning, as long as we know that the primary conditions (*causes* or *necessary conditions*) are true and that there is no information that the secondary conditions (*enabling conditions* or *possible conditions*) are false, then we can deduce that effects will follow. The theory is described in terms of modal logic, with one basic modal operator (□ or *necessity*) for specifying necessary conditions. and one auxiliary modal operator (◇ or *possibility*) for specifying possible conditions. The formal definition is as follows:

Definition 1 *A* **Causal Theory** *is a set of formulas of the following form*

$$\wedge_i \Box n_i a_i(t_{i1}, t_{i2}) \wedge_j \Diamond n_j b_j(t_{j1}, t_{j2}) \longrightarrow \Box c(t_1, t_2)$$

where n_i's are either $\neg$ or nothing, $t_2 > t_{i2}$ for all i's, $t_2 > t_{j2}$ for all j's, $n_i a_i$'s are necessary conditions (causes), and $n_j b_i$'s are possible conditions (enabling conditions). C is concluded iff all $n_i a_i$'s are true and none of $n_j b_j$'s are known to be false. [1]

From the standpoint of modeling commonsense knowledge, this model has some advantages, such as that it provides a simple and elegant formalism with efficient inference algorithms, that it is easily representable (and implementable), and that it has

[1]This process is formally described by a *minimization* principle in Shoham (1987).

compatibility with philosophical accounts of causality (Shoham 1987). On the other hand, the model ignores or discounts many aspects of commonsense causal reasoning; for example,

1. All propositions in this theory are binary: either true or false, and there is no sense of gradedness. Commonsense knowledge is certainly not limited to true/false only (Sun 1991, Hink & Woods 1987).
2. Beside the inexactness of individual concepts, reasoning processes in reality are also inexact and evidential. Specifically, the evidential combination process is cumulative (as observed in protocol data; Sun 1991, 1991a); that is, it "adds up" various pieces of evidence to reach a conclusion, with a confidence that is determined from the "sum" of the confidences of the different pieces of evidence. Moreover, different pieces of evidence are weighted, that is, each of them may have more or less impact, depending on its importance or saliency, on the reasoning process and the conclusion reached. We have to find a way of combining evidence from different sources cumulatively and with weights, without incurring too much computational overhead (such as in probabilistic reasoning or Dempster-Shafer calculus; cf. Pearl 1988).
3. Because of the lack of gradedness, the model will make projections too far along a chain of reasoning (or too far into the future; Sun 1991). An example from Shoham (1987):

$$\Box alive(t_0, t_0)$$

$$\Box shoot(t,t) \longrightarrow \Box\neg alive(t+1, t+1)$$

$$\Box alive(t,t) \Diamond\neg shoot(t,t) \Diamond\neg otherwise\text{-}killed(t,t) \longrightarrow \Box alive(t+1,t+1)$$

   which means that if one is alive at time t_0, one will continue to be alive as long as not being shot or otherwise killed. So if there is nothing known about "shoot" and "otherwise-killed", then according to the minimal model approach, we will predict that

$$\Box alive(t,t) \quad where\ t \longrightarrow \infty$$

   This is certainly not true. The problem is that, along a chain of inference (as well as in temporal projections), the confidence for the conclusions reached should weaken. We can weaken confidence along the way only when gradedness is reinstated into causal theories.
4. The clear-cut *necessity* and *possibility* is a problem, because in reality there is little, if any, qualitative difference between *causes* and *enabling conditions*. The difference is more quantitative (as will be illustrated later), and sometimes the two are interchangeable; for example, "He is shot dead" is expressed in CT as

$$\Box shoot(t,t) \wedge \Diamond\neg wearing\text{-}bullet\text{-}proof\text{-}vest(t,t) \ldots\ldots \longrightarrow \Box dead(t+1,t+1)$$

   and "His failure to wear the bullet-proof vest caused his tragic death" is expressed as

$$\Box\neg wearing\text{-}bullet\text{-}proof\text{-}vest(t,t) \wedge \Diamond shoot(t,t) \ldots\ldots \longrightarrow \Box dead(t+1,t+1)$$

   So one fact can be both a cause and an enabling condition.
5. Although the model does distinguish two different types of conditions, it does not explain why some conditions are necessary, and some conditions need only to be possible.
6. According to the model, it is necessary to list all causes and all enabling conditions, in order to guarantee correct results. This could be hard to do, because the number of enabling conditions could be infinite.
7. The causal connection between events in the left hand side of an implication and events in the right-hand side of the same implication may not be deterministic. It could be probabilistic, or otherwise uncertain (Suppes 1970).

For reviews of other accounts of causality, see Sun (1991).

Defining FEL

FEL (*Fuzzy Evidential Logic*) is aimed at resolving the problems inherent in existing logical accounts of causality. Like Shoham's formalism, FEL is defined around rules; however, FEL encodes rules with the weighted-sum computation. This formalism is meant to capture, among other things, the gradedness and evidentiality of commonsense reasoning, in a cognitively motivated way. Formal definitions follow (cf. Zadeh 1988):

Definition 2 *A* **Fact** *is an atom or its negation, represented by a letter (with or without a negation symbol) and having a value between l and u. The value of an atom is related to the value of its negation by a specific method, so that knowing the value of an atom results in immediately knowing the value of its negation, or vice versa.* [2]

Now we can define rules and their related weighting schemes:

[2]We will adopt a generic confidence measure as the value of a fact.

Definition 3 *A* **Rule** *is a structure composed of two parts: a left-hand side (LHS), which consists of one or more facts, and a right-hand side (RHS), which consists of one fact. When facts in LHS get assigned values, the fact in RHS can be assigned a value according to a weighting scheme* [3].

Definition 4 *A* **Weighting Scheme** *is a way of assigning a weight to each fact in LHS of a rule, with the total weights (i.e., the sum of the absolute values of all the weights) less than or equal to 1, and of determining the value of the fact in RHS of a rule by thresholded (if thresholds are used) weighted-sum of the values of the facts in LHS (or inner-products of weight vectors and vectors of values of LHS facts). When the range of values is continuous, then the weighted-sum is passed on if its absolute value is greater than the threshold, or 0 if otherwise. When the range of values is binary (or bipolar), then the result will be one or the other depending on whether the weighted-sum (or the absolute value of it) is greater than the threshold or not (usually the result will be 1 if the weighted-sum is greater than the threshold, 0 or -1 if otherwise).* [4]

Definition 5 *A* **Conclusion** *in FEL is a value associated with a fact, calculated from rules and facts by doing the following:*

(1) for each rule having that conclusion in its RHS, obtain conclusions of all facts in its LHS (if any fact is unobtainable, assume it to be zero); and then calculate the value of the conclusion in question using the weighting scheme;

(2) take the MAX of all these values associated with that conclusion calculated from different rules or given in initial input.

Definition 6 *A rule set is said to be* **Hierarchical**, *if the graph depicting the rule set is acyclic; the graph is constructed by drawing a unidirectional link from each fact (atom) in LHS of a rule to the fact (atom) in RHS of a rule.*

Making a rule set hierarchical avoids circular reasoning.

Now FEL can be defined as follows:

Definition 7 *A* **Fuzzy Evidential Logic** *(FEL) is a 6-tuple:* $< A, R, W, T, I, C >$, *where A is a set of facts (the values of which are assumed to be zero initially), R is a set of rules, W is a weighting scheme for R, T is a set of thresholds each of which is for one rule, I is a set of elements of the form* (f, v) *(where f is a fact, and v is a value associated with f), and C is a procedure for deriving conclusions (i.e. computing values of facts in RHS of a rule in R, based on the initial condition I).*

We want differentiate FEL into two versions: FEL_1 and FEL_2, which differ in their respective ranges for values associated with facts.

Definition 8 FEL_1 *is FEL when the range of values is restricted to between 0 and 1 (i.e.* l=0 *and* u=1*), and the way the value of a fact is related to the value of its negation is:*

$$a = 1 - \neg a$$

for any fact a.

Definition 9 FEL_2 *is FEL when the range of values is restricted to between -1 and 1 (i.e.* l=-1 *and* u=1*), and the way the value of a fact related to the value of its negation is:*

$$a = -\neg a$$

for any fact a.

As an illustration of its capability and correctness, we want to show that FEL can implement Horn clause logic as a special case (we will only deal with the propositional version here, and extensions to first order cases is dealt with in Sun 1991). Let us define Horn clause logic first (cf. Chang & Lee 1973):

Definition 10 **Horn clause logic** *is a logic in which all formulas are in the forms of*

$$p$$

or

$$p_1 p_2 \ldots\ldots p_n \longrightarrow q$$

where p's and q are propositions.

Definition 11 *A* **Binary FEL** *is a reduced version of FEL (either* FEL_1 *or* FEL_2*), in which values associated with facts are binary (or bipolar), total weights of each rule sum to 1, and all thresholds are set to 1.*

Here is the theorem for the equivalence (see Sun 1991 for proofs):

Theorem 1 *The binary FEL is sound and complete with respect to Horn clause logic.*

We want to show that FEL can simulate Shoham's Causal Theory, to further explore the logical capability of FEL. (We will only consider a non-temporal version of CT, that is, we strip away all

[3]When the value of a fact in LHS is unknown, assign a zero as its value.

[4]This weighting scheme can be generalized, as will be discussed later on.

temporal notations.) We have to find a mapping between truth values of formulas in Causal Theory and values of facts in FEL. Since in CT and in FEL, there is no logical OR and there is only a (implicit) logical AND in the LHS of a rule, which can be taken care of by a weighting scheme as will be discussed later, we do not have to worry about these two connectives in the mapping now. Therefore, we can use a mapping as follows, which can be easily verified to be consistent with regard to logical equivalence (without AND and OR; for example, □ a = ¬◇¬ a, etc.):

(1) M(a= true) = 'a=1'
(2) M(¬ a= true) = 'a=-1'
(3) M(□ a= true) = 'a=1'
(4) M(□¬ a= true) = 'a=-1'
(5) M(◇ a= true) = 'a=0'
(6) M(◇¬ a= true) = 'a=0'
(7) M(¬□ a= true) = 'a=0'
(8) M(¬□¬ a= true) = 'a=0'
(9) M(¬◇ a= true) = 'a=-1'
(10) M(¬◇¬ a= true) = 'a=1'

With the mapping in hand, we can proceed to find a weighting scheme to enable FEL to simulate Causal Theory. The problem is that in FEL we have nodes only for atoms such as a, b, m, n, etc. but not for □a or ◇b, etc. We have two ways of dealing with this:

1. Extending and making more complex the weighting scheme,
2. Adding nodes that can be used to represent atoms with modal operators.

We will adopt the first approach here (the second approach will also work — the difference is insignificant). For a formula in Causal Theory

$$\wedge_i \Box n_i a_i \wedge_j \Diamond n_j b_j \longrightarrow \Box nc$$

we can assign arbitrary weights to atoms: a_i's and b_i's (if there is a negation, the corresponding weight is negative; otherwise, weights are positive), as long as their absolute values sum to 1. However, for b_i's, we will also apply the following function to the link between b_i and c:

$$f_j(b_j) = \begin{cases} 1 & \text{if } b_j = 1 \\ 1 & \text{if } b_j = 0 \text{ and } n_j \neq \neg \\ -1 & \text{if } b_j = 0 \text{ and } n_j = \neg \\ -1 & \text{if } b_j = -1 \end{cases}$$

We will call this function the *elevation* function because it turns all 0's into 1's or -1's. We have thresholds equal to 1 for all rules. We restrict the possible values of facts to -1 or 1.

Now it is easy to verify that a rule in FEL with this specific weighting scheme and thresholds is equivalent to a corresponding formula in Causal Theory: e.g., suppose we have the following formula in CT:

$$\Box a \Box b \Diamond a \Diamond b \longrightarrow \Box e$$

It can be translated into FEL as follows:

$$abc'd' \longrightarrow e \quad (w_1 w_2 w_3 w_4)$$

where $c' = f_c(c)$ and $d' = f_d(d)$ and $\sum_i w_i = 1$, and the threshold equal to 1 for the rule. The equivalence can be verified case by case.

To find a full correspondence between FEL and Causal Theory, we also need a proof procedure that enables the derivation of all correct results (theorems). Here is a proof procedure for CT:

Given a Causal Theory CT, and a set of initial conditions (true events) I:

— For all a ∈ I, infer a, □a, and ◇a.
— Repeat:
for $\wedge_i \Box n_i a_i \wedge_j \Diamond n_j b_j \longrightarrow \Box c$ where $n_i a_i$'s are inferred, and $\neg n_j b_j$'s are non-inferable, [5]
infer c, □c, and ◇c.

It is easy to see the correctness of this procedure (see Sun 1991 for all the proofs):

Theorem 2 *The above proof procedure is sound and complete for Causal Theory as defined above.*

We can have a similar proof procedure for FEL:

Given a FEL theory, and a set of initial conditions (true facts) I:

— For all $(a, v_a) \in$ I, infer a with v_a.
— Repeat:
for $\wedge_i n_i a_i \longrightarrow c$ where each $n_i a_i$ is inferred with a certain value, or is non-inferable (and therefore a value zero is assumed), [6]
infer c with v_c, where v_c is calculated according to the weighting scheme used.

It is easy to see the correctness of this procedure for FEL, and the correspondence between the two proof procedures:

Theorem 3 *The above proof procedure is sound and complete for hierarchical FEL*

[5] They are not in the RHS of any rule and not in I, or in order to infer it, we have to use a rule which has a fact as a necessary condition in its LHS that is not inferable. Since CT is hierarchical, this is easy to detect. We can preconstruct a "dependency graph" which depicts inferability relations.

[6] According to the weighting scheme used to simulate CT, if a fact is inferred, it must be inferred with a value 1 or -1; if a fact is non-inferable, then its value is 0. When other weighting schemes are used, the results will be different.

Theorem 4 *The proof procedure for FEL carries out exactly the proof procedure for Causal Theory when Causal Theory is implemented in FEL in the aforementioned way.*

Therefore,

Theorem 5 *For every hierarchical, non-temporal Causal Theory, there is a FEL such that CT: w $\models$ a iff FEL: c $\models$ 'a=1', where w is a set of initial conditions for Causal Theory CT, and c is the set of initial conditions for FEL mapped over from w in CT.*

Accounting for Commonsense Causality

Now we are ready to show that FEL extends justifiably Shoham's Causal Theory and solves the problems identified earlier. To extend the FEL version of Shoham's Causal Theory, we first notice that the causes need not be known with absolute certainty, i.e. we should allow a confidence measure associated with each *necessary* fact (i.e. the one with $\Box$), because of the gradedness, uncertainty and fuzziness of our knowledge. By the same token, the conclusions need not be binary either, so that uncertain causes can generate uncertain effects. Moreover, even facts (causes) of absolutely certainty may not guarantee the expected effects (i.e. the idea of uncertain causality; Suppes 1970). Therefore, we will associate a confidence measure with each of the causes (i.e. the facts in LHS of a rule) between -1 and 1, and a confidence measure also with the effect (i.e. the fact in RHS of a rule). We can use weights to create a mapping between confidence measures of causes (i.e. values of the corresponding facts) and confidence measures of effects (i.e. values of the corresponding facts), so from a set of causes and their confidence measures (i.e. a set of facts and their values) we can deduce a confidence measure for an effect (or a value for a fact in RHS). Moreover, the set of weights associated with facts in LHS of a rule should reflect their relative importance: more important causes should have a larger weight associated with them, and since the total weights sum to 1, the value of a weight for a particular fact (condition) reflects its relative importance against a background of all other conditions.

Another issue to consider is how to handle *possible condition* facts (i.e. those with $\Diamond$). As explained before, there is a special function associated with them, which elevates 0 to 1 or -1, according to whether positive or negative forms appear in the causal rule. Since we now extend the binary (or bipolar) space for truth values into a graded, continuous space, there is no more need for that elevation function. It follows from the fact that when a *possible condition* fact is unknown (i.e. its value is 0), the conclusion can still be reached, albeit with a smaller value (in confidence level). Now that we no longer require a binary (or bipolar) outcome, it is fine to have a smaller value for a conclusion when some enabling conditions are unknown. When one of these enabling conditions become known, the value will become higher; that is, we will have more confidence in the conclusion. Normally the weights associated with those enabling conditions will be relatively small anyway, because they are non-essential and close to "*don't care*" conditions. So it is advantageous to remove the elevation functions in the FEL version of Shoham's Causal Theory and assign weights instead.

An alternative perspective of viewing the extension is that of "fuzzifying" the necessity function and the possibility function. Once fuzzified, these new functions wind up to be identity functions. Therefore, combining the above two perspectives, *causes* are those conditions that have *high* weights, and *enabling conditions* are those conditions that have *low* weights.

We can now easily map the FEL terminology into the causal terminology as follows:

Events *are facts in FEL.*
Causal Statements *are rules in FEL.*
Causes *are those conditions of a rule that have high weights associated with them according to some particular weighting scheme.*
Enabling Conditions *are those conditions of a rule that have low weights associated with them according to some particular weighting scheme.*
Effects *are facts in the RHS of a rule.*

Let us go back to the issues we raised before:

- The gradedness is readily taken care of in FEL by the confidence values associated with each fact.
- Because of the introduction of the gradedness and uncertain rules (i.e. total weights sum to less than 1), the confidence we have in the conclusions will weaken along the way in a chaining. For example, here is a FEL rule stating that if one is alive at time t, one will be alive at time $t+1$:

  $$alive(t) \longrightarrow alive(t+1)$$

  Suppose the weight is equal to 0.99, then if given *alive(0)=1*, we will have *alive(1)=0.99*, *alive(2)=0.98*, *alive(3)=0.97*, and so on.
- There is no more need to tell exactly which condition is necessary and which condition is possi-

ble: they are graded and the difference is only quantitative.

- There is no more need to list all conditions (the total number of which might be infinite), as long as we leave room in the weight distribution (by keeping total weights less than 1). We can list only those conditions that we care about, and by doing so, the sum of weights will then be less than 1, accommodating possible roles of other unlisted conditions in determining the causal outcome.
- The indeterminate or probabilistic nature of causality is readily captured in the weighting scheme: the weights do not have to sum to 1, and not all conditions have to be known for certain in order to deduce a plausible conclusion.

Let us look back to the shooting example. Instead of having two separate causal statements in CT as before,

$$\Box shoot(t,t) \wedge \Diamond\neg wearing\text{-}bullet\text{-}proof\text{-}vest(t,t)......$$

$$\longrightarrow \Box dead(t+1, t+1)$$

and

$$\Box\neg wearing\text{-}bullet\text{-}proof\text{-}vest(t,t) \wedge \Diamond shoot(t,t)......$$

$$\longrightarrow \Box dead(t+1, t+1)$$

we will have in FEL one single causal statement for all the situations:

$$\neg wearing\text{-}bullet\text{-}proof\text{-}vest \wedge shoot......$$

$$\longrightarrow dead \quad (w_1, w_2, ...)$$

and weights are assigned to each fact in LHS (Sun 1991). We assume the values of the unknown facts are zero and calculate the value of the conclusion by inner-products of the weights and the values of the facts in the LHS of the rule.

The Neural Net Connection

An implementation of FEL is a network of elements connected via links, where each element represents an atom and its negation and links represent rules, going from elements representing facts in LHS of a rule to elements representing facts in RHS of a rule; an element is a structure that has multiple sites each of which receives a group of links that represents one single rule, and the weighted-sum computation is carried out for computing and propagating activations. This implementation of FEL is clearly a connectionist network (Sun 1989, 1992).

Concluding Remarks

In this paper, Shoham's modal logic formalism for causal reasoning is critically analyzed; knowing its weakness in expressing graded concepts and other problems resulting from this, we proceed to define a different formalism, FEL, that utilizes weighted-sum computation and corresponds directly to neural net models. We prove that FEL can implement Shoham's logic as a special case as well as Horn clause logic, and furthermore that FEL is a justified extension of Shoham's logic. This work serves to justify a particular connectionist architecture proposed by the author, *CONSYDERR*, in its capabilities for coding rules and for performing commonsense causal reasoning (Sun 1991a).

REFERENCES

Chang, C. and Lee, R.C. 1973. *Symbolic Logic and Mechanical Theorem Proving.* Academic Press.

de Kleer, J. and Brown, J. 1986. Theories of Causal Ordering. *Artificial Intelligence*, 29:33-61.

Hink, R. and Woods, D. 1987. How Human Process Uncertain Knowledge. *AI Magazine*, 8:41-53.

Iwasaki, Y. and Simon, H. 1986. Causality in device behavior. *Artificial Intelligence*, 29:3-32.

Pearl, J. 1988. *Probabilistic Reasoning in Intelligent Systems.* Morgan Kaufman.

Shoham, Y. 1987. *Reasoning about Change.* Ph.D Dissertation, Yale University.

Sun, R. 1989. A discrete neural network model for conceptual representation and reasoning. *Proc.11th Cognitive Science Society Conference*, 916-923. Erlbaum.

Sun, R. and Waltz, D. 1991. Neurally Inspired Massively Parallel Model of Rule-Based Reasoning. in B. Soucek ed. *Neural and Intelligent System Integration*, John Wiley & Sons.

Sun, R. 1991. *Integrating Rules and Connectionism for Robust Reasoning.* Ph.D Dissertation, Brandeis University.

Sun, R. 1991a. Connectionist Models of Rule-Based Reasoning. *Proc.13th Annual Conference of Cognitive Science Society*, 437-442. Erlbaum.

Sun, R. 1992. Beyond associative memories: Logics and varaibles in connectionist models. *Information Sciences.* forthcoming.

Suppes, P. 1970. *A Probabilistic Theory of Causation.* North Holland.

Waltz, D. and Feldman, J. (eds.) 1986. *Connectionist Models and Their implications.* Ablex.

Zadeh, L. 1988. Fuzzy Logic. *Computer*, 21:83-93.

Exemplar competition: A variation on category learning in the Competition Model*

Roman Taraban
Department of Psychology
tirmt@ttacs.ttu.edu
J. Marcos Palacios
Computer Science
cjejp@ttacs.ttu.edu
Texas Tech University
Lubbock, TX 79409

Abstract

Two cue validity models for category learning were compared to the exemplar model of Medin & Schaffer (1978). The cue validity models tested for the use of two cue validity measures from the Competition Model of Bates & MacWhinney (1982, 1987, 1989) ("reliability" and "overall validity"); one of these models additionally tested for "rote" associations between items and categories. Twenty-four undergraduate subjects learned to classify pseudowords into two categories over 40 blocks of trials. The overall fit of the cue validity model without rote associations was poor, but the fit of the model that included these was nearly identical to the exemplar model (R^2 = .89 *vs* .90). However, both cue validity models failed to capture differences predicted by exemplar similarity, but not cue validity, that were apparent as early as the first block of learning trials. The critical parameters in the Medin-Schaffer model were fit as a logarithmic function of the learning block to provide a uniform account of learning across the 40 blocks of trials. The evidence that we provide suggests that competition at the level of exemplars should be considered as a possible extension of the Competition Model.

*This paper is based in part upon work supported by the Texas Advanced Research Program under Grant No. 0216-44-5829 to the first author.

Models of category learning have appeared in at least two distinct guises. *Independent-cue* models (Anderson, 1991; Beach, 1964; Reed, 1972; Rosch & Mervis, 1975) posit the summing of weighted "evidence" for a category derived from information provided by individual cues or features. *Exemplar* models (Kruschke, 1992; Medin & Schaffer, 1978; Nosofsky, 1984) usually require the analysis of exemplars into simpler components, but compute the evidence for a category on the basis of between-*item* similarity.

The *Competition Model* of Bates and MacWhinney (1982, 1987, 1989) is an independent cue model that has been quite successful in accounting for the learning of natural language categories. An important thesis in this model is that children and adults weight cues differently depending on their level of learning. These differences are described through various cue validity measures that assess the relative contribution of a cue to category selection. Taraban, McDonald, & MacWhinney, 1989), for instance, used human and computer simulation data to argue that *overall validity* provides the best characterization of cue weights early in learning; later in learning the weights are best described by *reliability* and then by a least-mean squares solution. McDonald & MacWhinney (1991) have provided evidence for early use of *overall validity* and later reliance on *conflict validity.*

Although the Competition Model provides the best current account for learning linguistic categories, the exemplar view has not been explored and it is still not known whether the Competition Model could benefit from an exemplar approach. In this paper we are not concerned with the standard Competition Model questions that focus on shifts in weights of independent cues. Instead we set up contrasting predictions for independent-cue models and an exemplar model in a learning experiment to test whether the exemplar model provides a better fit to performance at *any* stage in learning. In an experimental setting, it is difficult to systematically explore language learning with natural language materials, so in some of these studies the experimenters have resorted to using artificial materials (e.g. McDonald & MacWhinney, 1991). We have adopted the same approach in the present study using a very simple set of pseudowords for which subjects learned category labels over the course of a single, long, experimental session.

Three models: Cue Validity, Cue + Rote, Exemplar

Reliability is closely related to formulations in Beach (1964) and Reed (1972). For any given cue and a category X, reliability corresponds to the conditional probability $P(X|cue)$. In the *Cue Validity* model, we fit one parameter for each letter position in the pseudoword stimuli to allow for differences in attention to cue reliabilities in those positions. As indicated in (1), the "evidence" for some category X given a test item t is a weighted sum of cue reliabilities. *Overall validity* corresponds to the product of the overall frequency of a cue and its reliability. In the context of the present study, it is important to point out that the overall frequency of each cue was 0.5. Thus, a fitted *overall validity* model differs from a *reliability* model by a constant factor – i.e. we could fit the *overall validity* model directly from (1) by simply multiplying each fitted parameter by 2. This means that (1) should give a good account of a substantial part of learning performance, based on current Competition Model thinking.

$$E_{X|t} \equiv \sum a_i * \mathsf{reliability}_i \quad (1)$$

Is a weighted model like (1) sufficient for describing category learning? Clearly it is not, particularly if the categories are "non-linearly" separable, a condition which by definition precludes complete learning. MacWhinney, Leinbach, Taraban, & McDonald (1989) discuss the possibility that cue-to-category associations like those represented in (1) are supplemented by "rote" associations of items to their respective category. The *Cue + Rote* model discussed in this paper is identical to (1), except that the sum includes an additional product ($a_i * \mathsf{item}$) that estimates the strength of association of pseudowords to their respective categories, with the value of item equal to 1 for its association to its own category, and 0 for its association to the competing category. Does adding a parameter for rote associations render the reliabilities superfluous? The answer is "no." If subjects simply learned "paired associations" there would be no between-item differences in fit to a category (viz. typicality), which is, in general, unlikely for categories and not the case for our stimuli, as described later.

The *Exemplar* model presented in (2) is the one used in Medin & Schaffer (1978). In this paper, (2) computes the ***overall similarity*** of an item t to a category X. $Similarity(t,x) = \prod s_i$, with an s_i fitted for each letter position, computes the similarity of an item t to a particular category member x. As in Medin & Schaffer, $s_i = 1$ if $letter_i$ in x and in t match, and $0 \leq s_i \leq 1$ if they mismatch. In the tests done by Medin & Schaffer (1978), independent-cue models that did include item-level (rote) information generally did not appear to do more poorly than the exemplar model, motivating a further examination here of both types of models.

$$E_{X|t} \equiv \frac{\sum_{x\in X} Sim(t,x)}{\sum_{x\in X} Sim(t,x) + \sum_{y\in Y} Sim(t,y)} \quad (2)$$

In order to compare the models, we chose to use an instantiation of Type V stimuli in Shepard, Hovland, & Jenkins (1961). This set was important since cue validity and exemplar similarity predict different patterns of performance

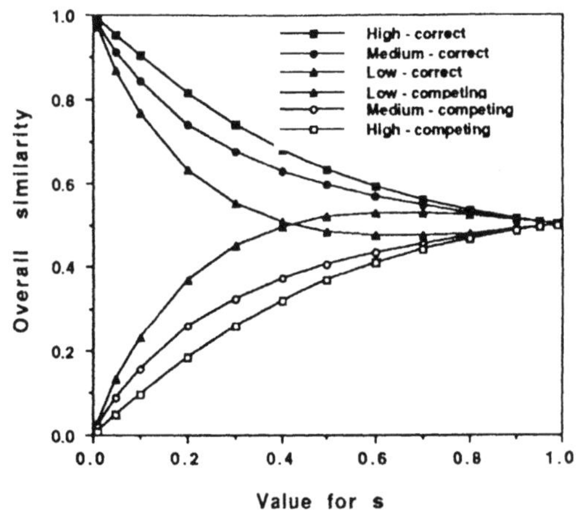

Figure 1: Overall similarity values for stimuli in Table 1, using (2).

across the learning trials. First, as shown in Figure 1, similarity calculations for the stimuli in Table 1 result in three groups, which we will term the *high-*, *medium-*, and *low-similarity* groups. The stimuli fall into these three groups for any value of s between 0 and 1, where s is the parameter estimated for (2) above. A sample set of similarities is shown in Table 1 for $s = \frac{1}{e}$. On the other hand, the sum of cue validities for each item in Table 1, for $a_i = 0.33$, shows that cue validities result in only two distinct groups. This is true whether the cue validity measure is "reliability" or "overall validity," as explained above.

Pseudo-word	*Category Label*	*Overall Sim*	$\sum$*Cue Validity*
zub	*Jets*	.70 (.30)	.58 (.42)
zud	*Jets*	.64 (.36)	.58 (.42)
zob	*Jets*	.64 (.36)	.58 (.42)
vod	*Jets*	.51 (.49)	.42 (.58)
vub	*Sharks*	.70 (.30)	.58 (.42)
vud	*Sharks*	.64 (.36)	.58 (.42)
vob	*Sharks*	.64 (.36)	.58 (.42)
zod	*Sharks*	.51 (.49)	.42 (.58)

Table 1. The *overall similarities*, using (2) and $s = \frac{1}{e}$), and *cue validities*, (using $a_i = 0.33$), are for the item's category; the value for the competing category is shown in parentheses.

The crucial comparison in this experiment was between the *high similarity* (zub, vub) and *medium similarity* (zud, zob, vud, vob) groups. Using the estimates shown in Table 1, the *Exemplar* model predicts a difference between these groups, based on their relative *similarities*. Neither the *Cue Validity* model nor the *Cue + Rote* model predicts a difference, and, in fact, there is no set of parameters for these two models that could separate the items into the *high* and *medium* subsets. In this experiment we tested to see whether the exemplar model provided a better fit to the data than either of the cue validity models at any point in learning.

Method

Subjects. Twenty-four undergraduates participated in this experiment for course credit.

Stimuli. The stimuli are shown in Table 1. Each category consisted of 4 three-letter pseudowords, which were presented to subjects as codenames for gang members in the Jets and the Sharks.

Procedure. Each subject was presented with 40 blocks of trials on an IBM AT clone, with the pseudowords appearing in random order within each block. Subjects used a rating scale of 0-9 to indicate membership for both gangs – i.e. subjects rated the pseudoword twice on each trial. The order of ratings was random. Feedback was provided after each trial to indicate the correct gang. Subjects were warned that early on in the experiment they would know little about the gang membership, so they should avoid extreme ratings.

Results

Since subjects were instructed to use whole number ratings, a middle rating (4.5), important in the early trials, was not available to them, and subjects tended to begin with ratings of 5. In order to convert the ratings to the range 0-1, to correct for the artifact of the rating scale, and to assure that the sum of residuals in the analyses was 0, each rating was divided by 9 and then 0.069 was subtracted.

In the current experiment, items should elicit high ratings for the item's own correct category and low ratings for the competing category. An examination of Table 2 shows higher ratings for high- vs medium- vs low-similarity items for the items' correct category; similarly, lower ratings for high- vs medium- vs low-similarity items for the items' competing category. An ANOVA using Similarity (high, medium, low), Rating Type (either for its own category or for the competing category), and Block showed a significant effect for the crucial 2-way interaction in these data: Similarity X Rating Type [$F(2,46) = 6.58$, $p < .004$, by subjects; $F(2,5) = 6.69$, $p < .04$, by items]. Importantly, the effect of the 3-way interaction was non-significant [F-values < 1, by subjects and items]. This suggests that there was a significant difference between the ***high, medium,*** and ***low*** items and that the effect did not vary significantly across the blocks of trials. One-*df* F-tests were used to verify that there was a significant difference between the mean ***high*** and mean ***medium*** ratings for items' own category (.70 vs .66: $F(1,23) = 13.94$, $p < .002$, by subjects; $F(1,4) = 9.14$, $p < .04$, by items), and between the mean ***high*** and mean ***medium*** ratings for items' competing category (.29 vs .33: $F(1,23) = 7.61$, $p < .02$, by subjects; $F(1,4) = 6.03$, $p = .07$, by items). As is evident in Table 2, the differences between ratings for ***high*** and ***medium*** similarity items clearly emerges in block 1, at least for items' own category. (Subjects' mean ratings for all the blocks are shown in Figures 3A and 3B.)

Correct category	*High*	*Med*	*Low*
Overall	.70	.66	.62
Block 1	.55	.50	.41
Block 2	.63	.47	.46
Competing category			
Overall	.29	.33	.37
Block 1	.47	.47	.59
Block 2	.37	.53	.52

Table 2: Mean ratings. (High, medium, and low groups are based on the overall similarity estimates in Figure 1.)

Fit to models. Each of the models was first assessed on a block-by-block basis – basically, 40 regression analyses for each model – using the models specified at (1), (2) above. This was to allow for the most liberal fit of parameters for each model and was equivalent to 40 hypothetical experiments for which testing would simply occur at the n-th block after $n - 1$ blocks of training. A comparison of the three models is shown in Figure 2 in terms of the residual error in the analyses done for each model at each block. The general result here is that all three models were quite close early on. After the first 5 blocks, the Cue Validity model began showing a clear disadvantage, and generally, the Exemplar model showed a slight advantage over the Cue + Rote model.

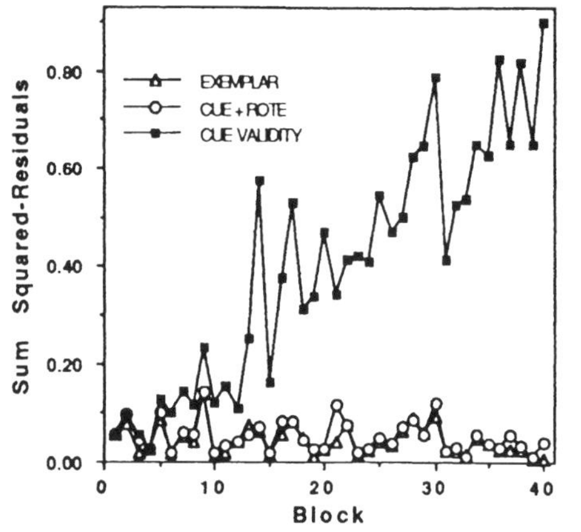

Figure 2: Fit of the three models.

To provide a uniform account of the **learning** that took place, we fit the data from all 40 blocks of trials by reinterpreting each s_i from the Exemplar model and each a_i from the Cue + Rote model as the logarithmic function in (3), with $constant_i$ defining the starting value for the redefined variable, s_i or a_i, and $lrate_i$ specifying how quickly it changes over the 40 blocks of trials. Figures 3A and 3B show the fitted Exemplar model, with (3) substituted for the s_is, superimposed on the human data. The overall fit of the model was excellent, with $R^2 = .90$. The overall fit of the Cue + Rote model (not shown) was similarly very good, with $R^2 = .89$. Figures 3C and 3D show how the reinterpreted s_i and a_i parameters change over the 40 blocks of trials. An important diagnostic characteristic of the Cue + Rote model is that it made identical predictions for *high* and *medium* similarity items, for each of the 40 blocks of trials. An examination of *human performance* in Figures 3A and 3B shows that this is a major flaw of the Cue + Rote model; the Exemplar Model correctly distinguishes between all three levels of similarity.

$$s_i, a_i = constant_i + lrate_i * \ln(\mathsf{block}) \quad (3)$$

Discussion

A major focus of recent work in categorization has been on learning, and a compelling insight has concerned between-item similarity, as first described by Medin & Schaffer (1978). As learning proceeds, the s parameter in the Exemplar model goes to 0. This reflects a reduction in the contribution of stored items that are "similar" to the test item on the categorization outcome. In the limit, the influence of other items is nil. The Cue + Rote model helps us to distinguish between the process in the Exemplar model and the buildup of rote associations. If they were similar, we might expect the two models to converge at some point in learning, but they clearly do not when one uses the high- and medium-similarity items to monitor the behavior of the models.

A question that has interested us is how the three s values that we fit in Figure 3C contribute to the categorization rating. A cursory examination of the distribution of the letter values in the second and third positions shows the reliability (conditional probability) of these letter values to be 0.5 – i.e. they are distributed equally in both categories. The first letter position is the only informative one. Interestingly, when we computed the *predicted ratings* using only the fitted Exemplar model parameters for the second and third letter positions, they were uniformly 0.5 for each item in each block. This means that the work in the Exemplar model is being done by the first letter position. This is somewhat striking, since it shows that the Exemplar model is fully consistent with predictions about cue informativeness that would be made based on cue validities. Yet, it is not simply cue validities, as tested in the Cue Validity model, that are being computed. Rather, the Exemplar model goes deeper to uncover something about the human representations that cue validities cannot capture.

At this point, it is not clear how relevant these results will be to the Competition Model, which is meant to account for children's natural language learning. It could indeed be the case that children do tend to pick up independent cues and over time organize these into a dominance hierarchy, as suggested recently by McDonald & MacWhinney (1991). Given the present result, though, it would seem worthwhile to consider the notion of competition from the perspective presented here.

The Exemplar model provides a mathematical formulation for category learning. It provides some insight into the characteristics of a process model, however, nothing nearly as complete as a blueprint. At this point it would be important to look at available models that have in recent tests demonstrated an excellent ability to model category learning problems of the sort presented here. Two models that we have in mind are the "backpropagation" model of MacWhinney, et al. (1989) and Kruschke's ALCOVE (1992). From our current perspective we can only speculate that the ability of models in this class to effectively model human data may depend crucially on the characteristics of "hidden units"–i.e. that part of the model that plays a major role in internal representations that the model processes.

Acknowledgments

We are indebted to Jerry Myers and Steve Dopkins for some of the original ideas for this research and to Yiannis Vourtsanis, and Vir Phoha for helpful discussions. We would also like to thank Sandra Douglas, Chris McGee, Mukesh Rohatgi, and Mark Stephan for help in organizing and running the experiment and in analyzing the data. Finally, our thanks to Bob Bell, Brian MacWhinney, Janet McDonald, Jerry Myers, Glenn Nakamura, and two anonymous conference reviewers for helpful comments on an earlier draft of this paper.

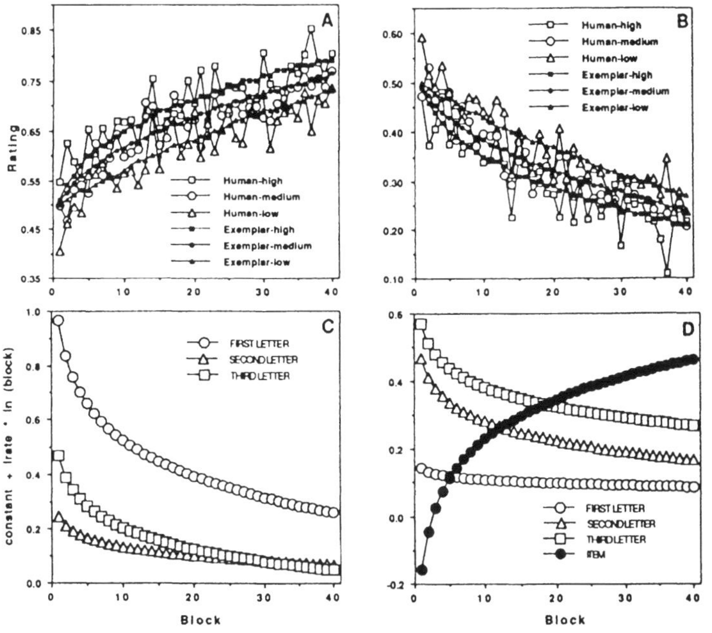

Figure 3: A: Data and model for correct category; B: data and model for competing category; C: plot of changes in parameter values for Exemplar Model; D: plot of parameters for Cue + Rote Model.

References

Anderson, J. R. (1991). The adaptive nature of human categorization. *Psychological Review, 98*, 409-429.

Bates, E., & MacWhinney, B. (1982). Functionalist approaches to grammar. In E. Wanner & L. Gleitman (Eds.), *Language acquisition: The state of the art.* Cambridge, UK: Cambridge University Press.

Bates, E. & MacWhinney, B. (1987). Competition, variation, and language learning. In B. MacWhinney (Ed.), *Mechanisms of language acquisition.* Hillsdale, NJ: Erlbaum.

Bates, E. & MacWhinney, B. (1989). Functionalism and the Competition Model. In B. MacWhinney & E. Bates (Eds.), *The crosslinguistic study of sentence processing.* New York: Cambridge University Press.

Beach, L. (1964). Cue probabilism and inference behavior. *Psychological Monographs, 78*, 21-37.

Kruschke, J. (1992). ALCOVE: An exemplar-based connectionist model of category learning. *Psychological Review, 99*, 22-44.

MacWhinney, B., Leinbach, J., Taraban, R., & McDonald, J. (1989). Language learning: Cues or rules? *Journal of Memory and Language, 28*, 255-277.

McDonald, J. & MacWhinney, B. (1991). Levels of learning: A comparison of concept formation and language acquisition. *Journal of Memory and Language, 30*, 407-430.

Medin, D. & Schaffer, M. (1978). Context theory of classification learning. *Psychological Review, 85*, 207-238.

Nosofsky, R. (1984). Choice, similarity, and the context theory of classification. *Journal of Experimental Psychology: Learning, Memory, and Cognition, 10*, 104-114.

Reed, S. (1972). Pattern recognition and categorization. *Cognitive Psychology, 4*, 382-407.

Rosch, E. & Mervis, C. (1975). Family resemblance studies in the internal structure of categories. *Cognitive Psychology, 7*, 573-605

Shepard, R., Hovland, C., & Jenkins, H. (1961). Learning and memorization of classifications. *Psychological Monographs: General and Applied, 75*, Whole No. 517.

Taraban, R., McDonald, J., & MacWhinney, B. (1989). Category learning in a connectionist model: Learning to decline the German definite article. In R. Corrigan, F. Eckman, & M. Noonan (Eds.), *Linguistic categorization* (pp. 163-193). Philadelphia: Benjamins.

A View of Diagnostic Reasoning as a Memory-directed Task

Roy M. Turner
Department of Computer Science
University of New Hampshire
Durham, NH 03824
rmt@cs.unh.edu

Abstract

Diagnostic reasoning underlies many intelligent activities, including (but not limited to) situation assessment/context recognition, natural language understanding, scene recognition, interpretation of scientific observations, and, of course, medical diagnosis and other forms of fault-finding. In this paper, we present a memory-directed, schema-based approach to diagnostic reasoning. Features of the problem are used to "evoke" one or more possible diagnoses, stored as schemas. Schemas contain information about their "manifestations" that be used to confirm or deny the diagnosis and, in some applications, information that can be used to take action based on the diagnosis. Potential advantages of the approach include cognitive plausibility, rule exception handling via (generalized) case-based reasoning, applicability to multiple domains, extensibility from experience, and a natural way to organize knowledge about what to do after a diagnosis is made.

Introduction

Diagnosis is central to much of intelligence. We can view all of the following as instances of the process of diagnosing features present in a reasoner's input: assessing its current situation (context recognition); understanding natural language; interpreting scientific observations by determining which theories they fit; and, of course, diagnosing medical problems and other kinds of faults.

We are interested in developing a computer-based model of diagnostic reasoning that supports our work in situation assessment and context-sensitive reactive reasoning in the autonomous underwater vehicle (AUV) domain, as part of the ORCA project (Turner & Stevenson, 1991). Such an approach would take features of the current problem-solving environment as input (derived ultimately from sensors or from information from other agents) and produce as output an assessment of the current problem-solving situation. This assessment would then provide the reasoner with the information necessary to assure that its behavior is appropriate for the context.

To be useful in a variety of real-world applications, a computer-based approach to diagnostic reasoning should have several properties. It should be general, that is, not tailored to one particular domain such as medical diagnosis. It should facilitate knowledge acquisition, both initially and throughout the problem solver's lifetime. This includes facilitating learning from the reasoner's own experience. Since complete knowledge is impossible to obtain in almost any domain, the reasoner should be able not only to handle usual situations, but also exceptions that follow no known rules. Finally, we should keep in mind that diagnostic reasoning is often only a means to an end. Consequently, a diagnostic reasoner should facilitate additional reasoning based on its diagnoses, for example, by allowing treatment information (for medical systems) or context-specific actions (for, e.g., reactive planners) to be associated with its representation of the diagnosis (a disease or context).

We are developing an approach to diagnostic reasoning that is based on the view of diagnosis as a memory-directed task; consequently, our approach is called MD, for Memory-based Diagnosis. This does *not* mean that memory is the sole mechanism underlying diagnosis in our approach; indeed, we have argued elsewhere (Turner, 1989a; 1989b) for a view of diagnosis as a planning task. Instead, we view the overall process as being *guided* by the organization of knowledge in the reasoner's memory.

It should be stressed that this is not completed work; rather, we build on some early work done in the domain of medical diagnosis during the MEDIC project (Turner, 1989a; 1989b; 1989c), and current work is ongoing in the area of situation assessment and context-specific reactive reasoning for AUVs. We believe that this approach shows promise not only for our domain but for diagnostic reasoning in general.

Diagnostic Reasoning

Diagnosis is fundamentally an abductive reasoning task (Reggia *et al.*, 1985): from a set of features of a problem or situation ("symptoms"), hypothesize a cause ("diagnosis") based on *a priori* knowledge linking causes to effects and effects to likely causes. The kind of reasoning done by diagnosticians (e.g., physicians) to implement this abductive inference is *hypotheticodeductive reasoning*: hypothesize a diagnosis, then test predictions made should that hypothesis be true.

Though most researchers agree on this much, differences of opinion arise with respect to a reasonable computational model of diagnosis. The prevalent model is backward chaining rule-based reasoning (Buchanan & Shortliffe, 1984). Chandrasekaran *et al.* (1979) describe diagnosis as *classification*: the process of assigning a problem to a particular location in a taxonomy of problems. Reggia and colleagues (1985), in perhaps the most formal treatment, view diagnosis as abduction, in particular parsimonious set covering. As mentioned above, the author has elsewhere argued for a view of the act of diagnosis as a planning task, view that attempts to tease out *how* diagnosis is done rather than what it is *per se*. The mechanism described in this paper is one part of the latter approach.

Most successful approaches to diagnosis have been rule-based, for example MYCIN (Buchanan & Shortliffe, 1984) and its descendants. Such systems, however, seem to miss an essential flavor of the hypotheticodeductive reasoning style as seen in humans, in which the reasoner is essentially "reminded" of possible diagnoses based on features in the problem.[1] Humans seem to be able to relate symptoms to diagnoses with relatively little cognitive effort, and with an accuracy that increases with the human's experience. This characteristic would be highly desirable in a computer-based diagnostic reasoner as well. To capture this aspect of diagnostic reasoning in a computer model would seem to require a memory of diagnoses, linked together (organized) to facilitate retrieval of diagnoses at appropriate times.

One system that begins to address this is INTERNIST-1/CADUCEUS (Miller *et al.*, 1982; Pople, Jr., 1982) (hereafter referred to as INTERNIST), a diagnostic reasoner in the domain of internal medicine. This system directly approaches diagnosis as an abductive task. Symptoms are stored with information about what diseases they *evoke*—that is, that they should bring to mind for an expert diagnostician. Unfortunately, INTERNIST's knowledge is relatively unstructured and the program has no facility to restructure its knowledge based on experience, nor does it provide a means of easily associating additional knowledge (e.g., treatments) with diagnoses. Another system, MDX (Chandrasekaran *et al.*, 1979) organizes diagnoses in a hierarchy that is traversed (by using rules) to find more specific diseases. However, the links between diseases in MDX's memory are static, and contain little or no information to help focus the reasoner's attention on one disease over another. Two other systems, CASEY (Koton, 1988) and MEDIC (Turner, 1989a; 1989b; 1989c), respectively a case-based reasoner (e.g., (Kolodner *et al.*, 1985)) and a schema-based reasoner (a generalization of a case-based reasoner), are more similar to the approach presented here. The kind of memory used by these programs allows a degree of interconnectedness not present in MDX, while at the same time providing mechanisms to allow updating from experience. However, though both of these systems use a content-addressable, flexible memory, they both also rely solely on mechanisms external to the memory to decide the appropriateness of evoked diagnoses.

What is desirable is an approach that can use the flexible, content-addressable memory style of CASEY or MEDIC, augmented with information such as INTERNIST has to provide a measure of likelihood of the diagnoses the reasoner is reminded of by the memory. Such a scheme would provide the reasoner with a metric, potentially updatable from experience, to allow it to effectively direct the diagnostic process.

Memory-based Diagnosis

Memory-based Diagnosis (MD) is based on using a content-addressable memory of schemas representing diagnoses to perform the hypothesis generation phase of hypotheticodeductive reasoning. The memory structures are organized to facilitate retrieval based on cues in the diagnostic problem, and they contain information useful in the context of the diagnoses they represent.

Before proceeding to describe the MD approach, a brief word on terminology is in order. Our terminology is a melange of terms borrowed from INTERNIST, the memory and case-based reasoning literature, and psychological research on the process of diagnosis. "Symptom" and "manifestation" will be used synonymously to mean a feature of a given situation. A "diagnosis" is a conclusion of a fault or disease causing the problem or a characterization of the current situation. A symptom's "evoking strength" (Miller *et al.*, 1982) for a diagnosis is a measure of how strongly its presence indicates the presence of that diagnosis. A diagnosis' "manifestation strength" for a symptom is a measure of how strongly the symptom is expected, given the diagnosis. The two terms "diagnostic task" (Miller *et al.*, 1982) and "logical competitor set" (LCS) (Feltovich *et al.*, 1984) are used synonymously to mean a set of diseases that are in competition to explain the same subset of manifestations; a reasoner seeks to "solve" an LCS by finding the best diagnosis from among its members.

In the MD approach, information about known diagnoses are represented as schematic memory struc-

[1] This is not to say that all remindings are based on surface features; other, deeper, features as they are discovered may also form the basis for hypotheses.

tures. Elsewhere, we have referred to similar memory structures as *contextual schemas* (c-schemas) (Turner, 1989a; 1989b); we will continue to use that terminology here, though the reader should be aware that we are extending our prior definition of c-schema slightly to accommodate diagnoses such as diseases. Figure 1 shows two example c-schemas.

Note that c-schemas contain additional information apart from the manifestation list (list of those manifestations that are expected). The actual contents vary depending on domain, but, in general, c-schemas can contain any information useful in the context of the diagnosis. In the case of a disease, this might be treatment information. In the case of situation assessment for an autonomous vehicle, this might include context-specific ways to handle unanticipated events, to order goals, to set parameters, or to select action, as discussed elsewhere (Turner, 1989c; 1990). The point is that by making the diagnosis, the information needed to deal with the diagnosis is immediately at hand.

The overall MD process can be described as follows:

1. Find candidate diagnoses.
2. Evaluate their "fit" to the current situation.
3. Form logical competitor sets (LCSs): sets of diagnoses in which each diagnosis competes to explain the same subset of the features in the problem.[2]
4. Solve the LCSs, obtaining one or more non-competing diagnoses to explain the problem.
5. Merge the diagnoses into an overall picture of the situation.

Finding candidate diagnoses. C-schemas are organized in a memory to facilitate their retrieval as well as their modification from future experience. The memory used in MEDIC is modeled after the CYRUS memory program (Kolodner, 1984), which is essentially a set of highly-interconnected discrimination nets. Such memories have been used with good results in case-based reasoners (CBR) and a schema-based reasoner (MEDIC) to find needed cases and schemas based on features of the current problem-solving environment. They are often referred to as "MOP memories", since the primary memory structures are memory organization packets (MOPs). MOPs are schemas that encode information about a class of episodes (in CYRUS); their primary function, as their name suggests, is to organize more specialized MOPs, which organize more specialized ones, etc., until the level of individual episodes ("cases", in the CBR literature) is reached. C-schemas are a kind of MOP; consequently, they serve as organizing points in the memory, with more general c-schemas indexing ones that are more specialized. We use the term "case" for the individual episodes in memory: cases of diseases that have been seen, situations that have been encountered, and so forth.

We should point out that our approach does not exclude case-based reasoning—far from it. MD, which is a kind of schema-based reasoning, can be validly viewed as generalized CBR: that is, CBR that uses generalized cases instead of or in addition to "real" cases of problem solving. We have argued elsewhere (Turner, 1989a) for the utility of this view.

C-schemas are linked together by *indices*, each of which is a feature type/value pair. The feature types present in a c-schema's indices correspond to types of features expected to be present in the diagnosis represented by the c-schema. For example, a medical reasoner's c-schemas might contain such feature types as: symptoms, patient characteristics, medical history, etc. The "value" half of an index is an indication of how the indexed c-schema differs from its parent. For example, a c-schema representing "lung disease" will index one representing "sarcoid", a common lung problem in inner-city hospitals in some regions of the country. Some feature type/value pairs we might expect as indices from "lung disease" to "sarcoid" would be: "patient home/Atlanta", "patient race/black", etc. Values are chosen that are different from values of the corresponding feature type in the parent c-schema; these are considered *predictive* differences by Kolodner; we can also think of them as *evocative* differences, since their presence *evokes* the indexed c-schema—that is, it "brings to mind" the diagnosis it contains. Figure 2 shows a portion of a hypothetical c-schema memory.

Traversal of a MOP memory entails first comparing features of the current problem to the features predicted by some subset of its top-level memory structures, then selecting potential indices based on salient (evocative) differences.[3] These potential indices are then compared to the c-schema's actual indices to produce a set of one or more c-schemas. Traversal continues in this manner using these structures until no further c-schemas are found. The c-schemas (and/or possibly episodes) found are, in our approach, the set of most specific possible diagnoses for the problem.

One difference in the MD approach from typical MOP memory applications is that we are interested in how strongly the current situation evokes a given diagnosis: this provides one measure of how well the diagnosis fits the problem, in the same manner as in INTERNIST. To provide for this, we augment

[2]This may, in turn, require looping to earlier steps as new information is uncovered; see, e.g., (Miller *et al.*, 1982).

[3]This process is actually quite a bit more complex than this, as described fully by Kolodner (1984).

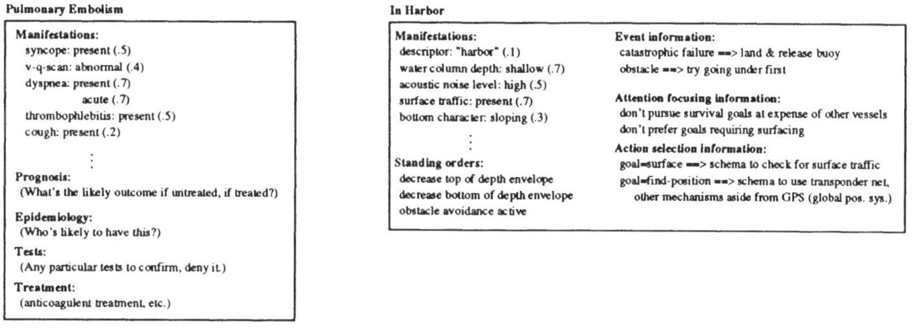

Figure 1: Contextual schemas from two different domains: medical diagnosis (left) and AUV control (right). Numbers refer to manifestation strengths.

the indices of c-schemas with an *evoking strength*, a measure of how strongly the reasoner should be reminded of a c-schema, given its parent c-schema and the current situation. An important property of MOP memories is that they can be updated from experience, and, as new episodes are added, the memory reorganizes its indices appropriately as well as updates the generalized information in its MOPs. In our scheme, the evoking strengths would also be updated when new information is added. This would theoretically allow the reasoner's "remindings" of possible diagnoses to become more accurate as it gains experience.

Of course, how strongly a diagnosis is evoked by a particular set of features in the problem description is not solely a function of the indices from the c-schema's parent. The strength with which the parent is evoked is also important. In addition, a c-schema can be reached through possibly many paths in a MOP memory, including multiple indices from the same parent; these different paths should contribute to the overall assessment of evoking strength. While we have not yet worked out the details of the combination function for evoking strengths, we can offer some initial observations. First, a c-schema should pass evoking strength to its children in proportion to how strongly it is evoked, yet we would not want simple multiplication of probability-like numbers, since this would result in a progressive decrease in evoking strength the deeper memory is traversed—the counter-intuitive situation of c-schemas that are better and better matches being less and less evoked! Second, when a c-schema is indexed more than once by a c-schema, or by more than one c-schema, its evoking strength should be a function of the evoking strengths of all the ways it is indexed, since these represent potentially different ways the c-schema matches the current situation. However, care will have to be taken to ensure that, as can happen in a MOP memory, the c-schema is not reached by two or more paths that are essentially the same, just permutations on the order in which features were examined.

The traversal process just described can determine a set of c-schemas comprising the potential diagnoses in one of at least two ways. As described above, memory traversal will create a set of all c-schemas encountered that the traversal process was not able to find children of, given the current problem features. The obvious drawback of this scheme is that the size of the set may grow quite large, and it may contain c-schemas that have very little to do with the actual problem (i.e., they were found solely due to slight resemblances). This means that the reasoner will have more hypotheses to select from, but will have to do more work to determine which hypothesis or small set of hypotheses is actually the correct diagnosis. A second approach would be to use information during memory traversal to narrow the search for hypotheses. The evoking strengths could be used during memory search to prune the search, in the manner of either beam search or a best-first search method. In this case, when considering memory traversal from some c-schema in memory, only the best n children might be considered next, where "best" is determined by evoking strength. With appropriate care, the memory could return either a single likely diagnosis or a set of likely diagnoses for the reasoner to consider.

These methods of finding a set of potential diagnoses are not mutually exclusive. It may be that different techniques are appropriate in different situations. For example, when in a hurry, the latter technique may be used, potentially sacrificing com-

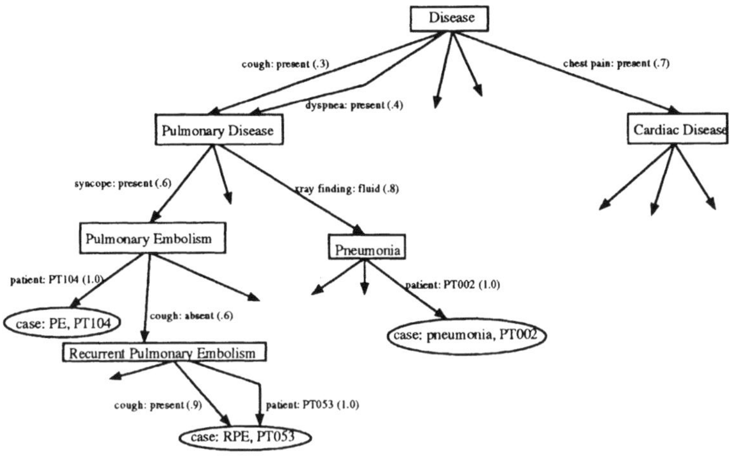

Figure 2: A portion of a c-schema memory. Numbers are evoking strengths.

pleteness for speed. In addition, it may be that if a reasoner is able to learn from its experiences, it can change the mechanism it uses. When a novice, a reasoner may painstakingly elucidate all possible diagnoses and carefully examine each one. With more experience, the reasoner's memory may become well-tuned, both in terms of its organization and the accuracy of its evoking strengths. In this case, the reasoner may use the latter strategy and immediately "home in" on the best diagnosis.

Evaluate fit of diagnoses. For this portion of the process, we turn to other work on diagnostic reasoning. We can, for example, use a mechanism patterned after INTERNIST's and formulate an overall score for each potential diagnosis from summing a positive component, based on the evoking strength, and a negative component, based on both manifestations predicted but not present and on unexplained manifestations. A better version of this scheme may be to consider manifestations that *are* explained as part of the positive component.

Logical competitor set formation and solution. The mechanism for these two steps can also be borrowed from INTERNIST. LCSs are useful for a diagnostician because, unlike rule-based reasoning, LCSs guide the strategies selected for information gathering. This allows the reasoner to play one disease off against others. For a more complete discussion of this, see (Miller *et al.*, 1982), (Feltovich *et al.*, 1984), or, in the context of a memory-based reasoner, (Turner, 1989a).

Merging diagnoses. For medical diagnosis and related tasks, when several complementary diagnoses are found, they can simply be output as a diagnosis list.[4] For other diagnostic tasks, however, it is not so simple. For example, in situation assessment for an autonomous underwater vehicle controller, the "diagnoses" are classes of situations; at any one time, the vehicle may be in several: operating with low power, under ice, within a harbor, etc. In domains such as this, the reasoner must merge the diagnoses to form an overall picture of its situation. This is an intended area for future research.

One exciting extension of the idea of representing diagnoses as c-schemas is the possibility of using a c-schema to represent a group of diagnoses that often occur together or that are often considered as a group during diagnosis. The former case should drastically improve the reasoner's ability to quickly diagnose a situation in which several different diagnoses are present, especially since the presence of one diagnosis often masks the presence of another (see, e.g., Sticklen (1987)). The latter case essentially means storing logical competitor sets as c-schemas; since LCS formation can be difficult, this, too, can improve the reasoner's performance, as suggested by Feltovich *et al.* (1984).

Conclusion

In this paper, we have briefly sketched memory-based diagnosis, an approach for using a dynamic conceptual memory to aid the process of diagnos-

[4]See Sticklen (1987), however, for a different view.

tic reasoning. There are several potential benefits of memory-based diagnosis. MD allows a reasoner to be "reminded" of appropriate diagnoses based on features in the problem being solved in such a way that the reasoner has an initial estimate of the accuracy of the diagnoses. Since the memory is of a type that can reorganize with experience, there is the possibility that as the reasoner gains experience, its memory will return better hypotheses more quickly. MD also allows the reasoner to store frequently-occurring logical competitor sets and clusters of diagnoses, both of which can improve diagnostic ability. Since additional information can reside in the memory structures, retrieving a diagnosis automatically retrieves knowledge about how to act on the diagnosis (e.g., treatment for a disease, appropriate behavior for a kind of situation, etc.). As the memory organizes diagnoses hierarchically and allows case-based reasoning, situations which do not follow usual rules can be stored and retrieved easily. And, finally, to the extent that the underlying memory is cognitively plausible (e.g., (Kolodner, 1984)), so is MD; indeed, there is some psychological evidence supporting the schematic representation of context (Chi *et al.*, 1982) and LCSs (Feltovich *et al.*, 1984) in expert problem solvers.

As we have said, this work is just beginning. Earlier work in the MEDIC project was encouraging, and we are currently investigating the approach in the domain of AUV control as part of the ORCA project. We believe that memory-based diagnosis is a promising approach to diagnostic reasoning, a kind of reasoning that underlies much behavior we consider intelligent.

References

Buchanan, B. G. & Shortliffe, E. H. (1984). *Rule-Based Expert Systems: The MYCIN Experiments of the Stanford Heuristic Programming Project.* Addison-Wesley Publishing Company, Reading, Massachusetts.

Chandrasekaran, B., Gomez, F., Mittal, S., & Smith, J. (1979). An approach to medical diagnosis based on conceptual structures. In *Proceedings of the Sixth International Joint Conference on Artificial Intelligence*, Stanford, California.

Chi, M. T. H., Glaser, R., & Rees, E. (1982). Expertise in problem solving. In Sternberg, R. J., editor, *Advances in the Psychology of Human Intelligence*, volume 1, pages 7–75. Erlbaum, Hillsdale, NJ.

Feltovich, P. J., Johnson, P. E., Moller, J. A., & Swanson, D. B. (1984). LCS: The role and development of medical knowledge and diagnostic expertise. In Clancey, W. J. & Shortliffe, E. H., editors, *Readings in Medical Artificial Intelligence*, pages 275–319. Addison-Wesley Publishing Company, Reading, Massachusetts.

Kolodner, J. L. (1984). *Retrieval and Organizational Strategies in Conceptual Memory.* Lawrence Erlbaum Associates, Publishers, Hillsdale, New Jersey.

Kolodner, J. L., Simpson, R. L., & Sycara-Cyranski, K. (1985). A process model of case-based reasoning in problem-solving. In *Proceedings of the International Joint Conference on Artificial Intelligence*, Los Angeles, California.

Koton, P. (1988). Reasoning about evidence in causal explanations. In *Proceedings of the DARPA Case-Based Reasoning Workshop*, pages 260–270, Clearwater Beach, Florida.

Miller, R. A., Pople, Jr., H. E., & Myers, J. D. (1982). INTERNIST–1, an experimental computer-based diagnostic consultant for general internal medicine. *New England Journal of Medicine*, 307:468–476.

Pople, Jr., H. E. (1982). Heuristic methods for imposing structure on ill-structured problems: The structuring of medical diagnostics. In Szolovits, P., editor, *Artificial Intelligence in Medicine*, pages 119–189. Westview Press, Boulder, Colorado.

Reggia, J. A., Nau, D. S., & Peng, Y. (1985). A formal model of diagnostic inference. I. Problem formulation and decomposition. *Information Sciences*, 37:227–256.

Sticklen, J. (1987). *MDX2: An Integrated Medical Diagnostic System.* PhD thesis, Department of Computer and Information Science, The Ohio State University.

Turner, R. M. (1989a). *A Schema-based Model of Adaptive Problem Solving.* PhD thesis, School of Information and Computer Science, Georgia Institute of Technology. Technical report GIT-ICS-89/42.

Turner, R. M. (1989b). Using schemas for diagnosis. *Computer Methods and Programs in Biomedicine*, 30(2/3):199–208.

Turner, R. M. (1989c). When reactive planning is not enough: Using contextual schemas to react appropriately to environmental change. In *Proceedings of the Eleventh Annual Conference of the Cognitive Science Society*, pages 940–947, Detroit, MI.

Turner, R. M. (1990). A mechanism for context-sensitive reasoning. Technical Report 90-68, Department of Computer Science, University of New Hampshire, Durham, NH 03824.

Turner, R. M. & Stevenson, R. A. G. (1991). ORCA: An adaptive, context-sensitive reasoner for controlling AUVs. In *Proceedings of the 7th International Symposium on Unmanned Untethered Underwater Submersible Technology (AUV '91).*

Defining the Action Selection Problem

Toby Tyrrell

email: lrtt@uk.ac.ed.cns
Centre for Cognitive Science
University of Edinburgh
Edinburgh EH8 9LW
Great Britain

Abstract

There has been a lack of progress in the field of action selection due to an incomplete understanding of the problem being faced. The differing nature of constituent parts of the action selection/'time-allocation' problem has not been properly appreciated. Some common sub-problems, such as obtaining food and avoiding predators, are described in terms of the demands they make on an animal's time. The significant differences between these sub-problems are highlighted and a *classificatory scheme* is proposed, with which sub-problems can be categorised. The need to take into account the full range of different sub-problems is demonstrated with a few examples. A particular shortcoming shared by all of the more well-known action selection mechanisms, from both robotics and animal behaviour, is described.

Introduction

Action selection is the choosing of the most appropriate action out of a set of possible candidates. The term *action* here refers to one of a set of mutually exclusive entities at the level of the behavioural final common path. That is, their demands on the effectors of the animal are such that only one of them may be executed at any one time. But how should the term 'most appropriate' be defined? The most appropriate action is that which maximises the number of copies of the animal's genes in future generations (assuming that an animal is just a carrier/propagator of its genes, as argued in (Dawkins, 89)).

There are two ways in which an animal can try to maximise the number of copies of its genes in future generations: (1) it can try to reproduce as often as possible (and so create new individuals with many of its genes), and (2) it can try to bring about the reproduction of other individuals which share many of its genes (e.g. offspring, siblings, parents).

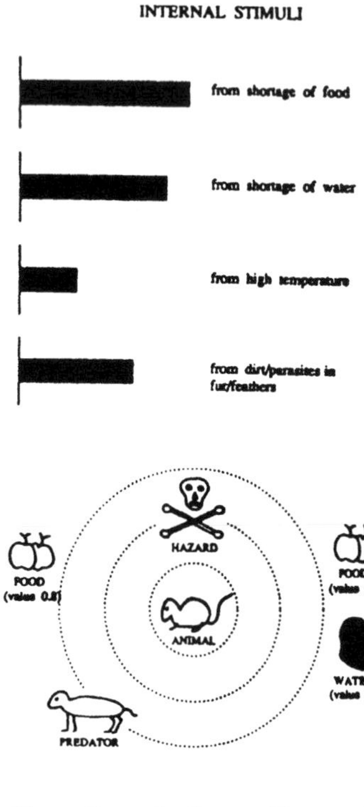

Figure 1: The action selection problem – what is the most appropriate action?

Obviously (1) engenders a further goal of surviving so as to be able to reproduce in the future, and (2) engenders a further goal of helping close genetic relatives to survive, so that they too can reproduce in the future. Therefore the 'purpose' of the mechanism selecting an animal's actions (and indeed of the rest of the animal as well) is to allow it to survive and to cause it to mate successfully as often as possible, and to help closely-related conspecifics to do the same.

This fairly abstract, high-level problem can be split into several *sub-problems*. For instance, the problem of survival leads to sub-problems such as obtaining enough food, avoiding predators, regulating body temperature, etc.

The following list contains some of the action selection sub-problems that commonly occur for different animals: obtaining food, obtaining water, regulating body temperature, avoiding predators, avoiding hazards in the environment, cleaning/preening, sleeping somewhere safe at night, and mating.

The problem of action selection can be thought of as that of allocating the animal's available time to different sub-problems. The nature of the time-allocation problem as a whole depends on the form of the individual demands on the animal's time made by each sub-problem. In this paper a description is given of the nature of some of these sub-problems. The different sub-problems are compared and a set of terms with which to describe them is presented. The usefulness of this tentative classificatory scheme is then demonstrated by showing that some proposed mechanisms for selecting actions are not able to deal successfully with some different types of sub-problem, or with some combinations of the different types.

In the rest of this paper the abbreviation AS will be used for action selection.

Classifying Sub-Problems of the Action Selection Problem

Several common sub-problems of the AS problem have been implemented in a complex simulated environment (Tyrrell & Mayhew, 91), and different AS mechanisms have been tested to see if they can cope with the different sub-problems present. Experience with this simulated environment has given the author some familiarity with the different types of sub-problem and with the problems of designing a mechanism to deal with them.

Some well-known mechanisms are (1) Maes (Maes, 91), a spreading activation network with two 'waves' of input from perception and motivations respectively and with a node for each behaviour; (2) Tinbergen (Tinbergen, 50 & Tinbergen, 51), a hierarchical model in which decisions are made at progressively lower levels until an action has been selected; (3) Lorenz (Lorenz, 50), the hydraulic model which uses an analogy with a reservoir of water to reproduce many of the phenomena of AS; and (4) the drive model (see (Hull, 43) or the description in (McFarland, 85)) which calculates a 'drive' for each sub-problem and then selects an action to satisfy the sub-problem with the highest drive. More explanation of these or other mechanisms (e.g. (Brooks, 86), (Ludlow, 80), (Halperin, 91)) cannot be given here.

Explicit planning systems are not considered here because (1) they presuppose a rather high degree of intelligence on the part of the animal, and (2) they are computationally and intellectually intractable in complex environments about which they only receive incomplete and unreliable information. It should also be noted that many of the mechanisms presented here are able to perform *implicit planning*. They can produce sequences of appetitive and consummatory actions in order to satisfy needs, but can interrupt these if urgent alternative actions are required.

Figure 2 shows some graphs of activations over time for the six example sub-problems now considered:

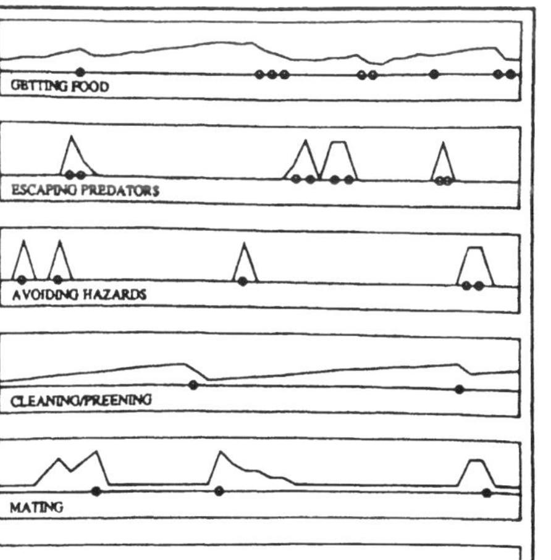

Figure 2: 'Longitudinal Profiles' for the six example sub-problems (after (Maes, 91)). The graphs show the 'urgency' of the sub-problem over time. The circles denote instances when the sub-problem in question determines the animal's action.

1. Trying to obtain enough food is the most commonly covered sub-problem in the ethological literature. The likelihood of the animal trying to obtain food should depend on both external stimuli (whether it sees food in the vicinity) and on internal stimuli (whether it is short of food). It is also a recurring, continual problem. An animal will need to eat a certain average amount each day. The need to obtain food will in general place a low-urgency demand on the animal's time, unless the animal is particularly short of food, since the animal is not likely to die suddenly or become injured if it does something else. Finally, food intake is a homeostatic problem.
2. A second common sub-problem is the need to escape predators. This is usually urgent and overriding, in that if the animal does not attend to it immediately then the consequences could be fatal. It is also highly dependent on external stimuli but not at all on internal stimuli. The priority which the animal should assign to trying to escape from a predator is highly dependent on whether the animal senses any predators, and if so how close they are. This sub-problem is a non-periodic, non-continual sort in that there is no pattern to how often the animal will need to attend to it. An animal may need to escape from a predator twice in one day and then not need to do so again for many days. There is no homeostatic aspect to this sub-problem.
3. A third sub-problem is that of avoiding hazards in the environment - places where an animal will endanger itself if it goes there (e.g. cliffs, streams). When an animal is near to one of these it is important for it that it does not move towards it. The demand on the animal's actions is *proscriptive* ('ruling-out'), rather than *prescriptive* ('specifying'), as is the case for other sub-problems. A proscriptive sub-problem specifies that certain actions should *not* be chosen (e.g. do *not* move towards a hazard) rather than that they should be chosen (e.g. eat food). This sub-problem will be urgent. There is no homeostatic or periodic or continual aspect to it.
4. Another common sub-problem is cleaning, preening or grooming. Most animals need to spend some time every so often to remove dirt/parasites from their fur, clean and oil their feathers, or whatever. This will not be an urgent activity, since it will not be crucial to the animal to pay attention to it at any particular moment in time. It will tend to occur most frequently at moments when no other activity is urgently required. It is continual in that the need for it will recur frequently. It is dependent on internal stimuli but not on external ones.
5. A fifth sub-problem is that of mating. External stimuli are important n that the animal should attach more priority to this sub-problem when a potential mate is perceived (assuming an animal which makes occasional matings with different mates and which forms no long-term partnerships). It is sometimes periodic and related to internal stimuli (e.g. menstrual cycles) and sometimes not. This sub-problem will be prescriptive, non-continual and non-homeostatic. It will probably be fairly urgent but the level of urgency in relation to the other sub-problems will depend on factors such as how often opportunities for mating arise and how much longer the animal can expect to live.
6. A final sub-problem is that of the animal needing to return to its den and sleep there at night. This is periodic (since it will occur every 24 hours). It will have increasing urgency as nightfall approaches. It is non-homeostatic and prescriptive.

Six different common sub-problems that compete for a 'share' of the animal's time have now been described. It does not matter so much that some of this set may not be completely general or that the description of them may not be completely accurate. The important point is that the competing demands on an animal's time vary in their nature. Some of the ways in which they vary have been highlighted by the preceding discussion. Past discussions of AS, and past proposals for AS mechanisms, have not taken fully into account the variety of different types of sub-problems. The study of AS/time-allocation has been held back by a lack of understanding of the different possible sub-problems and a vocabulary to describe them. A tentative vocabulary is now proposed here.

- **Homeostatic *v.* non-homeostatic** - a homeostatic sub-problem contains an internal variable which has a desired 'set-point' (optimal value), or at least a desired range of values. The behaviour of the animal will always act so as to return the value of the variable towards the set-point or range of values (Toates, 80).
- **External stimulus dependent *v.* external stimulus independent** - the urgency with which certain sub-problems should 'demand' the animal's attention is dependent on the appearance of certain external cues (e.g. getting water on

the stimulus of a water source, escaping predators on the appearance of a predator). Other sub-problems are independent of external cues (e.g. cleaning).

- **Internal stimulus dependent *v.* internal stimulus independent** - as for above, except that the important factor is an internal cue (e.g. body heat is too high, not enough food in the animal's stomach).
- **Periodic *v.* non-periodic** - some sub-problems such as sleeping at night are highly periodic, with the desirability of paying attention to them rising and falling with a regular rhythm.
- **Continual *v.* occasional** - some sub-problems need to be attended to frequently and the need for them keeps recurring (e.g. cleaning, getting food/water). They will need to be carried out several times each day. They are often internal stimulus dependent. Others only occur very occasionally and are usually external stimulus dependent (e.g. escaping predators, mating).
- **Degree of urgency** - some external stimulus dependent sub-problems arise only occasionally but are extremely urgent and over-ridingly important when they do occur (e.g. escaping predators, maybe mating). There will be significant consequences for the animal (in terms of future expected genetic fitness) if the sub-problem is not allowed to influence the action the animal selects. Some sub-problems generally have a fairly low urgency (e.g. cleaning), and tend to take over only when none of the more urgent sub-problems are relevant.
- **Prescriptive *v.* proscriptive** - most sub-problems require a certain set of actions to be carried out (e.g. find food, approach it then eat it), whereas others (e.g. avoid hazards) only require that certain actions should *not* be carried out.

This list is almost certainly incomplete in that there are other ways in which sub-problems can vary, but the most important differences are contained here.

Application of the New Classificatory Scheme

The previous section developed a list of terms that can be used to describe the sub-problems of an AS/time-allocation problem. Some well-known AS mechanisms will now be considered which do not produce optimal selections for some of the types of sub-problems outlined (although they perform satisfactorily in most respects). This is because they were designed without a full awareness of the differences that could occur in sub-problem parts of the AS problem.

Example 1 – Improper Combination of Proscriptive and Prescriptive Needs

Most of the proposed mechanisms do not take account of the fact that an AS sub-problem which only proscribes certain actions (e.g. the need to avoid a hazard prohibits movement towards that hazard) still leaves a wide range of actions which can be performed to the advantage of the animal. It can be seen in figure 3(a) that if the animal moves to the top left then it will both *not move towards* the hazard, and *move towards* the food.

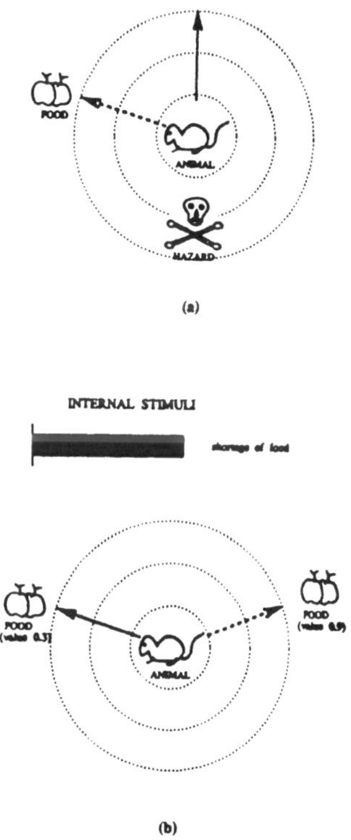

Figure 3: Example situations in which certain mechanisms may select the wrong action. In both (a) and (b) the dashed arrow indicates the optimal action and the solid arrow indicates the sub-optimal choice that might be made.

Most mechanisms cannot generate the optimal action because they select a particular sub-problem as the most urgent/appropriate, and *then* decide which action is most relevant for that sub-problem only. Mechanisms which have this fault are Maes, Tinbergen, Lorenz and the drive model.

Example 2 – Improper Selection between External Stimulus Dependent Sub-Problems

Consider the situation shown in figure 3(b). Here the animal is short of food, and there are two food sources with differing utility to the animal. Some mechanisms, e.g. Tinbergen and Maes, do not allow the *sizes* of external stimuli to affect the priority of a sub-problem. The information from the environment is expressed only in terms of logical conditions about the environment which can be either true or false (the environmental propositions of Maes, or the Innate Releasing Mechanisms of Tinbergen). Because the only information coming from the environment in their mechanisms is, in effect, along binary links, the two mechanisms cannot respond to varying sizes of external stimuli. Therefore they will not always be able to choose the most appropriate action in situations such as shown in figure 3(b).

Example 3 – Problems with Non-Periodic, Internal Stimulus Dependent Sub-Problems

The regulation of temperature is prescriptive, usually non-urgent, *non-periodic*, occasional, homeostatic and *internal stimulus dependent*. Consider an occasion on which the external temperature is fairly hot and the animal has recently been undertaking some strenuous activity. In this case, the animal's body temperature will rise and it should receive a high internal stimulus from its temperature receptors. Therefore there should be a high likelihood that its actions will result in a reduction of body temperature. Lorenz's hydraulic model (Lorenz, 50) is not able to reproduce this since the only source of internal, or endogenous activation energy is from his 'dripping tap' which flows at a constant rate. Since Lorenz's mechanism only allows internal variables which increase slowly with time and are then decreased by execution of the sub-problem, it cannot work for the case of non-periodic, internal stimulus dependent sub-problems.

Discussion – Making Decisions at the Wrong Level

One point to arise out of the previous section concerns the *level* at which decisions should be made. Lorenz's mechanism, Maes's mechanism, the drive model and Tinbergen's mechanism share a common attribute. Initially they all make a decision as to which sub-problem of the AS problem is the most appropriate (usually by calculating a value for each candidate sub-problem and picking the sub-problem with the highest value), and then later they make a decision as to what action is most appropriate for that sub-problem only. There is a two-stage process: (i) select a *sub-problem* (e.g. obtain food) to attend to, and (ii) select the most appropriate *action* for that sub-problem (e.g. move in a certain direction, eat food). While the mechanisms are fine in most respects, this common attribute would seem to be incorrect.

As described in the first example of the previous section, this approach leads to a shortcoming in that the needs of prescriptive and proscriptive sub-problems cannot be combined. The problem is more general than that though. No 'compromise' actions can be selected, whether the compromise is between prescriptive and proscriptive sub-problems, between two prescriptive sub-problems, or between two proscriptive sub-problems. This is because only one sub-problem is taken into account when selecting the best action. So, for instance, in the situation in figure 1 (ignoring the predator), the animal would not be able to choose the best action, that of moving to the right (medium-valued water, medium-valued food and not moving to the hazard). Instead, since it would only consider the most urgent need – to avoid the hazard – it would move directly away from that.

One AS mechanism which addresses the above problem and seems as if it would be able to cope with the whole range of AS sub-problems is (Rosenblatt & Payton, 89). This mechanism deserves further attention but cannot be considered here.

Conclusions

In the introduction to this paper the terms *action selection problem*, *sub-problem*, and *action selection mechanism* were defined. In §2 some of the most common sub-problems of an animal's action selection problem were described. This led on naturally to a set of descriptors for action selection sub-problems. The usefulness of this *classificatory scheme* was then shown in §3, in which several shortcomings with various action selection mechanisms were described using the new vocabulary.

The importance of the classificatory scheme presented in this paper is not just that it gives some convenient labels for describing action selection sub-problems. Rather, the importance lies in the whole way of thinking that it engenders. Action selection is the problem of dividing an animal's time amongst a number of sub-problems of differing nature. Past suggestions for action selection mechanisms have not come to terms with the wide range of differing sub-problems. Any valid action selection mechanism needs to be able to cope with all of them, and to interweave their demands on the animal's time/actions efficiently. Progress in the study of action selection has been hampered because this point has not been fully appreciated. It is to be hoped that the theory presented in this paper will enable more progress to be made.

Acknowledgements I would like to thank David Willshaw and John Hallam for their help and advice during the time I was carrying out this work. I would also like to thank Janet Halperin and Bridget Hallam for commenting on drafts of this paper. This project was supported by SERC (grant no. 89310818).

References

Brooks, Rodney A. 1986. A Robust Layered Control System for a Mobile Robot. *IEEE Journal of Robotics and Automation* 2:14–23.

Dawkins, Richard. 1989. *The Selfish Gene* (2nd edition). Oxford University Press.

Halperin, J. R. P. 1991. Machine Motivation. In Proceedings of the First International Conference on Simulation of Adaptive Behaviour. MIT Press/Bradford Books.

Hull, Clark L. 1943. *Principles of Behaviour: An Introduction to Behaviour Theory*. D. Appleton–Century Company.

Lorenz, Konrad. 1950. The Comparative Method in Studying Innate Behaviour Patterns. *Symposia of the Society for Experimental Biology* 4:221–268.

Lorenz, Konrad. 1985. *Foundations of Ethology*. Heidelberg: Springer-Verlag.

Ludlow, A. R. 1980. The Evolution and Simulation of a Decision Maker. In *Analysis of Motivational Processes*. Toates, F. M. & Halliday, T. R. eds. Academic Press.

Maes, Pattie. 1990. How To Do The Right Thing. *Connection Science Journal* (special issue on hybrid systems). 1(3).

Maes, Pattie. 1991. Bottom-Up Mechanism for Behaviour Selection In An Artificial Creature. In Proceedings of the First International Conference on Simulation of Adaptive Behaviour. MIT Press/Bradford Books.

McFarland, David. 1974. Time-Sharing as a Behavioural Phenomena. In *Advances in the Study of Behaviour (4)*. Lehrman D. S., Rosenblatt J. S., Hinde R. A. & Shaw E. eds. New York: Academic Press.

McFarland, David. 1985. *Animal Behaviour*. Pitman.

Rosenblatt, Kenneth J & Payton, David W. 1989. A Fine-Grained Alternative to the Subsumption Architecture for Mobile Robot Control. In Proceedings of the IEEE/INNS International Joint Conference on Neural Networks.

Tinbergen, Nikko. 1950. The Hierarchical Organisation of Nervous Mechanisms Underlying Instinctive Behaviour. *Sympos. Soc. Exper. Biol.* 4:305–312.

Tinbergen, Nikko. 1951. *The Study of Instinct*. Clarendon Press.

Tyrrell, T. & Mayhew, J. E. W. 1991. Computer Simulation of an Animal Environment. In Proceedings of the First International Conference on Simulation of Adaptive Behaviour. MIT Press/Bradford Books.

Analogical versus Rule-based Classification[1]

William D. Wattenmaker, Heather L. McQuaid, and Stephanie J. Schwertz
University of Pittsburgh
Department of Psychology/
Learning Research and Development Center
3939 O'Hara St.
Pittsburgh, PA 15260
wdw@vms.cis.pitt.edu

Abstract

Classification models have implicitly assumed that the nature of the representation that emerges from encoding will determine the type of classification strategy that will be used. These experiments, however, demonstrate that differences in classification performance can occur even when different transfer strategies operate on identical representations. Specifically, a series of examples was presented under incidental concept learning conditions. When the encoding task was completed, subjects were induced to make transfer decisions by analogy to stored information or to search for and apply rules. Across four experiments, an analogical transfer mode was found to be more effective than a rule-based transfer mode for preserving co-occurring features in classification decisions. This result held across a variety of category structures and stimulus materials. It was difficult for subjects who adopted an analytic transfer strategy to test hypotheses and identify regularities that were embedded in stored instances. Alternatively, subjects who adopted an analogical strategy preserved feature covariations as an indirect result of similarity-based retrieval and comparison processes.

Introduction

There are a variety of different strategies that people can use to learn concepts. Research that has examined the influences of learning strategies has found that different strategies produce clear differences in classification performance (e.g., Medin & Smith, 1981; Nosofsky, Clark, & Shin, 1989; Wattenmaker, 1991). Although this research has examined influences of alternative encoding strategies, it seems likely that there are also a variety of transfer or *postencoding* classification strategies that can be adopted. All research that has investigated relationships between strategies and classification performance, however, has only manipulated encoding strategies. In contrast, the present experiments investigated the influences that different transfer strategies have on classification performance.

In particular, the contrast between analytic and nonanalytic encoding strategies (e.g., Brooks, 1978; 1987) was extended to transfer strategies. In *nonanalytic transfer* conditions subjects were encouraged to make decisions by analogy to known instances (e.g., Medin & Edelson, 1988; Gentner, 1983; Holyoak & Koh, 1987). In *analytic transfer* conditions, however, subjects were encouraged to make transfer decisions by developing and applying rules. There is evidence that analogy can be effective for preserving complex regularities in decisions (e.g., Brooks, 1978, 1987; Wattenmaker, 1991), but there is very little research that has examined the ability of people to detect regularities that are embedded in stored information.

The general procedure in the experiments was to have subjects memorize a set of instances (short descriptions of hypothetical people) under incidental

[1]This research was supported by NIMH Grant MH45585 to William D. Wattenmaker.

concept learning conditions. After the examples had been memorized, the subjects were induced to make decisions by analogy to stored instances or to analyze the stored examples and develop rules. The analogy and analytic tasks were concealed until memorization had been completed. Thus all comparisons and analyses were heavily dependent on retrieval processes. Many features co-occurred in the memorized examples, and the central question was whether these co-occurrences would be preserved in classification judgments in the analogical and rule-based transfer conditions.

One possibility was that an analogical transfer mode would preserve co-occurrences as an indirect result of similarity-based retrieval mechanisms (Medin, Altom, Edelson, & Freko, 1982; Medin, 1983; Ross, 1989; Wattenmaker, Nakamura, & Medin, 1988), but that limitations associated with retrieving and analyzing stored information would make it difficult to recover co-occurrences in the rule condition (Wattenmaker, in press). In this case, an analogical transfer mode would be more effective than a rule-based transfer mode for preserving co-occurrences. A major way that encoding strategies influence classification performance, however, is by influencing *what* information is encoded (Medin & Smith, 1981; Medin, 1986). Thus, it is not clear that any differences in classification should be expected when different transfer strategies operate on identical representations.

Experiment 1

To examine the ability of analogical and rule-based transfer strategies to preserve correlated features, subjects in both conditions initially memorized descriptions of hypothetical people that had co-occurring features. When the descriptions had been memorized, a set of transfer tests was presented, and subjects were induced (through instructional manipulations) to make decisions by analogy to stored examples or to analyze the stored examples and develop rules. Performance on the transfer tests was examined to determine the extent to which decisions preserved the feature co-occurrences.

Method

Subjects. The subjects were 50 undergraduates from the University of Pittsburgh who participated in the experiment to fulfill course requirements.

Stimuli and procedure. Eight descriptions of hypothetical people (e.g., *likes diet pepsi, has a blue car, was born in July, likes apples, and has dark hair*) were memorized by subjects in both conditions. Four of the examples belonged to one category (Category A) and four of the examples belonged to another category (Category B). Each example had one feature on each of the following dimensions: beverage preference (diet coke or diet pepsi), color of car (blue or green), month of birth (July or August), fruit preference (apples or peaches), and hair color (light or dark). In terms of an abstract binary notation, the Category A examples were 11111, 11010, 00111, and 00001, whereas the Category B examples were 10100, 10010, 01100, and 01001. To construct the examples from this notation, the five stimulus dimensions (e.g., beverage preference) and features (e.g., diet coke vs. diet pepsi) were randomly assigned to the 1's and 0's. Thus, for one subject the pattern 11111 might have been represented by a description such as: *green car, July, apples, diet coke, and dark hair*. For another subject, however, this pattern might have been represented by the description: *diet pepsi, blue car, July, dark hair, and apples*.

The values on the first two dimensions were perfectly correlated in that the feature combinations 11--- and 00--- only occurred in Category A, whereas the feature combinations 10--- and 01--- only occurred in Category B. The assignment of features to values was counterbalanced, but if drink preference was on the first dimension and car preference was on the second dimension, then in Category A whenever *diet pepsi* occurred *blue car* also occurred, but whenever *diet coke* occurred *green car* occurred. Alternatively, in Category B, *diet pepsi* always occurred with *green car*, and *diet coke* always occurred with *blue car*. On the fourth and fifth dimensions, the value 1 was typical of Category A whereas the value 0 was typical of Category B.

After the eight examples had been memorized, subjects in the rule condition were instructed to analyze the stored examples and attempt to develop a rule or set of rules that would separate the descriptions in Category A from the descriptions in Category B. Subjects in the analogy condition, however, were discouraged from looking for rules. Instead of searching for rules, these subjects were instructed to make classification decisions by attempting to decide if the transfer items seemed more like the descriptions in Category A or the descriptions in Category B. The category assignments of the examples were not revealed until after the memorization phase of the experiment. Thus, all analyses of the categories were based on stored information.

The transfer tests also consisted of descriptions of hypothetical people (e.g., likes diet pepsi, has a blue car, born in August, likes apples, and has dark hair). Twenty-four different transfer test were presented. All of the transfer tests were new items but the values on the correlated dimensions were either consistent with Category A or Category B. With the transfer test 11011, for example, the values on the correlated dimensions (i.e., 11---) favored a Category A response. The results were analyzed in terms of the proportion of responses that were consistent with the correlated features. Classifying 11011 as a member of Category A, for example, would be consistent with the correlated features (see Wattenmaker, McQuaid, & Schwertz, 1992, for additional details).

Results and Discussion

Across the twenty-four transfer tests, there were significantly more decisions that were consistent with the co-occurring features in the analogy than the rule condition (.65 vs. .54), $F(1,38) = 4.71$, $p<.05$, $MSe=14.93$. Examination of the pattern of errors across the transfer tests revealed that subjects in the analogy condition preserved co-occurrences as a by-product of using similarity relations. The transfer items appeared to be placed in the category that had learning examples that were perceived to be highly similar to the transfer items. This process of retrieving similar learning examples, and making classification decisions by analogy to the retrieved examples, indirectly preserved the correlations. Subjects in the rule condition, however, had difficulty identifying the co-occurrences. Instead, they developed simple rules that were either inaccurate or had little generality. The retrieval and computational processes that are required with postencoding analyses appeared to make it difficult to develop accurate rules that involved features from multiple dimensions.

Experiment 2

Although little sensitivity to co-occurrences was observed in the rule condition of Experiment 1, the presence of simple rules (i.e., typical features) might have prevented subjects from entertaining more complex hypotheses. Thus, this experiment was identical to Experiment 1, except that the category structure was altered so that there were no regularities in the categories other than the co-occurrences. (In terms of the abstract notation the Category A examples were 11100, 11011, 00001 and 00110, whereas the Category B examples were 10001, 10110, 01100, and 01011).

Results and Discussion

Again an analogical transfer mode was very effective for preserving co-occurrences, but very few subjects in the rule condition were able to detect the co-occurrences. Overall, when the co-occurrences were consistent with similarity relations, 86% of the decisions in the analogy condition preserved the co-occurrences whereas 72% of the decisions in the rule condition preserved the co-occurrences, $F(1,42) = 5.08$, $MSe = 7.52$, $p<.05$. As in Experiment 1, subjects in the analogy condition appeared to make decisions by analogy to highly similar learning examples, and the feature co-occurrences were preserved as a by-product of this process. It was still difficult to capture the co-occurrences with postencoding analyses, however, as only four subjects accurately reported the correlation. Indeed, eleven subjects in the rule condition reported that they were unable to find a rule, and ten of these subjects spontaneously adopted an analogical strategy.

Experiment 3

Experiment 3 was designed to produce conditions that would be especially conducive to rule-based transfer. Specifically, the materials in the first two experiments consisted of lengthy lists of unrelated features, and the examples were highly similar, unfamiliar, and poorly integrated (e.g., blue car, July, diet coke, light hair, and apples). Materials of this type make it difficult to retrieve examples and to keep retrieved examples active in working memory. Thus, to increase the accessibility of the examples and the features, an attempt was made to use distinct, familiar, and well-integrated examples.

As in Experiments 1 and 2, the subjects were shown exemplars from two categories, and within each of these categories specific pairs of features were perfectly correlated with each other. The following features were used to construct the examples: male vs. female, politician vs. entertainer, and active career vs. inactive career. However, rather than presenting these features in list form, a single expression that integrated the features was presented. For example, if the underlying features of an example were *male, politician, and inactive*, then rather than presenting this list of features, the name of a well-known person who represented this combination was presented (e.g., *Winston Churchill).* Thus, the feature co-occurrences that were present in the underlying features (e.g., that male co-occurred with inactive) were preserved in the specific examples.

These familiar, well-integrated, and distinct examples should facilitate rule-based transfer by increasing the ease and accuracy of retrieval, by making it easier to keep retrieved features active in working memory, and by minimizing confusions between exemplars during retrieval and analysis. To eliminate potential problems in identifying the relevant features, participants were provided with the relevant features before the start of the transfer phase.

Method

Participants in both conditions initially memorized name-number associations (e.g., Winston Churchill-12). There were sixteen famous people (eight in each category) and the categories were distinguished by feature co-occurrences. In terms of the abstract notation, the examples in Category A were 110, 111, 111, 111, 000, 001, 001, and 001, whereas the examples in Category B were 101, 100, 100, 100, 011, 010, 010, and 010. Although some of the abstract patterns re-occurred within a category, each occurrence was represented by a different person. For instance, if the pattern 111 corresponded to the features *male, political, and inactive*, then in one case this pattern might be represented by Abraham Lincoln, in a second case by Thomas Jefferson, and in a third case by Winston Churchill.

The features on the first two dimensions co-occurred. These co-occurrences were of the same form as the co-occurrences in Experiment 1 and 2. Although the correlations were never directly presented, they were preserved in the exemplars. Consider, for example, a case where the Category A examplars were Ted Kennedy, John Adams, Thomas Jefferson, Winston Churchill, Barbra Streisand, Judy Garland, Greta Garbo, and Rita Hayworth, whereas the Category B examples were Eleanor Roosevelt, Margaret Thatcher, Geraldine Ferraro, Sandra O'Connor, Charlie Chaplin, Paul Newman, Robert DeNiro, and Michael Douglas. Notice that in Category A all the males were connected to politics whereas all the females were connected to entertainment. In Category B, however, all the females were connected to politics whereas all the males were connected to entertainment. Thus, all of the co-occurrences were implicitly represented in the exemplars.

Following the paired-associate learning procedure, the numbers (but not the names) were used to reveal the categories, and participants were induced to make transfer decisions by analogy or to search for rules. Before the transfer tests were presented, subjects in all conditions were given the features that had been used to construct the names. Unlike the learning examples, the transfer items consisted of lists of relevant features (e.g., male, political, active), and the task was to place the items in Category A or B. Again the results were analyzed in terms of the proportions of the responses that were consistent with the correlation. In terms of the categories that were illustrated above, for example, classifying the description *political, male, active* as a member of Category A would be consistent with the correlation. Ratings were collected to ensure that the famous people were perceived to have the correct combinations of underlying features.

Results and Discussion

Although very different stimulus materials were used in this experiment, exactly the same results that were observed in Experiments 1 and 2 were obtained: participants who used an analogical transfer mode were significantly more likely to make classification decisions that were consistent with the co-occurring features than subjects who used a rule-based transfer mode (.78 vs .64), $F(1,62) = 5.02$, $p<.05$, $MSe=3.81$. Even though the stimulus materials were designed to facilitate hypothesis testing in the rule condition, very few subjects in the rule condition were able to detect the co-occurrences. Instead, these subjects tended to develop rules that were inaccurate or that had very little generality.

Experiment 4

The feature co-occurrences were implicitly represented in the examples that were presented in Experiment 3. Although these materials were designed to facilitate postencoding rule abstraction, materials of this type have rarely been used and it is possible that intra-dimensional correlations of the type used in Experiment 3 are difficult to detect. To examine this possibility, the same materials that were used in Experiment 3 were presented to subjects under rule-based *encoding* conditions rather than rule-based memory conditions (i.e., the examples were visible rather than stored in memory during analysis).

Results and Discussion

When the names were visible during rule-seeking activity the vast majority of subjects detected the co-occurrences. Indeed, 90% of the classification decisions in the on-line rule condition preserved the co-occurrences. Thus, the failure of subjects in the rule condition of Experiment 3 to detect the co-occurrences appears to reflect basic limitations associated with analyzing information in memory.

Experiment 5

An analogical transfer mode was very effective for preserving co-occurrences in the first three experiments. In these experiments, however, the transfer tests were lists of features, the co-occurring features were expressed directly in this list, and no irrelevant properties were included in the transfer items. For example, although the learning examples in Experiment 3 were unique (e.g., Winston Churchill), the transfer tests consisted of lists of features that included the co-occurrences (e.g., "male, political, and inactive" where the co-occurrence was between male and inactive). To see if correlated features would be preserved as a by-product of analogy in noisier retrieval environments, the same materials and category structures that were used in Experiment 3 were used in this experiment but rather than presenting the underlying features on the transfer tests, new names of famous people (e.g., *Abraham Lincoln*) were used as transfer items. These names contained implicit correlations that were either associated with Category A or Category B. Clearly, when learning and test items both possess unique or irrelevant features, there are a multitude of idiosyncratic properties that can be ascribed to the items (nationality, political views, personality attributes, etc.). Thus, to see if an analogical transfer mode would still be effective under these conditions, learning and transfer items that possessed a wealth of irrelevant properties were presented. The procedure in this experiment was identical to Experiment 3, except that only an analogical condition was tested.

Results and Discussion

Even under more difficult retrieval conditions, an analogical transfer mode was effective for preserving co-occurring properties. Indeed, for those subjects who used analogy, 73% of the time they selected the category that was consistent with the co-occurring features, $t(39)= 2.23$, $p< .05$. The co-occurrences appeared to constrain retrieval and similarity calculations, and the influence of these implicit co-occurrences was strong enough to override possible influences of irrelevant features.

General Discussion

In all of the experiments, an analogical transfer mode was more effective than a rule-based transfer mode for preserving feature co-occurrences. This result held across a variety of category structures and stimulus materials and was highly consistent within as well as between condition: whenever subjects adopted a rule-based transfer mode they were significantly less likely to preserve correlated features in classification decisions than subjects who relied on analogy.

Although there was no direct awareness of the co-occurrences in the analogy condition, the process of making decisions by analogy to retrieved examples indirectly preserved the co-occurrences. This mechanism was effective even when the co-occurrences were not explicitly presented in the examples, and when both learning and transfer items had irrelevant features.

Subjects who used an analytic transfer mode, however, had difficulty detecting feature co-occurrences. Even when the category structures (Experiment 2) and stimulus materials (Experiment 3) were designed to facilitate postencoding analyses, little sensitivity to correlated features was observed. Results that are very similar to the current pattern of results have been observed with memory-based category construction tasks (Wattenmaker, in press). The results of these two sets of experiments suggest that when people attempt to induce rules from stored instances they will have a tendency to develop simplistic or inaccurate generalizations at the cost of missing more complex regularities that exist in stored information. In general, the retrieval and computational processes that are required with postencoding analyses appear to make it difficult to develop accurate rules that involve features from multiple dimensions. Many of these difficulties appear to be due to limitations in working memory capacity. An analogical transfer mode is adaptive because it provides a way for relational properties such as co-occurrences to be retrieved and to influence classification decisions without overwhelming processing and memorial capacities.

References

Brooks, L. (1978). Nonanalytic concept formation and memory for instances. In E. Rosch and B. C. Lloyd (Eds.), *Cognition and Categorization* (pp. 169-215). New York: Lawrence Erlbaum Associates.

Brooks, L. (1987). Decentralized control of categorization: The role of prior processing episodes. In U. Neisser (Ed.), *Concepts and Conceptual Development: Ecological and Intellectual Factors in Categorization* (pp. 141-174). Cambridge: Cambridge University Press.

Gentner, D. (1983) Structure-mapping: A theoretical framework for analogy. *Cognitive Science, 7,* 155-170.

Holyoak, K. J. & Koh, K. (1987). Surface and structural similarity in analogical transfer. *Memory & Cognition, 15,* 332-440.

Medin, D. L., & Smith, E. E. (1981). Strategies and classification learning. *Journal of Experimental Psychology: Human Learning and Memory, 7,* 241-253.

Medin, D. L., Altom, M. W., Edelson, S. M. & Freko, D. (1982). Correlated symptoms and simulated medical classification. *Journal of Experimental Psychology: Human Learning and Memory, 8,* 37-50.

Medin, D. L. (1983). Structural principles in categorization. In Tighe, T. J. & Shepp, B. E. (Eds.). *Perception, Cognition, and Development: Interactional Analyses,* (pp. 203-230). Hillsdale, NJ: Lawrence Erlbaum.

Medin, D. L. (1986). Comment on "Memory storage and retrieval processes in category learning." *Journal of Experimental Psychology: General, 115,* 373-381.

Medin, D. L., & Edelson, S. (1988). Problem structure and the use of base rate information from experience. *Journal of Experimental Psychology: General, 1,* 68-85.

Nosofsky, R. M., Clark, S. E., & Shin, H. J. (1989). Rules and exemplars in categorization, identification and recognition. *Journal of Experimental Psychology: Learning, Memory and Cognition, 15,* 282-304.

Ross, B. H. (1989). Remindings in learning and instruction. In S. Vosniadou & A. Ortony (Eds.). *Similarity and Analogical Reasoning* (pp. 438-469). Cambridge: Cambridge University Press.

Wattenmaker, W. D., Nakamura, G. N., & Medin, D. L. (1988). Relationships between similarity- based and explanation-based categorization. In D. Hilton (Ed.), *Science and Natural Explanation: Common Sense Conceptions of Causality* (pp. 204-240). Brighton, England: Harvester Press.

Wattenmaker, W. D. (1991). Learning modes, feature correlations, and memory-based categorization. *Journal of Experimental Psychology: Learning, Memory, and Cognition, 17,* 908-923.

Wattenmaker, W. D. (In press). Relational properties and memory-based category construction. *Journal of Experimental Psychology: Learning, Memory, and Cognition.*

Wattenmaker, W. D., McQuaid, H. L., & Schwertz, S J. (1992). Rules, analogy, and memory-based classification. Under review.

A Simple recurrent network model of serial conditioning: Implications for temporal event representation

Michael E. Young
University of Minnesota
317 Elliott Hall
Minneapolis, MN 55455
young@turtle.psych.umn.edu

Abstract

Elman (1990) proposed a connectionist architecture for the representation of temporal relationships. This approach is applied to the modeling of serial conditioning. Elman's basic simple recurrent network (SRN) was modified to focus its attention on the prediction of important events (Unconditioned Stimuli, or USs) by limiting the connection weights for other events (the Conditioned Stimuli, or CSs). With this modification, the model exhibited blocking and serial conditioning to sequential stimulus compounds. An exploration of the underlying mechanisms suggests that event terminations (CS offsets) were used in predicting US occurrences following simple trace conditioning and event beginnings (CS onsets) were more important following serial conditioning. The results held true under a series of learning rate and momentum values.

Introduction

The study of classical conditioning, beginning with Pavlov's demonstration of dogs salivating to bells in the early 1900s, is perhaps the most mature area of contemporary psychology. The Rescorla-Wagner model (Rescorla & Wagner, 1972) is one of the best known attempts to explain and predict classically conditioned behavior. Their model focused on the behavior of subjects at the *trial level*, which differs from recent models (e.g. Desmond, 1990; Grossberg & Levine, 1987; Grossberg & Schmajuk, 1987; Lee, 1991; Sutton & Barto, 1981, 1990) where the focus is on *intratrial* stimulus relationships. To demonstrate the distinction, consider the case of *blocking*. Here we have two concurrent events (CS_1 and CS_2) that consistently precede a third event (the US). However, the subject has received prior training in which CS_1 consistently preceded the US. This pretraining retards or *blocks* the subsequent learning of the CS_2-US relationship. Trial level theories can account for this observation, but they are silent on the role of intratrial variables like inter-stimulus interval (ISI - the temporal distance between CS onset and US onset), the temporal relationships among multiple CSs, and the duration of the CSs and USs. Intratrial models like the present one are designed to address these variables.

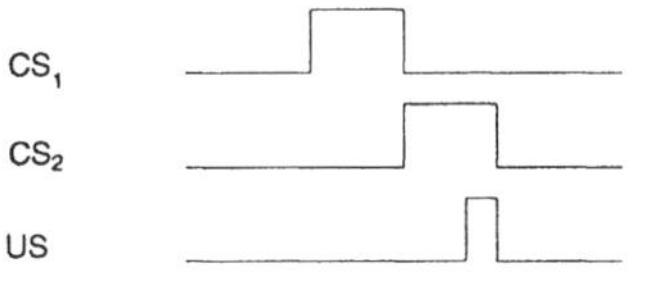

Figure 1: Serial conditioning

The particular classical conditioning paradigm investigated here is that of *serial* or *sequential* conditioning. In serial conditioning, multiple CSs precede the US, but unlike simple compound conditioning where the CSs co-occur, serial CSs are sequentially ordered (see Figure 1) where the first CS is presented in a *trace* relationship with the US. *Trace conditioning* is the term used to describe a situation where the CS is no longer present at the time of occurrence of the US. The second CS serves to provide a mechanism by which the first event acts on the US. In fact, earlier work (e.g. Bolles, Collier, Bouton & Marlin, 1978; Kehoe, Gibbs, Garcia & Gormezano, 1979) has shown that an intervening CS facilitates learned responding to a CS presented in a trace relationship to the US. In essence, it "bridges the gap."

Neural modelers have used a variety of approaches to capture temporal relationships among inputs. The approach adopted in this article is based on the work of Elman (1990). Elman used a simple recurrent network (SRN) to encode time where inputs can have sustained effect via a recurrent delay loop (see Figure 2).

This research supported by the Center for Research in Learning, Perception and Cognition and the National Institute of Child Health and Human Development (HD-07151).

The activations of the hidden units are fed back as input to themselves during the following time step. This permits the model to have some memory of its previous state. Elman trained this type of network to predict the input on the next time cycle: the values presented at the output layer at time *t* are identical to the inputs at time *t+1*. The primary goal of this study is to evaluate the promise of SRNs for modeling intratrial relationships. The secondary goal is to determine the mechanisms underlying the encoding of these temporal relationships.

The Model

Earlier work with SRNs demonstrated their ability to model basic excitatory conditioning and phenomena such as blocking (Young, unpublished data). To show blocking it was necessary to treat the two classes of events, the CSs and the USs, differently. Historically, there have been two theoretical approaches to capturing this difference and the relationship between the CSs and the US. Mackintosh (1975) assumes that concurrent CSs must compete for attention. Blocking occurs because the subject learns to attend to the pretrained CS_1 thus interfering with later attending to CS_2. Alternatively, Rescorla & Wagner (1972) assume a competition among the CSs for US associative strength. Blocking occurs because CS_1 has captured most of the associative strength available from the US. The US is no longer a surprising event (being predicted by CS_1) and thus does not require any additional predictors.

The software (*tlearn*) used in these simulations was developed at the University of California - San Diego's Center for Research in Language. Tlearn provided a mechanism to encourage CS competition. Figure 2 illustrates the architecture used. The weighted connections between the internal representations (the hidden layer) and the non-US portion of the output layer were limited in value. The limits were chosen as a result of the earlier work with blocking and represent one of the free parameters within this model. These limits constrain the degree to which errors in CS prediction can affect the learning process.

As the weights increase, hidden nodes will be more sensitive to US errors than to CS errors. If the US is being adequately predicted, less error will be propagated back to the internal connections. This process is analogous to the competition among CSs for US associative strength. When any error in US prediction is reintroduced (e.g. a change in salience), the model will be sensitive to these changes, thus allowing the CSs to compete for the ability to predict the "new" US.

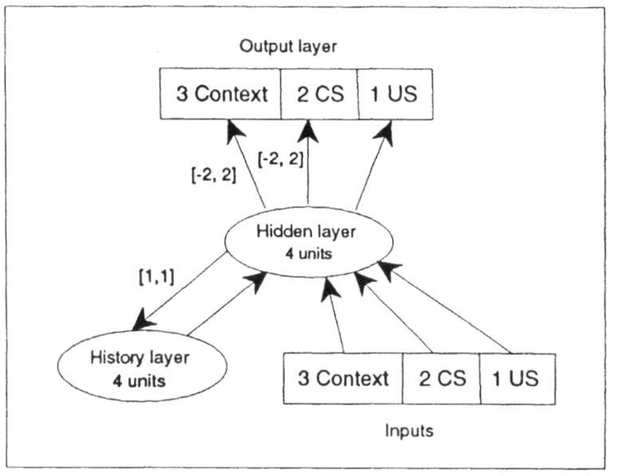

Figure 2: The SRN architecture used in the simulations. Labeled arrows have a limit on connection weights: [lower limit, upper limit]. The [1,1] connections are the copy back links discussed in Elman (1990).

In the first set of simulations, the author tested the model's performance during serial conditioning (see Figure 1). Learning of the CS_1->US relationship should be facilitated by the presence of the intervening CS_2 as compared to a control without CS_2 (Bolles *et al*, 1978). In the second set of simulations the roles of the onset and offset of the trace CS following both serial and trace conditioning were tested. The results were compared *qualitatively* to previous empirical research. Hence, no claims as to the correspondence between time in the model and real time will be made. The simulations demonstrated the performance of the model under different parameter settings (learning rate and momentum) to examine their effect on the qualitative results.

The output of the US node is the dependent variable of interest. This is a measure of the model's US expectancy for the following time step on a scale of [0,1]. Most previous modeling work uses the conditioned response (CR) as a dependent variable. Since I am not prepared to deal with the issues of learning vs. performance, I opted for a measure of the model's learning and suggest that the CR is a function of the US expectancy. For comparison purposes, it may be assumed that measures of CR and US expectancy are correlated.

Simulation 1

The first set of simulations were run to examine performance of the model during serial conditioning.

Method

Tlearn was trained on 2 different training sets. One training set represented the serial paradigm and consisted of 2 sequential CSs preceding the US. The first CS (CS_1) was three time steps long and was immediately followed by the second CS (CS_2), also three time steps long. During the third time step of CS_2, the US was presented and lasted for one time step, overlapping with CS_2 (Figure 1). The ITI (inter-trial interval) alternated between 5 and 7 time steps. On the average, 12 time steps corresponded to one trial or CS-US pairing. The presence of a CS or US was signalled by a 1 at the corresponding input value. The trained output values, as in Elman's (1990) model, were the input values for the subsequent time step. Learning was accomplished via backpropagation (Rumelhart, Hinton & Williams, 1986). The model also contained 3 additional inputs/outputs that were present for future work regarding the effect of contextual cues on conditioning. For the current simulations, these values were constant with values of [.5, 1, .5]. The second training set represented a control in which the intervening CS_2 was absent. Previous empirical work (e.g. Bolles *et al*, 1978) suggests that learning of the CS_1->US relationship should be slower following trace as compared to serial conditioning.

The model was run six times for each of three sets of parameter settings. Learning rate, designated r, and momentum, m, were set to the following: 1) r=.1, m=0, 2) r=.2, m=0 and 3) r=.1, m=.3. After training, performance was measured in response to CS_1, CS_2 and the CS_1->CS_2 compound *in the absence of the US.* No learning was permitted during this phase, thus preventing any extinction. The ITI between the end of the last US and the first of the test CSs was longer than that present during training to encourage the model to flush its temporal memory of previously occurring stimuli. The average ITI during training was 6 while that during testing was 12.

Results

The general results are presented in Figure 3 which represent the model's performance under the r=.2, m=0 settings. Qualitative results for the other settings were quite similar and will be described below. Figure 3 illustrates the US expectancy as a function of time since CS onset. Note that for optimal prediction, the peak of US expectancy should occur on the time step *before* presentation of the US. This was true for CS_1 under all of the parameter settings and after both serial and trace conditioning. Regarding the facilitation of learning the CS_1 ->US relationship, the results are mixed. The expectancy appeared to grow faster under serial conditioning during the first few 1000 time steps. This qualitative result is clearest with the r=.1, m=0 settings.

Given the small sample size (n=6), the only statistically significant difference between serial and trace condition peaks (at 4 time steps following CS onset) was at 30,000 time steps for r=.1, m=0 ($t(10)=2.408$, $p=.037$). Most of the other apparent differences at step 4 had p-values $< .2$.

With further training, peak expectancy to CS_1 under both conditions reached approximately the same asymptote (after 40K time steps, p=.168 for r=.1, m=0; p=.225 for r=.1, m=.3; p=.741 for r=.2, m=0).. However, note that US expectancy was significantly greater during the immediately preceding time steps under the serial paradigm.

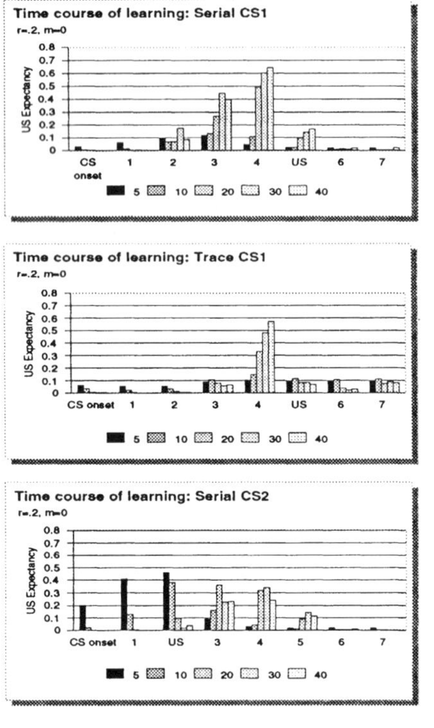

Figure 3: Time course of learning. The bars represent the US expectancy after training the model for 5, 10, 20, 30 and 40 thousand simulation time steps. The x-axis represents time steps during testing, indexed against the onset of the test CS. Onset of US during training (not testing) relative to the CS is noted on each graph.

Regarding CS_2, the peak expectancy tended to occur later than optimal as training progressed. In the r=.1, m=.3 case, the peak expectancies occurred two steps later than expected considering the CS_2->US relationship. There was also a trend toward later expectancies as the strength of the CS_1->US relationship grew. The earliest US expectancy peaks (and highest under all but the r=.1, m=.3 setting) occurred during the first five to twenty thousand time steps (depending on the settings). Latency grew longer with more training.

Discussion

The first thing to note was the common qualitative results for all three of the parameter settings. There were some differences in degree, but the trends were similar. Given the small sample size, the only conclusion regarding US expectancy peak that can be made is that the mean differences between the two conditions grew smaller with more training. A similar result was observed by Bolles *et al* (1978) in their animal subjects. They compared delay, trace and filler conditions where the filler condition represented serial conditioning. They observed facilitation (as measured by suppression ratio) in the filler condition after 16 CS-US pairings. However, after 64 pairings there were no significant differences among the three groups.

One of the more interesting simulation results is that serial conditioning did result in shorter latencies to CS_1 under all three parameter settings. If the goal of the system is to accurately predict the occurrence of the US, then the trace procedure was more accurate over the long run. However, in an adaptive sense, having a bit more forewarning of the USs' occurrence is beneficial. By that criteria, serial conditioning was superior, although the mechanism for this is unclear.

The graphs illustrate average performance over a number of runs. This conceals a couple of the interesting strategies adopted by some of the networks. In two runs (one at r=.2, m=0; one at r=.1, m=.3), the system was observed to develop no US expectancy to presentation of CS_1 or CS_2 alone but showed a normal expectancy to the compound (with a peak of approximately .8 to .9). This is evidence of configural learning where the compound is treated differently than the sum of its elements. Configural learning in animals usually results when there is differential reinforcement of the compound and its elements. However, in an experiment involving simple compound (non-serial) conditioning, Kehoe (1986) has found low levels of responding to the elements following conditioning of the compound only. In a similar vein, the model twice (once at r=.2, m=0; once at r=.1, m=0) showed US expectancy following CS_1, no expectancy following CS_2, but more expectancy to the compound than to CS_1 alone. Similar empirical results have also been observed (e.g. Kehoe, 1979).

The tendency for CS_2 latencies to grow longer with more training may be the result of generalization from the earlier CS_1. This type of generalization and the degree of supremacy of the first element in a serial compound has been extensively studied by Kehoe & Napier (1991) with eye blink conditioning in rabbits. Using serial pulse stimuli, Kehoe observed that the CR topography during test of later elements of the compound was very similar to the topography expected and observed to the first element of the compound. In Experiment 2, an A->B->C->D compound was presented where the ISI from A to the US was 400 msec. Observed CR peaks to all of the singly presented elements occurred after 400 to 450 msec despite the fact that the ISIs of B, C and D during training were 300, 200 and 100 msec respectively.

In simulation 2, I was interested in exploring the variables that drive the model's US expectancy. Moore, Desmond and Belthier's (1989) model relied on both CS onset and CS offset for its responses. The next simulation investigates the SRN model's dependence on these two variables following both trace and serial conditioning.

Simulation 2

In this set of simulations, the duration of the test CS was systematically manipulated. If, after equating for CS onset, the latency of US expectancy was the same for all durations of the test CS, then CS onset is determining expectancy. However, if the US latency systematically covaried with the changes in duration (and thus offset), then US expectancy is based on CS offset. The data from delay conditioning in the Kehoe & Napier (1991) studies indicate that the earliest part of a sequence of pulse stimuli commands substantial responding (the *temporal primacy effect*). This might generalize to apply to the earliest part of a single CS. A different result is suggested by Boyd & Levis (1976). Their results demonstrated a greater reliance on the *later* stimuli in the compound following avoidance conditioning. However, there is a significant difference in the CS durations (and hence ISIs) in the two studies. In Boyd & Levis (1976), the CSs were 6 sec. long. Given that they were using a three component compound, the ISI from CS_1 to the US was 18 sec. My hypothesis was that the durations being used in the present simulations would be better approximated by those used by Kehoe & Napiers (1991) rather than those of Boyd & Levis (1976). Note that this begs the question of optimal ISIs for the two different paradigms, NMR in the former and avoidance responding in the latter.

Method

The training method was identical to that used in Simulation 1. The testing phase consisted of a systematic variation of the duration of CS_1 including time step lengths of 1, 2, 3 (the one on which it was trained) and 6. The ITI between the last training US and the first testing CS was also systematically varied to investigate any effect on the system's performance. Testing was performed following forty thousand training steps (approximately 3300 trials) for each of the parameter settings used. Two runs at each of the settings were conducted.

Results

The results are shown in Figure 4 collapsed across parameter settings (there were no significant differences among results for different parameter values). Note that in the graphs, CS onset is equated across test CS durations. Offsets vary with duration (e.g. duration 6 is *on* through 5 on the graph and *off* at 6). Given the training set, a US expectancy that depends on the CS onset should peak at 4 on the graph. A US expectancy that depends on offset should peak at 2 for duration 1, 3 for duration 2, 4 for duration 3, and 7 for duration 6.

The variation of ITIs did have an impact on performance for some networks. Specifically the shortest test ITI (length 7) resulted in the poorest performance while test ITIs of 11 and 16 performed equally well.

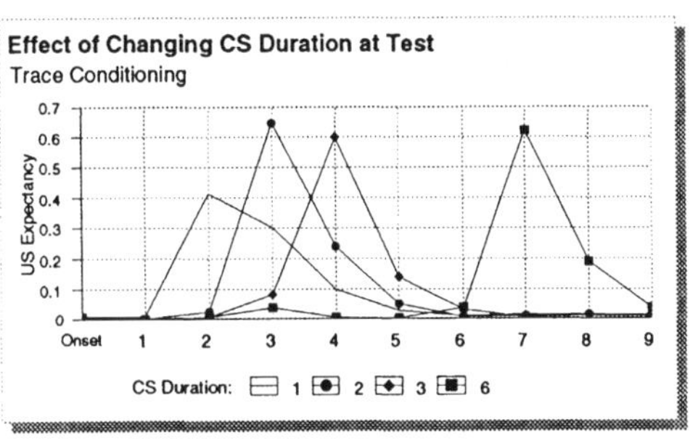

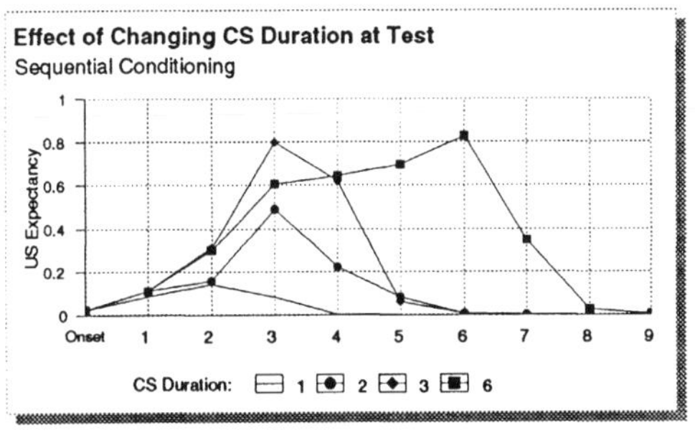

Figure 4: Effect of changing the duration of CS_1 at test. Time since CS onset represented on x-axis.

Discussion

It is apparent that the model depends primarily on CS offset for predicting US occurrence during trace conditioning. However during serial conditioning, the CS onset played a major role. All durations longer than 1 resulted in significant expectancy at time steps 3 & 4.

Following serial conditioning, the CS offset played a larger role in signalling when to stop expecting the US rather than in initiating expectancy. The longer the CS was on, the more sustained the expectancy. Hence, the offset tends to attenuate expectancies at time 3, 4, 5, and 8 for durations of 1, 2, 3, and 6 respectively. This helps to explain the lack of a peak at time 3 or 4 for the CS of length 1 and the drop in expectancy from 3 to 4 for the CS of length 2. The apparent drop at 4 for the CS of length 3 was not significant.

Note that the observed dependence on CS onset vs. offset will likely change for different trained CS durations. Longer CSs will drive the system to use the nearer CS offset for US initiation while shorter CSs lessen the burden of reliance on the farther onset.

The fact that the test ITI of 7 performed worst was a surprise considering that the model was trained using an ITI that varied between 5 and 7. The initial reason for testing this independent variable was to insure that the system was not learning about the regularity of US occurrence. Hence, it was a surprise that matching the test ITI to training ITI resulted in *worse* performance. The short training ITI may have actually retarded the network's learning.

General Discussion

The model produced very different event representations as the result of serial vs. trace conditioning. Other intratrial models of conditioning (e.g. Grossberg & Levine, 1987; Grossberg & Schmajuk,1987, 1989;

Klopf, 1988; Lee, 1991; Sutton & Barto, 1981, 1990) have been tested on a wide variety of conditioning paradigms. The SRN model's performance on serial conditioning demonstrates promise and it should be compared to that of the other models. Empirical work can then be planned to resolve the theoretical differences. As a model of conditioning, the current model is not comprehensive. Motivation, drive, habituation, and instrumental training have yet to be explored.

References

Blazis, D.E.J. & Moore, J.W. (1991). Conditioned stimulus duration in classical trace conditioning: Test of a real-time neural network model. *Behavioral Brain Research*, **43**, 73-78.

Bolles, R.C., Collier, A.C., Bouton, M.E. & Marlin, N.A. (1978). Some tricks for ameliorating the trace-conditioning deficit. *Bulletin of the Psychonomic Society*, **11**, 403-406.

Boyd, T.L. & Levis, D.J. (1976). The effects of single-component extinction of a three-component serial CS on resistance to extinction of the conditioned avoidance response. *Learning & Motivation*, **7**, 517-531.

Desmond, J.E. (1990). Temporally adaptive responses in neural models: The stimulus trace. In M. Gabriel & J. Moore (Eds.), *Learning and Computational Neuroscience: Foundations of Adaptive Networks.* Cambridge, MA: MIT Press.

Elman, J.L. (1990). Finding structure in time. *Cognitive Science*, **14**, 179–211.

Grossberg, S. & Levine, D.S. (1987). Neural dynamics of attentionally-modulated Pavlovian conditioning: Blocking, inter-stimulus interval, and secondary reinforcement. *Applied Optics*, **26**, 5015–5030.

Grossberg, S. & Schmajuk, N.A. (1989). Neural dynamics of adaptive timing and temporal discrimination during associative learning. *Neural Networks*, **2**, 79–102.

Kehoe, E.J. (1979). The role of CS-US contiguity in classical conditioning of the rabbit's nictitating membrane response to serial stimuli. *Learning & Motivation*, **10**, 23–38.

Kehoe, E.J. (1986). Summation and configuration in conditioning of the rabbit's nictitating membrane response. *Journal of Experimental Psychology: Animal Behavior Processes*, **8**, 313-328.

Kehoe, E.J., Gibbs, C.M., Garcia, E. & Gormezano, I. (1979). Associative transfer and stimulus selection in classical conditioning of the rabbit's nictitating membrane response to serial compound CSs. *Journal of Experimental Psychology: Animal Behavior Processes*, **5**, 1–18.

Kehoe, E.J. & Napier, R.M. (1991) Real-time factors in the rabbit's nictitating membrane response to pulsed and serial conditioned stimuli. *Animal Learning & Behavior,* **19**, 195-206.

Klopf, A.H. (1988) A neuronal model of classical conditioning. *Psychobiology*, **16**, 85–125.

Lee, C. (1991). Modeling the behavioral substrates of associate learning and memory: Adaptive neural models. *IEEE Transactions on Systems, Man, and Cybernetics*, **21**, 510–520.

Mackintosh, N.J. (1975). A theory of attention: Variations in the associability of stimuli with reinforcement. *Psychological Review*, **82**, 276–298.

Moore, J.W., Desmond, J.E., & Belthier, N.E. (1989). Adaptively timed conditioned responses and the cerebellum: A neural network approach. *Biological Cybernetics*, **62**, 17–28.

Rescorla, R.A. & Wagner, A.R. (1972). A theory of Pavlovian conditioning: Variations in the effectiveness of reinforcement and nonreinforcement. In A.H. Black & W.F. Prokasy (Eds.), *Classical Conditioning II: Current Research and Theory.* New York: Appleton-Century-Crofts.

Rumelhart, D.E., Hinton, G.E. & Williams, R.J. (1986). Learning internal representations by error propagation. In D.E. Rumelhart & J.L. McClelland (Eds.), *Parallel Distributed Processing: Explorations in the Microstructure of Cognition* (Vol. 1, pp. 318-362). Cambridge, MA: MIT Press.

Sutton, R.S. & Barto, A.G. (1981). Toward a modern theory of adaptive networks: Expectation and prediction. *Psychological Review*, **88**, 135-170.

Sutton, R.S. & Barto, A.G. (1990). Time-derivative models of Pavlovian reinforcement. In M. Gabriel & J. Moore (Eds.), *Learning and Computational Neuroscience: Foundations of Adaptive Networks.* Cambridge, MA: MIT Press.

The Figural Effect and a Graphical Algorithm for Syllogistic Reasoning*

Peter Yule
Centre for Cognitive Science
Edinburgh University
U.K.
pgy%cogsci.ed.ac.uk@nsfnet-relay.ac.uk

Keith Stenning
Human Communication Research Centre
Edinburgh University
U.K.
keith%cogsci.ed.ac.uk@nsfnet-relay.ac.uk

Abstract

Theories of syllogistic reasoning based on Euler Circles have foundered on a combinatorial explosion caused by an inappropriate interpretation of the diagrams. A new interpretation is proposed, allowing single diagrams to abstract over multiple logical models of premisses, permitting solution by a simple rule, which involves the identification of individuals whose existence is entailed by the premisses. This solution method suggests a performance model, which predicts some of the phenomena of the Figural Effect, a tendency for subjects to prefer conclusions in which the terms preserve their grammatical status from the premisses (Johnson-Laird & Steedman 1978). 21 students were asked to identify the necessary individuals for each of the 64 pairs of premisses. The order in which the three terms specifying the individuals were produced was shown to be as predicted by the performance model, but contrary to the presumed predictions of Mental Models theory.

Introduction

Syllogisms are arguments from two premisses to a conclusion. Both premisses and conclusion are statements of one of four types: "All of the A are B" (A), "Some of the A are B" (I), "None of the A are B" (E) and "Some of the A are not B" (O). Each statement in the premisses contains two terms: one term, the middle term (b), occurs in both premisses, while the other two (a and c) are known as the end terms. The arrangement of the end and middle terms in each of the premisses gives rise to a four-way classification, known as the figure of the syllogism (see Table 1). It should be noted that our terminology in this matter follows the usage of Johnson-Laird (1983), and that like Johnson-Laird, we take the term "syllogism" to refer to any of the 64 pairs of premisses, rather than a pair of premisses plus a conclusion.

Psychological theories of syllogistic reasoning attempt to account for subjects' errors and response biases in solving syllogisms. Although early theories such as the Atmosphere Hypothesis (Woodworth and Sells 1935) account for some common errors, they cannot account for correct performance in many cases. Theories which do explain correct performance are often based on the logical model theory of the syllogism. Most serious theories of this type are based on the method of Euler Circles, which uses circles to represent sets of entities.

Unfortunately the term "method" is perhaps a misnomer for the use of Euler Circles, since the exact method of their employment is at best implicit in logic textbooks. Psychological theories typically have used each diagram to represent a single logical model of the premisses, resulting in a many-many mapping between statement types and diagrams (see Fig. 1). Two diagrams, one for each premiss, are integrated to form a *registration diagram* by superimposing the middle term circles from each. To solve a syllogism this way, all possible combinations of all possible pairs of diagrams have to be considered. A valid conclusion is then one which is modelled by all of the possible registration diagrams for a problem. The theories of Erickson (1974) and Sternberg (e.g. Guyote and Sternberg 1981) both make this assumption. Unfortunately the number of registration diagrams can be inordinately large, and the methods employed by these theories to cut down the search space make them incomplete and

Premiss	*Figure* 1	2	3	4
1	ab	ba	ab	ba
2	bc	cb	cb	bc

Table 1: The four figures of the syllogism.

*The support of the Economic and Social Research Council UK (ESRC) is gratefully acknowledged, both for their funding of the Human Communication Research Centre (HCRC), and the support given to Peter Yule in the form of an ESRC Research Studentship Award.

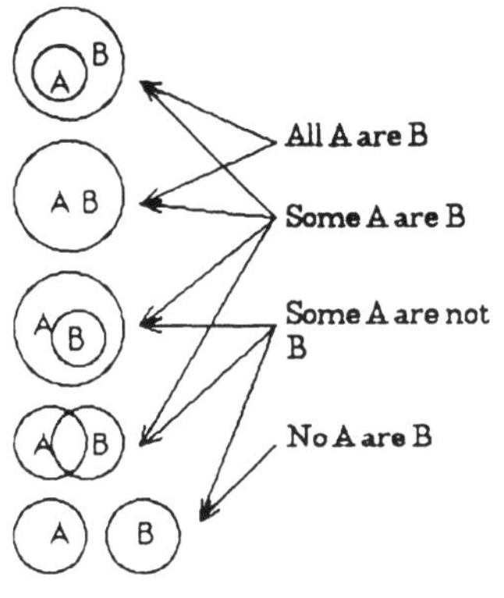

Fig. 1: The five Gergonne relations between a pair of circles, and how they model each of the four statement types.

consequently unable to account for correct performance (Johnson-Laird 1983).

Johnson-Laird's rejection of all methods based on Euler circles rests on a tacit assumption that Euler circles must be 'primitively' interpreted, so that one diagram always stands for one logical model. His proposed alternative is the theory of Mental Models, in which individual mental representations stand for more than one logical model, by devices such as the representation of optional elements (Johnson-Laird 1983), or a convention that makes it explicit when a set has been exhaustively represented, constraining the ways in which a skeletal model can be "fleshed out" (Byrne and Johnson-Laird 1991). Either way, each mental model abstracts over one or more logical models, so that an explosion of mental representations is avoided.

But Euler circles can also be interpreted in such a way that single diagrams abstract over multiple logical models. Each statement type has a *maximal model*, which contains all individuals which are compatible with the statement, and a *minimal model*, which contains only individuals whose existence is entailed by it. Representing the minimal model as a shaded region within the diagram which represents the maximal model, we obtain just one *characteristic diagram* per statement (see Fig. 2).

Each region within the diagram represents a different type of individual. A simple notation for individuals is the *type descriptor*. This is just a feature structure. Features are constructed from the terms appearing in the premisses, prefixed by "+" or "−", which indicate whether the individual concerned is or is not a member of the set denoted by the term. For example "+sentient-creature −Martian", describes a sentient creature which is not Martian.

The set of individuals whose existence is entailed by the premisses (*necessary individuals*) can be determined by forming a registration diagram from the two characteristic diagrams, overlapping the circles representing the end terms if this is consistent with the premisses (this is equivalent to forming a maximal model of the premiss pair). The necessary individuals correspond to shaded regions from the characteristic diagrams which are not bisected during the formation of the registration diagram. Fig. 3 shows an example registration diagram, for a problem which establishes a necessary individual, and Fig. 4 shows the registration diagram for a problem which does not establish any necessary individuals, and so lacks valid conclusions. In all there are 21 distinct registration diagram types, the full set of which can be found in Stenning (1992).

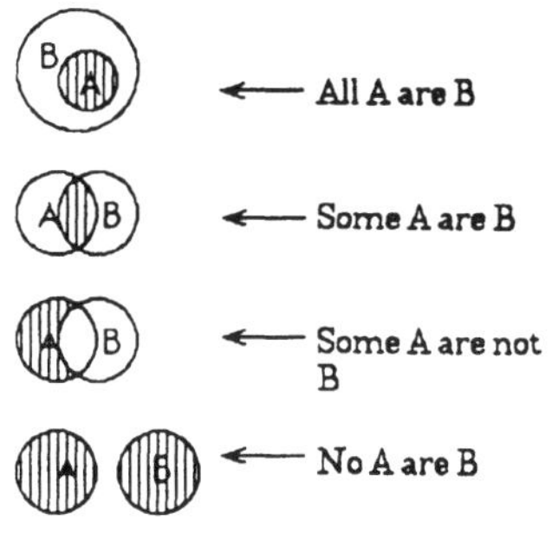

Fig. 2: Characteristic diagrams for each statement type, with shaded regions representing minimal models.

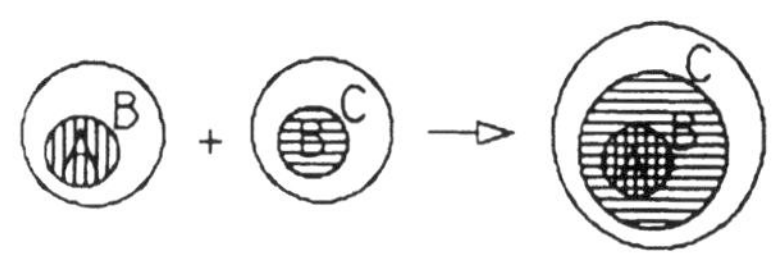

Fig. 3: Registration diagram for the problem "All of the a's are b's, All of the b's are c's". The central, unbisected circle corresponds to the individual +a+b+c.

Necessary individuals can form the basis for quantified conclusions. Particular (i.e. existential) conclusions can be drawn immediately by dropping the middle term from the type descriptor and picking a + term as the subject of the conclusion (e.g.

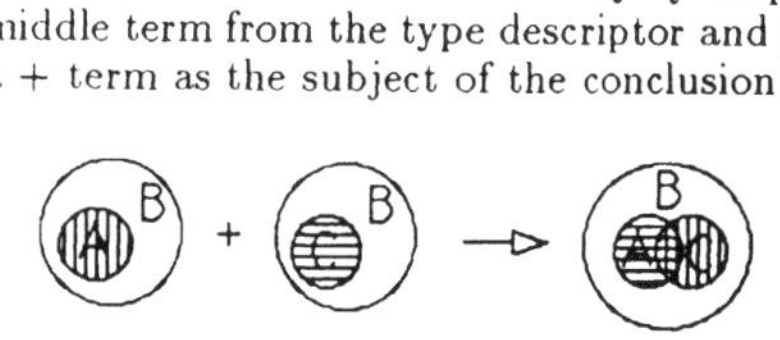

Fig. 4: Registration diagram for the problem "All of the a's are b's, All of the c's are b's", which lacks valid conclusions. Note that the shaded end-term circles overlap, bisecting each other.

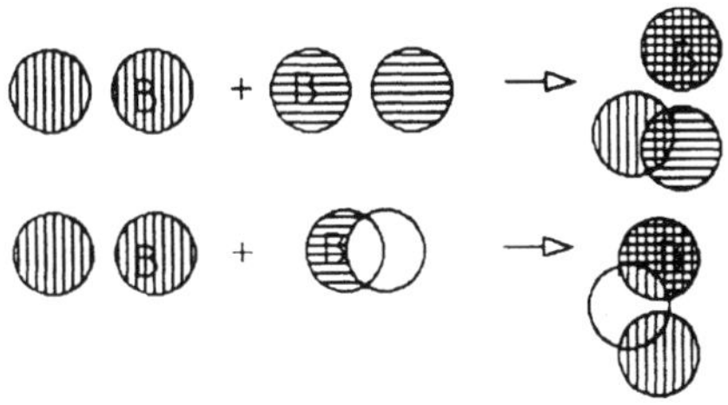

Fig. 5: Registration diagrams for the U-valid problem types, which establish necessary individuals but lack quantified conclusions.

"Some sentient creatures are not Martian"). For universal conclusions with subject X, drop the middle term feature and pick a feature +X such that X corresponds to an unbisected circular shaded region in the registration diagram. Note that although the "double negative" problems corresponding to the diagrams in Fig. 5 establish necessary individuals, they lack quantified conclusions because neither of the end-term features is +. It would be necessary to have another statement type "Some of the not A are not B", which we call type U, to express these conclusions.

This method constitutes a decision procedure for the categorial syllogism, and it has been implemented in Prolog. The interpretation of the method, and its relation to the theory of Mental Models, is detailed in Stenning and Oberlander (1992) and Stenning (1992).

The method can be adapted to provide a range of performance models of syllogisers. Our approach is to adapt the competence model to give a model of correct performance, which permits breakdowns to account for subjects' errors. Erroneous conclusions can be accommodated by assuming that subjects use sub-maximal diagrams, or register diagrams sub-maximally. This is equivalent to assuming subjects fail to consider all possible logical models. We assume the inference process involves the identification of a shaded region in the characteristic diagram for one of the premisses, followed by a test using the information from the other premiss to find out if the shaded region is bisected by registration or not. For optimal performance this process must be iterated for each shaded region, and failure to do so may result in an erroneous "No valid conclusion" response. We call the premiss which provides the shaded region the *source premiss*. We make no claims about the way in which subjects find the source premiss; it could be systematic or random, exhaustive or not. But by hypothesis, production of a valid conclusion entails that the subject has successfully identified the source.

We claim that the Figural Effect described by Johnson-Laird *et al* (Johnson-Laird & Steedman 1978, Johnson-Laird 1983, Johnson-Laird & Bara 1984, Byrne & Johnson-Laird 1991), is a consequence of the assymmetry between the roles of the two premisses in establishing necessary individuals. Basically, the Figural Effect is the tendency for terms to preserve their grammatical status from premisses to conclusion. So in the problem *Some of the Artists are Beekeepers, All of the Beekeepers are Chemists,* both of the conclusions *Some of the Artists are Chemists* and *Some of the Chemists are Artists* are valid, although the vast majority of subjects produce only the former. In figure 1, then, the preferred conclusions are ac ones, in figure 2 they are ca ones, while in the other two figures both types are equally common. Johnson-Laird's account of this effect is touched on briefly below, but is described in detail in the sources cited above.

We assume that the type descriptor is built up incrementally during the reasoning process, so the terms from the source premiss should precede the end term from the other premiss. Given that quantified conclusions will tend to preserve this ordering if possible, response biases toward conclusions with one or other end-term ordering are explicable in terms of which of the premisses is the source. This tends to vary with the figure of the syllogism. Consider the problem AabAbc (see Fig. 3). Although the characteristic diagrams for both the premisses contain shaded areas, only the shaded area from the first, corresponding to the term a, is unbisected by registration. So we predict conclusions in which a is the subject, namely Aac or Iac, but not Ica, although all three are valid. There are three problems in figure 1 that permit valid conclusions with both possible term orders (ac and ca). For two of these problems, the only possible source premiss is Premiss 1, and for the other, both premisses are potential sources (see Fig. 6 below). Therefore the theory predicts figural effects for the first two, but does not predict a figural effect for the third (but see below). Similarly in figure 2, there are three problems with free term order in valid conclusions, two of which have source Premiss 2, and the other has two potential sources. In figures 3 and 4, we can also predict term orderings in individual problems, and the predicted numbers of ac and ca orderings are equal.[1]

Our approach therefore offers an extension of the traditional Eigural Effect, since we can make specific predictions in figures 3 and 4, but it is at present limited to predictions for valid conclusions only. This is because only in these cases is there a principled basis on which to decide what the source premiss is, but in principle the theory should be ex-

[1] Figure 2 problems are equivalent to figure 1 problems when the premiss order is inverted, and in the remaining two figures each problem has an equivalent inverted problem in the same figure.

tensible to handle the Figural Effect in invalid conclusions. To do this, it is necessary to identify the representations used by subjects: then the source premiss can be identified *post hoc*, providing a basis on which to predict term orders. We intend to attempt this work in future.

Although Mental Models theory can at present offer an account of the Figural Effect in invalid conclusions as well as in valid conclusions, the present approach has the advantage that the Figural Effect as we treat it follows from logical rather than physical features of the representations used in inference. Consequently it is not so dependent on the assumption that all subjects follow the same procedure when solving syllogisms, which is rather implausible considering the great differences in experience between different subject groups.

The traditional syllogism task has a "degrees-of-freedom" problem, because for many syllogisms the order of the end terms in valid conclusions is not free to vary, except in I and E conclusions. However, in a task in which subjects are asked to identify the individuals entailed by a pair of premisses, the order of mention of the terms is completely independent of the validity of the conclusion. Additionally, there are valid conclusions for more problems than in the traditional task, since some of the double-negative problems establish individuals but none of the conclusions A, I, E or O. These are the problems which establish 'U-conclusions'. Our theory maintains that the identification of necessary individuals is a prerequisite to drawing a quantified conclusion, so this task is relevant to the traditional task.

With this task, we also need to be able to predict the position of the b term, which can be done by specifying the order in which the terms from the source premiss are mentioned. In the case of universal premisses, each candidate individual is uniquely identified by a single positive term, for example, the minimal model of *All of the a's are b's* is just the set a, all of whose members are implied to be b, so the predicted ordered type descriptor is +a+b. Particular premisses are more troublesome, but for the present we assume that the subject feature precedes the predicate feature. Type descriptors for necessary individuals are then composed of the ordered type descriptor for the candidate individual, followed by the feature corresponding to the remaining end term.

Although the present theory makes predictions about the ordering of end terms in valid quantified conclusions rather similar to those made by Mental Models theory, the predictions of the two theories diverge when we consider the necessary individuals task. According to Mental Models theory, subjects create a mental model in which tokens representing instances of the middle term occur *between* tokens representing instances of the end terms, in a two-

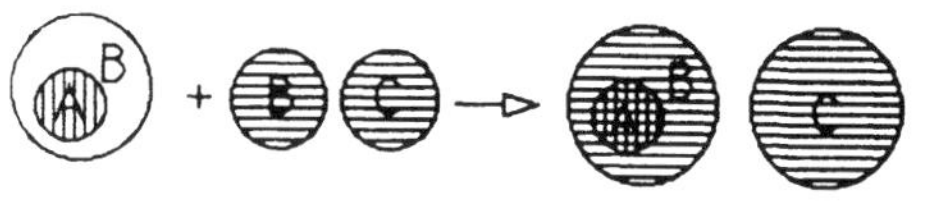

Fig. 6: Registration diagram for the problem "All of the a's are b's, None of the b's are c's". Note that there are two unbisected shaded areas.

dimensional spatial array, and then read off conclusions from one end of the model to the other, so the only predicted orders would presumably be abc and cba. However, the present theory additionally predicts occurrences of the orders bac and bca.

Finally, this task allows us to make predictions concerning some of the problems whose source can be either premiss. The problems in question are those which have an A premiss whose subject is an end term, and an E premiss (e.g. AabEbc - Fig. 6 shows the registration diagram). Problems of this type establish 2 individuals, each of which has a different source premiss. According to the theory, each of these individuals supports a different quantified conclusion, since each has only one positive feature corresponding to an end term (e.g. +a+b−c supports Eac and +c−b−c supports Eca). Since we cannot predict which premiss will be source here we cannot predict a figural effect, but we can test for an association between the source and term order, since the source can be determined *post hoc* on the basis of the conclusion that is drawn by the subject. Specifically, answers whose source is the A premiss should have the end term from that premiss before that from the other, and answers whose source is the E premiss should have the opposite end term order.

Method

Design Each subject produced one answer for each of the 64 syllogisms, which were presented in random order.

Subjects 21 subjects, either Psychology undergraduates or postgraduates in Cognitive Science, were each paid £5 to take part.

Materials Each subject received a set of 64 slips of paper, on each of which was printed a different pair of premisses. The vocabulary used was selected from sets of nouns denoting nationalities, professions and interests, for example *None of the musicians are chessplayers. All of the musicians are Italians.* Each vocabulary item appeared in two syllogisms, and two different random assignments of vocabulary to syllogisms were used.

Procedure Subjects were instructed to imagine that the premisses described a group of people at a party, and to decide whether any kind of person who could be described with certainty, in terms of

End term order	Source Premiss 1	2	Both	Overall
ac	77.9	28.0	65.7	55.8
N	213	218	134	565

Table 2: Percentages of correct valid conclusions with end-term ordering ac in problems with source premiss first, second or either.

	Figure 1	2	3	4	Overall
Predicted	86.2	79.5	85.2	62.2	75.4
N	138	122	88	217	565

Table 3: Percentages of predicted term orderings in correct responses to problems with valid individual conclusions in each figure.

either positive or negative values of all three features, had to be present in the room. They were instructed to assume that some people corresponding to each of the three terms existed. They were asked to describe the individual on the slip of paper, or if there was no individual, to write "No valid conclusion". Subjects worked alone in quiet surroundings, and were given as much time as they needed to finish all the problems.

Results

Table 2 shows how the order in which the end terms are mentioned, in type descriptors for correct answers to problems with valid conclusions, varies with source premiss. The variables are strongly associated ($\chi^2(2) = 116.0$, $p < 0.0001$), such that a tends to precede c when the first premiss is the source, and c tends to precede a when the second premiss is the source. Note that there is some overall tendency for a to be mentioned before c, particularly in problems where either premiss can function as the source.

Table 3 shows the percentages of correct responses to problems with valid conclusions which occur in the predicted orders in each figure of the syllogism. Overall, 75.4% of responses were as predicted, and there was a majority of predicted responses in each figure. However, the effect is not entirely independent of figure ($\chi^2(3) = 34.77$, $p < .0001$), the main divergence from the overall trend being in figure 4, where a substantial minority of responses occurred in unpredicted orders.

Table 4 shows the frequencies of responses having each of the possible term orderings. It is clear that, against the presumed predictions of Mental Models theory, responses with bac and bca orders actually outnumber those with abc and cba orders. As predicted by both accounts, there are very few cases in which the b term occurs last.

	Order abc	cba	bac	bca	cab	acb
N	194	148	236	156	31	19
Total	342		392		50	

Table 4: Frequencies of responses with each possible term ordering (total N=784).

End-term order	Source Premiss A	E	Total
AE	64	0	64
EA	4	10	14
Total	68	10	78

Table 5: Association between end-term order and source premiss for valid conclusions to problems which establish two individuals. End-term order is said to be AE if the end term from the A premiss precedes that from the E premiss.

Finally, Table 5 shows the relation between end-term order and source premiss for all of the the problems which establish two individuals (e.g. Fig. 6). On the basis of the response, we can determine whether the A or E premiss is the source, and we predict that the end-term from the source premiss should precede the end-term from the other. As the Table shows, there are more responses with the A premiss as source than with the E premiss as source, and the end-term order is strongly predicted by the source premiss ($Yates'\ \chi^2(1) = 46.24$, $p < .0001$).

Discussion

The results of the experiment show that the source premiss for a problem is strongly associated with the order of mention of the end terms, in both determinate and indeterminate problems. The three-term order predictions are also confirmed, and with the exception of the anomalous result for figure 4 (discussed below), the effect is uniform across figures, so the figural effect is explained by the distribution of different problem types among the figures. So our model can offer a uniform account of order effects in both this task and the traditional syllogism task. Although Mental Models theory handles the phenomena of the traditional task well, it fails to predict the large number of responses in this task in which the b term occurs first, so it cannot at present offer a uniform account of the two tasks.

Mental Models theory, of course, purports to account for other aspects of performance in the traditional task. Johnson-Laird (e.g. 1983) has argued that the number of mental models which need to be considered to solve a problem predicts its empirical difficulty quite well. However, Ardin (1991) has shown that in this experiment's data, although one-

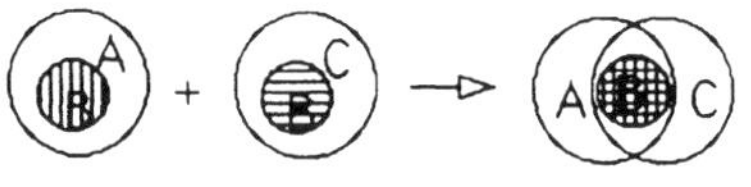

Fig. 7: Registration diagram for the problem "All of the b's are a's, Some of the b's are c's".

model problems are easier than the others, there is no significant difference between two- and three-model problems. Our theory does not rely on the construction of different models, so it cannot account for differences in difficulty in such terms, but it may be that difficulty can be predicted by the number of candidate individuals which have to be considered. This has yet to be investigated.

The results of the experiment showed that a substantial minority of correct responses to figure 4 problems did not occur in the predicted orders. There are indications that this is due to another strategy which can occur when the non-source premiss is of type A. Subjects sometimes produce the terms from this premiss before those from the "true" source premiss, but only under special conditions, when the shaded region from the A premiss is bisected by registration, but one of the halves is itself an unbisected shaded region, and so represents a necessary individual. Ordinarily, we would expect subjects who have detected bisection of a region either to give a "no valid conclusion" response, or start afresh with a shaded region from the other premiss to find the unbisected shaded region, but in this case it appears that they can detect the critical region on the first pass, and so produce the terms in the order which would be predicted if the A premiss was the source. An example is AbaIbc (see Fig. 7). There are similar problems in all the figures, but most are in figure 4, where the effect is most pronounced. The tendency for subjects to consider the A premiss first, or the "A-effect", has previously been noted by Lee (1987).

The results of the analysis of the Fig. 6-type problems also suggest the presence of an A-effect, insofar as most of the valid conclusions to these problems are only accessible from the A source rather than the E source. This, along with the core theory, can explain the figural effect for these problems, as follows. For the problem AabEbc, the A-source individual conclusion is +a+b−c, which supports only the quantified conclusion Eac, since the positive end-term feature is a, which must therefore form the subject of the conclusion, so the conclusion is figural. Similar arguments hold for the figure 2 problem, as well as the two in figure 3, one of which supports Eac and one of which supports Eca. Subjects' preferences concerning which premiss to consider first are independent of the central claims of this paper, but it appears that augmenting the theory by including the A-effect would successfully account for some figural phenomena on which the core theory remains neutral.

In conclusion, it is clear that a method for solving syllogisms using isomorphs of Euler Circles can not only avoid large numbers of representations, but can also explain some classic psychological results using minimal auxiliary assumptions. We hope soon to produce evidence that the theory can also account for the Figural Effect in invalid conclusions.

Acknowledgements

Thanks to Cath Ardin and Morten Christiansen for work on the preparation and running of the experiment, and to Jon Oberlander for preparing some of the diagrams.

References

Ardin, C. 1991. Mental Representations Underlying Syllogistic Reasoning. Ph.D Thesis, Centre for Cognitive Science, Edinburgh University.

Byrne, R. and Johnson-Laird, P.N. 1991. *Deduction*. Hove: LEA.

Erickson, J.R. 1974. A set analysis theory of behaviour in formal syllogistic reasoning tasks. In R.Solso (ed) *Theories in cognitive psychology: the Loyola Symposium*. Potomac: LEA.

Guyote, M.J. and Sternberg, R.J. 1981. A transitive-chain theory of syllogistic reasoning. *Cognitive Psychology* 13:461-525.

Johnson-Laird, P.N. 1983. *Mental Models*. Cambridge: CUP.

Johnson-Laird, P.N. and Bara, B. 1984. Syllogistic Inference. *Cognition* 16:1-61.

Johnson-Laird, P.N. and Steedman, M.J. 1978. The psychology of syllogisms. *Cognitive Psychology* 3:395-400.

Lee, J. 1987. Metalogic and the psychology of reasoning. Ph.D Thesis, Centre for Cognitive Science, Edinburgh University.

Stenning, K. 1992. Spatial inclusion as an analogy for set membership: a case study of analogy at work. To appear in Holyoak, K. and Barnden, J. (eds) *Analogical Connections*.

Stenning, K. and Oberlander, J. 1992. A cognitive theory of graphical and linguistic reasoning: logic and implementation. Research Paper RP20, H.C.R.C., Edinburgh University.

Woodworth, R.S. and Sells, S.B. 1935. An atmosphere effect in formal syllogistic reasoning. *Journal of Experimental Psychology* 18:451-460.

A Computational Best-Examples Model

Jianping Zhang[1]
Department of Computer Science
Utah State University
Logan, Utah 84322-4205 USA
Email: jianping@zhang.cs.usu.edu

Abstract

In the past, several machine learning algorithms were developed based on the exemplar view. However, none of the algorithms implemented the best-examples model in which the concept representation is restricted to exemplars that are typical of the concept. This paper describes a computational best-examples model and empirical evaluations on the algorithm. In this algorithm, typicalities of instances are first measured, then typical instances are selected to store as concept descriptions. The algorithm is also able to handle irrelevant attributes by learning attribute relevancies for each concept. The experimental results empirically showed that the best-examples model recorded lower storage requirements and higher classification accuracies than three other algorithms on several domains.

1. Introduction

Smith and Medin (1981) proposed the exemplar view for concept representation and category classification. Specifically, two cognitive models of the exemplar view, the proximity model and the best-examples model, were took up. In the proximity model, each concept is represented by all of its instances that have been encountered. The best-examples model assumes that the representation is restricted to exemplars that are typical of the concept. It seems impossible for an adult to remember all instances for each concept. The best-examples model strongly supports human concept formation. People tend to remember those most often encountered instances and forget those rarely encountered instances. Concepts involved in real world applications usually possess graded structures (Barsalou, 1985). Instead of being equivalent, instances of a concept may be characterized by a degree of typicality in representing the concept. Typical instances of a concept better characterize the concept than atypical instances. Typical instances represent the central tendency of a concept, so concepts described by typical instances are more human understandable than those described by atypical instances and also easier for human to capture the basic principles underlying these concepts.

In the past, several machine learning algorithms were developed based on the exemplar view, these learning algorithms are called instance-based learning algorithms, e.g., Protos (Bareiss, et al., 1990), IBL (Aha, et al., 1991), and Each (Salzberg, 1991). Although all these algorithms restricted the number of stored instances, none of them truly implemented the idea of the best-examples model. These algorithms selected misclassified instances which were proved to be near-boundary instances by Aha et. al (1991). Salzberg (1991) developed a method which assigned a weight to each stored instance. In his approach, typical instances got smaller weights than near-boundary instances, so they played more important role than near-boundary instances. However, this approach did not restrict stored instances to typical instances.

This paper presents a computational best-examples model developed from the cognitive best-examples model proposed in (Smith and Medin, 1981). Several problems were addressed in the computational best-examples model. First, an algorithm was developed to measure typicalities of instances. Second, an approach was designed to learn the weights of attributes for each class. Finally, an algorithm was proposed to select typical instances of a concept to store in memory. The computational model has been implemented and tested on both artificial and practical domains, and compared with three different instance-based learning algorithms: storing all instances, storing only incorrectly classified instances, and storing near-boundary instances. The empirical results showed that the

[1] This research was supported by the Department of Computer Science at Utah State University and the Utah State University Faculty Research Grant SCS-11107. The author would like to thank Steven Salzberg for providing the datasets of the malignant tumor classification and diabetes in Pima Indians.

computational best-examples model recorded lower storage requirements and higher classification accuracies than previous instance-based algorithms.

2. Learning Attribute Weights

Relevancies of attributes have a great impact on the performance of instance-based learning algorithms. Not all attributes chosen to describe a problem are relevant, even they do, the degrees of their relevancies differ. Different concepts in a problem may have different set of relevant attributes. For instance, an attribute that well distinguishes Concept1 from Concept2 may not do well to distinguish Concept2 from Concept3. In our model, the relevancies of attributes not only affect the classification of an instance, but also the typicality measured for each instance.

Both Aha (1989) and Salzberg (1991) assigned a weight to each attribute as its relevancy. Aha (1989) also assigned a different weight to the same attribute for different concepts. This is the approach used in our algorithm, but the weights were computed differently. In both (Aha, 1989) and (Salzberg, 1991), weights are computed incrementally. That is, each time a new instance was seen, the weights of attributes were modified based on the classification of the new instance made by the current descriptions. Weights were calculated during the process of instance selection in their algorithms. In our model, instances are selected according to their typicalities. Weights of attributes are used in measuring typicalities of instances, so we need the weights before selecting instances. Therefore, Aha's and Salzberg's methods cannot be used in our model. We use a statistical method to calculate the weights of attributes.

In our method, the weight of the attribute A with respect to the concept C is computed based on the difference of the distribution of the positive examples of C on all values of A and the distribution of the negative examples of C on all values of A. If the two distributions are very similar, the attribute A does not distinguish the concept C from other concepts well. In this situation, the difference of the two distributions is very small so the attribute gets a low weight (close to 0). If the two distributions do not intersect each other, the attribute A completely distinguishes C from other concepts. The difference of the two distributions in this situation reaches the maximum value so the attribute gets the largest weight. Generally, a more relevant attribute has a less intersection and a larger difference between the two distributions so it gets a larger weight.

Specifically, the weight of the attribute A which takes a value from {0, 1, ..., n} with respect to the concept C is computed by the following formula:

$$\frac{1}{2}\sum_{i=0}^{n} ABS\left(\frac{|\{e|A(e)=i \wedge e\in P\}|}{|P|} - \frac{|\{e|A(e)=i \wedge e\in N\}|}{|N|}\right)$$

where P and N are the sets of positive and negative examples of the concept C, respectively. $|\{e|A(e)=i \wedge e\in P\}|$ and $|\{e|A(e)=i \wedge e\in N\}|$ are the numbers of positive and negative examples whose value of the attribute A is i, respectively. The weight ranges from 0 to 1. if $\frac{|\{e|A(e)=i \wedge e\in P\}|}{|P|} = \frac{|\{e|A(e)=i \wedge e\in N\}|}{|N|}$ for all i ($0 \le i \le n$), The weight is 0 with respect to the concept C. IF one of $\frac{|\{e|A(e)=i \wedge e\in P\}|}{|P|}$ and $\frac{|\{e|A(e)=i \wedge e\in N\}|}{|N|}$ is 0 for all i ($0 \le i \le n$), The weight equals to 1 and the attribute A completely distinguishes C from other concepts.

3. Measuring Instance Typicalities

In our model, the typicality of an instance is measured based on its family resemblance (Rosch and Mervis, 1975), where family resemblance is defined as an instance's average similarity to other concept instances (intra-concept similarity) and its average similarity to instances of contrast concepts (inter-concept similarity). The more similar an instance is to other concept instances and the less similar it is to instances of contrast concepts, the higher its family resemblance, and the more typical it is of its concept. In other words, typical instances have higher intra-concept similarity and lower inter-concept similarity than atypical instances. The typicality of an instance is measured as the ratio of its intra-concept similarity to its inter-concept similarity. Thus, a larger intra-concept similarity implies a larger typicality, and a larger inter-concept similarity implies a smaller typicality. Generally, the typicalities of typical instances are much larger than 1, boundary instances have typicalities close to 1, and the instances with typicalities less than 1 are either noise or exceptions.

The intra-concept similarity of an instance of a concept C is computed as the average of the similarities of the instance to all other instances of C with respect to C, and the inter-concept similarity of an instance of a concept C is computed as the average of the similarities of the instance to all instances of contrast concepts (negative examples of C) with respect to C. The similarity of instances e^1 to e^2 with respect to C $sim(C, e^1, e^2)$ is the opposite of the distance of e^1 to e^2 with respect to C:

$$sim(C, e^1, e^2) = 1 - dis(C, e^1, e^2)$$

$dis(C, e^1, e^2)$ is computed by measuring the weighted Euclidean distance of the instance e^1 to the instance e^2. Specifically,

$$\mathrm{dis}(C,e^1,e^2) = \frac{\sqrt{\sum_{i=1}^{m} W(i,C) * \left(\frac{e^1{}_i - e^2{}_i}{\max_i - \min_i}\right)^2}}{\sqrt{\sum_{i=1}^{m} W(i,C)}}$$

where $e^j{}_i$ (j = 1, 2) is the value of the ith attribute on example e^j, $\max_i$ and $\min_i$ are respectively the maximum and minimum values of the ith attribute, and m is the number of attributes. W(i, C) is the weight of the attribute i with respect to the concept C. When the ith attribute is symbolic-valued, $e^1{}_i - e^2{}_i = 1$ if they are different, $e^1{}_i - e^2{}_i = 0$ otherwise. For missing values, $e^1{}_i - e^2{}_i = 0.5$. The distance of a linear attribute is normalized to the range of 0 to 1. The distance between two instances is also normalized to the range of 0 to 1.

4. Selecting Typical Instances

In the sections 2 and 3, we discussed the algorithms for computing weights of attributes and typicalities of instances. In this section, we shall first introduce the method for instance selection and classification, then present the complete instance-based algorithm in the computational best-examples model. The nearest neighbor algorithm stores all instances as concept descriptions. Aha et al. (1991) and Salzberg (1991) developed storage reduction instance-based learning algorithms in which only incorrectly classified instances were stored. Aha et al. (1991) empirically demonstrated that their storage reduction algorithm IB2 significantly reduced the storage requirements, and only slightly degraded classification accuracies. As indicated by Aha et al (1991), majority of stored instances by IB2 were near-boundary instances.

Similar to many IBL algorithms, the instance-based learning algorithm in our model stores a subset of training instances in its memory, and uses a distance measure to decide the distance between new instances and those stored. New instances are classified according to their closest neighbor's classification. The distance measure used is the one introduced in the section 3 with respect to the concept to which the stored instance belongs. Each time a new instance is incorrectly classified, our algorithm does not store the incorrectly classified instance itself, instead it stores the most typical instance which correctly classifies the new instance. That is, the algorithm finds the most typical instance such that after the instance is stored into the memory, the new instance can be correctly classified.

Similar to *Each* (Salzberg, 1991) and PEBLS (Cost and Salzberg, 1991), each stored instance is associated with a weight. The weight is used in measuring the distance between a new instance and the stored instance. The distance between a stored instance X of a concept C and a new instance Y is:

$$D(C, X, Y) = W_X * \mathrm{dis}(C, X, Y)$$

where dis(C, X, Y) is the distance measure introduced in section 3, W_X is the weight of X, and C is the concept to which X belongs. Each stored instance covers an area in the instance space. The area covered by an instance depends on the distribution of all stored instances and the weight assigned to the instance. Generally, the smaller the weight of an instance, the larger the area covered by the instance. By changing the weight, one can change the area that the instance covers. Detailed discussion about the issue can be found in (Cost and Salzberg, 1991). The weight of an instance in our algorithm is simply the reciprocal of its typicality. The rationale for this is that a typical instance is more reliable than a boundary instance and should cover a larger area. Namely, it should have a smaller weight. An exceptional case should cover only a small area so it should have a large weight.

Specifically, our computational best-examples model is described as follows:

1. Compute weights of all attributes with respect to each concept,
2. Compute typicalities for all instances,
3. CD = null,
4. pick up the most typical incorrectly classified instance x, find the most typical instance y which correctly classifies x,
5. compute the weight of y: $\mathrm{weight}(y) = \frac{1}{\mathrm{typicaity}(y)}$,
6. add y to CD,
7. repeat the step 4, 5 and 6 until all instances are correctly classified.

We have implemented the algorithm in a system TIBL (Typical-Instance-Base Learning). To compare with other instance-based learning algorithms, we have also implemented three other instance-based learning algorithms, BIBL (Boundary-Instance-Based Learning), SRIBL (Storage Reduction Instance-Based Learning), and IBL (Instance-Based Learning). BIBL algorithm stores the lest typical instances, that is, exceptional and boundary instances. This algorithm repeats the process of finding the incorrectly classified instance with the smallest typicality and storing it until all instances are correctly classified. SRIBL is similar to IB2 (Aha, et al., 1991). It repeats the process of finding an incorrectly classified instance and storing it until all instances are correctly covered.

IBL is the 1-nearest neighbor algorithm and stores all training instances.

5. Empirical Evaluation

To empirically evaluate the typical-instance-based learning algorithm, we have conducted two kinds of experiments with TIBL. The first kind of experiments was designed to evaluate the algorithm in comparison with other instance-based learning algorithms, while the second kind of experiments was to evaluate the effect of learning attribute relevancies. The performance was evaluated on two aspects: classification accuracy and storage requirement. Classification accuracy was measured as the percentage of correct classifications made by the concept description on a set of randomly selected test instances. Storage requirement was measured by the number of instances stored in descriptions. All results reported in this section were averaged over 10 trials.

We applied the four instance-based learning algorithms: TIBL (Typical-Instance-Based Learning), BIBL (Boundary-Instance-Based Learning), SRIBL (Storage Reduction Instance-Based Learning), and IBL (Instance-Based Learning) to five domains: classification of n-of-m concept, classification of congressional voting recording, malignant tumor classification, diagnosis of diabetes in Pima Indians, and diagnosis of heart disease. In these experiments, TIBL was applied without attribute weight learning. Table 1 summaries the characteristics of the five domains and table 2 reports the experimental results of the four different instance-based learning algorithms on the five domains. Test sets were disjoint with training sets except for the n-of-m concept on which test set was the whole instance space. In table 2, ACC and #ins represent accuracy and the number of instances, respectively.

Domain	Training Set Size	Test Set Size	Number of Attributes
n-of-m	400	1024	10
Voting	200	235	16
Tumor	150	219	9
Diabetes	200	568	8
Heart	100	203	13

Table 1: Summary of Domain Characteristics

	TIBL		BIBL		SRIBL		IBL	
Domains	ACC(%)	#ins	ACC(%)	#ins	ACC(%)	#ins	ACC(%)	#ins
n-of-m	99.5	10.8	76.0	182.8	80.3	219.4	85.5	400.0
Voting	90.4	31.6	92.0	59.5	92.4	51.9	93.4	200.0
Tumor	93.1	19.5	90.4	29.4	91.2	28.8	93.7	150.0
Diabetes	70.2	105.6	66.5	106.9	65.5	105.3	69.9	200.0
Heart	82.0	33.7	73.9	46.6	75.6	45.2	77.8	100.0

Table 2: Experimental Results of 4 IBL Algorithms on 5 Domains

The n-of-m concept is an artificial domain and contains 10 binary attributes and 2 concepts, C1 and C2. If 5 or more of the 10 attributes of an instance are 1, then the instance belongs to C1, otherwise it belongs to C2. TIBL significantly improved both accuracy and storage requirements over BIBL, SRIBL and IBL. The reason for such a large improvement is that the n-of-m concept has a very clear graded structure. When the two most typical instances, 1111111111 and 0000000000, appeared in the training set, they were the only two instances chosen by TIBL, 1111111111 for C1 and 0000000000 for C2. These two instances were weighted differently, 1111111111 had a slightly smaller weight than 0000000000 so that 1111111111 covered larger area than 0000000000. The concept C1 did cover a larger area than C2. 100% accuracy was achieved by these descriptions. Following is an example of such descriptions.

1111111111: weight = 0.483

0000000000: weight = 0.523

The congressional Voting database contains the voting records of the members of the United States House of Representatives during the second session of 1984. It is described by 16 binary attributes and has 288 missing values among its 435 instances. TIBL's classification accuracy is slightly lower than BIBL's, SRIBL's and IBL's, but TIBL saved much fewer instances. An interesting result is that almost all descriptions generated by TIBL included only one or two instances with very high typicalities plus a number of instances with very low typicalities. Very few instances with medium typicalities (1.2 to 3) were included in the descriptions. This is because that these instances were correctly classified by the typical instances stored. The typical instances of a description represented the central tendency, while the instances with low typicalities were exceptions which could not be correctly classified by any typical instances. The lower TIBL accuracy may be due to the fact that the test set included some exceptions which were not correctly classified by the typical instances stored.

The malignant tumor classification domain includes a set of 369 breast cancer patients, of which 201 have no malignancy and the remainder have confirmed malignancies (Wolberg and Mangasarian, 1989). The problem is to determine whether the tumors were benign or malignant from these cancer patients. Each patient is described by nine real-valued features. Mangasarian et al. (1989) applied a new linear programming technique to this domain, and good results have been achieved. Although the accuracy of IBL was slightly better than TIBL, it stored about 7 times more instances. TIBL outperformed both BIBL and SRIBL in terms of both accuracy and storage requirement, but the accuracy improvement is not significant. BIBL and SRIBL performed similarly. Similar to those obtained in the congressional voting records, the concept descriptions generated consisted of a few typical instances and a number of exceptional instances. These exceptional instances can be removed without degrading the accuracy.

The Diabetes in Pima Indians data set contains 768 instances, of which 500 (65%) have no diabetes, and 268 are diabetes patients. Each instance is described by 8 linear attributes. The problem is to diagnose who has diabetes and who has no. The accuracy of TIBL was consistently better than those of BIBL and SRIBL. The storage requirement of TIBL is about the same as those of BIBL and SRIBL. TIBL performed equally well as IBL in accuracy, while it reduced the storage by half.

The heart disease data set contains 303 instances, each instance is represented as 13 numeric attributes plus a classification: presence or absence of heart disease. 164 of the 303 instances have no heart disease. The goal is to learn to distinguish presence of heart disease from absence. Excellent results were obtained by TIBL on this domain. TIBL's classification accuracy was over 80% and higher than previously published results. Aha et al. (1991) reported 75.7% accuracy for standard nearest neighbor and 78% for a variant of NN that discards apparently noisy instances. They also reported that the C4 decision tree learning algorithm (Quinlan, 1987) achieved 75.5% accuracy. In our experiments, TIBL showed a significant accuracy improvement over the other three methods BIBL, SRIBL and IBL. It stored fewer instances than BIBL and SRIBL.

TIBL reduced the storage requirements dramatically on the datasets on which high accuracy were achieved by learning systems, e.g.,. congressional voting records and malignant tumor. This result was partially caused by the fact that high quality datasets enabled our algorithm to better distinguish typical instances from atypical ones. Another reason was that instances in high quality datasets are very concentrated and constitute few peaks which are well represented by a few typical instances.

To evaluate the effect of attribute weight learning, TIBL has been run on two domains, n-of-m concept and congressional voting, with and without attribute weight learning. The congressional voting dataset was the same as the one used in the experiments reported above. The n-of-m concept was modified by adding 5 irrelevant attributes. Table 3 presents the experimental results. Descriptions of n-of-m were tested on 2000 examples and descriptions of congressional voting records were tested 335 and 235 examples for the training sizes 100 and 200, respectively.

Domain	Training Set Size	With Attribute Relevancy		No Attribute Relevancy	
		ACC	#ins	ACC	#ins
n-of-m	200	95.6%	12.4	85.3%	60.4
	400	99.4%	7.5	89.0%	96.3
Voting	100	90.1%	12.6	88.7%	14.2
	200	91.3%	27.8	90.4%	31.6

Table 3: Experimental Results with and Without Learning Attribute Relevancies

Significant improvements on both classification accuracy and storage requirement were achieved on the domain of n-of-m concept. Although the improvements on congressional voting records were minor, they were stable. Improvement on accuracy was obtained on 19 of the 20 trials made over the two training set sizes and the improvement on storage requirement was observed for all 20 trials. These improvements were due to the attribute weight learning. Attribute weights not only helped TIBL in classifying new instances, but also in identifying typical instances, because attribute weights were used to compute typicalities of instances. For example, the typicalities of the most typical instances of n-of-m concept were around 1.25 without using attribute weights and were around 2.5 with using attribute weights. The typicalities of the most typical instances of congressional voting records were around 2.8 without using attribute weights and were around 4.0 with using attribute weights.

6. Summary and Future Work

The main contribution of the work described in this paper is the development of a computational best-examples model from the cognitive best-examples model proposed by Smith and Medin (1981). Three algorithms, attribute weight learning algorithm, instance typicality measuring algorithm and instance selection algorithm, were developed in this computational model. This model was empirically evaluated and compared with other instance-based learning algorithms. The results confirmed that the best-examples model can be adopted in developing instance-based learning systems. The results showed that when concepts have graded structures instances-based learning systems developed best-examples model may outperform other instance-based learning systems. The computational model may also help cognitive researchers to better understand the best-examples model.

One of the limitations of the computational model is the way to compute the distance of instances when attributes are symbolic-valued. In this case, distance in TIBL is computed by counting the attribute values that match. As indicated in (Cost and Salzberg, 1991), this approach for computing distance may not perform well when the domains are complex. In the future, we shall implement a more complicated method called Value Difference Metric (VDM) (Stanfill and Waltz, 1986; Cost and Salzberg, 1991) which takes into account the overall similarity of classification of all instances for each possible value of each attribute. In this method, a matrix defining the distance between all values of an attribute is derived statistically, based on the examples in the training set. Other future work includes developing a method for learning weights of linear attributes, especially continuous attributes. The problem of classifying new instances with degrees of membership should be addressed in the future too.

References

Aha, D., "Incremental, instance-based learning of independent and graded concept descriptions," Proceedings of the Sixth International Workshop on Machine Learning, Ithaca, NY, 1989.

Aha, D. and Kibler, D., "Noise-tolerant instance-based learning algorithms," Proceedings of the Eleventh International Joint Conference on Artificial Intelligence, Detroit, MI, 1989.

Aha, D., Kibler, D, and Albert, M., "Instance-Based Learning Algorithm," *Machine Learning* 6, 1991

Bareiss, E. R., Porter, B. W., and Wier, C. C., "Protos: An Exemplar-Based Learning Apprentice," *Machine Learning: An Artificial Intelligence Approach V III*, 1990.

Barsalou, L., "Ideals, central tendency, and frequency of instantiation as determinants of graded structure in categories," in Journal of Experimental Psychology: Learning, Memory and Cognition, 11. 1985.

Cost, S., and Salzberg, S., "Q weighted nearest neighbor algorithm for learning with symbolic features," Technique report, Department of Computer Science, The Johns Hopkins University, 1991.

Mangasarian, O., Setiono, R., and Wolberg, W., "Pattern recognition via linear programming: theory and application to medical diagnosis," Technical Report #878, Computer Science Department, University of Wisconsin-Madison, 1989.

Quinlan, J. R., "Simplifying decision trees." In *International Journal of Man-Machine Studies*, vol. 27, 1987.

Rosch, E. and Mervis, C. B., "Family Resemblances: Studies in the Internal Structure of Categories." In *Cognitive Psychology*, vol. 7, 1975.

Salzberg, S., "A nearest hyperrectangle learning method," Machine Learning, 6:3, 1991.

Smith, E. E., Medin, D. L., *Categories and Concepts*. Harvard University Press, 1981.

Stanfill, C. and Waltz, D., "Toward memory-based reasoning," Communications of the ACM, 29:12, 1986.

Wolberg, W. and Mangasarian, O., "Multisurface method of pattern separation applied to breast cytology diagnosis," Manuscript, Department of Surgery, Clinical Science Center, University of Wisconsin, Madison, WI, 1989.

INDEX

9780415651066